Distribution of Bird Species of the World

The Clements Checklist of Birds of the World

SIXTH EDITION

The Clements Checklist
of Birds of the World

SIXTH EDITION

James F. Clements

Forewords by Jared Diamond and Anthony W. White
Preface by John W. Fitzpatrick

*The Official World Checklist of the American Birding Association published
in collaboration with the Cornell Laboratory of Ornithology*

COMSTOCK PUBLISHING ASSOCIATES
a division of
CORNELL UNIVERSITY PRESS
ITHACA, NEW YORK

Sixth Edition copyright © 2007 by Cornell University
Previous editions copyright © James F. Clements
Fifth Edition published as *Birds of the World: A Checklist* by Ibis
   Publishing Company
Photograph of James Clements courtesy of Karen Clements

First published 2007 by Cornell University Press

Printed in the United States of America

Library of Congress Cataloging-in-Publication Data

Clements, James F.
    The Clements checklist of birds of the world / James F. Clements;
foreword by Jared Diamond; foreword by Anthony W. White;
preface by John W. Fitzpatrick. --6th ed.
        p. cm.
    "Published in association with the Cornell Laboratory of
Ornithology and the American Birding Association."
    Previous ed. published under the title: Birds of the world.
ISBN-13: 978-0-8014-4501-9 (cloth: alk. paper)
1. Birds.   I. Clements, James F. Birds of the world.   II. Cornell
University. Laboratory of Ornithology.   III. American Birding
Association.   IV. Title.
QL673.C53 2007
598--dc22   2006102996

Cornell University Press strives to use environmentally responsible
suppliers and materials to the fullest extent possible in the publishing of its books. Such materials include vegetable-based, low-VOC
inks and acid-free papers that are recycled, totally chlorine-free, or
partly composed of nonwood fibers. For further information, visit
our website at www.cornellpress.cornell.edu.

Cloth printing   10 9 8 7 6 5 4 3 2 1

Contents

Foreword: In Memoriam by Jared Diamond   vii

Foreword: The American Birding Association and the
Clements Checklist by Anthony W. White   ix

Preface: The Cornell Laboratory of Ornithology and the
Clements Checklist by John W. Fitzpatrick   xi

Abbreviations   xiii

Orders and Families   xv

Part I: Non-Passerines   1

Part II: Passerines   263

Extinct Species   699

Appendix A: Distribution of Bird Species of the World   703

Appendix B: Distribution of Endemic Bird Species of the World   707

Major Family References   711

Bibliography   721

Index of Scientific Names   728

Index of English Names   793

Foreword: In Memoriam

James Franklin Clements (1927–2005)

My dictionary defines "enthusiast" as someone with "intense or eager interest." The word conjures up for us the image of someone who is fascinated by a subject, continues to be fascinated by it, and remains insatiable. We also associate the word with youth: it's especially young people who are enthusiastic, while older people learn to temper their enthusiasm.

All his life, Jim Clements remained enthusiastic about birds. He was fascinated by all of them. Of course he waxed eloquent about big rare distinctive famous birds like the Kagu of New Caledonia, but he was equally enthusiastic about inconspicuous ones. I recall a discus-sion with him about New Guinea's Yellow-legged Flycatcher, an obscure uncommon small flycatcher confined to New Guinea, about which no one other than Jim and I and a handful of other New Guinea bird addicts has ever been enthusiastic. After returning from New Guinea, Jim told me with pleasure and pride not only about the birds of paradise that he had seen but also about identifying a Yellow-legged Flycatcher in a mountain range where it had never before been recorded. He tried to see as many species as possible, and he would have liked sooner or later to have observed every bird species in the world.

Every year or so, Jim visited me for breakfast at my house in Los Angeles. Whatever else we talked about—friends, marriage, health, travel—the conversation always turned to birds. A non–bird-watching friend of mine coined the sentence "This bird does this, that bird does that," as a description of what bird-watchers talk about when they get together. Yes, Jim and I did talk about what this bird does and what that bird does, and we enjoyed it whatever this bird and that bird happened to be at the particular moment.

Jim had the inexhaustible enthusiasm of a young person who has just discovered the object of his enthusiasm. For Jim, it was always as if birds were a new discovery for him. They kept him young, happy, and optimistic even in the face of health setbacks later in his life.

The word "enthusiast" is also sometimes associated with pejorative connotations, such as the connotation of superficiality. My dictionary gives a second definition of enthusiast: "a person who is carried away by his feelings for a cause." By that definition, Jim was the opposite of an enthusiast. To put it more accurately, he was indeed an enthusiast in the sense of the first positive definition of someone with "intense or eager interest," but he combined that interest with the opposite of superficiality and succumbing to feelings: he maintained a remarkable mastery of organization and detail, as demonstrated by every page of this new sixth edition of the authoritative checklist of the birds of the world that now bears his name, *The Clements Checklist of Birds of the World.*

In the months between our breakfast meetings, Jim and I corresponded about the editions of *Birds of the World: A Checklist*, starting with the first edition of 1972, and continuing up to this, the sixth edition. At the time when the first edition appeared, there was no comprehensive reference to the world's birds. The closest thing to it was Peters' *Birds of the World*, whose sixteenth and last volume would eventually be completed in 1987. The Peters volumes were scholarly, sound, and detailed, however, as of 1972 they weren't even a list of the birds of the world, because four of the volumes had not yet been published. The whole set cost hundreds of dollars, virtually confining it to libraries and specialists. Its restriction to Latin names and lack of vernacular names reinforced the point that it was intended for professional scientists, not for people who enjoyed birds nonprofessionally. It lacked a unified approach, because each volume or even each family was compiled by a different expert, and ornithological standards changed greatly over the course of the fifty-six years between the publication of the first and last volume of Peters. In short, Peters was essential to scientists but of little use to birders.

Jim's *Birds of the World: A Checklist* revolutionized world birding and gave tour leaders the opportunity to sharpen their talents, just as did Jim's 1989 worldwide birdathon for the Museum of Natural History in Los Angeles. Both motivated birders to travel the world over in pursuit of the object of their enthusiasm. His checklist was simultaneously a volume to whet their appetite, a notebook in which to check off their finds, a travel resource with which to plan the next trip, bedtime reading to browse at home, and a grand framework within which to see the birds of the world in a whole broad context. With the appearance of Jim's checklist, birding became more than a matter of "This bird does this, that bird does that."

The checklist fueled an explosion of international bird tours. That's not just a major contribution to the happiness of millions of birders but also to an explosion of knowledge about birds around the world, and a source of support for efforts to conserve birds. The result of the 1972 checklist was five future editions—as well as books by others similarly aiming to list all of the world's birds. Each of those subsequent books by other authors differs from each other and from Jim's, but this work of Jim's remains unique. It isn't a work by committee, but the work of one man, digesting information from hundreds of other people.

Jim's achievement reminds me of a line from a poem of Horace, the greatest Roman poet, whose odes and epodes rank as the highest achievement of Roman literature. Horace concluded the thirtieth and last ode of his third and last book of odes with the line

"Exegi monumentum aere perennius,"

meaning "I have erected a monument more lasting than bronze." Horace was thereby predicting that the memory of his poems would outlast his own life, and that they would represent a contribution to humanity at least as important as Rome's statues and buildings.

This checklist is the monument to Jim Clements: enthusiast, master of organization, master of detail, lover of birds, and lover of his friends and dear ones.

—Jared Diamond
Los Angeles, California

Foreword: The American Birding Association and the Clements Checklist

Bird-watching, or "birding" as the aficionados call it, fledged after World War II. New cars, cheap un-rationed gas, better optics, and the new Peterson field guide made birding easy and inexpensive. At first most birders only took day trips around home. An overnight trip to the coast was considered a major outing. As birding grew more popular, adventurous birders traveled farther from home to see the exciting birds illustrated in their field guides. In 1969 a group of "600" birders, those who had seen 600 species in North America, formed the American Birding Association (ABA), a forum to exchange information and compare bird lists. Today the ABA is an organization of roughly 17,500 members with a variety of programs and publications promoting birding and supporting birders and their hobby.

Back in 1969 it soon became apparent that a set of rules and standards had to be adopted to make the comparisons of bird lists meaningful. These standards included a checklist of acceptable species in North America or the "ABA Area" as it was called. ABA members were also expanding their horizons and birding outside the ABA Area frequently. If birding within North America needed a standard list on which all could rely, the need for a standard world list of birds was far greater. James Clements actually beat the ABA to the punch, publishing his *Birds of the World: A Checklist* in 1974, a year before ABA published its first checklist. Clements' book was written and designed with birders in mind; after each species it had space for birders to enter their sighting data. Some professional ornithologists disparaged the idea of an amateur bird-watcher compiling and publishing a list of the world's birds, but Clements persisted. He corrected the inevitable errors and

omissions in succeeding editions—a second edition in 1978 and a third in 1981. Each new edition also contained the latest taxonomic changes.

In 1983, shortly after the third edition was published, the ABA accepted Clements' *Birds of the World* as the standard for birds outside the ABA and American Ornithologists' Union (AOU) Areas. In reality this meant most of the world. James Clements continued to update his world list, publishing a fourth edition in 1991 and a fifth in 2000. He posted changes regularly on his website and licensed software providers to incorporate his list in their products. This further enhanced the utility of his checklist to birders and other users. In 2004 the ABA conducted a review of recent world checklists. Several books reflected the highest standards of research and excellence, but Clements' birder-friendly format, the regular promulgation of changes over the Internet, and the frequent publication of updated editions made his book an obvious choice to remain the ABA standard. Today over 1,000 birders submit lists to ABA for publication and comparison. James Clements' book is the standard for most of the world, but it is only one of many standards and rules the ABA birders are asked to follow.

ABA is seen by some as a group of hard-core, irresponsible birders, people who would endanger a bird in order to add it to their life list. Nothing could be further from the truth. ABA requires that members who submit their lists comply with a code of birding ethics—a code that is summarized by the statement, "In any conflict of interest between birds and birders, the welfare of the birds and their environment comes first." Here is the ABA Code of Birding Ethics and Recording Rules.

ABA Code of Birding Ethics
1. Promote the welfare of birds and their environment.
2. Respect the law and the rights of others.
3. Ensure that feeders, nest structures, and other artificial bird environments are safe.
4. Group birding, whether organized or impromptu, requires special care.

ABA Recording Rules
1. The bird must have been within the prescribed area and time period when encountered.
2. The bird must have been a species currently accepted by the ABA Checklist Committee for lists within its area, or by the AOU Checklist for lists outside the ABA Area and within the AOU Area, or by Clements for all other areas.
3. The bird must have been alive, wild, and unrestrained when encountered.
4. Diagnostic field marks for the bird, sufficient to identify to species, must have been seen and/or

documented by the recorder at the time of the encounter.
5. The bird must have been encountered under conditions that conform to the ABA Code of Birding Ethics.

James Clements' checklist and the ABA attract people to birding, an outdoor nature-oriented pursuit, and the developing birders in turn become conservationists and support the preservation of our natural heritage. James Clements' spirit and commitment live on after his untimely death in this posthumous publication of a sixth edition and in the Cornell Laboratory of Ornithology's plans to continue updating and promulgating changes to the Clements Checklist. The American Birding Association is proud to endorse this effort and is confident that it will benefit birders and ornithologists around the world, but, more importantly, the birds themselves.

—Anthony W. White
Chair, ABA Recording Standards and Ethics Committee

Preface: The Cornell Laboratory of Ornithology and the Clements Checklist

We birders might be called a peculiar lot if there weren't so many of us. Estimates of our population size in the United States alone vary among surveys, but numbers like 25 million, 54 million, even 70 million have emerged from recent, scholarly efforts to measure our burgeoning pastime. We vary from home naturalists content with our bird-feeders (like my late mother) to globe-trotting ecotourists like the late Jim Clements, eager to plan the next trip's potential bird list even while unpacking from the last one. Some of us even converted this peculiar passion into a profession and are fortunate enough to devote our lives to fostering appreciation, study, and conservation of the world's birds.

As a member of the last category, I am privileged to represent the Cornell Laboratory of Ornithology in honoring the lifelong contributions of Jim Clements to birding, exploration, and ornithology. As a sponsor of this Sixth Edition of Clements' monumental checklist, the Cornell Lab of Ornithology has agreed to assume official responsibility for updating it on a regular basis. I personally admired Jim Clements enormously and felt privileged to know him. Besides his boundless energy and enthusiasm for birds and travel, I respected his audacity to straddle the line between scientific ornithology and bird watching in producing and maintaining his pioneering Checklist. As were so many others, I was deeply saddened by Jim's premature departure. When Karen Clements asked if Cornell could step in to ensure publication of his nearly complete Sixth Edition, it was difficult to refuse despite the enormity of the undertaking. Thankfully, Cornell University Press accepted the significant up-front challenge of producing this handsome book. But to honor Jim Clements

genuinely, we recognized the need to commit to its regular updating.

We do not make this commitment lightly. Few tasks could be as Sisyphean as publishing an accurate, up-to-date list of common and scientific names, and their respective distributions, for a worldwide group of organisms as popular, mobile, and closely studied as birds. Jim Clements was a brave man to tackle such a task, and he did it so brilliantly that his Checklist has become an essential tool for serious birders the world over. Clements had the foresight to recognize that we birders are awed by the global diversity of these wonderful creatures, and we love our lists. He saw the need for a scholarly compilation of bird diversity as a reference tool, but he also kept sight of four specific needs of real birders: (1) the list must remain easy to use, pared of any but the most essential words in presenting the names and approximate ranges of all birds; (2) the list must be consistent with the findings of scientific professionals, who study bird biology and publish their decisions in peer-reviewed technical journals; (3) where differences of opinion exist among professionals, the list must be decisive, but biologically consistent and clear about its choice and hierarchy of scholarly references; (4) the list must "stay alive," remaining as synchronous as possible with a constantly changing body of information about both names and distributions. Jim kept abreast of these changes admirably, and the Internet became his vehicle for communicating regular updates to birders.

Beginning with his fifth edition, Jim Clements took another one of his signature steps, at once both monumental and risky. He elected to provide birders entry into the biologically fascinating world of within-species diversity. His ambitious cataloging of all currently recognized subspecies clearly was spurred by recognition that these names represent the grist for most future taxonomic splits. The proliferation of excellent behavioral studies all over the world continues to yield a steady stream of such changes, and the rate at which these changes appear will no doubt increase now that DNA sequencing has become a mainstream scientific tool. The sixth edition is largely Jim's own work and continues this ambitious effort to provide a comprehensive list of currently recognized subspecies for each bird species. Through our updates in this edition and our future updates, which will be posted regularly on our website at http://www.birds.cornell.edu/clementschecklist, we will do our best to keep these names current with the recognized authorities worldwide. The recent English names suggested by Frank Gill and Minturn Wright on behalf of the International Ornithological Congress in *Birds of the World: Recommended English Names* will undoubtedly spur much conversation on this topic. Our first electronic update, available soon after publication of the sixth edition of the *Clements Checklist*, will include a comprehensive list of species having different common names in the Clements and IOC lists. As in the past, future updates to the *Clements Checklist* will reflect official name changes as they are approved by the American Ornithologists' Union (for the AOU checklist area) and other professional organizations (as ap-

propriate elsewhere around the world.) Birders with suggested taxonomic changes, suggestions, or questions are invited to submit their queries to cornellbirds@cornell.edu. Please write "checklist" in the subject line.

Jim Clements was an energetic man of the world, who might have been called larger-than-life had he lacked his genuine warmth and infectious, friendly grin. He embraced big challenges, especially if they provided an opportunity to make a difference in the world. He was an explorer, and a communicator, who opened up the world of birds and birding to countless legions of people in all walks of life and on every continent. I have no illusion that anyone, or even any team, could fill the shoes of the late Jim Clements. But his unique Checklist, straddling as it does the interface between birding and ornithology, fits squarely within the mission of the Cornell Lab of Ornithology. For this reason, and with appreciation in advance to all who will help us live up to Jim's legacy, we accept the challenge to try.

Good birding!

—John W. Fitzpatrick
Director, Cornell Laboratory of Ornithology

Abbreviations

adjacent	adj.	Lake	L.
Afghanistan	Afghan.	Madagascar	Madag.
Amazonian	Amaz.	Massachusetts	Mass.
America	Am.	Mediterranean	Medit.
Archipelago	Arch.	Montana	Mont.
Arizona	Ariz.	Mountain	Mt.
Atlantic Ocean	Atl. Oc.	Mountains	Mts.; mts.
British	Br.	Netherlands Antilles	Neth. Ant.
British Columbia	Br. Col.	New Hampshire	N Hamp.
California	Calif.	New Mexico	N Mex.
Cape Province	C. Prov.	New South Wales	NSW
Central	C; c, cent.	New York	NY
Central African	Cent. African	New Zealand	NZ
Republic	Rep.	North	N; n
circa	ca	North America	N Am
Colorado	Colo.	North Carolina	N Car.
Connecticut	Conn.	N. Dakota	N Dak.
County	Co.	Northeast	NE; ne
District	Dist.	Northwest	NW; nw
Dominican Republic	Dom. Rep.	Northwest Territory	NWT
East	E; e	New York	NY
El Salvador	El Sal.	Pacific Ocean	Pac. Oc.
Extinct	†	Papua New Guinea	Papua NG
Federal	Fed.	Peninsula	Pen.
French	Fr.	Queensland	Queens.
Greater	Gr.	Republic	Rep.
Indian Ocean	Ind. Oc.	River	R.
Introduced	Introd.	Siberia	Sib.
Isla	I.	South	S; s
Island	I.	South America	S Am.
Islands	Is.; is.	South Carolina	S Car.

Southeast	SE; se	Tropical	Trop.; trop.
Southwest	SW; sw	United Kingdom	UK
Strait	Str.	United States	US
Subtropical	Subtrop.	Washington	Wash.
Synonymous	Syn.; syn.	West	W; w
Tasmania	Tas.	Wyoming	Wyo.
Tierra del Fuego	T. D. Fuego	>	Winters or disperses to

Orders and Families

PART I: Non-Passerines

English name	Order	Family	Number of species	Page
Ostrich	Struthioniformes	Struthionidae	1	2
Rheas	Struthioniformes	Rheidae	2	2
Cassowaries	Struthioniformes	Casuariidae	3	2
Emu	Struthioniformes	Dromaiidae	1	2
Kiwis	Struthioniformes	Apterygidae	5	2
Tinamous	Tinamiformes	Tinamidae	47	3 6
Penguins	Sphenisciformes	Spheniscidae	17	6–7
Loons	Gaviiformes	Gaviidae	5	7
Grebes	Podicipediformes	Podicipedidae	19	7–9
Albatrosses	Procellariiformes	Diomedeidae	13	9–10
Shearwaters and Petrels	Procellariiformes	Procellariidae	77	10–13
Storm-Petrels	Procellariiformes	Hydrobatidae	20	13–14
Diving-Petrels	Procellariiformes	Pelecanoididae	4	14–15
Tropicbirds	Pelecaniformes	Phaethontidae	3	15
Pelicans	Pelecaniformes	Pelecanidae	8	15
Boobies and Gannets	Pelecaniformes	Sulidae	10	15–16
Cormorants	Pelecaniformes	Phalacrocoracidae	39	16–18
Anhinga	Pelecaniformes	Anhingidae	2	18
Frigatebirds	Pelecaniformes	Fregatidae	5	18
Herons, Egrets and Bitterns	Ciconiiformes	Ardeidae	63	19–23
Hamerkop	Ciconiiformes	Scopidae	1	23
Storks	Ciconiiformes	Ciconiidae	19	23–24
Shoebill	Ciconiiformes	Balaenicipitidae	1	24
Ibises and Spoonbills	Ciconiiformes	Threskiornithidae	33	24–25
Flamingos	Phoenicopteriformes	Phoenicopteridae	6	25–26
Screamers	Anseriformes	Anhimidae	3	26
Ducks, Geese and Swans	Anseriformes	Anatidae	159	26–33
New World Vultures	Falconiformes	Cathartidae	7	34
Osprey	Falconiformes	Pandionidae	1	34

Hawks, Eagles and Kites	Falconiformes	Accipitridae	240	34–49
Secretary-bird	Falconiformes	Sagittariidae	1	49
Falcons and Caracaras	Falconiformes	Falconidae	64	49–53
Megapodes	Galliformes	Megapodiidae	21	53–54
Guans, Chachalacas and Curassows	Galliformes	Cracidae	50	54–57
Turkeys	Galliformes	Meleagriddae	2	57
Grouse, Ptarmigans and Prairie-chickens	Galliformes	Tetraonidae	19	57–60
New World Quail	Galliformes	Odontophoridae	31	60–63
Pheasants and Partridges	Galliformes	Phasianidae	155	63–74
Guineafowl	Galliformes	Numididae	6	74
Hoatzin	Gruiformes	Opisthocomidae	1	74
Mesites	Gruiformes	Mesitornithidae	3	75
Buttonquail	Gruiformes	Turnicidae	16	75–76
Cranes	Gruiformes	Gruidae	15	76–77
Limpkin	Gruiformes	Aramidae	1	77
Trumpeters	Gruiformes	Psophiidae	3	77
Rails, Gallinules and Coots	Gruiformes	Rallidae	135	78–86
Finfoots	Gruiformes	Heliornithidae	3	86
Kagu	Gruiformes	Rhynochetidae	1	86
Sunbittern	Gruiformes	Eurypygidae	1	86
Seriemas	Gruiformes	Cariamidae	2	87
Bustards	Gruiformes	Otididae	26	87–88
Jacanas	Charadriiformes	Jacanidae	8	88–89
Painted-Snipes	Charadriiformes	Rostratulidae	3	89
Crab Plover	Charadriiformes	Dromadidae	1	89
Oystercatchers	Charadriiformes	Haematopodidae	11	89
Ibisbill	Charadriiformes	Ibidorhynchidae	1	90
Avocets and Stilts	Charadriiformes	Recurvirostridae	10	90
Thick-knees	Charadriiformes	Burhinidae	9	90–91
Pratincoles and Coursers	Charadriiformes	Glareolidae	17	91–92
Plovers and Lapwings	Charadriiformes	Charadriidae	66	92–95
Magellanic Plover	Charadriiformes	Pluvianellidae	1	95
Sandpipers and Allies	Charadriiformes	Scolopacidae	89	95–99
Plains-wanderer	Charadriiformes	Pedionomidae	1	100
Seedsnipes	Charadriiformes	Thinocoridae	4	100
Sheathbills	Charadriiformes	Chionididae	2	100
Gulls	Charadriiformes	Laridae	56	100–103
Terns	Charadriiformes	Sternidae	44	103–106
Skimmers	Charadriiformes	Rynchopidae	3	106
Jaegers and Skuas	Charadriiformes	Stercorariidae	7	106–107
Auks, Murres and Puffins	Charadriiformes	Alcidae	23	107–108
Sandgrouse	Pterocliformes	Pteroclidae	16	108–109
Pigeons and Doves	Columbiformes	Columbidae	308	109–129
Cockatoos	Psittaciformes	Cacatuidae	21	129–131
Parrots	Psittaciformes	Psittacidae	347	131–151
Turacos	Cuculiformes	Musophagidae	23	151–152
Cuckoos	Cuculiformes	Cuculidae	141	152–160
Barn-Owls	Strigiformes	Tytonidae	16	160–162
Owls	Strigiformes	Strigidae	199	162–176
Oilbird	Caprimulgiformes	Steatornithidae	1	176

Owlet-Nightjars	Caprimulgiformes	Aegothelidae	9	176–177
Frogmouths	Caprimulgiformes	Podargidae	12	177
Potoos	Caprimulgiformes	Nyctibiidae	7	177–178
Nightjars and Allies	Caprimulgiformes	Caprimulgidae	91	178–184
Swifts	Apodiformes	Apodidae	100	184–191
Treeswifts	Apodiformes	Hemiprocnidae	4	191
Hummingbirds	Apodiformes	Trochilidae	339	191–211
Mousebirds	Coliiformes	Coliidae	6	211–212
Trogons and Quetzals	Trogoniformes	Trogonidae	40	212–215
Kingfishers	Coraciiformes	Alcedinidae	93	215–223
Todies	Coraciiformes	Todidae	5	223–224
Motmots	Coraciiformes	Momotidae	10	224–225
Bee-eaters	Coraciiformes	Meropidae	26	225–226
Rollers	Coraciiformes	Coraciidae	12	226–227
Ground-Rollers	Coraciiformes	Brachypteraciidae	5	227
Cuckoo-Roller	Coraciiformes	Leptosomatidae	1	228
Hoopoes	Coraciiformes	Upupidae	2	228
Woodhoopoes and Scimitar-bills	Coraciiformes	Phoeniculidae	8	228–229
Hornbills	Coraciiformes	Bucerotidae	57	229–231
Jacamars	Galbuliformes	Galbulidae	18	231–232
Puffbirds	Galbuliformes	Bucconidae	35	232–235
Barbets	Piciformes	Capitonidae	83	235–240
Toucans	Piciformes	Ramphastidae	40	241–243
Honeyguides	Piciformes	Indicatoridae	17	243–244
Woodpeckers	Piciformes	Picidae	219	244–261

PART II: Passerines

Broadbills	Passeriformes	Eurylaimidae	15	264–265
Asities or False-Sunbirds	Passeriformes	Philepittidae	4	265
Sapayoa	Passeriformes	Sapayaoidae	1	265
Pittas	Passeriformes	Pittidae	32	265–267
Ovenbirds and Woodcreepers	Passeriformes	Furnariidae	294	268–289
Typical Antbirds	Passeriformes	Thamnophilidae	212	290–303
Antthrushes and Antpittas	Passeriformes	Formicariidae	63	303–308
Gnateaters	Passeriformes	Conopophagidae	8	308
Tapaculos	Passeriformes	Rhinocryptidae	58	308–311
Cotingas	Passeriformes	Cotingidae	71	311–314
Manakins	Passeriformes	Pipridae	57	314–318
Tyrant Flycatchers	Passeriformes	Tyrannidae	435	318–347
New Zealand Wrens	Passeriformes	Acanthisittidae	3	347
Scrub-birds	Passeriformes	Atrichornithidae	2	347
Lyrebirds	Passeriformes	Menuridae	2	348
Larks	Passeriformes	Alaudidae	96	348–357
Swallows	Passeriformes	Hirundinidae	83	357–363
Wagtails and Pipits	Passeriformes	Motacillidae	66	363–368
Cuckoo-shrikes	Passeriformes	Campephagidae	82	368–375
Bulbuls	Passeriformes	Pycnonotidae	130	375–385
Kinglets	Passeriformes	Regulidae	6	385–386

Leafbirds	Passeriformes	Chloropseidae	8	386
Ioras	Passeriformes	Aegithinidae	4	386–387
Silky-flycatchers	Passeriformes	Ptilogonatidae	4	387
Waxwings	Passeriformes	Bombycillidae	3	387
Hypocolius	Passeriformes	Hypocoliidae	1	387
Palmchat	Passeriformes	Dulidae	1	387
Dippers	Passeriformes	Cinclidae	5	388
Wrens	Passeriformes	Troglodytidae	80	388–397
Mockingbirds and Thrashers	Passeriformes	Mimidae	35	397–399
Accentors	Passeriformes	Prunellidae	13	400
Thrushes and Allies	Passeriformes	Turdidae	176	401–414
Cisticolas and Allies	Passeriformes	Cisticolidae	111	414–425
Old World Warblers	Passeriformes	Sylviidae	291	425–445
Gnatcatchers	Passeriformes	Polioptilidae	15	445–446
Old World Flycatchers	Passeriformes	Muscicapidae	275	446–466
Wattle-eyes	Passeriformes	Platysteiridae	31	466–468
Fantails	Passeriformes	Rhipiduridae	45	468–471
Monarch Flycatchers	Passeriformes	Monarchidae	100	471–479
Australasian Robins	Passeriformes	Petroicidae	43	479–482
Whistlers and Allies	Passeriformes	Pachycephalidae	57	483–488
Rockfowl	Passeriformes	Picathartidae	2	488
Babblers	Passeriformes	Timaliidae	274	488–511
Pseudo-babblers	Passeriformes	Pomatostomidae	5	511
Parrotbills	Passeriformes	Paradoxornithidae	20	511–513
Logrunners	Passeriformes	Orthonychidae	3	513
Whipbirds and Quail-thrushes	Passeriformes	Eupetidae	15	513–514
Long-tailed Tits	Passeriformes	Aegithalidae	8	514–515
Fairywrens	Passeriformes	Maluridae	27	515–517
Thornbills and Allies	Passeriformes	Acanthizidae	65	517–522
Australian Chats	Passeriformes	Epthianuridae	5	522
Sittellas	Passeriformes	Neosittidae	2	522–523
Australasian Treecreepers	Passeriformes	Climacteridae	7	523
Chickadees and Tits	Passeriformes	Paridae	59	523–529
Nuthatches	Passeriformes	Sittidae	24	529–531
Wallcreeper	Passeriformes	Tichidromidae	1	531
Creepers	Passeriformes	Certhiidae	8	531–532
Philippine Creepers	Passeriformes	Rhabdornithidae	3	532–533
Penduline Tits	Passeriformes	Remizidae	13	533–534
Sunbirds and Spiderhunters	Passeriformes	Nectariniidae	131	534–544
Berrypeckers and Longbills	Passeriformes	Melanocharitidae	10	544–545
Tit Berrypecker and Crested Berrypecker	Passeriformes	Paramythiidae	2	545
Flowerpeckers	Passeriformes	Dicaeidae	44	545–549
Pardalotes	Passeriformes	Pardalotidae	4	549
White-eyes	Passeriformes	Zosteropidae	96	549–556
Sugarbirds	Passeriformes	Promeropidae	2	556
Honeyeaters	Passeriformes	Meliphagidae	174	556–567
Old World Orioles	Passeriformes	Oriolidae	30	567–569
Fairy-bluebirds	Passeriformes	Irenidae	2	569
Shrikes	Passeriformes	Laniidae	30	570–572
Bushshrikes and Allies	Passeriformes	Malaconotidae	46	572–575

Helmetshrikes and Allies	Passeriformes	Prionopidae	12	576
Vangas	Passeriformes	Vangidae	15	577
Drongos	Passeriformes	Dicruridae	24	577–580
Wattlebirds	Passeriformes	Callaeidae	2	580
Mudnest Builders	Passeriformes	Grallinidae	2	580
White-winged Chough and Apostlebird	Passeriformes	Corcoracidae	2	580
Woodswallows	Passeriformes	Artamidae	11	580–581
Bristlehead	Passeriformes	Pityriaseidae	1	581
Bellmagpies and Allies	Passeriformes	Cracticidae	13	581–582
Birds-of-paradise	Passeriformes	Paradisaeidae	44	582–585
Bowerbirds	Passeriformes	Ptilonorhynchidae	20	585–586
Crows, Jays and Magpies	Passeriformes	Corvidae	119	586–595
Starlings	Passeriformes	Sturnidae	114	596–602
Old World Sparrows	Passeriformes	Passeridae	38	602–604
Weavers and Allies	Passeriformes	Ploceidae	116	604–611
Waxbills and Allies	Passeriformes	Estrildidae	141	611–620
Indigobirds	Passeriformes	Viduidae	20	620–621
Vireos and Allies	Passeriformes	Vireonidae	52	621–625
Siskins, Crossbills and Allies	Passeriformes	Fringillidae	176	625–638
Hawaiian Honeycreepers	Passeriformes	Drepanididae	21	638–639
Olive Warbler	Passeriformes	Peucedramidae	1	639
New World Warblers	Passeriformes	Parulidae	118	639–647
Bananaquit	Passeriformes	Coerebidae	1	647
Tanagers and Allies	Passeriformes	Thraupidae	226	648–664
Przewalski's Rosefinch	Passeriformes	Urocynchramidae	1	664
Buntings, Sparrows, Seedeaters and Allies	Passeriformes	Emberizidae	329	664–688
Saltators, Cardinals and Allies	Passeriformes	Cardinalidae	43	688–691
Troupials and Allies	Passeriformes	Icteridae	101	691–698

Part I

Non-Passerines

ORDER: STRUTHIONIFORMES
FAMILY: STRUTHIONIDAE (Ostrich—1)

☐ **Ostrich** *Struthio camelus*

_____	*S. c. camelus*	Sahel of North Africa and the Sudan
_____	*S. c. syriacus†*	Formerly Syrian and Arabian deserts. Extinct ca 1966
_____	*S. c. molybdophanes*	S Ethiopia to Somalia and adjacent ne Kenya
_____	*S. c. massaicus*	S Kenya and e Tanzania
_____	*S. c. australis*	Southern Africa

ORDER: STRUTHIONIFORMES
FAMILY: RHEIDAE (Rheas—2)

☐ **Greater Rhea** *Rhea americana*

_____	*R. a. americana*	Campos of n and e Brazil
_____	*R. a. intermedia*	Extreme se Brazil (Rio Grande do Sul) and Uruguay
_____	*R. a. nobilis*	E Paraguay (east of the Río Paraguay)
_____	*R. a. araneipes*	*Chaco* of Paraguay to Bolivia and Brazilian Mato Grosso
_____	*R. a. albescens*	Plains of Argentina south to Río Negro

☐ **Lesser Rhea** *Rhea pennata*

_____	*R. p. garleppi*	Desert *puna* of se Peru, sw Bolivia and nw Argentina
_____	*R. p. tarapacensis*	*Puna* of n Chile (Arica to Atacama)
_____	*R. p. pennata*	Patagonian steppes of s Argentina and Magellanic Chile

ORDER: STRUTHIONIFORMES
FAMILY: CASUARIIDAE (Cassowaries—3)

☐ **Southern Cassowary** *Casuarius casuarius*

Humid forests of ne Australia, New Guinea and Aru Islands

☐ **Dwarf Cassowary** *Casuarius bennetti*

New Guinea, Yapen I. and New Britain

☐ **Northern Cassowary** *Casuarius unappendiculatus*

Lowlands of New Guinea, Yapen I. and w Papuan islands

ORDER: STRUTHIONIFORMES
FAMILY: DROMAIIDAE (Emu—1)

☐ **Emu** *Dromaius novaehollandiae*

_____	*D. n. novaehollandiae*	Australia
_____	*D. n. diemenensis†*	Formerly Tasmania. Extinct ca 1865

ORDER: STRUTHIONIFORMES
FAMILY: APTERYGIDAE (Kiwis—5)

☐ **Southern Brown Kiwi** *Apteryx australis*

_____	*A. a. australis*	Southwest South I. (New Zealand)
_____	*A. a. lawryi*	Stewart I. (New Zealand)

☐ **Okarito Brown Kiwi** *Apteryx rowi*

Okarito Forest (west coast of South I. (New Zealand)

☐ **North Island Brown Kiwi** *Apteryx mantelli*

North Island (New Zealand)

☐ **Little Spotted Kiwi** *Apteryx owenii*

Kapiti I., South I. and formerly North I. (New Zealand)

☐ **Great Spotted Kiwi** *Apteryx haastii*

Forests of South I. (New Zealand)

ORDER: TINAMIFORMES
FAMILY: TINAMIDAE (Tinamous—47)

☐ **Gray Tinamou** *Tinamus tao*
____ *T. t. larensis*   Montane forests of central Colombia and nw Venezuela
____ *T. t. kleei*   S-c Colombia to e Ecuador, e Peru, e Bolivia and w Brazil
____ *T. t. septentrionalis*   NE Venezuela and (?) nw Guyana
____ *T. t. tao*   N-central Brazil to borders of e Peru and Bolivia

☐ **Solitary Tinamou** *Tinamus solitarius*

E Brazil to se Paraguay and extreme ne Argentina (Misiones)

☐ **Black Tinamou** *Tinamus osgoodi*
____ *T. o. hershkovitzi*   Andes of s-central Colombia
____ *T. o. osgoodi*   Andes of se Peru

☐ **Great Tinamou** *Tinamus major*
____ *T. m. robustus*   Lowlands of se Mexico to Guatemala and n Nicaragua
____ *T. m. percautus*   SE Mexico (Yucatán Pen.) to Petén of Guatemala and Belize
____ *T. m. fuscipennis*   N Nicaragua to Costa Rica and w Panama
____ *T. m castaneiceps*   SW Costa Rica and w Panama
____ *T. m. brunneiventris*   S-central Panama
____ *T. m. saturatus*   Pacific slope of e Panama and nw Colombia
____ *T. m. latifrons*   SW Colombia and w Ecuador
____ *T. m. zuliensis*   Tropical ne Colombia and w Venezuela
____ *T. m. peruvianus*   SE Colombia east of the Andes to Bolivia and extreme w Brazil
____ *T. m. serratus*   Extreme s Venezuela and adjacent nw Brazil
____ *T. m. major*   E Venezuela to the Guianas and ne Brazil
____ *T. m. olivascens*   Amazonian Brazil

☐ **White-throated Tinamou** *Tinamus guttatus*

SE Colombia and s Venezuela to n Bolivia and Amaz. Brazil

☐ **Highland Tinamou** *Nothocercus bonapartei*
____ *N h frantzii*   Highlands of Costa Rica and w Panama
____ *N. b. intercedens*   W Andes of Colombia
____ *N. b. discrepans*   Base of Eastern Andes of Colombia (Tolima and Meta)
____ *N. b. bonapartei*   Central and E Andes of Colombia and w Venezuela
____ *N. b. plumbeiceps*   Andes of e Ecuador to extreme n Peru

☐ **Tawny-breasted Tinamou** *Nothocercus julius*

Andes of central Colombia and extreme w Venezuela; s-c Peru

☐ **Hooded Tinamou** *Nothocercus nigrocapillus*
____ *N. n. cadwaladeri*   Andes of nw Peru
____ *N. n. nigrocapillus*   Andes of central Peru to Bolivia

☐ **Berlepsch's Tinamou** *Crypturellus berlepschi*

Tropical forests of nw Colombia to nw Ecuador

☐ **Cinereous Tinamou** *Crypturellus cinereus*

SE Colombia to Guianas, s Venezuela, n Bolivia and ne Brazil

☐ **Red-legged Tinamou** *Crypturellus erythropus*
____ *C. e. columbianus*   Tropical north-central Colombia
____ *C. e. saltuarius*   NE Colombia (Sierra de Ocaña)
____ *C. e. idoneus*   NE Colombia and adjacent nw Venezuela
____ *C. e. cursitans*   N Colombia east of the Andes and nw Venezuela
____ *C. e. spencei*   N Venezuela
____ *C. e. margaritae*   Margarita I. (off Venezuela)
____ *C. e. erythropus*   E Venezuela to Guyana, Suriname and ne Brazil

☐ **Little Tinamou** *Crypturellus soui*
____ *C. s. meserythrus*   Tropical s Mexico to Belize, Honduras and se Nicaragua
____ *C. s. modestus*   Costa Rica and w Panama

____ *C. s. capnodes*	Humid lowlands of nw Panama
____ *C. s. poliocephalus*	Pacific slope of Panama (Veraguas to Canal Zone)
____ *C. s. panamensis*	Pacific and Caribbean slopes of Panama
____ *C. s. harterti*	Pacific slope of Colombia and Ecuador
____ *C. s. caucae*	Magdalena Valley of n-central Colombia
____ *C. s. mustelinus*	NE Colombia and extreme nw Venezuela
____ *C. s. soui*	E Colombia to the Guianas and ne Brazil
____ *C. s. caquetae*	SE Colombia (Meta to Caquetá)
____ *C. s. andrei*	Coastal n Venezuela (Falcón to Monagas); Trinidad
____ *C. s. nigriceps*	Tropical e Ecuador and ne Peru
____ *C. s. inconspicuus*	Central and e Peru and n Bolivia
____ *C. s. albigularis*	N and e Brazil

☐ **Tepui Tinamou** *Crypturellus ptaritepui*

Tepuis of s Venezuela (Ptari-tepui and Sororopán-tepui)

☐ **Brown Tinamou** *Crypturellus obsoletus*

____ *C. o. castaneus*	Tropical e Colombia to e Ecuador and n Peru
____ *C. o. ochraceiventris*	Subtropical central Peru (Huánuco to Cuzco)
____ *C. o. traylori*	Subtropical se Peru (Marcapata Valley of Cuzco)
____ *C. o. punensis*	Extreme se Peru and *yungas* of n Bolivia
____ *C. o. cerviniventris*	N Venezuela
____ *C. o. knoxi*	Subtropical nw Venezuela
____ *C. o. griseiventris*	N-central Brazil (Santarém region along Rio Tapajós)
____ *C. o. hypochraceus*	SW Brazil (upper Río Madeira in Rondônia)
____ *C. o. obsoletus*	E Paraguay to se Brazil and extreme ne Argentina (Misiones)

☐ **Undulated Tinamou** *Crypturellus undulatus*

____ *C. u. manapiare*	S Venezuela (upper Río Ventuari in Amazonas)
____ *C. u. simplex*	SW Guyana and immediately adjacent Brazil
____ *C. u. yapura*	SE Colombia to e Ecuador, e Peru and nw Brazil
____ *C. u. vermiculatus*	E Brazil (s Maranhão to Mato Grosso)
____ *C. u. adspersus*	Brazil south of the Amazon (Rio Madeira to Rio Tapajós)
____ *C. u. undulatus*	SE Peru to n Argentina

☐ **Pale-browed Tinamou** *Crypturellus transfasciatus*

Tropical forests of w Ecuador to nw Peru

☐ **Brazilian Tinamou** *Crypturellus strigulosus*

Tropical s Amazonian Brazil, adjacent e Peru and nw Bolivia

☐ **Gray-legged Tinamou** *Crypturellus duidae*

Tropical forests of e-central Colombia to s Venezuela

☐ **Yellow-legged Tinamou** *Crypturellus noctivagus*

____ *C. n. zabele*	Lowlands of ne Brazil (Piauí to Bahia and Minas Gerais)
____ *C. n. noctivagus*	Coastal se Brazil (Minas Gerais to Rio Grande do Sul)

☐ **Black-capped Tinamou** *Crypturellus atrocapillus*

____ *C. a. atrocapillus*	Lowlands of se Peru
____ *C. a. garleppi*	Lowlands of n Bolivia

☐ **Slaty-breasted Tinamou** *Crypturellus boucardi*

____ *C. b. boucardi*	Gulf-Caribbean lowlands of se Mexico to nw Honduras
____ *C. b. costaricensis*	Caribbean slope of Honduras to n Costa Rica

☐ **Choco Tinamou** *Crypturellus kerriae*

Humid foothills of extreme se Panama and nw Colombia

☐ **Variegated Tinamou** *Crypturellus variegatus*

Colombia to Venezuela, the Guianas, n Bolivia and Amaz. Brazil

☐ **Thicket Tinamou** *Crypturellus cinnamomeus*

____ *C. c. occidentalis*	Coastal w Mexico (Sinaloa to Guerrero)
____ *C. c. soconuscensis*	Pacific slope of s Mexico (Oaxaca and Chiapas)

____	*C. c. mexicanus*	Atlantic coast of Mexico (Tamaulipas to Puebla)
____	*C. c. sallaei*	S Mexico (Puebla to s Veracruz, Oaxaca and Chiapas)
____	*C. c. goldmani*	SE Mexico (Yucatán Peninsula) to n Guatemala and n Belize
____	*C. c. vicinior*	Highlands of s Mexico (Chiapas) to Guatemala and c Honduras
____	*C. c. cinnamomeus*	Coastal se Mexico (Chiapas) to El Salvador and Honduras
____	*C. c. delattrii*	Pacific lowlands of Nicaragua
____	*C. c. praepes*	Lowlands of nw Costa Rica

☐ **Rusty Tinamou** *Crypturellus brevirostris*

French Guiana and extreme ne Brazil; e Peru and nw Brazil

☐ **Bartlett's Tinamou** *Crypturellus bartletti*

Tropical w Amazonian Brazil, e Peru and n Bolivia

☐ **Small-billed Tinamou** *Crypturellus parvirostris*

Patchily distributed Amazon basin south to ne Argentina

☐ **Barred Tinamou** *Crypturellus casiquiare*

Extreme e Colombia and adjacent s Venezuela

☐ **Tataupa Tinamou** *Crypturellus tataupa*

____	*C. t. inops*	Marañón Valley of nw Peru
____	*C. t. peruvianus*	W-central Peru (Chanchamayo Valley of Junín)
____	*C. t. lepidotus*	NE Brazil (Maranhão, Ceará, Piauí, Pernambuco and Bahia)
____	*C. t. tataupa*	E Bolivia to Paraguay, s Brazil and n Argentina

☐ **Red-winged Tinamou** *Rhynchotus rufescens*

____	*R. r. catingae*	Central and ne Brazil
____	*R. r. rufescens*	SE Peru to Bolivia, e Paraguay, se Brazil and ne Argentina
____	*R. r. maculicollis*	Andes of nw Bolivia to nw Argentina
____	*R. r. pallescens*	N Argentina (e Formosa to Río Negro)

☐ **Ornate Tinamou** *Nothoprocta ornata*

____	*N. o. branickii*	*Puna* of central Peru (Ancash to Apurímac)
____	*N. o. ornata*	Andes of se Peru to Bolivia and extreme n Chile
____	*N. o. rostrata*	Andes of nw Argentina (Jujuy and La Rioja)

☐ **Chilean Tinamou** *Nothoprocta perdicaria*

____	*N. p. perdicaria*	Semiarid grasslands of n-central Chile (Atacama to Ñuble)
____	*N. p. sanborni*	S-central Chile (Maule to Llanquihue) and adjacent Argentina

☐ **Brushland Tinamou** *Nothoprocta cinerascens*

____	*N. c. cinerascens*	SE Bolivia to nw Paraguay and central Argentina
____	*N. c. parvimaculata*	Arid nw Argentina (e La Rioja)

☐ **Andean Tinamou** *Nothoprocta pentlandii*

____	*N. p. ambigua*	Andes of s Ecuador and nw Peru
____	*N. p. oustaleti*	W slope of Andes of central and s Peru
____	*N. p. niethammeri*	Coastal central Peru
____	*N. p. fulvescens*	Andes of se Peru
____	*N. p. pentlandii*	Andes of w Bolivia to nw Argentina and extreme n Chile
____	*N. p. doeringi*	Mountains of central Argentina (San Luis and Córdoba)
____	*N. p. mendozae*	Mountains of w-central Argentina (n Neuquén and Mendoza)

☐ **Curve-billed Tinamou** *Nothoprocta curvirostris*

____	*N. c. curvirostris*	Andes of central Ecuador to n Peru (Cordillera del Condor)
____	*N. c. peruviana*	Andes of n and central Peru (south to Huánuco)

☐ **Taczanowski's Tinamou** *Nothoprocta taczanowskii*

Andes of s-central Peru (Junín to Puno)

☐ **Kalinowski's Tinamou** *Nothoprocta kalinowskii*

Pacific Andes of Peru

☐ **White-bellied Nothura** *Nothura boraquira*

NE Brazil to e Bolivia and ne Paraguay

☐ **Lesser Nothura** *Nothura minor*

Semiarid grasslands and scrub of interior se Brazil

☐ **Darwin's Nothura** *Nothura darwinii*
____ *N. d. peruviana* Highlands of s Peru (Urubamba Valley of Cuzco)
____ *N. d. agassizii* Altiplano of extreme se Peru and w Bolivia
____ *N. d. boliviana* Highlands of w Bolivia (Cochabamba to Tarija)
____ *N. d. salvadorii* Semiarid subtropical w Argentina
____ *N. d. darwinii* Patagonian steppes of s-central Argentina

☐ **Spotted Nothura** *Nothura maculosa*
____ *N. m. cearensis* NE Brazil (s Ceará)
____ *N. m. major* Interior e-c Brazil (Minas Gerais, Goiás and adjacent Bahia)
____ *N. m. paludivaga* Central Paraguay and n-central Argentina
____ *N. m. maculosa* SE Brazil to e Paraguay, Uruguay and ne Argentina
____ *N. m. pallida* Moist *chaco* grasslands of nw Argentina
____ *N. m. annectens* Moist grasslands of e Argentina
____ *N. m. submontana* Andean foothills of sw Argentina (Neuquén to Chubut)
____ *N. m. nigroguttata* Plains of s-central Argentina (Río Negro to se Neuquén)

☐ **Chaco Nothura** *Nothura chacoensis*

Chaco of nw Paraguay and n-central Argentina

☐ **Dwarf Tinamou** *Taoniscus nanus*

Interior se Brazil and ne Argentina (Misiones)

☐ **Elegant Crested Tinamou** *Eudromia elegans*
____ *E. e. intermedia* Andes of nw Argentina (Salta to Catamarca)
____ *E. e. magnistriata* Andes of nw Argentina (Tucumán and n Córdoba)
____ *E. e. riojana* Andes of nw Argentina (La Rioja and San Juan)
____ *E. e. albida* Dry savanna of w Argentina (San Juan)
____ *E. e. wetmorei* Andean foothills in w Argentina (n-central Mendoza)
____ *E. e. devia* SW Argentina at base of Andes (Neuquén)
____ *E. e. numida* Dry grasslands of central Argentina
____ *E. e. multiguttata* Dry grasslands of e-central Argentina
____ *E. e. elegans* Central Argentina (Río Negro and Neuquén)
____ *E. e. patagonica* S Argentina (Neuquén to Santa Cruz) and adjacent s Chile

☐ **Quebracho Crested Tinamou** *Eudromia formosa*
____ *E. f. mira* Arid *chaco* of Paraguay and (?) possibly adjacent n Argentina
____ *E. f. formosa* Arid *quebracho* woodlands of n-central Argentina

☐ **Puna Tinamou** *Tinamotis pentlandii*

Andes of Peru to n Bolivia, Chile and nw Argentina

☐ **Patagonian Tinamou** *Tinamotis ingoufi*

Savanna of sw Argentina and s Chile

ORDER: SPHENISCIFORMES
FAMILY: SPHENISCIDAE (Penguins—17)

☐ **King Penguin** *Aptenodytes patagonicus*
____ *A. p. patagonicus* Staten I., South Georgia I. and Falkland Islands
____ *A. p. halli* Macquarie, Kerguelen, Crozet and Marion islands

☐ **Emperor Penguin** *Aptenodytes forsteri*

Antarctic continent and seas to edge of ice pack

☐ **Gentoo Penguin** *Pygoscelis papua*
____ *P. p. papua* Subantarctic regions south to ca 60°S
____ *P. p. ellsworthi* Antarctic Peninsula to South Sandwich Islands

☐ **Adelie Penguin** *Pygoscelis adeliae*

Circumpolar Antarctic seas to edge of ice pack

☐ **Chinstrap Penguin** *Pygoscelis antarcticus*

Circumpolar Antarctic seas and adjacent islands

☐ **Fiordland Penguin** *Eudyptes pachyrhynchus*

South I. (New Zealand) and adjacent subantarctic islands

☐ **Snares Penguin** *Eudyptes robustus*

Snares I. and adjacent waters off New Zealand

☐ **Erect-crested Penguin** *Eudyptes sclateri*

Subantarctic New Zealand and Australian waters

☐ **Rockhopper Penguin** *Eudyptes chrysocome*
_____ *E. c. chrysocome*
_____ *E. c. filholi*
_____ *E. c. moseleyi*

Cape Horn Archipelago and Falkland Islands
Kerguelen Islands and subantarctic New Zealand islands
Tristan da Cunha, Gough, St. Paul and Amsterdam islands

☐ **Royal Penguin** *Eudyptes schlegeli*

Macquarie I. and adjacent islets

☐ **Macaroni Penguin** *Eudyptes chrysolophus*

Subantarctic islands in s Atlantic Ocean and s Indian Ocean

☐ **Yellow-eyed Penguin** *Megadyptes antipodes*

Auckland I., Stewart I., Campbell I. and South I. (New Zealand)

☐ **Little Penguin** *Eudyptula minor*
_____ *E. m. novaehollandiae*
_____ *E. m. iredalei*
_____ *E. m. variabilis*
_____ *E. m. albosignata*
_____ *E. m. minor*
_____ *E. m. chathamensis*

Southern Australia and Tasmania
Northern North I. (New Zealand)
Southern North I. and Cook Strait (New Zealand)
Eastern South I. (New Zealand)
Western and southern South I. and Stewart I. (New Zealand)
Chatham Islands

☐ **Jackass Penguin** *Spheniscus demersus*

Coasts and islands off Namibia, South Africa and adj. waters

☐ **Humboldt Penguin** *Spheniscus humboldti*

Humboldt Current region of coastal n Peru to s Chile

☐ **Magellanic Penguin** *Spheniscus magellanicus*

Patagonian coasts, Staten, Falkland and Juan Fernández islands

☐ **Galapagos Penguin** *Spheniscus mendiculus*

Galapagos Islands (Fernandina and Isabela)

ORDER: GAVIIFORMES
FAMILY: GAVIIDAE (Loons—5)

☐ **Red-throated Loon** *Gavia stellata*

N Eurasia and n N Am.; winters to Caspian and Mediterranean

☐ **Arctic Loon** *Gavia arctica*

Arctic Eurasia and w Alaska; winters to s Palearctic region

☐ **Pacific Loon** *Gavia pacifica*

Coastal e Siberia and n N America; winters to Japan, s Baja

☐ **Common Loon** *Gavia immer*

W Palearctic and N America; winters to s US and s Palearctic

☐ **Yellow-billed Loon** *Gavia adamsii*

N Eurasia and n North America; winters to n Baja California

ORDER: PODICIPEDIFORMES
FAMILY: PODICIPEDIDAE (Grebes—19)

☐ **Little Grebe** *Tachybaptus ruficollis*
_____ *T. r. ruficollis*
_____ *T. r. iraquensis*
_____ *T. r. capensis*
_____ *T. r. poggei*

Europe east to Ural Mountains and nw Africa
Iraq and sw Iran
Caucasus to Myanmar and Sri Lanka; Africa s of the Sahara
SE to ne Asia, Hainan, Taiwan, Japan and s Kuril Islands

____	*T. r. philippensis*	N Philippine Islands
____	*T. r. cotabato*	Mindanao (s Philippines)
____	*T. r. tricolor*	Sulawesi to New Guinea and Lesser Sundas
____	*T. r. vulcanorum*	Java to Timor
____	*T. r. collaris*	NE New Guinea to Bougainville (Solomon Islands)

☐ **Australasian Grebe** *Tachybaptus novaehollandiae*

____	*T. n. novaehollandiae*	S New Guinea to Australia, Tasmania and New Zealand
____	*T. n. leucosternos*	Vanuatu and New Caledonia
____	*T. n. renellianus*	Rennell (Solomon Islands)
____	*T. n. javanicus*	Java
____	*T. n. timorensis*	Timor (e Lesser Sundas)
____	*T. n. fumosus*	Sangihi I. and Talaud Islands (off ne Sulawesi)
____	*T. n. incola*	N New Guinea

☐ **Madagascar Grebe** *Tachybaptus pelzelnii*

Madagascar

☐ **Least Grebe** *Tachybaptus dominicus*

____	*T. d. brachypterus*	S Texas to w-central Mexico and Panama
____	*T. d. bangsi*	W Mexico (southern half of Baja California and s Sonora)
____	*T. d. dominicus*	Cozumel I., Bahamas, Greater Antilles and Virgin Islands
____	*T. d. speciosus*	Tropical n South America to s Brazil and n Argentina

☐ **Pied-billed Grebe** *Podilymbus podiceps*

____	*P. p. podiceps*	Alaska to Panama and Cuba
____	*P. p. antillarum*	Greater and Lesser Antilles
____	*P. p. antarcticus*	Northern South America to s Argentina

☐ **White-tufted Grebe** *Rollandia rolland*

____	*R. r. morrisoni*	Andes of central Peru (Lake Junín)
____	*R. r. chilensis*	S Peru and s Brazil to Tierra del Fuego and Cape Horn Arch.
____	*R. r. rolland*	Falkland Islands

☐ **Short-winged Grebe** *Rollandia microptera*

Andes of s Peru and w Bolivia (Lake Titicaca basin)

☐ **Hoary-headed Grebe** *Poliocephalus poliocephalus*

Australia, Tasmania and South I. (New Zealand)

☐ **New Zealand Grebe** *Poliocephalus rufopectus*

North I. (New Zealand)

☐ **Great Grebe** *Podiceps major*

____	*P. m. major*	S Brazil to s Argentina and central Chile; coastal Peru
____	*P. m. navasi*	S Argentina and s Chile

☐ **Red-necked Grebe** *Podiceps grisegena*

____	*P. g. grisegena*	Locally in Eurasia
____	*P. g. holboellii*	North America and ne Asia

☐ **Great Crested Grebe** *Podiceps cristatus*

____	*P. c. cristatus*	Palearctic region and s Oriental region
____	*P. c. infuscatus*	Afrotropical region
____	*P. c. australis*	Australia, Tasmania and South I. (New Zealand)

☐ **Horned Grebe** *Podiceps auritus*

____	*P. a. auritus*	Locally in Palearctic region
____	*P. a. cornutus*	Locally in North America

☐ **Eared Grebe** *Podiceps nigricollis*

____	*P. n. nigricollis*	Locally in Eurasia
____	*P. n. gurneyi*	Africa south of the Sahara
____	*P. n. californicus*	Canada to Mexico; winters to Guatemala

☐ **Silvery Grebe** *Podiceps occipitalis*
_____ *P. o. juninensis*   Locally in Andes of Colombia to n Chile and Argentina
_____ *P. o. occipitalis*   Andes of c Argentina to Tierra del Fuego and Falkland Islands

☐ **Junin Grebe** *Podiceps taczanowskii*
         Andes of central Peru (Lake Junín)

☐ **Hooded Grebe** *Podiceps gallardoi*
         SW Argentina (w Santa Cruz); winter range unknown

☐ **Western Grebe** *Aechmophorus occidentalis*
_____ *A. o. occidentalis*   W North America (se Alaska to n Mexico and Baja Calif.)
_____ *A. o. ephemeralis*   Western Mexico

☐ **Clark's Grebe** *Aechmophorus clarkii*
_____ *A. c. clarkii*   W North America (se Alaska to n Mexico)
_____ *A. c. transitionalis*   Coastal w Mexico (Nayarit) and Mexican plateau

ORDER: PROCELLARIIFORMES
FAMILY: DIOMEDEIDAE (Albatrosses—13)

☐ **Wandering Albatross** *Diomedea exulans*
_____ *D. e. exulans*   Southern oceans in South Georgia area
_____ *D. e. dabbenena*   Tristan da Cunha and Gough islands
_____ *D. e. antipodensis*   Antipodes Islands
_____ *D. e. gibsoni*   Marion and Crozet islands
_____ *D. e. amsterdamensis*   Breeds Amsterdam I. (French subantarctic islands)

☐ **Royal Albatross** *Diomedea epomophora*
_____ *D. e. epomophora*   Campbell and Auckland islands; ranges circumpolar s oceans
_____ *D. e. sanfordi*   Chatham Islands and New Zealand; ranges circumpolar s oceans

☐ **Short-tailed Albatross** *Phoebastria albatrus*
         Breeds Torishima (Izu Islands); ranges at sea through n Pacific

☐ **Waved Albatross** *Phoebastria irrorata*
         Breeds Hood (Galapagos Is.) and Isla La Plata off Ecuador

☐ **Laysan Albatross** *Phoebastria immutabilis*
         Breeds w Hawaiian and Revillagigedo islands; ranges n Pacific

☐ **Black-footed Albatross** *Phoebastria nigripes*
         W Hawaiian, Izu, Bonin and s Ryukyu islands

☐ **Gray-headed Albatross** *Thalassarche chrysostoma*
         Circumpolar high s latitudes; ranges s oceans north to 35°S

☐ **Black-browed Albatross** *Thalassarche melanophris*
_____ *T. m. impavida*   Campbell Islands and adjacent islands off New Zealand
_____ *T. m. melanophris*   Cape Horn Archipelago to Antipodes Islands

☐ **Buller's Albatross** *Thalassarche bulleri*
         Breeds islands off N. Zealand; ranges s. Pacific

☐ **Shy Albatross** *Thalassarche cauta*
_____ *T. c. cauta*   Breeds Tasmania and adjacent islands
_____ *T. c. steadi*   Breeds Auckland Islands
_____ *T. c. eremita*   Chatham Islands
_____ *T. c. salvini*   Crozet, Snares and Bounty islands

☐ **Yellow-nosed Albatross** *Thalassarche chlororhynchos*
_____ *T. c. chlororhynchos*   Tristan da Cunha and Gough islands.; ranges southern oceans
_____ *T. c. bassi (carteri)*   Breeds on s Indian Ocean islands; ranges southern oceans

☐ **Sooty Albatross** *Phoebetria fusca*
         S Atlantic Ocean and Indian Ocean north to about 30°S

☐ **Light-mantled Albatross** *Phoebetria palpebrata*

Circumpolar subantarctic islands; ranges north to 35°S

ORDER: PROCELLARIIFORMES
FAMILY: PROCELLARIIDAE (Shearwaters and Petrels—77)

☐ **Antarctic Giant Petrel** *Macronectes giganteus*

Circumpolar southern oceans south to the pack ice

☐ **Hall's Giant Petrel** *Macronectes halli*

Southern oceans, generally north of Antarctic convergence

☐ **Northern Fulmar** *Fulmarus glacialis*
_____ *F. g. glacialis*
_____ *F. g. auduboni*
_____ *F. g. rodgersii*

Breeds high Arctic regions of North Atlantic; ranges widely
Breeds low Arctic and boreal North Atlantic; ranges widely
Breeds coasts of e Siberia and Alaskan peninsula; ranges widely

☐ **Southern Fulmar** *Fulmarus glacialoides*

Antarctic circumpolar; ranges widely southern oceans

☐ **Antarctic Petrel** *Thalassoica antarctica*

Breeds Antarctic islands and coasts; ranges southern oceans

☐ **Cape Petrel** *Daption capense*
_____ *D. c. capense*
_____ *D. c. australe*

Breeds circumpolar subantarctic islands; ranges southern oceans
Breeds New Zealand subantarctic islands; ranges southern oceans

☐ **Snow Petrel** *Pagodroma nivea*
_____ *P. n. nivea*
_____ *P. n. confusa*

South Georgia and adj. islands, Scotia Arc and Antarctic Pen.
South Sandwich Islands and Géologie Archipelago

☐ **Great-winged Petrel** *Pterodroma macroptera*
_____ *P. m. macroptera*
_____ *P. m. gouldi*

Breeds and ranges islands and seas in southern oceans
Breeds on islands off North I. (New Zealand) and sw Australia

☐ **Mascarene Petrel** *Pterodroma aterrima*

Réunion (Mascarene Islands); ranges adjacent Indian Ocean

☐ **Tahiti Petrel** *Pterodroma rostrata*
_____ *P. r. rostrata*
_____ *P. r. becki*
_____ *P. r. trouessarti*

Breeds Marquesas and Society islands; confined to trop. Pacific
Known from two 1928 specimens from Rendova (Solomon Is.)
Breeds New Caledonia; ranges s Pacific Ocean

☐ **White-headed Petrel** *Pterodroma lessonii*

Islands in s Indian Ocean and s Pacific Ocean

☐ **Black-capped Petrel** *Pterodroma hasitata*
_____ *P. h. hasitata*
_____ *P. h. caribbaea†*

Cuba, Hispaniola, Guadeloupe and Dominica ; ranges w Atlantic
Formerly Jamaica. Extinct ca 1936

☐ **Bermuda Petrel** *Pterodroma cahow*

Breeds Nonsuch I. (Bermuda); disperses to Gulf Stream

☐ **Atlantic Petrel** *Pterodroma incerta*

Tristan da Cunha and Gough islands; ranges s Atlantic Ocean

☐ **Phoenix Petrel** *Pterodroma alba*

Breeds French Polynesia to Kermadec I.; ranges s Pacific waters

☐ **Mottled Petrel** *Pterodroma inexpectata*

Breeds Stewart I., Snares Islands and sw South I. (New Zealand)

☐ **Providence Petrel** *Pterodroma solandri*

Breeds Lord Howe I. and Philip I.; ranges to nw Pacific

☐ **Murphy's Petrel** *Pterodroma ultima*

Tuamotu Archipelago, Austral Islands and Pitcairn I.

☐ **Kermadec Petrel** *Pterodroma neglecta*
_____ *P. n. neglecta*
_____ *P. n. juana*

Breeds South Pacific islands from New Zealand to Easter I.
Juan Fernández, San Ambrosio and San Félix islands (off Chile)

☐ **Magenta Petrel** *Pterodroma magentae*

Chatham Islands (population ±50 birds in 1992)

☐ **Herald Petrel** *Pterodroma arminjoniana*
_____ *P. a. arminjoniana*
_____ *P. a. heraldica*

Breeds Trindade I. and Martín Vaz I. (s Atlantic Ocean)
Breeds Raine I., Tonga and French Polynesia to Easter I.

☐ **Henderson Petrel** *Pterodroma atrata*

Breeds Henderson I. (se Pacific); >unknown

☐ **Soft-plumaged Petrel** *Pterodroma mollis*
_____ *P. m. mollis*
_____ *P. m. dubia*

Gough, Tristan da Cunha and Antipodes islands
Marion, Crozet, Kerguelen and Amsterdam islands

☐ **Cape Verde (Fea's) Petrel** *Pterodroma feae*

Cape Verde and Desertas islands; ranges e Atlantic Ocean

☐ **Madeira (Zino's) Petrel** *Pterodroma madeira*

Highlands of Madeira I.; ranges e Atlantic Ocean

☐ **Barau's Petrel** *Pterodroma baraui*

Breeds Réunion I. and Rodrigues I.; pelagic range unknown

☐ **Galapagos Petrel** *Pterodroma phaeopygia*

Breeds Galapagos Islands; ranges Clipperton I. to n Peru

☐ **Hawaiian Petrel** *Pterodroma sandwichensis*

Hawaiian Islands; ranges south to Polynesia

☐ **Juan Fernandez Petrel** *Pterodroma externa*

Alejandro Selkirk I. (Juan Fernández Islands off Chile)

☐ **White-necked Petrel** *Pterodroma cervicalis*

Breeds Kermadec Islands; ranges s Pacific Ocean

☐ **Cook's Petrel** *Pterodroma cookii*

Breeds islands off New Zealand; ranges to e and n Pacific

☐ **Defilippe's Petrel** *Pterodroma defilippiana*

Juan Fernández, San Ambrosio and San Félix islands (off Chile)

☐ **Gould's Petrel** *Pterodroma leucoptera*
_____ *P. l. leucoptera*
_____ *P. l. caledonica*
_____ *P. l. brevipes*

Breeds Cabbage Tree Is. (off e Australia); ranges s Pacific Ocean
New Caledonia
Fiji and Cook Islands

☐ **Bonin Petrel** *Pterodroma hypoleuca*

Volcano, Bonin and w Hawaiian islands; ranges to Polynesia

☐ **Black-winged Petrel** *Pterodroma nigripennis*

Breeds sw Pacific; ranges s-central Pacific Ocean

☐ **Chatham Petrel** *Pterodroma axillaris*

Rangitira I. (Chatham Islands) and adjacent seas

☐ **Stejneger's Petrel** *Pterodroma longirostris*

Alejandro Selkirk I. (off Chile); ranges e and n Pacific

☐ **Pycroft's Petrel** *Pterodroma pycrofti*

Breeds small islands off New Zealand coast; ranges to n Pacific

☐ **Fiji Petrel** *Pterodroma macgillivrayi*

Breeds Gau I. (Fiji). Status uncertain

☐ **Vanuatu Petrel** *Pterodroma occulta*

Presumed to breed on Banks Is. off n Vanuatu

☐ **Blue Petrel** *Halobaena caerulea*

Islands in subantarctic southern oceans and islands off Cape Horn

☐ **Broad-billed Prion** *Pachyptila vittata*

Breeds islands off New Zealand and Tristan da Cunha group

☐ **Salvin's Prion** *Pachyptila salvini*
_____ *P. s. salvini*
_____ *P. s. macgillivrayi*

Prince Edward and Crozet islands
Amsterdam and St. Paul islands

☐ **Antarctic Prion** *Pachyptila desolata*
_____ *P. d. desolata*

Crozet, Kerguelen and Macquarie islands

____	*P. d. altera*	Auckland and Heard islands
____	*P. d. banksi*	Scotia Arc, South Georgia, South Sandwich and Scott islands

☐ **Slender-billed Prion** *Pachyptila belcheri*

Crozet, Kerguelen and Falkland islands; Noir I. (off s Chile)

☐ **Fulmar Prion** *Pachyptila crassirostris*

____	*P. c. eatoni*	Breeds Heard and Auckland islands (New Zealand)
____	*P. c. crassirostris*	Snares, Bounty and Chatham islands (New Zealand)

☐ **Fairy Prion** *Pachyptila turtur*

Breeds scattered subtropical and subantarctic islands

☐ **Bulwer's Petrel** *Bulweria bulwerii*

Azores to Cape Verde, Johnston and nw Hawaiian islands

☐ **Jouanin's Petrel** *Bulweria fallax*

NW Indian Ocean and s Arabian Sea; breeding grounds unknown

☐ **Gray Petrel** *Procellaria cinerea*

Breeds and ranges circumpolar subantarctic seas

☐ **White-chinned Petrel** *Procellaria aequinoctialis*

____	*P. a. aequinoctialis*	Circumpolar subantarctic islands
____	*P. a. conspicillata*	Inaccessible I. (Tristan da Cunha)

☐ **Parkinson's Petrel** *Procellaria parkinsoni*

Little and Great Barrier is. (New Zealand); ranges to S America

☐ **Westland Petrel** *Procellaria westlandica*

South I. (New Zealand); disperses to Australia and w S. America

☐ **Kerguelen Petrel** *Aphrodroma brevirostris*

Tristan da Cunha, Gough, Prince Edward, Crozet and Kerguelen is.

☐ **Streaked Shearwater** *Calonectris leucomelas*

Breeds coastal islands off Japan and China; ranges to s Pacific

☐ **Cory's Shearwater** *Calonectris diomedea*

____	*C. d. diomedea*	Breeds Mediterranean islands
____	*C. d. borealis*	Breeds Azores, Madeira, Canary and Berlenga islands

☐ **Cape Verde Shearwater** *Calonectris edwardsii*

Breeds Cape Verde Islands

☐ **Pink-footed Shearwater** *Puffinus creatopus*

Mocha and Juan Fernández islands off Chile; ranges to n Pacific

☐ **Flesh-footed Shearwater** *Puffinus carneipes*

S Indian and sw Pacific Oceans; winters to Arabian Sea, nw Pacific

☐ **Greater Shearwater** *Puffinus gravis*

Islands in s Atlantic Ocean; ranges n Atlantic to Arctic Circle

☐ **Wedge-tailed Shearwater** *Puffinus pacificus*

Widespread tropical Pacific and Indian oceans

☐ **Buller's Shearwater** *Puffinus bulleri*

Breeds islands off New Zealand; wide transpacific dispersal

☐ **Sooty Shearwater** *Puffinus griseus*

S S. Am., New Zealand, se Australia; winters to n Pacific, n Atlantic

☐ **Short-tailed Shearwater** *Puffinus tenuirostris*

S Australia and Tasmania; winters to n Pacific

☐ **Christmas Shearwater** *Puffinus nativitatis*

Widespread throughout tropical central Pacific Ocean

☐ **Manx Shearwater** *Puffinus puffinus*

Breeds n Atlantic; ranges to Argentina and s African waters

☐ **Levantine Shearwater** *Puffinus yelkouan*

Breeds central and e Mediterranean islands

☐ **Balearic Shearwater** *Puffinus mauretanicus*

Breeds Balearic Islands; ranges to coastal n Europe

☐ **Hutton's Shearwater** *Puffinus huttoni*

Breeds ne South I. (New Zealand); ranges to Australia

☐ **Black-vented Shearwater** *Puffinus opisthomelas*

Islands off w coast of Baja Calif.; disperses adj. Mexican waters

☐ **Townsend's Shearwater** *Puffinus auricularis*

____ *P. a. newelli* — Breeds Kauai (Hawaiian Islands); dispersal unknown
____ *P. a. auricularis* — Breeds Revillagigedo Islands (off w Mexico); disperses to 8°N

☐ **Fluttering Shearwater** *Puffinus gavia*

Breeds is. off New Zealand; ranges to se Australia and Vanuatu

☐ **Little Shearwater** *Puffinus assimilis*

____ *P. a. baroli* — Azores, Desertas, Salvage and Canary islands
____ *P. a. boydi* — Cape Verde Islands
____ *P. a. tunneyi* — Islands off sw Australia (Abrolhos Islands to Récherche Arch.)
____ *P. a. assimilis* — Norfolk and Lord Howe islands
____ *P. a. kermadecensis* — Kermadec Islands
____ *P. a. haurakiensis* — Islets off ne coast of North I. (New Zealand)
____ *P. a. elegans* — Tristan da Cunha, Gough, Chatham and Antipodes islands
____ *P. a. myrtae* — Rapa I. (Austral Islands)

☐ **Audubon's Shearwater** *Puffinus lherminieri*

____ *P. l. lherminieri* — Breeds Bahamas and West Indies; formerly Bermuda
____ *P. l. loyemilleri* — Islets in sw Caribbean
____ *P. l. subalaris* — Breeds Galapagos Islands
____ *P. l. dichrous* — Islands throughout central Pacific (Samoa to Marquesas Islands)
____ *P. l. gunax* — Breeds Banks Group (Vanuatu)
____ *P. l. bannermani* — Breeds Bonin and Volcano islands (off Japan)
____ *P. l. bailloni* — Mascarene Islands
____ *P. l. nicolae* — NW Indian Ocean (Aldabra to Seychelles and Maldives)
____ *P. l. temptator* — Mohéli I. (Comoro Islands)

☐ **Persian Shearwater** *Puffinus persicus*

Breeds and ranges Arabian Sea and adjacent waters

☐ **Heinroth's Shearwater** *Puffinus heinrothi*

New Britain and Solomon Islands

☐ **Mascarene Shearwater** *Puffinus atrodorsalis*

Southwestern Indian Ocean

ORDER: PROCELLARIIFORMES
FAMILY: HYDROBATIDAE (Storm-Petrels—20)

☐ **Gray-backed Storm-Petrel** *Garrodia nereis*

Circumpolar subantarctic waters, north to about 35°S

☐ **Wilson's Storm-Petrel** *Oceanites oceanicus*

____ *O. o. oceanicus* — Subantarctic islands from Cape Horn to Kerguelen Islands
____ *O. o. exasperatus* — South Shetland Is., South Sandwich Is. and adj. Antarctic coast

☐ **White-vented Storm-Petrel** *Oceanites gracilis*

____ *O. g. galapagoensis* — Galapagos Islands (breeding grounds unknown)
____ *O. g. gracilis* — Coast of Ecuador to Chile (breeding grounds unknown)

☐ **White-faced Storm-Petrel** *Pelagodroma marina*

____ *P. m. hypoleuca* — Salvage I. (North Atlantic Ocean)
____ *P. m. eadesi* — Cape Verde Islands
____ *P. m. marina* — Tristan da Cunha and Gough islands
____ *P. m. dulciae* — Breeds islands off s Australia and Tasmania
____ *P. m. maoriana* — Stewart, Auckland, Chatham and islands off New Zealand
____ *P. m. albiclunis* — Kermadec Islands

☐ **Black-bellied Storm-Petrel** *Fregetta tropica*
_____ *F. t. tropica* — Subantarctic circumpolar islands; ranges north to tropics
_____ *F. t. melanoleuca* — Tristan da Cunha and Gough islands

☐ **White-bellied Storm-Petrel** *Fregetta grallaria*
_____ *F. g. leucogaster* — Tristan da Cunha, Gough, Amsterdam and St. Paul islands
_____ *F. g. grallaria* — Lord Howe and Kermadec islands
_____ *F. g. segethi* — Juan Fernández Islands (off Chile)
_____ *F. g. titan* — Rapa I. (Austral Islands)

☐ **Polynesian Storm-Petrel** *Nesofregetta fuliginosa* — Tropical central and w Pacific Ocean

☐ **European Storm-Petrel** *Hydrobates pelagicus* — E Atlantic and Mediterranean; ranges to Indian Ocean

☐ **Least Storm-Petrel** *Oceanodroma microsoma* — Breeds islands off Baja Calif.; ranges south to extreme n Peru

☐ **Wedge-rumped Storm-Petrel** *Oceanodroma tethys*
_____ *O. t. tethys* — Galapagos Islands (Pitt, Tower and Redonda)
_____ *O. t. kelsalli* — Islas Pescadores and San Gallán (off coast of Peru)

☐ **Band-rumped Storm-Petrel** *Oceanodroma castro* — Breeds and ranges tropical Atlantic and Pacific Oceans

☐ **Leach's Storm-Petrel** *Oceanodroma leucorhoa*
_____ *O. l. leucorhoa* — N Atlantic; Japan to Aleutians and islands off n Mexico
_____ *O. l. willeti* — Coronados Islands (off w Mexico)
_____ *O. l. chapmani* — San Benito Islands (off w Mexico)
_____ *O. l. socorroensis* — Summer breeder on Guadalupe I. (off w Mexico)
_____ *O. l. cheimomnestes* — Winter breeder on Guadalupe I. (off w Mexico)

☐ **Swinhoe's Storm-Petrel** *Oceanodroma monorhis* — Breeds islands off Japan; disperses to n Indian Ocean

☐ **Tristram's Storm-Petrel** *Oceanodroma tristrami* — Breeds and ranges Hawaii, Izu and Volcano islands

☐ **Markham's Storm-Petrel** *Oceanodroma markhami* — Breeds Paracas Peninsula (Peru); ranges to s Mexico and Chile

☐ **Matsudaira's Storm-Petrel** *Oceanodroma matsudairae* — Breeds Volcano Islands (Japan); disperses to Indian Ocean

☐ **Black Storm-Petrel** *Oceanodroma melania* — Breeds islands off California and Baja; ranges to Peru

☐ **Ashy Storm-Petrel** *Oceanodroma homochroa* — Breeds islands off California; ranges to s Baja California

☐ **Ringed Storm-Petrel** *Oceanodroma hornbyi* — Ranges coastal Ecuador to central Chile

☐ **Fork-tailed Storm-Petrel** *Oceanodroma furcata*
_____ *O. f. furcata* — N Kuril, Komandorskiye and Aleutian islands
_____ *O. f. plumbea* — Islands off s Alaska to n California

ORDER: PROCELLARIIFORMES
FAMILY: PELECANOIDIDAE (Diving-Petrels—4)

☐ **Peruvian Diving-Petrel** *Pelecanoides garnotii* — Arid coasts of Peru and n Chile

☐ **Magellanic Diving-Petrel** *Pelecanoides magellani* — Islands and fiords of s Chile, s Argentina and extreme se Brazil

☐ **South Georgia Diving-Petrel** *Pelecanoides georgicus* — Subantarctic circumpolar regions

☐ **Common Diving-Petrel** *Pelecanoides urinatrix*
_____ *P. u. berard* — Falkland Islands

___	*P. u. dacunhae*	Tristan da Cunha and Gough islands
___	*P. u. exsul*	South Georgia I. east to Antipodes Islands
___	*P. u. urinatrix*	Tasmania, New Zealand and islands in Bass Strait
___	*P. u. chathamensis*	Chatham and Snares islands (off New Zealand)
___	*P. u. copperingeri*	Southern Chile

ORDER: PELECANIFORMES
FAMILY: PHAETHONTIDAE (Tropicbirds—3)

☐ **Red-billed Tropicbird** *Phaethon aethereus*

___	*P. a. mesonauta*	Subtropical and tropical e Pacific, Caribbean and e Atlantic
___	*P. a. aethereus*	Fernando de Noronha, Ascension and St. Helena is. (s Atlantic)
___	*P. a. indicus*	Red Sea, Persian Gulf and Gulf of Aden

☐ **Red-tailed Tropicbird** *Phaethon rubricauda*

___	*P. r. melanorhynchos*	Breeds and disperses widely in tropical Pacific Ocean
___	*P. r. roseotinctus*	Breeds sw Pacific islands
___	*P. r. rubricauda*	Breeds islands in w Indian Ocean
___	*P. r. westralis*	Islands in e Indian Ocean and Easter I.

☐ **White-tailed Tropicbird** *Phaethon lepturus*

___	*P. l. lepturus*	Islands in Indian Ocean
___	*P. l. fulvus*	Christmas I. (Indian Ocean)
___	*P. l. europae*	Europa I. (s Mozambique Channel)
___	*P. l. dorothea*	Islands in tropical w Pacific (Hawaii to New Caledonia)
___	*P. l. catesbyi*	Breeds islands in tropical Atlantic Ocean
___	*P. l. ascensionis*	Fernando de Noronha and Ascension islands

ORDER: PELECANIFORMES
FAMILY: PELECANIDAE (Pelicans—8)

☐ **Great White Pelican** *Pelecanus onocrotalus*

Locally in s-central Eurasia, s Asia and Africa

☐ **Pink-backed Pelican** *Pelecanus rufescens*

Locally in Africa south of the Sahara and Madagascar

☐ **Spot-billed Pelican** *Pelecanus philippensis*

Lowlands of India to SE Asia and Philippines

☐ **Dalmatian Pelican** *Pelecanus crispus*

Breeds s Eurasia; winters to India

☐ **Australian Pelican** *Pelecanus conspicillatus*

Australia and Tasmania; winters to New Guinea region

☐ **American White Pelican** *Pelecanus erythrorhynchos*

S Canada to s US; winters to Costa Rica

☐ **Peruvian Pelican** *Pelecanus thagus*

Pacific coast of s Ecuador to s Chile

☐ **Brown Pelican** *Pelecanus occidentalis*

___	*P. o. occidentalis*	West Indies and Caribbean to islands off Venezuela
___	*P. o. carolinensis*	Locally on Atlantic coast of tropical America
___	*P. o. californicus*	Anacapa I. and islands off Baja and in Gulf of California
___	*P. o. murphyi*	Pacific coast of nw South America from Colombia to Peru
___	*P. o. urinator*	Galapagos Islands

ORDER: PELECANIFORMES
FAMILY: SULIDAE (Gannets and Boobies—10)

☐ **Northern Gannet** *Morus bassanus*

Breeds n Atlantic coasts; ranges to coastal nw Africa

☐ **Cape Gannet** *Morus capensis*

Breeds islands off s Africa; disperses to coastal Mozambique

☐ **Australian Gannet** *Morus serrator*

Breeds is. off se Australia, Tasmania, coasts and is. off N. Zealand

☐ **Abbott's Booby** *Sula abbotti*

Christmas I. (e Indian Ocean)

☐ **Blue-footed Booby** *Sula nebouxii*
____ *S. n. nebouxii*　　Pacific coast of Mexico to Peru
____ *S. n. excisa*　　Galapagos Islands

☐ **Peruvian Booby** *Sula variegata*

Coastal sw Colombia to s Chile

☐ **Masked Booby** *Sula dactylatra*
____ *S. d. personata*　　Islands in central and w Pacific to islands off w Australia
____ *S. d. fullagari*　　Islands in n Tasman Sea
____ *S. d. dactylatra*　　Breeds islands in Caribbean and sw Atlantic Ocean
____ *S. d. melanops*　　Breeds islands in w Indian Ocean

☐ **Nazca Booby** *Sula granti*

Breeds islands in Galapagos and off w Mexico

☐ **Red-footed Booby** *Sula sula*
____ *S. s. sula*　　Breeds islands in Caribbean and off Brazil
____ *S. s. rubripes*　　Breeds islands in tropical Pacific and Indian oceans
____ *S. s. websteri*　　Islands off w Mexico, Central America and Galapagos Islands

☐ **Brown Booby** *Sula leucogaster*
____ *S. l. brewsteri*　　Islands in Gulf of California and off w Mexico
____ *S. l. etesiaca*　　Islands off Central America and Colombia
____ *S. l. leucogaster*　　Islands in Gulf of Mexico, Caribbean and tropical Atlantic
____ *S. l. plotus*　　Islands in Red Sea, tropical Indian Ocean and s China Sea

ORDER: PELECANIFORMES
FAMILY: PHALACROCORACIDAE (Cormorants—39)

☐ **Little Black Cormorant** *Phalacrocorax sulcirostris*

Australasian region to Malay Archipelago

☐ **Double-crested Cormorant** *Phalacrocorax auritus*
____ *P. a. cincinatus*　　Aleutian Islands across Gulf of Alaska to Yakutat Peninsula
____ *P. a. albociliatus*　　SW British Columbia to Gulf of California
____ *P. a. auritus*　　Gulf of St. Lawrence to Cape Cod and locally west to Utah
____ *P. a. floridanus*　　Coastal North Carolina to Florida, Bahamas and Cuba

☐ **Indian Cormorant** *Phalacrocorax fuscicollis*

Lowlands of India and SE Asia

☐ **Neotropic Cormorant** *Phalacrocorax brasilianus*
____ *P. b. mexicanus*　　Extreme s US to Nicaragua, Bahamas, Cuba and Isle of Pines
____ *P. b. brasilianus*　　Costa Rica s through South America to Tierra del Fuego

☐ **Great Cormorant** *Phalacrocorax carbo*
____ *P. c. carbo*　　N Europe and n N America; winters to Gulf coast, nw Africa
____ *P. c. sinensis*　　N-central Europe to s China; winters to SE Asia and Indonesia
____ *P. c. hanedae*　　Honshu I. (Japan)
____ *P. c. maroccanus*　　Coastal nw Africa (Morocco to Mauritania)
____ *P. c. lucidus*　　Africa south of the Sahara and Cape Verde Islands
____ *P. c. novaehollandiae*　　Australia, Tasmania, New Zealand and Chatham Islands

☐ **Cape Cormorant** *Phalacrocorax capensis*

Coastal sw Namibia and s South Africa

16

☐ **Socotra Cormorant** *Phalacrocorax nigrogularis*

Seacoasts and islands in Persian Gulf

☐ **Bank Cormorant** *Phalacrocorax neglectus*

Coastal sw Africa (Namibia to sw Cape Province)

☐ **Japanese Cormorant** *Phalacrocorax capillatus*

Rocky seacoasts and islands of ne Asia

☐ **Brandt's Cormorant** *Phalacrocorax penicillatus*

Coastal s Alaska to Baja California

☐ **European Shag** *Phalacrocorax aristotelis*
____ *P. a. aristotelis*
____ *P. a. desmarestii*
____ *P. a. riggenbachi*

Iceland and n Scandinavia south to Iberian Peninsula
Mediterranean coasts and islands
W coast of Morocco (Casablanca to Puerto Cansado)

☐ **Pelagic Cormorant** *Phalacrocorax pelagicus*
____ *P. p. pelagicus*
____ *P. p. resplendens*

Coastal ne Asia, Bering Sea and Arctic Ocean islands
Coastal sw British Columbia to s Baja California

☐ **Red-faced Cormorant** *Phalacrocorax urile*

Islands off n Japan to coastal s Alaska

☐ **Rock Shag** *Phalacrocorax magellanicus*

Coasts of Chile, Argentina and Falkland Islands

☐ **Guanay Cormorant** *Phalacrocorax bougainvillii*

Seacoasts and islands off Peru and Chile

☐ **Pied Cormorant** *Phalacrocorax varius*
____ *P. v. hypoleucos*
____ *P. v. varius*

Coastal and interior Australia; rare vagrant to Tasmania
Coastal New Zealand and Stewart I.

☐ **Black-faced Cormorant** *Phalacrocorax fuscescens*

Coastal s Australia, Tasmania and islands in Bass Strait

☐ **Rough-faced Shag** *Phalacrocorax carunculatus*

Islands in Cook Strait (New Zealand)

☐ **Bronze Shag** *Phalacrocorax chalconotus*

Coasts of Otago, Stewart and South islands (New Zealand)

☐ **Chatham Islands Shag** *Phalacrocorax onslowi*

Chatham Islands

☐ **Auckland Islands Shag** *Phalacrocorax colensoi*

Auckland Islands

☐ **Campbell Islands Shag** *Phalacrocorax campbelli*

Campbell Islands

☐ **Bounty Islands Shag** *Phalacrocorax ranfurlyi*

Bounty Islands (New Zealand); vagrant to Antipodes Islands

☐ **Antarctic Shag** *Phalacrocorax bransfieldensis*

South Shetland Islands and Antarctic Peninsula

☐ **South Georgia Shag** *Phalacrocorax georgianus*

South Georgia, South Sandwich and South Orkney islands

☐ **Imperial Shag** *Phalacrocorax atriceps*
____ *P. a. atriceps*
____ *P. a. albiventer*

Islands and coasts of s Argentina and Chile
Falkland Islands

☐ **Heard Island Shag** *Phalacrocorax nivalis*

Heard I. (s Indian Ocean)

☐ **Crozet Shag** *Phalacrocorax melanogenis*

Prince Edward, Marion and Crozet islands

☐ **Kerguelen Shag** *Phalacrocorax verrucosus*

Kerguelen I. (s Indian Ocean)

☐ **Macquarie Shag** *Phalacrocorax purpurascens*

Macquarie I. and adjacent Bishop and Clerk Rocks

☐ **Red-legged Cormorant** *Phalacrocorax gaimardi*

Coastal Peru and Chile; isolated population in s Argentina

☐ **Spotted Shag** *Phalacrocorax punctatus*
_____ *P. p. punctatus* — North I. and South I. (New Zealand)
_____ *P. p. oliveri* — Stewart I. and adjacent w coast of South I. (New Zealand)

☐ **Pitt Island Shag** *Phalacrocorax featherstoni*

Chatham Islands (New Zealand)

☐ **Little Pied Cormorant** *Phalacrocorax melanoleucos*
_____ *P. m. melanoleucos* — Lesser Sundas to Solomon Islands, Australia and Tasmania
_____ *P. m. brevicauda* — Rennel I. (Solomon Islands)
_____ *P. m. brevirostris* — New Zealand, Stewart and Campbell islands

☐ **Long-tailed Cormorant** *Phalacrocorax africanus*
_____ *P. a. africanus* — Africa south of the Sahara
_____ *P. a. pictilis* — Madagascar

☐ **Crowned Cormorant** *Phalacrocorax coronatus*

Coastal sw Africa (Angola to South Africa)

☐ **Little Cormorant** *Phalacrocorax niger*

Lowlands of India to SE Asia and n Java

☐ **Pygmy Cormorant** *Phalacrocorax pygmaeus*

Inland lakes and rivers of se Europe to central Asia

☐ **Flightless Cormorant** *Phalacrocorax harrisi*

Galapagos Islands (coasts of Fernandina and Isabela)

ORDER: PELECANIFORMES
FAMILY: ANHINGIDAE (Anhingas—2)

☐ **Anhinga** *Anhinga anhinga*
_____ *A. a. leucogaster* — SE US to Panama, Cuba and Isle of Pines
_____ *A. a. anhinga* — Trinidad, Tobago and n South America to n Argentina

☐ **Darter** *Anhinga melanogaster*
_____ *A. m. rufa* — Africa south of the Sahara and Middle East
_____ *A. m. vulsini* — Madagascar
_____ *A. m. melanogaster* — India to SE Asia, Malay Archipelago and Philippine Islands
_____ *A. m. novaehollandiae* — Australia to Lesser Sundas, Moluccas and New Guinea

ORDER: PELECANIFORMES
FAMILY: FREGATIDAE (Frigatebirds—5)

☐ **Ascension Island Frigatebird** *Fregata aquila*

Breeds Ascension I.; ranges to w African coast

☐ **Christmas Island Frigatebird** *Fregata andrewsi*

Christmas I.; ranges to s China Sea and Australia

☐ **Magnificent Frigatebird** *Fregata magnificens*

Tropical w Atlantic and e Pacific oceans

☐ **Great Frigatebird** *Fregata minor*
_____ *F. m. palmerstoni* — Breeds islands in w and central Pacific
_____ *F. m. ridgwayi* — Breeds e Pacific on Revillagigedo, Cocos and Galapagos islands
_____ *F. m. nicolli* — Breeds Trindade I. and Martín Vaz I.; ranges to Brazil
_____ *F. m. aldabrensis* — Breeds w Indian Ocean on Aldabra and adjacent islands
_____ *F. m. minor* — Cocos and Christmas is. (Indian Ocean); Paracel Is. (S China Sea)

☐ **Lesser Frigatebird** *Fregata ariel*
_____ *F. a. ariel* — Islands in Indian and Pacific oceans
_____ *F. a. trinitatis* — Trindade I. and Martín Vaz I.; wanders to coastal Brazil
_____ *F. a. iredalei* — Mascarene Islands; disperses to coasts of India and Somalia

ORDER: CICONIIFORMES
FAMILY: ARDEIDAE (Herons, Egrets and Bitterns—63)

☐ **Whistling Heron** *Syrigma sibilatrix*
_____ *S. s. fostersmithi* E Colombia and Venezuela
_____ *S. s. sibilatrix* Wet grasslands of Bolivia to se Brazil and ne Argentina

☐ **Capped Heron** *Pilherodius pileatus*

Lowlands of e Panama to the Guianas, Brazil and n Paraguay

☐ **Gray Heron** *Ardea cinerea*
_____ *A. c. cinerea* Eurasia to Manchuria, India, Africa and Comoro Islands
_____ *A. c. jouyi* Japan, China, Indochina, Malaya, Sumatra and Java
_____ *A. c. firasa* Madagascar
_____ *A. c. monicae* Islands off Banc d'Arguin (Mauritania)

☐ **Great Blue Heron** *Ardea herodias*
_____ *A. h. fannini* SE Alaska to coastal Washington
_____ *A. h. herodias* S Canada to s Baja California and Central America
_____ *A. h. wardi* S-central US to Gulf Coast and Florida
_____ *A. h. occidentalis* S Florida through West Indies to islands off Venezuela
_____ *A. h. cognata* Galapagos Islands

☐ **Cocoi Heron** *Ardea cocoi*

Widespread South America (excluding the Andes)

☐ **Pacific Heron** *Ardea pacifica*

Wetlands of Australia and Tasmania; straggler to s New Guinea

☐ **Black-headed Heron** *Ardea melanocephala*

Grasslands and savanna of Africa south of the Sahara

☐ **Humblot's Heron** *Ardea humbloti*

Aquatic lowlands of Madagascar and Comoro Islands

☐ **White-bellied Heron** *Ardea insignis*

Himalayan foothills (Nepal to ne India and Myanmar)

☐ **Great-billed Heron** *Ardea sumatrana*
_____ *A. s. sumatrana* Coasts of SE Asia, Indonesia, Philippines and New Guinea
_____ *A. s. mathewsae* Coasts of tropical n Australia

☐ **Goliath Heron** *Ardea goliath*

Locally in Africa, Iraq and Iran; casual to India and Sri Lanka

☐ **Purple Heron** *Ardea purpurea*
_____ *A. p. purpurea (bournei)* SW Palearctic to Iran, Africa s of the Sahara and Cape Verde Is.
_____ *A. p. madagascariensis* Madagascar
_____ *A. p. manilensis* Southern and e Asia, Indonesia and Philippine Islands

☐ **Great Egret** *Ardea alba*
_____ *A. a. egretta* S Canada to Tierra del Fuego and West Indies
_____ *A. a. alba* Central Europe to central Asia (south to Iran)
_____ *A. a. melanorhyncha* Africa south of the Sahara and Madagascar
_____ *A. a. modesta* Southern and e Asia to Indonesia, Australia and New Zealand

☐ **Reddish Egret** *Egretta rufescens*
_____ *E. r. rufescens* S US, Bahamas and West Indies; winters to nw South America
_____ *E. r. dickeyi* S Baja California; winters to Guatemala and El Salvador

☐ **Pied Heron** *Egretta picata*

Locally from New Guinea to Indonesia and n Australia

☐ **Slaty Egret** *Egretta vinaceigula*

Swamps and reedbeds of s-central Africa

☐ **Black Heron** *Egretta ardesiaca*

Locally in Africa south of the Sahara and Madagascar

☐ **Tricolored Heron** *Egretta tricolor*
_____ *E. t. ruficollis (occidentalis)* — Tropical s US to Colombia, nw Venezuela and West Indies
_____ *E. t. tricolor* — NE Venezuela and the Guianas to s Peru and ne Brazil; Trinidad

☐ **Intermediate Egret** *Egretta intermedia*
_____ *E. i. intermedia* — Japan to s India and Greater Sundas
_____ *E. i. plumifera* — New Guinea, eastern Indonesia and Australia
_____ *E. i. brachyrhyncha* — Africa south of the Sahara

☐ **White-faced Heron** *Egretta novaehollandiae*
_____ *E. n. novaehollandiae* — Indonesia and Australasian region
_____ *E. n. parryi* — NW Australia

☐ **Little Blue Heron** *Egretta caerulea* — US to s Brazil, Uruguay and West Indies

☐ **Snowy Egret** *Egretta thula*
_____ *E. t. thula* — Western US to Baja California and coastal nw Mexico
_____ *E. t. brewsteri* — Locally from US to central Argentina and West Indies

☐ **Little Egret** *Egretta garzetta*
_____ *E. g. garzetta* — Widespread Eurasia, east and South Africa
_____ *E. g. nigripes* — Java and Philippines to New Guinea
_____ *E. g. immaculata* — N and e Australia; occasional New Zealand
_____ *E. g. dimorpha* — Madagascar, Aldabra and Assumption I.

☐ **Western Reef-Heron** *Egretta gularis*
_____ *E. g. gularis* — Coastal w Africa to Gulf of Guinea islands and Gabon
_____ *E. g. schistacea* — Coastal e Africa to Red Sea, Persian Gulf and se India

☐ **Chinese Egret** *Egretta eulophotes* — E Asia; winters to SE Asia, Philippines and Indonesia

☐ **Pacific Reef-Heron** *Egretta sacra*
_____ *E. s. sacra* — Coastal SE Asia, Malay Archipelago, Oceania and Australasia
_____ *E. s. albolineata* — New Caledonia and Loyalty Islands

☐ **Squacco Heron** *Ardeola ralloides* — Locally in s Palearctic region, Africa and Madagascar

☐ **Indian Pond-Heron** *Ardeola grayii* — Persian Gulf to India, Myanmar, Andaman and Nicobar islands

☐ **Chinese Pond-Heron** *Ardeola bacchus* — Lowlands of s Asia; winters to Greater Sundas

☐ **Javan Pond-Heron** *Ardeola speciosa*
_____ *A. s. continentalis* — Central Thailand to s Indochina
_____ *A. s. speciosa* — West and central Indonesian Archipelago

☐ **Madagascar Pond-Heron** *Ardeola idae* — Madagascar and Aldabra; post-breeding dispersal to c Africa

☐ **Rufous-bellied Heron** *Ardeola rufiventris* — Locally in e and se Africa

☐ **Cattle Egret** *Bubulcus ibis*
_____ *B. i. ibis* — W Palearctic, Africa, North and South America
_____ *B. i. coromandus* — S and E Asia to Indian subcontinent, Australia and New Zealand
_____ *B. i. seychellarum* — Seychelles

☐ **Striated Heron** *Butorides striata*
_____ *B. s. striata* — E Panama and all South America to n Argentina and Chile
_____ *B. s. sundevalli* — Coasts and mangroves of Galapagos Islands
_____ *B. s. atricapilla* — Africa south of the Sahara and islands in Gulf of Guinea
_____ *B. s. rutenbergi* — Madagascar

____	*B. s. brevipes*	Red Sea environs and n Somalia
____	*B. s. crawfordi*	Aldabra and Amirante islands
____	*B. s. rhizophorae*	Comoro Islands
____	*B. s. degens*	Seychelles
____	*B. s. albolimbata*	Diego Garcia, Chagos and Maldive islands
____	*B. s. chloriceps*	Indian subcontinent, Sri Lanka and Laccadive Islands
____	*B. s. javanica*	Myanmar and Thailand to Greater Sundas and Mascarene Islands
____	*B. s. amurensis*	Manchuria to ne China, Japan, Ryukyu and Bonin islands
____	*B. s. actophila*	E China to n Vietnam and n Myanmar
____	*B. s. spodiogaster*	Andaman Islands, Nicobar Islands and islands off w Sumatra
____	*B. s. carcinophila*	Taiwan, Philippines and Sulawesi
____	*B. s. steini*	Lesser Sundas
____	*B. s. moluccarum*	Moluccas
____	*B. s. papuensis*	Aru Islands and nw New Guinea
____	*B. s. idenburgi*	N-central New Guinea
____	*B. s. rogersi*	NW Western Australia
____	*B. s. cinerea*	NE Western Australia
____	*B. s. stagnatilis*	N-central Australia
____	*B. s. littleri*	South-central New Guinea and ne Queensland
____	*B. s. macrorhyncha*	E Queensland, New Caledonia and Loyalty Islands
____	*B. s. solomonensis*	Melanesia (New Hanover to w Fiji)
____	*B. s. patruelis*	Tahiti (Society Islands)

☐ **Green Heron** *Butorides virescens*

____	*B. v. virescens*	Central US and e Canada to Panama and Caribbean
____	*B. v. anthonyi*	Western US and n Baja California
____	*B. v. frazari*	S Baja California
____	*B. v. bahamensis*	Bahamas

☐ **Agami Heron** *Agamia agami*

Tropical s Mexico to n Bolivia and w Amazonian Brazil

☐ **Black-crowned Night-Heron** *Nycticorax nycticorax*

____	*N. n. nycticorax*	Eurasia south to Indonesia, Africa and Madagascar
____	*N. n. hoactli*	S Canada to n Argentina and Chile
____	*N. n. obscurus*	N Chile and n-central Argentina to Tierra del Fuego
____	*N. n. falklandicus*	Falkland Islands

☐ **Rufous Night-Heron** *Nycticorax caledonicus*

____	*N. c. manillensis*	Philippines, e Borneo and Sulawesi
____	*N. c. hilli*	Australia and New Guinea, west to Java
____	*N. c. mandibularis*	Bismarck Archipelago to Solomon Islands
____	*N. c. pelewensis*	Palau and Caroline islands
____	*N. c. caledonicus*	New Caledonia

☐ **Yellow-crowned Night-Heron** *Nyctanassa violacea*

____	*N. v. violacea*	Central and e US to e Mexico and Honduras
____	*N. v. bancrofti*	Baja California and w Mexico to El Salvador and West Indies
____	*N. v. gravirostris*	Socorro I. (Revillagigedo Islands off w Mexico)
____	*N. v. calignis*	Panama to Peru
____	*N. v. pauper*	Galapagos Islands
____	*N. v. cayennensis*	Colombia to e Brazil

☐ **White-backed Night-Heron** *Gorsachius leuconotus*

Locally in Africa south of the Sahara

☐ **White-eared Night-Heron** *Gorsachius magnificus*

Highlands of Hainan and se China

☐ **Japanese Night-Heron** *Gorsachius goisagi*

S Japan; winters se China to Ryukyu Islands and Indonesia

☐ **Malayan Night-Heron** *Gorsachius melanolophus*

Humid forests of s Asia and Malay Archipelago

☐ **Boat-billed Heron** *Cochlearius cochlearius*
____   *C. c. zeledoni*   W-central Mexico
____   *C. c. phillipsi*   Tropical e Mexico and Belize
____   *C. c. ridgwayi*   Tropical s Mexico to w Honduras and El Salvador
____   *C. c. panamensis*   Costa Rica and Panama
____   *C. c. cochlearius*   E Panama to the Guianas, Amazon basin and ne Argentina

☐ **Bare-throated Tiger-Heron** *Tigrisoma mexicanum*

      Wet lowlands of Mexico to nw Colombia

☐ **Fasciated Tiger-Heron** *Tigrisoma fasciatum*
____   *T. f. salmoni*   Costa Rica to Venezuela and n Bolivia
____   *T. f. fasciatum*   SE Brazil to ne Argentina
____   *T. f. pallescens*   NW Argentina

☐ **Rufescent Tiger-Heron** *Tigrisoma lineatum*
____   *T. l. lineatum*   SE Mexico to Amazonian Brazil and n Argentina
____   *T. l. marmoratum*   Central Bolivia to e Brazil and ne Argentina

☐ **Forest Bittern** *Zonerodius heliosylus*

      Lowlands of New Guinea, Salawati I. and Aru Islands

☐ **White-crested Bittern** *Tigriornis leucolopha*

      Sierra Leone to Cameroon, Gabon, Zaire and Cent. African Rep.

☐ **Zigzag Heron** *Zebrilus undulatus*

      Locally in ponds and streams of Amazon basin

☐ **Stripe-backed Bittern** *Ixobrychus involucris*

      Colombia to the Guianas, s Venezuela, c Argentina and c Chile

☐ **Least Bittern** *Ixobrychus exilis*
____   *I. e. exilis*   S Canada to Central America and West Indies
____   *I. e. pullus*   NW Mexico
____   *I. e. erythromelas*   E Panama to the Guianas, se Brazil and Paraguay
____   *I. e. bogotensis*   Central Colombia (declining due to habitat destruction)
____   *I. e. peruvianus*   W-central Peru

☐ **Yellow Bittern** *Ixobrychus sinensis*

      S Asia, Malay Archipelago, New Guinea region and s Oceania

☐ **Little Bittern** *Ixobrychus minutus*
____   *I. m. minutus*   Central and s Europe to Siberia; North Africa
____   *I. m. payesii*   Africa south of the Sahara
____   *I. m. podiceps*   Madagascar
____   *I. m. dubius*   SW and e Australia and New Guinea

☐ **Schrenck's Bittern** *Ixobrychus eurhythmus*

      E Asia; winters to SE Asia, Philippines and Greater Sundas

☐ **Cinnamon Bittern** *Ixobrychus cinnamomeus*

      India to SE Asia, Philippines and Indonesia

☐ **Dwarf Bittern** *Ixobrychus sturmii*

      Africa south of the Sahara

☐ **Black Bittern** *Ixobrychus flavicollis*
____   *I. f. flavicollis*   India and SE Asia to Indonesia and Philippines
____   *I. f. australis*   Moluccas, New Guinea and Bismarck Arch. to w, n and e Australia
____   *I. f. woodfordi*   Solomon Islands

☐ **Pinnated Bittern** *Botaurus pinnatus*
____   *B. p. caribaeus*   Lowlands of e Mexico
____   *B. p. pinnatus*   SE Nicaragua to Ecuador, the Guianas, n Argentina and Brazil

☐ **American Bittern** *Botaurus lentiginosus*

      Alaska to Mexico; winters to Panama and West Indies

☐ **Great Bittern** *Botaurus stellaris*
_____ *B. s. stellaris*   Palearctic and n Afrotropical region; winters to Philippines
_____ *B. s. capensis*   Southern Africa

☐ **Australasian Bittern** *Botaurus poiciloptilus*

       S Australia, Tasmania, New Zealand and New Caledonia

ORDER: CICONIIFORMES
FAMILY: SCOPIDAE (Hamerkop—1)

☐ **Hamerkop** *Scopus umbretta*
_____ *S. u. umbretta*   Tropical Africa, Madagascar and sw Arabia
_____ *S. u. minor*   Coastal w Africa (Sierra Leone to w Cameroon)

ORDER: CICONIIFORMES
FAMILY: CICONIIDAE (Storks—19)

☐ **Wood Stork** *Mycteria americana*

       S US to n Argentina, Brazil, Cuba and Hispaniola

☐ **Milky Stork** *Mycteria cinerea*

       Lowlands of Malaya, Indochina, Greater Sundas and Sulawesi

☐ **Yellow-billed Stork** *Mycteria ibis*

       Africa south of the Sahara and Madagascar

☐ **Painted Stork** *Mycteria leucocephala*

       Lowlands of Indian subcontinent to s China and SE Asia

☐ **Asian Openbill** *Anastomus oscitans*

       Lowlands of Indian subcontinent to SE Asia

☐ **African Openbill** *Anastomus lamelligerus*
_____ *A. l. lamelligerus*   Africa south of the Sahara
_____ *A. l. madagascariensis*   Madagascar

☐ **Black Stork** *Ciconia nigra*

       Central and s Eurasia; s Africa; winters to c Africa and India

☐ **Abdim's Stork** *Ciconia abdimii*

       Sub-Saharan Africa and sw Arabia

☐ **Woolly-necked Stork** *Ciconia episcopus*
_____ *C. e. microscelis*   Tropical Africa
_____ *C. e. episcopus*   India to Indochina, n Malay Peninsula and Philippines
_____ *C. e. neglecta*   Java and Wallacea

☐ **Storm's Stork** *Ciconia stormi*

       Lowlands of Borneo, e Sumatra and Malay Peninsula

☐ **Maguari Stork** *Ciconia maguari*

       Tropical plains and marshes of South America east of the Andes

☐ **White Stork** *Ciconia ciconia*
_____ *C. c. ciconia*   W Palearctic and w Asia: winters to tropical and South Africa
_____ *C. c. asiatica*   Turkestan; winters to Iran and India

☐ **Oriental Stork** *Ciconia boyciana*

       Siberia, Manchuria and Korea; winters to s China and n India

☐ **Black-necked Stork** *Ephippiorhynchus asiaticus*
_____ *E. a. asiaticus*   Indian subcontinent and SE Asia
_____ *E. a. australis*   N and e Australia and s New Guinea

☐ **Saddle-billed Stork** *Ephippiorhynchus senegalensis*

       Locally in Africa south of the Sahara

23

☐ **Jabiru** *Jabiru mycteria*

☐ **Lesser Adjutant** *Leptoptilos javanicus*

Tropical s Mexico through South America to ne Argentina

☐ **Marabou Stork** *Leptoptilos crumeniferus*

India and Sri Lanka to s China, Indochina and Indonesia

☐ **Greater Adjutant** *Leptoptilos dubius*

Tropical Africa south of the Sahara

NE India (population ±300 birds 1992)

ORDER: CICONIIFORMES
FAMILY: BALAENICIPIDIDAE (Shoebill—1)

☐ **Shoebill** *Balaeniceps rex*

Dense swamps of central Africa

ORDER: CICONIIFORMES
FAMILY: THRESKIORNITHIDAE (Ibis and Spoonbills—33)

☐ **Sacred Ibis** *Threskiornis aethiopicus*
____ *T. a. aethiopicus* — Africa south of the Sahara and se Iraq; formerly Egypt
____ *T. a. bernieri* — Madagascar
____ *T. a. abbotti* — Aldabra I.

☐ **Black-headed Ibis** *Threskiornis melanocephalus*

India to SE Asia; winters to e China, Sumatra and Philippines

☐ **Australian Ibis** *Threskiornis molucca*
____ *T. m. molucca* — Australia to s New Guinea, s Moluccas and e Lesser Sundas
____ *T. m. pygmaeus* — Solomon Islands (Rennell and Bellona)

☐ **Straw-necked Ibis** *Threskiornis spinicollis*

☐ **Red-naped Ibis** *Pseudibis papillosa*

Australia; ranges to n Tasmania and s New Guinea

☐ **White-shouldered Ibis** *Pseudibis davisoni*

Semiarid lowlands of India and Pakistan

☐ **Giant Ibis** *Pseudibis gigantea*

SW China to Myanmar, peninsular Thailand and Indochina

☐ **Waldrapp** *Geronticus eremita*

Lowlands of Thailand, Cambodia, Laos and s Vietnam

☐ **Bald Ibis** *Geronticus calvus*

Patchily distributed mts. of Morocco, Syria, and Red Sea area

☐ **Crested Ibis** *Nipponia nippon*

Mountains of inland regions of South Africa

China (±50 birds in Shaanxi Province in 1998)

☐ **Olive Ibis** *Bostrychia olivacea*
____ *B. o. olivacea* — Lowland forests of Sierra Leone and Liberia
____ *B. o. cupreipennis* — Lowland forests of Cameroon, Gabon, Congo and Zaire
____ *B. o. rothschildi†* — Formerly Príncipe I. (Gulf of Guinea). Extinct ca 1901
____ *B. o. bocagei* — São Tomé I. (Gulf of Guinea)
____ *B. o. akleyorum* — Montane forests of Kenya and Tanzania

☐ **Spot-breasted Ibis** *Bostrychia rara*

Liberia to Cameroon, Gabon, Zaire and extreme ne Angola

☐ **Hadada Ibis** *Bostrychia hagedash*
____ *B. h. brevirostris* — Senegal to Kenya and south to Zambezi Valley
____ *B. h. nilotica* — Sudan and Ethiopia to ne Zaire, Uganda and nw Tanzania
____ *B. h. hagedash* — Southern Africa (south of the Zambezi Valley)

☐ **Wattled Ibis** *Bostrychia carunculata*

Highlands of Ethiopia

☐ **Plumbeous Ibis** *Theristicus caerulescens*

Lowlands of s Brazil to Bolivia, Paraguay and ne Argentina

☐ **Buff-necked Ibis** *Theristicus caudatus*
___ *T. c. caudatus*
___ *T. c. hyperorius*

E Colombia to Venezuela, Guianas and sw Brazil (Mato Grosso)
E Bolivia to se Brazil, Paraguay, Uruguay and n Argentina

☐ **Andean Ibis** *Theristicus branickii*

Andes of Ecuador to extreme n Chile

☐ **Black-faced Ibis** *Theristicus melanopis*

Coastal Peru, n Chile and Argentina south to Tierra del Fuego

☐ **Sharp-tailed Ibis** *Cercibis oxycerca*

Llanos of e Colombia to the Guianas and w Amazonian Brazil

☐ **Green Ibis** *Mesembrinibis cayennensis*

Lowlands of Costa Rica to ne Argentina and Brazil

☐ **Bare-faced Ibis** *Phimosus infuscatus*
___ *P. i. berlepschi*
___ *P. i. nudifrons*
___ *P. i. infuscatus*

E Colombia to the Guianas, Suriname and adjacent nw Brazil
Brazil south of the Amazon
E Bolivia to Paraguay, ne Argentina and Uruguay

☐ **White Ibis** *Eudocimus albus*

Southern US to se Brazil, Bahamas and Greater Antilles

☐ **Scarlet Ibis** *Eudocimus ruber*

Coastal Colombia to the Guianas and ne Brazil; Trinidad

☐ **Glossy Ibis** *Plegadis falcinellus*

Locally in e N and s S America, Africa, Eurasia to Australasia

☐ **White-faced Ibis** *Plegadis chihi*

Great Basin of w US to sw Brazil and central Argentina

☐ **Puna Ibis** *Plegadis ridgwayi*

High Andes of central Peru to nw Argentina and n Chile

☐ **Madagascar Ibis** *Lophotibis cristata*
___ *L. c. cristata*
___ *L. c. urschi*

Forests of e Madagascar
Forests of w Madagascar

☐ **Eurasian Spoonbill** *Platalea leucorodia*
___ *P. l. leucorodia*
___ *P. l. balsaci*
___ *P. l. archeri*

S Palearctic to India; winters to central Africa and se China
Banc d'Arguin (off coast of Mauritania)
Coasts of Red Sea and Somalia

☐ **Royal Spoonbill** *Platalea regia*

Australia, New Zealand, Indonesia, New Guinea and Solomon Is.

☐ **African Spoonbill** *Platalea alba*

Africa south of the Sahara and Madagascar

☐ **Black-faced Spoonbill** *Platalea minor*

Breeds ne China and Korea; winters to SE Asia

☐ **Yellow-billed Spoonbill** *Platalea flavipes*

Australia; vagrant to Tasmania and New Zealand

☐ **Roseate Spoonbill** *Platalea ajaja*

S US to n Argentina, Brazil and West Indies

ORDER: PHOENICOPTERIFORMES
FAMILY: PHOENICOPTERIDAE (Flamingos—6)

☐ **Greater Flamingo** *Phoenicopterus roseus*

S Europe and Medit. basin to S Africa and Indian subcontinent

☐ **Caribbean Flamingo** *Phoenicopterus ruber*

Locally from Caribbean to ne Brazil; Galapagos Islands

☐ **Chilean Flamingo** *Phoenicopterus chilensis*

Andes of s South America; pampas of s Brazil to s Argentina

☐ **Lesser Flamingo** *Phoenicopterus minor*

Locally from Africa and Madagascar to nw India

☐ **Andean Flamingo** *Phoenicopterus andinus*

High Andes of s Peru to nw Argentina and n Chile

☐ **Puna Flamingo** *Phoenicopterus jamesi*

High Andes of s Peru to nw Argentina and n Chile

ORDER: ANSERIFORMES
FAMILY: ANHIMIDAE (Screamers—3)

☐ **Horned Screamer** *Anhima cornuta*

Lowlands of Venezuela to n Bolivia and Amazonian Brazil

☐ **Northern Screamer** *Chauna chavaria*

N Colombia and nw Venezuela

☐ **Southern Screamer** *Chauna torquata*

Wet lowlands of se Peru to n Argentina and s Brazil

ORDER: ANSERIFORMES
FAMILY: ANATIDAE (Ducks, Geese and Swans—159)

☐ **Magpie Goose** *Anseranas semipalmata*

Coastal n Australia and Trans-Fly savanna of s New Guinea

☐ **Spotted Whistling-Duck** *Dendrocygna guttata*

Sulawesi to New Guinea, Bismarck Arch. and s Philippines

☐ **Plumed Whistling-Duck** *Dendrocygna eytoni*

Lowlands of n and e Australia

☐ **Fulvous Whistling-Duck** *Dendrocygna bicolor*

S US to Argentina; e Africa, Madagascar and s Asia

☐ **Wandering Whistling-Duck** *Dendrocygna arcuata*
____ *D. a. arcuata*
____ *D. a. australis*
____ *D. a. pygmaea*

Philippines to Indonesia
S New Guinea and n and e Australia
New Britain I. (Bismarck Archipelago)

☐ **Lesser Whistling-Duck** *Dendrocygna javanica*

Indian subcontinent to SE Asia and Greater Sundas

☐ **White-faced Whistling-Duck** *Dendrocygna viduata*

Costa Rica to Brazil; Africa, Madagascar and Comoro Islands

☐ **West Indian Whistling-Duck** *Dendrocygna arborea*

Bahamas, Greater Antilles and n Lesser Antilles

☐ **Black-bellied Whistling-Duck** *Dendrocygna autumnalis*

Extreme s Texas to n Argentina (mainly east of Andes)

☐ **White-backed Duck** *Thalassornis leuconotus*
____ *T. l. leuconotus*
____ *T. l. insularis*

Locally in Africa south of the Sahara
Locally in aquatic lowlands of Madagascar

☐ **Mute Swan** *Cygnus olor*

Palearctic region; winters to India and se China

☐ **Black Swan** *Cygnus atratus*

Australia and Tasmania; introduced New Zealand

☐ **Black-necked Swan** *Cygnus melanocoryphus*

S Brazil to Tierra del Fuego and Falkland Islands

☐ **Trumpeter Swan** *Cygnus buccinator*

Western North America

☐ **Whooper Swan** *Cygnus cygnus*

Palearctic; winters to India and se China

☐ **Tundra Swan** *Cygnus columbianus*
____ *C. c. bewickii* — Kola Peninsula to arctic n Siberia; winters w Europe to s Asia
____ *C. c. columbianus* — Tundra of arctic North America; winters to w and coastal e US

☐ **Coscoroba Swan** *Coscoroba coscoroba*
— S Brazil to Paraguay, Uruguay, Tierra del Fuego and Falkland Is.

☐ **Swan Goose** *Anser cygnoides*
— N-central Asia (s-central Siberia to n China)

☐ **Bean Goose** *Anser fabalis*
____ *A. f. fabalis* — *Taiga* of Scandinavia to Ural Mountains
____ *A. f. johanseni* — *Taiga* and wooded tundra of Ural Mountains to Lake Baikal
____ *A. f. middendorffii* — *Taiga* of e Siberia (east of Lake Baikal)
____ *A. f. rossicus* — Tundra of n Russia and nw Siberia
____ *A. f. serrirostris* — Tundra of ne Siberia

☐ **Pink-footed Goose** *Anser brachyrhynchus*
— Breeds Greenland, Iceland and Spitzbergen; winters nw Europe

☐ **Greater White-fronted Goose** *Anser albifrons*
____ *A. a. albifrons* — N Russia and Siberia; winters to Mediterranean and n India
____ *A. a. frontalis* — E Siberia to n Canada; winters w US, n Mexico and China
____ *A. a. flavirostris* — Breeds w coast of Greenland; winters mainly in Ireland
____ *A. a. gambelli* — *Taiga* of nw Canada and w Alaska; winters Gulf Coast
____ *A. a. elgasi* — *Taiga* south of Alaskan tundra; winters Sacramento Valley

☐ **Lesser White-fronted Goose** *Anser erythropus*
— Arctic Eurasia; winters to s Europe, India and China

☐ **Greylag Goose** *Anser anser*
____ *A. a. anser* — Breeds nw Eurasia; winters to North Africa, Turkey and Iran
____ *A. a. rubrirostris* — NE Eurasia; winters to Asia Minor, India and n Indochina

☐ **Bar-headed Goose** *Anser indicus*
— Alpine lakes in central Asia; winters to India and Myanmar

☐ **Snow Goose** *Chen caerulescens*
____ *C. c. caerulescens* — Siberia and Alaska; winters to California and Gulf Coast
____ *C. c. atlantica* — NW Greenland and islands in Baffin Bay; winters to ne Mexico

☐ **Ross' Goose** *Chen rossii*
— Tundra of Arctic Canada; winters to s US

☐ **Emperor Goose** *Chen canagica*
— NE Siberia to w Alaska; winters s Alaska to n California

☐ **Brant** *Branta bernicla*
____ *B. b. bernicla* — N-central Siberia; winters coastal England and nw Europe
____ *B. b. orientalis* — NE Siberia
____ *B. b. hrota* — E Arctic Canada, Greenland and Spitzbergen; winters e N Am.
____ *B. b. nigricans* — Extreme ne Siberia to n Canada; winters to n Mexico and China

☐ **Barnacle Goose** *Branta leucopsis*
— Greenland to Novaya Zemlya; winters to n Mediterranean

☐ **Cackling Goose** *Branta hutchinsii*
____ *B. h. hutchinsii* — Breeds n-c Canada and Greenland; winters Texas and Mexico
____ *B. h. leucopareia* — Buldir I. (w Alaska). Formerly Kuril Is. and Aleutian Is.
____ *B. h. taverneri* — Central Alaska to Mackenzie River delta; winters to Mexico
____ *B. h. minima* — Coastal w Alaska to Mackenzie delta; winters to s California
____ *B. h. asiatica* — Bering Sea†

☐ **Canada Goose** *Branta canadensis*
____ *B. c. occidentalis* — SW Alaska (Prince William Sound to Copper River Delta)
____ *B. c. fulva* — Coastal s Alaska to British Columbia; occasional n California
____ *B. c. parvipes* — Central Alaska to Canadian prairie provinces; winters s US

____	*B. c. moffitti*	N Great Plains and s Canada; disperses southward in winter
____	*B. c. maxima*	Formerly Great Plains; now only on wildlife reserves
____	*B. c. interior*	Breeds ne Canada; winters to Florida and Louisiana
____	*B. c. canadensis*	Breeds Labrador and Newfoundland; winters to Florida

☐ **Hawaiian Goose** *Branta sandvicensis*

Upland lava flows of Hawaii; introduced to Maui

☐ **Red-breasted Goose** *Branta ruficollis*

Siberian tundra; winters Black, Caspian and Aral seas

☐ **Cape Barren Goose** *Cereopsis novaehollandiae*

Islands in Bass Strait, adjacent Australia and Tasmania

☐ **Freckled Duck** *Stictonetta naevosa*

Locally in se and extreme sw Australia

☐ **Blue-winged Goose** *Cyanochen cyanoptera*

Highlands of Ethiopia

☐ **Andean Goose** *Chloephaga melanoptera*

Andes of s Peru to nw Argentina and central Chile

☐ **Upland Goose** *Chloephaga picta*

____ *C. p. picta* — Mountains of central Argentina and Chile to Tierra del Fuego
____ *C. p. leucoptera* — Falkland Islands

☐ **Kelp Goose** *Chloephaga hybrida*

____ *C. h. hybrida* — Coastal s Argentina and Chile to Tierra del Fuego
____ *C. h. malvinarum* — Falkland Islands

☐ **Ashy-headed Goose** *Chloephaga poliocephala*

S Argentina and s Chile to Tierra del Fuego

☐ **Ruddy-headed Goose** *Chloephaga rubidiceps*

Tierra del Fuego and Falkland Islands

☐ **Orinoco Goose** *Neochen jubata*

Orinoco and Amazon River basins to nw Argentina

☐ **Egyptian Goose** *Alopochen aegyptiaca*

Africa south of the Sahara and Nile Valley

☐ **Ruddy Shelduck** *Tadorna ferruginea*

S Mediterranean basin to e Asia

☐ **South African Shelduck** *Tadorna cana*

Karoo of s Africa

☐ **Australian Shelduck** *Tadorna tadornoides*

Patchily distributed sw and se Australia and Tasmania

☐ **Paradise Shelduck** *Tadorna variegata*

North, South and Stewart islands (New Zealand)

☐ **Common Shelduck** *Tadorna tadorna*

Palearctic region; winters to Near East, India and Myanmar

☐ **Radjah Shelduck** *Tadorna radjah*

____ *T. r. radjah* — Moluccas to New Guinea and adjacent islands
____ *T. r. rufitergum* — Coastal n and e tropical Australia

☐ **Flightless Steamerduck** *Tachyeres pteneres*

S South America (Tierra del Fuego and Cape Horn Archipelago)

☐ **White-headed Steamerduck** *Tachyeres leucocephalus*

S Argentina (s coast of Chubut Province)

☐ **Falkland Steamerduck** *Tachyeres brachypterus*

Falkland Islands

☐ **Flying Steamerduck** *Tachyeres patachonicus*

Coastal s Chile, Argentina and Falkland Islands

☐ **Spur-winged Goose** *Plectropterus gambensis*

____ *P. g. gambensis* — Gambia to Ethiopia and south to the Zambesi River
____ *P. g. niger* — Namibia and Zimbabwe to Cape Province

☐ **Muscovy Duck** *Cairina moschata*

Lowlands of s Mexico to ne Argentina and Brazil

☐ **White-winged Duck** *Cairina scutulata*

India to SE Asia, Sumatra and Java

☐ **Comb Duck** *Sarkidiornis melanotos*
____ *S. m. melanotos*
____ *S. m. sylvicola*

Tropical Africa and Madagascar; India to s China
Tropical South America (east of the Andes) to n Argentina

☐ **Hartlaub's Duck** *Pteronetta hartlaubii*

Locally in forest streams of equatorial w Africa

☐ **Green Pygmy-goose** *Nettapus pulchellus*

Sulawesi to Moluccas, New Guinea and tropical n Australia

☐ **Cotton Pygmy-goose** *Nettapus coromandelianus*
____ *N. c. coromandelianus*
____ *N. c. albipennis*

Lowlands of India, s Asia, Indonesia and New Guinea
Lowlands of coastal ne Australia

☐ **African Pygmy-goose** *Nettapus auritus*

Africa south of the Sahara (except sw Africa) and Madagascar

☐ **Ringed Teal** *Callonetta leucophrys*

S Brazil to Bolivia, Paraguay, Uruguay and ne Argentina

☐ **Wood Duck** *Aix sponsa*

Inland waters of Canada to n Mexico, Cuba and Bahamas

☐ **Mandarin Duck** *Aix galericulata*

Wooded ponds, swamps and streams of ne Asia

☐ **Maned Duck** *Chenonetta jubata*

Australia (except driest regions); visitor to Tasmania

☐ **Brazilian Teal** *Amazonetta brasiliensis*
____ *A. b. brasiliensis*
____ *A. b. ipecutiri*

Colombia and Venezuela south to Brazil
S Brazil to e Bolivia, Uruguay and Argentina

☐ **Blue Duck** *Hymenolaimus malacorhynchos*
____ *H. m. malacorhynchos*
____ *H. m. hymenolaimus*

Mountain streams of w South I. (New Zealand)
Mountain streams of central North I. (New Zealand)

☐ **Torrent Duck** *Merganetta armata*
____ *M. a. colombiana*
____ *M. a. leucogenis*
____ *M. a. turneri*
____ *M. a. garleppi*
____ *M. a. berlepschi*
____ *M. a. armata*

Andes of Colombia, adjacent nw Venezuela and Ecuador
Andes of central and s Ecuador and Peru
Andes of s Peru (Cuzco and Arequipa)
Andes of Bolivia
Andes of nw Argentina and n Chile
Andes of Chile and adjacent Argentina south to Tierra del Fuego

☐ **Salvadori's Teal** *Salvadorina waigiuensis*

Mountain streams of New Guinea

☐ **African Black Duck** *Anas sparsa*
____ *A. s. leucostigma*
____ *A. s. sparsa*

W equatorial Africa; Ethiopia and Sudan to Zimbabwe
Southern Africa (south of Zimbabwe)

☐ **Eurasian Wigeon** *Anas penelope*

N and central Eurasia; winters to Africa and s Asia

☐ **American Wigeon** *Anas americana*

Alaska to s US; winters to nw South America

☐ **Chiloe Wigeon** *Anas sibilatrix*

S South America and Falkland Islands; winters to se Brazil

☐ **Falcated Duck** *Anas falcata*

E Siberia and Mongolia to n Japan; winters to India

☐ **Gadwall** *Anas strepera*
____ *A. s. strepera*
____ *A. s. couesi†*

Widespread Palearctic and Nearctic regions
Formerly Fanning Islands (central Pacific). Extinct ca 1874

☐ **Baikal Teal** *Anas formosa*

E Siberia to Kamchatka; winters to India, Myanmar and Japan

☐ **Eurasian Teal** *Anas crecca*
_____ *A. c. crecca*
_____ *A. c. nimia*

Palearctic region; winters to Africa, India and SE Asia
Aleutian Islands

☐ **Green-winged Teal** *Anas carolinensis*

Breeds North America; winters to Mexico and West Indies

☐ **Speckled Teal** *Anas flavirostris*
_____ *A. f. altipetens*
_____ *A. f. andium*
_____ *A. f. oxyptera*
_____ *A. f. flavirostris*

Andes of Colombia to nw Venezuela
Andes of Colombia and n Ecuador
Andes of central Peru to n Chile and Argentina
N Argentina to Tierra del Fuego, South Georgia and Falkland Is.

☐ **Cape Teal** *Anas capensis*

Locally from Sudan and Ethiopia to Namibia and South Africa

☐ **Bernier's Teal** *Anas bernieri*

Lowlands of w Madagascar (population ±20 birds 1993)

☐ **Sunda Teal** *Anas gibberifrons*
_____ *A. g. gibberifrons*
_____ *A. g. remissa†*

Java and Sulawesi to e Lesser Sundas (Timor and Wetar)
Formerly Rennell (Solomon Islands). Extinct ca 1959

☐ **Andaman Teal** *Anas albogularis*

Andaman Islands and Great Coco I.

☐ **Gray Teal** *Anas gracilis*

New Guinea to Australia, New Caledonia and New Zealand

☐ **Chestnut Teal** *Anas castanea*

Swamps and marshes of sw and se Australia and Tasmania

☐ **Auckland Islands Teal** *Anas aucklandica*

Islets off Auckland Islands

☐ **Campbell Islands Teal** *Anas nesiotis*

Rediscovered 1975 on Campbell Islands after considered extinct

☐ **Brown Teal** *Anas chlorotis*

New Zealand and offshore islands

☐ **Mallard** *Anas platyrhynchos*
_____ *A. p. platyrhynchos*
_____ *A. p. conboschas*
_____ *A. p. maculosa*
_____ *A. p. oustaletti†*
_____ *A. p. diazi*

Holarctic; winters to Mexico, North Africa, India and Borneo
Coastal sw Greenland
Atlantic s US to Mexico
Formerly Marianas Archipelago. Extinct ca 1974
S Texas, New Mexico and Arizona south to central Mexico

☐ **Laysan Duck** *Anas laysanensis*

Laysan I. (nw Hawaiian Islands)

☐ **Hawaiian Duck** *Anas wyvilliana*

Hawaiian Islands (Kauai and Oahu)

☐ **Mottled Duck** *Anas fulvigula*

Florida and coastal s Texas to se Mexico

☐ **American Black Duck** *Anas rubripes*

NE North America; winters to Bahamas and Gulf Coast

☐ **Yellow-billed Duck** *Anas undulata*
_____ *A. u. undulata*
_____ *A. u. rueppelli*

Locally from Kenya and Uganda to Angola and South Africa
Ethiopia (upper Blue Nile region) to n Kenya and s Sudan

☐ **Meller's Duck** *Anas melleri*

High plateau and e Madagascar

☐ **Spot-billed Duck** *Anas poecilorhyncha*
_____ *A. p. poecilorhyncha*
_____ *A. p. haringtoni*
_____ *A. p. zonorhyncha*

Indian subcontinent and Sri Lanka
Myanmar and Assam to extreme s China and Laos
Breeds ne Asia; winters to s China, Taiwan and Philippines

☐ **Pacific Black Duck** *Anas superciliosa*
_____ *A. s. superciliosa* — New Zealand, Auckland, Campbell and Macquarie islands
_____ *A. s. pelewensis* — New Guinea to Solomon Islands and French Polynesia
_____ *A. s. rogersi* — Sundas to s New Guinea, Australia and Tasmania

☐ **Philippine Duck** *Anas luzonica*
Philippine Islands

☐ **Spectacled Duck** *Anas specularis*
Mainly forested regions of s Argentina and Chile

☐ **Crested Duck** *Anas specularioides*
_____ *A. s. alticola* — Andes of Peru and Bolivia to nw Argentina and Chile
_____ *A. s. specularioides* — Central Chile and Argentina to Tierra del Fuego and Falkland Is.

☐ **Northern Pintail** *Anas acuta*
Palearctic and N America; winters to s Eurasia and n S America

☐ **Eaton's Pintail** *Anas eatoni*
_____ *A. e. eatoni* — Kerguelen Islands
_____ *A. e. drygalskii* — Crozet I.

☐ **Yellow-billed Pintail** *Anas georgica*
_____ *A. g. spinicauda* — Highlands of s Colombia to Tierra del Fuego and Falkland Islands
_____ *A. g. georgica* — South Georgia I.
_____ *A. g. nicefori†* — Formerly Andes of Colombia. Extinct ca 1952

☐ **White-cheeked Pintail** *Anas bahamensis*
_____ *A. b. bahamensis* — Locally in West Indies and n South America
_____ *A. b. rubrirostris* — S Brazil and Bolivia to Argentina and Chile
_____ *A. b. galapagensis* — Galapagos Islands

☐ **Red-billed Duck** *Anas erythrorhyncha*
Locally in e and s Africa and Madagascar

☐ **Puna Teal** *Anas puna*
Andes of Peru to nw Argentina and n Chile

☐ **Silver Teal** *Anas versicolor*
_____ *A. v. versicolor* — S Bolivia, Paraguay and s Brazil to Tierra del Fuego
_____ *A. v. fretensis* — S Chile, s Argentina and Falkland Islands

☐ **Hottentot Teal** *Anas hottentota*
Locally in Africa south of the Sahara and Madagascar

☐ **Garganey** *Anas querquedula*
Palearctic; winters to s Africa and Australasian region

☐ **Blue-winged Teal** *Anas discors*
North America; winters s US to central Argentina

☐ **Cinnamon Teal** *Anas cyanoptera*
_____ *A. c. septentrionalium* — British Columbia to nw Mexico; winters to nw South America
_____ *A. c. tropica* — Cauca and Magdalena valleys of Colombia
_____ *A. c. borreroi* — E Andes of Colombia
_____ *A. c. orinomus* — Altiplano of Peru and Bolivia to n Chile
_____ *A. c. cyanoptera* — S Peru and s Brazil to Tierra del Fuego and Falkland Islands

☐ **Red Shoveler** *Anas platalea*
S Peru and s Brazil to Tierra del Fuego and Falkland Islands

☐ **Cape Shoveler** *Anas smithii*
Locally in southern Africa

☐ **Australian Shoveler** *Anas rhynchotis*
_____ *A. r. rhynchotis* — Discontinuously distributed sw and se Australia and Tasmania
_____ *A. r. variegata* — New Zealand

☐ **Northern Shoveler** *Anas clypeata*
Holarctic; winters to Africa, n South America and Malay Arch.

☐ **Pink-eared Duck** *Malacorhynchus membranaceus*

☐ **Marbled Teal** *Marmaronetta angustirostris*

Nomadic throughout Australia (except driest areas)

☐ **Red-crested Pochard** *Netta rufina*

Canary Islands and Mediterranean basin to extreme sw China

☐ **Rosy-billed Pochard** *Netta peposaca*

Locally from Mediterranean basin to central Asia

☐ **Southern Pochard** *Netta erythrophthalma*

Lowlands of se Brazil to s Argentina and Chile

_____ *N. e. erythrophthalma*
_____ *N. e. brunnea*

Northern half of South America
Patchily distributed Sudan and Ethiopia to South Africa

☐ **Common Pochard** *Aythya ferina*

Palearctic; winters to tropical Africa, India and SE Asia

☐ **Canvasback** *Aythya valisineria*

Breeds North America; winters to s Mexico

☐ **Redhead** *Aythya americana*

Alaska to s US; winters to Guatemala and Greater Antilles

☐ **Ring-necked Duck** *Aythya collaris*

Breeds Alaska to s US; winters to Panama and s Lesser Antilles

☐ **Ferruginous Pochard** *Aythya nyroca*

Discontinuous Palearctic; winters to India, SE Asia and e China

☐ **Madagascar Pochard** *Aythya innotata*

Lake Alaotra (Madagascar). On verge of extinction

☐ **Baer's Pochard** *Aythya baeri*

NE Eurasia; winters to India, SE Asia and se China

☐ **White-eyed Duck** *Aythya australis*

Australia (except driest areas), Tasmania and sw Oceania

☐ **Tufted Duck** *Aythya fuligula*

N Palearctic region; winters to n Africa and s Asia

☐ **New Zealand Scaup** *Aythya novaeseelandiae*

New Zealand

☐ **Greater Scaup** *Aythya marila*

_____ *A. m. marila*
_____ *A. m. mariloides*

N Eurasia; winters to Mediterranean region and India
N Asia and n North America; winters to s US and China

☐ **Lesser Scaup** *Aythya affinis*

Alaska to s US; winters n South America and Hawaiian Islands

☐ **Common Eider** *Somateria mollissima*

_____ *S. m. mollissima*
_____ *S. m. faeroeensis*
_____ *S. m. v-nigra*
_____ *S. m. borealis*
_____ *S. m. sedentaria*
_____ *S. m. dresseri*

Coast of nw Eurasia; winters to coastal s France
Faeroe Islands
Arctic coasts of ne Siberia to Alaska and s British Columbia
Arctic coast of e Canada and Greenland; winters to Long Island
Coasts and islands of Hudson Bay to James Bay
Coastal Labrador to Maine; winters to Long Island

☐ **King Eider** *Somateria spectabilis*

Arctic Eurasia and n North America

☐ **Spectacled Eider** *Somateria fischeri*

Coastal n Siberia east to n Alaska

☐ **Steller's Eider** *Polysticta stelleri*

Arctic Siberia (Taymyr Peninsula) east to n Alaska

☐ **Harlequin Duck** *Histrionicus histrionicus*

_____ *H. h. histrionicus*
_____ *H. h. pacificus*

Greenland, Baffin I. and n Labrador; winters to Long Island
N and e Palearctic and Bering Sea islands to central California

☐ **Long-tailed Duck** *Clangula hyemalis*

Coasts of Holarctic region

☐ **Black Scoter** *Melanitta nigra*
_____ *M. n. nigra* — N Eurasia; winters w Europe to Mediterranean and Caspian Sea
_____ *M. n. americana* — Breeds n Siberia to Alaska; winters to n US

☐ **Surf Scoter** *Melanitta perspicillata* — Northern North America; winters to Baja California and s US

☐ **White-winged Scoter** *Melanitta fusca*
_____ *M. f. fusca* — N Eurasia; winters Norway to Spain and Caspian Sea
_____ *M. f. stejnegeri* — Breeds ne Asia; winters coastal e Asia to Japan and China
_____ *M. f. deglandi* — Northern North America; winters to coastal s US
_____ *M. f. dixoni* — Breeds nw Alaska; winters Aleutians and Alaska Pen. to Baja

☐ **Common Goldeneye** *Bucephala clangula*
_____ *B. c. clangula* — N Eurasia; winters to Mediterranean, Persian Gulf and s China
_____ *B. c. americana* — Breeds n North America; winters to California and Florida

☐ **Barrow's Goldeneye** *Bucephala islandica* — Disjunct populations in w Palearctic and n North America

☐ **Bufflehead** *Bucephala albeola* — Breeds n North America; winters to Mexico and Greater Antilles

☐ **Smew** *Mergellus albellus* — N Eurasia; winters to North Africa, India and e China

☐ **Hooded Merganser** *Lophodytes cucullatus* — Breeds n North America; winters to Mexico and West Indies

☐ **Brazilian Merganser** *Mergus octosetaceus* — S-central South America (Paraguai-Paraná drainage system)

☐ **Red-breasted Merganser** *Mergus serrator* — N Palearctic and n N America; winters s Palearctic and Mexico

☐ **Common Merganser** *Mergus merganser*
_____ *M. m. merganser* — Palearctic region; winters Mediterranean to n India and China
_____ *M. m. orientalis* — Afghanistan to Tibet and s China; winters to India and sw China
_____ *M. m. americanus* — Widespread North America

☐ **Scaly-sided Merganser** *Mergus squamatus* — Manchuria and extreme se Siberia; winters to s China

☐ **Black-headed Duck** *Heteronetta atricapilla* — Lowlands of s South America

☐ **Masked Duck** *Nomonyx dominica* — S Texas to n Argentina and Brazil; West Indies

☐ **Ruddy Duck** *Oxyura jamaicensis*
_____ *O. j. jamaicensis* — Interior nw N America (sw Canada to Mexico); West Indies
_____ *O. j. andina* — Lakes and marshes of Central and E Andes of Colombia

☐ **Andean Duck** *Oxyura ferruginea* — Locally from Andes of s Colombia to s Argentina and s Chile

☐ **White-headed Duck** *Oxyura leucocephala* — Patchily distributed Mediterranean basin to central Asia

☐ **Maccoa Duck** *Oxyura maccoa* — Locally in highlands of e and s Africa

☐ **Lake Duck** *Oxyura vittata* — S Argentina and Chile; winters north to s Brazil and Paraguay

☐ **Blue-billed Duck** *Oxyura australis* — Patchily distributed sw and se Australia and Tasmania

☐ **Musk Duck** *Biziura lobata* — Lakes and swamps of sw and se Australia and Tasmania

ORDER: FALCONIFORMES
FAMILY: CATHARTIDAE (New World Vultures—7)

☐ **Black Vulture** *Coragyps atratus*

____ *C. a. atratus*	Extreme s US and n Mexico
____ *C. a. brasiliensis*	Central America to n and e South America
____ *C. a. foetens*	Western South America

☐ **Turkey Vulture** *Cathartes aura*

____ *C. a. aura*	W North America south to Costa Rica; Greater Antilles
____ *C. a. septentrionalis*	E North America
____ *C. a. ruficollis*	Central America and lowlands of South America; Trinidad
____ *C. a. jota*	Pacific coast of Ecuador to Tierra del Fuego and Falkland Islands

☐ **Lesser Yellow-headed Vulture** *Cathartes burrovianus*

____ *C. b. burrovianus*	S Mexico to central Colombia and nw Venezuela
____ *C. b. urubitinga*	S America to Argentina and Brazil (east of the Andes)

☐ **Greater Yellow-headed Vulture** *Cathartes melambrotus*

Guianas and s Venezuela to n Bolivia and n Brazil

☐ **California Condor** *Gymnogyps californianus*

Formerly s California. ±132 birds extant in 1998

☐ **Andean Condor** *Vultur gryphus*

Andes and coasts of Colombia to Tierra del Fuego

☐ **King Vulture** *Sarcoramphus papa*

S Mexico to n Argentina and Brazil

ORDER: FALCONIFORMES
FAMILY: PANDIONIDAE (Osprey—1)

☐ **Osprey** *Pandion haliaetus*

____ *P. h. haliaetus*	Palearctic; winters to South Africa, India and Philippines
____ *P. h. carolinensis*	Canada to s US; winters to Peru and Brazil
____ *P. h. ridgwayi*	Caribbean (including Bahamas, Cuba and Belize)
____ *P. h. cristatus*	Sulawesi and Java to New Guinea, Australia and New Caledonia

ORDER: FALCONIFORMES
FAMILY: ACCIPITRIDAE (Hawks, Eagles and Kites—240)

☐ **African Cuckoo-Hawk** *Aviceda cuculoides*

____ *A. c. cuculoides*	Senegal to sw Ethiopia and n Zaire
____ *A. c. batesi*	Sierra Leone to Uganda and n Angola
____ *A. c. verreauxii*	Kenya to Namibia and South Africa

☐ **Madagascar Cuckoo-Hawk** *Aviceda madagascariensis*

Woodlands and scrub of Madagascar

☐ **Jerdon's Baza** *Aviceda jerdoni*

____ *A. j. ceylonensis*	SW India and Sri Lanka
____ *A. j. jerdoni*	NE India to Myanmar, s China and n Malay Peninsula
____ *A. j. borneensis*	Borneo
____ *A. j. magnirostris*	Philippine Islands
____ *A. j. celebensis*	Sulawesi, Banggai and Sula islands

☐ **Pacific Baza** *Aviceda subcristata*

____ *A. s. timorlaoensis*	Lesser Sundas and islands off Sulawesi
____ *A. s. pallida*	Seram Laut (Manawoka and Gorong) and Kai Islands
____ *A. s. reinwardtii*	S Moluccas (Boano, Seram, Ambon and Haruku)

____ *A. s. stresemanni*	Buru (central Moluccas)
____ *A. s. rufa*	Moluccas (Morotai, Halmahera, Ternate, Tidore, Bacan and Obi)
____ *A. s. waigeuensis*	Waigeo I. (off n New Guinea)
____ *A. s. obscura*	Biak I. (off n New Guinea)
____ *A. s. stenozoma*	Aru Islands and w New Guinea
____ *A. s. megala*	Eastern New Guinea
____ *A. s. coultasi*	Admiralty Islands
____ *A. s. bismarckii*	Bismarck Archipelago
____ *A. s. gurneyi*	Solomon Islands
____ *A. s. subcristata*	N and e Australia

☐ **Black Baza** *Aviceda leuphotes*

____ *A. l. wolfei*	W-central China (Sichuan)
____ *A. l. syama*	NE India to s China; winters to SE Asia and Sumatra
____ *A. l. leuphotes*	SW India to s Myanmar and w Thailand
____ *A. l. andamanica*	Andaman Islands

☐ **Gray-headed Kite** *Leptodon cayanensis*

____ *L. c. cayanensis*	SE Mexico to w Ecuador, the Guianas and Amazonia; Trinidad
____ *L. c. monachus*	Central Brazil to e Bolivia, n Argentina and Paraguay

☐ **White-collared Kite** *Leptodon forbesi*

NE Brazil (Pernambuco and Alagoas)

☐ **Hook-billed Kite** *Chondrohierax uncinatus*

____ *C. u. uncinatus*	S US and w Mexico to Brazil and n Argentina
____ *C. u. wilsonii*	E Cuba
____ *C. u. mirus*	Grenada (Lesser Antilles)

☐ **Long-tailed Honey-buzzard** *Henicopernis longicauda*

New Guinea, Aru, Yapen and w Papuan islands

☐ **Black Honey-buzzard** *Henicopernis infuscatus*

New Britain (Bismarck Archipelago)

☐ **European Honey-buzzard** *Pernis apivorus*

W Palearctic; winters s Europe and Iran to s Africa

☐ **Barred Honey-buzzard** *Pernis celebensis*

____ *P. c. celebensis*	Sulawesi, Muna, Butung and Peleng islands
____ *P. c. steerei*	Philippine Islands (except Palawan)

☐ **Oriental Honey-buzzard** *Pernis ptilorhynchus*

____ *P. p. orientalis*	S Siberia to Manchuria and Japan; winters to Greater Sundas
____ *P. p. ruficollis*	India and Sri Lanka to Myanmar and extreme sw China
____ *P. p. philippensis*	N and e Philippine Islands
____ *P. p. palawanensis*	S Philippines (Palawan and Calauit)
____ *P. p. torquatus*	Malay Peninsula, Sumatra and Borneo
____ *P. p. ptilorhynchus*	Java

☐ **Square-tailed Kite** *Lophoictinia isura*

Locally in open woodland and scrub throughout Australia

☐ **Black-breasted Kite** *Hamirostra melanosternon*

N and interior plains and scrub of Australia

☐ **Swallow-tailed Kite** *Elanoides forficatus*

____ *E. f. forficatus*	Lowlands of coastal se US to n Mexico
____ *E. f. yetapa*	S Mexico (except Yucatán Peninsula) to Brazil and ne Argentina

☐ **Bat Hawk** *Macheiramphus alcinus*

____ *M. a. alcinus*	S Myanmar to Malay Peninsula, Sumatra, Borneo and Sulawesi
____ *M. a. papuanus*	E New Guinea
____ *M. a. anderssoni*	Africa south of the Sahara and Madagascar

☐ **Pearl Kite** *Gampsonyx swainsonii*

____ *G. s. leonae*	Nicaragua; n South America south to the Amazon

____ *G. s. swainsonii*   Brazil south of the Amazon to e Peru, Bolivia and ne Argentina

____ *G. s. magnus*   Coastal w Colombia to Ecuador and n Peru

☐ **Black-shouldered Kite** *Elanus caeruleus*

____ *E. c. caeruleus*   SW Iberian Peninsula, Africa and sw Arabia

____ *E. c. vociferus*   Pakistan to e China, Indochina and Malay Peninsula

____ *E. c. hypoleucus*   Greater and Lesser Sundas, Sulawesi and Philippines

____ *E. c. wahgiensis*   New Guinea

☐ **Australian Kite** *Elanus axillaris*

Open woodlands, savanna and grasslands throughout Australia

☐ **White-tailed Kite** *Elanus leucurus*

____ *E. l. majusculus*   W and s US to w Panama

____ *E. l. leucurus*   E Panama to Brazil, central Argentina and central Chile

☐ **Letter-winged Kite** *Elanus scriptus*

Interior of Australia; disperses irregularly to coasts

☐ **Scissor-tailed Kite** *Chelictinia riocourii*

Savanna of sub-Saharan Africa

☐ **Snail Kite** *Rostrhamus sociabilis*

____ *R. s. plumbeus*   Freshwater marshes of Florida, Cuba and Isle of Pines

____ *R. s. major*   E Mexico and Petén of n Guatemala

____ *R. s. sociabilis*   Honduras and Nicaragua to Brazil and ne Argentina

☐ **Slender-billed Kite** *Rostrhamus hamatus*

E Panama to the Guianas, n Bolivia and Amazonian Brazil

☐ **Double-toothed Kite** *Harpagus bidentatus*

____ *H. b. fasciatus*   SE Mexico to w Colombia and w Ecuador

____ *H. b. bidentatus*   E Colombia and Ecuador through Amazonia to se Brazil

☐ **Rufous-thighed Kite** *Harpagus diodon*

Lowlands of the Guianas to n Argentina and all of Brazil

☐ **Mississippi Kite** *Ictinia mississippiensis*

S US (Arizona to Florida); winters in South America

☐ **Plumbeous Kite** *Ictinia plumbea*

E Mexico to ne Argentina and all of Brazil

☐ **Red Kite** *Milvus milvus*

____ *M. m. milvus*   Locally in western Palearctic region

____ *M. m. fasciicauda*   Cape Verde Islands

☐ **Black Kite** *Milvus migrans*

____ *M. m. migrans*   NW Africa and Europe to s-central Asia; winters to s Africa

____ *M. m. lineatus*   Siberia to n India, China and Ryukyu Is.; winters to Iraq, SE Asia

____ *M. m. formosanus*   Taiwan and Hainan (s China)

____ *M. m. govinda*   Indian subcontinent to Indochina and Malay Peninsula

____ *M. m. affinis*   Sulawesi to Moluccas, New Guinea, Solomons and Australia

____ *M. m. aegyptius*   Egypt, sw Arabia and coastal ne Africa

____ *M. m. parasitus*   Africa s of the Sahara, Madagascar, Cape Verde and Comoro is.

☐ **Whistling Kite** *Haliastur sphenurus*

Australia, New Guinea and New Caledonia

☐ **Brahminy Kite** *Haliastur indus*

____ *H. i. indus*   Indian subcontinent and SE Asia to s China

____ *H. i. intermedius*   Malay Peninsula, Indonesian Archipelago and Philippine Islands

____ *H. i. girrenera*   Moluccas, New Guinea, Bismarck Arch. and coastal n Australia

____ *H. i. flavirostris*   Solomon Islands

☐ **White-bellied Sea-Eagle** *Haliaeetus leucogaster*

Coasts and islands of s Asia to Philippines and Australia

☐ **Solomon Sea-Eagle** *Haliaeetus sanfordi*

Solomon Islands

☐ **African Fish-Eagle** *Haliaeetus vocifer*

Africa south of the Sahara

☐ **Madagascar Fish-Eagle** *Haliaeetus vociferoides*

Locally along coasts, lakes and rivers of nw Madagascar

☐ **Pallas' Fish-Eagle** *Haliaeetus leucoryphus*

Central Asia to India, Myanmar and s-central China (Sichuan)

☐ **White-tailed Eagle** *Haliaeetus albicilla*

Locally in Palearctic, sw Greenland, w Iceland and w Alaska

☐ **Bald Eagle** *Haliaeetus leucocephalus*
____ *H. l. washingtoniensis*
____ *H. l. leucocephalus*

Locally in Aleutian Islands, Alaska, Canada and n US
Locally from southern US to nw Mexico

☐ **Steller's Sea-Eagle** *Haliaeetus pelagicus*

Coastal e Siberia; winters to China, Korea, Japan and Ryukyu Is.

☐ **Lesser Fish-Eagle** *Ichthyophaga humilis*
____ *I. h. plumbea*
____ *I. h. humilis*

Kashmir to SE Asia and Hainan (s China)
Malaya to Borneo, Sumatra, Sulawesi, Banggai and Sula islands

☐ **Gray-headed Fish-Eagle** *Ichthyophaga ichthyaetus*

Lowlands of India to SE Asia, Borneo, Java and Philippines

☐ **Palm-nut Vulture** *Gypohierax angolensis*

Oil-palm forests and savanna of tropical Africa

☐ **Hooded Vulture** *Necrosyrtes monachus*

Savanna of Africa south of the Sahara

☐ **Lammergeier** *Gypaetus barbatus*
____ *G. b. barbatus*
____ *G. b. meridionalis*

Mountains of s Europe and nw Africa to central and ne China
Locally in e and s Africa and sw Arabia

☐ **Egyptian Vulture** *Neophron percnopterus*
____ *N. p. percnopterus*
____ *N. p. majorensis*
____ *N. p. ginginianus*

Africa, s Europe to nw India; Cape Verde Is.
Canary Islands (Fuerteventura I.)
Nepal and India (except northwest)

☐ **White-backed Vulture** *Gyps africanus*

Open plains and savanna of Africa south of the Sahara

☐ **White-rumped Vulture** *Gyps bengalensis*

Lowlands of Iran to India, sw China and SE Asia

☐ **Indian Vulture** *Gyps indicus*

Pakistan and India (south of the Ganges River)

☐ **Slender-billed Vulture** *Gyps tenuirostris*

Kashmir through base of Himalayas to SE Asia

☐ **Rueppell's Griffon** *Gyps rueppellii*
____ *G. r. rueppellii*
____ *G. r. erlangeri*

SW Mauritania to e Sudan, Uganda, Kenya and Tanzania
Ethiopia, Eritrea and nw Somalia; s Arabia?

☐ **Himalayan Griffon** *Gyps himalayensis*

Himalayas from nw India to Tibet and w-central China

☐ **Eurasian Griffon** *Gyps fulvus*
____ *G. f. fulvus*
____ *G. f. fulvescens*

NW Africa and Iberian Peninsula to Middle East
Afghanistan, Pakistan and n India to Assam

☐ **Cape Griffon** *Gyps coprotheres*

Open plains and mountains of southern Africa

☐ **Cinereous Vulture** *Aegypius monachus*

Mediterranean basin environs to e Asia

☐ **Lappet-faced Vulture** *Torgos tracheliotus*
____ *T. t. tracheliotus*

SW Morocco and Africa south of the Sahara

____ *T. t. nubicus*	Egypt and n Sudan	
____ *T. t. negevensis*	S Israel and Arabian Peninsula	

☐ **White-headed Vulture** *Trigonoceps occipitalis*

Thornscrub and deserts of Africa south of the Sahara

☐ **Red-headed Vulture** *Sarcogyps calvus*

India and SE Asia

☐ **Short-toed Eagle** *Circaetus gallicus*

W Palearctic to c Asia, Indian subcontinent and Lesser Sundas

☐ **Beaudouin's Snake-Eagle** *Circaetus beaudouini*

Senegal and Mauritania to s Sudan, n Uganda and nw Kenya

☐ **Black-breasted Snake-Eagle** *Circaetus pectoralis*

E Sudan and Ethiopia to South Africa

☐ **Brown Snake-Eagle** *Circaetus cinereus*

Senegambia to n Ethiopia and south to South Africa

☐ **Fasciated Snake-Eagle** *Circaetus fasciolatus*

Mainly coastal districts of s Somalia to e South Africa

☐ **Banded Snake-Eagle** *Circaetus cinerascens*

Savanna and thornscrub of Africa south of the Sahara

☐ **Bateleur** *Terathopius ecaudatus*

Savanna and thornscrub of Africa south of the Sahara

☐ **Nicobar Serpent-Eagle** *Spilornis klossi*

Great Nicobar I. (Nicobar Islands)

☐ **Sulawesi Serpent-Eagle** *Spilornis rufipectus*

____ *S. r. rufipectus*	Sulawesi and adjacent islands
____ *S. r. sulaensis*	Banggai and Sula islands (off e Sulawesi)

☐ **Mountain Serpent-Eagle** *Spilornis kinabaluensis*

Mountains of n Borneo

☐ **Crested Serpent-Eagle** *Spilornis cheela*

____ *S. c. cheela*	N India and Nepal
____ *S. c. melanotis*	Indo-Gangetic plain
____ *S. c. spilogaster*	Sri Lanka
____ *S. c. burmanicus*	Myanmar to sw China, Thailand and Indochina
____ *S. c. davisoni*	Andaman Islands
____ *S. c. minimus*	Nicobar Islands
____ *S. c. ricketti*	Southern China and n Vietnam
____ *S. c. perplexus*	S Ryukyu Islands
____ *S. c. hoya*	Taiwan
____ *S. c. rutherfordi*	Hainan (s China)
____ *S. c. palawanensis*	Palawan (sw Philippines)
____ *S. c. pallidus*	Lowlands of n Borneo
____ *S. c. richmondi*	S Borneo
____ *S. c. natunensis*	Natunas and Belitung islands (off Borneo)
____ *S. c. malayensis*	Malay Peninsula, n Sumatra and Anambas Islands
____ *S. c. batu*	S Sumatra and Batu Islands
____ *S. c. abbotti*	Simeulue I. (off w Sumatra)
____ *S. c. asturinus*	Nias I. (off w Sumatra)
____ *S. c. sipora*	Mentawai Archepelago (off w Sumatra)
____ *S. c. bido*	Java and Bali
____ *S. c. baweanus*	Bawean I. (off n Java)

☐ **Philippine Serpent-Eagle** *Spilornis holospilus*

Forests of larger Philippine Islands (except Palawan)

☐ **Andaman Serpent-Eagle** *Spilornis elgini*

Forests of Andaman Islands

☐ **Congo Serpent-Eagle** *Dryotriorchis spectabilis*

____ *D. s. spectabilis*	Sierra Leone to Nigeria and nw Cameroon
____ *D. s. batesi*	S Cameroon to Uganda, Gabon and n Angola

☐ **Madagascar Serpent-Eagle** *Eutriorchis astur*

Rainforests of ne Madagascar (on verge of extinction)

☐ **Western Marsh-Harrier** *Circus aeruginosus*
_____ *C. a. aeruginosus*
_____ *C. a. harterti*

W and central Palearctic; winters to SE Asia and Greater Sundas
Morocco to Tunisia

☐ **African Marsh-Harrier** *Circus ranivorus*

Marshes and grasslands of e and s Africa

☐ **Eastern Marsh-Harrier** *Circus spilonotus*
_____ *C. s. spilonotus*
_____ *C. s. spilothorax*

E Asia; winters to SE Asia, Philippines and Indonesia
Central and e New Guinea

☐ **Swamp Harrier** *Circus approximans*

Australasian region and sw Oceania

☐ **Reunion Harrier** *Circus maillardi*
_____ *C. m. maillardi*
_____ *C. m. macrosceles*

Réunion (w Indian Ocean)
Madagascar and Comoro Islands

☐ **Long-winged Harrier** *Circus buffoni*

Grasslands of tropical South America, Trinidad and Tobago

☐ **Spotted Harrier** *Circus assimilis*

Sulawesi, Sula Islands, e Lesser Sundas and Australia

☐ **Black Harrier** *Circus maurus*

Arid grasslands of s Namibia, sw Botswana and South Africa

☐ **Cinereous Harrier** *Circus cinereus*

Andes of Colombia to Tierra del Fuego and Falkland Islands

☐ **Northern Harrier** *Circus cyaneus*
_____ *C. c. cyaneus*
_____ *C. c. hudsonius*

Widespread Eurasia
Widespread North America; winters to n South America

☐ **Pallid Harrier** *Circus macrourus*

Central Eurasia; winters to s Africa, India and Myanmar

☐ **Pied Harrier** *Circus melanoleucos*

E Asia; winters to s Asia, Philippines and Greater Sundas

☐ **Montagu's Harrier** *Circus pygargus*

N Palearctic; winters to s Africa, Iran and India

☐ **African Harrier-Hawk** *Polyboroides typus*
_____ *P. t. pectoralis*
_____ *P. t. typus*

Senegambia to w Sudan, Niger and Zaire
E Sudan to Eritrea, Angola and South Africa

☐ **Madagascar Harrier-Hawk** *Polyboroides radiatus*

Woodlands and savanna of Madagascar

☐ **Lizard Buzzard** *Kaupifalco monogrammicus*
_____ *K. m. monogrammicus*
_____ *K. m. meridionalis*

Senegambia to Ethiopia, Uganda and Kenya
S Kenya to Angola, n Namibia and n South Africa

☐ **Dark Chanting-Goshawk** *Melierax metabates*
_____ *M. m. theresae*
_____ *M. m. neumanni*
_____ *M. m. ignoscens*
_____ *M. m. metabates*
_____ *M. m. mechowi*

SW Morocco
Mali east to n Sudan
SW Arabian Peninsula
Senegambia to Ethiopia and south to Zaire and n Tanzania
SE Gabon to Angola, s Tanzania s to n Namibia and ne S Africa

☐ **Eastern Chanting-Goshawk** *Melierax poliopterus*

SE Ethiopia and Somalia to e Uganda and n Tanzania

☐ **Pale Chanting-Goshawk** *Melierax canorus*
_____ *M. c. argentior*
_____ *M. c. canorus*

S Angola to Zimbabwe and ne South Africa
Southern South Africa

☐ **Gabar Goshawk** *Micronisus gabar*

    ____ *M. g. niger* — Senegambia to Sudan, n Ethiopia and sw Arabia

    ____ *M. g. aequatorius* — Highlands of Ethiopia to Zaire, Zambia and n Mozambique

    ____ *M. g. gabar* — S Angola to Zambia, Mozambique and South Africa

☐ **Gray-bellied Goshawk** *Accipiter poliogaster*

Locally in humid forests of South America east of the Andes

☐ **Crested Goshawk** *Accipiter trivirgatus*

    ____ *A. t. indicus* — India and Nepal to s China, Indochina and Malay Peninsula

    ____ *A. t. peninsulae* — SW India

    ____ *A. t. layardi* — Sri Lanka

    ____ *A. t. formosae* — Taiwan

    ____ *A. t. trivirgatus* — Sumatra

    ____ *A. t. niasensis* — Nias I. (off w Sumatra)

    ____ *A. t. javanicus* — Java (vagrant to Bali)

    ____ *A. t. microstictus* — Borneo

    ____ *A. t. palawanus* — SW Philippines (Palawan and Calamianes)

    ____ *A. t. extimus* — SE Philippine Islands

    ____ *A. t. castroi* — Polillo (off Luzon in n Philippines)

☐ **Red-chested Goshawk** *Accipiter toussenelii*

    ____ *A. t. macroscelides* — Rainforests of Senegambia to w Cameroon

    ____ *A. t. toussenelii* — Lower Zaire River basin (s Cameroon to Gabon)

    ____ *A. t. canescens* — Upper Zaire River basin

    ____ *A. t. lopezi* — Bioko (Gulf of Guinea)

☐ **Sulawesi Goshawk** *Accipiter griseiceps*

Sulawesi, Muna, Butung and Togian islands

☐ **African Goshawk** *Accipiter tachiro*

    ____ *A. t. unduliventer* — Highlands of Ethiopia

    ____ *A. t. croizati* — SW Ethiopia

    ____ *A. t. sparsimfasciatus* — Somalia to n Zaire, Angola, Zambia and Mozambique

    ____ *A. t. pembaensis* — Pemba I. (off Tanzania)

    ____ *A. t. tachiro* — S Angola to Mozambique and South Africa

☐ **Chestnut-flanked Sparrowhawk** *Accipiter castanilius*

Dense forests of Nigeria to Zaire River basin

☐ **Shikra** *Accipiter badius*

    ____ *A. b. cenchroides* — Azerbaijan to Kazakstan, Iran and nw India

    ____ *A. b. dussumieri* — Central India and Bangladesh

    ____ *A. b. badius* — SW India and Sri Lanka

    ____ *A. b. poliopsis* — N India to s China, Thailand and Vietnam

    ____ *A. b. sphenurus* — Senegambia to sw Arabia, n Zaire and n Tanzania

    ____ *A. b. polyzonoides* — S Zaire and s Tanzania to n South Africa

☐ **Nicobar Sparrowhawk** *Accipiter butleri*

    ____ *A. b. butleri* — Car Nicobar (north Nicobar Islands)

    ____ *A. b. obsoletus* — Central Nicobar Islands (Katchall and Camorta)

☐ **Levant Sparrowhawk** *Accipiter brevipes*

Balkans to Russia; winters to ne Africa and Arabian Peninsula

☐ **Chinese Goshawk** *Accipiter soloensis*

China and Korea; winters to SE Asia, Philippines and Indonesia

☐ **Frances' Goshawk** *Accipiter francesii*

    ____ *A. f. francesii* — Madagascar

    ____ *A. f. griveaudi* — Grand Comoro I. (Comoro Islands)

    ____ *A. f. pusillus* — Anjouan (Comoro Islands)

    ____ *A. f. brutus* — Mayotte (Comoro Islands)

☐ **Spot-tailed Goshawk** *Accipiter trinotatus*

Sulawesi, Talisei, Muna and Butung islands

☐ **Variable Goshawk** *Accipiter hiogaster*
_____ *A. h. sylvestris* — Lesser Sundas
_____ *A. n. polionotus* — Banda and Tanimbar islands (e Indonesian Archipelago)
_____ *A. h. albiventris* — Tayandu and Kai islands (e Indonesian Archipelago)
_____ *A. h. obiensis* — Obi I. (central Moluccas)
_____ *A. h. griseogularis* — N Moluccas (Halmahera, Ternate, Tidore and Bacan)
_____ *A. h. mortyi* — Morotai I. (n Moluccas)
_____ *A. h. hiogaster* — South Moluccas
_____ *A. h. pallidiceps* — Buru I. (s Moluccas)
_____ *A. h. leucosomus* — New Guinea
_____ *A. h. pallidimas* — D'Entrecasteaux Archipelago
_____ *A. h. manusi* — Admiralty Islands
_____ *A. h. bougainvillei* — Bougainville (n Solomon Islands)
_____ *A. h. rufoschistaceus* — Solomon Islands (Choiseul, Santa Isabel and Florida Group)
_____ *A. h. rubianae* — Central Solomon Islands
_____ *A. h. pulchellus* — Guadalcanal (sw Solomon Islands)
_____ *A. h. malaitae* — Malaita (se Solomon Islands)
_____ *A. h. misulae* — Louisiade Archipelago
_____ *A. h. misoriensis* — Biak I. (New Guinea)
_____ *A. h. dampieri* — New Britain (Bismarck Archipelago)
_____ *A. h. lavongai* — Bismarck Archipelago (New Hanover and New Ireland)
_____ *A. h. lihirensis* — Bismarck Archipelago (Lihir and Tanga)
_____ *A. h. matthiae* — St. Matthias I. (Bismarck Archipelago)

☐ **Gray Goshawk** *Accipiter novaehollandiae*

N and e Australia and Tasmania

☐ **Brown Goshawk** *Accipiter fasciatus*
_____ *A. f. natalis* — Christmas I. (Indian Ocean)
_____ *A. f. tjendanae* — Sumba (Lesser Sundas)
_____ *A. f. wallacii* — Lombok, Sumbawa, Flores and adjacent Lesser Sundas
_____ *A. f. stresemanni* — Tanahjampea, Kalao, Bonerate, Kalaotoa, Madu and Tukanbesi is.
_____ *A. f. hellmayri* — Lesser Sundas (Timor, Alor, Semau and Roti)
_____ *A. f. savu* — Sawu (Lesser Sundas)
_____ *A. f. polycryptus* — E New Guinea
_____ *A. f. dogwa* — S New Guinea
_____ *A. f. didimus (buruensis)* — Buru (s Moluccas) and n Australia
_____ *A. f. fasciatus* — Australia, Tasmania, Solomon Is. (Rennell, Bellona) and Timor
_____ *A. f. vigilax* — New Caledonia, Loyalty Islands and Vanuatu

☐ **Black-mantled Goshawk** *Accipiter melanochlamys*
_____ *A. m. melanochlamys* — W New Guinea (Vogelkop Mountains)
_____ *A. m. schistacinus* — Montane forests of central and e New Guinea

☐ **Pied Goshawk** *Accipiter albogularis*
_____ *A. a. eichhorni* — Feni I. (Bismarck Archipelago)
_____ *A. a. woodfordi* — Solomon Islands
_____ *A. a. albogularis* — Solomon Islands (San Cristóbal and Santa Anna)
_____ *A. a. gilvus* — Central Solomon Islands
_____ *A. a. sharpei* — Lowlands of Santa Cruz Group (Solomon Islands)

☐ **New Caledonia Goshawk** *Accipiter haplochrous*

New Caledonia

☐ **Fiji Goshawk** *Accipiter rufitorques*

Fiji Islands

☐ **Moluccan Goshawk** *Accipiter henicogrammus*

N Moluccas (Bacan, Halmahera, Ternate and Morotai)

☐ **Slaty-mantled Goshawk** *Accipiter luteoschistaceus*

Mountains of New Britain (Bismarck Archipelago)

☐ **Imitator Sparrowhawk** *Accipiter imitator*

Solomon Islands (Bougainville, Choiseul and Santa Isabel)

☐ **Gray-headed Goshawk** *Accipiter poliocephalus*

New Guinea, Aru Is., D'Entrecasteaux Arch. and Louisiade Arch.

☐ **New Britain Goshawk** *Accipiter princeps*

Highlands of New Britain (Bismarck Archipelago)

☐ **Tiny Hawk** *Accipiter superciliosus*
    ___ *A. s. fontanieri* — Nicaragua to w Colombia and Ecuador
    ___ *A. s. superciliosus* — S America east of the Andes to extreme ne Argentina and Brazil

☐ **Semicollared Hawk** *Accipiter collaris*

Mountains of Colombia to sw Venezuela, Ecuador and s Peru

☐ **Red-thighed Sparrowhawk** *Accipiter erythropus*
    ___ *A. e. erythropus* — Senegambia to Nigeria
    ___ *A. e. zenkeri* — Cameroon to w Uganda, central Zaire, Gabon and n Angola

☐ **Little Sparrowhawk** *Accipiter minullus*

Forests and thornscrub of e and s Africa

☐ **Japanese Sparrowhawk** *Accipiter gularis*
    ___ *A. g. sibiricus* — Mongolia to e China and Taiwan; winters to India and Indonesia
    ___ *A. g. gularis* — Sakhalin, Kuril Is. and Japan; winters to Philippines and Indonesia
    ___ *A. g. iwasakii* — S Ryukyu Islands (Iriomote and Ishigaki)

☐ **Small Sparrowhawk** *Accipiter nanus*

Montane forests of Sulawesi

☐ **Besra** *Accipiter virgatus*
    ___ *A. v. affinis* — N India and Nepal to central China and Indochina
    ___ *A. v. fuscipectus* — Mountains of Taiwan
    ___ *A. v. besra* — S India and Sri Lanka
    ___ *A. v. abdulali* — Andaman and Nicobar islands
    ___ *A. v. nisoides* — Myanmar, Thailand and n Malay Peninsula
    ___ *A. v. confusus* — Philippines (Luzon, Mindoro, Negros and Catanduanes)
    ___ *A. v. quagga* — Philippines (Cebu, Bohol, Leyte, Samar, Siquijor and Mindanao)
    ___ *A. v. rufotibialis* — N Borneo
    ___ *A. v. vanbemmeli* — Sumatra
    ___ *A. v. virgatus* — Java and Bali
    ___ *A. v. quinquefasciatus* — Flores (Lesser Sundas)

☐ **Rufous-necked Sparrowhawk** *Accipiter erythrauchen*
    ___ *A. e. erythrauchen* — N Moluccas (Morotai, Halmahera, Bacan and Obi)
    ___ *A. e. ceramensis* — S Moluccas (Buru, Ambon and Seram)

☐ **Collared Sparrowhawk** *Accipiter cirrocephalus*
    ___ *A. c. papuanus* — New Guinea, w Papuan islands and Aru Islands
    ___ *A. c. rosselianus* — Rossell (Louisiade Archipelago)
    ___ *A. c. cirrocephalus* — Australia and Tasmania

☐ **New Britain Sparrowhawk** *Accipiter brachyurus*

New Britain (Bismarck Archipelago)

☐ **Vinous-breasted Sparrowhawk** *Accipiter rhodogaster*
    ___ *A. r. rhodogaster* — Sulawesi
    ___ *A. r. butonensis* — Muna and Butung islands (off Sulawesi)
    ___ *A. r. sulaensis* — Banggai and Sula islands (off Sulawesi)

☐ **Madagascar Sparrowhawk** *Accipiter madagascariensis*

Scrub and savanna of Madagascar

☐ **Ovampo Sparrowhawk** *Accipiter ovampensis*

Locally in savanna and thornscrub of Africa south of the Sahara

☐ **Eurasian Sparrowhawk** *Accipiter nisus*
    ___ *A. n. granti* — Madeira and Canary Islands

____	*A. n. nisus*	Europe to Asia Minor and Siberia; winters to Africa
____	*A. n. wolterstorffi*	Corsica and Sardinia
____	*A. n. punicus*	NW Africa (Morocco to Tunisia)
____	*A. n. nisosimilis*	Central and e Asia; winters to India, Sri Lanka and Indochina
____	*A. n. melaschistos*	Himalayas and mountains of central Asia

☐ **Rufous-chested Sparrowhawk** *Accipiter rufiventris*

____	*A. r. perspicillaris*	Highland forests of Ethiopia
____	*A. r. rufiventris*	Montane forests of Kenya and e Zaire to South Africa

☐ **Sharp-shinned Hawk** *Accipiter striatus*

____	*A. s. perobscurus*	Queen Charlotte Is. and (?) adjacent coastal British Columbia
____	*A. s. velox*	Alaska and Canada to s US; winters to Panama
____	*A. s. suttoni*	Locally from extreme s New Mexico to se Mexico (Veracruz)
____	*A. s. madrensis*	W Mexico (Guerrero and w Oaxaca)
____	*A. s. chionogaster*	Oak-pine highlands of s Mexico (Chiapas) to Nicaragua
____	*A. s. fringilloides*	Cuba
____	*A. s. striatus*	Hispaniola
____	*A. s. venator*	Puerto Rico

☐ **Plain-breasted Hawk** *Accipiter ventralis*

Andes of Colombia and w Venezuela to w Bolivia

☐ **Rufous-thighed Hawk** *Accipiter erythronemius*

Bolivia and Paraguay to n Argentina and s Brazil

☐ **Cooper's Hawk** *Accipiter cooperii*

Woodlands of s Canada and US; winters to Central America

☐ **Gundlach's Hawk** *Accipiter gundlachi*

____	*A. g. gundlachi*	Lowland forests of w and central Cuba
____	*A. g. wileyi*	Lowland forests of e Cuba

☐ **Bicolored Hawk** *Accipiter bicolor*

____	*A. b. fidens*	Lowlands of s Mexico (Oaxaca, Veracruz and Yucatán Peninsula)
____	*A. b. bicolor*	SE Mexico (Yucatán Pen.) to the Guianas, Brazil and nw Peru
____	*A. b. pileatus*	Brazil south of the Amazon to ne Argentina
____	*A. b. guttifer*	Bolivia to Paraguay, sw Brazil (Mato Grosso) and n Argentina

☐ **Chilean Hawk** *Accipiter chilensis*

Andes of central Chile and Argentina to Tierra del Fuego

☐ **Black Goshawk** *Accipiter melanoleucus*

____	*A. m. temminckii*	Senegambia to Gabon, Congo and Central African Republic
____	*A. m. melanoleucus*	E Sudan and nw Ethiopia; Kenya to Angola and South Africa

☐ **Henst's Goshawk** *Accipiter henstii*

Forests and savanna of Madagascar

☐ **Northern Goshawk** *Accipiter gentilis*

____	*A. g. gentilis*	Europe and extreme nw Africa
____	*A. g. arrigonii*	Corsica and Sardinia
____	*A. g. buteoides*	N Eurasia (Sweden to Lena River); winters to central Asia
____	*A. g. albidus*	NE Siberia to Kamchatka Peninsula
____	*A. g. schvedowi*	NE Asia to central China; winters to n Indochina
____	*A. g. fujiyamae*	Japan
____	*A. g. atricapillus*	North America south to s US and w Mexico
____	*A. g. laingi*	SW Canada (Queen Charlotte Islands and Vancouver I.)

☐ **Meyer's Goshawk** *Accipiter meyerianus*

Moluccas to New Guinea, New Britain and Solomon Islands

☐ **Chestnut-shouldered Goshawk** *Erythrotriorchis buergersi*

Montane forests of n and e New Guinea

☐ **Red Goshawk** *Erythrotriorchis radiatus*

Sparsely distributed in open woodlands of n and e Australia

☐ **Doria's Goshawk** *Megatriorchis doriae*

New Guinea and Batanta I.

☐ **Long-tailed Hawk** *Urotriorchis macrourus*

Rainforests of Liberia to w Uganda and s Zaire

☐ **Grasshopper Buzzard** *Butastur rufipennis*

Savanna and grasslands of sub-Saharan Africa

☐ **White-eyed Buzzard** *Butastur teesa*

Woodlands and plains of se Iran to India, s Tibet and Myanmar

☐ **Rufous-winged Buzzard** *Butastur liventer*

Savanna and woodlands of SE Asia, Java and Sulawesi

☐ **Gray-faced Buzzard** *Butastur indicus*

NE Asia; winters SE Asia to Philippines and Indonesia

☐ **Crane Hawk** *Geranospiza caerulescens*
____ *G. c. livens* — NW Mexico
____ *G. c. nigra* — Northern Mexico to central Panama
____ *G. c. balzarensis* — E Panama to w Colombia, Ecuador and extreme nw Peru
____ *G. c. caerulescens* — Guianas and Amazonian Brazil to e Colombia and Peru
____ *G. c. gracilis* — NE Brazil (Maranhão, Ceará and Piauí to Bahia)
____ *G. c. flexipes* — S Brazil to *chaco* of Paraguay, Bolivia and n Argentina

☐ **Plumbeous Hawk** *Leucopternis plumbeus*

E Panama to w Colombia, Ecuador and extreme nw Peru

☐ **Slate-colored Hawk** *Leucopternis schistaceus*

Tropical s Venezuela to e Bolivia and Amazonian Brazil

☐ **Barred Hawk** *Leucopternis princeps*

Montane forests of Costa Rica to Colombia and n Ecuador

☐ **Black-faced Hawk** *Leucopternis melanops*

SE Colombia to s Venezuela, ne Ecuador and n Amaz. Brazil

☐ **White-browed Hawk** *Leucopternis kuhli*

Rainforests of e Peru, n Bolivia and s Amazonian Brazil

☐ **White-necked Hawk** *Leucopternis lacernulatus*

Lowland forests of coastal se Brazil

☐ **Semiplumbeous Hawk** *Leucopternis semiplumbeus*

Humid forests of Honduras to w Colombia and nw Ecuador

☐ **White Hawk** *Leucopternis albicollis*
____ *L. a. ghiesbreghti* — Tropical forests of s Mexico to Guatemala and Belize
____ *L. a. costaricensis* — Honduras to Panama and w Colombia
____ *L. a. williaminae* — NW Colombia to extreme nw Venezuela
____ *L. a. albicollis* — Humid forests of the Guianas and Amazonian basin; Trinidad

☐ **Gray-backed Hawk** *Leucopternis occidentalis*

Montane forests of w Ecuador and adjacent nw Peru

☐ **Mantled Hawk** *Leucopternis polionotus*

Tropical forests of e Brazil to e Uruguay and e Paraguay

☐ **Rufous Crab-Hawk** *Buteogallus aequinoctialis*

NE Venezuela (Orinoco delta) to e Brazil (Paraná)

☐ **Common Black-Hawk** *Buteogallus anthracinus*
____ *B. a. anthracinus* — SW US to n South America, St. Vincent and Trinidad
____ *B. a. gundlachii* — Cuba and Isle of Pines
____ *B. a. utilensis* — Cancún, Cozumel I. and islands in Gulf of Honduras

☐ **Mangrove Black-Hawk** *Buteogallus subtilis*
____ *B. s. rhizophorae* — Pacific coast of El Salvador and Honduras
____ *B. s. bangsi* — Pacific coast of Costa Rica and Panama; Pearl Islands
____ *B. s. subtilis* — Pacific coast of Colombia, Ecuador and extreme n Peru

☐ **Great Black-Hawk** *Buteogallus urubitinga*
____ *B. u. ridgwayi* — Lowlands of n Mexico to w Panama
____ *B. u. urubitinga* — E Panama through South America to n Argentina

☐ **Savanna Hawk** *Buteogallus meridionalis*

Savanna and marshes of w Panama to Brazil and n Argentina

☐ **Harris' Hawk** *Parabuteo unicinctus*
____ *P. u. harrisi*
____ *P. u. unicinctus*

Arid sw US to Pacific slope of Colombia, Ecuador and Peru
E Colombia and Venezuela to Brazil, s Argentina and s Chile

☐ **Black-collared Hawk** *Busarellus nigricollis*
____ *B. n. nigricollis*
____ *B. n. leucocephalus*

Lowlands of central Mexico to Amazonian Brazil and e Bolivia
Paraguay, Uruguay and n Argentina

☐ **Black-chested Buzzard-Eagle** *Geranoaetus melanoleucus*
____ *G. m. australis*
____ *G. m. melanoleucus*

Andes of w Venezuela to Tierra del Fuego
SE Brazil to Paraguay, Uruguay and ne Argentina

☐ **Solitary Eagle** *Harpyhaliaetus solitarius*
____ *H. s. sheffleri*
____ *H. s. solitarius*

Locally in montane forests of s Mexico to Panama
Locally in montane forests of Venezuela to nw Argentina

☐ **Crowned Eagle** *Harpyhaliaetus coronatus*

Savanna of s Brazil, Paraguay and Bolivia to n Argentina

☐ **Roadside Hawk** *Buteo magnirostris*
____ *B. m. griseocauda*
____ *B. m. conspectus*
____ *B. m. gracilis*
____ *B. m. sinushonduri*
____ *B. m. petulans*
____ *B. m. alius*
____ *B. m. magnirostris*
____ *B. m. occiduus*
____ *B. m. saturatus*
____ *B. m. nattereri*
____ *B. m. magniplumis*
____ *B. m. pucherani*

Mexico to nw Costa Rica and w Panama
SE Mexico (Tabasco and Yucatán Peninsula) to n Belize
Cozumel and Holbox islands (off Yucatán Peninsula)
Bonacca and Roatán islands (Honduras)
SW Costa Rica and w Panama and adjacent islands
Pearl Islands (San José and San Miguel) in Gulf of Panama
Colombia and w Ecuador to the Guianas and Amazonian Brazil
W Amazonian Brazil, e Peru and n Bolivia
SW Brazil to Paraguay, Bolivia and w Argentina
NE Brazil (south to Bahia)
S Brazil to ne Argentina (Misiones) and adjacent Paraguay
Uruguay and ne Argentina (south to Buenos Aires Province)

☐ **Red-shouldered Hawk** *Buteo lineatus*
____ *B. l. elegans*
____ *B. l. lineatus*
____ *B. l. texanus*
____ *B. l. alleni*
____ *B. l. extimus*

S Oregon to n Baja California
E North America (s Canada to central US)
S Texas to se Mexico (Veracruz)
S-central Texas to South Carolina and n Florida
Florida and Florida Keys

☐ **Ridgway's Hawk** *Buteo ridgwayi*

Lowlands of Hispaniola and satellite islands

☐ **Broad-winged Hawk** *Buteo platypterus*
____ *B. p. platypterus*
____ *B. p. cubanensis*
____ *B. p. brunnescens*
____ *B. p. insulicola*
____ *B. p. rivierei*
____ *B. p. antillarum*

Central and s Canada to s US; winters to Brazil and Bolivia
Cuba
Puerto Rico
Antigua (Lesser Antilles)
Lesser Antilles (Dominica, Martinique and St. Lucia)
Lesser Antilles (St. Vincent and Grenada) to Tobago

☐ **Gray Hawk** *Buteo nitidus*
____ *B. n. plagiatus*
____ *B. n. costaricensis*
____ *B. n. nitidus*
____ *B. n. pallidus*

Lowlands of sw US to nw Costa Rica
SW Costa Rica to n Colombia and w Ecuador
E Colombia and Ecuador to the Guianas and Amazonian Brazil
S-central Brazil to e Bolivia, Paraguay and n Argentina

☐ **White-rumped Hawk** *Buteo leucorrhous*

Forests of Venezuela to n Argentina and s Brazil

☐ **Short-tailed Hawk** *Buteo brachyurus*
____ *B. b. fuliginosus*
____ *B. b. brachyurus*

S Florida; e Mexico to Panama
N South America to Brazil, Bolivia, Paraguay and n Argentina

☐ **White-throated Hawk** *Buteo albigula*

Andes of Venezuela to nw Argentina and Chile

☐ **Swainson's Hawk** *Buteo swainsoni*

W N America; winters to n Argentina, s Brazil and Paraguay

☐ **White-tailed Hawk** *Buteo albicaudatus*
____ *B. a. hypospodius*
____ *B. a. colonus*
____ *B. a. albicaudatus*

S Texas and nw Mexico to n Colombia and nw Venezuela
Netherlands Antilles, n South America and Amazon basin
SE Peru to Bolivia, Paraguay, se Brazil, Uruguay, n Argentina

☐ **Galapagos Hawk** *Buteo galapagoensis*

Galapagos Islands

☐ **Red-backed Hawk** *Buteo polyosoma*
____ *B. p. polyosoma*
____ *B. p. exsul*

Andes of sw Colombia to Tierra del Fuego and Falkland Islands
Juan Fernández Islands (off Chile)

☐ **Puna Hawk** *Buteo poecilochrous*

High Andes of Ecuador to nw Argentina

☐ **Zone-tailed Hawk** *Buteo albonotatus*

Arid sw US to n Bolivia, Paraguay and Brazil

☐ **Hawaiian Hawk** *Buteo solitarius*

Forests of Hawaii (Hawaiian Islands)

☐ **Red-tailed Hawk** *Buteo jamaicensis*
____ *B. j. alascensis*
____ *B. j. harlani*
____ *B. j. calurus*
____ *B. j. borealis*
____ *B. j. kriderii*
____ *B. j. fuertesi*
____ *B. j. hadropus*
____ *B. j. kemsiesi*
____ *B. j. costaricensis*
____ *B. j. fumosus*
____ *B. j. socorroensis*
____ *B. j. umbrinus*
____ *B. j. solitudinis*
____ *B. j. jamaicensis*

SE Alaska and coastal British Columbia
Interior Alaska to sw Yukon and n British Columbia
W North America (west of the Great Plains)
North America (east of the Great Plains)
Plains of s-central Canada to n-central US
Texas to n Mexico
Highlands of central Mexico
S Mexico (Chiapas) to n Nicaragua
Costa Rica
Tres Marías Islands (off w Mexico)
Socorro I. (Revillagigedo Islands off w Mexico)
Florida
Bahamas and Cuba
Jamaica, Hispaniola, Puerto Rico and n Lesser Antilles

☐ **Rufous-tailed Hawk** *Buteo ventralis*

Patagonian forests of s Chile and Argentina

☐ **Eurasian Buzzard** *Buteo buteo*
____ *B. b. buteo*
____ *B. b. arrigonii*
____ *B. b. rothschildi*
____ *B. b. insularum*
____ *B. b. bannermani*
____ *B. b. vulpinus*
____ *B. b. menetriesi*
____ *B. b. japonicus*
____ *B. b. refectus*
____ *B. b. toyoshimai*
____ *B. b. oshiroi*

W Palearctic region and Madeira; winters to w Africa
Corsica and Sardinia
Azores
Canary Islands
Cape Verde Islands
N Palearctic; winters to s Asia and Africa south of the Sahara
S Crimea and Caucasus to n Iran
Central Asia to Japan and Tibet; winters India to Japan
Himalayas and w China
Izu Islands and Bonin Islands
Daito Islands

☐ **Mountain Buzzard** *Buteo oreophilus*
____ *B. o. oreophilus*
____ *B. o. trizonatus*

Highlands of Ethiopia to Tanzania and Malawi
Montane forests of Transvaal and South Africa

☐ **Madagascar Buzzard** *Buteo brachypterus*

Woodlands and savanna of Madagascar

☐ **Long-legged Buzzard** *Buteo rufinus*
____ *B. r. rufinus*
____ *B. r. cirtensis*

SE Europe to Mongolia and India; winters to Africa
Mauritania to Egypt and Arabian Peninsula

☐ **Upland Buzzard** *Buteo hemilasius*

Open steppes and montane slopes of e Asia

☐ **Ferruginous Hawk** *Buteo regalis*

Prairies and plains of s Canada to n Mexico

☐ **Rough-legged Hawk** *Buteo lagopus*
____ *B. l. lagopus*
____ *B. l. menzbieri*
____ *B. l. kamtschatkensis*
____ *B. l. sanctijohannis*

Tundra and grasslands of n Eurasia; winters central Eurasia
NE Asia; winters to central Asia, n China and Japan
Kamchatka Peninsula; winters in e-central Asia
Alaska and n Canada; winters to s US

☐ **Red-necked Buzzard** *Buteo auguralis*

Sub-Saharan and w-central Africa; disperses to Sahel

☐ **Augur Buzzard** *Buteo augur*

Highlands of Ethiopia to Zimbabwe, Angola and Namibia

☐ **Archer's Buzzard** *Buteo archeri*

Highlands of n Somalia

☐ **Jackal Buzzard** *Buteo rufofuscus*

Savanna of s Africa

☐ **Crested Eagle** *Morphnus guianensis*

Forests of e Guatemala to ne Argentina and s Brazil

☐ **Harpy Eagle** *Harpia harpyja*

Forests of Central America to ne Argentina and s Brazil

☐ **New Guinea Eagle** *Harpyopsis novaeguineae*

Forests of New Guinea

☐ **Great Philippine Eagle** *Pithecophaga jefferyi*

Philippines (Luzon, Leyte, Samar and Mindanao)

☐ **Black Eagle** *Ictinaetus malayensis*
____ *I. m perniger*
____ *I. m. malayensis*

N India and Nepal; s India and Sri Lanka
Myanmar to s China, SE Asia and Indonesia

☐ **Lesser Spotted Eagle** *Aquila pomarina*

Locally in e Europe to Caspian lowlands; winters to s Africa

☐ **Indian Spotted Eagle** *Aquila hastata*

India, Bangladesh and n Myanmar

☐ **Greater Spotted Eagle** *Aquila clanga*

C Eurasia and s Asia; winters to Africa, China and Indochina

☐ **Tawny Eagle** *Aquila rapax*
____ *A. r. belisarius*
____ *A. r. rapax*
____ *A. r. vindhiana*

Morocco and Algeria; s Arabia and w Africa to n Kenya
S Kenya and Zaire to Angola, Namibia and South Africa
Locally in Pakistan, India and s Nepal

☐ **Steppe Eagle** *Aquila nipalensis*
____ *A. n. orientalis*
____ *A. n. nipalensis*

Central Eurasia; winters to Middle East, Arabia and s Africa
Altai Mts. to Tibet and Manchuria; winters India to se China

☐ **Spanish Eagle** *Aquila adalberti*

Locally in Iberian Peninsula; formerly Morocco

☐ **Imperial Eagle** *Aquila heliaca*

C Europe to Mongolia; winters to Africa, n India and China

☐ **Wahlberg's Eagle** *Aquila wahlbergi*

Savanna of Africa south of the Sahara

☐ **Gurney's Eagle** *Aquila gurneyi*

Moluccas to New Guinea, w Papuan islands and Aru Islands

☐ **Golden Eagle** *Aquila chrysaetos*
    ____ *A. c. homeyeri*   Iberian Peninsula; nw Africa to Arabia and Iran
    ____ *A. c. chrysaetos*   Western Palearctic region to Siberia and Altai Mountains
    ____ *A. c. daphanea*   Turkestan to Manchuria, Pakistan, Himalayas and sw China
    ____ *A. c. japonica*   Korea and Japan
    ____ *A. c. kamtschatica*   Siberia and Altai Mountains to Kamchatka Peninsula
    ____ *A. c. canadensis*   Alaska to w-central Mexico and coastal ne US

☐ **Wedge-tailed Eagle** *Aquila audax*
    ____ *A. a. audax*   Plains, deserts, open woodlands of Australia and s New Guinea
    ____ *A. a. fleayi*   Tasmania

☐ **Verreaux's Eagle** *Aquila verreauxii*

    Locally in Africa south of the Sahara and s Arabian Peninsula

☐ **Bonelli's Eagle** *Aquila fasciatus*
    ____ *A. f. fasciatus*   Mediterranean basin to India, s China and Indochina
    ____ *A. f. renschi*   Lesser Sundas

☐ **African Hawk-Eagle** *Aquila spilogaster*

    Woodlands and savanna of Africa south of the Sahara

☐ **Booted Eagle** *Aquila pennatus*

    S Palearctic and Africa s of the Sahara; winters to SE Asia, India

☐ **Little Eagle** *Aquila morphnoides*
    ____ *A. m. weiskei*   New Guinea; vagrant to Moluccas (Halmahera and Seram)
    ____ *A. m. morphnoides*   Open woodlands and savannas of Australia

☐ **Ayres' Hawk-Eagle** *Aquila ayresii*

    Woodlands and savanna of Africa south of the Sahara

☐ **Rufous-bellied Eagle** *Aquila kienerii*
    ____ *A. k. kienerii*   NE India and Nepal; sw India (Western Ghats) and Sri Lanka
    ____ *A. k. formosus*   Myanmar to Indochina, Malay Pen., Indonesia and Philippines

☐ **Martial Eagle** *Polemaetus bellicosus*

    Savanna and thornbush of Africa south of the Sahara

☐ **Black-and-white Hawk-Eagle** *Spizastur melanoleucus*

    Forests of s Mexico to n Argentina and Brazil

☐ **Long-crested Eagle** *Lophaetus occipitalis*

    Savanna of Africa south of the Sahara

☐ **Cassin's Hawk-Eagle** *Spizaetus africanus*

    Humid forests of w and central Africa

☐ **Changeable Hawk-Eagle** *Spizaetus cirrhatus*
    ____ *S. c. limnaeetus*   N India to Indochina, Malaya, Greater Sundas and Philippines
    ____ *S. c. cirrhatus*   Peninsular India
    ____ *S. c. ceylanensis*   Sri Lanka
    ____ *S. c. andamanensis*   Andaman Islands
    ____ *S. c. vanheurni*   Simeulue I. (off w Sumatra)

☐ **Flores Hawk-Eagle** *Spizaetus floris*

    Lesser Sundas (Sumbawa, Komodo, Flores and Paloe)

☐ **Mountain Hawk-Eagle** *Spizaetus nipalensis*
    ____ *S. n. orientalis*   Japan
    ____ *S. n. nipalensis*   India to e China, Taiwan, Indochina and Malay Peninsula
    ____ *S. n. kelaarti*   SW India (Western Ghats) and Sri Lanka

☐ **Blyth's Hawk-Eagle** *Spizaetus alboniger*

    Forests of s Myanmar to Malay Peninsula, Sumatra and Borneo

☐ **Javan Hawk-Eagle** *Spizaetus bartelsi*

    Wooded hills of w Java

☐ **Sulawesi Hawk-Eagle** *Spizaetus lanceolatus*

    Forests of Sulawesi, Banggai and Sula islands

☐ **Philippine Hawk-Eagle** *Spizaetus philippensis*

Philippine Islands

☐ **Wallace's Hawk-Eagle** *Spizaetus nanus*
____ *S. n. nanus* — S Myanmar and Thailand to Malay Pen., Sumatra and Borneo
____ *S. n. stresemanni* — Nias I. (off w Sumatra)

☐ **Black Hawk-Eagle** *Spizaetus tyrannus*
____ *S. t. serus* — Forests of s Mexico to ne Argentina and Brazil; Trinidad
____ *S. t. tyrannus* — E and s Brazil to extreme ne Argentina (Misiones)

☐ **Ornate Hawk-Eagle** *Spizaetus ornatus*
____ *S. o. vicarius* — Humid forests of s Mexico to w Colombia and w Ecuador
____ *S. o. ornatus* — Humid tropical n South America to n Argentina and Brazil

☐ **Crowned Hawk-Eagle** *Stephanoaetus coronatus*

Forests of Africa south of the Sahara

☐ **Black-and-chestnut Eagle** *Oroaetus isidori*

Coastal mountains of Venezuela to Andes of nw Argentina

ORDER: FALCONIFORMES
FAMILY: SAGITTARIIDAE (Secretary-bird—1)

☐ **Secretary-bird** *Sagittarius serpentarius*

Savanna and grasslands of Africa south of the Sahara

ORDER: FALCONIFORMES
FAMILY: FALCONIDAE (Falcons and Caracaras—64)

☐ **Black Caracara** *Daptrius ater*

Guianas and s Venezuela to n Bolivia and Amazonian Brazil

☐ **Red-throated Caracara** *Ibycter americanus*

Extreme s Mexico to n Bolivia and s Brazil

☐ **Carunculated Caracara** *Phalcoboenus carunculatus*

Páramo of sw Colombia and Ecuador

☐ **Mountain Caracara** *Phalcoboenus megalopterus*

Páramo of n Peru to nw Argentina and central Chile

☐ **White-throated Caracara** *Phalcoboenus albogularis*

Andes of s Argentina and s Chile to Tierra del Fuego

☐ **Striated Caracara** *Phalcoboenus australis*

Tierra del Fuego, Staten I., Navarino I. and Falkland Islands

☐ **Crested Caracara** *Caracara cheriway*
____ *C. c. pallidus* — Tres Marías Islands (off w Mexico)
____ *C. c. audubonii* — S US to w Panama, Cuba and Isle of Pines
____ *C. c. cheriway* — E Panama and n S Am. to n Peru, Brazil, Aruba and Trinidad

☐ **Southern Caracara** *Caracara plancus*

Amazon basin to e Peru, Tierra del Fuego and Falkland Islands

☐ **Yellow-headed Caracara** *Milvago chimachima*
____ *M. c. cordata* — Savanna of sw Costa Rica to Brazil n of the Amazon; Trinidad
____ *M. c. chimachima* — Brazil s of the Amazon to e Bolivia, Paraguay and n Argentina

☐ **Chimango Caracara** *Milvago chimango*
____ *M. c. chimango* — Forests of s Brazil and Paraguay to c Argentina and c Chile
____ *M. c. temucoensis* — S Chile and s Argentina to Tierra del Fuego and Cape Horn Arch.

☐ **Laughing Falcon** *Herpetotheres cachinnans*
____ *H. c. chapmani* — Lowlands of n Mexico to Honduras

_____ *H. c. cachinnans*   Nicaragua to Colombia, the Guianas, Peru and central Brazil
_____ *H. c. queribundus*   E Bolivia and Brazil to Paraguay and n Argentina

☐ **Barred Forest-Falcon** *Micrastur ruficollis*

_____ *M. r. guerilla*   Humid forests of s Mexico to Nicaragua
_____ *M. r. interstes*   Costa Rica and Panama to w Colombia and w Ecuador
_____ *M. r. zonothorax*   E Andean foothills of Colombia and Venezuela to Bolivia
_____ *M. r. concentricus*   S Venezuela to the Guianas and Amazonian Brazil
_____ *M. r. ruficollis*   S Brazil to Paraguay and n Argentina
_____ *M. r. olrogi*   Subtropical forests of nw Argentina

☐ **Plumbeous Forest-Falcon** *Micrastur plumbeus*

Humid forests of sw Colombia and nw Ecuador

☐ **Lined Forest-Falcon** *Micrastur gilvicollis*

Guianas and s Venezuela to Bolivia and Amazonian Brazil

☐ **Cryptic Forest-Falcon** *Micrastur mintoni*

Rainforests of Brazil and northeastern Bolivia

☐ **Slaty-backed Forest-Falcon** *Micrastur mirandollei*

Humid Costa Rica to Bolivia and Amazonian Brazil

☐ **Collared Forest-Falcon** *Micrastur semitorquatus*

_____ *M. s. naso*   N-central Mexico to Ecuador
_____ *M. s. semitorquatus*   Rainforests of n South America to Brazil and n Argentina

☐ **Buckley's Forest-Falcon** *Micrastur buckleyi*

Humid Amazon forests of e Ecuador and e Peru

☐ **Spot-winged Falconet** *Spiziapteryx circumcincta*

Chaco of n Argentina, w Paraguay and e Bolivia

☐ **Pygmy Falcon** *Polihierax semitorquatus*

Discontinuously distributed acacia thornscrub of ne and s Africa

☐ **White-rumped Falcon** *Polihierax insignis*

_____ *P. i. insignis*   Myanmar (valley of Irrawaddy River)
_____ *P. i. cinereiceps*   S Myanmar and Thailand
_____ *P. i. harmandi*   S Indochina

☐ **Collared Falconet** *Microhierax caerulescens*

_____ *M. c. caerulescens*   Himalayas of India and Nepal to Assam
_____ *M. c. burmanicus*   Myanmar to s Indochina

☐ **Black-thighed Falconet** *Microhierax fringillarius*

Forests of Myanmar to Malay Peninsula and Greater Sundas

☐ **White-fronted Falconet** *Microhierax latifrons*

Forests of n Borneo

☐ **Philippine Falconet** *Microhierax erythrogenys*

_____ *M. e. erythrogenys*   N Philippines (Luzon, Mindoro, Negros and Bohol)
_____ *M. e. meridionalis*   S Philippines (Samar, Leyte and Cebu to Mindanao)

☐ **Pied Falconet** *Microhierax melanoleucos*

Forests of ne India to s China and n Indochina

☐ **Lesser Kestrel** *Falco naumanni*

Mediterranean basin to e China; winters to s Asia and s Africa

☐ **Eurasian Kestrel** *Falco tinnunculus*

_____ *F. t. tinnunculus*   N Africa, Europe and Middle East to Siberia
_____ *F. t. interstinctus*   Tibet to China and Japan; winters to India, Malaya, Philippines
_____ *F. t. objurgatus*   S India (Western and Eastern Ghats) and Sri Lanka
_____ *F. t. canariensis*   Madeira and w Canary Islands
_____ *F. t. dacotiae*   E Canary Islands
_____ *F. t. neglectus*   N Cape Verde Islands
_____ *F. t. alexandri*   SE Cape Verde Islands
_____ *F. t. rupicolaeformis* NE Africa and Arabia

____ *F. t. archerii*	Somalia, coastal Kenya and Socotra
____ *F. t. rufescens*	West Africa to Ethiopia, Tanzania and n Angola
____ *F. t. rupicolus*	N Angola to s Zaire, s Tanzania and South Africa

☐ **Madagascar Kestrel** *Falco newtoni*

Madagascar and Aldabra

☐ **Mauritius Kestrel** *Falco punctatus*

Dense forests of sw Mauritius (w Indian Ocean)

☐ **Seychelles Kestrel** *Falco araea*

Locally in Seychelles

☐ **Spotted Kestrel** *Falco moluccensis*

____ *F. m. moluccensis*	N and s Moluccas
____ *F. m. microbalius*	Java to Lesser Sundas, Sulawesi and Tanimbar Islands

☐ **Australian Kestrel** *Falco cenchroides*

____ *F. c. cenchroides*	Australian region; winters New Guinea to Java and Moluccas
____ *F. c. baru*	Montane forests of w-central New Guinea

☐ **Greater Kestrel** *Falco rupicoloides*

____ *F. r. fieldi*	Ethiopia and Somalia
____ *F. r. arthuri*	Kenya and ne Tanzania
____ *F. r. rupicoloides*	Acacia steppes of southern Africa

☐ **American Kestrel** *Falco sparverius*

____ *F. s. sparverius*	North America (Alaska to Newfoundland, south to w Mexico)
____ *F. s. paulus*	Coastal s US to Florida
____ *F. s. peninsularis*	W Mexico (s Baja California, Sonora and Sinaloa)
____ *F. s. tropicalis*	S Mexico to n Honduras
____ *F. s. nicaraguensis*	Savanna of Honduras and Nicaragua
____ *F. s. sparverioides*	Bahamas, Cuba and Isle of Pines
____ *F. s. dominicensis*	Hispaniola
____ *F. s. caribaearum*	West Indies (Puerto Rico to Grenada)
____ *F. s. brevipennis*	Netherlands Antilles (Aruba, Curaçao and Bonaire)
____ *F. s. ochraceus*	Mountains of e Colombia and nw Venezuela
____ *F. s. caucae*	Mountains of w Colombia
____ *F. s. isabellinus*	Venezuela to n Brazil
____ *F. s. aequatorialis*	Subtropical n Ecuador
____ *F. s. peruvianus*	Subtropical sw Ecuador, Peru and n Chile
____ *F. s. fernandensis*	Robinson Crusoe I. (Juan Fernández Islands off Chile)
____ *F. s. cinnamominus*	SE Peru, Chile and Argentina to Tierra del Fuego
____ *F. s. cearae*	Tablelands of ne Brazil to e Bolivia

☐ **Fox Kestrel** *Falco alopex*

Rocky hills and gorges of sub-Saharan Africa

☐ **Gray Kestrel** *Falco ardosiaceus*

Savanna and woodlands of Africa south of the Sahara

☐ **Dickinson's Kestrel** *Falco dickinsoni*

Palms and savanna of e-central and ne South Africa

☐ **Banded Kestrel** *Falco zoniventris*

Humid lowlands and subdesert of Madagascar

☐ **Red-necked Falcon** *Falco chicquera*

____ *F. c. chicquera*	Iran to India, Nepal and Bangladesh
____ *F. c. ruficollis*	Senegambia to Ethiopia, Somalia, Zambia and n Mozambique
____ *F. c. horsbrughi*	Zimbabwe and s Mozambique to Angola and n South Africa

☐ **Red-footed Falcon** *Falco vespertinus*

Central Eurasia; winters in Africa south of the Sahara

☐ **Amur Falcon** *Falco amurensis*

Steppes of ne Asia; winters from Malawi to South Africa

☐ **Eleonora's Falcon** *Falco eleonorae*

Mediterranean islands and coasts; winters to Madagascar

☐ **Sooty Falcon** *Falco concolor*

Rocky areas of ne Africa; winters to s Africa and Madagascar

☐ **Aplomado Falcon** *Falco femoralis*
  ____ *F. f. septentrionalis*
  ____ *F. f. femoralis*
  ____ *F. f. pichinchae*

Savanna and woodlands of n Mexico and Guatemala
Nicaragua and Belize through S America to Tierra del Fuego
Temperate Colombia to n Chile and nw Argentina

☐ **Merlin** *Falco columbarius*
  ____ *F. c. subaesalon*
  ____ *F. c. aesalon*
  ____ *F. c. insignis*
  ____ *F. c. pacificus*
  ____ *F. c. pallidus*
  ____ *F. c. lymani*
  ____ *F. c. suckleyi*
  ____ *F. c. columbarius*
  ____ *F. c. richardsoni*

Iceland
N Eurasia (Faeroes to central Siberia); winters to N Africa
Siberia (Yenisey River to Kolyma River)
NE Asia and Sakhalin
Steppes of Asia (Aral Sea to Altai Mountains)
Mountains of central Asia
Alaska and British Columbia to n Washington
North America (except for Pacific coast and Great Plains)
Great Plains of North America (central Alberta to Wyoming)

☐ **Bat Falcon** *Falco rufigularis*
  ____ *F. r. petoensis*
  ____ *F. r. rufigularis*
  ____ *F. r. ophryophanes*

Humid lowlands of n Mexico to s Ecuador (west of the Andes)
Lowlands of n S America to s Brazil and n Argentina; Trinidad
Tableland of Brazil, adjacent Bolivia, Paraguay and Argentina

☐ **Orange-breasted Falcon** *Falco deiroleucus*

Locally from s Mexico to n Argentina and Brazil

☐ **Eurasian Hobby** *Falco subbuteo*
  ____ *F. s. subbuteo*
  ____ *F. s. streichi*

Palearctic; winters to s Africa, s Eurasia and Greater Sundas
S and e China to Myanmar and n Indochina

☐ **African Hobby** *Falco cuvierii*

Savanna of Africa south of the Sahara

☐ **Oriental Hobby** *Falco severus*

S Asia, Malay Arch. and New Guinea region; winters to s India

☐ **Australian Hobby** *Falco longipennis*
  ____ *F. l. hanieli*
  ____ *F. l. longipennis*

Lesser Sundas (Lombok to Timor)
Australia and Tasmania; winters to New Guinea and Moluccas

☐ **New Zealand Falcon** *Falco novaeseelandiae*

Locally in New Zealand, Stewart I. and Auckland Islands

☐ **Brown Falcon** *Falco berigora*
  ____ *F. b. novaeguineae*
  ____ *F. b. berigora*
  ____ *F. b. occidentalis*

E and central New Guinea and coastal n Australia
Australia (except southwest) and Tasmania
SW and w-central Australia

☐ **Gray Falcon** *Falco hypoleucos*

Sparse in arid and semiarid areas of Australia with scattered trees

☐ **Black Falcon** *Falco subniger*

Woodlands and grasslands of Australia (seldom near coasts)

☐ **Lanner Falcon** *Falco biarmicus*
  ____ *F. b. feldeggii*
  ____ *F. b. erlangeri*
  ____ *F. b. tanypterus*
  ____ *F. b. abyssinicus*
  ____ *F. b. biarmicus*

Sicily and s Italy to Armenia, Azerbaijan and Lebanon
Mauritania to Morocco and Tunisia
Egypt and Sudan to Arabia, Israel and Iraq
Senegal and Ghana to Ethiopia, Somalia, Uganda and n Zaire
Angola to s Zaire, Kenya and South Africa

☐ **Laggar Falcon** *Falco jugger*

SE Iran to India and Myanmar

☐ **Saker Falcon** *Falco cherrug*
____ *F. c. cherrug*   S-cent. Eurasia to Altai Mts.; winters to ne Africa and nw India
____ *F. c. milvipes*   NE Asia; winters to Iran, nw India, Tibet and central China

☐ **Gyrfalcon** *Falco rusticolus*
    Mountains and tundra of n Palearctic region and n North America

☐ **Prairie Falcon** *Falco mexicanus*
    Arid North America (British Columbia to n Mexico)

☐ **Barbary Falcon** *Falco pelegrinoides*
    Canary Islands; locally from n Africa (Morocco) to w Iran

☐ **Taita Falcon** *Falco fasciinucha*
    Locally in highlands of sw Ethiopia to sw Mozambique

☐ **Peregrine Falcon** *Falco peregrinus*
____ *F. p. tundrius*   Arctic tundra of North America (Alaska to Greenland)
____ *F. p. anatum*   North America (south of tundra) to n Mexico
____ *F. p. pealei*   Coastal w North America (Aleutian Islands to Washington)
____ *F. p. cassini*   W South America (Ecuador to Tierra del Fuego and Falkland Is.)
____ *F. p. japonensis*   NE Siberia to Kamchatka Peninsula and Japan
____ *F. p. furuitii*   Volcano Islands and Bonin Islands
____ *F. p. calidus*   Tundra of Eurasia (Lapland to ne Siberia)
____ *F. p. peregrinus*   N Eurasia (south of the tundra)
____ *F. p. brookei*   Mediterranean basin east to the Caucasus Mountains
____ *F. p. babylonicus*   E Iran to Mongolia
____ *F. p. madens*   Cape Verde Islands
____ *F. p. minor*   Morocco, Mauritania and Africa south of the Sahara
____ *F. p. radama*   Madagascar and Comoro Islands
____ *F. p. peregrinator*   Pakistan, India and Sri Lanka to se China
____ *F. p. ernesti*   Philippines to New Guinea, Bismarck Arch. and Indonesia
____ *F. p. nesiotes*   Vanuatu and New Caledonia
____ *F. p. macropus*   Australia (expect for sw part)
____ *F. p. submelanogenys*   SW Australia

ORDER: GALLIFORMES
FAMILY: MEGAPODIIDAE (Megapodes—21)

☐ **Australian Brush-turkey** *Alectura lathami*
____ *A. l. purpureicollis*   NE Australia (n Cape York Peninsula)
____ *A. l. lathami*   E Australia (Cape York Peninsula to n New South Wales)

☐ **Wattled Brush-turkey** *Aepypodius arfakianus*
____ *A. a. arfakianus*   High mountains of New Guinea and Yapen I.
____ *A. a. misoliensis*   Mountains of Misool I. (off nw New Guinea)

☐ **Bruijn's Brush-turkey** *Aepypodius bruijnii*
    Waigeo I. (off New Guinea)

☐ **Red-billed Brush-turkey** *Talegalla cuvieri*
    Lowlands of nw New Guinea, Misool and Salawati islands

☐ **Black-billed Brush-turkey** *Talegalla fuscirostris*
____ *T. f. occidentis*   Lowlands of sw New Guinea and Aru Islands
____ *T. f. fuscirostris*   Lowlands of se New Guinea

☐ **Brown-collared Brush-turkey** *Talegalla jobiensis*
____ *T. j. jobiensis*   Lowland forests of n-central New Guinea and Yapen I.
____ *T. j. longicauda*   Lowland forests of e New Guinea

☐ **Malleefowl** *Leipoa ocellata*
    Scrub and heath of sw and s Australia

☐ **Maleo** *Macrocephalon maleo*
    Sulawesi, Bangka, Lembeh and Butung islands

☐ **Moluccan Scrubfowl** *Megapodius wallacei*

Moluccas and Misool I.

☐ **Niaufoou Scrubfowl** *Megapodius pritchardii*

Forests of Niauafo'ou I. (n Tonga)

☐ **Micronesian Scrubfowl** *Megapodius laperouse*
____ *M. l. laperouse*   Locally in n Mariana Islands
____ *M. l. senex*   Locally on Palau Is. (w Caroline Islands)

☐ **Nicobar Scrubfowl** *Megapodius nicobariensis*
____ *M. n. nicobariensis*   Lowlands of central and n Nicobar Islands
____ *M. n. abbotti*   Lowlands of Great Nicobar I. and Little Nicobar I.

☐ **Tabon Scrubfowl** *Megapodius cumingii*
____ *M. c. pusillus*   N and e Philippine Islands
____ *M. c. cumingii*   N Borneo, Palawan and Sulu Archipelago
____ *M. c. gilberti*   Sulawesi, Talisei, Tendila, Lembeh and Togian islands
____ *M. c. talautensis*   Talaud Islands (n Moluccas)
____ *M. c. sanghirensis*   Sangihe, Siau, Tahulandang and Ruang islands (off Sulawesi)

☐ **Sula Scrubfowl** *Megapodius bernsteinii*

Lowlands of Banggai and Sula islands (off Sulawesi)

☐ **Tanimbar Scrubfowl** *Megapodius tenimberensis*

Tanimbar I. (Banda Sea)

☐ **Dusky Scrubfowl** *Megapodius freycinet*
____ *M. f. freycinet*   N Moluccas and west Papuan islands (off New Guinea)
____ *M. f. geelvinkianus*   Islands in Geelvink Bay (n New Guinea)

☐ **Forsten's Scrubfowl** *Megapodius forstenii*

Moluccas (Ambon, Seram, Haruku, Gorang and Buru)

☐ **Melanesian Scrubfowl** *Megapodius eremita*

Admiralty Is., New Britain, New Ireland and Solomon Is.

☐ **Vanuatu Scrubfowl** *Megapodius layardi*

Forests of Vanuatu and Banks Group

☐ **New Guinea Scrubfowl** *Megapodius affinis*

N New Guinea and adjacent islands

☐ **Orange-footed Scrubfowl** *Megapodius reinwardt*
____ *M. r. buruensis*   Buru I. (s Moluccas)
____ *M. r. reinwardt*   Lesser Sundas, se Moluccas, Aru Is., s and se New Guinea
____ *M. r. macgillivrayi*   D'Entrecasteaux Islands and Louisiade Archipelago
____ *M. r. tumulus*   N Australia
____ *M. r. yorki*   NE Australia (Cape York Peninsula and adjacent islands)
____ *M. r. castanonotus*   E-c Queensland (Cooktown to Yeppoon and offshore islands)

ORDER: GALLIFORMES
FAMILY: CRACIDAE (Guans, Chachalacas and Curassows—50)

☐ **Plain Chachalaca** *Ortalis vetula*
____ *O. v. mccallii*   Extreme s Texas to ne Mexico (n Veracuz)
____ *O. v. vetula*   SE Mexico (s Veracruz) to nw Costa Rica
____ *O. v. pallidiventris*   SE Mexico (n Yucatán Peninsula)
____ *O. v. deschauenseei*   Utila I. (off n Honduras)

☐ **Gray-headed Chachalaca** *Ortalis cinereiceps*

Tropical e Honduras to nw Colombia

☐ **Chestnut-winged Chachalaca** *Ortalis garrula*

Woodlands and scrub of n Colombia

☐ **Rufous-vented Chachalaca** *Ortalis ruficauda*
____ *O. r. ruficrissa*   Tropical n Colombia and nw Venezuela
____ *O. r. ruficauda*   NE Colombia to n Venezuela, Tobago and Isla Margarita

☐ **Rufous-headed Chachalaca** *Ortalis erythroptera*

W Ecuador and extreme nw Peru

☐ **Rufous-bellied Chachalaca** *Ortalis wagleri*

Semiarid nw Mexico (Sonora to Jalisco)

☐ **West Mexican Chachalaca** *Ortalis poliocephala*

Semiarid w Mexico (c Jalisco and Michoacán to Chiapas)

☐ **Chaco Chachalaca** *Ortalis canicollis*
____ *O. c. canicollis*
____ *O. c. pantanalensis*

Chaco of e Bolivia to w Paraguay and n Argentina
W Brazil (sw Mato Grosso)

☐ **White-bellied Chachalaca** *Ortalis leucogastra*

Pacific slope of s Mexico (Chiapas) to nw Costa Rica

☐ **Little Chachalaca** *Ortalis motmot*
____ *O. m. motmot*
____ *O. m. ruficeps*

Guianas to s Venezuela and n Amazonian Brazil
N-central Brazil (south of the Amazon)

☐ **Speckled Chachalaca** *Ortalis guttata*
____ *O. g. columbiana*
____ *O. g. guttata*
____ *O. g. subaffinis*
____ *O. g. araucuan*
____ *O. g. squamata*

N and central Colombia
E Colombia to Ecuador, Peru, Bolivia and adjacent w Brazil
E and ne Bolivia and adjacent Brazil
E Brazil
SE Brazil

☐ **Buff-browed Chachalaca** *Ortalis superciliaris*

NE Brazil south of the Amazon (Pará to Piauí and n Goiás)

☐ **Band-tailed Guan** *Penelope argyrotis*
____ *P. a. albicauda*
____ *P. a. colombiana*
____ *P. a. argyrotis*

Sierra de Perijá (Colombia/ Venezuela border)
Santa Marta Mountains (ne Colombia)
Montane forests of n Colombia and n Venezuela

☐ **Bearded Guan** *Penelope barbata*

Locally in Western Andes of sw Ecuador and nw Peru

☐ **Baudo Guan** *Penelope ortoni*

W slope of Andes of Colombia and Ecuador

☐ **Andean Guan** *Penelope montagnii*
____ *P. m. montagnii*
____ *P. m. atrogularis*
____ *P. m. brooki*
____ *P. m. plumosa*
____ *P. m. sclateri*

N and central Colombia and nw Venezuela
W slope of Andes of s Colombia and Ecuador
E slope of Andes of s Colombia and Ecuador
E slope of Andes of Peru
Yungas of Bolivia

☐ **Marail Guan** *Penelope marail*
____ *P. m. marail*
____ *P. m. jacupeba*

Tropical Guianas and e Venezuela (south of the Orinoco)
SE Venezuela and n Amazonian Brazil

☐ **Rusty-margined Guan** *Penelope superciliaris*
____ *P. s. superciliaris*
____ *P. s. jacupemba*
____ *P. s. major*

Amazonian Brazil
Central and s Brazil to e Bolivia
Extreme s Brazil to e Paraguay and ne Argentina

☐ **Red-faced Guan** *Penelope dabbenei*

Humid Andes of se Bolivia to extreme nw Argentina

☐ **Crested Guan** *Penelope purpurascens*
____ *P. p. purpurascens*
____ *P. p. aequatorialis*
____ *P. p. brunnescens*

Humid forests of Mexico to Honduras and Nicaragua
S Honduras and Nicaragua to nw Colombia and se Ecuador
N Colombia to e Venezuela

☐ **Cauca Guan** *Penelope perspicax*

Subtropical W and C Andes of Colombia

☐ **White-winged Guan** *Penelope albipennis*

Dry forests of nw Peru (Tumbes, Piura and Lambayeque)

☐ **Spix's Guan** *Penelope jacquacu*
____ *P. j. granti* — S Venezuela to the Guianas and n Amazonian Brazil
____ *P. j. orienticola* — SE Venezuela and nw Brazil north of the Amazon
____ *P. j. jacquacu* — E Colombia to Bolivia and adjacent w Amazonian Brazil
____ *P. j. speciosa* — E Bolivia

☐ **Dusky-legged Guan** *Penelope obscura*
____ *P. o. bronzina* — E Brazil (Espíritu Santo to Santa Catarina)
____ *P. o. obscura* — Extreme se Brazil to se Paraguay, Uruguay and ne Argentina
____ *P. o. bridgesi* — E slope of Andes of Bolivia to nw Argentina

☐ **White-crested Guan** *Penelope pileata* — Amazonian Brazil (lower Rio Madeira to Rio Tapajós)

☐ **Chestnut-bellied Guan** *Penelope ochrogaster* — Lowlands of e Brazil (west to Mato Grosso)

☐ **White-browed Guan** *Penelope jacucaca* — Interior ne Brazil (Ceará and Paraíba to Bahia)

☐ **Blue-throated Piping-Guan** *Pipile cumanensis*
____ *P. c. cumanensis* — E Colombia to Venezuela, the Guianas, w Brazil and Peru
____ *P. c. grayi* — SW Brazil to Bolivia, ne Paraguay and se Peru

☐ **Trinidad Piping-Guan** *Pipile pipile* — Forests of Trinidad (seriously endangered)

☐ **Red-throated Piping-Guan** *Pipile cujubi*
____ *P. c. cujubi* — Forests of w Amazonian Brazil (Rio Madeira to n Pará)
____ *P. c. nattereri* — W Amazonian Brazil to extreme ne Bolivia

☐ **Black-fronted Piping-Guan** *Pipile jacutinga* — E Brazil (Bahia) to se Paraguay and ne Argentina

☐ **Wattled Guan** *Aburria aburri* — Montane forests of w Venezuela to n Peru

☐ **Black Guan** *Chamaepetes unicolor* — Montane forests of Costa Rica and w Panama

☐ **Sickle-winged Guan** *Chamaepetes goudotii*
____ *C. g. goudotii* — Andes of n Colombia
____ *C. g. sanctaemarthae* — Santa Marta Mountains (ne Colombia)
____ *C. g. fagani* — W slope of Andes of sw Colombia and Ecuador
____ *C. g. tschudii* — E slope of Andes of s Colombia, Ecuador and n Peru
____ *C. g. rufiventris* — E slope of Andes of central Peru
____ *C. g. ssp.* — Undescribed race from mountains of Bolivia (La Paz)

☐ **Highland Guan** *Penelopina nigra* — Humid montane forests of s Mexico to Nicaragua

☐ **Horned Guan** *Oreophasis derbianus* — Humid forests of s Mexico (Chiapas) and Guatemala

☐ **Nocturnal Curassow** *Nothocrax urumutum* — Humid s Venezuela to ne Peru and w Amazonian Brazil

☐ **Crestless Curassow** *Mitu tomentosum* — Guyana to s Venezuela, e Colombia and adjacent nw Brazil

☐ **Salvin's Curassow** *Mitu salvini* — Humid lowlands of s Colombia to e Ecuador and ne Peru

☐ **Razor-billed Curassow** *Mitu tuberosum* — Tropical se Colombia to n Bolivia and Amazonian Brazil

☐ **Alagoas Curassow** *Mitu mitu* — Coastal e Brazil (Alagoas); extinct in wild

☐ **Helmeted Curassow** *Pauxi pauxi*
____ *P. p. pauxi* — Montane forests of Venezuela and adjacent ne Colombia
____ *P. p. gilliardi* — Sierra de Perijá (Colombia/Venezuela border)

☐ **Horned Curassow** *Pauxi unicornis*
_____ *P. u. koepckeae* — Andes of se Peru (Cerros del Sira)
_____ *P. u. unicornis* — E slope of Andes of central Bolivia

☐ **Great Curassow** *Crax rubra*
_____ *C. r. rubra* — Humid forests of e Mexico to w Ecuador
_____ *C. r. griscomi* — Cozumel I. (off Yucatán coast of Mexico). Possibly extirpated

☐ **Blue-knobbed Curassow** *Crax alberti*
— Humid forests of n Colombia

☐ **Yellow-knobbed Curassow** *Crax daubentoni*
— *Llanos* of ne Colombia and adjacent n Venezuela

☐ **Black Curassow** *Crax alector*
_____ *C. a. erythrognatha* — E Colombia and Venezuela (south of the Orinoco River)
_____ *C. a. alector* — Extreme e Venezuela to Guianas and Brazil (n of the Amazon)

☐ **Bare-faced Curassow** *Crax fasciolata*
_____ *C. f. pinima* — NE Brazil
_____ *C. f. fasciolata* — Lowlands of Brazil to Paraguay and ne Argentina
_____ *C. f. grayi* — E Bolivia

☐ **Wattled Curassow** *Crax globulosa*
— Humid se Colombia to n Bolivia and w Amazonian Brazil

☐ **Red-billed Curassow** *Crax blumenbachii*
— Lowland forests of se Brazil (on verge of extinction)

ORDER: GALLIFORMES
FAMILY: MELEAGRIDIDAE (Turkeys—2)

☐ **Wild Turkey** *Meleagris gallopavo*
_____ *M. g. silvestris* — Central and e US
_____ *M. g. osceola* — Locally in Florida
_____ *M. g. intermedia* — N Texas to e-central Mexico
_____ *M. g. merriami* — Western US
_____ *M. g. mexicana* — Mountains west of central plateau of Mexico
_____ *M. g. gallopavo* — S Mexico (Jalisco to Veracruz and south to Guerrero)

☐ **Ocellated Turkey** *Meleagris ocellata*
— SE Mexico (Yucatán Pen.) to n Guatemala (Petén) and Belize

ORDER: GALLIFORMES
FAMILY: TETRAONIDAE (Grouse, Ptarmigans and Prairie-chickens—19)

☐ **Spruce Grouse** *Falcipennis canadensis*
_____ *F. c. osgoodi* — Alaska and Yukon to Great Slave Lake and Lake Athabasca
_____ *F. c. atratus* — S Alaska (Bristol Bay) to Prince William Sound and Kodiak I.
_____ *F. c. franklinii* — Coniferous forests of extreme se Alaska to nw US
_____ *F. c. canadensis* — Coniferous forests of Canada (central Alberta to Labrador)
_____ *F. c. canace* — SE Canada and adjacent US (Minnesota to Maine)
_____ *F. c. torridus* — Nova Scotia

☐ **Blue Grouse** *Dendragapus obscurus*
_____ *D. o. richardsoni* — Alaska and s Yukon to Idaho, w Montana and nw Wyoming
_____ *D. o. pallidus* — S-central British Columbia to e Washington and ne Oregon
_____ *D. o. oreinus* — Mountains of ne Nevada and adjacent Utah
_____ *D. o. obscurus* — Mts. of Wyoming to Colorado, Arizona and New Mexico

☐ **Siberian Grouse** *Dendragapus falcipennis*

Coniferous forests of ne Asia

☐ **Sooty Grouse** *Dendragapus fuliginosus*
____ *D. f. fuliginosus* — Yukon/Alaska border to nw California; Vancouver I.
____ *D. f. sitkensis* — SE Alaska to Queen Charlotte Islands
____ *D. f. sierrae* — Cascades from Washington to California and Nevada
____ *D. f. howardi* — S California (s Sierra Nevada and Tehachapi Mountains)

☐ **White-tailed Ptarmigan** *Lagopus leucura*
____ *L. l. peninsularis* — Mountains of s-central Alaska to Glacier Bay and White Pass
____ *L. l. leucura* — N Yukon, w Br. Columbia and w Alberta to n border of US
____ *L. l. saxatilis* — Higher peaks of Vancouver I.
____ *L. l. rainierensis* — Alpine summits of Cascades of Washington
____ *L. l. altipetens* — Rocky Mountains (Montana to New Mexico)

☐ **Willow Ptarmigan** *Lagopus lagopus*
____ *L. l. scotica* — British Isles
____ *L. l. variegata* — Coastal Norway (islands off Trondheim Fjord)
____ *L. l. lagopus* — Scandinavia and n Russia
____ *L. l. rossica* — Baltic countries to central Russia
____ *L. l. birulai* — New Siberian Islands
____ *L. l. koreni* — Siberia to Kamchatka Peninsula
____ *L. l. kamtschatkensis* — Kamchatka Peninsula and Kuril Islands
____ *L. l. maior* — Steppes of sw Siberia and n Kazakstan
____ *L. l. brevirostris* — Altai Mts. and Sayan Mts.
____ *L. l. kozlowae* — W Mongolia (Tanmu-Ola, Khangai and Kentei Mountains)
____ *L. l. sserebrowsky* — E Siberia (Lake Baikal to Sea of Okhotsk and Sikhote Alin Mts.)
____ *L. l. okadai* — Sakhalin I.
____ *L. l. muriei* — E Aleutian Islands and Kodiak I.
____ *L. l. alexandrae* — Alaskan Peninsula to nw British Columbia
____ *L. l. alascensis* — Alaska
____ *L. l. leucoptera* — Arctic islands of n Canada and adjacent mainland to s Baffin I.
____ *L. l. alba* — Tundra of n Yukon and c Br. Columbia to Gulf of St. Lawrence
____ *L. l. ungavus* — N Quebec and n Labrador
____ *L. l. alleni* — Newfoundland

☐ **Rock Ptarmigan** *Lagopus muta*
____ *L. m. hyperborea* — Svalbard, Franz Josef Land and Bear I.
____ *L. m. muta* — Norway, n Sweden, n Finland and Kola Peninsula
____ *L. m. millaisi* — Scotland
____ *L. m. pyrenaica* — Pyrénées
____ *L. m. helvetica* — Alps (Savoie to central Austria)
____ *L. m. komensis* — N Ural Mountains
____ *L. m. pleskei* — N Siberia (Taymyr Peninsula to Chukotsk Peninsula)
____ *L. m. macrorhyncha* — Tarbagatay Mountains (Russia)
____ *L. m. ssp.* — Undescribed race from Pamir Alaï Mts. (Tajikistan)
____ *L. m. nadezdae* — Mountains of s Siberia and Mongolia
____ *L. m. transbaicalica* — SE Siberia (Lake Baikal to Sea of Okhotsk)
____ *L. m. kraschennikovi* — Kamchatka Peninsula
____ *L. m. ridgwayi* — Komandorskiye Islands
____ *L. m. kurilensis* — Kuril Islands
____ *L. m. japonica* — Honshu I. (Japan)
____ *L. m. evermanni* — Attu I. (Aleutian Islands)
____ *L. m. townsendi* — Aleutian Islands (Kiska and Little Kiska)
____ *L. m. gabrielsoni* — Aleutian Islands (Amchitka, Little Sitkin and Rats)
____ *L. m. sanfordi* — Aleutian Islands (Tanaga and Kanaga)
____ *L. m. chamberlaini* — Adak I. (Aleutian Islands)
____ *L. m. atkhensis* — Atka I. (Aleutian Islands)
____ *L. m. yunaskensis* — Yunaska I. (Aleutian Islands)
____ *L. m. nelsoni* — Aleutian Islands (Unimak, Unalaska and Amaknak)
____ *L. m. dixoni* — Coasts and mountains of Glacier Bay to nw British Columbia

____	*L. m. kelloggae*	Alaska and n Yukon
____	*L. m. rupestris*	Tundra of n North America
____	*L. m. saturata*	NW Greenland
____	*L. m. capta*	E Greenland
____	*L. m. reinhardti*	SW Greenland
____	*L. m. welchi*	Newfoundland
____	*L. m. islandorum*	Iceland

☐ **Black-billed Capercaillie** *Tetrao parvirostris*

____	*T. p. parvirostris*	E Siberia to n Manchuria, Ussuriland and Sakhalin I.
____	*T. p. kamschaticus*	Kamchatka Peninsula
____	*T. p. stegmanni*	Lake Baikal region, Sayan Mountains and n Mongolia

☐ **Eurasian Capercaillie** *Tetrao urogallus*

____	*T. u. cantabricus*	Cantabrian Mountains (nw Spain)
____	*T. u. aquitanicus*	Pyrénées
____	*T. u. major*	Germany to sw Baltic countries and Balkan Peninsula
____	*T. u. rudolfi*	Carpathian Mountains and Rhodope Mountains
____	*T. u. urogallus*	Scandinavia
____	*T. u. lonnbergi*	Kola Peninsula
____	*T. u. karelicus*	Finland and n Russian (Karelia)
____	*T. u. pleskei*	Belarus, n Ukraine and European Russia
____	*T. u. obsoletus*	N Russia and n Siberia to upper Lena River
____	*T. u. volgensis*	Central and se Russia
____	*T. u. uralensis*	S Ural Mountains and sw Siberia
____	*T. u. taczanowskii*	Central Siberia to Altai Mountains and nw Mongolia

☐ **Black Grouse** *Tetrao tetrix*

____	*T. t. britannicus*	N England, Scotland and Inner Hebrides
____	*T. t. tetrix*	Scandinavia to France and n Italy east to Siberia
____	*T. t. viridanus*	SE Russia to Siberian steppes
____	*T. t. tschusii*	S Siberia south to nw Altai and Sayan mountains
____	*T. t. baikalensis*	SE Siberia to n Mongolia and nw Manchuria
____	*T. t. mongolicus*	Russian Altai to Chinese Turkestan
____	*T. t. ussuriensis*	SE Siberia (Lake Baikal) to nw Korea and n Mongolia

☐ **Caucasian Grouse** *Tetrao mlokosiewiczi*

Caucasus Mountains to ne Turkey and nw Iran

☐ **Hazel Grouse** *Bonasa bonasia*

____	*B. b. styriaca*	Jura Mountains, Alps, Hungary, Slovakia and s Poland
____	*B. b. rhenana*	NE France, Luxembourg, Belgium and West Germany
____	*B. b. rupestris*	S Germany, Bohemia and Sudety Mountains
____	*B. b. schiebeli*	Balkan Peninsula
____	*B. b. volgensis*	Poland and Ukraine to central European Russia
____	*B. b. bonasia*	S Scandinavia, Finland and n European Russia to Ural Mts.
____	*B. b. griseonota*	N Sweden
____	*B. b. sibirica*	Siberia to Altai Mountains, Sayan Mountains and n Mongolia
____	*B. b. kolymensis*	Extreme e Siberia to Sea of Okhotsk
____	*B. b. amurensis*	S Amurland and Little Khingan Mountains to n Korea
____	*B. b. yamashinai*	Sakhalin (Russia)
____	*B. b. vicinitas*	Hokkaido (n Japan)

☐ **Severtzov's Grouse** *Bonasa sewerzowi*

Mts. of w China (Gansu to e Tibet, nw Yunnan and n Sichuan)

☐ **Ruffed Grouse** *Bonasa umbellus*

____	*B. u. yukonensis*	W Alaska and Yukon to s Mackenzie and nw Saskatchewan
____	*B. u. umbelloides*	SE Alaska to British Columbia and east to Quebec
____	*B. u. labradorensis*	Labrador
____	*B. u. sabini*	Coastal sw British Columbia to nw California
____	*B. u. brunnescens*	Vancouver I. and adjacent mainland
____	*B. u. castanea*	Olympic Peninsula (nw Washington)

____	*B. u. affinis*	Inland British Columbia to Washington and central Oregon
____	*B. u. phaios*	SE Br. Columbia and e Washington to Rocky Mts. of s Idaho
____	*B. u. incana*	SE Idaho, Wyoming, N Dakota to Colorado and w S Dakota
____	*B. u. mediana*	Minnesota and s Wisconsin
____	*B. u. togata*	S Ontario and s Quebec to n Wisconsin, c Michigan and c NY
____	*B. u. thayeri*	Nova Scotia
____	*B. u. umbellus*	NY and Massachusetts to e Pennsylvania and New Jersey
____	*B. u. monticola*	S Michigan, Ohio and Pennsylvania to n Georgia

☐ **Greater Sage-Grouse** *Centrocercus urophasianus*

Prairies and sage grasslands of w Canada to sw US

☐ **Gunnison Sage-Grouse** *Centrocercus minimus*

Gunnison Basin and sw Colorado adj. se Utah

☐ **Sharp-tailed Grouse** *Tympanuchus phasianellus*

____	*T. p. caurus*	N Alaska to s Yukon, n British Columbia and n Alberta
____	*T. p. kennicotti*	Mackenzie River to Great Slave Lake
____	*T. p. phasianellus*	N Manitoba to n Ontario and w-central Quebec
____	*T. p. campestris*	S Manitoba to n Michigan, Minnesota and Wisconsin
____	*T. p. jamesi*	N-c Alberta and Saskatchewan to Colorado and Nebraska
____	*T. p. columbianus*	N-c British Columbia to e Oregon, n Utah and w Colorado

☐ **Greater Prairie-Chicken** *Tympanuchus cupido*

____	*T. c. cupido†*	Formerly ne US; extirpated ca 1932
____	*T. c. pinnatus*	S-central Canada to ne Oklahoma and n-central Tennessee
____	*T. c. attwateri*	Coastal se Texas

☐ **Lesser Prairie-Chicken** *Tympanuchus pallidicinctus*

Arid grasslands of s-central US

ORDER: GALLIFORMES
FAMILY: ODONTOPHORIDAE (New World Quail—31)

☐ **Bearded Wood-Partridge** *Dendrortyx barbatus*

Humid montane forests of ne Mexico

☐ **Long-tailed Wood-Partridge** *Dendrortyx macroura*

____	*D. m. macroura*	S Mexico (valley of México and Veracruz)
____	*D. m. diversus*	SW Mexico (montane oak-pine forests of nw Jalisco)
____	*D. m. griseipectus*	W slope of mts. of Distrito Federal, México and Morelos
____	*D. m. striatus*	SW Mexico (s Jalisco to Michoacán and Guerrero)
____	*D. m. inesperatus*	S Mexico (Chilpancingo area of Guerrero)
____	*D. m. oaxacae*	S Mexico (montane oak-pine forests w Oaxaca)

☐ **Buffy-crowned Wood-Partridge** *Dendrortyx leucophrys*

____	*D. l. leucophrys*	Mountains of s Mexico (Chiapas) to Nicaragua
____	*D. l. hypospodius*	Mountains of n Costa Rica

☐ **Mountain Quail** *Oreortyx pictus*

____	*O. p. pictus*	Cascades of Washington to coastal mountains of c California
____	*O. p. plumifera*	S Washington to w Nevada and central California
____	*O. p. russelli*	S California (Little San Bernardino Mountains)
____	*O. p. palmeri*	S Washington to c California (nw San Luis Obispo County)
____	*O. p. eremophilus*	Sierra Nevada of s California to n Baja, extreme sw Nevada
____	*O. p. confinis*	Mts. of n Baja California (Sierra Juárez and San Pedro Mártir)

☐ **Scaled Quail** *Callipepla squamata*

____	*C. s. hargravi*	SE Colorado to Oklahoma, sw Kansas and nw Texas
____	*C. s. pallida*	S Arizona to w Texas, n Sonora and Chihuahua
____	*C. s. squamata*	N Mexico (n Sonora and Tamaulipas s to valley of México)
____	*C. s. castanogastris*	S Texas to n Mexico (Tamaulipas, Nuevo León, e Coahuila)

☐ **Elegant Quail** *Callipepla douglasii*

____ *C. d. bensoni*	Arid nw Mexico (Sonora)
____ *C. d. languens*	Arid nw Mexico (w Chihuahua)
____ *C. d. douglasii*	W Mexico (extreme s Sonora to Sinaloa and nw Durango)
____ *C. d. impedita*	Arid w Mexico (Nayarit)
____ *C. d. teres*	Arid w Mexico (nw Jalisco)

☐ **California Quail** *Callipepla californica*

____ *C. c. californica*	N Oregon and w Nevada to s California and Coronados Islands
____ *C. c. orecta*	SE Oregon (Warner Valley) and extreme n California
____ *C. c. brunnescens*	Extreme n coastal California to s Santa Cruz County
____ *C. c. catalinensis*	Santa Catalina I. (off s California)
____ *C. c. canfieldae*	Owens Valley of e-central California
____ *C. c. plumbea*	San Diego County south through nw Baja California
____ *C. c. decolorata*	Baja California between latitude 25°N and 30°N
____ *C. c. achrustera*	S Baja California

☐ **Gambel's Quail** *Callipepla gambelii*

____ *C. g. gambelii*	Utah and Nevada to Colorado, Mohave deserts and ne Baja
____ *C. g. sana*	Arid scrub of w Colorado
____ *C. g. ignoscens*	S New Mexico and extreme w Texas
____ *C. g. pembertoni*	Isla Tiburón (Gulf of California)
____ *C. g. fulvipectus*	SE Arizona and sw New Mexico to nw Mexico (s Sonora)
____ *C. g. stephensi*	W Mexico (s Sonora adjacent to Sinaloa border)
____ *C. g. friedmanni*	W Mexico (coastal Sonora from Río Fuerte to Río Culiacán)

☐ **Banded Quail** *Philortyx fasciatus*

	Arid w Mexico (sw Jalisco to se Guerrero, Morelos and Puebla)

☐ **Northern Bobwhite** *Colinus virginianus*

____ *C. v. marilandicus*	NE US (se Maine to Pennsylvania and central Virginia)
____ *C. v. virginianus*	Atlantic coast (Virginia to n Florida and se Alabama)
____ *C. v. floridanus*	Peninsular Florida
____ *C. v. cubanensis*	Cuba and Isle of Pines
____ *C. v. mexicanus*	E US west of Atlantic seaboard to Great Plains
____ *C. v. taylori*	S Dakota to n Texas, w Missouri and nw Arkansas
____ *C. v. texanus*	SW Texas to n Mexico (Coahuila, Nuevo León, Tamaulipas)
____ *C. v. ridgwayi*	NW Mexico (n-central Sonora); extirpated in Arizona
____ *C. v. maculatus*	E Mexico (c Tamaulipas to n Veracruz and se San Luis Potosí)
____ *C. v. aridus*	E Mexico (c and w-central Tamaulipas to se San Luis Potosí)
____ *C. v. graysoni*	W-c Mexico (s Nayarit to Morelos, s Hidalgo, San Luis Potosí)
____ *C. v. nigripectus*	E Mexico (Puebla, Morelos and México)
____ *C. v. pectoralis*	SE Mexico (e slopes of mountains of central Veracruz)
____ *C. v. godmani*	Lowlands of se Mexico (Veracruz)
____ *C. v. minor*	SE Mexico (ne Chiapas and adjacent Tabasco)
____ *C. v. atriceps*	S Mexico (interior of w Oaxaca)
____ *C. v. thayeri*	S Mexico (ne Oaxaca)
____ *C. v. harrisoni*	S Mexico (sw Oaxaca)
____ *C. v. coyolcos*	Pacific coast of s Mexico (Oaxaca and Chiapas)
____ *C. v. salvini*	S Mexico (coastal southern Chiapas)
____ *C. v. insignis (nelsoni)*	S Mexico (s Chiapas) and adjacent Guatemala

☐ **Black-throated Bobwhite** *Colinus nigrogularis*

____ *C. n. persiccus*	SE Mexico (Progresso area of Yucatán Peninsula)
____ *C. n. caboti*	SE Mexico (n Campeche, Yucatán and n Quintana Roo)
____ *C. n. nigrogularis*	Belize and adjacent n Guatemala
____ *C. n. segoviensis*	E Honduras and ne Nicaragua

☐ **Crested Bobwhite** *Colinus cristatus*

____ *C. c. incanus*	S Guatemala
____ *C. c. hypoleucus*	W El Salvador and adjacent Guatemala

____	*C. c. leucopogon*	SE El Salvador and w Honduras
____	*C. c. leylandi*	NW Honduras
____	*C. c. sclateri*	Central Honduras to nw Nicaragua
____	*C. c. dickeyi*	NW and central Costa Rica
____	*C. c. mariae*	Savanna of sw Costa Rica and e Panama (Chiriquí)
____	*C. c. panamensis*	Lowlands of Pacific slope of Panama
____	*C. c. decoratus*	Caribbean coast of Colombia
____	*C. c. cristatus*	NE Colombia and nw Venezuela
____	*C. c. continentis*	Coastal nw Venezuela, Aruba and Curaçao
____	*C. c. littoralis*	N base of Santa Marta Mountains (ne Colombia)
____	*C. c. badius*	Cauca Valley to Pacific slope of Western Andes of Colombia
____	*C. c. leucotis*	N Colombia (Magdalena and Sinú valleys)
____	*C. c. bogotensis*	E Andes of Colombia (Boyacá and Cundinamarca)
____	*C. c. parvicristatus*	E slope of e Andes of Colombia and adjacent Venezuela
____	*C. c. horvathi*	Andes of nw Venezuela (Mérida)
____	*C. c. barnesi*	W-central Venezuela (Portuguesa and Barinas)
____	*C. c. mocquerysi*	NE Venezuela (Sucre, n Monagas and n Anzoátegui)
____	*C. c. sonnini*	Coastal n Venezuela to the Guianas and extreme n Brazil

☐ **Marbled Wood-Quail** *Odontophorus gujanensis*

____	*O. g. castigatus*	SW Costa Rica and (?) w Panama
____	*O. g. marmoratus*	E Panama to n Colombia and nw Venezuela
____	*O. g. gujanensis*	SE Venezuela to the Guianas, Brazil and extreme ne Paraguay
____	*O. g. medius*	S Venezuela to nw Brazil
____	*O. g. buckleyi*	Base of Eastern Andes of Colombia to e Ecuador and n Peru
____	*O. g. rufogularis*	NE Peru (upper Río Javarí)
____	*O. g. pachyrhynchus*	E-central Peru (Junín and Ayacucho)
____	*O. g. simonsi*	Tropical e Bolivia

☐ **Spot-winged Wood-Quail** *Odontophorus capueira*

____	*O. c. plumbeicollis*	Tropical ne Brazil (Ceará and Alagoas)
____	*O. c. capueira*	Tropical e Brazil to e Paraguay and ne Argentina

☐ **Black-eared Wood-Quail** *Odontophorus melanotis*

____	*O. m. verecundus*	Humid Caribbean lowlands of Honduras
____	*O. m. melanotis*	SE Honduras to Nicaragua, Costa Rica and Panama

☐ **Rufous-fronted Wood-Quail** *Odontophorus erythrops*

____	*O. e. parambae*	Tropical Colombia and w Ecuador
____	*O. e. erythrops*	Tropical sw Ecuador

☐ **Black-fronted Wood-Quail** *Odontophorus atrifrons*

____	*O. a. atrifrons*	Santa Marta Mountains (ne Colombia)
____	*O. a. variegatus*	E Andes of ne Colombia
____	*O. a. navai*	Sierra de Perijá (Colombia/Venezuela border)

☐ **Chestnut Wood-Quail** *Odontophorus hyperythrus*

Western and Central Andes of Colombia

☐ **Dark-backed Wood-Quail** *Odontophorus melanonotus*

Montane forests of nw Ecuador and adjacent sw Colombia

☐ **Rufous-breasted Wood-Quail** *Odontophorus speciosus*

____	*O. s. soederstroemii*	Tropical forests of e Ecuador
____	*O. s. speciosus*	Tropical e-central Peru
____	*O. s. loricatus*	Tropical se Peru and e Bolivia

☐ **Tacarcuna Wood-Quail** *Odontophorus dialeucos*

Extreme e Panama and adjacent nw Colombia (Chocó)

☐ **Gorgeted Wood-Quail** *Odontophorus strophium*

Temp. E Andes of Colombia (Santander and Cundinamarca)

☐ **Venezuelan Wood-Quail** *Odontophorus columbianus*

W Venezuela (sw Táchira) and coastal cordillera e to Miranda

☐ **Black-breasted Wood-Quail** *Odontophorus leucolaemus*

Tropical and subtropical forests of Costa Rica and w Panama

☐ **Stripe-faced Wood-Quail** *Odontophorus balliviani*

Andes of se Peru and n Bolivia

☐ **Starred Wood-Quail** *Odontophorus stellatus*

Tropical forests of w Amazon basin

☐ **Spotted Wood-Quail** *Odontophorus guttatus*

Forests of se Mexico to extreme w Panama

☐ **Singing Quail** *Dactylortyx thoracicus*

____ *D. t. pettingilli*	E Mexico (sw Tamaulipas and se San Luis Potosí)
____ *D. t. thoracicus*	E Mexico (ne Puebla and central Veracruz)
____ *D. t. devius*	W Mexico (Jalisco)
____ *D. t. molodus*	W Mexico (central Guerrero)
____ *D. t. ginetensis*	S Mexico (Chiapas/Oaxaca border region)
____ *D. t. edwardsi*	S Mexico (mountains of Chiapas adjacent to Oaxaca border)
____ *D. t. chiapensis*	S Mexico (central Chiapas)
____ *D. t. moorei*	S Mexico (mountains of central Chiapas)
____ *D. t. dolichonyx*	S Mexico (Sierra Madre del Sur of Chiapas)
____ *D. t. sharpei*	Campeche, Yucatán and Quintana Roo to Petén of Guatemala
____ *D. t. paynteri*	S Mexico (s-central Quintana Roo)
____ *D. t. calophonus*	Pacific Cordillera of Guatemala
____ *D. t. salvadoranus*	El Salvador (Volcán de San Miguel)
____ *D. t. taylori*	El Salvador (Mt. Cacaguatique region)
____ *D. t. fuscus*	Honduras (Tegucigalpa region)
____ *D. t. rufescens*	Honduras (San Juancito Mountains)
____ *D. t. conoveri*	Honduras (Department of Olancho)

☐ **Montezuma Quail** *Cyrtonyx montezumae*

____ *C. m. mearnsi*	W Texas to central Arizona and n Mexico (n Coahuila)
____ *C. m. montezumae*	E Mexico (Tamaulipas to Hidalgo, Puebla and Oaxaca)
____ *C. m. merriami*	SE Mexico (Mt. Orizaba area of Veracruz)
____ *C. m. sallei*	S Mexico (s Michoacán to Guerrero and w Oaxaca)
____ *C. m. rowleyi*	S Mexico (Sierra de Miahuatlán of Guerrero and Oaxaca)

☐ **Ocellated Quail** *Cyrtonyx ocellatus*

Oak-pine forests of s Mexico to n Nicaragua

☐ **Tawny-faced Quail** *Rhynchortyx cinctus*

____ *R. c. pudiobundus*	Caribbean lowlands of ne Honduras and e Nicaragua
____ *R. c. cinctus*	Caribbean coast of Costa Rica and Panama
____ *R. c. australis*	Pacific coast of Colombia and nw Ecuador

ORDER: GALLIFORMES
FAMILY: PHASIANIDAE (Pheasants and Partridges—155)

☐ **Snow Partridge** *Lerwa lerwa*

Himalayas of e Afghanistan to s Tibet and sw China

☐ **Verreaux's Partridge** *Tetraophasis obscurus*

Alpine mountain slopes of ne Tibet

☐ **Szechenyi's Partridge** *Tetraophasis szechenyii*

Mountains of e Tibet, sw China and extreme ne India

☐ **Caucasian Snowcock** *Tetraogallus caucasicus*

Rocky heights of Caucasus Mountains

☐ **Caspian Snowcock** *Tetraogallus caspius*

____ *T. c. caspius*	Mountains of e Turkey, s Russia and w Iran
____ *T. c. semenowtianschanskii*	Zagros Mountains (sw Iran)

☐ **Altai Snowcock** *Tetraogallus altaicus*

Mountains of sw Siberia and w Mongolia

63

□ **Tibetan Snowcock** *Tetraogallus tibetanus*

____	*T. t. tibetanus*	Pamir Mountains to w Tibet and Ladakh
____	*T. t. przewalskii*	NE India to w-central China and nw Sichuan
____	*T. t. aquilonifer*	W Nepal to Bhutan
____	*T. t. henrici*	E Tibet to nw Sichuan

□ **Himalayan Snowcock** *Tetraogallus himalayensis*

____	*T. h. sewerzowi*	Tien Shan Mountains to nw China (e Xinjiang)
____	*T. h. incognitus*	Mountains of s Tajikistan and n Afghanistan
____	*T. h. himalayensis*	E Afghanistan to nw India and Nepal
____	*T. h. grombczewskii*	W China (Kunlun Mountains) to n Tibet and s Xinjiang
____	*T. h. koslowi*	W China (Nam Shan and Ching Hai Ku Mountains)

□ **Rock Partridge** *Alectoris graeca*

____	*A. g. saxatilis*	Alps (France to Austria and w Yugoslavia) and Appennines
____	*A. g. graeca*	SE Yugoslavia to Greece and Bulgaria
____	*A. g. whitakeri*	Sicily

□ **Chukar** *Alectoris chukar*

____	*A. c. cypriotes*	SE Bulgaria to s Syria, Crete, Rhodes, and Cyprus
____	*A. c. sinaica*	N Syrian Desert south to Sinai Peninsula
____	*A. c. kurdestanica*	Caucasus Mountains to Iran
____	*A. c. werae*	E Iraq and sw Iran
____	*A. c. koroviakovi*	E Iran to Pakistan
____	*A. c. subpallida*	Tajikistan (Kyzl Kum and Kara Kum mountains)
____	*A. c. falki*	N-central Afghanistan to Pamirs and w China (w Xinjiang)
____	*A. c. dzungarica*	NW Mongolia to Russian Altai and e Tibet
____	*A. c. pallescens*	NE Afghanistan to Ladakh and w Tibet
____	*A. c. pallida*	NW China (Tarim basin of w Xinjiang)
____	*A. c. fallax*	NW China (e and s Tien Shan Mountains of Xinjiang)
____	*A. c. chukar*	E Afghanistan to e Nepal
____	*A. c. pubescens*	Inner Mongolia to nw Sichuan and e Qinghai
____	*A. c. potanini*	W Mongolia

□ **Philby's Partridge** *Alectoris philbyi*

Rocky mountains of sw Saudi Arabia and n Yemen

□ **Przevalski's Partridge** *Alectoris magna*

Desolate regions of n-central China (Qinghai and Gansu)

□ **Barbary Partridge** *Alectoris barbara*

____	*A. b. koenigi*	NW Morocco; introduced to Canary Islands and s Spain
____	*A. b. barbara*	N Morocco and n Algeria; Sardinia (introduced?)
____	*A. b. spatzi*	S Morocco to central Algeria and s Tunisia
____	*A. b. barbata*	Libya and nw Egypt

□ **Red-legged Partridge** *Alectoris rufa*

____	*A. r. rufa*	France, nw Italy, Elba and Corsica
____	*A. r. hispanica*	N and w Iberian Peninsula
____	*A. r. intercedens*	E and s Iberian Peninsula and Balearic Islands

□ **Arabian Partridge** *Alectoris melanocephala*

Arid regions of s Arabian Peninsula

□ **See-see Partridge** *Ammoperdix griseogularis*

Arid se Turkey and Syria to sw Russia and Pakistan

□ **Sand Partridge** *Ammoperdix heyi*

____	*A. h. heyi*	Jordan Valley to Sinai Peninsula and w Saudi Arabia
____	*A. h. nicolli*	N Egypt east of the Nile
____	*A. h. cholmleyi*	Central Egypt east of the Nile to n Sudan
____	*A. h. intermedius*	S Arabian Peninsula

☐ **Black Francolin** *Francolinus francolinus*
___ *F. f. francolinus*   Cyprus and Asia Minor to Iraq and Iran
___ *F. f. arabistanicus*   S Iraq and w Iran
___ *F. f. bogdanovi*   S Iran and Afghanistan to s Pakistan
___ *F. f. henrici*   S Pakistan to w India
___ *F. f. asiae*   N India
___ *F. f. melanonotus*   E India to Sikkim and Bangladesh

☐ **Painted Francolin** *Francolinus pictus*
___ *F. p. pallidus*   N-central India
___ *F. p. pictus*   Central and s India
___ *F. p. watsoni*   Sri Lanka

☐ **Chinese Francolin** *Francolinus pintadeanus*
___ *F. p. phayrei*   Dry scrub of ne India to Myanmar and Indochina
___ *F. p. pintadeanus*   SE China and Hainan

☐ **Gray Francolin** *Francolinus pondicerianus*
___ *F. p. mecranensis*   Arid se Iran and s Pakistan
___ *F. p. interpositus*   NW India and Pakistan
___ *F. p. pondicerianus*   S India and Sri Lanka

☐ **Swamp Francolin** *Francolinus gularis*
                            Terai of n India to s Nepal and Bangladesh

☐ **Coqui Francolin** *Francolinus coqui*
___ *F. c. spinetorum*   Mali to Nigeria and Angola
___ *F. c. maharao*   Ethiopia to s Uganda, Kenya and n Tanzania
___ *F. c. hubbardi*   W and s Kenya to central Tanzania
___ *F. c. coqui*   Kenya to Zaire, Botswana, Natal and n Namibia

☐ **White-throated Francolin** *Francolinus albogularis*
___ *F. a. albogularis*   Senegambia to Ivory Coast
___ *F. a. buckleyi*   E Ivory Coast to Cameroon
___ *F. a. dewittei*   SE Zaire to e Angola and nw Zambia

☐ **Schlegel's Francolin** *Francolinus schlegelii*
                            Savanna of Cameroon and s Chad to sw Sudan

☐ **Forest Francolin** *Francolinus lathami*
___ *F. l. lathami*   Sierra Leone to Gabon, nw Zaire and Angola
___ *F. l. schubotzi*   W Zaire to extreme sw Sudan, w Uganda and nw Tanzania

☐ **Crested Francolin** *Francolinus sephaena*
___ *F. s. grantii*   Ethiopia to s Sudan, Uganda and n-central Tanzania
___ *F. s. spilogaster*   E Ethiopia to Somalia and ne Kenya
___ *F. s. rovuma*   Coastal Kenya and Tanzania to n Mozambique
___ *F. s. sephaena*   E Zimbabwe to se Botswana and ne South Africa
___ *F. s. zambesiae*   W-central Mozambique to nw Namibia and s Angola

☐ **Ring-necked Francolin** *Francolinus streptophorus*
                            Southwest Cameroon; w Uganda to w Kenya and nw Tanzania

☐ **Finsch's Francolin** *Francolinus finschi*
                            Brachystegia belt of nw Angola, Gabon and sw Zaire

☐ **Red-winged Francolin** *Francolinus levaillantii*
___ *F. l. kikuyuensis*   Angola to e Zaire, w-central Kenya and Zambia
___ *F. l. levaillantii*   Malawi and ne Zambia to e South Africa

☐ **Gray-winged Francolin** *Francolinus africanus*
                            Grasslands of Lesotho and w-central South Africa

☐ **Moorland Francolin** *Francolinus psilolaemus*
    ____ *F. p. psilolaemus* — Montane moorlands of central and s Ethiopia
    ____ *F. p. elgonensis* — Montane moorlands of e Uganda to central Kenya

☐ **Shelley's Francolin** *Francolinus shelleyi*
    ____ *F. s. shelleyi* — S Uganda and sw Kenya to ne South Africa
    ____ *F. s. whytei* — SE Zaire to n Zambia and n Malawi

☐ **Orange River Francolin** *Francolinus levaillantoides*
    ____ *F. l. gutturalis* — N Ethiopia
    ____ *F. l. lorti* — Uganda to Sudan, s Ethiopia and Somalia
    ____ *F. l. jugularis* — Kalahari Desert of n Namibia to sw Angola
    ____ *F. l. levaillantoides* — S Botswana to e Namibia, Lesotho and n South Africa

☐ **Scaly Francolin** *Francolinus squamatus* — Equatorial Africa from s-c Nigeria to s Ethiopia and Malawi

☐ **Ahanta Francolin** *Francolinus ahantensis* — Discontinuous in lowlands of Senegambia to sw Nigeria

☐ **Gray-striped Francolin** *Francolinus griseostriatus* — Locally in escarpment of w Angola. Last recorded 1954

☐ **Nahan's Francolin** *Francolinus nahani* — Humid forests of w Uganda and ne Zaire

☐ **Hartlaub's Francolin** *Francolinus hartlaubi* — Rocky hill country of sw Angola to central Namibia

☐ **Double-spurred Francolin** *Francolinus bicalcaratus*
    ____ *F. b. ayesha* — W Morocco (Rabat to Essaouira)
    ____ *F. b. bicalcaratus* — Senegambia to Central African Republic

☐ **Heuglin's Francolin** *Francolinus icterorhynchus* — Central African Republic to s Sudan, n Zaire and w Uganda

☐ **Clapperton's Francolin** *Francolinus clappertoni* — Sub-Saharan Africa (Mali to Sudan and Ethiopia)

☐ **Harwood's Francolin** *Francolinus harwoodi* — Highlands of central Ethiopia

☐ **Red-billed Francolin** *Francolinus adspersus* — S Angola to Namibia, Botswana, sw Zambia and w Zimbabwe

☐ **Cape Francolin** *Francolinus capensis* — South Africa (riverine scrub of s and w Cape Province)

☐ **Natal Francolin** *Francolinus natalensis* — Zambia and Mozambique to Cape Province

☐ **Hildebrandt's Francolin** *Francolinus hildebrandti* — Kenya to Tanzania, se Zaire, ne Zambia and s Malawi

☐ **Yellow-necked Francolin** *Francolinus leucoscepus* — SE Sudan to Ethiopia, Somalia, Kenya and n Tanzania

☐ **Gray-breasted Francolin** *Francolinus rufopictus* — NW Tanzania (Lake Victoria to Serengeti and s to Wembere)

☐ **Red-necked Francolin** *Francolinus afer*
    ____ *F. a. cranchii* — W Congo to e Uganda and w Kenya, south to ne Zambia
    ____ *F. a. harterti* — N shore of Lake Tanganyika (Burundi, ne Tanzania, e Zaire)
    ____ *F. a. leucoparaeus* — Coastal Kenya (Tana River to Tanzania border)
    ____ *F. a. afer* — W Angola and extreme nw Namibia
    ____ *F. a. melanogaster* — Mozambique n of Zambezi R. to e Tanzania and e Zambia
    ____ *F. a. swynnertoni* — Interior Mozambique s of Zambezi River to se Zimbabwe
    ____ *F. a. castaneiventer* — South Africa (s and e Cape Province)

☐ **Swainson's Francolin** *Francolinus swainsonii*
    ____ *F. s. swainsonii* — SE Angola to n Namibia, s Botswana and ne South Africa
    ____ *F. s. lundazi* — N and w Zimbabwe to s Mozambique

☐ **Jackson's Francolin** *Francolinus jacksoni* — Montane forests of w and central Kenya

☐ **Handsome Francolin** *Francolinus nobilis*

Montane forests of sw Uganda, e Zaire and Rwanda

☐ **Cameroon Francolin** *Francolinus camerunensis*

Montane forests of Cameroon Mountain

☐ **Swierstra's Francolin** *Francolinus swierstrai*

Montane forests of w Angola (Cuanza Sul to Huila)

☐ **Chestnut-naped Francolin** *Francolinus castaneicollis*
___ *F. c. castaneicollis*
___ *F. c. atrifrons*

Mountains of ne Ethiopia and Somalia to Kenya border
S Ethiopia and extreme n Kenya

☐ **Erckel's Francolin** *Francolinus erckelii*

Eritrea to n Ethiopia and e Sudan (Red Sea Province)

☐ **Djibouti Francolin** *Francolinus ochropectus*

Montane forests of Forêt du Day (Djibouti)

☐ **Gray Partridge** *Perdix perdix*
___ *P. p. perdix (italica)*
___ *P. p. sphangnetorum*
___ *P. p. armoricana*
___ *P. p. hispaniensis*
___ *P. p. lucida*
___ *P. p. robusta*
___ *P. p. canescens*

British Isles and s Scandinavia to Alps, Italy and Balkans
Moors of n Holland and nw Germany
Locally in France
Central Pyrénées (ne Portugal and n Spain)
Finland east to Ural Mts. and s to Black Sea and n Caucasus
Ural Mountains to sw Siberia and nw China
Turkey east to the Caucasus, Transcaucasia and nw Iran

☐ **Daurian Partridge** *Perdix dauurica*
___ *P. d. dauurica*
___ *P. d. suschkini*

Steppes of Mongolia and n China (Xinjiang)
Steppes of Manchuria to w China (Gansu and Qinghai)

☐ **Tibetan Partridge** *Perdix hodgsoniae*
___ *P. h. sifanica*
___ *P. h. caraganae*
___ *P. h. hodgsoniae*

E Tibet to w-central China
E Kashmir to extreme e Tibet
Himalayas (w Nepal to Assam and e Tibet)

☐ **Long-billed Partridge** *Rhizothera longirostris*
___ *R. l. longirostris*
___ *R. l. dulitensis*

Myanmar, Malay Pen. and s Thailand to Sumatra and Borneo
N Borneo (Buto Song Mountains and Sarawak)

☐ **Madagascar Partridge** *Margaroperdix madagascarensis*

Madagascar; introduced Réunion

☐ **Black Partridge** *Melanoperdix niger*
___ *M. n. niger*
___ *M. n. borneensis*

Lowland forests of Malay Peninsula and Sumatra
Lowland forests of Borneo

☐ **Japanese Quail** *Coturnix japonica*

E Palearctic; winters to SE Asia and e China

☐ **Common Quail** *Coturnix coturnix*

Widespread Palearctic region; e and s Africa

☐ **Harlequin Quail** *Coturnix delegorguei*
___ *C. d. delegorguei*
___ *C. d. histrionica*
___ *C. d. arabica*

Grasslands of Africa south of the Sahara and Madagascar
São Tomé (Gulf of Guinea)
Known from some old specimens from sw Arabia

☐ **Rain Quail** *Coturnix coromandelica*

Pakistan to Myanmar and w Thailand

☐ **Stubble Quail** *Coturnix pectoralis*

Grasslands of Australia and Tasmania

☐ **Brown Quail** *Coturnix ypsilophora*
___ *C. y. raaltenii*
___ *C. y. pallidior*
___ *C. y. saturatior*

Lesser Sundas (Flores, Timor and adjacent islands)
Lesser Sundas (Sumba and Sawu)
Lowlands of n New Guinea

___	_C. y. lamonti_	Mid-montane central highlands of New Guinea
___	_C. y. dogwa_	Lowlands of s New Guinea
___	_C. y. plumbea_	Lowlands of e New Guinea
___	_C. y. monticola_	Alpine grasslands of se New Guinea
___	_C. y. mafulu_	S slopes of mountains of se New Guinea
___	_C. y. australis_	Moist areas of Australia
___	_C. y. ypsilophora_	Tasmania

☐ **Blue-breasted Quail** _Coturnix chinensis_

___	_C. c. chinensis_	India to Sri Lanka, Malaya, Indochina, se China and Taiwan
___	_C. c. trinkutensis_	Andaman Islands and Nicobar Islands
___	_C. c. palmeri_	Sumatra and Java
___	_C. c. lineata_	Philippines, Borneo, Sulawesi and Sula Islands
___	_C. c. lineatula_	Lesser Sundas (Lombok to Sumba, Flores and Timor)
___	_C. c. novaeguineae_	Montane forests of New Guinea
___	_C. c. papuensis_	SE New Guinea
___	_C. c. lepida_	Bismarck Archipelago
___	_C. c. colletti_	N Australia (Northern Territory)
___	_C. c. victoriae_	E Australia (Queensland to Victoria)

☐ **Blue Quail** _Coturnix adansonii_

Wet grasslands of Africa south of the Sahara

☐ **Snow Mountain Quail** _Anurophasis monorthonyx_

W New Guinea (alpine grasslands of Snow Mountains)

☐ **Jungle Bush-Quail** _Perdicula asiatica_

___	_P. a. punjabi_	NW India (Kashmir to Uttar Pradesh)
___	_P. a. vidali_	W India
___	_P. a. asiatica_	Central and ne India (Gujarat to Bihar)
___	_P. a. ceylonensis_	Sri Lanka

☐ **Rock Bush-Quail** _Perdicula argoondah_

___	_P. a. meinertzhageni_	NW India (south to Rann of Kutch and Madhya Pradesh)
___	_P. a. argoondah_	Peninsular India south to Madras
___	_P. a. salimali_	S India (stony lateritic soils of e-central Mysore)

☐ **Painted Bush-Quail** _Perdicula erythrorhyncha_

| ___ | _P. e. erythrorhyncha_ | W India (Western Ghats) |
| ___ | _P. e. blewitti_ | Central and eastern India |

☐ **Manipur Bush-Quail** _Perdicula manipurensis_

| ___ | _P. m. manipurensis_ | Manipur and Assam hills (south of the Brahmaputra River) |
| ___ | _P. m. inglisi_ | W Bengal and Assam (north of the Brahmaputra River) |

☐ **Udzungwa Partridge** _Xenoperdix udzungwensis_

| ___ | _X. u. udzungwensis_ | S Tanzania (Udzungwa Mountains) |
| ___ | _X. u. obscurata_ | S Tanzania (Maf-wemiro forest in Rubeho highlands) |

☐ **Hill Partridge** _Arborophila torqueola_

___	_A. t. millardi_	W Himalayas (Himanchal Pradesh to w Nepal)
___	_A. t. torqueola_	E Himalayas (Nepal to Tibet and n Myanmar)
___	_A. t. batemani_	N Myanmar to sw China (w Yunnan and sw Sichuan)
___	_A. t. griseata_	NW Vietnam

☐ **Sichuan Partridge** _Arborophila rufipectus_

SW China (hills of s-central Sichuan)

☐ **Chestnut-breasted Partridge** _Arborophila mandellii_

Coniferous foothill forests of Sikkim to se Tibet

☐ **White-necklaced Partridge** _Arborophila gingica_

Montane forests of se China

☐ **Rufous-throated Partridge** *Arborophila rufogularis*

____ *A. r. rufogularis* — N India (Uttar Pradesh to Assam) and Nepal
____ *A. r. intermedia* — NE India to Myanmar and extreme nw Yunnan
____ *A. r. tickelli* — E Myanmar to Thailand and sw Laos
____ *A. r. euroa* — S China (se Yunnan) to n Laos
____ *A. r. guttata* — Central Vietnam and n Laos
____ *A. r. vietnamenis* — S Vietnam (Langbian Plateau region)

☐ **White-cheeked Partridge** *Arborophila atrogularis*

Humid forests of ne India to Myanmar and sw China (Yunnan)

☐ **Taiwan Partridge** *Arborophila crudigularis*

Montane forests of Taiwan

☐ **Hainan Partridge** *Arborophila ardens*

Montane forests of Hainan I. (s China)

☐ **Chestnut-bellied Partridge** *Arborophila javanica*

____ *A. j. javanica* — Mountains of w Java
____ *A. j. bartelsi* — Mountains of w-central Java
____ *A. j. lawuana* — Mountains of e-central Java

☐ **Gray-breasted Partridge** *Arborophila orientalis*

____ *A. o. campbelli* — Mountains of Malay Peninsula
____ *A. o. rolli* — NW Sumatra (Batak highlands)
____ *A. o. sumatrana* — Mountains of central Sumatra
____ *A. o. orientalis* — Mountains of e Java

☐ **Bar-backed Partridge** *Arborophila brunneopectus*

____ *A. b. brunneopectus* — SW China (sw Yunnan) to e Myanmar, n Laos and w Thailand
____ *A. b. henrici* — N and central Vietnam
____ *A. b. albigula* — S-central Vietnam

☐ **Orange-necked Partridge** *Arborophila davidi*

S Vietnam and adjacent Cambodia

☐ **Chestnut-headed Partridge** *Arborophila cambodiana*

____ *A. c. diversa* — Tropical forests of se Thailand
____ *A. c. cambodiana* — Tropical forests of sw Cambodia
____ *A. c. chandamonyi* — Southwest Cambodia (Cardamom Mts.)

☐ **Red-breasted Partridge** *Arborophila hyperythra*

____ *A. h. hyperythra* — Montane forests of n-central Borneo
____ *A. h. erythrophrys* — Mt. Kinabalu (n Borneo)

☐ **Red-billed Partridge** *Arborophila rubrirostris*

Montane forests of Sumatra

☐ **Scaly-breasted Partridge** *Arborophila chloropus*

____ *A. c. chloropus* — Extreme sw China (Yunnan) to Myanmar and w Thailand
____ *A. c. peninsularis* — SW Thailand
____ *A. c. tonkinensis* — N Vietnam
____ *A. c. olivacea* — Laos and Cambodia
____ *A. c. cognacqi* — S Vietnam

☐ **Vietnam Partridge** *Arborophila merlini*

____ *A. m. merlini* — Interior of central Vietnam
____ *A. m. vivida* — Coastal hills of central Vietnam

☐ **Chestnut-necklaced Partridge** *Arborophila charltonii*

____ *A. c. charltonii* — S Thailand to s Myanmar and Malay Peninsula
____ *A. c. atjenensis* — N Sumatra (Aceh Province)
____ *A. c. graydoni* — N Borneo (Sabah)

☐ **Ferruginous Partridge** *Caloperdix oculeus*
___ *C. o. oculeus* — SE Myanmar and sw Thailand to Malay Peninsula
___ *C. o. ocellatus* — Sumatra
___ *C. o. borneensis* — Borneo

☐ **Crimson-headed Partridge** *Haematortyx sanguiniceps*

Montane forests of n Borneo

☐ **Crested Partridge** *Rollulus rouloul*

Malay Pen., Sumatra, Borneo, Banka and Belitung islands

☐ **Stone Partridge** *Ptilopachus petrosus*
___ *P. p. petrosus* — Senegambia to s Sudan, n Uganda and n Kenya
___ *P. p. major* — Rocky areas of nw Ethiopia

☐ **Mountain Bamboo-Partridge** *Bambusicola fytchii*
___ *B. f. hopkinsoni* — NE India to Bangladesh and n Myanmar
___ *B. f. fytchii* — SW China (Sichuan and Yunnan) to Myanmar and n Vietnam

☐ **Chinese Bamboo-Partridge** *Bambusicola thoracicus*
___ *B. t. thoracicus* — Arid bush of s and central China
___ *B. t. sonorivox* — Taiwan

☐ **Red Spurfowl** *Galloperdix spadicea*
___ *G. s. caurina* — W India (Arvalli Hills of s Rajasthan)
___ *G. s. spadicea* — N India (Uttar Pradesh) and *terai* of w Nepal to s India
___ *G. s. stewarti* — S India (Kerala coast)

☐ **Painted Spurfowl** *Galloperdix lunulata*

Semiarid steppes of peninsular India

☐ **Ceylon Spurfowl** *Galloperdix bicalcarata*

Humid forests of s Sri Lanka

☐ **Blood Pheasant** *Ithaginis cruentus*
___ *I. c. cruentus* — N Nepal to nw Bhutan
___ *I. c. affinis* — Sikkim
___ *I. c. tibetanus* — E Bhutan and s Tibet
___ *I. c. kuseri* — NE India (upper Assam) and se Tibet
___ *I. c. geoffroyi* — W China (w Sichuan) and se Tibet
___ *I. c. marionae* — Mountains of sw China (nw Yunnan) and ne Myanmar
___ *I. c. rocki* — SW China (Mekong Valley of nw Yunnan)
___ *I. c. holoptilus* — SW China (Likiang District of Yunnan)
___ *I. c. clarkei* — SW China (Likiang Mountains of nw Yunnan)
___ *I. c. michaelis* — N-central China (Nan Shan Mountains of nw Gansu)
___ *I. c. beicki* — N-central China (ne Qinghai and adjacent Gansu)
___ *I. c. berezowskii* — Mountains of central China (s Gansu and n Sichuan)
___ *I. c. annae* — Mountains of sw China (nw Sichuan)
___ *I. c. sinensis* — C China (Tsinling Mountains of s Shensi and sw Hunan)

☐ **Western Tragopan** *Tragopan melanocephalus*

Himalayas of n Pakistan to nw India and adjacent sw Tibet

☐ **Satyr Tragopan** *Tragopan satyra*

Oak-rhododendron forests of n India to Nepal and se Tibet

☐ **Blyth's Tragopan** *Tragopan blythii*
___ *T. b. blythii* — Himalayas of ne India to sw China and adjacent Myanmar
___ *T. b. molesworthi* — Known from 3 specimens from e Bhutan

☐ **Temminck's Tragopan** *Tragopan temminckii*

Mts. of ne India to central China, n Myanmar and nw Tonkin

☐ **Cabot's Tragopan** *Tragopan caboti*
___ *T. c. caboti* — Foothill forests of se China
___ *T. c. guangxiensis* — SE China (ne Guangxi Zhuangu Autonomous Region)

☐ **Koklass Pheasant** *Pucrasia macrolopha*
____ *P. m. castanea*   Mountains of e Afghanistan and adjacent Pakistan
____ *P. m. biddulphi*   Himalayas of Kashmir
____ *P. m. bethelae*   NW India (Kulu Valley)
____ *P. m. macrolopha*   W Himalayas (Kashmir to Kumaon)
____ *P. m. nipalensis*   Mountains of w Nepal
____ *P. m. meyeri*   Mountains of s-central China (w Sichuan to nw Yunnan)
____ *P. m. ruficollis*   Mountains of central China (s Gansu, Shaanxi and w Sichuan)
____ *P. m. xanthospila*   N Shaanxi to Inner Mongolia, w Liaoning and sw Manchuria
____ *P. m. joretiana*   Mountains of e-central China (sw Anhui)
____ *P. m. darwini*   Mountains of c China (Hubei and se Sichuan to Fujian)

☐ **Himalayan Monal** *Lophophorus impejanus*

Himalayas of Afghanistan to s Tibet, sw China and ne Myanmar

☐ **Sclater's Monal** *Lophophorus sclateri*
____ *L. s. sclateri*   Himalayas of ne India to sw China
____ *L. s. arunachalensis*   N India (mts. of w Arunachal Pradesh)

☐ **Chinese Monal** *Lophophorus lhuysii*

Mountains of sw China (s Gansu, nw Sichuan and Yunnan)

☐ **Red Junglefowl** *Gallus gallus*
____ *G. g. murghi*   N India and adjacent Nepal and Bangladesh
____ *G. g. spadiceus*   Myanmar to sw Yunnan, Malay Peninsula and n Sumatra
____ *G. g. jabouillei*   N Vietnam to s China (se Yunnan, Guangxi and Hainan I.)
____ *G. g. gallus*   N Indochina to e Thailand
____ *G. g. bankiva*   S Sumatra, Java and Bali

☐ **Gray Junglefowl** *Gallus sonneratii*

Peninsular India

☐ **Ceylon Junglefowl** *Gallus lafayetii*

Sri Lanka

☐ **Green Junglefowl** *Gallus varius*

Java, Bali, Lombok, Sumbawa, Flores and Alor islands

☐ **Kalij Pheasant** *Lophura leucomelanos*
____ *L. l. hamiltoni*   Western Himalayas (Indus River to w Nepal)
____ *L. l. leucomelanos*   Subtropical pine, *sal*, and moist temperate forests of Nepal
____ *L. l. melanota*   Sikkim and w Bhutan
____ *L. l. moffitti*   Range unknown; possibly central Bhutan
____ *L. l. lathami*   E Bhutan and n India to Myanmar
____ *L. l. williamsi*   W Myanmar (east to Irrawaddy River)
____ *L. l. oatesi*   S Myanmar (Arakan Yoma Mountains)
____ *L. l. lineata*   S Myanmar (east of Irrawaddy River) to nw Thailand
____ *L. l. crawfurdi*   SE Myanmar (Tenasserim) and peninsular Thailand

☐ **Imperial Pheasant** *Lophura imperialis*

Limestone mountains of central Vietnam

☐ **Edwards' Pheasant** *Lophura edwardsi*

Lowlands of central Vietnam (on verge of extinction in wild)

☐ **Vietnamese Pheasant** *Lophura hatinhensis*

Lowlands of n-central Vietnam

☐ **Swinhoe's Pheasant** *Lophura swinhoii*

Montane forests of central Taiwan

☐ **Salvadori's Pheasant** *Lophura inornata*
____ *L. i. inornata*   Montane forests of s Sumatra
____ *L. i. hoogerwerfi*   NW Sumatra (known from two E specimens ca 1939)

☐ **Silver Pheasant** *Lophura nycthemera*
____ *L. n. occidentalis*   S-central China (nw Yunnan) and ne Myanmar
____ *L. n. rufipes*   Highlands of n Myanmar (Northern Shan States)
____ *L. n. ripponi*   Highlands of n Myanmar (Southern Shan States)

____	*L. n. jonesi*	Myanmar to sw China (sw Yunnan) and central Thailand
____	*L. n. omeiensis*	S-central China (s Sichuan)
____	*L. n. rongjiangensis*	S-central China (se Guizhou)
____	*L. n. beaulieui*	S-central China (se Yunnan) to n Laos and n Vietnam
____	*L. n. nycthemera*	S China (Guangdong and Guangxi) to n Vietnam
____	*L. n. whiteheadi*	Hainan (s China)
____	*L. n. fokiensis*	SE China (nw Fujian and (?) Zhejiang
____	*L. n. berliozi*	Central Vietnam (w slope of Annamitic Mountains)
____	*L. n. beli*	Central Vietnam (e slope of Annamitic Mountains)
____	*L. n. engelbachi*	S Laos (Bolavens Plateau)
____	*L. n. lewisi*	Mountains of sw Cambodia and se Thailand
____	*L. n. annamensis*	Montane forests of s Vietnam

☐ **Crestless Fireback** *Lophura erythrophthalma*

____	*L. e. erythrophthalma*	Lowland forests of Malay Peninsula and Sumatra
____	*L. e. pyronota*	Lowland forests of n Borneo

☐ **Crested Fireback** *Lophura ignita*

____	*L. i. rufa*	Malay Peninsula and Sumatra (except for range of *macartneyi*)
____	*L. i. macartneyi*	SE Sumatra
____	*L. i. ignita*	Kalimantan (Borneo) and Banka I. (off se Sumatra)
____	*L. i. nobilis*	N Borneo (Sarawak and Sabah)

☐ **Siamese Fireback** *Lophura diardi*

Lowlands of e Myanmar, Thailand and Indochina

☐ **Bulwer's Pheasant** *Lophura bulweri*

Submontane forests of interior Borneo

☐ **White Eared-Pheasant** *Crossoptilon crossoptilon*

____	*C. c. harmani*	Rhododendron forests of s Tibet and adjacent ne India
____	*C. c. drouynii*	Montane forests of e Tibet
____	*C. c. dolani*	W-central China (s Qinghai)
____	*C. c. crossoptilon*	SW China (w Sichuan) to se Tibet and extreme ne India
____	*C. c. lichiangense*	S-central China (nw Yunnan)

☐ **Brown Eared-Pheasant** *Crossoptilon mantchuricum*

Montane forests of ne China (Liaoning and Shanxi)

☐ **Blue Eared-Pheasant** *Crossoptilon auritum*

Montane forests of n-central China

☐ **Cheer Pheasant** *Catreus wallichi*

Montane forests of e Afghanistan to central Nepal

☐ **Elliot's Pheasant** *Syrmaticus ellioti*

Montane bamboo forests of se China

☐ **Hume's Pheasant** *Syrmaticus humiae*

____	*S. h. humiae*	Montane forests of extreme ne India and n Myanmar
____	*S. h. burmanicus*	SW China (sw Yunnan) to Myanmar and nw Thailand

☐ **Mikado Pheasant** *Syrmaticus mikado*

Montane forests of central Taiwan

☐ **Copper Pheasant** *Syrmaticus soemmerringii*

____	*S. s. scintillans*	Japan (coniferous forests of n and central Honshu)
____	*S. s. intermedius*	Japan (coniferous forests of sw Honshu and Shikoku)
____	*S. s. subrufus*	Japan (coniferous forests of s Honshu and sw Shikoku)
____	*S. s. soemmerringii*	Japan (coniferous forests of n and central Kyushu)
____	*S. s. ijimae*	Japan (coniferous forests of se Kyushu)

☐ **Reeves' Pheasant** *Syrmaticus reevesii*

Low altitude deciduous forests of n-central China

☐ **Ring-necked Pheasant** *Phasianus colchicus*

____	*P. c. septentrionalis*	N Caucasus
____	*P. c. colchicus*	E Georgia to ne Azerbaijan, s Armenia and nw Iran

____	*P. c. talischensis*	SE Transcaucasia
____	*P. c. persicus*	SW Transcaspia
____	*P. c. bergii*	Islands in Aral Sea
____	*P. c. turcestanicus*	Kazakstan (Valley of River Syrdar'ya)
____	*P. c. mongolicus*	NE Russian Turkestan
____	*P. c. principalis*	S Russian Turkestan and n Afghanistan
____	*P. c. chrysomelas*	Turkestan (upper River Amudar'ya)
____	*P. c. zerafschanicus*	S Uzbekistan (Bukhara and Zerafshan Valley)
____	*P. c. zarudnyi*	Turkestan (valleys of central Amudar'ya)
____	*P. c. bianchii*	Turkestan (Amudar'ya delta)
____	*P. c. shawii*	Chinese Turkestan
____	*P. c. tarimensis*	E-central Chinese Turkestan
____	*P. c. hagenbecki*	NW Mongolia
____	*P. c. edzinensis*	S-central Mongolia
____	*P. c. satschuensis*	N-central China (extreme w Gansu)
____	*P. c. vlangallii*	N-central China (n Qinghai)
____	*P. c. alashanicus*	N-central China (foothills of Alaschan Mountains)
____	*P. c. sohokhotensis*	N-central China (Sohokhoto Oasis and Qilian Shan)
____	*P. c. pallasi*	SE Siberia and ne China
____	*P. c. karpowi*	NE China (s Manchuria and n Liaoning) to Korea
____	*P. c. kiangsuensis*	NE China (n Shanxi and Shaanxi) to se Mongolia
____	*P. c. strauchi*	Central China (s Shaanxi and s Gansu)
____	*P. c. suehschanensis*	W-central China (nw Sichuan)
____	*P. c. elegans*	W-central China (w Sichuan)
____	*P. c. decollatus*	Central China (Sichuan to Liaoning, ne Yunnan and Guizhou)
____	*P. c. torquatus*	E China (Shandong) to Vietnam border
____	*P. c. rothschildi*	SW China (e Yunnan) and n Vietnam
____	*P. c. takatsukasae*	S China and n Vietnam
____	*P. c. formosanus*	Taiwan

☐ **Green Pheasant** *Phasianus versicolor*

____	*P. v. versicolor*	Japan (sw Honshu and Kyushu)
____	*P. v. tanensis*	Japan (central Honshu) and Izu Islands
____	*P. v. robustipes*	Japan (nw Honshu and Sado I.)

☐ **Golden Pheasant** *Chrysolophus pictus*

Mountain slopes of central and s China

☐ **Lady Amherst's Pheasant** *Chrysolophus amherstiae*

Mountains of se Tibet and sw China to n Myanmar

☐ **Bronze-tailed Peacock-Pheasant** *Polyplectron chalcurum*

____	*P. c. scutulatum*	Mountains of n Sumatra
____	*P. c. chalcurum*	Mountains of s Sumatra

☐ **Mountain Peacock-Pheasant** *Polyplectron inopinatum*

Montane forests of Malay Peninsula

☐ **Germain's Peacock-Pheasant** *Polyplectron germaini*

Humid forests of s Vietnam

☐ **Gray Peacock-Pheasant** *Polyplectron bicalcaratum*

____	*P. b. bakeri*	Humid forests of ne India and Bhutan
____	*P. b. bailyi*	Patchily distributed w Assam and adjacent e Himalayas
____	*P. b. bicalcaratum*	NE Assam and Myanmar to sw Thailand and central Laos
____	*P. b. ghigii*	Central and n Vietnam to e Tonkin and central Laos
____	*P. b. katsumatae*	Hainan (s China)

☐ **Malayan Peacock-Pheasant** *Polyplectron malacense*

S Myanmar to s Thailand and Malay Peninsula

☐ **Bornean Peacock-Pheasant** *Polyplectron schleiermacheri*

Lowland primary forests of Borneo

☐ **Palawan Peacock-Pheasant** *Polyplectron napoleonis*

Humid forests of Palawan (sw Philippines)

☐ **Crested Argus** *Rheinardia ocellata*

___ *R. o. ocellata* — Mountains of central Vietnam and e Laos

___ *R. o. nigrescens* — Mts. of central Malay Peninsula (Taman Nagara Nat. Park)

☐ **Great Argus** *Argusianus argus*

___ *A. a. argus* — Malay Peninsula and Sumatra

___ *A. a. grayi* — Borneo

☐ **Indian Peafowl** *Pavo cristatus*

Forests and scrub of e Pakistan, India and Sri Lanka

☐ **Green Peafowl** *Pavo muticus*

___ *P. m. spificer* — NE India and se Bangladesh to nw Myanmar

___ *P. m. imperator* — Myanmar to Thailand, s China and Indochina

___ *P. m. muticus* — Locally in Java; formerly Malay Peninsula

☐ **Congo Peacock** *Afropavo congensis*

Locally in humid forests of central Zaire

ORDER: GALLIFORMES
FAMILY: NUMIDIDAE (Guineafowl—6)

☐ **White-breasted Guineafowl** *Agelastes meleagrides*

SE Sierra Leone to Ivory Coast and w Ghana

☐ **Black Guineafowl** *Agelastes niger*

SE Nigeria to n Angola and extreme ne Zaire

☐ **Helmeted Guineafowl** *Numida meleagris*

___ *N. m. sabyi* — NW Morocco

___ *N. m. galeatus* — W Africa to s Chad, central Zaire and n Angola

___ *N. m. meleagris* — E Chad to Ethiopia, n Zaire, Uganda and n Kenya

___ *N. m. somaliensis* — NE Ethiopia and Somalia

___ *N. m. reichenowi* — Kenya and central Tanzania

___ *N. m. mitratus* — Tanzania to e Mozambique, Zambia and n Botswana

___ *N. m. marungensis* — S Congo basin to w Angola and Zambia

___ *N. m. damarensis* — S Angola to Botswana and Namibia

___ *N. m. coronatus* — E South Africa

☐ **Plumed Guineafowl** *Guttera plumifera*

___ *G. p. plumifera* — S Cameroon to Congo basin, n Gabon and n Angola

___ *G. p. schubotzi* — N Zaire to Rift Valley and forests west of Lake Tanganyika

☐ **Crested Guineafowl** *Guttera pucherani*

___ *G. p. verreauxi* — Guinea-Bissau to w Kenya, Angola and Zambia

___ *G. p. sclateri* — NW Cameroon

___ *G. p. pucherani* — Somalia to Tanzania, Zanzibar and Tumbatu I.

___ *G. p. barbata* — SE Tanzania to e Mozambique and Malawi

___ *G. p. edouardi* — E Zambia to Mozambique and ne South Africa

☐ **Vulturine Guineafowl** *Acryllium vulturinum*

Arid acacia scrub of s Ethiopia and Somalia to ne Tanzania

ORDER: GRUIFORMES
FAMILY: OPISTHOCOMIDAE (Hoatzin—1)

☐ **Hoatzin** *Opisthocomus hoazin*

Amazon and Orinoco basin lowlands and the Guianas

ORDER: GRUIFORMES
FAMILY: MESITORNITHIDAE (Mesites—3)

☐ **White-breasted Mesite** *Mesitornis variegatus*

Deciduous dry forests of w Madagascar

☐ **Brown Mesite** *Mesitornis unicolor*

Rainforests of e Madagascar

☐ **Subdesert Mesite** *Monias benschi*

Coastal subdeserts of sw Madagascar

ORDER: GRUIFORMES
FAMILY: TURNICIDAE (Buttonquail—16)

☐ **Small Buttonquail** *Turnix sylvaticus*

____	*T. s. sylvaticus*	S Iberian Peninsula, n Morocco, Algeria and Tunisia
____	*T. s. lepuranus (alleni)*	Africa south of the Sahara and extreme s Arabian Peninsula
____	*T. s. dussumier*	Extreme e Iran to India and Myanmar
____	*T. s. davidi (mikado)*	Peninsular Thailand to s China, n Indochina and Taiwan
____	*T. s. whiteheadi*	Luzon I. (n Philippines)
____	*T. s. nigrorum*	Negros I. (Philippines)
____	*T. s. celestinoi*	S Philippines (Bohol and Mindanao)
____	*T. s. suluensis*	Sulu Archipelago
____	*T. s. bartelsorum*	Java and Bali

☐ **Red-backed Buttonquail** *Turnix maculosus*

____	*T. m. beccarii*	Sulawesi, Muna and Tomia I. (Tukangbesi Islands)
____	*T. m. kinneari*	Peleng I. (Banggai Islands off e Sulawesi)
____	*T. m. obiensis*	Obi I., Kai Kecil I. (Kai Islands) and Babar I.
____	*T. m. sumbanus*	Sumba I. (Lesser Sundas)
____	*T. m. floresianus*	Lesser Sundas (Sumbawa, Komodo, Padar, Flores and Alor)
____	*T. m. maculosus*	Lesser Sundas (Roti, Semau, Timor, Wetar, Moa and Kisar)
____	*T. m. savuensis*	Sawu I. (Lesser Sundas)
____	*T. m. saturatus*	Bismarck Archipelago (New Britain and Duke of York)
____	*T. m. furvus*	Huon Peninsula (ne New Guinea)
____	*T. m. giluwensis*	E-central New Guinea
____	*T. m. horsbrughi*	S New Guinea
____	*T. m. mayri*	Louisiade Archipelago
____	*T. m. salamonis*	Guadalcanal I. (Solomon Islands)
____	*T. m. melanotus (yorki, pseutes)*	N and e Australia

☐ **Hottentot Buttonquail** *Turnix hottentottus*

____	*T. h. nanus*	Ghana to Kenya, Uganda and se Cape Province
____	*T. h. hottentottus*	Mts. of South Africa (sw Cape Province to Port Elizabeth)

☐ **Yellow-legged Buttonquail** *Turnix tanki*

____	*T. t. tanki*	Indian subcontinent, Andaman and Nicobar Islands
____	*T. t. blanfordii*	Manchuria to Myanmar, s China and Indochina

☐ **Spotted Buttonquail** *Turnix ocellatus*

____	*T. o. benguetensis*	Mountains of n Luzon I. (n Philippines)
____	*T. o. ocellatus*	S and central Luzon I. (n Philippines)

☐ **Barred Buttonquail** *Turnix suscitator*

____	*T. s. taigoor*	India
____	*T. s. leggei*	Sri Lanka
____	*T. s. plumbipes*	Nepal, Sikkim and Bangladesh to n Myanmar
____	*T. s. bengalensis*	NE India (lower w Bengal)
____	*T. s. okinavensis*	S Kyushu I. and Makenoshima I. south to Ryukyu Islands
____	*T. s. rostratus*	Taiwan

____ *T. s. blakistoni*	Myanmar to s China, n Indochina and Hainan I.
____ *T. s. pallescens*	S-central Myanmar
____ *T. s. thai*	Central Thailand
____ *T. s. interrumpens*	Peninsular Myanmar and Thailand
____ *T. s. atrogularis*	Peninsular Malaysia
____ *T. s. suscitator (machetes, kuiperi)*	Sumatra, Belitung I. and Bangka I. to Java and Bali
____ *T. s. baweanus*	Bawean I. (off Java)
____ *T. s. fasciatus*	N Philippines (Luzon to Mindoro, Sibuyan and Masbate)
____ *T. s. haynaldi*	SW Philippines (Palawan and Calamian Is.)
____ *T. s. nigrescens*	Philippines (Negros, Cebu and Panay)
____ *T. s. rufilatus*	Sulawesi
____ *T. s. powelli*	Lesser Sundas

☐ **Madagascar Buttonquail** *Turnix nigricollis*

Madagascar; introduced (?) Mauritius, Réunion and Glorieuses

☐ **Black-breasted Buttonquail** *Turnix melanogaster*

Coastal e Australia (se Queensland and n New South Wales)

☐ **Chestnut-backed Buttonquail** *Turnix castanotus*

Locally in coastal n Australia and offshore islands

☐ **Buff-breasted Buttonquail** *Turnix olivii*

N Australia (Cape York Peninsula of n Queensland)

☐ **Painted Buttonquail** *Turnix varius*

____ *T. v. novaecaledoniae*	New Caledonia
____ *T. v. scintillans*	Houtman Abrolhos Islands (off sw Australia)
____ *T. v. varius*	SW, e and se Australia and Tasmania

☐ **Luzon Buttonquail** *Turnix worcesteri*

Known from four specimens from Luzon I. (n Philippines)

☐ **Sumba Buttonquail** *Turnix everetti*

Sumba I. (Lesser Sundas)

☐ **Red-chested Buttonquail** *Turnix pyrrhothorax*

Savanna and scrub of n and e Australia

☐ **Little Buttonquail** *Turnix velox*

Grasslands and woodlands throughout Australia

☐ **Quail-plover** *Ortyxelos meiffrenii*

Discontinuously distributed sahel of sub-Saharan Africa

ORDER: GRUIFORMES
FAMILY: GRUIDAE (Cranes—15)

☐ **Gray Crowned-Crane** *Balearica regulorum*

____ *B. r. gibbericeps*	Uganda and Kenya to n Zimbabwe and n Mozambique
____ *B. r. regulorum*	S Angola and n Namibia to Zimbabwe and e South Africa

☐ **Black Crowned-Crane** *Balearica pavonina*

____ *B. p. pavonina*	Sub-Saharan Africa (Senegambia to Lake Chad)
____ *B. p. ceciliae*	Sub-Saharan Africa (Chad to Ethiopia and Kenya)

☐ **Demoiselle Crane** *Anthropoides virgo*

Palearctic; winters in ne Africa and s Asia

☐ **Blue Crane** *Anthropoides paradiseus*

Locally in n Namibia, s Zimbabwe and South Africa

☐ **Wattled Crane** *Bugeranus carunculatus*

Patchily distributed ne and s Africa

☐ **Siberian Crane** *Grus leucogeranus*

Breeds Arctic Siberia; winters to n India and China

☐ **Sandhill Crane** *Grus canadensis*

___ *G. c. canadensis*	Arctic N America and e Siberia; winters sw US and n Mexico
___ *G. c. rowani*	British Columbia to n Ontario; winters to n Mexico
___ *G. c. tabida*	Mid-continental North America; winters s US and n Mexico
___ *G. c. pulla*	Gulf Coast of s US
___ *G. c. pratensis*	Georgia and Florida
___ *G. c. nesiotes*	Cuba and Isle of Pines

☐ **Sarus Crane** *Grus antigone*

___ *G. a. antigone*	N India to Nepal and (formerly?) Bangladesh
___ *G. a. sharpii*	Cambodia and s Laos; winters in Vietnam
___ *G. a. gilliae*	Spottily distributed coastal n Australia (mainly Queensland)

☐ **Brolga** *Grus rubicunda*

N and e Australia and Trans-Fly of s New Guinea

☐ **White-naped Crane** *Grus vipio*

Siberia and Manchuria; winters to s China, Korea and Japan

☐ **Common Crane** *Grus grus*

Breeds n Eurasia; winters to n Africa, s India and SE Asia

☐ **Hooded Crane** *Grus monacha*

Siberia and nw Manchuria; winters to e China, Korea, and Japan

☐ **Whooping Crane** *Grus americana*

Breeds n Canada; winters coastal se Texas

☐ **Black-necked Crane** *Grus nigricollis*

Breeds Tibetan plateau; winters to ne India and s China

☐ **Red-crowned Crane** *Grus japonensis*

Siberia, Hokkaido and Mongolia; winters e China and Korea

ORDER: GRUIFORMES
FAMILY: ARAMIDAE (Limpkin—1)

☐ **Limpkin** *Aramus guarauna*

___ *A. g. pictus*	Florida, Cuba and Jamaica
___ *A. g. elucus*	Hispaniola and Puerto Rico
___ *A. g. dolosus*	SE Mexico to Panama
___ *A. g. guarauna*	S America (except for arid w coast, Andes and extreme south)

ORDER: GRUIFORMES
FAMILY: PSOPHIIDAE (Trumpeters—3)

☐ **Gray-winged Trumpeter** *Psophia crepitans*

___ *P. c. napensis*	SE Colombia to ne Peru and extreme nw Brazil
___ *P. c. crepitans*	SE Colombia to Venezuela, the Guianas and n Brazil

☐ **Dark-winged Trumpeter** *Psophia viridis*

___ *P. v. viridis*	Brazil s of the Amazon between Rio Madeira and Rio Tapajós
___ *P. v. dextralis*	E Brazil s of Amazon between Rio Tapajós and Rio Tocantins
___ *P. v. obscura*	NE Brazil south of the Amazon (ne Pará east of Rio Tocantins)

☐ **Pale-winged Trumpeter** *Psophia leucoptera*

___ *P. l. ochroptera*	NW Brazil north of the Amazon and west of Rio Negro
___ *P. l. leucoptera*	E Peru to central Brazil and ne Bolivia

ORDER: GRUIFORMES
FAMILY: RALLIDAE (Rails, Gallinules and Coots—135)

☐ **White-spotted Flufftail** *Sarothrura pulchra*
　____ *S. p. pulchra* — S Senegal to n Cameroon
　____ *S. p. zenkeri* — Extreme se Nigeria, coastal Cameroon and n Gabon
　____ *S. p. batesi* — S Cameroon
　____ *S. p. centralis* — Congo to s Sudan, w Kenya, nw Tanzania and n Angola

☐ **Buff-spotted Flufftail** *Sarothrura elegans*
　____ *S. e. reichenovi* — W Africa to Uganda and Angola
　____ *S. e. elegans* — Ethiopia to Somalia, e Kenya and south to South Africa

☐ **Red-chested Flufftail** *Sarothrura rufa*
　____ *S. r. bonapartii* — Sierra Leone to Gabon and Congo
　____ *S. r. elizabethae* — C African Rep. to ne Zaire, Ethiopia, Uganda and w Kenya
　____ *S. r. rufa* — Central Kenya to s Zaire, Angola and South Africa

☐ **Chestnut-headed Flufftail** *Sarothrura lugens*
　____ *S. l. lugens* — Swamps of Cameroon to Zaire and w Tanzania
　____ *S. l. lynesi* — Angola to Zambia and Zimbabwe

☐ **Streaky-breasted Flufftail** *Sarothrura boehmi* — Locally in wet grasslands of central Africa

☐ **Striped Flufftail** *Sarothrura affinis*
　____ *S. a. antonii* — Montane grasslands of extreme s Sudan to e Zimbabwe
　____ *S. a. affinis* — Montane grasslands of South Africa

☐ **Madagascar Flufftail** *Sarothrura insularis* — Humid forests of e and nw Madagascar

☐ **White-winged Flufftail** *Sarothrura ayresi* — Highlands of Ethiopia and e South Africa

☐ **Slender-billed Flufftail** *Sarothrura watersi* — Highlands of e Madagascar

☐ **Nkulengu Rail** *Himantornis haematopus* — Humid forests of Sierra Leone to ne Zaire and Gabon

☐ **Gray-throated Rail** *Canirallus oculeus* — Rainforests of Sierra Leone to w Uganda and e Zaire

☐ **Madagascar Wood-Rail** *Canirallus kioloides*
　____ *C. k. kioloides* — Humid rainforests of eastern and high plateau of Madagascar
　____ *C. k. berliozi* — Locally in nw Madagascar (Sambirano district)

☐ **Swinhoe's Rail** *Coturnicops exquisitus* — Siberia and n Manchuria; winters to s China and Ryukyu Is.

☐ **Yellow Rail** *Coturnicops noveboracensis*
　____ *C. n. noveboracensis* — Disjunct in marshes of Canada and n US; winters to s US
　____ *C. n. goldmani* — Locally in marshes near Río Lerma (central Mexico)

☐ **Speckled Rail** *Coturnicops notatus* — Locally in lowlands of South America east of the Andes

☐ **Ocellated Crake** *Micropygia schomburgkii*
　____ *M. s. schomburgkii* — Locally in Guyana and French Guiana
　____ *M. s. chapmani* — E Brazil (Bahia) to Mato Grosso and n Bolivia

☐ **Chestnut Forest-Rail** *Rallina rubra*
　____ *R. r. rubra* — Arfak Mountains of w New Guinea
　____ *R. r. klossi* — New Guinea (Weyland Mts. to Oranje Mts.)
　____ *R. r. telefolminensis* — E New Guinea (Victor Emanuel and Hindenberg mountains)

☐ **White-striped Forest-Rail** *Rallina leucospila* — Montane forests of w New Guinea (Vogelkop Peninsula)

☐ **Forbes' Rail** *Rallina forbesi*
_____ *R. f. steini*   New Guinea (Weyland Mts. to Bismarck Mts.)
_____ *R. f. parva*   NE New Guinea (Mt. Mengam in Adelbert Range)
_____ *R. f. dryas*   SE New Guinea (Huon Peninsula)
_____ *R. f. forbesi*   SE New Guinea (Herzog Mts. to Owen Stanley Mts.)

☐ **Mayr's Rail** *Rallina mayri*
_____ *R. m. mayri*   W New Guinea (Cyclops Mountains)
_____ *R. m. carmichaeli*   NW Papus New Guinea (Torricelli and Bewani mountains)

☐ **Red-necked Crake** *Rallina tricolor*
                                      New Guinea to Bismarck Arch., Lesser Sundas and ne Australia

☐ **Andaman Crake** *Rallina canningi*
                                      Andaman Islands

☐ **Red-legged Crake** *Rallina fasciata*
                                      Lowlands of SE Asia, Malay Archipelago and Philippines

☐ **Slaty-legged Crake** *Rallina eurizonoides*
_____ *R. e. amauroptera*   Pakistan and India to Assam; winters to Sri Lanka
_____ *R. e. telmatophila*   Myanmar to n Thailand, Sumatra and Java
_____ *R. e. sepiaria*   Ryukyu Islands
_____ *R. e. formosana*   Taiwan and Lan-yü I.
_____ *R. e. eurizonoides*   Philippines; vagrant to Palau Islands (w Micronesia)
_____ *R. e. alvarezi*   Batan Islands (n Philippines)
_____ *R. e. minahasa*   Sulawesi and Sula Islands

☐ **Chestnut-headed Crake** *Anurolimnas castaneiceps*
_____ *A. c. coccineipes*   Tropical forests of sw Colombia and ne Ecuador
_____ *A. c. castaneiceps*   E Ecuador to e Peru and extreme nw Bolivia (Pando)

☐ **Russet-crowned Crake** *Anurolimnas viridis*
_____ *A. v. brunnescens*   E Colombia (middle Magdalena Valley)
_____ *A. v. viridis*   S Venezuela to Guianas, Amaz. Brazil, e Peru and n Bolivia

☐ **Black-banded Crake** *Anurolimnas fasciatus*
                                      Marshes of se Colombia to e Peru and w Amazonian Brazil

☐ **Rufous-sided Crake** *Laterallus melanophaius*
_____ *L. m. oenops*   Tropical se Colombia to e Peru and extreme w Brazil
_____ *L. m. melanophaius*   S Venezuela to the Guianas, Brazil, Bolivia and n Argentina

☐ **Rusty-flanked Crake** *Laterallus levraudi*
                                      Locally in wetlands of n Venezuela north of the Orinoco River

☐ **Ruddy Crake** *Laterallus ruber*
                                      Lowlands of s Mexico to nw Costa Rica (n Guanacaste)

☐ **White-throated Crake** *Laterallus albigularis*
_____ *L. a. cinereiceps*   SE Honduras and Caribbean slope of Nicaragua to nw Panama
_____ *L. a. albigularis*   Pacific lowlands of Costa Rica to w Colombia and w Ecuador
_____ *L. a. cerdaleus*   E Colombia (Córdoba to Santa Marta)

☐ **Gray-breasted Crake** *Laterallus exilis*
                                      Guatemala and Belize to n Bolivia, Amazonian and e Brazil

☐ **Junin Rail** *Laterallus tuerosi*
                                      Andes of central Peru (Lake Junín)

☐ **Black Rail** *Laterallus jamaicensis*
_____ *L. j. coturniculus*   Coastal central California south to n Baja California
_____ *L. j. jamaicensis*   E US to Belize and Cuba; winters to C America and W Indies
_____ *L. j. murivagans*   Arid littoral of Peru
_____ *L. j. salinasi*   Central Chile (Atacama to Malleco) and extreme w Argentina

☐ **Galapagos Rail** *Laterallus spilonotus*
                                      Galapagos Islands

☐ **Red-and-white Crake** *Laterallus leucopyrrhus*

Marshes of se Brazil to Uruguay, Paraguay and n Argentina

☐ **Rufous-faced Crake** *Laterallus xenopterus*

Marshes of se Paraguay, Bolivia (Beni) and adjacent se Brazil

☐ **Woodford's Rail** *Nesoclopeus woodfordi*
- ____ *N. w. tertius* — Bougainville I. (Solomon Islands)
- ____ *N. w. immaculatus* — Locally on Santa Isabel I. (Solomon Islands)
- ____ *N. w. woodfordi* — Guadalcanal I. (Solomon Islands)

☐ **Weka** *Gallirallus australis*
- ____ *G. a. greyi* — North I. (New Zealand)
- ____ *G. a. australis* — Western region of South I. (New Zealand)
- ____ *G. a. hectori* — Formerly South I. (New Zealand); introduced Chatham Islands
- ____ *G. a. scotti* — Stewart, Solander and Codfish islands (off New Zealand)

☐ **New Caledonian Rail** *Gallirallus lafresnayanus*

Forests of New Caledonia (possibly extinct)

☐ **Lord Howe Rail** *Gallirallus sylvestris*

Highlands of Lord Howe I. (on verge of extinction)

☐ **Okinawa Rail** *Gallirallus okinawae*

Swamps of n Okinawa I. (s Ryukyu Islands)

☐ **Buff-banded Rail** *Gallirallus philippensis*
- ____ *G. p. andrewsi* — Cocos Islands (Bay of Bengal)
- ____ *G. p. philippensis* — Philippine Islands
- ____ *G. p. pelewensis* — Palau Is. (w Caroline Islands)
- ____ *G. p. xerophilus* — Gunungapi I. (Banda Sea)
- ____ *G. p. wilkinsoni* — Flores I. (Lesser Sundas)
- ____ *G. p. lacustris* — N New Guinea
- ____ *G. p. reductus (wahgiensis)* — Central highlands, coastal ne New Guinea and Long I.
- ____ *G. p. anachoretae* — Anchorite Is. (Admiralty Islands)
- ____ *G. p. admiralitatis* — Admiralty Islands
- ____ *G. p. praedo* — Skoki I. (Admiralty Islands)
- ____ *G. p. lesouefi* — Bismarck Arch. (New Hanover, New Ireland, Tabar and Tanga)
- ____ *G. p. meyeri* — Bismarck Archipelago (Witu Islands and New Britain)
- ____ *G. p. christophori* — Solomon Islands
- ____ *G. p. mellori (randi, norfolkensis, australis)* — S New Guinea, Australia and Norfolk I.
- ____ *G. p. assimilis* — New Zealand
- ____ *G. p. tounelierie (yorki)* — Coral Sea islets (se New Guinea to n New Caledonia)
- ____ *G. p. swindellsi* — New Caledonia and Loyalty Islands
- ____ *G. p. sethsmithi* — Fiji Islands and Vanuatu
- ____ *G. p. ecaudatus* — Tonga Islands
- ____ *G. p. goodsoni* — Samoa Islands and Niue I.
- ____ *G. p. macquariensis†* — Macquarie I. Extinct

☐ **New Britain Rail** *Gallirallus insignis*

Forests of New Britain I. (Bismarck Archipelago)

☐ **Guam Rail** *Gallirallus owstoni*

Forests of Guam (s Mariana Islands). On verge of extinction

☐ **Barred Rail** *Gallirallus torquatus*
- ____ *G. t. torquatus* — Philippine Islands
- ____ *G. t. celebensis* — Sulawesi, Muna and adjacent islands
- ____ *G. t. sulcirostris* — Peleng I. (Banggai Is.) and Sula Is. (Tabiabu, Mangole, Sanana)
- ____ *G. t. kuehni* — Tukangbesi Islands (Binongka and Kaledupa)
- ____ *G. t. limarius* — Salawati I. and nw New Guinea

☐ **Calayan Rail** *Gallirallus calayanensis*

Northern Philippines (Calayan I.)

☐ **Roviana Rail** *Gallirallus rovianae*

New Georgia Group (Solomon Islands)

☐ **Slaty-breasted Rail** *Gallirallus striatus*
- ____ *G. s. albiventer* — India and Sri Lanka to s China (Yunnan) and Thailand

____	*G. s. obscurior*	Andaman and Nicobar islands
____	*G. s. jouyi*	Coastal s China and Hainan I.
____	*G. s. taiwanus*	Taiwan
____	*G. s. gularis*	Malaysia to Indochina, Sumatra, Java and s Borneo
____	*G. s. striatus*	Philippines, Sulu Archipelago, n Borneo and Sulawesi
____	*G. s. paratermus*	Samar I. (Philippines)

☐ **Clapper Rail** *Rallus longirostris*

____	*R. l. obsoletus*	N California (Humboldt Bay to Monterey Bay)
____	*R. l. levipes*	S Calif. (Santa Barbara) to Baja Calif. (Scammons Lagoon)
____	*R. l. yumanensis*	Salton Sea and Colorado River basin to w Mexico (Nayarit)
____	*R. l. beldingi*	W Mexico (s Baja from Magdalena Bay to Espírito Santo I.)
____	*R. l. crepitans*	Atlantic coast (Connecticut to ne North Carolina)
____	*R. l. waynei*	Coastal Atlantic salt marshes (se North Carolina to e Florida)
____	*R. l. saturatus*	Gulf Coast (sw Alabama to Texas and Tamaulipas)
____	*R. l. scotti*	Coastal Florida (Pensacola to Cape Sable and Jupiter)
____	*R. l. insularum*	Mangrove swamps of Florida Keys
____	*R. l. coryi*	Mangrove swamps of Bahamas
____	*R. l. pallidus*	Mangroves of se Mexico (coastal n Yucatán Peninsula)
____	*R. l. grossi*	SE Mexico (islands on Chinchorro Bank off Quintana Roo)
____	*R. l. belizensis*	Belize (Ycacos Lagoon)
____	*R. l. leucophaeus*	Isle of Pines
____	*R. l. caribaeus*	Cuba, Hispaniola and Puerto Rico to Antigua and n Antilles
____	*R. l. cypereti*	Coastal sw Colombia to Ecuador and nw Peru (Tumbes)
____	*R. l. phelpsi*	Extreme ne coastal Colombia and extreme nw Venezuela
____	*R. l. margaritae*	Margarita I. (Venezuela)
____	*R. l. pelodramus*	Trinidad
____	*R. l. longirostris*	Coasts of Guyana, Suriname and French Guiana
____	*R. l. crassirostris*	Coastal e Brazil (Amazon estuary to Santa Catarina)

☐ **King Rail** *Rallus elegans*

____	*R. e. elegans*	E Canada and ne US; winters to e Mexico
____	*R. e. tenuirostris*	Central Mexico
____	*R. e. ramsdeni*	Cuba and Isle of Pines

☐ **Plain-flanked Rail** *Rallus wetmorei*

Coastal swamps of nw Venezuela

☐ **Virginia Rail** *Rallus limicola*

____	*R. l. limicola*	S Canada and US; winters to Baja and Guatemala
____	*R. l. friedmanni*	SE Mexico (Puebla, México, Veracruz and Chiapas)
____	*R. l. aequatorialis*	Locally in mountains of sw Colombia, Ecuador and Peru

☐ **Bogota Rail** *Rallus semiplumbeus*

____	*R. s. semiplumbeus*	E Andes of Colombia (Boyacá and Cundinamarca)
____	*R. s. peruvianus*	One 1886 record from an unknown location in Peru

☐ **Austral Rail** *Rallus antarcticus*

Marshes of central Chile and Argentina to Tierra del Fuego

☐ **Water Rail** *Rallus aquaticus*

____	*R. a. hibernans*	Iceland
____	*R. a. aquaticus*	W Palearctic
____	*R. a. korejewi*	Iran to nw China; winters to India and s China
____	*R. a. indicus*	E Siberia to Japan; winters to SE Asia and Borneo

☐ **African Rail** *Rallus caerulescens*

Swamps and reedbeds of e and s Africa

☐ **Madagascar Rail** *Rallus madagascariensis*

Humid forests of e Madagascar

☐ **Luzon Rail** *Lewinia mirificus*

Wetlands of Luzon I. (n Philippines). Status unknown

☐ **Lewin's Rail** *Lewinia pectoralis*

____ *L. p. exsul* — Known from 4 specimens from w Flores I. (Lesser Sundas)

____ *L. p. mayri* — W New Guinea (Arfak and Weyland mountains)

____ *L. p. captus* — Central Highlands of New Guinea

____ *L. p. insulsus* — E New Guinea (Herzog Mountains)

____ *L. p. alberti* — Mountains of s-central Papua New Guinea

____ *L. p. clelandi†* — SW Australia. Extinct

____ *L. p. pectoralis* — Coastal e Queensland to Victoria and South Australia

____ *L. p. brachipus* — Tasmania

☐ **Auckland Islands Rail** *Lewinia muelleri*

Auckland Islands (Adams and Disappointment)

☐ **White-throated Rail** *Dryolimnas cuvieri*

____ *D. c. cuvieri* — Lowlands of Madagascar; formerly Mauritius

____ *D. c. abbotti†* — Formerly Assumption I. Extinct

____ *D. c. aldabranus* — Aldabra I.

☐ **African Crake** *Crecopsis egregia*

Africa south of the Sahara

☐ **Corn Crake** *Crex crex*

Palearctic; winters Mediterranean to Africa and Madagascar

☐ **Rouget's Rail** *Rougetius rougetii*

Highlands of Eritrea and Ethiopia

☐ **Platen's Rail** *Aramidopsis plateni*

Sulawesi

☐ **Inaccessible Island Rail** *Atlantisia rogersi*

Inaccessible I. (Tristan da Cunha)

☐ **Little Wood-Rail** *Aramides mangle*

Coastal e Brazil (Maranhão to Rio de Janeiro)

☐ **Rufous-necked Wood-Rail** *Aramides axillaris*

Coastal nw Mexico to extreme nw Peru, Suriname, Trinidad

☐ **Gray-necked Wood-Rail** *Aramides cajanea*

____ *A. c. mexicanus* — S Mexico (Tamaulipas to Chiapas)

____ *A. c. albiventris* — Yucatán Peninsula, Cozumel I., Belize and adj. n Guatemala

____ *A. c. vanrossemi* — S Mexico (Oaxaca) to sw Guatemala and w El Salvador

____ *A. c. pacificus* — Caribbean slope of Honduras and Nicaragua

____ *A. c. plumbeicollis* — Caribbean lowlands of ne Costa Rica

____ *A. c. latens* — San Miguel Islands and Pearl Islands (Panama)

____ *A. c. morrisoni* — Pearl Islands (San José and Pedro González)

____ *A. c. cajanea* — Costa Rica to n Argentina, Uruguay, Brazil and the Guianas

☐ **Brown Wood-Rail** *Aramides wolfi*

W Colombia to sw Ecuador and extreme n Peru

☐ **Giant Wood-Rail** *Aramides ypecaha*

Marshes of se Brazil to Paraguay, Uruguay and ne Argentina

☐ **Slaty-breasted Wood-Rail** *Aramides saracura*

Forests of se Brazil to Paraguay and ne Argentina (Misiones)

☐ **Red-winged Wood-Rail** *Aramides calopterus*

E Ecuador to ne Peru (Loreto) and w Amazonian Brazil

☐ **Uniform Crake** *Amaurolimnas concolor*

____ *A. c. concolor†* — Formerly Jamaica. Extinct

____ *A. c. guatemalensis* — S Mexico to Ecuador

____ *A. c. castaneus* — Venezuela to the Guianas, Brazil, e Peru and Bolivia

☐ **Bare-faced Rail** *Gymnocrex rosenbergii*

Sulawesi and Peleng I.

☐ **Bare-eyed Rail** *Gymnocrex plumbeiventris*

____ *G. p. plumbeiventris* — N Moluccas, New Guinea, Misool, Karkar and New Ireland

____ *G. p. hoeveni* — Aru Islands and Trans-Fly lowlands of s New Guinea

☐ **Talaud Rail** *Gymnocrex talaudensis*

Karekelong I. (Talaud Archipelago)

☐ **Brown Crake** *Amaurornis akool*
____ *A. a. akool*　　　　India to Bangladesh and w Myanmar
____ *A. a. coccineipes*　　SE China to ne Vietnam

☐ **Isabelline Bush-hen** *Amaurornis isabellina*

Lowlands of Sulawesi

☐ **Plain Bush-hen** *Amaurornis olivacea*

Philippine Islands (except Palawan)

☐ **White-breasted Waterhen** *Amaurornis phoenicurus*
____ *A. p. phoenicurus*　　S Asia, Malay Archipelago and Philippine Islands
____ *A. p. insularis*　　　Andaman and Nicobar islands
____ *A. p. midnicobarica*　Central Nicobar Islands
____ *A. p. leucomelana*　　Sulawesi, w Moluccas and Lesser Sundas

☐ **Talaud Bush-hen** *Amaurornis magnirostris*

Talaud Islands (n Moluccas)

☐ **Rufous-tailed Bush-hen** *Amaurornis moluccana*
____ *A. m. moluccana*　　Sangihe I., Moluccas, Misool I., w and n New Guinea
____ *A. m. nigrifrons*　　Bismarck Archipelago and Solomon Islands
____ *A. m. ultima*　　　　E Solomon Islands
____ *A. m. ruficrissa*　　S and e New Guinea, and n and e Australia

☐ **Black Crake** *Amaurornis flavirostra*

Africa south of the Sahara

☐ **Sakalava Rail** *Amaurornis olivieri*

Sakalava region of nw Madagascar (status unknown)

☐ **Black-tailed Crake** *Amaurornis bicolor*

NE India and Myanmar to sw China and n SE Asia

☐ **Little Crake** *Porzana parva*

S Palearctic region; winters Mediterranean to Africa and India

☐ **Baillon's Crake** *Porzana pusilla*
____ *P. p. intermedia (obscura)*　Europe to Asia Minor, e and s Africa and Madagascar
____ *P. p. pusilla*　　　Central and e Asia; winters to India, Malaya and Philippines
____ *P. p. mira*　　　　Known from a 1912 specimen from Borneo
____ *P. p. mayri*　　　Known from 4 specimens from New Guinea
____ *P. p. palustris*　E New Guinea, s Australia (winters to n); accidental Tasmania
____ *P. p. affinis*　　New Zealand and Chatham Islands

☐ **Spotted Crake** *Porzana porzana*

Palearctic; winters Mediterranean to s Africa and SE Asia

☐ **Australian Crake** *Porzana fluminea*

Moist areas of se and sw Australia and Tasmania

☐ **Sora** *Porzana carolina*

S Alaska to n Baja and s US; winters to W Indies and n S Am.

☐ **Dot-winged Crake** *Porzana spiloptera*

S Uruguay and n Argentina (status unknown)

☐ **Ash-throated Crake** *Porzana albicollis*
____ *P. a. olivacea*　　N Colombia to Venezuela, the Guianas and Suriname; Trinidad
____ *P. a. albicollis*　E Brazil to Paraguay, e Bolivia and extreme n Argentina

☐ **Ruddy-breasted Crake** *Porzana fusca*
____ *P. f. fusca*　　　Pakistan and India to Malaysia, Indonesia and Philippines
____ *P. f. zeylonica*　W peninsular India and Sri Lanka
____ *P. f. erythrothorax*　Japan, e China, Manchuria, Indochina and Taiwan
____ *P. f. phaeopyga*　Ryukyu Islands

☐ **Band-bellied Crake** *Porzana paykullii*

NE Asia; winters in SE Asia and Greater Sundas

☐ **Spotless Crake** *Porzana tabuensis*
 ____ *P. t. tabuensis*   Philippines, Australasian region and Oceania
 ____ *P. t. edwardi*   Central highlands of Papua New Guinea
 ____ *P. t. richardsoni*   W New Guinea (Jayawijaya Mountains)
 ____ *P. t. plumbea*   SE and sw Australia, Tasmania, New Zealand and Chatham Is.

☐ **Henderson Island Crake** *Porzana atra*

 Henderson I. (Pitcairn Archipelago)

☐ **Yellow-breasted Crake** *Porzana flaviventer*
 ____ *P. f. gossii*   Cuba and Jamaica
 ____ *P. f. hendersoni*   Hispaniola and Puerto Rico
 ____ *P. f. woodi*   S Mexico to nw Costa Rica
 ____ *P. f. flaviventer*   Panama to the Guianas, e Brazil, Paraguay and n Argentina
 ____ *P. f. bangsi*   Tropical n Colombia

☐ **White-browed Crake** *Porzana cinerea*

 Malaysia and Philippines to coastal n Australia and sw Oceanea

☐ **Striped Crake** *Aenigmatolimnas marginalis*

 Locally in Africa south of the Sahara

☐ **Zapata Rail** *Cyanolimnas cerverai*

 Zapata Swamp of sw Cuba

☐ **Colombian Crake** *Neocrex colombiana*
 ____ *N. c. ripleyi*   Caribbean lowlands of Panama and adjacent nw Colombia
 ____ *N. c. colombiana*   Santa Marta Mountains of ne Colombia to coastal nw Ecuador

☐ **Paint-billed Crake** *Neocrex erythrops*
 ____ *N. e. olivascens*   W Panama to Venezuela, the Guianas, nw Argentina and Brazil
 ____ *N. e. erythrops*   Galapagos Islands; coastal Peru (Lima to Lambayeque)

☐ **Spotted Rail** *Pardirallus maculatus*
 ____ *P. m. insolitus*   Locally from s Mexico to Costa Rica
 ____ *P. m. maculatus*   Cuba, Trinidad and Tobago; Venezuela to Argentina and Peru

☐ **Blackish Rail** *Pardirallus nigricans*
 ____ *P. n. caucae*   Cauca Valley of Colombia (status unknown)
 ____ *P. n. nigricans*   E Ecuador to e Peru, e Brazil, Paraguay and ne Argentina

☐ **Plumbeous Rail** *Pardirallus sanguinolentus*
 ____ *P. s. simonsi*   Arid littoral of Peru to n Chile
 ____ *P. s. tschudii*   Temperate Peru (upper Río Marañón) to Lake Titicaca
 ____ *P. s. zelebori*   SE Brazil
 ____ *P. s. sanguinolentus*   Extreme se Brazil to Uruguay, Paraguay and n Argentina
 ____ *P. s. landbecki*   Central Chile (Atacama to Llanquihue) and adjacent Argentina
 ____ *P. s. luridus*   Tierra del Fuego and Cape Horn Archipelago

☐ **Invisible Rail** *Habroptila wallacii*

 Halmahera I. (n Moluccas)

☐ **Chestnut Rail** *Eulabeornis castaneoventris*
 ____ *E. c. sharpei*   Aru Islands
 ____ *E. c. castaneoventris*   Coastal n Australia (n Western Australia to nw Queensland)

☐ **New Guinea Flightless Rail** *Megacrex inepta*
 ____ *M. i. pallida*   Coastal n New Guinea (Idenburg River to Sepik River)
 ____ *M. i. inepta*   Trans-Fly lowlands of se New Guinea

☐ **Watercock** *Gallicrex cinerea*

 Lowlands of s and e Asia and Malay Archipelago

☐ **Purple Swamphen** *Porphyrio porphyrio*
 ____ *P. p. porphyrio*   Iberian Peninsula and nw Africa
 ____ *P. p. madagascariensis*   Egypt, Africa south of the Sahara and Madagascar

_____ *P. p. caspius* — Caspian Sea to nw Iran and Turkey
_____ *P. p. seistanicus* — Iraq and s Iran to Afghanistan, Pakistan and nw India
_____ *P. p. poliocephalus* — India to Sri Lanka, s China, n Thailand, Andaman, Nicobar is.
_____ *P. p. viridis* — S Myanmar to s Thailand, s China, Malay Pen. and Indochina
_____ *P. p. indicus* — Sumatra, Java, Bali, Borneo and Sulawesi
_____ *P. p. pulverulentus* — Karekelong I. (Talaud Islands) and Philippine Islands
_____ *P. p. pelewensis* — Palau Islands (Koror and Anguar)
_____ *P. p. melanopterus* — Moluccas and Lesser Sundas to Aru Islands and New Guinea
_____ *P. p. bellus* — Extreme sw Australia
_____ *P. p. melanotus (chathamensis)* — Australia, New Zealand, Kermadec and Chatham Is.
_____ *P. p. samoensis* — Admiralty Is. to Samoa, New Caledonia, Solomon Is. and Fiji

☐ **Takahe** *Porphyrio mantelli*
_____ *P. m. mantelli†* — Formerly North I. (New Zealand). Extinct ca 1894
_____ *P. m. hochstetteri* — Mts. of s South I. (New Zealand). On verge of extinction

☐ **Allen's Gallinule** *Porphyrio alleni*
Africa south of the Sahara, Madagascar and Comoro Islands

☐ **Purple Gallinule** *Porphyrio martinica*
Locally from s US to n Argentina and West Indies

☐ **Azure Gallinule** *Porphyrio flavirostris*
E Colombia to s Venezuela, the Guianas, n Argentina and Brazil

☐ **San Cristobal Moorhen** *Gallinula silvestris*
Known from a 1929 specimen from San Cristobal I. (Solomon Is.)

☐ **Tristan Moorhen** *Gallinula nesiotis*
_____ *G. n. nesiotis†* — Formerly Tristan da Cunha. Extinct
_____ *G. n. comeri* — Gough I. (South Atlantic Ocean)

☐ **Common Moorhen** *Gallinula chloropus*
_____ *G. c. chloropus (correina, indica)* — Palearctic; winters to Arabia and s China
_____ *G. c. meridionalis* — Africa south of the Sahara and St. Helena I.
_____ *G. c. pyrrhorrhoa* — Madagascar, Réunion, Mauritius and Comoro Islands
_____ *G. c. orientulis* — Seychelles, Andamans, Malay Pen., Indonesia and Philippines
_____ *G. c. guami* — N Marianas (Guam, Saipan, Tinian and Pagan)
_____ *G. c. sandvicensis* — Hawaiian Islands
_____ *G. c. cachinnans* — SE Canada to w Panama, Bermuda and Galapagos Islands
_____ *G. c. cerceris* — Greater and Lesser Antilles
_____ *G. c. barbadensis* — Barbados
_____ *G. c. pauxilla* — E Panama to n and w Colombia, arid w Ecuador and nw Peru
_____ *G. c. garmani* — Andes of Peru to Chile, Bolivia and nw Argentina
_____ *G. c. galeata* — Guianas to n Argentina, Uruguay and Brazil; Trinidad

☐ **Dusky Moorhen** *Gallinula tenebrosa*
_____ *G. t. frontata* — SE Borneo to Sulawesi, s Moluccas, L Sundas, se New Guinea
_____ *G. t. neumanni* — N New Guinea
_____ *G. t. tenebrosa* — Locally in sw and e Australia and Tasmania

☐ **Lesser Moorhen** *Gallinula angulata*
Aquatic habitats of Africa south of the Sahara

☐ **Spot-flanked Gallinule** *Gallinula melanops*
_____ *G. m. bogotensis* — Temperate Eastern Andes of Colombia
_____ *G. m. melanops* — Peru to e Bolivia, e Brazil to Uruguay and Paraguay
_____ *G. m. crassirostris* — Argentina and Chile (except extreme south)

☐ **Black-tailed Native-hen** *Gallinula ventralis*
Aquatic habitats of Australia (except n and e coastal areas)

☐ **Tasmanian Native-hen** *Gallinula mortierii*
Tasmania

☐ **Red-knobbed Coot** *Fulica cristata*
S Spain and Morocco; e and s Africa and Madagascar

☐ **Eurasian Coot** *Fulica atra*

____ *F. a. atra*	Palearctic; winters to Africa, Indonesia and Philippines
____ *F. a. lugubris*	Mountains of Java and nw New Guinea
____ *F. a. novaeguinea*	Mountains of central New Guinea
____ *F. a. australis*	Wetlands of Australia, Tasmania, New Zealand and Buru I.

☐ **Hawaiian Coot** *Fulica alai*

Main Hawaiian Islands (except Lanai)

☐ **American Coot** *Fulica americana*

____ *F. a. americana*	Alaska to nw Costa Rica, Cuba, Jamaica and Grand Cayman I.
____ *F. a. columbiana*	Andes of Colombia to n Ecuador

☐ **Caribbean Coot** *Fulica caribaea*

S Bahamas to Greater and Lesser Antilles and extreme n S Am.

☐ **White-winged Coot** *Fulica leucoptera*

Extreme se Brazil to e Bolivia and south to Tierra del Fuego

☐ **Slate-colored Coot** *Fulica ardesiaca*

Andes of Ecuador to nw Argentina and n Chile

☐ **Red-gartered Coot** *Fulica armillata*

Paraguay to Uruguay, se Brazil and Tierra del Fuego

☐ **Red-fronted Coot** *Fulica rufifrons*

Paraguay to Uruguay, se Brazil, s Peru and Tierra del Fuego

☐ **Giant Coot** *Fulica gigantea*

Andes of s Peru to n Chile and nw Argentina

☐ **Horned Coot** *Fulica cornuta*

High Andean lakes of sw Bolivia to n Chile and nw Argentina

ORDER: GRUIFORMES
FAMILY: HELIORNITHIDAE (Finfoots—3)

☐ **African Finfoot** *Podica senegalensis*

____ *P. s. senegalensis*	Senegal to e Zaire, Uganda, nw Tanzania and Ethiopia
____ *P. s. somereni*	Kenya and ne Tanzania
____ *P. s. camerunensis*	S Cameroon to Gabon, Congo and n Zaire
____ *P. s. petersii*	Angola to se Zaire, Zambia, Mozambique and e South Africa

☐ **Masked Finfoot** *Heliopais personatus*

Bangladesh and ne India to Indochina; winters to Sumatra

☐ **Sungrebe** *Heliornis fulica*

Tropical lowlands of s Mexico to ne Argentina and Brazil

ORDER: GRUIFORMES
FAMILY: RHYNOCHETIDAE (Kagu—1)

☐ **Kagu** *Rhynochetos jubatus*

New Caledonia

ORDER: GRUIFORMES
FAMILY: EURYPYGIDAE (Sunbittern—1)

☐ **Sunbittern** *Eurypyga helias*

____ *E. h. major*	Extreme s Mexico and Guatemala to w Ecuador
____ *E. h. meridionalis*	S-central Peru (Junín and Cuzco)
____ *E. h. helias*	Colombia to Venezuela, Guianas, Amaz. Brazil and e Bolivia

ORDER: GRUIFORMES
FAMILY: CARIAMIDAE (Seriemas—2)

☐ **Red-legged Seriema** *Cariama cristata*

C and e Brazil to Bolivia, Paraguay and central Argentina

☐ **Black-legged Seriema** *Chunga burmeisteri*

Chaco of Paraguay and adjacent Bolivia to central Argentina

ORDER: GRUIFORMES
FAMILY: OTIDIDAE (Bustards—26)

☐ **Great Bustard** *Otis tarda*

_____ *O. t. tarda*	S Palearctic
_____ *O. t. dybowskii*	SE Russia to Mongolia and ne China

☐ **Arabian Bustard** *Ardeotis arabs*

_____ *A. a. lynesi*	W Morocco (probably extinct)
_____ *A. a. stieberi*	SW Mauritania and Senegambia to ne Sudan
_____ *A. a .butleri*	S Sudan; single record for nw Kenya
_____ *A. a. arabs*	Ethiopia to nw Somalia, sw Saudi Arabia and w Yemen

☐ **Kori Bustard** *Ardeotis kori*

_____ *A. k. struthiunculus*	Ethiopia to nw Somalia, se Sudan, ne Uganda and n Tanzania
_____ *A. k. kori*	S Angola to Namibia, Botswana, s Zimbabwe and Mozambique

☐ **Indian Bustard** *Ardeotis nigriceps*

Semiarid grasslands of nw and central India

☐ **Australian Bustard** *Ardeotis australis*

Lowlands of Australia (except sw and se) and s New Guinea

☐ **Houbara Bustard** *Chlamydotis undulata*

_____ *C. u. fuertaventurae*	E Canary Islands (Fuerteventura and Lanzarote)
_____ *C. u. undulata*	North Africa (Morocco to western Nile Valley)

☐ **Macqueen's Bustard** *Chlamydotis macqueenii*

Nile Valley of Egypt to Arabian Peninsula and Pakistan

☐ **Ludwig's Bustard** *Neotis ludwigii*

Extreme sw Angola to Namibia, sw Botswana and South Africa

☐ **Stanley Bustard** *Neotis denhami*

_____ *N. d. denhami*	SW Mauritania and Senegambia to n Uganda and Ethiopia
_____ *N. d. jacksoni*	Kenya and w Tanzania to Botswana, Zimbabwe and s Angola
_____ *N. d. stanleyi*	Swaziland and South Africa

☐ **Heuglin's Bustard** *Neotis heuglinii*

Savanna of Eritrea to s Ethiopia, n Somalia and n Kenya

☐ **Nubian Bustard** *Neotis nuba*

Sahel from w Mauritania to e Sudan

☐ **White-bellied Bustard** *Eupodotis senegalensis*

_____ *E. s. senegalensis*	SW Mauritania to Guinea, Central African Rep. and s Sudan
_____ *E. s. canicollis*	Ethiopia to Kenya and ne Tanzania
_____ *E. s. erlangeri*	S Kenya and w Tanzania
_____ *E. s. mackenziei*	E Gabon to s Zaire, e Angola and w Zambia
_____ *E. s. barrowii*	Botswana to Transvaal, Swaziland and e Cape Province

☐ **Blue Bustard** *Eupodotis caerulescens*

Acacia and grasslands of South Africa

☐ **Karoo Bustard** *Eupodotis vigorsii*

_____ *E. v. namaqua*	Arid s Namibia and nw Cape Province
_____ *E. v. vigorsii*	Orange Free State to s Cape Province

☐ **Rueppell's Bustard** *Eupodotis rueppellii*
___ *E. r. rueppellii* — Arid coastal sw Angola (Benguela) to nw Namibia
___ *E. r. fitzsimonsi* — S Namibia (Maltahöhe to Windhoek)

☐ **Little Brown Bustard** *Eupodotis humilis* — Acacia and thornbush of e Ethiopia and n Somalia

☐ **Savile's Bustard** *Eupodotis savilei* — SW Mauritania and Senegal to Nigeria, Chad and s Sudan

☐ **Buff-crested Bustard** *Eupodotis gindiana* — SE Sudan to s Ethiopia, Somalia, Kenya and n Tanzania

☐ **Red-crested Bustard** *Eupodotis ruficrista* — S Angola to ne Namibia, Botswana, s Zambia and n S Africa

☐ **Black Bustard** *Eupodotis afra* — Grasslands of sw Africa (Cape Province)

☐ **White-quilled Bustard** *Eupodotis afraoides*
___ *E. a. etoschae* — NW Namibia and n Botswana
___ *E. a. damarensis* — Namibia and central Botswana
___ *E. a. afraoides* — SE Botswana to ne South Africa and Lesotho

☐ **Black-bellied Bustard** *Lissotis melanogaster*
___ *L. m. melanogaster* — Senegal to Ethiopia, s Angola and Mozambique
___ *L. m. notophila* — SE Africa s of the Zambezi to w Zimbabwe and South Africa

☐ **Hartlaub's Bustard** *Lissotis hartlaubii* — E Sudan to Ethiopia, Somalia, ne Uganda and n Tanzania

☐ **Bengal Florican** *Houbaropsis bengalensis*
___ *H. b. bengalensis* — Grasslands of e and n India and *terai* of Nepal
___ *H. b. blandini* — Cambodia to Cochinchina and sw Vietnam

☐ **Lesser Florican** *Sypheotides indicus* — Locally in grasslands of Indian subcontinent

☐ **Little Bustard** *Tetrax tetrax* — S Palearctic region

ORDER: CHARADRIIFORMES
FAMILY: JACANIDAE (Jacanas—8)

☐ **Lesser Jacana** *Microparra capensis* — Lakes, ponds and marshes of Africa south of the Sahara

☐ **African Jacana** *Actophilornis africanus* — Lakes, ponds and marshes of Africa south of the Sahara

☐ **Madagascar Jacana** *Actophilornis albinucha* — Lakes, ponds and marshes of Madagascar

☐ **Comb-crested Jacana** *Irediparra gallinacea*
___ *I. g. gallinacea* — S Borneo, Sulawesi, Mindanao, Moluccas and Lesser Sundas
___ *I. g. novaeguinae* — N and central New Guinea, Misool I. and Aru Islands
___ *I. g. novaehollandiae* — S New Guinea, D'Entrecasteaux Arch. and n and e Australia

☐ **Pheasant-tailed Jacana** *Hydrophasianus chirurgus* — Swamps and marshes of India, SE Asia and Philippine Islands

☐ **Bronze-winged Jacana** *Metopidius indicus* — Lowlands of India to sw China, SE Asia, Sumatra and Java

☐ **Northern Jacana** *Jacana spinosa*
___ *J. s. gymnostoma* — N Mexico to Chiapas, Yucatán Peninsula and Cozumel I.
___ *J. s. spinosa* — Belize and Guatemala to w Panama
___ *J. s. violacea* — Cuba, Isle of Pines, Jamaica and Hispaniola

☐ **Wattled Jacana** *Jacana jacana*
___ *J. j. hypomelaena* — W-central Panama to n Colombia

_____ *J. j. melanopygia*	W Colombia to w Venezuela
_____ *J. j. jacana*	SE Colombia to the Guianas, Brazil, Uruguay and n Argentina
_____ *J. j. intermedia*	N and central Venezuela
_____ *J. j. scapularis*	Lowlands of w Ecuador and nw Peru
_____ *J. j. peruviana*	NE Peru (lower Río Ucayalí) and adjacent nw Brazil

ORDER: CHARADRIIFORMES
FAMILY: ROSTRATULIDAE (Painted-snipes—3)

☐ **Greater Painted-snipe** *Rostratula benghalensis*

Locally in Africa, Madagascar and Oriental region

☐ **Australian Painted Snipe** *Rostratula australis*

E and n Australia; vagrant to Tasmania

☐ **American Painted-snipe** *Rostratula semicollaris*

Lowlands of se Brazil to Paraguay, cent. Argentina and Chile

ORDER: CHARADRIIFORMES
FAMILY: DROMADIDAE (Crab Plover—1)

☐ **Crab Plover** *Dromas ardeola*

Coastal n Indian Ocean; ranges to Madagascar and Andaman Is.

ORDER: CHARADRIIFORMES
FAMILY: HAEMATOPODIDAE (Oystercatchers—11)

☐ **Magellanic Oystercatcher** *Haematopus leucopodus*

S-c Chile and Argentina to Cape Horn Arch. and Falkland Is.

☐ **Blackish Oystercatcher** *Haematopus ater*

N Peru to Tierra del Fuego and Falkland Is.; winters to Uruguay

☐ **Black Oystercatcher** *Haematopus bachmani*

W Aleutians to central Baja Calif. and Los Coronados Islands

☐ **American Oystercatcher** *Haematopus palliatus*

_____ *H. p. palliatus*	Seacoasts and islands of US to s S America and West Indies
_____ *H. p. galapagensis*	Galapagos Islands

☐ **African Oystercatcher** *Haematopus moquini*

Coasts of s Africa (n Namibia to e Cape Province)

☐ **Eurasian Oystercatcher** *Haematopus ostralegus*

_____ *H. o. ostralegus*	Iceland and Scandinavia to s Europe; winters to Africa
_____ *H. o. longipes*	Russia to Siberia and south to Caspian Sea and Aral Sea
_____ *H. o. osculans*	Kamchatka Peninsula and North Korea; winters e China

☐ **Pied Oystercatcher** *Haematopus longirostris*

Coastal Australia, Tasmania and s New Guinea, Aru and Kai Is.

☐ **South Island Oystercatcher** *Haematopus finschi*

Highlands of South I. (New Zealand); winters to North I.

☐ **Chatham Oystercatcher** *Haematopus chathamensis*

Chatham Islands (Chatham, Mangere, Rangatira and Pitt)

☐ **Variable Oystercatcher** *Haematopus unicolor*

Coasts and islands of New Zealand

☐ **Sooty Oystercatcher** *Haematopus fuliginosus*

_____ *H. f. opthalmicus*	Coasts and islands of n Australia (Shark Bay to Lady Elliot I.)
_____ *H. f. fuliginosus*	Coasts and islands of s Australia (n to Brisbane); Tasmania

ORDER: CHARADRIIFORMES
FAMILY: IBIDORHYNCHIDAE (Ibisbill—1)

☐ **Ibisbill** *Ibidorhyncha struthersii*

Rocky mountain streams and rivers of central Asia

ORDER: CHARADRIIFORMES
FAMILY: RECURVIROSTRIDAE (Avocets and Stilts—10)

☐ **Black-winged Stilt** *Himantopus himantopus*

Mediterranean and sub-Saharan Africa to SE Asia and Taiwan

☐ **Pied Stilt** *Himantopus leucocephalus*

Indonesia to Australia and New Zealand; winters to Philippines

☐ **Black Stilt** *Himantopus novaezelandiae*

MacKenzie Basin (South I., New Zealand); winters to North I.

☐ **Black-necked Stilt** *Himantopus mexicanus*
_____ *H. m. mexicanus* — W and s US to e Ecuador, sw Peru and ne Brazil; West Indies
_____ *H. m. knudseni* — Hawaiian Islands

☐ **White-backed Stilt** *Himantopus melanurus*

N Chile and e-central Peru to se Brazil and c Argentina

☐ **Banded Stilt** *Cladorhynchus leucocephalus*

Inland salt lakes of sw and s-central Australia

☐ **Pied Avocet** *Recurvirostra avosetta*

N Africa and Eurasia; winters to South Africa and s Asia

☐ **American Avocet** *Recurvirostra americana*

S Canada to n Mexico; winters s US to Costa Rica and Cuba

☐ **Red-necked Avocet** *Recurvirostra novaehollandiae*

Locally in Australia; vagrant to Tasmania and New Zealand

☐ **Andean Avocet** *Recurvirostra andina*

Andes of s Peru to nw Argentina and n Chile

ORDER: CHARADRIIFORMES
FAMILY: BURHINIDAE (Thick-knees—9)

☐ **Water Thick-knee** *Burhinus vermiculatus*
_____ *B. v. buettikoferi* — Liberia to Nigeria and Gabon; vagrant to Senegambia
_____ *B. v. vermiculatus* — Zaire to Somalia and South Africa

☐ **Eurasian Thick-knee** *Burhinus oedicnemus*
_____ *B. o. distinctus* — W Canary Islands
_____ *B. o. insularum* — E Canary Islands
_____ *B. o. saharae* — N Africa, Mediterranean is., Greece, Turkey to Iraq and Iran
_____ *B. o. oedicnemus* — S Britain and Iberian Peninsula to n Balkans and Caucasus
_____ *B. o. harterti* — Volga River to Turkestan, Pakistan and extreme nw India
_____ *B. o. indicus* — India and Sri Lanka to Indochina; winters to Africa and Arabia

☐ **Senegal Thick-knee** *Burhinus senegalensis*

Sandy lake and river banks of sub-Saharan Africa

☐ **Spotted Thick-knee** *Burhinus capensis*
_____ *B. c. maculosus* — Senegal to Somalia, Uganda and Kenya
_____ *B. c. dodsoni* — Coastal Somalia and Saudi Arabia
_____ *B. c. capensis* — Kenya to South Africa and west from Zambia to Angola
_____ *B. c. damarensis* — Namibia to Botswana and Cape Province

☐ **Double-striped Thick-knee** *Burhinus bistriatus*
_____ *B. b. bistriatus* — Arid s Mexico to nw Costa Rica

____	*B. b. dominicensis*	Hispaniola
____	*B. b. pediacus*	Savanna and pastures of n Colombia
____	*B. b. vocifer*	Venezuela to Guyana and extreme n Brazil

☐ **Peruvian Thick-knee** *Burhinus superciliaris*

Arid littoral of sw Ecuador to s Peru; one record from n Chile

☐ **Bush Thick-knee** *Burhinus grallarius*

Australia (except arid areas), s New Guinea; vagrant to Tasmania

☐ **Great Thick-knee** *Burhinus recurvirostris*

SE Iran to Indian subcontinent, Indochina and Hainan I.

☐ **Beach Thick-knee** *Burhinus magnirostris*

Andaman Is. and Malay Pen. to Philippines and Australasia

ORDER: CHARADRIIFORMES
FAMILY: GLAREOLIDAE (Pratincoles and Coursers—17)

☐ **Egyptian Plover** *Pluvianus aegyptius*

Sub-Saharan Africa south to n Zaire and extreme n Angola

☐ **Cream-colored Courser** *Cursorius cursor*

____	*C. c. bogolubovi*	SE Turkey to Iran, Afghanistan, s Pakistan and nw India
____	*C. c. cursor*	Canary Islands, North Africa, Arabian Pen. and Socotra I.
____	*C. c. exsul*	Cape Verde Islands
____	*C. c. somalensis*	Eritrea to Ethiopia and Somalia
____	*C. c. littoralis*	Extreme se Sudan to n Kenya and s Somalia

☐ **Burchell's Courser** *Cursorius rufus*

Semi-deserts of sw Angola to Namibia, Botswana and S Africa

☐ **Temminck's Courser** *Cursorius temminckii*

Savanna of sub-Saharan Africa to n South Africa

☐ **Indian Courser** *Cursorius coromandelicus*

Patchily distributed Pakistan to India and Sri Lanka

☐ **Double-banded Courser** *Smutsornis africanus*

____	*S. a. raffertyi*	Eritrea to Ethiopia and Djibouti
____	*S. a. hartingi*	SE Ethiopia (Ogaden Depression) and Somalia
____	*S. a. gracilis*	Kenya and Tanzania
____	*S. a. bisignatus*	SW Angola
____	*S. a. traylori*	Namibia (Etosha region) to Botswana (Makgadikgadi area)
____	*S. a. sharpei*	Central Namibia
____	*S. a. africanus*	Central Kalahari and s Namibia to n Cape Province
____	*S. a. granti*	W Cape Province and Karoo of South Africa

☐ **Three-banded Courser** *Rhinoptilus cinctus*

____	*R. c. cinctus*	SE Sudan to e Ethiopia, Somalia and n Kenya
____	*R. c. emini*	S Kenya to Tanzania and n Zambia
____	*R. c. seebohmi*	S Angola and n Namibia to Zimbabwe and n South Africa

☐ **Bronze-winged Courser** *Rhinoptilus chalcopterus*

Open woodlands of Africa south of the Sahara

☐ **Jerdon's Courser** *Rhinoptilus bitorquatus*

SE India (Andhra Pradesh). On verge of extinction

☐ **Australian Pratincole** *Stiltia isabella*

Australia; winters to s New Guinea and Greater Sundas

☐ **Collared Pratincole** *Glareola pratincola*

____	*G. p. pratincola*	S Europe to Pakistan; winters sub-Saharan Africa n of 5°N
____	*G. p. erlangeri*	Coastal plains of s Somalia and n Kenya
____	*G. p. fuelleborni*	Senegal to s Kenya, Zaire, Namibia and e South Africa

☐ **Oriental Pratincole** *Glareola maldivarum*

E Asia; winters India to SE Asia, Philippines and Australasia

☐ **Black-winged Pratincole** *Glareola nordmanni*

Romania to sw Russia and n Kazakstan; winters to S Africa

☐ **Madagascar Pratincole** *Glareola ocularis*

Madagascar; winters coastal Somalia to n Mozambique

☐ **Rock Pratincole** *Glareola nuchalis*
    ____ *G. n. liberiae*
    ____ *G. n. nuchalis*

Sierra Leone to w Cameroon
Chad to Ethiopia and s to Zambia, Namibia and Mozambique

☐ **Gray Pratincole** *Glareola cinerea*

Rivers of Mali to Cameroon, w Zaire and nw Angola

☐ **Small Pratincole** *Glareola lactea*

E Afghanistan and Pakistan to India and Indochina

ORDER: CHARADRIIFORMES
FAMILY: CHARADRIIDAE (Plovers and Lapwings—66)

☐ **Northern Lapwing** *Vanellus vanellus*

Palearctic; winters to n Africa, India, Myanmar and s China

☐ **Long-toed Lapwing** *Vanellus crassirostris*
    ____ *V. c. crassirostris*
    ____ *V. c. leucopterus*

S Sudan to e Zaire, n Malawi and w Angola
Tanzania to Zaire, Angola, n Botswana and ne South Africa

☐ **Blacksmith Plover** *Vanellus armatus*

Lakes and marshes of eastern and southern Africa

☐ **Spur-winged Plover** *Vanellus spinosus*

Senegambia to Red Sea, ne Tanzania and e Mediterranean

☐ **River Lapwing** *Vanellus duvaucelii*

Rivers of India and Nepal to sw China and Indochina

☐ **Yellow-wattled Lapwing** *Vanellus malabaricus*

Lowlands of s Pakistan, India, Bangladesh and Sri Lanka

☐ **Black-headed Lapwing** *Vanellus tectus*
    ____ *V. t. tectus*
    ____ *V. t. latifrons*

Senegambia to Ethiopia, Kenya and Uganda
S Somalia to e Kenya

☐ **White-headed Lapwing** *Vanellus albiceps*

Sandy riverbanks of Africa south of the Sahara

☐ **Senegal Lapwing** *Vanellus lugubris*

Savanna of Africa south of the Sahara

☐ **Black-winged Lapwing** *Vanellus melanopterus*
    ____ *V. m. melanopterus*
    ____ *V. m. minor*

Extreme s Sudan to Ethiopia; sw Kenya to central Tanzania
E Cape Province to ne Transvaal; winters to s Mozambique

☐ **Crowned Lapwing** *Vanellus coronatus*
    ____ *V. c. coronatus*
    ____ *V. c. demissus*

Arid brush and deserts of Ethiopia to South Africa
Arid brush and deserts of Somalia

☐ **Wattled Lapwing** *Vanellus senegallus*
    ____ *V. s. senegallus*
    ____ *V. s. major*
    ____ *V. s. lateralis*

Senegambia to s Sudan, ne Zaire and n Uganda
Eritrea and Ethiopia
S Congo and Angola to Mozambique and ne South Africa

☐ **Spot-breasted Lapwing** *Vanellus melanocephalus*

Montane grasslands of Ethiopia

☐ **Brown-chested Lapwing** *Vanellus superciliosus*

Ghana to Cameroon and Zaire; winters in e Africa

☐ **Gray-headed Lapwing** *Vanellus cinereus*

Breeds ne China and Japan; winters to India and SE Asia

☐ **Red-wattled Lapwing** *Vanellus indicus*
    ____ *V. i. aigneri*   SE Turkey to Pakistan
    ____ *V. i. indicus*   E Pakistan, India, Nepal and Bangladesh
    ____ *V. i. lankae*   Sri Lanka
    ____ *V. i. atronuchalis*   NE India and Myanmar to n Malaysia and Indochina

☐ **Sunda Lapwing** *Vanellus macropterus*
     Formerly Sumatra and Java. Last recorded ca 1920

☐ **Banded Lapwing** *Vanellus tricolor*
     Southern Australia and Tasmania

☐ **Masked Lapwing** *Vanellus miles*
    ____ *V. m. miles*   New Guinea and n Australia; visitor to s Wallacea
    ____ *V. m. novaehollandiae*   E and se Australia, Tasmania and New Zealand

☐ **Sociable Lapwing** *Vanellus gregarius*
     S-central Russia and Kazakstan; winters ne Africa to India

☐ **White-tailed Lapwing** *Vanellus leucurus*
     SE Turkey to Afghanistan; winters ne Africa to India

☐ **Pied Lapwing** *Vanellus cayanus*
     S America east of the Andes to se Brazil and ne Argentina

☐ **Southern Lapwing** *Vanellus chilensis*
    ____ *V. c. cayennensis*   N South America north of the Amazon
    ____ *V. c. lampronotus*   Amazonia s of the Amazon to s Brazil, n Chile and n Argentina
    ____ *V. c. chilensis*   Argentina (Comodoro Rivadavia) to Chile (Chiloé I.)
    ____ *V. c. fretensis*   S Argentina and s Chile

☐ **Andean Lapwing** *Vanellus resplendens*
     Andes of sw Colombia to nw Argentina and n Chile

☐ **Red-kneed Dotterel** *Erythrogonys cinctus*
     Australia (except arid areas) and s New Guinea

☐ **Pacific Golden-Plover** *Pluvialis fulva*
     Siberia and w Alaska; winters to Africa, s Asia and Australasia

☐ **American Golden-Plover** *Pluvialis dominica*
     Breeds Arctic North America; winters s South America

☐ **Eurasian Golden-Plover** *Pluvialis apricaria*
    ____ *P. a. albifrons*   E-central Greenland, Iceland and Faeroes to Taymyr Peninsula
    ____ *P. a. apricaria*   Br. Isles to Baltic Pen.; winters Mediterranean and Persian Gulf

☐ **Black-bellied Plover** *Pluvialis squatarola*
     Holarctic; almost cosmopolitan post-breeding dispersal

☐ **Red-breasted Dotterel** *Charadrius obscurus*
    ____ *C. o. aquilonius*   North I. (New Zealand)
    ____ *C. o. obscurus*   Stewart I. (New Zealand)

☐ **Common Ringed Plover** *Charadrius hiaticula*
    ____ *C. h. hiaticula*   NE Canada and Greenland to Scandinavia; winters to Africa
    ____ *C. h. tundrae*   Russia and Siberia; winters Caspian Sea, sw Asia to S Africa

☐ **Semipalmated Plover** *Charadrius semipalmatus*
     Breeds n N America; winters to s S America and Hawaiian Is.

☐ **Long-billed Plover** *Charadrius placidus*
     Breeds e Asia; winters to India and Indochina

☐ **Little Ringed Plover** *Charadrius dubius*
    ____ *C. d. curonicus*   Palearctic; winters to Africa, Arabia, e China and Indonesia
    ____ *C. d. jerdoni*   India and SE Asia
    ____ *C. d. dubius*   Philippines to New Guinea and Bismarck Archipelago

☐ **Wilson's Plover** *Charadrius wilsonia*
    ____ *C. w. wilsonia*   Coastal e US to Belize and West Indies; winters to e Brazil

_____ *C. w. beldingi* — Pacific coast of Baja California to s Peru
_____ *C. w. cinnamominus* — Colombia to French Guiana; Netherlands Antilles

☐ **Killdeer** *Charadrius vociferus*
_____ *C. v. vociferus* — Canada, US and Mexico; winters to nw South America
_____ *C. v. ternominatus* — Greater Antilles
_____ *C. v. peruvianus* — Peru and nw Chile

☐ **Piping Plover** *Charadrius melodus* — E Canada and US; winters se US, Bahamas and Gr. Antilles

☐ **Madagascar Plover** *Charadrius thoracicus* — Coastal sw Madagascar

☐ **Kittlitz's Plover** *Charadrius pecuarius* — Africa south of the Sahara, ne Egypt and Madagascar

☐ **St. Helena Plover** *Charadrius sanctaehelenae* — St. Helena I. (s Atlantic Ocean)

☐ **Three-banded Plover** *Charadrius tricollaris*
_____ *C. t. tricollaris* — Ethiopia to Tanzania, Gabon, Chad and South Africa
_____ *C. t. bifrontatus* — Madagascar

☐ **Forbes' Plover** *Charadrius forbesi* — Grasslands and rocky hillsides of west and central Africa

☐ **White-fronted Plover** *Charadrius marginatus*
_____ *C. m. mechowi* — Africa s of Sahara to n Angola, Botswana and Mozambique
_____ *C. m. marginatus* — S Angola to sw Cape Province
_____ *C. m. arenaceus* — S Mozambique to s Cape Province
_____ *C. m. tenellus* — Madagascar

☐ **Chestnut-banded Plover** *Charadrius pallidus*
_____ *C. p. venustus* — Rift Valley soda lakes on Kenya/Tanzania border
_____ *C. p. pallidus* — Locally in southern Africa

☐ **Snowy Plover** *Charadrius alexandrinus*
_____ *C. a. alexandrinus* — W Palearctic to ne China; winters to Africa, s Asia, Indonesia
_____ *C. a. dealbatus* — S Japan, Ryukyu Is. and e China; winters to Philippines, Borneo
_____ *C. a. seebohmi* — SE India and Sri Lanka
_____ *C. a. nivosus* — US to Mexico and West Indies; winters to Panama
_____ *C. a. occidentalis* — Coastal Peru to s-central Chile

☐ **Javan Plover** *Charadrius javanicus* — Coastal lowlands of Java, Bali and Kangean Islands

☐ **Red-capped Plover** *Charadrius ruficapillus* — Australia (except arid areas) and Tasmania

☐ **Malaysian Plover** *Charadrius peronii* — Sandy coasts of SE Asia to Philippines and Indonesia

☐ **Collared Plover** *Charadrius collaris* — Mexico to n Argentina and central Chile

☐ **Puna Plover** *Charadrius alticola* — Andes of Peru to nw Argentina and n Chile

☐ **Two-banded Plover** *Charadrius falklandicus* — S Chile, Argentina and Falkland Islands; winters to s Brazil

☐ **Double-banded Plover** *Charadrius bicinctus*
_____ *C. b. bicinctus* — New Zealand and Chatham Islands; winters to Australasia
_____ *C. b. exilis* — Auckland Islands

☐ **Lesser Sandplover** *Charadrius mongolus*
_____ *C. m. pamirensis* — Pamirs to w China (w Xinjiang); winters to Africa and w India
_____ *C. m. atrifrons* — Himalayas and s Tibet: winters from India to Sumatra
_____ *C. m. schaeferi* — E Tibet to s Mongolia; winters Thailand to Greater Sundas

____ *C. m. mongolus*	E Siberia and Russian Far East; winters Taiwan to Australia
____ *C. m. stegmanni*	Kamchatka to Chukotsk Peninsula; winters to Australia
☐ **Greater Sandplover** *Charadrius leschenaultii*	
____ *C. l. columbinus*	Turkey to s Afghanistan; winters se Mediterranean and Red Sea
____ *C. l. crassirostris*	Transcaspia to se Kazakstan; winters to South Africa
____ *C. l. leschenaultii*	W China to s Mongolia and s Siberia; winters Australasia
☐ **Caspian Plover** *Charadrius asiaticus*	
	Caspian Sea to extreme w China; winters in e and s Africa
☐ **Oriental Plover** *Charadrius veredus*	
	Siberia to Manchuria and Mongolia; winters to Australasia
☐ **Eurasian Dotterel** *Charadrius morinellus*	
	W Alaska and n Palearctic; winters North Africa to w Iran
☐ **Rufous-chested Dotterel** *Charadrius modestus*	
	Tierra del Fuego and Falkland Islands; winters to se Brazil
☐ **Mountain Plover** *Charadrius montanus*	
	Great Plains of w Canada to sw US; winters to n Mexico
☐ **Hooded Plover** *Thinornis cucullatus*	
	Coastal s Australia, Tasmania and islands in Bass Strait
☐ **Shore Plover** *Thinornis novaeseelandiae*	
	Rangitara I. (Chatham Islands off New Zealand)
☐ **Black-fronted Dotterel** *Elseyornis melanops*	
	Australia (except arid areas), Tasmania and New Zealand
☐ **Inland Dotterel** *Peltohyas australis*	
	Arid interior of s Australia
☐ **Wrybill** *Anarhynchus frontalis*	
	Breeds n South I. (New Zealand); winters North I.
☐ **Diademed Sandpiper-Plover** *Phegornis mitchellii*	
	High elevation bogs of s Peru to w Argentina and Chile
☐ **Tawny-throated Dotterel** *Oreopholus ruficollis*	
____ *O. r. pallidus*	Arid littoral of sw Ecuador and n Peru
____ *O. r. ruficollis*	Coastal central Peru to Tierra del Fuego; winters to se Brazil

ORDER: CHARADRIIFORMES
FAMILY: PLUVIANELLIDAE (Magellanic Plover—1)

☐ **Magellanic Plover** *Pluvianellus socialis*	
	Lagoons and ponds of extreme s Argentina and Chile

ORDER: CHARADRIIFORMES
FAMILY: SCOLOPACIDAE (Sandpipers and Allies—89)

☐ **Eurasian Woodcock** *Scolopax rusticola*	
	Locally in moist woodlands and bogs of Eurasia
☐ **Amami Woodcock** *Scolopax mira*	
	Amami-O-Shima, Tokuno-Shima, Okinawa, Tokashiki-Shima
☐ **Bukidnon Woodcock** *Scolopax bukidnonensis*	
	Philippines (central and n Luzon and mts. of Mindanao)
☐ **Dusky Woodcock** *Scolopax saturata*	
____ *S. s. saturata*	Mountains of Sumatra and w Java
____ *S. s. rosenbergii*	Mountains of New Guinea
☐ **Sulawesi Woodcock** *Scolopax celebensis*	
	Montane forests of Sulawesi
☐ **Moluccan Woodcock** *Scolopax rochussenii*	
	Known from 8 specimens from n Moluccas (Obi and Bacan)

☐ **American Woodcock** *Scolopax minor*

Breeds e Canada and US; winters to Gulf Coast

☐ **Chatham Islands Snipe** *Coenocorypha pusilla*

Chatham Islands (Rangitara, Mangere and Star Keys)

☐ **Subantarctic Snipe** *Coenocorypha aucklandica*
- ____ *C. a. iredalei* — Islands off Stewart I. (New Zealand)
- ____ *C. a. huegeli* — Snares Islands (Northeast, Broughton and Alert Stack)
- ____ *C. a. aucklandica* — Auckland Islands
- ____ *C. a. meinertzhagenae* — Antipodes Islands
- ____ *C. a. ssp.* — Undescribed form from Campbell Islands

☐ **Jack Snipe** *Lymnocryptes minimus*

Scandinavia to Siberia; winters tropical Africa and SE Asia

☐ **Solitary Snipe** *Gallinago solitaria*
- ____ *G. s. solitaria* — High mts. of central Asia; winters to Pakistan and n India
- ____ *G. s. japonica* — Sakhalin and ne China; winters Korea, Japan and e China

☐ **Latham's Snipe** *Gallinago hardwickii*

Sakhalin and Japan; winters e Australia, Tasmania, New Guinea

☐ **Wood Snipe** *Gallinago nemoricola*

Breeds Himalayas and Tibet; winters in India and SE Asia

☐ **Pintail Snipe** *Gallinago stenura*

Siberia; winters India to SE Asia, Indonesia and Philippines

☐ **Swinhoe's Snipe** *Gallinago megala*

Siberia; winters India to SE Asia, Philippines and n Australia

☐ **African Snipe** *Gallinago nigripennis*
- ____ *G. n. aequatorialis* — Ethiopia to e Zaire, Tanzania, Malawi and Mozambique
- ____ *G. n. angolensis* — Angola and Namibia to Zambia and w Zimbabwe
- ____ *G. n. nigripennis* — S Mozambique and South Africa

☐ **Madagascar Snipe** *Gallinago macrodactyla*

Marshes of e and central massif of Madagascar

☐ **Great Snipe** *Gallinago media*

Breeds n Palearctic region; winters in sub-Saharan Africa

☐ **Common Snipe** *Gallinago gallinago*
- ____ *G. g. faeroeensis* — Iceland, Faeroe, Orkney and Sheltand is.; winters in British Is.
- ____ *G. g. gallinago* — N Palearctic and Aleutians; winters to Africa, India, Indonesia

☐ **Wilson's Snipe** *Gallinago delicata*

Aleutian Is. and Alaska to s US; winters to n South America

☐ **South American Snipe** *Gallinago paraguaiae*
- ____ *G. p. paraguaiae* — E Colombia to the Guianas, Brazil, n Argentina and Trinidad
- ____ *G. p. magellanica* — C Chile and Argentina to Tierra del Fuego and Falkland Islands

☐ **Puna Snipe** *Gallinago andina*

Andes of n Peru to nw Argentina and n Chile

☐ **Noble Snipe** *Gallinago nobilis*

Andes of Colombia to sw Venezuela, Ecuador and n Peru

☐ **Giant Snipe** *Gallinago undulata*
- ____ *G. u. undulata* — Colombia to Venezuela, the Guianas and adjacent n Brazil
- ____ *G. u. gigantea* — E Bolivia to Paraguay, se Brazil and ne Argentina

☐ **Fuegian Snipe** *Gallinago stricklandii*

Andes of s Argentina and s Chile to Tierra del Fuego

☐ **Andean Snipe** *Gallinago jamesoni*

Andes of Colombia to w Venezuela and e Bolivia

☐ **Imperial Snipe** *Gallinago imperialis*

Locally in Andes of Colombia to e Peru

☐ **Short-billed Dowitcher** *Limnodromus griseus*
    ____ *L. g. caurinus* — S Alaska and s Yukon; winters coastal w US to s Peru
    ____ *L. g. hendersoni* — Plains of central Canada; winters se US to Panama
    ____ *L. g. griseus* — Quebec and Labrador; winters coastal e US to Brazil

☐ **Long-billed Dowitcher** *Limnodromus scolopaceus* — Breeds Siberia and Alaska; winters s US to Panama

☐ **Asian Dowitcher** *Limnodromus semipalmatus* — Siberia and Manchuria; winters to s Asia and n Australia

☐ **Black-tailed Godwit** *Limosa limosa*
    ____ *L. l. islandica* — Iceland, Faeroe Is. and Shetland Is.; winters to sw Europe
    ____ *L. l. limosa* — W Palearctic; winters to sub-Saharan Africa and India
    ____ *L. l. melanuroides* — E Palearctic; winters to SE Asia, Philippines and Australia

☐ **Hudsonian Godwit** *Limosa haemastica* — Canadian Arctic; winters Atlantic coast of s South America

☐ **Bar-tailed Godwit** *Limosa lapponica*
    ____ *L. l. lapponica* — Lapland to Taymyr Peninsula; winters to Africa and India
    ____ *L. l. menzbieri* — N Siberia; winters SE Asia to coastal Australia, Tasmania
    ____ *L. l. baueri* — NE Siberia to w Alaska; winters China to New Zealand

☐ **Marbled Godwit** *Limosa fedoa*
    ____ *L. f. beringiae* — Alaskan Peninsula; winters coastal Washington to California
    ____ *L. f. fedoa* — Great Plains of North America; winters to Argentina and Chile

☐ **Eskimo Curlew** *Numenius borealis* — Canadian Arctic; winters to s South America (possibly extinct)

☐ **Little Curlew** *Numenius minutus* — Siberia; winters to Philippines, Indonesia and n Australia

☐ **Whimbrel** *Numenius phaeopus*
    ____ *N. p. phaeopus* — NW Palearctic; winters to Africa and India
    ____ *N. p. alboaxillaris* — Steppes n of Caspian Sea; winters coastal w Indian Ocean
    ____ *N. p. variegatus* — Siberia; winters to India, Philippines, Indonesia and Australia
    ____ *N. p. hudsonicus* — Alaska to n Canada; winters to s South America

☐ **Bristle-thighed Curlew** *Numenius tahitiensis* — Breeds w Alaska; winters Hawaii and Micronesia to Polynesia

☐ **Slender-billed Curlew** *Numenius tenuirostris* — Breeds sw Siberia and n Kazakstan; winters nw Africa

☐ **Eurasian Curlew** *Numenius arquata*
    ____ *N. a. arquata* — British Isles to Ural Mts.; winters to nw Africa and India
    ____ *N. a. orientalis* — E Russia, Manchuria; winters to Africa, SE Asia and Indonesia

☐ **Long-billed Curlew** *Numenius americanus*
    ____ *N. a. parvus* — SW Canada to California; winters to Mexico
    ____ *N. a. americanus* — W US; winters to Central America

☐ **Far Eastern Curlew** *Numenius madagascariensis* — Breeds ne Asia; winters to Philippines, Indonesia and Australia

☐ **Upland Sandpiper** *Bartramia longicauda* — Breeds Alaska to s US; winters in South America

☐ **Terek Sandpiper** *Xenus cinereus* — N Eurasia; winters to s Africa, s Asia, Philippines, n Australia

☐ **Common Sandpiper** *Actitis hypoleucos* — Palearctic; winters to s Africa, c Asia, Philippines and Australia

☐ **Spotted Sandpiper** *Actitis macularius* — Breeds North America; winters to s South America

☐ **Green Sandpiper** *Tringa ochropus* — N Eurasia; winters to s Africa, s Asia, Philippines and Australia

☐ **Solitary Sandpiper** *Tringa solitaria*
_____ *T. s. cinnamomea*   Alaska and w Canada; winters n South America to Argentina
_____ *T. s. solitaria*   E British Columbia to Labrador; winters Cent. and S America

☐ **Gray-tailed Tattler** *Tringa brevipes*

     Mountains of Siberia; winters to SE Asia and Australasia

☐ **Wandering Tattler** *Tringa incana*

     Siberia and Alaska; winters to n S Am., Hawaii and sw Oceania

☐ **Spotted Redshank** *Tringa erythropus*

     Breeds n Eurasia; winters Mediterranean region to SE Asia

☐ **Greater Yellowlegs** *Tringa melanoleuca*

     Alaska and Canada; winters to s South America

☐ **Common Greenshank** *Tringa nebularia*

     Palearctic; winters to s Africa, s Asia, Philippines and Australia

☐ **Nordmann's Greenshank** *Tringa guttifer*

     E Siberia; winters SE Asia to Philippines and Indonesia

☐ **Willet** *Tringa semipalmatus*
_____ *T. s. inornatus*   Central Canada to Nebraska and Colorado; winters to n Chile
_____ *T. s. semipalmatus*   SE Canada to Gulf Coast and West Indies; winters to s Brazil

☐ **Lesser Yellowlegs** *Tringa flavipes*

     Alaska and Canada; winters to Tierra del Fuego and Galapagos

☐ **Marsh Sandpiper** *Tringa stagnatilis*

     Palearctic; winters to s Africa, s Asia and Australasia

☐ **Wood Sandpiper** *Tringa glareola*

     Breeds n Eurasia; winters to s Africa, s Asia and Australia

☐ **Common Redshank** *Tringa totanus*
_____ *T. t. robusta*   Iceland, Faeroes and Scotland; winters Br. Isles and w Europe
_____ *T. t. totanus*   Scandinavia to Iberia; winters to Africa, India and Indonesia
_____ *T. t. ussuriensis*   Siberia and Mongolia to e Russia; winters to Africa and India
_____ *T. t. terrignotae*   S Manchuria; winters in SE Asia
_____ *T. t. craggi*   NW China (nw Xinjiang); winter grounds unknown
_____ *T. t. eurhinus*   Pamir Mountains to n India and Tibet; winters in India

☐ **Tuamotu Sandpiper** *Prosobonia cancellata*

     Isolated islands in Tuamotu Archipelago (French Polynesia)

☐ **Ruddy Turnstone** *Arenaria interpres*
_____ *A. i. interpres*   Alaska, n N Am. and n Eurasia, winters to Africa, Australasia
_____ *A. i. morinella*   NE Alaska and Arctic Canada; winters se US to s S America

☐ **Black Turnstone** *Arenaria melanocephala*

     Breeds coastal Alaska; winters se Alaska to nw Mexico

☐ **Surfbird** *Aphriza virgata*

     Breeds Alaska and Yukon; winters to Straits of Magellan

☐ **Great Knot** *Calidris tenuirostris*

     NE Siberia; winters to India, SE Asia, Philippines and Australia

☐ **Red Knot** *Calidris canutus*
_____ *C. c. canutus*   Siberia; winters to South Africa and Australasia
_____ *C. c. piersmai*   New Siberian Archipelago; winters to Australasia
_____ *C. c. rogersi*   Chukotsk Peninsula (Russia); winters Australasia
_____ *C. c. roselaari*   Wrangel I. (Russia) and nw Alaska; winters to n Venezuela
_____ *C. c. rufa*   Canadian low Arctic; winters to s South America
_____ *C. c. islandica*   Canadian high Arctic and n Greenland; winters in w Europe

☐ **Sanderling** *Calidris alba*

     Breeds Holarctic; worldwide coastal post-breeding dispersal

☐ **Semipalmated Sandpiper** *Calidris pusilla*

     Breeds Arctic North America; winters to s South America

☐ **Western Sandpiper** *Calidris mauri*

     Breeds Siberia and Alaska; winters to n South America

☐ **Red-necked Stint** *Calidris ruficollis*

Breeds Siberia and Alaska; disperses to s Asia and Australasia

☐ **Little Stint** *Calidris minuta*

N Palearctic; winters to Africa and Indian subcontinent

☐ **Temminck's Stint** *Calidris temminckii*

N Palearctic; winters to Africa, Indonesia and Philippines

☐ **Long-toed Stint** *Calidris subminuta*

NE Palearctic; winters SE Asia to Philippines and Australia

☐ **Least Sandpiper** *Calidris minutilla*

Breeds n N America; winters to s S America and Hawaiian Is.

☐ **White-rumped Sandpiper** *Calidris fuscicollis*

Breeds Arctic North America; winters to Tierra del Fuego

☐ **Baird's Sandpiper** *Calidris bairdii*

Siberia, n Alaska to Greenland; winters in Andes to s S America

☐ **Pectoral Sandpiper** *Calidris melanotos*

Arctic N America and Siberia; winters to s S America, Australia

☐ **Sharp-tailed Sandpiper** *Calidris acuminata*

Breeds ne Siberia; winters in Australasia and Polynesia

☐ **Curlew Sandpiper** *Calidris ferruginea*

Arctic Siberia; winters to s Africa, SE Asia and Australasia

☐ **Dunlin** *Calidris alpina*
____ *C. a. arctica* — NE Greenland; winters mainly nw Africa
____ *C. a. schinzii* — Greenland and Iceland to s Scandinavia; winters to nw Africa
____ *C. a. alpina* — Scandinavia to e Russia; winters to Mediterranean and India
____ *C. a. sakhalina* — Russia to Chukotsk Pen.; winters China, Japan and Taiwan
____ *C. a. actites* — N Sakhalin; wintering grounds unknown
____ *C. a. kistchinskii* — Sea of Okhotsk to Kuril Islands; wintering grounds unknown
____ *C. a. arcticola* — NW Alaska and nw Canada; winters e China, Korea and Japan
____ *C. a. pacifica* — SW Alaska; winters in w US and w Mexico
____ *C. a. hudsonia* — Central Canada; winters se US and e Mexico

☐ **Purple Sandpiper** *Calidris maritima*

Holarctic tundra; winters coastal e US and nw Europe

☐ **Rock Sandpiper** *Calidris ptilocnemis*
____ *C. p. quarta* — Kuril Islands, s Kamchatka Pen. and Komandorskiye Islands
____ *C. p. tschuktschorum* — Chukotsk Pen. to w Alaska; winters nw N Am. and e Japan
____ *C. p. ptilocnemis* — Pribilof, St. Matthew and Hall islands; winters Alaska Pen.
____ *C. p. couesi* — Aleutian Islands and Alaska Peninsula

☐ **Stilt Sandpiper** *Calidris himantopus*

Breeds n North America; winters sw US to s South America

☐ **Spoon-billed Sandpiper** *Eurynorhynchus pygmeus*

Breeds ne Siberia; winters in SE Asia

☐ **Broad-billed Sandpiper** *Limicola falcinellus*
____ *L. f. falcinellus* — Scandinavia and nw Russia; winters to s Africa and India
____ *L. f. sibirica* — N Russia; winters to India, SE Asia, Philippines and Australia

☐ **Buff-breasted Sandpiper** *Tryngites subruficollis*

Breeds Arctic North America; winters in s South America

☐ **Ruff and Reeve** *Philomachus pugnax*

Breeds n Palearctic; winters to s Africa, s Asia and Australia

☐ **Wilson's Phalarope** *Phalaropus tricolor*

Breeds Canada and US; winters in w and s South America

☐ **Red-necked Phalarope** *Phalaropus lobatus*

Holarctic circumpolar; winters at sea in southern hemisphere

☐ **Red Phalarope** *Phalaropus fulicarius*

Holarctic circumpolar; winters at sea in southern hemisphere

ORDER: CHARADRIIFORMES
FAMILY: PEDIONOMIDAE (Plains-wanderer—1)

☐ **Plains-wanderer** *Pedionomus torquatus*

Locally in sparse grasslands of inland se Australia

ORDER: CHARADRIIFORMES
FAMILY: THINOCORIDAE (Seedsnipes—4)

☐ **Rufous-bellied Seedsnipe** *Attagis gayi*
_____ *A. g. latreillii* — Andes of n Ecuador
_____ *A. g. simonsi* — Andes of Peru to n Chile and extreme nw Argentina
_____ *A. g. gayi* — Andes of Chile and Argentina to Tierra del Fuego

☐ **White-bellied Seedsnipe** *Attagis malouinus*

S Argentina, s Chile, Cape Horn Archipelago and Staten I.

☐ **Gray-breasted Seedsnipe** *Thinocorus orbignyianus*
_____ *T. o. ingae* — Andes of n Peru to n Chile and nw Argentina
_____ *T. o. orbignyianus* — Andes of n-central Chile and Argentina to Tierra del Fuego

☐ **Least Seedsnipe** *Thinocorus rumicivorus*
_____ *T. r. pallidus* — Arid littoral of sw Ecuador and extreme nw Peru
_____ *T. r. cuneicauda* — Arid littoral of Peru
_____ *T. r. bolivianus* — Altiplano of s Peru to n Chile and extreme nw Argentina
_____ *T. r. rumicivorus* — Patagonia to Tierra del Fuego; winters to c Argentina and Chile

ORDER: CHARADRIIFORMES
FAMILY: CHIONIDIDAE (Sheathbills—2)

☐ **Snowy Sheathbill** *Chionis albus*

S Argentina, s Chile, Antarctic Peninsula and Falkland Islands

☐ **Black-faced Sheathbill** *Chionis minor*
_____ *C. m. marionensis* — Marion and Prince Edward islands
_____ *C. m. crozettensis* — Crozet Islands
_____ *C. m. minor* — Kerguelen Islands
_____ *C. m. nasicornis* — Heard and McDonald islands

ORDER: CHARADRIIFORMES
FAMILY: LARIDAE (Gulls—56)

☐ **Dolphin Gull** *Larus scoresbii*

Coasts of s Chile, Argentina and Falkland Islands

☐ **Pacific Gull** *Larus pacificus*
_____ *L. p. georgii* — Coastal Western Australia to South Australia and Kangaroo I.
_____ *L. p. pacificus* — SE Australia (Victoria) and Tasmania; casual to Queensland

☐ **Belcher's Gull** *Larus belcheri*

Humboldt Current of Peru and Chile; disperses to Ecuador

☐ **Olrog's Gull** *Larus atlanticus*

Atlantic coast of ne Argentina; winters north to Uruguay

☐ **Black-tailed Gull** *Larus crassirostris*

Coastal Siberia, Kuril Islands, Korea, China and Japan

☐ **Gray Gull** *Larus modestus*

Inland nitrate deserts of Peru and Chile; ranges to Ecuador

☐ **Heermann's Gull** *Larus heermanni*

Coastal w Mexico; winters British Columbia to Guatemala

☐ **White-eyed Gull** *Larus leucophthalmus*

Red Sea and Gulf of Aqaba to Gulf of Aden

☐ **Sooty Gull** *Larus hemprichii*

Red Sea and Persian Gulf area to s Pakistan and n Kenya

☐ **Mew Gull** *Larus canus*

____ *L. c. canus* — Iceland and British Isles to White Sea; winters to N Africa
____ *L. c. heinei* — W Russia to Siberia; winters to Black Sea and Caspian Sea
____ *L. c. kamtschatschensis* — NE Siberia; winters SE Asia
____ *L. c. brachyrhynchus* — Alaska to Br. Columbia and Saskatchewan; winters to California

☐ **Audouin's Gull** *Larus audouinii*

Mediterranean basin; winters to Senegambia

☐ **Ring-billed Gull** *Larus delawarensis*

N America; winters to s Mexico, Bahamas and Gr. Antilles

☐ **Kelp Gull** *Larus dominicanus*

____ *L. d. dominicanus* — Coastal s S America, Falklands, S. Georgia, N .Zealand, Australia
____ *L. d. vetula* — Coastal southern Africa and Namibia
____ *L. d. austrinus* — Antarctica and Antarctic islands
____ *L. d. judithae* — Subantarctic Indian Ocean islands
____ *L. d. melisandae* — Coasts of sw and s Madagascar

☐ **California Gull** *Larus californicus*

____ *L. c. albertaensis* — S Mackenzie, Alberta and w Manitoba to South Dakota
____ *L. c. californicus* — E Washington to Wyoming; winters to s Mexico

☐ **Great Black-backed Gull** *Larus marinus*

Palearctic and ne N Am.; winters to W Indies and Iberian Pen.

☐ **Glaucous-winged Gull** *Larus glaucescens*

Bering Sea to nw Oregon; winters to Japan and nw Mexico

☐ **Western Gull** *Larus occidentalis*

____ *L. o. occidentalis* — Pacific coast of British Columbia to central California
____ *L. o. wymani* — Central California (Monterey Bay) to s Baja California

☐ **Yellow-footed Gull** *Larus livens*

Islands in Gulf of California; ranges north to Salton Sea

☐ **Glaucous Gull** *Larus hyperboreus*

____ *L. h. hyperboreus* — Jan Mayen and Spitsbergen east to Taymyr Peninsula
____ *L. h. pallidissimus* — Taymyr Peninsula east to Bering Sea and Pribilof Islands
____ *L. h. barrovianus* — Alaska to w Canada (nw Mackenzie)
____ *L. h. leuceretes* — E Mackenzie and n Canadian Arch. to Greenland and Iceland

☐ **Iceland Gull** *Larus glaucoides*

____ *L. g. kumlieni* — NE Canada (Baffin I. and nw Ungava); winters to n US
____ *L. g. glaucoides* — S and w Greenland; winters to n Europe

☐ **Thayer's Gull** *Larus thayeri*

Hudson Bay to w Greenland; winters British Columbia to Baja

☐ **European Herring Gull** *Larus argentatus*

____ *L. a. argenteus* — Northwest Europe; winters to n Iberia
____ *L. a. argentatus* — Scandinavia to Kola Peninsula; winters n and w Europe

☐ **Lesser Black-backed Gull** *Larus fuscus*

____ *L. f. graellsii* — Iceland, Faeroes, Br. Isles, France, Portugal; winters to w Africa
____ *L. f. fuscus* — N Norway, Sweden, Finland to White Sea; > Africa, sw Asia
____ *L. f. intermedius* — Denmark to w Norway, locally s to ne Spain; winters to w Africa

☐ **Heuglin's Gull** *Larus heuglini*

NW Russia; winters to Middle East, South Africa and se Asia

☐ **East Siberian Gull** *Larus vegae*
_____ *L. v. vegae* — NE Siberia; winters south to China
_____ *L. v. mongolicus* — SE Altai and Lake Baikal to Mongolia; winters s Asia

☐ **American Herring Gull** *Larus smithsonianus*
N North America (Alaska to Atlantic coast); winters to C America

☐ **Caspian Gull** *Larus cachinnans*
Black Sea to Kazakstan; winters to s Asia and ne Africa

☐ **Armenian Gull** *Larus armenicus*
Lakes of Caucasus to e Turkey and Iran; winters to Red Sea

☐ **Steppe Gull** *Larus barabensis*
Steppes of central Asia; winters mainly in sw Asia east to India

☐ **Yellow-legged Gull** *Larus michahellis*
_____ *L. m. atlantis* — Azores Islands
_____ *L. m. michahellis* — Macaronesian Is. and nw Africa east through Mediterranean

☐ **Great Black-headed Gull** *Larus ichthyaetus*
S-central Asia; winters Mediterranean to SE Asia

☐ **Slaty-backed Gull** *Larus schistisagus*
Breeds ne Siberia to Japan; winters south to Taiwan

☐ **Brown-headed Gull** *Larus brunnicephalus*
Mts. of s-central Asia; winters to Arabia, India and SE Asia

☐ **Gray-headed Gull** *Larus cirrocephalus*
_____ *L. c. cirrocephalus* — Ecuador and Peru; coastal central Brazil to Argentina
_____ *L. c. poiocephalus* — Coasts and rivers of sub-Saharan Africa and Madagascar

☐ **Hartlaub's Gull** *Larus hartlaubii*
Coastal sw Namibia to sw South Africa

☐ **Silver Gull** *Larus novaehollandiae*
_____ *L. n. forsteri* — N Australia, New Caledonia and Loyalty Islands
_____ *L. n. novaehollandiae* — S Australia and Tasmania

☐ **Red-billed Gull** *Larus scopulinus*
New Zealand, Chatham, Auckland and adjacent islands

☐ **Black-billed Gull** *Larus bulleri*
South I. (New Zealand); winters to North I.

☐ **Brown-hooded Gull** *Larus maculipennis*
Lakes, rivers and coasts of s South America and Falkland Is.

☐ **Black-headed Gull** *Larus ridibundus*
N Palearctic; winters to Africa, s Asia and e North America

☐ **Slender-billed Gull** *Larus genei*
Mediterranean basin to nw India; winters to ne Africa

☐ **Bonaparte's Gull** *Larus philadelphia*
N North America; winters to Mexico and Greater Antilles

☐ **Saunders' Gull** *Larus saundersi*
Coastal e China; winters South Korea and s Japan to n Vietnam

☐ **Andean Gull** *Larus serranus*
Andean lakes of Ecuador to n Argentina and central Chile

☐ **Mediterranean Gull** *Larus melanocephalus*
North Sea to Mediterranean and Black Sea; winters to n Africa

☐ **Relict Gull** *Larus relictus*
Kazakstan to Mongolia; winters to eastern China Sea

☐ **Lava Gull** *Larus fuliginosus*
Galapagos Islands

☐ **Laughing Gull** *Larus atricilla*
_____ *L. a. megalopterus* — SE California to w Mexico; Maine to C Am.; winters to Peru
_____ *L. a. atricilla* — West Indies to Trinidad; winters to n Brazil

☐ **Franklin's Gull** *Larus pipixcan*
W-central N America; winters Pacific coasts of S America

☐ **Little Gull** *Larus minutus*

N Eurasia and ne North America

☐ **Ivory Gull** *Pagophila eburnea*

Arctic circumpolar (mainly associated with ice pack)

☐ **Ross' Gull** *Rhodostethia rosea*

Locally in Siberia and Arctic North America

☐ **Sabine's Gull** *Xema sabini*
____ *X. s. palaearctica* — Spitsbergen east to Taymyr Peninsula and Lena Delta
____ *X. s. tschuktschorum* — Chukotsk Peninsula (Russia)
____ *X. s. woznesenskii* — NE Siberia to Alaska
____ *X. s. sabini* — N Canada to Greenland; winters sw Africa and nw S America

☐ **Swallow-tailed Gull** *Creagrus furcatus*

Galapagos Islands; winters coastal Colombia to Chile

☐ **Red-legged Kittiwake** *Rissa brevirostris*

Komandorskiye, Aleutian and Pribilof islands; winters n Pacific

☐ **Black-legged Kittiwake** *Rissa tridactyla*
____ *R. t. tridactyla* — Circumpolar n Atlantic; winters to Saragossa Sea and w Africa
____ *R. t. pollicaris* — Circumpolar n Pacific; winters to e China Sea and nw Mexico

ORDER: CHARADRIIFORMES
FAMILY: STERNIDAE (Terns—44)

☐ **Brown Noddy** *Anous stolidus*
____ *A. s. plumbeigularis* — S Red Sea and Gulf of Aden
____ *A. s. pileatus* — Seychelles and Madagascar to Australia, Polynesia and Hawaii
____ *A. s. galapagensis* — Galapagos Islands
____ *A. s. ridgwayi* — W Mexico (Revillagigedo Islands) to Costa Rica (Cocos I.)
____ *A. s. stolidus* — Caribbean and s Atlantic islands; Gulf of Guinea to Cameroon

☐ **Black Noddy** *Anous minutus*
____ *A. m. worcesteri* — Cavilli I. and Tubbataha Reef (Sulu Sea)
____ *A. m. minutus* — NE Australia and New Guinea to Tuamotu Archipelago
____ *A. m. marcusi* — Marcus I. and Wake I. through Micronesia to Caroline Islands
____ *A. m. melanogenys* — Hawaiian Islands
____ *A. m. diamesus* — Clipperton I. (off w Mexico) and Cocos I. (off Costa Rica)
____ *A. m. americanus* — Islands off Central America and Venezuela; Lesser Antilles
____ *A. m. atlanticus* — St. Helena and adjacent s Atlantic islands to Gulf of Guinea

☐ **Lesser Noddy** *Anous tenuirostris*
____ *A. t. tenuirostris* — Seychelles, Mascarene and Maldive islands
____ *A. t. melanops* — Houtman Abrolhos Islands (w Australia); formerly Indonesia

☐ **Blue Noddy** *Procelsterna cerulea*
____ *P. c. saxatilis* — Marcus I. and n Marshall Islands to nw Hawaiian Islands
____ *P. c. nebouxi* — Ellice I. to Phoenix Islands, Fiji and Western Samoa
____ *P. c. cerulea* — Christmas I. (Line Islands) and Marquesas Islands
____ *P. c. teretirostris* — Tuamotu Archipelago, Cook, Austral and Society islands
____ *P. c. murphyi* — Gambier Islands

☐ **Gray Noddy** *Procelsterna albivitta*
____ *P. a. albivitta* — Lord Howe I., Norfolk I., Kermadec Islands and Tonga
____ *P. a. skottsbergii* — Henderson I., Easter I. and Sala y Gómez I. (Chile)
____ *P. a. imitatrix* — Desaventurados Is. off Chile (San Ambrosio and San Félix)

☐ **White Tern** *Gygis alba*
____ *G. a. alba* — Caroline Is. to Hawaii, Clipperton, Cocos and s Atlantic islands
____ *G. a. candida* — Seychelles and Mascarene islands to s-central Pacific

____	*G. a. leucopes*	Henderson I. and Pitcairn I.
____	*G. a. microrhyncha*	Marquesas, Phoenix and Line islands

☐ **Sooty Tern** *Onychoprion fuscatus*

____	*O. f. fuscatus*	Gulf of Mexico, e Mexico and W Indies; Gulf of Guinea islands
____	*O. f. nubilosus*	S Red Sea and Indian Ocean to Ryukyu Is. and Philippines
____	*O. f. infuscatus*	Central Indonesia
____	*O. f. serratus*	New Guinea, Australia and New Caledonia
____	*O. f. kermadeci*	Kermadec Islands
____	*O. f. oahuensis*	Bonin Islands to Hawaii and South Pacific islands
____	*O. f. crissalis*	Islands off w Mexico and Central America to Galapagos Is.
____	*O. f. luctuosus*	Juan Fernández Islands (off Chile)

☐ **Gray-backed Tern** *Onychoprion lunatus*

Islands in tropical Pacific Ocean

☐ **Bridled Tern** *Onychoprion anaethetus*

____	*O. a. melanopterus*	Coastal w Africa
____	*O. a. fuligulus*	Red Sea and East Africa to India
____	*O. a. antarcticus*	Madagascar, Aldabra, Seychelles, Mascarene and Andaman is.
____	*O. a. anaethetus*	Ryukyu Is., Taiwan, Philippines, Indonesia and Australia
____	*O. a. nelsoni*	W coast of Mexico and Central America
____	*O. a. recognitus*	West Indies, Belize and islands off Venezuela

☐ **Aleutian Tern** *Onychoprion aleuticus*

Alaska and Siberia; winters to Singapore and Indonesia

☐ **Little Tern** *Sternula albifrons*

____	*S. a. albifrons*	Europe to w Asia and w Indian Ocean; winters Africa to India
____	*S. a. guineae*	Ghana to Gabon
____	*S. a. innominata*	Islands in Persian Gulf
____	*S. a. pusilla*	NE India, Sri Lanka, Myanmar and islands off Sumatra and Java
____	*S. a. sinensis*	SE Russia to Japan, SE Asia, Philippines and New Guinea
____	*S. a. placens*	E Australia and e Tasmania

☐ **Least Tern** *Sternula antillarum*

____	*S. a. browni*	S California to Baja and w Mexico; winters to Central America
____	*S. a. athalassos*	N Great Plains to Louisiana and Texas; winters to n Brazil
____	*S. a. antillarum*	E US to Honduras, Caribbean and Guianas; winters to n Brazil

☐ **Yellow-billed Tern** *Sternula superciliaris*

Rivers and lakes of South America east of the Andes

☐ **Fairy Tern** *Sternula nereis*

____	*S. n. horni*	Western Australia
____	*S. n. nereis*	South Australia, Victoria and Tasmania
____	*S. n. exsul*	New Caledonia
____	*S. n. davisae*	N North I. (New Zealand)

☐ **Peruvian Tern** *Sternula lorata*

Arid Humboldt Current coasts of Ecuador to n Chile

☐ **Saunders' Tern** *Sternula saundersi*

Red Sea to India and Sri Lanka; winters to Malay Peninsula

☐ **Damara Tern** *Sternula balaenarum*

Coastal Namibia to Cape Province; winters to n Angola

☐ **Large-billed Tern** *Phaetusa simplex*

____	*P. s. simplex*	E Colombia to e Brazil and Amazonia; Trinidad; w Ecuador
____	*P. s. chloropoda*	Basins of Río Paraguay and Río Paraná to n Argentina

☐ **Gull-billed Tern** *Gelochelidon nilotica*

____	*G. n. nilotica*	Palearctic; winters tropical Africa and Persian Gulf to India
____	*G. n. addenda*	Transbaikalia to Manchuria and e China; winters SE Asia

____	*G. n. macrotarsa*	Australia (except arid areas)
____	*G. n. arenea*	E US to Gr. Antilles and Yucatán; winters to Brazil and Peru
____	*G. n. vanrossemi*	California to n Baja and nw Mexico; winters to Ecuador
____	*G. n. groenvoldi*	Coasts and rivers of French Guiana to ne Argentina

☐ **Caspian Tern** *Hydroprogne caspia*

Cosmopolitan—wide distribution worldwide

☐ **Inca Tern** *Larosterna inca*

Islands off Peru and Chile; ranges rarely n to Ecuador

☐ **Black Tern** *Chlidonias niger*

____	*C. n. niger*	W Palearctic; winters in Africa
____	*C. n. surinamensis*	N North America; winters Central America and n S America

☐ **White-winged Tern** *Chlidonias leucopterus*

Palearctic; winters to Africa, s Asia and Australasia

☐ **Whiskered Tern** *Chlidonias hybrida*

____	*C. h. hybrida*	SW Europe to Kazakstan; winters Africa and sw Asia
____	*C. h. swinhoei*	Transbaikalia to e China and Taiwan
____	*C. h. indicus*	E Iran and Pakistan to n India
____	*C. h. javanicus*	NE India and Sri Lanka; winters Malaysia and Indonesia
____	*C. h. sclateri*	South Africa and Madagascar
____	*C. h. delelandii*	Kenya to Tanzania
____	*C. h. fluviatilis*	Australia; disperses to New Guinea and Moluccas

☐ **Black-fronted Tern** *Chlidonias albostriatus*

South I.; disperses to Stewart I. and North I. (New Zealand)

☐ **Roseate Tern** *Sterna dougallii*

____	*S. d. dougallii*	Coastal e N America to W Indies; Azores, Europe and Africa
____	*S. d. arideensis*	Seychelles to Madagascar and Rodrigues I.
____	*S. d. korustes*	Sri Lanka, Andaman Islands and Mergui Archipelago
____	*S. d. bangsi*	Arabian Sea; e China to New Guinea, Solomons and Ryukyus
____	*S. d. gracilis*	Moluccas and n and w Australia

☐ **White-fronted Tern** *Sterna striata*

____	*S. s. incerta*	Flinders I. and Cape Barren I. (off Tasmania)
____	*S. s. striata*	Breeds New Zealand; winters coastal se Australia
____	*S. s. aucklandorna*	Chatham Islands, Auckland Islands and (?) Snares Islands

☐ **Black-naped Tern** *Sterna sumatrana*

____	*S. s. sumatrana*	Andaman and Nicobar is. to Japan, Malaysia and Australasia
____	*S. s. mathewsi*	Aldabra, Amirante, Chagos and Maldive islands

☐ **South American Tern** *Sterna hirundinacea*

Coasts and islands of s South America and Falkland Islands

☐ **Antarctic Tern** *Sterna vittata*

____	*S. v. tristanensis*	Tristan da Cunha, Gough I.; Amsterdam I. and (?) St. Paul I.
____	*S. v. georgiae*	South Georgia I.; possibly S Orkney and S Sandwich islands
____	*S. v. gaini*	South Shetland Islands
____	*S. v. vittata*	Prince Edward, Marion, Crozet and Kerguelen islands
____	*S. v. bethunei*	Stewart, Snares, Auckland, Bounty, Antipodes, Campbell is.
____	*S. v. macquariensis*	Macquarie I.

☐ **Arctic Tern** *Sterna paradisaea*

Arctic circumpolar; winters sub-Antarctic and Antarctic seas

☐ **Common Tern** *Sterna hirundo*

____	*S. h. hirundo*	N Am., S Am., Atlantic islands, Europe and w Africa to China
____	*S. h. minussensis*	C Asia to n Mongolia and s Tibet; winters Indian Ocean
____	*S. h. tibetana*	W Mongolia to Kashmir and Tibet; winters e Indian Ocean
____	*S. h. longipennis*	NE Siberia to ne China; winters SE Asia to Australia

☐ **Forster's Tern** *Sterna forsteri*

N America; winters s US to Costa Rica and Greater Antilles

☐ **Snowy-crowned Tern** *Sterna trudeaui*

Lagoons and marshes of se Brazil and Uruguay to Patagonia

☐ **Black-bellied Tern** *Sterna acuticauda*

Pakistan, Nepal and India to sw China, Myanmar and SE Asia

☐ **River Tern** *Sterna aurantia*

Pakistan to s India, Sri Lanka, Nepal and sw China

☐ **White-cheeked Tern** *Sterna repressa*

Islands and coasts of Red Sea and Persian Gulf to India

☐ **Kerguelen Tern** *Sterna virgata*

Marion, Crozet, and Kerguelen islands

☐ **Royal Tern** *Thalasseus maximus*
_____ *T. m. albididorsalis* — Coastal Mauritania to Guinea; winters to Namibia
_____ *T. m. maximus* — Coastal US to W Indies, Guianas, Brazil; winters to Argentina

☐ **Great Crested Tern** *Thalasseus bergii*
_____ *T. b. bergii* — Namibia to South Africa; disperses to Mozambique
_____ *T. b. enigmus* — Islands off Mozambique, Zambezi River delta and Madagascar
_____ *T. b. thalassinus* — Tanzania, Seychelles, Aldabra and Rodriques I.
_____ *T. b. velox* — Red Sea and nw Somalia to Maldives, Myanmar and Sri Lanka
_____ *T. b. cristatus* — Malaysia to Philippines and Ryukyus; e Australia to Society Is.
_____ *T. b. gwendolenae* — W and nw Australia

☐ **Sandwich Tern** *Thalasseus sandvicensis*
_____ *T. s. sandvicensis* — Europe to Caspian Sea; winters to S Africa, India, Sri Lanka
_____ *T. s. acuflavidus* — E N America to s Caribbean; winters to s Peru and Uruguay
_____ *T. s. eurygnathus* — Islands off Venezuela, the Guianas, e Brazil and n Argentina

☐ **Elegant Tern** *Thalasseus elegans*

S California to w Mexico (Nayarit); winters Guatemala to Chile

☐ **Lesser Crested Tern** *Thalasseus bengalensis*
_____ *T. b. emigrate* — Libya; winters off w African coast
_____ *T. b. bengalensis* — Red Sea, Pakistan, Laccadives and Maldives; winters S Africa
_____ *T. b. torresii* — Persian Gulf; Sulawesi to New Guinea and n Australia

☐ **Chinese Crested Tern** *Thalasseus bernsteini*

Coastal e China; winters to Thailand, Borneo and Moluccas

ORDER: CHARADRIIFORMES
FAMILY: RYNCHOPIDAE (Skimmers—3)

☐ **Black Skimmer** *Rynchops niger*
_____ *R. n. niger* — Coastal US and Mexico; winters to Panama
_____ *R. n. cinerascens* — Coasts and rivers of n S America to Bolivia and nw Argentina
_____ *R. n. intercedens* — E Brazil to Paraguay, Uruguay and ne Argentina

☐ **African Skimmer** *Rynchops flavirostris*

Major rivers, lakes and coasts of Africa south of the Sahara

☐ **Indian Skimmer** *Rynchops albicollis*

Rivers and lakes of Indian subcontinent and Mekong Delta

ORDER: CHARADRIIFORMES
FAMILY: STERCORARIIDAE (Jaegers and Skuas—7)

☐ **Chilean Skua** *Stercorarius chilensis*

Coasts of s Chile and s Argentina; ranges north to tropics

☐ **South Polar Skua** *Stercorarius maccormicki*

☐ **Brown Skua** *Stercorarius antarcticus*   Antarctica; ranges to n Atlantic, n Pacific and Indian oceans

____ *S. a. antarcticus*   Falkland Is. and se Argentina; winters off se South America
____ *S. a. hamiltoni*   Tristan da Cunha and Gough I.
____ *S. a. lonnbergi*   Antarctic Pen. and circumpolar subantarctic s ocean islands

☐ **Great Skua** *Stercorarius skua*

                                                   Arctic nw Eurasia; ranges to Mediterranean and n S America

☐ **Pomarine Jaeger** *Stercorarius pomarinus*

                                                   Circumpolar Arctic tundra; winters at sea in southern oceans

☐ **Parasitic Jaeger** *Stercorarius parasiticus*

                                                   Circumpolar Arctic tundra; winters at sea in southern oceans

☐ **Long-tailed Jaeger** *Stercorarius longicaudus*

____ *S. l. longicaudus*   N Scandinavia and Russia; winters to s S America and S Africa
____ *S. l. pallescens*   Arctic N Am and Siberia; winters to s S America and S Africa

ORDER: CHARADRIIFORMES
FAMILY: ALCIDAE (Auks, Murres and Puffins—23)

☐ **Dovekie** *Alle alle*

____ *A. a. alle*   Baffin I., Greenland and Iceland to Novaya Zemlya
____ *A. a. polaris*   Frans Josef Land to St. Lawrence I.; winters to Maine

☐ **Common Murre** *Uria aalge*

____ *U. a. aalge*   E N America, Greenland and Iceland to Norway and Baltic Sea
____ *U. a. albionis*   British Isles to w Iberian Peninsula; Helgoland
____ *U. a. hyperborea*   Svalbard and n Norway to Murmansk and Novaya Zemlya
____ *U. a. inornata*   Korea, Japan and Kamchatka to Bering Sea and Br. Columbia
____ *U. a. californica*   N Washington to s California

☐ **Thick-billed Murre** *Uria lomvia*

____ *U. l. lomvia*   Gulf of St. Lawrence to Greenland and Novaya Zemlya
____ *U. l. eleonorae*   E Taymyr Peninsula to New Siberian Islands (Russia)
____ *U. l. heckeri*   Wrangel I., Herald I. and n Chukotsk Peninsula
____ *U. l. arra*   N Japan and Aleutian Islands to se Alaska

☐ **Razorbill** *Alca torda*

____ *A. t. torda*   NE North America; Scandinavia to Murmansk and White Sea
____ *A. t. islandica*   Iceland and Br. Isles to France; winters to Mediterranean

☐ **Black Guillemot** *Cepphus grylle*

____ *C. g. mandtii*   Arctic e North America and Arctic n Palearctic
____ *C. g. arcticus*   Subarctic e N America and s Greenland to Br. Isles, Scandinavia
____ *C. g. islandicus*   Iceland
____ *C. g. faeroeensis*   Faeroe Islands
____ *C. g. grylle*   Baltic Sea

☐ **Pigeon Guillemot** *Cepphus columba*

____ *C. c. columba*   NE Siberia to Bering Sea and w Alaska
____ *C. c. snowi*   Kuril Islands
____ *C. c. kaiurka*   Komandorskiye Islands to w-central Aleutian Islands
____ *C. c. adiantus*   Central Aleutian Islands south to Washington
____ *C. c. eureka*   Oregon and California

☐ **Spectacled Guillemot** *Cepphus carbo*

                                                   Kamchatka and Sea of Okhotsk to Korea, Kuril Is. and n Japan

☐ **Marbled Murrelet** *Brachyramphus marmoratus*

                                                   W Aleutian Islands and Alaska to central California

☐ **Long-billed Murrelet** *Brachyramphus perdix*

☐ **Kittlitz's Murrelet** *Brachyramphus brevirostris*

Kamchatka Peninsula and Sea of Okhotsk to Hokkaido

☐ **Xantus' Murrelet** *Synthliboramphus hypoleucus*

Bering Sea to Gulf of Alaska; winters to Glacier Bay

_____ *S. h. scrippsi*

_____ *S. h. hypoleucus*

Channel Islands and islands off w coast of Baja California

San Benito I. and Guadalupe I. (off Baja California)

☐ **Craveri's Murrelet** *Synthliboramphus craveri*

Islands in Gulf of California; disperses north to s California

☐ **Ancient Murrelet** *Synthliboramphus antiquus*

_____ *S. a. antiquus*

_____ *S. a. microrhynchos*

E Asia, Aleutians and s Alaska; winters to California

Komandorskiye Islands; winters to Ryukyu Islands

☐ **Japanese Murrelet** *Synthliboramphus wumizusume*

Coasts and islands off e and s Japan and South Korea

☐ **Cassin's Auklet** *Ptychoramphus aleuticus*

_____ *P. a. aleuticus*

_____ *P. a. australis*

Aleutian Islands and Alaska to n Baja California

S Baja (San Benito to Asunción and San Roque islands)

☐ **Parakeet Auklet** *Aethia psittacula*

N Pacific and Bering Sea; winters to Japan and s California

☐ **Crested Auklet** *Aethia cristatella*

Breeds w Alaska and e Siberia; winters south to Japan

☐ **Whiskered Auklet** *Aethia pygmaea*

Breeds e Siberia and Aleutian Islands; winters south to Japan

☐ **Least Auklet** *Aethia pusilla*

Breeds e Siberia and w Alaska; winters south to n Japan

☐ **Rhinoceros Auklet** *Cerorhinca monocerata*

E Asia and w North America; winters to Baja California

☐ **Atlantic Puffin** *Fratercula arctica*

_____ *F. a. naumanni*

_____ *F. a. arctica*

_____ *F. a. grabae*

N Canada and Greenland to Spitsbergen and n Novaya Zemlya

Baffin I. to Maine, Scandinavia and s Novaya Zemlya

Faeroe Islands, s Scandinavia and British Isles to nw France

☐ **Horned Puffin** *Fratercula corniculata*

E Siberia and nw N America; winters to Japan and s California

☐ **Tufted Puffin** *Fratercula cirrhata*

Coasts and islands of ne Asia to Aleutians and s California

ORDER: PTEROCLIFORMES
FAMILY: PTEROCLIDAE (Sandgrouse—16)

☐ **Tibetan Sandgrouse** *Syrrhaptes tibetanus*

Inhospitable wastes of e Afghanistan to Tibetan plateau

☐ **Pallas' Sandgrouse** *Syrrhaptes paradoxus*

Sandy steppes of central Asia

☐ **Pin-tailed Sandgrouse** *Pterocles alchata*

_____ *P. a. alchata*

_____ *P. a. caudacutus*

Spain and s France; formerly Portugal

W Sahara to Middle East and w India

☐ **Namaqua Sandgrouse** *Pterocles namaqua*

Dry grasslands of southern Africa

☐ **Chestnut-bellied Sandgrouse** *Pterocles exustus*

_____ *P. e. exustus*

_____ *P. e. floweri†*

_____ *P. e. ellioti*

_____ *P. e. olivascens*

_____ *P. e. erlangeri*

Mauritania and Senegambia east to Sudan

Formerly Egypt. Extinct

SE Sudan to Eritrea, n Ethiopia and Somalia

S Ethiopia to Somalia, Kenya and n Tanzania

W and s Arabian Peninsula

____ *P. e. hindustan*   SE Iran to Pakistan and India

☐ **Spotted Sandgrouse** *Pterocles senegallus*

Deserts of sub-Saharan Africa and Arabian Pen. to w India

☐ **Black-bellied Sandgrouse** *Pterocles orientalis*
____ *P. o. orientalis*   Fuerteventura I., Iberian Peninsula and Morocco to w Iran
____ *P. o. arenarius*   Kazakstan to s Iran, Afghanistan, and nw China (nw Xinjiang)

☐ **Yellow-throated Sandgrouse** *Pterocles gutturalis*
____ *P. g. saturatior*   Ethiopia to Kenya, Tanzania and extreme n Zambia
____ *P. g. gutturalis*   S Zambia and Botswana to Transvaal and Cape Province

☐ **Crowned Sandgrouse** *Pterocles coronatus*
____ *P. c. coronatus*   Central Sahara to Mediterranean and Morocco east to Red Sea
____ *P. c. vastitas*   Sinai Peninsula and deserts of s Israel and Jordan
____ *P. c. atratus*   S Arabia, Iraq and s Iran to w Pakistan and Afghanistan
____ *P. c. saturatus*   Mountains of interior of Oman
____ *P. c. ladas*   N Pakistan (Kashmir) and adjacent India

☐ **Black-faced Sandgrouse** *Pterocles decoratus*
____ *P. d. ellenbecki*   NE Uganda to n Kenya, s Ethiopia and s Somalia
____ *P. d. decoratus*   Savanna and coastal dunes of se Kenya and e Tanzania
____ *P. d. loverridgei*   W Kenya and w Tanzania

☐ **Madagascar Sandgrouse** *Pterocles personatus*

Arid lowlands of Madagascar

☐ **Lichtenstein's Sandgrouse** *Pterocles lichtensteinii*
____ *P. l. targius*   Sahara and Sahel from Morocco and Mauritania to Chad
____ *P. l. lichtensteinii*   S Israel, Sinai, se Egypt to n Ethiopia, n Somalia and Socotra I.
____ *P. l. sukensis*   SE Sudan and s Ethiopia to central Kenya
____ *P. l. ingramsi*   S Yemen (Hadramaut)
____ *P. l. arabicus*   S Arabia to s Iran, s Afghanistan and Pakistan

☐ **Double-banded Sandgrouse** *Pterocles bicinctus*
____ *P. b. ansorgei*   SW Angola
____ *P. b. bicinctus*   Namibia, Botswana and nw Cape Province
____ *P. b. multicolor*   Mozambique, Malawi and Zambia to Transvaal

☐ **Four-banded Sandgrouse** *Pterocles quadricinctus*

Savanna and thornscrub of sub-Saharan Africa

☐ **Painted Sandgrouse** *Pterocles indicus*

Arid rocky lowlands of e Pakistan and peninsular India

☐ **Burchell's Sandgrouse** *Pterocles burchelli*

SE Angola to Namibia, Botswana and n Cape Province

ORDER: COLUMBIFORMES
FAMILY: COLUMBIDAE (Pigeons and Doves—308)

☐ **Rock Pigeon** *Columba livia*
____ *C. l. livia*   Western Palearctic
____ *C. l. atlantis*   Madeira, Azores and Cape Verde Islands
____ *C. l. canariensis*   Canary Islands and islands off Morocco
____ *C. l. gymnocycla*   Mauritania, Mali and Ghana; coastal Senegambia and Guinea
____ *C. l. targia*   Central Sahara to central Sudan
____ *C. l. dakhlae*   Egypt (Dakhla and Kharga oases)
____ *C. l. butleri*   Red Sea Province and Egyptian Sudan
____ *C. l. schimperi*   Nile Valley to Khartoum; Red Sea Hills of e Egypt to n Eritrea
____ *C. l. palaestinae*   Palestine, Sinai and Arabia to Aden and Oman

____ *C. l. gaddi*	Iran to Azerbaijan, Transcaspia, Afghanistan and Uzbekistan
____ *C. l. neglecta*	Mountains of central Asia
____ *C. l. intermedia*	Peninsular India and Sri Lanka
____ *C. l. nigricans*	Mongolia and n China (Shanxi, Jilin and Gansu)

☐ **Hill Pigeon** *Columba rupestris*

____ *C. r. turkestanica*	Altai Mts. to Turkestan, Tibet and n Himalayas
____ *C. r. rupestris*	W Mongolia to Mongolia, e Tibet, s China and Korea

☐ **Snow Pigeon** *Columba leuconota*

____ *C. l. leuconota*	Himalayas from w Afghanistan to Sikkim
____ *C. l. gradaria*	Mts. of e Tibet to sw China (Yunnan) and extreme n Myanmar

☐ **Speckled Pigeon** *Columba guinea*

____ *C. g. guinea*	Senegambia to Ethiopia, Somalia, Uganda, Kenya and Tanzania
____ *C. g. phaeonota*	SW Angola to Zimbabwe and Cape Province

☐ **White-collared Pigeon** *Columba albitorques*

Highlands of Ethiopia and Eritrea

☐ **Stock Dove** *Columba oenas*

____ *C. o. oenas*	W Europe and nw Africa to Caspian Sea and Kazakstan
____ *C. o. yarkandensis*	Uzbekistan and Tajikistan to Tien Shan and e Xinjiang

☐ **Pale-backed Pigeon** *Columba eversmanni*

NE Iran to Siberia, nw India and extreme w China

☐ **Somali Pigeon** *Columba oliviae*

Arid coastal hills of n Somalia

☐ **Common Wood-Pigeon** *Columba palumbus*

____ *C. p. azorica*	E and central Azores
____ *C. p. maderensis†*	Formerly mountains of Madeira. Probably extinct
____ *C. p. excelsa*	Morocco, Algeria and Tunisia
____ *C. p. palumbus*	Europe to w Siberia, e Turkey and Iraq; winters to n Africa
____ *C. p. iranica*	S Transcaspia to Iran
____ *C. p. casiotis*	Kazakstan to n Afghanistan, n Pakistan, nw India and Nepal

☐ **Trocaz Pigeon** *Columba trocaz*

Laurel forests of Madeira

☐ **Bolle's Pigeon** *Columba bollii*

W Canary Islands (La Palma, Gomera, El Hierro and Tenerife)

☐ **Afep Pigeon** *Columba unicincta*

Equatorial forests of w and central Africa

☐ **Laurel Pigeon** *Columba junoniae*

W Canary Islands (La Palma, La Gomera and Tenerife)

☐ **Rameron Pigeon** *Columba arquatrix*

Ethiopia to e Zaire, Tanzania and South Africa; w Angola

☐ **Cameroon Pigeon** *Columba sjostedti*

Highland forests of e Nigeria and sw Cameroon

☐ **Maroon Pigeon** *Columba thomensis*

Forests of São Tomé I. (Gulf of Guinea)

☐ **Delegorgue's Pigeon** *Columba delegorguei*

____ *C. d. sharpei*	SE Sudan to Uganda, Kenya, Tanzania and Zanzibar
____ *C. d. delegorguei*	Malawi to e Zimbabwe, Mozambique and South Africa

☐ **Bronze-naped Pigeon** *Columba iriditorques*

Sierra Leone to nw Angola, Zaire, sw Uganda and Rwanda

☐ **São Tomé Pigeon** *Columba malherbii*

São Tomé, Príncipe and Pagalu (Gulf of Guinea)

☐ **Lemon Dove** *Columba larvata*

____ *C. l. hypoleuca*	Sierra Leone to Liberia, se Nigeria, Cameroon, Gabon; Bioko
____ *C. l. principalis*	Príncipe (Gulf of Guinea)

_____ *C. l. bronzina*   Ethiopia and se Sudan (Boma Hills)
_____ *C. l. larvata*   S Sudan to Uganda, w Tanzania, Malawi and South Africa

☐ **Forest Dove** *Columba simplex*

São Tomé (Gulf of Guinea)

☐ **Comoro Pigeon** *Columba pollenii*

Mainly high elevation evergreen forests of Comoro Islands

☐ **Speckled Wood-Pigeon** *Columba hodgsonii*

Montane forests of Kashmir to India, w-c China and Myanmar

☐ **White-naped Pigeon** *Columba albinucha*

W Cameroon; e Zaire, w Uganda and w Rwanda

☐ **Ashy Wood-Pigeon** *Columba pulchricollis*

India to w China, Tibet, Myanmar, nw Thailand and Taiwan

☐ **Nilgiri Wood-Pigeon** *Columba elphinstonii*

SW India (Western Ghats)

☐ **Ceylon Wood-Pigeon** *Columba torringtoni*

Sri Lanka

☐ **Pale-capped Pigeon** *Columba punicea*

S Tibet to e India, Myanmar, Thailand and Hainan I. (s China)

☐ **Silvery Wood-Pigeon** *Columba argentina*

Sumatra, Borneo and adjacent islands

☐ **Andaman Wood-Pigeon** *Columba palumboides*

Andaman Islands and Nicobar Islands

☐ **Japanese Wood-Pigeon** *Columba janthina*
_____ *C. j. janthina*   Small islands sw of South Korea to Ryukyu Islands
_____ *C. j. nitens*   Ogasawara (Bonin Islands) and Iwo (Volcano Islands)

☐ **Metallic Pigeon** *Columba vitiensis*
_____ *C. v. griseogularis*   Philippines, Sulu Archipelago and islands off n Borneo
_____ *C. v. anthracina*   Palawan, Calauit and islands off north Borneo
_____ *C. v. metallica*   Lesser Sundas
_____ *C. v. halmaheira*   Banggai, Sulas, Kai, Moluccas to New Guinea and Solomons
_____ *C. v. leopoldi*   Vanuatu
_____ *C. v. hypoenochroa*   New Caledonia, Isle of Pines and Loyalty Islands
_____ *C. v. vitiensis*   Fiji Islands
_____ *C. v. castaneiceps*   W Samoa (Savai'i, Apolima, Manono and Upolu)

☐ **White-headed Pigeon** *Columba leucomela*

Coasts and forested islands of e Australia

☐ **Yellow-legged Pigeon** *Columba pallidiceps*

Bismarck Archipelago and Solomon Islands

☐ **White-crowned Pigeon** *Patagioenas leucocephala*

S Florida and offshore islands from West Indies to nw Panama

☐ **Scaly-naped Pigeon** *Patagioenas squamosa*

Greater Antilles, Lesser Antilles and Netherlands Antilles

☐ **Scaled Pigeon** *Patagioenas speciosa*

Tropical s Mexico to Brazil and ne Argentina

☐ **Picazuro Pigeon** *Patagioenas picazuro*
_____ *P. p. marginalis*   NE Brazil (Piauí, Bahia and Goiás)
_____ *P. p. picazuro*   E Brazil (Pernambuco) to Bolivia and s-central Argentina

☐ **Bare-eyed Pigeon** *Patagioenas corensis*

Arid coastal ne Colombia to n Venezuela and adjacent islands

☐ **Spot-winged Pigeon** *Patagioenas maculosa*
_____ *P. m. albipennis*   S Peru to w Bolivia and extreme nw Argentina
_____ *P. m. maculosa*   S Bolivia to Paraguay, se Brazil, Uruguay and s-c Argentina

☐ **Band-tailed Pigeon** *Patagioenas fasciata*
_____ *P. f. fasciata*   W North America from sw Canada to Nicaragua
_____ *P. f. monilis*   San Pedro Mártir of n Baja California

____ *P. f. vioscae*   Mountains of extreme s Baja California (Sierra de la Laguna)
____ *P. f. letonai*   Honduras and El Salvador (including Volcán de San Miguel)
____ *P. f. parva*   N Nicaragua
____ *P. f. crissalis*   Costa Rica and w Panama
____ *P. f. roraimae*   *Tepuis* of s Venezuela (Mt. Roraima)
____ *P. f. albilinea*   Colombia to Venezuela, nw Brazil, e Bolivia and n Argentina

☐ **Chilean Pigeon** *Patagioenas araucana*

Araucaria woodlands of c and s Chile and adjacent Argentina

☐ **Ring-tailed Pigeon** *Patagioenas caribaea*

Jamaica

☐ **Pale-vented Pigeon** *Patagioenas cayennensis*
____ *P. c. pallidicrissa*   Gulf lowlands of se Mexico to n Colombia
____ *P. c. andersoni*   SE Colombia and e Ecuador to Brazil north of the Amazon
____ *P. c. tobagensis*   Trinidad and Tobago
____ *P. c. cayennensis*   Guyana, Suriname and French Guiana
____ *P. c. sylvestris*   E Peru to Brazil s of the Amazon, Paraguay and n Argentina

☐ **Red-billed Pigeon** *Patagioenas flavirostris*
____ *P. f. restricta*   W Mexico (s Sonora to Sinaloa)
____ *P. f. madrensis*   Tres Marías Islands (off w Mexico)
____ *P. f. flavirostris*   S Texas (Rio Grande Valley) to e Costa Rica
____ *P. f. minima*   Lowlands of Costa Rica (Gulf of Nicoya region)

☐ **Peruvian Pigeon** *Patagioenas oenops*

N Peru (subtropical Marañón Valley) and adjacent s Ecuador

☐ **Plain Pigeon** *Patagioenas inornata*
____ *P. i. inornata*   Cuba, Isle of Pines and Hispaniola
____ *P. i. exigua*   Jamaica
____ *P. i. wetmorei*   Puerto Rico

☐ **Plumbeous Pigeon** *Patagioenas plumbea*
____ *P. p. delicata*   E Colombia to Venezuela, the Guianas, n Brazil and n Bolivia
____ *P. p. chapmani*   NW Ecuador
____ *P. p. pallescens*   Small tributaries of Amazon from Rio Purús to Pará
____ *P. p. baeri*   Central Brazil (Goiás and nw Minas Gerais)
____ *P. p. plumbea*   SE Brazil and Paraguay

☐ **Short-billed Pigeon** *Patagioenas nigrirostris*

Lowland rainforests of se Mexico to nw Colombia (Chocó)

☐ **Ruddy Pigeon** *Patagioenas subvinacea*
____ *P. s. subvinacea*   Subtropical Costa Rica and Panama
____ *P. s. berlepschi*   Pacific slope of se Panama to sw Ecuador
____ *P. s. zuliae*   NE Colombia and w Venezuela
____ *P. s. peninsularis*   NE Venezuela (Paría Peninsula)
____ *P. s. purpureotincta*   SE Colombia to Venezuela and the Guianas
____ *P. s. bogotensis*   Andes of Colombia to ne Bolivia and Amazonian Brazil

☐ **Dusky Pigeon** *Patagioenas goodsoni*

Lowland rainforests of w Colombia to nw Ecuador

☐ **Pink Pigeon** *Nesoenas mayeri*

Forests of sw Mauritius (on verge of extinction)

☐ **Eurasian Turtle-Dove** *Streptopelia turtur*
____ *S. t. turtur*   Azores, Canary Is. and Europe to w Siberia and Kazakstan
____ *S. t. arenicola*   Balearic Islands and nw Africa to Iran and extreme w China
____ *S. t. hoggara*   S Sahara (Aïr Massif and Hoggar Mountains)
____ *S. t. rufescens* (*isabellina*)   Egypt (Dakhla and Kharga oases) and n Sudan (Faiyûm)

☐ **Dusky Turtle-Dove** *Streptopelia lugens*

Montane forests of e Africa and sw Arabia

☐ **Adamawa Turtle-Dove** *Streptopelia hypopyrrha*

	Highlands of e Nigeria, n Cameroon and extreme sw Chad
☐ **Oriental Turtle-Dove** *Streptopelia orientalis*	
____ S. o. meena	SW Siberia to Iran, Afghanistan, Kashmir and Nepal
____ S. o. orientalis	Central Siberia to China, Korea, Japan and Kuril Islands
____ S. o. stimpsoni	Ryukyu Islands
____ S. o. orii	Taiwan
____ S. o. erythrocephala	Peninsular India
____ S. o. agricola	NE India to Myanmar and s-c China (w Yunnan and Hainan I.)
☐ **Island Collared-Dove** *Streptopelia bitorquata*	
____ S. b. dusumieri	Philippines and Sulu Archipelago; vagrant to n Borneo
____ S. b. bitorquata	Java, Bali and Lombok to Sumbawa, Flores, Solor and Timor
☐ **Eurasian Collared-Dove** *Streptopelia decaocto*	
____ S. d. decaocto	Europe to Middle East, India, Sri Lanka, w China and Korea
____ S. d. xanthocycla	Myanmar (Shan States) to s China (Yunnan) and e China
☐ **African Collared-Dove** *Streptopelia roseogrisea*	
____ S. r. roseogrisea	SW Mauritania and Senegambia to s Sudan and w Ethiopia
____ S. r. arabica	Coastal Eritrea, Ethiopia and Somalia to Arabia
☐ **White-winged Collared-Dove** *Streptopelia reichenowi*	
	Borassus palm regions of se Ethiopia and sw Somalia
☐ **African Mourning Dove** *Streptopelia decipiens*	
____ S. d. shelleyi	Mauritania and Senegambia to s Niger and central Nigeria
____ S. d. logonensis	Lake Chad basin to s Sudan, e Zaire and n and w Uganda
____ S. d. decipiens	E Sudan (Darfur) to Ethiopia and nw Somalia
____ S. d. elegans	S Ethiopia to s Somalia and e Kenya
____ S. d. perspicillata	W Kenya and central Tanzania
____ S. d. ambigua	E Angola to se Zaire, Zambia, Malawi and Limpopo Valley
☐ **Red-eyed Dove** *Streptopelia semitorquata*	
	Africa south of the Sahara and sw Arabia
☐ **Ring-necked Dove** *Streptopelia capicola*	
____ S. c. electa	W Ethiopia
____ S. c. somalica	E Ethiopia, Somalia and n Kenya south to Uaso Nyiro River
____ S. c. tropica	Cent. Kenya to Angola, Zimbabwe, South Africa and Zanzibar
____ S. c. onguati	SW Angola and n Namibia
____ S. c. damarensis	Namibia, Botswana and sw Zimbabwe
____ S. c. capicola	W Cape Province
☐ **Vinaceous Dove** *Streptopelia vinacea*	
	Dry savanna of sub-Saharan Africa
☐ **Red Collared-Dove** *Streptopelia tranquebarica*	
____ S. t. tranquebarica	Sind, Punjab and w Nepal south through peninsular India
____ S. t. humilis	Tibet to Myanmar, Thailand, SE Asia and n Philippines
☐ **Madagascar Turtle-Dove** *Streptopelia picturata*	
____ S. p. rostrata	Seychelles Islands
____ S. p. aldabrana	Amirante Islands
____ S. p. copperingi	Aldabra, Cosmoledo Atoll and Îles Glorieuses
____ S. p. comorensis	Comoro Islands
____ S. p. picturata	Madagascar
☐ **Spotted Dove** *Streptopelia chinensis*	
____ S. c. suratensis	Pakistan, Nepal and India to Sri Lanka, Bhutan and Assam
____ S. c. chinensis	Myanmar to e China and Taiwan
____ S. c. tigrina	N India to Malaya, Indochina, Philippines, Gr. and Lesser Sundas
☐ **Laughing Dove** *Streptopelia senegalensis*	

____	*S. s. phoenicophila*	Oases south of Atlas Mts. in Morocco, Algeria and Tunisia
____	*S. s. aegyptiaca*	Nile Valley (Suez Canal and delta south to Wadi Halfa)
____	*S. s. sokotrae*	Socotra (off ne Somalia)
____	*S. s. senegalensis (thome)*	Africa south of the Sahara and Arabia
____	*S. s. cambayensis*	E Arabia to s Iran, Indian subcontinent, w China and Andamans

☐ **Barred Cuckoo-Dove** *Macropygia unchall*

____	*M. u. tusalia*	Himalayas (Kashmir to Assam, sw China and Myanmar)
____	*M. u. minor*	Mts. of se China to Vietnam, Laos, n Thailand and Hainan
____	*M. u. unchall*	Mts. of Malay Peninsula, Sumatra, Java, Lombok and Flores

☐ **Brown Cuckoo-Dove** *Macropygia phasianella*

Coastal e Australia (Cape York to e Victoria)

☐ **Dusky Cuckoo-Dove** *Macropygia magna*

____	*M. m. macassariensis*	SW Sulawesi, Tanakeke and Salayar islands
____	*M. m. longa*	E Lesser Sundas (Tanahjampea and Kalatoa)
____	*M. m. magna*	Timor, Wetar and adjacent e Lesser Sundas
____	*M. m. timorlaoensis*	Tanimbar Islands (Yamdena, Larat and Selaru)

☐ **Slender-billed Cuckoo-Dove** *Macropygia amboinensis*

____	*M. a. sanghirensis*	Sangihe, Siau, Tahulandang, Ruang and Talaud islands
____	*M. a. albicapilla*	Sulawesi, Banggai, Tukangbesi and adjacent islands
____	*M. a. batchianensis*	N Moluccas
____	*M. a. amboinensis*	S Moluccas (Buru, Seram, Ambon and Seram Laut)
____	*M. a. keyensis*	Kai Islands (se Moluccas)
____	*M. a. doreya*	NW New Guinea and w Papuan islands
____	*M. a. maforensis*	Numfor I. (Geelvink Bay off n New Guinea)
____	*M. a. griseinucha*	Meos Num I. (Geelvink Bay off n New Guinea)
____	*M. a. kerstingi*	N New Guinea (Mamberano to Astrolabe Bay) and Yapen I.
____	*M. a. goldiei*	Coastal s New Guinea (Merauke region to Milne Bay)
____	*M. a. meeki*	Manam I. (off ne New Guinea)
____	*M. a. carteretia*	Bismarck Archipelago (except New Hanover) and Lihir Is.
____	*M. a. huskeri*	New Hanover (Bismarck Archipelago)
____	*M. a. cinereiceps*	D'Entrecasteaux Archipelago
____	*M. a. cunctata*	Louisiade Archipelago

☐ **Andaman Cuckoo-Dove** *Macropygia rufipennis*

Andaman and Nicobar islands

☐ **Philippine Cuckoo-Dove** *Macropygia tenuirostris*

Philippines, Taiwan and Lan-yü I.

☐ **Ruddy Cuckoo-Dove** *Macropygia emiliana*

____	*M. e. borneensis*	N Borneo
____	*M. e. hypopercna*	Simeulue I. (off nw Sumatra)
____	*M. e. modiglianii*	Nias I. (off w Sumatra)
____	*M. e. elassa*	Mentawi Islands (Siberut, Sipura and Pagai)
____	*M. e. cinnamomea*	Enggano I. (off w Sumatra)
____	*M. e. emiliana*	Krakatau, Java, Lombok, Sumbawa, Flores and Paloe islands
____	*M. e. megala*	Kangean Islands (off ne Java)

☐ **Black-billed Cuckoo-Dove** *Macropygia nigrirostris*

New Guinea, Bismarck Arch. and D'Entrecasteaux Arch.

☐ **Mackinlay's Cuckoo-Dove** *Macropygia mackinlayi*

____	*M. m. goodsoni*	Admiralty Is., St. Matthias Is. and w New Britain
____	*M. m. krakari*	Karkar I. (Papua New Guinea)
____	*M. m. arossi*	Solomon Islands
____	*M. m. mackinlayi*	Santa Cruz Islands, Banks Group and Vanuatu

☐ **Little Cuckoo-Dove** *Macropygia ruficeps*

____	*M. r. assimilis*	S Myanmar, nw Thailand and extreme sw China (Yunnan)
____	*M. r. engelbachi*	NW Vietnam (w Tonkin and n Laos)

____	*M. r. malayana*	Malay Peninsula
____	*M. r. simalurensis*	Simeulue I. (off Sumatra)
____	*M. r. sumatrana*	Sumatra
____	*M. r. nana*	Borneo and Sibatik I.
____	*M. r. ruficeps*	Java and Bali
____	*M. r. orientalis*	Lombok, Sumbawa, Komodo, Flores, Sumba, Pantar and Timor

☐ **Great Cuckoo-Dove** *Reinwardtoena reinwardtii*

____	*R. r. reinwardtii*	Moluccas
____	*R. r. griseotincta*	W Papuan islands, New Guinea and D'Entrecasteaux Arch.
____	*R. r. brevis*	Biak I. (off n New Guinea)

☐ **Pied Cuckoo-Dove** *Reinwardtoena browni*

Admiralty Islands and Bismarck Archipelago

☐ **Crested Cuckoo-Dove** *Reinwardtoena crassirostris*

Solomon Islands (Bougainville to San Cristobal)

☐ **White-faced Cuckoo-Dove** *Turacoena manadensis*

Sulawesi, Togian, Butung, Banggai and Sula islands

☐ **Slaty Cuckoo-Dove** *Turacoena modesta*

E Lesser Sundas (Timor and Wetar)

☐ **Emerald-spotted Wood-Dove** *Turtur chalcospilos*

Woodland and thornscrub of east and southern Africa

☐ **Black-billed Wood-Dove** *Turtur abyssinicus*

Arid scrub and woodlands of sub-Saharan Africa

☐ **Blue-spotted Wood-Dove** *Turtur afer*

Africa south of the Sahara, Zanzibar and Pemba I.

☐ **Tambourine Dove** *Turtur tympanistria*

Africa south of the Sahara, Bioko and Comoro Islands

☐ **Blue-headed Wood-Dove** *Turtur brehmeri*

____	*T. b. infelix*	Coastal Guinea and Sierra Leone to Cameroon
____	*T. b. brehmeri*	S Cameroon to n Congo, e Zaire and extreme nw Angola

☐ **Namaqua Dove** *Oena capensis*

____	*O. c. capensis*	Africa south of the Sahara, Socotra and Arabia
____	*O. c. aliena*	Madagascar

☐ **Emerald Dove** *Chalcophaps indica*

____	*C. i. indica*	India to Malaysia, Philippines, Indonesia and w Papuan islands
____	*C. i. robinsoni*	Sri Lanka
____	*C. i. natalis*	Christmas I. (Indian Ocean)
____	*C. i. minima*	Numfor, Biak and Mios Num islands (n New Guinea)
____	*C. i. augusta*	Nicobar Islands
____	*C. i. chrysochlora*	E Lesser Sundas, e Australia, New Guinea and adj. islands
____	*C. i. longirostris*	N Australia (n Western Australia and Northern Territory)
____	*C. i. sandwichensis*	Santa Cruz Is., Banks Is., Vanuatu and New Caledonia

☐ **Stephan's Dove** *Chalcophaps stephani*

____	*C. s. wallacei*	Sulawesi and Sula Islands (Taliabu I.)
____	*C. s. stephani*	Kai Is., e New Guinea, Admiralty Is. and Bismarck Arch.
____	*C. s. mortoni*	Solomon Is. (Bougainville to San Cristobal and Santa Anna)

☐ **New Guinea Bronzewing** *Henicophaps albifrons*

____	*H. a. albifrons*	W Papuan islands, New Guinea and Yapen I.
____	*H. a. schlegeli*	Aru Islands

☐ **New Britain Bronzewing** *Henicophaps foersteri*

S Bismarck Archipelago (Umboi, New Britain and Lolobau)

☐ **Common Bronzewing** *Phaps chalcoptera*

Australia and Tasmania

☐ **Brush Bronzewing** *Phaps elegans*

| ____ | *P. e. occidentalis* | SW Australia (Dongara to Point Culver) |
| ____ | *P. e. elegans* | SE Queensland to s-central Australia and Tasmania |

☐ **Flock Bronzewing** *Phaps histrionica*

NW Australia to w Queensland and nw New South Wales

☐ **Crested Pigeon** *Geophaps lophotes*
____ *G. l. whitlocki* — Arid woodlands and plains of western Australia
____ *G. l. lophotes* — Central and e Australia

☐ **Spinifex Pigeon** *Geophaps plumifera*
____ *G. p. ferruginea* — W Australia (De Grey R. to Gascoyne R. and Carnarvon Range)
____ *G. p. plumifera* — Western Australia (Edgar Range) to w Northern Territory
____ *G. p. leucogaster* — Central Australia to nw Queensland and ne South Australia

☐ **Squatter Pigeon** *Geophaps scripta*
____ *G. s. peninsulae* — NE Queensland (Cape York Peninsula to Burdekin River)
____ *G. s. scripta* — Central Queensland and extreme n New South Wales

☐ **Partridge Pigeon** *Geophaps smithii*
____ *G. s. blaauwi* — NE Western Australia (Kimberley Region)
____ *G. s. smithii* — NE Western Australia (Cockatoo Springs) and n N Territory

☐ **Chestnut-quilled Rock-Pigeon** *Petrophassa rufipennis*

W escarpment of Arnhem Land (n Northern Territory)

☐ **White-quilled Rock-Pigeon** *Petrophassa albipennis*
____ *P. a. albipennis* — NE Western Australia and n Northern Territory
____ *P. a. boothi* — N Northern Territory (Stokes Range to upper Baines River)

☐ **Diamond Dove** *Geopelia cuneata*

Arid interior of Australia

☐ **Zebra Dove** *Geopelia striata*

S Myanmar to Malaysia, Sumatra and Java

☐ **Peaceful Dove** *Geopelia placida*
____ *G. p. papua* — Savanna of s New Guinea (Merauke to Port Moresby)
____ *G. p. placida* — N and e Australia
____ *G. p. clelandi* — N-central Western Australia (Pilbara region)

☐ **Barred Dove** *Geopelia maugei*

SE Moluccas and Lesser Sundas

☐ **Bar-shouldered Dove** *Geopelia humeralis*
____ *G. h. gregalis* — Lowlands of coastal se New Guinea
____ *G. h. humeralis* — Lowlands of n and e Australia
____ *G. h. headlandi* — N-central Western Australia (Pilbara region)

☐ **Wonga Pigeon** *Leucosarcia melanoleuca*

Coastal forests of e Australia (cent. Queensland to se Victoria)

☐ **Mourning Dove** *Zenaida macroura*
____ *Z. m. marginella* — Br. Columbia to Baja California, w US and s-central Mexico
____ *Z. m. carolinensis* — E US, Bahamas and Bermuda
____ *Z. m. macroura* — Cuba, Isle of Pines, Hispaniola, Puerto Rico and Jamaica
____ *Z. m. clarionensis* — Isla Clarión (Revillagigedo Islands off w Mexico)
____ *Z. m. turturilla* — Costa Rica and w Panama

☐ **Socorro Dove** *Zenaida graysoni*

Formerly Socorro I. (off w Mexico); small captive population

☐ **Eared Dove** *Zenaida auriculata*
____ *Z. a. rubripes* — L Antilles, Trinidad and c Colombia to Venezuela and n Brazil
____ *Z. a. hypoleuca* — Arid littoral of w Ecuador and w Peru
____ *Z. a. caucae* — W Colombia (Cauca Valley)
____ *Z. a. antioquiae* — N-central Andes of Colombia (Antioquia)

____	*Z. a. ruficauda*	E Andes of Colombia to w Venezuela (Mérida)
____	*Z. a. vinaceorufa*	Netherlands Antilles (Curaçao, Aruba and Bonaire)
____	*Z. a. jessieae*	Bank of lower Amazon near Santarém
____	*Z. a. marajoensis*	Marajó and Mexiana islands in estuary of the Amazon
____	*Z. a. noronha*	NE Brazil (Maranhão, Piauí, Bahia); Fernando de Noronha I.
____	*Z. a. virgata*	Bolivia to c Brazil, Uruguay and Argentina to Tierra del Fuego
____	*Z. a. auriculata*	Central Chile (Atacama to Llanquihue) and w-c Argentina

☐ **Zenaida Dove** *Zenaida aurita*

____	*Z. a. salvadorii*	Coastal n Yucatán Pen., Cozumel, Holbox and Isla Mujeres
____	*Z. a. zenaida*	Bahamas, Greater Antilles and Virgin Islands
____	*Z. a. aurita*	Lesser Antilles (Anguilla to Grenada)

☐ **Galapagos Dove** *Zenaida galapagoensis*

____	*Z. g. galapagoensis*	Drier parts of Galapagos Islands (except range of *exsul*)
____	*Z. g. exsul*	Galapagos Islands (Culpepper and Wenman)

☐ **White-winged Dove** *Zenaida asiatica*

____	*Z. a. mearnsi*	Arid s California, N Mex., Arizona, w Mexico; Tres Marías Is.
____	*Z. a. asiatica*	Arid s Texas to Nicaragua; W Indies
____	*Z. a. australis*	W Costa Rica to w Panama

☐ **Pacific Dove** *Zenaida meloda*

Arid tropical sw Ecuador to n Chile (Coquimbo)

☐ **Common Ground-Dove** *Columbina passerina*

____	*C. p. passerina*	Coastal se US (South Carolina to Florida and se Texas)
____	*C. p. pallescens*	Arid sw US to Guatemala and Belize
____	*C. p. socorroensis*	Socorro I. (Revillagigedo Islands off w Mexico)
____	*C. p. neglecta*	Honduras to Costa Rica and Panama
____	*C. p. bahamensis*	Bermuda and Bahamas (except Inagua I.)
____	*C. p. exigua*	Great Inagua I. (s Bahamas) and Mona I. (Puerto Rico)
____	*C. p. insularis*	Cuba, Isle of Pines, Cayman Is., Hispaniola and adj. islands
____	*C. p. jamaicensis*	Jamaica
____	*C. p. navassae*	Navassa I. (off sw Hispaniola)
____	*C. p. portoricensis*	Puerto Rico, Mona, Culebra and Virgin is. (except St. Croix)
____	*C. p. nigrirostris*	St. Croix and n Lesser Antilles
____	*C. p. trochila*	Martinique (Lesser Antilles)
____	*C. p. antillarum*	S Lesser Antilles (St. Lucia and Barbados to Grenada)
____	*C. p. albivitta*	N Colombia, n Venezuela, Netherlands Antilles and Trinidad
____	*C. p. parvula*	Central Colombia (upper Magdalena Valley)
____	*C. p. nana*	W Colombia (Cauca Valley and arid upper Dagua Valley)
____	*C. p. quitensis*	Central Ecuador (Río Guaillabamba to Riobamba)
____	*C. p. griseola*	Extreme s Venezuela to the Guianas and e Brazil

☐ **Plain-breasted Ground-Dove** *Columbina minuta*

____	*C. m. interrupta*	SE Mexico to Belize, Guatemala and Nicaragua
____	*C. m. elaeodes*	Costa Rica to w-central Colombia
____	*C. m. minuta*	E Colombia to Venezuela, Guianas, s Brazil and ne Argentina
____	*C. m. amazilia*	Arid coastal sw Ecuador and Peru (south to Lima)

☐ **Ecuadorian Ground-Dove** *Columbina buckleyi*

Arid littoral of w Ecuador and extreme nw Peru (Tumbes)

☐ **Ruddy Ground-Dove** *Columbina talpacoti*

____	*C. t. eluta*	Coastal w Mexico (n Sinaloa to Chiapas)
____	*C. t. rufipennis*	SE Mexico to Colombia, n Venezuela, Trinidad and Tobago
____	*C. t. caucae*	W Colombia (Cauca Valley)
____	*C. t. talpacoti*	Guianas and Brazil to e Bolivia, Paraguay and c Argentina

☐ **Picui Ground-Dove** *Columbina picui*

_____ *C. p. strepitans* NE Brazil (Maranhão, Piauí, Ceará and Bahia)

_____ *C. p. picui* E Peru to Bolivia, Paraguay, s Argentina, Chile and s Brazil

☐ **Croaking Ground-Dove** *Columbina cruziana*

Arid w Ecuador, Peru and n Chile

☐ **Blue-eyed Ground-Dove** *Columbina cyanopis*

Cerrado of s-central Brazil

☐ **Inca Dove** *Columbina inca*

Semiarid sw US to nw Costa Rica

☐ **Scaled Dove** *Columbina squammata*

_____ *C. s. ridgwayi* Coastal ne Colombia to Venezuela, Margarita I. and Trinidad

_____ *C. s. squammata* C and e Brazil to Bolivia, Paraguay and ne Argentina

☐ **Blue Ground-Dove** *Claravis pretiosa*

SE Mexico (San Luis Potosí) to n Argentina and s Brazil

☐ **Purple-winged Ground-Dove** *Claravis godefrida*

SE Brazil (s Bahia) to e Paraguay and ne Argentina (Misiones)

☐ **Maroon-chested Ground-Dove** *Claravis mondetoura*

_____ *C. m. ochoterena* Mountains of se Mexico (Veracruz to Chiapas)

_____ *C. m. salvini* Guatemala, El Salvador and Honduras

_____ *C. m. umbrina* Costa Rica

_____ *C. m. pulchra* W Panama

_____ *C. m. mondetoura* Andes of Colombia to n Venezuela and e Ecuador

_____ *C. m. inca* Andes of Peru and w-central Bolivia

☐ **Bare-faced Ground-Dove** *Metriopelia ceciliae*

_____ *M. c. ceciliae* Andes of w Peru

_____ *M. c. obsoleta* N Peru (upper Marañón Valley)

_____ *M. c. zimmeri* Andes of extreme s Peru to Bolivia, nw Argentina and n Chile

☐ **Bare-eyed Ground-Dove** *Metriopelia morenoi*

Andes of nw Argentina

☐ **Black-winged Ground-Dove** *Metriopelia melanoptera*

_____ *M. m. saturatior* Andes of sw Colombia and Ecuador

_____ *M. m. melanoptera* Andes of Peru to s Argentina, s Chile and Tierra del Fuego

☐ **Golden-spotted Ground-Dove** *Metriopelia aymara*

Andes of s Peru to nw Argentina and n Chile

☐ **Long-tailed Ground-Dove** *Uropelia campestris*

W-c Brazil (Mato Grosso) and adjacent Bolivia

☐ **White-tipped Dove** *Leptotila verreauxi*

_____ *L. v. capitalis* Tres Marías Islands (off w Mexico)

_____ *L. v. angelica* S Texas and coastal Mexico south to Guerrero and Veracruz

_____ *L. v. fulviventris* SE Mexico and Yucatán Pen. to e Guatemala and Belize

_____ *L. v. bangsi* W Guatemala, El Salvador, Nicaragua and w Honduras

_____ *L. v. nuttingi* W shore of Lake Nicaragua and Isla de Ométepe

_____ *L. v. riottei* Caribbean slope of Costa Rica

_____ *L. v. verreauxi* Extreme sw Nicaragua to Colombia, Venezuela and offshore is.

_____ *L. v. zapluta* Trinidad

_____ *L. v. tobagensis* Tobago

_____ *L. v. decolor* W Andes of Colombia to n Peru (Marañón Valley and Trujillo)

_____ *L. v. brasiliensis* The Guianas and n Brazil south to north bank of the Amazon

_____ *L. v. approximans* NE Brazil (Piauí and Ceará to n Bahia)

_____ *L. v. decipiens* Lowlands of e Peru, e Bolivia and w Brazil s of the Amazon

_____ *L. v. chalcauchenia* S Bolivia to Paraguay, s Brazil, Uruguay and n-c Argentina

☐ **White-faced Dove** *Leptotila megalura*

Andes of Bolivia and nw Argentina

☐ **Gray-fronted Dove** *Leptotila rufaxilla*

_____ *L. r. pallidipectus* Tropical e Colombia and adjacent w Venezuela

_____ *L. r. dubusi* SE Colombia and e Ecuador to *tepuis* of Venezuela and Brazil

____ *L. r. rufaxilla*	Venezuela to Guianas and n Brazil (R. Madeira to n Maranhão)
____ *L. r. hellmayri*	NE Venezuela (María Peninsula) and Trinidad
____ *L. r. bahiae*	Central Brazil (s Mato Grosso to Bahia)
____ *L. r. reichenbachii*	Mato Grosso to Espírito Santo, Paraguay and ne Argentina

☐ **Gray-headed Dove** *Leptotila plumbeiceps*

____ *L. p. plumbeiceps*	SE Mexico (s Tamaulipas), w Costa Rica and w Colombia
____ *L. p. notia*	Caribbean slope of w Panama

☐ **Pallid Dove** *Leptotila pallida*

Tropical lowlands of w Colombia and w Ecuador

☐ **Brown-backed Dove** *Leptotila battyi*

____ *L. b. malae*	Pacific slope of Panama (s Veraguas and w Herrera); Cébaco I.
____ *L. b. battyi*	Isla Coiba (off s-central Panama)

☐ **Grenada Dove** *Leptotila wellsi*

Arid scrub of s Grenada (s Lesser Antilles)

☐ **Caribbean Dove** *Leptotila jamaicensis*

____ *L. j. gaumeri*	N Yucatán, Mujeres, Holbox, Cozumel and is. off Honduras
____ *L. j. collaris*	Cayman Islands
____ *L. j. jamaicensis*	Jamaica
____ *L. j. neoxena*	Isla San Andrés (Caribbean Sea off Nicaragua)

☐ **Gray-chested Dove** *Leptotila cassini*

____ *L. c. cerviniventris*	Caribbean lowlands of Guatemala to w Panama
____ *L. c. rufinucha*	SW Costa Rica to nw Panama (Chiriquí)
____ *L. c. cassini*	Panama (Canal Zone) to n Colombia (Cauca-Magdalena area)

☐ **Ochre-bellied Dove** *Leptotila ochraceiventris*

SW Ecuador (Manabí) to extreme nw Peru (Tumbes and Piura)

☐ **Tolima Dove** *Leptotila conoveri*

E slope of Central Andes of Colombia (Tolima to Huila)

☐ **Purplish-backed Quail-Dove** *Geotrygon lawrencii*

Costa Rica to e Panama (Darién)

☐ **Tuxtla Quail-Dove** *Geotrygon carrikeri*

SE Mexico (Sierra de Tuxtla in se Veracruz)

☐ **Buff-fronted Quail-Dove** *Geotrygon costaricensis*

Humid montane forests of Costa Rica and w Panama

☐ **Russet-crowned Quail-Dove** *Geotrygon goldmani*

____ *G. g. oreas*	E Panama (Cerro Chucantí in e Panamá Province)
____ *G. g. goldmani*	E Panama (e Darién and extreme nw Colombia)

☐ **Sapphire Quail-Dove** *Geotrygon saphirina*

____ *G. s. purpurata*	Humid forests of nw Colombia to w Ecuador
____ *G. s. saphirina*	Trop. e Ecuador to se Peru and extreme w Amazonian Brazil
____ *G. s. rothschildi*	SE Peru (Marcapata Valley)

☐ **Gray-fronted Quail-Dove** *Geotrygon caniceps*

Cuba (on verge of extinction)

☐ **White-fronted Quail-Dove** *Geotrygon leucometopia*

Dominican Republic (on verge of extinction)

☐ **Crested Quail-Dove** *Geotrygon versicolor*

Montane forests of Jamaica

☐ **Chiriqui Quail-Dove** *Geotrygon chiriquensis*

Humid montane forests of Costa Rica and w Panama

☐ **Olive-backed Quail-Dove** *Geotrygon veraguensis*

Caribbean lowlands of Costa Rica to nw Ecuador

☐ **White-faced Quail-Dove** *Geotrygon albifacies*

Mountains of se Mexico (San Luis Potosí) to nw Nicaragua

☐ **Lined Quail-Dove** *Geotrygon linearis*

□ **White-throated Quail-Dove** *Geotrygon frenata*

Mountains of Colombia, Venezuela, Trinidad and Tobago

_____ *G. f. bourcieri* Andes of w Colombia and w Ecuador
_____ *G. f. erythropareia* Andes of e Ecuador
_____ *G. f. frenata* Andes of n Peru to central Bolivia
_____ *G. f. margaritae* Andes of s Bolivia to nw Argentina

□ **Key West Quail-Dove** *Geotrygon chrysia*

Bahamas, Cuba, Isle of Pines, Hispaniola and sw Puerto Rico

□ **Bridled Quail-Dove** *Geotrygon mystacea*

Puerto Rico and Virgin Is. to Lesser Antilles (s to St. Lucia)

□ **Violaceous Quail-Dove** *Geotrygon violacea*

_____ *G. v. albiventer* Nicaragua to n Colombia and w Venezuela
_____ *G. v. violacea* Suriname to e Brazil, Bolivia, Paraguay and ne Argentina

□ **Ruddy Quail-Dove** *Geotrygon montana*

_____ *G. m. montana* Trop. s Mexico and Gr. Antilles to s Brazil and n Argentina
_____ *G. m. martinica* Lesser Antilles

□ **Blue-headed Quail-Dove** *Starnoenas cyanocephala*

Lowlands of Cuba (in danger of extinction)

□ **Nicobar Pigeon** *Caloenas nicobarica*

_____ *C. n. nicobarica* Malay Arch. to New Guinea, Philippines and Solomon Islands
_____ *C. n. pelewensis* Palau Islands

□ **Luzon Bleeding-heart** *Gallicolumba luzonica*

_____ *G. l. griseolateralis* N Luzon (n Philippines)
_____ *G. l. luzonica* N Philippines (central and s Luzon and Polillo)
_____ *G. l. rubiventris* NE Philippines (Vigo-Gigmoto watershed on Catanduanes)

□ **Mindanao Bleeding-heart** *Gallicolumba crinigera*

_____ *G. c. leytensis* Philippines (Samar, Leyte and Bohol)
_____ *G. c. crinigera* S Philippines (Mindanao and Dinagat)
_____ *G. c. bartletti* Basilan I. (se Philippines)

□ **Mindoro Bleeding-heart** *Gallicolumba platenae*

Forests of Mindoro (central Philippines). Status unknown

□ **Negros Bleeding-heart** *Gallicolumba keayi*

Forests of Negros (central Philippines). Status unknown

□ **Sulu Bleeding-heart** *Gallicolumba menagei*

Tawitawi I. (Sulu Archipelago). Status unknown

□ **Cinnamon Ground-Dove** *Gallicolumba rufigula*

_____ *G. r. rufigula* W Papuan islands (New Guinea)
_____ *G. r. septentrionalis* N New Guinea (east to Huon Gulf)
_____ *G. r. helviventris* Aru Is. and s New Guinea (Waitakwa River to Fly River)
_____ *G. r. alaris* S New Guinea east to Karimui (Chimbu Province)
_____ *G. r. orientalis* SE New Guinea (w to Mambare River and Angabunga River)

□ **Sulawesi Ground-Dove** *Gallicolumba tristigmata*

_____ *G. t. tristigmata* Humid forests of n and n-central Sulawesi
_____ *G. t. bimaculata* S Sulawesi
_____ *G. t. auripectus* S-central and se Sulawesi

□ **White-bibbed Ground-Dove** *Gallicolumba jobiensis*

_____ *G. j. jobiensis* New Guinea, Bismarck Arch. and D'Entrecasteaux Arch.
_____ *G. j. chalconota* Solomon Islands (Guadalcanal and Vella Lavella)

□ **Caroline Islands Ground-Dove** *Gallicolumba kubaryi*

Montane forests of e Caroline Islands (Truk and Pohnpei)

□ **Polynesian Ground-Dove** *Gallicolumba erythroptera*

□ **White-throated Ground-Dove** *Gallicolumba xanthonura*

□ **Friendly Ground-Dove** *Gallicolumba stairi*
_____ *G. s. stairi*
_____ *G. s. vitiensis*

□ **Santa Cruz Ground-Dove** *Gallicolumba sanctaecrucis*

□ **Thick-billed Ground-Dove** *Gallicolumba salamonis*

□ **Marquesas Ground-Dove** *Gallicolumba rubescens*

□ **Bronze Ground-Dove** *Gallicolumba beccarii*
_____ *G. b. beccarii*
_____ *G. b. johannae*
_____ *G. b. eichhorni*
_____ *G. b. admiralitatis*
_____ *G. b. intermedia*
_____ *G. b. solomonensis*

□ **Palau Ground-Dove** *Gallicolumba canifrons*

□ **Wetar Ground-Dove** *Gallicolumba hoedtii*

□ **Thick-billed Ground-Pigeon** *Trugon terrestris*
_____ *T. t. terrestris*
_____ *T. t. mayri*
_____ *T. t. leucopareia*

□ **Pheasant Pigeon** *Otidiphaps nobilis*
_____ *O. n. nobilis*
_____ *O. n. aruensis*
_____ *O. n. cervicalis*
_____ *O. n. insularis*

□ **Western Crowned-Pigeon** *Goura cristata*
_____ *G. c. cristata*
_____ *G. c. minor*

□ **Southern Crowned-Pigeon** *Goura scheepmakeri*
_____ *G. s. sclaterii*
_____ *G. s. scheepmakeri*

□ **Victoria Crowned-Pigeon** *Goura victoria*
_____ *G. v. victoria*
_____ *G. v. beccarii*

□ **Tooth-billed Pigeon** *Didunculus strigirostris*

□ **White-eared Dove** *Phapitreron leucotis*
_____ *P. l. leucotis*
_____ *P. l. nigrorum*
_____ *P. l. brevirostris (albifrons)*
_____ *P. l. occipitalis*

□ **Amethyst Dove** *Phapitreron amethystinus*
_____ *P. a. amethystinus*
_____ *P. a. imeldae*
_____ *P. a. maculipectus*

Uninhabited atolls of Tuamotu Archipelago

Yap (Caroline Islands) and Mariana Islands

Wallis and Futuna Islands and Samoa
Fiji and Tonga (Vava'u, Ha'apai and Nomuka group)

Santa Cruz Islands (Tinakula and Utupua) and Vanuatu

Solomon Islands (Ramos and San Cristobal). Possibly extinct

Marquesas Islands (Fatuhuku and Hatuta'a)

Mountains of New Guinea
Bismarck Archipelago (Karkar and Nissan)
St. Matthias Islands (Mussau and Emira)
Manus I. (Admiralty Islands)
W Solomon Is. (Bougainville, Gizo and New Georgia Group)
E Solomon Is. (Guadalcanal, San Cristobal, Santa Ana, Rennell)

Palau Islands (Babelthuap south to Angaur)

E Lesser Sundas (Timor and Wetar)

Salawati I. and nw New Guinea e to Geelvink and Etna bays
N-central New Guinea (Mamberamo River to Humboldt Bay)
S New Guinea (Setekwa River to Milne Bay)

Mountains of w New Guinea, Batanta and Waigeo islands
Aru Islands (New Guinea)
Mts. of e and se New Guinea (Saruwaged, Sepik and Kuper)
Fergusson I. (D'Entrecasteaux Archipelago)

NW New Guinea (Vogelkop to Etna Bay and Siriwo River)
W Papuan islands (Misool, Salawati, Batanta and Waigeo)

S New Guinea (Mimika River to Fly River)
Coastal se New Guinea (Hall Sound to Orangerie Bay)

Yapen I. and Biak I. (New Guinea)
N New Guinea

Mountains of Western Samoa (Upolu and Savai'i)

N Philippines (Catanduanes, Luzon, Mindoro and adj. islands)
Philippines (Cebu, Guimaras, Masbate, Negros, Panay, adj. is.)
Leyte, Mindoro, Bohol, Dinagat, Samar, Mindanao and Siquijor
S Philippines (Basilan) and Sulu Archipelago

Luzon, Mindanao and adjacent Philippines
Marinduque (n-central Philippines)
Montane forests of Negros (e-central Philippines)

____ *P. a. frontalis*　　　　　　　　　　Cebu (e-central Philippines). Probably extinct

☐ **Dark-eared Dove** *Phapitreron cinereiceps*
____ *P. c. brunneiceps*　　　　　　　　SE Philippines (Mindanao and Basilan)
____ *P. c. cinereiceps*　　　　　　　　Tawitawi (Sulu Archipelago)

☐ **Little Green-Pigeon** *Treron olax*
　　　　　　　　　　　　　　　　　　Lowlands of s Thailand, Malay Peninsula and Greater Sundas

☐ **Pink-necked Pigeon** *Treron vernans*
　　　　　　　　　　　　　　　　　　SE Asia to Philippines and Indonesia

☐ **Cinnamon-headed Pigeon** *Treron fulvicollis*
____ *T. f. fulvicollis*　　　　　　　　Malay Peninsula, Sumatra and adjacent islands
____ *T. f. melopogenys*　　　　　　　Nias I. (off w Sumatra)
____ *T. f. oberholseri*　　　　　　　　Natuna Islands (off nw Borneo)
____ *T. f. baramensis*　　　　　　　　N Borneo and adjacent islands off north coast

☐ **Orange-breasted Pigeon** *Treron bicinctus*
____ *T. b. bicinctus*　　　　　　　　Indian subcontinent and SE Asia
____ *T. b. leggei*　　　　　　　　　　Sri Lanka
____ *T. b. domvilii*　　　　　　　　Hainan (s China)
____ *T. b. javanus*　　　　　　　　Java and Bali

☐ **Pompadour Green-Pigeon** *Treron pompadora*
____ *T. p. pompadora*　　　　　　　India to sw China, Thailand, Laos and s Vietnam
____ *T. p. chloropterus*　　　　　　Andaman and Nicobar islands
____ *T. p. phayrei*　　　　　　　　Myanmar to sw China (Yunnan), Thailand, Laos and s Vietnam
____ *T. p. amadoni*　　　　　　　　N Luzon (n Philippines)
____ *T. p. axillaris*　　　　　　　　S Luzon, Polillo, Alabat, Catanduanes, Lubang and Mindoro
____ *T. p. canescens*　　　　　　　E Philippines (Mandate to Cebu, Basilan and Mindanao)
____ *T. p. everetti*　　　　　　　　Sulu Archipelago (Bongao, Jolo, Sibutu and Tawitawi)
____ *T. p. aromaticus*　　　　　　S Moluccas (Buru), Tanahjampea, Kalao and Kalaotoa islands

☐ **Thick-billed Pigeon** *Treron curvirostra*
　　　　　　　　　　　　　　　　　　Lowlands forests of s Asia and Malay Archipelago

☐ **Gray-cheeked Pigeon** *Treron griseicauda*
____ *T. g. sangirensis*　　　　　　Talaud Islands and Sangihe I.
____ *T. g. wallacei*　　　　　　　　Sulawesi, Banggai and Sula islands
____ *T. g. griseicauda*　　　　　　Java and Bali
____ *T. g. vordermani*　　　　　　Kangean Islands (Java Sea)

☐ **Sumba Green-Pigeon** *Treron teysmannii*
　　　　　　　　　　　　　　　　　　Lowlands of Sumba I. (w Lesser Sundas)

☐ **Flores Green-Pigeon** *Treron floris*
　　　　　　　　　　　　　　　　　　Lowlands of Lombok, Flores and adjacent w Lesser Sundas

☐ **Timor Green-Pigeon** *Treron psittaceus*
　　　　　　　　　　　　　　　　　　Lowlands of e Lesser Sundas (Timor, Roti and Semau)

☐ **Large Green-Pigeon** *Treron capellei*
　　　　　　　　　　　　　　　　　　Malay Peninsula to n Sumatra, Borneo, Java and adj. islands

☐ **Yellow-footed Pigeon** *Treron phoenicopterus*
____ *T. p. phillipsi*　　　　　　　　Sri Lanka
____ *T. p. chlorigaster*　　　　　　Peninsular India south of the Gangetic Plain
____ *T. p. phoenicopterus*　　　　E Pakistan and n India to Assam and Bangladesh
____ *T. p. viridifrons*　　　　　　Extreme sw China (Yunnan) to Myanmar and nw Thailand
____ *T. p. annamensis*　　　　　　E Thailand to s Laos and s Vietnam

☐ **Bruce's Green-Pigeon** *Treron waalia*
　　　　　　　　　　　　　　　　　　Savanna of sub-Saharan Africa, sw Arabia and Socotra I.

☐ **Madagascar Green-Pigeon** *Treron australis*
____ *T. a. griveaudi*　　　　　　　Mohéli (Comoro Islands)
____ *T. a. xenius*　　　　　　　　　W Madagascar

____ *T. a. australis*   Madagascar (east of the high plateau)

☐ **Pemba Green-Pigeon** *Treron pembaensis*

Pemba I. (off ne Tanzania)

☐ **São Tomé Green-Pigeon** *Treron sanctithomae*

São Tomé (extirpated on adjacent Ilha das Rôlas)

☐ **Yellow-vented Pigeon** *Treron seimundi*
____ *T. s. seimundi*   Mountains of Malay Peninsula
____ *T. s. modestus*   Mountains of Laos and Vietnam (Annam and Cochinchina)

☐ **Pin-tailed Pigeon** *Treron apicauda*
____ *T. a. apicauda*   Foothills of ne India to sw China and s Myanmar (Tenasserim)
____ *T. a. lowei*   Mountains of Thailand, central Laos and central Vietnam
____ *T. a. laotinus*   Mountains of Laos and n Vietnam

☐ **African Green-Pigeon** *Treron calvus*
____ *T. c. nudirostris*   Senegal to Gambia and Guinea-Bissau
____ *T. c. sharpei*   Sierra Leone to s Nigeria and n Cameroon
____ *T. c. calvus*   E Nigeria to ne Zaire, central Angola and Príncipe I.
____ *T. c. poensis*   Bioko I. (Gulf of Guinea)
____ *T. c. uellensis*   N Zaire to s Sudan and Uganda
____ *T. c. brevicerus*   SW Ethiopia to n Tanzania e of Rift Valley (except for coast)
____ *T. c. salvadorii*   Uganda, Rwanda and Burundi to e Zaire
____ *T. c. granviki*   W Kenya and nw Tanzania
____ *T. c. wakefieldii*   Coastal Kenya and nw Tanzania
____ *T. c. granti*   Lowlands of e Tanzania and Zanzibar
____ *T. c. orientalis*   S Tanzania to Mozambique and lower Zambezi Valley
____ *T. c. schalowi*   S Zaire and Zambia to Victoria Falls
____ *T. c. chobiensis*   SW Zimbabwe and n Botswana
____ *T. c. ansorgei*   W Angola (south of Cuanza River)
____ *T. c. vylderi*   NW Namibia (east to Grootfontein)
____ *T. c. damarensis*   NE Namibia and nw Botswana
____ *T. c. delalandii*   Coastal Kenya (Mombasa) to e Cape Province

☐ **Green-spectacled Pigeon** *Treron oxyurus*

Montane forests of Sumatra and w Java

☐ **Wedge-tailed Pigeon** *Treron sphenurus*
____ *T. s. sphenurus*   Mts. of Kashmir to sw China, Myanmar, n Thailand and Laos
____ *T. s. robinsoni*   Mountains of Malay Peninsula and central Vietnam
____ *T. s. korthalsi*   High mountains of Sumatra, Java, Bali and Lombok

☐ **White-bellied Pigeon** *Treron sieboldii*
____ *T. s. sieboldii*   Japan, mountains of Taiwan and e China (Jiangsu and Fujian)
____ *T. s. fopingensis*   Lowlands and foothills of c China (e Sichuan and s Shaanxi)
____ *T. s. murielae*   Extreme sw China to n Thailand, central Vietnam and Hainan

☐ **Whistling Green-Pigeon** *Treron formosae*
____ *T. f. permagnus*   N Ryukyu Islands (Yakushima, Amani-Oshima and Okinawa)
____ *T. f. medioximus*   S Ryukyu Islands (Ishigaki, Iriomote and Yonaguni)
____ *T. f. formosae*   Mountains of Taiwan and Botel Tobago
____ *T. f. filipinus*   N Philippines (Batan, Calayan, Camiguin Norte and Sabtang)

☐ **Black-backed Fruit-Dove** *Ptilinopus cinctus*
____ *P. c. baliensis*   Bali
____ *P. c. albocinctus*   Lesser Sundas (Lombok, Sumbawa and Flores)
____ *P. c. everetti*   Lesser Sundas (Pantar and Alor)
____ *P. c. cinctus*   Lesser Sundas (Timor, Wetar and Romang)
____ *P. c. lettiensis*   Lesser Sundas (Leti, Moa, Luang, Sermata and Teun)
____ *P. c. ottonis*   Lesser Sundas (Damar, Babar and Nila)

☐ **Black-banded Fruit-Dove** *Ptilinopus alligator*

☐ **Red-naped Fruit-Dove** *Ptilinopus dohertyi*

☐ **Pink-headed Fruit-Dove** *Ptilinopus porphyreus*

☐ **Yellow-breasted Fruit-Dove** *Ptilinopus occipitalis*
_____ *P. o. occipitalis*
_____ *P. o. incognitus*

☐ **Flame-breasted Fruit-Dove** *Ptilinopus marchei*

☐ **Cream-breasted Fruit-Dove** *Ptilinopus merrilli*
_____ *P. m. faustinoi*
_____ *P. m. merrilli*

☐ **Red-eared Fruit-Dove** *Ptilinopus fischeri*
_____ *P. f. fischeri*
_____ *P. f. centralis*
_____ *P. f. meridionalis*

☐ **Jambu Fruit-Dove** *Ptilinopus jambu*

☐ **Maroon-chinned Fruit-Dove** *Ptilinopus subgularis*
_____ *P. s. epia*
_____ *P. s. subgularis*
_____ *P. s. mangoliensis*

☐ **Black-chinned Fruit-Dove** *Ptilinopus leclancheri*
_____ *P. l. longialis*
_____ *P. l. leclancheri*
_____ *P. l. gironieri*

☐ **Scarlet-breasted Fruit-Dove** *Ptilinopus bernsteinii*
_____ *P. b. bernsteinii*
_____ *P. b. micrus*

☐ **Wompoo Fruit-Dove** *Ptilinopus magnificus*
_____ *P. m. alaris*
_____ *P. m. puella*
_____ *P. m. interpositus*
_____ *P. m. septentrionalis*
_____ *P. m. poliurus*
_____ *P. m. assimilis*
_____ *P. m. keri*
_____ *P. m. magnificus*

☐ **Pink-spotted Fruit-Dove** *Ptilinopus perlatus*
_____ *P. p. perlatus*
_____ *P. p. plumbeicollis*
_____ *P. p. zonurus*

☐ **Ornate Fruit-Dove** *Ptilinopus ornatus*
_____ *P. o. ornatus*
_____ *P. o. gestroi*

☐ **Tanna Fruit-Dove** *Ptilinopus tannensis*

☐ **Orange-fronted Fruit-Dove** *Ptilinopus aurantiifrons*

☐ **Wallace's Fruit-Dove** *Ptilinopus wallacii*

N-central Australia (western escarpment of Arnhem Land)

Forests of Sumba (w Lesser Sundas)

Montane forests of s Sumatra, Java and Bali

Lowland forests of n and central Philippines
Mountains of Mindanao (se Philippines)

Montane forests of Luzon (n Philippines)

N Luzon (n Philippines)
N Philippines (s Luzon, Polillo and Catanduanes)

Montane forests of n Sulawesi
Montane forests of central and se Sulawesi
SW Sulawesi (Lompobattang Massif)

Peninsular Thailand, Malay Pen, Sumatra, Borneo and w Java

Lowland forests of Sulawesi
Banggai Islands (Peleng and Banggai)
Sula Islands (Taliabu, Seho and Mangole)

Lan-yü I. (Taiwan); Batan, Calayan and Camiguin Norte is.
Philippines (except Palawan, Basilan and Sulu Archipelago)
Palawan (sw Philippines)

N Moluccas (Halmahera, Ternate and Bacan)
Obi I. (n-central Moluccas)

W Papuan islands (Waigeo, Misool, Batanta and Salawati)
NW New Guinea (Vogelkop Mountains)
W-central and sw New Guinea
N and ne New Guinea, Yapen, Manam and Karkar islands
SE New Guinea (west to Huon Gulf and Edrich River)
NE Australia (Cape York Peninsula)
NE Australia (Bellenden Ker Range of ne Queensland)
E Australia (s Queensland to n New South Wales)

W Papuan islands, nw New Guinea and Yapen I.
NE New Guinea (Astrolabe Bay to Huon Gulf)
Aru Islands, s New Guinea and D'Entrecasteaux Archipelago

NW New Guinea (Arfak Mountains and coastal Vogelkop)
New Guinea (west to Cyclops Mountains and Onin Peninsula)

Vanuatu and Banks Islands

W Papuan, Yapen, Aru is., New Guinea, D'Entrecasteaux Arch.

☐ **Superb Fruit-Dove** *Ptilinopus superbus*

Lowlands of sw New Guinea, s Moluccas, Kai and Aru islands

_____ *P. s. temminckii*
_____ *P. s. superbus*

Sulawesi and Sula Islands
Moluccas to Bismarck Arch., Solomon Is. and ne Australia

☐ **Many-colored Fruit-Dove** *Ptilinopus perousii*

_____ *P. p. perousii*
_____ *P. p. mariae*

Samoa (Savai'i, Upolu, Tutuila, Ofu and Tau)
Fiji and Tonga

☐ **Crimson-crowned Fruit-Dove** *Ptilinopus porphyraceus*

_____ *P. p. ponapensis*
_____ *P. p. hernsheimi*
_____ *P. p. porphyraceus (graeffei)*
_____ *P. p. fasciatus*

Caroline Islands (Truk and Pohnpei)
Kosrae (e Caroline Islands)
Small islands of Tonga, Fiji and Niue
Samoa

☐ **Palau Fruit-Dove** *Ptilinopus pelewensis*

Palau Islands (Babelthuap to Angaur)

☐ **Cook Islands Fruit-Dove** *Ptilinopus rarotongensis*

_____ *P. r. rarotongensis*
_____ *P. r. goodwini*

Rarotonga (Cook Islands)
Atiu (s Cook Islands)

☐ **Mariana Fruit-Dove** *Ptilinopus roseicapilla*

Mariana Islands (Saipan, Tinian, Agiguan, Rota and Guam)

☐ **Rose-crowned Fruit-Dove** *Ptilinopus regina*

_____ *P. r. flavicollis*
_____ *P. r. roseipileum*
_____ *P. r. xanthogaster*
_____ *P. r. ewingii*
_____ *P. r. regina*

Lesser Sundas (Flores, Roti, Sawu, Semau and w Timor)
Lesser Sundas (e Timor, Wetar, Romang, Kissar, Moa and Leti)
Banda, Kai, Damar, Sermata, Babar, Tanimbar and Aru islands
N Australia (Kimberley region to n N Territory and Melville I.)
Cape York Pen. to s New S Wales and islands in Torres Strait

☐ **Silver-capped Fruit-Dove** *Ptilinopus richardsii*

_____ *P. r. richardsii*
_____ *P. r. cyanopterus*

E Solomon Islands (Ugi and Santa Anna)
SE Solomon Islands (Rennell and Bellona)

☐ **Gray-green Fruit-Dove** *Ptilinopus purpuratus*

_____ *P. p. chrysogaster*
_____ *P. p. frater*
_____ *P. p. purpuratus*

W Society Islands (Bora Bora, Tahaa, Huahine and Maupiti)
Moorea (e Society Islands)
Tahiti (e Society Islands)

☐ **Makatea Fruit-Dove** *Ptilinopus chalcurus*

Makatea I. (w Tuamotu Archipelago)

☐ **Atoll Fruit-Dove** *Ptilinopus coralensis*

Larger islands in Tuamotu Archipelago (except Makatea)

☐ **Red-bellied Fruit-Dove** *Ptilinopus greyii*

E Solomons to Santa Cruz, Banks, New Caledonia, Isle of Pines

☐ **Rapa Fruit-Dove** *Ptilinopus huttoni*

Rapa I. (Austral Archipelago). Seriously endangered

☐ **White-capped Fruit-Dove** *Ptilinopus dupetithouarsii*

_____ *P. d. viridior*
_____ *P. d. dupetithouarsii*

N Marquesas Islands (Nukuhiva, Uahuka and Uapou)
S Marquesas Is. (Hivaoa, Tahuata, Mohotani and Fatuhiva)

☐ **Henderson Island Fruit-Dove** *Ptilinopus insularis*

Henderson I. (Pitcairn Archipelago)

☐ **Coroneted Fruit-Dove** *Ptilinopus coronulatus*

_____ *P. c. trigeminus*
_____ *P. c. geminus*
_____ *P. c. quadrigeminus*
_____ *P. c. huonensis*
_____ *P. c. coronulatus*

Salawati I. and w coast of Vogelkop Pen. (nw New Guinea)
N New Guinea (head of Geelvink Bay to Takar) and Yapen I.
N New Guinea (Humboldt to Astrolabe Bay) and Manam I.
N coastal New Guinea (Huon Bay to Goodenough Bay)
Aru Is. and s coastal New Guinea (Mimika River to Milne Bay)

☐ **Beautiful Fruit-Dove** *Ptilinopus pulchellus*

____	*P. p. pulchellus*	W Papuan islands and New Guinea
____	*P. p. decorus*	N New Guinea (east shore of Geelvink Bay to Astrolabe Bay)

☐ **Blue-capped Fruit-Dove Ptilinopus monacha**

N Moluccas (Halmahera, Ternate, Bacan and adjacent islands)

☐ **White-breasted Fruit-Dove Ptilinopus rivoli**

____	*P. r. prasinorrhous*	Moluccas, Aru Is., w Papuan is. and islands in Geelvink Bay
____	*P. r. bellus*	Mountains of New Guinea, Karkar I. and Goodenough I.
____	*P. r. miquelii*	Yapen I. and Meos Num I. (n New Guinea)
____	*P. r. rivoli*	Bismarck Archipelago
____	*P. r. strophium*	Egum Atoll (Trobriand Islands) and Louisiade Archipelago

☐ **Yellow-bibbed Fruit-Dove Ptilinopus solomonensis**

____	*P. s. speciosus*	Numfor, Biak and Traitor's islands (n New Guinea)
____	*P. s. johannis*	Admiralty Islands, St. Matthias Group and New Hanover
____	*P. s. meyeri*	New Britain and satellite islands
____	*P. s. neumanni*	Nissan I. (w Solomon Islands)
____	*P. s. bistictus*	Solomon Islands (Bougainville and Buka)
____	*P. s. vulcanorum*	SW Solomon Islands
____	*P. s. ocularis*	Guadalcanal (se Solomon Islands)
____	*P. s. ambiguus*	Malaita I. (e Solomon Islands)
____	*P. s. solomonensis*	Solomon Islands (San Cristóbal and Ugi)

☐ **Claret-breasted Fruit-Dove Ptilinopus viridis**

____	*P. v. viridis*	S Moluccas (Buru, Seram, Ambon and adjacent islands)
____	*P. v. pectoralis*	W Papuan islands and nw New Guinea
____	*P. v. geelvinkianus*	Numfor, Biak and Meos Num islands (n New Guinea)
____	*P. v. salvadorii*	Yapen I. and n New Guinea (Mamberamo River to Madang)
____	*P. v. vicinus*	Trobriand Islands and D'Entrecasteaux Archipelago
____	*P. v. lewisii*	Manus I., Lihir Is., and Nissan I. to w Solomon islands

☐ **White-headed Fruit-Dove Ptilinopus eugeniae**

E Solomon Islands (San Cristóbal, Malaupaina and Ugi)

☐ **Orange-bellied Fruit-Dove Ptilinopus iozonus**

____	*P. i. iozonus*	Aru Islands
____	*P. i. humeralis*	W Papuan islands and lowlands of nw New Guinea
____	*P. i. jobiensis*	Yapen I., n New Guinea, Manam and adjacent islands
____	*P. i. pseudohumeralis*	Central New Guinea (upper Fly River region)
____	*P. i. finschii*	Huon Peninsula and Fly River to se New Guinea

☐ **Knob-billed Fruit-Dove Ptilinopus insolitus**

____	*P. i. insolitus*	Bismarck Archipelago
____	*P. i. inferior*	St. Matthias Group (Mussau and Emira)

☐ **Gray-headed Fruit-Dove Ptilinopus hyogastrus**

N Moluccas (Morotai, Halmahera, Bacan, Tidore and Ternate)

☐ **Carunculated Fruit-Dove Ptilinopus granulifrons**

Obi (n-central Moluccas)

☐ **Black-naped Fruit-Dove Ptilinopus melanospilus**

____	*P. m. bangueyensis*	S Philippines and islands off n Borneo
____	*P. m. xanthorrhous*	Talaud Islands, Sangihe I. (Sulawesi) and Doi I. (n Moluccas)
____	*P. m. melanospilus*	Sulawesi, Talisei, Bangka, Lembeh and Togian islands
____	*P. m. chrysorrhous*	Banggai Is., Sula Is. and s Moluccas (Obi and Seram)
____	*P. m. melanauchen*	Matasiri I., Java, Lesser Sundas and islands south of Sulawesi

☐ **Dwarf Fruit-Dove Ptilinopus nanus**

____	*P. n. minimus*	W Papuan islands (Waigeo, Batanta, Salawati and Misol)
____	*P. n. nanus*	S New Guinea (Lobo Bay to Port Moresby)

☐ **Negros Fruit-Dove** *Ptilinopus arcanus*

Negros (known from a 1953 specimen from Mt. Canlaon)

☐ **Orange Dove** *Ptilinopus victor*
____ *P. v. victor* — N Fiji (Vanua Levu, Rabi, Kooa and Taveuni)
____ *P. v. aureus* — NE Fiji (Qamea and Laucala)

☐ **Golden Dove** *Ptilinopus luteovirens*

W Fiji (Waya Group, Viti Levu, Beqa, Ovalau and Gau)

☐ **Velvet Dove** *Ptilinopus layardi*

Fiji (Kandavu and Ono)

☐ **Cloven-feathered Dove** *Drepanoptila holosericea*

New Caledonia and Isle of Pines

☐ **Madagascar Blue-Pigeon** *Alectroenas madagascariensis*

Humid forests of e Madagascar

☐ **Comoro Blue-Pigeon** *Alectroenas sganzini*
____ *A. s. minor* — Humid forests of Aldabra
____ *A. s. sganzini* — Comoro Islands (Grand Comoro, Anjouan and Mayotte)

☐ **Seychelles Blue-Pigeon** *Alectroenas pulcherrima*

Seychelles (Praslin, Mahé, Félicité and Silhouette)

☐ **Pink-bellied Imperial-Pigeon** *Ducula poliocephala*

Highlands of Philippine Islands

☐ **White-bellied Imperial-Pigeon** *Ducula forsteni*

Mts. of Sulawesi and Sula Islands (Taliabu and Mangole)

☐ **Mindoro Imperial-Pigeon** *Ducula mindorensis*

Highlands of Mindoro (central Philippines)

☐ **Gray-headed Imperial-Pigeon** *Ducula radiata*

Montane forests of Sulawesi

☐ **Spotted Imperial-Pigeon** *Ducula carola*
____ *D. c. carola* — N Philippines (Luzon, Mindoro and Sibuyan)
____ *D. c. nigrorum* — Philippines (Negros and Siquijor)
____ *D. c. mindanensis* — S Philippines (Mindoro and Mindanao)

☐ **Green Imperial-Pigeon** *Ducula aenea*
____ *D. a. sylvatica (andamanica)* — N India to Nepal, Thailand, Indochina and Andaman Islands
____ *D. a. pusilla* — S India and Sri Lanka
____ *D. a. nicobarica* — Nicobar Islands
____ *D. a. mista* — Simeulue I. (off nw Sumatra)
____ *D. a. babiensis* — Babi and Lasia islands (off se coast of Simeulue I.)
____ *D. a. consobrina* — Nias I. (off w Sumatra)
____ *D. a. vicina* — Batu and Mentawi islands (off w Sumatra)
____ *D. a. aneothorax* — Enggano I. (off w Sumatra)
____ *D. a. aenea (polia)* — Malay Peninsula, Sumatra and Borneo to Bali and Philippines
____ *D. a. palawanensis* — Palawan, adjacent s Philippines and Banggai Islands
____ *D. a. fugaensis* — N Philippines (Calayan, Camiguin Norte and Fuga)
____ *D. a. nuchalis* — N Luzon (n Philippines)
____ *D. a. paulina (intermedia, pallidinucha, sulana)* — Sulawesi, Sangihe, Talaud, Togian, Sula and adjacent islands

☐ **White-eyed Imperial-Pigeon** *Ducula perspicillata*
____ *D. p. perspicillata* — N Moluccas and Kofiau I. (w Papuan islands)
____ *D. p. neglecta* — S Moluccas (Boano, Seram, Ambon and Saparua)

☐ **Elegant Imperial-Pigeon** *Ducula concinna*

Small islands off Moluccas to e Lesser Sundas

☐ **Pacific Imperial-Pigeon** *Ducula pacifica*
____ *D. p. sejuncta* — Small islands off n New Guinea, Ninigo Group and Hermit Is.
____ *D. p. pacifica* — Louisiade Arch. to Solomons, Samoa, Tonga, Niue and Cook Is.

☐ **Red-knobbed Imperial-Pigeon** *Ducula rubricera*
____ *D. r. rubricera* — Bismarck Archipelago
____ *D. r. rufigula* — Solomon Islands (except Rennell)

127

☐ **Micronesian Imperial-Pigeon** *Ducula oceanica*

____ *D. o. monacha* — Palau Islands and Yap I. (w Caroline Islands)
____ *D. o. teraokai* — Truk (Caroline Islands)
____ *D. o. townsendi* — Pohnpei (Caroline Islands)
____ *D. o. oceanica* — Kosrae (e Caroline Islands)
____ *D. o. ratakensis* — Marshall Islands (Wotje, Ailinglaplap, Arno and Jaluit)

☐ **Polynesian Imperial-Pigeon** *Ducula aurorae*

____ *D. a. aurorae* — Makatea I. (Tuamotu Archipelago)
____ *D. a. wilkesii* — Tahiti (Society Islands)

☐ **Marquesas Imperial-Pigeon** *Ducula galeata*

Nukuhiva I. (Marquesas Islands)

☐ **Spice Imperial-Pigeon** *Ducula myristicivora*

____ *D. m. myristicivora* — Widi I. (off Halmahera) and w Papuan islands (New Guinea)
____ *D. m. geelvinkiana* — Islands in Geelvink Bay (Meos Num, Numfor and Biak)

☐ **Purple-tailed Imperial-Pigeon** *Ducula rufigaster*

____ *D. r. rufigaster* — W Papuan is., Vogelkop and s New Guinea e to Orangerie Bay
____ *D. r. uropygialis* — Yapen I. and n New Guinea (east to Huon Gulf)

☐ **Cinnamon-bellied Imperial-Pigeon** *Ducula basilica*

____ *D. b. basilica* — N Moluccas (Morotai, Halmahera, Ternate, Kasiruta, Bacan)
____ *D. b. obiensis* — Obi I. (central Moluccas)

☐ **Finsch's Imperial-Pigeon** *Ducula finschii*

Bismarck Archipelago

☐ **Rufescent Imperial-Pigeon** *Ducula chalconota*

____ *D. c. chalconota* — Vogelkop Mountains (nw New Guinea)
____ *D. c. smaragdina* — Montane forests of New Guinea (except Vogelkop)

☐ **Island Imperial-Pigeon** *Ducula pistrinaria*

____ *D. p. rhodinolaema* — Admiralty Is., New Hanover and small is. off n New Guinea
____ *D. p. vanwyckii* — Bismarck Arch. (New Britain, New Ireland and Witu)
____ *D. p. postrema* — Misima I., D'Entrecasteaux and Louisiade archipelagos
____ *D. p. pistrinaria* — Solomon Islands and Lihir Group

☐ **Pink-headed Imperial-Pigeon** *Ducula rosacea*

Lesser Sundas and islands in Flores Sea and Java Seas

☐ **Christmas Island Imperial-Pigeon** *Ducula whartoni*

Inland plateau of Christmas I. (e Indian Ocean)

☐ **Gray Imperial-Pigeon** *Ducula pickeringii*

____ *D. p. pickeringii* — Sulu Archipelago and small islands off n and ne Borneo
____ *D. p. langhornei* — Sulu Islands (Bolod and Loran)
____ *D. p. palmasensis* — Miangas and Talaud islands (off Sulawesi)

☐ **Peale's Imperial-Pigeon** *Ducula latrans*

Forests of larger Fiji Islands

☐ **Chestnut-bellied Imperial-Pigeon** *Ducula brenchleyi*

E Solomon Islands (Guadalcanal, Malaita and San Cristobal)

☐ **Baker's Imperial-Pigeon** *Ducula bakeri*

Banks Group and n Vanuatu

☐ **New Caledonian Imperial-Pigeon** *Ducula goliath*

Montane forests of New Caledonia and Isle of Pines

☐ **Pinon Imperial-Pigeon** *Ducula pinon*

____ *D. p. pinon* — Aru Islands, w Papuan islands and sw New Guinea
____ *D. p. jobiensis* — Yapen I.; n New Guinea e to Huon Gulf and offshore islands
____ *D. p. rubiensis* — Central and s New Guinea
____ *D. p. salvadorii* — D'Entrecasteaux and Louisiade archipelagos

☐ **Bismarck Imperial-Pigeon** *Ducula melanochroa*

Bismarck Archipelago

☐ **Collared Imperial-Pigeon** *Ducula mullerii*
____ *D. m. aurantia* — Lowlands of n New Guinea (Geelvink to Astrolabe Bay)
____ *D. m. mullerii* — Aru Islands, s New Guinea, Boigu and Daru islands

☐ **Zoe Imperial-Pigeon** *Ducula zoeae*

Lowland forests of New Guinea and larger satellite islands

☐ **Mountain Imperial-Pigeon** *Ducula badia*
____ *D. b. cuprea* — SW India (Western Ghats from Goa to Kerala)
____ *D. b. insignis* — Himalayan foothills (w Nepal to Sikkim and Bhutan)
____ *D. b. griseicapilla* — Myanmar to sw China, Hainan, Thailand and Indochina
____ *D. b. badia* — Malay Pen. and Mergui Arch. to Sumatra, Borneo and w Java

☐ **Dark-backed Imperial-Pigeon** *Ducula lacernulata*
____ *D. l. lacernulata* — Montane forests of w and central Java
____ *D. l. williami* — Montane forests of e Java and Bali
____ *D. l. sasakensis* — W Lesser Sundas (Lombok, Sumbawa and Flores)

☐ **Timor Imperial-Pigeon** *Ducula cineracea*
____ *D. c. cineracea* — Montane forests of Timor (e Lesser Sundas)
____ *D. c. schistacea* — Montane forests of Wetar (e Lesser Sundas

☐ **Pied Imperial-Pigeon** *Ducula bicolor*
____ *D. b. bicolor* — Widespread SE Asia and Malay Archipelago
____ *D. b. melanura* — Moluccas, Tanimbar and Kai islands

☐ **Torresian Imperial-Pigeon** *Ducula spilorrhoa*
____ *D. s. subflavescens* — Bismarck Archipelago and Admiralty Islands
____ *D. s. spilorrhoa* — Aru Is., New Guinea and adjacent islands to n and ne Australia

☐ **White Imperial-Pigeon** *Ducula luctuosa*

Sulawesi subregion and Sula Islands

☐ **Topknot Pigeon** *Lopholaimus antarcticus*

Coastal e Australia (Cape York to s New South Wales)

☐ **New Zealand Pigeon** *Hemiphaga novaeseelandiae*
____ *H. n. novaeseelandiae* — Forests of New Zealand and larger offshore islands
____ *H. n. chathamensis* — Chatham Islands
____ *H. n. spadicea†* — Formerly Norfolk I. Extinct

☐ **Sombre Pigeon** *Cryptophaps poecilorrhoa*

Humid montane forests of Sulawesi

☐ **Papuan Mountain-Pigeon** *Gymnophaps albertisii*
____ *G. a. exsul* — Montane forests of Bacan (n Moluccas)
____ *G. a. albertisii* — Yapen I., New Guinea and Bismarck Archipelago

☐ **Long-tailed Mountain-Pigeon** *Gymnophaps mada*
____ *G. m. mada* — Montane forests of Buru (s Moluccas)
____ *G. m. stalkeri* — Montane forests of Seram (s Moluccas)

☐ **Pale Mountain-Pigeon** *Gymnophaps solomonensis*

Solomon Islands (Bougainville to Guadalcanal and Malaita)

ORDER: PSITTACIFORMES
FAMILY: CACATUIDAE (Cockatoos—21)

☐ **Palm Cockatoo** *Probosciger aterrimus*
____ *P. a. stenolophus* — Yapen I. and nw New Guinea
____ *P. a. goliath* — W Papuan islands and w and central New Guinea
____ *P. a. aterrimus* — Aru Is., Misool I., s New Guinea and ne Australia (Cape York)

☐ **Red-tailed Black-Cockatoo** *Calyptorhynchus banksii*
- ____ *C. b. banksii* — Tropical northern Australia
- ____ *C. b. macrorhynchus* — N-central and ne Australia
- ____ *C. b. samueli* — W-central to e-central Australia
- ____ *C. b. naso* — Forests of sw Australia
- ____ *C. b. graptogyne* — Forests of se South Australia and sw Victoria

☐ **Glossy Black-Cockatoo** *Calyptorhynchus lathami*
- ____ *C. l. erebus* — E Australia (coastal e-central Queensland)
- ____ *C. l. lathami* — Inland and coastal e Australia
- ____ *C. l. halmaturinus* — Kangaroo I. (South Australia)

☐ **Yellow-tailed Black-Cockatoo** *Calyptorhynchus funereus*
- ____ *C. f. funereus* — E Australia (e-central Queensland to e Victoria)
- ____ *C. f. whiteae* — SE Australia (s Victoria to Eyre Peninsula) and Kangaroo I.
- ____ *C. f. xanthonotus* — Tasmania and islands in Bass Strait

☐ **Slender-billed Black-Cockatoo** *Calyptorhynchus latirostris*

Woodlands and scrub of sw Australia

☐ **White-tailed Black-Cockatoo** *Calyptorhynchus baudinii*

Extreme sw Australia (south of the Murchison River)

☐ **Gang-gang Cockatoo** *Callocephalon fimbriatum*

Coastal se Australia (c New South Wales to se S Australia)

☐ **Galah** *Eolophus roseicapilla*
- ____ *E. r. kuhli* — N Australia (Northern Territory)
- ____ *E. r. roseicapilla* — Western and w-central Australia
- ____ *E. r. albiceps* — E-central and e Australia south to Tasmania

☐ **Long-billed Corella** *Cacatua tenuirostris*

Woodlands and forests of se Australia

☐ **Western Corella** *Cacatua pastinator*
- ____ *C. p. derbyi* — Western Australia (Dongara to Moora and Quairading)
- ____ *C. p. pastinator* — SW Western Australia (Lake Muir and Unicup region)

☐ **Little Corella** *Cacatua sanguinea*
- ____ *C. s. transfreta* — Lowlands of s New Guinea
- ____ *C. s. sanguinea* — NW Western Australia and Northern Territory
- ____ *C. s. westralensis* — Western Australia (Murchison River region)
- ____ *C. s. gymnopsis* — Inland central and e Australia
- ____ *C. s. normantoni* — NE Australia (western Cape York Peninsula)

☐ **Tanimbar Corella** *Cacatua goffiniana*

Coastal lowlands of Tanimbar Islands (e Lesser Sundas)

☐ **Philippine Cockatoo** *Cacatua haematuropygia*

Forests and scrub of Philippine Islands and Palawan

☐ **Yellow-crested Cockatoo** *Cacatua sulphurea*
- ____ *C. s. sulphurea* — Sulawesi, Muna, Butung, Tanahjampea and adjacent islands
- ____ *C. s. abbotti* — Masalembu Besar I. (Java Sea)
- ____ *C. s. parvula* — Lesser Sundas (Sumbawa to Timor)
- ____ *C. s. citrinocristata* — Sumba (Lesser Sundas)

☐ **Ducorps' Cockatoo** *Cacatua ducorpsii*

E Solomon Islands (Bougainville to Malaita and Guadalcanal)

☐ **Pink Cockatoo** *Cacatua leadbeateri*

Arid and semiarid interior and s Australia

☐ **Sulphur-crested Cockatoo** *Cacatua galerita*
- ____ *C. g. triton* — New Guinea and adjacent islands
- ____ *C. g. eleonora* — Aru Islands
- ____ *C. g. fitzroyi* — N Australia (Fitzroy River to Gulf of Carpenteria)
- ____ *C. g. galerita* — E and se Australia (Cape York to Adelaide, King I., Tasmania)

☐ **Blue-eyed Cockatoo** *Cacatua ophthalmica*

☐ **Salmon-crested Cockatoo** *Cacatua moluccensis*

☐ **White Cockatoo** *Cacatua alba*

☐ **Cockatiel** *Nymphicus hollandicus*

	Bismarck Archipelago (New Britain and New Ireland)
	S Moluccas (Seram, Ambon, Saparua and Haruku)
	N Moluccas (Bacan, Halmahera, Ternate, Tidore, adj. islands)
	Irregularly abundant near water in interior of Australia

ORDER: PSITTACIFORMES
FAMILY: PSITTACIDAE (Parrots—347)

☐ **Black Lory** *Chalcopsitta atra*
_____ *C. a. bernsteini* — Misool I. (off w New Guinea)
_____ *C. a. atra* — W New Guinea (w Vogelkop Pen.), Batanta and Salawati is.
_____ *C. a. insignis* — W New Guinea (e Vogelkop and Onin Pen.) and Amberpon I.

☐ **Brown Lory** *Chalcopsitta duivenbodei*
_____ *C. d. duivenbodei* — Coastal n New Guinea (Geelvink Bay to Aitape region)
_____ *C. d. syringanuchalis* — Coastal lowlands of New Guinea (Aitape to Astrolabe Bay)

☐ **Yellow-streaked Lory** *Chalcopsitta sintillata*
_____ *C. s. rubrifrons* — Aru Islands
_____ *C. s. sintillata* — S New Guinea (Triton Bay to lower Fly River)
_____ *C. s. chloroptera* — Upper Fly River to se Papua New Guinea

☐ **Cardinal Lory** *Chalcopsitta cardinalis*

Lowlands of Solomon Islands and Bismarck Archipelago

☐ **Red-and-blue Lory** *Eos histrio*
_____ *E. h. talautensis* — Talaud Islands (Karakelang, Salibabu and Kabaruan)
_____ *E. h. histrio* — Sangihi, Siau and Ruang islands (n of Sulawesi)

☐ **Violet-necked Lory** *Eos squamata*
_____ *E. s. riciniata* — N Moluccas and Widi I.
_____ *E. s. obiensis* — N Moluccas (Obi and Bisa)
_____ *E. s. squamata* — W Papuan islands and Schildpad I.

☐ **Red Lory** *Eos bornea*
_____ *E. b. bornea* — S Moluccas and Kai Islands
_____ *E. b. cyanonothorus* — Buru (s Moluccas)

☐ **Blue-streaked Lory** *Eos reticulata*

Tanimbar Islands (Arafura Sea)

☐ **Black-winged Lory** *Eos cyanogenia*

Islands in Geelvink Bay (off nw New Guinea)

☐ **Blue-eared Lory** *Eos semilarvata*

Montane forests of Seram (s Moluccas)

☐ **Dusky Lory** *Pseudeos fuscata*

New Guinea, Salawati and Yapen islands

☐ **Ornate Lorikeet** *Trichoglossus ornatus*

Sulawesi and larger satellite islands

☐ **Rainbow Lorikeet** *Trichoglossus haematodus*
_____ *T. h. mitchellii* — Bali and Lombok
_____ *T. h. forsteni* — Sumbawa (Lesser Sundas)
_____ *T. h. djampeanus* — Tanahjampea I. (Flores Sea)
_____ *T. h. stresemanni* — Kalaotoa I. (Flores Sea)
_____ *T. h. fortis* — Sumba (Lesser Sundas)
_____ *T. h. weberi* — Flores (Lesser Sundas)
_____ *T. h. capistratus* — Timor (Lesser Sundas)
_____ *T. h. flavotectus* — E Lesser Sundas (Wetar and Romang)

_____ *T. h. rosenbergii*	Biak I. (off n New Guinea)
_____ *T. h. intermedius*	N New Guinea (Sepik River to Astrolabe Bay) and Manam I.
_____ *T. h. haematodus*	S Moluccas, w Papuan islands and w New Guinea
_____ *T. h. nigrogularis (caeruleiceps)*	E Kai Islands, Aru Islands and s New Guinea
_____ *T. h. brooki*	Known from two cage birds from Trangan I. (Aru Islands)
_____ *T. h. micropteryx*	New Guinea e of Huon Pen.; Kimuta and adjacent islands
_____ *T. h. nesophilus*	Admiralty Islands (Ninigo and Hermit groups)
_____ *T. h. flavicans*	New Hanover and Admiralty Islands
_____ *T. h. massena*	Bismarck Archipelago, Solomon Islands and Vanuatu
_____ *T. h. deplanchii*	New Caledonia and Loyalty Islands
_____ *T. h. moluccanus*	E and se Australia (Cape York to Eyre Pen., South Australia)
_____ *T. h. rubritorquis*	N Australia (Kimberley region to Gulf of Carpenteria)

☐ **Olive-headed Lorikeet** *Trichoglossus euteles*

Lesser Sundas (Timor and adj. islands from Lomblen to Babar)

☐ **Yellow-and-green Lorikeet** *Trichoglossus flavoviridis*

_____ *T. f. meyeri*	Montane forests of Sulawesi
_____ *T. f. flavoviridis*	Sula Islands (Taliabu, Seho and Mangole)

☐ **Mindanao Lorikeet** *Trichoglossus johnstoniae*

Montane forests of Mindanao (s Philippines)

☐ **Pohnpei Lorikeet** *Trichoglossus rubiginosus*

Lowlands of Pohnpei (e Caroline Islands)

☐ **Scaly-breasted Lorikeet** *Trichoglossus chlorolepidotus*

Coastal e Australia (n Queensland to cent. New South Wales)

☐ **Varied Lorikeet** *Psitteuteles versicolor*

N Australia (Kimberley Division to Cape York Peninsula)

☐ **Iris Lorikeet** *Psitteuteles iris*

_____ *P. i. iris*	W Timor (e Lesser Sundas)
_____ *P. i. rubripileum*	E Timor (e Lesser Sundas)
_____ *P. i. wetterensis*	Wetar (e Lesser Sundas)

☐ **Goldie's Lorikeet** *Psitteuteles goldiei*

Mts. of New Guinea (Weyland Mts. to Owen Stanley Range)

☐ **Chattering Lory** *Lorius garrulus*

_____ *L. g. morotaianus*	N Moluccas (Morotai and Rau)
_____ *L. g. garrulus*	N Moluccas (Halmahera, Widi and Ternate)
_____ *L. g. flavopalliatus*	N Moluccas (Kasiruta, Bacan, Obi and Mandiole)

☐ **Purple-bellied Lory** *Lorius hypoinochrous*

_____ *L. h. devittatus*	SE New Guinea, Bismarck Archipelago and adjacent islands
_____ *L. h. hypoinochrous*	Louisiade Archipelago (Misima and Tagula)
_____ *L. h. rosselianus*	Rossel I. (Louisiade Archipelago)

☐ **Purple-naped Lory** *Lorius domicella*

S Moluccas (Seram and Ambon)

☐ **Black-capped Lory** *Lorius lory*

_____ *L. l. lory*	W Papuan islands and Vogelkop Peninsula (w New Guinea)
_____ *L. l. cyanauchen*	Biak I. (n New Guinea)
_____ *L. l. jobiensis*	Yapen I. and Mios Num I. (n New Guinea)
_____ *L. l. viridicrissalis*	N New Guinea (Humboldt Bay to Mamberamo River)
_____ *L. l. salvadorii*	NE New Guinea (Aitape area to Astrolabe Bay)
_____ *L. l. erythrothorax*	S and e New Guinea (except for range of *somu*)
_____ *L. l. somu*	Papua New Guinea (Fly River to Purari River)

☐ **White-naped Lory** *Lorius albidinucha*

New Ireland (Bismarck Archipelago)

☐ **Yellow-bibbed Lory** *Lorius chlorocercus*

E Solomon Islands

☐ **Collared Lory** *Phigys solitarius*

Coastal lowland forests of Fiji Islands

132

☐ **Blue-crowned Lorikeet** *Vini australis*

☐ **Kuhl's Lorikeet** *Vini kuhlii*

Samoa, Lau Arch., Tonga and adjacent islands in s-c Polynesia

☐ **Stephen's Lorikeet** *Vini stepheni*

N Line Islands (Rimitara, Kiritimati, Tabuaeran and Teraina)

☐ **Blue Lorikeet** *Vini peruviana*

Henderson I. (Pitcairn Islands)

☐ **Ultramarine Lorikeet** *Vini ultramarina*

Society Islands, Cook Islands and w Tuamotu Archipelago

☐ **Musk Lorikeet** *Glossopsitta concinna*

Marquesas Islands (montane forests of Uapou and Nukuhiva)

☐ **Little Lorikeet** *Glossopsitta pusilla*

SE Queensland to Victoria, Tasmania and Eyre Pen., S Australia

☐ **Purple-crowned Lorikeet** *Glossopsitta porphyrocephala*

E and se Australia (Atherton Tableland to se South Australia)

☐ **Palm Lorikeet** *Charmosyna palmarum*

Semiarid lowlands of sw and se Australia, including Kangaroo I.

☐ **Red-chinned Lorikeet** *Charmosyna rubrigularis*

Vanuatu, Duff, Santa Cruz and Banks islands (sw Pacific)

☐ **Meek's Lorikeet** *Charmosyna meeki*

Bismarck Arch. (New Britain and New Ireland) and Karkar I.

☐ **Blue-fronted Lorikeet** *Charmosyna toxopei*

Montane forests of Solomon Islands (including Bougainvile)

☐ **Striated Lorikeet** *Charmosyna multistriata*

Buru (s Moluccas)

☐ **Pygmy Lorikeet** *Charmosyna wilhelminae*

Mountains of w New Guinea (Snow Mts. to Chimbu Province)

☐ **Red-fronted Lorikeet** *Charmosyna rubronotata*

Mountains of New Guinea (Vogelkop to Owen Stanley Range)

____ *C. r. rubronotata*

____ *C. r. kordoana*

Salawati I. and nw New Guinea (Vogelkop to Adelbert Mts.)
Biak I. (off nw New Guinea)

☐ **Red-flanked Lorikeet** *Charmosyna placentis*

____ *C. p. intensior*
____ *C. p. placentis*
____ *C. p. ornata*
____ *C. p. subplacens*
____ *C. p. pallidior*

N Moluccas and Gebe I. (w Papuan islands)
S Moluccas, Aru islands and s New Guinea
W Papuan islands and nw New Guinea
E New Guinea
Bismarck Arch. and Solomon Islands (Bougainville and Fead)

☐ **Red-throated Lorikeet** *Charmosyna amabilis*

Fiji (Viti Levu, Vanua Levu, Ovalau and Taveuni)

☐ **Duchess Lorikeet** *Charmosyna margarethae*

Montane forests of Solomon Islands (including Bougainville)

☐ **Fairy Lorikeet** *Charmosyna pulchella*

____ *C. p. rothschildi*
____ *C. p. pulchella (bella)*

Cyclops Mountains and montane slopes above Idenburg River
Montane forests of New Guinea

☐ **Josephine's Lorikeet** *Charmosyna josefinae*

____ *C. j. josefinae*
____ *C. j. cyclopum*
____ *C. j. sepikiana*

Mts. of w New Guinea (Vogelkop to Snow Mountains)
Cyclops Mountains (w New Guinea)
New Guinea (Sepik River and W Highlands to Mt. Bosavi)

☐ **Papuan Lorikeet** *Charmosyna papou*

____ *C. p. papou*
____ *C. p. goliathina*
____ *C. p. wahnesi*
____ *C. p. stellae*

W New Guinea (montane forests of Vogelkop Peninsula)
Weyland Mts. to Eastern Highlands of Papua New Guinea
Mountains of Huon Peninsula (ne New Guinea)
Mts. of se New Guinea (Herzog Mts. to Owen Stanley Range)

☐ **Plum-faced Lorikeet** *Oreopsittacus arfaki*

____ *O. a. arfaki*

Vogelkop Mountains (w New Guinea)

_____ *O. a. major* Snow Mountains (w New Guinea)
_____ *O. a. grandis* Central mountains of Papua New Guinea

☐ **Yellow-billed Lorikeet** *Neopsittacus musschenbroekii*
_____ *N. m. musschenbroekii* Vogelkop Mountains (w New Guinea)
_____ *N. m. major* Snow Mts. to Huon Pen. and Owen Stanley Range

☐ **Orange-billed Lorikeet** *Neopsittacus pullicauda*
_____ *N. p. alpinus* New Guinea (Snow Mountains to Mt. Capella)
_____ *N. p. socialis* SE New Guinea (mountains of Huon Pen. and Herzog Mts.)
_____ *N. p. pullicauda* Mts. of se New Guinea (Mt. Capella to Owen Stanley Range)

☐ **Pesquet's Parrot** *Psittrichas fulgidus*
Patchily distributed mountains of New Guinea

☐ **Kea** *Nestor notabilis*
Mountains of South I. (New Zealand)

☐ **New Zealand Kaka** *Nestor meridionalis*
_____ *N. m. septentrionalis* New Zealand (North I. and adjacent offshore islands)
_____ *N. m. meridionalis* South I., Stewart I. and larger New Zealand offshore islands

☐ **Kakapo** *Strigops habroptila*
Nothofagus forests of New Zealand (on verge of extinction)

☐ **Yellow-capped Pygmy-Parrot** *Micropsitta keiensis*
_____ *M. k. keiensis* Kai Islands and Aru Islands
_____ *M. k. chloroxantha* W Papuan islands, Vogelkop and Onin peninsulas
_____ *M. k. viridipectus* S New Guinea (Mimika River to Fly River)

☐ **Geelvink Pygmy-Parrot** *Micropsitta geelvinkiana*
_____ *M. g. geelvinkiana* Numfor I. (Geelvink Bay off w New Guinea)
_____ *M. g. misoriensis* Biak I. (Geelvink Bay off n New Guinea)

☐ **Buff-faced Pygmy-Parrot** *Micropsitta pusio*
_____ *M. p. beccarii* N New Guinea, Manam, Karkar, Bagabag and Rook islands
_____ *M. p. pusio* SE New Guinea and Bismarck Archipelago
_____ *M. p. harterti* Fergusson I. (D'Entrecasteaux Archipelago)
_____ *M. p. stresemanni* Louisiade Archipelago (Misima and Tagula)

☐ **Red-breasted Pygmy-Parrot** *Micropsitta bruijnii*
_____ *M. b. pileata* S Moluccas (Seram and Buru)
_____ *M. b. bruijnii* Mts. of New Guinea (Vogelkop to Owen Stanley Range)
_____ *M. b. necopinata* Bismarck Archipelago (New Britain and New Ireland)
_____ *N. b. rosea* Solomon Is. (Bougainville, Guadalcanal and Kulambangra)

☐ **Meek's Pygmy-Parrot** *Micropsitta meeki*
_____ *M. m. meeki* Admiralty Islands
_____ *M. m. proxima* Bismarck Archipelago (St. Matthias and Squally Islands)

☐ **Finsch's Pygmy-Parrot** *Micropsitta finschii*
_____ *M. f. viridifrons* Bismarck Arch. (New Hanover, New Ireland and Lihir Group)
_____ *M. f. nanina* Solomon Islands (Bougainville, Choiseul and Santa Isabel)
_____ *M. f. tristrami* Vella Lavella, Kulambangra, Rendova and adj. Solomon Is.
_____ *M. f. aolae* E-c Solomon Islands (Russel Is., Guadalcanal and Malaita)
_____ *M. f. finschii* SE Solomon Islands (Ugi, San Cristóbal and Rennell)

☐ **Orange-breasted Fig-Parrot** *Cyclopsitta gulielmitertii*
_____ *C. g. melanogenia* Aru Islands
_____ *C. g. gulielmitertii* W New Guinea (Salawati I. and w Vogelkop Peninsula)
_____ *C. g. nigrifrons* N New Guinea
_____ *C. g. ramuensis* NE New Guinea (Ramu River district)
_____ *C. g. fuscifrons* S New Guinea

____ *C. g. amabilis*	NE New Guinea (Huon Peninsula to Milne Bay)
____ *C. g. suavissima*	SE Papua New Guinea

☐ **Double-eyed Fig-Parrot** *Cyclopsitta diophthalma*

____ *C. d. diophthalma*	W Papuan islands and w New Guinea
____ *C. d. aruensis*	Aru Islands and extreme s New Guinea
____ *C. d. coccineifrons*	E New Guinea east of Astrolabe Bay and Central Highlands
____ *C. d. virago*	D'Entrecasteaux Archipelago (Goodenough and Fergusson)
____ *C. d. inseparabilis*	Tagula I. (Louisiade Archipelago)
____ *C. d. marshalli*	Extreme n Queensland (Cape York Peninsula)
____ *C. d. macleayana*	NE Queensland (Atherton Tableland to Townsville)
____ *C. d. coxeni*	E Australia (se Queensland and ne New South Wales)

☐ **Large Fig-Parrot** *Psittaculirostris desmarestii*

____ *P. d. blythii*	Misool I. (w Papuan Islands)
____ *P. d. occidentalis*	W Vogelkop Peninsula, Salawati and Batanta islands
____ *P. d. desmarestii*	W New Guinea (e regions of Vogelkop Peninsula)
____ *P. d. intermedia*	W New Guinea (Onin Peninsula)
____ *P. d. godmani*	S New Guinea (se Irian Jaya to Fly River)
____ *P. d. cervicalis*	SE New Guinea (Fly River to extreme e Papua New Guinea)

☐ **Edwards' Fig-Parrot** *Psittaculirostris edwardsii*

Lowlands of ne New Guinea (Humboldt Bay to Huon Gulf)

☐ **Salvadori's Fig-Parrot** *Psittaculirostris salvadorii*

NW New Guinea (east shore of Geelvink Bay to Cyclops Mts.)

☐ **Guaiabero** *Bolbopsittacus lunulatus*

____ *B. l. lunulatus*	Luzon (n Philippines)
____ *B. l. callainipictus*	Samar (central Philippines)
____ *B. l. intermedius*	N Philippines (Leyte and Panaon)
____ *B. l. mindanensis*	Mindanao (s Philippines)

☐ **Crimson Shining-Parrot** *Prosopeia splendens*

SW Fiji (Kandavu and Ono)

☐ **Red Shining-Parrot** *Prosopeia tabuensis*

____ *P. t. tabuensis*	Fiji (Vanua Levu, Kioa, Koro and Gau); 'Eua I. (Tonga)
____ *P. t. taviuensis*	Fiji (Taveuni and Ngamea)

☐ **Masked Shining-Parrot** *Prosopeia personata*

Viti Levu I. (Fiji); extirpated on Ovalau and Mbau

☐ **Horned Parakeet** *Eunymphicus cornutus*

____ *E. c. cornutus*	New Caledonia
____ *E. c. uvaeensis*	Uvéa (Loyalty Islands). Population ±617 birds 1997

☐ **Antipodes Parakeet** *Cyanoramphus unicolor*

Locally in Antipodes Islands

☐ **Red-fronted Parakeet** *Cyanoramphus novaezelandiae*

____ *C. n. cyanurus*	Kermadec Islands
____ *C. n. novaezelandiae*	North I., South I., Stewart I. and Auckland Is. (New Zealand)
____ *C. n. chathamensis*	Chatham Islands
____ *C. n. hochstetteri*	Antipodes Islands
____ *C. n. erythrotis†*	Formerly Macquarie I. Extinct

☐ **New Caledonian Parakeet** *Cyanoramphus saissetti*

New Caledonia

☐ **Norfolk Island Parakeet** *Cyanoramphus cookii*

Norfolk I.

☐ **Yellow-fronted Parakeet** *Cyanoramphus auriceps*

North I., South I., Stewart I. and Auckland Is. (New Zealand)

☐ **Chatham Islands Parakeet** *Cyanoramphus forbesi*

Chatham Islands (Mangere and Little Mangare)

☐ **Malherbe's Parakeet** *Cyanoramphus malherbi*

Nothofagus forests of n South I. (New Zealand)

☐ **Red-capped Parrot** *Purpureicephalus spurius*

Lowlands of extreme sw Australia

☐ **Port Lincoln Parrot** *Barnardius zonarius*

____ *B. z. semitorquatus* — Extreme w Western Australia
____ *B. z. occidentalis* — SW Western Australia
____ *B. z. zonarius* — W Australia to s-c Northern Territory and s-c South Australia

☐ **Mallee Ringneck** *Barnardius barnardi*

____ *B. b. macgillivrayi* — N Australia (e Northern Territory and adj. nw Queensland)
____ *B. b. whitei* — South Australia (Flinders Range)
____ *B. b. barnardi* — Interior of se Australia (except in range of *whitei*)

☐ **Green Rosella** *Platycercus caledonicus*

Tasmania and larger islands in Bass Strait

☐ **Crimson Rosella** *Platycercus elegans*

____ *P. e. nigrescens* — E Australia (coastal ne Queensland)
____ *P. e. elegans* — E Australia (se Queensland to se South Australia)
____ *P. e. melanopterus* — Kangaroo I.

☐ **Yellow Rosella** *Platycercus flaveolus*

Interior se Australia (Murray-Murrumbidgee river systems)

☐ **Adelaide Rosella** *Platycercus adelaidae*

____ *P. a. subadelaidae* — S South Australia (s Flinders Range)
____ *P. a. adelaidae* — S South Australia (Mt. Lofty Range to Fleurieu Peninsula)

☐ **Northern Rosella** *Platycercus venustus*

Kimberley Range to Northern Territory/Queensland border

☐ **Eastern Rosella** *Platycercus eximius*

____ *P. e. cecilae* — E Australia (se Queensland and ne New South Wales)
____ *P. e. eximius* — SE New South Wales, Victoria and se South Australia
____ *P. e. diemenensis* — Tasmania

☐ **Pale-headed Rosella** *Platycercus adscitus*

____ *P. a. adscitus* — E Australia (extreme n Cape York Peninsula south to Cairns)
____ *P. a. palliceps* — N Queensland south of Mitchell River to n New South Wales

☐ **Western Rosella** *Platycercus icterotis*

____ *P. i. icterotis* — Coastal areas of extreme sw corner of Australia
____ *P. i. xanthogenys* — Drier interior of extreme sw corner of Australia

☐ **Mulga Parrot** *Psephotus varius*

Interior of s Australia (w W Australia to c New South Wales)

☐ **Red-rumped Parrot** *Psephotus haematonotus*

____ *P. h. caeruleus* — South Australia (Lake Eyre region) and adjacent Queensland
____ *P. h. haematonotus* — Scrub and riverine woodlands of interior se Australia

☐ **Hooded Parrot** *Psephotus dissimilis*

N Australia (ne Northern Territory)

☐ **Golden-shouldered Parrot** *Psephotus chrysopterygius*

NE Australia (interior s Cape York Peninsula)

☐ **Bluebonnet** *Northiella haematogaster*

____ *N. h. haematorrhous* — E Australia (interior s Queensland and n New South Wales)
____ *N. h. haematogaster* — W and s New South Wales, nw Victoria and se S Australia
____ *N. h. pallescens* — Inland South Australia
____ *N. h. narethae* — SE Western Australia to sw South Australia

☐ **Bourke's Parrot** *Neophema bourkii*

Locally in *Acacia* scrub of interior of s Australia

☐ **Blue-winged Parrot** *Neophema chrysostoma*

SE Australia and Tasmania

☐ **Elegant Parrot** *Neophema elegans*

Disjunct in sw and s-central Australia, including Kangaroo I.

☐ **Rock Parrot** *Neophema petrophila*

Coasts and islands of w and s Australia

☐ **Orange-bellied Parrot** *Neophema chrysogaster*

Breeds sw Tasmania; winters to coastal se Australia

☐ **Turquoise Parrot** *Neophema pulchella*

SE Australia (se Queensland to n Victoria)

☐ **Scarlet-chested Parrot** *Neophema splendida*

Interior of s Australia

☐ **Swift Parrot** *Lathamus discolor*

Tasmania and Flinders I.; winters e and se Australia

☐ **Budgerigar** *Melopsittacus undulatus*

Widespread and locally abundant in interior of Australia

☐ **Ground Parrot** *Pezoporus wallicus*
_____ *P. w. flaviventris* — Coastal sw Australia
_____ *P. w. wallicus* — Tasmania, islands in Bass Strait and coastal se Australia

☐ **Night Parrot** *Geopsittacus occidentalis*

Arid interior of w and c Australia (on verge of extinction)

☐ **Blue-rumped Parrot** *Psittinus cyanurus*
_____ *P. c. cyanurus* — S Thailand to s Myanmar, Malay Pen., Sumatra and Borneo
_____ *P. c. abbotti* — Simeulue and Siumat islands (off w coast of Sumatra)
_____ *P. c. pontius* — Mentawi Is. (Siberut, Sipura, North Pagai and South Pagai)

☐ **Painted Tiger-Parrot** *Psittacella picta*
_____ *P. p. lorentzi* — W New Guinea (Snow Mountains)
_____ *P. p. excelsa* — Mountains of Central Highlands of Papua New Guinea
_____ *P. p. picta* — SE New Guinea (Wharton and Owen Stanley mountains)

☐ **Brehm's Tiger-Parrot** *Psittacella brehmii*
_____ *P. b. brehmii* — NW New Guinea (montane forests of Vogelkop Peninsula)
_____ *P. b. intermixta* — W New Guinea (Snow Mts., Weyland Mts. and Mt. Goliath)
_____ *P. b. harterti* — E New Guinea (mountains of Huon Peninsula)
_____ *P. b. pallida* — Central mountains of Papua New Guinea

☐ **Modest Tiger-Parrot** *Psittacella modesta*
_____ *P. m. modesta* — W New Guinea (Vogelkop Mountains)
_____ *P. m. subcollaris* — W New Guinea (n slope of Snow Mts. e to Hindenburg Range)
_____ *P. m. collaris* — W New Guinea (s slopes of Snow Mountains)

☐ **Madarasz's Tiger-Parrot** *Psittacella madaraszi*
_____ *P. m. major* — W New Guinea (Weyland Mts. and n slope of Snow Mts.)
_____ *P. m. hallstromi* — Central Highlands and Hindenburg Range of New Guinea
_____ *P. m. huonensis* — NE New Guinea (mountains of Huon Peninsula)
_____ *P. m. madaraszi* — Mountains of se New Guinea

☐ **Red-cheeked Parrot** *Geoffroyus geoffroyi*
_____ *G. g. cyanicollis* — N Moluccas (Morotai, Halmahera and Bacan)
_____ *G. g. obiensis* — Central Moluccas (Obi and Bisa)
_____ *G. g. rhodops* — S Moluccas (Buru, Seram, Ambon and adjacent islands)
_____ *G. g. explorator* — Seram Laut I. (s Moluccas)
_____ *G. g. keyensis* — Kai Islands
_____ *G. g. floresianus* — W Lesser Sundas (Lombok, Sumbawa, Flores, Besar, Sumba)
_____ *G. g. geoffroyi* — E Lesser Sundas (Timor, Samau and Wetar)
_____ *G. g. timorlaoensis* — Tanimbar Islands (Arafura Sea)
_____ *G. g. pucherani* — W Papuan islands and nw New Guinea east to Etna Bay
_____ *G. g. minor* — N New Guinea (Mamberamo River to Astrolabe Bay)
_____ *G. g. jobiensis* — Yapen I. and Meos Num I. (n New Guinea)

_____ *G. g. mysoriensis*	Biak I. and Numfor I. (n New Guinea)
_____ *G. g. orientalis*	NE New Guinea (Huon Peninsula)
_____ *G. g. sudestiensis*	Louisiade Archipelago (Misima and Tagula)
_____ *G. g. cyanicarpus*	Rossel I. (Louisiade Archchipelago)
_____ *G. g. aruensis*	Aru Is., s New Guinea, Louisiade Arch. and n Queensland

☐ **Blue-collared Parrot** *Geoffroyus simplex*
_____ *G. s. simplex*	W New Guinea (Vogelkop Mountains)
_____ *G. s. buergersi*	Snow Mountains to Owen Stanley Range (New Guinea)

☐ **Singing Parrot** *Geoffroyus heteroclitus*
_____ *G. h. heteroclitus*	Bismarck Archipelago and Solomon Islands (except Rennell)
_____ *G. h. hyancinthinus*	Rennell (Solomon Islands)

☐ **Luzon Racquet-tail** *Prioniturus montanus*

Mountains of Luzon (n Philippines)

☐ **Mindanao Racquet-tail** *Prioniturus waterstradti*

Mountains of Mindanao (s Philippines)

☐ **Blue-headed Racquet-tail** *Prioniturus platenae*

S Philippines (Balabac, Palawan, Calamian and adj. islands)

☐ **Green Racquet-tail** *Prioniturus luconensis*

N Philippines (lowlands of Luzon and Marinduque)

☐ **Blue-crowned Racquet-tail** *Prioniturus discurus*
_____ *P. d. whiteheadi*	Philippines (Negros, Bohol, Samar, Leyte, Masbate, Cebu)
_____ *P. d. mindorensis*	Mindoro (n-central Philippines)
_____ *P. d. discurus*	Mindanao, Basilan and islands in Sulu Archipelago

☐ **Blue-winged Racquet-tail** *Prioniturus verticalis*

Sulu Archipelago

☐ **Yellowish-breasted Racquet-tail** *Prioniturus flavicans*

Lowlands of n Sulawesi, Bangka, Lembeh and Togian islands

☐ **Golden-mantled Racquet-tail** *Prioniturus platurus*
_____ *P. p. talautensis*	Talaud Islands (n Moluccas)
_____ *P. p. platurus*	Sulawesi, Togian, Banggai and adjacent islands
_____ *P. p. sinerubris*	Sula Islands (Taliabu and Mangole)

☐ **Buru Racquet-tail** *Prioniturus mada*

Montane forests of Buru (s Moluccas)

☐ **Black-lored Parrot** *Tanygnathus gramineus*

Montane forests of Buru (s Moluccas)

☐ **Great-billed Parrot** *Tanygnathus megalorynchos*
_____ *T. m. megalorynchos*	Sulawesi and adjacent islands to Moluccas and w Papuan is.
_____ *T. m. affinis*	S Moluccas (Buru, Seram, Ambon, Haruku and Seram Laut)
_____ *T. m. sumbensis*	Sumba (e Lesser Sundas)
_____ *T. m. hellmayri*	E Lesser Sundas (Roti, Semau and sw Timor)
_____ *T. m. subaffinis*	Babar and Tanimbar Islands (Yamdena and Larat)

☐ **Blue-naped Parrot** *Tanygnathus lucionensis*
_____ *T. l. lucionensis*	N Philippines (Luzon and Mindoro)
_____ *T. l. hybridus*	Polillo (n Philippines)
_____ *T. l. salvadorii*	S Philippines, Sulu Archipelago and islands off north Borneo
_____ *T. l. talautensis*	Talaud Islands (n Moluccas)

☐ **Azure-rumped Parrot** *Tanygnathus sumatranus*
_____ *T. s. duponti*	Luzon (n Philippines)
_____ *T. s. freeri*	Polillo (n Philippines)
_____ *T. s. everetti*	Philippines (Visayan Islands and Mindanao)
_____ *T. s. burbidgii*	Sulu Archipelago
_____ *T. s. sangirensis*	Sangihi I. and Talaud Islands
_____ *T. s. sumatranus*	Sulawesi, Togian, Sula, Muna, Buton is. and Banggai Arch.

☐ **Eclectus Parrot** *Eclectus roratus*

____	*E. r. vosmaeri*	Larger islands in n and central Moluccas
____	*E. r. roratus*	S Moluccas (Buru, Seram, Ambon, Saparua and Haruku)
____	*E. r. cornelia*	Sumba I. (Lesser Sundas)
____	*E. r. riedeli*	Tanimbar Islands (Arafura Sea)
____	*E. r. aruensis*	Aru Islands (New Guinea)
____	*E. r. biaki*	Biak I. (off nw New Guinea)
____	*E. r. polychloros*	New Guinea, adjacent islands and archipelagos
____	*E. r. solomonensis*	Admiralty Islands, Bismarck Arch. and Solomon Islands
____	*E. r. macgillivrayi*	NE Australia (extreme n Queensland)

☐ **Australian King-Parrot** *Alisterus scapularis*

____	*A. s. minor*	N Australia (ne Queensland)
____	*A. s. scapularis*	Coastal e Australia (n Queensland to s Victoria)

☐ **Moluccan King-Parrot** *Alisterus amboinensis*

____	*A. a. hypophonius*	Halmahera I. (n Moluccas)
____	*A. a. sulaensis*	Sula Islands (Taliabu, Seho and Mangole)
____	*A. a. versicolor*	Peleng I. (Banggai Islands)
____	*A. a. buruensis*	Buru (s Moluccas)
____	*A. a. amboinensis*	S Moluccas (Boano, Ambon and Seram)
____	*A. a. dorsalis*	W Papuan islands and nw New Guinea

☐ **Papuan King-Parrot** *Alisterus chloropterus*

____	*A. c. moszkowskii*	N New Guinea (Geelvink Bay to Aitape district)
____	*A. c. callopterus*	Central New Guinea (Weyland Mountains to Fly River)
____	*A. c. chloropterus*	E New Guinea (Huon Gulf to Hall Sound)

☐ **Olive-shouldered Parrot** *Aprosmictus jonquillaceus*

____	*A. j. wetterensis*	Wetar (e Lesser Sundas)
____	*A. j. jonquillaceus*	E Lesser Sundas (Timor and Roti)

☐ **Red-winged Parrot** *Aprosmictus erythropterus*

____	*A. e. coccineopterus*	Trans-Fly lowlands of s New Guinea and n Australia
____	*A. e. erythropterus*	Interior e Australia

☐ **Superb Parrot** *Polytelis swainsonii*

Interior se Australia (New South Wales and n Victoria)

☐ **Regent Parrot** *Polytelis anthopeplus*

____	*P. a. anthopeplus*	SW Australia
____	*P. a. monarchoides*	Interior western part of se Australia

☐ **Alexandra's Parrot** *Polytelis alexandrae*

Dry eucalyptus forests of interior central and w Australia

☐ **Alexandrine Parakeet** *Psittacula eupatria*

____	*P. e. nipalensis*	E Afghanistan to Pakistan, n India and Bangladesh
____	*P. e. eupatria*	S India and Sri Lanka
____	*P. e. magnirostris*	Andaman Islands
____	*P. e. avensis*	N Myanmar and adjacent ne India
____	*P. e. siamensis*	Thailand to Laos, Cambodia and Vietnam

☐ **Rose-ringed Parakeet** *Psittacula krameri*

____	*P. k. krameri*	Mauritania to Senegal, Guinea, w Uganda and s Sudan
____	*P. k. parvirostris*	E Sudan (Sennar) to Eritrea, Ethiopia, Djibouti and nw Somalia
____	*P. k. borealis*	NW Pakistan to n India, Nepal, se China and c Myanmar
____	*P. k. manillensis*	S peninsular India and Sri Lanka

☐ **Mauritius Parakeet** *Psittacula echo*

Montane forests of Mauritius (on verge of extinction)

☐ **Slaty-headed Parakeet** *Psittacula himalayana*

Himalayas (Afghanistan to n India, Nepal and w Assam)

☐ **Gray-headed Parakeet** *Psittacula finschii*

N India (w Bengal) to s China, Myanmar and Indochina

☐ **Plum-headed Parakeet** *Psittacula cyanocephala*

Indian subcontinent and Sri Lanka

☐ **Blossom-headed Parakeet** *Psittacula roseata*

____ *P. r. roseata*
____ *P. r. juneae*

N India (w Bengal) to Bhutan, Bangladesh and n Myanmar
S Myanmar and Thailand to Laos, Cambodia and Vietnam

☐ **Malabar Parakeet** *Psittacula columboides*

SW India (Western Ghats)

☐ **Layard's Parakeet** *Psittacula calthropae*

Sri Lanka

☐ **Derbyan Parakeet** *Psittacula derbiana*

Extreme sw China to se Tibet and ne Assam

☐ **Red-breasted Parakeet** *Psittacula alexandri*

____ *P. a. fasciata*
____ *P. a. abbotti*
____ *P. a. cala*
____ *P. a. major*
____ *P. a. alexandri*
____ *P. a. kangeanensis*
____ *P. a. dammermani*

N India to Nepal, Myanmar, Thailand, Indochina and Hainan
Andaman Islands
Simeulue I. (off w Sumatra)
Lasia I. and Babi I. (off Sumatra)
Java, Bali and extreme s Borneo
Kangean Islands (Java Sea)
Karimunjawa Islands (Java Sea)

☐ **Nicobar Parakeet** *Psittacula caniceps*

Nicobar Islands

☐ **Long-tailed Parakeet** *Psittacula longicauda*

____ *P. l. tytleri*
____ *P. l. nicobarica*
____ *P. l. longicauda*
____ *P. l. modesta*
____ *P. l. defontainei*

Andaman Islands
Nicobar Islands
S Malay Pen., Borneo, Sumatra, Nias, Bangka and Anambas is.
Enggano I. (off sw Sumatra)
Natuna Islands (off w Borneo)

☐ **Vernal Hanging-Parrot** *Loriculus vernalis*

NE and sw India to s China, SE Asia and Andaman Islands

☐ **Ceylon Hanging-Parrot** *Loriculus beryllinus*

Sri Lanka

☐ **Philippine Hanging-Parrot** *Loriculus philippensis*

____ *L. p. philippensis*
____ *L. p. mindorensis*
____ *L. p. bournsi*
____ *L. p. regulus*
____ *L. p. chrysonotus†*
____ *L. p. worcesteri*
____ *L. p. siquijorensis*
____ *L. p. apicalis*
____ *L. p. dohertyi*
____ *L. p. bonapartei*

Philippines (Banton, Catanduanes, Luzon, Polillo, Marinduque)
Mindoro (Philippines)
Sibuyan (Philippines)
Guimaras, Masbate, Negros, Panay, Tablas, Ticao and Romblon
Formerly Cebu (Philippines). Extinct
Philippines (Bohol, Leyte and Samar)
Formerly Siquijor (Philippines). Probably extinct
S Philippines (Mindanao, Basol and Dinagat)
Basilan (Philippines)
Sulu Archipelago (Bongao, Jolo and Tawitawi)

☐ **Camiguin Hanging-Parrot** *Loriculus camiguinensis*

Camiguin (Philippines)

☐ **Blue-crowned Hanging-Parrot** *Loriculus galgulus*

S Thailand, Malay Pen., Sumatra, Borneo and adjacent islands

☐ **Sulawesi Hanging-Parrot** *Loriculus stigmatus*

Sulawesi and adjacent islands

☐ **Sula Hanging-Parrot** *Loriculus sclateri*

Sula Islands and Banggai Islands

☐ **Moluccan Hanging-Parrot** *Loriculus amabilis*

N Moluccas (Halmahera and Bacan)

☐ **Sangihe Hanging-Parrot** *Loriculus catamene*

Sangihe I. (n of Sulawesi)

☐ **Papuan Hanging-Parrot** *Loriculus aurantiifrons*
____ *L. a. aurantiifrons* — Misool I. (w Papuan islands)
____ *L. a. batavorum* — Waigeo I. and coastal nw New Guinea
____ *L. a. meeki* — E New Guinea, Fergusson I., Goodenough I. and Karkar I.

☐ **Green-fronted Hanging-Parrot** *Loriculus tener* — Bismarck Archipelago

☐ **Pygmy Hanging-Parrot** *Loriculus exilis* — Sulawesi

☐ **Yellow-throated Hanging-Parrot** *Loriculus pusillus* — Java and Bali

☐ **Wallace's Hanging-Parrot** *Loriculus flosculus* — Flores (w Lesser Sundas)

☐ **Gray-headed Lovebird** *Agapornis canus*
____ *A. c. canus* — W and e Madagascar (except range of *ablectaneus*)
____ *A. c. ablectaneus* — Arid sw Madagascar

☐ **Red-headed Lovebird** *Agapornis pullarius*
____ *A. p. pullarius* — Sierra Leone to Guinea, Sudan, Angola and Zaire; São Tomé I.
____ *A. p. ugandae* — Ethiopia to Uganda, extreme e Zaire, Rwanda and Tanzania

☐ **Black-winged Lovebird** *Agapornis taranta* — Highland forests of Ethiopia

☐ **Black-collared Lovebird** *Agapornis swindernianus*
____ *A. s. swindernianus* — Patchily distributed Liberia, Ivory Coast and Ghana
____ *A. s. zenkeri* — Cameroon to Gabon, s Central African Republic and w Zaire
____ *A. s. emini* — Lowland forests of Zaire and w Uganda

☐ **Rosy-faced Lovebird** *Agapornis roseicollis*
____ *A. r. catumbellus* — Subdeserts of sw Angola
____ *A. r. roseicollis* — Subdeserts of Namibia to n Cape Province

☐ **Fischer's Lovebird** *Agapornis fischeri* — N Tanzania (south and east of Lake Victoria)

☐ **Yellow-collared Lovebird** *Agapornis personatus* — Tanzania

☐ **Lilian's Lovebird** *Agapornis lilianae* — S Tanzania to Malawi, Zambia, Zimbabwe and Mozambique

☐ **Black-cheeked Lovebird** *Agapornis nigrigenis* — S Zambia and extreme n Zimbabwe

☐ **Vasa Parrot** *Coracopsis vasa*
____ *C. v. comorensis* — Comoro Islands (Grand Comoro, Mohéli and Anjouan)
____ *C. v. drouhardi* — W and s Madagascar
____ *C. v. vasa* — Savanna and forests of e Madagascar

☐ **Black Parrot** *Coracopsis nigra*
____ *C. n. sibilans* — Comoro Islands (Grand Comoro and Anjouan)
____ *C. n. libs* — Drier areas of w Madagascar
____ *C. n. nigra* — Forests of e Madagascar
____ *C. n. barklyi* — Seychelles (Praslin and Curieuse)

☐ **Gray Parrot** *Psittacus erithacus*
____ *P. e. timneh* — S Guinea to Sierra Leone, Liberia, Mali and w Ivory Coast
____ *P. e. erithacus* — Ivory Coast to Kenya, Tanzania, Príncipe, São Tomé and Bioko

☐ **Brown-necked Parrot** *Poicephalus robustus*
____ *P. r. fuscicollis* — Senegambia to Nigeria and n Angola
____ *P. r. suahelicus* — C Tanzania to ne Transvaal, se Zaire, Angola and Namibia
____ *P. r. robustus* — Extreme se Africa

☐ **Red-fronted Parrot** *Poicephalus gulielmi*
    ____ *P. g. fantiensis*
    ____ *P. g. gulielmi*
    ____ *P. g. massaicus*

Liberia to Ivory Coast and Ghana	
Cameroon to n Angola e Zaire and w Uganda	
W Kenya and n Tanzania	

☐ **Meyer's Parrot** *Poicephalus meyeri*
    ____ *P. m. meyeri*
    ____ *P. m. saturatus*
    ____ *P. m. matschiei*
    ____ *P. m. reichenowi*
    ____ *P. m. damarensis*
    ____ *P. m. transvaalensis*

N Cameroon to s Chad, n Zaire, s Sudan and Ethiopia
W Kenya to Uganda, e Zaire, Rwanda, Burunda, nw Tanzania
SE Zaire to Tanzania, e Angola, n Zambia and n Malawi
W Angola
Extreme s Angola to n Namibia and n-central Botswana
S Zambia to n Mozambique, e Botswana and n South Africa

☐ **Rueppell's Parrot** *Poicephalus rueppellii*

Arid sw Angola to central Namibia

☐ **Brown-headed Parrot** *Poicephalus cryptoxanthus*
    ____ *P. c. tanganyikae*
    ____ *P. c. cryptoxanthus*

SE Kenya to Malawi, n Mozambique, Zanzibar and Pemba I.
SE Zimbabwe and Mozambique (s of Save R.) to ne S Africa

☐ **Niam-Niam Parrot** *Poicephalus crassus*

SW Chad to Central African Republic and extreme sw Sudan

☐ **Red-bellied Parrot** *Poicephalus rufiventris*
    ____ *P. r. pallidus*
    ____ *P. r. rufiventris*

Dry thornbush of e Ethiopia and Somalia
Central Ethiopia to n Tanzania

☐ **Senegal Parrot** *Poicephalus senegalus*
    ____ *P. s. senegalus*
    ____ *P. s. versteri*

Gambia and Guinea-Bissau to s Niger, n Cameroon, sw Chad
NW Ivory Coast to sw Nigeria (south of range of *senegalus*)

☐ **Yellow-fronted Parrot** *Poicephalus flavifrons*

Montane forests of Ethiopia

☐ **Hyacinth Macaw** *Anodorhynchus hyacinthinus*

Interior s Brazil, extreme nw Paraguay and adjacent e Bolivia

☐ **Lear's Macaw** *Anodorhynchus leari*

Caatinga of e Brazil (n Bahia). On verge of extinction

☐ **Spix's Macaw** *Cyanopsitta spixii*

Palm groves of ne Brazil (n Bahia). On verge of extinction

☐ **Blue-and-yellow Macaw** *Ara ararauna*

Tropical e Panama to e Peru, n Bolivia, Paraguay and e Brazil

☐ **Blue-throated Macaw** *Ara glaucogularis*

Chaco of e Bolivia (Beni and Santa Cruz)

☐ **Military Macaw** *Ara militaris*
    ____ *A. m. mexicanus*
    ____ *A. m. militaris*
    ____ *A. m. bolivianus*

Arid w Mexico (Sonora to Isthmus of Tehuántepec)
Tropical Colombia to nw Venezuela, Ecuador and n Peru
Tropical Bolivia and extreme nw Argentina

☐ **Great Green Macaw** *Ara ambiguus*
    ____ *A. a. ambiguus*
    ____ *A. a. guayaquilensis*

E Honduras to nw Colombia
W Ecuador and adjacent sw Colombia

☐ **Scarlet Macaw** *Ara macao*
    ____ *A. m. cyanopterus*
    ____ *A. m. macao*

SE Mexico to Nicaragua
Costa Rica to Colombia, the Guianas, Brazil, Peru and Bolivia

☐ **Red-and-green Macaw** *Ara chloropterus*

Humid e Panama to Brazil, e Peru, ne Bolivia and Paraguay

☐ **Red-fronted Macaw** *Ara rubrogenys*

Andean valleys of central Bolivia

☐ **Chestnut-fronted Macaw** *Ara severus*

Tropical e Panama to the Guianas, n Bolivia and Amaz. Brazil

☐ **Red-bellied Macaw** *Orthopsittaca manilata*

SE Colombia to the Guianas, Trinidad, n Bolivia, Amaz. Brazil

☐ **Blue-headed Macaw** *Primolius couloni*

E Peru to n Bolivia and extreme w Brazil

☐ **Blue-winged Macaw** *Primolius maracana*

E Brazil to Paraguay and extreme ne Argentina

☐ **Golden-collared Macaw** *Primolius auricollis*

NE Bolivia to Paraguay, sw Brazil and n Argentina

☐ **Red-shouldered Macaw** *Diopsittaca nobilis*
____ *D. n. nobilis*
____ *D. n. cumanensis*
____ *D. n. longipennis*

E Venezuela to the Guianas and n Brazil north of the Amazon
N Brazil south of lower Amazon to ne Brazil
SE Peru and ne Bolivia to central and se Brazil

☐ **Thick-billed Parrot** *Rhynchopsitta pachyrhyncha*

Mountains of w Mexico (Sierra Madre Occidental)

☐ **Maroon-fronted Parrot** *Rhynchopsitta terrisi*

Mountains of e Mexico (Sierra Madre Oriental)

☐ **Yellow-eared Parrot** *Ognorhynchus icterotis*

Andes of Colombia and n Ecuador

☐ **Golden Parakeet** *Guarouba guarouba*

NE Brazil (n Maranhão and Pará)

☐ **Blue-crowned Parakeet** *Aratinga acuticaudata*
____ *A. a. koenigi*
____ *A. a. neoxena*
____ *A. a. haemorrhous*
____ *A. a. neumanni*
____ *A. a. acuticaudata*

NE Colombia and n Venezuela
Isla Margarita (Venezuela)
Interior ne Brazil
Highlands of e Bolivia
E Bolivia to Paraguay, s Brazil, w Uruguay and n Argentina

☐ **Green Parakeet** *Aratinga holochlora*
____ *A. h. brewsteri*
____ *A. h. holochlora*

Mountains of nw Mexico (Sonora, Sinaloa and Chihuahua)
Open woodlands and pine forests of s Mexico

☐ **Pacific Parakeet** *Aratinga strenua*

Arid lowlands of se Mexico to n Nicaragua

☐ **Socorro Parakeet** *Aratinga brevipes*

Socorro I. (Revillagigedo Islands off w Mexico)

☐ **Red-throated Parakeet** *Aratinga rubritorquis*

Highlands of e Guatemala to n Nicaragua

☐ **Scarlet-fronted Parakeet** *Aratinga wagleri*
____ *A. w. wagleri*
____ *A. w. transilis*
____ *A. w. frontata*
____ *A. w. minor*

N Colombia (south to n Nariño) and extreme nw Venezuela
Extreme e Colombia to n Venezuela
W Ecuador and w Peru (south to Arequipa)
C and s central Peru (Marañón Valley south to Ayacucho)

☐ **Hocking's Parakeet** *Aratinga hockingi*

E Andes, n Peru

☐ **Mitred Parakeet** *Aratinga mitrata*
____ *A. m. mitrata*
____ *A. m. chlorogenys*
____ *A. m. tucumana*

Andes of central Peru to Bolivia and nw Argentina
E Andes, n Peru
Tucumán and Córdoba, Argentina

☐ **Chapman's Mitred Parakeet** *Aratinga alticola*

Andes of se Peru (Cuzco)

☐ **Red-masked Parakeet** *Aratinga erythrogenys*

Arid littoral of w Ecuador and nw Peru

☐ **Crimson-fronted Parakeet** *Aratinga finschi*

Humid lowlands of se Nicaragua, Costa Rica and w Panama

☐ **White-eyed Parakeet** *Aratinga leucophthalma*
____ *A. l. nicefori*
____ *A. l. callogenys*
____ *A. l. leucophthalma (propinqua)*

Known from one specimen from e Colombia (Meta)
SE Colombia to e Ecuador, nw Peru and extreme nw Brazil
Venezuela to Guianas, Brazil, Bolivia, Paraguay, n Argentina

☐ **Cuban Parakeet** *Aratinga euops*

SW Cuba (Zapata Swamp); formerly Isle of Pines

☐ **Hispaniolan Parakeet** *Aratinga chloroptera*
_____ *A. c. chloroptera*
_____ *A. c. maugei†*

Hispaniola
Formerly Mona I. (off Puerto Rico). Extinct

☐ **Sulphur-breasted Parakeet** *Aratinga pintoi*

Brazil (north bank of lower Amazon River in Pará)

☐ **Sun Parakeet** *Aratinga solstitialis*

Guyana, Suriname and n Amazonian Brazil

☐ **Jandaya Parakeet** *Aratinga jandaya*

Lowlands of ne Brazil (e Pará and Goiás to Alagoas)

☐ **Golden-capped Parakeet** *Aratinga auricapillus*
_____ *A. a. auricapillus*
_____ *A. a. aurifrons*

E central Brazil (n and central Bahia)
SE Brazil (s Bahia to s Paraná)

☐ **Dusky-headed Parakeet** *Aratinga weddellii*

SE Colombia to n Bolivia and adjacent w Amazonian Brazil

☐ **Brown-throated Parakeet** *Aratinga pertinax*
_____ *A. p. ocularis*
_____ *A. p. aeruginosa*
_____ *A. p. griseipecta*
_____ *A. p. lehmanni*
_____ *A. p. arubensis*
_____ *A. p. pertinax*
_____ *A. p. xanthogenia*
_____ *A. p. tortugensis*
_____ *A. p. margaritensis*
_____ *A. p. venezuelae*
_____ *A. p. surinama*
_____ *A. p. chrysophrys*
_____ *A. p. chrysogenys*
_____ *A. p. paraensis*

Pacific lowlands of sw Costa Rica and Panama
N Colombia to nw Venezuela
NE Colombia (Sinú River Valley)
Llanos of e Colombia (possibly adjacent w Venezuela)
Aruba (Netherlands Antilles)
Curaçao (Netherlands Antilles)
Bonaire (Netherlands Antilles)
Isla la Tortuga (off n Venezuela)
Isla Margarita and Islas Los Frailes (off n Venezuela)
Generally distributed throughout Venezuela
NE Venezuela and the Guianas
Tepuis of se Venezuela and adjacent Brazil
NW Brazil (Rio Negro region)
N Amazonian Brazil (Rio Tapajós and Rio Cururu)

☐ **Olive-throated Parakeet** *Aratinga nana*
_____ *A. n. vicinalis*
_____ *A. n. astec*
_____ *A. n. nana*

NE Mexico (Tamaulipas to ne Veracruz)
Caribbean slope of se Mexico to extreme w Panama
Jamaica

☐ **Orange-fronted Parakeet** *Aratinga canicularis*
_____ *A. c. clarae*
_____ *A. c. eburnirostrum*
_____ *A. c. canicularis*

W Mexico (Sinaloa to Colima, Durango and Michoacán)
SW Mexico (e Michoacán to Guerrero and Oaxaca)
Arid tropical w Mexico (Chiapas) to w Costa Rica

☐ **Peach-fronted Parakeet** *Aratinga aurea*

Suriname to s Brazil, se Peru, e Bolivia, Paraguay, n Argentina

☐ **Caatinga Parakeet** *Aratinga cactorum*
_____ *A. c. caixana*
_____ *A. c. cactorum*

Caatinga of ne Brazil (Pará to nw Bahia)
Inland *caatinga* of ne Brazil (Bahia and adj. Minas Gerais)

☐ **Nanday Parakeet** *Nandayus nenday*

Pantanal of se Bolivia, sw Brazil, Paraguay and n Argentina

☐ **Golden-plumed Parakeet** *Leptosittaca branickii*

Humid Andes of s Colombia to Ecuador and central Peru

☐ **Burrowing Parrot** *Cyanoliseus patagonus*
_____ *C. p. andinus*
_____ *C. p. conlara*
_____ *C. p. patagonus*
_____ *C. p. bloxami*

NW Argentina (Salta to San Luis)
W-central Argentina (San Luis and Córdoba)
C to se Argentina; winters to n Argentina and Uruguay
Central Chile (Atacama to Valdivia)

☐ **Blue-throated Parakeet** *Pyrrhura cruentata*

Lowlands of se Brazil (Bahia to Rio de Janeiro)

☐ **Blaze-winged Parakeet** *Pyrrhura devillei*

Forests of n Paraguay and sw Brazil (sw Mato Grosso)

☐ **Maroon-bellied Parakeet** *Pyrrhura frontalis*
____ *P. f. frontalis*
____ *P. f. chiripepe*

E Brazil (Bahia to Rio de Janeiro and n São Paulo)
SE Brazil to se Paraguay and n Argentina

☐ **Crimson-bellied Parakeet** *Pyrrhura perlata*

Brazil (w Pará, e Amazonas and Mato Grosso) to n Bolivia

☐ **Pearly Parakeet** *Pyrrhura lepida*
____ *P. l. lepida*
____ *P. l. anerythra*
____ *P. l. coerulescens*

N-central Brazil (ne Pará and nw Maranhão)
N-central Brazil (e Pará)
N-central Brazil (w and central Maranhão)

☐ **Green-cheeked Parakeet** *Pyrrhura molinae*
____ *P. m. phoenicura*
____ *P. m. molinae*
____ *P. m. restricta*
____ *P. m. sordida*
____ *P. m. australis*

NE Bolivia and w Brazil (w Mato Grosso)
Highlands of e Bolivia
Lowlands of e Bolivia (Palmarito)
Extreme e Bolivia and sw Brazil (s Mato Grosso)
S Bolivia (Tarija) to nw Argentina

☐ **Painted Parakeet** *Pyrrhura picta*
____ *P. p. picta*
____ *P. p. microtera*
____ *P. p. pantchenkoi*

SE Venezuela to the Guianas and n Amazonian Brazil (Amapá)
N-central Brazil south of the Amazon (Pará to n Goiás)
Sierra de Perijá (Colombia/Venezuela border)

☐ **Sinu Parakeet** *Pyrrhura subandina*

NW Colombia (lower Sinú River Valley)

☐ **Todd's Parakeet** *Pyrrhura caeruleiceps*

W slope of Eastern Andes of n Colombia and Sierra de Perijá

☐ **Azuero Parakeet** *Pyrrhura eisenmanni*

S-central Panama (Azuero Peninsula)

☐ **Red-crowned Parakeet** *Pyrrhura roseifrons*

Disjunct in w Amazonia (w Brazil; n-central Peru and Bolivia)

☐ **Deville's Parakeet** *Pyrrhura lucianii*

Brazil (w-central Amazonia along Rio Solimões at Tefe and Rio Purús)

☐ **Hellmayr's Parakeet** *Pyrrhura amazonum*

N-central Brazil (Pará, n Mato Grosso, n Goiás and Maranhão)

☐ **Madeira Parakeet** *Pyrrhura snethlage*

Drainage of Rio Madeira in s Brazil and n Bolivia

☐ **Wavy-breasted Parakeet** *Pyrrhura peruviana*

Disjunct in se Ecuador and n Peru; s-central Peru (Junín)

☐ **Fiery-shouldered Parakeet** *Pyrrhura egregia*
____ *P. e. egregia*
____ *P. e. obscura*

Tepuis of s Venezuela (Mt. Roraima) and adjacent Guyana
Tepuis of se Venezuela and extreme ne Brazil

☐ **White-eared Parakeet** *Pyrrhura leucotis*
____ *P. l. emma*
____ *P. l. auricularis*
____ *P. l. pfrimeri*
____ *P. l. griseipectus*
____ *P. l. anca*
____ *P. l. leucotis*

Patchily distributed humid forests of coastal n Venezuela
Coastal ne Venezuela (Sucre, Anzoátegui and Monagas)
Known from the type locality in central Brazil (Goiás)
NE Brazil (Ceará)
Coastal e Brazil (s Bahia to São Paulo)
E Brazil (Bahia to São Paulo)

☐ **Santa Marta Parakeet** *Pyrrhura viridicata*

Santa Marta Mountains (ne Colombia)

☐ **Maroon-tailed Parakeet** *Pyrrhura melanura*
____ *P. m. pacifica*
____ *P. m. chapmani*

W slope of Andes of sw Colombia (Nariño) and nw Ecuador
Subtropical e slope of Central Andes of s Colombia

____ *P. m. melanura*		SE Colombia to e Ecuador, ne Peru, s Venezuela and nw Brazil
____ *P. m. souancei*		S-central Colombia (Macarena Mountains)
____ *P. m. berlepschi*		E slope of Andes of se Ecuador and n Peru

☐ **El Oro Parakeet** *Pyrrhura orcesi*

Andean foothills of sw Ecuador (El Oro and Azuay)

☐ **Black-capped Parakeet** *Pyrrhura rupicola*
____ *P. r. rupicola* — Humid forests of e-central Peru
____ *P. r. sandiae* — Tropical se Peru, n Bolivia and extreme w Amazonian Brazil

☐ **White-necked Parakeet** *Pyrrhura albipectus*

Humid lowlands of se Ecuador

☐ **Flame-winged Parakeet** *Pyrrhura calliptera*

E Andes of Colombia (Boyacá and Cundinamarca)

☐ **Red-eared Parakeet** *Pyrrhura hoematotis*
____ *P. h. immarginata* — Known only from the type locality in nw Venezuela (Lara)
____ *P. h. hoematotis* — Montane forests of n Venezuela (Aragua to Miranda)

☐ **Rose-headed Parakeet** *Pyrrhura rhodocephala*

Andes of w Venezuela (Mérida, Táchira, and Trujillo)

☐ **Sulphur-winged Parakeet** *Pyrrhura hoffmanni*
____ *P. h. hoffmanni* — Highlands of s Costa Rica
____ *P. h. gaudens* — Mts. of w Panama and Caribbean slope of Bocas del Toro

☐ **Austral Parakeet** *Enicognathus ferrugineus*
____ *E. f. minor* — S Chile (Colchagua to Aysén) and adj. Andes of sw Argentina
____ *E. f. ferrugineus* — *Nothofagus* forests of extreme s Chile and s Argentina

☐ **Slender-billed Parakeet** *Enicognathus leptorhynchus*

Lowlands of Chile (Aconcagua to n Aysén and Chiloé I.)

☐ **Monk Parakeet** *Myiopsitta monachus*
____ *M. m. cotorra* — S Bolivia to Paraguay, s Brazil and nw Argentina
____ *M. m. monachus* — SE Brazil (Rio Grande do Sul), Uruguay and ne Argentina
____ *M. m. calita* — W Argentina (Salta to w Córdoba, Mendoza and La Pampa)

☐ **Cliff Parakeet** *Myiopsitta luchsi*

Xeric intermontane valleys of central Bolivia

☐ **Andean Parakeet** *Bolborhynchus orbygnesius*

Andes of Peru and w Bolivia

☐ **Barred Parakeet** *Bolborhynchus lineola*
____ *B. l. lineola* — Humid montane forests of s Mexico to w Panama
____ *B. l. tigrinus* — Mountains of nw Venezuela and Colombia to s Peru

☐ **Rufous-fronted Parakeet** *Bolborhynchus ferrugineifrons*

Central Andes of w Colombia (Tolima and Cauca)

☐ **Gray-hooded Parakeet** *Psilopsiagon aymara*

Andes of Bolivia to nw Argentina and n Chile

☐ **Mountain Parakeet** *Psilopsiagon aurifrons*
____ *P. a. robertsi* — N central Peru (Marañón Valley)
____ *P. a. aurifrons* — Coastal regions and adj. w slopes of Andes of central Peru
____ *P. a. margaritae* — Andes of s Peru to Bolivia, n Chile and extreme nw Argentina
____ *P. a. rubrirostris* — Andes of nw Argentina (Catamarca to Córdoba) and adj. Chile

☐ **Mexican Parrotlet** *Forpus cyanopygius*
____ *F. c. cyanopygius* — Arid w Mexico (se Sonora to Sinaloa, w Durango and Colima)
____ *F. c. insularis* — Tres Marías Islands (off w Mexico)

☐ **Green-rumped Parrotlet** *Forpus passerinus*
____ *F. p. cyanophanes* — Arid tropical n Colombia
____ *F. p. viridissimus* — NE Colombia to n Venezuela and Trinidad

___	*F. p. passerinus*	Guyana, Suriname and French Guiana
___	*F. p. cyanochlorus*	Extreme n Brazil (upper Rio Branco region of Roraima)
___	*F. p. deliciosus*	Lower Amazonian Brazil

☐ **Blue-winged Parrotlet *Forpus xanthopterygius***

___	*F. x. spengeli*	N Colombia
___	*F. x. xanthopterygius (olallae)*	SE Colombia to e Ecuador, n Peru and w Brazil
___	*F. x. flavescens*	SE Peru and e Bolivia
___	*F. x. flavissimus*	NE Brazil (Maranhão, Ceará and Paraíba to n Bahia)
___	*F. x. vividus*	E and se Brazil to Paraguay and n Argentina

☐ **Spectacled Parrotlet *Forpus conspicillatus***

___	*F. c. conspicillatus*	Tropical e Panama to n-central Colombia
___	*F. c. metae*	E slope of E Andes of Colombia to extreme w Venezuela
___	*F. c. caucae*	SW Colombia w of Andes (Cauca and Nariño); w Ecuador?

☐ **Dusky-billed Parrotlet *Forpus sclateri***

___	*F. s. eidos*	Extreme e Colombia to Venezuela, the Guianas and n Brazil
___	*F. s. sclateri*	SE Colombia to n Bolivia and Amazonian Brazil

☐ **Pacific Parrotlet *Forpus coelestis***

Arid littoral of w Ecuador and nw Peru

☐ **Yellow-faced Parrotlet *Forpus xanthops***

N Peru (dry scrub of upper Marañón Valley)

☐ **Plain Parakeet *Brotogeris tirica***

E Brazil (Bahia and s Goiás to Rio Grande do Sul)

☐ **Canary-winged Parakeet *Brotogeris versicolurus***

Lowlands of se Colombia to e Peru and Amazonian Brazil

☐ **Yellow-chevroned Parakeet *Brotogeris chiriri***

___	*B. c. behni*	Central and s Bolivia
___	*B. c. chiriri*	N Bolivia to Paraguay, se Brazil and n Argentina

☐ **Gray-cheeked Parakeet *Brotogeris pyrrhoptera***

Arid scrub of w Ecuador and extreme nw Peru

☐ **Orange-chinned Parakeet *Brotogeris jugularis***

___	*B. j. jugularis*	Tropical sw Mexico to n Colombia and nw Venezuela
___	*B. j. exsul*	E Colombia and w Venezuela

☐ **Cobalt-winged Parakeet *Brotogeris cyanoptera***

___	*B. c. cyanoptera*	SE Colombia to s Venezuela, e Ecuador, e Peru and w-c Brazil
___	*B. c. gustavi*	N Peru (upper Río Huallaga Valley)
___	*B. c. beniensis*	N Bolivia (Beni)

☐ **Tui Parakeet *Brotogeris sanctithomae***

___	*B. s. sanctithomae*	SE Colombia through Amaz. Brazil to se Peru and ne Bolivia
___	*B. s. takatsukasae*	Lower Amazon basin of n-central Brazil

☐ **Golden-winged Parakeet *Brotogeris chrysoptera***

___	*B. c. chrysoptera*	NE Venezuela to the Guianas and adjacent n Brazil
___	*B. c. tenuifrons*	N Brazil (upper Rio Negro in Amazonas)
___	*B. c. solimoensis*	N Brazil (Codajas and Manaus regions)
___	*B. c. tuipara*	Coastal n Brazil (Rio Tapajós to ne Maranhão)
___	*B. c. chrysosema*	W Brazil (Rio Madeira to n Mato Grosso)

☐ **Tepui Parrotlet *Nannopsittaca panychlora***

Tepuis of s Venezuela, s Guyana and extreme n Brazil

☐ **Amazonian Parrotlet *Nannopsittaca dachilleae***

Tropical se Peru and nw Bolivia (probably adjacent w Brazil)

☐ **Lilac-tailed Parrotlet *Touit batavicus***

N Venezuela to the Guianas, Trinidad and Tobago

☐ **Scarlet-shouldered Parrotlet** *Touit huetii*

Patchily distributed n South America to n Bolivia and n Brazil

☐ **Red-fronted Parrotlet** *Touit costaricensis*

Humid foothill forests of se Costa Rica and w Panama

☐ **Blue-fronted Parrotlet** *Touit dilectissimus*

Humid e Panama to nw Venezuela and nw Ecuador

☐ **Sapphire-rumped Parrotlet** *Touit purpuratus*
____ *T. p. viridiceps*
____ *T. p. purpuratus*

SE Colombia to s Venezuela, e Ecuador and ne Peru
S Venezuela to the Guianas and n Amazonian Brazil

☐ **Brown-backed Parrotlet** *Touit melanonotus*

SE Brazil (s Bahia to Rio de Janeiro and São Paulo)

☐ **Golden-tailed Parrotlet** *Touit surdus*

E Brazil (Pernambuco to São Paulo)

☐ **Spot-winged Parrotlet** *Touit stictopterus*

Locally in Andes of s Colombia to Ecuador and ne Peru

☐ **Black-headed Parrot** *Pionites melanocephalus*
____ *P. m. melanocephalus*
____ *P. m. pallidus*

SE Colombia to Venezuela, the Guianas and n Brazil
S Colombia to e Ecuador and ne Peru

☐ **White-bellied Parrot** *Pionites leucogaster*
____ *P. l. xanthomerius*
____ *P. l. xanthurus*
____ *P. l. leucogaster*

E Peru and n Bolivia to w Brazil south of the Amazon
Brazil s of the Amazon (R. Purús and R. Juruá to R. Madeira)
N Brazil south of the Amazon (Rio Madeira to Maranhão)

☐ **Vulturine Parrot** *Pionopsitta vulturina*

Humid forests of e Amazonian Brazil south of the Amazon

☐ **Bald Parrot** *Pionopsitta aurantiocephala*

Amazonian Brazil (middle Rio Tapajós and possibly lower Rio Madeira)

☐ **Brown-hooded Parrot** *Pionopsitta haematotis*
____ *P. h. haematotis*
____ *P. h. coccinicollaris*

Humid forests of s Caribbean Mexico to w Panama
E Panama (east of Canal Zone) and nw Colombia

☐ **Rose-faced Parrot** *Pionopsitta pulchra*

Humid forests of w Colombia and nw Ecuador

☐ **Orange-cheeked Parrot** *Pionopsitta barrabandi*
____ *P. b. barrabandi*
____ *P. b. aurantiigena*

SE Colombia and s Venezuela to Brazil north of the Amazon
E Ecuador to e Peru, n Bolivia and Brazil south of the Amazon

☐ **Saffron-headed Parrot** *Pionopsitta pyrilia*

Extreme e Panama to n Colombia and nw Venezuela

☐ **Caica Parrot** *Pionopsitta caica*

SE Venezuela to the Guianas and Brazil north of the Amazon

☐ **Pileated Parrot** *Pionopsitta pileata*

Humid forests of se Brazil to e Paraguay and ne Argentina

☐ **Black-winged Parrot** *Hapalopsittaca melanotis*
____ *H. m. peruviana*
____ *H. m. melanotis*

Andes of central and s Peru
Andes of Bolivia (La Paz and Cochabamba)

☐ **Rusty-faced Parrot** *Hapalopsittaca amazonina*
____ *H. a. velezi*
____ *H. a. amazonina*
____ *H. a. theresae*

Central Andes of Colombia
E Andes of Colombia
Andes of extreme e Colombia and nw Venezuela

☐ **Indigo-winged Parrot** *Hapalopsittaca fuertesi*

Central Andes of Colombia

☐ **Red-faced Parrot** *Hapalopsittaca pyrrhops*

Andes of sw Ecuador and adjacent nw Peru

☐ **Short-tailed Parrot** *Graydidascalus brachyurus*

SE Colombia to e Ecuador, e Peru and Brazil n of the Amazon

☐ **Blue-headed Parrot** *Pionus menstruus*
_____ *P. m. rubrigularis*   Tropical n Costa Rica to w Colombia and w Ecuador
_____ *P. m. menstruus*   E Colombia to the Guianas, Trinidad, n Brazil and Bolivia
_____ *P. m. reichenowi*   Coastal ne Brazil (Alagoas to Espírito Santo)

☐ **Red-billed Parrot** *Pionus sordidus*
_____ *P. s. antelius*   Mountains of ne Venezuela (Anzoátegui, Sucre and Monagas)
_____ *P. s. sordidus*   Mountains of n Venezuela (Lara and Falcón to Caracas)
_____ *P. s. ponsi*   Mountains of ne Colombia to Sierra de Perijá (nw Venezuela)
_____ *P. s. saturatus*   Santa Marta Mountains (ne Colombia)
_____ *P. s. corallinus*   E Andes of Colombia to e Ecuador, e Peru and n Bolivia
_____ *P. s. mindoensis*   Mountains of w Ecuador

☐ **Scaly-headed Parrot** *Pionus maximiliani*
_____ *P. m. maximiliani*   NE Brazil (Ceará to Espírito Santo and s Goiás)
_____ *P. m. siy*   SE Bolivia to Paraguay, w Brazil (Mato Grosso), n Argentina
_____ *P. m. melanoblepharus*   E Paraguay to se Brazil and ne Argentina (Misiones)
_____ *P. m. lacerus*   NW Argentina (Tucumán, Catamarca and s Salta)

☐ **White-crowned Parrot** *Pionus senilis*

    Caribbean slope of se Mexico to w Panama

☐ **Speckle-faced Parrot** *Pionus tumultuosus*
_____ *P. t. seniloides*   Andes of Colombia to nw Venezuela and extreme n Peru
_____ *P. t. tumultuosus*   Andes of Peru and w Bolivia

☐ **Bronze-winged Parrot** *Pionus chalcopterus*

    Humid forests of ne Colombia to w Venezuela and nw Peru

☐ **Dusky Parrot** *Pionus fuscus*

    Humid ne Colombia to the Guianas and n Amazonian Brazil

☐ **Cuban Parrot** *Amazona leucocephala*
_____ *A. l. bahamensis*   Bahamas (Great Inagua, Abaco and Acklins)
_____ *A. l. leucocephala (palmarum)*   Cuba and Isle of Pines
_____ *A. l. caymanensis*   Grand Cayman I.
_____ *A. l. hesterna*   Cayman Brac I. and (formerly) Little Cayman I.

☐ **Yellow-billed Parrot** *Amazona collaria*

    Humid forests of Jamaica

☐ **Hispaniolan Parrot** *Amazona ventralis*

    Hispaniola and satellite islands

☐ **Puerto Rican Parrot** *Amazona vittata*
_____ *A. v. vittata*   Montane forests of e Puerto Rico (critically endangered)
_____ *A. v. gracilipes†*   Formerly Culebra I. off e Puerto Rico. Extinct

☐ **Yellow-lored Parrot** *Amazona xantholora*

    SE Mexico (Yucatán Peninsula and Cozumel I.) to Belize

☐ **White-fronted Parrot** *Amazona albifrons*
_____ *A. a. saltuensis*   Arid nw Mexico (s Sonora and w Durango to Sinaloa)
_____ *A. a. albifrons*   Arid w Mexico (Nayarit) to sw Guatemala
_____ *A. a. nana*   SE Mexico (extreme se Veracruz) to nw Costa Rica

☐ **Black-billed Parrot** *Amazona agilis*

    Jamaica

☐ **Tucuman Parrot** *Amazona tucumana*

    Montane forests of se Bolivia and nw Argentina

☐ **Red-spectacled Parrot** *Amazona pretrei*

    SE Brazil (Rio Grande do Sul); vagrant to Paraguay, Argentina

☐ **Red-crowned Parrot** *Amazona viridigenalis*

    Lowlands of ne Mexico (Nuevo León to n Veracruz)

☐ **Lilac-crowned Parrot** *Amazona finschi*

    W Mexico (s Sonora, Chihuahua and Durango to Oaxaca)

☐ **Red-lored Parrot** *Amazona autumnalis*
____ *A. a. autumnalis* — Caribbean slope of e Mexico to n Nicaragua and Bay Islands
____ *A. a. salvini* — N Nicaragua to sw Colombia and extreme nw Venezuela
____ *A. a. lilacina* — W Ecuador (north of Gulf of Guayaquil)
____ *A. a. diadema* — NW Brazil (lower Rio Negro and n bank of the upper Amazon)

☐ **Blue-cheeked Parrot** *Amazona dufresniana* — SE Venezuela (Gran Sabana) to the Guianas and (?) adj. Brazil

☐ **Red-browed Parrot** *Amazona rhodocorytha* — E Brazil (Alagoas and Bahia to Rio de Janeiro and São Paulo)

☐ **Red-tailed Parrot** *Amazona brasiliensis* — SE Brazil (se São Paulo and Paraná)

☐ **Festive Parrot** *Amazona festiva*
____ *A. f. bodini* — E Colombia to Orinoco basin of Venezuela
____ *A. f. festiva* — SE Colombia to e Ecuador, e Peru and w Amazonian Brazil

☐ **Yellow-faced Parrot** *Amazona xanthops* — Interior Brazil (Maranhão) to n Paraguay and e Bolivia

☐ **Yellow-shouldered Parrot** *Amazona barbadensis* — Coastal n Venezuela, Bonaire, La Blanquilla and Margarita is.

☐ **Blue-fronted Parrot** *Amazona aestiva*
____ *A. a. aestiva* — E Brazil (Maranhão and Pará to Rio Grande do Sul)
____ *A. a. xanthopteryx* — Bolivia to sw Brazil, Paraguay and n Argentina

☐ **Yellow-headed Parrot** *Amazona oratrix*
____ *A. o. oratrix* — Trop. Pacific slope of s, ne and Gulf lowlands of Mexico
____ *A. o. tresmariae* — Tres Marías Islands (off w Mexico)
____ *A. o. belizensis* — Belize
____ *A. o. caribaea* — Bay Islands off Honduras (Isla Barbareta and Isla Guanaja)
____ *A. o. parvipes* — Coastal ne Honduras to ne Nicaragua
____ *A. o. hondurensis* — N Honduras (Sula Valley)

☐ **Yellow-crowned Parrot** *Amazona ochrocephala*
____ *A. o. panamensis* — Pearl Islands and w Panama to nw Colombia
____ *A. o. ochrocephala* — E Colombia to Venezuela, the Guianas, Trinidad and n Brazil
____ *A. o. xantholaema* — Ilha de Marajó (off n Brazil)
____ *A. o. nattereri* — S Colombia to e Ecuador, e Peru, n Bolivia and w Brazil

☐ **Yellow-naped Parrot** *Amazona auropalliata* — Tropical e Mexico (Oaxaca) to nw Costa Rica

☐ **Kawall's Parrot** *Amazona kawalli* — Amazonian Brazil

☐ **Orange-winged Parrot** *Amazona amazonica*
____ *A. a. amazonica* — E Colombia to Venezuela, the Guianas, n Bolivia and e Brazil
____ *A. a. tobagensis* — Trinidad and Tobago

☐ **Scaly-naped Parrot** *Amazona mercenaria*
____ *A. m. canipalliata* — Andes of Colombia to nw Venezuela and Ecuador
____ *A. m. mercenaria* — Andes of n Peru to n Bolivia; single record from Argentina

☐ **Mealy Parrot** *Amazona farinosa*
____ *A. f. guatemalae* — Caribbean slope of se Mexico to nw Honduras
____ *A. f. virenticeps* — Honduras (Sula Valley) to extreme w Panama
____ *A. f. farinosa (inornata, chapmani)* — E Panama to Colombia, the Guianas, ne Bolivia and e Brazil

☐ **Vinaceous Parrot** *Amazona vinacea* — E Brazil (Bahia) to e Paraguay and n Argentina

☐ **St. Lucia Parrot** *Amazona versicolor* — Montane forests of St. Lucia (Lesser Antilles)

☐ **Red-necked Parrot** *Amazona arausiaca* — Montane forests of Dominica (Lesser Antilles)

150

☐ **St. Vincent Parrot** *Amazona guildingii*

Montane forests of St. Vincent (Lesser Antilles)

☐ **Imperial Parrot** *Amazona imperialis*

Montane forests of Dominica (Lesser Antilles)

☐ **Red-fan Parrot** *Deroptyus accipitrinus*
____ *D. a. accipitrinus* SE Colombia to Venezuela, the Guianas, ne Peru and n Brazil
____ *D. a. fuscifrons* Brazil s of the Amazon (Pará to n Mato Grosso); adj. Bolivia?

☐ **Blue-bellied Parrot** *Triclaria malachitacea*

Lowlands of se Brazil (s Bahia to Rio Grande do Sul)

ORDER: CUCULIFORMES
FAMILY: MUSOPHAGIDAE (Turacos—23)

☐ **Great Blue Turaco** *Corythaeola cristata*

Lowland rainforests of w and central Africa

☐ **Guinea Turaco** *Tauraco persa*
____ *T. p. buffoni* Senegambia to Liberia
____ *T. p. persa* Ivory Coast to Ghana and Cameroon
____ *T. p. zenkeri* S Cameroon to Gabon, n Angola, Congo and nw Zaire

☐ **Livingstone's Turaco** *Tauraco livingstonii*
____ *T. l. reichenowi* Tanzania (Nguru and Uluguru Mts. to Njombe Highlands)
____ *T. l. cabanisi* Coastal lowlands of Tanzania to Mozambique and ne Zululand
____ *T. l. livingstonii* Highlands of Malawi to Mozambique and e Zimbabwe

☐ **Schalow's Turaco** *Tauraco schalowi*

Humid forests of s-central Africa

☐ **Knysna Turaco** *Tauraco corythaix*
____ *T. c. phoebus* Humid forests of Transvaal and nw Swaziland
____ *T. c. corythaix* Natal to w Zululand, s Swaziland and e Cape Province

☐ **Black-billed Turaco** *Tauraco schuettii*
____ *T. s. emini* S Sudan to e Zaire, Uganda, w Kenya, nw Tanzania, Burundi
____ *T. s. schuettii* Zaire east to Ituri basin and south to n Angola

☐ **White-crested Turaco** *Tauraco leucolophus*

Extreme se Nigeria to n Uganda, sw Sudan and w Kenya

☐ **Fischer's Turaco** *Tauraco fischeri*
____ *T. f. fischeri* Coastal forests of s Somalia, coastal Kenya and ne Tanzania
____ *T. f. zanzibaricus* Zanzibar

☐ **Yellow-billed Turaco** *Tauraco macrorhynchus*
____ *T. m. macrorhynchus* Lowland rainforests of Sierra Leone to Ghana
____ *T. m. verreauxii* Nigeria to Cameroon, Gabon, Congo and n Angola; Bioko I.

☐ **Bannerman's Turaco** *Tauraco bannermani*

SW Cameroon (Bamenda-Banso highlands)

☐ **Red-crested Turaco** *Tauraco erythrolophus*

Locally in woodlands and savanna of w Angola

☐ **Hartlaub's Turaco** *Tauraco hartlaubi*

Highlands of Kenya, adjacent Uganda and ne Tanzania

☐ **White-cheeked Turaco** *Tauraco leucotis*
____ *T. l. leucotis* *Podocarpus* forests of Eritrea, Ethiopia and se Sudan
____ *T. l. donaldsoni* S-central Ethiopia south of Rift Valley and extreme w Somalia

☐ **Prince Ruspoli's Turaco** *Tauraco ruspolii*

Juniper forests of s Ethiopia

☐ **Purple-crested Turaco** *Tauraco porphyreolophus*
    ____ *T. p. chlorochlamys* — SE Kenya and sw Uganda to Tanzania and n Mozambique
    ____ *T. p. porphyreolophus* — Zimbabwe and Mozambique to e Transvaal and Natal

☐ **Ruwenzori Turaco** *Ruwenzorornis johnstoni*
    ____ *R. j. johnstoni* — Ruwenzori Mountains (ne Zaire and sw Uganda)
    ____ *R. j. bredoi* — E Zaire (Mt. Kabobo)
    ____ *R. j. kivuensis* — Highlands of e Zaire, Rwanda, Burundi and sw Uganda

☐ **Violet Turaco** *Musophaga violacea* — S Senegambia to nw Cameroon, s Chad and Cent. African Rep.

☐ **Ross' Turaco** *Musophaga rossae* — Riparian forests and woodlands of central Africa

☐ **Bare-faced Go-away-bird** *Corythaixoides personatus*
    ____ *C. p. personatus* — Rift Valley of Ethiopia
    ____ *C. p. leopoldi* — S Uganda to Rwanda, Burundi, sw Kenya, Malawi and Zambia

☐ **Gray Go-away-bird** *Corythaixoides concolor*
    ____ *C. c. molybdophanes* — NE Angola to s Zaire, Zambia, s Tanzania and n Mozambique
    ____ *C. c. pallidiceps* — W Angola to s Namibia and w Botswana
    ____ *C. c. bechuanae (chobiensis)* — S Angola to ne Namibia, Botswana, s Zambia and w Transvaal
    ____ *C. c. concolor* — S Malawi and n Mozambique to e Transvaal and e Zululand

☐ **White-bellied Go-away-bird** *Corythaixoides leucogaster* — *Acacia* savanna of Somalia and Ethiopia to ne Tanzania

☐ **Western Plantain-eater** *Crinifer piscator* — *Acacia* savanna of Senegambia to C African Rep. and w Zaire

☐ **Eastern Plantain-eater** *Crinifer zonurus* — *Acacia* savanna of central and east Africa

ORDER: CUCULIFORMES
FAMILY: CUCULIDAE (Cuckoos—141)

☐ **Pied Cuckoo** *Clamator jacobinus*
    ____ *C. j. pica* — Sub-Saharan Africa; nw India to Nepal and Myanmar
    ____ *C. j. serratus* — South Africa
    ____ *C. j. jacobinus* — S India and Sri Lanka

☐ **Levaillant's Cuckoo** *Clamator levaillantii* — Africa south of the Sahara

☐ **Chestnut-winged Cuckoo** *Clamator coromandus* — India to SE Asia; winters to Greater Sundas

☐ **Great Spotted Cuckoo** *Clamator glandarius* — SW Palearctic and Africa south of the Sahara

☐ **Thick-billed Cuckoo** *Pachycoccyx audeberti*
    ____ *P. a. brazzae* — Sierra Leone to Ghana, Nigeria, Cameroon and w Zaire
    ____ *P. a. validus* — E Zaire to se Kenya, Tanzania, Mozambique and e Transvaal
    ____ *P. a. audeberti* — NE Madagascar

☐ **Sulawesi Hawk-Cuckoo** *Cuculus crassirostris* — Mountains of n and central Sulawesi

☐ **Large Hawk-Cuckoo** *Cuculus sparverioides*
    ____ *C. s. sparverioides* — N Pakistan to India, s China, Myanmar, Thailand and Indochina
    ____ *C. s. bocki* — Mountains of Malay Peninsula, Sumatra and Borneo

☐ **Common Hawk-Cuckoo** *Cuculus varius*
    ____ *C. v. varius* — India to Nepal, Bangladesh and Myanmar
    ____ *C. v. ciceliae* — Sri Lanka

☐ **Moustached Hawk-Cuckoo** *Cuculus vagans*

S Myanmar, Mergui Arch., Malay Pen., Sumatra and Borneo

☐ **Hodgson's Hawk-Cuckoo** *Cuculus nisicolor*

Nepal and e Himalayas to Myanmar, Thailand and Hainan I.

☐ **Northern Hawk-Cuckoo** *Cuculus hyperythrus*

NE China to Korea, lower Yangtze Valley and s Japan

☐ **Malaysian Hawk-Cuckoo** *Cuculus fugax*

S Thailand, Malay Peninsula and Greater Sundas

☐ **Philippine Hawk-Cuckoo** *Cuculus pectoralis*

Philippine Islands

☐ **Red-chested Cuckoo** *Cuculus solitarius*

Intra-African migrant south of the Sahara

☐ **Black Cuckoo** *Cuculus clamosus*
    ____ *C. c. gabonensis* — Liberia to Ghana, Nigeria, s Sudan, Uganda and w Kenya
    ____ *C. c. clamosus* — Highlands of Ethiopia and Somalia to e South Africa

☐ **Indian Cuckoo** *Cuculus micropterus*
    ____ *C. m. micropterus* — India and Myanmar to SE Asia; winters to Greater Sundas
    ____ *C. m. concretus* — Vietnam to s Thailand, Malay Pen., Sumatra, Java and Borneo

☐ **Common Cuckoo** *Cuculus canorus*
    ____ *C. c. bangsi* — Iberian Pen., Balearic Is. and nw Africa; winters in Africa
    ____ *C. c. canorus* — Europe, Siberia to Kamchatka and Japan; winters to s Africa
    ____ *C. c. subtelephonus* — Turkestan to s Mongolia; winters to s Asia and Africa
    ____ *C. c. bakeri* — W China to n India, Nepal, Myanmar, nw Thailand and s China

☐ **African Cuckoo** *Cuculus gularis*

Senegambia to Somalia and South Africa

☐ **Himalayan Cuckoo** *Cuculus saturatus*

S Himalayas to s China and Taiwan; winters to Indonesia

☐ **Oriental Cuckoo** *Cuculus optatus*

Russia to Siberia, n China, Korea, Japan; winters to Australia

☐ **Sunda Cuckoo** *Cuculus lepidus*

Malay Peninsula, Greater and Lesser Sundas

☐ **Lesser Cuckoo** *Cuculus poliocephalus*

South and east Asia; winters pen. India, Sri Lanka and e Africa

☐ **Madagascar Cuckoo** *Cuculus rochii*

Madagascar; winters in e Africa

☐ **Pallid Cuckoo** *Cuculus pallidus*

Australia and Tasmania; winters north to Wallacea

☐ **Dusky Long-tailed Cuckoo** *Cercococcyx mechowi*

Sierra Leone to nw Angola and e Zaire

☐ **Olive Long-tailed Cuckoo** *Cercococcyx olivinus*

Liberia to Zaire, w Uganda, nw Zambia and n Angola

☐ **Barred Long-tailed Cuckoo** *Cercococcyx montanus*
    ____ *C. m. montanus* — Montane forests of sw Uganda, e Zaire and Rwanda
    ____ *C. m. patulus* — Montane forests of Kenya to s Zaire, Zambia and Mozambique

☐ **Banded Bay Cuckoo** *Cacomantis sonneratii*
    ____ *C. s. waiti* — Sri Lanka
    ____ *C. s. sonneratii* — India to Nepal, Myanmar, Thailand and s Indochina
    ____ *C. s. malayanus* — Malay Peninsula
    ____ *C. s. schlegeli* — Sumatra, Borneo and Palawan (sw Philippines)
    ____ *C. s. musicus* — Java

☐ **Plaintive Cuckoo** *Cacomantis merulinus*
    ____ *C. m. passerinus* — India and Pakistan; winters to Sri Lanka
    ____ *C. m. querulus* — E Himalayas to s China, Myanmar, Malay Pen. and Indochina
    ____ *C. m. threnodes* — S Malay Peninsula, Sumatra and Borneo
    ____ *C. m. lanceolatus* — Java, Sulawesi and Togian Islands
    ____ *C. m. merulinus* — Philippine Islands

☐ **Brush Cuckoo** *Cacomantis variolosus*

____	*C. v. sepulcralis*	S Thailand, Malay Pen., Gr. and Lesser Sundas and Philippines
____	*C. v. everetti*	Sulu Archipelago (Jolo, Basilan, Tawitawi and adjacent islands)
____	*C. v. virescens*	Sulawesi, Butung, Tukangbesi and Banggai Islands
____	*C. v. infaustus*	N Moluccas to n and central New Guinea
____	*C. v. aeruginosus*	S Moluccas (Buru, Ambon and Seram) and Sula Islands
____	*C. v. oreophilus*	Highlands of e and s New Guinea
____	*C. v. blandus*	Admiralty Islands
____	*C. v. macrocercus*	Bismarck Archipelago (New Britain, New Ireland and Tabar)
____	*C. v. websteri*	New Hanover (Bismarck Archipelago)
____	*C. v. addendus*	Solomon Islands
____	*C. v. variolosus*	N and e Australia; winters to Moluccas and New Guinea

☐ **Moluccan Cuckoo** *Cacomantis heinrichi*

N Moluccas (Halmahera and Bacan)

☐ **Chestnut-breasted Cuckoo** *Cacomantis castaneiventris*

____	*C. c. arfakianus*	W Papuan islands and nw New Guinea
____	*C. c. weiskei*	Central and e New Guinea
____	*C. c. castaneiventris*	Aru Islands and ne Australia (Cape York Peninsula)

☐ **Fan-tailed Cuckoo** *Cacomantis flabelliformis*

____	*C. f. excitus*	Montane forests of New Guinea
____	*C. f. flabelliformis*	SW, e and se Australia (Cape York to se S Australia), Tasmania
____	*C. f. pyrrophanus*	New Caledonia and Loyalty Islands
____	*C. f. schistaceigularis*	Vanuatu
____	*C. f. simus*	Fiji Islands

☐ **Black-eared Cuckoo** *Chrysococcyx osculans*

Australia

☐ **Horsfield's Bronze-Cuckoo** *Chrysococcyx basalis*

Australia and Tasmania; winters to Java

☐ **Shining Bronze-Cuckoo** *Chrysococcyx lucidus*

____	*C. l. harterti*	Solomon Islands (Rennell and Bellona)
____	*C. l. layardi (aeneus)*	New Caledonia, Loyalty Is., Vanuatu, Banks and Santa Cruz is.
____	*C. l. lucidus*	Australia, Tasmania and New Zealand; winters to n Melanesia

☐ **Rufous-throated Bronze-Cuckoo** *Chrysococcyx ruficollis*

Humid montane forests of New Guinea

☐ **White-eared Bronze-Cuckoo** *Chrysococcyx meyeri*

Montane forests of New Guinea and Batanta I.

☐ **Little Bronze-Cuckoo** *Chrysococcyx minutillus*

____	*C. m. peninsularis*	Extreme s Thailand and Malay Peninsula
____	*C. m. albifrons*	N Sumatra and w Java
____	*C. m. cleis*	N and e Borneo
____	*C. m. aheneus*	SE Borneo and s Philippine Islands
____	*C. m. jungei*	Sulawesi, Madu I. and Flores
____	*C. m. rufomerus*	Lesser Sundas (Romang, Kisar, Leti, Moa, Sermata and Damar)
____	*C. m. ssp.*	Undescribed race from Timor (e Lesser Sundas)
____	*C. m. crassirostris*	Moluccas (Tayandu, Kai Is.) and Tanimbar Is. (Yamdena, Larat)
____	*C. m. salvadorii*	Known from one specimen from Tepa I. (Babar Islands)
____	*C. m. misoriensis*	Lowlands of coastal n New Guinea and adjacent islands
____	*C. m. poecilurus*	Lowlands of coastal s New Guinea and adjacent islands
____	*C. m. minutillus*	Moluccas, Lesser Sundas, n Australia and Melville I.
____	*C. m. russatus*	NE Australia (n and e Queensland)
____	*C. m. barnardi*	E Australia (se Queensland to ne New South Wales)

☐ **Asian Emerald Cuckoo** *Chrysococcyx maculatus*

India to s China, SE Asia, Sumatra, Andaman and Nicobar is.

☐ **Violet Cuckoo** *Chrysococcyx xanthorhynchus*

____	*C. x. xanthorhynchus*	NE India to SE Asia, Greater Sundas and Palawan
____	*C. x. amethystinus*	Philippine Islands

☐ **Yellow-throated Cuckoo** *Chrysococcyx flavigularis*

Sierra Leone to s Cameroon, sw Sudan, w Uganda and e Zaire

☐ **Klaas' Cuckoo** *Chrysococcyx klaas*

Widespread Africa south of the Sahara and Bioko

☐ **African Emerald Cuckoo** *Chrysococcyx cupreus*
____ *C. c. cupreus*   Africa south of the Sahara
____ *C. c. intermedius*   Bioko (Gulf of Guinea)
____ *C. c. insularum*   São Tomé, Príncipe and Pagalu (Gulf of Guinea)

☐ **Dideric Cuckoo** *Chrysococcyx caprius*

Intra-African migrant Africa south of the Sahara and s Arabia

☐ **Long-billed Cuckoo** *Rhamphomantis megarhynchus*
____ *R. m. megarhynchus*   Aru Islands and New Guinea
____ *R. m. sanfordi*   Waigeo I. (off n New Guinea)

☐ **Asian Drongo-Cuckoo** *Surniculus lugubris*
____ *S. l. dicruroides*   N India to s China and Indochina; winters to Indonesia
____ *S. l. lugubris*   Coastal sw India, Sri Lanka, Java and Bali
____ *S. l. brachyurus*   Malaysia, Sumatra, Bangka I., Borneo and sw Philippines
____ *S. l. musschenbroeki*   Sulawesi, Butung, Halmahera, Bacan and Obi islands

☐ **Philippine Drongo-Cuckoo** *Surniculus velutinus*

Philippine Islands and Sulu Archipelago

☐ **White-crowned Koel** *Caliechthrus leucolophus*

New Guinea and Salawati I.

☐ **Dwarf Koel** *Microdynamis parva*
____ *M. p. grisecens*   N New Guinea (Humboldt Bay to Kumusi River)
____ *M. p. parva*   Locally in s New Guinea and D'Entrecasteaux Archipelago

☐ **Black-billed Koel** *Eudynamys melanorhynchus*
____ *E. m. melanorhynchus*   Sulawesi
____ *E. m. facialis*   Sula Islands

☐ **Asian Koel** *Eudynamys scolopaceus*
____ *E. s. scolopaceus*   Nepal to Pakistan, India, Sri Lanka, Laccadives and Maldives
____ *E. s. chinensis*   S China and Indochina; winters to Borneo
____ *E. s. harterti*   Hainan (s China)
____ *E. s. malayanus (dolosa)* NE India to Thailand, Malaya, Sumatra, Borneo and L Sundas
____ *E. s. simalurensis*   Simeulue I. (off w Sumatra)
____ *E. s. frater*   N Philippines (Calayan and Fuga)
____ *E. s. mindanensis*   Philippines, Palawan, Sulu Arch., Sangihe I. and Talaud Islands
____ *E. s. corvinus*   N Moluccas (Morotai, Halmahera, Ternate, Tidore and Bacan)
____ *E. s. orientalis*   S Moluccas (Buru, Manipa, Kelang, Seram, Ambon, Watubela)
____ *E. s. picatus*   Kai Islands and Sumba to Timor and Roma
____ *E. s. rufiventer*   New Guinea
____ *E. s. salvadorii*   Bismarck Archipelago
____ *E. s. alberti*   Solomon Islands

☐ **Australian Koel** *Eudynamys cyanocephalus*
____ *E. c. subcyanocephalus* NW Australia to nw Queensland; winters to s Moluccas
____ *E. c. cyanocephalus*   N Queensland to s New South Wales; winters to Moluccas

☐ **Long-tailed Koel** *Eudynamys taitensis*

New Zealand; winters to Polynesia and Bismarck Archipelago

☐ **Channel-billed Cuckoo** *Scythrops novaehollandiae*

Sulawesi to n and e Australia; winters to Moluccas, New Guinea

☐ **Yellowbill** *Ceuthmochares aereus*
____ *C. a. flavirostris*   Gambia to Nigeria (west of the Niger River)
____ *C. a. aereus*   Nigeria to s Sudan, w Kenya, Zaire, Angola, n Zambia; Bioko I.
____ *C. a. australis*   Ethiopia to Kenya, Tanzania, Mozambique and e South Africa

☐ **Black-bellied Malkoha** *Phaenicophaeus diardi*

    ____ *P. d. diardi*   S Myanmar, s Thailand, Malay Peninsula and Sumatra

    ____ *P. d. borneensis*   Borneo

☐ **Chestnut-bellied Malkoha** *Phaenicophaeus sumatranus*

                                            S Myanmar, s Thailand, Malay Pen., Sumatra and Borneo

☐ **Blue-faced Malkoha** *Phaenicophaeus viridirostris*

                                            S peninsular India and Sri Lanka

☐ **Green-billed Malkoha** *Phaenicophaeus tristis*

    ____ *P. t. tristis*   N India to Nepal, Sikkim, Bhutan, Assam and Bangladesh

    ____ *P. t. saliens*   N Myanmar to n Thailand, n Indochina and sw China (Yunnan)

    ____ *P. t. hainanus*   Hainan I. (s China)

    ____ *P. t. longicaudatus*   S Myanmar, s Thailand, s Indochina and Malaysia

    ____ *P. t. elongatus*   Sumatra

    ____ *P. t. kangeangensis*   Kangean Islands (Java Sea)

☐ **Sirkeer Malkoha** *Phaenicophaeus leschenaultii*

    ____ *P. l. sirkee*   Pakistan and nw India

    ____ *P. l. infuscatus*   Sub-Himalayas (Kumaon to Nepal, w Assam and Bangladesh)

    ____ *P. l. leschenaultii*   S India and Sri Lanka

☐ **Raffles' Malkoha** *Phaenicophaeus chlorophaeus*

    ____ *P. c. chlorophaeus*   S Myanmar to s Thailand, Malay Pen., Sumatra and Borneo

    ____ *P. c. fuscigularis*   NW Borneo (Sarawak and nw Kalimantan)

☐ **Red-billed Malkoha** *Phaenicophaeus javanicus*

                                            S Myanmar, Malay Peninsula, Greater Sundas and Natuna Is.

☐ **Yellow-billed Malkoha** *Phaenicophaeus calyorhynchus*

    ____ *P. c. calyorhynchus*   Sulawesi and Togian Islands

    ____ *P. c. meridionalis*   Central and s Sulawesi

    ____ *P. c. rufiloris*   Butung I. (off Sulawesi)

☐ **Chestnut-breasted Malkoha** *Phaenicophaeus curvirostris*

    ____ *P. c. singularis*   S Myanmar to s Thailand, Malay Peninsula and Sumatra

    ____ *P. c. oeneicaudus*   Mentawi Islands (off sw Sumatra)

    ____ *P. c. curvirostris*   W and central Java

    ____ *P. c. deningeri*   E Java and Bali

    ____ *P. c. microrhinus*   Borneo and Bangka I.

    ____ *P. c. harringtoni*   S Philippines (Palawan, Balabac, Busuanga, Culion, Calauit)

☐ **Red-faced Malkoha** *Phaenicophaeus pyrrhocephalus*

                                            Sri Lanka

☐ **Red-crested Malkoha** *Phaenicophaeus superciliosus*

    ____ *P. s. cagayanensis*   N Luzon (Cagayan Province)

    ____ *P. s. superciliosus*   Luzon (south of range of *cagayanensis*)

☐ **Scale-feathered Malkoha** *Phaenicophaeus cumingi*

                                            N Philippines (Luzon, Marinduque and Catanduanes)

☐ **Sumatran Ground-Cuckoo** *Carpococcyx viridis*

                                            Lowlands and foothills of sw Sumatra (Basiran Mts.)

☐ **Bornean Ground-Cuckoo** *Carpococcyx radiatus*

                                            Borneo

☐ **Coral-billed Ground-Cuckoo** *Carpococcyx renauldi*

                                            SE Thailand, Laos, Cambodia and Vietnam

☐ **Giant Coua** *Coua gigas*

                                            Thinly distributed in forests and savanna of w and s Madagascar

☐ **Coquerel's Coua** *Coua coquereli*

                                            Humid and dry deciduous forests of w Madagascar

☐ **Red-breasted Coua** *Coua serriana*

                                            Humid forests of ne Madagascar

☐ **Red-fronted Coua** *Coua reynaudii*

Humid forests of n and e Madagascar

☐ **Red-capped Coua** *Coua ruficeps*
  ____ *C. r. ruficeps*
  ____ *C. r. olivaceiceps*

Lowlands of nw Madagascar
Lowlands of sw Madagascar

☐ **Running Coua** *Coua cursor*

Semiarid lowland forests of sw Madagascar

☐ **Crested Coua** *Coua cristata*
  ____ *C. c. cristata*
  ____ *C. c. dumonti*
  ____ *C. c. pyropyga*
  ____ *C. c. maxima*

N and e Madagascar (south to Mahajanga)
W Madagascar (Mahajanga to Morondava)
SW Madagascar (Morondava and Toliara to Amboasary)
SE Madagascar (Tolagnaro region)

☐ **Verreaux's Coua** *Coua verreauxi*

Locally in subdeserts of sw Madagascar

☐ **Blue Coua** *Coua caerulea*

Rainforests of nw and e Madagascar

☐ **Bay Coucal** *Centropus celebensis*
  ____ *C. c. celebensis*
  ____ *C. c. rufescens*

N Sulawesi and Togian Islands
Central and s Sulawesi, Labuan Blanda, Muna and Butung is.

☐ **Rufous Coucal** *Centropus unirufus*

N Philippines (Luzon, Polillo and Catanduanes)

☐ **Black-faced Coucal** *Centropus melanops*
  ____ *C. m. banken*
  ____ *C. m. melanops*

S Philippines (Bohol, Leyte, Samar and Biliran)
S Philippines (Basilan, Mindanao, Nipa, Dinagat and Siargao)

☐ **Sunda Coucal** *Centropus nigrorufus*

Lowlands of Java

☐ **Buff-headed Coucal** *Centropus milo*
  ____ *C. m. albidiventris*
  ____ *C. m. milo*

Solomon Is. (Vellalavella, Kulambangra, Gizo and Rendova)
S Solomon Islands (Guadalcanal and Florida Group)

☐ **Goliath Coucal** *Centropus goliath*

N Moluccas (Morotai, Halmahera, Tidore, Bacan and Obi)

☐ **Violaceous Coucal** *Centropus violaceus*

Bismarck Archipelago (New Ireland and New Britain)

☐ **Greater Black Coucal** *Centropus menbeki*
  ____ *C. m. menbeki*
  ____ *C. m. jobiensis*
  ____ *C. m. aruensis*

New Guinea, w Papuan islands and Numfor I.
Yapen I. (Geelvink Bay off n New Guinea)
Aru Islands

☐ **Pied Coucal** *Centropus ateralbus*

Bismarck Archipelago (New Ireland and New Britain)

☐ **Pheasant Coucal** *Centropus phasianinus*
  ____ *C. p. mui*
  ____ *C. p. propinquus*
  ____ *C. p. nigricans*
  ____ *C. p. thierfelderi*
  ____ *C. p. melanurus*
  ____ *C. p. phasianinus*

Known from one specimen from Timor (e Lesser Sundas)
N New Guinea (Mamberamo River to Astrolabe Bay)
SE New Guinea and Yule I.
S New Guinea and islands in nw Torres Strait
N and nw Australia
Coastal e Australia (n Queensland to n New South Wales)

☐ **Kai Coucal** *Centropus spilopterus*

Kai Islands (se Moluccas)

☐ **Lesser Black Coucal** *Centropus bernsteini*
  ____ *C. b. bernsteini*
  ____ *C. b. manam*

W and central New Guinea
Manam I. (off ne New Guinea)

☐ **Biak Coucal** *Centropus chalybeus*

Biak I. (Geelvink Bay off n New Guinea)

☐ **Short-toed Coucal** *Centropus rectunguis*

Malay Peninsula, Sumatra and Borneo

☐ **Black-hooded Coucal** *Centropus steerii*

Forests of Mindoro (n-central Philippines)

☐ **Greater Coucal** *Centropus sinensis*

____ *C. s. sinensis* — Pakistan to n India and s China
____ *C. s. parroti* — S peninsular India and Sri Lanka
____ *C. s. intermedius* — Bangladesh to Myanmar, s Thailand, Indochina and Malay Pen.
____ *C. s. bubutus* — Greater Sundas and adjacent islands to sw Philippines
____ *C. s. anonymus* — S Philippines (Jolo, Tawitawi, Basilan and Sanga Sanga)
____ *C. s. kangeanensis* — Kangean Islands (Java Sea)

☐ **Andaman Coucal** *Centropus andamanensis*

Andaman Islands and adj. Table, Great and Little Coco islands

☐ **Philippine Coucal** *Centropus viridis*

____ *C. v. major* — Babuyanes Islands (n Philippines)
____ *C. v. viridis* — Widespread throughout Philippine Islands
____ *C. v. mindorensis* — N-central Philippines (Mindoro and Semirara)
____ *C. v. carpenteri* — Batanas Islands north of Luzon (n Batan, Sabtang and Ibuhos)

☐ **Madagascar Coucal** *Centropus toulou*

____ *C. t. toulou* — Madagascar
____ *C. t. insularis* — Aldabra
____ *C. t. assumptionis†* — Formerly Assumption I. Extinct

☐ **Black Coucal** *Centropus grillii*

Intra-African migrant south of the Sahara

☐ **Green-billed Coucal** *Centropus chlororhynchus*

Locally in wet zone of sw Sri Lanka

☐ **Lesser Coucal** *Centropus bengalensis*

____ *C. b. bengalensis* — India and Nepal to Bangladesh, Myanmar and Indochina
____ *C. b. lignator* — S and se China, Hainan I. and Taiwan
____ *C. b. javanensis* — Malay Pen., Sumatra, Java, Borneo, Palawan and Philippines
____ *C. b. sarasinorum* — Sulawesi and Lesser Sundas
____ *C. b. medius* — Moluccas (Indonesia)

☐ **Black-throated Coucal** *Centropus leucogaster*

____ *C. l. leucogaster* — S Senegal and Guinea-Bissau to se Nigeria
____ *C. l. efulenensis* — Lowland forests of sw Cameroon and Gabon
____ *C. l. neumanni* — Locally in ne Zaire (Ituri Forest)

☐ **Gabon Coucal** *Centropus anselli*

Lowland swamps of s Cameroon to nw Angola and c Zaire

☐ **Blue-headed Coucal** *Centropus monachus*

____ *C. m. fischeri (verheyeni)* — Ivory Coast to w Kenya, s Sudan, Ethiopia and n Angola
____ *C. m. monachus* — Ethiopia to central Kenya

☐ **Coppery-tailed Coucal** *Centropus cupreicaudus*

Angola to s Zaire, Zambia, Zimbabwe, Tanzania and Malawi

☐ **Senegal Coucal** *Centropus senegalensis*

____ *C. s. aegyptius* — Lower Egypt (Nile River to El Minya)
____ *C. s. senegalensis* — Senegambia to Somalia and south to Zaire
____ *C. s. flecki* — E Angola to n Botswana, Zambia, Malawi and sw Tanzania

☐ **White-browed Coucal** *Centropus superciliosus*

____ *C. s. sokotrae* — Socotra I. and sw Arabia
____ *C. s. superciliosus* — E Sudan to Ethiopia, w Somalia, Kenya, ne Uganda, ne Tanzania
____ *C. s. loandae* — Uganda to sw Kenya, n Zimbabwe, Zambia, Botswana, Angola
____ *C. s. burchellii (fasciipygialis)* — E Botswana to s Zimbabwe, Mozambique and South Africa

☐ **Dwarf Cuckoo** *Coccyzus pumilus*

Tropical n Colombia to ne Venezuela; one record from n Brazil

☐ **Ash-colored Cuckoo** *Coccyzus cinereus*

S Brazil to n Argentina, Paraguay, Bolivia and extreme se Peru

☐ **Black-billed Cuckoo** *Coccyzus erythropthalmus*

Breeds e North America; winters to Bolivia

☐ **Yellow-billed Cuckoo** *Coccyzus americanus*

Canada to Mexico and West Indies; winters to n Argentina

☐ **Pearly-breasted Cuckoo** *Coccyzus euleri*

Locally from n South America to s Brazil and ne Argentina

☐ **Mangrove Cuckoo** *Coccyzus minor*

Locally from s Florida to coastal n Brazil

☐ **Cocos Island Cuckoo** *Coccyzus ferrugineus*

Cocos I. (off w Costa Rica)

☐ **Dark-billed Cuckoo** *Coccyzus melacoryphus*

Venezuela to the Guianas, Brazil, n Argentina and Galapagos

☐ **Gray-capped Cuckoo** *Coccyzus lansbergi*

N Colombia and n Venezuela; migrates to w Peru

☐ **Great Lizard-Cuckoo** *Coccyzus merlini*
____ *C. m. bahamensis (andrina)*
____ *C. m. santamariae*
____ *C. m. merlini*
____ *C. m. decolor*

Bahamas (Andros, New Providence, and Eleuthera)
Islands off n-central Cuba
Cuba
Isle of Pines

☐ **Puerto Rican Lizard-Cuckoo** *Coccyzus vieilloti*

Puerto Rico

☐ **Jamaican Lizard-Cuckoo** *Coccyzus vetula*

Jamaica

☐ **Hispaniolan Lizard-Cuckoo** *Coccyzus longirostris*
____ *C. l. longirostris*
____ *C. l. petersi*

Hispaniola and Saona I.
La Mohotiere I. and Gonâve I. (off w Haiti)

☐ **Chestnut-bellied Cuckoo** *Coccyzus pluvialis*

Jamaica

☐ **Bay-breasted Cuckoo** *Coccyzus rufigularis*

Hispaniola and Gonâve I.

☐ **Squirrel Cuckoo** *Piaya cayana*
____ *P. c. Mexicana*
____ *P. c. thermophila*
____ *P. c. nigricrissa*
____ *P. c. mehleri*
____ *P. c. mesura*
____ *P. c. circe*
____ *P. c. cayana*
____ *P. c. insulana*
____ *P. c. obscura*
____ *P. c. hellmayri*
____ *P. c. pallescens*
____ *P. c. cabinisi*
____ *P. c. macroura*
____ *P. c. mogenseni*

Pacific slope of Mexico (Sinaloa to Isthmus of Tehuántepec)
E Mexico to e Panama, nw Colombia and offshore islands
W Colombia and w Ecuador to central Peru
NE Colombia and coastal n Venezuela east to París Peninsula
Colombia east of the Andes and e Ecuador
W Venezuela (region south of Lake Maracaibo)
Orinoco Valley of Venezuela to the Guianas and n Brazil
Trinidad
Brazil south of the Amazon (Rio Juruá to Rio Tapajós)
Brazil south of the Amazon (Santarém to Amazon delta)
E Brazil (Piauí, Pernambuco, n Bahia and adjacent e Goiás)
S-central Brazil (central Mato Grosso and adjacent Goiás)
SE Brazil to Paraguay, Uruguay and ne Argentina
S Bolivia and adjacent nw Argentina

☐ **Black-bellied Cuckoo** *Piaya melanogaster*

Guianas and Venezuela to e Ecuador, Peru and w Brazil

☐ **Little Cuckoo** *Coccycua minuta*
____ *C. m. panamensis*
____ *C. m. barinensis*
____ *C. m. gracilis*
____ *C. m. minuta*
____ *C m. chaparensis*

Lowlands of e Panama and n Colombia (west of Gulf of Urabá)
Extreme e Colombia and adjacent w Venezuela
Colombia west of the Andes and w Ecuador
E Colombia to Venezuela, the Guianas, Amaz. Brazil and Peru
N Bolivia (Río Chaparé region)

☐ **Greater Ani** *Crotophaga major*

E Panama and S America e of Andes to n Argentina; Trinidad

☐ **Smooth-billed Ani** *Crotophaga ani*

Tropical s US to Brazil, n Argentina and West Indies

☐ **Groove-billed Ani** *Crotophaga sulcirostris*

S Baja; s Texas to n Chile, n Argentina, Trinidad and Curaçao

☐ **Guira Cuckoo** *Guira guira*

NE Brazil to Bolivia, Paraguay, Uruguay and central Argentina

☐ **Striped Cuckoo** *Tapera naevia*
____ *T. n. excellens*　　　　Tropical se Mexico to Panama
____ *T. n. naevia*　　　　N S America to Brazil, Argentina, Trinidad and Isla Margarita

☐ **Pheasant Cuckoo** *Dromococcyx phasianellus*

Lowlands of s Mexico to Brazil, Paraguay and ne Argentina

☐ **Pavonine Cuckoo** *Dromococcyx pavoninus*

Tropical South America east of the Andes to ne Argentina

☐ **Lesser Ground-Cuckoo** *Morococcyx erythropygus*
____ *M. e. mexicanus*　　　Arid w Mexico (Sinaloa to Isthmus of Tehuántepec)
____ *M. e. erythropygus*　　S Mexico (Isthmus of Tehuántepec to n Costa Rica)

☐ **Greater Roadrunner** *Geococcyx californianus*

Arid sw US to s Mexico

☐ **Lesser Roadrunner** *Geococcyx velox*

Arid w Mexico (s Sonora to n Nicaragua)

☐ **Scaled Ground-Cuckoo** *Neomorphus squamiger*

Brazil south of the Amazon (lower Rio Tapajós region)

☐ **Rufous-vented Ground-Cuckoo** *Neomorphus geoffroyi*
____ *N. g. salvini*　　　　Humid lowlands of Nicaragua to Pacific coast of Colombia
____ *N. g. aequatorialis*　　Tropical se Colombia to e Ecuador and n Peru
____ *N. g. australis*　　　S Peru and nw Bolivia
____ *N. g. geoffroyi*　　　Brazil south of the Amazon (Pará)
____ *N. g. dulcis*　　　　E Brazil (Espírito Santo to Rio de Janeiro)

☐ **Banded Ground-Cuckoo** *Neomorphus radiolosus*

Humid lowlands of sw Colombia and nw Ecuador

☐ **Rufous-winged Ground-Cuckoo** *Neomorphus rufipennis*

S Venezuela, Guyana and n Brazil (Roraima)

☐ **Red-billed Ground-Cuckoo** *Neomorphus pucheranii*
____ *N. p. pucheranii*　　　Amazonian Peru and w Brazil north of the Amazon
____ *N. p. lepidophanes*　　Amazonian Peru and Brazil south of the Amazon

ORDER: STRIGIFORMES
FAMILY: TYTONIDAE (Barn-Owls—16)

☐ **Greater Sooty-Owl** *Tyto tenebricosa*
____ *T. t. arfaki*　　　　New Guinea and Yapen I.
____ *T. t. tenebricosa*　　Coastal e Australia (central Queensland to s Victoria)

☐ **Lesser Sooty-Owl** *Tyto multipunctata*

Rainforests of ne Australia (ne Queensland)

☐ **Australian Masked-Owl** *Tyto novaehollandiae*
____ *T. n. calabyi*　　　　Trans-Fly lowlands of s New Guinea and Daru I.
____ *T. n. melvillensis*　　N Australia (Melville I. and Bathurst I.)
____ *T. n. galei*　　　　NE Queensland (ne Cape York Peninsula)
____ *T. n. kimberli*　　　N Australia (Yampi Peninsula to Atherton Tableland)
____ *T. n. novaehollandiae (perplexa)*　　SW W Australia to Victoria and ne Queensland (Townsville)
____ *T. n. castanops*　　　Tasmania, Maria I. and Maatsuyker I.

☐ **New Britain Masked-Owl** *Tyto aurantia*

New Britain (Bismarck Archipelago)

☐ **Lesser Masked-Owl** *Tyto sororcula*
____ *T. s. cayelii* — S Moluccas (Buru and Seram)
____ *T. s. sororcula* — Tanimbar Islands (Yamdena and Larat)

☐ **Manus Owl** *Tyto manusi*

Manus I. (Admiralty Islands)

☐ **Taliabu Owl** *Tyto nigrobrunnea*

Taliabu I. (Sula Islands)

☐ **Minahassa Owl** *Tyto inexspectata*

Hill forests of n and n-central Sulawesi

☐ **Sulawesi Owl** *Tyto rosenbergii*
____ *T. r. rosenbergii* — Rainforests of Sulawesi and Sangihe Island
____ *T. r. pelengensis* — Peleng I. (Banggai Islands)

☐ **Australasian Grass-Owl** *Tyto longimembris*
____ *T. l. longimembris (walleri)* — India to Indochina, Sulawesi, Lesser Sundas, n and e Australia
____ *T. l. chinensis (melli)* — SE China (se Yunnan to Jiangsu) and Vietnam
____ *T. l. pithecops* — Taiwan
____ *T. l. amauronota* — Philippine Islands
____ *T. l. baliem* — W New Guinea
____ *T. l. papuensis* — Montane grasslands of e New Guinea

☐ **African Grass-Owl** *Tyto capensis*

Wet grasslands of Africa south of the Sahara

☐ **Ashy-faced Owl** *Tyto glaucops*
____ *T. g. glaucops* — Hispaniola and Tortue I.
____ *T. g. nigrescens* — Dominica (Lesser Antilles)
____ *T. g. insularis* — St. Vincent, Bequia, Union, Carriacou and Grenada

☐ **Madagascar Red Owl** *Tyto soumagnei*

E Madagascar (on verge of extinction)

☐ **Barn Owl** *Tyto alba*
____ *T. a. alba* — W and s Europe; w Canary Islands and North Africa
____ *T. a. guttata* — Central Europe east to sw European USSR and ne Greece
____ *T. a. ernesti* — Corsica and Sardinia
____ *T. a. erlangeri* — Crete and Cyprus to sw Iran, ne Egypt and s Arabian Peninsula
____ *T. a. schmitzi* — Madeira and Porto Santo I.
____ *T. a. gracilirostris* — E Canary Islands (Fuerteventura, Lanzarote and Alegranza)
____ *T. a. detorta* — Cape Verde Islands
____ *T. a. affinis (hypermetra)* — Sub-Saharan Africa, Zanzibar, Pemba, Madagascar, Comoros
____ *T. a. poensis* — Bioko (Gulf of Guinea)
____ *T. a. thomensis* — São Tomé (Gulf of Guinea)
____ *T. a. stertens* — Indian subcontinent to n Sri Lanka, sw China and s Thailand
____ *T. a. deroepstorffi* — S Andaman Islands
____ *T. a. javanica* — Malay Peninsula to Greater Sundas
____ *T. a. sumbaensis* — Sumba (Lesser Sundas)
____ *T. a. meeki* — E New Guinea, Manam and Karkar islands
____ *T. a. delicatula (everetti, kuehni, bellonae, lulu)* — Timor to Australia, Solomon Is., Loyalty Is., and Samoa
____ *T. a. crassirostris* — Tanga I. (Bismarck Archipelago)
____ *T. a. interposita* — N Vanuatu, Santa Cruz Islands and Banks Group
____ *T. a. pratincola (lucayana)* — S Canada to n Mexico, Bermuda, Bahamas and Hispaniola
____ *T. a. guatemalae (subandeana)* — W Guatemala to Panama, Pearl Islands and Colombia
____ *T. a. bondi* — Bay Islands off n Honduras (Roatán and Guanaja)
____ *T. a. furcata* — Cuba, Cayman Islands and Jamaica
____ *T. a. niveicauda* — Isle of Pines (off Cuba)
____ *T. a. bargei* — Netherlands Antilles (Curaçao and Bonaire)
____ *T. a. punctatissima* — Galapagos Islands
____ *T. a. contempta* — W Colombia to Venezuela, Ecuador and Peru
____ *T. a. hellmayri* — Guianas to n Brazil, Margarita I., Trinidad and Tobago
____ *T. a. tuidara* — Brazil s of the Amazon to Tierra del Fuego and Falkland Is.

☐ **Oriental Bay-Owl** *Phodilus badius*

___	*P. b. saturatus*	Sikkim and ne India to s China, Myanmar, Thailand, Indochina
___	*P. b. ripleyi*	SW India (Anaimalai-Nelliamathy Hills)
___	*P. b. assimilis*	Sri Lanka
___	*P. b. badius*	Malay Peninsula, Borneo, Sumatra, Java and Nias I.
___	*P. b. arixuthus*	Bunguran I. (Natuna Islands)
___	*P. b. parvus*	Belitung I. (off sw Borneo)

☐ **Congo Bay-Owl** *Phodilus prigoginei*

Second specimen in 50 years banded in mts. of e Zaire in 1996

ORDER: STRIGIFORMES
FAMILY: STRIGIDAE (Owls—199)

☐ **White-fronted Scops-Owl** *Otus sagittatus*

S Myanmar, s Thailand and Malay Peninsula

☐ **Andaman Scops-Owl** *Otus balli*

Andaman Islands

☐ **Reddish Scops-Owl** *Otus rufescens*

___	*O. r. malayensis*	S peninsular Thailand and Malay Peninsula
___	*O. r. rufescens*	Sumatra, Bangka I., Java and Borneo

☐ **Serendib Scops-Owl** *Otus thilohoffmanni*

Lowland rainforests of southwest Sri Lanka

☐ **Sandy Scops-Owl** *Otus icterorhynchus*

___	*O. i. icterorhynchus*	Rainforests of Liberia, Ivory Coast and Ghana
___	*O. i. holerythrus*	S Cameroon to n Congo, n and e Zaire and (?) Gabon

☐ **Sokoke Scops-Owl** *Otus ireneae*

E Kenya (Sokoke Forest) and ne Tanzania (Usambara Mts.)

☐ **Flores Scops-Owl** *Otus alfredi*

Flores (Lesser Sundas)

☐ **Mountain Scops-Owl** *Otus spilocephalus*

___	*O. s. huttoni*	Western Himalayas (n Pakistan to central Nepal)
___	*O. s. spilocephalus*	Himalayas (cent. Nepal to Arunachal Pradesh and Myanmar)
___	*O. s. latouchi*	N Thailand and Laos to se China and Hainan
___	*O. s. hambroecki*	Taiwan
___	*O. s. siamensis*	Mountains of s Thailand to s Vietnam
___	*O. s. vulpes*	Mountains of Malay Peninsula
___	*O. s. vandewateri*	Mountains of Sumatra
___	*O. s. luciae*	Mountains of Borneo

☐ **Rajah Scops-Owl** *Otus brookii*

___	*O. b. solokensis*	Montane forests of Sumatra
___	*O. b. brookii*	Montane forests of Borneo

☐ **Javan Scops-Owl** *Otus angelinae*

Mountains of Java

☐ **Mentawai Scops-Owl** *Otus mentawi*

Mentawi Islands (off w Sumatra)

☐ **Indian Scops-Owl** *Otus bakkamoena*

___	*O. b. plumipes*	W Himalayas (n Pakistan to w Nepal)
___	*O. b. deserticolor*	S Pakistan and (?) se Iran
___	*O. b. gangeticus*	NW India to lowlands of Nepal
___	*O. b. marathae*	Central India
___	*O. b. bakkamoena*	S India and Sri Lanka

☐ **Collared Scops-Owl** *Otus lettia*

___	*O. l. lettia*	E Nepal to Bangladesh, Myanmar, Thailand and Indochina
___	*O. l. erythrocampe*	SE China

____ O. l. ussuriensis	Sakhalin I., Ussuriland and ne China
____ O. l. glabripes	Taiwan
____ O. l. umbratilis	Hainan I. (s China)

☐ **Sunda Scops-Owl** *Otus lempiji*

____ O. l. condorensis	S peninsular Thailand (south of Isthmus of Kra)
____ O. l. lempiji	Malay Pen., s Sumatra, Bangka, Belitung, Java, Borneo and Bali
____ O. l. cnephaeus	S Malay Peninsula
____ O. l. hypnodes	N and central Sumatra
____ O. l. lemurum	N Borneo
____ O. l. kangeanus	Kangean Islands (Java Sea)

☐ **Japanese Scops-Owl** *Otus semitorques*

____ O. s. semitorques	S Kuril Islands and Hokkaido south to Yakushima I.
____ O. s. pryeri	Hachijo I. (s Izu Is.) and s Ryukyu Is. (Okinawa to Iriomote)

☐ **Wallace's Scops-Owl** *Otus silvicola*

Lesser Sundas (Sumbawa and Flores)

☐ **Palawan Scops-Owl** *Otus fuliginosus*

Palawan (sw Philippines)

☐ **Philippine Scops-Owl** *Otus megalotis*

____ O. m. megalotis	Philippines (Luzon, Catanduanes and Marinduque)
____ O. m. everetti	Philippines (Samar, Biliran, Leyte, Mindando and Basilan)
____ O. m. nigrorum	Negros (Philippines)
____ O. m. boholensis	Bohol (Philippines)

☐ **Mindanao Scops-Owl** *Otus mirus*

Montane rainforests of Mindanao (s Philippines)

☐ **Luzon Scops-Owl** *Otus longicornis*

Montane forests of Luzon (n Philippines)

☐ **Mindoro Scops-Owl** *Otus mindorensis*

Montane forests of Mindoro (Philippines)

☐ **Pallid Scops-Owl** *Otus brucei*

____ O. b. brucei	E Aral Sea to Kyrgystan and Tajikistan
____ O. b. obsoletus	S Turkey to n Syria, n Iraq, Uzbekistan and n Afghanistan
____ O. b. semenowi	S Tajikistan to w China, e Afghanistan and n Pakistan
____ O. b. exiguus	Israel to Iraq, s Iran, Oman, s Afghanistan and w Pakistan

☐ **African Scops-Owl** *Otus senegalensis*

____ O. s. senegalensis	Widespread sub-Saharan Africa
____ O. s. pamelae	S Saudi Arabia
____ O. s. socotranus	Socotra
____ O. s. feae	Pagulu (Gulf of Guinea)
____ O. s. nivosus	SE Kenya (lower Tana River to Lali Hills)

☐ **European Scops-Owl** *Otus scops*

____ O. s. scops	France and Medit. is. to Volga R., n Greece and Transcaucasia
____ O. s. pulchellus	Volga R. to Lake Baikal and south to Altai and Tien Shan Mts.
____ O. s. mallorcae	Iberian Pen., Balearic Is., n Morocco, Algeria and Tunisia
____ O. s. cycladum	S Greece and Crete, s Asia Minor, Israel, s Turkey and Jordan
____ O. s. cyprius	Cyprus
____ O. s. turanicus	Iraq through Iran and s Transcaspia to nw Pakistan

☐ **Oriental Scops-Owl** *Otus sunia*

____ O. s. sunia	N Pakistan to Bangladesh and n India
____ O. s. rufipennis	S India
____ O. s. leggei	Sri Lanka
____ O. s. modestus (nicobaricus, distans)	Assam to Myanmar, Thailand, Indochina; Andaman, Nicobar is.
____ O. s. malayanus	S China (Yunnan to e Guangdong)
____ O. s. stictonotus	SE Siberia to ne China, Sakhalin I. and n Korea
____ O. s. japonicus	Japan

☐ **Flammulated Owl** *Otus flammeolus*

S Br. Columbia to sw US and s Mexico; winters to Guatemala

☐ **Moluccan Scops-Owl** *Otus magicus*

____ *O. m. morotensis*	N Moluccas (Morotai and Ternate)
____ *O. m. leucospilus*	N Moluccas (Halmahera, Kasiruta and Bacan)
____ *O. m. obira*	Obi I. (central Moluccas)
____ *O. m. magicus*	S Moluccas (Seram and Ambon)
____ *O. m. bouruensis*	Buru (s Moluccas)
____ *O. m. albiventris*	Lesser Sundas (Lombok, Sumbawa, Flores, Besar, Lomblen)
____ *O. m. tempestatis*	Wetar (Lesser Sundas)

☐ **Mantanani Scops-Owl** *Otus mantananensis*

____ *O. m. romblonis*	Philippines (Romblon, Tablas, Sibuyan, Banton and Semirara)
____ *O. m. cuyensis*	SW Philippines (Cuyo, Dicabaito and Linapacan)
____ *O. m. mantananensis*	Mantanani I. (off Borneo); Rasa and Ursula is. (off Palawan)
____ *O. m. sibutuensis (steerei)*	SW Sulu Islands (Sibutu and Tumindao)

☐ **Ryukyu Scops-Owl** *Otus elegans*

____ *O. e. elegans*	Ryukyu Islands (s Japan)
____ *O. e. interpositus*	Daito Islands (s Japanese Archipelago)
____ *O. e. botelensis*	Lan-yü I. (off se Taiwan)
____ *O. e. calayensis*	N Philippines (Batan, Sabtang and Calayan)

☐ **Sulawesi Scops-Owl** *Otus manadensis*

____ *O. m. siaoensis*	Siau I. (off n Sulawesi)
____ *O. m. manadensis*	Sulawesi
____ *O. m. mendeni*	Banggai Islands (Peleng and Labobo)
____ *O. m. sulaensis*	Sula Islands (Taliabu, Seho, Mangole and Sanana)
____ *O. m. kalidupae*	Kaledupa I. (Tukangbesi Islands)

☐ **Sangihe Scops-Owl** *Otus collari*

Sangihe I. (north of Sulawesi)

☐ **Biak Scops-Owl** *Otus beccarii*

Biak I. (off nw New Guinea)

☐ **Seychelles Scops-Owl** *Otus insularis*

Highland forests of Mahé (Seychelles)

☐ **Simeulue Scops-Owl** *Otus umbra*

Simeulue I. (off nw Sumatra)

☐ **Enggano Scops-Owl** *Otus enganensis*

Enggano I. (off sw Sumatra)

☐ **Nicobar Scops-Owl** *Otus alius*

Great Nicobar I. (s Nicobar Islands)

☐ **Pemba Scops-Owl** *Otus pembaensis*

Pemba I. (off n Tanzania)

☐ **Comoro Scops-Owl** *Otus pauliani*

Mt. Karthala on Grand Comoro I. (Comoro Islands)

☐ **Anjouan Scops-Owl** *Otus capnodes*

Anjouan (Comoro Islands)

☐ **Moheli Scops-Owl** *Otus moheliensis*

Mohéli (Comoro Islands)

☐ **Malagasy Scops-Owl** *Otus rutilus*

Rainforests of e Madagascar

☐ **Torotoroka Scops-Owl** *Otus madagascariensis*

W Madagascar

☐ **Mayotte Scops-Owl** *Otus mayottensis*

Mayotte (Comoro Islands)

☐ **São Tomé Scops-Owl** *Otus hartlaubi*

Highlands of São Tomé (Gulf of Guinea)

☐ **Western Screech-Owl** *Megascops kennicottii*

____ *M. k. kennicottii (saturatus)*	Coastal s Alaska to nw Canada and nw California

____ *M. k. bendirei (macfarlanei, brewsteri)*	E Washington and Montana to se California
____ *M. k. aikeni (myochophilus, inyoensis, cineraceus)*	SW US (California) to w Oklahoma, s to n Mexico (Sonora)
____ *M. k. cardonensis (quercinus, clazus, gilmani)*	S California and n Baja California
____ *M. k. xantusi*	Cape district of s Baja California
____ *M. k. yumanensis*	SE California and sw Arizona to n Mexico (nw Sonora)
____ *M. k. suttoni (sortilegus)*	SW Texas to Mexican Plateau
____ *M. k. vinaceus (sinaloensis)*	N Mexico (s Sonora and w Chihuahua to n Sinaloa)

☐ **Balsas Screech-Owl** *Megascops seductus*

SW Mexico (s Jalisco and Colima to w Guerrero)

☐ **Pacific Screech-Owl** *Megascops cooperi*

____ *M. c. lambi*	Coastal s Mexico (Pacific slope of Oaxaca)
____ *M. c. chiapensis*	Coastal s Mexico (Chiapas)
____ *M. c. cooperi*	Extreme s Chiapas to nw Costa Rica (Guanacaste Peninsula)

☐ **Whiskered Screech-Owl** *Megascops trichopsis*

____ *M. t. aspersus (pinosus, ridgwayi, guerrerensis)*	SE Arizona to nw Mexico (Sonora and Chihuahua)
____ *M. t. trichopsis*	Highlands of c Mexico (Durango to Veracruz and Chiapas)
____ *M. t. mesamericanus (pumilus)*	SE Mexico (Chiapas) to n-central Nicaragua

☐ **Eastern Screech-Owl** *Megascops asio*

____ *M. a. maxwelliae (swenki)*	S-central Canada and n-central US
____ *M. a. naevius*	SE Canada and ne US (south to North Carolina)
____ *M. a. asio*	Oklahoma to South Carolina and Georgia
____ *M. a. hasbroucki*	Central Oklahoma to Texas
____ *M. a. floridanus*	Louisiana to Florida
____ *M. a. mccallii (semplei)*	S Texas to ne Mexico (Nuevo León and Tamaulipas)

☐ **Tropical Screech-Owl** *Megascops choliba*

____ *M. c. luctisonus*	Costa Rica to nw Colombia; Pearl Islands (Panama)
____ *M. c. margaritae*	Isla Margarita (off nw Venezuela)
____ *M. c. duidae*	Duida Mountains (s Venezuela)
____ *M. c. crucigerus (montanus, kelsoi, alticola, caucae)*	E Colombia to Venezuela, the Guianas, e Peru and ne Brazil
____ *M. c. surutus*	Bolivia
____ *M. c. decussatus (caatingensis)*	S-central and e Brazil
____ *M. c. choliba (chapadensis)*	S Brazil (s Mato Grosso and São Paulo) to e Paraguay
____ *M. c. wetmorei (alilucoco)*	W Paraguay and n Argentina
____ *M. c. uruguaiensis*	SE Brazil to Uruguay and ne Argentina

☐ **Koepcke's Screech-Owl** *Megascops koepckeae*

Disjunct in Andes of nw Peru and w-central Bolivia (La Paz)

☐ **West Peruvian Screech-Owl** *Megascops roboratus*

____ *M. r. pacificus*	SW Ecuador and nw Peru (south to Lambayeque)
____ *M. r. roboratus*	Extreme s Ecuador and nw Peru (between W and C Andes)

☐ **Bare-shanked Screech-Owl** *Megascops clarkii*

Montane forests of Costa Rica to extreme nw Colombia

☐ **Bearded Screech-Owl** *Megascops barbarus*

Montane forests of s Mexico (Chiapas) and c highlands

☐ **Rufescent Screech-Owl** *Megascops ingens*

____ *M. i. venezuelanus*	Andes of n Colombia to nw Venezuela
____ *M. i. ingens (minimus)*	Andes of n Ecuador to Peru and w-central Bolivia

☐ **Colombian Screech-Owl** *Megascops colombianus*

W slope of Andes of Colombia to nw Ecuador

☐ **Cinnamon Screech-Owl** *Megascops petersoni*

Cloud forests of s Ecuador to n Peru

☐ **Cloud-forest Screech-Owl** *Megascops marshalli*

Cloud forests of central Peru (Pasco and Cuzco)

☐ **Tawny-bellied Screech-Owl** *Megascops watsonii*
____ *M. w. watsonii*   E Colombia to ne Peru, Venezuela, Guianas and Amaz. Brazil
____ *M. w. usta*   E Peru and s Amaz. Brazil to n Mato Grosso and n Bolivia

☐ **Guatemalan Screech-Owl** *Megascops guatemalae*
____ *M. g. tomlini*   NW Mexico (se Sonora and sw Chihuahua to Sinaloa)
____ *M. g. hastatus (pettingilli)*   W Mexico (sw Sinaloa to Oaxaca)
____ *M. g. cassini*   E Mexico (s Tamaulipas and n Veracruz)
____ *M. g. fuscus*   Mountains of e Mexico (central Veracruz)
____ *M. g. thompsoni*   Yucatán Peninsula and Cozumel I.
____ *M. g. guatemalae (peteni)*   Mts. of se Mexico (se Veracruz and ne Oaxaca) to Honduras
____ *M. g. dacrysistactus*   Mountains of n Nicaragua

☐ **Vermiculated Screech-Owl** *Megascops vermiculatus*
     Lowlands of Costa Rica to nw Colombia and n Venezuela

☐ **Roraima Screech-Owl** *Megascops roraimae*
     Tepuis of se Venezuela and adjacent n Brazil

☐ **Rio Napo Screech-Owl** *Megascops napensis*
____ *M. n. napensis*   Tropical e Ecuador and e Colombia
____ *M. n. helleri*   Tropical e Peru
____ *M. n. bolivianus*   Tropical n Bolivia (Cochabamba)

☐ **Hoy's Screech-Owl** *Megascops hoyi*
     Montane forests of s Bolivia and nw Argentina

☐ **Variable Screech-Owl** *Megascops atricapilla*
     SE Brazil (s Bahia) to se Paraguay and ne Argentina (Misiones)

☐ **Long-tufted Screech-Owl** *Megascops sanctaecatarinae*
     Foothills of se Brazil, Uruguay and ne Argentina (Misiones)

☐ **Puerto Rican Screech-Owl** *Megascops nudipes*
____ *M. n. nudipes*   Puerto Rico
____ *M. n. newtoni*   Vieques I., Culebra I. and Virgin Islands

☐ **White-throated Screech-Owl** *Megascops albogularis*
____ *M. a. obscurus*   Sierra de Perijá (Colombia/Venezuela border)
____ *M. a. macabrum*   Western and Central Andes of Colombia and Ecuador to n Peru
____ *M. a. albogularis*   Eastern Andes of Colombia and n Ecuador
____ *M. a. meridensis*   Andes of w Venezuela
____ *M. a. aequatorialis*   Andes of e Ecuador
____ *M. a. remotus*   Andes of e Peru to w Bolivia (Cochabamba)

☐ **Palau Owl** *Pyrroglaux podarginus*
     Lowlands of Palau Islands (w Caroline Islands)

☐ **Bare-legged Owl** *Gymnoglaux lawrencii*
____ *G. l. exsul*   Western Cuba and Isle of Pines
____ *G. l. lawrencii*   Central and e Cuba

☐ **Northern White-faced Owl** *Ptilopsis leucotis*
     Senegambia to Somalia, n Zaire, n Uganda and central Kenya

☐ **Southern White-faced Owl** *Ptilopsis granti*
     SE Gabon to s Zaire, sw Kenya, Namibia and n Cape Province

☐ **Mindanao Eagle-Owl** *Mimizuku gurneyi*
     S Philippines (Mindanao, Dinagat and Siargao)

☐ **Great Horned Owl** *Bubo virginianus*
____ *B. v. lagophonus*   Central Alaska to ne Oregon and Montana; winters to Texas
____ *B. v. saturatus*   Coastal sw Alaska (Cook Inlet) to coastal central California
____ *B. v. pacificus*   Coastal California to nw Baja California
____ *B. v. elachistus*   S Baja California and Isla Espírito Santo
____ *B. v. subarcticus (occidentalis)*   Mackenzie and nw Br. Col. to Hudson Bay, Wyoming, N Dak.
____ *B. v. pallescens*   Deserts of cent. and se Calif. to Kansas and s Mexico (Oaxaca)
____ *B. v. heterocnemis*   NE Canada south to Great Lakes region

_____	*B. v. virginianus*	Minnesota to Nova Scotia, s to Kansas, e Texas and Florida
_____	*B. v. mayensis*	SE Mexico (Yucatán Peninsula)
_____	*B. v. mesembrinus*	S Mexico (Isthmus of Tehuántepec) to w Panama
_____	*B. v. nigrescens*	Andes of Colombia to Ecuador and nw Peru
_____	*B. v. nacurutu*	E Colombia to the Guianas, ne Brazil, Bolivia and Argentina

☐ **Magellanic Horned Owl** *Bubo magellanicus*

C Peru to w Bolivia, w Argentina, Tierra del Fuego, Cape Horn

☐ **Eurasian Eagle-Owl** *Bubo bubo*

_____	*B. b. hispanus*	Iberian Peninsula; formerly Atlas Mts. of n Africa (extinct?)
_____	*B. b. bubo*	Scandinavia and Spain through w Europe to w Russia
_____	*B. b. ruthenus*	Cent. European Russia to Ural Mts. and lower Volga basin
_____	*B. b. interpositus*	Turkey and nw Iran to s Ukraine, Romania and Bulgaria
_____	*B. b. sibiricus (baschkiricus)*	Western foothills of Ural Mountains to Ob River and w Altai
_____	*B. b. yenisseensis (zaissanensis)*	Central Siberia to n Mongolia
_____	*B. b. turcomanus (tarimensis)*	Lower Volga R. and Ural R. to nw China and w Mongolia
_____	*B. b. omissus*	Turkmenistan to extreme w China
_____	*B. b. hemachalanus (tibetanus)*	Pamirs and n Tien Shan south to w Himalayas and w Tibet
_____	*B. b. nikolskii*	E Iraq to Iran, Afghanistan and w Pakistan
_____	*B. b. jakutensis*	NE Siberia (Lena River to Sea of Okhotsk)
_____	*B. b. ussuriensis (dauricus, borissowi)*	SE Siberia to ne China, Sakhalin I., n Hokkaido and s Kuril Is.
_____	*B. b. kiautschensis (jarlandi, setschuanus, inexpectatus)*	W and central China (s to Yunnan and Sichuan) to Korea
_____	*B. b. swinhoei*	SE China

☐ **Rock Eagle-Owl** *Bubo bengalensis*

Indian subcontinent to Himalayan foothills and w Myanmar

☐ **Pharaoh Eagle-Owl** *Bubo ascalaphus*

_____	*B. a. ascalaphus*	NW Africa and n Egypt to w Iraq
_____	*B. a. desertorum*	Sahara to Mauritania, Niger, Ethiopia, Arabia and s Iraq

☐ **Cape Eagle-Owl** *Bubo capensis*

_____	*B. c. dillonii*	Highlands of s Eritrea and Ethiopia
_____	*B. c. mackinderi*	Kenya and Uganda to Zimbabwe, Mozambique and Malawi
_____	*B. c. capensis*	Extreme s Namibia and South Africa

☐ **Spotted Eagle-Owl** *Bubo africanus*

_____	*B. a. milesi*	SW Arabia, Yemen and Oman
_____	*B. a. africanus*	Gabon to Zaire, s Uganda, central Kenya and s to the Cape
_____	*B. a. tanae*	SE Kenya (central and lower Tana River and Lali Hills)

☐ **Grayish Eagle-Owl** *Bubo cinerascens*

Senegambia to Ethiopia, Somalia, n Uganda and n Kenya

☐ **Fraser's Eagle-Owl** *Bubo poensis*

Rainforests of Liberia to w Uganda and nw Angola; Bioko

☐ **Usambara Eagle-Owl** *Bubo vosseleri*

NE Tanzania (Usambara and Uluguru Mts.)

☐ **Spot-bellied Eagle-Owl** *Bubo nipalensis*

_____	*B. n. nipalensis*	Himalayas to India, sw China (Yunnan), Myanmar and Vietnam
_____	*B. n. blighi*	Sri Lanka

☐ **Barred Eagle-Owl** *Bubo sumatranus*

_____	*B. s. sumatranus*	S Myanmar, peninsular Thailand, Malay Pen., Sumatra, Bangka I.
_____	*B. s. strepitans*	Borneo, Java and Bali

☐ **Shelley's Eagle-Owl** *Bubo shelleyi*

Sierra Leone and Liberia to Cameroon, ne Zaire and Gabon

☐ **Verreaux's Eagle-Owl** *Bubo lacteus*

Savanna and woodlands of Africa south of the Sahara

☐ **Dusky Eagle-Owl** *Bubo coromandus*

_____	*B. c. coromandus*	Pakistan to central India, s Nepal, Assam and Bangladesh
_____	*B. c. klossii*	Extreme s China to s Myanmar and w Thailand

167

☐ **Akun Eagle-Owl** *Bubo leucostictus*

Sierra Leone and Liberia to Cameroon, Zaire and nw Angola

☐ **Philippine Eagle-Owl** *Bubo philippensis*
　____ 　*B. p. philippensis*　　　　N Philippines (Luzon and Catanduanes)
　____ 　*B. p. mindanensis*　　　　S Philippines (Mindanao, Samar, Leyte and Bohol)

☐ **Snowy Owl** *Bubo scandiacus*

Arctic circumpolar; irregular southern post-breeding irruptions

☐ **Blakiston's Fish-Owl** *Ketupa blakistoni*
　____ 　*K. b. piscivora*　　　　W Manchuria (west of Great Khingan Mountains)
　____ 　*K. b. doerriesi*　　　　SE Siberia and extreme ne China to Korea
　____ 　*K. b. karafutonis*　　　　Sakhalin I.
　____ 　*K. b. blakistoni*　　　　N Japan (s Kuril Islands and Hokkaido)

☐ **Brown Fish-Owl** *Ketupa zeylonensis*
　____ 　*K. z. semenowi*　　　　Extreme se Turkey to Israel, n Syria and nw India
　____ 　*K. z. leschenaultii*　　　　India south of Himalayas to Myanmar and Thailand
　____ 　*K. z. zeylonensis*　　　　Sri Lanka
　____ 　*K. z. orientalis*　　　　NE Myanmar to se China, Malay Pen., Indochina and Hainan

☐ **Tawny Fish-Owl** *Ketupa flavipes*

Himalayas to s China, ne Myanmar, s Indochina and Taiwan

☐ **Buffy Fish-Owl** *Ketupa ketupu*
　____ 　*K. k. aagaardi*　　　　S Assam to s Thailand and Vietnam
　____ 　*K. k. ketupu*　　　　Malay Pen., Riau Arch., Sumatra, Java, Bali, Borneo and Bangka
　____ 　*K. k. minor*　　　　Nias I. (off nw Sumatra)
　____ 　*K. k. pageli*　　　　NW Borneo

☐ **Pel's Fishing-Owl** *Scotopelia peli*

Locally in riverine forests of Africa south of the Sahara

☐ **Rufous Fishing-Owl** *Scotopelia ussheri*

Rainforests of Sierra Leone, Liberia, Ivory Coast and Ghana

☐ **Vermiculated Fishing-Owl** *Scotopelia bouvieri*

S Cameroon to Cent. African Rep., Zaire, Gabon and n Angola

☐ **Spotted Wood-Owl** *Strix seloputo*
　____ 　*S. s. seloputo*　　　　S Myanmar, Malay Pen., Thailand, Sumatra and Java
　____ 　*S. s. baweana*　　　　Bawean I. (Java Sea off n Java)
　____ 　*S. s. wiepkeni*　　　　S Philippines (Palawan and Calamian Islands)

☐ **Mottled Wood-Owl** *Strix ocellata*
　____ 　*S. o. grisescens*　　　　Base of Himalayas (Pakistan to Rajasthan and Bihar)
　____ 　*S. o. grandis*　　　　W India (Kathiawar Peninsula of s Gujarat)
　____ 　*S. o. ocellata*　　　　Peninsular India

☐ **Brown Wood-Owl** *Strix leptogrammica*
　____ 　*S. l. newarensis*　　　　Himalayas (Jammu and Kashmir to ne India)
　____ 　*S. l. ticehursti (orientalis, shahensis)*　　　　Myanmar to se China, Thailand, n Laos and n Vietnam
　____ 　*S. l. caligata*　　　　Hainan and Taiwan
　____ 　*S. l. laotiana*　　　　S Laos and central Vietnam (Annam)
　____ 　*S. l. indranee (connectens)*　　　　Peninsular India
　____ 　*S. l. ochrogenys*　　　　Sri Lanka
　____ 　*S. l. maingayi*　　　　S Myanmar, s Thailand and Malay Peninsula
　____ 　*S. l. myrtha*　　　　Sumatra
　____ 　*S. l. nyctiphasma*　　　　Banyak I. (off nw Sumatra)
　____ 　*S. l. niasensis*　　　　Nias I. (off nw Sumatra)
　____ 　*S. l. chaseni*　　　　Belitung I. (Java Sea off se Sumatra)
　____ 　*S. l. vaga*　　　　N Borneo
　____ 　*S. l. leptogrammica*　　　　Central and s Borneo
　____ 　*S. l. bartelsi*　　　　Java

☐ **Tawny Owl** *Strix aluco*
- ____ *S. a. aluco* — N and e Europe to Ukraine, Crimea, Balkans and Black Sea
- ____ *S. a. siberiae* — Ural Mountains to w Siberia
- ____ *S. a. sylvatica* — Britain, France, Iberia, s Italy, Greece, w and central Turkey
- ____ *S. a. mauritanica* — Morocco, Algeria and Tunisia
- ____ *S. a. willkonskii* — NE Turkey, Caucasus and nw Iran to Turkmenistan
- ____ *S. a. sanctinicolai* — Zagros Mountains (ne Iraq and w Iran)
- ____ *S. a. harmsi* — Turkestan
- ____ *S. a. biddulphi* — Pakistan and nw India
- ____ *S. a. nivicola* — Nepal to se China, n Myanmar and n Indochina
- ____ *S. a. ma* — NE China and Korea
- ____ *S. a. yamadae* — Mountains of s Taiwan

☐ **Hume's Owl** *Strix butleri*

Deserts of Syria, Jordan, Arabia, Sinai Pen. and far se Egypt

☐ **Spotted Owl** *Strix occidentalis*
- ____ *S. o. caurina* — Temperate forests from s British Columbia to n California
- ____ *S. o. occidentalis* — Mountains of s California to n Baja (San Pedro Mártir)
- ____ *S. o. lucida* — Mountains of sw US to c Mexico (Michoacán and Guanajuato)

☐ **Barred Owl** *Strix varia*
- ____ *S. v. varia* — SE Alaska to se Canada and e-central US
- ____ *S. v. georgica* — SE US (Arkansas to e Texas, the Gulf Coast and s Florida)
- ____ *S. v. helveola* — S-central Texas
- ____ *S. v. sartorii* — Mountains of n Mexico (Durango) to Veracruz and Oaxaca

☐ **Fulvous Owl** *Strix fulvescens*

Mountains of s Mexico (e Oaxaca and Chiapas) to El Salvador

☐ **Rusty-barred Owl** *Strix hylophila*

Paraguay to se Brazil (Minas Gerais) and extreme ne Argentina

☐ **Rufous-legged Owl** *Strix rufipes*
- ____ *S. r. rufipes* — Central Chile and w-central Argentina to Tierra del Fuego
- ____ *S. r. sanborni* — Chiloe I. (Chile)

☐ **Chaco Owl** *Strix chacoensis*

Chaco of s Bolivia, w Paraguay and n Argentina

☐ **Ural Owl** *Strix uralensis*
- ____ *S. u. liturata* — N Europe to nw Russia, n Poland, Belarus and middle Volga
- ____ *S. u. uralensis* — E European Russia to Sea of Okhotsk
- ____ *S. u. macroura (carpathaca)* — Carpathian Mountains to Bulgaria and w Balkans
- ____ *S. u. yenisseensis* — Central Siberian plateau
- ____ *S. u. nikolskii (daurica, tatibanai, coreensis)* — Transbaikalia to Sakhalin, ne China and Korea
- ____ *S. u. japonica* — Hokkaido (n Japan)
- ____ *S. u. hondoensis (momiyamae)* — N and central Honshu (Japan)
- ____ *S. u. fuscescens* — S Honshu south to Kyushu (Japan)

☐ **Père David's Owl** *Strix davidi*

Mountains of central China (se Qinghai and Sichuan)

☐ **Great Gray Owl** *Strix nebulosa*
- ____ *S. n. nebulosa* — Boreal forests of n North America
- ____ *S. n. lapponica (elisabethae)* — Boreal forests of n Europe, n Asia and Sakhalin

☐ **African Wood-Owl** *Strix woodfordii*
- ____ *S. w. nuchalis* — Senegambia to s Sudan, Uganda, n Angola, w Zaire; Bioko
- ____ *S. w. umbrina* — Ethiopia and se Sudan
- ____ *S. w. nigricantior* — S Somalia to Kenya, Tanzania, Zanzibar and e Zaire
- ____ *S. w. woodfordii* — S Angola to s Zaire, Botswana, sw Tanzania and South Africa

☐ **Mottled Owl** *Ciccaba virgata*
- ____ *C. v. squamulata* — W Mexico (Sonora to Guerrero, Guanajuato and Morelos)

____ *C. v. tamaulipensis*	NE Mexico (s Nuevo León and Tamaulipas)
____ *C. v. centralis*	SE Mexico (Oaxaca and Veracruz) to w Panama
____ *C. v. virgata (minuscula)*	E Panama to Colombia, Venezuela and Ecuador; Trinidad
____ *C. v. macconnelli*	N South America (the Guianas)
____ *C. v. superciliaris*	N-central and ne Amazonian Brazil
____ *C. v. borelliana*	E Paraguay to se Brazil and ne Argentina (Misiones)

☐ **Black-and-white Owl** *Ciccaba nigrolineata*

Lowlands of s Mexico to w Ecuador and extreme nw Peru

☐ **Black-banded Owl** *Ciccaba huhula*

____ *C. h. huhula*	E Colombia to the Guianas, ne Brazil, e Peru and nw Argentina
____ *C. h. albomarginata*	E Paraguay to se Brazil and ne Argentina (Misiones)

☐ **Rufous-banded Owl** *Ciccaba albitarsis*

Andes of Colombia to Venezuela, Ecuador, Peru and Bolivia

☐ **Crested Owl** *Lophostrix cristata*

____ *L. c. stricklandi*	S Mexico (Veracruz) to w Panama and w Colombia
____ *L. c. wedeli*	E Panama to ne Colombia and nw Venezuela
____ *L. c. cristata*	S Colombia to n Bolivia, s Venezuela, Guianas, w Amaz. Brazil

☐ **Maned Owl** *Jubula lettii*

Patchily distributed in humid forests of Liberia to e Zaire

☐ **Spectacled Owl** *Pulsatrix perspicillata*

____ *P. p. saturata*	S Mexico (Veracruz and Oaxaca) to w Panama (Chiriquí)
____ *P. p. chapmani*	E Costa Rica and Panama to Colombia, w Ecuador and nw Peru
____ *P. p. perspicillata*	E Colombia to Venezuela, the Guianas, Brazil and n Bolivia
____ *P. p. trinitatis*	Trinidad
____ *P. p. boliviana*	S Bolivia and n Argentina
____ *P. p. pulsatrix*	Paraguay to e Brazil (Bahia) and ne Argentina (Misiones)

☐ **Tawny-browed Owl** *Pulsatrix koeniswaldiana*

E Paraguay to se Brazil and ne Argentina (Misiones)

☐ **Band-bellied Owl** *Pulsatrix melanota*

____ *P. m. melanota*	Humid forests of se Colombia to e Ecuador and se Peru
____ *P. m. philoscia*	*Yungas* of w-central Bolivia

☐ **Northern Hawk-Owl** *Surnia ulula*

____ *S. u. caparoch*	Alaska to Canada, Newfoundland and extreme n US
____ *S. u. ulula*	N Eurasia
____ *S. u. tianschanica*	Central Asia to nw and ne China and n Mongolia

☐ **Eurasian Pygmy-Owl** *Glaucidium passerinum*

____ *G. p. passerinum*	Scandinavia, mts. of Europe to Siberia, Sakhalin and ne China
____ *G. p. orientale*	Central and e Siberia to Manchuria

☐ **Collared Owlet** *Glaucidium brodiei*

____ *G. b. brodiei*	Pakistan to s China, se Tibet, n Indochina and Malay Peninsula
____ *G. b. pardalotum*	Taiwan
____ *G. b. peritum*	Sumatra
____ *G. b. borneense*	Borneo

☐ **Pearl-spotted Owlet** *Glaucidium perlatum*

____ *G. p. perlatum*	Senegambia to w Sudan
____ *G. p. licua*	E Sudan and Ethiopia to n S Africa, Angola and Namibia

☐ **Northern Pygmy-Owl** *Glaucidium californicum*

____ *G. c. grinnelli*	Coniferous forests of se Alaska to n California
____ *G. c. swarthi*	Vancouver I.
____ *G. c. californicum*	Central British Columbia to sw US and nw Mexico
____ *G. c. pinicola*	Rocky Mountains (w-central US)

☐ **Mountain Pygmy-Owl** *Glaucidium gnoma*

SE Arizona to highlands of Mexico (Chihuahua to Oaxaca)

☐ **Guatemalan Pygmy-Owl** *Glaucidium cobanense*

Mountains of s Mexico (Chiapas) to Guatemala and Honduras

☐ **Cape Pygmy-Owl** *Glaucidium hoskinsii*

Mountains of s Baja California

☐ **Costa Rican Pygmy-Owl** *Glaucidium costaricanum*

Mountains of central Costa Rica to w Panama

☐ **Cloud-forest Pygmy-Owl** *Glaucidium nubicola*

Pacific slope of Western Andes of Colombia and Ecuador

☐ **Andean Pygmy-Owl** *Glaucidium jardinii*

Mountains of n Colombia to w Venezuela, Ecuador and Peru

☐ **Colima Pygmy-Owl** *Glaucidium palmarum*
_____ *G. p. oberholseri*   NW Mexico (Sonora to Sinaloa)
_____ *G. p. palmarum*   W Mexico (Nayarit to Oaxaca)
_____ *G. p. griscomi*   Central Mexico (sw Morelos and ne Guerrero)

☐ **Tamaulipas Pygmy-Owl** *Glaucidium sanchezi*

NE Mexico (sw Tamaulipas, e San Luis Potosí and n Hidalgo)

☐ **Central American Pygmy-Owl** *Glaucidium griseiceps*
_____ *G. g. occultum*   SE Mexico (se Veracruz, n Oaxaca and Chiapas)
_____ *G. g. griseiceps*   Guatemala, Belize and Honduras
_____ *G. g. rarum*   Costa Rica and Panama

☐ **Subtropical Pygmy-Owl** *Glaucidium parkeri*

East slope of Andes of Ecuador, Peru and extreme n Bolivia

☐ **Yungas Pygmy-Owl** *Glaucidium bolivianum*

E slope of Andes of se Peru, w-c Bolivia and nw Argentina

☐ **Amazonian Pygmy-Owl** *Glaucidium hardyi*

SE Venezuela to the Guianas, Amaz. Brazil, se Peru and Bolivia

☐ **Least Pygmy-Owl** *Glaucidium minutissimum*

E Paraguay to se Brazil and ne Argentina (Misiones)

☐ **Pernambuco Pygmy-Owl** *Glaucidium mooreorum*

Eastern Brazil (southern Pernambuco)

☐ **Ferruginous Pygmy-Owl** *Glaucidium brasilianum*
_____ *G. b. cactorum*   SE Arizona and w Mexico (Sonora to Oaxaca)
_____ *G. b. saturatum*   S Mexico (Chiapas) and Guatemala
_____ *G. b. ridgwayi*   S Texas (lower Rio Grande Valley) to Panama (Canal Zone)
_____ *G. b. medianum*   Tropical lowlands of n Colombia
_____ *G. b. margaritae*   Isla Margarita (Venezuela)
_____ *G. b. phaloenoides*   Tropical n Venezuela, Trinidad and the Guianas
_____ *G. b. duidae*   *Tepuis* of s Venezuela (Mt. Duida)
_____ *G. b. olivaceum*   *Tepuis* of s Venezuela (Mt. Auyan-Tepuí)
_____ *G. b. ucayalae*   E base of Andes of se Colombia to Peru and n Bolivia
_____ *G. b. brasilianum*   S Amazonian Brazil to e Paraguay, Uruguay and ne Argentina
_____ *G. b. pallens*   *Chaco* of e Bolivia, w Paraguay and n Argentina
_____ *G. b. stranecki*   S Uruguay to central Argentina

☐ **Tucuman Pygmy-Owl** *Glaucidium tucumanum*

Subtropical w Argentina (Salta and Tucumán to Córdoba)

☐ **Peruvian Pygmy-Owl** *Glaucidium peruanum*

Lowlands and foothills of sw Ecuador and w Peru

☐ **Austral Pygmy-Owl** *Glaucidium nanum*

Andes of s Chile and s Argentina; winters to n Argentina

☐ **Cuban Pygmy-Owl** *Glaucidium siju*
_____ *G. s. siju*   Cuba
_____ *G. s. vittatum*   Isle of Pines

☐ **Red-chested Owlet** *Glaucidium tephronotum*
_____ *G. t. tephronotum*   Equatorial forests of Liberia, Ivory Coast and Ghana

____	*G. t. pycrafti*	Forests of s Cameroon
____	*G. t. medje (elgonense)*	Congo River basin to Zaire, e Uganda and w Kenya

☐ **Sjostedt's Owlet** *Glaucidium sjostedti*

Lowlands of sw Cameroon to Gabon, n Congo and nw Zaire

☐ **Asian Barred Owlet** *Glaucidium cuculoides*

____	*G. c. cuculoides*	Himalayas (ne Pakistan and Kashmir to w Sikkim)
____	*G. c. austerum*	E Sikkim to Bhutan, ne Assam and nw Myanmar
____	*G. c. rufescens*	NE India, Bangladesh and n Myanmar
____	*G. c. bruegeli*	S Myanmar and s Thailand
____	*G. c. delacouri*	N Indochina
____	*G. c. deignani*	SE Thailand and s Indochina
____	*G. c. whitelyi*	Sichuan, Yunnan and se China s of Yangtze to ne Vietnam
____	*G. c. persimile*	Hainan (s China)

☐ **Javan Owlet** *Glaucidium castanopterum*

Java and Bali

☐ **Jungle Owlet** *Glaucidium radiatum*

____	*G. r. radiatum*	Himalayas to Bhutan, India, w Myanmar and Sri Lanka
____	*G. r. malabaricum*	SW peninsular India

☐ **Chestnut-backed Owlet** *Glaucidium castanonotum*

Forests of wet zone of Sri Lanka

☐ **African Barred Owlet** *Glaucidium capense*

____	*G. c. scheffleri*	Extreme s coastal Somalia, e Kenya and ne Tanzania
____	*G. c. ngamiense*	Central Tanzania to e Zaire, Angola, Mozambique and Mafia I.
____	*G. c. capense*	S Mozambique to Natal and e Cape Province

☐ **Chestnut Owlet** *Glaucidium castaneum*

____	*G. c. etchecopari*	Patchily distributed in Liberia and Ivory Coast
____	*G. c. castaneum*	NE Zaire (Semliki Valley) and sw Uganda (Bwamba Forest)

☐ **Albertine Owlet** *Glaucidium albertinum*

Known from 5 specimens from ne Zaire and n Rwanda

☐ **Long-whiskered Owlet** *Xenoglaux loweryi*

Cloud forests of n Peru (Amazonas and San Martín)

☐ **Elf Owl** *Micrathene whitneyi*

____	*M. w. whitneyi*	Arid sw US and adjacent nw Mexico (Sonora)
____	*M. w. idonea*	S Texas (lower Rio Grand Valley) to central Mexico
____	*M. w. sanfordi*	Lower Baja California (south of latitude 23°40'N)
____	*M. w. graysoni†*	Formerly Socorro I. (Revillagigedo Is. off w Mexico). Extinct

☐ **Burrowing Owl** *Athene cunicularia*

____	*A. c. hypugaea*	SW Canada to El Salvador
____	*A. c. rostrata*	Isla Clarión (Revillagigedo Islands off w Mexico)
____	*A. c. floridana*	Prairies of central and s Florida, Bahamas, Cuba, Isle of Pines
____	*A. c. guantanamensis*	Coastal n Cuba (Guantánamo Province)
____	*A. c. troglodytes*	Hispaniola, Gonâve and Beata islands
____	*A. c. amaura†*	Formerly Nevis and Antigua (West Indies). Extinct
____	*A. c. guadeloupensis†*	Formerly Guadeloupe (West Indies). Extinct
____	*A. c. arubensis*	Aruba (Netherlands Antilles)
____	*A. c. brachyptera*	Isla Margarita (off n Venezuela)
____	*A. c. apurensis*	N-central Venezuela
____	*A. c. minor*	S Guyana and adjacent extreme n Brazil (Roraima)
____	*A. c. carrikeri*	E Colombia
____	*A. c. tolimae*	W Colombia (Tolima)
____	*A. c. pichinchae*	W Ecuador (except for arid littoral)
____	*A. c. punensis*	Arid littoral of sw Ecuador and nw Peru
____	*A. c. intermedia*	Coastal w Peru (Paita to Pacasmayo)
____	*A. c. nanodes*	Arid littoral of w Peru (Trujillo to Arequipa)

____	*A. c. juninensis*	Andes of central Peru (Junín) to w Bolivia and nw Argentina
____	*A. c. boliviana*	Bolivia
____	*A. c. grallaria*	E Brazil (Maranhão to Mato Grosso and Paraná)
____	*A. c. partridgei*	N Argentina (Corrientes Province)
____	*A. c. cunicularia*	S Bolivia and s Brazil to Paraguay and Tierra del Fuego

☐ **Spotted Owlet** *Athene brama*

____	*A. b. albida*	S Iran and s Pakistan
____	*A. b. indica*	N and central peninsular India
____	*A. b. brama*	Southern India
____	*A. b. pulchra*	Myanmar, extreme sw China, s Laos, Cambodia and s Vietnam

☐ **Forest Owlet** *Athene blewitti*

Rediscovered in central India in 1998 after a 100-year absence

☐ **Little Owl** *Athene noctua*

____	*A. n. vidalii*	S Baltic to Iberian Pen., Balearic Is., Poland and nw Russia
____	*A. n. noctua*	Sardinia, Corsica, Italy and Yugoslavia to Carpathian Mts.
____	*A. n. indigena*	Balkans to Turkey, s Russia, Transcaucasia and sw Siberia
____	*A. n. glaux*	N Africa and coastal Israel (north to Haifa)
____	*A. n. saharae (solitudinis)*	Morocco to Tunisia s of Atlas, Libya coast, w Egypt and Arabia
____	*A. n. spilogastra*	Red Sea coast of e Sudan and n Ethiopia
____	*A. n. somaliensis*	E Ethiopia and Somalia
____	*A. n. lilith*	Cyprus; inland Middle East from se Turkey to s Sinai
____	*A. n. bactriana*	Azerbaijan to Iraq, Iran, Afghanistan and Lake Balkhash
____	*A. n. orientalis*	Extreme nw China and adjacent Siberia
____	*A. n. impasta*	W-central China (Kokonor and w Gansu)
____	*A. n. ludlowi*	S-central China and Tibet to n Himalayas
____	*A. n. plumipes*	NE China, Mongolia and Ussuriland

☐ **Boreal Owl** *Aegolius funereus*

____	*A. f. funereus*	N Scandinavia to Pyrénées and Urals (except for Caucasus Mts.)
____	*A. f. caucasicus*	N Caucasus Mountains
____	*A. f. pallens*	W Siberia, Tien Shan and s Siberia east to Sakhalin
____	*A. f. magnus*	NE Siberia (Kolyma to Kamchatka Peninsula)
____	*A. f. beickianus*	Extreme nw India (Lahul) to sw China (Qinghai)
____	*A. f. richardsoni*	Central Alaska and n Canada to n US

☐ **Northern Saw-whet Owl** *Aegolius acadicus*

____	*A. a. acadicus*	Mixed woodlands of s Alaska to s Mexico
____	*A. a. brooksi*	Queen Charlotte Islands (off British Columbia)

☐ **Unspotted Saw-whet Owl** *Aegolius ridgwayi*

____	*A. r. tacanensis*	Oak-pine woodlands of s Mexico (Chiapas)
____	*A. r. rostratus*	Locally from Guatemala to nw El Salvador
____	*A. r. ridgwayi*	Costa Rica and w Panama

☐ **Buff-fronted Owl** *Aegolius harrisii*

____	*A. h. harrisii*	Patchily distributed Colombia to Ecuador, Peru and Venezuela
____	*A. h. iheringi*	E Bolivia to Paraguay, e Brazil, Uruguay and ne Argentina
____	*A. h. dabbenei*	W Bolivia and nw Argentina (Tucumán, Salta and Jujuy)

☐ **Rufous Owl** *Ninox rufa*

____	*N. r. humeralis (aruensis)*	New Guinea, Aru Islands and Waigeo I.
____	*N. r. rufa*	Coastal n Australia (Kimberleys and n Northern Territory)
____	*N. r. meesi*	Coastal Cape York Pen. south to Endeavor R. and Mitchell R.
____	*N. r. queenslandica*	Coastal e Queensland (Endeavour River to Rockhamton)

☐ **Powerful Owl** *Ninox strenua*

E Australia (se Queensland to Victoria and se South Australia)

☐ **Barking Owl** *Ninox connivens*

____	*N. c. rufostrigata*	N Moluccas (Halmahera, Morotai, Bacan and Obi)

____ *N. c. assimilis*	E New Guinea, Manam I. and Karkar I.
____ *N. c. peninsularis (occidentalis)*	Coastal n Australia and islands in Torres Strait
____ *N. c. connivens*	SW Australia; widespread e Australia to base of Cape York Peninsula

☐ **Sumba Boobook** *Ninox rudolfi*

Lowlands of Sumba I. (w Lesser Sundas)

☐ **Andaman Hawk-Owl** *Ninox affinis*

Andaman Islands and Nicobar Islands

☐ **Morepork** *Ninox novaeseelandiae*

____ *N. n. leucopsis*	Tasmania and islands in Bass Strait
____ *N. n. albaria†*	Lord Howe I. Extinct
____ *N. n. undulata*	Norfolk I.
____ *N. n. novaeseelandiae*	New Zealand and offshore islands

☐ **Southern Boobook** *Ninox boobook*

____ *N. b. rotiensis*	Roti (Lesser Sundas)
____ *N. b. fusca*	Timor (e Lesser Sundas)
____ *N. b. plesseni*	Alor (north of Timor)
____ *N. b. moae*	Moa, Leti and Romang islands (e of Timor)
____ *N. b. cinnamomina*	Babar I. (e of Timor)
____ *N. b. remigialis*	Kai Islands
____ *N. b. pusilla*	S New Guinea
____ *N. b. ocellata*	Sawu (Lesser Sundas) and Australia (except east coast)
____ *N. b. lurida*	NE Australia (ne Queensland between Cooktown and Paluma)
____ *N. b. boobook*	Coastal e Australia (north to s Queensland)

☐ **Little Sumba Hawk-Owl** *Ninox sumbaensis*

Newly described species from Sumba I. (Lesser Sundas)

☐ **Brown Hawk-Owl** *Ninox scutulata*

____ *N. s. lugubris*	N India to w Assam and central peninsular India
____ *N. s. hirsuta*	S India and Sri Lanka
____ *N. s. obscura*	Andaman Islands and Nicobar Islands
____ *N. s. burmanica*	E Assam to s Yunnan, n Malay Pen., Thailand and Indochina
____ *N. s. palawanensis*	Palawan (sw Philippines)
____ *N. s. scutulata (malaccensis)*	S Malay Peninsula, Riau Archipelago, Sumatra and Bangka I.
____ *N. s. javanensis*	W Java
____ *N. s. borneensis*	Borneo and North Natuna Islands

☐ **Northern Boobook** *Ninox japonica*

____ *N. j. japonica*	SE Siberia to se Manchuria, e China, Korea and Japan; winters S Asia
____ *N. j. totogo*	Ryukyu Islands and Taiwan

☐ **Chocolate Boobook** *Ninox randi*

Philippine Islands, Sulu Archipelago and Talaud Archipelado

☐ **White-browed Owl** *Ninox superciliaris*

Forests and semiarid scrub of ne, sw and s Madagascar

☐ **Philippine Hawk-Owl** *Ninox philippensis*

____ *N. p. philippensis*	Leyte, Luzon, Marinduque, Samar, Polillo and Catanduanes
____ *N. p. mindorensis*	Mindoro (n Philippines)
____ *N. p. spilonota*	Philippines (Cebu, Sibuyan, Tablas and Camiguin Sur)
____ *N. p. proxima (ticaoensis)*	Philippines (Masbate and Ticao)
____ *N. p. centralis*	Philippines (Guimaras, Negros, Panay and Siquijor)
____ *N. p. spilocephala*	Philippines (Basilan, Mindanao, Siargao and Dinagat)
____ *N. p. reyi (everetti)*	Jolo, Tawitawi and adjacent islands in Sulu Archipelago

☐ **Ochre-bellied Hawk-Owl** *Ninox ochracea*

Humid forests of Sulawesi and Butung I.

☐ **Togian Hawk-Owl** *Ninox burhani*

Togian Islands off Sulawesi

☐ **Cinnabar Hawk-Owl** *Ninox ios*

NE Sulawesi (Bogani Nani Wartabone National Park)

☐ **Moluccan Hawk-Owl** *Ninox squamipila*

____	*N. s. hypogramma*	N Moluccas (Halmahera, Ternate and Bacan)
____	*N. s. hantu*	Buru (s Moluccas)
____	*N. s. squamipila*	Seram (s Moluccas)
____	*N. s. forbesi*	Tanimbar Islands (Arafura Sea)

☐ **Christmas Island Hawk-Owl** *Ninox natalis*

Christmas I. (Indian Ocean south of Java)

☐ **Jungle Hawk-Owl** *Ninox theomacha*

____	*N. t. hoedtii*	Waigeo and Misool islands (off w New Guinea)
____	*N. t. theomacha*	New Guinea
____	*N. t. goldii*	D'Entrecasteaux Arch. (Goodenough, Fergusson, Normanby)
____	*N. t. rosseliana*	Louisiade Archipelago (Tagula and Rossel)

☐ **Manus Hawk-Owl** *Ninox meeki*

Manus (Admiralty Islands)

☐ **Speckled Hawk-Owl** *Ninox punctulata*

Sulawesi, Kabaena, Muna and Butung islands

☐ **Bismarck Hawk-Owl** *Ninox variegata*

____	*N. v. superior*	New Hanover (Bismarck Archipelago)
____	*N. v. variegata*	New Ireland (Bismarck Archipelago)

☐ **New Britain Hawk-Owl** *Ninox odiosa*

New Britain (Bismarck Archipelago)

☐ **Solomon Hawk-Owl** *Ninox jacquinoti*

____	*N. j. eichhorni*	N Solomon Islands (Buka, Bougainville and Choiseul)
____	*N. j. jacquinoti*	Central Solomon Islands (Santa Isabel and San Jorge)
____	*N. j. granti*	Guadalcanal (s Solomon Islands)
____	*N. j. mono*	Mono (nw Solomon Islands)
____	*N. j. floridae*	Florida Group (central Solomon Islands)
____	*N. j. malaitae*	Malaita (s Solomon Islands)
____	*N. j. roseoaxillaris*	S Solomon Islands (Bauro and San Cristóbal)

☐ **Papuan Hawk-Owl** *Uroglaux dimorpha*

Sparsely distributed New Guinea and Yapen I.

☐ **Jamaican Owl** *Pseudoscops grammicus*

Woodlands of Jamaica

☐ **Striped Owl** *Pseudoscops clamator*

____	*P. c. forbesi*	Tropical s Mexico to Panama
____	*P. c. clamator*	Colombia to Venezuela, e Peru and central and ne Brazil
____	*P. c. oberi*	Trinidad and Tobago
____	*P. c. midas*	E Bolivia to Paraguay, s Brazil, Uruguay and n Argentina

☐ **Stygian Owl** *Asio stygius*

____	*A. s. lambi*	Highlands of w Mexico (sw Chihuahua to Jalisco)
____	*A. s. robustus*	S Mexico (Guerrero and Veracruz) to Venezuela and Ecuador
____	*A. s. siguapa*	Cuba and Isle of Pines
____	*A. s. noctipetens*	Hispaniola and Gonâve I.
____	*A. s. stygius*	E Bolivia to n and se Brazil and ne Argentina
____	*A. s. barberoi*	Paraguay and n Argentina

☐ **Northern Long-eared Owl** *Asio otus*

____	*A. o. tuftsi*	W Canada to nw Baja, s Texas and n Mexico (Nuevo León)
____	*A. o. wilsonianus*	S-central and se Canada to s-central US
____	*A. o. otus*	Europe, Asia and North Africa
____	*A. o. canariensis*	Canary Islands

☐ **African Long-eared Owl** *Asio abyssinicus*

____	*A. a. abyssinicus*	Highlands of Eritrea and Ethiopia
____	*A. a. graueri*	Mt. Kenya and Ruwenzori Mts. to Mt. Kabobo (e Zaire)

☐ **Madagascar Long-eared Owl** *Asio madagascariensis*

Forests of Madagascar

☐ **Short-eared Owl** *Asio flammeus*

____	*A. f. flammeus*	North America, Europe, n Asia and North Africa
____	*A. f. ponapensis*	Pohnpei (e Caroline Islands)
____	*A. f. sandwichensis*	Hawaiian Islands
____	*A. f. domingensis*	Hispaniola and (?) Cuba
____	*A. f. portoricensis*	Puerto Rico
____	*A. f. pallidicaudus*	N Venezuela and Guyana
____	*A. f. bogotensis*	Andes of Colombia, Ecuador and nw Peru
____	*A. f. galapagoensis*	Galapagos Islands
____	*A. f. suinda*	S Peru to Bolivia, se Brazil and Tierra del Fuego
____	*A. f. sanfordi*	Falkland Islands

☐ **Marsh Owl** *Asio capensis*

____	*A. c. tingitanus*	NW Morocco
____	*A. c. capensis*	Patchily distributed Senegambia to Ethiopia and South Africa
____	*A. c. hova*	Grasslands and marshes of Madagascar

☐ **Fearful Owl** *Nesasio solomonensis*

Solomon Islands (Bougainville, Santa Isabel and Choiseul)

ORDER: CAPRIMULGIFORMES
FAMILY: STEATORNITHIDAE (Oilbird—1)

☐ **Oilbird** *Steatornis caripensis*

Locally from Panama and n S Am. to w Bolivia; Trinidad

ORDER: CAPRIMULGIFORMES
FAMILY: AEGOTHELIDAE (Owlet-Nightjars—9)

☐ **Feline Owlet-Nightjar** *Aegotheles insignis*

Mountains of New Guinea (Vogelkop to se New Guinea)

☐ **Spangled Owlet-Nightjar** *Aegotheles tatei*

S New Guinea (lower elevations of Fly River headwaters)

☐ **Moluccan Owlet-Nightjar** *Aegotheles crinifrons*

N Moluccas (Halmahera, Kasiruta and Bacan)

☐ **Wallace's Owlet-Nightjar** *Aegotheles wallacii*

____	*A. w. wallacii*	W New Guinea and Aru Islands
____	*A. w. gigas*	Weyland Mountains (w-central New Guinea)
____	*A. w. manni*	N coastal mountains of New Guinea (Mt. Menawa and Mt. Turu)

☐ **Archbold's Owlet-Nightjar** *Aegotheles archboldi*

Central New Guinea (Wissel Lakes region)

☐ **Mountain Owlet-Nightjar** *Aegotheles albertisi*

____	*A. a. albertisi*	Arfak Mountains (nw New Guinea)
____	*A. a. wondiwoi*	Wandammen Peninsula (New Guinea)
____	*A. a. salvadorii*	Mountains of New Guinea (Weyland Mts. to se New Guinea)

☐ **New Caledonian Owlet-Nightjar** *Aegotheles savesi*

Rediscovered 2000 on New Caledonia after a 119-year absence

☐ **Barred Owlet-Nightjar** *Aegotheles bennettii*

____	*A. b. affinis*	Arfak Mountains (nw New Guinea)
____	*A. b. wiedenfeldi*	N New Guinea (Idenburg River to Holnicote Bay)
____	*A. b. terborghi*	E highlands of Papua New Guinea (Karimui basin region)
____	*A. b. bennettii*	Coastal se New Guinea (Koembe River to Milne Bay)
____	*A. b. plumiferus*	D'Entrecasteaux Archipelago (Fergusson and Goodenough)

☐ **Australian Owlet-Nightjar** *Aegotheles cristatus*
_____ *A. c. cristatus (major, leucogaster)* — SE New Guinea and Australia
_____ *A. c. tasmanicus* — Tasmania

ORDER: CAPRIMULGIFORMES
FAMILY: PODARGIDAE (Frogmouths—12)

☐ **Tawny Frogmouth** *Podargus strigoides*
_____ *P. s. phalaenoides (lilae, gouldi)* — N Australia (north of latitude 20°S)
_____ *P. s. brachypterus* — Mainland Australia (west of Great Dividing Range)
_____ *P. s. strigoides* — Australia (east of Great Dividing Range) and Tasmania

☐ **Marbled Frogmouth** *Podargus ocellatus*
_____ *P. o. ocellatus* — New Guinea, w Papuan is., Aru Is. and islands in Geelvink Bay
_____ *P. o. intermedius* — Trobriand Islands and D'Entrecasteaux Archipelago
_____ *P. o. meeki* — Tagula I. (Louisiade Archipelago)
_____ *P. o. inexpectatus* — N Solomon Islands (Bougainville, Choiseul and Santa Isabel)
_____ *P. o. marmoratus* — NE Australia (Cape York Peninsula)
_____ *P. o. plumiferus* — Coastal e Australia (se Queensland to ne New South Wales)

☐ **Papuan Frogmouth** *Podargus papuensis*
New Guinea, Aru, w Papuan is., and ne Australia (south to Townsville)

☐ **Large Frogmouth** *Batrachostomus auritus*
S Thailand, Malaysia, Sumatra, Borneo, N Natuna and Labuan is.

☐ **Dulit Frogmouth** *Batrachostomus harterti*
Mountains of n and central Borneo

☐ **Philippine Frogmouth** *Batrachostomus septimus*
_____ *B. s. microrhynchus* — N Philippines (mountains of Luzon and Catanduanes)
_____ *B. s. menagei* — Central Philippines (Negros and Panay)
_____ *B. s. septimus* — Philippines (Basilan, Mindanao, Leyte, Bohol and Samar)

☐ **Gould's Frogmouth** *Batrachostomus stellatus*
Pen. Thailand to Malay Pen., Sumatra, Borneo and adj. islands

☐ **Ceylon Frogmouth** *Batrachostomus moniliger*
Humid forests of sw India and Sri Lanka

☐ **Hodgson's Frogmouth** *Batrachostomus hodgsoni*
NE India to Myanmar, sw China, nw Thailand, Laos and Annam

☐ **Short-tailed Frogmouth** *Batrachostomus poliolophus*
_____ *B. p. poliolophus* — Montane forests of Sumatra
_____ *B. p. mixtus* — Montane forests of Borneo

☐ **Javan Frogmouth** *Batrachostomus javensis*
_____ *B. j. continentalis* — S Myanmar to Thailand, s Laos and central Vietnam
_____ *B. j. affinis (chaseni)* — SE peninsular Thailand to Sumatra, Borneo and Palawan
_____ *B. j. javensis* — Lowlands of w and central Java

☐ **Sunda Frogmouth** *Batrachostomus cornutus*
_____ *B. c. cornutus* — Sumatra, Borneo, Bangka, Belitung and Banggi islands
_____ *B. c. longicaudatus* — Kangean Islands (ne of Java)

ORDER: CAPRIMULGIFORMES
FAMILY: NYCTIBIIDAE (Potoos—7)

☐ **Great Potoo** *Nyctibius grandis*
Extreme s Mexico to n Bolivia, Paraguay and se Brazil

☐ **Long-tailed Potoo** *Nyctibius aethereus*
_____ *N. a. chocoensis* — Locally in w Colombia (Chocó)
_____ *N. a. longicaudatus* — Tropical e Ecuador to Peru and the Guianas
_____ *N. a. aethereus* — SE Paraguay to se Brazil and ne Argentina

☐ **Northern Potoo** *Nyctibius jamaicensis*

____	*N. j. lambi*	Pacific slope of w Mexico
____	*N. j. mexicanus*	E and s Mexico to El Salvador, Honduras and Roatán I.
____	*N. j. costaricensis*	Pacific slope of Nicaragua, Costa Rica and adj. w Panama
____	*N. j. jamaicensis*	Jamaica
____	*N. j. abbotti*	Hispaniola and Gonâve I.

☐ **Andean Potoo** *Nyctibius maculosus*

Locally in Andes of e Colombia to w Venezuela and w Bolivia

☐ **Common Potoo** *Nyctibius griseus*

____	*N. g. panamensis*	Nicaragua and sw Costa Rica to nw Venezuela and w Ecuador
____	*N. g. griseus (cornutus)*	Colombia to Guianas, Trinidad, Tobago, Brazil and n Argentina

☐ **White-winged Potoo** *Nyctibius leucopterus*

N Amazonian Brazil; e Brazil (Bahia)

☐ **Rufous Potoo** *Nyctibius bracteatus*

Tropical e Ecuador and e Peru to n Brazil and Guyana

ORDER: CAPRIMULGIFORMES
FAMILY: CAPRIMULGIDAE (Nightjars and Allies—91)

☐ **Short-tailed Nighthawk** *Lurocalis semitorquatus*

____	*L. s. stonei*	SE Mexico to Guatemala, n Honduras and ne Nicaragua
____	*L. s. noctivagus*	Costa Rica to Panama, coastal w Colombia and nw Ecuador
____	*L. s. semitorquatus*	NE Colombia to the Guianas, n Brazil, Trinidad and Tobago
____	*L. s. schaeferi*	N Venezuela
____	*L. s. nattereri*	E Ecuador to e Peru, Brazil s of the Amazon and ne Argentina

☐ **Rufous-bellied Nighthawk** *Lurocalis rufiventris*

Andes of sw Colombia to w Venezuela, Ecuador, Peru and Bolivia

☐ **Least Nighthawk** *Chordeiles pusillus*

____	*C. p. septentrionalis*	E Colombia to s Venezuela, the Guianas and adjacent Brazil
____	*C. p. esmeraldae*	SE Colombia to s Venezuela and extreme nw Brazil
____	*C. p. xerophilus*	Extreme ne Brazil (Paraíba and Pernambuco)
____	*C. p. novaesi*	NE Brazil (Maranhão and Piauí)
____	*C. p. pusillus*	E Brazil (Tocantins, Bahia and Goiás)
____	*C. p. saturatus*	Extreme e Bolivia and w-central Brazil

☐ **Sand-colored Nighthawk** *Chordeiles rupestris*

____	*C. r. xyostictus*	Sandbars and river banks of central Colombia
____	*C. r. rupestris*	SE Colombia to Venezuela, c Brazil, ne Peru and c Bolivia

☐ **Lesser Nighthawk** *Chordeiles acutipennis*

____	*C. a. texensis*	SW US to central Mexico; winters to n Colombia
____	*C. a. micromeris*	N Yucatán Pen., Isla Mujeres and Belize; winters to Panama
____	*C. a. littoralis*	S Mexico to Costa Rica
____	*C. a. acutipennis*	Tropical n South America to n Bolivia, Paraguay and Brazil
____	*C. a. crissalis*	SW Colombia
____	*C. a. aequatorialis*	W Colombia, w Ecuador and adjacent nw Peru
____	*C. a. exilis*	W Peru to extreme n Chile

☐ **Common Nighthawk** *Chordeiles minor*

____	*C. m. minor*	Central and s Canada to n and ne US; winters to n Argentina
____	*C. m. hesperis*	SW Canada and w US; winters n South America
____	*C. m. sennetti*	S-central Canada and n-central US ; winters to South America
____	*C. m. howelli*	W-central and s-central US; winters to South America
____	*C. m. henryi*	SW US and n-central Mexico; winters to Colombia
____	*C. m. asserriensis*	S-central US to extreme n Mexico (n Tamaulipas)
____	*C. m. chapmani*	SE US; winters to Argentina
____	*C. m. panamensis*	E Honduras, Belize and Nicaragua to Panama; winters to S Am.

☐ **Antillean Nighthawk** *Chordeiles gundlachii*

____ *C. g. vicinus*	S Florida and the Bahamas
____ *C. g. gundlachii*	Cuba, Isle of Pines, Jamaica, Hispaniola, Puerto Rico, Virgin Is.

☐ **Nacunda Nighthawk** *Podager nacunda*

____ *P. n. minor*	Colombia to Venezuela, Trinidad, the Guianas and n Brazil
____ *P. n. nacunda*	E Peru and Brazil s of Amazon to Paraguay and c Argentina

☐ **Band-tailed Nighthawk** *Nyctiprogne leucopyga*

____ *N. l. pallida*	Tropical ne Colombia to central Venezuela
____ *N. l. leucopyga*	E Venezuela to the Guianas and n Brazil
____ *N. l. exigua*	Tropical e Colombia and s Venezuela
____ *N. l. latifascia*	Extreme s Venezuela
____ *N. l. majuscula*	NE Peru to e Bolivia and central Brazil

☐ **Bahia Nighthawk** *Nyctiprogne vielliardi*

E Brazil (xeric *caatinga* of n Bahia and n Minas Gerais)

☐ **Spotted Nightjar** *Eurostopodus argus*

Australia (except e coast); winters to Aru Is. and Lesser Sundas

☐ **White-throated Nightjar** *Eurostopodus mystacalis*

____ *E. m. nigripennis*	N and central Solomon Islands
____ *E. m. exul*	New Caledonia
____ *E. m. mystacalis (gilberti)*	E Australia; winters to New Guinea

☐ **Diabolical Nightjar** *Eurostopodus diabolicus*

Rediscovered in 1998 after 60-year absence in central Sulawesi

☐ **Papuan Nightjar** *Eurostopodus papuensis*

Lowlands of New Guinea and Salawati I.

☐ **Archbold's Nightjar** *Eurostopodus archboldi*

Highlands of New Guinea

☐ **Malaysian Nightjar** *Eurostopodus temminckii*

S Thailand to Malaysia, Sumatra, Borneo and adjacent islands

☐ **Great Eared-Nightjar** *Eurostopodus macrotis*

____ *E. m. bourdilloni*	SW India
____ *E. m. cerviniceps*	Bangladesh and ne India to s China, Indochina and n Malay Peninsula
____ *E. m. jacobsoni*	Simeulue I. (off nw Sumatra)
____ *E. m. macrotis*	N and e Philippine Islands
____ *E. m. macropterus*	Sulawesi, Talaud Is., Sangihe I., Banggai Is. and Sula Is.

☐ **Pauraque** *Nyctidromus albicollis*

____ *N. a. insularis*	Tres Marías Islands (off w Mexico)
____ *N. a. merrilli*	Lower Rio Grande Valley to Tamaulipas; winters to Puebla
____ *N. a. yucatanensis*	Tropical n Mexico to Belize, Cozumel I. and Guatemala
____ *N. a. intercedens*	S Guatemala to Costa Rica and w Panama
____ *N. a. gilvus*	Panama and n Colombia
____ *N. a. albicollis*	E Colombia to Venezuela, Guianas, ne Brazil and n Bolivia
____ *N. a. derbyanus*	Central and s Brazil to ne Argentina

☐ **Common Poorwill** *Phalaenoptilus nuttallii*

____ *P. n. nuttallii*	S Br. Columbia to w US and n Mexico; winters to c Mexico
____ *P. n. californicus*	California (west of the Sierra Nevada) to n Baja
____ *P. n. hueyi*	Lower Colorado River of California; n Baja and sw Arizona
____ *P. n. dickeyi*	S Baja California (south of latitude 30°N)
____ *P. n. adustus*	Extreme s Arizona to n Mexico (central Sonora)

☐ **Least Poorwill** *Siphonorhis brewsteri*

Semiarid lowlands of Hispaniola and Gonâve I.

☐ **Eared Poorwill** *Nyctiphrynus mcleodii*

____ *N. m. mcleodii*	W Mexico (Chihuahua and s Sonora to Jalisco and Colima)
____ *N. m. rayi*	Oak-pine woodlands of w-central Mexico (Guerrero)

☐ **Yucatan Poorwill** *Nyctiphrynus yucatanicus*

S Mexico (Yucatán Pen.) to n Belize and Petén of n Guatemala

☐ **Choco Poorwill** *Nyctiphrynus rosenbergi*

W Colombia (Chocó) and extreme nw Ecuador

☐ **Ocellated Poorwill** *Nyctiphrynus ocellatus*
____ *N. o. lautus*
____ *N. o. ocellatus*

E Honduras to Nicaragua, Costa Rica and (?) w Panama
Colombia to e Ecuador, Peru, Brazil, Paraguay, ne Argentina

☐ **Chuck-will's-widow** *Caprimulgus carolinensis*

E US; winters se US to Greater Antilles and n South America

☐ **Rufous Nightjar** *Caprimulgus rufus*
____ *C. r. minimus*
____ *C. r. otiosus*
____ *C. r. rufus (noctivigularus)*
____ *C. r. rutilus (ornatus, cortapau)*
____ *C. r. saltarius*

SE Costa Rica and Panama to Colombia and Venezuela; Coiba I.
St. Lucia (Lesser Antilles)
S Venezuela to the Guianas and n-central Brazil
S Brazil to e Bolivia, Paraguay and ne Argentina
NW Argentina and se Bolivia

☐ **Greater Antillean Nightjar** *Caprimulgus cubanensis*
____ *C. c. cubanensis*
____ *C. c. insulaepinorum*

Cuba
Isle of Pines and Cayo Coco

☐ **Hispaniolan Nightjar** *Caprimulgus ekmani*

Hispaniola

☐ **Tawny-collared Nightjar** *Caprimulgus salvini*

Tropical e Mexico (Nuevo León to Oaxaca and Chiapas)

☐ **Yucatan Nightjar** *Caprimulgus badius*

Yucatán Pen. and Cozumel I.; winters to Belize and n Honduras

☐ **Silky-tailed Nightjar** *Caprimulgus sericocaudatus*
____ *C. s. mengeli*
____ *C. s. sericocaudatus*

N Peru to nw Bolivia and n Brazil
SE Brazil to e Paraguay and ne Argentina

☐ **Buff-collared Nightjar** *Caprimulgus ridgwayi*
____ *C. r. ridgwayi*
____ *C. r. troglodytes*

SE Arizona and w Mexico (Sonora to Oaxaca)
Central Guatemala to Honduras and central Nicaragua

☐ **Whip-poor-will** *Caprimulgus vociferus*
____ *C. v. vociferus*
____ *C. v. arizonae*
____ *C. v. setosus*
____ *C. v. oaxacae*
____ *C. v. chiapensis*
____ *C. v. vermiculatus*

S Canada and e US; winters to Cuba and w Panama
SW US to central Mexico
E Mexico
SW Mexico
SE Mexico and highlands of Guatemala
Highlands of Honduras and El Salvador

☐ **Puerto Rican Nightjar** *Caprimulgus noctitherus*

Dry lowland forests of sw Puerto Rico

☐ **Dusky Nightjar** *Caprimulgus saturatus*

Montane forests of Costa Rica and w Panama

☐ **Band-winged Nightjar** *Caprimulgus longirostris*
____ *C. l. ruficervix*
____ *C. l. roraimae*
____ *C. l. atripunctatus*
____ *C. l. decussatus*
____ *C. l. bifasciatus*
____ *C. l. longirostris*
____ *C. l. patagonicus*

Andes of Colombia to w Venezuela and Ecuador
Tepuis of s Venezuela
Andes of Peru to Bolivia, nw Argentina and n Chile
Arid littoral of w Peru and extreme n Chile
Chile and w Argentina
SE Brazil to Paraguay, Uruguay and ne Argentina
Central and s Argentina

☐ **White-winged Nightjar** *Caprimulgus candicans*

Interior of n Bolivia, s-central Brazil and e Paraguay

☐ **Pygmy Nightjar** *Caprimulgus hirundinaceus*
____ *C. h. cearae*

E Brazil (Ceará to extreme n Bahia)

_____	_C. h. hirundinaceus_	NE Brazil (s Piauí to Bahia and Alagoas)
_____	_C. h. vielliardi_	E Brazil (Espírito Santo)

☐ **Little Nightjar** _Caprimulgus parvulus_

_____	_C. p. heterurus_	N Colombia to central Venezuela
_____	_C. p. parvulus_	E Peru to Brazil s of the Amazon, Uruguay and n Argentina

☐ **Spot-tailed Nightjar** _Caprimulgus maculicaudus_

Locally from se Mexico to n Bolivia, e Paraguay and se Brazil

☐ **White-tailed Nightjar** _Caprimulgus cayennensis_

_____	_C. c. albicauda_	Savanna of se Costa Rica to nw Colombia
_____	_C. c. apertus_	W Colombia to extreme n Ecuador
_____	_C. c. insularis_	NE Colombia, nw Venezuela, Isla Margarita and adj. islands
_____	_C. c. manati_	Martinique (Lesser Antilles)
_____	_C. c. leopetes_	Trinidad, Tobago, Bocas Islands and Little Tobago
_____	_C. c. cayennensis_	E Colombia to Venezuela, the Guianas and extreme n Brazil

☐ **Scrub Nightjar** _Caprimulgus anthonyi_

Lowlands of sw Ecuador and nw Peru

☐ **Cayenne Nightjar** _Caprimulgus maculosus_

Known from a 1917 specimen from French Guiana

☐ **Blackish Nightjar** _Caprimulgus nigrescens_

E Colombia to s Venezuela, the Guianas, Bolivia and Brazil

☐ **Roraiman Nightjar** _Caprimulgus whitelyi_

Tepuis of se Venezuela, adjacent n Brazil and Guyana

☐ **Brown Nightjar** _Caprimulgus binotatus_

Locally from Liberia to n Gabon and central Zaire

☐ **Red-necked Nightjar** _Caprimulgus ruficollis_

_____	_C. r. ruficollis_	Arid Iberian Peninsula and n Morocco
_____	_C. r. desertorum_	NE Morocco, n Algeria and n Tunisia; winters mainly in Mali

☐ **Gray Nightjar** _Caprimulgus indicus_

_____	_C. i. hazarae_	NE Pakistan to Bangladesh, s China, Myanmar and Malay Pen.
_____	_C. i. indicus_	Peninsular India south of the Himalayas
_____	_C. i. kelaarti_	Sri Lanka
_____	_C. i. jotaka_	SE Siberia to e China, Japan and Korea; winters to Gr. Sundas
_____	_C. i. phalaena_	Palau Islands (w Caroline Islands)

☐ **Eurasian Nightjar** _Caprimulgus europaeus_

_____	_C. e. europaeus_	N and c Europe to n Asia and L. Baikal area; winters to Africa
_____	_C. e. meridionalis_	Mediterranean basin to nw Iran and Caspian Sea
_____	_C. e. sarudnyi_	E side of Caspian Sea (Kazakstan) to Altai Mountains
_____	_C. e. unwini_	Iraq and Iran to w Tien Shan, Turkmenistan and Uzbekistan
_____	_C. e. plumipes_	NW China to w Mongolia
_____	_C. e. dementievi_	NE Mongolia and s Transbaikalia

☐ **Sombre Nightjar** _Caprimulgus fraenatus_

Ethiopia to nw Somalia, sw Kenya and ne Tanzania

☐ **Rufous-cheeked Nightjar** _Caprimulgus rufigena_

_____	_C. r. damarensis_	Coastal w Angola to Namibia, Botswana and nw South Africa
_____	_C. r. rufigena_	Zimbabwe and s Zambia to South Africa

☐ **Egyptian Nightjar** _Caprimulgus aegyptius_

_____	_C. a. saharae_	Morocco to Nile Delta; winters in western Sahel
_____	_C. a. aegyptius (arenicolor)_	NE Egypt and Arabia to w China, w Pakistan and se Iran

☐ **Nubian Nightjar** _Caprimulgus nubicus_

_____	_C. n. tamaricis_	Israel to Jordan, sw Saudi Arabia and Yemen
_____	_C. n. nubicus_	Central Sudan
_____	_C. n. torridus (taruensis)_	Central Ethiopia to Somalia, Kenya and ne Uganda
_____	_C. n. jonesi_	Socotra

☐ **Sykes' Nightjar** *Caprimulgus mahrattensis*

SE Iran to s Afghanistan, Pakistan and nw India

☐ **Vaurie's Nightjar** *Caprimulgus centralasicus*

W China (known from a 1960 specimen from w Xinjiang)

☐ **Golden Nightjar** *Caprimulgus eximius*
____ *C. e. simplicior*
____ *C. e. eximius*

S Mauritania and n Senegal to central Chad
Central Sudan

☐ **Large-tailed Nightjar** *Caprimulgus macrurus*
____ *C. m. albonotatus*
____ *C. m. bimaculatus (ambiguus, aequabilis, hainanus)*
____ *C. m. johnsoni*
____ *C. m. salvadorii*
____ *C. m. macrurus*
____ *C. m. schlegelii (yorki, meeki)*

NE Pakistan and n India to Bhutan and Bangladesh
NE India to s China, Sumatra and Riau Archipelago
S Philippines (Palawan, Busuanga and Culion)
N Borneo, Labuan, Balambangan, Banguey and s Sula islands
Java and Bali
Wallacea, New Guinea, New Britain and coastal n Australia

☐ **Andaman Nightjar** *Caprimulgus andamanicus*

Andaman Islands

☐ **Mees's Nightjar** *Caprimulgus meesi*

Lesser Lundas (Sumba and Flores)

☐ **Jerdon's Nightjar** *Caprimulgus atripennis*
____ *C. a. atripennis*
____ *C. a. aequabilis*

S peninsular India (Western Ghats and Eastern Ghats)
Sri Lanka

☐ **Philippine Nightjar** *Caprimulgus manillensis*

Philippine Islands (except Palawan)

☐ **Sulawesi Nightjar** *Caprimulgus celebensis*
____ *C. c. celebensis*
____ *C. c. jungei*

Sulawesi and Butung I.
Sula Islands (Taliabu and Mangole)

☐ **Donaldson-Smith's Nightjar** *Caprimulgus donaldsoni*

Ethiopia and Somalia to se Sudan and ne Tanzania

☐ **Black-shouldered Nightjar** *Caprimulgus nigriscapularis*

Senegambia to se Sudan, w Kenya and sw Zaire

☐ **Fiery-necked Nightjar** *Caprimulgus pectoralis*
____ *C. p. shelleyi*
____ *C. p. fervidus*
____ *C. p. crepusculans*
____ *C. p. pectoralis*

Angola to s Zaire, se Kenya and sw Tanzania
S Angola to n Namibia, Botswana, Zimbabwe and ne S Africa
SE Zimbabwe to Mozambique, Swaziland and e South Africa
S Transvaal, Natal and Cape Province

☐ **Abyssinian Nightjar** *Caprimulgus poliocephalus*

SW Saudi Arabia to Ethiopia, ne Uganda and n Tanzania

☐ **Montane Nightjar** *Caprimulgus ruwenzorii*
____ *C. r. ruwenzorii*
____ *C. r. guttifer*

SW Uganda to e Zaire; isolated population in w Angola
SW Tanzania to n Malawi, ne Zambia; isolated population ne Tanzania

☐ **Indian Nightjar** *Caprimulgus asiaticus*
____ *C. a. asiaticus*
____ *C. a. eidos*
____ *C. a. siamensis*

SE Pakistan to India, s Thailand and s Indochina
Sri Lanka
N Thailand

☐ **Madagascar Nightjar** *Caprimulgus madagascariensis*
____ *C. m. aldabrensis*
____ *C. m. madagascariensis*

Aldabra
Madagascar and Nosy Boraha

☐ **Swamp Nightjar** *Caprimulgus natalensis*
____ *C. n. natalensis*
____ *C. n. accrae*

E Gambia to Sudan, sw Ethiopia, e Tanzania and n S Africa
Coastal n-central Sierra Leone to w Cameroon

☐ **Plain Nightjar** *Caprimulgus inornatus*

Thornscrub of sub-Saharan Africa and sw Arabian Peninsula

☐ **Star-spotted Nightjar** *Caprimulgus stellatus*

SE Sudan to Ethiopia, Djibouti, Somalia and central Kenya

☐ **Nechisar Nightjar** *Caprimulgus solala*

Described from one wing from a road corpse found in Ethiopia

☐ **Savanna Nightjar** *Caprimulgus affinis*

____ *C. a. monticolus*	NE Pakistan to India, Myanmar, s Thailand, Cambodia, Vietnam
____ *C. a. amoyensis*	SE China and n Vietnam
____ *C. a. stictomus*	Taiwan
____ *C. a. griseatus*	N Philippines
____ *C. a. mindanensis*	S Philippines (Mindanao); sight record from Jolo (Sulu Arch.)
____ *C. a. affinis*	Greater and Lesser Sundas
____ *C. a. propinquus*	N-central and s Sulawesi
____ *C. a. undulatus*	W Lesser Sundas (Sumbawa, Komodo and Flores)
____ *C. a. kasuidori*	Central Lesser Sundas (Sumba and Sawu)
____ *C. a. timorensis*	E Lesser Sundas (Alor, Timor, Roti and Kisar)

☐ **Freckled Nightjar** *Caprimulgus tristigma*

____ *C. t. sharpei*	Guinea to Togo, Cameroon and Central African Republic
____ *C. t. pallidogriseus*	Nigeria
____ *C. t. tristigma*	NE Zaire to s Sudan, Burundi, Ethiopia and n Tanzania
____ *C. t. lentiginosus*	W Angola to Namibia and w Cape Province
____ *C. t. granosus*	SE Zaire to Zambia, s Tanzania and e Cape Province

☐ **Bonaparte's Nightjar** *Caprimulgus concretus*

Lowlands of Sumatra, Borneo and Belitung I.

☐ **Salvadori's Nightjar** *Caprimulgus pulchellus*

____ *C. p. pulchellus*	Sumatra
____ *C. p. bartelsi*	Java

☐ **Itombwe Nightjar** *Caprimulgus prigoginei*

Single 1955 specimen from e Zaire (Itombwe Forest)

☐ **Collared Nightjar** *Caprimulgus enarratus*

Humid forests of c Madagascar

☐ **Bates' Nightjar** *Caprimulgus batesi*

Congo basin (s Cameroon to n Gabon, e Zaire and w Uganda)

☐ **Long-tailed Nightjar** *Caprimulgus climacurus*

____ *C. c. climacurus*	Mauritania to Sudan, w Ethiopia and e Zaire
____ *C. c. sclateri*	Guinea to nw Uganda
____ *C. c. nigricans*	E Sudan (White Nile region) Nile Valley

☐ **Slender-tailed Nightjar** *Caprimulgus clarus*

SE Sudan to Ethiopia, Djibouti, Somalia and n Tanzania

☐ **Square-tailed Nightjar** *Caprimulgus fossii*

____ *C. f. fossii*	Grassy savanna of Gabon and sw Congo
____ *C. f. welwitschii*	S Zaire to Angola, Natal; Zanzibar and Pemba I.
____ *C. f. griseoplurus*	Kalahari Desert and extreme n South Africa

☐ **Pennant-winged Nightjar** *Macrodipteryx vexillarius*

Brachystegia woodlands of southern Africa

☐ **Standard-winged Nightjar** *Macrodipteryx longipennis*

Senegambia to sw Sudan, n Uganda, Ethiopia and Somalia

☐ **Lyre-tailed Nightjar** *Uropsalis lyra*

____ *U. l. lyra*	Andes of w Colombia to w Venezuela and central Ecuador
____ *U. l. peruana*	Andes of Peru to w Bolivia
____ *U. l. argentina*	Andes of s Bolivia and adjacent n Argentina

☐ **Swallow-tailed Nightjar** *Uropsalis segmentata*

____ *U. s. segmentata*	Andes of Colombia and n Ecuador
____ *U. s. kalinowskii*	E slope of Andes of central Peru to w Bolivia

☐ **Ladder-tailed Nightjar** *Hydropsalis climacocerca*

____ *H. c. schomburgki*	Riverine habitats of e Venezuela to the Suriname
____ *H. c. climacocerca*	Mainly along rivers from se Colombia to n Bolivia
____ *H. c. intercedens*	Central Peru (Obidos region in w Pará)
____ *H. c. pallidior*	N-central Brazil (Santarém region of w Pará)
____ *H. c. canescens*	N-central Brazil (lower Rio Tapajós region of w Pará)

☐ **Scissor-tailed Nightjar** *Hydropsalis torquata*

____ *H. t. torquata*	S Suriname to Amazonian and e Brazil and e Peru
____ *H. t. furcifer*	S Peru to e Bolivia, s Brazil, Uruguay, Paraguay, c Argentina

☐ **Long-trained Nightjar** *Macropsalis forcipata*

SE Brazil and adjacent ne Argentina (Misiones)

☐ **Sickle-winged Nightjar** *Eleothreptus anomalus*

Swamps of se Brazil to s Paraguay and n Argentina

ORDER: APODIFORMES
FAMILY: APODIDAE (Swifts—100)

☐ **Tepui Swift** *Cypseloides phelpsi*

Tepuis of s Venezuela, adj. nw Guyana and n Brazil (Roraima)

☐ **Black Swift** *Cypseloides niger*

____ *C. n. borealis*	Mainly mountains of se Alaska to sw US
____ *C. n. costaricensis*	Highlands of central Mexico to Costa Rica
____ *C. n. niger*	West Indies and Trinidad; Guyana (Merumé Mountains)

☐ **White-chested Swift** *Cypseloides lemosi*

SW Colombia (mts. of upper Cauca Valley), n Ecuador

☐ **Rothschild's Swift** *Cypseloides rothschildi*

Andes of nw Argentina; rarely adj. s Bolivia and Peru (Cuzco)

☐ **Sooty Swift** *Cypseloides fumigatus*

E Bolivia to se Brazil, ne Argentina and adjacent Paraguay

☐ **White-fronted Swift** *Cypseloides storeri*

Mountains of sw Mexico (Jalisco, Guerrero and Michoacán)

☐ **Spot-fronted Swift** *Cypseloides cherriei*

Mts. of Costa Rica; Colombia to w Ecuador and n Venezuela

☐ **White-chinned Swift** *Cypseloides cryptus*

Locally in mts. of nw S America; scattered records C America

☐ **Great Dusky Swift** *Cypseloides senex*

Central and s Brazil to e Paraguay and ne Argentina

☐ **Chestnut-collared Swift** *Streptoprocne rutila*

____ *S. r. griseifrons*	W Mexico (Nayarit to Jalisco, s Durango and w Zacatecas)
____ *S. r. brunnitorques*	SE Mexico to w Bolivia
____ *S. r. rutila*	Venezuela to Guyana and Trinidad

☐ **White-naped Swift** *Streptoprocne semicollaris*

Mts. of w Mexico (Chihuahua to Nayarit, Hidalgo and Morelos)

☐ **White-collared Swift** *Streptoprocne zonaris*

____ *S. z. mexicana*	Highlands of s Mexico to Belize and El Salvador
____ *S. z. bouchellii*	Nicaragua to Panama
____ *S. z. pallidifrons*	Greater Antilles and locally in Lesser Antilles
____ *S. z. subtropicalis*	Mountains of Colombia to w Venezuela (Mérida) and Peru
____ *S. z. altissima*	Andes of Colombia and Ecuador
____ *S. z. minor*	Cordillera of coastal n Venezuela and Trinidad
____ *S. z. albicincta*	Tropical s Venezuela and the Guianas
____ *S. z. kuenzeli*	Andes of Bolivia and nw Argentina
____ *S. z. zonaris*	Lowlands of s Brazil, Bolivia, Paraguay and n Argentina

☐ **Biscutate Swift** *Streptoprocne biscutata*

_____ *S. b. seridoensis*   NE Brazil (Seridó region of Paraíba)

_____ *S. b. biscutata*   SE Brazil (Minas Gerais) to Paraguay and ne Argentina

☐ **Waterfall Swift** *Hydrochous gigas*

Mts. of peninsular Malaysia, Borneo, Sumatra and w Java

☐ **Glossy Swiftlet** *Collocalia esculenta*

_____ *C. e. affinis*   Andaman and Nicobar Islands

_____ *C. e. elachyptera*   Mergui Archipelago (off Myanmar)

_____ *C. e. cyanoptila*   Malay Pen., Sumatra and satellite islands and lowland Borneo

_____ *C. e. vanderbilti*   Nias I. (off w Sumatra)

_____ *C. e. oberholseri*   Batu Islands and Mentawi Islands (off w Sumatra)

_____ *C. e. natalis*   Christmas I. (Indian Ocean south of Java)

_____ *C. e. septentrionalis*   Philippines (Calayan, Camiguin Norte, Babuyan, Claro, Fuga)

_____ *C. e. isonota*   Luzon (n Philippines)

_____ *C. e. marginata*   N and w Philippines (s Luzon to Palawan and Bohol)

_____ *C. e. bagobo*   S Philippines (Mindanao, Mindoro and Sulu Archipelago)

_____ *C. e. spilura*   N Moluccas

_____ *C. e. manadensis*   N Sulawesi, Sangihe, Siau, Talasea and Talaud islands

_____ *C. e. esculenta*   S Moluccas, s Sulawesi, Banggai and Sula islands

_____ *C. e. minuta*   Salayar, Bonerate, Tanahjampea and Kalao is. (n Flores Sea)

_____ *C. e. sumbawae*   W Lesser Sundas (Sumbawa, Sumba, Flores and Besar)

_____ *C. e. perneglecta*   Alor, Sawu, Wetar, Kisar, Romang, Damar and Tanimbar is.

_____ *C. e. neglecta*   E Lesser Sundas (Roti, Dao, Semau, Timor and Jaco)

_____ *C. e. amethystina*   Waigeo I. (off nw New Guinea)

_____ *C. e. erwini*   High mountains of w New Guinea

_____ *C. e. numforensis*   Numfor I. (off nw New Guinea)

_____ *C. e. nitens*   Lowlands of New Guinea and w Papuan islands

_____ *C. e. misimae*   Louisiade Archipelago (Misima and Rossel)

_____ *C. e. stresemanni*   Admiralty Islands (Manus, Rambutyo, Nauna, Los Negros)

_____ *C. e. kalili*   Bismarck Arch. (New Ireland, New Hanover and Dyaul)

_____ *C. e. spilogaster*   Bismarck Archipelago (Lihir Group and Tatau Islands)

_____ *C. e. hypogrammica*   Bismarck Archipelago (Nissan and Green)

_____ *C. e. tametamele*   Bismarck Arch. (New Britain and Witu); Bougainville

_____ *C. e. becki*   Central and ne Solomon Islands; single record from Malaita

_____ *C. e. makirensis*   San Cristóbal I. (se Solomon Islands)

_____ *C. e. desiderata*   Rennell (s Solomon Islands)

_____ *C. e. uropygialis*   Santa Cruz Is. and Vanuatu (including Torres and Banks Is.)

_____ *C. e. albidior*   New Caledonia and Loyalty Islands

☐ **Cave Swiftlet** *Collocalia linchi*

_____ *C. l. ripleyi*   Barisan Mountains (Sumatra)

_____ *C. l. dodgei*   High elevations on Mt. Kinabalu (n Borneo)

_____ *C. l. linchi*   Java, Madura, Nusa Penida and Bawean islands

_____ *C. l. dedii*   Bali and Lombok

☐ **Pygmy Swiftlet** *Collocalia troglodytes*

Philippines and Palawan (absent in Sulu Archipelago)

☐ **Seychelles Swiftlet** *Aerodramus elaphrus*

Seychelles (Mahé, Praslin and La Digue)

☐ **Mascarene Swiftlet** *Aerodramus francicus*

W Mascarene Islands (Mauritius and Réunion)

☐ **Indian Swiftlet** *Aerodramus unicolor*

SW pen. India, Sri Lanka and small islands off Malabar coast

☐ **Moluccan Swiftlet** *Aerodramus infuscatus*

_____ *A. i. sororum*   Sulawesi, Sangihe, Siau and Taliabu islands

_____ *A. i. infuscatus*   N Moluccas (Halmahera, Ternate and Morotai)

_____ *A. i. ceramensis*   S Moluccas (Buru, Boano, Seram and Ambon)

☐ **Philippine Swiftlet** *Aerodramus mearnsi*

Lowlands of Philippine Islands

☐ **Mountain Swiftlet** *Aerodramus hirundinaceus*
- _____ *A. h. baru* — Yapen I.
- _____ *A. h. excelsus* — Central New Guinea (Snow Mountains and Mt. Carstenz)
- _____ *A. h. hirundinaceus* — New Guinea, Dampier and Goodenough islands

☐ **White-rumped Swiftlet** *Aerodramus spodiopygius*
- _____ *A. s. delichon* — Manus I. (Admiralty Islands)
- _____ *A. s. eichhorni* — Bismarck Archipelago (Mussau I. in St. Matthias group)
- _____ *A. s. noonaedanae* — Bismarck Archipelago (New Ireland and New Britain)
- _____ *A. s. reichenowi* — Southern and e Solomon Islands
- _____ *A. s. desolatus* — Solomon Islands (Duff, Swallow and Santa Cruz)
- _____ *A. s. epiensis* — Banks Islands and n Vanuatu to Epi Islands
- _____ *A. s. ingens* — S Vanuatu
- _____ *A. s. leucopygius* — Loyalty Islands and New Caledonia
- _____ *A. s. assimilis* — Fiji Islands
- _____ *A. s. townsendi* — Tonga
- _____ *A. s. spodiopygius* — Samoa

☐ **Australian Swiftlet** *Aerodramus terraereginae*
- _____ *A. t. terraereginae* — Coastal n Queensland), Dunk, Hinchinbrook and Family Is.
- _____ *A. t. chillagoensis* — NE Australia (inland Queensland w of Great Dividing Range)

☐ **Himalayan Swiftlet** *Aerodramus brevirostris*
- _____ *A. b. brevirostris* — Himalayas to Nepal, ne India, Myanmar and Thailand
- _____ *A. b. innominatus* — E-central China to n Vietnam; winters to Malay Peninsula

☐ **Indochinese Swiftlet** *Aerodramus rogersi*
- Mountains of e Myanmar, w Thailand, n Laos and Vietnam

☐ **Volcano Swiftlet** *Aerodramus vulcanorum*
- Mountains of Java

☐ **Whitehead's Swiftlet** *Aerodramus whiteheadi*
- _____ *A. w. whiteheadi* — N Philippines (Mt. Data on n Luzon)
- _____ *A. w. origenis* — S Philippines (Mt. Apo on Mindanao)

☐ **Bare-legged Swiftlet** *Aerodramus nuditarsus*
- Mountains of central and e New Guinea

☐ **Mayr's Swiftlet** *Aerodramus orientalis*
- _____ *A. o. leletensis* — Bismarck Archipelago (Lelet Plateau in central New Ireland)
- _____ *A. o. orientalis* — Guadalcanal (Solomon Islands)

☐ **Palawan Swiftlet** *Aerodramus palawanensis*
- Palawan (sw Philippines)

☐ **Uniform Swiftlet** *Aerodramus vanikorensis*
- _____ *A. v. amelis* — Philippines (Luzon, Mindoro, Cebu, Bohol and Mindanao)
- _____ *A. v. aenigma* — Central and se Sulawesi and Muna I.
- _____ *A. v. heinrichi* — S Sulawesi
- _____ *A. v. moluccarum* — S Moluccas (Seram, Ambon, Banda, Gorong, Tayandu, Kai Is.)
- _____ *A. v. waigeuensis* — N Moluccas (Morotai and Halmahera) and w Papuan islands
- _____ *A. v. steini* — Biak and Numfor islands (off nw New Guinea)
- _____ *A. v. yorki* — Aru Islands, New Guinea and D'Entrecasteaux Archipelago
- _____ *A. v. tagulae* — Louisiade Archipelago, Trobriand Islands and Woodlark I.
- _____ *A. v. coultasi* — Admiralty Is. (Manus, Rambutyo, Los Negros); St. Matthias Is.
- _____ *A. v. pallens* — New Britain, New Ireland, Dyaul I. and New Hanover
- _____ *A. v. lihirensis* — Lihir, Feni, Tabar, Nuguria and Hibernian is. (e of New Ireland)
- _____ *A. v. lugubris* — Solomon Islands
- _____ *A. v. vanikorensis* — Santa Cruz Islands (including Duff and Swallow) and Vanuatu

☐ **Mossy-nest Swiftlet** *Aerodramus salangana*
- _____ *A. s. natunae* — N Borneo, Natuna Islands and (?) Sumatra
- _____ *A. s. maratua* — Maratua Archipelago (off ne Borneo)

| ____ | *A. s. aerophilus* | Nias and adjacent islands off w Sumatra |
| | *A. s. salangana* | Java; single Philippine record from Basilan |

☐ **Palau Swiftlet** *Aerodramus pelewensis*

Lowlands of Palau Islands (w Caroline Islands)

☐ **Mariana Swiftlet** *Aerodramus bartschi*

S Mariana Islands (Saipan, Tinian, Aguijan and Guam)

☐ **Caroline Islands Swiftlet** *Aerodramus inquietus*

____	*A. i. rukensis*	Central Caroline Islands (Yap and Truk)
	A. i. ponapensis	Pohnpei (e Caroline Islands)
	A. i. inquietus	Kosrae (Caroline Islands)

☐ **Atiu Swiftlet** *Aerodramus sawtelli*

Atiu (s Cook Archipelago)

☐ **Polynesian Swiftlet** *Aerodramus leucophaeus*

E Society Islands (Tahiti and Moorea)

☐ **Marquesan Swiftlet** *Aerodramus ocistus*

| ____ | *A. o. ocistus* | N Marquesas Islands (Eiao, Nukuhiva and Uahuka) |
| | *A. o. gilliardi* | S Marquesas Islands (Uapou, Hivaoa and Tahuata) |

☐ **Black-nest Swiftlet** *Aerodramus maximus*

____	*A. m. maximus*	S Myanmar to s Malay Peninsula, se Vietnam and w Java
	A. m. lowi	Sumatra, Nias I., and Borneo; one 1887 Palawan specimen
	A. m. tichelmani	SE Borneo

☐ **Edible-nest Swiftlet** *Aerodramus fuciphagus*

____	*A. f. inexpectatus*	Andaman and Nicobar islands
	A. f. vestitus	Sumatra, Belitung I. and Borneo
	A. f. perplexus	Maratua Archipelago (off e Borneo)
	A. f. fuciphagus	Java, Kangean Is. and Bali to w L Sundas and Tanahjampea I.
	A. f. dammermani	Flores (w Lesser Sundas)
	A. f. micans	E Lesser Sundas (Sumba, Sawu and Timor)

☐ **German's Swiftlet** *Aerodramus germani*

| ____ | *A. g. germani* | Coasts of Malay Peninsula, n Borneo and s Philippines |
| | *A. g. amechanus* | Anambas Islands (South China Sea) |

☐ **Papuan Swiftlet** *Aerodramus papuensis*

N New Guinea (Idenburg River to Huon Peninsula)

☐ **Scarce Swift** *Schoutedenapus myoptilus*

____	*S. m. poensis*	Bioko (Gulf of Guinea)
	S. m. chapini	Highlands of e Zaire, Rwanda and sw Uganda
	S. m. myoptilus	Highlands of Ethiopia to Zimbabwe and w Mozambique

☐ **Schouteden's Swift** *Schoutedenapus schoutedeni*

Known from five specimens (1956-59) from e Zaire

☐ **Philippine Needletail** *Mearnsia picina*

Philippines (Cebu, Leyte, Mindanao, Negros, Biliran, Samar)

☐ **Papuan Needletail** *Mearnsia novaeguineae*

| ____ | *M. n. buergersi* | N New Guinea (Sepik River area) |
| | *M. n. novaeguineae* | S New Guinea |

☐ **Malagasy Spinetail** *Zoonavena grandidieri*

| ____ | *Z. g. grandidieri* | Madagascar |
| | *Z. g. mariae* | Grand Comoro I. (Comoro Islands) |

☐ **São Tomé Spinetail** *Zoonavena thomensis*

Mountains of São Tomé and Príncipe (Gulf of Guinea)

☐ **White-rumped Needletail** *Zoonavena sylvatica*

India south of the Himalayas to w Myanmar

☐ **Mottled Spinetail** *Telacanthura ussheri*

____	*T. u. ussheri*	Senegambia to Nigeria
____	*T. u. sharpei*	Cameroon to Gabon, Zaire and Uganda
____	*T. u. stictilaema*	Coastal s Kenya to n Tanzania, Zanzibar and Pemba I.
____	*T. u. benguellensis*	W Angola to Mozambique

☐ **Black Spinetail** *Telacanthura melanopygia*

Sierra Leone to s Ghana, Gabon, ne Zaire and ne Angola

☐ **Silver-rumped Needletail** *Rhaphidura leucopygialis*

S Myanmar to Malay Peninsula, Greater Sundas and Bangka I.

☐ **Sabine's Spinetail** *Rhaphidura sabini*

Rainforests of s Guinea to e Zaire, w Uganda and w Kenya

☐ **Cassin's Spinetail** *Neafrapus cassini*

Sierra Leone and Liberia to nw Angola and w Uganda; Bioko

☐ **Bat-like Spinetail** *Neafrapus boehmi*

____	*N. b. boehmi*	Angola to s Zaire, w Tanzania and n Zambia
____	*N. b. sheppardi*	SE Kenya to s Tanzania, Mozambique and ne South Africa

☐ **White-throated Needletail** *Hirundapus caudacutus*

____	*H. c. caudacutus*	Siberia to Japan and Kuril Islands.; winters to Australia
____	*H. c. nudipes*	Himalayas to sw China; winters to India and Myanmar

☐ **Silver-backed Needletail** *Hirundapus cochinchinensis*

____	*H. c. rupchandi*	Central Nepal; winters to Malay Peninsula, Sumatra and Java
____	*H. c. cochinchinensis*	E Himalayas to SE Asia; winters to Sumatra and Java
____	*H. c. formosanus*	Taiwan

☐ **Brown-backed Needletail** *Hirundapus giganteus*

____	*H. g. indicus*	SW India and Sri Lanka; Bangladesh to SE Asia, Andaman Is.
____	*H. g. giganteus*	Malay Peninsula, Greater Sundas and Palawan

☐ **Purple Needletail** *Hirundapus celebensis*

N Sulawesi and Philippines (absent from Palawan)

☐ **Band-rumped Swift** *Chaetura spinicaudus*

____	*C. s. aetherodroma*	E Panama to s Ecuador and extreme n Peru
____	*C. s. latirostris*	S Venezuela to Brazilian border and Delta Amacuro
____	*C. s. spinicaudus*	E Venezuela, the Guianas and n Brazil (n Amapá)
____	*C. s. aethalea*	Central Brazil south of the Amazon

☐ **Costa Rican Swift** *Chaetura fumosa*

Southwest Costa Rica and w Panama

☐ **Lesser Antillean Swift** *Chaetura martinica*

Guadeloupe, Dominica, Martinique, St. Lucia and St. Vincent

☐ **Gray-rumped Swift** *Chaetura cinereiventris*

____	*C. c. phaeopygos*	Caribbean slope of e Nicaragua to Panama
____	*C. c. occidentalis*	W Colombia to w Ecuador and extreme nw Peru
____	*C. c. schistacea*	E Colombia to w Venezuela (Mérida and Táchira)
____	*C. c. lawrencei*	Grenada, Trinidad, Tobago, Isla Margarita and n Venezuela
____	*C. c. guianensis*	*Tepuis* of e Venezuela and w Guyana
____	*C. c. sclateri*	S Colombia to s Venezuela, nw Brazil, e Peru and nw Bolivia
____	*C. c. cinereiventris*	E Brazil to Paraguay and ne Argentina (Misiones)

☐ **Pale-rumped Swift** *Chaetura egregia*

E Ecuador to e Peru, n Bolivia and w Amazonian Brazil

☐ **Chimney Swift** *Chaetura pelagica*

E North America; winters to Brazil and n Chile

☐ **Vaux's Swift** *Chaetura vauxi*

____	*C. v. vauxi*	Locally from se Alaska to sw US; winters to Guatemala
____	*C. v. tamaulipensis*	E Mexico (sw Tamaulípas and se San Luis Potosí)
____	*C. v. gaumeri*	SE Mexico (Yucatán Peninsula) and Cozumel I.
____	*C. v. richmondi*	S Mexico to Costa Rica and extreme w Panama (Chiriquí)

_____ *C. v. ochropygia*	E Panama
_____ *C. v. aphanes*	N Venezuela

☐ **Chapman's Swift** *Chaetura chapmani*
_____ *C. c. chapmani*	Panama to Colombia, Venezuela, Guianas, ne Brazil; Trinidad
_____ *C. c. viridipennis*	Tropical e Peru to e Bolivia and w Amazonian Brazil

☐ **Short-tailed Swift** *Chaetura brachyura*
_____ *C. b. brachyura*	Panama to the Guianas, Trinidad, w-c Brazil and n Bolivia
_____ *C. b. praevelox*	S Lesser Antilles (Grenada, St. Vincent and Tobago)
_____ *C. b. cinereocauda*	N-central Brazil

☐ **Tumbes Swift** *Chaetura ocypetes*
	Locally in sw Ecuador and extreme nw Peru

☐ **Ashy-tailed Swift** *Chaetura andrei*
_____ *C. a. andrei*	Orinoco Valley of e-central Venezuela
_____ *C. a. meridionalis*	Colombia to the Guianas, Brazil, n Paraguay and nw Argentina

☐ **White-throated Swift** *Aeronautes saxatalis*
_____ *A. s. saxatalis*	S British Columbia to Baja Calif. and sw Mexico (Oaxaca)
_____ *A. s. nigrior*	S Mexico (Chiapas) to central Honduras

☐ **White-tipped Swift** *Aeronautes montivagus*
_____ *A. m. montivagus*	Locally in mts. of Colombia to n Venezuela and w Bolivia
_____ *A. m. tatei*	*Tepuis* of s Venezuela and extreme n Brazil

☐ **Andean Swift** *Aeronautes andecolus*
_____ *A. a. parvulus*	Andes of w Peru to extreme n Chile
_____ *A. a. peruvianus*	Andes of se Peru
_____ *A. a. andecolus*	Andes of Bolivia to w Argentina (Río Negro)

☐ **Antillean Palm-Swift** *Tachornis phoenicobia*
_____ *T. p. iradii*	Cuba and Isle of Pines
_____ *T. p. phoenicobia*	Jamaica, Hispaniola, Saona, Beata and Île-á-Vache

☐ **Pygmy Swift** *Tachornis furcata*
_____ *T. f. furcata*	Lowlands of ne Colombia and nw Venezuela
_____ *T. f. nigrodorsalis*	Lowlands of w Venezuela

☐ **Fork-tailed Palm-Swift** *Tachornis squamata*
_____ *T. s. semota*	E Colombia to s Venezuela, e Ecuador, ne Peru and nw Brazil
_____ *T. s. squamata*	Trinidad and the Guianas to Amazonian and e Brazil

☐ **Great Swallow-tailed Swift** *Panyptila sanctihieronymi*
	Mountains of s Mexico to s Honduras and (rarely) n Costa Rica

☐ **Lesser Swallow-tailed Swift** *Panyptila cayennensis*
_____ *P. c. veraecrucis*	Humid lowlands of se Mexico (Veracruz) to n Honduras
_____ *P. c. cayennensis*	S Honduras to n Bolivia and se Brazil; Trinidad and Tobago

☐ **Asian Palm-Swift** *Cypsiurus balasiensis*
_____ *C. b. balasiensis*	Indian subcontinent and Sri Lanka
_____ *C. b. infumatus*	Myanmar to Indochina, Malay Peninsula, Sumatra and Borneo
_____ *C. b. bartelsorum*	Java and Bali
_____ *C. b. pallidior*	Philippine Islands

☐ **African Palm-Swift** *Cypsiurus parvus*
_____ *C. p. parvus*	Senegambia to s Sudan, Ethiopia and sw Arabia
_____ *C. p. brachypterus*	Sierra Leone to ne Zaire, Angola and Gulf of Guinea islands
_____ *C. p. myochrous*	Higher elevations from s Sudan to ne South Africa
_____ *C. p. laemostigma*	Coastal lowlands of s Somalia to Mozambique
_____ *C. p. hyphaenes*	N Namibia and n Botswana

___	C. p. celer	Mozambique to Natal
___	C. p. griveaudi	Comoro Islands
___	C. p. gracilis	Madagascar

☐ **Alpine Swift** *Tachymarptis melba*

___	T. m. melba	S Europe to Asia Minor and nw Iran; winters African tropics
___	T. m. tuneti	E Morocco to Middle East, Iran, Kazakstan and w Pakistan
___	T. m. archeri	N Somalia to sw Arabia and Dead Sea depression
___	T. m. africanus	Ethiopia to Cape Province and sw Angola
___	T. m. maximus	Ruwenzori Mountains of ne Zaire and Uganda
___	T. m. marjoriae	N-central Namibia to nw Cape Province
___	T. m. willsi	Madagascar
___	T. m. nubifugus	Himalayas; winters in central India
___	T. m. dorabtatai	Mountains of w peninsular India
___	T. m. bakeri	Sri Lanka

☐ **Mottled Swift** *Tachymarptis aequatorialis*

___	T. a. lowei	Sierra Leone to Nigeria
___	T. a. furensis	W Sudan (Darfur region)
___	T. a. aequatorialis (schubotzi, bamendae)	Eritrea and Ethiopia to Cameroon, Angola and Mozambique
___	T. a. gelidus	SW Zimbabwe

☐ **Alexander's Swift** *Apus alexandri*

Cape Verde Islands

☐ **Common Swift** *Apus apus*

___	A. a. apus	W Palearctic east to Lake Baikal and Iran; winters to s Africa
___	A. a. pekinensis	Iran to Himalayas, Mongolia and n China; winters to s Africa

☐ **Plain Swift** *Apus unicolor*

Madeira and w Canary Islands; winters in n Africa

☐ **Nyanza Swift** *Apus niansae*

___	A. n. niansae	Eritrea to Ethiopia, e Uganda, w Kenya and n Tanzania
___	A. n. somalicus	N Somalia and adjacent Ethiopia

☐ **Pallid Swift** *Apus pallidus*

___	A. p. brehmorum	Madeira, Canary Islands, s Europe to Turkey, coastal n Africa
___	A. p. illyricus	Dalmatian coast of Adriatic Sea; winters in Sahel
___	A. p. pallidus	Mauritania (Banc d'Arguin), Sahara hills and Egypt to Pakistan

☐ **African Swift** *Apus barbatus*

___	A. b. glanvillei	Known from two specimens from Sierra Leone (Rokupr)
___	A. b. sladeniae	Locally in se Nigeria, w Cameroon, Bioko I. and w Angola
___	A. b. serlei	W Cameroon (Bamenda Plateau)
___	A. b. roehli	E Ethiopia to ne Uganda, Kenya, Malawi, e Zaire; ne Angola
___	A. b. hollidayi	Victoria Falls area on Zambia/Zimbabwe border
___	A. b. oreobates	Zimbabwe and Mozambique (Mt. Gorongoza)
___	A. b. barbatus	South Africa

☐ **Forbes-Watson's Swift** *Apus berliozi*

___	A. b. berliozi	Breeds Socotra, winters to Africa
___	A. b. bensoni	Coastal e Somalia; winters to coastal Kenya

☐ **Bradfield's Swift** *Apus bradfieldi*

___	A. b. bradfieldi	Deserts and arid savanna of sw Angola and Namibia
___	A. b. deserticola	South Africa (n Cape Province)

☐ **Madagascar Swift** *Apus balstoni*

___	A. b. balstoni	Madagascar
___	A. b. mayottensis	Comoro Islands

☐ **Fork-tailed Swift** *Apus pacificus*

___ *A. p. pacificus*	Siberia to Kamchatka, n China and s Japan; winters to Australia
___ *A. p. kanoi*	Southeast Tibet to e China and Taiwan; winters to Indonesia
___ *A. p. leuconyx*	Outer Himalayas and Assam Hills; winters in India
___ *A. p. cooki*	Thailand, Myanmar and Indochina; Hainan I. and Lan-yü I.

☐ **Dark-rumped Swift** *Apus acuticauda*

Breeds ne India >Bhutan, Myanmar, Thailand and Nepal

☐ **Little Swift** *Apus affinis*

___ *A. a. galilejensis*	N and sub-Saharan Africa east to Pakistan
___ *A. a. aerobates*	SW Mauritania to Ethiopia, Somalia, c Angola and S Africa
___ *A. a. bannermani*	Bioko, São Tomé and Príncipe (Gulf of Guinea)
___ *A. a. theresae*	W and s Angola to s Zambia and South Africa
___ *A. a. affinis*	S Somalia to n Mozambique, Pemba I. and Zanzibar to India
___ *A. a. singalensis*	S India and Sri Lanka

☐ **House Swift** *Apus nipalensis*

___ *A. n. nipalensis*	Nepal to se China, Myanmar, Thailand, Indochina, Philippines
___ *A. n. subfurcatus*	Malay Peninsula to Borneo, Sumatra and adjacent islands
___ *A. n. furcatus*	Java and Bali
___ *A. n. kuntzi*	Taiwan

☐ **Horus Swift** *Apus horus*

___ *A. h. horus (toulsoni)*	Widespread intra-African migrant south of the Sahara
___ *A. h. fuscobrunneus*	SW Angola

☐ **White-rumped Swift** *Apus caffer*

S Iberian Peninsula, central Morocco and sub-Saharan Africa

☐ **Bates' Swift** *Apus batesi*

Humid forests of w Cameroon to n Gabon and e Zaire

ORDER: APODIFORMES
FAMILY: HEMIPROCNIDAE (Treeswifts—4)

☐ **Crested Treeswift** *Hemiprocne coronata*

India to sw China, Myanmar and Indochina

☐ **Gray-rumped Treeswift** *Hemiprocne longipennis*

___ *H. l. harterti*	S Myanmar to sw Thailand, Malaysia, Sumatra and Borneo
___ *H. l. perlonga*	Islands off w Sumatra (Simeulue to Enggano)
___ *H. l. longipennis*	Java and Bali to Lombok and Kangean Islands
___ *H. l. wallacii*	Sulawesi, Banggai, Sula and adjacent islands

☐ **Whiskered Treeswift** *Hemiprocne comata*

___ *H. c. comata (stresemanni)*	S Myanmar and pen. Thailand to Sumatra, Borneo and adj. is.
___ *H. c. major (nakamurai)*	Philippine Islands and Sulu Archipelago (absent from Palawan)

☐ **Moustached Treeswift** *Hemiprocne mystacea*

___ *H. m. confirmata*	Moluccas and Aru Islands
___ *H. m. mystacea*	New Guinea and w Papuan islands
___ *H. m. aeroplanes*	Bismarck Archipelago
___ *H. m. macrura*	Admiralty Islands
___ *H. m. woodfordiana*	Solomon Islands, Feni I. and Bougainville
___ *H. m. carbonaria*	San Cristóbal (Solomon Islands)

ORDER: APODIFORMES
FAMILY: TROCHILIDAE (Hummingbirds—339)

☐ **Saw-billed Hermit** *Ramphodon naevius*

Lowlands of se Brazil (s Minas Gerais Santa Catarina)

☐ **White-tipped Sicklebill** *Eutoxeres aquila*
_____ *E. a. salvini (mundus)* — Humid foothills of e Costa Rica to w Colombia
_____ *E. a. heterurus* — Western Andes (sw Colombia to w Ecuador)
_____ *E. a. aquila* — Eastern Andes (Colombia to n Peru)

☐ **Buff-tailed Sicklebill** *Eutoxeres condamini*
_____ *E. c. condamini* — Eastern Andes (se Colombia to n Peru)
_____ *E. c. gracilis* — Eastern Andes (Peru to nw Bolivia)

☐ **Hook-billed Hermit** *Glaucis dohrnii*
Coastal se Brazil (Bahia and Espírito Santo)

☐ **Rufous-breasted Hermit** *Glaucis hirsutus*
_____ *G. h. insularum* — Grenada, Trinidad and Tobago
_____ *G. h. hirsutus* — Panama to w Colombia, Venezuela, Guianas, Brazil, n Bolivia

☐ **Bronzy Hermit** *Glaucis aeneus*
E Honduras to w Panama; w Colombia to nw Ecuador

☐ **Band-tailed Barbthroat** *Threnetes ruckeri*
_____ *T. r. ventosus* — Tropical e Guatemala and Belize to w Panama
_____ *T. r. ruckeri* — N and w Colombia to w Ecuador
_____ *T. r. venezuelensis* — NW Venezuela (region sw of Lake Maracaibo)

☐ **Pale-tailed Barbthroat** *Threnetes niger*
_____ *T. n. cervinicauda* — E Colombia to e Ecuador, ne Peru and adj. w Amaz. Brazil
_____ *T. n. rufigastra* — Central Peru to n Bolivia
_____ *T. n. leucurus* — S Venezuela to Suriname, Amazonian Brazil and n Bolivia
_____ *T. n. niger* — French Guiana and adjacent Brazil (n Amapá)
_____ *T. n. loehkeni* — NE Brazil north of the Amazon (Amapá)
_____ *T. n. medianus* — NE Brazil south of the Amazon (e Pará and n Maranhão)

☐ **Broad-tipped Hermit** *Anopetia gounellei*
Lowlands of e Brazil (Piauí, Ceará and Bahia)

☐ **White-whiskered Hermit** *Phaethornis yaruqui*
Tropical Pacific Colombia and w Ecuador

☐ **Green Hermit** *Phaethornis guy*
_____ *P. g. coruscus* — Mainly subtropical Costa Rica, Panama and nw Colombia
_____ *P. g. emiliae* — Major river valleys of w-central Colombia
_____ *P. g. apicalis* — E slope of Andes (n Colombia to nw Venezuela and e Peru)
_____ *P. g. guy* — NE Venezuela and Trinidad

☐ **White-bearded Hermit** *Phaethornis hispidus*
E Colombia to s Venezuela, n Bolivia and w Amazonian Brazil

☐ **Western Long-tailed Hermit** *Phaethornis longirostris*
_____ *P. l. griseoventer* — W Mexico (Nayarit to Colima)
_____ *P. l. mexicanus* — SW Mexico (w Guerrero to se Oaxaca)
_____ *P. l. longirostris (veracrucis)* — S Mexico (n Oaxaca and Chiapas) to Belize and n Honduras
_____ *P. l. cephalus (cassinii)* — E Honduras to nw Colombia
_____ *P. l. sussurus* — Santa Marta Mountains (ne Colombia)
_____ *P. l. baroni* — W Ecuador and nw Peru

☐ **Eastern Long-tailed Hermit** *Phaethornis superciliosus*
_____ *P. s. superciliosus* — S Venezuela to Suriname and nw Brazil
_____ *P. s. muelleri* — N Brazil south of Amazon (Pará and Maranhão)

☐ **Great-billed Hermit** *Phaethornis malaris*
_____ *P. m. insolitus* — E Colombia to s Venezuela and adjacent n Brazil
_____ *P. m. malaris* — Suriname and French Guiana to adjacent n Brazil (Amapá)
_____ *P. m. margarettae* — Coastal e Brazil (Pernambuco to Espírito Santo)
_____ *P. m. moorei* — E Colombia to e Ecuador and n Peru
_____ *P. m. ochraceiventris* — NE Peru and w Brazil to lower Rio Madeira (s of the Amazon)
_____ *P. m. bolivianus* — SE Peru to Bolivia and w Brazil (to west bank of Rio Tapajós)

☐ **Tawny-bellied Hermit** *Phaethornis syrmatophorus*
____ *P. s. syrmatophorus*
____ *P. s. columbianus (huallagae)*

W Andes of Colombia to sw Ecuador
E Andes of Colombia to n Peru

☐ **Koepcke's Hermit** *Phaethornis koepckeae*

Foothills of e slope of Andes of Peru

☐ **Needle-billed Hermit** *Phaethornis philippii*

Trop. e Peru, n Bolivia and w Amaz. Brazil s of the Amazon

☐ **Straight-billed Hermit** *Phaethornis bourcieri*
____ *P. b. bourcieri (whitelyi)*
____ *P. b. major*

E Colombia to s Venezuela, the Guianas, n Brazil and n Peru
Brazil south of the Amazon (e bank of lower Rio Tapajós)

☐ **Pale-bellied Hermit** *Phaethornis anthophilus*
____ *P. a. hyalinus*
____ *P. a. anthophilus (fuliginosus)*

Pearl Islands (Bay of Panama)
Central Panama to Colombia and n Venezuela

☐ **Scale-throated Hermit** *Phaethornis eurynome*
____ *P. e. paraguayensis*
____ *P. e. eurynome*

E Paraguay and ne Argentina (Misiones)
SE Brazil (Bahia to Rio Grande do Sul)

☐ **Planalto Hermit** *Phaethornis pretrei*

E Brazil to e Bolivia, e Paraguay and n Argentina

☐ **Sooty-capped Hermit** *Phaethornis augusti*
____ *P. a. curiosus*
____ *P. a. augusti (vicarius)*
____ *P. a. incanescens*

Santa Marta Mountains (ne Colombia)
Colombia (E Andes and Macarene Mts.) to mts. of n Venezuela
Tepuis of se Venezuela and adjacent Guyana

☐ **Buff-bellied Hermit** *Phaethornis subochraceus*

Lowlands of e Bolivia and adjacent Brazil (w Mato Grosso)

☐ **Dusky-throated Hermit** *Phaethornis squalidus*

SE Brazil (s Minas Gerais and Espírito Santo to Santa Catarina)

☐ **Streak-throated Hermit** *Phaethornis rupurumii*
____ *P. r. rupurumii*
____ *P. r. amazonicus*

Extreme e Colombia to Venezuela, w Guyana and adj. w Brazil
Valley of lower Amazon in n-central Brazil

☐ **Little Hermit** *Phaethornis longuemareus*

NE Venezuela to French Guiana; Trinidad

☐ **Minute Hermit** *Phaethornis idaliae*

Lowlands of se Brazil (Bahia to Rio de Janeiro)

☐ **Cinnamon-throated Hermit** *Phaethornis nattereri*

E Bolivia and adj. sw Brazil; ne Brazil (Maranhão to Ceará)

☐ **Reddish Hermit** *Phaethornis ruber*
____ *P. r. episcopus*
____ *P. r. ruber*
____ *P. r. nigricinctus*
____ *P. r. longipennis*

Central and e Venezuela, Guyana and adjacent Brazil
Suriname to French Guiana, Brazil, se Peru and n Bolivia
Extreme e Colombia to sw Venezuela and n Peru
S Peru

☐ **White-browed Hermit** *Phaethornis stuarti*

Foothills of se Peru to central Bolivia

☐ **Black-throated Hermit** *Phaethornis atrimentalis*
____ *P. a. atrimentalis*
____ *P. a. riojae*

E Andes of Colombia, Ecuador and n Peru
Central Peru

☐ **Stripe-throated Hermit** *Phaethornis striigularis*
____ *P. s. saturatus*
____ *P. s. subrufescens*
____ *P. s. striigularis*
____ *P. s. ignobilis*

S Mexico (s Veracruz) to nw Colombia
W Colombia and w Ecuador
N Colombia (Magdalena Valley) and adjacent w Venezuela
N Venezuela

☐ **Gray-chinned Hermit** *Phaethornis griseogularis*
____ *P. g. griseogularis*

Andes of Colombia to n Peru, s Venezuela and adjacent Brazil

_____ *P. g. zonura*   NW Peru (Cajamarca and adjacent Amazonas)

_____ *P. g. porcullae*   Andes of n Peru (Tumbes, Piura and Lambayeque)

☐ **Tooth-billed Hummingbird** *Androdon aequatorialis*

                                 Humid e Panama to w Colombia (Chocó) and nw Ecuador

☐ **Green-fronted Lancebill** *Doryfera ludovicae*

_____ *D. l. veraguensis*   Humid montane forests of Costa Rica and w Panama

_____ *D. l. ludovicae (rectirostris, grisea)*   Mts. of e Panama to Colombia, w Venezuela and nw Bolivia

☐ **Blue-fronted Lancebill** *Doryfera johannae*

_____ *D. j. johannae*   E slope of Andes of se Colombia and e Ecuador to ne Peru

_____ *D. j. guianensis*   *Tepuis* of s Venezuela, s Guyana and adjacent n Brazil

☐ **Scaly-breasted Hummingbird** *Phaeochroa cuvierii*

_____ *P. c. roberti*   Extreme se Mexico to Belize, Guatemala and ne Costa Rica

_____ *P. c. maculicauda*   Pacific slope of Costa Rica

_____ *P. c. saturatior*   Coiba I.

_____ *P. c. cuvierii*   E and central Panama

_____ *P. c. berlepschi*   Coastal n Colombia (Cartagena to Barranquilla)

☐ **Wedge-tailed Sabrewing** *Campylopterus curvipennis*

_____ *C. c. curvipennis*   S Mexico (se San Luis Potosí and sw Tamaulipas to n Oaxaca)

_____ *C. c. pampa*   Yucatán Peninsula to n Guatemala, Belize and n Honduras

☐ **Long-tailed Sabrewing** *Campylopterus excellens*

                                 Sierra de Tuxtla (se Mexico)

☐ **Gray-breasted Sabrewing** *Campylopterus largipennis*

_____ *C. l. aequatorialis*   E Colombia to Ecuador, Peru, n Bolivia and nw Brazil

_____ *C. l. largipennis*   E Venezuela, the Guianas and Rio Negro region of nw Brazil

_____ *C. l. obscurus*   NE Brazil (e Pará and Maranhão)

_____ *C. l. diamantinensis*   SE Brazil (Serra Espinhaço in Minas Gerais)

☐ **Rufous Sabrewing** *Campylopterus rufus*

                                 Highlands of s Mexico (Oaxaca and Chiapas) to El Salvador

☐ **Violet Sabrewing** *Campylopterus hemileucurus*

_____ *C. h. hemileucurus*   Patchily distributed highlands of s Mexico to s-c Nicaragua

_____ *C. h. mellitus*   Costa Rica and w Panama

☐ **Rufous-breasted Sabrewing** *Campylopterus hyperythrus*

                                 Tepuis of se Venezuela and adjacent nw Brazil

☐ **White-tailed Sabrewing** *Campylopterus ensipennis*

                                 Mountains of ne Venezuela and Tobago

☐ **Lazuline Sabrewing** *Campylopterus falcatus*

                                 Mountains of Colombia, Venezuela and e Ecuador

☐ **Santa Marta Sabrewing** *Campylopterus phainopeplus*

                                 Santa Marta Mountains (ne Colombia)

☐ **Napo Sabrewing** *Campylopterus villaviscensio*

                                 Foothills of s Colombia to e Ecuador and adjacent ne Peru

☐ **Buff-breasted Sabrewing** *Campylopterus duidae*

_____ *C. d. guaiquinimae*   *Tepuis* of s Venezuela (Mt. Guaiquinima)

_____ *C. d. duidae*   *Tepuis* of s Venezuela (Mt. Duida) and adjacent n Brazil

☐ **Sombre Hummingbird** *Campylopterus cirrochloris*

                                 Forest and scrub of e and central Brazil

☐ **Swallow-tailed Hummingbird** *Campylopterus macrourus*

_____ *C. m. macrourus*   Guianas to n, central and se Brazil and Paraguay

_____ *C. m. simoni*   NE Brazil (s Maranhão, Piauí and Ceará to Minas Gerais)

_____ *C. m. cyanoviridis*   SE Brazil (Serra do Mar in s São Paulo)

_____ *C. m. hirundo*   E Peru (Huiro)

_____ *C. m. bolivianus*   Savanna of nw Bolivia (Beni)

☐ **White-necked Jacobin** *Florisuga mellivora*
   _____ *F. m. mellivora*
   _____ *F. m. flabellifera*

Tropical s Mexico to n Bolivia and Amazonian Brazil; Trinidad
Tobago (s Lesser Antilles)

☐ **Black Jacobin** *Florisuga fuscus*

E Brazil to Uruguay and ne Argentina

☐ **Brown Violet-ear** *Colibri delphinae*

Belize and Guatemala to the Guianas, Brazil and Bolivia

☐ **Green Violet-ear** *Colibri thalassinus*
   _____ *C. t. thalassinus (minor)*
   _____ *C. t. cabanidis*
   _____ *C. t. cyanotus*
   _____ *C. t. crissalis*

Open mountain slopes of s Mexico to n-central Nicaragua
Highlands of Costa Rica and w Panama
Mountains of Colombia, Venezuela and Ecuador
Andes of Peru, Bolivia and extreme nw Argentina

☐ **Sparkling Violet-ear** *Colibri coruscans*
   _____ *C. c. germanus (rostratus)*
   _____ *C. c. coruscans*

Tepuis of s Venezuela, e Guyana and adjacent n Brazil
Mountains of Colombia and Venezuela to nw Argentina

☐ **White-vented Violet-ear** *Colibri serrirostris*

Savanna of e Bolivia to Paraguay, s Brazil and n Argentina

☐ **Green-throated Mango** *Anthracothorax viridigula*

NE Venezuela to the Guianas and n Brazil; Trinidad

☐ **Green-breasted Mango** *Anthracothorax prevostii*
   _____ *A. p. prevostii*
   _____ *A. p. gracilirostris*
   _____ *A. p. hendersoni (pinchoti)*
   _____ *A. p. viridicordatus*
   _____ *A. p. iridescens*

E Mexico to Guatemala, Belize and El Salvador
El Salvador and Honduras to central Costa Rica
Isla Providéncia and Isla San Andrés (off e Nicaragua)
Extreme ne Colombia (Guajira Pen.) and coastal n Venezuela
W Colombia (Cauca Valley) to w Ecuador and nw Peru

☐ **Black-throated Mango** *Anthracothorax nigricollis*

Panama and Colombia e of Andes to ne Argentina and Brazil

☐ **Veraguan Mango** *Anthracothorax veraguensis*

W Panama (Chiriquí to s Coclé) and adjacent islands

☐ **Antillean Mango** *Anthracothorax dominicus*
   _____ *A. d. dominicus*
   _____ *A. d. aurulentus*

Hispaniola, Île-á-Vache, Tortue, Gonâve and Beata islands
Puerto Rico, Culebra I., Vieques I. and Virgin Islands

☐ **Green Mango** *Anthracothorax viridis*

Puerto Rico

☐ **Jamaican Mango** *Anthracothorax mango*

Jamaica

☐ **Fiery-tailed Awlbill** *Avocettula recurvirostris*

SE Venezuela to the Guianas and n-central Brazil; e Ecuador

☐ **Fiery Topaz** *Topaza pyra*
   _____ *T. p. pyra*
   _____ *T. p. amaruni*
   _____ *T. p. pamprepta*

SE Colombia to e Ecuador, extreme ne Peru and s Venezuela
W Amazonian Peru and Ecuador (ríos Napo and Corrientes)
E Ecuador (Río Napo and Río Suno region)

☐ **Crimson Topaz** *Topaza pella*
   _____ *T. p. pella (smaragdula)*
   _____ *T. p. microrhyncha*

S Venezuela to Suriname and n Brazil (Amapá)
NE Brazil (south bank of lower Amazon near Belém)

☐ **Purple-throated Carib** *Eulampis jugularis*

Montane forests of Lesser Antilles

☐ **Green-throated Carib** *Eulampis holosericeus*
   _____ *E. h. holosericeus*
   _____ *E. h. chlorolaemus*

Puerto Rico, Virgin Is. and Lesser Antilles (except Grenada)
Grenada

☐ **Ruby-topaz Hummingbird** *Chrysolampis mosquitus*

Trop. e Panama to Colombia, Venezuela, e Bolivia and Brazil

☐ **Antillean Crested Hummingbird** *Orthorhyncus cristatus*
_____ *O. c. exilis* — E Puerto Rico, Virgin Islands and Lesser Antilles to St. Lucia
_____ *O. c. ornatus* — St. Vincent (Lesser Antilles)
_____ *O. c. cristatus* — Barbados (Lesser Antilles)
_____ *O. c. emigrans* — Lesser Antilles (Grenadines and Grenada)

☐ **Violet-headed Hummingbird** *Klais guimeti*
_____ *K. g. merritti* — E Honduras to e Panama
_____ *K. g. guimeti* — E Colombia to Venezuela, Brazil, e Ecuador and n Peru
_____ *K. g. pallidiventris* — E Peru to w-central Bolivia

☐ **Plovercrest** *Stephanoxis lalandi*
_____ *S. l. lalandi* — E Brazil (s Minas Gerais to Espírito Santo and ne São Paulo)
_____ *S. l. loddigesii* — E Paraguay and ne Argentina (Misiones) to s Brazil

☐ **Emerald-chinned Hummingbird** *Abeillia abeillei*
_____ *A. a. abeillei* — Mountains of se Mexico to n Honduras
_____ *A. a. aurea* — Mountains of s Honduras and n Nicaragua

☐ **Tufted Coquette** *Lophornis ornatus* — E Venezuela, Trinidad and the Guianas to n Brazil

☐ **Dot-eared Coquette** *Lophornis gouldii* — Lowlands of n-central Brazil to e Bolivia (Santa Cruz)

☐ **Frilled Coquette** *Lophornis magnificus* — Forests and scrub of e-central Brazil

☐ **Short-crested Coquette** *Lophornis brachylophus* — S Mexico (Sierra Madre del Sur of Guerrero)

☐ **Rufous-crested Coquette** *Lophornis delattrei*
_____ *L. d. lessoni* — Locally from sw Costa Rica to Andes of central Colombia
_____ *L. d. delattrei* — Locally from s Ecuador to e Peru and n Bolivia

☐ **Spangled Coquette** *Lophornis stictolophus* — Andes of e Colombia to w Venezuela and n Peru

☐ **Festive Coquette** *Lophornis chalybeus*
_____ *L. c. verreauxii* — E Colombia to e Ecuador, e Peru, nw Brazil and c Bolivia
_____ *L. c. klagesi* — SE Venezuela
_____ *L. c. chalybeus* — SE Brazil (Espírito Santo, Minas Gerais and Santa Catarina)

☐ **Peacock Coquette** *Lophornis pavoninus*
_____ *L. p. pavoninus* — *Tepuis* of se Venezuela to Guyana (Merumé Mountains)
_____ *L. p. duidae* — *Tepuis* of se Venezuela (Mt. Duida) and adjacent smaller *tepuis*

☐ **Black-crested Coquette** *Lophornis helenae* — Gulf slope of s Mexico (Veracruz) to e Costa Rica

☐ **White-crested Coquette** *Lophornis adorabilis* — Costa Rica to w Panama

☐ **Wire-crested Thorntail** *Popelairia popelairii* — E Colombia to e Ecuador and ne Peru

☐ **Black-bellied Thorntail** *Popelairia langsdorffi*
_____ *P. l. melanosternon* — SE Colombia to s Venezuela, e Ecuador, e Peru and w Brazil
_____ *P. l. langsdorffi* — E Brazil (Bahia, Espírito Santo and Rio de Janeiro)

☐ **Coppery Thorntail** *Popelairia letitiae* — Known from three specimens ca 1852 labeled from "Bolivia"

☐ **Green Thorntail** *Discosura conversii* — Costa Rica to Panama, w Colombia and w Ecuador

☐ **Racket-tailed Coquette** *Discosura longicaudus* — Tropical s Venezuela to the Guianas and e Brazil

☐ **Red-billed Streamertail** *Trochilus polytmus* — Jamaica (except range of *scitulus*)

☐ **Black-billed Streamertail** *Trochilus scitulus*

Extreme ne Jamaica (Portland Parish)

☐ **Blue-chinned Sapphire** *Chlorostilbon notatus*
_____ *C. n. notatus (cyanogenys)*
_____ *C. n. puruensis*
_____ *C. n. obsoletus*

N Colombia to Venezuela, Trinidad, Tobago, Guianas, e Brazil
SE Colombia to ne Peru and nw Brazil
NE Peru (lower Río Ucayali near mouth of Río Napo)

☐ **Blue-tailed Emerald** *Chlorostilbon mellisugus*
_____ *C. m. pumilus*
_____ *C. m. melanorhynchus*
_____ *C. m. gibsoni*
_____ *C. m. chrysogaster*
_____ *C. m. nitens*
_____ *C. m. caribaeus*
_____ *C. m. duidae*
_____ *C. m. subfurcatus*
_____ *C. m. mellisugus*
_____ *C. m. phoeopygus (napensis)*
_____ *C. m. peruanus*

Pacific slope of w Colombia and w Ecuador
Andes of sw Colombia (Nariño) and w Ecaudor
Colombia (upper Magdalena Valley)
Lowlands of n Colombia (Cartagena to Santa Marta)
Arid coast of extreme ne Colombia and nw Venezuela
NE Venezuela, Curaçao, Aruba, Bonaire, Trinidad, Margarita I.
Tepuis of s Venezuela (Mt. Duida)
S Venezuela to Guyana and nw Brazil (Rio Branco region)
Suriname, French Guiana and lower Amazonian Brazil
Upper Amazon and tributaries from Colombia to Bolivia
E Peru and e Bolivia

☐ **Golden-crowned Emerald** *Chlorostilbon auriceps*

W Mexico (s Sinaloa to Durango, Guerrero and Oaxaca)

☐ **Cozumel Emerald** *Chlorostilbon forficatus*

SE Mexico (Cozumel I. and rarely Isla Mujeres)

☐ **Canivet's Emerald** *Chlorostilbon canivetii*
_____ *C. c. canivetii*
_____ *C. c. osberti*
_____ *C. c. salvini*

SE Mexico (Tamaulipas) to Belize, n Guatemala and Nicaragua
SE Mexico (se Chiapas) to Honduras, Holbox, Bay and Hog is.
Highlands of Pacific slope of nw Costa Rica

☐ **Garden Emerald** *Chlorostilbon assimilis*

Pacific slope of sw Costa Rica and Panama; Coiba and Pearl is.

☐ **Glittering-bellied Emerald** *Chlorostilbon aureoventris*
_____ *C. a. pucherani*
_____ *C. a. aureoventris*
_____ *C. a. igneus*
_____ *C. a. berlepschi*

E Brazil (Maranhão and Ceará to Paraná)
E Bolivia to Paraguay and w-central Brazil (Mato Grosso)
NW Argentina (Jujuy and Chaco to Mendoza and San Luis)
S Brazil (Rio Grande do Sul) to Uruguay and ne Argentina

☐ **Chiribiquete Emerald** *Chlorostilbon olivaresi*

SE Colombia (Sierra de Chiribiquete)

☐ **Cuban Emerald** *Chlorostilbon ricordii*

Cuba, Isle of Pines, Grand Bahama, Great Abaco and Andros

☐ **Hispaniolian Emerald** *Chlorostilbon swainsonii*

Primarily montane forests of Hispaniola

☐ **Puerto Rican Emerald** *Chlorostilbon maugaeus*

Puerto Rico

☐ **Coppery Emerald** *Chlorostilbon russatus*

Highlands of ne Colombia and extreme nw Venezuela

☐ **Narrow-tailed Emerald** *Chlorostilbon stenurus*
_____ *C. s. stenurus*
_____ *C. s. ignotus*

Andes of ne Colombia to nw Venezuela and ne Ecuador
Coastal mts. of n Venezuela to highlands of extreme se Lara

☐ **Green-tailed Emerald** *Chlorostilbon alice*

Mts. of n Venezuela (Falcón to Lara, Sucre and n Monagas)

☐ **Short-tailed Emerald** *Chlorostilbon poortmani*
_____ *C. p. euchloris*
_____ *C. p. poortmani*

Humid montane forests of central Colombia
E slope of Eastern Andes of Colombia and nw Venezuela

☐ **Fiery-throated Hummingbird** *Panterpe insignis*
_____ *P. i. eisenmanni*
_____ *P. i. insignis*

NW Costa Rica (Cordillera de Guanacaste)
N-c Costa Rica (Cordillera de Tilarán) to extreme w Panama

☐ **White-tailed Emerald** *Elvira chionura*

Pacific slope of s Costa Rica to central Panama

☐ **Coppery-headed Emerald** *Elvira cupreiceps*

Highlands of n and central Costa Rica

☐ **Blue-capped Hummingbird** *Eupherusa cyanophrys*

S Mexico (Sierra de Miahuatlán of Oaxaca)

☐ **White-tailed Hummingbird** *Eupherusa poliocerca*

S Mexico (Sierra Madre del Sur from Guerrero to w Oaxaca)

☐ **Stripe-tailed Hummingbird** *Eupherusa eximia*
_____ *E. e. nelsoni*
_____ *E. e. eximia*
_____ *E. e. egregia*

Humid rainforests of se Mexico (Veracruz and Oaxaca)
S Mexico (Chiapas) to Belize and n Nicaragua
Highlands of Costa Rica and w Panama

☐ **Black-bellied Hummingbird** *Eupherusa nigriventris*

Montane forests of central Costa Rica and extreme w Panama

☐ **Rufous-cheeked Hummingbird** *Goethalsia bella*

Highlands of e Panama (Darién) and extreme nw Colombia

☐ **Violet-capped Hummingbird** *Goldmania violiceps*

Highlands of e Panama and extreme nw Colombia

☐ **Dusky Hummingbird** *Cynanthus sordidus*

Mountains of central Mexico (Michoacán to Oaxaca)

☐ **Broad-billed Hummingbird** *Cynanthus latirostris*
_____ *C. l. magicus*
_____ *C. l. latirostris*
_____ *C. l. lawrencei*
_____ *C. l. propinquus*
_____ *C. l. doubledayi (nitidus)*

Arid sw US to nw Mexico (Nayarit)
E Mexico (San Luis Potosí and Tamaulipas to n Veracruz)
Tres Marías Islands (off w Mexico)
Central Mexico (Guanajuato to Michoacán)
S Mexico (Guerrero, Oaxaca and Chiapas)

☐ **Blue-headed Hummingbird** *Cyanophaia bicolor*

Lesser Antilles (mountains of Dominica and Martinique)

☐ **Violet-crowned Woodnymph** *Thalurania colombica*
_____ *T. c. townsendi*
_____ *T. c. venusta*
_____ *T. c. colombica*
_____ *T. c. rostrifera*

Belize and e Guatemala to se Honduras
E Nicaragua to w Panama
N Colombia and nw Venezuela
NW Venezuela (sw Táchira)

☐ **Mexican Woodnymph** *Thalurania ridgwayi*

Pacific slope of w Mexico (s Nayarit, Jalisco and Guerrero)

☐ **Green-crowned Woodnymph** *Thalurania fannyi*
_____ *T. f. fannyi*
_____ *T. f. subtropicalis*
_____ *T. f. verticeps*
_____ *T. f. hypochlora*

E Panama to w Colombia
W-c Colombia (Cauca Valley and adjacent W and C Andes)
Pacific slope of W Andes of sw Colombia and nw Ecuador
Pacific lowlands of Ecuador to extreme n Peru

☐ **Fork-tailed Woodnymph** *Thalurania furcata*
_____ *T. f. refulgens*
_____ *T. f. furcata*
_____ *T. f. fissilis (orenocensis)*
_____ *T. f. nigrofasciata*
_____ *T. f. viridipectus (taczanowskii)*
_____ *T. f. jelskii*
_____ *T. f. simoni*
_____ *T. f. balzani*
_____ *T. f. furcatoides*
_____ *T. f. boliviana*
_____ *T. f. baeri*
_____ *T. f. eriphile*

NE Venezuela (Paría Peninsula and Sierra de Cumaná)
Extreme e Venezuela, Guianas and ne Brazil n of the Amazon
E Venezuela, adjacent extreme w Guyana and ne Brazil
SE Colombia to extreme s Venezuela and nw Brazil
E slope of Andes and lowlands of e Colombia to ne Peru
Tropical e Peru and adjacent Brazil
Amazonia s of the Amazon in extreme e Peru and w Brazil
N-central Brazil south of the Amazon
Lower Amazon region of e Brazil south of the Amazon
Andean foothills and adj. lowlands of se Peru and ne Bolivia
NE and central Brazil to se Bolivia and n Argentina
SE Brazil, adjacent Paraguay and ne Argentina (Misiones)

☐ **Long-tailed Woodnymph** *Thalurania watertonii*

Coastal e Brazil (e Pará to Pernambuco and Bahia)

☐ **Violet-capped Woodnymph** *Thalurania glaucopis*

SE Brazil to e Paraguay and ne Argentina

☐ **Violet-bellied Hummingbird** *Damophila julie*
____ *D. j. panamensis*
____ *D. j. julie*
____ *D. j. feliciana*

Humid forests of central Panama (east to Darién)
Tropical n Colombia
Tropical sw Colombia to w Ecuador and extreme nw Peru

☐ **Sapphire-throated Hummingbird** *Lepidopyga coeruleogularis*
____ *L. c. coeruleogularis*
____ *L. c. confinis*
____ *L. c. coelina*

Pacific slope of w Panama (Chiriquí to Canal Zone)
Caribbean slope of e Panama (Darién) and adj. nw Colombia
N Colombia (n Chocó to Santa Marta region)

☐ **Sapphire-bellied Hummingbird** *Lepidopyga lilliae*

Coastal n-central Colombia

☐ **Shining-green Hummingbird** *Lepidopyga goudoti*
____ *L. g. luminosa*
____ *L. g. goudoti*
____ *L. g. zuliae*
____ *L. g. phaeochroa*

Coastal lowlands of n Colombia
N-central Colombia (middle and upper Magdalena Valley)
N and w area of Lake Maracaibo basin (Colombia/Venezuela)
NW Venezuela (south and east area of Lake Maracaibo basin)

☐ **Blue-throated Goldentail** *Hylocharis eliciae*
____ *H. e. eliciae*
____ *H. e. earina*

Lowlands of se Mexico and Belize to s Costa Rica
W Panama, Coiba I., Bay of Panama islands and nw Colombia

☐ **Rufous-throated Sapphire** *Hylocharis sapphirina*

E Colombia to Guianas, s Venezuela, se Brazil and ne Argentina

☐ **White-chinned Sapphire** *Hylocharis cyanus*
____ *H. c. viridiventris*
____ *H. c. rostrata*
____ *H. c. conversa*
____ *H. c. cyanus*
____ *H. c. griseiventris*

Colombia to the Guianas, s Venezuela and n Brazil
E Peru to ne Bolivia and w Brazil (Mato Grosso)
E Bolivia to n Paraguay and sw Brazil (Mato Grosso do Sul)
Coastal e Brazil (Pernambuco to Rio de Janeiro)
Coastal se Brazil (São Paulo) to ne Argentina (Buenos Aires)

☐ **Gilded Sapphire** *Hylocharis chrysura*

Bolivia to Paraguay, Uruguay, se Brazil and n Argentina

☐ **Blue-headed Sapphire** *Hylocharis grayi*

W Colombia and n Ecuador

☐ **Humboldt's Sapphire** *Hylocharis humboldtii*

Extreme se Panama to Colombia and nw Ecuador

☐ **Xantus' Hummingbird** *Hylocharis xantusii*

Arid scrub of s Baja California

☐ **White-eared Hummingbird** *Hylocharis leucotis*
____ *H. l. borealis*
____ *H. l. leucotis*
____ *H. l. pygmaea*

Mountains of se Arizona and n Mexico
Highland pine forests of central and s Mexico to Guatemala
Highlands of El Salvador, Honduras and Nicaragua

☐ **Golden-tailed Sapphire** *Chrysuronia oenone*
____ *C. o. oenone (longirostris)*
____ *C. o. josephinae*
____ *C. o. alleni*

E Colombia to e Venezuela, e Ecuador, ne Peru and w Brazil
Tropical e Amazonian Peru
N Bolivia

☐ **White-throated Hummingbird** *Leucochloris albicollis*

E Bolivia to e Paraguay, n Argentina and se Brazil

☐ **White-tailed Goldenthroat** *Polytmus guainumbi*
____ *P. g. andinus*
____ *P. g. guainumbi*
____ *P. g. thaumantias*

E Colombia (south to Meta and Vichada)
Venezuela to the Guianas, n Brazil and Trinidad
E Bolivia to e Paraguay, c and e Brazil and ne Argentina

☐ **Tepui Goldenthroat** *Polytmus milleri*

Tepuis of s Venezuela and adj. Brazil (Roraima)

☐ **Green-tailed Goldenthroat** *Polytmus theresiae*
____ *P. t. theresiae* — The Guianas and n-c Brazil (Amazonas, Pará and Amapá)
____ *P. t. leucorrhous* — Savanna of Colombia to s Venezuela, nw Brazil and e Peru

☐ **Buffy Hummingbird** *Leucippus fallax*
— Coastal n Colombia, Venezuela, La Tortuga I. and Margarita I.

☐ **Tumbes Hummingbird** *Leucippus baeri*
— Arid littoral of extreme sw Ecuador and adjacent nw Peru

☐ **Spot-throated Hummingbird** *Leucippus taczanowskii*
— W slope of Central Andes of n and central Peru

☐ **Olive-spotted Hummingbird** *Leucippus chlorocercus*
— Extreme se Colombia to e Ecuador, ne Peru and nw Brazil

☐ **White-bellied Hummingbird** *Leucippus chionogaster*
____ *L. c. chionogaster* — N and central Peru
____ *L. c. hypoleucus* — SE Peru to Bolivia, Paraguay, nw Argentina and Mato Grosso

☐ **Green-and-white Hummingbird** *Leucippus viridicauda*
— E slope of Andes of Peru (s Huánuco to s Puno)

☐ **Many-spotted Hummingbird** *Leucippus hypostictus*
— Andes of e Ecuador to se Bolivia and nw Argentina; sw Brazil

☐ **Rufous-tailed Hummingbird** *Amazilia tzacatl*
____ *A. t. tzacatl* — E Mexico (s Tamaulipas) to central Panama
____ *A. t. handleyi* — Isla Escudo de Veraguas (off Caribbean coast of nw Panama)
____ *A. t. fuscicaudata* — N and w Colombia to w Venezuela
____ *A. t. brehmi* — Colombia (Nariño)
____ *A. t. jucunda* — SW Colombia and w Ecuador; Isla Gorgona (off w Colombia)

☐ **Chestnut-bellied Hummingbird** *Amazilia castaneiventris*
— W slope of Eastern Andes of n-central Colombia

☐ **Amazilia Hummingbird** *Amazilia amazilia*
____ *A. a. dumerilii* — Pacific slope of Ecuador to n Peru; e slope in Zamora Valley
____ *A. a. leucophaea* — NW Peru
____ *A. a. amazilia* — Arid littoral of w Peru
____ *A. a. caeruleigularis* — Foothills of sw Peru (Nazca Valley in Ica)

☐ **Loja Hummingbird** *Amazilia alticola*
— Andes of s Ecuador

☐ **Buff-bellied Hummingbird** *Amazilia yucatanensis*
____ *A. y. chalconota* — S Texas (lower Rio Grande Valley) to ne Mexico
____ *A. y. cerviniventris* — S Mexico (Veracruz, Puebla, Oaxaca and Chiapas)
____ *A. y. yucatanensis* — Yucatán Peninsula, Petén of n Guatemala and n Belize

☐ **Cinnamon Hummingbird** *Amazilia rutila*
____ *A. r. diluta* — Coastal nw Mexico (Sinaloa and Nayarit)
____ *A. r. graysoni* — Tres Marías Islands (off w Mexico)
____ *A. r. rutila* — W Mexico (Jalisco to Oaxaca)
____ *A. r. corallirostris* — SE Mexico (Chiapas) to w Costa Rica; Holbox I., Isla Mujeres

☐ **Plain-bellied Emerald** *Agyrtria leucogaster*
____ *A. l. leucogaster* — E Venezuela to the Guianas and n Brazil
____ *A. l. bahiae* — E Brazil (Pernambuco to Bahia)

☐ **Versicolored Emerald** *Agyrtria versicolor*
____ *A. v. millerii* — Tropical e Colombia to s Venezuela, e Peru and n Brazil
____ *A. v. hollandi* — Tropical se Venezuela and (?) adjacent Guyana
____ *A. v. nitidifrons* — NE Brazil
____ *A. v. versicolor* — SE Brazil
____ *A. v. kubtcheki* — NE Bolivia to e Paraguay, sw Brazil and extreme ne Argentina

☐ **Rondonia Emerald** *Agyrtria rondoniae*
— N Bolivia and w-c Brazil (Rondônia)

☐ **White-chested Emerald** *Agyrtria brevirostris*
_____ *A. b. chionopectus* | Trinidad
_____ *A. b. brevirostris* | E Venezuela to Guyana, Suriname and extreme n-c Brazil
_____ *A. b. orienticola* | Coastal French Guiana

☐ **Andean Emerald** *Agyrtria franciae*
_____ *A. f. franciae (veneta)* | Subtropical Andes of nw and central Colombia
_____ *A. f. viridiceps* | Tropical sw Colombia and w Ecuador
_____ *A. f. cyanocollis* | E slope of Andes of n Peru

☐ **White-bellied Emerald** *Agyrtria candida*
_____ *A. c. genini* | Caribbean slope of se Mexico
_____ *A. c. pacifica* | Pacific slope of se Mexico (Chiapas) to s Guatemala
_____ *A. c. candida* | Humid se Mexico (Yucatán Pen.) to Belize and Nicaragua

☐ **Azure-crowned Hummingbird** *Agyrtria cyanocephala*
_____ *A. c. cyanocephala* | Oak-pine woodlands of se Mexico to n-central Nicaragua
_____ *A. c. chlorostephana* | Mosquito coast of e Honduras and ne Nicaragua

☐ **Violet-crowned Hummingbird** *Agyrtria violiceps*
_____ *A. v. ellioti* | Semiarid sw US to w Mexico (Jalisco and Guanajuato)
_____ *A. v. violiceps* | SW Mexico

☐ **Green-fronted Hummingbird** *Agyrtria viridifrons*
_____ *A. v. viridifrons* | S Mexico (e Guerrero to w Oaxaca; e Oaxaca to c Chiapas)
_____ *A. v. wagneri* | S Mexico (central and s Oaxaca)

☐ **Sapphire-spangled Emerald** *Polyerata lactea*
_____ *P. l. zimmeri* | *Tepuis* of se Venezuela (Mt. Auyan-tepui)
_____ *P. l. bartletti* | E Peru to n Bolivia
_____ *P. l. lactea* | Central and s Brazil (s Bahia to São Paulo)

☐ **Glittering-throated Emerald** *Polyerata fimbriata*
_____ *P. f. elegantissima* | Extreme ne Colombia to n Venezuela
_____ *P. f. distans* | Andes of w Venezuela (Táchira). Probable hybrid
_____ *P. f. fimbriata (maculicauda, alia)* | NE Venezuela to Guianas and Brazil north of the Amazon
_____ *P. f. apicalis* | Colombia east of the Andes
_____ *P. f. fluviatilis* | SE Colombia and e Ecuador
_____ *P. f. laeta* | NE Peru
_____ *P. f. nigricauda* | E Bolivia and central Brazil south of the Amazon
_____ *P. f. tephrocephala* | Coastal se Brazil (Espírito Santo to Rio Grande do Sul)

☐ **Blue-chested Hummingbird** *Polyerata amabilis*
| NE Nicaragua to Colombia and Ecuador (west of the Andes)

☐ **Charming Hummingbird** *Polyerata decora*
| Pacific slope of sw Costa Rica and extreme w Panama

☐ **Purple-chested Hummingbird** *Polyerata rosenbergi*
| Pacific lowlands of w Colombia and nw Ecuador

☐ **Mangrove Hummingbird** *Polyerata boucardi*
| W coast of Costa Rica (Gulf of Nicoya to Gulf of Dulce)

☐ **Honduran Emerald** *Polyerata luciae*
| Arid interior valleys of n and central Honduras

☐ **Steely-vented Hummingbird** *Saucerottia saucerrottei*
_____ *S. s. hoffmanni* | Semiarid w Nicaragua to central Costa Rica
_____ *S. s. warscewiczi* | N Colombia and extreme nw Venezuela
_____ *S. s. saucerrottei (australis)* | Colombia (W slope of Western Andes and Cauca Valley)
_____ *S. s. braccata* | Andes of w Venezuela (Mérida and Trujillo)

☐ **Indigo-capped Hummingbird** *Saucerottia cyanifrons*
| Lowlands and foothills of n and central Colombia

☐ **Alfaros Hummingbird** *Saucerottia alfaroana*

Costa Rica (Volcán Miravalles). Possible hybrid; probably extinct

☐ **Snowy-bellied Hummingbird** *Saucerottia edward*
____ *S. e. niveoventer* Extreme sw Costa Rica and w Panama; Isla Coiba
____ *S. e. edward* Arid tropical Panama (Canal Zone to w Darién)
____ *S. e. collata* Central Panama
____ *S. e. margaritarum (crosbyi)* Urabá, Taboga, Taboguilla, Pearl is.; e Panama to sw Darién

☐ **Blue-tailed Hummingbird** *Saucerottia cyanura*
____ *S. c. guatemalae* Pacific slope of s Mexico (se Chiapas) to s Guatemala
____ *S. c. cyanura* S Honduras to e El Salvador and nw Nicaragua
____ *S. c. impatiens* NW and central Costa Rica

☐ **Berylline Hummingbird** *Saucerottia beryllina*
____ *S. b. viola* Oak-pine woods of nw Mexico (Sonora to Guerrero)
____ *S. b. beryllina* E and central Mexico (Veracruz to Chiapas)
____ *S. b. lichtensteini* Montane slopes of s Mexico (w Chiapas)
____ *S. b. sumichrasti* Coastal mountains of s Mexico (central and s Chiapas)
____ *S. b. devillei* Highlands of Guatemala to Honduras

☐ **Green-bellied Hummingbird** *Saucerottia viridigaster*
____ *S. v. viridigaster* E slope of Eastern Andes of Colombia
____ *S. v. pacaraimae* Mts. of s Venezuela (Sierra de Pacaraima)
____ *S. v. iodura* Andes of w Venezuela (Táchira)

☐ **Copper-tailed Hummingbird** *Saucerottia cupreicauda*
____ *S. c. cupreicauda* *Tepuis* of s Venezuela, Guyana and extreme n Brazil (Roraima)
____ *S. c. duidae* *Tepuis* of s Venezuela (Mt. Duida)
____ *S. c. laireti* *Tepuis* of s Venezuela (Cerro de la Neblina)

☐ **Copper-rumped Hummingbird** *Saucerottia tobaci*
____ *S. t. tobaci* Tobago
____ *S. t. erythronotos* Trinidad
____ *S. t. aliciae* Arid littoral of ne Venezuela and Isla Margarita
____ *S. t. monticola* NW Venezuela
____ *S. t. feliciae (apurensis)* Coastal ranges and arid littoral of n Venezuela
____ *S. t. caudata* NE Venezuela
____ *S. t. caurensis* E and se Venezuela (Orinoco Valley to the *tepuis*)

☐ **Snowcap** *Microchera albocoronata*
____ *M. a. parvirostris* Caribbean slope of s Honduras to s Costa Rica and (?) Panama
____ *M. a. albocoronata* Atlantic and Pacific slopes of w-central Panama

☐ **Blossomcrown** *Anthocephala floriceps*
____ *A. f. floriceps* Santa Marta Mountains (ne Colombia)
____ *A. f. berlepschi* E slope of Central Andes of Colombia (Magdalena Valley)

☐ **White-vented Plumeleteer** *Chalybura buffonii*
____ *C. b. micans* Central Panama to Pacific slope of nw Colombia
____ *C. b. buffonii* N-central Colombia (Magdalena Valley) and nw Venezuela
____ *C. b. aeneicauda* N Colombia (Santa Marta region) and n Venezuela
____ *C. b. caeruleogaster* E slope of Eastern Andes of se Colombia
____ *C. b. intermedia* Subtropical sw Ecuador to nw Peru (San Martín)

☐ **Bronze-tailed Plumeleteer** *Chalybura urochrysia*
____ *C. u. melanorrhoa* Caribbean slope of Nicaragua and Costa Rica
____ *C. u. isaurae (incognita)* Caribbean slope of w Panama to nw Colombia; e Panama
____ *C. u. urochrysia* Extreme se Panama to w Colombia and nw Ecuador

☐ **Blue-throated Hummingbird** *Lampornis clemenciae*
_____ *L. c. bessophilus* — Mts. of sw US to nw Mexico (e Sonora and w Chihuahua)
_____ *L. c. clemenciae* — W Texas (Chisos Mountains) to s Mexico (s Oaxaca)

☐ **Amethyst-throated Hummingbird** *Lampornis amethystinus*
_____ *L. a. amethystinus* — Mountains of e Mexico (Nayarit to s Tamaulipas and e Oaxaca)
_____ *L. a. margaritae (brevirostris)* — Mountains of sw Mexico (Michoacán, Guerrero and w Oaxaca)
_____ *L. a. salvini* — Highlands of s Mexico (Chiapas) to Guatemala, El Salvador
_____ *L. a. nobilis* — Highlands of Honduras

☐ **Green-throated Mountain-gem** *Lampornis viridipallens*
_____ *L. v. amadoni* — S Mexico (Cerro Baúl in Oaxaca)
_____ *L. v. ovandensis* — Highlands of s Mexico (Chiapas) and nw Guatemala
_____ *L. v. viridipallens (connectans)* — Highlands of Guatemala, n El Salvador and w Honduras
_____ *L. v. nubivagus* — Upper tropical El Salvador

☐ **Green-breasted Mountain-gem** *Lampornis sybillae* — Humid montane forests of e Honduras and n-c Nicaragua

☐ **White-bellied Mountain-gem** *Lampornis hemileucus* — Caribbean slope of n-central Costa Rica to w Panama

☐ **White-throated Mountain-gem** *Lampornis castaneoventris* — Montane forests of extreme w Panama

☐ **Purple-throated Mountain-gem** *Lampornis calolaemus*
_____ *L. c. pectoralis* — Mountains of Nicaragua and nw Costa Rica
_____ *L. c. calolaemus* — Mountains of Costa Rica and w-central Panama

☐ **Gray-tailed Mountain-gem** *Lampornis cinereicauda*
_____ *L. c. cinereicauda* — S Costa Rica (Cordillera de Talamanca)
_____ *L. c. homogenes* — Pacific slope of extreme s Costa Rica and w Panama

☐ **Garnet-throated Hummingbird** *Lamprolaima rhami* — Mts. of s Mexico to Guatemala, El Salvador and Honduras

☐ **Ecuadorian Piedtail** *Phlogophilus hemileucurus* — Locally in foothills of s Colombia, e Ecuador and n Peru

☐ **Peruvian Piedtail** *Phlogophilus harterti* — Subtropical se Peru (Huánuco, Pasco, Puno and Cuzco)

☐ **Speckled Hummingbird** *Adelomyia melanogenys*
_____ *A. m. cervina* — W and Central Andes of Colombia
_____ *A. m. melanogenys* — E Andes of Colombia and w Venezuela to s-central Peru
_____ *A. m. connectens* — S Colombia (Huila)
_____ *A. m. debellardiana* — Mts. of Venezuela (Lara, Trujillo, Mérida, Táchira, Perijá)
_____ *A. m. aeneosticta* — Mountains of central and n Venezuela
_____ *A. m. maculata* — Andes of Ecuador and n Peru
_____ *A. m. chlorospila* — Andes of se Peru
_____ *A. m. inornata* — *Yungas* of Bolivia and adj. nw Argentina (Jujuy and Salta)

☐ **Brazilian Ruby** *Clytolaema rubricauda* — SE Brazil (Goiás and Minas Gerais to Rio Grande do Sul)

☐ **Gould's Jewelfront** *Heliodoxa aurescens* — E Colombia to n Bolivia, s Venezuela and w Amaz. Brazil

☐ **Fawn-breasted Brilliant** *Heliodoxa rubinoides*
_____ *H. r. rubinoides* — Central and Eastern Andes of Colombia
_____ *H. r. aequatorialis* — W slope of Andes of Colombia and w Ecuador
_____ *H. r. cervinigularis* — E slope of Andes of e Ecuador and ne Peru

☐ **Violet-fronted Brilliant** *Heliodoxa leadbeateri*
_____ *H. l. leadbeateri* — Coastal mts. of n Venezuela (Falcón to Carabobo and Miranda)
_____ *H. l. parvula* — Andes of Colombia and w Venezuela
_____ *H. l. sagitta* — Andes of e Ecuador and n Peru
_____ *H. l. otero* — Andes of central Peru to nw Bolivia

☐ **Velvet-browed Brilliant** *Heliodoxa xanthogonys*
    ____ *H. x. xanthogonys* — *Tepuis* of s Venezuela and adjacent Guyana
    ____ *H. x. willardi* — S Venezuela, n Brazil (Serranía de la Neblina and Sierra Imeri)

☐ **Black-throated Brilliant** *Heliodoxa schreibersii*
    ____ *H. s. schreibersii* — SE Colombia to Ecuador, ne Peru and nw Amazonian Brazil
    ____ *H. s. whitelyana* — Tropical e Peru

☐ **Pink-throated Brilliant** *Heliodoxa gularis* — Trop. s Colombia to ne Ecuador, ne Peru and extreme nw Brazil

☐ **Rufous-webbed Brilliant** *Heliodoxa branickii* — Andean foothills of e Peru and (?) nw Bolivia

☐ **Empress Brilliant** *Heliodoxa imperatrix* — Pacific slope of sw Colombia to nw Ecuador

☐ **Green-crowned Brilliant** *Heliodoxa jacula*
    ____ *H. j. henryi* — Humid montane forests of Costa Rica and w Panama
    ____ *H. j. jacula* — Mountains of extreme e Panama and Andes of Colombia
    ____ *H. j. jamesoni* — Andes of sw Colombia and w Ecuador

☐ **Magnificent Hummingbird** *Eugenes fulgens*
    ____ *E. f. fulgens* — Mts. of s Arizona, New Mexico and Texas to ne Nicaragua
    ____ *E. f. spectabilis* — Montane forests of Costa Rica and w Panama

☐ **Scissor-tailed Hummingbird** *Hylonympha macrocerca* — NE Venezuela (mountains of Paría Peninsula)

☐ **Violet-chested Hummingbird** *Sternoclyta cyanopectus* — Andes of w Venezuela and coastal cordillera east to Miranda

☐ **White-tailed Hillstar** *Urochroa bougueri*
    ____ *U. b. bougueri* — Andes of sw Colombia and nw Ecuador
    ____ *U. b. leucura* — Andes of s Colombia (Nariño) to e Ecuador and ne Peru

☐ **Chestnut-breasted Coronet** *Boissonneaua matthewsii* — Andes of extreme se Colombia, Ecuador and e Peru

☐ **Buff-tailed Coronet** *Boissonneaua flavescens*
    ____ *B. f. flavescens* — Andes of Colombia and w Venezuela (Mérida)
    ____ *B. f. tinochlora* — Andes of sw Colombia and adjacent nw Ecuador

☐ **Velvet-purple Coronet** *Boissonneaua jardini* — Western Andes of sw Colombia and nw Ecuador

☐ **Shining Sunbeam** *Aglaeactis cupripennis*
    ____ *A. c. cupripennis (parvula, ruficauda)* — Andes of Colombia to Ecuador and central Peru
    ____ *A. c. caumatonota* — Andes of s-c Peru (Junín, Apurímac, Ayacucho and Cuzco)

☐ **White-tufted Sunbeam** *Aglaeactis castelnaudii*
    ____ *A. c. regalis* — Andes of central Peru (Huánuco, Pasco and Junín)
    ____ *A. c. castelnaudii* — Andes of s Peru (Huancavelica, Ayacucho, Apurímac, Cuzco)

☐ **Purple-backed Sunbeam** *Aglaeactis aliciae* — E slope of Andes of n Peru (La Libertad and Ancash)

☐ **Black-hooded Sunbeam** *Aglaeactis pamela* — Andes of w Bolivia (La Paz and Cochabamba)

☐ **Andean Hillstar** *Oreotrochilus estella*
    ____ *O. e. estella* — Andes of sw Peru to w Bolivia, n Chile and nw Argentina
    ____ *O. e. bolivianus* — Andes of central Bolivia (Cochabamba)

☐ **Chimborazo Hillstar** *Oreotrochilus chimborazo*
    ____ *O. c. jamesoni* — Andes of extreme s Colombia and n Ecuador
    ____ *O. c. soderstroemi* — Andes of central Ecuador (Mt. Quilotoa)
    ____ *O. c. chimborazo* — Andes of central Ecuador (Mt. Chimborazo and Mt. Azuay)

☐ **Green-headed Hillstar** *Oreotrochilus stolzmanni*

Andes of n and c Peru (Cajamarca, Huánuco) and adj. s Ecuador

☐ **White-sided Hillstar** *Oreotrochilus leucopleurus*

Andes of s Bolivia to s Argentina and s-c Chile (Bío-Bío)

☐ **Black-breasted Hillstar** *Oreotrochilus melanogaster*

Andes of Peru (Junín, Huancavelica, Ancash, Lima, Ayacucho)

☐ **Wedge-tailed Hillstar** *Oreotrochilus adela*

Andes of Bolivia and adj. Argentina

☐ **Mountain Velvetbreast** *Lafresnaya lafresnayi*
___ *L. l. liriope* — Santa Marta Mountains (ne Colombia)
___ *L. l. greenewalti* — Andes of w Venezuela (ne Táchira, Mérida and s Trujillo)
___ *L. l. lafresnayi* — Central and Eastern Andes of Colombia
___ *L. l. tamae* — Extreme w Venezuela (Páramo de Tamá in s Táchira)
___ *L. l. saul (orestes)* — Andes of sw Colombia, Ecuador and n Peru
___ *L. l. rectirostris* — Temperate Andes of n and central Peru

☐ **Bronzy Inca** *Coeligena coeligena*
___ *C. c. coeligena (zuloagae)* — Coastal mountains of n Venezuela (Lara to Miranda)
___ *C. c. zuliana* — Sierra de Perijá (Colombia/Venezuela border)
___ *C. c. columbiana* — E and Central Andes of Colombia and nw Venezuela
___ *C. c. ferruginea* — W and Central Andes of Colombia
___ *C. c. obscura* — Andes of extreme s Colombia, Ecuador and Peru
___ *C. c. boliviana* — Andes of central and se Bolivia

☐ **Brown Inca** *Coeligena wilsoni*

Andes of sw Colombia and w Ecuador

☐ **Black Inca** *Coeligena prunellei*

W slope of Central Andes and Eastern Andes of Colombia

☐ **Gould's Inca** *Coeligena inca*
___ *C. i. omissa* — Andes of se Peru (Urubamba to Puno)
___ *C. i. inca* — Andes of Bolivia (La Paz and Cochabamba)

☐ **Collared Inca** *Coeligena torquata*
___ *C. t. conradii* — E Andes of Colombia to nw Venezuela (Trujillo and Mérida)
___ *C. t. torquata* — Andes of Colombia to nw Venezuela (Táchira) and n Peru
___ *C. t. fulgidigula* — Andes of w Ecuador
___ *C. t. margaretae* — Andes of n Peru (Chachapoyas)
___ *C. t. insectivora* — Andes of central Peru
___ *C. t. eisenmanni* — S Peru (Cordillera Vilcabamba)

☐ **White-tailed Starfrontlet** *Coeligena phalerata*

Santa Marta Mountains (ne Colombia)

☐ **Golden Starfrontlet** *Coeligena eos*

Andes of w Venezuela (Trujillo, Barinas, Mérida and Táchira)

☐ **Golden-bellied Starfrontlet** *Coeligena bonapartei*
___ *C. b. consita* — Sierra de Perijá (Colombia/Venezuela border)
___ *C. b. bonapartei* — E Andes of Colombia (Boyacá to Bogotá)
___ *C. b. orina* — N Colombia (single specimen from Páramo de Frontino)

☐ **Blue-throated Starfrontlet** *Coeligena helianthea*
___ *C. h. helianthea* — Andes of ne Colombia (Sierra de Perijá to Bogotá)
___ *C. h. tamai* — W Venezuela (Páramo de Tamá in Táchira)

☐ **Buff-winged Starfrontlet** *Coeligena lutetiae*

Central Andes of Colombia to Ecuador and extreme nw Peru

☐ **Violet-throated Starfrontlet** *Coeligena violifer*
___ *C. v. dichroura* — Andes of s Ecuador (Loja) to Peru (Junín, Huánuco and Lima)
___ *C. v. albicaudata* — Andes of s Peru (Cuzco, Apurímac and Ayacucho)
___ *C. v. osculans* — Andes of se Peru (Cuzco)
___ *C. v. violifer* — Andes of nw Bolivia (La Paz and Cochabamba)

☐ **Rainbow Starfrontlet** *Coeligena iris*

____	*C. i. hesperus*	Andes of s-central Ecuador (Cuenca)
____	*C. i. iris*	Andes of s Ecuador (Loja) to n Peru (Piura)
____	*C. i. aurora*	Andes of n Peru (Cutervo, Cerros de Amachonga)
____	*C. i. fulgidiceps (hypocrita)*	Andes of n Peru (east of Río Marañón in Amazonas)
____	*C. i. flagrans*	Andes of nw Peru (Cajamarca)
____	*C. i. eva*	N Peru (west of Río Marañón in Cajamarca and La Libertad)

☐ **Sword-billed Hummingbird** *Ensifera ensifera*

Andes of w Venezuela to ne Bolivia

☐ **Great Sapphirewing** *Pterophanes cyanopterus*

____	*P. c. cyanopterus*	E Andes of n-c Colombia (Santander and Cundinamarca)
____	*P. c. caeruleus*	C Andes of Colombia (Tolima) to extreme sw Andes (Nariño)
____	*P. c. peruvianus*	Andes of Ecuador, Peru and n Bolivia

☐ **Giant Hummingbird** *Patagona gigas*

____	*P. g. peruviana*	Andes of sw Colombia to n Chile and nw Argentina
____	*P. g. gigas*	Central and s Chile to w-central Argentina

☐ **Green-backed Firecrown** *Sephanoides sephaniodes*

C Chile and Argentina to Tierra del Fuego; Juan Fernandez Is.

☐ **Juan Fernandez Firecrown** *Sephanoides fernandensis*

____	*S. f. fernandensis*	Robinson Crusoe I. (Juan Fernández Islands off Chile)
____	*S. f. leyboldi†*	Alejandro Selkirk I. (Juan Fernández Islands off Chile). Extinct

☐ **Longuemare's Sunangel** *Heliangelus clarisse*

____	*H. c. violiceps*	Sierra de Perijá (Colombia/Venezuela border)
____	*H. c. spencei*	Andes of nw Venezuela (Mérida)
____	*H. c. clarisse*	Eastern Andes of Colombia and adjacent w Venezuela

☐ **Orange-throated Sunangel** *Heliangelus mavors*

Andes of ne Colombia to nw Venezuela (s Lara to Táchira)

☐ **Amethyst-throated Sunangel** *Heliangelus amethysticollis*

____	*H. a. laticlavius*	Andes of s Ecuador and n Peru
____	*H. a. decolor*	E Andes of central Peru (south of Río Marañón)
____	*H. a. amethysticollis*	Andes of s Peru to nw Bolivia

☐ **Gorgeted Sunangel** *Heliangelus strophianus*

Andes of sw Colombia (Nariño) and nw Ecuador

☐ **Tourmaline Sunangel** *Heliangelus exortis*

Andes of Colombia and e slope of Andes of nw Ecuador

☐ **Little Sunangel** *Heliangelus micraster*

____	*H. m. micraster*	Andes of se Ecuador and adjacent n Peru
____	*H. m. cutervensis*	Andes of nw Peru (Cajamarca)

☐ **Purple-throated Sunangel** *Heliangelus viola*

Andes of s Ecuador and n Peru

☐ **Royal Sunangel** *Heliangelus regalis*

Andes of n Peru (Cajamarca and San Martín)

☐ **Black-breasted Puffleg** *Eriocnemis nigrivestis*

Andes of n Ecuador (Pichincha, Atacazo and Imbabura)

☐ **Glowing Puffleg** *Eriocnemis vestita*

____	*E. v. paramillo*	N end of Western and Central Andes of Colombia
____	*E. v. vestita*	E Andes of Colombia and nw Venezuela
____	*E. v. smaragdinipectus*	Central Andes of Colombia and Ecuador
____	*E. v. arcosi*	Andes of s Ecuador and extreme n Peru

☐ **Black-thighed Puffleg** *Eriocnemis derbyi*

Central Andes of Colombia and nw Ecuador

☐ **Turquoise-throated Puffleg** *Eriocnemis godini*

Andes of nw Ecuador and (?) sw Colombia

☐ **Sapphire-vented Puffleg** *Eriocnemis luciani*
_____ *E. l. luciani* — Andes of sw Colombia (Nariño) and n Ecuador
_____ *E. l. baptistae* — Andes of central and s Ecuador
_____ *E. l. meridae* — Andes of w Venezuela (Mérida)

☐ **Coppery-bellied Puffleg** *Eriocnemis cupreoventris*

E Andes of Colombia and nw Venezuela (Mérida)

☐ **Coppery-naped Puffleg** *Eriocnemis sapphiropygia*
_____ *E. s. catharina* — E Andes of n Peru (Utcubamba Valley)
_____ *E. s. sapphiropygia* — E Andes of central and s Peru (Pasco and Junín to Puno)

☐ **Golden-breasted Puffleg** *Eriocnemis mosquera*

SW and Central Andes of Colombia to nw Ecuador

☐ **Blue-capped Puffleg** *Eriocnemis glaucopoides*

Andes of s Bolivia and nw Argentina

☐ **Colorful Puffleg** *Eriocnemis mirabilis*

W slope of Western Andes of Colombia (Cauca)

☐ **Emerald-bellied Puffleg** *Eriocnemis alinae*
_____ *E. a. alinae* — S-central and Eastern Andes of Colombia to Ecuador
_____ *E. a. dybowskii* — Eastern Andes of n and central Peru

☐ **Greenish Puffleg** *Haplophaedia aureliae*
_____ *H. a. caucensis* — Mts. of se Panama to Western and Central Andes of Colombia
_____ *H. a. aureliae* — Eastern Andes of Colombia and (?) e slope of Central Andes
_____ *H. a. russata* — E slope of Andes of Ecuador
_____ *H. a. cutucuensis* — Mts. of e Ecuador (Cordillera de Cutucú)

☐ **Buff-thighed Puffleg** *Haplophaedia assimilis*
_____ *H. a. affinis* — Eastern Andes of n and central Peru
_____ *H. a. assimilis* — Andes of se Peru (Puno) and w Bolivia

☐ **Hoary Puffleg** *Haplophaedia lugens*

Pacific slope of sw Colombia (Nariño) and nw Ecuador

☐ **Purple-bibbed Whitetip** *Urosticte benjamini*

Pacific slope of w Colombia to sw Ecuador and nw Peru

☐ **Rufous-vented Whitetip** *Urosticte ruficrissa*

E slope of Andes in s Colombia and e Ecuador

☐ **Booted Racket-tail** *Ocreatus underwoodii*
_____ *O. u. polystictus* — Coastal mountains of n Venezuela (Carabobo to Miranda)
_____ *O. u. discifer* — Colombia to nw Venezuela (Zulia to Falcón, Táchira, w Barinas)
_____ *O. u. underwoodii* — Eastern Andes of Colombia
_____ *O. u. incommodus (ambiguus)* — Western and Central Andes of Colombia
_____ *O. u. melanantherus* — Andes of Ecuador
_____ *O. u. peruanus* — E Ecuador and ne Peru
_____ *O. u. annae* — Andes of central and s Peru
_____ *O. u. addae* — *Yungas* of Bolivia (La Paz to Santa Cruz and Chuquisaca)

☐ **Black-tailed Trainbearer** *Lesbia victoriae*
_____ *L. v. victoriae (aequatorialis)* — Andes of s Colombia (Nariño) and Ecuador
_____ *L. v. juliae* — Andes of n and central Peru
_____ *L. v. berlepschi* — Andes of se Peru

☐ **Green-tailed Trainbearer** *Lesbia nuna*
_____ *L. n. gouldii* — Andes of Colombia; old record from w Venezuela (Mérida)
_____ *L. n. gracilis* — Andes of Ecuador
_____ *L. n. pallidiventris* — Andes of n Peru
_____ *L. n. eucharis (chlorura)* — Andes of central Peru (Huánuco)
_____ *L. n. nuna (boliviana)* — Andes of sw Peru and n Bolivia

☐ **Red-tailed Comet** *Sappho sparganura*
    ____ *S. s. sparganura*   Andes of n and central Bolivia; accidental s Peru (Puno)
    ____ *S. s. sapho*   Andes of s Bolivia to w Argentina and n Chile

☐ **Bronze-tailed Comet** *Polyonymus caroli*
         Pacific slope of Andes of Peru

☐ **Purple-backed Thornbill** *Ramphomicron microrhynchum*
    ____ *R. m. andicola*   Andes of w Venezuela (Mérida)
    ____ *R. m. microrhynchum*   Andes of Colombia, Ecuador and nw Peru
    ____ *R. m. albiventre*   Andes of c Peru (Huánuco to Junín, Cuzco and Apurímac)
    ____ *R. m. bolivianum*   Andes of w Bolivia (Cochabamba)

☐ **Black-backed Thornbill** *Ramphomicron dorsale*
         Santa Marta Mountains (ne Colombia)

☐ **Bearded Mountaineer** *Oreonympha nobilis*
    ____ *O. n. albolimbata*   Andes of w-central Peru (Huancavelica)
    ____ *O. n. nobilis*   Andes of s Peru (Cuzco and Apurímac)

☐ **Bearded Helmetcrest** *Oxypogon guerinii*
    ____ *O. g. cyanolaemus*   Santa Marta Mountains (ne Colombia)
    ____ *O. g. lindenii*   Andes of nw Venezuela (Mérida and Trujillo)
    ____ *O. g. guerinii*   *Páramo* of E Andes of Colombia (south to Cundinamarca)
    ____ *O. g. stuebelii*   W Andes of Colombia (Volcán de Ruiz)

☐ **Tyrian Metaltail** *Metallura tyrianthina*
    ____ *M. t. districta*   Santa Marta Mountains (ne Colombia)
    ____ *M. t. chloropogon*   Coastal mountains of n Venezuela
    ____ *M. t. oreopola*   Andes of w Venezuela (Lara, Trujillo and Mérida)
    ____ *M. t. tyrianthina*   Andes of Colombia, Venezuela (Táchira), Ecuador and n Peru
    ____ *M. t. quitensis*   Andes of nw Ecuador
    ____ *M. t. septentrionalis*   Andes of n Peru (west of Río Marañón)
    ____ *M. t. smaragdinicollis (peruviana)*   Andes of e Peru and n Bolivia

☐ **Perija Metaltail** *Metallura iracunda*
         Sierra de Perijá (Colombia/Venezuela border)

☐ **Scaled Metaltail** *Metallura aeneocauda*
    ____ *M. a. aeneocauda*   Andes of se Peru (Cuzco and Puno) to nw Bolivia (La Paz)
    ____ *M. a. malagae*   *Yungas* of Bolivia (Cochabamba and w Santa Cruz)

☐ **Fire-throated Metaltail** *Metallura eupogon*
         E Andes of Peru (Huánuco, Junín, Apurímac and Ayacucho)

☐ **Coppery Metaltail** *Metallura theresiae*
    ____ *M. t. parkeri*   Andes of n Peru (Cordillera de Colán)
    ____ *M. t. theresiae*   Andes of n Peru (Amazonas to Huánuco)

☐ **Neblina Metaltail** *Metallura odomae*
         Andes of s Ecuador (Loja) and n Peru (Piura and Cajamarca)

☐ **Violet-throated Metaltail** *Metallura baroni*
         Páramo of Andes of s-central Ecuador (Azuay)

☐ **Viridian Metaltail** *Metallura williami*
    ____ *M. w. recisa*   Andes of n-c Colombia (Páramo de Frontino in Antioquia)
    ____ *M. w. williami*   Central Andes of Colombia
    ____ *M. w. primolinus*   Eastern Andes of s Colombia (Nariño) and n Ecuador
    ____ *M. w. atrigularis*   Andes of s Ecuador (Cordillera de Chilla in Azuay and Loja)

☐ **Black Metaltail** *Metallura phoebe*
         Andes of n Peru (Cajamarca) to extreme n Chile

☐ **Rufous-capped Thornbill** *Chalcostigma ruficeps*
         Andes of se Ecuador to e Peru and *yungas* of w Bolivia

☐ **Olivaceous Thornbill** *Chalcostigma olivaceum*
_____ *C. o. pallens* — Locally in Andes of central Peru
_____ *C. o. olivaceum* — Locally in Eastern Andes of se Peru and w Bolivia

☐ **Blue-mantled Thornbill** *Chalcostigma stanleyi*
_____ *C. s. stanleyi* — *Páramo* of Andes of Ecuador
_____ *C. s. versigulare* — Andes of Peru (east of Río Marañón) to Carpish Mountains
_____ *C. s. vulcani* — Andes of e Peru (south of Río Huallaga) to w Bolivia

☐ **Bronze-tailed Thornbill** *Chalcostigma heteropogon*
— *Páramo* of ne Colombia to extreme w Venezuela

☐ **Rainbow-bearded Thornbill** *Chalcostigma herrani*
_____ *C. h. tolimae* — Central Andes of Colombia (Volcán de Tolima)
_____ *C. h. herrani* — W Andes of s Colombia to extreme n Peru (Piura)

☐ **Mountain Avocetbill** *Opisthoprora euryptera*
— C Andes of Colombia to n Peru (Amazonas to La Libertad)

☐ **Gray-bellied Comet** *Taphrolesbia griseiventris*
— Andes of nw Peru (s Cajamarca to w Huánuco)

☐ **Long-tailed Sylph** *Aglaiocercus kingi*
_____ *A. k. margarethae* — Mountains of n-central and coastal Venezuela
_____ *A. k. caudatus* — Andes of n Colombia and w Venezuela
_____ *A. k. emmae* — Central and Western Andes of Colombia to nw Ecuador
_____ *A. k. kingi* — Eastern Andes of Colombia
_____ *A. k. mocoa* — Central Andes of s Colombia to Ecuador and n Peru
_____ *A. k. smaragdinus* — E Andes of Peru and *yungas* of w Bolivia

☐ **Violet-tailed Sylph** *Aglaiocercus coelestis*
_____ *A. c. coelestis (pseudocoelestis)* — Pacific slope of Western Andes of Colombia and n Ecuador
_____ *A. c. aethereus* — Andes of sw Ecuador

☐ **Venezuelan Sylph** *Aglaiocercus berlepschi*
— Mountains of ne Venezuela (Sucre and Monagas)

☐ **Hyacinth Visorbearer** *Augastes scutatus*
_____ *A. s. scutatus* — High montane forests of se Brazil (central and e Minas Gerais)
_____ *A. s. ilseae* — Mid-montane forests of se Brazil (central and e Minas Gerais)
_____ *A. s. soaresi* — SE Brazil (Rio Paricicaba basin in s-central Minas Gerais)

☐ **Hooded Visorbearer** *Augastes lumachella*
— Locally in ne Brazilian plateaux (Bahia and Minas Gerais)

☐ **Wedge-billed Hummingbird** *Augastes geoffroyi*
_____ *A. g. geoffroyi* — Andes of e Colombia to n Venezuela and e Peru
_____ *A. g. albogularis* — Western and Central Andes of Colombia and w Ecuador
_____ *A. g. chapmani* — Andes of central Bolivia (Cochabamba)

☐ **Purple-crowned Fairy** *Heliothryx barroti*
— Gulf slope of se Mexico to sw Ecuador (El Oro)

☐ **Black-eared Fairy** *Heliothryx auritus*
_____ *H. a. auritus* — Tropical se Colombia to Venezuela, the Guianas and n Brazil
_____ *H. a. phainolaemus* — Amazonian Brazil south of the Amazon (Pará and Maranhão)
_____ *H. a. auriculatus* — Tropical e Peru, n Bolivia, Amazonian and s Brazil

☐ **Horned Sungem** *Heliactin bilophus*
— Extreme s Suriname to interior Brazil and adjacent e Bolivia

☐ **Marvelous Spatuletail** *Loddigesia mirabilis*
— Andes of n Peru (along east bank of Río Utcubamba)

☐ **Plain-capped Starthroat** *Heliomaster constantii*
_____ *H. c. pinicola* — Arid nw Mexico (Sonora to Jalisco)
_____ *H. c. leocadiae (surdus)* — Arid tropical w Mexico (Nayarit) to w Guatemala
_____ *H. c. constantii* — El Salvador and Nicaragua to sw Costa Rica

209

☐ **Long-billed Starthroat** *Heliomaster longirostris*
   ____ *H. l. pallidiceps*   Tropical s Mexico to Nicaragua
   ____ *H. l. longirostris*   E Costa Rica to Bolivia and Brazil; Trinidad
   ____ *H. l. albicrissa*   Tropical w Ecuador and nw Peru

☐ **Stripe-breasted Starthroat** *Heliomaster squamosus*

E Brazil (Pernambuco to Bahia, Goiás and São Paulo)

☐ **Blue-tufted Starthroat** *Heliomaster furcifer*

Bolivia to Paraguay, c and s Brazil, Uruguay and n Argentina

☐ **Oasis Hummingbird** *Rhodopis vesper*
   ____ *R. v. koepckeae*   Arid nw Peru (Cerro Illescas in sw Piura)
   ____ *R. v. vesper (tertia)*   Arid w Peru to extreme n Chile (Tacna and Tarapacá)
   ____ *R. v. atacamensis*   Arid n Chile (Atacama)

☐ **Peruvian Sheartail** *Thaumastura cora*

Arid w Peru to extreme n Chile (Arica)

☐ **Sparkling-tailed Hummingbird** *Tilmatura dupontii*

Oak-pine highlands of w Mexico to n Nicaragua

☐ **Slender Sheartail** *Doricha enicura*

Highlands of s Mexico (Chiapas) to w Honduras, El Salvador

☐ **Mexican Sheartail** *Doricha eliza*

Disjunct in arid s Mexico (Veracruz and Yucatán Peninsula)

☐ **Bahama Woodstar** *Calliphlox evelynae*
   ____ *C. e. evelynae*   Bahamas
   ____ *C. e. lyrura*   Great Inagua I. (Bahamas)

☐ **Magenta-throated Woodstar** *Calliphlox bryantae*

Highlands of n Costa Rica and w Panama

☐ **Purple-throated Woodstar** *Calliphlox mitchellii*

Tropical e Panama (Darién) to w Colombia and sw Ecuador

☐ **Amethyst Woodstar** *Calliphlox amethystina*

Widespread trop. South America to ne Argentina and s Brazil

☐ **Slender-tailed Woodstar** *Microstilbon burmeisteri*

Arid scrub of e Bolivia and nw Argentina

☐ **Lucifer Hummingbird** *Calothorax lucifer*

Arid mountains of sw US and n Mexico; winters to s Mexico

☐ **Beautiful Hummingbird** *Calothorax pulcher*

Arid scrub of s Mexico (Guerrero and s Puebla to e Oaxaca)

☐ **Vervain Hummingbird** *Mellisuga minima*
   ____ *M. m. minima*   Jamaica
   ____ *M. m. vieilloti*   Hispaniola, Gonâve, Tortue, Saona, Catalina and Île-à-Vache

☐ **Bee Hummingbird** *Mellisuga helenae*

Cuba and Isle of Pines

☐ **Ruby-throated Hummingbird** *Archilochus colubris*

E Canada and US; winters Mexico to Panama

☐ **Black-chinned Hummingbird** *Archilochus alexandri*

Arid sw British Columbia to nw Mexico; winters to s Mexico

☐ **Anna's Hummingbird** *Calypte anna*

Arid sw British Columbia to nw Baja; winters to n Mexico

☐ **Costa's Hummingbird** *Calypte costae*

Arid sw US to s Baja California and nw Mexico

☐ **Bumblebee Hummingbird** *Atthis heloisa*
   ____ *A. h. margarethae*   Mts. of nw Mexico (se Sinaloa and sw Chihuahua to Jalisco)
   ____ *A. h. heloisa*   Highlands of Mexico (c Tamaulipas to Guerrero and Oaxaca)

☐ **Wine-throated Hummingbird** *Atthis ellioti*
   ____ *A. e. ellioti*   Montane forests of s Mexico (Chiapas) and Guatemala
   ____ *A. e. selasphoroides*   Humid montane forests of Honduras

☐ **Calliope Hummingbird** *Stellula calliope*

Mts. of Br. Columbia to sw US and n Baja; winters to s Mexico

☐ **Purple-collared Woodstar** *Myrtis fanny*
____ *M. f. fanny* | W and se Ecuador to w Peru (Piura to Arequipa)
____ *M. f. megalura* | N Peru (Cajabamba to La Libertad and extreme nw Huánuco)

☐ **Chilean Woodstar** *Eulidia yarrellii*

Arid s Peru (Tacna and Moquegua) to n Chile (Arica)

☐ **Short-tailed Woodstar** *Myrmia micrura*

Arid scrub of sw Ecuador and nw Peru

☐ **White-bellied Woodstar** *Chaetocercus mulsant*

Andes of Colombia to w Bolivia (Cochabamba)

☐ **Little Woodstar** *Chaetocercus bombus*

Andes of extreme sw Colombia (Nariño) to n Peru

☐ **Gorgeted Woodstar** *Chaetocercus heliodor*
____ *C. h. heliodor* | Andes of Colombia and w Ecuador to nw Venezuela (Mérida)
____ *C. h. cleavesi* | Andes of ne Ecuador

☐ **Santa Marta Woodstar** *Chaetocercus astreans*

Santa Marta Mountains (ne Colombia)

☐ **Esmeraldas Woodstar** *Chaetocercus berlepschi*

Lowlands of w Ecuador (Esmeraldas, Manabí and Guayas)

☐ **Rufous-shafted Woodstar** *Chaetocercus jourdanii*
____ *C. j. jourdanii* | Mountains of ne Venezuela (Cumaná); Trinidad
____ *C. j. rosae* | Mountains of n Venezuela (Zulia to Distrito Federal)
____ *C. j. andinus* | Andes of ne Colombia to w Venezuela (Lara to Táchira)

☐ **Scintillant Hummingbird** *Selasphorus scintilla*

Humid montane forests of Costa Rica and w Panama

☐ **Glow-throated Hummingbird** *Selasphorus ardens*

Mountains of w Panama (Chiriquí and Veraguas)

☐ **Volcano Hummingbird** *Selasphorus flammula*
____ *S. f. simoni* | Costa Rica (Volcán Poás, Volcán Barba and Cerros de Escazú)
____ *S. f. flammula* | Costa Rica (Volcán Irazú and Volcán Turrialba)
____ *S. f. torridus* | Sierra de Talamanca (Costa Rica) and Volcán Barú (w Panama)

☐ **Broad-tailed Hummingbird** *Selasphorus platycercus*

Mountains of sw US to Mexico and Guatemala

☐ **Rufous Hummingbird** *Selasphorus rufus*

Alaska to nw US; winters to s Mexico

☐ **Allen's Hummingbird** *Selasphorus sasin*
____ *S. s. sasin* | S Oregon to s California; winters to central Mexico
____ *S. s. sedentarius* | Channel Islands (off s California)

ORDER: COLIIFORMES
FAMILY: COLIIDAE (Mousebirds—6)

☐ **Speckled Mousebird** *Colius striatus*
____ *C. s. nigricollis* | Nigeria to Cameroon, Gabon and sw Congo
____ *C. s. striatus* | S Cape Province east to Great Kei River (South Africa)
____ *C. s. minor* | Natal to sw Zululand, Swaziland, e and n Transvaal
____ *C. s. integralis* | E Transvaal to Zululand, s Mozambique and se Zimbabwe
____ *C. s. simulans* | Lower Zambezi valley in Mozambique and Malawi
____ *C. s. rhodesiae* | Highlands of e Zimbabwe and adjacent Mozambique
____ *C. s. affinis* | Zimbabwe to Malawi, n Mozambique, s and coastal Tanzania
____ *C. s. mombassicus* | S Somalia to coastal Kenya and ne Tanzania (south to Amani)
____ *C. s. cinerascens* | W and s Tanzania
____ *C. s. berlepschi* | N Malawi to ne Zambia and sw Tanzania
____ *C. s. kikuyensis* | Cent. Kenya and high rainfall areas of peripheral n Tanzania

____ *C. s. kiwuensis*	Uganda to nw Tanzania, Rwanda and e Zaire
____ *C. s. congicus*	E Angola to s Zaire and nw Zambia
____ *C. s. jebelensis*	N border of Uganda and s Sudan
____ *C. s. leucotis*	E Sudan to w and sw Ethiopia
____ *C. s. hilgerti*	NE Ethiopia, sw Djibouti and nw Somalia
____ *C. s. leucophthalmus*	NE Zaire to s Sudan and se Central African Republic

☐ **White-headed Mousebird** *Colius leucocephalus*

____ *C. l. leucocephalus*	Arid s Somalia, Kenya (except n) and extreme ne Tanzania
____ *C. l. turneri*	N Kenya (Lake Turkana to Mt. Kenya and Isiolo)

☐ **Red-backed Mousebird** *Colius castanotus*

Euphorbia savanna of Angola; formerly sw Gabon

☐ **White-backed Mousebird** *Colius colius*

Dry savanna of Namibia, Botswana and South Africa

☐ **Blue-naped Mousebird** *Urocolius macrourus*

____ *U. m. macrourus*	Senegambia to e Ethiopia
____ *U. m. laeneni*	Aïr region of Niger
____ *U. m. pulcher*	SE Sudan to s Somalia, Kenya, Uganda and Tanzania
____ *U. m. abyssinicus*	Central and s Ethiopia to nw Somalia
____ *U. m. griseogularis*	SW Ethiopia to Sudan, Uganda, e Zaire, Rwanda, nw Tanzania
____ *U. m. massaicus*	Central to coastal Tanzania

☐ **Red-faced Mousebird** *Urocolius indicus*

____ *U. i. indicus*	South Africa (s Cape Province east to Great Kei River)
____ *U. i. pallidus*	Coastal se Tanzania and ne Mozambique
____ *U. i. lacteifrons*	Angola to Namibia and w Botswana
____ *U. i. mossambicus*	E Angola to Zambia, se Zaire, Malawi and sw Tanzania
____ *U. i. transvaalensis*	S Mozambique to Zimbabwe, sw Zambia and n Cape Province

ORDER: TROGONIFORMES
FAMILY: TROGONIDAE (Trogons and Quetzals—40)

☐ **Narina Trogon** *Apaloderma narina*

____ *A. n. narina*	Highlands of Ethiopia to Angola and South Africa
____ *A. n. littoralis*	Coastal lowlands from Somalia to Natal; Zanzibar
____ *A. n. brachyurum*	SE Nigeria and s Cameroon to Zaire and Uganda
____ *A. n. constantia*	Sierra Leone, Liberia and s Guinea (Mt. Nimba) to Ghana

☐ **Bare-cheeked Trogon** *Apaloderma aequatoriale*

Lowland rainforests of Cameroon, Gabon and ne Zaire

☐ **Bar-tailed Trogon** *Apaloderma vittatum*

____ *A. v. vittatum*	Tanzania, Malawi and Mozambique
____ *A. v. camerunensis*	Nigeria and Cameroon to Angola, w Zaire, w Uganda; Bioko

☐ **Cuban Trogon** *Priotelus temnurus*

____ *P. t. temnurus*	Cuba
____ *P. t. vescus*	Isle of Pines

☐ **Hispaniolan Trogon** *Priotelus roseigaster*

Montane forests of Hispaniola; locally in coastal mangroves

☐ **Black-headed Trogon** *Trogon melanocephalus*

____ *T. m. melanocephalus*	Gulf-Caribbean slope of e Mexico to ne Costa Rica
____ *T. m. illaetabilis*	W Costa Rica

☐ **Citreoline Trogon** *Trogon citreolus*

____ *T. c. citreolus*	Pacific slope of w Mexico (Sinaloa to w Oaxaca)
____ *T. c. sumichrasti*	Pacific slope of s Mexico (central Oaxaca to Chiapas)

☐ **White-tailed Trogon** *Trogon viridis*

____ *T. v. chionurus*	Lowlands of e Panama to w Colombia and w Ecuador
____ *T. v. viridis*	Colombia east of the Andes to n Bolivia and Brazil; Trinidad
____ *T. v. melanopterus*	Tropical se Brazil (Bahia to São Paulo)

☐ **Baird's Trogon** *Trogon bairdii*

	Pacific slope of sw Costa Rica and w Panama

☐ **Violaceous Trogon** *Trogon violaceus*

____ *T. v. braccatus*	Tropical s Mexico (Oaxaca) to Costa Rica
____ *T. v. concinnus*	Tropical Panama to Colombia, w Ecuador and nw Peru
____ *T. v. caligatus*	N Colombia to w Venezuela (Maracaibo basin)
____ *T. v. violaceus*	Venezuela, the Guianas and adjacent n Brazil; Trinidad
____ *T. v. ramonianus*	Upper Amazonia (w Brazil and e Peru)
____ *T. v. crissalis*	E Brazil, n Peru, e Ecuador, Colombia and s Venezuela

☐ **Mountain Trogon** *Trogon mexicanus*

____ *T. m. clarus*	Oak-pine woodlands of nw Mexico (Sinaloa to Durango)
____ *T. m. mexicanus*	Oak-pine woodlands of central Mexico to w Guatemala
____ *T. m. lutescens*	Highlands of El Salvador and Honduras

☐ **White-eyed Trogon** *Trogon comptus*

	Humid forests of w Colombia and nw Ecuador

☐ **Collared Trogon** *Trogon collaris*

____ *T. c. puella*	Tropical and subtropical central Mexico to w Panama
____ *T. c. extimus*	Subtropical e Panama
____ *T. c. heothinus*	Eastern Panama (Darién)
____ *T. c. virginalis*	W Colombia to w Ecuador and nw Peru
____ *T. c. subtropicalis*	Central Colombia
____ *T. c. collaris*	E Colombia to Bolivia, Venezuela and the Guianas
____ *T. c. castaneus*	Tropical e Colombia to nw Brazil, e Peru and n Bolivia
____ *T. c. exoptatus*	N Colombia, n Venezuela, Trinidad and Tobago
____ *T. c. eytoni*	E Brazil

☐ **Elegant Trogon** *Trogon elegans*

____ *T. e. canescens*	Oak-pine woodlands of s Arizona to nw Mexico
____ *T. e. goldmani*	Tres Marías Islands (off w Mexico)
____ *T. e. ambiguus*	Extreme s Texas to e and central Mexico
____ *T. e. elegans*	SE Guatemala
____ *T. e. lubricus*	Honduras, Nicaragua and nw Costa Rica

☐ **Orange-bellied Trogon** *Trogon aurantiiventris*

____ *T. a. underwoodi*	Humid montane forests of nw Costa Rica
____ *T. a. aurantiiventris*	Humid montane forests of central Costa Rica to w Panama
____ *T. a. flavidior*	W Panama (extreme e Chiriquí on Cerro Flores)

☐ **Masked Trogon** *Trogon personatus*

____ *T. p. sanctaemartae*	Santa Marta Mountains (ne Colombia)
____ *T. p. ptaritepui*	*Tepuis* of s Venezuela
____ *T. p. personatus*	Subtropical mountains of w Venezuela, e Colombia and e Peru
____ *T. p. assimilis*	Subtropical w Ecuador and nw Peru
____ *T. p. temperatus*	Andes of central Colombia, Ecuador and Peru
____ *T. p. heliothrix*	Montane forests of Peru
____ *T. p. submontanus*	Foothills of Andes of s Peru and Bolivia
____ *T. p. duidae*	*Tepuis* of s Venezuela (Mt. Duida)
____ *T. p. roraimae*	Auyan-tepui and Mt. Roraima (Venezuela/Guyana border)

☐ **Black-throated Trogon** *Trogon rufus*

____ *T. r. tenellus*	Tropical se Honduras to extreme nw Colombia
____ *T. r. cupreicauda*	Tropical w Colombia and w Ecuador
____ *T. r. rufus*	E Venezuela to the Guianas and n Brazil (Rio Negro region)

____ *T. r. sulphureus*   E Colombia, e Ecuador, ne Peru, s Venezuela and w Brazil
____ *T. r. amazonicus*   S Venezuela and ne Brazil
____ *T. r. chrysochloros*   S Brazil to Paraguay and ne Argentina

☐ **Surucua Trogon** *Trogon surrucura*

____ *T. s. aurantius*   E Brazil (Minas Gerais to Rio de Janeiro and n São Paulo)
____ *T. s. surrucura*   S Brazil to Paraguay, Uruguay and n Argentina

☐ **Blue-crowned Trogon** *Trogon curucui*

____ *T. c. peruvianus (bolivianus)*   S-central Colombia to Ecuador, Peru and e Brazil
____ *T. c. curucui*   Humid e Brazil, Paraguay and Bolivia
____ *T. c. behni*   E Bolivia to s Brazil, Paraguay and n Argentina

☐ **Black-tailed Trogon** *Trogon melanurus*

____ *T. m. macroura*   E Panama (Canal Zone) to n Colombia
____ *T. m. melanurus*   E Colombia to the Guianas, n Bolivia and e Brazil
____ *T. m. mesurus*   W Ecuador and nw Peru
____ *T. m. eumorphus*   S Colombia to Ecuador, Peru, Bolivia and Amazonian Brazil
____ *T. m. occidentalis*   SE Brazil (São Paulo region)

☐ **Slaty-tailed Trogon** *Trogon massena*

____ *T. m. massena*   Humid lowland forests of se Mexico to Nicaragua
____ *T. m. hoffmanni*   Costa Rica and Panama to extreme nw Colombia
____ *T. m. australis*   Coastal Pacific Colombia to nw Ecuador

☐ **Lattice-tailed Trogon** *Trogon clathratus*

Caribbean coast of Costa Rica and w Panama

☐ **Eared Quetzal** *Euptilotis neoxenus*

Montane forests of w Mexico; occasional sw Arizona

☐ **Resplendent Quetzal** *Pharomachrus mocinno*

____ *P. m. mocinno*   Montane forests of s Mexico to n Nicaragua
____ *P. m. costaricensis*   Montane forests of Costa Rica to w Panama

☐ **Crested Quetzal** *Pharomachrus antisianus*

Andes of w Venezuela to central Bolivia

☐ **White-tipped Quetzal** *Pharomachrus fulgidus*

____ *P. f. festatus*   Santa Marta Mountains (ne Colombia)
____ *P. f. fulgidus*   Mountains of n Venezuela

☐ **Golden-headed Quetzal** *Pharomachrus auriceps*

____ *P. a. hargitti*   Mts. of e Colombia and w Venezuela
____ *P. a. auriceps*   Mts. of e Panama to n Bolivia

☐ **Pavonine Quetzal** *Pharomachrus pavoninus*

____ *P. p. heliactin*   Subtropical w Ecuador
____ *P. p. pavoninus*   Upper Amazonia from se Colombia to ne Peru (Río Negro)
____ *P. p. viridiceps*   Lower Amazonian Brazil (Rio Tapajós)

☐ **Javan Trogon** *Harpactes reinwardtii*

Montane forests of w Java

☐ **Sumatran Trogon** *Harpactes mackloti*

Barisan Mts. of Sumatra

☐ **Malabar Trogon** *Harpactes fasciatus*

____ *H. f. malabaricus*   Forests of w and s India
____ *H. f. legerli*   E-central India
____ *H. f. fasciatus (parvus)*   Sri Lanka

☐ **Red-naped Trogon** *Harpactes kasumba*

____ *H. k. kasumba*   Lowland forests of Malay Peninsula and Sumatra
____ *H. k. impavidus*   Lowlands of Borneo

☐ **Diard's Trogon** *Harpactes diardii*

____ *H. d. neglectus*	N Peninsular Thailand and Malay Peninsula
____ *H. d. sumatranus*	Sumatra
____ *H. d. diardii*	Borneo, Bangka I. and Lingga Archipelago

☐ **Philippine Trogon** *Harpactes ardens*

____ *H. a. herberti*	N Philippines (ne Luzon and Marinduque
____ *H. a. luzoniensis*	N Philippines (s and central Luzon)
____ *H. a. minor*	Polillo I. (n Philippines)
____ *H. a. linae*	Central Philippines (Bohol, Leyte and Samar)
____ *H. a. ardens*	S Philippines (Basilan, Dinagat and Mindanao)

☐ **Whitehead's Trogon** *Harpactes whiteheadi*

Mountains of n Borneo

☐ **Cinnamon-rumped Trogon** *Harpactes orrhophaeus*

____ *H. o. orrhophaeus*	Lowlands of Malay Peninsula and Sumatra
____ *H. o. vidua*	Lowlands of n Borneo

☐ **Scarlet-rumped Trogon** *Harpactes duvaucelii*

Lowland forests of Malay Peninsula, Sumatra and Borneo

☐ **Red-headed Trogon** *Harpactes erythrocephalus*

____ *H. e. hodgsoni*	NE India to e Nepal and Bhutan
____ *H. e. erythrocephalus*	E Himalayas of Myanmar and nw Thailand
____ *H. e. helenae*	S China (w Yunnan) and n Myanmar
____ *H. e. yamakanensis (rosa)*	SE China (Sichuan, Fujian and n Guangdong)
____ *H. e. intermedius*	N Laos, n Vietnam and s China (Yunnan)
____ *H. e. annamensis*	E Thailand, s Laos and Vietnam
____ *H. e. klossi*	Banthat Mountains (w Cambodia and extreme sw Thailand)
____ *H. e. chaseni*	Malay Peninsula
____ *H. e. hainanus*	Hainan I. (s China)
____ *H. e. flagrans*	Mountains of Sumatra

☐ **Orange-breasted Trogon** *Harpactes oreskios*

____ *H. o. stellae*	Lowlands of sw China (sw Yunnan), s Myanmar and Indochina
____ *H. o. uniformis*	S Thailand, Malay Peninsula and Sumatra
____ *H. o. oreskios*	Java
____ *H. o. dulitensis*	Mountains of nw Borneo
____ *H. o. nias*	Nias I. (off nw Sumatra)

☐ **Ward's Trogon** *Harpactes wardi*

E Bhutan to s China and nw Vietnam

ORDER: CORACIIFORMES
FAMILY: ALCEDINIDAE (Kingfishers—93)

☐ **Blyth's Kingfisher** *Alcedo hercules*

Sikkim to sw China, Myanmar, n Laos, n Vietnam and Hainan

☐ **Common Kingfisher** *Alcedo atthis*

____ *A. a. atthis*	SE Europe and North Africa to nw India
____ *A. a. ispida*	British Isles to w Russia, Iberian and Baltic peninsulas
____ *A. a. bengalensis*	L Baikal and n India through e and SE Asia mainland and islands
____ *A. a. taprobana*	S India and Sri Lanka
____ *A. a. floresiana*	Lesser Sundas (Bali to Timor and Wetar)
____ *A. a. hispidoides*	Sulawesi to Moluccas, New Guinea and Bismarck Arch.
____ *A. a. solomonensis*	Bougainville and Solomon Islands (east to San Cristóbal)

☐ **Half-collared Kingfisher** *Alcedo semitorquata*

____ *A. s. heuglini*	Ethiopia
____ *A. s. semitorquata*	S Mozambique and South Africa
____ *A. s. tephria*	Angola to Tanzania and Mozambique

☐ **Shining-blue Kingfisher** *Alcedo quadribrachys*
_____ *A. q. quadribrachys*   Senegambia to w-central Nigeria
_____ *A. q. guentheri*   Coastal sw Nigeria to extreme s Sudan, Kenya and Zambia

☐ **Blue-eared Kingfisher** *Alcedo meninting*
_____ *A. m. coltarti*   Lowlands of Nepal and n India to s Myanmar and Laos
_____ *A. m. laubmanni*   E India
_____ *A. m. phillipsi*   SW India (Kerala) and Sri Lanka
_____ *A. m. scintillans*   Peninsular Myanmar and peninsular Thailand
_____ *A. m. rufigastra*   Andaman Islands
_____ *A. m. verreauxii*   Malay Peninsula, Sumatra, Borneo and Sulu Archipelago
_____ *A. m. proxima*   Pagai Islands (Mentawi Archipelago off w Sumatra)
_____ *A. m. subviridis*   Nias I. (off nw Sumatra)
_____ *A. m. callima*   Batu Islands (off w Sumatra)
_____ *A. m. meninting*   Java and Bali to Lombok, Sulawesi, Sula Is., w Lesser Sundas
_____ *A. m. amadoni*   S Philippines (Balabac, Busuanga and Culion)

☐ **Azure Kingfisher** *Alcedo azurea*
_____ *A. a. affinis*   N Moluccas (Morotai, Halmahera and Bacan)
_____ *A. a. wallaceanus*   Aru Islands
_____ *A. a. lessonii*   Lowlands of New Guinea, w Papuan islands and Fergusson I.
_____ *A. a. ochrogaster*   N New Guinea and islands in Geelvink Bay
_____ *A. a. yamdenae*   Tanimbar Islands (s Banda Sea)
_____ *A. a. ruficollaris*   Coastal n Australia and major offshore outlying islands
_____ *A. a. azurea*   Coastal e and se Australia
_____ *A. a. diemenensis*   Tasmania

☐ **Bismarck Kingfisher** *Alcedo websteri*
         Lowlands of Bismarck Archipelago

☐ **Blue-banded Kingfisher** *Alcedo euryzona*
_____ *A. e. peninsulae*   S Myanmar, Malay Peninsula, Sumatra and Borneo
_____ *A. e. euryzona*   Java

☐ **Indigo-banded Kingfisher** *Alcedo cyanopectus*
_____ *A. c. cyanopectus*   Luzon, Marinduque, Masbate, Mindoro, Sibuyan, Polillo, Ticao
_____ *A. c. nigrirostris*   Central Philippines (Cebu, Negros and Panay)

☐ **Silvery Kingfisher** *Alcedo argentata*
_____ *A. a. flumenicola*   Central Philippines (Bohol, Samar and Leyte)
_____ *A. a. argentata*   S Philippines (Basilan, Dinagat and Mindanao)

☐ **Malachite Kingfisher** *Alcedo cristata*
_____ *A. c. galerita*   Africa s of Sahara south to Limpopo River and Mozambique
_____ *A. c. cristata*   S Angola, sw Zambia and Limpopo R. south to Cape Province
_____ *A. c. thomensis*   São Tomé (Gulf of Guinea)
_____ *A. c. robertsi*   N Namibia and s Angola to nw Zimbabwe and sw Zambia

☐ **Malagasy Kingfisher** *Alcedo vintsioides*
_____ *A. v. vintsioides*   Madagascar
_____ *A. v. johannae*   Comoro Islands

☐ **White-bellied Kingfisher** *Alcedo leucogaster*
_____ *A. l. bowdleri*   Tropical Guinea to Mali and Ghana
_____ *A. l. leucogaster (batesi)*   Nigeria to s Cameroon, Gabon and nw Angola; Bioko
_____ *A. l. nais*   Príncipe (Gulf of Guinea)
_____ *A. l. leopoldi*   Congo basin (e Congo to s Uganda and nw Zambia)

☐ **Small Blue Kingfisher** *Alcedo coerulescens*
         S Sumatra, Java, Bali, Lombok, Sumbawa and Kangean Islands

☐ **Little Kingfisher** *Alcedo pusilla*
_____ *A. p. pusilla*   New Guinea, Aru, w Papuan is., n Moluccas and Kai Is.

____ *A. p. laetior*	N New Guinea (Geelvink Bay to Astrolabe Bay)
____ *A. p. ramsayi*	Coastal n Australia (Anson Bay to w Cape York Peninsula)
____ *A. p. halli*	NE Australia (coastal ne Queensland and Hinchinbrook I.)
____ *A. p. masauji*	Bismarck Arch. (New Britain, New Ireland and New Hanover)
____ *A. p. bougainvillei*	Solomon Islands (Bougainville, Santa Isabel and Choiseul)
____ *A. p. halmaherae*	N Moluccas (Halmahera and Bacan)
____ *A. p. richardsi*	Central Solomon Islands between Vellalavella and Vangunu
____ *A. p. aolae*	Known from the type specimen from Guadalcanal

☐ **Black-backed Kingfisher** *Ceyx erithaca*

____ *C. e. erithaca*	India and Sri Lanka to se China, Indochina and Sumatra
____ *C. e. macrocarus*	Nicobar Islands and (?) Andaman Islands
____ *C. e. motleyi*	Borneo and adjacent northern offshore islands
____ *C. e. captus*	Nias I. (off nw Sumatra)
____ *C. e. jungei*	Batu I. and Simeulue I. (off nw Sumatra)

☐ **Philippine Kingfisher** *Ceyx melanurus*

____ *C. m. melanurus*	N Philippines (Luzon, Polillo, Alabat and Catanduanes)
____ *C. m. samarensis*	Central Philippines (Samar and Leyte)
____ *C. m. mindanensis (platenae)*	S Philippines (Mindanao and Basilan)

☐ **Sulawesi Kingfisher** *Ceyx fallax*

____ *C. f. sangirensis*	Sangihi I. (ne of Sulawesi)
____ *C. f. fallax*	Sulawesi and Lembeh I.

☐ **Rufous-backed Kingfisher** *Ceyx rufidorsa*

	Malay Peninsula, Greater and Lesser Sundas to w Philippines

☐ **Variable Kingfisher** *Ceyx lepidus*

____ *C. l. margarethae*	Central and s Philippines
____ *C. l. wallacii*	Sula Islands (Taliabu, Seho, Mangole and Sanana)
____ *C. l. uropygialis*	Obi, Bisa, Bacan, Ternate, Halmahera, Tidore and Morotai is.
____ *C. l. lepidus*	Ambon, Ambelau, Seram, Saparua, Seram Laut, Watubela is.
____ *C. l. cajeli*	Buru I. (s Moluccas)
____ *C. l. solitarius*	New Guinea, Aru, w Papuan islands and D'Entrecasteaux Arch.
____ *C. l. dispar*	Admiralty Islands
____ *C. l. mulcatus*	Bismarck Arch. (New Hanover, New Ireland and Lihir Islands)
____ *C. l. sacerdotis*	Bismarck Archipelago (New Britain and Umboi)
____ *C. l. pallidus*	Solomon Islands (Bougainville and Buka)
____ *C. l. collectoris*	Solomon Is. (Choiseul, Vellalavella, New Georgia and Rendova)
____ *C. l. meeki*	Solomon Islands (Choiseul and Santa Isabel)
____ *C. l. malaitae*	Known from the type specimen from Malaita (Solomon Islands)
____ *C. l. nigromaxilla*	Guadalcanal (Solomon Islands)
____ *C. l. gentianus*	San Cristóbal (e Solomon Islands)

☐ **Madagascar Pygmy-Kingfisher** *Ispidina madagascariensis*

____ *I. m. madagascariensis*	Lowland forests of Madagascar
____ *I. m. diluta*	SW Madagascar (Sakaraha region)

☐ **African Pygmy-Kingfisher** *Ispidina picta*

____ *I. p. picta*	Senegambia to Ethiopia, Uganda, s Mozambique and Pemba I.
____ *I. p. ferruginea*	Rainforests of Sierra Leone to Congo Basin and w Uganda
____ *I. p. natalensis*	Angola to s Mozambique and South Africa

☐ **Dwarf Kingfisher** *Ispidina lecontei*

	Sierra Leone to w Uganda, n Zaire and n Angola

☐ **Banded Kingfisher** *Lacedo pulchella*

____ *L. p. amabilis*	Lowland forests of s Myanmar, Thailand and s Vietnam
____ *L. p. deignani*	S Thailand
____ *L. p. pulchella*	Malay Pen., Sumatra, Java, Riau Arch. and North Natuna Islands
____ *L. p. melanops*	Borneo and Bangka I.

☐ **Laughing Kookaburra** *Dacelo novaeguineae*
____ *D. n. minor* — NE Australia (Cape York Peninsula south to Cooktown)
____ *D. n. novaeguineae* — E and se Australia; introduced Tasmania and sw Australia

☐ **Blue-winged Kookaburra** *Dacelo leachii*
____ *D. l. superflua* — SW Irian Jaya (Mimika River to Merauke District)
____ *D. l. intermedia* — S New Guinea (Amazon Bay to Bensbach River)
____ *D. l. cliftoni* — NW Australia (Hamersley and Pilbara regions)
____ *D. l. kempi* — Islands in Torres Strait and ne Australia
____ *D. l. cervina* — Melville I. and adjacent humid coastal Northern Territory
____ *D. l. leachii* — NW Australia (Kimberley Division) to se Queensland)

☐ **Spangled Kookaburra** *Dacelo tyro*
____ *D. t. archboldi* — Trans-Fly savanna of s-central New Guinea
____ *D. t. tyro* — Aru Islands

☐ **Rufous-bellied Kookaburra** *Dacelo gaudichaud*

Lowlands of New Guinea, Aru, Yapen and w Papuan islands

☐ **Shovel-billed Kookaburra** *Clytoceyx rex*
____ *C. r. rex* — Arfak and Snow mountains to se Papua New Guinea
____ *C. r. imperator* — Irian Jaya (between Lorentz River and Mt. Goliath)

☐ **Lilac Kingfisher** *Cittura cyanotis*
____ *C. c. sanghirensis* — Rainforests of Sangihe and Siau islands (ne of Sulawesi)
____ *C. c. cyanotis* — Sulawesi and Lembeh I.

☐ **Brown-winged Kingfisher** *Pelargopsis amauroptera*

Coastal areas from e India to Myanmar and n Malay Peninsula

☐ **Stork-billed Kingfisher** *Pelargopsis capensis*
____ *P. c. capensis* — Nepal to India, Sri Lanka and nw Myanmar
____ *P. c. burmanica* — Myanmar to Thailand, Indochina and south to Isthmus of Kra
____ *P. c. intermedia* — Nicobar Islands
____ *P. c. osmastoni* — Andaman Islands
____ *P. c. malaccensis* — S Malay Peninsula, Riau Archipelago and Lingga Archipelago
____ *P. c. cyanopteryx* — Sumatra, Bangka and Billiton islands
____ *P. c. simalurensis* — Simeulue I. (off nw Sumatra)
____ *P. c. sodalis* — Banyak I. (off nw Sumatra)
____ *P. c. nesoeca* — Nias I. and Batu Islands (off w Sumatra)
____ *P. c. isoptera* — Mentawi Archipelago (Pagai, Siberut and Sipura)
____ *P. c. inominata* — Borneo
____ *P. c. floresiana* — Lesser Sundas (Bali, Lombok, Sumbawa and Flores)
____ *P. c. javana* — Java
____ *P. c. gouldi* — Philippines (Balabac, Culion, Lubang, Mindoro, Palawan, Calauit)
____ *P. c. gigantea (smithi)* — Central and s Philippines

☐ **Black-billed Kingfisher** *Pelargopsis melanorhyncha*
____ *P. m. melanorhyncha* — Sulawesi
____ *P. m. dichrorhyncha* — Peleng I. and Banggai Islands (off Sulawesi)
____ *P. m. eutreptorhyncha* — Sula Islands (e of Sulawesi)

☐ **Ruddy Kingfisher** *Halcyon coromanda*
____ *H. c. major (ochrothorectis)* — Japan to Korea, ne China; winters to Taiwan, Philippines, Borneo
____ *H. c. coromanda* — E Himalayas to n Myanmar and sw China; winters to Sumatra
____ *H. c. mizorhina* — Andaman Islands; questionable Nicobar I. specimen
____ *H. c. minor* — Riau Arch., Mentawi Arch., s Malay Pen. and Greater Sundas
____ *H. c. bangsi* — Ryukyu Islands: winters Philippines and Talaud Islands
____ *H. c. linae* — Palawan (sw Philippines)
____ *H. c. claudiae* — Sulu Archipelago (Tawitawi, Bulubuk and Sanga Sanga)
____ *H. c. pelingensis* — Peleng I. (off e Sulawesi)

____ *H. c. rufa*	S Sulawesi, Sangihe, Muna and Butung islands
____ *H. c. sulana*	Sula Islands (e of Sulawesi)

☐ **Chocolate-backed Kingfisher** *Halcyon badia*

____ *H. b. badia*	Sierra Leone to w Uganda, e Zaire and n Angola
____ *H. b. lopezi*	Bioko I. (Gulf of Guinea)

☐ **White-throated Kingfisher** *Halcyon smyrnensis*

____ *H. s. smyrnensis*	Arabian Peninsula to Caucasus Mountains and nw India
____ *H. s. fusca*	W India and Sri Lanka
____ *H. s. saturatior*	Andaman Islands
____ *H. s. perpulchra*	Myanmar to Malay Peninsula and Indochina
____ *H. s. fokiensis*	S and e China; Hainan and Taiwan
____ *H. s. gularis*	Philippine Islands

☐ **Gray-headed Kingfisher** *Halcyon leucocephala*

____ *H. l. acteon*	Cape Verde Islands (Santiago, Fogo and Brava)
____ *H. l. semicaerulea*	Red Sea coast of extreme s Arabian Peninsula
____ *H. l. leucocephala*	Senegambia to Ethiopia, nw Somalia and ne Zaire
____ *H. l. hyacinthina*	S Somalia and coastal Kenya to Mozambique
____ *H. l. pallidiventris*	Congo, Zaire, Rwanda and nw Tanzania to n South Africa

☐ **Black-capped Kingfisher** *Halcyon pileata*

Manchuria, Korea and China to s India, Malaya and Sulu Arch.

☐ **Javan Kingfisher** *Halcyon cyanoventris*

Lowlands of Java and Bali

☐ **Woodland Kingfisher** *Halcyon senegalensis*

____ *H. s. senegalensis*	Senegambia to Ethiopia and n Tanzania
____ *H. s. fuscopileus*	Forests of Sierra Leone to s Nigeria and Congo basin
____ *H. s. cyanoleuca*	NW Tanzania to Angola, Botswana and Natal

☐ **Mangrove Kingfisher** *Halcyon senegaloides*

____ *H. s. rantvorus*	Coastal Somalia to Tanzania, Zanzibar, Pemba and Mafia islands
____ *H. s. senegaloides*	Coastal se Africa (Mozambique to South Africa)

☐ **Blue-breasted Kingfisher** *Halcyon malimbica*

____ *H. m. malimbica*	Riverine woodlands of Cameroon to Uganda and Zambia
____ *H. m. torquata*	S Senegambia and Guinea-Bissau to extreme w Mali
____ *H. m. forbesi*	Sierra Leone to e Nigeria and extreme w Cameroon; Bioko I.
____ *H. m. dryas*	Príncipe I. and (formerly) São Tomé I. (Gulf of Guinea)

☐ **Brown-hooded Kingfisher** *Halcyon albiventris*

____ *H. a. albiventris*	Southwest Cape Province to Natal; winters to se Zimbabwe
____ *H. a. vociferans*	E Botswana to s Natal, s Mozambique and Orange Free State
____ *H. a. orientalis*	Coastal Somalia to n Botswana and Mozambique
____ *H. a. prentissgrayi*	SE Kenya to Congo, Zaire, Angola and Zambia

☐ **Striped Kingfisher** *Halcyon chelicuti*

____ *H. c. chelicuti*	S Mauritania and Senegambia to Ethiopia and South Africa
____ *H. c. eremogiton*	Sahara edge (central Mali) to e Sudan (White Nile region)

☐ **Blue-black Kingfisher** *Todiramphus nigrocyaneus*

____ *T. n. nigrocyaneus*	Lowlands of w New Guinea, Salawati and Batanta islands
____ *T. n. quadricolor*	Yapen I. and n New Guinea (Geelvink Bay to Astrolabe Bay)
____ *T. n. stictolaemus*	S New Guinea (se Irian Jaya to Owen Stanley Range)

☐ **Rufous-lored Kingfisher** *Todiramphus winchelli*

____ *T. w. nigrorum*	Bohol, Cebu, Negros, Samar, Siquijor, Leyte, Biliran
____ *T. w. nesydrionetes*	Central Philippines (Romblon, Sibuyan and Tablas)

____	*T. w. mindanensis*	Mindanao (s Philippines)
____	*T. w. winchelli*	Basilan (s Philippines)
____	*T. w. alfredi*	Sulu Archipelago (Bongao, Jolo, Papahag and Tawitawi)

☐ **Blue-and-white Kingfisher** *Todiramphus diops*

Halmahera, Ternate, Morotai, Bacan, Obi and adjacent Moluccas

☐ **Lazuli Kingfisher** *Todiramphus lazuli*

S Moluccas (Ambon, Seram and Haruku)

☐ **Forest Kingfisher** *Todiramphus macleayii*

____	*T. m. elizabeth*	E New Guinea
____	*T. m. macleayii (insularis)*	N Australia (n N Territory); winters to Sermata I. (Lesser Sundas)
____	*T. m. incinctus*	E Queensland to NSW; winters to e New Guinea and Kai Is.

☐ **New Britain Kingfisher** *Todiramphus albonotatus*

Coastal lowlands of New Britain (Bismarck Archipelago)

☐ **Ultramarine Kingfisher** *Todiramphus leucopygius*

Eastern and central Solomon Is.

☐ **Chestnut-bellied Kingfisher** *Todiramphus farquhari*

Vanuatu (Espíritu Santo, Malo and Malakula)

☐ **Red-backed Kingfisher** *Todiramphus pyrrhopygius*

Australia (except sw and e, from s Queensland to se S Australia)

☐ **Flat-billed Kingfisher** *Todiramphus recurvirostris*

W Samoa (Apolima, Upolu and Savai'i)

☐ **Micronesian Kingfisher** *Todiramphus cinnamominus*

____	*T. c. miyakoensis*	Known from one specimen from Miyako-Jima (Ryukyu Islands)
____	*T. c. cinnamominus†*	Formerly Guam (Mariana Is.). Extinct
____	*T. c. pelewensis*	Palau Islands (w Caroline Islands)
____	*T. c. reichenbachii*	Pohnpei (e Caroline Islands)

☐ **Collared Kingfisher** *Todiramphus chloris*

____	*T. c. abyssinicus*	W coast of Red Sea to head of Gulf of Aden
____	*T. c. kalbaensis*	S coast of Arabian Peninsula to extreme nw Oman
____	*T. c. vidali*	Peninsular India (Ratnagiri District)
____	*T. c. davisoni*	Andaman Islands and Cocos Islands (Indian Ocean)
____	*T. c. occipitalis*	Nicobar Islands
____	*T. c. humii*	NW India to Malay Pen., Thailand, Myanmar, Mergui Arch.
____	*T. c. armstrongi*	S Thailand and Myanmar
____	*T. c. chloropterus*	Islands off w Sumatra (except Enggano)
____	*T. c. azelus*	Enggano I. (off sw Sumatra)
____	*T. c. palmeri*	Java, Bali and adjacent islands in Java Sea
____	*T. c. laubmannianus*	S Sumatra, Borneo and adjacent islands
____	*T. c. collaris*	Philippines, Sulu Archipelago and Palawan
____	*T. c. chloris*	Sulawesi to nw New Guinea and Lesser Sundas
____	*T. c. pilbara*	W Australia (Exmouth Gulf to mouth of Turner River)
____	*T. c. sordidus*	S New Guinea, Aru Islands and coastal n Australia
____	*T. c. colonus*	Islands off se Papua New Guinea and Louisiade Archipelago
____	*T. c. teraokai*	Palau Islands (w Caroline Islands)
____	*T. c. owstoni*	N Mariana Islands (Asuncion, Pagan, Almagan and Agrihan)
____	*T. c. albicilla*	S Mariana Islands (Saipan, Tinian and Aguiguan)
____	*T. c. orii*	Rota I. (s Mariana Islands)
____	*T. c. matthiae*	St. Matthias Islands (Papua New Guinea)
____	*T. c. stresemanni*	Witu, Umboi and adjacent islands in Dampier Straits
____	*T. c. nusae*	New Ireland (except sw), New Hanover and Feni Islands
____	*T. c. novaehiberniae*	New Ireland (Bismarck Archipelago)
____	*T. c. bennetti*	Nissan I. (e Papua New Guinea)
____	*T. c. tristrami*	New Britain (Bismarck Archipelago)
____	*T. c. alberti*	Buka, Bougainville and Solomon Islands (east to Guadalcanal)
____	*T. c. mala*	Malaita (e Solomon Islands)
____	*T. c. sororum*	S Solomon Islands (Malaupaina and Malaulalo)
____	*T. c. pavuvu*	Pavuvu I. (Russel Group in central Solomon Islands)

____	*T. c. solomonis*	Solomon Islands (Uki Ni Masi, San Cristóbal and Santa Anna)
____	*T. c. amoenus*	Solomon Islands (Rennell and Bellona)
____	*T. c. brachyurus*	Reef Islands (Fenualoa and Lomlon)
____	*T. c. vicina*	Duff Group (e Solomon Islands)
____	*T. c. ornatus*	E Solomon Islands (Santa Cruz and Tinakula)
____	*T. c. utupuae*	Utupua I. (Santa Cruz Group in e Solomon Islands)
____	*T. c. melanoderus*	Vanikolo I. (Santa Cruz Group in e Solomon Islands)
____	*T. c. torresianus*	Torres Group (Toga, Loh and Hiu)
____	*T. c. santoensis*	Banks Group to Espíritu Santo and Malo (n Vanuatu)
____	*T. c. juliae*	N and central Vanuatu (Maewo and Aoba islands to Efate)
____	*T. c. erromangae*	Erromango I. (s Vanuatu)
____	*T. c. tannensis*	Tanna I. (s Vanuatu)
____	*T. c. vitiensis*	Fiji (Ngau, Ovalau, Koro, Viti Levu, Vanua Levu and Taveuni)
____	*T. c. eximius*	Fiji (Kandavu, Ono and Vanua Kula)
____	*T. c. marinus*	Lau Archipelago (e Fiji)
____	*T. c. sacer*	Tonga
____	*T. c. regina*	Futuna (Wallis and Futuna, central Polynesia)
____	*T. c. pealei*	Tutuila I. (American Samoa)
____	*T. c. manuae*	American Samoa (Ofu, Olosega and Tau)

☐ **Sombre Kingfisher** *Todiramphus funebris*

Halmahera (n Moluccas)

☐ **Talaud Kingfisher** *Todiramphus enigma*

Talaud Islands (Karakelong and Salebabu)

☐ **Beach Kingfisher** *Todiramphus saurophagus*

____	*T. s. saurophagus*	Moluccas to Bismarck Archipelago and Solomon Islands
____	*T. s. admiralitatis* (*anchoretus*)	Admiralty Islands (Anchorite, Hermit and Ninigo)

☐ **Cinnamon-banded Kingfisher** *Todiramphus australasia*

____	*T. a. australasia*	Lesser Sundas (Lombok, Sumba, Timor, Romang and Wetar)
____	*T. a. dammerianus*	E Lesser Sundas (Moa, Leti, Babar and Damar)
____	*T. a. odites*	Tanimbar Islands (Yamdena and Larat)

☐ **Sacred Kingfisher** *Todiramphus sanctus*

____	*T. s. sanctus*	Australia to Solomon Islands; winters to New Guinea
____	*T. s. vagans* (*norfolkiensis, adamsi*)	New Zealand, Norfolk, Lord Howe and Kermadec islands
____	*T. s. canacorum*	New Caledonia and Isle of Pines
____	*T. s. macmillani*	Loyalty Islands

☐ **Tahiti Kingfisher** *Todiramphus veneratus*

____	*T. v. veneratus*	Tahiti (French Polynesia)
____	*T. v. youngi*	Moorea (French Polynesia)

☐ **Mangaia Kingfisher** *Todiramphus ruficollaris*

Mangroves of Mangaia I. (s Cook Islands)

☐ **Chattering Kingfisher** *Todiramphus tutus*

____	*T. t. tutus*	Borabora, Maupiti, Raiatea, Huahine, Tahaa and Tahiti
____	*T. t. atiu*	Atiu I. (e Cook Islands)
____	*T. t. mauke*	Mauke I. (e Cook Islands)

☐ **Marquesas Kingfisher** *Todiramphus godeffroyi*

S Marquesas Islands (Hivaoa, Tahuata and Fatuhiva)

☐ **Tuamotu Kingfisher** *Todiramphus gambieri*

____	*T. g. gambieri*†	Formerly Mangareva I. (e Tuamotu Archipelago). Extinct
____	*T. g. gertrudae*	Niau I. (nw Tuamotu Archipelago)

☐ **White-rumped Kingfisher** *Caridonax fulgidus*

Lesser Sundas (Lombok, Sumbawa, Flores and Besar)

☐ **Hook-billed Kingfisher** *Melidora macrorrhina*

____	*M. m. waigiuensis*	Waigeo I. (w Papuan islands)

221

____ *M. m. macrorrhina*	New Guinea, Misool, Salawati and Batanta islands
____ *M. m. jobiensis*	Yapen I. and n New Guinea (Geelvink to Astrolabe Bay)

☐ **Moustached Kingfisher** *Actenoides bougainvillei*

____ *A. b. bougainvillei*	Bougainville (Solomon Islands)
____ *A. b. excelsus*	Guadalcanal (Solomon Islands)

☐ **Rufous-collared Kingfisher** *Actenoides concretus*

____ *A. c. concretus*	S Myanmar, Malay Pen., Sumatra, Bangka and Belitung islands
____ *A. c. peristephes*	S Myanmar and peninsular Thailand
____ *A. c. borneanus*	Borneo

☐ **Spotted Kingfisher** *Actenoides lindsayi*

____ *A. l. lindsayi*	N Philippines (Luzon, Marinduque and Catanduanes)
____ *A. l. moseleyi*	Forests of Negros (central Philippines)

☐ **Blue-capped Kingfisher** *Actenoides hombroni*

	Forests of Mindanao (s Philippines)

☐ **Green-backed Kingfisher** *Actenoides monachus*

____ *A. m. monachus*	N and central Sulawesi, Manadotua and Lembeh islands
____ *A. m. capucinus*	E, se and s Sulawesi

☐ **Scaly Kingfisher** *Actenoides princeps*

____ *A. p. princeps*	Humid montane forests of ne Sulawesi
____ *A. p. erythrorhamphus*	Humid montane forests of nw and central Sulawesi
____ *A. p. regalis*	Humid montane forests of se Sulawesi

☐ **Yellow-billed Kingfisher** *Syma torotoro*

____ *S. t. torotoro (tentelare, pseustes, brevirostris, meeki)*	Lowlands of New Guinea, w Papuan Is., Yapen I. and Aru Is.
____ *S. t. flavirostris*	Cape York Pen. of ne Australia (s to Weipa and Massy Creek)
____ *S. t. ochracea*	D'Entrecasteaux Archipelago

☐ **Mountain Kingfisher** *Syma megarhyncha*

____ *S. m. wellsi*	W New Guinea (Snow and Weyland Mts.)
____ *S. m. sellamontis*	NE New Guinea (Mts. of Huon Peninsula)
____ *S. m. megarhyncha*	C and se New Guinea (Sudirman range to Owen Stanley range)

☐ **Little Paradise-Kingfisher** *Tanysiptera hydrocharis*

	Aru Islands and s New Guinea (Merauke to Fly River)

☐ **Common Paradise-Kingfisher** *Tanysiptera galatea*

____ *T. g. doris*	Morotai (n Moluccas)
____ *T. g. emiliae*	Rau (Moluccas)
____ *T. g. browningi*	Halmahera (n Moluccas)
____ *T. g. brunhildae*	Doi (Moluccas)
____ *T. g. sabrina*	Kayoa (n Moluccas)
____ *T. g. margarethae*	Bacan (n Moluccas)
____ *T. g. obiensis*	Central Moluccas (Obi and Bisa)
____ *T. g. acis*	Buru (s Moluccas)
____ *T. g. boanensis*	Boano (Moluccas)
____ *T. g. nais*	S Moluccas (Ambon, Manipa, Seram, Manawoka, Gorong)
____ *T. g. galatea*	NW New Guinea and w Papuan islands
____ *T. g. meyeri*	N New Guinea (Mamberamo River to Jimi River Valley)
____ *T. g. minor*	S New Guinea (Digul River to Kumusi River) and Darnley I.
____ *T. g. vulcani*	Manam I. (off Papua New Guinea)
____ *T. g. rosseliana*	Rossel I. (Louisiade Archipelago)

☐ **Kofiau Paradise-Kingfisher** *Tanysiptera ellioti*

	Kofiau I. (w Papuan islands)

☐ **Biak Paradise-Kingfisher** *Tanysiptera riedelii*

	Biak I. (off w New Guinea)

☐ **Numfor Paradise-Kingfisher** *Tanysiptera carolinae*

Numfor I. (off w New Guinea)

☐ **Red-breasted Paradise-Kingfisher** *Tanysiptera nympha*

Locally in mangroves and forests of New Guinea

☐ **Brown-headed Paradise-Kingfisher** *Tanysiptera danae*

SE New Guinea (Aroa River to Waria River)

☐ **Buff-breasted Paradise-Kingfisher** *Tanysiptera sylvia*
_____ *T. s. leucura*
_____ *T. s. nigriceps*
_____ *T. s. salvadoriana*
_____ *T. s. sylvia*

Umboi I. (Bismarck Archipelago)
Bismarck Archipelago (New Britain and Duke of York)
SE New Guinea (Hall Sound to Kemp Welch River)
Breeds ne Australia (n Queensland); winters in New Guinea

☐ **Giant Kingfisher** *Megaceryle maximus*
_____ *M. m. maximus*
_____ *M. m. giganteus*

Senegambia to Ethiopia and South Africa
Rainforests of Liberia to w Tanzania and n Angola

☐ **Crested Kingfisher** *Megaceryle lugubris*
_____ *M. l. continentalis*
_____ *M. l. guttulatus*
_____ *M. l. pallidus*
_____ *M. l. lugubris*

W Himalayas (Kashmir to central Bhutan)
E Himalayas to China, Myanmar and Thailand; Hainan
S Kuril Islands and Hokkaido
Honshu, Shikoku and Kyushu

☐ **Belted Kingfisher** *Ceryle alcyon*

Cent. and e Canada to n S America; W Indies; Trinidad; Tobago

☐ **Ringed Kingfisher** *Ceryle torquatus*
_____ *C. t. torquatus*
_____ *C. t. stictipennis*
_____ *C. t. stellatus*

Extreme s Texas to n Argentina; Trinidad; Isla Margarita
Lesser Antilles (Guadeloupe, Martinique, Dominica, Grenada)
S Chile and Argentina to Tierra del Fuego; > to ne Argentina

☐ **Pied Kingfisher** *Ceryle rudis*
_____ *C. r. rudis*
_____ *C. r. leucomelanurus*
_____ *C. r. travancoreensis*
_____ *C. r. insignis*

Africa south of the Sahara, Egypt to Iran and Turkey
Kashmir and ne Afghanistan to India and Indochina
Extreme sw India (Cape Comorin to n Kerala)
E China (south of Yangtze River Valley) and Hainan

☐ **Amazon Kingfisher** *Chloroceryle amazona*

S Mexico to Colombia, Venezuela and e of Andes to Argentina

☐ **Green Kingfisher** *Chloroceryle americana*
_____ *C. a. hachisukai*
_____ *C. a. septentrionalis (isthmica)*
_____ *C. a. americana (bottomeana, croteta)*
_____ *C. a. cabanisii (hellmayri, ecuadorensis)*
_____ *C. a. mathewsii*

Extreme s Arizona to w-central Texas and nw Mexico
S-central Texas to s Colombia and w Venezuela
Trop. S America (primarily e of Andes); Trinidad; Tobago
W Colombia and w Ecuador w of Andes to n Chile
S Brazil and Bolivia to n Argentina

☐ **Green-and-rufous Kingfisher** *Chloroceryle inda*

SE Nicaragua to n Bolivia, e Paraguay and s Brazil

☐ **American Pygmy Kingfisher** *Chloroceryle aenea*
_____ *C. a. stictoptera*
_____ *C. a. aenea*

S Mexico and Yucatán Peninsula to n Costa Rica
Central Costa Rica to n Bolivia, Guianas and Brazil; Trinidad

ORDER: CORACIIFORMES
FAMILY: TODIDAE (Todies—5)

☐ **Cuban Tody** *Todus multicolor*

Forests and woodlands of Cuba and Isle of Pines

☐ **Broad-billed Tody** *Todus subulatus*

Arid scrub and woodlands of Hispaniola and Gonâve I.

☐ **Narrow-billed Tody** *Todus angustirostris*

Humid montane forests of Hispaniola

☐ **Jamaican Tody** *Todus todus*

Hills and mountains of Jamaica

☐ **Puerto Rican Tody** *Todus mexicanus*

Woodlands and scrub of Puerto Rico

ORDER: CORACIIFORMES
FAMILY: MOMOTIDAE (Motmots—10)

☐ **Tody Motmot** *Hylomanes momotula*
____ *H. m. chiapensis*   Pacific coast of s Mexico (Chiapas)
____ *H. m. momotula*   Caribbean slope of s Mexico (Veracruz) to Honduras
____ *H. m. obscurus*   NW Costa Rica to extreme nw Colombia

☐ **Blue-throated Motmot** *Aspatha gularis*

Oak-pine highlands of s Mexico (Oaxaca) to Honduras

☐ **Russet-crowned Motmot** *Momotus mexicanus*
____ *M. m. vanrossemi*   NW Mexico (s Sonora to adjacent Chihuahua and Sinaloa)
____ *M. m. mexicanus*   N-central and central Mexico
____ *M. m. saturatus*   Southwest Mexico (Oaxaca and Chiapas)
____ *M. m. castaneiceps*   Arid Guatemala (Zacapa Plains and Motagua Valley)

☐ **Blue-crowned Motmot** *Momotus momota*
____ *M. m. coeruliceps*   NE Mexico (Nuevo León and Tamaulipas to n Veracruz)
____ *M. m. goldmani*   Tropical se Mexico (Veracruz) to Petén of n Guatemala
____ *M. m. exiguus*   Tropical s Mexico (Campeche and Yucatán)
____ *M. m. lessonii*   Tropical s Mexico (Chiapas) to w Panama
____ *M. m. conexus*   Panama (Canal Zone) to n Colombia (lower Río Cauca)
____ *M. m. reconditus*   E Panama to n Colombia (Atrato Valley)
____ *M. m. spatha*   E Colombia (Serrania Macuira in e Guajira)
____ *M. m. olivaresi*   E Colombia
____ *M. m. subrufescens*   Caribbean coast of n Colombia and Venezuela
____ *M. m. osgoodi*   Humid forests of w Venezuela (Lake Maracaibo region)
____ *M. m. bahamensis*   Trinidad and Tobago
____ *M. m. microstephanus*   Colombia and Ecuador (east of Andes) and adjacent nw Brazil
____ *M. m. momota*   Tropical e Venezuela to the Guianas and n Brazil
____ *M. m. argenticinctus*   Tropical w Ecuador and nw Peru
____ *M. m. ignobilis*   E Amazonian Peru and w Brazil
____ *M. m. nattereri*   Tropical base of Andes of ne Bolivia
____ *M. m. simplex*   Brazil/Peru border e to Rio Tapajós and s to n Mato Grosso
____ *M. m. cametensis*   N-central Brazil between Rio Tapajós and Rio Tocantins
____ *M. m. parensis*   E Brazil (Rio Tocantins to Maranhão and Piauí)
____ *M. m. pilcomajensis*   S Bolivia to s Brazil and nw Argentina

☐ **Highland Motmot** *Momotus aequatorialis*
____ *M. a. aequatorialis*   Subtropical Andes of Colombia and e Ecuador
____ *M. a. chlorolaemus*   Subtropical e Peru

☐ **Rufous Motmot** *Baryphthengus martii*

Caribbean slope of Honduras to n Bolivia and Amaz. Brazil

☐ **Rufous-capped Motmot** *Baryphthengus ruficapillus*

Lowlands of e Brazil to e Paraguay and ne Argentina

☐ **Keel-billed Motmot** *Electron carinatum*

Caribbean slope of se Mexico to ne Costa Rica

☐ **Broad-billed Motmot** *Electron platyrhynchum*
____ *E. p. minus*   E Honduras to n Colombia (lower Cauca Valley)
____ *E. p. platyrhynchum*   W Colombia and w Ecuador
____ *E. p. pyrrholaemum*   E Colombia to e Ecuador, e Peru and n Bolivia
____ *E. p. colombianum*   N Colombia (humid lowlands north of the Andes)

| ____ | *E. p. orienticola* | W Brazil (Río Purús region) |
| ____ | *E. p. chlorophrys* | Brazil (Mato Grosso, Pará and Goiás) |

☐ **Turquoise-browed Motmot** *Eumomota superciliosa*

____	*E. s. bipartita*	Gulf slope of s Mexico to Pacific slope of Guatemala
____	*E. s. superciliosa*	SE Mexico (Tabasco, Campeche, n Yucatán and Cozumel I.)
____	*E. s. vanrossemi*	Arid interior Guatemala (Río Negro and Motagua valleys)
____	*E. s. sylvestris*	Caribbean lowlands of e Guatemala
____	*E. s. apiaster*	El Salvador to w Honduras and nw Nicaragua
____	*E. s. euroaustris*	Arid Caribbean slope of n Honduras
____	*E. s. australis*	Pacific slope of nw Costa Rica

ORDER: CORACIIFORMES
FAMILY: MEROPIDAE (Bee-eaters—26)

☐ **Red-bearded Bee-eater** *Nyctyornis amictus*

Malay Peninsula, Sumatra and Borneo

☐ **Blue-bearded Bee-eater** *Nyctyornis athertoni*

| ____ | *N. a. athertoni* | India to sw China and SE Asia |
| ____ | *N. a. brevicaudatus* | Hainan I. (s China) |

☐ **Purple-bearded Bee-eater** *Meropogon forsteni*

Humid forests of Sulawesi

☐ **Black Bee-eater** *Merops gularis*

| ____ | *M. g. gularis* | Rainforests of Sierra Leone to Nigeria (Cross River) |
| ____ | *M. g. australis* | Nigeria (Cross River) to Uganda and n Angola |

☐ **Blue-headed Bee-eater** *Merops muelleri*

| ____ | *M. m. mentalis* | Mali and Sierra Leone to Cameroon (Douala) |
| ____ | *M. m. muelleri* | Congo basin (Cameroon to Kenya and Zaire) |

☐ **Red-throated Bee-eater** *Merops bulocki*

| ____ | *M. b. bulocki* | Senegambia to Chad (Chari River) and Central African Republic |
| ____ | *M. b. frenatus* | Sub-Saharan Sudan and adj. Zaire to nw Uganda and w Ethiopia |

☐ **White-fronted Bee-eater** *Merops bullockoides*

| ____ | *M. b. bullockoides* | Semiarid tropical savannas of central and s Africa |
| ____ | *M. b. randorum* | Highlands of s Tanzania |

☐ **Little Bee-eater** *Merops pusillus*

____	*M. p. pusillus*	Savannas from Senegambia to Cameroon, Sudan and ne Zaire
____	*M. p. ocularis*	E Zaire to n Uganda, s Sudan and Red Sea coast
____	*M. p. cyanostictus*	E Ethiopia to Somalia and e Kenya
____	*M. p. meridionalis*	Congo basin to e Zaire, Uganda and w Kenya south to Natal
____	*M. p. argutus*	SW Angola to sw Zambia and Botswana

☐ **Blue-breasted Bee-eater** *Merops variegatus*

____	*M. v. bangweoloensis*	E Angola to se Zaire, Zambia and extreme w Tanzania
____	*M. v. loringi*	SE Nigeria and Cameroon to Uganda and Kenya
____	*M. v. variegatus*	Gabon, Rio Muni and sw Cameroon to Zaire and n Angola
____	*M. v. lafresnayii*	Ethiopia to Sudan (Boma Hills)

☐ **Cinnamon-chested Bee-eater** *Merops oreobates*

Highland forests of e Africa

☐ **Swallow-tailed Bee-eater** *Merops hirundineus*

____	*M. h. chrysolaimus*	Senegal to s Chad and nw Central African Republic
____	*M. h. heuglini*	S Sudan and adjacent Ethiopia, Uganda and Zaire
____	*M. h. furcatus*	S Zaire to Zambia, s Zimbabwe, Mozambique and Tanzania
____	*M. h. hirundineus*	Angola and Namibia to Orange R., Transvaal and Zimbabwe

☐ **Black-headed Bee-eater** *Merops breweri*

Coastal Gabon and locally along Congo River

☐ **Somali Bee-eater** *Merops revoilii*

Arid thornscrub of e Ethiopia, Somalia and Kenya

☐ **White-throated Bee-eater** *Merops albicollis*

Subdesert steppes of sub-Saharan Africa and sw Arabia

☐ **Green Bee-eater** *Merops orientalis*

____ *M. o. viridissimus*	Savanna of Senegal to Eritrea, Ethiopia and w Sudan
____ *M. o. flavoviridis*	Subdesert steppes of Chad to Red Sea coast of Sudan
____ *M. o. cleopatra*	Nile Valley (Lake Nasser to delta)
____ *M. o. cyanophrys*	Arabian Peninsula
____ *M. o. najdanus*	Central Arabian plateau
____ *M. o. beludschicus*	North end of Persian Gulf to Baluchistan and w India
____ *M. o. orientalis*	Rann of Kutch to Bangladesh and Sri Lanka
____ *M. o. ferrugeiceps*	Assam and Myanmar to Vietnam

☐ **Boehm's Bee-eater** *Merops boehmi*

Humid forests of se Africa (Malawi and adjacent countries)

☐ **Blue-throated Bee-eater** *Merops viridis*

____ *M. v. viridis*	S China, Thailand and Indochina to Sumatra, Borneo and Java
____ *M. v. americanus*	Philippine Islands

☐ **Blue-cheeked Bee-eater** *Merops persicus*

____ *M. p. chrysocercus*	NW Africa (south of Atlas Mts.); Senegambia to Lake Chad
____ *M. p. persicus*	Egypt to Lake Balkhash and Hindu Kush; winters to s Africa

☐ **Madagascar Bee-eater** *Merops superciliosus*

____ *M. s. superciliosus*	East Africa, Comoro Islands and Madagascar
____ *M. s. alternans*	Arid littoral of w Angola and nw Namibia

☐ **Blue-tailed Bee-eater** *Merops philippinus*

India to s China, SE Asia, New Guinea and Indonesia

☐ **Rainbow Bee-eater** *Merops ornatus*

Australia to New Guinea, Solomons, Gr. and Lesser Sundas

☐ **European Bee-eater** *Merops apiaster*

S Palearctic; winters to sub-Saharan Africa and w India

☐ **Chestnut-headed Bee-eater** *Merops leschenaulti*

____ *M. l. leschenaulti*	Sri Lanka and sw India to Thailand, Indochina and Malay Pen.
____ *M. l. andamanensis*	Andaman Islands
____ *M. l. quinticolor*	Java and Bali

☐ **Rosy Bee-eater** *Merops malimbicus*

Humid forests of w and central Africa

☐ **Northern Carmine Bee-eater** *Merops nubicus*

Savanna and grasslands of sub-Saharan Africa

☐ **Southern Carmine Bee-eater** *Merops nubicoides*

Acacia savanna of central and s Africa

ORDER: CORACIIFORMES
FAMILY: CORACIIDAE (Rollers—12)

☐ **European Roller** *Coracias garrulus*

____ *C. g. garrulous*	N Africa, Europe to Iran and sw Siberia; winters to s Africa
____ *C. g. semenowi*	Iraq to w Xinjiang and s Kazakstan; winters to s Africa

☐ **Abyssinian Roller** *Coracias abyssinicus*

Senegambia to Ethiopia, Somalia and Kenya; central Arabia

☐ **Lilac-breasted Roller** *Coracias caudatus*
    ____ *C. c. lorti* — Ethiopia south to Lake Turkana, Somalia and ne Kenya
    ____ *C. c. caudatus* — Southern and e Africa north to Uganda and Kenya

☐ **Racket-tailed Roller** *Coracias spatulatus*
    Tropical sw Angola to ne Tanzania and s Mozambique

☐ **Rufous-crowned Roller** *Coracias noevius*
    ____ *C. n. noevius* — Senegambia to Ethiopia, Somalia and n Tanzania
    ____ *C. n. mosambicus* — Angola and Namibia to Zambia and n South Africa

☐ **Indian Roller** *Coracias benghalensis*
    ____ *C. b. benghalensis* — E Arabia to ne India
    ____ *C. b. indicus* — S India and Sri Lanka
    ____ *C. b. affinis* — NE India to s-central China, n Malay Peninsula and Indochina

☐ **Purple-winged Roller** *Coracias temminckii*
    Sulawesi, Lembeh, Bangka, Manterawu, Muna and Butung is.

☐ **Blue-bellied Roller** *Coracias cyanogaster*
    Senegambia to Cent. African Rep., extreme s Sudan and ne Zaire

☐ **Broad-billed Roller** *Eurystomus glaucurus*
    ____ *E. g. afer* — Senegal to Sudan, coastal w Africa and n Zaire
    ____ *E. g. aethiopicus* — Sudan and w Ethiopia
    ____ *E. g. suahelicus* — S Somalia to central Zaire, ne Zambia, Angola and Natal
    ____ *E. g. glaucurus* — Madagascar; non-breeding migrant to e Africa

☐ **Blue-throated Roller** *Eurystomus gularis*
    ____ *E. g. gularis* — Rainforests of Guinea and w Cameroon
    ____ *E. g. neglectus* — S Nigeria to Uganda and Angola

☐ **Dollarbird** *Eurystomus orientalis*
    ____ *E. o. abundus* — Himalayas to China, Manchuria and Korea; winters to Indonesia
    ____ *E. o. deignani* — N Thailand; winters to Malaysia, Sumatra, Borneo and Java
    ____ *E. o. orientalis* — S Himalayas to SE Asia, Ryukyu Islands and Indonesian Arch.
    ____ *E. o. gigas* — S Andaman Islands
    ____ *E. o. oberholseri* — Simeulue I. (off Sumatra)
    ____ *E. o. connectens* — S Sulawesi, Sula Islands and Lesser Sundas
    ____ *E. o. latouchei* — NE China
    ____ *E. o. waigiouensis* — New Guinea, w Papuan is., D'Entrecasteaux and Louisiade Arch.
    ____ *E. o. pacificus* — N and e Australia; winters to New Guinea, s Moluccas, adj. islands
    ____ *E. o. crassirostris* — Bismarck Archipelago
    ____ *E. o. solomonensis* — Feni I. and Solomon Islands

☐ **Purple Roller** *Eurystomus azureus*
    N Moluccas

ORDER: CORACIIFORMES
FAMILY: BRACHYPTERACIIDAE (Ground-Rollers—5)

☐ **Short-legged Ground-Roller** *Brachypteracias leptosomus*
    Dense rainforests of central and ne Madagascar

☐ **Scaly Ground-Roller** *Brachypteracias squamiger*
    Dense rainforests of central and ne Madagascar

☐ **Pitta-like Ground-Roller** *Atelornis pittoides*
    Humid forests of Madagascar

☐ **Rufous-headed Ground-Roller** *Atelornis crossleyi*
    Rainforests of central and ne Madagascar

☐ **Long-tailed Ground-Roller** *Uratelornis chimaera*
    Subdesert of sw Madagascar

ORDER: CORACIIFORMES
FAMILY: LEPTOSOMATIDAE (Cuckoo-Roller—1)

☐ **Cuckoo-Roller** *Leptosomus discolor*

____ *L. d. gracilis*	Grand Comoro I. (Comoro Islands)
____ *L. d. intermedius*	Anjouan I. (Comoro Islands)
____ *L. d. discolor*	Mohéli, Mayotte and Madagascar

ORDER: CORACIIFORMES
FAMILY: UPUPIDAE (Hoopoes—2)

☐ **Eurasian Hoopoe** *Upupa epops*

____ *U. e. epops*	W Palearctic to Baluchistan; winters to s India and Africa
____ *U. e. major*	Egypt (Nile Valley to Suez Canal)
____ *U. e. senegalensis*	Senegambia to Ethiopia, Somalia and Uganda
____ *U. e. waibeli*	Chad to Cameroon (Adamawa Plateau), n Uganda and Kenya
____ *U. e. africana*	S Zaire to Uganda, Kenya and Cape Province
____ *U. e. orientalis*	NW India
____ *U. e. ceylonensis*	Central and s India and Sri Lanka
____ *U. e. saturata*	E Siberia, Manchuria and n China to extreme sw China
____ *U. e. longirostris*	Assam to Malay Peninsula, Indochina and Sumatra

☐ **Madagascar Hoopoe** *Upupa marginata*

	Madagascar

ORDER: CORACIIFORMES
FAMILY: PHOENICULIDAE (Woodhoopoes and Scimitar-bills—8)

☐ **Green Woodhoopoe** *Phoeniculus purpureus*

____ *P. p. guineensis*	Senegambia to n Ghana, Nigeria, Chad and Central African Rep.
____ *P. p. senegalensis*	S Senegal and Gambia to s Ghana
____ *P. p. niloticus*	Sudan to w Ethiopia and ne Zaire
____ *P. p. marwitzi*	E Uganda and Kenya to central and e Natal; 2 records Somalia
____ *P. p. angolensis*	Angola to w Zambia, w Zimbabwe, Namibia and Botswana
____ *P. p. purpureus (erythrorhynchos)*	South Africa (Cape Province and Transkei)

☐ **Violet Woodhoopoe** *Phoeniculus damarensis*

____ *P. d. damarensis*	Angola and Namibia
____ *P. d. granti*	Ethiopia and Kenya

☐ **Black-billed Woodhoopoe** *Phoeniculus somaliensis*

____ *P. s. somaliensis*	SE Ethiopia to w Somalia and ne Kenya
____ *P. s. neglectus*	Arid thornscrub of central Ethiopia
____ *P. s. abyssinicus*	N Ethiopia and Eritrea

☐ **White-headed Woodhoopoe** *Phoeniculus bollei*

____ *P. b. jacksoni*	Humid forests of e Zaire to Sudan and Kenya
____ *P. b. bollei*	Liberia to Central African Republic
____ *P. b. okuensis*	Cameroon (Lake Oku region)

☐ **Forest Woodhoopoe** *Phoeniculus castaneiceps*

____ *P. c. castaneiceps*	Liberia to Nigeria
____ *P. c. brunneiceps*	Cameroon to Zaire, Uganda and Kenya

☐ **Black Scimitar-bill** *Rhinopomastus aterrimus*

____ *R. a. aterrimus*	Senegambia to w Sudan
____ *R. a. emini*	Central Sudan to ne Zaire and Uganda
____ *R. a. notatus*	Ethiopia
____ *R. a. anchietae*	Zaire to w Zambia, Angola and se Gabon

☐ **Common Scimitar-bill** *Rhinopomastus cyanomelas*
____ *R. c. cyanomelas* — Angola and Namibia to Transvaal
____ *R. c. schalowi (intermedius)* — Somalia to Zambia and w Natal

☐ **Abyssinian Scimitar-bill** *Rhinopomastus minor*
____ *R. m. minor* — Ethiopia to Somalia and n Kenya
____ *R. m. cabanisi* — Sudan to s Ethiopia, Kenya and Tanzania

ORDER: CORACIIFORMES
FAMILY: BUCEROTIDAE (Hornbills—57)

☐ **White-crested Hornbill** *Tockus albocristatus*
____ *T. a. albocristatus* — Humid forests of Guinea to w Ivory Coast
____ *T. a. macrourus* — E Ivory Coast and Ghana
____ *T. a. cassini* — Nigeria through Congo basin to w Uganda and n Angola

☐ **Black Dwarf Hornbill** *Tockus hartlaubi*
____ *T. h. hartlaubi* — Sierra Leone to Zaire (west of Congo River)
____ *T. h. granti* — Congo basin of Central African Republic to Zaire and Uganda

☐ **Red-billed Dwarf Hornbill** *Tockus camurus*

S Sierra Leone to extreme s Sudan, w Uganda and n Angola

☐ **Monteiro's Hornbill** *Tockus monteiri*

Arid savanna of s Angola and n Namibia

☐ **Red-billed Hornbill** *Tockus erythrorhynchus*
____ *T. e. kempi* — Senegambia to (Mali) inner Niger delta
____ *T. e. erythrorhynchus* — Sierra Leone to Somalia and south to Tanzania
____ *T. e. rufirostris* — Angola and n Namibia to Mozambique and e South Africa
____ *T. e. damarensis* — Central Namibia (s Damaraland)

☐ **Eastern Yellow-billed Hornbill** *Tockus flavirostris*

Ethiopia and Somalia to n Uganda and ne Tanzania

☐ **Southern Yellow-billed Hornbill** *Tockus leucomelas*

Arid savanna of s-central and s Africa

☐ **Jackson's Hornbill** *Tockus jacksoni*

Arid savanna of ne Uganda and nw Kenya

☐ **Von der Decken's Hornbill** *Tockus deckeni*

Ethiopia and s Somalia to Kenya, ne Uganda and Tanzania

☐ **Crowned Hornbill** *Tockus alboterminatus*

Savanna of e and s Africa, Zanzibar and Pemba I.

☐ **Bradfield's Hornbill** *Tockus bradfieldi*

NW Zimbabwe to sw Zambia, n Botswana and s Angola

☐ **African Pied Hornbill** *Tockus fasciatus*
____ *T. f. semifasciatus* — Senegambia to just east of Niger River
____ *T. f. fasciatus* — Nigeria (east of Niger River) to Angola, Zaire and Uganda

☐ **Hemprich's Hornbill** *Tockus hemprichii*

Rocky *Euphorbia* areas of Ethiopia to nw Kenya and Uganda

☐ **African Gray Hornbill** *Tockus nasutus*
____ *T. n. nasutus (forskali)* — Senegambia to Ethiopia, Kenya and Uganda; Arabian Peninsula
____ *T. n. epirhinus (dorsalis)* — S Uganda and se Kenya to n South Africa

☐ **Pale-billed Hornbill** *Tockus pallidirostris*
____ *T. p. pallidirostris* — Angola to s Zaire, Zambia, Mozambique and s Tanzania
____ *T. p. neumanni* — E Zambia to Mozambique, Malawi and Tanzania

☐ **Malabar Gray Hornbill** *Ocyceros griseus*

Moist deciduous forests of sw India

☐ **Ceylon Gray Hornbill** *Ocyceros gingalensis*

Moist deciduous forests of Sri Lanka

☐ **Indian Gray Hornbill** *Ocyceros birostris*

NE Pakistan, India and nw Bangladesh

☐ **Malabar Pied-Hornbill** *Anthracoceros coronatus*

Mixed forests of s India and Sri Lanka

☐ **Oriental Pied-Hornbill** *Anthracoceros albirostris*
____ *A. a. albirostris*
____ *A. a. convexus*

India to Assam, Nepal, Myanmar, s China and Indochina
S Thailand, Malay Pen., Greater Sundas and adjacent islands

☐ **Black Hornbill** *Anthracoceros malayanus*

Malay Peninsula, Sumatra and Borneo

☐ **Palawan Hornbill** *Anthracoceros marchei*

SW Philippines (Palawan, Balabac, Busuanga and Calauit)

☐ **Sulu Hornbill** *Anthracoceros montani*

Sulu Archipelago (Jolo, Tawitawi and Sanga Sanga)

☐ **Rhinoceros Hornbill** *Buceros rhinoceros*
____ *B. r. rhinoceros*
____ *B. r. borneoensis*
____ *B. r. silvestris*

Lowlands of s Malay Peninsula and Sumatra
Borneo
Java

☐ **Great Hornbill** *Buceros bicornis*

Lowlands of India to sw China, SE Asia and Sumatra

☐ **Rufous Hornbill** *Buceros hydrocorax*
____ *B. h. hydrocorax*
____ *B. h. mindanensis*
____ *B. h. semigaleatus*

Philippines (Luzon and Marinduque)
Philippines (Mindanao, Basilan, Dinagat and Siargao)
Samar, Leyte, Bohol, Panaon, Buad, Calicoan and Biliran

☐ **Helmeted Hornbill** *Buceros vigil*

Myanmar to s Thailand, Malay Peninsula, Sumatra and Borneo

☐ **Brown Hornbill** *Anorrhinus austeni*

Assam, Myanmar and sw China to Thailand and Indochina

☐ **Rusty-cheeked Hornbill** *Anorrhinus tickelli*

S Myanmar and se Thailand

☐ **Bushy-crested Hornbill** *Anorrhinus galeritus*

S Myanmar, Malay Peninsula, Sumatra, Borneo and Penang I.

☐ **Luzon Hornbill** *Penelopides manillae*

Philippines (Luzon, Marinduque and Catanduanes)`

☐ **Mindoro Hornbill** *Penelopides mindorensis*

Mindoro (central Philippines)

☐ **Tarictic Hornbill** *Penelopides panini*
____ *P. p. panini*
____ *P. p. ticaensis*

Panay, Masbate, Guimaras, Negros, Pan de Azucar and Sicogon
Ticao I. (central Philippines)

☐ **Samar Hornbill** *Penelopides samarensis*

Philippines (Samar, Leyte, Calicoan and Bohol)

☐ **Mindanao Hornbill** *Penelopides affinis*
____ *P. a. affinis*
____ *P. a. basilanicus*

S Philippines (Mindanao, Dinagat and Siargao)
Basilan (s Philippines)

☐ **Sulawesi Hornbill** *Penelopides exarhatus*
____ *P. e. exarhatus*
____ *P. e. sanfordi*

N Sulawesi and Lembeh I.
Central and s Sulawesi, Muna, Butung and Togian islands

☐ **White-crowned Hornbill** *Aceros comatus*

Extreme s Myanmar, Malay Peninsula, Sumatra and Borneo

☐ **Rufous-necked Hornbill** *Aceros nipalensis*

Himalayan foothills to sw China and n SE Asia

☐ **Wrinkled Hornbill** *Aceros corrugatus*

Malay Peninsula, Sumatra and Borneo

230

☐ **Writhe-billed Hornbill** *Aceros waldeni*

☐ **Writhed Hornbill** *Aceros leucocephalus*
> Central Philippines (Panay, Guimaras and Negros)

☐ **Knobbed Hornbill** *Aceros cassidix*
> S Philippines (Camiguin, Dinagat and Mindanao)

☐ **Wreathed Hornbill** *Aceros undulatus*
> Sulawesi, Lembeh, Togian Islands, Muna and Butung

☐ **Narcondam Hornbill** *Aceros narcondami*
> NE India to sw China, SE Asia and Greater Sundas.

☐ **Sumba Hornbill** *Aceros everetti*
> Lowland forests of Narcondam I. (Andaman Islands)

☐ **Plain-pouched Hornbill** *Aceros subruficollis*
> Sumba (Lesser Sundas)

☐ **Blyth's Hornbill** *Aceros plicatus*
> S Myanmar, sw and s Thailand and n Malaysia

    ____ *A. p. ruficollis*
    ____ *A. p. plicatus*
    ____ *A. p. jungei*
    ____ *A. p. dampieri*
    ____ *A. p. harterti*
> N Moluccas and w New Guinea
> S Moluccas (Kelang, Seram and Ambon)
> E New Guinea; vagrant to Fergusson I. (D'Entrecasteaux Arch.)
> Bismarck Arch. (New Hanover, New Ireland and New Britain)
> Solomon Is. (Buka, Bougainville, Fauro and Shortland Islands)

☐ **Trumpeter Hornbill** *Ceratogymna bucinator*

☐ **Piping Hornbill** *Ceratogymna fistulator*
> Forests of s-central and se Africa

    ____ *C. f. fistulator*
    ____ *C. f. sharpii*
    ____ *C. f. duboisi*
> Mangroves and humid forests of Senegambia to Niger River
> Niger River to Cameroon, Gabon and n Angola
> Cameroon to n Angola, Central African Republic and Uganda

☐ **Silvery-cheeked Hornbill** *Ceratogymna brevis*

☐ **Black-and-white-casqued Hornbill** *Ceratogymna subcylindrica*
> Montane and coastal forests of Ethiopia to s Mozambique

    ____ *C. s. subcylindrica*
    ____ *C. s. subquadrata*
> Ivory Coast to Nigeria (west of Niger River)
> Nigeria (east of Niger River) to w Kenya, Burundi and Angola

☐ **Brown-cheeked Hornbill** *Ceratogymna cylindrica*

☐ **White-thighed Hornbill** *Ceratogymna albotibialis*
> Humid forests of Sierra Leone to Ghana

☐ **Black-casqued Hornbill** *Ceratogymna atrata*
> Humid forests of Benin to Angola and Uganda

☐ **Yellow-casqued Hornbill** *Ceratogymna elata*
> Patchily distributed forests of w and central Africa; Bioko I.

☐ **Abyssinian Ground-Hornbill** *Bucorvus abyssinicus*
> Patchily distributed forests of Senegambia to Cameroon

☐ **Southern Ground-Hornbill** *Bucorvus leadbeateri*
> Senegambia to Ethiopia, n Uganda and ne Kenya

> Savanna of e and s Africa

ORDER: GALBULIFORMES
FAMILY: GALBULIDAE (Jacamars—18)

☐ **White-eared Jacamar** *Galbalcyrhynchus leucotis*

☐ **Purus Jacamar** *Galbalcyrhynchus purusianus*
> Colombia east of the Andes to ne Peru and w Amazonian Brazil

☐ **Dusky-backed Jacamar** *Brachygalba salmoni*
> Tropical e Peru, n Bolivia and w Amazonian Brazil

☐ **Pale-headed Jacamar** *Brachygalba goeringi*
> Tropical e Panama and nw Colombia

> Tropical ne Colombia and nw Venezuela

☐ **Brown Jacamar** *Brachygalba lugubris*

____ *B. l. fulviventris*	E base of E Andes of Colombia (Buenavista and Villavicencio)
____ *B. l. caquetae*	E base of Eastern Andes of Colombia (s of Caquetá) to e Peru
____ *B. l. lugubris*	E and s Venezuela to the Guianas and ne Brazil
____ *B. l. obscuriceps*	Extreme s Venezuela and nw Brazil
____ *B. l. naumburgi*	NE Brazil (Maranhão and Piauí)
____ *B. l. melanosterna*	E Bolivia; central and sw Brazil
____ *B. l. phaeonota*	Central Brazil (known from one specimen from Rio Solimões)

☐ **White-throated Jacamar** *Brachygalba albogularis*

Patchily distributed se Peru to n Bolivia and sw Amaz. Brazil

☐ **Three-toed Jacamar** *Jacamaralcyon tridactyla*

Lowlands of se Brazil (Minas Gerais to Paraná)

☐ **Yellow-billed Jacamar** *Galbula albirostris*

____ *G. a. albirostris*	E Colombia to Venezuela, the Guianas and n Brazil
____ *G. a. chalcocephala*	SE Colombia to Ecuador, ne Peru; w Brazil (upper Rio Negro)

☐ **Blue-cheeked Jacamar** *Galbula cyanicollis*

Humid lowlands of ne Peru and Brazil south of the Amazon

☐ **Green-tailed Jacamar** *Galbula galbula*

E Colombia to s Venezuela, the Guianas, n and central Brazil

☐ **Rufous-tailed Jacamar** *Galbula ruficauda*

____ *G. r. melanogenia*	Lowlands of se Mexico (Veracruz) to w Ecuador
____ *G. r. pallens*	Arid tropical n Colombia
____ *G. r. ruficauda*	C Colombia to the Guianas and n Brazil; Trinidad and Tobago
____ *G. r. brevirostris*	NE Colombia and nw Venezuela (Lake Maracaibo region)
____ *G. r. rufoviridis*	Brazil south of Amazon to n Bolivia, Paraguay and ne Argentina
____ *G. r. heterogyna*	Bolivia east of the Andes and sw Brazil (w Mato Grosso)

☐ **Coppery-chested Jacamar** *Galbula pastazae*

Foothills of se Colombia and e Ecuador

☐ **White-chinned Jacamar** *Galbula tombacea*

____ *G. t. tombacea*	Amazonian Colombia to e Ecuador, ne Peru and w Brazil
____ *G. t. mentalis*	Central and w Amazonian Brazil

☐ **Bluish-fronted Jacamar** *Galbula cyanescens*

Riverine forests of e Peru to n Bolivia and w Amazonian Brazil

☐ **Purplish Jacamar** *Galbula chalcothorax*

S Colombia to e Peru and sw Amazonian Brazil

☐ **Bronzy Jacamar** *Galbula leucogastra*

S Venezuela to the Guianas, w Amazonian Brazil and n Bolivia

☐ **Paradise Jacamar** *Galbula dea*

____ *G. d. dea*	S Venezuela to Guianas and Brazil n of Amazon (e of Rio Negro)
____ *G. d. amazonum*	N Bolivia (Río Beni) and Brazil s of the Amazon (east to Pará)
____ *G. d. brunneiceps*	Extreme e Colombia, e Peru and w Brazil
____ *G. d. phainopepla*	W Brazil south of the Amazon and west of Rio Madeira)

☐ **Great Jacamar** *Jacamerops aureus*

____ *J. a. penardi*	Caribbean slope of Costa Rica to w Colombia
____ *J. a. aureus*	E Colombia to Venezuela and the Guianas
____ *J. a. ridgwayi*	Lower Amazonian Brazil (Rio Negro and Rio Tapajós eastward)
____ *J. a. isidori*	E Ecuador to e Peru, w Brazil and n Bolivia

ORDER: GALBULIFORMES
FAMILY: BUCCONIDAE (Puffbirds—35)

☐ **White-necked Puffbird** *Notharchus macrorhynchos*

____ *N. m. hyperrynchus*	Semiarid s Mexico to nw South America
____ *N. m. cryptoleucus*	Pacific lowlands of El Salvador and nw Nicaragua

____	*N. m. macrorhynchos*	Guianas and Brazil north of the Amazon
____	*N. m. paraensis*	Lower Amazon east of Rio Tapajós (Pará and Maranhão)

☐ **Buff-bellied Puffbird** *Notharchus swainsoni*

SE Brazil to e Paraguay and ne Argentina

☐ **Black-breasted Puffbird** *Notharchus pectoralis*

Humid forests of e Panama to w Colombia and nw Ecuador

☐ **Brown-banded Puffbird** *Notharchus ordii*

S Venezuela to n Amazonian Brazil, se Peru and n Bolivia

☐ **Pied Puffbird** *Notharchus tectus*

____	*N. t. subtectus*	Caribbean coast of Costa Rica to cent. Colombia and sw Ecuador
____	*N. t. picatus*	E Ecuador and e Peru
____	*N. t. tectus*	S Venezuela to Guianas and Amazonian Brazil (e to Maranhão)

☐ **Chestnut-capped Puffbird** *Bucco macrodactylus*

SE Colombia to Venezuela, n Bolivia and w Amazonian Brazil

☐ **Spotted Puffbird** *Bucco tamatia*

____	*B. t. pulmentum*	E Colombia to e Ecuador, ne Peru, w Brazil and ne Bolivia
____	*B. t. tamatia*	Extreme e Colombia to Venezuela, Guianas and n Brazil
____	*B. t. hypneleus*	Amazonian Brazil east of Rio Tapajós and delta islands

☐ **Sooty-capped Puffbird** *Bucco noanamae*

Coastal w Colombia (Gulf of Urabá to Río San Juan)

☐ **Collared Puffbird** *Bucco capensis*

SE Colombia to e Peru, Guianas, Amaz. Brazil and Mato Grosso

☐ **Barred Puffbird** *Nystalus radiatus*

Humid forests of w Panama to w Ecuador

☐ **White-eared Puffbird** *Nystalus chacuru*

____	*N. c. uncirostris*	SE Peru to ne Bolivia and adjacent w Brazil
____	*N. c. chacuru*	Campos of s Brazil, e Paraguay and ne Argentina

☐ **Striolated Puffbird** *Nystalus striolatus*

____	*N. s. striolatus*	E Ecuador to e Peru, Bolivia and sw Amazonian Brazil
____	*N. s. torridus*	E Brazil south of the Amazon (Pará)

☐ **Spot-backed Puffbird** *Nystalus maculatus*

____	*N. m. maculatus (parvirostris)*	NE and central Brazil (to sw Mato Grosso)
____	*N. m. striatipectus (pallidigula)*	S Bolivia and s-central Brazil to Paraguay and nw Argentina

☐ **Russet-throated Puffbird** *Hypnelus ruficollis*

____	*H. r. ruficollis*	N Colombia and nw Venezuela (w Lake Maracaibo region)
____	*H. r. decolor (striaticollis)*	Extreme ne Colombia (Guajira Pen.) and nw Venezuela (Falcón)
____	*H. r. coloratus*	W Venezuela (south of Lake Maracaibo)

☐ **Two-banded Puffbird** *Hypnelus bicinctus*

____	*H. b. bicinctus*	Llanos of interior ne Colombia and n Venezuela
____	*H. b. stoicus*	Isla Margarita (off n Venezuela)

☐ **Crescent-chested Puffbird** *Malacoptila striata*

____	*M. s. minor*	Humid lowlands of ne Brazil (ne Maranhão)
____	*M. s. striata*	SE Brazil (s Bahia and e Minas Gerais to Santa Catarina)

☐ **White-chested Puffbird** *Malacoptila fusca*

SE Colombia to e Peru, the Guianas and w Amazonian Brazil

☐ **Semicollared Puffbird** *Malacoptila semicincta*

Humid lowlands of se Peru, extreme w Brazil and nw Bolivia

☐ **Black-streaked Puffbird** *Malacoptila fulvogularis*

E slope of Andes from central Colombia to nw Bolivia

☐ **Rufous-necked Puffbird** *Malacoptila rufa*
_____ *M. r. rufa*   E Ecuador to e Peru and w Brazil (s of Amazon to Rio Madeira)
_____ *M. r. brunnescens*   Humid forests of Amazonian Brazil

☐ **White-whiskered Puffbird** *Malacoptila panamensis*
_____ *M. p. inornata (fuliginosa)*   Caribbean slope of se Mexico to w Panama
_____ *M. p. panamensis*   Pacific slope of sw Costa Rica to nw Colombia
_____ *M. p. poliopis*   Tropical sw Colombia and w Ecuador
_____ *M. p. magdalenae*   W-central Colombia (Magdalena Valley)

☐ **Moustached Puffbird** *Malacoptila mystacalis*
Andes of Colombia and n Venezuela

☐ **Lanceolated Monklet** *Micromonacha lanceolata*
E Costa Rica to Ecuador, ne Peru, extreme w Brazil, n Bolivia

☐ **Rusty-breasted Nunlet** *Nonnula rubecula*
_____ *N. r. simulatrix*   SE Colombia and nw Brazil (between Río Negro and Amazon)
_____ *N. r. duidae*   E Venezuela in s Amazonas (n of Río Orinoco)
_____ *N. r. interfluvialis*   S Venezuela (s of Río Orinoco) s to Río Negro in n Brazil
_____ *N. r. cineracea*   NE Ecuador, ne Peru and w Brazil s of Amazon (s to Rondônia)
_____ *N. r. tapanahoniensis*   S Guianas and n Brazil (n bank of lower Amazon)
_____ *N. r. simplex*   N Brazil along s bank of lower Amazon
_____ *N. r. rubecula*   E and se Brazil to e Paraguay and ne Argentina (Misiones)

☐ **Fulvous-chinned Nunlet** *Nonnula sclateri*
Trop. se Peru to n Bolivia w Brazil south of the Amazon

☐ **Brown Nunlet** *Nonnula brunnea*
Humid lowlands of se Colombia, e Ecuador and ne Peru

☐ **Gray-cheeked Nunlet** *Nonnula frontalis*
_____ *N. f. stulta*   Central Panama to extreme nw Colombia
_____ *N. f. pallescens*   Caribbean lowlands of n Colombia
_____ *N. f. frontalis*   Interior n Colombia

☐ **Rufous-capped Nunlet** *Nonnula ruficapilla*
_____ *N. r. rufipectus*   NE Peru (north of the Amazon)
_____ *N. r. ruficapilla*   E Peru and w Brazil s of the Amazon
_____ *N. r. nattereri*   S Amazon basin (Mato Grosso, n Bolivia and n Brazil in w Pará)
_____ *N. r. inundata*   E Brazil (e Pará) on left bank of Rio Tocantins

☐ **Chestnut-headed Nunlet** *Nonnula amaurocephala*
NW Brazil n of Amazon and w of Río Negro

☐ **White-faced Nunbird** *Hapaloptila castanea*
Locally in Andean foothills of w Colombia to nw Peru

☐ **Black Nunbird** *Monasa atra*
Extreme e Colombia to Venezuela, the Guianas and n Brazil

☐ **Black-fronted Nunbird** *Monasa nigrifrons*
_____ *M. n. nigrifrons*   Lowlands of se Colombia to e Peru and much of Brazil
_____ *M. n. canescens*   Tropical Bolivia e of Andes

☐ **White-fronted Nunbird** *Monasa morphoeus*
_____ *M. m. grandior*   Caribbean slope of Honduras and Nicaragua to w Panama
_____ *M. m. fidelis*   Caribbean e Panama and nw Colombia (e to s Córdoba)
_____ *M. m. pallescens*   SE Panama to w Colombia (s to upper Río San Juan)
_____ *M. m. sclateri*   N and cent. Colombia (Magdalena Valley s to n Tolima)
_____ *M. m. peruana*   SE Colombia to s Venezuela, ne Bolivia and upper Amaz. Brazil
_____ *M. m. rikeri*   Amazonian Brazil from Río Tapajós e to Piauí
_____ *M. m. morphoeus*   E Brazil (Bahia south to Rio de Janeiro)

☐ **Yellow-billed Nunbird** *Monasa flavirostris*
Amazon basin (se Colombia to n Bolivia and w Amaz. Brazil)

☐ **Swallow-wing** *Chelidoptera tenebrosa*

____ *C. t. tenebrosa*	E Colombia to the Guianas, n Bolivia and Brazil (except se)
____ *C. t. brasiliensis*	Coastal se Brazil
____ *C. t. pallida*	Northwest Venezuela

ORDER: PICIFORMES
FAMILY: CAPITONIDAE (Barbets—83)

☐ **Yellow-billed Barbet** *Trachyphonus purpuratus*

____ *T. p. goffinii*	Sierra Leone east to Ghana/Togo border
____ *T. p. togoensis*	Extreme e Ghana to sw Nigeria
____ *T. p. purpuratus*	SE Nigeria to Cent. African Rep., n Angola and central Zaire
____ *T. p. elgonensis*	S Sudan e Zaire east to w Kenya

☐ **Crested Barbet** *Trachyphonus vaillantii*

____ *T. v. suahelicus*	N-c Angola and Zaire to Tanzania and w Mozambique
____ *T. v. vaillantii*	S Angola, ne Namibia to Botswana, Mozambique and e S Africa

☐ **Red-and-yellow Barbet** *Trachyphonus erythrocephalus*

____ *T. e. shelleyi*	SE Ethiopia to nw and s Somalia
____ *T. e. versicolor*	NE Uganda to se Sudan, Ethiopia and n Kenya
____ *T. e. erythrocephalus*	Central Kenya to ne and n-central Tanzania

☐ **Yellow-breasted Barbet** *Trachyphonus margaritatus*

____ *T. m. margaritatus*	E Mauritania to Chad, Sudan, n Ethiopia and Eritrea
____ *T. m. somalicus*	E Ethiopia and Djibouti to n Somalia

☐ **D'Arnaud's Barbet** *Trachyphonus darnaudii*

____ *T. d. darnaudii*	SE Sudan and sw Ethiopia to ne Uganda and w-central Kenya
____ *T. d. boehmi*	S and e Ethiopia to s Somalia, e Kenya and ne Tanzania
____ *T. d. usambiro*	SW Kenya to n-central Tanzania
____ *T. d. emini*	N-central Tanzania (e to Dar es Salaam suburbs)

☐ **Gray-throated Barbet** *Gymnobucco bonapartei*

____ *G. b. bonapartei*	W Cameroon and Gabon east to sw and e-central Zaire
____ *G. b. cinereiceps*	C Afr. Rep., s Sudan to e Zaire, Angola, w Kenya and Tanzania

☐ **Sladen's Barbet** *Gymnobucco sladeni*

	Forests of central and e Zaire; recorded from sw C African Rep.

☐ **Bristle-nosed Barbet** *Gymnobucco peli*

	Sierra Leone to s Cameroon, w Zaire, Gabon and n Angola

☐ **Naked-faced Barbet** *Gymnobucco calvus*

____ *G. c. calvus*	Lowland forests of Sierra Leone to Cameroon and Gabon
____ *G. c. congicus*	W Congo (Brazzaville) and w Zaire to nw Angola)
____ *G. c. vernayi*	Uplands of w-central Angola

☐ **White-eared Barbet** *Stactolaema leucotis*

____ *S. l. kilimensis*	Locally in highlands of central Kenya to ne Tanzania
____ *S. l. leucogrammica*	Highlands of s-central Tanzania (Uluguru Mts. to Mahenge)
____ *S. l. leucotis*	Malawi and Mozambique s to Swaziland and Natal

☐ **Whyte's Barbet** *Stactolaema whytii*

____ *S. w. terminata*	Highlands of s Tanzania (Iringa region)
____ *S. w. stresemanni*	SW corner of Tanzania and adjacent ne Zambia
____ *S. w. sowerbyi*	E Zimbabwe to extreme s Malawi
____ *S. w. whytii*	S-central Tanzania to se Malawi and Mozambique
____ *S. w. angoniensis*	E Zambia to sw Malawi (west of Shire River)
____ *S. w. buttoni*	N-central Zambia (along Zaire border); probably s Zaire

☐ **Anchieta's Barbet** *Stactolaema anchietae*

____ *S. a. rex*	W-central Angola
____ *S. a. anchietae*	Highlands of s-central Angola to w Zambia border
____ *S. a. katangae*	NE Angola, s Zaire and Zambia

☐ **Green Barbet** *Stactolaema olivacea*

____ *S. o. olivacea*	SE Kenya and ne Tanzania
____ *S. o. howelli*	Cent. Tanzania (Udzungwe and Mahenge mountains)
____ *S. o. woodwardi*	SE Tanzania (Rondo Plateau) and Zululand (Ngoye Forest)
____ *S. o. rungweensis*	Highlands of sw Tanzania to n Malawi
____ *S. o. belcheri*	S Malawi (Mt. Thyolo) and n Mozambique (Mt. Namuli)

☐ **Speckled Tinkerbird** *Pogoniulus scolopaceus*

____ *P. s. scolopaceus*	Sierra Leone and se Guinea to s Nigeria
____ *P. s. stellatus*	Bioko I. (Gulf of Guinea)
____ *P. s. flavisquamatus*	S Cameroon to extreme w Kenya , n Angola and s Zaire

☐ **Green Tinkerbird** *Pogoniulus simplex*

	Coastal forests of se Kenya to Tanzania, Malawi, s Mozambique

☐ **Moustached Tinkerbird** *Pogoniulus leucomystax*

	Extreme e Uganda and c Kenya s to Malawi (w of Rift Valley)

☐ **Western Tinkerbird** *Pogoniulus coryphaeus*

____ *P. c. coryphaeus*	E Nigeria (Obudu Plateau) and adjacent sw Cameroon
____ *P. c. hildamariae*	E Zaire, sw Uganda and w Rwanda
____ *P. c. angolensis*	W-central Angola (Mt. Moco and Mombolo highlands)

☐ **Red-rumped Tinkerbird** *Pogoniulus atroflavus*

	Senegambia to w Uganda, Congo R. mouth and Ruwenzori Mts.

☐ **Yellow-throated Tinkerbird** *Pogoniulus subsulphureus*

____ *P. s. chrysopygus*	Sierra Leone and se Guinea to s Ghana
____ *P. s. flavimentum*	Togo to s Central African Republic, s Uganda and e Zaire
____ *P. s. subsulphureus*	Bioko I. (Gulf of Guinea)

☐ **Yellow-rumped Tinkerbird** *Pogoniulus bilineatus*

____ *P. b. leucolaimus*	Senegambia to s Sudan, Uganda, se Zaire and n Angola
____ *P. b. poensis*	Highlands of Bioko I. (Gulf of Guinea)
____ *P. b. mfumbiri*	Extreme sw Uganda to e Zaire, w Tanzania and Zambia
____ *P. b. jacksoni*	Highlands of e Uganda, c Kenya, Rwanda and n Tanzania
____ *P. b. fischeri*	Coastal Kenya to ne Tanzania; Zanzibar and Mafia I.
____ *P. b. bilineatus*	E Zambia and s Tanzania to Mozambique and e South Africa

☐ **Red-fronted Tinkerbird** *Pogoniulus pusillus*

____ *P. p. uropygialis*	Eritrea to n and central Ethiopia and n Somalia
____ *P. p. affinis*	SE Sudan to se Ethiopia, s Somalia, Kenya, Uganda, se Tanzania
____ *P. p. pusillus*	S Mozambique to e South Africa

☐ **Yellow-fronted Tinkerbird** *Pogoniulus chrysoconus*

____ *P. c. chrysoconus*	SW Mauritania and Senegambia to Ethiopia and nw Tanzania
____ *P. c. xanthostictus*	Highlands of central and s Ethiopia
____ *P. c. extoni*	Angola to s Zaire, Tanzania, Namibia, Botswana, s Mozambique

☐ **Yellow-spotted Barbet** *Buccanodon duchaillui*

	Sierra Leone to s Sudan, Zaire, Uganda, Rwanda and w Kenya

☐ **Hairy-breasted Barbet** *Tricholaema hirsuta*

____ *T. h. hirsuta*	Sierra Leone to s-central Nigeria
____ *T. h. flavipunctata*	SE Nigeria and Cameroon to n and central Gabon
____ *T. h. angolensis*	S Gabon to s Zaire, s Congo and nw Angola
____ *T. h. ansorgii*	SE Cameroon to e Zaire, Uganda, w Kenya and nw Tanzania

☐ **Red-fronted Barbet** *Tricholaema diademata*

_____ *T. d. diademata* — SE Sudan and n-cent. Ethiopia to se Uganda and central Kenya
_____ *T. d. massaica* — S-central Kenya to central and sw Tanzania

☐ **Miombo Barbet** *Tricholaema frontata*

S Zaire to sw Tanzania, central Angola, s Zambia and w Malawi

☐ **Pied Barbet** *Tricholaema leucomelas*

_____ *T. l. centralis* — Angola, extreme sw Zambia and w Zimbabwe to n Cape Prov.
_____ *T. l. affinis* — E Zimbabwe and sw Mozambique to e Cape Prov. and n Natal
_____ *T. l. leucomelas* — South Africa (central, southern and sw Cape Province)

☐ **Spot-flanked Barbet** *Tricholaema lacrymosa*

_____ *T. l. lacrymosa* — Extreme ne Zaire to s Sudan, n Kenya, Uganda and ne Tanzania
_____ *T. l. radcliffei* — E Zaire and s Uganda to c Kenya, ne Zambia and sw Tanzania

☐ **Black-throated Barbet** *Tricholaema melanocephala*

_____ *T. m. melanocephala* — N and central Ethiopia, Djibouti and nw Somalia
_____ *T. m. stigmatothorax* — SE Sudan, s Ethiopia and s Somalia to Kenya and Tanzania
_____ *T. m. blandi* — N-central and ne Somalia
_____ *T. m. flavibuccalis* — N-central Tanzania (Seronera region and Wembere steppes)

☐ **Banded Barbet** *Lybius undatus*

_____ *L. u. thiogaster* — Highlands of Eritrea and ne Ethiopia
_____ *L. u. undatus* — Highlands of nw to central Ethiopia
_____ *L. u. leucogenys* — Highlands of sw and s-central Ethiopia
_____ *L. u. salvadorii* — Highlands of se Ethiopia

☐ **Vieillot's Barbet** *Lybius vieilloti*

_____ *L. v. buchanani* — S Mauritania and Mali to Niger, n Nigeria and s Chad
_____ *L. v. rubescens* — Sahel fringe from Gambia and Sierra Leone to Zaire and Nigeria
_____ *L. v. vieilloti* — Central Sudan (Khartoum) to ne Zaire and w Ethiopia

☐ **White-headed Barbet** *Lybius leucocephalus*

_____ *L. l. leucocephalus* — E Cent. African Rep. and s Sudan to Kenya and nw Tanzania
_____ *L. l. adamauae* — N Nigeria to s Chad, w Central African Republic and nw Zaire
_____ *L. l. albicauda* — SW and s Kenya to n Tanzania
_____ *L. l. senex* — Highlands of central and s-central Kenya
_____ *L. l. lynesi* — Central Tanzania
_____ *L. l. leucogaster* — Highlands of sw Angola

☐ **Chaplin's Barbet** *Lybius chaplini*

Miombo woodlands of s-central Zambia

☐ **Red-faced Barbet** *Lybius rubrifacies*

Acacia belt of sw Uganda to e Rwanda and nw Tanzania

☐ **Black-billed Barbet** *Lybius guifsobalito*

E Sudan to w Ethiopia, ne Zaire, Uganda, w Kenya, n Tanzania

☐ **Black-collared Barbet** *Lybius torquatus*

_____ *L. t. pumilio* — E Zaire to n Zambia, e Rwanda, w Tanzania, nw Mozambique
_____ *L. t. irroratus* — Coastal e Kenya (Lamu and Tana River) to central Tanzania
_____ *L. t. zombae* — SE Tanzania to s-central Malawi and central Mozambique
_____ *L. t. congicus* — N Angola to nw Zambia and s-central Zaire
_____ *L. t. vivacens* — S-central and w Mozambique to e Zimbabwe and s Malawi
_____ *L. t. bocagei* — S Angola to n Namibia, n Botswana, sw Zambia, w Zimbabwe
_____ *L. t. torquatus* — SE Botswana to e Cape Province, Swaziland and Transvaal

☐ **Brown-breasted Barbet** *Lybius melanopterus*

Savanna of s Somalia to Kenya, se Malawi and n Mozambique

☐ **Black-backed Barbet** *Lybius minor*

_____ *L. m. minor* — S Gabon and w Zaire to w Angola
_____ *L. m. macclounii* — S-central Zaire to c Angola, n Zambia, Malawi and w Tanzania

☐ **Double-toothed Barbet** *Lybius bidentatus*
- _____ *L. b. bidentatus* — Guinea-Bissau, Guinea, Sierra Leone and Liberia to c Cameroon
- _____ *L. b. aequatorialis* — E Cameroon to Ethiopia, nw Angola, n Zaire and nw Tanzania

☐ **Bearded Barbet** *Lybius dubius* — W Sahel (n Senegambia to nw Central African Republic)

☐ **Black-breasted Barbet** *Lybius rolleti* — S Chad to ne Cent. African Rep., s Sudan, ne Zaire and n Uganda

☐ **Brown Barbet** *Calorhamphus fuliginosus*
- _____ *C. f. detersus* — S Burma and adjacent peninsular Thailand (s to Trang)
- _____ *C. f. hayii* — Malay Peninsula and Sumatra
- _____ *C. f. tertius* — N Borneo
- _____ *C. f. fuliginosus* — Borneo (except northern part)

☐ **Fire-tufted Barbet** *Psilopogon pyrolophus* — Montane forests of Malay Pen. and Sumatra; recorded s Thailand

☐ **Great Barbet** *Megalaima virens*
- _____ *M. v. marshallorum* — NE Pakistan and nw India to w Nepal
- _____ *M. v. magnifica* — E Nepal to central Assam
- _____ *M. v. clamator (mayri)* — NE Assam, n Burma, sw China (w Yunnan) and nw Thailand
- _____ *M. v. virens (indochinensis)* — C Burma and se China to n-c Thailand and n Vietnam

☐ **Red-vented Barbet** *Megalaima lagrandieri*
- _____ *M. l. rothschildi* — N Vietnam and n Laos
- _____ *M. l. lagrandieri* — S Laos and s Vietnam

☐ **Brown-headed Barbet** *Megalaima zeylanica*
- _____ *M. z. inornata* — W-central and sw India (Maharashtra, Goa and Karnataka)
- _____ *M. z. caniceps (kangrae)* — SW Nepal and n India
- _____ *M. z. zeylanica* — S India (Kerala and s Tamil Nandu) and Sri Lanka

☐ **Lineated Barbet** *Megalaima lineata*
- _____ *M. l. hodgsoni* — NW India and Nepal to s China, Burma, Vietnam, n Malay Pen.
- _____ *M. l. lineata* — Java and Bali

☐ **White-cheeked Barbet** *Megalaima viridis* — Lowlands and foothills of w-central and sw India

☐ **Green-eared Barbet** *Megalaima faiostricta*
- _____ *M. f. praetermissa* — N Thailand, Laos, n Vietnam and s China
- _____ *M. f. faiostricta* — Central and s Thailand, Cambodia and s Vietnam

☐ **Brown-throated Barbet** *Megalaima corvina* — Montane forests of w Java

☐ **Gold-whiskered Barbet** *Megalaima chrysopogon*
- _____ *M. c. laeta* — SW Thailand and Malay Peninsula; possibly extreme s Burma
- _____ *M. c. chrysopogon* — Sumatra
- _____ *M. c. chrysopsis* — Borneo

☐ **Red-crowned Barbet** *Megalaima rafflesii* — S Burma and pen. Thailand Sumatra, Borneo, Bangka, Belitung

☐ **Red-throated Barbet** *Megalaima mystacophanos*
- _____ *M. m. mystacophanos (humii)* — S Burma, s pen. Thailand, Malay Peninsula, Sumatra and Borneo
- _____ *M. m. ampala* — Batu Islands (off w Sumatra)

☐ **Black-banded Barbet** *Megalaima javensis* — Lowlands of Java and Bali

☐ **Yellow-fronted Barbet** *Megalaima flavifrons* — Sri Lanka

☐ **Golden-throated Barbet** *Megalaima franklinii*
- _____ *M. f. franklinii (tonkinensis)* — Nepal to se Tibet, n Burma, s China, n Thailand and n Vietnam
- _____ *M. f. ramsayi (trangensis, minor)* — Mts. of central Burma and nw Thailand s through Malay Pen.
- _____ *M. f. auricularis* — Mts. of e Laos and s Vietnam

☐ **Black-browed Barbet** *Megalaima oorti*
- _____ *M. o. oorti* — Montane forests of Malay Peninsula and Sumatra
- _____ *M. o. annamensis* — S and e Laos and Vietnam
- _____ *M. o. sini* — Montane forests of s China (Guangxi)
- _____ *M. o. faber* — Hainan I. (s China)
- _____ *M. o. nuchalis* — Taiwan

☐ **Blue-throated Barbet** *Megalaima asiatica*
- _____ *M. a. asiatica* — NE Pakistan to central Burma and s China (w Yunnan)
- _____ *M. a. davisoni* — SE Burma to s China (s Yunnan) and central Vietnam
- _____ *M. a. chersonesus* — Mountains of e peninsular Thailand

☐ **Moustached Barbet** *Megalaima incognita*
- _____ *M. i. incognita* — S Burma and w Thailand
- _____ *M. i. elbeli* — NE and e Thailand, n Laos and n Vietnam
- _____ *M. i. euroa* — SE Thailand, Cambodia, s Laos and s Vietnam

☐ **Mountain Barbet** *Megalaima monticola* — Mountains of n Borneo

☐ **Yellow-crowned Barbet** *Megalaima henricii*
- _____ *M. h. henricii* — S peninsular Thailand south to Malay Peninsula and Sumatra
- _____ *M. h. brachyrhyncha* — Borneo

☐ **Flame-fronted Barbet** *Megalaima armillaris*
- _____ *M. a. armillaris* — Java
- _____ *M. a. baliensis* — Bali

☐ **Golden-naped Barbet** *Megalaima pulcherrima* — Mountains of n Borneo

☐ **Blue-eared Barbet** *Megalaima australis*
- _____ *M. a. cyanotis* — SE Nepal to Bangladesh, ne India, s China, Burma, Malay Pen.
- _____ *M. a. orientalis* — E Thailand, Cambodia, Laos and Vietnam
- _____ *M. a. duvaucelii* — Malay Peninsula, Sumatra, Bangka I. and Borneo
- _____ *M. a. gigantorhina* — Nias I. (off nw Sumatra)
- _____ *M. a. tanamassae* — Batu Islands (off w Sumatra)
- _____ *M. a. australis (hebereri)* — Java and Bali

☐ **Bornean Barbet** *Megalaima eximia*
- _____ *M. e. eximia* — Mountains of n Borneo
- _____ *M. e. cyanea* — N Borneo (Mt. Kinabalu)

☐ **Crimson-fronted Barbet** *Megalaima rubricapillus*
- _____ *M. r. malabarica* — SW India (Goa south to Kerala and w Tamil Nadu)
- _____ *M. r. rubricapillus* — Lowlands and foothills of Sri Lanka

☐ **Coppersmith Barbet** *Megalaima haemacephala*
- _____ *M. h. indica (confusa, lutea)* — NE Pakistan to s China, s to Sri Lanka, Singapore and Vietnam
- _____ *M. h. delica* — Sumatra
- _____ *M. h. rosea* — Java and Bali
- _____ *M. h. haemacephala* — N Philippines (Luzon and Mindoro)
- _____ *M. h. intermedia* — Philippines (Guimaras, Negros, Panay, Calagayan, Pan de Azucar)
- _____ *M. h. celestinoi* — Philippines (Samar, Leyte, Biliran and Catanduanes)
- _____ *M. h. mindanensis* — Mindanao (s Philippines)
- _____ *M. h. cebuensis* — Cebu (central Philippines)
- _____ *M. h. homochroa* — Philippines (Masbate, Romblon and Tablas); possibly Palawan

☐ **Scarlet-crowned Barbet** *Capito aurovirens*

SE Colombia to e Peru and w Amazonian Brazil (e to Río Negro)

☐ **Scarlet-banded Barbet** *Capito wallacei*

N-central Peru e of Andes nw of Contamana at 1350-1500 m

☐ **Spot-crowned Barbet** *Capito maculicoronatus*

____ *C. m. maculicoronatus* — W Panama (Veraguas) east to Canal Zone)
____ *C. m. rubrilateralis (pirrensis, melas)* — E Panama to nw Colombia (east to Antioquia and south to Valle)

☐ **Orange-fronted Barbet** *Capito squamatus*

Extreme sw Colombia (se Nariño) and w Ecuador (s to El Oro)

☐ **White-mantled Barbet** *Capito hypoleucus*

____ *C. h. hypoleucus* — Central Andes of nw Colombia (Bolivar to Antioquia)
____ *C. h. carrikeri* — N-central Colombia (Botero area of Río Porce in Antioquia)
____ *C. h. extinctus* — Colombia (Magdalena Valley in Caldas, Cundinamarca, Tolima)

☐ **Black-girdled Barbet** *Capito dayi*

W Amazonian Brazil south to e Bolivia and w-c Mato Grosso

☐ **Brown-chested Barbet** *Capito brunneipectus*

Brazil s of Amazon between lower Río Madeira and Río Tapajós

☐ **Black-spotted Barbet** *Capito niger*

____ *C. n. aurantiicinctus (intermedius)* — Venezuela (upper Orinoco region, w Bolivar and Amazonas)
____ *C. n. niger* — E Venezuela, the Guianas, and ne Brazil n of the Amazon
____ *C. n. orosae* — E Peru (Río Orosa to Rio Javari, s to extreme w Brazil in Acre)
____ *C. n. amazonicus (novaolindae, arimae)* — W Brazil s of Rio Solimões from upper Rio Jurúa e to Rio Purús
____ *C. n. nitidior (transilens)* — Extreme e Colombia and s Venezuela to Peru/Brazil border
____ *C. n. hypochondriacus* — N Brazil (Roraima along Rio Branco to Rio Negro and Solimões)

☐ **Gilded Barbet** *Capito auratus*

____ *C. a. punctatus* — S-central Colombia along lower E Andes to central Peru (Junín)
____ *C. a. auratus* — NE Peru (mouth of Río Napo s along Amazon and Río Ucayali)
____ *C. a. insperatus* — SE Peru, n Bolivia and w Brazil (Calama)

☐ **Five-colored Barbet** *Capito quinticolor*

W Colombia (Chocó) to extreme nw Ecuador (Esmeraldas)

☐ **Lemon-throated Barbet** *Eubucco richardsoni*

____ *E. r. richardsoni (granadensis)* — SE Colombia, e Ecuador and n Peru (west of Iquitos)
____ *E. r. nigriceps* — NE Peru (lower Río Napo to extreme nw Brazil in w Amazonas)
____ *E. r. aurantiicollis (coccineus)* — E Peru s of R. Marañón and w Brazil s of Amazon to nw Bolivia
____ *E. r. purusianus* — W Brazil s of Amazon (Rio Juruá to upper Rio Madeira)

☐ **Red-headed Barbet** *Eubucco bourcierii*

____ *E. b. salvini* — Highlands of Costa Rica and w Panama
____ *E. b. anomalus* — E Panama and possibly adjacent nw Colombia
____ *E. b. occidentalis* — Western Andes of Colombia
____ *E. b. bourcierii* — Andes of central Colombia to w Venezuela
____ *E. b. aequatorialis* — Coastal mts. and Western Andes of Ecuador
____ *E. b. orientalis* — E Andes of Ecuador and n Peru (Cajamarca and Amazonas)

☐ **Scarlet-hooded Barbet** *Eubucco tucinkae*

Humid e Peru, n Bolivia and w Amazonian Brazil

☐ **Versicolored Barbet** *Eubucco versicolor*

____ *E. v. steerii* — Andes of n Peru (Amazonas to n Huánuco)
____ *E. v. glaucogularis* — Andes of central Peru (e Huánuco to n Cusco)
____ *E. v. versicolor* — Andes of s Peru (Cusco and Puno) to n-c Bolivia (Cochabamba)

☐ **Prong-billed Barbet** *Semnornis frantzii*

Humid montane forests of Costa Rica and w Panama

☐ **Toucan Barbet** *Semnornis ramphastinus*

____ *S. r. caucae* — W Andes of sw Colombia (Valle to Nariño)
____ *S. r. ramphastinus* — Andes of nw and w-central Ecuador

ORDER: PICIFORMES
FAMILY: RAMPHASTIDAE (Toucans—40)

☐ **Wagler's Toucanet** *Aulacorhynchus wagleri*

Sierra Madre del Sur of sw Mexico (Guerrero and sw Oaxaca)

☐ **Emerald Toucanet** *Aulacorhynchus prasinus*

____ *A. p. prasinus* — SE Mexico (Veracruz, adj. San Luis Potosí and Oaxaca)
____ *A. p. warneri* — Mts. of se Mexico (Sierra de Los Tuxtlas in s Veracruz)
____ *A. p. virescens* — SE Mexico (Chiapas) to Honduras and Nicaragua
____ *A. p. stenorhabdus* — Subtropical s Mexico to w Guatemala and n El Salvador
____ *A. p. chiapensis* — Mts. of extreme s Mexico (Mt. Ovando, Chiapas)
____ *A. p. volcanius* — Eastern El Salvador (Volcán San Miguel)

☐ **Blue-throated Toucanet** *Aulacorhynchus caeruleogularis*

____ *A. c. maxillaris* — Highlands of Costa Rica and w Panama
____ *A. c. caeruleogularis* — Highlands of e-central Panama (Chiriquí and Veraguas)

☐ **Violet-throated Toucanet** *Aulacorhynchus cognatus*

Mts. of e Panama (Darién) and adjacent Colombia

☐ **Santa Marta Toucanet** *Aulacorhynchus lautus*

Santa Marta Mountains (ne Colombia)

☐ **Andean Toucanet** *Aulacorhynchus albivitta*

____ *A. a. albivitta* — E and Central Andes of Colombia, e Ecuador and w Venezuela
____ *A. a. griseigularis* — N end of W Andes and w slope of Central Andes of Colombia
____ *A. a. phaeolaemus (petax)* — Subtropical W Andes of Colombia

☐ **Black-throated Toucanet** *Aulacorhynchus atrogularis*

____ *A. a. atrogularis* — Humid Andes of e Peru and n Bolivia
____ *A. a. dimidiatus* — Subtropical Andes of n Peru
____ *A. a. cyanolaemus* — Subtropical Andes of se Ecuador and n Peru

☐ **Groove-billed Toucanet** *Aulacorhynchus sulcatus*

____ *A. s. calorhynchus* — Santa Marta Mts. (ne Colombia) and Sierra de Perijá (w Venezuela)
____ *A. s. sulcatus* — Coastal cordillera of n Venezuela (Falcón to Miranda)
____ *A. s. erythrognathus* — Mountains of ne Venezuela

☐ **Chestnut-tipped Toucanet** *Aulacorhynchus derbianus*

____ *A. d. derbianus (nigrirostris)* — Extreme s Colombia along e slope of Andes to Bolivia
____ *A. d. duidae* — Mts. of s Venezuela (Amazonas and w Bolivar) and adj. n Brazil
____ *A. d. whitelianus* — Mts. of s Venezuela (se Bolivar) and n Guyans
____ *A. d. osgoodi* — S Guyana (Acary Mts.) and Suriname (Wilhelmina Mts.)

☐ **Crimson-rumped Toucanet** *Aulacorhynchus haematopygus*

____ *A. h. sexnotatus* — Extreme sw Colombia (Nariño) and Andes of w Ecuador
____ *A. h. haematopygus* — Andes of Colombia and Sierra de Perijá (w Venezuela)

☐ **Yellow-browed Toucanet** *Aulacorhynchus huallagae*

Locally in mountains of n Peru (San Martín and La Libertad)

☐ **Blue-banded Toucanet** *Aulacorhynchus coeruleicinctis*

Andes of e Peru (Huánuco) to s Bolivia (Santa Cruz, Chuquisaca)

☐ **Guianan Toucanet** *Selenidera culik*

Extreme se Venezuela to the Guianas and n Brazil n of Amazon

☐ **Tawny-tufted Toucanet** *Selenidera nattereri*

Extreme e Colombia and s Venezuela to nw Brazil

☐ **Golden-collared Toucanet** *Selenidera reinwardtii*

____ *S. r. reinwardtii* — SE Colombia (along Brazil border) to e Ecuador and ne Peru
____ *S. r. langsdorffii* — Tropical e Peru to nw Bolivia and w Amazonian Brazil

☐ **Gould's Toucanet** *Selenidera gouldii*

Central and e Brazil s of Amazon to e Bolivia and Mato Grosso

☐ **Spot-billed Toucanet** *Selenidera maculirostris*

Atlantic forests of se Brazil to e Paraguay and ne Argentina

☐ **Yellow-eared Toucanet** *Selenidera spectabilis*

Humid forests of Honduras to extreme nw Ecuador (Esmeraldas)

☐ **Gray-breasted Mountain-Toucan** *Andigena hypoglauca*

____ *A. h. hypoglauca*   Central Andes of Colombia to e Ecuador
____ *A. h. lateralis*   Andes of e Ecuador and e Peru

☐ **Hooded Mountain-Toucan** *Andigena cucullata*

Andes of se Peru (Puno) and w Bolivia (La Paz, Cochabamba)

☐ **Plate-billed Mountain-Toucan** *Andigena laminirostris*

Andes of sw Colombia to s Ecuador

☐ **Black-billed Mountain-Toucan** *Andigena nigrirostris*

____ *A. n. occidentalis*   Western Andes of Colombia
____ *A. n. spilorhyncha*   C Andes of s Colombia and ne Ecuador; recorded from n Peru
____ *A. n. nigrirostris*   Eastern Andes of Colombia and w Venezuela

☐ **Saffron Toucanet** *Baillonius bailloni*

Humid forests of se Brazil to ne Argentina and e Paraguay

☐ **Green Araçari** *Pteroglossus viridis*

E Venezuela to the Guianas and n Brazil

☐ **Lettered Araçari** *Pteroglossus inscriptus*

____ *P. i. humboldti*   SE Colombia and w Brazil to n-central Bolivia
____ *P. i. inscriptus*   N-central Brazil s of the Amazon to e Bolivia and Mato Grosso

☐ **Red-necked Araçari** *Pteroglossus bitorquatus*

____ *P. b. sturmii*   N-central Brazil s of Amazon to e Bolivia and Mato Grosso
____ *P. b. reichenowi*   Brazil s of Amazon (between Rio Tapajós and Rio Tocantins)
____ *P. b. bitorquatus*   NE Brazil s of Amazon e of Rio Tocantins to Maranhão

☐ **Ivory-billed Araçari** *Pteroglossus azara*

____ *P. a. flavirostris*   SE Colombia to Ecuador, ne Peru, s Venezuela and nw Brazil
____ *P. a. azara*   W Brazil in Amazonas (between Rio Negro and Rio Solimões)
____ *P. a. mariae*   E Peru and w Brazil s of Amazon to n-central Bolivia

☐ **Black-necked Araçari** *Pteroglossus aracari*

____ *P. a. atricollis (roraimae)*   E Venezuela and Guianas to n Brazil (s to Amazon)
____ *P. a. aracari*   Disjunct populations in n-central, east and se Brazil
____ *P. a. vergens*   E Brazil (Minas Gerais, São Paulo, Paraná and Santa Catarina)

☐ **Chestnut-eared Araçari** *Pteroglossus castanotis*

____ *P. c. castanotis*   S and e Colombia to e Ecuador, se Peru and nw Brazil
____ *P. c. australis*   E Bolivia to w and se Brazil, e Paraguay and ne Argentina

☐ **Many-banded Araçari** *Pteroglossus pluricinctus*

NE Colombia to s Venezuela, ne Peru and nw Amazonian Brazil

☐ **Collared Araçari** *Pteroglossus torquatus*

____ *P. t. torquatus (esperanzae)*   Tropical s Mexico to nw Colombia
____ *P. t. erythrozonus*   Yucatán Peninsula to Belize and Petén of ne Guatemala
____ *P. t. nuchalis (pectoralis)*   NE Colombia (Santa Marta Mts.) and mts. of n Venezuela
____ *P. t. sanguineus*   E Panama and n Colombia south to nw Ecuador
____ *P. t. erythropygius*   W Ecuador; recently recorded in extreme n Peru (Tumbes)

☐ **Fiery-billed Araçari** *Pteroglossus frantzii*

Pacific slope of Costa Rica and w Panama (w Chiriquí)

☐ **Curl-crested Araçari** *Pteroglossus beauharnaesii*

E Peru to n Bolivia and w Amazonian Brazil s of the Amazon

☐ **Red-breasted Toucan** *Ramphastos dicolorus*

Humid forests of se Brazil to e Paraguay and ne Argentina

☐ **Keel-billed Toucan** *Ramphastos sulfuratus*

____	*R. s. sulfuratus*	Lowlands of se Mexico to Belize, Guatemala and Honduras
____	*R. s. brevicarinatus*	SE Guatemala to n Colombia and extreme nw Venezuela

☐ **Choco Toucan** *Ramphastos brevis*

Pacific slope of w Colombia to sw Ecuador

☐ **Channel-billed Toucan** *Ramphastos vitellinus*

____	*R. v. citreolaemus*	Humid forests of n Colombia and nw Venezuela
____	*R. v. culminatus*	Lowlands of w Venezuela to n Bolivia and w Amaz Brazil
____	*R. v. vitellinus (aurantiirostris)*	Venezuela to the Guianas and Brazil n of the Amazon; Trinidad
____	*R. v. ariel*	Tropical Brazil south of the Amazon
____	*R. v. pintoi*	SE Brazil (s Goiás and w São Paulo)
____	*R. v. theresae*	NE Brazil (Piauí and Maranhão)

☐ **Toco Toucan** *Ramphastos toco*

____	*R. t. toco*	The Guianas and ne Brazil; recently recorded in se Peru
____	*R. t. albogularis*	E and s Brazil to Paraguay, n Bolivia and n Argentina

☐ **Red-billed Toucan** *Ramphastos tucanus*

____	*R. t. cuvieri*	SE Colombia to Venezuela, w Amazonian Brazil and n Bolivia
____	*R. t. tucanus*	E Venezuela to the Guianas and n Brazil
____	*R. t. inca*	N and central Bolivia

☐ **Black-mandibled Toucan** *Ramphastos ambiguus*

____	*R. a. swainsonii*	Humid forests of n Honduras to w Ecuador
____	*R. a. abbreviatus*	E slope of Andes of Colombia to w Venezuela and e Peru
____	*R. a. ambiguus*	Northern section of upper Amazon basin

ORDER: PICIFORMES
FAMILY: INDICATORIDAE (Honeyguides—17)

☐ **Cassin's Honeyguide** *Prodotiscus insignis*

____	*P. i. flavodorsalis*	Disjunct from Sierra Leone to sw Nigeria
____	*P. i. insignis*	SE Nigeria to n Angola, Zaire, s Sudan, Uganda and cent. Kenya

☐ **Green-backed Honeyguide** *Prodotiscus zambesiae*

____	*P. z. zambesiae*	C Angola to s Zaire, s Tanzania, Namibia and Mozambique
____	*P. z. ellenbecki*	S Ethiopia to Kenya and n Tanzania

☐ **Wahlberg's Honeyguide** *Prodotiscus regulus*

____	*P. r. regulus*	E-c Sudan to s Zaire, Angola, ne Namibia and e Cape Province
____	*P. r. camerunensis*	Disjunct from Guinea to Cameroon and Central African Republic

☐ **Zenker's Honeyguide** *Melignomon zenkeri*

S Cameroon to ne Zaire, C African Rep., Gabon and sw Uganda

☐ **Yellow-footed Honeyguide** *Melignomon eisentrauti*

Sierra Leone to Liberia, sw Ghana and sw Cameroon

☐ **Dwarf Honeyguide** *Indicator pumilio*

Mts. of e Zaire, w Rwanda, w Burundi and sw Uganda

☐ **Willcock's Honeyguide** *Indicator willcocksi*

____	*I. w. ansorgei*	Locally in humid forests of Guinea-Bissau
____	*I. w. willcocksi*	Sierra Leone to s Nigeria, s Cameroon, Zaire and w Uganda
____	*I. w. hutsoni*	N-central Nigeria to s Chad and sw Sudan

☐ **Pallid Honeyguide** *Indicator meliphilus*

____	*I. m. meliphilus*	E Uganda and central Kenya to central Tanzania and sw Sudan
____	*I. m. angolensis*	C Angola to s Zaire, s Malawi, Zimbabwe and c Mozambique

☐ **Least Honeyguide** *Indicator exilis*

____	*I. e. exilis*	Senegal to e Cent. African Rep., ne Zaire, n Angola, nw Zambia

____ *I. e. poensis*	Humid forests of Bioko I. (Gulf of Guinea)
____ *I. e. pachyrhynchus*	SW Sudan and e Zaire to w Kenya, nw Tanzania and Rwanda

☐ **Thick-billed Honeyguide** *Indicator conirostris*

____ *I. c. ussheri*	Disjunct from se Sierra Leone to s Ghana
____ *I. c. conirostris*	S Nigeria to w Kenya, Zaire and n Angola

☐ **Lesser Honeyguide** *Indicator minor*

____ *I. m. senegalensis*	Senegambia and Guinea to n Cameroon, Chad and w Sudan
____ *I. m. riggenbachi*	C Cameroon to sw Sudan, w Uganda, Burundi and ne Zaire
____ *I. m. diadematus*	Central Sudan to n Somalia
____ *I. m. damarensis*	S Angola and n Namibia
____ *I. m. teitensis*	SE Sudan to Somalia, ne Namibia, Zimbabwe and c Mozambique
____ *I. m. minor*	S Namibia, se Botswana and s Mozambique to South Africa

☐ **Spotted Honeyguide** *Indicator maculatus*

____ *I. m. maculatus*	Gambia to Nigeria
____ *I. m. stictithorax*	S Cameroon to sw Sudan, n Angola, e Zaire and sw Uganda

☐ **Scaly-throated Honeyguide** *Indicator variegatus*

Savanna and riparian woodlands of east and southern Africa

☐ **Yellow-rumped Honeyguide** *Indicator xanthonotus*

____ *I. x. xanthonotus (fulvus)*	NE Pakistan and Himalayas of nw India to extreme w Nepal
____ *I. x. radcliffi*	Nepal to ne India, se Tibet and n Burma

☐ **Malaysian Honeyguide** *Indicator archipelagicus*

S pen. Thailand, adj. Burma, Malay Pen., Sumatra and Borneo

☐ **Greater Honeyguide** *Indicator indicator*

Widespread Africa south of the Sahara

☐ **Lyre-tailed Honeyguide** *Melichneutes robustus*

Guinea to Cent. African Rep., s Uganda, e Zaire and nw Angola

ORDER: PICIFORMES
FAMILY: PICIDAE (Woodpeckers and Allies—219)

☐ **Eurasian Wryneck** *Jynx torquilla*

____ *J. t. torquilla*	W Europe to SE Asia and Japan; winters to Africa and s Asia
____ *J. t. tschusii*	Italy, Sardinia, Corsica and e Adriatic coast; winters in Africa
____ *J. t. mauretanica*	Northwest Africa
____ *J. t. himalayana*	NW Himalayas; winters to s India at lower elevations

☐ **Rufous-necked Wryneck** *Jynx ruficollis*

____ *J. r. ruficollis*	SE Gabon to s Zaire, e Uganda, s Sudan, n Angola and e S Africa
____ *J. r. pulchricollis*	SE Nigeria and Cameroon to nw Zaire, s Sudan and nw Uganda
____ *J. r. aequatorialis*	Highlands of Ethiopia

☐ **Speckled Piculet** *Picumnus innominatus*

____ *P. i. innominatus*	NE Afghanistan and n Pakistan to n India, Nepal and se Tibet
____ *P. i. malayorum*	Pen. and ne India to sw China, Indochina, Sumatra and n Borneo
____ *P. i. chinensis*	Central, e and s China (from Sichuan to Jiangsu)

☐ **Bar-breasted Piculet** *Picumnus aurifrons*

____ *P. a. aurifrons*	N Mato Grosso of Brazil (upper Rio Madeira to Rio Tapajós)
____ *P. a. transfasciatus*	E-central Brazil (Rio Tapajós to Rio Tocantins)
____ *P. a. borbae*	Central Brazil (lower Rio Tapajós to lower Rio Madeira)
____ *P. a. wallacii*	Brazil (lower Rio Madeira to lower Rio Purús)
____ *P. a. purusianus*	W Brazil (upper Rio Purús)
____ *P. a. flavifrons*	NE Peru and w Brazil (along Rio Solimões)
____ *P. a. juruanus*	E Peru to w Brazil (upper Rio Juruá)

☐ **Orinoco Piculet** *Picumnus pumilus*

E Colombia, adjacent s Venezuela (s Amazonas) and nw Brazil

☐ **Lafresnaye's Piculet** *Picumnus lafresnayi*
____ *P. l. lafresnayi*
____ *P. l. punctifrons*
____ *P. l. taczanowskii*
____ *P. l. pusillus*

Tropical se Colombia to e Ecuador and n Peru
Tropical e Peru
NE and n-central Peru (Huambo-Inayabamba-Huánuco region)
N-central Brazil (middle Rio Amazon east to Rio Negro)

☐ **Golden-spangled Piculet** *Picumnus exilis*
____ *P. e. clarus*
____ *P. e. undulatus*
____ *P. e. buffoni*
____ *P. e. pernambucensis*
____ *P. e. alegriae*
____ *P. e. exilis*

E Venezuela (e Bolívar and Delta Amacuro)
E Colombia to se Venezuela, n Brazil (Roraima) and w Guyana
E Guyana to ne Brazil (Amapá)
Coastal e Brazil (Pernambuco and Alagoas)
Coastal forests of ne Brazil (ne Pará and nw Maranhão)
Coastal e Brazil (Bahia to Espírito Santo)

☐ **Black-spotted Piculet** *Picumnus nigropunctatus*

NE Venezuela (Sucre to Monagas and Delta Amacuro)

☐ **Ecuadorian Piculet** *Picumnus sclateri*
____ *P. s. parvistriatus*
____ *P. s. sclateri*
____ *P. s. porcullae*

Arid scrub of w Ecuador (Manabi to Guayas)
Arid scrub of sw Ecuador (El Oro, Loja) and extreme nw Peru
Arid scrub of nw Peru (central Piura to n Lambayeque)

☐ **Scaled Piculet** *Picumnus squamulatus*
____ *P. s. roehli*
____ *P. s. squamulatus*
____ *P. s. lovejoyi*
____ *P. s. apurensis*
____ *P. s. obsoletus*

NE Colombia (Boyacá and Santa Marta region) to n Venezuela
NE and central Colombia (Arauca to Huila and ne Meta)
Extreme nw Venezuela (nw Zulia)
N-c Venezuela (Apure, Guárico and Anzoátegui)
Extreme ne Venezuela (e Sucre)

☐ **White-bellied Piculet** *Picumnus spilogaster*
____ *P. s. orinocensis*
____ *P. s. spilogaster*
____ *P. s. pallidus*

Central Venezuela (se Apure to Delta Amacuro)
N Guianas and n Brazil (Roraima)
NE Brazil (Belém area of e Pará)

☐ **Guianan Piculet** *Picumnus minutissimus*

Coastal lowlands from Guyana to Suriname and French Guiana

☐ **Spotted Piculet** *Picumnus pygmaeus*

E Brazil (c Maranhão and Piauí to s Bahia and n Minas Gerais)

☐ **Speckle-chested Piculet** *Picumnus steindachneri*

Andes of ne Peru in se Amazonas and nw San Martín

☐ **Varzea Piculet** *Picumnus varzeae*

Amazonian Brazil (lower Rio Madeira to extreme w Pará)

☐ **White-barred Piculet** *Picumnus cirratus*
____ *P. c. macconnelli*
____ *P. c. confusus*
____ *P. c. cirratus*
____ *P. c. pilcomayensis*
____ *P. c. tucumanus*
____ *P. c. thamnophiloides*

NE Brazil (e Amazonian basin west to lower Rio Tapajós)
SW Guyana, extreme n Brazil (e Roraima) and French Guiana
SE Brazil (Minas Gerais to Paraná) to e Paraguay
SE Bolivia and Paraguay to n Argentina
N Argentina (w Salta to La Rioja)
Andes of se Bolivia (Chuquisaca) to nw Argentina (n Salta)

☐ **Ocellated Piculet** *Picumnus dorbignyanus*
____ *P. d. jelskii*
____ *P. d. dorbignyanus*

Andes of e Peru
Andes of Bolivia and extreme nw Argentina

☐ **Ochre-collared Piculet** *Picumnus temminckii*

Lowlands of se Brazil to ne Argentina and e Paraguay

☐ **White-wedged Piculet** *Picumnus albosquamatus*
____ *P. a. albosquamatus*
____ *P. a. guttifer*

N Bolivia to sw Brazil (Mato Grosso) and adjacent n Paraguay
Brazil (e Mato Grosso to Pará, Maranhão, Goiás and Minas Gerais)

245

☐ **Rusty-necked Piculet** *Picumnus fuscus*

☐ **Rufous-breasted Piculet** *Picumnus rufiventris*

NE Bolivia (Beni) and w-central Brazil (extreme w Mato Grosso)

_____ *P. r. rufiventris*
_____ *P. r. grandis*
_____ *P. r. brunneifrons*

E Colombia to e Ecuador, ne Peru and w Amazonian Brazil
E Peru (Huánuco and Junín) and adjacent w Amazonian Brazil
N Bolivia (Pando, Beni and Cochabamba)

☐ **Tawny Piculet** *Picumnus fulvescens*

E Brazil (e Piauí to s Ceará, Paraíba, Pernambuco and Alagoas)

☐ **Ochraceous Piculet** *Picumnus limae*

Lowlands of e Brazil (n-central Ceará)

☐ **Mottled Piculet** *Picumnus nebulosus*

SE Brazil to ne Argentina, Uruguay and (?) e Paraguay

☐ **Plain-breasted Piculet** *Picumnus castelnau*

E slope of Andes of se Colombia and ne Peru

☐ **Fine-barred Piculet** *Picumnus subtilis*

Foothills of e Peru (Loreto to Cusco and Madre de Dios)

☐ **Olivaceous Piculet** *Picumnus olivaceus*

_____ *P. o. dimotus*
_____ *P. o. flavotinctus*
_____ *P. o. olivaceus*
_____ *P. o. harterti*
_____ *P. o. eisenmanni*
_____ *P. o. tachirensis*

Lowlands of e Guatemala, n Honduras and e Nicaragua
Costa Rica to Panama and extreme nw Colombia (Chocó)
W Colombia (Sucre) s in Andes to Cauca Valley and e to Huila
SW Colombia and w Ecuador; recorded from nw Peru (Tumbes)
W Venezuela (Sierra de Perijá) and adjacent Colombia
E Andes of Colombia and adjacent w Venezuela (sw Táchira)

☐ **Grayish Piculet** *Picumnus granadensis*

_____ *P. g. antioquensis*
_____ *P. g. granadensis*

W Andes of Colombia (Antioquia to upper Río San Juan)
N Colombia (central Cauca Valley to upper Patía Valley)

☐ **Chestnut Piculet** *Picumnus cinnamomeus*

_____ *P. c. cinnamomeus*
_____ *P. c. perijanus*
_____ *P. c. persaturatus*
_____ *P. c. venezuelensis*

Coastal n Colombia, s locally to Cauca and Magdalena valleys
NW Venezuela (n portions of Lake Maracaibo basin)
N-central Colombia (Serranía de San Jerónimo in Bolívar)
W Venezuela (s and e shores of Lake Maracaibo)

☐ **African Piculet** *Sasia africana*

S Cameroon to e Zaire, sw Uganda and nw Angola

☐ **Rufous Piculet** *Sasia abnormis*

_____ *S. a. abnormis*
_____ *S. a. magnirostris*

S Burma and adj. sw Thailand to Sumatra, Borneo and Java
Nias I. (off nw Sumatra)

☐ **White-browed Piculet** *Sasia ochracea*

_____ *S. o. ochracea*
_____ *S. o. reichenowi*
_____ *S. o. kinneari*

N India (Uttar Pradesh) and c Nepal to Thailand and Vietnam
S Burma and adjacent sw Thailand south to Isthmus of Kra
S China (Yunnan and Guangxi) and extreme n Vietnam

☐ **Antillean Piculet** *Nesoctites micromegas*

_____ *N. m. micromegas*
_____ *N. m. abbotti*

Hispaniola
Gonâve I. (off w Haiti)

☐ **White Woodpecker** *Melanerpes candidus*

Dry forests of South America east of the Andes to n Argentina

☐ **Lewis' Woodpecker** *Melanerpes lewis*

Oak-pine woodlands of w North America; winters to n Mexico

☐ **Guadeloupe Woodpecker** *Melanerpes herminieri*

Guadeloupe (Lesser Antilles)

☐ **Puerto Rican Woodpecker** *Melanerpes portoricensis*

Puerto Rico and Vieques I.

☐ **Red-headed Woodpecker** *Melanerpes erythrocephalus*

E North America from s Canada to Gulf of Mexico and Florida

☐ **Acorn Woodpecker** *Melanerpes formicivorus*
_____ *M. f. bairdi* — Oak-pine woodlands of nw Oregon to n Baja California
_____ *M. f. angustifrons* — Cape region of s Baja California
_____ *M. f. formicivorus (aculeatus)* — Arizona, New Mexico and w Texas to se Mexico (w of Chiapas)
_____ *M. f. albeolus* — S Mexico (e Chiapas) to ne Guatemala and Belize
_____ *M. f. lineatus* — S Mexico (Chiapas) to Guatemala and n Nicaragua
_____ *M. f. striatipectus* — Nicaragua to w Panama
_____ *M. f. flavigula* — Andes of Colombia

☐ **Black-cheeked Woodpecker** *Melanerpes pucherani* — S Mexico (Veracruz, Chiapas) to w Colombia and w Ecuador

☐ **Golden-naped Woodpecker** *Melanerpes chrysauchen* — SW Costa Rica and adjacent w Panama

☐ **Beautiful Woodpecker** *Melanerpes pulcher* — N Colombia (Magdalena Valley)

☐ **Yellow-tufted Woodpecker** *Melanerpes cruentatus* — E Colombia to the Guianas, e Bolivia, ne Brazil and Mato Grosso

☐ **Yellow-fronted Woodpecker** *Melanerpes flavifrons* — E and se Brazil to e Paraguay and ne Argentina (Misiones)

☐ **White-fronted Woodpecker** *Melanerpes cactorum* — Arid se Peru (Puno?) to n Argentina, Paraguay and sw Brazil

☐ **Hispaniolan Woodpecker** *Melanerpes striatus* — Hispaniola and Beata I.

☐ **Jamaican Woodpecker** *Melanerpes radiolatus* — Jamaica

☐ **Golden-cheeked Woodpecker** *Melanerpes chrysogenys*
_____ *M. c. chrysogenys* — Coastal lowlands of nw Mexico (s Sinaloa to Nayarit)
_____ *M. c. flavinuchus* — W Mexico (Jalisco to se Pueble and e Oaxaca)

☐ **Gray-breasted Woodpecker** *Melanerpes hypopolius* — Interior sw Mexico (n Guerrero and Morelos to e-central Oaxaca)

☐ **Yucatan Woodpecker** *Melanerpes pygmaeus*
_____ *M. p. rubricomus* — Yucatán Peninsula south to central Belize
_____ *M. p. pygmaeus* — Cozumel I.
_____ *M. p. tysoni* — Guanaja I. (off n Honduras)

☐ **Red-crowned Woodpecker** *Melanerpes rubricapillus*
_____ *M. r. rubricapillus* — SW Costa Rica to Colombia and the Guianas; Tobago
_____ *M. r. subfusculus* — Coiba I. (Panama)
_____ *M. r. seductus* — San Miguel del Rey I. (se Panama)
_____ *M. r. paraguanae* — NW Venezuela (Paraguaná Peninsula)

☐ **Gila Woodpecker** *Melanerpes uropygialis*
_____ *M. u. uropygialis (fuscescens)* — Arid lowlands of sw US to central Mexico
_____ *M. u. cardonensis* — N Baja California
_____ *M. u. brewsteri* — S Baja California

☐ **Red-bellied Woodpecker** *Melanerpes carolinus* — E North America (s Canada to Texas and Florida Keys)

☐ **West Indian Woodpecker** *Melanerpes superciliaris*
_____ *M. s. nyeanus* — Grand Bahama and San Salvador I.
_____ *M. s. blakei* — Great Abaco I. (n Bahamas)
_____ *M. s. superciliaris* — Cuba, Cantiles Keys and adjacent islands
_____ *M. s. murceus* — Isle of Pines, Cayo Largo and Cayo Real
_____ *M. s. caymanensis* — Grand Cayman I.

☐ **Golden-fronted Woodpecker** *Melanerpes aurifrons*
_____ *M. a. aurifrons* — Lowlands of sw Oklahoma and nw Texas to s-central Mexico
_____ *M. a. polygrammus* — Arid sw Mexico (sw Oaxaca through interior Chiapas)
_____ *M. a. grateloupensis* — E Mexico (Tamaulipas and San Luís Potosí to Puebla, Veracruz)

____ *M. a. veraecrucis*	E Mexico (Atlantic slope from Veracruz to ne Guatemala)
____ *M. a. dubius*	S Mexico (Yucatán Peninsula) to ne Guatemala and Belize
____ *M. a. leei*	Cozumel I. (off e Mexico)
____ *M. a. turneffensis*	Turneffe Islands (off Belize)
____ *M. a. santacruzi*	SE Chiapas to El Salvador, sw Honduras and n-central Nicaragua
____ *M. a. hughlandi*	Guatemala (upper Río Negro and upper Motagua Valley)
____ *M. a. pauper*	Coastal n Honduras
____ *M. a. insulanus*	Utila I. (off n Honduras)
____ *M. a. canescens*	Roatán I. and Barbareta I. (off n Honduras)

☐ **Hoffmann's Woodpecker** *Melanerpes hoffmannii*

Pacific slope of s Honduras to Nicaragua and Costa Rica

☐ **Yellow-bellied Sapsucker** *Sphyrapicus varius*

N America (Yukon, Canada, US); winters se US to Panama

☐ **Red-naped Sapsucker** *Sphyrapicus nuchalis*

SW Canada to sw US; winters to n Mexico and s Baja California

☐ **Red-breasted Sapsucker** *Sphyrapicus ruber*
____ *S. r. ruber*	Coastal s Alaska to w Oregon
____ *S. r. daggetti*	SW Oregon to Sierra Nevada of s California and w Nevada

☐ **Williamson's Sapsucker** *Sphyrapicus thyroideus*
____ *S. t. thyroideus*	Mts. of s British Columbia to n Baja; winters to n Mexico
____ *S. t. nataliae*	SE British Columbia to Rocky Mts. and Great Basin ranges

☐ **Cuban Woodpecker** *Xiphidiopicus percussus*
____ *X. p. percussus*	Cuba
____ *X. p. insulaepinorum (gloriae, marthae)*	Isle of Pines, Cantiles Keys and Archipelago Jardines de la Reina

☐ **Fine-spotted Woodpecker** *Campethera punctuligera*
____ *C. p. punctuligera*	SW Mauritania to Senegambia, Cameroon, sw Sudan, ne Zaire
____ *C. p. balia*	S Sudan to extreme ne Zaire

☐ **Nubian Woodpecker** *Campethera nubica*
____ *C. n. nubica*	Sudan to Ethiopia, Kenya, Uganda, ne Zaire and sw Tanzania
____ *C. n. pallida*	S Somalia and coastal Kenya

☐ **Bennett's Woodpecker** *Campethera bennettii*
____ *C. b. bennettii*	Angola to se Zaire, Tanzania, Malawi and ne South Africa
____ *C. b. capricorni*	S Angola to n Namibia, n Botswana and sw Zambia

☐ **Reichenow's Woodpecker** *Campethera scriptoricauda*

Central and e Tanzania to se Malawi and n Mozambique

☐ **Golden-tailed Woodpecker** *Campethera abingoni*
____ *C. a. chrysura*	Senegambia to s Sudan, ne Zaire and w Uganda
____ *C. a. kavirondensis*	E Rwanda to n Tanzania and sw Kenya
____ *C. a. suahelica*	N Tanzania to e Zimbabwe, Mozambique and n Swaziland
____ *C. a. abingoni*	W Zaire to w Tanzania, ne Namibia, nw Zambia and n Transvaal
____ *C. a. anderssoni*	SW Angola to Namibia, sw Botswana, n Cape Prov., s Transvaal
____ *C. a. constricta*	S Swaziland, extreme s Mozambique and Natal

☐ **Mombasa Woodpecker** *Campethera mombassica*

S Somalia to coastal Kenya and ne Tanzania

☐ **Knysna Woodpecker** *Campethera notata*

Coastal e South Africa (extreme s Natal to Cape Province)

☐ **Little Green Woodpecker** *Campethera maculosa*

Senegal and Guinea-Bissau to central Ghana and Liberia

☐ **Green-backed Woodpecker** *Campethera cailliautii*
____ *C. c. permista*	E Ghana to sw Sudan, sw Uganda, n Angola and central Zaire
____ *C. c. nyansae*	SW Ethiopia to sw Kenya, nw Tanzania, e Zaire and ne Angola

____ *C. c. cailliautii*   Coastal s Somalia through Kenya to ne Tanzania; Zanzibar
____ *C. c. loveridgei*   Central Tanzania to extreme e Zimbabwe and Mozambique

□ **Tullberg's Woodpecker**   *Campethera tullbergi*
____ *C. t. tullbergi*   Mountains of se Nigeria and w Cameroon; Bioko I.
____ *C. t. taeniolaema*   E Zaire to Rwanda, Uganda, Burundi, w Tanzania and w Kenya
____ *C. t. hausburgi*   Highlands of Kenya (east of Rift Valley) and extreme n Tanzania

□ **Buff-spotted Woodpecker**   *Campethera nivosa*
____ *C. n. nivosa*   Senegambia to Gabon, w Zaire, nw Zambia and nw Angola
____ *C. n. poensis*   Bioko I. (Gulf of Guinea)
____ *C. n. herberti*   Central African Rep. to w Kenya, sw Sudan, Uganda and e Zaire
____ *C. n. maxima*   Known from two specimens from n Ivory Coast

□ **Brown-eared Woodpecker**   *Campethera caroli*
____ *C. c. arizela*   Guinea-Bissau and Sierra Leone to Ghana
____ *C. c. caroli*   Benin to Nigeria, Cameroon, Angola, Kenya and nw Tanzania

□ **Ground Woodpecker**   *Geocolaptes olivaceus*
____ *G. o. olivaceus*   South Africa (western/Cape Province)
____ *G. o. petrobates*   Highlands of ne Cape Province, adj. Transkei to Natal
____ *G. o. prometheus*   Orange Free State to interior Natal, Sande Transvaal

□ **Little Gray Woodpecker**   *Dendropicos elachus*
       Sahel (sw Mauritania to Senegambia, Chad and w-central Sudan)

□ **Speckle-breasted Woodpecker**   *Dendropicos poecilolaemus*
       Extreme se Nigeria to sw Sudan, Uganda and w Kenya

□ **Abyssinian Woodpecker**   *Dendropicos abyssinicus*
       Savanna and juniper highlands of Ethiopia and Eritrea

□ **Cardinal Woodpecker**   *Dendropicos fuscescens*
____ *D. f. fuscescens*   N-c Namibia through South Africa to s Transvaal and w Natal
____ *D. f. intermedius*   Natal and Transvaal to Mozambique (lower Zambesi River)
____ *D. f. centralis*   E Angola to n Namibia, s Zaire, w Tanzania and Zambia
____ *D. f. hartlaubii*   Coastal s Kenya to Tanzania, Malawi, Zambia and Mozambique
____ *D. f. lafresnayi*   Senegambia and Sierra Leone to Nigeria
____ *D. f. sharpii*   Cameroon to s Sudan, w Zaire and n Angola
____ *D. f. lepidus*   E Zaire to highlands of Ethiopia to Uganda, Kenya, nw Tanzania
____ *D. f. massaicus*   S Ethiopia, inland and w Kenya and n-central Tanzania
____ *D. f. hemprichii*   Lower elevations of Ethiopia to Somalia and e Kenya

□ **Melancholy Woodpecker**   *Dendropicos lugubris*
       Sierra Leone to se Ghana, s Nigeria and sw Cameroon

□ **Gabon Woodpecker**   *Dendropicos gabonensis*
____ *D. g. reichenowi*   S Nigeria and extreme sw Cameroon
____ *D. g. gabonensis*   S Cameroon to Cent. Afr. Rep., Zaire, w Uganda and n Angola

□ **Stierling's Woodpecker**   *Dendropicos stierlingi*
       S Tanzania to n Mozambique and s Malawi; e Zambia?

□ **Bearded Woodpecker**   *Dendropicos namaquus*
____ *D. n. namaquus*   W Cent. Afr. Rep. to s Sudan, e Zaire, Tanzania and Angola
____ *D. n. schoensis*   Ethiopia to Somalia and n Kenya
____ *D. n. coalescens*   Central and s Mozambique to e South Africa

□ **Fire-bellied Woodpecker**   *Dendropicos pyrrhogaster*
       Sierra Leone and s Guinea to s Nigeria and sw Cameroon

□ **Golden-crowned Woodpecker**   *Dendropicos xantholophus*
       SW Cameroon to s Sudan, Uganda, w Kenya and nw Angola

□ **Elliot's Woodpecker**   *Dendropicos elliotii*
____ *D. e. elliotii (gabela)*   SW Cameroon to w Angola, Gabon, Zaire and e Uganda
____ *D. e. johnstoni*   Montane forests of se Nigeria and sw Cameroon; Bioko I.

☐ **Gray Woodpecker** *Dendropicos goertae*

____	*D. g. abessinicus*	E Sudan to w Ethiopia
____	*D. g. goertae*	Senegambia to s Sudan, Zaire, w Kenya and nw Tanzania
____	*D. g. koenigi*	Sahel of Sahara (e Mali to Niger, Chad and w Sudan)
____	*D. g. meridionalis*	S Gabon to nw Angola and s-central Zaire

☐ **Gray-headed Woodpecker** *Dendropicos spodocephalus*

____	*D. s. rhodeogaster*	Highlands of central Kenya to n-central Tanzania
____	*D. s. spodocephalus*	E Sudan and highlands of central and s Ethiopia

☐ **Olive Woodpecker** *Dendropicos griseocephalus*

____	*D. g. ruwenzori*	Namibia to Zaire, Uganda, Tanzania, Zimbabwe and Malawi
____	*D. g. kilimensis*	Humid montane forests of Tanzania
____	*D. g. griseocephalus*	Extreme s Mozambique to Natal, e Transvaal to Cape Province

☐ **Brown-backed Woodpecker** *Dendropicos obsoletus*

____	*D. o. obsoletus*	Senegambia to s Cameroon, ne Zaire, w Sudan and Uganda
____	*D. o. heuglini*	E Sudan to n Ethiopia and Eritrea
____	*D. o. ingens*	Cent. and s Ethiopia to ne Uganda, Kenya and extreme n Tanzania
____	*D. o. crateri*	N Tanzania (Crater Highlands south to Nou Forest)

☐ **Sulawesi Woodpecker** *Dendrocopos temminckii*

	Sulawesi, Butung I. and Togian Islands

☐ **Philippine Woodpecker** *Dendrocopos maculatus*

____	*D. m. validirostris*	Philippines (Catanduanes, Lubang, Luzon, Marinduque, Mindoro)
____	*D. m. maculatus*	Sibuyan, Cebu, Guimaras, Negros, Panay and Gigantes
____	*D. m. fulvifasciatus*	Samar, Calicoan, Leyte, Bohol, Basilan, Mindanao and Dinagat)
____	*D. m. ramsayi*	Sulu Arch. (Bongao, Jolo, Tawitawi, Sanga Sanga and Sibutu)

☐ **Brown-capped Woodpecker** *Dendrocopos moluccensis*

____	*D. m. nanus*	N and central peninsular India
____	*D. m. cinereigula*	S India (Madras and Kerala)
____	*D. m. gymnophthalmus*	Sri Lanka
____	*D. m. moluccensis*	Malay Peninsula to Borneo, Sumatra, Java and Riau Archipelago
____	*D. m. grandis*	Lesser Sundas (Lombok, Lomblen, Sumbawa, Flores, Besar, Alor)

☐ **Gray-capped Woodpecker** *Dendrocopos canicapillus*

____	*D. c. doerriesi*	E Siberia (Ussuriland) to Korea and e Manchuria
____	*D. c. scintilliceps*	E and c China (Liaoning to Sichuan and Zhejiang)
____	*D. c. kaleensis*	SW China (Sichuan to Fujian) to n Burma, n Indochina; Taiwan
____	*D. c. swinhoei*	Hainan I. (s China)
____	*D. c. semicoronatus*	Extreme e Nepal to w Assam
____	*D. c. mitchellii*	W Nepal to nw India and n Pakistan
____	*D. c. canicapillus*	E Assam, Bangladesh to central and s Burma, Thailand and Laos
____	*D. c. delacouri*	SE Thailand, Cambodia and Cochinchina
____	*D. c. auritus*	S Thailand and Malay Peninsula
____	*D. c. volzi*	Sumatra, Riau Archipelago and Nias I.
____	*D. c. aurantiiventris*	Borneo

☐ **Pygmy Woodpecker** *Dendrocopos kizuki*

____	*D. k. ijimae*	SE Siberia and Sakhalin to n Korea, n Japan and Kuril Is.
____	*D. k. seebohmi*	Korea (except ne), Cheju-Do Islands and Honshu
____	*D. k. amamii*	N Ryukyu Islands (Amami-O-Shima and Tokuno-Shima)
____	*D. k. kizuki*	NE China, s Japan, Okinawa, s Ryukyu and Izu Islands

☐ **Lesser Spotted Woodpecker** *Dendrocopos minor*

____	*D. m. comminutus*	England and Wales
____	*D. m. hortorum*	France to Poland, Switzerland, Hungary and n Romania
____	*D. m. buturlini*	Iberia, s France and Italy to Romania, Bulgaria and n Greece
____	*D. m. minor*	N Europe (Scandinavia to Ural Mountains)

____	*D. m. amurensis*	Lower Amur River and Sakhalin to ne Korea and n Japan
____	*D. m. kamtschatkensis*	Ural Mts. to Anadyr River and Kamchatka
____	*D. m. colchicus*	Caucasus and Transcaucasia
____	*D. m. danfordi*	Greece and Turkey
____	*D. m. quadrifasciatus*	SE Transcaucasia (Lenkoran region)
____	*D. m. morgani*	Zagros Mountains and nw Iran
____	*D. m. ledouci*	NW Africa (ne Algeria and nw Tunisia)

☐ **Brown-fronted Woodpecker** *Dendrocopos auriceps*

N Baluchistan and w Himalayas (Afghanistan to n India, Nepal)

☐ **Fulvous-breasted Woodpecker** *Dendrocopos macei*

____	*D. m. westermanni*	N Pakistan, nw India (Himachal Pradesh) and w Nepal
____	*D. m. macei*	C Nepal to n Burma and e pen. India (Orissa and Andra Pradesh)
____	*D. m. longipennis*	S Burma to Thailand, Laos, Annam, Cambodia and Cochinchina
____	*D. m. andamanensis*	Andaman Islands
____	*D. m. analis*	S Sumatra, Java and Bali

☐ **Stripe-breasted Woodpecker** *Dendrocopos atratus*

NE India to Burma, sw China, Thailand, Cambodia and s Laos

☐ **Yellow-crowned Woodpecker** *Dendrocopos mahrattensis*

____	*D. m. mahrattensis*	India to Burma, sw China, e Cambodia, s Laos; Sri Lanka
____	*D. m. pallescens*	E Pakistan (e of Indus River) and nw India

☐ **Arabian Woodpecker** *Dendrocopos dorae*

Southwest Arabian Peninsula

☐ **Rufous-bellied Woodpecker** *Dendrocopos hyperythrus*

____	*D. h. marshalli*	NE Pakistan to Kashmir and n India
____	*D. h. hyperythrus*	Nepal to se Tibet, sw China, Burma and nw Thailand
____	*D. h. subrufinus*	Manchuria and Ussuriland; winters to s China
____	*D. h. annamensis*	E Thailand, Cambodia and s Vietnam

☐ **Crimson-breasted Woodpecker** *Dendrocopos cathpharius*

____	*D. c. cathpharius*	E Himalayas (Nepal to n Assam)
____	*D. c. ludlowi*	SE Tibet
____	*D. c. pyrrhothorax*	Hills south of Brahmaputra River and adjacent n Burma
____	*D. c. tenebrosus*	N Burma to Thailand, Laos, n Vietnam and Yunnan
____	*D. c. pernyii*	W China (nw Yunnan, Sichuan and Xinjiang north to Gansu)
____	*D. c. innixus*	E-central China (central Hubei)

☐ **Darjeeling Woodpecker** *Dendrocopos darjellensis*

Nepal to se Tibet, ne India, n Burma, sw China and nw Tonkin

☐ **Middle Spotted Woodpecker** *Dendrocopos medius*

____	*D. m. medius*	NW Spain to France, Estonia, w Russia, Ukraine, Italy, Balkans
____	*D. m. caucasicus*	N Turkey to Caucasus and Transcaucasia; nw Iran?
____	*D. m. anatoliae*	W and s Asia Minor
____	*D. m. sanctijohannis*	Zagros Mountains (sw Iran)

☐ **White-backed Woodpecker** *Dendrocopos leucotos*

____	*D. l. leucotos (uralensis)*	N, central and e Europe to ne Asia, Korea and Sakhalin
____	*D. l. lilfordi*	Pyrenees to Asia Minor, Caucasus and Transcaucasia
____	*D. l. tangi*	W China (Sichuan)
____	*D. l. subcirris*	N Japan (Hokkaido)
____	*D. l. stejnegeri*	Japan (n Honshu)
____	*D. l. namiyei*	S Honshu, Kyushu, Shikoku and Cheju-Do Islands
____	*D. l. takahashii*	Ullung I. (off e Korea)
____	*D. l. owstoni*	Amami-O-Shima I. (n Ryukyus)
____	*D. l. fohkiensis*	Mountains of se China (Fujian)
____	*D. l. insularis*	Taiwan

☐ **Great Spotted Woodpecker** *Dendrocopos major*

____	*D. m. major*	Scandinavia and w Siberia to Ural Mts., n Poland and n Ukraine

____ *D. m. pinetorum*	Britain, France and c Europe to Italy, Balkans, Turkey, Caucasus
____ *D. m. harterti*	Sardinia and Corsica
____ *D. m. hispanus*	Iberian Peninsula
____ *D. m. canariensis*	Tenerife (Canary Islands)
____ *D. m. thanneri*	Gran Canaria I. (Canary Islands)
____ *D. m. mauritanus*	Morocco
____ *D. m. numidus*	N Algeria and Tunisia
____ *D. m. poelzami*	Transcaucasia and s Caspian region
____ *D. m. brevirostris*	W Siberia to lower Amur River, Manchuria and Mongolia
____ *D. m. kamtschaticus*	Kamchatka Peninsula and n coast of Sea of Okhotsk
____ *D. m. japonicus*	E Manchuria, Sakhalin, Kuril Islands, Korea and n Japan
____ *D. m. cabanisi*	S Manchuria to se China, e Burma and Indochina; Hainan
____ *D. m. stresemanni*	W China to Burma, se Tibet and ne India

☐ **Syrian Woodpecker** *Dendrocopos syriacus*

SE Europe, Transcaucasia, Turkey and Iran to Israel and Jordan

☐ **White-winged Woodpecker** *Dendrocopos leucopterus*

Aral Sea to Lake Balkhash, ne Afghanistan and extreme w China

☐ **Sind Woodpecker** *Dendrocopos assimilis*

Arid woodlands and desert scrub of se Iran to Pakistan

☐ **Himalayan Woodpecker** *Dendrocopos himalayensis*

____ *D. h. albescens*	NE Afghanistan (Safed Koh) and n Pakistan to Himachal Pradesh
____ *D. h. himalayensis*	N Pakistan (e Himachal Pradesh) to w Nepal

☐ **Striped Woodpecker** *Picoides lignarius*

Arid highlands of Bolivia to s Chile and s Argentina

☐ **Checkered Woodpecker** *Picoides mixtus*

____ *P. m. mixtus*	N Argentina (Paraná R. drainage and Buenos Aires Province)
____ *P. m. berlepschi*	Central Argentina (Córdoba south to Neuquén and Río Negro)
____ *P. m. malleator*	*Chaco* of n Argentina, Paraguay and se Bolivia
____ *P. m. cancellatus*	Extreme e Bolivia to e Brazil, ne Paraguay and w Uruguay

☐ **Nuttall's Woodpecker** *Picoides nuttallii*

Chaparral and oak woodlands of n California to nw Baja

☐ **Ladder-backed Woodpecker** *Picoides scalaris*

____ *P. s. cactophilus*	Arid sw US to ne Baja California and central Mexico
____ *P. s. eremicus*	N Baja California
____ *P. s. lucasanus*	S Baja California
____ *P. s. graysoni*	Tres Marías Islands (off w Mexico)
____ *P. s. sinaloensis*	Coastal w Mexico (s Sonora to Guerrero, sw Puebla, w Oaxaca)
____ *P. s. scalaris*	S Mexico (Veracruz and Chiapas)
____ *P. s. parvus*	N Yucatán Peninsula, Cozumel I. and Holbox I.
____ *P. s. leucoptilurus*	Belize, Guatemala and El Salvador to ne Nicaragua

☐ **Downy Woodpecker** *Picoides pubescens*

____ *P. p. pubescens*	Kansas to N Carolina, south to e Texas and Florida
____ *P. p. medianus*	C Alaska to Newfoundland and c US (east of Rocky Mountains)
____ *P. p. leucurus*	Rocky Mountains (se Alaska to sw US)
____ *P. p. glacialis*	Coastal se Alaska
____ *P. p. gairdnerii*	Coastal w British Columbia to nw California
____ *P. p. turati*	Interior Washington, Oregon and California

☐ **Red-cockaded Woodpecker** *Picoides borealis*

Patchily distributed pine forests of se US

☐ **Strickland's Woodpecker** *Picoides stricklandi*

Coniferous forests of e Mexico (Michoacán to w-c Veracruz)

☐ **Arizona Woodpecker** *Picoides arizonae*

____ *P. a. arizonae*	SE Arizona to nw Mexico (n Sinaloa and adjacent Durango)
____ *P. a. fraterculus*	W Mexico (s Sinaloa and adjacent Durango to Michoacán)

□ **Hairy Woodpecker** *Picoides villosus*

____	*P. v. septentrionalis*	W North America (Alaska to n New Mexico)
____	*P. v. villosus*	E N Dak. to s Quebec, Nova Scotia, c Texas, Missouri, Virginia
____	*P. v. terraenovae*	Newfoundland
____	*P. v. sitkensis*	Coastal se Alaska and n British Columbia
____	*P. v. picoideus*	Queen Charlotte Islands (off British Columbia)
____	*P. v. harrisi*	Coastal s British Columbia to nw California
____	*P. v. audubonii*	S Illinois to se Virginia, e Texas and Gulf Coast
____	*P. v. hyloscopus*	W California to n Baja California
____	*P. v. orius*	Cascade Mts. of Br. Columbia to se California and w Texas
____	*P. v. icastus*	SE Arizona and New Mexico through w Mexico to Jalisco
____	*P. v. jardinii*	Central and e Mexico to Jalisco, Guerrero and Oaxaca
____	*P. v. sanctorum*	S Mexico (Chiapas) and Guatemala to w Panama
____	*P. v. piger*	N Bahamas (Abaco, Mores and Grand Bahama)
____	*P. v. maynardi*	S Bahamas (Andros and New Providence)

□ **White-headed Woodpecker** *Picoides albolarvatus*

____	*P. a. albolarvatus*	Montane coniferous forests of s British Columbia to sw US
____	*P. a. gravirostris*	SW California (mountains of Los Angeles and San Diego region)

□ **Eurasian Three-toed Woodpecker** *Picoides tridactylus*

____	*P. t. tridactylus*	N Europe to n Mongolia, Manchuria and Sakhalin
____	*P. t. crissoleucus*	N *taiga* from Ural Mountains to Sea of Okhotsk
____	*P. t. albidior*	Kamchatka Peninsula
____	*P. t. alpinus*	Mountains of Europe to ne Korea and n Japan (Hokkaido)
____	*P. t. funebris*	SW China to Tibet

□ **American Three-toed Woodpecker** *Picoides dorsalis*

____	*P. d. fasciatus*	Alaska and Yukon south to Oregon, n Idaho and w Montana
____	*P. d. bacatus*	Alberta, Labrador and Newfoundland s to Minnesota and ne US
____	*P. d. dorsalis*	Rocky Mountains from Montana to Nevada and nw New Mexico

□ **Black-backed Woodpecker** *Picoides arcticus*

Boreal forests of n North America

□ **Scarlet-backed Woodpecker** *Veniliornis callonotus*

____	*V. c. callonotus*	SW Colombia (Nariño) and nw Ecuador (s to Guayas, El Oro)
____	*V. c. major*	SW Ecuador (El Oro and Loja) and nw Peru

□ **Yellow-vented Woodpecker** *Veniliornis dignus*

____	*V. d. dignus*	Humid montane forests of extreme sw Venezuela and n Ecuador
____	*V. d. baezae*	E slope of Andes of Ecuador
____	*V. d. valdizani*	E slope of Andes of Peru

□ **Bar-bellied Woodpecker** *Veniliornis nigriceps*

____	*V. n. equifasciatus*	Andes of n Colombia and Ecuador
____	*V. n. pectoralis*	Andes of Peru
____	*V. n. nigriceps*	Andes of Bolivia (Cochabamba and w Santa Cruz)

□ **Smoky-brown Woodpecker** *Veniliornis fumigatus*

____	*V. f. oleagineus*	Lowlands and foothills of e Mexico
____	*V. f. sanguinolentus*	Central and s Mexico to w Panama
____	*V. f. fumigatus*	E Panama to w Bolivia and nw Argentina (Jujuy)
____	*V. f. obscuratus*	SW Ecuador to nw Peru
____	*V. f. reichenbachi*	N Venezuela

□ **Little Woodpecker** *Veniliornis passerinus*

____	*V. p. fidelis*	E Colombia to w Venezuela
____	*V. p. modestus*	NE Venezuela
____	*V. p. passerinus*	The Guianas and ne Brazil
____	*V. p. diversus*	N Brazil

____ *V. p. agilis*	E Ecuador to n Bolivia and w Brazil
____ *V. p. olivinus*	S Brazil to Paraguay, Bolivia and n Argentina
____ *V. p. taenionotus*	E Brazil
____ *V. p. tapajozensis*	Central Brazil
____ *V. p. insignis*	W-central Brazil

☐ **Dot-fronted Woodpecker** *Veniliornis frontalis*

Humid Andean slopes of s Bolivia and nw Argentina

☐ **White-spotted Woodpecker** *Veniliornis spilogaster*

SE Brazil to se Paraguay, Uruguay and ne Argentina

☐ **Blood-colored Woodpecker** *Veniliornis sanguineus*

Humid coastal lowlands of the Guianas

☐ **Red-rumped Woodpecker** *Veniliornis kirkii*

____ *V. k. neglectus*	SW Costa Rica and w Panama; Coiba I.
____ *V. k. cecilii*	E Panama to w Colombia, w Ecuador and extreme n Peru
____ *V. k. kirkii*	Trinidad and Tobago; Paría Peninsula of ne Venezuela
____ *V. k. continentalis*	N and w Venezuela
____ *V. k. monticola*	*Tepuis* of se Venezuela (Mt. Roraima and Cerro Uei-tepui)

☐ **Choco Woodpecker** *Veniliornis chocoensis*

Humid forests of nw Colombia to nw Ecuador

☐ **Golden-collared Woodpecker** *Veniliornis cassini*

E and se Venezuela, the Guianas and ne Brazil (n of Amazon)

☐ **Red-stained Woodpecker** *Veniliornis affinis*

____ *V. a. orenocensis*	SE Colombia to s Venezuela and n Brazil
____ *V. a. hilaris*	E Ecuador through e Peru to n Bolivia and w Mato Grosso
____ *V. a. ruficeps*	Central and ne Brazil (south to Mato Grosso)
____ *V. a. affinis*	E Brazil (Alagoas and e Bahia)

☐ **Yellow-eared Woodpecker** *Veniliornis maculifrons*

Lowlands of se Brazil (se Bahia to Rio de Janeiro)

☐ **Rufous-winged Woodpecker** *Piculus simplex*

Caribbean slope of Honduras to Costa Rica and w Panama

☐ **Stripe-cheeked Woodpecker** *Piculus callopterus*

Humid lowlands and foothills on both slopes of Panama

☐ **Lita Woodpecker** *Piculus litae*

Humid lowlands of w Colombia to nw Ecuador (Pichincha)

☐ **White-throated Woodpecker** *Piculus leucolaemus*

S Colombia to e Ecuador, se Peru, n Bolivia, w Amaz. Brazil

☐ **Yellow-throated Woodpecker** *Piculus flavigula*

____ *P. f. flavigula*	Extreme Colombia to Venezuela, the Guianas, n Amaz. Brazil
____ *P. f. magnus*	SE Colombia to n Bolivia and ne Brazil
____ *P. f. erythropis*	E Brazil

☐ **Golden-green Woodpecker** *Piculus chrysochloros*

____ *P. c. aurosus*	Tropical e Panama
____ *P. c. xanthochlorus*	N Colombia and nw Venezuela
____ *P. c. capistratus*	SE Colombia to nw Brazil and Suriname
____ *P. c. guianensis*	French Guiana
____ *P. c. paraensis*	NE Brazil
____ *P. c. laemostictus*	NW Brazil
____ *P. c. hypochryseus*	W Brazil to n Bolivia
____ *P. c. chrysochloros*	Central and s Brazil to Bolivia, w Paraguay and n Argentina
____ *P. c. polyzonus*	SE Brazil (Espírito Santo and Rio de Janeiro)

☐ **Yellow-browed Woodpecker** *Piculus aurulentus*

Lowlands of se Brazil to ne Argentina and e Paraguay

☐ **Gray-crowned Woodpecker** *Piculus auricularis*

Pacific slope of Mexico (s Sonora to central Oaxaca)

☐ **Golden-olive Woodpecker** *Piculus rubiginosus*

____ P. r. aeruginosus	E Mexico (Tamaulipas to Veracruz)
____ P. r. yucatanensis	S Mexico (Oaxaca) to w Panama
____ P. r. alleni	Santa Marta Mountains (ne Colombia)
____ P. r. buenavistae	Andean slopes of e Colombia and e Ecuador
____ P. r. tobagensis	Tobago
____ P. r. trinitatis	Trinidad
____ P. r. meridensis	NW Venezuela
____ P. r. deltanus	NE Venezuela (Delta Amacuro)
____ P. r. guianae	E Venezuela and adjacent Guyana
____ P. r. paraquensis	Mountains of s-central Venezuela
____ P. r. rubiginosus	Mountains of n-central and ne Venezuela
____ P. r. viridissimus	*Tepuis* of s Venezuela (high plateau of Auyán-tepui)
____ P. r. nigriceps	Acari Mountains (s Guyana and adjacent s Suriname)
____ P. r. gularis	Colombia (Central and Western Andes)
____ P. r. rubripileus	Extreme sw Colombia to w Ecuador and nw Peru
____ P. r. coloratus	Extreme se Ecuador and n-central Peru
____ P. r. chrysogaster	Central Peru
____ P. r. canipileus	Central and se Bolivia
____ P. r. tucumanus	S Bolivia to nw Argentina (s to Tucumán)

☐ **Crimson-mantled Woodpecker** *Piculus rivolii*

____ P. r. rivolii	Andes of e-central Colombia to nw Venezuela
____ P. r. meridae	Andes of w Venezuela (Mérida and Táchira)
____ P. r. quindiuna	Andes of n-central Colombia
____ P. r. brevirostris	Andes of sw Colombia to central Peru
____ P. r. atriceps	Andes of se Peru and s Bolivia

☐ **Black-necked Woodpecker** *Colaptes atricollis*

____ C. a. atricollis	Xeric w slopes of Andes of Peru (La Libertad to w Arequipa)
____ C. a. peruvianus	N Peru (xeric slopes of Marañón Valley)

☐ **Spot-breasted Woodpecker** *Colaptes punctigula*

____ C. p. ujhelyii	E Pamana (d Darién) and n Colombia
____ C. p. striatigularis	W-central Colombia
____ C. p. zuliae	NW Venezuela
____ C. p. punctipectus	E Colombia and Venezuela (except for range of *zuliae*)
____ C. p. punctigula	The Guianas
____ C. p. guttatus	Amazonian Ecuador to nw Bolivia, ne Brazil and n Mato Grosso

☐ **Green-barred Woodpecker** *Colaptes melanochloros*

____ C. m. melanochloros	SE Brazil to se Paraguay, ne Argentina and Uruguay
____ C. m. nattereri	NE Brazil to Bolivia (Santa Cruz)
____ C. m. melanolaimus	Arid upland valleys of central and s Bolivia
____ C. m. nigroviridis	S Bolivia to w Paraguay, n Argentina and w Uruguay
____ C. m. leucofrenatus	NW and w-c Argentina (s to Neuquén and w Río Negro)

☐ **Northern Flicker** *Colaptes auratus*

____ C. a. luteus	C Alaska to s Labrador, Newfoundland, Montana and ne US
____ C. a. auratus	SE US
____ C. a. cafer	S Alaska and British Columbia to n California
____ C. a. collaris	SW US to nw Baja and w Mexico (Durango)
____ C. a. mexicanus	W Mexico (Durango) to San Luis Potosí and Oaxaca)
____ C. a. mexicanoides	Highlands of s Mexico (Chiapas) to Nicaragua
____ C. a. chrysocaulosus	Cuba
____ C. a. gundlachi	Grand Cayman I.

☐ **Gilded Flicker** *Colaptes chrysoides*

____ C. c. mearnsi	Extreme se California to Arizona and nw Mexico (n Sonora)
____ C. c. tenebrosus	NW Mexico (n Sonora to n Sinaloa)

____	*C. c. brunnescens*	N and central Baja California
____	*C. c. chrysoides*	S Baja California

☐ **Fernandina's Flicker** *Colaptes fernandinae*

Locally in palm groves of Cuba

☐ **Chilean Flicker** *Colaptes pitius*

Humid forests and scrub of central and s Chile and sw Argentina

☐ **Andean Flicker** *Colaptes rupicola*

____	*C. r. cinereicapillus*	Andean grasslands of extreme s Ecuador and n Peru
____	*C. r. puna*	Central and s Peru
____	*C. r. rupicola*	Bolivia to n Chile and nw Argentina

☐ **Campo Flicker** *Colaptes campestris*

____	*C. c. campestris*	S Suriname to e Brazil, Bolivia and central Paraguay
____	*C. c. campestroides*	S Paraguay to se Brazil, Uruguay and n Argentina

☐ **Rufous Woodpecker** *Celeus brachyurus*

____	*C. b. humei*	NW India (Himachal Pradesh) to extreme w Nepal
____	*C. b. jerdonii*	Central and s peninsular India and Sri Lanka
____	*C. b. phaioceps*	C Nepal and e India to se Tibet, Burma, s China and s Thailand
____	*C. b. squamigularis*	S pen. Thailand to Sumatra, Bangka, Belitung and Nias islands
____	*C. b. brachyurus*	Java
____	*C. b. badiosus*	Borneo and North Natuna Islands
____	*C. b. fokiensis*	S China and n Vietnam
____	*C. b. holroydi*	Hainan I. (s China)
____	*C. b. annamensis*	Laos, Cambodia and s Vietnam

☐ **Cinnamon Woodpecker** *Celeus loricatus*

____	*C. l. diversus*	Humid forests of e Nicaragua to w Panama
____	*C. l. mentalis*	Panama to extreme nw Colombia
____	*C. l. loricatus*	W Colombia to sw Ecuador (Guayas)
____	*C. l. innotatus*	N Colombia

☐ **Scaly-breasted Woodpecker** *Celeus grammicus*

____	*C. g. grammicus*	SE Colombia and s Venezuela to ne Peru and w Brazil
____	*C. g. verreauxii*	Tropical e Ecuador and adjacent ne Peru
____	*C. g. subcervinus*	W Amazonian Brazil and n Mato Grosso
____	*C. g. latifasciatus*	SE Peru to n Bolivia and w Brazil (upper Rio Madeira)

☐ **Waved Woodpecker** *Celeus undatus*

____	*C. u. undatus*	E Venezuela to the Guianas and ne Brazil
____	*C. u. amacurensis*	NE Venezuela (Delta Amacuro)
____	*C. u. multifasciatus*	NE Brazil s of the Amazon (Pará east to Rio Tocantins)

☐ **Chestnut-colored Woodpecker** *Celeus castaneus*

Gulf-Caribbean slope of s Mexico (s Veracruz) to w Panama

☐ **Chestnut Woodpecker** *Celeus elegans*

____	*C. e. leotaudi*	Trinidad
____	*C. e. hellmayri*	E Venezuela to Guyana and Suriname
____	*C. e. deltanus*	NE Venezuela (Delta Amacuro)
____	*C. e. elegans*	French Guiana, adj. Suriname and ne Brazil north of the Amazon
____	*C. e. jumanus*	E Colombia to sw Venezuela, n Bolivia and s Amazonian Brazil
____	*C. e. citreopygius*	E Ecuador and e Peru

☐ **Pale-crested Woodpecker** *Celeus lugubris*

____	*C. l. lugubris*	Dry lowlands of e Bolivia and sw Brazil (w Mato Grosso)
____	*C. l. kerri*	Paraguay and s Mato Grosso to ne Argentina

☐ **Blond-crested Woodpecker** *Celeus flavescens*

____	*C. f. ochraceus*	Lower Amazonian and e Brazil south to e Bahia

____	*C. f. intercedens*	E Brazil (w Bahia to Goiás and Minas Gerais)
____	*C. f. flavescens*	E Brazil (Rio de Janeiro) to e Paraguay and ne Argentina

☐ **Cream-colored Woodpecker** *Celeus flavus*

____	*C. f. flavus*	E Colombia toVenezuela, the Guianas, w Brazil and n Bolivia
____	*C. f. peruvianus*	Tropical e Peru
____	*C. f. subflavus*	E Brazil (Alagoas; Bahia to Espírito Santo)
____	*C. f. tectricialis*	NE Brazil (Maranhão)

☐ **Rufous-headed Woodpecker** *Celeus spectabilis*

____	*C. s. spectabilis*	Lowlands of e Ecuador and adjacent ne Peru
____	*C. s. exsul*	Tropical se Peru, extreme w Brazil (w Acre) and n Bolivia
____	*C. s. obrieni*	E Brazil (w Piauí)

☐ **Ringed Woodpecker** *Celeus torquatus*

____	*C. t. torquatus*	E Venezuela to the Guianas and ne Brazil (Pará)
____	*C. t. occidentalis*	SE Colombia to n Bolivia, Amazonian Brazil and Mato Grosso
____	*C. t. tinnunculus*	E Brazil (Bahia and Espírito Santo)

☐ **Helmeted Woodpecker** *Dryocopus galeatus*

Humid forests of se Brazil to e Paraguay and ne Argentina

☐ **Pileated Woodpecker** *Dryocopus pileatus*

____	*D. p. abieticola*	S British Columbia to central California and ne US
____	*D. p. pileatus*	SE US from Illinois to Virginia, Texas and Florida

☐ **Lineated Woodpecker** *Dryocopus lineatus*

____	*D. l. scapularis*	W Mexico (s Sonora to Oaxaca)
____	*D. l. similis*	E Mexico (Tamaulipas) to nw Costa Rica
____	*D. l. lineatus*	E Costa Rica to w Colombia, e Peru, n Paraguay and e Brazil
____	*D. l. fuscipennis*	Arid littoral of w Ecuador and nw Peru
____	*D. l. erythrops*	SE Brazil to e Paraguay and ne Argentina

☐ **Black-bodied Woodpecker** *Dryocopus schulzi*

Dry forests of s Bolivia to cent. Paraguay and n-cent. Argentina

☐ **White-bellied Woodpecker** *Dryocopus javensis*

____	*D. j. hodgsonii*	Peninsular India
____	*D. j. feddeni*	Thailand, Burma and Indochina
____	*D. j. javensis*	S Thailand and Malay Pen. to Gr. Sundas and offshore islands
____	*D. j. parvus*	Simeulue I. (off nw Sumatra)
____	*D. j. forresti*	Montane forests of n Burma and adjacent sw China
____	*D. j. richardsi*	Korea; extinct on Tsushima I. (Japan)
____	*D. j. confusus*	Luzon (n Philippines)
____	*D. j. pectoralis*	Philippines (Leyte, Samar, Panaon, Calicoan and Bohol)
____	*D. j. philippinensis*	Philippines (Panay, Negros, Masbate and Guimaras)
____	*D. j. mindorensis*	Mindoro (Philippines)
____	*D. j. cebuensis†*	Formerly Cebu (central Philippines). Extinct
____	*D. j. multilunatus*	S Philippines (Basilan, Dinagat and Mindanao)
____	*D. j. suluensis*	Sulu Archipelago
____	*D. j. hargitti*	Palawan (sw Philippines)

☐ **Andaman Woodpecker** *Dryocopus hodgei*

High evergreen forests of Andaman Islands

☐ **Black Woodpecker** *Dryocopus martius*

____	*D. m. martius*	Coniferous and beech forests of Eurasia
____	*D. m. khamensis*	Tibet and sw China

☐ **Powerful Woodpecker** *Campephilus pollens*

____	*C. p. pollens*	Andes of n Colombia to w Venezuela and Ecuador
____	*C. p. peruvianus*	Andes of n Peru (south to Pasco)

☐ **Crimson-bellied Woodpecker** *Campephilus haematogaster*

    ____ *C. h. splendens*   Humid forests of Panama to w Ecuador

    ____ *C. h. haematogaster*   Humid forests of e Colombia to e Peru

☐ **Red-necked Woodpecker** *Campephilus rubricollis*

    ____ *C. r. rubricollis*   E Colombia and e Ecuador to s Venezuela, Guianas and n Brazil

    ____ *C. r. trachelopyrus*   NE Peru to n Bolivia (La Paz) and w Brazil s of the Amazon

    ____ *C. r. olallae*   Brazil south of the Amazon to Bolivia (Cochabamba)

☐ **Robust Woodpecker** *Campephilus robustus*

                            Lowlands of se Brazil to e Paraguay and ne Argentina (Misiones)

☐ **Pale-billed Woodpecker** *Campephilus guatemalensis*

    ____ *C. g. nelsoni*   Lowlands of w Mexico (s Sonora to Oaxaca)

    ____ *C. g. regius*   Lowlands of ne Mexico (Tamaulipas to Veracruz)

    ____ *C. g. guatemalensis*   Lowlands of s Mexico (Veracruz) to w Panama

☐ **Crimson-crested Woodpecker** *Campephilus melanoleucos*

    ____ *C. m. malherbii*   W Panama to n and central Colombia

    ____ *C. m. melanoleucos*   S America east of Andes to ne Argentina and Brazil; Trinidad

☐ **Guayaquil Woodpecker** *Campephilus gayaquilensis*

                            Arid Pacific lowlands of sw Colombia to nw Peru (Cajamarca)

☐ **Cream-backed Woodpecker** *Campephilus leucopogon*

                            Dry woodlands of n-cent. Bolivia to n Argentina and se Brazil

☐ **Magellanic Woodpecker** *Campephilus magellanicus*

                            Nothofagus forests of sw Argentina and s Chile

☐ **Imperial Woodpecker** *Campephilus imperialis*

                            Sierra Madre Occidental of w Mexico. Probably extinct

☐ **Ivory-billed Woodpecker** *Campephilus principalis*

    ____ *C. p. principalis*   Big Woods Arkansas; once considered extinct from se US

    ____ *C. p. bairdii*   Formerly Cuba (last recorded 1987). Probably extinct

☐ **Banded Woodpecker** *Picus mineaceus*

    ____ *P. m. perlutus*   S Burma and peninsular Thailand

    ____ *P. m. malaccensis*   Malay Peninsula, Sumatra, Borneo, Bangka and Belitung islands

    ____ *P. m. niasensis*   Nias I. (off nw Sumatra)

    ____ *P. m. mineaceus*   Java

☐ **Lesser Yellownape** *Picus chlorolophus*

    ____ *P. c. chlorolophus*   E Nepal to Burma and n Vietnam

    ____ *P. c. simlae*   N India (Himachal Pradesh) to w Nepal

    ____ *P. c. annamensis*   SE Thailand to s Vietnam

    ____ *P. c. chlorigaster*   Peninsular India

    ____ *P. c. wellsi*   Sri Lanka

    ____ *P. c. citrinocristatus*   N Vietnam (Tonkin) and se China (Fujian)

    ____ *P. c. longipennis*   Hainan I. (s China)

    ____ *P. c. rodgeri*   Highlands of w Malaysia

    ____ *P. c. vanheysti*   Highlands of Sumatra

☐ **Crimson-winged Woodpecker** *Picus puniceus*

    ____ *P. p. observandus*   S Burma and pen. Thailand to Sumatra, Borneo and Bangka I.

    ____ *P. p. soligae*   Nias I. (off nw Sumatra)

    ____ *P. p. puniceus*   Java

☐ **Greater Yellownape** *Picus flavinucha*

    ____ *P. f. flavinucha*   N and e India to Burma, sw China and n Vietnam

    ____ *P. f. styani*   Hainan I. and immediately adjacent mainland China

    ____ *P. f. ricketti*   N Vietnam (Tonkin) to se China (Fujian)

    ____ *P. f. pierrei*   SE Thailand to s Vietnam

____	*P. f. mystacalis*	N Sumatra
____	*P. f. korinchi*	SW Sumatra
____	*P. f. wrayi*	Highlands of Malay Peninsula

□ **Checker-throated Woodpecker Picus mentalis**

____	*P. m. humii*	S Burma and s pen. Thailand to Sumatra, Borneo and Bangka I.
____	*P. m. mentalis*	Evergreen and moss forests of w Java

□ **Streak-breasted Woodpecker Picus viridanus**

Lowlands of Burma and sw Thailand to extreme nw Malay Pen.

□ **Laced Woodpecker Picus vittatus**

SE Asia, Sumatra, Java, Bali and Kangean Islands

□ **Streak-throated Woodpecker Picus xanthopygaeus**

Indian subcontinent to Burma, sw China and SE Asia

□ **Scaly-bellied Woodpecker Picus squamatus**

____	*P. s. squamatus*	NE Afghanistan to n India (Darjeeling)
____	*P. s. flavirostris*	Afghanistan (except ne) and w Pakistan

□ **Japanese Woodpecker Picus awokera**

Japan (Honshu to Tanegashima and offshore islands)

□ **Green Woodpecker Picus viridis**

____	*P. v. viridis*	Britain south to France, Alps, n Yugoslavia and Romania
____	*P. v. karelini*	SE Europe to Asia Minor, n Iran and sw Turkmenistan
____	*P. v. innominatus*	Zagros Mountains (sw Iran)
____	*P. v. sharpei*	Iberian Peninsula and the Pyrénées

□ **Levaillant's Woodpecker Picus vaillantii**

Mountains of Morocco, Algeria and Tunisia

□ **Red-collared Woodpecker Picus rabieri**

SW China (extreme s Yunnan) to Laos and Vietnam

□ **Black-headed Woodpecker Picus erythropygius**

____	*P. e. nigrigenis*	Burma and w Thailand
____	*P. e. erythropygius*	NE Thailand and Indochina

□ **Gray-faced Woodpecker Picus canus**

____	*P. c. canus*	Europe (s Scandinavia and France) to w Siberia
____	*P. c. jessoensis*	E Siberia to ne China, Korea, Sakhalin and Hokkaido
____	*P. c. guerini*	N-central China (central Sichuan to Yangtze River basin)
____	*P. c. sobrinus*	SE China (Guangxi to Fujian) and ne Vietnam
____	*P. c. tancolo*	Hainan I. and Taiwan
____	*P. c. kogo*	Central China (Shaanxi to Qinghai and n Sichuan)
____	*P. c. sordidior*	SE Tibet to sw China (Sichuan and Yunnan) and ne Burma
____	*P. c. hessei*	Nepal and n India to Burma, s China, Thailand and Vietnam
____	*P. c. sanguiniceps*	NE Pakistan to n India and extreme w Nepal
____	*P. c. robinsoni*	Malaysia (Gunung Tahan and Cameron Highlands)
____	*P. c. dedemi*	Highlands of Sumatra

□ **Olive-backed Woodpecker Dinopium rafflesii**

____	*D. r. rafflesii*	S Burma, pen. Thailand, Malay Pen., Sumatra and Bangka I.
____	*D. r. dulitense*	Borneo

□ **Himalayan Flameback Dinopium shorii**

____	*D. s. shorii*	Himalayas (nw India to n Bangladesh); w India (Western Ghats)
____	*D. s. anguste*	W Burma and adjacent ne India (Assam)

□ **Common Flameback Dinopium javanense**

____	*D. j. malabaricum*	Wet woodlands of w India
____	*D. j. intermedium*	Bangladesh and Assam to Burma, sw China and Indochina
____	*D. j. javanense*	Peninsular Thailand to Sumatra, Riau Arch., w Java and Borneo

____	*D. j. exsul*	E Java and Bali
____	*D. j. raveni*	Eraban I. and adjacent ne Borneo
____	*D. j. everetti*	S Philippines (Balabac, Palawan and Calamian Islands)

☐ **Black-rumped Flameback** *Dinopium benghalense*

____	*D. b. benghalense*	N India to Assam and sw Burma
____	*D. b. dilutum*	Pakistan
____	*D. b. puncticolle*	Central and s India; n Sri Lanka
____	*D. b. psarodes*	Central and s Sri Lanka

☐ **Greater Flameback** *Chrysocolaptes lucidus*

____	*C. l. guttacristatus*	NW India and Nepal to sw China (Yunnan), Thailand, Indochina
____	*C. l. socialis*	Coastal w India
____	*C. l. stricklandi*	Sri Lanka
____	*C. l. chersonesus*	Malay Peninsula to Sumatra and w Java
____	*C. l. andrewsi*	NE Borneo
____	*C. l. strictus*	E Java
____	*C. l. kangeanensis*	Coastal e Java, Bali and Kangean Islands
____	*C. l. haematribon (grandis)*	N Philippines (Luzon, Polillo, Catanduanes and Marinduque)
____	*C. l. xanthocephalus*	Philippines (Negros, Guimaras, Panay, Masbate and Ticao)
____	*C. l. rufopunctatus*	Philippines (Bohol, Leyte, Samar, Biliran and Panaon)
____	*C. l. erythrocephalus*	Philippines (Balabac, Palawan, Busuanga and Calamian Is.)
____	*C. l. lucidus*	S Philippines (Zamboanga Pen. of w Mindando and Basilan)
____	*C. l. montanus*	Mindanao (except Zamboanga Peninsula) and Samal

☐ **White-naped Woodpecker** *Chrysocolaptes festivus*

____	*C. f. festivus*	Lowlands and hills of sw Nepal and most of India
____	*C. f. tantus*	Sri Lanka

☐ **Pale-headed Woodpecker** *Gecinulus grantia*

____	*G. g. grantia*	E Nepal to Burma and s China (w Yunnan)
____	*G. g. indochinensis*	SW China (sw Yunnan) to nw Thailand, Laos and Vietnam
____	*G. g. viridanus*	SE China (Fujian and n Guangdong)

☐ **Bamboo Woodpecker** *Gecinulus viridis*

Lowlands of e Burma, adjacent Thailand, n Laos and Malay Pen.

☐ **Okinawa Woodpecker** *Sapheopipo noguchii*

Okinawa (Yambaru Moutains). On verge of extinction

☐ **Maroon Woodpecker** *Blythipicus rubiginosus*

S Burma, sw Thailand, Malay Peninsula, Sumatra and Borneo

☐ **Bay Woodpecker** *Blythipicus pyrrhotis*

____	*B. p. pyrrhotis*	Nepal to s China (Sichuan, Yunnan), Laos and n Vietnam
____	*B. p. sinensis*	SE China (Guizhou and Guangxi to Fujian)
____	*B. p. annamensis*	Highlands of s Vietnam
____	*B. p. hainanus*	Mountains of Hainan I. (s China)
____	*B. p. cameroni*	Highlands of Malay Peninsula

☐ **Orange-backed Woodpecker** *Reinwardtipicus validus*

____	*R. v. xanthopygius*	Extreme s Thailand to Sumatra, Borneo, Riau Arch., n Natuna Is.
____	*R. v. validus*	Java

☐ **Buff-rumped Woodpecker** *Meiglyptes tristis*

____	*M. t. grammithorax*	S Burma and pen. Thailand to Sumatra, Borneo and adj. islands
____	*M. t. tristis*	Java

☐ **Black-and-buff Woodpecker** *Meiglyptes jugularis*

Humid forests of Burma, Thailand and Indochina

☐ **Buff-necked Woodpecker** *Meiglyptes tukki*

____	*M. t. tukki*	S Burma, Malay Pen., Sumatra, n Borneo and adjacent islands

_____	*M. t. percnerpes*	S Borneo
_____	*M. t. batu*	Batu Islands (off w Sumatra)
_____	*M. t. pulonis*	Banggi I. (off n Borneo)
_____	*M. t. infuscatus*	Nias I. (off nw Sumatra)

☐ **Gray-and-buff Woodpecker** *Hemicircus concretus*

_____	*H. c. sordidus*	S Burma, pen. Thailand to Sumatra, Borneo and adj. islands
_____	*H. c. concretus*	W and central Java

☐ **Heart-spotted Woodpecker** *Hemicircus canente*

Humid forests of pen. India, Burma, Thailand and Indochina

☐ **Ashy Woodpecker** *Mulleripicus fulvus*

_____	*M. f. fulvus*	N Sulawesi, Bangka, Lembeh, Manterawu and Togian islands
_____	*M. f. wallacei*	S Sulawesi, Muna and Butung islands

☐ **Sooty Woodpecker** *Mulleripicus funebris*

_____	*M. f. funebris*	Philippines (central and s Luzon, Catanduanes and Marinduque)
_____	*M. f. mayri*	N Luzon (n Philippines)
_____	*M. f. parkesi*	Polillo I. (Philippines)
_____	*M. f. fuliginosus*	Philippines (Samar, Leyte and Mindanao)

☐ **Great Slaty Woodpecker** *Mulleripicus pulverulentus*

_____	*M. p. harterti*	N India and Nepal to n Burma, sw China and Indochina
_____	*M. p. pulverulentus*	Malay Pen. to Borneo, Sumatra, Java, Balabac and Palawan

Part II

Passerines: Perching Birds
Order: Passeriformes

FAMILY: EURYLAIMIDAE (Broadbills—15)

☐ **African Broadbill** *Smithornis capensis*
____	*S. c. delacouri*	Sierra Leone to Ghana
____	*S. c. camarunensis*	Cameroon to Gabon and C African Republic
____	*S. c. albigularis*	Angola to Zaire, n Zambia, w Tanzania and n Malawi
____	*S. c. medianus*	Highlands of s Kenya and n Tanzania
____	*S. c. meinertzhageni*	Highlands of ne Zaire, Uganda and w Kenya
____	*S. c. medianus*	C Kenya and ne Tanzania
____	*S. c. suahelicus*	SE Kenya, e Tanzania and ne Mozambique
____	*S. c. conjunctus*	Caprivi Strip and middle Zambezi Valley to nw Mozambique
____	*S. c. cryptoleucus*	S Malawi and se Tanzania to ne Transvaal and Mozambique
____	*S. c. capensis*	South Africa (coastal Natal and s Zululand)

☐ **Gray-headed Broadbill** *Smithornis sharpei*
____	*S. s. zenkeri*	Extreme se Nigeria, w Cameroon, n Gabon to sw CAR
____	*S. s. sharpei*	Bioko (Gulf of Guinea)
____	*S. s. eurylaemus*	E Zaire

☐ **Rufous-sided Broadbill** *Smithornis rufolateralis*
____	*S. r. rufolateralis*	E Sierra Leone to w Zaire, sw Congo and n Angola
____	*S. r. budongoensis*	E-c and ne Zaire to w Uganda

☐ **Green Broadbill** *Calyptomena viridis*
____	*C. v. viridis*	Borneo, Sumatra, Nias, Batu, Lingga and n Natuna islands
____	*C. v. caudacuta*	S Myanmar, sw Thailand, Malay Peninsula and adjacent islands
____	*C. v. siberu*	Mentawai Is. (Siberut, North Pagai and South Pagai)

☐ **Hose's Broadbill** *Calyptomena hosii*

Patchily distributed submontane forests of n and c Borneo

☐ **Whitehead's Broadbill** *Calyptomena whiteheadi*

Montane forests of n and c Borneo

☐ **Black-and-red Broadbill** *Cymbirhynchus macrorhynchos*
____	*C. m. affinis*	Southwest Myanmar
____	*C. m. siamensis*	S Myanmar, s Thailand, Cambodia, s Laos and s Vietnam
____	*C. m. malaccensis*	Extreme s Thailand and Malay Peninsula
____	*C. m. macrorhynchos (lemniscatus)*	Sumatra, Borneo, Bangka, Belitung and Palau Laut

☐ **Long-tailed Broadbill** *Psarisomus dalhousiae*
____	*P. d. dalhousiae*	Himalayas to ne India, Myanmar, sw China and Vietnam
____	*P. d. cyanicauda*	Southeast Thailand and Cambodia
____	*P. d. divinus*	S Vietnam (s Annam)
____	*P. d. psittacinus*	Malay Peninsula and Sumatra
____	*P. d. borneensis*	N Borneo

☐ **Silver-breasted Broadbill** *Serilophus lunatus*
____	*S. l. rubropygius*	NE India, Bhutan and e Bangladesh to ne Myanmar
____	*S. l. atrestus*	Cent. Myanmar to s China, ne Thaiiland, Laos and nw Vietnam
____	*S. l. elisabethae*	SE China (se Yunnan and sw Guangxi) and e Tonkin
____	*S. l. polionotus*	Mountains of Hainan I.
____	*S. l. lunatus*	S-c and s Myanmar and adjacent nw Thailand
____	*S. l. stolidus*	S Myanmar and peninsular Thailand (except extreme south)
____	*S. l. aphobus*	SE Thailand and Cambodia
____	*S. l. impavidus*	S Laos (Bolaven Plateau)
____	*S. l. rothschildi*	Malay Peninsula and extreme s peninsular Thailand
____	*S. l. intensus*	Sumatra

☐ **Banded Broadbill** *Eurylaimus javanicus*
____	*E. j. friedmanni*	SE Myanmar, Thailand and Indochina
____	*E. j. pallidus*	S Thailand (Isthmus of Kra) and Malay Peninsula

_____ *E. j. harterti*	Sumatra, Riau Archipelago, Bangka I. and Belitung I.
_____ *E. j. brookei*	Borneo and North Natuna Islands
_____ *E. j. javanicus*	Java

☐ **Black-and-yellow Broadbill** *Eurylaimus ochromalus*

S Myanmar to Thailand, Malay Pen., Sumatra, Borneo, adj. is.

☐ **Wattled Broadbill** *Eurylaimus steerii*

_____ *E. s. steerii*	Philippines (Basilan, Malamaui and Mindanao (Zamboanga Pen.)
_____ *E. s. mayri*	S Philippines (Dinagat, Poneas, Siargao and Mindanao)

☐ **Visayan Broadbill** *Eurylaimus samarensis*

C Philippines (Leyte, Samar and Bohol)

☐ **Dusky Broadbill** *Corydon sumatranus*

_____ *C. s. laoensis*	S Myanmar to n Thailand and Indochina
_____ *C. s. sumatranus*	Extreme s Thailand, Malay Peninsula, Sumatra and Penang I.
_____ *C. s. brunnescens*	Northwest Borneo (Sarawak) and North Natuna Islands
_____ *C. s. orientalis*	Borneo (except northwest)

☐ **Grauer's Broadbill** *Pseudocalyptomena graueri*

E Zaire and sw Uganda

FAMILY: PHILEPITTIDAE (Asities or False-Sunbirds—4)

☐ **Velvet Asity** *Philepitta castanea*

Humid montane slopes of e Madagascar

☐ **Schlegel's Asity** *Philepitta schlegeli*

Patchily distributed dense forests of w Madagascar

☐ **Sunbird Asity** *Neodrepanis coruscans*

Highland forests of e Madagascar

☐ **Yellow-bellied Asity** *Neodrepanis hypoxantha*

Rain forests of e-c Madagascar

FAMILY: SAPAYAOIDAE (Sapayoa—1)

☐ **Broad-billed Sapayoa** *Sapayoa aenigma* | Tropical central Panama to Colombia and extreme nw Ecuador

FAMILY: PITTIDAE (Pittas—32)

☐ **Eared Pitta** *Pitta phayrei*

Bangladesh to Myanmar, Thailand, s China, s Cambodia, c Annam

☐ **Blue-naped Pitta** *Pitta nipalensis*

_____ *P. n. nipalensis (nuchalis)*	Himalayas (c Nepal to se Tibet, ne India and Myanmar)
_____ *P. n. hendeei*	S China (se Yunnan, sw Guangxi) to n Vietnam and n Laos

☐ **Blue-rumped Pitta** *Pitta soror*

_____ *P. s. tonkinensis*	S China (Guangxi) and n Vietnam (c Tonkin)
_____ *P. s. petersi*	SE Tonkin, n Annam and c Laos
_____ *P. s. douglasi*	Hainan I. (Seven Finger Mountains)
_____ *P. s. soror (annamensis)*	SE Laos and Vietnam (s Annam and Cochinchina)
_____ *P. s. flynnstonei*	SE Thailand and sw Cambodia

☐ **Rusty-naped Pitta** *Pitta oatesi*

_____ *P. o. oatesi*	Myanmar to ne Laos and sw Thailand
_____ *P. o. castaneiceps*	S China (s Yunnan) to c Laos and nw Vietnam
_____ *P. o. bolovenensis*	S Laos and s Annam
_____ *P. o. deborah*	C Peninsular Malaysia

☐ **Schneider's Pitta** *Pitta schneideri*

Batak Mountains of n Sumatra (critically endangered)

☐ **Giant Pitta** *Pitta caerulea*
- ____ *P. c. caerulea*
- ____ *P. c. hosei*

S Myanmar, s Thailand, Malay Peninsula and Sumatra
Borneo

☐ **Blue Pitta** *Pitta cyanea*
- ____ *P. c. cyanea*
- ____ *P. c. aurantiaca*
- ____ *P. c. willoughbyi*

E Bangladesh, ne India, s China, Myanmar, Thailand, Indochina
Mts. of sw Cambodia and se Thailand
Mts. of c Laos to s Annam

☐ **Banded Pitta** *Pitta guajana*
- ____ *P. g. ripleyi*
- ____ *P. g. irena*
- ____ *P. g. schwaneri*
- ____ *P. g. guajana (affinis)*

S Peninsular Thailand
Malay Peninsula and Sumatra
Borneo
Java and Bali

☐ **Bar-bellied Pitta** *Pitta elliotii*

Extreme se Thailand, Laos, Vietnam and Cambodia

☐ **Gurney's Pitta** *Pitta gurneyi*

Peninsular Thailand and s Myanmar (critically endangered)

☐ **Blue-headed Pitta** *Pitta baudii*

Primary lowland forests of Borneo

☐ **Superb Pitta** *Pitta superba*

Manus I. (Admiralty Islands)

☐ **Ivory-breasted Pitta** *Pitta maxima*
- ____ *P. m. maxima*
- ____ *P. m. morotaiensis*

N Moluccas (Halmahera, Bacan, Kasiruta and Obi)
Morotai I. (n Moluccas)

☐ **Blue-banded Pitta** *Pitta arquata*

Primary submontane forests of n half of Borneo

☐ **Garnet Pitta** *Pitta granatina*
- ____ *P. g. granatina*
- ____ *P. g. coccinea*

Borneo (except north)
Malay Peninsula, s peninsular Thailand, s Myanmar and Sumatra

☐ **Black-headed Pitta** *Pitta ussheri*

Lowlands of n Borneo

☐ **Black-crowned Pitta** *Pitta venusta*

Highlands of Sumatra

☐ **African Pitta** *Pitta angolensis*
- ____ *P. a. pulih*
- ____ *P. a. angolensis*
- ____ *P. a. longipennis*

Sierra Leone and se Guinea to w Cameroon
Southwest Cameroon to Gabon, nw Angola and w Zaire
Intratropical migrant (se Zaire to e South Africa)

☐ **Green-breasted Pitta** *Pitta reichenowi*

Cameroon to n Gabon, sw CAR, Zaire and sw Uganda

☐ **Azure-breasted Pitta** *Pitta steerii*
- ____ *P. s. coelestis*
- ____ *P. s. steerii*

Philippines (Samar, Leyte and Bohol)
S Philippines (Mindanao)

☐ **Hooded Pitta** *Pitta sordida*
- ____ *P. s. cucullata*
- ____ *P. s. abbotti*
- ____ *P. s. mulleri*
- ____ *P. s. bangkana*
- ____ *P. s. sordida*
- ____ *P. s. palawanensis*
- ____ *P. s. sanghirana*
- ____ *P. s. forsteni*
- ____ *P. s. novaeguineae*

Foothills of n India to s China (Yunnan) and Indochina
Nicobar Islands
Extreme s Thailand, n Malay Pen., Greater Sundas and w Sulu Is.
Bangka and Belitung islands (off Sumatra)
Philippine Islands (except Palawan group)
W Philippines (Palawan, Culion, Balabac, Calauit and Busuanga)
Sangihe I. (ne of Sulawesi)
N Sulawesi (Minahassa Peninsula)
W Papuan islands, New Guinea and Karkar I.

____	*P. s. mefoorana*	Numfor I. (off nw New Guinea)
____	*P. s. rosenbergii*	Biak I. (off nw New Guinea)
____	*P. s. goodfellowi*	Aru Islands (off s New Guinea)

☐ **Whiskered Pitta** *Pitta kochi*

Mountains of n Luzon (n Philippines)

☐ **Red-bellied Pitta** *Pitta erythrogaster*

____	*P. e. erythrogaster (thompsoni)*	Philippine Islands (except Palawan group)
____	*P. e. propinqua*	SW Philippines (Palawan and Balabac)
____	*P. e. caeruleitorques*	Sangihe I. (ne of Sulawesi)
____	*P. e. inspeculata*	Talaud Islands
____	*P. e. palliceps*	Siau and Tahulandang islands (Celebes Sea)
____	*P. e. celebensis*	Sulawesi, Manterawu I. and Togian Islands
____	*P. e. rufiventris (obbiensis)*	Morotai, Halmahera, Kasiruta, Bacan, Moti, Damar and Obi islands
____	*P. e. cyanonota*	Ternate I. (North Moluccas)
____	*P. e. bernsteini*	Gebe I. (east of Halmahera)
____	*P. e. rubrinucha*	Buru I. (South Moluccas)
____	*P. e. piroensis*	Seram I. (South Moluccas)
____	*P. e. mackloti (kuehni, digglesi)*	W and s New Guinea, w Papuan Is., n and e Cape York Peninsula
____	*P. e. habenichti*	N New Guinea (Weyland Mts. to Astrolabe Bay)
____	*P. e. aruensis*	Aru Islands (off sw New Guinea)
____	*P. e. oblita*	Papua New Guinea (upper montane Aroa River region)
____	*P. e. loriae*	Extreme se New Guinea (e from Kumusi River and Cloudy Bay)
____	*P. e. extima*	New Hanover (Bismarck Archipelago)
____	*P. e. novaehibernicae*	New Ireland (Bismarck Archipelago)
____	*P. e. splendida*	Tabar I. (e of New Ireland)
____	*P. e. gazellae*	S Bismarck Archipelago (New Britain and adjacent islands)
____	*P. e. finschii*	D'Entrecasteaux Archipelago (Fergusson and Goodenough is.)
____	*P. e. meeki*	Rossel I. (Louisiade Archipelago)

☐ **Sula Pitta** *Pitta dohertyi*

Banggai Is. (Peleng, Banggai) and Sula Is. (Taliabu, Seho, Mangole)

☐ **Indian Pitta** *Pitta brachyura*

Widespread Indian subcontinent

☐ **Fairy Pitta** *Pitta nympha*

S Japan to Korea and se China; winters to SE Asia and Borneo

☐ **Blue-winged Pitta** *Pitta moluccensis*

SE Asia to Borneo; winters to Gr. Sundas; vagrant to nw Australia

☐ **Mangrove Pitta** *Pitta megarhyncha*

S Bangladesh to Malay Pen, Sumatra, Bangka I. and Riau Arch.

☐ **Elegant Pitta** *Pitta elegans*

____	*P. e. concinna (hutzi, everetti)*	Lombok, Sumbawa, Flores, Adonara, Lomblen and Alor islands
____	*P. e. maria*	Lesser Sundas (Sumba and s Flores)
____	*P. e. virginalis (kalaonensis, plesseni)*	Tanahjampea, Kalaotoa and Kalao islands (off Sulawesi)
____	*P. e. elegans*	Sangihe I., Sula Islands and c Moluccas
____	*P. e. vigorsii*	Seram, Watubela, Banda, Tayandu, Kai and Tanimbar islands

☐ **Noisy Pitta** *Pitta versicolor*

____	*P. v. simillima*	NE Queensland (e Cape York Pen. and islands in Torres Strait)
____	*P. v. versicolor*	E Australia (Gladstone, Queensland to Sydney area, NSW)

☐ **Black-faced Pitta** *Pitta anerythra*

____	*P. a. pallida*	Bougainville (n Solomon Islands)
____	*P. a. nigrifrons*	Choiseul (c Solomon Islands)
____	*P. a. anerythra*	Santa Isabel (c Solomon Islands)

☐ **Rainbow Pitta** *Pitta iris*

____	*P. i. johnstoneiana*	NW Western Australia (nw Kimberley region)
____	*P. i. iris*	N Australia (Arnhem Land, Melville I. and Groote Eylandt)

FAMILY: FURNARIIDAE (Ovenbirds and Woodcreepers—294)

☐ **Campo Miner** *Geobates poecilopterus*

S-c Brazil (Minas Gerais to São Paulo) and adjacent ne Bolivia

☐ **Common Miner** *Geositta cunicularia*

    ____ *G. c. juninensis* — Highlands of Peru (Junín and Huancavelica)
    ____ *G. c. titicacae* — High plateau of s Peru to extreme n Chile and nw Argentina
    ____ *G. c. frobeni* — Arid Andes of s Peru (Arequipa and Tacna)
    ____ *G. c. georgei* — *Lomas* of coastal s Peru
    ____ *G. c. deserticolor* — Arid littoral of sw Peru (Arequipa) to n Chile (Atacama)
    ____ *G. c. fissirostris* — C Chile (s Atacama to Llanquihué)
    ____ *G. c. contrerasi* — W-c Argentina (Sierra Grandes in Córdoba)
    ____ *G. c. hellmayri* — W Argentina
    ____ *G. c. cunicularia* — Extreme s Brazil to Uruguay, Argentina and Tierra del Fuego

☐ **Slender-billed Miner** *Geositta tenuirostris*

    ____ *G. t. tenuirostris* — Andes of n Ecuador to nw Argentina and extreme n Chile
    ____ *G. t. kalimayae* — Andes of n Ecuador (w slopes of Volcán Iliniza)

☐ **Short-billed Miner** *Geositta antarctica*

Breeds s Chile and s Argentina; winters north to Mendoza

☐ **Grayish Miner** *Geositta maritima*

Arid littoral of Peru (Ancash) to n Chile (Atacama)

☐ **Coastal Miner** *Geositta peruviana*

    ____ *G. p. paytae* — Arid littoral of Peru (Tumbes to Ancash)
    ____ *G. p. peruviana* — Arid littoral of c Peru (Lima)
    ____ *G. p. rostrata* — Arid littoral of s-c Peru (Ica)

☐ **Dark-winged Miner** *Geositta saxicolina*

Andes of c Peru (w Junín, Huancavelica, Lima and Pasco)

☐ **Puna Miner** *Geositta punensis*

Andes of extreme s Peru to w Bolivia, n Chile and nw Argentina

☐ **Rufous-banded Miner** *Geositta rufipennis*

    ____ *G. r. fasciata* — W Bolivia and Pacific slope of n and c Chile
    ____ *G. r. harrisoni* — N Chile (sw Antofogasta)
    ____ *G. r. rufipennis* — Northwest Argentina (Jujuy to San Juan)
    ____ *G. r. fragai* — Northwest Argentina (Cerro Famatina in La Rioja)
    ____ *G. r. ottowi* — W-c Argentina (Sierra de Córdoba)
    ____ *G. r. hoyi* — W Argentina (Mendosa to Neuquén) and s Chile (s Aysén)
    ____ *G. r. giaii* — Southwest Argentina (s Neuquén to Chubut)

☐ **Creamy-rumped Miner** *Geositta isabellina*

Andes of c Chile and adjacent w Argentina

☐ **Thick-billed Miner** *Geositta crassirostris*

W slope of Andes of Peru (Lima south to Arequipa)

☐ **Scale-throated Earthcreeper** *Upucerthia dumetaria*

    ____ *U. d. peruana* — Andes of s Peru (Puno)
    ____ *U. d. hypoleuca (hallinani)* — W Bolivia, c Chile and w Argentina
    ____ *U. d. saturatior* — C Chile and sw Argentina
    ____ *U. d. dumetaria* — S Chile and s and c Argentina to Tierra del Fuego

☐ **White-throated Earthcreeper** *Upucerthia albigula*

W slope of Andes of sw Peru and extreme n Chile (Arica)

☐ **Plain-breasted Earthcreeper** *Upucerthia jelskii*

    ____ *U. j. saturata* — W Andes of c Peru (Ancash, Huánuco and nw Pasco)
    ____ *U. j. jelskii* — Andes of c Peru (Lima, Junín and Huancavelica)
    ____ *U. j. pallida* — Andes of s Peru, n Chile, w Bolivia and nw Argentina

☐ **Buff-breasted Earthcreeper** *Upucerthia validirostris*

Andes of extreme s Bolivia (Potosí) and nw Argentina

☐ **Striated Earthcreeper** *Upucerthia serrana*
    ____   *U. s. serrana*   Andes of Peru (Cajamarca to Lima)
    ____   *U. s. huancavelicae*   Andes of sw Peru (Huancavelica)

☐ **Rock Earthcreeper** *Upucerthia andaecola*

Andes of w Bolivia to nw Argentina and adjacent n Chile

☐ **Straight-billed Earthcreeper** *Upucerthia ruficaudus*
    ____   *U. r. montana*   *Puna* of s Peru (Arequipa and Tacna)
    ____   *U. r. ruficaudus*   *Puna* of w Bolivia to n Chile and nw Argentina
    ____   *U. r. famatinae*   N Argentina

☐ **Bolivian Earthcreeper** *Ochetorhynchus harterti*

Andes of s Bolivia (Cochabamba, w Santa Cruz and Chuquisaca)

☐ **Chaco Earthcreeper** *Ochetorhynchus certhioides*
    ____   *O. c. estebani*   *Chaco* of se Bolivia, w Paraguay and adjacent n Argentina
    ____   *O. c. luscinia*   *Chaco* of w Argentina (La Rioja and Mendoza)
    ____   *O. c. certhioides*   *Chaco* of n and c Argentina

☐ **Band-tailed Earthcreeper** *Eremobius phoenicurus*

S Argentina (Neuquén to s Santa Cruz) and extreme s Chile

☐ **Crag Chilia** *Chilia melanura*
    ____   *C. m. atacamae*   Mountains of n-c Chile (Atacama and Coquimbo)
    ____   *C. m. melanura*   Mountains of c Chile (south to Colchagua)

☐ **Stout-billed Cinclodes** *Cinclodes excelsior*
    ____   *C. e. columbianus*   N Colombia (Santa Marta Mountains and Nevado de Tolima)
    ____   *C. e. excelsior*   Andes of sw Colombia and Andes of Ecuador

☐ **Royal Cinclodes** *Cinclodes aricomae*

Andes of s Peru and extreme w Bolivia (La Paz)

☐ **Bar-winged Cinclodes** *Cinclodes fuscus*
    ____   *C. f. oreobates (paramo)*   Santa Marta Mts., C and Eastern Andes of n Colombia
    ____   *C. f. heterurus*   Andes of w Venezuela (Mérida, Trujillo and s Lara)
    ____   *C. f. albidiventris*   Andes of n Ecuador to n Peru (Piura, Cajamarca)
    ____   *C. f. albiventris (longipennis, rivularis)*   Andes of n Peru, Bolivia, n Chile and nw Argentina
    ____   *C. f. tucumanus*   Northwest Argentina (Tucumán)
    ____   *C. f. rufus*   Andes of nw Argentina Campo de Arenal in Catamarca)
    ____   *C. f. yzurietae*   Andes of nw Argentina (Sierra de Manchao in se Catamarca)
    ____   *C. f. riojanus*   Andes of nw Argentina (Sierra de Famatina in La Rioja)
    ____   *C. f. fuscus*   S Chile and s Argentina; winters to se Brazil, Uruguay, Paraguay

☐ **Comechingones Cinclodes** *Cinclodes comechingonus*

Andes of n-c Argentina (w Córdoba to e Tucumán)

☐ **Long-tailed Cinclodes** *Cinclodes pabsti*

SE Brazil (se Santa Catarina and adjacent ne Rio Grande do Sul)

☐ **Olrog's Cinclodes** *Cinclodes olrogi*

Andes of n-c Argentina (w Córdoba and ne San Luis)

☐ **Gray-flanked Cinclodes** *Cinclodes oustaleti*
    ____   *C. o. oustaleti*   C and s Chile and adjacent Argentina; Chiloé I.
    ____   *C. o. baeckstroemii*   Juan Fernández Islands (off Chile)
    ____   *C. o. hornensis*   Tierra del Fuego and Cape Horn Archipelago

☐ **Dark-bellied Cinclodes** *Cinclodes patagonicus*
    ____   *C. p. chilensis*   C Chile and w Argentina; Chiloé I.
    ____   *C. p. patagonicus*   S Chile and s Argentina to Tierra del Fuego and adjacent islands

☐ **Peruvian Seaside Cinclodes** *Cinclodes taczanowskii*

Rocky coasts of Peru (Ancash south to Tacna)

☐ **Chilean Seaside Cinclodes** *Cinclodes nigrofumosus*

Rocky coasts of Chile (Arica south to Valdivia)

☐ **Blackish Cinclodes** *Cinclodes antarcticus*
____ *C. a. maculirostris*   Tierra del Fuego and Cape Horn Archipelago
____ *C. a. antarcticus*   Falkland Islands

☐ **White-winged Cinclodes** *Cinclodes atacamensis*
____ *C. a. atacamensis*   Andes of Peru to w Bolivia, n Chile and n Argentina
____ *C. a. schocolatinus*   C Argentina (Sierra de Córdoba and ne San Luis)

☐ **White-bellied Cinclodes** *Cinclodes palliatus*

Andes of c Peru (Junín, Lima and Huancavelica)

☐ **Pale-legged Hornero** *Furnarius leucopus*
____ *F. l. longirostris*   Arid coastal n Colombia and nw Venezuela
____ *F. l. endoecus*   N Colombia (Magdalena Valley) and w Venezuela (s Zulia)
____ *F. l. leucopus*   Interior sw Guyana and adjacent n Brazil
____ *F. l. cinnamomeus*   SW Ecuador and nw Peru
____ *F. l. tricolor*   Amazonian Peru, adjacent w Brazil and n Bolivia (Beni)
____ *F. l. araguaiae*   S-c Brazil (e Mato Grosso and w Goiás)
____ *F. l. assimilis*   E and s Brazil to extreme se Bolivia (se Santa Cruz)

☐ **Bay Hornero** *Furnarius torridus*

Along Amazon R. (s Colombia, ne Ecuador, ne Peru and w Brazil)

☐ **Lesser Hornero** *Furnarius minor*

Amazonian Brazil, s Colombia, e Ecuador and ne Peru

☐ **Rufous Hornero** *Furnarius rufus*
____ *F. r. albogularis*   SE Brazil (Goiás and Bahia to São Paulo)
____ *F. r. commersoni*   W Brazil (Mato Grosso) and adjacent Bolivia
____ *F. r. schuhmacheri*   N and e Bolivia (La Paz and Beni to Tarija)
____ *F. r. paraguayae*   Paraguay and n Argentina
____ *F. r. rufus*   S Brazil and Uruguay to c Argentina

☐ **Crested Hornero** *Furnarius cristatus*

Chaco of se Bolivia to w Paraguay and n Argentina

☐ **Tail-banded Hornero** *Furnarius figulus*
____ *F. f. pileatus*   Lower Amazonian Brazil
____ *F. f. figulus*   E Brazil (south to Minas Gerais and Espírito Santo)

☐ **Curve-billed Reedhaunter** *Limnornis curvirostris*

Extreme s Brazil to s Uruguay and e Argentina

☐ **Straight-billed Reedhaunter** *Limnornis rectirostris*

Extreme s Brazil to s Uruguay and e Argentina

☐ **Wren-like Rushbird** *Phleocryptes melanops*
____ *P. m. brunnescens*   Coastal w Peru (Trujillo to Pisco)
____ *P. m. schoenobaenus (juninensis)*   Highlands of s Peru, w Bolivia and nw Argentina
____ *P. m. loaensis*   N Chile (Tarapacá)
____ *P. m. melanops*   S Brazil to c Chile, c Argentina, Paraguay and Uruguay

☐ **Thorn-tailed Rayadito** *Aphrastura spinicauda*
____ *A. s. spinicauda*   C Chile and adjacent w Argentina south to Tierra del Fuego
____ *A. s. bullocki*   Mocha I. (Chile)
____ *A. s. fulva*   Chiloé I. (Chile)

☐ **Masafuera Rayadito** *Aphrastura masafuerae*

Alejandro Selkirk I. (Juan Fernández Islands off Chile)

☐ **Brown-capped Tit-Spinetail** *Leptasthenura fuliginiceps*
____ *L. f. fuliginiceps*   Andes of w Bolivia (north to La Paz)
____ *L. f. paranensis*   Andes of w Argentina and mountains of Córdoba and San Luis

☐ **Tawny Tit-Spinetail** *Leptasthenura yanacensis*

Locally in Andes of n Peru and w Bolivia

☐ **Tufted Tit-Spinetail** *Leptasthenura platensis*

S Paraguay; extreme sw Brazil, Uruguay and Argentina

☐ **Plain-mantled Tit-Spinetail** *Leptasthenura aegithaloides*
    ____ *L. a. grisescens* — Coastal arid s Peru and n Chile
    ____ *L. a. berlepschi* — Altiplano of s Peru, Bolivia, n Chile and nw Argentina
    ____ *L. a. aegithaloides* — Lowlands of Chile (s Coquimbo to n Aysén)
    ____ *L. a. pallida* — Lowlands of w and s Argentina to Tierra del Fuego

☐ **Striolated Tit-Spinetail** *Leptasthenura striolata*
SE Brazil (Paraná south to n Rio Grande do Sul)

☐ **Rusty-crowned Tit-Spinetail** *Leptasthenura pileata*
    ____ *L. p. latistriata* — Andes of w Peru (Ayacucho)
    ____ *L. p. cajabambae* — Andes of Peru (Cajamarca to Junín and Huancavelica)
    ____ *L. p. pileata* — Cordillera of w Peru (Lima)

☐ **White-browed Tit-Spinetail** *Leptasthenura xenothorax*
Andes of s Peru (Apurímac and Cuzco)

☐ **Streaked Tit-Spinetail** *Leptasthenura striata*
    ____ *L. s. superciliaris* — W slope of coastal cordillera of Peru (Ancash to Lima)
    ____ *L. s. albigularis* — Andes of w Peru (Huancavelica)
    ____ *L. s. striata* — Andes of sw Peru (Arequipa and Tacna) to n Chile (Tarapacá)

☐ **Andean Tit-Spinetail** *Leptasthenura andicola*
    ____ *L. a. extima* — Santa Marta Mountains (ne Colombia)
    ____ *L. a. exterior* — E Andes of Colombia (Boyacá)
    ____ *L. a. andicola* — C Andes of Colombia and Ecuador
    ____ *L. a. certhia* — Andes of w Venezuela (Mérida and Trujillo)
    ____ *L. a. peruviana* — Andes of Peru to n Bolivia (La Paz)

☐ **Araucaria Tit-Spinetail** *Leptasthenura setaria*
SE Brazil (Rio de Janeiro to Rio Grande do Sul) to ne Argentina

☐ **Bay-capped Wren-Spinetail** *Spartonoica maluroides*
Extreme s Brazil to Uruguay and c Argentina

☐ **Des Murs' Wiretail** *Sylviorthorhynchus desmursii*
C and s Chile and adjacent Argentina

☐ **Perija Thistletail** *Schizoeaca perijana*
Sierra de Perijá (Colombia/Venezuela border)

☐ **Ochre-browed Thistletail** *Schizoeaca coryi*
Andes of w Venezuela (n Táchira, Mérida and Trujillo)

☐ **White-chinned Thistletail** *Schizoeaca fuliginosa*
    ____ *S. f. fumigata* — C and Western Andes of Colombia
    ____ *S. f. fuliginosa* — Andes of e Colombia, extreme w Venezuela and n Ecuador
    ____ *S. f. peruviana* — Andes of n Peru (Amazonas and San Martín)
    ____ *S. f. plengei* — Andes of Peru (Cordillera Carpish in Huánuco)

☐ **Mouse-colored Thistletail** *Schizoeaca griseomurina*
Andes of s Ecuador and extreme n Peru

☐ **Eye-ringed Thistletail** *Schizoeaca palpebralis*
Andes of c Peru (Junín)

☐ **Vilcabamba Thistletail** *Schizoeaca vilcabambae*
    ____ *S. v. vilcabambae* — S Peru (Cordillera Vilcabamba in Cuzco)
    ____ *S. v. ayacuchensis* — Andes of s Peru (Ayacucho)

☐ **Puna Thistletail** *Schizoeaca helleri*
Andes of s Peru (s Cuzco and Puno)

☐ **Black-throated Thistletail** *Schizoeaca harterti*
Andes of w Bolivia (La Paz, Cochabamba and w Santa Cruz)

☐ **Itatiaia Thistletail** *Oreophylax moreirae*
High mountains of se Brazil (s Espírito Santo to ne São Paulo)

☐ **Chotoy Spinetail** *Schoeniophylax phryganophilus*
    ____ *S. p. phryganophilus* — N and e Bolivia to s Brazil, Uruguay, Paraguay and n Argentina
    ____ *S. p. petersi* — Interior e Brazil (n Minas Gerais and w Bahia)

☐ **Rufous-capped Spinetail** *Synallaxis ruficapilla*

E Paraguay to se Brazil and ne Argentina (Misiones)

☐ **Bahia Spinetail** *Synallaxis whitneyi*

E Brazil (interior and s Bahia and ne Minas Gerais)

☐ **Pinto's Spinetail** *Synallaxis infuscata*

NE Brazil (ne Maranhão, e Pernambuco and adjacent Alagoas)

☐ **Gray-bellied Spinetail** *Synallaxis cinerascens*

E Paraguay to se Brazil, n Uruguay and ne Argentina

☐ **Silvery-throated Spinetail** *Synallaxis subpudica*

E Andes of n Colombia (Boyacá and Cundinamarca)

☐ **Sooty-fronted Spinetail** *Synallaxis frontalis*
_____ *S. f. frontalis* — E and c Brazil to Paraguay, Uruguay and n Argentina
_____ *S. f. fuscipennis* — E Bolivia and nw Argentina

☐ **Azara's Spinetail** *Synallaxis azarae*
_____ *S. a. elegantior* — E Andes of Colombia and w Venezuela
_____ *S. a. media* — W and C Andes of Colombia and n Ecuador
_____ *S. a. ochracea* — Subtropical sw Ecuador and nw Peru
_____ *S. a. fruticicola* — N Peru (La Libertad, Cajamarca, San Martín and Amazonas)
_____ *S. a. infumata* — N-c Peru (San Martín, Huánuco and Junín)
_____ *S. a. urubambae* — SE Peru (Cuzco)
_____ *S. a. carabayae* — Andes of se Peru (Puno) and n Bolivia (La Paz)
_____ *S. a. azarae* — Andes of n Bolivia (Cochabamba)
_____ *S. a. samaipatae* — Andes of s Bolivia (Santa Cruz, Chuquisaca and Tarija)
_____ *S. a. superciliosa* — Andes of nw Argentina (Jujuy and Tucumán)

☐ **Apurimac Spinetail** *Synallaxis courseni*

Andes of s Peru (Bosque de Ampay region of Apurímac)

☐ **Pale-breasted Spinetail** *Synallaxis albescens*
_____ *S. a. latitabunda* — SW Costa Rica
_____ *S. a. hypoleuca* — Pacific coast of s Panama and nw Colombia
_____ *S. a. insignis* — Andes of Colombia
_____ *S. a. occipitalis* — E Colombia and nw Venezuela
_____ *S. a. littoralis* — Caribbean littoral of n Colombia
_____ *S. a. perpallida* — NE Colombia (Guajira Peninsula) and nw Venezuela
_____ *S. a. nesiotis* — Santa Marta Mts. (ne Colombia) to n Venezuela; Isla Margarita
_____ *S. a. trinitatis* — E Venezuela and Trinidad
_____ *S. a. josephinae* — S Venezuela, Guyana, Suriname and adjacent n Brazil
_____ *S. a. inaequalis* — French Guiana
_____ *S. a. griseonota* — C Brazil (lower Rio Tapajós region)
_____ *S. a. albescens* — C and e Brazil to e Paraguay and ne Argentina
_____ *S. a. australis* — E Bolivia to w Paraguay and nw Argentina

☐ **Dark-breasted Spinetail** *Synallaxis albigularis*
_____ *S. a. rodolphei* — S Colombia
_____ *S. a. albigularis* — SE Colombia to e Ecuador, e Peru and w Amazonian Brazil

☐ **Chicli Spinetail** *Synallaxis spixi*

SE Brazil to e Paraguay, Uruguay and ne Argentina

☐ **Cinereous-breasted Spinetail** *Synallaxis hypospodia*

Locally from se Peru and n Bolivia to c and ne Brazil

☐ **Ruddy Spinetail** *Synallaxis rutilans*
_____ *S. r. caquetensis* — Tropical se Colombia to e Ecuador and ne Peru
_____ *S. r. confinis* — NW Brazil (between Rio Negro and Rio Solimões)
_____ *S. r. dissors* — E Colombia to the Guianas and Brazil north of the Amazon
_____ *S. r. amazonica* — E Peru to n Bolivia and w Amazonian Brazil
_____ *S. r. rutilans* — Brazil s of the Amazon (Rio Tapajós to Rio Tocantins)
_____ *S. r. omissa* — Brazil south of the Amazon (Rio Tocantins to n Maranhão)
_____ *S. r. tertia* — NE Bolivia (n La Paz) and adjacent w Brazil (Mato Grosso)

□ **Chestnut-throated Spinetail** *Synallaxis cherriei*
___ *S. c. napoensis* — Extreme s Ecuador to se Colombia and e Peru (San Martín)
___ *S. c. cherriei* — S Amazonian Brazil

□ **Rufous Spinetail** *Synallaxis unirufa*
___ *S. u. unirufa* — Andes of Colombia and e Ecuador
___ *S. u. munoztebari* — Sierra de Perijá (e Colombia) and Andes of w Venezuela
___ *S. u. meridana* — Extreme e Colombia and Andes of w Venezuela
___ *S. u. ochrogaster* — Andes of Peru south to Cordillera Vilcabamba (n Cuzco)

□ **Black-throated Spinetail** *Synallaxis castanea*
Coastal mountains of n Venezuela (Aragua to Distrito Federal)

□ **Rusty-headed Spinetail** *Synallaxis fuscorufa*
Santa Marta Mountains (ne Colombia)

□ **Rufous-breasted Spinetail** *Synallaxis erythrothorax*
___ *S. e. furtiva* — Lowlands of se Mexico (Veracruz, Oaxaca and Tabasco)
___ *S. e. erythrothorax* — SE Mexico (Yucatán Peninsula) to nw Honduras
___ *S. e. pacifica* — Pacific lowlands of s Mexico (Chiapas) to El Salvador

□ **Slaty Spinetail** *Synallaxis brachyura*
___ *S. b. nigrofumosa* — Caribbean slope of Honduras to Panama
___ *S. b. chapmani* — SW Costa Rica to extreme nw Peru (Tumbes)
___ *S. b. caucae* — Tropical and subtropical Colombia (Cauca Valley)
___ *S. b. brachyura* — Magdalena Valley and n Western Andes of Colombia

□ **Black-faced Spinetail** *Synallaxis tithys*
Arid tropical sw Ecuador to extreme nw Peru (Tumbes)

□ **White-bellied Spinetail** *Synallaxis propinqua*
River islands in Amazon basin (Guianas to Bolivia)

□ **McConnell's Spinetail** *Synallaxis macconnelli*
Tepuis of s Venezuela, adjacent Brazil, Guyana and Suriname

□ **Dusky Spinetail** *Synallaxis moesta*
___ *S. m. moesta* — Tropical and subtropical Eastern Andes of Colombia
___ *S. m. obscura* — Tropical se Colombia
___ *S. m. brunneicaudalis* — E Ecuador and ne Peru (San Martín and Loreto)

□ **Cabanis' Spinetail** *Synallaxis cabanisi*
___ *S. c. cabanisi* — E base of Andes of Peru (north to Huánuco)
___ *S. c. fulviventris* — E base of Andes of Bolivia (south to s Beni and Cochabamba)

□ **Marañón Spinetail** *Synallaxis maranonica*
S Ecuador (Zamora-Chinchipe) to nw Peru (n Cajamarca)

□ **Plain-crowned Spinetail** *Synallaxis gujanensis*
___ *S. g. columbiana* — Tropical and subtropical Eastern Andes of Colombia
___ *S. g. gujanensis* — Venezuela to the Guianas and n Brazil
___ *S. g. huallagae* — NE Peru (Loreto)
___ *S. g. canipileus* — SE Peru (Cuzco and Puno)
___ *S. g. inornata* — Brazil south of the Amazon and adjacent Bolivia
___ *S. g. certhiola* — Bolivia (Beni, La Paz and Santa Cruz)
___ *S. g. simoni* — C Brazil (along Rio Araguaia in Goiás)

□ **White-lored Spinetail** *Synallaxis albilora*
___ *S. a. albilora* — E Bolivia (e Santa Cruz) to interior sw Brazil and n Paraguay
___ *S. a. simoni* — Pantanal of w Brazil (Araguaia Valley)

□ **Ochre-cheeked Spinetail** *Synallaxis scutata*
___ *S. s. scutata* — E and c Brazil
___ *S. s. whitii* — E Bolivia to sw Brazil (Mato Grosso) and nw Argentina
___ *S. s. teretiala* — E Brazil (Serra dos Carajás in s Pará)

273

☐ **White-whiskered Spinetail** *Synallaxis candei*
  ____ *S. c. candei* — Arid Caribbean littoral of n Colombia and w Venezuela
  ____ *S. c. atrigularis* — N Colombia (Magdalena Valley)
  ____ *S. c. venezuelensis* — NE Colombia (Guajira Peninsula) and arid nw Venezuela

☐ **Hoary-throated Spinetail** *Synallaxis kollari*
  Locally in campos of n Brazil (Roraima) and (?) adjacent Guyana

☐ **Stripe-breasted Spinetail** *Synallaxis cinnamomea*
  ____ *S. c. cinnamomea* — E Andes of Colombia to nw Venezuela (Sierra de Perijá)
  ____ *S. c. carri* — Trinidad
  ____ *S. c. terrestris* — Tobago
  ____ *S. c. aveledoi* — W Venezuela (Falcón, Lara and n Táchira)
  ____ *S. c. bolivari* — Coastal cordillera of n Venezuela
  ____ *S. c. striatipectus* — Mountains of ne Venezuela (Sucre, Anzoátegui and Monagas)
  ____ *S. c. pariae* — Subtropical mountains of ne Venezuela (Paría Peninsula)

☐ **Russet-bellied Spinetail** *Synallaxis zimmeri*
  W slope of Andes of c Peru (Ancash)

☐ **Necklaced Spinetail** *Synallaxis stictothorax*
  ____ *S. s. stictothorax* — Arid littoral of sw Ecuador and Isla Puná
  ____ *S. s. maculata* — Arid nw Peru (Tumbes, Lambayeque, Piura and La Libertad)

☐ **Chinchipe Spinetail** *Synallaxis chinchipensis*
  NW Peru (upper Marañón and Chinchipe valleys)

☐ **Great Spinetail** *Siptornopsis hypochondriaca*
  NW Peru (upper arid Río Marañón Valley)

☐ **Red-shouldered Spinetail** *Gyalophylax hellmayri*
  Arid ne Brazil (Piauí, n Bahia and w Pernambuco)

☐ **White-browed Spinetail** *Hellmayrea gularis*
  ____ *H. g. gularis* — Andes of Colombia to n Ecuador and w Venezuela
  ____ *H. g. brunneidorsalis* — Sierra de Perijá (Colombia/Venezuela border)
  ____ *H. g. cinereiventris* — Andes of w Venezuela (Mérida and Trujillo)
  ____ *H. g. rufiventris* — Andes of c Peru (Junín)

☐ **Marcapata Spinetail** *Cranioleuca marcapatae*
  ____ *C. m. marcapatae* — Andes of se Peru (Cuzco)
  ____ *C. m. weskei* — SE Peru (cloud forests of Cordillera Vilcabamba in Cuzco)

☐ **Light-crowned Spinetail** *Cranioleuca albiceps*
  ____ *C. a. albiceps* — Andes of extreme s Peru (s Puno) and w Bolivia (La Paz)
  ____ *C. a. discolor* — Andes of Bolivia (Cochabamba and Santa Cruz)

☐ **Rusty-backed Spinetail** *Cranioleuca vulpina*
  ____ *C. v. dissita* — Coiba I. (Panama)
  ____ *C. v. apurensis* — W Venezuela (w Apure)
  ____ *C. v. vulpina (alopecias, solimonensis)* — NE Colombia to Venezuela, Brazil and extreme e Bolivia
  ____ *C. v. foxi* — C Bolivia (e Cochabamba)
  ____ *C. v. reiseri* — NE Brazil (Piauí, w Pernambuco and w Bahia)

☐ **Parker's Spinetail** *Cranioleuca vulpecula*
  River islands in Amazon system from e Brazil to e Peru

☐ **Sulphur-bearded Spinetail** *Cranioleuca sulphurifera*
  S Brazil (Rio Grande do Sul) to Uruguay and e Argentina

☐ **Crested Spinetail** *Cranioleuca subcristata*
  ____ *C. s. fuscivertex* — Sierra de Perijá (Colombia/Venezuela border)
  ____ *C. s. subcristata* — E Andes of Colombia and mountains of n Venezuela

☐ **Stripe-crowned Spinetail** *Cranioleuca pyrrhophia*
  ____ *C. p. rufipennis* — Known from type locality in La Paz, Bolivia
  ____ *C. p. striaticeps* — E Bolivia (Cochabamba, Santa Cruz and Chuquisaca)
  ____ *C. p. pyrrhophia* — S Bolivia to Uruguay, Paraguay, n Argentina and extreme s Brazil

☐ **Bolivian Spinetail** *Cranioleuca henricae*

Andes of n Bolivia (Río La Paz drainage system)

☐ **Olive Spinetail** *Cranioleuca obsoleta*

E Paraguay to se Brazil (São Paulo) and ne Argentina

☐ **Pallid Spinetail** *Cranioleuca pallida*

Mts. of se Brazil (Brasília to Espírito Santo and se São Paulo)

☐ **Gray-headed Spinetail** *Cranioleuca semicinerea*
_____ *C. s. semicinerea* — NE Brazil (Ceará, Alagoas, s Bahia and n Minas Gerais)
_____ *C. s. goyana* — E Brazil (s-c Goiás)

☐ **Creamy-crested Spinetail** *Cranioleuca albicapilla*
_____ *C. a. albicapilla* — Andes of Peru (Junín, Huancavelica, Apurímac and Ayacucho)
_____ *C. a. albigula* — Andes of se Peru (Cuzco)

☐ **Red-faced Spinetail** *Cranioleuca erythrops*
_____ *C. e. rufigenis* — Highlands of Costa Rica and w Panama (Chiriquí)
_____ *C. e. griseigularis* — W Andes and w slope of C Andes of Colombia
_____ *C. e. erythrops* — Andes of w Ecuador (sw Manabí and w Guayas)

☐ **Tepui Spinetail** *Cranioleuca demissa*

Tepuis of s Venezuela, adjacent Guyana and n Brazil (n Roraima)

☐ **Streak-capped Spinetail** *Cranioleuca hellmayri*

Santa Marta Mountains (ne Colombia)

☐ **Ash-browed Spinetail** *Cranioleuca curtata*
_____ *C. c. curtata* — Subtropical Eastern Andes of Colombia
_____ *C. c. cisandina* — Subtrop. Magdalena Valley of c Colombia to n Peru
_____ *C. c. debilis* — Andes of Peru (Junín) to w Bolivia

☐ **Line-cheeked Spinetail** *Cranioleuca antisiensis*
_____ *C. a. antisiensis* — Andes of sw Ecuador (Azuay, El Oro and Loja)
_____ *C. a. furcata* — Andes of nw Peru (San Martín)
_____ *C. a. palamblae* — Andes of nw Peru (south to n Cajamarca and Lambayeque)

☐ **Baron's Spinetail** *Cranioleuca baroni*
_____ *C. b. baroni* — Andes of Peru (Amazonas and Cajamarca)
_____ *C. b. capitalis* — Andes of Peru (Huánuco)
_____ *C. b. zaratensis* — Andes of s Peru (Pasco and Lima)

☐ **Speckled Spinetail** *Cranioleuca gutturata*

S Colombia to n Bolivia, the Guianas and n Amazonian Brazil

☐ **Scaled Spinetail** *Cranioleuca muelleri*

Lower Amazonian Brazil and Mexiana I.

☐ **Yellow-chinned Spinetail** *Certhiaxis cinnamomeus*
_____ *C. c. fuscifrons* — N Colombia (Río Atrato to Santa Marta region)
_____ *C. c. marabinus* — NE Colombia and nw Venezuela
_____ *C. c. valencianus* — W-c Venezuela
_____ *C. c. orenocensis* — E Venezuela (lower Orinoco Valley)
_____ *C. c. cinnamomeus* — NE Venezuela to the Guianas and ne Brazil; Trinidad
_____ *C. c. pallidus* — Extreme se Colombia and w and c Amazonian Brazil
_____ *C. c. cearensis* — E Brazil (s Maranhão, Ceará, Piauí, Pernambuco and n Bahia)
_____ *C. c. russeolus* — E Bolivia to se Brazil, Paraguay, n Argentina and Uruguay

☐ **Red-and-white Spinetail** *Certhiaxis mustelinus*

Extreme se Colombia to e Peru and Amazonian Brazil

☐ **Striated Softtail** *Thripophaga macroura*

Locally in e Brazil (s Bahia to Espírito Santo and n Rio de Janeiro)

☐ **Orinoco Softtail** *Thripophaga cherriei*

N Venezuela (upper Orinoco River in nw Amazonas)

☐ **Plain Softtail** *Thripophaga fusciceps*
_____ *T. f. dimorpha* — Locally in ne Ecuador and ne Peru; se Peru (Pasco south to Puno)

____	*T. f. obidensis*	E Amazonian Brazil (mouth of Rio Madeira to Rio Tapajós)
____	*T. f. fusciceps*	N Bolivia (Beni, e La Paz and n Cochabamba)

☐ **Russet-mantled Softtail** *Thripophaga berlepschi*

Andes of n Peru (Amazonas to La Libertad)

☐ **Canyon Canastero** *Asthenes pudibunda*

____	*A. p. neglecta*	Andes of w Peru (Ancash)
____	*A. p. pudibunda*	*Polylepis* zone of Peru (La Libertad to Tacna)

☐ **Rusty-fronted Canastero** *Asthenes ottonis*

Andes of c Peru (Huancavelica to Cuzco)

☐ **Iquico Canastero** *Asthenes heterura*

Andes of w Bolivia (La Paz and Cochabamba)

☐ **Cordilleran Canastero** *Asthenes modesta*

____	*A. m. proxima*	C and se Peru (Junín and Cuzco)
____	*A. m. modesta*	*Puna* of s Peru to w Bolivia, n Chile and nw Argentina
____	*A. m. rostrata*	Andes of n Bolivia (Cochabamba)
____	*A. m. serrana*	C Argentina (south to Santa Cruz)
____	*A. m. australis*	Andes of Chile (Atacama to Colchagua) and adjacent Argentina

☐ **Cactus Canastero** *Asthenes cactorum*

____	*A. c. cactorum*	Pacific slope of Andes of w Peru (Lima, Ica and Arequipa)
____	*A. c. lachayensis*	C coast of Peru (Lomas de Lachay)

☐ **Streak-throated Canastero** *Asthenes humilis*

____	*A. h. cajamarcae*	Arid temperate w Andes of Peru (Cajamarca)
____	*A. h. humilis*	Arid Andes of Peru (Ancash to Junín and Huancavelica)
____	*A. h. robusta*	Andes of se Peru (Puno) and w Bolivia (La Paz)

☐ **Streak-backed Canastero** *Asthenes wyatti*

____	*A. w. wyatti*	Arid Andes of n Colombia (Santander)
____	*A. w. sanctaemartae*	Santa Marta Mountains (ne Colombia)
____	*A. w. mucuchiesi*	Andes of nw Venezuela (Mérida)
____	*A. w. perijana*	Sierra de Perijá (Colombia/Venezuela border)
____	*A. w. aequatorialis*	W Andes of c Ecuador
____	*A. w. azuay*	Andes of s Ecuador (Azuay)
____	*A. w. graminicola*	Andes of Peru (Junín, Cuzco, Huancavelica and Puno)

☐ **Puna Canastero** *Asthenes sclateri*

____	*A. s. punensis*	Andes of extreme se Peru (Puno) and w Bolivia (La Paz)
____	*A. s. cuchacanchae*	Andes of sw Bolivia (Cochabamba and Potosí)
____	*A. s. lilloi*	Andes of nw Argentina (Tucumán, Catamarca and La Rioja)
____	*A. s. sclateri*	Andes of n-c Argentina (Sierra de Córdoba)

☐ **Austral Canastero** *Asthenes anthoides*

Extreme s Chile and s Argentina to Tierra del Fuego and Staten I.

☐ **Hudson's Canastero** *Asthenes hudsoni*

Extreme s Brazil to Uruguay and e Argentina (s to Río Negro)

☐ **Line-fronted Canastero** *Asthenes urubambensis*

____	*A. u. huallagae*	Andes of n Peru (La Libertad to Huánuco and Pasco)
____	*A. u. urubambensis*	Andes of se Peru (Cuzco) to n Bolivia (La Paz and Cochabamba)

☐ **Many-striped Canastero** *Asthenes flammulata*

____	*A. f. multostriata*	E Andes of Colombia
____	*A. f. quindiana*	C Andes of Colombia (Nevado de Tolima)
____	*A. f. flammulata*	Andes of s Colombia (Nariño) and adjacent Ecuador
____	*A. f. pallida*	Andes of n Peru (La Libertad and Cajamarca)
____	*A. f. taczanowskii*	Andes of n c Peru (La Libertad to Junín)

☐ **Junin Canastero** *Asthenes virgata*

Very local in Andes of c and s Peru

276

☐ **Scribble-tailed Canastero** *Asthenes maculicauda*

Locally in Andes of s Peru (Puno) to w Bolivia and nw Argentina

☐ **Lesser Canastero** *Asthenes pyrrholeuca*
____ *A. p. affinis* — C Argentina and Chile; winters to Bolivia and Paraguay
____ *A. p. pyrrholeuca* — NW Argentina
____ *A. p. sordida* — Chile (Aconcagua to Aysén); w Argentina (Lake Nahuel Huapí)
____ *A. p. flavogularis* — E and s Argentina (Buenos Aires to Santa Cruz)

☐ **Dusky-tailed Canastero** *Asthenes humicola*
____ *A. h. goodalli* — Andes of n Chile (sw Antofogasta)
____ *A. h. humicola* — N and c Chile (Atacama to n Maule)
____ *A. h. polysticta* — S Chile (s Maule, Concepción and Malleco)

☐ **Creamy-breasted Canastero** *Asthenes dorbignyi*
____ *A. d. consobrina* — Andes of nw Bolivia (La Paz)
____ *A. d. dorbignyi* — Andes of Bolivia and nw Argentina

☐ **Dark-winged Canastero** *Asthenes arequipae*

Andes of w Peru (Lima and Ayacucho) to w Bolivia and n Chile

☐ **Pale-tailed Canastero** *Asthenes huancavelicae*
____ *A. h. usheri* — Arid Andes of s-c Peru (Ancash to Apurímac)
____ *A. h. huancavelicae* — Arid Andes of sw Peru (Huancavelica)

☐ **Berlepsch's Canastero** *Asthenes berlepschi*

Andes of w Bolivia (Nevado Illampu region in La Paz)

☐ **Steinbach's Canastero** *Asthenes steinbachi*

Andes of w Argentina (w Salta to Mendoza)

☐ **Short-billed Canastero** *Asthenes baeri*
____ *A. b. chacoensis* — NW Paraguay (Puerto Casado)
____ *A. b. baeri* — W Paraguay to extreme se Brazil, Uruguay and c Argentina

☐ **Cipo Canastero** *Asthenes luizae*

Interior se Brazil (Serra do Cipó in Minas Gerais)

☐ **Patagonian Canastero** *Asthenes patagonica*

S Argentina (Mendoza south to n Santa Cruz)

☐ **Common Thornbird** *Phacellodomus rufifrons*
____ *P. r. inornatus* — *Llanos* of ne Colombia and Venezuela
____ *P. r. castilloi* — S Venezuela (n Bolívar)
____ *P. r. peruvianus* — NW Peru (Río Marañón Valley) and adjacent s Ecuador
____ *P. r. specularis* — NE Brazil (Pernambuco)
____ *P. r. rufifrons* — E Brazil (Piauí, Bahia and Minas Gerais)
____ *P. r. fargoi* — SW Brazil (Mato Grosso) and n Paraguay
____ *P. r. sincipitalis* — Bolivia (Santa Cruz and Tarija) and nw Argentina

☐ **Little Thornbird** *Phacellodomus sibilatrix*

Chaco of w Paraguay to s Bolivia, w Uruguay and n Argentina

☐ **Streak-fronted Thornbird** *Phacellodomus striaticeps*
____ *P. s. griseipectus* — Andes of s Peru (Cuzco, Apurímac and Puno)
____ *P. s. striaticeps* — Andes of w Bolivia and nw Argentina

☐ **Freckle-breasted Thornbird** *Phacellodomus striaticollis*

SE Brazil (e Paraná) to Uruguay and ne Argentina

☐ **Chestnut-backed Thornbird** *Phacellodomus dorsalis*

N Peru (upper Marañón Valley)

☐ **Greater Thornbird** *Phacellodomus ruber*

N Brazil to n Bolivia, Paraguay and n Argentina

☐ **Spot-breasted Thornbird** *Phacellodomus maculipectus*

Andes of s Bolivia and nw Argentina (south to La Rioja)

☐ **Red-eyed Thornbird** *Phacellodomus erythrophthalmus*
____ *P. e. erythrophthalmus* — Coastal e Brazil (s Bahia to São Paulo)
____ *P. e. ferrugineigula* — Coastal se Brazil (São Paulo to s Rio Grande do Sul)

☐ **Canebrake Groundcreeper** *Clibanornis dendrocolaptoides*

E Paraguay to se Brazil and ne Argentina

☐ **Firewood-gatherer** *Anumbius annumbi*

Savanna of Paraguay to s Brazil, Uruguay and c Argentina

☐ **Lark-like Brushrunner** *Coryphistera alaudina*
_____ *C. a. campicola*
_____ *C. a. alaudina*

Chaco of se Bolivia and adjacent w Paraguay
S Bolivia to n Argentina and extreme s Brazil

☐ **Spectacled Prickletail** *Siptornis striaticollis*

Locally in Andes of s Colombia to extreme n Peru

☐ **Orange-fronted Plushcrown** *Metopothrix aurantiaca*

Tropical se Colombia to n Bolivia and w Amazonian Brazil

☐ **Double-banded Graytail** *Xenerpestes minlosi*
_____ *X. m. minlosi*
_____ *X. m. umbraticus*

Tropical e Panama and Caribbean lowlands of Colombia
Pacific lowlands of Colombia

☐ **Equatorial Graytail** *Xenerpestes singularis*

Locally in foothills of e Ecuador and n Peru

☐ **Pink-legged Graveteiro** *Acrobatornis fonsecai*

E Brazil (cocoa-growing region of se Bahia)

☐ **Rusty-winged Barbtail** *Premnornis guttuligera*
_____ *P. g. guttuligera*
_____ *P. g. venezuelanus*

Andes of Colombia and Ecuador to s Peru
Andes of extreme nw Venezuela (sw Táchira)

☐ **Spotted Barbtail** *Premnoplex brunnescens*
_____ *P. b. brunneicauda*
_____ *P. b. distinctus*
_____ *P. b. mnionophilus*
_____ *P. b. albescens*
_____ *P. b. coloratus*
_____ *P. b. brunnescens*
_____ *P. b. stictonotus*
_____ *P. b. rostratus*

Subtropical highlands of Costa Rica and w Panama
Subtropical mountains of c Panama (Veraguas)
Mountains of Panama (w San Blas)
Subtropical mountains of e Panama (Cerro Tacarcuna)
Subtropical Santa Marta Mountains (ne Colombia)
Andes of Colombia, Venezuela, Ecuador and n Peru
SE Peru (Puno) and w Bolivia (La Paz and Cochabamba)
Mts. of n Venezuela (Lara, Aragua, Carabobo and Miranda)

☐ **White-throated Barbtail** *Premnoplex tatei*
_____ *P. t. tatei*
_____ *P. t. pariae*

Mountains of n Venezuela (Sucre, Anzoátegui and n Monagas)
NE Venezuela (mountains of Paría Peninsula)

☐ **Roraiman Barbtail** *Roraimia adusta*
_____ *R. a. obscurodorsalis*
_____ *R. a. duidae*
_____ *R. a. adusta*

Tepuis of se Venezuela (Bolívar and Amazonas)
Tepuis of s Venezuela (Mt. Duida)
Tepuis of e Venezuela, adjacent w Guyana and extreme n Brazil

☐ **Ruddy Treerunner** *Margarornis rubiginosus*
_____ *M. r. rubiginosus*
_____ *M. r. boultoni*

Highlands of Costa Rica and w Panama (w Chiriquí)
Highlands of c Panama (e Chiriquí and Veraguas)

☐ **Fulvous-dotted Treerunner** *Margarornis stellatus*

W Andes of Colombia and nw Ecuador

☐ **Beautiful Treerunner** *Margarornis bellulus*

Humid montane forests of e Panama (Mt. Pirre)

☐ **Pearled Treerunner** *Margarornis squamiger*
_____ *M. s. perlatus*
_____ *M. s. peruvianus*
_____ *M. s. squamiger*

Andes of Colombia to n Peru and w Venezuela
Andes of Peru (Cajamarca to Huánuco, Junín and Cuzco)
Andes of se Peru (Puno) and w Bolivia (La Paz and Cochabamba)

☐ **Gray-crested Cacholote** *Pseudoseisura unirufa*

E Bolivia, *chaco* of n Paraguay and sw Brazil (Mato Grosso)

☐ **Caatinga Cacholote** *Pseudoseisura cristata*

Arid ne Brazil (Pernambuco, Piauí, Bahia and Minas Gerais)

☐ **Brown Cacholote** *Pseudoseisura lophotes*

SE Bolivia to w Paraguay, s Brazil, Uruguay and n Argentina

☐ **White-throated Cacholote** *Pseudoseisura gutturalis*

W and c Argentina

☐ **Buffy Tuftedcheek** *Pseudocolaptes lawrencii*

___ *P. l. lawrencii (panamensis)*   Highlands of Costa Rica and w Panama (w Chiriquí)
___ *P. l. johnsoni*   W slope of Western Andes of sw Colombia and w Ecuador

☐ **Streaked Tuftedcheek** *Pseudocolaptes boissonneautii*

___ *P. b. boissonneautii*   Andes of Colombia and n Ecuador
___ *P. b. striaticeps*   Coastal cordillera of n Venezuela
___ *P. b. meridae*   Andes of w Venezuela (Táchira, Mérida and Trujillo)
___ *P. b. oberholseri*   S Colombia
___ *P. b. orientalis*   Andes of s Ecuador
___ *P. b. intermedianus*   W Andes of Peru (Piura)
___ *P. b. pallidus*   Andes of nw Peru (Cajamarca)
___ *P. b. medianus*   Andes of n Peru (Cajamarca to La Libertad)
___ *P. b. auritus*   Andes of c Peru (Huánuco, Cuzco and Puno)
___ *P. b. carabayae*   Andes of se Peru (Puno) and w Bolivia

☐ **Point-tailed Palmcreeper** *Berlepschia rikeri*

SE Colombia to e Peru, nw Bolivia, Guianas and Amaz. Brazil

☐ **Scaly-throated Foliage-gleaner** *Anabacerthia variegaticeps*

___ *A. v. variegaticeps*   S Mexico (Guerrero and Veracruz) to w Panama
___ *A. v. temporalis*   W slope of Western Andes of Colombia and Ecuador

☐ **Montane Foliage-gleaner** *Anabacerthia striaticollis*

___ *A. s. striaticollis*   Subtropical Andes of Colombia and Venezuela
___ *A. s. anxia*   Santa Marta Mountains (ne Colombia)
___ *A. s. perijana*   Sierra de Perijá (Colombia/Venezuela border)
___ *A. s. venezuelana*   Coastal cordillera of n Venezuela
___ *A. s. montana*   Subtropical e Ecuador and e Peru
___ *A. s. yungae*   Andes of se Peru (Cuzco and Puno) and w Bolivia

☐ **White-browed Foliage-gleaner** *Anabacerthia amaurotis*

SE Paraguay to se Brazil (s Espírito Santo) and ne Argentina

☐ **Guttulated Foliage-gleaner** *Syndactyla guttulata*

___ *S. g. guttulata*   Mountains of n Venezuela (Yaracuy to Distrito Federal)
___ *S. g. pallida*   Mountains of ne Venezuela (Anzoátegui, Sucre and n Monagas)

☐ **Lineated Foliage-gleaner** *Syndactyla subalaris*

___ *S. s. lineata*   Subtropical Costa Rica and w Panama
___ *S. s. tacarcunae*   Mountains of e Panama (Darién)
___ *S. s. subalaris*   Western and C Andes of Colombia and w Ecuador
___ *S. s. striolata*   Eastern Andes of Colombia and Andes of w Venezuela
___ *S. s. mentalis*   Subtropical e Ecuador
___ *S. s. colligata*   Subtropical nw Peru (Cajamarca)
___ *S. s. ruficrissa*   Subtropical c Peru (Junín)

☐ **Buff-browed Foliage-gleaner** *Syndactyla rufosuperciliata*

___ *S. r. similis*   Andes of extreme s Ecuador and n Peru (Cajamarca)
___ *S. r. cabanisi*   E slope of Andes of Peru and Bolivia
___ *S. r. oleaginea*   SE Bolivia and ne Argentina
___ *S. r. rufosuperciliata*   SE Brazil (Minas Gerais to São Paulo and Paraná)
___ *S. r. acrita*   S Brazil to Paraguay, Uruguay and ne Argentina

☐ **Rufous-necked Foliage-gleaner** *Syndactyla ruficollis*

___ *S. r. celicae*   Andes of extreme sw Ecuador (Loja)
___ *S. r. ruficollis*   Andes of nw Peru (Piura, Cajamarca and Lambayeque)

☐ **Peruvian Recurvebill** *Simoxenops ucayalae*

Tropical se Peru, extreme nw Bolivia and Amazonian Brazil

☐ **Bolivian Recurvebill** *Simoxenops striatus*

Yungas of w Bolivia (La Paz, Cochabamba and w Santa Cruz)

☐ **Chestnut-winged Hookbill** *Ancistrops strigilatus*

_____ *A. s. strigilatus* — W Amazonia from se Colombia to nw Bolivia and w Brazil
_____ *A. s. cognitus* — C Brazil (lower Rio Tapajós)

☐ **Striped Woodhaunter** *Hyloctistes subulatus*

_____ *H. s. nicaraguae* — Caribbean lowlands of e Nicaragua
_____ *H. s. virgatus* — Lowlands of Costa Rica and w Panama
_____ *H. s. assimilis* — Extreme e Panama (Darién) to w Colombia and w Ecuador
_____ *H. s. cordobae* — NW Colombia
_____ *H. s. lemae* — S Venezuela (Amazonas and Bolívar)
_____ *H. s. subulatus* — Tropical se Colombia to n Bolivia and w Amazonian Brazil

☐ **Rufous-tailed Foliage-gleaner** *Philydor ruficaudatum*

_____ *P. r. ruficaudatum* — Tropical se Colombia to n Bolivia and w Amazonian Brazil
_____ *P. r. flavipectus* — S Venezuela to the Guianas and n Amazonian Brazil

☐ **Slaty-winged Foliage-gleaner** *Philydor fuscipenne*

_____ *P. f. fuscipenne* — Lowlands of Panama (Veraguas, Coclé, Colón and Canal Zone)
_____ *P. f. erythronotum* — Lowlands of e Panama to nw Colombia and w Ecuador

☐ **Rufous-rumped Foliage-gleaner** *Philydor erythrocercum*

_____ *P. e. subfulvum* — Tropical se Colombia to e Ecuador and n Peru
_____ *P. e. erythrocercum* — Guianas and Brazil north of the Amazon (east of Rio Negro)
_____ *P. e. ochrogaster* — C and se Peru and Bolivia
_____ *P. e. lyra* — Tropical e Peru, n Bolivia and Brazil south of the Amazon
_____ *P. e. subalare* — W Brazil (north bank of Rio Solimões)

☐ **Chestnut-winged Foliage-gleaner** *Philydor erythropterum*

_____ *P. e. erythropterum* — SE Colombia to s Venezuela, n Bolivia and w Amazonian Brazil
_____ *P. e. diluviale* — N-c Brazil (lower Rio Tapajós)

☐ **Ochre-breasted Foliage-gleaner** *Philydor lichtensteini*

E Paraguay to e and se Brazil and ne Argentina (Misiones)

☐ **Alagoas Foliage-gleaner** *Philydor novaesi*

Montane forests of ne Brazil (Alagoas)

☐ **Black-capped Foliage-gleaner** *Philydor atricapillus*

E Paraguay to se Brazil (s Bahia) and ne Argentina

☐ **Buff-fronted Foliage-gleaner** *Philydor rufum*

_____ *P. r. panerythrum* — Highlands of Costa Rica to Eastern Andes of Colombia
_____ *P. r. riveti* — W Andes of Colombia and nw Ecuador (Pichincha and El Oro)
_____ *P. r. columbianum* — Coastal cordillera of n Venezuela
_____ *P. r. cuchiverus* — Subtropical s Venezuela (Bolívar)
_____ *P. r. bolivianum* — E Peru and Bolivia (La Paz and Santa Cruz)
_____ *P. r. chapadense* — SW Brazil (Mato Grosso)
_____ *P. r. rufum* — E Brazil (Goiás and Bahia) to e Paraguay and ne Argentina

☐ **Cinnamon-rumped Foliage-gleaner** *Philydor pyrrhodes*

SE Colombia to s Venezuela, Guianas, n Bolivia, e Amaz. Brazil

☐ **Russet-mantled Foliage-gleaner** *Philydor dimidiatum*

_____ *P. d. dimidiatum* — SW Brazil (s Mato Grosso)
_____ *P. d. baeri* — SE Brazil (Goiás) to extreme ne Paraguay (Concepción)

☐ **Crested Foliage-gleaner** *Anabazenops dorsalis*

SE Colombia to extreme nw Bolivia and w Amazonian Brazil

☐ **White-collared Foliage-gleaner** *Anabazenops fuscus*

SE Brazil (Minas Gerais and Espírito Santo to e Santa Catarina)

☐ **Pale-browed Treehunter** *Cichlocolaptes leucophrus*

E Brazil (s Bahia to ne Santa Catarina)

☐ **Uniform Treehunter** *Thripadectes ignobilis*

Humid Western Andes of Colombia and w Ecuador (s to El Oro)

☐ **Streak-breasted Treehunter** *Thripadectes rufobrunneus*

C highlands of Costa Rica and w Panama

☐ **Black-billed Treehunter** *Thripadectes melanorhynchus*
___ *T. m. striaticeps*
___ *T. m. melanorhynchus*

E Andes of Colombia (single record from w Meta)
E slope of Andes of Ecuador and e Peru (south to Puno)

☐ **Striped Treehunter** *Thripadectes holostictus*
___ *T. h. striatidorsus*
___ *T. h. holostictus*
___ *T. h. moderatus*

Andes of sw Colombia (Nariño) and w Ecuador
Andes of Colombia to sw Venezuela, e Ecuador and n Peru
Subtropical e Peru (Junín and Cuzco) and w Bolivia

☐ **Streak-capped Treehunter** *Thripadectes virgaticeps*
___ *T. v. sclateri*
___ *T. v. magdalenae*
___ *T. v. klagesi*
___ *T. v. tachirensis*
___ *T. v. virgaticeps*
___ *T. v. sumaco*

W Andes of Colombia (Cauca and Nariño)
N Colombia (subtropical Magdalena Valley in Huila)
Coastal mountains of n Venezuela (Carabobo to Distrito Federal)
Andes of w Venezuela (sw Lara and Táchira)
Subtropical nw Ecuador (south to Pichincha)
Subtropical e Ecuador (w Napo)

☐ **Flammulated Treehunter** *Thripadectes flammulatus*
___ *T. f. flammulatus*
___ *T. f. bricenoi*

Andes of Colombia, Ecuador and extreme n Peru
Andes of w Venezuela (Mérida)

☐ **Buff-throated Treehunter** *Thripadectes scrutator*

E slope of Andes of Peru and w Bolivia (Cochabamba)

☐ **Buff-throated Foliage-gleaner** *Automolus ochrolaemus*
___ *A. o. cervinigularis*
___ *A. o. amusos*
___ *A. o. hypophaeus*
___ *A. o. exsertus*
___ *A. o. pallidigularis*
___ *A. o. turdinus*
___ *A. o. ochrolaemus*
___ *A. o. auricularis*

Gulf-Caribbean slope of s Mexico to Belize and Nicaragua
Tropical se Guatemala to Honduras
Caribbean slope of e Nicaragua to nw Panama
Pacific slope of sw Costa Rica to w Panama (Chiriquí)
E Panama to Colombia and nw Ecuador
Tropical n and w Amazon basin
Tropical e Peru to n Bolivia and w Brazil
NE Bolivia (Rio Beni) and w Brazil (Rio Purús to Rio Tapajós)

☐ **Olive-backed Foliage-gleaner** *Automolus infuscatus*
___ *A. i. infuscatus*
___ *A. i. badius*
___ *A. i. cervicalis*
___ *A. i. perusianus*

Tropical se Colombia to e Ecuador, e Peru and nw Bolivia
E Colombia to s Venezuela and nw Brazil
E Venezuela to Guyana, Suriname and Brazil (n of the Amazon)
W Brazil (Rio Purús to Rio Solimões)

☐ **Pará Foliage-gleaner** *Automolus paraensis*

Brazil south of the Amazon (Rio Madeira to Rio Tocantins)

☐ **White-eyed Foliage-gleaner** *Automolus leucophthalmus*
___ *A. l. lammi*
___ *A. l. leucophthalmus*
___ *A. l. sulphurascens*

E Brazil (Paraíba, Pernambuco and Alagoas)
E Brazil (Bahia to Rio Grande do Sul)
S Brazil (se Mato Grosso) to e Paraguay and ne Argentina

☐ **Brown-rumped Foliage-gleaner** *Automolus melanopezus*

SE Colombia to se Peru, extreme nw Bolivia and w Amaz. Brazil

☐ **White-throated Foliage-gleaner** *Automolus roraimae*
___ *A. r. paraquensis*
___ *A. r. duidae*
___ *A. r. albigularis*
___ *A. r. roraimae*

Tepuis of s Venezuela (Mt. Paraque, Parú and Ptari-tepui)
Tepuis of s Venezuela (Mt. Duida and Mt. Yaví)
Subtropical mountains of se Venezuela (Gran Sabana)
Tepuis of extreme n Brazil (Mt. Roraima)

☐ **Ruddy Foliage-gleaner** *Automolus rubiginosus*

_____ *A. r. guerrerensis*	Subtropical sw Mexico (Guerrero and w Oaxaca)
_____ *A. r. rubiginosus*	Subtropical e Mexico (Veracruz)
_____ *A. r. veraepacis*	Subtropical n Guatemala (Alta Verapaz)
_____ *A. r. umbrinus*	Subtropical s Mexico (Chiapas) to n Nicaragua
_____ *A. r. fumosus*	Subtropical w Panama (Chiriquí)
_____ *A. r. saturatus*	Tropical e Panama and nw Colombia (Antioquia)
_____ *A. r. sasaimae*	N Colombia (upper tropical Magdalena Valley)
_____ *A. r. nigricauda*	W Colombia (Baudó Mountains) to w Ecuador
_____ *A. r. rufipectus*	Santa Marta Mountains (ne Colombia)
_____ *A. r. cinnamomeigula*	Tropical base of Eastern Andes of Colombia
_____ *A. r. caquetae*	SE Colombia (Caquetá and Putumayo)
_____ *A. r. venezuelanus*	*Tepuis* of s Venezuela
_____ *A. r. obscurus*	S Venezuela, the Guianas and n Amazonian Brazil
_____ *A. r. brunnescens*	E Ecuador (Río Suno region) and adjacent ne Peru
_____ *A. r. moderatus*	Tropical ne Peru (San Martín and Loreto)
_____ *A. r. watkinsi*	SE Peru (Huánuco) to w Bolivia (La Paz)

☐ **Chestnut-crowned Foliage-gleaner** *Automolus rufipileatus*

_____ *A. r. consobrinus*	Tropical e Colombia to the Guianas, n Bolivia and w Brazil
_____ *A. r. rufipileatus*	Brazil south of the Amazon (Rio Purús to n Maranhão)

☐ **Henna-hooded Foliage-gleaner** *Hylocryptus erythrocephalus*

_____ *H. e. erythrocephalus*	SW Ecuador (Loja) and nw Peru (south to Lambayeque)
_____ *H. e. palamblae*	W Andes of Peru (Piura)

☐ **Chestnut-capped Foliage-gleaner** *Hylocryptus rectirostris*

	Campos of interior e and s-c Brazil to ne Paraguay

☐ **Tawny-throated Leaftosser** *Sclerurus mexicanus*

_____ *S. m. mexicanus*	Tropical se Mexico (Veracruz and Chiapas) to Honduras
_____ *S. m. pullus*	Costa Rica to w Panama; e Panama (Mt. Tacarcuna)
_____ *S. m. andinus*	Tropical e Panama to Colombia
_____ *S. m. obscurior*	W Andes of Colombia and w Ecuador
_____ *S. m. peruvianus*	Tropical w Amazon basin
_____ *S. m. macconnelli*	S Venezuela to the Guianas and e Amazonian Brazil
_____ *S. m. bahiae*	E Brazil (Alagoas to ne São Paulo)

☐ **Short-billed Leaftosser** *Sclerurus rufigularis*

_____ *S. r. fulvigularis*	SE Colombia to s Venezuela, the Guianas and n Brazil
_____ *S. r. brunnescens (furfurosus)*	Amazonian Brazil (north of Rio Solimões)
_____ *S. r. rufigularis*	S Amazonian Brazil to ne Peru and n Bolivia

☐ **Scaly-throated Leaftosser** *Sclerurus guatemalensis*

_____ *S. g. guatemalensis*	Tropical s Mexico (Veracruz) to e Panama
_____ *S. g. salvini*	W Colombia (Chocó) to w Ecuador (south to Guayas)
_____ *S. g. ennosiphyllus*	Colombia (Río Magdalena Valley in Santander)

☐ **Black-tailed Leaftosser** *Sclerurus caudacutus*

_____ *S. c. caudacutus*	The Guianas
_____ *S. c. insignis*	S Venezuela (Amazonas and Bolívar) and adjacent n Brazil
_____ *S. c. brunneus*	Tropical se Colombia to Peru and w Amazonian Brazil
_____ *S. c. olivascens*	E Peru (Ayacucho) to extreme n Bolivia (Pando)
_____ *S. c. pallidus*	N Brazil south of the Amazon (Rio Madeira to Rio Capím)
_____ *S. c. umbretta*	Coastal e Brazil (Alagoas to Espírito Santo)

☐ **Gray-throated Leaftosser** *Sclerurus albigularis*

_____ *S. a. canigularis*	Montane forests of Costa Rica and w Panama (Chiriquí)
_____ *S. a. propinquus*	Santa Marta Mountains (ne Colombia)
_____ *S. a. albigularis*	Tropical e Colombia to Venezuela, Trinidad and Tobago
_____ *S. a. kunanensis*	NE Venezuela (Paría Peninsula)
_____ *S. a. zamorae*	E Ecuador, e Peru and sw Amazonian Brazil

____ *S. a. kempffi*	NE Bolivia (Serrania Huanchaca)
____ *S. a. albicollis*	N Bolivia (south to w Santa Cruz)

☐ **Rufous-breasted Leaftosser** *Sclerurus scansor*

____ *S. s. cearensis*	NE Brazil (Ceará and n Bahia)
____ *S. s. scansor*	E Brazil (Goiás) to e Paraguay and ne Argentina

☐ **Sharp-tailed Streamcreeper** *Lochmias nematura*

____ *L. n. nelsoni*	Extreme e Panama (Darién)
____ *L. n. sororius*	Andes of Colombia to ne Peru; coastal mts. of n Venezuela
____ *L. n. chimantae*	Mountains of se Venezuela (Gran Sabana)
____ *L. n. castanonotus*	*Tepuis* of s Venezuela (Bolívar and Amazonas)
____ *L. n. obscuratus*	E-c Peru and w Bolivia
____ *L. n. nematura*	SE Brazil to e Paraguay, Uruguay and ne Argentina

☐ **Sharp-billed Treehunter** *Heliobletus contaminatus*

	E Paraguay to se Brazil (Espírito Santo) and ne Argentina

☐ **Rufous-tailed Xenops** *Xenops milleri*

	SE Colombia to s Venezuela, Guianas, e Peru and Amaz. Brazil

☐ **Slender-billed Xenops** *Xenops tenuirostris*

____ *X. t. acutirostris*	SE Colombia to s Venezuela, e Ecuador and ne Peru
____ *X. t. hellmayri*	French Guiana, Suriname and extreme n Brazil (Roraima)
____ *X. t. tenuirostris*	S Venezuela to e Peru, n Bolivia and Amazonian Brazil

☐ **Plain Xenops** *Xenops minutus*

____ *X. m. mexicanus*	Tropical s Mexico to Honduras
____ *X. m. ridgwayi*	Tropical Nicaragua to Costa Rica and w Panama
____ *X. m. littoralis*	Tropical e Panama to w Ecuador (El Oro)
____ *X. m. neglectus*	N Colombia and n Venezuela
____ *X. m. remoratus*	Tropical e Colombia to s Venezuela and n Brazil
____ *X. m. ruficaudus*	Extreme e Colombia to Venezuela, the Guianas and n Brazil
____ *X. m. olivaceus*	Lowlands of ne Colombia
____ *X. m. obsoletus*	Tropical e Ecuador to e Peru, n Bolivia and w Brazil
____ *X. m. genibarbis*	N Brazil south of the Amazon (Rio Madeira to Maranhão)
____ *X. m. minutus*	E Brazil (Pernambuco) to e Paraguay and ne Argentina

☐ **Streaked Xenops** *Xenops rutilans*

____ *X. r. septentrionalis*	Highlands of Costa Rica and w Panama
____ *X. r. heterurus*	E Panama to ne Ecuador and Venezuela; Trinidad
____ *X. r. incomptus*	E Panama (Darién)
____ *X. r. perijanus*	Sierra de Perijá (Colombia/Venezuela border)
____ *X. r. phelpsi*	Santa Marta Mountains (ne Colombia)
____ *X. r. guayae*	Tropical w Ecuador and extreme nw Peru (Piura)
____ *X. r. peruvianus*	Tropical e Ecuador and e Peru
____ *X. r. purusianus*	Brazil south of the Amazon (Rio Purús, Madeira and Tapajós)
____ *X. r. connectens*	E Bolivia and nw Argentina
____ *X. r. chapadensis*	SW Brazil (Mato Grosso) and n Bolivia (Río Beni)
____ *X. r. rutilans*	SE Brazil to e Paraguay and ne Argentina

☐ **Great Xenops** *Megaxenops parnaguae*

	Arid *caatinga* of interior ne and e-c Brazil

☐ **White-throated Treerunner** *Pygarrhichas albogularis*

	C Chile and adjacent Argentina south to Tierra del Fuego

☐ **Tyrannine Woodcreeper** *Dendrocincla tyrannina*

____ *D. t. tyrannina*	Andes of Colombia to e Peru (Cordillera Vilcabamba)
____ *D. t. hellmayri*	Andes of e Colombia and w Venezuela (sw Táchira)

☐ **Plain-brown Woodcreeper** *Dendrocincla fuliginosa*

____ *D. f. ridgwayi*	Tropical se Honduras to w Colombia, w Ecuador and nw Peru
____ *D. f. lafresnayei*	N and e Colombia and adjacent nw Venezuela
____ *D. f. meruloides*	Coastal n Venezuela, Trinidad and Tobago

283

____ *D. f. barinensis*	Llanos of n Colombia and w-c Venezuela
____ *D. f. deltana*	Northeast Venezuela in Orinoco River delta
____ *D. f. phaeochroa*	Amazonian e Colombia to e Ecuador, e Peru and nw Brazil
____ *D. f. neglecta*	W Amazonia from e Ecuador and e Peru to w Brazil
____ *D. f. fuliginosa*	Southeast Venezuela to the Guianas and adjacent n Brazil
____ *D. f. atritrostris*	Southeast Peru to n and e Bolivia and sw Brazil (Mato Grosso)
____ *D. f. rufoolivacea*	E Amazonian Brazil (Rio Tapajós to n Maranhão)
____ *D. f. trumaii*	Locally in s Amazonian Brazil (upper Rio Xingu)
____ *D. f. taunayi*	NE Brazil (e Pernambuco and e Alagoas)

☐ **Thrush-like Woodcreeper** *Dendrocincla turdina*

	E Paraguay to se Brazil (Bahia) and ne Argentina (Misiones)

☐ **Tawny-winged Woodcreeper** *Dendrocincla anabatina*

____ *D. a. anabatina*	Gulf-Caribbean slope of se Mexico to ne Nicaragua
____ *D. a. typhla*	Southeast Mexico (Yucatán Peninsula)
____ *D. a. saturata*	Pacific slope of Costa Rica and w Panama (w Chiriquí)

☐ **White-chinned Woodcreeper** *Dendrocincla merula*

____ *D. m. bartletti*	SE Colombia to s Venezuela, e Peru and w Brazil
____ *D. m. merula*	The Guianas and adjacent n Brazil
____ *D. m. obidensis*	Lower Amazonian Brazil (Faro and Obidos regions)
____ *D. m. remota*	E-c Bolivia (Santa Cruz)
____ *D. m. olivascens*	Brazil south of the Amazon (Rio Madeira to Rio Tapajós)
____ *D. m. castanoptera*	Brazil south of the Amazon (Rio Tapajós to Rio Tocantins)
____ *D. m. badia*	Brazil south of the Amazon (Rio Tocantins to Rio Guamá)

☐ **Ruddy Woodcreeper** *Dendrocincla homochroa*

____ *D. h. homochroa*	S Mexico to Guatemala and Honduras
____ *D. h. acedesta*	SW Nicaragua and w Costa Rica to extreme w Panama
____ *D. h. ruficeps*	E Panama to nw Venezuela (Lara, Mérida, Barinas and Apure)
____ *D. h. meridionalis*	Sierra de Perijá (Colombia/Venezuela border)

☐ **Long-tailed Woodcreeper** *Deconychura longicauda*

____ *D. l. typica*	SE Honduras to w Panama
____ *D. l. darienensis*	E Panama (Darién)
____ *D. l. minor*	Tropical n Colombia
____ *D. l. longicauda*	The Guianas and Brazil north of the Amazon
____ *D. l. connectens*	NW Amazonian Brazil to e Ecuador and Peru
____ *D. l. zimmeri*	E Brazil (Pará)
____ *D. l. pallida*	SE Peru to n Bolivia and w Brazil (Mato Grosso)

☐ **Spot-throated Woodcreeper** *Deconychura stictolaema*

____ *D. s. secunda*	W Amazonia (s Colombia to ne Peru and nw Brazil)
____ *D. s. clarior*	NE Amazonia n of Amazon in Brazil, Venezuela and Fr. Guiana
____ *D. s. stictolaema*	S Amazonian Brazil s of Amazon

☐ **Olivaceous Woodcreeper** *Sittasomus griseicapillus*

____ *S. g. jaliscensis (harrisoni)*	Mexico (Nayarit and San Luis Potosí to Isthmus of Tehuantepec)
____ *S. g. sylvioides (levis, veraguensis)*	S Mexico to nw Colombia
____ *S. g. gracileus*	SE Mexico (Yucatán Peninsula) n Belize and adj. n Guatemala
____ *S. g. perijanus*	NE Colombia and extreme nw Venezuela (Sierra de Perijá)
____ *S. g. tachirensis*	N Colombia and w Venezuela (sw Táchira)
____ *S. g. griseus*	E Andes and coastal ranges of n Venezuela; Tobago
____ *S. g. aequatorialis*	W Ecuador (w Esmeraldas) to extreme nw Peru (Tumbes)
____ *S. g. amazonus*	Tropical e Colombia, Ecuador, Peru and Amazonian Brazil
____ *S. g. axillaris*	Tropical se Venezuela to Guianas and extreme n Brazil
____ *S. g. transitivus*	SE Amazonian Brazil to ne Mato Grosso
____ *S. g. viridis (viridior)*	Amazonian Bolivia (La Paz, Beni, Cochabamba and Santa Cruz)
____ *S. g. griseicapillus*	W Brazil to Paraguay, n Argentina and s Bolivia
____ *S. g. reiseri*	NE Brazil (Maranhão and Piauí to n Goiás and w Bahia)

____ *S. g. olivaceus*	Coastal e Brazil (se Bahia)
____ *S. g. sylviellus*	SE Brazil to ne Argentina, se Paraguay and ne Uruguay

☐ **Wedge-billed Woodcreeper** *Glyphorynchus spirurus*

____ *G. s. pectoralis (sublestus)*	S Mexico (s Veracruz) to Costa Rica and w Panama
____ *G. s. pallidulus*	E Panama and adjacent nw Colombia (n Chocó)
____ *G. s. subrufescens*	Pacific coast of se Panama to w Colombia and w Ecuador
____ *G. s. integratus*	N Colombia and w Venezuela
____ *G. s. rufigularis*	Tropical e Colombia to s Venezuela, ne Ecuador and nw Brazil
____ *G. s. amacurensis*	NE Venezuela (Delta Amacuro)
____ *G. s. spirurus*	NE Venezuela, the Guianas and adjacent n Brazil
____ *G. s. coronobscurus*	S Venezuela (Cerro de la Neblina in sw Amazonas)
____ *G. s. castelnaudii*	Tropical e Peru and w Amazonian Brazil
____ *G. s. albigularis*	SE Peru (Puno) to n Bolivia (La Paz and Cochabamba)
____ *G. s. inornatus*	S Amazonian Brazil (Rio Madeira to Tapajós and s Mato Grosso)
____ *G. s. pararensis*	SE Amazonian Brazil s of Amazon (R Tapajos to n Maranhão)
____ *G. s. cuneatus*	Coastal e Brazil (n Bahia to n Espírito Santo)

☐ **Scimitar-billed Woodcreeper** *Drymornis bridgesii*

S Bolivia to c Argentina, Uruguay, w Paraguay and sw Brazil

☐ **Long-billed Woodcreeper** *Nasica longirostris*

E Colombia to sw Venezuela, n Bolivia, Amaz. Brazil and Fr. Guiana

☐ **Cinnamon-throated Woodcreeper** *Dendrexetastes rufigula*

____ *D. r. devillei*	W Amazonian basin (se Colombia to ne Bolivia)
____ *D. r. rufigula*	The Guianas and n Brazil
____ *D. r. monileger*	W Brazil (Rio Madeira to Amazonas/Mato Grosso border)
____ *D. r. paraensis*	NE Brazil (south of the Amazon in Pará)

☐ **Red-billed Woodcreeper** *Hylexetastes perrotii*

SE Venezuela to the Guianas and Brazil north of the Amazon

☐ **Uniform Woodcreeper** *Hylexetastes uniformis*

C Amazonian Brazil to ne Bolivia

☐ **Brigida's Woodcreeper** *Hylexetastes brigidai*

Brazil (Pará and Mato Grosso)

☐ **Bar-bellied Woodcreeper** *Hylexetastes stresemanni*

____ *H. s. insignis*	W Amazonian Brazil (Rio Uaupés region)
____ *H. s. stresemanni*	NW Brazil (lower Rio Negro to Rio Solimões)
____ *H. s. undulatus*	Tropical se Peru to extreme nw Bolivia and w Brazil

☐ **Strong-billed Woodcreeper** *Xiphocolaptes promeropirhynchus*

____ *X. p. omiltemensis*	Subtropical sw Mexico (Sierra Madre del Sur of Guerrero)
____ *X. p. sclateri*	SE Mexico (se San Luis Potosí, w Veracruz and n Oaxaca)
____ *X. p. emigrans*	S Mexico (Chiapas) to n-c Nicaragua
____ *X. p. costaricensis*	Highlands of c Costa Rica and sw Panama (Chiriquí)
____ *X. p. panamensis*	Mountains of Pacific slope of s Panama (Veraguas)
____ *X. p. sanctaemartae*	N Colombia (Santa Marta Mts.)
____ *X. p. rostratus*	Lowlands of n Colombia (Córdoba and Bolívar)
____ *X. p. fortis*	Known from a single specimen of unknown location
____ *X. p. virgatus*	W slope of C Andes of Colombia east to Río Magdalena
____ *X. p. promeropirhynchus*	W slope of E Andes of Colombia and Andes of w Venezuela
____ *X. p. procerus*	Mountains of n and c Venezuela
____ *X. p. macarenae*	W-c Colombia (Macarena Mts. and foothills of Cent. Andes)
____ *X. p. neblinae*	*Tepuis* of s Venezuela (Cerro de la Neblina and [?] adj. Brazil)
____ *X. p. tenebrosus*	*Tepuis* of se Venezuela and adj. Guyana
____ *X. p. ignotus*	Subtropical and temperate Andes of Ecuador
____ *X. p. crassirostris*	Andean foothills of sw Ecuador (El Oro and Loja) to nw Peru
____ *X. p. compressirostris*	Temperate Andes of n Peru (Cajamarca, Amazonas, San Martín)
____ *X. p. phaeopygus*	Temperate Andes of c Peru (Junín)
____ *X. p. lineatocephalus*	Andes of se Peru (Cusco) to c Bolivia
____ *X. p. solivagus*	Upper trop. e Cordillera of Peru (Junín and n Húanuco))

____	*X. p. orenocensis*	Trop. Colombia to Venezuela, e Ecuador, e Peru, adj. nw Brazil
____	*X. p. berlepschi*	Trop. w Brazil s of Rio Solimões e to Rio Madeira)
____	*X. p. paraensis*	C Amazonian Brazil (Rio Madeira s of Amazon to n Mato Grosso)
____	*X. p. obsoletus*	Lowlands of n and e Bolivia (La Paz to Santa Cruz)

☐ **Carajás Woodcreeper** *Xiphocolaptes carajaensis*

Amazonian Brazil between Xingú and Tocantins/Araguaia

☐ **White-throated Woodcreeper** *Xiphocolaptes albicollis*

____	*X. a. villanovae*	Known only from type locality in ne Brazil (ne Bahia)
____	*X. a. bahiae*	E Brazil (e and c Bahia)
____	*X. a. albicollis*	SE and s Brazil (Goiás and Bahia) to Paraguay and ne Argentina

☐ **Moustached Woodcreeper** *Xiphocolaptes falcirostris*

____	*X. f. falcirostris*	NE Brazil (Maranhão and Ceará to w Paraíba and nw Bahia)
____	*X. f. franciscanus*	E Brazil (w Bahia and n and nw Minas Gerais)

☐ **Great Rufous Woodcreeper** *Xiphocolaptes major*

____	*X. m. castaneus*	SW Brazil to Bolivia and nw Argentina (se Jujuy and n Salta)
____	*X. m. remoratus*	SW Brazil (sw Mato Grosso)
____	*X. m. major*	W and c Paraguay to n Argentina
____	*X. m. estebani*	NW Argentina (Tucumán)

☐ **Northern Barred-Woodcreeper** *Dendrocolaptes sanctithomae*

____	*D. s. scheffleri*	Pacific slope of sw Mexico (Guerrero and Oaxaca)
____	*D. s. sanctithomae*	S Mexico (Veracruz) to n and w Colombia
____	*D. s. hesperius*	Pacific slope of sw Costa Rica and adjacent w Panama
____	*D. s. punctipectus (hyleorus)*	N Colombia and nw Venezuela (Zulia and n Mérida)

☐ **Amazonian Barred-Woodcreeper** *Dendrocolaptes certhia*

____	*D. c. radiolatus*	W Amazonia from se Colombia to ne Peru and nw Brazil
____	*D. c. certhia*	Extreme e Colombia to the Guianas and n Brazil
____	*D. c. juruanus*	SW Amazonia from e Peru and w Brazil to n Bolivia
____	*D. c. concolor*	S Amazonian Brazil s of Amazon to Mato Grosso and ne Bolivia
____	*D. c. medius*	SE Amazonia s of Amazon (Rio Tocantins to nw Maranhão)
____	*D. c. polyzonus*	N Bolivia (La Paz, Cochabamba and Santa Cruz)

☐ **Hoffmann's Woodcreeper** *Dendrocolaptes hoffmannsi*

S Amazonian Brazil s of Amazon to Rondônia and s Mato Grosso

☐ **Black-banded Woodcreeper** *Dendrocolaptes picumnus*

____	*D. p. puncticollis*	Highlands of s Mexico (Chiapas) to Guatemala and w Honduras
____	*D. p. costaricensis (veraguensis)*	Highlands of Costa Rica and Pacific slope of w Panama
____	*D. p. multistrigatus*	Andes of Colombia and n and w Venezuela
____	*D. p. seilerni*	Santa Marta Mts. (ne Colombia) and coastal mts. of n Venezuela
____	*D. p. validus*	Lowlands and foothills of w Amazonia
____	*D. p. picumnus*	Trop. e Venezuela to the Guianas and Brazil n of the Amazon
____	*D. p. transfasciatus*	Amazonian Brazil s of lower Amazon to n Mato Grosso
____	*D. p. olivaceus*	Foothills of Bolivian Andes (La Paz, Cochabamba, Santa Cruz)
____	*D. p. pallescens (extimus)*	Chaco of e Bolivia, s Brazil and w Paraguay
____	*D. p. casaresi*	Andean foothills of nw Argentina (Jujuy, Salta and Tucumán)

☐ **Planalto Woodcreeper** *Dendrocolaptes platyrostris*

____	*D. p. intermedius*	NE Brazil (Pará and Bahia) to e Paraguay)
____	*D. p. platyrostris*	E and se Brazil (s Bahia) to e Paraguay and ne Argentina

☐ **Lesser Woodcreeper** *Xiphorhynchus fuscus*

____	*X. f. atlanticus*	NE Brazil (Ceará and Paraíba to Alagoas)
____	*X. f. brevirostris*	Arid ne Brazil in w Bahia
____	*X. f. tenuirostris*	Coastal e Brazil (cent. Bahia to Espírito Santo north of Rio Doce)
____	*X. f. fuscus*	SE Brazil (s Goiás) to se Paraguay and ne Argentina

☐ **Tschudi's Woodcreeper** *Xiphorhynchus chunchotambo*

____ *X. c. napensis* — Tropical se Colombia to e Ecuador and ne Peru

____ *X. c. chunchotambo* — Trop., subtrop. e Peru (Loreto, Amazonas, Huánuco and Junín)

____ *X. c. brevirostris* — Tropical se Peru and ne Bolivia (La Paz, w Beni, w Santa Cruz)

☐ **Ocellated Woodcreeper** *Xiphorhynchus ocellatus*

____ *X. o. lineatocapilla* — Range uncertain: possibly along Río Orinoco in n Venezuela

____ *X. o. weddellii* — NW Amazonia (E Colombia to ne Peru, s Venezuela, nw Brazil)

____ *X. o. perplexus* — W Amazonia (ne Peru and adjacent w Brazil)

____ *X. o. ocellatus* — S Amazonian Brazil s of Amazon (Rio Purús to Rio Tocantins)

☐ **Chestnut-rumped Woodcreeper** *Xiphorhynchus pardalotus*

____ *X. p. caurensis* — Tepuis of se Venezuela and adj. n Brazil to w Guyana

____ *X. p. pardalotus* — NE Amazonia n of Amazon (Guianas and n Brazil)

☐ **Spix's Woodcreeper** *Xiphorhynchus spixii*

SE Amazonian Brazil south of the Amazon

☐ **Elegant Woodcreeper** *Xiphorhynchus elegans*

____ *X. e. buenavistae* — E slope of E Andes of Colombia in upper Orinoco drainage

____ *X. e. ornatus* — Tropical se Colombia to e Ecuador, ne Peru and adj. nw Brazil

____ *X. e. insignis* — Foothills of e-cent. Peru (s of Río Marañón)

____ *X. e. elegans* — Brazil south of the Amazon and n Bolivia

☐ **Juruá Woodcreeper** *Xiphorhynchus juruanus*

SE Peru to ne Bolivia and w Brazil (w of Rio Madeira)

☐ **Striped Woodcreeper** *Xiphorhynchus obsoletus*

____ *X. o. notatus* — E Colombia to s Venezuela and adj. nw Brazil

____ *X. o. caicarae* — Middle Orinoco Valley in cent. Venezuela (nw Bolívar)

____ *X. o. palliatus* — W Amazonia (se Colombia to ne Bolivia and w Amazonian Brazil)

____ *X. o. obsoletus* — E Amazonia (e Venezuela, the Guianas and n Brazil)

☐ **Buff-throated Woodcreeper** *Xiphorhynchus guttatus*

____ *X. g. polystictus* — E Colombia to the Guianas and extreme n Brazil

____ *X. g. connectens* — NW Amazonian Brazil n of the Amazon (Manaus to Amapá)

____ *X. g. guttatus* — Coastal e Brazil (Paraíba to Espírito Santo)

☐ **Lafresnaye's Woodcreeper** *Xiphorhynchus guttatoides*

____ *X. g. eytoni* — S Amazonian Brazil s of the Amazon (Rio Madeira to Maranhão)

____ *X. g. guttatoides* — Tropical w Amazonian basin

____ *X. g. vicinalis* — Brazil south of the Amazon (Rio Madeira to Rio Tapajós)

____ *X. g. dorbignyanus* — Tropical ne Bolivia to e-c Brazil

☐ **Cocoa Woodcreeper** *Xiphorhynchus susurrans*

____ *X. s. confinis* — Caribbean slope of e Guatemala and n Honduras

____ *X. s. costaricensis* — SE Honduras to Nicaragua, Costa Rica and w Panama

____ *X. s. marginatus* — E Panama (e Chiriquí, Veraguas and w Azuero Peninsula)

____ *X. s. nanus (demonstratus)* — Both slope of e Panama to n Colombia and w Venezuela

____ *X. s. rosenbergi* — W Colombia (upper Cauca Valley in Valle)

____ *X. s. jardinei* — NE Venezuela (ne Anzoátegui, Sucre and n Monagas)

____ *X. s. margaritae* — Isla Margarita (off n Venezuela)

____ *X. s. susurrans* — Trinidad and Tobago; one record from ne Venezuela (se Sucre)

☐ **Ivory-billed Woodcreeper** *Xiphorhynchus flavigaster*

____ *X. f. tardus* — Tropical nw Mexico (extreme se Sonora and n Sinaloa)

____ *X. f. mentalis* — W Mexico (Sinaloa and Durango to Jalisco and Michoacán)

____ *X. f. flavigaster* — SW Mexico (Guerrero and s Oaxaca)

____ *X. f. saltuarius* — NE Mexico (s Tamaulipas, se San Luis Potosí and n Veracruz)

____ *X. f. ascensor* — Caribbean slope of s Mexico (s Veracruz to Oaxaca and Tabasco)

____ *X. f. yucatanensis* — SE Mexico (Yucatán Peninsula to n Belize); Meco I.)

____ *X. f. eburneirostris* — Pacific slope of s Mexico to Belize, Honduras and nw Costa Rica

____ *X. f. ultimus* — NW Costa Rica (Nicoya Peninsula)

☐ **Black-striped Woodcreeper** *Xiphorhynchus lachrymosus*

_____ *X. l. lachrymosus*	E Nicaragua to Pacific coast of Colombia and nw Ecuador
_____ *X. l. eximius*	Pacific slope of sw Costa Rica and adj. w Panama
_____ *X. l. alarum*	Tropical n Colombia (Sinú, Cauca and Magdalena valleys)

☐ **Spotted Woodcreeper** *Xiphorhynchus erythropygius*

_____ *X. e. erythropygius*	Highlands of s Mexico (San Luis Potosí to Oaxaca and Guerrero)
_____ *X. e. parvus*	Highlands of s Mexico (se Oaxaca, Chiapas) to n-cent. Nicaragua
_____ *X. e. punctigula*	Tropical se Nicaragua to cent. Panama (Veraguas)
_____ *X. e. insolitus*	Cent. and e Panama to nw Colombia (Atrato and Truando rivers)
_____ *X. e. aequatorialis*	Tropical Pacific slope of w Colombia to sw Ecuador

☐ **Olive-backed Woodcreeper** *Xiphorhynchus triangularis*

_____ *X. t. hylodromus*	Coastal and interior mts. of n Venezuela
_____ *X. t. triangularis*	Andes of Colombia, e Ecuador, n Peru and w Venezuela
_____ *X. t. intermedius*	Andes of c Peru (Pasco and Junín)
_____ *X. t. bangsi*	Andes of se Peru to Bolivia (La Paz, Cochabamba and Santa Cruz)

☐ **Straight-billed Woodcreeper** *Dendroplex picus*

_____ *D. p. extimus*	C and e Panama to nw Colombia
_____ *D. p. dugandi*	NW Colombia (Chocó, Bolívar, Atlántico and Santa Marta Mts.)
_____ *D. p. picirostris*	Coastal n Colombia (Santa Marta Mts.) to extreme nw Venezuela
_____ *D. p. parguanae*	NW Venezuela (Falcón, n Lara)
_____ *D. p. choicus*	N Venezuela (e Falcón east to Miranda)
_____ *D. p. longirostris*	Isla Margarita (Venezuela)
_____ *D. p. phalara*	*Llanos* of interior n Venezuela
_____ *D. p. saturatior*	Tropical e Colombia to w Venezuela (cent. and s Maracaibo basin)
_____ *D. p. duidae*	E Colombia (e Vichada), s Venezuela and adj. w Brazil
_____ *D. p. altirostris*	Trinidad
_____ *D. p. deltanus*	NE Venezuela (Delta Amacuro)
_____ *D. p. picus (rufescens, bahiae)*	E Colombia to Venezuela, the Guianas and Amazonian Brazil
_____ *D. p. peruvianus (borreroi)*	SW Amazonia (e Peru, adjacent w Brazil and n Bolivia)

☐ **Zimmer's Woodcreeper** *Dendroplex kienerii*

	Extreme se Colombia to ne Ecuador, ne Peru and adj. Brazil

☐ **Narrow-billed Woodcreeper** *Lepidocolaptes angustirostris*

_____ *L. a. griseiceps*	Suriname (Sipaliwini savanna)
_____ *L. a. coronatus*	NE Brazil (Maranhão and Piauí to Goiás and nw Bahia)
_____ *L. a. bahiae*	NE Brazil (Piauí, Ceará and interior Bahia)
_____ *L. a. bivittatus*	E Bolivia and Brazilian plateau
_____ *L. a. hellmayri*	Andean foothills of Bolivia (Cochabamba, Santa Cruz and Tarija)
_____ *L. a. certhiolus*	C Bolivia, *chaco* of w Paraguay and nw Argentina
_____ *L. a. angustirostris*	E Paraguay to sw Brazil (sw Mato Grosso) and n Argentina
_____ *L. a. praedatus (dabbenei)*	W Uruguay to n and cent. Argentina and extreme s Brazil

☐ **White-striped Woodcreeper** *Lepidocolaptes leucogaster*

_____ *L. l. umbrosus*	NW Mexico (se Sonora to Durango, Nayarit and Jalisco)
_____ *L. l. leucogaster*	SW Mexico (Jalisco to Colima, Oaxaca, Puebla and Veracruz)

☐ **Lineated Woodcreeper** *Lepidocolaptes albolineatus*

_____ *L. a. albolineatus*	E Venezuela (Bolívar) to the Guianas and n Brazil
_____ *L. a. duidae*	E Colombia, s Venezuela (Amazonas) and nw Brazil
_____ *L. a. fuscicapillus*	Amazonian Ecuador to Peru, e Bolivia and s Amazonian Brazil
_____ *L. a. madeirae*	S-cent. Amazonian Brazil south of the Amazon
_____ *L. a. layardi*	SE Amazonian Brazil south of the Amazon

☐ **Spot-crowned Woodcreeper** *Lepidocolaptes affinis*

_____ *L. a. lignicida*	NE Mexico (sw Tamaulipas and e San Luis Potosí)
_____ *L. a. affinis*	S Mexico (se San Luis Potosí and w Guerrero) to n Nicaragua
_____ *L. a. neglectus*	Mts. of Costa Rica and w Panama (Chiriquí highlands)

☐ **Montane Woodcreeper** *Lepidocolaptes lacrymiger*

____	*L. l. sanctaemartae*	Santa Marta Mts. of n Colombia
____	*L. l. lacrymiger*	E Andes of Colombia and Sierra de Perijá of adj. w Venezuela
____	*L. l. lafresnayi*	Coastal cordillera of n Venezuela
____	*L. l. sneiderni*	Andes of Colombia
____	*L. l. frigidus*	E slope of Andes of s Colombia (Nariño)
____	*L. l. aequatorialis*	Pacific slope of Andes of sw Colombia (Nariño) and Ecuador
____	*L. l. warscewiczi*	E slope of Andes of se Ecuador and Peru (Cajamarca to Junín)
____	*L. l. carabayae*	E slope of Andes of se Peru (Cuzco and Puno)
____	*L. l. bolivianus*	E slope of Andes of Bolivia (La Paz, Cochabamba, w Santa Cruz)

☐ **Streak-headed Woodcreeper** *Lepidocolaptes souleyetii*

____	*L. s. guerrerensis*	W Mexico (Sierra Madre del Sur of Guerrero and sw Oaxaca)
____	*L. s. compressus (insignis)*	S Mexico (Veracruz and Chiapas) to w Panama
____	*L. s. lineaticeps*	Cent. and e Panama to n Colombia and w Venezuela
____	*L. s. littoralis*	Tropical nw Colombia to Guyana and adj. n Brazil; Trinidad
____	*L. s. uaireni*	Extreme se Venezuela (along Rio Uairén in se Bolívar)
____	*L. s. esmeraldae*	Tropical sw Colombia (Nariño) and w Ecuador (s to El Oro)
____	*L. s. souleyetii*	Tropical sw Ecuador and nw Peru (south to Lambayeque)

☐ **Scaled Woodcreeper** *Lepidocolaptes squamatus*

____	*L. s. wagleri*	NE Brazil (s Piauí, w Bahia and n Minas Gerais)
____	*L. s. squamatus*	E and se Brazil (Bahia, Minas Gerais, Rio de Janeiro and n São Paulo)

☐ **Scalloped Woodcreeper** *Lepidocolaptes falcinellus*

SE Brazil to ne Argentina, ne Uruguay and se Paraguay

☐ **Greater Scythebill** *Campylorhamphus pucherani*

Locally in Western Andes of Colombia to se Peru (Cuzco)

☐ **Black-billed Scythebill** *Campylorhamphus falcularius*

SE Brazil (cent. Bahia) to e Paraguay and ne Argentina

☐ **Curve-billed Scythebill** *Campylorhamphus procurvoides*

____	*C. p. sanus*	Trop. se Colombia to ne Peru, Venezuela, n Guyana and n Brazil
____	*C. p. procurvoides*	Suriname, French Guiana and adj. Brazil north of the Amazon
____	*C. p. successor*	SW Amazonia s of the Amazon in w Brazil
____	*C. p. probatus*	S-cent. Amazonian Brazil (Rio Madeira to Rio Tapajós)
____	*C. p. multostriatus*	SE Amazonian Brazil (Rio Tapajós to Rio Tocantins)

☐ **Red-billed Scythebill** *Campylorhamphus trochilirostris*

____	*C. t. brevipennis*	Tropical e Panama (Canal Zone to Darién) and nw Colombia
____	*C. t. venezuelensis*	Locally in n Colombia and n and cent. Venezuela
____	*C. t. thoracicus*	Coastal sw Colombia (sw Nariño) and w Ecuador
____	*C. t. zarumillanus*	Coastal extreme nw Peru (Tumbes and Piura)
____	*C. t. napensis*	W Amazonia (e Ecuador and e Peru)
____	*C. t. notabilis*	W Amazonian Brazil s of the Amazon (Rio Purús to Rio Madeira)
____	*C. t. snethlageae*	Cent. Amazonian Brazil (Rio Madeira to Rio Tapajós)
____	*C. t. devius*	SW Amazonia (n Bolivia in La Paz and Cochabamba)
____	*C. t. lafresnayanus*	E Bolivia to sw Brazil (w Mato Grosso) and *chaco* of w Paraguay
____	*C. t. hellmayri*	SW Paraguay (Ñeembucú) and n Argentina
____	*C. t. major (omissus, guttistriatus)*	Interior e and s Brazil (Piauí to Minas Gerais and w Paraná)
____	*C. t. trochilirostris*	Coastal e Brazil (Pernambuco to se Bahia)

☐ **Brown-billed Scythebill** *Campylorhamphus pusillus*

____	*C. p. borealis*	Costa Rica and w Panama (w Chiriquí and Bocas del Toro)
____	*C. p. olivaceus*	C and e Panama (Veraguas to e Darién)
____	*C. p. tachirensis*	Andes of extreme ne Colombia and nw Venezuela (Zulia, Táchira)
____	*C. p. pusillus*	N and cent. Andes of Colombia to w Ecuador and n Peru
____	*C. p. guapiensis*	Coastal lowlands of sw Colombia (Cauca)

FAMILY: THAMNOPHILIDAE (Typical Antbirds—212)

☐ **Fasciated Antshrike** *Cymbilaimus lineatus*
____ *C. l. fasciatus* | Extreme se Honduras and Nicaragua to nw Ecuador
____ *C. l. intermedius* | Tropical s Colombia to s Venezuela, n Bolivia, Amazonian Brazil
____ *C. l. lineatus* | E Venezuela to French Guiana and ne Amazonian Brazil

☐ **Bamboo Antshrike** *Cymbilaimus sanctaemariae*
Local from se Peru to nw Bolivia and sw Amazonian Brazil

☐ **Spot-backed Antshrike** *Hypoedaleus guttatus*
SE Brazil to e Paraguay and extreme ne Argentina (Misiones)

☐ **Giant Antshrike** *Batara cinerea*
____ *B. c. excubitor* | E slope of Andes of c Bolivia (w Santa Cruz)
____ *B. c. argentina* | E Bolivia to w Paraguay and nw Argentina
____ *B. c. cinerea* | SE Brazil (s Espírito Santo) to extreme ne Argentina (Misiones)

☐ **Large-tailed Antshrike** *Mackenziaena leachii*
SE Brazil (Minas Gerais) to e Paraguay and extreme ne Argentina

☐ **Tufted Antshrike** *Mackenziaena severa*
SE Brazil to e Paraguay and extreme ne Argentina (Misiones)

☐ **Black-throated Antshrike** *Frederickena viridis*
SE Venezuela to the Guianas and ne Amazonian Brazil

☐ **Undulated Antshrike** *Frederickena unduligera*
____ *F. u. unduligera* | NW Amazonian Brazil on upper Rio Negro
____ *F. u. fulva* | SE Colombia to e Ecuador and e Peru (n of Río Marañón)
____ *F. u. diversa* | E Peru (s of Río Marañón) and nw Bolivia (s La Paz, s Beni)
____ *F. u. pallida* | SW Amazonian Brazil s of the Amazon; n Bolivia (?)

☐ **Great Antshrike** *Taraba major*
____ *T. m. melanocrissus* | Caribbean slope of se Mexico (San Luis Potosí) to w Panama
____ *T. m. obscurus* | W Costa Rica to Panama and n Colombia
____ *T. m. transandeanus* | Coastal sw Colombia to w Ecuador and nw Peru (Tumbes)
____ *T. m. granadensis* | Caribbean slope of n Colombia to nw Venezuela
____ *T. m. semifasciatus* | Extreme e Colombia to s Venezuela, the Guianas, n and e Brazil
____ *T. m. duidae* | *Tepuis* of se Venezuela (Mt. Duida)
____ *T. m. melanurus* | SE Colombia to e Ecuador, e Peru and sw Amazonian Brazil)
____ *T. m. borbae* | S-c Amazonian Brazil (Rio Purús to Rio Madeira)
____ *T. m. stagurus* | NE Brazil (e Maranhão to Pernambuco and Espírito Santo)
____ *T. m. major* | E Bolivia to s-cent. Brazil, w Paraguay and n Argentina

☐ **Black-crested Antshrike** *Sakesphorus canadensis*
____ *S. c. pulchellus (phainoleucus, paraguanae)* | Caribbean slope of n Colombia and extreme nw Venezuela
____ *S. c. intermedius* | E Colombia and Venezuela north of Río Orinoco
____ *S. c. trinitatis* | NE and s Venezuela, Trinidad and Guyana
____ *S. c. canadensis* | Suriname and coastal French Guiana
____ *S. c. fumosus* | SW Venezuela (s Amazonas) and extreme n Brazil
____ *S. c. loretoyacuensis* | Extreme se Colombia to ne Peru

☐ **Silvery-cheeked Antshrike** *Sakesphorus cristatus*
Arid ne Brazil (Piauí and Ceará to extreme n Minas Gerais)

☐ **Collared Antshrike** *Sakesphorus bernardi*
____ *S. b. bernardi (piurae, cajamarcae)* | Arid tropical sw Ecuador to n-cent. Peru; Isla Puná
____ *S. b. shumbae* | N-cent. Peru (Río Marañón drainage in Cajamarca and Amazonas)

☐ **Black-backed Antshrike** *Sakesphorus melanonotus*
Caribbean slope of n Colombia and nw Venezuela (e to Miranda)

☐ **Band-tailed Antshrike** *Sakesphorus melanothorax*
Suriname, French Guiana and ne Amazonian Brazil

☐ **Glossy Antshrike** *Sakesphorus luctuosus*
____ *S. l. luctuosus* | Central and e Amazonian Brazil
____ *S. l. araguayae* | Central Brazil (s R. Araguaia drainage)

☐ **White-bearded Antshrike** *Biatas nigropectus*

SE Brazil (se Minas Gerais) to extreme ne Argentina (Misiones)

☐ **Barred Antshrike** *Thamnophilus doliatus*
____ *T. d. intermedius (yucatanensis, pacificus)*
____ *T. d. nigricristatus*
____ *T. d. eremnus*
____ *T. d. nesiotes*
____ *T. d. albicans*
____ *T. d. nigrescens*
____ *T. d. tobagensis*
____ *T. d. doliatus (fraterculus)*
____ *T. d. radiatus (subradiatus, signatus, novus)*
____ *T. d. cadwaladeri*
____ *T. d. difficilis*
____ *T. d. capistratus*

E Mexico (Tamaulipas) to Belize, Guatemala and w Panama
C Panama (e Chiriquí and s Veraguas to w San Blas)
Coiba I. (Panama)
Pearl Islands (Gulf of Panama)
Caribbean slope of Colombia and s in Magdalena Valley to Huila
N-cent. Colombia e of Andes and nw Venezuela n of Andes
Tobago
NE Colombia to the Guianas and n Amazonian Brazil; Trinidad
Extreme se Colombia to e Peru, Bolivia, Paraguay, n Argentina
S Bolivia (Tarija)
E-cent. Brazil (e Maranhão to e Mato Grosso, Goiás and w Bahia)
E Brazil (Ceará to extreme n Minas Gerais and cent. Bahia)

☐ **Chapman's Antshrike** *Thamnophilus zarumae*
____ *T. z. zarumae*
____ *T. z. palamblae*

SW Ecuador (El Oro and Loja) and nw Peru (Tumbes, ne Piura)
NW Peru (se Piura and e Lambayeque)

☐ **Bar-crested Antshrike** *Thamnophilus multistriatus*
____ *T. m. brachyurus*
____ *T. m. selvae*
____ *T. m. multistriatus*
____ *T. m. oecotonophilus*

W Colombia in Western Andes and w slope of Central Andes
W slope of Western Andes of Colombia (upper Río San Juan)
Colombia (e slope of Cent. Andes and w slope of E and S Andes)
W slope of E Andes of Colombia and Sierra de Perijá (w Venezuela)

☐ **Lined Antshrike** *Thamnophilus tenuepunctatus*
____ *T. t. tenuepunctatus*
____ *T. t. tenuifasciatus*
____ *T. t. berlepschi*

E slope of Eastern Andes of n-cent. Colombia
E slope of s-cent. Colombia (Putumayo) and e Ecuador
E slope of se Ecuador (s Zamora-Chinchipe) and ne Peru

☐ **Chestnut-backed Antshrike** *Thamnophilus palliatus*
____ *T. p. similis*
____ *T. p. puncticeps*
____ *T. p. palliatus*
____ *T. p. vestitus*

E slope of Andes of central Peru (Huánuco and Junín)
SE Peru (Cusco and Puno) to n Bolivia and s Amazonian Brazil
Brazil s of R. Amazon and coastal ne Brazil (Paraíba to n Bahia)
Coastal e Brazil (s Bahia to Rio de Janeiro)

☐ **Rufous-winged Antshrike** *Thamnophilus torquatus*

Lowlands of ne Bolivia to ne Paraguay and e and central Brazil

☐ **Rufous-capped Antshrike** *Thamnophilus ruficapillus*
____ *T. r. jaczewskii*
____ *T. r. marcapatae*
____ *T. r. subfasciatus*
____ *T. r. cochabambae*
____ *T. r. ruficapillus*

N Peru (Amazonas s of Río Marañón, Cajamarca and San Martín)
SE Peru (Cusco and Puno)
Yungas of nw Bolivia (w Cochabamba and La Paz)
S Bolivia to nw Argentina (Jujuy, Salta, Tucumán)
SE Brazil to ne Argentina, e Paraguay and Uruguay

☐ **Black-hooded Antshrike** *Thamnophilus bridgesi*

Pacific slope of sw Costa Rica and w Panama

☐ **Black Antshrike** *Thamnophilus nigriceps*

Tropical e Panama (Darién) and n Colombia

☐ **Cocha Antshrike** *Thamnophilus praecox*

Tropical e Ecuador (along upper Río Napo and tributaries)

☐ **Blackish-gray Antshrike** *Thamnophilus nigrocinereus*
____ *T. n. cinereoniger*
____ *T. n. tschudii*
____ *T. n. huberi*
____ *T. n. nigrocinereus*
____ *T. n. kulczynskii*

NE Colombia, sw Venezuela and nw Amazonian Brazil
W-cent. Brazil (e Amazonas along lower R. Madeira)
E-cent. Brazil (w Pará along lower R. Tapajós)
E Brazil (lower Amazon River to Amapá and estuary islands)
E French Guiana and adjacent Brazil (extreme n Amapá)

☐ **Castelnau's Antshrike** *Thamnophilus cryptoleucus*

NE Ecuador to n Peru and w Amazonian Brazil

291

☐ **White-shouldered Antshrike** *Thamnophilus aethiops*

_____	*T. a. wetmorei*	Foothills of se Colombia (w Meta to e Cauca and w Putumayo)
_____	*T. a. aethiops*	Foothills of e Ecuador and ne Peru
_____	*T. a. polionotus*	S and e Venezuela and nw Brazil
_____	*T. a. kapouni*	E and se Peru, n Bolivia and w Brazil (R. Solimões)
_____	*T. a. juruanus*	W Brazil (south of R. Solimões from R. Juruá to R. Purús)
_____	*T. a. injunctus*	Brazil south of R. Amazon (R. Madeira to R. Purús)
_____	*T. a. punctuliger*	Central Brazil north of R. Amazon
_____	*T. a. atriceps*	Brazil south of R. Amazon (R. Tapajós to R. Xingú)
_____	*T. a. incertus*	NE Brazil (R. Tocantins to Maranhão)
_____	*T. a. distans*	Coastal ne Brazil (Pernambuco and Alagoas)

☐ **Uniform Antshrike** *Thamnophilus unicolor*

_____	*T. u. grandior*	Colombia to e Ecuador and n Peru (south to n San Martín)
_____	*T. u. unicolor*	Pacific slope of Ecuador
_____	*T. u. caudatus*	E slope of Andes of Peru (s San Martín south locally to Cusco)

☐ **Plain-winged Antshrike** *Thamnophilus schistaceus*

_____	*T. s. heterogynus*	Extreme e Colombia to s-central Amazonian Brazil
_____	*T. s. capitalis*	E base of Andes of se Colombia, e Ecuador and ne Peru
_____	*T. s. schistaceus (dubius, inornatus)*	Tropical se Peru to n Bolivia and s Amazonian Brazil

☐ **Mouse-colored Antshrike** *Thamnophilus murinus*

_____	*T. m. cayennensis*	French Guiana and ne Amazonian Brazil (Amapá and n Pará)
_____	*T. m. murinus*	Extreme e Colombia to Guians, Suriname and Brazil n of Amazon
_____	*T. m. canipennis*	Extreme se Colombia, e Ecuador, e Peru, w Brazil and n Bolivia

☐ **Upland Antshrike** *Thamnophilus aroyae*

Lower Andean slopes of extreme se Peru (Puno) and nw Bolivia

☐ **Western Slaty-Antshrike** *Thamnophilus atrinucha*

_____	*T. a. atrinucha*	S Belize and ne Guatemala to nw Peru and nw Venezuela
_____	*T. a. gorgonae*	Gorgona I. (off w Colombia)

☐ **Northern Slaty-Antshrike** *Thamnophilus punctatus*

_____	*T. p. punctatus*	Extreme e Venezuela to the Guianas and Brazil n of R. Amazon
_____	*T. p. interpositus*	E base of Andes of w Venezuela and adjacent Colombia

☐ **Marañón Slaty-Antshrike** *Thamnophilus leucogaster*

_____	*T. l. huallagae*	N Peru (Río Huallaga drainage in San Martín)
_____	*T. l. leucogaster*	Río Marañón drainage in extreme s Ecuador and n Peru

☐ **Natterer's Slaty-Antshrike** *Thamnophilus stictocephalus*

_____	*T. s. stictocephalus*	Locally in Brazil south of R. Amazon and extreme n Bolivia
_____	*T. s. parkeri*	Extreme ne Bolivia (Serranía de Huanchaca in n Santa Cruz)

☐ **Bolivian Slaty-Antshrike** *Thamnophilus sticturus*

Central and e Bolivia, immediately adj. Brazil and n Paraguay

☐ **Planalto Slaty-Antshrike** *Thamnophilus pelzelni*

Plateau region of central and e Brazil

☐ **Sooretama Slaty-Antshrike** *Thamnophilus ambiguus*

Coastal se Brazil (s Bahia and Espírito Santo to Rio de Janeiro)

☐ **Amazonian Antshrike** *Thamnophilus amazonicus*

_____	*T. a. cinereiceps*	E-central Colombia to sw Venezuela and nw Brazil
_____	*T. a. divaricatus*	Extreme e Venezuela to the Guianas and ne Amazonian Brazil
_____	*T. a. amazonicus*	SE Colombia to n Bolivia and w Brazil (s of R. Amazon)
_____	*T. a. obscurus*	S-cent. Amaz. Brazil (s Pará between R. Tapajós and Tocantins)
_____	*T. a. paraensis*	Brazil (e of R. Tocantins in e Pará, w Maranhão and n Tocantins)

☐ **Streak-backed Antshrike** *Thamnophilus insignis*

_____	*T. i. nigrifrontalis*	Extreme sw Venezuela (Macizo de Sipapo in w Amazonas)
_____	*T. i. insignis*	*Tepuis* of s Venezuela and adjacent w Guyana and n Brazil

☐ **Variable Antshrike** *Thamnophilus caerulescens*

_____ *T. c. subandinus (melanochrous)*	E slope of Andes of Peru (Amazonas s of Río Marañón to n Puno)
_____ *T. c. aspersiventer*	SE Peru (Puno) to w-c Bolivia (La Paz and Cochabamba)
_____ *T. c. dinellii (connectens)*	Central and s Bolivia to nw Argentina
_____ *T. c. paraguayensis*	SE Bolivia, n Paraguay and s Brazil (Mato Grosso do Sul)
_____ *T. c. gilvigaster*	Extreme se Brazil to Uruguay and ne Argentina
_____ *T. c. caerulescens (albonotatus)*	SE Brazil (Minas Gerais) to se Paraguay and ne Argentina
_____ *T. c. ochraceiventer*	E-central Brazil (s Tocantins, Goiás, Distrito Federal, s-c Bahia)
_____ *T. c. cearensis (pernambucensis)*	NE Brazil (Ceará, Pernambuco and Alagoas)

☐ **Acre Antshrike** *Thamnophilus divisorius*

W Amazonian Brazil in Serra do Divisor National Park

☐ **Spot-winged Antshrike** *Pygiptila stellaris*

_____ *P. s. occipitalis*	Extreme e Colombia to n Venezuela, Suriname and n Brazil
_____ *P. s. stellaris (maculipennis, purusiana)*	Brazil south of R. Amazon and nw Mato Grosso

☐ **Black Bushbird** *Neoctantes niger*

SE Colombia to w Amazonian Brazil, e Ecuador and se Peru

☐ **Recurve-billed Bushbird** *Clytoctantes alixii*

N Colombia and nw Venezuela (Sierra de Perijá)

☐ **Rondonia Bushbird** *Clytoctantes atrogularis*

Locally in sw Amazonian Brazil (e Rondônia)

☐ **Pearly Antshrike** *Megastictus margaritatus*

SE Colombia to s Venezuela, e Peru and w-c Amazonian Brazil

☐ **Speckled Antshrike** *Xenornis setifrons*

E Panama (Colón) to extreme nw Colombia (Chocó)

☐ **Russet Antshrike** *Thamnistes anabatinus*

_____ *T. a. anabatinus*	Atlantic slope of se Mexico to Guatemala, Belize and Honduras
_____ *T. a. saturatus*	Atlantic slope of Nicaragua to Costa Rica and extreme w Panama
_____ *T. a. coronatus*	Central and e Panama (Veraguas to Darién) and nw Colombia
_____ *T. a. intermedius*	Pacific slope of w Colombia and Ecuador
_____ *T. a. gularis*	Extreme nw Venezuela (Táchira) and (? adj. ne Colombia)
_____ *T. a. aequatorialis*	Foothills of se Colombia, e Ecuador and extreme n Peru
_____ *T. a. rufescens*	E Peru south of the Río Marañón (Amazonas) to w Bolivia

☐ **Spot-breasted Antvireo** *Dysithamnus stictothorax*

SE Brazil (e Minas Gerais to Paraná) and ne Argentina (Misiones)

☐ **Plain Antvireo** *Dysithamnus mentalis*

_____ *D. m. septentrionalis*	Atlantic slope of s Mexico (Campeche, Chiapas) to w Panama
_____ *D. m. suffusus*	E Panama (Darién) and nw Colombia (n Chocó and n Antioquia)
_____ *D. m. extremus*	W Andes and w slope of Central Andes of Colombia
_____ *D. m. aequatorialis*	Pacific slope of w Ecuador and extreme nw Peru (Tumbes)
_____ *D. m. viridis*	Mts. of n Colombia and n Venezuela
_____ *D. m. cumbreanus*	Coastal mts. of n Venezuela (Falcón and Lara to n Sucre)
_____ *D. m. oberi*	Tobago
_____ *D. m. andrei*	NE Venezuela (s Sucre to ne Bolívar); Trinidad
_____ *D. m. ptaritepui*	*Tepuis* of s Venezuela (Ptari-tepui and Sororopán-tepui)
_____ *D. m. spodionotus*	S Venezuela (s Bolívar, Amazonas) and n Brazil (Roraima)
_____ *D. m. semicinereus*	Andes of w-central Colombia
_____ *D. m. napensis*	Extreme s Colombia to extreme n Peru (n Amazonas)
_____ *D. m. tambillanus*	E slope of Andes of n and central Peru
_____ *D. m. olivaceus*	E slope of Andes of Peru (Pasco to Cusco and w Madre de Dios)
_____ *D. m. tavarae`*	SE Peru (se Madre de Dios) to central Bolivia
_____ *D. m. emiliae*	NE Brazil (se Pará, n Maranhão, Ceará and Pernambuco, Alagoas)
_____ *D. m. affinis*	Extreme ne Bolivia and central Brazil
_____ *D. m. mentalis*	SE Brazil (Bahia) to e Paraguay and ne Argentina

☐ **Streak-crowned Antvireo** *Dysithamnus striaticeps*

Extreme se Honduras, e Nicaragua and Costa Rica

☐ **Spot-crowned Antvireo** *Dysithamnus puncticeps*

SE Costa Rica to w Colombia and nw Ecuador

☐ **Rufous-backed Antvireo** *Dysithamnus xanthopterus*

Coastal mts. of se Brazil (Rio de Janeiro to Paraná)

☐ **White-streaked Antvireo** *Dysithamnus leucostictus*
_____ *D. l. tucuyensis*
_____ *D. l. leucostictus*

Coastal mts. of n Venezuela (Falcón and Lara to Monagas)
E slope of Andes of Colombia to extreme n Peru (Cajamarca)

☐ **Plumbeous Antvireo** *Dysithamnus plumbeus*

SE Brazil (s Bahia to e Minas Gerais and n Rio de Janeiro)

☐ **Bicolored Antvireo** *Dysithamnus occidentalis*
_____ *D. o. occidentalis*
_____ *D. o. punctitectus*

Pacific slope of Andes of sw Colombia and n Ecuador (Carchi)
E slope of Andes of Ecuador (w Napo and Morona-Santiago)

☐ **Dusky-throated Antshrike** *Thamnomanes ardesiacus*
_____ *T. a. ardesiacus*
_____ *T. a. obidensis*

SE Colombia to e Peru, ne Bolivia and adjacent Brazil (se Acre)
E Colombia to e Venezuela, the Guianas and n Amazonian Brazil

☐ **Saturnine Antshrike** *Thamnomanes saturninus*
_____ *T. s. huallagae*
_____ *T. s. saturninus*

NE Peru n of R. Amazon and sw Amazonian Brazil
S-central Brazil and extreme ne Bolivia (ne Santa Cruz)

☐ **Cinereous Antshrike** *Thamnomanes caesius*
_____ *T. c. glaucus*
_____ *T. c. persimilis*
_____ *T. c. simillimus*
_____ *T. c. hoffmannsi*
_____ *T. c. caesius*

E Colombia to the Guianas, ne Peru and Brazil n of R. Amazon
Cent. Brazil south of R. Amazon and extreme ne Bolivia
S-cent. Amazonian Brazil along middle R. Purús
E-c Brazil s of Amazon (R. Tapajós to n Maranhão, ne Mato Grosso)
Coastal e Brazil (Pernambuco Rio de Janeiro, inland to Minas Gerais)

☐ **Bluish-slate Antshrike** *Thamnomanes schistogynus*
_____ *T. s. intermedius*
_____ *T. s. schistogynus*

Cent. Peru s of R. Amazon and e of R. Huallaga (s to Junín)
W Brazil s of R. Amazon to se Peru and nw Bolivia

☐ **Pygmy Antwren** *Myrmotherula brachyura*

E Colombia to the Guianas, Amazonian Brazil and n Bolivia

☐ **Moustached Antwren** *Myrmotherula ignota*
_____ *M. i. ignota*
_____ *M. i. obscura*

E Panama to w Colombia and nw Ecuador
S-cent. and e Colombia to ne Peru and w Amazonian Brazil

☐ **Guianan Antwren** *Myrmotherula surinamensis*

S Venezuela to the Guianas and n Amazonian Brazil

☐ **Amazonian Antwren** *Myrmotherula multostriata*

E Colombia to n Bolivia and Brazil south of R. Amazon

☐ **Pacific Antwren** *Myrmotherula pacifica*

Panama (Veraguas) to coastal Pacific Colombia and w Ecuador

☐ **Cherrie's Antwren** *Myrmotherula cherriei*

SE Colombia, sw Venezuela, ne Peru and nw Brazil

☐ **Klages' Antwren** *Myrmotherula klagesi*

Central Amazonian Brazil

☐ **Stripe-chested Antwren** *Myrmotherula longicauda*
_____ *M. l. soderstromi*
_____ *M. l. pseudoaustralis*
_____ *M. l. longicauda*
_____ *M. l. australis*

E slope of Andes of s Colombia and n Ecuador (Napo)
E slope of n Ecuador (Zamora-Chinchipe) and n Peru (s to Pasco)
Central Peru (Junín)
SE Peru (Cusco, Madre de Dios and Puno) to nw Bolivia

☐ **Sclater's Antwren** *Myrmotherula sclateri*

E Peru s of Amazon, n Bolivia and s Amazonian Brazil

☐ **Yellow-throated Antwren** *Myrmotherula ambigua*

Extreme e Colombia, sw Venezuela and nw Amazonian Brazil

☐ **Rufous-bellied Antwren** *Myrmotherula guttata*

S Venezuela to the Guianas and ne Amazonian Brazil

☐ **Plain-throated Antwren** *Myrmotherula hauxwelli*
_____ *M. h. suffusa* — SE Colombia to ne Peru and extreme nw Amazonian Brazil
_____ *M. h. hauxwelli (clarior)* — E Peru (drainage of R. Huallaga), s Amazonian Brazil, n Bolivia
_____ *M. h. hellmayri* — NE Brazil (e of R. Xingú in e Pará and w Maranhão)

☐ **Star-throated Antwren** *Myrmotherula gularis*

SE Brazil (s Bahia to w Paraná and n Rio Grande do Sul)

☐ **Brown-bellied Antwren** *Myrmotherula gutturalis*

E Venezuela to the Guianas and ne Amazonian Brazil

☐ **Checker-throated Antwren** *Myrmotherula fulviventris*

Honduras to Colombia (lower Cauca Valley) and w Ecuador

☐ **White-eyed Antwren** *Myrmotherula leucophthalma*
_____ *M. l. leucophthalma* — E Peru, w Brazil s of R. Amazon and n Bolivia
_____ *M. l. dissita* — SE Peru (Puno) and adjacent n Bolivia (La Paz)
_____ *M. l. phaeonota* — S Amazonian Brazil (lower R. Madeira to lower R. Tapajós)
_____ *M. l. sordida* — E Brazil south of R. Amazon (R. Tapajós east to R. Tocantins)

☐ **Foothill Antwren** *Myrmotherula spodionota*
_____ *M. s. spodionota* — E slope of Andes of s Colombia south to n Peru (n Amazonas)
_____ *M. s. sororia* — E Peru (south of R. Marañón from San Martín to Madre de Dios)

☐ **Stipple-throated Antwren** *Myrmotherula haematonota*
_____ *M. h. pyrrhonota* — SE Colombia to s Venezuela, ne Peru and nw Brazil
_____ *M. h. haematonota* — E Peru (Loreto to Madre de Dios) and w Brazil (Amazonas, Acre)
_____ *M. h. amazonica* — S-central Amazonian Brazil and n Bolivia (Pando)

☐ **Brown-backed Antwren** *Myrmotherula fjeldsaai*

SE Ecuador and extreme n-central Peru

☐ **Ornate Antwren** *Myrmotherula ornata*
_____ *M. o. ornata* — Andean foothills of central Colombia (Meta)
_____ *M. o. saturata* — S-cent. Colombia to e Ecuador and ne Peru (n of R. Marañón)
_____ *M. o. atrogularis* — E-cent. Peru and extreme sw Amazonian Brazil
_____ *M. o. meridionalis* — SE Peru (Madre de Dios, Puno), adj. Brazil and nw Bolivia
_____ *M. o. hoffmannsi* — SE Amazonian Brazil

☐ **Rufous-tailed Antwren** *Myrmotherula erythrura*
_____ *M. e. erythrura* — SE Colombia to ne Peru and extreme nw Brazil
_____ *M. e. septentrionalis* — E Peru (Loreto to Puno) and w-central Amazonian Brazil

☐ **White-flanked Antwren** *Myrmotherula axillaris*
_____ *M. a. albigula* — SE Honduras to w Colombia and w Ecuador
_____ *M. a. melaena* — E Colombia to ne Peru, w Venezuela and n Amazonian Brazil
_____ *M. a. axillaris* — E Venezuela to the Guianas, e Amaz. Brazil, ne Bolivia; Trinidad
_____ *M. a. heterozyga* — E Peru (Ucayali to Madre de Dios) and sw Amazonian w Brazil
_____ *M. a. fresnayana* — Extreme se Peru (Puno) and nw Bolivia
_____ *M. a. luctuosa* — Coastal e Brazil (Paraíba to Rio de Janeiro)

☐ **Slaty Antwren** *Myrmotherula schisticolor*
_____ *M. s. schisticolor* — Extreme se Mexico (Chiapas) to w Ecuador (south to Loja)
_____ *M. s. sanctaemartae* — NE Colombia (Santa Marta Mts.) and mts. of n Venezuela
_____ *M. s. interior* — Andean slopes from e Colombia to s Peru (Puno)

☐ **Rio Suno Antwren** *Myrmotherula sunensis*
_____ *M. s. sunensis* — E slope of Andes of s-cent. Colombia and Ecuador
_____ *M. s. yessupi* — E-central Peru (Huánuco, Pasco) and sw Amazonian Brazil

☐ **Salvadori's Antwren** *Myrmotherula minor*

Coastal se Brazil (se Bahia to extreme ne Santa Catarina)

☐ **Ihering's Antwren** *Myrmotherula iheringi*
_____ *M. i. heteroptera* — SW Amazonian Brazil to se Peru (Madre de Dios) and nw Bolivia
_____ *M. i. iheringi* — S-central Amazonian Brazil

☐ **Rio de Janeiro Antwren** *Myrmotherula fluminensis*

SE Brazil (known from a 1988 specimen from Serra dos Órgãos)

☐ **Plain-winged Antwren** *Myrmotherula behni*
____ *M. b. yavii*
____ *M. b. camanii*
____ *M. b. inornata*
____ *M. b. behni*

S Venezuela (nw Bolívar, Amazonas) and adj. n Brazil (Amazonas)
S Venezuela (Cerro Camani in n Amazonas)
SE Venezuela (se Bolívar) and adjacent Brazil (n Roraima)
E slope of Andes of s-central Colombia to e Ecuador

☐ **Ashy Antwren** *Myrmotherula grisea*

Andean foothills of w Bolivia (La Paz, Cochabamba, Santa Cruz)

☐ **Unicolored Antwren** *Myrmotherula unicolor*

SE Brazil (n Rio de Janeiro to extreme n Rio Grande do Sul)

☐ **Alagoas Antwren** *Myrmotherula snowi*

NE Brazil (Alagoas near Murici on R. Pedra Branca)

☐ **Long-winged Antwren** *Myrmotherula longipennis*
____ *M. l. longipennis*
____ *M. l. zimmeri*
____ *M. l. garbei*
____ *M. l. ochrogyna*
____ *M. l. paraensis*
____ *M. l. transitiva*

SE Colombia to the Guianas, n Peru and n Amazonian Brazil
E Ecuador (s of R. Napo) and ne Peru
E Peru (s of R. Marañón), sw Amazonian Brazil and nw Bolivia
Brazil s of R. Amazon (lower R. Madeira to R. Tapajós)
Brazil s of R. Amazon (R. Tapajós to w Maranhão, Mato Grosso)
S-cent. Brazil (Rondônia, sw Mato Grosso)

☐ **Band-tailed Antwren** *Myrmotherula urosticta*

Coastal se Brazil (central Bahia to n Rio de Janeiro)

☐ **Gray Antwren** *Myrmotherula menetriesii*
____ *M. m. pallida*
____ *M. m. cinereiventris*
____ *M. m. menetriesii*
____ *M. m. berlepschi*
____ *M. m. omissa*

E Colombia to sw Venezuela, ne Peru and nw Brazil
SE Venezuela to the Guianas and ne Amazonian Brazil
E Peru s of R. Amazon to nw Bolivia and sw Amazonian Brazil
S-central Amazonian Brazil (R. Madeira to R. Tapajós), n Bolivia
NE Brazil (R. Tapajós to w Maranhão)

☐ **Leaden Antwren** *Myrmotherula assimilis*

River islands of R. Amazon and its major tributaries

☐ **Banded Antbird** *Dichrozona cincta*

SE Colombia to e Peru, nw Bolivia, nw and s Amazonian Brazil

☐ **Stripe-backed Antbird** *Myrmorchilus strigilatus*
____ *M. s. strigilatus*
____ *M. s. suspicax*

NE Brazil (e Piauí, Ceará and Pernambuco to n Minas Gerais)
SE Bolivia to w Brazil (w Mato Grosso), w Paraguay, n Argentina

☐ **Spot-backed Antwren** *Herpsilochmus dorsimaculatus*

Extreme e Colombia, s Venezuela and nw Amazonian Brazil

☐ **Caatinga Antwren** *Herpsilochmus sellowi*

Caatinga of n-central and e Brazil

☐ **Pileated Antwren** *Herpsilochmus pileatus*

Coastal e Brazil (Bahia south of Salvador)

☐ **Black-capped Antwren** *Herpsilochmus atricapillus*

NE Brazil to e Bolivia, nw Paraguay and extreme nw Argentina

☐ **Creamy-bellied Antwren** *Herpsilochmus motacilloides*

E slope of Andes of Peru (Huánuco to Cusco)

☐ **Ash-throated Antwren** *Herpsilochmus parkeri*

E slope of Andes of n-central Peru (San Martín)

☐ **Dugand's Antwren** *Herpsilochmus dugandi*

Locally in extreme se Colombia, e Ecuador and ne Peru

☐ **Spot-tailed Antwren** *Herpsilochmus sticturus*

E Venezuela to the Guianas and ne Amazonian Brazil

☐ **Roraiman Antwren** *Herpsilochmus roraimae*
____ *H. r. kathleenae*
____ *H. r. roraimae*

Tepuis of sw Venezuela and adj. Brazil (n Amazonas)
Tepuis of se Venezuela, adj. n Brazil (n Roraima) and w-c Guyana

☐ **Yellow-breasted Antwren** *Herpsilochmus axillaris*
____ *H. a. senex*
____ *H. a. aequatorialis*

Tropical and subtropical Western Andes of sw Colombia
E slope of Andes of se Colombia and e Ecuador

_____ *H. a. puncticeps*   E slope of Andes of n Peru (Loreto to Huánuco)
_____ *H. a. axillaris*   E slope of Andes of s Peru (Junín, Cusco and Puno)

☐ **Ancient Antwren *Herpsilochmus gentryi***

Sandy soil forests in extreme se Ecuador and n Peru

☐ **Todd's Antwren *Herpsilochmus stictocephalus***

Extreme e Venezuela to the Guianas and extreme n Brazil

☐ **Large-billed Antwren *Herpsilochmus longirostris***

Brazilian plateau (Piauí to Mato Grosso do Sul) and ne Bolivia

☐ **Pectoral Antwren *Herpsilochmus pectoralis***

Locally in ne Brazil (Maranhão, Rio Grande do Norte and Bahia)

☐ **Rufous-winged Antwren *Herpsilochmus rufimarginatus***
_____ *H. r. exiguus*   Pacific slope of e Panama (e Darién)
_____ *H. r. frater*   Colombia and Venezuela to Bolivia and e Brazil (Maranhão)
_____ *H. r. scapularis*   E Brazil (Pernambuco to Espírito Santo and Minas Gerais)
_____ *H. r. rufimarginatus*   SE Brazil (Rio de Janeiro) to e Paraguay and ne Argentina

☐ **Dot-winged Antwren *Microrhopias quixensis***
_____ *M. q. boucardi*   Tropical se Mexico to Belize, e Guatemala, n Honduras
_____ *M. q. virgatus*   SE Honduras and e Nicaragua to Costa Rica and Panama
_____ *M. q. consobrinus*   E Panama to w Colombia and w Ecuador
_____ *M. q. microstictus*   S Guyana, Suriname, French Guiana and ne Amazonian Brazil
_____ *M. q. quixensis*   S Colombia to e Ecuador and ne Peru
_____ *M. q. intercedens*   Lowlands of central Peru and sw Amazonian Brazil
_____ *M. q. nigriventris*   E slope of Andes of central Peru (San Martín to n Cusco)
_____ *M. q. albicauda*   SE Peru (Cusco, Madre de Dios, Puno) and adj. n Bolivia (Pando)
_____ *M. q. bicolor*   S-central Amazonian Brazil
_____ *M. q. emiliae*   Amazonian Brazil (R. Tapajós to R. Tocantins, n Mato Grosso)

☐ **Narrow-billed Antwren *Neorhopias iheringi***

Interior e Brazil (e Bahia and ne Minas Gerais)

☐ **White-fringed Antwren *Formicivora grisea***
_____ *F. g. alticincta*   Pearl Islands (Bay of Panama)
_____ *F. g. hondae*   NW Colombia
_____ *F. g. intermedia*   N Colombia, nw Venezuela, Margarita I. and Chacachacare I.
_____ *F. g. tobagensis*   Tobago
_____ *F. g. fumosa*   N base of Andes of ne Colombia and w Venezuela
_____ *F. g. rufiventris*   Extreme e Colombia and s Venezuela (w Amazonas)
_____ *F. g. orenocensis*   S Venezuela s of R. Orinoco (Bolívar andextreme n Amazonas)
_____ *F. g. grisea*   Guyana, coastal Suriname, French Guiana, n and e Brazil
_____ *F. g. deluzae*   Known from a single specimen from se Brazil (Rio de Janeiro)

☐ **Serra Antwren *Formicivora serrana***
_____ *F. s. serrana*   Mts. of se Brazil (e Minas Gerais and adj. Espírito Santo)
_____ *F. s. interposita*   SE Brazil (nw Rio de Janeiro and adjacent se Minas Gerais)

☐ **Restinga Antwren *Formicivora littoralis***

Coastal se Brazil (Rio de Janeiro) and adjacent offshore islands

☐ **Black-bellied Antwren *Formicivora melanogaster***
_____ *F. m. bahiae*   NE Brazil (e Maranhão to n Bahia and w Pernambuco)
_____ *F. m. melanogaster*   Central Brazil to se Bolivia and extreme n Paraguay

☐ **Rusty-backed Antwren *Formicivora rufa***
_____ *F. r. urubambae*   Locally in foothills of e Peru (San Martín and Cusco)
_____ *F. r. chapmani*   S Suriname and e-central Brazil
_____ *F. r. rufa*   Amazonian Brazil to se Peru, e Bolivia and Paraguay

☐ **Black-hooded Antwren *Formicivora erythronotos***

Locally in lowlands of se Brazil (e São Paulo and Rio de Janeiro)

☐ **Parana Antwren *Formicivora acutirostris***

Coastal se Brazil (Paraná and extreme ne Santa Catarina)

297

☐ **Ferruginous Antbird** *Drymophila ferruginea*

SE Brazil (se Bahia to n Santa Catarina, Minas Gerais, São Paulo)

☐ **Bertoni's Antbird** *Drymophila rubricollis*

SE Brazil to extreme e Paraguay and ne Argentina (Misiones)

☐ **Rufous-tailed Antbird** *Drymophila genei*

SE Brazil (se Minas Gerais, s Espírito Santo and Rio de Janeiro)

☐ **Ochre-rumped Antbird** *Drymophila ochropyga*

SE Brazil (se Bahia to Espírito Santo, Minas Gerais, São Paulo)

☐ **Dusky-tailed Antbird** *Drymophila malura*

SE Brazil to se Paraguay and ne Argentina (Misiones)

☐ **Scaled Antbird** *Drymophila squamata*
_____ *D. s. squamata* — E Brazil (e Alagoas and e Bahia)
_____ *D. s. stictocorypha* — SE Brazil (e Minas Gerais to ne Santa Catarina)

☐ **Striated Antbird** *Drymophila devillei*
_____ *D. d. devillei* — S-cent. Colombia to e Peru, n Bolivia, sw Amazonian Brazil
_____ *D. d. subochracea* — S-cent. Amazonian Brazil and ne Bolivia (ne Santa Cruz)

☐ **Long-tailed Antbird** *Drymophila caudata*
_____ *D. c. klagesi* — Mts. of n Colombia and n Venezuela
_____ *D. c. aristeguietana* — Mts. of nw Venezuela (Sierra de Perijá)
_____ *D. c. hellmayri* — NE Colombia (Santa Marta Mts.) and e slope of E Andes
_____ *D. c. caudata* — Andes of Colombia, Peru and nw Bolivia (La Paz)

☐ **Streak-capped Antwren** *Terenura maculata*

SE Brazil (se Bahia) to e Paraguay and ne Argentina (Misiones)

☐ **Orange-bellied Antwren** *Terenura sicki*

E Brazil (e Alagoas and e Pernambuco)

☐ **Rufous-rumped Antwren** *Terenura callinota*
_____ *T. c. callinota* — Costa Rica to Ecuador and n Peru
_____ *T. c. venezuelana* — NW Venezuela (Sierra de Perijá), Mérida and Barinas
_____ *T. c. guianensis* — S Guyana (Acari Mountains) and adjacent central Suriname
_____ *T. c. peruviana* — Central Peru (San Martín to Junín and Cusco)

☐ **Chestnut-shouldered Antwren** *Terenura humeralis*

E Ecuador to nw Bolivia and sw Amazonian Brazil

☐ **Yellow-rumped Antwren** *Terenura sharpei*

E slope of Andes of se Peru (Cusco, Puno) and w Bolivia

☐ **Ash-winged Antwren** *Terenura spodioptila*
_____ *T. s. signata* — SE Colombia ne Peru and nw Amazonian Brazil (R. Negro region)
_____ *T. s. spodioptila (elaopteryx)* — S Venezuela to the Guianas and ne Amazonian Brazil
_____ *T. s. meridionalis* — S-cent. Amazonian Brazil (R. Madeira to lower R. Tapajós)

☐ **Blackish Antbird** *Cercomacra nigrescens*
_____ *C. n. nigrescens* — Coastal Suriname, French Guiana, adj. Brazil (Amazonas, Pará)
_____ *C. n. aequatorialis* — E Andean slopes s Colombia to ne Peru (Amazonas, San Martín)
_____ *C. n. notata* — E Andean slope of central Peru (w Ucayali to Junín)
_____ *C. n. fuscicauda* — SE Colombia to ne Bolivia and sw Amazonian Brazil
_____ *C. n. approximans* — S-cent. Amazonian Brazil and e Bolivia (Beni, Santa Cruz)
_____ *C. n. ochrogyna* — E-cent. Amazonian Brazil

☐ **Willis' Antbird** *Cercomacra laeta*
_____ *C. l. waimiri* — N-cent. Amazonian Brazil and extreme s Guyana
_____ *C. l. laeta* — SE Amazonian Brazil (e Pará, w Maranhão)
_____ *C. l. sabinoi* — Coastal ne Brazil (Pernambuco and Alagoas)

☐ **Dusky Antbird** *Cercomacra tyrannina*
_____ *C. t. crepera* — SE Mexico to w Panama
_____ *C. t. tyrannina (rufiventris)* — Cent. Panama to e Colombia, s Venezuela and nw Brazil
_____ *C. t. vicina* — E slope of Eastern Andes of n Colombia and nw Venezuela
_____ *C. t. saturatior* — The Guianas and ne Amazonian Brazil

☐ **Parker's Antbird** *Cercomacra parkeri*

Andes of w-central Colombia

☐ **Black Antbird** *Cercomacra serva*

S Colombia e of Andes to nw Bolivia and sw Amazonian Brazil

☐ **Gray Antbird** *Cercomacra cinerascens*
_____ *C. c. cinerascens*
_____ *C. c. immaculata*
_____ *C. c. sclateri*
_____ *C. c. iterata*

SE Colombia to ne Peru, s Venezuela and nw Amazonian Brazil
E Venezuela to the Guianas and ne Amazonian Brazil
E Peru to nw Bolivia and sw Amazonian Brazil
SE Amazonian Brazil and ne Bolivia (n Santa Cruz)

☐ **Rio de Janeiro Antbird** *Cercomacra brasiliana*

SE Brazil (s Bahia and e Minas Gerais to Rio de Janeiro)

☐ **Rio Branco Antbird** *Cercomacra carbonaria*

Extreme n Brazil (along R. Branco) and adjacent Guyana

☐ **Jet Antbird** *Cercomacra nigricans*

E Panama (including Pearl Islands) to w Ecuador and e Venezuela

☐ **Mato Grosso Antbird** *Cercomacra melanaria*

E Bolivia to s-cent. Brazil (Mato Grosso) and n Paraguay

☐ **Manu Antbird** *Cercomacra manu*

Locally in se Peru, nw Bolivia and s Amazonian Brazil

☐ **Bananal Antbird** *Cercomacra ferdinandi*

Cent. Brazil (Bananal I. and right bank of R. Javaés, Tocantins)

☐ **White-backed Fire-eye** *Pyriglena leuconota*
_____ *P. l. pacifica*
_____ *P. l. castanoptera*
_____ *P. l. picea*
_____ *P. l. marcapatensis*
_____ *P. l. hellmayri*
_____ *P. l. maura*
_____ *P. l. similis*
_____ *P. l. interposita*
_____ *P. l. leuconota*
_____ *P. l. pernambucensis*

W Ecuador and extreme nw Peru (Tumbes)
E Anden slope of s Colombia to e Ecuador and n Peru
E Andean slope of Peru (s Amazonas to Junín and Ayacucho)
SE Peru (Madre de Dios, Cusco and Puno)
W-cent. Bolivia (Beni, La Paz, Cochabamba and w Santa Cruz)
E Bolivia to s-cent. Brazil (Mato Grosso) and n Paraguay
S-cent. Amazonian Brazil (R. Tapajós to R. Xingú)
E-central Brazil (R. Xingú to R. Tocantins)
E Brazil (e Pará e of R. Tocantins and n Maranhão)
NE Brazil (e Pernambuco and e Alagoas)

☐ **Fringe-backed Fire-eye** *Pyriglena atra*

Lowlands of coastal e Brazil (s Sergipe and coastal ne Bahia)

☐ **White-shouldered Fire-eye** *Pyriglena leucoptera*

E Brazil (e Bahia) to e Paraguay and ne Argentina (Misiones)

☐ **Slender Antbird** *Rhopornis ardesiacus*

Highlands of e Brazil (se Bahia and ne Minas Gerais)

☐ **White-browed Antbird** *Myrmoborus leucophrys*
_____ *M. l. angustirostris*
_____ *M. l. erythrophrys*
_____ *M. l. koenigorum*
_____ *M. l. leucophrys (griseigula)*

S Venezuela to the Guianas and Brazil north of R. Amazon
E slope of Eastern Andes of Colombia and nw Venezuela
Central Peru (upper Río Huallaga Valley in Huánuco)
S Colombia to e Peru, ne Bolivia and w Amazonian Brazil

☐ **Ash-breasted Antbird** *Myrmoborus lugubris*
_____ *M. l. berlepschi*
_____ *M. l. stictopterus*
_____ *M. l. femininus*
_____ *M. l. lugubris*

NE Peru (Loreto) and extreme w Amazonian Brazil
Cent. Amazonian Brazil (lower R. Negro and adj. R. Solimões)
S-cent. Amazonian Brazil (lower R. Madeira)
Central Brazil (along both banks of R. Amazon)

☐ **Black-faced Antbird** *Myrmoborus myotherinus*
_____ *M. m. elegans (napensis)*
_____ *M. m. myotherinus*
_____ *M. m. incanus*
_____ *M. m. ardesiacus*
_____ *M. m. proximus*
_____ *M. m. sororius*
_____ *M. m. ochrolaema*

SE Colombia to s Venezuela, cent. Peru and nw Amazonian Brazil
Extreme e Peru to nw Bolivia and sw Amazonian Brazil
NW Amazonian Brazil (along north bank of R. Solimões)
W Brazil (R. Japurá to lower R. Negro)
W Brazil (south bank of R. Amazon from R. Purús to R. Madeira)
S-cent. Amazonian Brazil (se Amazonas, Rondônia)
Brazil s of R. Amazon (R. Madeira to Tocantins, Mato Grosso)

☐ **Black-tailed Antbird** *Myrmoborus melanurus*

Tropical ne Peru and adjacent Brazil (extreme w Amazonas)

☐ **Warbling Antbird** *Hypocnemis cantator*
____ *H. c. cantator (notaea)* — E Venezuela (ne Bolívar) to the Guianas and ne Amazonian Brazil
____ *H. c. flavescens* — E-cent. Colombia to s Venezuela and extreme nw Amaz. Brazil
____ *H. c. saturata* — S Colombia to ne Peru and w-cent. Amazonian Brazil
____ *H. c. peruviana* — Trop. e Peru s of R. Amazon to n Bolivia and sw Amaz. Brazil
____ *H. c. subflava* — Andean foothills of e Peru (Huánuco to Cusco)
____ *H. c. collinsi* — S Peru (Ucayali to Puno) to w-cent. Bolivia and sw Amaz. Brazil
____ *H. c. ochrogyna* — E Bolivia (e Beni, Santa Cruz) and s-cent. Amazonian Brazil
____ *H. c. implicata* — S-cent. Amazonian Brazil (lower R. Madeira to R. Tapajós)
____ *H. c. striata* — Brazil s of R. Amazon (R. Tapajós to R. Xingú, n Mato Grosso)
____ *H. c. affinis* — Brazil s of R. Amazon (R. Xingú to R. Tocantins, e Mato Grosso)

☐ **Yellow-browed Antbird** *Hypocnemis hypoxantha*
____ *H. h. hypoxantha* — SE Colombia to e Ecuador, ne Peru and w Amazonian Brazil
____ *H. h. ochraceiventris* — SE Amazonian Brazil (R. Tapajós to R. Xingú)

☐ **Black-chinned Antbird** *Hypocnemoides melanopogon*
____ *H. m. melanopogon (occidentalis)* — E Colombia to the Guianas and n Brazil n of R. Amazon
____ *H. m. minor* — S-central Amazonian Brazil

☐ **Band-tailed Antbird** *Hypocnemoides maculicauda*

E Peru to n Bolivia and s Amazonian Brazil

☐ **Black-and-white Antbird** *Myrmochanes hemileucus*

Amazon river islands (ne Ecuador to n Bolivia and w Brazil)

☐ **Bare-crowned Antbird** *Gymnocichla nudiceps*
____ *G. n. chiroleuca* — Caribbean slope of e Guatemala and Belize to w Panama
____ *G. n. erratilis* — Pacific slope of Costa Rica and w Panama (Chiriquí)
____ *G. n. nudiceps* — E Panama and Pacific slope of nw Colombia
____ *G. n. sanctamartae* — N Colombia (Santa Marta region and Magdalena Valley)

☐ **Silvered Antbird** *Sclateria naevia*
____ *S. n. naevia* — E Venezuela to the Guianas and ne Amazonian Brazil; Trinidad
____ *S. n. diaphora* — S-cent. Venezuela (lower R. Caura drainage in nw Bolívar)
____ *S. n. toddi* — S-cent. Amazonian Brazil (lower R. Madeira to R. Tocantins)
____ *S. n. argentata* — SE Colombia to n Bolivia, sw Venezuela and w Amazonian Brazil

☐ **Black-headed Antbird** *Percnostola rufifrons*
____ *P. r. jensoni* — NE Peru (Quebrada Orán in Loreto)
____ *P. r. minor* — E Colombia to sw Venezuela and nw Amazonian Brazil
____ *P. r. subcristata* — N Brazil (lower R. Negro to R. Trombetas)
____ *P. r. rufifrons* — Guyana, Suriname, French Guiana and ne Amazonian Brazil

☐ **Allpahuayo Antbird** *Percnostola arenarum*

NE Peru north of R. Marañón (R. Morona to R. Nanay drainage)

☐ **Slate-colored Antbird** *Percnostola schistacea*

SE Colombia to ne Ecuador, e Peru and w-cent. Amazonian Brazil

☐ **Spot-winged Antbird** *Percnostola leucostigma*
____ *P. l. obscura* — Tepuis of se Venezuela (e Bolívar) and adj. Brazil
____ *P. l. leucostigma* — E Venezuela (e Bolívar) to the Guianas and ne Amazonian Brazil
____ *P. l. infuscata* — E Colombia to s Venezuela and nw Amazonian Brazil
____ *P. l. subplumbea* — E Colombia to extreme w Venezuela, ne Peru and adj. w Brazil
____ *P. l. intensa* — Central Peru (Huánuco, Pasco, Junín and s Ucayali)
____ *P. l. brunneiceps* — S Peru (Cusco, Madre de Dios, Puno) and extreme w-cent. Bolivia
____ *P. l. humaythae* — SW and cent. Amazonian Brazil and extreme n Bolivia (Pando)
____ *P. l. rufifacies* — Brazil s of R. Amazon (R. Madeira to R. Tocantins, n Rondônia)

☐ **Roraiman Antbird** *Percnostola saturata*

Extreme se Venezuela and adj. Guyana (Mt. Roraima environs)

Typical Antbirds

☐ **Caura Antbird** *Percnostola caurensis*
_____ *P. c. caurensis* — S Venezuela (w Bolívar and n Amazonas)
_____ *P. c. australis* — S Venezuela (s Amazonas) and immediately adjacent n Brazil

☐ **White-lined Antbird** *Percnostola lophotes*
— Extreme sw Amazonian Brazil to se Peru and nw Bolivia

☐ **Yapacana Antbird** *Myrmeciza disjuncta*
— Locally in e Colombia, sw Venezuela and nw Brazil (Amazonas)

☐ **Gray-headed Antbird** *Myrmeciza griseiceps*
— Andes of extreme sw Ecuador and nw Peru (Tumbes, Piura)

☐ **White-bellied Antbird** *Myrmeciza longipes*
_____ *M. l. panamensis* — E Panama and n Colombia
_____ *M. l. longipes* — Extreme ne Colombia and n Venezuela; Trinidad
_____ *M. l. boucardi* — Colombia (upper Magdalena Valley from Cundinamarca to Huila)
_____ *M. l. griseipectus* — SE Colombia to s Venezuela, the Guianas and ne Brazil

☐ **Chestnut-backed Antbird** *Myrmeciza exsul*
_____ *M. e. exsul* — Caribbean slope of Nicaragua to w Panama
_____ *M. e. occidentalis* — Pacific slope of Costa Rica and w Panama
_____ *M. e. niglarus* — E Panama and adjacent extreme nw Colombia (n Chocó)
_____ *M. e. cassini* — Extreme se Panama (Darién) and n Colombia
_____ *M. e. maculifer* — Pacific slope of w Colombia and w Ecuador

☐ **Ferruginous-backed Antbird** *Myrmeciza ferruginea*
_____ *M. f. ferruginea* — Extreme e Venezuela to the Guianas and ne Amazonian Brazil
_____ *M. f. eluta* — S-cent. Amazonian Brazil (lower R. Tapajós to lower R. Madeira)

☐ **Scalloped Antbird** *Myrmeciza ruficauda*
_____ *M. r. soror* — Coastal ne Brazil (Paraíba south to s Alagoas)
_____ *M. r. ruficauda* — E Brazil (se Bahia, extreme e Minas Gerais and Espírito Santo)

☐ **White-bibbed Antbird** *Myrmeciza loricata*
— E Brazil (Bahia, Minas Gerais, Espírito Santo, Rio de Janeiro)

☐ **Squamate Antbird** *Myrmeciza squamosa*
— SE Brazil (Rio de Janeiro to ne Rio Grande do Sul)

☐ **Dull-mantled Antbird** *Myrmeciza laemosticta*
_____ *M. l. laemosticta* — Caribbean slope of e Costa Rica and both slope of Panama
_____ *M. l. palliata (bolivari, venezuelae)* — N slope of Andes of Colombia and nw Venezuela (Zulia, Mérida)

☐ **Esmeraldas Antbird** *Myrmeciza nigricauda*
— Pacific slope of w Colombia and w Ecuador (south to El Oro)

☐ **Stub-tailed Antbird** *Myrmeciza berlepschi*
— Pacific slope of w Colombia and nw Ecuador (Esmeraldas)

☐ **Gray-bellied Antbird** *Myrmeciza pelzelni*
— Extreme e Colombia to sw Venezuela and extreme nw Brazil

☐ **Southern Chestnut-tailed Antbird** *Myrmeciza hemimelaena*
_____ *M. h. hemimelaena* — E Peru (s of R. Marañón) to nw Bolivia and sw Amazonian Brazil
_____ *M. h. pallens* — Central Brazil s of R. Amazon to ne Bolivia (e Santa Cruz)

☐ **Zimmer's Antbird** *Myrmeciza castanea*
_____ *M. c. centunculorum* — Extreme s Colombia to e Ecuador and ne Peru (Loreto)
_____ *M. c. castanea* — SE Ecuador (Zamora-Chinchipe) and n Peru (San Martín)

☐ **Plumbeous Antbird** *Myrmeciza hyperythra*
— S Colombia to e Peru, nw Bolivia and sw Amazonian Brazil

☐ **Goeldi's Antbird** *Myrmeciza goeldii*
— SE Peru to extreme nw Bolivia and extreme sw Amazonian Brazil

☐ **White-shouldered Antbird** *Myrmeciza melanoceps*
— SE Colombia to ne Peru and w Amazonian Brazil

☐ **Sooty Antbird** *Myrmeciza fortis*

____ *M. f. fortis*	SE Colombia to extreme nw Bolivia and w Amazonian Brazil
____ *M. f. incanescens*	Brazil (known only from the type locality on R. Solimões)

☐ **Immaculate Antbird** *Myrmeciza immaculata*

____ *M. i. zeledoni*	Extreme s Nicaragua (San Juan) to w Panama
____ *M. i. macrorhyncha*	E Panama (Darién) to w Colombia and w Ecuador
____ *M. i. brunnea*	Mountains of nw Venezuela (Sierra de Perijá)
____ *M. i. immaculata*	Locally in central and E Andes of Colombia to w Venezuela

☐ **Black-throated Antbird** *Myrmeciza atrothorax*

____ *M. a. metae*	Central Colombia (Meta, w Guaviare)
____ *M. a. atrothorax*	E Colombia to s Venezuela, the Guianas and n Amaz. Brazil
____ *M. a. tenebrosa*	E Ecuador n of R. Amazon, ne Peru and n Brazil
____ *M. a. maynana*	N-cent. Peru (south of Río Marañón and west of Río Huallaga)
____ *M. a. melanura (stictothorax, obscurata, griseiventris)*	E Peru s of R. Amazon to n Bolivia and w and central Brazil

☐ **Spotted Antbird** *Hylophylax naevioides*

____ *H. n. capnitis*	Caribbean slope of e Honduras to w Panama
____ *H. n. naevioides (subsimilis)*	E Panama to w Colombia and w Ecuador (south to e Guayas)

☐ **Spot-backed Antbird** *Hylophylax naevius*

____ *H. n. naevius (consobrinus, obscurus)*	SE Colombia to s Venezuela, the Guianas n Peru and n Brazil
____ *H. n. theresae*	S Ecuador to n Peru (Loreto) and w Amazonian Brazil
____ *H. n. peruvianus*	Foothills of n Peru (Amazonas and San Martín)
____ *H. n. inexpectatus*	SE Peru to extreme sw Amaz. Brazil and nw Bolivia
____ *H. n. ochraceus*	SE Amaz. Brazil (R. Tapajós to R. Tocantins, s to n Mato Grosso)

☐ **Dot-backed Antbird** *Hylophylax punctulatus*

	SE Colombia to s Venezuela, n Bolivia and Amazonian Brazil

☐ **Scale-backed Antbird** *Hylophylax poecilinotus*

____ *H. p. poecilinotus*	S Venezuela to the Guianas and ne Amazonian Brazil
____ *H. p. duidae*	E Colombia to sw Venezuela and nw Brazil (lower R. Negro)
____ *H. p. lepidonotus*	SE Colombia to e Ecuador and e Peru (south to Cusco)
____ *H. p. gutturalis*	NE Peru and adjacent w Brazil (east to lower R. Juruá)
____ *H. p. griseiventris*	E-cent. and se Peru to n Bolivia and sw Amazonian Brazil
____ *H. p. nigrigula*	S-central Amazonian Brazil
____ *H. p. vidua*	E Amazonian Brazil s of R. Amazon (R. Xingú to w Maranhão)

☐ **Wing-banded Antbird** *Myrmornis torquata*

____ *M. t. stictoptera*	Caribbean lowlands of e Nicaragua to extreme nw Colombia
____ *M. t. torquata*	E Colombia to the Guianas, ne Peru and Amazonian Brazil

☐ **White-plumed Antbird** *Pithys albifrons*

____ *P. a. peruvianus (brevibarba)*	E Colombia to w Venezuela, n-c Peru and nw Amazonian Brazil
____ *P. a. albifrons*	S Venezuela to the Guianas and Brazil n of R. Amazon

☐ **White-masked Antbird** *Pithys castaneus*

	NE Peru (lower R. Pastaza and lower R. Morona)

☐ **Bicolored Antbird** *Gymnopithys leucaspis*

____ *G. l. olivascens*	Caribbean slope of Honduras to w Panama
____ *G. l. bicolor*	E Panama and nw Colombia (Pacific slope in Chocó)
____ *G. l. daguae*	Pacific slope of w Colombia (s Chocó to Cauca)
____ *G. l. aequatorialis*	Tropical sw Colombia (Nariño) and w Ecuador
____ *G. l. ruficeps*	Tropical central Colombia (Antioquia)
____ *G. l. leucaspis*	Tropical e Colombia
____ *G. l. castaneus*	Tropical e Ecuador and ne Peru (Loreto)
____ *G. l. lateralis*	NW Amazonian Brazil
____ *G. l. peruanus*	N Peru (Marañón Valley)

☐ **Rufous-throated Antbird** *Gymnopithys rufigula*
_____ *G. r. pallidus* | S Venezuela (Bolívar and Amazonas)
_____ *G. r. pallidigula* | *Tepuis* of extreme s Venezuela (Pica Yavita-Pimichín environs)
_____ *G. r. rufigula* | Extreme e Venezuela to the Guianas and E Brazil n of R. Amazon

☐ **White-throated Antbird** *Gymnopithys salvini*

E Peru to sw Amazonian Brazil and central Bolivia

☐ **Lunulated Antbird** *Gymnopithys lunulatus*

Locally in e Ecuador and e Peru (w of R. Napo and R. Ucayali)

☐ **Bare-eyed Antbird** *Rhegmatorhina gymnops*

SE Amazonian Brazil (R. Tapajós to R. Iriri, s to Mato Grosso)

☐ **Harlequin Antbird** *Rhegmatorhina berlepschi*

E Amazonian Brazil (near west bank of lower R. Tapajós)

☐ **White-breasted Antbird** *Rhegmatorhina hoffmannsi*

Central Amazonian Brazil to w Mato Grosso

☐ **Hairy-crested Antbird** *Rhegmatorhina melanosticta*
_____ *R. m. melanosticta* | Base of Andes of se Colombia to e Ecuador and ne Peru
_____ *R. m. brunneiceps* | Cent. Peru s of R. Marañón (San Martín to n Ayacucho)
_____ *R. m. purusiana (badia)* | E Peru to sw Amazonian Brazil and nw Bolivia (Pando, La Paz)

☐ **Chestnut-crested Antbird** *Rhegmatorhina cristata*

SE Colombia and adjacent nw Brazil

☐ **Pale-faced Antbird** *Skutchia borbae*

S-cent. Amaz. Brazil (R. Tapajós to R. Madeira, s to R. Aripuña)

☐ **Black-spotted Bare-eye** *Phlegopsis nigromaculata*
_____ *P. n. nigromaculata* | SE Colombia to e Peru, n Bolivia and sw Amazonian Brazil
_____ *P. n. bowmani* | S-central Amazonian Brazil and ne Bolivia (n Santa Cruz)
_____ *P. n. confinis* | N Amazonian Brazil (R. Xingú to R. Tocantins and R. Araguaia)
_____ *P. n. paraensis* | E Brazil south of R. Amazon (R. Tocantins to w Maranhão)

☐ **Reddish-winged Bare-eye** *Phlegopsis erythroptera*
_____ *P. e. erythroptera* | SE Colombia to ne Peru, s Venezuela and nw Amazonian Brazil
_____ *P. e. ustulata* | E Peru to sw Amaz. Brazil and extreme nw Bolivia (nw Pando)

☐ **Ocellated Antbird** *Phaenostictus mcleannani*
_____ *P. m. saturatus* | Tropical e Honduras to Costa Rica and extreme w Panama
_____ *P. m. mcleannani (chocoanus)* | Central and e Panama to nw Colombia
_____ *P. m. pacificus* | Extreme sw Colombia (Nariño) to nw Ecuador (Esmeraldas)

FAMILY: FORMICARIIDAE (Antthrushes and Antpittas—63)

☐ **Rufous-capped Antthrush** *Formicarius colma*
_____ *F. c. colma* | E Colombia to s Venezuela, the Guianas, Brazil n of R. Amazon
_____ *F. c. nigrifrons* | E Ecuador to e Peru, n Bolivia and Brazil s of R. Amazon
_____ *F. c. amazonicus* | Brazil s of R. Amazon (R. Madeira to nw Maranhão, s Mato Grosso)
_____ *F. c. ruficeps* | Coastal e and se Brazil (Pernambuco to Rio Grande do Sul)

☐ **Mexican Antthrush** *Formicarius moniliger*
_____ *F. m. moniliger* | Caribbean slope of s Mexico and e Guatemala (except Petén)
_____ *F. m. pallidus* | SE Mexico (Yucatán Peninsula) and Petén of n Guatemala
_____ *F. m. intermedius* | E Guatemala, Belize and nw Honduras

☐ **Black-faced Antthrush** *Formicarius analis*
_____ *F. a. umbrosus* | Caribbean slope of Honduras to w Panama
_____ *F. a. hoffmanni* | Lowlands of sw Costa Rica to w Panama (w Chiriquí)
_____ *F. a. panamensis* | E Panama (Chocó and Darién) and adjacent nw Colombia
_____ *F. a. virescens* | West base of Santa Marta Mts. (ne Colombia)
_____ *F. a. griseoventris* | Mts. of n Colombia and nw Venezuela in w Maracaibo basin
_____ *F. a. saturatus* | N Colombia to nw Venezuela; Trinidad

____	*F. a. connectens*	E Colombia east of the Andes
____	*F. a. crissalis*	Extreme e Venezuela to the Guianas and adjacent ne Brazil
____	*F. a. zamorae (olivaceus)*	E Ecuador to ne Peru and w Brazil (n of R. Solimões)
____	*F. a. analis*	Amazonian Peru s of R. Amazon to w Brazil and n Bolivia
____	*F. a. paraensis*	E Brazil (R. Tapajós to Belém and w Maranhão)

☐ **Black-headed Antthrush** *Formicarius nigricapillus*

____	*F. n. nigricapillus*	Caribbean slope of e Costa Rica and both slopes of Panama
____	*F. n. destructus*	Tropical Pacific Colombia and w Ecuador

☐ **Rufous-fronted Antthrush** *Formicarius rufifrons*

SE Peru (Madre de Dios), adjacent w Brazil and nw Bolivia

☐ **Rufous-breasted Antthrush** *Formicarius rufipectus*

____	*F. r. rufipectus*	E Costa Rica and both slopes of w Panama to Darién
____	*F. r. lasallei*	Mts. of nw Venezuela (Sierra de Perijá and sw Táchira)
____	*F. r. carrikeri*	W and Central Andes of Colombia and w Ecuador
____	*F. r. thoracicus*	Subtropical e Ecuador to se Peru (south to Cusco)

☐ **Short-tailed Antthrush** *Chamaeza campanisona*

____	*C. c. venezuelana*	Mts. of n Venezuela (east to Distrito Federal and Aragua)
____	*C. c. yavii*	*Tepuis* of south-central Venezuela (Cerro Yaví)
____	*C. c. huachamacarii*	Tepuis of s Venezuela (Cerro Huachamacari)
____	*C. c. obscura*	Subtropical mountains of se Venezuela (Bolívar and s Amazonas)
____	*C. c. fulvescens*	*Tepuis* of se Venezuela (Mt. Roraima) and adjacent Guyana
____	*C. c. columbiana*	E slope of Andes of Colombia
____	*C. c. punctigula*	E Ecuador (Napo and Pastaza) to n Peru (north of R. Marañón)
____	*C. c. olivacea*	Tropical e-central Peru (south to Madre de Dios)
____	*C. c. berlepschi*	SE Peru (Cusco) and extreme w Bolivia
____	*C. c. boliviana*	Andes of Bolivia (La Paz, Cochabamba and Santa Cruz)
____	*C. c. campanisona*	E and se Brazil (Ceará, Alagoas, and s Bahia to Santa Catarina)
____	*C. c. tshororo*	E Paraguay, s Brazil and ne Argentina

☐ **Striated Antthrush** *Chamaeza nobilis*

____	*C. n. rubida*	SE Colombia to e Ecuador, ne Peru and adjacent w Brazil
____	*C. n. nobilis*	E Peru to nw Bolivia and w Amazonian Brazil
____	*C. n. fulvipectus*	N-central Amaz. Brazil (left bank of Tapajós near Santarém, Pará)

☐ **Brazilian Antthrush** *Chamaeza ruficauda*

SE Brazil (Minas Gerais to n Rio Grande do Sul) and ne Argentina

☐ **Such's Antthrush** *Chamaeza meruloides*

SE Brazil (Minas Gerais and Espírito Santo to ne Santa Catarina)

☐ **Schwartz's Antthrush** *Chamaeza turdina*

____	*C. t. chionogaster*	Mts. of n Venezuela
____	*C. t. turdina*	Andes of Colombia (upper Magdalena and middle Cauca valleys)

☐ **Barred Antthrush** *Chamaeza mollissima*

____	*C. m. mollissima*	Locally in Andes of Colombia, Ecuador and n Peru
____	*C. m. yungae*	Locally in Andes of se Peru (Cusco) and n Bolivia (Cochabamba)

☐ **Spotted Antpitta** *Hylopezus macularius*

____	*H. m. macularius*	NE Venezuela to the Guianas and adjacent n Brazil
____	*H. m. diversus*	Extreme se Colombia to s Venezuela and ne Peru
____	*H. m. paraensis*	Amazonian Brazil (south to Rondônia and east to e Pará)

☐ **Masked Antpitta** *Hylopezus auricularis*

N Bolivia (se Pando and n Beni)

☐ **Streak-chested Antpitta** *Hylopezus perspicillatus*

____	*H. p. intermedius*	Caribbean slope of ne Honduras to w Panama
____	*H. p. lizanoi*	Pacific slope of sw Costa Rica
____	*H. p. perspicillatus*	E Panama to nw Colombia (n Chocó)

____	*H. p. pallidior*	Colombia (lower Cauca, upper Sinú and Magdalena valleys)
____	*H. p. periophthalmicus*	Pacific coast of w Colombia and nw Ecuador

☐ **Fulvous-bellied Antpitta** *Hylopezus dives*
____	*H. d. dives*	Caribbean slope of ne Honduras to Costa Rica
____	*H. d. flammulatus*	Caribbean slope of w Panama (Bocas del Toro)
____	*H. d. barbacoae*	Lowlands of e Panama (Darién) and w Colombia (Nariño)

☐ **White-lored Antpitta** *Hylopezus fulviventris*
____	*H. f. caquetae*	Lowlands of extreme se Colombia (w Caquetá)
____	*H. f. fulviventris*	Tropical e Ecuador and extreme n Peru (south to Iquitos)

☐ **Amazonian Antpitta** *Hylopezus berlepschi*
____	*H. b. yessupi*	E and central Peru and immediately adjacent w Brazil
____	*H. b. berlepschi*	S Amazonian Brazil to se Peru and n-central Bolivia

☐ **White-browed Antpitta** *Hylopezus ochroleucus*

Interior ne Brazil (Piauí and Ceará to extreme n Minas Gerais)

☐ **Speckle-breasted Antpitta** *Hylopezus nattereri*

SE Brazil to ne Argentina (Misiones) and extreme e Paraguay

☐ **Thrush-like Antpitta** *Myrmothera campanisona*
____	*M. c. modesta*	Base of Eastern Andes of se Colombia (south from Meta)
____	*M. c. dissors*	E Colombia to s Venezuela and nw Amazonian Brazil
____	*M. c. campanisona*	SE Venezuela to the Guianas and adjacent n Brazil
____	*M. c. signata*	Tropical e Ecuador and ne Peru (north of R. Amazon)
____	*M. c. minor*	Tropical e Peru to extreme nw Bolivia and w Amazonian Brazil
____	*M. c. subcanescens*	N Brazil south of R. Amazon (R. Madeira to R. Tapajós)

☐ **Tepui Antpitta** *Myrmothera simplex*
____	*M. s. duidae*	*Tepuis* of s Venezuela (erros of Yaví, Duida and Neblina)
____	*M. s. guaiquinimae*	*Tepuis* of se Venezuela (cerros Guaiquinima and Paurai-tepui)
____	*M. s. simplex*	SE Venezuela (Gran Sabana and Mt. Roraima) and adj. Guyana
____	*M. s. pacaraimae*	SE Venezuela (Pacaraima Mts.) and immediately adj. n Brazil

☐ **Black-crowned Antpitta** *Pittasoma michleri*
____	*P. m. zeledoni*	Caribbean slope of Costa Rica and w Panama
____	*P. m. michleri*	E Panama and extreme nw Colombia (Chocó)

☐ **Rufous-crowned Antpitta** *Pittasoma rufopileatum*
____	*P. r. rosenbergi*	W Andes of Colombia and Baudó Mountains (south to Chocó)
____	*P. r. harterti*	Lowlands and foothills of Colombia (Cauca and w Nariño)
____	*P. r. rufopileatum*	Pacific lowlands of nw Ecuador

☐ **Great Antpitta** *Grallaria excelsa*
____	*G. e. excelsa*	W Venezuela (Sierra de Perijá) and Andes of Lara to Táchira
____	*G. e. phelpsi*	Coastal mts. n Venezuela (Colonia Tovar in Aragua)

☐ **Giant Antpitta** *Grallaria gigantea*
____	*G. g. lehmanni*	Andes of sw Colombia (head of Magdalena Valley)
____	*G. g. hylodroma*	SW Colombia (Nariño) and W slope of Andes of Ecuador
____	*G. g. gigantea*	E slope of Andes of Ecuador (Carchi and Napo s to Tungurahua)

☐ **Undulated Antpitta** *Grallaria squamigera*
____	*G. s. squamigera*	Andes of Colombia, Ecuador and w Venezuela
____	*G. s. canicauda*	Andes of se Ecuador and e Peru (Cajamarca) to w Bolivia

☐ **Variegated Antpitta** *Grallaria varia*
____	*G. v. cinereiceps*	Extreme s Venezuela, adj. nw Brazil and ne Peru
____	*G. v. varia*	The Guianas and Brazil n of R. Amazon (w from lower R. Napo)
____	*G. v. distincta*	Brazil s of R. Amazon (R. Madeira to R. Tapajós, s to Mato Grosso)

____ *G. v. intercedens*	E Brazil (Pernambuco and s Bahia to se Minas Gerais)
____ *G. v. imperator*	SE Brazil (s Minas Gerais) to e Paraguay and ne Argentina

☐ **Scaled Antpitta** *Grallaria guatimalensis*

____ *G. g. ochraceiventris*	Mts. of sw Mexico (Jalisco to Hidalgo, Guerrero and Morelos)
____ *G. g. guatimalensis*	S Mexico (nw Veracruz and Oaxaca) to n Nicaragua
____ *G. g. princeps*	Subtropical Costa Rica to w Panama (Veraguas)
____ *G. g. chocoensis*	Mts. of e Panama (Darién) and nw Colombia (Chocó)
____ *G. g. carmelitae*	Santa Marta Mts. (ne Colombia) and Sierra de Perijá
____ *G. g. aripoensis*	Trinidad
____ *G. g. regulus*	Andes of Colombia, e Ecuador and Peru (south to Cusco)
____ *G. g. sororia*	S Peru (Cusco) to cent. Bolivia (Santa Cruz)
____ *G. g. roraimae*	*Tepuis* of s Venezuela, adj. n Brazil and w Guyana

☐ **Moustached Antpitta** *Grallaria alleni*

____ *G. a. alleni*	W slope of Central Andes of Colombia (Risaralda in Cauca)
____ *G. a. andaquiensis*	Andes of Colombia (Magdalena) to Ecuador (Napo, Cotopaxi)

☐ **Tachira Antpitta** *Grallaria chthonia*

	Andes of w Venezuela (sw Táchira)

☐ **Plain-backed Antpitta** *Grallaria haplonota*

____ *G. h. haplonota*	Mts. of n Venezuela (Lara, Carabobo, Aragua and Miranda)
____ *G. h. pariae*	NE Venezuela (subtropical mts. of Pária Peninsula)
____ *G. h. parambae*	Pacific slope of w Colombia (Risaralda) to s Ecuador (El Oro)
____ *G. h. chaplinae*	E slope of Andes of Colombia, Ecuador and extreme n Peru

☐ **Tawny Antpitta** *Grallaria quitensis*

____ *G. q. alticola*	E Andes of Colombia
____ *G. q. quitensis*	Central Andes of Colombia , Ecuador and extreme n Peru
____ *G. q. atuensis*	Andes of n Peru (s Amazonas and e La Libertad)

☐ **Rufous-faced Antpitta** *Grallaria erythrotis*

	Yungas of w Bolivia (La Paz, Cochabamba and Santa Cruz)

☐ **Brown-banded Antpitta** *Grallaria milleri*

	W slope of Cent. Andes of Colombia (Caldas, Risaralda, Quindío)

☐ **Bicolored Antpitta** *Grallaria rufocinerea*

____ *G. r. rufocinerea*	Cent. Andes of Colombia (s Antioquia to w Huila)
____ *G. r. romeroana*	S Colombia to Andes of ne Ecuador (nw Sucumbíos)

☐ **Stripe-headed Antpitta** *Grallaria andicolus*

____ *G. a. andicolus*	Andes of Peru, south locally to Ayacucho
____ *G. a. punensis*	Andes of s Peru (Cusco, Puno) and w Bolivia (w La Paz)

☐ **Cundinamarca Antpitta** *Grallaria kaestneri*

	E Andes of Colombia (Cundinamarca and Meta)

☐ **Santa Marta Antpitta** *Grallaria bangsi*

	Santa Marta Mountains (ne Colombia)

☐ **White-throated Antpitta** *Grallaria albigula*

	Andes of se Peru and w Bolivia to extreme nw Argentina

☐ **Red-and-white Antpitta** *Grallaria erythroleuca*

	Andes of se Peru in Cusco (Vilcabamba and Vilcanota Mts.)

☐ **Bay Antpitta** *Grallaria capitalis*

	E slope of Andes of Peru (Junín, Pasco and Huánuco)

☐ **Ochre-striped Antpitta** *Grallaria dignissima*

	Extreme se Colombia (w Putumayo) to e Ecuador and ne Peru

☐ **Elusive Antpitta** *Grallaria eludens*

	Lowlands of e Peru (Ucayali, Madre de Dios), adj. Brazil (Acre)

☐ **Chestnut-crowned Antpitta** *Grallaria ruficapilla*

____ *G. r. ruficapilla*	Andes of Colombia and n Ecuador
____ *G. r. perijana*	Sierra de Perijá (Colombia/Venezuela border)
____ *G. r. avilae*	Coastal cordillera of n Venezuela and Andes of Lara

306

____	*G. r. nigrolineata*	Andes of w Venezuela (Táchira, Mérida and Trujillo)
____	*G. r. connectens*	Subtropical Andes of sw Ecuador
____	*G. r. albiloris*	S Ecuador (Loja) and nw Peru (south to Lambayeque)
____	*G. r. interior*	Central Cordillera of n Peru (Amazonas and San Martín)

☐ **Watkins' Antpitta** *Grallaria watkinsi*

Lowlands and Andes of sw Ecuador and nw Peru (Tumbes)

☐ **Chestnut-naped Antpitta** *Grallaria nuchalis*

____	*G. n. ruficeps*	Colombia (Central Andes and w slope of Eastern Andes)
____	*G. n. obsoleta*	W slope of Andes of nw Ecuador (Imbabura and Pichincha)
____	*G. n. nuchalis*	E slope of Andes of Ecuador and extreme n Peru (Piura)

☐ **Jocotoco Antpitta** *Grallaria ridgelyi*

Andes of s Ecuador (upper Río Chinchipe drainage)

☐ **Pale-billed Antpitta** *Grallaria carrikeri*

E slope of Andes of n Peru (Amazonas and La Libertad)

☐ **Yellow-breasted Antpitta** *Grallaria flavotincta*

W slope of Western Andes of Colombia and nw Ecuador

☐ **White-bellied Antpitta** *Grallaria hypoleuca*

____	*G. h. hypoleuca*	W slope of Eastern Andes of Colombia
____	*G. h. castanea*	Central Andes of Colombia to extreme n Peru (Piura)

☐ **Rusty-tinged Antpitta** *Grallaria przewalskii*

E slope of Andes of n Peru (Amazonas to e La Libertad)

☐ **Gray-naped Antpitta** *Grallaria griseonucha*

____	*G. g. tachirae*	Andes of w Venezuela (ne Táchira)
____	*G. g. griseonucha*	Andes of w Venezuela (e Mérida)

☐ **Rufous Antpitta** *Grallaria rufula*

____	*G. r. spatiator*	Santa Marta Mountains (ne Colombia)
____	*G. r. saltuensis*	Sierra de Perijá (Colombia/Venezuela border)
____	*G. r. rufula*	Andes of Colombia to w Venezuela (Táchira) and Ecuador
____	*G. r. cajamarcae*	Andes of n Peru (Cajamarca)
____	*G. r. obscura*	Andes of central Peru (Huánuco and Junín)
____	*G. r. occabambae*	Andes of se Peru (Cusco)
____	*G. r. cochabambae*	Andes of w Bolivia (La Paz and Cochabamba)

☐ **Chestnut Antpitta** *Grallaria blakei*

Andes of central Peru (Amazonas, Huánuco and Pasco)

☐ **Ochre-breasted Antpitta** *Grallaricula flavirostris*

____	*G. f. costaricensis*	Mountains of Costa Rica and w Panama (east to Veraguas)
____	*G. f. brevis*	E Panama (upper slopes of Mt. Pirre)
____	*G. f. ochraceiventris*	Locally in Western Andes of Colombia
____	*G. f. flavirostris*	Subtropical Eastern Andes of Colombia and e Ecuador
____	*G. f. mindoensis*	Subtropical n Ecuador (Pichincha)
____	*G. f. zarumae*	Subtropical sw Ecuador (El Oro)
____	*G. f. similis*	Subtropical Andes of e Peru
____	*G. f. boliviana*	Andes of n Bolivia (La Paz and Cochabamba)

☐ **Rusty-breasted Antpitta** *Grallaricula ferrugineipectus*

____	*G. f. rara*	E Andes of Colombia and nw Venezuela (Sierra de Perijá)
____	*G. f. ferrugineipectus*	Santa Marta Mts. (ne Colombia) and mountains of n Venezuela
____	*G. f. leymebambae*	Andes of Peru and w Bolivia (La Paz)

☐ **Scallop-breasted Antpitta** *Grallaricula loricata*

Coastal mountains of n Venezuela (Yaracuy to Distrito Federal)

☐ **Hooded Antpitta** *Grallaricula cucullata*

____	*G. c. cucullata*	Central and Eastern Andes of Colombia
____	*G. c. venezuelana*	Andes of central Colombia and nw Venezuela

☐ **Peruvian Antpitta** *Grallaricula peruviana*

E slope of Andes of se Ecuador and extreme n Peru

☐ **Ochre-fronted Antpitta** *Grallaricula ochraceifrons*

Andes of n Peru (San Martín and Amazonas)

☐ **Slate-crowned Antpitta** *Grallaricula nana*

___ *G. n. nana (occidentalis)*	Andes of Colombia, Ecuador, n Peru and w Venezuela
___ *G. n. olivascens*	Coastal mountains of n Venezuela (Aragua and Distrito Federal)
___ *G. n. cumanensis*	Coastal mts. of n Venezuela (Anzoátegui, Sucre and Monagas)
___ *G. n. pariae*	NE Venezuela (subtropical mountains of Paría Peninsula)
___ *G. n. kukenamensis*	*Tepuis* of se Venezuela (e Bolívar) and (?) adjacent Guyana

☐ **Crescent-faced Antpitta** *Grallaricula lineifrons*

Locally in Andes of s Colombia and Ecuador

FAMILY: CONOPOPHAGIDAE (Gnateaters—8)

☐ **Chestnut-belted Gnateater** *Conopophaga aurita*

___ *C. a. aurita*	Guyana to French Guiana and n Brazil (Manaus to Amapá)
___ *C. a. inexpectata*	SE Colombia and adjacent nw Brazil
___ *C. a. occidentalis*	Tropical ne Ecuador and ne Peru (e of R. Napo)
___ *C. a. australis*	NE Peru (s to Ucayali) and w Amazonian Brazil
___ *C. a. snethlageae*	Brazil south of R. Amazon (lower R. Tapajós to cent. Pará)
___ *C. a. pallida*	Cent. Brazil (cent. Pará to w bank of R. Tocantins)

☐ **Black-cheeked Gnateater** *Conopophaga melanops*

___ *C. m. nigrifrons*	Coastal ne Brazil (Paraíba to Alagoas)
___ *C. m. perspicillata*	E Brazil (coastal Bahia and (?) adj. Sergipe)
___ *C. m. melanops*	SE Brazil (Espírito Santo to Santa Catarina)

☐ **Rufous Gnateater** *Conopophaga lineata*

___ *C. l. cearae*	NE Brazil (Serra de Baturité in n Ceará
___ *C. l. lineata*	E Brazil (Pernambuco, s Bahia, Goiás and Mato Grosso do Sul)
___ *C. l. vulgaris*	SE Brazil to e Paraguay, ne Argentina and e Uruguay

☐ **Hooded Gnateater** *Conopophaga roberti*

NE Brazil (e Pará to w Ceará and s Piauí)

☐ **Ash-throated Gnateater** *Conopophaga peruviana*

E Ecuador to e Peru, n Bolivia and w Amazonian Brazil

☐ **Slaty Gnateater** *Conopophaga ardesiaca*

___ *C. a. saturata*	E slope of Andes of s Peru (south from Cusco)
___ *C. a. ardesiaca*	Andes of se Peru (Puno) to s Bolivia (Tarija)

☐ **Chestnut-crowned Gnateater** *Conopophaga castaneiceps*

___ *C. c. chocoensis*	Colombia (w slope of W Andes and Baudó Mts. in w Chocó)
___ *C. c. castaneiceps*	Central and E Andes of Colombia to cent. Ecuador
___ *C. c. chapmani*	E slope of Andes of s Ecuador to n Peru (San Martín)
___ *C. c. brunneinucha*	E slope of Andes of cent. Peru (Huánuco to Cusco)

☐ **Black-bellied Gnateater** *Conopophaga melanogaster*

Lower Amaz. Brazil s of R. Amazon; single record for nw Bolivia

FAMILY: RHINOCRYPTIDAE (Tapaculos—58)

☐ **Chestnut-throated Huet-huet** *Pteroptochos castaneus*

Cent. Chile (Colchagua to R. Bío Bío) and adj. Argentina

☐ **Black-throated Huet-huet** *Pteroptochos tarnii*

S Chile (R. Bío Bío) and adj. sw Argentina

☐ **Moustached Turca** *Pteroptochos megapodius*
_____ *P. m. atacamae* — Arid n Chile (Atacama)
_____ *P. m. megapodius* — Central Chile (Coquimbo to Concepción)

☐ **White-throated Tapaculo** *Scelorchilus albicollis*
_____ *S. a. atacamae* — N Chile (sw Antofogasta and Atacama to n Coquimbo)
_____ *S. a. albicollis* — Central Chile (s Coquimbo to Curicó)

☐ **Chucao Tapaculo** *Scelorchilus rubecula*
_____ *S. r. rubecula* — S Chile (Bío Bío to Aysén) and adjacent w Argentina
_____ *S. r. mochae* — Mocha I. (off Chile)

☐ **Crested Gallito** *Rhinocrypta lanceolata*
_____ *R. l. saturata* — SE Bolivia and w Paraguay
_____ *R. l. lanceolata* — N and cent. Argentina (s to Río Negro and s Buenos Aires)

☐ **Sandy Gallito** *Teledromas fuscus* — Andean slope of Argentina (sw Salta to Río Negro)

☐ **Rusty-belted Tapaculo** *Liosceles thoracicus*
_____ *L. t. dugandi* — SE Colombia and adj. w Brazil (R. Solimões)
_____ *L. t. erithacus* — E Ecuador and e Peru (south to mouth of Río Urubamba)
_____ *L. t. thoracicus* — SE Peru and w Amazonian Brazil (s to R. Tapajós)

☐ **Ocellated Tapaculo** *Acropternis orthonyx*
_____ *A. o. orthonyx* — Cent. and E Andes of Colombia and Andes of nw Venezuela
_____ *A. o. infuscatus* — Andes of e Ecuador and extreme nw Peru (s Amazonas)

☐ **Ochre-flanked Tapaculo** *Eugralla paradoxa* — S-central Chile (Santiago to Chiloé) and adj. Argentina; Mocha I.

☐ **Wetland Tapaculo** *Scytalopus iraiensis* — SW Brazil (river basins in Iguaçu and Tibagi, e Paraná)

☐ **Mouse-colored Tapaculo** *Scytalopus speluncae* — Mts. of se Brazil to ne Argentina (Misiones) and (?) se Paraguay

☐ **Planalto Tapaculo** *Scytalopus pachecoi* — S Brazil and ne Argentina

☐ **Brasilia Tapaculo** *Scytalopus novacapitalis* — S Brazil (s Goiás, Distrito Federal and w Minas Gerais)

☐ **White-breasted Tapaculo** *Scytalopus indigoticus* — SE Brazil (e Bahia to Santa Catarina and ne Rio Grande do Sul)

☐ **Bahia Tapaculo** *Scytalopus psychopompus* — Coastal lowlands of e Brazil (se Bahia)

☐ **Bolivian Tapaculo** *Scytalopus bolivianus* — Andes of se Peru (Ayacucho, Puno) to s Bolivia (Chuquisaca)

☐ **White-crowned Tapaculo** *Scytalopus atratus*
_____ *S. a. nigricans* — Andes of w Venezuela (Táchira, Mérida and Sierra de Perijá)
_____ *S. a. confusus* — Cent. Andes and e slope of W Andes of Colombia
_____ *S. a. atratus* — E slope of Andes of Colombia, Ecuador and Peru (s to Cusco)

☐ **Santa Marta Tapaculo** *Scytalopus sanctaemartae* — Santa Marta Mountains (ne Colombia)

☐ **Rufous-vented Tapaculo** *Scytalopus femoralis* — E slope of Andes of e Peru (Amazonas to Junín)

☐ **Long-tailed Tapaculo** *Scytalopus micropterus* — Andes of Colombia to Ecuador and extreme n Peru

☐ **Nariño Tapaculo** *Scytalopus vicinior* — Pacific slope of sw Colombia (Risaralda) to Ecuador (Cotopaxi)

☐ **Upper Magdalena Tapaculo** *Scytalopus rodriguezi* — Upper Magdalena Valley, Colombia

☐ **Stiles' Tapaculo** *Scytalopus stilesi* — N Central Andes of Colombia

☐ **Ecuadorian Tapaculo** *Scytalopus robbinsi*

Pacific slope of sw Ecuador (Azuay and El Oro)

☐ **Choco Tapaculo** *Scytalopus chocoensis*

Pacific slope of e Panama (Cerro Pirre) to nw Ecuador

☐ **Pale-throated Tapaculo** *Scytalopus panamensis*

E Panama and adj. n Colombia (Cerro Tacarcuna massif)

☐ **Silvery-fronted Tapaculo** *Scytalopus argentifrons*
____ *S. a. argentifrons*
____ *S. a. chiriquensis*

Montane forests of Costa Rica and w Panama (Volcán Chiriquí)
Montane forests of w Panama (e Chiriquí and e Veraguas)

☐ **Caracas Tapaculo** *Scytalopus caracae*

Coastal mts. of n Venezuela (Aragua to Miranda and w Sucre)

☐ **Merida Tapaculo** *Scytalopus meridanus*

Andes of w Venezuela (Mérida and Táchira)

☐ **Brown-rumped Tapaculo** *Scytalopus latebricola*

Santa Marta Mountains (ne Colombia)

☐ **Spillman's Tapaculo** *Scytalopus spillmanni*

Central Andes of Colombia to central Ecuador

☐ **Chusquea Tapaculo** *Scytalopus parkeri*

Andes of s Ecuador to extreme n Peru (e Piura, n Cajamarca)

☐ **Trilling Tapaculo** *Scytalopus parvirostris*

Andes of nw Peru to e-cent. Bolivia (Santa Cruz)

☐ **Tschudi's Tapaculo** *Scytalopus acutirostris*

Central Andes of Peru (e La Libertad to Junín)

☐ **Unicolored Tapaculo** *Scytalopus unicolor*

Andes of n Peru (s Cajamarca and La Libertad) and adj. Ecuador

☐ **Lara Tapaculo** *Scytalopus fuscicauda*

Andes of Venezuela (s Lara andTrujillo)

☐ **Matorral Tapaculo** *Scytalopus griseicollis*
____ *S. g. infasciatus*
____ *S. g. griseicollis*

E Andes of Colombia (Páramo de Beltrán)
E Andes of Colombia (Cundinamarca and Boyacá)

☐ **Paramo Tapaculo** *Scytalopus canus*
____ *S. c. canus*
____ *S. c. opacus*

W Andes of Colombia (Antioquia)
Cent. Andes of Colombia to Ecuador and extreme n Peru

☐ **Ancash Tapaculo** *Scytalopus affinis*

Western Andes of Peru (s Cajamarca to Ancash)

☐ **Neblina Tapaculo** *Scytalopus altirostris*

Central Andes of Peru (s Amazonas to Huánuco)

☐ **Vilcabamba Tapaculo** *Scytalopus urubambae*

Andes of e-central Peru (s Cordillera Vilcabamba)

☐ **Diademed Tapaculo** *Scytalopus schulenbergi*

Andes of se Peru (Vilcanota Mts.) to cent. Bolivia (Cochabamba)

☐ **Puna Tapaculo** *Scytalopus simonsi*

Andes of se Peru (Vilcanota Mts.) to cent. Bolivia (Cochabamba)

☐ **Zimmer's Tapaculo** *Scytalopus zimmeri*

Andes of s Bolivia (Chuquisaca and Tarija)

☐ **White-browed Tapaculo** *Scytalopus superciliaris*
____ *S. s. superciliaris*
____ *S. s. santabarbarae*

Andes of nw Argentina (Jujuy to Cacamarca)
Andes of nw Argentina (Santa Barbara Mts.)

☐ **Magellanic Tapaculo** *Scytalopus magellanicus*

Andes of central Chile and Argentina to Tierra del Fuego

☐ **Dusky Tapaculo** *Scytalopus fuscus*

Andes of central Chile (Atacama to Bío-Bío)

☐ **Blackish Tapaculo** *Scytalopus latrans*
____ *S. l. latrans*
____ *S. l. subcinereus*
____ *S. l. intermedius*

Andes of Colombia to w Venezuela, e Ecuador and n Peru
Pacific slope of sw Ecuador (Azuay) to nw Peru (Cajamarca)
Central Andes of n Peru (s Amazonas)

☐ **Large-footed Tapaculo** *Scytalopus macropus*

E slope of Central Andes of Peru (s Amazonas to Junín)

☐ **Ash-colored Tapaculo** *Myornis senilis*

Andes of Colombia, Ecuador and n Peru (s to Pasco)

☐ **Stresemann's Bristlefront** *Merulaxis stresemanni*

Rare and local in coastal e Brazil (Bahia)

☐ **Slaty Bristlefront** *Merulaxis ater*

SE Brazil (s Bahia and Espírito Santo to e Paraná, Santa Catarina)

☐ **Spotted Bamboowren** *Psilorhamphus guttatus*

SE Brazil (Minas Gerais) to ne Argentina and (?) se Paraguay

☐ **Collared Crescent-chest** *Melanopareia torquata*
- ____ *M. t. bitorquata* — *Cerrado* of e Bolivia (ne Santa Cruz)
- ____ *M. t. rufescens* — *Cerrado* of central Brazil and extreme ne Paraguay
- ____ *M. t. torquata* — *Cerrado* of interior e Brazil (s Piauí and w Bahia)

☐ **Olive-crowned Crescent-chest** *Melanopareia maximiliani*
- ____ *M. m. maximiliani* — Yungas of w Bolivia (La Paz)
- ____ *M. m. argentina* — Central Bolivia to nw Argentina (w Córdoba, n San Luis)
- ____ *M. m. pallida* — SE Bolivia and chaco of w Paraguay and n Argentina

☐ **Marañón Crescent-chest** *Melanopareia maranonica*

Extreme s Ecuador (s Zamora-Chinchipe) and adj. n Peru

☐ **Elegant Crescent-chest** *Melanopareia elegans*
- ____ *M. e. elegans* — Arid sw Ecuador (north to Manabí and extreme s Pichincha)
- ____ *M. e. paucalensis* — Arid nw Peru (s to La Libertad)

FAMILY: COTINGIDAE (Cotingas—71)

☐ **Sharpbill** *Oxyruncus cristatus*
- ____ *O. c. frater* — Discontinuously distributed Costa Rica and w Panama
- ____ *O. c. brooksi* — Mountains of e Panama
- ____ *O. c. hypoglaucus* — Mountains of Guyana, Suriname and se Venezuela
- ____ *O. c. phelpsi* — Mountains of Venezuela (Bolívar and Amazonas) and adj. Guyana
- ____ *O. c. tocantinsi* — Central Brazil (Goiás to Pará and Amapá)
- ____ *O. c. cristatus* — Disjunct ranges in se Brazil, s Paraguay, e Bolivia and e Peru

☐ **Peruvian Plantcutter** *Phytotoma raimondii*

Coastal nw Peru (Tumbes south to n Lima)

☐ **White-tipped Plantcutter** *Phytotoma rutila*
- ____ *P. r. angustirostris* — Highlands of w Bolivia and nw Argentina
- ____ *P. r. rutila* — *Chaco* of w Paraguay, w Uruguay, n Argentina and extreme s Brazil

☐ **Rufous-tailed Plantcutter** *Phytotoma rara*

Central and s Chile and Argentina; accidental Falkland Islands

☐ **Red-crested Cotinga** *Ampelion rubrocristatus*

Andes of Colombia to w Venezuela and w Bolivia

☐ **Chestnut-crested Cotinga** *Ampelion rufaxilla*
- ____ *A. r. antioquiae* — Subtropical Andes of w Colombia and (?) n Ecuador
- ____ *A. r. rufaxilla* — E slope of Andes of s Ecuador, Peru and w Bolivia

☐ **Bay-vented Cotinga** *Doliornis sclateri*

E slope of Andes of central Peru

☐ **Chestnut-bellied Cotinga** *Doliornis remseni*

Eastern Andes of Ecuador and extreme n Peru

☐ **White-cheeked Cotinga** *Zaratornis stresemanni*

Andes of w Peru (La Libertad to Ayacucho)

☐ **Shrike-like Cotinga** *Laniisoma elegans*
- ____ *L. e. venezuelense* — Locally from tropical ne Colombia to nw Venezuela
- ____ *L. e. buckleyi* — E slope of Andes of e Ecuador and e Peru

311

____ L. e. elegans	Tropical se Brazil (Espírito Santo to e Paraná)
____ L. e. cadwaladeri	Tropical nw Bolivia (La Paz)

☐ **Swallow-tailed Cotinga** *Phibalura flavirostris*

____ P. f. flavirostris	SE Brazil (Goiás) to e Paraguay and ne Argentina
____ P. f. boliviana	Foothills of w Bolivia (La Paz)

☐ **Hooded Berryeater** *Carpornis cucullata*

SE Brazil (Espírito Santo to Rio Grande do Sul)

☐ **Black-headed Berryeater** *Carpornis melanocephala*

Coastal se Brazil (Alagoas and e Bahia to ne Paraná)

☐ **Green-and-black Fruiteater** *Pipreola riefferii*

____ P. r. occidentalis	Andes of sw Colombia and w Ecuador
____ P. r. riefferii	Andes of e Colombia to nw Venezuela (Táchira) and e Ecuador
____ P. r. melanolaema	Andes of w Venezuela (north to s Lara)
____ P. r. confusa	Andes of e Ecuador to Andes of Peru (n Amazonas)
____ P. r. chachapoyas	Subtropical Central Andes of n Peru (San Martín)
____ P. r. tallmanorum	Andes of central Peru (Huánuco)

☐ **Band-tailed Fruiteater** *Pipreola intermedia*

____ P. i. intermedia	Subtropical Andes of Peru (La Libertad to Junín)
____ P. i. signata	Andes of se Peru (Cusco and Puno) to w Bolivia

☐ **Barred Fruiteater** *Pipreola arcuata*

____ P. a. arcuata	Andes of Colombia, w Venezuela, Ecuador, n Peru and w Bolivia
____ P. a. viridicauda	Andes of central Peru (Junín) to w Bolivia

☐ **Golden-breasted Fruiteater** *Pipreola aureopectus*

____ P. a. decora	Santa Marta Mountains (ne Colombia)
____ P. a. aureopectus	Subtropical Andes of e Colombia and w Venezuela
____ P. a. festiva	Coastal mountains of n Venezuela

☐ **Orange-breasted Fruiteater** *Pipreola jucunda*

W slope of Western Andes of Colombia and w Ecuador

☐ **Black-chested Fruiteater** *Pipreola lubomirskii*

Andes of s Colombia to extreme n Peru (Cajamarca)

☐ **Masked Fruiteater** *Pipreola pulchra*

Andes of e Peru (Amazonas to Cordillera Vilcabamba of Cusco)

☐ **Scarlet-breasted Fruiteater** *Pipreola frontalis*

____ P. f. squamipectus	Andes of se Ecuador (w Napo) and n Peru (San Martín)
____ P. f. frontalis	Andes of se Peru (Huánuco) to w Bolivia

☐ **Fiery-throated Fruiteater** *Pipreola chlorolepidota*

Andean foothills of se Colombia to Ecuador and c Peru (Pasco)

☐ **Handsome Fruiteater** *Pipreola formosa*

____ P. f. formosa	Coastal cordillera of n Venezuela
____ P. f. rubidior	Mts. of ne Venezuela (Anzoátegui, Monagas and Sucre)
____ P. f. pariae	NE Venezuela (Paría Peninsula)

☐ **Red-banded Fruiteater** *Pipreola whitelyi*

____ P. w. kathleenae	*Tepuis* of s Venezuela (se Bolívar)
____ P. w. whitelyi	*Tepuis* of w Guyana (Mt. Twek-quay) and adj. n Brazil (Roraima)

☐ **Scaled Fruiteater** *Ampelioides tschudii*

Andes of Colombia and w Venezuela to w Bolivia (La Paz)

☐ **Dusky Purpletuft** *Iodopleura fusca*

SE Venezuela to the Guianas and extreme n Brazil (Roraima)

☐ **White-browed Purpletuft** *Iodopleura isabellae*

____ I. i. isabellae	SE Colombia to e Ecuador, e Peru, n Bolivia and n Brazil
____ I. i. paraensis	NE Brazil (R. Tocantins to Pará and Goiás)

☐ **Buff-throated Purpletuft** *Iodopleura pipra*
    ____ *I. p. leucopygia*
    ____ *I. p. pipra*

Coastal ne Brazil (Paraíba, Pernambuco, Alagoas and Bahia)
Coastal e Brazil (Espírito Santo, Rio de Janeiro and São Paulo)

☐ **Kinglet Calyptura** *Calyptura cristata*

SE Brazil (Rio de Janeiro area)

☐ **Gray-tailed Piha** *Snowornis subalaris*

E slope of Andes of s Colombia, e Ecuador and e Peru

☐ **Olivaceous Piha** *Snowornis cryptolophus*
    ____ *S. c. mindoensis*
    ____ *S. c. cryptolophus*

Andes of sw Colombia and w Ecuador
Andes of e Colombia, e Ecuador and e Peru (south to Huánuco)

☐ **Dusky Piha** *Lipaugus fuscocinereus*

Andes of Colombia to Ecuador and extreme n Peru (Piura)

☐ **Chestnut-capped Piha** *Lipaugus weberi*

N slope of Central Cordillera of Andes of Colombia

☐ **Scimitar-winged Piha** *Lipaugus uropygialis*

S Peru (Cordillera Apolobamba of Puno) and w Bolivia

☐ **Screaming Piha** *Lipaugus vociferans*

E Colombia to s Venezuela, Guianas, n Bolivia, Amaz. and e Brazil

☐ **Rufous Piha** *Lipaugus unirufus*
    ____ *L. u. unirufus*
    ____ *L. u. castaneotinctus*

Gulf lowlands of s Mexico to n Colombia
Tropical sw Colombia to nw Ecuador

☐ **Cinnamon-vented Piha** *Lipaugus lanioides*

SE Brazil (Espírito Santo and Minas Gerais to ne Santa Catarina)

☐ **Rose-collared Piha** *Lipaugus streptophorus*

Tepuis of s Venezuela, adjacent w Guyana and extreme n Brazil

☐ **Black-and-gold Cotinga** *Tijuca atra*

SE Brazil (Rio de Janeiro, s Minas Gerais to extreme e São Paulo)

☐ **Gray-winged Cotinga** *Tijuca condita*

Cloud forests of se Brazil in Rio de Janeiro (Serra dos Órgãos)

☐ **Purple-throated Cotinga** *Porphyrolaema porphyrolaema*

Tropical se Colombia to se Peru and w Amazonian Brazil

☐ **Lovely Cotinga** *Cotinga amabilis*

Gulf lowlands of s Mexico (Veracruz) to se Costa Rica

☐ **Turquoise Cotinga** *Cotinga ridgwayi*

Pacific slope of sw Costa Rica to w Panama (w Chiriquí)

☐ **Blue Cotinga** *Cotinga nattererii*

Central Panama to Colombia, nw Ecuador and w Venezuela

☐ **Plum-throated Cotinga** *Cotinga maynana*

Lowlands of se Colombia to n Bolivia and w Amazonian Brazil

☐ **Purple-breasted Cotinga** *Cotinga cotinga*

E Colombia to s Venezuela, Guianas, n and e Amazonian Brazil

☐ **Banded Cotinga** *Cotinga maculata*

SE Brazil (s Bahia and Minas Gerais to Rio de Janeiro)

☐ **Spangled Cotinga** *Cotinga cayana*

S Venezuela and the Guianas to n Bolivia and Amazonian Brazil

☐ **Pompadour Cotinga** *Xipholena punicea*

E Colombia to s Venezuela, Guianas, ne Bolivia and Amaz. Brazil

☐ **White-tailed Cotinga** *Xipholena lamellipennis*

Lower Amazonian Brazil (R. Tapajós to n Maranhão)

☐ **White-winged Cotinga** *Xipholena atropurpurea*

Coastal e Brazil (Paraíba to n Rio de Janeiro)

☐ **Snowy Cotinga** *Carpodectes nitidus*

Caribbean slope of n Honduras to extreme w Panama

☐ **Yellow-billed Cotinga** *Carpodectes antoniae*

Pacific lowlands of sw Costa Rica and extreme w Panama

☐ **Black-tipped Cotinga** *Carpodectes hopkei*

Humid lowlands of e Panama to w Colombia and nw Ecuador

☐ **Black-faced Cotinga** *Conioptilon mcilhennyi*

Tropical se Peru (s Ucayali and s Madre de Dios)

☐ **Bare-necked Fruitcrow** *Gymnoderus foetidus*

S Venezuela and the Guianas to n Bolivia and Amazonian Brazil

☐ **Crimson Fruitcrow** *Haematoderus militaris*

Extreme s Venezuela to the Guianas and n Amazonian Brazil

☐ **Purple-throated Fruitcrow** *Querula purpurata*

Trop. Costa Rica to n Bolivia, the Guianas and Amazonian Brazil

☐ **Red-ruffed Fruitcrow** *Pyroderus scutatus*
___ *P. s. occidentalis* — Andes of w Colombia and nw Ecuador
___ *P. s. granadensis* — E Andes of Colombia and w Venezuela
___ *P. s. orenocensis* — Upper tropical Venezuela (ne Bolívar) and n Guyana
___ *P. s. masoni* — Subtropical e Peru (s Amazonas to Junín)
___ *P. s. scutatus* — Tropical se Brazil to e Paraguay and ne Argentina

☐ **Bare-necked Umbrellabird** *Cephalopterus glabricollis*

Humid highlands of Costa Rica and w Panama

☐ **Amazonian Umbrellabird** *Cephalopterus ornatus*

S Venezuela to Guyana, n Bolivia and Amazonian Brazil

☐ **Long-wattled Umbrellabird** *Cephalopterus penduliger*

W Andes of sw Colombia and w Ecuador (south to El Oro)

☐ **Capuchinbird** *Perissocephalus tricolor*

E Colombia to s Venezuela, the Guianas and ne Amazonian Brazil

☐ **Three-wattled Bellbird** *Procnias tricarunculatus*

Humid montane forests of Nicaragua to w Panama

☐ **White Bellbird** *Procnias albus*
___ *P. a. albus* — SE Venezuela to the Guianas and e Amazonian Brazil
___ *P. a. wallacei* — NE Brazil (Carajás, se Pará)

☐ **Bearded Bellbird** *Procnias averano*
___ *P. a. carnobarba* — NE Colombia to n Venezuela, w Guyana and adj. Brazil; Trinidad
___ *P. a. averano* — NE Brazil (Maranhão and Piauí to Pernambuco and Alagoas)

☐ **Bare-throated Bellbird** *Procnias nudicollis*

SE Brazil (Alagoas) to e Paraguay and ne Argentina (Misiones)

☐ **Guianan Red-Cotinga** *Phoenicircus carnifex*

Extreme s Venezuela to the Guianas and lower Amazonian Brazil

☐ **Black-necked Red-Cotinga** *Phoenicircus nigricollis*

SE Colombia to sw Venezuela, ne Peru and w and c Amaz. Brazil

☐ **Guianan Cock-of-the-rock** *Rupicola rupicola*

E Colombia to s Venezuela, the Guianas and n Amazonian Brazil

☐ **Andean Cock-of-the-rock** *Rupicola peruvianus*
___ *R. p. sanguinolentus* — Andes of w Colombia and nw Ecuador
___ *R. p. aequatorialis* — Andes of e Colombia to w Venezuela, e Ecuador and e Peru
___ *R. p. peruvianus* — Andes of central Peru (San Martín to Junín)
___ *R. p. saturatus* — Andes of se Peru (Cusco and Puno) and w Bolivia

FAMILY: PIPRIDAE (Manakins—57)

☐ **Crimson-hooded Manakin** *Pipra aureola*
___ *P. a. aureola* — Tropical ne Venezuela to the Guianas and ne Brazil
___ *P. a. aurantiicollis* — Brazil on R. Tapajós in w Pará and adj. n bank of R. Amazon
___ *P. a. flavicollis* — Lower Amazonian Brazil in e Amazonas and w Pará
___ *P. a. borbae* — Brazil along lower and middle R. Madeira in e Amazonas

☐ **Band-tailed Manakin** *Pipra fasciicauda*
___ *P. f. calamae* — W-c Brazil near upper Rio Madeira and extreme nw Mato Grosso
___ *P. f. saturata* — E slope of Central Andes of Peru (San Martín and s Loreto)
___ *P. f. purusiana* — Tropical e Peru (south to Cusco) and adjacent w Amazonian Brazil

____	*P. f. fasciicauda*	Tropical se Peru (Puno) and ne Bolivia
____	*P. f. scarlatina*	N Bolivia to se Brazil, e Paraguay and ne Argentina

☐ **Wire-tailed Manakin** *Pipra filicauda*

____	*P. f. subpallida*	Tropical e Colombia and nw Venezuela
____	*P. f. filicauda*	E Ecuador to ne Peru, s Venezuela and w Amazonian Brazil

☐ **Golden-headed Manakin** *Pipra erythrocephala*

____	*P. e. erythrocephala*	E Panama to the Guianas and Brazil north of R. Amazon; Trinidad
____	*P. e. berlepschi*	Tropical se Colombia to ne Peru and w Amazonian Brazil
____	*P. e. flammiceps*	E Colombia (Santander)

☐ **Red-headed Manakin** *Pipra rubrocapilla*

Tropical e Peru to n Bolivia, s Amazonian and e Brazil

☐ **Red-capped Manakin** *Pipra mentalis*

____	*P. m. mentalis*	Gulf lowlands of se Mexico to e Costa Rica; Isla de Mujeres
____	*P. m. ignifera*	Tropical w Costa Rica and w Panama
____	*P. m. minor*	Extreme e Panama to w Colombia and nw Ecuador

☐ **Round-tailed Manakin** *Pipra chloromeros*

Lowlands and foothills of e Peru, adj. w Brazil and n Bolivia

☐ **Scarlet-horned Manakin** *Pipra cornuta*

Tepuis of s Venezuela, w Guyana and extreme n Brazil

☐ **Blue-crowned Manakin** *Lepidothrix coronata*

____	*L. c. velutina*	Humid sw Costa Rica to w Panama
____	*L. c. minuscula*	E Panama to nw Ecuador (south to Pichincha)
____	*L. c. caquetae*	Tropical se Colombia (east of Andes in w Meta and w Caquetá)
____	*L. c. carbonata*	SE Colombia to n Peru, s Venezuela and n Amazonian Brazil
____	*L. c. coronata*	Tropical e Ecuador to ne Peru and w Amazonian Brazil
____	*L. c. exquisita*	Tropical e Peru (s Loreto to Junín)
____	*L. c. caelestipileata*	SE Peru (Puno) and adjacent w Amazonian Brazil
____	*L. c. regalis*	Tropical n Bolivia south to Cochabamba

☐ **Blue-rumped Manakin** *Lepidothrix isidorei*

____	*L. i. isidorei*	E slope of Andes of Colombia (w Meta) and e Ecuador
____	*L. i. leucopygia*	Andean foothills of n Peru (San Martín and n Huánuco)

☐ **Cerulean-capped Manakin** *Lepidothrix coeruleocapilla*

E slope of Andes of s Peru (s Huánuco to Puno)

☐ **Snow-capped Manakin** *Lepidothrix nattereri*

____	*L. n. nattereri*	Central Brazil south of the Amazon
____	*L. n. gracilis*	S-central Brazil (upper Rio Madeira) to extreme ne Bolivia

☐ **Golden-crowned Manakin** *Lepidothrix vilasboasi*

E Amazonian Brazil (near headwaters of R. Tapajós in sw Pará)

☐ **Opal-crowned Manakin** *Lepidothrix iris*

____	*L. i. iris*	E Amazonian Brazil (Belém region of e Pará to n Maranhão)
____	*L. i. eucephala*	Brazil south of R. Amazon (east bank of lower R. Tapajós)

☐ **Tepui Manakin** *Lepidothrix suavissima*

Tepuis of s Venezuela to n Guyana and extreme n Brazil

☐ **White-fronted Manakin** *Lepidothrix serena*

SE Guyana, Suriname, French Guiana and n Amazonian Brazil

☐ **White-crowned Manakin** *Dixiphia pipra*

____	*D. p. anthracina*	Caribbean slope of Costa Rica and w Panama
____	*D. p. bolivari*	NW Colombia (upper Sinú Valley)
____	*D. p. coracina*	E Andes of Colombia, e Ecuador and n Peru
____	*D. p. minima*	W slope of Western Andes of Colombia (Cauca)
____	*D. p. unica*	Subtropical Colombia (Antioquia and Huila)
____	*D. p. pipra*	Extreme e Colombia to Venezuela, the Guianas and n Brazil

____ *D. p. discolor*	Tropical ne Peru (along Río Napo in ne Loreto)
____ *D. p. occulta*	E slope of Central Andes of Peru (San Martín and Huánuco)
____ *D. p. pygmaea*	Tropical e Peru (lower Río Huallaga in Loreto)
____ *D. p. microlopha*	Tropical e Peru (south of Río Marañón) and w Amazonian Brazil
____ *D. p. comata*	Tropical e Peru (s Pasco, Junín and n Cusco)
____ *D. p. separabilis*	Amazonian Brazil (R. Tapajós to Belém)
____ *D. p. cephaleucos*	Coastal se Brazil (s Bahia to Rio de Janeiro)

☐ **Helmeted Manakin** *Antilophia galeata*

Tableland of interior s Brazil to ne Bolivia and ne Paraguay

☐ **Araripe Manakin** *Antilophia bokermanni*

E Brazil (Chapada do Araripe in Ceará)

☐ **Long-tailed Manakin** *Chiroxiphia linearis*

____ *C. l. linearis*	Pacific lowlands of s Mexico (Oaxaca) and Guatemala
____ *C. l. fastuosa*	El Salvador to w Nicaragua and nw Costa Rica

☐ **Lance-tailed Manakin** *Chiroxiphia lanceolata*

SW Costa Rica to n Colombia, n Venezuela and Isla Margarita

☐ **Blue-backed Manakin** *Chiroxiphia pareola*

____ *C. p. atlantica*	Tobago
____ *C. p. napensis*	Tropical se Colombia (east of the Andes) to e Ecuador and e Peru
____ *C. p. pareola*	E Venezuela, the Guianas, n and e Amazonian Brazil
____ *C. p. regina*	Tropical ne Peru (south of R. Amazon) and w Amazonian Brazil

☐ **Yungas Manakin** *Chiroxiphia boliviana*

SE Peru (Cusco, s Madre de Dios and Puno) and w Bolivia

☐ **Blue Manakin** *Chiroxiphia caudata*

Lowlands of se Brazil (s Bahia) to e Paraguay and ne Argentina

☐ **Pin-tailed Manakin** *Ilicura militaris*

SE Brazil (Espírito Santo to Paraná and e Santa Catarina)

☐ **Golden-winged Manakin** *Masius chrysopterus*

____ *M. c. bellus*	W slope of W Andes to w side of Central Andes of Colombia
____ *M. c. pax*	Subtropical se Colombia (e Nariño) and e Ecuador
____ *M. c. coronulatus*	W Andes of Colombia (sw Cauca and Nariño) and w Ecuador
____ *M. c. chrysopterus*	Central and Eastern Andes of Colombia and nw Venezuela
____ *M. c. peruvianus*	Subtropical n Peru (n Cajamarca and n San Martín)

☐ **White-ruffed Manakin** *Corapipo altera*

____ *C. a. altera*	E Honduras (Olancho) to nw Costa Rica
____ *C. a. heteroleuca*	Tropical sw Costa Rica to nw Colombia (Chocó)
____ *C. a. leucorrhoa*	Foothills of w Colombia and w Venezuela

☐ **White-throated Manakin** *Corapipo gutturalis*

S Venezuela to the Guianas and ne Amazonian Brazil

☐ **White-collared Manakin** *Manacus candei*

Gulf lowlands of s Mexico to e Costa Rica

☐ **Orange-collared Manakin** *Manacus aurantiacus*

Pacific slope of s Costa Rica and w Panama (east to Azuero Pen.)

☐ **Golden-collared Manakin** *Manacus vitellinus*

____ *M. v. vitellinus*	Lowlands of Panama to nw Colombia (south to sw Cauca)
____ *M. v. amitinus*	Escudo de Veraguas I. (off n Panama)
____ *M. v. milleri*	N Colombia in Sinú and lower Cauca valleys
____ *M. v. viridiventris*	W Colombia w of W Andes and e side in upper Cauca Valley

☐ **White-bearded Manakin** *Manacus manacus*

____ *M. m. milleri*	Tropical n Colombia (Sinú, Cauca and Magdalena valleys)
____ *M. m. abditivus*	Santa Marta, lower Cauca and middle Magdalena valleys
____ *M. m. flaveolus*	N Colombia (tropical upper Magdalena Valley)
____ *M. m. bangsi*	Tropical sw Colombia and extreme nw Ecuador
____ *M. m. interior*	Colombia (e of the Andes) to e Ecuador, n Peru and nw Brazil

____	*M. m. trinitatis*	Trinidad
____	*M. m. umbrosus*	Tropical s Venezuela (Amazonas)
____	*M. m. manacus*	S Venezuela to the Guianas and n Brazil
____	*M. m. leucochlamys*	Tropical nw Ecuador (Esmeraldas, Manabí and Guayas)
____	*M. m. maximus*	Tropical sw Ecuador (El Oro and w Loja)
____	*M. m. expectatus*	Tropical ne Peru (Loreto) and adjacent w Brazil
____	*M. m. longibarbatus*	Lower Amazonian Brazil (R. Xingú to R. Tocantins)
____	*M. m. purissimus*	E Brazil (R. Tocantins to se Pará and n Maranhão)
____	*M. m. gutturosus*	SE Brazil (Alagoas) to e Paraguay and ne Argentina
____	*M. m. purus*	N Brazil (R. Madeira to R. Tapajós and sw Pará)
____	*M. m. subpurus*	S-cent. Brazil (se Amazona, e Rondônia and nw Mato Grosso)

☐ **Fiery-capped Manakin** *Machaeropterus pyrocephalus*

____	*M. p. pallidiceps*	S Venezuela (nw Bolívar) and extreme n Brazil (Roraima)
____	*M. p. pyrocephalus*	E Peru (San Martín) to n Bolivia and w Amazonian Brazil

☐ **Western Striped Manakin** *Machaeropterus striolatus*

____	*M. s. antioquiae*	Eastern Andes of w and central Colombia
____	*M. s. striolatus*	Tropical e Colombia to e Ecuador, ne Peru and w Amaz. Brazil
____	*M. s. obscurostriatus*	Tropical nw Venezuela (Mérida)
____	*M. s. zulianus*	Tropical w Venezuela (w Zulia, Táchira and n Barinas)
____	*M. s. aureopectus*	Tropical se Venezuela (ne Amazonas and se Bolívar)

☐ **Eastern Striped Manakin** *Machaeropterus regulus*

Coastal se Brazil (Bahia to Rio de Janeiro)

☐ **Club-winged Manakin** *Machaeropterus deliciosus*

W slope of Andes of sw Colombia and w Ecuador

☐ **Black Manakin** *Xenopipo atronitens*

E Colombia to s Venezuela, ne Bolivia and s Amazonian Brazil

☐ **Jet Manakin** *Chloropipo unicolor*

E slope of Andes from e Ecuador (Napo) to s Peru (Puno)

☐ **Olive Manakin** *Chloropipo uniformis*

____	*C. u. duidae*	*Tepuis* of s Venezuela (s Bolívar and Amazonas)
____	*C. u. uniformis*	*Tepuis* of w Guyana and extreme n Brazil (Roraima)

☐ **Green Manakin** *Chloropipo holochlora*

____	*C. h. suffusa*	Extreme e Panama (Darién) and adjacent nw Colombia
____	*C. h. litae*	Tropical w Colombia to nw Ecuador (south to Pichincha)
____	*C. h. holochlora*	Tropical se Colombia, e Ecuador and e Peru (south to Junín)
____	*C. h. viridior*	Foothills of se Peru (Cusco and nw Puno)

☐ **Yellow-headed Manakin** *Chloropipo flavicapilla*

Western and Central Andes of Colombia and ne Ecuador

☐ **Flame-crested Manakin** *Heterocercus linteatus*

S Amazonian Brazil to ne Peru and extreme ne Bolivia

☐ **Yellow-crested Manakin** *Heterocercus flavivertex*

E Colombia to sw Venezuela and n Amazonian Brazil

☐ **Orange-crested Manakin** *Heterocercus aurantiivertex*

Locally in e Ecuador and adjacent ne Peru (sw Loreto)

☐ **Saffron-crested Tyrant-Manakin** *Neopelma chrysocephalum*

E Colombia to s Venezuela, the Guianas and n Amazonian Brazil

☐ **Sulphur-bellied Tyrant-Manakin** *Neopelma sulphureiventer*

Tropical e Peru to n Bolivia and adjacent w Amazonian Brazil

☐ **Pale-bellied Tyrant-Manakin** *Neopelma pallescens*

Lower Amazonian Brazil; central and e Brazil to ne Bolivia

☐ **Wied's Tyrant-Manakin** *Neopelma aurifrons*

Coastal se Brazil (s Bahia, Espírito Santo and Minas Gerais)

☐ **Serra Tyrant-Manakin** *Neopelma chrysolophum*

SE Brazil (s Minas Gerais, Rio de Janeiro and e São Paulo)

☐ **Dwarf Tyrant-Manakin** *Tyranneutes stolzmanni*

E Colombia to s Venezuela, n Bolivia and Amazonian Brazil

☐ **Tiny Tyrant-Manakin** *Tyranneutes virescens*

SE Venezuela to the Guianas and ne Amazonian Brazil

☐ **Gray-headed Piprites** *Piprites griseiceps*

Caribbean lowlands of e Guatemala to Costa Rica

☐ **Wing-barred Piprites** *Piprites chloris*
____ *P. c. antioquiae* | Central Andes of Colombia (Antioquia)
____ *P. c. perijana* | Sierra de Perijá (e Colombia) and Andes of w Venezuela
____ *P. c. tschudii* | Tropical se Colombia to central Peru and nw Brazil
____ *P. c. chlorion* | Tropical se Venezuela, the Guianas and n Brazil
____ *P. c. grisescens* | N Brazil (e Pará)
____ *P. c. boliviana* | Tropical n Bolivia and sw Amazonian Brazil
____ *P. c. chloris* | SE Brazil (Espírito Santo) to e Paraguay and ne Argentina

☐ **Black-capped Piprites** *Piprites pileata*

Locally from se Brazil (Rio de Janeiro) to ne Argentina

☐ **Greater Schiffornis** *Schiffornis major*
____ *S. m. major* | Tropical se Colombia to n Bolivia and Amazonian Brazil
____ *S. m. duidae* | Tropical se Venezuela (s Amazonas)

☐ **Thrush-like Schiffornis** *Schiffornis turdina*
____ *S. t. veraepacis* | Gulf-Caribbean slope of se Mexico to s Costa Rica
____ *S. t. dumicola* | Tropical w Panama (Chiriquí, Veraguas and n Coclé)
____ *S. t. panamensis* | Tropical e Panama and nw Colombia
____ *S. t. acrolophites* | E Panama (Cerro Mali and Tacarcuna), adj. Colombia (Chocó)
____ *S. t. rosenbergi* | Trop. w Colombia (Chocó to Nariño) and w Ecuador (s to Loja)
____ *S. t. stenorhyncha* | Tropical ne Colombia and n Venezuela
____ *S. t. olivacea* | Tropical se Venezuela (Bolívar) and adjacent Guyana
____ *S. t. aenea* | E Andes of Ecuador and adjacent n Peru (Piura and Cajamarca)
____ *S. t. amazona* | S Venezuela to e Peru and w Amazonian Brazil
____ *S. t. wallacii* | Tropical French Guiana, Suriname and ne Brazil
____ *S. t. steinbachi* | Upper tropical se Peru (Junín) to n Bolivia
____ *S. t. intermedia* | Campos of e Brazil (Alagoas and Paraíba)
____ *S. t. turdina* | SE Brazil (s Bahia, e Minas Gerais and e Espírito Santo)

☐ **Greenish Schiffornis** *Schiffornis virescens*

SE Brazil (s Bahia) to e Paraguay and ne Argentina

FAMILY: TYRANNIDAE (Tyrant Flycatchers—435)

☐ **Planalto Tyrannulet** *Phyllomyias fasciatus*
____ *P. f. cearae* | E Brazil (Ceará and e Pernambuco)
____ *P. f. fasciatus* | E Brazil (Maranhão to se Mato Grosso and s Goiás)
____ *P. f. brevirostris* | SE Brazil (Minas Gerais) to e Paraguay and ne Argentina

☐ **Rough-legged Tyrannulet** *Phyllomyias burmeisteri*
____ *P. b. zeledoni* | Mountains of Costa Rica and w Panama
____ *P. b. leucogonys* | E Andes from Colombia to e Ecuador and se Peru
____ *P. b. wetmorei* | Sierra de Perijá (Colombia/Venezuela border)
____ *P. b. viridiceps* | Coastal mountains of n Venezuela
____ *P. b. bunites* | *Tepuis* of s Venezuela in Bolívar (Cerro Chimantá-tepui)
____ *P. b. burmeisteri* | E slope of Andes of e Bolivia, n Argentina, e Paraguay, se Brazil

☐ **Greenish Tyrannulet** *Phyllomyias virescens*

SE Brazil (Espírito Santo) to e Paraguay and ne Argentina

☐ **Reiser's Tyrannulet** *Phyllomyias reiseri*

Interior e Brazil (Piauí) to ne Paraguay

☐ **Urich's Tyrannulet** *Phyllomyias urichi*

Extreme ne Venezuela (Sucre, Monagas and Anzoátegui)

☐ **Sclater's Tyrannulet** *Phyllomyias sclateri*
　____ *P. s. subtropicalis* — E slope of Andes of se Peru (Cuzco and Puno)
　____ *P. s. sclateri* — Andean foothills of Bolivia to nw Argentina (s to Tucumán)

☐ **Gray-capped Tyrannulet** *Phyllomyias griseocapilla*
　SE Brazil (Minas Gerais and Espírito Santo to Santa Catarina)

☐ **Sooty-headed Tyrannulet** *Phyllomyias griseiceps*
　E Panama to Colombia, Ecuador, Guyana, e Peru and n Brazil

☐ **Plumbeous-crowned Tyrannulet** *Phyllomyias plumbeiceps*
　Andes of Colombia to Ecuador and s Peru (Cuzco)

☐ **Black-capped Tyrannulet** *Phyllomyias nigrocapillus*
　____ *P. n. flavimentum* — Santa Marta Mountains (ne Colombia)
　____ *P. n. nigrocapillus* — Andes of Colombia and Ecuador to s Peru (Cuzco)
　____ *P. n. aureus* — Andes of w Venezuela (n Táchira to s Lara)

☐ **Ashy-headed Tyrannulet** *Phyllomyias cinereiceps*
　Andes of Colombia to Ecuador and s Peru (Puno)

☐ **Tawny-rumped Tyrannulet** *Phyllomyias uropygialis*
　Andes of Colombia and w Venezuela to w Bolivia (s to Tarija)

☐ **Yellow-crowned Tyrannulet** *Tyrannulus elatus*
　SW Costa Rica to n Bolivia, the Guianas and Amazonian Brazil

☐ **Forest Elaenia** *Myiopagis gaimardii*
　____ *M. g. macilvainii* — Tropical e Panama and Caribbean coast of Colombia
　____ *M. g. trinitatis* — Trinidad
　____ *M. g. bogotensis* — Tropical ne Colombia and n Venezuela
　____ *M. g. guianensis* — Extreme e Colombia to se Venezuela, the Guianas and n Brazil
　____ *M. g. gaimardii (subcinerea)* — Tropical s Ecuador to e Peru, n Bolivia, sw and e Brazil

☐ **Gray Elaenia** *Myiopagis caniceps*
　____ *M. c. absita* — Extreme e Panama (Darién)
　____ *M. c. parambae* — Tropical w Colombia and nw Ecuador (south to w Cañar)
　____ *M. c. cinerea* — E Colombia to e Ecuador, ne Peru, s Venezuela and nw Brazil
　____ *M. c. caniceps* — Tropical se Brazil to Paraguay, s Bolivia and n Argentina

☐ **Foothill Elaenia** *Myiopagis olallai*
　Disjunct in foothills on e slope of Andes of Ecuador and Peru

☐ **Pacific Elaenia** *Myiopagis subplacens*
　W Ecuador (w Esmeraldas and Manabí) to nw Peru

☐ **Yellow-crowned Elaenia** *Myiopagis flavivertex*
　S Venezuela to the Guianas, ne Peru and w Amazonian Brazil

☐ **Jamaican Elaenia** *Myiopagis cotta*
　Jamaica

☐ **Greenish Elaenia** *Myiopagis viridicata*
　____ *M. v. jaliscensis* — Tropical w Mexico (Sinaloa and Durango to Guerrero)
　____ *M. v. minima* — Tres Marías Islands (off w Mexico)
　____ *M. v. placens* — Tropical se Mexico (Tamaulipas) to Honduras; Cozumel I.
　____ *M. v. pacifica* — Pacific lowlands of s Mexico (Chiapas) to w Honduras
　____ *M. v. accola* — Tropical Nicaragua to Panama, n Colombia and w Venezuela
　____ *M. v. pallens* — Colombia (Cundinamarca, Huila and Santa Marta region)
　____ *M. v. restricta* — Tropical s Venezuela
　____ *M. v. zuliae* — Sierra de Perijá (Colombia/Venezuela border)
　____ *M. v. implacens* — Tropical sw Colombia (Nariño) and w Ecuador
　____ *M. v. viridicata* — SE Peru to e Bolivia, e Paraguay, n Argentina, e and se Brazil

☐ **Caribbean Elaenia** *Elaenia martinica*
　____ *E. m. remota* — Islands off e Mexico (Cozumel, Meco, Mujeres, Holbox)
　____ *E. m. chinchorrensis* — Great Cay I. off Quintana Roo (e Mexico)
　____ *E. m. cinerascens* — San Andrés, Providéncia, Santa Catalina Is. (off Honduras)
　____ *E. m. caymanensis* — Cayman Islands
　____ *E. m. riisii* — Puerto Rico to Anguilla, se to Antigua; Aruba, Curaçao, Bonaire

____	*E. m. martinica*	Lesser Antilles (south to Grenada)
____	*E. m. barbadensis*	Barbados

☐ **Yellow-bellied Elaenia** *Elaenia flavogaster*

____	*E. f. subpagana*	S Mexico (Oaxaca and Veracruz) to Costa Rica; Coiba I.
____	*E. f. pallididorsalis*	Panama; Cébaco I. and Pearl Islands
____	*E. f. flavogaster*	Colombia to the Guianas, Brazil, Paraguay, Argentina; Trinidad
____	*E. f. semipagana*	Tropical w Ecuador (El Oro and Loja) and nw Peru; Isla Puná

☐ **Large Elaenia** *Elaenia spectabilis*

Extreme se Colombia to n Argentina and Brazil

☐ **Noronha Elaenia** *Elaenia ridleyana*

Ilha Fernando de Noronha (off ne Brazil)

☐ **White-crested Elaenia** *Elaenia albiceps*

____	*E. a. griseigularis*	Andes of sw Colombia (Nariño) to Ecuador and nw Peru
____	*E. a. diversa*	Central Andes of Peru (Cajamarca to Huánuco)
____	*E. a. urubambae*	Subtropical se Peru (Cuzco)
____	*E. a. modesta*	Arid tropical w Peru (La Libertad) to nw Chile
____	*E. a. albiceps*	Extreme se Peru (Puno) and nw Bolivia
____	*E. a. chilensis*	Andes of Bolivia to Tierra del Fuego; winters north to Brazil

☐ **Small-billed Elaenia** *Elaenia parvirostris*

S Brazil to Bolivia and c Argentina; winters n to Colombia

☐ **Olivaceous Elaenia** *Elaenia mesoleuca*

SE Brazil (Goiás) to e Paraguay and ne Argentina

☐ **Slaty Elaenia** *Elaenia strepera*

Andes of s Bolivia and nw Argentina; winters n to Venezuela

☐ **Mottle-backed Elaenia** *Elaenia gigas*

Tropical se Colombia to e Ecuador, e Peru and w Bolivia

☐ **Brownish Elaenia** *Elaenia pelzelni*

River islands of Amazon system

☐ **Plain-crested Elaenia** *Elaenia cristata*

____	*E. c. cristata*	Venezuela and Guianas to Amaz. and e Brazil and ne Bolivia
____	*E. c. alticola*	*Tepuis* of s Venezuela (se Bolívar) and adjacent n Brazil

☐ **Lesser Elaenia** *Elaenia chiriquensis*

____	*E. c. chiriquensis*	NW Costa Rica to Panama; Coiba, Cébaco and Pearl islands
____	*E. c. brachyptera*	Pacific slope of sw Colombia (Nariño) and nw Ecuador
____	*E. c. albivertex*	Colombia to the Guianas, Brazil and n Argentina; Trinidad

☐ **Rufous-crowned Elaenia** *Elaenia ruficeps*

SE Colombia to the Guianas, s Venezuela and n Amaz. Brazil

☐ **Mountain Elaenia** *Elaenia frantzii*

____	*E. f. ultima*	Mountains of Guatemala, El Salvador, Honduras and Nicaragua
____	*E. f. frantzii*	Mountains of Costa Rica and w Panama (Chiriquí and Veraguas)
____	*E. f. pudica*	Andes of Colombia and coastal mountains of n Venezuela
____	*E. f. browni*	N Colombia (Santa Marta Mountains and Sierra de Perijá)

☐ **Greater Antillean Elaenia** *Elaenia fallax*

____	*E. f. fallax*	Humid montane forests of Jamaica
____	*E. f. cherriei*	Humid montane forests of Hispaniola

☐ **Highland Elaenia** *Elaenia obscura*

____	*E. o. obscura*	E slope of Andes of s Ecuador to Bolivia and nw Argentina
____	*E. o. sordida*	SE Brazil (Rio de Janeiro) to e Paraguay and ne Argentina

☐ **Great Elaenia** *Elaenia dayi*

____	*E. d. tyleri*	*Tepuis* of s Venezuela (Duida, Huachamacare and Parú)
____	*E. d. auyantepui*	*Tepuis* of s Venezuela (Auyan-tepui)
____	*E. d. dayi*	*Tepuis* of s Venezuela (Roraima, Kukenamk and Ptari-tepui)

☐ **Sierran Elaenia** *Elaenia pallatangae*
_____ *E. p. pallatangae* — Andes of s Colombia and Ecuador
_____ *E. p. davidwillardi* — *Tepuis* of s Venezuela
_____ *E. p. olivina* — *Tepuis* of s Venezuela, adjacent Guyana and extreme n Brazil
_____ *E. p. intensa* — Andes of Peru
_____ *E. p. exsul* — Andes of Bolivia (La Paz and Cochabamba)

☐ **Yellow-bellied Tyrannulet** *Ornithion semiflavum*

Humid lowlands of s Mexico (Oaxaca) to ne Costa Rica

☐ **Brown-capped Tyrannulet** *Ornithion brunneicapillus*

Costa Rica to w Colombia, nw Ecuador and nw Venezuela

☐ **White-lored Tyrannulet** *Ornithion inerme*

S Venezuela to the Guianas, n Bolivia, Amazonian and e Brazil

☐ **Northern Beardless-Tyrannulet** *Camptostoma imberbe*

SE Arizona to s Texas, w Mexico, n Costa Rica; Cozumel I.

☐ **Southern Beardless-Tyrannulet** *Camptostoma obsoletum*
_____ *C. o. flaviventre* — Pacific coast of sw Costa Rica and Panama
_____ *C. o. orphnum* — Coiba and Cébaco islands (Panama)
_____ *C. o. majus* — Pearl Islands (Panama)
_____ *C. o. caucae (bogotensis)* — Colombia (Western Andes, Cauca and Magdalena valleys)
_____ *C. o. pusillum (venezuelae)* — Caribbean coast of n Colombia and n Venezuela; Trinidad
_____ *C. o. napaeum* — Extreme s-central Venezuela to the Guianas and n Brazil
_____ *C. o. maranonicum* — N Peru (middle Marañón Valley)
_____ *C. o. olivaceum* — SE Colombia to e Ecuador, ne Peru and w Brazil (w Amazonas)
_____ *C. o. sclateri* — Tropical w Ecuador and extreme nw Peru (Tumbes and n Piura)
_____ *C. o. griseum* — Arid littoral of w Peru (Lambayeque to Lima)
_____ *C. o. bolivianum* — Central Bolivia to nw Argentina (Tucumán)
_____ *C. o. cinerascens* — E Brazil (Maranhão to Ceará and Mato Grosso) and w Bolivia
_____ *C. o. obsoletum* — SE Brazil to Uruguay, Paraguay and n Argentina

☐ **Suiriri Flycatcher** *Suiriri suiriri*
_____ *S. s. affinis* — *Chaco* of Suriname; e Brazil (Pará) to nw Bolivia
_____ *S. s. bahiae* — E Brazil (Paraíba, Pernambuco, ne Bahia and e Piauí)
_____ *S. s. suiriri* — SW Brazil to Paraguay, Uruguay, Bolivia and n Argentina

☐ **Chapada Flycatcher** *Suiriri islerorum*

Cerrado region of Brazil and adjacent eastern Bolivia

☐ **White-tailed Tyrannulet** *Mecocerculus poecilocercus*

Andes of Colombia to Ecuador and s Peru (Cuzco)

☐ **Buff-banded Tyrannulet** *Mecocerculus hellmayri*

E slope of Andes of s Peru (Puno) to w Bolivia and nw Argentina

☐ **White-banded Tyrannulet** *Mecocerculus stictopterus*
_____ *M. s. stictopterus* — Andes of Colombia, Ecuador and n Peru
_____ *M. s. albocaudatus* — Andes of nw Venezuela (Trujillo, Mérida and Táchira)
_____ *M. s. taeniopterus* — Andes of e Peru to w Bolivia (La Paz and Cochabamba)

☐ **White-throated Tyrannulet** *Mecocerculus leucophrys*
_____ *M. l. notatus* — W and central Andes of Colombia (south to Cauca)
_____ *M. l. setophagoides (gularis)* — E Andes of Colombia and nw Venezuela
_____ *M. l. rufomarginatis* — S Colombia (Nariño) to w Ecuador and nw Peru (Piura)
_____ *M. l. nigriceps (montensis, palliditergum)* — Mts. of n Venezuela
_____ *M. l. chapmani* — *Tepuis* of s Venezuela
_____ *M. l. roraimae* — Subtropical central Venezuela (Amazonas and Bolívar)
_____ *M. l. parui* — *Tepuis* of s Venezuela (Cerro Parú in Amazonas)
_____ *M. l. brunneomarginatus* — Andes of Peru (La Libertad and Cajamarca to Cuzco)
_____ *M. l. pallidior* — Humid temperate Western Andes of Peru (w Ancash)
_____ *M. l. leucophrys* — Temperate Andes of se Peru to Bolivia and nw Argentina

☐ **Rufous-winged Tyrannulet** *Mecocerculus calopterus*

Andes of Ecuador and nw Peru (s to Lambayeque and La Libertad)

321

☐ **Sulphur-bellied Tyrannulet** *Mecocerculus minor*

Andes of Colombia and w Venezuela to Ecuador and e Peru

☐ **Black-crested Tit-Tyrant** *Anairetes nigrocristatus*

Andes of extreme s Ecuador and w Peru (south to Pasco)

☐ **Pied-crested Tit-Tyrant** *Anairetes reguloides*

_____ *A. r. albiventris* Arid littoral of w Peru (Ancash to Ica and w Ayacucho)
_____ *A. r. reguloides* Arid sw Peru (s Ayacucho) to extreme n Chile (Tarapacá)

☐ **Ash-breasted Tit-Tyrant** *Anairetes alpinus*

_____ *A. a. alpinus* Andes of Peru (n Ancash. Apurímac and Cuzco)
_____ *A. a. bolivianus* *Yungas* of w Bolivia (La Paz)

☐ **Yellow-billed Tit-Tyrant** *Anairetes flavirostris*

_____ *A. f. huancabambae* Western and Central Andes of nw Peru south to Huánuco
_____ *A. f. arequipae* Andes of sw Peru (Lima) to nw Chile
_____ *A. f. cuzcoensis* Andes of se Peru (Cuzco)
_____ *A. f. flavirostris* Andes of s Peru (Puno) to Bolivia, n Chile and w Argentina

☐ **Tufted Tit-Tyrant** *Anairetes parulus*

_____ *A. p. aequatorialis* Andes of s Colombia to n Argentina (Salta and Jujuy)
_____ *A. p. patagonicus* W Argentina (s Mendoza to n Santa Cruz)
_____ *A. p. parulus* Andes of Chile and sw Argentina to Tierra del Fuego

☐ **Juan Fernandez Tit-Tyrant** *Anairetes fernandezianus*

Robinson Crusoe I. (Juan Fernández Islands off s Chile)

☐ **Agile Tit-Tyrant** *Anairetes agilis*

Andes of Colombia to extreme w Venezuela and Ecuador

☐ **Unstreaked Tit-Tyrant** *Anairetes agraphia*

_____ *A. a. plengei* E Peru (Cordillera de Colán in Amazonas)
_____ *A. a. squamigera* Central Peru (Carpish Mountains in e La Libertad and Huánuco)
_____ *A. a. agraphia* SE Peru (Cordillera Vilcanota and Urubamba Valley of Cuzco)

☐ **Torrent Tyrannulet** *Serpophaga cinerea*

_____ *S. c. grisea* Rocky mountain streams of Costa Rica and w Panama
_____ *S. c. cinerea* Mountain streams of Colombia and Venezuela to w Bolivia

☐ **River Tyrannulet** *Serpophaga hypoleuca*

_____ *S. h. hypoleuca* SE Colombia to e Ecuador, e Peru, n Bolivia and w Amaz. Brazil
_____ *S. h. venezuelana* Tropical Venezuela (Apure, Anzoátegui and n Bolívar)
_____ *S. h. pallida* Amazonian Brazil south of the Amazon

☐ **Sooty Tyrannulet** *Serpophaga nigricans*

SE Bolivia to c Argentina, e Paraguay, Uruguay and s Brazil

☐ **White-crested Tyrannulet** *Serpophaga subcristata*

_____ *S. s. straminea* SE Brazil (s Piauí and Bahia) to Uruguay
_____ *S. s. subcristata* E Bolivia to sw Brazil (Mato Grosso), Paraguay and c Argentina

☐ **White-bellied Tyrannulet** *Serpophaga munda*

W Bolivia to w Paraguay, extreme sw Brazil and w Argentina

☐ **Mouse-colored Tyrannulet** *Phaeomyias murina*

_____ *P. m. eremonoma* Pacific lowlands of Panama (Chiriquí to e Panama Province)
_____ *P. m. incomta* Colombia to Venezuela, the Guianas and n Brazil; Trinidad
_____ *P. m. tumbezana* Arid tropical sw Ecuador to nw Peru (south to ne Lambayeque)
_____ *P. m. inflava* Arid tropical nw Peru (Lambayeque to n Lima)
_____ *P. m. maranonica* N Peru (arid tropical Marañón Valley)
_____ *P. m. wagae* Tropical e Peru to nw Bolivia and w Amazonian Brazil
_____ *P. m. murina (ignobilis)* S Bolivia to Paraguay, nw Argentina and se Brazil

☐ **Cocos Island Flycatcher** *Nesotriccus ridgwayi*

Cocos I. (off Costa Rica)

☐ **Yellow Tyrannulet** *Capsiempis flaveola*

____ *C. f. semiflava* — Tropical s Nicaragua to e-central Panama; Coiba I.
____ *C. f. leucophrys* — Colombia (Magdalena Valley) to nw Venezuela
____ *C. f. cerula (amazonus)* — E Colombia to ne Ecuador, the Guianas and n Brazil
____ *C. f. magnirostris* — SW Ecuador (Pichincha to El Oro)
____ *C. f. flaveola* — SE Brazil to ne Bolivia, e Paraguay, ne Argentina and se Peru

☐ **Sharp-tailed Tyrant** *Culicivora caudacuta*

E Bolivia to s-central Brazil, e Paraguay and ne Argentina

☐ **Bearded Tachuri** *Polystictus pectoralis*

____ *P. p. bogotensis* — W Colombia (Cauca Valley and Cundinamarca)
____ *P. p. brevipennis* — NE Colombia to s Venezuela, the Guianas and extreme n Brazil
____ *P. p. pectoralis* — S Brazil to e Bolivia, Uruguay, Paraguay and n Argentina

☐ **Gray-backed Tachuri** *Polystictus superciliaris*

SE Brazil (central Bahia to Minas Gerais and n São Paulo)

☐ **Dinelli's Doradito** *Pseudocolopteryx dinelliana*

Locally in n Argentina; > to s Bolivia, w Paraguay and sw Brazil

☐ **Crested Doradito** *Pseudocolopteryx sclateri*

S Venezuela to Guyana, ne Argentina and s Brazil; Trinidad

☐ **Subtropical Doradito** *Pseudocolopteryx acutipennis*

Andes of Colombia to nw Argentina (south to La Rioja)

☐ **Warbling Doradito** *Pseudocolopteryx flaviventris*

Extreme s Brazil to e Bolivia, Uruguay, Argentina and c Chile

☐ **Bronze-olive Pygmy-Tyrant** *Pseudotriccus pelzelni*

____ *P. p. berlepschi* — Mountains of extreme e Panama and adjacent nw Colombia
____ *P. p. annectens* — Andes of sw Colombia and nw Ecuador (south to El Oro)
____ *P. p. pelzelni* — E Colombia (Meta) and e Ecuador (Napo-Pastaza)
____ *P. p. peruvianus* — Andes of e Peru (San Martín to Cuzco)

☐ **Hazel-fronted Pygmy-Tyrant** *Pseudotriccus simplex*

E slope of Andes of s Peru (Puno) and w Bolivia

☐ **Rufous-headed Pygmy-Tyrant** *Pseudotriccus ruficeps*

Andes of Colombia to w Bolivia (La Paz and Cochabamba)

☐ **Ringed Antpipit** *Corythopis torquatus*

____ *C. t. sarayacuensis* — SE Colombia to e Ecuador and ne Peru
____ *C. t. torquatus (subtorquatus)* — E Peru to n Bolivia and w Amaz. Brazil
____ *C. t. anthoides* — S Venezuela to the Guianas and n Amazonian Brazil

☐ **Southern Antpipit** *Corythopis delalandi*

S Brazil to e Bolivia, e Paraguay and ne Argentina

☐ **Tawny-crowned Pygmy-Tyrant** *Euscarthmus meloryphus*

____ *E. m. paulus* — NE Colombia (Santa Marta region) and n Venezuela
____ *E. m. fulviceps* — Tropical sw Ecuador and w Peru (south to La Libertad)
____ *E. m. meloryphus* — SE Brazil to n Uruguay, e Bolivia, e Paraguay and n Argentina

☐ **Rufous-sided Pygmy-Tyrant** *Euscarthmus rufomarginatus*

S Suriname and e Brazil to ne Bolivia and ne Paraguay

☐ **Gray-and-white Tyrannulet** *Pseudelaenia leucospodia*

Arid scrub of sw Ecuador and nw Peru (Tumbes to La Libertad)

☐ **Lesser Wagtail-Tyrant** *Stigmatura napensis*

____ *S. n. napensis* — Amazon system is. (se Colombia, e Ecuador, ne Peru and w Brazil)
____ *S. n. bahiae* — River islands in e Brazil (Pernambuco and Bahia)

☐ **Greater Wagtail-Tyrant** *Stigmatura budytoides*

____ *S. b. budytoides* — S Bolivia (Cochabamba south to Tarija)
____ *S. b. inzonata* — SE Bolivia to w Paraguay and n Argentina
____ *S. b. flavocinerea* — Central Argentina
____ *S. b. gracilis* — NE Brazil (Pernambuco and n Bahia)

☐ **Paltry Tyrannulet** *Zimmerius vilissimus*
____ *Z. v. vilissimus*
____ *Z. v. parvus*

S Mexico (Chiapas) to El Salvador
Honduras to Panama and extreme nw Colombia (Chocó)

☑ **Venezuelan Tyrannulet** *Zimmerius improbus*
____ *Z. i. improbus*
____ *Z. i. tamae*
____ *Z. i. petersi*

Andes of n Colombia and Sierra de Perijá (w Venezuela)
Santa Marta Mountains (ne Colombia)
Coastal cordillera of n Venezuela (s Lara east to Miranda)

☐ **Bolivian Tyrannulet** *Zimmerius bolivianus*

Andes of se Peru (Huánuco to Puno) to w Bolivia

☐ **Red-billed Tyrannulet** *Zimmerius cinereicapilla*

Foothills of e Ecuador and e Peru (south to Madre de Dios)

☐ **Mishana Tyrannulet** *Zimmerius villarejoi*

White sand *varillal* forests of ne Peru in Iquitos region

☐ **Slender-footed Tyrannulet** *Zimmerius gracilipes*
____ *Z. g. gracilipes*
____ *Z. g. acer*
____ *Z. g. gilvus*

SE Colombia to s Venezuela, the Guianas and ne Brazil
NE Brazil (Alagoas)
E Ecuador to e Peru, n Bolivia and Amazonian Brazil

☐ **Golden-faced Tyrannulet** *Zimmerius chrysops*
____ *Z. c. minimus*
____ *Z. c. cumanensis*
____ *Z. c. albigularis*
____ *Z. c. flavidifrons*
____ *Z. c. chrysops*

Santa Marta Mountains (ne Colombia)
Coastal mts. of Venezuela (Anzoátegui, Sucre and Monagas)
SW Colombia (Nariño) and w Ecuador (south to sw Guayas)
SW Ecuador (se Guayas to w Loja and El Oro) to ext. n Peru
S Colombia to Ecuador, n Peru and nw Venezuela

☐ **Peruvian Tyrannulet** *Zimmerius viridiflavus*

E slope of Andes of Peru (Huánuco to Junín)

☐ **Mottle-cheeked Tyrannulet** *Phylloscartes ventralis*
____ *P. v. angustirostris*
____ *P. v. tucumanus*
____ *P. v. ventralis*

E slope of Andes of Peru (San Martín) to n Bolivia
Andes of nw Argentina (Jujuy to Tucumán and Catamarca)
SE Brazil (Minas Gerais) to Uruguay, e Paraguay, ne Argentina

☐ **Restinga Tyrannulet** *Phylloscartes kronei*

Coastal se Brazil (se São Paulo to ne Santa Catarina)

☐ **Bahia Tyrannulet** *Phylloscartes beckeri*

Montane forests of se Brazil (s Bahia near Boa Nova)

☐ **Yellow-green Tyrannulet** *Phylloscartes flavovirens*

Pacific lowlands of Panama (Canal Zone to e Darién)

☐ **Olive-green Tyrannulet** *Phylloscartes virescens*

Lowlands of Guianas and adjacent n Brazil (Amazonas)

☐ **Ecuadorian Tyrannulet** *Phylloscartes gualaquizae*

E slope of Andes of e Ecuador and n Peru (San Martín)

☐ **Black-fronted Tyrannulet** *Phylloscartes nigrifrons*

Tepuis of s Venezuela (s Bolívar and Amazonas)

☐ **Alagoas Tyrannulet** *Phylloscartes ceciliae*

Highlands of ne Brazil (Alagoas)

☐ **Rufous-browed Tyrannulet** *Phylloscartes superciliaris*
____ *P. s. superciliaris*
____ *P. s. palloris*
____ *P. s. griseocapillus*

Mountains of Costa Rica to w Panama (Veraguas)
E Panama (Darién) and adjacent Colombia (Chocó)
Sierra de Perijá to se Ecuador and extreme n Peru

☐ **Rufous-lored Tyrannulet** *Phylloscartes flaviventris*

Coastal mts. of n Venezuela

☐ **Cinnamon-faced Tyrannulet** *Phylloscartes parkeri*

Andean foothills of central and se Peru and adjacent n Bolivia

☐ **Minas Gerais Tyrannulet** *Phylloscartes roquettei*

Known from a 1926 specimen from e Brazil (n Minas Gerais)

☐ **Bay-ringed Tyrannulet** *Phylloscartes sylviolus*

SE Brazil (Espírito Santo) to e Paraguay and ne Argentina

☐ **São Paulo Tyrannulet** *Phylloscartes paulista*

SE Brazil (Espírito Santo) to e Paraguay and ne Argentina

☐ **Oustalet's Tyrannulet** *Phylloscartes oustaleti*

SE Brazil (Espírito Santo to e Santa Catarina)

☐ **Serra do Mar Tyrannulet** *Phylloscartes difficilis*

Coastal mts. of se Brazil (Espírito Santo to n Rio Grande do Sul)

☐ **Variegated Bristle-Tyrant** *Pogonotriccus poecilotis*

Andes of n Colombia to w Venezuela, e Ecuador and s Peru

☐ **Chapman's Bristle-Tyrant** *Pogonotriccus chapmani*
____ *P. c. chapmani*
____ *P. c. duidae*

Tepuis of s Venezuela (s Bolívar and ne Amazonas)
Tepuis of se Venezuela (Cerro de la Neblina) and (?) adj. Brazil

☐ **Marble-faced Bristle-Tyrant** *Pogonotriccus ophthalmicus*
____ *P. o. ophthalmicus*
____ *P. o. purus*
____ *P. o. ottonis*

Andes of Colombia to nw Venezuela, e Ecuador and n Peru
Coastal cordillera of n Venezuela (Yaracuy to Distrito Federal)
Andes of se Peru (Puno) and w Bolivia

☐ **Southern Bristle-Tyrant** *Pogonotriccus eximius*

SE Brazil (Espírito Santo) to e Paraguay and ne Argentina

☐ **Spectacled Bristle-Tyrant** *Pogonotriccus orbitalis*

Andes of extreme s Colombia to Ecuador, Peru and w Bolivia

☐ **Venezuelan Bristle-Tyrant** *Pogonotriccus venezuelanus*

Coastal cordillera of n Venezuela (Carabobo to Distrito Federal)

☐ **Antioquia Bristle-Tyrant** *Pogonotriccus lanyoni*

Central Andes of n Colombia (Antioquia and e Caldas)

☐ **Sepia-capped Flycatcher** *Leptopogon amaurocephalus*
____ *L. a. pileatus (faustus)*
____ *L. a. idius*
____ *L. a. diversus*
____ *L. a. orenocensis (obscuritergum)*
____ *L. a. peruvianus*
____ *L. a. amaurocephalus*

Tropical s Mexico to Costa Rica and Panama
Coiba I. (Panama)
Santa Marta and Magdalena valleys of n Colombia and w Zulia
Tropical Venezuela to e Brazil (Amapá)
Tropical e Colombia to n Bolivia
Tropical se Brazil to e Paraguay, n Argentina and e Bolivia

☐ **Slaty-capped Flycatcher** *Leptopogon superciliaris*
____ *L. s. superciliaris*
____ *L. s. albidiventer*

Mts. of Costa Rica to Venezuela, n Brazil and w Bolivia
SE Peru (Cuzco and Puno) and *yungas* of w Bolivia

☐ **Rufous-breasted Flycatcher** *Leptopogon rufipectus*

Andes of Colombia to e Ecuador, w Venezuela and n Peru

☐ **Inca Flycatcher** *Leptopogon taczanowskii*

E slope of Andes of Peru (Amazonas to Cuzco)

☐ **Streak-necked Flycatcher** *Mionectes striaticollis*
____ *M. s. columbianus (selvae)*
____ *M. s. viridiceps*
____ *M. s. palamblae*
____ *M. s. striaticollis (poliocephalus)*

Andes of e Colombia and e Ecuador
Extreme tropical sw Colombia (w Nariño) and w Ecuador
Andes of n Peru (Piura south to Huánuco)
Andes of central Peru to w Bolivia

☐ **Olive-striped Flycatcher** *Mionectes olivaceus*
____ *M. o. olivaceus*
____ *M. o. hederaceus*
____ *M. o. galbinus*
____ *M. o. fasciaticollis*
____ *M. o. venezuelensis (pallidus, meridae)*

Lowlands and foothills of e Costa Rica and w Panama
E Panama to w Colombia and w Ecuador
Santa Marta Mountains (ne Colombia)
Andes of s Colombia to s Peru and extreme n Bolivia
Andes of s Colombia and n Venezuela; Trinidad

☐ **Ochre-bellied Flycatcher** *Mionectes oleagineus*
____ *M. o. assimilis (obscurus, dyscolus, lutescens)*
____ *M. o. parcus*
____ *M. o. pacificus*

Tropical s Mexico to e Costa Rica and w Panama
Tropical e Panama to n Colombia and nw Venezuela
Tropical sw Colombia to w Ecuador

____ *M. o. abdominalis*	Tropical n Venezuela (Distrito Federal and Miranda)
____ *M. o. pallidiventris*	Tropical ne Venezuela; Trinidad and Tobago
____ *M. o. dorsalis*	*Tepuis* of se Venezuela (Gran Sabana of Bolívar)
____ *M. o. oleagineus*	E Colombia to the Guianas, e Brazil, e Peru and n Bolivia

☐ **McConnell's Flycatcher** *Mionectes macconnelli*

____ *M. m. roraimae (mercedesfosteri)*	Tepuis of s Venezuela and Guyana
____ *M. m. macconnelli (amazonus)*	E Venezuela to the Guianas, ne Brazil, se Peru and ne Bolivia
____ *M. m. peruanus*	Central Peru (Junín)

☐ **Gray-hooded Flycatcher** *Mionectes rufiventris*

SE Brazil (Espírito Santo) to e Paraguay and ne Argentina

☐ **Northern Scrub-Flycatcher** *Sublegatus arenarum*

____ *S. a. arenarum*	Costa Rica and w Panama
____ *S. a. atrirostris*	Pearl Islands (off s Panama) and n Colombia
____ *S. a. glaber*	Coastal n Venezuela, n Guianas, Trinidad, Margarita, Patos I.
____ *S. a. pallens*	Curaçao
____ *S. a. tortugensis*	Isla La Tortuga (off n Venezuela)
____ *S. a. orinocensis*	Extreme e Colombia, Orinoco Valley of s Venezuela, adj. Brazil

☐ **Amazonian Scrub-Flycatcher** *Sublegatus obscurior*

E Colombia to s Venezuela, Guianas, n Bolivia, Amaz. Brazil

☐ **Southern Scrub-Flycatcher** *Sublegatus modestus*

____ *S. m. modestus*	Tropical e Peru to e Brazil (Maranhão, Pernambuco and Paraná)
____ *S. m. brevirostris*	E Bolivia to Paraguay, Uruguay and central Argentina

☐ **Slender-billed Tyrannulet** *Inezia tenuirostris*

Arid ne Colombia to nw Venezuela (Zulia, Falcón and n Lara)

☐ **Plain Tyrannulet** *Inezia inornata*

SE Peru to nw Argentina, Paraguay, sw and Amazonian Brazil

☐ **Amazonian Tyrannulet** *Inezia subflava*

____ *I. s. obscura*	Extreme e Colombia to sw Venezuela and adjacent nw Brazil
____ *I. s. subflava*	S Amazonian Brazil and ne Bolivia (ne Beni and ne Santa Cruz)

☐ **Pale-tipped Tyrannulet** *Inezia caudata*

____ *I. c. intermedia*	Caribbean lowlands of ne Colombia and n Venezuela
____ *I. c. caudata*	S Venezuela to the Guianas and extreme n Brazil

☐ **Ornate Flycatcher** *Myiotriccus ornatus*

____ *M. o. ornatus*	W slope of Eastern and Central Andes of Colombia
____ *M. o. stellatus*	Western Andes of Colombia and Ecuador
____ *M. o. phoenicurus*	Eastern Andes of se Colombia to e Ecuador and n Peru
____ *M. o. aureiventris*	Andes of se Peru (Huánuco to Puno)

☐ **Many-colored Rush-Tyrant** *Tachuris rubrigastra*

____ *T. r. libertatis*	Marshes of coastal w Peru (La Libertad to n Ica)
____ *T. r. alticola*	Andes of se Peru to w Bolivia and nw Argentina
____ *T. r. rubrigastra*	SE Brazil to Paraguay, Uruguay, n Argentina and w Chile
____ *T. r. loaensis*	N Chile (Antofagasta)

☐ **Eared Pygmy-Tyrant** *Myiornis auricularis*

____ *M. a. cinereicollis*	E Brazil (se Bahia, Minas Gerais and Espírito Santo)
____ *M. a. auricularis*	SE Paraguay to ne Argentina and se Brazil

☐ **White-bellied Pygmy-Tyrant** *Myiornis albiventris*

E slope of Andes of Peru (Huánuco) to w Bolivia (Santa Cruz)

☐ **Black-capped Pygmy-Tyrant** *Myiornis atricapillus*

Humid lowlands of Costa Rica to w Colombia and nw Ecuador

☐ **Short-tailed Pygmy-Tyrant** *Myiornis ecaudatus*

____ *M. e. miserabilis*	E Colombia to the Guianas and n Brazil; Trinidad
____ *M. e. ecaudatus*	Amazonian Brazil to e Ecuador, Peru and n Bolivia

☐ **Northern Bentbill** *Oncostoma cinereigulare*

Lowlands of s Mexico (Veracruz) to w Panama

☐ **Southern Bentbill** *Oncostoma olivaceum*

Tropical e Panama and n Colombia (east to Santa Marta region)

☐ **Scale-crested Pygmy-Tyrant** *Lophotriccus pileatus*
  ____ *L. p. luteiventris*
  ____ *L. p. santaeluciae*
  ____ *L. p. squamaecrista*
  ____ *L. p. pileatus*
  ____ *L. p. hypochlorus*

Highlands of Costa Rica to e Panama (Darién)
Andes of Colombia and nw Venezuela (Zulia to Táchira)
Andes of Colombia and w Ecuador
Andes of e Ecuador and Peru (south to n Cuzco)
Andes of se Peru (Cuzco and Puno)

☐ **Long-crested Pygmy-Tyrant** *Lophotriccus eulophotes*

Extreme sw Amaz. Brazil, adj. se Peru and extreme nw Bolivia

☐ **Double-banded Pygmy-Tyrant** *Lophotriccus vitiosus*
  ____ *L. v. affinis*
  ____ *L. v. guianensis*
  ____ *L. v. congener*
  ____ *L. v. vitiosus*

SE Colombia to ne Peru and nw Amazonian Brazil
The Guianas and ne Brazil (Amapá and Pará); Mato Grosso?
W Amazonian Brazil (Rio Juruá in sw Amazonas)
E Peru (e San Martín, s Loreto and Huánuco)

☐ **Helmeted Pygmy-Tyrant** *Lophotriccus galeatus*

E Colombia to s Venezuela, the Guianas and n Amaz. Brazil

☐ **Pale-eyed Pygmy-Tyrant** *Atalotriccus pilaris*
  ____ *A. p. wilcoxi*
  ____ *A. p. pilaris*
  ____ *A. p. venezuelensis*
  ____ *A. p. griseiceps*

Arid tropical Pacific coast of Panama
N Colombia and adjacent Venezuela (Zulia and Táchira)
N Venezuela south to n Amazonas and n Bolívar
E Colombia to Venezuela, w Guyana and extreme n Brazil

☐ **Snethlage's Tody-Tyrant** *Hemitriccus minor*
  ____ *H. m. minor*
  ____ *H. m. pallens*
  ____ *H. m. snethlageae*

E Amazonian Brazil in central central Pará
W-central Brazil (Amazonas) east to Rio Negro and Rio Madeira
Central Amazonian Brazil to extreme ne Bolivia

☐ **Yungas Tody Tyrant** *Hemitriccus spodiops*

Yungas of Bolivia (La Paz, Cochabamba and s Beni)

☐ **Drab-breasted Bamboo-Tyrant** *Hemitriccus diops*

SE Brazil (se Bahia) to e Paraguay and ne Argentina

☐ **Brown-breasted Bamboo-Tyrant** *Hemitriccus obsoletus*
  ____ *H. o. obsoletus*
  ____ *H. o. zimmeri*

Mountains of se Brazil (Rio de Janeiro and São Paulo)
Mountains of se Brazil (Paraná and Rio Grande do Sul)

☐ **Flammulated Bamboo-Tyrant** *Hemitriccus flammulatus*
  ____ *H. f. flammulatus*
  ____ *H. f. olivascens*

Tropical e Peru, adjacent sw Brazil and *yungas* of n Bolivia
Tropical n Bolivia (south to n Santa Cruz)

☐ **Boat-billed Tody-Tyrant** *Hemitriccus josephinae*

S Guyana to Suriname, French Guiana and n Brazil (Amapá)

☐ **White-eyed Tody-Tyrant** *Hemitriccus zosterops*
  ____ *H. z. zosterops*
  ____ *H. z. flaviviridis*
  ____ *H. z. naumburgae*

S Colombia to s Venezuela, the Guianas and nw Brazil
N Peru (central Amazonas and n San Martín)
NE Brazil (Paraíba to Alagoas)

☐ **White-bellied Tody-Tyrant** *Hemitriccus griseipectus*

SE Peru (Cuzco) to n Bolivia and central Amazonian Brazil

☐ **Eye-ringed Tody-Tyrant** *Hemitriccus orbitatus*

Lowlands of se Brazil (s Minas Gerais to ne Rio Grande do Sul)

☐ **Johannes' Tody-Tyrant** *Hemitriccus iohannis*

SE Colombia to e Peru, n Bolivia and w Amazonian Brazil

☐ **Stripe-necked Tody-Tyrant** *Hemitriccus striaticollis*
  ____ *H. s. striaticollis*
  ____ *H. s. griseiceps*

NE Colombia and ne Peru to n Bolivia, central and e Brazil
W-central Brazil (lower Rio Tapajós in e Pará)

☐ **Hangnest Tody-Tyrant** *Hemitriccus nidipendulus*
    ____ *H. n. nidipendulus* — E-central Brazil (Bahia)
    ____ *H. n. paulistus* — SE Brazil (Espírito Santo and Minas Gerais to São Paulo)

☐ **Pearly-vented Tody-Tyrant** *Hemitriccus margaritaceiventer*
    ____ *H. m. impiger* — Arid tropical ne Colombia and n Venezuela; Isla Margarita
    ____ *H. m. septentrionalis* — S Colombia (arid tropical upper Magdalena Valley)
    ____ *H. m. chiribiquetensis* — S Colombia (Sierra de Chiribiquete)
    ____ *H. m. duidae* — *Tepuis* of s Venezuela (Mt. Duida)
    ____ *H. m. auyantepui* — Subtropical *tepuis* of se Venezuela (se Bolívar)
    ____ *H. m. breweri* — Subtropical se Venezuela (Jaua massif)
    ____ *H. m. rufipes* — Tropical Peru east of the Andes (Cuzco) to nw Bolivia
    ____ *H. m. margaritaceiventer* — E Bolivia to n Argentina, e Paraguay and w Brazil
    ____ *H. m. wuchereri* — E Brazil (Maranhão to Ceará, Pernambuco and Bahia)

☐ **Pelzeln's Tody-Tyrant** *Hemitriccus inornatus*
    NW Brazil (locally on Rio Negro and tributaries)

☐ **Zimmer's Tody-Tyrant** *Hemitriccus minimus*
    E Amazonian Brazil to ne Bolivia (n Beni and ne Santa Cruz)

☐ **Black-throated Tody-Tyrant** *Hemitriccus granadensis*
    ____ *H. g. lehmanni* — Santa Marta Mountains (ne Colombia)
    ____ *H. g. granadensis* — Andes of Colombia and ne Ecuador
    ____ *H. g. andinus* — E Andes of Colombia and w Venezuela (Táchira)
    ____ *H. g. intensus* — Sierra de Perijá (Colombia/Venezuela border)
    ____ *H. g. federalis* — Montane forests of coastal n Venezuela (Distrito Federal)
    ____ *H. g. pyrrhops* — Andes of se Ecuador and Peru (south to Cuzco)
    ____ *H. g. caesius* — Andes of se Peru (Puno) and adjacent n Bolivia

☐ **Buff-breasted Tody-Tyrant** *Hemitriccus mirandae*
    Lowlands of e Brazil (Ceará, Pernambuco and Alagoas)

☐ **Cinnamon-breasted Tody-Tyrant** *Hemitriccus cinnamomeipectus*
    S Ecuador (Cordillera del Cóndor) to n Peru (San Martín)

☐ **Kaempfer's Tody-Tyrant** *Hemitriccus kaempferi*
    SE Brazil (e Santa Catarina)

☐ **Buff-throated Tody-Tyrant** *Hemitriccus rufigularis*
    Foothills on e slope of Andes in Ecuador, Peru and w Bolivia

☐ **Fork-tailed Tody-Tyrant** *Hemitriccus furcatus*
    SE Brazil (Rio de Janeiro, s Minas Gerais and ne São Paulo)

☐ **Black-chested Tyrant** *Taeniotriccus andrei*
    ____ *T. a. andrei* — Tropical se Venezuela and nw Amazonian Brazil
    ____ *T. a. klagesi* — E Amazonian Brazil (along Rio Tapajós and Rio Xingú)

☐ **Rufous-crowned Tody-Tyrant** *Poecilotriccus ruficeps*
    ____ *P. r. melanomystax* — Central Andes of Colombia and w Venezuela
    ____ *P. r. ruficeps* — E Andes of Colombia, s-central Ecuador and sw Venezuela
    ____ *P. r. rufigenis* — Andes of sw Colombia and nw Ecuador
    ____ *P. r. peruvianus* — Andes of s Ecuador and extreme n Peru (Piura and Cajamarca)

☐ **Lulu's Tody-Tyrant** *Poecilotriccus luluae*
    N Peru (Cordillera de Colán) and adjacent Eastern Andes

☐ **White-cheeked Tody-Tyrant** *Poecilotriccus albifacies*
    SE Peru (s Madre de Dios and adjacent ne Cuzco)

☐ **Black-and-white Tody-Tyrant** *Poecilotriccus capitalis*
    SE Colombia to e Ecuador, ne Peru and sw Brazil (Rondônia)

☐ **Buff-cheeked Tody-Flycatcher** *Poecilotriccus senex*
    W Brazil (known from an 1830 specimen from Amazonas)

☐ **Ruddy Tody-Flycatcher** *Poecilotriccus russatus*
    Tepuis of se Venezuela and n Brazil (Cerro Uei-tepui in Roraima)

☐ **Ochre-faced Tody-Flycatcher** *Poecilotriccus plumbeiceps*

____	*P. p. obscurus*	E slope of Andes of se Peru (Puno) to n Bolivia
____	*P. p. viridiceps*	S Bolivia to nw Argentina (Jujuy and Salta)
____	*P. p. plumbeiceps*	SE Brazil (São Paulo) to e Paraguay and ne Argentina
____	*P. p. cinereipectus*	SE Brazil (Espírito Santo and se Minas Gerais); Alagoas

☐ **Smoky-fronted Tody-Flycatcher** *Poecilotriccus fumifrons*

____	*P. f. penardi*	Suriname to French Guiana and lower Amazonian Brazil
____	*P. f. fumifrons*	Coastal ne Brazil (Paraíba to ne Bahia)

☐ **Rusty-fronted Tody-Flycatcher** *Poecilotriccus latirostris*

____	*P. l. mituensis*	SE Colombia (Mitú region of Vaupés)
____	*P. l. caniceps*	Tropical se Colombia to e Ecuador and e Peru
____	*P. l. latirostris*	Central Brazil (Rio Juruá to Rio Purús in Amazonas)
____	*P. l. mixtus*	SE Peru (n Puno) to n Bolivia
____	*P. l. ochropterus*	S Brazil (n São Paulo) to e Bolivia and w Brazil (Mato Grosso)
____	*P. l. austroriparius*	E Brazil (right bank of Rio Tapajós near Santarém in w Pará)
____	*P. l. senectus*	Amazonian Brazil (ne Amazonas and nw Pará)

☐ **Slate-headed Tody-Flycatcher** *Poecilotriccus sylvia*

____	*P. s. schistaceiceps*	Gulf slope of s Mexico (Veracruz) to Panama (Canal Zone)
____	*P. s. superciliaris*	Tropical n Colombia (Cauca and Magdalena valleys)
____	*P. s. griseolus*	Extreme e Colombia to nw Venezuela (east to n Bolívar)
____	*P. s. sylvia*	Guianas and n Brazil (along Rio Branco)
____	*P. s. schulzi*	NE Brazil (se Pará to n Piauí)

☐ **Golden-winged Tody-Flycatcher** *Poecilotriccus calopterus*

	SE Colombia (Putumayo) to e Ecuador and ne Peru (Loreto)

☐ **Black-backed Tody-Flycatcher** *Poecilotriccus pulchellus*

	Humid foothills of se Peru (Cuzco and Puno)

☐ **Spotted Tody-Flycatcher** *Todirostrum maculatum*

____	*T. m. signatum*	SE Colombia to ne Ecuador, e Peru, n Bolivia and w Brazil
____	*T. m. amacurense*	Extreme ne Venezuela and n Guyana; Trinidad
____	*T. m. maculatum*	French Guiana to Suriname and ne Amazonian Brazil
____	*T. m. diversum*	Central Amazonian Brazil
____	*T. m. annectens*	N-central Brazil (Rio Branco to Rio Negro)

☐ **Yellow-lored Tody-Flycatcher** *Todirostrum poliocephalum*

	Lowlands of se Brazil (s Minas Gerais to e Santa Catarina)

☐ **Common Tody-Flycatcher** *Todirostrum cinereum*

____	*T. c. virididorsale*	Tropical s Mexico (Veracruz and n Oaxaca)
____	*T. c. finitimum*	Tropical s Mexico (Tabasco and Chiapas) to nw Costa Rica
____	*T. c. wetmorei*	Tropical central and e Costa Rica and Panama
____	*T. c. sclateri*	SW Colombia (Nariño) to w Ecuador and nw Peru
____	*T. c. cinereum*	Tropical s Colombia to s Venezuela, the Guianas and n Brazil
____	*T. c. peruanum*	E Ecuador and e Peru (south to Cuzco)
____	*T. c. coloreum*	SE Brazil (Espírito Santo) to n Paraguay and n Bolivia
____	*T. c. cearae*	NE Brazil (e Pará to Piauí, Ceará, Alagoas and n Bahia)

☐ **Short-tailed Tody-Flycatcher** *Todirostrum viridanum*

	Coastal nw Venezuela (Zulia and Falcón)

☐ **Painted Tody-Flycatcher** *Todirostrum pictum*

	S Venezuela to the Guianas and n Brazil (east of Rio Negro)

☐ **Yellow-browed Tody-Flycatcher** *Todirostrum chrysocrotaphum*

____	*T. c. guttatum*	SE Colombia to ne Peru, extreme sw Venezuela and nw Brazil
____	*T. c. neglectum*	E Peru to n Bolivia and sw Brazil
____	*T. c. chrysocrotaphum*	E Peru (south of Río Marañón) and w Amazonian Brazil
____	*T. c. simile*	NE Brazil (along lower Rio Tapajós in w Pará)
____	*T. c. illigeri*	NE Brazil (along Rio Tapajós from w Pará to n Maranhão)

☐ **Black-headed Tody-Flycatcher** *Todirostrum nigriceps*

Caribbean slope of Costa Rica to w Ecuador and nw Venezuela

☐ **Eye-ringed Flatbill** *Rhynchocyclus brevirostris*

____ *R. b. brevirostris* — S Mexico (e Oaxaca and Veracruz) to w Panama
____ *R. b. pallidus* — Pacific coast of s Mexico (Guerrero to Oaxaca)
____ *R. b. hellmayri* — Mountains of e Panama (Darién) and extreme nw Colombia

☐ **Olivaceous Flatbill** *Rhynchocyclus olivaceus*

____ *R. o. bardus* — E Panama and nw Colombia (n Chocó to s Bolívar)
____ *R. o. mirus* — NW Colombia (lower Atrato Valley and inland from coast)
____ *R. o. flavus* — N and c Colombia and n Venezuela
____ *R. o. jelambianus* — NE Venezuela (Sucre and n Monagas)
____ *R. o. tamborensis* — Central Colombia (Rio Lebrija region of Santander)
____ *R. o. aequinoctialis* — S-cent. and se Colombia to e Ecuador, e Peru and n-cent. Bolivia
____ *R. o. guianensis* — S Venezuela , the Guianas and n Amazonian Brazil
____ *R. o. sordidus* — Brazil s of the Amazon (Rio Tapajós to n Maranhão)
____ *R. o. olivaceus* — N-c and e Brazil (e Pará; Pernambuco to Rio de Janeiro)

☐ **Pacific Flatbill** *Rhynchocyclus pacificus*

W Colombia (Chocó) to nw Ecuador (south to s Pichincha)

☐ **Fulvous-breasted Flatbill** *Rhynchocyclus fulvipectus*

Andes of Colombia and extreme w Venezuela to w Bolivia

☐ **Yellow-olive Flycatcher** *Tolmomyias sulphurescens*

____ *T. s. cinereiceps* — Tropical s Mexico (Oaxaca) to Costa Rica
____ *T. s. flavoolivaceus* — W Panama (Chiriquí and Darién) to Colombia (sw Bolívar)
____ *T. s. asemus* — W Colombia (Cauca and Magdalena valleys)
____ *T. s. confusus* — E Andes of Colombia to sw Venezuela and ne Ecuador
____ *T. s. exortivus* — Santa Marta region of n Colombia to n Venezuela
____ *T. s. berlepschi* — Trinidad
____ *T. s. cherriei* — S Venezuela to the Guianas and n Amazonian Brazil
____ *T. s. duidae* — *Tepuis* of se Venezuela and adjacent nw Brazil
____ *T. s. aequatorialis* — W Ecuador and nw Peru (Tumbes and Piura)
____ *T. s. peruvianus* — SE Ecuador (Loja) and n Peru (south to Junín)
____ *T. s. insignis* — NE Peru (Loreto) and adjacent w Amazonian Brazil
____ *T. s. mixtus* — NE Brazil (e Pará to nw Maranhão)
____ *T. s. inornatus* — Subtropical se Peru (n Puno)
____ *T. s. pallescens* — E Brazil (Minas Gerais) to n Bolivia and n Argentina
____ *T. s. grisescens* — Paraguay and n Argentina (e Chaco, Formosa and n Santa Fe)
____ *T. s. sulphurescens* — SE Brazil to e Paraguay and ne Argentina

☐ **Orange-eyed Flycatcher** *Tolmomyias traylori*

Tropical s Colombia and ne Peru north of the Amazon

☐ **Yellow-margined Flycatcher** *Tolmomyias assimilis*

____ *T. a. flavotectus* — E Costa Rica to w Colombia and nw Ecuador
____ *T. a. neglectus* — E Colombia to sw Venezuela and nw Amazonian Brazil
____ *T. a. examinatus* — SE Venezuela to the Guianas and ne Brazil (Pará and Amapá)
____ *T. a. obscuriceps* — SE Colombia (Meta) to ne Ecuador and ne Peru (e Loreto)
____ *T. a. clarus* — Peru immediately north of Río Marañón south to n Puno
____ *T. a. assimilis* — Cent. Brazil (e Amazonas to Rio Tapajós in w Pará)
____ *T. a. paraensis* — NE Brazil (e Pará and nw Maranhão)
____ *T. a. calamae* — Tropical n Bolivia and sw Brazil (se Amazonas)

☐ **Gray-crowned Flycatcher** *Tolmomyias poliocephalus*

____ *T. p. poliocephalus* — SE Colombia to e Peru, s Venezuela and w Amazonian Brazil
____ *T. p. klagesi* — Tropical e Venezuela (n Amazonas to Delta Amacuro)
____ *T. p. sclateri* — Tropical Guianas, Amazonian and e Brazil and nw Bolivia

☐ **Yellow-breasted Flycatcher** *Tolmomyias flaviventris*

____ *T. f. aurulentus (collingwoodi)* — N Colombia to the Guianas and ne Brazil; Trinidad and Tobago

____	*T. f. viridiceps*	SE Colombia to e Ecuador, e Peru and upper Amazonian Brazil
____	*T. f. dissors*	SW Venezuela (Amazonas) and lower Amazonian Brazil
____	*T. f. zimmeri*	N-central Peru (San Martín to Junín)
____	*T. f. subsimilis*	SE Peru (n Puno) to nw Bolivia and sw Brazil
____	*T. f. flaviventris*	E Brazil (Maranhão to Espírito Santo and Mato Grosso)

☐ **Cinnamon-crested Spadebill** *Platyrinchus saturatus*

____	*P. s. saturatus*	E Colombia to s Venezuela, the Guianas and n Amaz. Brazil
____	*P. s. pallidiventris*	Brazil on s bank of the Amazon (Rio Tapajós to n Maranhão)

☐ **Stub-tailed Spadebill** *Platyrinchus cancrominus*

Tropical s Mexico to w Costa Rica and nw Panama

☐ **White-throated Spadebill** *Platyrinchus mystaceus*

____	*P. m. neglectus*	E Costa Rica to e Colombia and extreme nw Venezuela
____	*P. m. perijanus*	Subtropical Sierra de Perijá (Colombia/Venezuela border)
____	*P. m. insularis*	N Venezuela; Trinidad and Tobago
____	*P. m. imatacae*	S Venezuela (Sierra de Imataca in Bolívar)
____	*P. m. ventralis*	S Venezuela (Cerro de la Neblina) and adjacent Brazil
____	*P. m. duidae*	*Tepuis* of se Venezuela and adjacent n Brazil
____	*P. m. ptaritepui*	*Tepuis* of se Venezuela (se Bolívar)
____	*P. m. albogularis*	Colombia (Western Andes and Cauca Valley) to w Ecuador
____	*P. m. zamorae*	Andes of e Ecuador and n Peru (south to Junín)
____	*P. m. mystaceus*	SE Brazil to e Paraguay and ne Argentina
____	*P. m. partridgei*	*Yungas* of Bolivia (Cochabamba and Santa Cruz)
____	*P. m. bifasciatus*	S Brazil (central Mato Grosso to central Goiás)
____	*P. m. cancromus*	E Brazil (Maranhão to Ceará, n Bahia and e Paraná)
____	*P. m. niveigularis*	Coastal ne Brazil (Paraíba to Alagoas)

☐ **Golden-crowned Spadebill** *Platyrinchus coronatus*

____	*P. c. superciliaris*	Caribbean slope of Honduras to Colombia and nw Ecuador
____	*P. c. coronatus*	Extreme se Colombia to n Bolivia, s Venezuela and w Brazil
____	*P. c. gumia*	SE Venezuela to the Guianas and n Amazonian Brazil

☐ **Yellow-throated Spadebill** *Platyrinchus flavigularis*

____	*P. f. flavigularis*	Andes of Colombia and w Venezuela to se Peru (Cuzco)
____	*P. f. vividus*	Sierra de Perijá (Colombia/Venezuela border)

☐ **White-crested Spadebill** *Platyrinchus platyrhynchos*

____	*P. p. platyrhynchos*	Extreme e Colombia to s Venezuela, the Guianas and n Brazil
____	*P. p. senex*	E Ecuador to e Peru (Loreto), nw Bolivia and extreme w Brazil
____	*P. p. nattereri*	W Brazil (Rio Purús to Rio Madeira and Rio Jiparaná)
____	*P. p. amazonicus*	Amazonian Brazil (Rio Tapajós east to Pará)

☐ **Russet-winged Spadebill** *Platyrinchus leucoryphus*

SE Brazil (Espírito Santo) to e Paraguay and ne Argentina

☐ **Northern Royal Flycatcher** *Onychorhynchus mexicanus*

Gulf lowlands of se Mexico (Veracruz) to e Panama (Darién)

☐ **Amazonian Royal Flycatcher** *Onychorhynchus coronatus*

____	*O. c. fraterculus*	NE Colombia to e Venezuela (w Zulia and w Barinas)
____	*O. c. castelnaui*	SE Colombia to n Bolivia and w Brazil (Amazonas)
____	*O. c. coronatus*	E Venezuela to the Guianas and n Amazonian Brazil
____	*O. c. swainsoni*	SE Brazil (Minas Gerais, Rio de Janeiro, São Paulo and Paraná)

☐ **Pacific Royal Flycatcher** *Onychorhynchus occidentalis*

W Ecuador (Esmeraldas) to extreme n Peru (Tumbes)

☐ **Brownish Flycatcher** *Cnipodectes subbrunneus*

____	*C. s. subbrunneus (panamensis)*	Lowlands of e Panama to w Colombia and w Ecuador
____	*C. s. minor*	SE Colombia to e Ecuador, e Peru, nw Bolivia and w Amaz. Brazil

☐ **Flavescent Flycatcher** *Myiophobus flavicans*

    ____ *M. f. flavicans*   Andes of Colombia to Ecuador and Peru (north of Río Marañón)

    ____ *M. f. perijanus*   NW Venezuela (Sierra de Perijá and Páramo de Tamá)

    ____ *M. f. venezuelanus*   Mts. of n Venezuela (Táchira to Distrito Federal and Miranda)

    ____ *M. f. caripensis*   Coastal cordillera of ne Venezuela (Monagas and Sucre)

    ____ *M. f. superciliosus*   Central Andes of Peru (s Amazonas to Cuzco)

☐ **Orange-crested Flycatcher** *Myiophobus phoenicomitra*

    ____ *M. p. litae*   W slope of Andes of Colombia (s Chocó) and Ecuador

    ____ *M. p. phoenicomitra*   E slope of Andes of e Ecuador to n Peru (San Martín)

☐ **Unadorned Flycatcher** *Myiophobus inornatus*

    E slope of Andes of s Peru (Cuzco) to w Bolivia

☐ **Roraiman Flycatcher** *Myiophobus roraimae*

    ____ *M. r. roraimae*   SE Colombia to e Ecuador, s Venezuela, w Guyana and w Brazil

    ____ *M. r. sadiecoatsae*   *Tepuis* of s Venezuela (Bolívar and Amazonas)

    ____ *M. r. rufipennis*   Locally in se Peru (San Martín to Puno)

☐ **Handsome Flycatcher** *Myiophobus pulcher*

    ____ *M. p. pulcher*   W Andes of sw Colombia and nw Ecuador (south to Pichincha)

    ____ *M. p. bellus*   Central and Eastern Andes of Colombia and ne Ecuador

    ____ *M. p. oblitus*   Andes of se Peru (Cuzco and n Puno)

☐ **Orange-banded Flycatcher** *Myiophobus lintoni*

    E slope of Andes of s Ecuador to extreme n Peru (Piura)

☐ **Ochraceous-breasted Flycatcher** *Myiophobus ochraceiventris*

    E slope of Andes of e Peru (Amazonas) to w Bolivia (La Paz)

☐ **Olive-chested Flycatcher** *Myiophobus cryptoxanthus*

    Tropical e Ecuador and ne Peru (San Martín)

☐ **Bran-colored Flycatcher** *Myiophobus fasciatus*

    ____ *M. f. furfurosus*   SW Costa Rica and w Panama; Pearl Islands

    ____ *M. f. fasciatus*   Colombia to n Venezuela, the Guianas and adj. Brazil; Trinidad

    ____ *M. f. crypterythrus*   Tropical sw Colombia to w Ecuador and extreme nw Peru

    ____ *M. f. rufescens*   Arid w Peru (La Libertad) to extreme n Chile (Tarapacá)

    ____ *M. f. saturatus*   E Peru (San Martín to central Cuzco)

    ____ *M. f. auriceps*   SE Peru (Cuzco) to n Bolivia, n Argentina and w Paraguay

    ____ *M. f. flammiceps*   E Brazil (e Pará) to Uruguay, e Paraguay and ne Argentina

☐ **Tawny-breasted Flycatcher** *Myiobius villosus*

    ____ *M. v. villosus*   E Panama (Cerro Tacarcuna) to w Colombia and w Ecuador

    ____ *M. v. schaeferi*   E Andes of n Colombia and extreme w Venezuela (Táchira)

    ____ *M. v. clarus*   Foothills of e Ecuador and e Peru (south to Junín)

    ____ *M. v. peruvianus*   Foothills of se Peru (Puno) and nw Bolivia (La Paz)

☐ **Sulphur-rumped Flycatcher** *Myiobius sulphureipygius*

    ____ *M. s. sulphureipygius*   Tropical se Mexico to Honduras

    ____ *M. s. aureatus*   S Honduras to w Colombia and w Ecuador

☐ **Whiskered Flycatcher** *Myiobius barbatus*

    ____ *M. b. semiflavus*   E-central Colombia (Nechí region of Antioquia)

    ____ *M. b. barbatus*   SE Colombia to n Peru, s Venezuela, the Guianas and n Brazil

    ____ *M. b. amazonicus*   E Peru (south of Río Marañón) east to Rio Madeira (Brazil)

    ____ *M. b. insignis*   NE Brazil (south of the Amazon from Rio Tapajós to Pará)

☐ **Yellow-rumped Flycatcher** *Myiobius mastacalis*

    SE Brazil (s Goiás, Paraíba and Bahia to Santa Catarina)

☐ **Black-tailed Flycatcher** *Myiobius atricaudus*

    ____ *M. a. atricaudus*   Tropical sw Costa Rica to Panama and w Colombia

    ____ *M. a. portovelae*   W Ecuador and extreme nw Peru (Tumbes)

_____ *M. a. modestus*   E Venezuela (n Bolívar)
_____ *M. a. adjacens*   S Colombia (Putumayo) to e Ecuador, e Peru and w Brazil
_____ *M. a. connectens*   NE Brazil (Rio Tapajós to n Maranhão)
_____ *M. a. snethlagei*   NE Brazil (Maranhão, Piauí, Ceará and w Bahia to se Goiás)
_____ *M. a. ridgwayi*   SE Brazil (Espírito Santa and s Minas Gerais to ne Paraná)

☐ **Ruddy-tailed Flycatcher**   *Terenotriccus erythrurus*

_____ *T. e. fulvigularis*   Tropical se Mexico to Colombia, w Ecuador and Venezuela
_____ *T. e. signatus*   E Colombia to ne Peru (north of Río Marañón)
_____ *T. e. venezuelensis*   Extreme e Colombia to s Venezuela and nw Brazil
_____ *T. e. brunneifrons*   E Peru (south of Río Marañón) to n Bolivia and sw Brazil
_____ *T. e. erythrurus*   S Venezuela (Bolívar) to the Guianas and ne Brazil
_____ *T. e. purusianus*   Amazonian Brazil (middle Rio Purús)
_____ *T. e. amazonus*   Amazonian Brazil (Rio Purús to Rio Tapajós)
_____ *T. e. hellmayri*   NE Brazil (along lower Rio Tocantins east to Maranhão)

☐ **Cinnamon Tyrant**   *Neopipo cinnamomea*

_____ *N. c. helenae*   Extreme s Venezuela to the Guianas and n Brazil (Amapá)
_____ *N. c. cinnamomea*   Extreme e Colombia to e Ecuador, e Peru and w Amaz. Brazil

☐ **Cinnamon Flycatcher**   *Pyrrhomyias cinnamomeus*

_____ *P. c. assimilis*   Santa Marta Mountains (ne Colombia)
_____ *P. c. pyrrhopterus*   Andes of Colombia to nw Venezuela, Ecuador and n Peru
_____ *P. c. vieillotioides*   Coastal mountains of nw Venezuela (Lara to Miranda)
_____ *P. c. spadix*   Coastal mountains of ne Venezuela (Anzoátegui to w Sucre)
_____ *P. c. pariae*   NE Venezuela (Cerro Azul and Cerro Humo on Paría Peninsula)
_____ *P. c. cinnamomeus*   Subtropical e Peru (San Martín) to nw Argentina

☐ **Cliff Flycatcher**   *Hirundinea ferruginea*

_____ *H. f. sclateri*   E Andes of Colombia to w Venezuela and e Peru (s to Cuzco)
_____ *H. f. ferruginea*   Extreme e Colombia to se Venezuela, the Guianas and nw Brazil
_____ *H. f. bellicosa*   S and e Brazil to e Paraguay, Uruguay and ne Argentina
_____ *H. f. pallidior*   N and e Bolivia to w Paraguay and nw Argentina

☐ **Euler's Flycatcher**   *Lathrotriccus euleri*

_____ *L. e. flaviventris†*   Grenada (Lesser Antilles). Extinct
_____ *L. e. lawrencei*   Extreme e Colombia to n Venezuela and Suriname; Trinidad
_____ *L. e. bolivianus*   E Ecuador to n Bolivia, s Venezuela and Amazonian Brazil
_____ *L. e. argentinus*   E Bolivia to Paraguay and n Argentina; winters to e Brazil
_____ *L. e. euleri*   SE Brazil to ne Argentina; winters to Peru, Bolivia and Brazil

☐ **Gray-breasted Flycatcher**   *Lathrotriccus griseipectus*

Arid tropical sw Ecuador and nw Peru

☐ **Tawny-chested Flycatcher**   *Aphanotriccus capitalis*

E Nicaragua to n Costa Rica (south to Puerto Limón)

☐ **Black-billed Flycatcher**   *Aphanotriccus audax*

Tropical e Panama and n Colombia

☐ **Belted Flycatcher**   *Xenotriccus callizonus*

Highlands of s Mexico (Chiapas), Guatemala and El Salavdor

☐ **Pileated Flycatcher**   *Xenotriccus mexicanus*

C Mexico (Michoacán to Oaxaca) and adj. Guatemala

☐ **Fuscous Flycatcher**   *Cnemotriccus fuscatus*

_____ *C. f. cabanisi*   N Colombia to ne Venezuela; Trinidad and Tobago
_____ *C. f. duidae*   S Venezuela and nw Brazil
_____ *C. f. fumosus*   The Guianas and ne Brazil
_____ *C. f. fuscatior*   SW Venezuela, se Colombia, e Ecuador, e Peru and c Brazil
_____ *C. f. beniensis*   N Bolivia
_____ *C. f. bimaculatus*   C Bolivia, s and e Brazil, Paraguay and n Argentina
_____ *C. f. fuscatus*   SE Brazil and ne Argentina

☐ **Yellow-bellied Flycatcher** *Empidonax flaviventris*

Breeds e Canada and US; winters Mexico to Panama

☐ **Acadian Flycatcher** *Empidonax virescens*

Breeds e US; winters Nicaragua to Ecuador and Venezuela

☐ **Willow Flycatcher** *Empidonax traillii*
_____ *E. t. brewsteri*
_____ *E. t. adastus*
_____ *E. t. extimus*
_____ *E. t. traillii*

Pacific northwest (s British Columbia) to mts. of cent. California
Great Basin and central Rocky Mts. south to Utah and Colorado
Breeds s Calif. e to New Mexico and (?) formerly w Texas
Great Plains to ne US and se Canada; winters to nw S America

☐ **Alder Flycatcher** *Empidonax alnorum*

Breeds Alaska to ne US; winters Colombia to Bolivia

☐ **White-throated Flycatcher** *Empidonax albigularis*
_____ *E. a. timidus*
_____ *E. a. albigularis*
_____ *E. a. australis*

Highlands of nw Mexico (Chihuahua to s Durango)
Highlands of e Mexico to Guatemala, El Salvador and Honduras
Highlands of Nicaragua to w Panama (Chiriquí)

☐ **Least Flycatcher** *Empidonax minimus*

Breeds Canada and US; winters n Mexico to Panama

☐ **Hammond's Flycatcher** *Empidonax hammondii*

Breeds w North America; winters to Nicaragua

☐ **Gray Flycatcher** *Empidonax wrightii*

Breeds w North America; winters to s Mexico

☐ **Dusky Flycatcher** *Empidonax oberholseri*

Breeds w North America; winters to s Mexico

☐ **Pine Flycatcher** *Empidonax affinis*
_____ *E. a. pulverius*
_____ *E. a. trepidus*
_____ *E. a. affinis*
_____ *E. a. bairdi*
_____ *E. a. vigensis*

Oak-pine forests of nw Mexico (Sinaloa to Jalisco)
N Mexico (Coahuila and Tamaulipas); winters to Guatemala
Pine forests of Mexican plateau (Michoacán to Puebla)
Pine forests of s Mexico (Guerrero, Oaxaca and Chiapas)
Oak-pine forests of e Mexico (Veracruz)

☐ **Pacific-slope Flycatcher** *Empidonax difficilis*
_____ *E. d. difficilis*
_____ *E. d. insulicola*
_____ *E. d. cineritius*

W North America (se Alaska to n Baja); winters to s Mexico
Channel Islands off s California
Cape district of Baja California

☐ **Cordilleran Flycatcher** *Empidonax occidentalis*
_____ *E. o. hellmayri*
_____ *E. o. occidentalis*

Woodlands of sw Canada to n Mexico; winters to s Mexico
Highlands of Mexico

☐ **Yellowish Flycatcher** *Empidonax flavescens*
_____ *E. f. imperturbatus*
_____ *E. f. salvini*
_____ *E. f. flavescens*

S Mexico (Sierra de Tuxtla in se Veracruz)
Highlands of se Mexico (Oaxaca) to Nicaragua
Highlands of Costa Rica and w Panama (east to Veraguas)

☐ **Buff-breasted Flycatcher** *Empidonax fulvifrons*
_____ *E. f. pygmaeus*
_____ *E. f. fulvifrons*
_____ *E. f. rubicundus*
_____ *E. f. brodkorbi*
_____ *E. f. fusciceps*
_____ *E. f. inexpectatus*

SW US and nw Mexico (Sonora to Coahuila)
Mountains of ne Mexico (Tamaulipas and San Luis Potosí)
Mexico (Chihuahua and Durango to Guerrero and Veracruz)
S Mexico (Río Molino area of s Oaxaca)
SE Mexico (Chiapas) to Guatemala and El Salvador
Central and s Honduras

☐ **Black-capped Flycatcher** *Empidonax atriceps*

Mountains of Costa Rica and w Panama

☐ **Olive-sided Flycatcher** *Contopus cooperi*

Breeds temperate North America; winters south to Bolivia

334

☐ **Greater Pewee** *Contopus pertinax*
  ____ *C. p. pertinax (pallidiventris)* — SW US to s Mexico; winters to Belize and Guatemala
  ____ *C. p. minor* — Highlands of Belize, Honduras, El Salvador and n Nicaragua

☐ **Dark Pewee** *Contopus lugubris*

Montane forests of Costa Rica and extreme w Panama (Chiriquí)

☐ **Smoke-colored Pewee** *Contopus fumigatus*
  ____ *C. f. ardosiacus* — Colombia to ne Venezuela, e Ecuador and ne Peru
  ____ *C. f. cineraceus* — Subtropical n Venezuela (Yaracuy to Miranda)
  ____ *C. f. duidae* — *Tepuis* of s Venezuela (s Bolívar and Amazonas), adj. Guyana
  ____ *C. f. zarumae* — SW Ecuador (Nariño) to n Peru (sw Cajamarca)
  ____ *C. f. fumigatus* — SE Peru (Puno) and w Bolivia (La Paz and Cochabamba)
  ____ *C. f. brachyrhynchus* — SE Bolivia (Santa Cruz and Tarija) to nw Argentina (Tucumán)

☐ **Ochraceous Pewee** *Contopus ochraceus*

Montane forests of Costa Rica and extreme w Panama (Chiriquí)

☐ **Western Wood-Pewee** *Contopus sordidulus*
  ____ *C. s. saturatus* — SE Alaska to w Oregon; winters to n South America
  ____ *C. s. veliei (siccicola, amplus)* — E Alaska to Texas and n Mexico; winters to South America
  ____ *C. s. peninsulae* — S Baja California; winters to nw South America
  ____ *C. s. sordidulus (griscomi)* — Highlands of Mexico to Honduras; winters to Peru

☐ **Eastern Wood-Pewee** *Contopus virens*

Breeds e North America; winters to n Bolivia and w Brazil

☐ **Tropical Pewee** *Contopus cinereus*
  ____ *C. c. brachytarsus* — Tropical se Mexico (Oaxaca and Veracruz) to Panama
  ____ *C. c. rhizophorus* — Arid Pacific littoral of w Costa Rica (Guanacaste)
  ____ *C. c. aithalodes* — Isla Coiba (Panama)
  ____ *C. c. bogotensis* — N Colombia to n Venezuela and nw Brazil; Trinidad
  ____ *C. c. surinamensis* — S Venezuela to the Guianas and ne Brazil
  ____ *C. c. punensis* — Andes of Ecuador and n Peru (s to Junín)
  ____ *C. c. pallescens* — Extreme se Peru to e Brazil, Bolivia and nw Argentina
  ____ *C. c. cinereus* — SE Brazil (Bahia to Paraná) to e Paraguay and ne Argentina

☐ **Cuban Pewee** *Contopus caribaeus*
  ____ *C. c. bahamensis* — Grand Bahama, Abaco, Andros, New Providence and Eleuthera
  ____ *C. c. caribaeus* — Cuba and Isle of Pines
  ____ *C. c. morenoi* — S Cuba (Zapata Swamp) and adjacent offshore cays
  ____ *C. c. nerlyi* — Islands off s Camagüey (Cuba)

☐ **Jamaican Pewee** *Contopus pallidus*

Jamaica

☐ **Hispaniolan Pewee** *Contopus hispaniolensis*
  ____ *C. h. hispaniolensis* — Hispaniola
  ____ *C. h. tacitus* — Gonâve I. (Haiti)

☐ **Lesser Antillean Pewee** *Contopus latirostris*
  ____ *C. l. blancoi* — Montane forests of Puerto Rico
  ____ *C. l. brunneicapillus* — Montane forests of Dominica, Guadeloupe and Martinique
  ____ *C. l. latirostris* — Montane forests of St. Lucia

☐ **White-throated Pewee** *Contopus albogularis*

Suriname to French Guiana and extreme ne Brazil (Amapá)

☐ **Blackish Pewee** *Contopus nigrescens*
  ____ *C. n. nigrescens* — E slope of Andes of Ecuador (Napo-Pastaza and Santiago-Zamora)
  ____ *C. n. canescens* — E Peru (s to Cuzco); s Guyana (Acari Mts.) and e Amaz. Brazil

☐ **Tufted Flycatcher** *Mitrephanes phaeocercus*
  ____ *M. p. tenuirostris* — Mts. of w Mexico (Sonora and Chihuahua to Jalisco)

____ *M. p. phaeocercus (burleighi, nicaraguae)*	Mountains of e Mexico to El Salvador and ne Nicaragua
____ *M. p. aurantiiventris (vividus)*	Highlands of Costa Rica and w Panama
____ *M. p. berlepschi (eminulus)*	Extreme e Panama to nw Colombia and nw Ecuador

☐ **Olive-tufted Flycatcher** *Mitrephanes olivaceus*

E slope of Andes of ne Peru (Piura) to nw Bolivia

☐ **Black Phoebe** *Sayornis nigricans*

____ *S. n. semiater*	W US (Oregon) to Baja California and w Mexico (Nayarit)
____ *S. n. nigricans*	Highlands of ne Mexico (Tamaulipas) to n Chiapas
____ *S. n. aquaticus*	S Mexico (highlands of s Chiapas) to Guatemala and Nicaragua
____ *S. n. amnicola*	Highlands of Costa Rica and w Panama (Chiriquí)
____ *S. n. angustirostris*	E Panama to Colombia, Ecuador, s Peru and n Venezuela
____ *S. n. latirostris*	Andes of Bolivia and nw Argentina

☐ **Eastern Phoebe** *Sayornis phoebe*

Breeds e Canada and US; winters to se Mexico

☐ **Say's Phoebe** *Sayornis saya*

____ *S. s. saya (pallidus)*	Arid scrub of North America (Alaska to s Mexico)
____ *S. s. quiescens*	Northern half of Baja California and Isla de Cedros

☐ **Vermilion Flycatcher** *Pyrocephalus rubinus*

____ *P. r. flammeus*	Arid sw US to Baja California and nw Mexico (Nayarit)
____ *P. r. mexicanus*	Arid sw Texas to Guerrero, Oaxaca, Puebla and Veracruz
____ *P. r. blatteus*	SE Mexico (s Veracruz) to Guatemala and Honduras
____ *P. r. pinicola*	Lowland pine savanna of ne Nicaragua
____ *P. r. nanus*	Galapagos Islands (except Chatham I.)
____ *P. r. dubius*	Chatham I. (Galapagos Islands)
____ *P. r. saturatus*	NE Colombia to n Venezuela, Guyana and n Brazil
____ *P. r. piurae*	Colombia (west of Eastern Andes) to w Ecuador and nw Peru
____ *P. r. ardens*	N Peru (Cajamarca, Amazonas and extreme e Piura)
____ *P. r. obscurus*	W Peru (Lima)
____ *P. r. cocachacrae*	SW Peru (Ica to Tacna) and adjacent n Chile
____ *P. r. major* (?)	SE Peru (Cuzco and Puno)
____ *P. r. rubinus*	Extreme se Brazil to se Bolivia, Paraguay, Uruguay, ne Argentina

☐ **Austral Negrito** *Lessonia rufa*

C Chile and Argentina to Tierra del Fuego; winters to se Brazil

☐ **Andean Negrito** *Lessonia oreas*

Andes of Peru (Huánuco) to n Chile and nw Argentina

☐ **Spectacled Tyrant** *Hymenops perspicillatus*

____ *H. p. perspicillatus*	SW Brazil to Uruguay, Paraguay and n Argentina
____ *H. p. andinus*	C Chile and c Argentina; winters to n Bolivia and sw Brazil

☐ **Cinereous Tyrant** *Knipolegus striaticeps*

E Bolivia to n Argentina, w Paraguay and extreme sw Brazil

☐ **Hudson's Black-Tyrant** *Knipolegus hudsoni*

Central Argentina; winters to Bolivia, sw Brazil and se Peru

☐ **Amazonian Black-Tyrant** *Knipolegus poecilocercus*

S Venezuela and w Guyana to e Ecuador and Amazonian Brazil

☐ **Andean Tyrant** *Knipolegus signatus*

____ *K. s. signatus*	E slope of Andes of Peru (Cajamarca to Junín)
____ *K. s. cabanisi*	Andes of se Peru (Cuzco) to w Bolivia and n Argentina

☐ **Blue-billed Black-Tyrant** *Knipolegus cyanirostris*

SE Brazil to Uruguay, e Paraguay and ne Argentina

☐ **Rufous-tailed Tyrant** *Knipolegus poecilurus*

____ *K. p. poecilurus*	Andes of Colombia and w Venezuela
____ *K. p. venezuelanus*	Coastal cordillera of n Venezuela (Distrito Federal)
____ *K. p. paraquensis*	Subtropical s Venezuela (Cerro Paraque area of Amazonas)

_____ *K. p. salvini*   *Tepuis* of s Venezuela, Guyana and extreme n Brazil

_____ *K. p. peruanus*   Andes of se Ecuador, e Peru and n Bolivia

☐ **Riverside Tyrant** *Knipolegus orenocensis*

_____ *K. o. orenocensis*   Extreme se Colombia and Orinoco system of Venezuela

_____ *K. o. xinguensis*   E Brazil (Rio Xingú, Rio Tapajós and Rio Araguaia)

_____ *K. o. sclateri*   Tropical ne Peru (Loreto) and central Amazonian Brazil

☐ **White-winged Black-Tyrant** *Knipolegus aterrimus*

_____ *K. a. heterogyna*   N Peru (Marañón Valley in Cajamarca, La Libertad and Ancash)

_____ *K. a. anthracinus*   Andes of Peru (Ayacucho, Cuzco, Puno) and w Bolivia (La Paz)

_____ *K. a. aterrimus*   Andes of e Bolivia, w Argentina and *chaco* of Paraguay

☐ **Caatinga Black-Tyrant** *Knipolegus franciscanus*

E-central Brazil (Minas Gerais and Rio São Francisco in Bahia)

☐ **Crested Black-Tyrant** *Knipolegus lophotes*

S-central and se Brazil to Uruguay and ne Paraguay

☐ **Velvety Black-Tyrant** *Knipolegus nigerrimus*

E Brazil (ne Bahia and Alagoas to ne Rio Grande do Sul)

☐ **Tumbes Tyrant** *Tumbezia salvini*

Arid nw Peru (Tumbes to La Libertad)

☐ **Crowned Chat-Tyrant** *Ochthoeca frontalis*

_____ *O. f. albidiadema*   Eastern Andes of Colombia (north to Norte de Santander)

_____ *O. f. frontalis (orientalis)*   Central Andes of Colombia to Andes of Ecuador and n Peru

☐ **Peruvian Chat-Tyrant** *Ochthoeca spodionota*

_____ *O. s. spodionota*   Andes of Peru (Junín, Ayacucho and Cordillera Vilcabamba)

_____ *O. s. boliviana*   Andes of Peru (Huánuco and Cuzco) to w Bolivia

☐ **Golden-browed Chat-Tyrant** *Ochthoeca pulchella*

_____ *O. p. similis*   E slope of Andes of Peru (s Amazonas to Junín)

_____ *O. p. pulchella*   Andes of se Peru (Cuzco and Ayacucho) to w Bolivia

☐ **Yellow-bellied Chat-Tyrant** *Ochthoeca diadema*

_____ *O. d. jesupi*   Santa Marta Mountains (ne Colombia)

_____ *O. d. rubellula*   NE Colombia (Sierra de Perijá) and nw Venezuela (Zulia)

_____ *O. d. tovarensis*   Coastal cordillera of n Venezuela (Aragua and Distrito Federal)

_____ *O. d. diadema (meridana)*   E Andes of Colombia and w Venezuela

_____ *O. d. gratiosa (cajamarcae)*   Andes of Colombia, Ecuador and n Peru

☐ **Jelski's Chat-Tyrant** *Ochthoeca jelskii*

Andes of sw Ecuador and nw Peru (s to Lima and Huánuco)

☐ **Slaty-backed Chat-Tyrant** *Ochthoeca cinnamomeiventris*

_____ *O. c. cinnamomeiventris*   Andes of Colombia to n Ecuador and sw Venezuela

_____ *O. c. nigrita*   Andes of w Venezuela (Mérida, w Barinas and Táchira)

_____ *O. c. angustifasciata*   Andes of n Peru (s Amazonas, San Martín and Cajamarca)

☐ **Maroon-chested Chat-Tyrant** *Ochthoeca thoracica*

Andes of se Peru (Pasco) to w Bolivia (Cochabamba)

☐ **Rufous-breasted Chat-Tyrant** *Ochthoeca rufipectoralis*

_____ *O. r. poliogastra*   Santa Marta Mountains (ne Colombia)

_____ *O. r. rubicundula*   Sierra de Perijá (Colombia/Venezuela border)

_____ *O. r. obfuscata*   Central and W Andes of Colombia to Peru (nw San Martín)

_____ *O. r. rufopectus*   Eastern Andes of Colombia (south to Bogotá)

_____ *O. r. centralis*   Andes of Peru (s La Libertad, Ancash and Huánuco)

_____ *O. r. tectricialis*   W slope of Eastern Andes of Peru (Pasco to Cuzco)

_____ *O. r. rufipectoralis*   Andes of se Peru (Cuzco and Puno) to w Bolivia

☐ **Brown-backed Chat-Tyrant** *Ochthoeca fumicolor*

_____ *O. f. fumicolor*   E Andes of Colombia and w Venezuela (w Táchira)

____	*O. f. ferruginea*	Central and Western Andes of Colombia (Antioquia)
____	*O. f. superciliosa*	Andes of w Venezuela (Trujillo, Mérida and e Táchira)
____	*O. f. brunneifrons*	Central and Western Andes of Colombia to central Peru
____	*O. f. berlepschi*	Andes of se Peru (Cuzco and Puno) to w Bolivia

☐ **D'Orbigny's Chat-Tyrant** *Ochthoeca oenanthoides*

____	*O. o. polionota*	Andes of Peru (north to La Libertad and w Huánuco)
____	*O. o. oenanthoides*	Andes of Bolivia to extreme n Chile and nw Argentina

☐ **White-browed Chat-Tyrant** *Ochthoeca leucophrys*

____	*O. l. dissors*	N Peru (upper Marañón Valley)
____	*O. l. interior*	Andes of central Peru (Huánuco and Pasco)
____	*O. l. urubambae*	Andes of s Peru (Junín to ne Ayacucho and Cuzco)
____	*O. l. leucometopa*	W slope of Andes of Peru (Ancash) to extreme nw Chile
____	*O. l. leucophrys*	Andes of w Bolivia
____	*O. l. tucumana*	Andes of nw Argentina (Salta to San Juan)

☐ **Piura Chat-Tyrant** *Ochthoeca piurae*

Locally in Andes of nw Peru (Piura to Ancash)

☐ **Patagonian Tyrant** *Colorhamphus parvirostris*

S Chile (Valdivia) and adj. Argentina south to Tierra del Fuego

☐ **Drab Water-Tyrant** *Ochthornis littoralis*

S Guyana to s Venezuela, n Bolivia and Amazonian Brazil

☐ **Yellow-browed Tyrant** *Satrapa icterophrys*

Extreme se Peru to n Argentina, e and s Brazil; Venezuela

☐ **Pied Water-Tyrant** *Fluvicola pica*

E Panama to Venezuela, the Guianas, extreme n Brazil; Trinidad

☐ **Black-backed Water-Tyrant** *Fluvicola albiventer*

Amazonian and e Brazil to e Bolivia, Paraguay and n Argentina

☐ **Masked Water-Tyrant** *Fluvicola nengeta*

____	*F. n. atripennis*	SW Ecuador to nw Peru (Tumbes)
____	*F. n. nengeta*	E Brazil (Maranhão to Minas Gerais and ne São Paulo)

☐ **White-headed Marsh-Tyrant** *Arundinicola leucocephala*

Colombia to the Guianas, s Brazil and n Argentina; Trinidad

☐ **Cock-tailed Tyrant** *Alectrurus tricolor*

Locally in e Bolivia, ne Argentina, ne Paraguay and s Brazil

☐ **Strange-tailed Tyrant** *Alectrurus risora*

Locally in e Paraguay, s Brazil, Uruguay and n Argentina

☐ **Streak-throated Bush-Tyrant** *Myiotheretes striaticollis*

____	*M. s. striaticollis*	Andes of Colombia to w Venezuela, Ecuador and central Peru
____	*M. s. pallidus*	Andes of e Peru (Cuzco) to n Bolivia and nw Argentina

☐ **Santa Marta Bush-Tyrant** *Myiotheretes pernix*

Santa Marta Mountains (ne Colombia)

☐ **Smoky Bush-Tyrant** *Myiotheretes fumigatus*

____	*M. f. olivaceus*	Sierra de Perijá (Colombia/Venezuela border)
____	*M. f. fumigatus*	Andes of Colombia and n Ecuador
____	*M. f. lugubris*	Andes of w Venezuela (Táchira, Trujillo and Mérida)
____	*M. f. cajamarcae*	Andes of s Ecuador (Cañar) and Peru (south to Cuzco)

☐ **Rufous-bellied Bush-Tyrant** *Myiotheretes fuscorufus*

E slope of Andes of s Peru (Pasco) to w Bolivia

☐ **Red-rumped Bush-Tyrant** *Cnemarchus erythropygius*

____	*C. e. orinomus*	Santa Marta Mountains and Eastern Andes of n Colombia
____	*C. e. erythropygius*	Andes of s Colombia to Ecuador, e Peru and w Bolivia

☐ **Chocolate-vented Tyrant** *Neoxolmis rufiventris*

Extreme s Argentina and Chile; winters n to extreme se Brazil

☐ **Rufous-webbed Tyrant** *Polioxolmis rufipennis*
____ *P. r. rufipennis* — Andes of Peru to extreme nw Argentina and (?) adjacent n Chile
____ *P. r. bolivianus* — Andes of central Bolivia

☐ **Fire-eyed Diucon** *Xolmis pyrope*
____ *X. p. pyrope* — Andes of central Chile and adj. Argentina to Tierra del Fuego
____ *X. p. fortis* — Chiloé I. (Chile)

☐ **Gray Monjita** *Xolmis cinereus*
____ *X. c. cinereus* — Suriname; Amazonian and e Brazil, Uruguay and ne Argentina
____ *X. c. pepoaza* — SE Peru (Madre de Dios) to e Bolivia, Paraguay and n Argentina

☐ **Black-crowned Monjita** *Xolmis coronatus* — S Argentina; winters to s Bolivia, Paraguay and extreme s Brazil

☐ **White-rumped Monjita** *Xolmis velatus* — Savanna and campos of e Brazil, ne Paraguay and e Bolivia

☐ **White Monjita** *Xolmis irupero*
____ *X. i. niveus* — E Brazil (Ceará and Pernambuco to Bahia and Minas Gerais)
____ *X. i. irupero* — SE Brazil to Paraguay, Uruguay, Bolivia and n Argentina

☐ **Salinas Monjita** *Xolmis salinarum* — Arid *salinas* of nw Argentina

☐ **Rusty-backed Monjita** *Xolmis rubetra* — W Argentina (Mendoza to Santa Cruz); winters to w Uruguay

☐ **Black-and-white Monjita** *Xolmis dominicanus* — SE Brazil (Paraná) to Paraguay, Uruguay and ne Argentina

☐ **Black-billed Shrike-Tyrant** *Agriornis montanus*
____ *A. m. solitarius* — Andes of Colombia and Ecuador
____ *A. m. insolens* — Andes of Peru
____ *A. m. intermedius* — Andes of w Bolivia (La Paz and Oruro) to n Chile (Tarapacá)
____ *A. m. montanus* — Andes of e and s Bolivia and nw Argentina (south to La Rioja)
____ *A. m. maritimus (leucurus)* — Andes of Chile and s Argentina

☐ **White-tailed Shrike-Tyrant** *Agriornis andicola*
____ *A. a. andicola* — Andes of Ecuador (north to Imbabura)
____ *A. a. albicauda* — Andes of Peru to w Bolivia, n Chile and nw Argentina

☐ **Great Shrike-Tyrant** *Agriornis lividus*
____ *A. l. lividus* — Coast and mountains of Chile (Atacama to Valdivia)
____ *A. l. fortis* — S Chile (Aysén) and s Argentina to Tierra del Fuego

☐ **Gray-bellied Shrike-Tyrant** *Agriornis micropterus*
____ *A. m. andecola* — Andes of s Peru to Bolivia, nw Argentina and n Chile
____ *A. m. micropterus* — Andes of s Argentina; winters to w Paraguay and s Uruguay

☐ **Lesser Shrike-Tyrant** *Agriornis murinus* — S Argentina; winters north to w Paraguay and s Bolivia

☐ **Little Ground-Tyrant** *Muscisaxicola fluviatilis* — SE Colombia to e Ecuador, e Peru, n Bolivia and sw Amaz. Brazil

☐ **Spot-billed Ground-Tyrant** *Muscisaxicola maculirostris*
____ *M. m. niceforoi* — Eastern Andes of Colombia (Boyacá and Cundinamarca)
____ *M. m. rufescens* — Andes of Ecuador (Pichincha to Azuay)
____ *M. m. maculirostris* — Andes of Peru to w Bolivia, w Argentina and Chile

☐ **Taczanowski's Ground-Tyrant** *Muscisaxicola griseus* — *Páramo* of Peru and w Bolivia (La Paz to Cochabamba)

☐ **Puna Ground-Tyrant** *Muscisaxicola juninensis* — Andes of Peru (Junín and Lima) to n Chile and nw Argentina

☐ **Cinereous Ground-Tyrant** *Muscisaxicola cinereus*
____ *M. c. cinereus* — Andes of s Peru (Puno) to w Bolivia, n Chile and n Argentina
____ *M. c. argentinus* — Andes of nw Argentina (Jujuy to Catamarca and Tucumán)

☐ **White-fronted Ground-Tyrant** *Muscisaxicola albifrons*

Andes of s Peru (Ancash) to w Bolivia and extreme n Chile

☐ **Ochre-naped Ground-Tyrant** *Muscisaxicola flavinucha*
_____ *M. f. flavinucha*
_____ *M. f. brevirostris*

Andes of Chile and Argentina; winters to n Peru (La Libertad)
S Chile (Aysén) south to Tierra del Fuego

☐ **Rufous-naped Ground-Tyrant** *Muscisaxicola rufivertex*
_____ *M. r. occipitalis*
_____ *M. r. pallidiceps*
_____ *M. r. rufivertex*

Andes of Peru and n Bolivia (La Paz and Cochabamba)
Andes of sw Peru to nw Argentina and n Chile
Andes of s Chile and s Argentina; winters north to Antofagasta

☐ **Dark-faced Ground-Tyrant** *Muscisaxicola maclovianus*
_____ *M. m. mentalis*
_____ *M. m. maclovianus*

S Argentina and Chile; winters on coast to n Peru and Uruguay
Falkland Islands

☐ **White-browed Ground-Tyrant** *Muscisaxicola albilora*

Andes of c and s Chile and adj. Argentina; winters to s Ecuador

☐ **Plain-capped Ground-Tyrant** *Muscisaxicola alpinus*
_____ *M. a. columbianus*
_____ *M. a. quesadea*
_____ *M. a. alpinus*
_____ *M. a. griseus*

Central Andes of Colombia (Nevado del Ruiz to Cauca)
E Andes of Colombia (Boyacá and Cundinamarca)
Páramo of Ecuador
Páramo of Peru and w Bolivia (La Paz to Cochabamba)

☐ **Cinnamon-bellied Ground-Tyrant** *Muscisaxicola capistratus*

Breeds s Chile and s Argentina; winters north in Andes to s Peru

☐ **Black-fronted Ground-Tyrant** *Muscisaxicola frontalis*

Andes of Chile and Argentina; winters n to s Peru (Arequipa)

☐ **Streamer-tailed Tyrant** *Gubernetes yetapa*

E Bolivia to Paraguay, s-central and se Brazil and ne Argentina

☐ **Shear-tailed Gray Tyrant** *Muscipipra vetula*

SE Brazil (Minas Gerais) to e Paraguay and ne Argentina

☐ **Long-tailed Tyrant** *Colonia colonus*
_____ *C. c. leuconota*
_____ *C. c. fuscicapilla*
_____ *C. c. poecilonota*
_____ *C. c. niveiceps*
_____ *C. c. colonus*

SE Honduras and e Nicaragua to w Colombia and w Ecuador
E Andes of Colombia, n Ecuador and extreme ne Peru
SE Venezuela (e Bolívar) and the Guianas
SE Ecuador, Peru (San Martín to n Puno) and n Bolivia
C and e Brazil (s Maranhão) to e Paraguay and ne Argentina

☐ **Short-tailed Field-Tyrant** *Muscigralla brevicauda*

Arid sw Ecuador to extreme n Chile; Gorgona I. (Colombia)

☐ **Cattle Tyrant** *Machetornis rixosa*
_____ *M. r. flavigularis*
_____ *M. r. obscurodorsalis*
_____ *M. r. rixosa*

Caribbean coast of n Colombia and n Venezuela
Llanos of e Colombia and sw Venezuela
Interior e Brazil to e Bolivia, Paraguay, Uruguay and n Argentina

☐ **Piratic Flycatcher** *Legatus leucophaius*
_____ *L. l. variegatus*
_____ *L. l. leucophaius*

Tropical se Mexico (San Luis Potosí) to Honduras
Nicaragua to n Argentina and s Brazil; Trinidad and Tobago

☐ **Rusty-margined Flycatcher** *Myiozetetes cayanensis*
_____ *M. c. hellmayri*
_____ *M. c. rufipennis*
_____ *M. c. cayanensis*
_____ *M. c. erythropterus*

E Panama to Colombia, e Ecuador and extreme nw Venezuela
E Colombia to n Venezuela and e Ecuador
S Venezuela, the Guianas and Amazonian Brazil to n Bolivia
SE Brazil (e Minas Gerais to Rio de Janeiro)

☐ **Social Flycatcher** *Myiozetetes similis*
_____ *M. s. primulus*
_____ *M. s. hesperis*
_____ *M. s. texensis*
_____ *M. s. columbianus*

W Mexico (s Sonora to n Sinaloa)
W Mexico (s Sinaloa to s Zacatecas, sw Puebla and Oaxaca)
E Mexico (s Tamaulipas) to n Costa Rica
Tropical sw Costa Rica to n Colombia and n Venezuela

_____ _M. s. similis_   E Colombia to n Bolivia, Venezuela and n Amazonian Brazil
_____ _M. s. grandis_   W Ecuador (Esmeraldas) to extreme nw Peru (Tumbes)
_____ _M. s. pallidiventris_   E Brazil (Pará) to e Paraguay and ne Argentina

☐ **Gray-capped Flycatcher** _Myiozetetes granadensis_
_____ _M. g. granadensis_   Caribbean slope of e Honduras to central highlands of Panama
_____ _M. g. occidentalis_   E Panama (Darién) to w Ecuador and extreme nw Peru (Tumbes)
_____ _M. g. obscurior_   E Colombia to s Venezuela, n Bolivia and w Amazonian Brazil

☐ **Dusky-chested Flycatcher** _Myiozetetes luteiventris_
_____ _M. l. luteiventris_   SE Colombia to se Venezuela, n Bolivia and w Amaz. Brazil
_____ _M. l. septentrionalis_   Suriname (Marowijne area) and adjacent Brazil (Amapá)

☐ **White-bearded Flycatcher** _Phelpsia inornata_
                                    Llanos of Venezuela

☐ **Great Kiskadee** _Pitangus sulphuratus_
_____ _P. s. texanus_   S Texas (Rio Grande Valley) to se Mexico (Veracruz)
_____ _P. s. derbianus_   Arid w Mexico (s Sonora to Isthmus of Tehuántepec)
_____ _P. s. guatimalensis_   SE Mexico (Nuevo León) to central Panama
_____ _P. s. trinitatis_   Extreme e Colombia to e Venezuela and nw Brazil; Trinidad
_____ _P. s. caucensis_   W and s Colombia (sw Bolívar, Cauca and Magdalena valleys)
_____ _P. s. rufipennis_   Coastal n Colombia and n Venezuela
_____ _P. s. sulphuratus_   Tropical se Colombia to se Peru, the Guianas and n Brazil
_____ _P. s. maximiliani_   Amazonian Brazil to e Bolivia and _chaco_ of Paraguay
_____ _P. s. bolivianus_   Highlands of e Bolivia (Cochabamba to Tarija)
_____ _P. s. argentinus_   Extreme se Brazil to e Paraguay, Uruguay and central Argentina

☐ **Lesser Kiskadee** _Philohydor lictor_
_____ _P. l. panamensis_   E Panama to n Colombia
_____ _P. l. lictor_   E Colombia to the Guianas, e Bolivia, Amazonian and e Brazil

☐ **White-ringed Flycatcher** _Conopias albovittatus_
_____ _C. a. albovittatus_   Lowlands of c Honduras to w Colombia and nw Ecuador
_____ _C. a. distinctus_   Lower Caribbean slopes of Costa Rica

☐ **Yellow-throated Flycatcher** _Conopias parvus_
                                    E Colombia to s Venezuela, the Guianas, n Brazil, n Peru

☐ **Three-striped Flycatcher** _Conopias trivirgatus_
_____ _C. t. berlepschi_   S Venezuela (Bolívar) to ne Peru and lower Amazonian Brazil
_____ _C. t. trivirgatus_   SE Brazil (se Bahia to Paraná) to e Paraguay and ne Argentina

☐ **Lemon-browed Flycatcher** _Conopias cinchoneti_
_____ _C. c. icterophrys_   Andes of Colombia to nw Venezuela (Sierra de Perijá)
_____ _C. c. cinchoneti_   Andes of e Ecuador and e Peru (south to Cuzco)

☐ **Golden-bellied Flycatcher** _Myiodynastes hemichrysus_
                                    Montane forests of Costa Rica and w Panama (e to Veraguas)

☐ **Golden-crowned Flycatcher** _Myiodynastes chrysocephalus_
_____ _M. c. minor_   Extreme e Panama (Darién) and Colombia south to Ecuador
_____ _M. c. cinerascens_   Andes of n Colombia and coastal cordillera of n Venezuela
_____ _M. c. chrysocephalus_   E slope of Andes of Peru to Bolivia and nw Argentina

☐ **Baird's Flycatcher** _Myiodynastes bairdii_
                                    Arid sw Ecuador to nw Peru (Lima)

☐ **Sulphur-bellied Flycatcher** _Myiodynastes luteiventris_
                                    SE Arizona to Costa Rica; winters e Ecuador to n Bolivia

☐ **Streaked Flycatcher** _Myiodynastes maculatus_
_____ _M. m. insolens_   Gulf slope of Mexico to Honduras; winters to n South America
_____ _M. m. difficilis_   Costa Rica to Colombia and Venezuela; Coiba I. and Cébaco I.
_____ _M. m. nobilis_   Caribbean coast of ne Colombia to w slope of Sierra de Perijá

____	*M. m. chapmani*	Pacific Colombia (Chocó) to nw Peru (Piura)
____	*M. m. maculatus*	Venezuela and the Guianas to ne Peru and n Amazonian Brazil
____	*M. m. tobagensis*	N Venezuela and Guyana; Trinidad and Tobago
____	*M. m. solitarius*	S Peru to Paraguay, Uruguay, central Argentina and Brazil

☐ **Boat-billed Flycatcher** *Megarynchus pitangua*

____	*M. p. tardiusculus*	Foothills of nw Mexico (s Sinaloa to Nayarit)
____	*M. p. caniceps*	SW Mexico (s Jalisco)
____	*M. p. mexicanus*	E Mexico (Tamaulipas) to nw Colombia; Cébaco I. (Panama)
____	*M. p. deserticola*	Central Guatemala (valley of Río Negro)
____	*M. p. pitangua*	Tropical n and central South America to n Argentina; Trinidad
____	*M. p. chrysogaster*	Pacific slope of w Ecuador and nw Peru (Tumbes and n Piura)

☐ **Sulphury Flycatcher** *Tyrannopsis sulphurea*

S Venezuela and Guianas to n Bolivia and Amaz. Brazil; Trinidad

☐ **Variegated Flycatcher** *Empidonomus varius*

____	*E. v. rufinus*	E Venezuela to the Guianas, n and w Amazonian Brazil
____	*E. v. varius*	SE Brazil to Paraguay, Uruguay, n Argentina, e Peru, e Bolivia

☐ **Crowned Slaty Flycatcher** *Griseotyrannus aurantioatrocristatus*

____	*G. a. pallidiventris*	E Brazil (Rio Tapajós to n Goiás and Piauí)
____	*G. a. aurantioatrocristatus*	Bolivia to n Argentina and s Brazil; winters n to w Amazonia

☐ **Snowy-throated Kingbird** *Tyrannus niveigularis*

Extreme sw Colombia to w Ecuador and nw Peru (s to Ancash)

☐ **White-throated Kingbird** *Tyrannus albogularis*

S Venezuela and the Guianas to n Bolivia and Amazonian Brazil

☐ **Tropical Kingbird** *Tyrannus melancholicus*

____	*T. m. satrapa*	S Arizona to n Colombia and n Venezuela; Trinidad and Tobago
____	*T. m. despotes*	NE Brazil (Amapá, Maranhão and Ceará to Bahia)
____	*T. m. melancholicus*	Tropical n South America to central Argentina and Brazil

☐ **Couch's Kingbird** *Tyrannus couchii*

Extreme s Texas to n Guatemala and Belize

☐ **Cassin's Kingbird** *Tyrannus vociferans*

____	*T. v. vociferans*	Arid w US to central Mexico; winters to Honduras
____	*T. v. xenopterus*	Highlands of sw Mexico (Guerrero)

☐ **Thick-billed Kingbird** *Tyrannus crassirostris*

____	*T. c. pompalis*	Extreme se Arizona south along w coast of Mexico to Colima
____	*T. c. crassirostris*	SW Mexico (Guerrero to Oaxaca); winters to Guatemala

☐ **Western Kingbird** *Tyrannus verticalis*

Breeds w N America; > s Mexico to Costa Rica and (?) Panama

☐ **Eastern Kingbird** *Tyrannus tyrannus*

E North America; winters mainly w Amazon basin

☐ **Gray Kingbird** *Tyrannus dominicensis*

____	*T. d. dominicensis*	Coastal se US to Colombia and Venezuela; Trinidad; Neth. Ant.
____	*T. d. vorax*	Lesser Antilles; winters to Trinidad and the Guianas

☐ **Loggerhead Kingbird** *Tyrannus caudifasciatus*

____	*T. c. bahamensis*	Grand Bahama, Abaco, Andros and New Providence islands
____	*T. c. caudifasciatus*	Cuba
____	*T. c. flavescens*	Isle of Pines
____	*T. c. caymanensis*	Cayman Islands
____	*T. c. jamaicensis*	Jamaica
____	*T. c. taylori*	Puerto Rico and Vieques I.
____	*T. c. gabbii*	Hispaniola

☐ **Giant Kingbird** *Tyrannus cubensis*

Cuba and Isle of Pines; formerly Great Inagua and Caicos islands

☐ **Scissor-tailed Flycatcher** *Tyrannus forficatus*

Breeds s-central US; winters to w Panama

☐ **Fork-tailed Flycatcher** *Tyrannus savana*
____ *T. s. monachus*
____ *T. s. sanctaemartae*
____ *T. s. circumdatus*
____ *T. s. savana*

S Mexico (Veracruz) to Colombia, the Guianas and n Brazil
Caribbean coast of n Colombia and nw Venezuela (Guajira Pen.)
Lower Amazonian Brazil (west to Manaus area)
Central and s S America and Falkland Is.; winters to West Indies

☐ **Grayish Mourner** *Rhytipterna simplex*
____ *R. s. frederici*
____ *R. s. simplex*

Colombia (e of Andes) to the Guianas, n Bolivia and Amaz. Brazil
SE Brazil (Alagoas to São Paulo and Rio de Janeiro)

☐ **Pale-bellied Mourner** *Rhytipterna immunda*

E Colombia to Suriname, French Guiana and Amazonian Brazil

☐ **Rufous Mourner** *Rhytipterna holerythra*
____ *R. h. holerythra*
____ *R. h. rosenbergi*

Gulf lowlands of se Mexico (Veracruz) to n Colombia
W Colombia (s Chocó and Nariño) to nw Ecuador

☐ **Sirystes** *Sirystes sibilator*
____ *S. s. albogriseus*
____ *S. s. albocinereus*
____ *S. s. subcanescens*
____ *S. s. sibilator*
____ *S. s. atimastus*

E Panama (Veraguas) to nw Colombia and nw Ecuador
SE Colombia to e Ecuador, e Peru, n Bolivia and w Brazil
S Suriname and ne Brazil
E Brazil (Goiás and Bahia) to e Paraguay and ne Argentina
SW Brazil (Mato Grosso)

☐ **Rufous Casiornis** *Casiornis rufus*

SE Peru to e Bolivia, Paraguay, n Argentina, Amaz. and e Brazil

☐ **Ash-throated Casiornis** *Casiornis fuscus*

Caatinga and scrub of ne Brazil

☐ **Rufous Flycatcher** *Myiarchus semirufus*

Arid coastal nw Peru (Tumbes to n Lima)

☐ **Yucatan Flycatcher** *Myiarchus yucatanensis*
____ *M. y. yucatanensis*
____ *M. y. lanyoni*
____ *M. y. navai*

E Mexico (extreme e Tabasco and n and cent. Yucatán Pen.)
Cozumel I. (off Yucatán Peninsula of e Mexico)
S Mexico (s Quintana Roo) to n Belize and Guatemala

☐ **Sad Flycatcher** *Myiarchus barbirostris*

Open woodlands and montane forests of Jamaica

☐ **Dusky-capped Flycatcher** *Myiarchus tuberculifer*
____ *M. t. olivascens*
____ *M. t. lawrenceii*
____ *M. t. querulous (tresmariae)*
____ *M. t. platyrhynchus*
____ *M. t. manens*
____ *M. t. connectens*
____ *M. t. littoralis*
____ *M. t. nigricapillus*
____ *M. t. brunneiceps*
____ *M. t. pallidus*
____ *M. t. tuberculifer*
____ *M. t. nigriceps*
____ *M. t. atriceps*

SW US and nw Mexico; winters to Oaxaca
E Mexico (Nuevo León) to highlands of Guatemala
SW Mexico (s Sinaloa to Oaxaca); Tres Marias Islands
Cozumel I. (off Yucatán Peninsula of e Mexico)
SE Mexico (s Yucatán Peninsula)
Guatemala to n Nicaragua
Pacific coast of se Honduras to nw Costa Rica
Extreme se Nicaragua to Costa Rica and w Panama
Tropical e Panama and w Colombia
N Colombia to n and w Venezuela
E Colombia to Suriname and Amaz. Brazil; se Brazil; Trinidad
SW Colombia to w Ecuador and nw Peru
E slope of Andes of Ecuador to e Peru, Bolivia and nw Argentina

☐ **Swainson's Flycatcher** *Myiarchus swainsoni*
____ *M. s. phaeonotus*
____ *M. s. pelzelni*
____ *M. s. ferocior*
____ *M. s. swainsoni*

S Venezuela and the Guianas to lower Amazonian Brazil
SE Peru to n Bolivia and e Brazil (Mato Grosso to Pará)
SE Bolivia to w Paraguay and c Argentina; winters to Colombia
SE Brazil to e Paraguay and ne Argentina; winters to Trinidad

☐ **Venezuelan Flycatcher** *Myiarchus venezuelensis*

Caribbean lowlands of n Colombia and n Venezuela; Tobago

☐ **Panama Flycatcher** *Myiarchus panamensis*
_____ *M. p. actiosus* — Pacific coast of nw Costa Rica
_____ *M. p. panamensis* — Pacific slope of sw Costa Rica to sw Colombia and nw Venezuela

☐ **Short-crested Flycatcher** *Myiarchus ferox*
_____ *M. f. brunnescens* — *Llanos* of e Colombia to Venezuela and Guyana
_____ *M. f. ferox* — Tropical se Colombia to n Bolivia, the Guianas and n Brazil
_____ *M. f. australis* — S Brazil to e Bolivia, e Paraguay and ne Argentina

☐ **Apical Flycatcher** *Myiarchus apicalis* — Arid scrub of Colombia (west of the Eastern Andes)

☐ **Pale-edged Flycatcher** *Myiarchus cephalotes*
_____ *M. c. caribbaeus* — Mountains of n Venezuela (Trujillo and Lara to Sucre)
_____ *M. c. cephalotes* — Andes of Colombia and w Venezuela to w Bolivia

☐ **Sooty-crowned Flycatcher** *Myiarchus phaeocephalus*
_____ *M. p. phaeocephalus* — Arid w Ecuador and nw Peru (south to Lambayeque))
_____ *M. p. interior* — E slope of Andes of nw Peru and adjacent Ecuador

☐ **Ash-throated Flycatcher** *Myiarchus cinerascens*
_____ *M. c. cinerascens* — Semiarid w US and w Mexico; winters to n Costa Rica
_____ *M. c. pertinax* — S Baja California (south of latitude 29°)

☐ **Nutting's Flycatcher** *Myiarchus nuttingi*
_____ *M. n. inquietus* — Semiarid w Mexico (Sonora to Chiapas) and central Mexico
_____ *M. n. nuttingi* — Arid interior montane valleys of Chiapas to nw Costa Rica
_____ *M. n. flavidior* — Pacific lowlands of s Mexico (Chiapas) to nw Costa Rica

☐ **Great Crested Flycatcher** *Myiarchus crinitus* — Breeds se Canada to Gulf States; winters to nw South America

☐ **Brown-crested Flycatcher** *Myiarchus tyrannulus*
_____ *M. t. magister* — SW US through w Mexico to Oaxaca; Tres Marías Islands
_____ *M. t. cooperi* — S Texas through e Mexico to Guatemala, Belize and Honduras
_____ *M. t. cozumelae* — Cozumel I. (off Yucatán Peninsula of e Mexico)
_____ *M. t. insularum* — Utila, Roatán and Bonaca islands (off Honduras)
_____ *M. t. brachyurus* — Pacific coast of El Salvador to nw Costa Rica
_____ *M. t. tyrannulus* — E Colombia to the Guianas, n Argentina and n Brazil; Trinidad
_____ *M. t. bahiae* — N and e Brazil (Amapá to São Paulo) and ne Argentina

☐ **Galapagos Flycatcher** *Myiarchus magnirostris* — Galapagos Islands

☐ **Grenada Flycatcher** *Myiarchus nugator* — S Lesser Antilles (St. Vincent, the Grenadines and Grenada)

☐ **Rufous-tailed Flycatcher** *Myiarchus validus* — Wooded hills and mountains of Jamaica

☐ **La Sagra's Flycatcher** *Myiarchus sagrae*
_____ *M. s. lucaysiensis* — Bahamas
_____ *M. s. sagrae* — Cuba, Isle of Pines and Grand Cayman I.

☐ **Stolid Flycatcher** *Myiarchus stolidus*
_____ *M. s. dominicensis* — Hispaniola, Gonâve, Tortue, Beata and Grand Cayemite islands
_____ *M. s. stolidus* — Jamaica

☐ **Puerto Rican Flycatcher** *Myiarchus antillarum* — Puerto Rico, Vieques I., Culebra I. and Virgin Islands

☐ **Lesser Antillean Flycatcher** *Myiarchus oberi*
_____ *M. o. oberi* — Lesser Antilles (Dominica and Guadeloupe)
_____ *M. o. sanctaeluciae* — St. Lucia (Lesser Antilles)
_____ *M. o. berlepschii* — Lesser Antilles (St. Kitts, Barbuda and Nevis)
_____ *M. o. sclateri* — Martinique (Lesser Antilles)

☐ **Flammulated Flycatcher** *Deltarhynchus flammulatus*

Pacific lowlands of w Mexico (Sinaloa to w Chiapas)

☐ **Large-headed Flatbill** *Ramphotrigon megacephalum*
_____ *R. m. venezuelense*
_____ *R. m. pectorale*
_____ *R. m. bolivianum*
_____ *R. m. megacephalum*

NW Venezuela (and probably adjacent Colombia)
SE Colombia to e Ecuador and s Venezuela; adj. Brazil and Peru?
Tropical e Peru (Loreto) to n Bolivia and w Amazonian Brazil
SE Brazil (e Minas Gerais) to se Paraguay and ne Argentina

☐ **Rufous-tailed Flatbill** *Ramphotrigon ruficauda*

SE Colombia to s Venezuela, Guianas, n Bolivia, Amaz. Brazil

☐ **Dusky-tailed Flatbill** *Ramphotrigon fuscicauda*

Extreme se Colombia to n Bolivia and sw Brazil (Mato Grosso)

☐ **Rufous-tailed Attila** *Attila phoenicurus*

Amazonian and se Brazil to e Paraguay and ne Argentina

☐ **Cinnamon Attila** *Attila cinnamomeus*

S Venezuela and the Guianas to n Bolivia and Amazonian Brazil

☐ **Ochraceous Attila** *Attila torridus*

Extreme sw Colombia to w Ecuador and nw Peru (Tumbes)

☐ **Citron-bellied Attila** *Attila citriniventris*

E Colombia to s Venezuela, ne Peru (Loreto) and w Amaz. Brazil

☐ **Dull-capped Attila** *Attila bolivianus*
_____ *A. b. nattereri*
_____ *A. b. bolivianus*

Extreme se Colombia to ne Peru and w Amazonian Brazil
Tropical sw Brazil (Amazonas and Mato Grosso) and n Bolivia

☐ **Gray-hooded Attila** *Attila rufus*
_____ *A. r. hellmayri*
_____ *A. r. rufus*

Tropical e Brazil (central Bahia)
Tropical se Brazil (Minas Gerais to Rio Grande do Sul)

☐ **Bright-rumped Attila** *Attila spadiceus*
_____ *A. s. pacificus*
_____ *A. s. cozumelae*
_____ *A. s. gaumeri*
_____ *A. s. flammulatus*
_____ *A. s. salvadorensis*
_____ *A. s. citreopyga*
_____ *A. s. sclateri*
_____ *A. s. caniceps*
_____ *A. s. parvirostris*
_____ *A. s. parambae*
_____ *A. s. spadiceus*
_____ *A. s. uropygiatus*

Coastal nw Mexico (extreme s Sonora to w Oaxaca)
Cozumel I. (off Yucatán Peninsula of e Mexico)
Tropical n Yucatán Pen.; Holbox I., Meco I. and Isla Mujeres
Tropical se Mexico (Veracruz) to El Salvador
Tropical El Salvador to nw Nicaragua
Tropical se Honduras and Nicaragua to w Panama
Tropical e Panama and nw Colombia (upper Sinú Valley)
Trop. n Colombia (middle Magdalena and lower Sinú valleys)
Santa Marta Mountains (ne Colombia) and nw Venezuela
W Colombia (Río Atrato to Nariño) and nw Ecuador
E Colombia to the Guianas, ne Peru, n Bolivia, n Brazil; Trinidad
SE Brazil (Alagoas and Bahia to Rio de Janeiro)

☐ **Cinereous Mourner** *Laniocera hypopyrra*

E Colombia to Venezuela, e Brazil, n Bolivia and Amaz. Brazil

☐ **Speckled Mourner** *Laniocera rufescens*
_____ *L. r. rufescens*
_____ *L. r. griseigula*
_____ *L. r. tertia*

Gulf lowlands of s Mexico to nw Colombia (Chocó)
Tropical n Colombia (Córdoba, Antioquia and Santander)
Tropical sw Colombia and nw Ecuador (Esmeraldas)

☐ **Masked Tityra** *Tityra semifasciata*
_____ *T. s. hannumi*
_____ *T. s. griseiceps*
_____ *T. s. personata*
_____ *T. s. deses*
_____ *T. s. costaricensis*
_____ *T. s. columbiana*
_____ *T. s. nigriceps*
_____ *T. s. semifasciata*
_____ *T. s. fortis*

Arid tropical nw Mexico (se Sonora and ne Sinaloa)
Pacific coast of Mexico (Sinaloa and w Durango to Oaxaca)
Arid tropical e Mexico (Tamaulipas) to n Nicaragua
SE Mexico (Yucatán Peninsula)
SE Honduras to Nicaragua, Costa Rica and w Panama
Tropical e Panama to Colombia and w Venezuela
Tropical sw Colombia (Nariño) and nw Ecuador
Amazonian Brazil south of the Amazon
SE Colombia to se Peru, n Bolivia and w Amazonian Brazil

☐ **Black-crowned Tityra** *Tityra inquisitor*

____	*T. i. fraserii*	Tropical se Mexico (San Luis Potosí) to central Panama
____	*T. i. albitorques*	Tropical e Panama to nw Bolivia and w Amazonian Brazil
____	*T. i. buckleyi*	Tropical se Colombia (Caquetá) and e Ecuador (Napo-Pastaza)
____	*T. i. erythrogenys*	Tropical e Colombia to Venezuela, the Guianas and n Brazil
____	*T. i. pelzelni*	Tropical ne Bolivia, Mato Grosso and Brazil s of the Amazon
____	*T. i. inquisitor*	Tropical se Brazil (s Piauí) to e Paraguay and ne Argentina

☐ **Black-tailed Tityra** *Tityra cayana*

____	*T. c. cayana*	E Colombia to n Bolivia, the Guianas and n Brazil; Trinidad
____	*T. c. braziliensis*	E Brazil (Maranhão) to e Paraguay and ne Argentina

☐ **White-naped Xenopsaris** *Xenopsaris albinucha*

____	*X. a. minor*	Locally in w and c Venezuela and extreme n Brazil (Roraima)
____	*X. a. albinucha*	Interior ne Brazil to e Bolivia, w Paraguay and n Argentina

☐ **Green-backed Becard** *Pachyramphus viridis*

____	*P. v. griseigularis*	SE Venezuela (e Bolívar) and lower Amazonian Brazil
____	*P. v. viridis*	E Bolivia to n Argentina, e Uruguay, Paraguay and e Brazil

☐ **Yellow-cheeked Becard** *Pachyramphus xanthogenys*

____	*P. x. xanthogenys*	E slope of Andes of e Ecuador (south to Zamora-Chinchipe)
____	*P. x. peruanus*	E slope of Andes of central Peru (Huánuco and Junín)

☐ **Barred Becard** *Pachyramphus versicolor*

____	*P. v. costaricensis*	Humid montane forests of Costa Rica and w Panama (Chiriquí)
____	*P. v. versicolor*	E and Central Andes of Colombia to w Venezuela and Ecuador
____	*P. v. meridionalis*	E slope of Andes of s Ecuador to e Peru and w Bolivia

☐ **Slaty Becard** *Pachyramphus spodiurus*

Pacific lowlands of w Ecuador and extreme nw Peru

☐ **Cinereous Becard** *Pachyramphus rufus*

____	*P. r. rufus*	E Panama to Colombia, the Guianas, Amazonian and e Brazil
____	*P. r. juruanus*	SE Ecuador to e Peru (Loreto) and w Brazil (sw Amazonas)

☐ **Chestnut-crowned Becard** *Pachyramphus castaneus*

____	*P. c. saturatus*	Trop. se Colombia to e Ecuador, n Peru and w Amazonian Brazil
____	*P. c. intermedius*	Tropical n Venezuela (Falcón to Sucre and Monagas)
____	*P. c. parui*	Tropical s Venezuela (Cerro Parú in Amazonas)
____	*P. c. amazonus*	Lower Amazonian Brazil (e Amazonas and Pará)
____	*P. c. castaneus*	SE Brazil (Bahia and se Goiás) to e Paraguay and ne Argentina

☐ **Cinnamon Becard** *Pachyramphus cinnamomeus*

____	*P. c. fulvidior*	SE Mexico (Oaxaca and Chiapas) to extreme w Panama
____	*P. c. cinnamomeus*	Tropical e Panama to Colombia and Ecuador (south to El Oro)
____	*P. c. magdalenae*	N Colombia to nw Venezuela (Maracaibo basin)
____	*P. c. badius*	W Venezuela (s Táchira)

☐ **White-winged Becard** *Pachyramphus polychopterus*

____	*P. p. similis*	Caribbean slope of Guatemala to extreme n Colombia (Chocó)
____	*P. p. cinereiventris*	N Colombia (east to Santa Marta region)
____	*P. p. dorsalis*	Tropical and subtropical sw Colombia and nw Ecuador
____	*P. p. tenebrosus*	Tropical se Colombia (Caquetá) to e Ecuador and ne Peru
____	*P. p. tristis*	NE Colombia to Guianas and ne Brazil; Trinidad and Tobago
____	*P. p. nigriventris*	Trop. se Colombia (Meta) to s Venezuela, n Bolivia, w Brazil
____	*P. p. polychopterus*	E Brazil (Piauí and Ceará to Alagoas and Bahia)
____	*P. p. spixii*	S Brazil to Paraguay, Uruguay, e Bolivia and n Argentina

☐ **Black-capped Becard** *Pachyramphus marginatus*

____	*P. m. nanus*	Colombia to the Guianas, n Bolivia and Amazonian Brazil
____	*P. m. marginatus*	Coastal e Brazil (Pernambuco to se São Paulo)

☐ **Black-and-white Becard** *Pachyramphus albogriseus*

____	*P. a. ornatus*	Costa Rica to w Panama (Chiriquí and Veraguas)
____	*P. a. coronatus*	Santa Marta Mts. (n Colombia) and nw Venezuela (Zulia)
____	*P. a. albogriseus*	Subtropical E Andes of n Colombia (Boyacá) and n Venezuela
____	*P. a. guayaquilensis*	W Ecuador (Guayaquil basin) and Isla Puná
____	*P. a. salvini*	E slope of Andes of e Ecuador and Peru (south to Ayacucho)

☐ **Gray-collared Becard** *Pachyramphus major*

____	*P. m. uropygialis*	W Mexico (Sonora, Sinaloa, Durango, Michoacán and Guerrero)
____	*P. m. major*	E Mexico (Nuevo León and San Luis Potosí to Chiapas)
____	*P. m. matudai*	Pacific slope of s Mexico (Chiapas) and n Guatemala
____	*P. m. itzensis*	SE Mexico (Campeche, Yucatán, Quintana Roo) and Belize
____	*P. m. australis*	Guatemala to n-central Nicaragua

☐ **Glossy-backed Becard** *Pachyramphus surinamus*

	Suriname, French Guiana and lower Amazonian Brazil

☐ **Rose-throated Becard** *Pachyramphus aglaiae*

____	*P. a. albiventris*	SE Arizona and w Mexico (south to Guerrero and Zacatecas)
____	*P. a. gravis*	S Texas and ne Mexico (Tamaulipas to San Luis Potosí)
____	*P. a. yucatanensis*	SE Mexico (Yucatán, Campeche and Quintana Roo)
____	*P. a. insularis*	Tres Marías Islands (off w Mexico)
____	*P. a. aglaiae*	Coastal s Mexico (Guerrero to Oaxaca)
____	*P. a. sumichrasti*	Lowlands of se Mexico (Veracruz) to w Guatemala
____	*P. a. hypophaeus*	Belize and Honduras to w-central Costa Rica
____	*P. a. latirostris*	Pacific slope of n El Salvador to nw Costa Rica

☐ **Jamaican Becard** *Pachyramphus niger*

	Wooded hills of Jamaica

☐ **One-colored Becard** *Pachyramphus homochrous*

____	*P. h. homochrous*	Tropical central Panama to nw Peru (Tumbes and Piura)
____	*P. h. quimarinus*	NW Colombia (Sinú Valley)
____	*P. h. canescens*	Caribbean coast of n Colombia and nw Venezuela

☐ **Pink-throated Becard** *Pachyramphus minor*

	S Venezuela and the Guianas to e Bolivia and Amazonian Brazil

☐ **Crested Becard** *Pachyramphus validus*

____	*P. v. audax*	S Peru (Ayacucho) to Bolivia and nw Argentina
____	*P. v. validus*	Tropical e Bolivia to ne Argentina, Paraguay and e Brazil

FAMILY: ACANTHISITTIDAE (New Zealand Wrens—3)

☐ **Rifleman** *Acanthisitta chloris*

____	*A. c. granti*	North, Great and Little Barrier islands (New Zealand)
____	*A. c. chloris (citrina)*	South, Stewart and Codfish islands (New Zealand)

☐ **Bush Wren** *Xenicus longipes*

____	*X. l. stokesii†*	North I. (New Zealand). Extinct ca 1850
____	*X. l. longipes*	Montane forests of South I. (New Zealand). Probably extinct
____	*X. l. variabilis*	Small islands off Stewart I. (New Zealand). Probably extinct

☐ **South Island Wren** *Xenicus gilviventris*

	High mts. of South I. (New Zealand)

FAMILY: ATRICHORNITHIDAE (Scrub-birds—2)

☐ **Rufous Scrub-bird** *Atrichornis rufescens*

____	*A. r. rufescens*	E Australia (extreme se Queensland to Gibraltar Range, NSW)
____	*A. r. ferrieri*	E New South Wales (Dorrigo Plateau to Barrington Tops)

347

☐ **Noisy Scrub-bird** *Atrichornis clamosus*

SW Western Australia (coastal heaths east of Albany)

FAMILY: MENURIDAE (Lyrebirds—2)

☐ **Albert's Lyrebird** *Menura alberti*

Rainforests of e Australia (se Queensland and extreme ne NSW

☐ **Superb Lyrebird** *Menura novaehollandiae*
_____ *M. n. edwardi*
_____ *M. n. novaehollandiae*
_____ *M. n. victoriae*

E Australia (extreme se Queensland to Hunter River, NSW)
SE Australia (c New South Wales to Victoria border)
SE NSW to Dandenong Range, Victoria; introduced s Tasmania

FAMILY: ALAUDIDAE (Larks—96)

☐ **Australasian Bushlark** *Mirafra javanica*
_____ *M. j. williamsoni (beaulieui)*
_____ *M. j. philippinensis*
_____ *M. j. mindanensis*
_____ *M. j. javanica*
_____ *M. j. parva*
_____ *M. j. timorensis*
_____ *M. j. aliena (sepikiana)*
_____ *M. j. woodwardi*
_____ *M. j. halli*
_____ *M. j. forresti (subrufescens)*
_____ *M. j. melvillensis*
_____ *M. j. soderbergi*
_____ *M. j. rufescens*
_____ *M. j. athertonensis*
_____ *M. j. horsfieldii (keasti)*
_____ *M. j. secunda*

Local in s China, Myanmar, Thailand, Indochina and s Vietnam
N Philippines (Luzon and Mindoro)
S Philippines (Mindanao)
S Borneo, Java and Bali
Lesser Sundas (Lombok, Sumbawa, Sumba and Flores)
E Lesser Sundas (Sawu and Timor)
N, ne and s New Guinea
Extreme nw Western Australia
N Western Australia
NE Western Australia
Melville and Bathurst Islands (Northern Territory)
Northern Territory (N Northern Territory)
E Northern Territory to nw Queensland and ne South Australia
NE Queensland (Atherton-Evelyn Tablelands)
Southeast South Australia)
S South Australia

☐ **Jerdon's Bushlark** *Mirafra affinis*

E and s India and Sri Lanka

☐ **Indian Bushlark** *Mirafra erythroptera*

India and se Pakistan

☐ **Indochinese Bushlark** *Mirafra erythrocephala*

S Myanmar, Thailand, Cambodia, Laos and s Vietnam

☐ **Bengal Bushlark** *Mirafra assamica*

N India to s Nepal, Bhutan, Bangladesh and extreme w Myanmar.

☐ **Burmese Bushlark** *Mirafra microptera*

C Myanmar

☐ **Madagascar Lark** *Mirafra hova*

Grasslands of Madagascar

☐ **Singing Bushlark** *Mirafra cantillans*
_____ *M. c. marginata*
_____ *M. c. chadensis*
_____ *M. c. simplex*
_____ *M. c. cantillans*

S Sudan to Eritrea, ne Ethiopia, Somalia and Kenya
Senegal and Mali to Sudan and w Ethiopia
W Arabia
N India

☐ **White-tailed Lark** *Mirafra albicauda*

W Chad to s Sudan, Ethiopia, Kenya and Tanzania

☐ **Kordofan Lark** *Mirafra cordofanica*

Senegambia to Mauritania, Niger and Sudan (Darfur and Kordofan)

☐ **Williams' Lark** *Mirafra williamsi*

Black lava deserts of n Kenya (Marsabit region)

☐ **Rusty Lark** *Mirafra rufa*
_____ *M. r. nigriticola*

E Mali to n Togo and Niger

_____ *M. r. rufa*   Chad to w Sudan (Darfur)

_____ *M. r. lynesi*   C Sudan (Kordofan)

☐ **Monotonous Lark** *Mirafra passerina*

SW Angola to s Zambia, n Namibia, Zimbabwe, n Cape Province

☐ **Latakoo Lark** *Mirafra cheniana*

Botswana to Zimbabwe and interior ne South Africa

☐ **Friedmann's Lark** *Mirafra pulpa*

SW Ethiopia and n Kenya

☐ **Ash's Lark** *Mirafra ashi*

Arid coastal s Somalia

☐ **Somali Long-billed Lark** *Mirafra somalica*

_____ *M. s. somalica*   N and ne Somalia and possibly extreme e Ethiopia

_____ *M. s. rochei*   Coastal c Somalia

☐ **Red-winged Lark** *Mirafra hypermetra*

_____ *M. h. kidepoensis*   S Sudan and ne Uganda

_____ *M. h. kathangorensis*   Extreme se Sudan

_____ *M. h. gallarum*   E and s Ethiopia

_____ *M. h. hypermetra*   Somalia to Kenya and n Tanzania

☐ **Rufous-naped Lark** *Mirafra africana*

_____ *M. a. henrici*   Guinea to Liberia

_____ *M. a. batesi*   Niger to Nigeria

_____ *M. a. bamendae*   W Cameroon

_____ *M. a. stresemanni*   N Cameroon (Ngaoundéré)

_____ *M. a. kurrae*   Sudan (Kurra and Darfur provinces)

_____ *M. a. ruwenzoria*   E Zaire to sw Uganda

_____ *M. a. tropicalis*   S Uganda to w Kenya and n Tanzania

_____ *M. a. sharpii*   NW Somalia (Silo Plain, Tuyo Plain and Bankisah)

_____ *M. a. athi*   C Kenya (Nairobi and Nakuru) to ne Tanzania

_____ *M. a. harterti*   E Kenya (Ukamba)

_____ *M. a. occidentalis*   W Angola (Huila escarpment north to Kisama)

_____ *M. a. gomesi*   E Angola (Macondo) to nw Zambia (Kabompo)

_____ *M. a. kabalii*   NE Angola (Luiacana) and w Zambia (Balovale)

_____ *M. a. pallida*   Namibia (Windhoek north to Kaokoveld and Ovamboland)

_____ *M. a. ghansiensis*   Namibia and w Botswana

_____ *M. a. malbranti*   C and s Zaire (Djambala, Petianga and Kasai) to se Gabon

_____ *M. a. chapini*   S Zaire and nw Zambia

_____ *M. a. nigrescens*   NE Zambia (Lundazi) and s Tanzania (Ukinga and Njombe)

_____ *M. a. transvaalensis*   SE Botswana to Zimbabwe, Tanzania, Malawi and South Africa

_____ *M. a. nyikae*   Nyika Plateau (e Zambia and Malawi)

_____ *M. a. isolata*   Malawi (Mangochi District)

_____ *M. a. grisescens*   W Zambia to nw Zimbabwe and n Botswana

_____ *M. a. africana*   S Natal to e Cape Province

☐ **Angola Lark** *Mirafra angolensis*

_____ *M. a. angolensis*   N and w-c Angola

_____ *M. a. antonii (minyanyae)*   E Angola to s Zaire and nw Zambia

_____ *M. a. marungensis*   SE Zaire (Marungu Plateau)

☐ **Flappet Lark** *Mirafra rufocinnamomea*

_____ *M. r. buckleyi*   S Mauritania and Senegal to n Cameroon

_____ *M. r. serlei*   SE Nigeria

_____ *M. r. tigrina*   E Cameroon to n Zaire

_____ *M. r. furensis*   W-c Sudan (w Darfur)

_____ *M. r. sobatensis*   C Sudan (confluence of White Nile and Sobat rivers)

_____ *M. r. torrida*   SE Sudan to s Ethiopia, Uganda, c Kenya and Tanzania

_____ *M. r. rufocinnamomea*   NW and c Ethiopia

_____ *M. r. omoensis*   SW Ethiopia (Omo to Madji and Baro rivers)

____ *M. r. kawirondensis*	E Zaire to w Uganda and w Kenya
____ *M. r. fischeri*	E Angola to s Somalia, e Kenya, Tanzania and n Mozambique
____ *M. r. schoutedeni*	Gabon to C African Republic, w Zaire and ne Angola
____ *M. r. lwenarum*	N Zambia (Balovale)
____ *M. r. smithersi*	Zambia to Zimbabwe, ne Botswana and n Transvaal
____ *M. r. mababiensis*	W Zambia to c Botswana
____ *M. r. pintoi*	E Transvaal to ne Natal, Swaziland and s Mozambique

☐ **Cape Clapper Lark** *Mirafra apiata*

____ *M. a. apiata (algoensis)*	SW Namibia and w coastal plain and interior South Africa
____ *M. a. marjoriae*	South Africa (Coastal plain of Western Cape)

☐ **Eastern Clapper Lark** *Mirafra fasciolata*

____ *M. f. reynoldsi*	SE Zambia, n Namibia and n Bostwana
____ *M. f. jappi*	W Zambia
____ *M. f. nata*	NE Botswana
____ *M. f. deserti*	E-c Namibia to w, e and c Botswana
____ *M. f. fasciolata (hewitti)*	C and eastern South Africa

☐ **Collared Lark** *Mirafra collaris*

Arid acacia of se Ethiopia, Somalia and ne Kenya

☐ **Gillett's Lark** *Mirafra gilletti*

____ *M. g. gilletti*	E Ethiopia and nw Somalia
____ *M. g. arorihensis*	C Somalia to ne Kenya

☐ **Degodi Lark** *Mirafra degodiensis*

SE Ethiopia

☐ **Archer's Lark** *Heteromirafra archeri*

Highlands of nw Somalia

☐ **Sidamo Lark** *Heteromirafra sidamoensis*

Known from two specimens from s Ethiopia

☐ **Rudd's Lark** *Heteromirafra ruddi*

High altitude grasslands of e South Africa

☐ **Pink-breasted Lark** *Calendulauda poecilosterna*

Savanna of s Ethiopia to Kenya, e Uganda and n Tanzania

☐ **Sabota Lark** *Calendulauda sabota*

____ *C. s. plebeja*	Coastal n Angola (Cabinda)
____ *C. s. ansorgei*	Coastal Angola (Moçamedes to Novo Redondo)
____ *C. s. sabota*	Coastal Angola to Namibia, Botswana, Natal and Cape Province
____ *C. s. naevia*	NW Namibia
____ *C. s. waibeli*	N Namibia (Etosha and Ovamboland) to n Botswana
____ *C. s. herero*	S and e Namibia to nw Cape Province
____ *C. s. sabotoides*	W Zimbabwe to s Botswana and nw Transvaal
____ *C. s. suffusca*	SW Zimbabwe to s Mozambique, ne Natal, e Transvaal, Swaziland
____ *C. s. bradfieldi*	South Africa (n, c and e Cape Province)

☐ **Fawn-colored Lark** *Calendulauda africanoides*

____ *C. a. trapnelli*	SE Angola and sw Zambia
____ *C. a. makarikari*	SE Angola to n Namibia, w Zambia and n and c Botswana
____ *C. a. harei (omaruru, isseli, rubidior)*	NW and c Namibia
____ *C. a. sarwensis*	W Botswana
____ *C. a. africanoides (transvaalensis, austinrobertsi)*	S Namibia, s and e Botswana, w Zimbabwe and n South Africa
____ *C. a. vincenti*	C Zimbabwe to s Mozambique

☐ **Foxy Lark** *Calendulauda alopex*

____ *C. a. alopex*	N Somalia and extreme e Ethiopia
____ *C. a. intercedens (macdonaldi, longonotensis)*	E and s Ethiopia, adjacent Somalia, e Uganda, Kenya, n Tanzania

☐ **Ferruginous Lark** *Calendulauda burra*

South Africa (northwest Northern Cape)

☐ **Karoo Lark** *Calendulauda albescens*
_____ *C. a. albescens*
_____ *C. a. codea*
_____ *C. a. guttata*
_____ *C. a. karruensis*

SW Cape Province (Cape Town to Berg River)
Coastal South Africa (Saldanha Bay to Port Nolloth)
Cape Province (Clanwilliam to Little Namaqualand)
Cape Province (Calvinia and Williston to Oudtshoorn)

☐ **Barlow's Lark** *Calendulauda barlowi*
_____ *C. b. barlowi*
_____ *C. b. patae*
_____ *C. b. cavei*

SW Namibia from s of R. Koichab (inland of Lüderitz) e to Aus
Coastal sw Namibia (Lüderitz) to extreme nw South Africa
Inland coastal plain of sw Namibia to nw South Africa (R. Holgat)

☐ **Dune Lark** *Calendulauda erythrochlamys*

W-c Namibia between R. Kuiseb (Walvis Bay) and R. Koichab

☐ **Rufous-rumped Lark** *Pinarocorys erythropygia*

Arid thornscrub of sub-Saharan Africa

☐ **Dusky Lark** *Pinarocorys nigricans*
_____ *P. n. occidentis*
_____ *P. n. nigricans*

SW Zaire and n Angola; >Namibia and Mozambique
S Zaire, nw Zambia and sw Tanzania; >Namibia and Mozambique

☐ **Gray's Lark** *Ammomanopsis grayi*
_____ *A. g. hoeschi*
_____ *A. g. grayi*

Sandy deserts of extreme sw Angola to n Namibia (Cape Cross)
W Namibia (Cape Cross to southern edge of Namib Desert)

☐ **Spike-heeled Lark** *Chersomanes albofasciata*
_____ *C. a. obscurata*
_____ *C. a. boweni*
_____ *C. a. erikssoni*
_____ *C. a. arenaria*
_____ *C. a. kalahariae (bathoeni)*
_____ *C. a. barlowi*
_____ *C. a. albofasciata (baddeleyi)*
_____ *C. a. garrula (bushmanensis, meinertzhageni)*
_____ *C. a. macdonaldi*
_____ *C. a. alticola (subpallida)*

SW, c and e Angola
NW Namibia (Cunene to Swakopmund, Usakos and Karibib)
N Namibia (Ovamboland and Outjo)
S Namibia to n Cape Province (Van Wyksvlei)
W and s Botswana to n Cape Province (Stella and Vryburg)
E Botswana
S Botswana to e Cape Province
South Africa (w and n Cape Province)
South Africa (s and e Karoo)
South Africa (s and c Transvaal)

☐ **Beesley's Lark** *Chersomanes beesleyi*

N Tanzania (area just north of Mt. Meru)

☐ **Cape Lark** *Certhilauda curvirostris*
_____ *C. c. falcirostris*
_____ *C. c. curvirostris*

South Africa (nw Cape Province)
South Africa (s and w Cape Province)

☐ **Algulhas Lark** *Certhilauda brevirostris*

South Africa (Algulhas Plain of coastal w Cape Province)

☐ **Eastern Long-billed Lark** *Certhilauda semitorquata*
_____ *C. s. transvaalensis*
_____ *C. s. semitorquata*
_____ *C. s. algida*

E South Africa
South Africa (e Karoo grasslands)
South Africa (southern part of e Cape Province)

☐ **Karoo Long-billed Lark** *Certhilauda subcoronata*
_____ *C. c. bradshawi*
_____ *C. s. damarensis*
_____ *C. s. subcoronata*
_____ *C. s. gilli*

S Namibia and n Cape Province
W-c Namibia
Karoo Plains of s Namibia and w Cape Province
South Africa (s-c Cape Province)

☐ **Benguela Lark** *Certhilauda benguelensis*
_____ *C. b. benguelensis*
_____ *C. b. kaokoensis*

Extreme coastal sw Angola and n Namibia
Brandberg Mts. (s Angola and n Namibia)

☐ **Short-clawed Lark** *Certhilauda chuana*

E Botswana to Transvaal and w Orange Free State

☐ **Greater Hoopoe-Lark** *Alaemon alaudipes*

   ____ *A. a. boavistae*

   ____ *A. a. alaudipes*

   ____ *A. a. desertorum*

   ____ *A. a. doriae*

☐ **Lesser Hoopoe-Lark** *Alaemon hamertoni*

   ____ *A. h. hamertoni*

   ____ *A. h. tertius*

   ____ *A. h. alter*

☐ **Dupont's Lark** *Chersophilus duponti*

   ____ *C. d. duponti*

   ____ *C. d. margaritae*

☐ **Black-eared Sparrow-Lark** *Eremopterix australis*

☐ **Chestnut-backed Sparrow-Lark** *Eremopterix leucotis*

   ____ *E. l. melanocephalus*

   ____ *E. l. leucotis*

   ____ *E. l. madaraszi*

   ____ *E. l. smithi*

   ____ *E. l. hoeschi*

☐ **Black-crowned Sparrow-Lark** *Eremopterix nigriceps*

   ____ *E. n. nigriceps*

   ____ *E. n. albifrons*

   ____ *E. n. melanauchen (forbeswatsoni)*

   ____ *E. n. affinis*

☐ **Gray-backed Sparrow-Lark** *Eremopterix verticalis*

   ____ *E. v. damarensis*

   ____ *E. v. harti*

   ____ *E. v. khama*

   ____ *E. v. verticalis*

☐ **Chestnut-headed Sparrow-Lark** *Eremopterix signatus*

   ____ *E. s. signatus*

   ____ *E. s. harrisoni*

☐ **Fischer's Sparrow-Lark** *Eremopterix leucopareia*

☐ **Ashy-crowned Sparrow-Lark** *Eremopterix griseus*

☐ **Bar-tailed Lark** *Ammomanes cinctura*

   ____ *A. c. cinctura*

   ____ *A. c. arenicolor (pallens)*

   ____ *A. c. zarudnyi*

☐ **Rufous-tailed Lark** *Ammomanes phoenicura*

☐ **Desert Lark** *Ammomanes deserti*

   ____ *A. d. payni (monodi)*

   ____ *A. d. algeriensis (mirei)*

   ____ *A. d. mya*

   ____ *A. d. geyri (janeti, bensoni)*

   ____ *A. d. whitakeri*

   ____ *A. d. kollmanspergeri*

   ____ *A. d. isabellina*

   ____ *A. d. deserti*

Cape Verde Islands
Deserts of North Africa (Morocco, Algeria and Tunisia)
Coastal Sudan (Port Sudan) to nw Somalia and Aden
E Arabia to Iraq, Iran and nw India

Deserts of n Somalia (south of latitude 7°N)
Deserts of n Somalia (west of longitude 47°E)
Deserts of n Somalia (longitude 47°E to 49°E)

Iberian Peninsula, Morocco, n Algeria and nw Tunisia
Algeria (s slopes of Atlas Mts.) to se Tunisia, n Libya and nw Egypt

---

Grassy plains of s Namibia to s Botswana and n South Africa

Senegambia to Nile River
E and s Sudan to Eritrea and Ethiopia
NE Uganda to Kenya and n Tanzania; n Malawi and Mozambique
Zambia to s Malawi, Zimbabwe, e Botswana and South Africa
S Angola to n Namibia, ne Botswana and w Zimbabwe

Cape Verde Islands
S Morocco to Mauritania, Mali, Chad and w Sudan
E Sudan to Ethiopia, Somalia, s Iraq, s Iran, s Pakistan and nw India
Peninsular India and Sri Lanka

Coastal Angola to Namibia, Zambia and nw Botswana
Zambia (Liuwa Plain, Kalabo, Senanga and Siloana Plains)
NE Botswana (Makgadikgadi) to Zambia and Zimbabwe
Zambia to Botswana, w Transvaal, Zimbabwe and Cape Province

Extreme se Sudan to se Ethiopia and Somalia
SE Sudan to nw Kenya (west of Lake Turkana)

NE Uganda to Kenya, Tanzania, n Zambia and Malawi

Grasslands and stony plains of Indian subcontinent

Cape Verde Islands
Deserts of North Africa to Sinai Peninsula and Arabia
Deserts of e Iran to s Afghanistan and s Pakistan

NE Pakistan and most of India

Morocco south of the High Atlas and adj. sw Algeria
N Algeria to Tunisia and nw Libya
C Algerian Sahara (between 27°N and 30°N)
Mauritania to s Algeria and nw Niger (Aïr Massif)
SE Algeria to sw Libya and nw Chad (Tibesti Mts.)
NE Chad (Ennedi Mts.) and w Sudan (Darfur)
N Egypt w of Nile Valley to Israel, s Jordan, nw Saudi Arabia, Iraq
E Egypt (e of Nile to Red Sea) s to Sudan

_____ A. d. erythrochroa	W Chad (Ndjamena) to n Sudan (Dongola to Kordofan)
_____ A. d. samharensis	Red Sea coast of e Sudan, Eritrea and Arabia Pen. (s to n Yemen)
_____ A. d. assabensis	S Eritrea, Ethiopia and nw Somalia
_____ A. d. akeleyi	Highlands of n Somalia
_____ A. d. coxi	S Turkey (Birecik), Syria and n Iraq
_____ A. d. annae	Black lava deserts of Jordan (Azraq area) and extreme s Syria
_____ A. d. azizi	NE Saudi Arabia (Al Hufuf area)
_____ A. d. saturata	Black lava deserts of s Arabia (n Hijaz to Aden)
_____ A. d. insularis	Bahrain (Persian Gulf)
_____ A. d. taimuri	Oman (Muscat area)
_____ A. d. cheesmani	Iraq (e of Tigris River) to w Iran (w of Zagros Mts. to Persian Gulf
_____ A. d. darica	S Zagros Mts. (sw Iran)
_____ A. d. parvirostris	W Turkmenistan (Kara-Bogaz-Gol to Kopet Dagh and Atrak Basin)
_____ A. d. orientalis	NE Iran, Turkestan and n Afghanistan
_____ A. d. iranica	E Iran to sw Afghanistan and w Pakistan (Baluchistan)
_____ A. d. phoenicuroides	SE Afghanistan, e Pakistan and nw India (extreme w Rajasthan)

☐ **Thick-billed Lark** _Ramphocoris clotbey_

Stony deserts of nw Africa to Jordan and nw Arabia

☐ **Calandra Lark** _Melanocorypha calandra_

_____ M. c. calandra	Mediterranean basin to e Turkey, nw Iran, Transcaucasia and Urals
_____ M. c. hebraica	S-c Turkey, adj. nw Syria to Israel, Palestine and w Jordan
_____ M. c. gaza	SE Turkey to e Syria, Iraq and sw Iran
_____ M. c. psammochroa	N Iraq and n Iran to Turkmenistan and e Kazakhstan

☐ **Bimaculated Lark** _Melanocorypha bimaculata_

Turkey to Iran and Afghanistan; >ne Africa and India

☐ **Tibetan Lark** _Melanocorypha maxima_

Tibetan plateau from nw India to c China

☐ **Mongolian Lark** _Melanocorypha mongolica_

High steppes of Mongolia and n China; >c China

☐ **White-winged Lark** _Melanocorypha leucoptera_

Arid steppes of c Eurasia

☐ **Black Lark** _Melanocorypha yeltoniensis_

Steppes of s Russia and sw Siberia; >Black Sea region

☐ **Greater Short-toed Lark** _Calandrella brachydactyla_

_____ C. b. brachydactyla	Mediterranean environs and islands
_____ C. b. hungarica	Hungary
_____ C. b. rubiginosa	N Africa (Morocco, Algeria and Tunisia); Malta
_____ C. b. woltersi	NW Syria and adjacent s Turkey
_____ C. b. hermonensis	Sinai Peninsula to extreme s Turkey and e Syria
_____ C. b. artemisiana	Asia Minor, Transcaucasia and nw Iran
_____ C. b. longipennis (orientalis)	Caucasus to Ukraine, n Mongolia and ne China; >e Africa
_____ C. b. dukhunensis	Tibet to ne China

☐ **Red-capped Lark** _Calandrella cinerea_

_____ C. c. saturatior	Nigeria to Zaire, Uganda, Tanzania, Angola, n Zambia, Malawi
_____ C. c. williamsi	S Kenya and n Tanzania
_____ C. c. spleniata (ongumaensis)	SW Angola and nw Namibia
_____ C. c. fulvida	S Angola, s Zambia and Zimbabwe
_____ C. c. alluvia	Coastal s Mozambique; disperses inland to n Botswana
_____ C. c. millardi	S Botswana
_____ C. c. cinerea (witputzi, anderssoni)	S Namibia and w South Africa
_____ C. c. niveni	E South Africa

☐ **Blanford's Lark** _Calandrella blanfordi_

_____ C. b. eremica	SW Arabia
_____ C. b. blanfordi	Highlands of Ethiopia
_____ C. b. daaroodensis	N and c Somalia

☐ **Erlanger's Lark** *Calandrella erlangeri*

Highlands of Ethiopia

☐ **Hume's Lark** *Calandrella acutirostris*

____ *C. a. acutirostris*   Mts. of ne Iran to n Afghanistan, e Turkestan and w China
____ *C. a. tibetana*   Himalayas of n Pakistan to n India and e Tibet; >India

☐ **Somali Short-toed Lark** *Calandrella somalica*

____ *C. s. megaensis*   S Ethiopia
____ *C. s. somalica*   E Ethiopia to n Somalia
____ *C. s. perconfusa*   C and w plateau of n Somalia
____ *C. s. athensis*   Kenya and ne Tanzania

☐ **Lesser Short-toed Lark** *Calandrella rufescens*

____ *C. r. rufescens*   Tenerife I. (Canary Islands)
____ *C. r. polatzeki*   E Canary Islands (Gran Canaria, Fuerteventura and Lanzarote)
____ *C. r. apetzii*   S Iberian Peninsula
____ *C. r. minor*   North Africa to n Sinai and s Turkey and e Iraq
____ *C. r. nicolli*   N Egypt (northern Nile Delta)
____ *C. r. pseudobaetica*   Highlands of Armenia and sw shores of Caspian Sea
____ *C. r. persica (aharonii)*   S Iraq and Iran to se Afghanistan
____ *C. r. heinei*   SE Russia n to Volga basin and sw Siberia; >Asia Minor
____ *C. r. cheleensis (obscura)*   E Transbaikalia and n Mongolia to ne China
____ *C. r. leucophaea*   Turkestan (south of Kyzyl Kum Desert); >Asia Minor
____ *C. r. niethammeri*   C Turkey (Anatolia high plateau)
____ *C. r. kukunoorensis*   SW Mongolia (Lake Kokonor basin)
____ *C. r. seebohmi*   NW Mongolia and w China (Xinjiang e to Khotan River)
____ *C. r. beicki*   W China (s Gobi Desert to nw Gansu)
____ *C. r. stegmanni*   Arid w China (ne Gansu)
____ *C. r. tangutica*   Mts. of w China (s Kokonor) to ne Tibet

☐ **Sand Lark** *Calandrella raytal*

____ *C. r. raytal*   Coastal se Iran to Afghanistan, n India and s Myanmar
____ *C. r. krishnarkumarsinhji*   NW India (Kathiawar Peninsula)
____ *C. r. adamsi*   Coastal nw India

☐ **Horned Lark** *Eremophila alpestris*

____ *E. a. arcticola*   N Alaska to mts. of British Columbia and n Washington
____ *E. a. alpina*   Arctic-alpine summits of nw US (Mt. Rainier and Mt. St. Helens)
____ *E. a. hoyti*   Arctic coast of North America to s Canada; >n US
____ *E. a. alpestris*   Arctic ne Canada to Newfoundland; >coastal se US
____ *E. a. leucolaema*   S Canada to sw US and nw Texas; >nw Mexico
____ *E. a. enthymia*   Great Plains of c Canada to c US; >n Mexico
____ *E. a. praticola*   SE Canada to c and e-c US
____ *E. a. strigata*   Humid coastal belt of sw Br. Col. and nw US w of the Cascades
____ *E. a. merrilli*   E slope of Cascades and adj. lowlands from Br. Col. to ne Calif.
____ *E. a. lamprochroma*   SE Oregon to sw Idaho, ne California and w Nevada
____ *E. a. utahensis*   S-c Idaho to e-c Nevada and w-c Utah
____ *E. a. sierrae*   Mts. of ne California (s Cascades and n Sierra Nevada)
____ *E. a. rubea*   C California (Sacramento Valley)
____ *E. a. actia*   Coastal range of s California (Humboldt Co.) to n Baja California
____ *E. a. insularis*   Channel Islands (off s California)
____ *E. a. ammophila*   Deserts of sw Nevada and se California; >nw Mexico
____ *E. a. leucansiptila*   Colorado Desert (sw Nevada, w Arizona, ne Baja and nw Sonora)
____ *E. a. occidentalis*   N and cent. Arizona to n-cent. New Mexico; >n Mexico
____ *E. a. adusta*   S Arizona (s of Tucson) to extreme sw New Mexico and n Sonora
____ *E. a. giraudi*   Coastal prairie region of se Texas to e Mexico (ne Tamaulipas)
____ *E. a. enertera*   W-c Baja California and coastal islands s of Magdalena Bay
____ *E. a. aphrasta*   NW Mexico (Chihuahua and Durango)
____ *E. a. lactea*   NE Mexico (Coahuila)
____ *E. a. diaphora*   NE Mexico (se Coahuila to s Tamaulipas, Hidalgo and ne Puebla)

____	*E. a. chrysolaema*	S Mexican Plateau (Jalisco to Michoacán, Puebla and Veracruz)
____	*E. a. oaxacae*	S Mexico (e Oaxaca)
____	*E. a. peregrina*	E Andes of Colombia
____	*E. a. flava*	N Palearctic region
____	*E. a. balcanica*	Mts. of se Europe (Yugoslavia, Bulgaria and n Greece)
____	*E. a. kumerloevei*	West and c Asia Minor
____	*E. a. atlas*	High plateaux of Morocco
____	*E. a. bicornis*	W Turkey (Taurus Mts.) to Lebanon and Palestine
____	*E. a. penicillata*	Mts. of Asia Minor, the Caucasus and w Iran
____	*E. a. albigula*	Mts. of n and e Iran to Pamirs, Afghanistan and w China
____	*E. a. brandti*	Steppes of c Asia to mts. of w Mongolia and n China
____	*E. a. longirostris*	NW Himalayas
____	*E. a. teleschowi*	Mts. of w China (extreme se Xinjiang)
____	*E. a. khamensis*	SW China (Kham region of w and s Sichuan)
____	*E. a. przewalskii*	W China (nw Qinghai)
____	*E. a. nigrifrons*	W China (Kokonor to w Gansu)
____	*E. a. argalea*	W China (extreme sw Xinjiang) to nw India (Kashmir to Ladakh)
____	*E. a. elwesi*	W China (s Qinghai and s Tibet) to n Sikkim

☐ **Temminck's Lark** *Eremophila bilopha*

Deserts of North Africa to n Arabia and sw Iraq

☐ **Dunn's Lark** *Eremalauda dunni*

____	*E. d. dunni*	S edge of Sahara (Mauritania to Mali, Chad and c Sudan)
____	*E. d. eremodites*	SW Arabia

☐ **Stark's Lark** *Spizocorys starki*

Extreme sw Angola to Namibia, sw Botswana and nw South Africa

☐ **Pink-billed Lark** *Spizocorys conirostris*

____	*S. c. damarensis*	NW Namibia (Swakop River to Ovamboland)
____	*S. c. barlowi*	S Namibia to s Botswana and nw Cape Province
____	*S. c. harti*	SW Zambia (Matabele Plain)
____	*S. c. makawai*	Extreme sw Zambia (Liuwa and Mutala plains)
____	*S. c. crypta*	NE Botswana (Makgadıkgadı Pan)
____	*S. c. conirostris (transiens)*	South Africa and nw Lesotho

☐ **Botha's Lark** *Spizocorys fringillaris*

High altitude grasslands of se Transvaal and Orange Free State

☐ **Obbia Lark** *Spizocorys obbiensis*

Deserts of coastal Somalia (Obbia to Hal Hambo)

☐ **Masked Lark** *Spizocorys personata*

____	*S. p. personata*	E Ethiopia
____	*S. p. yavelloensis*	S Ethiopia and n Kenya (south to Dida Galgallu Desert)
____	*S. p. mcchesneyi*	N Kenya (Marsabit Plateau)
____	*S. p. intensa*	C Kenya

☐ **Sclater's Lark** *Spizocorys sclateri*

Deserts of s Namibia and Cape Province

☐ **Short-tailed Lark** *Pseudalaemon fremantlii*

____	*P. f. fremantlii*	SE Ethiopia and Somalia
____	*P. f. megaensis*	S Ethiopia to n Kenya
____	*P. f. delamerei*	S Kenya to n Tanzania

☐ **Crested Lark** *Galerida cristata*

____	*G. c. cristata*	C Europe to Slovenia, Belarus, n Hungary, n Ukraine
____	*G. c. pallida*	Iberian Peninsula
____	*G. c. neumanni*	Italy (Toscana s to Rome area)
____	*G. c. apuliae*	S peninsular Italy and Sicily
____	*G. c. tenuirostris*	E Hungary and Romania to s Russia and w Kazakhstan
____	*G. c. meridionalis*	S Yugoslavia to mainland Greece, Ionian Is., Crete and w Turkey
____	*G. c. caucasica*	E Aegean Is., n Turkey, s Caucasus and w Transcaucasia

____	*G. c. subtaurica*	C Turkey to s Transcaucasia, nw Iran, w Turkmenistan and e Iraq
____	*G. c. cypriaca*	Kárpathos, Rhodes and Cyprus
____	*G. c. zion*	S Turkey, Syria, e Lebanon, and e Israel (s to Jerusalem)
____	*G. c. cinnamomina*	W Lebanon (w from Beirut) and nw Israel (Mt. Carmel and Haifa)
____	*G. c. magna*	S Kazakhstan to e Mongolia and n China (Inner Mongolia)
____	*G. c. leautungensis*	Manchuria and ne China
____	*G. c. coreensis*	Korea
____	*G. c. iwanowi*	C Turkmenistan to Iran, Tadjikistan, Afghanistan and nw Pakistan
____	*G. c. lynesi*	N Kashmir (Gilgit Valley)
____	*G. c. kleinschmidti*	NW Morocco (e to Rif Mts. and s to Middle Atlas)
____	*G. c. riggenbachi*	W Morocco (Casablanca to Sous Valley)
____	*G. c. carthaginis*	Coastal ne Morocco to n Tunisia (e to Sousse)
____	*G. c. randoni*	Hauts Plateaux of e Morocco and nw Algeria
____	*G. c. macrorhyncha*	S Morocco and nw Algeria s of Atlas Saharien to w-c Mauritania
____	*G. c. balsaci*	Coastal Mauritania
____	*G. c. arenicola*	NE Algerian Sahara to s Tunisia and nw Libya
____	*G. c. helenae*	SE Algeria and immediately adjacent sw Libya
____	*G. c. festae*	Coastal ne Libya (Benghazi to Tobruq)
____	*G. c. brachyura*	NE Libya to coastal n Egypt, n Sinai, n Saudi Arabia and s Iraq
____	*G. c. nigricans*	N Egypt (Nile Delta)
____	*G. c. maculata*	Egypt (Nile Valley from Cairo to Aswan and El Faiyum)
____	*G. c. halfae*	Egypt (Nile Valley s of Aswan) to extreme n Sudan (Wadi Halfa)
____	*G. c. senegallensis*	S Mauritania, Senegambia and Guinea-Bissau to Niger
____	*G. c. jordansi*	N Niger (Aïr Mts.)
____	*G. c. alexanderi (zalingei)*	N Nigeria to w Sudan and ne Zaire
____	*G. c. isabellina*	C Sudan (Kordofan to Nile Valley)
____	*G. c. altirostris*	E Sudan (e of Nile Valley) and Eritrea
____	*G. c. somaliensis*	N Somalia, s Ethiopia and n Kenya
____	*G. c. tardinata*	S Arabia
____	*G. c. chendoola*	Foothills of s Kashmir to e Pakistan, w and n India and s Nepal

☐ **Thekla Lark** *Galerida theklae*

____	*G. t. theklae*	E and s Portugal, Spain, Balearic Is., and extreme s France
____	*G. t. erlangeri*	N Morocco (e to Algerian border, s to Middle Atlas)
____	*G. t. ruficolor (aguirrei)*	NE and c Morocco, coastal Algeria and n Tunisia
____	*G. t. theresae*	SW Morocco (s from Anti-Atlas Mts.) and Western Sahara
____	*G. t. superflua*	NE Morocco (w to Moulouya River) to n Algeria and Tunisia
____	*G. t. carolinae (deichleri)*	N Sahara (extreme e Morocco to ne Libya and extreme nw Egypt)
____	*G. t. praetermissa*	Highlands of s Eritrea to c Ethiopia
____	*G. t. huei*	S-c Ethiopia (Bale Mts., Arussi)
____	*G. t. huriensis*	S Ethiopia and n Kenya (Huri Hills s to Marsabit)
____	*G. t. ellioti*	N and c Somalia
____	*G. t. harrarensis*	E Ethiopia (Jigjiga and Harar)
____	*G. t. mallablensis*	Coastal s Somalia

☐ **Malabar Lark** *Galerida malabarica*

Arid w peninsular India

☐ **Sun Lark** *Galerida modesta*

____	*G. m. nigrita*	Senegambia to Sierra Leone and Mali
____	*G. m. modesta (giffardi)*	Burkina Faso to n Nigeria, n Cameroon, Chad and w Sudan
____	*G. m. struempelli*	Cameroon (Foumban, Tibati and Ngaoundéré)
____	*G. m. bucolica*	N Zaire to w Uganda

☐ **Tawny Lark** *Galerida deva*

Semiarid c plateau of India

☐ **Large-billed Lark** *Galerida magnirostris*

____	*G. m. sedentaria*	Extreme sw Namibia and w South Africa (e to Griqualand West)
____	*G. m. magnirostris*	SW South Africa
____	*G. m. harei (montivaga)*	W grasslands of South Africa and Lesotho

☐ **Eurasian Skylark** *Alauda arvensis*

____ *A. a. arvensis*	Azores; Europe from Wales to Norway, Ural Mts. and Alps
____ *A. a. scotica*	Ireland, nw England, Scotland and Faeroe Islands
____ *A. a. guillelmi*	N Portugal and nw Spain
____ *A. a. sierrae*	C and s Portugal to s Spain
____ *A. a. harterti*	Mts. of nw Africa
____ *A. a. cantarella*	S Europe to Balkans, Crimea and Iran; >N Africa
____ *A. a. armenica*	Transcaucasia and e Turkey to sw Iran (Zagros and Elburz mts.)
____ *A. a. dulcivox*	SE Russia to Yenisey basin and Afghanistan; >nw India
____ *A. a. kiborti*	S Siberia to n Mongolia, Manchuria and Korea; >e China
____ *A. a. intermedia*	SE Siberia to lower Amur River and ne Manchuria; >e China
____ *A. a. pekinensis*	NE Siberia to Sea of Okhotsk, Kamchatka Pen. and Kuril Is.
____ *A. a. lonnbergi*	Shantar and Sakhalin islands (Sea of Okhotsk); >Japan
____ *A. a. japonica*	Major islands in Japanese Archipelago; >Ryukyu Islands

☐ **Oriental Skylark** *Alauda gulgula*

____ *A. g. inconspicua*	Transcaspia to Turkmenistan, e Iran, Afghanistan and nw India
____ *A. g. lhamarum*	Pamir Mts. and w Himalayas (Kashmir to n Punjab)
____ *A. g. weigoldi*	E China (Shandong to s Shaanxi and c Sichuan)
____ *A. g. inopinata*	Tibetan plateau, e Qinghai, Gansu and sw Inner Mongolia
____ *A. g. vernayi*	E Himalayas and adjacent China (ae Xizang and w Yunnan)
____ *A. g. gulgula (herberti, australis)*	E India to Sri Lanka and Indochina
____ *A. g. coelivox (sala)*	SE and s China to n Vietnam; Hainan I.
____ *A. g. wattersi (wolfei)*	Taiwan and Philippine Islands

☐ **Razo Skylark** *Alauda razae*

	Razo I. (Cape Verde Islands)

☐ **Wood Lark** *Lullula arborea*

____ *L. a. arborea*	N Europe to Portugal, n Spain, n Italy, n Yugoslavia and Ukraine
____ *L. a. pallida*	S Europe to Crimea, Caucasus, Iran and Turkmenistan

FAMILY: HIRUNDINIDAE (Swallows—83)

☐ **African River Martin** *Pseudochelidon eurystomina*

	Large rivers of Zaire; >coastal Gabon

☐ **White-eyed River Martin** *Pseudochelidon sirintarae*

	Last recorded ca 1980 from c Thailand. Probably extinct

☐ **Square-tailed Sawwing** *Psalidoprocne nitens*

____ *P. n. nitens*	S Guinea to Cameroon, Gabon, extreme w Zaire and nw Angola
____ *P. n. cis*	NW Zaire (Tshuapa to Semliki Valley)

☐ **Mountain Sawwing** *Psalidoprocne fuliginosa*

	Mts. of se Nigeria and sw Cameroon; Bioko

☐ **White-headed Sawwing** *Psalidoprocne albiceps*

____ *P. a. albiceps*	S Sudan to w Kenya, e Zaire, Tanzania, Zambia and Malawi
____ *P. a. suffusa*	N Angola to extreme sw Zaire

☐ **Black Sawwing** *Psalidoprocne pristoptera*

____ *P. p. petiti*	SE Nigeria and sw Cameroon to Cabinda and lower Congo River
____ *P. p. chalybea*	Extreme se Nigeria to Cameroon, C African Rep. and Zaire
____ *P. p. pristoptera*	Highlands of Eritrea to n Ethiopia and Somalia
____ *P. p. blanfordi*	Highlands of w-c Ethiopia
____ *P. p. antinorii*	Highlands of ne, s and e-c Ethiopia (s to Lake Turkana)
____ *P. p. oleaginea*	SW Ethiopia (Maji region)
____ *P. p. mangbettorum*	Extreme ne Zaire to extreme sw Sudan
____ *P. p. ruwenzori*	Ruwenzori Mts. (e Zaire and sw Uganda)
____ *P. p. reichenowi*	SW Gabon to c Angola, s Zaire and w Zambia
____ *P. p. massaica*	Mts. of w and c Kenya to Uluguru and Usambara Mts. (Tanzania)

____ *P. p. orientalis*	SE Tanzania to ne Zambia, Zimbabwe, Malawi and Mozambique
____ *P. p. holomelas*	Highlands of Kenya and Tanzania to South Africa

☐ **Fanti Sawwing** *Psalidoprocne obscura*

Senegambia to e Nigeria, sw Cameroon and C African Rep.

☐ **Gray-rumped Swallow** *Pseudhirundo griseopyga*

____ *P. g. melbina*	Liberia and Guinea-Bissau to Gabon and lower Congo River
____ *P. g. griseopyga (andrewi)*	S Ethiopia and Sudan to Kenya, Uganda, Cameroon and n S Africa

☐ **White-backed Swallow** *Cheramoeca leucosterna*

Southern Australia (except sw Western Australia and s Victoria)

☐ **Mascarene Martin** *Phedina borbonica*

____ *P. b. borbonica*	Mauritius and Réunion (w Mascarene Islands)
____ *P. b. madagascariensis*	Madagascar and Pemba I.; wanders to e Africa

☐ **Brazza's Martin** *Phedina brazzae*

Locally along rivers of sw Zaire and ne Angola (Lunda)

☐ **Plain Martin** *Riparia paludicola*

____ *R. p. mauritanica*	W Morocco (Oued Oum R'bia and Oued Sous)
____ *R. p. minor (paludibula)*	Senegambia to Sudan and ne Ethiopia
____ *R. p. schoensis*	Highlands of Ethiopia
____ *R. p. newtoni*	Mts. of e Nigeria and w Cameroon (Bamenda highlands)
____ *R. p. ducis*	E Zaire (Kivu) to Uganda, Kenya and c Tanzania
____ *R. p. paludicola*	Angola to Zambia, s Tanzania and South Africa
____ *R. p. cowani*	Madagascar
____ *R. p. chinensis*	Afghanistan and Pakistan to n India, Myanmar and SE Asia
____ *R. p. tantilla*	N Philippines (Luzon and [?] Negros)

☐ **Congo Martin** *Riparia congica*

Zaire (middle and lower Congo River and lower Ubangi River)

☐ **Bank Swallow** *Riparia riparia*

____ *R. r. riparia*	Breeds widely in Holarctic regions; >in tropics
____ *R. r. ijimae*	Kamchatka Pen. and Kuril Is. to Amur River and Hokkaido
____ *R. r. shelleyi*	Lower Egypt and Suez Canal region
____ *R. r. eilata*	S Israel
____ *R. r. innominata*	SE Kazakhstan; may winter in Africa or s Asia

☐ **Pale Sand Martin** *Riparia diluta*

____ *R. d. gavrilovi*	C Siberia to River Lena, south to Altai and Tuva Republic
____ *R. d. transbaykalica*	Transbaikalia
____ *R. d. diluta*	South and se Kazakhstan; >nw India and Nepal
____ *R. d. indica*	Pakistan and n India
____ *R. d. fohkienensis*	C and e China
____ *R. d. tibetana*	Southwest China (Xizang-Qinghai plateau)

☐ **Banded Martin** *Riparia cincta*

____ *R. c. erlangeri*	Ethiopia and Sudan (upper White Nile River)
____ *R. c. suahelica*	S Sudan to Kenya, Uganda, Zambia, Zimbabwe, Mozambique
____ *R. c. parvula*	N Angola to sw Zaire and nw Zambia; >Cameroon
____ *R. c. xerica*	Kalahari of n Botswana and Namibia to Angola and w Zambia
____ *R. c. cincta*	Zimbabwe and Kwazulu-Natal to Cape Province

☐ **Tree Swallow** *Tachycineta bicolor*

Breeds Alaska to s US; >n South America

☐ **Violet-green Swallow** *Tachycineta thalassina*

____ *T. t. thalassina (lepida)*	Alaska and Canada to c Baja and s Mexico; >Costa Rica
____ *T. t. brachyptera*	Mts. of c and s Baja; coastal w Mexico (Sonora to Sinaloa)

☐ **Golden Swallow** *Tachycineta euchrysea*

____ *T. e. euchrysea*	Mts. of Jamaica
____ *T. e. sclateri*	Mts. of Hispaniola

☐ **Bahama Swallow** *Tachycineta cyaneoviridis*

Pine forests of n Bahamas; >in Bahamas and e Cuba

☐ **Tumbes Swallow** *Tachycineta stolzmanni*

Coastal nw Peru (Tumbes to La Libertad) and adj. sw Ecuador

☐ **Mangrove Swallow** *Tachycineta albilinea*

Lowlands of n Mexico to Panama

☐ **White-winged Swallow** *Tachycineta albiventer*

Guianas and Venezuela to se Brazil and n Argentina; Trinidad

☐ **White-rumped Swallow** *Tachycineta leucorrhoa*

Bolivia to Paraguay, se Brazil and n Argentina; >s Peru

☐ **Chilean Swallow** *Tachycineta meyeni*

Breeds s Chile and Argentina; >n to Bolivia and Brazil

☐ **Purple Martin** *Progne subis*
_____ *P. s. subis*
_____ *P. s. arboricola*
_____ *P. s. hesperia*

S Canada to highlands of c Mexico; >Brazil
Mts. of w North America
SW Arizona to nw Mexico (Sonora), s Baja Calif. and Isla Tiburón

☐ **Cuban Martin** *Progne cryptoleuca*

Cuba and Isle of Pines; casual in s Florida

☐ **Caribbean Martin** *Progne dominicensis*

West Indies (except Cuba and Isle of Pines); Tobago

☐ **Sinaloa Martin** *Progne sinaloae*

Breeds Sierra Madre Occidental of w Mexico; winter range unknown

☐ **Gray-breasted Martin** *Progne chalybea*
_____ *P. c. warneri*
_____ *P. c. chalybea*
_____ *P. c. macrorhamphus*

Coastal western Mexico (Sinaloa to Chiapas)
E Mexico (s Tamaulipas) to n Argentina and Brazil
E Bolivia and e Brazil to Paraguay, Uruguay and ne Argentina

☐ **Galapagos Martin** *Progne modesta*

Galapagos Islands

☐ **Peruvian Martin** *Progne murphyi*

Coastal Peru (Piura south to Ica), rarely to n Chile (Arica)

☐ **Southern Martin** *Progne elegans*

Bolivia to Uuruguay and Argentina; >north to Colombia

☐ **Brown-chested Martin** *Progne tapera*
_____ *P. t. tapera*
_____ *P. t. fusca*

Trop. e Colombia to Bolivia, the Guianas and Amazonian Brazil
SE Brazil to Paraguay, e Bolivia, Uruguay and n Argentina

☐ **Brown-bellied Swallow** *Notiochelidon murina*
_____ *N. m. murina*
_____ *N. m. meridensis*
_____ *N. m. cyanodorsalis*

Andes of Colombia to s Peru (Arequipa and Cuzco)
Andes of w Venezuela (Mérida and Trujillo)
Cordillera of w Bolivia and possibly adjacent Peru (Puno)

☐ **Blue-and-white Swallow** *Notiochelidon cyanoleuca*
_____ *N. c. cyanoleuca*
_____ *N. c. peruviana*
_____ *N. c. patagonica*

Highlands of Costa Rica to Venezuela, Brazil and n Argentina
Coastal Peru (La Libertad to Arequipa)
C Chile and Argentina to Tierra del Fuego

☐ **Pale-footed Swallow** *Notiochelidon flavipes*

Andes of Colombia and w Venezuela to w Bolivia

☐ **Black-capped Swallow** *Notiochelidon pileata*

Highlands of s Mexico (Chiapas) to w Honduras

☐ **Andean Swallow** *Haplochelidon andecola*
_____ *H. a. andecola*
_____ *H. a. oroyae*

Andes of s Peru (Cuzco, Puno and Arequipa) to n Bolivia and Chile
Puna of c Peru

☐ **White-banded Swallow** *Atticora fasciata*

E Colombia to Guianas, s Venezuela, n Bolivia and Amazonian Brazil

☐ **Black-collared Swallow** *Atticora melanoleuca*

Rapids on rivers of Guianas, s Venezuela and Amazonian Brazil

☐ **White-thighed Swallow** *Neochelidon tibialis*

_____	*N. t. minima*	Extreme e Panama (Darién) to w Colombia and w Ecuador
_____	*N. t. griseiventris*	S Colombia to se Venezuela, n Bolivia and w Amazonian Brazil
_____	*N. t. tibialis*	SE Brazil (Espírito Santo to Rio de Janeiro and São Paulo)

☐ **Northern Rough-winged Swallow** *Stelgidopteryx serripennis*

_____	*S. s. serripennis*	SE Alaska to Arizona, New Mexico, e Texas and Gulf States
_____	*S. s. psammochroa*	S California to Baja, s Texas and s Mexico (Oaxaca)
_____	*S. s. fulvipennis*	S Mexico (Oaxaca and Veracruz) to highlands of Costa Rica
_____	*S. s. stuarti*	S Veracruz, Oaxaca and Chiapas
_____	*S. s. ridgwayi*	SE Mexico (n Yucatán Peninsula)
_____	*S. s. burleighi*	Belize and Guatemala (s Yucatán Peninsula)

☐ **Southern Rough-winged Swallow** *Stelgidopteryx ruficollis*

_____	*S. r. uropygialis*	Caribbean lowlands of Honduras and Nicaragua to nw Peru
_____	*S. r. decolor*	Pacific coast of Costa Rica and Panama
_____	*S. r. aequalis*	N Colombia to e and s Venezuela
_____	*S. r. ruficollis*	SE Colombia to the Guianas, Brazil and n Argentina

☐ **Tawny-headed Swallow** *Alopochelidon fucata*

Venezuela and n Brazil; Brazil s of the Amazon to n Argentina

☐ **Barn Swallow** *Hirundo rustica*

_____	*H. r. erythrogaster*	Alaska and Canada to s Mexico; >s Argentina
_____	*H. r. rustica*	Europe, w Asia and n Africa; >sub-Saharan Africa and s Asia
_____	*H. r. savignii*	Egyptian delta (south to Luxor)
_____	*H. r. transitiva*	Lebanon, Syria, Israel and w Jordan
_____	*H. r. tytleri*	S Siberia to n Inner Mongolia; >India and SE Asia
_____	*H. r. gutturalis*	E Himalayas to ne Myanmar, Japan, Korea; >n Australia
_____	*H. r. mandschurica*	Northeast China; >SE Asia

☐ **Red-chested Swallow** *Hirundo lucida*

_____	*H. l. lucida (clara)*	Sahel and Senegal to Ghana, Togo and extreme w Nigeria
_____	*H. l. rothschildi*	C and sw Ethiopia
_____	*H. l. subalaris*	E Zaire (lower and upper valley of the Congo River) and n Gabon

☐ **Angola Swallow** *Hirundo angolensis*

Uganda to w Kenya, Tanzania, Zambia, Malawi, Angola, Gabon

☐ **Pacific Swallow** *Hirundo tahitica*

_____	*H. t. domicola*	S India and Sri Lanka
_____	*H. t. javanica (abbotti, mallopega)*	Andamans and Myanmar to Indochina, Sundas, Wallacea, Philippines
_____	*H. t. namiyei*	Ryukyu Islands and Taiwan
_____	*H. t. frontalis*	N and w New Guinea
_____	*H. t. albescens*	S and e New Guinea
_____	*H. t. ambiens*	New Britain (Bismarck Archipelago)
_____	*H. t. subfusca*	New Ireland to Solomons, New Caledonia ,Vanuatu , Fiji and Tonga
_____	*H. t. tahitica*	Society Islands (Moorea and Tahiti)

☐ **Welcome Swallow** *Hirundo neoxena*

_____	*H. n. carteri*	SW Western Australia (North West Cape to about Eyre); >n
_____	*H. n. neoxena (parsonsi)*	NE Australia (ne Queensland to s S Australia, Tasmania); >n

☐ **White-throated Swallow** *Hirundo albigularis*

Angola to se Zaire, Botswana, Zambia, Malawi and South Africa

☐ **Wire-tailed Swallow** *Hirundo smithii*

_____	*H. s. smithii*	Widespread Africa south of the Sahara
_____	*H. s. filifera*	Afghanistan and Baluchistan to India, Myanmar and Indochina

☐ **White-throated Blue Swallow** *Hirundo nigrita*

Sierra Leone to s Zaire, Gabon and n Angola

☐ **Ethiopian Swallow** *Hirundo aethiopica*
_____ *H. a. aethiopica*   Senegambia to se Sudan, w Kenya, Uganda and Tanzania
_____ *H. a. amadoni*   E Ethiopia and Somalia to e Kenya

☐ **Pied-winged Swallow** *Hirundo leucosoma*

     Savanna of w Africa (Senegambia to extreme w Cameroon)

☐ **White-tailed Swallow** *Hirundo megaensis*

     Arid highlands of s Ethiopia (Sidamo Province)

☐ **Pearl-breasted Swallow** *Hirundo dimidiata*
_____ *H. d. marwitzi*   Angola to Zaire, Zambia, Zimbabwe, sw Tanzania and Malawi
_____ *H. d. dimidiata*   Zimbabwe to Transvaal, Orange Free State and Cape Province

☐ **Blue Swallow** *Hirundo atrocaerulea*

     E South Africa and w Swaziland; >n to Kenya and s Uganda

☐ **Black-and-rufous Swallow** *Hirundo nigrorufa*

     Savanna of Angola to s Zaire and n Zambia

☐ **Eurasian Crag-Martin** *Ptyonoprogne rupestris*

     S Palearctic to c Asia; >Arabia and India

☐ **Rock Martin** *Ptyonoprogne fuligula*
_____ *P. f. presaharica*   S Morocco, Algeria (except s) and n Mauritania
_____ *P. f. spatzi*   S Algeria, sw Libya and n Chad
_____ *P. f. buchanani*   S-c Sahara (Aïr Massif of n Niger)
_____ *P. f. obsoleta*   Egypt to Arabia and Iran
_____ *P. f. perpallida*   S Iraq and ne Saudi Arabia
_____ *P. f. pallida (peloplasta)*   E Iran, s Afghanistan and Pakistan
_____ *P. f. arabica*   NE Chad, n Sudan, sw Arabia, Eritrea, n Somalia and Socotra I.
_____ *P. f. pusilla*   S Mali to Eritrea and Ethiopia
_____ *P. f. bansoensis (birwae)*   Sierra Leone to Nigeria and Cameroon
_____ *P. f. fusciventris (rufigula)*   S Chad, CAR, w and s Sudan, sw Ethiopia to n Mozambique
_____ *P. f. anderssoni*   N and sw Angola and n and c Namibia
_____ *P. f. fuligula*   S Namibia, Botswana and w South Africa
_____ *P. f. pretoriae*   SW Zimbabwe and s Mozambique to e South Africa

☐ **Dusky Crag-Martin** *Ptyonoprogne concolor*
_____ *P. c. concolor*   Himalayan foothills (nw India to w Bengal)
_____ *P. c. sintaungensis*   SW China (s Yunnan) to e Myanmar, n Thailand and Indochina

☐ **House Martin** *Delichon urbicum*
_____ *D. u. urbicum*   Europe to c Asia, w and se Africa
_____ *D. u. meridionale*   Mediterranean basin to North Africa, Iran and n India
_____ *D. u. lagopodum*   Siberia to Mongolia and Manchuria; >s China, Thailand

☐ **Asian Martin** *Delichon dasypus*
_____ *D. d. dasypus*   Siberia to Kuril Is., Japan; >Greater Sundas, Philippines
_____ *D. d. cashmiriensis*   Himalayas (se Afghanistan to India and w China)
_____ *D. d. nigrimentale*   SE China (Fujian and Guangxi) and Taiwan

☐ **Nepal Martin** *Delichon nipalense*
_____ *D. n. nipalense*   Himalayas from Nepal and se Tibet to w and e Myanmar
_____ *D. n. cuttingi*   NE Myanmar to s China (swYunnan), n Thailand, n Laos, w Tonkin

☐ **Greater Striped-Swallow** *Cecropis cucullata*

     Breeds southern Africa; >north to Tanzania

☐ **Lesser Striped-Swallow** *Cecropis abyssinica*
_____ *C. a. abyssinica*   E Sudan to Eritrea, Ethiopia and Somalia
_____ *C. a. unitatis*   S Sudan to Kenya, w Uganda, Gabon and e Cape Province
_____ *C. a. puella*   Senegambia to ne Nigeria and n Cameroon
_____ *C. a. maxima*   SE Nigeria and s Cameroon to sw C African Republic
_____ *C. a. bannermani*   NE C African Republic to sw Sudan (Darfur Province)
_____ *C. a. ampliformis*   S Angola to n Namibia, w Zambia and nw Zimbabwe

☐ **Rufous-chested Swallow** *Cecropis semirufa*

____ *C. s. gordoni* Senegal to s Sudan, n Angola, sw Kenya and nw Tanzania

____ *C. s. semirufa* Botswana to Malawi, Mozambique and e Cape Province

☐ **Mosque Swallow** *Cecropis senegalensis*

____ *C. s. senegalensis* Mauritania and Senegambia to s Chad and s Sudan

____ *C. s. saturatior* S Ghana to Gabon, Ethiopia, Uganda and Kenya

____ *C. s. monteiri* Angola to Zaire, Zambia, Malawi and Mozambique

☐ **Red-rumped Swallow** *Cecropis daurica*

____ *C. d. daurica (gephyra)* S Siberia to Amur River, n Mongolia, w China and Transbaikalia

____ *C. d. japonica* Korea, e and c China and Japan; migrant to coastal n Australia

____ *C. d. nipalensis* C Himalayas to sw China (Yunnan), n India and n Myanmar

____ *C. d. erythropygia* N India (base of Himalayas to Nilgiri)

____ *C. d. hyperythra* Sri Lanka

____ *C. d. rufula* Iberian Peninsula to N Africa, Iran, Afghanistan and nw India

____ *C. d. domicella* Senegambia to s Sudan and extreme nw Uganda

____ *C. d. kumboensis (disjuncta)* Sierra Leone (Birwa Plateau) and Cameroon (Bamenda highlands)

____ *C. d. emini* S Sudan to e Zaire, Uganda, Kenya, Tanzania and Malawi

____ *C. d. melanocrissus* Highlands of Ethiopia

☐ **Striated Swallow** *Cecropis striolata*

____ *C. s. mayri* NW India to n Myanmar and nw Thailand

____ *C. s. stanfordi* NE Myanmar to sw China (s Yunnan), n Thailand and n Laos

____ *C. s. vernayi* Thailand/Tenasserim border and w Thailand

____ *C. s. striolata* Greater and Lesser Sundas to the Philippines and Taiwan

☐ **Rufous-bellied Swallow** *Cecropis badia* Malay Peninsula

☐ **Red-throated Swallow** *Petrochelidon rufigula* Gabon and s-c Zaire to Angola and nw Zambia

☐ **Preuss' Swallow** *Petrochelidon preussi* Guinea-Bissau, Sierra Leone and Mali to Cameroon and ne Zaire

☐ **Red Sea Swallow** *Petrochelidon perdita* Known from one specimen from Red Sea coast of e Sudan

☐ **South African Swallow** *Petrochelidon spilodera* S Zimbabwe and South Africa; >lower Congo basin

☐ **Forest Swallow** *Petrochelidon fuliginosa* Lowland forests of e Nigeria, s Cameroon and Gabon

☐ **Streak-throated Swallow** *Petrochelidon fluvicola* NE Afghanistan to n Pakistan and peninsular India

☐ **Fairy Martin** *Petrochelidon ariel* Watercourses of mainland Australia; >New Guinea

☐ **Tree Martin** *Petrochelidon nigricans*

____ *P. n. timoriensis* Lesser Sundas (Timor, Alor and Flores); >Solomon Islands

____ *P. n. nigricans* S New Guinea; Tasmania, King and Flinders Islands; >n

____ *P. n. neglecta* Treed areas of mainland Australia and islands of s Torres Strait

☐ **Cliff Swallow** *Petrochelidon pyrrhonota*

____ *P. p. pyrrhonota (hypopolia)* Breeds Alaska and US; >South America

____ *P. p. tachina* SW US to nw Mexico and Baja; >South America

____ *P. p. melanogaster* SE Arizona and New Mexico to Oaxaca; >South America

____ *P. p. ganieri* S US w of Appalachians (Tennessee to Texas); >S America

☐ **Cave Swallow** *Petrochelidon fulva*

____ *P. f. pallida (pelodoma)* N Arizona to New Mexico, s Texas and ne Mexico (Tamaulipas)

____ *P. f. citata* S Mexico (n Yucatán Peninsula and interior valley of Chiapas)

____ *P. f. cavicola* Cuba and Isle of Pines

____ *P. f. poeciloma* Jamaica

____	*P. f. fulva*	Hispaniola and Gonâve I.
____	*P. f. puertoricensis*	Puerto Rico

☐ **Chestnut-collared Swallow** *Petrochelidon rufocollaris*

____	*P. r. aequatorialis*	Pacific coast of s Ecuador (Loja and Guayaquil)
____	*P. r. rufocollaris*	Pacific coast of n and c Peru (south to Lima)

FAMILY: MOTACILLIDAE (Wagtails and Pipits—66)

☐ **Australasian Pipit** *Anthus novaeseelandiae*

____	*A. n. exiguus*	Grasslands of c New Guinea (Mt. Hagen to upper Watut River)
____	*A. n. rogersi*	N Australia (Northern Territory to Gulf of Carpenteria)
____	*A. n. bilbali*	SW Australia
____	*A. n. australis (subaustralis)*	Queensland to Victoria and e South Australia
____	*A. n. bistriatus*	Tasmania and islands in Bass Strait
____	*A. n. novaeseelandiae (reischeki)*	South I. (New Zealand)
____	*A. n. chathamensis*	Chatham Islands
____	*A. n. aucklandicus*	Auckland Islands
____	*A. n. steindachneri*	Antipodes Islands

☐ **Richard's Pipit** *Anthus richardi*

____	*A. r. richardi*	SW Siberia and ne Kazakhstan to L Baikal; >mainly sw Asia
____	*A. r. dauricus*	Transbaikalia to Sea of Okhotsk, n Mongolia; ne China; >s Asia
____	*A. r. centralasiae*	E Kazakhstan to w and s Mongolia and n China; >s Asia
____	*A. r. ussuriensis*	SE Russia to e China (probably Korea); >SE Asia
____	*A. r. sinensis*	SE China (south of the Yangtze R.)

☐ **Oriental Pipit** *Anthus rufulus*

____	*A. r. waitei*	NW Indian subcontinent
____	*A. r. rufulus*	Indian subcontinent (except range of *watei*), to s China, Indochina
____	*A. r. malayensis*	Extreme sw India, Sri Lanka, Malay Pen., Gr. Sundas, s Indochina
____	*A. r. lugubris*	Philippine, Palawan and (?) n Borneo
____	*A. r. albidus*	Sulawesi, Bali and w Lesser Sundas (Lombok to Sumba)
____	*A. r. medius*	E Lesser Sundas (Sawu, Timor, Roti, Kisar, Leti, Moa, Sermata)

☐ **African Pipit** *Anthus cinnamomeus*

____	*A. c. lynesi*	SE Nigeria to Cameroon and w Sudan (Darfur)
____	*A. c. camaroonensis*	Cameroon (Mt. Cameroon and Mt. Manenguba)
____	*A. c. stabilis*	C and se Sudan
____	*A. c. cinnamomeus*	Highlands of w and se Ethiopia
____	*A. c. eximius*	Yemen
____	*A. c. annae*	Eritrea to Ethiopia, Djibouti, Somalia, e Kenya and ne Tanzania
____	*A. c. itombwensis*	E Zaire (Itombwe Highlands and Mt. Kabobo)
____	*A. c. lacuum*	Kenya to se Uganda and c Tanzania
____	*A. c. lichenya*	NE Angola to s Zaire, w Uganda, w Tanzania and Mozambique
____	*A. c. spurius*	NE Namibia to n Botswana, s Malawi and s Mozambique
____	*A. c. bocagei*	Angola to Botswana, Zimbabwe, s Mozambique and n Cape Prov.
____	*A. c. grotei*	Salt pans of n Namibia and n Botswana
____	*A. c. rufuloides*	South Africa (except nw), Swaziland and Lesotho lowlands

☐ **Mountain Pipit** *Anthus hoeschi*

South Africa (Drakensberg Mts. of s Lesotho)

☐ **Jackson's Pipit** *Anthus latistriatus*

Highlands of sw Uganda, nw Tanzania and w Kenya

☐ **Woodland Pipit** *Anthus nyassae*

____	*A. n. schoutedeni*	SE Gabon and s PRCongo to s DRCongo and s Angola
____	*A. n. nyassae*	Zambia to s Tanzania, Malawi and nw Mozambique
____	*A. n. chersophilus*	NE Namibia and n Botswana
____	*A. n. frondicolus*	Zimbabwe and adjacent s Mozambique

☐ **Long-billed Pipit** *Anthus similis*

____	*A. s. asbenaicus*	S Sahara in c and e Mali and c Niger
____	*A. s. bannermani*	Mts. of sw Mali, Guinea, Sierra Leone, n Liberia to w Cameroon
____	*A. s. captus*	Lebanon, Syria, Israel, Palestine and w Jordan
____	*A. s. jebelmarrae*	Mts. of w and c Sudan
____	*A. s. nivescens*	Mts. of se Egypt; coastal Red Sea in ne Sudan to n Kenya
____	*A. s. hararensis (chyuluensis)*	Highlands of Eritrea, Ethiopia, Kenya and Tanzania
____	*A. s. dewittei (hallae)*	Highlands of e DRCongo, sw Uganda, Rwanda, Burundo, Angola
____	*A. s. dewittei (moco, hallae)*	Highlands of se Zaire to Rwanda, Burundi and sw Uganda
____	*A. s. palliditinctus*	Extreme sw Angola and nw Namibia
____	*A. s. leucocraspedon*	W and s Namibia and sw South Africa
____	*A. s. nicholsoni*	SE Botswana and ne South Africa
____	*A. s. petricolus*	Lesotho and e South Africa
____	*A. s. primarius*	S South Africa (s Western Cape to n Eastern Cape)
____	*A. s. arabicus*	SW, s and se Arabian Peninsula
____	*A. s. sokotrae*	Socotra I.
____	*A. s. decaptus*	S Iran to w Pakistan; >nw India
____	*A. s. jerdoni*	Mts. of e Afghanistan to w Nepal; >n-c India, Bangladesh
____	*A. s. similis*	Peninsular India (Bombay to Karnataka and w Tamil Nadu)
____	*A. s. travancoriensis*	SW India
____	*A. s. yamethini*	C Myanmar

☐ **Kimberley Pipit** *Anthus pseudosimilis*

Kimberley and adjacent regions of South Africa

☐ **Blyth's Pipit** *Anthus godlewskii*

S Siberia to Mongolia, China, Tibet and ne India

☐ **Tawny Pipit** *Anthus campestris*

____	*A. c. campestris (boehmi)*	Europe to North Africa, Middle East and Asia; >Africa and India
____	*A. c. kastschenkoi*	S Siberia to nw Mongolia; >s Asia
____	*A. c. griseus*	SW Kazakhstan to ne Iran, Afghanistan, nw China; >sw & s Asia

☐ **Plain-backed Pipit** *Anthus leucophrys*

____	*A. l. ansorgei*	S Mauritania to Senegambia and Guinea-Bissau
____	*A. l. zenkeri*	S Mali to s Sudan, n Zaire, w Uganda, w Kenya and nw Tanzania
____	*A. l. gouldii*	Sierra Leone to Liberia and Ivory Coast
____	*A. l. omoensis*	Extreme e Sudan and Ethiopia
____	*A. l. bohndorffi*	Angola to s Zaire, se Gabon, Zambia, n Malawi and Tanzania
____	*A. l. tephridorsus*	S Angola to sw Zambia, ne Namibia and nw Botswana
____	*A. l. leucophrys (enunciator)*	Mozambique to Swaziland, Lesotho and South Africa

☐ **Long-tailed Pipit** *Anthus longicaudatus*

Breeding range unknown; >in South Africa (Kimberly area)

☐ **Buffy Pipit** *Anthus vaalensis*

____	*A. v. saphiroi*	SE Ethiopia and nw Somalia
____	*A. v. goodsoni*	C and sw Kenya to extreme n Tanzania
____	*A. v. neumanni (muhingae)*	Plateau of Angola; disperses to Namibia and Botswana
____	*A. v. chobiensis*	NE Namibia to s Zaire, w Tanzania, n Botswana and Mozambique
____	*A. v. marungensis*	N Zambia to s Tanzania
____	*A. v. namibicus*	NE and c Namibia
____	*A. v. exasperatus*	Salt pans of ne Botswana; wanders to w Zimbabwe
____	*A. v. vaalensis (daviesi, clanceyi)*	S Botswana to s Mozambique, nevSouth Africa and w Lesotho

☐ **Long-legged Pipit** *Anthus pallidiventris*

____	*A. p. pallidiventris*	Cameroon and Equatorial Guinea to Gabon and nw Angola
____	*A. p. esobe*	C Zaire (upper Congo River)

☐ **Nilgiri Pipit** *Anthus nilghiriensis*

Montane grasslands of sw India (w Tamil Nadu and Kerala)

☐ **Upland Pipit** *Anthus sylvanus*

Himalayas (e Afghanistan to Nepal and se China)

☐ **Berthelot's Pipit** *Anthus berthelotii*
- ____ *A. b. madeirensis* — Madeira Arch. (Madeira, Desertas, Porto Santo and Baixo Is.)
- ____ *A. b. berthelotii* — Selvagens and Canary Islands

☐ **Malindi Pipit** *Anthus melindae*
- ____ *A. m. mallablensis* — Somalia (ne of Mogadishu in Mallable region)
- ____ *A. m. melindae* — Coastal s Somalia and se Kenya (s to Mombassa)

☐ **Striped Pipit** *Anthus lineiventris*
- ____ *A. l. stygium* — W Angola to w Tanzania, se Kenya and coastal e South Africa
- ____ *A. l. lineiventris* — SE Botswana to ne South Africa and w Swaziland

☐ **Yellow-tufted Pipit** *Anthus crenatus*

South Africa (montane slopes of e Cape Province to w Swaziland)

☐ **Alpine Pipit** *Anthus gutturalis*
- ____ *A. g. wollastoni* — High alpine grasslands of w-c New Guinea
- ____ *A. g. rhododendri* — Alpine grasslands of e-c New Guinea and Huon Peninsula
- ____ *A. g. gutturalis* — High alpine grasslands of se New Guinea

☐ **Sprague's Pipit** *Anthus spragueii*

Prairies and plains of North America; >s Mexico

☐ **Short-billed Pipit** *Anthus furcatus*
- ____ *A. f. brevirostris* — *Puna* of Andes of Peru and Bolivia
- ____ *A. f. furcatus* — Extreme se Brazil to Paraguay, Uruguay and n Argentina

☐ **Yellowish Pipit** *Anthus lutescens*
- ____ *A. l. parvus* — Savanna of w Panama
- ____ *A. l. lutescens* — Savanna of e Colombia to Venezuela, Guianas, Brazil, Argentina
- ____ *A. l. peruvianus* — Coastal n Peru (Lambayeque) to extreme n Chile (Tacna)

☐ **Chaco Pipit** *Anthus chacoensis*

Locally in *chaco* of e Paraguay and n Argentina

☐ **Correndera Pipit** *Anthus correndera*
- ____ *A. c. calcaratus* — Arid *páramo* of Peru (Junín, Cuzco and Puno)
- ____ *A. c. catamarcae* — *Páramo* of Bolivia (Potosí) to n Chile and extreme nw Argentina
- ____ *A. c. chilensis* — S Chile and s Argentina
- ____ *A. c. correndera* — Coastal se Brazil to Uruguay, Paraguay and n Argentina
- ____ *A. c. grayi* — Falkland Islands

☐ **South Georgia Pipit** *Anthus antarcticus*

Grasslands of South Georgia I.

☐ **Ochre-breasted Pipit** *Anthus nattereri*

Locally in se Brazil, se Paraguay and ne Argentina

☐ **Hellmayr's Pipit** *Anthus hellmayri*
- ____ *A. h. hellmayri* — Andes of s Peru (Puno) to Bolivia and nw Argentina (Tucumán)
- ____ *A. h. dabbenei* — Andes of w Argentina (w Neuquén and w Chubut) and adj. Chile
- ____ *A. h. brasilianus* — SE Brazil (São Paulo, Rio de Janeiro) to Uruguay and n Argentina

☐ **Paramo Pipit** *Anthus bogotensis*
- ____ *A. b. bogotensis* — E Andes of Colombia and Ecuador
- ____ *A. b. meridae* — Andes of nw Venezuela (Mérida, Táchira and Trujillo)
- ____ *A. b. immaculatus* — Andes of s Peru (Junín) to n Bolivia (La Paz and Cochabamba)
- ____ *A. b. shiptoni* — Andes of s Bolivia (Cochabamba) to nw Argentina (Tucumán)

☐ **Meadow Pipit** *Anthus pratensis*
- ____ *A. p. whistleri* — Iceland and Faroes to Scotland, Ireland and England
- ____ *A. p. pratensis* — SE Greenland to Europe and w Siberia; >N Africa, Iran

☐ **Rosy Pipit** *Anthus roseatus*

E Afghanistan to se Tibet and sw China; >SE Asia

☐ **Red-throated Pipit** *Anthus cervinus*

Tundra of n Palearctic, Alaska; >Africa and Indonesia

☐ **Olive-backed Pipit** *Anthus hodgsoni*

_____ *A. h. yunnanensis* — NE Eurasia; >India, Myanmar, Philippines and Borneo
_____ *A. h. hodgsoni (berezowskii)* — Himalayas to China and Japan; >SE Asia

☐ **Tree Pipit** *Anthus trivialis*

_____ *A. t. trivialis* — Europe to L. Baikal and n Iran; >Africa and India
_____ *A. t. schlueteri* — Kazakhstan to nw China, Afghanistan; >in cent. China
_____ *A. t. haringtoni* — NW Himalayas (Kashmir to Garhwal); >in Indian subcont.

☐ **Pechora Pipit** *Anthus gustavi*

_____ *A. g. gustavi* — N Eurasia; >Philippines, n Borneo and Wallacea
_____ *A. g. stejnegeri (commandorensis)* — Commander Is.; >E Asia
_____ *A. g. menzbieri* — Russian Far East and extreme ne China; winter area unknown

☐ **Water Pipit** *Anthus spinoletta*

_____ *A. s. spinoletta* — Mts. of c and sw Europe (Iberia to Balkans and nw Turkey)
_____ *A. s. coutellii* — E Turkey to Caucasus, n Iran and Turkmenistan
_____ *A. s. blakistoni* — NE Afghanistan to Transbaikalia and Nan Shan Mts.

☐ **Rock Pipit** *Anthus petrosus*

_____ *A. p. kleinschmidti* — Faeroes, Shetlands, Orkneys, Fair Isla and St. Kilda I.
_____ *A. p. petrosus (meinertzhageni)* — Coastal Ireland, British Isles, nw France and Channel Islands
_____ *A. p. littoralis* — Fennoscandia and nw Russia; >s Spain and nw Africa

☐ **American Pipit** *Anthus rubescens*

_____ *A. r. japonicus (harmsi)* — E Siberia (s of *rubescens*); >s China, India and Myanmar
_____ *A. r. pacificus* — Alaska to Oregon; >w Mexico
_____ *A. r. alticola* — Mts. of s British Columbia to California; >Mexico
_____ *A. r. rubescens* — N and e Canada, w Greenland, extreme ne US; >Cent. Am.

☐ **Short-tailed Pipit** *Anthus brachyurus*

_____ *A. b. leggei (eludens)* — SE Gabon to Angola, Zaire, w Uganda and Zambia to e S Africa
_____ *A. b. brachyurus* — Grasslands of s Mozambique and e South Africa

☐ **Bush Pipit** *Anthus caffer*

_____ *A. c. australoabyssinicus* — Highlands of extreme s Ethiopia
_____ *A. c. blayneyi* — Highlands of Kenya and Tanzania
_____ *A. c. mzimbaensis* — N Zambia to ne Botswana, Zimbabwe plateau and w Malawi
_____ *A. c. caffer* — SE Botswana to sw Zimbabwe, Transvaal, w Swaziland, n Natal
_____ *A. c. traylori* — S Mozambique, adjacent e Transvaal and extreme ne Natal

☐ **Sokoke Pipit** *Anthus sokokensis*

Coastal forests of se Kenya and ne Tanzania

☐ **Golden Pipit** *Tmetothylacus tenellus*

Arid scrub of se Sudan to Ethiopia, Somalia, Kenya and n Tanzania

☐ **Yellow-breasted Pipit** *Hemimacronyx chloris*

S Transvaal to w Natal, Orange Free State, Lesotho, e Cape Prov.

☐ **Sharpe's Longclaw** *Hemimacronyx sharpei*

Grassy highlands of Kenya

☐ **Orange-throated Longclaw** *Macronyx capensis*

_____ *M. c. stabilior* — Zimbabwe (cent. Plateau to Inyanga Mts.) and adj. Mozambique
_____ *M. c. colletti (latimerae)* — SE Botswana to e South Africa, Swaziland and Lesotho
_____ *M. c. capensis* — Coastal belt of sw and s South Africa

☐ **Yellow-throated Longclaw** *Macronyx croceus*

Widespread grasslands of sub-Saharan Africa

☐ **Fuelleborn's Longclaw** *Macronyx fuelleborni*

_____ *M. f. fuelleborni* — Grasslands and *brachystegia* woodlands of extreme s Tanzania
_____ *M. f. ascensi* — Grasslands and *brachystegia* woodlands of c Africa

☐ **Abyssinian Longclaw** *Macronyx flavicollis*

High grasslands of Ethiopia

☐ **Pangani Longclaw** *Macronyx aurantiigula*

Arid savanna of s Somalia, Kenya and ne Tanzania

☐ **Rosy-throated Longclaw** *Macronyx ameliae*

_____ *M. a. wintoni* C and sw Kenya and n Tanzania
_____ *M. a. altanus* E and s Angola to sw Tanzania, n Botswana and nw Mozambique
_____ *M. a. ameliae* Coastal s Mozambique and ne South Africa

☐ **Grimwood's Longclaw** *Macronyx grimwoodi*

Wet grasslands of e Angola to sw Zaire and extreme nw Zambia

☐ **Forest Wagtail** *Dendronanthus indicus*

Russian Far East to se China and s Japan; >s and SE Asia

☐ **White Wagtail** *Motacilla alba*

_____ *M. a. alba* SE Greenland, Iceland to Faeroes, Europe, s Urals, Asia Minor
_____ *M. a. yarrellii* Britain, Ireland and adj. coastal w continental Europe; >nw Africa
_____ *M. a. dukhunensis* C Russia to Caucasus, nw Iran, Altai Mts.; >Middle East to India
_____ *M. a. ocularis* N Siberia and ne Alaska; >Indian subcontinent and s Asia
_____ *M. a. subpersonata* W Morocco
_____ *M. a. persica* N-c and w Iran (Elburz and Zagros Mts.)
_____ *M. a. personata* C Siberia to nw Mongolia, n Iran, Afghanistan and n India
_____ *M. a. baicalensis* S-c Siberia to Mongolia and ne China; >ne India to c Indochina
_____ *M. a. lugens* SE Russia to n Korea and Japan.; >Myanmar to se China
_____ *M. a. leucopsis* C & e China, Russian Far East, Korea, sw Japan; >n India, s Asia
_____ *M. a. alboides* Himalayas, s China, n Myanmar, n Laos, n Vietnam; >to n Thailand

☐ **White-browed Wagtail** *Motacilla madaraspatensis*

Himalayas (n Pakistan to Bangladesh) and peninsular India

☐ **Japanese Wagtail** *Motacilla grandis*

Breeds Japanese Islands; >China, Korea and Taiwan

☐ **Mekong Wagtail** *Motacilla samveasnae*

Lower Mekong of Cambodia and Laos

☐ **African Pied Wagtail** *Motacilla aguimp*

_____ *M. a. vidua* Sierra Leone to s Sudan, Ethiopia, Kenya and South Africa
_____ *M. a. aguimp* Namibia to Orange Free State, Lesotho and sw Transvaal

☐ **Yellow Wagtail** *Motacilla flava*

_____ *M. f. flavissima* Britain and adj. coastal Europe; >Africa
_____ *M. f. thunbergi* Scandinavia to nw Siberia; >sub-Saharan Africa, s and SE Asia
_____ *M. f. flava* S Scandinavia to c Europe, and Ural Mts.; >sub-Saharan Africa
_____ *M. f. iberiae* SW France, Iberian Pen. and nw Africa; >w and n-c Africa
_____ *M. f. cinereocapilla* Italy, Sicily, Corsica, Sardinia, Slovenia; >w-c Africa
_____ *M. f. pygmaea* Egypt (delta and s along Nile River)
_____ *M. f. feldegg (melanogrisea)* Balkans to Turkey, Iraq, Iran and Afghanistan; >e Africa, s Asia
_____ *M. f. lutea* Lower Volga R basin to Kazakhstan; >Africa, Indian subcontinent
_____ *M. f. beema* Russia to w Siberia, n Kazakhstan and Altai; >e Africa, India
_____ *M. f. taivana* SE Siberia to Sea of Okhotsk and n Japan; >s Asia, Indonesia
_____ *M. f. leucocephala* NW Mongolia to nw China and adj. USSR; >mostly in India
_____ *M. f. macronyx* Ussuriland to ne Mongolia and c Manchuria; >SE Asia, se China

☐ **Eastern Yellow Wagtail** *Motacilla tschutschensis*

_____ *M. t. plexa* N Siberia; >India and se Asia
_____ *M. t. tschutschensis* NE Siberia and extreme nw N America; >se Asia, Indonesia
_____ *M. t. angarensis* S Siberia, w Transbaikalia to n Mongolia; >Myanmar to se China
_____ *M. t. simillima (zaissanensis)* Kamchatka, Commander Is. and n Kurile Is.; >to Australia

☐ **Citrine Wagtail** *Motacilla citreola*

_____ *M. c. citreola* NE Russia to Siberia, Mongolia and Manchuria; >to India
_____ *M. c. werae* Russian and Siberian steppes to e Iran, Afghanistan and India
_____ *M. c. calcarata* E Iran to n Afghanistan, Tibet, s China and Myanmar

☐ **Cape Wagtail** *Motacilla capensis*

____ *M. c. simplicissima*	Angola to se Zaire, Zambia, Caprivi Strip, Botswana and Zimbabwe
____ *M. c. capensis (bradfieldi)*	Namibia to Zimbabwe and Mozambique
____ *M. c. wellsi*	Highlands of e Zaire to Uganda and Kenya

☐ **Madagascar Wagtail** *Motacilla flaviventris*

Open country and streams throughout Madagascar

☐ **Gray Wagtail** *Motacilla cinerea*

____ *M. c. cinerea*	Eurasia; N. Africa, se Asia, New Guinea
____ *M. c. patriciae*	Azores (Furnas and São Miguel)
____ *M. c. schmitzi*	Madeira

☐ **Mountain Wagtail** *Motacilla clara*

____ *M. c. chapini*	Sierra Leone to Gabon, Zaire and w Uganda
____ *M. c. torrentium*	E Uganda to Kenya, Rwanda, Angola and South Africa
____ *M. c. clara*	Ethiopia

FAMILY: CAMPEPHAGIDAE (Cuckoo-shrikes—82)

☐ **Ground Cuckoo-shrike** *Coracina maxima*

Interior of mainland Australia

☐ **Large Cuckoo-shrike** *Coracina macei*

____ *C. m. macei*	India south of the Himalayas (Garhwal to Travancore)
____ *C. m. nipalensis*	Lower Himalayas of India to Nepal, Sikkim and w Assam
____ *C. m. rexpineti*	SE China (Fujian, Guangdong and Yunnan) to n Laos and Taiwan
____ *C. m. layardi*	Sri Lanka
____ *C. m. andamana*	Andaman Islands
____ *C. m. siamensis*	SW China (se Yunnan) to Myanmar, pen. Thailand, s Indochina
____ *C. m. larutensis*	N Malaya
____ *C. m. larvivora*	Hainan (s China)

☐ **Sunda Cuckoo-shrike** *Coracina larvata*

____ *C. l. melanocephala*	Mts. of Sumatra
____ *C. l. normani*	Mts. of Borneo
____ *C. l. larvata*	Mts. of Java

☐ **Javan Cuckoo-shrike** *Coracina javensis*

Malay Peninsula, Java and Bali

☐ **Slaty Cuckoo-shrike** *Coracina schistacea*

Banggai and Sula islands (off Sulawesi)

☐ **Wallacean Cuckoo-shrike** *Coracina personata*

____ *C. p. pollens*	Kai Islands (Kai Kecil, Kai Besar and Add)
____ *C. p. floris*	Lesser Sundas (Sumbawa, Komodo, Rinca, Besar and Flores)
____ *C. p. alfrediana*	Lesser Sundas (Lomblen and Alor)
____ *C. p. sumbensis*	Sumba (Lesser Sundas)
____ *C. p. personata*	Lesser Sundas (Roti, Semau, Timor, Wetar, Leti, Moa, Sermata)
____ *C. p. unimoda*	Tanimbar Islands (Yamdena, Larat and Loetoe)

☐ **Melanesian Cuckoo-shrike** *Coracina caledonica*

____ *C. c. bougainvillei*	Bougainville (Solomon Islands)
____ *C. c. kulambangrae*	Kulambangra (Solomon Islands)
____ *C. c. welchmani*	Santa Isabel (Solomon Islands)
____ *C. c. amadonis*	Guadalcanal (Solomon Islands)
____ *C. c. thilenii*	Vanuatu (Espíritu Santo, Malo and Malakula)
____ *C. c. seiuncta*	Erromango (Vanuatu)
____ *C. c. lifuensis*	Lifou (Loyalty Islands)
____ *C. c. caledonica*	New Caledonia

☐ **Black-faced Cuckoo-shrike** *Coracina novaehollandiae*

____ *C. n. melanops (lettiensis)*	Australia; >New Guinea, Sundas, w Solomon Is.

____	*C. n. novaehollandiae*	Tasmania, King and Flinders Islands; >e Australia
____	*C. n. subpallida*	C Western Australia (Pilbara region); >Kai Islands

☐ **Stout-billed Cuckoo-shrike** *Coracina caeruleogrisea*

____	*C. c. strenua*	Mts. of w and c New Guinea and Yapen Island
____	*C. c. caeruleogrisea*	Aru Islands and Trans-Fly lowlands of s New Guinea
____	*C. c. adamsoni*	SE New Guinea (Astrolabe Bay to Hall Sound)

☐ **Bar-bellied Cuckoo-shrike** *Coracina striata*

____	*C. s. dobsoni*	Andaman Islands
____	*C. s. sumatrensis*	S peninsular Thailand, Malaysia, Sumatra and Borneo
____	*C. s. bungurensis*	Anambas and Natuna islands (South China Sea)
____	*C. s. simalurensis*	Simeulue I. (off Sumatra)
____	*C. s. babiensis*	Babi I. (off Sumatra)
____	*C. s. kannegieteri*	Nias I. (off Sumatra)
____	*C. s. enganensis*	Enggano I. (off Sumatra)
____	*C. s. vordermani*	Kangean Islands (Java Sea)
____	*C. s. striata*	Philippines (Luzon, Polillo and Lubang)
____	*C. s. mindorensis*	Philippines (Mindoro, Libagao and Tablas)
____	*C. s. panayensis*	Philippines (Guimaras, Masbate, Panay, Ticao and Negros)
____	*C. s. boholensis*	Philippines (Bohol, Leyte, Panaon, Calicoan and Samar)
____	*C. s. cebuensis*	Cebu (Philippines). Probably extinct
____	*C. s. kochii*	S Philippines (Mindanao, Nipa and Basilan)
____	*C. s. difficilis*	S Philippines (Palawan, Busuanga and Balabac)
____	*C. s. guillemardi*	Sulu Archipelago

☐ **Pied Cuckoo-shrike** *Coracina bicolor*

Sulawesi, Sangihe, Manterawu, Bangka, Muna, Butung, Togian is.

☐ **Moluccan Cuckoo-shrike** *Coracina atriceps*

____	*C. a. magnirostris*	N Moluccas (Ternate, Halmahera, Bacan and Kasiruta)
____	*C. a. atriceps*	Seram (s Moluccas)

☐ **Buru Cuckoo-shrike** *Coracina fortis*

Forests of Buru (s Moluccas)

☐ **Cerulean Cuckoo-shrike** *Coracina temminckii*

____	*C. t. temminckii*	Montane forests of n Sulawesi
____	*C. t. rileyi*	Montane forests of c and se Sulawesi
____	*C. t. tonkeana*	Montane forests of e Sulawesi

☐ **Yellow-eyed Cuckoo-shrike** *Coracina lineata*

____	*C. l. axillaris*	Mts. of c New Guinea and Waigeo I.
____	*C. l. maforensis*	Numfor I. (New Guinea)
____	*C. l. sublineata*	Bismarck Archipelago (New Ireland and New Britain)
____	*C. l. nigrifrons*	Solomon Islands (Bougainville, Choiseul and Santa Isabel)
____	*C. l. ombriosa*	Solomon Is. (Kulambangra, New Georgia Group and Rendova)
____	*C. l. pusilla*	Guadalcanal (Solomon Islands)
____	*C. l. malaitae*	Malaita (Solomon Islands)
____	*C. l. makirae*	San Cristóbal (Solomon Islands)
____	*C. l. gracilis*	Rennell (Solomon Islands)
____	*C. l. lineata*	NE Australia (e Cape York Peninsula to ne New South Wales)

☐ **Boyer's Cuckoo-shrike** *Coracina boyeri*

____	*C. b. boyeri*	Misool, Yapen, Salawati islands and w New Guinea
____	*C. b. subalaris*	S New Guinea

☐ **White-rumped Cuckoo-shrike** *Coracina leucopygia*

Lowlands of Sulawesi and adjacent islands

☐ **White-bellied Cuckoo-shrike** *Coracina papuensis*

____	*C. p. melanolora*	N Moluccas
____	*C. p. papuensis*	Yapan, Batanta, Salawati islands and w New Guinea
____	*C. p. intermedia*	S New Guinea (Mimika River to Lorentz River)

____	*C. p. oriomo*	Lowlands of se New Guinea and ne Australia (n Queensland)
____	*C. p. angustifrons*	SE New Guinea (Huon Gulf to Hall Sound)
____	*C. p. louisiadensis*	Tagula I. (Louisiade Archipelago)
____	*C. p. ingens*	Admiralty Islands (Manus and Los Negros)
____	*C. p. sclaterii*	Bismarck Archipelago
____	*C. p. perpallida*	Solomon Is. (Bougainville, Choiseul, Santa Isabel and Florida)
____	*C. p. elegans*	Solomon Is. (New Georgia Group, Rendova and Guadalcanal)
____	*C. p. eyerdami*	Malaita (Solomon Islands)
____	*C. p. timorlaoensis*	Tanimbr Islands (Banda Sea)
____	*C. p. hypoleuca*	Kai, Tanimbar and Aru Is.; Kimberly to nw Queensland
____	*C. p. apsleyi*	Melville and Bathurst Islands (Northern Territory)
____	*C. p. oriomo*	N Queensland (islands in s Torres Strait and Cape York Peninsula)
____	*C.p. artamoides*	C and eastern Queensland to n New South Wales)
____	*C.p. robusta*	N-c and e NSW to sw Victoria and adj. se South Australia

☐ **Hooded Cuckoo-shrike *Coracina longicauda***

____	*C. l. grisea*	Jayawaijaya Mts. (c New Guinea)
____	*C. l. longicauda*	C Highlands and mts. of Huon Peninsula (ne New Guinea)

☐ **Halmahera Cuckoo-shrike *Coracina parvula***

Halmahera (n Moluccas)

☐ **Pygmy Cuckoo-shrike *Coracina abbotti***

High montane forests of n, c and se Sulawesi

☐ **New Caledonian Cuckoo-shrike *Coracina analis***

New Caledonia

☐ **White-breasted Cuckoo-shrike *Coracina pectoralis***

Brachystegia woodlands of Africa south of the Sahara

☐ **Blue Cuckoo-shrike *Coracina azurea***

Sierra Leone to Gabon, Zaire and sw Uganda; Bioko I.

☐ **Gray Cuckoo-shrike *Coracina caesia***

____	*C. c. pura*	Sudan and Ethiopia to Uganda and Malawi
____	*C. c. preussi*	Humid montane forests of se Nigeria, Cameroon and Bioko I.
____	*C. c. caesia*	Zimbabwe and Mozambique to e South Africa

☐ **Grauer's Cuckoo-shrike *Coracina graueri***

Montane forests of e Zaire and adjacent sw Uganda

☐ **Ashy Cuckoo-shrike *Coracina cinerea***

____	*C. c. cucullata*	Comoro Islands (Grand Comoro and Mohéli)
____	*C. c. cinerea*	Coastal n and e Madagascar
____	*C. c. pallida*	Arid c, w and sw Madagascar

☐ **Mauritius Cuckoo-shrike *Coracina typica***

Native forests of Mauritius (w Mascarene Islands)

☐ **Reunion Cuckoo-shrike *Coracina newtoni***

Forests of nw Réunion (w Mascarene Islands)

☐ **Cicadabird *Coracina tenuirostris***

____	*C. t. edithae*	S Sulawesi
____	*C. t. emancipata*	Tanahjampea I. (Flores Sea)
____	*C. t. kalaotuae*	Kalaotoa I. (Flores Sea)
____	*C. t. pererrata*	Tukangbesi Islands (Kaledupa and Tomea)
____	*C. t. pelingi*	Banggai Islands (Peleng and Banggai)
____	*C. t. grayi*	N Moluccas (Morotai, Halmahera, Ternate, Tidore and Bacan)
____	*C. t. obiensis*	S Moluccas (Obi and Bisa)
____	*C. t. timoriensis*	E Lesser Sundas (Lomblen and Timor)
____	*C. t. aruensis*	Aru Islands and Trans-Fly lowlands of s New Guinea
____	*C. t. nehrkorni*	Waigeo I. (New Guinea)
____	*C. t. muellerii*	Kofiau and Misool is., New Guinea and D'Entrecasteaux Arch.
____	*C. t. numforana*	Numfor I. (New Guinea)
____	*C. t. meyerii*	Biak I. (New Guinea)
____	*C. t. tagulana*	Louisiade Archipelago (Tagula and Misima)

____	*C. t. rooki*	Umboi (Bismarck Archipelago)
____	*C. t. remota*	Bismarck Arch. (New Ireland, New Hanover, Dyaul and Feni Is.)
____	*C. t. heinrothi*	New Britain and Duke of York I. (Bismarck Archipelago)
____	*C. t. matthiae*	St. Matthias Islands (Bismarck Archipelago)
____	*C. t. ultima*	Bismarck Archipelago (Tabar, Lihir and Tanga)
____	*C. t. admiralitatis*	Manus I. (Admiralty Islands)
____	*C. t. rostrata*	Rossel (Solomon Islands)
____	*C. t. erythropygia*	Solomon Islands (Guadalcanal, Malaita, Florida and Savo)
____	*C. t. salomonis*	San Cristóbal (Solomon Islands)
____	*C. t. saturatior*	N and c Solomon Islands
____	*C. t. nisoria*	Solomon Islands (Pavuvu and Russell Group)
____	*C. t. monacha*	Palau Islands (w Caroline Islands)
____	*C. t. nesiotis*	Yap (w Caroline Islands)
____	*C. t. insperata*	Pohnpei (Caroline Islands)
____	*C. t. melvillensis*	N Australia (Arnhem Land, N Territory to Cape York Peninsula)
____	*C. t. tenuirostris*	E Australia (ne Queensland to se Victoria)

☐ **Blackish Cuckoo-shrike** *Coracina coerulescens*

____	*C. c. coerulescens*	N Philippines (lowlands of Luzon and Catanduanes)
____	*C. c. deschauenseei*	Marinduque (n Philippines)
____	*C. c. altera†*	Formerly Cebu (c Philippines). Extinct

☐ **Sumba Cuckoo-shrike** *Coracina dohertyi*

Lesser Sundas (Sumbawa, Flores and Sumba)

☐ **Sula Cuckoo-shrike** *Coracina sula*

Lowlands of Sula Islands (Talibau, Seho, Mangole and Sanana)

☐ **Kai Cuckoo-shrike** *Coracina dispar*

Lowlands of Kai Islands (s Ceram Sea)

☐ **Black-bibbed Cuckoo-shrike** *Coracina mindanensis*

____	*C. m. lecroyae*	Luzon (n Philippines)
____	*C. m. elusa*	Mindoro (Philippines)
____	*C. m. ripleyi*	Philippines (Bohol, Samar, Diliran and Leyte)
____	*C. m. mindanensis*	S Philippines (Mindanao and Basilan)
____	*C. m. everetti*	Sulu Archipelago (Bongao, Jolo, Lapac and Tawitawi)

☐ **Sulawesi Cuckoo-shrike** *Coracina morio*

____	*C. m. talautensis*	Talaud Islands (Salebabu, Karakelong and Kaburuang)
____	*C. m. salvadorii*	Sangihe I. (off n Sulawesi)
____	*C. m. morio*	Sulawesi, Lembeh, Muna, Tomea, Kabaena and Butung islands

☐ **Pale-gray Cuckoo-shrike** *Coracina ceramensis*

____	*C. c. ceramensis*	S Moluccas (Seram, Buru and Boano)
____	*C. c. hoogerwerfi*	Obi (s Moluccas)

☐ **Papuan Cuckoo-shrike** *Coracina incerta*

Montane forests of New Guinea, Yapen and w Papuan islands

☐ **Gray-headed Cuckoo-shrike** *Coracina schisticeps*

____	*C. s. schisticeps*	Misool I. and Salawati I. and nw New Guinea
____	*C. s. reichenowi*	N New Guinea (Geelvink Bay to Sepik flood-plain)
____	*C. s. poliopsa*	S New Guinea (Kapare River to Astrolabe Mts.)
____	*C. s. vittata*	D'Entrecasteaux Archipelago (Fergusson and Goodenough)

☐ **New Guinea Cuckoo-shrike** *Coracina melas*

____	*C. m. waigeuensis*	Waigeo I. (New Guinea)
____	*C. m. tommasonis*	Yapen I. (New Guinea)
____	*C. m. melas*	Salawati I. and w New Guinea
____	*C. m. meeki*	E New Guinea
____	*C. m. goodsoni*	Aru Islands (New Guinea)
____	*C. m. batantae*	Batanta I. (New Guinea)

☐ **Black-bellied Cuckoo-shrike** *Coracina montana*
    ____ *C. m. montana*   Montane forests of New Guinea
    ____ *C. m. bicinia*   Montane forests in Sepik River region of New Guinea

☐ **Solomon Islands Cuckoo-shrike** *Coracina holopolia*
    ____ *C. h. holopolia*   Bougainville, Choiseul, Buka, Guadalcanal and Santa Isabel
    ____ *C. h. pygmaea*   Solomon Islands (Kulambangra and Vangunu)
    ____ *C. h. tricolor*   Malaita (Solomon Islands)

☐ **McGregor's Cuckoo-shrike** *Coracina mcgregori*
       Mts. of Mindanao (s Philippines)

☐ **Indochinese Cuckoo-shrike** *Coracina polioptera*
    ____ *C. p. jabouillei*   N Vietnam
    ____ *C. p. indochinensis*   Myanmar to c Thailand, c Laos and s Vietnam
    ____ *C. p. polioptera*   S Myanmar to s Thailand, Cambodia and s Laos

☐ **White-winged Cuckoo-shrike** *Coracina ostenta*
       C Philippines (Guimaras, Negros and Panay)

☐ **Black-winged Cuckoo-shrike** *Coracina melaschistos*
    ____ *C. m. melaschistos*   N Pakistan, Himalayas and ne Indian subcontinent
    ____ *C. m. avensis*   W China to c and s Myanmar, n Thailand and n Vietnam
    ____ *C. m. intermedia*   C and s China and Taiwan; >Indochina
    ____ *C. m. saturata*   N Vietnam and Hainan; >Indochina

☐ **Lesser Cuckoo-shrike** *Coracina fimbriata*
    ____ *C. f. neglecta*   S Myanmar and peninsular Thailand
    ____ *C. f. culminata*   S Malay Peninsula
    ____ *C. f. schierbrandii*   Sumatra and Borneo
    ____ *C. f. compta*   Simeulue and Siberut islands (off w Sumatra)
    ____ *C. f. fimbriata*   Java and Bali

☐ **Black-headed Cuckoo-shrike** *Coracina melanoptera*
    ____ *C. m. melanoptera*   N India (n Punjab and United Provinces); >Myanmar
    ____ *C. m. sykesi*   S India and Sri Lanka

☐ **Golden Cuckoo-shrike** *Campochaera sloetii*
    ____ *C. s. sloetii*   W New Guinea (Arfak Mts. to Idenberg River)
    ____ *C. s. flaviceps*   SE New Guinea (Mimika River to Port Moresby)

☐ **Black-and-white Triller** *Lalage melanoleuca*
    ____ *L. m. melanoleuca*   N Philippines (Luzon and Mindoro)
    ____ *L. m. minor*   S Philippines (Samar, Leyte and Mindanao)

☐ **Pied Triller** *Lalage nigra*
    ____ *L. n. davisoni*   Nicobar Islands
    ____ *L. n. nigra*   Malay Peninsula, Sumatra, Java and offshore islands
    ____ *L. n. chilensis*   Borneo and Philippine Islands

☐ **White-rumped Triller** *Lalage leucopygialis*
       Sulawesi subregion and Sula Islands

☐ **White-shouldered Triller** *Lalage sueurii*
       E Java, Bali, Sulawesi subregion and Lesser Sundas

☐ **White-winged Triller** *Lalage tricolor*
       Port Moresby area (New Guinea); s Torres Strait is.; mainland Australia

☐ **Rufous-bellied Triller** *Lalage aurea*
       N Moluccas (Morotai, Ternate, Halmahera, Bacan, Kasiruta, Obi)

☐ **White-browed Triller** *Lalage moesta*
       Tanimbar Islands (Arafura Sea)

☐ **Varied Triller** *Lalage leucomela*
    ____ *L. l. keyensis*   Kai Islands (Kai Kecil, Kai Besar and Add)

____	*L. l. polygrammica*	Aru Islands and e New Guinea
____	*L. l. obscurior*	D'Entrecasteaux Archipelago
____	*L. l. trobriandi*	Trobriand Islands
____	*L. l. pallescens*	Louisiade Archipelago (Misima and Tagula)
____	*L. l. falsa*	Bismarck Archipelago (New Britain, Umboi and Duke of York)
____	*L. l. karu*	New Ireland (Bismarck Archipelago)
____	*L. l. albidior*	New Hanover (Bismarck Archipelago)
____	*L. l. ottomeyeri*	Lihir Islands (Bismarck Archipelago)
____	*L. l. tabarensis*	Tabar I. (Bismarck Archipelago)
____	*L. l. conjuncta*	St. Matthias I. (Bismarck Archipelago)
____	*L. l. sumunae*	Dyaul I. (Bismarck Archipelago)
____	*L. l. macrura*	NW Western Australia (w Kimberley Division)
____	*L. l. rufiventris*	N Northern Territory (Melville I., Arnhem Land, Groote Eylandt)
____	*L. l. yorki*	N Queensland (islands in s Torres Strait and Cape York Pen.)
____	*L. l. leucomela*	E Australia (c Queensland to c New South Wales)

☐ **Black-browed Triller** *Lalage atrovirens*

____	*L. a. atrovirens*	Misool, Salawati and Waigeo islands and n New Guinea
____	*L. a. leucoptera*	Biak I. (n New Guinea)

☐ **Samoan Triller** *Lalage sharpei*

____	*L. s. sharpei*	Highlands of Upolu (Western Samoa)
____	*L. s. tenebrosa*	Highlands of Savai'i (Western Samoa)

☐ **Polynesian Triller** *Lalage maculosa*

____	*L. m. modesta*	N and c Vanuatu
____	*L. m. ultima*	Efate (Vanuatu)
____	*L. m. melanopygia*	Santa Cruz and Utupua (Solomon Is.)
____	*L. m. vanikorensis*	Vanikoro (Solomon Is.)
____	*L. m. soror*	Kandavu (Fiji)
____	*L. m. pumila*	Viti Levu (Fiji)
____	*L. m. mixta*	Oavlau and adjacent Fiji Is.
____	*L. m. woodi*	Vanua Levu, Taveuni and Oamea (Fiji Is.)
____	*L. m. rotumae*	Rotuma (Fiji)
____	*L. m. nesophila*	Lau Archipelago (Fiji)
____	*L. m. vauana*	Vava'u Group (Fiji)
____	*L. m. tabuensis*	Ha'apai, Nomuka and Tongatapu (Tonga)
____	*L. m. keppeli*	Niuatoputapu and Tafahi (extreme n Tonga)
____	*L. m. futunae*	Futuna and Alofi islands (Wallis and Futuna Islands)
____	*L. m. whitmeei*	Niue I. (c Pacific)
____	*L. m. maculosa*	Western Samoa (Upolu and Savai'i)

☐ **Long-tailed Triller** *Lalage leucopyga*

____	*L. l. affinis*	Solomon Islands (San Cristóbal and Ugi)
____	*L. l. deficiens*	Vanuatu (Torres I. and Banks Group)
____	*L. l. albiloris*	C and n Vanuatu
____	*L. l. simillima*	S Vanuatu and Loyalty Islands
____	*L. l. montrosieri*	New Caledonia
____	*L. l. leucopyga†*	Norfolk I. Extinct

☐ **Petit's Cuckoo-shrike** *Campephaga petiti*

SE Nigeria and Cameroon to ne Zaire, w Uganda and w Kenya

☐ **Black Cuckoo-shrike** *Campephaga flava*

Angola to Kenya, Tanzania, Mozambique and South Africa

☐ **Red-shouldered Cuckoo-shrike** *Campephaga phoenicea*

Senegambia to s Sudan, Ethiopia, w Kenya and n Angola

☐ **Purple-throated Cuckoo-shrike** *Campephaga quiscalina*

____	*C. q. quiscalina*	Guinea and Sierra Leone to Cameroon, Zambia and n Angola
____	*C. q. martini*	E Zaire to Uganda and c Kenya
____	*C. q. muenzneri*	Highlands of e Tanzania

☐ **Ghana Cuckoo-shrike** *Campephaga lobata*

E Sierra Leone to Liberia, Ivory Coast and s Ghana

☐ **Oriole Cuckoo-shrike** *Campephaga oriolina*

S Cameroon to sw C African Republic, e Zaire and Gabon

☐ **Rosy Minivet** *Pericrocotus roseus*
 ____ *P. r. roseus*
 ____ *P. r. stanfordi*

Himalayas (Afghanistan to sw China, Myanmar and n India)
S China to s Thailand and s Laos

☐ **Brown-rumped Minivet** *Pericrocotus cantonensis*

Breeds c and se China; >Thailand and Indochina

☐ **Ashy Minivet** *Pericrocotus divaricatus*

NE Asia; >Philippines and Indonesia

☐ **Small Minivet** *Pericrocotus cinnamomeus*
 ____ *P. c. peregrinus*
 ____ *P. c. pallidus*
 ____ *P. c. malabaricus*
 ____ *P. c. cinnamomeus*
 ____ *P. c. vividus*
 ____ *P. c. thai*
 ____ *P. c. sacerdos*
 ____ *P. c. separatus*
 ____ *P. c. saturatus*

Himalayas and n India
Pakistan (Indus River valley from Rann of Kutch to Punjab)
W India (Western Ghats from Belgaum to Kerala)
S peninsular India and Sri Lanka
Andaman Islands
Myanmar to n Thailand and Laos
Cambodia and s Vietnam
S Myanmar (Mergui District) and s peninsular Thailand
Java and Bali

☐ **Ryukyu Minivet** *Pericrocotus tegimae*

S Kyushu and Ryukyu Islands (s Japan)

☐ **Fiery Minivet** *Pericrocotus igneus*
 ____ *P. i. igneus*
 ____ *P. i. trophis*

Myanmar, Malaya, Sumatra, Borneo and adj. islands; Palawan
Simeulue I. (off Sumatra)

☐ **Flores Minivet** *Pericrocotus lansbergei*

W Lesser Sundas (Sumbawa and Flores)

☐ **White-bellied Minivet** *Pericrocotus erythropygius*
 ____ *P. e. erythropygius*
 ____ *P. e. albifrons*

Peninsular India (Punjab and Rajasthan to Bihar and Mysore)
Plains of c Myanmar

☐ **Long-tailed Minivet** *Pericrocotus ethologus*
 ____ *P. e. favillaceus*
 ____ *P. e. laetus*
 ____ *P. e. ethologus*
 ____ *P. e. yvettae (mariae)*
 ____ *P. e. ripponi*
 ____ *P. e. annamensis*

Himalayas of Afghanistan and Kashmir to Nepal
E Nepal to Sikkim, Bengal and w Assam (Khasi Hills)
S China and extreme ne India; >n Indochina
Myanmar and adjacent sw China (Yunnan)
SW China (Yunnan), e Myanmar (s Shan States) and nw Thailand
S Vietnam (Langbian Plateau)

☐ **Short-billed Minivet** *Pericrocotus brevirostris*
 ____ *P. b. brevirostris*
 ____ *P. b. affinis*
 ____ *P. b. neglectus*
 ____ *P. b. anthoides*

E Himalayas to se Tibet, Nepal and w Assam
SW China (s Sichuan to w Yunnan) to e Assam and nw Myanmar
N Thailand to n Tenasserim and n Laos
SW China (se Yunnan, Guangxi and Guangdong) to n Vietnam

☐ **Sunda Minivet** *Pericrocotus miniatus*

Highlands of Sumatra and Java

☐ **Scarlet Minivet** *Pericrocotus flammeus*
 ____ *P. f. speciosus*
 ____ *P. f. flammeus*
 ____ *P. f. fohkiensis*
 ____ *P. f. fraterculus*
 ____ *P. f. elegans*
 ____ *P. f. semiruber*
 ____ *P. f. flammifer*
 ____ *P. f. xanthogaster*

Himalayas (Kashmir to e Assam); >n India
Peninsular India and Sri Lanka
SE China (Hunan, Fujian, Guangdong and Guangxi)
Hainan (s China)
SW China (nw Yunnan) to ne India, n Myanmar and n Indochina
SE India (E Ghats) to s Myanmar, Thailand and n Indochina
S Myanmar to sw Thailand and n Malay Peninsula
S Malaya, Sumatra, Bangka and Belitung islands

____	*P. f. andamanensis*	Andaman Islands
____	*P. f. minythomelas*	Simeulue I. (off Sumatra)
____	*P. f. modiglianii*	Enggano I. (off Sumatra)
____	*P. f. insulanus*	Borneo
____	*P. f. novus*	N Philippines (Luzon and Negros)
____	*P. f. leytensis*	C Philippines (Samar and Leyte)
____	*P. f. johnstoniae (gonzalesi)*	Mindanao (s Philippines)
____	*P. f. marchesae*	Jolo Group (Sulu Archipelago)
____	*P. f. siebersi*	Java and Bali
____	*P. f. exul*	Lombok (Lesser Sundas)

☐ **Gray-chinned Minivet** *Pericrocotus solaris*

____	*P. s. solaris*	E Himalayas (Nepal to nw Myanmar)
____	*P. s. rubrolimbatus*	S Myanmar and n Thailand
____	*P. s. montpellieri*	SW China (nw and c Yunnan)
____	*P. s. griseogularis*	SE China to n Indochina and Taiwan
____	*P. s. deignani*	A Vietnam (Langbian Plateau)
____	*P. s. nassovicus*	Mts. of se Thailand and w Cambodia
____	*P. s. montanus*	Mts. of Malaya and w Sumatra
____	*P. s. cinereigula*	Mts. of n Borneo

☐ **Bar-winged Flycatcher-shrike** *Hemipus picatus*

____	*H. p. capitalis*	Himalayas to n Burma, sw China, n Thailand and n Indochina
____	*H. p. picatus*	Peninsular India to s Burma, s Thailand and s Indochina
____	*H. p. intermedius*	Peninsular Thailand to nw Malaysia, Sumatra and ne Borneo
____	*H. p. leggei*	Sri Lanka

☐ **Black-winged Flycatcher-shrike** *Hemipus hirundinaceus*

Malaysia, Sumatra, Borneo, Java, Bali and adjacent islands

FAMILY: PYCNONOTIDAE (Bulbuls—130)

☐ **Crested Finchbill** *Spizixos canifrons*

____	*S. c. canifrons*	S Assam (south of the Brahmaputra) to hills of w Burma
____	*S. c. ingrami*	E Burma to sw China (Yunnan), nw Thailand and n Indochina

☐ **Collared Finchbill** *Spizixos semitorques*

____	*S. s. semitorques*	Mts. of s China to n Vietnam (nw Tonkin)
____	*S. s. cinereicapillus*	Taiwan

☐ **Straw-headed Bulbul** *Pycnonotus zeylanicus*

Burma to Malay Peninsula, Sumatra, Nias I., Java and Borneo

☐ **Striated Bulbul** *Pycnonotus striatus*

____	*P. s. striatus*	Himalayas of Nepal to sw China, Assam and w Burma
____	*P. s. arctus*	NE Assam (Mishmi Hills)
____	*P. s. paulus*	Burma to sw China, n Thailand, n Laos and n Vietnam

☐ **Cream-striped Bulbul** *Pycnonotus leucogrammicus*

Highland forests of w Sumatra

☐ **Spot-necked Bulbul** *Pycnonotus tympanistrigus*

Foothills of w Sumatra

☐ **Black-and-white Bulbul** *Pycnonotus melanoleucus*

Peninsular Thailand, Malaysia, Sumatra, Siberut I. and Borneo

☐ **Gray-headed Bulbul** *Pycnonotus priocephalus*

SW pen. India (s Maharashtra and Goa to w Mysore and Kerala)

☐ **Black-headed Bulbul** *Pycnonotus atriceps*

____	*P. a. atriceps*	NE India to sw China, SE Asia, Bali, Borneo and Palawan

____	*P. a. fuscoflavescens*	Andaman Islands
____	*P. a. hyperemnus*	Sumatra, Simeulue, Nias, Mentawai, Bangka and Belitung islands
____	*P. a. baweanus*	Bawean I. (Java Sea)
____	*P. a. hodiernus*	Maratua Islands (Celebes Sea)

☐ **Black-crested Bulbul** *Pycnonotus melanicterus*

____	*P. m. flaviventris*	Himalayas (Punjab) to ne India, n Burma and sw China
____	*P. m. gularis*	Hills of sw India (w Mysore to Kerala and Tamil Nadu)
____	*P. m. melanicterus*	Sri Lanka
____	*P. m. vantynei*	S Burma to n Thailand, n Laos, Tonkin and n Annam
____	*P. m. xanthops*	SE Burma to n Thailand
____	*P. m. negatus*	S Burma and adjacent sw Thailand
____	*P. m. auratus*	N plateau of ne Thailand and adjacent w Laos
____	*P. m. johnsoni*	S plateau of Thailand to s Laos, Cambodia and Vietnam
____	*P. m. elbeli*	Islets off coast of se Thailand
____	*P. m. caecilii*	N Malay Peninsula
____	*P. m. dispar*	Sumatra and Java
____	*P. m. montis*	Highlands of n Borneo

☐ **Styan's Bulbul** *Pycnonotus taivanus*

Coastal lowland forests of e and s Taiwan

☐ **Scaly-breasted Bulbul** *Pycnonotus squamatus*

____	*P. s. weberi*	S Burma to peninsular Thailand, Malaya and Sumatra
____	*P. s. squamatus*	W and central Java
____	*P. s. borneensis*	Borneo

☐ **Gray-bellied Bulbul** *Pycnonotus cyaniventris*

____	*P. c. cyaniventris*	S Burma to Thailand, Malaya, Sumatra and Sipura I.
____	*P. c. paroticalis*	Borneo

☐ **Red-whiskered Bulbul** *Pycnonotus jocosus*

____	*P. j. fuscicaudatus*	W India (Tapiti River to Kerala and n Madras)
____	*P. j. abuensis*	W India (n Bombay to sw Rajasthan)
____	*P. j. pyrrhotis*	Valley of Nepal and n India (e Punjab to Bihar)
____	*P. j. emeria*	Lowlands of e India to Burma and sw Thailand
____	*P. j. whistleri*	Andaman Islands
____	*P. j. monticola*	E Himalayas from Sikkim to n Burma and sw China (Yunnan)
____	*P. j. pattani*	Thailand to n Malaya and s Indochina
____	*P. j. hainanensis*	N Vietnam and se China (s Guangdong); Naozhou I.
____	*P. j. jocosus*	S China (Guizhou to Guangxi, e Guangdong and Hong Kong)

☐ **Brown-breasted Bulbul** *Pycnonotus xanthorrhous*

____	*P. x. xanthorrhous*	Himalayas of Tibet to ne Burma, sw China and n Vietnam
____	*P. x. andersoni*	S China (Sichuan to n Guangdong and nw Fujian)

☐ **Light-vented Bulbul** *Pycnonotus sinensis*

____	*P. s. hoyi*	Middle Yangtze River Valley (Sichuan, Hubei and Hunan)
____	*P. s. sinensis*	S China (lower Yangtze River Valley and maritime provinces)
____	*P. s. hainanus*	S China (sw Guangdong and s Guangxi); n Vietnam; Hainan
____	*P. s. formosae*	Taiwan
____	*P. s. orii*	S Ryukyu Islands (Yonaguni and Ishigaki)

☐ **Common Bulbul** *Pycnonotus barbatus*

____	*P. b. barbatus*	Morocco, Algeria and Tunisia
____	*P. b. inornatus*	Senegal to Ghana, n Niger, n Nigeria, n Cameroon and w Chad
____	*P. b. gabonensis*	Central Nigeria and central Cameroon to Gabon and s Congo
____	*P. b. arsinoe*	E Chad to Egypt and Sudan (s to Darfur, Kordofan, Nile Valley)
____	*P. b. schoanus*	Eritrea and e Ethiopia to extreme se Sudan (Boma Hills)
____	*P. b. somaliensis*	Djibouti to nw Somalia and se Ethiopia
____	*P. b. spurius*	S Ethiopia (s Bale to n Sidamo-Borama)

____	*P. b. dodsoni*	S Somalia and adjacent Ethiopia to n Kenya
____	*P. b. tricolor (minor, fayi, ngamii)*	E Cameroon to Zaire, s Sudan, Angola, Namibia and Zambia
____	*P. b. layardi (micrus, tenebrior, naumanni)*	SE Kenya to e Tanzania, Zambia, ne Botswana and S Africa

☐ **Black-fronted Bulbul** *Pycnonotus nigricans*
- ____ *P. n. nigricans (grisescentior)* — Arid s Angola to Namibia, s Botswana and n Transvaal
- ____ *P. n. superior* — S Transvaal to Lesotho, Orange Free State and ne Cape Province

☐ **Cape Bulbul** *Pycnonotus capensis*

South Africa (s and sw Cape Province)

☐ **White-spectacled Bulbul** *Pycnonotus xanthopygos*

Coastal s Turkey to Near East, Sinai Peninsula and Arabia

☐ **White-eared Bulbul** *Pycnonotus leucotis*

E Iraq to s Iran, n Arabia, s Afghanistan and w India

☐ **White-cheeked Bulbul** *Pycnonotus leucogenys*
- ____ *P. l. mesopotamiae* — Iraq, Arabia and s Iran (valleys of the Tigris and Euphrates)
- ____ *P. l. dactylus* — Persian Gulf coast of e Arabia
- ____ *P. l. humii* — NW Pakistan
- ____ *P. l. leucogenys* — Himalayas of ne Afghanistan to e Assam (n of the Brahmaputra)

☐ **Red-vented Bulbul** *Pycnonotus cafer*
- ____ *P. c. intermedius* — Himalayas (w Pakistan to w Uttar Pradesh)
- ____ *P. c. humayuni* — W Pakistan (Salt Range) to nw India
- ____ *P. c. bengalensis* — E Himalayas (e Uttar Pradesh) to ne India, Nepal and Bhutan
- ____ *P. c. wetmorei* — NE peninsular India
- ____ *P. c. primrosei* — S Assam (s of the Brahmaputra), Bangladesh and West Bengal
- ____ *P. c. pusillus* — S India (Bombay, Madhya Pradesh and Andhra to Kerala)
- ____ *P. c. cafer* — Sri Lanka
- ____ *P. c. stanfordi* — N Burma to extreme sw China (w Yunnan)
- ____ *P. c. melanchimus* — S-central Burma (Mandalay to Rangoon)

☐ **Sooty-headed Bulbul** *Pycnonotus aurigaster*
- ____ *P. a. chrysorrhoides* — S China (Fujian, e Guangdong and Hong Kong)
- ____ *P. a. resurrectus* — S China (Guangdong and Naozhou I.); n Vietnam
- ____ *P. a. dolichurus* — Central Vietnam (Quangtri and Thuathien provinces)
- ____ *P. a. latouchei* — SW China to n Thailand, n Laos and n Vietnam
- ____ *P. a. klossi* — SE Burma to n Thailand
- ____ *P. a. schauenseei* — S Burma to sw Thailand
- ____ *P. a. thais* — S Thailand
- ____ *P. a. germani* — SE Thailand to s Indochina
- ____ *P. a. aurigaster* — Java and Bali; introduced Singapore, Sumatra and s Sulawesi

☐ **Puff-backed Bulbul** *Pycnonotus eutilotus*

S Burma, Malaya, Sumatra, Bangka I. and Borneo

☐ **Blue-wattled Bulbul** *Pycnonotus nieuwenhuisii*
- ____ *P. n. inexpectatus* — Rediscovered in nw Sumatra and Borneo
- ____ *P. n. nieuwenhuisii* — Known from a 1901 specimen from ne Borneo

☐ **Yellow-wattled Bulbul** *Pycnonotus urostictus*
- ____ *P. u. ilokensis* — N Luzon (n Philippines)
- ____ *P. u. urostictus* — N Philippines (Luzon and Polillo); formerly Catanduanes
- ____ *P. u. atricaudatus* — Philippines (Bohol, Samar, Panaon, Biliran and Leyte)
- ____ *P. u. philippensis* — S Philippines (Dinagat, Siargao and Mindanao)
- ____ *P. u. basilanicus* — S Philippiness (Basilan and Zamboanga area of Mindanao)

☐ **Orange-spotted Bulbul** *Pycnonotus bimaculatus*
- ____ *P. b. snouckaerti* — Mts. of nw Sumatra
- ____ *P. b. barat* — Mts. of sw Sumatra, w and central Java
- ____ *P. b. bimaculatus* — Mts. of e Java and Bali

□ **Stripe-throated Bulbul** *Pycnonotus finlaysoni*
____ *P. f. davisoni*	S Burma (delta of the Irrawaddy River)
____ *P. f. eous*	SE Burma to sw China (s Yunnan), Thailand and s Indochina
____ *P. f. finlaysoni*	Malay Peninsula (s Burma and Isthmus of Kra to Malacca)

□ **Yellow-throated Bulbul** *Pycnonotus xantholaemus*

Thornscrub of s India

□ **Yellow-eared Bulbul** *Pycnonotus penicillatus*

Highlands of Sri Lanka

□ **Flavescent Bulbul** *Pycnonotus flavescens*
____ *P. f. flavescens*	Hills of s Assam (south of the Brahmaputra) to w Burma
____ *P. f. vividus*	NE Burma to sw China (Yunnan), Thailand and n Indochina
____ *P. f. sordidus*	S Indochina
____ *P. f. leucops*	N Borneo (Mt. Kinabalu to Mt. Mulu and Mt. Murud)

□ **White-browed Bulbul** *Pycnonotus luteolus*
____ *P. l. luteolus*	Arid scrub of coastal s peninsular India
____ *P. l. insulae*	Lowlands of Sri Lanka

□ **Yellow-vented Bulbul** *Pycnonotus goiavier*
____ *P. g. jambu*	Lowlands of se Thailand to s Indochina
____ *P. g. personatus*	Malay Peninsula, Riau Arch., Sumatra, Bangka and Belitung is.
____ *P. g. analis*	Java and Bali; probably introduced to Lombok and s Sulawesi
____ *P. g. gourdini*	Borneo and Maratua Islands
____ *P. g. goiavier*	N and central Philippine Islands
____ *P. g. samarensis*	Philippines (Bohol, Cebu, Leyte, Samar, Ticao and Biliran)
____ *P. g. suluensis*	S Philippines (Mindanao, Basilan, Camiguin Sur and Sulu Arch.)

□ **Olive-winged Bulbul** *Pycnonotus plumosus*
____ *P. p. plumosus*	Malay Peninsula, Riau Arch., e Sumatra and Java
____ *P. p. porphyreus*	W Sumatra, Nias, Batu, Banyak and Mentawi islands
____ *P. p. billitonis*	W and s Borneo and Belitung I.
____ *P. p. hutzi*	N and e Borneo
____ *P. p. chiroplethis*	Anambas Islands (South China Sea)
____ *P. p. hachisukae*	Banggai and adjacent islands off ne Borneo; Cagayan Sulu
____ *P. p. cinereifrons*	SW Philippines (Palawan, Culion and Busuanga)
____ *P. p. sibergi*	Bawean I. (Java Sea)

□ **Streak-eared Bulbul** *Pycnonotus blanfordi*
____ *P. b. blanfordi*	Central and s Burma
____ *P. b. conradi*	Thailand to n Malaysia and s Indochina
____ *P. b. robinsoni*	Central Malaysia

□ **Cream-vented Bulbul** *Pycnonotus simplex*
____ *P. s. simplex*	S Thailand to Malaya, Sumatra, Riau, Lingga, Nias and Batu is.
____ *P. s. prillwitzi*	Java
____ *P. s. perplexus*	N and e Borneo; Balembangan I.
____ *P. s. oblitus*	S and w Borneo; Bangka, Belitung and South Natuna islands
____ *P. s. halizonus*	Anambas Islands and North Natuna Islands

□ **Red-eyed Bulbul** *Pycnonotus brunneus*
____ *P. b. brunneus*	Lowlands of Malay Peninsula, Sumatra, Borneo and adj. islands
____ *P. b. zapolius*	Anambas Islands (South China Sea)

□ **Spectacled Bulbul** *Pycnonotus erythropthalmos*
____ *P. e. erythropthalmos*	Malay Peninsula, Belitung I., Sumatra and adjacent w islands
____ *P. e. salvadorii*	Borneo

□ **Cameroon Mountain Greenbul** *Andropadus montanus*

Montane forests of Nigeria and Cameroon

☐ **Shelley's Greenbul** *Andropadus masukuensis*

____	*A. m. kakamegae*	E Zaire to Uganda, w Kenya and w Tanzania
____	*A. m. roehli*	Highlands of e Tanzania
____	*A. m. masukuensis*	SW Tanzania (Rungwe Mts.) to n Malawi (Masuku Mts.)

☐ **Little Greenbul** *Andropadus virens*

____	*A. v. erythropterus*	Lowlands of Gambia to s Nigeria
____	*A. v. virens*	Cameroon to Gabon, Angola, s Sudan and w Kenya; Bioko
____	*A. v. holochlorus*	Lowland forests of w Uganda
____	*A. v. hallae*	Single specimen from e Zaire (probable melanistic *A. v. virens*)
____	*A. v. zombensis*	SE Zaire to e Kenya, Tanzania and Mozambique; Mafia I.
____	*A. v. marwitzi*	Kenya/Tanzania border (coast to Usambara, Kilimanjaro Mts.)
____	*A. v. zanzibaricus*	Zanzibar (Tanzania)

☐ **Gray Greenbul** *Andropadus gracilis*

____	*A. g. extremus*	Sierra Leone to sw Nigeria
____	*A. g. gracilis*	SE Nigeria to s Cameroon, Gabon, nw Angola and central Zaire
____	*A. g. ugandae*	E Zaire to Uganda and w Kenya

☐ **Ansorge's Greenbul** *Andropadus ansorgei*

____	*A. a. ansorgei*	Sierra Leone to Gabon, n Angola, n Zaire and w Uganda
____	*A. a. kavirondensis*	W Kenya (n Kavirondo to Mt. Elgon area)

☐ **Plain Greenbul** *Andropadus curvirostris*

____	*A. c. leoninus*	Sierra Leone to central Ghana
____	*A. c. curvirostris*	S Ghana to n Angola, s Sudan, Uganda and w Kenya; Bioko

☐ **Slender-billed Greenbul** *Andropadus gracilirostris*

____	*A. g. gracilirostris*	Guinea to extreme s Sudan, w Zaire, w Kenya and Angola
____	*A. g. percivali*	Highlands of central Kenya to extreme w Tanzania

☐ **Sombre Greenbul** *Andropadus importunus*

____	*A. i. insularis (somaliensis, subularis, fricki)*	S Somalia to Kenya and n Tanzania; Manda I.
____	*A. i. hypoxanthus (loquax)*	S Tanzania to Malawi, Zambia, e Zimbabwe and Mozambique
____	*A. i. oleaginus*	S Mozambique to n Kwazulu-Natal and ne Transvaal
____	*A. i. importunus (errolius, noomei)*	Natal to s Zululand, Transvaal, Swaziland and Cape Province

☐ **Yellow-whiskered Greenbul** *Andropadus latirostris*

____	*A. l. congener*	Senegal to sw Nigeria
____	*A. l. latirostris*	S Nigeria to n Angola, e Zaire, Kenya and Tanzania; Bioko

☐ **Western Mountain-Greenbul** *Andropadus tephrolaemus*

____	*A. t. bamendae*	Mts. of se Nigeria and adjacent w Cameroon
____	*A. t. tephrolaemus*	SW Cameroon (Mt. Cameroon); Bioko

☐ **Eastern Mountain-Greenbul** *Andropadus nigriceps*

____	*A. n. kikuyuensis*	Mts. of e Zaire to w Uganda and central Kenya
____	*A. n. nigriceps*	Mts. of s Kenya (Nguruman Hills) to n Tanzania
____	*A. n. kungwensis*	W Tanzania (Mt. Kungwe area)
____	*A. n. usambarae*	SE Kenya (Taita Hills) to ne Tanzania (s Pare, w Usambara mts.)
____	*A. n. neumanni*	NE Tanzania (Uluguru Mts.)
____	*A. n. chlorigula*	E Tanzania (Nguru Mts. and highlands of Iringa District)
____	*A. n. fusciceps*	Mts. of sw Tanzania, ne Zambia, Malawi and ne Mozambique

☐ **Stripe-cheeked Greenbul** *Andropadus milanjensis*

____	*A. m. striifacies*	Highlands of se Kenya to n Tanzania (Kilimanjaro to Iringa)
____	*A. m. olivaceiceps*	Highlands of sw Tanzania to n Malawi and n Mozambique
____	*A. m. milanjensis*	Malawi (Mt. Milanje) to w Mozambique and Zimbabwe

☐ **Golden Greenbul** *Calyptocichla serina*

Sierra Leone to Gabon, Angola and ne Zaire; Bioko

☐ **Honeyguide Greenbul** *Baeopogon indicator*
_____ *B. i. leucurus (togoensis)*
_____ *B. i. indicator (chlorosaturatus)*

Sierra Leone to Liberia, Ivory Coast, Ghana and Togo
S Nigeria to e Zaire, s Sudan, Uganda, w Kenya and nw Zambia

☐ **Sjostedt's Greenbul** *Baeopogon clamans*

Locally in Cameroon, Equatorial Guinea, w Gabon and ne Zaire

☐ **Spotted Greenbul** *Ixonotus guttatus*

Liberia to s Cameroon, Gabon, Zaire, w Uganda, nw Tanzania

☐ **Simple Greenbul** *Chlorocichla simplex*

Guinea-Bissau to ne Angola, e Zaire and extreme s Sudan

☐ **Yellow-throated Greenbul** *Chlorocichla flavicollis*
_____ *C. f. flavicollis*
_____ *C. f. adamauae*
_____ *C. f. simplicicolor*
_____ *C. f. soror*
_____ *C. f. flavigula (pallidigula)*

Senegal to e Nigeria and n Cameroon
N Cameroon (Adamawa Plateau)
E Cameroon (Uam region)
N-central Cameroon to Gabon, Zaire, s Sudan and Ethiopia
Angola to Zaire, w Uganda, w Kenya, Zambia and nw Tanzania

☐ **Yellow-necked Greenbul** *Chlorocichla falkensteini*
_____ *C. f. viridescentior*
_____ *C. f. falkensteini*

S Cameroon (River Ja region)
S Central African Republic to Rio Muni, sw Zaire and n Angola

☐ **Yellow-bellied Greenbul** *Chlorocichla flaviventris*
_____ *C. f. centralis*
_____ *C. f. occidentalis (zambesiae, ortiva)*
_____ *C. f. flaviventris*

Somalia to Kenya, e Tanzania and n Mozambique
W Tanzania to Zaire, Angola, Botswana, Transvaal, Mozambique
Natal and n Mozambique

☐ **Joyful Greenbul** *Chlorocichla laetissima*
_____ *C. l. laetissima*
_____ *C. l. schoutedeni*

Montane forests of e Zaire to s Sudan, Uganda and nw Kenya
Montane forests of se Zaire to sw Tanzania and ne Zambia

☐ **Prigogine's Greenbul** *Chlorocichla prigoginei*

Submontane forests of e Zaire (Lendu Plateau and Butembo)

☐ **Swamp Greenbul** *Thescelocichla leucopleura*

Sierra Leone to Cameroon, e Zaire and w Uganda

☐ **Leaf-love** *Phyllastrephus scandens*
_____ *P. s. scandens*
_____ *P. s. orientalis (acedis, upembae)*

Senegal to n Nigeria and n Cameroon
S Cameroon to s Zaire, s Sudan and extreme w Tanzania

☐ **Cabanis' Greenbul** *Phyllastrephus cabanisi*
_____ *P. c. placidus*
_____ *P. c. sucosus*
_____ *P. c. cabanisi*
_____ *P. c. nandensis*
_____ *P. c. ngurumanensis*

N Kenya to Tanzania, ne Zambia, Malawi and ne Mozambique
E Zaire to Rwanda, s Sudan, Uganda, w Kenya and nw Tanzania
Highlands of w Angola to se Zaire, Zambia and sw Tanzania
W Kenya (n Nandi Hills)
SW Kenya (Nguruman Hills)

☐ **Fischer's Greenbul** *Phyllastrephus fischeri*

Extreme se Somalia to Kenya, Tanzania and n Mozambique

☐ **Terrestrial Brownbul** *Phyllastrephus terrestris*
_____ *P. t. suahelicus (bensoni)*
_____ *P. t. intermedius (katangae) (robertsi)*
_____ *P. t. terrestris*

E Kenya to e Tanzania and n Mozambique
S Angola to s Zaire, Zambia, e Zululand and s Mozambique
W Zululand, Swaziland, Transvaal and Natal to Cape Province

☐ **Northern Brownbul** *Phyllastrephus strepitans*

Extreme s Sudan to n Uganda, s Ethiopia, Kenya and e Tanzania

☐ **Pale-olive Greenbul** *Phyllastrephus fulviventris*

Riparian vegetation of extreme w Zaire, Cabinda and w Angola

☐ **Gray-olive Greenbul** *Phyllastrephus cerviniventris*

S Kenya to Tanzania, Zambia, s Zaire, Malawi and Mozambique

☐ **Baumann's Greenbul** *Phyllastrephus baumanni*

Lowland forests of Sierra Leone and Liberia to s Nigeria

☐ **Toro Olive-Greenbul** *Phyllastrephus hypochloris*

Forests of ne Zaire to extreme se Sudan, Uganda and w Kenya

☐ **Cameroon Olive-Greenbul** *Phyllastrephus poensis*

Montane forests of se Nigeria and sw Cameroon; Bioko

☐ **Sassi's Greenbul** *Phyllastrephus lorenzi*

Primary forests of e Zaire and immediately adjacent Uganda

☐ **Yellow-streaked Bulbul** *Phyllastrephus flavostriatus*

____ *P. f. tenuirostris*	Coastal se Kenya to Tanzania and ne Mozambique
____ *P. f. kungwensis*	W Tanzania (Kungwe-Mahari Mts.)
____ *P. f. uzungwensis*	E Tanzania (Udzungwa Mts.)
____ *P. f. alfredi*	SW Tanzania to e Zambia and n Malawi
____ *P. f. graueri*	E Zaire (highlands west of lakes Kivu, Edward and Albert)
____ *P. f. olivaceogriseus*	Highlands of e Zaire (nw of Lake Tanganyika) to sw Uganda
____ *P. f. vincenti*	Highlands of se Malawi and adjacent n Mozambique
____ *P. f. flavostriatus (distans, dendrophilus, dryobates)*	E Zimbabwe and Mozambique (s of Zambezi River) to Natal

☐ **Gray-headed Greenbul** *Phyllastrephus poliocephalus*

Montane rainforests of se Nigeria and sw Cameroon

☐ **Tiny Greenbul** *Phyllastrephus debilis*

____ *P. d. rabai*	Lowlands of coastal se Kenya to se Tanzania (Rufiji River)
____ *P. d. albigula*	E Tanzania (Nguru and Usambara mts.)
____ *P. d. debilis*	SE Tanzania to e Zimbabwe and s Mozambique

☐ **White-throated Greenbul** *Phyllastrephus albigularis*

____ *P. a. albigularis*	SW Senegal to Cameroon, Gabon, s Sudan and Uganda
____ *P. a. viridiceps*	NW Angola

☐ **Icterine Greenbul** *Phyllastrephus icterinus*

Sierra Leone to Gabon, Zaire and extreme w Uganda; Bioko

☐ **Liberian Greenbul** *Phyllastrephus leucolepis*

Humid forests of se Liberia

☐ **Xavier's Greenbul** *Phyllastrephus xavieri*

____ *P. x. serlei*	Lowlands and foothills of Mt. Cameroon
____ *P. x. xavieri (sethsmithi)*	Cameroon to n Zaire, w Uganda and nw Tanzania

☐ **Long-billed Greenbul** *Phyllastrephus madagascariensis*

____ *P. m. madagascariensis*	Forests of e Madagascar
____ *P. m. inceleber*	Forests of n and w Madagascar

☐ **Spectacled Greenbul** *Phyllastrephus zosterops*

____ *P. z. fulvescens*	Humid forests of extreme n Madagascar (Mt. d'Ambre)
____ *P. z. andapae*	NE Madagascar (Andapa region)
____ *P. z. zosterops*	E Madagascar
____ *P. z. ankafanae*	Highlands of se Madagascar (Fianarantsoa region)

☐ **Appert's Greenbul** *Phyllastrephus apperti*

SW Madagascar (Zombitse and Vohibasia forests)

☐ **Dusky Greenbul** *Phyllastrephus tenebrosus*

Rainforests of e-central Madagascar

☐ **Gray-crowned Greenbul** *Phyllastrephus cinereiceps*

Rainforests of e Madagascar

☐ **Common Bristlebill** *Bleda syndactylus*

____ *B. s. syndactylus (multicolor)*	Sierra Leone to Gabon, nw Angola, w Zaire and Zambia
____ *B. s. woosnami*	E Zaire to s Sudan, Kenya, Uganda and Zambia
____ *B. s. nandensis*	W Kenya (Nandi Hills)

☐ **Green-tailed Bristlebill** *Bleda eximius*

Guinea to Sierra Leone and Ghana

□ **Lesser Bristlebill** *Bleda notatus*
_____ *B. n. notatus* — SE Nigeria to Central African Republic; Bioko
_____ *B. n. ugandae* — NE Zaire to s Sudan and Uganda

□ **Gray-headed Bristlebill** *Bleda canicapillus*
_____ *B. c. morelorum* — Senegal and Gambia
_____ *B. c. canicapillus* — Guinea-Bissau to Nigeria

□ **Yellow-spotted Nicator** *Nicator chloris* — Senegal to Zaire, Uganda, extreme s Sudan and w Tanzania

□ **Eastern Nicator** *Nicator gularis* — S Somalia to Kenya, Tanzania, Zambia, Mozambique and Natal

□ **Yellow-throated Nicator** *Nicator vireo* — S Cameroon to Gabon, n Angola and extreme w Uganda

□ **Red-tailed Greenbul** *Criniger calurus*
_____ *C. c. verreauxi* — Senegambia to sw Nigeria
_____ *C. c. calurus* — S Nigeria (Benin) to extreme w Zaire (lower Congo); Bioko
_____ *C. c. emini* — W-central Zaire to ne Angola, Uganda and Tanzania

□ **Western Bearded-Greenbul** *Criniger barbatus*
_____ *C. b. barbatus* — Sierra Leone to Togo
_____ *C. b. ansorgeanus* — S Nigeria (lower Niger River delta)

□ **Eastern Bearded-Greenbul** *Criniger chloronotus* — Cameroon to e Zaire, n Angola and extreme w Uganda

□ **Yellow-bearded Greenbul** *Criniger olivaceus* — Senegambia to Ghana and Ivory Coast

□ **White-bearded Greenbul** *Criniger ndussumensis* — SE Nigeria and Cameroon to Gabon, Angola and n Zaire

□ **Finsch's Bulbul** *Alophoixus finschii* — Lowlands of s peninsular Thailand, Malaya, Sumatra and Borneo

□ **White-throated Bulbul** *Alophoixus flaveolus*
_____ *A. f. flaveolus* — Himalayas (Nepal to ne Burma)
_____ *A. f. burmanicus* — SW China (w Yunnan) to se Burma to w Thailand

□ **Puff-throated Bulbul** *Alophoixus pallidus*
_____ *A. p. griseiceps* — Pegu Yoma Mts. (Burma)
_____ *A. p. robinsoni* — S Burma (Amherst District of Tenasserim)
_____ *A. p. henrici* — SW China (w Yunnan to s Guangxi) to n Thailand, n Indochina
_____ *A. p. pallidus* — Hainan (s China)
_____ *A. p. isani* — NW part of eastern plateau of Thailand
_____ *A. p. annamensis* — Central Indochina
_____ *A. p. khmerensis* — S Laos to Cambodia and s Vietnam

□ **Ochraceous Bulbul** *Alophoixus ochraceus*
_____ *A. o. hallae* — S Vietnam
_____ *A. o. cambodianus* — SE Thailand and sw Cambodia (Chaine de l'Éléphant)
_____ *A. o. ochraceus* — S Burma to sw Thailand
_____ *A. o. sordidus* — Malay Peninsula and Mergui Archipelago
_____ *A. o. sacculatus* — Highlands of s Malaya (n Perak to Negri Sembilan and Pahang)
_____ *A. o. sumatranus* — Highlands of w Sumatra
_____ *A. o. fowleri* — Mountain and hill forests of Borneo (except range of *ruficrissus*)
_____ *A. o. ruficrissus* — Highlands of n Borneo (Mt Kinabalu)

□ **Gray-cheeked Bulbul** *Alophoixus bres*
_____ *A. b. tephrogenys* — S Burma, Malay Peninsula and lowlands of e Sumatra
_____ *A. b. bres* — W and central Java
_____ *A. b. balicus* — E Java and Bali
_____ *A. b. gutturalis* — Borneo
_____ *A. b. frater* — SW Philippines (Balabac, Busuanga, Calamianes and Palawan)

☐ **Yellow-bellied Bulbul** *Alophoixus phaeocephalus*

____	*A. p. phaeocephalus*	Malay Pen., Sumatra, Bangka, Belitung and North Natuna Is.
____	*A. p. connectens*	NE Borneo
____	*A. p. sulphuratus*	Central, east and s Borneo
____	*A. p. diardi*	W Borneo

☐ **Golden Bulbul** *Alophoixus affinis*

____	*A. a. platenae*	Sangihe I. (off Sulawesi)
____	*A. a. aureus*	Togian Islands (off Sulawesi)
____	*A. a. harterti*	Banggai Islands (Peleng, Banggai, Labobo and Banda)
____	*A. a. longirostris*	Sula Islands (Taliabu, Mangole and Sanana)
____	*A. a. chloris*	N Moluccas (Morotai, Halmahera, Bacan and Kasiruta)
____	*A. a. lucasi*	Obi (n Moluccas)
____	*A. a. mystacalis*	Buru (s Moluccas)
____	*A. a. affinis*	Seram (s Moluccas)
____	*A. a. flavicaudus*	Ambon (s Moluccas)

☐ **Hook-billed Bulbul** *Setornis criniger*

Lowland forests of Borneo, e Sumatra and Bangka I.

☐ **Hairy-backed Bulbul** *Tricholestes criniger*

____	*T. c. criniger*	Malay Peninsula, Tioman I. and e Sumatra
____	*T. c. sericeus*	W Sumatra, Batu Islands, Lingga Archipelago and Musala I.
____	*T. c. viridis*	Borneo and North Natuna Islands

☐ **Olive Bulbul** *Iole virescens*

Forests of Bangladesh and ne India to Burma and w Thailand

☐ **Gray-eyed Bulbul** *Iole propinqua*

____	*I. p. propinqua*	E Burma to sw China, n Thailand, n Laos and n Vietnam
____	*I. p. lekhakuni*	S Burma to sw Thailand
____	*I. p. simulator*	SE Thailand to s Laos, Cambodia and n Vietnam
____	*I. p. cinnamomeoventris*	Malay Pen. (Mergui District and Isthmus of Kra to Trang)
____	*I. p. aquilonis*	S China (sw Guangxi) and n Vietnam
____	*I. p. innectens*	S Vietnam

☐ **Buff-vented Bulbul** *Iole olivacea*

Malay Peninsula, Sumatra, Borneo and adjacent islands

☐ **Yellow-browed Bulbul** *Iole indica*

____	*I. i. icterica*	W India (w Ghats from s Maharashtra to Belgaum and Goa)
____	*I. i. indica*	SW India and Sri Lanka (except for southwest)
____	*I. i. guglielmi*	Southwestern Sri Lanka

☐ **Sulphur-bellied Bulbul** *Ixos palawanensis*

Mts. of Palawan (sw Philippines)

☐ **Philippine Bulbul** *Ixos philippinus*

____	*I. p. philippinus*	Luzon, Marinduque, Catanduanes, Polillo and adj. small islands
____	*I. p. parkesi*	Burias (Philippines)
____	*I. p. guimarasensis*	Philippines (Guimaras, Masbate, Panay, Negros, Ticao, Verde)
____	*I. p. mindorensis*	Philippines (Mindoro and Semirara)
____	*I. p. saturatior*	Samar, Leyte, Cebu, Bohol and Mindanao and adj. small islands

☐ **Zamboanga Bulbul** *Ixos rufigularis*

S Philippines (Basilan and Zamboanga Pen. of w Mindanao)

☐ **Streak-breasted Bulbul** *Ixos siquijorensis*

____	*I. s. cinereiceps*	Philippines (Romblon and Tablas)
____	*I. s. monticola†*	Cebu (Philippines). Extinct
____	*I. s. siquijorensis*	Siquijor (Philippines)

☐ **Brown-eared Bulbul** *Ixos amaurotis*

____	*I. a. hensoni*	Breeds sw Hokkaido; winters to s Korea, n Japan and se China
____	*I. a. matchiae*	S Japan (Hachijo-jima, Tanegashima and Yakushima)

383

____ *I. a. amaurotis*	Central Japanese is. (Honshu to Kyushu); Cheju-Do I. (Korea)
____ *I. a. squamiceps*	Bonin Islands (Mukojima, Chichijima and Hahajima)
____ *I. a. magnirostris*	Volcano Islands (Japan)
____ *I. a. borodinonis*	Daito Islands (Japan)
____ *I. a. ogawae*	N Ryukyu Islands (Amami-O-Shima and Tokuno-Shima)
____ *I. a. pryeri*	Central Ryukyu Islands (Ihiya, Okinawa, Zamami and Kume)
____ *I. a. insignis*	Miyako-Jima (s Ryukyu Islands)
____ *I. a. stejnegeri*	Ryukyu Islands (Ishigaki, Iriomote and Yonaguni)
____ *I. a. harterti*	S Taiwan and Lan-yü I.
____ *I. a. batanensis*	Philippines (Babuyan Claro, Batan, Ivojos and Sabtang)
____ *I. a. fugensis*	Philippines (Calayan, Fuga and Dalupiri)
____ *I. a. camiguinensis*	Camiguin Norte (n Philippines)

☐ **Yellowish Bulbul** *Ixos everetti*

____ *I. e. everetti*	Philippines (Dinagat, Mindanao, Panaon, Biliran and Siargao)
____ *I. e. samarensis*	Philippines (Samar and Leyte)
____ *I. e. haynaldi*	Sulu Arch. (Bongao, Jolo, Sibutu, Tawitawi and Sanga Sanga)
____ *I. e. catarmanensis*	Camiguin Sur (s Philippines)

☐ **Mountain Bulbul** *Ixos mcclellandii*

____ *I. m. mcclellandii*	E Himalayas (w Uttar Pradesh to e Assam)
____ *I. m. ventralis*	SW Burma (Chin Hills and Arakan Yoma Mts.)
____ *I. m. tickelli*	E Burma (n Shan States) to nw Thailand
____ *I. m. similis*	NE Burma (Kachin) to sw China (Yunnan) and n Indochina
____ *I. m. holtii*	S China (Sichuan to Fujian and Guangdong)
____ *I. m. loquax*	N and e Thailand to s Laos (Bolavens Plateau)
____ *I. m. griseiventer*	S Vietnam (Langbian Plateau)
____ *I. m. canescens*	SE Thailand (Khao Kuap region)
____ *I. m. peracensis*	Peninsular Thailand to n Malaysia (Selangor and Pahang)

☐ **Sunda Bulbul** *Ixos virescens*

____ *I. v. sumatranus*	Montane forests of w Sumatra
____ *I. v. virescens*	Montane forests of Java

☐ **Streaked Bulbul** *Ixos malaccensis*

	S Thailand, Malay Pen., Sumatra, Borneo and adj. islands

☐ **Ashy Bulbul** *Hemixos flavala*

____ *H. f. flavala*	E Himalayas (Garhwal to Nepal); w Yunnan and sw Burma
____ *H. f. bourdellei*	SW China (s Yunnan) to e Thailand and n Laos
____ *H. f. remotus*	S Vietnam (Bolavens and Langbian plateaux)
____ *H. f. hildebrandti*	N Burma to nw Thailand
____ *H. f. davisoni*	Central Burma to sw Thailand
____ *H. f. cinereus*	Malay Peninsula and Sumatra
____ *H. f. connectens*	Highlands of n Borneo

☐ **Chestnut Bulbul** *Hemixos castanonotus*

____ *H. c. canipennis*	S China (Hunan, Guangxi, Fujian and Guangdong)
____ *H. c. castanonotus*	N Vietnam (Tonkin) and Hainan

☐ **Madagascar Bulbul** *Hypsipetes madagascariensis*

____ *H. m. madagascariensis*	Madagascar
____ *H. m. grotei*	Îles Glorieuses (Indian Ocean off Réunion)
____ *H. m. rostratus*	Aldabra

☐ **Seychelles Bulbul** *Hypsipetes crassirostris*

	Seychelles (Mahé, Praslin and Félicité)

☐ **Comoro Bulbul** *Hypsipetes parvirostris*

	Comoro Islands (highlands of Grand Comoro and Mohéli)

☐ **Reunion Bulbul** *Hypsipetes borbonicus*

	Evergreen forests of Réunion (w Mascarene Islands)

☐ **Mauritius Bulbul** *Hypsipetes olivaceus*

Maccabe forest of sw Mauritius (w Mascarene Islands)

☐ **Black Bulbul** *Hypsipetes leucocephalus*

____	*H. l. psaroides*	Himalayas of n Afghanistan to e Assam and se Tibet
____	*H. l. nigrescens*	E Assam (s of the Brahmaputra) to w Burma
____	*H. l. ganeesa*	SW India (Western Ghats)
____	*H. l. humii*	Sri Lanka
____	*H. l. ambiens*	NE Burma to sw China (w Yunnan in Irrawaddy watershed)
____	*H. l. concolor*	E Burma to sw China, e Thailand, Laos and s Vietnam
____	*H. l. sinensis*	E Burma to sw China; winters to Thailand and s Laos
____	*H. l. stresemanni*	SW China (Likiang Mts. of nw Yunnan) to Thailand and s Laos
____	*H. l. leucothorax*	E China (Sichuan to Shaanxi and Hebei); winters to s Laos
____	*H. l. leucocephalus*	Maritime provinces of se China
____	*H. l. perniger*	Hainan (s China)
____	*H. l. nigerrimus*	Taiwan

☐ **Nicobar Bulbul** *Hypsipetes virescens*

Forests of Nicobar Islands

☐ **White-headed Bulbul** *Hypsipetes thompsoni*

Montane forests of s Burma to nw Thailand

☐ **Black-collared Bulbul** *Neolestes torquatus*

Savanna of Gabon to Angola, se Zaire and nw Zambia

FAMILY: REGULIDAE (Kinglets—6)

☐ **Golden-crowned Kinglet** *Regulus satrapa*

____	*R. s. olivaceus*	SE Alaska to Oregon (w of Cascades); winters to s California
____	*R. s. amoenus*	Kenai Pen. and central Yukon to Rocky Mts.; winters to sw US
____	*R. s. satrapa*	Labrador and Newfoundland to e US; winters to Gulf Coast
____	*R. s. apache*	Mts. of s Arizona; winters to s Texas and New Mexico
____	*R. s. aztecus*	Mts. of Mexico (Michoacán to Hidalgo, Puebla and Guerrero)
____	*R. s. clarus*	Mts. of s Mexico (Chiapas) and s Guatemala

☐ **Ruby-crowned Kinglet** *Regulus calendula*

____	*R. c. grinnelli*	Coastal Alaska and British Columbia; winters to s California
____	*R. c. calendula*	N and e Canada to ne US; winters to Guatemala, Cuba, Bahamas
____	*R. c. obscurus*	Guadalupe I. (off w Mexico)

☐ **Goldcrest** *Regulus regulus*

____	*R. r. regulus (anglorum, interni)*	Europe to Asia Minor and w Siberia; winters to Mediterranean
____	*R. r. azoricus*	São Miguel I. (Azores)
____	*R. r. sanctaemariae*	Santa Maria I. (Azores)
____	*R. r. inermis*	W Azores (Flores, Faial, Pico, São Jorge and Terciera)
____	*R. r. buturlini*	Crimea, Caucasus and Azerbaijan; winters to n Iran
____	*R. r. hyrcanus*	E Turkey (Elburz Mts.) to n Iran (s Caspian District)
____	*R. r. coatsi*	W Siberia to Altai Mts.; winters to s Nan Shan Mts.
____	*R. r. tristis*	Mts. of central Asia; winters to Transcaspia and w Iran
____	*R. r. himalayensis*	Himalayas of Afghanistan to Pakistan and Nepal
____	*R. r. sikkimensis*	Himalayas of Nepal to se Tibet and w China
____	*R. r. japonensis*	Mts. of Manchuria to n and e China, Korea and Japan
____	*R. r. yunnanensis*	Mts. of w China (s Gansu and Shaanxi to Sichuan and Yunnan)

☐ **Canary Islands Kinglet** *Regulus teneriffae*

Coniferous and mixed forests of Canary Islands

☐ **Flamecrest** *Regulus goodfellowi*

Montane forests of central Taiwan

☐ **Firecrest** *Regulus ignicapilla*

____	*R. i. madeirensis*	Madeira (e Atlantic Ocean)
____	*R. i. ignicapilla*	England and w Europe to Mediterranean and Asia Minor
____	*R. i. balearica*	Balearic Islands and North Africa (Morocco to n Tunisia)

FAMILY: CHLOROPSEIDAE (Leafbirds—8)

☐ **Philippine Leafbird** *Chloropsis flavipennis*

Philippines (Cebu, Leyte and Mindanao)

☐ **Yellow-throated Leafbird** *Chloropsis palawanensis*

Philippines (Balabac, Busuanga, Palawan and Calamian)

☐ **Greater Green Leafbird** *Chloropsis sonnerati*
____ *C. s. zosterops* | Burma, peninsular Thailand, Malaysia, Sumatra and Borneo
____ *C. s. parvirostris* | Nias I. (off Sumatra)
____ *C. s. sonnerati* | Java

☐ **Lesser Green Leafbird** *Chloropsis cyanopogon*
____ *C. c. cyanopogon* | S Burma, n pen. Thailand, Malaysia, Sumatra and Borneo
____ *C. c. septentrionalis* | S peninsular Thailand

☐ **Blue-winged Leafbird** *Chloropsis cochinchinensis*
____ *C. c. jerdoni* | Peninsular India and Sri Lanka
____ *C. c. kinneari* | SW China (s Yunnan) to e Thailand and n Indochina
____ *C. c. cochinchinensis* | SE Thailand and s Indochina
____ *C. c. serithai* | Peninsular Thailand south to Isthmus of Kra
____ *C. c. moluccensis* | S Thailand and Malay Peninsula
____ *C. c. icterocephala* | Sumatra
____ *C. c. natunensis* | Natuna Islands (China Sea)
____ *C. c. viridinucha* | Borneo
____ *C. c. billitonis* | Belitung I. (off Borneo)
____ *C. c. nigricollis* | Java

☐ **Golden-fronted Leafbird** *Chloropsis aurifrons*
____ *C. a. aurifrons* | Himalayas and ne India to Burma
____ *C. a. frontalis* | Peninsular India
____ *C. a. insularis* | SW India (Travancore) and Sri Lanka
____ *C. a. pridii* | SW China (s Yunnan) to s Burma, n Thailand and n Laos
____ *C. a. inornata* | Central and se Thailand to Cambodia and s Vietnam
____ *C. a. incompta* | SW Thailand and s Indochina
____ *C. a. media* | Sumatra

☐ **Orange-bellied Leafbird** *Chloropsis hardwickii*
____ *C. h. hardwickii* | E Himalayas to sw China, Burma, n Thailand and n Vietnam
____ *C. h. melliana* | S China (Guangxi, Fujian and Guangdong) to n Vietnam
____ *C. h. lazulina* | Hainan (s China)
____ *C. h. malayana* | Malay Peninsula

☐ **Blue-masked Leafbird** *Chloropsis venusta*

Foothills of Sumatra

FAMILY: AEGITHINIDAE (Ioras—4)

☐ **Common Iora** *Aegithina tiphia*
____ *A. t. multicolor* | S India and Sri Lanka
____ *A. t. deignani* | Peninsular India to n and central Burma
____ *A. t. humei* | Central India (south of the Ganges River)
____ *A. t. tiphia* | NE India (Kumaon to Bengal and Assam)
____ *A. t. septentrionalis* | Pakistan and nw India (Punjab)
____ *A. t. philipi* | SW China to central Burma, n Thailand, Laos and n Vietnam
____ *A. t. cambodiana* | Cambodia to se Thailand and s Vietnam
____ *A. t. horizoptera* | S Burma to Thailand, Malaysia, Sumatra and adj. islands
____ *A. t. scapularis* | Java and Bali
____ *A. t. viridis* | S Borneo
____ *A. t. aequanimis* | N Borneo, adjacent northern islands and Palawan

☐ **White-tailed Iora** *Aegithina nigrolutea*

Lowlands of n Pakistan and nw India

☐ **Green Iora** *Aegithina viridissima*
    ____ *A. v. viridissima* — S Burma, peninsular Thailand, Malaya, Sumatra and Borneo
    ____ *A. v. thapsina* — Anambas Islands (South China Sea)

☐ **Great Iora** *Aegithina lafresnayei*
    ____ *A. l. lafresnayei* — S Thailand and Malaysia
    ____ *A. l. innotata* — SW China (s Yunnan) to Burma, Thailand and n Indochina
    ____ *A. l. xanthotis* — Cambodia and s Indochina

FAMILY: PTILOGONATIDAE (Silky-flycatchers—4)

☐ **Black-and-yellow Silky-flycatcher** *Phainoptila melanoxantha*
    ____ *P. m. parkeri* — Mts. of n Costa Rica (Cordilleras de Guanacaste and Tilarán)
    ____ *P. m. melanoxantha* — Cordillera of central Costa Rica to w Panama (e to Veraguas)

☐ **Gray Silky-flycatcher** *Ptilogonys cinereus*
    ____ *P. c. otofuscus* — Sierra Madre Occidental of w Mexico
    ____ *P. c. cinereus* — Highlands of central and e Mexico
    ____ *P. c. pallescens* — Highlands of sw Mexico (e Michoacán and Guerrero)
    ____ *P. c. molybdophanes* — Highlands of s Mexico (Chiapas) and w Guatemala

☐ **Long-tailed Silky-flycatcher** *Ptilogonys caudatus*

Mts. of Costa Rica and w Panama (Volcán de Chiriquí)

☐ **Phainopepla** *Phainopepla nitens*
    ____ *P. n. lepida* — Arid sw US to Baja and nw Mexico (Sonora and Chihuahua)
    ____ *P. n. nitens* — S Texas to s Mexican plateau

FAMILY: BOMBYCILLIDAE (Waxwings—3)

☐ **Bohemian Waxwing** *Bombycilla garrulus*
    ____ *B. g. pallidiceps* — NW North America; highly nomadic in winter
    ____ *B. g. garrulus* — Fenno-Scandia to w Siberia; winters to central Europe
    ____ *B. g. centralasiae* — Central Siberia to Sea of Okhotsk; winters to s China and Japan

☐ **Cedar Waxwing** *Bombycilla cedrorum*

North America; winters to n S America and Greater Antilles

☐ **Japanese Waxwing** *Bombycilla japonica*

SE Siberia and n Manchuria; winters to s China and Ryukyu Is.

FAMILY: HYPOCOLIIDAE (Hypocolius—1)

☐ **Hypocolius** *Hypocolius ampelinus*

Iraq (Tigris-Euphrates valleys) to Turkmenia; winters Arabia

FAMILY: DULIDAE (Palmchat—1)

☐ **Palmchat** *Dulus dominicus*

Hispaniola, Gonâve I. and Saona I.

FAMILY: CINCLIDAE (Dippers—5)

☐ **White-throated Dipper** *Cinclus cinclus*

_____ *C. c. hibernicus*	Ireland, Outer Hebrides and w coast of Scotland
_____ *C. c. gularis*	Orkney Islands, c and e Scotland, w and c England and Wales
_____ *C. c. cinclus*	Fenno-Scandia to s coast of White Sea and Kaliningrad region
_____ *C. c. aquaticus*	Central and s Europe to Balkan Peninsula
_____ *C. c. minor*	Mts. of Morocco, Tunisia and Algeria
_____ *C. c. olympicus†*	Formerly Cyprus. Extinct
_____ *C. c. caucasicus*	Caucasus Mts. to nw Iran; winters to Iraq and Pakistan
_____ *C. c. rufiventris*	Anti-Lebanon Mts.
_____ *C. c. persicus*	SW Iran (Zagros and Bakhtiari mts.)
_____ *C. c. uralensis*	Ural Mts.
_____ *C. c. leucogaster*	Mts. of central Asia
_____ *C. c. cashmeriensis*	Himalayas (w Kashmir to Sikkim)
_____ *C. c. przewalskii*	Mts. of s Tibet and w China

☐ **Brown Dipper** *Cinclus pallasii*

_____ *C. p. tenuirostris*	Mts. of central Asia and Himalayas
_____ *C. p. dorjei*	Mts. of e Sikkim, Assam, e Tibet, n Burma and n Thailand
_____ *C. p. pallasii*	Mts. of ne Asia, Japan, w China, n Thailand, n Vietnam
_____ *C. p. marila*	NE India (Khasi Hills)

☐ **American Dipper** *Cinclus mexicanus*

_____ *C. m. unicolor*	Aleutian Islands to Alaska, w Canada and w US
_____ *C. m. mexicanus*	Highlands of n and central Mexico
_____ *C. m. anthonyi*	Mts. of s Mexico (Chiapas) to Guatemala and Honduras
_____ *C. m. ardesiacus*	Mts. of Costa Rica and w Panama

☐ **White-capped Dipper** *Cinclus leucocephalus*

_____ *C. l. rivularis*	Santa Marta Mts. (ne Colombia)
_____ *C. l. leuconotus*	Mts. of Colombia to w Venezuela and Ecuador
_____ *C. l. leucocephalus*	Mts. of Peru and Bolivia

☐ **Rufous-throated Dipper** *Cinclus schulzi*

	E slope of Andes of extreme nw Argentina and se Bolivia

FAMILY: TROGLODYTIDAE (Wrens—80)

☐ **Black-capped Donacobius** *Donacobius atricapilla*

_____ *D. a. brachypterus*	Tropical e Panama (Darién) to n Colombia
_____ *D. a. nigrodorsalis*	SE Colombia to e Ecuador and se Peru (Madre de Dios)
_____ *D. a. atricapilla*	Venezuela to Guianas, Amazonian and e Brazil and ne Argentina
_____ *D. a. albovittatus*	E Bolivia (Beni, Cochabamba and Santa Cruz); adjacent Brazil?

☐ **White-headed Wren** *Campylorhynchus albobrunneus*

_____ *C. a. albobrunneus*	Humid lowlands of central and e Panama
_____ *C. a. harterti*	E Panama (Darién) and w Colombia (south to Valle)
_____ *C. a. aenigmaticus*	SW Colombia (Nariño); probable hybrid *albobrunneus* x *zonatus*

☐ **Band-backed Wren** *Campylorhynchus zonatus*

_____ *C. z. zonatus*	E Mexico (e San Luis Potosí and n Veracruz to n Puebla)
_____ *C. z. restrictus*	S Mexico (s Veracruz and n Oaxaca) to Belize and Guatemala
_____ *C. z. vulcanius*	S Mexico (Chiapas) to central Nicaragua
_____ *C. z. costaricensis*	Caribbean slope of Costa Rica and nw Panama
_____ *C. z. panamensis*	W Panama (Veraguas)
_____ *C. z. curvirostris*	N Colombia (tropical base of Santa Marta Mts.)
_____ *C. z. brevirostris*	N Colombia to nw Ecuador

☐ **Gray-barred Wren** *Campylorhynchus megalopterus*
___	*C. m. megalopterus*	Coniferous forests of Mexican plateau
___	*C. m. nelsoni*	Mts. of s Mexico (e Puebla, sw Veracruz and Oaxaca)

☐ **Giant Wren** *Campylorhynchus chiapensis*

Humid Pacific lowlands of s Mexico (Chiapas)

☐ **Rufous-naped Wren** *Campylorhynchus rufinucha*
___	*C. r. humilis*	Arid lowlands of sw Mexico (Colima and s Jalisco to w Chiapas)
___	*C. r. rufinucha*	Lowlands of e Mexico (Veracruz and adjacent Oaxaca)
___	*C. r. nigricaudatus*	S Mexico (coastal Chiapas) to w Guatemala
___	*C. r. xerophilum*	Interior of Guatemala
___	*C. r. nicaraguae*	Interior of w Nicaragua
___	*C. r. castaneus*	W-central Honduras (Sula Valley)
___	*C. r. capistratus (nicoyae)*	E Guatemala, coastal Honduras to Nicaragua and nw Costa Rica

☐ **Spotted Wren** *Campylorhynchus gularis*

Oak-pine woodlands of w and central Mexico

☐ **Boucard's Wren** *Campylorhynchus jocosus*

Arid oak-pine forests of s Mexican plateau

☐ **Yucatan Wren** *Campylorhynchus yucatanicus*

Arid coastal lowlands of se Mexico (n Yucatán Peninsula)

☐ **Cactus Wren** *Campylorhynchus brunneicapillus*
___	*C. b. anthonyi (couesi)*	Arid sw US to n Baja, nw Mexico and extreme n Tamaulipas
___	*C. b. sandiegensis*	Arid San Diego County (s California) and adjacent nw Baja
___	*C. b. bryanti*	Pacific slope of w Baja California between 31º and 29º 30'N
___	*C. b. purus*	Coastal central Baja California between 29º and 25ºN
___	*C. b. seri*	Isla Tiburón (Sea of Cortés)
___	*C. b. affinis*	S Baja California (south of 25º N to Cabo San Lucas)
___	*C. b. brunneicapillus*	NW Mexico (central Sonora to n Sinaloa)
___	*C. b. guttatus*	S and central Texas; central plateau of Mexico

☐ **Bicolored Wren** *Campylorhynchus griseus*
___	*C. g. albicilius*	Tropical n Colombia to extreme nw Venezuela
___	*C. g. zimmeri*	Colombia (Huila and Tolima)
___	*C. g. bicolor*	Colombia (upper Río Magdalena, Santander and Boyacá)
___	*C. g. minor*	Tropical e Colombia to n Venezuela
___	*C. g. pallidus*	S Venezuela (Amazonas)
___	*C. g. griseus*	E Venezuela (n Amazonas) to w Guyana and extreme n Brazil

☐ **Thrush-like Wren** *Campylorhynchus turdinus*
___	*C. t. hypostictus (chanchamayoensis)*	SE Colombia to Ecuador, Peru, Bolivia, Brazil s of the Amazon
___	*C. t. turdinus*	E-central Brazil (Maranhão to Goiás, Bahia and Espírito Santo)
___	*C. t. unicolor*	Tropical e Bolivia to sw Brazil (Mato Grosso) and n Paraguay

☐ **Stripe-backed Wren** *Campylorhynchus nuchalis*
___	*C. n. pardus*	Arid tropical n Caribbean Colombia s to lower Magdalena Valley
___	*C. n. brevipennis*	Coastal n Venezuela (Miranda to Carabobo and Guarico)
___	*C. n. nuchalis*	Central and e Venezuela (Barinas to Bolívar and Sucre)

☐ **Fasciated Wren** *Campylorhynchus fasciatus*
___	*C. f. pallescens*	Arid sw Ecuador to nw Peru (Tumbes, Piura and Lambayeque)
___	*C. f. fasciatus*	Arid w Peru (s Piura to Huánuco and n Lima)

☐ **Gray-mantled Wren** *Odontorchilus branickii*
___	*O. b. branickii*	Trop. and subtrop. e Colombia, Ecuador, Peru and ne Bolivia
___	*O. b. minor*	W Andes of s Colombia and extreme n Ecuador

☐ **Tooth-billed Wren** *Odontorchilus cinereus*

Lowlands of Amazonian Brazil s of the Amazon and e Bolivia

☐ **Rock Wren** *Salpinctes obsoletus*

____	*S. o. obsoletus*	SW Canada and w US to n and central Mexico
____	*S. o. guadeloupensis*	Guadalupe I. (off w Mexico)
____	*S. o. tenuirostris*	San Benito Islands (off s Baja California)
____	*S. o. exsul†*	San Benedicto I. (Revillagigedo Islands off s Baja California)
____	*S. o. neglectus*	Highlands of se Mexico (Chiapas) to Guatemala and c Honduras
____	*S. o. guttatus*	Highlands of El Salvador to Nicaragua and nw Costa Rica

☐ **Canyon Wren** *Catherpes mexicanus*

____	*C. m. conspersus (griseus, punctulatus)*	S Br. Col. through w US to nw Mexico (Sonora)
____	*C. m. albifrons*	SW Texas and n Mexico (s to Zacatecas and San Luis Potosí)
____	*C. m. mexicanus*	Mexico (s Chihuahua to Isthmus of Tehuantepec)

☐ **Sumichrast's Wren** *Hylorchilus sumichrasti*

Lowlands of s Mexico (Veracruz, Puebla and adjacent n Oaxaca)

☐ **Nava's Wren** *Hylorchilus navai*

Lowlands of s Mexico (Chiapas and extreme e Veracruz)

☐ **Rufous Wren** *Cinnycerthia unirufa*

____	*C. u. unirufa*	Andes of ne Colombia and extreme w Venezuela (Táchira)
____	*C. u. unibrunnea*	Central Andes of Colombia to Ecuador and extreme n Peru
____	*C. u. chakei*	Perijá Mts. (Colombia/Venezuela border)

☐ **Sharpe's Wren** *Cinnycerthia olivascens*

____	*C. o. bogotensis*	W slope of Eastern Andes of Colombia
____	*C. o. olivascens*	Central and Western Andes of Colombia to extreme n Peru

☐ **Peruvian Wren** *Cinnycerthia peruana*

Eastern Andes of Peru (Amazonas to Ayacucho)

☐ **Fulvous Wren** *Cinnycerthia fulva*

____	*C. f. fitzpatricki*	E Peru (Cordillera Vilcabamba of Cusco)
____	*C. f. fulva*	Eastern Andes of central Peru (s Cusco)
____	*C. f. gravesi*	Andes of s Peru (Puno) to n Bolivia (Cochabamba and La Paz)

☐ **Black-throated Wren** *Thryothorus atrogularis*

Caribbean lowlands of e Nicaragua to w Panama

☐ **Sooty-headed Wren** *Thryothorus spadix*

____	*T. s. xerampelinus*	Pacific slope of e Panama (Darién)
____	*T. s. spadix*	W Colombia

☐ **Black-bellied Wren** *Thryothorus fasciatoventris*

____	*T. f. melanogaster*	Pacific lowlands of s Costa Rica to w Panama
____	*T. f. albigularis*	E Panama (Canal Zone) to w Colombia (Chocó)
____	*T. f. fasciatoventris*	Tropical n Colombia to Río Magdalena Valley

☐ **Inca Wren** *Thryothorus eisenmanni*

Andes of s Peru (Cusco)

☐ **Whiskered Wren** *Thryothorus mystacalis*

____	*T. m. saltuensis*	Western and Central Andes of Colombia
____	*T. m. yananchae*	SW Colombia (Yanachá region of Nariño)
____	*T. m. mystacalis*	S Colombia to w Ecuador (south to El Oro)
____	*T. m. macrurus*	Colombia (known from one specimen of unknown provenance)
____	*T. m. amaurogaster*	Subtropical Eastern Andes of Colombia
____	*T. m. consobrinus*	Western Venezuela (Lara and Mérida)
____	*T. m. ruficaudatus*	N Venezuela (Carabobo and Distrito Federal)
____	*T. m. tachirensis*	Andes of nw Venezuela (Páramo de Tama in Táchira)

☐ **Plain-tailed Wren** *Thryothorus euophrys*

____	*T. e. euophrys*	Andes of extreme s Colombia (w Nariño) and n Ecuador
____	*T. e. longipes*	Temperate e slope of Andes of Ecuador
____	*T. e. atriceps*	Andes of nw Peru (north of Río Marañón)
____	*T. e. schulenbergi*	E slope of Andes of n Peru (south of Río Marañón)

☐ **Moustached Wren** *Thryothorus genibarbis*

    ____ *T. g. genibarbis*   E Brazil (Maranhão to Espirito Santo, west to Rio Madeira

    ____ *T. g. juruanus*   W Amazonian Brazil to se Peru (Ucayali) and nw Bolivia

    ____ *T. g. intercedens*   Central Brazil (Goiás to Minas Gerais and Mato Grosso)

    ____ *T. g. bolivianus*   Lowlands of Bolivia (La Paz, Cochabamba and Santa Cruz)

☐ **Coraya Wren** *Thryothorus coraya*

    ____ *T. c. griseipectus*   E Ecuador, adjacent Brazil and ne Peru n of the Marañón

    ____ *T. c. caurensis*   E Venezuela (Caura Valley)

    ____ *T. c. barrowcloughianus*   *Tepuis* of s Venezuela (Mt. Roraima and Mt. Cuquenam)

    ____ *T. c. ridgwayi*   Mts. of e Venezuela (Gran Sabana) to w Guyana

    ____ *T. c. obscurus*   *Tepuis* of se Venezuela in Bolívar (Auyan-tepui)

    ____ *T. c. coraya*   The Guianas and adjacent Brazil n of the Amazon

    ____ *T. c. herberti*   N Brazil s of the Amazon

    ____ *T. c. albiventris*   E Peru (e slope of Andes in San Martín)

    ____ *T. c. amazonicus*   Tropical e Peru south of Río Marañón (Loreto and Huánuco)

    ____ *T. c. cantator*   Subtropical mts. of se Peru (Junín and Cusco)

☐ **Happy Wren** *Thryothorus felix*

    ____ *T. f. sonorae*   Pacific slope of nw Mexico (s Sonora to n Sinaloa)

    ____ *T. f. pallidus*   W Mexico (c Sinaloa and w Durango to Jalisco and Michoacán)

    ____ *T. f. lawrencii*   María Madre I. (Tres Marías Islands off w Mexico)

    ____ *T. f. magdalenae*   María Magdalena I. (Tres Marías Islands off w Mexico)

    ____ *T. f. felix*   W Mexico (se Jalisco to Michoacán, Guerrero and w Oaxaca)

    ____ *T. f. grandis*   S Mexico (upper Río Balsas drainage to sw Puebla, n Guerrero)

☐ **Spot-breasted Wren** *Thryothorus maculipectus*

    ____ *T. m. microstictus*   NE Mexico (e Nuevo León, San Luis Potosí and Tamaulipas)

    ____ *T. m. maculipectus*   Gulf-Caribbean slope of e Mexico (Veracruz to n Oaxaca)

    ____ *T. m. umbrinus*   S Mexico (Tabasco and Chiapas) and s Belize

    ____ *T. m. varians*   S Mexico (Pacific slope of Chiapas), Guatemala and El Salvador

    ____ *T. m. canobrunneus*   SE Mexico (Yucatán Pen. to Petén of Guatemala and n Belize

    ____ *T. m. petersi*   N Honduras to n Costa Rica

☐ **Rufous-breasted Wren** *Thryothorus rutilus*

    ____ *T. r. hyperythrus*   Pacific slope of sw Costa Rica to e Panama (Chiriqui to Panama)

    ____ *T. r. laetus*   N Colombia (Santa Marta Mts.)

    ____ *T. r. hypospodius*   Eastern Andes of Colombia and extreme w Venezuela (Táchira)

    ____ *T. r. interior*   Colombia (valley of Río Lebrija in Santander)

    ____ *T. r. intensus*   W Venezuela (Zulia)

    ____ *T. r. rutilus*   Mts. of n Venezuela (Sucre and Monagua to Lara); Trinidad

    ____ *T. r. tobagensis*   Tobago

☐ **Speckle-breasted Wren** *Thryothorus sclateri*

    ____ *T. s. columbianus*   Andes of Colombia (Valle and Cundinamarca)

    ____ *T. s. paucimaculatus*   Tropical w Ecuador (Manabí to Loja)

    ____ *T. s. sclateri*   S Ecuador and adjacent n Peru (Cajamarca)

☐ **Riverside Wren** *Thryothorus semibadius*

    Pacific lowlands of sw Costa Rica and extreme w Panama

☐ **Bay Wren** *Thryothorus nigricapillus*

    ____ *T. n. costaricensis*   Caribbean lowlands of se Nicaragua, Costa Rica and nw Panama

    ____ *T. n. castaneus*   Panama (Veraguas to Canal Zone)

    ____ *T. n. odicus*   Panama (Escudo de Veraguas I., Bocas del Toro)

    ____ *T. n. schotti*   Pacific slope of e Panama (Darién) and nw Colombia

    ____ *T. n. reditus*   Caribbean slope of e Panama (e Colón, San Blas)

    ____ *T. n. connectens*   SW Colombia (Cauca and Nariño) to ne Ecuador

    ____ *T. n. nigricapillus*   Tropical w Ecuador (Esmeraldas to El Oro)

☐ **Stripe-breasted Wren** *Thryothorus thoracicus*

    Caribbean lowlands of e Nicaragua to w Panama

□ **Stripe-throated Wren** *Thryothorus leucopogon*
_____ *T. l. grisescens* — Caribbean coast of e Panama and adjacent Colombia
_____ *T. l. leucopogon* — Pacific coast of e Panama (Darién), w Colombia and nw Ecuador

□ **Banded Wren** *Thryothorus pleurostictus*
_____ *T. p. nisorius* — W Mexico (Morelos, Puebla and Michoacán)
_____ *T. p. oaxacae* — SW Mexico (coastal central Guerrero to Oaxaca)
_____ *T. p. acaciarum* — S Mexico (Chiapas)
_____ *T. p. oblitus* — Pacific lowlands of e Chiapas to Guatemala and w El Salvador
_____ *T. p. pleurostictus* — Guatemala (Gualán region of Zacapa)
_____ *T. p. lateralis* — Lowlands of El Salvador and w Honduras
_____ *T. p. ravus* — Pacific lowlands of Nicaragua to nw Costa Rica

□ **Carolina Wren** *Thryothorus ludovicianus*
_____ *T. l. ludovicianus* — SE Canada to Texas and se US
_____ *T. l. miamensis* — Peninsular Florida
_____ *T. l. nesophilus* — Dog I. (off nw Florida)
_____ *T. l. burleighi* — Cat I., Ship I. and Horn I. (off Mississippi)
_____ *T. l. lomitensis* — Texas (lower Rio Grande Valley) and ne Mexico (n Tamaulipas)
_____ *T. l. berlandieri* — Mts. of e Mexico (e Coahuila, Nuevo León and sw Tamaulipas)
_____ *T. l. tropicalis* — Tropical ne Mexico (s Tamaulipas and e San Luis Potosí)
_____ *T. l. albinucha* — SE Mexico (Yucatán Peninsula) to Petén of n Guatemala
_____ *T. l. subfulvus* — Arid interior of Guatemala to nw Nicaragua

□ **Rufous-and-white Wren** *Thryothorus rufalbus*
_____ *T. r. transfinis* — S Mexico (Pacific slope of extreme sw Chiapas)
_____ *T. r. rufalbus* — Highlands of Guatemala and El Salvador
_____ *T. r. castanonotus* — Pacific slope of Nicaragua to Costa Rica and w Panama
_____ *T. r. cumanensis* — Caribbean coast of n Colombia to ne Venezuela (Pária Peninsula)
_____ *T. r. minlosi* — Tropical e Colombia to nw Venezuela

□ **Sinaloa Wren** *Thryothorus sinaloa*
_____ *T. s. cinereus* — NW Mexico (se Sonora, sw Chihuahua and n Sinaloa)
_____ *T. s. sinaloa* — W Mexico (c Sinaloa to w Durango, Nayarit, Jalisco and Colima)
_____ *T. s. russeus* — SW Mexico (central Guerrero to extreme sw Oaxaca)

□ **Plain Wren** *Thryothorus modestus*
_____ *T. m. modestus (pullus)* — S Mexico (Oaxaca) to Guatemala and n Nicaragua
_____ *T. m. roberti* — Caribbean lowlands of Honduras
_____ *T. m. vanrossemi* — Pacific lowlands of El Salvador
_____ *T. m. zeledoni* — E Nicaragua to e Costa Rica and nw Panama (w Bocas del Toro)
_____ *T. m. elutus* — W Panama (Chiriquí to Canal Zone)

□ **Buff-breasted Wren** *Thryothorus leucotis*
_____ *T. l. galbraithii* — E Panama and nw Colombia (n Chocó and n Antioquia)
_____ *T. l. conditus* — Pearl Islands and Coiba I. (Gulf of Panama)
_____ *T. l. leucotis* — N Colombia (w slope of Santa Marta Mts. to Magdalena Valley)
_____ *T. l. collinus* — N Colombia (n Guajira Peninsula in Serranía de Macuira)
_____ *T. l. venezuelanus* — N tropical Colombia and nw Venezuela
_____ *T. l. zuliensis* — E Colombia (Norte de Santander) to w Venezuela
_____ *T. l. peruanus* — SE Colombia to e Ecuador, e Peru, n Bolivia and w Amaz. Brazil
_____ *T. l. bogotensis* — *Llanos* of e Colombia to central Venezuela
_____ *T. l. hypoleucus* — *Llanos* of n central Venezuela
_____ *T. l. albipectus* — NE Venezuela to the Guianas, ne Brazil and n Mato Grosso
_____ *T. l. rufiventris* — E Brazil (s Maranhão to Piauí, Goiás, Minas Gerais and São Paulo)

□ **Niceforo's Wren** *Thryothorus nicefori*
W slope of E Andes of n Colombia (near San Gil on Río Fonce)

□ **Superciliated Wren** *Thryothorus superciliaris*
_____ *T. s. superciliaris* — Arid coastal Ecuador (Manabi to Guayas); Isla Puná
_____ *T. s. baroni* — Arid s Ecuador (El Oro) to nw Peru (Ancash)

☐ **Fawn-breasted Wren** *Thryothorus guarayanus*

N Bolivia, adj. sw Brazil (sw Mato Grosso) and Paraguay

☐ **Long-billed Wren** *Thryothorus longirostris*
___ *T. l. bahiae*
___ *T. l. longirostris*

NE Brazil (Ceará and e Piauí to n Bahia and Alagoas)
Coastal se Brazil (e Minas Gerais to s Bahia and n Paraná)

☐ **Gray Wren** *Thryothorus griseus*

W Amazonian Brazil (sw Amazonas)

☐ **Bewick's Wren** *Thryomanes bewickii*
___ *T. b. calophonus*
___ *T. b. drymoecus*
___ *T. b. atrestus*
___ *T. b. marinensis*
___ *T. b. spilurus*
___ *T. b. correctus*
___ *T. b. eremophilus*
___ *T. b. leucophrys†*
___ *T. b. charienturus*
___ *T. b. nesophilus*
___ *T. b. catalinae*
___ *T. b. cerroensis*
___ *T. b. magdalenensis*
___ *T. b. brevicauda†*
___ *T. b. bewickii*
___ *T. b. altus*
___ *T. b. cryptus*
___ *T. b. bairdii*
___ *T. b. percnus*

SW British Columbia to w Washington and w Oregon
SW Oregon to California (Sacramento and n San Joaquin valleys)
S-central Oregon to ne California and w-central Nevada
Coastal California (Del Norte County to Marin County)
Coastal central California (San Francisco to Monterey Bay)
Coastal California (San Benito County to San Diego)
E California to Utah, Wyoming, w Texas and w-central Mexico
San Clemente I. (off s California). Extinct
N Baja California (south to 30°N)
Santa Cruz, Santa Rosa and Anacapa Islands off s California
Catalina Island (off Southern California)
W-central Baja California (30° to 26°N) and Isla Cedros
S Baja California south of 26°N
Formerly Guadalupe I. (off Baja California). Extinct ca 1903
N-central US to Kansas, Nebraska and Mississippi
E North America (s Ontario to central Georgia)
Kansas and Oklahoma to s Texas and ne Mexico (n Tamaulipas)
SE Mexico (Oaxaca, Veracruz and s Puebla)
W Mexico (Jalisco to s Zacatecas)

☐ **Zapata Wren** *Ferminia cerverai*

SW Cuba (dense vegetation of Zapata Swamp)

☐ **Winter Wren** *Troglodytes troglodytes*
___ *T. t. islandicus*
___ *T. t. borealis*
___ *T. t. zetlandicus*
___ *T. t. hebridensis*
___ *T. t. fridariensis*
___ *T. t. hirtensis*
___ *T. t. indigenus*
___ *T. t. troglodytes*
___ *T. t. kabylorum*
___ *T. t. koenigi*
___ *T. t. cypriotes*
___ *T. t. hyrcanus (zagrossiensis)*
___ *T. t. juniperi*
___ *T. t. tianschanicus*
___ *T. t. pallescens*
___ *T. t. kurilensis*
___ *T. t. fumigatus*
___ *T. t. mosukei*
___ *T. t. ogawae*
___ *T. t. taivanus*
___ *T. t. dauricus*
___ *T. t. idius*
___ *T. t. szetschuanus*
___ *T. t. talifuensis*
___ *T. t. subpallidus*
___ *T. t. neglectus*
___ *T. t. nipalensis*
___ *T. t. magrathi*
___ *T. t. alascensis*

Iceland
Faeroe Islands (n Atlantic Ocean)
Shetland Islands (Scotland)
Outer Hebrides Islands (Scotland)
Fair Isle (Scotland)
St. Kilda I. (Scotland)
Ireland, Inner Hebrides, Orkneys, Scotland and England
Continental Europe and Asia Minor
Balearic Islands, s Spain and nw Africa (Morocco to Tunisia)
Corsica and Sardinia
Crete, Rhodes, Cyprus and Near East
Crimean Peninsula to Caucasus Mts., n Iraq and Iran
NW Libya
NE Iran and s Transcaspia to n Afghanistan and Turkestan
Kamchatka Peninsula and Komandorskiye Islands
N Kuril Islands (Shasukotan and Ushichi)
S Kuril Islands and Japan
Izu Islands and Daito Islands
S Japanese Archipelago (Tanegashima and Yakushima)
Taiwan
E Siberia to Sakhalin, Manchuria and Korea
N China (s Hebei to Shandong)
SW China (s Shaanxi and Sichuan east to Hupei)
W China (s Sichuan to w Yunnan) and ne Burma
Himalayas of Afghanistan
W Himalayas (Gilgit to w Nepal)
Himalayas of Nepal to ne Assam and s Tibet
Mts. on borders of Pakistan and Afghanistan
Pribilof Islands (St. George, St. Paul and Otter)

____	*T. t. kiskensis*	W Aleutians (Kiska, Little Kiska, Amchitka, Ogliuga)
____	*T. t. meligerus*	W Aleutians (Attu, Agattu, Alaid, Nitzi and Buldir)
____	*T. t. tanagensis*	Central Aleutians (Tanaga, Adak and Atka))
____	*T. t. seguamensis*	Central Aleutians (Seguam, Amutka and Yunaska)
____	*T. t. stevensoni*	W Alaska Peninsula, Amak and Amagat islands
____	*T. t. petrophilus*	E Aleutians (Unalaska, Amaknak and Akutan)
____	*T. t. semidiensis*	SE Alaska (Semidi Islands)
____	*T. t. helleri*	S Alaska (Kodiak, Afognak and Raspberry islands)
____	*T. t. pacificus*	Coastal sw Alaska to California
____	*T. t. salebrosus*	Interior British Columbia to we Montana and e Oregon
____	*T. t. hiemalis*	E North America from s Canada to s Georgia
____	*T. t. pullus*	Appalachian Mts. (e W Virginia and w Virginia to ne Georgia)
____	*T. t. muiri*	Coastal n California (south to Marin County)
____	*T. t. obscurior*	Coastal central Calif. (San Francisco to San Luis Obispo Co.)

☐ **House Wren** *Troglodytes aedon*

____	*T. a. parkmanii*	SW Canada to central and w US and n Baja California
____	*T. a. aedon*	SE Canada and e US
____	*T. a. baldwini*	S-central Canada to s US
____	*T. a. cahooni*	Mts. of se Arizona to central Mexican plateau
____	*T. a. compositus*	Mts. of e Mexico (Coahuila and Nuevo León to Puebla)
____	*T. a. brunneicollis*	Central and s Mexico (San Luis Potosí and Hidalgo to Oaxaca)
____	*T. a. nitidus*	S Mexico (Mt. Zempoaltepec in Oaxaca)
____	*T. a. intermedius*	S Mexico (se Oaxaca and e Tabasco) to Costa Rica
____	*T. a. beani*	E Mexico (Cozumel I. off Quintana Roo)
____	*T. a. inquietus*	Extreme s Costa Rica, Panama and Pearl Islands
____	*T. a. carychrous*	Coiba I. (Panama)
____	*T. a. rufescens*	Dominica (Lesser Antilles)
____	*T. a. martinicensis*	Martinique (Lesser Antilles)
____	*T. a. mesoleucus*	St. Lucia (Lesser Antilles)
____	*T. a. guadeloupensis*	Guadeloupe (Lesser Antilles)
____	*T. a. musicus*	St. Vincent (Lesser Antilles)
____	*T. a. grenadensis*	Grenada (Lesser Antilles)
____	*T. a. clarus*	Trinidad, the Guianas, Venezuela, Brazil, n Peru, Colombia
____	*T. a. atopus*	N Colombia (Santa Marta region)
____	*T. a. striatulus*	W and Central Andes of Colombia
____	*T. a. columbae*	E Colombia and w Venezuela
____	*T. a. albicans*	SW Colombia and w Ecuador
____	*T. a. tobagensis*	Tobago
____	*T. a. audax*	Arid littoral of w Peru (Cajamarca to n Ica)
____	*T. a. puna*	*Puna* of n Peru to nw Bolivia (La Paz)
____	*T. a. rex*	Central and e Bolivia
____	*T. a. carabayae*	Central and s Peru (Junín, Cusco and Puno)
____	*T. a. tecellatus*	Coastal s Peru (Arequipa) to n Chile (Tarapacá)
____	*T. a. atacamensis*	N Chile (Antofagasta, Atacama and n Coquimbo)
____	*T. a. musculus*	Central and s Brazil to e Paraguay and ne Argentina (Misiones)
____	*T. a. bonariae*	S Brazil, Uruguay and ne Argentina
____	*T. a. chilensis*	S Chile and s Argentina to Tierra del Fuego

☐ **Socorro Wren** *Troglodytes sissonii*

Socorro I. (Revillagigedo Islands off w Mexico)

☐ **Clarion Wren** *Troglodytes tanneri*

Isla Clarión (Revillagigedo Islands off w Mexico)

☐ **Cobb's Wren** *Troglodytes cobbi*

Falkland Islands

☐ **Rufous-browed Wren** *Troglodytes rufociliatus*

____	*T .r. chiapensis*	Highlands of s Mexico (Chiapas)
____	*T. r. rufociliatus*	Highlands of e Guatemala and n El Salvador
____	*T. r. nannoides*	Highlands of w El Salvador (Volcán Santa Ana)
____	*T. r. rehni*	Highlands of Honduras to nw Nicaragua

☐ **Ochraceous Wren** *Troglodytes ochraceus*
- ____ *T. o. ochraceus* — Highlands of Costa Rica
- ____ *T. o. ligea* — Highlands of w Panama (Chiriquí)
- ____ *T. o. festinus* — E Panama (Mt. Pirre) and possibly adjacent Colombia

☐ **Santa Marta Wren** *Troglodytes monticola*

Santa Marta Mts. (ne Colombia)

☐ **Mountain Wren** *Troglodytes solstitialis*
- ____ *T. s. solitarius* — Andes of Colombia and w Venezuela
- ____ *T. s. solstitialis* — Andes of s Colombia to Ecuador and n Peru (Cajamarca)
- ____ *T. s. macrourus* — Andes of e-central Peru (s Amazonas to Cusco)
- ____ *T. s. frater* — Andes of extreme se Peru (Puno) to Bolivia
- ____ *T. s. auricularis* — Andes of nw Argentina (south to Tucumán and Catamarca)

☐ **Tepui Wren** *Troglodytes rufulus*
- ____ *T. r. rufulus* — Subtrop. Mt. Roraima and Vei-Tepui (Venezuela/Guyana border)
- ____ *T. r. fulvigularis* — *Tepuis* of se Venezuela (Ptari-tepui, Sororopón and Auyan-tepui)
- ____ *T. r. yavii* — *Tepuis* of se Venezuela (Cerro Yaví and Cerro Sarisariñama)
- ____ *T. r. duidae* — *Tepuis* of s Venezuela (Duida, Parú and Paraque)
- ____ *T. r. wetmorei* — *Tepuis* of se Venezuela (Cerro de la Neblina)

☐ **Sedge Wren** *Cistothorus platensis*
- ____ *C. p. stellaris* — E Canada to e US; winters Florida to ne Mexico
- ____ *C. p. tinnulus* — W Mexico (Nayarit to Michoacán, México and Distrito Federal)
- ____ *C. p. potosinus* — N-central Mexico (San Luis Potosí)
- ____ *C. p. jalapensis* — E Mexico (interior central Veracruz to Orizaba region)
- ____ *C. p. warneri* — Tropical s Mexico (Veracruz, Tabasco and w Chiapas)
- ____ *C. p. elegans* — S-central Guatemala
- ____ *C. p. russelli* — Pine ridge region of Belize
- ____ *C. p. graberi* — E Honduras to ne Nicaragua
- ____ *C. p. lucidus* — Subtropical central Costa Rica to w Panama (Chiriquí)
- ____ *C. p. alticola* — Mts. of n Colombia to n Venezuela; s Guyana
- ____ *C. p. tamae* — E Andes of Colombia and sw Venezuela
- ____ *C. p. tolimae* — Central Andes of Colombia (Tolima and Caldas)
- ____ *C. p. aequatorialis* — Central and Western Andes of s Colombia and Ecuador
- ____ *C. p. graminicola* — Andes of s Peru (Junín to Cusco) to nw Bolivia
- ____ *C. p. minimus (boliviae)* — Andes of s Peru (Puno) and adjacent Bolivia
- ____ *C. p. polyglottus* — SE Brazil (Goiás and Minas Gerais) to Paraguay and ne Argentina
- ____ *C. p. tucumanus* — NW Argentina (Jujuy to Catamarca and Tucumán)
- ____ *C. p. platensis* — Central and e Argentina to Córdoba and Mendoza
- ____ *C. p. hornensis* — S Argentina (Neuquén) and Chile (Coquimbo) to Tierra del Fuego
- ____ *C. p. falklandicus* — Falkland Islands

☐ **Apolinar's Wren** *Cistothorus apolinari*
- ____ *C. a. apolinari* — E Andes of Colombia (Boyacá and Cundinamarca)
- ____ *C. a. hernandezi* — Paramo, central Colombia

☐ **Paramo Wren** *Cistothorus meridae*

Andes of nw Venezuela (Trujillo and Mérida)

☐ **Marsh Wren** *Cistothorus palustris*
- ____ *C. p. browningi* — Coastal marshes of sw British Columbia to central Washington
- ____ *C. p. paludicola* — SW Washington to nw Oregon
- ____ *C. p. pulverius* — Cent. Br. Columbia and Idaho to ne California and nw Nevada
- ____ *C. p. plesius* — SE Idaho to Colorado and New Mexico; winters to c Mexico
- ____ *C. p. laingi* — W-central Canada to Montana; winters to s Mexico
- ____ *C. p. iliacus* — W-central Canada to w-central US; winters to Gulf Coast
- ____ *C. p. dissaeptus* — S-central Canada to n-central US; winters to ne Mexico
- ____ *C. p. clarkae* — Coastal s California (Los Angeles Co. to San Diego County)
- ____ *C. p. aestuarinus* — Inland valleys of s California, s Nevada and sw Arizona
- ____ *C. p. deserticola* — Deserts of s California

____ *C. p. palustris* — Coastal marshes of New England to Virginia
____ *C. p. waynei* — Coastal marshes of s Virginia to North Carolina
____ *C. p. griseus* — Coastal marshes of ne South Carolina to e-central Florida
____ *C. p. marianae (thryophilus)* — Coastal marshes of extreme e Texas to sw Florida
____ *C. p. tolucensis* — Central Mexico (Río Lerma marshes to Hidalgo and w Puebla)

☐ **White-bellied Wren** *Uropsila leucogastra*

____ *U. l. leucogastra*	Gulf lowlands of e Mexico (s Tamaulipas to n Oaxaca)
____ *U. l. pacifica*	Disjunct in sw Mexico (Colima, Michoacán and Guerrero)
____ *U. l. musica*	Coastal plain of s Mexico (Tabasco and n Chiapas), n Guatemala
____ *U. l. brachyura*	Yucatán Pen. and Belize
____ *U. l. hawkinsi*	Honduras (Yoro)

☐ **Timberline Wren** *Thryorchilus browni*

____ *T. b. ridgwayi*	Mts. of central Costa Rica (Volcán Turialba and Volcán Irazú)
____ *T. b. basultoi*	Mts. of sw Costa Rica (Cerros de Dota)
____ *T. b. browni*	Mts. of w Panama (Volcán Barú and Volcan de Chiriquí)

☐ **White-breasted Wood-Wren** *Henicorhina leucosticta*

____ *H. l. decolorata*	SE Mexico
____ *H. l. prostheleuca*	Trop. e Mexico (San Luis Potosí) to Belize and w Guatemala
____ *H. l. tropaea*	Honduras and Nicaragua
____ *H. l. smithei*	Petén of Guatemala
____ *H. l. costaricensis*	Central Costa Rica (Cartago and Limón)
____ *H. l. pittieri*	SW Costa Rica and Panama (east to Canal Zone)
____ *H. l. darienensis*	E Panama (Canal Zone) to nw Colombia (south to Chocó)
____ *H. l. albilateralis*	Tropical and subtropical n and nw Colombia
____ *H. l. eucharis*	Subtropical western Colombia
____ *H. l. inornata*	Pacific lowlands of s Colombia and extreme nw Ecuador
____ *H. l. hauxwelli*	Trop. s Colombia to e Ecuador and central Peru e of Andes
____ *H. l. leucosticta*	E Venezuela, Guyana, Suriname and n Brazil

☐ **Gray-breasted Wood-Wren** *Henicorhina leucophrys*

____ *H. l. festiva*	W Mexico (cloud forests of w Michoacán and Guerrero)
____ *H. l. mexicana*	E Mexico (San Luis Potosí to Puebla, Veracruz and n Oaxaca)
____ *H. l. castanea*	Extreme s Mexico (Chiapas) and n Guatemala
____ *H. l. capitalis*	S Mexico (w Chiapas) to w Guatemala and El Salvador
____ *H. l. collina*	Highlands of Costa Rica and w Panama (Chiriquí and Veraguas)
____ *H. l. anachoreta*	Temp. and upper subtrop. Santa Marta Mts. (ne Colombia)
____ *H. l. bangsi*	Subtrop. and upper trop. Santa Marta Mts. (ne Colombia)
____ *H. l. leucophrys*	Andes of Colombia to Ecuador and Peru
____ *H. l. brunneiceps*	Andes of sw Colombia to extreme n Ecuador (Imbabura)
____ *H. l. hilaris*	Subtropical mts. of sw Ecuador
____ *H. l. venezuelensis*	Subtropical coastal cordillera of n Venezuela (Lara to Miranda)
____ *H. l. meridana*	E slope of Andes of Colombia and w Venezuela
____ *H. l. boliviana*	Subtrop. mts. of w Bolivia (Cochabamba, La Paz and Santa Cruz)

☐ **Munchique Wood-Wren** *Henicorhina negreti*

Western Andes of Colombia

☐ **Bar-winged Wood-Wren** *Henicorhina leucoptera*

Andes of n Peru and immediately adjacent Ecuador

☐ **Nightingale Wren** *Microcerculus philomela*

Humid s Mexico (n Chiapas) to central Costa Rica

☐ **Scaly-breasted Wren** *Microcerculus marginatus*

____ *M. m. luscinia*	Central Costa Rica to e Panama and extreme nw Colombia
____ *M. m. taeniatus*	Tropical w Ecuador
____ *M. m. marginatus*	E Colombia to n Bolivia and w Amazonian Brazil
____ *M. m. squamulatus*	NE Colombia and nw Venezuela
____ *M. m. corrasus*	Santa Marta region of n Colombia
____ *M. m. occidentalis*	W Colombia and nw Ecuador

☐ **Flutist Wren** *Microcerculus ustulatus*

____ *M. u. duidae*	Mts. of s Venezuela (Duida, Yaví, Paraque and Sierra de Curupira)
____ *M. u. lunatipectus*	*Tepuis* of s Venezuela (Mt. Guaiquinima)
____ *M. u. obscurus*	*Tepuis* of se Venezuela (Ptari-tepui, Uei-tepui, Sororopán-tepui)
____ *M. u. ustulatus*	*Tepuis* of se Venezuela, n Brazil and w Guyana (Mt. Twek-quay)

☐ **Wing-banded Wren** *Microcerculus bambla*

____ *M. b. albigularis*	Tropical e Ecuador and w Amazonian Brazil; se Peru
____ *M. b. caurensis*	Tropical e Venezuela (Amazonas and w Bolívar)
____ *M. b. bambla*	Trop. se Venezuela (Auyan-tepui) to the Guianas and ne Brazil

☐ **Song Wren** *Cyphorhinus phaeocephalus*

____ *C. p. richardsoni*	Caribbean lowlands of se Honduras to Nicaragua
____ *C. p. infuscatus*	Caribbean lowlands of Costa Rica and extreme nw Panama
____ *C. p. lawrencii*	Lowlands of e Panama to nw Colombia
____ *C. p. propinquus*	Tropical lowlands of nw Colombia (Bolívar and Santander)
____ *C. p. phaeocephalus*	Pacific lowlands of sw Colombia to w Ecuador (s to El Oro)

☐ **Chestnut-breasted Wren** *Cyphorhinus thoracicus*

____ *C. t. dichrous*	Central and Western Andes of Colombia to Peru (San Martín)
____ *C. t. thoracicus*	Trop. and subtrop. e slope of Andes of se Peru (Huánuco to Puno)

☐ **Musician Wren** *Cyphorhinus arada*

____ *C. a. transfluvialis*	SE Colombia (Caquetá) to n Brazil (e Amazonas)
____ *C. a. salvini*	SE Colombia to e Ecuador and ne Peru (Loreto)
____ *C. a. urbanoi*	Brazil (Pará in vicinity of Faro and Obidos)
____ *C. a. arada*	S Venezuela (Gran Sabana) to the Guianas and adjacent ne Brazil
____ *C. a. faroensis*	N Brazil
____ *C. a. griseolateralis*	N Amazonian Brazil in drainage of Rio Tapajós
____ *C. a. interpositus*	Brazil (west of Rio Tapajós to n Mato Grosso)
____ *C. a. modulator*	Tropical e Peru, n Bolivia and w Amazonian Brazil

FAMILY: MIMIDAE (Mockingbirds and Thrashers—35)

☐ **Gray Catbird** *Dumetella carolinensis*

S Br. Columbia to Gulf States; winters to West Indies and Panama

☐ **Black Catbird** *Melanoptila glabrirostris*

Coastal e Mexico (Yucatán Pen. and adj. islands) to n Honduras

☐ **Bahama Mockingbird** *Mimus gundlachii*

____ *M. g. gundlachii*	Bahamas, cays off n Cuba, Great Inagua and Caicos islands
____ *M. g. hillii*	Arid coastal lowlands of s Jamaica

☐ **Northern Mockingbird** *Mimus polyglottos*

____ *M. p. leucopterus*	SW Canada to s Baja California and sw Mexico (Oaxaca)
____ *M. p. polyglottos*	E Canada to central, e and se US
____ *M. p. orpheus*	Bahamas and Greater Antilles

☐ **Tropical Mockingbird** *Mimus gilvus*

____ *M. g. gracilis*	S Mexico (Oaxaca) to Guatemala, Honduras and El Salvador
____ *M. g. leucophaeus*	Humid tropical se Mexico, Cozumel I., Isla Mujeres and Belize
____ *M. g. antillarum*	Martinique, St. Lucia, St. Vincent, the Grenadines and Grenada
____ *M. g. tobagensis*	Trinidad and Tobago
____ *M. g. rostratus*	Netherlands Antilles and adjacent islands off n coast of Venezuela
____ *M. g. magnirostris*	Isla San Andrés (w Caribbean Sea)
____ *M. g. tolimensis*	W and central Colombia
____ *M. g. melanopterus*	Coastal n Colombia to Venezuela, Guyana and extreme n Brazil
____ *M. g. gilvus*	French Guiana and Suriname
____ *M. g. antelius*	Coastal e Brazil (Pará to Rio de Janeiro)

☐ **Socorro Mockingbird** *Mimus graysoni*

Socorro I. (Revillagigedo Islands off w Mexico)

☐ **Chalk-browed Mockingbird** *Mimus saturninus*
    ____ *M. s. saturninus* — S Suriname and n Brazil (Amapá to se Pará)
    ____ *M. s. arenaceus* — NE Brazil (Paraíba, Alagoas and Bahia)
    ____ *M. s. frater* — N Bolivia to ne and sw Brazil (Mato Grosso)
    ____ *M. s. modulator* — SE Bolivia to s Brazil, Uruguay, Paraguay and n Argentina

☐ **Patagonian Mockingbird** *Mimus patagonicus*

Central and s Argentina and s Chile

☐ **Brown-backed Mockingbird** *Mimus dorsalis*

Arid montane scrub of Bolivia and extreme nw Argentina

☐ **White-banded Mockingbird** *Mimus triurus*

Central Argentina and Bolivia (Beni); winters to sw Brazil

☐ **Long-tailed Mockingbird** *Mimus longicaudatus*
    ____ *M. l. platensis* — Isla La Plata (off w Ecuador)
    ____ *M. l. albogriseus* — Arid sw Ecuador (Manabi to s Loja) and extreme n Peru (Piura)
    ____ *M. l. longicaudatus* — W Peru (La Libertad to Ica)
    ____ *M. l. maranonicus* — N Peru (upper Marañón Valley)

☐ **Chilean Mockingbird** *Mimus thenca*

Coastal Chile (Atacama to Valdivia)

☐ **Galapagos Mockingbird** *Nesomimus parvulus*
    ____ *N. p. parvulus* — Main Galapagos Islands except extreme eastern islands
    ____ *N. p. barringtoni* — Barrington (Galapagos Islands)
    ____ *N. p. personatus* — Galapagos Islands (Abingdon, Bindloe, James and Jervis)
    ____ *N. p. wenmani* — Wenman (Galapagos Islands)
    ____ *N. p. hulli* — Culpepper (Galapagos Islands)
    ____ *N. p. bauri* — Tower (Galapagos Islands)

☐ **Charles Mockingbird** *Nesomimus trifasciatus*

Galapagos Islands (Champion and Gardner)

☐ **Hood Mockingbird** *Nesomimus macdonaldi*

Hood and adjacent se Galapagos Islands

☐ **San Cristobal Mockingbird** *Nesomimus melanotis*

San Cristóbal (Galapagos Islands)

☐ **Sage Thrasher** *Oreoscoptes montanus*

Arid s British Columbia to Baja California and central Mexico

☐ **Brown Thrasher** *Toxostoma rufum*
    ____ *T. r. rufum* — SE Canada to Gulf States (mainly east of Rocky Mts.)
    ____ *T. r. longicauda* — S-central Canada to e Colorado and Kansas; winters to se US

☐ **Long-billed Thrasher** *Toxostoma longirostre*
    ____ *T. l. sennetti* — Arid s Texas to ne Mexico
    ____ *T. l. longirostre* — E Mexico (ne Querétaro to n Puebla and central Veracruz)

☐ **Cozumel Thrasher** *Toxostoma guttatum*

Cozumel I. (se Mexico off Quintana Roo)

☐ **Gray Thrasher** *Toxostoma cinereum*
    ____ *T. c. mearnsi* — Desert scrub of w Baja California (latitude 31°N to 28°N)
    ____ *T. c. cinereum* — Cape District of s Baja California (south of latitude 28°N)

☐ **Bendire's Thrasher** *Toxostoma bendirei*
    ____ *T. b. bendirei* — Arid sw US to nw Mexico (n Sonora)
    ____ *T. b. candidum* — Sonoran Desert of w Mexico (w Sonora)
    ____ *T. b. rubricatum* — Central and s interior of se Sonora and coast near Isla Tiburón

☐ **Ocellated Thrasher** *Toxostoma ocellatum*
    ____ *T. o. ocellatum* — Highlands of c Mexico (San Luis Potosí to Hidalgo and México)
    ____ *T. o. villai* — Oak-pine highlands of s Mexico (Puebla to Oaxaca)

☐ **Curve-billed Thrasher** *Toxostoma curvirostre*

_____ *T. c. palmeri*   Arid s Arizona to w Mexico (central Sonora)
_____ *T. c. celsum*   SE Arizona to w Texas and n Mexico (ne Sonora to w Coahuila)
_____ *T. c. oberholseri*   S Texas to ne Mexico (e Coahuila, Nuevo León and Tamaulipas)
_____ *T. c. maculatum*   NW Mexico (s Sonora to n Sinaloa and sw Chihuahua)
_____ *T. c. insularum*   Islands in Sea of Cortés (San Estéban and Tiburón)
_____ *T. c. occidentale*   NW Mexico (s Sinaloa, Nayarit, nw Jalisco and w Durango)
_____ *T. c. curvirostre*   S Mexico (se Jalisco to Guerrero, México, Puebla and Oaxaca)

☐ **California Thrasher** *Toxostoma redivivum*

_____ *T. r. sonomae*   Chaparral belt of n California (south to Monterey)
_____ *T. r. redivivum*   Chaparral belt of s California and nw Baja

☐ **Crissal Thrasher** *Toxostoma crissale*

_____ *T. c. coloradense*   Arid s California and Arizona to ne Baja and nw Sonora
_____ *T. c. crissale*   Arid sw US to w Texas and n Mexico (Sonora to nw Coahuila)
_____ *T. c. trinitatis*   N Baja California (Valle de La Trinidad)
_____ *T. c. dumosum*   N Mexico (Zacatecas, s Coahuila, San Luis Potosí and Hidalgo)

☐ **Le Conte's Thrasher** *Toxostoma lecontei*

_____ *T. l. lecontei*   Arid sw US to n Baja California and nw Mexico
_____ *T. l. macmillanorum*   Inland central California (s San Joaquin Valley)

☐ **Vizcaino Thrasher** *Toxostoma arenicola*

                                    Vizcaino desert of central Baja California (29°N to 26°N)

☐ **White-breasted Thrasher** *Ramphocinclus brachyurus*

_____ *R. b. brachyurus*   Martinique (Lesser Antilles)
_____ *R. b. sanctaeluciae*   St. Lucia (Lesser Antilles)

☐ **Blue Mockingbird** *Melanotis caerulescens*

_____ *M. c. longirostris*   Tres Marías Islands (off w Mexico)
_____ *M. c. caerulescens*   Oak-pine zone of w Mexico (s Sonora to Isthmus of Tehuántepec)

☐ **Blue-and-white Mockingbird** *Melanotis hypoleucus*

                                    Humid highlands of s Mexico (Chiapas) to w Honduras

☐ **Gray Trembler** *Cinclocerthia gutturalis*

_____ *C. g. gutturalis*   Martinique (Lesser Antilles)
_____ *C. g. macrorhyncha*   St. Lucia (Lesser Antilles)

☐ **Brown Trembler** *Cinclocerthia ruficauda*

_____ *C. r. pavida*   Lesser Antilles (Montserrat and adjacent nw Leeward Islands)
_____ *C. r. tremula*   Guadeloupe (Lesser Antilles)
_____ *C. r. ruficauda*   Dominica (Lesser Antilles)
_____ *C. r. tenebrosa*   St. Vincent (Lesser Antilles)

☐ **Scaly-breasted Thrasher** *Allenia fusca*

_____ *A. f. atlanticus*   Barbados
_____ *A. f. hypenemus*   N Lesser Antilles
_____ *A. f. schwartzi*   St. Lucia (Lesser Antilles)
_____ *A. f. vincenti*   St. Vincent (Lesser Antilles)
_____ *A. f. fuscus*   Lesser Antilles (Dominica to Grenada)

☐ **Pearly-eyed Thrasher** *Margarops fuscatus*

_____ *M. f. fuscatus*   S Bahamas, Hispaniola, Puerto Rico, Lesser and Neth. Antilles
_____ *M. f. densirostris*   Lesser Antilles (Guadeloupe, Dominica and Martinique)
_____ *M. f. klinikowskii*   St. Lucia (Lesser Antilles)
_____ *M. f. bonairensis*   Bonaire and Horquilla (Los Hermanos Archipelago off Venezuela)

FAMILY: PRUNELLIDAE (Accentors—13)

☐ **Alpine Accentor** *Prunella collaris*
_____ *P. c. collaris* — Mts. of Europe to Carpathians, n Yugoslavia and nw Africa
_____ *P. c. subalpina* — Mts. of se Europe to Crete and w Turkey
_____ *P. c. montana* — Caucasus Mts. to n Iraq and s Iran
_____ *P. c. rufilata* — Tajikistan to n Afghanistan, w China (w Xinjiang) and se Tibet
_____ *P. c. whymperi* — W Himalayas (Kashmir to Garhwal and Kumaon)
_____ *P. c. nipalensis* — E Himalayas to sw China (e Xinjiang to n Yunnan) and se Tibet
_____ *P. c. tibetana (berezowski)* — NW China (n Xinjiang, s Qinghai and Gansu) to e Tibet
_____ *P. c. erythropygia* — Altai Mts. to n China, Sea of Okhotsk, Korea and Japan
_____ *P. c. fennelli* — Mts. of Taiwan

☐ **Himalayan Accentor** *Prunella himalayana*

Mts. of central Asia to Afghanistan, s Tibet and nw India

☐ **Robin Accentor** *Prunella rubeculoides*
_____ *P. r. rubeculoides* — Himalayas of n Pakistan to Nepal, Sikkim and se Tibet
_____ *P. r. fusca* — Mts. of e Tibet and w China (Xinjiang, Gansu and Shaanxi)

☐ **Rufous-breasted Accentor** *Prunella strophiata*
_____ *P. s. jerdoni* — Mts. of e Afghanistan to n India (Kashmir and Kumaon)
_____ *P. s. strophiata* — Himalayas of Nepal to se Tibet, sw China and n Burma

☐ **Siberian Accentor** *Prunella montanella*
_____ *P. m. montanella* — Ural Mts. to Altai, Lake Baikal and Sikhote Alin Mts.
_____ *P. m. badia* — NE Siberia (lower Lena River to Sea of Okhotsk)

☐ **Radde's Accentor** *Prunella ocularis*

Mts. of Turkey, Armenia, n Georgia and Iran

☐ **Yemen Accentor** *Prunella fagani*

High mts. of Yemen; winters to s Arabia

☐ **Brown Accentor** *Prunella fulvescens*
_____ *P. f. fulvescens* — Tien Shan Mts. to Afghanistan and Pakistan
_____ *P. f. dahurica* — Altai Mts. to Mongolia
_____ *P. f. dresseri* — Mts. of w China (sw Xinjiang to w Gansu and n Tibet)
_____ *P. f. nanschanica* — Montane forests of w China on Qinghai/Gansu border
_____ *P. f. khamensis* — Mts. of w China (Xinjiang to s Qinghai, Gansu and ne Tibet)
_____ *P. f. sushkini* — Treeline mountain slopes of s Tibet

☐ **Black-throated Accentor** *Prunella atrogularis*
_____ *P. a. atrogularis* — N Ural Mts.; winters to Afghanistan and Iran
_____ *P. a. huttoni (lucens)* — Russian Altai (Dzhungarski Mts.); winters to Pakistan, ne India

☐ **Mongolian Accentor** *Prunella koslowi*

Mts. of Mongolia and immediately adjacent n China (Ningxia)

☐ **Dunnock** *Prunella modularis*
_____ *P. m. hebridium* — Ireland, Outer Hebrides, Inner Hebrides and w Scotland
_____ *P. m. occidentalis* — E Scotland, England, Wales and w France
_____ *P. m. modularis* — N and cent. Europe; winters to w Mediterranean is. and N Africa
_____ *P. m. mabbotti* — SW France to Pyrénées, Iberian Peninsula and Apennine Mts.
_____ *P. m. meinertzhageni* — S Yugoslavia and Bulgaria
_____ *P. m. fuscata* — Mountains of Crimean Peninsula
_____ *P. m. euxina* — N Turkey to w Caucasus Mountains
_____ *P. m. obscura* — Caucasus and e Turkey to n Iran; winters to mts. of Lebanon

☐ **Japanese Accentor** *Prunella rubida*
_____ *P. r. rubida* — S Kuril Islands, Shikoku and Hokkaido
_____ *P. r. fervida* — Honshu (Japan); winters to Kyushu

☐ **Maroon-backed Accentor** *Prunella immaculata*

Nepal to se Tibet and sw China; winters to ne Myanmar

FAMILY: TURDIDAE (Thrushes and Allies—176)

☐ **Rufous Flycatcher-Thrush** *Neocossyphus fraseri*
_____ *N. f. rubicundus* — Nigeria to Central African Republic, w Zaire, Gabon and Angola
_____ *N. f. vulpine* — S Sudan to Uganda, ne Zaire, nw Zambia and nw Tanzania
_____ *N. f. fraseri* — Bioko (Gulf of Guinea)

☐ **Finsch's Flycatcher-Thrush** *Neocossyphus finschii*
Lowlands of Sierra Leone and Liberia to Ghana and s Nigeria

☐ **Red-tailed Ant-Thrush** *Neocossyphus rufus*
_____ *N. r. gabunensis* — Lowlands of se Cameroon to Gabon, ne Zaire and w Uganda
_____ *N. r. rufus* — Coastal n Kenya to n Tanzania (Tana River to Mikindani); Zanzibar

☐ **White-tailed Ant-Thrush** *Neocossyphus poensis*
_____ *N. p. poensis* — Sierra Leone to Cameroon, Gabon and s Congo; Bioko
_____ *N. p. praepectoralis* — N Angola to Central African Republic, w Zaire and Uganda
_____ *N. p. kakamegoes* — W Kenya (Kakamega Forest)
_____ *N. p. nigridorsalis* — W Kenya (n Nandi Hills)
_____ *N. p. pallidigularis* — NW Angola (Canzele)

☐ **Forest Rock-Thrush** *Pseudocossyphus sharpei*
Humid e-central plateau of Madagascar

☐ **Benson's Rock-Thrush** *Pseudocossyphus bensoni*
Montane rocky areas of sw Madagascar (Isalo Massif)

☐ **Littoral Rock-Thrush** *Pseudocossyphus imerinus*
_____ *P. i. erythronotus* — Arid littoral of n Madagascar
_____ *P. i. salomonseni* — Coastal lowlands of e Madagascar
_____ *P. i. imerinus* — Coastal lowlands of se Madagascar

☐ **Cape Rock-Thrush** *Monticola rupestris*
SE Botswana to s Mozambique, Swaziland and Cape Province

☐ **Sentinel Rock-Thrush** *Monticola explorator*
_____ *M. e. explorator* — Lowlands of Natal to Transvaal and sw Cape Province
_____ *M. e. tenebriformis* — Swaziland (Lebombo Mts.); winters north to s Mozambique

☐ **Short-toed Rock-Thrush** *Monticola brevipes*
_____ *M. b. niveiceps* — W Angola (Huila escarpment)
_____ *M. b. brevipes (leucocapilla)* — S Angola to Namibia, Botswana and Cape Province
_____ *M. b. pretoriae* — Mountains of se Botswana to w Transvaal

☐ **Miombo Rock-Thrush** *Monticola angolensis*
_____ *M. a. angolensis (niassae)* — Angola to s Zaire, Rwanda, Zambia and Tanzania
_____ *M. a. hylophila* — S Zambia to Zimbabwe, w Malawi and w Mozambique

☐ **Rufous-tailed Rock-Thrush** *Monticola saxatilis*
Rocky regions of s Palearctic region; winters to e Africa

☐ **Little Rock-Thrush** *Monticola rufocinereus*
_____ *M. r. rufocinereus* — Mts. of se Sudan to Ethiopia, e Uganda, w Kenya and ne Tanzania
_____ *M. r. sclateri* — W Saudi Arabia

☐ **Blue-capped Rock-Thrush** *Monticola cinclorhynchus*
Mts. of e Afghanistan, n Pakistan and India; winters to Myanmar

☐ **White-throated Rock-Thrush** *Monticola gularis*
SE Siberia to ne China and Korea; winters to SE Asia

☐ **Chestnut-bellied Rock-Thrush** *Monticola rufiventris*
Pakistan to se Tibet, sw China, Myanmar, n Laos and n Vietnam

☐ **Blue Rock-Thrush** *Monticola solitarius*
_____ *M. s. solitarius* — S Europe, nw Africa and Middle East; winters to central Africa
_____ *M. s. longirostris* — N Iraq and Iran to Pakistan; winters to n India and ne Africa
_____ *M. s. pandoo* — Central Asia and Himalayas; winters to Malaysia and Indonesia
_____ *M. s. philippensis* — SE Siberia to China, Japan and Lan-yü I.; winters to Indonesia
_____ *M. s. madoci* — Malay Peninsula and Sumatra

☐ **Ceylon Whistling-Thrush** *Myophonus blighi*

Mountain ravines of Sri Lanka

☐ **Shiny Whistling-Thrush** *Myophonus melanurus*

Montane moss forests of Sumatra

☐ **Javan Whistling-Thrush** *Myophonus glaucinus*

Montane forests of Java and Bali

☐ **Chestnut-winged Whistling-Thrush** *Myophonus castaneus*

Foothill forests of Sumatra

☐ **Bornean Whistling-Thrush** *Myophonus borneensis*

Foothill and montane forests of Borneo

☐ **Malayan Whistling-Thrush** *Myophonus robinsoni*

Moist montane forests of central Malaya

☐ **Malabar Whistling-Thrush** *Myophonus horsfieldii*

Swift flowing rocky streams of peninsular India

☐ **Formosan Whistling-Thrush** *Myophonus insularis*

Mountain streams of Taiwan

☐ **Blue Whistling-Thrush** *Myophonus caeruleus*
  ____ *M. c. temminckii*   Central Asia to n India, Pakistan, se Tibet and Myanmar
  ____ *M. c. eugenei*   NE Assam to s Myanmar, n Thailand, sw China and Indochina
  ____ *M. c. caeruleus*   W China (Sichuan); winters to s China and n Indochina
  ____ *M. c. crassirostris* Peninsular and extreme se Thailand to n Malaysia
  ____ *M. c. dichrorhynchus* Central and s Malaysia; foothills of w Sumatra
  ____ *M. c. flavirostris*   Foothill and montane forests of Java

☐ **Geomalia** *Geomalia heinrichi*

Mountains of Sulawesi

☐ **Slaty-backed Thrush** *Zoothera schistacea*

E Lesser Sundas (Yamdena and Larat)

☐ **Buru Thrush** *Zoothera dumasi*

Mid-mountain forests of Buru (s Moluccas)

☐ **Seram Thrush** *Zoothera joiceyi*

Mid-mountain forests of Seram (s Moluccas)

☐ **Chestnut-capped Thrush** *Zoothera interpres*

S Thailand, Malaysia, Greater and Lesser Sundas and s Philippines

☐ **Enggano Thrush** *Zoothera leucolaema*

Enggano I. (off w Sumatra)

☐ **Chestnut-backed Thrush** *Zoothera dohertyi*

Lesser Sundas (Lombok, Sumbawa, Sumba, Flores) and w Timor

☐ **Rusty-backed Thrush** *Zoothera erythronota*
  ____ *Z. e. erythronota*   Lowland forests of Sulawesi
  ____ *Z. e. subspecies?*   Taliabu I. (Sula Islands)
  ____ *Z. e. kabaena*   Kabaena I. (Flores Sea)

☐ **Red-and-black Thrush** *Zoothera mendeni*

Peleng I. (Banggai Islands)

☐ **Pied Thrush** *Zoothera wardii*

Himalayas of n India from Nepal to Assam; winters to Sri Lanka

☐ **Ashy Thrush** *Zoothera cinerea*

N Philippines (Luzon and Mindoro)

☐ **Orange-banded Thrush** *Zoothera peronii*
  ____ *Z. p. peronii*   E Lesser Sundas (Roti and w Timor)
  ____ *Z. p. audacis*   E Lesser Sundas (e Timor, Wetar, Romang, Damar and Babar)

☐ **Orange-headed Thrush** *Zoothera citrina*
  ____ *Z. c. citrina*   W Pakistan to n Myanmar; winters to Sri Lanka
  ____ *Z. c. cyanota*   Peninsular India (north to Gujarat and Andhra)
  ____ *Z. c. innotata*   S Myanmar to s China and Indochina; winters to Malaysia
  ____ *Z. c. melli*   SE China (Fujian and n Guangdong)
  ____ *Z. c. courtoisi* E China (Anhui Province)

____	*Z. c. aurimacula*	S Vietnam and Hainan
____	*Z. c. andamanensis*	Andaman Islands
____	*Z. c. albogularis*	Nicobar Islands
____	*Z. c. gibsonhilli*	Central Malay Peninsula (s Myanmar to s Thailand)
____	*Z. c. aurata*	Mountains of n Borneo
____	*Z. c. rubecula*	W Java
____	*Z. c. orientis*	E Java and Bali

☐ **Everett's Thrush** *Zoothera everetti*

High mountains of n Borneo (Sarawak and Sabah)

☐ **Siberian Thrush** *Zoothera sibirica*

____	*Z. s. sibirica*	NE Asia; winters to SE Asia, Sumatra and Java
____	*Z. s. davisoni*	Sakhalin I. and n Japan; winters to s China, SE Asia and Sumatra

☐ **Abyssinian Ground-Thrush** *Zoothera piaggiae*

____	*Z. p. piaggiae*	Ethiopia to se Sudan, n Kenya, sw Uganda and mts. of e Zaire
____	*Z. p. hadii*	SE Sudan (Imatong and Dongotona mountains)
____	*Z. p. ruwenzorii*	Ruwenzori Mountains (Zaire/Uganda border)
____	*Z. p. kilimensis*	Kenya (east of Rift Valley) to n Tanzania (Mt. Kilimanjaro)
____	*Z. p. rowei*	N Tanzania (Loliondo and Magaidu forests)

☐ **Kivu Ground-Thrush** *Zoothera tanganjicae*

Montane forests of e Zaire, Rwanda, n Burundi and sw Uganda

☐ **Crossley's Ground-Thrush** *Zoothera crossleyi*

____	*Z. c. crossleyi*	Montane forests of se Nigeria, Cameroon, Congo and w Zaire
____	*Z. c. pilettei*	NE Zaire (Semliki Vallei and w slope of Itombwe highlands)

☐ **Orange Ground-Thrush** *Zoothera gurneyi*

____	*Z. g. otomitra*	Angola (Mt. Moco) to Zaire, Tanzania and n Malawi
____	*Z. g. chuka*	Mt. Kenya and Kikuyu escarpment
____	*Z. g. raineyi (chyulu)*	Montane forests of se Kenya (Taita and Chyulu Hills)
____	*Z. g. disruptans*	Central Malawi to Mozambique, e Zimbabwe and n Transvaal
____	*Z. g. gurneyi*	South Africa (Natal and e Cape Province)

☐ **Black-eared Ground-Thrush** *Zoothera camaronensis*

____	*Z. c. camaronensis*	Lowland forests of Cameroon and Gabon
____	*Z. c. graueri*	Lowland forests of ne Zaire and w Uganda

☐ **Gray Ground-Thrush** *Zoothera princei*

____	*Z. p. princei*	Dense lowland forests of Liberia to Ivory Coast and Ghana
____	*Z. p. batesi*	Coastal s Cameroon to Gabon, ne Zaire and extreme w Uganda

☐ **Oberlaender's Ground-Thrush** *Zoothera oberlaenderi*

Lowlands of e Zaire and w Uganda (Bwamba Forest)

☐ **Spotted Ground-Thrush** *Zoothera guttata*

____	*Z. g. maxis*	S Sudan
____	*Z. g. fischeri*	Coastal e Kenya and Tanzania
____	*Z. g. lippensi*	E Zaire
____	*Z. g. belcheri*	S Malawi (Mt. Thyolo)
____	*Z. g. guttata*	S Malawi to Natal and Cape Province

☐ **Spot-winged Thrush** *Zoothera spiloptera*

Montane forests of Sri Lanka

☐ **Sunda Thrush** *Zoothera andromedae*

Patchily distributed Philippines to Sumatra, Java, Lesser Sundas

☐ **Plain-backed Thrush** *Zoothera mollissima*

____	*Z. m. whiteheadi*	W Himalayas (n Pakistan to w Nepal)
____	*Z. m. mollissima*	E Himalayas to se Tibet; winters to Myanmar and Indochina
____	*Z. m. griseiceps*	SW China (Sichuan and Yunnan) to n Vietnam (Tonkin)

403

☐ **Long-tailed Thrush** *Zoothera dixoni*

Himalayas to sw China and se Tibet; winters to SE Asia

☐ **Scaly Thrush** *Zoothera dauma*
    ____ *Z. d. aurea*   Siberia to Manchuria and Korea; winters to s China and Indochina
    ____ *Z. d. toratugumi*   Manchuria and Japan; winters in Taiwan and Lan-yü I.
    ____ *Z. d. dauma*   Pakistan to Myanmar, s China and Thailand; winters to s India
    ____ *Z. d. neilgherriensis*   S peninsular India (Mysore, Madras and Kerala)
    ____ *Z. d. imbricata*   Highlands of Sri Lanka
    ____ *Z. d. hancii*   Peninsular Thailand to s Vietnam; s Ryukyu Islands and Taiwan
    ____ *Z. d. major*   Amami-O-Shima (n Ryukyu Islands)
    ____ *Z. d. horsfieldi*   Mountains of Sumatra, Java, Bali, Lombok and Sumbawa

☐ **Fawn-breasted Thrush** *Zoothera machiki*

Tanimbar Islands (Yamdena and Larat)

☐ **Olive-tailed Thrush** *Zoothera lunulata*
    ____ *Z. l. cuneata*   Montane ne Queensland (Windsor and Atherton Tablelands)
    ____ *Z. l. lunulata*   SE Australia (se Queensland to s Victoria and Tasmania)
    ____ *Z. l. halmaturina*   South Australia (Mt. Lofty and Flinders Ranges; Kangaroo I.)

☐ **Russet-tailed Thrush** *Zoothera heinei*

E Australia (Clarke Range, Queensland to about Sydney, NSW)

☐ **New Britain Thrush** *Zoothera talaseae*

Bismarck Archipelago (New Britain and Umboi) and Bougainville

☐ **San Cristobal Thrush** *Zoothera margaretae*
    ____ *Z. m. margaretae*   Mountains of San Cristóbal (Solomon Islands)
    ____ *Z. m. turipavae*   Mountains of Guadalcanal (Solomon Islands)

☐ **Long-billed Thrush** *Zoothera monticola*
    ____ *Z. m. monticola*   Himalayas of n India to Nepal and ne Myanmar
    ____ *Z. m. atrata*   N Vietnam (nw Tonkin)

☐ **Dark-sided Thrush** *Zoothera marginata*

Nepal to n India, sw China, Myanmar, Thailand and n Indochina

☐ **Sulawesi Thrush** *Cataponera turdoides*
    ____ *C. t. abditiva*   N-central Sulawesi
    ____ *C. t. tenebrosa*   S-central Sulawesi (Latimojong Mountains)
    ____ *C. t. turdoides*   S Sulawesi (Lompobattang Mountains)
    ____ *C. t. heinrichi*   SE Sulawesi (Mekonga Mountains)

☐ **Tristan Thrush** *Nesocichla eremita*
    ____ *N. e. eremita*   Tristan da Cunha I. (Atlantic Ocean)
    ____ *N. e. gordoni*   Inaccessible I. (Atlantic Ocean)
    ____ *N. e. procax*   Nightingale I. (Atlantic Ocean)

☐ **Forest Thrush** *Cichlherminia lherminieri*
    ____ *C. l. lherminieri*   Guadeloupe (Lesser Antilles)
    ____ *C. l. lawrencii*   Montserrat (Lesser Antilles)
    ____ *C. l. dominicensis* Dominica (Lesser Antilles)
    ____ *C. l. sanctaeluciae* St. Lucia (Lesser Antilles)

☐ **Eastern Bluebird** *Sialia sialis*
    ____ *S. s. sialis*   S-central and e Canada to se US; winters to n Mexico and Cuba
    ____ *S. s. grata*   S peninsular Florida
    ____ *S. s. episcopus*   S coastal Texas (Rockport) to e Mexico (s Tamaulipas)
    ____ *S. s. fulva*   Mts. of s-c Arizona to s Mexico (Guerrero); winters to Guatemala
    ____ *S. s. guatemalae*   Mts. of se Mexico (s Tamaulipas to Chiapas) and Guatemala
    ____ *S. s. meridionalis* Mountains of El Salvador to ne Nicaragua

☐ **Western Bluebird** *Sialia mexicana*
    ____ *S. m. occidentalis* S British Columbia to s California and w Nevada
    ____ *S. m. bairdi*   SW US to nw Mexico (Sonora and Chihuahua)

____	*S. m. anabelae*	Mountains of n Baja Calif. (Sierra Juárez and San Pedro Mártir)
____	*S. m. amabilis*	Sierra Madre Occidental of Mexico (s Chihuahua to Zacatecas)
____	*S. m. mexicana*	Plateau of ne Mexico (Coahuila to Nuevo León and Tamaulipas)
____	*S. m. australis*	S plateau of Mexico (Jalisco to Morelos, Puebla and Veracruz)

☐ **Mountain Bluebird *Sialia currucoides***

W North America (Alaska to c Mexico); winters to Baja Calif.

☐ **Townsend's Solitaire *Myadestes townsendi***

____	*M. t. townsendi*	Central Alaska to w US, Baja California and nw Mexico
____	*M. t. calophonus*	N Mexico (s Chihuahua to Durango, Jalisco and Zacatecas)

☐ **Brown-backed Solitaire *Myadestes occidentalis***

____	*M. o. cinereus*	Mts. of Mexico (se Sonora to Sinaloa, Chihuahua and Durango)
____	*M. o. occidentalis*	Mts. of w Mexico (Nayarit to Guerrero, w Oaxaca and Morelos)
____	*M. o. insularis*	Tres Marías Islands (off w Mexico)
____	*M. o. deignani*	Mountains of s Mexico (Oaxaca and s Chiapas)
____	*M. o. oberholseri*	Mts. of s Mexico to Guatemala, El Salvador and central Honduras

☐ **Cuban Solitaire *Myadestes elisabeth***

____	*M. e. elisabeth*	Locally in mountains of Cuba
____	*M. e. retrusus†*	Formerly Isle of Pines (Cuba). Extinct

☐ **Rufous-throated Solitaire *Myadestes genibarbis***

____	*M. g. solitarius*	Montane forests of Jamaica
____	*M. g. montanus*	Montane forests of Hispaniola
____	*M. g. dominicanus*	Montane forests of Dominica
____	*M. g. genibarbis*	Montane forests of Martinique
____	*M. g. sanctaeluciae*	Montane forests of St. Lucia
____	*M. g. sibilans*	Montane forests of St. Vincent

☐ **Black-faced Solitaire *Myadestes melanops***

Mountains of Costa Rica and w Panama (e to Veraguas)

☐ **Varied Solitaire *Myadestes coloratus***

Highlands of extreme e Panama (Darién) and adj. nw Colombia

☐ **Slate-colored Solitaire *Myadestes unicolor***

____	*M. u. unicolor*	Humid montane forests of s Mexico to Guatemala and n Honduras
____	*M. u. pallens*	Humid montane forests of n-central Nicaragua

☐ **Andean Solitaire *Myadestes ralloides***

____	*M. r. plumbeiceps*	W and central Andes of Colombia and w Ecuador
____	*M. r. candelae*	N-central Colombia (Magdalena Valley)
____	*M. r. venezuelensis*	E Andes of Colombia to n Venezuela, e Ecuador and n Peru
____	*M. r. ralloides*	Central Peru (La Libertad and Huánuco) to n Bolivia

☐ **Kamao *Myadestes myadestinus***

Kauai (Alakai Swamp)

☐ **Olomao *Myadestes lanaiensis***

____	*M. l. rutha*	Montane forests of Molokai (Mt. Olokui). On verge of extinction
____	*M. l. lanaiensis†*	Formerly Lanai. Extinct

☐ **Omao *Myadestes obscurus***

Highlands of Hawaii

☐ **Puaiohi *Myadestes palmeri***

Kauai (Alakai Swamp)

☐ **Rufous-brown Solitaire *Cichlopsis leucogenys***

____	*C. l. chubbi*	W slope of Andes of sw Colombia (w Valle) and nw Ecuador
____	*C. l. peruvianus*	E slope of Andes of central Peru (Junín and Huánuco)
____	*C. l. gularis*	*Tepuis* of se Venezuela (Bolívar) and adjacent Guyana
____	*C. l. leucogenys*	Coastal se Brazil (s Bahia and Espírito Santo)

☐ **White-eared Solitaire** *Entomodestes leucotis*

E slope of Andes of Peru and w Bolivia (La Paz and Cochabamba)

☐ **Black Solitaire** *Entomodestes coracinus*

W slope of Western Andes of Colombia (Chocó) and nw Ecuador

☐ **Orange-billed Nightingale-Thrush** *Catharus aurantiirostris*

____ *C. a. aenopennis*	Highlands of nw Mexico (n Sinaloa to sw Chihuahua)
____ *C. a. clarus*	N central Mexico (s Sinaloa to w Puebla and sw Tamaulipas)
____ *C. a. melpomene*	S Mexico (Veracruz, ne Puebla, Oaxaca and Chiapas)
____ *C. a. bangsi*	Guatemala to El Salvador and Honduras
____ *C. a. costaricensis*	Nicaragua to nw Costa Rica
____ *C. a. russatus*	Mountains of sw Costa Rica and w Panama
____ *C. a. griseiceps*	Mountains of w Panama (e Chiriquí and Veraguas)
____ *C. a. insignis*	N Colombia (upper Magdalena Valley)
____ *C. a. aurantiirostris*	Mountains of ne Colombia and nw Venezuela
____ *C. a. sierrae*	Santa Marta Mountains (ne Colombia)
____ *C. a. inornatus*	W slope of e Andes of Colombia (Santander)
____ *C. a. phaeoplurus*	Central Colombia (Cauca, upper Patía and Guáitara valleys)
____ *C. a. birchalli*	Mountains of ne Venezuela (Sucre); Trinidad
____ *C. a. barbaritoi*	W Venezuela (Sierra de Perijá) and valley of upper Río Negro

☐ **Slaty-backed Nightingale-Thrush** *Catharus fuscater*

____ *C. f. hellmayri*	Mountains of Costa Rica and w Panama (Chiriquí and Veraguas)
____ *C. f. mirabilis*	Extreme e Panama (Mt. Pirre)
____ *C. f. sanctaemartae*	Santa Marta Mountains (ne Colombia)
____ *C. f. fuscater*	E Panama (Mt. Tacarcuna); e Andes of Venezuela to Ecuador
____ *C. f. opertaneus*	W Andes of Colombia (Antioquia)
____ *C. f. caniceps*	Andes of n and central Peru
____ *C. f. mentalis*	Andes of extreme se Peru (Puno) and nw Bolivia (La Paz)

☐ **Russet Nightingale-Thrush** *Catharus occidentalis*

____ *C. o. olivascens*	Mts. of nw Mexico (extreme s Sonora, n Sinaloa, w Chihuahua)
____ *C. o. durangensis*	Mountains of nw Mexico (nw Durango)
____ *C. o. lambi*	Mountains of e Mexico (n Puebla)
____ *C. o. fulvescens*	Mts. of central Mexico (Jalisco to s Tamaulipas and w Puebla)
____ *C. o. occidentalis*	Mts. of se Mexico (e San Luis Potosí to Puebla and s Oaxaca)

☐ **Black-billed Nightingale-Thrush** *Catharus gracilirostris*

____ *C. g. gracilirostris*	Humid montane forests of Costa Rica
____ *C. g. accentor*	Humid montane forests of w Panama

☐ **Ruddy-capped Nightingale-Thrush** *Catharus frantzii*

____ *C. f. frantzii*	Mountains of w Jalisco to c Michoacán, c México and Morelos
____ *C. f. confusus*	Mts. of se San Luis Potosí to ne Hidalgo, ne Puebla and n Oaxaca
____ *C. f. nelsoni*	Mountains of s Mexico (sw Guerrero to se Oaxaca)
____ *C. f. chiapensis*	Mountains of s Mexico (central Chiapas)
____ *C. f. juancitonis*	Mountains of s Mexico (s Chiapas) to Guatemala and Honduras
____ *C. f. waldroni*	Mountains of n Nicaragua
____ *C. f. wetmorei*	Mountains of Costa Rica and w Panama (Chiriquí)

☐ **Black-headed Nightingale-Thrush** *Catharus mexicanus*

____ *C. m. mexicanus*	Mountains of e Mexico (Tamaulipas to Veracruz and w Chiapas)
____ *C. m. cantator*	Highlands of s Mexico (Chiapas) to e Guatemala and Honduras
____ *C. m. fumosus*	Highlands of Nicaragua to Costa Rica and w Panama

☐ **Spotted Nightingale-Thrush** *Catharus dryas*

____ *C. d. harrisoni*	Highlands of se Mexico (Oaxaca)
____ *C. d. ovandensis*	Highlands of s Mexico (Chiapas)
____ *C. d. dryas*	W Guatemala (Sierra de las Minas) to Honduras; w Ecuador
____ *C. d. maculatus*	E slope of Andes of Colombia to e Ecuador, e Peru and n Bolivia
____ *C. d. ecuadoreanus*	Andes of w Ecuador
____ *C. d. blakei*	Andes of extreme n Argentina (Jujuy and Salta)

☐ **Veery** *Catharus fuscescens*

____ *C. f. fuscescens*	E Canada and e US; winters to Amazonian Brazil
____ *C. f. fuliginosus*	SW Newfoundland to s-central Quebec; winters to South America
____ *C. f. salicicola*	W Canada and w US; winters to w Brazil (Mato Grosso)
____ *C. f. subpallidus*	N Washington to ne Oregon, w Montana and Colorado

☐ **Gray-cheeked Thrush** *Catharus minimus*

____ *C. m. minimus*	NE Siberia and Canada; winters to n South America
____ *C. m. aliciae*	SE Canada; winters to West Indies

☐ **Bicknell's Thrush** *Catharus bicknelli*

Newfoundland and adj. Canada to ne US; winters in Hispaniola

☐ **Swainson's Thrush** *Catharus ustulatus*

____ *C. u. almae*	S Alaska and w Canada; winters to Gulf Coast of s US
____ *C. u. ustulatus*	Coastal se Alaska to s Oregon; winters to w Mexico
____ *C. u. oedicus*	N Washington to s California; winters to s Mexico
____ *C. u. swainsoni*	E Canada to e US; winters to West Indies and n Argentina

☐ **Hermit Thrush** *Catharus guttatus*

____ *C. g. guttatus*	Alaskan Peninsula to sw Canada; winters to central Mexico
____ *C. g. nanus (verecundus)*	Coastal se Alaska and w Br. Columbia; winters to Baja California
____ *C. g. vaccinius*	Vancouver I.; winters to coastal central California
____ *C. g. slevini (oromelus, jewetti)*	Washington and Oregon to c California; winters to nw Mexico
____ *C. g. sequoiensis*	Sierra Nevada of California and w Nevada; winters to n Mexico
____ *C. g. polionotus*	E Washington to e-central California, Nevada and sw Utah
____ *C. g. auduboni*	Rocky Mts. of sw US to New Mexico; winters to Guatemala
____ *C. g. faxoni*	Yukon and n Canada to e US; winters to Florida
____ *C. g. crymophilus*	Newfoundland; winters to se US

☐ **Wood Thrush** *Hylocichla mustelina*

Breeds e North America; winters s Texas to Panama

☐ **Pale-eyed Thrush** *Platycichla leucops*

Andes of Colombia to w Bolivia; *tepuis* and mts. of w Venezuela

☐ **Yellow-legged Thrush** *Platycichla flavipes*

____ *P. f. venezuelensis*	N Colombia to n and w Venezuela
____ *P. f. melanopleura*	NE Venezuela, Isla Margarita and Trinidad
____ *P. f. xanthoscela*	Tobago
____ *P. f. polionota*	S Venezuela (Bolívar) and Guyana
____ *P. f. flavipes*	SE Brazil (s Bahia) to ne Paraguay and ne Argentina

☐ **Groundscraper Thrush** *Psophocichla litsipsirupa*

____ *P. l. simensis*	Highlands of Eritrea and Ethiopia
____ *P. l. stierlingi*	N Angola to se Zaire, w Tanzania, w Malawi and Mozambique
____ *P. l. pauciguttatus*	S Angola to n Namibia and nw Botswana
____ *P. l. litsipsirupa*	C Namibia to Botswana, Zimbabwe, Mozambique and S Africa

☐ **Yemen Thrush** *Turdus menachensis*

Mountains of sw Arabian Peninsula

☐ **Olive Thrush** *Turdus olivaceus*

____ *T. o. ludoviciae*	Mountains of n Somalia
____ *T. o. oldeani*	Mountains of n-central Tanzania
____ *T. o. roehli*	Mountains of ne Tanzania (Pare and Usambara mountains)
____ *T. o. helleri*	SE Kenya (Taita Hills and Mt. Kasigau)
____ *T. o. deckeni*	N Tanzania (Longido and Ketumbeine to Mt. Kilimanjaro)
____ *T. o. abyssinicus*	Highlands of Ethiopia, se Sudan, n Uganda, Kenya and n Tanzania
____ *T. o. baraka*	Mountains of e Zaire and w Uganda (Ruwenzori Mountains)
____ *T. o. bambusicola*	Highlands of Burundi, Rwanda, sw Uganda, nw Tanzania, e Zaire
____ *T. o. nyikae*	Tanzania (Nguru and Uluguru mts.), n Malawi and ne Zambia
____ *T. o. milanjensis*	Mountains of s Malawi and Mozambique
____ *T. o. swynnertoni*	Montane forests of e Zimbabwe and adjacent Mozambique

____	*T. o. culminans*	Natal (Drakensberg to Nkandhla, Qudeni and Ngorne forests)
____	*T. o. transvaalensis*	N and e Transvaal and w Swaziland
____	*T. o. smithi*	S Namibia to se Botswana, sw Transvaal and n Cape Province
____	*T. o. pondoensis*	Natal and Swaziland to Transkei and e Cape Province
____	*T. o. olivaceus*	South Africa (sw Cape Province)

☐ **Olivaceous Thrush** *Turdus olivaceofuscus*

____	*T. o. olivaceofuscus*	São Tomé (Gulf of Guinea)
____	*T. o. xanthorhynchus*	Príncipe (Gulf of Guinea)

☐ **Comoro Thrush** *Turdus bewsheri*

____	*T. b. comorensis*	Grand Comoro I.
____	*T. b. moheliensis*	Mohéli (Comoro Islands)
____	*T. b. bewsheri*	Anjouan (Comoro Islands)

☐ **Kurrichane Thrush** *Turdus libonyanus*

____	*T. l. verreauxi*	S Zaire to Angola, n Namibia, Zambia, w Zimbabwe, n Botswana
____	*T. l. libonyanus*	E Botswana to Transvaal and n Cape Province
____	*T. l. peripheris*	S Mozambique (Maputo) to Natal and se Swaziland
____	*T. l. tropicalis*	SE Zaire to Tanzania, Malawi, Zimbabwe and Mozambique

☐ **African Thrush** *Turdus pelios*

____	*T. p. chiguancoides*	Senegal to Gambia, Guinea, Sierra Leone, Liberia and n Ghana
____	*T. p. saturatus*	W Ghana to Cameroon, w Congo and Gabon
____	*T. p. pelios*	E Cameroon to Chad, s Sudan, Eritrea and Ethiopia
____	*T. p. adamauae*	N Cameroon (Adamawa Plateau)
____	*T. p. nigrilorum*	Highlands of Mt. Cameroon
____	*T. p. poensis*	Bioko (Gulf of Guinea)
____	*T. p. centralis*	N Zaire to s Sudan, sw Ethiopia, Uganda, Kenya and Tanzania
____	*T. p. graueri*	Extreme e Zaire to Rwanda, Burundi and w Tanzania
____	*T. p. bocagei*	W Angola (Benguela highlands) to w Zaire
____	*T. p. stormsi*	SE Zaire (Shaba) to ne Angola and nw Zambia

☐ **African Bare-eyed Thrush** *Turdus tephronotus*

Arid lowlands of Ethiopia and Somalia to Kenya and ne Tanzania

☐ **Gray-backed Thrush** *Turdus hortulorum*

Breeds e Siberia, Manchuria and n Korea; winters to SE Asia

☐ **Tickell's Thrush** *Turdus unicolor*

Himalayas of Kashmir to Nepal; winters to s peninsular India

☐ **Black-breasted Thrush** *Turdus dissimilis*

Mountains of Assam to sw China, n Myanmar and n SE Asia

☐ **Japanese Thrush** *Turdus cardis*

Central China and Japan; winters to s China and Indochina

☐ **White-collared Blackbird** *Turdus albocinctus*

Himalayas of n India to se Tibet, sw China and nw Myanmar

☐ **Ring Ouzel** *Turdus torquatus*

____	*T. t. torquatus*	Scandinavia, Britain, Ireland and coastal w France
____	*T. t. alpestris*	Mts. of central and s Europe; winters to Asia Minor and N Africa
____	*T. t. amicorum*	Caucasus and e Turkey to n Iran; winters to s Iran

☐ **Gray-winged Blackbird** *Turdus boulboul*

Himalayas of w Pakistan to n Myanmar, s China and n SE Asia

☐ **Eurasian Blackbird** *Turdus merula*

____	*T. m. merula*	W Europe; introd. se Australia, Tasmania, Norfolk, Lord Howe is.
____	*T. m. azorensis*	Azores
____	*T. m. cabrerae*	Madeira and w Canary Islands
____	*T. m. mauritanicus*	North Africa (Morocco to Tunisia)
____	*T. m. aterrimus (insularum)*	SE Europe to Crete, Rhodes, Caucasus, Transcaucasia and n Iran
____	*T. m. syriacus*	S Turkey to Syria, n Iraq and s Iran
____	*T. m. intermedius*	C Asia to ne Afghanistan, Pamirs and Xinjiang; winters to s Iraq
____	*T. m. maximus*	W Pakistan and India to Sikkim, Bhutan and se Tibet

____	*T. m. sowerbyi*	SW China (Sichuan)
____	*T. m. mandarinus*	W-central China (Guizhou)
____	*T. m. nigropileus*	W Ghats (Gujarat to Mysore) and Nilgiri Plateau of s central India
____	*T. m. spencei*	E Ghats (Madhya Pradesh to Seshachalam Hills) of e India
____	*T. m. simillimus*	SW India (Mysore and w Madras)
____	*T. m. bourdilloni*	SW India (Kerala)
____	*T. m. kinnisii*	Hills of Sri Lanka

☐ **Island Thrush** *Turdus poliocephalus*

____	*T. p. erythropleurus*	Christmas I. (Indian Ocean)
____	*T. p. loeseri*	Mountains of n Sumatra
____	*T. p. indrapurae*	Mountains of sw Sumatra
____	*T. p. biesenbachi*	W Java (Mt. Papandajan region)
____	*T. p. fumidus*	W Java (Mt. Gedeh region)
____	*T. p. stresemanni*	W Java (Mt. Lawoe region)
____	*T. p. javanicus*	Central Java
____	*T. p. whiteheadi*	Mountains of e Java
____	*T. p. seebohmi*	N Borneo (Mt. Kinabalu and Trus Madi)
____	*T. p. niveiceps*	Mountains of Taiwan and Lan-yü I.
____	*T. p. thomassoni*	Mountains of n Luzon (n Philippines)
____	*T. p. mayonensis*	Mountains of s Luzon (n Philippines)
____	*T. p. mindorensis*	Mountains of Mindoro (Philippines)
____	*T. p. nigrorum*	Mountains of Negros (Philippines)
____	*T. p. malindangensis*	S Philippines (Mt. Malindang region of nw Mindanao)
____	*T. p. katanglad*	S Philippines (Mt. Katanglad region of central Mindanao)
____	*T. p. kelleri*	S Philippines (Mt. Apo and adjacent mountains of se Mindanao)
____	*T. p. hygroscopus*	S-central Sulawesi (Latimojong Mountains)
____	*T. p. celebensis*	SW Sulawesi (Bonthain Peak and Wawa Kareng)
____	*T. p. schlegelii*	E Lesser Sundas (Mount Mutis on w Timor)
____	*T. p. sterlingi*	E Lesser Sundas (Mount Ramelan on e Timor)
____	*T. p. deningeri*	S Moluccas (Mt. Binaia on Seram)
____	*T. p. versteegi*	W New Guinea (Jayawijaya Mountains)
____	*T. p. carbonarius*	New Guinea (Bismarck Mountains)
____	*T. p. keysseri*	NE New Guinea (Saruwaged Mountains of Huon Peninsula)
____	*T. p. papuensis*	Mountains of se New Guinea
____	*T. p. tolokiwae*	Tolokiwa I. (Bismarck Archipelago)
____	*T. p. heinrothi*	St. Matthias I. (Bismarck Archipelago)
____	*T. p. canescens*	Goodenough I. (D'Entrecasteaux Archipelago)
____	*T. p. bougainvillei*	Bougainville (Solomon Islands)
____	*T. p. kulambangrae*	Kulambangra (Solomon Islands)
____	*T. p. sladeni*	Guadalcanal (Solomon Islands)
____	*T. p. rennellianus*	Rennell (Solomon Islands)
____	*T. p. vanikorensis*	Vanuatu (Vanikoro, Santa Cruz and Espíritu Santo)
____	*T. p. placens*	Banks Group (Ureparapara and Vanua Lava)
____	*T. p. whitneyi*	Gau I. (Banks Group)
____	*T. p. malekulae*	Vanuatu (Pentecost, Malakulu and Ambrim)
____	*T. p. becki*	Vanuatu (Paama, Lopevi, Epi and Mai)
____	*T. p. efatensis*	Vanuatu (Efate and Nguna)
____	*T. p. albifrons*	Erromango (Vanuatu)
____	*T. p. xanthopus*	New Caledonia
____	*T. p. pritzbueri*	Loyalty Islands (Tanna and Lifu)
____	*T. p. mareensis*	Maré (Loyalty Islands).
____	*T. p. poliocephalus†*	Norfolk I. Extinct
____	*T. p. vinitinctus†*	Lord Howe I. Extinct
____	*T. p. layardi*	Fiji (Viti Levu, Ovalau, Yasawa and Koro)
____	*T. p. ruficeps*	Kandavu (Fiji)
____	*T. p. vitiensis*	Vanua Levu (Fiji)
____	*T. p. hades*	Ngau (Fiji)
____	*T. p. tempesti*	Taveuni (Fiji)
____	*T. p. samoensis*	Western Samoa (Savai'i and Upolu)

☐ **Chestnut Thrush** *Turdus rubrocanus*
   ____ *T. r. rubrocanus*   Himalayas of Afghanistan to Nepal, Sikkim and Bhutan
   ____ *T. r. gouldi*   Himalayas of se Tibet to sw China and n Myanmar

☐ **White-backed Thrush** *Turdus kessleri*

               Himalayas of w China (Gansu to Sichuan); winters to n India

☐ **Gray-sided Thrush** *Turdus feae*

               Mountains of ne China (Liaoning); winters to India and Myanmar

☐ **Eyebrowed Thrush** *Turdus obscurus*

               Siberia, Mongolia and Japan; winters to Indonesia and Philippines

☐ **Pale Thrush** *Turdus pallidus*

               NE Siberia to Kuril Is. and Japan; winters to SE Asia and Sumatra

☐ **Brown-headed Thrush** *Turdus chrysolaus*
   ____ *T. c. orii*   Kuril Islands; winters to main Japanese islands and Ryukyu Is.
   ____ *T. c. chrysolaus*   Sakhalin I. (Russia) to n Japan; winters to s China and Philippines

☐ **Izu Thrush** *Turdus celaenops*

               Izu Islands and Yakushima (Ryukyu Islands)

☐ **Dark-throated Thrush** *Turdus ruficollis*
   ____ *T. r. atrogularis*   E Russia to w Siberia; winters to n India and China
   ____ *T. r. ruficollis*   E Siberia to n Manchuria; winters to w China, Myanmar, ne India

☐ **Dusky Thrush** *Turdus naumanni*
   ____ *T. n. eunomus*   N Siberia to Kamchatka; winters to Japan, s China and Myanmar
   ____ *T. n. naumanni*   C Siberia to n Manchuria, Amurland, Sakhalin; winters to Korea

☐ **Fieldfare** *Turdus pilaris*

               N Palearctic; winters to n Africa, Mediterranean and Near East

☐ **Redwing** *Turdus iliacus*
   ____ *T. i. coburni*   Iceland and Faeroes; winters to nw Europe
   ____ *T. i. iliacus*   N Europe to central Asia; winters to North Africa and Near East

☐ **Song Thrush** *Turdus philomelos*
   ____ *T. p. hebridensis*   Outer Hebrides and Isle of Skye
   ____ *T. p. clarkei*   British Isles and w Europe; winters to n Mediterranean basin
   ____ *T. p. philomelos*   N and e Europe to central Asia; winters to North Africa and Iran
   ____ *T. p. nataliae*   Sayan Mountains to Lake Baikal and n Iran; winters to s Iran

☐ **Chinese Thrush** *Turdus mupinensis*

               W China (s Gansu to Shaanxi, Hubei, Sichuan and nw Yunnan)

☐ **Mistle Thrush** *Turdus viscivorus*
   ____ *T. v. viscivorus*   Western Palearctic (except range of *deichleri*) to w Siberia
   ____ *T. v. deichleri*   Northwest Africa, Corsica and Sardinia
   ____ *T. v. bonapartei*   E Siberia to central Asia and the Himalayas; winters to n India

☐ **Red-legged Thrush** *Turdus plumbeus*
   ____ *T. p. plumbeus*   N Bahamas
   ____ *T. p. schistaceus*   E Cuba
   ____ *T. p. rubripes*   Central and w Cuba and Isle of Pines; formerly Swan I.
   ____ *T. p. coryi*   Cayman Brac (Greater Antilles)
   ____ *T. p. ardosiaceus*   Hispaniola, Puerto Rico, Gonâve I. and Tortue I.
   ____ *T. p. albiventris*   Dominica (Lesser Antilles)

☐ **Chiguanco Thrush** *Turdus chiguanco*
   ____ *T. c. conradi*   Andes of s Ecuador and central Peru
   ____ *T. c. chiguanco*   Coastal Peru; nw Bolivia (La Paz)
   ____ *T. c. anthracinus*   W Bolivia to ne Chile (Atacama) and w Argentina

☐ **Sooty Robin** *Turdus nigrescens*

               Mountains of Costa Rica and w Panama (extreme w Chiriquí)

☐ **Great Thrush** *Turdus fuscater*

  ____ *T. f. cacozelus*   Santa Marta Mountains (ne Colombia)
  ____ *T. f. clarus*   Sierra de Perijá (Colombia/Venezuela border)
  ____ *T. f. quindio*   Central and Western Andes of Colombia to n Ecuador
  ____ *T. f. gigas*   E Andes of Colombia to w Venezuela (Mérida and Táchira to Lara)
  ____ *T. f. gigantodes*   S Ecuador to n Peru (Junín)
  ____ *T. f. ockendeni*   Andes of se Peru (Cuzco and Puno)
  ____ *T. f. fuscater*   Andes of w Bolivia (La Paz and Cochabamba)

☐ **Black Robin** *Turdus infuscatus*

  Humid montane forests of s Mexico to nw Honduras

☐ **Glossy-black Thrush** *Turdus serranus*

  ____ *T. s. cumanensis*   NE Venezuela (Anzoátegui, Sucre and Monagas)
  ____ *T. s. atrosericeus*   NE Colombia (Páramo de Tamá) to Andes of n Venezuela
  ____ *T. s. fuscobrunneus*   Mountains of central and s Colombia to Ecuador
  ____ *T. s. serranus*   Mountains of Peru, Bolivia and nw Argentina (Salta and Jujuy)

☐ **Andean Slaty-Thrush** *Turdus nigriceps*

  Andes of s Ecuador to Peru, Bolivia and nw Argentina

☐ **Eastern Slaty-Thrush** *Turdus subalaris*

  S Brazil (Goiás, Mato Grosso, Paraná) to Paraguay, ne Argentina

☐ **Black-hooded Thrush** *Turdus olivater*

  ____ *T. o. sanctaemartae*   Santa Marta Mountains (ne Colombia)
  ____ *T. o. olivater*   E Colombia and coastal mountains of n Venezuela
  ____ *T. o. caucae*   SW Colombia (Cauca Valley)
  ____ *T. o. paraquensis*   *Tepuis* of s Venezuela (Cerro Paraque)
  ____ *T. o. kemptoni*   *Tepuis* of s Venezuela (Cerro de la Neblina)
  ____ *T. o. duidae*   *Tepuis* of s Venezuela (Mt. Duida)
  ____ *T. o. roraimae*   *Tepuis* of s Venezuela (s Bolívar), s Guyana and adjacent n Brazil
  ____ *T. o. ptaritepui*   *Tepuis* of se Venezuela (Mt. Ptari-tepui)

☐ **Plumbeous-backed Thrush** *Turdus reevei*

  Arid scrub of sw Ecuador and nw Peru

☐ **Marañón Thrush** *Turdus maranonicus*

  N Peru (upper Marañón Valley) and adjacent Ecuador

☐ **Chestnut-bellied Thrush** *Turdus fulviventris*

  E Andes of Colombia to nw Venezuela and extreme n Peru

☐ **Rufous-bellied Thrush** *Turdus rufiventris*

  ____ *T. r. juensis*   NE Brazil (Piauí and Ceará to Pernambuco and w Bahia)
  ____ *T. r. rufiventris*   S Brazil (s Bahia) to Uruguay, Paraguay, Bolivia and n Argentina

☐ **Austral Thrush** *Turdus falcklandii*

  ____ *T. f. pembertoni*   S-central Argentina (Río Negro and Neuquén)
  ____ *T. f. magellanicus*   S Chile and s Argentina to Tierra del Fuego; Juan Fernández Is.
  ____ *T. f. falcklandii*   Falkland Islands

☐ **Pale-breasted Thrush** *Turdus leucomelas*

  ____ *T. l. albiventer*   N Colombia to Venezuela, the Guianas and n Brazil
  ____ *T. l. cautor*   NE Colombia (Guajira Peninsula)
  ____ *T. l. leucomelas*   S Brazil to e Paraguay, n Bolivia, ne Argentina; e Peru (San Martín)

☐ **Creamy-bellied Thrush** *Turdus amaurochalinus*

  E Peru to central Argentina, Paraguay, Uruguay and Brazil

☐ **Mountain Robin** *Turdus plebejus*

  ____ *T. p. differens*   Mountains of se Mexico (extreme se Chiapas) and Guatemala
  ____ *T. p. rafaelensis*   Mountains of El Salvador and Nicaragua
  ____ *T. p. plebejus*   Mts. of Costa Rica and w Panama (Bocas del Toro and Chiriquí)

☐ **Black-billed Thrush** *Turdus ignobilis*

  ____ *T. i. ignobilis*   E and Central Andes of Colombia
  ____ *T. i. goodfellowi*   Colombia (Cauca Valley and west slope of Western Andes)

____ *T. i. debilis*   E Colombia to w Venezuela, nw Brazil, e Peru and n Bolivia

____ *T. i. murinus*   SE Venezuela (Amazonas and Bolívar) and Guyana

____ *T. i. arthuri*   SE Venezuela (Mt. Duida), adjacent Guyana and French Guiana

☐ **Lawrence's Thrush** *Turdus lawrencii*

SE Colombia to n Bolivia, s Venezuela and w Amazonian Brazil

☐ **Cocoa Thrush** *Turdus fumigatus*

____ *T. f. personus*   Lesser Antilles (St. Vincent and Grenada)

____ *T. f. aquilonalis*   Coastal ne Colombia to n Venezuela; Trinidad

____ *T. f. orinocensis*   E Colombia (e Vichada and Meta) and w Venezuela

____ *T. f. fumigatus*   The Guianas to n and e Brazil, Mato Grosso and e Bolivia

☐ **Pale-vented Thrush** *Turdus obsoletus*

____ *T. o. obsoletus*   Caribbean slope of Costa Rica to Panama and nw Colombia

____ *T. o. parambanus*   Pacific coast of Colombia and w Ecuador

____ *T. o. colombianus*   E slope of Western Andes of Colombia

☐ **Hauxwell's Thrush** *Turdus hauxwelli*

SE Colombia to n Bolivia, s Venezuela and w Amazonian Brazil

☐ **Clay-colored Robin** *Turdus grayi*

____ *T. g. tamaulipensis*   Tropical e Mexico (s Tamaulipas to Yucatán Pen. and n Chiapas)

____ *T. g. microrhynchus*   E Mexico (Santa María del Río region of San Luis Potosí)

____ *T. g. grayi*   E Mexico (Sierra Madre Oriental) to Guatemala

____ *T. g. megas*   W Guatemala to Nicaragua

____ *T. g. casius*   Costa Rica to nw Colombia (nw Chocó)

____ *T. g. incomptus*   Coastal n Colombia (Barranquilla to Santa Marta Peninsula)

☐ **Bare-eyed Thrush** *Turdus nudigenis*

____ *T. n. nudigenis*   S Lesser Antilles; Trinidad; Colombia to Guianas and n Brazil

____ *T. n. extimus*   N Brazil (s bank of lower Amazon in Santarém region)

☐ **Ecuadorian Thrush** *Turdus maculirostris*

Coastal w Ecuador (Esmeraldas) to extreme nw Peru (Tumbes)

☐ **Unicolored Thrush** *Turdus haplochrous*

Rare and local in n Bolivia (Beni and Santa Cruz)

☐ **White-eyed Thrush** *Turdus jamaicensis*

Montane forests and wooded hills of Jamaica

☐ **White-throated Thrush** *Turdus assimilis*

____ *T. a. calliphthongus*   Highlands of nw Mexico (se Sonora to ne Sinaloa and Chihuahua)

____ *T. a. lygrus*   W Mexico (s Sinaloa to w Oaxaca and sw Chiapas)

____ *T. a. assimilis*   E Mexico (s Tamaulipas to e México, n Oaxaca and w Veracruz)

____ *T. a. oaxacae*   Highlands of Oaxaca

____ *T. a. leucauchen*   S Mexico (s Veracruz) to Honduras

____ *T. a. rubicundus*   Pacific slope of w Guatemala and El Salvador

____ *T. a. atrotinctus*   Caribbean highlands of e Nicaragua

____ *T. a. oblitus*   Highlands of n and central Costa Rica

____ *T. a. cnephosus*   Highlands of sw Costa Rica to w Panama (Chiriquí and Veraguas)

____ *T. a. coibensis*   Coiba I. (w Panama)

____ *T. a. daguae*   E Panama (Darién) to w Colombia and nw Ecuador

☐ **White-necked Thrush** *Turdus albicollis*

____ *T. a. phaeopygoides*   NE Colombia to n Venezuela; Trinidad and Tobago

____ *T. a. phaeopygus*   Extreme e Colombia to the Guianas and n Amazonian Brazil

____ *T. a. spodiolaemus*   E Ecuador to e Peru, n Bolivia and w Brazil

____ *T. a. contemptus*   *Yungas* of Bolivia (La Paz, Santa Cruz and Tarija)

____ *T. a. crotopezus*   E Brazil (Bahia, Espírito Santo and Alagoas)

____ *T. a. albicollis*   SE Brazil (Rio de Janeiro to Rio Grande do Sul)

____ *T. a. paraguayensis*   SW Brazil (Mato Grosso) to Paraguay and ne Argentina

☐ **Rufous-backed Robin** *Turdus rufopalliatus*

____ *T. r. rufopalliatus*   Arid w Mexico (Sonora to w Puebla and Oaxaca)

____ *T. r. graysoni*   Tres Marías Islands and adjacent coastal Nayarit (w Mexico)

☐ **Rufous-collared Robin** *Turdus rufitorques*

Highlands of s Mexico (Chiapas) to central Honduras

☐ **American Robin** *Turdus migratorius*
_____ *T. m. migratorius* N Alaska and n Canada to central US; winters e Mexico and Cuba
_____ *T. m. caurinus* SE Alaska and w Canada to nw Oregon; winters to California
_____ *T. m. nigrideus* E Canada (n Quebec and Labrador) to Gulf Coast of US
_____ *T. m. achrusterus* S-central US; winters to se Mexico
_____ *T. m. propinquus* E British Columbia to sw US and sw Mexico; winters to Guatemala
_____ *T. m. phillipsi* E Mexico (sw Tamaulipas to Puebla, Guerrero and Oaxaca)
_____ *T. m. confinis* Mountains of s Baja California (Sierra de la Laguna)

☐ **La Selle Thrush** *Turdus swalesi*
_____ *T. s. dodae* Humid montane forests of central Dominican Republic
_____ *T. s. swalesi* Humid forests of Haiti (Morne La Selle)

☐ **White-chinned Thrush** *Turdus aurantius*

Wooded hills and mountains of Jamaica

☐ **Varied Thrush** *Ixoreus naevius*
_____ *I. n. naevius* SE Alaska to coastal nw California; winters to s California
_____ *I. n. meruloides* N Alaska and nw Canada to nw US; winters to w-central US
_____ *I. n. carlottae* Queen Charlotte Islands (off sw Canada)

☐ **Aztec Thrush** *Ridgwayia pinicola*

Oak-pine forests of w and central Mexico (Sonora to Oaxaca)

☐ **Fruit-hunter** *Chlamydochaera jefferyi*

Patchily distributed mountains of n Borneo (Sabah)

☐ **Rusty-bellied Shortwing** *Brachypteryx hyperythra*

NE India (Sikkim to Arunachal Pradesh) to sw China (Yunnan)

☐ **Gould's Shortwing** *Brachypteryx stellata*
_____ *B. s. stellata* Himalayas (Nepal to Bhutan, se Tibet, sw China and ne Myanmar)
_____ *B. s. fusca* Mountains of n Vietnam (nw Tonkin)

☐ **White-bellied Shortwing** *Brachypteryx major*
_____ *B. m. major* Peninsular India (Nilgiri Hills of Mysore and w Madras)
_____ *B. m. albiventris* Peninsular India (sw Madras to Kerala)

☐ **Lesser Shortwing** *Brachypteryx leucophrys*
_____ *B. l. nipalensis* Himalayas of n India to sw China (Yunnan) and Myanmar
_____ *B. l. carolinae* S China to n Thailand and n Indochina
_____ *B. l. langbianensis* Mountains of s Laos (Langbian Plateau) and s Vietnam
_____ *B. l. wrayi* Mountains of Malaysia
_____ *B. l. leucophrys* Mts. of Sumatra, Java, Bali, Lombok, Sumbawa, Alor and Timor

☐ **White-browed Shortwing** *Brachypteryx montana*
_____ *B. m. cruralis* E Himalayas to n Myanmar and w China; winters to n Indochina
_____ *B. m. sinensis* Mountains of se China (nw Fujian and Guangxi)
_____ *B. m. goodfellowi* Mountains of Taiwan
_____ *B. m. poliogyna* Mountains of n Luzon (n Philippines)
_____ *B. m. andersoni* Mountains of s Luzon (n Philippines)
_____ *B. m. mindorensis* Highlands of Mindoro (Philippines)
_____ *B. m. brunneiceps* Mountains of Negros (Philippines)
_____ *B. m. mindanensis* S Philippines (Mt. Apo and Mt. Matutum on Mindanao)
_____ *B. m. malindangensis* S Philippines (Mt. Malindang region of nw Mindanao)
_____ *B. m. sillimani* Mountains of s Palawan (s Philippines)
_____ *B. m. erythrogyna* Mountains of n Borneo
_____ *B. m. saturata* Mountains of Sumatra
_____ *B. m. montana* Mountains of Java
_____ *B. m. floris* Mountains of Flores (w Lesser Sundas)

☐ **Great Shortwing** *Heinrichia calligyna*
_____ *H. c. simplex* Tentolo-Matinan Mountains (ne peninsula of Sulawesi)

____ *H. c. calligyna*	Mount Latimojong (s central Sulawesi)
____ *H. c. picta*	Mekonga Mountains (se Sulawesi)

☐ **Brown-chested Alethe** *Alethe poliocephala*

____ *A. p. poliocephala*	Sierra Leone to Ghana
____ *A. p. hallae*	Angola (Gabela escarpment of Cuanza Sul)
____ *A. p. giloensis*	S Sudan
____ *A. p. carruthersi*	Extreme s Sudan to ne Zaire, Uganda and w Kenya
____ *A. p. compsonota*	Nigeria and Cameroon to nw Angola (Quicolungo); Bioko
____ *A. p. vandeweghei*	Rwanda and Burundi
____ *A. p. nandensis*	W Kenya (Nandi Hills)
____ *A. p. akeleyae*	Kenya east of the Rift Valley
____ *A. p. kungwensis*	W Tanzania (Kungwe-Mahari Mountains)
____ *A. p. ufipae*	SW Tanzania (Ufipa Plateau) and se Zaire

☐ **Red-throated Alethe** *Alethe poliophrys*

____ *A. p. poliophrys*	Montane forests of ne Zaire, Rwanda, Burundi and sw Uganda
____ *A. p. kaboboensis*	Montane forests of e Zaire (Mt. Kabobo)

☐ **Cholo Alethe** *Alethe choloensis*

____ *A. c. choloensis*	Montane forests of s Malawi (east of Rift Valley)
____ *A. c. namuli*	NW Mozambique (Namuli massif)

☐ **White-chested Alethe** *Alethe fuelleborni*

	Mountains of e Tanzania to n Malawi and s-central Mozambique

☐ **Fire-crested Alethe** *Alethe diademata*

____ *A. d. diademata*	Senegambia to Sierra Leone, Liberia, Ghana and Togo
____ *A. d. castanea*	S Nigeria to Cameroon, Gabon and Zaire; Bioko
____ *A. d. woosnami*	E Zaire to sw Sudan, w Uganda and nw Tanzania

FAMILY: CISTICOLIDAE (Cisticolas and Allies—111)

☐ **Red-faced Cisticola** *Cisticola erythrops*

____ *C. e. erythrops*	Senegambia to Central African Republic
____ *C. e. sylvia*	NE Zaire and s Sudan to central Tanzania and s Zaire
____ *C. e. nyasa*	Extreme se Zaire to s Tanzania, n Botswana and Natal
____ *C. e. lepe*	Angola
____ *C. e. pyrrhomitra*	SE Sudan to Ethiopia
____ *C. e. niloticus*	Sudan (upper Blue Nile)

☐ **Singing Cisticola** *Cisticola cantans*

____ *C. c. swanzii*	Senegambia to s Nigeria
____ *C. c. concolor*	N Nigeria to s Sudan
____ *C. c. adamauae*	Cameroon to Congo and nw Zaire
____ *C. c. belli*	Central African Republic to ne Zaire, Uganda and nw Tanzania
____ *C. c. cantans*	S Eritrea to s Ethiopia
____ *C. c. pictipennis*	Kenya and n Tanzania (south to Nguru Mountains)
____ *C. c. muenzneri*	S Tanzania (Uluguru Mountains) to Zimbabwe

☐ **Whistling Cisticola** *Cisticola lateralis*

____ *C. l. lateralis*	Senegambia to Nigeria and Cameroon
____ *C. l. antinorii*	Central African Republic to w Kenya
____ *C. l. modestus (vincenti)*	Gabon to n Angola and s Zaire

☐ **Chattering Cisticola** *Cisticola anonymus*

	Nigeria to Cameroon, Gabon, Congo, Zaire and nw Angola

☐ **Trilling Cisticola** *Cisticola woosnami*

____ *C. w. woosnami*	NE Zaire to Uganda, Burundi and n Tanzania (south to Iringa)
____ *C. w. lufira*	SW Tanzania (Kigoma and Rukwa)

☐ **Bubbling Cisticola** *Cisticola bulliens*
___ *C. b. septentrionalis* — Cabinda and lower Congo River to n Angola (south to Gabela)
___ *C. b. bulliens* — S Angola (Benguela escarpment)

☐ **Chubb's Cisticola** *Cisticola chubbi*
___ *C. c. adametzi* — SE Nigeria to Cameroon
___ *C. c. discolor* — S Cameroon (Mt. Cameroon)
___ *C. c. chubbi* — Zaire to Rwanda, Burundi, Uganda and Kenya
___ *C. c. marungensis* — SE Zaire (Marungu Plateau)

☐ **Hunter's Cisticola** *Cisticola hunteri*
High mountains of w Kenya, Uganda and n Tanzania

☐ **Black-lored Cisticola** *Cisticola nigriloris*
Highlands of n Malawi, ne Zambia and s Tanzania

☐ **Rock-loving Cisticola** *Cisticola aberrans*
___ *C. a. admiralis* — Guinea to Sierra Leone, Mali and s Ghana
___ *C. a. petrophilus* — N Nigeria to w Cameroon, ne Zaire and sw Sudan
___ *C. a. bailunduensis* — Central Angola
___ *C. a. emini (teitensis)* — S Kenya to n Tanzania
___ *C. a. nyika* — SW Tanzania to Zambia, Malawi, Zimbabwe and Mozambique
___ *C. a. lurio* — Malawi (east of Rift Valley) and adjacent nw Mozambique
___ *C. a. aberrans* — Botswana to Transvaal, w Swaziland and Natal
___ *C. a. minor* — Lowlands of s Mozambique to Natal and e Cape Province

☐ **Boran Cisticola** *Cisticola bodessa*
___ *C. b. bodessa* — Juniper woodlands of s Sudan to Eritrea, s Ethiopia and n Kenya
___ *C. b. kaffensis* — W Ethiopia (Kaffa Province)

☐ **Rattling Cisticola** *Cisticola chiniana*
___ *C. c. simplex* — S Sudan to n Uganda
___ *C. c. fricki* — S Ethiopia to n Kenya
___ *C. c. humilis* — W Kenya and ne Uganda; n Tanzania (Loliondo region)
___ *C. c. ukamba* — Highlands of e Kenya and ne Tanzania
___ *C. c. victoria* — Lake Victoria basin (sw Kenya and adjacent Tanzania)
___ *C. c. heterophrys* — Coastal Kenya and Tanzania
___ *C. c. fischeri* — N-central Tanzania (south to Tabora region)
___ *C. c. keithi* — S-central Tanzania (Dodoma to Iringa)
___ *C. c. mbeya* — S Tanzania (Mbeya to Chimala)
___ *C. c. emendatus* — SE Tanzania to n Mozambique, Malawi and extreme e Zambia
___ *C. c. procerus* — Extreme s Malawi (Chiromo) and n Mozambique (Tete)
___ *C. c. chiniana (vulpiniceps)* — Zimbabwe to w Mozambique, Transvaal and se Botswana
___ *C. c. fortis* — N Angola to Zambia, s Congo, Gabon and s Zaire
___ *C. c. frater* — Central Namibia
___ *C. c. bensoni* — S Zambia
___ *C. c. smithersi* — W Zimbabwe to nw Botswana, sw Zambia, n Namibia and s Angola
___ *C. c. campestris* — S coastal Mozambique to Swaziland and Natal

☐ **Ashy Cisticola** *Cisticola cinereolus*
___ *C. c. cinereolus* — NE Ethiopia and s Somalia
___ *C. c. schillingsi* — S Ethiopia and extreme se Sudan to n Tanzania

☐ **Red-pate Cisticola** *Cisticola ruficeps*
___ *C. r. guinea* — Senegal to Nigeria and Cameroon (Adamawa Plateau)
___ *C. r. ruficeps* — Chad to s Sudan (Kordofan and Bahr-el-Ghazal)
___ *C. r. scotopterus* — Central Sudan (White and Blue Nile valleys) to Eritrea
___ *C. r. mongalla* — S Sudan (upper White Nile) to n Uganda

☐ **Dorst's Cisticola** *Cisticola dorsti*
Grassy steppes of nw Nigeria, n Cameroon and s Chad

☐ **Gray Cisticola** *Cisticola rufilatus*
_____ *C. r. rufilatus* — S Angola and n Namibia to Botswana, Zimbabwe and n Cape Prov.
_____ *C. r. ansorgei (venustulus)* — SE Gabon to e Angola, s Zaire, n Zambia and Malawi
_____ *C. r. vicinior* — Plateau of Zimbabwe

☐ **Red-headed Cisticola** *Cisticola subruficapilla*
_____ *C. s. newtoni* — SW Angola to Namibia (Kaokoveld)
_____ *C. s. windhoekensis* — Central Namibia (Waterberg Mts. to Naukluft Mts.)
_____ *C. s. karasensis* — S Namibia (Great Namaqualand) to nw Cape Province
_____ *C. s. namaqua* — NW Cape Prov. (Oliphants River to Orange River and w Karoo)
_____ *C. s. subruficapilla* — SW Cape Province (east to Knysna and Oliphants)
_____ *C. s. jamesi* — SE Cape Prov. (Port Elizabeth to e Karoo and Orange Free State)

☐ **Wailing Cisticola** *Cisticola lais*
_____ *C. l. namba* — Highlands of w Angola
_____ *C. l. distinctus* — Highlands of e Uganda and central Kenya
_____ *C. l. semifasciatus* — S Tanzania (Iringa Plateau) to Malawi, Zambia, n Mozambique
_____ *C. l. mashona* — S Mozambique to Zimbabwe, n Transvaal and Swaziland
_____ *C. l. oreobates* — Central Mozambique (Mt. Gorongoza)
_____ *C. l. monticola* — S Transvaal
_____ *C. l. lais* — SE Transvaal to Natal, e Orange Free State, Lesotho, e Cape Prov.
_____ *C. l. maculatus* — S Cape Province (east to Port Elizabeth)

☐ **Tana River Cisticola** *Cisticola restrictus*
NE Kenya (lower Tana River basin)

☐ **Churring Cisticola** *Cisticola njombe*
_____ *C. n. njombe* — Highlands of s Tanzania
_____ *C. n. mariae* — NE Zambia and w Malawi (Nyika Plateau)

☐ **Winding Cisticola** *Cisticola galactotes*
_____ *C. g. amphilectus (griseus)* — Senegal to Ghana, coastal Nigeria and w Congo basin
_____ *C. g. zalingei* — N Nigeria to s Sudan (Darfur)
_____ *C. g. marginatus* — S Sudan (upper White Nile) to n Uganda
_____ *C. g. lugubris* — Ethiopia
_____ *C. g. haematocephala* — Coastal s Somalia to Kenya and n Tanzania
_____ *C. g. nyansae* — Central and e Zaire to Uganda and w Kenya
_____ *C. g. suahelicus (isodactylus)* — Central Tanzania to se Zaire, n Zambia, Malawi and Mozambique
_____ *C. g. galactotes* — SE Zimbabwe and South Africa
_____ *C. g. luapula* — N Zambia (Lake Mweru and Lake Bangweulu basin)
_____ *C. g. schoutedeni* — W Zambia
_____ *C. g. stagnans* — Caprivi Strip and adjacent s Zambia and w Zimbabwe

☐ **Chirping Cisticola** *Cisticola pipiens*
_____ *C. p. pipiens* — W Angola
_____ *C. p. congo* — E Angola to Zambia, Zaire, Tanzania and Burundi
_____ *C. p. arundicola* — SE Angola (Cuando Cubango) to Botswana and Zimbabwe

☐ **Carruthers' Cisticola** *Cisticola carruthersi*
Zaire to Uganda, w Kenya, Rwanda, Burundi and nw Tanzania

☐ **Tinkling Cisticola** *Cisticola tinniens*
_____ *C. t. perpullus* — Angola to s Zaire and nw Zambia
_____ *C. t. oreophilus* — Highlands of w and central Kenya
_____ *C. t. shiwae* — NE Zambia to se Zaire and extreme sw Tanzania
_____ *C. t. dyleffi* — Mountains nw of Lake Tanganyika (west of Ruzizi Valley)
_____ *C. t. tinniens* — Zimbabwe to South Africa

☐ **Stout Cisticola** *Cisticola robustus*
_____ *C. r. santae* — Highlands of e Nigeria and w Cameroon
_____ *C. r. schraderi* — Central Eritrea and adjacent n Ethiopia (Adigrat)
_____ *C. r. robustus* — N Ethiopian plateau and Harrar

____	_C. r. omo_	S Ethiopian plateau
____	_C. r. nuchalis (ambiguus)_	NE Zaire to Kenya and n Tanzania
____	_C. r. angolensis_	S Zaire (w Shaba) to nw Zambia and Angola
____	_C. r. awemba_	SE Zaire (e Shaba) to sw Tanzania and ne Zambia

☐ **Croaking Cisticola** *Cisticola natalensis*

____	_C. n. strangei (validus, kapitensis, littoralis)_	Senegal to Cabinda, Sudan, Rwanda, Burundi, Uganda and Kenya
____	_C. n. tonga_	Sudan (White Nile and Blue Nile regions)
____	_C. n. inexpectatus_	Highlands of Ethiopia
____	_C. n. argenteus_	S Ethiopia and se Somalia to n Kenya
____	_C. n. natalensis (vigilax)_	Interior Tanzania to se Zambia, Mozambique and South Africa
____	_C. n. holubii_	Extreme n Botswana, extreme w Zimbabwe and adj. s Zambia
____	_C. n. katanga_	SW Tanzania, adj. n Malawi, ne Zambia, se Zaire and ne Angola

☐ **Piping Cisticola** *Cisticola fulvicapilla*

____	_C. f. dispar_	W Zaire to se Gabon, central plateau of Angola and nw Zambia
____	_C. f. hallae_	S Angola to sw Zambia, nw Zimbabwe, n Botswana, n Namibia
____	_C. f. muelleri_	NW Zambia to e Tanzania, Mozambique and ne Zimbabwe
____	_C. f. dexter_	Plateau of Zimbabwe to Transvaal and extreme e Botswana
____	_C. f. lebombo_	Mozambique to Transvaal, n Zululand, Swaziland and w Natal
____	_C. f. ruficapilla_	Highlands of Transvaal to w Orange Free State and ne Cape Prov.
____	_C. f. fulvicapilla_	Interior e Cape Prov. to Drakensberg escarpment and w Lesotho
____	_C. f. dumicola_	W Zululand to Natal, coastal Transkei and e Cape Province
____	_C. f. silberbaueri_	South Africa (winter rainfall area of sw Cape Province)

☐ **Aberdare Cisticola** *Cisticola aberdare*

W-central Kenya (Aberdare Mountains)

☐ **Tabora Cisticola** *Cisticola angusticauda*

SE Uganda to sw Kenya, Rwanda, Tanzania, se Zaire and Zambia

☐ **Slender-tailed Cisticola** *Cisticola melanurus*

Locally in ne Angola, s Zaire and extreme w Zambia

☐ **Siffling Cisticola** *Cisticola brachypterus*

____	_C. b. brachypterus_	Gambia to Central African Republic, Sudan, n Zaire and n Angola
____	_C. b. zedlitzi_	Eritrea and Ethiopian plateau
____	_C. b. reichenowi_	Extreme s Somalia to coastal Kenya and Tanzania
____	_C. b. loanda_	SE Zaire to w Zambia and interior Angola
____	_C. b. hypoxanthus_	NE Zaire to n Uganda and se Sudan
____	_C. b. ankole_	S Uganda to Rwanda, Burundi, adjacent e Zaire and nw Tanzania
____	_C. b. katonae_	Interior of Kenya and n Tanzania
____	_C. b. kericho_	SW Kenya (Kericho)
____	_C. b. isabellinus (tenebricosus)_	Central Tanzania to Mozambique, Zimbabwe and e Zambia

☐ **Rufous Cisticola** *Cisticola rufus*

Grasslands of Gambia to Lake Chad and Central African Republic

☐ **Foxy Cisticola** *Cisticola troglodytes*

____	_C. t. troglodytes_	Central African Republic to s Sudan (White Nile) and w Kenya
____	_C. t. ferrugineus_	W Ethiopia and adjacent e Sudan (Blue Nile)

☐ **Tiny Cisticola** *Cisticola nana*

SE Sudan to Ethiopia, Somalia, Kenya and n Tanzania

☐ **Zitting Cisticola** *Cisticola juncidis*

____	_C. j. cisticola_	Coastal w France to Iberian Pen., Balearic Islands and nw Africa
____	_C. j. juncidis_	S France to Corsica, Sardinia, Balkans, Turkey, Syria and Israel
____	_C. j. neuroticus_	Cyprus, Levant, Iraq and w Iran
____	_C. j. uropygialis (perrenius)_	Senegal to s Nigeria, Sudan, Rwanda and n Tanzania; Mafia I.
____	_C. j. terrestris_	Rio Muni to central Zaire, Burundi and s Tanzania
____	_C. j. cursitans_	E Afghanistan to Pakistan, Nepal, n Myanmar, India, Sri Lanka
____	_C. j. salimalii_	SW India (Kerala)
____	_C. j. omalurus_	Sri Lanka

____ C. j. malaya	S Myanmar to Thailand, Malaysia, Sumatra and w Java
____ C. j. brunniceps	Japan (Honshu to Ryukyu, Izu and Cheju-Do is.) to n Philippines
____ C. j. tinnabulans	S China to Indochina, Hainan, Taiwan and Philippines
____ C. j. nigrostriatus	SW Philippines (Culion and Palawan)
____ C. j. fuscicapilla	E Java, Kangean Islands and Lesser Sundas
____ C. j. constans	Sulawesi, Togian Is., Muna I., Tukangbesi Is. and Peleng I.
____ C. j. normani	Coastal s New Guinea and coastal n Queensland
____ C. j. leanyeri	Disjunct in coastal n Australia to western Gulf of Carpenteria
____ C. j. laveryi	Coastal ne Queensland (Cape York Pen. south to Keppel I.)

□ **Socotra Cisticola** *Cisticola haesitatus*

Socotra I.

□ **Madagascar Cisticola** *Cisticola cherina*

Grasslands of Madagascar and Aldabra

□ **Desert Cisticola** *Cisticola aridulus*

____ C. a. aridulus	Mali, Niger and n Nigeria to s Sudan
____ C. a. lavendulae	Coastal Eritrea to Ethiopia and Somalia
____ C. a. tanganyika	Kenya and Tanzania
____ C. a. lobito	Coastal Angola
____ C. a. traylori	E Angola and extreme w Zambia
____ C. a. eremicus	S Angola to n Namibia, s Zambia, Zimbabwe and n Botswana
____ C. a. perplexus	N Zambia (Bangweulu swamps)
____ C. a. kalahari	Central Namibia to s Botswana and South Africa
____ C. a. caliginus	Natal to Swaziland, s Mozambique and adjacent ne Transvaal

□ **Cloud Cisticola** *Cisticola textrix*

____ C. t. bulubulu	Highlands of s Angola
____ C. t. anselli	Highlands of e Angola and nw Zambia
____ C. t. marleyi	S Mozambique to ne Zululand and coastal Natal
____ C. t. major	Transvaal to w Natal, w Swaziland and e Cape Province
____ C. t. textrix	S Cape Province (Cape Town to Port Elizabeth)

□ **Black-necked Cisticola** *Cisticola eximius*

____ C. e. occidens	S Senegal to Sierra Leone, s Mali and Nigeria
____ C. e. winneba	Coastal Ghana (Winneba)
____ C. e. eximius	N Zaire to Ethiopia, extreme w Kenya and s Central African Rep.

□ **Cloud-scraping Cisticola** *Cisticola dambo*

____ C. d. dambo	SE Gabon to ne Angola, s Zaire and nw Zambia
____ C. d. kasai	S Zaire (nw Kasai)

□ **Pectoral-patch Cisticola** *Cisticola brunnescens*

____ C. b. mbangensis	N Cameroon (Adamawa Plateau)
____ C. b. brunnescens	Ethiopia and nw Somalia
____ C. b. wambera	NW Ethiopia (Wambera Plateau)
____ C. b. nakuruensis	Highlands of Kenya and n Tanzania (west of Rift Valley)
____ C. b. hindii	Highlands of Kenya and n Tanzania (east of Rift Valley)
____ C. b. lynesi	Highlands of w Cameroon

□ **Pale-crowned Cisticola** *Cisticola cinnamomeus*

____ C. c. midcongo	SE Gabon (Teke Plateau)
____ C. c. cinnamomeus	S Tanzania to se Zaire, Zambia, Zimbabwe, Botswana, Angola
____ C. c. egregius (taciturnus)	S Mozambique to South Africa

□ **Wing-snapping Cisticola** *Cisticola ayresii*

____ C. a. gabun	Gabon to Congo and nw Zaire
____ C. a. imatong	S Sudan (Imatong Mountains)
____ C. a. entebbe	Extreme e Zaire to Rwanda, Uganda, nw Tanzania and w Kenya
____ C. a. itombwensis	Mountains ne of Lake Tanganyika

____	*C. a. mauensis*	Highlands of Kenya
____	*C. a. ayresii*	S Tanzania to se Zaire, e Zambia, Mozambique and South Africa

☐ **Golden-headed Cisticola** *Cisticola exilis*

____	*C. e. tytleri*	Foothills of Nepal to ne India, n Myanmar and sw China (Yunnan)
____	*C. e. erythrocephalus*	S India (s Mysore, w Tamil Nadu and Kerala)
____	*C. e. equicaudatus*	E Myanmar to central Thailand, Cambodia and s Vietnam
____	*C. e. courtoisi*	SE China (se Yunnan to s Hunan, Jiangxi and Fujian)
____	*C. e. volitans*	Taiwan
____	*C. e. semirufus*	Philippine Islands and Sulu Archipelago
____	*C. e. rusticus*	Sulawesi subregion and s Moluccas
____	*C. e. lineocapilla*	Java, Bali, Lesser Sundas and n Australia (Northern Territory)
____	*C. e. diminutus*	New Guinea to Solomon Islands and islands in Torres Strait
____	*C. e. polionotus*	Bismarck Archipelago
____	*C. e. alexandrae*	N Australia (Pilbara region to Mackenzie River, Queensland)
____	*C. e. exilis*	E Australia (c Queensland to s Victoria and adj. South Australia)

☐ **White-browed Chinese Warbler** *Rhopophilus pekinensis*

____	*R. p. albosuperciliaris*	NW China (Xinjiang) from Tarim Basin to Lop Nor
____	*R. p. leptorhynchus*	N-central China (ne Qinghai, Shaanxi and Gansu)
____	*R. p. pekinensis*	S Manchuria to Korea and ne China

☐ **Socotra Warbler** *Incana incanus*

Socotra (off e Somalia)

☐ **Streaked Scrub-Warbler** *Scotocerca inquieta*

____	*S. i. saharae (harterti)*	Morocco to Tunisia, Algeria and Libya
____	*S. i. theresae*	Mauritania and s Morocco
____	*S. i. inquieta*	Deserts of e Egypt and n Arabia
____	*S. i. grisea*	W Saudi Arabia (Taif Plateau), e South Yemen and Oman
____	*S. i. buryi*	SW Saudi Arabia, North Yemen and Hadramaut
____	*S. i. striata*	Iran to Baluchistan, Pakistan and nw India
____	*S. i. platyura*	Transcaspia to s Uzbekistan, n Turkmenistan and w Tajikistan
____	*S. i. montana*	Mts. of s Turkmenistan to w Tajikistan and n Afghanistan

☐ **Rufous-vented Prinia** *Prinia burnesii*

Elephant and *sarkhan* grass of Pakistan to nw India (w Punjab)

☐ **Swamp Prinia** *Prinia cinerascens*

Wet grasslands of ne India (Assam) and n Bangladesh

☐ **Striated Prinia** *Prinia crinigera*

____	*P. c. striatula*	Foothills of ne Afghanistan and w Pakistan
____	*P. c. crinigera*	Foothills of Pakistan and Kashmir to Arunachal Pradesh
____	*P. c. catharia*	NE India (Assam) to s China and w Myanmar (Chin Hills)
____	*P. c. parvirostris*	SW China (se Yunnan)
____	*P. c. parumstriata*	Coastal provinces of se China and Yangtze River drainage
____	*P. c. striata*	Taiwan

☐ **Brown Prinia** *Prinia polychroa*

____	*P. p. bangsi*	S China (se Yunnan and w Jiangxi) and Taiwan
____	*P. p. cooki*	Central Myanmar to e Thailand, Laos and Cambodia
____	*P. p. rocki*	Laos (Langbian Plateau)
____	*P. p. polychroa*	Java

☐ **Hill Prinia** *Prinia atrogularis*

____	*P. a. atrogularis*	E Nepal to Sikkim, Bhutan, se Tibet and Arunachal Pradesh
____	*P. a. khasiana*	NE India (Assam) to Bangladesh, w Myanmar and s China
____	*P. a. erythropleura*	Myanmar (s Shan States, Kayah and Tenasserim) to n Thailand
____	*P. a. superciliaris*	Hills of e Myanmar to s China (Yunnan), n Laos and n Vietnam
____	*P. a. waterstradti*	Highlands of e Malaysia (Gunong Tahan)
____	*P. a. klossi*	High plateaus of s Laos and s Vietnam
____	*P. a. dysancrita*	Hills of w Sumatra

☐ **Gray-crowned Prinia** *Prinia cinereocapilla*

Himalayan foothills of n India (Kashmir to Bhutan and s Assam)

☐ **Rufous-fronted Prinia** *Prinia buchanani*

Thornscrub of India and Pakistan (Indus River plain)

☐ **Rufescent Prinia** *Prinia rufescens*

_____ *P. r. rufescens* — Nepal to Bhutan, se Tibet, s China, Bangladesh, Myanmar and India
_____ *P. r. beavani* — SE Myanmar to sw Thailand and n Indochina (Laos and Vietnam)
_____ *P. r. peninsularis* — S Myanmar and peninsular Thailand (Isthmus of Kra to Trang)
_____ *P. r. objurgans* — SE Thailand
_____ *P. r. extrema* — S Peninsular Thailand and Malay Peninsula
_____ *P. r. dalatensis* — S Vietnam

☐ **Gray-breasted Prinia** *Prinia hodgsonii*

_____ *P. h. rufula* — Kashmir to Assam, sw China (nw Yunnan) and n Myanmar
_____ *P. h. hodgsonii* — India to w Myanmar
_____ *P. h. albogularis* — SW peninsular India (E Ghats to s Mysore and Kerala)
_____ *P. h. leggei* — Sri Lanka
_____ *P. h. erro* — E Myanmar (Shan States) to Thailand and s Indochina
_____ *P. h. confusa* — S China (s Sichuan and w Yunnan) to ne Laos and n Vietnam

☐ **Bar-winged Prinia** *Prinia familiaris*

_____ *P. f. prinia* — Lowlands of Sumatra, w Java and Karimunjawa Islands
_____ *P. f. familiaris* — E Java and Bali

☐ **Graceful Prinia** *Prinia gracilis*

_____ *P. g. akyildizi* — Coastal s Turkey (Antalya to Adana)
_____ *P. g. irakensis* — NE Syria to Iraq and sw Iran (foothills of Zagros Mountains)
_____ *P. g. palaestinae* — E Sinai to s Israel, Lebanon, Syria, Jordan and nw Arabia
_____ *P. g. deltae* — Egypt (Nile Delta) to Sinai and w Israel
_____ *P. g. gracilis* — Nile Valley (Cairo to n Sudan) and n Egypt (El Faiyum)
_____ *P. g. natronensis* — N Egypt (Wadi el Natrun)
_____ *P. g. yemenensis* — Coastal Arabia and Yemen (Mecca to Aden and Hadramaut)
_____ *P. g. carlo* — S Sudan to Eritrea, Ethiopia, Djibouti and s Somalia
_____ *P. g. hufufae* — E Saudi Arabia (Hufuf Oasis) and Bahrain
_____ *P. g. carpenteri* — Coastal Oman
_____ *P. g. lepida* — S coastal Iran to Afghanistan, Pakistan and n India
_____ *P. g. stevensi* — S Nepal to ne India (Assam and Arunachal Pradesh)

☐ **Jungle Prinia** *Prinia sylvatica*

_____ *P. s. insignis* — NW India (Rann of Kutch and Gujarat to w Rajasthan)
_____ *P. s. gangetica* — *Terai* of Nepal to n India and Bangladesh
_____ *P. s. mahendrae* — NE India (Orissa)
_____ *P. s. sylvatica* — Peninsular India (north to Madhya Pradesh and Mahjarashtra)
_____ *P. s. valida* — Sri Lanka

☐ **Yellow-bellied Prinia** *Prinia flaviventris*

_____ *P. f. sindiana* — Pakistan (Indus River system) to nw India
_____ *P. f. flaviventris* — Nepal to Bhutan, ne India, Bangladesh and n Myanmar
_____ *P. f. delacouri* — SE Myanmar to central Thailand and Indochina
_____ *P. f. sonitans* — SE China (n Guangxi, Guangdong, Fujian), Hainan and Taiwan
_____ *P. f. rafflesi* — S Myanmar, peninsular Thailand, Malaya, Sumatra and Java
_____ *P. f. halistona* — Nias I. (off Sumatra)
_____ *P. f. latrunculus* — Borneo

☐ **Ashy Prinia** *Prinia socialis*

_____ *P. s. stewarti* — N Pakistan (upper Indus River) to Nepal and n India
_____ *P. s. inglisi* — NE India to Sikkim, Bhutan, Assam and Bangladesh
_____ *P. s. socialis* — Peninsular India
_____ *P. s. brevicauda* — Sri Lanka

☐ **Tawny-flanked Prinia** *Prinia subflava*

____ *P. s. subflava*	Senegal to s Sudan, adj. Uganda, s-central Ethiopia and s Eritrea
____ *P. s. pallescens*	Mali to Sudan, Ethiopia and nw Eritrea
____ *P. s. melanorhyncha*	Sierra Leone to Cameroon, n Zaire, Kenya and nw Tanzania
____ *P. s. graueri*	E Zaire (Kivu) to Rwanda and highlands of Angola
____ *P. s. affinis*	E Zaire to sw Tanzania, Zambia, e Botswana and s Mozambique
____ *P. s. bechuanae (ovampensis)*	SW Angola to n Namibia, n Botswana, sw Zambia, w Zimbabwe
____ *P. s. mutatrix*	S Tanzania to Malawi, e Zambia, e Zimbabwe and Mozambique
____ *P. s. kasokae*	W Zambia (west of Zambezi River) and adjacent e Angola
____ *P. s. tenella*	Coastal e Africa (Somalia to s Tanzania)
____ *P. s. pondoensis*	S Mozambique to Natal, e Swaziland and e Cape Province

☐ **Plain Prinia** *Prinia inornata*

____ *P. i. terricolor*	E Baluchistan to Pakistan and nw India
____ *P. i. inornata*	Central and peninsular India (south to s Madras)
____ *P. i. franklinii*	S India (sw Mysore, Kerala and hills of w and s Madras)
____ *P. i. insularis*	Sri Lanka
____ *P. i. fusca*	Nepal to Sikkim, Bhutan, Assam and Bangladesh
____ *P. i. extensicauda*	S China to n Laos, n Vietnam and Hainan
____ *P. i. blanfordi*	Myanmar and n Thailand
____ *P. i. herberti*	S Myanmar and s Thailand to s Laos, Cambodia and s Vietnam
____ *P. i. flavirostris*	Taiwan
____ *P. i. blythi*	Java

☐ **Pale Prinia** *Prinia somalica*

____ *P. s. erlangeri*	SE Sudan to Ethiopia, Uganda, Kenya and Somalia
____ *P. s. somalica*	Extreme n Somalia

☐ **River Prinia** *Prinia fluviatilis*

	Locally in Niger, Chad and n Cameroon (status unknown)

☐ **Black-chested Prinia** *Prinia flavicans*

____ *P. f. ansorgei*	Coastal Angola and Namibia (Namib Desert to Walvis Bay)
____ *P. f. bihe*	Angola highlands to Zambia
____ *P. f. flavicans*	Namibia to Botswana and nw Cape Province
____ *P. f. nubilosa*	E Botswana to sw Zambia, sw Zimbabwe and Transvaal
____ *P. f. ortleppi*	SW Transvaal to w Orange Free State and ne Cape Province

☐ **Karoo Prinia** *Prinia maculosa*

____ *P. m. maculosa*	S Namibia to s Orange Free State and Cape Province
____ *P. m. psammophila*	SW Namibia to w Cape Province
____ *P. m. exultans*	South Africa (Lesotho, adjacent ne Cape Province and w Natal)

☐ **Drakensberg Prinia** *Prinia hypoxantha*

	Transvaal to Natal, Lesotho and e Cape Province

☐ **Namaqua Prinia** *Prinia substriata*

____ *P. s. confinis*	S Namibia (arid lower Orange River)
____ *P. s. substriata*	Orange Free State, Karoo and s Little Namaqualand

☐ **São Tomé Prinia** *Prinia molleri*

	São Tomé (Gulf of Guinea)

☐ **Roberts' Prinia** *Prinia robertsi*

	Highland forests of e Zimbabwe and adjacent sw Mozambique

☐ **Banded Prinia** *Prinia bairdii*

____ *P. b. bairdii*	SE Nigeria to Cabinda, ne Zaire and w Uganda
____ *P. b. heinrichi*	NW Angola (Cuanza Norte)
____ *P. b. obscura*	Highlands of e Zaire to w Uganda, Rwanda and Burundi
____ *P. b. melanops*	Kenya (west of Rift Valley)

☐ **Red-winged Prinia** *Prinia erythroptera*

____ *P. e. erythroptera*	Senegal to s Mali and n Cameroon

____	*P. e. jodoptera*	Central and s Cameroon to s Sudan
____	*P. e. major*	Ethiopia
____	*P. e. rhodoptera*	Kenya to Mozambique

☐ **Sierra Leone Prinia** *Schistolais leontica*

Montane ravines of e Sierra Leone, s Guinea and sw Ivory Coast

☐ **White-chinned Prinia** *Schistolais leucopogon*

____	*P. l. leucopogon*	E Nigeria to middle Ubangi River and w side of Lake Tanganyika
____	*P. l. reichenowi*	Zaire to Uganda, Rwanda, Burundi, Kenya and Tanzania

☐ **Rufous-eared Warbler** *Malcorus pectoralis*

____	*M. p. etoshae*	N Namibia (Etosha Pan to n Damaraland)
____	*M. p. ocularius*	Namibia to Botswana, sw Transvaal and n Cape Province
____	*M. p. pectoralis*	South Africa (w Cape Province to sw Orange Free State)

☐ **Red-winged Gray Warbler** *Drymocichla incana*

E Nigeria and Cameroon to se Sudan and nw Uganda

☐ **Green Longtail** *Urolais epichlorus*

____	*U. e. epichlorus (cinderella)*	Montane forests of Nigeria (Obudu Plateau) and s Cameroon
____	*U. e. mariae*	Bioko I. (Gulf of Guinea)

☐ **Cricket Longtail** *Spiloptila clamans*

Mauritania and Senegal to Mali, central Sudan and n Ethiopia

☐ **Black-collared Apalis** *Apalis pulchra*

____	*A. p. pulchra*	Mts. of se Nigeria and Cameroon; se Sudan to Uganda and Kenya
____	*A. p. murphyi*	Extreme e Zaire (Marungu Plateau)

☐ **Ruwenzori Apalis** *Apalis ruwenzorii*

Montane forests of e Zaire, sw Uganda, Rwanda and Burundi

☐ **Bar-throated Apalis** *Apalis thoracica*

____	*A. t. fuscigularis*	SE Kenya (Taita Hills)
____	*A. t. griseiceps (iringae)*	SE Kenya (Chyulu Hills) and highlands of Tanzania
____	*A. t. pareensis*	N Tanzania (South Pare Mountains)
____	*A. t. uluguru*	NE Tanzania (Uluguru Mountains)
____	*A. t. murina*	NE Tanzania to n Malawi and adjacent Zambia
____	*A. t. youngi*	SW Tanzania to ne Malawi and adjacent Zambia
____	*A. t. whitei*	E Zambia to s Malawi and adjacent Mozambique (Zobue)
____	*A. t. flavigularis*	SE Malawi (e of Nyasa-Shire Rift) and adjacent Mozambique
____	*A. t. quarta*	NE Zimbabwe (Mt. Nyangani) and Mozambique (Mt. Gorongoza)
____	*A. t. arnoldi*	E Zimbabwe and adjacent Mozambique
____	*A. t. lynesi*	N Mozambique (Mt. Namuli)
____	*A. t. rhodesiae*	Zimbabwe plateau and ne Botswana
____	*A. t. flaviventris*	SE Botswana to n and w Transvaal
____	*A. t. spelonkensis*	E and n Transvaal
____	*A. t. lebomboensis*	NE Zululand (Lebombo Mts.) to e Swaziland and s Mozambique
____	*A. t. venusta (darglensis)*	Zululand to Natal, e Griqualand and Great Kei River
____	*A. t. drakensbergensis*	South Africa (Drakensberg Mountains to w Swaziland)
____	*A. t. thoracica*	SE Cape Province (Great Kei and Gamtoos River to Umtata)
____	*A. t. claudei*	S Cape Province (Knysna to Humansdorp and Beaufort West)
____	*A. t. capensis*	S and sw Cape Province (Paarl to Oudtshoorn and Mossel Bay)
____	*A. t. griseopyga*	Coastal w Cape Province (Lamberts Bay to Cape Town)

☐ **Black-capped Apalis** *Apalis nigriceps*

____	*A. n. nigriceps*	Sierra Leone to Gabon and Central African Republic; Bioko
____	*A. n. collaris*	E Zaire to sw Uganda

☐ **Black-throated Apalis** *Apalis jacksoni*

____	*A. j. bambuluensis*	Highlands of Nigeria and Cameroon
____	*A. j. minor (albimentalis)*	Lowlands of Cameroon and Zaire
____	*A. j. jacksoni*	Zaire to Angola, Sudan, Uganda and Kenya

☐ **White-winged Apalis** *Apalis chariessa*
____ *A. c. macphersoni* | Montane forests of se Kenya to n Mozambique
____ *A. c. chariessa* | Kenya (lower Tana River)

☐ **Masked Apalis** *Apalis binotata*

| SW Cameroon to ne Gabon, e Zaire, e Uganda and nw Tanzania

☐ **Black-faced Apalis** *Apalis personata*
____ *A. p. personata* | Montane forests of e Zaire, w Uganda, Rwanda and Burundi
____ *A. p. marungensis* | SE Zaire (Marungu Plateau)

☐ **Yellow-breasted Apalis** *Apalis flavida*
____ *A. f. caniceps* | Gambia to n Angola and w Kenya
____ *A. f. flavocincta (malensis)* | SE Sudan to n Uganda, s Ethiopia, Somalia, Kenya, ne Tanzania
____ *A. f. viridiceps* | N Somalia, adjacent Ethiopia and n Kenya
____ *A. f. abyssinica* | Highlands of sw Ethiopia
____ *A. f. pugnax* | Highlands of s Kenya
____ *A. f. golzi* | SE Kenya (Taita Hills) to interior Tanzania and Rwanda
____ *A. f. flavida* | W Angola to n Namibia, n Botswana and sw Zambia
____ *A. f. neglecta (renata, lucidigula, niassae, tenerrima)* | E Angola to se Kenya, Tanzania, Mozambique and n Natal
____ *A. f. florisuga* | Central Natal to South Africa

☐ **Rudd's Apalis** *Apalis ruddi*
____ *A. r. caniviridis* | S Malawi
____ *A. r. ruddi* | Coastal Mozambique (Save River to lower Incomati River)
____ *A. r. fumosa* | Mozambique (Maputo District) to coastal Natal

☐ **Sharpe's Apalis** *Apalis sharpii*

| Humid forests of Sierra Leone, Ivory Coast and Ghana

☐ **Buff-throated Apalis** *Apalis rufogularis*
____ *A. r. sanderi* | SW Nigeria (Lagos to Ife and Niger River)
____ *A. r. rufogularis* | E Nigeria to Cameroon, Gabon and Central African Rep.; Bioko
____ *A. r. angolensis* | NW Angola
____ *A. r. brauni* | W Angola (Cuanza Sul escarpment)
____ *A. r. argenteus (eidos)* | E Zaire to Rwanda, Burundi and w Tanzania
____ *A. r. nigrescens* | SW Sudan to Zaire, Zambia, ne Angola, Uganda and nw Tanzania
____ *A. r. kigezi* | SW Uganda (Bwindi-Impenetrable Forest)

☐ **Bamenda Apalis** *Apalis bamendae*

| Montane forests of w Cameroon and Adamawa Plateau

☐ **Gosling's Apalis** *Apalis goslingi*

| Montane forests of s Cameroon to Gabon, ne Angola and Zaire

☐ **Chestnut-throated Apalis** *Apalis porphyrolaema*
____ *A. p. kaboboensis* | Montane forests of e Zaire (Mt. Kabobo)
____ *A. p. affinis* | Mountains of sw Uganda
____ *A. p. porphyrolaema* | Montane forests of Kenya to n Tanzania

☐ **Chapin's Apalis** *Apalis chapini*
____ *A. c. strausae* | SW Tanzania to ne Zambia and w Malawi
____ *A. c. chapini* | Montane forests of e Tanzania (Uluguru Mountains)

☐ **Black-headed Apalis** *Apalis melanocephala*
____ *A. m. melanocephala* | S Somalia to coastal Kenya and coastal ne Tanzania
____ *A. m. nigrodorsalis* | Highlands of Kenya
____ *A. m. moschi* | S Kenya (Taita Hills) and highlands of e Tanzania
____ *A. m. muhuluensis* | SE Tanzania (Mahenge and Songea)
____ *A. m. adjacens* | SE Malawi
____ *A. m. fuliginosa* | SE Malawi (Mulanje and Thyolo mountains)
____ *A. m. lightoni* | W Mozambique to se Zimbabwe
____ *A. m. tenebricosa* | N Mozambique (Njesi Plateau, Mt. Chiperoni and Mt. Namuli)
____ *A. m. addenda* | S Mozambique

☐ **Chirinda Apalis** *Apalis chirindensis*

＿＿＿ *A. c. vumbae*   Zimbabwe (Nyanga Highlands to Bvumba Mountains)
＿＿＿ *A. c. chirindensis*   W-central Mozambique (Mt. Gorongoza) and adj. e Zimbabwe

☐ **Gray Apalis** *Apalis cinerea*

＿＿＿ *A. c. funebris*   Montane forests of Nigeria and Cameroon
＿＿＿ *A. c. sclateri*   Mt. Cameroon; Bioko (Gulf of Guinea)
＿＿＿ *A. c. grandis*   W Angola
＿＿＿ *A. c. cinerea*   Zaire to s Sudan, ne Uganda, Rwanda, Kenya and nw Tanzania

☐ **Brown-headed Apalis** *Apalis alticola*

＿＿＿ *A. a. alticola*   SW Kenya to Tanzania, n Malawi, s Zaire and n Zambia
＿＿＿ *A. a. dowsetti*   S Zaire (Marungu Plateau)

☐ **Karamoja Apalis** *Apalis karamojae*

＿＿＿ *A. k. karamojae*   Mountains of n Uganda
＿＿＿ *A. k. stronachi*   Highlands of ne Tanzania (Nzega district)

☐ **Red-fronted Warbler** *Urorhipis rufifrons*

＿＿＿ *U. r. rufifrons*   Chad to n Sudan, ne Ethiopia, Djibouti and nw Somalia
＿＿＿ *U. r. smithi*   S Sudan to se Ethiopia, Somalia, Uganda, Kenya and Tanzania
＿＿＿ *U. r. rufidorsalis*   SE Kenya (Tsavo)

☐ **Oriole Warbler** *Hypergerus atriceps*

Gallery forests and palms of Senegal to Central African Republic

☐ **Gray-capped Warbler** *Eminia lepida*

SE Sudan to n Zaire, Rwanda, Burundi, Uganda, Kenya, Tanzania

☐ **Green-backed Camaroptera** *Camaroptera brachyura*

＿＿＿ *C. b. brevicaudata*   Senegal and Sierra Leone to c Sudan and lowlands of Ethiopia
＿＿＿ *C. b. tincta*   Liberia to w Kenya, nw Angola, nw Zambia and w Tanzania
＿＿＿ *C. b. abessinica*   S Sudan to ne Zaire, Ethiopia, n Uganda, n Kenya and w Somalia
＿＿＿ *C. b. insulata*   Ethiopia (Ghere region and Kaffa Province)
＿＿＿ *C. b. erlangeri (albiventris)*   S Somalia to coastal Kenya and ne Tanzania
＿＿＿ *C. b. aschani*   Highlands of Kenya to extreme sw Uganda and e Zaire (Kivu)
＿＿＿ *C. b. griseigula*   W Kenya to e Uganda and n Tanzania
＿＿＿ *C. b. pileata*   S Kenya to se Tanzania; Mafia I. and Zanzibar
＿＿＿ *C. b. fugglescouchmani*   E Tanzania to Zambia and Malawi
＿＿＿ *C. b. bororensis*   S Tanzania to s Malawi and n Mozambique
＿＿＿ *C. b. harterti*   Escarpment of se Gabon and nw Angola
＿＿＿ *C. b. intercalata*   S Zaire (Shaba) to w Tanzania, e Angola and e Zambia
＿＿＿ *C. b. sharpei (noomei)*   S Angola to Namibia, Zambia, Malawi, Zimbabwe, w Transvaal
＿＿＿ *C. b. transitiva*   SE Botswana to Zimbabwe
＿＿＿ *C. b. beirensis (marleyi)*   Mozambique (n of Save River) to e Zimbabwe and ne Zululand
＿＿＿ *C. b. brachyura*   Natal and w Zululand to South Africa
＿＿＿ *C. b. constans*   E Zululand to e Swaziland, Transvaal, se Zimbabwe, Mozambique

☐ **Yellow-browed Camaroptera** *Camaroptera superciliaris*

Guinea and Sierra Leone to Zaire, Uganda and nw Angola

☐ **Olive-green Camaroptera** *Camaroptera chloronota*

＿＿＿ *C. c. kelsalli*   Senegal to Ghana
＿＿＿ *C. c. chloronota*   Togo to s Cameroon, Gabon and Congo
＿＿＿ *C. c. granti*   Bioko (Gulf of Guinea)
＿＿＿ *C. c. toroensis*   Central African Rep. to sw Sudan, Uganda, w Kenya and Tanzania
＿＿＿ *C. c. kamitugaensis*   E Zaire (Itombwe region)

☐ **Miombo Wren-Warbler** *Calamonastes undosus*

＿＿＿ *C. u. cinereus*   Congo to w Zaire, Angola and nw Zambia
＿＿＿ *C. u. katangae*   S Zaire (Shaba) to n Zambia
＿＿＿ *C. u. huilae*   W-central Angola

____ C. u. stierlingi (buttoni, neglectus)	SE Angola to ne Namibia, s Zambia, e Tanzania and Mozambique
____ C. u. undosus	Kenya to Rwanda and Tanzania
____ C. u. olivascens	S coastal Tanzania to Malawi and coastal Mozambique
____ C. u. irwini	Extreme e Zambia to Botswana, Zimbabwe and Mozambique
____ C. u. pintoi	Transvaal to Swaziland, Natal and Zululand

☐ **Gray Wren-Warbler** *Calamonastes simplex*

SE Sudan to Ethiopia, Somalia, Kenya, ne Uganda, ne Tanzania

☐ **Barred Wren-Warbler** *Calamonastes fasciolatus*

____ C. f. pallidior	SW Angola
____ C. f. fasciolatus	Central Namibia to Botswana, Zimbabwe and n Cape Province
____ C. f. europhilus	SE Botswana to Zimbabwe (s Matabeleland) and Transvaal

☐ **Kopje Warbler** *Euryptila subcinnamomea*

____ E. s. petrophila	Namibia and extreme nw Cape Province
____ E. s. subcinnamomea	South Africa (Cape Province)

FAMILY: SYLVIIDAE (Old World Warblers—291)

☐ **Chestnut-headed Tesia** *Tesia castaneocoronata*

____ T. c. castaneocoronata	E Himalayas to Bangladesh, n Myanmar, s Tibet and nw Thailand
____ T. c. ripleyi	SE Tibet to sw China (Sichuan and Yunnan)
____ T. c. abadiei	N Vietnam (nw Tonkin)

☐ **Javan Tesia** *Tesia superciliaris*

Montane forests of Java

☐ **Slaty-bellied Tesia** *Tesia olivea*

NE India to s China, Myanmar, nw Thailand, n Laos and Tonkin

☐ **Gray-bellied Tesia** *Tesia cyaniventer*

Nepal to s China (w Yunnan and Guangxi), SE Asia and Java

☐ **Russet-capped Tesia** *Tesia everetti*

____ T. e. sumbawana	Sumbawa (w Lesser Sundas)
____ T. e. everetti	Flores (w Lesser Sundas)

☐ **Timor Stubtail** *Urosphena subulata*

____ U. s. subulata	E Lesser Sundas (Timor and Wetar)
____ U. s. advena	Babar (e Lesser Sundas)

☐ **Bornean Stubtail** *Urosphena whiteheadi*

Mountains of n Borneo (Kinabalu to Liang Kubung)

☐ **Asian Stubtail** *Urosphena squameiceps*

Breeds ne Asia; winters to SE Asia and Taiwan

☐ **Manchurian Bush-Warbler** *Cettia canturians*

E Siberia to Manchuria, China and Korea; winters to s China

☐ **Pale-footed Bush-Warbler** *Cettia pallidipes*

____ C. p. pallidipes	Himalayan foothills to n Myanmar
____ C. p. laurentei	S China; winters to nw Thailand, n Laos and n Vietnam
____ C. p. osmastoni	Andaman Islands

☐ **Japanese Bush-Warbler** *Cettia diphone*

____ C. d. borealis	Manchuria to Korea and s China; winters to Taiwan
____ C. d. viridis	S Sakhalin and s Kuril Islands; winters to se China
____ C. d. cantans	Main Japanese islands south to Cheju-Do Islands
____ C. d. riukiuensis	Ryukyu Islands
____ C. d. restricta	Daito Islands
____ C. d. diphone	S Izu Islands, Bonin Islands and Volcano Islands

☐ **Philippine Bush-Warbler** *Cettia seebohmi*

Montane forests of n Luzon (n Philippines)

☐ **Palau Bush-Warbler** *Cettia annae*

Palau Islands (Babelthaup, Koror, Garakayo, Peleliu and Ngabad)

☐ **Shade Warbler** *Cettia parens*

Montane forests of San Cristóbal I. (s Solomon Islands)

☐ **Odeni** *Cettia haddeni*

Bougainville I.

☐ **Fiji Bush-Warbler** *Cettia ruficapilla*

____ *C. r. ruficapilla* — Kandavu (Fiji Islands)
____ *C. r. badiceps* — Viti Levu (Fiji Islands)
____ *C. r. castaneoptera* — Vanua Levu (Fiji Islands)
____ *C. r. funebris* — Taveuni (Fiji Islands)

☐ **Tanimbar Bush-Warbler** *Cettia carolinae*

Yamdena (Tanimbar Islands)

☐ **Brownish-flanked Bush-Warbler** *Cettia fortipes*

____ *C. f. pallida* — NE Himalayas (Kashmir to w Nepal)
____ *C. f. fortipes* — Himalayas of e Nepal to Bhutan, se Tibet, ne India and Myanmar
____ *C. f. davidiana* — Mountains of s China to n Laos and n Vietnam
____ *C. f. robustipes* — Mountains of Taiwan

☐ **Sunda Bush-Warbler** *Cettia vulcania*

____ *C. v. sepiaria* — Mountains of n Sumatra
____ *C. v. flaviventris* — Mountains of central and s Sumatra
____ *C. v. vulcania* — Java, Bali, Lombok and Sumbawa
____ *C. v. oreophila* — N Borneo (Mt. Kinabalu)
____ *C. v. banksi* — Mountains of n Borneo (Sabah and Sarawak)
____ *C. v. palawana* — Mountains of Palawan (sw Philippines)
____ *C. v. everetti* — Timor (e Lesser Sundas)

☐ **Chestnut-crowned Bush-Warbler** *Cettia major*

____ *C. m. major* — Himalayas of Nepal to ne India, se Tibet, Myanmar and sw China
____ *C. m. vafra* — NE India (Meghalaya and Cachar hills of Assam)

☐ **Aberrant Bush-Warbler** *Cettia flavolivacea*

____ *C. f. flavolivacea* — Himalayas of Garhwal to Nepal, Arunachal Pradesh and se Tibet
____ *C. f. intricata* — NE Myanmar to nw Thailand and sw China (s Shanxi, e Sichuan)
____ *C. f. dulcivox* — S China (s Sichuan to s Yunnan)
____ *C. f. stresemanni* — NE India (Garo and Khasi hills of Assam)
____ *C. f. weberi* — W Myanmar (Chin Hills)
____ *C. f. alexanderi* — Extreme ne India (Manipur) and adjacent Myanmar
____ *C. f. oblita* — N Laos and n Vietnam

☐ **Yellowish-bellied Bush-Warbler** *Cettia acanthizoides*

____ *C. a. acanthizoides* — Himalayas of n India to s Tibet, s China and e Myanmar
____ *C. a. brunnescens* — SE Tibet (Tsangpo Valley)
____ *C. a. concolor* — Mountains of Taiwan

☐ **Gray-sided Bush-Warbler** *Cettia brunnifrons*

____ *C. b. whistleri* — NW Himalayas (Kashmir to Garhwal)
____ *C. b. brunnifrons* — E Himalayas (Garhwal to Nepal, Sikkim, Bhutan and se Tibet)
____ *C. b. muroides* — S Tibet to s China (Sichuan and Yunnan west of Mekong River)
____ *C. b. umbratica* — N Myanmar to extreme ne India (Assam) and sw China

☐ **Cetti's Warbler** *Cettia cetti*

____ *C. c. cetti* — S Europe to Asia Minor and North Africa
____ *C. c. orientalis* — Turkey to Crimea, n Iran (Zagros Mts.) and n Afghanistan
____ *C. c. albiventris* — Iran to Kazakstan, Afghanistan, Pakistan and w Xinjiang

☐ **African Bush-Warbler** *Bradypterus baboecala*

____ *B. b. centralis* — Nigeria to s Cameroon, ne Zaire, Burundi, Rwanda, sw Uganda
____ *B. b. chadensis* — W Chad

____	*B. b. sudanensis*	S Sudan
____	*B. b. abyssinicus*	Ethiopia
____	*B. b. elgonensis*	Highlands of w and central Kenya and se Uganda
____	*B. b. tongensis (moreaui)*	SE Kenya to Tanzania, Zambia, Malawi, Mozambique and Natal
____	*B. b. benguellensis*	W Angola
____	*B. b. msiri (bedfordi)*	E Angola to Zambia, se Zaire (Shaba) and ne Botswana
____	*B. b. transvaalensis*	C Zimbabwe to w Swaziland, Transvaal, Lesotho and w Natal
____	*B. b. baboecala*	S South Africa (east to Great Kei River)

☐ **Ja River Scrub-Warbler** *Bradypterus grandis*

Swamps and reedbeds of s Cameroon and Gabon

☐ **White-winged Scrub-Warbler** *Bradypterus carpalis*

Lowland papyrus swamps of ne Zaire, Uganda and Rwanda

☐ **Grauer's Scrub-Warbler** *Bradypterus graueri*

Highland papyrus swamps of e Zaire and adj. Rwanda and Uganda

☐ **Bamboo Scrub-Warbler** *Bradypterus alfredi*

____	*B. a. alfredi*	W Ethiopia to w Uganda and w Zaire
____	*B. a. kungwensis*	W Tanzania and nw Zambia

☐ **Knysna Scrub-Warbler** *Bradypterus sylvaticus*

____	*B. s. pondoensis*	Coastal scrub of e Cape Province and Natal coast
____	*B. s. sylvaticus*	South Africa (Cape Peninsula to Port Elizabeth)

☐ **Cameroon Scrub-Warbler** *Bradypterus lopezi*

____	*B. l. camerunensis*	Mt. Cameroon
____	*B. l. lopezi*	Bioko (Gulf of Guinea)
____	*B. l. barakae*	Zaire, Rwanda and sw Uganda
____	*B. l. boultoni*	Angola
____	*B. l. mariae*	Kenya to ne Tanzania
____	*B. l. usambarae*	S Kenya (Taita Hills) to sw Tanzania, n Malawi, n Mozambique
____	*B. l. ufipae*	SW Tanzania to se Zaire and n Zambia
____	*B. l. granti*	Malawi (south of Nyika) to n Mozambique (Mt. Chiperone)

☐ **African Scrub-Warbler** *Bradypterus barratti*

____	*B. b. priesti*	E Zimbabwe and adjacent s Mozambique
____	*B. b. barratti*	SW Mozambique to e Transvaal, Swaziland, n Zululand, n Natal
____	*B. b. cathkinensis*	Drakensberg Mts. (e Griqualand to Natal/Transvaal border)
____	*B. b. godfreyi (wilsoni)*	Natal and Cape Province

☐ **Bangwa Scrub-Warbler** *Bradypterus bangwaensis*

E Nigeria (Obudu Plateau) and adjacent s Cameroon

☐ **Cinnamon Bracken-Warbler** *Bradypterus cinnamomeus*

____	*B. c. cinnamomeus*	Ethiopia to Kenya, Uganda, Burundi, Rwanda and n Tanzania
____	*B. c. cavei*	SE Sudan
____	*B. c. mildbreadi*	Ruwenzori Mountains (Zaire/Uganda border)
____	*B. c. nyassae*	N Tanzania to se Zaire, ne Zambia and Malawi

☐ **Victorin's Scrub-Warbler** *Bradypterus victorini*

Mountains of South Africa (s Cape Province)

☐ **Spotted Bush-Warbler** *Bradypterus thoracicus*

____	*B. t. suschkini*	N Altai Mountains to sw Transbaikalia and ne Baikal
____	*B. t. davidi*	SE Transbaikalia and w Amurland to Manchuria and n Liaoning
____	*B. t. kashmirensis*	NW Himalayas (Kashmir to Kumaon)
____	*B. t. thoracicus*	Himalayas (Nepal to Assam), se Tibet and sw China
____	*B. t. przevalskii*	Mountains of w China to se Tibet and n Myanmar
____	*B. t. shanensis*	Mountains of n Myanmar; winters to lowlands of Thailand

☐ **Long-billed Bush-Warbler** *Bradypterus major*

____	*B. m. major*	Mountains of Uzbekistan and Xinjiang to w Himalayas (Kashmir)
____	*B. m. innae*	Extreme w China (Kunlun Shan Mountains)

☐ **Chinese Bush-Warbler** *Bradypterus tacsanowskius*

E Siberia to se Tibet and s China; winters to n Indochina

☐ **Russet Bush-Warbler** *Bradypterus seebohmi*

____ *B. s. idoneus*   SE Tibet to n Thailand and s Vietnam
____ *B. s. melanorhynchus*   Mountains of se China (n Guangdong, nw Fujian) and Taiwan
____ *B. s. seebohmi*   Mountains of n Luzon (n Philippines)
____ *B. s. montis*   Mountains of e Java
____ *B. s. timoriensis*   Mountains of Timor (e Lesser Sundas)

☐ **Brown Bush-Warbler** *Bradypterus luteoventris*

____ *B. l. luteoventris*   Himalayas of e Nepal to ne India (Assam), se Tibet and sw China
____ *B. l. ticehursti*   S Myanmar and n Thailand

☐ **Taiwan Bush-Warbler** *Bradypterus alishanensis*

Montane forests of Taiwan

☐ **Ceylon Bush-Warbler** *Bradypterus palliseri*

Dwarf bamboo montane forests of Sri Lanka

☐ **Friendly Bush-Warbler** *Bradypterus accentor*

Montane forests of n Borneo (Mt. Kinabalu and Mt. Trus Madi)

☐ **Long-tailed Bush-Warbler** *Bradypterus caudatus*

____ *B. c. caudatus*   Mountains of n Luzon (n Philippines)
____ *B. c. malindangensis*   Mt. Malindang on nw Mindanao (s Philippines)
____ *B. c. unicolor*   Mt. Apo on s-central Mindanao (s Philippines)

☐ **Chestnut-backed Bush-Warbler** *Bradypterus castaneus*

____ *B. c. castaneus*   Mountains of Sulawesi
____ *B. c. disturbans*   Buru (s Moluccas)
____ *B. c. musculus*   Seram (s Moluccas)

☐ **Brown Emu-tail** *Dromaeocercus brunneus*

Humid rainforests of e Madagascar

☐ **Gray Emu-tail** *Dromaeocercus seebohmi*

Humid grassy swamps of e Madagascar

☐ **Black-capped Rufous-Warbler** *Bathmocercus cerviniventris*

Sierra Leone to se Guinea, Liberia, Ivory Coast and Ghana

☐ **Black-faced Rufous-Warbler** *Bathmocercus rufus*

____ *B. r. rufus*   S Cameroon to Gabon and Central African Republic
____ *B. r. vulpinus (jacksoni)*   E Zaire to extreme s Sudan, Uganda, Kenya and Tanzania

☐ **Mrs. Moreau's Warbler** *Bathmocercus winifredae*

Montane forests of e Tanzania (Uluguru Mountains)

☐ **Aldabra Brush-Warbler** *Nesillas aldabrana*

Formerly Aldabra (Comoro Islands). Possibly extinct

☐ **Anjouan Brush-Warbler** *Nesillas longicaudata*

Forest undergrowth of Anjouan (Comoro Islands)

☐ **Madagascar Brush-Warbler** *Nesillas typica*

____ *N. t. moheliensis*   Mohéli (Comoro Islands)
____ *N. t. obscura*   NW Madagascar
____ *N. t. typica*   Central and e Madagascar
____ *N. t. lantzii*   Subdesert of sw Madagascar

☐ **Grand Comoro Brush-Warbler** *Nesillas brevicaudata*

Forest undergrowth of Grand Comoro I.

☐ **Moheli Brush-Warbler** *Nesillas mariae*

Forests of Mohéli (Comoro Islands)

☐ **Thamnornis** *Thamnornis chloropetoides*

Subdesert of sw Madagascar

☐ **Moustached Grass-Warbler** *Melocichla mentalis*

____ *M. m. mentalis*   Senegambia to Gabon, s Zaire, Angola, Zambia and n Malawi

____ *M. m. amauroura (atricauca, chyulu, granviki)*	S Sudan to sw Ethiopia, Kenya, Tanzania and Zambia
____ *M. m. orientalis*	E Kenya to s Tanzania, s Malawi, Zambia, Zimbabwe, Mozambique
____ *M. m. incana*	NE Tanzania (Mt. Meru)
____ *M. m. luangwae*	Zambia

☐ **Cape Grassbird *Sphenoeacus afer***

____ *S. a. excisus*	E Zimbabwe and adjacent sw Mozambique
____ *S. a. natalensis*	Kwazulu-Natal to w Swaziland, n Lesotho and Transvaal
____ *S. a. intermedius*	Lesotho to Transkei and Port Elizabeth
____ *S. a. afer*	SW Cape Province

☐ **Lanceolated Warbler *Locustella lanceolata***

E Palearctic; winters to s Asia, Greater Sundas and Philippines

☐ **Grasshopper Warbler *Locustella naevia***

____ *L. n. naevia*	Europe to e Russia and Crimea Pen.; winters to n and w Africa
____ *L. n. obscurior*	Caucasus to Georgia and n Armenia; winters to ne Africa
____ *L. n. straminea*	W Siberia to w China (Tien Shan Mountains of w Xinjiang)
____ *L. n. mongolica*	Kazakstan to Afghanistan and w Mongolia; winters to n India

☐ **Pallas' Warbler *Locustella certhiola***

____ *L. c. rubescens*	N Siberia to Sea of Okhotsk and Kamchatka; winters to s India
____ *L. c. sparsimstriata*	S Siberia to n Altai Mts., Sayan Mts. and Transbaicalia
____ *L. c. certhiola (minor)*	S Siberia to Manchuria and Sea of Japan; winters to India
____ *L. c. centralasiae*	SE Siberia to ne China; winters to Andaman and Nicobar is.

☐ **Middendorff's Warbler *Locustella ochotensis***

____ *L. o. subcerthiola*	Kamchatka Peninsula and n Kuril Islands; winters to Philippines
____ *L. o. ochotensis*	E Siberia to n Japan; winters to Philippines, Borneo and Sulawesi

☐ **Pleske's Warbler *Locustella pleskei***

E Siberia to Korea, Kyushu and Izu Islands; winters in s China

☐ **Eurasian River Warbler *Locustella fluviatilis***

Central and e Europe to w Siberia; winters in e Africa

☐ **Savi's Warbler *Locustella luscinioides***

____ *L. l. luscinioides*	Central and e Europe to Iberian Pen. and N Africa; winters to Sudan
____ *L. l. sarmatica*	Ukraine and Sea of Azov to Volga and s Urals; winters ne Africa
____ *L. l. fusca*	Turkey and Jordan to central Asia; winters to Sudan and Ethiopia

☐ **Gray's Warbler *Locustella fasciolata***

NE Asia; winters to Philippines, Indonesia and w New Guinea

☐ **Sakhalin Warbler *Locustella amnicola***

Sakhalin, s Kuril Islands and Hokkaido; winters in Philippines

☐ **Moustached Warbler *Acrocephalus melanopogon***

____ *A. m. melanopogon*	Mediterranean basin (Europe and North Africa)
____ *A. m. mimicus*	W Turkey to s Russia, Iraq, Iran and Afghanistan
____ *A. m. albiventris*	Russia (e coast of Sea of Azov to lower Don River)

☐ **Aquatic Warbler *Acrocephalus paludicola***

Mainly open marshes of w Palearctic; winters in Africa

☐ **Sedge Warbler *Acrocephalus schoenobaenus***

Palearctic region; winters to s Africa

☐ **Streaked Reed-Warbler *Acrocephalus sorghophilus***

Breeds ne China (Liaoning to Hubei); winters in Philippines

☐ **Black-browed Reed-Warbler *Acrocephalus bistrigiceps***

____ *A. b. bistrigiceps*	E Siberia to n China and n Manchuria; winters to s Asia
____ *A. b. tangorum*	N Manchuria; winters to s China and Thailand

☐ **Paddyfield Warbler *Acrocephalus agricola***

Central Eurasia; winters in s Asia

☐ **Blunt-winged Warbler** *Acrocephalus concinens*
  _____ *A. c. haringtoni*
  _____ *A. c. stevensi*
  _____ *A. c. concinens*

Mountains of n Afghanistan to nw India (Kashmir) and Pakistan
Plains of Brahmaputra River (Assam) and adjacent Myanmar
N and central China; winters to s Myanmar and s Thailand

☐ **Eurasian Reed-Warbler** *Acrocephalus scirpaceus*
  _____ *A. s. scirpaceus*
  _____ *A. s. fuscus*

NW Africa and Europe to Crimea and Volga R; > w and c Africa
E Mediterranean and Caspian to Kazakstan; winters to s Africa

☐ **African Reed-Warbler** *Acrocephalus baeticatus*
  _____ *A. b. guiersi*
  _____ *A. b. cinnamomeus (fraterculus, hopsoni)*
  _____ *A. b. hallae*
  _____ *A. b. avicenniae*
  _____ *A. b. suahelicus*
  _____ *A. b. baeticatus*

N Senegal
Senegal to s Sudan, Ethiopia and Somalia south to Mozambique
SW Angola to Namibia, sw Botswana, sw Zambia and Malawi
Mangroves of coastal Sudan, Eritrea, Somalia and w Arabia
Coastal Tanzania to Mozambique and Natal
N Botswana to Transvaal, Natal, e and s Cape Province

☐ **Blyth's Reed-Warbler** *Acrocephalus dumetorum*

E Palearctic; winters in India and Sri Lanka

☐ **Marsh Warbler** *Acrocephalus palustris*

Europe to central Russia; winters to coastal se Africa (Natal)

☐ **Great Reed-Warbler** *Acrocephalus arundinaceus*
  _____ *A. a. arundinaceus*
  _____ *A. a. zarudnyi*

Europe to w Siberia, Turkey, n Iran and nw Africa; winters Africa
N Iraq and Iran to s Afghanistan, Altai, nw Mongolia and w China

☐ **Oriental Reed-Warbler** *Acrocephalus orientalis*

SE Siberia to n China; winters s Asia, n Australia and Philippines

☐ **Clamorous Reed-Warbler** *Acrocephalus stentoreus*
  _____ *A. s. stentoreus*
  _____ *A. s. brunnescens*
  _____ *A. s. amyae*
  _____ *A. s. meridionalis*
  _____ *A. s. harterti*
  _____ *A. s. celebensis*
  _____ *A. s. siebersi*
  _____ *A. s. lentecaptus*
  _____ *A. s. sumbae*

Egypt to Sinai Peninsula, Levant and Jordan
Arabia and Iran to Mongolia and India; winters to India
Plains of Brahmaputra River (Assam) to Myanmar and sw China
Sri Lanka
Philippines (Luzon, Mindoro, Leyte, Bohol and Mindanao)
S Sulawesi
W Java
SE Borneo, Java and w Lesser Sundas (Lombok and Sumbawa)
Buru I. (s Moluccas) and e Lesser Sundas (Sumba and Timor)

☐ **Large-billed Reed-Warbler** *Acrocephalus orinus*

Known from an 1867 specimen from n India (Himachal Pradesh)

☐ **Basra Reed-Warbler** *Acrocephalus griseldis*

S Iraq (Tigris and Euphrates valleys); winters Kenya to Malawi

☐ **Australian Reed-Warbler** *Acrocephalus australis*
  _____ *A. a. toxopei*
  _____ *A. a. carterae*
  _____ *A. a. australis*
  _____ *A. a. gouldi*

New Guinea, Bismarck Archipelago and Solomon Islands
NW Australia
Eastern half of Australia; winters to north
SW Western Australia(Pilbara to Esperance); winters to north

☐ **Nightingale Reed-Warbler** *Acrocephalus luscinius*
  _____ *A. l. luscinius*
  _____ *A. l. yamashinae*
  _____ *A. l. nijoi*

Mariana Islands (Guam, Agrihan, Alamagan and Saipan)
Pagan (n Mariana Islands)
Aguijan (n Mariana Islands)

☐ **Caroline Reed-Warbler** *Acrocephalus syrinx*

Caroline Is. (Woleai, Lamotrek, Truk, Pohnpei, Nukuoro, Kosrae)

☐ **Nauru Reed-Warbler** *Acrocephalus rehsei*

Nauru I. (Melanesia)

☐ **Millerbird** *Acrocephalus familiaris*
  _____ *A. f. kingi*
  _____ *A. f. familiaris†*

Nihoa (w Hawaiian Islands)
Laysan (w Hawaiian Islands). Extirpated ca 1923

☐ **Christmas Island Warbler** *Acrocephalus aequinoctialis*
____ *A. a. aequinoctialis*   Kiritimati (n Line Islands in central Pacific)
____ *A. a. pistor*   N Line Islands (Teraina and Tabuaeran)

☐ **Tahiti Reed-Warbler** *Acrocephalus caffer*
____ *A. c. caffer*   Tahiti (Society Islands)
____ *A. c. garretti*   Huahine (Society Islands). Possibly extinct
____ *A. c. longirostris*   Moorea (Society Islands).

☐ **Tuamotu Reed-Warbler** *Acrocephalus atyphus*
____ *A. a. atyphus*   Islands in nw Tuamotu Archipelago
____ *A. a. ravus*   Islands in se Tuamotu Archipelago
____ *A. a. palmarum*   Anaa I. (Tuamotu Archipelago)
____ *A. a. niauensis*   Niau I. (Tuamotu Archipelago)
____ *A. a. eremus*   Makatea I. (Tuamotu Archipelago)
____ *A. a. flavidus*   Napuka I. (Tuamotu Archipelago)

☐ **Rimitara Reed-Warbler** *Acrocephalus rimitarae*
                                Rimitara (Tubuai Islands)

☐ **Pitcairn Reed-Warbler** *Acrocephalus vaughani*
                                Pitcairn I. (s Polynesia)

☐ **Henderson Island Reed-Warbler** *Acrocephalus taiti*
                                Henderson I. (sw Pitcairn Islands)

☐ **Marquesan Reed-Warbler** *Acrocephalus mendanae*
____ *A. m. percernis*   Nukuhiva I. (Marquesas Islands)
____ *A. m. consobrina*   Mohotane I. (Marquesas Islands)
____ *A. m. mendanae*   Marquesas Islands (Hivaoa and Tahuata)
____ *A. m. fatuhivae*   Fatuhiva I. (Marquesas Islands)
____ *A. m. idae*   Uahuka I. (Marquesas Islands)
____ *A. m. dido*   Uapou I. (Marquesas Islands)
____ *A. m. aquilonis*   Eiao I. (Marquesas Islands)
____ *A. m. postremus*   Hatutu I. (Marquesas Islands)

☐ **Cook Islands Reed-Warbler** *Acrocephalus kerearako*
____ *A. k. kaoko*   Mitiaro I. (Cook Islands)
____ *A. k. kerearako*   Mangaia I. (Cook Islands)

☐ **Greater Swamp-Warbler** *Acrocephalus rufescens*
____ *A. r. senegalensis*   Senegal
____ *A. r. rufescens*   Ghana to Nigeria, Cameroon, Cabinda and nw Zaire; Bioko
____ *A. r. chadensis*   Lake Chad environs
____ *A. r. ansorgei (niloticus, foxi)*   S Sudan to Uganda, e Zaire, w Kenya, n Botswana and nw Angola

☐ **Cape Verde Swamp-Warbler** *Acrocephalus brevipennis*
                                Santiago (Cape Verde Is.). Extirpated on São Nicolau and Brava

☐ **Lesser Swamp-Warbler** *Acrocephalus gracilirostris*
____ *A. g. neglectus*   W Chad
____ *A. g. jacksoni*   S Sudan to w Kenya, Uganda and adjacent Zaire
____ *A. g. tsanae*   NW Ethiopia (Lake Tana)
____ *A. g. leptorhynchus*   Ethiopia to Kenya, Zambia, Zimbabwe, e Transvaal, coastal Natal
____ *A. g. parvus*   Highlands of sw Ethiopia to Kenya, n Tanzania, Rwanda, Burundi
____ *A. g. cunenensis*   SW Angola to n Namibia, n Botswana, sw Zambia, w Zimbabwe
____ *A. g. winterbottomi*   N and nw Zambia to e Angola and sw Tanzania
____ *A. g. gracilirostris (zuluensis)*   S Mozambique to se Zimbabwe and South Africa

☐ **Madagascar Swamp-Warbler** *Acrocephalus newtoni*
                                Aquatic habitats of Madagascar

☐ **Thick-billed Warbler** *Acrocephalus aedon*
____ *A. a. aedon*   S Siberia to w Mongolia; winters to Myanmar, Thailand, Indonesia
____ *A. a. stegmanni*   E Siberia to Mongolia; winters to se China, Thailand and Indochina

☐ **Rodrigues Brush-Warbler** *Acrocephalus rodericanus*

Rodrigues (e Mascarene Islands). Seriously endangered

☐ **Seychelles Brush-Warbler** *Acrocephalus sechellensis*

Cousin (Seychelles Islands)

☐ **Booted Warbler** *Hippolais caligata*

Breeds e Palearctic; winters to India and Sri Lanka

☐ **Sykes' Warbler** *Hippolais rama*

Arabia to Turkestan and w China; winters in s India and Sri Lanka

☐ **Eastern Olivaceous Warbler** *Hippolais pallida*
____ *H. p. elaeica*
____ *H. p. pallida*
____ *H. p. reiseri*
____ *H. p. laeneni*

SE Europe to Iran and sw Asia; winters to ne Africa
Egypt; winters to s Sudan and Ethiopia
Algerian Sahara, s Morocco, Mauritania and Libya
Niger, Chad and Nigeria to w Sudan

☐ **Western Olivaceous Warbler** *Hippolais opaca*

Iberian Peninsula, Morocco, Tunisia, Algeria and Libya

☐ **Upcher's Warbler** *Hippolais languida*

S-central Asia; winters in ne Africa and s Arabia

☐ **Olive-tree Warbler** *Hippolais olivetorum*

Balkan Peninsula and Asia Minor; winters in e and se Africa

☐ **Melodious Warbler** *Hippolais polyglotta*

SW Palearctic; winters in savanna of w Africa

☐ **Icterine Warbler** *Hippolais icterina*

Central Europe to w Siberia and n Iran; winters to s Africa

☐ **African Yellow Warbler** *Chloropeta natalensis*
____ *C. n. batesi*
____ *C. n. massaica*
____ *C. n. major*
____ *C. n. natalensis*

Nigeria to n Zaire and sw Sudan
SE Sudan and Ethiopia to e Zaire, Uganda, Kenya and s Tanzania
Gabon and Angola to s Zaire and n Zambia
S Zambia and s Tanzania to e South Africa

☐ **Mountain Yellow Warbler** *Chloropeta similis*

Mountains of e Zaire to se Sudan, Kenya, Tanzania and n Malawi

☐ **Papyrus Yellow Warbler** *Chloropeta gracilirostris*
____ *C. g. gracilirostris*
____ *C. g. bensoni*

E Zaire to Kenya, w Uganda and Burundi
NE Zambia (mouth of Luapula River)

☐ **Buff-bellied Warbler** *Phyllolais pulchella*

E Nigeria to Chad, s Sudan, Ethiopia, Zaire, Kenya and n Tanzania

☐ **African Tailorbird** *Orthotomus metopias*
____ *O. m. metopias (pallidus)*
____ *O. m. altus*

Montane forests of e Tanzania to nw Mozambique (Njesi Plateau)
E Tanzania (Uluguru Mountains)

☐ **Long-billed Tailorbird** *Orthotomus moreaui*
____ *O. m. moreaui*
____ *O. m. sousae*

NE Tanzania (Usambara Mountains)
W-central Mozambique (Njesi Plateau)

☐ **Mountain Tailorbird** *Orthotomus cuculatus*
____ *O. c. coronatus*
____ *O. c. thais*
____ *O. c. cuculatus*
____ *O. c. cinereicollis*
____ *O. c. philippinus*
____ *O. c. viridicollis*
____ *O. c. riedeli*
____ *O. c. stentor*
____ *O. c. meisei*
____ *O. c. hedymeles*
____ *O. c. batjanensis*
____ *O. c. dumasi*
____ *O. c. everetti*

Mts. of e Nepal to ne India, sw China, n Thailand, Laos, Vietnam
Mountains of peninsular Thailand (south of Isthmus of Kra)
Sumatra, Java and Bali
Mts. of ne Borneo (Kinabalu to Mulu and Tama Abu Range)
N Luzon (n Philippines)
Mountains of Palawan (sw Philippines)
N Sulawesi
N-central and se Sulawesi
S-central Sulawesi (Latimojong Mountains)
S Sulawesi (Mt. Lompobatang) and Taliabu I. (Sula Islands)
Mountains of Bacan I. (n Moluccas)
S Moluccas (Buru and Seram)
Flores (w Lesser Sundas)

☐ **Common Tailorbird** *Orthotomus sutorius*

____	*O. s. guzuratus*	Pakistan and peninsular India
____	*O. s. patia*	*Terai* of Nepal to ne India and Myanmar
____	*O. s. luteus*	NE India (ne Assam) to n Myanmar
____	*O. s. sutorius*	Plains and foothills of Sri Lanka
____	*O. s. fernandonis*	Central highlands of Sri Lanka
____	*O. s. inexpectatus*	E Myanmar, Laos, Yunnan and n Thailand
____	*O. s. longicauda*	SE China, Hainan and ne Indochina
____	*O. s. maculicollis*	SE Myanmar to s Malay Pen. and s Indochina
____	*O. s. edela*	Java

☐ **Rufous-headed Tailorbird** *Orthotomus heterolaemus*

Mountains of Mindanao (s Philippines)

☐ **Dark-necked Tailorbird** *Orthotomus atrogularis*

____	*O. a. nitidus*	NE India (Assam) to Myanmar, s China, Thailand and Indochina
____	*O. a. atrogularis*	Malaysia to Sumatra, Bangka I., Belitung I. and Borneo
____	*O. a. humphreysi*	N and e Borneo
____	*O. a. anambensis*	Tioman, Natuna and Anambas islands (South China Sea)
____	*O. a. chloronotos*	N Luzon (n Philippines)

☐ **Philippine Tailorbird** *Orthotomus castaneiceps*

____	*O. c. castaneiceps*	Philippines (Masbate, Panay, Guimaras, Bantayan and Ticao)
____	*O. c. rabori*	Negros (Philippines)

☐ **Rufous-fronted Tailorbird** *Orthotomus frontalis*

____	*O. f. frontalis*	S Philippines (Samar, Leyte, Dinagat, Bohol and Mindanao)
____	*O. f. mearnsi*	Basilan (s Philippines)

☐ **Gray-backed Tailorbird** *Orthotomus derbianus*

____	*O. d. derbianus*	Luzon (n Philippines); single specimen from Palawan
____	*O. d. nilesi*	Catanduanes (n Philippines)

☐ **Rufous-tailed Tailorbird** *Orthotomus sericeus*

____	*O. s. hesperius*	Myanmar to Thailand, Malaysia, Sumatra, Riau and Lingga archs.
____	*O. s. sericeus*	Borneo
____	*O. s. rubicundulus*	Sirhassen I. (South Natuna Islands)
____	*O. s. nuntius*	SW Philippines (Balabac, Palawan, Cagayan Sulu and Sulu Arch.)

☐ **Ashy Tailorbird** *Orthotomus ruficeps*

____	*O. r. cineraceus*	S Myanmar to Malaysia, Indochina, Sumatra and adjacent islands
____	*O. r. baeus*	Nias and Pagai islands (off w Sumatra)
____	*O. r. concinnus*	Siberut and Sipoura islands (off w Sumatra)
____	*O. r. ruficeps*	Coastal mangroves of Java
____	*O. r. palliolatus*	Kangean and Karimunjawa islands (Java Sea)
____	*O. r. baweanus*	Bawean I. (Java Sea)
____	*O. r. borneoensis*	Borneo
____	*O. r. cagayanensis*	Cagayan Sulu (sw Philippines)

☐ **Olive-backed Tailorbird** *Orthotomus sepium*

____	*O. s. sundaicus*	Panaitan I. (off w Java)
____	*O. s. sepium*	Lowlands of Java, Bali, Madura and Lombok

☐ **Yellow-breasted Tailorbird** *Orthotomus samarensis*

Philippines (Bohol, Leyte and Samar)

☐ **White-browed Tailorbird** *Orthotomus nigriceps*

S Philippines (Mindanao, Dinagat and Siargao)

☐ **White-eared Tailorbird** *Orthotomus cinereiceps*

____	*O. c. obscurior*	Mindanao (s Philippines)
____	*O. c. cinereiceps*	Basilan (s Philippines)

☐ **White-tailed Warbler** *Poliolais lopezi*
    ____ *P. l. manengubae*
    ____ *P. l. alexanderi*
    ____ *P. l. lopezi*

SE Nigeria and s Cameroon (Mt. Manenguba and Mt. Kupé)
Mt. Cameroon
Bioko (Gulf of Guinea)

☐ **Grauer's Warbler** *Graueria vittata*

Dense montane forests of e Zaire, sw Uganda and w Rwanda

☐ **Salvadori's Eremomela** *Eremomela salvadorii*

E Zaire to se Gabon, central plateau of Angola and w Zambia

☐ **Yellow-vented Eremomela** *Eremomela flavicrissalis*

S Ethiopia to s Somalia, ne Uganda and se Kenya

☐ **Yellow-bellied Eremomela** *Eremomela icteropygialis*
    ____ *E. i. alexanderi*
    ____ *E. i. griseoflava (karamojensis, crawfurdi)*
    ____ *E. i. abdominalis*
    ____ *E. i. polioxantha*
    ____ *E. i. puellula*
    ____ *E. i. icteropygialis*
    ____ *E. i. helenorae (viriditincta)*
    ____ *E. i. perimacha*
    ____ *E. i. saturatior (sharpei)*

Senegambia to Sudan (Darfur and Kordofan)
Ethiopia to Somalia, e Uganda, Rwanda, w Kenya and Tanzania
Kenya to n Tanzania
S Zaire to Tanzania, sw Zimbabwe, e Transvaal and Mozambique
SW Angola
Namibia and w Botswana
Caprivi Strip to Zimbabwe, sw Zambia and w Mozambique
S Botswana to w Transvaal and nw Cape Province)
South Africa (Orange Free State to Cape Province)

☐ **Senegal Eremomela** *Eremomela pusilla*

Senegambia to Cameroon, sw Chad and nw Central African Rep.

☐ **Green-backed Eremomela** *Eremomela canescens*
    ____ *E. c. canescens*
    ____ *E. c. elegans*
    ____ *E. c. abyssinica*
    ____ *E. c. elgonensis*

Central African Rep. to Chad, s Sudan, Uganda and w Kenya
Sudan (Darfur and Kordofan to Sennar)
Eritrea to Ethiopia and Sudan
W Kenya (Mt. Elgon to s Nandi Hills)

☐ **Greencap Eremomela** *Eremomela scotops*
    ____ *E. s. pulchra (extrema)*
    ____ *E. s. congensis (angolensis)*
    ____ *E. s. citriniceps*
    ____ *E. s. kikuyuensis*
    ____ *E. s. scotops (chlorochlamys, occipitalis)*

Angola , Zaire, Zambia, w Malawi, Transvaal, Natal
SE Gabon to Congo, nw Zaire and n Angola
Uganda to w Kenya and w Tanzania
Highlands of central Kenya
E Kenya to Tanzania, Botswana, Zimbabwe, Mozambique, Natal

☐ **Yellow-rumped Eremomela** *Eremomela gregalis*
    ____ *E. g. damarensis*
    ____ *E. g. gregalis (albigularis)*

Namibia (Oösop region on Swakop River)
Arid *karoo* of s Namibia and nw Cape Province

☐ **Rufous-crowned Eremomela** *Eremomela badiceps*
    ____ *E. b. fantiensis*
    ____ *E. b. badiceps*
    ____ *E. b. latukae*

Sierra Leone to w Nigeria
Nigeria to n Angola and w Uganda; Bioko
S Sudan

☐ **Turner's Eremomela** *Eremomela turneri*
    ____ *E. t. kalindei*
    ____ *E. t. turneri*

E-central Zaire and extreme sw Uganda (Nyondo Forest)
W Kenya (Mt. Elgon, Kakamega Forest and s Nandi Hills)

☐ **Black-necked Eremomela** *Eremomela atricollis*

Brachystegia woodlands of Angola to se Zaire and Zambia

☐ **Burnt-neck Eremomela** *Eremomela usticollis*
    ____ *E. u. rensi*
    ____ *E. u. usticollis (baumgarti)*

S Zambia to s Malawi and Mozambique (north of Save River)
S Angola to Namibia, Zimbabwe, Mozambique and South Africa

☐ **Rand's Warbler** *Randia pseudozosterops*

Rainforests of e Madagascar

☐ **Dark Newtonia** *Newtonia amphichroa*

Humid highland forests of e Madagascar

434

☐ **Common Newtonia** *Newtonia brunneicauda*
_____ *N. b. brunneicauda* — Wooded areas of Madagascar
_____ *N. b. monticola* — Central Madagascar (Mt. Ankarata)

☐ **Archbold's Newtonia** *Newtonia archboldi*

Subdesert of sw Madagascar

☐ **Red-tailed Newtonia** *Newtonia fanovanae*

Locally in rainforests of e Madagascar

☐ **Cryptic Warbler** *Cryptosylvicola randriansoloi*

Rainforests of e Madagascar

☐ **Green Crombec** *Sylvietta virens*
_____ *S. v. flaviventris* — Senegambia to sw Nigeria
_____ *S. v. virens* — SE Nigeria to Cameroon, Gabon and central Zaire
_____ *S. v. tando (meridionalis)* — Congo to s Zaire and nw Angola
_____ *S. v. baraka* — E Zaire to Kenya

☐ **Lemon-bellied Crombec** *Sylvietta denti*
_____ *S. d. hardyi* — Sierra Leone to Ghana (race in Gambia and Nigeria unknown)
_____ *S. d. denti* — S Cameroon to Zaire

☐ **White-browed Crombec** *Sylvietta leucophrys*
_____ *S. l. leucophrys* — W Kenya and w Uganda (Kibale Forest and Ruwenzori Mts.)
_____ *S. l. chloronota (arileuca)* — SW Uganda (Kigezi) to e Zaire and w Tanzania
_____ *S. l. chapini* — E Zaire (Lendu Plateau)

☐ **Northern Crombec** *Sylvietta brachyura*
_____ *S. b. brachyura* — Senegambia and Sierra Leone to Sudan and n Eritrea
_____ *S. b. carnapi (dilutior)* — Cameroon to Uganda and w Kenya
_____ *S. b. leucopsis* — S Eritrea to Ethiopia, se Sudan, Somalia, Kenya and Tanzania

☐ **Short-billed Crombec** *Sylvietta philippae*

Acacia steppes of nw Somalia and adjacent Ethiopia

☐ **Red-capped Crombec** *Sylvietta ruficapilla*
_____ *S. r. rufigenis* — Lower Congo inland to Kasai and se Gabon
_____ *S. r. schoutedeni* — E Zaire (L. Tanganyika to Marungu and Mt. Kabobo) to se Gabon
_____ *S. r. gephyra* — S Zaire (w Shaba) to Zambia and Zimbabwe
_____ *S. r. chubbi* — S Zaire (se Shaba) to Malawi and n Mozambique
_____ *S. r. makayii* — Interior of n Angola
_____ *S. r. ruficapilla* — Central Angola to s Zaire (sw Shaba)

☐ **Red-faced Crombec** *Sylvietta whytii*
_____ *S. w. loringi (abayensis)* — Sudan and Ethiopia to n Uganda, w Kenya and ne Tanzania
_____ *S. w. jacksoni* — S and e Uganda to sw Kenya, w Tanzania and n Malawi
_____ *S. w. minima* — Coastal e Kenya and Tanzania
_____ *S. w. whytii (nemorivaga)* — Coastal s Tanzania to Mozambique, Zimbabwe and s Malawi

☐ **Somali Crombec** *Sylvietta isabellina*

Dry acacia steppes of Ethiopia, Somalia and n Kenya

☐ **Cape Crombec** *Sylvietta rufescens*
_____ *S. r. adelphe* — Zaire to Zambia and n Malawi
_____ *S. r. ansorgei* — Coastal Angola (Benguela to Luanda)
_____ *S. r. flecki (mossamedes, ochrocara)* — S Angola to e Namibia, e Botswana, sw Zambia and Zimbabwe
_____ *S. r. pallida* — SE Zambia to Mozambique, Zimbabwe, Malawi and Transvaal
_____ *S. r. rufescens* — S Botswana to sw Transvaal and w Cape Province
_____ *S. r. diverga* — S Transvaal to Orange Free State and e Cape Province
_____ *S. r. resurga* — Natal

☐ **Neumann's Warbler** *Hemitesia neumanni*

Montane forests of e Zaire, sw Uganda and w Rwanda

☐ **Kemp's Longbill** *Macrosphenus kempi*
_____ *M. k. kempi*   Sierra Leone to sw Nigeria
_____ *M. k. flammeus*   SE Nigeria and w Cameroon

☐ **Yellow Longbill** *Macrosphenus flavicans*
_____ *M. f. flavicans*   SE Nigeria to Cameroon, Angola and w Zaire; Bioko
_____ *M. f. hypochondriacus*   E Zaire to Uganda, Central African Rep. and extreme sw Sudan

☐ **Gray Longbill** *Macrosphenus concolor*

S Guinea and Sierra Leone to extreme ne Angola and sw Uganda

☐ **Pulitzer's Longbill** *Macrosphenus pulitzeri*

Escarpment of w Angola (Vila Nova do Seles to Chingoroi area)

☐ **Kretschmer's Longbill** *Macrosphenus kretschmeri*
_____ *M. k. kretschmeri*   Extreme se Kenya to central Tanzania
_____ *M. k. griseiceps*   SE Tanzania to ne Mozambique

☐ **Bocage's Longbill** *Amaurocichla bocagei*

Locally in forests of s São Tomé

☐ **Green Hylia** *Hylia prasina*
_____ *H. p. prasina*   Senegambia to Angola, Zaire, s Sudan, w Kenya and nw Tanzania
_____ *H. p. poensis*   Bioko (Gulf of Guinea)

☐ **White-browed Tit-Warbler** *Leptopoecile sophiae*
_____ *L. s. sophiae*   Mountains of central Asia to Pakistan and nw India
_____ *L. s. stoliczkae*   S-central Asia to w Gobi Desert
_____ *L. s. major*   Kazakstan (e Tien Shan Mountains) to w China (Xinjiang)
_____ *L. s. obscura*   SE Tibet to s China (se Xinjiang, Sichuan and s Gansu)

☐ **Crested Tit-Warbler** *Leptopoecile elegans*

Coniferous forests of n-central China to Tibet and Sichuan

☐ **Red-faced Woodland-Warbler** *Phylloscopus laetus*
_____ *P. l. laetus*   E Zaire (Lendu Plateau and Ruwenzori Mountains) to Burundi
_____ *P. l. schoutedeni*   E Zaire (Mt. Kabobo)

☐ **Laura's Wood-Warbler** *Phylloscopus laurae*
_____ *P. l. laurae*   W Angola (Mt. Moco)
_____ *P. l. eustacei*   SE Zaire to nw Zambia and sw Tanzania

☐ **Yellow-throated Wood-Warbler** *Phylloscopus ruficapilla*
_____ *P. r. minullus*   SE Kenya and e Tanzania
_____ *P. r. ochrogularis*   W Tanzania (Kungwe-Mahale Mountains)
_____ *P. r. johnstoni*   Malawi to nw Mozambique, ne Zambia and s Tanzania
_____ *P. r. alacris*   E Zimbabwe and w Mozambique (Mt. Gorongoza)
_____ *P. r. quelimanensis*   N Mozambique (Mt. Namuli)
_____ *P. r. ruficapilla (ochraceiceps)*   South Africa (e Transvaal and Natal)
_____ *P. r. voelckeri*   South Africa (e and s Cape Province)

☐ **Uganda Wood-Warbler** *Phylloscopus budongoensis*

Primary forests of Gabon to ne Zaire, Uganda and w Kenya

☐ **Brown Woodland-Warbler** *Phylloscopus umbrovirens*
_____ *P. u. yemenensis*   SW Arabian Peninsula
_____ *P. u. umbrovirens*   Eritrea to Ethiopia and nw Somalia
_____ *P. u. omoensis*   W and s Ethiopia
_____ *P. u. williamsi*   N Somalia (Erigave district)
_____ *P. u. mackensianus*   S Sudan to e Uganda and central Kenya
_____ *P. u. dorcadichroa*   SE Kenya to n Tanzania
_____ *P. u. alpinus*   Ruwenzori Mountains
_____ *P. u. wilhelmi*   E Zaire to Rwanda and sw Uganda (Kivu Volcanoes)
_____ *P. u. fugglescouchmani*   E Tanzania (Uluguru Mountains)

□ **Black-capped Woodland-Warbler** *Phylloscopus herberti*
____ *P. h. camerunensis* — Highlands of se Nigeria and w Cameroon
____ *P. h. herberti* — Bioko (Gulf of Guinea)

□ **Willow Warbler** *Phylloscopus trochilus*
____ *P. t. acredula* — Scandinavia to Siberia (Yenisey River); winters to w Africa
____ *P. t. trochilus* — W Europe to s Poland and Romania; winters to w Africa
____ *P. t. yakutensis* — E Siberia (Taymyr Peninsula to Anadyr River); winters to s Africa

□ **Common Chiffchaff** *Phylloscopus collybita*
____ *P. c. abietinus* — Scandinavia to Urals, Caucasus, Transcaucasia and n Iran
____ *P. c. collybita* — Denmark to Pyrénées, Poland and Romania; winters to n Africa
____ *P. c. brevirostris* — Highlands of w Turkey and Black Sea coastlands of n Turkey
____ *P. c. caucasicus* — East of range of *brevirostris* at lower elevations south to Armenia
____ *P. c. menzbieri* — Mts. of ne Iran, e Elburz and Khorasan Mts. n to adj. Turkmenia
____ *P. c. tristis* — Ural Mountains to ne Iran, n India and Bangladesh

□ **Canary Islands Chiffchaff** *Phylloscopus canariensis*
____ *P. c. canariensis* — Canary Is. (La Palma, Hierro, Gomera, Tenerife and Gran Canaria)
____ *P. c. exsul* — Lanzarote (ne Canary Islands). Possibly extinct

□ **Iberian Chiffchaff** *Phylloscopus ibericus*
____ *P. i. biscayensis* — N Portugal and n Spain to extreme France Basque country
____ *P. i. ibericus* — Central and s Portugal to sw Spain (Andalucia)

□ **Mountain Chiffchaff** *Phylloscopus sindianus*
____ *P. s. lorenzii* — SW Asia (e Turkey to Caucasus, Transcaucasia and ne Iran)
____ *P. s. sindianus* — Extreme w China (sw Xinjiang) to n Pakistan and n India

□ **Plain Leaf-Warbler** *Phylloscopus neglectus* — Oak-juniper woodlands of Iran to Afghanistan and Kashmir

□ **Western Bonelli's Warbler** *Phylloscopus bonelli* — West and central Europe; winters to sahel of n Africa

□ **Eastern Bonelli's Warbler** *Phylloscopus orientalis* — Balkans, Turkey and Levant; winters to Sudan

□ **Wood Warbler** *Phylloscopus sibilatrix* — Breeds Europe and Russia; winters in tropical Africa

□ **Dusky Warbler** *Phylloscopus fuscatus*
____ *P. f. fuscatus* — Siberia to Mongolia and w China; winters to India and Indochina
____ *P. f. robustus* — N China (south of Gobi Desert) to n Sichuan; winters to Indochina
____ *P. f. weigoldi* — Mts. of s Tibet to e Himalayas and sw China; winters to ne India

□ **Smoky Warbler** *Phylloscopus fuligiventer*
____ *P. f. fuligiventer* — Himalayas of Nepal to Sikkim, Bhutan, sw Tibet and ne India
____ *P. f. tibetanus* — SE Tibet to s China (sw Xinjiang); winters to ne India

□ **Tickell's Leaf-Warbler** *Phylloscopus affinis* — Mts. of n India to s China, se Tibet, Myanmar and Thailand

□ **Buff-throated Warbler** *Phylloscopus subaffinis* — Alpine scrub of central and s China to se Tibet

□ **Sulphur-bellied Warbler** *Phylloscopus griseolus* — Alpine scrub of s Asia; winters in India

□ **Yellow-streaked Warbler** *Phylloscopus armandii*
____ *P. a. armandii* — Mts. of Mongolia to e China; winters to Myanmar and n Laos
____ *P. a. perplexus* — SE Tibet to n Myanmar and sw China (Sichuan, Hubei, Yunnan)

□ **Radde's Warbler** *Phylloscopus schwarzi* — Breeds ne Asia; winters in SE Asia

□ **Buff-barred Warbler** *Phylloscopus pulcher*
____ *P. p. kangrae* — Montane oak-rhododendron forests of nw Himalayas
____ *P. p. pulcher* — Nepal to Tibet, sw China and n Myanmar; winters to n Thailand

☐ **Ashy-throated Warbler** *Phylloscopus maculipennis*
_____ *P. m. virens*   N India (Kashmir and n Punjab to Arunachal Pradesh)
_____ *P. m. maculipennis*   E Himalayas to se Tibet, sw China, n Myanmar and n Indochina

☐ **Lemon-rumped Warbler** *Phylloscopus proregulus*

Coniferous forests and *taiga* of e Asia; > to Indochina

☐ **Gansu Leaf-Warbler** *Phylloscopus kansuensis*

Mountains of w China (Qinghai and Gansu)

☐ **Pale-rumped Warbler** *Phylloscopus chloronotus*
_____ *P. c. chloronotus*   Himalayas of Pakistan to se Tibet, ne India and s-central China
_____ *P. c. simlaensis*   NW Himalayas (Afghanistan to w Nepal)

☐ **Sichuan Leaf-Warbler** *Phylloscopus forresti*

Mts. of southern China

☐ **Chinese Leaf-Warbler** *Phylloscopus yunnanensis*

Mountains of central China (Sichuan, Liaoning and Shanxi)

☐ **Brooks' Leaf-Warbler** *Phylloscopus subviridis*

Coniferous forests of Turkestan, ne Afghanistan and n India

☐ **Yellow-browed Warbler** *Phylloscopus inornatus*

Ural Mts. to Sea of Okhotsk, Mongolia, Manchuria and Korea

☐ **Hume's Warbler** *Phylloscopus humei*
_____ *P. h. humei*   Sayan and Altai Mts. to nw Himalayas; winters to SE Asia
_____ *P. h. mandellii*   S Tibet to Sikkim, Myanmar and sw China; winters to n Thailand

☐ **Arctic Warbler** *Phylloscopus borealis*
_____ *P. b. borealis (talovka)*   Scandinavia to Chukotsk Pen; winters to se China and Philippines
_____ *P. b. transbaicalicus*   E Siberia to n Mongolia; winters to SE Asia
_____ *P. b. xanthodryas*   Sea of Okhotsk to Kamchatka Pen., Kuril Is., Hokkaido and Honshu
_____ *P. b. hylebata*   E Amurland to n Manchuria, Ussuriland and North Korea
_____ *P. b. kennicotti*   W Alaska; winters in Philippines

☐ **Greenish Warbler** *Phylloscopus trochiloides*
_____ *P. t. viridanus*   NE Europe to central Asia and Afghanistan; winters to s India
_____ *P. t. nitidus*   Caucasus to n Turkey, n Iran and nw Afghanistan; winters s India
_____ *P. t. plumbeitarsus*   S Siberia to Mongolia and Manchuria; winters to SE Asia
_____ *P. t. trochiloides*   Himalayas to Tibet and w China; winters n India to Indochina
_____ *P. t. ludlowi*   W Himalayas (Gilgit and Kashmir to Kumaon); winters to s India
_____ *P. t. obscuratus*   NW China to Tibet; winters to Myanmar, Thailand and Indochina

☐ **Pale-legged Leaf-Warbler** *Phylloscopus tenellipes*

Breeds river valleys of ne Asia; winters in SE Asia

☐ **Sakhalin Leaf-Warbler** *Phylloscopus borealoides*

Sakhalin, Kuril Islands and Hokkaido (n Japan)

☐ **Large-billed Leaf-Warbler** *Phylloscopus magnirostris*

Kashmir to s Tibet and s China; winters to India and Myanmar

☐ **Tytler's Leaf-Warbler** *Phylloscopus tytleri*

Coniferous forests of Pakistan and n India; winters to Myanmar

☐ **Western Crowned Leaf-Warbler** *Phylloscopus occipitalis*

Mountains of e Afghanistan and Kashmir; winters in India

☐ **Eastern Crowned Leaf-Warbler** *Phylloscopus coronatus*

Siberia and n China; winters in SE Asia and Greater Sundas

☐ **Ijima's Leaf-Warbler** *Phylloscopus ijimae*

Izu Islands (s Japanese Archipelago); > n Philippines

☐ **Blyth's Leaf-Warbler** *Phylloscopus reguloides*
_____ *P. r. kashmiriensis*   Himalayas of nw India (Kashmir to Garhwal)
_____ *P. r. reguloides*   Himalayas of ne India to Nepal, s Tibet and s China (sw Sichuan)
_____ *P. r. assamensis*   NE India (Assam) to n Myanmar and sw China (nw Yunnan)
_____ *P. r. claudiae*   Mts. of w China to e Tibet; > to s China and SE Asia
_____ *P. r. fokiensis*   S China (w Hubei, Guizhou, Guangxi, nw Fujian and Anhui)
_____ *P. r. goodsoni*   Breeding range unknown; > on Hainan (s China)
_____ *P. r. ticehursti*   S Vietnam (Langbian Plateau)

☐ **Hainan Leaf-Warbler** *Phylloscopus hainanus*

Hainan (s China)

☐ **Emei Leaf-Warbler** *Phylloscopus emeiensis*

SW China (Mt. Emei Shan in Sichuan)

☐ **White-tailed Leaf-Warbler** *Phylloscopus davisoni*
____ *P. d. davisoni*
____ *P. d. disturbans*
____ *P. d. ogilviegranti*
____ *P. d. intensior*
____ *P. d. klossi*

Extreme sw China to e Myanmar, n Thailand, n Laos and Tonkin
S China (Sichuan to se Yunnan, n Guizhou and se Hunan)
SE China (nw Fujian and adjacent Guangdong)
SE Thailand (Trat Province) to mountains of n Cambodia
Mountains of s Laos and s Vietnam

☐ **Yellow-vented Warbler** *Phylloscopus cantator*
____ *P. c. cantator*
____ *P. c. pernotus*

Mts. of Sikkim and ne India to Myanmar; winters to nw Thailand
N Laos

☐ **Sulphur-breasted Warbler** *Phylloscopus ricketti*

Mountains of s China; winters to Laos and s Vietnam

☐ **Lemon-throated Warbler** *Phylloscopus cebuensis*
____ *P. c. luzonensis*
____ *P. c. sorsogonensis*
____ *P. c. cebuensis*

N and central Luzon (n Philippines)
S Luzon (s Philippines)
Philippines (Cebu and Negros)

☐ **Mountain Warbler** *Phylloscopus trivirgatus*
____ *P. t. parvirostris*
____ *P. t. trivirgatus*
____ *P. t. kinabaluensis*
____ *P. t. sarawacensis*
____ *P. t. benguetensis*
____ *P. t. nigrorum*
____ *P. t. diuatae*
____ *P. t. mindanensis*
____ *P. t. malindangensis*
____ *P. t. flavostriatus*
____ *P. t. peterseni*

Malaya
Sumatra, Java, Bali, Lombok, Sumbawa and nw Borneo
Mountains of ne Borneo (Mt. Kinabalu)
Mountains of w Borneo (Poi Mountains of w Sarawak)
N Luzon (n Philippines)
Philippines (s Luzon, Negros and Mindoro)
Diuata Mountains of ne Mindanao (s Philippines)
Mindanao (Mt. Apo and Mt. Mayo)
Mindanao (Mt. Malindang and Zamboanga Peninsula)
Mindanao (Mt. Katanglad and mts. of Misamis Oriental Prov.)
Mountains of Palawan (sw Philippines)

☐ **Sulawesi Leaf-Warbler** *Phylloscopus sarasinorum*
____ *P. s. nesophilus*
____ *P. s. sarasinorum*

Mountains of n Sulawesi
Mountains of s Sulawesi

☐ **Timor Leaf-Warbler** *Phylloscopus presbytes*
____ *P. p. floris*
____ *P. p. presbytes*

Flores (w Lesser Sundas)
Timor (e Lesser Sundas)

☐ **Island Leaf-Warbler** *Phylloscopus poliocephalus*
____ *P. p. henrietta*
____ *P. p. waterstradti*
____ *P. p. everetti*
____ *P. p. ceramensis*
____ *P. p. avicola*
____ *P. p. maforensis*
____ *P. p. misoriensis*
____ *P. p. poliocephalus*
____ *P. p. albigularis*
____ *P. p. paniaiae*
____ *P. p. cyclopum*
____ *P. p. giulianettii*
____ *P. p. hamlini*
____ *P. p. matthiae*
____ *P. p. moorhousei*
____ *P. p. leletensis*
____ *P. p. becki*

N Moluccas (Halmahera and Ternate)
Moluccas (Bacan and Obi)
Buru (s Moluccas)
S Moluccas (Seram and Ambon)
Kai Besar I. (Kai Islands)
Numfor I. (n New Guinea)
Biak I. (n New Guinea)
NW New Guinea (Tamrau, Arfak and Wandammen mountains)
W-central New Guinea (Weyland Mountains)
W-central New Guinea (Wissel Lakes region)
N New Guinea (Cyclops Mountains)
New Guinea (Snow, Sepik, Saruwaged and Herzog mountains)
Goodenough I. (D'Entrecasteaux Archipelago)
St. Matthias I. (Bismarck Archipelago)
Bismarck Archipelago (New Britain and Umboi)
New Ireland (Bismarck Archipelago)
Solomon Islands (Guadalcanal, Santa Isabel and Malaita)

____	*P. p. bougainvillei*	Bougainville (Solomon Islands)
____	*P. p. pallescens*	Kulambangra (Solomon Islands)

☐ **Philippine Leaf-Warbler** *Phylloscopus olivaceus*

S Philippines (Samar, Leyte, Mindanao, Negros) and Sulu Arch.

☐ **San Cristobal Leaf-Warbler** *Phylloscopus makirensis*

San Cristóbal (s Solomon Islands)

☐ **Kulambangra Leaf-Warbler** *Phylloscopus amoenus*

Montane forests of Kulambangra (central Solomon Islands)

☐ **Golden-spectacled Warbler** *Seicercus burkii*

W Himalayas (Pakistan to Kashmir and Bhutan); winters India

☐ **Gray-crowned Warbler** *Seicercus tephrocephalus*

N Myanmar to sw China and n Thailand; winters to Indochina

☐ **Plain-tailed Warbler** *Seicercus soror*

W and s China; > to Thailand, Cambodia and n Vietnam

☐ **Whistler's Warbler** *Seicercus whistleri*

____	*S. w. whistleri*	Pakistan to n India
____	*S. w. nemoralis*	Nepal to sw China, Myanmar and n Thailand

☐ **Bianchi's Warbler** *Seicercus valentini*

____	*S. v. valentini*	Cent. and sw China; winters to s Yunnan, nw Thailand and n Laos
____	*S. v. latouchei*	Central and e China

☐ **Gray-hooded Warbler** *Seicercus xanthoschistos*

____	*S. x. xanthoschistos*	W Himalayas (nw Pakistan to Kashmir, Nepal and se Tibet)
____	*S. x. jerdoni*	E Himalayas (e Nepal to Sikkim, Bhutan and Arunachal Pradesh)
____	*S. x. tephrodiras*	NE India (Assam, Nagaland and Manipur) to Myanmar
____	*S. x. flavogularis*	NE India (Abor and Mishmi hills) to n Myanmar

☐ **White-spectacled Warbler** *Seicercus affinis*

____	*S. a. affinis*	Nepal to ne India, n Myanmar, sw China, n Laos and s Vietnam
____	*S. a. intermedius*	Mts. of se China (nw Fujian); winters to sw China and Indochina

☐ **Gray-cheeked Warbler** *Seicercus poliogenys*

Montane forests of n India to sw China, Myanmar and Indochina

☐ **Chestnut-crowned Warbler** *Seicercus castaniceps*

____	*S. c. castaniceps*	E Himalayas (Nepal to Sikkim, Bhutan and ne India)
____	*S. c. collinsi*	Myanmar (s Shan States) to nw Thailand
____	*S. c. laurentei*	SW China (se Yunnan)
____	*S. c. sinensis*	S China (Shaanxi, Sichuan, nw Fujian) to n Laos and n Vietnam
____	*S. c. stresemanni*	S Laos (Bolavens Plateau) and sw Cambodia
____	*S. c. youngi*	Mountains of peninsular Thailand (south of Isthmus of Kra)
____	*S. c. annamensis*	Mountains of Vietnam (Langbian and Da Lat plateaux)
____	*S. c. butleri*	Mountains of Malay Peninsula
____	*S. c. muelleri*	W Sumatra (Barisan Mountains)

☐ **Yellow-breasted Warbler** *Seicercus montis*

____	*S. m. davisoni*	High mountains of s Malay Peninsula
____	*S. m. inornatus*	Mountains of Sumatra
____	*S. m. montis*	Mountains of Borneo (Kinabalu to Poi Range)
____	*S. m. xanthopygius*	Mountains of Palawan (sw Philippines)
____	*S. m. floris*	Mountains of Flores (w Lesser Sundas)
____	*S. m. paulinae*	Mountains of Timor (e Lesser Sundas)

☐ **Sunda Warbler** *Seicercus grammiceps*

____	*S. g. sumatrensis*	Mountains of Sumatra
____	*S. g. grammiceps*	Mountains of Java and Bali

☐ **Rufous-faced Warbler** *Abroscopus albogularis*

____	*A. a. albogularis*	Nepal to Sikkim, ne India, Bangladesh, Yunnan and w Myanmar

____	*A. a. fulvifacies*	S China to n Laos and n Vietnam; Hainan
____	*A. a. hugonis*	NW Thailand

☐ **Yellow-bellied Warbler** *Abroscopus superciliaris*

____	*A. s. flaviventris*	Central Nepal to Sikkim, Bhutan, ne India and Bangladesh
____	*A. s. superciliaris*	S China (Yunnan) to Myanmar, w Thailand and nw Laos
____	*A. s. drasticus*	NE India (Arunachal Pradesh) to n Myanmar; winters sw Thailand
____	*A. s. smythiesi*	Myanmar (central Irrawaddy basin from Pakokkuu to Prome)
____	*A. s. euthymus*	N and central Vietnam
____	*A. s. bambusarum*	Peninsular Thailand (Isthmus of Kra to Phangnga)
____	*A. s. sakaiorum*	Malay Peninsula (s Thailand to Negeri Sembilan)
____	*A. s. papilio*	Sumatra
____	*A. s. schwaneri*	Borneo
____	*A. s. vordermani*	Java

☐ **Black-faced Warbler** *Abroscopus schisticeps*

____	*A. s. schisticeps*	Central Nepal to Sikkim and ne India (Darjiling)
____	*A. s. flavimentalis*	SE Tibet to ne India and Myanmar (Chin Hills and Mt. Victoria)
____	*A. s. ripponi*	S China (Sichuan and Yunnan) to e Myanmar and n Vietnam

☐ **Broad-billed Warbler** *Tickellia hodgsoni*

____	*T. h. hodgsoni*	Nepal to ne India and w Myanmar
____	*T. h. tonkinensis*	SW China (se Yunnan) to n Laos and nw Tonkin

☐ **Yellow-bellied Hyliota** *Hyliota flavigaster*

____	*H. f. flavigaster*	Senegal to s Sudan, w Ethiopia, Kenya and Tanzania
____	*H. f. barbozae (marginalis)*	Lake Victoria to Angola, Zambia, Malawi and n Mozambique

☐ **Southern Hyliota** *Hyliota australis*

____	*H. a. slatini*	W Cameroon; ne Zaire to w Uganda and w Kenya
____	*H. a. inornata (pallidipectus)*	Angola to s Zaire (Shaba), Zambia, Malawi and n Mozambique
____	*H. a. australis (rhodesiae)*	Zimbabwe and Mozambique

☐ **Usambara Hyliota** *Hyliota usambarae*

	NE Tanzania (Rubu River to Usambara Mountains)

☐ **Violet-backed Hyliota** *Hyliota violacea*

____	*H. v. nehrkorni*	Liberia to Ghana and Togo
____	*H. v. violacea*	Lowlands of Nigeria and Cameroon to Gabon and e Zaire

☐ **Marsh Grassbird** *Megalurus pryeri*

____	*M. p. sinensis*	Reedbeds of ne China (e Liaoning and ne Hebei); winters se China
____	*M. p. pryeri*	Honshu (Japan)

☐ **Tawny Grassbird** *Megalurus timoriensis*

____	*M. t. tweeddalei*	Philippines (Luzon, Panay, Tablas, Marinduque, Ticao, Negros)
____	*M. t. mindorensis*	Mindoro (Philippines)
____	*M. t. alopex*	Philippines (Bohol, Cebu and Leyte)
____	*M. t. crex*	S Philippines (Mindanao and Camiguin Sur)
____	*M. t. celebensis*	N-central Sulawesi
____	*M. t. amboinensis*	Ambon (s Moluccas)
____	*M. t. inquirendus*	Sumba (Lesser Sundas)
____	*M. t. timoriensis*	Timor (e Lesser Sundas)
____	*M. t. stresemanni*	NW New Guinea (Lake Giji, Arfak Mts. and Wissel Lakes)
____	*M. t. mayri*	N New Guinea (Lake Sentani, Humboldt Bay to Astrolabe Bay)
____	*M. t. wahgiensis*	Central Highlands of New Guinea
____	*M. t. montanus*	Central Highlands of New Guinea (Mt. Hagen and Mt. Wilhelm)
____	*M. t. macrurus*	SE New Guinea
____	*M. t. harterti*	E New Guinea (Huon Peninsula)
____	*M. t. alpinus*	Alpine grasslands of Snow Mountains to se New Guinea
____	*M. t. muscalis*	S New Guinea (Middle Fly River region)

____ *M. t. interscapularis*	Bismarck Arch. (New Britain, New Ireland and New Hanover)
____ *M. t. alisteri*	Coastal n and e Australia (King Sound to Sydney, NSW)

☐ **Little Grassbird** *Megalurus gramineus*

____ *M. g. papuensis*	W New Guinea (Wissel Lakes region)
____ *M. g. goulburni*	Eastern half of Australia (except Cape York Peninsula)
____ *M. g. gramineus*	Tasmania, King and Flinders Islands (Bass Strait)
____ *M. g. thomasi*	Southwest Western Australia (Shark Bay to Esperance)

☐ **Striated Grassbird** *Megalurus palustris*

____ *M. p. toklao*	Pakistan to India, s China, s Myanmar, Thailand and Indochina
____ *M. p. forbesi*	Philippines (Luzon, Mindoro, Panay, Samar and Mindanao)
____ *M. p. palustris*	Borneo and Java

☐ **Fly River Grassbird** *Megalurus albolimbatus*

SE New Guinea (Fly River lowlands)

☐ **Fernbird** *Megalurus punctatus*

____ *M. p. vealeae*	North I. (New Zealand)
____ *M. p. punctatus*	South I. (New Zealand)
____ *M. p. stewartianus*	Stewart I. (New Zealand)
____ *M. p. wilsoni*	Codfish I. (New Zealand)
____ *M. p. caudatus*	Snares I. (New Zealand)

☐ **Brown Songlark** *Cincloramphus cruralis*

Australia (except most north and east coast regions)

☐ **Rufous Songlark** *Cincloramphus mathewsi*

Australia (except n Cape York Pen. and interior W Australia)

☐ **Spinifex-bird** *Eremiornis carteri*

Spinifex regions of n-central Australia (Pilbara to nw Queensland)

☐ **Buff-banded Bushbird** *Buettikoferella bivittata*

Lowland scrub of Timor (e Lesser Sundas)

☐ **New Caledonian Grassbird** *Megalurulus mariei*

Grasslands and open heath of New Caledonia

☐ **Bismarck Thicketbird** *Megalurulus grosvenori*

Known from 2 specimens from New Britain (Bismarck Arch.)

☐ **Bougainville Thicketbird** *Megalurulus llaneae*

Mountains of Bougainville (n Solomon Islands)

☐ **Guadalcanal Thicketbird** *Megalurulus whitneyi*

____ *M. w. whitneyi*	Mountains of Espíritu Santo (Vanuatu)
____ *M. w. turipavae*	Mountains of Guadalcanal (se Solomon Islands)

☐ **Rusty Thicketbird** *Megalurulus rubiginosus*

Scrub of New Britain (Bismarck Archipelago)

☐ **Long-legged Warbler** *Trichocichla rufa*

____ *T. r. rufa*	Viti Levu (Fiji Islands). Possibly extinct
____ *T. r. cluniei*	Vanua Levu (Fiji Islands). Possibly extinct

☐ **Bristled Grassbird** *Chaetornis striata*

Grasslands of Indian subcontinent

☐ **Rufous-rumped Grassbird** *Graminicola bengalensis*

____ *G. b. bengalensis*	W Nepal to n India, Bangladesh and n Myanmar
____ *G. b. sinicus*	S China (e Guangxi and Guangdong)
____ *G. b. striatus*	S Myanmar to central Thailand, n Vietnam and Hainan

☐ **Broad-tailed Grassbird** *Schoenicola platyurus*

Grasslands of sw India (Western Ghats from Mysore to Kerala)

☐ **Fan-tailed Grassbird** *Schoenicola brevirostris*

____ *S. b. alexinae*	Guinea to Ethiopia and n Malawi
____ *S. b. brevirostris*	Malawi to South Africa

□ **Wrentit** *Chamaea fasciata*
____ *C. f. phaea*
____ *C. f. rufula*
____ *C. f. intermedia*
____ *C. f. fasciata*
____ *C. f. henshawi*
____ *C. f. canicauda*

Coastal nw Oregon (Columbia River to California border)
Chaparral of coastal n California (Del Norte to Marin counties)
Chaparral belt of central California (San Francisco region)
Coastal s California (Monterey to San Luis Obispo counties)
Chaparral belt of interior s Oregon to s California (San Diego)
Chaparral belt of nw Baja California

□ **Yemen Warbler** *Sylvia buryi*

Acacia scrub of s Yemen and sw Saudi Arabia

□ **Blackcap** *Sylvia atricapilla*
____ *S. a. gularis (atlantis)*
____ *S. a. heineken*
____ *S. a. atricapilla*
____ *S. a. pauluccii (koenigi)*
____ *S. a. dammholzi*

Cape Verde Islands and Azores
SW Spain, Portugal, Madeira and Canary Islands
Europe to w Siberia and nw Africa; winters to s Africa
Corsica, Sardinia, Balearic Islands, Tunisia, Italy and Sicily
Caucasus, Transcaucasia and n Iran; winters to ne Africa

□ **Garden Warbler** *Sylvia borin*
____ *S. b. borin*
____ *S. b. woodwardi (pallida)*

Br. Isles to Scandinavia, s Urals and Caucasus; winters to s Africa
N European Russia and w Siberia; winters to s Africa

□ **Barred Warbler** *Sylvia nisoria*
____ *S. n. nisoria*
____ *S. n. merzbacheri*

S Scandinavia and Europe to Ural Mountains; winters to e Africa
W Siberia to n Iran, Afghanistan and w China; winters to e Africa

□ **Small Whitethroat** *Sylvia minula*

Deserts of western China

□ **Margelanic Whitethroat** *Sylvia margelanica*

Uzbekistan and Kyrgyzstan to w China (Tien Shan Mountains)

□ **Hume's Whitethroat** *Sylvia althaea*

Mts. of e Iran to Afghanistan and n India; winters to s India

□ **Western Orphean Warbler** *Sylvia hortensis*

SW Europe and North Africa; winters s Mauritania to w Sudan

□ **Eastern Orphean Warbler** *Sylvia crassirostris*
____ *S. c. crassirostris*
____ *S. c. balchanica*
____ *S. c. jerdoni*

Slovenia and Balkan Pen. to Transcaucasia, Turkey and Levant
S Transcaspia, Turkmenistan and Iran
Baluchistan to Pakistan and n to Tadzhikistan and Kyrgyzstan

□ **Asian Desert Warbler** *Sylvia nana*

Caspian Sea to Mongolia and w China; winters to ne Africa, India

□ **Greater Whitethroat** *Sylvia communis*
____ *S. c. communis*
____ *S. c. volgensis*
____ *S. c. icterops*
____ *S. c. rubicola*

W Europe and Scandinavia to North Africa; winters to n Africa
E Europe to w Siberia
W Siberia to Iran and Asia Minor; winters to e Africa
NW China and w Mongolia to Kazakstan; winters to s Africa

□ **Lesser Whitethroat** *Sylvia curruca*
____ *S. c. curruca*
____ *S. c. blythi (affinis)*
____ *S. c. halimodendri*
____ *S. c. caucasica*
____ *S. c. telengitica*
____ *S. c. jaxartica (snigirewskii)*

W Europe to Caucasus and w Siberia; winters to central Africa
E Siberia to n Altai and n Mongolia
Plains of lower Volga to e Kazakstan (Lake Zaysan) and w Altai
Mountains of Balkan Peninsula to w Iran and Caucasus Mountains
Deserts of Russian Altai to w and s Mongolia
Plains of s Transcaspia

□ **African Desert Warbler** *Sylvia deserti*

Deserts of nw Africa to e Libya

□ **Red Sea Warbler** *Sylvia leucomelaena*
____ *S. l. blanfordi*
____ *S. l. somaliensis*

Red Sea coast of Egypt, ne Sudan and Eritrea
Djibouti and n Somalia

_____ *S. l. leucomelaena*   W Saudi Arabia to w Yemen and Oman

_____ *S. l. negevensis*   Arava Valley (Israel-Jordan border)

□ **Cyprus Warbler** *Sylvia melanothorax*

Arid scrub of Cyprus; winters to Near East and ne Egypt

□ **Ménétries' Warbler** *Sylvia mystacea*

_____ *S. m. mystacea*   Transcaucasia and ne Turkey to lower Volga; winters to ne Africa

_____ *S. m. rubescens (semenowi)*   Lebanon and se Turkey to Iraq and sw Iran; winters to ne Africa

_____ *S. m. turcmenica*   E Iran to n Afghanistan and w Tajikistan; winters ne Africa

□ **Spectacled Warbler** *Sylvia conspicillata*

_____ *S. c. orbitalis*   Madeira, Canary Islands and Cape Verde Islands

_____ *S. c. conspicillata*   W Mediterranean basin and nw Africa; winters to Senegal and Niger

□ **Tristram's Warbler** *Sylvia deserticola*

_____ *S. d. maroccana*   Morocco and w Algeria (Haut and Moyen Atlas mountains)

_____ *S. d. deserticola (ticehursti)*   Algeria (Atlas Saharien and Aurès) and adjacent Tunisia

□ **Dartford Warbler** *Sylvia undata*

_____ *S. u. dartfordiensis*   S England, w France, nw Spain and n Portugal

_____ *S. u. undata*   Medit. France, Corsica, Sardinia, Sicily, Balearic Is. and Italy

_____ *S. u. toni*   Iberian Peninsula and coastal Morocco, Algeria, Tunisia

□ **Marmora's Warbler** *Sylvia sarda*

Corsica, Sardinia, Montecristo, Pantelleria, Giannutri and Zembra

□ **Balearic Warbler** *Sylvia balearica*

Balearic Islands except Menorca

□ **Rueppell's Warbler** *Sylvia rueppelli*

Rocky slopes of e Mediterranean region; winters in ne Africa

□ **Subalpine Warbler** *Sylvia cantillans*

_____ *S. c. cantillans*   Mainly coastal and continental Europe, from Iberia east to Italy

_____ *S. c. moltonii*   W Mediterranean islands, including Balearics, Corsica, Sardinia

_____ *S. c. inornata*   Morocco to Tunisia and (?) nw Libya; winters Senegal to w Niger

_____ *S. c. albistriata*   Balkan Peninsula to w Turkey; winters Sahara oases and Arabia

□ **Sardinian Warbler** *Sylvia melanocephala*

_____ *S. m. melanocephala (leucogastra, pasiphae)*   S Europe, Canary and Mediterranean is., w Turkey and N Africa

_____ *S. m. norrisae†*   Formerly Egypt (Faiyum region). Extinct

_____ *S. m. momus*   Syria, Israel, Jordan and Sinai Peninsula; winters to ne Africa

□ **Layard's Warbler** *Parisoma layardi*

_____ *P. l. layardi*   Namibia to nw Cape Province, Namaqualand and middle Orange R.

_____ *P. l. aridicola*   Highlands of Namibia, Damaraland and n Cape Province

_____ *P. l. barnesi*   Highlands of Lesotho and ne Cape Province; w Cape Province

_____ *P. l. subsolanum*   Highlands of sw Cape Province to Orange Free State

□ **Rufous-vented Warbler** *Parisoma subcaeruleum*

_____ *P. s. ansorgei*   Coastal sw Angola

_____ *P. s. cinerascens*   Namibia

_____ *P. s. subcaeruleum*   Botswana to Cape Province and Orange Free State

_____ *P. s. orpheanum*   Zimbabwe to Transvaal, Natal, w Zululand and Lesotho

□ **Brown Warbler** *Parisoma lugens*

_____ *P. l. lugens*   Ethiopia

_____ *P. l. griseiventre*   S Ethiopia (Bale Mountains)

_____ *P. l. jacksoni*   Sudan and Uganda to Kenya, n Tanzania, e Zaire and Malawi

_____ *P. l. prigoginei*   E Zaire (Itombwe Mountains)

_____ *P. l. clarum*   Tanzania (Matengo Highlands)

□ **Banded Warbler** *Parisoma boehmi*

_____ *P. b. somalicum*   Ethiopia and Somalia

____ *P. b. marsabit*	N-central Kenya
____ *P. b. boehmi*	S Kenya and Tanzania

FAMILY: POLIOPTILIDAE (Gnatcatchers—15)

☐ **Collared Gnatwren** *Microbates collaris*

____ *M. c. collaris*	SE Colombia to the Guianas, Suriname and adjacent nw Brazil
____ *M. c. paraguensis*	S Venezuela (Bolívar and Amazonas)
____ *M. c. perlatus*	N Amazonian Brazil and ne Peru (Loreto and n San Martín)

☐ **Tawny-faced Gnatwren** *Microbates cinereiventris*

____ *M. c. semitorquatus*	Caribbean slope of se Nicaragua to w Panama
____ *M. c. magdalenae*	Caribbean slope of extreme e Panama to n Colombia
____ *M. c. cinereiventris*	Pacific coast of Colombia to sw Ecuador (Guayas)
____ *M. c. peruvianus*	Trop. se Colombia (Nariño) to e Ecuador and se Peru (Puno)

☐ **Long-billed Gnatwren** *Ramphocaenus melanurus*

____ *R. m. rufiventris*	Trop. se Mexico (Oaxaca) to Panama, Colombia and e Ecuador
____ *R. m. ardeleo*	SE Mexico (Yucatán Peninsula) and Petén of n Guatemala
____ *R. m. sanctaemarthae*	Caribbean coast of n Colombia to nw Venezuela (Zulia)
____ *R. m. griseodorsalis*	W-central Colombia (Antioquia south to Valle)
____ *R. m. pallidus*	NE Colombia e of Andes to n Venezuela (e Falcón to Miranda)
____ *R. m. trinitatis*	Tropical e Colombia (Meta) to ne Venezuela; Trinidad
____ *R. m. albiventris*	S Venezuela (e Bolívar) to the Guianas and ne Brazil
____ *R. m. duidae*	Tropical ne Ecuador to s Venezuela (Amazonas and Bolívar)
____ *R. m. badius*	SE Ecuador to ne Peru (north of Río Marañón)
____ *R. m. obscurus*	Tropical e Peru (Loreto) to n Bolivia (La Paz)
____ *R. m. amazonum*	E Peru (right bank of upper Río Ucayali) and adjacent nw Brazil
____ *R. m. sticturus*	SW Brazil (Mato Grosso)
____ *R. m. austerus*	E Brazil (e Pará and n Maranhão)
____ *R. m. melanurus*	Coastal ne Brazil (Pernambuco to São Paulo)

☐ **Blue-gray Gnatcatcher** *Polioptila caerulea*

____ *P. c. caerulea*	E and c US to Gulf Coast; winters to ne Mexico and West Indies
____ *P. c. amoenissima*	SW Oregon to n Baja and n Mexico; winters to s Mexico
____ *P. c. obscura*	S Baja California (28°N to Cape District)
____ *P. c. gracilis*	Foothills of nw Mexico (se Sonora)
____ *P. c. nelsoni*	S Mexico (Guerrero to Oaxaca and s Chiapas)
____ *P. c. deppei*	E Mexico (San Luis Potosí to Veracruz, Tabasco and n Chiapas)
____ *P. c. mexicana*	SE Mexico (Yucatán Peninsula)
____ *P. c. cozumelae*	Cozumel I. (off e Mexico)

☐ **Cuban Gnatcatcher** *Polioptila lembeyei*

	Semiarid coastal scrub of e Cuba and Cayo Coco

☐ **California Gnatcatcher** *Polioptila californica*

	SW California to s Baja, Santa Margarita I. and Espírito Santo I.

☐ **Black-tailed Gnatcatcher** *Polioptila melanura*

____ *P. m. lucida*	Arid sw US to ne Baja California and nw Mexico (Durango)
____ *P. m. melanura*	W Nevada to Texas and e Mexico (Tamaulipas, San Luis Potosí)
____ *P. m. pontilis*	Central Baja California from latitude 30°N to 27°N
____ *P. m. margaritae*	S Baja California; Santa Margarita I. and Espírito Santo I.
____ *P. m. curtata*	Isla Tiburón (Sea of Cortés)

☐ **Black-capped Gnatcatcher** *Polioptila nigriceps*

____ *P. n. restricta*	Extreme s Arizona to nw Mexico (Sonora and Chihuahua))
____ *P. n. nigriceps*	Arid w Mexico (n Sinaloa to Durango, Jalisco and Colima)

☐ **White-lored Gnatcatcher** *Polioptila albiloris*

____ *P. a. vanrossemi*	Arid w and s Mexico (Michoacán and Guerrero to s Chiapas)

____ *P. a. albiventris*	SE Mexico (extreme n Yucatán Peninsula)
____ *P. a. albiloris*	Interior of Guatemala to nw Costa Rica

☐ **Guianan Gnatcatcher** *Polioptila guianensis*

____ *P. g. facilis*	S Venezuela (Amazonas) to extreme ne Brazil (upper Rio Negro)
____ *P. g. guianensis*	Guyana, Suriname and French Guiana
____ *P. g. paraensis*	E Brazil (Manaus area and from lower Rio Tapajós to Belém)

☐ **Iquitos Gnatcatcher** *Polioptila clementsi*

White-sands forests adjacent to Iquitos, Peru

☐ **Tropical Gnatcatcher** *Polioptila plumbea*

____ *P. p. brodkorbi*	Lowlands of se Mexico (s Veracruz) to e Nicaragua
____ *P. p. superciliaris*	SE Mexico (Quintana Roo and Campeche) to Panama
____ *P. p. cinericia*	Panama (Coiba and Pearl islands)
____ *P. p. bilineata*	N Colombia to w Peru (n Lima)
____ *P. p. maranonica*	W Peru (Río Marañón Valley from Piura south to Lima)
____ *P. p. plumbiceps*	E slope of Andes of n Colombia to n Venezuela; Isla Margarita
____ *P. p. anteocularis*	N Colombia (upper Magdalena Valley)
____ *P. p. daguae*	Colombia (upper Río Dagua and upper Río Patía)
____ *P. p. innotata*	Extreme e Colombia to s Venezuela and extreme n Brazil
____ *P. p. plumbea*	The Guianas and ne Brazil (Rio Tapajós to n Maranhão)
____ *P. p. maior*	Trop. e Peru (upper Río Marañón) from Piura to La Libertad
____ *P. p. parvirostris*	Trop. e Peru (upper Amazon, Río Huallaga and Río Marañón)
____ *P. p. atricapilla*	NE Brazil (Maranhão to Piauí, Ceará, Pernambuco and Bahia)

☐ **Creamy-bellied Gnatcatcher** *Polioptila lactea*

Lowlands of se Brazil to e Paraguay and ne Argentina

☐ **Slate-throated Gnatcatcher** *Polioptila schistaceigula*

Tropical forests of e Panama to nw Ecuador

☐ **Masked Gnatcatcher** *Polioptila dumicola*

____ *P. d. berlepschi*	Interior e Brazil (n Goiás, se Pará to Mato Grosso) and e Bolivia
____ *P. d. dumicola*	Extreme s Brazil to Bolivia, Paraguay, Uruguay and n Argentina
____ *P. d. saturata*	Highlands of Bolivia (Cochabamba)

FAMILY: MUSCICAPIDAE (Old World Flycatchers—275)

☐ **Silverbird** *Empidornis semipartitus*

S Sudan to n Ethiopia, Uganda. w Kenya and w Tanzania

☐ **Pale Flycatcher** *Bradornis pallidus*

____ *B. p. pallidus*	Senegambia to n Zaire, s Sudan and w Ethiopia
____ *B. p. parvus*	SW Ethiopia to e Sudan, e Zaire and nw Uganda
____ *B. p. bowdleri*	Eritrea to central Ethiopia
____ *B. p. bafirawari*	S Ethiopia and ne Kenya
____ *B. p. duyerali*	NE Ethiopia (Duyer Ali) to central Somalia (El Bur)
____ *B. p. subalaris*	Coastal e Kenya to ne Tanzania
____ *B. p. erlangeri*	S Somalia (Bardera and Serenli to Hanole)
____ *B. p. modestus (nigeriae)*	Guinea to se Mali and Central African Republic
____ *B. p. murinus*	Gabon to Congo, Angola, w Kenya, n Botswana, nw Zimbabwe
____ *B. p. aquaemontis*	Central Namibia (Waterberg Plateau)
____ *B. p. griseus*	SE Kenya to c Tanzania, e Zambia, e Zimbabwe and n Malawi
____ *B. p. divisus*	SE Zambia to Mozambique, n Transvaal and ne Swaziland
____ *B. p. sibilans*	Mozambique (south of Sul do Save) to n Natal

☐ **Chat Flycatcher** *Bradornis infuscatus*

____ *B. i. benguellensis*	Arid coastal sw Angola (Benguela) to nw Namibia (Kaokoveld)
____ *B. i. namaquensis*	Namibia
____ *B. i. placidus*	Botswana to w Transvaal, nw Orange Free State and n Cape Prov.
____ *B. i. seimundi*	South Africa (n Cape Province to sw Orange Free State)
____ *B. i. infuscatus*	SW Namibia to sw Cape Province

☐ **Mariqua Flycatcher** *Bradornis mariquensis*
____ *B. m. acaciae* — Savanna of s Angola to sw Botswana and n Cape Province
____ *B. m. mariquensis* — S Botswana to w Zimbabwe and w Transvaal
____ *B. m. territinctus* — NE Namibia and nw Botswana

☐ **African Gray Flycatcher** *Bradornis microrhynchus*
____ *B. m. neumanni* — SE Sudan to s Ethiopia, central Somalia, n Kenya and ne Uganda
____ *B. m. pumilus* — Central Ethiopia to n Somalia
____ *B. m. burae* — E Kenya to se Somalia
____ *B. m. microrhynchus* — SW Kenya to w Tanzania and ne Zambia
____ *B. m. taruensis* — SE Kenya

☐ **Angola Slaty-Flycatcher** *Melaenornis brunneus*
____ *M. b. brunneus* — Northern end of western escarpment of Angola
____ *M. b. bailunduensis* — Angola (Mt. Moco and central highlands)

☐ **White-eyed Slaty-Flycatcher** *Melaenornis fischeri*
____ *M. f. fischeri* — Mountains of se Sudan to Uganda, Kenya and ne Tanzania
____ *M. f. toruensis* — Highlands of sw Uganda and e Zaire to Rwanda and Burundi
____ *M. f. nyikensis (ufipae)* — E Zaire (Marungu Highlands) to Tanzania and Malawi
____ *M. f. semicinctus* — Highlands of e Zaire (west of Lake Albert)

☐ **Abyssinian Slaty-Flycatcher** *Melaenornis chocolatinus*
____ *M. c. chocolatinus* — High plateau of s Eritrea to w and central Ethiopia
____ *M. c. reichenowi* — Highlands of w Ethiopia (Wallegha to Gimirra)

☐ **Northern Black-Flycatcher** *Melaenornis edolioides*
____ *M. e. edolioides* — Savanna of Senegambia to Mali, Sierra Leone and Cameroon
____ *M. e. lugubris* — E Cameroon to w Ethiopia, nw Zaire, Uganda, w Kenya, Tanzania
____ *M. e. schistaceus* — Eritrea and e Ethiopia to n Kenya (Moyale)

☐ **Southern Black-Flycatcher** *Melaenornis pammelaina*
____ *M. p. ater* — SE Botswana to Malawi, e Zimbabwe, Mozambique, e S Africa
____ *M. p. pammelaina* — S Tanzania to se Malawi and Mozambique (n of Sul do Save)
____ *M. p. diabolicus* — S Angola to n Namibia and nw Botswana
____ *M. p. tropicalis* — Zaire to Uganda,, Rwanda, Kenya and w Tanzania
____ *M. p. poliogygna* — Angola to Caprivi Strip, Zimbabwe, nw Malawi and sw Tanzania

☐ **Yellow-eyed Black-Flycatcher** *Melaenornis ardesiacus*
— Mountains of e Zaire to Rwanda, Burundi and sw Uganda

☐ **Nimba Flycatcher** *Melaenornis annamarulae*
— Humid forests of e Sierra Leone to Liberia and s Ivory Coast

☐ **African Forest-Flycatcher** *Fraseria ocreata*
____ *F. o. kelsalli* — Humid forests of Sierra Leone
____ *F. o. prosphora* — Liberia to Ghana
____ *F. o. ocreata* — Nigeria and Cameroon to Angola, Zaire and w Uganda; Bioko

☐ **White-browed Forest-Flycatcher** *Fraseria cinerascens*
____ *F. c. cinerascens* — Senegal and Gambia to Ghana
____ *F. c. ruthae* — S Nigeria to Cameroon, Zaire and Cabinda

☐ **Fiscal Flycatcher** *Sigelus silens*
— SE Botswana and South Africa; > to Mozambique

☐ **Buru Jungle-Flycatcher** *Rhinomyias additus*
— Forests of Buru (s Moluccas)

☐ **Flores Jungle-Flycatcher** *Rhinomyias oscillans*
____ *R. o. oscillans* — Flores and Sumbawa (w Lesser Sundas)
____ *R. o. stresemanni* — Sumba (w Lesser Sundas)

☐ **Brown-chested Jungle-Flycatcher** *Rhinomyias brunneatus*
____ *R. b. brunneatus* — Breeds se China; > to Malaysia, Thailand and Nicobar Is.

447

_____ _R. b. nicobaricus_   S China (Guangxi); > to Nicobar Islands

☐ **Gray-chested Jungle-Flycatcher** _Rhinomyias umbratilis_

S pen. Thailand, Malaysia, Sumatra, Borneo, Java; N Natuna Is.

☐ **Fulvous-chested Jungle-Flycatcher** _Rhinomyias olivaceus_
_____ _R. o. olivaceus_   N Myanmar, peninsular Thailand, Sumatra, Java and Borneo
_____ _R. o. perolivaceus_   North Natuna Islands

☐ **Chestnut-tailed Jungle-Flycatcher** _Rhinomyias ruficauda_
_____ _R. r. samarensis_   Philippines (Leyte, Samar and e Mindanao)
_____ _R. r. boholensis_   Bohol (Philippines)
_____ _R. r. zamboanga_   W Mindanao (s Philippines)
_____ _R. r. ruficauda_   Basilan (s Philippines)
_____ _R. r. ocularis_   Sulu Archipelago (Pangamican and Tawitawi)
_____ _R. r. ruficrissa_   N Borneo (Mt. Kinabalu)
_____ _R. r. isola_   Montane forests of Sarawak (n Borneo)

☐ **Henna-tailed Jungle-Flycatcher** _Rhinomyias colonus_
_____ _R. c. subsolanus_   E Sulawesi
_____ _R. c. pelingensis_   Peleng I. (Banggai Islands off Sulawesi)
_____ _R. c. colonus_   Sula Islands (Taliabu, Seho, Mangole and Sanana)

☐ **Eyebrowed Jungle-Flycatcher** _Rhinomyias gularis_

Mountains of n Borneo (Mt. Kinabalu and adjacent mountains)

☐ **Rusty-flanked Jungle-Flycatcher** _Rhinomyias insignis_

Montane forests of n Luzon (n Philippines)

☐ **Negros Jungle-Flycatcher** _Rhinomyias albigularis_

Philippines (lowlands of Negros and Guimaras)

☐ **Mindanao Jungle-Flycatcher** _Rhinomyias goodfellowi_

Montane forests of Mt. Apo on Mindanao (s Philippines)

☐ **Spotted Flycatcher** _Muscicapa striata_
_____ _M. s. striata_   Europe to N Africa, Siberia and Asia Minor; > to s Africa
_____ _M. s. neumanni_   E Siberia to Caucasus, s China and s Asia; > to e Africa
_____ _M. s. balearica_   Balearic Islands; > to w and sw Africa
_____ _M. s. tyrrhenica_   Corsica and Sardinia
_____ _M. s. sarudnyi_   Caucasus Mts. to n Iran and Afghanistan; > to East Africa
_____ _M. s. inexpectata_   Crimean Peninsula
_____ _M. s. mongola_   SE Altai to n Mongolia to se Transbaikalia

☐ **Gambaga Flycatcher** _Muscicapa gambagae_

Semiarid s Mali to Ghana, Kenya, Somalia and sw Arabia

☐ **Gray-streaked Flycatcher** _Muscicapa griseisticta_

SE Siberia to ne China; > to New Guinea and Philippines

☐ **Dark-sided Flycatcher** _Muscicapa sibirica_
_____ _M. s. sibirica_   SE Siberia to Japan; > to Indochina and Greater Sundas
_____ _M. s. gulmergi_   W Himalayas (e Afghanistan to Kashmir and Garhwal)
_____ _M. s. cacabata_   E Himalayas to se Tibet and ne India; > to s Thailand
_____ _M. s. rothschildi_   Mts. of w China to n Myanmar; > to Malaysia and Indochina

☐ **Asian Brown Flycatcher** _Muscicapa dauurica_

Siberia to Japan, s China and India; > to Greater Sundas

☐ **Brown-streaked Flycatcher** _Muscicapa williamsoni_
_____ _M. w. williamsoni_   S Myanmar to pen. Thailand, Malaya, s Vietnam and Sumatra
_____ _M. w. siamensis_   N plateau of Thailand and Vietnam
_____ _M. w. umbrosa_   NE Borneo (Sabah)

☐ **Ashy-breasted Flycatcher** _Muscicapa randi_

N Philippines (Luzon and Negros)

☐ **Sumba Brown Flycatcher** _Muscicapa segregata_

Sumba (Lesser Sundas)

☐ **Rusty-tailed Flycatcher** *Muscicapa ruficauda*

Uzbekistan, Tajikistan and e Afghanistan to n India and Nepal

☐ **Brown-breasted Flycatcher** *Muscicapa muttui*

NE India to s China and n Vietnam; > to Sri Lanka

☐ **Ferruginous Flycatcher** *Muscicapa ferruginea*

Nepal to n India, s China and Taiwan; > to Indochina

☐ **Ussher's Flycatcher** *Muscicapa ussheri*

Sierra Leone to Ghana and Nigeria

☐ **Sooty Flycatcher** *Muscicapa infuscata*
_____ *M. i. infuscata (chapini)*   Nigeria to nw Angola, Central African Republic and w Zaire
_____ *M. i. minuscula*   NE Zaire to Uganda, nw Zambia and n Tanzania

☐ **Boehm's Flycatcher** *Muscicapa boehmi*

Angola to Zaire, Zambia, sw Tanzania and n Mozambique

☐ **Swamp Flycatcher** *Muscicapa aquatica*
_____ *M. a. aquatica*   Gambia to sw Sudan and n Zaire
_____ *M. a. infulata (ruandae)*   S Sudan to e Zaire, w Kenya, nw Tanzania and ne Zambia
_____ *M. a. lualabae*   SE Zaire (swamps along Lualaba River)
_____ *M. a. grimwoodi*   S Zambia (Kabwe district, Suye Lake and Lukanga Swamp)

☐ **Olivaceous Flycatcher** *Muscicapa olivascens*
_____ *M. o. olivascens*   Ghana and Nigeria to Zaire
_____ *M. o. nimbae*   Liberia and Ivory Coast

☐ **Chapin's Flycatcher** *Muscicapa lendu*
_____ *M. l. lendu*   Mts. of ne Zaire to sw Uganda and w Kenya (Kakamega Forest)
_____ *M. l. itombwensis*   E Zaire (Itombwe Mountains)

☐ **African Dusky Flycatcher** *Muscicapa adusta*
_____ *M. a. poensis (obscura, albiventris, kumboensis, okuensis)*   Highlands of Cameroon; Bioko (Gulf of Guinea)
_____ *M. a. pumila (grotei, subtilis, interposita, chyulu)*   Mountains of s Sudan to Cameroon, Uganda and n Tanzania
_____ *M. a. minima*   Highlands of Eritrea and ne Ethiopia
_____ *M. a. marsabit*   N Kenya (Mt. Marsabit region)
_____ *M. a. murina (roehli)*   Mountains of se Kenya (Taita Hills) to nw Tanzania
_____ *M. a. fuelleborni*   Highlands of s and central Tanzania
_____ *M. a. subadusta (angolensis)*   Angola to s Zaire, nw Zambia, Zimbabwe and Mozambique
_____ *M. a. mesica*   Zimbabwe (except for eastern highlands)
_____ *M. a. fuscula*   Coastal Transkei to Swaziland, Natal and e Cape Province
_____ *M. a. adusta*   N and e Transvaal to Natal, Swaziland and Cape Province

☐ **Little Gray Flycatcher** *Muscicapa epulata*

Lowlands of se Guinea to Liberia, Gabon and ne Zaire

☐ **Yellow-footed Flycatcher** *Muscicapa sethsmithi*

S Nigeria and Cameroon to Gabon, Zaire and w Uganda; Bioko

☐ **Dusky-blue Flycatcher** *Muscicapa comitata*
_____ *M. c. aximensis*   Sierra Leone and se Guinea to s Nigeria
_____ *M. c. camerunensis*   Mt. Cameroon
_____ *M. c. comitata (stuhlmanni)*   Cameroon to nw Angola, Zaire, Uganda and sw Sudan

☐ **Tessmann's Flycatcher** *Muscicapa tessmanni*

Lowlands of Ivory Coast to s Cameroon and ne Zaire

☐ **Cassin's Flycatcher** *Muscicapa cassini*

Sierra Leone to Angola, w Uganda and extreme n Zambia

☐ **Ashy Flycatcher** *Muscicapa caerulescens*
_____ *M. c. nigrorum*   SE Guinea to Sierra Leone, Ghana and Togo
_____ *M. c. brevicauda*   SE Nigeria to nw Angola, e Zaire, s Sudan and Uganda
_____ *M. c. cinereola*   S Somalia to e Kenya and e Tanzania
_____ *M. c. impavida*   S Zaire to sw Angola, Namibia, w Tanzania and n Mozambique
_____ *M. c. vulturna*   S Malawi to s Zimbabwe, e Transvaal and n Swaziland
_____ *M. c. caerulescens*   Extreme s Mozambique to Natal and e Cape Province

☐ **Gray-throated Tit-Flycatcher** *Myioparus griseigularis*
____ *M. g. parelii* — Liberia (Mt. Nimba) to Ivory Coast and Ghana
____ *M. g. griseigularis* — SE Nigeria to nw Angola, e Zaire, w Uganda and nw Tanzania

☐ **Gray Tit-Flycatcher** *Myioparus plumbeus*
____ *M. p. plumbeus* — Senegambia to nw Angola, s Ethiopia, Uganda and nw Tanzania
____ *M. p. orientalis* — Lowlands of e Kenya to e Tanzania, Mozambique and Natal
____ *M. p. catoleucum (grandior)* — Angola plateau to Namibia, s Zaire, Botswana, Malawi and Natal

☐ **Fairy Flycatcher** *Stenostira scita*
____ *S. s. scita* — W Cape Province; > to s Namibia
____ *S. s. rudebecki* — Lesotho; > to Transvaal
____ *S. s. saturatior* — S Africa (Great and Little Karoo); > to Orange Free State

☐ **Grand Comoro Flycatcher** *Humblotia flavirostris*
Grand Comoro I. (Mt. Karthala)

☐ **European Pied Flycatcher** *Ficedula hypoleuca*
____ *F. h. hypoleuca* — British Isles and n Europe to w Siberia; > to tropical Africa
____ *F. h. iberiae* — Iberian Peninsula; > in west Africa
____ *F. h. sibirica (syn. tomensis)* — *Taiga* of w Siberia (Ural Mts. to Yenisey R.); > to e Africa

☐ **Atlas Flycatcher** *Ficedula speculigera*
Morocco (south to Middle Atlas Mts.), n Algeria and n Tunisia

☐ **Collared Flycatcher** *Ficedula albicollis*
E France to Balkans and Ukraine; > in tropical and s Africa

☐ **Semicollared Flycatcher** *Ficedula semitorquata*
Montane forests of Balkan Pen. to nw Iran; > in e Africa

☐ **Korean Flycatcher** *Ficedula zanthopygia*
Mountains of ne Asia; > in SE Asia and Greater Sundas

☐ **Narcissus Flycatcher** *Ficedula narcissina*
____ *F. n. elisae* — Mts. of ne China (n Liaoning); > to s China and Hainan
____ *F. n. narcissina* — Sakhalin to Japan; > to Philippines and Borneo
____ *F. n. owstoni* — S Ryukyu Islands

☐ **Beijing Flycatcher** *Ficedula beijingnica*
New species described in 1999 from mountains of w Beijing

☐ **Mugimaki Flycatcher** *Ficedula mugimaki*
SE Siberia, Sakhalin and ne China; > to SE Asia, Indonesia

☐ **Slaty-backed Flycatcher** *Ficedula hodgsonii*
Himalayas (Nepal to n Myanmar and sw China); > SE Asia

☐ **Rufous-gorgeted Flycatcher** *Ficedula strophiata*
____ *F. s. strophiata* — Himalayas to s China and n Thailand; > to n Indochina
____ *F. s. fuscogularis* — S Laos (Langbian Plateau)

☐ **Red-breasted Flycatcher** *Ficedula parva*
N Europe to s Urals, Balkans and s Caspian; > to s Asia

☐ **Taiga Flycatcher** *Ficedula albicilla*
Siberia to Kamchatka Pen. and n Mongolia; > India to SE Asia

☐ **Kashmir Flycatcher** *Ficedula subrubra*
Himalayas of n India (Kashmir); > to s India and Sri Lanka

☐ **Snowy-browed Flycatcher** *Ficedula hyperythra*
____ *F. h. hyperythra* — E Himalayas to s China, Myanmar, nw Thailand and n Vietnam
____ *F. h. annamensis* — S China (sw Yunnan) and n Laos (Langbian Plateau)
____ *F. h. sumatrana* — Malay Peninsula, Sumatra and Borneo
____ *F. h. mjobergi* — W Borneo (Poi Mountains)
____ *F. h. vulcani* — Java, Bali and w Lesser Sundas (Lombok, Sumbawa and Flores)
____ *F. h. innexa* — Taiwan
____ *F. h. luzoniensis* — Luzon (n Philippines)
____ *F. h. mindorensis* — Mindoro (Philippines)
____ *F. h. calayensis* — Calayan (Philippines)

450

____ *F. h. nigrorum*	Negros (Philippines)
____ *F. h. montigena*	Mountains of central Mindanao (s Philippines)
____ *F. h. daggayana*	N Mindanao (s Philippines)
____ *F. h. malindangensis*	S Philippines (Mt. Malindang region of nw Mindanao)
____ *F. h. rara*	Palawan (sw Philippines)
____ *F. h. annalisa*	N peninsula of Sulawesi
____ *F. h. jugosae*	Central and s Sulawesi and Taliabu I. (Sula Islands)
____ *F. h. pallidipectus*	Bacan (s Moluccas)
____ *F. h. alifura*	Buru (s Moluccas)
____ *F. h. negroides*	Seram (s Moluccas)
____ *F. h. clarae*	Timor (e Lesser Sundas)
____ *F. h. audacis*	Babar (e Lesser Sundas)

☐ **White-gorgeted Flycatcher** *Ficedula monileger*

____ *F. m. monileger*	Himalayas of Nepal to Bhutan and ne India (Arunachal Pradesh)
____ *F. m. leucops*	NE India to s China (Yunnan), Myanmar, Thailand and n Vietnam
____ *F. m. gularis*	Myanmar (Arakan Yoma Mountains)

☐ **Rufous-browed Flycatcher** *Ficedula solitaris*

____ *F. s. submonileger*	Mountains of se Myanmar to peninsular Thailand and s Vietnam
____ *F. s. malayana*	Mountains of Malay Peninsula
____ *F. s. solitaris*	Mountains of Sumatra

☐ **Rufous-chested Flycatcher** *Ficedula dumetoria*

____ *F. d. muelleri*	Peninsular Thailand, Malaysia, Sumatra and Borneo
____ *F. d. dumetoria*	Java, Bali and w Lesser Sundas (Lombok, Sumbawa and Flores)
____ *F. d. riedeli*	Tanimbar Islands (Larat and Yamdena)

☐ **Rufous-throated Flycatcher** *Ficedula rufigula*

	Lowland rainforests of Sulawesi

☐ **Cinnamon-chested Flycatcher** *Ficedula buruensis*

____ *F. b. buruensis*	Buru (s Moluccas)
____ *F. b. ceramensis*	Seram (s Moluccas)
____ *F. b. siebersi*	Kai Besar (Kai Islands)

☐ **Little Slaty Flycatcher** *Ficedula basilanica*

____ *F. b. samarensis*	Central Philippines (Leyte and Samar)
____ *F. b. basilanica*	S Philippines (Basilan, Dinagat and Mindanao)

☐ **Damar Flycatcher** *Ficedula henrici*

	Damar I. (eastern Lesser Sundas)

☐ **Sumba Flycatcher** *Ficedula harterti*

	Lowlands of Sumba (w Lesser Sundas)

☐ **Palawan Flycatcher** *Ficedula platenae*

	Lowlands of Palawan (sw Philippines)

☐ **Russet-tailed Flycatcher** *Ficedula crypta*

	Submontane forests of Mindanao (s Philippines)

☐ **Furtive Flycatcher** *Ficedula disposita*

	Submontane forests of Luzon (n Philippines)

☐ **Lompobattang Flycatcher** *Ficedula bonthaina*

	SW Sulawesi (Lompobattang massif)

☐ **Little Pied Flycatcher** *Ficedula westermanni*

____ *F. w. collini*	Himalayas (Nepal to Sikkim); > to plains of India
____ *F. w. australorientis*	Himalayas (Bhutan to s China, n Myanmar, Thailand, Indochina)
____ *F. w. langbianis*	S Laos (Langbian Plateau) and Vietnam
____ *F. w. westermanni*	S Thailand, Malaysia, Sumatra, Borneo, Sulawesi and Mindanao
____ *F. w. hasselti*	S Sumatra, Java, Bali, Lombok, Sumbawa, Flores and s Sulawesi
____ *F. w. mayri*	E Lesser Sundas (Timor and Wetar)
____ *F. w. rabori*	N Philippines (Luzon, Negros and Panay)
____ *F. w. palawanensis*	Mountains of s Palawan (sw Philippines)

☐ **Ultramarine Flycatcher** *Ficedula superciliaris*
　____　*F. s. superciliaris*　　　Himalayas (n Pakistan to Nepal and Sikkim); > to c India
　____　*F. s. aestigma*　　　　　Himalayas (Bhutan to se Tibet, sw China, Myanmar and n India)

☐ **Slaty-blue Flycatcher** *Ficedula tricolor*
　____　*F. t. tricolor*　　　　　Himalayas (Kashmir to central Nepal)
　____　*F. t. minuta*　　　　　　Himalayas (e Nepal to se Tibet) and ne India (Arunachal Pradesh)
　____　*F. t. cerviniventris*　　　N India (Manipur Hills) to Myanmar (Chin Hills)
　____　*F. t. diversa*　　　　　　Mountains of s-central China; > to n Indochina

☐ **Black-and-rufous Flycatcher** *Ficedula nigrorufa*
　　　　　　　　　　　　　　　　Mountains of sw India (w Maharashtra south to Kerala)

☐ **Sapphire Flycatcher** *Ficedula sapphira*
　____　*F. s. sapphira*　　　　　Himalayas (e Nepal to se Tibet, sw China and ne India)
　____　*F. s. tienchuanensis*　　　Mountains of w-central China (Sichuan to s Shaanxi)
　____　*F. s. laotiana*　　　　　　Mountains of nw Thailand, n Laos and n Vietnam

☐ **Black-banded Flycatcher** *Ficedula timorensis*
　　　　　　　　　　　　　　　　Timor (e Lesser Sundas)

☐ **Blue-and-white Flycatcher** *Cyanoptila cyanomelana*
　____　*C. c. cumatilis*　　　　　NE Asia; > to Philippines, Indochina and Greater Sundas
　____　*C. c. cyanomelana*　　　　Japan and Korea; > to Myanmar, Thailand and Gr. Sundas

☐ **Verditer Flycatcher** *Eumyias thalassinus*
　____　*E. t. thalassinus*　　　　N Pakistan to s China and Indochina; > to pen. India
　____　*E. t. thalassoides*　　　　Peninsular Thailand, Malaya, Sumatra and (rarely) Borneo

☐ **Dull-blue Flycatcher** *Eumyias sordidus*
　　　　　　　　　　　　　　　　Forested uplands of Sri Lanka

☐ **Island Flycatcher** *Eumyias panayensis*
　____　*E. p. septentrionalis*　　　Montane forests of n, central and se Sulawesi
　____　*E. p. meridionalis*　　　　Montane forests of s Sulawesi
　____　*E. p. sanghirensis*　　　　Talaud Islands., Sangihe, Siau, Tahulandang, Ruang and Biaro is.
　____　*E. p. subspecies?*　　　　Taliabu (Sula Islands)
　____　*E. p. obiensis*　　　　　　Montane forests of Obi (s Moluccas)
　____　*E. p. harterti*　　　　　　Montane forests of Seram (s Moluccas)
　____　*E. p. nigrimentalis*　　　　N Philippines (montane forests of Luzon and Mindoro)
　____　*E. p. panayensis*　　　　　Central Philippines (montane forests of Negros and Panay)
　____　*E. p. nigriloris*　　　　　S Philippines (montane forests of Mindanao)

☐ **Nilgiri Flycatcher** *Eumyias albicaudatus*
　　　　　　　　　　　　　　　　Foothill forests of sw peninsular India (Mysore south to Kerala)

☐ **Indigo Flycatcher** *Eumyias indigo*
　____　*E. i. ruficrissa*　　　　　High montane forests of Sumatra
　____　*E. i. indigo*　　　　　　　High montane forests of Java
　____　*E. i. cerviniventris*　　　High montane forests of Borneo

☐ **Large Niltava** *Niltava grandis*
　____　*N. g. grandis*　　　　　　Himalayas (Nepal to n Myanmar, s China and n Indochina)
　____　*N. g. griseiventris*　　　　S China (se Yunnan)
　____　*N. g. decipiens*　　　　　Peninsular Thailand to Malaysia and Sumatra
　____　*N. g. decorata*　　　　　　S Laos (Langbian Plateau)

☐ **Small Niltava** *Niltava macgrigoriae*
　____　*N. m. macgrigoriae*　　　Himalayas of Nepal to s China and ne India (Darjiling)
　____　*N. m. signata*　　　　　　E Himalayas to Bhutan, n Myanmar, nw Thailand and Indochina

☐ **Fujian Niltava** *Niltava davidi*
　　　　　　　　　　　　　　　　Central and s China; > to se Thailand and Indochina

☐ **Rufous-bellied Niltava** *Niltava sundara*
　____　*N. s. whistleri*　　　　　W Himalayas from Pakistan to n India (Kumaon)

_____	*N. s. sundara*	E Himalayas to se Tibet, s China (n Yunnan) and n Laos
_____	*N. s. denotata*	SW China to n Myanmar; > to n Thailand and n Laos

☐ **Rufous-vented Niltava** *Niltava sumatrana*

Mountains of Malay Peninsula and Sumatra

☐ **Vivid Niltava** *Niltava vivida*

_____	*N. v. oatesi*	SE Tibet to sw China, ne India, Myanmar and n Vietnam
_____	*N. v. vivida*	Taiwan

☐ **Matinan Flycatcher** *Cyornis sanfordi*

Mountains of n Sulawesi

☐ **Blue-fronted Flycatcher** *Cyornis hoevelli*

Montane forests of central and se Sulawesi

☐ **Timor Blue-Flycatcher** *Cyornis hyacinthinus*

_____	*C. h. hyacinthinus*	E Lesser Sundas (Roti, Semau and Timor)
_____	*C. h. kuehni*	Wetar (e Lesser Sundas)

☐ **White-tailed Flycatcher** *Cyornis concretus*

_____	*C. c. cyaneus*	Mountains of ne India to s China, Myanmar and n Thailand
_____	*C. c. concretus*	Mountains of s Malay Peninsula and Sumatra
_____	*C. c. everetti*	Mountains of n Borneo

☐ **Rueck's Blue-Flycatcher** *Cyornis ruckii*

Known from four specimens ca 1919 from ne Sumatra

☐ **Blue-breasted Flycatcher** *Cyornis herioti*

_____	*C. h. herioti*	N Philippines (mountains of n and central Luzon)
_____	*C. h. camarinensis*	N Philippines (mountains of s Luzon and Catanduanes)

☐ **Hainan Blue-Flycatcher** *Cyornis hainanus*

Mts. of s China to s Myanmar, Thailand and Indochina; Hainan

☐ **White-bellied Blue-Flycatcher** *Cyornis pallipes*

Uplands of sw India (w Maharashtra to Kerala)

☐ **Pale-chinned Blue-Flycatcher** *Cyornis poliogenys*

_____	*C. p. poliogenys*	Himalayas (central Nepal to ne India, Bhutan and w Myanmar)
_____	*C. p. cachariensis*	E Himalayas (Assam) to nw Myanmar and s China (nw Yunnan)
_____	*C. p. laurentei*	SW China (se Yunnan)
_____	*C. p. vernayi*	E India (Eastern Ghats from n Orissa to Andhra Pradesh)

☐ **Pale Blue-Flycatcher** *Cyornis unicolor*

_____	*C. u. unicolor*	Himalayas (Garhwal) to n India, n Myanmar, s China and n Laos
_____	*C. u. diaoluoensis*	Hainan (s China)
_____	*C. u. harterti*	Malay Pen. (south of Isthmus of Kra), Sumatra, Java and Borneo

☐ **Blue-throated Flycatcher** *Cyornis rubeculoides*

_____	*C. r. rubeculoides*	Kashmir to n India and n Myanmar; > to Sri Lanka
_____	*C. r. dialilaemus*	E Myanmar to n and sw Thailand
_____	*C. r. rogersi*	Myanmar (Arakan Yoma and lower Chindwin River area)
_____	*C. r. glaucicomans*	S China (Sichuan, Guizhou, w Hubei and Shaanxi)
_____	*C. r. klossi*	E Thailand to s Laos and s Vietnam

☐ **Hill Blue-Flycatcher** *Cyornis banyumas*

_____	*C. b. magnirostris*	Himalayas (Nepal to Bangladesh); > to pen. Thailand
_____	*C. b. whitei*	NE Myanmar to s China, ne Thailand, n Laos and n Vietnam
_____	*C. b. lekhakuni*	Eastern plateau of Thailand
_____	*C. b. deignani*	SE Thailand
_____	*C. b. coerulifrons*	Malay Peninsula (south of Isthmus of Kra)
_____	*C. b. ligus*	E Java
_____	*C. b. banyumas*	Central Java
_____	*C. b. coeruleatus*	Borneo

☐ **Long-billed Blue-Flycatcher** *Cyornis caerulatus*
_____ *C. c. albiventer* Lowland forests of Sumatra
_____ *C. c. rufifrons* Lowland forests of w Borneo
_____ *C. c. caerulatus* Lowland forests of n, e and s Borneo

☐ **Malaysian Blue-Flycatcher** *Cyornis turcosus*
_____ *C. t. rupatensis* Lowland forests of Malaysia, Sumatra and w Borneo
_____ *C. t. turcosus* Lowland forests of central and e Borneo

☐ **Palawan Blue-Flycatcher** *Cyornis lemprieri*
  SW Philippines (Balabac, Calamian and Palawan)

☐ **Bornean Blue-Flycatcher** *Cyornis superbus*
  Montane forests of Borneo

☐ **Tickell's Blue-Flycatcher** *Cyornis tickelliae*
_____ *C. t. tickelliae* India to sw China (s Yunnan), n Myanmar and Bangladesh
_____ *C. t. jerdoni* Sri Lanka
_____ *C. t. sumatrensis* S peninsular Thailand to Malaysia and ne Sumatra
_____ *C. t. indochina* S Myanmar to Thailand and Indochina
_____ *C. t. lamprus* Anambas Islands (South China Sea)

☐ **Mangrove Blue-Flycatcher** *Cyornis rufigastra*
_____ *C. r. rufigastra* Coastal lowlands of s Thailand, Malaya, Sumatra and Borneo
_____ *C. r. lepidulus* Karimunjawa Islands (Java Sea)
_____ *C. r. rhizophorae* W Java and Sebesi I. (Sunda Strait)
_____ *C. r. karimatensis* Karimata I. (off sw Borneo)
_____ *C. r. blythi* N Philippines (Luzon and Polillo)
_____ *C. r. marinduquensis* Philippines (Marinduque)
_____ *C. r. mindorensis* Philippines (Mindoro)
_____ *C. r. philippensis (litoralis)* Central and s Philippines, Palawan and Sulu Archipelago

☐ **Sulawesi Blue-Flycatcher** *Cyornis omissus*
_____ *C. o. omissus* Sulawesi
_____ *C. o. peromissus* Salayar I. (Flores Sea)
_____ *C. o. djampeanus* Tanahjampea I. (Flores Sea)
_____ *C. o. kalaoensis* Kalao I. (Flores Sea)

☐ **Pygmy Blue-Flycatcher** *Muscicapella hodgsoni*
_____ *M. h. hodgsoni* Himalayas of Nepal to ne India, Bhutan, Myanmar and Thailand
_____ *M. h. sondaica* Mountains of Thailand, Malaya, Sumatra and Borneo

☐ **Gray-headed Canary-Flycatcher** *Culicicapa ceylonensis*
_____ *C. c. calochrysea* Pakistan to n India, s China, Myanmar, Malaysia and Indochina
_____ *C. c. ceylonensis* S India and Sri Lanka to Sumatra, Java, Borneo and Palawan
_____ *C. c. sejuncta* W Lesser Sundas (Lombok and Flores)
_____ *C. c. connectens* Sumba (w Lesser Sundas)

☐ **Citrine Canary-Flycatcher** *Culicicapa helianthea*
_____ *C. h. septentrionalis* N Philippines (nw Luzon)
_____ *C. h. zimmeri* Philippines (Luzon and Catanduanes)
_____ *C. h. panayensis* Panay, Negros, Cebu, Leyte, Mindanao, Biliran and Palawan
_____ *C. h. mayri* Sulu Archipelago (Bongao and Tawitawi)
_____ *C. h. helianthea* Sulawesi, Salayar, Banggai and Sula islands

☐ **White-starred Robin** *Pogonocichla stellata*
_____ *P. s. ruwenzorii (friedmanni)* NE Zaire (Kivu region) to sw Uganda, Rwanda and Burundi
_____ *P. s. guttifer* Mt. Kilimanjaro (n Tanzania)
_____ *P. s. elgonensis* Mt. Elgon region on Kenya/Uganda border
_____ *P. s. pallidiflava* S Sudan (Imatong Mountains)
_____ *P. s. macarthuri* SE Kenya (Chyulu Mountains)
_____ *P. s. helleri* Kenya (Taita Hills) and Tanzania (Pare Mountains)

____	*P. s. orientalis*	Tanzania to Zambia, Malawi and central Mozambique
____	*P. s. transvaalensis (lebombo)*	Highlands of e Zimbabwe, ne Transvaal and Mozambique
____	*P. s. stellata*	Zululand and Natal to Orange Free State and s Cape Province
____	*P. s. intensa*	Sudan to Kenya and n Tanzania

☐ **Swynnerton's Robin** *Swynnertonia swynnertoni*

____	*S. s. rodgersi*	E Tanzania (Udzungwa Mountains)
____	*S. s. swynnertoni (umbratica)*	Mountains of e Zimbabwe and w Mozambique

☐ **Forest Robin** *Stiphrornis erythrothorax*

____	*S. e. erythrothorax*	Sierra Leone to s Nigeria
____	*S. e. gabonensis*	Coastal Cameroon and Gabon; Bioko (Gulf of Guinea)
____	*S. e. xanthogaster (mabirae)*	E Cameroon to e Zaire, w Uganda and extreme s Sudan

☐ **Bocage's Akalat** *Sheppardia bocagei*

____	*S. b. granti*	Mountains of se Nigeria and w Cameroon
____	*S. b. poensis (insulana)*	Bioko (Gulf of Guinea)
____	*S. b. kaboboensis*	E Zaire (Mt. Kabobo)
____	*S. b. schoutedeni*	E Zaire (mountains west of Lake Edward to Kivu))
____	*S. b. bocagei*	Western highlands of Angola
____	*S. b. kungwensis*	W Tanzania (Kungwe-Mahari Mountains)
____	*S. b. ilyai*	W Tanzania (east of Mt. Kungwe)
____	*S. b. chapini*	SE Zaire and n-central Zambia

☐ **Lowland Akalat** *Sheppardia cyornithopsis*

____	*S. c. houghtoni*	Lowlands of Guinea to Sierra Leone, Liberia and Ivory Coast
____	*S. c. cyornithopsis*	Lowland forests of s Cameroon and Gabon
____	*S. c. lopezi*	Lowland forests of e Zaire to w Uganda and nw Tanzania

☐ **Equatorial Akalat** *Sheppardia aequatorialis*

____	*S. a. aequatorialis*	Mts. of e Zaire to sw Uganda, w Rwanda, Burundi and w Kenya
____	*S. a. acholiensis*	S Sudan (Imatong Mountains)

☐ **Sharpe's Akalat** *Sheppardia sharpei*

____	*S. s. usambarae*	Tanzania (Usambara Mountains and Nguru Mountains)
____	*S. s. sharpei*	Montane forests of sw Tanzania to Zambia and n Malawi

☐ **East Coast Akalat** *Sheppardia gunningi*

____	*S. g. sokokensis*	Coastal and riverine forests of se Kenya to e Tanzania
____	*S. g. alticola*	E Tanzania (Nguu Mountains)
____	*S. g. bensoni*	Lowlands of nw Malawi
____	*S. g. gunningi*	SE Mozambique

☐ **Gabela Akalat** *Sheppardia gabela*

Angola (humid montane forests of Gabela escarpment)

☐ **Usambara Akalat** *Sheppardia montana*

NE Tanzania (w Usambara Mountains)

☐ **Iringa Akalat** *Sheppardia lowei*

Dry montane forests of s-central Tanzania

☐ **Rubeho Akalat** *Sheppardia aurantiithorax*

Montane forests of e Tanzania (Rubehos and Ukagura Mts.)

☐ **European Robin** *Erithacus rubecula*

____	*E. r. melophilus*	British Isles and Scandinavia
____	*E. r. rubecula*	W Europe, nw Morocco, Azores, Madeira and w Canary Islands
____	*E. r. superbus*	Central Canary Islands (Teneriffe and Gran Canaria)
____	*E. r. witherbyi (sardus)*	S Spain, Corsica, Sardinia, ne Morocco, Algeria and Tunisia
____	*E. r. balcanicus*	Balkan Peninsula to w Turkey
____	*E. r. valens*	Crimean Peninsula
____	*E. r. hyrcanus*	SE Transcaucasia to n Iran
____	*E. r. tataricus*	W Siberia (Ural Mountains to Semipalatinsk); > to Iran

☐ **Japanese Robin** *Erithacus akahige*

    ____ *E. a. akahige* — S Kuril and Sakhalin is. to n Japanese Arch.; > to s China

    ____ *E. a. rishirensis* — Rishiri I. (off nw Hokkaido)

    ____ *E. a. tanensis* — S Japanese Arch. (Izu, Tanegashima and Yakushima islands)

☐ **Ryukyu Robin** *Erithacus komadori*

    ____ *E. k. komadori* — Ryukyu Is. (Tanega-Shima, Amami-O-Shima, Tokuno-Shima)

    ____ *E. k. namiyei* — Okinawa (central Ryukyu Islands)

    ____ *E. k. subrufus* — S Ryukyu Islands (Ishigaki, Iriomote and Yonaguni)

☐ **Rufous-tailed Robin** *Luscinia sibilans* — Breeds s Siberia to Sea of Okhotsk; > s China to SE Asia

☐ **Thrush Nightingale** *Luscinia luscinia* — N Eurasia; > to e and s Africa

☐ **Common Nightingale** *Luscinia megarhynchos*

    ____ *L. m. megarhynchos* — W Europe, N Africa and Asia Minor; > in tropical Africa

    ____ *L. m. africana* — Caucasus and e Turkey to sw Iran and Iraq; > to E Africa

    ____ *L. m. hafizi* — Aral Sea to Mongolia; > coastal e Africa

☐ **Siberian Rubythroat** *Luscinia calliope* — Siberia to Japan; > to SE Asia, Philippines and Palau Is.

☐ **White-tailed Rubythroat** *Luscinia pectoralis*

    ____ *L. p. pectoralis* — Mountains of Turkestan to Afghanistan (Pamirs) and n Pakistan

    ____ *L. p. confusa* — E Himalayas (Nepal to Bhutan); > to ne India

    ____ *L. p. ballioni* — W China (w Xinjiang from Tien Shan to Kashgar)

    ____ *L. p. tschebaiewi* — E Ladakh to nw China (Gansu), se Tibet and extreme n Myanmar

☐ **Bluethroat** *Luscinia svecica*

    ____ *L. s. svecica (gaetkei, robusta)* — Scandinavia across Siberia to w Alaska; > N Africa, s Asia

    ____ *L. s. namnetum* — Western France

    ____ *L. s. cyanecula* — Central Europe and Spain; > to North Africa

    ____ *L. s. volgae* — NE Ukraine to middle Volga River

    ____ *L. s. magna* — Caucasus area, e Turkey and Iran; > to Sudan and Ethiopia

    ____ *L. s. luristanica* — Armenia to sw Iran; > to Iraq and the Sudan

    ____ *L. s. pallidogularis (saturatior, altaica)* — SW Siberia to Turkmenistan, Altai Mts. and upper Yenisey

    ____ *L. s. tianschanica* — Pamir Mountains and Tien Shan Mountains

    ____ *L. s. abbotti* — W Pakistan and nw India

    ____ *L. s. przewalskii* — Inner Mongolia to w China (Qinghai) and s Tibet

    ____ *L. s. kobdensis* — W China (Xinjiang)

☐ **Rufous-headed Robin** *Luscinia ruficeps* — N-central China (Tsingling Mountains of Shaanxi and Sichuan)

☐ **Black-throated Blue Robin** *Luscinia obscura* — W cent. China (se Gansu, sw Shaanxi, n Sichuan and n Yunnan)

☐ **Firethroat** *Luscinia pectardens* — Mts. of se Tibet to w-central China; > to ne Myanmar

☐ **Indian Blue Robin** *Luscinia brunnea*

    ____ *L. b. brunnea* — Himalayas (Pakistan to Bhutan and se Tibet); > to Sri Lanka

    ____ *L. b. wickhami* — Myanmar (Chin Hills)

☐ **Siberian Blue Robin** *Luscinia cyane*

    ____ *L. c. cyane* — S Siberia (Altai Mts. to Sea of Okhotsk); > to Indonesia

    ____ *L. c. bochaiensis* — E Siberia to ne China, Korea and Japan; > to Malaysia

☐ **Red-flanked Bluetail** *Tarsiger cyanurus*

    ____ *T. c. cyanurus* — N Russia to n Japan; > to s China, Taiwan and Gr. Sundas

    ____ *T. c. pallidior* — Afghanistan to Nepal; > to Myanmar, Thailand and Laos

    ____ *T. c. rufilatus* — Nepal to ne India and sw China; > to Indochina

☐ **Golden Bush-Robin** *Tarsiger chrysaeus*

    ____ *T. c. whistleri* — N Pakistan to Kashmir and nw India

____ *T. c. chrysaeus*   Nepal to ne India, n Myanmar and w China; > to n Vietnam

☐ **White-browed Bush-Robin** *Tarsiger indicus*
- ____ *T. i. indicus*   Himalayas (Nepal to Bhutan, se Tibet and Assam)
- ____ *T. i. yunnanensis*   SW China (Yunnan); > to n Myanmar and n Tonkin
- ____ *T. i. formosanus*   Mountains of Taiwan

☐ **Rufous-breasted Bush-Robin** *Tarsiger hyperythrus*

Himalayas (ne India, se Tibet and s China); > to Myanmar

☐ **Collared Bush-Robin** *Tarsiger johnstoniae*

Mountains of Taiwan

☐ **White-throated Robin** *Irania gutturalis*

Turkey to Iraq, Iran and s Turkestan; > in East Africa

☐ **White-bellied Robin-Chat** *Cossyphicula roberti*
- ____ *C. r. roberti*   Montane forests of se Nigeria and w Cameroon; Bioko
- ____ *C. r. rufescentior*   Montane forests of e Zaire, Rwanda and sw Uganda

☐ **Mountain Robin-Chat** *Cossypha isabellae*
- ____ *C. i. batesi*   Montane forests of se Nigeria
- ____ *C. i. isabellae*   Montane forests of Mt. Cameroon

☐ **Archer's Robin-Chat** *Cossypha archeri*
- ____ *C. a. archeri (albimentalis)*   Montane forests of e Zaire to Rwanda, Burundi and sw Uganda
- ____ *C. a. kimbutui*   Montane forests of se Zaire (Mt. Kabobo)

☐ **Olive-flanked Robin-Chat** *Cossypha anomala*
- ____ *C. a. grotei*   Highlands of e Tanzania
- ____ *C. a. mbuluensis*   N-central Tanzania (Mbulu region)
- ____ *C. a. macclounii*   S Tanzania (Tukuyu District) to n Malawi (Viphya Plateau)
- ____ *C. a. anomala*   N-central Malawi (Mt. Mulanje region)
- ____ *C. a. gurue*   Montane forests of n-central Mozambique

☐ **Cape Robin-Chat** *Cossypha caffra*
- ____ *C. c. iolaema*   Mts. of extreme s Sudan to Kenya, Zambia and Mozambique
- ____ *C. c. kivuensis*   E Zaire (Kivu highlands) and sw Uganda
- ____ *C. c. drakensbergi*   Natal-Transvaal border to e Transvaal
- ____ *C. c. vespera*   Highlands of e Zimbabwe
- ____ *C. c. namaquensis*   S Namibia to Orange Free State and w Transvaal
- ____ *C. c. caffra*   Natal to Swaziland and Cape Province

☐ **White-throated Robin-Chat** *Cossypha humeralis*

E Botswana to s Mozambique, e Transvaal and n Natal

☐ **Blue-shouldered Robin-Chat** *Cossypha cyanocampter*
- ____ *C. c. cyanocampter*   Guinea to Mali, Sierra Leone, Cameroon and Gabon
- ____ *C. c. bartteloti (pallidiventris)*   NE Zaire to s Sudan, Uganda and w Kenya

☐ **Gray-winged Robin-Chat** *Cossypha polioptera*
- ____ *C. p. nigriceps*   Sierra Leone to n Cameroon
- ____ *C. p. tessmanni*   E Cameroon
- ____ *C. p. polioptera (grimwoodi)*   S Sudan to Uganda, nw Tanzania, n Angola and nw Zambia

☐ **Rueppell's Robin-Chat** *Cossypha semirufa*
- ____ *C. s. semirufa*   SE Sudan (Boma Hills) to Eritrea, Ethiopia and n Kenya
- ____ *C. s. donaldsoni*   E and se Ethiopia (Harrar and e Gallaland)
- ____ *C. s. intercedens*   S-central highlands of Kenya to n Tanzania (Mt. Kilimanjaro)

☐ **White-browed Robin-Chat** *Cossypha heuglini*
- ____ *C. h. heuglini*   Zaire to s Sudan, s Ethiopia, Angola, Zimbabwe and e Transvaal
- ____ *C. h. subrufescens*   Gabon to n Angola and extreme w Zaire
- ____ *C. h. intermedia*   Coastal e Somalia to e Kenya, Tanzania and n Natal

☐ **Red-capped Robin-Chat** *Cossypha natalensis*
    ____ *C. n. larischi*   Nigeria to n Angola
    ____ *C. n. intensa (garguensis) (tennenti)*   S Somalia and Sudan to Angola, e Transvaal and Mozambique
    ____ *C. n. natalensis*   Coastal Natal to e Cape Province

☐ **Chorister Robin-Chat** *Cossypha dichroa*
    ____ *C. d. dichroa*   Forests of e and s South Africa and w Swaziland
    ____ *C. d. mimica*   NE Transvaal (Zoutpansberg and Woodbush)

☐ **White-headed Robin-Chat** *Cossypha heinrichi*
                                 Savanna and forests of nw Angola and adjacent w Zaire

☐ **Snowy-crowned Robin-Chat** *Cossypha niveicapilla*
    ____ *C. n. niveicapilla*   Senegal to s Sudan, sw Ethiopia, Uganda, Kenya and Tanzania
    ____ *C. n. melanonota*   Lake Victoria basin

☐ **White-crowned Robin-Chat** *Cossypha albicapilla*
    ____ *C. a. albicapilla*   Senegal to Guinea
    ____ *C. a. giffardi*   Ghana to Nigeria, s Chad and n Cameroon
    ____ *C. a. omoensis*   Extreme se Sudan to sw Ethiopia

☐ **Angola Cave-Chat** *Xenocopsychus ansorgei*
                                 Rocky caves and gorges of w Angola

☐ **Collared Palm-Thrush** *Cichladusa arquata*
                                 Coastal Kenya to Mozambique, Caprivi Strip and se Angola

☐ **Rufous-tailed Palm-Thrush** *Cichladusa ruficauda*
                                 Scrub and palms of s Gabon to coastal Angola and n Namibia

☐ **Spotted Morning-Thrush** *Cichladusa guttata*
    ____ *C. g. guttata*   S Sudan to w Uganda, Zaire and nw Kenya (w of Lake Turkana)
    ____ *C. g. intercalans*   SW Ethiopia to Kenya, Tanzania to e Zaire
    ____ *C. g. rufipennis*   Littoral of s Somalia to e Kenya and e Tanzania (Dar-es-Salaam)

☐ **Forest Scrub-Robin** *Cercotrichas leucosticta*
    ____ *C. l. leucosticta*   Ghana
    ____ *C. l. colstoni*   Sierra Leone (Kambui Hills) and Liberia
    ____ *C. l. collsi*   Central African Republic to ne Zaire and w Uganda
    ____ *C. l. reichenowi*   N Angola (Huila escarpment)

☐ **Bearded Scrub-Robin** *Cercotrichas quadrivirgata*
    ____ *C. q. quadrivirgata*   S Somalia to Kenya, Tanzania, Botswana, Zambia and ne S Africa
    ____ *C. q. greenwayi*   Zanzibar and Mafia I. (off Tanzania)

☐ **Miombo Scrub-Robin** *Cercotrichas barbata*
                                 Angola to Zaire, Burundi, Zambia, sw Tanzania, n Mozambique

☐ **Brown Scrub-Robin** *Cercotrichas signata*
    ____ *C. s. tongensis*   Extreme s Mozambique to e Swaziland and n Natal
    ____ *C. s. signata (reclusa, oatleyi)*   South Africa (Transvaal and s Natal to se Cape Province)

☐ **Brown-backed Scrub-Robin** *Cercotrichas hartlaubi*
                                 Locally in s Cameroon; nw Angola to w Kenya and nw Tanzania

☐ **Red-backed Scrub-Robin** *Cercotrichas leucophrys*
    ____ *C. l. leucoptera*   S Sudan to ne Uganda, s Ethiopia, n Somalia and n Kenya
    ____ *C. l. zambesiana*   S Sudan to e Kenya, n Mozambique and e Zambia
    ____ *C. l. eluta*   S Somalia (Juba River area) and adjacent ne Kenya
    ____ *C. l. brunneiceps*   Central Kenya to ne Tanzania (w of Kilimanjaro to Loliondo)
    ____ *C. l. vulpina*   S-central Kenya to extreme ne Tanzania
    ____ *C. l. sclateri*   Central Tanzania
    ____ *C. l. munda*   Congo River to central Angola and Zaire (Katanga)
    ____ *C. l. ovamboensis*   S Angola to sw Zambia, n Botswana, n Namibia, w Zimbabwe
    ____ *C. l. leucophrys*   S Zimbabwe and Transvaal to s Cape Province

□ **Rufous-tailed Scrub-Robin** *Cercotrichas galactotes*

____	*C. g. galactotes*	W Mediterranean basin and North Africa; > to s Sahara
____	*C. g. syriacus*	E Mediterranean basin and Middle East; > to e Africa
____	*C. g. familiaris*	S Caucasus to Iran and Pakistan; > to s Arabia and Kenya

□ **Kalahari Scrub-Robin** *Cercotrichas paena*

____	*C. p. benguellensis*	SW Angola (Benguela Province)
____	*C. p. damarensis*	Namibia
____	*C. p. paena*	Botswana to Zimbabwe, w Transvaal and n Cape Province
____	*C. p. oriens*	W Orange Free State to s Transvaal and s Zimbabwe

□ **African Scrub-Robin** *Cercotrichas minor*

____	*C. m. minor*	Senegambia to Sudan, Eritrea, Ethiopia and n Somalia
____	*C. m. hamertoni*	E Somalia (Beira and Wagar mountains)

□ **Karoo Scrub-Robin** *Cercotrichas coryphaeus*

____	*C. c. coryphaeus*	S Namibia to Botswana and w Cape Province
____	*C. c. cinerea*	SE Namibia to sw Cape Province
____	*C. c. eurina*	Orange Free State to Lesotho

□ **Black Scrub-Robin** *Cercotrichas podobe*

____	*C. p. podobe*	Mauritania to Chad, Sudan, Eritrea, Ethiopia and n Somalia
____	*C. p. melanoptera*	W Saudi Arabia, Yemen and Aden

□ **Herero Chat** *Namibornis herero*

Rocky bush country of extreme sw Angola to w Namibia

□ **Madagascar Magpie-Robin** *Copsychus albospecularis*

____	*C. a. albospecularis*	N Madagascar
____	*C. a. inexpectatus*	E Madagascar
____	*C. a. pica*	W Madagascar
____	*C. a. winterbottomi*	SW Madagascar

□ **Oriental Magpie-Robin** *Copsychus saularis*

____	*C. s. saularis*	Lowlands of Pakistan to n and w India
____	*C. s. ceylonensis*	SE India and Sri Lanka
____	*C. s. erimelas*	NE India to Myanmar, Thailand and Indochina
____	*C. s. andamanensis*	Andaman Islands
____	*C. s. prosthopellus*	S China (Sichuan to mouth of Yangtze River and Hainan)
____	*C. s. musicus*	Peninsular Thailand, Malaysia and Sumatra
____	*C. s. nesiotes*	SE Sumatra, Rhio Archipelago, Belitung and Bangka islands
____	*C. s. zacnecus*	Simeulue I. (off Sumatra)
____	*C. s. nesiarchus*	Nias I. (off Sumatra)
____	*C. s. masculus*	Batu Islands (Pini, Tello and Tana Massa)
____	*C. s. pagiensis*	Mentawi Archipelago, Siberut and Sipoura islands (off Sumatra)
____	*C. s. javensis*	W Java
____	*C. s. amoenus*	E Java and Bali
____	*C. s. problematicus*	SW and w Borneo
____	*C. s. adamsi*	N Borneo, Banggi and adjacent islands
____	*C. s. pluto*	E Borneo and Maratua Islands
____	*C. s. deuteronymus*	N Philippines (Luzon, Lubang and Palaui)
____	*C. s. mindanensis*	S Philippines and Sulu Archipelago

□ **White-rumped Shama** *Copsychus malabaricus*

____	*C. m. malabaricus*	S peninsular India
____	*C. m. leggei*	Sri Lanka
____	*C. m. indicus*	Nepal to Assam and ne India
____	*C. m. albiventris*	Andaman Islands
____	*C. m. interpositus*	SW China to Myanmar, Thailand, Indochina and Mergui Arch.
____	*C. m. minor*	Hainan (s China)

____	*C. m. mallopercnus*	Malay Peninsula, Riau Archipelago and Lingga Archipelago
____	*C. m. tricolor*	Sumatra, w Java, Banka, Belitung and Karimata islands
____	*C. m. mirabilis*	Prinsen I. (Sunda Strait)
____	*C. m. melanurus*	Islands off nw Sumatra
____	*C. m. opisthopelus*	Islands off sw Sumatra
____	*C. m. javanus*	Central Java
____	*C. m. omissus*	E Java
____	*C. m. ochroptilus*	Anambas Islands (South China Sea)
____	*C. m. abbotti*	Bangka and Belitung islands (off Borneo)
____	*C. m. eumesus*	Natuna Islands (off Borneo)
____	*C. m. suavis*	Borneo (except northern part)
____	*C. m. nigricauda*	Kangean Islands and Matasiri I. (Java Sea)
____	*C. m. stricklandii*	Lowlands of n Borneo, Labuan, Balembangan and Banggi islands
____	*C. m. barbouri*	Maratua Islands (off n Borneo)

☐ **Seychelles Magpie-Robin** *Copsychus sechellarum*

Frégate (Seychelles). Seriously endangered

☐ **White-browed Shama** *Copsychus luzoniensis*

____	*C. l. luzoniensis*	N Philippines (Luzon and Catanduanes)
____	*C. l. parvimaculatus*	Polillo (Philippines)
____	*C. l. shemleyi*	Marinduque (Philippines)
____	*C. l. superciliaris*	Philippines (Ticao, Masbate, Panay and Negros)

☐ **White-vented Shama** *Copsychus niger*

S Philippines (Balabac, Busuanga, Culion, Bantac and Palawan)

☐ **Black Shama** *Copsychus cebuensis*

Cebu (s-central Philippines)

☐ **Rufous-tailed Shama** *Trichixos pyrropygus*

S Peninsular Thailand, Malaya, Sumatra and Borneo

☐ **Indian Robin** *Saxicoloides fulicatus*

____	*S. f. cambaiensis*	Pakistan to n and w India and lowlands of Nepal
____	*S. f. erythrurus*	NE India (plains of Bihar and West Bengal)
____	*S. f. intermedius*	Central India (Bombay to Hyderabad and Krishna River)
____	*S. f. fulicatus*	S India (Bombay to Mysore and Kerala)
____	*S. f. leucopterus*	Lowlands of Sri Lanka

☐ **Ala Shan Redstart** *Phoenicurus alaschanicus*

Montane coniferous forests of n China to ne Tibet

☐ **Rufous-backed Redstart** *Phoenicurus erythronotus*

Montane forests of central Asia; > to Iran and n India

☐ **Blue-capped Redstart** *Phoenicurus caeruleocephala*

Montane juniper and pine forests of s Asia

☐ **Black Redstart** *Phoenicurus ochruros*

____	*P. o. gibraltariensis*	W and central Europe to Crimea and North Africa
____	*P. o. aterrimus*	Iberian Peninsula
____	*P. o. ochruros*	Mountains of e Turkey to n Iran; > to Iraq
____	*P. o. semirufus*	Hills of Syria and Lebanon; > to Israel and Sinai Peninsula
____	*P. o. phoenicuroides*	Tien Shan Mts. to n Mongolia; > to ne Africa and India
____	*P. o. rufiventris*	Himalayas (Tibet to nw China); > to India and n Myanmar
____	*P. o. xerophilus*	W China (Astin Tagh Mountains to w Gansu and Qinghai)

☐ **Common Redstart** *Phoenicurus phoenicurus*

____	*P. p. phoenicurus*	Europe and North Africa to central Asia; > to trop. Africa
____	*P. p. samamisicus*	Crimea and Caucasus to w Afghanistan; > to ne Africa

☐ **Hodgson's Redstart** *Phoenicurus hodgsoni*

Himalayas of w-central China; > to India and Myanmar

☐ **White-throated Redstart** *Phoenicurus schisticeps*

Nepal to s Tibet and w-central China; > to Myanmar

☐ **Daurian Redstart** *Phoenicurus auroreus*
    ____ *P. a. auroreus* — S Siberia to Mongolia; > to Japan and Ryukyu Islands
    ____ *P. a. leucopterus* — W China to se Tibet and nw Thailand; > to n Myanmar

☐ **Moussier's Redstart** *Phoenicurus moussieri* — Bare plateaux of s Morocco, n Algeria and n Tunisia

☐ **White-winged Redstart** *Phoenicurus erythrogastrus*
    ____ *P. e. erythrogastrus* — Caucasus Mountains to s Caspian region of Iran
    ____ *P. e. grandis* — Central Asia to se Tibet, s China, Pakistan and n India

☐ **Blue-fronted Redstart** *Phoenicurus frontalis* — Himalayas of s Asia; > to SE Asia

☐ **White-capped Redstart** *Chaimarrornis leucocephalus* — Himalayas of s Asia; > to SE Asia

☐ **Plumbeous Redstart** *Rhyacornis fuliginosa*
    ____ *R. f. fuliginosa* — Himalayas (Pakistan to Myanmar, se Tibet, w China, n Vietnam)
    ____ *R. f. affinis* — Taiwan

☐ **Luzon Redstart** *Rhyacornis bicolor* — Rocky streams of n Luzon (n Philippines)

☐ **White-bellied Redstart** *Hodgsonius phaenicuroides*
    ____ *H. p. phaenicuroides* — Himalayas of Pakistan to se Tibet, sw China and n Myanmar
    ____ *H. p. ichangensis* — Himalayas of w China; > to n Vietnam and Laos

☐ **White-tailed Robin** *Cinclidium leucurum*
    ____ *C. l. leucurum* — Mts. of Nepal to Myanmar, sw China, Indochina and Malaysia
    ____ *C. l. cambodianum* — S Cambodia (Chaine de l'Éléphant)
    ____ *C. l. montium* — Mountains of Taiwan

☐ **Sunda Robin** *Cinclidium diana*
    ____ *C. d. sumatranum* — Mountains of n and w-central Sumatra
    ____ *C. d. diana* — Mountains of Java

☐ **Blue-fronted Robin** *Cinclidium frontale*
    ____ *C. f. frontale* — Himalayan foothills of e Nepal, Sikkim and sw China
    ____ *C. f. orientale* — Mountains of nw Thailand, n Vietnam (Tonkin) and Laos

☐ **Grandala** *Grandala coelicolor* — Alpine meadows of Kashmir to se Tibet, Myanmar and w China

☐ **Little Forktail** *Enicurus scouleri*
    ____ *E. s. scouleri* — Mountains of se Russia to the Himalayas, n India and sw China
    ____ *E. s. fortis* — Mountain streams of Taiwan

☐ **Sunda Forktail** *Enicurus velatus*
    ____ *E. v. sumatranus* — Rocky mountain streams of Sumatra
    ____ *E. v. velatus* — Rocky mountain streams of Java

☐ **Chestnut-naped Forktail** *Enicurus ruficapillus* — S Myanmar to s Thailand, Malaya, Sumatra and Borneo

☐ **Black-backed Forktail** *Enicurus immaculatus* — Rocky streams of n India to Myanmar, sw China, nw Thailand

☐ **Slaty-backed Forktail** *Enicurus schistaceus* — Rocky mountain streams of n India to s China and SE Asia

☐ **White-crowned Forktail** *Enicurus leschenaulti*
    ____ *E. l. indicus* — Himalayas of ne India to Myanmar, n Thailand and Indochina
    ____ *E. l. sinensis* — W and s China; Hainan
    ____ *E. l. frontalis* — Malaysia, Sumatra, Nias I. and lowlands of Borneo
    ____ *E. l. chaseni* — Tanahmasa I. (Batu Islands off w Sumatra)
    ____ *E. l. leschenaulti* — Java and Bali
    ____ *E. l. borneensis* — Mountains of n Borneo

☐ **Spotted Forktail** *Enicurus maculatus*
____ *E. m. maculatus*   Himalayas of n Afghanistan to Kashmir, Nepal and s Tibet
____ *E. m. guttatus*   Himalayas of extreme e Nepal to sw China and Myanmar
____ *E. m. bacatus*   Mts. of s China (se Yunnan and nw Fujian) to n Vietnam
____ *E. m. robinsoni*   South Vietnam (Da Lat Plateau)

☐ **Purple Cochoa** *Cochoa purpurea*

Montane forests of n India to sw China, Myanmar and Indochina

☐ **Green Cochoa** *Cochoa viridis*

Montane forests of n India to sw China, Myanmar and Indochina

☐ **Sumatran Cochoa** *Cochoa beccarii*

Highlands of w Sumatra

☐ **Javan Cochoa** *Cochoa azurea*

Mountains of w and central Java

☐ **Whinchat** *Saxicola rubetra*

W Palearctic; > tropical and s Africa

☐ **White-browed Bushchat** *Saxicola macrorhynchus*

Arid plains of s Afghanistan, e Pakistan and nw India

☐ **White-throated Bushchat** *Saxicola insignis*

Rocky alpine meadows of central Asia; > to n India

☐ **Canary Island Stonechat** *Saxicola dacotiae*
____ *S. d. dacotiae*   Fuerteventura (Canary Islands)
____ *S. d. murielae†*   Formerly Canary Is. (Montaña Clara and Allegranza). Extinct

☐ **European Stonechat** *Saxicola rubicola*
____ *S. r. hibernans*   Britain, Ireland, coastal w France and w coast of Iberian Pen.
____ *S. r. rubicola*   W and s Europe and North Africa; > to Middle East

☐ **Siberian Stonechat** *Saxicola maurus*
____ *S. m. variegatus*   Steppes of lower Volga and mouth of Ural River to e Caucasus
____ *S. m. armenicus*   Mountains of e Turkey to Transcaucasia and Iran
____ *S. m. maurus*   E Russia to central Asia; > to Iran, Iraq and n India
____ *S. m. indicus*   Himalayas (Kashmir to Sikkim and Assam); > to India
____ *S. m. stejnegeri*   E Siberia to Japan and Korea; > to s China and Indochina
____ *S. m. przewalskii*   Mountains of w China; > to Myanmar and n India

☐ **African Stonechat** *Saxicola torquatus*
____ *S. t. felix*   Mountains of sw Arabia and Yemen
____ *S. t. albofasciatus*   Highlands of Ethiopia, se Sudan and ne Uganda
____ *S. t. jebelmarrae*   W Sudan (Darfur region)
____ *S. t. moptanus*   Inner Niger delta, Mali, n Senegal and Senegal delta
____ *S. t. nebularum*   Highlands of Sierra Leone, Guinea, Liberia and w Ivory Coast
____ *S. t. adamauae*   Highlands of n and w Cameroon
____ *S. t. pallidigula*   Cameroon Mt. and Bioko
____ *S. t. axillaris*   Highlands of e Zaire to Rwanda, Uganda, Kenya and n Tanzania
____ *S. t. promiscuus*   Highlands of e Tanzania
____ *S. t. salax*   SE Nigeria to Cameroon, Gabon, lower Congo and Angola
____ *S. t. stonei*   Angola to w Tanzania, s Mozambique and n Cape Province
____ *S. t. clanceyi*   Coastal w Namibia to nw Cape Province
____ *S. t. torquatus*   SW Cape Province to Natal and Transvaal
____ *S. t. oreobates*   Highlands of Lesotho; > to e Zimbabwe and s Mozambique
____ *S. t. altivagus*   Highlands of s Malawi and adj. Mozambique to n Transvaal
____ *S. t. sibilla*   Madagascar
____ *S. t. voeltzkowi*   Grand Comoro I.

☐ **Reunion Stonechat** *Saxicola tectes*

Réunion (w Mascarene Islands)

☐ **White-tailed Stonechat** *Saxicola leucurus*

E Pakistan to Nepal, n India and Myanmar

☐ **Pied Bushchat** *Saxicola caprata*

____	*S. c. rossorum*	Transcaspia to e Iran, Afghanistan and n Kashmir
____	*S. c. bicolor*	Pakistan to Baluchistan and Kashmir; > to central India
____	*S. c. burmanicus*	Central India to sw China, Myanmar, n Thailand and Indochina
____	*S. c. nilgiriensis*	S India (w Madras and Kerala)
____	*S. c. atratus*	Sri Lanka
____	*S. c. caprata*	N Philippines (Luzon, Lubang and Mindoro)
____	*S. c. randi*	Philippines (Negros, Bohol, Masbate, Ticao, Cebu, Siquijor)
____	*S. c. anderseni*	S Philippines (Mindanao, Camiguin Sur, Leyte and Biliran)
____	*S. c. fruticola*	Java, Bali. Lombok, Sumbawa, Flores, Lomblen and Alor
____	*S. c. pyrrhonotus*	Lesser Sundas (Kisar, Wetar, Sawu, Semau, Roti and Timor)
____	*S. c. francki*	Sumba I. (Lesser Sundas)
____	*S. c. cognatus*	Babar I. (Lesser Sundas)
____	*S. c. albonotatus*	Sulawesi, Salayar and Butung islands
____	*S. c. aethiops*	N New Guinea and New Britain (Bismarck Archipelago)
____	*S. c. belensis*	Central mts. of New Guinea (Wissel Lakes to Snow Mountains)
____	*S. c. wahgiensis*	Central highlands of New Guinea to Huon Pen. and se mountains

☐ **Jerdon's Bushchat** *Saxicola jerdoni*

____	*S. j. harringtoni*	Afghanistan to Nepal, Myanmar, s China and n Indochina
____	*S. j. jerdoni*	S Tibet to sw China (s Yunnan)

☐ **Gray Bushchat** *Saxicola ferreus*

N Pakistan to s Tibet, se China, Myanmar and n Indochina

☐ **Timor Bushchat** *Saxicola gutturalis*

____	*S. g. gutturalis*	E Lesser Sundas (Timor and Roti)
____	*S. g. luctuosus*	Semau (e Lesser Sundas)

☐ **Buff-streaked Bushchat** *Saxicola bifasciatus*

Rocky montane areas of Natal to Transvaal and Cape Province

☐ **White-tailed Wheatear** *Oenanthe leucopyga*

____	*O. l. aegra*	Rocky deserts of Mauritania to Tunisia
____	*O. l. leucopyga*	Rocky deserts of Mali to Chad, Sudan, Eritrea and Ethiopia
____	*O. l. ernesti*	Deserts of e Egypt to Dead Sea, Saudi Arabia, Iraq and sw Iran

☐ **Hooded Wheatear** *Oenanthe monacha*

Rocky desert ravines of ne Sudan and Egypt to s Pakistan

☐ **Hume's Wheatear** *Oenanthe albonigra*

Bare rocky hills of e Arabia to s Iran, Afghanistan and Pakistan

☐ **Black Wheatear** *Oenanthe leucura*

____	*O. l. leucura*	Iberian Peninsula to coastal s France, Italy, Sardinia and Sicily
____	*O. l. syenitica*	Extreme nw Mauritania to Morocco, Tunisia, Algeria and Libya

☐ **Mountain Wheatear** *Oenanthe monticola*

____	*O. m. albipileata*	Coastal Angola (Benguela escarpment)
____	*O. m. nigricauda*	Angola (highlands of Huambo and s Cuanza Sul)
____	*O. m. atmorii*	N Namibia (south to Damaraland)
____	*O. m. monticola (griseiceps)*	S Botswana and Transvaal to Swaziland and Natal

☐ **Somali Wheatear** *Oenanthe phillipsi*

Mountains of se Ethiopia and n Somalia

☐ **Northern Wheatear** *Oenanthe oenanthe*

____	*O. o. leucorhoa*	NE Canada to Greenland and Iceland; > to w Africa
____	*O. o. oenanthe*	British Isles to Mediterranean and Siberia; > to c Africa
____	*O. o. libanotica*	S Spain and Balearic Is. to Iran, Kazakstan and Mongolia
____	*O. o. seebohmi*	Morocco to ne Algeria; > to Mauritania

☐ **Mourning Wheatear** *Oenanthe lugens*

____	*O. l. halophila*	N Sahara (e Morocco to n Libya and nw Egypt)

_____	_O. l. lugens_	Egypt (east of the Nile) to Israel, Syria, Jordan and n Iraq
_____	_O. l. lugentoides_	W Arabia (Taif to Yemen)
_____	_O. l. boscaweni_	S Arabia (Hadramaut)
_____	_O. l. vauriei_	NE Somalia
_____	_O. l. lugubris_	Highlands of n and central Ethiopia
_____	_O. l. schalowi_	Highlands of s Kenya to ne Tanzania
_____	_O. l. persica_	S Iran; wanders to s Egypt, n Sudan and s Israel

☐ **Finsch's Wheatear** _Oenanthe finschii_

_____	_O. f. finschii_	Turkey to Israel, n Arabia and s Iran; > to Cyprus and Egypt
_____	_O. f. barnesi_	E Turkey to e Caucasus, n Iran, Afghanistan and w Pakistan

☐ **Variable Wheatear** _Oenanthe picata_

Iran to n Baluchistan and Pakistan; > to s Iran and n India

☐ **Red-rumped Wheatear** _Oenanthe moesta_

_____	_O. m. moesta_	Extreme nw Mauritania to Morocco and coastal nw Egypt
_____	_O. m. brooksbanki_	S Syria to Jordan, nw Saudi Arabia and sw Iraq

☐ **Pied Wheatear** _Oenanthe pleschanka_

Stony s-central Eurasia; > Arabia and Iran to ne Africa

☐ **Cyprus Wheatear** _Oenanthe cypriaca_

Stony areas of Cyprus; > in ne Africa

☐ **Black-eared Wheatear** _Oenanthe hispanica_

_____	_O. h. hispanica_	S Europe and North Africa; > Senegal to Mali
_____	_O. h. melanoleuca (xanthomeleana)_	SE Europe to Caspian and Iran; > to ne and w Africa

☐ **Red-tailed Wheatear** _Oenanthe xanthoprymna_

_____	_O. x. chrysopygia (kingi)_	Transcaucasia to Afghanistan; > to Pakistan and nw India
_____	_O. x. xanthoprymna_	Mts. of se Turkey to sw Iran; > Sinai, e Egypt and Sudan

☐ **Desert Wheatear** _Oenanthe deserti_

_____	_O. d. homochroa_	Deserts of North Africa (Western Sahara to w Egypt)
_____	_O. d. deserti_	Levant
_____	_O. d. atrogularis_	Transcaucasia and Iran to Afghanistan and Mongolia
_____	_O. d. oreophila_	W China to Kashmir and Tibet; > to Pakistan and ne Africa

☐ **Capped Wheatear** _Oenanthe pileata_

_____	_O. p. neseri_	S Angola and n Namibia to w Botswana
_____	_O. p. livingstonii_	E Angola to Zaire, Kenya, Tanzania, Malawi and Mozambique
_____	_O. p. pileata_	S Namibia and South Africa

☐ **Isabelline Wheatear** _Oenanthe isabellina_

S-central Eurasia; > ne Africa, Arabia and India

☐ **Red-breasted Wheatear** _Oenanthe bottae_

_____	_O. b. bottae_	Highlands of sw Arabia (Mecca to Yemen)
_____	_O. b. frenata_	Highlands of Eritrea and Ethiopia

☐ **Heuglin's Wheatear** _Oenanthe heuglini_

Mauritania to Mali, Cameroon, Sudan, Ethiopia and nw Kenya

☐ **Sicklewing Chat** _Cercomela sinuata_

_____	_C. s. ensifera_	S Namibia to Transvaal, Orange Free State and n Cape Province
_____	_C. s. hypernephela_	South Africa (arid regions of Lesotho); > to Natal
_____	_C. s. sinuata_	South Africa (s Cape Province)

☐ **Karoo Chat** _Cercomela schlegelii_

_____	_C. s. benguellensis_	Coastal sw Angola (Benguella escarpment)
_____	_C. s. schlegelii_	Coastal Namibia (w Damaraland to Erongo Mountains)
_____	_C. s. namaquensis_	S Namibia to nw Cape Province
_____	_C. s. kobosensis_	Namibia (Great Namaqualand)
_____	_C. s. pollux_	South Africa (w Orange Free State and Cape Province)

☐ **Tractrac Chat** *Cercomela tractrac*

____	*C. t. hoeschi*	Coastal deserts of sw Angola and nw Namibia
____	*C. t. albicans*	Coastal n Namibia (w Damaraland and n Great Namaqualand)
____	*C. t. barlowi*	Namibia (central and s Great Namaqualand)
____	*C. t. nebulosa*	Coastal sand dunes of sw Namibia and w South Africa
____	*C. t. tractrac*	South Africa (Karoo to Aliwal)

☐ **Familiar Chat** *Cercomela familiaris*

____	*C. f. falkensteini*	Ghana to sw Sudan, n Ethiopia, Uganda and Kenya
____	*C. f. omoensis*	SE Sudan (Boma Hills) to sw Ethiopia
____	*C. f. modesta*	NE Angola to Malawi, Zambia, Zimbabwe and Mozambique
____	*C. f. angolensis*	W Angola to n Namibia
____	*C. f. galtoni*	E Namibia to w Botswana and n Cape Province
____	*C. f. hellmayri*	SE Botswana to Zimbabwe, Transvaal and Orange Free State
____	*C. f. actuosa*	Drakensberg, Transkei, w Natal and Lesotho
____	*C. f. familiaris*	S Mozambique to Natal and s Cape Province

☐ **Brown-tailed Chat** *Cercomela scotocerca*

____	*C. s. furensis*	W Sudan (Darfur)
____	*C. s. scotocerca*	Red Sea coast of e Sudan to n Ethiopia
____	*C. s. turkana*	SW Ethiopia to nw Kenya and Uganda
____	*C. s. spectatrix*	Ethiopia (Awash Valley) and n Somalia
____	*C. s. validior*	Somalia (Run region)

☐ **Indian Chat** *Cercomela fusca*

Rocky hills and cliffs of ne Pakistan and n India

☐ **Sombre Chat** *Cercomela dubia*

Montane deserts of central Ethiopia and nw Somalia

☐ **Blackstart** *Cercomela melanura*

____	*C. m. ultima*	Mali and s Niger
____	*C. m. airensis*	E Niger (Aïr Massif) to Chad and w Sudan (Kordofan)
____	*C. m. lypura*	W coast of Red Sea from se Egypt to e Sudan and Eritrea
____	*C. m. aussae*	E Ethiopia (Donakil Depression), Djibouti and adjacent Somalia
____	*C. m. melanura*	Dead Sea depression of Egypt and Jordan to Saudi Arabia
____	*C. m. neumanni*	W Saudi Arabia to Yemen, Aden and Hadramaut

☐ **Moorland Chat** *Cercomela sordida*

____	*C. s. sordida*	High altitude moorlands of Ethiopia
____	*C. s. rudolfi*	N Kenya (Mt. Elgon moorlands) and adjacent e Uganda
____	*C. s. ernesti*	N Kenya (Mt. Kenya and Aberdare Mountains)
____	*C. s. olimotiensis*	High altitude moorlands of n Tanzania (Crater Highlands)
____	*C. s. hypospodia*	N Tanzania (Mt. Kilimanjaro moorlands)

☐ **Congo Moorchat** *Myrmecocichla tholloni*

Gabon to s Angola, Central African Republic and w Zaire

☐ **Northern Anteater-Chat** *Myrmecocichla aethiops*

____	*M. a. aethiops*	Senegambia to Chad, n Nigeria and n Cameroon
____	*M. a. sudanensis*	W Sudan (Darfur and Kordofan)
____	*M. a. cryptoleuca*	Highlands of n Kenya to n Tanzania

☐ **Southern Anteater-Chat** *Myrmecocichla formicivora*

Namibia to Botswana and South Africa

☐ **Sooty Chat** *Myrmecocichla nigra*

Nigeria to Angola, extreme s Sudan, Tanzania and Zambia

☐ **Rueppell's Chat** *Myrmecocichla melaena*

High plateau of Eritrea and n Ethiopia

☐ **White-fronted Black-Chat** *Myrmecocichla albifrons*

____	*M. a. frontalis*	Extreme s Mauritania to Senegal, Nigeria, Chad and Cameroon
____	*M. a. limbata*	E Cameroon to Central African Republic (Ubangi-Shari region)
____	*M. a. clericalis*	S Sudan (west of the Nile) to ne Zaire and n Uganda

____	*M. a. albifrons*	Eritrea and n Ethiopia
____	*M. a. pachyrhyncha*	SW Ethiopia

☐ **White-headed Black-Chat** *Myrmecocichla arnotti*

____	*M. a. harterti*	Angola
____	*M. a. arnotti*	SW Zaire to Namibia, Zambia, Tanzania, Malawi and n Transvaal

☐ **Mocking Cliff-Chat** *Thamnolaea cinnamomeiventris*

____	*T. c. bambarae*	Mali (Mandingo Mountains)
____	*T. c. coronata*	Togo and se Burkina Faso to n Cameroon and w Sudan (Darfur)
____	*T. c. kordofanensis*	Central Sudan (Kordofan and Nuba Hills)
____	*T. c. albiscapulata*	S Ethiopia
____	*T. c. subrufipennis*	Extreme se Sudan to sw Ethiopia, Zambia and Malawi
____	*T. c. odica*	E Zimbabwe
____	*T. c. cinnamomeiventris*	E Transvaal to Orange Free State, Natal and e Cape Province
____	*T. c. autochthones*	S Mozambique to e Transvaal, e Swaziland and n Natal

☐ **White-winged Cliff-Chat** *Thamnolaea semirufa*

Highlands of Eritrea and Ethiopia

☐ **Boulder Chat** *Pinarornis plumosus*

SE Zambia to Botswana, s Malawi, Zimbabwe and Mozambique

FAMILY: PLATYSTEIRIDAE (Wattle-eyes—31)

☐ **African Shrike-flycatcher** *Megabyas flammulatus*

____	*M. f. flammulatus*	Sierra Leone to Cameroon, Gabon and w Zaire; Bioko
____	*M. f. aequatorialis (carolathi)*	NW Angola to Zaire, Uganda, w Kenya and extreme s Sudan

☐ **Black-and-white Shrike-flycatcher** *Bias musicus*

____	*B. m. musicus (feminina, pallidiventris)*	Sierra Leone to n Angola, Zaire, Uganda and nw Tanzania
____	*B. m. changamwensis*	Kenya to e Tanzania
____	*B. m. clarens*	S Malawi to e Zimbabwe and Mozambique

☐ **Ward's Flycatcher** *Pseudobias wardi*

Rainforests of e Madagascar

☐ **Brown-throated Wattle-eye** *Platysteira cyanea*

____	*P. c. cyanea*	Senegal to Gabon, Angola, Central African Republic and Zaire
____	*P. c. aethiopica*	SE Sudan (Boma) and Ethiopia
____	*P. c. nyansae*	S Sudan to n Zaire, Kenya, Uganda and nw Tanzania

☐ **White-fronted Wattle-eye** *Platysteira albifrons*

Lowlands and escarpment of w Angola and adjacent sw Zaire

☐ **Black-throated Wattle-eye** *Platysteira peltata*

____	*P. p. cryptoleuca*	Somalia to e Zimbabwe and n Mozambique; Mafia I.
____	*P. p. mentalis*	Angola to s Zaire, Zambia, Uganda, Kenya and w Tanzania
____	*P. p. peltata*	Zambia to Malawi, Mozambique, e Zimbabwe and e S Africa

☐ **Banded Wattle-eye** *Platysteira laticincta*

Bamenda Mountains (w Cameroon)

☐ **Chestnut Wattle-eye** *Platysteira castanea*

____	*P. c. hormophora*	Sierra Leone to Togo
____	*P. c. castanea*	S Nigeria to se Sudan, Uganda, Kenya, n Tanzania, n Angola

☐ **White-spotted Wattle-eye** *Platysteira tonsa*

Forests of s Ivory Coast to Nigeria, Gabon and n Zaire

☐ **Red-cheeked Wattle-eye** *Platysteira blissetti*

Humid forests of Guinea and s Sierra Leone to s Cameroon

☐ **Black-necked Wattle-eye** *Platysteira chalybea*

Humid forests of s Cameroon and Gabon; Bioko

☐ **Jameson's Wattle-eye** *Platysteira jamesoni*

E Zaire to Uganda, se Sudan, w Kenya and nw Tanzania

☐ **Yellow-bellied Wattle-eye** *Platysteira concreta*
_____ *P. c. concreta*
_____ *P. c. ansorgei*
_____ *P. c. graueri (kumbaensis, harterti, silvae)*
_____ *P. c. kungwensis*

Sierra Leone to Ghana
Escarpment of w Angola (Cuanza Norte to n Huila)
Nigeria to Gabon, Zaire and w Kenya
Extreme w Tanzania (Mt. Nkungwe and Mt. Mahari)

☐ **Boulton's Batis** *Batis margaritae*
_____ *B. m. margaritae*
_____ *B. m. kathleenae*

W-central Angola (Mt. Moco)
Mountains of nw Zambia and adjacent extreme se Zaire

☐ **Short-tailed Batis** *Batis mixta*
_____ *B. m. ultima*
_____ *B. m. mixta*

Coastal se Kenya
Highlands of s Kenya to n Tanzania and n Malawi

☐ **Ruwenzori Batis** *Batis diops*

E Zaire to w Uganda, Rwanda, Burundi and nw Tanzania

☐ **Cape Batis** *Batis capensis*
_____ *B. c. reichenowi*
_____ *B. c. sola*
_____ *B. c. dimorpha*
_____ *B. c. erythrophthalma*
_____ *B. c. kennedyi*
_____ *B. c. hollidayi*
_____ *B. c. capensis*

SE Tanzania (Mikindani to Lindi)
N Malawi (Mwantjati to Nyika)
Mountains of s Malawi and adjacent Mozambique
E highlands of Zimbabwe and adjacent w Mozambique
SW Zimbabwe (Mopoto Hills region)
Zululand, Swaziland and Mozambique (Lebombo Range)
S Natal to Orange Free State and Cape Province

☐ **Woodward's Batis** *Batis fratrum*

S Malawi to e Zimbabwe, Mozambique and Natal

☐ **Chinspot Batis** *Batis molitor*
_____ *B. m. pintoi*
_____ *B. m. puella*
_____ *B. m. palliditergum*
_____ *B. m. molitor*

SE Gabon to Angola, sw Zaire and nw Zambia
E Zaire to Uganda, w Kenya and w Tanzania
S Zaire to Namibia, Botswana, Malawi and n Cape Province
S Mozambique to e Cape Province

☐ **Pale Batis** *Batis soror*

Lowlands of se Kenya to Malawi and Mozambique; Zanzibar

☐ **Pririt Batis** *Batis pririt*
_____ *B. p. affinis*
_____ *B. p. pririt*

Arid coastal sw Angola to n Namibia and w Botswana
Central Botswana to sw Transvaal and w Cape Province

☐ **Senegal Batis** *Batis senegalensis*

Senegambia to s Mauritania, s Niger and Cameroon

☐ **Gray-headed Batis** *Batis orientalis*
_____ *B. o. bella (somaliensis)*
_____ *B. o. orientalis*
_____ *B. o. chadensis*
_____ *B. o. lynesi*

N Eritrea and e Ethiopia to Djibouti, n Somalia and n Kenya
Central Eritrea to n Ethiopia
NE Nigeria to Zaire, Chad, Sudan and w Ethiopia
N Sudan (e Red Sea Province)

☐ **Black-headed Batis** *Batis minor*
_____ *B. m. erlangeri (congoensis, nyansae, batesi)*
_____ *B. m. minor*
_____ *B. m. suahelicus*

Ethiopian Plateau and Somalia to Cameroon and Angola
S Somalia
Kenya and Tanzania

☐ **Pygmy Batis** *Batis perkeo*

Arid s Ethiopia, Sudan, Somalia, Kenya to extreme ne Tanzania

☐ **Verreaux's Batis** *Batis minima*

Humid forests of s Cameroon and w Gabon

☐ **Ituri Batis** *Batis ituriensis*

Humid forests of Zaire and adjacent w Uganda

☐ **Fernando Po Batis** *Batis poensis*

Bioko (Gulf of Guinea)

☐ **West African Batis** *Batis occulta*

Humid forests of Sierra Leone to s Cameroon and Gabon

☐ **Angola Batis** *Batis minulla*

Forests of se Gabon to w Angola, Cabinda and adjacent Zaire

☐ **White-tailed Shrike** *Lanioturdus torquatus*

Angola to escarpment of central Namibia

FAMILY: RHIPIDURIDAE (Fantails—45)

☐ **Yellow-bellied Fantail** *Rhipidura hypoxantha*

E Himalayas to se Tibet, s China, Myanmar, Thailand, Tonkin

☐ **Blue Fantail** *Rhipidura superciliaris*
___ *R. s. superciliaris*	Philippines (Basilan and Zamboanga Peninsula of n Mindanao)
___ *R. s. apo*	Philippines (se Mindanao)
___ *R. s. samarensis*	Philippines (Bohol, Leyte and Samar)

☐ **Blue-headed Fantail** *Rhipidura cyaniceps*
___ *R. c. pinicola*	Philippines (highlands of n Luzon)
___ *R. c. cyaniceps*	Philippines (Luzon and Catanduanes)
___ *R. c. sauli*	Philippines (Tablas)
___ *R. c. albiventris*	Philippines (Guimaras, Masbate, Negros, Panay and Ticao)

☐ **Rufous-tailed Fantail** *Rhipidura phoenicura*

Montane forests of Java

☐ **Black-and-cinnamon Fantail** *Rhipidura nigrocinnamomea*
___ *R. n. hutchinsoni*	Philippines (montane forests of n Mindanao)
___ *R. n. nigrocinnamomea*	Philippines (montane forests of se Mindanao)

☐ **White-throated Fantail** *Rhipidura albicollis*
___ *R. a. canescens*	W Himalayas (Pakistan and Kashmir to w Nepal)
___ *R. a. albicollis*	Himalayas (w Nepal and Sikkim)
___ *R. a. orissae*	NE India
___ *R. a. stanleyi*	E Himalayas to Assam and Myanmar
___ *R. a. vernayi*	SE India
___ *R. a. celsa*	SE Tibet to s China, Hainan, w Thailand and n Indochina
___ *R. a. atrata*	S Thailand to Malaya and Sumatra
___ *R. a. sarawacensis*	N Borneo (Poi Mountains)
___ *R. a. kinabalu*	Mountains of n Borneo (Kinabalu to Murud and Mulu)
___ *R. a. cinerascens*	S Indochina

☐ **Spot-breasted Fantail** *Rhipidura albogularis*

S and c India (north to Rajasthan, Madhya Pradesh and Orissa)

☐ **White-bellied Fantail** *Rhipidura euryura*

Montane forests of Java

☐ **White-browed Fantail** *Rhipidura aureola*
___ *R. a. aureola*	N India
___ *R. a. compressirostris*	S peninsular India and Sri Lanka
___ *R. a. burmanica*	Assam to w Yunnan, Myanmar, pen. Thailand and Indochina

☐ **Northern Fantail** *Rhipidura rufiventris*
___ *R. r. obiensis*	Obi I. (n Moluccas)
___ *R. r. bouruensis*	Buru I. (s Moluccas)
___ *R. r. cinerea*	S Moluccas (Seram, Ambon and Boano)
___ *R. r. finitima*	Tayandu Islands (Taam, Kilsuin and Kur)
___ *R. r. perneglecta*	Watubela Islands (s Moluccas)
___ *R. r. assimilis*	Kai Islands (Kai Kecil and Kai Besar)
___ *R. r. tenkatei*	Roti (e Lesser Sundas)

____	*R. r. rufiventris*	E Lesser Sundas (Semau, Timor and Jaco)
____	*R. r. pallidiceps*	Wetar (e Lesser Sundas)
____	*R. r. buttikoferi*	E Lesser Sundas (Sermata, Moa, Leti, Romang and Damar)
____	*R. r. gularis*	New Guinea, Yapen and w Papuan islands
____	*R. r. kordensis*	Biak I. (New Guinea)
____	*R. r. vidua*	Kofiau I. (New Guinea)
____	*R. r. nigromentalis*	Louisiade Archipelago (Tagula and Misima)
____	*R. r. gigantea*	Bismarck Archipelago (Lihir and Tabar Groups)
____	*R. r. tangenensis*	Bismarck Archipelago (Boang and Tanga)
____	*R. r. mussai*	St. Matthias Islands (Bismarck Archipelago)
____	*R. r. setosa*	Bismarck Archipelago (New Ireland, New Hanover and Dyaul)
____	*R. r. finschii*	Bismarck Archipelago (New Britain and Duke of York)
____	*R. r. niveiventris*	Admiralty Islands
____	*R. r. tenkatei*	Roti I. (e Lesser Sundas)
____	*R. r. gularis*	Islands of n Torres Strait (Boigu I.)
____	*R. r. isura*	Western Australia (Broome) to Cape York and s to Burdekin R.

□ **Pied Fantail** *Rhipidura javanica*

____	*R. j. longicauda*	SE Asia to Sumatra, Borneo and adjacent islands
____	*R. j. javanica*	Java and Bali; single record from Lombok
____	*R. j. nigritorquis*	Philippine Islands and Sulu Archipelago

□ **Spotted Fantail** *Rhipidura perlata*

S peninsular Thailand, Malaya, Sumatra and Borneo

□ **Willie-wagtail** *Rhipidura leucophrys*

____	*R. l. picata*	N Australia (Broome, W Australia to Cape York Peninsula)
____	*R. l. leucophrys*	Mainland Australia, except north
____	*R. l. melaleuca*	Moluccas, New Guinea, Bismarck Arch. and Solomon Islands

□ **Brown-capped Fantail** *Rhipidura diluta*

____	*R. d. sumbawensis*	Sumbawa (w Lesser Sundas)
____	*R. d. diluta*	W Lesser Sundas (Flores and Lomblen)

□ **Cinnamon-tailed Fantail** *Rhipidura fuscorufa*

Tanimbar Is. (Larat, Yamdena, Lutu, Mutu, Selaru) and Babar

□ **White-winged Fantail** *Rhipidura cockerelli*

____	*R. c. cockerelli*	Solomon Islands (Guadalcanal)
____	*R. c. coultasi*	Solomon Islands (Malaita)
____	*R. c. septentrionalis*	Solomon Islands (Buka, Bougainville and Shortland)
____	*R. c. interposita*	Solomon Islands (Choiseul and Santa Isabel)
____	*R. c. floridana*	Solomon Islands (Florida and Tulagi)
____	*R. c. lavellae*	Solomon Islands (Vellalavella and Ranongga)
____	*R. c. albina*	Solomon Islands (Kulambangra and Rendova)

□ **Friendly Fantail** *Rhipidura albolimbata*

____	*R. a. albolimbata*	Mountains of nw New Guinea to Huon Peninsula
____	*R. a. lorentzi*	Snow Mountains and Central Highlands of New Guinea

□ **Chestnut-bellied Fantail** *Rhipidura hyperythra*

____	*R. h. hyperythra*	Aru Islands (New Guinea)
____	*R. h. mulleri*	Yapen I. and w New Guinea (Astrolabe Bay to Lake Kutubu)
____	*R. h. castaneothorax*	SE New Guinea (Saruwaged Mtns. to Angabunga River)

□ **Sooty Thicket-Fantail** *Rhipidura threnothorax*

____	*R. t. threnothorax*	New Guinea, Aru, Waigeo, Salawati and Misool islands
____	*R. t. fumosa*	Yapen I. (New Guinea)

□ **Black Thicket-Fantail** *Rhipidura maculipectus*

Lowlands of New Guinea, Aru, Batanta and Salawati islands

☐ **White-bellied Thicket-Fantail** *Rhipidura leucothorax*
_____ *R. l. leucothorax* — NW New Guinea (Astrolabe Bay to Port Moresby)
_____ *R. l. clamosa* — Karimui Basin and adjacent e-central New Guinea
_____ *R. l. episcopalis* — SE New Guinea (Kapa Kapa to Astrolabe Bay)

☐ **Black Fantail** *Rhipidura atra*
_____ *R. a. atra* — Mountains of New Guinea and Waigeo I.
_____ *R. a. vulpes* — N New Guinea (Cyclops Mountains)

☐ **Mangrove Fantail** *Rhipidura phasiana*

SE New Guinea (Trans-Fly lowlands) and n Australia

☐ **Brown Fantail** *Rhipidura drownei*
_____ *R. d. drownei* — Montane forests of Bougainville (Solomon Islands)
_____ *R. d. ocularis* — Montane forests of Guadalcanal (Solomon Islands)

☐ **Dusky Fantail** *Rhipidura tenebrosa*

Mountains of San Cristóbal (Solomon Islands)

☐ **Rennell Fantail** *Rhipidura rennelliana*

Forests of Rennell (se Solomon Islands)

☐ **Gray Fantail** *Rhipidura albiscapa*
_____ *R. a. bulgeri* — New Caledonia and Lifou I.
_____ *R. a. brenchleyi* — Vanuatu and Banks Islands; San Cristóbal (s Solomon Islands)
_____ *R. a. keasti* — NE Queensland (Cooktown to Clarke Range)
_____ *R. a. pelzelni* — Norfolk Island
_____ *R. a. alisteri* — S-central Qld. to s Victoria and Eyre Pen., South Australia
_____ *R. a. albiscapa* — Tasmania, King and Flinders Islands (Bass Strait); > to n
_____ *R. a. preissi* — Southwest Western Australia; > to north
_____ *R. a. albicauda* — Interior of s-central Western Australia and s Northern Territory

☐ **New Zealand Fantail** *Rhipidura fuliginosa*
_____ *R. f. fuliginosa* — Stewart I. and South I. (New Zealand)
_____ *R. f. placabilis* — North I. (New Zealand)
_____ *R. f. penitus* — Chatham Islands
_____ *R. f. cervina†* — Formerly Lord Howe I. Extinct

☐ **Streaked Fantail** *Rhipidura spilodera*
_____ *R. s. spilodera* — Vanuatu and Banks Group
_____ *R. s. layardi* — Fiji (Ovalau and Viti Levu)
_____ *R. s. erythronota* — Fiji (Yanganga and Vanua Levu)
_____ *R. s. rufilateralis* — Taveuni (Fiji)
_____ *R. s. verreauxi* — New Caledonia, Lifou and Maré islands

☐ **Kandavu Fantail** *Rhipidura personata*

Dense riparian thickets of Kandavu (sw Fiji)

☐ **Samoan Fantail** *Rhipidura nebulosa*
_____ *R. n. nebulosa* — Mountains of Upolu (Western Samoa)
_____ *R. n. altera* — Mountains of Savai'i (Western Samoa)

☐ **Dimorphic Fantail** *Rhipidura brachyrhyncha*
_____ *R. b. brachyrhyncha* — NW New Guinea (Arfak Mountains)
_____ *R. b. devisi* — Mountains of se New Guinea and Huon Peninsula

☐ **Rusty-flanked Fantail** *Rhipidura teysmanni*
_____ *R. t. toradja* — Mountains of n, central and se Sulawesi
_____ *R. t. teysmanni* — Mt. Lompobatang (sw Sulawesi)
_____ *R. t. sulaensis* — Taliabu I. (Sula Islands)

☐ **Cinnamon-backed Fantail** *Rhipidura superflua*

Montane forests of Buru (s Moluccas)

☐ **Streaky-breasted Fantail** *Rhipidura dedemi*

Seram (s Moluccas)

☐ **Long-tailed Fantail** *Rhipidura opistherythra*

Tanimbar Islands (Larat, Yamdena and Maru)

☐ **Palau Fantail** *Rhipidura lepida*

Palau Islands (Babelthuap to Peleliu)

☐ **Rufous-backed Fantail** *Rhipidura rufidorsa*

_____ *R. r. rufidorsa*   NW New Guinea, Misool and Yapen islands
_____ *R. r. kumusi*   N coast of se New Guinea (Kumusi River to Collingwood Bay)
_____ *R. r. kubuna*   S coast of se New Guinea

☐ **Matthias Fantail** *Rhipidura matthiae*

St. Matthias Group (n Bismarck Archipelago)

☐ **Bismarck Fantail** *Rhipidura dahli*

_____ *R. d. dahli*   Mountains of New Britain (Bismarck Archipelago)
_____ *R. d. antonii*   Mountains of New Ireland (Bismarck Archipelago)

☐ **Malaita Fantail** *Rhipidura malaitae*

Montane forests of Malaita (se Solomon Islands)

☐ **Manus Fantail** *Rhipidura semirubra*

Manus, San Miguel, Tong and adjacent Admiralty Islands

☐ **Rufous Fantail** *Rhipidura rufifrons*

_____ *R. r. torrida*   N Moluccas (Halmahera, Ternate, Bacan and Obi)
_____ *R. r. louisiadensis*   D'Entrecasteaux and Louisiade archipelagos
_____ *R. r. uraniae*   Guam (Mariana Islands)
_____ *R. r. saipanensis*   Mariana Islands (Saipan and Tinian)
_____ *R. r. mariae*   Mariana Islands (Agiguan and Rota)
_____ *R. r. versicolor*   Yap (Caroline Islands)
_____ *R. r. kubaryi*   Pohnpei I. (Caroline Islands)
_____ *R. r. melaenolaema*   Vanikoro (Santa Cruz Islands)
_____ *R. r. agilis*   Santa Cruz Group (Solomon Islands)
_____ *R. r. utupuae*   Utupua (Solomon Islands)
_____ *R. r. commoda*   Bougainville, Choiseul and adjacent Solomon Islands
_____ *R. r. rufofronta*   Guadalcanal (Solomon Islands)
_____ *R r. granti*   Central Solomon Islands
_____ *R. r. russata*   San Cristóbal (Solomon Islands)
_____ *R. r. brunnea*   Malaita (Solomon Islands)
_____ *R. r. kuperi*   Santa Anna (Solomon Islands)
_____ *R. r. ugiensis*   Ugi (Solomon Islands)
_____ *R. r. intermedia*   E Queensland (Cooktown to NSW border); > to north
_____ *R. r. rufifrons*   SE Australia (ne NSW to s and cent. Victoria); > to north

☐ **Arafura Fantail** *Rhipidura dryas*

_____ *R. d. celebensis*   Tanahjampea and Lalao islands (Flores Sea)
_____ *R. d. mimosae*   Kalaotoa I. (Flores Sea)
_____ *R. d. sumbaensis*   Sumba (Lesser Sundas)
_____ *R. d. semicollaris*   Flores to Timor and adjacent Lesser Sundas
_____ *R. d. elegantula*   E Lesser Sundas (Leti, Moa, Romang and Damar)
_____ *R. d. reichenowi*   Babar I. (e Lesser Sundas)
_____ *R. d. hamadryas*   Tanimbar Islands (Arafura Sea)
_____ *R. d. squamata*   Banda, Seram Laut, Tayandu Is. and Kai Is.
_____ *R. d. henrici*   Aru Islands
_____ *R. d. streptophora*   S New Guinea
_____ *R. d. dryas*   Coastal n Australia (Kimberley to w Cape York Pen.)

☐ **Pohnpei Fantail** *Rhipidura kubaryi*

Pohnpei (e Caroline Islands)

FAMILY: MONARCHIDAE (Monarch Flycatchers—100)

☐ **Chestnut-capped Flycatcher** *Erythrocercus mccallii*

_____ *E. m. nigeriae*   Dense forests of Sierra Leone to Guinea and sw Nigeria

____ *E. m. mccallii*	SE Nigeria to Cameroon, Gabon and Zaire
____ *E. m. congicus*	E and s Zaire to w Uganda

☐ **Yellow Flycatcher** *Erythrocercus holochlorus*

Coastal s Somalia to se Kenya and ne Tanzania

☐ **Livingstone's Flycatcher** *Erythrocercus livingstonei*

____ *E. l. thomsoni*	S Tanzania to Malawi and n Mozambique
____ *E. l. livingstonei*	Zambia to Zimbabwe and nw Mozambique
____ *E. l. francisi*	S Malawi and Mozambique (south to Limpopo River)

☐ **African Blue-Flycatcher** *Elminia longicauda*

____ *E. l. longicauda*	Senegal to Gambia and Nigeria
____ *E. l. teresita*	Cameroon to s Sudan, w Kenya, nw Tanzania and Angola

☐ **White-tailed Blue-Flycatcher** *Elminia albicauda*

Angola to sw Uganda, Tanzania and n Mozambique

☐ **Dusky Crested-Flycatcher** *Elminia nigromitrata*

____ *E. n. colstoni*	Liberia to Nigeria
____ *E. n. nigromitrata*	Cameroon to s Sudan, Kenya, Uganda and Tanzania

☐ **White-bellied Crested-Flycatcher** *Elminia albiventris*

____ *E. a. albiventris*	Montane forests of se Nigeria and s Cameroon; Bioko
____ *E. a. toroensis*	Montane forests of e Zaire, Rwanda and sw Uganda

☐ **White-tailed Crested-Flycatcher** *Elminia albonotata*

____ *E. a. albonotata*	Zaire to s Ethiopia, Kenya, Tanzania, n Malawi and Zambia
____ *E. a. subcaerulea*	E Tanzania to Malawi and Mozambique (north of the Zambezi)
____ *E. a. swynnertoni*	E Zimbabwe and Mozambique (south of the Zambezi)

☐ **Blue-headed Crested-Flycatcher** *Trochocercus nitens*

____ *T. n. reichenowi*	Guinea and Sierra Leone to Togo
____ *T. n. nitens*	Nigeria and Cameroon to Gabon, n Angola, e Zaire and s Sudan

☐ **African Crested-Flycatcher** *Trochocercus cyanomelas*

____ *T. c. vivax*	Uganda and w Tanzania to s Zaire and Zambia
____ *T. c. bivittatus*	Somalia to Kenya, e Tanzania and Zanzibar
____ *T. c. megalolophus*	Malawi and n Mozambique to Zimbabwe and n Natal
____ *T. c. segregus*	E Transvaal to Natal and w Zululand
____ *T. c. cyanomelas*	South Africa (w Transkei to sw Cape Province)

☐ **Short-crested Monarch** *Hypothymis helenae*

____ *H. h. personata*	N Philippines (Camiguin Norte)
____ *H. h. helenae*	N Philippines (Luzon, Samar and Polillo)
____ *H. h. agusanae*	S Philippines (Mindanao, Dinagat and Siargao)

☐ **Black-naped Monarch** *Hypothymis azurea*

____ *H. a. styani*	India and Nepal to s China and Indochina; Hainan
____ *H. a. oberholseri*	Taiwan
____ *H. a. forrestia*	Mergui Archipelago (s Myanmar)
____ *H. a. montana*	N and central Thailand
____ *H. a. galerita*	Peninsular Thailand
____ *H. a. prophata*	S Thailand, Malaysia, Sumatra and Borneo
____ *H. a. tytleri*	Andaman Islands and Cocos Islands (Bay of Bengal)
____ *H. a. ceylonensis*	Sri Lanka
____ *H. a. idiochroa*	Car Nicobar I.
____ *H. a. nicobarica*	Nicobar Islands
____ *H. a. opisthocyanea*	Anambas Islands (South China Sea)
____ *H. a. javana*	Java and Bali
____ *H. a. penidae*	Penida I. (Lesser Sundas)
____ *H. a. karimatensis*	Karimata I. (off w Borneo)
____ *H. a. gigantoptera*	Natunas Islands (South China Sea)

____ *H. a. aeria*	Maratua Islands (off Borneo)
____ *H. a. consobrina*	Simeulue I. (off Sumatra)
____ *H. a. leucophila*	Mentawi Archipelago (off Sumatra)
____ *H. a. richmondi*	Enggano I. (Sumatra)
____ *H. a. abbotti*	Babi and Masia islands (Malaysia)
____ *H. a. symmixta*	Lesser Sundas
____ *H. a. azurea*	Philippine Islands
____ *H. a. catarmanensis*	S Philippines (Camiguin Sur)

☐ **Pale-blue Monarch** *Hypothymis puella*

____ *H. p. puella*	Sulawesi and adjacent islands
____ *H. p. blasii*	Banggai and Sula islands

☐ **Celestial Monarch** *Hypothymis coelestis*

Philippine Islands

☐ **Cerulean Paradise-Flycatcher** *Eutrichomyias rowleyi*

Rediscovered 1995 on Sangihe I. after considered extinct

☐ **Black-headed Paradise-Flycatcher** *Terpsiphone rufiventer*

____ *T. r. rufiventer*	Senegal and Gambia to Guinea-Bissau and w Guinea
____ *T. r. nigriceps*	Sierra Leone to Togo and sw Benin
____ *T. r. fagani*	Benin and sw Nigeria
____ *T. r. tricolor*	Bioko (Gulf of Guinea)
____ *T. r. smithii*	Pagalu (Gulf of Guinea)
____ *T. r. neumanni*	SE Nigeria to s Cameroon, Gabon, Cabinda and n Angola
____ *T. r. schubotzi*	SE Cameroon and sw Central African Republic
____ *T. r. mayombe*	Congo and w Zaire (Lukolela, Mayombe and Ubangi)
____ *T. r. somereni*	Forests of w Uganda
____ *T. r. emini*	SE Uganda to extreme w Kenya and nw Tanzania
____ *T. r. ignea*	E Central African Rep. to s Zaire, ne Angola and nw Zambia

☐ **Bedford's Paradise-Flycatcher** *Terpsiphone bedfordi*

E Zaire (ne Ituri and region west of Itombwe and Kahuzi Mts.)

☐ **Rufous-vented Paradise-Flycatcher** *Terpsiphone rufocinerea*

SE Nigeria and s Cameroon to Gabon to n Angola

☐ **Bates' Paradise-Flycatcher** *Terpsiphone batesi*

____ *T. b. batesi*	S Cameroon, Rio Muni and Gabon to e Zaire
____ *T. b. bannermani*	N Angola to Zaire (lower Congo River) and Congo

☐ **African Paradise-Flycatcher** *Terpsiphone viridis*

____ *T. v. viridis*	Senegal and Gambia to Sierra Leone
____ *T. v. speciosa*	S Cameroon to e Zaire, s Sudan and Gabon
____ *T. v. ferreti*	Mali and Ivory Coast to ne Zaire, Sudan, Kenya and Tanzania
____ *T. v. restricta*	Lake Victoria region of w Kenya and Uganda
____ *T. v. kivuensis*	SW Uganda to e Zaire, Rwanda, Burundi and nw Tanzania
____ *T. v. suahelica*	Highlands of w Kenya and Tanzania
____ *T. v. ungujaensis*	E Tanzania to Zambia; Zanzibar, Pemba I. and Mafia I.
____ *T. v. plumbeiceps*	S Angola to w Zaire, sw Tanzania and ne South Africa
____ *T. v. granti*	Natal to sw Cape Province; > to s Tanzania
____ *T. v. harterti*	S Arabian peninsula

☐ **Sáo Tomé Paradise-Flycatcher** *Terpsiphone atrochalybeia*

Sáo Tomé (Gulf of Guinea)

☐ **Madagascar Paradise-Flycatcher** *Terpsiphone mutata*

____ *T. m. mutata*	Forests of e Madagascar
____ *T. m. singetra*	Forests of w Madagascar
____ *T. m. pretiosa*	Mayotte (Indian Ocean)
____ *T. m. vulpina*	Anjouan (Comoro Islands)
____ *T. m. voeltzkowiana*	Mohéli (Comoro Islands)
____ *T. m. comoroensis*	Grand Comoro I.

☐ **Seychelles Paradise-Flycatcher** *Terpsiphone corvina*

Forests of La Digue (Seychelles)

☐ **Mascarene Paradise-Flycatcher** *Terpsiphone bourbonnensis*

_____ *T. b. bourbonnensis* Réunion (Mascarene Islands)
_____ *T. b. desolata* Mauritius (Mascarene Islands)

☐ **Japanese Paradise-Flycatcher** *Terpsiphone atrocaudata*

_____ *T. a. atrocaudata* Japan (Honshu, Shikoku and Kyushu); > to SE Asia
_____ *T. a. illex* Ryukyu Islands and Taiwan
_____ *T. a. periophthalmica* Philippines (Luzon, Batan, Mindoro and Palawan)

☐ **Blue Paradise-Flycatcher** *Terpsiphone cyanescens*

SW Philippines (Calamian Group and Palawan)

☐ **Rufous Paradise-Flycatcher** *Terpsiphone cinnamomea*

_____ *T. c. unirufa* N Philippines (Luzon to Negros)
_____ *T. c. cinnamomea* S Philippines (Mindanao, Basilan and islands in Sulu Arch.)
_____ *T. c. talautensis* Talaud Islands (Karakelong, Salebabu and Kaburuang)

☐ **Asian Paradise-Flycatcher** *Terpsiphone paradisi*

_____ *T. p. leucogaster* Mountains of Afghanistan, Pakistan and w India
_____ *T. p. paradisi* Central and s India; > to Sri Lanka
_____ *T. p. ceylonensis* Sri Lanka
_____ *T. p. incei* China, Manchuria and Japan; > to Malaysia and Sumatra
_____ *T. p. saturatior* E Himalayas, Assam and Bangladesh; > to Malaysia
_____ *T. p. myanmare* Central and s Myanmar
_____ *T. p. indochinensis* S China (s Yunnan) to s Thailand and Indochina
_____ *T. p. affinis* Malaya, Sumatra, Riau and Lingga arch., Bangka, Belitung is.
_____ *T. p. nicobarica* Andaman and Nicobar islands
_____ *T. p. madzoedi* N Sumatra
_____ *T. p. australis* S Sumatra and Java
_____ *T. p. borneensis* Borneo
_____ *T. p. procera* Simeulue I. (off Sumatra)
_____ *T. p. insularis* Nias I. (off Sumatra)
_____ *T. p. sumbaensis* Sumba (w Lesser Sundas)
_____ *T. p. floris* W Lesser Sundas (Sumbawa, Alor, Besar, Lomblen and Flores)

☐ **Elepaio** *Chasiempis sandwichensis*

_____ *C. s. sandwichensis* Drier areas of Hawaii (Hawaiian Islands)
_____ *C. s. bryani* *Mamame-naio* forests on Mauna Kea (Hawaii)
_____ *C. s. ridgwayi* Wetter areas of Hawaii (Hawaiian Islands)
_____ *C. s. sclateri* Kauai (Hawaiian Islands)
_____ *C. s. gayi* Oahu (Hawaiian Islands)

☐ **Rarotonga Monarch** *Pomarea dimidiata*

Forest undergrowth of Rarotonga (sw Cook Islands)

☐ **Tahiti Monarch** *Pomarea nigra*

_____ *P. n. nigra* Highlands of Tahiti (Society Islands)
_____ *P. n. pomarea†* Formerly highlands of Maupiti I. (Society Islands). Extinct

☐ **Iphis Monarch** *Pomarea iphis*

_____ *P. i. iphis* Uahuka (n Marquesas Islands)
_____ *P. i. fluxa†* Formerly Eioa (n Marquesas Islands). Extinct

☐ **Marquesas Monarch** *Pomarea mendozae*

Marquesas (Nukuhiva, Uapou, Hivaoa, Tahuata and Motane)

☐ **Fatuhiva Monarch** *Pomarea whitneyi*

Fatuhiva (s Marquesas Islands)

☐ **Ogea Monarch** *Mayrornis versicolor*

Forests of Ogea Levu (se Fiji)

☐ **Slaty Monarch** *Mayrornis lessoni*
    ____   *M. l. lessoni*   Northwest Fiji Islands and Lau Archipelago
    ____   *M. l. orientalis*   E Fiji Islands

☐ **Vanikoro Monarch** *Mayrornis schistaceus*

    Vanikoro (Santa Cruz Islands)

☐ **Buff-bellied Monarch** *Neolalage banksiana*

    Vanuatu and Banks Group (se Melanesia)

☐ **Southern Shrikebill** *Clytorhynchus pachycephaloides*
    ____   *C. p. pachycephaloides*   New Caledonia
    ____   *C. p. grisescens*   Vanuatu, Banks and Torres groups (se Melanesia)

☐ **Rennell Shrikebill** *Clytorhynchus hamlini*

    Rennell (se Solomon Islands)

☐ **Fiji Shrikebill** *Clytorhynchus vitiensis*
    ____   *C. v. powelli*   American Samoa (Tau, Ofu, Olosega and Manua)
    ____   *C. v. compressirostris*   Fiji (Kandavu, Ono and Vanuakula)
    ____   *C. v. vitiensis*   W Fiji Islands
    ____   *C. v. buensis*   Fiji (Vanua Levu and Kioa)
    ____   *C. v. layardi*   Taveuni (Fiji)
    ____   *C. v. pontifex*   W Fiji Islands (Ngamea and Rambi)
    ____   *C. v. wiglesworthi*   Rotuma I. (Fiji)
    ____   *C. v. vatuanus*   N Lau Archipelago (e Fiji Islands)
    ____   *C. v. nesiotes*   S Lau Archipelago (e Fiji Islands)
    ____   *C. v. heinei*   Central Tonga Islands
    ____   *C. v. fortunae*   Futuna and Alifi islands (Wallis and Futuna Group)
    ____   *C. v. keppeli*   Keppel and Boscawen islands (between Tonga and Samoa)

☐ **Black-throated Shrikebill** *Clytorhynchus nigrogularis*
    ____   *C. n. nigrogularis*   Mountains of larger Fiji Islands
    ____   *C. n. sanctaecrucis*   Mountains of Santa Cruz Group (Solomon Islands*)*

☐ **Truk Monarch** *Metabolus rugensis*

    Truk (e Caroline Islands)

☐ **Black Monarch** *Monarcha axillaris*
    ____   *M. a. axillaris*   NW New Guinea (Arfak, Weyland and Wandammen mts.)
    ____   *M. a. fallax*   Mountains of se New Guinea and Goodenough I.

☐ **Rufous Monarch** *Monarcha rubiensis*

    Lowlands of w New Guinea (absent from Vogelkop Peninsula)

☐ **Island Monarch** *Monarcha cinerascens*
    ____   *M. c. commutatus*   Sangihe and Siau islands (north of Sulawesi)
    ____   *M. c. jabobii*   Talaud Islands (Moluccas)
    ____   *M. c. disjunctus*   Lesser Sundas
    ____   *M. c. intercedens*   Sulawesi, Tukangbesi, Peleng, Banggai and Sula islands
    ____   *M. c. cinerascens*   Lesser Sundas (Timor, Wetar and Romang)
    ____   *M. c. kisserensis*   Lesser Sundas (Kisar, Damar, Kai and Tanimbar Islands)
    ____   *M. c. harterti*   N and s Moluccas
    ____   *M. c. brunneus*   Great Banda I. (s Moluccas)
    ____   *M. c. inornatus*   NW New Guinea, Aru, Waigeo, Salawati and Misool islands
    ____   *M. c. steini*   Numfor I. (New Guinea)
    ____   *M. c. geelvinkianus*   Yapen and Biak islands (New Guinea)
    ____   *M. c. fuscescens*   Islands in Geelvink Bay (n New Guinea)
    ____   *M. c. nigrirostris*   Coastal n New Guinea (Dagua to Huon Gulf)
    ____   *M. c. fulviventris*   Ninigo, Hermit, Anchorite and Admiralty islands
    ____   *M. c. perpallidus*   St. Matthias Group and w Bismarck Archipelago
    ____   *M. c. impediens*   E Bismarck Archipelago to Solomon Islands
    ____   *M. c. rosselianus*   D'Entrecasteaux, Bismarck and Louisiade archipelagos

☐ **Black-winged Monarch** *Monarcha frater*
   ____ *M. f. frater*
   ____ *M. f. kunupi*
   ____ *M. f. periophthalmicus*
   ____ *M. f. canescens*

Vogelkop Mountains (nw New Guinea)
Weyland Mountains (central New Guinea)
Mountains of e and se New Guinea
S Torres Strait islands and e Cape York Pen. s to Cape Flattery

☐ **Black-faced Monarch** *Monarcha melanopsis*

E Australia (Cooktown to ne Victoria); > to New Guinea

☐ **Bougainville Monarch** *Monarcha erythrostictus*

Rainforests of Bougainville (Solomon Islands)

☐ **Chestnut-bellied Monarch** *Monarcha castaneiventris*
   ____ *M. c. castaneiventris*
   ____ *M. c. obscurior*
   ____ *M. c. megarhyncha*
   ____ *M. c. ugiensis*

Solomons (Guadalcanal, Malaita, Santa Isabel, Florida, Choiseul)
Rossel (Solomon Islands)
San Cristóbal (Solomon Islands)
Ugi (Solomon Islands)

☐ **White-capped Monarch** *Monarcha richardsii*

Forests of central Solomon Islands

☐ **White-naped Monarch** *Monarcha pileatus*
   ____ *M. p. pileatus*
   ____ *M. p. buruensis*

Halmahera (n Moluccas)
Buru (n Moluccas)

☐ **Loetoe Monarch** *Monarcha castus*

Tanimbar Is. (Larat, Yamdena, Selaru); Tayandu Is. (Kilsuin)

☐ **White-eared Monarch** *Monarcha leucotis*

NE Australia (McIlwraith Range, Qld. to far northeast NSW)

☐ **Spot-winged Monarch** *Monarcha guttulus*

New Guinea, w Papuan islands and Louisiade Archipelago

☐ **Black-bibbed Monarch** *Monarcha mundus*

E Lesser Sundas (Babar, Damar, Larat, Yamdena and Selaru)

☐ **Spectacled Monarch** *Monarcha trivirgatus*
   ____ *M. t. bimaculatus*
   ____ *M. t. diadematus*
   ____ *M. t. boanensis*
   ____ *M. t. nigrimentum*
   ____ *M. t. wellsi*
   ____ *M. t. trivirgatus*
   ____ *M. t. bernsteinii*
   ____ *M. t. albiventris*
   ____ *M. t. melanorrhoa*
   ____ *M. t. gouldii*
   ____ *M. t. melanopterus*

N Moluccas (Morotai, Halmahera and Bacan)
N Moluccas (Obi and Bisa)
Boano I. (s Moluccas)
S Moluccas (Seram and Ambon)
Seram Laut Is. (Gorong, Manawoka) and Watubela Is. (Kasiui)
Lesser Sundas
Salawati I. (New Guinea)
S New Guinea; Torres Straits is.; Cape York Pen to Burdekin R.
E Queensland (Cooktown to Burdekin River)
E Australia (Clarke Range, Queensland to near Sydney, NSW)
Louisiade Archipelago

☐ **Flores Monarch** *Monarcha sacerdotum*

SW Flores (w Lesser Sundas)

☐ **White-tipped Monarch** *Monarcha everetti*

Tanahjampea I. (Flores Sea)

☐ **Black-tipped Monarch** *Monarcha loricatus*

Lowlands of Buru (s Moluccas)

☐ **Black-chinned Monarch** *Monarcha boanensis*

Boano I. (s Moluccas)

☐ **White-tailed Monarch** *Monarcha leucurus*

Kai Islands (Kai Kecil, Kai Besar and Baer)

☐ **Black-backed Monarch** *Monarcha julianae*

Known from a 1959 specimen from Kofiau I. (New Guinea)

☐ **Hooded Monarch** *Monarcha manadensis*

Patchily distributed lowland forests of New Guinea

☐ **Biak Monarch** *Monarcha brehmii*

Biak I. off nw New Guinea

☐ **Manus Monarch** *Monarcha infelix*
_____ *M. i. infelix* — Manus I. (n Bismarck Archipelago)
_____ *M. i. coultasi* — Rambutyo I. (n Bismarck Archipelago)

☐ **White-breasted Monarch** *Monarcha menckei*
— Mussau I. (St. Matthias Group in Bismarck Archipelago)

☐ **Black-tailed Monarch** *Monarcha verticalis*
_____ *M. v. ateralbus* — Dyaul I. (Bismarck Archipelago)
_____ *M. v. verticalis* — New Britain, New Ireland and adj. islands in Bismarck Arch.

☐ **Kulambangra Monarch** *Monarcha browni*
_____ *M. b. browni* — Kulambangra (New Georgia Group of Solomon Islands)
_____ *M. b. ganongae* — Ranongga (New Georgia Group of Solomon Islands)
_____ *M. b. nigrotectus* — Vellalavella (New Georgia Group of Solomon Islands)
_____ *M. b. meeki* — Rendova and Tetipari (New Georgia Group of Solomon Islands)

☐ **White-collared Monarch** *Monarcha viduus*
— San Cristóbal (s Solomon Islands)

☐ **Black-and-white Monarch** *Monarcha barbatus*
_____ *M. b. barbatus* — Solomons (Bougainville, Guadalcanal, Choiseul and Santa Isabel)
_____ *M. b. malaitae* — Malaita (Solomon Islands)

☐ **Yap Monarch** *Monarcha godeffroyi*
— Forests of Yap (nw Caroline Islands)

☐ **Tinian Monarch** *Monarcha takatsukasae*
— Mariana Islands (Tinian and Agiguan)

☐ **Golden Monarch** *Monarcha chrysomela*
_____ *M. c. aurantiacus* — N New Guinea (Geelvink to Astrolabe Bay and Ramu River)
_____ *M. c. melanonotus* — NW New Guinea and w Papuan islands
_____ *M. c. kordensis* — Biak and Misool islands (New Guinea)
_____ *M. c. nitidus* — E and se New Guinea and Louisiade Archipelago
_____ *M. c. aruensis* — S New Guinea and Aru Islands
_____ *M. c. pulcherrimus* — Dyaul I. (Bismarck Archipelago)
_____ *M. c. chrysomela* — Bismarck Archipelago (New Hanover and New Ireland)
_____ *M. c. whitneyorum* — Lihir Group (Bismarck Archipelago)
_____ *M. c. tabarensis* — Tabar Group (Bismarck Archipelago)

☐ **Frilled Monarch** *Arses telescophthalmus*
_____ *A. t. telescophthalmus* — NW New Guinea, Salawati and Misool islands
_____ *A. t. batantae* — Batanta and Waigeo islands (New Guinea)
_____ *A. t. aruensis* — Aru Islands (New Guinea)
_____ *A. t. lauterbachi* — N coast of se New Guinea (Milne Bay to Huon Peninsula)
_____ *A. t. harterti* — S New Guinea (Mimika River to Purari River); Boigu I.
_____ *A. t. henkei* — Coastal se New Guinea (Hall Sound to Orangerie Bay)

☐ **Frill-necked Monarch** *Arses lorealis*
— N Queensland (coastal Cape York Pen. south to Cooktown)

☐ **Rufous-collared Monarch** *Arses insularis*
— Lowlands of n New Guinea and Yapen I.

☐ **Pied Monarch** *Arses kaupi*
_____ *A. k. terraereginae* — NE Queensland (Annan River to Mossman)
_____ *A. k. kaupi* — NE Queensland (Mossman to Seaview Range)

☐ **Guam Flycatcher** *Myiagra freycineti*
— Guam (s Mariana Islands). On verge of extinction

☐ **Palau Flycatcher** *Myiagra erythrops*
— Mangroves and lowlands of Palau (w Caroline Islands)

☐ **Pohnpei Flycatcher** *Myiagra pluto*
— Pohnpei (e Caroline Islands)

☐ **Oceanic Flycatcher** *Myiagra oceanica*
— Truk (w Caroline Islands)

☐ **Biak Flycatcher** *Myiagra atra*

Numfor and Biak islands (New Guinea)

☐ **Moluccan Flycatcher** *Myiagra galeata*

____ *M. g. galeata*   N Moluccas (Obi, Bacan, Ternate, Halmahera, Bisa, Morotai)
____ *M. g. buruensis*   Buru (s Moluccas)
____ *M. g. seramensis*   S Moluccas (Seram, Ambon and Boano)
____ *M. g. goramensis*   Seram Laut and Kai Islands (Kai Cecil)

☐ **Leaden Flycatcher** *Myiagra rubecula*

____ *M. r. papuana*   Savanna of s New Guinea and Boigu I. (n Torres Strait)
____ *M. r. concinna*   N Australia (King Sound, Western Australia to nw Queensland)
____ *M. r. okyri*   N Queensland (Cape York Peninsula)
____ *M. r. yorki*   E Australia (Burdekin River, Qld. to far ne New South Wales)
____ *M. r. rubecula*   SE Australia (ne NSW to central and s Victoria); > to north
____ *M. r. sciurorum*   Louisiade and D'Entrecasteaux archipelagos

☐ **Steel-blue Flycatcher** *Myiagra ferrocyanea*

____ *M. f. ferrocyanea* Solomon Islands (Santa Isabel, Choiseul and Guadalcanal)
____ *M. f. feminina*   Kulambangra (New Georgia Group in Solomon Islands)
____ *M. f. cinerea*   Bougainville (Solomon Islands)
____ *M. f. malaitae*   Malaita (Solomon Islands)

☐ **Ochre-headed Flycatcher** *Myiagra cervinicauda*

Lowlands of San Cristóbal (s Solomon Islands)

☐ **Melanesian Flycatcher** *Myiagra caledonica*

____ *M. c. caledonica*   New Caledonia
____ *M. c. melanura*   Maré (Loyalty Islands) and Vanuatu (Tanna and Erromanga)
____ *M. c. viridinitens* Loyalty Islands (Lifou and Ovéa)
____ *M. c. marinae*   N and central Vanuatu, Banks and Torres groups
____ *M. c. occidentalis* Rennell (se Solomon Islands)

☐ **Vanikoro Flycatcher** *Myiagra vanikorensis*

____ *M. v. vanikorensis* Vanikoro (Santa Cruz Islands)
____ *M. v. rufiventris*   W Fiji Islands
____ *M. v. kandavensis*   Kandavu and adjacent w Fiji Islands
____ *M. v. dorsalis*   S-central Fiji Islands and n Lau Archipelago
____ *M. v. townsendi*   S Lau Archipelago (Fiji)

☐ **Samoan Flycatcher** *Myiagra albiventris*

Western Samoa (Savai'i and Upolu)

☐ **Blue-crested Flycatcher** *Myiagra azureocapilla*

____ *M. a. azureocapilla*   Fiji (montane forests of Taveuni)
____ *M. a. castaneigularis* Fiji (montane forests of Vanua Levu and Kambara)
____ *M. a. whitneyi*   Fiji (montane forests of Viti Levu)

☐ **Broad-billed Flycatcher** *Myiagra ruficollis*

____ *M. r. ruficollis*   Lesser Sundas and small islands in Flores Sea
____ *M. r. fulviventris* Tanimbar Islands (Larat, Yamdena and Selaru)
____ *M. r. mimikae*   New Guinea; nw W Australia to Cape York Pen.; Keppel Bay

☐ **Satin Flycatcher** *Myiagra cyanoleuca*

SE Queensland to se S Australia, Tasmania; > to New Guinea

☐ **Restless Flycatcher** *Myiagra inquieta*

N Queensland to Victoria and s S Australia; sw W Australia

☐ **Paperbark Flycatcher** *Myiagra nana*

N Australia, extreme s New Guinea, islands of n Torres Strait

☐ **Shining Flycatcher** *Myiagra alecto*

____ *M. a. alecto*   N Moluccas
____ *M. a. longirostris*   Tanimbar Islands (Arafura Sea)
____ *M. a. chalybeocephala* New Guinea, w Papuan islands and Bismarck Archipelago

____	_M. a. manumudari_	Manam I. (New Guinea)
____	_M. a. wardelli_	Trans-Fly of se New Guinea and n Queensland
____	_M. a. lucida_	Louisiade Archipelago and D'Entrecasteaux Archipelago
____	_M. a. melvillensis_	King Sound (W Australia) to Arnhem Land and nw Queensland
____	_M. a. wardelli_	Torres Strait islands and coastal e Queensland s to Moreton Bay

□ **Dull Flycatcher** *Myiagra hebetior*

____	_M. h. hebetior_	St. Matthias Group (Bismarck Archipelago)
____	_M. h. eichhorni_	Bismarck Arch. (New Hanover, New Ireland and New Britain)
____	_M. h. cervinicolor_	Dyaul I. (Bismarck Archipelago)

□ **Silktail** *Lamprolia victoriae*

____	_L. v. victoriae_	Fiji (mountains of Taveuni)
____	_L. v. kleinschmidti_	Fiji (mountains of Vanua Levu)

□ **Black-breasted Boatbill** *Machaerirhynchus nigripectus*

____	_M. n. nigripectus_	NW New Guinea (Vogelkop Mountains)
____	_M. n. saturatus_	Montane forests of central New Guinea
____	_M. n. harterti_	Mountains of Huon Peninsula and se New Guinea

□ **Yellow-breasted Boatbill** *Machaerirhynchus flaviventer*

____	_M. f. albifrons_	Waigeo I. (New Guinea)
____	_M. f. albigula_	W Papuan islands and w New Guinea
____	_M. f. novus_	N coast of se New Guinea
____	_M. f. xanthogenys_	Aru Islands and s New Guinea (Mimika River to Milne Bay)
____	_M. f. flaviventer_	N Queensland (n Cape York Peninsula)
____	_M. f. secundus_	NE Queensland (Cooktown to about Townsville)

FAMILY: PETROICIDAE (Australasian Robins—43)

□ **Greater Ground-Robin** *Amalocichla sclateriana*

____	_A. s. occidentalis_	Snow Mountains (w New Guinea)
____	_A. s. sclateriana_	Owen Stanley Mountains (se New Guinea)

□ **Lesser Ground-Robin** *Amalocichla incerta*

____	_A. i. incerta_	Arfak Mountains of w New Guinea
____	_A. i. olivascentior_	Wandamman and Weyland to Snow Mts. (w New Guinea)
____	_A. i. brevicauda_	Mountains of e and se New Guinea

□ **Torrent Flycatcher** *Monachella muelleriana*

____	_M. m. muelleriana_	Swift flowing streams of New Guinea
____	_M. m. coultasi_	New Britain (Bismarck Archipelago)

□ **Jacky-winter** *Microeca fascinans*

____	_M. f. zimmeri_	Papua New Guinea (Port Moresby area)
____	_M. f. pallida (barcoo)_	N and e Australia (nw Western Australia to nw NSW)
____	_M. f. fascinans_	E Australia (Cooktown to s Victoria and se South Australia)
____	_M. f. assimilis_	Southern Australia (Shark Bay, Western Australia to sw NSW)

□ **Golden-bellied Flyrobin** *Microeca hemixantha*

	Tanimbar Islands (Larat, Yamdena and Lutu)

□ **Lemon-bellied Flycatcher** *Microeca flavigaster*

____	_M. f. tarara_	E New Guinea
____	_M. f. laeta_	New Guinea (Wandamman and Victor Emanual mountains)
____	_M. f. terraereginae_	SE New Guinea
____	_M. f. tormenti_	Mangroves of nw Western Australia (Broome to Kimberley)
____	_M. f. flavigaster_	Northern Territory (Arnhem Land, Melville I. , Groote Eylandt
____	_M. f. flavissima_	N Queensland (Cape York Peninsula)
____	_M. f. laetissima_	E Queensland (Rockingham Bay to Broad Sound)

☐ **Yellow-legged Flycatcher** *Microeca griseoceps*
_____ *M. g. occidentalis* — Mountains of nw New Guinea
_____ *M. g. griseoceps* — Mountains of se New Guinea
_____ *M. g. kempi* — Coastal ne Queensland (e Cape York Pen. to Silver Plains)

☐ **Olive Flyrobin** *Microeca flavovirescens*
_____ *M. f. flavovirescens* — S New Guinea (Wassi Kussa to Fly River) and Aru Islands
_____ *M. f. cuicui* — New Guinea, Yapen I. and w Papuan islands

☐ **Canary Flycatcher** *Microeca papuana*
Montane forests of New Guinea

☐ **Garnet Robin** *Eugerygone rubra*
_____ *E. r. rubra* — NW New Guinea (Arfak Mountains)
_____ *E. r. saturatior* — Mountains of central and se New Guinea

☐ **Alpine Robin** *Petroica bivittata*
_____ *P. b. caudata* — Central New Guinea (Snow and Nassau mountains)
_____ *P. b. bivittata* — High mountains of se New Guinea

☐ **Snow Mountain Robin** *Petroica archboldi*
New Guinea (Mt. Wilhelmina and Mt. Carstensz)

☐ **Scarlet Robin** *Petroica multicolor*
_____ *P. m. pusilla* — Western Samoa (Upolu and Savai'i)
_____ *P. m. feminina* — Vanuatu (Efate and Mai)
_____ *P. m. similis* — Vanuatu (Tanna and Aneityum)
_____ *P. m. cognata* — Vanuatu (Erromanga I.)
_____ *P. m. ambrynensis* — Vanuatu and Banks Group
_____ *P. m. soror* — Vanua Lava (Banks Group)
_____ *P. m. kleinschmidti* — Fiji (Viti Levu and Vanua Levu)
_____ *P. m. taveunensis* — Taveuni (Fiji)
_____ *P. m. becki* — Kandavu (Fiji)
_____ *P. m. polymorpha* — San Cristóbal (Solomon Islands)
_____ *P. m. septentrionalis* — Bougainville (Solomon Islands)
_____ *P. m. kulambangrae* — Kulambangra (Solomon Islands)
_____ *P. m. dennisi* — Guadalcanal (Solomon Islands)
_____ *P. m. campbelli* — SW Western Australia and s Eyre Peninsula (South Australia)
_____ *P. m. leggii* — Tasmania and islands of Furneaux Group (Bass Strait)
_____ *P. m. boodang* — SE Australia (se Queensland to se South Australia
_____ *P. m. multicolor* — Norfolk Island

☐ **Tomtit** *Petroica macrocephala*
_____ *P. m. toitoi* — North I. and adjacent offshore islands (New Zealand)
_____ *P. m. macrocephala* — South I. and Stewart Island (New Zealand)
_____ *P. m. marrineri* — Auckland Islands
_____ *P. m. chathamensis* — Chatham Islands
_____ *P. m. dannefaerdi* — Snares Islands

☐ **Red-capped Robin** *Petroica goodenovii*
Interior of s Australia, to southern and western coasts

☐ **Flame Robin** *Petroica phoenicea*
SE Australia (ne NSW to Victoria and Tasmania)

☐ **Rose Robin** *Petroica rosea*
SE Australia (ne NSW to sw Victoria)

☐ **Pink Robin** *Petroica rodinogaster*
_____ *P. r. rodinogaster* — Tasmania and islands in Bass Strait
_____ *P. r. inexpectata* — SE Australia (se NSW to s Victoria); > to north and west

☐ **New Zealand Robin** *Petroica australis*
_____ *P. a. longipes* — North I., Little Barrier I. and Kapiti I. (New Zealand)
_____ *P. a. australis* — South I. (New Zealand)
_____ *P. a. rakiura* — Stewart I. (New Zealand)

☐ **Chatham Robin** *Petroica traversi*

Chatham Islands (Mangere I. and South East I). Endangered

☐ **Hooded Robin** *Melanodryas cucullata*
- _____ *M. c. melvillensis* — Melville and Bathurst Islands (Northern Territory)
- _____ *M. c. picata* — Broome (Western Australia) to Arnhem Land and w NSW
- _____ *M. c. cucullata* — SE Australia (se Queensland to Victoria and se South Australia)
- _____ *M. c. westralensis* — S Western Australia, w South Australia and sw N Territory

☐ **Dusky Robin** *Melanodryas vittata*
- _____ *M. v. vittata* — Tasmania and Flinders I. (Bass Strait)
- _____ *M. v. kingi* — King I. (Bass Strait)

☐ **White-faced Robin** *Tregellasia leucops*
- _____ *T. l. leucops* — NW New Guinea (mountains of Vogelkop Peninsula)
- _____ *T. l. mayri* — New Guinea (Wandammen and Weyland mountains)
- _____ *T. l. nigroorbitalis* — New Guinea (Nassau and Snow mountains)
- _____ *T. l. heurni* — New Guinea (Weyland Mts. and upper Mamberano River)
- _____ *T. l. nigriceps* — New Guinea (Victor Emanuel and Snow mountains)
- _____ *T. l. melanogenys* — N New Guinea (Cyclops Mountains to Aicora River)
- _____ *T. l. wahgiensis* — Mountains of e New Guinea
- _____ *T. l. albifacies* — Mountains of se New Guinea
- _____ *T. l. auricularis* — Lowlands of s New Guinea (Orimo River region)
- _____ *T. l. albigularis* — Coastal ne Queensland (Cape York to McIlwraith Range)

☐ **Pale-yellow Robin** *Tregellasia capito*
- _____ *T. c. nana* — Rainforests of n Queensland (Cooktown to Mt. Spec)
- _____ *T. c. capito* — Rainforests of e Australia (se Queensland to Hunter R., NSW)

☐ **Yellow Robin** *Eopsaltria australis*
- _____ *E. a. chrysorrhos* — E Australia (Cooktown, Queensland to Hunter River, NSW)
- _____ *E. a. australis* — SE Australia (central NSW to Victoria and se South Australia)

☐ **Gray-breasted Robin** *Eopsaltria griseogularis*
- _____ *E. g. griseogularis* — Southern Australia (Shark Bay to Eyre Pen., South Australia)
- _____ *E. g. rosinae* — Extreme sw Western Australia

☐ **Yellow-bellied Robin** *Eopsaltria flaviventris*

Forests of New Caledonia

☐ **White-breasted Robin** *Eopsaltria georgiana*

SW Western Australia (Geraldton to Esperance)

☐ **Mangrove Robin** *Eopsaltria pulverulenta*
- _____ *E. p. pulverulenta* — Coastal lowlands of New Guinea
- _____ *E. p. leucura* — Aru Is. and coastal n Qld. (Cape York Pen. to Prosperine)
- _____ *E. p. cinereiceps* — Coastal W. Australia (North West Cape to Cambridge Gulf)
- _____ *E. p. alligator* — Coastal N Territory (Arnhem Land, Melville I., Groote Eylandt)

☐ **Black-chinned Robin** *Poecilodryas brachyura*
- _____ *P. b. brachyura* — W New Guinea (Vogelkop, Wandammen and Weyland mts.)
- _____ *P. b. albotaeniata* — New Guinea (Geelvink Bay region) and Yapen I.
- _____ *P. b. dumasi* — N New Guinea (Humboldt Bay to Sepik River)

☐ **Black-sided Robin** *Poecilodryas hypoleuca*
- _____ *P. h. steini* — Waigeo I. (New Guinea)
- _____ *P. h. hypoleuca* — W New Guinea to Port Moresby area and w Papuan islands
- _____ *P. h. hermani* — N New Guinea (Mamberamo River to upper Watut River)

☐ **White-browed Robin** *Poecilodryas superciliosa*
- _____ *P. s. superciliosa* — NE Queensland (Cape York Peninsula to Sarina)
- _____ *P. s. cervviniventris* — N Australia (Fitzroy R., Western Australia to nw Queensland)

☐ **Olive-yellow Robin** *Poecilodryas placens*

Patchily distributed New Guinea and Batanta I.

☐ **Black-throated Robin** *Poecilodryas albonotata*

_____ *P. a. albonotata*   W New Guinea (Vogelkop Mountains)

_____ *P. a. griseiventris*   Central Highlands of New Guinea

_____ *P. a. correcta*   New Guinea (mountains of Huon Peninsula and se peninsula)

☐ **White-winged Robin** *Peneothello sigillata*

_____ *P. s. quadrimaculata*   W New Guinea (Nassau and Snow mountains)

_____ *P. s. saruwagedi*   New Guinea (mountains of Huon Peninsula)

_____ *P. s. hagenensis*   C New Guinea (highlands of Mt. Hagen and Star Mountains)

_____ *P. s. sigillata*   Central Highlands and mountains of se New Guinea

☐ **Smoky Robin** *Peneothello cryptoleuca*

_____ *P. c. cryptoleuca*   NW New Guinea (Tamrau and Arfak mountains)

_____ *P. c. albidior*   New Guinea (Weyland, Gauttier and Nassau mountains)

_____ *P. c. maxima*   W New Guinea (Kumawa Mountains)

☐ **White-rumped Robin** *Peneothello bimaculata*

_____ *P. b. bimaculata*   Mountains of nw New Guinea

_____ *P. b. vicaria*   Mountains of se New Guinea and Huon Peninsula

☐ **Blue-gray Robin** *Peneothello cyanus*

_____ *P. c. cyanus*   NW New Guinea (Vogelkop Mountains)

_____ *P. c. atricapilla*   Mountains of ne and central New Guinea

_____ *P. c. subcyana*   Central highlands and mts. of se New Guinea and Huon Pen.

☐ **Gray-headed Robin** *Heteromyias albispecularis*

_____ *H. a. albispecularis*   NW New Guinea (Tamrau and Arfak mountains)

_____ *H. a. atricapilla*   NE New Guinea (mountains of Huon Peninsula)

_____ *H. a. rothschildi*   New Guinea (Weyland and Snow mountains)

_____ *H. a. centralis*   Central Highlands of New Guinea

_____ *H. a. armiti*   Herzog Mountains and mountains of se New Guinea

_____ *H. a. cinereifrons*   Rainforests of n Queensland (Cooktown to Mt. Spec)

☐ **Green-backed Robin** *Pachycephalopsis hattamensis*

_____ *P. h. hattamensis*   NW New Guinea (Tamrau and Arfak mountains)

_____ *P. h. ernesti*   NW New Guinea (Wandammen Mountains)

_____ *P. h. axillaris*   New Guinea (Weyland, Nassau and Snow mountains)

_____ *P. h. insulanus*   Mountains of New Guinea

_____ *P. h. lecroyae*   Mountains of e-central New Guinea

☐ **White-eyed Robin** *Pachycephalopsis poliosoma*

_____ *P. p. idenburgi*   N slopes of central ranges of n New Guinea

_____ *P. p. hypopolia*   NE New Guinea (mountains of Huon Peninsula)

_____ *P. p. albigularis*   New Guinea (Weyland and Victor Emanuel mountains)

_____ *P. p. balim*   Central New Guinea (valleys of the Bele and Balim rivers)

_____ *P. p. approximans*   Central New Guinea (south slopes of Snow Mountains)

_____ *P. p. hunsteini*   New Guinea (mountains along upper Sepik River)

_____ *P. p. poliosoma*   Herzog Mountains and mountains of se New Guinea

☐ **Northern Scrub-Robin** *Drymodes superciliaris*

_____ *D. s. beccarii*   NW New Guinea (Arfak and Wandammen mountains)

_____ *D. s. nigriceps*   New Guinea (Cyclops Mts. and n slope of Snow Mountains)

_____ *D. s. brevirostris*   S and se New Guinea and Aru Islands

_____ *D. s. superciliaris*   N Queensland (rainforests of coastal Cape York Peninsula)

☐ **Southern Scrub-Robin** *Drymodes brunneopygia*

Southern Australia (Shark Bay, W Australia to w and c NSW)

FAMILY: PACHYCEPHALIDAE (Whistlers and Allies—57)

☐ **Whitehead** *Mohoua albicilla*

Little and Great Barrier is., s North, Arid and Kapiti islands

☐ **Yellowhead** *Mohoua ochrocephala*

Forests of South I. and Stewart I. (New Zealand)

☐ **Pipipi** *Mohoua novaeseelandiae*

Forests of South I. and Stewart I. (New Zealand)

☐ **Crested Shrike-tit** *Falcunculus frontatus*
- ____ *F. f. frontatus* — SE Australia (c Qld. to s Victoria and se South Australia)
- ____ *F. f. leucogaster* — Southwest Western Australia
- ____ *F. f. whitei* — N Australia (Kimberley to Arnhem Land and far nw Qld.)

☐ **Crested Bellbird** *Oreoica gutturalis*
- ____ *O. g. gutturalis* — Central Australia (to coast in central Western Australia)
- ____ *O. g. pallescens* — SW Western Australia to nw Victoria and central Queensland

☐ **Mottled Whistler** *Rhagologus leucostigma*
- ____ *R. l. leucostigma* — NW New Guinea (Arfak and Tamrau mountains)
- ____ *R. l. novus* — N New Guinea (Weyland and Nassau mountains)
- ____ *R. l. obscurus* — Mountains of central and se New Guinea and Huon Peninsula

☐ **Dwarf Whistler** *Pachycare flavogriseum*
- ____ *P. f. flavogriseum* — W New Guinea (Vogelkop and Wandammen mountains)
- ____ *P. f. subaurantium* — Mountains of central New Guinea
- ____ *P. f. randi* — New Guinea (Snow Mountains in Idenburg River area)
- ____ *P. f. subpallidum* — SE New Guinea (Herzog and Saruwaged mountains)

☐ **Olive-flanked Whistler** *Hylocitrea bonensis*
- ____ *H. b. bonensis* — Mountains of n, central and se Sulawesi
- ____ *H. b. bonthaina* — S Sulawesi (Mt. Lompobattang)

☐ **Maroon-backed Whistler** *Coracornis raveni*

Mountains of Sulawesi

☐ **Rufous-naped Whistler** *Aleadryas rufinucha*
- ____ *A. r. rufinucha* — NW New Guinea (Volgelkop Mountains)
- ____ *A. r. niveifrons* — Mountains of central New Guinea
- ____ *A. r. lochmia* — NE New Guinea (mountains of Huon Peninsula)
- ____ *A. r. gamblei* — Herzog Mountains and mountains of se New Guinea

☐ **Olive Whistler** *Pachycephala olivacea*
- ____ *P. o. macphersoniana* — E Australia (far se Queensland to central New South Wales)
- ____ *P. o. olivacea* — SE Australia (se New South Wales to central and e Victoria)
- ____ *P. o. bathychroa* — S Victoria (Otway Peninsula and Strzelecki Range)
- ____ *P. o. apatetes* — Tasmania, King and Flinders Islands (Bass Strait)
- ____ *P. o. hesperus* — Far southeast South Australia and adjacent sw Victoria

☐ **Red-lored Whistler** *Pachycephala rufogularis*

SE South Australia, adj. nw Victoria and sw and central NSW

☐ **Gilbert's Whistler** *Pachycephala inornata*

S Western Australia to n Victoria and sw and central NSW

☐ **Mangrove Whistler** *Pachycephala grisola*

India to Myanmar, Andaman Is., Greater Sundas and Palawan

☐ **Green-backed Whistler** *Pachycephala albiventris*

N Philippines (Luzon and Mindoro)

☐ **White-vented Whistler** *Pachycephala homeyeri*

Central and s Philippines and Sulu Archipelago

☐ **Island Whistler** *Pachycephala phaionota*

Moluccas, Kai, Aru and w Papuan islands

☐ **Rusty Whistler** *Pachycephala hyperythra*

____ *P. h. hyperythra* — W New Guinea (Vogelkop Mountains)

____ *P. h. sepikiana* — New Guinea (Sepik Mts. to mts. south of Mamberamo River)

____ *P. h. reichenowi* — NE New Guinea (Saruwaged Mountains)

____ *P. h. salvadorii* — Mountains of se New Guinea

☐ **Brown-backed Whistler** *Pachycephala modesta*

____ *P. m. modesta* — Herzog Mountains and mountains of se New Guinea

____ *P. m. hypoleuca* — New Guinea (Sepik and Saruwaged mountains)

____ *P. m. telefolminensis* — Central New Guinea (Victor Emanuel and Hindenburg mts.)

☐ **Bornean Whistler** *Pachycephala hypoxantha*

____ *P. h. hypoxantha* — Mountains of n Borneo (Kinabalu to n Sarawak)

____ *P. h. sarawacensis* — N Borneo (Poi Mountains)

☐ **Sulphur-bellied Whistler** *Pachycephala sulfuriventer*

Montane forests of Sulawesi

☐ **Vogelkop Whistler** *Pachycephala meyeri*

NW New Guinea (Arfak, Tamrau and [?] Foya Mountains)

☐ **Yellow-bellied Whistler** *Pachycephala philippinensis*

____ *P. p. fallax* — Calayan (n Philippines)

____ *P. p. illex* — Camiguin Norte (n Philippines)

____ *P. p. philippinensis* — N Philippines (Luzon and Catanduanes)

____ *P. p. siquijorensis* — Siquijor (Philippines)

____ *P. p. apoensis* — Philippines (Dinagat, Samar, Leyte, Biliran and Mindanao)

____ *P. p. basilanica* — Basilan (Philippines)

____ *P. p. boholensis* — Bohol (Philippines)

☐ **Gray-headed Whistler** *Pachycephala griseiceps*

____ *P. g. rufipennis* — Kai Islands (Kai Kecil and Kai Besar)

____ *P. g. gagiensis* — Gagi I. (New Guinea)

____ *P. g. waigeuensis* — Waigeo and Gebe islands (New Guinea)

____ *P. g. miosnomensis* — Meos Num I. (New Guinea)

____ *P. g. griseiceps* — Aru Islands and ne New Guinea

____ *P. g. jobiensis* — N New Guinea and Yapen I.

____ *P. g. perneglecta* — S New Guinea

____ *P. g. dubia* — SE New Guinea and D'Entrecasteaux Archipelago

____ *P. g. sudestensis* — Tagula I. (Louisiade Archipelago)

____ *P. g. peninsulae* — NE Queensland s to Rockingham Bay; Hinchinbrook I.

☐ **Fawn-breasted Whistler** *Pachycephala orpheus*

E Lesser Sundas (Semau, Timor, Jaco and Wetar)

☐ **Gray Whistler** *Pachycephala simplex*

____ *P. s. simplex* — N Northern Territory (Melville and Bathhurst Is., Arnhem Land)

____ *P. s. peninsulae* — Coastal n Queensland (Cape York Pen. s to about Townsville)

☐ **Golden Whistler** *Pachycephala pectoralis*

____ *P. p. javana* — E Java and Bali

____ *P. p. teysmanni* — Salayar I. (Flores Sea)

____ *P. p. everetti* — Tanahjampea, Kalaotoa and Madu islands (Flores Sea)

____ *P. p. pelengensis* — Banggai Islands (Peleng and Banggai)

____ *P. p. clio* — Sula Islands (Taliabu, Seho, Mangole and Sanana)

____ *P. p. mentalis* — N Moluccas (Bacan, Halmahera and Morotai)

____ *P. p. tidorensis* — N Moluccas (Tidore and Ternate)

____ *P. p. obiensis* — S Moluccas (Obi and Bisa)

____ *P. p. buruensis* — Buru (s Moluccas)

____ *P. p. macrorhyncha* — S Moluccas (Ambon and Seram)

____ *P. p. fulvotincta* — E Lesser Sundas

____ *P. p. fulviventris* — Sumba (Lesser Sundas)

____ *P. p. calliope* — E Lesser Sundas (Roti, Timor, Semau and Wetar)

____	*P. p. compar*	E Lesser Sundas (Leti and Moa)
____	*P. p. par*	Romang (e Lesser Sundas)
____	*P. p. dammeriana*	Damar (e Lesser Sundas)
____	*P. p. sharpei*	Babar (Lesser Sundas)
____	*P. p. fuscoflava*	Tanimbar Islands (Larat and Yamdena)
____	*P. p. tabarensis*	Tabar I. (Papua New Guinea)
____	*P. p. ottomeyeri*	Lihir Islands (Bismarck Archipelago)
____	*P. p. goodsoni*	Admiralty Islands (Bismarck Archipelago)
____	*P. p. citreogaster*	Bismarck Arch. (New Hanover, New Britain and New Ireland)
____	*P. p. sexuvaria*	St. Matthias Islands (Bismarck Archipelago)
____	*P. p. fergussonis*	Fergusson I. (D'Entrecasteaux Archipelago)
____	*P. p. pectoralis*	E Australia (Cooktown, Queensland to Hunter River, NSW
____	*P. p. youngi*	SE Australia (central NSW to sw Victoria); > to north
____	*P. p. glaucura*	Tasmania and Flinders I. (Bass Strait)
____	*P. p. fuliginosa*	SW Western Australia to se South Australia and w Victoria
____	*P. p. collaris*	Louisiade Archipelago
____	*P. p. rosseliana*	Rossel I. (Louisiade Archipelago)
____	*P. p. misimae*	Misima I. (Louisiade Archipelago)
____	*P. p. whitneyi*	Shortland I. (Solomon Islands)
____	*P. p. bougainvillei*	Solomon Islands (Buka and Bougainville)
____	*P. p. orioloides*	Solomon Islands (Choiseul, Santa Isabel and Florida)
____	*P. p. cinnamomea*	Solomon Islands (Guadalcanal and Beagle)
____	*P. p. sanfordi*	Malaita (Solomon Islands)
____	*P. p. pavuvu*	Pavuvu Islands (Solomon Islands)
____	*P. p. centralis*	E New Georgia Group (Solomon Islands)
____	*P. p. feminina*	Rennell (se Solomon Islands)
____	*P. p. melanoptera*	S New Georgia Group (Solomon Islands)
____	*P. p. melanonota*	Solomon Islands (Ranongga and Vellalavella)
____	*P. p. christophori*	Solomon Islands (Santa Anna and San Cristóbal)
____	*P. p. utupuae*	Utupua I. (Solomon Islands)
____	*P. p. littayei*	New Caledonia and Loyalty Islands (Lifou and Maré)
____	*P. p. cucullata*	Aneityum (Vanuatu)
____	*P p chlorura*	Erromango (Vanuatu)
____	*P. p. intacta*	Vanuatu and Banks Group
____	*P. p. vanikorensis*	Vanikoro and Santa Cruz Islands
____	*P. p. ornata*	N Santa Cruz Islands
____	*P. p. kandavensis*	Kandavu Islands (Fiji)
____	*P. p. lauana*	S Lau Archipelago (Fiji)
____	*P. p. vitiensis*	Ngau (Fiji)
____	*P. p. koroana*	Karo (Fiji)
____	*P. p. torquata*	Taveuni (Fiji)
____	*P. p. ambigua*	Fiji (Rambi and Kioa)
____	*P. p. optata*	Fiji (Ovalu and se Viti Levu)
____	*P. p. graeffii*	Fiji (Wala and Viti Levu)
____	*P. p. aurantiiventris*	Fiji (Yanganga and Vanua Levu)
____	*P. p. bella*	Vanua Lava (Banks Group)
____	*P. p. contempta*	Lord Howe I.
____	*P. p. xanthoprocta*	Norfolk I.

☐ **Sclater's Whistler** *Pachycephala soror*

____	*P. s. soror*	NW New Guinea (Vogelkop Mountains)
____	*P. s. klossi*	Mountains of central and e New Guinea
____	*P. s. bartoni*	Mountains of se New Guinea and Goodenough I.
____	*P. s. octogenarii*	New Guinea (Kumawa Mountains)
____	*P. s. remota*	Mountains of Goodenough I. (D'Entrecasteaux Archipelago)

☐ **Lorentz's Whistler** *Pachycephala lorentzi*

New Guinea (Snow, Victor Emanuel and Hindenburg mts.)

☐ **Black-tailed Whistler** *Pachycephala melanura*

____	*P. m. balim*	N New Guinea (Balim and Bele valleys)

____ *P. m. dahli* — Islands off se New Guinea and Bismarck Archipelago
____ *P. m. melanura* — Mangroves of nw Western Australia (Pilbara and w Kimberley)
____ *P. m. robusta* — Mangroves of n Australia (Arnhem Land to central Queensland)
____ *P. m. spinicaudus* — Islands of north Torres Strait

☐ **New Caledonian Whistler** *Pachycephala caledonica*

Forests of New Caledonia

☐ **Samoan Whistler** *Pachycephala flavifrons*

W Samoa (Savai'i and Upolu)

☐ **Tongan Whistler** *Pachycephala jacquinoti*

Low scrub of Vava'u (n Tonga)

☐ **Regent Whistler** *Pachycephala schlegelii*
____ *P. s. schlegelii* — New Guinea (Vogelkop and Wandammen mountains)
____ *P. s. obscurior* — Mountains of central and e New Guinea
____ *P. s. cyclopum* — New Guinea (Cyclops Mountains)

☐ **Bare-throated Whistler** *Pachycephala nudigula*
____ *P. n. ilsa* — Montane forests of Sumbawa (Lesser Sundas)
____ *P. n. nudigula* — Montane forests of Flores (Lesser Sundas)

☐ **Hooded Whistler** *Pachycephala implicata*
____ *P. i. implicata* — Montane forests of Guadalcanal (Solomon Islands)
____ *P. i. richardsi* — Montane forests of Bougainville (Solomon Islands)

☐ **Golden-backed Whistler** *Pachycephala aurea*

Locally from Weyland and Snow Mountains to se New Guinea

☐ **Drab Whistler** *Pachycephala griseonota*
____ *P. g. lineolata* — Sula Islands (Taliabu, Seho and Sanana)
____ *P. g. cinerascens* — N Moluccas (Morotai, Halmahera, Ternate, Tidore and Bacan)
____ *P. g. johni* — Obi (n Moluccas)
____ *P. g. examinata* — Buru (s Moluccas)
____ *P. g. griseonota* — Seram (s Moluccas)
____ *P. g. kuehni* — Kai Islands (Kai Kecil and Kai Besar)

☐ **Wallacean Whistler** *Pachycephala arctitorquis*
____ *P. a. tianduana* — Tayandu Islands (Banda Sea)
____ *P. a. kebirensis* — E Lesser Sundas (Moa, Romang, Babar, Wedan and Damar)
____ *P. a. arctitorquis* — Tanimbar Islands (Yamdena, Larat, Lutu and Mutu)

☐ **Black-headed Whistler** *Pachycephala monacha*

Mountains of central New Guinea and Aru Islands

☐ **White-bellied Whistler** *Pachycephala leucogastra*
____ *P. l. dorsalis* — Mountains of central and e New Guinea
____ *P. l. leucogastra* — Coastal se New Guinea (Hall Sound to Port Moresby)
____ *P. l. meeki* — Rossel (Louisiade Archipelago)

☐ **Rufous Whistler** *Pachycephala rufiventris*
____ *P. r. minor* — Melville and Bathurst Islands (Northern Territory)
____ *P. r. falcata* — N Australia (Broome, Western Australia to ne N Territory)
____ *P. r. pallida* — N and nw Queensland (Nicholson River to Cape York Pen.)
____ *P. r. rufiventris* — Mainland Australia (except north and arid treeless areas)
____ *P. r. xanthetraea* — New Caledonia

☐ **White-breasted Whistler** *Pachycephala lanioides*
____ *P. l. carnaroni* — Coastal central Western Australia (Pilbara region)
____ *P. l. lanioides* — Coastal nw Western Australia (w Kimberley region)
____ *P. l. fretorum* — Coastal far nw W Australia to sw Cape York Peninsula

☐ **Sooty Shrike-Thrush** *Colluricincla umbrina*
____ *C. u. atra* — N New Guinea
____ *C. u. umbrina* — S New Guinea

☐ **Rufous Shrike-Thrush** *Colluricincla megarhyncha*

____	*C. m. affinis*	Waigeo I. (New Guinea)
____	*C. m. batantae*	Batanta I. (New Guinea)
____	*C. m. misoliensis*	Misool I. (New Guinea)
____	*C. m. aruensis*	Aru Islands (New Guiinea)
____	*C. m. obscura*	Yapen I. (New Guinea)
____	*C. m. melanorhyncha*	Biak I. (New Guinea)
____	*C. m. idenburgi*	N New Guinea (slopes south of Idenburg River)
____	*C. m. hybridus*	N New Guinea (Humboldt Bay to Mamberamo River)
____	*C. m. tappenbecki*	NE New Guinea (Astrolabe Bay to lower Sepik River)
____	*C. m. maeandrina*	NE New Guinea (upper Sepik River and Victor Emanuel Mts.)
____	*C. m. megarhyncha*	W New Guinea (Vogelkop to Onin Peninsula)
____	*C. m. ferruginea*	Head of Geelvink Bay (nw New Guinea)
____	*C. m. neos*	Herzog Mts., s coast of Huon Gulf and upper Watut River
____	*C. m. madaraszi*	NE New Guinea (Huon Peninsula)
____	*C. m. goodsoni*	S New Guinea (Merauke District)
____	*C. m. wuroi*	S New Guinea (Oriomo River to Morehead River)
____	*C. m. palmeri*	S New Guinea (Trans-Fly lowlands)
____	*C. m. despecta*	S coast of se New Guinea (Milne Bay to Hall Sound)
____	*C. m. superflua*	N coast of se New Guinea (Collingwood Bay to Aicora River)
____	*C. m. fortis*	D'Entrecasteaux Archipelago
____	*C. m. trobriandi*	Trobriand Islands (Solomon Sea)
____	*C. m. discolor*	Tagula I. (Louisiade Archipelago)
____	*C. m. parvula*	NW Australia (Kimberley, Western Australia to Arnhem Land)
____	*C. m. aelptes*	Northern Territory (coastal Gulf of Carpenteria)
____	*C. m. normani*	Coastal n Queensland (w and e Cape York Peninsula)
____	*C. m. griseata*	Rainforests of ne Queensland (Cooktown to Ingham)
____	*C. m. synaptica*	Rainforests of c Queensland (Burdekin R. to Dawson R. basin)
____	*C. m. gouldi*	E Queensland (Connors Range to Dawes Range)
____	*C. m. rufigaster*	E Australia (se Queensland to ne New South Wales)

☐ **Sangihe Shrike-Thrush** *Colluricincla sanghirensis*

Sangihe I. (north of Sulawesi)

☐ **Bower's Shrike-Thrush** *Colluricincla boweri*

Montane rainforests of ne Queensland (Cooktown to Ingham)

☐ **Sandstone Shrike-Thrush** *Colluricincla woodwardi*

N Australia (Kimberley to Arnhem Land and nw Queensland)

☐ **Gray Shrike-Thrush** *Colluricincla harmonica*

____	*C. h. tachycrypta*	Coastal se New Guinea
____	*C. h. brunnea*	N Australia (Fitzroy R., Western Australia to nw Queensland)
____	*C. h. superciliosa*	N Queensland (Torres Strait islands and n Cape York Pen)
____	*C. h. harmonica*	E Australia (n and c Qld. to Victoria and e South Australia)
____	*C. h. strigata*	Tasmania, King and Flinders Islands (Bass Strait)
____	*C. h. rufiventris*	S Australia (c and s Western Australia to Flinders Ranges)

☐ **Morningbird** *Colluricincla tenebrosa*

Palau Islands (Babelthuap to Peleliu)

☐ **Hooded Pitohui** *Pitohui dichrous*

____	*P. d. dichrous*	Mountains of n New Guinea and Yapen I.
____	*P. d. monticola*	Mountains of central New Guinea

☐ **White-bellied Pitohui** *Pitohui incertus*

Lowlands of s New Guinea (Lorentz River to upper Fly River)

☐ **Rusty Pitohui** *Pitohui ferrugineus*

____	*P. f. leucorhynchus*	Waigeo I. (New Guinea)
____	*P. f. fuscus*	Batanta I. (New Guinea)
____	*P. f. brevipennis*	Aru Islands (New Guinea)
____	*P. f. ferrugineus*	NW New Guinea, Misool and Salawati islands
____	*P. f. holerythrus*	N New Guinea and Yapen I.
____	*P. f. clarus*	SE New Guinea

☐ **Crested Pitohui** *Pitohui cristatus*

____	*P. c. cristatus*	W New Guinea (Arfak Mountains)
____	*P. c. arthuri*	New Guinea (Orimo River, Cyclops and Sepik mountains)
____	*P. c. kodonophonos*	Mountains and lowlands of se New Guinea

☐ **Variable Pitohui** *Pitohui kirhocephalus*

____	*P. k. kirhocephalus*	Coastal ne New Guinea (Vogelkop to Geelvink Bay)
____	*P. k. salvadorii*	NW New Guinea (Geelvink Bay region)
____	*P. k. dohertyi*	NW New Guinea (islands and peninsulas of Wandammen area)
____	*P. k. rubiensis*	NW New Guinea (head of Geelvink Bay)
____	*P. k. tibialis*	NW New Guinea (western half of Vogelkop Peninsula)
____	*P. k. stramineipectus*	SW New Guinea (Triton Bay region)
____	*P. k. decipiens*	SW New Guinea (Onin Peninsula)
____	*P. k. adiensis*	Adi Island (off s coast of Onin Peninsula, sw New Guinea)
____	*P. k. carolinae*	SW New Guinea (Etna Bay region)
____	*P. k. brunneivertex*	W New Guinea (se coast of Geelvink Bay)
____	*P. k. jobiensis*	Kurudu I. and Yapen I. (New Guinea)
____	*P. k. meyeri*	Coastal n New Guinea (Mamberamo River to Tami River)
____	*P. k. senex*	N New Guinea (upper Sepik Valley)
____	*P. k. brunneicaudus*	N New Guinea (lower Sepik River to upper Ramu River)
____	*P. k. nigripectus*	S New Guinea
____	*P. k. meridionalis*	SE New Guinea (Chads Bay to Yule I.)
____	*P. k. brunneiceps*	S New Guinea (Fly River to Gulf of Papua)
____	*P. k. aruensis*	Aru Islands (New Guinea)
____	*P. k. uropygialis*	Salawati and Misool islands (New Guinea)
____	*P. k. pallidus*	Sagewin and Batanta islands (New Guinea)
____	*P. k. cervineiventris*	Waigeo and Gemien islands (New Guinea)

☐ **Black Pitohui** *Pitohui nigrescens*

____	*P. n. nigrescens*	NW New Guinea (Arfak and Tamrau mountains)
____	*P. n. wandamensis*	New Guinea (Wandammen Peninsula)
____	*P. n. buergersi*	New Guinea (Sepik, Hindenburg and Hagen mountains)
____	*P. n. meeki*	Central New Guinea (Weyland, Nassau and Snow mountains)
____	*P. n. harterti*	NE New Guinea (Saruwaged Mountains of Huon Peninsula)
____	*P. n. schistaceus*	Herzog Mountains and mountains of se New Guinea

☐ **Wattled Ploughbill** *Eulacestoma nigropectus*

____	*E. n. clara*	Central Highlands of New Guinea
____	*E. n. nigropectus*	Mountains of se New Guinea

FAMILY: PICATHARTIDAE (Rockfowl—2)

☐ **White-necked Rockfowl** *Picathartes gymnocephalus*

Locally in Guinea, Liberia, Sierra Leone, Ivory Coast and Ghana

☐ **Gray-necked Rockfowl** *Picathartes oreas*

Forests of se Nigeria to s Cameroon and ne Gabon; Bioko

FAMILY: TIMALIIDAE (Babblers—274)

☐ **Malia** *Malia grata*

____	*M. g. recondita*	Montane forests of peninsular n Sulawesi
____	*M. g. stresemanni*	Central and se peninsular Sulawesi
____	*M. g. grata*	S Sulawesi (Mt. Lompobattang)

☐ **Ashy-headed Laughingthrush** *Garrulax cinereifrons*

Humid forests of sw Sri Lanka

☐ **Sunda Laughingthrush** *Garrulax palliatus*

____	*G. p. palliatus*	Montane forests of w Sumatra
____	*G. p. schistochlamys*	Montane forests of n Borneo

☐ **Rufous-fronted Laughingthrush** *Garrulax rufifrons*
_____ *G. r. rufifrons* — Montane forests of w Java
_____ *G. r. slamatensis* — Central Java (Mt. Slamet area)

☐ **Masked Laughingthrush** *Garrulax perspicillatus*
— Lowland scrub of s China to Indochina

☐ **White-throated Laughingthrush** *Garrulax albogularis*
_____ *G. a. whistleri* — Kashmir to Pakistan and nw India (Uttar Pradesh)
_____ *G. a. albogularis* — Himalayas (w Nepal to e Bhutan)
_____ *G. a. eous* — SW China (Qinghai, s Shaanxi, s Sichuan, n Yunnan) to nw Tonkin
_____ *G. a. ruficeps* — Montane forests of Taiwan

☐ **White-crested Laughingthrush** *Garrulax leucolophus*
_____ *G. l. leucolophus* — W Himalayas to Nepal, Sikkim, Bhutan and Assam (Mishmi Hills)
_____ *G. l. patkaicus* — S Assam (s of the Brahmaputra) to n Myanmar and nw Yunnan
_____ *G. l. belangeri* — S Myanmar and sw Thailand (valley of Mekong River)
_____ *G. l. diardi* — SE Myanmar to sw Yunnan, peninsular Thailand and Indochina
_____ *G. l. bicolor* — Mountains of w Sumatra

☐ **Lesser Necklaced Laughingthrush** *Garrulax monileger*
_____ *G. m. monileger* — Himalayas from Nepal to ne Myanmar and s China (sw Yunnan)
_____ *G. m. badius* — NE Assam (Mishmi Hills)
_____ *G. m. stuarti* — SE Myanmar to nw Thailand
_____ *G. m. fuscatus* — Central Myanmar to sw Thailand
_____ *G. m. mouhoti* — SE Thailand to Cambodia and s Vietnam
_____ *G. m. pasquieri* — Central Vietnam (Thuatien and Quangtri provinces)
_____ *G. m. schauenseei* — E Myanmar to sw Yunnan, ne plateau of Thailand and n Laos
_____ *G. m. tonkinensis* — S China (Guangxi and se Yunnan) to n Vietnam
_____ *G. m. melli* — SE China (Fujian and Hunan to n Guangdong)
_____ *G. m. schmackeri* — Hainan (s China)

☐ **Greater Necklaced Laughingthrush** *Garrulax pectoralis*
_____ *G. p. pectoralis* — Himalayas of Nepal to s China (s Yunnan w of the Mekong R.)
_____ *G. p. melanotis* — Himalayas (Sikkim to Assam, n Myanmar and s China)
_____ *G. p. pingi* — S China (w Yunnan south of range of *melanotis*)
_____ *G. p. subfusus* — SE Myanmar to w Thailand and nw Laos
_____ *G. p. robini* — S China (s Yunnan e of the Mekong) to n Vietnam and ne Laos
_____ *G. p. picticollis* — E China (Anhui to Shaanxi, Fujian and Hunan to Guangdong)
_____ *G. p. semitorquatus* — Hainan (s China)

☐ **Black Laughingthrush** *Garrulax lugubris*
_____ *G. l. lugubris* — Highlands of Malay Peninsula and w Sumatra
_____ *G. l. calvus* — Highlands of ne Borneo

☐ **Striated Laughingthrush** *Garrulax striatus*
_____ *G. s. striatus* — NW Himalayas (East Punjab to Kumaon)
_____ *G. s. vibex* — Himalayas (w and central Nepal to s Tibet)
_____ *G. s. sikkimensis* — Himalayas (e Nepal to se Tibet, sw China, Sikkim and Bhutan)
_____ *G. s. cranbrooki* — Bhutan to Assam, w Myanmar and s China (nw Yunnan)

☐ **White-necked Laughingthrush** *Garrulax strepitans*
_____ *G. s. strepitans* — Myanmar to sw China (sw Yunnan), w Thailand and nw Laos
_____ *G. s. terrarius* — SE Thailand

☐ **Black-hooded Laughingthrush** *Garrulax milleti*
_____ *G. m. milleti* — Montane forests of n and s Vietnam and s Laos
_____ *G. m. sweeti* — Central highlands of Vietnam (Kon Tum Province)

☐ **Gray Laughingthrush** *Garrulax maesi*
_____ *G. m. grahami* — SW China (sw Sichuan to se Guangxi and ne Yunnan)
_____ *G. m. maesi* — Mountains of sw China (Guangxi) and n Tonkin

____ *G. m. varennei*	NE and central Laos (Chiang Khwang and Thakkek provinces)
____ *G. m. castanotis*	Hainan (s China)

☐ **Rufous-necked Laughingthrush *Garrulax ruficollis***

Mixed forests of e Nepal to sw China, ne India and Myanmar

☐ **Chestnut-backed Laughingthrush *Garrulax nuchalis***

Lowlands of ne India (Arunachal Pradesh) to n Myanmar

☐ **Black-throated Laughingthrush *Garrulax chinensis***

____ *G. c. lochmius*	S China (sw Yunnan) to se Myanmar, n Thailand and n Laos
____ *G. c. chinensis*	S China (se Yunnan, s Guangxi and s Guangdong) to n Laos
____ *G. c. propinquus*	S Myanmar to sw Thailand
____ *G. c. germaini*	S Vietnam (Phantiet and Phanrang provinces)
____ *G. c. monachus*	Hainan (s China)

☐ **White-cheeked Laughingthrush *Garrulax vassali***

Montane forests of Vietnam and s Laos

☐ **Yellow-throated Laughingthrush *Garrulax galbanus***

____ *G. g. galbanus*	SE Assam (Manipur and Lushai Hills) to w Myanmar
____ *G. g. courtoisi*	E-central China (n Jiangxi Province)
____ *G. g. simaoensis*	S China (Yunnan)

☐ **Wynaad Laughingthrush *Garrulax delesserti***

SW India (W Ghats from Goa to Kerala and w Tamil Nadu)

☐ **Rufous-vented Laughingthrush *Garrulax gularis***

E Bhutan to ne India, n Myanmar and n Laos

☐ **Père David's Laughingthrush *Garrulax davidi***

____ *G. d. chinganicus*	N Manchuria (Khingan Mountains)
____ *G. d. davidi*	N China (Inner Mongolia to Gansu, e Qinghai and Liaoning)
____ *G. d. experrectus*	N Gansu (north spur of Nan Shan Mountains)
____ *G. d. concolor*	N Sichuan (Sungpan region)

☐ **Sukatschev's Laughingthrush *Garrulax sukatschewi***

SW China (montane forests of s Gansu and adjacent Sichuan)

☐ **Moustached Laughingthrush *Garrulax cineraceus***

____ *G. c. cineraceus*	S Assam (s of the Brahmaputra) to w Myanmar (Chin Hills)
____ *G. c. strenuus*	NE Myanmar to s China (se Sichuan and nw Yunnan)
____ *G. c. cinereiceps*	S China (w Sichuan to Anhui, Guandong and Zhejiang)

☐ **Rufous-chinned Laughingthrush *Garrulax rufogularis***

____ *G. r. occidentalis*	W Himalayas (Pakistan to nw Uttar Pradesh)
____ *G. r. grosvenori*	Himalayas of w Nepal
____ *G. r. rufogularis*	Himalayas (Nepal to Bhutan and n Assam)
____ *G. r. assamensis*	NE Assam
____ *G. r. rufitinctus*	Assam south of the Brahmaputra (Khasi Hills)
____ *G. r. rufiberbis*	N Myanmar
____ *G. r. intensior*	N Vietnam (Tonkin)

☐ **Chestnut-eared Laughingthrush *Garrulax konkakinhensis***

Central Vietnam (Mount Kon Ka Kinh)

☐ **Spotted Laughingthrush *Garrulax ocellatus***

____ *G. o. griseicauda*	Himalayas from nw India (nw Uttar Pradesh) to w Nepal
____ *G. o. ocellatus*	Himalayas (central Nepal to Bhutan and se Tibet)
____ *G. o. maculipectus*	S China (nw Yunnan) to ne Myanmar
____ *G. o. artemisiae*	S China (s Gansu to Sichuan and ne Yunnan)

☐ **Barred Laughingthrush *Garrulax lunulatus***

S-central China (s Gansu, s Shaanxi and w Sichuan)

☐ **Biet's Laughingthrush *Garrulax bieti***

Mountains of s China (sw Sichuan and nw Yunnan)

☐ **Giant Laughingthrush** *Garrulax maximus*

Mts. of se Tibet to s China (s Gansu, w Sichuan and nw Yunnan)

☐ **Gray-sided Laughingthrush** *Garrulax caerulatus*
_____ *G. c. caerulatus*
_____ *G. c. subcaerulatus*
_____ *G. c. livingstoni*
_____ *G. c. kaurensis*
_____ *G. c. latifrons*

Himalayas from Nepal to Bhutan and Assam (n of Brahmaputra)
S Assam south of the Brahmaputra (Khasi Hills)
E Assam (Naga Hills and Manipur) to nw Myanmar
N Myanmar (Kachin State)
NE Myanmar (Myitkyina District) and adj. s China (nw Yunnan)

☐ **Rusty Laughingthrush** *Garrulax poecilorhynchus*
_____ *G. p. ricinus*
_____ *G. p. berthemyi*
_____ *G. p. poecilorhynchus*

Mts. of s China (Gansu to s Sichuan and extreme nw Yunnan)
Mountains of se China (s Anhui to Zhejiang and nw Fujian)
Mountains of Taiwan

☐ **Chestnut-capped Laughingthrush** *Garrulax mitratus*
_____ *G. m. major*
_____ *G. m. mitratus*
_____ *G. m. damnatus*
_____ *G. m. treacheri*
_____ *G. m. griswoldi*

Highlands of Malay Pen. (n Perak to s Selangor and Pahang)
Highlands of w Sumatra
Mountains of e Sarawak (Mt. Dulit, Mt. Derian, Kelabit Plateau)
N Borneo (Mt. Kinabalu)
Highlands of central Borneo (Schwaner and Müller mountains)

☐ **Spot-breasted Laughingthrush** *Garrulax merulinus*
_____ *G. m. merulinus*
_____ *G. m. obscurus*
_____ *G. m. annamensis*

S China (w Yunnan) to n Myanmar and s Assam
S China (se Yunnan) to n Laos and nw Tonkin
Laos (Langbian Plateau)

☐ **Hwamei** *Garrulax canorus*
_____ *G. c. canorus*
_____ *G. c. owstoni*
_____ *G. c. taewanus*

S China (Yangtze Valley) to Tonkin, n Annam and n Laos
Mountains of Hainan (s China)
Taiwan

☐ **White-browed Laughingthrush** *Garrulax sannio*
_____ *G. s. albosuperciliaris*
_____ *G. s. comis*
_____ *G. s. sannio*
_____ *G. s. oblectans*

NE India (Naga Hills and Manipur in e Assam)
S China (Yunnan) to ne Myanmar, n Laos, n Annam and Tonkin
S China (Guangxi, Guandong, Fujian, Jiangxi, Hunan) to Tonkin
W central China (sw Hubei, n Guizhou and Sichuan)

☐ **Rufous-breasted Laughingthrush** *Garrulax cachinnans*

SW Peninsular India (Nilgiri Hills in w Tamil Nadu)

☐ **Gray-breasted Laughingthrush** *Garrulax jerdoni*
_____ *G. j. jerdoni*
_____ *G. j. fairbanki*
_____ *G. j. meridionalis*

Hill forests of sw India (Western Ghats in Coorg region)
S India (Palni and Anaimalai hills and n Kerala)
Hill forests of sw India (s Kerala)

☐ **Streaked Laughingthrush** *Garrulax lineatus*
_____ *G. l. bilkevitchi*
_____ *G. l. gilgit*
_____ *G. l. lineatus*
_____ *G. l. setafer*
_____ *G. l. imbricatus*

Tajikistan and e Afghanistan to nw Pakistan
NE Pakistan (Gilgit region of Kashmir)
Himalayas (central Kashmir to nw Uttar Pradesh and sw Tibet)
Nepal to Sikkim and w Bengal (Darjiling)
Bhutan and se Tibet

☐ **Striped Laughingthrush** *Garrulax virgatus*

Mountains of ne India (Assam) and sw Myanmar (Chin Hills)

☐ **Scaly Laughingthrush** *Garrulax subunicolor*
_____ *G. s. subunicolor*
_____ *G. s. griseatus*
_____ *G. s. fooksi*

Himalayas (Nepal to Sikkim, Bhutan, e Assam and se Tibet)
Extreme ne Myanmar (Kachin State) to s China (nw Yunnan)
Mountains of nw Tonkin

☐ **Brown-capped Laughingthrush** *Garrulax austeni*
_____ *G. a. austeni*
_____ *G. a. victoriae*

Montane forests of s Assam (south of the Brahmaputra)
W Myanmar (Mt. Victoria)

☐ **Blue-winged Laughingthrush** *Garrulax squamatus*

Montane forests of Nepal to sw China, Myanmar and nw Tonkin

☐ **Elliot's Laughingthrush** *Garrulax elliotii*
____ *G. e. przewalskii*
____ *G. e. elliotii*

Montane forests of s-central China (Gansu and e Qinghai)
Mts. of s China (s Shaanxi, w Hubei, Sichuan and nw Yunnan)

☐ **Variegated Laughingthrush** *Garrulax variegatus*
____ *G. v. similis*
____ *G. v. variegatus*

Himalayas (e Afghanistan to w Pakistan and w Kashmir)
Himalayas of nw India (Himachal Pradesh) to Nepal

☐ **Prince Henry's Laughingthrush** *Garrulax henrici*

Semiarid montane scrub of w China (sw Xinjiang and se Tibet)

☐ **Black-faced Laughingthrush** *Garrulax affinis*
____ *G. a. affinis*
____ *G. a. bethelae*
____ *G. a. oustaleti*
____ *G. a. muliensis*
____ *G. a. blythii*
____ *G. a. saturatus*

Mountains of w and central Nepal
Himalayas from e Nepal to e Bhutan and se Tibet
S China (nw Yunnan) to ne Assam and ne Myanmar
SW China (Yangtze River Valley of se Qinghai and nw Yunnan)
Mountains of w-central China (n Sichuan in Moupin region)
N Tonkin (Fan Si Pan Mountains)

☐ **White-whiskered Laughingthrush** *Garrulax morrisonianus*

Montane forests of Taiwan

☐ **Chestnut-crowned Laughingthrush** *Garrulax erythrocephalus*
____ *G. e. erythrocephalus*
____ *G. e. kali*
____ *G. e. nigrimentum*
____ *G. e. imprudens*
____ *G. e. chrysopterus*
____ *G. e. godwini*
____ *G. e. erythrolaemus*
____ *G. e. woodi*
____ *G. e. connectens*
____ *G. e. subconnectens*
____ *G. e. schistaceus*
____ *G. e. melanostigma*
____ *G. e. ramsayi*
____ *G. e. peninsulae*

Himalayas of w India (Himachal Pradesh to Uttar Pradesh)
W and central Nepal
Himalayas of Sikkim, Bhutan and se Tibet
Hill forests of Assam (north and east of the Brahmaputra)
Hill forests of s Assam (south of the Brahmaputra)
Hill forests of se Assam (Barail Mountains)
E Manipur and sw Myanmar (Chin Hills and Arakan Yoma Mts.)
NE Myanmar (Kachin and N Shan States) and adj. sw Yunnan
Mountains of nw Tonkin and ne Laos
Mountains of ne Thailand (Doi Phu Kha)
Mountains of e Myanmar and nw Thailand
SE Myanmar (s Shan States) to high mountains of nw Thailand
S Myanmar (Karenni State and Tavoy District)
High mountains of peninsular Thailand and Malay Peninsula

☐ **Golden-winged Laughingthrush** *Garrulax ngoclinhensis*

Western highlands of Vietnam (Mount Ngoc Linh)

☐ **Collared Laughingthrush** *Garrulax yersini*

S Laos (Langbian Plateau). Status unknown

☐ **Red-winged Laughingthrush** *Garrulax formosus*
____ *G. f. formosus*
____ *G. f. greenwayi*

SW China (sw Sichuan, ne Yunnan and s Guangxi)
N Tonkin (Fan Si Pan Mountains)

☐ **Red-tailed Laughingthrush** *Garrulax milnei*
____ *G. m. sharpei*
____ *G. m. vitryi*
____ *G. m. sinianus*
____ *G. m. milnei*

E Myanmar to s China (s Yunnan), nw Thailand and n Indochina
S Laos (Bolavens Plateau)
SE China (Guizhou and Guangxi)
Mountains of se China (nw Fujian)

☐ **Gray-faced Liocichla** *Liocichla omeiensis*

Mountains of sw China (central Sichuan on Mt. Omei Shan)

☐ **Bugun Liocichla** *Liocichla bugunorum*

Arunachal Pradesh, India

☐ **Steere's Liocichla** *Liocichla steerii*

Montane forests of Taiwan

☐ **Red-faced Liocichla** *Liocichla phoenicea*
____ *L. p. phoenicea*

Himalayas from Nepal to Bhutan and Assam (Mishmi Hills)

____	*L. p. bakeri*	S Assam (s of the Brahmaputra) to nw Myanmar and nw Yunnan
____	*L. p. ripponi*	E Myanmar (Kachin State to s Shan States) and nw Thailand
____	*L. p. wellsi*	S China (se Yunnan) to n Laos and n Tonkin

☐ **Spot-throat** *Modulatrix stictigula*
____	*M. s. stictigula*	NE Tanzania (Usambara and Nguru mountains)
____	*M. s. pressa*	E Tanzania (Ukaguru Mts.) to n Malawi (Masuku Mts.)

☐ **Dapple-throat** *Arcanator orostruthus*
____	*A. o. armani*	NE Tanzania (Usambara Mountains)
____	*A. o. sanjei*	NE Tanzania (Udzungwa Mountains)
____	*A. o. orostruthus*	N Mozambique (Mt. Namuli)

☐ **White-chested Babbler** *Trichastoma rostratum*
____	*T. r. rostratum*	Malay Pen., Sumatra, Belitung I., Riau and Lingga archipelagos
____	*T. r. macropterum*	Borneo and Banggai I.

☐ **Sulawesi Babbler** *Trichastoma celebense*
____	*T. c. celebense*	N peninsular Sulawesi, Bangka, Lembeh and Manterawu islands
____	*T. c. rufofuscum*	N-central, s-central and se Sulawesi and Butung I.
____	*T. c. finschi*	S Sulawesi
____	*T. c. togianense*	Togian Islands

☐ **Ferruginous Babbler** *Trichastoma bicolor*

Lowlands of Malay Peninsula, Sumatra, Bangka I. and Borneo

☐ **Bagobo Babbler** *Trichastoma woodi*

S Philippines (montane forests of Mindanao)

☐ **Abbott's Babbler** *Malacocincla abbotti*
____	*M. a. abbotti (rufescentior)*	S Myanmar to Thailand, nw Malay Pen. and Mergui Archipelago
____	*M. a. krishnarajui*	E India (Eastern Ghats in n Andhra Pradesh)
____	*M. a. williamsoni*	Thailand (e part of sw plateau) and nw Cambodia
____	*M. a. obscurior*	Coastal se Thailand (Chon Buri Province to Trat); Ko Kut I.
____	*M. a. altera*	Central Laos and central Vietnam
____	*M. a. olivacea*	Peninsular Thailand and Malay Peninsula to e Sumatra
____	*M. a. sirense*	Borneo, Matasiri and Belitung islands
____	*M. a. baweana*	Bawean I. (Java Sea)

☐ **Horsfield's Babbler** *Malacocincla sepiaria*
____	*M. s. tardinata*	Malay Peninsula (Pattani to Selangor and Pahang)
____	*M. s. liberalis*	Highlands of nw Sumatra
____	*M. s. barussana*	Highlands of sw Sumatra
____	*M. s. sepiaria (minor)*	Java and Bali
____	*M. s. harterti*	N and e Borneo
____	*M. s. rufiventris*	W and s Borneo

☐ **Short-tailed Babbler** *Malacocincla malaccensis*
____	*M. m. malaccensis*	Malay Pen., Sumatra, Anambas, N Natuna, Lingga and Riau arch.
____	*M. m. saturata*	W Borneo, Bangka and Belitung islands
____	*M. m. poliogenys*	E Borneo
____	*M. m. feriata*	Extreme ne Sarawak (Mt. Mulu)

☐ **Ashy-headed Babbler** *Malacocincla cinereiceps*

S Philippines (Balabac and Palawan)

☐ **Brown-capped Babbler** *Pellorneum fuscocapillus*
____	*P. f. babaulti*	Arid lowlands of n and e Sri Lanka
____	*P. f. fuscocapillus*	Wet zone of sw Sri Lanka
____	*P. f. scortillum*	Humid forests of sw Sri Lanka

☐ **Marsh Babbler** *Pellorneum palustre*

NE India (Arunachal Pradesh to Cachar, Khasi, Chittagong Hills)

☐ Buff-breasted Babbler *Pellorneum tickelli*

____	*P. t. assamense*	NE India (Arunachal Pradesh to Bangladesh and Manipur)
____	*P. t. grisescens*	SW Myanmar (Arakan Yoma Mountains)
____	*P. t. fulvum*	S China (sw Yunnan) to ne Myanmar, n Thailand and Indochina
____	*P. t. annamense*	Central and s Vietnam to s Laos (Bolavens Plateau)
____	*P. t. tickelli*	Central and s Malay Peninsula and adjacent Thailand
____	*P. t. buettikoferi*	S Sumatra and Belitung I.

☐ Temminck's Babbler *Pellorneum pyrrogenys*

____	*P. p. pyrrogenys*	W Java
____	*P. p. besuki*	E Java
____	*P. p. erythrote*	W Sarawak (Mt. Poi and Mt. Penrissen)
____	*P. p. longstaffi*	Montane forests of Sarawak
____	*P. p. canicapillus*	Highlands of n Borneo

☐ Spot-throated Babbler *Pellorneum albiventre*

____	*P. a. ignotum*	Hill forests of ne Assam (Mishmi Hills)
____	*P. a. albiventre (nagaense)*	Bhutan/Assam border to w Myanmar (Chin Hills)
____	*P. a. cinnamomeum*	Central Myanmar to nw Thailand, s Laos and s Annam
____	*P. a. pusillum*	Eastern regions of n Laos and w Tonkin

☐ Puff-throated Babbler *Pellorneum ruficeps*

____	*P. r. olivaceum*	SW India (Kerala)
____	*P. r. ruficeps*	Coastal lowlands and hills of w and central India
____	*P. r. punctatum*	W Himalayas (Kangra to Garhwal)
____	*P. r. mandellii*	Nepal to Sikkim, Bhutan and ne India (Darjiling District)
____	*P. r. chamelum*	S Assam south of the Brahmaputra (Garo Hills to Naga Hills)
____	*P. r. pectorale*	NE Assam (Mishmi Hills)
____	*P. r. ripleyi*	NE Assam south of the Brahmaputra (Lakhimpur District)
____	*P. r. vocale*	NE India (valley of central Manipur)
____	*P. r. victoriae*	N Myanmar (Chin Hills)
____	*P. r. stageri*	NE Myanmar (Myitkyina and Bhamo districts)
____	*P. r. shanense*	Central Myanmar (N and S Shan States) to s China (sw Yunnan)
____	*P. r. hilarum*	Arid zone of central Myanmar
____	*P. r. minus*	S Myanmar (lower Irrawaddy River)
____	*P. r. subochraceum*	S Myanmar and adjacent sw Thailand
____	*P. r. insularum*	S Myanmar (Mergui Archipelago)
____	*P. r. acrum*	Central plains of Thailand and n Malay Peninsula
____	*P. r. chthonium*	N plateau of Thailand
____	*P. r. indictinctum*	Mekong River drainage of n plateau of Thailand
____	*P. r. oreum*	S China (s Yunnan between Mekong and Salween river)
____	*P. r. vividum*	S Yunnan (Red River Valley) to extreme n Tonkin and c Annam
____	*P. r. elbeli*	Northwest part of e plateau of Thailand
____	*P. r. ubonense*	E part of e plateau of Thailand and adjacent s Laos
____	*P. r. deignani*	S Vietnam
____	*P. r. dilloni*	S Indochina
____	*P. r. euroum*	Central plains of Thailand (e of Chao Phaya) to w Cambodia
____	*P. r. smithi*	Islets off coastal se Thailand and Cambodia

☐ Black-capped Babbler *Pellorneum capistratum*

____	*P. c. nigrocapitatum*	Malay Peninsula to Singapore, North Natuna Is. and Belitung I.
____	*P. c. nyctilampis*	Sumatra and Bangka I.
____	*P. c. capistratoides*	W and s Borneo
____	*P. c. morrelli*	N and e Borneo and Banggai Islands
____	*P. c. capistratum*	Java

☐ Palawan Babbler *Malacopteron palawanense*

SW Philippines (Balabac and Palawan)

☐ Moustached Babbler *Malacopteron magnirostre*

____	*M. m. magnirostre (flavum)*	S Myanmar and s Thailand to Sumatra and adjacent islands
____	*M. m. cinereocapilla*	N Borneo

☐ **Sooty-capped Babbler** *Malacopteron affine*
_____ *M. a. affine*   SE peninsular Thailand to Malaya, Singapore and Sumatra
_____ *M. a. notatum*   Banyak I. (off Sumatra)
_____ *M. a. phoeniceum*   Borneo

☐ **Scaly-crowned Babbler** *Malacopteron cinereum*
_____ *M. c. indochinense*   SE Thailand to Cambodia and s Laos
_____ *M. c. cinereum*   Malay Peninsula to Sumatra, Borneo and adjacent islands
_____ *M. c. niasense*   Nias I. (off Sumatra)
_____ *M. c. rufifrons*   Java
_____ *M. c. bungurense*   North Natuna Islands (off n Borneo)

☐ **Rufous-crowned Babbler** *Malacopteron magnum*
_____ *M. m. magnum*   S Myanmar, Malay Pen., Sumatra, Borneo and North Natuna Is.
_____ *M. m. saba*   NE Borneo

☐ **Gray-breasted Babbler** *Malacopteron albogulare*
_____ *M. a. albogulare*   Malay Peninsula, ne Sumatra, Batu Islands and Lingga Arch.
_____ *M. a. moultoni*   NW Borneo

☐ **Damara Rockjumper** *Chaetops pycnopygius*
_____ *C. p. pycnopygius*   Rocky regions of sw Angola and n Namibia
_____ *C. p. spadix*   Escarpment of sw highlands of Angola (Huila and adj. Namibia)

☐ **Rufous Rockjumper** *Chaetops frenatus*
                             South Africa (Western and Eastern Cape Province)

☐ **Orange-breasted Rockjumper** *Chaetops aurantius*
                             Rocky montane slopes of Lesotho, Natal and e Cape Province

☐ **Blackcap Illadopsis** *Illadopsis cleaveri*
_____ *I. c. johnsoni*   Sierra Leone and Liberia (Ivory Coast?)
_____ *I. c. cleaveri*   Ghana
_____ *I. c. marchanti*   S Nigeria
_____ *I. c. batesi*   SE Nigeria to Cent. African Rep. and Congo
_____ *I. c. poensis*   Bioko (Gulf of Guinea)

☐ **Scaly-breasted Illadopsis** *Illadopsis albipectus*
                             NW Angola to Zaire, Uganda, w Kenya, nw Tanaznia, se Sudan

☐ **Rufous-winged Illadopsis** *Illadopsis rufescens*
                             Senegal to Ghana; single record from Togo

☐ **Puvel's Illadopsis** *Illadopsis puveli*
_____ *I. p. puveli*   Senegal to Togo
_____ *I. p. strenuipes*   Nigeria to Sudan and Uganda

☐ **Pale-breasted Illadopsis** *Illadopsis rufipennis*
_____ *I. r. extrema*   Sierra Leone to Ghana
_____ *I. r. rufipennis (bocagei)*   S Nigeria to Angola, Kenya and nw Tanzania
_____ *I. r. distans*   S Kenya to ne Tanzania and Zanzibar I.
_____ *I. r. pugensis*   E Tanzania (Pugu Hills)

☐ **Brown Illadopsis** *Illadopsis fulvescens*
_____ *I. f. gularis*   Sierra Leone to w Ghana
_____ *I. f. moloneyana*   E Ghana and Togo
_____ *I. f. iboensis*   SW Nigeria to w Cameroon
_____ *I. f. fulvescens*   Cameroon to Congo and w Zaire
_____ *I. f. ugandae*   Central and e Zaire to Sudan, Kenya and Tanzania
_____ *I. f. dilutior*   N Angola

☐ **Mountain Illadopsis** *Illadopsis pyrrhoptera*
_____ *I. p. nyasae*   N Malawi
_____ *I. p. pyrrhoptera*   Montane forests of Zaire to Kenya and Tanzania

☐ **Gray-chested Illadopsis** *Kakamega poliothorax*

SE Nigeria to s Cameroon, e Zaire and w Kenya; Bioko

☐ **African Hill Babbler** *Pseudoalcippe abyssinica*

____ *P. a. abyssinica*	Ethiopia and Sudan to e Uganda, sw Tanzania, se Zaire, Angola
____ *P. a. monachus*	Mt. Cameroon
____ *P. a. claudei*	Bioko (Gulf of Guinea)
____ *P. a. atriceps*	Nigeria to Cameroon, e Zaire, Rwanda and w Uganda
____ *P. a. stierlingi*	E and s Tanzania
____ *P. a. stictigula*	Malawi and n Mozambique

☐ **Thrush Babbler** *Ptyrticus turdinus*

____ *P. t. harterti*	Grasslands of central Cameroon
____ *P. t. turdinus*	S Sudan and ne Zaire
____ *P. t. upembae*	S Zaire, ne Angola and nw Zambia

☐ **Large Scimitar-Babbler** *Pomatorhinus hypoleucos*

____ *P. h. hypoleucos*	NE India (Assam) to Bangladesh and w Myanmar
____ *P. h. tickelli*	S China (s Yunnan) to s Myanmar, Thailand and n Indochina
____ *P. h. brevirostris*	S Indochina
____ *P. h. wrayi*	Malay Peninsula
____ *P. h. hainanus*	Hainan (s China)

☐ **Spot-breasted Scimitar-Babbler** *Pomatorhinus erythrocnemis*

____ *P. e. ferrugilatus*	Montane forests of Kashmir to central Nepal
____ *P. e. haringtoni*	Himalayas (Sikkim to Bhutan)
____ *P. e. mcclellandi*	Assam south of the Brahmaputra to w Myanmar (Chin Hills)
____ *P. e. odicus*	Mts. of ne Myanmar to s China (Yunnan), n Laos and nw Tonkin
____ *P. e. decarlei*	Mountains of sw China (Qinghai, s Sichuan and nw Yunnan)
____ *P. e. dedekeni*	Mountains of sw China (e Qinghai to nw Yunnan)
____ *P. e. gravivox*	Mountains of sw China (nw Sichuan to s Gansu and s Shaanxi)
____ *P. e. sowerbyi*	Central China (central Shaanxi)
____ *P. e. cowensae*	S China (e Sichuan to sw Hubei and n Guizhou)
____ *P. e. swinhoei*	E China (Anhui to ne Jiangxi and Fujian)
____ *P. e. abbreviatus*	SE China (s Hunan, Guangxi and n Guandgong)
____ *P. e. erythrocnemis*	Taiwan

☐ **Rusty-cheeked Scimitar-Babbler** *Pomatorhinus erythrogenys*

Himalayas of central Myanmar and nw Thailand

☐ **Indian Scimitar-Babbler** *Pomatorhinus horsfieldii*

____ *P. h. obscurus*	NW India (Aravalli Mountains of Rajasthan)
____ *P. h. horsfieldii*	W India (Western Ghats from Satpura Range to Goa)
____ *P. h. maderaspatensis*	E central India (Eastern Ghats from Andhra to Salem District)
____ *P. h. travancoreensis*	SW India (Western Ghats from North Kanara to Kerala)
____ *P. h. melanurus*	Sri Lanka

☐ **White-browed Scimitar-Babbler** *Pomatorhinus schisticeps*

____ *P. s. leucogaster*	Himalayas (Himachal Pradesh to nw Uttar Pradesh)
____ *P. s. schisticeps*	Himalayas (w Nepal to Bhutan, Assam and nw Myanmar)
____ *P. s. salimalii*	NE Assam (Mishmi Hills)
____ *P. s. cryptanthus*	NE Assam (Lakhimpur District) and (?) adjacent ne Myanmar
____ *P. s. mearsi*	W Myanmar (lower Chindwin District to Arakan Yoma Mts.)
____ *P. s. ripponi*	E Myanmar to n Thailand and adjacent nw Laos
____ *P. s. nuchalis*	W Myanmar (Southern Shan State and Karenni State)
____ *P. s. difficilis*	Mountains of nw Thailand and s Myanmar (Amherst District)
____ *P. s. olivaceus*	Lowlands of s Myanmar and peninsular Thailand
____ *P. s. humilis*	E Thailand (Nan Province) to s Laos and central Vietnam
____ *P. s. klossi*	SE Thailand and sw Cambodia
____ *P. s. annamensis*	S Vietnam (Langbian Plateau)
____ *P. s. fastidiosus*	Malay Peninsula (s Myanmar and Isthmus of Kra to Trang)

☐ **Chestnut-backed Scimitar-Babbler** *Pomatorhinus montanus*

____ P. m. occidentalis	Malay Peninsula, Sumatra and Bangka I.
____ P. m. montanus	W and central Java
____ P. m. ottolanderi	E Java and Bali
____ P. m. bornensis	Borneo

☐ **Streak-breasted Scimitar-Babbler** *Pomatorhinus ruficollis*

____ P. r. ruficollis	W and central Nepal
____ P. r. godwini	E Himalayas (e Nepal to Sikkim, Bhutan, se Tibet and n Assam)
____ P. r. bakeri	Hill forests of se Assam (s of the Brahmaputra) to w Myanmar
____ P. r. bhamoensis	N Myanmar (Bhamo District)
____ P. r. similis	NE Myanmar to s China (nw Yunnan)
____ P. r. albipectus	S China (sw Yunnan) and adjacent n Laos
____ P. r. beaulieui	N Laos
____ P. r. laurentei	S China (Kunming region of Yunnan)
____ P. r. reconditus (laurenti)	S China (se Yunnan to n Vietnam)
____ P. r. stridulus	Hill forests of se China (Guandong, Fujian and Jiangxi)
____ P. r. intermedius	Central China (se Hubei, Hunan, Guangxi and Guizhou)
____ P. r. eidos	SW China (s Sichuan)
____ P. r. nigrostellatus	Hainan (s China)
____ P. r. musicus	Taiwan

☐ **Red-billed Scimitar-Babbler** *Pomatorhinus ochraceiceps*

____ P. o. stenorhynchus	NE Assam (Mishmi Hills) to n Myanmar
____ P. o. austeni	Hill forests of e Assam (Naga Hills to Barail Mts. and Manipur)
____ P. o. ochraceiceps	Myanmar to mountains of n Thailand, Tonkin and n Laos
____ P. o. alius	Plateau of ne Thailand to s Indochina

☐ **Coral-billed Scimitar-Babbler** *Pomatorhinus ferruginosus*

____ P. f. ferruginosus	Himalayas (e Nepal to e Assam north of the Brahmaputra)
____ P. f. formosus	Hill forests of Assam (south of the Brahmaputra and Manipur)
____ P. f. phayrei	Hill forests of sw Myanmar (Chin Hills and Arakan Yoma Mts.)
____ P. f. stanfordi	NE Myanmar (Kachin State)
____ P. f. mariae	Central Myanmar
____ P. f. albogularis	E Myanmar to nw Thailand
____ P. f. orientalis	N Indochina (Tonkin and n Laos)
____ P. f. dickinsoni	Highlands of c Vietnam and Bolavens Plateau of s Laos

☐ **Slender-billed Scimitar-Babbler** *Xiphirhynchus superciliaris*

____ X. s. superciliaris	Himalayas (e Nepal to Sikkim and Bhutan)
____ X. s. intextus	Hill forests of s Assam (s and e of the Brahmaputra)
____ X. s. forresti	Mountains of ne Myanmar and sw China (nw Yunnan)
____ X. s. rothschildi	Montane forests of n Vietnam (Fan Si Pan Mountains)

☐ **Short-tailed Scimitar-Babbler** *Jabouilleia danjoui*

____ J. d. danjoui	Central Vietnam (Langbian Plateau)
____ J. d. parvirostris	Central Vietnam (Col des Nuages)

☐ **Naung Mung Scimitar-babbler** *Jabouilleia naungmungensis*

	Sub-Himalayan Myanmar

☐ **Long-billed Wren-Babbler** *Rimator malacoptilus*

____ R. m. malacoptilus	E Himalayas (Sikkim to e Assam and ne Myanmar)
____ R. m. pasquieri	N Vietnam (Fan Si Pan Mountains)
____ R. m. albostriatus	Highlands of w Sumatra

☐ **Bornean Wren-Babbler** *Ptilocichla leucogrammica*

	Patchily distributed lowlands of Borneo

☐ **Striated Wren-Babbler** *Ptilocichla mindanensis*

____ P. m. minuta	N Philippines (Leyte and Samar)
____ P. m. fortichi	Central Philippines (Bohol)

____	*P. m. mindanensis*	S Philippines (Mindanao)
____	*P. m. basilanica*	S Philippines (Basilan)

☐ **Falcated Wren-Babbler** *Ptilocichla falcata*

SW Philippines (Balabac and Palawan)

☐ **Striped Wren-Babbler** *Kenopia striata*

S peninsular Thailand, Malay Peninsula, Sumatra and Borneo

☐ **Large Wren-Babbler** *Napothera macrodactyla*

____	*N. m. macrodactyla*	SW Peninsular Thailand and Malay Peninsula
____	*N. m. beauforti*	NE Sumatra
____	*N. m. lepidopleura*	Java

☐ **Rusty-breasted Wren-Babbler** *Napothera rufipectus*

Montane forests of w Sumatra

☐ **Black-throated Wren-Babbler** *Napothera atrigularis*

Patchily distributed forests of Borneo

☐ **Marbled Wren-Babbler** *Napothera marmorata*

____	*N. m. grandior*	Montane forests of central Malaya (Selangor/Pahang border)
____	*N. m. marmorata*	Highlands of w Sumatra

☐ **Limestone Wren-Babbler** *Napothera crispifrons*

____	*N. c. crispifrons*	Limestone hills of n Thailand to s Myanmar
____	*N. c. calcicola*	Limestone hills of central Thailand (Sathani Hin Lap)
____	*N. c. annamensis*	Limestone hills of n Indochina

☐ **Streaked Wren-Babbler** *Napothera brevicaudata*

____	*N. b. striata*	Hill forests of s Assam (s of the Brahmaputra) to sw Myanmar
____	*N. b. venningi*	S China (w Yunnan) to ne Myanmar
____	*N. b. brevicaudata*	N Thailand south to s Myanmar
____	*N. b. stevensi*	S China (sw Guangxi) and n Indochina
____	*N. b. griseigularis*	SE Thailand and sw Cambodia
____	*N. b. proxima*	Central Vietnam and s Laos
____	*N. b. rufiventer*	S Vietnam (Langbian Plateau)
____	*N. b. leucosticta*	N Malay Peninsula

☐ **Mountain Wren-Babbler** *Napothera crassa*

High mountains of n Borneo (ne Sarawak and n Sabah)

☐ **Luzon Wren-Babbler** *Napothera rabori*

____	*N. r. rabori*	N Philippines (Ilicos Norte and Cagayan provinces of Luzon)
____	*N. r. mesoluzonica*	N Philippines (Laguna Province of Luzon)
____	*N. r. sorsogonensis*	N Philippines (Sorsogon and Camarines Sur provinces of Luzon)

☐ **Eyebrowed Wren-Babbler** *Napothera epilepidota*

____	*N. e. guttaticollis*	Hill forests of n Assam (north of the Brahmaputra)
____	*N. e. roberti*	Hill forests of s Assam (s of the Brahmaputra) to nw Myanmar
____	*N. e. bakeri*	Central Myanmar (Southern Shan and Karenni states)
____	*N. e. davisoni*	N Thailand south to s Myanmar
____	*N. e. amyea*	N Indochina
____	*N. e. delacouri*	S China (s Yunnan and Yao Shan region of Guangxi)
____	*N. e. hainana*	Hainan (s China)
____	*N. e. clara*	S Vietnam (Langbian Plateau)
____	*N. e. granti*	N Malay Peninsula
____	*N. e. diluta (lucilleae)*	Highlands of n and w Sumatra
____	*N. e. mendeni*	Highlands of sw Sumatra
____	*N. e. epilepidota*	Highlands of w and central Java
____	*N. e. exsul*	Highlands of n Borneo

☐ **Scaly-breasted Wren-Babbler** *Pnoepyga albiventer*

____	*P. a. pallidior*	Himalayas (East Punjab to w Nepal)
____	*P. a. albiventer*	Himalayas of Nepal to Assam, n Myanmar, s China and n Tonkin

☐ **Immaculate Wren-Babbler** *Pnoepyga immaculata*

NE India and Himalayas of Nepal; > in *terai* lowlands

☐ **Pygmy Wren-Babbler** *Pnoepyga pusilla*
____ *P. p. pusilla* — Nepal to Assam, n Myanmar, se Tibet, sw China and n Thailand
____ *P. p. formosana* — Highlands of Taiwan
____ *P. p. annamensis* — S Indochina (Langbian Plateau) and s Laos (Bolavens Plateau)
____ *P. p. harterti* — Highlands of Malay Peninsula
____ *P. p. lepida* — Highlands of w Sumatra
____ *P. p. rufa* — Highlands of Java
____ *P. p. everetti* — Highlands of Flores (e Lesser Sundas)
____ *P. p. timorensis* — Highlands of Timor (e Lesser Sundas)

☐ **Rufous-throated Wren-Babbler** *Spelaeornis caudatus*

Mountains of e Nepal to Sikkim, Darjiling and Bhutan

☐ **Mishmi Wren-Babbler** *Spelaeornis badeigularis*

NE India (Mishmi Hills)

☐ **Bar-winged Wren-Babbler** *Spelaeornis troglodytoides*
____ *S. t. sherriffi* — E Bhutan
____ *S. t. souliei* — SE Tibet to s China (nw Yunnan) and ne Myanmar
____ *S. t. rocki* — S China (nw Yunnan east of the Mekong River)
____ *S. t. troglodytoides* — W-central China (Qinghai to nw Sichuan)
____ *S. t. halsueti* — W-central China (Tsingling Mts. in s Shaanxi and adj. Gansu)

☐ **Spotted Wren-Babbler** *Spelaeornis formosus*

Humid forests of Sikkim to se China and w Myanmar

☐ **Long-tailed Wren-Babbler** *Spelaeornis chocolatinus*
____ *S. c. chocolatinus* — Hill forests of Assam (south of the Brahmaputra) and Manipur
____ *S. c. oatesi* — N Myanmar (Mt. Victoria)
____ *S. c. reptatus* — NE Myanmar to s China (sw Yunnan)
____ *S. c. kinneari* — N Vietnam (nw Tonkin)

☐ **Tawny-breasted Wren-Babbler** *Spelaeornis longicaudatus*

Oak-rhododendron forests of ne India (Meghalaya and Manipur)

☐ **Wedge-billed Wren-Babbler** *Sphenocichla humei*
____ *S. h. humei* — Himalayas from Sikkim to n Assam (n of the Brahmaputra)
____ *S. h. roberti* — Hill forests of s Assam (s of the Brahmaputra) to ne Myanmar

☐ **Common Jery** *Neomixis tenella*
____ *N. t. tenella* — Savanna of n Madagascar
____ *N. t. decaryi* — Savanna of w Madagascar
____ *N. t. orientalis* — Humid forests of central and s Madagascar
____ *N. t. debilis* — Arid subdesert of sw Madagascar

☐ **Green Jery** *Neomixis viridis*
____ *N. v. delacouri* — Humid highland forests of ne Madagascar
____ *N. v. viridis* — Humid highland forests of se Madagascar

☐ **Stripe-throated Jery** *Neomixis striatigula*
____ *N. s. sclateri* — Humid forests of ne Madagascar
____ *N. s. pallidior* — Arid subdesert of sw Madagascar
____ *N. s. striatigula* — Humid forests of se Madagascar

☐ **Wedge-tailed Jery** *Hartertula flavoviridis*

Rainforests of e Madagascar (Sianaka Forest s to Vondrozo)

☐ **Deignan's Babbler** *Stachyris rodolphei*

Montane bamboo forests of nw Thailand (Doi Luang Chiang)

☐ **Buff-chested Babbler** *Stachyris ambigua*
____ *S. a. ambigua* — Himalayas (Sikkim to Bhutan and Assam s of the Brahmaputra)
____ *S. a. planicola* — NE Myanmar (Kachin State) to nw Yunnan (Salween Valley)
____ *S. a. adjuncta* — N and e Thailand to n Laos and nw Tonkin
____ *S. a. insuspecta* — S Laos (Bolavens Plateau)

☐ **Rufous-fronted Babbler** *Stachyris rufifrons*
____   *S. r. pallescens*   Hill forests of sw Myanmar (Chin Hills and Arakan Yoma Mts.)
____   *S. r. rufifrons*   SE Myanmar to w Thailand
____   *S. r. obscura*   S Myanmar (Mergui District) and central peninsular Thailand
____   *S. r. poliogaster*   W Malaya (s Perak to Johore) to Sumatra
____   *S. r. sarawacensis*   N Borneo (Mt. Poi in w Sarawak)

☐ **Rufous-capped Babbler** *Stachyris ruficeps*
____   *S. r. ruficeps*   E Himalayas (Sikkim to Bhutan and n Assam n of Brahmaputra)
____   *S. r. rufipectus*   Hill forests of ne Assam to nw Myanmar
____   *S. r. davidi*   Central and s China (Yangtze River Valley) to n Indochina
____   *S. r. bhamoensis*   NE Myanmar (Kachin and N Shan States) to nw Yunnan
____   *S. r. goodsoni*   Hainan (s China)
____   *S. r. praecognita*   Taiwan
____   *S. r. pagana*   S Vietnam

☐ **Black-chinned Babbler** *Stachyris pyrrhops*
Himalayas (Kashmir to central Nepal)

☐ **Golden Babbler** *Stachyris chrysaea*
____   *S. c. chrysaea*   Nepal to Sikkim, Assam, Bhutan, sw China and n Myanmar
____   *S. c. binghami*   SE Assam to sw Myanmar (Chin Hills and Arakan Yoma Mts.)
____   *S. c. aurata*   S Myanmar (s Shan State) to extreme n Thailand and n Indochina
____   *S. c. assimilis*   Central Myanmar (west of the Salween River) to nw Thailand
____   *S. c. chrysops*   Hills of Malay Peninsula
____   *S. c. frigida*   Highlands of w Sumatra

☐ **Pygmy Babbler** *Stachyris plateni*
____   *S. p. pygmaea*   Philippines (Leyte and Samar)
____   *S. p. plateni*   S Philippines (Mindanao)

☐ **Golden-crowned Babbler** *Stachyris dennistouni*
N Philippines (Sierra Madre Mountains of n Luzon)

☐ **Black-crowned Babbler** *Stachyris nigrocapitata*
____   *S. n. affinis*   Philippines (s Sierra Madre Mountains of s Luzon)
____   *S. n. nigrocapitata*   N Philippines (Leyte and Samar)
____   *S. n. boholensis*   Central Philippines (Bohol)

☐ **Rusty-crowned Babbler** *Stachyris capitalis*
____   *S. c. capitalis*   S Philippines (Dinagat)
____   *S. c. euroaustralis*   S Philippines (Mindanao, excluding Zamboanga Peninsula)
____   *S. c. isabelae*   S Philippines (Basilan and Zamboanga Peninsula of Mindanao)

☐ **Flame-templed Babbler** *Stachyris speciosa*
____   *S. s. speciosa*   Central Philippines (Negros)
____   *S. s. ssp.*   Central Philippines (Panay). Undescribed subspecies

☐ **Chestnut-faced Babbler** *Stachyris whiteheadi*
____   *S. w. whiteheadi*   N Philippines (n Luzon)
____   *S. w. sorsogonensis*   N Philippines (s Luzon)

☐ **Luzon Striped-Babbler** *Stachyris striata*
N Philippines (n Luzon)

☐ **Panay Striped-Babbler** *Stachyris latistriata*
Central Philippines (montane forests of Panay)

☐ **Negros Striped-Babbler** *Stachyris nigrorum*
Philippines (montane forests of Negros)

☐ **Palawan Striped-Babbler** *Stachyris hypogrammica*
SW Philippines (montane forests of Palawan)

☐ **White-breasted Babbler** *Stachyris grammiceps*
Locally in forests of w Java

☐ **Sooty Babbler** *Stachyris herberti*

Rediscovered in 1994 in central Laos after 74-year absence

☐ **Gray-throated Babbler** *Stachyris nigriceps*

____	*S. n. nigriceps*	Himalayas (central Nepal to Sikkim, Bhutan and Assam)
____	*S. n. coei*	E Assam (Mishmi Hills)
____	*S. n. coltarti*	E Assam (Naga Hills) to n Myanmar and s China (w Yunnan)
____	*S. n. spadix*	S Assam (s of the Brahmaputra) to s Myanmar and nw Thailand
____	*S. n. yunnanensis*	E Myanmar to n Thailand, sw China and n Indochina
____	*S. n. rileyi*	S Vietnam
____	*S. n. dipora*	Malay Peninsula (Mergui District and Isthmus of Kra to Trang)
____	*S. n. davisoni*	Malay Peninsula (Pattani Province to Negri Sembilan)
____	*S. n. larvata*	Sumatra and Lingga Archipelago
____	*S. n. natunensis*	N Natuna Islands
____	*S. n. tionis*	Tioman I. (South China Sea)
____	*S. n. hartleyi*	Highlands of n Borneo (w Sarawak)
____	*S. n. borneensis*	Highlands of Borneo

☐ **Gray-headed Babbler** *Stachyris poliocephala*

Peninsular Thailand, Malaya, Sumatra, Lingga Arch. and Borneo

☐ **Snowy-throated Babbler** *Stachyris oglei*

Mountains of ne India (ne Assam and se Arunachal Pradesh)

☐ **Spot-necked Babbler** *Stachyris striolata*

____	*S. s. tonkinensis*	S China (Guangxi and s Yunnan) to n Indochina
____	*S. s. swinhoei*	Hainan (s China)
____	*S. s. helenae*	N plateau of Thailand (Nan Province) to n Laos
____	*S. s. guttata*	S Myanmar and adjacent w Thailand (Tak Province)
____	*S. s. nigrescentior*	Peninsular Thailand (Isthmus of Kra to Trang Province)
____	*S. s. umbrosa*	NE Sumatra
____	*S. s. striolata*	Highlands of w Sumatra

☐ **White-necked Babbler** *Stachyris leucotis*

____	*S. l. leucotis*	Peninsular Thailand and Malaya
____	*S. l. sumatrensis*	NE Sumatra (Aceh Province)
____	*S. l. obscurata*	N Borneo (Mt. Mulu)

☐ **Black-throated Babbler** *Stachyris nigricollis*

Forests of peninsular Thailand, Malaya, e Sumatra and Borneo

☐ **White-bibbed Babbler** *Stachyris thoracica*

____	*S. t. thoracica*	Foothill forests of s Sumatra, w and central Java
____	*S. t. orientalis*	E Java

☐ **Chestnut-rumped Babbler** *Stachyris maculata*

____	*S. m. pectoralis*	Lowlands of peninsular Thailand and Malaya
____	*S. m. maculata*	Sumatra, Borneo and Riau Archipelago
____	*S. m. banjakensis*	Banyak I. (off Sumatra)
____	*S. m. hypopyrrha*	Batu Islands (off Sumatra)

☐ **Chestnut-winged Babbler** *Stachyris erythroptera*

____	*S. e. erythroptera*	Malay Pen. (Isthmus of Kra to Singapore) and North Natuna Is.
____	*S. e. pyrrhophaea (apega)*	Sumatra, Bangka, Belitung and Batu islands
____	*S. e. fulviventris*	Banyak I. (off Sumatra)
____	*S. e. bicolor*	N and e Borneo and Banggai Islands
____	*S. e. rufa*	SW Borneo

☐ **Crescent-chested Babbler** *Stachyris melanothorax*

____	*S. m. melanothorax*	W Java (Mt. Gedeh and Mt. Pangerango)
____	*S. m. albigula*	W Java (Mt. Papandayan)
____	*S. m. mendeni*	W Java (Mt. Ciremay)
____	*S. m. intermedia*	E Java (Mt. Raung)
____	*S. m. baliensis*	Lowlands of Bali

☐ **Tawny-bellied Babbler** *Dumetia hyperythra*

____	*D. h. hyperythra*	Lowlands of sw Nepal to n and central India
____	*D. h. albogularis*	S India (Aravelli Mountains to Western and Eastern Ghats)
____	*D. h. navarroi*	W India
____	*D. h. phillipsi*	Sri Lanka

☐ **Dark-fronted Babbler** *Rhopocichla atriceps*

____	*R. a. atriceps*	Central India (Western Ghats from Bombay to Nilgiri Hills)
____	*R. a. bourdilloni*	Hill forests of sw India (Kerala)
____	*R. a. siccata*	Arid north, east and central hills of Sri Lanka
____	*R. a. nigrifrons*	Wet lowlands of sw Sri Lanka

☐ **Striped Tit-Babbler** *Macronous gularis*

____	*M. g. rubicapilla*	Lowlands of e Nepal to ne India and extreme n Myanmar
____	*M. g. ticehursti*	W Myanmar (Upper Chindwin District to Arakan)
____	*M. g. sulphureus*	S China (sw Yunnan) to e Myanmar and n plateau of Thailand
____	*M. g. lutescens*	S China (se Yunnan) to n and e Thailand, Laos and Tonkin
____	*M. g. saraburiensis*	E Thailand and w Cambodia
____	*M. g. kinneari*	Central Vietnam
____	*M. g. versuricola*	E Cambodia and s Vietnam
____	*M. g. connectens*	Coastal Gulf of Siam (Isthmus of Kra to Cambodia)
____	*M. g. inveteratus*	Coastal islets off se Thailand and Cambodia
____	*M. g. condorensis*	Pulau Kundur (South China Sea)
____	*M. g. archipelagicus*	Mergui Archipelago (off sw Myanmar)
____	*M. g. chersonesophilus*	Malay Peninsula (Isthmus of Kra to Perak and Trengganu)
____	*M. g. gularis*	S Malay Pen., Sumatra, Banyak, Batu, Lingga and Riau islands
____	*M. g. zopherus*	Anambas Islands
____	*M. g. zaperissus*	North Natuna Islands
____	*M. g. everetti*	Pulau Bunguran (North Natuna Islands)
____	*M. g. ruficoma*	Bangka and Belitung islands
____	*M. g. montanus*	NE Borneo
____	*M. g. bornensis*	Borneo
____	*M. g. argenteus*	Banggai Islands (off n Borneo)
____	*M. g. cagayanensis*	Cagayan Sulu (Sulu Sea)
____	*M. g. woodi*	SW Philippines (Balabac and Palawan)

☐ **Gray-cheeked Tit-Babbler** *Macronous flavicollis*

____	*M. f. javanicus*	Lowlands of w and central Java
____	*M. f. flavicollis*	Lowlands of e Java
____	*M. f. prillwitzi*	Kangean Islands (Java Sea)

☐ **Gray-faced Tit-Babbler** *Macronous kelleyi*

Forests of s Laos, central and s Vietnam and Cochinchina

☐ **Brown Tit-Babbler** *Macronous striaticeps*

____	*M. s. mindanensis*	S Philippines (Samar, Leyte, Bohol and Mindanao)
____	*M. s. alcasidi*	S Philippines (Dinagat and Siargao)
____	*M. s. striaticeps*	S Philippines (Basilan and Malamaui)
____	*M. s. kettlewelli*	Sulu Archipelago (Bongao, Jolo and Tawitawi islands)

☐ **Fluffy-backed Tit-Babbler** *Macronous ptilosus*

____	*M. p. ptilosus*	Peninsular Thailand and Malaya
____	*M. p. trichorrhos*	Sumatra and Batu Islands
____	*M. p. sordidus*	Bangka and Belitung islands
____	*M. p. reclusus*	Borneo

☐ **Miniature Tit-Babbler** *Micromacronus leytensis*

____	*M. l. leytensis*	S Philippines (Leyte and Samar)
____	*M. l. sordidus*	S Philippines (Mindanao)

☐ **Chestnut-capped Babbler** *Timalia pileata*

____	*T. p. bengalensis*	Submontane Himalayas (Nepal to Assam and nw Myanmar)

____ *T. p. smithi*	N Myanmar to s China, n Thailand and n Indochina
____ *T. p. intermedia*	Central and s Myanmar to sw Thailand
____ *T. p. patriciae*	W portion of central plains of Thailand
____ *T. p. dictator*	E and se Thailand to s Indochina
____ *T. p. pileata*	Java

☐ **Yellow-eyed Babbler** *Chrysomma sinense*

____ *C. s. hypoleucum*	E Pakistan to peninsular India, Bangladesh and w Myanmar
____ *C. s. nasale*	Sri Lanka
____ *C. s. saturatius*	Himalayas from Sikkim to Assam (north of the Brahmaputra)
____ *C. s. sinense*	S China to Myanmar, Thailand and Indochina

☐ **Jerdon's Babbler** *Chrysomma altirostre*

____ *C. a. scindica*	Grasslands of extreme s Pakistan (Mangrani region)
____ *C. a. griseigularis*	Base of Himalayas (Bhutan to s Assam and ne Myanmar)
____ *C. a. altirostre†*	S-c Myanmar (Irawaddy-Sittang grasslands). Extinct ca 1941

☐ **Rufous-tailed Babbler** *Chrysomma poecilotis*

Mts. of sw China (se Qinghai to nw Sichuan and nw Yunnan)

☐ **Spiny Babbler** *Turdoides nipalensis*

Himalayas of w and central Nepal

☐ **Iraq Babbler** *Turdoides altirostris*

SE Iraq and sw Iran (reed beds of lower Tigris-Euphrates Valley)

☐ **Common Babbler** *Turdoides caudata*

____ *T. c. salvadorii*	SE Iraq to sw Iran
____ *T. c. huttoni*	E Iran to s Afghanistan and s Pakistan
____ *T. c. eclipes*	N Pakistan (Fort Sandeman to Kashmir border)
____ *T. c. caudata*	Peninsular India, Laccadive Islands and Pamean I.

☐ **Striated Babbler** *Turdoides earlei*

____ *T. e. sonivius*	Pakistan (Rann of Kutch and Indus River Valley) to nw India
____ *T. e. earlei*	Grasslands of ne India to Assam and Myanmar

☐ **White-throated Babbler** *Turdoides gularis*

Dry grassy plains of central and s Myanmar

☐ **Slender-billed Babbler** *Turdoides longirostris*

Grasslands of Nepal to Assam and nw Myanmar

☐ **Large Gray Babbler** *Turdoides malcolmi*

Arid lowland scrub of peninsular India

☐ **Arabian Babbler** *Turdoides squamiceps*

____ *T. s. squamiceps*	Arabian Peninsula (Dead Sea depression to sw Saudi Arabia)
____ *T. s. yemensis*	Yemen and Aden
____ *T. s. muscatensis*	Arabian coast of Gulf of Oman

☐ **Fulvous Chatterer** *Turdoides fulva*

____ *T. f. maroccana (billypayni)*	S Morocco, adjacent Algeria and sw Libya
____ *T. f. fulva*	N Algeria to Tunisia and nw Libya
____ *T. f. buchanani*	SE Algeria and Mali to Niger and central Chad (Senegal?)
____ *T. f. acaciae*	N Chad and Sudan to n Eritrea

☐ **Scaly Chatterer** *Turdoides aylmeri*

____ *T. a. aylmeri*	Somalia and e Ethiopia
____ *T. a. boranensis*	S Ethiopia and n Kenya
____ *T. a. keniana (loveridgei)*	SE Kenya and ne Tanzania
____ *T. a. mentalis*	Interior ne Tanzania and adjacent s Kenya border

☐ **Rufous Chatterer** *Turdoides rubiginosa*

____ *T. r. rubiginosa*	SE Sudan, Ethiopia, n Uganda and Kenya
____ *T. r. sharpii*	SE Ethiopia (Dolo, Unsi) and adjacent Somalia
____ *T. r. heuglini*	S Somalia, se Kenya and ne Tanzania
____ *T. r. emini*	N-central Tanzania

☐ **Rufous Babbler** *Turdoides subrufa*

    _____ *T. s. subrufa*   SW India (Western Ghats to n Kerala, Madras and Nilgiri Hills)
    _____ *T. s. hyperythra*   SW India (sw Madras and Kerala)

☐ **Jungle Babbler** *Turdoides striata*

    _____ *T. s. sindiana*   Pakistan and nw India
    _____ *T. s. striata*   Himalayan foothills (n India to e Assam)
    _____ *T. s. orientalis*   Central and s India
    _____ *T. s. somervillei*   Coastal w India (Surat Dangs to Goa)
    _____ *T. s. malabarica*   SW India (Goa to Kerala)

☐ **Orange-billed Babbler** *Turdoides rufescens*

     Humid forests of Sri Lanka

☐ **Yellow-billed Babbler** *Turdoides affinis*

    _____ *T. a. affinis*   Arid lowlands and foothills of s India
    _____ *T. a. taprobana*   Sri Lanka

☐ **Blackcap Babbler** *Turdoides reinwardtii*

    _____ *T. r. reinwardtii*   Senegal to Sierra Leone and Mali
    _____ *T. r. stictilaema*   Ghana to Nigeria, Cameroon, Central African Rep. and n Zaire

☐ **Dusky Babbler** *Turdoides tenebrosa*

     NE Cent. African Rep. to nw Uganda and sw Ethiopia

☐ **Black-lored Babbler** *Turdoides sharpei*

    _____ *T. s. sharpei*   Lake Turkana to w Rift Valley and n Tanzania (Lake Rukwa)
    _____ *T. s. vepres*   Central Kenya (Nanyuki area)

☐ **Hartlaub's Babbler** *Turdoides hartlaubii*

    _____ *T. h. hartlaubii (ater)*   E Zaire and Rwanda to Zambia, n Botswana and Angola
    _____ *T. h. griseosquamata*   N Botswana to immediately adjacent Zimbabwe and Zambia

☐ **Black-faced Babbler** *Turdoides melanops*

    _____ *T. m. melanops*   SW Angola and n Namibia
    _____ *T. m. querula*   NW Botswana, ne Namibia and adjacent se Angola

☐ **Scaly Babbler** *Turdoides squamulata*

    _____ *T. s. carolinae*   Webi Shabeele River, s Somalia and se Ethiopia
    _____ *T. s. squamulata*   Kenya coast and Tana River and extreme s Somalia
    _____ *T. s. jubaensis*   Jubba River, s Somalia and se Ethiopia
    _____ *T. s. subsp.?*   Lake Bor watercourse and Daua River (Ethiopia/Kenya border)
    _____ *T. s. subsp.?*   Webi Gestro River, Ethiopia

☐ **White-rumped Babbler** *Turdoides leucopygia*

    _____ *T. l. limbata*   W Eritrea and nw Ethiopia
    _____ *T. l. leucopygia*   E Eritrea and nw Ethiopia
    _____ *T. l. omoensis (clarkei)*   S and sw Ethiopia to se Sudan
    _____ *T. l. lacuum*   Central Ethiopian Rift Valley
    _____ *T. l. smithii*   NW Somalia to e and se Ethiopia

☐ **Southern Pied-Babbler** *Turdoides bicolor*

     Namibia to Botswana, w Zimbabwe and nw South Africa

☐ **Northern Pied-Babbler** *Turdoides hypoleuca*

    _____ *T. h. hypoleuca*   Central and s Kenya and n Tanzania
    _____ *T. h. rufuensis*   N and ne Tanzania

☐ **Hinde's Pied-Babbler** *Turdoides hindei*

     Locally in foothill scrub of e-central Kenya

☐ **Cretzschmar's Babbler** *Turdoides leucocephala*

     Thornscrub of e Sudan to nw Ethiopia and n Eritrea

☐ **Brown Babbler** *Turdoides plebejus*

_____ *T. p. platycirca (togoensis)*	Senegal to w Nigeria
_____ *T. p. plebejus*	N Nigeria to Cameroon, s Chad and central Sudan (Kardofan)
_____ *T. p. cinerea (gularis)*	SE Nigeria to s Sudan, sw Ethiopia and w Kenya

☐ **Arrow-marked Babbler** *Turdoides jardineii*

_____ *T. j. emini*	S Kenya, nw Tanzania, s Uganda, Rwanda, Burundi, adj. Zaire
_____ *T. j. kirkii*	SE Kenya to e Tanzania, Malawi, e Zambia and Mozambique
_____ *T. j. hyposticta*	W Angola to w and s Zaire
_____ *T. j. tanganjicae*	E Angola to n Zambia and se Zaire
_____ *T. j. tamalakanae*	N Botswana, sw Zambia and s Angola
_____ *T. j. jardineii (convergens)*	N and e S Africa, s Mozambique, Zimbabwe, c and nw Zambia

☐ **Bare-cheeked Babbler** *Turdoides gymnogenys*

Arid bush of sw Angola and nw Namibia

☐ **Chinese Babax** *Babax lanceolatus*

_____ *B. l. lanceolatus*	Mountains of se Tibet to se Assam, sw China and ne Myanmar
_____ *B. l. woodi*	SE Assam (Lushai Hills) to w Myanmar (Chin Hills)
_____ *B. l. latouchei*	Montane forests of se China

☐ **Giant Babax** *Babax waddelli*

_____ *B. w. lumsdeni*	NE Tibet (on border with Qinghai)
_____ *B. w. waddelli*	SE Tibet (Lhasa, Loti, Chushul, Dzong and Chaksam)
_____ *B. w. jomo*	S-central Tibet (Gyangtse region)

☐ **Tibetan Babax** *Babax koslowi*

_____ *B. k. yuguensis*	Himalayas of se Tibet
_____ *B. k. koslowi*	Himalayas of s-central China (s Qinghai and nw Sichuan)

☐ **Silver-eared Mesia** *Leiothrix argentauris*

_____ *L. a. argentauris*	Himalayas (Garhwal to Nepal, Sikkim, Bhutan and n Assam)
_____ *L. a. aureigularis*	S Assam (s of the Brahmaputra) and sw Myanmar (Chin Hills)
_____ *L. a. vernayi*	NE Assam to n Myanmar and s China (w Yunnan)
_____ *L. a. galbana*	E Myanmar to n Thailand
_____ *L. a. ricketti*	S China (se Yunnan) to n Indochina
_____ *L. a. rubrogularis*	S China (se Yunnan and Guangxi)
_____ *L. a. cunhaci*	S Laos (Bolavens Plateau) and s Annam
_____ *L. a. tahanensis*	Mountains of Malay Peninsula
_____ *L. a. rookmakeri*	Highlands of nw Sumatra (Aceh District)
_____ *L. a. laurinae*	Highlands of w Sumatra

☐ **Red-billed Leiothrix** *Leiothrix lutea*

_____ *L. l. kumaiensis*	Himalayas (Kashmir to nw Uttar Pradesh)
_____ *L. l. calipyga*	W Nepal to Sikkim, Bhutan, e Assam and se Tibet
_____ *L. l. luteola*	S Assam to sw Myanmar (Chin Hills and Arakan Yoma Mts.)
_____ *L. l. yunnanensis*	NE Myanmar to sw China (se Qinghai and nw Yunnan)
_____ *L. l. kwangtungensis*	S China (se Yunnan, Guangxi and Guangdong) to ne Tonkin
_____ *L. l. lutea*	Central and se China

☐ **Cutia** *Cutia nipalensis*

_____ *C. n. nipalensis*	Nepal to e Assam, w Myanmar, s Sichuan and nw Yunnan
_____ *C. n. melanchima*	E Myanmar to s Yunnan, nw Thailand, n Laos and nw Tonkin
_____ *C. n. cervinicrissa*	Highlands of Malay Peninsula (s Perak to s Selangor)
_____ *C. n. hoae*	Central Vietnam (Mt. Ngoc Linh, Kon Tum Province)
_____ *C. n. legalleni*	S Vietnam (Langbian Plateau)

☐ **Black-headed Shrike-Babbler** *Pteruthius rufiventer*

_____ *P. r. rufiventer*	Nepal to Bhutan, Assam, n Myanmar, Sichuan and nw Yunnan
_____ *P. r. delacouri*	NW Tonkin (Fan Si Pan Mountains)

☐ **White-browed Shrike-Babbler** *Pteruthius flaviscapis*

____	*P. f. validirostris*	Himalayas (n Pakistan to s China, Assam and nw Myanmar)
____	*P. f. ricketti*	NE Myanmar to s China, n Thailand and n Indochina
____	*P. f. lingshuiensis*	Hainan (s China)
____	*P. f. annamensis*	S Vietnam (Langbian Plateau)
____	*P. f. aeralatus*	Mountains of e Myanmar to nw Thailand and (?) w Cambodia
____	*P. f. schauenseei*	Mountains of s Thailand and s Myanmar to Isthmus of Kra
____	*P. f. cameranoi*	Highlands of Malaya (s Perak to s Selangor) and w Sumatra
____	*P. f. flaviscapis*	Highlands of Java
____	*P. f. robinsoni*	Highlands of n Borneo

☐ **Green Shrike-Babbler** *Pteruthius xanthochlorus*

____	*P. x. occidentalis*	Himalayas (Kashmir to w Nepal)
____	*P. x. xanthochlorus*	Cent. Nepal to Sikkim, Bhutan, n Assam, se Tibet and w Sichuan
____	*P. x. hybrida*	Assam (Lushai and Naga Hills) to w Myanmar (Chin Hills)
____	*P. x. pallidus*	NE Myanmar to sw Qinghai, w Sichuan, Yunnan and nw Fujian

☐ **Black-eared Shrike-Babbler** *Pteruthius melanotis*

____	*P. m. melanotis*	Nepal to n Myanmar, s Yunnan, n Thailand and n Indochina
____	*P. m. tahanensis*	Highlands of Malaya (s Perak to Selangor and n Pahang)

☐ **Chestnut-fronted Shrike-Babbler** *Pteruthius aenobarbus*

____	*P. a. aenobarbulus*	Assam (Garo Hills)
____	*P. a. intermedius*	E Myanmar to s Yunnan, nw Thailand, Laos and nw Tonkin
____	*P. a. yaoshanensis*	SE China (Yao Shan region of Guangxi)
____	*P. a. indochinensis*	S Vietnam (Langbian Plateau)
____	*P. a. aenobarbus*	Highlands of w Java

☐ **White-hooded Babbler** *Gampsorhynchus rufulus*

____	*G. r. rufulus*	Nepal to Sikkim, Bhutan, Assam, sw Myanmar and w Yunnan
____	*G. r. torquatus*	SE Myanmar to se Yunnan, Thailand and n Indochina
____	*G. r. saturatior*	Highlands of Malaya (s Perak to s Selangor)

☐ **Rusty-fronted Barwing** *Actinodura egertoni*

____	*A. e. egertoni*	Nepal to Sikkim, Bhutan, n Assam and se Tibet
____	*A. e. lewisi*	NE Assam (Mishmi Hills)
____	*A. e. khasiana*	Hill forests of s Assam (south of the Brahmaputra)
____	*A. e. ripponi*	S China (w Yunnan and sw Guangxi) to sw Myanmar

☐ **Spectacled Barwing** *Actinodura ramsayi*

____	*A. r. radcliffei*	NE Myanmar (Ruby Mines district) to sw Yunnan and n Laos
____	*A. r. yunnanensis*	Mountains of s China (se Yunnan and s Guangxi) to n Vietnam
____	*A. r. ramsayi*	Mountains of extreme se Myanmar and adjacent nw Thailand

☐ **Black-crowned Barwing** *Actinodura sodangorum*

Western highlands of Vietnam

☐ **Hoary-throated Barwing** *Actinodura nipalensis*

____	*A. n. nipalensis*	Oak-rhododendron forests of w and central Nepal
____	*A. n. vinctura*	E Nepal to se Tibet (Pome District), Sikkim and Bhutan

☐ **Streak-throated Barwing** *Actinodura waldeni*

____	*A. w. daflaensis*	N Assam (north of the Brahmaputra) and se Tibet
____	*A. w. waldeni*	SE Assam (Naga Hills and Manipur) to nw Myanmar
____	*A. w. poliotis*	N Myanmar (Mt. Victoria)
____	*A. w. saturatior*	NE Myanmar (Kachin State) and s China (nw Yunnan)

☐ **Streaked Barwing** *Actinodura souliei*

____	*A. s. souliei*	Oak-rhododendron forests of s China (s Sichuan to nw Yunnan)
____	*A. s. griseinucha*	Oak-rhododendron forests of nw Tonkin

☐ **Taiwan Barwing** *Actinodura morrisoniana*

Montane evergreen forests of Taiwan

☐ **Blue-winged Minla** *Minla cyanouroptera*
_____ *M. c. cyanouroptera*   Himalayas (Uttar Pradesh to Nepal, Sikkim, Bhutan, e Assam)
_____ *M. c. aglae*   Hill forests of se Assam and w Myanmar
_____ *M. c. sordida*   E and s Myanmar to nw Thailand
_____ *M. c. wingatei*   NE Myanmar to n Thailand, s China and n Indochina; Hainan
_____ *M. c. croizati*   SW China (se Sichuan in Ipin region)
_____ *M. c. rufodorsalis*   Mountains of se Thailand and sw Cambodia
_____ *M. c. orientalis*   S Vietnam (Langbian Plateau)
_____ *M. c. sordidior*   Mountains of peninsular Thailand and n Malaya

☐ **Chestnut-tailed Minla** *Minla strigula*
_____ *M. s. simlaensis*   Himalayas (Kashmir to w Nepal)
_____ *M. s. strigula*   Himalayas (central Nepal to Bhutan, n Assam and se Tibet)
_____ *M. s. cinereigenae*   E Assam (Mt. Japvo)
_____ *M. s. yunnanensis*   E Assam to w Myanmar, sw China, n Laos and n Tonkin
_____ *M. s. castanicauda*   S Myanmar (Karenni Hills) to nw Thailand
_____ *M. s. malayana*   Highlands of Malaya (n Perak to s Selangor and Pahang)
_____ *M. s. traii*   Highlands of s Vietnam (Mt. Ngoc Linh, Kon Tum Province)

☐ **Red-tailed Minla** *Minla ignotincta*
_____ *M. i. ignotincta*   Nepal to Myanmar, se Tibet, Assam and s China (nw Yunnan)
_____ *M. i. mariae*   S China (se Yunnan) to n Tonkin
_____ *M. i. sini*   S China (Yao Shan region of Guangxi)
_____ *M. i. jerdoni (sini)*   S China (s Sichuan, s Hunan and Yao Shan region of Guangxi)

☐ **Golden-breasted Fulvetta** *Alcippe chrysotis*
_____ *A. c. chrysotis*   Himalayas (e Nepal to Sikkim, Bhutan and e Assam)
_____ *A. c. albilineata*   Hill forests of s Assam (south of the Brahmaputra)
_____ *A. c. forresti*   NE Myanmar to s China (nw Yunnan)
_____ *A. c. amoena*   S China (sc Yunnan) and nw Tonkin
_____ *A. c. robsoni*   Central Vietnam (Mt. Ngoc Linh, Kon Tum Province)
_____ *A. c. swinhoii*   S China (s Shaanxi, central Sichuan, ne Yunnan and nw Guangxi)

☐ **Gold-fronted Fulvetta** *Alcippe variegaticeps*

Montane forests of s China (e Sichuan and Guangxi)

☐ **Yellow-throated Fulvetta** *Alcippe cinerea*

Nepal to ne Myanmar, s China (Yunnan) and n Laos

☐ **Rufous-winged Fulvetta** *Alcippe castaneceps*
_____ *A. c. castaneceps*   Nepal to Assam, Myanmar, se Tibet and nw Thailand
_____ *A. c. exul*   S China (se Yunnan) to n Thailand plateau, Laos and nw Tonkin
_____ *A. c. soror*   Highlands of Malaya (n Perak to s Selangor and Pahang)
_____ *A. c. stepanyani*   Central Vietnam (Kon Tum Province and Gia Lai Province)
_____ *A. c. klossi*   S Vietnam (Langbian Plateau)

☐ **White-browed Fulvetta** *Alcippe vinipectus*
_____ *A. v. kangrae*   Himalayas (Kashmir to nw Uttar Pradesh)
_____ *A. v. vinipectus*   Montane forests of w and central Nepal
_____ *A. v. chumbiensis*   Montane forests of e Nepal to se Tibet, Sikkim and Bhutan
_____ *A. v. austeni*   Hill forests of s Assam (south of the Brahmaputra)
_____ *A. v. ripponi*   W Myanmar (highest regions of the Chin Hills)
_____ *A. v. perstriata*   NE Myanmar (Kachin State) to s China (w Yunnan)
_____ *A. v. valentinae*   N Vietnam (Fan Si Pan Mountains)
_____ *A. v. bieti*   S China (Sichuan to nw Yunnan and se Tibet)

☐ **Chinese Fulvetta** *Alcippe striaticollis*

Mts. of s-c China (sw Gansu to se Qinghai, Sichuan, nw Yunnan)

☐ **Spectacled Fulvetta** *Alcippe ruficapilla*
_____ *A. r. ruficapilla*   Central China (s Gansu to s Shaanxi and Sichuan)

_____ *A. r. sordidior*   S China (w Sichuan, Guizhou and n Yunnan)
_____ *A. r. danisi*   Northwest Tonkin and n Laos
_____ *A. r. bidoupensis*   S Vietnam (Da Lat Plateau)

☐ **Streak-throated Fulvetta** *Alcippe cinereiceps*

_____ *A. c. manipurensis*   Montane forests of ne India to ne Myanmar and nw Yunnan
_____ *A. c. tonkinensis*   S China (se Yunnan) to nw Tonkin and ne Laos
_____ *A. c. guttaticollis*   Highlands of se China (nw Fujian); > in n Guangdong
_____ *A. c. fucata (berliozi)*   Central China (Hubei to s Hunan and s Shaanxi)
_____ *A. c. cinereiceps*   W-central China (w Hubei to w Sichuan)
_____ *A. c. fessa*   W-central China (s Shaanxi and Gansu)
_____ *A. c. formosana*   Highlands of Taiwan

☐ **Ludlow's Fulvetta** *Alcippe ludlowi*

Himalayas (se Tibet to sw China, e Bhutan and extreme ne India)

☐ **Rufous-throated Fulvetta** *Alcippe rufogularis*

_____ *A. r. rufogularis*   Himalayas (Bhutan to n Assam north of the Brahmaputra)
_____ *A. r. collaris*   Hill forests of e Assam to Bangladesh
_____ *A. r. major*   NE Myanmar to n and e Thailand and n-central Laos
_____ *A. r. stevensi*   S China (sw Yunnan) and n Indochina
_____ *A. r. kelleyi*   Central Vietnam (Quangtri Province)
_____ *A. r. khmerensis*   SE Thailand and adjacent sw Cambodia

☐ **Dusky Fulvetta** *Alcippe brunnea*

_____ *A. b. mandellii*   S Assam (s of the Brahmaputra) to w Myanmar (Chin Hills)
_____ *A. b. genestieri*   SW China (se Qinghai, Yunnan and w Guizhou) to n Indochina
_____ *A. b. olivacea*   S-central China (Hubei, s Shaanxi and e Sichuan)
_____ *A. b. superciliaris*   Hills of se China (Anhui to Guangxi, Fujian and Guangdong)
_____ *A. b. intermedia*   E Myanmar (Kachin and Shan States) to s China (Yunnan)
_____ *A. b. arguta*   Hainan (s China)
_____ *A. b. brunnea*   Taiwan

☐ **Rusty-capped Fulvetta** *Alcippe dubia*

_____ *A. d. dubia*   Pine forests of s Myanmar and adj. China (Yunnan)
_____ *A. d. cui*   Central Vietnam (Mt. Ngoc Linh, Kon Tum Province)

☐ **Brown Fulvetta** *Alcippe brunneicauda*

_____ *A. b. brunneicauda*   S Thailand, Malaya, Sumatra, nw Borneo, N Natuna and Batu is.
_____ *A. b. eriphaea*   Borneo

☐ **Brown-cheeked Fulvetta** *Alcippe poioicephala*

_____ *A. p. poioicephala*   W India (W Ghats from s Mysore to Kerala and Palni Hills)
_____ *A. p. brucei*   Central and s peninsular India
_____ *A. p. fusca*   Assam (south of the Brahmaputra) to nw Myanmar
_____ *A. p. phayrei*   SW Myanmar (Chin Hills and Arakan Yoma Mountains)
_____ *A. p. haringtoniae*   NE Myanmar to s China (w Yunnan) and nw Thailand
_____ *A. p. alearis*   S China (s Yunnan) to n plateau of Thailand and n Indochina
_____ *A. p. karenni*   SE Myanmar (Karenni State) to sw Thailand
_____ *A. p. davisoni*   Malay Pen. (Isthmus of Kra to Trang) and Mergui Archipelago

☐ **Gray-cheeked Fulvetta** *Alcippe morrisonia*

_____ *A. m. yunnanensis*   S China (s Sichuan and e Yunnan) to ne Myanmar (Kachin)
_____ *A. m. fratercula*   S China (sw Yunnan) to se Myanmar and n Indochina
_____ *A. m. schaefferi*   S China (se Yunnan and Guangxi) to nw Vietnam
_____ *A. m. davidi*   S-central China (w Hubei to Hunan, Sichuan and ne Yunnan)
_____ *A. m. hueti*   Hill forests of se China (Guangdong to Anhui)
_____ *A. m. rufescentior*   Hainan (s China)
_____ *A. m. morrisonia*   Taiwan

☐ **Javan Fulvetta** *Alcippe pyrrhoptera*

Forests of w and central Java

508

☐ **Mountain Fulvetta** *Alcippe peracensis*
- ____ *A. p. grotei* — N and central Annam and adjacent Laos
- ____ *A. p. annamensis* — S Laos (Bolavens Plateau), s Annam and adjacent Cochinchina
- ____ *A. p. eremita* — SE Thailand
- ____ *A. p. peracensis* — Highlands of Malay Pen. (n Perak to s Selangor and Pahang)

☐ **Nepal Fulvetta** *Alcippe nipalensis*
- ____ *A. n. nipalensis* — Himalayas (Nepal to Sikkim, Bhutan and e Assam)
- ____ *A. n. commoda* — Assam (south and east of the Brahmaputra) to n Myanmar
- ____ *A. n. stanfordi* — Hill forests of sw Myanmar (Chin Hills and Arakan Yoma Mts.)

☐ **Bush Blackcap** *Lioptilus nigricapillus*

South Africa (Transvaal to Natal and e Cape Province)

☐ **White-throated Mountain-Babbler** *Kupeornis gilberti*

Montane forests of se Nigeria and sw Cameroon

☐ **Red-collared Mountain-Babbler** *Kupeornis rufocinctus*

Montane forests of e Zaire and sw Rwanda

☐ **Chapin's Mountain-Babbler** *Kupeornis chapini*
- ____ *K. c. chapini* — E Zaire (Lake Albert to Lake Edward)
- ____ *K. c. nyombensis* — E Zaire (Mt. Nyombe, Kivu)
- ____ *K. c. kalindei* — E Zaire (sw Itombwe highlands)

☐ **Dohrn's Thrush-Babbler** *Horizorhinus dohrni*

Príncipe I. (Gulf of Guinea)

☐ **Abyssinian Catbird** *Parophasma galinieri*

Highland juniper and *Hagenia* forests of Ethiopia

☐ **Capuchin Babbler** *Phyllanthus atripennis*
- ____ *P. a. atripennis* — Gambia to Ivory Coast
- ____ *P. a. haynesi* — Ivory Coast to s Nigeria
- ____ *P. a. bohndorffi* — NE Zaire and w Uganda

☐ **Gray-crowned Crocias** *Crocias langbianis*

Recorded after 57-year absence in s Vietnam (Langbian Plateau)

☐ **Spotted Crocias** *Crocias albonotatus*

Montane forests of w and central Java

☐ **Rufous-backed Sibia** *Heterophasia annectens*
- ____ *H. a. annectens* — Sikkim to Bhutan, Assam, nw Myanmar and w Yunnan
- ____ *H. a. saturata* — SE Myanmar to nw Thailand
- ____ *H. a. mixta* — SE Myanmar to sw Yunnan, n Thailand, n Laos and nw Tonkin
- ____ *H. a. roundi* — Central Vietnam (Kon Tum Province)
- ____ *H. a. eximia* — S Vietnam (Da Lat Plateau)

☐ **Rufous Sibia** *Heterophasia capistrata*
- ____ *H. c. capistrata* — W Himalayas (Pakistan to Garhwal)
- ____ *H. c. nigriceps* — Central Himalayas (Kumaon to central Nepal)
- ____ *H. c. bayleyi* — E Himalayas (e Nepal to Sikkim, Bhutan, s Tibet and n Assam)

☐ **Gray Sibia** *Heterophasia gracilis*

Montane forests of ne India, Myanmar and s China (w Yunnan)

☐ **Black-backed Sibia** *Heterophasia melanoleuca*
- ____ *H. m. melanoleuca* — E Myanmar to nw Thailand
- ____ *H. m. tonkinensis* — NW Tonkin
- ____ *H. m. engelbachi* — S Laos (Bolavens Plateau)
- ____ *H. m. kingi* — Central Vietnam (Mt. Ngoc Linh, Kon Tum Province)
- ____ *H. m. robinsoni* — S Vietnam (Langbian Plateau)
- ____ *H. m. castanoptera* — SE Myanmar (Southern Shan State and Karenni State)

☐ **Black-headed Sibia** *Heterophasia desgodinsi*

NE Myanmar to sw China (se Qinghai, s Sichuan, nw Yunnan)

☐ **White-eared Sibia** *Heterophasia auricularis*

Montane oak forests of Taiwan

☐ **Beautiful Sibia** *Heterophasia pulchella*

Mountains of ne India to se Tibet, sw China and ne Myanmar

☐ **Long-tailed Sibia** *Heterophasia picaoides*
____ *H. p. picaoides*
____ *H. p. cana*
____ *H. p. wrayi*
____ *H. p. simillima*

Nepal to Sikkim, Bhutan, Assam, nw Yunnan and ne Myanmar
S China (sw Yunnan) to s Myanmar, n Thailand and n Indochina
Highlands of Malay Pen. (n Perak to s Selangor and n Pahang)
Highlands of w Sumatra

☐ **Striated Yuhina** *Yuhina castaniceps*
____ *Y. c. rufigenis*
____ *Y. c. plumbeiceps*
____ *Y. c. castaniceps*
____ *Y. c. striata*
____ *Y. c. torqueola*

NE India (Darjiling) and Sikkim
N Assam to n Myanmar and s China (w Yunnan)
S Assam to sw Myanmar (Chin Hills and Arakan Yoma Mts.)
Mountains of e Myanmar to nw Thailand
N plateau of Thailand to s China and n Indochina

☐ **Chestnut-crested Yuhina** *Yuhina everetti*

Montane forests of n Borneo

☐ **White-naped Yuhina** *Yuhina bakeri*

Mts. of ne India (Assam) to Myanmar and s China (nw Yunnan)

☐ **Whiskered Yuhina** *Yuhina flavicollis*
____ *Y. f. albicollis*
____ *Y. f. flavicollis*
____ *Y. f. rouxi*
____ *Y. f. clarki*
____ *Y. f. constantiae*
____ *Y. f. rogersi*

W Himalayas (Kashmir to w Nepal)
Central Nepal to Bhutan, e Assam (Abor Hills) and se Tibet
S Assam to n Myanmar, s China (Yunnan) and n Indochina
Mountains of e Myanmar (S Shan State and Karenni State)
Mountains of n Laos (Chiang Khwang Province)
Extreme n Thailand (Doi Phu Kha)

☐ **Burmese Yuhina** *Yuhina humilis*

Mountains of se Myanmar and adjacent nw Thailand

☐ **Stripe-throated Yuhina** *Yuhina gularis*
____ *Y. g. vivax*
____ *Y. g. gularis*
____ *Y. g. uthaii*
____ *Y. g. omiensis*

W Himalayas (Garhwal to Kumaon)
Nepal to se Tibet, nw Yunnan, w Myanmar and nw Vietnam
Central Vietnam (Mt. Ngoc Linh, Kon Tum Province)
S China (Omei Shan Mountains of Sichuan to nw Yunnan)

☐ **White-collared Yuhina** *Yuhina diademata*

Mountains of sw China to se Myanmar and nw Tonkin

☐ **Rufous-vented Yuhina** *Yuhina occipitalis*
____ *Y. o. occipitalis*
____ *Y. o. obscurior*

E Himalayas (Nepal to se Tibet and n Assam)
NE Myanmar (Kachin State) to s China (nw Yunnan)

☐ **Taiwan Yuhina** *Yuhina brunneiceps*

Montane forests of Taiwan

☐ **Black-chinned Yuhina** *Yuhina nigrimenta*
____ *Y. n. nigrimenta*
____ *Y. n. intermedia*
____ *Y. n. pallida*

Himalayas (Garhwal to e Assam)
Myanmar to s China (Liaoning, Sichuan, Yunnan) and n Indochina
Highlands of se China (nw Fujian, Guangxi and Guangdong)

☐ **White-bellied Yuhina** *Yuhina zantholeuca*
____ *Y. z. zantholeuca*
____ *Y. z. tyrannula*
____ *Y. z. griseiloris*
____ *Y. z. sordida*
____ *Y. z. canescens*
____ *Y. z. interposita*
____ *Y. z. saani*
____ *Y. z. brunnescens*

E Himalayas to n Myanmar, s China (Yunnan) and w Thailand
NE Thailand to s China (se Yunnan), n Indochina and Hainan
SE China (Fujian, Guangdong, w Guangxi, se Yunnan); Taiwan
Extreme e plateau of Thailand to s Indochina
SE Thailand to w Cambodia
Malay Peninsula (Mergui District and Isthmus of Kra to Johore)
NW Sumatra
Borneo

☐ **Fire-tailed Myzornis** *Myzornis pyrrhoura*

Nepal to se Tibet, s China (Yunnan) and ne Myanmar

☐ **White-throated Oxylabes** *Oxylabes madagascariensis*

Dense humid forests of nw and e Madagascar

☐ **Yellow-browed Oxylabes** *Crossleyia xanthophrys*

Lowland rainforests of e-central Madagascar

☐ **Crossley's Babbler** *Mystacornis crossleyi*

Dense humid forests of e Madagascar

FAMILY: POMATOSTOMIDAE (Pseudo-babblers—5)

☐ **New Guinea Babbler** *Pomatostomus isidorei*
____ *P. i. calidus* — N New Guinea (Geelvink Bay to Astrolabe Bay)
____ *P. i. isidorei* — S New Guinea, Misool I. and Waigeo I.

☐ **Gray-crowned Babbler** *Pomatostomus temporalis*
____ *P. t. strepitans* — S New Guinea (Orimo River to Digul River)
____ *P. t. temporalis* — E Australia (Cape York Pen. to c Victoria and se S Australia)
____ *P. t. rubeculus* — N Australia (Shark Bay, Western Australia to nw Queensland)

☐ **White-browed Babbler** *Pomatostomus superciliosus*
____ *P. s. gilgandra* — E Australia (s Queensland to northern Victoria)
____ *P. s. superciliosus* — Southern Australia (Fortescue R. to w NSW and w Victoria)
____ *P. s. ashbyi* — Southwestern Western Australia
____ *P. s. centralis* — Central Western Australia to sw N Territory and nw S Australia

☐ **Hall's Babbler** *Pomatostomus halli*

Acacia woodlands of s-central Queensland and adj. nw NSW

☐ **Chestnut-crowned Babbler** *Pomatostomus ruficeps*

Arid woodlands of e-central Australia

FAMILY: PARADOXORNITHIDAE (Parrotbills—20)

☐ **Bearded Reedling** *Panurus biarmicus*
____ *P. b. biarmicus* — W Europe to Sweden, Poland, Italy, Balkans and Transcaucasia
____ *P. b. russicus* — C Europe (Austria to n Yugoslavia, Asia Minor, c Asia and China)
____ *P. b. kosswigi* — Formerly s Turkey (Amik Gölü). Probably extinct

☐ **Great Parrotbill** *Conostoma oemodium*

Bamboo forests of w Nepal to se Tibet, s China and ne Myanmar

☐ **Brown Parrotbill** *Paradoxornis unicolor*

Bamboo forests of Nepal to se Tibet, sw China and ne Myanmar

☐ **Gray-headed Parrotbill** *Paradoxornis gularis*
____ *P. g. gularis* — E Himalayas (Sikkim to Bhutan and n Assam)
____ *P. g. transfluvialis* — S Assam (s of the Brahmaputra) to Myanmar and nw Thailand
____ *P. g. rasus* — W Myanmar (Chin Hills)
____ *P. g. laotianus* — E Myanmar (Kengtung) to n Thailand, n Laos and nw Tonkin
____ *P. g. fokiensis* — Hill forests of s China (south of the Yangtze River)
____ *P. g. hainanus* — Hainan (s China)
____ *P. g. margaritae* — Highlands of s Vietnam (s Annam)

☐ **Three-toed Parrotbill** *Paradoxornis paradoxus*
____ *P. p. taipaiensis* — Central China (Tsingling Mountains of s Shaanxi)
____ *P. p. paradoxus* — S China (n Sichuan and sw Gansu)

☐ **Black-breasted Parrotbill** *Paradoxornis flavirostris*

Foothills of Nepal to ne Assam and w Myanmar (Chin Hills)

☐ **Spot-breasted Parrotbill** *Paradoxornis guttaticollis*

Hill forests of Assam to e Myanmar, s China and n Indochina

☐ **Spectacled Parrotbill** *Paradoxornis conspicillatus*
_____ *P. c. conspicillatus*   Mts. of c China (e Qinghai to e Sichuan, se Gansu, sw Shaanxi)
_____ *P. c. rocki*   Montane bamboo forests of ne China (w Liaoning)

☐ **Vinous-throated Parrotbill** *Paradoxornis webbianus*
_____ *P. w. suffusus*   Mts. of nw China (Shaanxi to w Sichuan, Jiangxi, Guangdong)
_____ *P. w. mantschuricus*   E Manchuria to ne China (e Liaoning)
_____ *P. w. fulvicauda*   NE China (s Liaoning) to s Korea
_____ *P. w. webbianus*   Coastal e China (s Jiangsu and n Zhejiang)
_____ *P. w. elisabethae*   Montane forests of s China (se Yunnan) and nw Tonkin
_____ *P. w. bulomachus*   Highlands of Taiwan

☐ **Brown-winged Parrotbill** *Paradoxornis brunneus*
_____ *P. b. ricketti*   S China (s Sichuan and n Yunnan)
_____ *P. b. styani*   S China (n Yunnan from Mekong Valley to Tali region)
_____ *P. b. brunneus*   S China (Yunnan from Tali region south to Tengyueh)

☐ **Ashy-throated Parrotbill** *Paradoxornis alphonsianus*
_____ *P. a. alphonsianus*   Mts. of w-central China (e Qinghai, e Sichuan and Guizhou)
_____ *P. a. yunnanensis*   Montane forests of s China (se Yunnan) and nw Tonkin

☐ **Gray-hooded Parrotbill** *Paradoxornis zappeyi*
                                 Mountains of s China (Washan and Omei Shan in e Sichuan)

☐ **Rusty-throated Parrotbill** *Paradoxornis przewalskii*
                                 Mts. of s-cent. China (se Qinghai to se Gansu and adj. Sichuan)

☐ **Fulvous Parrotbill** *Paradoxornis fulvifrons*
_____ *P. f. fulvifrons*   Montane forests of Nepal, Sikkim and Bhutan
_____ *P. f. chayulensis*   Montane forests of se Tibet and ne Myanmar
_____ *P. f. albifacies*   Montane forests of sw China (se Qinghai and nw Yunnan)
_____ *P. f. cyanophrys*   Mountains of sw China (w Sichuan, se Gansu and s Shaanxi)

☐ **Black-throated Parrotbill** *Paradoxornis nipalensis*
_____ *P. n. nipalensis*   Central Nepal (Kathmandu Valley)
_____ *P. n. humii*   Himalayas of e Nepal, Sikkim and w Bhutan
_____ *P. n. crocotius*   Mountains of se Tibet and e Bhutan
_____ *P. n. poliotis*   E Bhutan to Assam, ne Myanmar, se Tibet, s China (nw Yunnan)
_____ *P. n. partriciae*   SE Assam (Lushai Hills)
_____ *P. n. ripponi*   N Myanmar (Mt. Victoria)
_____ *P. n. feae*   Hill forests of se Myanmar (Karenni State) and nw Thailand
_____ *P. n. kamoli*   Central Vietnam (Mt. Ngoc Linh, Kon Tum Province)

☐ **Golden Parrotbill** *Paradoxornis verreauxi*
_____ *P. v. verreauxi*   Mountains of sw China (e Qinghai to Sichuan and n Yunnan)
_____ *P. v. craddocki*   S China (ne Guangxi), ne Tonkin (Fan Si Pan Mts.), w Myanmar
_____ *P. v. beaulieu*   Mountains of n Laos (Chiang Khwang Province)
_____ *P. v. pallidus*   Mountains of se China (nw Fujian)
_____ *P. v. morrisonianus*   Highlands of Taiwan

☐ **Short-tailed Parrotbill** *Paradoxornis davidianus*
_____ *P. d. davidianus*   Highlands of se China (s Zhejiang to central Fujian)
_____ *P. d. tonkinensis*   Highlands of n Vietnam (Bac Phan)
_____ *P. d. thompsoni*   S China (s Yunnan) to e Myanmar, e Thailand, nw Laos, n Tonkin

☐ **Black-browed Parrotbill** *Paradoxornis atrosuperciliaris*
_____ *P. a. oatesi*   NE India (Darjiling) to Sikkim
_____ *P. a. atrosuperciliaris* Assam to Myanmar, s China (w Yunnan), nw Thailand, n Laos

☐ **Rufous-headed Parrotbill** *Paradoxornis ruficeps*
_____ *P. r. ruficeps*   Nepal to Bhutan, n Assam, s China (nw Yunnan) and se Tibet
_____ *P. r. bakeri*   S Assam (south of the Brahmaputra) to n and e Myanmar
_____ *P. r. magnirostris*   Highlands of central Tonkin

☐ **Reed Parrotbill** *Paradoxornis heudei*
_____ *P. h. polivanovi* — Reedbeds of extreme se Siberia (Lake Khanka) and s Ussuriland
_____ *P. h. heudei* — Reedbeds of e China (s Heliongjiang, ne Zhejiang and Jiangsu)

FAMILY: ORTHONYCHIDAE (Logrunners—3)

☐ **Northern Logrunner** *Orthonyx novaeguineae*
_____ *O. n. novaeguineae* — W New Guinea (Arfak and Tamrau mountains)
_____ *O. n. dorsalis* — W New Guinea (Nassau and Snow mountains)
_____ *O. n. victorianus* — SE New Guinea (Herzog and Wharton mountains)

☐ **Southern Logrunner** *Orthonyx temminckii* — Upland rainforests of e Australia (Bunya Mts. To central NSW)

☐ **Chowchilla** *Orthonyx spaldingii*
_____ *O. s. melasmenus* — Rainforests of ne Queensland (Cooktown to Thornton Range)
_____ *O. s. spaldingii* — Rainforests of e Queensland (Herberton to Paluma Range)

FAMILY: EUPETIDAE (Whipbirds and Quail-thrushes—15)

☐ **Papuan Whipbird** *Androphobus viridis* — New Guinea (Snow and Weyland mountains)

☐ **Eastern Whipbird** *Psophodes olivaceus*
_____ *P. o. lateralis* — NE Queensland (Cooktown to Townsville)
_____ *P. o. olivaceus* — E Australia (central Queensland to s-central Victoria)

☐ **Western Whipbird** *Psophodes nigrogularis*
_____ *P. n. nigrogularis* — Extreme southwest Western Australia (Two Peoples Bay area)
_____ *P. n. leucogaster* — Mallee of s Eyre and s Yorke Pen. and adj. Victoria border
_____ *P. n. lashmari* — Kangaroo I. (South Australia)
_____ *P. n. oberon* — Southwest Western Australia

☐ **Chiming Wedgebill** *Psophodes occidentalis* — Arid central Australia, to west coast of Western Australia

☐ **Chirruping Wedgebill** *Psophodes cristatus* — Arid sw Qld. to ne South Australia and se Northern Territory

☐ **Spotted Quail-thrush** *Cinclosoma punctatum*
_____ *C. p. punctatum* — E Australia (c Queensland to s Victoria and se South Australia)
_____ *C. p. dovei* — Eastern Tasmania
_____ *C. p. anachoreta* — Mt. Lofty Range (South Australia). On verge of extinction.

☐ **Chestnut Quail-thrush** *Cinclosoma castanotum*
_____ *C. c. fordianum* — SW Western Australia to sw South Australia
_____ *C. c. clarum* — S Western Australia to sw Northern Territory and nw S Australia
_____ *C. c. castanotum* — Southeast South Australia, adjacent nw Victoria and sw NSW

☐ **Chestnut-breasted Quail-thrush** *Cinclosoma castaneothorax*
_____ *C. c. marginatum* — Arid c Western Australia and adjacent sw Northern Territory
_____ *C. c. castaneothorax* — S-central Queensland and adjacent nw New South Wales

☐ **Cinnamon Quail-thrush** *Cinclosoma cinnamomeum*
_____ *C. c. cinnamomeum* — SE Qld. to nw NSW, cent. and ne SA and se NT
_____ *C. c. alisteri* — SE Western Australia and adjacent sw South Australia

☐ **Painted Quail-thrush** *Cinclosoma ajax*
_____ *C. a. ajax* — Lowlands of w New Guinea (Geelvink Bay to Triton Bay)
_____ *C. a. muscale* — S-central New Guinea (upper Fly River Valley)
_____ *C. a. alare* — S-central New Guinea (Oriomo and lower Fly River valleys)
_____ *C. a. goldiei* — Lowlands of se New Guinea (Hall Sound to Milne Bay)

☐ **Spotted Jewel-babbler** *Ptilorrhoa leucosticta*

____	*P. l. leucosticta*	W New Guinea (Arfak and Tamrau mountains)
____	*P. l. mayri*	W New Guinea (Wandammen Mountains)
____	*P. l. centralis*	W New Guinea (Weyland, Nassau and Snow mountains)
____	*P. l. sibilans*	N New Guinea (Cyclops Mountains)
____	*P. l. menawa*	Coastal n New Guinea
____	*P. l. amabilis*	E New Guinea (Saruwaged Mountains)
____	*P. l. loriae*	Mountains of se New Guinea

☐ **Blue Jewel-babbler** *Ptilorrhoa caerulescens*

____	*P. c. caerulescens*	W New Guinea (Vogelkop to Etna Bay); Misool I.
____	*P. c. neumanni*	N New Guinea (Mamberamo River to Astrolabe Bay)
____	*P. c. nigricrissus*	S New Guinea (Etna Bay to Milne Bay)
____	*P. c. geislerorum*	E New Guinea (Huon Gulf to Collingwood Bay)

☐ **Chestnut-backed Jewel-babbler** *Ptilorrhoa castanonota*

____	*P. c. castanonota*	W New Guinea (mountains of Vogelkop Peninsula)
____	*P. c. saturata*	W New Guinea (Nassau Mountains)
____	*P. c. uropygialis*	W New Guinea (n slopes of Snow Mountains)
____	*P. c. buergersi*	Central New Guinea (Sepik Mountains)
____	*P. c. par*	E New Guinea (Saruwaged Mountains)
____	*P. c. pulchra*	Mountains of se New Guinea
____	*P. c. gilliardi*	Batanta I. and Yapen I.

☐ **Malaysian Rail-babbler** *Eupetes macrocerus*

____	*E. m. macrocerus*	Peninsular Thailand, Malaya, Sumatra and North Natuna Is.
____	*E. m. borneensis*	Mountains of n Borneo

☐ **Blue-capped Ifrita** *Ifrita kowaldi*

____	*I. k. brunnea*	Mountains of w-central New Guinea
____	*I. k. kowaldi*	New Guinea (Central Highlands and mountains of Huon Pen.)

FAMILY: AEGITHALIDAE (Long-tailed Tits—8)

☐ **Long-tailed Tit** *Aegithalos caudatus*

____	*A. c. rosaceus*	British Isles
____	*A. c. caudatus*	Scandinavia and ne Europe to Siberia, n China, Korea and Japan
____	*A. c. aremoricus*	W France, Channel Islands and Île d'Yeu
____	*A. c. taiti*	S and sw France to nw Spain and Portugal
____	*A. c. irbii*	S Spain, Portugal and Corsica
____	*A. c. europaeus*	France and Germany to n Italy, w Romania and n Bulgaria
____	*A. c. italiae*	Mainland Italy and Yugoslavia
____	*A. c. siculus*	Sicily
____	*A. c. macedonicus*	Albania, Yugoslavia, Greece and s Bulgaria
____	*A. c. tauricus*	S Crimean Peninsula
____	*A. c. tephronotus*	Asia Minor
____	*A. c. major*	Caucasus to w and central Transcaucasia
____	*A. c. alpinus*	SE Azerbaijan to n Iran and sw Turkmenistan
____	*A. c. passekii*	Zagros Mountains (sw Iran)
____	*A. c. glaucogularis*	Central China (mountains of w Sichuan to Yangtze delta)
____	*A. c. vinaceus*	N and w China (Liaoning to Gansu, Qinghai and n Yunnan)
____	*A. c. magnus*	S Korea and Tsushima Is. (Kamino-shima and Shimono-shima)
____	*A. c. trivirgatus*	Japan (Honshu, Awa-shima, Sado and Oki); Cheju-Do Is. (Korea)
____	*A. c. kiusiuensis*	S Japanese islands (Shikoku, Kyushu and Yakushima)

☐ **White-cheeked Tit** *Aegithalos leucogenys*

	Juniper and *ilex* scrub of w Kashmir, Afghanistan and n Pakistan

☐ **Black-throated Tit** *Aegithalos concinnus*

____	*A. c. iredalei*	Himalayas (ne Pakistan to n India and s Tibet)
____	*A. c. manipurensis*	NE India (se Arunachal Pradesh) to w Myanmar (Chin Hills)

____	*A. c. concinnus*	Central and e China and Taiwan
____	*A. c. talifuensis*	NE Myanmar to sw China, nw Vietnam and n Laos
____	*A. c. pulchellus*	E Myanmar (s Shan States and Kayah) to extreme nw Thailand
____	*A. c. annamensis*	S Laos (Bolavens Plateau) and Vietnam (central and s Annam)

☐ **White-throated Tit** *Aegithalos niveogularis*

Birch and pine forests of n Pakistan and nw India

☐ **Black-browed Tit** *Aegithalos iouschistos*

____	*A. i. iouschistos*	Himalayas (Nepal to n India and se Tibet
____	*A. i. bonvaloti*	S China (se Tibet to Sichuan and Yunnan) and ne Myanmar
____	*A. i. obscuratus*	Mountains of w China (Sungpan region of ne Sichuan)
____	*A. i. sharpei*	Montane forests of sw Myanmar (Mt. Victoria)

☐ **Sooty Tit** *Aegithalos fuliginosus*

Mts. of cent. China (Sichuan, s Gansu, s Shaanxi and sw Hubei)

☐ **Bushtit** *Psaltriparus minimus*

____	*P. m. saturatus*	S British Columbia, Puget Sound lowlands and Whidbey I.
____	*P. m. minimus*	Pacific coast west of the Cascades (n Oregon to s California)
____	*P. m. plumbeus*	E Oregon to Idaho, Wyoming, Arizona, New Mexico, w Texas
____	*P. m. melanurus*	Coastal s California (n San Diego County) to n Baja
____	*P. m. californicus*	Interior s Oregon to s California (Kern County)
____	*P. m. sociabilis*	S California (Little San Bernardino and Eagle mountains)
____	*P. m. grindae*	Mountains s Baja California (Sierra de la Laguna)
____	*P. m. dimorphicus*	Mountains of nw Mexico (Sonora to Sinaloa and n Coahuila)
____	*P. m. iulus*	Mts. of w Mexico (Durango to s Jalisco and w Tamaulipas)
____	*P. m. personatus*	Mts. of central Mexico (Michoacán to w Veracruz and Puebla)
____	*P. m. melanotis*	Mts. of s Mexico (Guerrero, Oaxaca and Chiapas) to Guatemala

☐ **Pygmy Tit** *Psaltria exilis*

Mountains of w and central Java

FAMILY: MALURIDAE (Fairywrens—27)

☐ **Orange-crowned Fairywren** *Clytomyias insignis*

____	*C. i. insignis*	NW New Guinea (Tamrau and Wandammen mountains)
____	*C. i. oorti*	New Guinea (Snow Mts. to mountains of Huon Peninsula)

☐ **Wallace's Fairywren** *Sipodotus wallacii*

Lowlands of New Guinea, Aru, Misool and Yapen islands

☐ **Broad-billed Fairywren** *Malurus grayi*

____	*M. g. grayi*	N New Guinea (Vogelkop to Sepik River) and Salawati I.
____	*M. g. campbelli*	SE New Guinea (middle Strickland River and Mt. Bosavi area)

☐ **White-shouldered Fairywren** *Malurus alboscapulatus*

____	*M. a. lorentzi*	S and southwestern New Guinea
____	*M. a. alboscapulatus*	NW New Guinea (Arfak and Tamrau mountains)
____	*M. a. naimii*	N and s lowlands and Central Highlands of New Guinea
____	*M. a. aida*	Northwestern New Guinea
____	*M. a. kutubu*	S highlands of central New Guinea
____	*M. a. moretoni*	N and s coasts of se New Guinea and Fergusson I.

☐ **Red-backed Fairywren** *Malurus melanocephalus*

____	*M. m. cruentatus*	N Australia (n Western Australia to Cape York Peninsula)
____	*M. m. melanocephalus*	E Australia (Burdekin R., Queensland to Hunter R., NSW)

☐ **White-winged Fairywren** *Malurus leucopterus*

____	*M. l. edouardi*	Barrow I. (Western Australia)
____	*M. l. leuconotus*	Interior of southern Australia (extending to w and s coasts)
____	*M. l. leucopterus*	Dirk Hartog I. (Western Australia)

☐ **Superb Fairywren** *Malurus cyaneus*

___	*M. c. cyaneus*	Tasmania
___	*M. c. samueli*	Flinders I. (Bass Strait)
___	*M. c. elizabethae*	King Island (Bass Strait)
___	*M. c. cyanochlamys*	SE Australia (central Queensland to s and central Victoria)
___	*M. c. leggei*	S Australia (s Eyre Pen, South Australia to extreme sw Victoria)
___	*M. c. ashbyi*	Kangaroo Island (South Australia)

☐ **Splendid Fairywren** *Malurus splendens*

___	*M. s. splendens*	S Western Australia (Shark Bay to Experance and Laverton)
___	*M. s. musgravi*	S-central Western Australia to Flinders Ranges, South Australia
___	*M. s. melanotus*	Semiarid se Australia (Flinders Ranges to west Darling basin)
___	*M. s. emmottorum*	Semiarid central Queensland

☐ **Variegated Fairywren** *Malurus lamberti*

___	*M. l. rogersi*	NE Northern Territory (Kimberley region)
___	*M. l. dulcis*	N Northern Territory (central Arnhem Land)
___	*M. l. lamberti*	SE Australia (Fitzroy R., Queensland to Bateman's Bay, NSW)
___	*M. l. assimilis*	Interior of Australia, extending to coasts in w Western Australia
___	*M. l. bernieri*	Bernier Island, Western Australia

☐ **Lovely Fairywren** *Malurus amabilis*

NE Queensland (coastal Cape York Pen. s to about Townsville

☐ **Red-winged Fairywren** *Malurus elegans*

Extreme sw Western Australia

☐ **Blue-breasted Fairywren** *Malurus pulcherrimus*

SW Western Australia and Eyre Peninsula, South Australia

☐ **Purple-crowned Fairywren** *Malurus coronatus*

___	*M. c. coronatus*	NW Australia) Kimberleys to Victoria R., Northern Territory
___	*M. c. macgillivrayi*	N Australia (coastal ne Northern Territory to nw Queensland)

☐ **Emperor Fairywren** *Malurus cyanocephalus*

___	*M. c. cyanocephalus*	Lowlands of w New Guinea, Salawati and Yapen islands
___	*M. c. mysorensis*	Lowlands of Biak I. (n New Guinea)
___	*M. c. bonapartii*	S New Guinea and Aru Islands

☐ **Southern Emuwren** *Stipiturus malachurus*

___	*S. m. malachurus*	SE Australia (s Queensland, coastal NSW to s Victoria)
___	*S. m. littleri*	N and w Tasmania
___	*S. m. polionotum*	SE South Australia and adjacent sw Victoria
___	*S. m. intermedius*	Mt. Lofty Range (South Australia)
___	*S. m. halmaturinus*	Kangaroo I. (South Australia)
___	*S. m. parimeda*	Southern tip of Eyre Peninsula (South Australia)
___	*S. m. westernensis*	Southwestern Western Australia
___	*S. m. hartogi*	Dirk Hartog I. (Western Australia)

☐ **Rufous-crowned Emuwren** *Stipiturus ruficeps*

Arid central Australia (Pilbara region to sw Queensland)

☐ **Mallee Emuwren** *Stipiturus mallee*

Mallee of se Australia and adjacent nw Victoria

☐ **Thick-billed Grasswren** *Amytornis textilis*

___	*A. t. textilis*	W Western Australia (Dirk Hartog I. & inland from Shark Bay)
___	*A. t. myall*	Gawler Range and n Eyre Peninsula (South Australia)
___	*A. t. modestus*	Central Australia (Alice Springs, N. Territory to central NSW)

☐ **Dusky Grasswren** *Amytornis purnelli*

Rocky spinifex hills of central Australia

☐ **Kalkadoon Grasswren** *Amytornis ballarae*

Spinifex stony hills of Selwyn Range (nw Queensland)

☐ **Black Grasswren** *Amytornis housei*

Sandstones of w Kimberley Division (Western Australia)

☐ **Striated Grasswren** *Amytornis striatus*

____ *A. s. rowleyi* — Forsyth Range (central Queensland)
____ *A. s. striatus* — Sand plains of central Australia to nw Victoria and c NSW
____ *A. s. whitei* — Stony hillsides of Pilbara region of Western Australia

☐ **White-throated Grasswren** *Amytornis woodwardi*

Sandstones of central Arnhem Land, Northern Territory

☐ **Carpentarian Grasswren** *Amytornis dorotheae*

Inland Gulf of Carpenteria (e N Territory to nw Queensland)

☐ **Short-tailed Grasswren** *Amytornis merrotsyi*

Stony hillsides of Flinders and Gawler Ranges (South Australia)

☐ **Gray Grasswren** *Amytornis barbatus*

____ *A. b. barbatus* — Lower Bulloo River (sw Queensland and adjacent nw NSW)
____ *A. b. diamantine* — Flood plains of lower Diamantina River and Coopers Creek

☐ **Eyrean Grasswren** *Amytornis goyderi*

Dunefields of Simpson-Strzelecki Deserts (central Australia)

FAMILY: ACANTHIZIDAE (Thornbills and Allies—65)

☐ **Western Bristlebird** *Dasyornis longirostris*

Coastal heaths of sw Western Australia (east of Albany)

☐ **Eastern Bristlebird** *Dasyornis brachypterus*

____ *D. b. monoides* — E Australia (se Queensland and ne New South Wales)
____ *D. b. brachypterus* — SE Australia (se New South Wales and ne Victoria)

☐ **Rufous Bristlebird** *Dasyornis broadbenti*

____ *D. b. caryochrous* — Otway Peninsula (southern Victoria)
____ *D. b. broadbenti* — SE South Australia and adjacent sw Victoria
____ *D. b. littoralis†* — Extreme sw Western Australia. Extinct.

☐ **Pilotbird** *Pycnoptilus floccosus*

____ *P. f. sandlandi* — SE Australia (subcoastal central NSW to central Victoria)
____ *P. f. floccosus* — SE Australia (Brindabella Range, NSW to Snowy Mts, Victoria)

☐ **Rock Warbler** *Origma solitaria*

Sandstone outcrops between Hunter R. and Bega, coastal NSW

☐ **Fernwren** *Oreoscopus gutturalis*

Montane wet tropical zone of ne Queensland

☐ **Rusty Mouse-Warbler** *Crateroscelis murina*

____ *C. m. murina* — Lowland forests of Salawati and Yapen islands (New Guinea)
____ *C. m. monacha* — Lowland forests of Aru Islands (New Guinea)
____ *C. m. pallida* — Trans-Fly lowlands of se New Guinea
____ *C. m. capitalis* — Lowland forests of Waigeo and Batanta islands (New Guinea)
____ *C. m. fumosa* — Lowland forests of Misool I. (New Guinea)

☐ **Bicolored Mouse-Warbler** *Crateroscelis nigrorufa*

____ *C. n. blissi* — New Guinea (Snow Mountains to Central Highlands)
____ *C. n. nigrorufa* — SE New Guinea (Owen Stanley Mountains)

☐ **Mountain Mouse-Warbler** *Crateroscelis robusta*

____ *C. r. peninsularis* — NW New Guinea (Arfak Mountains)
____ *C. r. ripleyi* — NW New Guinea (Tamrau Mountains)
____ *C. r. bastille* — Coastal n New Guinea (Bewani and Toricelli mountains)
____ *C. r. deficiens* — N New Guinea (Cyclops Mountains)
____ *C. r. sanfordi* — New Guinea (Weyland and Jayawijaya mountains)
____ *C. r. robusta* — SE New Guinea (Herzog and Owen Stanley mountains)

☐ **Yellow-throated Scrubwren** *Sericornis citreogularis*
_____ *S. c. cairnsi* — NE Queensland (Cooktown to Paluma Range)
_____ *S. c. intermedius* — E Australia (se Queensland to Clarence River, NSW)
_____ *S. c. citreogularis* — E New South Wales (Clarence River to Mt. Dromedary)

☐ **White-browed Scrubwren** *Sericornis frontalis*
_____ *S. f. laevigaster* — Disjunct in e Queensland (Atherton Tableland to Burnett River)
_____ *S. f. tweedi* — E New South Wales (Queensland border to Hunter River)
_____ *S. f. frontalis* — SE Australia (central NSW to s Victoria and se South Australia)
_____ *S. f. harterti* — Southern Victoria (Otway Peninsula to Strzelecki Range)
_____ *S. f. rosinae* — South Australia (Mt. Lofty Range to Fleurieu Peninsula)
_____ *S. f. ashbyi* — Kangaroo I. (South Australia)
_____ *S. f. mellori* — Southern Australia (Hopetoun, Western Australia to Adelaide)
_____ *S. f. maculatus* — SW Western Australia
_____ *S. f. balstoni* — Cent. W Australia (Geraldton to Shark Bay; Houtman Abrolhos)
_____ *S. f. flindersi* — Flinders I. and adjacent islands in Bass Strait

☐ **Tasmanian Scrubwren** *Sericornis humilis*
_____ *S. h. humilis* — Dense forest undergrowth of Tasmania
_____ *S. h. tregellasi* — King I. (Bass Strait)

☐ **Atherton Scrubwren** *Sericornis keri*
NE Queensland (Windsor Tablelands to Paluma Range)

☐ **Beccari's Scrubwren** *Sericornis beccarii*
_____ *S. b. wondiwoi* — NW New Guinea (Wondiwoi Mountains)
_____ *S. b. cyclopum* — N New Guinea (Cyclops Mountains)
_____ *S. b. weylandi* — New Guinea (Weyland Mountains)
_____ *S. b. imitator* — NE New Guinea (Arfak Mountains)
_____ *S. b. idenburgi* — New Guinea (Gauttier Mts. and slopes above Idenburg River)
_____ *S. b. boreonesioticus* — N New Guinea (Toricelli Mountains)
_____ *S. b. pontifex* — N New Guinea (Sepik Mountains)
_____ *S. b. randi* — S New Guinea (Trans-Fly lowlands)
_____ *S. b. beccarii* — Aru Islands
_____ *S. b. minimus* — NE Australia (n tip of Cape York Peninsula)
_____ *S. b. dubius* — NE Australia (n Queensland south to Cooktown)

☐ **Perplexing Scrubwren** *Sericornis virgatus*
_____ *S. v. virgatus* — Vogelkop Mts. and n slopes of Sepik-Ramu river drainage
_____ *S. v. jobiensis* — Yapen I. (New Guinea)

☐ **Large Scrubwren** *Sericornis nouhuysi*
_____ *S. n. cantans* — NW New Guinea (Vogelkop Mountains)
_____ *S. n. adelberti* — NE New Guinea (Adelbert Mountains)
_____ *S. n. nouhuysi* — New Guinea (Weyland, Nassau and Snow mountains)
_____ *S. n. stresemanni* — Central Highlands of New Guinea
_____ *S. n. oorti* — New Guinea (Herzog Mts. and mountains of Huon Peninsula)
_____ *S. n. monticola* — SE New Guinea (high Owen Stanley Mountains)

☐ **Large-billed Scrubwren** *Sericornis magnirostra*
_____ *S. m. viridior* — NE Australia (coastal and montane forests of Queensland)
_____ *S. m. magnirostra* — E Australia (Clarke Range, Queensland to far ne Victoria)
_____ *S. m. howei* — Victoria (w Gippsland and Strzelecki Range)

☐ **Vogelkop Scrubwren** *Sericornis rufescens*
W New Guinea (Arfak and Tamrau mts. and Bomberai Pen.)

☐ **Buff-faced Scrubwren** *Sericornis perspicillatus*
Humid montane forests of central and e New Guinea

☐ **Papuan Scrubwren** *Sericornis papuensis*
_____ *S. p. meeki* — W New Guinea (Jayawijaya Mountains)

____	*S. p. burgersi*	New Guinea (Weyland Mts. to Cent. Highlands and Sepik Mts.)
____	*S. p. papuensis*	Mountains of se New Guinea

☐ **Gray-green Scrubwren** *Sericornis arfakianus*

Humid montane forests of New Guinea

☐ **Pale-billed Scrubwren** *Sericornis spilodera*

____	*S. s. spilodera*	NW New Guinea (east to Astrolabe Bay) and Yapen I.
____	*S. s. granti*	W New Guinea (Snow Mountains)
____	*S. s. wuroi*	S New Guinea (Trans-Fly lowlands)
____	*S. s. guttatus*	Mountains of se New Guinea
____	*S. s. ferrugineus*	Waigeo I. (New Guinea)
____	*S. s. aruensis*	Aru Islands (New Guinea)
____	*S. s. batantae*	Batanta I. (New Guinea)

☐ **Scrubtit** *Acanthornis magna*

____	*A. m. magna*	Tasmania (except north)
____	*A. m. greenianus*	King Island (Bass Strait)

☐ **Redthroat** *Pyrrholaemus brunneus*

Arid s Australia

☐ **Speckled Warbler** *Pyrrholaemus sagittatus*

Central and e Queensland through NSW to e and cent. Victoria

☐ **Rufous Fieldwren** *Calamanthus campestris*

____	*C. c. winiam*	Sandplain heaths of se South Australia and adj. w Victoria
____	*C. c. campestris*	SE Western Australia through South Australia to nw Victoria
____	*C. c. rubiginosus*	Coastal areas of central Western Australia
____	*C. c. hartogi*	Dirk Hartog I. (Western Australia)
____	*C. c. dorrie*	Dorre I. (Western Australia)
____	*C. c. wayensis*	Salt lakes of central Western Australia
____	*C. c. montanellus*	S of line from Pt. Culver to Geraldton (except far sw corner)
____	*C. c. isabellinus*	North of L. Eyre-Frome basin of S. Australia and adj. w NSW

☐ **Striated Fieldwren** *Calamanthus fuliginosus*

____	*C. f. albiloris*	Southeast Australia (se New South Wales and se Victoria)
____	*C. f. bourneorum*	Southern Australia (se South Australia and sw Victoria)
____	*C. f. fuliginosus*	Eastern Tasmania
____	*C. f. diemenensis*	Western Tasmania

☐ **Chestnut-rumped Heathwren** *Hylacola pyrrhopygia*

____	*H. p. pyrrhopygia*	SE Australia (extreme se Queensland to se South Australia)
____	*H. p. parkeri*	Mt. Lofty Range (South Australia)
____	*H. p. pedleri*	Southern Flinders Ranges (South Australia)

☐ **Shy Heathwren** *Hylacola cauta*

____	*H. c. macrorhyncha*	Central New South Wales
____	*H. c. cauta*	Eyre Pen. and Flinders Ranges to nw Victoria and adj. NSW
____	*H. c. halmaturina*	Kangaroo I. (South Australia)
____	*H. c. whitlocki*	South-central Western Australia

☐ **Papuan Thornbill** *Acanthiza murina*

New Guinea (Snow Mountains to Owen Stanley range)

☐ **Buff-rumped Thornbill** *Acanthiza reguloides*

____	*A. r. squamata*	E Queensland (Atherton Tableland to Connors Range)
____	*A. r. nesa*	SE Queensland (Blackdown Tableland to NSW border)
____	*A. r. reguloides*	SE Australia (Queensland border to south and central Victoria)
____	*A. r. australis*	Mt. Lofty Range (South Australia) to sw Victoria

☐ **Western Thornbill** *Acanthiza inornata*

Southwestern Western Australia

☐ **Slender-billed Thornbill** *Acanthiza iredalei*

____	*A. i. hedleyi*	Salt marshes and heaths of se S Australia and adj. w Victoria
____	*A. i. rosinae*	South Australia (Adelaide ares to head of Gulf of St. Vincent)
____	*A. i. iredalei*	Southern Western Australia to eastern South Australia

☐ **Mountain Thornbill** *Acanthiza katherina*

Rainforests of ne Queensland (Cooktown to Paluma Range)

☐ **Brown Thornbill** *Acanthiza pusilla*

____	*A. p. dawsonensis*	E Queensland (Bowen to Dawes Range)
____	*A. p. pusilla*	E Australia (se Queensland to s Victoria and se South Australia)
____	*A. p. diemenensis*	Tasmania and Deal I. (Bass Strait)
____	*A. p. archibaldi*	King I. (Bass Strait). Probably extinct . . .
____	*A. p. zietzi*	Kangaroo I. (South Australia)

☐ **Tasmanian Thornbill** *Acanthiza ewingii*

____	*A. e. ewingii*	Tasmania and Flinders I. (Bass Strait)
____	*A. e. rufifrons*	King I. (Bass Strait)

☐ **Inland Thornbill** *Acanthiza apicalis*

____	*A. a. whitlocki*	Inland southern Australia, extending to coast in west
____	*A. a. apicalis*	Extreme sw Western Australia east to western Victoria
____	*A. a. albiventris*	Central Queensland and NSW to n Victoria and e South Australia
____	*A. a. cinerascens*	West-central Queensland

☐ **Yellow-rumped Thornbill** *Acanthiza chrysorrhoa*

____	*A. c. normantoni*	Interior of e-central Australia
____	*A. c. leighi*	SE Australia (se Queensland to Victoria and e South Australia)
____	*A. c. leachi*	Tasmania (except western areas)
____	*A. c. chrysorrhoa*	S Western Australia (Pilbara east to South Australian border)

☐ **Chestnut-rumped Thornbill** *Acanthiza uropygialis*

Arid and semiarid interior of s Australia; to coast in west

☐ **Slaty-backed Thornbill** *Acanthiza robustirostris*

Mulga areas of central Western Australia to s-cent. Queensland

☐ **Yellow Thornbill** *Acanthiza nana*

____	*A. n. flava*	NE Queensland (Atherton Tableland to Paluma Range)
____	*A. n. nana*	Coastal se Australia (Moreton Bay, Queensland, to Eden, NSW)
____	*A. n. modesta*	Central and e Queensland to s Victoria and se South Australia

☐ **Striated Thornbill** *Acanthiza lineata*

____	*A. l. alberti*	E Australia (se Queensland and adjacent ne New South Wales)
____	*A. l. lineata*	SE South Australia (ne New South Wales to e and s Victoria)
____	*A. l. clelandi*	SE South Australia and adjacent sw Victoria
____	*A. l. whitei*	Kangaroo I. (South Australia)

☐ **Weebill** *Smicrornis brevirostris*

____	*S. b. flavescens*	Northern and central Australia
____	*S. b. brevirostris*	E Australia (Burdekin R., Queensland to se South Australia)
____	*S. b. occidentalis*	Geraldton, Western Australia to nw Victoria and sw NSW
____	*S. b. ochrogaster*	W-central Australia (Pilbara region to Gibson Desert)

☐ **Mountain Gerygone** *Gerygone cinerea*

Arfak Mountains to se ranges of New Guinea

☐ **Green-backed Gerygone** *Gerygone chloronota*

____	*G. c. cinereiceps*	New Guinea
____	*G. c. aruensis*	Aru, Waigeo, Salawati and Batanta islands
____	*G. c. chloronota*	Northern Territory (Arnhem Land, Melville I., Groote Eylandt)
____	*G. c. darwini*	NW Western Australia (Kimberley Division)

☐ **Fairy Gerygone** *Gerygone palpebrosa*

    ____ *G. p. palpebrosa*   NW New Guinea, Aru, Waigeo, Misool and Salawati islands
    ____ *G. p. wahnesi*   Yapen I. and n New Guinea
    ____ *G. p. inconspicua*   SE New Guinea (west to upper Fly River)
    ____ *G. p. tarara*   S New Guinea (Moorhead River to mouth of Fly River)
    ____ *G. p. personata*   N Queensland (Cape York Peninsula)
    ____ *G. p. flavida*   E Queensland (Burdekin River to about Maryborough)

☐ **Biak Gerygone** *Gerygone hypoxantha*

    Biak I. (off nw New Guinea)

☐ **White-throated Gerygone** *Gerygone olivacea*

    ____ *G. o. cinerascens*   Lowlands of se New Guinea and n Queensland (Cape York Pen.)
    ____ *G. o. rogersi*   N Australia (Kimberley to Arnhem Land and nw Queensland)
    ____ *G. o. olivacea*   E Australia (ne Queensland through NSW to e Victoria)

☐ **Yellow-bellied Gerygone** *Gerygone chrysogaster*

    ____ *G. c. leucothorax*   Lowlands of w New Guinea (Geelvink Bay region)
    ____ *G. c. notata*   W New Guinea, Misool and Batanta islands
    ____ *G. c. neglecta*   Waigeo I. (New Guinea)
    ____ *G. c. dohertyi*   SW New Guinea (Onin Peninsula to Triton Bay)
    ____ *G. c. chrysogaster*   S and e New Guinea, Yapen I. and Aru Islands

☐ **Large-billed Gerygone** *Gerygone magnirostris*

    ____ *G. m. conspicillata*   NW New Guinea (Vogelkop region)
    ____ *G. m. affinis*   N New Guinea, Karkar, Manam and Yapen islands
    ____ *G. m. mimikae*   S New Guinea (Onin Peninsula to Port Moresby area)
    ____ *G. m. brunneipectus*   Aru Is. (New Guinea); Saibai and Boigu Is. (Torres Strait)
    ____ *G. m. cobana*   Waigeo, Batanta and Salawati islands (New Guinea)
    ____ *G. m. occasa*   Kofiau I. (New Guinea)
    ____ *G. m. proxima*   D'Entrecasteaux Arch. (Fergusson and Goodenough)
    ____ *G. m. onerosa*   Misima I. (Louisiade Archipelago)
    ____ *G. m. tagulana*   Tagula I. (Louisiade Archipelago)
    ____ *G. m. rosseliana*   Rossel I. (Louisiade Archipelago)
    ____ *G. m. magnirostris*   N Australia (Kimberley to Arnhem Land and Melville I.)
    ____ *G. m. cairnsensis*   NE Queensland (islands of s Torres Strait south to about Sarina)

☐ **Dusky Gerygone** *Gerygone tenebrosa*

    ____ *G. t. tenebrosa*   Mangroves of nw Western Australia (Broome to Kimberly)
    ____ *G. t. christophori*   Mangroves of c Western Australia (Shark Bay to De Grey R.)

☐ **Brown Gerygone** *Gerygone mouki*

    ____ *G. m. mouki*   NE Queensland (Cooktown to Paluma Range)
    ____ *G. m. amalia*   E Queensland (Mackay to Bowen)
    ____ *G. m. richmondi*   SE Australia (e-central Queensland to ne Victoria)

☐ **Golden-bellied Gerygone** *Gerygone sulphurea*

    ____ *G. s. flaveola*   Sulawesi, Salayar and Banggai Islands (Peleng and Banggai)
    ____ *G. s. sulphurea*   Malay Peninsula, Greater and Lesser Sundas
    ____ *G. s. simplex*   Lubang, Luzon, Mindoro, Verde, Negros, Bohol and Cebu
    ____ *G. s. rhizophorae*   S Philippines (Mindanao, Basilan and Sulu Archipelago)

☐ **Plain Gerygone** *Gerygone inornata*

    Lesser Sundas (Sawu, Roti, Timor and Wetar)

☐ **Rufous-sided Gerygone** *Gerygone dorsalis*

    ____ *G. d. senex*   Kalaotoa and Madu islands (Flores Sea)
    ____ *G. d. keyensis*   Tayandu and Kai Islands (Kai Kecil, Sawa and Ruin)
    ____ *G. d. fulvescens*   Lesser Sundas (Moa, Kisar, Leti, Sermata, Babar and Romang)
    ____ *G. d. kuehni*   Damar I. (Lesser Sundas)
    ____ *G. d. dorsalis*   Tanimbar Islands (Arafura Sea)

☐ **Brown-breasted Gerygone** *Gerygone ruficollis*

____ *G. r. ruficollis*	W New Guinea (Arfak Mountains)
____ *G. r. insperata*	New Guinea (C Highlands to Owen Stanley Mts. and Huon Pen.)

☐ **Western Gerygone** *Gerygone fusca*

____ *G. f. fusca*	SW Western Australia
____ *G. f. exsul*	N-central Queensland to s Victoria and e South Australia
____ *G. f. mungi*	Interior W Australia, N Territory and nw South Australia

☐ **Mangrove Gerygone** *Gerygone levigaster*

____ *G. l. levigaster*	Coastal n Australia (Broome, nw Australia to e Cape York Pen.)
____ *G. l. cantator*	Coastal e Australia (Halifax Bay, Qld. to Gosford, NSW)

☐ **Norfolk Island Gerygone** *Gerygone modesta*

Norfolk Island

☐ **Gray Gerygone** *Gerygone igata*

Mainland New Zealand and adjacent offshore islands

☐ **Chatham Island Gerygone** *Gerygone albofrontata*

Chatham Islands (off New Zealand)

☐ **Fan-tailed Gerygone** *Gerygone flavolateralis*

____ *G. f. flavolateralis*	New Caledonia and Maré
____ *G. f. lifuensis*	Lifou (Loyalty Islands)
____ *G. f. correiae*	N Vanuatu and Banks Group
____ *G. f. rouxi*	Uvéa (Loyalty Islands)
____ *G. f. citrina*	Rennell (se Solomon Islands)

☐ **Southern Whiteface** *Aphelocephala leucopsis*

____ *A. l. leucopsis*	Southeastern Australia to e Western Australia
____ *A. l. castaneiventris*	Interior of southern Western Australia

☐ **Chestnut-breasted Whiteface** *Aphelocephala pectoralis*

Elevated gibber plains of central South Australia

☐ **Banded Whiteface** *Aphelocephala nigricincta*

Interior of central Western Australia to far sw Queensland

FAMILY: EPTHIANURIDAE (Australian Chats—5)

☐ **Crimson Chat** *Epthianura tricolor*

Interior of Australia, to southern and western coasts

☐ **Orange Chat** *Epthianura aurifrons*

Interior of Australia, to southern and western coasts

☐ **Yellow Chat** *Epthianura crocea*

____ *E. c. tunneyi*	Northern Territory (w Arnhem Land)
____ *E. c. crocea*	Grassy swamps of n Western Australia to w Queensland
____ *E. c. macgregori*	E Queensland (Dawson and Mackenzie R. basins). Extinct?

☐ **White-fronted Chat** *Epthianura albifrons*

Southern Australia, Tasmania, King and Flinders Islands

☐ **Desert Chat** *Ashbyia lovensis*

Stony plains of central Australia

FAMILY: NEOSITTIDAE (Sittellas—2)

☐ **Black Sittella** *Neositta miranda*

____ *N. m. frontalis*	Mountains of nw New Guinea
____ *N. m. kuboriensis*	Central Highlands of ne New Guinea
____ *N. m. miranda*	Mountains of se New Guinea

☐ **Varied Sittella** *Neositta chrysoptera*

____	*N. c. papuensis*	W New Guinea (Arfak Mountains)
____	*N. c. wahgiensis*	New Guinea (Mt. Hagen region)
____	*N. c. toxopeusi*	NW New Guinea (Snow Mountains)
____	*N. c. intermedia*	New Guinea (Nassau Mountains)
____	*N. c. alba*	Central mountains of New Guinea
____	*N. c. albifrons*	Montane forests of se New Guinea
____	*N. c. leucoptera*	N Australia (Fitzroy R., Western Australia to nw Queensland
____	*N. c. striata*	N Queensland (Cape York Peninsula)
____	*N. c. leucocephala*	E Queensland (Repulse Bay to Brisbane)
____	*N. c. chrysoptera*	SE Australia (south-central Queensland to Victoria)
____	*N. c. pileata*	S Australia (North West Cape to w NSW and w Victoria)

FAMILY: CLIMACTERIDAE (Australasian Treecreepers—7)

☐ **Papuan Treecreeper** *Cormobates placens*

____	*C. p. placens*	NW New Guinea (Arfak and Tamrau mountains)
____	*C. p. steini*	W New Guinea (Weyland Mountains)
____	*C. p. inexpectata*	N New Guinea (n slope of Jayawijaya Mountains)
____	*C. p. meridionalis*	Mountains of se New Guinea

☐ **White-throated Treecreeper** *Cormobates leucophaea*

____	*C. l. minor*	NE Queensland (uplands of Cooktown, Atherton and Paluma)
____	*C. l. intermedius*	E Queensland (Clarke and Connors Ranges)
____	*C. l. metastasis*	E Australia (Fitzroy R., Queensland to Hunter R., NSW)
____	*C. l. leucophaea*	SE Australia (c NSW to s Victoria and se South Australia)
____	*C. l. grisescens*	South Australia (Mt. Lofty Range)

☐ **White-browed Treecreeper** *Climacteris affinis*

____	*C. a. affinis*	Arid s Australia (c Western Australia to Flinders Ranges)
____	*C. a. superciliosa*	Semiarid se Australia (Flinders Ranges to w Darling basin)

☐ **Red-browed Treecreeper** *Climacteris erythrops*

	SE Australia (se Queensland through e NSW to central Victoria)

☐ **Brown Treecreeper** *Climacteris picumnus*

____	*C. p. melanotus*	NE Queensland (s Cape York Peninsula)
____	*C. p. picumnus*	Central and e Queensland to c Victoria and e South Australia
____	*C. p. victoriae*	SE South Australia (se Qld. through e NSW to sw Victoria)

☐ **Black-tailed Treecreeper** *Climacteris melanurus*

____	*C. m. melanurus*	N Austalia (Kimberley to Arnhem Land and nw Queensland)
____	*C. m. wellsi*	W-central Western Australia (Pilbara region)

☐ **Rufous Treecreeper** *Climacteris rufus*

	S and s-central Western Australia to Eyre Pen., South Australia

FAMILY: PARIDAE (Chickadees and Tits—59)

☐ **Sombre Tit** *Poecile lugubris*

____	*P. l. lugubris (splendens, lugens)*	SE Europe south to n Greece; Crete
____	*P. l. anatoliae*	S Greece to Turkey, w Georgia, Armenia, n Iraq and nw Iran
____	*P. l. dubius*	W Iran (Zagros Mountains); winters to ne Iraq

☐ **Marsh Tit** *Poecile palustris*

____	*P. p. palustris (dresseri)*	Br. Isles, s Scandinavia and c Europe to Pyrénées and Balkans
____	*P. p. stagnatilis*	W Russia to n Turkey, e Yugoslavia and Bulgaria
____	*P. p. kabardensis (brandtii)*	Caucasus Mountains
____	*P. p. italicus*	French Alps and Italy

____ *P. p. brevirostris*   Siberia to Mongolia, ne China and Korea
____ *P. p. ernsti*   Sakhalin
____ *P. p. hensoni*   N Japan (s Kuril Islands and Hokkaido)
____ *P. p. hellmayri*   S and e China (Sichuan to Liaoning) and s Korea

☐ **Black-bibbed Tit** *Poecile hypermelaenus*

Central and e China to se Tibet and w Myanmar

☐ **Caspian Tit** *Poecile hyrcana*

____ *P. h. hyrcana*   N Iran (Elburz Mountains) and adjacent Azerbaijan
____ *P. h. kirmanensis*   S Iran (mountains of Kerman region)

☐ **Willow Tit** *Poecile montana*

____ *P. m. kleinschmidti*   Britain
____ *P. m. rhenana*   W Europe to w Germany and n Switzerland
____ *P. m. salicaria*   Central Europe to w Poland, sw Germany and nw Austria
____ *P. m. montana*   Central Europe
____ *P. m. borealis (colletti, loennbergi)*   Scandinavia to Baltic States, Carpathian Mts. and se Russia
____ *P. m. uralensis*   SE Russia to s Urals, sw Siberia and n Kazakstan
____ *P. m. baicalensis*   E Siberia to Sea of Okhotsk, Mongolia, Ussuriland and ne China
____ *P. m. anadyrensis*   NE Siberia to n coast of Sea of Okhotsk
____ *P. m. kamtschatkensis*   Kamchatka Peninsula
____ *P. m. sachalinensis*   Sakhalin; vagrant to Hokkaido
____ *P. m. restricta*   Japan

☐ **Songar Tit** *Poecile songara*

____ *P. s. songara*   Kazakstan (Tien Shan Mountains)
____ *P. s. weigoldica*   SW China (Sichuan to e Tibet, se Qinghai and nw Yunnan)
____ *P. s. affinis*   N-central China (Ningxia to s Gansu and ne Qinghai)
____ *P. s. stotzneri*   NE China (se Mongolia to Liaoning, Shaanxi and Henan)

☐ **Carolina Chickadee** *Poecile carolinensis*

____ *P. c. atricapilloides*   S-central US (Kansas to Oklahoma and Texas)
____ *P. c. agilis*   S US (sw Arkansas to w Louisiana and e Texas)
____ *P. c. extima*   E US (e Missouri to n New Jersey, Kentucky and Virginia)
____ *P. c. carolinensis*   SE US (e Arkansas to s Virginia, Mississippi and Florida)

☐ **Black-capped Chickadee** *Poecile atricapillus*

____ *P. a. turneri*   Alaska and adjacent nw Canada
____ *P. a. occidentalis*   Extreme sw Br. Columbia to nw California (west of Cascades)
____ *P. a. fortuita*   S interior British Columbia to nw Montanea and nw Idaho
____ *P. a. garrina*   Rocky Mts. (se Idaho to Wyoming, e Utah and New Mexico)
____ *P. a. nevadensis*   Great Basin of sw US (e Oregon to Idaho, Nevada and w Utah)
____ *P. a. bartletti*   Newfoundland and Miquelon
____ *P. a. atricapillus*   E Canada and ne US
____ *P. a. practica*   NE US (Appalachian Mountains region)
____ *P. a. septentrionalis*   W Canada and central US

☐ **Mountain Chickadee** *Poecile gambeli*

____ *P. g. abbreviata*   SW Canada and coastal ranges of n California
____ *P. g. gambeli*   Rocky Mts. (Montana to Wyoming, New Mexico and sw Texas)
____ *P. g. wasatchensis*   Coniferous forests of s Idaho and Utah
____ *P. g. inyoensis*   Great Basin (se Oregon, sw Idaho to w Utah and e-c California)
____ *P. g. baileyae*   Sierra Nevada of s California
____ *P. g. atrata*   Mountains of n Baja California

☐ **Mexican Chickadee** *Poecile sclateri*

____ *P. s. eidos*   Oak-pine forests of se Arizona and New Mexico to nw Mexico
____ *P. s. garzai*   N Mexico (se Coahuila/Nuevo León border)
____ *P. s. sclateri*   Mountains of w Mexico (se Sinaloa to Puebla and w Veracruz)
____ *P. s. rayi*   Mountains of w Mexico (s Jalisco to Guerrero and Oaxaca)

☐ **White-browed Tit** *Poecile superciliosa*

Mountains of w China (Qinghai to se Tibet, Gansu and Sichuan)

☐ **Père David's Tit** *Poecile davidi*

Mountains of w China (Hubei to Shaanxi, Gansu and Sichuan)

☐ **Chestnut-backed Chickadee** *Poecile rufescens*
_____ *P. r. rufescens* — Alaska to coastal central California and Sierra Nevada of n Calif.
_____ *P. r. neglecta* — Coastal California (sw Marin County)
_____ *P. r. barlowi* — Coastal s California (San Francisco Bay to Santa Barbara Co.)

☐ **Boreal Chickadee** *Poecile hudsonica*
_____ *P. h. stoneyi* — N Alaska to n Yukon and nw Mackenzie
_____ *P. h. hudsonica* — C Alaska and Yukon to Minnesota, Labrador and Newfoundland
_____ *P. h. columbiana* — S Alaska to s Yukon, Br. Columbia, Montana and Washington
_____ *P. h. farleyi* — NE British Columbia and Alberta to Saskatchewan and Manitoba
_____ *P. h. littoralis* — SE Canada (s Quebec) to Nova Scotia and ne US

☐ **Gray-headed Chickadee** *Poecile cincta*
_____ *P. c. lapponicus* — Fenno-Scandia, s Kola Peninsula and n European Russia
_____ *P. c. cincta* — N Ural Mountains to Bering Sea and ne Mongolia
_____ *P. c. sayanus* — Altai, Sayan and Tannu-Ola mountains to nw Mongolia
_____ *P. c. lathami* — Coniferous forests of Alaska and nw Canada

☐ **Coal Tit** *Periparus ater*
_____ *P. a. hibernicus* — Ireland (except extreme ne in County Down)
_____ *P. a. britannicus* — Britain and ne Ireland
_____ *P. a. ater* — Continental Europe to Siberia, Mongolia, Sakhalin and ne China
_____ *P. a. vieirae* — Iberian Peninsula
_____ *P. a. sardus* — Corsica and Sardinia
_____ *P. a. atlas* — N Morocco
_____ *P. a. ledouci* — North Africa (n Tunisia and n Algeria)
_____ *P. a. cypriotes* — Cyprus
_____ *P. a. moltchanovi* — Crimean Peninsula
_____ *P. a. derjugini* — Mts. of ne Turkey, w Georgia and Black Sea coast of Russia
_____ *P. a. michalowskii* — Caucasus and Transcaucasia
_____ *P. a. gaddi* — SE Azerbaijan and Caspian region of n Iran (east to Gorgan)
_____ *P. a. chorassanicus* — NE Iran and sw Turkmenistan
_____ *P. a. phaeonotus* — Zagros Mountains (sw Iran)
_____ *P. a. rufipectus* — Kazakstan (Tien Shan Mountains) to nw China (Xinjiang)
_____ *P. a. aemodius* — E Himalayas to ne Myanmar, Tibet and sw China
_____ *P. a. pekinensis* — NE China (s Liaoning to Shaanxi and Shantung Peninsula)
_____ *P. a. kuatunensis* — Montane forests of se China (Anhui, Fujian and Zhejiang)
_____ *P. a. ptilosus* — Montane forests of Taiwan
_____ *P. a. insularis* — S Kuril Islands, Japan and Cheju-Do Islands (Korea)

☐ **Black-breasted Tit** *Periparus rufonuchalis*

Oak-rhododendron forests of Turkestan to w China and Nepal

☐ **Rufous-vented Tit** *Periparus rubidiventris*
_____ *P. r. rubidiventris* — Rhododendron forests of nw India to central Nepal
_____ *P. r. beavani* — Central Nepal to ne Myanmar, nw India, s Tibet and sw China
_____ *P. r. saramatii* — NW Myanmar (Mount Sarameti in Naga Hills)

☐ **Black-crested Tit** *Periparus melanolophus*

Coniferous forests of e Afghanistan to w Nepal

☐ **Yellow-bellied Tit** *Pardaliparus venustulus*

Mixed woodlands of east, central and s China

☐ **Elegant Tit** *Pardaliparus elegans*
_____ *P. e. edithae* — N Philippines (Calayan and Camiguin Norte)
_____ *P. e. montigenus* — N Philippines (nw Luzon)
_____ *P. e. gilliardi* — N Philippines (Bataan Peninsula of central Luzon)
_____ *P. e. elegans* — Philippines (c and s Luzon, Panay, Mindoro and Catanduanes)

____ *P. e. albescens*	Philippines (Guimaras, Masbate, Negros and Ticao)
____ *P. e. visayanus*	Philippines (Cebu)
____ *P. e. mindanensis*	Philippines (Samar, Leyte, Biliran and Mindanao)
____ *P. e. suluensis*	Sulu Archipelago (Jolo, Tawitawi and Sanga Sanga)
____ *P. e. bongaoensis*	Bongao I. (Sulu Archipelago)

☐ **Palawan Tit** *Pardaliparus amabilis*

SW Philippines (Balabac, Calauit and Palawan)

☐ **Crested Tit** *Lophophanes cristatus*

____ *L. c. scoticus*	Scotland
____ *L. c. abadiei*	NW France (Bretagne)
____ *L. c. cristatus*	Fenno-Scandia to European Russia, n Urals and Ukraine
____ *L. c. mitratus*	Central and w Europe
____ *L. c. weigoldi*	Portugal and s Spain; rare vagrant to Morocco
____ *L. c. baschkirikus*	S Urals

☐ **Gray-crested Tit** *Lophophanes dichrous*

____ *L. d. kangrae*	Coniferous forests of Kashmir to n India (Uttar Pradesh)
____ *L. d. dichrous*	Coniferous forests of w Nepal to Arunachal Pradesh and s Tibet
____ *L. d. dichroides*	NE Tibet to sw China (n Sichuan, Qinghai, Gansu and Shaanxi)
____ *L. d. wellsi*	NE Myanmar and s China (w Sichuan and nw Yunnan)

☐ **White-winged Black-Tit** *Melaniparus leucomelas*

____ *M. l. leucomelas*	Central and se highlands of Ethiopia
____ *M. l. insignis*	Gabon to Zaire, Kenya, Zambia, Malawi and n Mozambique

☐ **White-shouldered Black-Tit** *Melaniparus guineensis*

Senegal to s Sudan, ne Zaire, sw Ethiopia, n Uganda, w Kenya

☐ **Southern Black-Tit** *Melaniparus niger*

____ *M. n. xanthostomus*	Angola to Namibia, Botswana, Zambia and s Tanzania
____ *M. n. ravidus*	E Zambia to Malawi, Zimbabwe and Mozambique
____ *M. n. niger*	S Mozambique to Natal, Zululand, Swaziland, e Cape Province

☐ **Carp's Tit** *Melaniparus carpi*

S Angola to nw Namibia

☐ **White-bellied Tit** *Melaniparus albiventris*

Disjunct in mts. of Nigeria/Cameroon; se Sudan to Tanzania

☐ **White-backed Black-Tit** *Melaniparus leuconotus*

Wooded mountain gorges of Ethiopia and adjacent Eritrea

☐ **Rufous-bellied Tit** *Melaniparus rufiventris*

____ *M. r. rufiventris*	N Angola to w Zaire, ne Namibia and central Zambia
____ *M. r. masukuensis*	Extreme se Zaire to Zambia, Malawi and Mozambique
____ *M. r. pallidiventris*	Tanzania, n Mozambique, Zimbabwe and extreme e Zambia

☐ **Dusky Tit** *Melaniparus funereus*

____ *M. f. funereus*	Guinea to Cameroon, Zaire, Uganda and Kenya
____ *M. f. gabela*	W-central Angola (Gabela escarpment)

☐ **Red-throated Tit** *Melaniparus fringillinus*

Acacia savanna of s Kenya and n Tanzania

☐ **Stripe-breasted Tit** *Melaniparus fasciiventer*

____ *M. f. fasciiventer*	Montane forests of e Zaire, sw Uganda, Rwanda and Burundi
____ *M. f. tanganjicae*	Montane forests of e Zaire (Itombwe Mts.)
____ *M. f. kaboboensis*	Montane forests of se Zaire (Mt. Kabobo)

☐ **Somali Tit** *Melaniparus thruppi*

____ *M. t. thruppi*	Dry acacia of Ethiopia and Somalia
____ *M. t. barakae*	Dry acacia of sw Somalia to Kenya, Uganda and ne Tanzania

☐ **Miombo Tit** *Melaniparus griseiventris*

Angola, Zambia, se Zaire, Tanzania, Zimbabwe, Mozambique

☐ **Ashy Tit** *Melaniparus cinerascens*
____ *M. c. benguelae* | SW Angola and extreme nw Namibia
____ *M. c. cinerascens* | Namibia to Botswana, Zimbabwe and South Africa

☐ **Gray Tit** *Melaniparus afer*
____ *M. a. afer* | Namibia and w Cape Province
____ *M. a. arens* | Lesotho to Natal, sw Orange Free State and e Cape Province

☐ **Great Tit** *Parus major*
____ *P. m. newtoni* | British Isles
____ *P. m. major* | Europe to nw Iran, Siberia, Lake Baikal, Altai and Sayan mts.
____ *P. m. kapustini* | NW China (nw Xinjiang) to Mongolia and e Siberia
____ *P. m. corsus* | Iberian Peninsula and Corsica
____ *P. m. mallorcae* | Balearic Islands
____ *P. m. ecki* | Sardinia
____ *P. m. excelsus* | NW Africa (Morocco to Tunisia)
____ *P. m. aphrodite* | S Italy, Sicily, s Greece, Mediterranean islands and Cyprus
____ *P. m. niethammeri* | Crete
____ *P. m. terraesanctae* | NW Syria, Lebanon, Israel and Jordan
____ *P. m. blanfordi (karelini)* | N Iraq and Iran
____ *P. m. intermedius* | NE Iran and adjacent sw Turkmenistan
____ *P. m. cashmirensis* | NE Afghanistan to n Pakistan and nw India
____ *P. m. ziaratensis* | S Afghanistan to n Baluchistan and Pakistan
____ *P. m. decolorans* | SE Afghanistan (east of Kabul and south of the Hindu Kush)
____ *P. m. nipalensis* | N India to Nepal, Bhutan, Bangladesh and w Myanmar
____ *P. m. vauriei* | E Assam (Lakhimpur district) and e Arunachal Pradesh
____ *P. m. stupae* | Central and peninsular India
____ *P. m. mahrattarum* | SW India (Kerala) and Sri Lanka
____ *P. m. templorum* | NE Thailand to s Laos and s Vietnam
____ *P. m. hainanus* | Hainan (s China)
____ *P. m. ambiguus* | SE Myanmar, peninsular Thailand, Malaya and Sumatra
____ *P. m. cinereus* | Java and w Lesser Sundas
____ *P. m. sarawacensis* | Borneo (w Sarawak)
____ *P. m. minor* | SE Russia to Japan, Korea, sw China and e Tibet
____ *P. m. tibetanus* | SW China to se Tibet; single record from Sikkim
____ *P. m. subtibetanus* | S China to se Tibet and nw Myanmar
____ *P. m. nubicolus* | SE Myanmar to n Thailand, n Laos and extreme w Tonkin
____ *P. m. dageletensis* | Ullung I. (South Korea)
____ *P. m. amamiensis* | N Ryukyu Islands (Amami-O-Shima and Tokuno-Shima)
____ *P. m. okinawae* | Central Ryukyu Islands (Okinawa and Yagachi)
____ *P. m. nigriloris* | S Ryukyu Islands (Ishigaki and Iriomote)
____ *P. m. commixtus* | S China (south of the Yangtze) to Hong Kong and e Tonkin

☐ **Turkestan Tit** *Parus bokharensis*
____ *P. b. bokharensis* | Russia to Tien Shan and Karatau mts. and nw Afghanistan
____ *P. b. ferghanensis* | S Kirgiz and w Tien Shan mts. to w Pamirs and Turkestan
____ *P. b. turkestanicus* | Lake Balkhash to w China (Xinjiang) and sw Mongolia

☐ **Green-backed Tit** *Parus monticolus*
____ *P. m. monticolus* | W Himalayas to w Nepal and se Tibet
____ *P. m. yunnanensis* | E Himalayas to ne India, Myanmar, w China and nw Vietnam
____ *P. m. legendrei* | S Vietnam (Da Lat Plateau)
____ *P. m. insperatus* | Montane forests of Taiwan

☐ **White-winged Tit** *Parus nuchalis*

Patchily distributed semiarid thorn forests of w and s India

☐ **Black-lored Tit** *Parus xanthogenys*
____ *P. x. xanthogenys* | Foothills of w Himalayas
____ *P. x. aplonotus* | N and e peninsular India
____ *P. x. tranvancoreensis* | S peninsular India

527

☐ **Yellow-cheeked Tit** *Parus spilonotus*
- ____ *P. s. spilonotus* — E Himalayas and adjacent ne Myanmar to s Tibet and sw China
- ____ *P. s. subviridis* — NE Indian hill states to Myanmar, n Thailand and nw Laos
- ____ *P. s. rex* — S China to nw Vietnam and ne Laos
- ____ *P. s. basileus* — S Laos (Bolavens Plateau) and s Vietnam (s Annam)

☐ **Yellow Tit** *Macholophus holsti*
- Montane forests of Taiwan

☐ **Eurasian Blue Tit** *Cyanistes caeruleus*
- ____ *C. c. obscurus* — Ireland, Britain and Channel Islands
- ____ *C. c. caeruleus* — Continental Europe to n Spain, Sicily, n Turkey and n Urals
- ____ *C. c. ogilastrae* — Portugal, s Spain, Corsica and Sardinia
- ____ *C. c. balearicus* — Majorca I. (Balearic Islands)
- ____ *C. c. calamensis* — S Greece, Pelopónnisos, Cyclades, Crete and Rhodes
- ____ *C. c. orientalis* — S European Russia (Volga River to central and s Ural Mts.)
- ____ *C. c. satunini* — Crimean Pen., Caucasus, Transcaucasia and nw Iran to e Turkey
- ____ *C. c. raddei* — N Iran
- ____ *C. c. persicus* — SW Iran (Zagros Mountains)

☐ **African Blue Tit** *Cyanistes teneriffae*
- ____ *C. t. ultramarinus* — NW Africa (Morocco to Tunisia) and Pantelleria I. (s Italy)
- ____ *C. t. cyrenaicae* — Libya
- ____ *C. t. palmensis* — La Palma I. (w Canary Islands)
- ____ *C. t. teneriffae* — Canary Islands (Gomera, Tenerife and Gran Canaria)
- ____ *C. t. ombriosus* — Hierro I. (sw Canary Islands)
- ____ *C. t. degener* — E Canary Islands (Fuerteventura and Lanzarote)

☐ **Azure Tit** *Cyanistes cyanus*
- ____ *C. c. cyanus* — European Russia to basin of middle Volga River
- ____ *C. c. hyperrhiphaeus* — E European Russia to w Siberia and n Kazakstan
- ____ *C. c. kotkalensis* — Lowlands of se Kazakstan (south of Lake Balkhash)
- ____ *C. c. tianschanicus* — SE Kazakstan to nw China, Manchuria and Pakistan
- ____ *C. c. yenisseensis* — SE Siberia to Sea of Japan and lower Amur River

☐ **Yellow-breasted Tit** *Cyanistes flavipectus*
- ____ *C. f. carruthersi* — W Pamir Mountains to Fergana basin and e Alayskiy Mountains
- ____ *C. f. flavipectus* — W Tien Shan Mountains to n Afghanistan and n Pakistan
- ____ *C. f. berezowskii* — N-central China (e Qinghai s of Kokonor on Upper Hwang Ho)

☐ **White-fronted Tit** *Sittiparus semilarvatus*
- ____ *S. s. snowi* — N Philippines (n Luzon)
- ____ *S. s. semilarvatus* — N Philippines (central and s Luzon)
- ____ *S. s. nehrkorni* — S Philippines (Mindanao)

☐ **Varied Tit** *Sittiparus varius*
- ____ *S. v. varius* — Kuril Islands, ne China, Korea and main Japanese islands
- ____ *S. v. sunsunpi* — Tanegashima (Ryukyu Islands)
- ____ *S. v. yakushimensis* — Yakushima (Ryukyu Islands)
- ____ *S. v. amamii* — Amami-O-Shima, Tokuno-Shima and Okinawa
- ____ *S. v. orii* — Daito Is. (Kita-Daito-jima, Minami-Daito-jima and Daito-jima)
- ____ *S. v. olivaceus* — Iriomote (s Ryukyu Islands)
- ____ *S. v. castaneoventris* — Taiwan
- ____ *S. v. namiyei* — N Izu Islands (To-shima, Nii-jima and Kozu-shima)
- ____ *S. v. owstoni* — S Izu Islands (Miyake-jima, Mikura-jima and Hachijo-jima)

☐ **Bridled Titmouse** *Baeolophus wollweberi*
- ____ *B. w. vandevenderi* — Oak-juniper forests of Arizona and New Mexico
- ____ *B. w. phillipsi* — SE Arizona (south of Gila River) to nw Mexico (Durango)
- ____ *B. w. wollweberi* — Central and s highlands of Mexico (Durango to Nuevo León)
- ____ *B. w. caliginosus* — W Mexico (Sierra Madre del Sur of Guerrero and Oaxaca)

☐ **Oak Titmouse** *Baeolophus inornatus*

____ *B. i. sequestriatus*	Interior coast ranges of sw Oregon and nw California
____ *B. i. inornatus (transpositus, kernensis, restrictus)*	W California (Mendocino to Santa Barbara e to Sierra Nevada)
____ *B. i. affabilis*	SW California (Ventura County) to n Baja California
____ *B. i. mohavensis*	Little San Bernardino Mountains of s California
____ *B. i. ridgwayi*	California (Modoc Plateau south to Providence Mountains)
____ *B. i. cineraceus*	Cape district of s Baja California

☐ **Juniper Titmouse** *Baeolophus ridgwayi*

____ *B. r. ridgwayi*	Interior w N Am. (Idaho to Nevada, se Calif., Arizona, ne Sonora)
____ *B. r. zaleptus*	S Oregon (e of Cascades) to Nevada and e Calif. (Inyo County)

☐ **Tufted Titmouse** *Baeolophus bicolor*

____ *B. b. bicolor*	S-central Canada (Ontario) to e, central and se US
____ *B. b. sennetti*	Central and s Texas (s to Brooks County, west to Terrell County)
____ *B. b. paloduro (dysleptus)*	Texas panhandle and sw Oklahoma; sw Texas and nw Coahuila

☐ **Black-crested Titmouse** *Baeolophus atricristatus*

	SW Oklahoma and Texas to se Mexico (s Veracruz)

☐ **Yellow-browed Tit** *Sylviparus modestus*

____ *S. m. simlaensis*	W Himalayas (Kashmir to Uttar Pradesh)
____ *S. m. modestus*	Nepal to ne India, n Myanmar, sw China, Thailand and n Laos
____ *S. m. klossi*	S Vietnam (Da Lat Plateau)

☐ **Sultan Tit** *Melanochlora sultanea*

____ *M. s. sultanea*	Nepal to Assam, Myanmar, n Thailand, n Laos and Vietnam
____ *M. s. flavocristata*	S Myanmar to s Thailand, Malay Pen., Sumatra and Hainan
____ *M. s. seorsa*	S China (central Fujian to s Guangxi) to Laos and ne Tonkin
____ *M. s. gayeti*	Central Annam (Col des Nuages) to s Laos (Bolavens Plateau)

☐ **Ground Tit** *Pseudopodoces humilis*

	Semiarid steppes of Tibetan plateau

FAMILY: SITTIDAE (Nuthatches—24)

☐ **Chestnut-bellied Nuthatch** *Sitta castanea*

____ *S. c. castanea*	Foothills of n and central India (Western Ghats)
____ *S. c. almorae*	Foothills of w Himalayas (Pakistan to e Nepal)
____ *S. c. cinnamoventris*	E Himalayas (e Nepal to nw Yunnan and Arunachal Pradesh)
____ *S. c. koelzi*	SE Arunachal Pradesh to Assam and adjacent nw Myanmar
____ *S. c. tonkinensis*	S Yunnan to n Thailand, n Vietnam, n and central Laos
____ *S. c. neglecta*	Myanmar to se Yunnan, Thailand, Laos, Cambodia, s Vietnam

☐ **Eurasian Nuthatch** *Sitta europaea*

____ *S. e. caesia*	Britain to Denmark, Carpathian Mts., Pyrénées and Balkan Pen.
____ *S. e. europaea*	Scandinavia and Russia to Volga and Vyatka basins and Ukraine
____ *S. e. asiatica*	E European Russia to Sea of Okhotsk, s Kuril Is. and n Japan
____ *S. e. arctica*	N-central Siberia to Anadyr River (e Russia)
____ *S. e. hispaniensis*	Iberian Peninsula
____ *S. e. atlas*	Morocco
____ *S. e. cisalpina*	S Switzerland, Italy, Sicily and coastal Yugoslavia
____ *S. e. levantina*	W Asia Minor, Levant and s Turkey (e to Euphrates River)
____ *S. e. caucasica*	N and ne Turkey, Caucasus region and Transcaucasia
____ *S. e. rubiginosa*	SE Transcaucasia (Talyshskiye and Gory mountains) to n Iran
____ *S. e. persica*	Zagros Mountains (sw Iran)
____ *S. e. seorsa*	NW China (e Tien Shan Mountains of n Xinjiang)
____ *S. e. sakhalinensis*	Sakhalin (Russia)
____ *S. e. albifrons*	Kamchatka Peninsula, Paramushir I. and n Kuril Islands
____ *S. e. amurensis*	SE Russia to ne China, Korea and Honshu (n Japan)

____ *S. e. roseillia*	S Japan (se Honshu, Shikoku and Kyushu)
____ *S. e. bedfordi*	Cheju-Do Islands (Korea)
____ *S. e. sinensis*	Central and e China and Taiwan
____ *S. e. nebulosa*	S China (lowlands of s Yunnan)

☐ **Chestnut-vented Nuthatch** *Sitta nagaensis*

____ *S. n. nagaensis (montium)*	N India to central and s China, Myanmar and Thailand
____ *S. n. grisiventris*	S Vietnam and sw Myanmar (Mt. Victoria)

☐ **Kashmir Nuthatch** *Sitta cashmirensis*

Mts. of ne Afghanistan to n Pakistan, nw India and nw Nepal

☐ **White-tailed Nuthatch** *Sitta himalayensis*

Himalayas from n India to se Tibet, sw China and Indochina

☐ **White-browed Nuthatch** *Sitta victoriae*

Alpine forests of sw Myanmar (Mt. Victoria)

☐ **Pygmy Nuthatch** *Sitta pygmaea*

____ *S. p. melanotis*	British Columbia to nw Mexico (Sonora and nw Coahuila)
____ *S. p. pygmaea*	W California (Mendocino County to San Luis Obispo County)
____ *S. p. leuconucha*	S California (San Jacinto and Laguna mts.) to n Baja California
____ *S. p. chihuahuae*	W Mexico (Sierra Madre Occidental of ne Sonora to n Jalisco)
____ *S. p. brunnescens*	SW Mexico (s Jalisco and Michoacán)
____ *S. p. flavinucha*	W Veracruz (Mt. Orizaba) to w Puebla, Morelos and México
____ *S. p. elii*	E Mexico (sw Nuevo León and se Coahuila)

☐ **Brown-headed Nuthatch** *Sitta pusilla*

____ *S. p. pusilla*	Pine forests of se US
____ *S. p. insularis*	Grand Bahama I.

☐ **Corsican Nuthatch** *Sitta whiteheadi*

Montane pine forests of Corsica

☐ **Algerian Nuthatch** *Sitta ledanti*

Montane oak-cedar forests of ne Algeria

☐ **Krueper's Nuthatch** *Sitta krueperi*

Pine, cedar and juniper forests of Turkey to Caucasus Mountains

☐ **Snowy-browed Nuthatch** *Sitta villosa*

____ *S. v. villosa*	Russia to ne China and Korea
____ *S. v. bangsi*	Central China (Qinghai to e Gansu)

☐ **Yunnan Nuthatch** *Sitta yunnanensis*

Mountains of w Sichuan to w Yunnan, se Tibet and nw Guizhou

☐ **Red-breasted Nuthatch** *Sitta canadensis*

Coniferous forests of North America

☐ **White-cheeked Nuthatch** *Sitta leucopsis*

____ *S. l. leucopsis*	Himalayas (ne Afghanistan to nw India and nw Nepal)
____ *S. l. przewalskii*	W China (Qinghai to s Sichuan, sw Gansu, ne Tibet and Yunnan)

☐ **White-breasted Nuthatch** *Sitta carolinensis*

____ *S. c. tenuissima*	British Columbia and Cascades to Sierra Nevada of n California
____ *S. c. aculeata*	W Washington to Oregon, California and n Baja (Sierra Juárez)
____ *S. c. nelsoni*	Rocky Mts. of w US to n Mexico (Sonora and n Chihuahua)
____ *S. c. carolinensis*	NE North America to Dakotas, Kansas, Oklahoma and e Texas
____ *S. c. alexandrae*	Mountains of n Baja California (San Pedro Mártir)
____ *S. c. lagunae*	Mountains of s Baja California (Sierra de la Laguna)
____ *S. c. oberholseri*	SW Texas (Chisos Mts.) to e Mexico (n Sierra Madre Oriental)
____ *S. c. mexicana*	Mountains of w Mexico (Sierra Madre Occidental)
____ *S. c. kinneari*	Mountains of w Mexico (Guerrero and Oaxaca)

☐ **Rock Nuthatch** *Sitta neumayer*

____ *S. n. neumayer*	Balkan Peninsula
____ *S. n. syriaca*	W and central Turkey to Syria, Lebanon and Israel

____	*S. n. rupicola*	Extreme e Turkey to n Iraq and n Iran
____	*S. n. tschitscherini*	Zagros Mountains (sw Iran)
____	*S. n. plumbea*	S-central Iran (mountains of s Kerman Province)

☐ **Persian Nuthatch** *Sitta tephronota*

____	*S. t. dresseri*	Extreme se Turkey to w Iran and Iraq
____	*S. t. obscura*	S Transcaucasia to n Iran and ne Turkey
____	*S. t. tephronota*	Extreme e Turkmenistan to Pamirs, Afghanistan and Pakistan
____	*S. t. iranica*	S Turkmenistan to Uzbekistan (Kyzylkum Desert)

☐ **Velvet-fronted Nuthatch** *Sitta frontalis*

____	*S. f. frontalis*	India to pen. Thailand to Sumatra, Lingga Arch. and Bangka I.
____	*S. f. saturatior*	S peninsular Thailand to Malaysia and n Sumatra; Simeulue I.
____	*S. f. corralipes*	Borneo and Maratua Islands
____	*S. f. palawana*	S Philippines (Palawan and Balabac)
____	*S. f. velata*	Java

☐ **Yellow-billed Nuthatch** *Sitta solangiae*

____	*S. s. solangiae*	N Vietnam (Fan Si Pan Mountains)
____	*S. s. fortior*	S-central Laos (Langbian Plateau)
____	*S. s. chienfengensis*	Hainan (s China)

☐ **Sulphur-billed Nuthatch** *Sitta oenochlamys*

____	*S. o. mesoleuca*	N Philippines (Cordillera Mountains of nw Luzon)
____	*S. o. isarog*	N Philippines (s and central Luzon)
____	*S. o. oenochlamys*	Philippines (Cebu, Guimaras, Panay and Negros)
____	*S. o. lilacea*	Philippines (Samar, Leyte and Biliran)
____	*S. o. apo*	S Philippines (e Mindanao)
____	*S. o. zamboanga*	S Philippines (w Mindanao, Basilan and East Blood)

☐ **Blue Nuthatch** *Sitta azurea*

____	*S. a. expectata*	Malay Peninsula and Sumatra
____	*S. a. nigriventer*	W Java
____	*S. a. azurea*	E Java

☐ **Giant Nuthatch** *Sitta magna*

____	*S. m. ligea*	SW China (extreme s Sichuan, nw Yunnan and sw Guizhou)
____	*S. m. magna*	S China (w Yunnan) to se Myanmar and nw Thailand

☐ **Beautiful Nuthatch** *Sitta formosa*

	Sikkim and Bhutan to Yunnan, Myanmar and nw Tonkin

FAMILY: TICHODROMIDAE (Wallcreeper—1)

☐ **Wallcreeper** *Tichodroma muraria*

____	*T. m. muraria*	Europe and sw Asia to n and w Iran
____	*T. m. nepalensis*	S-central Asia (Turkmenistan) and e Iran to China

FAMILY: CERTHIIDAE (Creepers—8)

☐ **Eurasian Treecreeper** *Certhia familiaris*

____	*C. f. britannica*	Great Britain and Ireland
____	*C. f. macrodactyla*	Western Europe to Oder River, Hungary and Yugoslavia
____	*C. f. corsa*	Corsica
____	*C. f. familiaris*	Scandinavia and e Europe to w Siberia
____	*C. f. daurica*	Siberia to Sea of Okhotsk, n Mongolia and ne China
____	*C. f. orientalis*	Amurland to ne China, Kuril Is., Sakhalin, Hokkaido and Korea
____	*C. f. japonica*	Japan (Honshu, Shikoku and Kyushu)

____	*C. f. persica*	Crimean Peninsula, Turkey, Caucasus, Transcaucasus and n Iran
____	*C. f. tianschanica*	Tien Shan Mts. (Kazakstan) to nw China
____	*C. f. bianchii*	W-central China (e Qinghai, Gansu, Shaanxi and Shanxi)
____	*C. f. khamensis*	S China (s Gansu) to s Tibet, Yunnan, ne Myanmar and Bhutan
____	*C. f. hodgsoni*	W Himalayas (east to Himachal Pradesh)
____	*C. f. mandellii*	Himalayas (Himachal Pradesh to extreme w Arunachal Pradesh)

☐ **Sichuan Treecreeper** *Certhia tianquanensis*

Mountains of w China (Sichuan Province)

☐ **Brown Creeper** *Certhia americana*

____	*C. a. alascensis*	S-central Alaska; winters to Arizona and New Mexico
____	*C. a. occidentalis*	Coastal se Alaska to central California
____	*C. a. stewarti*	Queen Charlotte Islands (British Columbia)
____	*C. a. montana*	British Columbia to Cascades, Dakotas, s Arizona and w Texas
____	*C. a. zelotes (phillipsi)*	S Oregon through Cascades and Coast ranges to s California
____	*C. a. americana*	N Saskatchewan to Newfoundland and ne US; winters to Mexico
____	*C. a. idahoensis*	N Idaho and nw Montana to central Alberta; winters to Arizona
____	*C. a. nigrescens*	Great Smoky Mountains of Tennessee and North Carolina
____	*C. a. albescens*	Mountains of se Arizona, sw New Mexico and nw Mexico
____	*C. a. alticola*	Mountains of s Mexico
____	*C. a. pernigra*	Mountains of extreme s Mexico (Chiapas) and n Guatemala
____	*C. a. extima*	Mountains of e Guatemala, Honduras and nw Nicaragua

☐ **Short-toed Treecreeper** *Certhia brachydactyla*

____	*C. b. megarhyncha*	W Europe (east to w Germany)
____	*C. b. brachydactyla*	Continental Europe (east of *megarhyncha*)
____	*C. b. mauritanica*	Morocco, Algeria and nw Tunisia
____	*C. b. dorotheae*	Cyprus
____	*C. b. harterti*	Asia Minor and the Caucasus Mountains

☐ **Bar-tailed Treecreeper** *Certhia himalayana*

____	*C. h. taeniura*	Central Asia and Afghanistan (north of the Hindu Kush)
____	*C. h. himalayana*	E Afghanistan, n Pakistan and Himalayas (east to central Nepal)
____	*C. h. yunnanensis*	SW China and adjacent Myanmar
____	*C. h. ripponi*	W Myanmar (Mt. Victoria)

☐ **Rusty-flanked Treecreeper** *Certhia nipalensis*

Mountains of w Nepal to ne Myanmar, se Tibet and sw China

☐ **Brown-throated Treecreeper** *Certhia discolor*

____	*C. d. discolor*	Himalayas of Nepal to s Tibet and nw India
____	*C. d. manipurensis*	NE India (south of the Brahmaputra) and adjacent sw Myanmar
____	*C. d. shanensis*	NE Myanmar to sw China (Yunnan), Thailand and nw Vietnam
____	*C. d. meridionalis*	S Vietnam (Da Lat Plateau)
____	*C. d. laotiana*	Laos (Tranninh Plateau)

☐ **Spotted Creeper** *Salpornis spilonotus*

____	*S. s. emini*	Gambia to ne Zaire and nw Uganda
____	*S. s. erlangeri*	Ethiopia
____	*S. s. salvadori*	Extreme e Uganda to w Kenya, Tanzania, Zambia and Malawi
____	*S. s. xylodromus*	Zimbabwe and adjacent Mozambique
____	*S. s. spilonotus*	Central and western India
____	*S. s. rajputanae*	W India (cent. and se Rajasthan from Sambhar to Mount Abu)

FAMILY: RHABDORNITHIDAE (Philippine Creepers—3)

☐ **Stripe-sided Rhabdornis** *Rhabdornis mysticalis*

____	*R. m. mysticalis*	Philippines (Luzon, Masbate, Negros, Catanduanes and Panay)
____	*R. m. minor*	Philippines (Basilan, Samar, Leyte, Bohol, Dinagat, Mindanao)

☐ **Long-billed Rhabdornis** *Rhabdornis grandis*

N Philippines (Sierra Madre Mountains of n Luzon)

☐ **Stripe-breasted Rhabdornis** *Rhabdornis inornatus*
_____ *R. i. inornatus*   Philippines (montane forests of Samar)
_____ *R. i. leytensis*   Philippines (Leyte and Biliran)
_____ *R. i. rabori*   Philippines (Negros)
_____ *R. i. alaris (zamboanga)*   S Philippines (Mindanao)

FAMILY: REMIZIDAE (Penduline Tits—13)

☐ **Verdin** *Auriparus flaviceps*
_____ *A. f. acaciarum*   Deserts of sw US to n Baja California and nw Mexico
_____ *A. f. ornatus*   SE Arizona to Oklahoma, Texas, s Coahuila and Tamaulipas
_____ *A. f. flaviceps (fraterculus)*   Central Baja California; ne Sonora to n Sinaloa and Tiburón I.
_____ *A. f. lamprocephalus*   S Baja California, San José and San Francisco islands
_____ *A. f. sinaloae*   NW Mexico (nw Sinaloa from Culiacán to Guamuchil)
_____ *A. f. hidalgensis*   W Mexico (ne Jalisco to s San Luis Potosí and e Hidalgo)

☐ **Eurasian Penduline-Tit** *Remiz pendulinus*
_____ *R. p. pendulinus*   Europe to Ural Mountains, Caucasus Mountains and w Turkey
_____ *R. p. menzbieri*   S and e Turkey to Armenia and nw Iran
_____ *R. p. caspius*   NW Kazakstan (Volga and Ural plains) to Caspian Sea
_____ *R. p. jaxarticus*   E Urals to w Siberia and n Kazakstan

☐ **Black-headed Penduline-Tit** *Remiz macronyx*
_____ *R. m. macronyx*   SW Kazakstan to Uzbekistan, Tajikistan and se Turkmenistan
_____ *R. m. neglectus*   SW Turkmenistan and n Iran (Atrak and Gorgan valleys)
_____ *R. m. nigricans*   SE Iran
_____ *R. m. ssaposhnikowi*   SE Kazakstan (region of lakes Balkhash, Sasykkol and Alakol)

☐ **White-crowned Penduline-Tit** *Remiz coronatus*

Deciduous woodlands of s-central Eurasia; winters to nw India

☐ **Chinese Penduline-Tit** *Remiz consobrinus*

Reedbeds and marshes of n China; winters to s China

☐ **Sennar Penduline-Tit** *Anthoscopus punctifrons*

S borders of Sahara; Mauritania to Eritrea

☐ **Mouse-colored Penduline-Tit** *Anthoscopus musculus*

Thornscrub of Ethiopia and Somalia to extreme n Tanzania

☐ **Yellow Penduline-Tit** *Anthoscopus parvulus*
_____ *A. p. senegalensis*   Senegal to Central African Republic
_____ *A. p. aureus*   N Ghana
_____ *A. p. parvulus*   Chad to s Sudan and ne Zaire

☐ **Forest Penduline-Tit** *Anthoscopus flavifrons*
_____ *A. f. waldroni*   Humid forests of Ghana to Ivory Coast and Liberia
_____ *A. f. flavifrons*   E Nigeria to Cameroon, Gabon and n Zaire
_____ *A. f. ruthae*   SE Zaire (se Kivu Province)

☐ **African Penduline-Tit** *Anthoscopus caroli*
_____ *A. c. ansorgei*   SE Gabon to Angola and sw Zaire
_____ *A. c. caroli*   N Namibia, s Angola, Botswana and sw Zambia
_____ *A. c. roccatii*   Uganda, ne Zaire, Rwanda, Burundi, ne Tanzania and w Kenya
_____ *A. c. rhodesiae*   SE Zaire to ne Zambia and sw Tanzania (Ufipa Plateau)
_____ *A. c. sylviella (rothschildi)*   S-central Kenya (e of Rift Valley) to central Tanzania
_____ *A. c. sharpei*   SW Kenya and n Tanzania
_____ *A. c. robertsi (taruensis)*   Mozambique, Malawi, se Zambia, e Tanzania and se Kenya
_____ *A. c. pallescens*   W Tanzania (Kigoma, Kungwe-Mahari)
_____ *A. c. winterbottomi*   NW Zambia and adjacent s Zaire (s Katanga)

____ *A. c. hellmayri*   E and s Zimbabwe, s Mozambique, S Africa and e Swaziland

____ *A. c. rankinei*   Zambesi River between ne Zimbabwe and nw Mozambique

☐ **Southern Penduline-Tit** *Anthoscopus minutus*

____ *A. m. damarensis*   Angola to n Namibia, Botswana, Zimbabwe and w Transvaal

____ *A. m. minutus*   W Namibia to w Cape Prov., Botswana and w Orange Free State

____ *A. m. gigi*   SE Cape Province (Little Karoo and s Great Karoo)

☐ **Fire-capped Tit** *Cephalopyrus flammiceps*

____ *C. f. flammiceps*   W Himalayas (Pakistan to Garhwal and w Nepal)

____ *C. f. olivaceus*   E Himalayas (e Nepal to se Tibet and central China)

☐ **Tit-hylia** *Pholidornis rushiae*

____ *P. r. ussheri*   Sierra Leone to Togo (and sw Nigeria?)

____ *P. r. rushiae*   SE Nigeria, sw Cameroon and Gabon

____ *P. r. bedfordi*   Bioko (Gulf of Guinea)v

____ *P. r. denti*   SE Cameroon to Uganda and Angola

FAMILY: NECTARINIIDAE (Sunbirds and Spiderhunters—131)

☐ **Ruby-cheeked Sunbird** *Chalcoparia singalensis*

____ *C. s. assamensis (rubigentis)*   E Nepal to Bangladesh, Assam, sw China, n Burma, n Thailand

____ *C. s. internota*   S Burma to s Thailand (Isthmus of Kra)

____ *C. s. koratensis (stellae)*   Plateau of e Thailand to Laos and Vietnam

____ *C. s. interposita*   S peninsular Thailand (south of Isthmus of Kra)

____ *C. s. singalensis*   Malay Peninsula from Perak southwards

____ *C. s. panopsia*   Islands off w Sumatra (Banyak, Nias and Tanahmasa)

____ *C. s. sumatrana*   Sumatra and Belitung I.

____ *C. s. pallida*   North Natuna Is.

____ *C. s. borneana*   Borneo and Banggi I.

____ *C. s. bantenensis*   Extreme w Java (Banten region)

____ *C. s. phoenicotis*   W-central and e Java

☐ **Scarlet-tufted Sunbird** *Deleornis fraseri*

____ *D. f. idius*   Guinea and Sierra Leone to Ghana and Togo

____ *D. f. cameroonensis*   S Nigeria to Central African Rep., nw Angola, Gabon and Congo

____ *D. f. fraseri*   Bioko (Gulf of Guinea)

☐ **Gray-headed Sunbird** *Deleornis axillaris*

     NE Zaire to w Uganda; single record from nw Tanzania

☐ **Plain-backed Sunbird** *Anthreptes reichenowi*

____ *A. r. yokanae*   Lowland coastal forests of se Kenya and ne Tanzania

____ *A. r. reichenowi*   Locally in e Zimbabwe, s Mozambique and ne South Africa

☐ **Anchieta's Sunbird** *Anthreptes anchietae*

     Angola to se Zaire, Zambia, sw Tanzania, Malawi, n Mozambique

☐ **Plain Sunbird** *Anthreptes simplex*

     S Burma to Malaya, Sumatra, Borneo, Nias I. and N Natuna Is.

☐ **Plain-throated Sunbird** *Anthreptes malacensis*

____ *A. m. malacensis*   S Burma to pen. Thailand, Indochina, Sumatra and s Borneo

____ *A. m. mjobergi*   Maratua Is. (off Borneo)

____ *A. m. borneensis*   N Borneo (Sabah, Brunei and adj. islands)

____ *A. m. birgitae*   N Philippines (Luzon, Mindoro and Catanduanes)

____ *A. m. chlorigaster*   Cebu, Masbate, Negros, Panay, Sibuyan, Tablas, Romblon, Ticao

____ *A. m. griseigularis*   Philippines (Samar, Leyte, Sakuyok, Camiguin Sur, ne Mindanao)

____ *A. m. heliolusius*   S Philippines (Basilan, w and central Mindanao and Talicod)

____ *A. m. cagayensis*   Cagayan Sulu I. (Sulu Sea)

____ *A. m. paraguae*   S Philippines (Balabac, Culion, Palawan and Calauit)

____ *A. m. wiglesworthi*   Sulu Archipelago (Bongao, Jolo, Tawitawi and Basbas)

_____ *A. m. iris*	S Philippines (Sibutu and Sitanki)
_____ *A. m. heliocalus (sanghirara)*	Sangihe and Siau islands (off Sulawesi)
_____ *A. m. celebensis (citrinus, nesophilus)*	Sulawesi and adjacent islands
_____ *A. m. extremus*	Banggai and Sula Islands
_____ *A. m. convergens*	W Lesser Sundas (Lombok to Pantar and Alor)
_____ *A. m. rubrigena*	Sumba I. (Lesser Sundas)
_____ *A. m. anambae*	Anambas Is. (South China Sea)

☐ **Red-throated Sunbird** *Anthreptes rhodolaemus*

Lowlands of s Burma, s pen. Thailand, Malaya, Sumatra, Borneo

☐ **Mouse-brown Sunbird** *Anthreptes gabonicus*

S Senegal to Cameroon, Gabon, n Angola and sw Zaire

☐ **Western Violet-backed Sunbird** *Anthreptes longuemarei*

_____ *A. l. longuemarei*	Senegal to Guinea-Bissau
_____ *A. l. haussarum*	Liberia to n Cameroon, s Sudan, n Uganda; vagrant to w Kenya
_____ *A. l. angolensis*	Gabon, s Zaire and Angola to w Tanzania and w Malawi
_____ *A. l. nyassae*	SE Tanzania to Malawi, e Mozambique and e Zimbabwe

☐ **Kenya Violet-backed Sunbird** *Anthreptes orientalis*

Somalia to se Sudan, Ethiopia, n Uganda, Kenya and ne Tanzania

☐ **Uluguru Violet-backed Sunbird** *Anthreptes neglectus*

Lowland coastal forests of se Kenya, Tanzania and n Mozambique

☐ **Violet-tailed Sunbird** *Anthreptes aurantium*

Cameroon to Gabon, ne Angola, Cent. African Rep. and ne Zaire

☐ **Little Green Sunbird** *Anthreptes seimundi*

_____ *A. s. kruensis*	Guinea and Sierra Leone to Ghana and Togo
_____ *A. s. seimundi*	Bioko (Gulf of Guinea)
_____ *A. s. minor*	Nigeria to Cameroon, n Angola, Zaire, s Sudan, Rwanda, Uganda

☐ **Green Sunbird** *Anthreptes rectirostris*

_____ *A. r. rectirostris*	Sierra Leone to Ghana
_____ *A. r. tephrolaemus*	S Nigeria to Angola, s Sudan, Uganda and w Kenya; Bioko

☐ **Banded Sunbird** *Anthreptes rubritorques*

NE Tanzania (Usambara, Nguru and Uluguru Mts.)

☐ **Collared Sunbird** *Hedydipna collaris*

_____ *H. c. subcollaris (nigeriae)*	Senegal to s Nigeria (Niger River delta)
_____ *H. c. hypodila*	Bioko I. (Gulf of Guinea)
_____ *H. c. somereni*	Extreme se Nigeria to n Zaire, nw Angola and sw Sudan
_____ *H. c. djamdjamensis*	SW Ethiopia (Alghe and Sagan river area)
_____ *H. c. garguensis*	E Angola to se Sudan, Uganda, Kenya, nw Tanzania and Zambia
_____ *H. c. zambesiana (chobiensis)*	Angola to Tanzania, Namibia, Zimbabwe and ne South Africa
_____ *H. c. zuluensis (patersonae, beverlyae)*	NE Natal to s Mozambique and Zimbabwe
_____ *H. c. elachior*	E Kenya to ne Tanzania, Manda I., Zanzibar and Mafia I.
_____ *H. c. collaris*	S Natal to w Swaziland, s Zululand and e Cape Province

☐ **Pygmy Sunbird** *Hedydipna platura*

SW Mauritania to n Nigeria, ne Zaire, s Sudan and n Uganda

☐ **Nile Valley Sunbird** *Hedydipna metallica*

N Egypt to Sudan, n Ethiopia, Somalia, sw Arabia, Yemen, Oman

☐ **Amani Sunbird** *Hedydipna pallidigaster*

Coastal se Kenya (Sokoke Forest) and ne Tanzania (rare)

☐ **Purple-naped Sunbird** *Hypogramma hypogrammicum*

_____ *H. h. lisettae*	S China (w Yunnan) to n Burma, n Thailand and c Indochina
_____ *H. h. mariae*	Cambodia and s Indochina
_____ *H. h. nuchale*	S Burma to s Thailand and Malay Peninsula
_____ *H. h. hypogrammicum*	Sumatra and Borneo
_____ *H. h. natunense*	North Natuna Islands (South China Sea)

☐ **Reichenbach's Sunbird** *Anabathmis reichenbachii*

Liberia to Ivory Coast, Nigeria, Cameroon, n Zaire and ne Angola

535

☐ **Principe Sunbird** *Anabathmis hartlaubii*

Príncipe I. (Gulf of Guinea)

☐ **Newton's Sunbird** *Anabathmis newtonii*

São Tomé I. (Gulf of Guinea)

☐ **São Tomé Sunbird** *Dreptes thomensis*

São Tomé I. (Gulf of Guinea)

☐ **Orange-breasted Sunbird** *Anthobaphes violacea*

Heathlands and *protea* shrubs of sw Cape Province (South Africa)

☐ **Green-headed Sunbird** *Cyanomitra verticalis*
____ *C. v. verticalis* — Senegal to Cameroon
____ *C. v. cyanocephala* — Coastal Equatorial Guinea and Gabon to mouth of Congo River
____ *C. v. boehndorffi* — S Cameroon to Cent. African Rep., n Angola and s Zaire
____ *C. v. viridisplendens* — E Zaire to s Sudan, w Kenya, Tanzania, Malawi and ne Zambia

☐ **Bannerman's Sunbird** *Cyanomitra bannermani*

N Angola to se Zaire and extreme nw Zambia

☐ **Blue-throated Brown Sunbird** *Cyanomitra cyanolaema*
____ *C. c. magnirostrata* — Sierra Leone and Liberia to Ivory Coast, Ghana and Togo
____ *C. c. octaviae* — Nigeria and Cameroon to Angola, Uganda and w Kenya
____ *C. c. cyanolaema* — Bioko I. (Gulf of Guinea)

☐ **Cameroon Sunbird** *Cyanomitra oritis*
____ *C. o. poensis* — Montane forests of Bioko I. (Gulf of Guinea)
____ *C. o. bansoensis* — Highlands of w Cameroon
____ *C. o. oritis* — Montane forests of Mt. Cameroon

☐ **Blue-headed Sunbird** *Cyanomitra alinae*
____ *C. a. marungensis* — E Zaire (Marungu Mts.)
____ *C. a. alinae* — Mts. of sw Uganda and nw Rwanda
____ *C. a. tanganjicae (vulcanorum)* — Montane forests of se Zaire, sw Rwanda and w Burundi
____ *C. a. derooi* — Highlands ne Zaire (west of Lake Albert and Lake Edward)
____ *C. a. kaboboensis* — E Zaire (montane forests of Mt. Kabobo)

☐ **Eastern Olive Sunbird** *Cyanomitra olivacea*
____ *C. o. neglecta* — Highlands of ne Kenya to ne Tanzania
____ *C. o. changamwensis (puguensis)* — S Somalia and coastal Kenya to ne Tanzania; Mafia I.
____ *C. o. alfredi* — S Tanzania to Malawi, se Zambia and n Mozambique
____ *C. o. olivacina* — Coastal s Tanzania to s Mozambique and ne Kawzulu-Natal
____ *C. o. olivacea* — South Africa (Pondoland to Natal and s Zululand)

☐ **Western Olive Sunbird** *Cyanomitra obscura*
____ *C. o. guineensis* — Guinea-Bissau to Togo
____ *C. o. cephaelis* — Ghana to Gabon, Zaire, n Angola and Congo Basin
____ *C. o. obscura* — Principe I. and Bioko I. (Gulf of Guinea)
____ *C. o. ragazzii (vicente, lowei)* — S Sudan to Uganda, w Kenya, w Tanzania, e Zaire and n Zambia
____ *C. o. granti* — Zanzibar and Pemba I.
____ *C. o. sclateri* — Mts. of e Zimbabwe and immediately adjacent Mozambique

☐ **Mouse-colored Sunbird** *Cyanomitra veroxii*
____ *C. v. fischeri* — S Somalia to e Kenya, Tanzania, Mozambique and n Natal
____ *C. v. zanzibarica* — Zanzibar
____ *C. v. veroxii* — E Natal to e Cape Province

☐ **Buff-throated Sunbird** *Chalcomitra adelberti*
____ *C. a. adelberti* — Sierra Leone to Ghana
____ *C. a. eboensis* — Togo to se Nigeria

☐ **Carmelite Sunbird** *Chalcomitra fuliginosa*
____ *C. f. aurea* — Sierra Leone to Liberia, Nigeria, Cameroon and coastal Gabon
____ *C. f. fuliginosa (nigrescens)* — Lower Congo River (s Zaire) to w Angola

☐ **Green-throated Sunbird** *Chalcomitra rubescens*

____ *C. r. stangerii* — Bioko I. (Gulf of Guinea)
____ *C. r. crossensis* — Nigeria and Cameroon
____ *C. r. rubescens* — E Cameroon to n Angola, se Sudan, Kenya, Tanzania and Zambia

☐ **Amethyst Sunbird** *Chalcomitra amethystina*

____ *C. a. kirkii (doggetti, kalkcreuthi)* — S Somalia to se Zaire, Kenya, Tanzania, e Zambia and Zimbabwe
____ *C. a. deminuta* — Zaire to Angola, Zambia, n Namibia and n Botswana
____ *C. a. amethystina (adjuncta)* — S Mozambique and Botswana to Cape Province

☐ **Scarlet-chested Sunbird** *Chalcomitra senegalensis*

____ *C. s. senegalensis* — Senegal to n Ghana and n Nigeria
____ *C. s. acik (adamauae)* — Cameroon to ne Zaire, Cent. African Rep., sw Sudan, nw Uganda
____ *C. s. cruentata* — SE Sudan (Boma Hills), Eritrea, Ethiopia and n Kenya
____ *C. s. lamperti (aequatorialis, erythrinae)* — E Zaire to s Sudan, Uganda, Kenya and Tanzania
____ *C. s. gutturalis (saturatior, inaestimata)* — SE Kenya to Angola, w Zambia, w Zaire, n Botswana, e S Africa

☐ **Hunter's Sunbird** *Chalcomitra hunteri*

____ *C. h. hunteri* — S Ethiopia to Kenya, ne Tanzania and Uganda
____ *C. h. siccata* — S Ethiopia to ne Kenya and plateau region of Somalia

☐ **Socotra Sunbird** *Chalcomitra balfouri*

Socotra I. and Abd el Kuri (off ne Somalia)

☐ **Purple-rumped Sunbird** *Leptocoma zeylonica*

____ *L. z. flaviventris (sola)* — Peninsular India, Bangladesh, and w Burma
____ *L. z. zeylonica* — Sri Lanka

☐ **Crimson-backed Sunbird** *Leptocoma minima*

India (Western Ghats to hills of s Kerala)

☐ **Purple-throated Sunbird** *Leptocoma sperata*

____ *L. s. brasiliana (hasseltii, phayrei)* — S Thailand to Burma, Malay Pen., Sumatra, Borneo and Java
____ *L. s. emmae* — Cambodia to s Laos and s Vietnam
____ *L. s. mecynorhyncha* — Simeulue, Banyak, Nias and Mentawi (off Sumatra)
____ *L. s. eumecis* — Anambas Islands (South China Sea)
____ *L. s. axantha* — North Natuna Islands (South China Sea)
____ *L. s. henkei* — Philippines (Luzon, Babuyan Claro, Calayan, Camiguin Norte)
____ *L. s. sperata (marinduquensis, trochilus, theresae)* — N Philippines (Luzon, Polillo, Catanduanes and Marinduque)
____ *L. s. juliae (davaoensis, theresae)* — S Philippines (w Mindanao, Basilan and Sulu Archipelago)
____ *L. s. oenopa* — Sumatra and Nias I.

☐ **Black Sunbird** *Leptocoma sericea*

____ *L. s. talautensis* — Talaud Is. (Karakelong, Salebabu, Kabruang and Sara)
____ *L. s. sangirensis* — Sangihe, Siau and Ruang islands
____ *L. s. grayi* — N Sulawesi, Bangka, Lembeh, Manadotua and Bangka islands
____ *L. s. porphyrolaema* — E and s Sulawesi, Muna, Labuan, Blanda, Butung and Togian is.
____ *L. s. auriceps* — Banggai Is., Sula Is. and Moluccas
____ *L. s. auricapilla* — Kayoa Is. (n Moluccas)
____ *L. s. proserpina* — Buru I. (s Moluccas)
____ *L. s. aspasioides (chlorocephala)* — S Moluccas (Seram, Ambon, Nusa Laut and Watubela); Aru Is.
____ *L. s. chlorolaema* — Kai Is.
____ *L. s. sericea* — Mainland New Guinea (except for southeast)
____ *L. s. vicina* — SE New Guinea
____ *L. s. mariae* — Kofiau I. (w New Guinea)
____ *L. s. cochrani* — Misol and Waigeo islands (w New Guinea)
____ *L. s. maforensis* — Numfor I. (Geelvink Bay off n New Guinea)
____ *L. s. salvadorii* — Yapen I. (Geelvink Bay off n New Guinea)
____ *L. s. mysorensis* — Biak I. (n New Guinea)
____ *L. s. nigriscapularis* — Meos Num and Rani islands (south of Biak I.)
____ *L. s. veronica* — Liki I. (off n coast of w New Guinea)
____ *L. s. cornelia* — Tarawai I. (n New Guinea off mouth of Sepik River)

_____ *L. s. christianae*   D'Entrecasteaux Archipelago and Louisiade Archipelago
_____ *L. s. caeruleogula*   Bismarck Archipelago (New Britain, Rook and Umboi)
_____ *L. s. corinna*   Bismarck Archipelago (except Feni I.)
_____ *L. s. eichhorni*   Feni I. (Bismarck Archipelago)

☐ **Copper-throated Sunbird** *Leptocoma calcostetha*

Mangroves and scrub of SE Asia, Palawan and Greater Sundas

☐ **Bocage's Sunbird** *Nectarinia bocagei*

Highlands of central Angola and sw Zaire

☐ **Purple-breasted Sunbird** *Nectarinia purpureiventris*

Montane forests of e Zaire, w Uganda, Rwanda and Burundi

☐ **Tacazze Sunbird** *Nectarinia tacazze*
_____ *N. t. tacazze*   Highlands of Eritrea and Ethiopia
_____ *N. t. jacksoni*   Mts. of se Sudan to e Uganda, w Kenya and ne Tanzania

☐ **Bronze Sunbird** *Nectarinia kilimensis*
_____ *N. k. gadowi*   Highlands of central Angola
_____ *N. k. kilimensis*   Highlands of e Zaire to Uganda, w Kenya and n Tanzania
_____ *N. k. arturi*   Highlands of s Tanzania to Malawi, ne Zambia and e Zimbabwe

☐ **Malachite Sunbird** *Nectarinia famosa*
_____ *N. f. cupreonitens (aeneigularis, centralis, subfamosa)*   E Zaire to se Sudan, Ethiopia, Uganda, Kenya, Tanzania, Malawi
_____ *N. f. famosa (major)*   Mts. of e Zimbabwe, Lesotho, w Swaziland and South Africa

☐ **Red-tufted Sunbird** *Nectarinia johnstoni*
_____ *N. j. johnstoni (idius)*   Highlands of Kenya and n Tanzania
_____ *N. j. dartmouthi*   Montane forests of e Zaire, sw Uganda and nw Rwanda
_____ *N. j. itombwensis*   E Zaire (Itombwe Mts.)
_____ *N. j. salvadorii*   Montane forests of s Tanzania to Zambia and n Malawi

☐ **Golden-winged Sunbird** *Drepanorhynchus reichenowi*
_____ *D. r. shelleyae*   Montane forests of e Zaire
_____ *D. r. lathburyi*   Montane forests of n Kenya (Mt. Kulal to Mt. Uraguess)
_____ *D. r. reichenowi*   Highlands of Kenya to se Uganda and ne Tanzania

☐ **Olive-bellied Sunbird** *Cinnyris chloropygius*
_____ *C. c. kempi*   Senegal to sw Nigeria
_____ *C. c. chloropygius (luehderi, insularis)*   SE Nigeria to Cameroon, nw Angola and cent. Zaire; Bioko I.
_____ *C. c. bineschensis*   Known only from type location in sw Ethiopia
_____ *C. c. orphogaster (uellensis)*   NE Angola to e Zaire, Ethiopia, Uganda, Kenya and w Tanzania

☐ **Tiny Sunbird** *Cinnyris minullus*

Sierra Leone to Cameroon, Gabon, e Zaire and w Uganda; Bioko

☐ **Miombo Sunbird** *Cinnyris manoensis*
_____ *C. m. pintoi*   Central Angola to se Zaire and w Zambia
_____ *C. m. manoensis*   S Tanzania to se Zambia, Zimbabwe and n Mozambique
_____ *C. m. amicorum*   Mozambique (Mt. Gorongoza)

☐ **Southern Double-collared Sunbird** *Cinnyris chalybeus*
_____ *C. c. subalaris (capricornensis)*   N Transvaal to Swaziland, Natal and e Cape Province
_____ *C. c. albilateralis*   South Africa (w Cape Province)
_____ *C. c. chalybeus*   Extreme s Namibia, Swaziland, Lesotho and South Africa

☐ **Neergaard's Sunbird** *Cinnyris neergaardi*

Coastal scrub of se Mozambique to extreme n Natal

☐ **Stuhlmann's Sunbird** *Cinnyris stuhlmanni*
_____ *C. s. stuhlmanni*   Ruwenzori Mts. (e Zaire and w Uganda)
_____ *C. s. schubotzi*   E Zaire (mts. w of Lake Kivu), adj. Rwanda and Burundi
_____ *C. s. chapini*   E Zaire and s Burundi (mts. west of Lake Edward to Mt. Kabobo)
_____ *C. s. graueri*   Kivu Volcanos, Rwanda and sw Uganda (Mt. Muhavura)

☐ **Prigogine's Sunbird** *Cinnyris prigoginei*

Montane forests of se Zaire and w Uganda

☐ **Montane Double-collared Sunbird** *Cinnyris ludovicensis*
_____ *C. l. ludovicensis*
_____ *C. l. whytei*

Montane forests of w Angola
Montane forests of ne Zambia and w Malawi (Nyika Plateau)

☐ **Northern Double-collared Sunbird** *Cinnyris reichenowi*
_____ *C. r. preussi (parvirostris, genderuensis)*
_____ *C. r. reichenowi (kikuyensis)*

SE Nigeria to Cameroon and w Cent. African Rep.; Bioko I.
Highlands of e Zaire to Uganda, se Sudan and w Kenya

☐ **Greater Double-collared Sunbird** *Cinnyris afer*
_____ *C. a. saliens*
_____ *C. a. afer*

E Cape and Natal to Transvaal and w Swaziland
South Africa (sw and s Cape Province)

☐ **Regal Sunbird** *Cinnyris regius*
_____ *C. r. regius (kivuensis)*
_____ *C. r. anderseni*

Albertine Rift Mts. (e Zaire, sw Uganda to Burundi; Mahare Mts.)
W Tanzania (Kungwe-Mahale Highlands)

☐ **Rockefeller's Sunbird** *Cinnyris rockefelleri*

Alpine moorlands of e Zaire (Albertine Rift Mts.)

☐ **Eastern Double-collared Sunbird** *Cinnyris mediocris*
_____ *C. m. mediocris*
_____ *C. m. usambaricus*
_____ *C. m. fuelleborni*
_____ *C. m. bensoni*

Highlands of Kenya and n Tanzania
Highlands of se Kenya (Taita Hills) and ne Tanzania
Highlands of Tanzania to n Malawi and ne Zambia
Highlands of Malawi and n Mozambique

☐ **Moreau's Sunbird** *Cinnyris moreaui*

Mts. of e Tanzania (Nguru, Ukaguru, Uvidunda and Udzungwa)

☐ **Loveridge's Sunbird** *Cinnyris loveridgei*

Mts. of e Tanzania (Uluguru Mts.)

☐ **Beautiful Sunbird** *Cinnyris pulchellus*
_____ *C. p. pulchellus (aegra, lucidipectus)*
_____ *C. p. melanogastrus*

Senegal to Mali, s Niger (Aïr Massif), Uganda and e Sudan
Eritrea to Ethiopia, ne Zaire, Uganda, Kenya and sw Tanzania

☐ **Mariqua Sunbird** *Cinnyris mariquensis*
_____ *C. m. osiris (hawkeri)*
_____ *C. m. suahelicus*
_____ *C. m. mariquensis (lucens, ovambensis)*

Extreme se Sudan to Eritrea, Ethiopia, n Kenya and n Uganda
E Zaire to central Uganda, Rwanda, Tanzania and ne Zambia
S Angola to n Namibia, sw Zambia, Zimbabwe and n S Africa

☐ **Shelley's Sunbird** *Cinnyris shelleyi*
_____ *C. s. shelleyi*
_____ *C. s. hofmanni*

SE Zaire to se Tanzania, e Zambia, Malawi and n Mozambique
E Tanzania (Ruvu and Pangani rivers to Morogoro region)

☐ **Congo Sunbird** *Cinnyris congensis*

Zaire and Congo (banks of upper Congo and Ubangi rivers)

☐ **Red-chested Sunbird** *Cinnyris erythrocercus*

S Sudan to e Zaire, Uganda, w Kenya and nw Tanzania

☐ **Black-bellied Sunbird** *Cinnyris nectarinioides*
_____ *C. n. erlangeri*
_____ *C. n. nectarinioides*

Savanna of se Ethiopia to s Somalia and extreme ne Kenya
Lowlands of interior n Kenya to extreme ne Tanzania

☐ **Purple-banded Sunbird** *Cinnyris bifasciatus*
_____ *C. b. bifasciatus*
_____ *C. b. microrhynchus (strophium)*

Gabon to lower Congo River and w Angola
E Zaire to Uganda, Kenya, Tanzania, Zambia and e South Africa

☐ **Tsavo Sunbird** *Cinnyris tsavoensis*

S Somalia to s Sudan, Ethiopia, e Kenya and ne Tanzania

☐ **Violet-breasted Sunbird** *Cinnyris chalcomelas*

Savanna of s Somalia and e Kenya

☐ **Pemba Sunbird** *Cinnyris pembae*

Pemba I. (off se Tanzania)

☐ **Orange-tufted Sunbird** *Cinnyris bouvieri*

Extreme se Nigeria to Cameroon, Gabon, n Angola and w Kenya

☐ **Palestine Sunbird** *Cinnyris osea*
_____ *C. o. decorsei (butleri)*
_____ *C. o. osea*

Lake Chad to s Sudan, extreme ne Zaire and nw Uganda
Syria, Israel to Turkey and Arabia (east to Oman)

☐ **Shining Sunbird** *Cinnyris habessinicus*
_____ *C. h. habessinicus*
_____ *C. h. alter*
_____ *C. h. turkanae*
_____ *C. h. kinneari*
_____ *C. h. hellmayri*

Red Sea Province of Sudan to Eritrea and w Ethiopia
N Somalia to Ethiopia
SE Sudan to s Ethiopia, sw Somalia, n Kenya and ne Uganda
W Saudi Arabia (Asir to Hejaz)
Saudi Arabia and Yemen

☐ **Splendid Sunbird** *Cinnyris coccinigastrus*

Senegal to sw Mali, Gabon, ne Zaire, sw Sudan and Uganda

☐ **Johanna's Sunbird** *Cinnyris johannae*
_____ *C. j. fasciatus*
_____ *C. j. johannae*

Sierra Leone to Benin and s Nigeria
S Nigeria and s Cameroon to e Zaire

☐ **Superb Sunbird** *Cinnyris superbus*
_____ *C. s. ashantiensis*
_____ *C. s. nigeriae*
_____ *C. s. superbus*

Sierra Leone to Togo and (?) Benin
S Nigeria
S Cameroon to n Angola, Zaire, extreme w Kenya and w Tanzania

☐ **Rufous-winged Sunbird** *Cinnyris rufipennis*

E Tanzania (eastern escarpment of Udzungwa Mts.)

☐ **Oustalet's Sunbird** *Cinnyris oustaleti*
_____ *C. o. oustaleti*
_____ *C. o. rhodesiae*

Central and s Angola (Huila to Cuanza Sul and n Bie)
NE Zambia to Malawi and extreme sw Tanzania

☐ **White-breasted Sunbird** *Cinnyris talatala*

SE Zaire to Angola, Tanzania, Zambia, n Natal and s Transvaal

☐ **Variable Sunbird** *Cinnyris venustus*
_____ *C. v. venustus*
_____ *C. v. falkensteini (niassae, kuanzae)*
_____ *C. v. igneiventris*
_____ *C. v. fazoglensis (sukensis)*
_____ *C. v. albiventris*

Senegal to n Cameroon and Ubangi-Shari
Gabon to n Angola, e Zaire, Kenya, Tanzania, Zambia, Zimbabwe
Highlands of extreme e Zaire, Rwanda and w Uganda
E-central Sudan to Eritrea, Ethiopia and nw Kenya
S Ethiopia to Somalia and n Kenya

☐ **Dusky Sunbird** *Cinnyris fuscus*
_____ *C. f. fuscus*
_____ *C. f. inclusus*

Namibia and Botswana to South Africa
Coastal sw Angola

☐ **Ursula's Sunbird** *Cinnyris ursulae*

Montane forests of Mt. Cameroon and Bioko I. (Gulf of Guinea)

☐ **Bates' Sunbird** *Cinnyris batesi*

Liberia to Nigeria, Gabon, s Zaire and nw Zambia; Bioko

☐ **Copper Sunbird** *Cinnyris cupreus*
_____ *C. c. cupreus*
_____ *C. c. chalceus*

Senegal to e Zaire, Uganda, Ethiopia, w Kenya and w Tanzania
Angola to se Zaire, w Zambia, Malawi and Zimbabwe

☐ **Purple Sunbird** *Cinnyris asiaticus*
_____ *C. a. brevirostris*
_____ *C. a. asiaticus*
_____ *C. a. intermedius*

NE Arabia and se Iran to Afghanistan, Pakistan and n India
India and Sri Lanka
Bangladesh to Assam, Burma, Thailand and Indochina

☐ **Olive-backed Sunbird** *Cinnyris jugularis*
_____ *C. j. andamanicus*
_____ *C. j. klossi (blanfordi)*
_____ *C. j. proselius*

Andaman Is. (Bay of Bengal)
N Nicobar Is. (Bay of Bengal)
Car Nicobar I. (Bay of Bengal)

____	*C. j. flammaxillaris*	Burma to Thailand, Cambodia and n Malay Pen. (s to Penang)
____	*C. j. ornatus*	S Malay Pen. to Sumatra, Borneo, Java, Lesser Sundas and adj. is.
____	*C. j. rhizophorae*	S China (s Yunnan, Guangxi, Guangdong, Hainan) and n Vietnam
____	*C. j. polyclystus*	Enggano I. (off w Sumatra)
____	*C. j. obscurior*	N Philippines (montane forests of n Luzon)
____	*C. j. jugularis*	S Luzon, central and s Philippine Is.
____	*C. j. aurora*	Agutaya, Balabac, Busuanga, Cagayancillo, Culion, Cuyo, Palawan
____	*C. j. woodi*	Sulu Archipelago (s Philippines)
____	*C. j. plateni*	Sulawesi, Talaud, Salayar and adjacent smaller islands
____	*C. j. infrenatus*	Tukangbesi Is. (off Sulawesi)
____	*C. j. robustirostris*	Banggai Is. and Sula Is.
____	*C. j. frenatus*	N Moluccas, Aru and w Papuan is., New Guinea and ne Queensland
____	*C. j. teysmanni*	Tanahjampea, Kalao, Bonerate, Kalaotoa and Madu islands
____	*C. j. buruensis*	Buru I. (s Moluccas)
____	*C. j. clementiae*	S Moluccas (Seram, Ambon and adjacent islands)
____	*C. j. keiensis*	Kai Is. (Kai Kecil, Kai Besar, Ohimas and Add)
____	*C. j. idenburgi*	Northwest New Guinea and Sepik Ramu
____	*C. j. flavigastra*	Bismarck Archipelago and Solomon Is.

☐ **Apricot-breasted Sunbird** *Cinnyris buettikoferi*

Lowlands of Sumba I. (w Lesser Sundas)

☐ **Flame-breasted Sunbird** *Cinnyris solaris*

____	*C. s. solaris*	Sumbawa, Flores, Besar, Lomblen, Alor, Semau, Roti and Timor
____	*C. s. exquisitus*	Wetar I. (e Lesser Sundas)

☐ **Souimanga Sunbird** *Cinnyris souimanga*

____	*C. s. aldabrensis*	Aldabra I.
____	*C. s. abbotti*	Assumption I. (Aldabra Archipelago)
____	*C. s. buchenorum*	Cosmoledo Atoll (Aldabra Archipelago)
____	*C. s. souimanga*	Îles Glorieuses and n Madagascar
____	*C. s. apolis*	Subdesert of sw Madagascar

☐ **Madagascar Sunbird** *Cinnyris notatus*

____	*C. n. moebii*	Grand Comoro I. (Comoro Is.)
____	*C. n. voeltzkowi*	Mohéli I. (Comoro Is.)
____	*C. n. notatus*	Madagascar

☐ **Seychelles Sunbird** *Cinnyris dussumieri*

Seychelles

☐ **Humblot's Sunbird** *Cinnyris humbloti*

____	*C. h. humbloti*	Grand Comoro I. (Comoro Is.)
____	*C. h. mohelicus*	Mohéli I. (Comoro Is.)

☐ **Anjouan Sunbird** *Cinnyris comorensis*

Anjouan I. (Comoro Is.)

☐ **Mayotte Sunbird** *Cinnyris coquerellii*

Mayotte I. (Comoro Is.)

☐ **Long-billed Sunbird** *Cinnyris lotenius*

____	*C. l. hindustanicus*	S peninsular India (north to Bombay and Andhra Pradesh)
____	*C. l. lotenius*	Sri Lanka

☐ **Gray-hooded Sunbird** *Aethopyga primigenia*

____	*A. p. primigenia*	S Philippines (montane forests of central and e Mindanao)
____	*A. p. diuatae*	S Philippines (Mt. Hilong Hilong on Mindanao)

☐ **Mount Apo Sunbird** *Aethopyga boltoni*

____	*A. b. boltoni*	S Philippines (montane forests of e Mindanao)
____	*A. b. malindangensis*	S Philippines (montane forests of Mt. Malindang on w Mindanao)
____	*A. b. tibolii*	S Philippines (mts. of s Mindanao)

☐ **Lina's Sunbird** *Aethopyga linaraborae*

S Philippines (montane forests of e Mindanao)

☐ **Flaming Sunbird** *Aethopyga flagrans*
____ *A. f. decolor*
____ *A. f. flagrans*
____ *A. f. guimarasensis*
____ *A. f. daphoenonota*

N Philippines (ne Luzon)
N Philippines (s Luzon and Catanduanes)
Philippines (Panay and Guimaras)
Philippines (Negros)

☐ **Metallic-winged Sunbird** *Aethopyga pulcherrima*
____ *A. p. jefferyi*
____ *A. p. pulcherrima*
____ *A. p. decorosa*

N Philippines (Luzon)
Basilan, Dinagat, Siargo, Biliran, Samar, Leyte and Mindanao
Philippines (Bohol)

☐ **Elegant Sunbird** *Aethopyga duyvenbodei*

Sangihe and Siau islands (off n Sulawesi)

☐ **Lovely Sunbird** *Aethopyga shelleyi*

S Philippines (Balabac, Busuanga, Culion and Palawan)

☐ **Handsome Sunbird** *Aethopyga bella*
____ *A. b. flavipectus*
____ *A. b. minuta*
____ *A. b. rubrinota*
____ *A. b. bella*
____ *A. b. bonita*
____ *A. b. arolasi*

N Philippines (n Luzon)
N Philippines (c Luzon, Mindoro, Polillo and Marinduque)
N Philippines (Lubang)
S Philippines (Samar, Leyte, Dinagat, Siargao and Mindanao)
Philippines (Ticao, Masbate, Panay, Negros and Cebu)
S Philippines (Sulu Archipelago)

☐ **Gould's Sunbird** *Aethopyga gouldiae*
____ *A. g. gouldiae*
____ *A. g. isolata*
____ *A. g. dabryii*
____ *A. g. annamensis*

Himalayas (Sutlej Valley to Arunachal Province and se Tibet)
S Assam to Bangladesh and nw Burma (Chin Hills)
W China (Xinjiang to Sichuan and Yunnan) to Burma, n Laos
S Laos (Bolavens Plateau) and s Vietnam (Langbian Plateau)

☐ **Green-tailed Sunbird** *Aethopyga nipalensis*
____ *A. n. horsfieldii*
____ *A. n. nipalensis*
____ *A. n. koelzi*
____ *A. n. victoriae*
____ *A. n. karenensis*
____ *A. n. angkanensis*
____ *A. n. australis*
____ *A. n. blanci*
____ *A. n. ezrai*

W Himalayas (Garhwal to w Nepal)
W Nepal to ne India (Darjiling) and Sikkim
Himalayas (Bhutan to ne Burma, sw China and nw Tonkin)
W Burma (Chin Hills)
SE Burma (Karen Hills)
High mts. of n Thailand (Doi Ang Ka)
High mts. of s peninsular Thailand south of Isthmus of Kra
High mts. of Laos
High mts. of s Vietnam

☐ **White-flanked Sunbird** *Aethopyga eximia*

Montane forests and scrub of Java

☐ **Fork-tailed Sunbird** *Aethopyga christinae*
____ *A. c. latouchii*
____ *A. c. christinae*
____ *A. c. sokolovi*

SE China (e Sichuan, s Hunan, Guandong, Fujian) to n Vietnam
Hainan I. (s China)
S Vietnam

☐ **Black-throated Sunbird** *Aethopyga saturata*
____ *A. s. saturata*
____ *A. s. assamensis*
____ *A. s. galenae*
____ *A. s. petersi*
____ *A. s. sanguinipectus*
____ *A. s. anomala*
____ *A. s. wrayi*
____ *A. s. ochra*
____ *A. s. cambodiana*
____ *A. s. johnsi*

Himalayas (Garhwal to Bhutan and se Tibet)
Bangladesh to Assam, n Burma and sw China (w Yunnan)
Mts. of nw Thailand
E Burma to n Thailand, Laos, n Vietnam and se Yunnan
Hills of se Burma (Karenni and n Tenasserim)
Hills of s peninsular Thailand (Phatthalung and Trang)
Mts. of Malay Peninsula (south of Trang)
S Laos (Bolavens Plateau) and central Vietnam (Dakto)
Mts. of sw Cambodia
Langbian Plateau (s Vietnam)

☐ **Western Crimson Sunbird** *Aethopyga vigorsii*

W India (Western Ghats from Narbada River to Kerala)

☐ **Eastern Crimson Sunbird** *Aethopyga siparaja*
____ *A. s. seheriae*
____ *A. s. labecula*
____ *A. s. owstoni*
____ *A. s. tonkinensis*
____ *A. s. mangini*
____ *A. s. insularis*
____ *A. s. cara*
____ *A. s. trangensis*
____ *A. s. siparaja*
____ *A. s. nicobarica*
____ *A. s. heliogona*
____ *A. s. natunae*
____ *A. s. magnifica*
____ *A. s. flavostriata*
____ *A. s. beccarii*

Nepal to Assam, Bangladesh, Burma, sw China and nw Thailand
E Himalayas (Bhutan to Arundal Pradesh, Assam and Bangladesh)
S China (Naochow I. off Luichow Peninsula)
S China (se Yunnan) and ne Vietnam
SE Thailand to central and s Indochina
Phu Quoc I. (off extreme s Cambodia)
S Burma, Thailand and Mergui Archipelago
Peninsular Thailand, n Malay Peninsula and adjacent Burma
Malay Peninsula, Sumatra, Borneo and adjacent offshore islands
Nicobar Islands
Java
North Natuna Islands
Philippines (Cebu, Negros, Panay, Sibuyan and Tablas)
N Sulawesi
Central, se and s Sulawesi; Butung, Muna and Kabaena islands

☐ **Scarlet Sunbird** *Aethopyga mystacalis*

Java

☐ **Temminck's Sunbird** *Aethopyga temminckii*

Extreme sw Thailand, Malay Peninsula, Sumatra and Borneo

☐ **Fire-tailed Sunbird** *Aethopyga ignicauda*
____ *A. i. ignicauda (exultans)*
____ *A. i. flavescens*

Himalayas (Garhwal to sw China and n Burma)
N Burma (Chin Hills)

☐ **Little Spiderhunter** *Arachnothera longirostra*
____ *A. l. longirostra*
____ *A. l. sordida*
____ *A. l. pallida (zharina)*
____ *A. l. cinireicollis*
____ *A. l. niasensis*
____ *A. l. prillwitzi*
____ *A. l. buettikoferi*
____ *A. l. atita*
____ *A. l. rothschildi*
____ *A. l. flammifera*
____ *A. l. randi*
____ *A. l. dilutior*

SW India; Nepal to Assam, w Yunnan, Burma and w Thailand
SW China (se Yunnan) to ne Thailand and n Indochina
SE Thailand and central Indochina
Pen. Thailand (s of Isthmus of Kra), Malay Pen. and Sumatra
Nias I. (off w Sumatra)
Java
Borneo
South Natuna Is. (South China Sea)
North Natuna Is. (South China Sea)
Philippines (Samar, Leyte, Bohol, Mindanao, Dinagat and Biliran)
S Philippines (Basilan I.)
SW Philippines (Palawan)

☐ **Thick-billed Spiderhunter** *Arachnothera crassirostris*

Forests of peninsular Thailand, Malay Pen., Sumatra and Borneo

☐ **Long-billed Spiderhunter** *Arachnothera robusta*
____ *A. r. robusta*
____ *A. r. armata*

Forests of s peninsular Thailand, Malay Pen., Sumatra and Borneo
Java

☐ **Spectacled Spiderhunter** *Arachnothera flavigaster*

Peninsular Thailand, Malaysia, Sumatra and Borneo

☐ **Yellow-eared Spiderhunter** *Arachnothera chrysogenys*
____ *A. c. chrysogenys*
____ *A. c. harrissoni*

S Burma, s Thailand, Malay Pen., Sumatra, Java and w Borneo
E Borneo

☐ **Naked-faced Spiderhunter** *Arachnothera clarae*
____ *A. c. luzonensis*
____ *A. c. philippensis*
____ *A. c. clarae*
____ *A. c. malindangensis*

N Philippines (Sierra Madre Mts. of e-central Luzon)
Philippines (Samar, Leyte and Biliran)
S Philippines (e Mindanao in Davao area)
S Philippines (central and w Mindanao); Basilan?

543

☐ **Gray-breasted Spiderhunter** *Arachnothera modesta*

____ A. m. caena	S Burma and n peninsular Thailand
____ A. m. modesta	S Thailand (s of Isthmus of Kra), Malay Peninsula and w Borneo
____ A. m. concolor	Sumatra and Mentawi Archipelago
____ A. m. pars	E Borneo

☐ **Streaky-breasted Spiderhunter** *Arachnothera affinis*

____ A. a. everetti	Mts. of n and central Borneo
____ A. a. affinis	Java and Bali

☐ **Streaked Spiderhunter** *Arachnothera magna*

____ A. m. magna	Himalayas (Garhwal to n Burma and sw China and Malay Pen.)
____ A. m. aurata	East and central Burma
____ A. m. musarum	SE Burma (s Shan States) to n Thailand and n Laos
____ A. m. pagodarum	S Burma and sw Thailand
____ A. m. remota	S Vietnam

☐ **Whitehead's Spiderhunter** *Arachnothera juliae*

	Montane forests of n Borneo

FAMILY: MELANOCHARITIDAE (Berrypeckers and Longbills—10)

☐ **Obscure Berrypecker** *Melanocharis arfakiana*

	Known from 2 specimens ca 1900 from mts. of se New Guinea

☐ **Black Berrypecker** *Melanocharis nigra*

____ M. n. pallida	Waigeu I. (off w New Guinea)
____ M. n. nigra	Misool I., Salawati I. and w New Guinea
____ M. n. unicolor	Yapen I., Meos Num I., n and e New Guinea
____ M. n. chloroptera	Aru Is. and s New Guinea (Mimika River to Fly River)

☐ **Lemon-breasted Berrypecker** *Melanocharis longicauda*

____ M. l. longicauda	Mts. of nw New Guinea (Vogelkop and Wandammen mts.)
____ M. l. umbrosa	NW New Guinea (slopes above Idenberg River)
____ M. l. chloris	W New Guinea (Weyland Mts. and s slopes of Jayawijaya Mts.)
____ M. l. captata	Mts. of cent. New Guinea (Huon Pen. and Central Highlands)
____ M. l. orientalis	Mts. of se New Guinea

☐ **Fan-tailed Berrypecker** *Melanocharis versteri*

____ M. v. versteri	NW New Guinea (Arfak Mts.)
____ M. v. meeki	W New Guinea (Weyland, Nassau, Orange and Hindenberg mts.)
____ M. v. virago	Mts. of n New Guinea, Central Highlands and Huon Peninsula
____ M. v. maculiceps	Herzog Mts. and Mts. of se New Guinea

☐ **Streaked Berrypecker** *Melanocharis striativentris*

____ M. s. axillaris	W New Guinea (Weyland Mts. and south slopes of Snow Mts.)
____ M. s. striativentris	Central Highlands and s slopes of Mts. of se New Guinea
____ M. s. chrysocome	E New Guinea (Mts. of Huon Peninsula)
____ M. s. prasina	N slopes of Mts. of se New Guinea

☐ **Spotted Berrypecker** *Melanocharis crassirostris*

____ M. c. crassirostris	New Guinea (Mts. of Vogelkop Peninsula and Central Highlands)
____ M. c. piperata	Mts. of se New Guinea
____ M. c. viridescens	SE New Guinea (Herzog Mts.)

☐ **Yellow-bellied Longbill** *Toxorhamphus novaeguineae*

____ T. n. novaeguineae	W Papuan islands, w New Guinea and islands in Geelvink Bay
____ T. n. flaviventris	Aru Is. and s New Guinea (Utakwa River to middle Fly River)

☐ **Slaty-chinned Longbill** *Toxorhamphus poliopterus*

____ *T. p. maximus*	N slopes of c mts. of New Guinea (Weyland to Jayawijaya mts.)
____ *T. p. poliopterus*	Mts. of se New Guinea

☐ **Dwarf Honeyeater** *Toxorhamphus iliolophus*

____ *T. i. cinerascens*	Waigeu I. (off w New Guinea)
____ *T. i. affine*	W New Guinea (Mts. of Vogelkop Peninsula)
____ *T. i. iliolophus*	Yapen and Meos Num islands and n New Guinea
____ *T. i. flavus*	S and se New Guinea
____ *T. i. fergussonis*	D'Entrecasteaux Arch. (Fergusson, Goodenough and Normanby)

☐ **Pygmy Honeyeater** *Toxorhamphus pygmaeum*

____ *T. p. waigeuense*	Waigeu I. (off w New Guinea)
____ *T. p. pygmaeum*	Misool I. and w New Guinea
____ *T. p. flavipectus*	S New Guinea (Etna Bay to Milne Bay)
____ *T. p. olivascens*	N coast of se New Guinea (Milne Bay to Huon Peninsula)
____ *T. p. meeki*	D'Entrecasteaux Archipelago (Fergusson and Goodenough)

FAMILY: PARAMYTHIIDAE (Tit Berrypecker and Crested Berrypecker—2)

☐ **Tit Berrypecker** *Oreocharis arfaki*

	Montane forests of New Guinea

☐ **Crested Berrypecker** *Paramythia montium*

____ *P. m. montium*	Central Highlands and Mts. of se New Guinea
____ *P. m. olivacea*	Weyland, Nassau and Jayawijaya Mts. (cent. New Guinea)
____ *P. m. alpina*	Upper slopes of Jayawijaya Mts. and n slopes of Nassau Mts.
____ *P. m. brevicauda*	SE New Guinea (Mts. of Huon Peninsula)

FAMILY: DICAEIDAE (Flowerpeckers—44)

☐ **Olive-backed Flowerpecker** *Prionochilus olivaceus*

____ *P. o. parsoni*	N Philippines (Sierra Madre Mts. of n Luzon); Catanduanes
____ *P. o. samarensis*	Central Philippines (s Luzon; Samar, Leyte and possibly Bohol)
____ *P. o. olivaceus*	S Philippines (Basilan, Dinagat and Mindanao)

☐ **Yellow-breasted Flowerpecker** *Prionochilus maculatus*

____ *P. m. septentrionalis*	S Burma and s peninsular Thailand
____ *P. m. oblitus*	Malay Peninsula
____ *P. m. maculatus*	Sumatra, Belitung I., Nias I. and Borneo
____ *P. m. natunensis*	North Natuna Islands

☐ **Crimson-breasted Flowerpecker** *Prionochilus percussus*

____ *P. p. ignicapilla*	Malaya, s Thailand, Sumatra, Borneo, Riau Arch., N Natuna Is.
____ *P. p. regulus*	Batu Is. (off w Sumatra)
____ *P. p. percussus*	Java

☐ **Palawan Flowerpecker** *Prionochilus plateni*

____ *P. p. plateni*	SW Philippines (Balabac and Palawan)
____ *P. p. culionensis*	SW Philippines (Busuanga, Culion, Calauit and Calamianes)

☐ **Yellow-rumped Flowerpecker** *Prionochilus xanthopygius*

	Lowlands of Borneo; single record from Bungarun (N Natuna Is.)

☐ **Scarlet-breasted Flowerpecker** *Prionochilus thoracicus*

	S Vietnam, Malay Pen., Sumatra, Riau and Lingga Arch., Borneo

☐ **Golden-rumped Flowerpecker** *Dicaeum annae*

____ *D. a. annae*	Flores (w Lesser Sundas)
____ *D. a. sumbawense*	Sumbawa (w Lesser Sundas)

☐ **Thick-billed Flowerpecker** *Dicaeum agile*

____ *D. a. agile*	NE Pakistan and peninsular India
____ *D. a. zeylonicum*	Sri Lanka
____ *D. a. modestum (remotum)*	S peninsular Thailand, Malay Peninsula and Borneo
____ *D. a. pallescens*	Bangladesh to Burma, n Thailand and n Vietnam
____ *D. a. sumatranum (atjehense)*	N Sumatra (Aceh, Utara and Selatan)
____ *D. a. finschi*	W Java
____ *D. a. tinctum*	Lesser Sundas (Sumba, Flores and Alor)
____ *D. a. obsoletum*	Timor (e Lesser Sundas)
____ *D. a. striatissimum*	N Philippines (Lubang, Luzon, Romblon, Sibuyan, Catanduanes)
____ *D. a. aeruginosum*	Cent. and s Philippines (Cebu, Negros, Mindoro and Mindanao)
____ *D. a. affine*	SW Philippines (Palawan)

☐ **Brown-backed Flowerpecker** *Dicaeum everetti*

____ *D. e. everetti (sordidum)*	S Malaya, Riau Archipelago, n Borneo and Labuan I.
____ *D. e. bungurense*	North Natuna Islands

☐ **Whiskered Flowerpecker** *Dicaeum proprium*

S Philippines (montane forests of Mindanao)

☐ **Yellow-vented Flowerpecker** *Dicaeum chrysorrheum*

____ *D. c. chrysochlore*	E Himalayas (Sikkim to sw China, Burma and n Indochina)
____ *D. c. chrysorrheum*	S peninsular Thailand to Malaya, Sumatra, Borneo, Java and Bali

☐ **Yellow-bellied Flowerpecker** *Dicaeum melanoxanthum*

E Himalayas (Nepal to ne India, sw China and Burma) to Vietnam

☐ **White-throated Flowerpecker** *Dicaeum vincens*

Wet lowlands of sw Sri Lanka

☐ **Yellow-sided Flowerpecker** *Dicaeum aureolimbatum*

____ *D. a. aureolimbatum*	Sulawesi, Bangka, Lembeh, Muna and Butung islands
____ *D. a. laterale*	Sangihe I. (off n Sulawesi)

☐ **Olive-capped Flowerpecker** *Dicaeum nigrilore*

____ *D. n. diuatae*	S Philippines (Diuata Mts. of ne Mindanao)
____ *D. n. nigrilore*	S Philippines (Mts. of Mindanao)

☐ **Flame-crowned Flowerpecker** *Dicaeum anthonyi*

____ *D. a. anthonyi*	N Philippines (montane forests of Luzon)
____ *D. a. kampalili*	S Philippines (montane forests of Mindanao)
____ *D. a. masawan*	S Philippines (Mt. Malindang on nw Mindanao)

☐ **Bicolored Flowerpecker** *Dicaeum bicolor*

____ *D. b. inexpectatum*	N Philippines (Luzon, Mindoro and Catanduanes)
____ *D. b. viridissimum*	Central Philippines (Negros and Guimaras)
____ *D. b. bicolor*	S Philippines (Bohol, Leyte, Dinagat, Mindanao and Samar)

☐ **Cebu Flowerpecker** *Dicaeum quadricolor*

Philippines (rediscovered 1992 after 85-year absence on Cebu)

☐ **Red-striped Flowerpecker** *Dicaeum australe*

Widespread throughout Philippine Islands

☐ **Red-keeled Flowerpecker** *Dicaeum haematostictum*

Philippines (Panay and Negros); probably extinct on Guimaras

☐ **Scarlet-collared Flowerpecker** *Dicaeum retrocinctum*

N Philippines (Mindoro); vagrant to Panay and Negros

☐ **Orange-bellied Flowerpecker** *Dicaeum trigonostigma*

____ *D. t. rubropygium*	NE India (Assam) to s Burma and peninsular Thailand
____ *D. t. trigonostigma*	Malay Peninsula to Singapore, Sumatra and satellite islands
____ *D. t. melanostigma*	Malaya, Sumatra, Bangka and Belitung islands
____ *D. t. antioproctum*	Simeulue I. (off Sumatra)
____ *D. t. megastoma*	Great Natuna I. (North Natuna Islands)
____ *D. t. flaviclunis*	Krakatoa I., Java and Bali

____ *D. t. dayakanum* — Borneo and adjacent offshore northern islands
____ *D. t. xanthopygium* — N Philippines (Luzon, Marinduque, Mindoro and Polillo)
____ *D. t. intermedium* — N Philippines (Romblon)
____ *D. t. cnecolaemum* — N Philippines (Tablas)
____ *D. t. sibuyanicum* — N Philippines (Sibuyan)
____ *D. t. dorsale* — Central Philippines (Masbate, Panay and Negros)
____ *D. t. besti* — S Philippines (Siquijor)
____ *D. t. cinereigulare* — Philippines (Mindanao, Samar, Leyte, Calicoan, Biliran and Bohol)
____ *D. t. pallidius* — Central Philippines (Cebu)
____ *D. t. isidroi* — S Philippines (Camiguin Sur)
____ *D. t. assimile* — Sulu Archipelago (Tawitawi, Jolo and Siasi)
____ *D. t. sibutuense* — Sulu Archipelago (Sibutu, Omapoy and Sipangkot)

☐ **White-bellied Flowerpecker** *Dicaeum hypoleucum*
____ *D. h. cagayanensis* — N Philippines (Sierra Madre Mts. of ne Luzon)
____ *D. h. obscurum (lagunae)* — N Philippines (central and s Luzon and Catanduanes)
____ *D. h. pontifex* — Philippines (Bohol, Samar, Leyte, Dinagat, Panaon, Mindanao)
____ *D. h. mindanense* — S Philippines (Zamboanga Peninsula of Mindanao)
____ *D. h. hypoleucum* — S Philippines (Basilan, Bongao, Jolo, Tawitawi and Siasi)

☐ **Pale-billed Flowerpecker** *Dicaeum erythrorhynchos*
____ *D. e. erythrorhynchos* — India to s Nepal, Bhutan, Bangladesh and w Burma
____ *D. e. ceylonense* — Sri Lanka

☐ **Plain Flowerpecker** *Dicaeum concolor*
____ *D. c. olivaceum* — Indian subcontinent, China, Burma, Thailand and n pen. Malaysia
____ *D. c. concolor (subflavum)* — Western Ghats and coastal sw India
____ *D. c. virescens* — South and Middle Andaman Islands
____ *D. c. minullum* — Hainan I. (s China)
____ *D. c. uchidai* — Montane forests of Taiwan
____ *D. c. borneanum* — Malay Peninsula, Sumatra, Borneo and North Natuna Is.
____ *D. c. sollicitans* — Java and Bali

☐ **Pygmy Flowerpecker** *Dicaeum pygmaeum*
____ *D. p. fugaensis* — Fuga I. (n Philippines off n Luzon)
____ *D. p. salomonseni* — N Philippines (Ilocos Norte Province of extreme n Luzon)
____ *D. p. pygmaeum* — Central and s Luzon and central Philippine Is.
____ *D. p. davao* — S Philippines (Mindanao and Camiguin Sur)
____ *D. p. palawanorum* — S Philippines (Balabac, Culion, Calauit and Palawan)

☐ **Crimson-crowned Flowerpecker** *Dicaeum nehrkorni*

Montane forests of Sulawesi

☐ **Flame-breasted Flowerpecker** *Dicaeum erythrothorax*
____ *D. e. schistaceiceps* — Moluccas (Morotai, Halmahera, Kasiruta, Bacan, Obi and Bisa)
____ *D. e. erythrothorax* — Buru (s Moluccas)

☐ **Ashy Flowerpecker** *Dicaeum vulneratum*

Moluccas (Boano, Seram, Ambon, Saparua, Gorong, Seram Laut)

☐ **Olive-crowned Flowerpecker** *Dicaeum pectorale*
____ *D. p. ignotum* — Gebe I. (Halmahera Sea off nw New Guinea)
____ *D. p. pectorale* — W Papuan islands and lowlands of w New Guinea

☐ **Red-capped Flowerpecker** *Dicaeum geelvinkianum*
____ *D. g. maforense* — Numfor I. (Geelvink Bay off n New Guinea)
____ *D. g. misoriense* — Biak I. (Geelvink Bay off n New Guinea)
____ *D. g. geelvinkianum* — Yapen and Kurudu islands (Geelvink Bay off n New Guinea)
____ *D. g. obscurifrons* — W New Guinea (Wissel Lakes region)
____ *D. g. setekwa* — SW New Guinea (s slopes of Snow Mts. to Lorentz River)
____ *D. g. diversum* — North coast of Irian Jaya)
____ *D. g. centrale* — Central New Guinea (Nassau and Jayawijaya Mts.)

____ *D. g. albopunctatum*   Lowlands of s-central New Guinea
____ *D. g. rubrigulare*   S New Guinea (Palmer Junction to mouth of Fly River)
____ *D. g. rubrocoronatum*   SE New Guinea
____ *D. g. violaceum*   D'Entrecasteaux Arch. (Fergusson, Goodenough and Dobu is.)

☐ **Louisiade Flowerpecker**   *Dicaeum nitidum*
____ *D. n. nitidum*   Louisiade Archipelago (Tagula and Misima islands)
____ *D. n. rosseli*   Rossel I. (Louisiade Archipelago)

☐ **Red-banded Flowerpecker**   *Dicaeum eximium*
____ *D. e. layardorum*   New Britain (Bismarck Archipelago)
____ *D. e. eximium*   Bismarck Archipelago (New Ireland and New Hanover)
____ *D. e. phaeopygium*   Dyaul I. (Bismarck Archipelago)

☐ **Midget Flowerpecker**   *Dicaeum aeneum*
____ *D. a. aeneum*   Bougainville, Choiseul, Ysabel and adj. n Solomon Is.
____ *D. a. becki*   Guadalcanal and Florida I. (Solomon Is.)
____ *D. a. malaitae*   Malaita I. (Solomon Is.)

☐ **Mottled Flowerpecker**   *Dicaeum tristrami*
     Makira I. (se Solomon Is.)

☐ **Black-fronted Flowerpecker**   *Dicaeum igniferum*
____ *D. i. igniferum*   Lesser Sundas (Sumbawa, Komodo, Flores, Lomblen and Besar)
____ *D. i. aetum*   Lesser Sundas (Pantar and Alor)

☐ **Red-chested Flowerpecker**   *Dicaeum maugei*
____ *D. m. splendidum*   Salayar and Tanahjampea islands (Flores Sea)
____ *D. m. neglectum*   W Lesser Sundas (Lombok and Penida)
____ *D. m. maugei*   E Lesser Sundas (Roti, Sawu, Semau, Timor, Romang and Damar)
____ *D. m. salvadorii*   E Lesser Sundas (Moa and Babar)

☐ **Fire-breasted Flowerpecker**   *Dicaeum ignipectum*
____ *D. i. ignipectum*   E Himalayas (Kashmir) to s-cent. China, Burma and Indochina
____ *D. i. dolichorhynchum*   Mts. of s peninsular Thailand and Malay Peninsula
____ *D. i. cambodianum*   Mts. of e Thailand and Cambodia
____ *D. i. formosum*   Montane forests of Taiwan
____ *D. i. luzoniense*   N Philippines (montane forests of n Luzon)
____ *D. i. bonga*   Central Philippines (Samar I.)
____ *D. i. apo*   Mts. of s Philippines (Negros and Mindanao)
____ *D. i. beccarii*   Montane forests of Sumatra

☐ **Black-sided Flowerpecker**   *Dicaeum monticolum*
     Mts. of n Borneo

☐ **Gray-sided Flowerpecker**   *Dicaeum celebicum*
____ *D. c. talautense*   Talaud Is. (n Moluccas)
____ *D. c. sanghirense*   Sangihe and Siau islands (off n Sulawesi)
____ *D. c. celebicum*   Manadotua, Bangka, Sulawesi, Lembeh, Togian, Muna and Butung Is.
____ *D. c. kuehni*   Tukangbesi Archipelago (off Sulawesi)
____ *D. c. sulaense*   Banggai Is. and Sula Is. (Taliabu, Mangole and Sanana)

☐ **Blood-breasted Flowerpecker**   *Dicaeum sanguinolentum*
____ *D. s. sanguinolentum*   Java and Bali; vagrant to Sumatra
____ *D. s. rhodopygiale*   Flores (w Lesser Sundas)
____ *D. s. wilhelminae*   Sumba (w Lesser Sundas)
____ *D. s. hanieli*   Timor (e Lesser Sundas)

☐ **Mistletoebird**   *Dicaeum hirundinaceum*
____ *D. h. kiense*   Southern Wallacea (Watubela, Tayandu and Kai islands)
____ *D. h. fulgidum*   Tanimbar Is. (Yamdena, Larat and Lutu)

548

____ *D. h. ignicolle*	Aru Islands
____ *D. h. hirundinaceum*	Islands in Torres Strait and treed areas of mainland Australia

☐ **Scarlet-backed Flowerpecker** *Dicaeum cruentatum*

____ *D. c. cruentatum (siamensis)*	E Nepal to ne India, s China, Burma, Thailand and Indochina
____ *D. c. ignitum*	Malay Peninsula
____ *D. c. sumatranum*	Sumatra and satellite islands
____ *D. c. batuense*	Mentawi Archipelago (off Sumatra)
____ *D. c. simalurense*	Simeulue I. (off w Sumatra)
____ *D. c. niasense*	Nias I. (off w Sumatra)
____ *D. c. nigrimentum (pryeri, hosei)*	Borneo and Karimata Is.

☐ **Scarlet-headed Flowerpecker** *Dicaeum trochileum*

____ *D. t. trochileum*	Java, Bali, Madura, se Borneo, Bawean and Kangean islands
____ *D. t. stresemanni*	Lombok (w Lesser Sundas)

FAMILY: PARDALOTIDAE (Pardalotes—4)

☐ **Spotted Pardalote** *Pardalotus punctatus*

____ *P. p. millitaris*	NE Queensland (Atherton Tablelands to about Townsville)
____ *P. p. punctatus*	SE Queensland to S Australia, Tasmania; sw South Australia
____ *P. p. xanthopyge*	SW Western Australia to nw Victoria and sc NSW; Kangaroo I.

☐ **Forty-spotted Pardalote** *Pardalotus quadragintus*

	Drier sclerohyll forests of e Tasmania and Flinders I.

☐ **Red-browed Pardalote** *Pardalotus rubricatus*

____ *P. r. yorki*	Cape York Peninsula (Queensland)
____ *P. r. rubricatus*	Northern and central Australia, to coasts in Western Australia

☐ **Striated Pardalote** *Pardalotus striatus*

____ *P. s. uropygialis*	N Australia (Fitzroy R., Western Australia to Cape York Pen.)
____ *P. s. melvillensis*	Melville and Bathurst islands (Northern Territory)
____ *P. s. melanocephalus*	E Queensland (about Ingham to New South Wales)
____ *P. s. ornatus*	NE New South Wales to e and s Victoria
____ *P. s. striatus*	Tasmania, King and Flinders islands; winters to n Queensland
____ *P. s. substriatus*	S and central Australia, except the most arid regions

FAMILY: ZOSTEROPIDAE (White-eyes—96)

☐ **Black-capped Speirops** *Speirops lugubris*

	São Tomé I. (Gulf of Guinea)

☐ **Cameroon Speirops** *Speirops melanocephalus*

	Montane forests of Mt. Cameroon

☐ **Fernando Po Speirops** *Speirops brunneus*

	Bioko I. (Pico Basilé)

☐ **Principe Speirops** *Speirops leucophoeus*

	Lowlands of Príncipe I. (Gulf of Guinea)

☐ **African Yellow White-eye** *Zosterops senegalensis*

____ *Z. s. senegalensis*	Senegal to Uganda, n Congo, n Eritrea and nw Ethiopia
____ *Z. s. demeryi*	Sierra Leone to Liberia and Ivory Coast
____ *Z. s. stenocricotus*	SE Nigeria to Cameroon and n Gabon; Bioko (Gulf of Guinea)
____ *Z. s. stuhlmanni*	NW Tanzania to w Kenya and Uganda
____ *Z. s. gerhardi*	S Sudan and ne Uganda border
____ *Z. s. reichenowi*	E Zaire (mts. nw of Lake Tanganyika and sw of Lake Kivu)
____ *Z. s. toroensis*	Lowlands of ne Zaire to w Uganda
____ *Z. s. jacksoni*	Highlands of Kenya and n Tanzania (Loliondo)
____ *Z. s. stierlingi*	Mts. of s Tanzania to e Zambia, Malawi and w Mozambique

____ *Z. s. kasaicus*   Central Zaire (Kasai) to ne Angola

____ *Z. s. heinrichi*   NW Angola (Cuanza Norte and Cuanza Sul)

____ *Z. s. quanzae*   Central Angola (Malanje to central highlands)

____ *Z. s. anderssoni*   S Angola to Zambia, s Tanzania, Malawi, Mozambique and Natal

____ *Z. s. tongensis*   SE Zimbabwe to ne Natal and s Mozambique

☐ **Broad-ringed White-eye** *Zosterops poliogastrus*

____ *Z. p. kaffensis*   Highlands of w and s Ethiopia (s of Lake Tana, w of Omo River)

____ *Z. p. poliogastrus*   Mts. of Eritrea and Ethiopia

____ *Z. p. kulalensis*   N Kenya (Mt. Kulal)

____ *Z. p. kikuyuensis*   Central Kenya (Aberdare Mts. and Mt. Kenya)

____ *Z. p. mbuluensis*   SE Kenya (Chyulu Mts.) and n Tanzania (North Pare Mts.)

____ *Z. p. eurycricotus*   N Tanzania (Mt. Kilimanjaro, Mt. Meru and Arusha regions)

____ *Z. p. winifredae*   NE Tanzania (South Pare Mts.)

____ *Z. p. silvanus*   SE Kenya (Taita Hills)

☐ **White-breasted White-eye** *Zosterops abyssinicus*

____ *Z. a. arabs*   S Arabian Peninsula (Yemen and extreme n Aden)

____ *Z. a. abyssinicus*   Lowlands of Eritrea and e Ethiopia to se Sudan

____ *Z. a. socotranus*   N Somalia and Socotra

____ *Z. a. omoensis*   SW Ethiopia

____ *Z. a. jubaensis*   SE Ethiopia to Somalia and extreme n Kenya

____ *Z. a. flavilateralis*   Extreme sw Ethiopia to e Kenya and e-central Tanzania

☐ **Cape White-eye** *Zosterops pallidus*

____ *Z. p. virens*   SW Mozambique to e Cape Province

____ *Z. p. caniviridis*   E Botswana to w Transvaal

____ *Z. p. atmorii*   Inland s Cape Province

____ *Z. p. capensis*   W Cap Province

____ *Z. p. pallidus*   S Namibia to Transvaal and nw Cape Province

____ *Z. p. sundevalli*   NE Cape Province

☐ **Pemba White-eye** *Zosterops vaughani*

Pemba I. (off Tanzania)

☐ **Mayotte White-eye** *Zosterops mayottensis*

Mayotte (se Comoro Is.); extirpated on Maria Anne I. ca 1940

☐ **Madagascar White-eye** *Zosterops maderaspatanus*

____ *Z. m. menaiensis*   Cosmoledo Atoll (Seychelles Is.)

____ *Z. m. aldabrensis*   Aldabra and Astove islands (Indian Ocean)

____ *Z. m. kirki*   Grand Comoro I.

____ *Z. m. anjouanensis*   Anjouan (Comoro Islands)

____ *Z. m. comorensis*   Mohéli (Comoro Islands)

____ *Z. m. maderaspatanus*   Madagascar and Isles Glorieuses

____ *Z. m. voeltzkowi*   Europa I. (s Mozambique Channel)

☐ **Comoro White-eye** *Zosterops mouroniensis*

Grand Comoro I. (montane heath of Mt. Karthala)

☐ **São Tomé White-eye** *Zosterops ficedulinus*

____ *Z. f. ficedulinus*   Príncipe (Gulf of Guinea)

____ *Z. f. feae*   São Tomé (Gulf of Guinea)

☐ **Annobon White-eye** *Zosterops griseovirescens*

Pagalu (Gulf of Guinea)

☐ **Mascarene White-eye** *Zosterops borbonicus*

____ *Z. b. borbonicus (alopekion, xerophilus)*   Réunion (w Mascarene Islands)

____ *Z. b. mauritianus*   Mauritius (w Mascarene Islands)

☐ **Reunion White-eye** *Zosterops olivaceus*

Forests of Réunion (w Mascarene Islands)

☐ **Mauritius White-eye** *Zosterops chloronothos*

Highlands of Mauritius (w Mascarene Islands)

☐ **Seychelles White-eye** *Zosterops modestus*

Forests of Mahé (Seychelles Is.). On verge of extinction

☐ **Ceylon White-eye** *Zosterops ceylonensis*

Montane forests of Sri Lanka

☐ **Chestnut-flanked White-eye** *Zosterops erythropleurus*

SE Siberia to sw China and ne Manchuria; winters to SE Asia

☐ **Oriental White-eye** *Zosterops palpebrosus*
- ____ *Z. p. egregius (occidentis)* — Afghanistan to Pakistan, India, Sri Lanka and Laccadive Is.
- ____ *Z. p. palpebrosus (siamensis)* — Nepal to se Tibet, sw China, Burma, n Thailand and Indochina
- ____ *Z. p. nilgiriensis* — S India (Nilgiri and Palani hills)
- ____ *Z. p. salimalii* — SE India (se Hyderabad)
- ____ *Z. p. nicobaricus* — Andaman and Nicobar islands
- ____ *Z. p. joannae* — SW China
- ____ *Z. p. williamsoni* — S Thailand and Malay Peninsula
- ____ *Z. p. auriventer* — S Burma to Malaysia, w Borneo, s Natuna and Bangka islands
- ____ *Z. p. sumatranus* — W Sumatra
- ____ *Z. p. buxtoni* — Mts. of e Sumatra and w Java
- ____ *Z. p. melanurus* — Mts. of central and e Java and Bali
- ____ *Z. p. unicus* — W Lesser Sundas (Sumbawa and Flores)

☐ **Japanese White-eye** *Zosterops japonicus*
- ____ *Z. j. yesoensis* — Hokkaido (n Japan)
- ____ *Z. j. japonicus* — Main Japanese islands (Honshu to Kyushu)
- ____ *Z. j. stejnegeri* — Izu Is. (s Japan); introduced to Bonin Is.
- ____ *Z. j. insularis* — Ryukyu Is. (Tanegashima and Yakushima)
- ____ *Z. j. loochooensis* — Iriomote (Ryukyu Is.)
- ____ *Z. j. alani* — Volcano Is. (Iwo Jima and Minami-iwo-Jima)
- ____ *Z. j. daitoensis* — Daito Is. (Philippine Sea)
- ____ *Z. j. simplex* — W China to Burma, n Vietnam and Taiwan; winters to Hainan
- ____ *Z. j. hainanus* — Hainan (s China)

☐ **Lowland White-eye** *Zosterops meyeni*
- ____ *Z. m. meyeni* — Philippines (Luzon, Banton, Calayan, Lubang, Verde, Caluya)
- ____ *Z. m. batanis* — Philippines (Batan, Sabtang, Ivojos, Itbayat and Y'Ami)

☐ **Enggano White-eye** *Zosterops salvadorii*

Enggano and Mega islands (off w Sumatra)

☐ **Bridled White-eye** *Zosterops conspicillatus*
- ____ *Z. c. rotensis* — Sabena Plateau on Rota (n Mariana Is.)
- ____ *Z. c. conspicillatus* — Guam (s Mariana Is.)
- ____ *Z. c. saypani* — SE Mariana Is. (Tinian, Agiguan and Saipan)

☐ **Caroline Islands White-eye** *Zosterops semperi*
- ____ *Z. s. semperi* — Palau Is. (Babelthuap, Koror, Garakayo and Palau)
- ____ *Z. s. owstoni* — Truk (Caroline Is.)
- ____ *Z. s. takasukasai* — Pohnpei (Caroline Is.)

☐ **Plain White-eye** *Zosterops hypolais*

Forest and scrub of Yap (nw Caroline Is.)

☐ **Black-capped White-eye** *Zosterops atricapilla*
- ____ *Z. a. viridicatus* — Mts. of n Sumatra
- ____ *Z. a. atricapilla* — Mts. of central and s Sumatra and n Borneo (Mt. Kinabalu)

☐ **Everett's White-eye** *Zosterops everetti*
- ____ *Z. e. tahanensis* — Peninsular Thailand to Malaysia and n Borneo
- ____ *Z. e. wetmorei* — S Thailand
- ____ *Z. e. everetti* — Philippines (Cebu)
- ____ *Z. e. basilanicus* — Philippines (Basilan, Dinagat, Mindanao, Siargao, Camiguin Sur)

_____ *Z. e. boholensis*	Philippines (Bohol, Leyte, Samar, Calicoan and Biliran)
_____ *Z. e. siquijorensis*	Philippines (Siquijor)
_____ *Z. e. mandibularis*	Sulu Archipelago (Sulu, Tawitawi, Jolo, Bongao, Sanga Sanga)
_____ *Z. e. babelo*	Talaud Is. (Karakelong and Salebabu) and n Sulawesi

☐ **Yellowish White-eye** *Zosterops nigrorum*

_____ *Z. n. meyleri*	N Philippines (Camiguin Norte)
_____ *Z. n. aureiloris*	N Philippines (Mts. of n Luzon)
_____ *Z. n. sierramadrensis*	N Philippines (Cagayan Province on s Luzon)
_____ *Z. n. luzonicus*	N Philippines (se Luzon and Catanduanes)
_____ *Z. n. nigrorum*	Philippines (Masbate, Negros, Ticao, Panay and Caluya)
_____ *Z. n. mindorensis*	Philippines (Mindoro)
_____ *Z. n. catamarensis*	S Philippines (Camiguin Sur)
_____ *Z. n. richmondi*	Cagayancillo I. (Sulu Archipelago)

☐ **Mountain White-eye** *Zosterops montanus*

_____ *Z. m. whiteheadi*	N Philippines (n Luzon)
_____ *Z. m. halconensis*	Philippines (Mindoro)
_____ *Z. m. gilli*	Philippines (Marinduque)
_____ *Z. m. pectoralis*	Philippines (n Negros)
_____ *Z. m. diuatae*	S Philippines (n Mindanao)
_____ *Z. m. vulcani*	S Philippines (Mt. Apo and Mt. Katanglad on Mindanao)
_____ *Z. m. parkesi*	SW Philippines (Palawan)
_____ *Z. m. obstinatus*	Moluccas (Ternate, Bacan and Seram)
_____ *Z. m. montanus*	Sulawesi, Sula Is., Buru, Lombok, Sumbawa, Flores, Timor
_____ *Z. m. difficilis*	S Sumatra

☐ **Christmas Island White-eye** *Zosterops natalis*

Christmas I. (e Indian Ocean). Introd. Cocos (Keeling) I.

☐ **Javan White-eye** *Zosterops flavus*

Mangroves and bamboo belt of coastal nw Java and se Borneo

☐ **Yellow-bellied White-eye** *Zosterops chloris*

_____ *Z. c. mentoris*	N-central Sulawesi
_____ *Z. c. intermedius*	S Sulawesi to Muna, Butung, Flores and Sumbawa
_____ *Z. c. flavissimus*	Tukangbesi Is. (Binongka, Kalidupa, Tomea and Wantjee)
_____ *Z. c. maxi*	Lombok, Nusa Penida and adjacent smaller w Lesser Sundas
_____ *Z. c. chloris*	Tayandu Is. and Kai Is. (Banda Sea)
_____ *Z. c. solombensis*	Solombo Besar I. (Java Sea)
_____ *Z. c. zachlorus*	Kalambau I. (Java Sea)

☐ **Ashy-bellied White-eye** *Zosterops citrinella*

_____ *Z. c. citrinella*	Lesser Sundas (Timor, Roti, Sawu and Sumba)
_____ *Z. c. albiventris*	Tanimbar Is. and islands off n Queensland south to Lizard I.
_____ *Z. c. harterti*	Alor (e Lesser Sundas)

☐ **Great Kai White-eye** *Zosterops grayi*

Kai Besar (Kai Islands)

☐ **Little Kai White-eye** *Zosterops uropygialis*

Kai Kecil (Kai Islands)

☐ **Sulawesi White-eye** *Zosterops consobrinorum*

SE peninsula of Sulawesi

☐ **Black-ringed White-eye** *Zosterops anomalus*

Hills of sw Sulawesi

☐ **Yellow-spectacled White-eye** *Zosterops wallacei*

Lesser Sundas (Sumbawa, Flores, Komodo, Sumba and Lomblen)

☐ **Black-crowned White-eye** *Zosterops atrifrons*

_____ *Z. a. atrifrons*	N, n-central and se Sulawesi and Peleng I. (Banggai Is.)
_____ *Z. a. sulaensis*	Sula Is. (Taliabu, Seho, Mangole and Sanana)
_____ *Z. a. surda*	Central Sulawesi
_____ *Z. a. subatrifrons*	Peleng I.

☐ **Sangihe White-eye** *Zosterops nehrkorni*

Sangihe I. (n of Sulawesi)

☐ **Seram White-eye** *Zosterops stalkeri*

Seram I. (s Moluccas)

☐ **Cream-throated White-eye** *Zosterops atriceps*
_____ *Z. a. dehaani* — Morotai (n Moluccas)
_____ *Z. a. fuscifrons* — Halmahera (n Moluccas)
_____ *Z. a. atriceps* — N Moluccas (Bacan and Obi)

☐ **Black-fronted White-eye** *Zosterops minor*
_____ *Z. m. chrysolaemus (tenuifrons)* — Mts. of nw New Guinea (Arfak Mts. and Onin Peninsula)
_____ *Z. m. minor* — Mts. of n New Guinea (Cyclops, Sepik and Snow Mts.); Yapen I.
_____ *Z. m. rothschildi* — Mts. of central New Guinea (Weyland Mts.)
_____ *Z. m. gregarius* — Mts. of se New Guinea (Huon Peninsula)
_____ *Z. m. delicatulus* — SE New Guinea (Herzog to Hydrographer mts. and se peninsula)
_____ *Z. m. pallidogularis* — D'Entrecasteaux Archipelago (Fergusson and Goodenough)

☐ **White-throated White-eye** *Zosterops meeki*

Uplands of Tagula I. (Louisiade Archipelago). Status unknown

☐ **Black-headed White-eye** *Zosterops hypoxanthus*
_____ *Z. h. hypoxanthus* — Bismarck Archipelago (New Britain, Uatom and Mioko)
_____ *Z. h. ultimus* — Bismarck Archipelago (New Hanover and New Ireland)
_____ *Z. h. admiralitatis* — Manus I. (Admiralty Islands)

☐ **Biak White-eye** *Zosterops mysorensis*

Mts. of Biak and Supiori islands (off nw New Guinea)

☐ **Capped White-eye** *Zosterops fuscicapilla*
_____ *Z. f. fuscicapilla* — Mts. of coastal n New Guinea
_____ *Z. f. crookshanki* — Mts. of Goodenough I. (D'Entrecasteaux Archipelago)

☐ **Buru White-eye** *Zosterops buruensis*

Mainly montane forests of Buru (s Moluccas)

☐ **Ambon White-eye** *Zosterops kuehni*

S Moluccas (Seram and Ambon)

☐ **New Guinea White-eye** *Zosterops novaeguineae*
_____ *Z. n. novaeguineae* — NW New Guinea (Vogelkop Mts.)
_____ *Z. n. aruensis* — Aru Islands
_____ *Z. n. magnirostris* — Coastal ne New Guinea (opposite Manam I.)
_____ *Z. n. wahgiensis* — C New Guinea (Wahgi Valley, Mt. Kubor and Bismarck Mts.)
_____ *Z. n. wuroi* — Coastal s New Guinea (west of Fly River mouth)
_____ *Z. n. crissalis* — Mts. of se New Guinea (Astrolabe Mts.)
_____ *Z. n. oreophilus* — Mts. of e New Guinea (Huon Peninsula)

☐ **Australian Yellow White-eye** *Zosterops luteus*
_____ *Z. l. balstoni* — Coastal Western Australia (Shark Bay to Kimberley Division)
_____ *Z. l. luteus* — Coastal n Australia (Kimberley to w Cape York Peninsula)

☐ **Louisiade White-eye** *Zosterops griseotinctus*
_____ *Z. g. pallidipes* — Rossel I. (Louisiade Archipelago)
_____ *Z. g. griseotinctus (aignani)* — Louisiade Arch. (Misima, Deboyne, Duchateau and Conflict)
_____ *Z. g. longirostris* — Bonvouloir, Heath and Alcester islands (Solomon Sea)
_____ *Z. g. eichhorni* — Bismarck Archipelago (Nauna, Nissan and Long islands)

☐ **Rennell White-eye** *Zosterops rennellianus*

Rennell I. (se Solomon Is.)

☐ **Banded White-eye** *Zosterops vellalavella*

Solomon Is. (Vella Lavella and Bagga)

☐ **Ganongga White-eye** *Zosterops splendidus*

Forests of Ganongga I. (w-central Solomon Is.)

☐ **Splendid White-eye** *Zosterops luteirostris*

Central Solomon Is. (Gizo and Ranongga)

☐ **Yellow-throated White-eye** *Zosterops metcalfii*
____ *Z. m. metcalfii*
____ *Z. m. floridanus*

Bougainville, Buka, Shortland, Choiseul, Santa Isabel, Molakobi
Florida I. (Solomon Is.)

☐ **Solomon Islands White-eye** *Zosterops rendovae*
____ *Z. r. rendovae*
____ *Z. r. tetiparius*
____ *Z. r. kulambangrae*

Rendova I. (Solomon Is.)
Tetipari I. (Solomon Is.)
Solomon Is. (Kulambangra to New Georgia group)

☐ **Kulambangra White-eye** *Zosterops murphyi*

Kulambangra I. (central Solomon Is.)

☐ **Gray-throated White-eye** *Zosterops ugiensis*
____ *Z. u. hamlini*
____ *Z. u. oblitus*
____ *Z. u. ugiensis*

Bougainville (Solomon Is.)
Guadalcanal (Solomon Is.)
San Cristóbal (Solomon Is.)

☐ **Malaita White-eye** *Zosterops stresemanni*

Malaita I. (se Solomon Is.)

☐ **Santa Cruz White-eye** *Zosterops santaecrucis*

Nendo I. (Santa Cruz Is. se of Solomon Is.)

☐ **Large Lifou White-eye** *Zosterops inornatus*

Forests of Lifou I. (Loyalty Is.)

☐ **Green-backed White-eye** *Zosterops xanthochroa*

Mts. of New Caledonia, Ile des Pins and Maré I.

☐ **Small Lifou White-eye** *Zosterops minutus*

Forest and scrub of Lifou (Loyalty Is.)

☐ **Lord Howe White-eye** *Zosterops tephropleurus*

Lord Howe I. (off New Zealand)

☐ **Slender-billed White-eye** *Zosterops tenuirostris*
____ *Z. t. tenuirostris*
____ *Z. t. stenurus†*

Norfolk I.
Lord Howe I. Extinct

☐ **White-chested White-eye** *Zosterops albogularis*

Norfolk I. Possibly extinct

☐ **Layard's White-eye** *Zosterops explorator*

Fiji (Viti Levu, Ovalau, Gau, Vanua Levu, Taveuni and Kandavu)

☐ **Silver-eye** *Zosterops lateralis*
____ *Z. l. vegetus*
____ *Z. l. cornwalli*
____ *Z. l. chlorocephalus*
____ *Z. l. westernensis*
____ *Z. l. tephropleurus*
____ *Z. l. lateralis*
____ *Z. l. ochrochrous*
____ *Z. l. pinarochrous*
____ *Z. l. chloronotus*
____ *Z. l. griseonota*
____ *Z. l. nigrescens*
____ *Z. l. melanops*
____ *Z. l. macmillani*
____ *Z. l. tropicus*
____ *Z. l. vatensis*
____ *Z. l. valuensis*
____ *Z. l. flaviceps*

NE Queensland (McIlwraith Range to Burdekin River)
E Australia (e-central Queensland to Hunter River, NSW)
E Queensland (islands of Capricorn coast)
SE Australia (southeast New South Wales to Victoria)
Lord Howe I.
Tasmania and Flinders I. (Bass Strait)
King I. (Bass Strait)
South Australia (Eyre Pen.) to nw Victoria and adj. NSW
SW Western Australia (Shark Bay to far sw South Australia)
New Caledonia
Loyalty Is. (Maré and Uvea)
Lifou (Loyalty Is.)
Vanuatu (Tanna and Aniwa islands)
Espíritu Santo I. (Vanuatu)
N Vanuatu, Banks Group and Torres Is.
Vanua Lava I. (Vanuatu)
Fiji Archipelago

☐ **Yellow-fronted White-eye** *Zosterops flavifrons*
____ *Z. f. gauensis*
____ *Z. f. perplexus*

Gau I. (n Vanuatu)
Vanua Lava I. (n Vanuatu)

____	*Z. f. brevicauda*	N Vanuatu (Malo and Espíritu Santo)
____	*Z. f. macgillivrayi*	Malekula I. (Vanuatu)
____	*Z. f. efatensis*	Vanuatu (Nguna, Efate and Erromanga)
____	*Z. f. flavifrons*	Tanna I. (Vanuatu)
____	*Z. f. majusculus*	Aneityum I. (Vanuatu)

☐ **Samoan White-eye** *Zosterops samoensis*

Mts. of Savai'i (Western Samoa)

☐ **Dusky White-eye** *Zosterops finschii*

Woodlands of Palau Archipelago (w Caroline Is.)

☐ **Gray White-eye** *Zosterops cinereus*

____	*Z. c. ponapensis*	Pohnpei (e Caroline Is.)
____	*Z. c. cinereus*	Kosrae (e Caroline Is.)

☐ **Yap White-eye** *Zosterops oleagineus*

Forests of Yap (nw Caroline Is.)

☐ **Truk White-eye** *Rukia ruki*

Forests and scrub of Truk (e Caroline Is.)

☐ **Long-billed White-eye** *Rukia longirostra*

Mts. of Pohnpei (e Caroline Is.)

☐ **Golden White-eye** *Cleptornis marchei*

S Mariana Is. (Saipan and Aguijan)

☐ **Rufescent White-eye** *Tephrozosterops stalkeri*

Montane epiphyte forests of Seram (s Moluccas)

☐ **Rufous-throated White-eye** *Madanga ruficollis*

Rediscovered after 75-year absence on Buru (s Moluccas)

☐ **Javan Gray-throated White-eye** *Lophozosterops javanicus*

____	*L. j. frontalis*	Mts. of w Java (Mt. Karang and Mt. Pangrango-Gedeh)
____	*L. j. javanicus*	Mts. of central and e Java
____	*L. j. elongatus*	Mts. of extreme e Java (Idjen Plateau) and Bali

☐ **Streak-headed White-eye** *Lophozosterops squamiceps*

____	*L. s. heinrichi*	NW Sulawesi (Tentolo-Matinan Mts.)
____	*L. s. stresemanni*	Mts. of ne Sulawesi
____	*L. s. striaticeps*	Mts. of n central Sulawesi
____	*L. s. stachyrinus*	S central Sulawesi (Latimojong Mts.)
____	*L. s. analogus*	SE Sulawesi (Mengkoka Mts.)
____	*L. s. squamiceps*	S Sulawesi (Lompobattang Massif)

☐ **Gray-hooded White-eye** *Lophozosterops pinaiae*

Tree-fern forests of Seram (s Moluccas)

☐ **Mindanao White-eye** *Lophozosterops goodfellowi*

____	*L. g. goodfellowi*	S Philippines (Mts. Apo, Matutum, Mayo, Katanglad on Mindanao)
____	*L. g. malindangensis*	S Philippines (Mt. Malindang on Mindanao)
____	*L. g. gracilis*	S Philippines (Mt. Hilong Hilong on ne Mindanao)

☐ **White-browed White-eye** *Lophozosterops superciliaris*

____	*L. s. hartertianus*	Sumbawa (w Lesser Sundas)
____	*L. s. superciliaris*	Flores (w Lesser Sundas)

☐ **Dark-crowned White-eye** *Lophozosterops dohertyi*

____	*L. d. dohertyi*	W Lesser Sundas (Mts. of Sumbawa and Satonda)
____	*L. d. subcristatus*	W Lesser Sundas (Mts. of Flores)

☐ **Pygmy White-eye** *Oculocincta squamifrons*

Montane moss forests of n Borneo (Sabah)

☐ **Flores White-eye** *Heleia crassirostris*

W Lesser Sundas (Flores and Sumbawa)

☐ **Timor White-eye** *Heleia muelleri*

Lowlands of w Timor (e Lesser Sundas)

☐ **Mountain Black-eye** *Chlorocharis emiliae*
____	*C. e. emiliae*	North Borneo (Mt. Kinabalu)
____	*C. e. trinitae*	North Borneo (Mt. Trus Madi)
____	*C. e. fusciceps*	Mts. of nw Borneo (ne Sarawak)
____	*C. e. moultoni*	Mts. of nw Borneo (w Sarawak)

☐ **Bare-eyed White-eye** *Woodfordia superciliosa*

Open woodlands of Rennell (se Solomon Is.)

☐ **Sanford's White-eye** *Woodfordia lacertosa*

Woodlands and scrub of Nendo (Santa Cruz Is.)

☐ **Giant White-eye** *Megazosterops palauensis*

Palau Archipelago (Babelthuap, Urukthapel and Peleliu)

☐ **Cinnamon White-eye** *Hypocryptadius cinnamomeus*

S Philippines (Mts. of Mindanao)

FAMILY: PROMEROPIDAE (Sugarbirds—2)

☐ **Gurney's Sugarbird** *Promerops gurneyi*
____	*P. g. ardens*	E Zimbabwe to w-central Mozambique n of Limpopo River
____	*P. g. gurneyi*	E Transvaal to Natal and e Cape Province s of Limpopo River

☐ **Cape Sugarbird** *Promerops cafer*

South Africa (Mts. of Cape Province)

FAMILY: MELIPHAGIDAE (Honeyeaters—174)

☐ **Olive Straightbill** *Timeliopsis fulvigula*
____	*T. f. fulvigula*	NW New Guinea (Arfak Mts.)
____	*T. f. montana*	Mts. of c New Guinea (Weyland Mts. to Wharton Range)
____	*T. f. meyeri*	Mts. of se New Guinea
____	*T. f. fuscicapilla*	E New Guinea (Mts. of Huon Peninsula)

☐ **Tawny Straightbill** *Timeliopsis griseigula*
____	*T. g. griseigula*	W New Guinea (Vogelkop Peninsula to Weyland Mts.)
____	*T. g. fulviventris*	Mts. of se New Guinea

☐ **Long-billed Honeyeater** *Melilestes megarhynchus*
____	*M. m. vagans*	Batanta and Waigeo islands (New Guinea)
____	*M. m. brunneus*	Mts. of nw New Guinea; Misool and Salawati islands
____	*M. m. megarhynchus*	S New Guinea and Aru Islands
____	*M. m. stresemanni*	N New Guinea (Astrolabe Bay to Geelvink Bay); Yapen I.

☐ **Bougainville Honeyeater** *Stresemannia bougainvillei*

Mts. of Bougainville (nw Solomon Is.)

☐ **Green-backed Honeyeater** *Glycichaera fallax*
____	*G. f. pallida*	Batanta and Waigeo islands (New Guinea)
____	*G. f. poliocephala*	Aru Islands, Misool I. and nw New Guinea
____	*G. f. fallax*	New Guinea (west to Geelvink Bay and Onin Pen.); Yapen I.
____	*G. f. claudi*	NE Queensland (McIlwraith Range)

☐ **Sunda Honeyeater** *Lichmera lombokia*

W Lesser Sundas (Lombok, Sumbawa and Flores)

☐ **Olive Honeyeater** *Lichmera argentauris*

Moluccas and w Papuan islands

☐ **Brown Honeyeater** *Lichmera indistincta*
____	*L. i. ocularis*	SE New Guinea and ne Australia (Cape York Pen. to n NSW)
____	*L. i. nupta*	Aru Islands
____	*L. i. indistincta*	SW Western Australia north and east to nw Queensland
____	*L. i. melvillensis*	Melville and Bathurst Islands (Northern Territory)

☐ **Indonesian Honeyeater** *Lichmera limbata*

Bali and Lesser Sundas to Flores and Timor

☐ **Dark-brown Honeyeater** *Lichmera incana*
- ___ *L. i. incana* — New Caledonia
- ___ *L. i. poliotis* — Loyalty Is. (Beautemps, Beaupré, Uvéa and Lifou)
- ___ *L. i. mareensis* — Maré (Loyalty Is.)
- ___ *L. i. griseoviridis* — Central Vanuatu islands
- ___ *L. i. flavotincta* — Erromango I. (Vanuatu)

☐ **White-tufted Honeyeater** *Lichmera squamata*

Lesser Sundas (Wetar to Tanimbar and Kai islands)

☐ **Silver-eared Honeyeater** *Lichmera alboauricularis*
- ___ *L. a. olivacea* — Lowlands of n New Guinea (Lake Sentani to Ramu River)
- ___ *L. a. alboauricularis* — South coast of se New Guinea, Doini and Rogeia islands

☐ **Buru Honeyeater** *Lichmera deningeri*

Mts. of Buru (s Moluccas)

☐ **Seram Honeyeater** *Lichmera monticola*

Mts. of Seram (s Moluccas)

☐ **Yellow-eared Honeyeater** *Lichmera flavicans*

Timor (e Lesser Sundas)

☐ **Black-chested Honeyeater** *Lichmera notabilis*

Wetar (Lesser Sundas). Status unknown

☐ **White-streaked Honeyeater** *Trichodere cockerelli*

NE Queensland (coastal n Cape York Peninsula)

☐ **Seram Myzomela** *Myzomela blasii*

S Moluccas (Seram, Ambon and Boano)

☐ **White-chinned Myzomela** *Myzomela albigula*
- ___ *M. a. albigula* — Rossell (Louisiade Archipelago)
- ___ *M. a. pallidior* — Louisiade Arch. (Misima, Bonvouloir, Deboyne and Conflict)

☐ **Red-throated Myzomela** *Myzomela eques*
- ___ *M. e. eques* — NW New Guinea, Waigeo, Salawati and Misool islands
- ___ *M. e. primitiva* — N New Guinea (Geelvink Bay to Astrolabe Bay)
- ___ *M. e. nymani* — S and e New Guinea
- ___ *M. e. karimuiensis* — E New Guinea

☐ **Ashy Myzomela** *Myzomela cineracea*

New Britain and Umboi islands (w Bismarck Archipelago)

☐ **Dusky Myzomela** *Myzomela obscura*
- ___ *M. o. aruensis* — Aru Islands (New Guinea)
- ___ *M. o. rubrobrunnea* — Biak I. (New Guinea)
- ___ *M. o. fumata* — S New Guinea (Vogelkop to Port Moresby)
- ___ *M. o. simplex* — N Moluccas (Halmahera, Damar, Ternate, Tidore and Bacan)
- ___ *M. o. rubrotincta* — N Moluccas (Obi and Bisa)
- ___ *M. o. mortyana* — Morotai (n Moluccas)
- ___ *M. o. obscura* — Northern Territory (Melville and Bathurst is. and Arnhem Land)
- ___ *M. o. harterti* — N and e Queensland (s Torres Straits islands south to Noosa)
- ___ *M. o. fumata* — Boigu and Saibai Islands (north Torres Straits)

☐ **Red Myzomela** *Myzomela cruentata*
- ___ *M. c. cruentata* — New Guinea and Yapen I.
- ___ *M. c. coccinea* — Bismarck Archipelago (New Britain and Duke of York)
- ___ *M. c. erythrina* — New Ireland (Bismarck Archipelago)
- ___ *M. c. lavongai* — New Hanover (Bismarck Archipelago)
- ___ *M. c. cantans* — Tabar I. (Bismarck Archipelago)
- ___ *M. c. vinacea* — Dyaul I. (Bismarck Archipelago)

☐ **Black Myzomela** *Myzomela nigrita*
- ___ *M. n. steini* — Waigeo I. (New Guinea)
- ___ *M. n. nigrita* — New Guinea and Aru Islands

557

____	*M. n. meyeri*	N New Guinea and Yapen I.
____	*M. n. nigerrima*	Long Island (ne New Guinea)
____	*M. n. pluto*	Meos Num I. (New Guinea)
____	*M. n. forbesi*	D'Entrecasteaux Archipelago
____	*M. n. louisiadensis*	Louisiade Archipelago
____	*M. n. hades*	St. Matthias Group (Bismarck Archipelago)
____	*M. n. ramsayi*	Tingwon Is. (Bismarck Archipelago)
____	*M. n. ernstmayri*	Manus I. (Admiralty Islands)

☐ **New Ireland Myzomela** *Myzomela pulchella*

New Ireland (Bismarck Archipelago)

☐ **Crimson-hooded Myzomela** *Myzomela kuehni*

Wetar (Lesser Sundas)

☐ **Red-headed Myzomela** *Myzomela erythrocephala*

____	*M. e. infuscata*	S New Guinea, Aru Is. and mangroves of Torres Straits islands
____	*M. e. erythrocephala*	Mangroves of n Australia (Broome to e Cape York Pen.)

☐ **Sumba Myzomela** *Myzomela dammermani*

Mangroves of Sumba (w Lesser Sundas)

☐ **Mountain Myzomela** *Myzomela adolphinae*

Patchily distributed mts. of New Guinea

☐ **Sulawesi Myzomela** *Myzomela chloroptera*

____	*M. c. chloroptera*	N Sulawesi (Mts. of Minahassa Peninsula)
____	*M. c. charlotta*	Mts. of central and se Sulawesi
____	*M. c. juga*	Mt. Lompobattang (sw Sulawesi)
____	*M. c. eva*	Salayar, Tanahjampea and Sula Islands (Taliabu)
____	*M. c. batjanensis*	Bacan (n Moluccas)

☐ **Wakolo Myzomela** *Myzomela wakoloensis*

____	*M. w. elisabethae*	Mts. of Seram (s Moluccas)
____	*M. w. wakoloensis*	Mts. of Buru (s Moluccas)

☐ **Banda Myzomela** *Myzomela boiei*

____	*M. b. boiei*	Banda (e Lesser Sundas)
____	*M. b. annabellae*	E Lesser Sundas (Babar, Yamdena and Selaru)

☐ **New Caledonian Myzomela** *Myzomela caledonica*

Mangroves and forests of New Caledonia

☐ **Scarlet Myzomela** *Myzomela sanguinolenta*

E Australia (Cooktown, Queensland to e Victoria)

☐ **Micronesian Myzomela** *Myzomela rubratra*

____	*M. r. rubratra*	Kosrae (Caroline Islands)
____	*M. r. dichromata*	Pohnpei (Caroline Islands)
____	*M. r. major*	Truk (Caroline Islands)
____	*M. r. kurodai*	Yap (Caroline Islands)
____	*M. r. kobayashii*	Palau Islands (Babelthuap to Angaur)
____	*M. r. asuncionis*	N Mariana Islands
____	*M. r. saffordi*	S Mariana Islands

☐ **Cardinal Myzomela** *Myzomela cardinalis*

____	*M. c. lifuensis*	Loyalty Islands
____	*M. c. cardinalis*	Southern Vanuatu
____	*M. c. tenuis*	Northern Vanuatu and Banks Group
____	*M. c. tucopiae*	Tikopia I. (ne of Banks Group)
____	*M. c. nigriventris*	Samoa (Upolu, Savai'i and Tutuila)
____	*M. c. santaecrucis*	Solomon Islands (Santa Cruz and Torres)
____	*M. c. sanfordi*	Rennell (se Solomon Islands)
____	*M. c. pulcherrima*	Solomon Islands (San Cristóbal and Ugi)

☐ **Rotuma Myzomela** *Myzomela chermesina*

Rotuma I. (nw of Fiji)

☐ **Scarlet-bibbed Myzomela** *Myzomela sclateri*

Karkar and small islands off ne New Guinea and New Britain

☐ **Ebony Myzomela** *Myzomela pammelaena*

Smaller islands in Bismarck Archipelago

☐ **Scarlet-naped Myzomela** *Myzomela lafargei*

Buka, Bougainville, Shortland, Fauro, Choiseul and Santa Isabel

☐ **Yellow-vented Myzomela** *Myzomela eichhorni*

____ *M. e. eichhorni* — New Georgia, Rendova, Vangunu and Kulambangra islands
____ *M. e. ganongae* — Ranongga (Solomon Islands)
____ *M. e. atrata* — Solomon Islands (Vellalavella and Baga)

☐ **Red-bellied Myzomela** *Myzomela malaitae*

Malaita (se Solomon Islands)

☐ **Black-headed Myzomela** *Myzomela melanocephala*

Solomon Islands (Florida, Savo and Guadalcanal)

☐ **Sooty Myzomela** *Myzomela tristrami*

Solomon Islands (San Cristóbal and Santa Anna)

☐ **Orange-breasted Myzomela** *Myzomela jugularis*

Fiji Islands, Yasawa Group and Lau Archipelago

☐ **Black-bellied Myzomela** *Myzomela erythromelas*

New Britain (Bismarck Archipelago)

☐ **Black-breasted Myzomela** *Myzomela vulnerata*

Timor (Lesser Sundas)

☐ **Red-collared Myzomela** *Myzomela rosenbergii*

____ *M. r. rosenbergii* — Mts. of nw New Guinea
____ *M. r. wahgiensis* — Mts. of w and central New Guinea
____ *M. r. longirostris* — Goodenough I. (D'Entrecasteaux Archipelago)

☐ **Banded Honeyeater** *Certhionyx pectoralis*

N Australia (Broome, Western Australia to Cape York Pen.)

☐ **Black Honeyeater** *Certhionyx niger*

Arid central Australia, to coasts in central Western Australia

☐ **Pied Honeyeater** *Certhionyx variegatus*

Arid central Australia, to coasts in central Western Australia

☐ **Forest Honeyeater** *Meliphaga montana*

____ *M. m. montana* — Mts. of nw New Guinea
____ *M. m. sepik* — N slope of Mts. of central New Guinea
____ *M. m. germanorum* — N New Guinea (Cyclops Mts.)
____ *M. m. huonensis* — NE New Guinea (Mts. of Huon Peninsula)
____ *M. m. aicora* — N slope of Mts. of se New Guinea
____ *M. m. steini* — Yapen I. (New Guinea)

☐ **Spot-breasted Meliphaga** *Meliphaga mimikae*

____ *M. m. rara* — N New Guinea (lowlands of upper Mamberamo River)
____ *M. m. mimikae* — Central Mts. of New Guinea
____ *M. m. bastille* — NE New Guinea (Hydrographic Range)
____ *M. m. granti* — Mts. of se New Guinea

☐ **Mountain Meliphaga** *Meliphaga orientalis*

____ *M. o. facialis* — Mts. of central New Guinea and Waigeo I.
____ *M. o. becki* — Mts. of ne New Guinea
____ *M. o. orientalis* — Mts. of se New Guinea
____ *M. o. citreola* — N New Guinea (n slope of Snow Mts.)

☐ **Scrub Honeyeater** *Meliphaga albonotata*

____ *M. a. setekwa* — Foothills of south-central New Guinea
____ *M. a. albonotata* — Foothills of s New Guinea

☐ **Puff-backed Honeyeater** *Meliphaga aruensis*

____ *M. a. sharpei* — New Guinea, w Papuan islands and D'Entrecasteaux Arch.
____ *M. a. aruensis* — S New Guinea and Aru Islands

☐ **Mimic Honeyeater** *Meliphaga analoga*

____ *M. a. papuae* — S New Guinea (Fly River to Hall Sound)

____ *M. a. analoga* — S New Guinea and w Papuan islands

____ *M. a. longirostris* — Aru Islands

____ *M. a. flavida* — Lowlands of n New Guinea, Yapen I. and Meos Num I.

____ *M. a. connectens* — Lowlands of n New Guinea (Wewak to Huon Gulf)

☐ **Tagula Honeyeater** *Meliphaga vicina*

Lowlands of Tagula I. (Louisiade Arch.). Status unknown

☐ **Graceful Honeyeater** *Meliphaga gracilis*

____ *M. g. stevensi* — N slope of Mts. of se New Guinea

____ *M. g. cinereifrons* — Lowlands of se New Guinea (west to Hall Sound); Sariba I.

____ *M. g. gracilis* — S New Guinea, Arus, is. in s Torres Strait and Cape York Pen.

____ *M. g. imitatrix* — NE Queensland (Mossman to Burdekin River)

☐ **Yellow-spotted Honeyeater** *Meliphaga notata*

____ *M. n. notata* — N Queensland (Cape York Pen. and islands in sw Torres Strait)

____ *M. n. mixta* — NE Queensland (Mossman to Burdekin River)

☐ **Yellow-gaped Honeyeater** *Meliphaga flavirictus*

____ *M. f. crockettorum* — New Guinea (except for range of flavirictus)

____ *M. f. flavirictus* — SE New Guinea (west to lower Fly River)

☐ **Lewin's Honeyeater** *Meliphaga lewinii*

____ *M. l. amphochlora* — NE Queensland (Lloyd Bay to Cooktown)

____ *M. l. mab* — E Queensland (Mossman to Fitzroy River)

____ *M. l. lewinii* — E Australia (se Queensland to s Victoria)

☐ **White-lined Honeyeater** *Meliphaga albilineata*

____ *M. a. albilineata* — N Australia (sandstones of w Arnhem Land, Northern Territory)

____ *M. a. fordiana* — NW Western Australia (sandstones of nw Kimberley)

☐ **Streak-breasted Honeyeater** *Meliphaga reticulata*

E Lesser Sundas (Timor and Semau)

☐ **Guadalcanal Honeyeater** *Guadalcanaria inexpectata*

Mts. of Guadacanal (se Solomon Islands)

☐ **Wattled Honeyeater** *Foulehaio carunculatus*

____ *F. c. carunculatus* — Samoa, e Fiji islands, Lau Archipelago and Tonga

____ *F. c. taviuensis* — Fiji (Taveuni and Vanua Levu)

____ *F. c. procerior* — Fiji (Viti Levu, Ovalau and islands in Yasawa Archipelago)

☐ **Black-throated Honeyeater** *Lichenostomus subfrenatus*

____ *L. s. subfrenatus* — NW New Guinea (Arfak Mts.)

____ *L. s. melanolaemus* — Central Mts. of New Guinea

____ *L. s. utakwensis* — New Guinea (Weyland and Snow Mts.)

____ *L. s. salvadorii* — Mts. of se New Guinea

☐ **Obscure Honeyeater** *Lichenostomus obscurus*

____ *L. o. viridifrons* — NE New Guinea (Mts. of Vogelkop Peninsula)

____ *L. o. obscurus* — New Guinea (Weyland, Snow and Sepik Mts.)

☐ **Bridled Honeyeater** *Lichenostomus frenatus*

Montane rainforests of ne Queensland (Cooktown to Paluma)

☐ **Eungella Honeyeater** *Lichenostomus hindwoodi*

Rainforests of central Queensland (Clarke Range)

☐ **Yellow-faced Honeyeater** *Lichenostomus chrysops*

____ *L. c. baroni* — NE Queensland (Blackdown Tableland to Atherton Tableland)

____ *L. c. chrysops* — E Australia (se Queensland to s Victoria and se South Australia)

____ *L. c. samueli* — South Australia (s Flinders Ranges to Mt. Lofty Range)

☐ **Varied Honeyeater** *Lichenostomus versicolor*

____ *L. v. sonoroides* — NW New Guinea (Vogelkop Peninsula) and w Papuan islands

____ *L. v. vulgaris* — Coastal n New Guinea, Yapen I. and Fergusson I.

____ *L. v. intermedia* — Samarai, Doini and Killerton islands (off e New Guinea)

____ *L. v. versicolor* — Coastal s New Guinea, Torres Strait islands and ne Queensland

☐ **Mangrove Honeyeater** *Lichenostomus fasciogularis*

Coastal e Australia (central Queensland to Hastings R., NSW)

☐ **Singing Honeyeater** *Lichenostomus virescens*

____ *L. v. cooperi* — N Australia (Arnhem Land, Melville I., Groote Eylandt)

____ *L. v. sonorus* — Cent. Queensland and NSW to s Victoria and se South Australia

____ *L. v. virescens* — Shark Bay, Western Australia to coastal w South Australia

____ *L. v. forresti* — Interior of Australia, to coast in w Western Australia

☐ **Yellow Honeyeater** *Lichenostomus flavus*

____ *L. f. flavus* — N Queensland (Cape York Peninsula)

____ *L. f. addensus* — Coastal central Queensland (Burdekin Rover to about Sarina)

☐ **White-gaped Honeyeater** *Lichenostomus unicolor*

N Australia (King Sound, W Australia, to Townsville, Qld.)

☐ **White-eared Honeyeater** *Lichenostomus leucotis*

____ *L. l. leucotis* — SE Australia (se Qld. to s Victoria and se South Australia)

____ *L. l. novaenorciae* — Cent. Queensland se South Australia and sw Western Australia

____ *L. l. thomasi* — Kargaroo I. (South Australia)

☐ **Yellow-throated Honeyeater** *Lichenostomus flavicollis*

Tasmania, King I. and Furneaux Group (Bass Strait)

☐ **Yellow-tufted Honeyeater** *Lichenostomus melanops*

____ *L. m. meltoni* — Cent. and e Queensland through NSW to far se S Australia

____ *L. m. melanops* — Eastern New South Wales (about Lismore to Jervis Bay)

____ *L. m. cassidix* — S-central Victoria (Yellingbo district of West Gippsland)

☐ **Purple-gaped Honeyeater** *Lichenostomus cratitius*

____ *L. c. occidentalis* — Disjunct in mallee from s-c Western Australia to cent. Victoria

____ *L. c. cratitius* — Kangaroo I. (South Australia)

☐ **Gray-headed Honeyeater** *Lichenostomus keartlandi*

N Australia (Pilbara, Western Australia to nw and c Queensland)

☐ **Yellow-tinted Honeyeater** *Lichenostomus flavescens*

____ *L. f. flavescens* — N Australia (Broome, Western Australia to w Cape York Pen.)

____ *L. f. melvillensis* — Melville and Bathurst Islands (Northern Territory)

Coastal n Australia and savanna of se New Guinea

☐ **Fuscous Honeyeater** *Lichenostomus fuscus*

____ *L. f. subgermanus* — E-central Queensland (Atherton Tablelands to about Sarina)

____ *L. f. fuscus* — E Australia (Rockhampton, Queensland to sw Victoria)

☐ **Gray-fronted Honeyeater** *Lichenostomus plumulus*

____ *L. p. plumulus* — Arid interior of Western Australia to central South Australia

____ *L. p. planasi* — N Australia (Fitzroy R., Western Australia to cent. Queensland)

____ *L. p. graingeri* — Semiarid central Queensland and NSW to e South Australia

☐ **Yellow-plumed Honeyeater** *Lichenostomus ornatus*

Mallee of s Austraia (sw Western Australia to central NSW)

☐ **White-plumed Honeyeater** *Lichenostomus penicillatus*

____ *L. p. penicillatus* — SE Australia (c Qld. to s Victoria and se South Australia)

____ *L. p. leilavalensis* — Eucalpytus-lined watercourses of central Australia

____ *L. p. carteri* — Central coast of Western Australia to sw Northern Territory

____ *L. p. calconi* — Broome, Western Australia to Victoria R., Northern Territory

☐ **Tawny-breasted Honeyeater** *Xanthotis flaviventer*

____ *X. f. fusciventris* — Waigeo and Batanta islands (New Guinea)

561

____ *X. f. flaviventer*	NW New Guinea, Misool and Salawati islands
____ *X. f. rubiensis*	West-central New Guinea
____ *X. f. saturatior*	Sabai, Boigu, Aru Islands and s New Guinea
____ *X. f. tararae*	Coastal s New Guinea (lower Digul River to Fly River)
____ *X. f. giulianettii*	SE New Guinea (Hall Sound to Port Moresby)
____ *X. f. visi*	SE New Guinea (Cloudy Bay to Milne Bay)
____ *X. f. kumusii*	N coast of se New Guinea (Collingwood Bay to Aicora River)
____ *X. f. madaraszi*	NE New Guinea (Huon Peninsula)
____ *X. f. philemon*	N New Guinea (Astrolabe Bay to Mamberamo River)
____ *X. f. meyeri*	Yapen I. (New Guinea)
____ *X. f. spilogaster*	Trobriand Islands and D'Entrecasteaux Archipelago
____ *X. f. filiger*	N Queensland (n Cape York Peninsula)

☐ **Spotted Honeyeater** *Xanthotis polygrammus*

____ *X. p. polygrammus*	Waigeo I. (New Guinea)
____ *X. p. keuhni*	Misool I. (New Guinea)
____ *X. p. poikilosternos*	Lower mountain slopes of w New Guinea and Salawati I.
____ *X. p. septentrionalis*	N New Guinea (Mamberamo River to upper Sepik River)
____ *X. p. lophotis*	Mts. of se New Guinea
____ *X. p. candidior*	S New Guinea (Trans-Fly lowlands)

☐ **Macleay's Honeyeater** *Xanthotis macleayanus*

Rainforests of ne Queensland (Cooktown to Townsville)

☐ **Kandavu Honeyeater** *Xanthotis provocator*

Forests and scrub of Kandavu (sw Fiji)

☐ **Orange-cheeked Honeyeater** *Oreornis chrysogenys*

W-central New Guinea (timberline zone of Snow Mts.)

☐ **Bonin Honeyeater** *Apalopteron familiare*

____ *A. f. familiare*	Muko-Shima (n Bonin Islands)
____ *A. f. hahasima*	Haha-Shima Group (s Bonin Islands)

☐ **White-naped Honeyeater** *Melithreptus lunatus*

____ *M. l. lunatus*	E Australia (ne Qld. to s Victoria and se South Australia)
____ *M. l. chloropsis*	SW Western Australia

☐ **Black-headed Honeyeater** *Melithreptus affinis*

Tasmania (except sw), King and Flinders Islands (Bass Strait)

☐ **White-throated Honeyeater** *Melithreptus albogularis*

____ *M. a. albogularis*	N Australia (Kimberley, Western Australia to Cape York Pen.)
____ *M. a. inopinatus*	E Australia (Cairns, Queensland to about Kempsey, NSW)

☐ **Black-chinned Honeyeater** *Melithreptus gularis*

____ *M. g. laetior*	N Australia (Pilbara region, W Australia to Cape York Pen.)
____ *M. g. gularis*	SE Australia (c Qld. to c Victoria and se South Australia)

☐ **Strong-billed Honeyeater** *Melithreptus validirostris*

Tasmania, King and Flinders Islands (Bass Strait)

☐ **Brown-headed Honeyeater** *Melithreptus brevirostris*

____ *M. b. brevirostris*	SE Australia (ne NSW to s Victoria and far se South Australia)
____ *M. b. wombeyi*	S Victoria (OtwayPeninsula and Gippsland)
____ *M. b. pallidiceps*	S-central Queensland to nw Victoria and se South Australia
____ *M. b. magnirostris*	Kangaroo I. (South Australia)
____ *M. b. leucogenys*	Southern Western Australia to Eyre Peninsula, South Australia

☐ **Stitchbird** *Notiomystis cincta*

____ *N. c. hautura*	Little Barrier I. (New Zealand)
____ *N. c. cincta†*	Formerly Great Barrier, North and Kapiti is. Extinct ca 1885

☐ **Plain Honeyeater** *Pycnopygius ixoides*

____ *P. i. ixoides*	NW New Guinea (east to Geelvink Bay)

____ *P. i. simplex*	N New Guinea (Mamberamo River to middle Sepik River)
____ *P. i. proximus*	N New Guinea (middle Sepik River to Astrolabe Bay)
____ *P. i. unicus*	NE New Guinea
____ *P. i. cinereifrons*	S New Guinea (Mimika River to upper Fly River)
____ *P. i. finschi*	N coast of se New Guinea (Kumusi River to Milne Bay)

☐ **Marbled Honeyeater *Pycnopygius cinereus***

____ *P. c. cinereus*	NW New Guinea (Mts. of Vogelkop Peninsula)
____ *P. c. dorsalis*	W New Guinea (Weyland and Nassau Mts.)
____ *P. c. marmoratus*	E New Guinea (Adelbert Mts. and Huon Peninsula Mts.)

☐ **Streak-headed Honeyeater *Pycnopygius stictocephalus***

Lowlands of New Guinea, Salawati I. and Aru Islands

☐ **White-streaked Friarbird *Melitograis gilolensis***

N Moluccas (Morotai, Halmahera, Kasiruta and Bacan)

☐ **Meyer's Friarbird *Philemon meyeri***

Lowlands of New Guinea (except for Vogelkop region)

☐ **Timor Friarbird *Philemon inornatus***

Timor (e Lesser Sundas)

☐ **Gray Friarbird *Philemon kisserensis***

E Lesser Sundas (Kisar, Leti and Moa)

☐ **Brass' Friarbird *Philemon brassi***

Unreported since 1940 discovery in nw New Guinea

☐ **Dusky Friarbird *Philemon fuscicapillus***

N Moluccas (Halmahera, Bacan and Morotai)

☐ **Little Friarbird *Philemon citreogularis***

____ *P. c. papuensis*	Trans-Fly savanna of s New Guinea
____ *P. c. sordidus*	N Australia (Broome, Western Australia to n Queensland)
____ *P. c. citreogularis*	E Australia (Cape York Pen. to n Victoria and s South Australia)

☐ **Black-faced Friarbird *Philemon moluccensis***

____ *P. m. moluccensis*	Buru (n Moluccas)
____ *P. m. plumigenis*	Tanimbar Is. (Larat, Yamdena); Kai Is. (Kai Kecil, Kai Besar)

☐ **Seram Friarbird *Philemon subcorniculatus***

Lowlands of Seram (s Moluccas)

☐ **Helmeted Friarbird *Philemon buceroides***

____ *P. b. novaeguineae*	New Guinea
____ *P. b. gordoni*	Northern Territory (Melville I. and coastal Arnhem Land)
____ *P. b. ammitophila*	N Australia (Interior of Arnhem Land, Northern Territory)
____ *P. b. yorki*	NE Queensland (island s of Torres Strait to Connors Range)
____ *P. b. buceroides*	E Lesser Sundas (Sawu, Roti, Timor, Semau and Wetar)
____ *P. b. neglectus*	W Lesser Sundas (Lombok and Sumba and Flores)

☐ **White-naped Friarbird *Philemon albitorques***

Manus I. (nw Bismarck Archipelago)

☐ **New Britain Friarbird *Philemon cockerelli***

____ *P. c. umboi*	Umboi I. (sw Bismarck Archipelago)
____ *P. c. cockerelli*	Bismarck Archipelago (New Britain and Duke of York)

☐ **New Ireland Friarbird *Philemon eichhorni***

New Ireland (e Bismarck Archipelago)

☐ **Silver-crowned Friarbird *Philemon argenticeps***

____ *P. a. argenticeps*	N Australia (Kimberley, Western Australia to nw Queensland)
____ *P. a. kempi*	N Queensland (Cape York Peninsula)

☐ **Noisy Friarbird *Philemon corniculatus***

____ *P. c. ellioti*	Trans-Fly lowlands of se New Guinea
____ *P. c. corniculatus*	N Queensland (Cape York Peninsula to Burdekin River)
____ *P. c. monachus*	E Australia (e and central Queensland to s and central Victoria)

☐ **New Caledonian Friarbird** *Philemon diemenensis*

Forests of New Caledonia, Maré and Lifou

☐ **Leaden Honeyeater** *Ptiloprora plumbea*
_____ *P. p. granti*
_____ *P. p. plumbea*

Central Highlands of New Guinea
Mts. of se New Guinea and Herzog Mts.

☐ **Olive-streaked Honeyeater** *Ptiloprora meekiana*
_____ *P. m. occidentalis*
_____ *P. m. meekiana*

Central New Guinea (Nassau and Snow Mts.)
Herzog Mts. and Mts. of se New Guinea

☐ **Rufous-sided Honeyeater** *Ptiloprora erythropleura*
_____ *P. e. erythropleura*
_____ *P. e. dammermani*

NW New Guinea (Mts. of Vogelkop Peninsula)
Central mountain ranges of New Guinea

☐ **Mayr's Honeyeater** *Ptiloprora mayri*

N New Guinea (Foya, Cyclops and Bewani Mts.)

☐ **Rufous-backed Honeyeater** *Ptiloprora guisei*
_____ *P. g. acrophila*
_____ *P. g. umbrosa*
_____ *P. g. guisei*

Coastal n New Guinea
N New Guinea (Sepik Mts.)
Herzog Mts. and Mts. of se New Guinea

☐ **Black-backed Honeyeater** *Ptiloprora perstriata*
_____ *P. p. praedicta*
_____ *P. p. incerta*
_____ *P. p. perstriata*

NW New Guinea (Wandammen Mts.)
Mts. of w-central New Guinea
Mts. of central and e New Guinea

☐ **Sooty Melidectes** *Melidectes fuscus*
_____ *M. f. occidentalis*
_____ *M. f. gilliardi*
_____ *M. f. fuscus*

Subalpine forests of central mountain ranges of New Guinea
E New Guinea (Bismarck Mts.)
SE New Guinea (Scratchley and Wharton Mts.)

☐ **Bismarck Melidectes** *Melidectes whitemanensis*

Mts. of central New Britain (Bismarck Archipelago)

☐ **Short-bearded Melidectes** *Melidectes nouhuysi*

W New Guinea (Mt. Wilhelmina in Snow Mts.)

☐ **Long-bearded Melidectes** *Melidectes princeps*

Alpine grasslands of e-central New Guinea

☐ **Cinnamon-browed Melidectes** *Melidectes ochromelas*
_____ *M. o. ochromelas*
_____ *M. o. batesi*
_____ *M. o. lucifer*

W New Guinea (Tamrau, Arfak and Wandammen Mts.)
SE New Guinea (Weyland and Nassau Mts.)
NE New Guinea (Mts. of Huon Peninsula)

☐ **Vogelkop Melidectes** *Melidectes leucostephes*

W New Guinea (Tamrau, Arfak, Fak Fak and Kumawa mts.)

☐ **Belford's Melidectes** *Melidectes belfordi*
_____ *M. b. brassi*
_____ *M. b. joiceyi*
_____ *M. b. kinneari*
_____ *M. b. belfordi*
_____ *M. b. schraderensis*

Mts. of nw New Guinea
W New Guinea (Weyland Mts.)
S New Guinea (Nassau and Snow Mts.)
Central Highlands and Mts. of se New Guinea
NE New Guinea (Schrader Mts.)

☐ **Yellow-browed Melidectes** *Melidectes rufocrissalis*
_____ *M. r. rufocrissalis*
_____ *M. r. thomasi*
_____ *M. r. gilliardi*

Sepik Mts. and central mountain ranges of New Guinea
Mts. of e New Guinea
Mts. of e-central New Guinea

☐ **Huon Melidectes** *Melidectes foersteri*

NE New Guinea (montane forests of Huon Peninsula)

☐ **Ornate Melidectes** *Melidectes torquatus*
- ____ *M. t. torquatus* — NW New Guinea (Mts. of Vogelkop Peninsula)
- ____ *M. t. nuchalis* — Central New Guinea (Weyland, Nassau and Snow Mts.)
- ____ *M. t. mixtus* — Central New Guinea (Snow Mts. to Victor Emanuel Mts.)
- ____ *M. t. cahni* — NE New Guinea (Mts. of Huon Peninsula)
- ____ *M. t. polyphonus* — NE New Guinea (Bismarck range to Herzog Mts.)
- ____ *M. t. emilii* — Mts. of se New Guinea

☐ **San Cristobal Melidectes** *Melidectes sclateri*

Mts. of San Cristóbal (Solomon Islands)

☐ **Arfak Honeyeater** *Melipotes gymnops*

NW New Guinea (Arfak, Tamrau and Wandammen Mts.)

☐ **Smoky Honeyeater** *Melipotes fumigatus*
- ____ *M. f. goliathi* — Mts. of central New Guinea
- ____ *M. f. fumigatus* — Mts. of se New Guinea
- ____ *M. f. kumawa* — Extreme nw New Guinea (Kumawa Mts.)

☐ **Spangled Honeyeater** *Melipotes ater*

NE New Guinea (montane forests of Huon Peninsula)

☐ **Dark-eared Honeyeater** *Myza celebensis*
- ____ *M. c. celebensis* — Mts. of n, central and se Sulawesi
- ____ *M. c. meriodionalis* — Mts. of s Sulawesi

☐ **Greater Streaked Honeyeater** *Myza sarasinorum*
- ____ *M. s. sarasinorum* — Mts. of n Sulawesi
- ____ *M. s. chionogenys* — Mts. of n-central and s-central Sulawesi
- ____ *M. s. pholidota* — Mts. of se Sulawesi

☐ **Giant Honeyeater** *Gymnomyza viridis*
- ____ *G. v. viridis* — Fiji (Taveuni and Vanua Levu)
- ____ *G. v. brunneirostris* — Fiji (Viti Levu)

☐ **Mao** *Gymnomyza samoensis*

Samoa (Mts. of Savai'i, Upolu and Tutuila)

☐ **Crow Honeyeater** *Gymnomyza aubryana*

New Caledonia

☐ **Kauai Oo** *Moho braccatus*

Alakai Swamp of Kauai (Hawaiian Is.). On verge of extinction

☐ **Bishop's Oo** *Moho bishopi*

Maui (Hawaiian Islands). On verge of extinction

☐ **Crescent Honeyeater** *Phylidonyris pyrrhopterus*
- ____ *P. p. pyrrhopterus* — Hunter R., NSW to s Victoria and se S Australia and Tasmania
- ____ *P. p. halmaturinus* — Mt. Lofty Range and Kangaroo I. (South Australia)

☐ **New Holland Honeyeater** *Phylidonyris novaehollandiae*
- ____ *P. n. novaehollandiae* — SE Queensland to Victoria and Eyre Pen. (South Australia)
- ____ *P. n. caudatus* — King I. and Flinders Islands (Bass Strait)
- ____ *P. n. canescens* — Tasmania
- ____ *P. n. campbelli* — Kangaroo I. (South Australia)
- ____ *P. n. longirostris* — SW Western Australia

☐ **White-cheeked Honeyeater** *Phylidonyris niger*
- ____ *P. n. niger* — E Australia (Atherton Tablelands to Wallaga Lake, se NSW)
- ____ *P. n. gouldii* — Coastal sw Western Australia

☐ **White-fronted Honeyeater** *Phylidonyris albifrons*

Woodlands of inland Australia, to southern and western coasts

☐ **Barred Honeyeater** *Phylidonyris undulatus*

Montane forests of New Caledonia

565

☐ **New Hebrides Honeyeater** *Phylidonyris notabilis*
_____ *P. n. notabilis*
_____ *P. n. superciliaris*

Banks Group and nw Vanuatu
N Vanuatu (Espiritu Santo to Epi)

☐ **Tawny-crowned Honeyeater** *Phylidonyris melanops*
_____ *P. m. melanops*
_____ *P. m. chelidonia*

NE NSW to se S Australia, e Tasmania and sw W Australia
Western Tasmania

☐ **Brown-backed Honeyeater** *Ramsayornis modestus*

Mangroves of s New Guinea, ne Queensland and adj. islands

☐ **Bar-breasted Honeyeater** *Ramsayornis fasciatus*

N Australia (Broome to Cape York Pen. and coastal c Qld.)

☐ **Striped Honeyeater** *Plectorhyncha lanceolata*

E Australia (central Qld. to nw Victoria and se South Australia)

☐ **Rufous-banded Honeyeater** *Conopophila albogularis*
_____ *C. a. mimikae*
_____ *C. a. albogularis*

N and s New Guinea and Aru Islands
Arnhem Land, Melville I., Cape York Pen. and Torres Strait is.

☐ **Rufous-throated Honeyeater** *Conopophila rufogularis*

N Australia (Fitzroy R., Western Australia to Noosa, Qld.)

☐ **Gray Honeyeater** *Conopophila whitei*

Mulga woodlands of inland western Australia

☐ **Painted Honeyeater** *Grantiella picta*

E Australia (Roper R., N. Territory to NSW and c Victoria)

☐ **Regent Honeyeater** *Xanthomyza phrygia*

SE Queensland to s and central Victoria; se South Australia

☐ **Eastern Spinebill** *Acanthorhynchus tenuirostris*
_____ *A. t. cairnsensis*
_____ *A. t. tenuirostris*
_____ *A. t. dubius*
_____ *A. t. halmaturinus*

Montane ne Queensland (Windsor Tablelands to Paluma Range)
E Australia (c Queensland to s Victoria and se South Australia)
Tasmania, King and Flinders Islands (Bass Strait)
SE South Australia (Mt. Lofty and Flinders Ranges; Kangaroo I.

☐ **Western Spinebill** *Acanthorhynchus superciliosus*

SW Western Australia

☐ **Blue-faced Honeyeater** *Entomyzon cyanotis*
_____ *E. c. harterti*
_____ *E. c. albipennis*
_____ *E. c. griseigularis*
_____ *E. c. cyanotis*

S New Guinea (Trans-Fly lowlands)
N Australia (Kimberley, Western Australia to nw Queensland)
N Queensland (Cape York Peninsula)
E Australia (e and cent. Qld. to Victoria and se South Australia)

☐ **Bell Miner** *Manorina melanophrys*

Coastal e Australia (se Queensland to s Victoria)

☐ **Noisy Miner** *Manorina melanocephala*
_____ *M. m. titaniota*
_____ *M. m. lepidota*
_____ *M. m. melanocephala*
_____ *M. m. leachi*

NE Queensland (interior of s Cape York Peninsula)
E Australia (Burdekin R., Queensland to sw New South Wales)
SE Australia (central NSW to s Victoria and se South Australia)
Eastern Tasmania

☐ **Yellow-throated Miner** *Manorina flavigula*
_____ *M. f. melvillensis*
_____ *M. f. lutea*
_____ *M. f. flavigula*
_____ *M. f. wayensis*
_____ *M. f. obscura*
_____ *M. f. melanotis*

Melville and Bathurst Islands (Northern Territory)
Fitzroy R., Western Australia to MacArthur R., Northern Territory
N-central Queensland to nw Victoria and se South Australia
Interior of Australia, excepts in east; to west and south coasts
Southwest Western Australia (except extreme southwest)
Murray mallee of se NSW, nw Victoria and e South Australia

☐ **Black-eared Miner** *Manorina melanotis*

Restricted to *mallee* of south-central Australia

☐ **New Zealand Bellbird** *Anthornis melanura*
_____ *A. m. obscura*
_____ *A. m. dumerilii*

Three Kings Islands (New Zealand)
North I. and adjacent offshore islands (New Zealand)

____	*A. m. oneho*	Central New Zealand
____	*A. m. melanocephala†*	Formerly Chatham Islands. Extinct ca 1906
____	*A. m. melanura*	South I. and Stewart I. (New Zealand)
____	*A. m. incoronata*	Auckland Islands

☐ **Spiny-cheeked Honeyeater** *Acanthagenys rufogularis*

Interior of Australia, to coasts in south and west

☐ **Red Wattlebird** *Anthochaera carunculata*

____	*A. c. carunculata*	SE Australia (se Queensland to s and central Victoria)
____	*A. c. clelandi*	Kangaroo I. (South Australia)
____	*A. c. woodwardi*	SW Western Australia to Eyre Peninsula (South Australia)

☐ **Brush Wattlebird** *Anthochaera chrysoptera*

____	*A. c. chrysoptera*	E Australia (se Queensland to s Victoria and se S Australia)
____	*A. c. halmaturina*	Kangaroo I. (South Australia)
____	*A. c. tasmanica*	E and northern Tasmania

☐ **Little Wattlebird** *Anthochaera lunulata*

SW Western Australia

☐ **Yellow Wattlebird** *Anthochaera paradoxa*

____	*A. p. paradoxa*	N and e Tasmania
____	*A. p. kingi*	King I. (Bass Strait)

☐ **Tui** *Prosthemadera novaeseelandiae*

____	*P. n. novaeseelandiae*	New Zealand, Stewart I. and Auckland Islands
____	*P. n. kermadecensis*	Kermadec Islands
____	*P. n. chathamensis*	Chatham Islands

FAMILY: ORIOLIDAE (Old World Orioles—30)

☐ **Timor Oriole** *Oriolus melanotis*

____	*O. m. melanotis*	E Lesser Sundas (Roti, Timor and Semau)
____	*O. m. finschi*	Wetar (e Lesser Sundas)

☐ **Buru Oriole** *Oriolus bouroensis*

____	*O. b. bouroensis*	Buru (s Moluccas)
____	*O. b. decipiens*	Tanimbar Islands (Arafura Sea)

☐ **Seram Oriole** *Oriolus forsteni*

Seram (s Moluccas)

☐ **Halmahera Oriole** *Oriolus phaeochromus*

Halmahera (n Moluccas)

☐ **Brown Oriole** *Oriolus szalayi*

New Guinea and w Papuan islands

☐ **Olive-backed Oriole** *Oriolus sagittatus*

____	*O. s. magnirostris*	Lowlands of s New Guinea
____	*O. s. affinis*	N Australia (Broome, Western Australia to nw Queensland)
____	*O. s. grisescens*	N Queensland (Cape York Pen. and islands of s Torres Strait)
____	*O. s. sagittatus*	E Australia (n Queensland to Victoria and se South Australia)

☐ **Green Oriole** *Oriolus flavocinctus*

____	*O. f. flavocinctus*	Aru Islands, s New Guinea, e Lesser Sundas and n Australia
____	*O. f. tiwi*	Melville and Bathhurst Islands (Northern Territory)
____	*O. f. flavotinctus*	N Queensland (Cape York Pen. and s Torres Straits islands)
____	*O. f. kingi*	NE Queensland (Atherton Tablelands s to about Townsville)

☐ **Dark-throated Oriole** *Oriolus xanthonotus*

____	*O. x. xanthonotus*	S Myanmar, Thailand, Malaysia, Sumatra, sw Borneo and Java
____	*O. x. consobrinus*	Borneo and adjacent islands

567

_____ *O. x. mentawi*   Mentawi Archipelago and adjacent islands off Sumatra

_____ *O. x. persuasus*   SW Philippines (Palawan and Culion)

☐ **White-lored Oriole** *Oriolus albiloris*

N Philippines (montane forests of Luzon)

☐ **Philippine Oriole** *Oriolus steerii*

_____ *O. s. samarensis*   Philippines (Leyte, e Mindanao and Samar)

_____ *O. s. steerii*   S Philippines (Basilan and w Mindanao)

_____ *O. s. nigrostriatus*   Philippines (Masbate and Negros)

_____ *O. s. assimilis†*   Philippines. Formerly Cebu (last recorded 1906)

_____ *O. s. cinereogenys*   Sulu Archipelago

☐ **Isabela Oriole** *Oriolus isabellae*

N Philippines (mountains of n Luzon)

☐ **Eurasian Golden Oriole** *Oriolus oriolus*

_____ *O. o. oriolus*   W Palearctic to e Siberia; > to Africa and nw India

_____ *O. o. kundoo*   W Siberia to Indian subcontinent

☐ **African Golden Oriole** *Oriolus auratus*

_____ *O. a. auratus*   Senegambia to Sudan, Uganda, s Ethiopia and s Somalia

_____ *O. a. notatus*   Angola to Tanzania, Mozambique and ne South Africa

☐ **Black-naped Oriole** *Oriolus chinensis*

_____ *O. c. invisus*   S Vietnam

_____ *O. c. diffusus*   E Asia; > to India, Malaysia and Indochina

_____ *O. c. andamanensis*   Andaman Islands

_____ *O. c. macrourus*   Nicobar Islands

_____ *O. c. chinensis*   Philippine Islands

_____ *O. c. suluensis*   Sulu Archipelago

_____ *O. c. melanisticus*   Talaud Islands (Karakelong and Salebabu)

_____ *O. c. sanghirensis*   Sangihe and Tabuken islands (off n Sulawesi)

_____ *O. c. formosus*   Siau, Tahulandang, Ruang, Biaro and Mayu is. (off Sulawesi)

_____ *O. c. celebensis*   Sulawesi, Bangka, Talisei, Lembeh, Togian Is., Muna, Butung

_____ *O. c. frontalis*   Banggai and Sula islands (off Sulawesi)

_____ *O. c. oscillans*   Tukangbesi Islands (off Sulawesi)

_____ *O. c. boneratensis*   Tanahjampea, Bonerate, Lalaotoa, Madu and Kayuadi islands

_____ *O. c. mundus*   Simeulue I. (off Sumatra)

_____ *O. c. sipora*   Sipura I. (off Sumatra)

_____ *O. c. richmondi*   Siberut and Pagi islands (off Sumatra)

_____ *O. c. insularis*   Kangean Islands (Java Sea)

_____ *O. c. broderipii*   Lesser Sundas (Lombok, Sumba, Sumbawa, Flores, Bisar, Alor)

_____ *O. c. maculatus*   Sumatra, Java, Borneo, Bali, Belitung and Nias islands

☐ **Slender-billed Oriole** *Oriolus tenuirostris*

E Himalayas to sw China and n SE Asia

☐ **Green-headed Oriole** *Oriolus chlorocephalus*

_____ *O. c. amani*   Montane forests of extreme se Kenya and neTanzania

_____ *O. c. chlorocephalus*   Malawi and Mozambique (Mt. Chiperone)

_____ *O. c. speculifer*   S Mozambique (Mt. Gorongoza)

☐ **São Tomé Oriole** *Oriolus crassirostris*

Forests of São Tomé I. (Gulf of Guinea)

☐ **Western Black-headed Oriole** *Oriolus brachyrhynchus*

_____ *O. b. brachyrhynchus*   Sierra Leone and Guinea to Benin

_____ *O. b. laetior*   Benin to Cameroon, s Uganda and extreme w Kenya

☐ **Dark-headed Oriole** *Oriolus monacha*

_____ *O. m. monacha*   Montane juniper forests of n Ethiopia and Eritrea

_____ *O. m. meneliki (permistus)*   Montane juniper forests of s Ethiopia

☐ **African Black-headed Oriole** *Oriolus larvatus*
____ *O. l. angolensis* | Angola to Namibia and Zambia
____ *O. l. additus* | SE Tanzania and Mozambique
____ *O. l. tibicen* | Coastal s Tanzania to s Mozambique
____ *O. l. reichenowi* | Coastal s Somalia to central Tanzania
____ *O. l. rolleti* | Ethiopia, Sudan, Uganda, Rwanda, Burundi (?) and e Zaire
____ *O. l. larvatus* | E South Africa, Mozambique and s Zimbabwe

☐ **Black-tailed Oriole** *Oriolus percivali*

E African and Albertine Rift montane forests

☐ **Black-winged Oriole** *Oriolus nigripennis*

S Guinea to se Sudan, Cameroon and Angola; Bioko I.

☐ **Black-hooded Oriole** *Oriolus xanthornus*
____ *O. x. xanthornus* | N India, Myanmar, Thailand, Malaysia, Indochina and Sumatra
____ *O. x. maderaspatanus* | S peninsular India
____ *O. x. andamanensis* | Andaman Islands
____ *O. x. ceylonensis* | Sri Lanka
____ *O. x. tanakae* | Coastal ne Borneo and adjacent offshore islands

☐ **Black Oriole** *Oriolus hosii*

Montane forests of n Sarawak (n Borneo)

☐ **Black-and-crimson Oriole** *Oriolus cruentus*
____ *O. c. malayanus* | Malay Peninsula
____ *O. c. consanguineus* | Sumatra
____ *O. c. vulneratus* | Mountains of n Borneo
____ *O. c. cruentus* | Java

☐ **Maroon Oriole** *Oriolus traillii*
____ *O. t. traillii* | Himalayan foothills to Myanmar, n Thailand and n Indochina
____ *O. t. robinsoni* | S Laos and s Vietnam
____ *O. t. nigellicauda* | N Vietnam and Hainan (s China)
____ *O. t. ardens* | Taiwan

☐ **Silver Oriole** *Oriolus mellianus*

Mountains of s China; > to s Thailand and Cambodia

☐ **Wetar Figbird** *Sphecotheres hypoleucus*

Lowlands of Wetar I. (e Lesser Sundas)

☐ **Green Figbird** *Sphecotheres viridis*
____ *S. v. viridis* | E Lesser Sundas (Roti, Semau and Timor)
____ *S. v. flaviventris* | W Australia, Northern Territory and Queensland

☐ **Australian Figbird** *Sphecotheres vieilloti*
____ *S. v. ashbyi* | N Australia (King Sound, Western Australia to Arnhem Land)
____ *S. v. flaviventris* | N Queensland (Cape York Pen. south to about Cairns); Kai Is.
____ *S. v. vieilloti* | Coastal se New Guinea and ne Australia (s to New South Wales)

FAMILY: IRENIDAE (Fairy-bluebirds—2)

☐ **Asian Fairy-bluebird** *Irena puella*
____ *I. p. puella* | India to Myanmar, Thailand and Indochina
____ *I. p. malayensis* | Malaysia
____ *I. p. crinigera* | Sumatra, Borneo and adjacent islands
____ *I. p. turcosa* | Java
____ *I. p. tweeddalei* | Philippines (Palawan, Busuanga, Balabac, Culion and Calamian)

☐ **Philippine Fairy-bluebird** *Irena cyanogastra*
____ *I. c. cyanogastra* | N Philippines (Luzon and Polillo)
____ *I. c. ellae* | Philippines (Bohol, Samar and Leyte)
____ *I. c. melanochlamys* | S Philippines (Basilan)
____ *I. c. hoogstraali* | S Philippines (Mindanao and Dinagat)

FAMILY: LANIIDAE (Shrikes—30)

☐ **Tiger Shrike** *Lanius tigrinus*

NE Asia; > to SE Asia, Greater Sundas and Philippines

☐ **Bull-headed Shrike** *Lanius bucephalus*
_____ *L. b. bucephalus*   E Asia, Japan, Korea and ne China
_____ *L. b. sicarius*   W-central China (Tao Valley of sw Gansu)

☐ **Red-backed Shrike** *Lanius collurio*

Widespread Palearctic region; > to South Africa

☐ **Rufous-tailed Shrike** *Lanius isabellinus*
_____ *L. i. phoenicuroides*   Iran to Pakistan, s Kazakstan and extreme w China (w Xinjiang)
_____ *L. i. isabellinus (speculigerus)*   SE Altai and n-central China; > to India, e and c Africa
_____ *L. i. arenarius*   NW China (Xinjiang); > to Iran, Pakistan and nw India
_____ *L. i. tsaidamensis*   N-central China (Qinghai); wintering grounds unknown

☐ **Brown Shrike** *Lanius cristatus*
_____ *L. c. cristatus*   E Siberia to nw Mongolia; > to India and Malay Pen.
_____ *L. c. confusus*   Manchuria and Amurland; > to Malay Pen. and Sumatra
_____ *L. c. superciliosus*   S Sakhalin and Japan; > to Sumatra and Lesser Sundas
_____ *L. c. lucionensis*   Korea and e China; > to Philippines, Borneo, Sulawesi

☐ **Burmese Shrike** *Lanius collurioides*
_____ *L. c. collurioides*   Mountains of Assam to s China, Myanmar and Indochina
_____ *L. c. nigricapillus*   S Vietnam

☐ **Emin's Shrike** *Lanius gubernator*

Savanna of Ivory Coast to s Sudan, n Uganda and ne Zaire

☐ **Souza's Shrike** *Lanius souzae*
_____ *L. s. souzae*   SE Gabon to s Congo, s Zaire, n Zambia and Angola
_____ *L. s. burigi*   NW Tanzania to e Zambia, e Rwanda and Burundi
_____ *L. s. tacitus*   SE Angola to n Namibia, s Zambia, Malawi and w Mozambique

☐ **Bay-backed Shrike** *Lanius vittatus*
_____ *L. v. vittatus*   Pakistan (Indus Plains) to w Bengal and s India (except for sw)
_____ *L. v. nargianus*   SE Iran, s Turkmenistan, Afghanistan, Baluchistan and Pakistan

☐ **Long-tailed Shrike** *Lanius schach*
_____ *L. s. erythronotus*   NE Iran to Pakistan and n India
_____ *L. s. caniceps*   W and s India and Sri Lanka
_____ *L. s. schach*   E and s China, Taiwan and Hainan
_____ *L. s. tricolor*   Nepal to n Myanmar, Yunnan, n Laos and n Thailand
_____ *L. s. bentet*   Malay Peninsula, Sumatra, Java, Borneo and Lesser Sundas
_____ *L. s. longicaudatus*   Central and se Thailand
_____ *L. s. nasutus*   Philippine Islands
_____ *L. s. suluensis*   Sulu Archipelago
_____ *L. s. stresemanni*   E New Guinea

☐ **Gray-backed Shrike** *Lanius tephronotus*
_____ *L. t. tephronotus*   Nepal to Sikkim, Bhutan, n India and w-central China
_____ *L. t. lahulensis*   N Kashmir to Ladakh and adjacent w Tibet

☐ **Gray-capped Shrike** *Lanius validirostris*
_____ *L. v. validirostris*   N Philippines (n Luzon)
_____ *L. v. tertius*   N Philippines (Mindoro)
_____ *L. v. hachisuka*   S Philippines (Mindanao)

☐ **Loggerhead Shrike** *Lanius ludovicianus*
_____ *L. l. gambeli*   W N America (sw Canada to sw US); > to w Mexico
_____ *L. l. excubitorides*   Great Plains region of North America; > to s Mexico

____ *L. l. migrans*	E North America (se Canada to e Texas); > to ne Mexico
____ *L. l. sonoriensis*	Arid sw US to nw Mexico (n Durango and s Sinaloa)
____ *L. l. anthonyi*	N Channel Islands (off s California)
____ *L. l. mearnsi*	San Clemente I. (off s California). ±13 birds in wild 1999
____ *L. l. grinnelli*	S California (San Diego County) and n Baja California
____ *L. l. nelsoni*	S Baja California
____ *L. l. ludovicianus*	Coastal se US (Virginia to Florida)
____ *L. l. miamensis*	S Florida
____ *L. l. mexicanus*	Central Mexico (s Tamaulipas and Nayarit to Oaxaca)

☐ **Northern Shrike** *Lanius excubitor*

____ *L. e. excubitor*	W and n Europe to w Siberia
____ *L. e. homeyeri*	Balkan Peninsula to s Ural Mountains and w Siberia
____ *L. e. leucopterus*	W Siberia to Yenisey River
____ *L. e. sibiricus*	E Siberia to n Mongolia and Kamchatka Peninsula
____ *L. e. bianchii*	Sakhalin and s Kuril Islands (n Japan)
____ *L. e. mollis*	Russian Altai and nw Mongolia
____ *L. e. funereus*	W China (Tien Shan Mountains)
____ *L. e. invictus*	N Alaska to extreme n British Columbia and Alberta
____ *L. e. borealis*	E Canada (Quebec and n Ontario); > to ne US

☐ **Southern Gray Shrike** *Lanius meridionalis*

____ *L. m. meridionalis*	Iberian Peninsula and s France; > to nw Africa
____ *L. m. koenigi*	Canary Islands
____ *L. m. algeriensis*	Morocco (n of Atlas Mountains), coastal n Algeria and Tunisia
____ *L. m. elegans*	N Sahara (Mauritania to Sinai Peninsula and Red Sea)
____ *L. m. leucopygos*	S Sahara (Mali to Nile River valley of the Sudan)
____ *L. m. aucheri*	W coast of Red Sea to s Iran and Arabian Peninsula
____ *L. m. theresae*	Galilee hills of n Israel and s Lebanon
____ *L. m. buryi*	Yemen; vagrant to Djibouti and Ethiopia
____ *L. m. uncinatus*	Socotra
____ *L. m. jebelmarrae*	W Sudan (Darfur)
____ *L. m. lahtora*	E Pakistan and n India
____ *L. m. pallidirostris*	Iran to arid steppes of w China (Xinjiang, Gansu and Ningsia)

☐ **Lesser Gray Shrike** *Lanius minor*

	Iberian Pen. to Siberia and central Asia; > to s Africa

☐ **Chinese Gray Shrike** *Lanius sphenocercus*

____ *L. s. sphenocercus*	Mountains of e Russia (Amur region) to ne and central China
____ *L. s. giganteus*	Mountains of e Tibet; > to se China

☐ **Gray-backed Fiscal** *Lanius excubitoroides*

____ *L. e. excubitoroides*	SE Mauritania to w Sudan, Mali, Chad, ne Zaire and Uganda
____ *L. e. intercedens*	Central Ethiopia to nw Uganda and w Kenya
____ *L. e. boehmi*	W Tanzania to Rwanda, sw Uganda and sw Kenya

☐ **Long-tailed Fiscal** *Lanius cabanisi*

	Dry thornscrub of s Somalia to Kenya and ne Tanzania

☐ **Taita Fiscal** *Lanius dorsalis*

	S Sudan to Somalia, Ethiopia and ne Tanzania

☐ **Somali Fiscal** *Lanius somalicus*

	Arid thornscrub of the Horn of Africa

☐ **Mackinnon's Shrike** *Lanius mackinnoni*

	Equatorial rainforests of central Africa

☐ **Common Fiscal** *Lanius collaris*

____ *L. c. smithii*	Guinea to s Sudan
____ *L. c. humeralis*	Eritrea and Ethiopia to s Tanzania and n Mozambique
____ *L. c. marwitzi*	NE and s-central Tanzania
____ *L. c. capelli*	SW Uganda to e Zaire, Angola, Zambia and Malawi
____ *L. c. collaris (predator)*	S Namibia, S Africa to s Transvaal, Natal and w Swaziland

____ *L. c. pyrrhostictus*	Zimbabwe to n Transvaal, Natal and e Botswana
____ *L. c. subcoronatus*	Namibia to n Cape Prov., Botswana and w Zimbabwe
____ *L. c. aridicolus*	Arid coastal nw Namibia and sw Angola

☐ **Newton's Fiscal** *Lanius newtoni*

São Tomé I. Rediscovered 1990 after 50-year absence

☐ **Masked Shrike** *Lanius nubicus*

Greece to sw Iran; > to central Africa and sw Arabia

☐ **Woodchat Shrike** *Lanius senator*

____ *L. s. senator*	Europe (Spain to w Turkey) and N Africa; > to c Africa
____ *L. s. badius*	W Mediterranean islands; > to central Africa
____ *L. s. niloticus*	E Turkey and Levant to Iran; > to central Africa

☐ **Yellow-billed Shrike** *Corvinella corvina*

____ *C. c. corvina*	S Mauritania and Senegal to Mali and nw Nigeria
____ *C. c. togoensis*	Guinea and Sierra Leone to s Chad and Sudan (Darfur)
____ *C. c. caliginosa*	Sudan (Bahr al Ghazal)
____ *C. c. affinis*	S Sudan to extreme ne Zaire, n Uganda and w Kenya

☐ **Magpie Shrike** *Corvinella melanoleuca*

____ *C. m. aequatorialis (angolensis)*	E Africa (north of the Zambezi River) north to Kenya
____ *C. m. expressa*	SE Zimbabwe to e Transvaal, s Mozambique and e Swaziland
____ *C. m. melanoleuca*	Africa (s of the Zambezi River) to e Angola and South Africa

☐ **White-rumped Shrike** *Eurocephalus rueppelli*

S Sudan to Somalia, Ethiopia, Kenya and Tanzania

☐ **White-crowned Shrike** *Eurocephalus anguitimens*

____ *E. a. anguitimens*	SW Angola to Botswana, Namibia and Zimbabwe
____ *E. a. niveus*	E Zambia to s Mozambique, Zimbabwe and Transvaal

FAMILY: MALACONOTIDAE (Bushshrikes and Allies—46)

☐ **Brubru** *Nilaus afer*

____ *N. a. afer*	Senegal to the Sudan, Eritrea and Ethiopia
____ *N. a. camerunensis*	S Cameroon and Central African Republic to e Zaire
____ *N. a. minor (brevialatus)*	SE Sudan to s Eritrea, Somalia, extreme n Kenya and Tanzania
____ *N. a. massaicus*	SW Kenya to n Tanzania, Rwanda and e Zaire
____ *N. a. nigritemporalis (occidentalis)*	E Angola, se Zaire, Tanzania, Zambia, Malawi, n Mozambique
____ *N. a. brubru*	S Angola to n Cape Province
____ *N. a. solivagus*	Local in Natal, Swaziland, Zimbabwe, extreme s Mozambique
____ *N. a. affinis*	Central highlands of w Angola and adjacent s Zaire
____ *N. a. miombensis*	Coastal Zululand and Mozambique (Sol do Save region)

☐ **Northern Puffback** *Dryoscopus gambensis*

____ *D. g. gambensis*	Senegal to Cameroon, Chad and n Gabon
____ *D. g. congicus*	SW Congo and w Zaire
____ *D. g. malzacii (erwini)*	E Cameroon to Cent. African Rep., s Uganda, w Kenya, e Zaire
____ *D. g. erythreae*	Extreme e Sudan to Eritrea and Ethiopia

☐ **Pringle's Puffback** *Dryoscopus pringlii*

Ethiopia and Somalia to Kenya and ne Tanzania

☐ **Black-backed Puffback** *Dryoscopus cubla*

____ *D. c. affinis*	Coastal Somalia and Kenya to n Tanzania, Zanzibar, Mafia I.
____ *D. c. nairobiensis*	Central Kenya to ne Tanzania
____ *D. c. hamatus (chapini)*	S Kenya to Mozambique, n Angola and ne South Africa
____ *D. c. okavangensis*	NW Cape Province to Namibia, s Angola, Botswana, Zambia
____ *D. c. cubla*	Natal to Cape Province

☐ **Red-eyed Puffback** *Dryoscopus senegalensis*

Lowlands of Senegambia to se Zaire, n Angola and Uganda

☐ **Pink-footed Puffback** *Dryoscopus angolensis*
____ *D. a. angolensis (boydi)*
____ *D. a. nandensis*
____ *D. a. kungwensis*

E Nigeria, Cameroon, Gabon, Congo, w Zaire and Angola
E Zaire to s Sudan, Uganda, w Rwanda and w Kenya
W Tanzania (Kungwe-Mahare region)

☐ **Large-billed Puffback** *Dryoscopus sabini*
____ *D. s. sabini*
____ *D. s. melanoleucus*

Sierra Leone to s Nigeria
Cameroon and Gabon to Angola and central Zaire (Ituri Forest)

☐ **Marsh Tchagra** *Tchagra minutus*
____ *T. m. minutus*
____ *T. m. reichenowi (remotus)*
____ *T. m. anchietae*

Sierra Leone to s Sudan, Ethiopia, w Kenya and nw Tanzania
E and s Tanzania, s Malawi, e Zimbabwe and Mozambique
S Zaire to Angola, n Zambia, sw Tanzania and n Malawi

☐ **Black-crowned Tchagra** *Tchagra senegalus*
____ *T. s. cucullatus*
____ *T. s. percivali*
____ *T. s. remigialis*
____ *T. s. nothus (timbuktanus)*
____ *T. s. senegalus (pallidus)*
____ *T. s. wardangliensis*
____ *T. s. habessinicus*
____ *T. s. armenus (camerunensis, sudanensis, rufofuscus)*
____ *T. s. orientalis (mozambicus, confusus)*
____ *T. s. kalahari*

Coastal Morocco to Algeria and Tunisia
S Arabian Peninsula
Central Chad to c Sudan (Darfur, Kordofan and Nile Valley)
Mali to n Nigeria and Lake Chad
Senegal and Sierra Leone to Mali, s Chad, Cent. African Rep.
N Somalia (Warsangli)
S Sudan (upper Nile Province) to Eritrea, Ethiopia and Somalia
S Cameroon to n Zaire, s Sudan, Uganda, Kenya, Mozambique
S Somalia to e Transvaal, Natal and e Cape Province
S Angola, n Namibia, sw Zambia, nw Zimbabwe to n S Africa

☐ **Brown-crowned Tchagra** *Tchagra australis*
____ *T. a. ussheri*
____ *T. a. emini (frater)*
____ *T. a. minor (littoralis, congener)*
____ *T. a. rhodesiensis*
____ *T. a. ansorgei*
____ *T. a. bocagei*
____ *T. a. souzae*
____ *T. a. australis*
____ *T. a. damarensis*

Sierra Leone to sw Nigeria
SE Nigeria to Zaire, s Sudan, Rwanda, w Kenya, nw Tanzania
SE Kenya, Tanzania, Zambia to n Zimbabwe and Mozambique
NE Namibia to ne Botswana, se Angola, extreme sw Zambia
W Angola (Luanda to Moçamedes)
Angola (central Cuando-Cubango)
SE Zaire to central Angola plateau and extreme n Zambia
E Botswana to Transvaal, se Zimbabwe and s Mozambique
Namibia to extreme sw Angola, Botswana and n South Africa

☐ **Three-streaked Tchagra** *Tchagra jamesi*
____ *T. j. jamesi*
____ *T. j. mandanus*

Extreme se Sudan to Somalia, Uganda, Ethiopia and n Kenya
Coastal e Kenya, ne Tanzania and adj. is. (Manda and Lamu)

☐ **Southern Tchagra** *Tchagra tchagra*
____ *T. t. tchagra*
____ *T. t. caffrariae*
____ *T. t. natalensis*

Southern Cape Province
SE Cape Province to Natal border
Natal to Swaziland, se Transvaal and e Cape Province

☐ **Red-naped Bushshrike** *Laniarius ruficeps*
____ *L. r. rufinuchalis*
____ *L. r. ruficeps*
____ *L. r. kismayensis*

Eritrea to Ethiopia, Djibouti and se Kenya
Extreme nw Somalia
Coastal s Somalia and n Kenya

☐ **Luehder's Bushshrike** *Laniarius luehderi*

E Nigeria to s Cameroon, s Sudan, w Kenya and sw Tanzania

☐ **Braun's Bushshrike** *Laniarius brauni*

NW Angola (Cuanza Norte escarpment)

☐ **Gabela Bushshrike** *Laniarius amboimensis*

W-central Angola (Gabela escarpment)

☐ **Bulo Burti Boubou** *Laniarius liberatus*

Known from a bird captured 1991 in central Somalia (hybrid?)

☐ **Turati's Boubou** *Laniarius turatii*

Rainforests of Guinea-Bissau, w Guinea and w Sierra Leone

☐ **Tropical Boubou** *Laniarius aethiopicus*

____	*L. a. major*	Sierra Leone to s Sudan, Kenya, Tanzania and Malawi
____	*L. a. aethiopicus*	Extreme e Sudan to Eritrea, Ethiopia, Somalia and n Kenya
____	*L. a. erlangeri*	S Somalia (lower Shabeelle and Jubba valleys)
____	*L. a. ambiguus*	Kenya and ne Tanzania
____	*L. a. sublacteus*	S Somalia, Kenya and ne Tanzania: Zanzibar
____	*L. a. limpopoensis*	SE Zimbabwe (Limpopo River Valley) to n Transvaal
____	*L. a. mossambicus*	NE Botswana to s Zambia, Malawi and extreme n South Africa

☐ **Gabon Boubou** *Laniarius bicolor*

____	*L. b. bicolor*	Mangroves of Cameroon to Gabon
____	*L. b. guttatus*	Lower Congo River (w Zaire and Angola)
____	*L. b. sticturus*	S Angola to w Zambia, ne Namibia, n Botswana, w Zimbabwe

☐ **Southern Boubou** *Laniarius ferrugineus*

____	*L. f. savensis*	SE Botswana to se Zimbabwe and s Mozambique
____	*L. f. tongensis*	Coastal s Mozambique and South Africa
____	*L. f. natalensis*	SE Cape Province and Zululand
____	*L. f. pondoensis*	Coastal Natal to e Cape Province
____	*L. f. transvaalensis*	Lowlands of n Zululand, e Swaziland and Transvaal
____	*L. f. ferrugineus*	SW and s Cape Province

☐ **Common Gonolek** *Laniarius barbarus*

____	*L. b. helenae*	Mangroves of coastal Sierra Leone
____	*L. b. barbarus*	Senegal to Nigeria, extreme n Cameroon and s Chad

☐ **Black-headed Gonolek** *Laniarius erythrogaster*

Acacia of Nigeria to Sudan, Ethiopia, Kenya and nw Tanzania

☐ **Crimson-breasted Gonolek** *Laniarius atrococcineus*

SW Angola and Namibia to Botswana and nw South Africa

☐ **Papyrus Gonolek** *Laniarius mufumbiri*

Papyrus swamps of Zaire to Uganda, Kenya and Tanzania

☐ **Yellow-breasted Boubou** *Laniarius atroflavus*

Highlands of se Nigeria and w Cameroon

☐ **Slate-colored Boubou** *Laniarius funebris*

____	*L. f. funebris (atrocaeruleus)*	NW Somalia to Ethiopia, Kenya and Tanzania
____	*L. f. degener*	SE Ethiopia to coastal s Somalia, Kenya and n Tanzania

☐ **Sooty Boubou** *Laniarius leucorhynchus*

Sierra Leone to Sudan, Uganda, Zaire and ne Angola

☐ **Mountain Sooty Boubou** *Laniarius poensis*

____	*L. p. camerunensis*	Cameroon and Nigeria
____	*L. p. poensis*	Bioko I. (Gulf of Guinea)
____	*L. p. holomelas*	E Zaire, Uganda, Rwanda and Burundi

☐ **Fuelleborn's Boubou** *Laniarius fuelleborni*

____	*L. f. usambaricus (ulugurensis)*	Mts. of w Tanzania
____	*L. f. fuelleborni*	SW Tanzania (Iringa Mts.), n Malawi and extreme e Zambia

☐ **Rosy-patched Bushshrike** *Rhodophoneus cruentus*

____	*R. c. cruentus*	Extreme se Egypt to Eritrea, Ethiopia and extreme se Sudan
____	*R. c. kordofanicus*	Sudan (w Kordofan)
____	*R. c. hilgerti*	E and s Ethiopia to Somalia and n Kenya
____	*R. c. cathemagmenus*	S Kenya (Tsavo and Lake Victoria) to ne Tanzania

☐ **Bokmakierie** *Telophorus zeylonus*

____	*T. z. phanus*	S Angola and nw Namibia
____	*T. z. thermophilus*	N Cape Province to w Transvaal, Botswana and Namibia

___ *T. z. restrictus*	Chimanimani Mts. (e Zimbabwe and w Mozambique)
___ *T. z. zeylonus*	Transvaal to s Cape Province

☐ **Gray-green Bushshrike** *Telophorus bocagei*

___ *T. b. bocagei (ansorgei)*	S Cameroon to n Angola
___ *T. b. jacksoni (andaryae)*	Central Zaire to Uganda and w Kenya

☐ **Sulphur-breasted Bushshrike** *Telophorus sulfureopectus*

___ *T. s. sulfureopectus*	Senegal to Gabon, n Zaire and w Uganda
___ *T. s. similis (terminus)*	S Sudan, Ethiopia, e Zaire, e Uganda, Kenya to South Africa

☐ **Olive Bushshrike** *Telophorus olivaceus*

___ *T. o. makawa*	C and s Malawi (w of Shire Valley) to e Zimbabwe, n Transvaal
___ *T. o. bertrandi*	S Malawi (east of Shire Valley)
___ *T. o. vitorum*	Coastal Mozambique (Save River to Algoa Bay)
___ *T. o. interfluvius*	Mozambique (except littoral)
___ *T. o. olivaceus (taylori, rubiginosus)*	S Mozambique to e Transvaal and e Cape Province

☐ **Many-colored Bushshrike** *Telophorus multicolor*

___ *T. m. multicolor*	SW Mali and Sierra Leone to Cameroon
___ *T. m. batesi*	S Cameroon and Gabon to Rwanda, w Uganda and n Angola
___ *T. m. graueri*	Highlands of e Zaire to sw Uganda and Rwanda

☐ **Black-fronted Bushshrike** *Telophorus nigrifrons*

___ *T. n. nigrifrons*	Mts. of central Kenya to Tanzania and n Malawi
___ *T. n. manningi*	SE Zaire to n Zambia and (?) e Angola
___ *T. n. sandgroundi*	SE Malawi to Zimbabwe, Mozambique and ne Transvaal

☐ **Mt. Kupe Bushshrike** *Telophorus kupeensis*

	Rainforests of Mt. Kupé (sw Cameroon)

☐ **Four-colored Bushshrike** *Telophorus viridis*

___ *T. v. viridis*	N Angola to s Zaire and nw Zambia
___ *T. v. nigricauda*	Coastal se Kenya and e Tanzania
___ *T. v. quartus*	S Malawi to Zimbabwe and Mozambique
___ *T. v. quadricolor*	E Cape to Swaziland and n Transvaal

☐ **Doherty's Bushshrike** *Telophorus dohertyi*

	Mts. of Rwanda, Burundi, w Uganda, e Zaire and w Kenya

☐ **Fiery-breasted Bushshrike** *Malaconotus cruentus*

	Guinea and Sierra Leone to e Zaire and extreme w Uganda

☐ **Lagden's Bushshrike** *Malaconotus lagdeni*

___ *M. l. lagdeni*	Montane forests of Sierra Leone, Liberia and Ivory Coast
___ *M. l. centralis*	Montane forests of e-central Zaire, Rwanda and w Uganda

☐ **Green-breasted Bushshrike** *Malaconotus gladiator*

	Montane forests of w Cameroon and adjacent se Nigeria

☐ **Gray-headed Bushshrike** *Malaconotus blanchoti*

___ *M. b. blanchoti*	Senegal to n Nigeria and n Cameroon
___ *M. b. catharoxanthus*	N Cameroon to Eritrea, n Ethiopia, Uganda and w Kenya
___ *M. b. interpositus*	Angola to w Zambia
___ *M. b. citrinipectus*	Namibia (n Ovamboland to Cuene Valley)
___ *M. b. approximans*	S Ethiopia to Somalia, Kenya and n Tanzania
___ *M. b. hypopyrrhus*	Rwanda to Tanzania, Malawi, Mozambique and ne S Africa
___ *M. b. extremus*	SE Cape Province

☐ **Monteiro's Bushshrike** *Malaconotus monteiri*

___ *M. m. perspicillatus*	Mt. Cameroon (rediscovered 1992)
___ *M. m. monteiri*	NW Angola (last recorded 1954)

☐ **Uluguru Bushshrike** *Malaconotus alius*

	Uluguru Mts. (central Tanzania)

FAMILY: PRIONOPIDAE (Helmetshrikes and Allies—12)

☐ **White Helmetshrike** *Prionops plumatus*
____ *P. p. plumatus*   Senegal to n Cameroon
____ *P. p. concinnatus (adamauae)*   Cent. Cameroon to Zaire, se Sudan, Ethiopia and n Uganda
____ *P. p. cristatus*   Eritrea and w Ethiopia to se Sudan, e Uganda and nw Kenya
____ *P. p. vinaceigularis (melanopterus)*   Somalia, Ethiopia, e Kenya and ne Tanzania
____ *P. p. poliocephalus (angolicus, talacomus)*   Cent. Kenya to s Uganda, se Zaire, Angola, Namibia, S Africa

☐ **Gray-crested Helmetshrike** *Prionops poliolophus*
                              Dry thornscrub of sw Kenya and w Tanzania

☐ **Yellow-crested Helmetshrike** *Prionops alberti*
                              Montane forests of e-central Zaire and (?) adjacent sw Uganda

☐ **Chestnut-bellied Helmetshrike** *Prionops caniceps*
____ *P. c. caniceps*   Guinea and Mali to Togo
____ *P. c. harterti*   Benin to w Cameroon

☐ **Rufous-bellied Helmetshrike** *Prionops rufiventris*
____ *P. r. rufiventris*   S Cameroon and Cent. African Rep. to Congo and nw Zaire
____ *P. r. mentalis*   Central and nw Zaire to Uganda

☐ **Retz's Helmetshrike** *Prionops retzii*
____ *P. r. graculinus (neumanni)*   S Somalia to e Kenya and ne Tanzania
____ *P. r. tricolor (intermedius)*   Tanzania to Malawi, Zambia, Mozambique and n Natal
____ *P. r. nigricans*   Angola to nw Zambia and se Zaire
____ *P. r. retzii*   Transvaal to Zimbabwe, s Zambia, Botswana, Namibia, Angola

☐ **Angola Helmetshrike** *Prionops gabela*
                              W Angola (Gabela escarpment)

☐ **Chestnut-fronted Helmetshrike** *Prionops scopifrons*
____ *P. s. keniensis*   Central Kenya
____ *P. s. kirki*   Coastal s Somalia to e Kenya (Lamu) and ne Tanzania
____ *P. s. scopifrons*   SE Tanzania to Mozambique and Zimbabwe

☐ **Large Woodshrike** *Tephrodornis gularis*
____ *T. g. sylvicola*   SW India (Western Ghats from Narbada River to Kerala)
____ *T. g. pelvicus*   E Himalayas (Nepal to Assam) to n Myanmar
____ *T. g. jugans*   S Myanmar and n Thailand
____ *T. g. vernayi*   SW Thailand
____ *T. g. annectens*   N peninsular Thailand and n Malaysia
____ *T. g. fretensis*   S peninsular Thailand, Malaysia and n Sumatra
____ *T. g. gularis*   Coastal sw and extreme s Sumatra and Java
____ *T. g. mekongensis*   E and s Thailand, Cambodia and s Indochina
____ *T. g. hainanus*   N Indochina and Hainan (s China)
____ *T. g. latouchei*   SE China (Fujian)
____ *T. g. frenatus*   Borneo

☐ **Common Woodshrike** *Tephrodornis pondicerianus*
____ *T. p. pallidus*   Pakistan and nw India
____ *T. p. pondicerianus*   E India to Myanmar, n Thailand and s Laos
____ *T. p. orientis*   Cambodia and s Vietnam
____ *T. p. affinis*   Sri Lanka

☐ **Rufous-winged Philentoma** *Philentoma pyrhoptera*
____ *P. p. pyrhoptera*   S Myanmar, Malaya, Sumatra, Borneo and offshore islands
____ *P. p. dubia*   Natuna Islands (China Sea)

☐ **Maroon-breasted Philentoma** *Philentoma velata*
____ *P. v. caesia*   S Myanmar, pen. Thailand, Malaya, Sumatra, Borneo and Java
____ *P. v. velata*   Lowland forests of Java

FAMILY: VANGIDAE (Vangas—15)

☐ **Red-tailed Vanga** *Calicalicus madagascariensis*

Forested regions of Madagascar

☐ **Red-shouldered Vanga** *Calicalicus rufocarpalis*

Southwestern desert of Madagascar

☐ **Rufous Vanga** *Schetba rufa*
____   *S. r. rufa*   Primary forests of e Madagascar
____   *S. r. occidentalis*   Primary forests of sw Madagascar

☐ **Hook-billed Vanga** *Vanga curvirostris*
____   *V. c. curvirostris*   Forests of n and e Madagascar
____   *V. c. cetera*   Forests of sw Madagascar

☐ **Lafresnaye's Vanga** *Xenopirostris xenopirostris*

Locally in arid subdeserts of sw Madagascar

☐ **Van Dam's Vanga** *Xenopirostris damii*

Dry woodlands of nw Madagascar

☐ **Pollen's Vanga** *Xenopirostris polleni*

Rainforests of nw coastal and e Madagascar

☐ **Sickle-billed Vanga** *Falculea palliata*

Savanna and mangroves of w and n Madagascar

☐ **White-headed Vanga** *Artamella viridis*
____   *A. v. viridis*   Forests of e Madagascar
____   *A. v. annae*   Forests of w Madagascar

☐ **Chabert Vanga** *Leptopterus chabert*
____   *L. c. chabert*   Forests and savanna of n and e Madagascar
____   *L. c. schistocercus*   Forests of sw Madagascar

☐ **Blue Vanga** *Cyanolanius madagascarinus*
____   *C. m. madagascarinus*   Forests of n and central Madagascar
____   *C. m. comorensis*   Grand Comoro I.
____   *C. m. bensoni*   Mohéli (Comoro Islands)

☐ **Bernier's Vanga** *Oriolia bernieri*

Humid forests of e Madagascar

☐ **Helmet Vanga** *Euryceros prevostii*

Dense forests of ne Madagascar

☐ **Tylas Vanga** *Tylas eduardi*
____   *T. e. eduardi*   Dense forests of e Madagascar
____   *T. e. albigularis*   Dense forests of w central Madagascar

☐ **Coral-billed Nuthatch** *Hypositta corallirostris*

Forests and plateaus of humid e Madagascar

FAMILY: DICRURIDAE (Drongos—24)

☐ **Papuan Drongo** *Chaetorhynchus papuensis*

Montane forests of New Guinea and Yule I.

☐ **Square-tailed Drongo** *Dicrurus ludwigii*
____   *D. l. sharpei*   Senegal to n Angola, Zaire, s Sudan, Uganda and w Kenya
____   *D. l. muenzneri*   S Somalia to e Kenya and s Tanzania
____   *D. l. saturnus*   Angola, Zambia and s Zaire (Katanga)
____   *D. l. ludwigii*   Swaziland and s Mozambique to e Cape Province
____   *D. l. tephrogaster*   S Malawi to Mozambique and e Zimbabwe

☐ **Shining Drongo** *Dicrurus atripennis*

Forests of Guinea and Sierra Leone to Gabon and e Zaire

☐ **Fork-tailed Drongo** *Dicrurus adsimilis*

_____ *D. a. adsimilis*   Africa south of the equatorial rain forests
_____ *D. a. fugax*   S Uganda and Kenya to n Natal and Zimbabwe
_____ *D. a. divaricatus*   Senegal to Chad, Sudan, Ethiopia, n Uganda and w Kenya
_____ *D. a. apivorus*   Angola, s Zaire, Zambia, Namibia, Botswana, and n Cape Prov.

☐ **Velvet-mantled Drongo** *Dicrurus modestus*

_____ *D. m. atactus*   Sierra Leone to s Nigeria and Cameroon; Bioko
_____ *D. m. coracinus*   S Nigeria to s Ghana and sw Nigeria
_____ *D. m. modestus*   Príncipe I. (Gulf of Guinea)

☐ **Aldabra Drongo** *Dicrurus aldabranus*

Aldabra (w Indian Ocean)

☐ **Comoro Drongo** *Dicrurus fuscipennis*

Grand Comoro I. (w Indian Ocean)

☐ **Crested Drongo** *Dicrurus forficatus*

_____ *D. f. forficatus*   Madagascar and Nosy Bé
_____ *D. f. potior*   Anjouan (Comoro Islands)

☐ **Mayotte Drongo** *Dicrurus waldenii*

Woodlands of Mayotte (Comoro Islands)

☐ **Black Drongo** *Dicrurus macrocercus*

_____ *D. m. albirictus*   SE Iran to Afghanistan and n India
_____ *D. m. macrocercus*   Peninsular India
_____ *D. m. minor*   Sri Lanka
_____ *D. m. cathoecus*   China, n Myanmar, n Thailand, Laos, n Vietnam and Malaysia
_____ *D. m. thai*   S Myanmar, s Thailand and s Vietnam
_____ *D. m. harterti*   Taiwan
_____ *D. m. javanus*   Java and Bali

☐ **Ashy Drongo** *Dicrurus leucophaeus*

_____ *D. l. longicaudatus*   E Afghanistan to Sikkim; > to s India and Sri Lanka
_____ *D. l. hopwoodi*   E Himalayas to Myanmar and s China; > to Indochina
_____ *D. l. mouhoti*   S Myanmar and n Thailand; > to Indochina
_____ *D. l. bondi*   S Thailand and Cambodia
_____ *D. l. nigrescens*   Extreme s Myanmar, s Thailand and Malaysia
_____ *D. l. leucogenis*   Manchuria and e China; > to Indochina
_____ *D. l. salangensis*   SE China and s Thailand; > to Hainan
_____ *D. l. innexus*   Hainan (s China)
_____ *D. l. stigmatops*   Mountains of n Borneo
_____ *D. l. batakensis*   N Sumatra
_____ *D. l. phaedrus*   S Sumatra
_____ *D. l. periophthalmicus*   Simeulue I. and adjacent Mentawi Islands (off Sumatra)
_____ *D. l. siberu*   Siberut I. (off Sumatra)
_____ *D. l. leucophaeus*   Java, Bali, Lombok, Palawan, Calamian and Balabac islands

☐ **White-bellied Drongo** *Dicrurus caerulescens*

_____ *D. c. caerulescens*   Foothills of s Nepal and peninsular India
_____ *D. c. insularis*   N Sri Lanka
_____ *D. c. leucopygialis*   S Sri Lanka

☐ **Crow-billed Drongo** *Dicrurus annectans*

E Himalayas to s China; > to SE Asia and Greater Sundas

☐ **Bronzed Drongo** *Dicrurus aeneus*

_____ *D. a. aeneus*   India to Myanmar, s China, Thailand, Malaysia and Indochina
_____ *D. a. malayensis*   S Malay Peninsula, Sumatra and Borneo
_____ *D. a. braunianus*   Taiwan

☐ **Lesser Racket-tailed Drongo** *Dicrurus remifer*

_____ *D. r. tectirostris*   Himalayas to s China, Myanmar, Thailand and Indochina
_____ *D. r. peracensis*   S Thailand to w Laos and Malay Peninsula

____ *D. r. lefoli*	S Cambodia (Chaine d'Éléphant and Cardamomes Mountains)
____ *D. r. remifer*	Sumatra and Java

☐ **Hair-crested Drongo** *Dicrurus hottentottus*

____ *D. h. hottentottus*	India to Myanmar, n Thailand and s Indochina
____ *D. h. brevirostris*	S China to n Myanmar, n Laos and n Vietnam
____ *D. h. viridinitens*	Mentawi Islands (off Sumatra)
____ *D. h. borneensis*	N Borneo, Maratua and Matasiri islands
____ *D. h. jentincki*	Bali and Kangean Islands
____ *D. h. leucops*	Sulawesi, adjacent islands and n Moluccas
____ *D. h. guillemardi*	Central Moluccas (Bisa and Obi)
____ *D. h. pectoralis*	Sula Islands (Taliabu, Mangola and Sanana)

☐ **Balicassiao** *Dicrurus balicassius*

____ *D. b. abraensis*	N Philippines (n Luzon)
____ *D. b. balicassius*	C and s Luzon, Lubang, Marinduque, Mindoro and Polillo
____ *D. b. mirabilis*	Philippines (Panay, Cebu, Negros, Guimaras, Ticao, Masbate)

☐ **Sulawesi Drongo** *Dicrurus montanus*

Hill and montane forests of Sulawesi

☐ **Sumatran Drongo** *Dicrurus sumatranus*

Humid lowlands of Sumatra

☐ **Wallacean Drongo** *Dicrurus densus*

____ *D. d. bimaensis*	Lesser Sundas (Lombok to Alor)
____ *D. d. renschi*	Sumbawa (Lesser Sundas)
____ *D. d. sumbae*	Sumba (Lesser Sundas)
____ *D. d. densus*	E Lesser Sundas (Roti, Semau, Timor, Wetar, Sermata, Luang)
____ *D. d. kuehni*	Tanimbar Islands (Arafura Sea)
____ *D. d. megalornis*	Seram Laut (Gorong, Manawoka), Watubela and Kai Islands

☐ **Ribbon-tailed Drongo** *Dicrurus megarhynchus*

New Ireland (e Bismarck Archipelago)

☐ **Spangled Drongo** *Dicrurus bracteatus*

____ *D. b. samarensis*	Philippines (Bohol, Leyte, Panaon, Samar and Calicoan)
____ *D. b. palawanensis*	Philippines (Palawan, Busuanga, Cagayan Sulu, Culion, Balabac)
____ *D. b. cuyensis*	Philippines (Cuyo and Semirara)
____ *D. b. menagei*	Philippines (Tablas)
____ *D. b. striatus*	S Philippines (Basilan, Mindanao and Nipa)
____ *D. b. suluensis*	Sulu Archipelago
____ *D. b. morotensis*	Morotai (Moluccas)
____ *D. b. atrocaeruleus*	N Moluccas (Bacan, Kasiruta and Halmahera)
____ *D. b. buruensis*	Buru (s Moluccas)
____ *D. b. amboinensis*	S Moluccas (Seram, Ambon, Haruku and Saparua)
____ *D. b. carbonarius*	Aru Is., New Guinea, Louisiade Arch., D'Entrecasteaux Arch.
____ *D. b. laemostictus*	New Britain (Bismarck Archipelago)
____ *D. b. meeki*	Guadalcanal (Solomon Islands)
____ *D. b. longirostris*	San Cristóbal (Solomon Islands)
____ *D. b. carbonarius*	Saibai, Boigu and Dauan Islands (north Torres Straits)
____ *D. b. baileyi*	N Australia (Kimberley to Arnhem Land and Melville I.)
____ *D. b. atrabectus*	Islands of s Torres Straits and Cape York Pen. s to Burdekin R.
____ *D. b. bracteatus*	E Australia (Burdekin R., Queensland to south coast of NSW)

☐ **Andaman Drongo** *Dicrurus andamanensis*

____ *D. a. andamanensis*	Andaman Islands
____ *D. a. dicruriformis*	S Myanmar and Cocos Islands (Bay of Bengal)

☐ **Greater Racket-tailed Drongo** *Dicrurus paradiseus*

____ *D. p. grandis*	N India to n Myanmar and n Vietnam
____ *D. p. rangoonensis*	Central India to s Myanmar, w Thailand and Indochina
____ *D. p. paradiseus*	S India to s Thailand and Indochina
____ *D. p. johni*	Hainan (s China)

____	*D. p. malayensis*	N Malaysia
____	*D. p. platurus*	S Malaysia, Sumatra and adjacent nw islands
____	*D. p. ceylonicus*	Sri Lanka
____	*D. p. lophorinus*	W Sri Lanka
____	*D. p. otiosus*	Andaman Islands
____	*D. p. nicobariensis*	Nicobar Islands
____	*D. p. banguey*	Islands off north Borneo
____	*D. p. brachyphorus*	Borneo
____	*D. p. microlophus*	North Natuna Islands
____	*D. p. formosus*	Java

FAMILY: CALLAEIDAE (Wattlebirds—2)

☐ **Kokako** *Callaeas cinereus*

____	*C. c. wilsoni*	North I. and Great Barrier I. (New Zealand)
____	*C. c. cinereus*	Formerly South I. and (?) Stewart I. (New Zealand)

☐ **Saddleback** *Philesturnus carunculatus*

____	*P. c. rufusater*	Hen and adjacent North Islands (New Zealand)
____	*P. c. carunculatus*	South Cape Islands (New Zealand)

FAMILY: GRALLINIDAE (Mudnest Builders—2)

☐ **Magpie-lark** *Grallina cyanoleuca*

____	*G. c. neglecta*	N Australia, extreme s New Guinea and Timor.
____	*G. c. cyanoleuca*	Mainland Australia, except north and the most arid areas

☐ **Torrent-lark** *Grallina bruijni*

	Swift-flowing mountain streams of New Guinea

FAMILY: CORCORACIDAE (White-winged Chough and Apostlebird—2)

☐ **White-winged Chough** *Corcorax melanorhamphos*

____	*C. m. melanorhamphos*	E-central Queensland to Victoria and se South Australia
____	*C. m. whiteae*	South Australia (Eyre Pen. and Mt. Lofty and Flinders Ranges) Forests and woodlands of e and se Australia

☐ **Apostlebird** *Struthidea cinerea*

____	*S. c. dalyi*	Interior of Northern Territory and n Queensland
____	*S. c. cinerea*	Central Queensland to n Victoria and se South Australia

FAMILY: ARTAMIDAE (Woodswallows—11)

☐ **Ashy Woodswallow** *Artamus fuscus*

	India and Sri Lanka to Myanmar, s China and SE Asia

☐ **Fiji Woodswallow** *Artamus mentalis*

	N Fiji (Yasawa and Viti Levu to Taveuni and Qamea)

☐ **White-backed Woodswallow** *Artamus monachus*

	Sulawesi, Lembeh I., Banggai Islands and Sula Islands

☐ **Great Woodswallow** *Artamus maximus*

	Mountains of New Guinea

☐ **White-breasted Woodswallow** *Artamus leucorynchus*

____	*A. l. pelewensis*	Babelthuap I. (Palau Islands)
____	*A. l. leucorhynchus*	Philippines, Palawan, Borneo and Natuna Islands
____	*A. l. amydrus*	Sumatra, Bangka, Belitung, Kangean islands, Java and Bali
____	*A. l. humei*	Andaman and Cocos islands (Bay of Bengal)

____ *A. l. albiventer*	Sulawesi and Lesser Sundas
____ *A. l. musschenbroeki*	Babar I. and Tanimbar Islands (Larat, Yamdena and Kirimoen)
____ *A. l. leucopygialis*	Moluccas, Kai Is., Aru Is., New Guinea, n and e Australia
____ *A. l. melaleucus*	New Caledonia, Maré and Lifou (Loyalty Islands)
____ *A. l. tenuis*	Vanuatu and Banks Islands

☐ **Bismarck Woodswallow** *Artamus insignis*

Bismarck Archipelago (New Britain and New Ireland)

☐ **Masked Woodswallow** *Artamus personatus*

Australian mainland (except most northern and eastern areas)

☐ **White-browed Woodswallow** *Artamus superciliosus*

Inland eastern and s-central Australia; > to north

☐ **Black-faced Woodswallow** *Artamus cinereus*

____ *A. c. normani*	N Queensland (se Cape York Peninsula)
____ *A. c. dealbatus*	E Queensland (Burdekin River to Burnett River)
____ *A. c. melanops*	Interior of c Western Australia to c Queensland and n Victoria
____ *A. c. cinereus*	Southwestern Western Australia
____ *A. c. perspicillatus*	E Lesser Sundas

☐ **Dusky Woodswallow** *Artamus cyanopterus*

____ *A. c. cyanopterus*	SE Australia (c Queensland to se South Australia; Tasmania)
____ *A. c. perthi*	SW Western Australia to Eyre Peninsula (South Australia)

☐ **Little Woodswallow** *Artamus minor*

____ *A. m. derbyi*	N and e Australia (Broome, Western Australia to n NSW)
____ *A. m. minor*	Pilbara region to sw N Territory and adj. nw South Australia

FAMILY: PITYRIASEIDAE (Bristlehead—1)

☐ **Bornean Bristlehead** *Pityriasis gymnocephala*

Lowland forests of Borneo

FAMILY: CRACTICIDAE (Bellmagpies and Allies—13)

☐ **Mountain Peltops** *Peltops montanus*

Mountains of New Guinea

☐ **Lowland Peltops** *Peltops blainvillii*

Lowlands of New Guinea, Waigeo, Misool and Salawati islands

☐ **Black-backed Butcherbird** *Cracticus mentalis*

____ *C. m. mentalis*	Savanna of se New Guinea (Merauke to Port Moresby)
____ *C. m. kempi*	Northern Queensland (Cape York Peninsula)

☐ **Gray Butcherbird** *Cracticus torquatus*

____ *C. t. torquatus*	SE Australia (ne New South Wales to s Victoria)
____ *C. t. leucopterus*	Southern Australia (Pilbara to s South Australia and e Qld.)
____ *C. t. cinereus*	Tasmania

☐ **Silver-backed Butcherbird** *Cracticus argenteus*

____ *C. a. argenteus*	NW Western Australia (King Sound to Cambridge Gulf)
____ *C. a. colletti*	Northwest Northern Territory (Daly River to Arnhem Land)

☐ **Hooded Butcherbird** *Cracticus cassicus*

____ *C. c. cassicus*	Lowlands of New Guinea, w Papuan, Aru, Yapen and Biak Is.
____ *C. c. hercules*	Trobriand Islands and D'Entrecasteaux Archipelago

☐ **Tagula Butcherbird** *Cracticus louisiadensis*

Tagula I. (Louisiade Archipelago)

☐ **Pied Butcherbird** *Cracticus nigrogularis*

____	*C. n. picatus*	Western Australia and Northern Territory to w Queensland
____	*C. n. nigrogularis*	E Australia (Cape York Pen. to n Victoria and se S Australia)

☐ **Black Butcherbird** *Cracticus quoyi*

____	*C. q. quoyi*	New Guinea, Yapen I. and w Papuan islands
____	*C. q. spaldingi*	Aru Islands; W Australia to Arnhem Land and adj. islands
____	*C. q. alecto*	Islands of North Torres Strait (Boigu and Saibai)
____	*C. q. jardini*	Coasts of w and e Cape York Peninsula south to Cooktown
____	*C. q. rufescens*	E Queensland (Cairns to about Sarina)

☐ **Australasian Magpie** *Gymnorhina tibicen*

____	*G. t. papuana*	SE New Guinea (Trans-Fly lowlands)
____	*G. t. eylandtensis*	Interior of Kimberley and Arnhem Land to nw Queensland
____	*G. t. terraereginae*	S Cape York Pen. to n Victoria and e South Australia
____	*G. t. tibicen*	SE Australia (se Queensland to Bega, New South Wales)
____	*G. t. tyrannica*	Southern Victoria and adjacent southeast South Australia
____	*G. t. hypoleuca*	E Tasmania, King and Flinders Islands (Bass Strait)
____	*G. t. telonucua*	Southern South Australia to s Eyre and Yorke Peninsulas
____	*G. t. dorsalis*	Southwest Western Australia
____	*G. t. longirostris*	Central Western Australia (Pilbara region north to King Sound)

☐ **Pied Currawong** *Strepera graculina*

____	*S. g. magnirostris*	NE Queensland (e Cape York Peninsula to Laura)
____	*S. g. robinsoni*	Uplands of ne Queensland (Cooktown to about Laura)
____	*S. g. graculina*	E Australia (central Queensland to Hunter River, NSW)
____	*S. g. nebuloas*	SE Australia (cent. NSW to s and e Victoria); > to north
____	*S. g. ashbyi*	Southwestern Victoria; > to northeast
____	*S. g. crissalis*	Lord Howe I.

☐ **Black Currawong** *Strepera fuliginosa*

____	*S. f. fuliginosa*	Tasmania (except northeast)
____	*S. f. pervior*	Flinders I. (Bass Strait)
____	*S. f. colei*	King I. (Bass Strait)

☐ **Gray Currawong** *Strepera versicolor*

____	*S. v. versicolor*	SE Australia (Hunter R., NSW to s and sw Victoria)
____	*S. v. arguta*	Eastern Tasmania
____	*S. v. melanoptera*	Southeast South Australia and adj. sw NSW and nw Victoria
____	*S. v. halmaturina*	Kangaroo I. (South Australia)
____	*S. v. intermedia*	S-central South Australia to Eyre and Yorke Peninsulas
____	*S. v. plumbea*	S Western Australia to nw and far sw South Australia

FAMILY: PARADISAEIDAE (Birds-of-paradise—44)

☐ **Loria's Bird-of-paradise** *Cnemophilus loriae*

____	*C. l. amethystina*	Western, Southern and Eastern Highlands of New Guinea
____	*C. l. inexpectata*	Weyland, Nassau, Jayawijaya, Hindenberg, Victor Emanuel Mts.
____	*C. l. loriae*	Mts. of se New Guinea (Herzog Mts. to Owen Stanley Range)

☐ **Crested Bird-of-paradise** *Cnemophilus macgregorii*

____	*C. m. sanguineus*	Central and Eastern Highlands of Papua New Guinea
____	*C. m. macgregorii*	Mountains of se Papua New Guinea

☐ **Yellow-breasted Bird-of-paradise** *Loboparadisea sericea*

____	*L. s. sericea*	Central New Guinea (Weyland Mts. to Victor Emanuel Mts.)
____	*L. s. aurora*	W New Guinea (Hertzog Mountains)

☐ **Macgregor's Bird-of-paradise** *Macgregoria pulchra*
____ *M. p. pulchra*   SE New Guinea (high summits of Owen Stanley Mountains)
____ *M. p. carolinae*   W New Guinea (Snow, Jayawijaya and Star mountains)

☐ **Paradise-crow** *Lycocorax pyrrhopterus*
____ *L. p. morotensis*   N Moluccas (Morotai and Rau)
____ *L. p. pyrrhopterus*   N Moluccas (Bacan, Kasiruta and Halmahera)
____ *L. p. obiensis*   N Moluccas (Obi and Bisa)

☐ **Glossy-mantled Manucode** *Manucodia ater*
____ *M. a. ater*   Western two-thirds of New Guinea
____ *M. a. subalter*   Aru, w Papuan islands and se peninsular Papua New Guinea
____ *M. a. alter*   Tagula I. (Louisiade Archipelago)

☐ **Jobi Manucode** *Manucodia jobiensis*
     Lowlands of n New Guinea and Yapen I.

☐ **Crinkle-collared Manucode** *Manucodia chalybatus*
     Hills and lower montane forests of New Guinea and Misool I.

☐ **Curl-crested Manucode** *Manucodia comrii*
____ *M. c. comrii*   D'Entrecasteaux Archipelago
____ *M. c. trobriandi*   Trobriand Islands (Kiriwina and Kaileuna)

☐ **Trumpet Manucode** *Manucodia keraudrenii*
____ *M. k. keraudrenii*   Vogelkop, Onin Peninsula and Weyland Mts. (w New Guinea)
____ *M. k. aruensis*   Aru Islands (w New Guinea)
____ *M. k. jamesii*   Lowlands of s New Guinea (Mimika River to Port Moresby)
____ *M. k. neumanni*   N escarpment of central range of New Guinea
____ *M. k. adelberti*   Adelbert Mountains (n Papua New Guinea)
____ *M. k. diamondi*   Southern watershed of Eastern Highlands of New Guinea
____ *M. k. purpureoviolaceus*   Highlands of se Papua New Guinea
____ *M. k. hunsteini*   D'Entrecasteaux Archipelago
____ *M. k. jamesi*   N Torres Straits islands (Saibai and Boigu)
____ *M. k. gouldii*   N Queensland (s Torres Straits islands and n Cape York Pen.)

☐ **Long-tailed Paradigalla** *Paradigalla carunculata*
     W New Guinea (Arfak and Farfak mountains)

☐ **Short-tailed Paradigalla** *Paradigalla brevicauda*
     Patchily distributed mountains of w and central New Guinea

☐ **Arfak Astrapia** *Astrapia nigra*
     W New Guinea (Arfak Mts.). Sight record from Tamrau Mts.

☐ **Splendid Astrapia** *Astrapia splendidissima*
____ *A. s. helios*   W New Guinea (C Range of Irian Jaya to Victor Emanuel Mts.)
____ *A. s. splendidissima*   W segment of central cordillera of New Guinea

☐ **Ribbon-tailed Astrapia** *Astrapia mayeri*
     Central cordillera of e New Guinea

☐ **Princess Stephanie's Astrapia** *Astrapia stephaniae*
____ *A. s. feminina*   Schrader and Bismarck ranges and Sepik-Wahgi divide
____ *A. s. stephaniae*   E New Guinea (Central Highlands to Owen Stanley Range)

☐ **Huon Astrapia** *Astrapia rothschildi*
     NE New Guinea (mountains of Huon Peninsula)

☐ **Western Parotia** *Parotia sefilata*
     W New Guinea (Arfak, Tamrau and Wondiwoi mountains)

☐ **Carola's Parotia** *Parotia carolae*
____ *P. c. carolae*   W New Guinea (Weyland Mountains east to Wissel Lakes)
____ *P. c. meeki*   New Guinea (Snow Mountains to Victor Emanuel Mountains)
____ *P. c. chalcothorax*   W New Guinea (Doorman Mountains)
____ *P. c. berlepschi*   W New Guinea (Van Rees and [?] Foya mountains)

____ *P. c. clelandiae*	Papua New Guinea border to s watershed of Eastern Highlands
____ *P. c. chrysenia*	E New Guinea (n escarpment of Central Highlands)

☐ **Lawes' Parotia** *Parotia lawesii*

____ *P. l. lawesii*	W and s highlands of Papua New Guinea
____ *P. l. helenae*	N watershed of Papua New Guinea (Waria to Milne Bay)

☐ **Wahnes' Parotia** *Parotia wahnesi*

NE New Guinea (n coastal ranges of Huon Peninsula)

☐ **King-of-Saxony Bird-of-paradise** *Pteridophora alberti*

Central cordillera of Papua New Guinea

☐ **Magnificent Riflebird** *Ptiloris magnificus*

____ *P. m. magnificus*	Lowlands of w and w-central New Guinea
____ *P. m. intercedens*	Lowlands of central and se Papua New Guinea
____ *P. m. alberti*	N Queensland (n Cape York Peninsula)

☐ **Paradise Riflebird** *Ptiloris paradiseus*

E Australia (Dawes Range, Queensland to Bellinger R., NSW)

☐ **Victoria's Riflebird** *Ptiloris victoriae*

Rainforests of ne Queensland (Cooktown to Paluma Range)

☐ **Superb Bird-of-paradise** *Lophorina superba*

____ *L. s. superba*	W New Guinea (Arfak and Tamrau mountains)
____ *L. s. niedda*	W New Guinea (Mt. Wondiwoi in Wandammen Peninsula)
____ *L. s. feminina*	W New Guinea (Weyland Mts. to Hindenberg Mts.)
____ *L. s. latipennis*	E New Guinea (Central and E Highlands to mts. of Huon Pen.)
____ *L. s. minor*	Mountains of se Papua New Guinea
____ *L. s. sphinx*	Mountains of extreme se New Guinea

☐ **Black Sicklebill** *Epimachus fastuosus*

____ *E. f. fastuosus*	W New Guinea (Tamrau and Arfak mountains)
____ *E. f. atratus*	New Guinea (Mts. of Wandammen Peninsula to Kratka Range)
____ *E. f. ultimus*	Coastal n Papua New Guinea (Mt. Menawa and Mt. Somoro)

☐ **Brown Sicklebill** *Epimachus meyeri*

____ *E. m. meyeri*	Mountains of extreme se New Guinea
____ *E. m. bloodi*	Central Highlands of New Guinea
____ *E. m. albicans*	Weyland Mountains to Hindenburg and Victor Emanuel Mts.

☐ **Black-billed Sicklebill** *Epimachus albertisi*

____ *E. a. albertisi*	New Guinea (mountains of Vogelkop and Huon Peninsula)
____ *E. a. cervinicauda*	Central cordillera of New Guinea

☐ **Pale-billed Sicklebill** *Epimachus bruijnii*

Lowlands of n Irian Jaya and nw Papua New Guinea

☐ **Magnificent Bird-of-paradise** *Cicinnurus magnificus*

____ *C. m. magnificus*	Extreme w Irian Jaya and Salawati I.
____ *C. m. chrysopterus*	W and central New Guinea and Yapen I.
____ *C. m. hunsteini*	E Papua New Guinea

☐ **Wilson's Bird-of-paradise** *Cicinnurus respublica*

W Papuan islands (Waigeo and Batanta)

☐ **King Bird-of-paradise** *Cicinnurus regius*

____ *C. r. regius*	S New Guinea, Aru and w Papuan islands
____ *C. r. coccineifrons*	N watershed of main body of New Guinea and Yapen I.

☐ **Wallace's Standardwing** *Semioptera wallacii*

____ *S. w. halmaherae*	Hill and montane forests of Halmahera (n Moluccas)
____ *S. w. wallacii*	Hill and montane forests of Bacan (n Moluccas)

☐ **Twelve-wired Bird-of-paradise** *Seleucidis melanoleucus*
_____ *S. m. melanoleucus* | Lowlands of s New Guinea and Salawati I.
_____ *S. m. auripennis* | N New Guinea (Mamberamo River to Ramu River)

☐ **Lesser Bird-of-paradise** *Paradisaea minor*
_____ *P. m. minor* | Misool I. and w New Guinea (e to Papua New Guinea border)
_____ *P. m. jobiensis* | Yapen I.
_____ *P. m. finschi* | N Papua New Guinea border to Gogol and upper Ramu River

☐ **Greater Bird-of-paradise** *Paradisaea apoda*
_____ *P. a. apoda* | Aru Islands
_____ *P. a. novaeguineae* | S New Guinea (Timika to Fly/Strickland watershed)

☐ **Raggiana Bird-of-paradise** *Paradisaea raggiana*
_____ *P. r. raggiana* | S watershed of se Papua New Guinea
_____ *P. r. salvadorii* | Lowlands of s Papua New Guinea
_____ *P. r. intermedia* | N coast of se Papua New Guinea
_____ *P. r. augustaevictoriae* | New Guinea (Huon Pen. to Waria and lower Mambare rivers)

☐ **Goldie's Bird-of-paradise** *Paradisaea decora* | D'Entrecasteaux Archipelago (Fergusson and Normanby)

☐ **Red Bird-of-paradise** *Paradisaea rubra* | W Papuan islands (Waigeo, Batanta, Gemien and Saonek)

☐ **Emperor Bird-of-paradise** *Paradisaea guilielmi* | NE New Guinea (lower mountains of Huon Peninsula)

☐ **Blue Bird-of-paradise** *Paradisaea rudolphi*
_____ *P. r. rudolphi* | Mountains of se Papua New Guinea
_____ *P. r. margaritae* | Mountains of central Papua New Guinea

☐ **Lesser Melampitta** *Melampitta lugubris*
_____ *M. l. lugubris* | W New Guinea (Arfak Mountains)
_____ *M. l. rostrata* | W New Guinea (Weyland and Nassau mountains)
_____ *M. l. longicauda* | New Guinea (Jayawijaya to Owen Stanley Mts. and Huon Pen.)

☐ **Greater Melampitta** *Melampitta gigantea* | Patchily distributed lower montane slopes of New Guinea

FAMILY: PTILONORHYNCHIDAE (Bowerbirds—20)

☐ **White-eared Catbird** *Ailuroedus buccoides*
_____ *A. b. geislerorum* | Yapen I. and n New Guinea
_____ *A. b. oorti* | W Papuan islands and w New Guinea
_____ *A. b. buccoides* | S New Guinea (Triton Bay to upper Fly River)
_____ *A. b. cinnamomeus* | S New Guinea (Mimika River to upper Fly River)
_____ *A. b. stonii* | S coastal New Guinea (Hall Sound to Port Moresby)

☐ **Spotted Catbird** *Ailuroedus melanotis*
_____ *A. m. melanotis* | Aru Islands and s New Guinea
_____ *A. m. misoliensis* | Misool I. (w Papuan islands)
_____ *A. m. melanocephalus* | Mountains of se New Guinea
_____ *A. m. facialis* | W New Guinea (Nassau and Jayawijaya mountains)
_____ *A. m. guttaticollis* | New Guinea (Sepik River area and Mt. Hagen region)
_____ *A. m. astigmaticus* | NE New Guinea (mountains of Huon Peninsula)
_____ *A. m. jobiensis* | New Guinea (Weyland and Adelbert mountains)
_____ *A. m. arfakianus* | W New Guinea (Arfak Mountains of Vogelkop Peninsula)
_____ *A. m. maculosus* | NE Australia (ne Queensland from Claudie River to Townsville)

☐ **Green Catbird** *Ailuroedus crassirostris*
_____ *A. c. joanae* | NE Queensland (McIlwraith Range)

585

____ *A. c. maculosus*	E Queensland (Atherton Tablelands to Burdekin River)
____ *A. c. crassirostris*	E Australia (Dawes Range, Queensland to Jervis Bay, NSW)

☐ **Tooth-billed Catbird** *Ailuroedus dentirostris*

Montane ne Queensland (Atherton Tablelands to Burdekin River)

☐ **Archbold's Bowerbird** *Archboldia papuensis*

W New Guinea (locally in Nassau Mountains)

☐ **Sanford's Bowerbird** *Archboldia sanfordi*

Humid Central Highlands of New Guinea

☐ **Vogelkop Bowerbird** *Amblyornis inornata*

NW New Guinea (Arfak, Tamrau and Wandammen mountains)

☐ **Macgregor's Bowerbird** *Amblyornis macgregoriae*

____ *A. m. mayri*	New Guinea (Weyland, Nassau and Jayawijaya mountains)
____ *A. m. macgregoriae*	New Guinea (Central, Huon and Adelbert mountains)
____ *A. m. germana*	NE New Guinea (mountains of Huon Peninsula)
____ *A. m. kombok*	Papua New Guinea (Mt. Hagen area)
____ *A. m. lecroyae*	NE Papua New Guinea (Mt. Bosavi area)
____ *A. m. nubicola*	SE Papua New Guinea (Owen Stanley Mountains)

☐ **Streaked Bowerbird** *Amblyornis subalaris*

SE Papua New Guinea (Owen Stanley Mountains)

☐ **Golden-fronted Bowerbird** *Amblyornis flavifrons*

N-central New Guinea (Foya Mountains)

☐ **Golden Bowerbird** *Prionodura newtoniana*

Highland rain forests of ne Australia (ne Queensland)

☐ **Flame Bowerbird** *Sericulus aureus*

____ *S. a. aureus*	Lowlands and foothill forests of w New Guinea
____ *S. a. ardens*	S New Guinea (Toricelli and Prince Alexander mountains)

☐ **Fire-maned Bowerbird** *Sericulus bakeri*

NE New Guinea (humid montane forests of Adelbert Mts.)

☐ **Regent Bowerbird** *Sericulus chrysocephalus*

E Australia (Clarke Range, Queensland to Gosford, NSW)

☐ **Satin Bowerbird** *Ptilonorhynchus violaceus*

____ *P. v. minor*	NE Queensland (Atherton Tablelands to Burdekin River)
____ *P. v. violaceus*	E Australia (central Queensland to e and s Victoria)

☐ **Western Bowerbird** *Chlamydera guttata*

____ *C. g. guttata*	North West Cape area of Western Australia
____ *C. g. carteri*	Western Australia (Pilbara region) to central Australia

☐ **Spotted Bowerbird** *Chlamydera maculata*

E Australia (c Queensland to w NSW and extreme nw Victoria)

☐ **Great Bowerbird** *Chlamydera nuchalis*

____ *C. n. nuchalis*	N Australia (Broome, Western Australia to far nw Queensland)
____ *C. n. orientalis*	N Queensland (Cape York Peninsula south to Clark Range)

☐ **Yellow-breasted Bowerbird** *Chlamydera lauterbachi*

____ *C. l. lauterbachi*	N-central New Guinea (upper Mamberamo River to Digul River)
____ *C. l. uniformis*	S-central New Guinea (Geelvink Bay to Digul River)

☐ **Fawn-breasted Bowerbird** *Chlamydera cerviniventris*

New Guinea and n Queensland (Cape York Peninsula)

FAMILY: CORVIDAE (Crows, Jays and Magpies—119)

☐ **Crested Jay** *Platylophus galericulatus*

____ *P. g. ardesiacus*	S Myanmar, peninsular Thailand and Malay Peninsula

_____ *P. g. coronatus*   Borneo and Sumatra
_____ *P. g. galericulatus*   Java

☐ **Black Magpie** *Platysmurus leucopterus*
_____ *P. l. leucopterus*   S Myanmar, peninsular Thailand, Malay Peninsula and Sumatra
_____ *P. l. aterrimus*   Borneo

☐ **Siberian Jay** *Perisoreus infaustus*
_____ *P. i. infaustus*   Lapland to n Norway, Sweden, Finland and Kola Peninsula
_____ *P. i. ostjakorum*   Ural Mountains and nw Siberia
_____ *P. i. yakutensis*   NE Asia
_____ *P. i. ruthenus*   Central Scandinavia to central Russia and w Siberia
_____ *P. i. opicus*   Altai and Sayan mountains
_____ *P. i. rogosowi (sibericus)*   Central Siberia to Irkutsk, middle Yenisey River and Outer Siberia
_____ *P. i. varnak*   Upper Amur River (north to Stanovoi Mountains)
_____ *P. i. sakhalinensis*   N Sakhalin and Shantar islands
_____ *P. i. maritimus*   Lower Amur River (Bureya River to ne Manchuria)

☐ **Sichuan Jay** *Perisoreus internigrans*
                              Mountains of w China (se Qinghai, sw Gansu and n Sichuan)

☐ **Gray Jay** *Perisoreus canadensis*
_____ *P. c. pacificus*   N-central Alaska, n Yukon and nw Mackenzie
_____ *P. c. canadensis*   Canada (central Yukon) to n US
_____ *P. c. nigricapillus*   E Canada (n Quebec to Newfoundland and Nova Scotia)
_____ *P. c. arcus*   SW Canada (coastal mountains of British Columbia)
_____ *P. c. albescens*   NE British Columbia to w South Dakota and nw Nebraska
_____ *P. c. bicolor*   SE British Columbia to e Washington, ne Oregon and w Montana
_____ *P. c. capitalis*   Rocky Mountains (Idaho and Montana to New Mexico)
_____ *P. c. griseus*   SW British Columbia and Vancouver I. to ne California
_____ *P. c. obscurus*   Coastal nw US (Washington to nw California)

☐ **Steller's Jay** *Cyanocitta stelleri*
_____ *C. s. stelleri*   S Alaska and coastal British Columbia to nw Oregon
_____ *C. s. carlottae*   Queen Charlotte Islands (British Columbia)
_____ *C. s. annectens*   Interior British Columbia to ne Oregon and nw Wyoming
_____ *C. s. frontalis*   Central Oregon to s California; > to extreme n Baja
_____ *C. s. carbonacea*   Coastal California (Marin and Contra Costa to Monterey counties)
_____ *C. s. macrolopha*   Rocky Mountains (Nevada and Utah to n Sonora)
_____ *C. s. diademata*   Sierra Madre Occidental (se Sonora to Chihuahua and Jalisco)
_____ *C. s. philippsi*   Central Mexico (San Luis Potosí, Guanajuato and Hidalgo)
_____ *C. s. coronata*   Highlands of central Mexico (San Luis Potosí to n Puebla)
_____ *C. s. purpurea*   SW Mexico (highlands of w and central Michoacán)
_____ *C. s. azteca*   Mts. of c Mexico (México, Morelos, Puebla and w-c Veracruz)
_____ *C. s. teotepecencis*   Mountains of s Mexico (central and s Guerrero)
_____ *C. s. restricta*   S Mexico (highlands of Oaxaca)
_____ *C. s. ridgwayi*   Highlands of s Mexico (Chiapas) to Guatemala and El Salvador
_____ *C. s. lazula*   Highlands of El Salvador
_____ *C. s. suavis*   Highlands of w Honduras to central Nicaragua

☐ **Blue Jay** *Cyanocitta cristata*
_____ *C. c. bromia*   S Canada (Alberta to Quebec) to central US; > to se US
_____ *C. c. cristata*   E-central and se US
_____ *C. c. semplei*   S Florida
_____ *C. c. cyanoptera*   SE Wyoming and Nebraska to w Kansas, Oklahoma and n Texas

☐ **Black-throated Magpie-Jay** *Calocitta colliei*
                              Pacific slope of nw Mexico (se Sonora to Jalisco and Colima)

☐ **White-throated Magpie-Jay** *Calocitta formosa*
_____ *C. f. formosa*   Coastal s Mexico (Colima, Michoacán and Puebla to Oaxaca)

_____ *C. f. azurea*
_____ *C. f. pompata*

Pacific slope of se Mexico (Chiapas) and Guatemala
Arid interior of s Mexico (Oaxaca) to nw Costa Rica

☐ **Tufted Jay** *Cyanocorax dickeyi*

Montane forests of nw Mexico (Sinaloa, Durango and Nayarit)

☐ **Black-chested Jay** *Cyanocorax affinis*
_____ *C. a. zeledoni*
_____ *C. a. affinis*

Humid forests of se Costa Rica and Panama
N Colombia and nw Venezuela

☐ **Green Jay** *Cyanocorax yncas*
_____ *C. y. speciosus*
_____ *C. y. vividus*
_____ *C. y. luxuosus*
_____ *C. y. centralis*
_____ *C. y. maya*
_____ *C. y. cozumelae*
_____ *C. y. galeatus*
_____ *C. y. cyanodorsalis*
_____ *C. y. andicolus*
_____ *C. y. guatimalensis*
_____ *C. y. yncas*
_____ *C. y. longirostris*

Pacific slope of w Mexico (Nayarit and Jalisco)
Pacific slope of s Mexico (Colima) to w Guatemala
S Texas (Rio Grande Valley) to e Mexico (Puebla and Veracruz)
SE Mexico (Tabasco) to Belize, e Guatemala and Honduras
SE Mexico (Yucatán Peninsula) and extreme s Quintana Roo
Cozumel I.
Subtropical central Colombia (west of Eastern Andes)
E Andes of Colombia and nw Venezuela
Mountains of n Venezuela
Coastal cordillera of n Venezuela
Subtropical sw Colombia to Ecuador, Peru and central Bolivia
N Peru (arid upper Marañón Valley)

☐ **Brown Jay** *Cyanocorax morio*
_____ *C. m. palliatus*
_____ *C. m. morio*
_____ *C. m. cyanogenys*
_____ *C. m. vociferus*

Gulf-Caribbean slope of se Texas to se Mexico (Veracruz)
SE Mexico (coastal c Veracruz to e Tabasco and n Chiapas)
SE Mexico (extreme e Tabasco and Campeche) to nw Panama
N Yucatán Peninsula (Campeche, Yucatán and Quintana Roo)

☐ **Bushy-crested Jay** *Cyanocorax melanocyaneus*
_____ *C. m. melanocyaneus*
_____ *C. m. chavezi*

Highlands of central Guatemala to Honduras
Highlands of s Honduras and n Nicaragua

☐ **San Blas Jay** *Cyanocorax sanblasianus*
_____ *C. s. nelsoni*
_____ *C. s. sanblasianus*

Coastal w Mexico (Nayarit, Jalisco, Colima, Michoacán, Guerrero)
Coastal sw Mexico (central Guerrero)

☐ **Yucatan Jay** *Cyanocorax yucatanicus*
_____ *C. y. yucatanicus*
_____ *C. y. rivularis*

SE Mexico (Yucatán Peninsula) to Belize and Petén of Guatemala
SE Mexico (Tabasco and sw Campeche)

☐ **Purplish-backed Jay** *Cyanocorax beecheii*

Lowlands of w Mexico (se Sonora and Chihuahua to Nayarit)

☐ **Purplish Jay** *Cyanocorax cyanomelas*

SE Peru to ne Argentina, Paraguay and sw Brazil

☐ **Azure Jay** *Cyanocorax caeruleus*

SE Brazil to e Paraguay and ne Argentina

☐ **Violaceous Jay** *Cyanocorax violaceus*
_____ *C. v. pallidus*
_____ *C. v. violaceus*

Caribbean littoral of n Venezuela (Anzoátegui)
E Colombia to Venezuela, Guianas, Brazil, Peru and n Bolivia

☐ **Curl-crested Jay** *Cyanocorax cristatellus*

S-central Brazil to e Bolivia and extreme ne Paraguay

☐ **Azure-naped Jay** *Cyanocorax heilprini*

SE Colombia to sw Venezuela (Amazonas) and extreme nw Brazil

☐ **Cayenne Jay** *Cyanocorax cayanus*

SE Venezuela (Bolívar) to the Guianas and ne Brazil

☐ **Plush-crested Jay** *Cyanocorax chrysops*
_____ *C. c. diesingii*

N Brazil south of the Amazon

_____ *C. c. chrysops* — SE Brazil to Paraguay, Uruguay, n Bolivia and ne Argentina
_____ *C. c. tucumanus* — NW Argentina

☐ **White-naped Jay** *Cyanocorax cyanopogon*

Tableland of e Brazil (Pará to Minas Gerais and e Mato Grosso)

☐ **White-tailed Jay** *Cyanocorax mystacalis*

Arid sw Ecuador and nw Peru (south to La Libertad)

☐ **Black-collared Jay** *Cyanolyca armillata*
_____ *C. a. armillata* — Andes of e Colombia and w Venezuela
_____ *C. a. quindiuna* — Andes of s Colombia and n Ecuador (e Carchi and nw Napo)
_____ *C. a. meridana* — Andes of nw Venezuela (north to Trujillo)

☐ **Turquoise Jay** *Cyanolyca turcosa*

Andes of se Colombia to n Peru (n Piura and nw Cajamarca)

☐ **White-collared Jay** *Cyanolyca viridicyanus*
_____ *C. v. jolyaea* — Andes of n Peru (Amazonas to Junín)
_____ *C. v. cyanolaema* — Andes of se Peru (Cuzco and Puno)
_____ *C. v. viridicyanus* — Andes of w Bolivia (La Paz and Cochabamba)

☐ **Azure-hooded Jay** *Cyanolyca cucullata*
_____ *C. c. mitrata* — E Mexico (San Luis Potosí to n-central Oaxaca)
_____ *C. c. guatemalae* — S Mexico (Chiapas) to central Guatemala
_____ *C. c. hondurensis* — W Honduras
_____ *C. c. cucullata* — Costa Rica and w Panama

☐ **Beautiful Jay** *Cyanolyca pulchra*

W Andes of sw Colombia and nw Ecuador (south to Pichincha)

☐ **Black-throated Jay** *Cyanolyca pumilo*

Montane forests of extreme s Mexico to nw Honduras

☐ **Dwarf Jay** *Cyanolyca nana*

Oak-pine forests of sw Mexico (Vera Cruz, Puebla and Oaxaca)

☐ **Silvery-throated Jay** *Cyanolyca argentigula*
_____ *C. a. albior* — Montane forests of Costa Rica
_____ *C. a. argentigula* — Montane forests of s Costa Rica and w Panama

☐ **White-throated Jay** *Cyanolyca mirabilis*

Oak-pine forests of sw Mexico (s Guerrero and s-central Oaxaca)

☐ **Florida Scrub-Jay** *Aphelocoma coerulescens*

Locally in central peninsular and Atlantic coastal Florida

☐ **Island Scrub-Jay** *Aphelocoma insularis*

Santa Cruz I. (Channel Islands off s California)

☐ **Western Scrub-Jay** *Aphelocoma californica*
_____ *A. c. inmanis* — Extreme sw Washington to w Oregon (Willamette Valley)
_____ *A. c. caurina* — Coastal sw Oregon to c Calif. (Trinity, Lake and Napa counties)
_____ *A. c. oocleptica* — S-central Oregon to San Francisco Bay and w Nevada
_____ *A. c. californica* — Coastal ranges of c Calif. (s San Mateo to sw Ventura County)
_____ *A. c. cana* — Arid s California (Eagle Mountain area of Riverside County)
_____ *A. c. obscura* — SW California and n Baja California (south to Todos Santos Bay)
_____ *A. c. cactophila* — Central Baja California (latitude 29°30' to Bahía Magdalena)
_____ *A. c. hypoleuca* — Cape District of Baja California
_____ *A. c. nevadae* — SE Oregon s through Great Basin to ne Sonora and nw Chihuahua
_____ *A. c. woodhouseii* — Rocky Mountains to w Oklahoma, w Texas and n Chihuahua
_____ *A. c. texana* — W-central Texas
_____ *A. c. grisea* — NW Mexico (e slopes of Sierra Madre Occidental)
_____ *A. c. cyanotis* — Mts. of e central Mexico (s Coahuila and Nuevo León to Hidalgo)
_____ *A. c. sumichrasti* — Highlands of s Mexican plateau (Veracruz to Puebla and Oaxaca)
_____ *A. c. remota* — SW Mexico (Sierra Madre del Sur of Guerrero)

☐ **Mexican Jay** *Aphelocoma ultramarina*
_____ *A. u. arizonae* — Mts. of Arizona and New Mexico to n Sonora and nw Chihuahua

____	*A. u. wollweberi*	Mts. of w Mexico (se Sonora to Durango, Zacatecas and n Jalisco)
____	*A. u. gracilis*	Mts. of w-central Mexico (e Nayarit and n Jalisco)
____	*A. u. couchii*	Mts. of extreme sw Texas to s Nuevo León and c Tamaulipas
____	*A. u. potosina*	Mts. of e-c Mexico (San Luis Potosí to Querétaro and c Hidalgo)
____	*A. u. ultramarina*	S Mexican Plateau (Jalisco to Michoacán, Puebla and w Veracruz)
____	*A. u. colimae*	Mountains w Mexico (nw Jalisco to ne Colima)

☐ **Unicolored Jay** *Aphelocoma unicolor*

____	*A. u. concolor*	Mountains of se Mexico (w-c Veracruz, e México and Puebla)
____	*A. u. guerrerensis*	Mountains of w Mexico (s-central Guerrero)
____	*A. u. oaxacae*	S Mexico (central highlands of Oaxaca)
____	*A. u. unicolor*	Mountains of se Mexico (Chiapas) and Guatemala
____	*A. u. griscomi*	Mountains of n El Salvador and w Honduras

☐ **Pinyon Jay** *Gymnorhinus cyanocephalus*

Pinyon-juniper woodlands of s Oregon to sw US and n Baja

☐ **Eurasian Jay** *Garrulus glandarius*

____	*G. g. rufitergum*	S Scotland, England, Wales and n France
____	*G. g. hibernicus*	Ireland
____	*G. g. glandarius*	N and central Europe
____	*G. g. severtzowi*	Scandinavia and w Russia
____	*G. g. lusitanicus*	N Portugal and n Spain
____	*G. g. fasciatus*	Southern, central and eastern Spain
____	*G. g. corsicanus*	Corsica
____	*G. g. albipectus*	Italy, Dalmatian coast of Yugoslavia, Albania and Ionian Islands
____	*G. g. jordansi*	Sicily
____	*G. g. ichnusae*	Sardinia
____	*G. g. graecus*	S Yugoslavia, s Bulgaria and Greece
____	*G. g. cretorum*	Crete
____	*G. g. glaszneri*	Cyprus
____	*G. g. fernandi*	SE Bulgaria to n Turkey (Istranca Mountains)
____	*G. g. atricapillus*	Lebanon to s Syria, Israel and w Jordan
____	*G. g. anatoliae*	W Turkey and e Aegean Sea to w Asia Minor, n Iraq and sw Iran
____	*G. g. samios*	Samos and Ikaria region of e Aegean Sea
____	*G. g. iphigenia*	Crimean Peninsula
____	*G. g. minor*	Atlas Mountains of Morocco and Algeria
____	*G. g. cervicalis*	NE Algeria and Tunisia
____	*G. g. whitakeri*	N Morocco and nw Algeria
____	*G. g. krynicki*	Caucasus, Transcaucasia and n Asia Minor
____	*G. g. hyrcanus*	N Iran (Elzburg Mountains and south shore of Caspian Sea)
____	*G. g. brandtii*	Ural Mountains to Siberia, Lake Baikal and Altai and Sayan mts.
____	*G. g. bambergi*	Mongolia to Sakhalin, s Kuril Islands, Hokkaido and Korea
____	*G. g. kansuensis*	Kazakstan (e Tien Shan) and w China (Gansu)
____	*G. g. pekingensis*	N China (Liaoning) and sw Manchuria
____	*G. g. sinensis*	W China to n Yunnan and ne Myanmar
____	*G. g. leucotis*	E Myanmar to s Yunnan, Thailand to central Vietnam
____	*G. g. oatesi*	Central Myanmar (upper Chindwin and Chin Hills)
____	*G. g. barringtoni*	Myanmar (Mt. Victoria in s Chin Hills)
____	*G. g. interstinctus*	E Himalayas and se Tibet
____	*G. g. persaturatus*	N India (Khasi Hills of Assam)
____	*G. g. bispecularis*	Himalayas (Kashmir to Nepal)
____	*G. g. japonicus*	Japan (Hondo, Shikoku and Kyushu)
____	*G. g. tokugawae*	Sado I. (Japan)
____	*G. g. hiugaensis*	Japan (Isu Peninsula of e Hondo, s Kyushu and Kagoshima)
____	*G. g. orii*	Yakushima I. (Ryukyu Islands)
____	*G. g. namiyei*	Tsushima Islands (sw Japan)
____	*G. g. taivanus*	Taiwan

☐ **Black-headed Jay** *Garrulus lanceolatus*

Himalayas (Afghanistan to n India and central Nepal)

☐ **Lidth's Jay** *Garrulus lidthi*

Ryukyu Islands (Amami-O-Shima and Tokuno-Shima)

☐ **Azure-winged Magpie** *Cyanopica cyanus*
_____ *C. c. cooki* — Iberian Peninsula
_____ *C. c. cyanus* — E-central Asia
_____ *C. c. pallescens* — Middle and lower Amur River region
_____ *C. c. koreensis* — Korea
_____ *C. c. stegmanni* — Manchuria
_____ *C. c. swinhoei* — E China (Liaoning to Fujian and Sichuan)
_____ *C. c. interposita* — N China (Shaanxi)
_____ *C. c. kansuensis* — W China (Gansu, Qinghai and nw Sichuan)
_____ *C. c. japonica* — Japan (Hondo and Kyushu)

☐ **Ceylon Magpie** *Urocissa ornata*

Dense evergreen hill forests of Sri Lanka

☐ **Formosan Magpie** *Urocissa caerulea*

Montane forests of Taiwan

☐ **Gold-billed Magpie** *Urocissa flavirostris*
_____ *U. f. cucullata* — W Himalayas (Hazara to e Nepal)
_____ *U. f. flavirostris* — E Himalayas to Assam, s Tibet and n Myanmar
_____ *U. f. schaferi* — W Myanmar (Chin Hills)
_____ *U. f. robini* — N Vietnam (nw Tonkin)

☐ **Blue Magpie** *Urocissa erythrorhyncha*
_____ *U. e. brevivexilla* — SW Manchuria and n China
_____ *U. e. erythrorhyncha* — Central China to s Yunnan, n Laos and n Vietnam
_____ *U. e. alticola* — SW China (n Yunnan) and ne Myanmar
_____ *U. e. occipitalis* — Himalayas (Punjab to Sikkim)
_____ *U. e. magnirostris* — Hills of Assam to Indochina

☐ **White-winged Magpie** *Urocissa whiteheadi*
_____ *U. w. xanthomelana* — Mountains of s China to central Laos and n Vietnam
_____ *U. w. whiteheadi* — Mountains of Hainan (s China)

☐ **Green Magpie** *Cissa chinensis*
_____ *C. c. chinensis* — E Himalayas to se Tibet, Myanmar, n Laos and n Vietnam
_____ *C. c. robinsoni* — Malaysia
_____ *C. c. klossi* — Central Indochina
_____ *C. c. margaritae* — S Vietnam (Langbian Mountains)
_____ *C. c. minor* — Sumatra and nw Borneo

☐ **Yellow-breasted Magpie** *Cissa hypoleuca*
_____ *C. h. jini* — SE China (Yaoshan Massif of Guangxi)
_____ *C. h. concolor* — N Vietnam
_____ *C. h. chauleti* — Central Vietnam
_____ *C. h. hypoleuca* — E Thailand and s Indochina
_____ *C. h. katsumatae* — Hainan (s China)

☐ **Short-tailed Magpie** *Cissa thalassina*
_____ *C. t. thalassina* — Java
_____ *C. t. jeffreyi* — Mountains of n Borneo

☐ **Rufous Treepie** *Dendrocitta vagabunda*
_____ *D. v. pallida* — W Himalayas and nw India
_____ *D. v. vagabunda* — Lower Himalayas and ne India (south to Hyderabad)
_____ *D. v. parvula* — SW India (s Kanara to Cape Comorin)
_____ *D. v. vernayi* — SE India
_____ *D. v. sclateri* — E Myanmar (upper Chindwin to Chin Hills and Arakan Yoma)
_____ *D. v. kinneari* — S Myanmar and nw Thailand

____	*D. v. saturatior*	Tenasserim and s Thailand
____	*D. v. sakeratensis*	E Thailand and Indochina

☐ **Gray Treepie** *Dendrocitta formosae*

____	*D. f. occidentalis*	W Himalayas (n Pakistan to Garhwal)
____	*D. f. himalayensis*	E Himalayas to Myanmar and n Laos
____	*D. f. sarkari*	E India (s Jaipur and n Madras)
____	*D. f. assimilis*	S Myanmar to Thailand and Andaman Islands
____	*D. f. sinica*	E and se China to n Vietnam
____	*D. f. sapiens*	S China on Mt. Omei (w Sichuan)
____	*D. f. formosae*	Taiwan
____	*D. f. insulae*	Hainan (s China)

☐ **Sumatran Treepie** *Dendrocitta occipitalis*

Mountains of Sumatra

☐ **Bornean Treepie** *Dendrocitta cinerascens*

Mountains of Borneo

☐ **White-bellied Treepie** *Dendrocitta leucogastra*

Lowlands of sw India (Western Ghats)

☐ **Collared Treepie** *Dendrocitta frontalis*

Himalayas of n India, n Myanmar and nw Vietnam

☐ **Andaman Treepie** *Dendrocitta bayleyi*

Dense forests of Andaman Islands

☐ **Racket-tailed Treepie** *Crypsirina temia*

S Myanmar to n Malaya, Indochina, Java and Bali

☐ **Hooded Treepie** *Crypsirina cucullata*

Arid lowlands of Myanmar

☐ **Ratchet-tailed Treepie** *Temnurus temnurus*

Forests of central Laos, n Vietnam and Hainan

☐ **Eurasian Magpie** *Pica pica*

____	*P. p. pica*	British Isles, s Scandinavia, central and e Europe to Asia Minor
____	*P. p. fennorum*	N Scandinavia and w Russia
____	*P. p. galliae*	W Europe to Balkans
____	*P. p. melanotos*	Iberian Peninsula
____	*P. p. mauretanica*	NE Mauritania to Morocco, Algeria and Tunisia
____	*P. p. asirensis*	Assir Mountains (sw Arabia)
____	*P. p. bactriana*	Central Russia to n India and w Tibet
____	*P. p. hemileucoptera*	W and s Siberia to Outer Mongolia
____	*P. p. leucoptera*	S Transbaicalia to e Mongolia and Altai Mountains
____	*P. p. camtschatica*	N shores of Sea of Okhotsk to Kamchatka Peninsula
____	*P. p. bottanensis*	E Himalayas to se Tibet and w China (Qinghai and Xinjiang)
____	*P. p. sericea*	S China to Myanmar, Indochina, Hainan and Taiwan

☐ **Black-billed Magpie** *Pica hudsonia*

Alaska and Yukon to w Canada and w US

☐ **Yellow-billed Magpie** *Pica nuttalli*

Interior, coastal valleys and foothills of central California

☐ **Stresemann's Bush-Crow** *Zavattariornis stresemanni*

S Ethiopia (Sidamo Province)

☐ **Mongolian Ground-Jay** *Podoces hendersoni*

Deserts of central Asia (Kazakstan to Outer and Inner Mongolia)

☐ **Xinjiang Ground-Jay** *Podoces biddulphi*

Deserts of nw China (w Xinjiang)

☐ **Turkestan Ground-Jay** *Podoces panderi*

Deserts from Turkmenistan to Kazakstan

☐ **Iranian Ground-Jay** *Podoces pleskei*

Desert steppes of central and e Iran

☐ **Clark's Nutcracker** *Nucifraga columbiana*

Rocky Mountains (sw Canada to n Baja); casual to nw Mexico

☐ **Eurasian Nutcracker** *Nucifraga caryocatactes*

____ *N. c. caryocatactes*	Scandinavia to n and e Europe; > to s Russia
____ *N. c. macrorhynchos*	N and ne Asia; irruptions to n Iran, Korea and n China
____ *N. c. rothschildi*	Tien Shan Mountains (Kazakstan)
____ *N. c. japonica*	Central and s Kuril Islands, Hokkaido and Hondo
____ *N. c. owstoni*	Taiwan
____ *N. c. interdicta*	Mountains of n China (Liaoning)
____ *N. c. multipunctata*	Pakistan and nw India
____ *N. c. hemispila*	Himalayas (w Nepal to s Kashmir)
____ *N. c. macella*	E Himalayas to s Tibet, w Nepal, n Myanmar and sw China
____ *N. c. yunnanensis*	SW China (Yunnan)

☐ **Red-billed Chough** *Pyrrhocorax pyrrhocorax*

____ *P. p. pyrrhocorax*	Locally in England, Wales, Isle of Man, Inner Hebrides and Ireland
____ *P. p. erythrorhamphos*	Alps, Pyrénées, Iberian Peninsula and Mediterranean islands
____ *P. p. barbarus*	La Palma (Canary Islands) and nw Africa (Morocco to Algeria)
____ *P. p. baileyi*	Highlands of Ethiopia
____ *P. p. docilis*	Crete and se Europe to n Arabia, n Iraq, Iran and Afghanistan
____ *P. p. centralis*	Central Asia (Tien Shan, Pamir and Altai mountains)
____ *P. p. himalayanus*	Himalayas and n India to w China
____ *P. p. brachypus*	Central and n China to Manchuria and Mongolia

☐ **Yellow-billed Chough** *Pyrrhocorax graculus*

____ *P. g. graculus*	Mountains of Europe, North Africa, Caucasus and s Caspian area
____ *P. g. digitatus*	Lebanon and Iran to s Tibet and Himalayas

☐ **Piapiac** *Ptilostomus afer*

Palm savanna of sub-Saharan Africa

☐ **Eurasian Jackdaw** *Corvus monedula*

____ *C. m. monedula*	Scandinavia; occasionally > to England and France
____ *C. m. spermologus*	W and central Europe; > to Canary Islands and Corsica
____ *C. m. soemmerringii*	E Europe, n and c Asia; > to Iran and nw India (Kashmir)
____ *C. m. cirtensis*	N Africa (Morocco and Algeria)

☐ **Daurian Jackdaw** *Corvus dauuricus*

S Siberia to Mongolia, n China and se Tibet; > to s China

☐ **House Crow** *Corvus splendens*

____ *C. s. zugmayeri*	Coastal s Iran to s Kashmir and nw India
____ *C. s. splendens*	India s of the Himalayas; ship-assisted immigrant to W Australia
____ *C. s. protegatus*	Sri Lanka
____ *C. s. maldevicius*	Laccadive and Maldive islands
____ *C. s. insolens*	S Myanmar to sw Thailand and sw China (w Yunnan)

☐ **New Caledonian Crow** *Corvus moneduloides*

Forests of New Caledonia

☐ **Slender-billed Crow** *Corvus enca*

____ *C. e. sierramadrensis*	N Philippines (Sierra Madre Mountains of Luzon)
____ *C. e. samarensis*	S Philippines (Samar and Mindanao)
____ *C. e. pusillus*	S Philippines (Balabac, Culion, Mindoro and Palawan)
____ *C. e. compilator*	Malaysia, Sumatra, Borneo, Riau Arch. and w Sumatran islands
____ *C. e. enca*	Java, Bali and Mentawi Archipelago
____ *C. e. celebensis*	Sulawesi, Talaud Is., Togian Is. and Tukangbesi Is.
____ *C. e. mangoli*	Sula Islands (Taliabu, Seho, Mangole and Sanana)
____ *C. e. violaceus*	S Moluccas (Seram, Buru and Ambon)

☐ **Piping Crow** *Corvus typicus*

Central and s Sulawesi, Muna and Butung islands

☐ **Flores Crow** *Corvus florensis*

Lowlands of Flores (w Lesser Sundas)

☐ **Mariana Crow** *Corvus kubaryi*

Mariana Islands (Rota and Guam). On verge of extinction

☐ **Long-billed Crow** *Corvus validus*

N Moluccas (Morotai, Halmahera, Kayoa, Bacan, Obi, Kasiruta)

☐ **Guadalcanal Crow** *Corvus woodfordi*
____   *C. w. woodfordi*
____   *C. w. vegetus*

Guadalcanal (Solomon Islands)
Solomon Islands (Choiseul and Santa Isabel)

☐ **Bougainville Crow** *Corvus meeki*

Solomon Islands (Bougainville and Shortland)

☐ **Brown-headed Crow** *Corvus fuscicapillus*
____   *C. f. megarhynchus*
____   *C. f. fuscicapillus*

Waigeo and Gemien islands (w New Guinea)
Aru Islands and lower Mamberamo River (New Guinea)

☐ **Gray Crow** *Corvus tristis*

New Guinea, Yapen, w Papuan islands and D'Entrecasteaux Arch.

☐ **Cape Crow** *Corvus capensis*
____   *C. c. capensis*
____   *C. c. kordofanensis*

Angola, Zambia and Zimbabwe to South Africa
Sudan, Eritrea and Somalia to Kenya, Uganda and Tanzania

☐ **Rook** *Corvus frugilegus*
____   *C. f. frugilegus*
____   *C. f. pastinator*

W Eurasia; > to North Africa and nw India
E Asia; > to Korea, Japan and se China

☐ **American Crow** *Corvus brachyrhynchos*
____   *C. b. hesperis*
____   *C. b. brachyrhynchos*
____   *C. b. paulus*
____   *C. b. pascuus*

N British Columbia to sw US and n Baja California
Central and e Canada to e-central US; > to se US
E and se US
Peninsular Florida

☐ **Northwestern Crow** *Corvus caurinus*

Kodiak I. and coastal s Alaska to sw Washington

☐ **Cuban Palm Crow** *Corvus minutus*

W Cuba

☐ **Hispaniolan Palm Crow** *Corvus palmarum*

Hispaniola

☐ **Cuban Crow** *Corvus nasicus*

Cuba, Isle of Pines and s Bahamas (Caicos Islands)

☐ **White-necked Crow** *Corvus leucognaphalus*

Hispaniola; casual on Gonâve and Saona

☐ **Jamaican Crow** *Corvus jamaicensis*

Locally in Jamaica (mainly in uplands)

☐ **Tamaulipas Crow** *Corvus imparatus*

S Texas (lower Rio Grande Valley) to se Mexico (n Veracruz)

☐ **Sinaloa Crow** *Corvus sinaloae*

Coastal lowlands of w Mexico (s Sonora to w Nayarit)

☐ **Fish Crow** *Corvus ossifragus*

Eastern US (New England to s Texas)

☐ **Hawaiian Crow** *Corvus hawaiiensis*

Montane forests of Hawaii (on verge of extinction)

☐ **Chihuahuan Raven** *Corvus cryptoleucus*

Arid sw US to central Mexico

☐ **Carrion Crow** *Corvus corone*
____   *C. c. corone*
____   *C. c. orientalis*

W Europe
Iran to n China, Korea and Japan

☐ **Hooded Crow** *Corvus cornix*
____   *C. c. cornix*
____   *C. c. sharpii*

N Europe to Yenisey Valley, Ukraine, Corsica and s Italy
Mainland Italy to Yugoslavia, Asia Minor, n Iran and Kazakstan

    C. c. pallescens   Coastal s Turkey to Levant, n Iraq and Egypt
    C. c. capellanus   S Iraq and adjacent sw Iran

☐ **Large-billed Crow** _Corvus macrorhynchos_
    C. m. japonensis   Sakhalin, Kuril Islands and n Japanese Archipelago
    C. m. connectens   S Ryukyu Islands (Amami-O-Shima, Okinawa and Miyako-Jima)
    C. m. osai   S Ryukyu Is. (Ishigaki, Iriomote, Kobama, Kuru and Aragusuku)
    C. m. mandschuricus   NE Asia
    C. m. colonorum   N China to n Indochina and Taiwan
    C. m. hainanus   Hainan (s China)
    C. m. mengtszensis   SW China (s Yunnan)
    C. m. tibetosinensis   E Himalayas to se Tibet, n Myanmar and w China
    C. m. intermedius   Extreme e Iran to nw India and w Himalayas
    C. m. culminatus   S India and Sri Lanka
    C. m. levaillantii   Indus River drainage of ne India, Sri Lanka and Andaman Islands
    C. m. macrorhynchos   Malaysia, s Indochina, Borneo, Sumatra, Java and Lesser Sundas
    C. m. philippinus   Philippine Islands

☐ **Torresian Crow** _Corvus orru_
    C. o. orru   N Moluccas, w Papuan is., New Guinea and D'Entrecasteaux Arch.
    C. o. insularis   Bismarck Archipelago
    C. o. latirostris   Tanimbar Islands and Babar Islands (Arafura Sea)
    C. o. ceciliae   Most of northern mainland Australia (except w Queensland)

☐ **Little Crow** _Corvus bennetti_

    Inland Australia (reaching coasts in west and south)

☐ **Australian Raven** _Corvus coronoides_
    C. c. coronoides   E Australia (Gulf of Carpenteria to c Australia and s Victoria)
    C. c. perplexus   SW Western Australia to far southwest South Australia

☐ **Little Raven** _Corvus mellori_

    SE Australia (n NSW to Victoria and Eyre Pen., South Australia)

☐ **Forest Raven** _Corvus tasmanicus_
    C. t. boreus   NE New South Wales (New England Tableland and adj. coast)
    C. t. tasmanicus   Tasmania to s Victoria and adjacent se South Australia

☐ **Collared Crow** _Corvus torquatus_

    Central and e China to n Vietnam; Hainan and Taiwan

☐ **Pied Crow** _Corvus albus_

    Africa s of the Sahara, Madagascar, Aldabra and Comoro Islands

☐ **Brown-necked Raven** _Corvus ruficollis_

    Cape Verde Archipelago; North Africa to w Pakistan

☐ **Somali Crow** _Corvus edithae_

    Eritrea, Ethiopia, Somalia, Kenya and se Sudan

☐ **Fan-tailed Raven** _Corvus rhipidurus_

    Sub-Saharan Africa, s Middle East and Arabian Peninsula

☐ **White-necked Raven** _Corvus albicollis_

    Rocks, cliffs and hills of e and se Africa

☐ **Thick-billed Raven** _Corvus crassirostris_

    Mts. of Ethiopia and Eritrea; vagrant to nw Somalia and se Sudan

☐ **Common Raven** _Corvus corax_
    C. c. principalis   Islands in Bering Sea, Alaska, Canada and n US
    C. c. sinuatus   W-central US to s Baja, Revillagigedo Islands and nw Nicaragua
    C. c. varius   Iceland and Faeroe Islands
    C. c. corax   Europe and Mediterranean islands to w Asia
    C. c. subcorax   SE Europe and Asia Minor to Pakistan
    C. c. tingitanus   Canary Islands; coastal Morocco to Egypt
    C. c. tibetanus   Mountains of central Asia and the Himalayas
    C. c. kamtschaticus   Siberia to Sea of Okhotsk, Sakhalin, Kuril Islands and n Japan

FAMILY: STURNIDAE (Starlings—114)

☐ **Metallic Starling** *Aplonis metallica*

____	*A. m. circumscripta*	Tanimbar Islands (Larat and Kirimoen) and Damar I.
____	*A. m. metallica*	Moluccas, Sula Is. and Aru Is. to New Guinea and ne Queensland
____	*A. m. inornata*	Biak I. and Numfor I. (n New Guinea)
____	*A. m. nitida*	Bismarck Archipelago
____	*A. m. purpureiceps*	Admiralty Islands

☐ **Yellow-eyed Starling** *Aplonis mystacea*

Lowland rainforests of s New Guinea

☐ **Singing Starling** *Aplonis cantoroides*

Aru Islands and New Guinea to Bismarck Arch. and Solomon Is.

☐ **Tanimbar Starling** *Aplonis crassa*

Tanimbar Islands (Larat and Yamdena)

☐ **Atoll Starling** *Aplonis feadensis*

____	*A. f. feadensis*	Ontong Java I. (Solomons); Nissan and Nuguria is. (New Guinea)
____	*A. f. heureka*	Bismarck Archipelago (Ninigo and Hermit Islands)

☐ **Rennell Starling** *Aplonis insularis*

Solomon Islands (Rennell and Bellona)

☐ **Long-tailed Starling** *Aplonis magna*

____	*A. m. magna*	Biak I. (off n New Guinea)
____	*A. m. brevicaudus*	Numfor I. (off nw New Guinea)

☐ **White-eyed Starling** *Aplonis brunneicapillus*

Solomon Is. (Bougainville, Guadalcanal, Choiseul and Rendova)

☐ **Brown-winged Starling** *Aplonis grandis*

____	*A. g. grandis*	Solomon Islands (Bougainville, Choiseul and Santa Isabel)
____	*A. g. macrura*	Guadalcanal (Solomon Islands)
____	*A. g. malaita*	Malaita (Solomon Islands)

☐ **San Cristobal Starling** *Aplonis dichroa*

Lowland forests of San Cristóbal (s Solomon Islands)

☐ **Rusty-winged Starling** *Aplonis zelandica*

____	*A. z. rufipennis*	Central and n Vanuatu and Banks Group
____	*A. z. maxwelli*	Santa Cruz I. (Vanuatu)
____	*A. z. zelandica*	Vanikoro I. (Vanuatu)

☐ **Striated Starling** *Aplonis striata*

____	*A. s. striata*	Forests of New Caledonia
____	*A. s. atronitens*	Loyalty Islands

☐ **Mountain Starling** *Aplonis santovestris*

Espíritu Santo (Vanuatu)

☐ **Asian Glossy Starling** *Aplonis panayensis*

____	*A. p. affinis*	Assam to Bangladesh and Myanmar (Arakan Yoma Mts.)
____	*A. p. strigata*	S Thailand to Malaysia, Sumatra, Java and w Borneo
____	*A. p. tytleri*	Andaman Islands and Car Nicobar I.
____	*A. p. albiris*	Great and Central Nicobar islands
____	*A. p. heterochlora*	Anambas and Natuna islands (off Borneo)
____	*A. p. eustathis*	E Borneo
____	*A. p. alipodis*	Panjang, Maratau and Derawan islands (off e Borneo)
____	*A. p. panayensis*	N Sulawesi and Philippine Islands
____	*A. p. sanghirensis*	Talaud, Sangihe, Siau, Tahjlandang, Ruang, and Biaro islands
____	*A. p. enganensis*	Enggano I. (off s Sumatra)
____	*A. p. altirostris*	Simuelue, Banyan and Nias islands (off w Sumatra)
____	*A. p. leptorrhyncha*	Batu I. (off w Sumatra)
____	*A. p. pachistorhina*	Mentawi Islands (off w Sumatra)
____	*A. p. gusti*	Bali

☐ **Moluccan Starling** *Aplonis mysolensis*
    ____ *A. m. sulaensis*   E Sulawesi, Banggai Islands and Sula Islands
    ____ *A. m. mysolensis*   Moluccas and w Papuan islands

☐ **Short-tailed Starling** *Aplonis minor*

          Sulawesi, Java, Bali, Lesser Sundas and s Philippines (Mindanao)

☐ **Micronesian Starling** *Aplonis opaca*
    ____ *A. o. opaca*   Kosrai (Caroline Islands)
    ____ *A. o. angus*   Truk, Ulithi, Fais, Wolea, Ifalik and adjacent Caroline Islands
    ____ *A. o. kurodae*   Yap (Caroline Islands)
    ____ *A. o. ponapensis*   Pohnpei (Caroline Islands)
    ____ *A. o. aeneus*   Mariana Islands (Alamagan, Pagan, Agrihan and Asuncion)
    ____ *A. o. guami*   Mariana Islands (Guam, Rota, Tinian and Saipan)
    ____ *A. o. orii*   Palau Islands

☐ **Pohnpei Starling** *Aplonis pelzelni*

          Rediscovered 1995 in mountains of Pohnpei after 50-year absence

☐ **Polynesian Starling** *Aplonis tabuensis*
    ____ *A. t. tabuensis*   S Tonga and Lau Archipelago
    ____ *A. t. tenebrosus*   Keppel and Boscawen islands (central Polynesia)
    ____ *A. t. nesiotes*   Niuafou I. (central Polynesia)
    ____ *A. t. brunnescens* Niue I. (Cook Isands)
    ____ *A. t. vitiensis*   Fiji Islands
    ____ *A. t. fortunae*   Futuna, Alofa, Uea and Horne islands (central Polynesia)
    ____ *A. t. rotumae*   Rotuma (Fiji)
    ____ *A. t. tucopiae*   Tukopia (Solomon Islands east of Santa Cruz Group)
    ____ *A. t. pachyrampha* Reef, Swallow and Tinakula islands (Santa Cruz Group)
    ____ *A. t. brevirostris*   Western Samoa
    ____ *A. t. tutuilae*   Tutuila (American Samoa)
    ____ *A. t. manuae*   Manua Islands (American Samoa)

☐ **Samoan Starling** *Aplonis atrifusca*

          Western and American Samoa

☐ **Rarotonga Starling** *Aplonis cinerascens*

          Rarotonga (sw Cook Islands). On verge of extinction

☐ **Yellow-faced Myna** *Mino dumontii*

          New Guinea, Aru and w Papuan Islands

☐ **Golden Myna** *Mino anais*
    ____ *M. a. anais*   Lowlands of nw New Guinea and Salawati I.
    ____ *M. a. orientalis*   N New Guinea (east to Huon Peninsula) and Yapen I.
    ____ *M. a. robertsoni*   S New Guinea (east to Milne Bay)

☐ **Long-tailed Myna** *Mino kreffti*

          Bismarck Archipelago, n and central Solomon Islands

☐ **Sulawesi Myna** *Basilornis celebensis*

          Sulawesi, Lembeh, Muna and Butung islands

☐ **Helmeted Myna** *Basilornis galeatus*

          Banggai Is. (Peleng and Banggai); Sula Is. (Taliabu and Mangole)

☐ **Long-crested Myna** *Basilornis corythaix*

          Forests of Seram (s Moluccas)

☐ **Apo Myna** *Basilornis mirandus*

          Montane forests of Mindanao (s Philippines)

☐ **Coleto** *Sarcops calvus*
    ____ *S. c. calvus*   N Philippine Islands
    ____ *S. c. melanonotus*   Mindanao, Cebu, Panay, Negros, Bohol, Samar and Ticao
    ____ *S. c. lowii*   Sulu Archipelago

☐ **White-necked Myna** *Streptocitta albicollis*
    ____ *S. a. torquata*   N Sulawesi, Lembeh I. and Togian Islands
    ____ *S. a. albicollis*   S Sulawesi, Muna and Butung islands

☐ **Bare-eyed Myna** *Streptocitta albertinae*

Sula Islands (Taliabu and Mangole)

☐ **Fiery-browed Myna** *Enodes erythrophris*

Mountains of Sulawesi

☐ **Finch-billed Myna** *Scissirostrum dubium*

Sulawesi, Bangka, Lembeh, Butung, Togian, and Banggai islands

☐ **Spot-winged Starling** *Saroglossa spiloptera*

Foothills of n India and Nepal; winters to Myanmar and Thailand

☐ **Madagascar Starling** *Saroglossa aurata*

Forests of Madagascar

☐ **Golden-crested Myna** *Ampeliceps coronatus*

NE India to Myanmar, n Malaya, Thailand and Indochina

☐ **Common Hill Myna** *Gracula religiosa*
 ____ *G. r. religiosa*
 ____ *G. r. batuensis*
 ____ *G. r. palawanensis*
 ____ *G. r. venerata*
 ____ *G. r. intermedia*
 ____ *G. r. peninsularis*
 ____ *G. r. andamanensis*

Malaysia, Sumatra, Java, Bali, Borneo and Bangka I.
Batu and Mentawi islands (off nw Sumatra)
Palawan (sw Philippines)
W Lesser Sundas (Sumbawa, Flores, Pantar, Lomblen and Alor)
N India to Myanmar, Thailand, Indochina and s China
NE peninsular India
Andaman and Nicobar islands

☐ **Southern Hill Myna** *Gracula indica*

SW India (Western Ghats) and s Sri Lanka

☐ **Enggano Myna** *Gracula enganensis*

Enggano I. (off s Sumatra)

☐ **Nias Myna** *Gracula robusta*

Nias, Pulan, Babi, Tuangku and Bangkaru islands (off Sumatra)

☐ **Ceylon Myna** *Gracula ptilogenys*

Humid forests of Sri Lanka

☐ **White-vented Myna** *Acridotheres grandis*

NE India to sw China, Myanmar, Thailand and Indochina

☐ **Crested Myna** *Acridotheres cristatellus*
 ____ *A. c. cristatellus*
 ____ *A. c. brevipennis*
 ____ *A. c. formosanus*

E Myanmar to n Indochina, se and central China
Hainan (s China)
Taiwan

☐ **Javan Myna** *Acridotheres javanicus*

Java and Bali

☐ **Pale-bellied Myna** *Acridotheres cinereus*

S peninsular Sulawesi (north to Rantepao)

☐ **Jungle Myna** *Acridotheres fuscus*
 ____ *A. f. fuscus*
 ____ *A. f. mahrattensis*
 ____ *A. f. fumidus*
 ____ *A. f. torquatus*

Himalayas (Pakistan to Assam, Rajasthan and n Orissa)
W peninsular India
NE Assam
Myanmar to n and central Malaysia

☐ **Collared Myna** *Acridotheres albocinctus*

NE India (Manipur) to n Myanmar and sw China (nw Yunnan)

☐ **Bank Myna** *Acridotheres ginginianus*

Foothills of e Pakistan to n Nepal and n-central India

☐ **Common Myna** *Acridotheres tristis*
 ____ *A. t. tristis*
 ____ *A. t. melanosternus*

SE Iran to India and SE Asia; introduced widely worldwide
Sri Lanka

☐ **Vinous-breasted Starling** *Acridotheres burmannicus*
 ____ *A. b. burmannicus*
 ____ *A. b. leucocephalus*

Myanmar to central Thailand and extreme sw China (Yunnan)
S Thailand to Cambodia and s Indochina

☐ **Black-winged Starling** *Acridotheres melanopterus*

Lowlands of Java, Bali and Lombok

☐ **Bali Myna** *Leucopsar rothschildi*

Coastal nw Bali (Bali Barat National Park). ±55 birds in 1993

598

☐ **Black-collared Starling** *Gracupica nigricollis*

Open country and scrub of s China and SE Asia

☐ **Asian Pied Starling** *Gracupica contra*
- ____ *G. c. contra* — N and central India
- ____ *G. c. sordidus* — N Assam
- ____ *G. c. superciliaris* — Manipur and Myanmar south to Tenasserim
- ____ *G. c. floweri* — S Myanmar to Thailand and Laos
- ____ *G. c. jalla* — Sumatra, Java and Bali

☐ **Daurian Starling** *Sturnia sturnina*

SE Siberia to n Mongolia and n Korea; winters to SE Asia

☐ **Chestnut-cheeked Starling** *Sturnia philippensis*

S Sakhalin, Kuril Is. and n Japan; winters to Philippines, e Indies

☐ **White-shouldered Starling** *Sturnia sinensis*

S China to Indochina; winters to SE Asia and n Philippines

☐ **Chestnut-tailed Starling** *Sturnia malabarica*

Peninsular India to Myanmar, sw China and n SE Asia

☐ **White-headed Starling** *Sturnia erythropygia*
- ____ *S. e. erythropygia* — Andaman Islands
- ____ *S. e. andamanensis* — Car Nicobar I.
- ____ *S. e. katchalensis* — Katchall I. (Nicobar Islands)

☐ **White-faced Starling** *Sturnia albofrontata*

Forests of wet zone of sw Sri Lanka

☐ **Brahminy Starling** *Temenuchus pagodarum*

E Afghanistan to Bangladesh, s Nepal, India and Sri Lanka

☐ **Rosy Starling** *Pastor roseus*

S-central Europe; winters primarily India and se Arabian Pen.

☐ **Red-billed Starling** *Sturnus sericeus*

Lowlands of s-central China; winters to SE Asia and Philippines

☐ **White-cheeked Starling** *Sturnus cineraceus*

NE Asia; winters in s China and Philippines

☐ **European Starling** *Sturnus vulgaris*
- ____ *S. v. granti* — Azores
- ____ *S. v. vulgaris* — Canary Is. and Iceland to Ural Mts., n Ukraine and se Europe
- ____ *S. v. faroensis* — Faeroes
- ____ *S. v. zetlandicus* — Shetland Islands
- ____ *S. v. tauricus* — E and s Ukraine, Crimea and Asia Minor
- ____ *S. v. purpurascens* — W Transcaucasia to Georgia and Armenia
- ____ *S. v. caucasicus* — Volga Delta and n Caucasus to Caspian Sea and s Iran
- ____ *S. v. nobilior* — Afghanistan, Transcaspia and Khorasan
- ____ *S. v. poltaratskyi* — E Ural Mountains to Lake Baikal, Kazakstan and w Mongolia
- ____ *S. v. porphyronotus* — S Dzungaria and Tien Shan Mts. to Pamir Mts. and Samarkand
- ____ *S. v. humii* — W Himalayas (Kashmir to Garhwal)
- ____ *S. v. minor* — Locally in w Pakistan (Sind)

☐ **Spotless Starling** *Sturnus unicolor*

Iberian Peninsula to Corsica, Sardinia, Sicily and nw Africa

☐ **Wattled Starling** *Creatophora cinerea*

Savanna of e and s Africa and Arabian Peninsula

☐ **Cape Glossy-Starling** *Lamprotornis nitens*

Acacia of w Gabon and Angola to South Africa

☐ **Greater Blue-eared Glossy-Starling** *Lamprotornis chalybaeus*
- ____ *L. c. chalybaeus* — Senegal to e Sudan
- ____ *L. c. cyaniventris* — Eritrea and Ethiopia to w Kenya and e Uganda
- ____ *L. c. scyobius* — SW Uganda to se Kenya, Tanzania, se Zaire and Mozambique
- ____ *L. c. nordmanni* — S Angola to Zimbabwe, s Zambia, Botswana and South Africa

☐ **Lesser Blue-eared Glossy-Starling** *Lamprotornis chloropterus*
- ____ *L. c. chloropterus* — Senegal to s Sudan, Eritrea, Ethiopia, n Uganda and w Kenya
- ____ *L. c. elisabeth* — SE Kenya and Tanzania to Zimbabwe

☐ **Bronze-tailed Glossy-Starling** *Lamprotornis chalcurus*
- ____ *L. c. chalcurus* — Senegal and Guinea Bissau to n Cameroon
- ____ *L. c. emini* — E Cameroon to s Sudan and w Kenya

☐ **Splendid Glossy-Starling** *Lamprotornis splendidus*
- ____ *L. s. chrysonotis* — Senegambia to Togo
- ____ *L. s. splendidus* — S Benin to nw Angola, Ethiopia and Zambia
- ____ *L. s. lessoni* — Bioko I. (Gulf of Guinea)
- ____ *L. s. bailundensis* — Cent. Angola to Zambia and se Zaire

☐ **Principe Glossy-Starling** *Lamprotornis ornatus* — Príncipe I. (Gulf of Guinea)

☐ **Emerald Starling** *Lamprotornis iris* — Wooded savanna of Guinea, Sierra Leone and Ivory Coast

☐ **Purple Glossy-Starling** *Lamprotornis purpureus*
- ____ *L. p. purpureus* — Senegal to Cameroon
- ____ *L. p. amethystinus* — Cameroon to we Kenya

☐ **Rueppell's Glossy-Starling** *Lamprotornis purpuroptera*
- ____ *L. p. aeneocephalus* — Eritrea, Sudan and n Ethiopia
- ____ *L. p. purpuroptera* — S Ethiopia to Somalia, s Sudan, Uganda, w Kenya and Tanzania

☐ **Long-tailed Glossy-Starling** *Lamprotornis caudatus* — Savanna of Senegal to s Sudan (Nile River region)

☐ **Golden-breasted Starling** *Lamprotornis regius* — Arid savanna of s Ethiopia to Somalia, e Kenya and ne Tanzania

☐ **Meves' Glossy-Starling** *Lamprotornis mevesii*
- ____ *L. m. mevesii* — Angola and n Namibia to Botswana, s Malawi and ne S Africa
- ____ *L. m. violacior* — N Namibia and sw Angola
- ____ *L. m. benguelensis* — *Mopane* woodlands of s end of w Angola escarpment

☐ **Burchell's Glossy-Starling** *Lamprotornis australis* — S Angola to w Zambia, Namibia, Mozambique and n South Africa

☐ **Sharp-tailed Glossy-Starling** *Lamprotornis acuticaudus*
- ____ *L. a. acuticaudus* — Central Angola to Zambia and sw Tanzania
- ____ *L. a. ecki* — N Namibia and adjacent s Angola

☐ **Black-bellied Glossy-Starling** *Lamprotornis corruscus*
- ____ *L. c. corruscus (mandanus, jombeni)* — Coastal bush of s Somalia to e Swaziland and Cape Province
- ____ *L. c. vaughani* — Pemba I. (off Tanzania)

☐ **Superb Starling** *Lamprotornis superbus* — SE Sudan and Ethiopia to Somalia, Uganda, Kenya and Tanzania

☐ **Hildebrandt's Starling** *Lamprotornis hildebrandti* — Acacia savanna of s Kenya and n Tanzania

☐ **Shelley's Starling** *Lamprotornis shelleyi* — Acacia of Somalia and s Ethiopia to se Sudan and se Kenya

☐ **Chestnut-bellied Starling** *Lamprotornis pulcher* — Savanna of sub-Saharan Africa (Senegal to n Ethiopia)

☐ **Purple-headed Glossy-Starling** *Lamprotornis purpureiceps* — Guinea to Cameroon, Gabon, n Zaire, Uganda and nw Kenya

☐ **Copper-tailed Glossy-Starling** *Lamprotornis cupreocauda* — Sierra Leone to Ghana

☐ **Violet-backed Starling** *Cinnyricinclus leucogaster*
- ____ *C. l. leucogaster* — Senegal to Ethiopia, Gabon, Zaire, Uganda, nw Kenya, Tanzania
- ____ *C. l. verreauxi* — Lower Congo to s Zaire, Uganda, Kenya, Namibia and S Africa
- ____ *C. l. arabicus* — N Ethiopia, Eritrea, e Sudan, nw Somalia and Arabian Pen.

☐ **African Pied Starling** *Spreo bicolor* — Extreme se Namibia to s Mozambique and South Africa

☐ **Fischer's Starling** *Spreo fischeri*

S Ethiopia and Somalia to e Kenya and ne Tanzania

☐ **Ashy Starling** *Spreo unicolor*

Acacia woodlands of s Kenya and Tanzania

☐ **White-crowned Starling** *Spreo albicapillus*
____ *S. a. albicapillus*
____ *S. a. horrensis*

Ethiopia and Somalia to Djibouti and extreme ne Kenya
Kenya (n edge of Dida Galgulu Desert north to Ethiopian border)

☐ **Red-winged Starling** *Onychognathus morio*
____ *O. m. morio (shelleyi)*
____ *O. m. rueppellii*

Savanna of Africa south of the Sahara
Eritrea to Ethiopia and extreme n Kenya

☐ **Slender-billed Starling** *Onychognathus tenuirostris*
____ *O. t. tenuirostris*
____ *O. t. theresae (raymondi)*

Mountains of Eritrea and Ethiopia
E Zaire to Kenya, Tanzania and n Malawi

☐ **Chestnut-winged Starling** *Onychognathus fulgidus*
____ *O. f. hartlaubii*
____ *O. f. fulgidus*
____ *O. f. intermedius*

Sierra Leone to w Sudan
Bioko I. and São Tomé I. (Gulf of Guinea)
Gabon, Congo, Zaire and Angola

☐ **Waller's Starling** *Onychognathus walleri*
____ *O. w. preussi*
____ *O. w. elgonensis*
____ *O. w. walleri*

Highlands of se Nigeria and Cameroon; Bioko
Kenya (west of Rift Valley) to Uganda, se Sudan and e Zaire
Kenya (e of Rift Valley) to Tanzania and n Malawi

☐ **Somali Starling** *Onychognathus blythii*

Eritrea and Ethiopia to n Somalia, Abd el Kuri I. and Socotra

☐ **Socotra Starling** *Onychognathus frater*

Rocky hills of Socotra I. (off ne Somalia)

☐ **Tristram's Starling** *Onychognathus tristramii*

Dead Sea Valley and w Arabia to Yemen and s Oman

☐ **Pale-winged Starling** *Onychognathus nabouroup*

Deserts of sw Angola to Namibia and arid interior South Africa

☐ **Bristle-crowned Starling** *Onychognathus salvadorii*

Extreme se Sudan to ne Uganda, s Ethiopia, Somalia and n Kenya

☐ **White-billed Starling** *Onychognathus albirostris*

Mountains of Eritrea and Ethiopia

☐ **Neumann's Starling** *Onychognathus neumanni*
____ *O. n. neumanni*
____ *O. n. modicus*

N Nigeria to Cameroon, Central African Republic and w Sudan
Senegal to Mali and w Niger

☐ **Narrow-tailed Starling** *Poeoptera lugubris*

Sierra Leone to n Angola, Zaire and w Uganda; Bioko I.

☐ **Stuhlmann's Starling** *Poeoptera stuhlmanni*

Mts. of s Ethiopia to s Sudan, Uganda, Tanzania and e Zaire

☐ **Kenrick's Starling** *Poeoptera kenricki*
____ *P. k. kenricki*
____ *P. k. bensoni*

Montane forests of ne Tanzania
Montane forests of central Kenya

☐ **Sharpe's Starling** *Pholia sharpii*

S Sudan to Ethiopia, n Tanzania, w Uganda, Rwanda and e Zaire

☐ **Abbott's Starling** *Pholia femoralis*

Mt. Kenya south to Mt. Meru and Mt. Kilimanjaro

☐ **White-collared Starling** *Grafisia torquata*

Cameroon to Gabon, Central African Republic and n Zaire

☐ **Magpie Starling** *Speculipastor bicolor*

Sudan to s Ethiopia, Somalia, ne Uganda, Kenya and Tanzania

☐ **Babbling Starling** *Neocichla gutturalis*
____ *N. g. gutturalis*
____ *N. g. angusta*

Brachystegia of interior sw Angola and w Zambia
E Zambia to Tanzania and n Malawi

☐ **Red-billed Oxpecker** *Buphagus erythrorhynchus*

Savanna of e and s Africa

☐ **Yellow-billed Oxpecker** *Buphagus africanus*
____ *B. a. africanus* — Savanna of Africa south of the Sahara
____ *B. a. langi* — Gabon, w Zaire and Congo

FAMILY: PASSERIDAE (Old World Sparrows—38)

☐ **Saxaul Sparrow** *Passer ammodendri*
____ *P. a. ammodendri* — Aral Sea to Kazakstan and Uzbekistan
____ *P. a. nigricans* — NW China (n Xinjiang to Manas River valley)
____ *P. a. stoliczkae* — W China (w Xinjiang to n Gansu) and s Mongolia

☐ **House Sparrow** *Passer domesticus*
____ *P. d. domesticus* — Europe to Mongolia, Amurland and n Manchuria
____ *P. d. balearoibericus* — Mediterranean Spain, Balearic Is., France, Balkans to Asia Minor
____ *P. d. tingitanus* — NW Africa (Morocco to Tunisia, Algeria and ne Libya)
____ *P. d. rufidorsalis* — Nile Valley of the Sudan
____ *P. d. niloticus* — NE Africa (Suez Canal region to n Sudan)
____ *P. d. biblicus* — Cyprus and Levant to Turkey, n Saudi Arabia, Iraq and w Iran
____ *P. d. indicus* — S Israel to n Saudi Arabia, s Iran, India, Sri Lanka and Myanmar
____ *P. d. hufufae* — NE Arabia (south to n Oman)
____ *P. d. bactrianus* — Transcaspia to Kazakstan, Afghanistan and nw Pakistan
____ *P. d. hyrcanus* — N Iran (south to Elburz Mountains) and adjacent Turkmenistan
____ *P. d. persicus* — Central Iran to sw Afghanistan and extreme w Pakistan
____ *P. d. parkini* — Himalayas (Pakistan to sw Tibet, Nepal and Sikkim)

☐ **Spanish Sparrow** *Passer hispaniolensis*
____ *P. h. hispaniolensis (maltae)* — Cape Verde Is., Canary Is., Madeira, s Europe and North Africa
____ *P. h. italiae* — Morocco, Algeria, Tunisia and Libya
____ *P. h. transcaspicus* — Iran and Transcaspia to e Kazakstan and Afghanistan

☐ **Sind Sparrow** *Passer pyrrhonotus*

Extreme se Iran to Pakistan and nw India

☐ **Somali Sparrow** *Passer castanopterus*
____ *P. c. castanopterus* — E Ethiopia and Somalia
____ *P. c. fulgens* — Extreme sw Ethiopia to n-central Kenya

☐ **Russet Sparrow** *Passer rutilans*
____ *P. r. rutilans* — Sakhalin to Japan, South Korea, e Manchuria, s China and Taiwan
____ *P. r. intensior* — SW China to n Myanmar, Laos and nw Tonkin
____ *P. r. cinnamomeus* — Himalayas of ne Afghanistan to ne India and se Tibet

☐ **Plain-backed Sparrow** *Passer flaveolus*

N Myanmar to Thailand, Malay Peninsula and s Vietnam

☐ **Dead Sea Sparrow** *Passer moabiticus*
____ *P. m. moabiticus* — Israel and Jordan
____ *P. m. mesoptamicus* — S Turkey to n Syria, Iraq and sw Iran; formerly Cyprus
____ *P. m. yatii* — E Iran and adjacent sw Afghanistan

☐ **Cape Verde Sparrow** *Passer iagoensis*

Locally in Cape Verde Islands (except Fogo I.)

☐ **Socotra Sparrow** *Passer insularis*

Socotra and Abd-al-Küri I. (off Somalia)

☐ **Great Rufous Sparrow** *Passer motitensis*
____ *P. m. benguellensis* — S Angola to Namibia
____ *P. m. motitensis* — Botswana to Transvaal and n Cape Province
____ *P. m. subsolanus* — S Zimbabwe to n Orange Free State and nw Swaziland

☐ **Kenya Rufous Sparrow** *Passer rufocinctus*

Kenya and Tanzania

☐ **Shelley's Rufous Sparrow** *Passer shelleyi*

SE Sudan to ne Uganda, e Ethiopia and nw Somalia

☐ **Kordofan Rufous Sparrow** *Passer cordofanicus*

E Chad to w-central Sudan

☐ **Cape Sparrow** *Passer melanurus*
____ *P. m. damarensis*
____ *P. m. melanurus*
____ *P. m. vicinus*

SW Angola to Namibia, Zimbabwe and n Cape Province
Cape Province and sw Free State
Cent. and s Transvaal to Swaziland, Lesotho and e Cape Province

☐ **Gray-headed Sparrow** *Passer griseus*
____ *P. g. griseus*
____ *P. g. laeneni*
____ *P. g. ugandae*

Senegal to s Chad, n Cameroon, n Gabon and s Sudan
Mali to extreme n Cameroon and w-central Sudan
Angola to s Sudan, n Ethiopia, Kenya and ne Tanzania

☐ **Swainson's Sparrow** *Passer swainsonii*

Extreme ne Sudan to Ethiopia, n Somalia and n-central Kenya

☐ **Parrot-billed Sparrow** *Passer gongonensis*

Extreme se Sudan to s Ethiopia, s Somalia, Kenya and ne Tanzania

☐ **Swahili Sparrow** *Passer suahelicus*

S Kenya to w Tanzania, Malawi and nw Mozambique

☐ **Southern Gray-headed Sparrow** *Passer diffusus*
____ *P. d. diffusus*
____ *P. d. luangwae*
____ *P. d. mosambicus*
____ *P. d. stygiceps*

Angola to Namibia, sw Zambia, w Zimbabwe and w South Africa
E Zambia (upper Luangwa Valley)
SE Malawi, n Mozambique and se Tanzania
SE Zambia, sw Malawi and s Mozambique to e South Africa

☐ **Desert Sparrow** *Passer simplex*
____ *P. s. saharae*
____ *P. s. simplex*
____ *P. s. zarudnyi*

Sahara of se Morocco to Algeria, s Tunisia and central Libya
S Sahara from Mali to n Niger (Aïr Massif), n Chad and c Sudan
Deserts of s-central Asia (Uzbekistan and Turkmenistan)

☐ **Eurasian Tree Sparrow** *Passer montanus*
____ *P. m. montanus*
____ *P. m. transcaucasicus*
____ *P. m. dilutus*
____ *P. m. dybowskii*
____ *P. m. kansuensis*
____ *P. m. iubilaeus*
____ *P. m. obscuratus*
____ *P. m. saturatus*
____ *P. m. malaccensis*

Europe to n Africa, n Mongolia, Manchuria and Sea of Okhotsk
S Caucasus (Black Sea coast of Georgia to n Iran)
Transcaspia to w Pakistan, Gobi Desert and w China (Xinjiang)
E Asia (lower Amur River to Manchuria and n Korea)
W China (Zaidam basin and n Gansu)
E China (Liaoning to lower Yangtze River and Shaanxi)
Nepal to ne India, Myanmar and w-c China (Sichuan to Hubei)
S Kuril Is., Japan, South Korea, Ryukyu Is., Taiwan and se China
Central Myanmar, Malaya, Hainan, Vietnam and w Indonesia

☐ **Sudan Golden Sparrow** *Passer luteus*

Mauritania to Senegal, Burkina Faso, Sudan and n Ethiopia

☐ **Arabian Golden Sparrow** *Passer euchlorus*

SW Saudi Arabia to South Yemen, adj. Ethiopia and n Somalia

☐ **Chestnut Sparrow** *Passer eminibey*

S Sudan and sw Ethiopia to Uganda, Somalia and n Tanzania

☐ **Yellow-spotted Petronia** *Petronia pyrgita*
____ *P. p. pallida*
____ *P. p. pyrgita*

S Mauritania and Senegal to Mali, Niger, Chad and c Sudan
SE Sudan to s Ethiopia, Somalia, ne Uganda and ne Tanzania

☐ **Chestnut-shouldered Petronia** *Petronia xanthocollis*
____ *P. x. transfuga*
____ *P. x. xanthocollis*

SE Turkey to Iraq, s Iran, s Pakistan and nw India
E Afghanistan to n Pakistan

☐ **Yellow-throated Petronia** *Petronia superciliaris*
____ *P. s. rufitergum*

S Zaire to sw Tanzania, n Malawi, Zambia, Angola, nw Botswana

____ *P. s. flavigula*	S Zambia, Zimbabwe, w Mozambique, e Botswana and n S Africa
____ *P. s. bororensis*	E Tanzania, s Malawi, n Mozambique to ne Natal and e Swaziland
____ *P. s. superciliaris*	South Africa (e Cape Province to Natal)

☐ **Bush Petronia** *Petronia dentata*

Senegambia to s Sudan, Ethiopia and s Arabian Peninsula

☐ **Rock Petronia** *Petronia petronia*

____ *P. p. petronia (madeirensis)*	Canary Is., Madeira and Europe to Bulgaria and w Asia Minor
____ *P. p. barbara*	N Africa (Morocco to Algeria, Tunisia and w Libya)
____ *P. p. puteicola*	S Turkey to s Syria, n Israel and Jordan
____ *P. p. exigua*	Central Turkey to n Caucasus, n Iraq and n Iran
____ *P. p. kirhizica*	Lower Volga River Valley to Turgay depression and Aral Sea
____ *P. p. intermedia*	Transcaspia to e Iran, n Afghanistan, Pamirs and w Kunlun Shan Mts.
____ *P. p. brevirostris*	E Siberia to Mongolia, nw Manchuria and sw China (n Sichuan)

☐ **Pale Rockfinch** *Carpospiza brachydactyla*

Arid Asia Minor; winters to sw Arabia and ne Africa

☐ **White-winged Snowfinch** *Montifringilla nivalis*

____ *M. n. nivalis*	Pyrénées and Alps to Italy, s Yugoslavia and n Greece
____ *M. n. leucura*	S and e Asia Minor
____ *M. n. alpicola*	Caucasus to n Iran, Afghanistan (Hindu Kush) and w Pamirs
____ *M. n. gaddi*	Zagros Mountains (sw Iran)
____ *M. n. tianshanica*	Alayskiy and Chatkal'skiy mountains to w Tien Shan Mts.
____ *M. n. groum-grzimaili*	E Tien Shan Mts. to w China (n Xinjiang) and w Mongolia
____ *M. n. kwenlunensis*	W China (w Kunlun Shan Mts. to s Tibet and e Qinghai)

☐ **Black-winged Snowfinch** *Montifringilla adamsi*

____ *M. a. adamsi*	Kashmir to Nepal, Sikkim, s Tibet and sw China
____ *M. a. xerophila*	W China (nw Nan Shan Mts. to Astin Tagh Mts.)

☐ **White-rumped Snowfinch** *Montifringilla taczanowskii*

Tibet to w China (Kokonor and Nan Shan Mts. to Sichuan)

☐ **Père David's Snowfinch** *Montifringilla davidiana*

____ *M. d. davidiana*	Inner Mongolia to w China (Gansu and Qinghai)
____ *M. d. potanini*	SE Altai Mountains to Outer Mongolia and ne China

☐ **Rufous-necked Snowfinch** *Montifringilla ruficollis*

____ *M. r. ruficollis*	W Tibet to Sikkim, Kokonor and s Nan Shan Mountains
____ *M. r. isabellina*	W China (s Xinjiang to nw Qinghai); winters to India

☐ **Blanford's Snowfinch** *Montifringilla blanfordi*

____ *M. b. ventorum*	Mountains of w-central China (se Xinjiang to w Qinghai)
____ *M. b. barbata*	Mountains of w China (ne Qinghai to Nan Shan Mts.)
____ *M. b. blanfordi*	Mountains of Tibet to Sikkim and w China; winters to India

☐ **Afghan Snowfinch** *Montifringilla theresae*

Hindu Kush Mountains (Afghanistan)

FAMILY: PLOCEIDAE (Weavers and Allies—116)

☐ **White-billed Buffalo-Weaver** *Bubalornis albirostris*

Senegambia to s Sudan, Ethiopia, n Uganda and nw Kenya

☐ **Red-billed Buffalo-Weaver** *Bubalornis niger*

____ *B. n. niger (militaris)*	Angola and Zambia to South Africa
____ *B. n. intermedius*	S Ethiopia to Somalia, Kenya and Tanzania

☐ **White-headed Buffalo-Weaver** *Dinemellia dinemelli*

____ *D. d. dinemelli*	Sudan to Somalia and Kenya
____ *D. d. boehmi*	SE Kenya to Tanzania

☐ **Speckle-fronted Weaver** *Sporopipes frontalis*
　____　*S. f. frontalis (pallidior)*　　　　Senegambia to Ethiopia and Eritrea
　____　*S. f. emini*　　　　S Sudan to ne Uganda, Kenya and n Tanzania

☐ **Scaly Weaver** *Sporopipes squamifrons*

　　　　SW Angola to Botswana, Zimbabwe, Transvaal and n Cape Prov.

☐ **White-browed Sparrow-Weaver** *Plocepasser mahali*
　____　*P. m. melanorhynchus*　　　　Sudan to Kenya
　____　*P. m. ansorgei*　　　　S Angola to extreme n Namibia
　____　*P. m. pectoralis*　　　　Tanzania, Zambia, Malawi, Mozambique, Zimbabwe, Botswsana
　____　*P. m. mahali (stentor, terricolor)*　　　　S Namibia, s Botswana, s Zimbabwe and South Africa

☐ **Chestnut-crowned Sparrow-Weaver** *Plocepasser superciliosus*

　　　　Senegal to Sudan, Ethiopia, n Uganda and nw Kenya

☐ **Chestnut-backed Sparrow-Weaver** *Plocepasser rufoscapulatus*

　　　　S Angola to se Zaire, Zambia and w Malawi

☐ **Donaldson-Smith's Sparrow-Weaver** *Plocepasser donaldsoni*

　　　　Savanna of sw Ethiopia, extreme s Somalia and n Kenya

☐ **Rufous-tailed Weaver** *Histurgops ruficauda*

　　　　Acacia savanna of n Tanzania se of Lake Victoria

☐ **Gray-headed Social-Weaver** *Pseudonigrita arnaudi*
　____　*P. a. arnaudi*　　　　SW Sudan to s Ethiopia, Kenya, Uganda and extreme n Tanzania
　____　*P. a. dorsalis*　　　　West and central Tanzania

☐ **Black-capped Social-Weaver** *Pseudonigrita cabanisi*

　　　　Thornscrub of Ethiopia to sw Somalia, Kenya and ne Tanzania

☐ **Social Weaver** *Philetairus socius*

　　　　Namibia and Botswana to n South Africa

☐ **Red-crowned Malimbe** *Malimbus coronatus*

　　　　S Cameroon to Gabon, sw Central African Republic and e Zaire

☐ **Black-throated Malimbe** *Malimbus cassini*

　　　　Lowlands of s Ghana and s Cameroon to Gabon and e Zaire

☐ **Ballman's Malimbe** *Malimbus ballmanni*

　　　　Humid forests of Sierra Leone, Liberia and sw Ivory Coast

☐ **Rachel's Malimbe** *Malimbus racheliae*

　　　　Lowland forests of se Nigeria to s Cameroon and Gabon

☐ **Red-vented Malimbe** *Malimbus scutatus*
　____　*M. s. scutatus*　　　　Sierra Leone to Ghana and Benin
　____　*M. s. scutopartitus*　　　　S Nigeria to sw Cameroon

☐ **Ibadan Malimbe** *Malimbus ibadanensis*

　　　　Savanna of sw Nigeria

☐ **Red-bellied Malimbe** *Malimbus erythrogaster*
　____　*M. e. erythrogaster*　　　　Nigeria and Cameroon to e-central Zaire, Sudan and Uganda
　____　*M. e. fagani*　　　　E Zaire (Ituri and Semliki forests)

☐ **Gray's Malimbe** *Malimbus nitens*

　　　　Senegambia to nw Angola, Zaire and w Uganda

☐ **Crested Malimbe** *Malimbus malimbicus*
　____　*M. m. nigrifrons*　　　　Sierra Leone to s Nigeria
　____　*M. m. malimbicus*　　　　S Cameroon lower Congo and n Angola
　____　*M. m. crassirostris*　　　　Zaire and Uganda

☐ **Red-headed Malimbe** *Malimbus rubricollis*
　____　*M. r. bartletti*　　　　W Guinea to Togo
　____　*M. r. nigeriae*　　　　Benin to s Nigeria (east to Niger delta)
　____　*M. r. rubricollis*　　　　SE Nigeria to lower Congo River and Angola (Cabinda)
　____　*M. r. centralis*　　　　E Zaire, Uganda and w Kenya
　____　*M. r. rufovelatus*　　　　Bioko (Gulf of Guinea)
　____　*M. r. praedi*　　　　Angola (Uige, Cuanza Norte and Cuanza Sul)

□ **Red-headed Weaver** *Anaplectes rubriceps*
_____ *A. r. leuconotus* — Senegambia to s Sudan, Ethiopia, Kenya, Tanzania and Malawi
_____ *A. r. jubaensis* — S Somalia and adjacent coastal Kenya
_____ *A. r. rubriceps* — S Angola to ne Namibia, s Zambia, Mozambique and ne S Africa

□ **Yellow-legged Weaver** *Ploceus flavipes* — NE Zaire (Ituri Forest)

□ **Bertram's Weaver** *Ploceus bertrandi* — Highlands of Tanzania, Zambia, Malawi and n Mozambique

□ **Baglafecht Weaver** *Ploceus baglafecht*
_____ *P. b. neumanni* — E Nigeria, Cameroon and Central African Republic
_____ *P. b. baglafecht* — Ethiopian highlands and s Eritrea
_____ *P. b. eremobius* — NE Zaire and sw Sudan
_____ *P. b. emini (budongoensis)* — SE Sudan and n Uganda
_____ *P. b. reichenowi* — Highlands of Kenya and n Tanzania
_____ *P. b. stuhlmanni* — E Zaire to s Uganda and w Tanzania
_____ *P. b. sharpii* — Montane forests of sw Tanzania
_____ *P. b. nyikae* — Nyika Plateau of Zambia and Malawi

□ **Black-chinned Weaver** *Ploceus nigrimentus* — SE Gabon (Lekoni) to central Angola and adj. Congo Republic

□ **Bannerman's Weaver** *Ploceus bannermani* — Montane forests of extreme e Nigeria and w Cameroon

□ **Bates' Weaver** *Ploceus batesi* — Lowland forests of s Cameroon

□ **Little Weaver** *Ploceus luteolus* — Senegambia to ne Zaire, s Sudan, Ethiopia, Kenya and Tanzania

□ **Slender-billed Weaver** *Ploceus pelzelni*
_____ *P. p. monacha* — Papyrus swamps of Liberia to w Zaire, Angola and Zambia
_____ *P. p. pelzelni (tuta)* — NE Zaire to Uganda, Rwanda, Burundi, Kenya and Tanzania

□ **Loango Weaver** *Ploceus subpersonatus* — Coastal s Gabon to mouth of Congo River

□ **Black-necked Weaver** *Ploceus nigricollis*
_____ *P. n. brachypterus (po)* — Senegal to Nigeria and w Cameroon; Bioko
_____ *P. n. nigricollis* — E Cameroon to n Angola, s Sudan, s Zaire, w Kenya, nw Tanzania
_____ *P. n. melanoxanthus (malensis)* — S Ethiopia to e Kenya and e Tanzania

□ **Spectacled Weaver** *Ploceus ocularis*
_____ *P. o. crocatus (tenuirostris)* — Cameroon to s Sudan, s Ethiopia, Angola and Zambia
_____ *P. o. suahelicus* — E Kenya to e Tanzania, e Zambia, Malawi and n Mozambique
_____ *P. o. ocularis (brevior)* — E South Africa, Swaziland and Mozambique s of the Limpopo R.

□ **Black-billed Weaver** *Ploceus melanogaster*
_____ *P. m. melanogaster* — Mountains of extreme se Nigeria and w Cameroon; Bioko
_____ *P. m. stephanophorus* — S Sudan to Uganda, e Zaire, Burundi, w Kenya and w Tanzania

□ **Strange Weaver** *Ploceus alienus* — Montane forests of e Zaire, w Uganda, Rwanda and Burundi

□ **Cape Weaver** *Ploceus capensis* — South Africa

□ **Bocage's Weaver** *Ploceus temporalis* — Riparian grasslands of Angola to se Zaire and nw Zambia

□ **African Golden-Weaver** *Ploceus subaureus*
_____ *P. s. aureoflavus* — Somalia to e Kenya, e Tanzania, Malawi and n Mozambique
_____ *P. s. subaureus (tongensis)* — S Mozambique to Natal, Zululand and e Cape Province

□ **Holub's Golden-Weaver** *Ploceus xanthops* — Gabon to Angola, Uganda, Kenya, Botswana and Mozambique

☐ **Orange Weaver** *Ploceus aurantius*
_____ *P. a. aurantius* — Sierra Leone to Cameroon, Gabon, ne Angola and Zaire
_____ *P. a. rex* — S Uganda (Lake Victoria region) to w Kenya and nw Tanzania

☐ **Golden Palm Weaver** *Ploceus bojeri* — Savanna of Ethiopia to s Somalia, e Kenya and ne Tanzania

☐ **Taveta Golden-Weaver** *Ploceus castaneiceps* — Riverine scrub of extreme se Kenya and ne Tanzania

☐ **Principe Golden-Weaver** *Ploceus princeps* — Príncipe I. (Gulf of Guinea)

☐ **Southern Brown-throated Weaver** *Ploceus xanthopterus*
_____ *P. x. castaneigula* — Extreme ne Namibia to sw Zambia, Caprivi Strip and n Botswana
_____ *P. x. xanthopterus* — Tanzania, Malawi, Mozambique and ne Zimbabwe
_____ *P. x. marleyi* — Extreme s coastal Mozambique to Natal and Zululand

☐ **Northern Brown-throated Weaver** *Ploceus castanops* — E Zaire to Uganda, Rwanda, w Kenya and nw Tanzania

☐ **Kilombero Weaver** *Ploceus burnieri* — SE Tanzania (floodplain of Kilombero River)

☐ **Northern Masked-Weaver** *Ploceus taeniopterus* — Extreme se Sudan to ne Zaire, n Uganda and s Ethiopia

☐ **Lesser Masked-Weaver** *Ploceus intermedius*
_____ *P. i. intermedius* — SE Sudan to s Ethiopia, Somalia, e Zaire, Kenya and Tanzania
_____ *P. i. beattyi* — Arid coastal w Angola
_____ *P. i. cabanisii* — Coastal Tanzania to Natal, n Namibia and interior s Angola

☐ **Southern Masked-Weaver** *Ploceus velatus*
_____ *P. v. velatus* — Western Cape, Northern Cape and Free State
_____ *P. v. nigrifrons* — E Cape, n and w Natal and w Swaziland
_____ *P. v. tahatali (peixotoi)* — Zimbabwe, se Botswana, sw Mozambique, e Swaziland, ne Natal
_____ *P. v. shelleyi* — Zambia, Malawi and Mozambique
_____ *P. v. caurinus* — S Angola, Namibia, Botswana and n Cape Province
_____ *P. v. finschi* — Coastal Angola

☐ **Vitelline Masked-Weaver** *Ploceus vitellinus*
_____ *P. v. vitellinus* — Senegal to w Sudan
_____ *P. v. uluensis* — Ethiopia to Kenya and n Tanzania

☐ **Tanganyika Masked-Weaver** *Ploceus reichardi* — Swamps of sw Tanzania (Karema to Rukwa) and Zambia

☐ **Katanga Masked-Weaver** *Ploceus katangae* — Spottily distributed e Zaire and n Zambia

☐ **Lake Lufira Masked-Weaver** *Ploceus ruweti* — SE Zaire (swamps in vicinity of Lake Lufira)

☐ **Heuglin's Masked-Weaver** *Ploceus heuglini* — Senegambia to sw Sudan, s Uganda and nw Kenya

☐ **Rueppell's Weaver** *Ploceus galbula* — E Sudan to Eritrea, n Ethiopia, Somalia and sw Arabia

☐ **Speke's Weaver** *Ploceus spekei* — S Ethiopia to Somalia, Kenya and n Tanzania

☐ **Fox's Weaver** *Ploceus spekeoides* — Swamps of e-central Uganda

☐ **Vieillot's Weaver** *Ploceus nigerrimus*
_____ *P. n. castaneofuscus* — Sierra Leone to e Nigeria
_____ *P. n. nigerrimus* — Extreme se Nigeria to Uganda, w Kenya, Angola and Zaire

☐ **Village Weaver** *Ploceus cucullatus*
_____ *P. c. cucullatus* — Mauretania to Chad, south to n Gabon and nw Zaire; Bioko
_____ *P. c. collaris* — Gabon to sw Zaire and n Angola
_____ *P. c. bohndorffi* — N Gabon to n Zaire, Uganda, w Kenya and sw Sudan

____	*P. c. frobenii*	S Zaire
____	*P. c. graueri*	E Zaire to Rwanda, Burundi and adjacent w Tanzania
____	*P. c. abyssinicus*	Ethiopia and se Sudan
____	*P. c. nigriceps (paroptus)*	S Somalia to Kenya, Tanzania and n Mozambique; São Tomé
____	*P. c. spilonotus (dilutescens)*	SE Botswana to s Mozambique, South Africa and Lesotho

☐ **Giant Weaver** *Ploceus grandis*

São Tomé (Gulf of Guinea)

☐ **Weyns' Weaver** *Ploceus weynsi*

Lowland forests of n Zaire, Uganda and extreme nw Tanzania

☐ **Clarke's Weaver** *Ploceus golandi*

Coastal se Kenya (Arabuko-Sokoke Forest)

☐ **Salvadori's Weaver** *Ploceus dichrocephalus*

Riparian vegetation of se Ethiopia, s Somalia and ne Kenya

☐ **Black-headed Weaver** *Ploceus melanocephalus*

____	*P. m. melanocephalus*	N Senegal to s Mauritania, Niger and Mali
____	*P. m. capitalis*	Guinea-Bissau to Niger, s Chad and Central African Republic
____	*P. m. duboisi*	Congo, s Central African Republic andn Zaire
____	*P. m. dimidiatus (fischeri)*	E Sudan to nw Ethiopia, Uganda, Kenya, Tanzania and Zambia

☐ **Golden-backed Weaver** *Ploceus jacksoni*

Extreme se Sudan to Uganda, Burundi, Kenya and n Tanzania

☐ **Chestnut Weaver** *Ploceus rubiginosus*

____	*P. r. trothae*	Acacia savanna of sw Angola, n Namibia and Botswana
____	*P. r. rubiginosus*	Extreme se Sudan to Ethiopia, Somalia, Kenya and n Tanzania

☐ **Cinnamon Weaver** *Ploceus badius*

Sudan (Nile tributaries south to Uganda birder)

☐ **Golden-naped Weaver** *Ploceus aureonucha*

NE Zaire (Ituri Forest north of Beni)

☐ **Yellow-mantled Weaver** *Ploceus tricolor*

____	*P. t. tricolor*	Sierra Leone to se Guinea, Cameroon, Gabon and n Angola
____	*P. t. interscapularis*	N Zaire to w Uganda, Kenya and extreme sw Sudan

☐ **Maxwell's Black Weaver** *Ploceus albinucha*

____	*P. a. albinucha*	Guinea to Ghana
____	*P. a. maxwelli*	Bioko (Gulf of Guinea)
____	*P. a. holomelas*	S Nigeria to Gabon, Cent. African Rep., n Zaire and w Uganda

☐ **Forest Weaver** *Ploceus bicolor*

____	*P. b. tephronotus (analogus)*	Extreme se Nigeria and Cameroon; Bioko
____	*P. b. amaurocephalus*	N Angola
____	*P. b. mentalis*	S Sudan (Imatong Hills) to ne Zaire, Uganda and w Kenya
____	*P. b. kigomaensis*	S Zaire to Zambia and extreme w Tanzania
____	*P. b. kersteni*	Extreme s Somalia to coastal e Kenya and e Tanzania
____	*P. b. stictifrons (lebomboensis, sclateri)*	SE Tanzania to e Zimbabwe, Mozambique and n KwaZulu Natal
____	*P. b. sylvanus*	E Zimbabwe and adjacent w Mozambique
____	*P. b. bicolor*	E Cape Province to KwaZulu Natal

☐ **Brown-capped Weaver** *Ploceus insignis*

SE Nigeria to n Angola, s Sudan, Kenya and w Tanzania; Bioko

☐ **Yellow-capped Weaver** *Ploceus dorsomaculatus*

S Cameroon to Gabon, Central African Republic and ne Zaire

☐ **Preuss' Weaver** *Ploceus preussi*

Guinea and Sierra Leone to s Cameroon, Gabon and ne Zaire

☐ **Olive-headed Weaver** *Ploceus olivaceiceps*

Mountains of s Tanzania, Zambia, Malawi and Mozambique

☐ **Usambara Weaver** *Ploceus nicolli*

____	*P. n. nicolli*	NE Tanzania (Usambara Mts.)
____	*P. n. anderseni*	NE Tanzania (Uluguru and Udzungwa Mts.)

☐ **Bar-winged Weaver** *Ploceus angolensis*

Angola to n Zambia and extreme se Zaire

☐ **São Tomé Weaver** *Ploceus sanctithomae*

Highlands of São Tomé (Gulf of Guinea)

☐ **Nelicourvi Weaver** *Ploceus nelicourvi*

Evergreen mossy forests of n, w and s Madagascar

☐ **Sakalava Weaver** *Ploceus sakalava*
_____ *P. s. sakalava* — Dry lowland forests of n and ne Madagascar
_____ *P. s. minor* — Subdeserts of w and sw Madagascar

☐ **Streaked Weaver** *Ploceus manyar*
_____ *P. m. flaviceps* — Lowlands of e Pakistan to w India and Sri Lanka
_____ *P. m. peguensis* — NE India (Assam) to Bangladesh and n Myanmar
_____ *P. m. williamsoni* — SW China (Yunnan) to Thailand and Vietnam
_____ *P. m. manyar* — Java, Bali and Bawean I.

☐ **Baya Weaver** *Ploceus philippinus*
_____ *P. p. philippinus* — Lowlands of se Pakistan to w India and Sri Lanka
_____ *P. p. travencoreensis* — SW India (Goa to Travancore and Kerala)
_____ *P. p. burmanicus* — NE India (Bengal) to Bangladesh, Assam and Myanmar
_____ *P. p. infortunatus* — Malay Peninsula to s Vietnam, Sumatra and Nias I.
_____ *P. p. angelorum* — Plains of central Thailand

☐ **Asian Golden Weaver** *Ploceus hypoxanthus*
_____ *P. h. hymenaicus* — S Myanmar to central Thailand, Cambodia and s Vietnam
_____ *P. h. hypoxanthus* — Sumatra and Java

☐ **Yellow Weaver** *Ploceus megarhynchus*
_____ *P. m. megarhynchus* — Foothills of s Himalayas
_____ *P. m. salimalii* — Foothills of ne India

☐ **Bengal Weaver** *Ploceus benghalensis*

Lowlands of Pakistan to peninsular India, Nepal and Bangladesh

☐ **Compact Weaver** *Pachyphantes superciliosus*

Sierra Leone to Angola, s Sudan, Ethiopia, Uganda and w Kenya

☐ **Cardinal Quelea** *Quelea cardinalis*

S Sudan to Ethiopia, Uganda, Kenya, e Zaire, Zambia, Zimbabwe

☐ **Red-headed Quelea** *Quelea erythrops*

Grasslands of Africa south of the Sahara

☐ **Red-billed Quelea** *Quelea quelea*
_____ *Q. q. quelea (intermedia)* — Senegal to Chad
_____ *Q. q. aethiopica* — Sudan to Somalia, ne Zaire, Uganda, Kenya and Tanzania
_____ *Q. q. lathami (spoliator)* — South Africa

☐ **Bob-tailed Weaver** *Brachycope anomala*

SE Cameroon and banks of Congo River system in Zaire

☐ **Red Fody** *Foudia madagascariensis*

Madagascar, Reunion, Mauritius and Chagos Arch.

☐ **Red-headed Fody** *Foudia eminentissima*
_____ *F. e. aldabrana* — Aldabra (w Indian Ocean)
_____ *F. e. consobrina* — Grand Comoro I. (Comoro Islands)
_____ *F. e. anjuanensis* — Anjouan (Comoro Islands)
_____ *F. e. eminentissima* — Mohéli (Comoro Islands)
_____ *F. e. algondae* — Mayotte (Comoro Islands)

☐ **Forest Fody** *Foudia omissa*

Forests of e Madagascar

☐ **Mauritius Fody** *Foudia rubra*

Uplands of sw Mauritius (w Mascarene Islands)

☐ **Seychelles Fody** *Foudia sechellarum*

Seychelles Islands (Cousin, Cousine and Frégate)

☐ **Rodrigues Fody** *Foudia flavicans*

Rodrigues (e Mascarene Islands). On verge of extinction

☐ **Orange Bishop** *Euplectes franciscanus*

Senegal to Sudan, e Zaire, Ethiopia, n Uganda and nw Kenya

☐ **Red Bishop** *Euplectes orix*

E Zaire to Uganda, Kenya, Tanzania, Malawi and Mozambique

☐ **Zanzibar Bishop** *Euplectes nigroventris*

SE Kenya to e Tanzania, Zanzibar, Pemba and n Mozambique

☐ **Black-winged Bishop** *Euplectes hordeaceus*

Widespread Africa south of the Sahara (except South Africa)

☐ **Black Bishop** *Euplectes gierowii*
____ *E. g. ansorgei* — Cameroon to Ethiopia and w Kenya
____ *E. g. friederichseni* — SW Kenya to n-central Tanzania
____ *E. g. gierowii* — N Angola to sw Zaire

☐ **Yellow-crowned Bishop** *Euplectes afer*
____ *E. a. afer (ladoensis, niassensis)* — S Mauritania to Chad, CAR, w Sudan, n Zaire and nw Angola
____ *E. a. strictus* — Highlands of Ethiopia
____ *E. a. taha* — S Sudan, s Ethiopia and Somalia to South Africa

☐ **Fire-fronted Bishop** *Euplectes diadematus*

SW Somalia to e Kenya and extreme ne Tanzania

☐ **Golden-backed Bishop** *Euplectes aureus*

Scrub of coastal Angola; São Tomé

☐ **Yellow Bishop** *Euplectes capensis*
____ *E. c. phoenicomerus* — Highlands of se Nigeria and Cameroon; Bioko
____ *E. c. crassirostris* — NE Transvaal to Uganda, Sudan and Kenya
____ *E. c. xanthomelas* — Ethiopia
____ *E. c. angolensis* — N and central highlands of Angola
____ *E. c. approximans (knysnae)* — E South Africa (Knysna to Gauteng Province)
____ *E. c. capensis (macrorhynchus)* — W South Africa to Knysna

☐ **White-winged Widowbird** *Euplectes albonotatus*
____ *E. a. eques* — CAR to Sudan, Ethiopia and s Tanzania
____ *E. a. asymmetrurus* — W Gabon to w Angola; São Tomé
____ *E. a. albonotatus* — S Tanzania to ne Botswana and South Africa

☐ **Yellow-shouldered Widowbird** *Euplectes macroura*
____ *E. m. macroura (soror, intermedia)* — Senegal to s Sudan, Zaire, Angola, Zambia and Malawi
____ *E. m. macrocerca* — Highlands of Ethiopia, Uganda and w Kenya
____ *E. m. conradsi* — NW Tanzania (Ukerewe Island in Lake Victoria)

☐ **Red-collared Widowbird** *Euplectes ardens*
____ *E. a. laticauda* — Highlands of se Sudan, Eritrea and Ethiopia
____ *E. a. suahelica* — Highlands of Kenya and ne Tanzania
____ *E. a. ardens (concolor)* — Sierra Leone to Uganda, sw Sudan and Tanzania to South Africa

☐ **Fan-tailed Widowbird** *Euplectes axillaris*
____ *E. a. bocagei (mechowi, quanzae)* — West Africa to Angola, sw Zaire, n Botswana and Namibia
____ *E. a. traversii* — Highlands of Ethiopia
____ *E. a. phoeniceus* — S Sudan to s Ethiopia, Uganda, w Kenya, w Tanzania and Zambia
____ *E. a. zanzibaricus* — Coastal Somalia, Kenya and Tanzania
____ *E. a. axillaris* — Malawi and Mozambique south to e South Africa

☐ **Marsh Widowbird** *Euplectes hartlaubi*
____ *E. h. humeralis* — Highlands of Nigeria to Cameroon, Zaire, Uganda and w Kenya
____ *E. h. hartlaubi* — Angola to s Zaire and Zambia

☐ **Buff-shouldered Widowbird** *Euplectes psammocromius*

Highlands of sw Tanzania to extreme ne Zambia and n Malawi

☐ **Long-tailed Widowbird** *Euplectes progne*
___ *E. p. delamerei* — Locally in highlands of Kenya
___ *E. p. progne* — Botswana to Zimbabwe, Mozambique and e Cape Province
___ *E. p. delacouri (definita)* — Zaire to Angola and Zambia

☐ **Jackson's Widowbird** *Euplectes jacksoni*

Grassy highlands of Kenya and n Tanzania

☐ **Grosbeak Weaver** *Amblyospiza albifrons*
___ *A. a. capitalba* — Senegal to sw Nigeria
___ *A. a. saturata* — SE Nigeria, Cameroon, n Gabon, n Congo, sw CAR, nw Zaire
___ *A. a. melanota* — Ethiopia, s Sudan, Uganda, w Kenya, ne Zaire, nw Tanzania
___ *A. a. unicolor* — S Somalia, coastal e Kenya, e Tanzania and extreme s Malawi
___ *A. a. montana* — Interior s Kenya and Tanzania to Malawi, Zambia and Zimbabwe
___ *A. a. tandae* — N Angola and extreme w Zaire
___ *A. a. kasaica* — S Zaire (Kasai)
___ *A. a. maxima* — SE Angola, Caprivi Strip, n Botswana, extreme nw Zimbabwe
___ *A. a. albifrons* — Transvaal, KwaZulu-Natal and e Cape Province
___ *A. a. woltersi* — Mozambique (s of Zambezi) and adj. e Zimbabwe to Zululand

FAMILY: ESTRILDIDAE (Waxbills and Allies—141)

☐ **Pale-fronted Negrofinch** *Nigrita luteifrons*
___ *N. l. luteifrons* — Ivory Coast to Central African Republic, n Angola and w Uganda
___ *N. l. alexanderi* — Bioko (Gulf of Guinea)

☐ **Gray-headed Negrofinch** *Nigrita canicapillus*
___ *N. c. emiliae* — Guinea and Sierra Leone to Ghana and Togo
___ *N. c. canicapillus* — S Nigeria to Gabon, lower and middle Congo and w Zaire
___ *N. c. angolensis* — SW Zaire to nw Angola
___ *N. c. schistaceus* — S Sudan to n Zaire, sw Uganda, w Kenya and n Tanzania
___ *N. c. diabolicus* — Central Kenya to n Tanzania
___ *N. c. candidus* — W Tanzania (Kungwe-Mahari Mountains)

☐ **Chestnut-breasted Negrofinch** *Nigrita bicolor*
___ *N. b. bicolor* — Sierra Leone to Ghana
___ *N. b. brunnescens* — Nigeria to Angola and Zaire; Príncipe I.
___ *N. b. saturatior* — E Zaire and Uganda

☐ **White-breasted Negrofinch** *Nigrita fusconotus*
___ *N. f. fusconotus* — SE Nigeria to Cameroon, Gabon, Angola and w Kenya; Bioko
___ *N. f. uropygialis* — SE Guinea and Liberia (Mt. Nimba) to s Ghana and sw Nigeria

☐ **Woodhouse's Antpecker** *Parmoptila woodhousei*
___ *P. w. woodhousei* — SE Nigeria to Cameroon, s Central African Rep. and cent. Zaire
___ *P. w. ansorgei* — NW Angola

☐ **Red-fronted Antpecker** *Parmoptila rubrifrons*

Liberia (Mt. Nimba) to s Mali and s-central Ghana

☐ **Jameson's Antpecker** *Parmoptila jamesoni*

E Congo and n Zaire to w Uganda and w Tanzania

☐ **Fernando Po Oliveback** *Nesocharis shelleyi*
___ *N. s. shelleyi* — S Cameroon (Mt. Cameroon); Bioko (Moka Highlands)
___ *N. s. bansoensis* — Highlands of Cameroon and se Nigeria (Obudu Plateau)

☐ **White-collared Oliveback** *Nesocharis ansorgei*

Mts. of e Zaire, w Uganda, n Rwanda, Burundi and nw Tanzania

☐ **Gray-headed Oliveback** *Nesocharis capistrata*

Guinea to s Mali, Cameroon, n Zaire, sw Sudan and w Uganda

☐ **Swee Waxbill** *Coccopygia melanotis*
_____ *C. m. melanotis* | SW Zimbabwe to extreme s Mozambique and n South Africa
_____ *C. m. bocagei* | Angola and Namibia

☐ **Yellow-bellied Waxbill** *Coccopygia quartinia*
_____ *C. q. quartinia* | Highlands of Eritrea and Ethiopia
_____ *C. q. kilimensis (nyanzae)* | SE Sudan, e Zaire, Uganda, Kenya to central Tanzania
_____ *C. q. stuartirwini* | S and e Tanzania to e Zimbabwe

☐ **Green-backed Twinspot** *Mandingoa nitidula*
_____ *M. n. schlegeli* | Sierra Leone to Angola, Zaire and Uganda
_____ *M. n. virginiae* | Bioko (Gulf of Guinea)
_____ *M. n. chubbi* | S Ethiopia to ne Zambia, n Malawi and n Mozambique
_____ *M. n. nitidula* | Zaire, nw Zambia, s Malawi and cent. Mozambique to S Africa

☐ **Shelley's Crimson-wing** *Cryptospiza shelleyi*

Mountains of e Zaire to sw Uganda, Rwanda and Burundi

☐ **Dusky Crimson-wing** *Cryptospiza jacksoni*

Mountains of e Zaire to sw Uganda, Rwanda and Burundi

☐ **Abyssinian Crimson-wing** *Cryptospiza salvadorii*
_____ *C. s. kilimensis* | S Sudan to e Uganda, se Kenya and n Tanzania
_____ *C. s. salvadorii* | S Ethiopia (Shoa Province) to n Kenya
_____ *C. s. ruwenzori* | E Zaire (Ruwenzori Mountains) to sw Uganda

☐ **Red-faced Crimson-wing** *Cryptospiza reichenovii*
_____ *C. r. reichenovii* | Cameroon to Angola; Bioko
_____ *C. r. ocularis* | Mts. of Albertine Rift
_____ *C. r. australis (homogenes)* | Tanzania to Zimbabwe and adjacent Mozambique

☐ **Lavender Waxbill** *Estrilda caerulescens*

Senegambia to sw Chad, n Cameroon and Central African Republic

☐ **Black-tailed Waxbill** *Estrilda perreini*
_____ *E. p. perreini* | Zaire, Angola, Zambia, Tanzania and extreme n Malawi
_____ *E. p. incana (poliogastra, torrida)* | S Malawi, se Zimbabwe, Mozambique and South Africa

☐ **Cinderella Waxbill** *Estrilda thomensis*

W Angola to extreme nw Namibia (Cunene River)

☐ **Fawn-breasted Waxbill** *Estrilda paludicola*
_____ *E. p. ochrogaster* | Highlands of extreme e Sudan
_____ *E. p. roseicrissa* | E Zaire, Rwanda, Burundi, sw Uganda and nw Tanzania
_____ *E. p. marwitzi* | SW Tanzania
_____ *E. p. paludicola* | S Sudan, ne Zaire, Uganda and w Kenya
_____ *E. p. ruthae* | Congo and immediately adjacent Gabon and w Zaire
_____ *E. p. benguellensis* | Angola, s Zaire and n Zambia

☐ **Anambra Waxbill** *Estrilda poliopareia*

S Nigeria (Niger River delta to Benin border)

☐ **Orange-cheeked Waxbill** *Estrilda melpoda*

Senegambia to Zaire, n Angola and n Zambia

☐ **Arabian Waxbill** *Estrilda rufibarba*

SW Arabian Peninsula

☐ **Crimson-rumped Waxbill** *Estrilda rhodopyga*
_____ *E. r. rhodopyga* | N Sudan to Eritrea, Ethiopia and Somalia
_____ *E. r. centralis* | S Sudan to n Malawi

☐ **Black-rumped Waxbill** *Estrilda troglodytes*

Senegambia to s Sudan, nw Ethiopia, sw Eritrea and w Kenya

☐ **Common Waxbill** *Estrilda astrild*
_____ *E. a. kempi* | Sierra Leone to s Guinea and Liberia

____	*E. a. occidentalis*	Ghana to CAR, middle Congo River and ne Zaire; Bioko
____	*E. a. rubriventris*	Coastal Gabon and lower sw Congo to nw Angola
____	*E. a. jagoensis*	Coastal w Angola (Benguela and Moçamedes)
____	*E. a. angolensis*	Plateau of w Angola
____	*E. a. niediecki (ngamiensis)*	Angola to sw Zambia, ne Namibia, n Botswana and w Zimbabwe
____	*E. a. damarensis*	Namibia and nw Cape Province
____	*E. a. peasei*	Ethiopia
____	*E. a. macmillani*	Central and s Sudan
____	*E. a. niansae*	Uganda to e Zaire, Rwanda, Burundi, w Kenya, nw Tanzania
____	*E. a. minor*	S Somalia to e Kenya, ne Tanzania, Zanzibar and Mafia I.
____	*E. a. massaica*	Kenya e of Rift Valley to interior n Tanzania
____	*E. a. cavendishi*	Tanzania, se Zaire, Zambia, Malawi to Zimbabwe, Mozambique
____	*E. a. astrild*	W South Africa and se Botswana
____	*E. a. tenebridorsa*	E South Africa, Lesotho and Swaziland

☐ **Black-faced Waxbill** *Estrilda nigriloris*

Swamps of se Zaire (no published records since 1950)

☐ **Black-crowned Waxbill** *Estrilda nonnula*

____	*E. n. nonnula*	Burkina Faso to se Nigeria, Uganda, w Kenya and nw Tanzania
____	*E. n. elizae*	Bioko (Gulf of Guinea)
____	*E. n. eisentrauti*	S Cameroon (Mt. Cameroon)

☐ **Black-headed Waxbill** *Estrilda atricapilla*

____	*E. a. atricapilla*	Extreme se Nigeria to Cameroon, Gabon and nw Zaire
____	*E. a. avakubi*	Central Zaire to extreme ne Angola
____	*E. a. marunguensis*	SE Zaire (Marungu Mts.)

☐ **Kandt's Waxbill** *Estrilda kandti*

Rift Valley mts. of e Zaïre, Uganda, Burundi, Rwanda and Kenya

☐ **Black-cheeked Waxbill** *Estrilda erythronotos*

____	*E. e. delamerei*	S Kenya to w Uganda and central Tanzania
____	*E. e. erythronotos (soligena)*	Angola, sw Zambia, w Zimbabwe, Namibia, Botswana, n S Africa

☐ **Red-rumped Waxbill** *Estrilda charmosyna*

____	*E. c. charmosyna (pallidior)*	Extreme s Sudan to Ethiopia, Somalia, ne Uganda and ne Kenya
____	*E. c. kiwanukae*	S Kenya

☐ **Grant's Bluebill** *Spermophaga poliogenys*

Primary forests of e Zaire and sw Uganda

☐ **Western Bluebill** *Spermophaga haematina*

____	*S. h. haematina*	Senegambia to s Mali, Sierra Leone, Liberia and Ghana
____	*S. h. togoensis*	Togo to sw Nigeria
____	*S. h. pustulata*	SE Nigeria to CAR., Congo, central Zaire and nw Angola

☐ **Red-headed Bluebill** *Spermophaga ruficapilla*

____	*S. r. ruficapilla*	S Sudan to Uganda, w Kenya, w Tanzania and nw Angola
____	*S. r. cana*	NE Tanzania (Usambara Mountains)

☐ **Black-bellied Seedcracker** *Pyrenestes ostrinus*

Ivory Coast to Gabon, n Angola, e Zaire, Tanzania, ne Zambia

☐ **Crimson Seedcracker** *Pyrenestes sanguineus*

Senegambia to s Mali, Sierra Leone, Liberia and Ivory Coast

☐ **Lesser Seedcracker** *Pyrenestes minor*

E Tanzania to s Malawi, e Zimbabwe and n Mozambique

☐ **Blue-breasted Cordonbleu** *Uraeginthus angolensis*

____	*U. a. angolensis (cyanopleurus)*	N Angola to s Zaire, nw Zambia and n Zimbabwe; São Tomé
____	*U. a. niassensis*	S Tanzania, s Zaire, Mozambique, Malawi, Zambia to S Africa

☐ **Red-cheeked Cordonbleu** *Uraeginthus bengalus*

____	*U. b. bengalus*	Senegal to Sudan, Eritrea, Ethiopia, Uganda and w Kenya

_____ *U. b. brunneigularis (littoralis)*	S Somalia and Kenya to Tanzania
_____ *U. b. ugogoensis*	S Kenya and Tanzania
_____ *U. b. katangae (semotus)*	Extreme ne Angola to s Zaire and n Zambia

☐ **Blue-capped Cordonbleu** *Uraeginthus cyanocephalus*

Extreme se Sudan to se Ethiopia, s Somalia, Kenya and n Tanzania

☐ **Purple Grenadier** *Granatina ianthinogaster*

Extreme se Sudan to s Ethiopia, Kenya, Uganda, Tanzania

☐ **Violet-eared Waxbill** *Granatina granatina*

Angola to Namibia, Zambia, s Zimbabwe, Botswana, n S Africa.

☐ **Dybowski's Twinspot** *Euschistospiza dybowskii*

Senegambia and Guinea to n Zaire and extreme s Sudan

☐ **Dusky Twinspot** *Euschistospiza cinereovinacea*

_____ *E. c. cinereovinacea*	Highlands of w and central Angola
_____ *E. c. graueri*	E Zaire to sw Uganda

☐ **Peters' Twinspot** *Hypargos niveoguttatus*

_____ *H. n. macrospilotus (interior)*	Angola to e Zaire, Kenya, Somalia, Tanzania, Zambia and Malawi
_____ *H. n. niveoguttatus*	E Zimbabwe and s Mozambique

☐ **Pink-throated Twinspot** *Hypargos margaritatus*

S Mozambique to e Transvaal, e Swaziland and n Natal

☐ **Brown Twinspot** *Clytospiza monteiri*

SE Nigeria to s Chad, s Sudan, w Uganda, w Kenya and n Angola

☐ **Red-faced Pytilia** *Pytilia hypogrammica*

NE Guinea and Sierra Leone to Cameroon and nw Zaire

☐ **Red-winged Pytilia** *Pytilia phoenicoptera*

_____ *P. p. phoenicoptera*	Senegambia to Burkina Faso, n Nigeria and Cameroon
_____ *P. p. emini*	Cameroon to n Zaire, n Uganda and extreme s Sudan

☐ **Red-billed Pytilia** *Pytilia lineata*

Highlands of w and central Ethiopia

☐ **Green-winged Pytilia** *Pytilia melba*

_____ *P. m. citerior (clanceyi)*	Senegal to Burkina Faso, s Chad, n Cameroon and w Sudan
_____ *P. m. jessei*	NE Sudan to Eritrea and nw Somalia
_____ *P. m. soudanensis (jubaensis)*	E Sudan to Ethiopia, Somalia, Kenya and ne Tanzania
_____ *P. m. flavicaudata*	Djibouti
_____ *P. m. grotei*	E Tanzania, Mozambique, Malawi and Zanzibar
_____ *P. m. belli*	W Uganda, Rwanda, Burundi, Zaire and w Tanzania
_____ *P. m. percivali*	Central and sw Kenya to n and central Tanzania
_____ *P. m. melba (thamnophila, damarensis)*	Congo and Zaire to sw Tanzania, w Malawi and South Africa
_____ *P. m. hygrophila*	N Zambia and n Malawi

☐ **Orange-winged Pytilia** *Pytilia afra*

Extreme s Sudan and Ethiopia to Angola and n Transvaal

☐ **Red-billed Firefinch** *Lagonosticta senegala*

_____ *L. s. senegala (guineensis)*	Cape Verde Islands; Senegal and Mali to Nigeria
_____ *L. s. rhodopsis (flavodorsalis)*	E Mali to Sudan and lowlands of w Eritrea and Ethiopia
_____ *L. s. ruberrima*	Uganda, Kenya, Tanzania, ne Zambia, Zaire and ne Angola
_____ *L. s. brunneiceps*	Highlands of Ethiopia, Eritrea and se Sudan
_____ *L. s. somaliensis*	SE Ethiopia and Somalia
_____ *L. s. rendalli (pallidicrissa, confidens)*	S Angola to s Tanzania, Natal, and n-central Cape Province

☐ **Bar-breasted Firefinch** *Lagonosticta rufopicta*

_____ *L. r. rufopicta*	Senegal to Nigeria, n Cameroon and Ubangi-Chari
_____ *L. r. lateritia*	NE Zaire, s Sudan, w Ethiopia, Uganda and Kenya

☐ **Brown Firefinch** *Lagonosticta nitidula*

Angola to Zaire, Zambia, s Tanzania, n Namibia and n Zimbabwe

☐ **Black-faced Firefinch** *Lagonosticta larvata*

____ *L. l. vinacea* — Senegal to Guinea and w Mali
____ *L. l. nigricollis* — Central and s Mali to Sudan and Uganda
____ *L. l. larvata* — W Ethiopia and e Sudan

☐ **Black-bellied Firefinch** *Lagonosticta rara*

____ *L. r. forbesi* — Senegal to Nigeria
____ *L. r. rara* — N Cameroon to n Zaire, s Sudan, n Uganda and w Kenya

☐ **African Firefinch** *Lagonosticta rubricata*

____ *L. r. polionota (neglecta)* — Guinea to Nigeria
____ *L. r. congica (ugandae)* — Cameroon to s Sudan, Angola, w Uganda and Zaire
____ *L. r. haematocephala (hildebrandti)* — Sudan, Ethiopia, Uganda, Kenya, Tanzania to Mozambique
____ *L. r. rubricata* — South Africa, Swaziland and Mozambique south of Save R.

☐ **Pale-billed Firefinch** *Lagonosticta landanae*

— Cabinda and lower Congo River to nw Angola and s Zaire

☐ **Jameson's Firefinch** *Lagonosticta rhodopareia*

____ *L. r. rhodopareia* — Ethiopia, Sudan and w Uganda to n Kenya
____ *L. r. jamesoni* — Kenya to South Africa
____ *L. r. ansorgei* — Angola, lower Zaire (Matadi) and n Namibia

☐ **Mali Firefinch** *Lagonosticta virata*

— SE Senegal and s Mali (upper Niger River to Guinea border)

☐ **Rock Firefinch** *Lagonosticta sanguinodorsalis*

— N Nigeria (Jos Plateau)

☐ **Reichenow's Firefinch** *Lagonosticta umbrinodorsalis*

— Dry grasslands of s Chad and adjacent n Cameroon

☐ **Cut-throat** *Amadina fasciata*

____ *A. f. fasciata* — Senegal to Sudan
____ *A. f. alexanderi* — Eritrea, Ethiopia, Somalia to se Sudan, Uganda, Kenya, Tanzania
____ *A. f. meridionalis* — Malawi to Mozambique, Angola, nc Namibia and n South Africa
____ *A. f. contigua* — S Zimbabwe, s Mozambique, South Africa and Swaziland

☐ **Red-headed Finch** *Amadina erythrocephala*

____ *A. e. erythrocephala* — NW Angola to sw Zimbabwe, Transvaal and n Cape Province
____ *A. e. dissita* — Arid w Free State and adj. sw Transvaal to n Cape Prov. and Karoo

☐ **Zebra Waxbill** *Sporaeginthus subflavus*

____ *S. s. subflavus* — Senegal to s Sudan, Ethiopia and w Kenya; South Yemen
____ *S. s. clarkei* — Gabon and w Angola to Kenya, Tanzania and South Africa

☐ **Green Avadavat** *Sporaeginthus formosus*

— Locally in grassland and scrub of peninsular India

☐ **Red Avadavat** *Amandava amandava*

____ *A. a. amandava* — Lowlands of Pakistan to India, s Nepal and Bangladesh
____ *A. a. flavidiventris* — SW China (Yunnan) to Myanmar, Malay Pen. and Lesser Sundas
____ *A. a. punicea* — SE China to se Thailand, Cambodia, Vietnam, Java and Bali

☐ **Black-faced Quailfinch** *Ortygospiza atricollis*

____ *O. a. atricollis (ugandae)* — Senegal to Cameroon
____ *O. a. ugandae* — S Sudan, nw Uganda, ne Zaire and w Kenya
____ *O. a. ansorgei* — Guinea-Bissau to Liberia, Ivory Coast and coastal Ghana

☐ **Red-billed Quailfinch** *Ortygospiza gabonensis*

____ *O. g. gabonensis* — Rio Muni to Gabon and central Zaire (Congo River)
____ *O. g. fuscata* — Angola to s Zaire and Zambia
____ *O. g. dorsostriata* — Extreme e Zaire to Uganda, Rwanda and nw Tanzania

☐ **African Quailfinch** *Ortygospiza fuscocrissa*
_____ *O. f. fuscocrissa* — Highlands of Ethiopia and Eritrea
_____ *O. f. muelleri (bradfieldi, miniscula)* — Kenya to Zambia, Angola, Namibia, n Cape Province, Transvaal
_____ *O. f. smithersi* — NE Zambia
_____ *O. f. pallida* — N Botswana and adjacent nw Zimbabwe
_____ *O. f. digressa* — E Zimbabwe, s Mozambique, South Africa, Lesotho, Swaziland

☐ **Locustfinch** *Paludipasser locustella*
_____ *P. l. locustella* — Angola to s Tanzania, n Mozambique and e Zimbabwe
_____ *P. l. uelensis* — W Congo to central Nigeria and w Kenya

☐ **Painted Firetail** *Emblema pictum*
Interior n-central Australia (Pilbara to nw Queensland)

☐ **Beautiful Firetail** *Stagonopleura bella*
_____ *S. b. bella* — SE Australia (Hunter R., NSW to s Victoria and Tasmania)
_____ *S. b. interposita* — SE South Australia and adjacent sw Victoria
_____ *S. b. samueli* — South Australia (Kangaroo I. and Mt. Lofty Range

☐ **Red-eared Firetail** *Stagonopleura oculata*
Southwest Western Australia east to about Esperance

☐ **Diamond Firetail** *Stagonopleura guttata*
SE Australia (e-central Qld. to s Victoria and se South Australia)

☐ **Mountain Firetail** *Oreostruthus fuliginosus*
_____ *O. f. pallidus* — W New Guinea (Snow Mountains)
_____ *O. f. hagenensis* — Central highlands of New Guinea
_____ *O. f. fuliginosus* — SE New Guinea (Owen Stanley Mountains)

☐ **Red-browed Firetail** *Neochmia temporalis*
_____ *N. t. minor* — NE Queensland (Cape York Peninsula)
_____ *N. t. temporalis* — E Australia (ne Queensland to s Victoria and se South Australia)

☐ **Crimson Finch** *Neochmia phaeton*
_____ *N. p. evangelinae* — S New Guinea and n Australia (Cape York Peninsula)
_____ *N. p. phaeton* — N Australia (King Sound to nw Queensland; n and e-central Qld.)

☐ **Star Finch** *Neochmia ruficauda*
_____ *N. r. subclarescens* — W and n Australia (disjunct Shark Bay to Gulf of Carpenteria)
_____ *N. r. clarescens* — N Queensland (southern Cape York Peninsula)
_____ *N. r. ruficauda* — Queensland (Burdekin R. to far northern NSW). Extinct?

☐ **Plum-headed Finch** *Neochmia modesta*
E Australia (Burdekin R., Queensland to about Sydney, NSW)

☐ **Zebra Finch** *Taeniopygia guttata*
Lesser Sundas (Lombok and Sumba to Timor and Sermata)

☐ **Chestnut-eared Finch** *Taeniopygia castanotis*
Mainland Australia (except n, s and e coastal regions)

☐ **Double-barred Finch** *Taeniopygia bichenovii*
_____ *T. b. annulosa* — N Australia (Kimberley to Gulf of Carpenteria, N Territory)
_____ *T. b. bichenovii* — E Australia (Cape York Peninsula to e Victoria)

☐ **Masked Finch** *Poephila personata*
_____ *P. p. personata* — N Australia (Kimberley, Western Australia to nw Queensland)
_____ *P. p. leucotis* — N Queensland (Cape York Peninsula)

☐ **Long-tailed Finch** *Poephila acuticauda*
_____ *P. a. hecki* — N Australia (Arnhem Land, Northern Territory to nw Queensland)
_____ *P. a. acuticauda* — NW Western Australia (Kimberley Division)

☐ **Black-throated Finch** *Poephila cincta*
_____ *P. c. atropygialis* — N Queensland (Cape York Peninsula)
_____ *P. c. cincta* — E Australia (Burdekin R., Queensland to far ne New South Wales)

☐ **Tawny-breasted Parrotfinch** *Erythrura hyperythra*

____	*E. h. malayana*	Mountains of Malay Peninsula
____	*E. h. brunneiventris*	Philippines (montane forests of Luzon and Mindoro)
____	*E. h. borneensis*	Borneo (Sarawak and Sabah)
____	*E. h. microrhyncha*	Montane forests of Sulawesi
____	*E. h. hyperythra*	Montane forests of Java
____	*E. h. intermedia*	W Lesser Sundas (Lombok, Sumbawa and Flores)

☐ **Pin-tailed Parrotfinch** *Erythrura prasina*

____	*E. p. prasina*	S Myanmar to Thailand, Laos, Malay Peninsula, Sumatra and Java
____	*E. p. coelica*	Borneo

☐ **Green-faced Parrotfinch** *Erythrura viridifacies*

N Philippines (Luzon and Negros)

☐ **Tricolored Parrotfinch** *Erythrura tricolor*

Lesser Sundas (Timor, Wetar, Babar, Damar, Romang, Tanimbar)

☐ **Blue-faced Parrotfinch** *Erythrura trichroa*

____	*E. t. sanfordi*	Sulawesi (Latimojong Mountains and Lore Lindu Nat. Park)
____	*E. t. modesta*	N Moluccas (Ternate, Tidore, Halmahera and Bacan)
____	*E. t. pinaiae*	S Moluccas (Seram and Buru)
____	*E. t. sigillifera*	New Guinea and Bismarck Archipelago
____	*E. t. macgillivrayi*	NE Queensland (Cape York to about Cairns)
____	*E. t. eichhorni*	St. Matthias Islands (Bismarck Archipelago)
____	*E. t. pelewensis*	Palau Islands (e Caroline Islands)
____	*E. t. clara*	Caroline Islands (Truk, Pohnpei and Kosrae)
____	*E. t. trichroa*	Kosrae (Caroline Islands)
____	*E. t. woodfordi*	Guadalcanal and Solomon Islands
____	*E. t. cyanofrons*	Vanuatu and Loyalty Islands (Lifou and Maré)

☐ **Red-eared Parrotfinch** *Erythrura coloria*

S Philippines (Mt. Katanglad and Mt. Apo on Mindanao)

☐ **Papuan Parrotfinch** *Erythrura papuana*

Mountains of New Guinea (Vogelkop Pen. to southeast ranges)

☐ **Red-throated Parrotfinch** *Erythrura psittacea*

New Caledonia

☐ **Fiji Parrotfinch** *Erythrura pealii*

Fiji (Kandavu, Viti Levu, Vanua Levu and Taveuni)

☐ **Red-headed Parrotfinch** *Erythrura cyaneovirens*

____	*E. c. cyaneovirens*	Western Samoa (Savai'i and Upolu)
____	*E. c. gaughrani*	Savai'i (Western Samoa)
____	*E. c. efatensis*	Efate (Vanuatu)
____	*E. c. serena*	Aneityum (Vanuatu)

☐ **Royal Parrotfinch** *Erythrura regia*

N Vanuatu and Banks Group

☐ **Pink-billed Parrotfinch** *Erythrura kleinschmidti*

Montane forests of Viti Levu (Fiji)

☐ **Gouldian Finch** *Chloebia gouldiae*

N Australia (Kimberley Division to w Cape York Peninsula)

☐ **Gray-headed Silverbill** *Odontospiza griseicapilla*

S Ethiopia to Kenya and central Tanzania

☐ **Bronze Mannikin** *Spermestes cucullatus*

____	*S. c. cucullatus*	Senegal to w Kenya and nw Angola; São Tomé and Príncipe
____	*S. c. scutatus (tessellates)*	Ethiopia and Sudan to Natal and e Cape Province; Comoro Is.

☐ **Black-and-white Mannikin** *Spermestes bicolor*

____	*S. b. bicolor*	Guinea-Bissau to s Nigeria
____	*S. b. poensis (stigmatophorus)*	S Cameroon to Ethiopia, Kenya, Tanzania and n Angola; Bioko
____	*S. b. nigriceps (minor)*	S Somalia to Kenya, Tanzania, Malawi, Zambia and S Africa
____	*S. b. woltersi*	SE Zaire to nw Zambia

☐ **Magpie Mannikin** *Spermestes fringilloides*

_____
Senegal to Kenya, south to Tanzania, Zanzibar, Zambia and Natal

☐ **African Silverbill** *Euodice cantans*
_____ *E. c. cantans*
_____ *E. c. orientalis (inornata)*

Senegal to central Sudan
Sudan to Ethiopia, Somalia, Kenya and n Tanzania; se Arabia

☐ **White-throated Munia** *Euodice malabarica*

_____
Arabian Peninsula to se Iran, India, Sri Lanka, Nepal and Sikkim

☐ **Madagascar Munia** *Lonchura nana*

Madagascar

☐ **White-rumped Munia** *Lonchura striata*
_____ *L. s. acuticauda*
_____ *L. s. striata*
_____ *L. s. fumigata*
_____ *L. s. semistriata*
_____ *L. s. swinhoei*
_____ *L. s. subsquamicollis*
_____ *L. s. explita*

SE Kashmir and n India to Nepal, Myanmar and n Thailand
S India and Sri Lanka
Andaman Islands
Nicobar Islands
S China and Taiwan
S peninsular Thailand and Malay Peninsula to Indochina
Sporadic but locally common on Sumatra and Bangka I.

☐ **Javan Munia** *Lonchura leucogastroides*

_____
Lowlands of s Sumatra, Java, Bali and Lombok

☐ **Dusky Munia** *Lonchura fuscans*

Borneo, Natuna, Banggi, Cagayan and Sulu islands

☐ **Black-faced Munia** *Lonchura molucca*

Lowlands of Wallacea (except for Lesser Sundas)

☐ **Black-throated Munia** *Lonchura kelaarti*
_____ *L. k. jerdoni*
_____ *L. k. kelaarti*

SW India (Kerala to w Tamil Nadu) and Eastern Ghats
Highlands of Sri Lanka

☐ **Nutmeg Mannikin** *Lonchura punctulata*
_____ *L. p. punctulata*
_____ *L. p. subundulata*
_____ *L. p. yunnanensis*
_____ *L. p. topela*
_____ *L. p. cabanisi*
_____ *L. p. fretensis*
_____ *L. p. nisoria*
_____ *L. p. sumbae*
_____ *L. p. blasii*
_____ *L. p. particeps*
_____ *L. p. baweana*
_____ *L. p. holmesi*

Nepal to Sikkim, India and Sri Lanka
NE India (Assam) to Bhutan and w Myanmar
NE Myanmar and sw China
S China to n Thailand, Indochina, Hainan and Taiwan
Philippines (Luzon, Mindoro, Panay, Cebu, Calauit and Palawan)
S Thailand and Malay Peninsula to Sumatra and adjacent islands
Java, Bali, Lombok and Sumbawa
Sumba (Lesser Sundas)
Flores, Timor, Tanimbar Islands and adj. Lesser Sundas
Sulawesi
Bawean I. (Java Sea)
Southeast Borneo (Kalimantan)

☐ **White-bellied Munia** *Lonchura leucogastra*
_____ *L. l. leucogastra*
_____ *L. l. everetti*
_____ *L. l. manueli*
_____ *L. l. palawana*
_____ *L. l. smythiesi*
_____ *L. l. castanonota*

SE Myanmar, peninsular Thailand and Malay Pen. to Sumatra
Luzon, Mindoro, Camiguin Norte, Catanduanes and Polillo
Widespread throughout Philippine Islands
Palawan, islands in Sulu Archipelago and Borneo
N Borneo (Kucing region of sw Sarawak)
S Borneo

☐ **Streak-headed Munia** *Lonchura tristissima*
_____ *L. t. tristissima*
_____ *L. t. hypomelaena*
_____ *L. t. calaminoros*
_____ *L. t. bigilalei*
_____ *L. t. leucosticta*
_____ *L. t. moresbyensis*

NW New Guinea (Vogelkop Mountains)
W New Guinea (Weyland Mountains)
Jayawijaya and Nassau mountains to Lorentz River and Karkar I.
SE Papua New Guinea
Lowlands of s New Guinea (Lorentz River to Turama River)
SE Papua New Guinea (Port Moresby)

618

□ **Black-headed Munia** *Lonchura malacca*
____ *L. m. rubronigra* — N India (Haryana to n Bihar) and lowlands of Nepal
____ *L. m. malacca* — Lowlands of s India (Tapi River to Raipur) and Sri Lanka

□ **Chestnut Munia** *Lonchura atricapilla*
____ *L. a. atricapilla* — SE Nepal and ne India to Myanmar and nw Yunnan
____ *L. a. deignani* — SW China (sw Yunnan) to Thailand, Laos and Vietnam
____ *L. a. sinensis* — Pen. Thailand, Malaya, Sumatra, Riau Arch. and Lingga Arch.
____ *L. a. batakana* — Mountains of n Sumatra
____ *L. a. formosana* — Taiwan and n Philippines
____ *L. a. jagori* — Philippines, Sulu Islands, Palawan, Borneo, Sulawesi, Muna and Butung
____ *L. a. obscura* — Borneo (Sampit region of Kalimantan)
____ *L. a. selimbauensis* — Borneo (Pontianak region of w Kalimantan)
____ *L. a. brunneiceps* — S Sulawesi (Makassar district)

□ **White-capped Munia** *Lonchura ferruginosa*
— Lowlands of Java and Bali

□ **Five-colored Munia** *Lonchura quinticolor*
____ *L. q. wallacii* — W Lesser Sundas (Lombok and Sumbawa)
____ *L. q. sumbae* — Lesser Sundas (Flores, Alor, Sumba, Roti and w Timor)
____ *L. q. quinticolor* — E Timor, Sermata I., Babar I. and Tanimbar Is. (Yamdena)

□ **White-headed Munia** *Lonchura maja*
____ *L. m. maja* — Peninsular Thailand, Malaya, Sumatra, Java and Bali
____ *L. m. vietnamensis* — Vietnam

□ **Pale-headed Munia** *Lonchura pallida*
____ *L. p. pallida* — Sulawesi and Lesser Sundas
____ *L. p. subcastanea* — N-central Sulawesi (lower Palu Valley)

□ **Grand Munia** *Lonchura grandis*
____ *L. g. ernesti* — N New Guinea (Ramu and Sepik Valley to Astrolabe Bay)
____ *L. g. destructa* — N New Guinea (Hollandia district)
____ *L. g. heurni* — N New Guinea (Idenberg and Hollandia regions)
____ *L. g. grandis* — SE New Guinea (Hall Sound to upper Watut River)

□ **Gray-banded Munia** *Lonchura vana*
— NW New Guinea (Arfak Mountains)

□ **Gray-crowned Munia** *Lonchura nevermanni*
— New Guinea (Kurik and Mapa areas to lower Fly River)

□ **Hooded Munia** *Lonchura spectabilis*
____ *L. s. wahgiensis* — Central and s highlands of New Guinea
____ *L. s. gajduseki* — New Guinea (Karimui basin and e highlands)
____ *L. s. mayri* — New Guinea (Cyclops Mountains and lowlands of Hollandia)
____ *L. s. sepikensis* — NE New Guinea (East Sepik Province)
____ *L. s. spectabilis* — Bismarck Archipelago (New Britain, Long and Umboi)

□ **Gray-headed Munia** *Lonchura caniceps*
____ *L. c. caniceps* — SE New Guinea (Yule I. and Hall Sound to Port Moresby)
____ *L. c. scratchleyana* — SE New Guinea (Malalaua and Kupriano Mountains)
____ *L. c. kumusii* — SE New Guinea (Kumusi River to upper Musa River)

□ **Mottled Munia** *Lonchura hunsteini*
— New Ireland (Bismarck Archipelago)

□ **New Ireland Munia** *Lonchura forbesi*
— New Ireland (Bismarck Archipelago)

□ **New Hanover Munia** *Lonchura nigerrima*
— New Hanover (Bismarck Archipelago)

☐ **Yellow-rumped Munia** *Lonchura flaviprymna*

N Australia (Cambridge Gulf, Western Australia to Arnhem Land)

☐ **Chestnut-breasted Munia** *Lonchura castaneothorax*

_____ *L. c. uropygialis*	NW New Guinea (Geelvink Bay environs)
_____ *L. c. sharpii*	New Guinea (Astrolabe Bay to Humboldt Bay and upper Watut R.)
_____ *L. c. boschmai*	W New Guinea (Wissel Lakes region)
_____ *L. c. ramsayi*	SE New Guinea and D'Entrecasteaux Archipelago
_____ *L. c. castaneothorax*	N and e Australia (King Sound, W Australia to s coastal NSW)

☐ **Black Munia** *Lonchura stygia*

S New Guinea (Trans-Fly lowlands)

☐ **Black-breasted Munia** *Lonchura teerinki*

_____ *L. t. mariae*	W New Guinea (Snow Mountains)
_____ *L. t. teerinki*	Mountains of w-central New Guinea

☐ **Snow Mountain Munia** *Lonchura montana*

W New Guinea (Snow Mts.) and Mt. Capella (Papua NG)

☐ **Alpine Munia** *Lonchura monticola*

SW New Guinea (Wharton and Owen Stanley mountains)

☐ **Bismarck Munia** *Lonchura melaena*

_____ *L. m. melaena*	New Britain (se Bismarck Archipelago)
_____ *L. m. bukaensis*	Buka (Solomon Islands)

☐ **Pictorella Munia** *Heteromunia pectoralis*

N Australia (Fitzroy River, Western Australia to nw Queensland)

☐ **Java Sparrow** *Padda oryzivora*

Java, Bali, Lombok, Sumbawa, Kangean Is.; introd. Christmas I.

☐ **Timor Sparrow** *Padda fuscata*

E Lesser Sundas (Timor, Roti and Semau)

FAMILY: VIDUIDAE (Indigobirds—20)

☐ **Pin-tailed Whydah** *Vidua macroura*

Senegambia to s Chad, s Sudan, s Somalia and South Africa

☐ **Northern Paradise-Whydah** *Vidua orientalis*

_____ *V. o. aucupum*	Senegal to nw Nigeria
_____ *V. o. orientalis*	Chad to Sudan and Eritrea

☐ **Long-tailed Paradise-Whydah** *Vidua interjecta*

Mali to s Nigeria, Cameroon, Central African Rep. and ne Zaire

☐ **Togo Paradise-Whydah** *Vidua togoensis*

Sierra Leone to Ivory Coast, Ghana, Togo; Cameroon

☐ **Broad-tailed Paradise-Whydah** *Vidua obtusa*

N Angola to Zaire, sw Uganda, Kenya and ne South Africa

☐ **Eastern Paradise-Whydah** *Vidua paradisaea*

Angola to se Sudan, Ethiopia and Kenya south to ne S Africa

☐ **Steel-blue Whydah** *Vidua hypocherina*

S Sudan to Ethiopia, Somalia, Uganda, Kenya and Tanzania

☐ **Straw-tailed Whydah** *Vidua fischeri*

Extreme se Sudan to Ethiopia, Somalia, Uganda and Tanzania

☐ **Shaft-tailed Whydah** *Vidua regia*

S Angola to Namibia, Zambia, Botswana, s Mozambique, Natal

☐ **Village Indigobird** *Vidua chalybeata*

_____ *V. c. chalybeata (aenea)*	Senegal to Timbuktu and Mali
_____ *V. c. neumanni*	E Mali and Burkina Faso to s Sudan and Eritrea
_____ *V. c. ultramarina*	Ethiopia
_____ *V. c. centralis*	E Zaire to Uganda, Rwanda, Kenya and w Tanzania
_____ *V. c. okavangoensis*	Angola to Botswana, n Namibia and w Zambia
_____ *V. c. amauropteryx*	Coastal e Africa from s Somalia to South Africa

☐ **Pale-winged Indigobird** *Vidua wilsoni*

Senegambia to n Zaire, s Sudan and nw Ethiopia

☐ **Quailfinch Indigobird** *Vidua nigeriae*

Discontinuous in Gambia, Mali, Nigeria, Cameroon and Sudan

☐ **Jos Plateau Indigobird** *Vidua maryae*

N Nigeria (Jos Plateau)

☐ **Jambandu Indigobird** *Vidua raricola*

Sierra Leone to n Nigeria, n Cameroon, n Zaire and s Sudan

☐ **Baka Indigobird** *Vidua larvaticola*

SE Senegal to Cameroon, n Zaire, Sudan and w Ethiopia

☐ **Cameroon Indigobird** *Vidua camerunensis*

Sierra Leone to e Cameroon, ne Zaire and s Sudan

☐ **Variable Indigobird** *Vidua funerea*
    ____ *V. f. nigerrima* — Kenya to Mozambique, Angola and Zimbabwe
    ____ *V. f. funerea* — South Africa, Swaziland, and Mozambique s of the Limpopo R.

☐ **Purple Indigobird** *Vidua purpurascens*

SW Angola to e Namibia, Kenya, Tanzania and Transvaal

☐ **Green Indigobird** *Vidua codringtoni*

Locally in Zambia, sw Tanzania, Malawi and Zimbabwe

☐ **Parasitic Weaver** *Anomalospiza imberbis*
    ____ *A. i. imberbis* — Ethiopia through East Africa, s Zaire and s-c Africa to S Africa
    ____ *A. i. butleri* — West Africa to southern Sudan and n Zaire

FAMILY: VIREONIDAE (Vireos and Allies—52)

☐ **Slaty Vireo** *Vireo brevipennis*

Oak-pine highlands of s-central Mexico

☐ **White-eyed Vireo** *Vireo griseus*
    ____ *V. g. noveboracensis* — Central and e US; winters to Bahamas, n Nicaragua and Cuba
    ____ *V. g. griseus* — SE US; winters to e Mexico, n Honduras and w Cuba
    ____ *V. g. maynardi* — Coastal and insular Florida
    ____ *V. g. bermudianus* — Bermuda
    ____ *V. g. micrus* — S Texas to e Mexico (south to Puebla and extreme n Veracruz)
    ____ *V. g. perquisitor* — E Mexico (ne Puebla and n-central Veracruz)

☐ **Thick-billed Vireo** *Vireo crassirostris*
    ____ *V. c. crassirostris* — Bahamas and Cayman Islands
    ____ *V. c. stalagmium* — Turks and Caicos Islands
    ____ *V. c. cubensis* — Cayo Paredón Grande in Sabana-Camegüey Arch. off n Cuba
    ____ *V. c. alleni* — Caiman Islands
    ____ *V. c. tortugae* — Tortue I. (off nw Haiti)
    ____ *V. c. approximans* — Isla Providéncia and Isla Santa Catalina (sw Caribbean Sea)

☐ **Mangrove Vireo** *Vireo pallens*
    ____ *V. p. paluster* — Mangroves of nw Mexico (extreme sw Sonora to Nayarit)
    ____ *V. p. ochraceus* — Mangroves of Pacific coast of Guatemala and El Salvador
    ____ *V. p. pallens* — Mangroves of w Honduras to w Nicaragua and w Costa Rica
    ____ *V. p. semiflavus* — SE Mexico (Yucatán Peninsula) to Belize and Nicaragua

☐ **Cozumel Vireo** *Vireo bairdi*

SE Mexico (Cozumel I. off Quintana Roo)

☐ **St. Andrew Vireo** *Vireo caribaeus*

Mangroves of Isla San Andrés (w Caribbean Sea)

☐ **Jamaican Vireo** *Vireo modestus*

Lowlands and mountains of Jamaica

☐ **Cuban Vireo** *Vireo gundlachii*

Woodlands of Cuba, Isle of Pines and adjacent islands

☐ **Puerto Rican Vireo** *Vireo latimeri*

Confined mainly to limestone hills of w and central Puerto Rico

☐ **Flat-billed Vireo** *Vireo nanus*

Semiarid lowland scrub of Hispaniola and Gonâve I.

☐ **Bell's Vireo** *Vireo bellii*
_____ *V. b. pusillus* — Arid s California; winters to s Baja
_____ *V. b. arizonae* — SW US; winters from Baja and central Sonora to Colima
_____ *V. b. medius* — SW Texas to s Durango and s Coahuila; winters to Oaxaca
_____ *V. b. bellii* — Central and s US; winters from s Mexico to n Nicaragua

☐ **Black-capped Vireo** *Vireo atricapilla*

Breeds Kansas to n Mexico (Coahuila); winters to Guerrero

☐ **Dwarf Vireo** *Vireo nelsoni*

Arid highlands of sw Mexico (se Jalisco to Oaxaca)

☐ **Gray Vireo** *Vireo vicinior*

SW US to n Baja; winters to n Mexico (Sonora)

☐ **Blue Mountain Vireo** *Vireo osburni*

Humid montane forests of nw Jamaica

☐ **Yellow-throated Vireo** *Vireo flavifrons*

E Canada to Gulf States; winters to Venezuela and West Indies

☐ **Plumbeous Vireo** *Vireo plumbeus*
_____ *V. p. plumbeus* — Rocky Mountains of n US to nw Mexico; winters to nw Mexico
_____ *V. p. pinicolus* — Mountains of nw Mexico (Sonora to Durango and Zacatecas)
_____ *V. p. repetens* — Oak-pinyon-juniper belt of central Mexico
_____ *V. p. montanus* — Mountains of s Mexico (Chiapas) to nw Honduras
_____ *V. p. notius* — Pine savanna of Belize

☐ **Cassin's Vireo** *Vireo cassinii*
_____ *V. c. cassinii* — S British Columbia to n Baja; winters to Guatemala
_____ *V. c. lucasanas* — Mountains of s Baja California (Sierra de la Laguna)

☐ **Blue-headed Vireo** *Vireo solitarius*
_____ *V. s. solitarius* — NE British Columbia to e US; winters to Cuba and Costa Rica
_____ *V. s. alticola* — S Appalachian Mountains to se US; winters to s Florida

☐ **Yellow-winged Vireo** *Vireo carmioli*

Montane forests of Costa Rica and w Panama (w Chiriquí)

☐ **Hutton's Vireo** *Vireo huttoni*
_____ *V. h. insularis* — Vancouver I. (British Columbia)
_____ *V. h. huttoni (mailliardorum)* — SW British Columbia (west of Cascades and Sierras) to n Baja
_____ *V. h. obscurus* — Inner coast ranges of California (south to Lake County)
_____ *V. h. parkesi* — Coastal California (Humboldt County to Marin County)
_____ *V. h. sierrae* — Sierra Nevada of California
_____ *V. h. unitti* — Santa Catalina I. (off s California)
_____ *V. h. oberholseri* — Inner Coast Ranges of California (Monterey Co. to s California)
_____ *V. h. cognatus* — Mountains of s Baja California (Sierra de la Laguna)
_____ *V. h. stephensi* — Mts. of se Arizona and New Mexico to nw Mexico (Sinaloa)
_____ *V. h. carolinae* — Chisos Mountains (Texas) to ne Mexico (sw Tamaulipas)
_____ *V. h. pacificus* — Coastal mountains of w Mexico (Nayarit to sw Oaxaca)
_____ *V. h. mexicanus* — Oak-pine forests of Mexican plateau
_____ *V. h. vulcani* — Mountains of s Mexico (Chiapas) and w Guatemala

☐ **Warbling Vireo** *Vireo gilvus*
_____ *V. g. swainsonii* — SE Alaska to nw Baja; winters to El Salvador
_____ *V. g. victoriae* — Mountains of s Baja California (Sierra de la Laguna)
_____ *V. g. leucopolius* — Great Basin of e Washington to se California and sw Utah
_____ *V. g. gilvus* — SW Canada to e-central US; winters to ne Costa Rica
_____ *V. g. brewsteri* — Sierra Madre Occidental of nw Mexico (s Sonora to Nayarit)
_____ *V. g. eleanorae* — Sierra Madre Oriental of ne Mexico (s Tamaulipas to Hidalgo)
_____ *V. g. bulli* — Mountains of se Mexico (Oaxaca)
_____ *V. g. amauronotus* — Mountains of e-central Mexico (Puebla and Veracruz)
_____ *V. g. connectens* — Mountains of s-central Mexico (Michoacán to Oaxaca)
_____ *V. g. strenuus* — Subtropical s Mexico (Chiapas) to Guatemala and Honduras

☐ **Brown-capped Vireo** *Vireo leucophrys*
     _____ *V. l. leucophrys*   Highlands of e Mexico (San Luis Potosí) to Nicaragua
     _____ *V. l. chiriquensis*   Subtropical Costa Rica and extreme w Panama
     _____ *V. l. disjunctus*   Subtropical Central and (?) Western Andes of Colombia
     _____ *V. l. mirandae*   Santa Marta Mts. (ne Colombia) and mts. of nw Venezuela
     _____ *V. l. dissors*   Western and Central Andes of Colombia
     _____ *V. l. josephae*   Central Andes of sw Colombia (Nariño) and w Ecuador
     _____ *V. l. maranonicus*   Western Andes of n Peru
     _____ *V. l. laetissimus*   Subtropical se Peru and n Bolivia (La Paz and Cochabamba)

☐ **Philadelphia Vireo** *Vireo philadelphicus*

        Breeds e North America; winters to n Colombia

☐ **Red-eyed Vireo** *Vireo olivaceus*
     _____ *V. o. olivaceus*   Canada, w-central and e US; winters to Cuba and c S America
     _____ *V. o. forreri*   W Mexico (Sonora to Jalisco) and Tres Marías Islands
     _____ *V. o. insulanus*   Pearl Islands (Gulf of Panama); winters to upper Amazonia
     _____ *V. o. caucae*   Tropical w Colombia
     _____ *V. o. griseobarbatus*   Tropical w Ecuador and nw Peru
     _____ *V. o. pectoralis*   N Peru (middle Marañón Valley)
     _____ *V. o. solimoensis*   W Amazonian Brazil to e Ecuador and ne Peru
     _____ *V. o. vividior*   Colombia to Venezuela, the Guianas and n Brazil; Trinidad
     _____ *V. o. tobagensis*   Tobago
     _____ *V. o. agilis*   Coastal ne Brazil (Pará to Rio de Janeiro)
     _____ *V. o. diversus*   SE Brazil (São Paulo) to Uruguay, Paraguay and ne Argentina
     _____ *V. o. chivi*   W Amazonian basin; winters north to Colombia and Venezuela

☐ **Choco Vireo** *Vireo masteri*

        Rainforests of Pacific slope of Western Andes of Colombia

☐ **Golden Vireo** *Vireo hypochryseus*
     _____ *V. h. nitidus*   Lowlands of w Mexico (extreme s Sonora)
     _____ *V. h. hypochryseus*   Tropical w Mexico (Sinaloa and w Durango to w Oaxaca)
     _____ *V. h. sordidus*   Tres Marías Islands (off w Mexico)

☐ **Yellow-green Vireo** *Vireo flavoviridis*

        S Texas to Panama; winters to w Brazil and n Bolivia

☐ **Noronha Vireo** *Vireo gracilirostris*

        Ilha Fernando de Noronha (off coastal ne Brazil)

☐ **Black-whiskered Vireo** *Vireo altiloquus*
     _____ *V. a. barbatulus*   Coastal s Florida, Cuba and Haiti; winters to Amazonian Brazil
     _____ *V. a. altiloquus*   Greater Antilles; winters to n South America
     _____ *V. a. barbadensis*   St. Croix and Barbados
     _____ *V. a. bonairensis*   Netherlands Antilles (Aruba, Curaçao and Bonaire); Margarita I.
     _____ *V. a. grandior*   Isla Providéncia and Isla Santa Catalina (w Caribbean Sea)
     _____ *V. a. canescens*   Isla San Andrés (w Caribbean Sea)

☐ **Yucatan Vireo** *Vireo magister*
     _____ *V. m. magister*   Yucatán Pen., Belize, Cozumel, Isla Mujeres and adj. islands
     _____ *V. m. caymanensis*   Grand Cayman I.

☐ **Rufous-crowned Greenlet** *Hylophilus poicilotis*

        N Bolivia to e Paraguay, ne Argentina and s Brazil

☐ **Gray-eyed Greenlet** *Hylophilus amaurocephalus*

        Lowlands of e Brazil (Piauí and Ceará to extreme n São Paulo)

☐ **Lemon-chested Greenlet** *Hylophilus thoracicus*
     _____ *H. t. aemulus*   E slope of Andes of Colombia to se Peru and n Bolivia
     _____ *H. t. griseiventris*   E Venezuela (Bolívar) to the Guianas and n Brazil
     _____ *H. t. thoracicus*   SE Brazil (Minas Gerais and Espírito Santo to Rio de Janeiro)

☐ **Gray-chested Greenlet** *Hylophilus semicinereus*
     _____ *H. s. viridiceps*   S Venezuela (Amazonas and Bolívar), the Guianas and n Brazil

_____ *H. s. semicinereus*   N Brazil south of lower Amazon (Maranhão to n Mato Grosso)

_____ *H. s. juruanus*   NW Brazil (south of Rio Solimões)

☐ **Ashy-headed Greenlet** *Hylophilus pectoralis*

Guianas and e Venezuela to n Bolivia and Amazonian Brazil

☐ **Tepui Greenlet** *Hylophilus sclateri*

Tepuis of s Venezuela, w Guyana and extreme n Brazil

☐ **Buff-cheeked Greenlet** *Hylophilus muscicapinus*

_____ *H. m. muscicapinus*   S Venezuela to the Guianas and Brazil north of the Amazon

_____ *H. m. griseifrons*   N Brazil south of the Amazon (Rio Madeira to Rio Tapajós)

☐ **Brown-headed Greenlet** *Hylophilus brunneiceps*

_____ *H. b. brunneiceps*   E Colombia to s Venezuela and nw Brazil

_____ *H. b. inornatus*   N Brazil south of the Amazon (Rio Tapajós to Rio Tocantins)

☐ **Dusky-capped Greenlet** *Hylophilus hypoxanthus*

_____ *H. h. hypoxanthus*   SE Colombia to extreme sw Venezuela and extreme nw Brazil

_____ *H. h. fuscicapillus*   E Ecuador and n Peru (south to Río Marañón)

_____ *H. h. flaviventris*   Tropical central Peru (San Martín to Ayacucho)

_____ *H. h. ictericus*   W Brazil, extreme se Peru (Puno) and n Bolivia

_____ *H. h. albigula*   N Brazil south of the Amazon (Rio Purús to Rio Xingú)

☐ **Rufous-naped Greenlet** *Hylophilus semibrunneus*

Andes of Colombia to extreme nw Venezuela and e Ecuador

☐ **Olivaceous Greenlet** *Hylophilus olivaceus*

Subtropical Andes of e Ecuador to central Peru (Junín)

☐ **Scrub Greenlet** *Hylophilus flavipes*

_____ *H. f. viridiflavus*   Tropical sw Costa Rica and w Panama

_____ *H. f. xuthus*   Isla Coiba (Panama)

_____ *H. f. flavipes*   Caribbean coast of n Colombia and Magdalena Valley

_____ *H. f. melleus*   N Colombia (e tip of Guajira Peninsula)

_____ *H. f. galbanus*   Tropical ne Colombia and nw Venezuela

_____ *H. f. acuticauda*   Tropical n Venezuela and Isla Margarita

_____ *H. f. insularis*   Tobago

☐ **Tawny-crowned Greenlet** *Hylophilus ochraceiceps*

_____ *H. o. ochraceiceps*   Gulf lowlands of s Mexico to Guatemala

_____ *H. o. pallidipectus*   Tropical Honduras to w Panama

_____ *H. o. nelsoni*   E Panama (Veraguas to Darién)

_____ *H. o. bulunensis*   Extreme e Panama to Pacific coast of Colombia and w Ecuador

_____ *H. o. ferrugineifrons*   SE Colombia to the Guianas, e Peru and nw Amazonian Brazil

_____ *H. o. viridior*   Tropical s Peru (Ayacucho and Cuzco) to n Bolivia

_____ *H. o. luteifrons*   Extreme e Venezuela, the Guianas and n Amazonian Brazil

_____ *H. o. lutescens*   N Brazil south of the Amazon (Rio Madeira to Rio Xingú)

_____ *H. o. rubrifrons*   NE Brazil (Rio Tocantins to Pará)

☐ **Golden-fronted Greenlet** *Hylophilus aurantiifrons*

_____ *H. a. aurantiifrons*   E Panama and Caribbean coast of n Colombia

_____ *H. a. helvinus*   Tropical nw Venezuela (Zulia to n Mérida and s Táchira)

_____ *H. a. saturatus*   Tropical e Colombia to n Venezuela; Trinidad

☐ **Lesser Greenlet** *Hylophilus decurtatus*

_____ *H. d. decurtatus*   E Mexico (San Luis Potosí) to Panama (Canal Zone)

_____ *H. d. dariensis*   Tropical e Panama and n Colombia

_____ *H. d. minor*   Trop. sw Colombia (Nariño) to w Ecuador and extreme n Peru

☐ **Chestnut-sided Shrike-Vireo** *Vireolanius melitophrys*

_____ *V. m. goldmani*   Oak forests of s central Mexico

_____ *V. m. melitophrys*   S Mexico (se Chiapas) to s Guatemala

☐ **Green Shrike-Vireo** *Vireolanius pulchellus*

_____ *V. p. pulchellus*	Gulf-Caribbean lowlands of se Mexico to Honduras
_____ *V. p. verticalis*	Caribbean slope of Nicaragua and Costa Rica
_____ *V. p. viridiceps*	Pacific slope of Costa Rica and w Panama

☐ **Yellow-browed Shrike-Vireo** *Vireolanius eximius*

_____ *V. e. mutabilis*	Extreme e Panama (Darién) and nw Colombia
_____ *V. e. eximius*	Tropical n Colombia and nw Venezuela (Zulia and Táchira)

☐ **Slaty-capped Shrike-Vireo** *Vireolanius leucotis*

_____ *V. l. mikettae*	W slope of Western Andes of Colombia and nw Ecuador
_____ *V. l. leucotis*	SE Colombia to e Ecuador, n Peru, the Guianas and nw Brazil
_____ *V. l. simplex*	Brazil (south of the Amazon) to s Peru (Huánuco to n Cuzco)
_____ *V. l. bolivianus*	SE Peru (Cuzco) to n Bolivia

☐ **Rufous-browed Peppershrike** *Cyclarhis gujanensis*

_____ *C. g. flaviventris*	E Mexico (San Luis Potosí) to e Guatemala and n Honduras
_____ *C. g. yucatanensis*	SE Mexico (Yucatán Peninsula) and Petén of n Guatemala
_____ *C. g. insularis*	Cozumel I. (se Mexico off Quintana Roo)
_____ *C. g. nicaraguae*	S Mexico (Chiapas) and Guatemala to Nicaragua
_____ *C. g. subflavescens*	Pacific slope of Costa Rica and w Panama
_____ *C. g. perrygoi*	W central Panama (Coclé to extreme e Veraguas)
_____ *C. g. flavens*	Coastal e Panama
_____ *C. g. coibae*	Isla Coiba (Panama)
_____ *C. g. cantica*	Tropical Caribbean coast of n Colombia and Magdalena Valley
_____ *C. g. flavipectus*	NE Venezuela (Paría Peninsula) and Trinidad
_____ *C. g. parva*	E slope of Eastern Andes of Colombia and n Venezuela
_____ *C. g. gujanensis*	E Colombia to the Guianas, Amaz. Brazil, e Peru and ne Bolivia
_____ *C. g. cearensis*	Tableland of e Brazil
_____ *C. g. ochrocephala*	SE Brazil to Uruguay, e Paraguay and ne Argentina
_____ *C. g. viridis*	*Chaco* of Paraguay and n Argentina
_____ *C. g. virenticeps*	Pacific slope of Ecuador and nw Peru
_____ *C. g. contrerasi*	Mountains of n Peru (south to La Libertad and San Martín)
_____ *C. g. saturata*	N Peru (upper Río Marañón Valley east of Western Andes)
_____ *C. g. pax*	*Yungas* of Bolivia (La Paz)
_____ *C. g. dorsalis*	Highlands of central Bolivia
_____ *C. g. tarijae*	Extreme se Bolivia (Tarija) and nw Argentina (Jujuy)

☐ **Black-billed Peppershrike** *Cyclarhis nigrirostris*

_____ *C. n. nigrirostris*	Andes of central Colombia and e Ecuador
_____ *C. n. atrirostris*	Andes of sw Colombia (Nariño) and w Ecuador

---

FAMILY: FRINGILLIDAE (Siskins, Crossbills and Allies—176)

☐ **Chaffinch** *Fringilla coelebs*

_____ *F. c. moreletti*	Azores
_____ *F. c. maderensis*	Madeira
_____ *F. c. canariensis*	Canary Islands (Gran Canaria, Gomera and Tenerife)
_____ *F. c. ombriosa*	Hierro (Canary Islands)
_____ *F. c. palmae*	La Palma (Canary Islands)
_____ *F. c. africana*	NW Africa (Morocco to nw Tunisia)
_____ *F. c. spodiogenys*	E Tunisia and nw Libya
_____ *F. c. coelebs*	Continental Europe to Siberia and Asia Minor; winters to Africa
_____ *F. c. gengleri*	British Isles, Orkneys and Outer Hebrides
_____ *F. c. balearica*	Iberian Peninsula and Balearic Islands
_____ *F. c. sarda*	Sardinia
_____ *F. c. schiebeli*	Crete
_____ *F. c. syriaca*	Cyprus and Levant

_____ *F. c. balearica*   Spain and Portugal
_____ *F. c. solomkoi*   Crimean Peninsula and Caucasus
_____ *F. c. alexandrovi*   N Iran
_____ *F. c. transcaspica*   NE Iran (s Transcaspia in Kopet Dagh and Khorasan)

☐ **Blue Chaffinch** *Fringilla teydea*
_____ *F. t. teydea*   Tenerife (Canary Islands)
_____ *F. t. polatzeki*   Gran Canaria (Canary Islands)

☐ **Brambling** *Fringilla montifringilla*

       N Eurasia; winters Mediterranean region and s Asia

☐ **Jamaican Euphonia** *Euphonia jamaica*

       Woodlands and scrub of Jamaica

☐ **Plumbeous Euphonia** *Euphonia plumbea*

       S Venezuela to Guyana, Suriname and n Amazonian Brazil

☐ **Scrub Euphonia** *Euphonia affinis*
_____ *E. a. godmani*   Arid tropical w Mexico (se Sonora to Guerrero)
_____ *E. a. olmecorum*   S Mexico (Oaxaca and Chiapas)
_____ *E. a. affinis*   Tropical e Mexico (Tamaulipas) to nw Costa Rica

☐ **Purple-throated Euphonia** *Euphonia chlorotica*
_____ *E. c. cynophora*   Tropical ne Colombia to s Venezuela and extreme n Brazil
_____ *E. c. chlorotica*   Tropical Guianas to ne Brazil
_____ *E. c. serrirostris*   SE Bolivia to e Paraguay, Uruguay, se Brazil, n Argentina
_____ *E. c. taczanowskii*   E Peru and n Bolivia
_____ *E. c. amazonica*   Amazonian Brazil

☐ **Yellow-crowned Euphonia** *Euphonia luteicapilla*

       Tropical e Nicaragua to Costa Rica and Panama

☐ **Trinidad Euphonia** *Euphonia trinitatis*

       Trop. n Caribbean Colombia to n Venezuela; Trinidad

☐ **Velvet-fronted Euphonia** *Euphonia concinna*

       W Colombia (upper Magdalena Valley)

☐ **Orange-crowned Euphonia** *Euphonia saturata*

       Andes of sw Colombia to Colombia and nw Peru (Tumbes)

☐ **Finsch's Euphonia** *Euphonia finschi*

       SE Venezuela to the Guianas and n Brazil (n Roraima)

☐ **Violaceous Euphonia** *Euphonia violacea*
_____ *E. v. rodwayi*   Tropical e Venezuela (Sucre to n Amazonas); Trinidad
_____ *E. v. violacea*   Tropical Guianas and n Brazil
_____ *E. v. auranticollis*   E Paraguay to se Brazil and ne Argentina (Misiones)

☐ **Thick-billed Euphonia** *Euphonia laniirostris*
_____ *E. l. crassirostris*   Costa Rica to n Colombia and n Venezuela
_____ *E. l. melanura*   E Colombia to e Ecuador, n Peru and w Amazonian Brazil
_____ *E. l. hypoxantha*   W Ecuador and nw Peru
_____ *E. l. zopholega*   Tropical e-central Peru (Junín and Cuzco)
_____ *E. l. laniirostris*   E Bolivia and adjacent sw Brazil

☐ **Yellow-throated Euphonia** *Euphonia hirundinacea*
_____ *E. h. hirundinacea (suttoni, russelli, caribbaea)*   Gulf-Caribbean lowlands of e Mexico to e Nicaragua
_____ *E. h. gnatho*   NW Nicaragua to Costa Rica and extreme w Panama

☐ **Green-chinned Euphonia** *Euphonia chalybea*

       Lowlands of e Paraguay to se Brazil and ne Argentina

☐ **Elegant Euphonia** *Euphonia elegantissima*
_____ *E. e. rileyi*   Montane forests of nw Mexico (se Sonora and ne Sinaloa)
_____ *E. e. elegantissima*   N Mexico (s Sinaloa and sw Tamaulipas) to nw Guatemala
_____ *E. e. vincens*   SE Guatemala to w Panama

☐ **Antillean Euphonia** *Euphonia musica*
- ____ *E. m. musica* — Hispaniola and Gonâve I.
- ____ *E. m. sclateri* — Puerto Rico
- ____ *E. m. flavifrons* — Lesser Antilles

☐ **Golden-rumped Euphonia** *Euphonia cyanocephala*
- ____ *E. c. pelzelni* — Andes of Colombia to nw Argentina, n Venezuela; Trinidad
- ____ *E. c. cyanocephala* — Tepuis of se Venezuela to Guyana and Suriname
- ____ *E. c. ssp.* — E Paraguay to se Brazil and ne Argentina

☐ **Spot-crowned Euphonia** *Euphonia imitans*

Humid sw Costa Rica and extreme w Panama

☐ **Fulvous-vented Euphonia** *Euphonia fulvicrissa*
- ____ *E. f. fulvicrissa* — Humid lowlands of central Panama to nw Colombia
- ____ *E. f. omissa* — Tropical central Colombia
- ____ *E. f. purpurascens* — Tropical sw Colombia (Nariño) to nw Ecuador (Esmeraldas)

☐ **Olive-backed Euphonia** *Euphonia gouldi*
- ____ *E. g. gouldi (loetscheri)* — Caribbean slope of e Mexico (Veracruz) to Honduras
- ____ *E. g. praetermissa* — Extreme e Honduras to w Panama (Bocas del Toro)

☐ **Bronze-green Euphonia** *Euphonia mesochrysa*
- ____ *E. m. mesochrysa* — Central Colombia (Magdalena Valley) to e Ecuador
- ____ *E. m. media* — Subtropical Andes of n and central Peru
- ____ *E. m. tavarae* — Andes of se Peru to nw Bolivia

☐ **White-lored Euphonia** *Euphonia chrysopasta*
- ____ *E. c. chrysopasta* — Tropical se Colombia to Bolivia and adjacent w Brazil
- ____ *E. c. nitida* — E Colombia to s Venezuela, the Guianas and n Brazil

☐ **White-vented Euphonia** *Euphonia minuta*
- ____ *E. m. humilis* — Gulf slope of se Mexico (se Chiapas) to w Ecuador
- ____ *E. m. minuta* — Colombia e of Andes to the Guianas, n Bolivia and w Brazil

☐ **Tawny-capped Euphonia** *Euphonia anneae*
- ____ *E. a. anneae* — Caribbean slope of Costa Rica and extreme w Panama
- ____ *E. a. rufivertex* — Tropical w Panama (Veraguas) to nw Colombia (Chocó)

☐ **Orange-bellied Euphonia** *Euphonia xanthogaster*
- ____ *E. x. chocoensis* — Humid e Panama (Darién) to nw Colombia and nw Ecuador
- ____ *E. x. exsul (badissima)* — Mountains of ne Colombia and n Venezuela
- ____ *E. x. dilutior* — Tropical se Colombia to ne Peru (Ucayali Valley)
- ____ *E. x. lecroyana* — W Venezuela (Táchira, Mérida, Lara, Barinas and Zulia)
- ____ *E. x. brevirostris* — E Colombia to Venezuela, Guianas, nw Brazil and e Peru
- ____ *E. x. quitensis* — Tropical and subtropical w Ecuador
- ____ *E. x. brunneifrons* — SE Peru (Cuzco and Puno)
- ____ *E. x. ruficeps* — W Bolivia (La Paz and Cochabamba)
- ____ *E. x. cyanonota* — W Brazil (Rio Juruá and Rio Purús regions)
- ____ *E. x. xanthogaster* — E Brazil (south to Rio de Janeiro)

☐ **Rufous-bellied Euphonia** *Euphonia rufiventris*

SE Colombia to s Venezuela, n Bolivia and w Amaz. Brazil

☐ **Golden-sided Euphonia** *Euphonia cayennensis*

Tropical se Venezuela, the Guianas and e Amazonian Brazil

☐ **Chestnut-bellied Euphonia** *Euphonia pectoralis*

E Brazil (e Alagoas) to e Paraguay and ne Argentina

☐ **Yellow-collared Chlorophonia** *Chlorophonia flavirostris*

Humid forests of sw Colombia and nw Ecuador

☐ **Blue-naped Chlorophonia** *Chlorophonia cyanea*
- ____ *C. c. psittacina* — Santa Marta Mountains (ne Colombia)

____ *C. c. intensa*	W slope of Western Andes of Colombia (Caldas and Valle)
____ *C. c. longipennis*	Andes of e Colombia to w Venezuela, e Peru and w Bolivia
____ *C. c. frontalis*	Mountains of n Venezuela (Falcón and Lara to Miranda)
____ *C. c. minuscula*	Coastal mountains of ne Venezuela
____ *C. c. roraimae*	S Venezuela to Guyana and extreme nw Brazil
____ *C. c. cyanea*	Paraguay to se Brazil (s Bahia) and ne Argentina (Misiones)

☐ **Chestnut-breasted Chlorophonia** *Chlorophonia pyrrhophrys*

Andes of Colombia to w Venezuela and e Peru (Huánuco)

☐ **Blue-crowned Chlorophonia** *Chlorophonia occipitalis*

Montane forests of se Mexico (Veracruz) to Nicaragua

☐ **Golden-browed Chlorophonia** *Chlorophonia callophrys*

Montane forests of Costa Rica and w Panama

☐ **São Tomé Grosbeak** *Neospiza concolor*

São Tomé (rediscovered in 1996 after 101-year absence)

☐ **Oriole Finch** *Linurgus olivaceus*

____ *L. o. olivaceus*	Mountains of se Nigeria and Cameroon; Bioko
____ *L. o. prigoginei*	Montane forests of e Zaire
____ *L. o. elgonensis*	SE Sudan (Imatog Mountains) to Kenya highlands
____ *L. o. kilimensis*	Highlands of Tanzania to n Malawi

☐ **Golden-winged Grosbeak** *Rhynchostruthus socotranus*

____ *R. s. percivali*	Mountains of sw Arabia
____ *R. s. socotranus*	Socotra (off Somalia)

☐ **Somali Grosbeak** *Rhynchostruthus louisae*

Mts. of n Somalia near Ethiopian border

☐ **Plain Mountain-Finch** *Leucosticte nemoricola*

____ *L. n. altaica*	Mts. of ne Afghanistan to sw China (w Xinjiang) and Altai Mts.
____ *L. n. nemoricola*	Himalayas (Nepal to w China); winters to n Myanmar

☐ **Black-headed Mountain-Finch** *Leucosticte brandti*

____ *L. b. margaritacea*	High barren plains of w Mongolia and se Altai
____ *L. b. brandti*	W Tien Shan Mts. to Kyrgystan and w China (w Xinjiang)
____ *L. b. pamirensis*	S Kyrgystan to Pamir Mountains and ne Afghanistan
____ *L. b. haematopygia (audreyana, walteri, intermedia)*	N Pakistan to Kashmir, Ladakhh, n Punjab, Tibet and w China
____ *L. b. pallidior*	Mountains of w China (sw Xinjiang to ne Qinghai)

☐ **Tawny-headed Mountain-Finch** *Leucosticte sillemi*

Mountains of western Tibet

☐ **Asian Rosy-Finch** *Leucosticte arctoa*

____ *L. a. arctoa*	Russian Altai
____ *L. a. cognata*	Sayan Mts. and adjacent mountains on Russia/Mongolia border
____ *L. a. sushkini*	N Mongolia (Khangai region)
____ *L. a. gigliolii*	Mountains north of Lake Baikal (east to Yablonovy Mts.)
____ *L. a. brunneonucha*	Mts. of e Siberia (Lena River to Kamchatka and Kuril Islands)

☐ **Gray-crowned Rosy-Finch** *Leucosticte tephrocotis*

____ *L. t. griseonucha*	Komandorskiye, Aleutian and Kodiak islands to Alaska
____ *L. t. umbrina*	St. Matthew I. and Pribilof Islands (Bering Sea)
____ *L. t. irvingi*	Brooks Range (n Alaska)
____ *L. t. littoralis*	Central Alaska to sw Yukon, w Canada and nw US
____ *L. t. tephrocotis*	Mts. of Yukon and w Alberta to se Br. Columbia and nw Montana
____ *L. t. wallowa*	Wallowa Mountains (ne Oregon); winters to w-central Nevada
____ *L. t. dawsoni*	E California (Sierra Nevada, White and Inyo mountains)

☐ **Black Rosy-Finch** *Leucosticte atrata*

Mts. of Idaho and Montana to Nevada and Utah; winters to Arizona

☐ **Brown-capped Rosy-Finch** *Leucosticte australis*

Mountains of se Wyoming, Colorado and n-central New Mexico

☐ **Pine Grosbeak** *Pinicola enucleator*
- ____ *P. e. enucleator* — N Scandinavia to Russia and w Siberia (Yenisey River)
- ____ *P. e. pacata* — Siberia (e of Yenisey R.) to Altai Mts., Mongolia and Manchuria
- ____ *P. e. kamschatkensis* — W Siberia (Anadyr River) to Kamchatka Peninsula
- ____ *P. e. sakhalinensis* — Sakhalin, Kuril Islands and high mountains of Hokkaido
- ____ *P. e. alascensis* — NW Alaska to nw Mackenzie and ne Br. Col.; winters to nw US
- ____ *P. e. flammula* — S Alaska to nw British Columbia; winters to nw US
- ____ *P. e. carlottae* — Islands and coasts from Queen Charlotte Islands to Vancouver I.
- ____ *P. e. montana* — Interior central British Columbia to Rocky Mountains of sw US
- ____ *P. e. californica* — Sierra Nevada Mountains (e California)
- ____ *P. e. leucura* — Central and e Canada; winters to ne US
- ____ *P. e. eschatosa* — Central Quebec and Newfoundland to New England states

☐ **Crimson-browed Finch** *Pinicola subhimachala*

Mountains of Nepal to se Tibet, sw China and ne Myanmar

☐ **Blanford's Rosefinch** *Carpodacus rubescens*

Coniferous forests of Nepal to se Tibet and sw China

☐ **Dark-breasted Rosefinch** *Carpodacus nipalensis*
- ____ *C. n. kangrae* — W Himalayas (Kashmir to Garhwal)
- ____ *C. n. nipalensis* — Himalayas (Kumaon to Nepal, Sikkim, Bhutan and se Tibet)
- ____ *C. n. intensicolor* — Mts. of sw China to s Tibet and n Vietnam; winters to s Myanmar

☐ **Common Rosefinch** *Carpodacus erythrinus*
- ____ *C. e. erythrinus* — E Europe to w Asia; winters to India and Indochina
- ____ *C. e. grebnitskii* — E Siberia to Bering Sea and Sea of Okhotsk; winters to se China
- ____ *C. e. kubanensis* — Caucasus to nw Iran; winters to sw Iran
- ____ *C. e. ferghanensis* — Mts. of ne Iran to n Afghanistan, Pakistan, and w Himalayas
- ____ *C. e. roseatus* — E Himalayas to Tibet and China; winters to s India and Indochina

☐ **Cassin's Finch** *Carpodacus cassinii*

Mountains of sw Canada to central Mexico and s Baja California

☐ **Purple Finch** *Carpodacus purpureus*
- ____ *C. p. purpureus* — E-central Canada to ne US; winters to Florida and Texas
- ____ *C. p. californicus* — SW British Columbia to sw US and n Baja California

☐ **House Finch** *Carpodacus mexicanus*
- ____ *C. m. frontalis* — SW Canada to w US, Baja California and nw Mexico
- ____ *C. m. clementis* — Channel Is. (off s California) and Los Coronados Is. (off n Baja)
- ____ *C. m. mcgregori†* — Formerly San Benito and Cedros is. (off Baja California). Extinct
- ____ *C. m. amplus* — Guadalupe I. (off Baja California)
- ____ *C. m. ruberrimus* — S Baja Calif. and nw Mexico (Sonora, s Sinaloa and sw Chihuahua)
- ____ *C. m. rhodopnus* — Arid tropical central Sinaloa
- ____ *C. m. coccineus* — Mts. of sw Mexico (s Nayarit and w Zacatecas to w Michoacán)
- ____ *C. m. potosinus* — S Texas (Rio Grande Valley) to Chihuahua and sw Tamaulipas
- ____ *C. m. centralis* — Central Mexican plateau (Guanajuato, Querétaro and adj. states)
- ____ *C. m. mexicanus* — S central Mexican plateau (e Michoacán to Hidalgo and Oaxaca)
- ____ *C. m. roseipectus* — S Mexico (s Puebla and Valley of Oaxaca)
- ____ *C. m. griscomi* — SW Mexico (Sierra Madre del Sur of Guerrero)

☐ **Beautiful Rosefinch** *Carpodacus pulcherrimus*
- ____ *C. p. pulcherrimus* — Himalayas (Himachal Pradesh to Nepal, Bhutan and se Tibet)
- ____ *C. p. waltoni* — SE Tibet to sw China (sw Xinjiang)
- ____ *C. p. argyrophrys* — W China (Xinjiang, Qinghai, Gansu and w Sichuan)
- ____ *C. p. davidianus* — Central Mongolia (south to sw Inner Mongolia)

☐ **Pink-rumped Rosefinch** *Carpodacus eos*

Alpine scrub and forests of central China and Tibet

☐ **Pink-browed Rosefinch** *Carpodacus rodochroa*

_____ Himalayan fir and birch forests from Kashmir to s Tibet

☐ **Vinaceous Rosefinch** *Carpodacus vinaceus*

_____ *C. v. vinaceus*   W China (s Gansu and n Shaanxi to Sichuan); winters to Myanmar
_____ *C. v. formosanus*   Mountains of Taiwan

☐ **Dark-rumped Rosefinch** *Carpodacus edwardsii*

_____ *C. e. edwardsii*   W China (se Tibet to w Sichuan and s Gansu)
_____ *C. e. rubicundus*   Himalayas (Nepal to se Tibet and sw China); winters to Myanmar

☐ **Pale Rosefinch** *Carpodacus synoicus*

_____ *C. s. synoicus*   Sinai Peninsula
_____ *C. s. salimalii*   NE Afghanistan
_____ *C. s. stoliczkae*   SW China (sw Xinjiang)
_____ *C. s. beicki*   W China (Sining Ho region of ne Qinghai to nw Gansu)

☐ **Pallas' Rosefinch** *Carpodacus roseus*

_____ *C. r. roseus*   Siberia and n Mongolia; winters to n China and Japan
_____ *C. r. portenkoi*   Sakhalin; winters to s Korea and n Japan

☐ **Three-banded Rosefinch** *Carpodacus trifasciatus*

_____ Coniferous forests of w China; winters to se Tibet

☐ **Spot-winged Rosefinch** *Carpodacus rhodopeplus*

_____ *C. r. rhodopeplus*   Himalayas (Garhwal to Nepal and Sikkim)
_____ *C. r. verreauxii*   SW China (extreme se Tibet to n Yunnan); winters to n Myanmar

☐ **White-browed Rosefinch** *Carpodacus thura*

_____ *C. t. blythi*   NE Afghanistan and Pakistan through w Himalayas to Kumaon
_____ *C. t. thura*   Central Himalayas (Nepal to Bhutan)
_____ *C. t. femininus*   SE Tibet to sw China (w Sichuan and n Yunnan)
_____ *C. t. dubius*   W China (se Qinghai to se Gansu, s Ningxia and n Sichuan)
_____ *C. t. deserticolor*   W China (ne Qinghai)

☐ **Tibetan Rosefinch** *Carpodacus roborowskii*

_____ Mountains of w-central China (central Qinghai)

☐ **Red-mantled Rosefinch** *Carpodacus rhodochlamys*

_____ *C. r. rhodochlamys*   Mts. of extreme e Uzbekistan to sw China, Altai and n Mongolia
_____ *C. r. kotschubeii*   Alai Mountains (Kyrgyzstan) to Pamirs and Tajikistan
_____ *C. r. grandis*   Mountains of nw Afghanistan, Pakistan, Ladakh and Kumaon

☐ **Streaked Rosefinch** *Carpodacus rubicilloides*

_____ *C. r. lucifer*   Mts. of Ladakh and se Kashmir to Nepal, Sikkim and sw Tibet
_____ *C. r. rubicilloides*   W China (Qinghai and Gansu) to se Tibet

☐ **Great Rosefinch** *Carpodacus rubicilla*

_____ *C. r. rubicilla*   Caucasus; winters to Transcaucasus
_____ *C. r. diabolicus*   NE Afghanistan (Sanglech region)
_____ *C. r. kobdensis*   Alai Mountains of w Mongolia to e and central Altai Mountains
_____ *C. r. severtzowi*   Kashmir to Nepal, Tibet and sw China

☐ **Red-fronted Rosefinch** *Carpodacus puniceus*

_____ *C. p. kilianensis*   Mts. of Kyrgystan and Tajikistan to sw Xinjiang and ne Ladakh
_____ *C. p. humii*   W Himalayas (Pakistan to Kumaon)
_____ *C. p. puniceus*   Central Himalayas (Nepal to se Tibet)
_____ *C. p. sikangensis*   SW China (Minya Konka to Mula region of extreme w Sichuan)
_____ *C. p. longirostris*   W China (e Qinghai to Gansu and nw Sichuan)

☐ **Parrot Crossbill** *Loxia pytyopsittacus*

_____ Coniferous forests of ne Europe to w Siberia

☐ **Scottish Crossbill** *Loxia scotica*

_____ Pine forests of n Scotland

☐ **Red Crossbill** *Loxia curvirostra*

_____ *L. c. curvirostra*   Coniferous forests of n Europe to e Siberia and n Mongolia
_____ *L. c. corsicana*   Coniferous forests of Corsica
_____ *L. c. balearica*   Balearic Islands
_____ *L. c. poliogyna*   Morocco, Algeria and Tunisia
_____ *L. c. guillemardi*   Troödos Mountains (Cyprus)
_____ *L. c. mariae*   SW Crimean Peninsula
_____ *L. c. altaiensis*   Altai Mountains
_____ *L. c. tianschanica*   Tien Shan Mountains (Kazakstan)
_____ *L. c. himalayensis*   Himalayas (Kashmir to Nepal, Sikkim, sw China and sw Tibet)
_____ *L. c. meridionalis*   Mountains of s Vietnam (Da Lat Plateau)
_____ *L. c. japonica*   Extreme ne Asia; winters to e-cent. China and s Japanese islands
_____ *L. c. luzoniensis*   N Philippines (mountains of n Luzon)
_____ *L. c. bendirei (neogaea)*   S Yukon and n Br. Col. to w US e of Cascades; winters to Baja
_____ *L. c. sitkensis*   Coastal s Alaska to nw California; winters to ne US
_____ *L. c. benti*   Mts. of se Montana and ne Wyoming to sw US; winters to s Texas
_____ *L. c. minor*   S-central Canada and n-central US; winters to se US
_____ *L. c. grinnelli*   Mountains of sw US
_____ *L. c. stricklandi*   Mts. of n Baja California, s Arizona and s New Mexico to Chiapas
_____ *L. c. mesamericana*   Montane pine forests from Guatemala to n Nicaragua
_____ *L. c. pusilla*   Newfoundland; winters to ne US

☐ **Hispaniola Crossbill** *Loxia megaplaga*

    Cordillera Central of Dominican Republic and adjacent Haiti

☐ **White-winged Crossbill** *Loxia leucoptera*

_____ *L. l. bifasciata*   Coniferous forests of n Eurasia
_____ *L. l. leucoptera*   N-central Alaska to Newfoundland, Canada and n US

☐ **Yellow-breasted Greenfinch** *Carduelis spinoides*

_____ *C. s. spinoides*   Pakistan to se Tibet, sw China, n India, Nepal, Sikkim and Bhutan
_____ *C. s. heinrichi*   NE India (se Assam and Manipur) to w Myanmar (Mt. Victoria)
_____ *C. s. monguilloti*   S Vietnam (Langbian Plateau)

☐ **Vietnamese Greenfinch** *Carduelis monguilloti*

    S Vietnam (Langbian Plateau)

☐ **European Greenfinch** *Carduelis chloris*

_____ *C. c. chloris*   British Isles, n Europe, Corsica and Sardinia; winters to s Europe
_____ *C. c. voousi*   Mts. of North Africa (Morocco and Algeria)
_____ *C. c. vanmarli*   Lowlands of n Morocco
_____ *C. c. aurantiiventris*   S Europe; winters North Africa
_____ *C. c. chlorotica*   Syria, Lebanon, Israel and Jordan; winters to Sinai Pen. and Egypt
_____ *C. c. turkestanica*   Crimea to Caucasus, n Iran and Turkmenistan; winters to s Iran

☐ **Black-headed Greenfinch** *Carduelis ambigua*

_____ *C. a. taylori*   Mountains of se Tibet and extreme w Sichuan
_____ *C. a. ambigua*   Mts. of se Tibet to sw China, n Myanmar, n Laos and nw Tonkin

☐ **Common Redpoll** *Carduelis flammea*

_____ *C. f. flammea*   N Eurasia and n North America
_____ *C. f. rostrata*   N Labrador, Baffin I. and s Greenland; winters to ne US, Br. Isles
_____ *C. f. islandica*   Iceland
_____ *C. f. cabaret*   British Isles, Alps and mountains of Czechoslovakia

☐ **Hoary Redpoll** *Carduelis hornemanni*

_____ *C. h. exilipes*   Tundra of n Eurasia and n North America
_____ *C. h. hornemanni*   Ellesmere I., Baffin I. and n Greenland; winters to n US, Br. Isles

☐ **Eurasian Siskin** *Carduelis spinus*

    N Palearctic; winters Mediterranean region, China and Ryukyu Is.

☐ **Pine Siskin** *Carduelis pinus*

_____ *C. p. pinus*   Coniferous forests of s Alaska to US; winters to central Mexico

631

____ *C. p. macroptera*   Coniferous forests of n Baja California, nw and central Mexico
____ *C. p. perplexa*   Mountains of s Mexico (Chiapas) to sw Guatemala

☐ **Black-capped Siskin** *Carduelis atriceps*

Coniferous forests of s Mexico (Chiapas) and w Guatemala

☐ **Black-headed Siskin** *Carduelis notata*

____ *C. n. notata*   Oak-pine forests of e Mexico (San Luis Potosí) to n Guatemala
____ *C. n. forreri*   Oak-pine belt of w Mexico (s Sonora and Chihuahua to Guerrero)
____ *C. n. oleacea*   Coniferous forests of Belize to Honduras and n Nicaragua

☐ **Andean Siskin** *Carduelis spinescens*

____ *C. s. spinescens (capitanea)*   Andes of Colombia to Venezuela and extreme nw Ecuador
____ *C. s. nigricauda*   Temperate Central and Western Andes of n Colombia

☐ **Yellow-faced Siskin** *Carduelis yarrellii*

N Venezuela; interior ne Brazil (Ceará to Bahia)

☐ **Thick-billed Siskin** *Carduelis crassirostris*

____ *C. c. amadoni*   Andes of se Peru (Tacna and Puno)
____ *C. c. crassirostris*   Andes of s Bolivia to central Chile and nw Argentina

☐ **Hooded Siskin** *Carduelis magellanica*

____ *C. m. capitalis*   Mountains of extreme s Colombia to Ecuador and nw Peru
____ *C. m. longirostris*   SE Venezuela to Guyana and n Brazil
____ *C. m. paula*   Tropical and subtrop. s Ecuador and w Peru (south to Arequipa)
____ *C. m. peruana*   Trop. and subtrop. central Peru (Huánuco to Ayacucho and Cuzco)
____ *C. m. urubambensis*   Temperate s Peru (Cuzco) to n Chile (Tarapacá)
____ *C. m. boliviana*   Temperate central and s Bolivia
____ *C. m. tucumana*   NW Argentina (Jujuy, Santiago and Santa Fe to Mendoza)
____ *C. m. santaecrucis*   E foothills of Andes of Bolivia (Santa Cruz)
____ *C. m. alleni*   SE Bolivia to Paraguay, ne Argentina and s Brazil
____ *C. m. icterica*   SE Brazil (Minas Gerais) to Paraguay and extreme ne Argentina
____ *C. m. magellanica*   Uruguay and e Argentina (s Corrientes to Río Negro)

☐ **Yellow-bellied Siskin** *Carduelis xanthogastra*

____ *C. x. xanthogastra*   Local in mts. of Costa Rica to Colombia, Ecuador and Venezuela
____ *C. x. stejnegeri*   SE Peru (Puno) and nw Bolivia (La Paz and Santa Cruz)

☐ **Saffron Siskin** *Carduelis siemiradzkii*

SW Ecuador, Isla Puná and extreme nw Peru (Tumbes)

☐ **Olivaceous Siskin** *Carduelis olivacea*

E slope of Andes of se Ecuador to Peru and w Bolivia

☐ **Red Siskin** *Carduelis cucullata*

Locally in ne Colombia and n Venezuela (on verge of extinction)

☐ **Antillean Siskin** *Carduelis dominicensis*

Montane pine forests of Hispaniola

☐ **Lesser Goldfinch** *Carduelis psaltria*

____ *C. p. hesperophilus*   W US to s Baja California and nw Mexico
____ *C. p. witti*   Tres Marías Islands (off w Mexico)
____ *C. p. psaltria*   S-central US to s Mexico (Guerrero, Veracruz and Oaxaca)
____ *C. p. jouyi*   SE Mexico (Yucatán Peninsula and n Quintana Roo)
____ *C. p. columbianus*   S Mexico (Chiapas) to Colombia, Ecuador, Venezuela and n Peru

☐ **Lawrence's Goldfinch** *Carduelis lawrencei*

S California and nw Baja; winters to Texas and n Sonora

☐ **Black-chinned Siskin** *Carduelis barbata*

Central Chile and s Argentina to Tierra del Fuego and Falkland Is.

☐ **Black Siskin** *Carduelis atrata*

Andes of central Peru to n Chile and Argentina

☐ **Yellow-rumped Siskin** *Carduelis uropygialis*

Andes of central Peru to nw Argentina and n Chile

☐ **American Goldfinch** *Carduelis tristis*

____	*C. t. jewetti*	SW British Columbia to sw Oregon (west of the Cascades)
____	*C. t. pallida*	S-cent. Br. Col. to w-cent. US; winters to Puebla and Veracruz
____	*C. t. salicamans*	California (west of Sierra Nevada) to nw Baja California
____	*C. t. tristis*	E Canada to e-central US; winters to se US and e Mexico

☐ **European Goldfinch** *Carduelis carduelis*

____	*C. c. britannica*	British Isles, Channel Islands and w Netherlands
____	*C. c. carduelis*	W and c Europe; winters to Mediterranean and Black Sea
____	*C. c. parva*	Azores, Madeira, Canary Islands and w Mediterranean region
____	*C. c. tschusii*	Corsica, Sardinia and Sicily
____	*C. c. balcanica*	S Yugoslavia to Bulgaria, Greece and Crete
____	*C. c. niediecki*	Rhodes, Karpathos, Cyprus; Egypt to Asia Minor, n Iraq, sw Iran
____	*C. c. brevirostris*	Crimean Peninsula to Caucasus and ne Turkey
____	*C. c. loudoni*	N Iran (Talysh Mountains to Elburz Mountains)
____	*C. c. paropanisi*	Iran to n Afghanistan and w China (Xinjiang)
____	*C. c. major*	SW Siberia (Ural Mountains to Yenisey River)
____	*C. c. subulata*	S-central Siberia to Lake Baikal and nw Mongolia
____	*C. c. caniceps*	W Himalayas (Kashmir to Nepal and w Tibet)

☐ **Oriental Greenfinch** *Carduelis sinica*

____	*C. s. sinica*	W China (Gansu) to s Manchuria
____	*C. s. chabarowi*	Inner Mongolia to n Manchuria
____	*C. s. ussuriensis*	E Manchuria to s Ussuriland and Korea
____	*C. s. kawarahiba*	Kamchatka, Kuril Is., Sakhalin and Hokkaido; winters to Japan
____	*C. s. minor*	S Japan (Honshu, Shikoku and Kyushu) and Korea (Cheju-Do Is.)
____	*C. s. kittlitzi*	S Japan (Bonin Islands and Volcano Islands)

☐ **Twite** *Carduelis flavirostris*

____	*C. f. flavirostris*	Norway, n Sweden, n Finland and Kola Pen.; winters to s Europe
____	*C. f. pipilans*	Shetland Is., Hebrides, Orkneys, Scotland, n England and Ireland
____	*C. f. brevirostris*	E Turkey and Caucasus to nw Iran
____	*C. f. korejevi*	S Ural Mts. to Caspian Sea, Kirghiz steppes and Tien Shan Mts.
____	*C. f. altaica*	Altai Mountains of central Russia to nw Outer Mongolia
____	*C. f. montanella*	Kyrgyzstan (Alai Mts.) to Pamirs and w China (w Xinjiang)
____	*C. f. miniakensis*	W China (e Xinjiang, Qinghai and Gansu) to se Tibet
____	*C. f. rufostrigata*	Himalayas (n Kashmir to n Nepal and sw Tibet)

☐ **Eurasian Linnet** *Carduelis cannabina*

____	*C. c. autochthona*	Scotland
____	*C. c. nana*	Madeira
____	*C. c. meadewaldoi*	Tenerife (w Canary Islands)
____	*C. c. harterti*	Lanzarote (e Canary Islands)
____	*C. c. mediterranea*	Iberian Peninsula and North Africa to Asia Minor
____	*C. c. cannabina*	Europe to w Siberia, Crimean Peninsula and North Africa
____	*C. c. bella*	Asia Minor to Caucasus, Afghanistan, sw China; winters to India

☐ **Yemen Linnet** *Carduelis yemenensis*

Mountains of sw Arabia

☐ **Warsangli Linnet** *Carduelis johannis*

Montane juniper forests of n Somalia

☐ **Ankober Serin** *Carduelis ankoberensis*

Montane scrub of central Ethiopia

☐ **Fire-fronted Serin** *Serinus pusillus*

Mountains of Asia Minor to w China; winters to Israel

☐ **European Serin** *Serinus serinus*

Open woodlands of s Palearctic region

☐ **Syrian Serin** *Serinus syriacus*

Mountains of Lebanon and Syria; winters to Turkey and nw Iraq

☐ **Island Canary** *Serinus canaria*

Madeira, Azores and w Canary Islands

☐ **Citril Finch** *Serinus citrinella*

Montane coniferous forests of s Europe

☐ **Corsican Finch** *Serinus corsicanus*

Corsica and Sardinia

☐ **Tibetan Serin** *Serinus thibetanus*

Coniferous forests of n India to Nepal, e Tibet and sw China

☐ **Mountain Serin** *Serinus estherae*

____ *S. e. vanderbilti* — Mountains of n Sumatra
____ *S. e. estherae* — Mountains of w Java
____ *S. e. orientalis* — Tengger Mountains (s-central Java)
____ *S. e. renatae* — Mountains of Sulawesi
____ *S. e. mindanensis* — S Philippines (Mt. Katanglad and Mt. Apo on Mindanao)

☐ **Yellow-crowned Canary** *Serinus flavivertex*

____ *S. f. flavivertex* — Eritrea to n Tanzania
____ *S. f. sassii* — SW Uganda; se Zaire to sw Tanzania, ne Zambia and n Malawi
____ *S. f. huillensis* — Highlands of central Angola (n Huambo, Huila and Bie)

☐ **Cape Canary** *Serinus canicollis*

____ *S. c. griseitergum* — Highlands of e Zimbabwe and adjacent Mozambique
____ *S. c. thompsonae* — Transvaal and Natal to e Griqualand
____ *S. c. canicollis* — W and sw Cape Province to e Cape and w Free State

☐ **Abyssinian Siskin** *Serinus nigriceps*

Alpine moorlands of n and central Ethiopia

☐ **African Citril** *Serinus citrinelloides*

____ *S. c. citrinelloides* — Highlands of Eritrea and Ethiopia
____ *S. c. kikuyensis* — Highlands of s and sw Kenya

☐ **Western Citril** *Serinus frontalis*

Highlands of e Zaire to Uganda and n Zambia

☐ **Southern Citril** *Serinus hyposticutus*

____ *S. h. brittoni* — S Sudan and w Kenya
____ *S. h. hypostictus* — SE Kenya and Tanzania to Malawi

☐ **Black-faced Canary** *Serinus capistratus*

____ *S. c. capistratus* — Gabon to central Angola, s Zaire and n Zambia
____ *S. c. hildegardae* — Central Angola

☐ **Papyrus Canary** *Serinus koliensis*

Papyrus swamps of e Zaire, Uganda, Rwanda, w Kenya, Tanzania

☐ **Forest Canary** *Serinus scotops*

____ *S. s. transvaalensis* — Highland evergreen forests of n and e Transvaal
____ *S. s. umbrosus* — SE Transvaal to w Zululand, Natal and coastal s Cape Province
____ *S. s. scotops* — Coastal s Natal and e Cape Province

☐ **White-rumped Seedeater** *Serinus leucopygius*

____ *S. l. riggenbachi (pallens)* — Seegal to w Sudan
____ *S. l. leucopygius* — Central and s Sudan to w Eritrea, w Ethiopia and nw Uganda

☐ **Olive-rumped Serin** *Serinus rothschildi*

Mountains of sw Arabia

☐ **Black-throated Canary** *Serinus atrogularis*

____ *S. a. somereni* — E Zaire, nw Tanzania, Uganda and w Kenya
____ *S. a. lwenarum (kasaicus)* — Congo, s Zaire, cent. Angola, Zambia and sw Tanzania
____ *S. a. deserti* — S Angola, Namibia and Cape Province
____ *S. a. semideserti* — S Angola to n Namibia, Botswana, s Zambia and nw Transvaal
____ *S. a. atrogularis* — Transvaal to adj. n Free State, se Botswana and Zimbabwe
____ *S. a. impiger* — NE Cape Province to Free State, Lesotho, se Transvaal, Natal
____ *S. a. seshekeensis* — SW Zimbabwe

☐ **Reichenow's Seedeater** *Serinus reichenowi*

SE Sudan to s Ethiopia, Somalia, ne Uganda, Kenya, ne Tanzania

☐ **Yellow-rumped Serin** *Serinus xanthopygius*

Savanna of Eritrea and n Ethiopia

☐ **Lemon-breasted Seedeater** *Serinus citrinipectus*

S Mozambique, s Malawi, se Zimbabwe, e Transvaal, ne Natal

☐ **Yellow-fronted Canary** *Serinus mozambicus*

____ *S. m. caniceps*	Senegal to Cameroon (south to Benue plain)
____ *S. m. punctigula*	Cameroon (north to Toukte, Grand Capitaine and Koum)
____ *S. m. tando (santhome)*	Gabon to n Angola and sw Zaire; introduced São Tomé
____ *S. m. vansoni*	Extreme se Angola and adj. Namibia to n Botswana, sw Zambia
____ *S. m. barbatus (pseudobarbatus)*	S Chad to Sudan, ne Zaire, w Kenya and central Tanzania
____ *S. m. samaliyae*	SE Zaire to sw Tanzania and adjacent Zambia
____ *S. m. grotei*	S Sudan (east of the Nile) to Eritrea and w Ethiopia
____ *S. m. gommaensis*	W Ethiopia (Lake Tana to Gomma)
____ *S. m. mozambicus*	Coastal Kenya and Mafia I. to Zambia, Mozambique, n Transvaal
____ *S. m. granti*	S Mozambique (s of Limpopo R.) to e Transvaal and e Cape Prov.

☐ **White-bellied Canary** *Serinus dorsostriatus*

____ *S. d. maculicollis*	Somalia, Ethiopia, Sudan, ne Uganda and n Kenya
____ *S. d. dorsostriatus*	SE Uganda, w Kenya and nw Tanzania
____ *S. d. taruensis*	Kenya to ne Tanzania

☐ **Yellow-throated Serin** *Serinus flavigula*

Savanna of Ethiopia (Shoa Province)

☐ **Salvadori's Serin** *Serinus xantholaemus*

Savanna of Ethiopia

☐ **Northern Grosbeak-Canary** *Serinus donaldsoni*

S Ethiopia, Somalia to n Kenya; isolated population in Taita Hills

☐ **Southern Grosbeak-Canary** *Serinus buchanani*

Savanna of s Kenya and ne Tanzania

☐ **Brimstone Canary** *Serinus sulphuratus*

____ *S. s. sharpii (frommi, shelleyi, loveridgei)*	Kenya, Uganda, e Zaire to Angola, Zambia and Mozambique
____ *S. s. wilsoni (languens)*	S Mozambique and S Africa (Transvaal to e Cape Province)
____ *S. s. sulphuratus*	SW and s Cape Province

☐ **Yellow Canary** *Serinus flaviventris*

____ *S. f. damarensis*	Angola, Namibia, Botswana and n Cape Province
____ *S. f. flaviventris (quintoni, hesperus)*	W, cent. and s Cape Province to extreme s Namibia
____ *S. f. marshalli*	Transvaal and Free State to Lesotho and ne Cape Province
____ *S. f. guillarmodi*	Highlands of Lesotho

☐ **White-throated Canary** *Serinus albogularis*

____ *S. a. crocopygius*	N Namibia and sw Angola
____ *S. a. sordahlae*	NE and n Cape Province and s Namibia
____ *S. a. albogularis (hewitti)*	W and sw Cape Province to e Cape and Griqualand
____ *S. a. orangensis*	NE Cape Province, Lesotho, Free State and sw Transvaal

☐ **Streaky Seedeater** *Serinus striolatus*

____ *S. s. striolatus (affinis)*	Eritrea, Ethiopia, s Sudan, e Uganda, Kenya and n Tanzania
____ *S. s. graueri*	E Zaire, sw Uganda and Rwanda

☐ **Yellow-browed Seedeater** *Serinus whytii*

Highlands of s Tanzania to Zambia and n Malawi

☐ **Thick-billed Seedeater** *Serinus burtoni*

____ *S. b. burtoni*	Nigeria, Cameroon and Bioko
____ *S. b. tanganjicae*	Highlands of w-central Angola; e Zaire and adjacent w Uganda
____ *S. b. kilimensis (gurneti)*	Montane forests of e Uganda to n Kenya and n Tanzania
____ *S. b. albifrons*	Montane forests of e Kenya (east of the Rift Valley)

☐ **Tanzania Seedeater** *Serinus melanochrous*

S Tanzania (Mt. Rungwe and Udzungwa Mts.)

☐ **Principe Seedeater** *Serinus rufobrunneus*
_____ *S. r. rufobrunneus*
_____ *S. r. thomensis*
_____ *S. r. fradei*

Príncipe (Gulf of Guinea)
São Tomé (Gulf of Guinea)
Caroo I. (Gulf of Guinea)

☐ **Protea Canary** *Serinus leucopterus*

Montane *protea* and heath of sw Cape Province (South Africa)

☐ **Black-eared Seedeater** *Serinus mennelli*

Brachystegia woodlands of e Angola to Mozambique

☐ **Streaky-headed Seedeater** *Serinus gularis*
_____ *S. g. canicapilla*
_____ *S. g. montanorum*
_____ *S. g. elgonensis*
_____ *S. g. benguellensis*
_____ *S. g. mendosus*
_____ *S. g. gularis*
_____ *S. g. endemion*
_____ *S. g. humilis*

Guinea to Central African Republic
Highlands of e Nigeria and w Cameroon
S Sudan to w Kenya
Highlands of central Angola to w Zambia
NE Botswana to w Zimbabwe and adj. highland Mozambique
N and w Transvaal, adj. Botswana to w Free State, n Cape Prov.
SE Transvaal, Swaziland, s Mozambique to Natal, Cape Prov.
SW and s Cape Province (east to Great Fish R.)

☐ **Reichard's Seedeater** *Serinus reichardi*
_____ *S. r. striatipectus*
_____ *S. r. reichardi*

Mts. of s Sudan to s Ethiopia and central highlands of Kenya
SE Zaire to Zambia, Tanzania, Malawi and n Mozambique

☐ **Brown-rumped Seedeater** *Serinus tristriatus*

Juniper woodlands of n Ethiopia, Eritrea and nw Somalia

☐ **Yemen Serin** *Serinus menachensis*

High altitude bushy hills of sw Saudi Arabia

☐ **Cape Siskin** *Pseudochloroptila totta*

Montane scrub of South Africa (s Cape Province)

☐ **Drakensberg Siskin** *Pseudochloroptila symonsi*

Montane scrub of w Natal and Lesotho (Drakensberg Mountains)

☐ **Black-headed Canary** *Alario alario*

South Africa and Lesotho

☐ **Damara Canary** *Alario leucolaemus*

Namibia, Damarland, Great Namaqualand, Botswana

☐ **Brown Bullfinch** *Pyrrhula nipalensis*
_____ *P. n. nipalensis*
_____ *P. n. ricketti*
_____ *P. n. victoriae*
_____ *P. n. waterstradti*
_____ *P. n. uchidae*

Himalayas (n Pakistan to n India, Nepal and Bhutan)
SE Tibet to s China, ne India and n Vietnam
Myanmar (s Chin Hills)
Mountains of Malay Peninsula (Perak, Pahang and Selangor)
Mountains of s Taiwan

☐ **White-cheeked Bullfinch** *Pyrrhula leucogenis*
_____ *P. l. leucogenis*
_____ *P. l. steerei (coriaria, apo)*

N Philippines (mountains of nw Luzon)
S Philippines (mountains of Mindanao)

☐ **Orange Bullfinch** *Pyrrhula aurantiaca*

Montane forests of n Pakistan to nw India (Himachal Pradesh)

☐ **Red-headed Bullfinch** *Pyrrhula erythrocephala*

Montane forests of n Pakistan, n India and se Tibet

☐ **Gray-headed Bullfinch** *Pyrrhula erythaca*
_____ *P. e. erythaca*
_____ *P. e. wilderi*
_____ *P. e. owstoni*

Himalayas (Sikkim to Bhutan, se Tibet, w China and n Myanmar)
NE China (Liaoning)
Taiwan (Mt. Ari Shan and Mt. Morrison)

☐ **Eurasian Bullfinch** *Pyrrhula pyrrhula*
_____ *P. p. pileata*

British Isles

___	*P. p. europoea*	Denmark, w Germany, Netherlands, Belgium and w France
___	*P. p. pyrrhula*	N Europe to w Mongolia; winters to s Europe and Iran
___	*P. p. iberiae*	Mountains of n Portugal to nw Spain (Pyrénées)
___	*P. p. murina*	San Miguel (Azores)
___	*P. p. rossikowi*	W Turkey to Caucasus and extreme ne Iran
___	*P. p. caspica*	NE and n Iran
___	*P. p. cineracea*	Sayan and n Altai Mts.; winters to Amur region and Manchuria
___	*P. p. cassinii*	Kamchatka Pen., Paramušir I. (n Kurils) and Sea of Okhotsk coast
___	*P. p. griseiventris*	E Manchuria to Sakhalin, Kuril Is. and Honshu; winters to Japan

☐ **Hawfinch** *Coccothraustes coccothraustes*

___	*C. c. coccothraustes*	England and n continental Europe to w Asia; winters to N Africa
___	*C. c. buvryi*	Mountains of Morocco, Algeria and Tunisia
___	*C. c. nigricans*	Ukraine to Crimean Peninsula, Caspian Sea and n Iran
___	*C. c. humii*	E Kazakstan to Kyrgystan, Tajikistan and w Afghanistan
___	*C. c. japonicus*	Sakhalin, Hokkaido and n Honshu; winters to e China and Bonin Is.

☐ **Evening Grosbeak** *Coccothraustes vespertinus*

___	*C. v. brooksi*	W-central Canada to mountains of w US; winters to Texas
___	*C. v. vespertinus*	E-cent. Canada (Alberta to New England); winters to se US
___	*C. v. montanus*	Mountains of se Arizona to s Mexico (Sierra Madre Occidental)

☐ **Hooded Grosbeak** *Coccothraustes abeillei*

___	*C. a. pallidus*	Sierra Madre Occidental of Mexico (Chihuahua, Sinaloa, Durango)
___	*C. a. saturatus*	Mountains of e Mexico (San Luis Potosí and sw Tamaulipas)
___	*C. a. abeillei*	Mountains of central and s Mexico
___	*C. a. cobanensis*	Highlands of s Mexico (Chiapas) to central Guatemala

☐ **Yellow-billed Grosbeak** *Eophona migratoria*

___	*E. m. migratoria*	E Manchuria to North Korea; winters to e China
___	*E. m. sowerbyi*	E China

☐ **Japanese Grosbeak** *Eophona personata*

___	*E. p. personata*	Japan (Hokkaido to Honshu); winters to s Japan, e China, Taiwan
___	*E. p. magnirostris*	E Manchuria to ne China (Shandong)

☐ **Black-and-yellow Grosbeak** *Mycerobas icterioides*

Montane forests of ne Afghanistan to central Nepal

☐ **Collared Grosbeak** *Mycerobas affinis*

Mountains of n Pakistan to se Tibet, ne Myanmar and c China

☐ **Spot-winged Grosbeak** *Mycerobas melanozanthos*

Mts. of n Pakistan to s Tibet, sw China, Myanmar and Thailand

☐ **White-winged Grosbeak** *Mycerobas carnipes*

___	*M. c. speculigerus*	Mts. of ne Iran and Transcaspia to n Afghanistan and Pakistan
___	*M. c. carnipes*	Montane forests of Turkestan to Pakistan and w China

☐ **Gold-naped Finch** *Pyrrhoplectes epauletta*

Montane forests of n India to sw China, se Tibet and n Myanmar

☐ **Spectacled Finch** *Callacanthis burtoni*

Himalayas (Kashmir to Nepal and Sikkim)

☐ **Crimson-winged Finch** *Rhodopechys sanguineus*

___	*R. s. alienus*	Mts. of Morocco and Algeria
___	*R. s. sanguineus*	Mts. of Turkey to Caucasus, Iran, n Afghanistan and nw India

☐ **Trumpeter Finch** *Bucanetes githagineus*

___	*B. g. amantum*	Canary Islands
___	*B. g. zedlitzi*	Mauritania to Morocco, Tunisia and Algeria
___	*B. g. githagineus*	S Egypt to n Sudan
___	*B. g. crassirostris*	Israel to Arabia, Iraq, Iran, Afghanistan and Pakistan

☐ **Mongolian Finch** *Bucanetes mongolicus*

Sparse montane vegetation of sw and central Asia

☐ **Desert Finch** *Rhodospiza obsoleta*

Locally in mountains of se Turkey to n China and n Pakistan

☐ **Long-tailed Rosefinch** *Uragus sibiricus*

    ____ *U. s. sibiricus*   S Siberia to n Mongolia and n Manchuria; winters to Turkestan

    ____ *U. s. ussuriensis*   Central Manchuria to N Korea; winters to ne China and S Korea

    ____ *U. s. sanguinolentus*   Sakhalin, s Kuril Is. and Hokkaido; winters to s Japanese islands

    ____ *U. s. lepidus*   W China (se Gansu and s Shaanxi)

    ____ *U. s. henrici*   Mountains of w China (se Tibet to w Sichuan and n Yunnan)

☐ **Scarlet Finch** *Haematospiza sipahi*

Himalayas (Nepal to sw China, Myanmar, n Laos and n Vietnam)

FAMILY: DREPANIDIDAE (Hawaiian Honeycreepers—21)

☐ **Laysan Finch** *Telespiza cantans*

Laysan I. (w Hawaiian Islands)

☐ **Nihoa Finch** *Telespiza ultima*

Nihoa I. (nw Hawaiian Islands)

☐ **Ou** *Psittirostra psittacea*

Humid montane forests of Hawaiian Islands (Hawaii and Kauai)

☐ **Palila** *Loxioides bailleui*

Montane *mamame-naio* forests of Hawaii

☐ **Maui Parrotbill** *Pseudonestor xanthophrys*

Maui (*Ohia* montane forests on e slopes of Haleakala Crater)

☐ **Hawaii Amakihi** *Hemignathus virens*

    ____ *H. v. wilsoni*   *Ohia* and *mamani* forests of Maui; formerly Molokai and Lanai

    ____ *H. v. virens*   *Ohia* and *mamani* forests of Hawaii

☐ **Oahu Amakihi** *Hemignathus flavus*

Humid montane forests of Oahu

☐ **Kauai Amakihi** *Hemignathus kauaiensis*

Humid montane *ohia/koa* forests of Kauai

☐ **Anianiau** *Hemignathus parvus*

Humid montane forests of Kauai

☐ **Nukupuu** *Hemignathus lucidus*

    ____ *H. l. lucidus*   Wet *ohia* forests of Hawaii

    ____ *H. l. affinis*   Wet *ohia* forests of Maui

    ____ *H. l. hanapepe*   Wet *ohia* forests of Kauai

☐ **Akiapolaau** *Hemignathus munroi*

Montane *koa* forests of Hawaii

☐ **Akikiki** *Oreomystis bairdi*

Montane *ohia* forests of Kauai

☐ **Hawaii Creeper** *Oreomystis mana*

Mixed *koa/ohia* forests of Hawaii

☐ **Oahu Alauahio** *Paroreomyza maculata*

Humid montane forests of Oahu

☐ **Maui Alauahio** *Paroreomyza montana*

Humid montane forests of e Maui

☐ **Akekee** *Loxops caeruleirostris*

Kauai (Kokee and Alakai Swamp environs)

☐ **Akepa** *Loxops coccineus*

    ____ *L. c. coccineus*   Montane *ohia* and *koa* forests of Hawaii

    ____ *L. c. wolstenholmei*†   Formerly montane *ohia* and *koa* forests on Oahu. Extinct ca 1900

    ____ *L. c. ochraceus*   Montane *ohia* and *koa* forests of Maui. Probably extinct

☐ Iiwi *Vestiaria coccinea*

Ohia and *mamame* forests of main Hawaiian Islands

☐ Akohekohe *Palmeria dolei*

Ohia forests of e Maui. Formerly on Molokai

☐ Apapane *Himatione sanguinea*

Wet *ohia* forests from Kauai to Hawaii (e Hawaiian Islands)

☐ Poo-uli *Melamprosops phaeosoma*

Maui (*Ohia* forests of Haleakala Crater). On verge of extinction

FAMILY: PEUCEDRAMIDAE (Olive Warbler—1)

☐ Olive Warbler *Peucedramus taeniatus*
___ *P. t. arizonae* — Mts. of sw US to n Mexico (n Chihuahua and n Coahuila)
___ *P. t. jaliscensis* — Mts. of nw Mexico (s Chihuahua to sw Jalisco and Colima)
___ *P. t. giraudi* — Mts. of central Mexico (Jalisco to n Puebla and w-c Veracruz)
___ *P. t. taeniatus (aurantiacus)* — Mts. of s Mexico (Guerrero, Oaxaca, Chiapas) to w Guatemala
___ *P. t. micrus* — Mts. of El Salvador, Honduras and n Nicaragua

FAMILY: PARULIDAE (New World Warblers—118)

☐ Bachman's Warbler *Vermivora bachmanii*

Formerly se US; winters to Cuba and Bahamas (probably extinct)

☐ Blue-winged Warbler *Vermivora pinus*

E US; winters se Mexico to Panama

☐ Golden-winged Warbler *Vermivora chrysoptera*

E N America; winters Guatemala to nw S Am. and Greater Antilles

☐ Tennessee Warbler *Vermivora peregrina*

SE Alaska and s Yukon to n US; winters s Mexico to nw S Am.

☐ Orange-crowned Warbler *Vermivora celata*
___ *V. c. lutescens* — SE Alaska to Br. Columbia and s Calif., winters to Guatemala
___ *V. c. celata* — Central Alaska to s Canada; winters to Guatemala and Bahamas
___ *V. c. orestera* — Rocky Mountains to sw US and w Texas; winters to s Mexico
___ *V. c. sordida* — Coastal s California and islands off sw California and Baja

☐ Nashville Warbler *Vermivora ruficapilla*
___ *V. r. ridgwayi* — S Br. Col. to California, Nevada and n Utah; winters to Mexico
___ *V. r. ruficapilla* — S-cent. Canada to e-central US; winters Mexico to central Honduras

☐ Virginia's Warbler *Vermivora virginiae*

Mountains of sw US; winters to sw Mexico

☐ Colima Warbler *Vermivora crissalis*

W Texas (Chisos Mountains) to central Mexico

☐ Lucy's Warbler *Vermivora luciae*

Arid sw US to ne Baja and ne Sonora; winters to sw Mexico

☐ Flame-throated Warbler *Parula gutturalis*

Humid montane forests of Costa Rica and w Panama

☐ Crescent-chested Warbler *Parula superciliosa*
___ *P. s. sodalis* — Sierra Madre Occidental of w Mexico
___ *P. s. mexicana* — Sierra Madre Oriental of e Mexico
___ *P. s. palliata* — SW Mexico (s Jalisco to w Michoacán and Guerrero)
___ *P. s. superciliosa* — Highlands of s Mexico (Chiapas) to Guatemala and w Honduras
___ *P. s. parva* — Highlands of central and e Honduras and Nicaragua

☐ Northern Parula *Parula americana*

E N Am. (s Canada to s US); winters to Nicaragua and West Indies

☐ Tropical Parula *Parula pitiayumi*
___ *P. p. graysoni* — Socorro I. (Revillagigedo Islands off s Baja California)
___ *P. p. insularis* — Tres Marías Islands (off w Mexico) and adjacent Nayarit

____	*P. p. pulchra*	Sierra Madre Occidental of nw Mexico (Sonora to Jalisco)
____	*P. p. nigrilora*	S Texas (lower Rio Grande Valley) to e Coahuila and n Veracruz
____	*P. p. inornata*	S Mexico (s Veracruz) to e Guatemala and n Honduras
____	*P. p. speciosa*	Tropical s Honduras to Nicaragua, Costa Rica and w Panama
____	*P. p. cirrha*	Coiba I. (Panama)
____	*P. p. nana*	E Panama (Darién) to nw Colombia (Córdoba)
____	*P. p. elegans*	Trop. n Colombia, n Venezuela and n Brazil; Trinidad and Tobago
____	*P. p. pacifica*	Tropical sw Colombia (Nariño) to w Ecuador and nw Peru
____	*P. p. roraimae*	*Tepuis* of s Venezuela and adjacent n Brazil
____	*P. p. alarum*	Subtrop. e Ecuador and n Peru e of the Andes (s to Huánuco)
____	*P. p. melanogenys*	S Peru (Junín) to w Bolivia (La Paz and Cochabamba)
____	*P. p. pitiayumi*	E Bolivia to Paraguay, Brazil, Uruguay and n Argentina

☐ **Yellow Warbler** *Dendroica petechia*

____	*D. p. amnicola*	Alaska, Canada and Newfoundland; winters to n South America
____	*D. p. rubiginosa*	S Alaska to w British Columbia; winters to s Baja and Panama
____	*D. p. aestiva*	S-central Canada and central US; winters to South America
____	*D. p. morcomi (brewsteri, ineditus)*	SE Br. Columbia, w US and n Baja; winters to n South America
____	*D. p. sonorana*	SW US to nw Mexico; winters to w Panama, Colombia and Ecuador
____	*D. p. castaneiceps (hueyi)*	Mangroves of coastal s Baja California (south of latitude 27°N)
____	*D. p. rhizophorae*	Mangroves of nw Mexico (Sonora to Nayarit); winters to Oaxaca
____	*D. p. dugesi*	Central plateau of Mexico
____	*D. p. oraria*	Mangroves of e Mexico (s Tamaulipas to w Tabasco)
____	*D. p. bryanti*	Mangroves of Yucatán Peninsula to Belize and Costa Rica
____	*D. p. rufivertex*	Cozumel I. (off Quintana Roo)
____	*D. p. xanthotera*	Pacific coast of w Guatemala to Costa Rica
____	*D. p. flavida*	Isla San Andrés (w Caribbean Sea)
____	*D. p. armouri*	Isla Providéncia (w Caribbean Sea)
____	*D. p. eoa*	Jamaica and Cayman Islands
____	*D. p. gundlachi*	Lower Florida Keys, Cuba, Isle of Pines and Bahamas
____	*D. p. albicollis*	Hispaniola, Gonâve and adjacent islands
____	*D. p. cruciana*	Puerto Rico and Virgin Islands
____	*D. p. bartholemica*	Montserrat and n Lesser Antilles
____	*D. p. melanoptera*	Guadeloupe, Dominica and central Lesser Antilles
____	*D. p. ruficapilla*	Martinique (Lesser Antilles)
____	*D. p. babad*	St. Lucia (Lesser Antilles)
____	*D. p. petechia*	Barbados (Lesser Antilles)
____	*D. p. alsiosa*	Grenadines (Lesser Antilles)
____	*D. p. rufopileata*	Netherlands Antilles (Aruba, Curaçao, Bonaire and adj. islands)
____	*D. p. obscura*	Islas Los Roques (off n Venezuela)
____	*D. p. chrysendeta*	NE Colombia (Guajira Peninsula) and nw Venezuela (Zulia)
____	*D. p. paraguanae*	NW Venezuela (Paraguaná Peninsula of Falcón)
____	*D. p. cienagae*	N Venezuela (coastal Carabobo and Aragua) and offshore islands
____	*D. p. aurifrons*	Coastal n-c Venezuela, Islas La Tortuga, Tortuguillas and Piritu
____	*D. p. aureola*	Cocos I. (off Costa Rica) and Galapagos Islands
____	*D. p. aequatorialis*	Pearl Islands and adjacent mainland Panama
____	*D. p. erithachorides*	Atlantic coast of Panama and Caribbean coast of n Colombia
____	*D. p. peruviana*	Extreme sw Colombia (Nariño) to w Ecuador and n Peru (Lima)

☐ **Chestnut-sided Warbler** *Dendroica pensylvanica*

E N America; winters Guatemala to Panama (casual n S America)

☐ **Magnolia Warbler** *Dendroica magnolia*

E North America; winters to Panama and West Indies

☐ **Cape May Warbler** *Dendroica tigrina*

Canada and ne US; winters mainly West Indies

☐ **Black-throated Blue Warbler** *Dendroica caerulescens*

____	*D. c. caerulescens*	SE Canada to ne US; winters to Bahamas and Greater Antilles
____	*D. c. cairnsi*	E-central US; winters to Bahamas and Greater Antilles

☐ **Yellow-rumped Warbler** *Dendroica coronata*

____	*D. c. coronata (hooveri)*	N Alaska, Canada and n US; winters to Panama and West Indies

____ *D. c. auduboni (memorabilis)* — SW Canada and w US; winters to w Honduras
____ *D. c. nigrifrons* — Sierra Madre Occidental of w Mexico (Chihuahua to Durango)
____ *D. c. goldmani* — High mountains of s Chiapas (Volcán Tacaná) and w Guatemala

☐ **Black-throated Gray Warbler** *Dendroica nigrescens*
____ *D. n. nigrescens* — SW Br. Columbia and w US; winters s California to n Guatemala
____ *D. n. halseii* — N Baja California, s Arizona, s New Mexico and n Sonora

☐ **Golden-cheeked Warbler** *Dendroica chrysoparia* — Texas (Edwards Plateau); winters to Nicaragua

☐ **Black-throated Green Warbler** *Dendroica virens* — E N America; winters Mexico to n South America and West Indies

☐ **Townsend's Warbler** *Dendroica townsendi* — Coniferous forests of w North America; winters to Costa Rica

☐ **Hermit Warbler** *Dendroica occidentalis* — Mountains of w US; winters to Nicaragua

☐ **Blackburnian Warbler** *Dendroica fusca* — E North America; winters Costa Rica to Bolivia

☐ **Yellow-throated Warbler** *Dendroica dominica*
____ *D. d. albilora* — E-central and se US; winters to Costa Rica, Cuba and Jamaica
____ *D. d. dominica* — New Jersey to cent. Florida; winters to Bahamas and Gr. Antilles
____ *D. d. stoddardi* — Extreme s Alabama and nw Florida
____ *D. d. flavescens* — Grand Bahama, Little Abaco and Great Abaco

☐ **Olive-capped Warbler** *Dendroica pityophila* — Pine barrens of e Cuba, Grand Bahama and Abaco

☐ **Grace's Warbler** *Dendroica graciae*
____ *D. g. graciae* — Mts. of sw US to n Mexico (n Sinaloa); winters to Michoacán
____ *D. g. yaegeri* — Mts. of w Mexico (s Sinaloa to Nayarit, w Zacatecas and w Jalisco)
____ *D. g. remota* — Mts. of s Mexico (Michoacán and Guerrero) to n Nicaragua
____ *D. g. decora* — Pine ridges of coastal Belize, e Honduras and ne Nicaragua

☐ **Adelaide's Warbler** *Dendroica adelaidae* — Puerto Rico and Vieques I.

☐ **Barbuda Warbler** *Dendroica subita* — Barbuda I. (Lesser Antilles)

☐ **St. Lucia Warbler** *Dendroica delicata* — St. Lucia I. (Lesser Antilles)

☐ **Pine Warbler** *Dendroica pinus*
____ *D. p. pinus* — SE Canada to se Texas and n Florida
____ *D. p. florida* — Peninsular Florida (Gainesville to Everglades National Park)
____ *D. p. achrustera* — Grand Bahama, Little and Great Abaco, Andros and New Providence
____ *D. p. chrysoleuca* — Hispaniola

☐ **Kirtland's Warbler** *Dendroica kirtlandii* — Jack-pine area of Michigan; winters in Bahamas

☐ **Prairie Warbler** *Dendroica discolor*
____ *D. d. discolor* — Central and e-central US; winters Bahamas and West Indies
____ *D. d. paludicola* — SE S Carolina to s Florida; winters to Gr. Antilles and El Salvador

☐ **Vitelline Warbler** *Dendroica vitellina*
____ *D. v. crawfordi* — Little Cayman I. and Cayman Brac
____ *D. v. vitellina* — Grand Cayman I.
____ *D. v. nelsoni* — Swan Islands (w Caribbean Sea)

☐ **Palm Warbler** *Dendroica palmarum*
____ *D. p. palmarum* — Cent. and e Canada to n US; winters to W Indies and Nicaragua
____ *D. p. hypochrysea* — SE Canada and ne US; winters to cent. Florida and n Gulf Coast

☐ **Bay-breasted Warbler** *Dendroica castanea* — Canada and ne US; winters Panama and nw South America

☐ **Blackpoll Warbler** *Dendroica striata*

☐ **Cerulean Warbler** *Dendroica cerulea*

Alaska and Canada; winters Colombia to Peru and w Amaz. Brazil

☐ **Plumbeous Warbler** *Dendroica plumbea*

E N America; winters mts. of Colombia to Venezuela and Bolivia

☐ **Arrow-headed Warbler** *Dendroica pharetra*

Dominica, Marie Galante, Guadeloupe and Terre-de-Haut I.

☐ **Elfin-woods Warbler** *Dendroica angelae*

Montane forests and wooded hills of Jamaica

☐ **Whistling Warbler** *Catharopeza bishopi*

Montane elfin forests of e Puerto Rico

☐ **Black-and-white Warbler** *Mniotilta varia*

Montane forests of St. Vincent (Lesser Antilles)

☐ **American Redstart** *Setophaga ruticilla*

Canada to Gulf states; winters Mexico to Peru and West Indies

☐ **Prothonotary Warbler** *Protonotaria citrea*

Alaska to s US; winters s US to n South America and West Indies

☐ **Worm-eating Warbler** *Helmitheros vermivorum*

E US; winters se Mexico to w Ecuador and Lesser Antilles

☐ **Swainson's Warbler** *Limnothlypis swainsonii*

E US; winters se Mexico to Panama and Greater Antilles

☐ **Ovenbird** *Seiurus aurocapilla*

SE US; winters se Mexico and Belize to Guatemala and W Indies

_____ *S. a. aurocapilla*
_____ *S. a. cinereus*
_____ *S. a. furvior*

Central and se Canada to e US; winters to n South America
Rocky Mts. of s Alberta to w-central US; winters to Costa Rica
Newfoundland; winters to Bahamas, Cuba and Panama

☐ **Northern Waterthrush** *Seiurus noveboracensis*

N North America; winters to West Indies and n South America

☐ **Louisiana Waterthrush** *Seiurus motacilla*

E US; winters s Florida to nw South America and West Indies

☐ **Kentucky Warbler** *Oporornis formosus*

E US; winters Mexico to nw South America

☐ **Connecticut Warbler** *Oporornis agilis*

N North America; winters S Am. e of Andes to Bolivia and Brazil

☐ **Mourning Warbler** *Oporornis philadelphia*

SE Canada and ne US; winters Nicaragua to nw South America

☐ **MacGillivray's Warbler** *Oporornis tolmiei*

W North America; winters highlands of Mexico to Panama

☐ **Common Yellowthroat** *Geothlypis trichas*

_____ *G. t. campicola*
_____ *G. t. arizela*
_____ *G. t. occidentalis*
_____ *G. t. sinuosa*
_____ *G. t. scirpicola*
_____ *G. t. trichas (brachidactylus)*
_____ *G. t. typhicola*
_____ *G. t. ignota*
_____ *G. t. insperata*
_____ *G. t. chryseola (riparia)*
_____ *G. t. modesta*
_____ *G. t. chapalensis*
_____ *G. t. melanops*

Yukon, w Canada and se Alaska to nw US; winters to n Mexico
Extreme se Alaska to s-c California; winters to s Baja, n Sonora
N Oregon to New Mexico and nw Texas; winters to Honduras
Saltwater marshes from San Francisco to San Diego
S California to Nevada, sw Utah, n Baja and extreme nw Sonora
SE Canada and e-cent. US; winters to West Indies, n S America
SE US; winters to se Mexico (Veracruz)
Coastal se S Carolina to s Florida, s Mississippi and se Louisiana
S Texas (Rio Grande Valley below Brownsville)
SE Arizona to s New Mexico, w Texas and nw Mexico
W Mexico (w-central Sonora south to Colima)
NW Mexico (Lake Chapala region of Jalisco)
Cent. Mexico (Zacatecas and n Jalisco to Oaxaca and Veracruz)

☐ **Belding's Yellowthroat** *Geothlypis beldingi*

_____ *G. b. goldmani*
_____ *G. b. beldingi*

Marshes of central Baja California (latitude 28°N to 26°N)
Marshes of Cape district of s Baja California

☐ **Altamira Yellowthroat** *Geothlypis flavovelata*

Coastal ne Mexico (s Tamaulipas, e San Luis Potosí, n Veracruz)

☐ **Bahama Yellowthroat** *Geothlypis rostrata*
_____ *G. r. coryi* — N Bahamas (Eleuthera and Cat I.)
_____ *G. r. tanneri* — Grand Bahama, Moranie Cay, Little and Great Abaco, Elbow Cay
_____ *G. r. rostrata* — W Bahamas (New Providence and Andros)

☐ **Olive-crowned Yellowthroat** *Geothlypis semiflava*
_____ *G. s. bairdi* — Tropical ne Honduras to Nicaragua, Costa Rica and nw Panama
_____ *G. s. semiflava* — Pacific lowlands of w Colombia to w Ecuador (El Oro)

☐ **Black-polled Yellowthroat** *Geothlypis speciosa* — Highlands of s-c Mexico (e Michoacán, s Guanajuato and México)

☐ **Masked Yellowthroat** *Geothlypis aequinoctialis*
_____ *G. a. chiriquensis* — Lowlands of sw Costa Rica and w Panama (w Chiriquí)
_____ *G. a. aequinoctialis* — NE Colombia to Venezuela, Guianas, Suriname, n Brazil; Trinidad
_____ *G. a. auricularis* — Pacific slope of w Ecuador to w Peru (south to Ica)
_____ *G. a. peruviana* — N Peru (upper Marañon Valley of Cajamarca and La Libertad)
_____ *G. a. velata* — SE Peru to Bolivia, Brazil, Paraguay, Uruguay and ne Argentina

☐ **Gray-crowned Yellowthroat** *Geothlypis poliocephala*
_____ *G. p. poliocephala (pontilis)* — Pacific slope of nw Mexico (n Sinaloa to extreme w Oaxaca)
_____ *G. p. ralphi* — Extreme s Texas to ne Mexico (Tamaulipas and San Luis Potosí)
_____ *G. p. palpebralis* — Caribbean slope of s Mexico (Veracruz) to Belize and Costa Rica
_____ *G. p. caninucha* — S Mexico (Oaxaca) to w Guatemala, El Salvador and se Honduras
_____ *G. p. icterotis* — Pacific slope of w Nicaragua and w Costa Rica
_____ *G. p. ridgwayi* — SW Costa Rica (Térraba Valley) to w Panama (Chiriquí)

☐ **Hooded Yellowthroat** *Geothlypis nelsoni*
_____ *G. n. nelsoni* — Sierra Madre Oriental of e Mexico (Coahuila to Veracruz, n Puebla)
_____ *G. n. karlenae* — Highlands of s Mexico (Distrito Fed. to Puebla and n-c Oaxaca)

☐ **Green-tailed Warbler** *Microligea palustris*
_____ *M. p. palustris* — Higher elevations of Hispaniola
_____ *M. p. vasta* — Xeric lowlands of sw Dominican Republic and Beata I.

☐ **Yellow-headed Warbler** *Teretistris fernandinae* — Forest undergrowth of w Cuba, Isle of Pines and Cayo Cantiles

☐ **Oriente Warbler** *Teretistris fornsi*
_____ *T. f. fornsi* — Humid mts. and semiarid coast of e Cuba and Camagüey Arch.
_____ *T. f. turquinensis* — Sierra Maestra of e Cuba in Pico Turquino region

☐ **Semper's Warbler** *Leucopeza semperi* — Montane forests of St. Lucia (Lesser Antilles). Possibly extinct

☐ **Hooded Warbler** *Wilsonia citrina* — E US; winters Mexico to Panama

☐ **Wilson's Warbler** *Wilsonia pusilla*
_____ *W. p. pileolata* — N Alaska and n Yukon to sw US; winters to w Panama
_____ *W. p. chryseola* — SW British Columbia to s California; winters to Panama
_____ *W. p. pusilla* — S and e Canada to ne US; winters to Costa Rica

☐ **Canada Warbler** *Wilsonia canadensis* — E North America; winters Panama and mts. of nw South America

☐ **Red-faced Warbler** *Cardellina rubrifrons* — Mts. of Ariz. and New Mexico to nw Mexico; winters to Honduras

☐ **Red Warbler** *Ergaticus ruber*
_____ *E. r. melanauris* — Sierra Madre Occidental of w Mexico (Chihuahua and Durango)
_____ *E. r. ruber* — Mts. of w Mexico (Jalisco and Michoacán to Veracruz and Oaxaca)
_____ *E. r. rowleyi* — S Mexico (mountains of Oaxaca in Lachao Nuevo region)

☐ **Pink-headed Warbler** *Ergaticus versicolor* — Mountains of extreme s Mexico (Chiapas) to Guatemala

☐ Painted Redstart *Myioborus pictus*
- ____ M. p. pictus — Mountains of sw US to s Mexico (Guerrero, Oaxaca and Veracruz)
- ____ M. p. guatemalae — Mountains of s Mexico (Chiapas) to n Nicaragua

☐ Slate-throated Redstart *Myioborus miniatus*
- ____ M. m. miniatus — Mts. of w Mexico (s Sonora to Guerrero, Oaxaca and w Chiapas)
- ____ M. m. molochinus — Mountains of e Mexico (Sierra de Tuxtla in se Veracruz)
- ____ M. m. intermedius — S Mexico (extreme e Oaxaca and Chiapas) to e Guatemala
- ____ M. m. hellmayri — Subtropical Pacific cordillera from Guatemala to sw El Salvador
- ____ M. m. connectens — Subtropical mountains of El Salvador and Honduras
- ____ M. m. comptus — Subtrop. mts. of Costa Rica (Cordillera Central and Guanacaste)
- ____ M. m. aurantiacus — Mts. of e Costa Rica (Cordillera de Talamanca) and w Panama
- ____ M. m. ballux — E Panama (Darién) to Colombia, w Venezuela and nw Ecuador
- ____ M. m. sanctaemartae — Santa Marta Mountains (ne Colombia)
- ____ M. m. pallidiventris — Coastal mountains of n Venezuela (e Falcón to Monagas)
- ____ M. m. subsimilis — W Andes of Ecuador (El Oro) to extreme nw Peru (Piura)
- ____ M. m. verticalis — SE Ecuador to Peru, Bolivia, se Venezuela, Guyana and nw Brazil

☐ Tepui Redstart *Myioborus castaneocapilla*
- ____ M. c. duidae — *Tepuis* of se Venezuela (Mt. Duida, Mt. Parú, Mt. Huachamacari)
- ____ M. c. castaneocapilla — *Tepuis* of e Venezuela (e Bolívar), w Guyana and extreme n Brazil
- ____ M. c. maguirei — *Tepuis* of se Venezuela (Cerro de la Neblina)

☐ Brown-capped Redstart *Myioborus brunniceps*

Andes of Bolivia (La Paz and Cochabamba) to nw Argentina

☐ Yellow-faced Redstart *Myioborus pariae*

NE Venezuela (humid cloud forests of Paría Peninsula)

☐ White-faced Redstart *Myioborus albifacies*

Tepuis of s Venezuela (Cerros Guany, Yaví and Paraque)

☐ Saffron-breasted Redstart *Myioborus cardonai*

Tepuis of s Venezuela (Cerro Guaiquinima in w-central Bolívar)

☐ Collared Redstart *Myioborus torquatus*

Humid montane forests of Costa Rica and w Panama

☐ Spectacled Redstart *Myioborus melanocephalus*
- ____ M. m. ruficoronatus — Andes of sw Colombia (Nariño) to s Ecuador (Loja)
- ____ M. m. griseonuchus — Western Andes of nw Peru (Piura and Cajamarca)
- ____ M. m. malaris — Central Andes of n Peru (north of Chachapoyas in Amazonas)
- ____ M. m. melanocephalus — Eastern Andes of Peru (Amazonas to Ayacucho)
- ____ M. m. bolivianus — Andes of s Peru (Cuzco) to w Bolivia (La Paz and Cochabamba)

☐ Golden-fronted Redstart *Myioborus ornatus*
- ____ M. o. chrysops — W and Cent. Andes of Colombia (Antioquia to Cauca and Huila)
- ____ M. o. ornatus — E Andes of Colombia and adjacent sw Venezuela (Táchira)

☐ White-fronted Redstart *Myioborus albifrons*

Andes of w Venezuela (Trujillo, Táchira and Mérida)

☐ Yellow-crowned Redstart *Myioborus flavivertex*

Santa Marta Mountains (ne Colombia)

☐ Fan-tailed Warbler *Euthlypis lachrymosa*

Tropical w Mexico (se Sonora) to n-central Nicaragua

☐ Gray-and-gold Warbler *Basileuterus fraseri*
- ____ B. f. ochraceicrista — Arid tropical w Ecuador (Manabí to El Oro) and Isla Puná
- ____ B. f. fraseri — Central Ecuador (Chimborazo) to nw Peru (Lambayeque)

☐ Two-banded Warbler *Basileuterus bivittatus*
- ____ B. b. roraimae — *Tepuis* of se Venezuela to Guyana and extreme n Brazil
- ____ B. b. bivittatus — E Andes of se Peru (Cuzco) to w Bolivia
- ____ B. b. argentinae — SE Bolivia (sw Santa Cruz) to nw Argentina (Jujuy and Salta)

☐ Golden-bellied Warbler *Basileuterus chrysogaster*

E slope of Andes of Peru (Huánuco and Junín to Puno)

☐ **Choco Warbler** *Basileuterus chlorophrys*

Tropical sw Colombia (Cauca and Nariño) to nw Ecuador

☐ **Pale-legged Warbler** *Basileuterus signatus*
- ____ *B. s. signatus* — Andes of central Peru (Junín to Cuzco)
- ____ *B. s. flavovirens* — Andes of se Peru (s Cuzco and Puno) to w Bolivia and nw Argentina

☐ **Citrine Warbler** *Basileuterus luteoviridis*
- ____ *B. l. luteoviridis* — E Andes of Colombia to sw Venezuela and e Ecuador
- ____ *B. l. quindianus* — Central Andes of Colombia (Caldas and Tolima)
- ____ *B. l. richardsoni* — Western Andes of Colombia (Cauca)
- ____ *B. l. striaticeps* — Andes of n Peru (Amazonas to Cuzco)
- ____ *B. l. euophrys* — Andes of sw Peru (Puno) to w Bolivia (La Paz and Cochabamba)

☐ **Black-crested Warbler** *Basileuterus nigrocristatus*

Andes of Colombia to n Venezuela, Ecuador and n Peru

☐ **Gray-headed Warbler** *Basileuterus griseiceps*

Cloud forests of coastal mountains of ne Venezuela

☐ **Santa Marta Warbler** *Basileuterus basilicus*

Santa Marta Mountains (ne Colombia)

☐ **Gray-throated Warbler** *Basileuterus cinereicollis*
- ____ *B. c. pallidulus* — Andes of ne Colombia and w Venezuela
- ____ *B. c. zuliensis* — Sierra de Perijá (Colombia/Venezuela border)
- ____ *B. c. cinereicollis* — Eastern Andes of Colombia (Santander del Norte to w Meta)

☐ **White-lored Warbler** *Basileuterus conspicillatus*

Santa Marta Mountains (ne Colombia)

☐ **Russet-crowned Warbler** *Basileuterus coronatus*
- ____ *B. c. regulus* — Andes of Colombia and w Venezuela
- ____ *B. c. elatus* — Andes of sw Colombia (Nariño) and w Ecuador
- ____ *B. c. orientalis* — E slope of Andes of Ecuador (Pichincha to Chimborazo)
- ____ *B. c. castaneiceps* — Andes of sw Ecuador to nw Peru (Tumbes and Piura)
- ____ *B. c. chapmani* — E slope of Western Andes of nw Peru (Cajamarca)
- ____ *B. c. inaequalis* — Central Andes of n Peru (Amazonas and San Martín)
- ____ *B. c. coronatus* — E Andes of se Peru (Junín) to w Bolivia
- ____ *B. c. notius* — E slope of Andes of Bolivia (La Paz and Cochabamba)

☐ **Golden-crowned Warbler** *Basileuterus culicivorus*
- ____ *B. c. flavescens* — W Mexico (Nayarit and w Jalisco)
- ____ *B. c. brasherii* — E Mexico (Nuevo León and Tamaulipas to Hidalgo and n Veracruz)
- ____ *B. c. culicivorus* — Tropical s Mexico (Puebla) to nw Costa Rica
- ____ *B. c. godmani* — S Costa Rica and w Panama (e to Veraguas)
- ____ *B. c. occultus* — Tropical and lower subtropical Andes of Colombia
- ____ *B. c. olivascens* — E slope of E Andes of Colombia, n Venezuela and Trinidad
- ____ *B. c. austerus* — E slope of E Andes of Colombia (Boyacá, Cundinamarca and Meta)
- ____ *B. c. indignus* — Santa Marta Mountains (ne Colombia)
- ____ *B. c. cabanisi* — Extreme ne Colombia (Santander del Norte) and nw Venezuela
- ____ *B. c. segrex* — SE Venezuela to w Guyana and adjacent nw Brazil
- ____ *B. c. auricapillus* — Tropical central Brazil
- ____ *B. c. azarae* — Paraguay to s Brazil, Uruguay and ne Argentina
- ____ *B. c. viridescens* — Tropical e Bolivia (Santa Cruz)

☐ **Three-banded Warbler** *Basileuterus trifasciatus*
- ____ *B. t. nitidior* — Andes of sw Ecuador (El Oro and Loja) to nw Peru (Tumbes)
- ____ *B. t. trifasciatus* — Andes of nw Peru (Piura, Cajamarca, Lambayeque, La Libertad)

☐ **White-bellied Warbler** *Basileuterus hypoleucus*

Lowlands of e Bolivia to ne Paraguay and s-central Brazil

☐ **Rufous-capped Warbler** *Basileuterus rufifrons*
- ____ *B. r. caudatus* — Sierra Madre Occidental of w Mexico (Sonora to Durango)
- ____ *B. r. dugesi* — Mts. of w and central Mexico (s Sinaloa to Guerrero and Oaxaca)
- ____ *B. r. jouyi* — Sierra Madre Oriental of e Mexico (Nuevo León to Veracruz)

___ *B. r. rufifrons*   Mts. of s Mexico (Puebla to Oaxaca and Chiapas) to n Guatemala

___ *B. r. salvini*   S Mexico (s Veracruz, Tabasco, Chiapas) to Belize and n Guatemala

___ *B. r. delattrii*   S Mexico (se Chiapas) to highlands of n Costa Rica

___ *B. r. mesochrysus*   SW Costa Rica to Panama, n Colombia and w Venezuela

___ *B. r. actuosus*   Coiba I. (off Pacific coast of Panama)

☐ **Golden-browed Warbler** *Basileuterus belli*

___ *B. b. bateli*   Highlands of w Mexico (se Sinaloa to Jalisco and Michoacán)

___ *B. b. belli*   Highlands of e Mexico (sw Tamaulipas to n Oaxaca)

___ *B. b. clarus*   Highlands of sw Mexico (s Morelos, Guerrero and w Oaxaca)

___ *B. b. scitulus*   Highlands of se Mexico (e Oaxaca and Chiapas) to nw El Salvador

___ *B. b. suboscurus*   High mountains of nw Honduras

☐ **Black-cheeked Warbler** *Basileuterus melanogenys*

___ *B. m. melanogenys*   Subtropical highlands of Costa Rica

___ *B. m. eximus*   Highlands of w Panama (Chiriquí)

___ *B. m. bensoni*   Highlands of w Panama (Veraguas)

☐ **Pirre Warbler** *Basileuterus ignotus*

Humid montane forests of e Panama and extreme nw Colombia

☐ **Three-striped Warbler** *Basileuterus tristriatus*

___ *B. t. chitrensis*   Mountains of Costa Rica and w Panama

___ *B. t. tacarcunae*   Highlands of e Panama (Darién) and nw Colombia (Chocó)

___ *B. t. daedalus*   Subtrop. Western and Central Andes of Colombia and w Ecuador

___ *B. t. auricularis*   Subtrop. E and Central Andes of Colombia and sw Venezuela

___ *B. t. meridanus*   Subtropical Andes of w Venezuela (Lara to Táchira)

___ *B. t. bessereri*   Subtropical mountains of n Venezuela (Yaracuy to Miranda)

___ *B. t. pariae*   NE Venezuela (subtropical mountains of Paría Peninsula)

___ *B. t. baezae*   Andes of e Ecuador (Pichincha to Chimborazo)

___ *B. t. tristriatus*   Andes of se Ecuador (Loja) to central Peru (Cuzco)

___ *B. t. inconspicuus*   Andes of se Peru (Puno) to nw Bolivia (La Paz)

___ *B. t. punctipectus*   Andes of central Bolivia (Cochabamba)

___ *B. t. canens*   E Bolivia (Santa Cruz)

☐ **White-rimmed Warbler** *Basileuterus leucoblepharus*

E Paraguay to se Brazil, Uruguay and ne Argentina

☐ **White-striped Warbler** *Basileuterus leucophrys*

Gallery forests of s-central Brazil

☐ **Flavescent Warbler** *Basileuterus flaveolus*

___ *B. f. pallidirostris*   Tropical ne Colombia to n Venezuela and s Guyana

___ *B. f. flaveolus*   E Bolivia to n Paraguay, interior s Brazil and n Argentina

☐ **Buff-rumped Warbler** *Basileuterus fulvicauda*

___ *B. f. leucopygia*   Tropical n-central Honduras to w Panama (Veraguas)

___ *B. f. veraguensis*   Tropical sw Costa Rica to central Panama (Canal Zone)

___ *B. f. semicervina*   Tropical e Panama (Darién) to nw Peru (Tumbes and Piura)

___ *B. f. motacilla*   N Colombia (upper Magdalena Valley in Tolima and Huila)

___ *B. f. fulvicauda*   Trop. e Colombia to Ecuador, Peru (Junín) and extreme w Brazil

___ *B. f. significans*   Trop. se Peru (Inambari and Tambopata drainages) to nw Bolivia

☐ **Neotropical River Warbler** *Basileuterus rivularis*

___ *B. r. mesoleuca*   E Venezuela to the Guianas and n Brazil

___ *B. r. rivularis*   E Paraguay to se Brazil and ne Argentina

___ *B. r. bolivianus*   Foothills of Bolivia (La Paz, Cochabamba, Santa Cruz and Tarija)

☐ **Wrenthrush** *Zeledonia coronata*

Humid montane forests of Costa Rica and w Panama

☐ **Yellow-breasted Chat** *Icteria virens*

___ *I. v. auricollis (tropicalis)*   S British Columbia to Baja and n Mexico; winters to Guatemala

___ *I. v. virens*   N-central and e US to Florida; winters to Panama

☐ **Red-breasted Chat** *Granatellus venustus*
> ____ *G. v. venustus (melanotis)* — Coastal w Mexico (n Sinaloa to Chiapas)
> ____ *G. v. francescae* — Isla María Madre (Tres Marías Islands off w Mexico)

☐ **Gray-throated Chat** *Granatellus sallaei*
> ____ *G. s. sallaei* — Coastal se Mexico (Veracruz, Tabasco, e Oaxaca and n Chiapas)
> ____ *G. s. boucardi* — SE Mexico (Yucatán Peninsula) to Belize and e Guatemala

☐ **Rose-breasted Chat** *Granatellus pelzelni*
> ____ *G. p. pelzelni* — Trop. se Venezuela to Guyana, Suriname, nw Brazil and n Bolivia
> ____ *G. p. paraensis* — Tropical n Brazil (e Pará)

☐ **White-winged Warbler** *Xenoligea montana*
> Montane forests of Hispaniola

FAMILY: COEREBIDAE (Bananaquit—1)

☐ **Bananaquit** *Coereba flaveola*
> ____ *C. f. mexicana* — SE Mexico (Veracruz) to w Panama (Veraguas) and Coiba I.
> ____ *C. f. caboti* — E Mexico (Cozumel I., Holbox I., Cancún and Cayo Culebra)
> ____ *C. f. tricolor* — Isla Providéncia (w Caribbean Sea)
> ____ *C. f. oblita* — Isla San Andrés (w Caribbean Sea)
> ____ *C. f. sharpei* — Grand Cayman, Little Cayman and Cayman Brac
> ____ *C. f. bahamensis* — Bahamas (Great Bahama and Little Abaco to Grand Turk)
> ____ *C. f. flaveola* — Jamaica
> ____ *C. f. bananivora* — Hispaniola, Gonâve, Petite Cayemite and Île-à-Vache
> ____ *C. f. nectarea* — Tortue I. (off n Haiti)
> ____ *C. f. portoricensis* — Puerto Rico
> ____ *C. f. sanctithomae* — Virgin Islands (including Vieques and Culebra)
> ____ *C. f. newtoni* — St. Croix (Virgin Islands)
> ____ *C. f. bartholemica* — N Lesser Antilles
> ____ *C. f. martinicana* — Martinique and St. Lucia (Lesser Antilles)
> ____ *C. f. barbadensis* — Barbados (Lesser Antilles)
> ____ *C. f. atrata* — St. Vincent (Lesser Antilles)
> ____ *C. f. aterrima* — Grenada and the Grenadines (Lesser Antilles)
> ____ *C. f. cerinoclunis* — Pearl Islands (Bay of Panama)
> ____ *C. f. columbiana* — Trop. e Panama to sw Colombia and s Venezuela (Amazonas)
> ____ *C. f. uropygialis* — Netherlands Antilles (Aruba and Curaçao)
> ____ *C. f. bonariensis* — Bonaire I. (Netherlands Antilles)
> ____ *C. f. melanornis* — N Venezuela (Cayo Sal off Chiririviche)
> ____ *C. f. lowii* — Islas Los Roques (off n Venezuela)
> ____ *C. f. ferryi* — Isla La Tortuga (off n Venezuela)
> ____ *C. f. frailensis* — Isla de Puerto Real and Morro El Fondeadero (off Venezuela)
> ____ *C. f. laurae* — Isla Testigo Grande and Isla Conejo (off n Venezuela)
> ____ *C. f. luteola* — Tropical n Colombia to n Venezuela; Trinidad and Tobago
> ____ *C. f. obscura* — Tropical ne Colombia and w Venezuela
> ____ *C. f. minima* — Trop. e Colombia to s Venezuela, the Guianas and n Brazil
> ____ *C. f. montana* — Highlands of w Venezuela (Mérida and Táchira)
> ____ *C. f. caucae* — W Colombia (upper Cauca Valley)
> ____ *C. f. gorgonae* — Gorgona I. (off w Colombia)
> ____ *C. f. intermedia* — SW Colombia to n Peru, sw Venezuela and w Brazil
> ____ *C. f. bolivari* — E Venezuela (lower Orinoco River Valley)
> ____ *C. f. guianensis* — Tropical e Venezuela (Bolívar) and adjacent Guyana
> ____ *C. f. roraimae* — *Tepuis* of se Venezuela, nw Brazil and adjacent sw Guyana
> ____ *C. f. pacifica* — Arid nw Peru (Lambayeque, w La Libertad and Ancash)
> ____ *C. f. magnirostris* — N Peru (upper Marañón Valley)
> ____ *C. f. dispar* — Central Peru (San Martín) to nw Bolivia (La Paz)
> ____ *C. f. chloropyga* — Trop. s Peru to Bolivia, Paraguay, w Brazil and ne Argentina
> ____ *C. f. alleni* — Plateau of central Brazil (Mato Grosso) and e Bolivia

FAMILY: THRAUPIDAE (Tanagers and Allies—226)

☐ **Chestnut-vented Conebill** *Conirostrum speciosum*

_____	*C. s. amazonum*	E Colombia to sw Venezuela, the Guianas, n Brazil and e Peru
_____	*C. s. guaricola*	Tropical cent. Venezuela (e Guárico and w Anzoátegui)
_____	*C. s. speciosum*	SE Peru (Puno) to e Brazil, Bolivia, Paraguay and n Argentina

☐ **White-eared Conebill** *Conirostrum leucogenys*

_____	*C. l. panamense*	Tropical e Panama to nw Colombia
_____	*C. l. leucogenys*	Tropical n Colombia and ne Venezuela
_____	*C. l. cyanochroum*	Andes of w Venezuela (Mérida) and Sierra de Perijá

☐ **Bicolored Conebill** *Conirostrum bicolor*

_____	*C. b. bicolor*	Coastal n Colombia to the Guianas and n Brazil; Trinidad
_____	*C. b. minus*	Trop. e Ecuador to extreme ne Peru (Loreto) and w Brazil

☐ **Pearly-breasted Conebill** *Conirostrum margaritae*

Tropical ne Peru (Loreto), w Amazonian Brazil and adj. Bolivia

☐ **Cinereous Conebill** *Conirostrum cinereum*

_____	*C. c. fraseri*	Central and E Andes of sw Colombia and Andes of Ecuador
_____	*C. c. littorale*	West slope of Western Andes of Peru to n Chile
_____	*C. c. cinereum*	Andes of se Peru and w Bolivia

☐ **Tamarugo Conebill** *Conirostrum tamarugense*

SW Peru (Arequipa and Tacna) to n Chile (Tarapacá)

☐ **White-browed Conebill** *Conirostrum ferrugineiventre*

Andes of e Peru (San Martín) and w Bolivia

☐ **Rufous-browed Conebill** *Conirostrum rufum*

Mts. of ne Colombia and extreme sw Venezuela (Táchira)

☐ **Blue-backed Conebill** *Conirostrum sitticolor*

_____	*C. s. intermedium*	Andes of w Venezuela (Mérida and Táchira)
_____	*C. s. sitticolor*	Andes of s Colombia, Ecuador and nw Peru
_____	*C. s. cyaneum*	Andes of Peru to w Bolivia (La Paz and Cochabamba)

☐ **Capped Conebill** *Conirostrum albifrons*

_____	*C. a. albifrons*	Central and E Andes of Colombia and w Venezuela (Táchira)
_____	*C. a. centralandium*	Central Andes of Colombia (Antioquia to Cauca)
_____	*C. a. cyanonotum*	Coastal mts. of n Venezuela (Aragua and Distrito Federal)
_____	*C. a. atrocyaneum*	Andes of sw Colombia, Ecuador and n Peru
_____	*C. a. sordidum*	Andes of s Peru (Junín) to w Bolivia (La Paz)
_____	*C. a. lugens*	*Yungas* of e Bolivia (Cochabamba and Santa Cruz)

☐ **Giant Conebill** *Oreomanes fraseri*

Polylepis woodlands of s Colombia to w Bolivia

☐ **Brown Tanager** *Orchesticus abeillei*

Foothills of se Brazil (s Bahia and Minas Gerais to Paraná)

☐ **Cinnamon Tanager** *Schistochlamys ruficapillus*

_____	*S. r. capistrata*	*Caatinga* of ne Brazil (s Pará to Maranhão and Bahia)
_____	*S. r. sicki*	SW Brazil (Rio das Mortes region of e Mato Grosso)
_____	*S. r. ruficapillus*	SE Brazil (s Minas Gerais to Paraná) and ne Argentina

☐ **Black-faced Tanager** *Schistochlamys melanopis*

_____	*S. m. aterrima*	NE Colombia to Venezuela, w Guyana and extreme n Brazil
_____	*S. m. melanopis*	E Guyana to Suriname, French Guiana and ne Brazil
_____	*S. m. grisea*	Subtropical Central Andes of Peru
_____	*S. m. olivina*	E Bolivia to Paraguay and s-central Brazil (Mato Grosso)
_____	*S. m. amazonica*	Amazonian and se Brazil

☐ **White-banded Tanager** *Neothraupis fasciata*

E Bolivia to ne Paraguay and campos of e and s Brazil

☐ **White-rumped Tanager** *Cypsnagra hirundinacea*
   ____ *C. h. pallidigula*
   ____ *C. h. hirundinacea*

Campos of Suriname and central Brazil to ne Bolivia
E Bolivia to ne Paraguay and campos of s Brazil

☐ **Black-and-white Tanager** *Conothraupis speculigera*

Locally from w Ecuador to nw Peru; e Peru to nw Bolivia

☐ **Cone-billed Tanager** *Conothraupis mesoleuca*

SW Brazil (known from a 1939 specimen from w Mato Grosso)

☐ **Magpie Tanager** *Cissopis leverianus*
   ____ *C. l. leverianus*
   ____ *C. l. major*

E Colombia to se Venezuela, Guianas, n Bolivia, Amaz. Brazil
Paraguay to se Brazil and adjacent ne Argentina (Misiones)

☐ **Red-billed Pied Tanager** *Lamprospiza melanoleuca*

Amazonian Brazil, adjacent n Bolivia, e Peru and Guianas

☐ **Grass-green Tanager** *Chlorornis riefferii*
   ____ *C. r. riefferii*
   ____ *C. r. dilutus*
   ____ *C. r. elegans*
   ____ *C. r. celatus*
   ____ *C. r. bolivianus*

Andes of Colombia and Ecuador
Central Andes of n Peru
Andes of central Peru (Junín)
Andes of extreme se Peru (Puno)
Andes of w Bolivia (La Paz and Cochabamba)

☐ **Scarlet-throated Tanager** *Compsothraupis loricata*

Arid interior of ne Brazil

☐ **White-capped Tanager** *Sericossypha albocristata*

Locally in Andes of Colombia and w Venezuela to se Peru

☐ **Puerto Rican Tanager** *Nesospingus speculiferus*

Highlands of Puerto Rico

☐ **Common Bush-Tanager** *Chlorospingus ophthalmicus*
   ____ *C. o. albifrons*
   ____ *C. o. wetmorei*
   ____ *C. o. persimilis*
   ____ *C. o. ophthalmicus*
   ____ *C. o. dwighti*
   ____ *C. o. postocularis*
   ____ *C. o. honduratius*
   ____ *C. o. regionalis*
   ____ *C. o. novicius*
   ____ *C. o. punctulatus*
   ____ *C. o. jaqueti*
   ____ *C. o. eminens*
   ____ *C. o. trudis*
   ____ *C. o. exitelus*
   ____ *C. o. flavopectus*
   ____ *C. o. macarenae*
   ____ *C. o. nigriceps*
   ____ *C. o. falconensis*
   ____ *C. o. venezuelanus*
   ____ *C. o. ponsi*
   ____ *C. o. phaeocephalus*
   ____ *C. o. cinereocephalus*
   ____ *C. o. hiaticolus*
   ____ *C. o. peruvianus*
   ____ *C. o. bolivianus*
   ____ *C. o. fulvigularis*
   ____ *C. o. argentinus*

Sierra Madre del Sur of sw Mexico (Guerrero and Oaxaca)
SE Mexico (Sierra de Tuxtla in Veracruz)
S Mexico (s Oaxaca)
SE Mexico (n Vera Cruz and s San Luis Potosí to w Chiapas)
Caribbean slope of s Mexico (Chiapas) and e Guatemala
Pacific slope of s Mexico (Chiapas) and w Guatemala
Subtropical El Salvador and Honduras
Subtropical Nicaragua and e Costa Rica
Subtropical sw Costa Rica and w Panama (Chiriquí)
Highlands of w Panama (Veraguas and Coclé)
W slope of Eastern Andes of ne Colombia and n Venezuela
E Andes of ne Colombia (Norte de Santander and Boyacá)
Colombia (w slope of Andes of Santander at La Pica)
Colombia (e slope of W Andes and C Andes in Antioquia)
E Andes of Colombia (s Santander and w Cundinamarca)
E Colombia (Macarena Mountains)
Subtropical and lower temperate Andes of Colombia
NW Venezuela (San Luis Mountains and Sierra de Aroa)
Andes of sw Venezuela (Lara, Mérida and Táchira)
Sierra de Perijá (Colombia/Venezuela border)
Subtropical mountains of e and w Ecuador
Subtropical central Peru (Junín)
Central Peru
Subtropical s Peru (Puno)
W-central Bolivia (Cochabamba and La Paz)
Central Bolivia (s Cordillera de Cochabamba)
Cent. Bolivia (upper Río Mizque) to n Argentina (Tucumán)

☐ **Tacarcuna Bush-Tanager** *Chlorospingus tacarcunae*

Montane forests of e Panama and extreme nw Colombia

☐ **Pirre Bush-Tanager** *Chlorospingus inornatus*

Humid montane forests of e Panama

☐ **Dusky Bush-Tanager** *Chlorospingus semifuscus*
    ____ *C. s. livingstoni* — Pacific slope of Western Andes of Colombia
    ____ *C. s. semifuscus* — Pacific slope of sw Colombia (Nariño) and w Ecuador

☐ **Sooty-capped Bush-Tanager** *Chlorospingus pileatus*
    ____ *C. p. pileatus* — Mountains of Costa Rica and w Panama (Volcán de Chiriquí)
    ____ *C. p. diversus* — Mountains of w Panama (e Chiriquí)

☐ **Short-billed Bush-Tanager** *Chlorospingus parvirostris*
    ____ *C. p. huallagae* — E Andes of s Colombia and n Peru
    ____ *C. p. medianus* — Andes of e-central Peru (Junín and Cuzco)
    ____ *C. p. parvirostris* — Andes of extreme se Peru (Puno) and nw Bolivia (La Paz)

☐ **Yellow-throated Bush-Tanager** *Chlorospingus flavigularis*
    ____ *C. f. hypophaeus* — Montane forests of w Panama (Volcán Chiriquí to Veraguas)
    ____ *C. f. marginatus* — W slope of Western Andes of sw Colombia and w Ecuador
    ____ *C. f. flavigularis* — Andes of central Colombia to e Ecuador and e Peru (Cuzco)

☐ **Yellow-green Bush-Tanager** *Chlorospingus flavovirens*
    Locally in humid Western Andes of Colombia and nw Ecuador

☐ **Ashy-throated Bush-Tanager** *Chlorospingus canigularis*
    ____ *C. c. olivaceiceps* — Subtropical Caribbean slope of w Costa Rica
    ____ *C. c. canigularis* — E Andes of Colombia (Cundinamarca) and sw Venezuela
    ____ *C. c. conspicillatus* — Western and Central Andes of Colombia
    ____ *C. c. paulus* — Subtropical Andes of sw Ecuador
    ____ *C. c. signatus* — Subtropical Andes of e Ecuador and Peru (south to Cuzco)

☐ **Gray-hooded Bush-Tanager** *Cnemoscopus rubrirostris*
    ____ *C. r. rubrirostris* — Andes of Colombia to sw Venezuela (Táchira) and e Ecuador
    ____ *C. r. chrysogaster* — Andes of Peru (Amazonas to Cordillera Vilcabamba in Cuzco)

☐ **Three-striped Hemispingus** *Hemispingus trifasciatus*
    Andes of se Peru to n Bolivia (La Paz and Cochabamba)

☐ **Black-capped Hemispingus** *Hemispingus atropileus*
    Andes of Colombia to sw Venezuela (Táchira) and Ecuador

☐ **White-browed Hemispingus** *Hemispingus auricularis*
    Andes of e Peru (Amazonas to Cuzco)

☐ **Orange-browed Hemispingus** *Hemispingus calophrys*
    Andes of se Peru (Puno) and w Bolivia (La Paz, Cochabamba)

☐ **Parodi's Hemispingus** *Hemispingus parodii*
    Andes of se Peru (Cuzco)

☐ **Superciliaried Hemispingus** *Hemispingus superciliaris*
    ____ *H. s. superciliaris* — E Andes of central Colombia (Cundinamarca)
    ____ *H. s. nigrifrons* — Central Andes of Colombia and Ecuador
    ____ *H. s. chrysophrys* — Andes of sw Venezuela (Trujillo, Mérida and Táchira)
    ____ *H. s. maculifrons* — Andes of extreme sw Ecuador and nw Peru
    ____ *H. s. insignis* — Highlands of n Peru (Utcubamba Valley east of Río Marañón)
    ____ *H. s. leucogastrus* — Temperate Andes of central Peru (Junín)
    ____ *H. s. urubambae* — Temperate Andes of s Peru (Cuzco) to w Bolivia (La Paz)

☐ **Black-headed Hemispingus** *Hemispingus verticalis*
    Andes of Colombia to sw Venezuela and extreme n Peru

☐ **Drab Hemispingus** *Hemispingus xanthophthalmus*
    Humid Andes of central Peru to nw Bolivia (Puno and La Paz)

☐ **Rufous-browed Hemispingus** *Hemispingus rufosuperciliaris*
    Andes of e Peru (Amazonas to La Libertad and Huánuco)

☐ **Slaty-backed Hemispingus** *Hemispingus goeringi*
    Andes of w Venezuela (Mérida and n Táchira)

☐ **Piura Hemispingus** *Hemispingus piurae*
    Locally in Andes of extreme sw Ecuador and nw Peru (Piura)

☐ **Oleaginous Hemispingus** *Hemispingus frontalis*

    ____   *H. f. frontalis*   Subtropical Andes of e Colombia, e Ecuador and e Peru

    ____   *H. f. ignobilis*   Andes of w Venezuela (s Lara, Trujillo, Mérida and Táchira)

    ____   *H. f. flavidorsalis*   Sierra de Perijá (Colombia/Venezuela border)

    ____   *H. f. hanieli*   Coastal mountains of n Venezuela (Aragua to Miranda)

    ____   *H. f. iteratus*   Coastal mountains of ne Venezuela (Monagas to Sucre)

☐ **Black-eared Hemispingus** *Hemispingus melanotis*

    ____   *H. m. melanotis*   Andes of Colombia to sw Venezuela and e Ecuador

    ____   *H. m. ochraceus*   W slope of Andes of sw Colombia (Nariño) and w Ecuador

    ____   *H. m. macrophrys*   Pacific slope of Andes of w Peru (La Libertad)

    ____   *H. m. berlepschi*   Subtropical Andes of central Peru (Junín)

    ____   *H. m. castaneicollis*   Andes of se Peru (Puno) and *yungas* of w Bolivia

☐ **Gray-capped Hemispingus** *Hemispingus reyi*

     Andes of sw Venezuela (Trujillo, Mérida and Táchira)

☐ **Chestnut-headed Tanager** *Pyrrhocoma ruficeps*

     E Paraguay to se Brazil and ne Argentina (Misiones)

☐ **Fulvous-headed Tanager** *Thlypopsis fulviceps*

    ____   *T. f. fulviceps*   E slope of Andes of ne Colombia and mountains of n Venezuela

    ____   *T. f. intensa*   E Andes of ne Colombia (s Magdalena)

    ____   *T. f. obscuriceps*   Sierra de Perijá (Colombia/Venezuela border)

    ____   *T. f. meridensis*   Andes of w Venezuela (Mérida)

☐ **Rufous-chested Tanager** *Thlypopsis ornata*

    ____   *T. o. ornata*   Andes of sw Colombia (Puracé) and w Ecuador

    ____   *T. o. media*   Andes of s Ecuador (Loja) to central Peru (Lima)

    ____   *T. o. macropteryx*   E slope of Andes of central and s Peru (Junín and Cuzco)

☐ **Brown-flanked Tanager** *Thlypopsis pectoralis*

     Andes of central Peru (Huánuco, Pasco and Junín)

☐ **Orange-headed Tanager** *Thlypopsis sordida*

    ____   *T. s. chrysopis*   Extreme s Colombia to e Ecuador, e Peru and w Brazil

    ____   *T. s. orinocensis*   Tropical e-central Venezuela (s Anzoátegui and n Bolívar)

    ____   *T. s. sordida*   E and s Brazil to e Bolivia, Paraguay and n Argentina

☐ **Buff-bellied Tanager** *Thlypopsis inornata*

     Andes of nw Peru; recently recorded in adjacent s Ecuador

☐ **Rust-and-yellow Tanager** *Thlypopsis ruficeps*

     Andes of se Peru to Bolivia and nw Argentina

☐ **Guira Tanager** *Hemithraupis guira*

    ____   *H. g. guirina*   W and central Colombia to e Ecuador and extreme nw Peru

    ____   *H. g. nigrigula*   N-central Colombia to n Venezuela, the Guianas and ne Brazil

    ____   *H. g. huambina*   Tropical se Colombia to e Ecuador, ne Peru and w Brazil

    ____   *H. g. roraimae*   *Tepuis* of se Venezuela and Guyana

    ____   *H. g. boliviana*   NE Bolivia to nw Argentina and adjacent w Brazil

    ____   *H. g. amazonica*   Brazil south of the Amazon (Rio Madeira to Rio Tapajós)

    ____   *H. g. guira*   E Brazil (Rio Tocantins to Ceará, Goiás and nw Bahia)

    ____   *H. g. forsteri*   E Paraguay to ne Argentina and interior se Brazil

☐ **Rufous-headed Tanager** *Hemithraupis ruficapilla*

    ____   *H. r. ruficapilla*   SE Brazil (s Minas Gerais and Espírito Santo to Santa Catarina)

    ____   *H. r. bahiae*   E Brazil (se Bahia)

☐ **Yellow-backed Tanager** *Hemithraupis flavicollis*

    ____   *H. f. ornata*   Tropical e Panama (Darién) and extreme nw Colombia

    ____   *H. f. albigularis*   Colombia (upper Sinú, lower Cauca, middle Magdalena valleys)

    ____   *H. f. peruana*   S-cent. Colombia to e Ecuador and ne Peru (n of Río Marañón)

    ____   *H. f. aurigularis*   Extreme se Colombia to s Venezuela and n Brazil

    ____   *H. f. hellmayri*   SE Venezuela (e Bolívar) to w Guyana (Merumé Mts.)

651

____ H. f. flavicollis	Suriname, French Guiana and adjacent Brazil n of the Amazon
____ H. f. sororia	N Peru (south of Río Marañón)
____ H. f. centralis	SE Peru to n Bolivia and central Brazil
____ H. f. obidensis	N Brazil (along north bank of the lower Amazon in Pará)
____ H. f. melanoxantha	E Brazil (Pernambuco and Bahia)
____ H. f. insignis	SE Brazil (Espírito Santo and Rio de Janeiro)

☐ **Black-and-yellow Tanager** *Chrysothlypis chrysomelas*

____ C. c. chrysomelas	Caribbean slope of e Costa Rica and w Panama
____ C. c. ocularis	Tropical e Panama (Darién)

☐ **Scarlet-and-white Tanager** *Chrysothlypis salmoni*

Pacific slope of Andes of Colombia and nw Ecuador

☐ **Hooded Tanager** *Nemosia pileata*

____ N. p. hypoleuca	Caribbean coast of n Colombia and n Venezuela
____ N. p. surinamensis	Guyana and Suriname
____ N. p. pileata	French Guiana, ne and Amazonian Brazil, extreme n Bolivia
____ N. p. interna	N Brazil (upper Rio Branco and lower Rio Negro)
____ N. p. nana	Tropical Amazon basin of ne Peru and adjacent w Brazil
____ N. p. caerulea	E Bolivia to Paraguay, e and s Brazil and n Argentina

☐ **Cherry-throated Tanager** *Nemosia rourei*

SE Brazil (rediscovered after 47-year absence in Espírito Santo)

☐ **Black-crowned Palm-Tanager** *Phaenicophilus palmarum*

Lowlands of Hispaniola (except s pen. Haiti) and Saona I.

☐ **Gray-crowned Palm-Tanager** *Phaenicophilus poliocephalus*

____ P. p. poliocephalus	S peninsular Haiti, Île-à-Vache and Grande Cayemite I.
____ P. p. coryi	Gonâve I. (e of Haiti)

☐ **Western Chat-Tanager** *Calyptophilus tertius*

Massifs of s Haiti and extreme sw Dominican Republic

☐ **Eastern Chat-Tanager** *Calyptophilus frugivorus*

____ C. f. frugivorus	W Dominican Republic (Benefactor to Sananá)
____ C. f. neibei	Central Dominican Republic
____ C. f. abbotti	Gonâve I. (e of Haiti)

☐ **Rosy Thrush-Tanager** *Rhodinocichla rosea*

____ R. r. schistacea	Tropical Pacific coast of Mexico (Sinaloa to Michoacán)
____ R. r. eximia	SW Costa Rica (Térraba Valley) to w Panama
____ R. r. harterti	W slope of E Andes of central Colombia (Bogotá to e Tolima)
____ R. r. beebei	Sierra de Perijá (Colombia/Venezuela border)
____ R. r. rosea	NW Venezuela (Falcón to Distrito Federal and Miranda)

☐ **Dusky-faced Tanager** *Mitrospingus cassinii*

____ M. c. costaricensis	Caribbean lowlands of e Costa Rica and extreme w Panama
____ M. c. cassinii	Tropical e Panama to w Colombia and w Ecuador

☐ **Olive-backed Tanager** *Mitrospingus oleagineus*

____ M. o. obscuripectus	SE Venezuela (Gran Sabana) and extreme n Brazil (Uei-tepui)
____ M. o. oleagineus	*Tepuis* of se Venezuela (Mt. Roraima) and adjacent Guyana

☐ **Olive Tanager** *Chlorothraupis carmioli*

____ C. c. carmioli	Tropical e Nicaragua to nw Panama (Almirante Bay)
____ C. c. magnirostris	Tropical w Panama (Veraguas)
____ C. c. lutescens	Tropical e Panama (San Blas and Darién) to nw Colombia
____ C. c. frenata	Tropical s Colombia to nw Bolivia (La Paz and Cochabamba)

☐ **Lemon-spectacled Tanager** *Chlorothraupis olivacea*

Tropical e Panama (Darién) to w Colombia and nw Ecuador

☐ **Ochre-breasted Tanager** *Chlorothraupis stolzmanni*

    _____ *C. s. dugandi*   W slope of Western Andes of sw Colombia

    _____ *C. s. stolzmanni*   Tropical w Ecuador (south to El Oro)

☐ **Olive-green Tanager** *Orthogonys chloricterus*

    Montane forests of se Brazil (Espírito Santo to Santa Catarina)

☐ **Gray-headed Tanager** *Eucometis penicillata*

    _____ *E. p. pallida*   Tropical se Mexico (Veracruz and Yucatán) to e Guatemala

    _____ *E. p. spodocephala*   Tropical Nicaragua and Pacific slope of Costa Rica

    _____ *E. p. stictothorax*   Tropical sw Costa Rica and w Panama (east to Veraguas)

    _____ *E. p. cristata*   E Panama to n Colombia and extreme w Venezuela

    _____ *E. p. penicillata*   SE Colombia east of Andes to the Guianas, e Peru and n Brazil

    _____ *E. p. affinis*   Tropical n Venezuela (Falcón to Miranda)

    _____ *E. p. albicollis*   E Bolivia to n Paraguay, ne Argentina and s-central Brazil

☐ **Fulvous Shrike-Tanager** *Lanio fulvus*

    _____ *L. f. peruvianus*   S Colombia (east of the Andes) to e Ecuador and ne Peru

    _____ *L. f. fulvus*   S Venezuela to the Guianas and n Amazonian Brazil

☐ **White-winged Shrike-Tanager** *Lanio versicolor*

    _____ *L. v. versicolor*   E Peru (s of Río Marañón) to n Bolivia and w Brazil

    _____ *L. v. parvus*   S Amazonian Brazil and n Mato Grosso

☐ **Black-throated Shrike-Tanager** *Lanio aurantius*

    Gulf-Caribbean lowlands of se Mexico to nw Honduras

☐ **White-throated Shrike-Tanager** *Lanio leucothorax*

    _____ *L. l. leucothorax*   Extreme e Honduras to e Nicaragua and e Costa Rica

    _____ *L. l. reversus*   NW Costa Rica (Nicoya Pen., Puntarenas and Las Agujas)

    _____ *L. l. melanopygius*   SW Costa Rica and Pacific slope of w Panama

    _____ *L. l. ictus*   Extreme nw Panama (Almirante Bay environs)

☐ **Rufous-crested Tanager** *Creurgops verticalis*

    Andes of Colombia to sw Venezuela and e Peru (Ayacucho)

☐ **Slaty Tanager** *Creurgops dentatus*

    Andes of se Peru to w Bolivia (La Paz and Cochabamba)

☐ **Sulphur-rumped Tanager** *Heterospingus rubrifrons*

    Humid lowlands of e Costa Rica and Panama

☐ **Scarlet-browed Tanager** *Heterospingus xanthopygius*

    _____ *H. x. xanthopygius*   Extreme e Panama and n Colombia

    _____ *H. x. berliozi*   Pacific slope of w Colombia and nw Ecuador

☐ **Flame-crested Tanager** *Tachyphonus cristatus*

    _____ *T. c. fallax*   S Colombia (se Nariño) to e Ecuador and ne Peru

    _____ *T. c. cristatellus*   Tropical se Colombia to ne Peru, s Venezuela and nw Brazil

    _____ *T. c. huarandosae*   N Peru (Chinchipe Valley in Río Marañón Valley)

    _____ *T. c. intercedens*   E Venezuela (e Bolívar), Guyana and Suriname

    _____ *T. c. cristatus*   French Guiana and ne Brazil (north of the Amazon)

    _____ *T. c. pallidigula*   NE Brazil (e Pará and lower Rio Tocantins)

    _____ *T. c. madeirae*   Brazil s of the Amazon (Teffe to Rio Xingú and Mato Grosso)

    _____ *T. c. nattereri*   SW Brazil (Mato Grosso)

    _____ *T. c. brunneus*   E Brazil (Pernambuco to São Paulo)

☐ **Yellow-crested Tanager** *Tachyphonus rufiventer*

    E Peru to nw Bolivia and adjacent w Amazonian Brazil

☐ **Fulvous-crested Tanager** *Tachyphonus surinamus*

    _____ *T. s. surinamus*   E and s Venezuela, the Guianas and Brazil n of the Amazon

    _____ *T. s. brevipes*   E Colombia to s Venezuela, nw Brazil, e Ecuador and ne Peru

    _____ *T. s. napensis*   Tropical e Peru (south of the Amazon) and nw Brazil

    _____ *T. s. insignis*   N Brazil south of the Amazon (lower Rio Madeira to Pará)

☐ **White-shouldered Tanager** *Tachyphonus luctuosus*

____	*T. l. axillaris*	Caribbean slope of e Honduras to w Panama
____	*T. l. nitidissimus*	Pacific slope of sw Costa Rica and extreme w Panama
____	*T. l. panamensis*	E Panama to w Colombia, w Ecuador and w Venezuela
____	*T. l. luctuosus*	SE Colombia e of Andes to Bolivia, the Guianas and n Brazil
____	*T. l. flaviventris*	Extreme ne Venezuela (Sucre) and Trinidad

☐ **Tawny-crested Tanager** *Tachyphonus delatrii*

Tropical e Honduras to w Ecuador; Isla Gorgona (off Colombia)

☐ **Ruby-crowned Tanager** *Tachyphonus coronatus*

E Paraguay to se Brazil and ne Argentina

☐ **White-lined Tanager** *Tachyphonus rufus*

Lowlands of Costa Rica to n Argentina and Amazonian Brazil

☐ **Red-shouldered Tanager** *Tachyphonus phoenicius*

Guianas and s Venezuela to ne Peru and Amazonian Brazil

☐ **Black-goggled Tanager** *Trichothraupis melanops*

Andes of Peru and w Bolivia; e Brazil, ne Argentina, e Paraguay

☐ **Red-crowned Ant-Tanager** *Habia rubica*

____	*H. r. holobrunnea*	Subtrop. e Mexico (s Tamaulipas to Veracruz and n Oaxaca)
____	*H. r. rosea*	Pacific slope of sw Mexico (Nayarit and Jalisco to Guerrero)
____	*H. r. affinis*	Pacific slope of s Mexico (Oaxaca)
____	*H. r. nelsoni*	SE Mexico (Yucatán Peninsula north of s Campeche)
____	*H. r. rubicoides*	S Mexico (Puebla and e Veracruz) to n Nicaragua
____	*H. r. alfaroana*	NW Costa Rica (Guanacaste Peninsula)
____	*H. r. vinacea*	Pacific slope of sw Costa Rica (Nicoya Peninsula) to e Panama
____	*H. r. rubra*	Trinidad
____	*H. r. crissalis*	Coastal mountains of ne Venezuela (Anzoátegui to Sucre)
____	*H. r. mesoptamia*	Venezuela (Río Yuruán region of e Bolívar)
____	*H. r. perijana*	Sierra de Perijá (Colombia/Venezuela border)
____	*H. r. coccinea*	E base of E Andes of n-central Colombia and w Venezuela
____	*H. r. rhodinolaema*	SE Colombia e of the Andes to ne Peru and extreme nw Brazil
____	*H. r. peruviana*	Tropical e Peru to central Bolivia and adjacent w Brazil
____	*H. r. hesterna*	Central Brazil south of the Amazon to n Mato Grosso
____	*H. r. bahiae*	Tropical e Brazil (Bahia)
____	*H. r. rubica*	SE Brazil (s Minas Gerais) to e Paraguay and ne Argentina

☐ **Red-throated Ant-Tanager** *Habia fuscicauda*

____	*H. f. salvini*	SE Mexico (San Luis Potosí and Veracruz) to El Salvador
____	*H. f. insularis*	Yucatán Peninsula, Meco I., Isla Mujeres and n Guatemala
____	*H. f. discolor*	Tropical Nicaragua
____	*H. f. fuscicauda*	Extreme s Nicaragua to extreme w Panama
____	*H. f. willisi*	Central Panama (ne Coclé and Colón to w San Blas)
____	*H. f. erythrolaema*	Caribbean coast of n Colombia

☐ **Sooty Ant-Tanager** *Habia gutturalis*

Foothills of Western Andes of n Colombia

☐ **Black-cheeked Ant-Tanager** *Habia atrimaxillaris*

Pacific lowlands of sw Costa Rica (Golfo Dulce)

☐ **Crested Ant-Tanager** *Habia cristata*

W Andes of Colombia (Antioquia to Cauca)

☐ **Rose-throated Tanager** *Piranga roseogularis*

____	*P. r. roseogularis*	SE Mexico (arid n Yucatán Peninsula)
____	*P. r. tincta*	Central and s Yucatán Peninsula, n Belize and n Guatemala
____	*P. r. cozumelae*	SE Mexico (Cozumel I. and Isla Mujeres)

☐ **Hepatic Tanager** *Piranga flava*

____	*P. f. hepatica*	SW US and w Mexico (south to Guerrero and Oaxaca)
____	*P. f. dextra*	SW US (e of Rockies) and e Mexico; winters to w Guatemala
____	*P. f. figlina*	Lowland pine savanna of e Guatemala and Belize
____	*P. f. savannarum*	Lowland pine savanna of e Honduras and ne Nicaragua

____ *P. f. albifacies*	Montane oak-pine belt of w Guatemala to n Nicaragua
____ *P. f. testacea*	Subtropical Costa Rica and Panama (east to Cape Gararchiné)
____ *P. f. desidiosa*	Upper tropical and subtropical sw Colombia
____ *P. f. lutea*	Extreme sw Colombia to w Ecuador, Peru and nw Bolivia
____ *P. f. haemalea*	Mountains of s Venezuela, w Guyana and extreme n Brazil
____ *P. f. faceta*	Mountains of n Colombia and n Venezuela; Trinidad
____ *P. f. toddi*	W slope of Eastern Andes of Colombia (Magdalena)
____ *P. f. macconnelli*	S Guyana and adjacent extreme n Brazil
____ *P. f. saira*	E Brazil (Amazon to Mato Grosso and Rio Grande do Sul)
____ *P. f. rosacea*	E Bolivia (Santa Cruz to Chiquitos)
____ *P. f. flava*	S Bolivia (Cochabamba) to Uruguay and n Argentina

☐ **Scarlet Tanager** *Piranga olivacea*

E Canada and US; winters mainly upper Amazon basin

☐ **Summer Tanager** *Piranga rubra*

____ *P. r. cooperi (ochracea)*	SW US and n Mexico; winters to s Baja and s Mexico
____ *P. r. rubra*	SE US; winters to Amazonian Brazil and n Bolivia

☐ **Western Tanager** *Piranga ludoviciana*

W North America; winters to w Panama

☐ **Flame-colored Tanager** *Piranga bidentata*

____ *P. b. bidentata*	W Mexico (Sonora and Chihuahua to Guerrero and Morelos)
____ *P. b. flammea*	Tres Marías Islands (off w Mexico)
____ *P. b. sanguinolenta*	E Mexico (Nuevo León and Tamaulipas) to El Salvador
____ *P. b. citrea*	Highlands of Costa Rica and w Panama

☐ **White-winged Tanager** *Piranga leucoptera*

____ *P. l. leucoptera*	Sierra Madre Oriental of e Mexico (Tamaulipas) to Nicaragua
____ *P. l. latifasciata*	Costa Rica and w Panama (east to Veraguas)
____ *P. l. venezuelae*	Andes of Colombia to Venezuela and n Brazil
____ *P. l. ardens*	Andes of extreme sw Colombia (Nariño) to nw Bolivia

☐ **Red-headed Tanager** *Piranga erythrocephala*

____ *P. e. candida*	Sierra Madre Occidental of w Mexico (se Sonora to nw Jalisco)
____ *P. e. erythrocephala*	Mountains of s Mexico (Jalisco and Guanajuato to Oaxaca)

☐ **Red-hooded Tanager** *Piranga rubriceps*

Andes of Colombia to ne Peru (Huánuco)

☐ **Vermilion Tanager** *Calochaetes coccineus*

Andes of se Colombia to Ecuador and e Peru (Cuzco)

☐ **Crimson-collared Tanager** *Ramphocelus sanguinolentus*

____ *R. s. sanguinolentus*	Tropical se Mexico (Veracruz) to Guatemala and Honduras
____ *R. s. apricus*	Caribbean slope of e Honduras to nw Panama (Almirante Bay)

☐ **Masked Crimson Tanager** *Ramphocelus nigrogularis*

Humid se Colombia to n Bolivia and w Amazonian Brazil

☐ **Crimson-backed Tanager** *Ramphocelus dimidiatus*

____ *R. d. isthmicus*	Tropical w and central Panama (east to Río Chepo)
____ *R. d. arestus*	Coiba I. (Panama)
____ *R. d. limatus*	Pearl Islands (Bay of Panama)
____ *R. d. dimidiatus*	Extreme e Panama (Darién) to n Colombia and w Venezuela
____ *R. d. molochinus*	N Colombia (upper Magdalena Valley)

☐ **Huallaga Tanager** *Ramphocelus melanogaster*

____ *R. m. melanogaster*	Highlands of n Peru (San Martín)
____ *R. m. transitus*	E-c Peru (upper Huallaga Valley in Huánuco and San Martín)

☐ **Silver-beaked Tanager** *Ramphocelus carbo*

____ *R. c. unicolor*	E base of E Andes of Colombia (Cundinamarca and Meta)
____ *R. c. magnirostris*	Trinidad; single specimen from ne Venezuela (Sucre)

_____	_R. c. carbo_	SE Colombia to the Guianas, e Peru and n Brazil
_____	_R. c. venezuelensis_	E Colombia and w Venezuela
_____	_R. c. capitalis_	NE Venezuela (ne Anzoátegui to se Monagas, Delta Amacuro)
_____	_R. c. connectens_	SE Peru (Cuzco) to nw Bolivia (Río Beni)
_____	_R. c. atrosericeus_	N and e Bolivia
_____	_R. c. centralis_	E-central Brazil and adjacent e Paraguay

☐ **Brazilian Tanager** _Ramphocelus bresilius_

_____	_R. b. bresilius_	NE Brazil (Paraíba to Bahia)
_____	_R. b. dorsalis_	SE Brazil (Minas Gerais and Espírito Santo to Santa Catarina)

☐ **Passerini's Tanager** _Ramphocelus passerinii_

S Mexico (se Veracruz and ne Oaxaca) to w Panama

☐ **Cherrie's Tanager** _Ramphocelus costaricensis_

Pacific slope of s Costa Rica (Puntarenas) and w Panama

☐ **Flame-rumped Tanager** _Ramphocelus flammigerus_

_____	_R. f. icteronotus_	Humid lowlands of Panama to w Colombia and w Ecuador
_____	_R. f. flammigerus_	W Colombia (middle Cauca Valley south to Nariño)

☐ **Western Spindalis** _Spindalis zena_

_____	_S. z. townsendi_	Grand Bahama I., the Abacos and Green Turtle Cay
_____	_S. z. zena_	Central Bahamas
_____	_S. z. pretrei_	Cuba, Isle of Pines and adjacent offshore cays
_____	_S. z. salvini_	Grand Cayman I.
_____	_S. z. benedicti_	Cozumel I. (off Yucatán coast of se Mexico)

☐ **Puerto Rican Spindalis** _Spindalis portoricensis_

Puerto Rico

☐ **Hispaniolan Spindalis** _Spindalis dominicensis_

Hispaniola and Gonâve I.

☐ **Jamaican Spindalis** _Spindalis nigricephala_

Jamaica

☐ **Blue-gray Tanager** _Thraupis episcopus_

_____	_T. e. cana_	SE Mexico (San Luis Potosí) to n Venezuela; Pearl Islands
_____	_T. e. caesita_	Caribbean coast of w Panama (Escudo de Veraguas)
_____	_T. e. cumatilis_	Coiba I. (off Pacific coast of Panama)
_____	_T. e. nesophilus_	Extreme e Colombia to e Venezuela; Trinidad
_____	_T. e. leucoptera_	E slope of Eastern Andes of central Colombia
_____	_T. e. quaesita_	Pacific slope of sw Colombia, w Ecuador and nw Peru
_____	_T. e. mediana_	SE Colombia to extreme n Bolivia and n Brazil
_____	_T. e. coelestis_	Tropical se Colombia to central Peru and w Amaz. Brazil
_____	_T. e. berlepschi_	Tobago
_____	_T. e. episcopus_	The Guianas and n Brazil
_____	_T. e. caerulea_	SE Ecuador and n Peru (south to Huánuco)
_____	_T. e. major_	Central Peru (Chanchamayo Valley of Ica)
_____	_T. e. urubambae_	SE Peru (Urubamba Valley and Amazonian drainage)

☐ **Glaucous Tanager** _Thraupis glaucocolpa_

Arid ne Colombia and n Venezuela; Isla Margarita

☐ **Sayaca Tanager** _Thraupis sayaca_

_____	_T. s. boliviana_	Tropical nw Bolivia (Río Beni to Río Mapiri)
_____	_T. s. obscura_	Central and s Bolivia to w Argentina
_____	_T. s. sayaca_	Paraguay to e Brazil, se Peru, Uruguay and ne Argentina

☐ **Azure-shouldered Tanager** _Thraupis cyanoptera_

Coastal se Brazil (Minas Gerais to Rio Grande do Sul)

☐ **Golden-chevroned Tanager** _Thraupis ornata_

SE Brazil (s Bahia to e Minas Gerais and Santa Catarina)

☐ **Blue-capped Tanager** _Thraupis cyanocephala_

_____	_T. c. cyanocephala_	Andes of w Ecuador to e Peru and n Bolivia

____	*T. c. annectens*	W and Central Andes of central Colombia
____	*T. c. auricrissa*	E Andes of n-central Colombia and w Venezuela
____	*T. c. margaritae*	Santa Marta Mountains (ne Colombia)
____	*T. c. hypophaea*	Subtropical nw Venezuela (Páramo de las Rosas in Lara)
____	*T. c. olivicyanea*	Coastal mountains of n Venezuela (Aragua to Miranda)
____	*T. c. subcinerea*	Coastal mountains of ne Venezuela (Sucre and Monagas)
____	*T. c. buesingi*	Mountains of ne Venezuela (Pariá Peninsula); Trinidad

☐ **Blue-and-yellow Tanager   *Thraupis bonariensis***

____	*T. b. darwinii*	Andes of Ecuador to n Chile
____	*T. b. composita*	Andes of e and central Bolivia
____	*T. b. schulzei*	Paraguay and nw Argentina (south to Mendoza and Lavalle)
____	*T. b. bonariensis*	S Brazil (Rio Grande do Sul) to Uruguay and n Argentina

☐ **Yellow-winged Tanager   *Thraupis abbas***

E Mexico to e Nicaragua

☐ **Palm Tanager   *Thraupis palmarum***

____	*T. p. atripennis*	E Nicaragua to n Colombia and extreme nw Venezuela
____	*T. p. violilavata*	Pacific slope of sw Colombia and w Ecuador
____	*T. p. melanoptera*	E Colombia to n Bolivia, the Guianas and Amazonian Brazil
____	*T. p. palmarum*	E and s Brazil to e Bolivia and Paraguay

☐ **Blue-backed Tanager   *Cyanicterus cyanicterus***

E Venezuela to the Guianas and adjacent ne Brazil

☐ **Blue-and-gold Tanager   *Bangsia arcaei***

____	*B. a. caeruleigularis*	Caribbean slope of Costa Rica
____	*B. a. arcaei*	Humid lowlands of w Panama

☐ **Black-and-gold Tanager   *Bangsia melanochlamys***

Central and Western Andes of Colombia

☐ **Golden-chested Tanager   *Bangsia rothschildi***

Andes of w Colombia to nw Ecuador (Esmeraldas)

☐ **Moss-backed Tanager   *Bangsia edwardsi***

Andean foothills of sw Colombia and nw Ecuador

☐ **Gold-ringed Tanager   *Bangsia aureocincta***

Pacific slope of Western Andes of Colombia

☐ **Hooded Mountain-Tanager   *Buthraupis montana***

____	*B. m. gigas*	Eastern Andes of Colombia and w Venezuela
____	*B. m. cucullata*	Western and Central Andes of Colombia and Ecuador
____	*B. m. cyanonota*	Andes of Peru (Amazonas to Junín)
____	*B. m. saturata*	Andes of se Peru (Cuzco and Puno)
____	*B. m. montana*	Andes of w Bolivia (La Paz and Cochabamba)

☐ **Black-chested Mountain-Tanager   *Buthraupis eximia***

____	*B. e. eximia*	Eastern Andes of Colombia and sw Venezuela
____	*B. e. zimmeri*	Western and Central Andes of Colombia
____	*B. e. chloronota*	E slope of Andes of se Colombia (Nariño) and nw Ecuador
____	*B. e. cyanocalyptra*	E slope of Andes of se Ecuador to n Peru

☐ **Golden-backed Mountain-Tanager   *Buthraupis aureodorsalis***

Andes of c Peru (e La Libertad to San Martín and Huánuco)

☐ **Masked Mountain-Tanager   *Buthraupis wetmorei***

Andes of sw Colombia to nw Peru

☐ **Orange-throated Tanager   *Wetmorethraupis sterrhopteron***

N Peru (Marañón Valley) and adjacent s Ecuador

☐ **Santa Marta Mountain-Tanager   *Anisognathus melanogenys***

Santa Marta Mountains (ne Colombia)

☐ **Lacrimose Mountain-Tanager   *Anisognathus lacrymosus***

____	*A. l. pallididorsalis*	Sierra de Perijá (Colombia/Venezuela border)
____	*A. l. melanops*	Andes of w Venezuela (Trujillo, Mérida and Táchira)

____ *A. l. tamae* — Mountains of n-central Colombia and sw Venezuela

____ *A. l. intensus* — E slope of W Andes of sw Colombia (Valle and Cauca)

____ *A. l. olivaceiceps* — N part of W and Central Andes of Colombia (s to Quindío)

____ *A. l. palpebrosus* — Andes of sw Colombia (Nariño) and e Ecuador

____ *A. l. caerulescens* — Mts. of s Ecuador (Loja) to n Peru (Cajamarca, Amazonas)

____ *A. l. lacrymosus* — Andes of central Peru (La Libertad to Junín and n Cuzco)

☐ **Scarlet-bellied Mountain-Tanager** *Anisognathus igniventris*

____ *A. i. lunulatus* — Andes of n-central Colombia and w Venezuela (Táchira)

____ *A. i. erythronotus* — Central Andes of s Colombia and Ecuador

____ *A. i. ignicrissa* — Andes of Peru (Cajamarca and Amazonas to Junín)

____ *A. i. igniventris* — Andes of se Peru (Cuzco) to nw Bolivia

☐ **Blue-winged Mountain-Tanager** *Anisognathus somptuosus*

____ *A. s. antioquiae* — Northern part of Western and Central Andes of Colombia

____ *A. s. victorini* — Andes of central Colombia to sw Venezuela (Táchira)

____ *A. s. cyanopterus* — W slope of Central Andes of sw Colombia and w Ecuador

____ *A. s. baezae* — E slope of Eastern Andes of s Colombia and e Ecuador

____ *A. s. venezuelanus* — Coastal mountains of n Venezuela (Yaracuy to Miranda)

____ *A. s. virididorsalis* — Subtropical n Venezuela (Golfo de Triste area)

____ *A. s. alamoris* — Subtropical sw Ecuador (Cuenca to Loja)

____ *A. s. somptuosus* — Extreme se Ecuador (Zamora) to e Peru (Junín)

____ *A. s. flavinuchus* — Subtropical se Peru (Cuzco) to nw Bolivia

☐ **Black-chinned Mountain-Tanager** *Anisognathus notabilis*

Humid Andes of w Colombia and w Ecuador

☐ **Diademed Tanager** *Stephanophorus diadematus*

Paraguay to se Brazil, Uruguay and ne Argentina

☐ **Purplish-mantled Tanager** *Iridosornis porphyrocephalus*

Humid Western Andes of Colombia and nw Ecuador

☐ **Yellow-throated Tanager** *Iridosornis analis*

Andes of se Colombia to Ecuador and se Peru (Puno)

☐ **Golden-collared Tanager** *Iridosornis jelskii*

____ *I. j. jelskii* — Temperate Andes of Peru (La Libertad to Junín)

____ *I. j. bolivianus* — Andes of se Peru (Cuzco) to w Bolivia (La Paz)

☐ **Golden-crowned Tanager** *Iridosornis rufivertex*

____ *I. r. caeruleoventris* — Western and Central Andes of nw Colombia

____ *I. r. ignicapillus* — Western and Central Andes of sw Colombia

____ *I. r. rufivertex* — E Andes of Colombia to sw Venezuela, e Ecuador, e Peru

____ *I. r. subsimilis* — W slope of Western Andes of Ecuador

☐ **Yellow-scarfed Tanager** *Iridosornis reinhardti*

E slope of Andes of Peru (s Amazonas to Cuzco)

☐ **Buff-breasted Mountain-Tanager** *Dubusia taeniata*

____ *D. t. carrikeri* — Santa Marta Mountains (ne Colombia)

____ *D. t. taeniata* — Andes of Colombia to w Venezuela and Ecuador

____ *D. t. stictocephala* — Andes of se Peru (Junín to Cuzco)

☐ **Chestnut-bellied Mountain-Tanager** *Delothraupis castaneoventris*

____ *D. c. peruviana* — Andes of e Peru (north to La Libertad)

____ *D. c. castaneoventris* — Andes of w Bolivia (La Paz, Cochabamba, w Santa Cruz)

☐ **Fawn-breasted Tanager** *Pipraeidea melanonota*

____ *P. m. venezuelensis* — Andes of Colombia to Venezuela and nw Argentina

____ *P. m. melanonota* — E Paraguay to se Brazil, Uruguay and ne Argentina

☐ **Glistening-green Tanager** *Chlorochrysa phoenicotis*

Andes of w Colombia and w Ecuador

☐ **Orange-eared Tanager** *Chlorochrysa calliparaea*
_____ *C. c. bourcieri* — W slope of Eastern Andes of Colombia to ne Peru
_____ *C. c. calliparaea* — Subtropical Andes of e-central Peru
_____ *C. c. fulgentissima* — Andes of se Peru (Puno) to w Bolivia

☐ **Multicolored Tanager** *Chlorochrysa nitidissima*

Locally in Andes of w Colombia

☐ **Plain-colored Tanager** *Tangara inornata*
_____ *T. i. rava* — Caribbean lowlands of se Costa Rica and w Panama
_____ *T. i. languens* — Tropical e Panama to extreme nw Colombia
_____ *T. i. inornata* — N Colombia (Sinú, lower Cauca, middle Magdalena valleys)

☐ **Turquoise Tanager** *Tangara mexicana*
_____ *T. m. media* — Extreme e Colombia to e Venezuela and nw Brazil
_____ *T. m. vieilloti* — Trinidad
_____ *T. m. boliviana* — E Colombia to n Bolivia and w Amazonian Brazil
_____ *T. m. mexicana* — Tropical zone of the Guianas
_____ *T. m. brasiliensis* — Coastal se Brazil (s Bahia to Rio de Janeiro)

☐ **Azure-rumped Tanager** *Tangara cabanisi*

Cloud forests of s Mexico (Chiapas) and adj. Guatemala

☐ **Gray-and-gold Tanager** *Tangara palmeri*

Humid foothills of e Panama to nw Ecuador (Pichincha)

☐ **Paradise Tanager** *Tangara chilensis*
_____ *T. c. coelicolor* — Colombia east of the Andes to s Venezuela and nw Brazil
_____ *T. c. chilensis* — SE Colombia to n Bolivia and w Amazonian Brazil
_____ *T. c. paradisea* — E Venezuela to the Guianas and n Brazil
_____ *T. c. chlorocorys* — N-central Peru (upper Huallaga Valley)

☐ **Seven-colored Tanager** *Tangara fastuosa*

Coastal ne Brazil (s Paraíba, e Pernambuco and Alagoas)

☐ **Green-headed Tanager** *Tangara seledon*

SE Paraguay to se Brazil and ne Argentina (Misiones)

☐ **Red-necked Tanager** *Tangara cyanocephala*
_____ *T. c. cearensis* — Forests of ne Brazil (Ceará)
_____ *T. c. corallina* — E Brazil (Pernambuco to Bahia)
_____ *T. c. cyanocephala* — SE Brazil (s Bahia) to e Paraguay and ne Argentina

☐ **Brassy-breasted Tanager** *Tangara desmaresti*

Coastal mountains of se Brazil (Espírito Santo to e Paraná)

☐ **Gilt-edged Tanager** *Tangara cyanoventris*

SE Brazil (s Bahia and se Minas Gerais to e São Paulo)

☐ **Blue-whiskered Tanager** *Tangara johannae*

Pacific lowlands of w Colombia and nw Ecuador

☐ **Green-and-gold Tanager** *Tangara schrankii*
_____ *T. s. anchicayae* — W slope of Western Andes of Colombia (Río Anchicayá)
_____ *T. s. schrankii* — SE Colombia to n Bolivia and w Amazonian Brazil
_____ *T. s. venezuelana* — Tropical s Venezuela (s Bolívar and e Amazonas)

☐ **Emerald Tanager** *Tangara florida*
_____ *T. f. florida* — Caribbean slope of Costa Rica and w Panama
_____ *T. f. auriceps* — Extreme e Panama (Darién) to Colombia and nw Ecuador

☐ **Golden Tanager** *Tangara arthus*
_____ *T. a. occidentalis* — W and Central Andes of Colombia (Antioquia to Nariño)
_____ *T. a. palmitae* — W slope of Eastern Andes of Colombia (s Magdalena)
_____ *T. a. sclateri* — Both slopes of Eastern Andes of Colombia
_____ *T. a. aurulenta* — Cent. Colombia (upper Magdalena Valley) to nw Venezuela
_____ *T. a. arthus* — Mts. of n Venezuela (Táchira to Lara, Falcón and Miranda)

____ *T. a. goodsoni*	Subtropical w Ecuador
____ *T. a. aequatorialis*	Subtropical e Ecuador and n Peru
____ *T. a. pulchra*	Central Peru (Chachapoyas to Chanchamayo)
____ *T. a. sophiae*	Tropical se Peru (Cuzco and Puno) to nw Bolivia

□ **Silver-throated Tanager** *Tangara icterocephala*

____ *T. i. frantzii*	Humid highlands of Costa Rica and w Panama
____ *T. i. oresbia*	Mountains of w-central Panama
____ *T. i. icterocephala*	E Panama (Darién) to w Colombia and extreme nw Peru

□ **Golden-eared Tanager** *Tangara chrysotis*

Andes of sw Colombia to nw Bolivia

□ **Saffron-crowned Tanager** *Tangara xanthocephala*

____ *T. x. venusta*	Andes of Colombia to w Venezuela and central Peru
____ *T. x. xanthocephala*	Subtropical Andes of central Peru (Chanchamayo region)
____ *T. x. lamprotis*	Andes of se Peru (Cuzco and Puno) to nw Bolivia

□ **Flame-faced Tanager** *Tangara parzudakii*

____ *T. p. parzudakii*	Andes of Colombia to sw Venezuela, e Ecuador and e Peru
____ *T. p. lunigera*	Pacific slope of Colombia and w Ecuador
____ *T. p. urubambae*	SE Peru (Cordillera Urubamba in Cuzco)

□ **Yellow-bellied Tanager** *Tangara xanthogastra*

____ *T. x. xanthogastra*	SE Colombia to s Venezuela and n Bolivia
____ *T. x. phelpsi*	S Venezuela and w Amazonian Brazil

□ **Spotted Tanager** *Tangara punctata*

____ *T. p. punctata*	Tropical s Venezuela, the Guianas and n Amazonian Brazil
____ *T. p. zamorae*	Tropical e Ecuador and n Peru
____ *T. p. perenensis*	Tropical and subtropical e Peru (Chanchamayo region)
____ *T. p. annectens*	Subtropical se Peru (Río Inambari region)
____ *T. p. punctulata*	*Yungas* of n Bolivia (La Paz and Cochabamba)

□ **Speckled Tanager** *Tangara guttata*

____ *T. g. eusticta*	Caribbean slope of Costa Rica and w Panama
____ *T. g. tolimae*	E slope of Central Andes of Colombia (Tolima)
____ *T. g. bogotensis*	Colombia (east of the Andes) and adjacent w Venezuela
____ *T. g. chrysophrys*	Venezuela and extreme nw Brazil (Sierra de Curupira)
____ *T. g. guttata*	SE Venezuela (s Bolívar) and extreme n Brazil (Roraima)
____ *T. g. trinitatis*	Mountains of n Trinidad

□ **Dotted Tanager** *Tangara varia*

S Venezuela to the Guianas and n Amazonian Brazil

□ **Rufous-throated Tanager** *Tangara rufigula*

Andes of w Colombia and w Ecuador (south to El Oro)

□ **Bay-headed Tanager** *Tangara gyrola*

____ *T. g. bangsi*	Humid tropical and subtropical Costa Rica and w Panama
____ *T. g. deleticia*	E Panama (Darién) and w Colombia
____ *T. g. nupera*	Extreme sw Colombia (Nariño) and w Ecuador
____ *T. g. toddi*	Mountains of n Colombia and nw Venezuela
____ *T. g. viridissima*	Coastal ne Venezuela; Trinidad
____ *T. g. catharinae*	E base of Eastern Andes of Colombia to central Bolivia
____ *T. g. parva*	SE Colombia to s Venezuela, ne Peru and nw Brazil
____ *T. g. gyrola*	S Venezuela to the Guianas and extreme n Brazil
____ *T. g. albertinae*	Brazil s of the Amazon (Rio Purús to Pará, n Mato Grosso)

□ **Rufous-winged Tanager** *Tangara lavinia*

____ *T. l. cara*	Honduras to Nicaragua and Costa Rica
____ *T. l. dalmasi*	Tropical w Panama (Chiriquí and Veraguas)
____ *T. l. lavinia*	E Panama to w Colombia and nw Ecuador; Isla Gorgona

☐ Burnished-buff Tanager *Tangara cayana*
_____ *T. c. fulvescens* — Eastern Andes of Colombia
_____ *T. c. cayana* — Colombia (e of the Andes) to Guianas, n Brazil and e Peru
_____ *T. c. huberi* — NE Brazil (Ilha Marajó region of Pará)
_____ *T. c. flava* — NE Brazil (Maranhão and n Goiás to extreme s Bahia)
_____ *T. c. sincipitalis* — Central Brazil (Goiás)
_____ *T. c. margaritae* — Central Brazil (Mato Grosso)
_____ *T. c. chloroptera* — Paraguay to se Brazil (Minas Gerais) and ne Argentina

☐ Black-backed Tanager *Tangara peruviana*
Lowlands of se Brazil (Rio de Janeiro to Santa Catarina)

☐ Lesser Antillean Tanager *Tangara cucullata*
_____ *T. c. versicolor* — St. Vincent (Lesser Antilles)
_____ *T. c. cucullata* — Grenada (Lesser Antilles)

☐ Chestnut-backed Tanager *Tangara preciosa*
E Paraguay to se Brazil, Uruguay and ne Argentina

☐ Scrub Tanager *Tangara vitriolina*
Arid scrub of w Colombia and nw Ecuador

☐ Green-capped Tanager *Tangara meyerdeschauenseei*
E slope of Andes of extreme se Peru (Puno)

☐ Rufous-cheeked Tanager *Tangara rufigenis*
Coastal mts. of n Venezuela (s Lara to Distrito Federal)

☐ Golden-naped Tanager *Tangara ruficervix*
_____ *T. r. ruficervix* — Andes of Colombia and Santa Marta Mountains
_____ *T. r. taylori* — SE Colombia (east of the Andes) and e Ecuador
_____ *T. r. leucotis* — Subtropical w Ecuador
_____ *T. r. amabilis* — Subtropical n Peru (south to Huánuco)
_____ *T. r. inca* — Subtropical s Peru (north to Junín)
_____ *T. r. fulvicervix* — *Yungas* of nw Bolivia (La Paz and Cochabamba)

☐ Metallic-green Tanager *Tangara labradorides*
_____ *T. l. labradorides* — Andes of w Colombia and w Ecuador
_____ *T. l. chaupensis* — Andes of ne Peru (south to San Martín)

☐ Blue-browed Tanager *Tangara cyanotis*
_____ *T. c. lutleyi* — Andes of s Colombia to e Ecuador and e Peru
_____ *T. c. cyanotis* — *Yungas* of nw Bolivia (La Paz and Cochabamba)

☐ Blue-necked Tanager *Tangara cyanicollis*
_____ *T. c. granadensis* — Andes of Colombia
_____ *T. c. caeruleocephala* — E Andes of central Colombia to e Ecuador and n Peru
_____ *T. c. hannahiae* — Colombia (east of Eastern Andes) and w Venezuela
_____ *T. c. cyanopygia* — Western Ecuador
_____ *T. c. cyanicollis* — E Peru (north to Huánuco) and e Bolivia
_____ *T. c. melanogaster* — W Amazonian Brazil (Amazon drainage of w Mato Grosso)
_____ *T. c. albotibialis* — E Brazil (s Pará and s Goiás)

☐ Golden-hooded Tanager *Tangara larvata*
_____ *T. l. larvata* — Tropical s Mexico (n Oaxaca and Tabasco) to n Costa Rica
_____ *T. l. centralis* — Caribbean slope of Costa Rica and w Panama
_____ *T. l. franciscae* — Pacific slope of Costa Rica and w Panama
_____ *T. l. fanny* — Pacific slope of Panama to Colombia and nw Ecuador

☐ Masked Tanager *Tangara nigrocincta*
SE Colombia to s Venezuela, Guyana, n Bolivia and w Brazil

☐ Spangle-cheeked Tanager *Tangara dowii*
Humid montane forests of Costa Rica and w Panama

☐ Green-naped Tanager *Tangara fucosa*
Highlands of extreme e Panama (Darién) and adj. Colombia

☐ **Beryl-spangled Tanager** *Tangara nigroviridis*
____ *T. n. cyanescens (consobrina)*   Andes of Colombia to n Venezuela and w Ecuador
____ *T. n. nigroviridis*   E slope of Eastern Andes of Colombia and e Ecuador
____ *T. n. lozanoana*   Mts. of Venezuela (Táchira, Mérida, Zulia and Lara)
____ *T. n. berlepschi*   Andes of e Peru to nw Bolivia (La Paz and Cochabamba)

☐ **Blue-and-black Tanager** *Tangara vassorii*
____ *T. v. vassorii*   Andes of Colombia to nw Venezuela, Ecuador and nw Peru
____ *T. v. branickii*   Andes of n and central Peru
____ *T. v. atrocaerulea*   Andes of s Peru (Huánuco) to w Bolivia

☐ **Black-capped Tanager** *Tangara heinei*
      Mountains of n Colombia to n Venezuela and n Ecuador

☐ **Sira Tanager** *Tangara phillipsi*
      Andes of e Peru (Cerros del Sira in e Huánuco)

☐ **Silver-backed Tanager** *Tangara viridicollis*
____ *T. v. fulvigula*   Andes of s Ecuador and n Peru
____ *T. v. viridicollis*   Andes of central and s Peru

☐ **Straw-backed Tanager** *Tangara argyrofenges*
____ *T. a. caeruleigularis*   Andes of extreme s Ecuador and n Peru (south to Junín)
____ *T. a. argyrofenges*   *Yungas* of w Bolivia (La Paz, Cochabamba and w Santa Cruz)

☐ **Black-headed Tanager** *Tangara cyanoptera*
____ *T. c. cyanoptera*   Mountains of n Colombia to n Venezuela
____ *T. c. whitelyi*   Tepuis of s Venezuela to Guyana and extreme n Brazil

☐ **Opal-rumped Tanager** *Tangara velia*
____ *T. v. iridina*   Colombia (east of the Andes) to n Bolivia and nw Brazil
____ *T. v. velia*   The Guianas and n Amazonian Brazil
____ *T. v. signata*   Tropical ne Brazil (south of the Amazon in Pará)
____ *T. v. cyanomelas*   Coastal se Brazil (Pernambuco to Rio de Janeiro)

☐ **Opal-crowned Tanager** *Tangara callophrys*
      SE Colombia to n Bolivia and w Amazonian Brazil

☐ **Golden-collared Honeycreeper** *Iridophanes pulcherrimus*
____ *I. p. pulcherrimus*   E slope of Andes of Colombia to e Ecuador and e Peru
____ *I. p. aureinucha*   Subtropical w Ecuador

☐ **Turquoise Dacnis-Tanager** *Pseudodacnis hartlaubi*
      Locally in Andes of Colombia

☐ **White-bellied Dacnis** *Dacnis albiventris*
      SE Colombia to s Venezuela, ne Peru and c Amaz. Brazil

☐ **Black-faced Dacnis** *Dacnis lineata*
____ *D. l. egregia*   Tropical central Colombia (Magdalena and Cauca valleys)
____ *D. l. lineata*   Colombia (e of the Andes) to n Bolivia and Amaz. Brazil
____ *D. l. aequatorialis*   Tropical w Ecuador (Esmeraldas to Chimbo)

☐ **Yellow-bellied Dacnis** *Dacnis flaviventer*
      SE Colombia to s Venezuela, n Bolivia and w Amaz. Brazil

☐ **Black-legged Dacnis** *Dacnis nigripes*
      SE Brazil (Minas Gerais to Santa Catarina)

☐ **Scarlet-thighed Dacnis** *Dacnis venusta*
____ *D. v. venusta*   Tropical Costa Rica and w Panama (Chiriquí)
____ *D. v. fuliginata*   E Panama (Caribbean slope of Darién) to nw Ecuador

☐ **Blue Dacnis** *Dacnis cayana*
____ *D. c. ultramarina*   Caribbean slope of ne Honduras to nw Colombia
____ *D. c. callaina*   W Costa Rica and w Panama (Chiriquí)
____ *D. c. napaea*   Tropical n Colombia

____ D. c. baudoana	Tropical sw Colombia (Baudó Mts.) to w Ecuador
____ D. c. coerbicolor	Central Colombia (Cauca and Magdalena valleys)
____ D. c. cayana	E Colombia to Venezuela, Guianas, n and c Brazil; Trinidad
____ D. c. glaucogularis	S Colombia to e Ecuador, e Peru and w Bolivia
____ D. c. paraguayensis	E Paraguay to e and s Brazil and ne Argentina

☐ **Viridian Dacnis** *Dacnis viguieri*

Lowlands of extreme e Panama and adjacent nw Colombia

☐ **Scarlet-breasted Dacnis** *Dacnis berlepschi*

Lowlands of extreme sw Colombia and nw Ecuador

☐ **Green Honeycreeper** *Chlorophanes spiza*

____ C. s. guatemalensis	S Mexico (Oaxaca) to Guatemala, Belize and Honduras
____ C. s. arguta	Extreme e Honduras to nw Colombia
____ C. s. exsul	Tropical sw Colombia to w Ecuador and extreme nw Peru
____ C. s. subtropicalis	Andes of Colombia and w Venezuela
____ C. s. caerulescens	SE Colombia to e Ecuador, e Peru and w Bolivia
____ C. s. spiza	E Colombia to Venezuela, the Guianas and n Brazil; Trinidad
____ C. s. axillaris	Coastal e Brazil (Pernambuco to Santa Catarina)

☐ **Short-billed Honeycreeper** *Cyanerpes nitidus*

SE Colombia to Venezuela, e Peru and w Amaz. Brazil

☐ **Shining Honeycreeper** *Cyanerpes lucidus*

____ C. l. lucidus	S Mexico (Chiapas) to Belize, Guatemala and n Nicaragua
____ C. l. isthmicus	Costa Rica to Panama and extreme nw Colombia

☐ **Purple Honeycreeper** *Cyanerpes caeruleus*

____ C. c. caeruleus	Extreme e Panama to Venezuela, the Guianas and ne Brazil
____ C. c. chocoanus	Tropical w Colombia and w Ecuador
____ C. c. microrhynchus	E Colombia to s Venezuela, n Bolivia and w Amaz. Brazil
____ C. c. longirostris	Trinidad
____ C. c. hellmayri	Highlands of Guyana

☐ **Red-legged Honeycreeper** *Cyanerpes cyaneus*

____ C. c. carneipes	Gulf slope of s Mexico to n Colombia; Coiba I. and Pearl Is.
____ C. c. gemmeus	N Colombia (Serranía de Macuire on Guajira Peninsula)
____ C. c. pacificus	Pacific coast of w Colombia and w Ecuador
____ C. c. gigas	Gorgona Islands (off Pacific coast of Colombia)
____ C. c. eximius	Tropical n Colombia to n Venezuela; Isla Margarita
____ C. c. dispar	E Colombia to s Venezuela, w Brazil and ne Peru
____ C. c. tobagensis	Tobago
____ C. c. cyaneus	Trop. se Venezuela to the Guianas and ne Brazil; Trinidad
____ C. c. brevipes	Central Amazonian Brazil
____ C. c. holti	E Brazil
____ C. c. violaceus	Central Bolivia and w Brazil

☐ **Tit-like Dacnis** *Xenodacnis parina*

____ X. p. bella	Locally in *polylepis* woodlands of sw Ecuador and n Peru
____ X. p. petersi	Cordillera Blanca (central Peru)
____ X. p. parina	Andes of s-central Peru (Junín, Ayacucho and Cuzco)

☐ **Swallow-Tanager** *Tersina viridis*

____ T. v. occidentalis	E Panama to Venezuela, the Guianas, n Bolivia and n Brazil
____ T. v. grisescens	Santa Marta Mountains (ne Colombia)
____ T. v. viridis	E Bolivia to Paraguay, e Brazil and ne Argentina

☐ **Plush-capped Finch** *Catamblyrhynchus diadema*

____ C. d. diadema	Andes of Colombia to nw Venezuela and s Ecuador
____ C. d. federalis	Coastal mts. of n Venezuela (Aragua and Distrito Federal)
____ C. d. citrinifrons	Andes of Peru to Bolivia and nw Argentina (Jujuy)

☐ **Tanager Finch** *Oreothraupis arremonops*

Pacific slope of w Andes of Colombia and nw Ecuador

☐ **Black-backed Bush-Tanager** *Urothraupis stolzmanni*

Andes of se Colombia and e Ecuador

☐ **Pardusco** *Nephelornis oneilli*

Andes of c Peru (San Martín, La Libertad and Huánuco)

FAMILY: UROCYNCHRAMIDAE (Przewalski's Rosefinch—1)

☐ **Przewalski's Rosefinch** *Urocynchramus pylzowi*

Mountains of w China and e Tibet

FAMILY: EMBERIZIDAE (Buntings, Sparrows, Seedeaters and Allies—329)

☐ **Crested Bunting** *Melophus lathami*

N Pakistan to se Tibet, s China, Laos and n Vietnam (Tonkin)

☐ **Slaty Bunting** *Latoucheornis siemsseni*

Mountains of central China (s Shaanxi, se Gansu and ne Sichuan)

☐ **Yellowhammer** *Emberiza citrinella*

____ *E. c. caliginosa*	Ireland, Scotland, Wales and n and w England
____ *E. c. citrinella*	SE England, n and w Europe to c Russia; > to North Africa
____ *E. c. erythrogenys*	E Europe to central Siberia; > to n Mongolia and Iraq

☐ **Pine Bunting** *Emberiza leucocephalos*

____ *E. l. leucocephalos*	Siberia to Sakhalin and Tibet; > to Iraq, India and China
____ *E. l. fronto*	NW China (Kokonor region of ne Qinghai to nw Gansu)

☐ **Cirl Bunting** *Emberiza cirlus*

____ *E. c. cirlus*	Wales and e England to Mediterranean and North Africa
____ *E. c. nigrostriata*	Corsica and Sardinia

☐ **Tibetan Bunting** *Emberiza koslowi*

W China (borders of arid Tibet, sw Qinghai and Sichuan)

☐ **Rock Bunting** *Emberiza cia*

____ *E. c. cia*	Iberian Peninsula and s Europe to w Asia Minor and N Africa
____ *E. c. hordei*	Greeca, central Asia Minor and Levant
____ *E. c. prageri*	Crimea, Caucasus, ne Turkey and nw Iran
____ *E. c. par*	N and central Iran to Pakistan, nw India and s Altai Mountains

☐ **Godlewski's Bunting** *Emberiza godlewskii*

____ *E. g. stracheyi*	W Himalayas (Chitral to Ladakh)
____ *E. g. decolorata*	W China (foothills of w Tarim Basin in Xinjiang)
____ *E. g. godlewskii*	Mongolia to nw China
____ *E. g. khamensis*	Tibet to w Sichuan and s Qinghai
____ *E. g. yunnanensis*	SE Tibet to n Myanmar, n Yunnan, ne Sichuan and w Hubei
____ *E. g. omissa*	S Mongolia to Hubei, Shaanxi and nw Sichuan
____ *E. g. flemingorum*	Nepal

☐ **Meadow Bunting** *Emberiza cioides*

____ *E. c. tarbagataica*	Mountains of central Asia; > in n Mongolia
____ *E. c. cioides*	NW Altai Mts. to Transbaikalia and mts. of Mongolia
____ *E. c. weigoldi*	E Transbaikalia to Manchuria, n Liaoning and n Korea
____ *E. c. castaneiceps*	S and central Korea and e China
____ *E. c. ciopsis*	S Kuril and Japanese islands; > from Honshu southward

☐ **Rufous-backed Bunting** *Emberiza jankowskii*

NE China and extreme ne Korea; > e-central China

☐ **Gray-hooded Bunting** *Emberiza buchanani*
_____ *E. b. cerrutii*   E Turkey, Russia s of the Caucasus to Iran; Mugodzhary Mts.
_____ *E. b. buchanani*   Afghanistan to w Pakistan; > to se India
_____ *E. b. neobscura*   Tajikistan to w Xinjiang, e Kazakstan and w Mongolia

☐ **Cinereous Bunting** *Emberiza cineracea*
_____ *E. c. cineracea*   Arid rocky slopes of w and s Turkey
_____ *E. c. semenowi*   Zagros Mountains (sw Iran); > to Yemen, Sudan and Eritrea

☐ **Ortolan Bunting** *Emberiza hortulana*

W Palearctic; > Mediterranean environs, Arabia, Africa

☐ **Chestnut-breasted Bunting** *Emberiza stewarti*

S Turkmenistan to n Afghanistan, n Pakistan and nw India

☐ **Cretzschmar's Bunting** *Emberiza caesia*

S Europe to Asia Minor; > to ne Africa and Arabia

☐ **House Bunting** *Emberiza striolata*
_____ *E. s. sahari (theresae)*   Mountains of Morocco, Algeria and Tunisia to Mali and se Niger
_____ *E. s. striolata*   NE Africa to Arabia, Iran, Pakistan and central India
_____ *E. s. sanghae*   S Mali (Mopi region)
_____ *E. s. saturatior*   Highlands of central Sudan, sw Ethiopia and nw Kenya
_____ *E. s. jebelmarrae*   Highlands of w-central Sudan (Jebel Marra)

☐ **Lark-like Bunting** *Emberiza impetuani*
_____ *E. i. eremica*   S Angola to n Namibia and nw Cape Province
_____ *E. i. impetuani*   Arid scrub of w Cape Province to Botswana and sw Zimbabwe
_____ *E. i. sloggetti*   W, s and cent. Cape Province to w Free State and s Transvaal

☐ **Cinnamon-breasted Bunting** *Emberiza tahapisi*
_____ *E. t. goslingi*   Gambia to Sudan west of the Nile and extreme ne Zaïre
_____ *E. t. tahapisi (nivenorum)*   Gabon to Uganda, s Sudan, Ethiopia, Somalia and South Africa
_____ *E. t. septemstriata*   Sudan east of the Nile to w and n Ethiopia
_____ *E. t. arabica*   S Arabia (Asir to Hadhramaut)
_____ *E. t, insularis*   Socotra I. (off ne Somalia)

☐ **Socotra Bunting** *Emberiza socotrana*

Highlands of Socotra I.

☐ **Cape Bunting** *Emberiza capensis*
_____ *E. c. nebularum*   SW Angola
_____ *E. c. bradfieldi*   N Namibia (Kaokoveld and highlands of Damaraland)
_____ *E. c. capensis*   Namibia to sw and w Cape Province
_____ *E. c. vinacea*   N Cape Province (Kaap Plateau and n Asbestos Mts.)
_____ *E. c. cinnamomea (media)*   Cape Province to s Transvaal and w Free State
_____ *E. c. limpopoensis*   SE Botswana to central and sw Transvaal
_____ *E. c. smithersii*   Mountains on e Zimbabwe/Mozambique border
_____ *E. c. plowesi*   Plateau of s Zimbabwe and adjacent ne Botswana
_____ *E. c. reidi*   N Lesotho, e Free State, KwaZulu-Natal and adj. Transvaal
_____ *E. c. basutoensis*   Highlands of Lesotho to e Griqualand and KwaZulu-Natal

☐ **Vincent's Bunting** *Emberiza vincenti*

Central Malawi to adjacent n Mozambique and e Zambia

☐ **Ochre-rumped Bunting** *Emberiza yessoensis*
_____ *E. y. continentalis*   SE Siberia and e Manchuria; > to s Korea and e China
_____ *E. y. yessoensis*   Japan (Hokkaido, Honshu and s Kuril Islands)

☐ **Tristram's Bunting** *Emberiza tristrami*

Siberia to ne China; > to s China and n Thailand

☐ **Chestnut-eared Bunting** *Emberiza fucata*
_____ *E. f. arcuata*   Himalayas (Pakistan to Bangladesh; n Yunnan); > to Myanmar
_____ *E. f. fucata*   Mountains of Mongolia and Manchuria to Korea and Japan
_____ *E. f. kuatunensis*   S China (Jiangsu to Guangdong and s Yunnan)

☐ **Little Bunting** *Emberiza pusilla*

Taiga of n Eurasia; > to India, SE Asia and Philippines

☐ **Yellow-browed Bunting** *Emberiza chrysophrys*

Taiga of central Siberia; > in central and se China

☐ **Rustic Bunting** *Emberiza rustica*

    ____ *E. r. rustica*   *Taiga* of n Eurasia; > to e China and Japan
    ____ *E. r. latifascia*   *Taiga* of ne Siberia (Yakutsk to Kamchatka)

☐ **Yellow-throated Bunting** *Emberiza elegans*

    ____ *E. e. elegans*   Manchuria and n Korea; > to s Korea, s Japan and e China
    ____ *E. e. ticehursti*   E Amurland; > s Manchuria to Shandong Province
    ____ *E. e. elegantula*   Mountains of sw China; > to ne Myanmar

☐ **Yellow-breasted Bunting** *Emberiza aureola*

    ____ *E. a. aureola*   Boreal forests of Finland to Bering Sea; > to Indochina
    ____ *E. a. ornata*   Amur River to Manchuria, N Korea, Kamchatka and Kuril Is.

☐ **Golden-breasted Bunting** *Emberiza flaviventris*

    ____ *E. f. flavigaster*   S edge of Sahara from Mauretania to Eritrea
    ____ *E. f. kalaharica (carychroa)*   S Angola to se Sudan, Kenya, Mozambique and n South Africa
    ____ *E. f. flaviventris*   S and e Cape Province to KwaZulu-Natal
    ____ *E. f. princeps*   N and nw Namibia to sw Angola

☐ **Somali Bunting** *Emberiza poliopleura*

Sudan to Ethiopia, Somalia, Uganda, Kenya and ne Tanzania

☐ **Brown-rumped Bunting** *Emberiza affinis*

    ____ *E. a. nigeriae*   Gambia to Nigeria and n Cameroon
    ____ *E. a. vulpecula*   Cameroon, s Chad and adjacent n Central African Republic
    ____ *E. a. affinis*   S Sudan, n Uganda and adjacent Zaïre
    ____ *E. a. omoensis*   S Ethiopia

☐ **Cabanis' Bunting** *Emberiza cabanisi*

    ____ *E. c. cabanisi*   Sierra Leone to s Sudan, ne Zaïre and nw Uganda
    ____ *E. c. orientalis (cognominata)*   S Zaïre to Tanzania, Angola, Zambia, Zimbabwe, n Mozambique

☐ **Chestnut Bunting** *Emberiza rutila*

Siberia to n Mongolia and ne China; > to India and SE Asia

☐ **Black-headed Bunting** *Emberiza melanocephala*

S-central Eurasia; > to India

☐ **Red-headed Bunting** *Emberiza bruniceps*

S-central Eurasia; > in Indian subcontinent

☐ **Yellow Bunting** *Emberiza sulphurata*

Honshu (Japan); > e China, n Philippines and Taiwan

☐ **Black-faced Bunting** *Emberiza spodocephala*

    ____ *E. s. spodocephala*   Central and e Asia; > to e China and Taiwan
    ____ *E. s. personata*   Sakhalin and s Kuril Is. to Honshu; > to Ryukyu Is.
    ____ *E. s. sordida*   W China; > to e India, n Myanmar and n Indochina

☐ **Gray Bunting** *Emberiza variabilis*

S Kamchatka, Sakhalin, Kuril Is. and n Japan; > to Ryukyu Is.

☐ **Pallas' Bunting** *Emberiza pallasi*

    ____ *E. p. polaris*   Siberia to Sea of Okhotsk; > Manchuria to e China
    ____ *E. p. pallasi*   Mts. of central and e Asia; > to w China and Mongolia
    ____ *E. p. lydiae*   Mountains of central Mongolia

☐ **Reed Bunting** *Emberiza schoeniclus*

    ____ *E. s. schoeniclus*   British Isles and nw Europe to c Russia; > to North Africa
    ____ *E. s. witherbyi*   S Portugal, coastal w Spain, France, Balearic Is. and Sardinia
    ____ *E. s. canetti (intermedia)*   Italy and Sicily to s Ukraine, Crimea and ne Turkey
    ____ *E. s. reiseri*   S Yugoslavia to sw Albania and n Greece
    ____ *E. s. caspia*   E Caucasus to w and s Iran, Syria, adj. se Turkey and ne Iraq

____	*E. s. korejewi*	E Iran
____	*E. s. pyrrhuloides*	Caspian Sea to Kazakstan, w Xinjiang and w Mongolia
____	*E. s. passerina*	NW Siberia; > to n Xinjiang, Mongolia and n Iran
____	*E. s. parvirostris*	Central Siberia and n Mongolia; > to n China
____	*E. s. pyrrhulina*	Transbaikalia to Kamchatka, Kuril Is., Sakhalin, Hokkaido
____	*E. s. pallidior*	SW Siberia; > Caucasus to nw India and Mongolia
____	*E. s. minor*	SE Siberia and adjacent Manchuria
____	*E. s. ukrainae*	S Russia to n Ukraine and Volga River; > to Caucasus
____	*E. s. incognita*	Russia e of Volga to s Urals, n Kazakstan; > to nw China
____	*E. s. zaidamensis*	W China (Tsaidam basin in n Qinghai)

☐ **Corn Bunting** *Emberiza calandra*

Grasslands and scrub of Palearctic region

☐ **Coal-crested Finch** *Charitospiza eucosma*

Campos of ne Bolivia to ne and central Brazil and ne Argentina

☐ **Black-masked Finch** *Coryphaspiza melanotis*

____	*C. m. marajoara*	Ilha de Marajó (e Brazil in Pará)
____	*C. m. melanotis*	SE Peru to n Bolivia, se Paraguay, se Brazil and ne Argentina

☐ **Many-colored Chaco-Finch** *Saltatricula multicolor*

SE Bolivia to w Paraguay, nw Uruguay and n Argentina

☐ **Pileated Finch** *Coryphospingus pileatus*

____	*C. p. rostratus*	Colombia (arid upper Magdalena Valley)
____	*C. p. brevicaudus*	N Colombia to n Venezuela; Isla Margarita
____	*C. p. pileatus*	E-cent. Brazil (Ceará and Piauí to Minas Gerais and Rio de Janeiro)

☐ **Red-crested Finch** *Coryphospingus cucullatus*

____	*C. c. cucullatus*	The Guianas and ne Brazil (east to Pará)
____	*C. c. rubescens*	E Paraguay to cent. and s Brazil, Uruguay and ne Argentina
____	*C. c. fargoi*	N Peru to Bolivia, w Paraguay and n Argentina

☐ **Crimson-breasted Finch** *Rhodospingus cruentus*

Arid scrub of w Ecuador and extreme nw Peru

☐ **Black-hooded Sierra-Finch** *Phrygilus atriceps*

Andes of sw Peru (Arequipa) to w Bolivia, n Chile, nw Argentina

☐ **Peruvian Sierra-Finch** *Phrygilus punensis*

____	*P. p. chloronotus*	Andes of s Peru (Cajamarca to Ayacucho and n Cuzco)
____	*P. p. punensis*	Andes of s Peru (Puno) to nw Bolivia (La Paz)

☐ **Gray-hooded Sierra-Finch** *Phrygilus gayi*

____	*P. g. gayi*	Andes of n Chile (Coquimbo to Colchagua)
____	*P. g. minor*	Coastal Chile (Atacama to Santiago)
____	*P. g. caniceps*	S Chile and Argentina to Tierra del Fuego

☐ **Patagonian Sierra-Finch** *Phrygilus patagonicus*

Central Argentina and Chile to Tierra del Fuego

☐ **Mourning Sierra-Finch** *Phrygilus fruticeti*

____	*P. f. peruvianus*	Andes of n Peru to Bolivia (La Paz and Cochabamba)
____	*P. f. coracinus*	Andes of Bolivia (w Oruro and Potosí)
____	*P. f. fruticeti*	Andes of sw Bolivia to s Chile and s Argentina

☐ **Plumbeous Sierra-Finch** *Phrygilus unicolor*

____	*P. u. nivarius*	NE Colombia (Santa Marta Mts.) and Andes of nw Venezuela
____	*P. u. geospizopsis*	E and Central Andes of Colombia and Ecuador
____	*P. u. inca*	Andes of Peru to w Bolivia (La Paz)
____	*P. u. unicolor*	Andes of sw Peru (Tacna) to s Chile and w Argentina
____	*P. u. tucumanus*	Andes of Bolivia and nw Argentina
____	*P. u. ultimus*	Mountains of s Argentina (Tierra del Fuego)

☐ **Red-backed Sierra-Finch** *Phrygilus dorsalis*

Andes of sw Bolivia to n Chile and nw Argentina

☐ **White-throated Sierra-Finch** *Phrygilus erythronotus*

High Andes of sw Peru, sw Bolivia and n Chile

☐ **Carbonated Sierra-Finch** *Phrygilus carbonarius*

Pampas of central Argentina; > north to Tucumán

☐ **Band-tailed Sierra-Finch** *Phrygilus alaudinus*
____ *P. a. bipartitus*   Andes of w Ecuador and Peru (south to Arequipa)
____ *P. a. humboldti*   Coastal s Ecuador and n Peru (south to Piura)
____ *P. a. bracki*   Central Peru (arid inter-Andean of upper Río Huallaga)
____ *P. a. excelsus*   Andes of s Peru (Puno) and Bolivia
____ *P. a. alaudinus*   Andes of Chile (Atacama to Valdivia)
____ *P. a. venturii*   Andes of Argentina (Jujuy and Salta to Tucumán and w Córdoba)

☐ **Ash-breasted Sierra-Finch** *Phrygilus plebejus*
____ *P. p. ocularis*   Andes of Ecuador and extreme n Peru (Tumbes and Piura)
____ *P. p. plebejus*   Andes of Peru to Bolivia, n Chile and nw Argentina

☐ **Canary-winged Finch** *Melanodera melanodera*
____ *M. m. princetoniana*   *Llanos* of s Chile and s Argentina to Tierra del Fuego
____ *M. m. melanodera*   Falkland Islands

☐ **Yellow-bridled Finch** *Melanodera xanthogramma*
____ *M. x. barrosi*   Mountains of Chile and w Argentina
____ *M. x. xanthogramma*   S Argentina (Tierra del Fuego and Cape Horn Archipelago)

☐ **Black-crested Finch** *Lophospingus pusillus*

Chaco of se Bolivia to w Paraguay and n Argentina

☐ **Gray-crested Finch** *Lophospingus griseocristatus*

Arid Andes of se Bolivia and nw Argentina (Jujuy and Salta)

☐ **Long-tailed Reed-Finch** *Donacospiza albifrons*

N Bolivia to se Paraguay, se Brazil, Uruguay and ne Argentina

☐ **Gough Island Finch** *Rowettia goughensis*

Gough I. (s Atlantic Ocean)

☐ **Nightingale Finch** *Nesospiza acunhae*
____ *N. a. acunhae*   Inaccessible I. (s Atlantic Ocean)
____ *N. a. questi*   Nightingale I. (s Atlantic Ocean)

☐ **Wilkins' Finch** *Nesospiza wilkinsi*
____ *N. w. dunnei*   Inaccessible I. (s Atlantic Ocean)
____ *N. w. wilkinsi*   Nightingale I. (s Atlantic Ocean)

☐ **White-winged Diuca-Finch** *Diuca speculifera*
____ *D. s. magnirostris*   Andes of central Peru (Ancash and Junín)
____ *D. s. speculifera*   Andes of s Peru to n Chile, nw Bolivia and nw Argentina

☐ **Common Diuca-Finch** *Diuca diuca*
____ *D. d. crassirostris*   Andes of n Chile and n Argentina
____ *D. d. diuca*   Andes of central Chile and central Argentina
____ *D. d. chiloensis*   Chiloé I. (off w Chile)
____ *D. d. minor*   Argentina (Córdoba to Santa Cruz); > to se Brazil

☐ **Short-tailed Finch** *Idiopsar brachyurus*

Andes of extreme s Peru (Puno) to n Bolivia and nw Argentina

☐ **Cinereous Finch** *Piezorhina cinerea*

Arid coastal nw Peru (Tumbes to La Libertad)

☐ **Slender-billed Finch** *Xenospingus concolor*

Coastal w Peru (Lima) to Andes of n Chile

☐ **Great Inca-Finch** *Incaspiza pulchra*

Arid Andes of Peru (Ancash to s Lima)

☐ **Rufous-backed Inca-Finch** *Incaspiza personata*

Arid Andes of w Peru (upper Marañón Valley drainage)

☐ **Gray-winged Inca-Finch** *Incaspiza ortizi*

Arid Andes of w Peru (upper Marañón Valley drainage)

☐ **Buff-bridled Inca-Finch** *Incaspiza laeta*

Arid Andes of w Peru (upper Marañón Valley drainage)

☐ **Little Inca-Finch** *Incaspiza watkinsi*

Foothills of arid w Peru (middle Marañón River drainage)

☐ **Bay-chested Warbling-Finch** *Poospiza thoracica*

Scrub of se Brazil (Minas Gerais to n Rio Grande do Sul)

☐ **Bolivian Warbling-Finch** *Poospiza boliviana*

Andes of Bolivia and adjacent nw Argentina (Jujuy)

☐ **Plain-tailed Warbling-Finch** *Poospiza alticola*

Andes of w Peru (s Cajamarca to La Libertad and e Ancash)

☐ **Rufous-sided Warbling-Finch** *Poospiza hypochondria*
____ *P. h. hypochondria*
____ *P. h. affinis*

Andes of w Bolivia (La Paz and Cochabamba)
Andes of nw Argentina (Jujuy to Mendoza)

☐ **Cinnamon Warbling-Finch** *Poospiza ornata*

Breeds c Argentina; > n to Salta, Tucumán and Córdoba

☐ **Rusty-browed Warbling-Finch** *Poospiza erythrophrys*
____ *P. e. cochabambae*
____ *P. e. erythrophrys*

Andes of w Bolivia (Cochabamba and Chuquisaca)
Andes of Bolivia (Tarija) and nw Argentina

☐ **Black-and-rufous Warbling-Finch** *Poospiza nigrorufa*

SE Paraguay to se Brazil, Uruguay and central Argentina

☐ **Black-and-chestnut Warbling-Finch** *Poospiza whitii*
____ *P. w. whitii*
____ *P. w. wagneri*

Andes of w Bolivia (La Paz and Cochabamba) to nw Argentina
Andes of Bolivia on Mt. Chulumaní (La Paz)

☐ **Red-rumped Warbling-Finch** *Poospiza lateralis*
____ *P. l. lateralis*
____ *P. l. cabanisi*

SE Brazil (sw Minas Gerais and Espírito Santo to n São Paulo)
SE Brazil (s São Paulo) to Uruguay, Paraguay and ne Argentina

☐ **Rufous-breasted Warbling-Finch** *Poospiza rubecula*

W slope of Andes of Peru (La Libertad to Ica)

☐ **Cochabamba Mountain-Finch** *Poospiza garleppi*

Andes of w Bolivia (s Cochabamba)

☐ **Tucuman Mountain-Finch** *Poospiza baeri*

Andes of nw Argentina (Jujuy to Tucumán, Salta and La Rioja)

☐ **Chestnut-breasted Mountain-Finch** *Poospiza caesar*

Andes of se Peru (Apurímac, Cuzco and Puno)

☐ **Collared Warbling-Finch** *Poospiza hispaniolensis*

Arid scrub of sw Ecuador and w Peru

☐ **Ringed Warbling-Finch** *Poospiza torquata*
____ *P. t. torquata*
____ *P. t. pectoralis*

Arid intermontane valleys of nw Bolivia
Lowlands of se Bolivia to w Paraguay and central Argentina

☐ **Black-capped Warbling-Finch** *Poospiza melanoleuca*

W Bolivia to n Argentina, Paraguay, Uruguay and sw Brazil

☐ **Cinereous Warbling-Finch** *Poospiza cinerea*

Locally in *campos* of s-central Brazil

☐ **Blue-black Grassquit** *Volatinia jacarina*
____ *V. j. splendens*
____ *V. j. peruviensis*
____ *V. j. jacarina*

N Mexico to n South America, Grenada, Trinidad and Tobago
Arid Pacific slope of Ecuador, Peru and extreme n Chile
SE Peru to e Bolivia, Paraguay, cent. and e Brazil and n Argentina

☐ **Buffy-fronted Seedeater** *Sporophila frontalis*

E Paraguay to se Brazil and ne Argentina

☐ **Temminck's Seedeater** *Sporophila falcirostris*

Lowlands of se Brazil, Paraguay and ne Argentina (Misiones)

☐ **Slate-colored Seedeater** *Sporophila schistacea*

_____ *S. s. subconcolor* — Belize and Guatemala to Nicaragua
_____ *S. s. schistacea* — SW Costa Rica to Panama and n Colombia
_____ *S. s. incerta* — Pacific slope of Andes of Colombia to Ecuador (Pichincha)
_____ *S. s. longipennis* — Trop. e Colombia to Venezuela, Guianas, n Brazil and nw Bolivia

☐ **Plumbeous Seedeater** *Sporophila plumbea*

_____ *S. p. colombiana* — N Colombia (Santa Marta Mts. and lower Magdalena Valley)
_____ *S. p. whiteleyana* — Llanos of e Colombia to s Venezuela, the Guianas and n Brazil
_____ *S. p. plumbea* — SE Peru to n Bolivia, e Paraguay, s Brazil and ne Argentina

☐ **Caqueta Seedeater** *Sporophila murallae*

SE Colombia to e Brazil, e Ecuador and ne Peru

☐ **Gray Seedeater** *Sporophila intermedia*

_____ *S. i. bogotensis (agustini)* — Colombia (Western Andes, Magdalena and Cauca valleys)
_____ *S. i. anchicayae* — Colombia (valley of the Río Anchicayá in Valle)
_____ *S. i. intermedia* — E Colombia to n Venezuela, w Guyana and ne Brazil; Trinidad

☐ **Wing-barred Seedeater** *Sporophila americana*

_____ *S. a. americana* — NE Venezuela to Guianas, ne Brazil; Tobago and Chacachacare I.
_____ *S. a. dispar* — W Amazonian Brazil

☐ **Variable Seedeater** *Sporophila corvina*

_____ *S. c. corvina (badiiventris)* — Caribbean slope of e Mexico (Veracruz) to Nicaragua
_____ *S. c. hoffmannii (collaris)* — S Costa Rica and w Panama
_____ *S. c. hicksii (chocoana)* — Pacific slope of e Panama (Darién) and w Colombia
_____ *S. c. ophthalmica* — SW Colombia w of Andes to Ecuador and n Peru (La Libertad)

☐ **White-collared Seedeater** *Sporophila torqueola*

_____ *S. t. sharpei* — S Texas (lower Rio Grande Valley) to e Mexico (Veracruz)
_____ *S. t. torqueola* — SW Mexico (Jalisco to Guanajuato, w Puebla and s Oaxaca)
_____ *S. t. morelleti* — Caribbean slope of s Mexico (Veracruz) to extreme w Panama
_____ *S. t. mutanda* — Pacific slope of s Mexico (Chiapas) to Guatemala and El Salvador

☐ **Rusty-collared Seedeater** *Sporophila collaris*

_____ *S. c. ochrascens* — E Bolivia (Beni) to w Brazil (n Mato Grosso and w São Paulo)
_____ *S. c. melanocephala* — Paraguay to w Brazil (sw Mato Grosso), Uruguay and n Argentina
_____ *S. c. collaris* — E Brazil (s Goiás, Minas Gerais, Espírito Santo and Rio de Janeiro)

☐ **Lesson's Seedeater** *Sporophila bouvronides*

_____ *S. b. bouvronides* — E Colombia to Guianas, Trinidad and Tobago; > to ne Peru
_____ *S. b. restricta* — Colombia (lower Magdalena Valley)

☐ **Lined Seedeater** *Sporophila lineola*

_____ *S. l. lineola* — Bolivia to Paraguay, se Brazil and n Argentina; > Amazonia
_____ *S. l. ssp.* — Caatinga of ne Brazil; > to Venezuela and the Guianas

☐ **Black-and-white Seedeater** *Sporophila luctuosa*

Andes of Colombia to w Venezuela, n Bolivia and sw Brazil

☐ **Yellow-bellied Seedeater** *Sporophila nigricollis*

_____ *S. n. nigricollis* — S Costa Rica to ne Argentina, Brazil, Trinidad, Tobago and Grenada
_____ *S. n. vivida* — SW Colombia (Nariño) and w Ecuador
_____ *S. n. inconspicua* — Andes of Peru (south to Cuzco)

☐ **Dubois' Seedeater** *Sporophila ardesiaca*

E Brazil (s Minas Gerais, Espírito Santo and Rio de Janeiro)

☐ **Hooded Seedeater** *Sporophila melanops*

Known from an 1870 specimen from se Brazil (possible hybrid)

☐ **Double-collared Seedeater** *Sporophila caerulescens*

_____ *S. c. caerulescens* — Bolivia to Paraguay, e Brazil (Pará), Uruguay and Argentina

_____ *S. c. yungae*   N Bolivia (La Paz, Cochabamba and Beni)

_____ *S. c. hellmayri*   E Brazil (Bahia)

☐ **White-throated Seedeater** *Sporophila albogularis*

NE Brazil (Piauí and Pernambuco to n Bahia)

☐ **Drab Seedeater** *Sporophila simplex*

Arid s Ecuador; upper Marañon Valley of n Peru south to Ica

☐ **White-bellied Seedeater** *Sporophila leucoptera*

_____ *S. l. mexianae*   S Suriname and Mexiana I. (off ne Brazil in Pará)

_____ *S. l. bicolor*   SE Peru (Madre de Dios) to nw Bolivia (Beni and Santa Cruz)

_____ *S. l. leucoptera*   E Paraguay to central and sw Brazil and n Argentina

_____ *S. l. cinereola*   E Brazil (s Maranhão to Rio de Janeiro)

☐ **Parrot-billed Seedeater** *Sporophila peruviana*

_____ *S. p. devronis*   Arid coastal sw Ecuador (Manabí) to n Peru (Tumbes)

_____ *S. p. peruviana*   Arid coast of Peru (La Libertad to Ica)

☐ **Black-and-tawny Seedeater** *Sporophila nigrorufa*

Extreme e Bolivia and sw Brazil (Mato Grosso)

☐ **Capped Seedeater** *Sporophila bouvreuil*

_____ *S. b. pileata*   E Paraguay to s Brazil (s Mato Grosso) and ne Argentina

_____ *S. b. bouvreuil*   Suriname; e Brazil (Amazon delta to ne São Paulo)

_____ *S. b. crypta*   SE Brazil (Farinha Sêca area near Rio de Janeiro)

_____ *S. b. saturata*   SE Brazil (São Paulo)

☐ **Ruddy-breasted Seedeater** *Sporophila minuta*

_____ *S. m. parva*   Arid Pacific lowlands of sw Mexico (Nayarit) to Nicaragua

_____ *S. m. centralis*   SW Costa Rica and Pacific slope of Panama

_____ *S. m. minuta*   Colombia to Venezuela, Guianas, Amaz. Brazil, Trinidad, Tobago

☐ **Tawny-bellied Seedeater** *Sporophila hypoxantha*

Bolivia to Paraguay, s Brazil and n Argentina

☐ **Dark-throated Seedeater** *Sporophila ruficollis*

N Bolivia to Paraguay, s Brazil, n Uruguay and n Argentina

☐ **Marsh Seedeater** *Sporophila palustris*

Locally in Paraguay, s Brazil, Uruguay and ne Argentina

☐ **Chestnut-bellied Seedeater** *Sporophila castaneiventris*

Guianas and s Venezuela to n Bolivia and Amazonian Brazil

☐ **Gray-and-chestnut Seedeater** *Sporophila hypochroma*

Locally in e Bolivia, Paraguay, sw Brazil and ne Argentina

☐ **Chestnut Seedeater** *Sporophila cinnamomea*

Paraguay to s-central Brazil, Uruguay and ne Argentina

☐ **Narosky's Seedeater** *Sporophila zelichi*

Lowlands of ne Argentina (Entre Ríos)

☐ **Black-bellied Seedeater** *Sporophila melanogaster*

SE Brazil (s Goiás and Minas Gerais to Rio Grande do Sul)

☐ **Chestnut-throated Seedeater** *Sporophila telasco*

Coastal lowlands of sw Colombia to extreme n Chile (Tarapacá)

☐ **Tumaco Seedeater** *Sporophila insulata*

SW Colombia (rediscovered after 82-year absence on Tumaco I.)

☐ **Nicaraguan Seed-Finch** *Oryzoborus nuttingi*

Caribbean lowlands of Nicaragua, Costa Rica and w Panama

☐ **Large-billed Seed-Finch** *Oryzoborus crassirostris*

_____ *O. c. crassirostris*   Colombia east of Andes to Venezuela, the Guianas and n Brazil

_____ *O. c. occidentalis*   Tropical se Colombia to e Ecuador, e Peru and n Bolivia

_____ *O. c. magnirostris*   E Venezuela (Delta Amacuro and n Bolívar) and Trinidad

_____ *O. c. ssp.*   Pacific slope of w Colombia to sw Ecuador

☐ **Black-billed Seed-Finch** *Oryzoborus atrirostris*

_____ *O. a. atrirostris*   E slope of Andes of n Peru (Loreto and San Martín)

_____ *O. a. gigantirostris*   N Bolivia (Beni)

☐ **Great-billed Seed-Finch** *Oryzoborus maximiliani*

SE Colombia to n Bolivia, the Guianas, Amazonian and se Brazil

☐ **Chestnut-bellied Seed-Finch** *Oryzoborus angolensis*
_____ *O. a. torridus* — S Colombia to Peru, Venezuela, the Guianas and Brazil; Trinidad
_____ *O. a. angolensis* — E Brazil to w Bolivia, Paraguay and extreme ne Argentina

☐ **Thick-billed Seed-Finch** *Oryzoborus funereus*

S Mexico (Veracruz) to w Colombia, w Ecuador; Coiba I.; Pearl Is.

☐ **Blackish-blue Seedeater** *Amaurospiza moesta*

Locally from se Paraguay to e Brazil and ne Argentina (Misiones)

☐ **Blue Seedeater** *Amaurospiza concolor*
_____ *A. c. relicta* — Mts. of s Mexico (s Jalisco to Guerrero, Morelos and Oaxaca)
_____ *A. c. concolor* — Mts. of s Mexico (Chiapas) to Nicaragua, Costa Rica and Panama
_____ *A. c. aequatorialis* — Mountains of sw Colombia (Nariño) to n Peru (Cajamarca)

☐ **Carrizal Seedeater** *Amaurospiza carrizalensis*

N Venezuela (lower Río Caroni in Bolívar)

☐ **Cuban Bullfinch** *Melopyrrha nigra*
_____ *M. n. nigra* — Cuba and Isle of Pines
_____ *M. n. taylori* — Grand Cayman I.

☐ **White-naped Seedeater** *Dolospingus fringilloides*

Savanna of e Colombia to s Venezuela and nw Amazonian Brazil

☐ **Band-tailed Seedeater** *Catamenia analis*
_____ *C. a. alpica* — Santa Marta Mountains (ne Colombia)
_____ *C. a. schistaceifrons* — Eastern Andes of central Colombia
_____ *C. a. soederstromi* — Andes of n Ecuador (Imbabura to Chimborazo)
_____ *C. a. insignis* — E slope of Andes of Peru (Cajamarca to Ancash)
_____ *C. a. analoides* — W slope of Andes of Peru (Piura to Ayacucho)
_____ *C. a. griseiventris* — Andes of se Peru (Cuzco)
_____ *C. a. analis (subinsignis)* — Andes of n Chile to central Bolivia and nw Argentina

☐ **Plain-colored Seedeater** *Catamenia inornata*
_____ *C. i. minor* — Andes of Colombia to w Venezuela, Ecuador and Peru (Junín)
_____ *C. i. mucuchiesi* — Andes of w Venezuela (Mérida)
_____ *C. i. inornata* — Andes of se Peru (Cuzco) to Bolivia, Chile and nw Argentina

☐ **Paramo Seedeater** *Catamenia homochroa*
_____ *C. h. homochroa* — Andes of Colombia to w Venezuela, Ecuador, Peru and n Bolivia
_____ *C. h. duncani* — *Tepuis* of s Venezuela (Bolívar and Amazonas) and ne Brazil

☐ **Dull-colored Grassquit** *Tiaris obscurus*
_____ *T. o. haplochroma* — NE Colombia (Santa Marta Mountains) to nw Venezuela
_____ *T. o. pauper* — Extreme s Colombia (Nariño) to Ecuador and nw Peru
_____ *T. o. obscurus* — E slope of Andes of central Peru to w Bolivia and nw Argentina
_____ *T. o. pacificus* — *Lomas* of coastal w Peru (n Lima to n Arequipa)

☐ **Cuban Grassquit** *Tiaris canorus*

Cuba and Isle of Pines; introduced Isla Providéncia

☐ **Yellow-faced Grassquit** *Tiaris olivaceus*
_____ *T. o. pusillus* — Gulf lowlands of e Mexico to Colombia and w Venezuela
_____ *T. o. intermedius* — Cozumel I. and Holbox I. (off Yucatán Peninsula)
_____ *T. o. ravidus* — Isla Coiba (Panama)
_____ *T. o. olivaceus* — Cuba, Isle of Pines, Jamaica and Cayman Islands
_____ *T. o. bryanti* — Puerto Rico

☐ **Black-faced Grassquit** *Tiaris bicolor*
_____ *T. b. bicolor* — Bahamas and cays off Las Villas Province (Cuba)
_____ *T. b. marchii* — Jamaica, Hispaniola and adjacent islands
_____ *T. b. omissus* — Puerto Rico, Tobago, Isla Margarita, n Colombia and n Venezuela

____	*T. b. huilae*	S-central Colombia (upper Magdalena Valley)
____	*T. b. grandior*	Islas Providéncia, Santa Catalina and San Andrés (w Caribbean Sea)
____	*T. b. johnstonei*	Isla La Blanquilla and Islas Los Hermanos (off Venezuela)
____	*T. b. sharpei*	Netherlands Antilles (Aruba, Curaçao and Bonaire)
____	*T. b. tortugensis*	Isla La Tortuga (off Venezuela)

☐ **Sooty Grassquit** *Tiaris fuliginosus*

Colombia to Venezuela, Guyana, e and sw Brazil

☐ **Yellow-shouldered Grassquit** *Loxipasser anoxanthus*

Hills and mountains of Jamaica

☐ **Orangequit** *Euneornis campestris*

Wooded highlands of Jamaica

☐ **St. Lucia Black Finch** *Melanospiza richardsoni*

St. Lucia (Lesser Antilles)

☐ **Puerto Rican Bullfinch** *Loxigilla portoricensis*

____	*L. p. portoricensis*	Puerto Rico
____	*L. p. grandis*	St. Kitts I. (unrecorded since 1920s)

☐ **Greater Antillean Bullfinch** *Loxigilla violacea*

____	*L. v. violacea*	Bahamas and Caicos Islands
____	*L. v. maurella*	Tortue, Gonâve and Saona islands (off Puerto Rico)
____	*L. v. affinis*	Hispaniola and Isla Catalina
____	*L. v. parishi*	Île-á-Vache and Beata I. (off Hispaniola)
____	*L. v. ruficollis*	Jamaica

☐ **Lesser Antillean Bullfinch** *Loxigilla noctis*

____	*L. n. coryi*	Lesser Antilles (St. Kitts and Montserrat)
____	*L. n. ridgwayi*	Lesser Antilles (Anguilla, St. Martin, Barbuda and Antigua)
____	*L. n. desiradensis*	Désirade I. (Lesser Antilles)
____	*L. n. dominicana*	Guadeloupe, Marie Galante, Dominica and Iles des Saintes
____	*L. n. noctis*	Martinique (Lesser Antilles)
____	*L. n. sclateri*	St. Lucia (Lesser Antilles)
____	*L. n. crissalis*	St. Vincent (Lesser Antilles)
____	*L. n. grenadensis*	Grenada (Lesser Antilles)

☐ **Barbados Bullfinch** *Loxigilla barbadensis*

Barbados (Lesser Antilles)

☐ **Cocos Island Finch** *Pinaroloxias inornata*

Forests of Cocos I. (off w Costa Rica)

☐ **Slaty Finch** *Haplospiza rustica*

____	*H. r. uniformis*	Highlands of s Mexico (Veracruz and Chiapas)
____	*H. r. barrilesensis*	Highlands of Honduras, Costa Rica and w Panama (Chiriquí)
____	*H. r. rustica*	N Colombia to n Venezuela, Ecuador, Peru and Bolivia
____	*H. r. arcana*	*Tepuis* of s Venezuela (Cerro Chimantá-tepui in Bolívar)

☐ **Uniform Finch** *Haplospiza unicolor*

E Paraguay to se Brazil and ne Argentina

☐ **Peg-billed Finch** *Acanthidops bairdii*

High volcanic peaks of Costa Rica and w Panama

☐ **Cinnamon-bellied Flowerpiercer** *Diglossa baritula*

____	*D. b. baritula*	Highlands of cent. Mexico (se Jalisco to Isthmus of Tehuántepec)
____	*D. b. montana*	Highlands of s Mexico (Chiapas) to Guatemala and El Salvador
____	*D. b. parva*	Highlands of e Guatemala to Honduras and n-central Nicaragua

☐ **Slaty Flowerpiercer** *Diglossa plumbea*

____	*D. p. plumbea*	Highlands of Costa Rica and extreme w Panama (Chiriquí)
____	*D. p. veraguensis*	Pacific slope of w Panama (Veraguas)

☐ **Rusty Flowerpiercer** *Diglossa sittoides*

____	*D. s. hyperythra*	N Colombia (Santa Marta Mts.) and coastal mts. of n Venezuela

____	*D. s. dorbignyi*	Andes of Colombia and w Venezuela (Lara, Mérida and Táchira)
____	*D. s. coelestis*	Sierra de Perijá (Colombia/Venezuela border)
____	*D. s. mandeli*	Subtropical mts. of ne Venezuela (Mt. Turumiquire in Sucre)
____	*D. s. decorata*	Subtropical Andes of Ecuador and Peru
____	*D. s. sittoides*	Subtropical Andes of w Bolivia and nw Argentina

☐ **Venezuelan Flowerpiercer** *Diglossa venezuelensis*

Mountains of ne Venezuela (nw Monagas and s Sucre)

☐ **Chestnut-bellied Flowerpiercer** *Diglossa gloriosissima*

Andes of w Colombia (Antioquia to Cauca)

☐ **White-sided Flowerpiercer** *Diglossa albilatera*

____	*D. a. federalis*	Coastal cordillera of n Venezuela (Aragua to Miranda)
____	*D. a. albilatera*	Santa Marta Mts. and Andes of Colombia to w Venezuela, Ecuador
____	*D. a. schistacea*	Andes of extreme sw Ecuador to nw Peru (Cajamarca)
____	*D. a. affinis*	Highlands of n-central Peru (above Río Utcubamba) to Cuzco

☐ **Glossy Flowerpiercer** *Diglossa lafresnayii*

____	*D. l. lafresnayii*	Andes of Colombia to w Venezuela, Ecuador and extreme n Peru
____	*D. l. unicincta*	Andes of n Peru (Cajamarca)

☐ **Moustached Flowerpiercer** *Diglossa mystacalis*

____	*D. m. pectoralis*	Andes of n and central Peru (Amazonas to Huánuco and Junín)
____	*D. m. albilinea*	Andes of se Peru (Cuzco and Puno)
____	*D. m. mystacalis*	Andes of nw Bolivia (La Paz, Cochabamba and w Santa Cruz)

☐ **Merida Flowerpiercer** *Diglossa gloriosa*

Andes of w Venezuela (Trujillo, Mérida and n Táchira)

☐ **Black Flowerpiercer** *Diglossa humeralis*

____	*D. h. nocticolor*	Santa Marta Mts. of n Colombia and Sierra de Perijá (w Venezuela)
____	*D. h. humeralis*	E Andes of Colombia and sw Venezuela (Páramo de Tamá)
____	*D. h. aterrima*	W and Central Andes of Colombia to Ecuador and nw Peru

☐ **Black-throated Flowerpiercer** *Diglossa brunneiventris*

Andes of n Colombia; Andes of Peru, nw Bolivia and n Chile

☐ **Gray-bellied Flowerpiercer** *Diglossa carbonaria*

Andes of w Bolivia (La Paz to Chuquisaca) and adj. nw Argentina

☐ **Scaled Flowerpiercer** *Diglossa duidae*

____	*D. d. hitchcocki*	*Tepuis* of s Venezuela (s Amazonas)
____	*D. d. duidae*	*Tepuis* of s Venezuela (c Amazonas) and adj. n Brazil (Roraima)
____	*D. d. georgebarrowcloughi*	*Tepuis* of s Venezuela (Cerro Jime)

☐ **Greater Flowerpiercer** *Diglossa major*

____	*D. m. gilliardi*	*Tepuis* of s Venezuela (Auyan-tepui in se Bolívar)
____	*D. m. disjuncta*	*Tepuis* of se Venezuela (Gran Sabana in Bolívar)
____	*D. m. chimantae*	*Tepuis* of se Venezuela (Chimantáa-tepui in Bolívar)
____	*D. m. major*	Mts. of se Venezuela and adj. n Brazil (Uei-tepui in Roraima)

☐ **Indigo Flowerpiercer** *Diglossopis indigoticus*

Andes of w Colombia and nw Ecuador (south to Pichincha)

☐ **Deep-blue Flowerpiercer** *Diglossopis glaucus*

____	*D. g. tyrianthinus*	E slope of Andes of sw Colombia (Nariño) and e Ecuador
____	*D. g. glaucus*	SE Peru (Junín) to yungas of w Bolivia (La Paz and Cochabamba)

☐ **Bluish Flowerpiercer** *Diglossopis caerulescens*

____	*D. c. saturatus*	Andes of Colombia and sw Venezuela
____	*D. c. ginesi*	Sierra de Perijá (Colombia/Venezuela border)
____	*D. c. caerulescens*	Coastal mts. of n Venezuela (Carabobo to Distrito Federal)
____	*D. c. medius*	Andes of s Ecuador (Loja) to nw Peru (Cajamarca and Amazonas)
____	*D. c. pallidus*	Andes of central Peru (La Libertad to Lima and Junín)
____	*D. c. mentalis*	Andes of se Peru and nw Bolivia (La Paz)

☐ **Masked Flowerpiercer** *Diglossopis cyaneus*
___ *D. c. cyaneus* — Andes of Colombia to Ecuador and w Venezuela
___ *D. c. tovarensis* — Coastal mountains of n Venezuela (Aragua and Distrito Federal)
___ *D. c. obscurus* — Sierra de Perijá (Colombia/Venezuela border)
___ *D. c. dispar* — Andes of sw Ecuador and nw Peru
___ *D. c. melanopis* — Andes of Peru and nw Bolivia

☐ **Puna Yellow-Finch** *Sicalis lutea*
Andes of s Peru (Cuzco) to w Bolivia and nw Argentina

☐ **Saffron Finch** *Sicalis flaveola*
___ *S. f. flaveola* — Tropical e Colombia to Venezuela and the Guianas; Trinidad
___ *S. f. valida* — Pacific lowlands of Ecuador to nw Peru (Ancash)
___ *S. f. brasiliensis* — Tropical ne Brazil (Maranhão, Minas Gerais and São Paulo)
___ *S. f. pelzelni* — E Bolivia to Paraguay, se Brazil, Uruguay and n Argentina

☐ **Grassland Yellow-Finch** *Sicalis luteola*
___ *S. l. chrysops* — S Mexico (Veracruz and Chiapas) to Guatemala and ne Honduras
___ *S. l. mexicana* — Pacific slope of s Mexico (sw Puebla and Morelos)
___ *S. l. eisenmanni* — Pacific slope of nw Costa Rica (Guanacaste) and Panama (Coclé)
___ *S. l. bogotensis* — E Andes of Colombia to Venezuela, Ecuador and s Peru
___ *S. l. luteola* — W Andes of Colombia to Venezuela, the Guianas and n Brazil
___ *S. l. flavissima* — Islands at mouth of Amazon River and adjacent Pará
___ *S. l. luteiventris* — Lowlands of s South America

☐ **Stripe-tailed Yellow-Finch** *Sicalis citrina*
___ *S. c. browni* — Highlands of Colombia to Venezuela, the Guianas and ne Brazil
___ *S. c. citrina* — E Brazil (s Pará to Goiás, Piauí, e Mato Grosso and Paraná)
___ *S. c. occidentalis* — Andes of extreme s Peru (Puno) and n Argentina (Tucumán)

☐ **Bright-rumped Yellow-Finch** *Sicalis uropygialis*
___ *S. u. sharpei* — Andes of n Peru (Cajamarca to Junín)
___ *S. u. connectens* — Andes of s Peru (above upper Urubamba Valley in Cuzco)
___ *S. u. uropygialis* — Andes of s Peru (Puno) to Bolivia, n Chile and nw Argentina

☐ **Citron-headed Yellow-Finch** *Sicalis luteocephala*
Highlands of w Bolivia and nw Argentina (Jujuy)

☐ **Greater Yellow-Finch** *Sicalis auriventris*
Andes of n and central Chile and adjacent Argentina

☐ **Greenish Yellow-Finch** *Sicalis olivascens*
___ *S. o. salvini* — Andes of n Peru (La Libertad to Huánuco, Junín and Ayacucho)
___ *S. o. chloris* — W slope of Andes of Peru (Ancash) to n Chile (Coquimbo)
___ *S. o. olivascens* — Andes of se Peru (Cuzco) to w Bolivia and nw Argentina
___ *S. o. mendozae* — Andes of w Argentina (Río Negro to Mendoza and San Luis)

☐ **Patagonian Yellow-Finch** *Sicalis lebruni*
Open plains of s Chile and s Argentina

☐ **Orange-fronted Yellow-Finch** *Sicalis columbiana*
___ *S. c. columbiana* — Extreme e Colombia to e Venezuela (Orinoco basin); Trinidad
___ *S. c. leopoldinae* — E Brazil (Piauí to w Bahia and Goiás)
___ *S. c. goeldii* — Amazonian Brazil (to w Pará)

☐ **Raimondi's Yellow-Finch** *Sicalis raimondii*
W Andes of Peru (Cajamarca to Arequipa and Moquegua)

☐ **Sulphur-throated Finch** *Sicalis taczanowskii*
Arid littoral of sw Ecuador and nw Peru

☐ **Wedge-tailed Grass-Finch** *Emberizoides herbicola*
___ *E. h. lucaris* — SW Costa Rica (Térraba area)
___ *E. h. hypochondriacus* — Foothills of w Panama (Volcán de Chiriquí) to Panama City
___ *E. h. floresae* — Mountains of w Panama (Cerro Flores in e Chiriquí)
___ *E. h. apurensis* — Tropical Colombia (east of Andes) and w Venezuela

_____ _E. h. sphenurus_ — N Colombia to Venezuela, the Guianas and n Brazil
_____ _E. h. herbicola_ — SE Peru to e Bolivia, Paraguay, ne Argentina, e and se Brazil

☐ **Duida Grass-Finch** _Emberizoides duidae_

Tepuis of s Venezuela (Cerro Duida in Amazonas)

☐ **Lesser Grass-Finch** _Emberizoides ypiranganus_

Marshes of e Paraguay to se Brazil and ne Argentina

☐ **Pale-throated Serra-Finch** _Embernagra longicauda_

Locally on _serras_ of e Brazil (interior c Bahia and Minas Gerais)

☐ **Great Pampa-Finch** _Embernagra platensis_
_____ _E. p. olivascens_ — SE Bolivia to sw Paraguay and nw Argentina
_____ _E. p. platensis_ — E Paraguay to se Brazil, Uruguay and central Argentina

☐ **Yellow Cardinal** _Gubernatrix cristata_

Extreme se Brazil and Uruguay to n Argentina

☐ **Red-crested Cardinal** _Paroaria coronata_

Bolivia to c Argentina, Uruguay, Paraguay and extreme se Brazil

☐ **Red-cowled Cardinal** _Paroaria dominicana_

Forests of ne Brazil (Maranhão to Minas Gerais)

☐ **Red-capped Cardinal** _Paroaria gularis_
_____ _P. g. nigrogenis_ — E Colombia and Venezuela; Trinidad
_____ _P. g. gularis_ — E Colombia to Venezuela, the Guianas, Peru and Amaz. Brazil
_____ _P. g. cervicalis_ — N Bolivia and adjacent Brazil (Mato Grosso)

☐ **Crimson-fronted Cardinal** _Paroaria baeri_
_____ _P. b. baeri_ — Central Brazil (w Goiás and adjacent ne Mato Grosso)
_____ _P. b. xinguensis_ — Central Brazil (upper Rio Xingú in n Mato Grosso)

☐ **Yellow-billed Cardinal** _Paroaria capitata_
_____ _P. c. capitata_ — SW Brazil (Mato Grosso) to Paraguay and n Argentina
_____ _P. c. fuscipes_ — SE Bolivia (Fortín Campero region in Tarija)

☐ **Sooty-faced Finch** _Lysurus crassirostris_

Humid montane forests of Costa Rica and w Panama

☐ **Olive Finch** _Lysurus castaneiceps_

Locally in Andes of Colombia to Ecuador and se Peru

☐ **Yellow-thighed Finch** _Pselliophorus tibialis_

Humid montane forests of Costa Rica and w Panama

☐ **Yellow-green Finch** _Pselliophorus luteoviridis_

Humid montane forests of w Panama

☐ **Large-footed Finch** _Pezopetes capitalis_

Humid montane forests of Costa Rica and w Panama

☐ **White-naped Brush-Finch** _Atlapetes albinucha_
_____ _A. a. albinucha_ — Highlands of s Mexico (Puebla and Veracruz to n Chiapas)
_____ _A. a. griseipectus_ — Pacific slope of s Mexico (Chiapas) to w Guatemala and El Salvador
_____ _A. a. fuscipygius_ — Highlands of Honduras to nw El Salvador and nw Nicaragua
_____ _A. a. parvirostris_ — Subtropical Costa Rica
_____ _A. a. brunnescens_ — Subtropical w Panama (w Chiriquí)
_____ _A. a. coloratus_ — Subtropical w Panama (e Chiriquí and Veraguas)
_____ _A. a. azuerensis_ — W Panama (Azuero Peninsula)
_____ _A. a. gutturalis_ — Upper tropical and subtropical n Colombia

☐ **Pale-naped Brush-Finch** _Atlapetes pallidinucha_
_____ _A. p. pallidinucha_ — E Andes of Colombia and sw Venezuela (Mérida and Táchira)
_____ _A. p. papallactae_ — Central Andes of Colombia to Ecuador and extreme nw Peru

☐ **Yellow-breasted Brush-Finch** _Atlapetes latinuchus_
_____ _A. l. phelpsi_ — Mts. of ne Colombia and western Venezuela (Perijá Mts.)
_____ _A. l. elaeoprorus_ — Subtropical n Central Andes of Colombia (Antioquia)

____	*A. l. simplex*	E Andes of Colombia (Bogotá)
____	*A. l. caucae*	Western and Central Andes of Colombia (Valle and Cauca)
____	*A. l. spodionotus*	Andes of s Colombia (Nariño) and n Ecuador
____	*A. l. comptus*	Andes of sw Ecuador (Cañar to w Loja) and n Peru (Piura)
____	*A. l. latinuchus*	Andes of se Ecuador (e Azuay and e Loja) to ne Peru (Amazonas)
____	*A. l. chugurensis*	Pacific slope of nw Peru (Cajamarca)
____	*A. l. baroni*	N Peru (upper Marañón Valley in Cajamarca and La Libertad)
____	*A. l. yariguierum*	Yariguies Mts. Colombia

☐ **Vilcabamba Brush-Finch** *Atlapetes terborghi*

Andes of se Peru (Cordillera Vilcabamba)

☐ **Black-faced Brush-Finch** *Atlapetes melanolaemus*

Subtropical Andes of se Peru (Cuzco and Puno)

☐ **Bolivian Brush-Finch** *Atlapetes rufinucha*

____	*A. r. rufinucha*	Subtropical Andes of w Bolivia (La Paz and Cochabamba)
____	*A. r. carrikeri*	Andes of e Bolivia (Santa Cruz)

☐ **Slaty Brush-Finch** *Atlapetes schistaceus*

____	*A. s. schistaceus*	W and Central Andes of Colombia and Ecuador
____	*A. s. tamae*	Andes of e Colombia and sw Venezuela (Táchira)
____	*A. s. fumidus*	Sierra de Perijá (Colombia/Venezuela border)
____	*A. s. castaneifrons*	Andes of w Venezuela (Trujillo, Mérida and e Táchira)
____	*A. s. taczanowskii*	Andes of central Peru (Huánuco and Junín)

☐ **Cusco Brush-Finch** *Atlapetes canigenis*

Andes of se Peru (Cuzco)

☐ **White-rimmed Brush-Finch** *Atlapetes leucopis*

Locally in Andes of s Colombia and e Ecuador

☐ **Rufous-capped Brush-Finch** *Atlapetes pileatus*

____	*A. p. dilutus*	Oak-pine forests of n Mexican plateau
____	*A. p. pileatus*	Oak-pine forests of s Mexican plateau

☐ **Santa Marta Brush-Finch** *Atlapetes melanocephalus*

Santa Marta Mountains (ne Colombia)

☐ **Yellow-headed Brush-Finch** *Atlapetes flaviceps*

Central Andes of Colombia

☐ **Dusky-headed Brush-Finch** *Atlapetes fuscoolivaceus*

Andes of Colombia (upper Magdalena Valley)

☐ **Tricolored Brush-Finch** *Atlapetes tricolor*

____	*A. t. crassus*	Andes of w Colombia and Ecuador (Pichincha and El Oro)
____	*A. t. tricolor*	Andes of central Peru (La Libertad and Junín)

☐ **Moustached Brush-Finch** *Atlapetes albofrenatus*

____	*A. a. albofrenatus*	E Andes of Colombia (Norte de Santander to Cundinamarca)
____	*A. a. meridae*	Andes of w Venezuela (Táchira and Mérida)

☐ **Bay-crowned Brush-Finch** *Atlapetes seebohmi*

____	*A. s. celicae*	Andes of extreme s Ecuador (w Loja)
____	*A. s. simonsi*	Andes of extreme s Ecuador (eastern and central Loja)
____	*A. s. seebohmi*	Pacific slope of Andes of nw Peru (La Libertad to Ancash)

☐ **Rusty-bellied Brush-Finch** *Atlapetes nationi*

____	*A. n. nationi*	Western Andes of Peru (Ancash to Arequipa)
____	*A. n. brunneiceps*	Andes of sw Peru (Ica and Ayacucho)

☐ **White-winged Brush-Finch** *Atlapetes leucopterus*

____	*A. l. leucopterus*	Andes of w Ecuador (south to Chimbo Valley)
____	*A. l. dresseri*	Andes of sw Ecuador (El Oro and Loja)
____	*A. l. paynteri*	N Peru (n Piura to Cajamarca border)

☐ **White-headed Brush-Finch** *Atlapetes albiceps*

Arid scrub of se Ecuador and nw Peru

☐ **Pale-headed Brush-Finch** *Atlapetes pallidiceps*

Arid Andes of sw Ecuador (Azuay)

☐ **Rufous-eared Brush-Finch** *Atlapetes rufigenis*

Andes of nw Peru (Cajamarca, La Libertad, Ancash and Huánuco)

☐ **Apurimac Brush-Finch** *Atlapetes forbesi*

Andes of s-central Peru (Apurímac, Cuzco and Puno)

☐ **Black-spectacled Brush-Finch** *Atlapetes melanopsis*

Andes of s Peru (Ayacucho and Huancavelica)

☐ **Ochre-breasted Brush-Finch** *Atlapetes semirufus*
- ____ *A. s. zimmeri* — E slope of E Andes of ne Colombia and w Venezuela (Táchira)
- ____ *A. s. majusculus* — E slope of Eastern Andes of Colombia (n Boyacá)
- ____ *A. s. semirufus* — E slope of Eastern Andes of Colombia (Cundinamarca)
- ____ *A. s. denisei* — Mts. of n Venezuela (Sucre and Monagas to Aragua and Carabobo)
- ____ *A. s. benedettii* — Mountains of n Venezuela (Falcón, Lara and Trujillo)
- ____ *A. s. albigula* — Andes of w Venezuela (ne Táchira)

☐ **Fulvous-headed Brush-Finch** *Atlapetes fulviceps*

Andes of Bolivia to nw Argentina

☐ **Tepui Brush-Finch** *Atlapetes personatus*
- ____ *A. p. personatus* — *Tepuis* of s Venezuela (Mt. Roraima and adjacent *tepuis*)
- ____ *A. p. collaris* — *Tepuis* of s Venezuela (Auyan-tepui in se Bolívar)
- ____ *A. p. duidae* — *Tepuis* of s Venezuela (Mt. Duida in Amazonas)
- ____ *A. p. parui* — *Tepuis* of s Venezuela (n Amazonas)
- ____ *A. p. paraquensis* — *Tepuis* of s Venezuela (Cerro Paraque and Cerro Yaví)
- ____ *A. p. jugularis* — Cerro de la Neblina (Venezuela/Brazil border in se Amazonas)

☐ **Yellow-striped Brush-Finch** *Atlapetes citrinellus*

Andes of nw Argentina

☐ **Chestnut-capped Brush-Finch** *Buarremon brunneinucha*
- ____ *B. b. suttoni* — Mountains of sw Mexico (Guerrero to central Oaxaca)
- ____ *B. b. nigrilatera* — Mountains of s Mexico (Oaxaca)
- ____ *B. b. brunneinucha* — Subtrop. e Mexico (San Luis Potosí and Veracruz to ne Oaxaca)
- ____ *B. b. apertus* — S Mexico (Sierra de Tuxtla of s Veracruz)
- ____ *B. b. macrourus* — Mountains of s Mexico (Chiapas) and sw Guatemala
- ____ *B. b. alleni* — Mountains of n El Salvador, Honduras and w Nicaragua
- ____ *B. b. elsae* — Mountains of Costa Rica to w and central Panama
- ____ *B. b. frontalis* — Mts. of extreme e Panama to Colombia, w Venezuela and s Peru
- ____ *B. b. allinornatus* — Mountains of nw Venezuela (Falcón and Yaracuy)
- ____ *B. b. inornatus* — Mts. of w-central Ecuador (Río Chimbo and Río Chanchan area)

☐ **Green-striped Brush-Finch** *Buarremon virenticeps*
- ____ *B. v. verecundus* — Mountains of w Mexico (s Sinaloa, n Nayarit and s Durango)
- ____ *B. v. virenticeps* — Mts. of w Mexico (Jalisco and Colima to Morelos and w Puebla)

☐ **Stripe-headed Brush-Finch** *Buarremon torquatus*
- ____ *B. t. costaricensis* — Highlands of sw Costa Rica and w Panama (Chiriquí)
- ____ *B. t. tacarcunae* — Mountains of e Panama
- ____ *B. t. atricapillus* — Central and Eastern Andes of n Colombia
- ____ *B. t. basilicus* — Santa Marta Mountains (ne Colombia)
- ____ *B. t. perijanus* — E slope of Andes of e Colombia and w Venezuela (Zulia)
- ____ *B. t. larensis* — Andes of w Venezuela (w Lara and w Táchira)
- ____ *B. t. phaeopleurus* — Coastal mts. of n Venezuela (Aragua, Miranda, Distrito Federal)
- ____ *B. t. phygas* — Mts. of ne Venezuela (Anzoátegui, Monagas and Sucre)
- ____ *B. t. assimilis* — Andes of Colombia to w Venezuela, Ecuador and nw Peru
- ____ *B. t. nigrifrons* — Andes of sw Ecuador (El Oro and Loja) and nw Peru
- ____ *B. t. poliophrys* — Andes of central and se Peru (Huánuco, Junín and Cuzco)
- ____ *B. t. torquatus* — Andes of w Bolivia (La Paz and w Cochabamba)
- ____ *B. t. fimbriatus* — Andes of Bolivia (e Cochabamba, w Santa Cruz and Chuquisaca)
- ____ *B. t. borelli* — Andes of s Bolivia (Tarija) and nw Argentina (Jujuy and Salta)

□ **Orange-billed Sparrow** *Arremon aurantiirostris*
____	*A. a. saturatus*	Caribbean slope of se Mexico (Veracruz) to Guatemala and Belize
____	*A. a. rufidorsalis*	Caribbean slope of Honduras to Nicaragua and Costa Rica
____	*A. a. aurantiirostris*	Pacific slope of Costa Rica and Panama
____	*A. a. strictocollaris*	Extreme e Panama and adjacent nw Colombia (Chocó)
____	*A. a. occidentalis*	Pacific slope of w Colombia and nw Ecuador
____	*A. a. erythrorhynchus*	N Colombia (middle Magdalena, lower Cauca, upper Sinú valleys)
____	*A. a. spectabilis*	SE Colombia (Putumayo) to e Ecuador and ne Peru (San Martín)
____	*A. a. santarosae*	SW Ecuador

□ **Pectoral Sparrow** *Arremon taciturnus*
____	*A. t. axillaris*	Colombia (east of the Andes) and adjacent w Venezuela
____	*A. t. taciturnus*	E Colombia to s Venezuela, the Guianas and Amazonian Brazil
____	*A. t. nigrirostris*	Tropical se Peru to n Bolivia and nw Argentina

□ **San Francisco Sparrow** *Arremon franciscanus*

Thick scrub caatinga of Bahia and Minas Gerais, Brazil

□ **Half-collared Sparrow** *Arremon semitorquatus*

SE Brazil (Rio de Janeiro to Rio Grande do Sul)

□ **Golden-winged Sparrow** *Arremon schlegeli*
____	*A. s. fratruelis*	N Colombia (Guajira Peninsula)
____	*A. s. canidorsum*	W slope of Eastern Andes of Colombia (Santander)
____	*A. s. schlegeli*	Caribbean coast of e Colombia and n Venezuela

□ **Black-capped Sparrow** *Arremon abeillei*
____	*A. a. abeillei*	Arid scrub of sw Ecuador (Manabí) to nw Peru (Cajamarca)
____	*A. a. nigriceps*	NW Peru (upper Marañón Valley)

□ **Saffron-billed Sparrow** *Arremon flavirostris*
____	*A. f. flavirostris*	E Brazil (Bahia to s Goiás, w Minas Gerais and se Mato Grosso)
____	*A. f. dorbignii*	Lowlands of e Bolivia and nw Argentina
____	*A. f. devillii*	E Bolivia (Santa Cruz) to se Brazil (s São Paulo and adj. Goiás)
____	*A. f. polionotus*	Paraguay, n Argentina and sw Brazil (w Paraná and Mato Grosso)

□ **Olive Sparrow** *Arremonops rufivirgatus*
____	*A. r. sinaloae*	Coastal w Mexico (Sinaloa to Nayarit)
____	*A. r. sumichrasti*	Coastal sw Mexico (Jalisco to Colima, Michoacán and Oaxaca)
____	*A. r. rufivirgatus*	S Texas to ne Mexico (Coahuila and Nuevo León to Tamaulipas)
____	*A. r. ridgwayi*	E Mexico (s Tamaulipas to San Luis Potosí and n Veracruz)
____	*A. r. crassirostris*	Coastal se Mexico (Veracruz to e Puebla and n Oaxaca)
____	*A. r. verticalis*	SE Mexico (Yucatán Pen.) to Petén of n Guatemala and n Belize
____	*A. r. rhypthothorax*	SE Mexico (Yucatán Peninsula)
____	*A. r. chiapensis*	S Mexico (central valley of Chiapas)
____	*A. r. superciliosus*	Pacific coast of nw Costa Rica

□ **Tocuyo Sparrow** *Arremonops tocuyensis*

Arid scrub of ne Colombia and nw Venezuela

□ **Green-backed Sparrow** *Arremonops chloronotus*
____	*A. c. chloronotus*	Gulf slope of se Mexico (Tabasco) to Belize and nw Honduras
____	*A. c. twomeyi*	Tropical n-central Honduras (Yoro and Olancho)

□ **Black-striped Sparrow** *Arremonops conirostris*
____	*A. c. richmondi*	Tropical e Honduras to Nicaragua, Costa Rica and w Panama
____	*A. c. striaticeps*	Central and e Panama to Pacific slope of Colombia and w Ecuador
____	*A. c. inexpectatus*	Colombia (arid upper Magdalena Valley)
____	*A. c. conirostris*	Caribbean coast of Colombia to n Venezuela and extreme n Brazil
____	*A. c. umbrinus*	E Colombia (Norte de Santander) to w Venezuela

□ **Rusty-crowned Ground-Sparrow** *Melozone kieneri*
____	*M. k. grisior*	Arid Pacific slope of nw Mexico (se Sonora and ne Sinaloa)
____	*M. k. kieneri*	Arid w Mexico (Sinaloa, w Durango, Nayarit, w Jalisco, Colima)
____	*M. k. rubricata (obscurior)*	Arid sw Mexico (Michoacán, Puebla, Morelos, Guerrero, Oaxaca)

☐ **Prevost's Ground-Sparrow** *Melozone biarcuata*

____ *M. b. hartwegi*	Highlands of s Mexico (Chiapas)
____ *M. b. biarcuata*	Highlands of Guatemala, El Salvador and w Honduras
____ *M. b. cabanisi*	Highlands of central Costa Rica

☐ **White-eared Ground-Sparrow** *Melozone leucotis*

____ *M. l. occipitalis*	Highlands of s Mexico (Chiapas) to Guatemala and El Salvador
____ *M. l. nigrior*	Highlands of n-central Nicaragua
____ *M. l. leucotis*	Highlands of central Costa Rica

☐ **Green-tailed Towhee** *Pipilo chlorurus*

Highlands of w US; > to central Mexico

☐ **Collared Towhee** *Pipilo ocai*

____ *P. o. alticola*	Mountains of w Mexico (w Jalisco and extreme ne Colima)
____ *P. o. nigrescens*	Mountains of w Mexico (n-central Michoacán)
____ *P. o. guerrerensis*	Mountains of sw Mexico (Sierra Madre del Sur of Guerrero)
____ *P. o. brunnescens*	Mountains of s Mexico (central Oaxaca)
____ *P. o. ocai*	Mountains of se Mexico (e Puebla and w-central Veracruz)

☐ **Socorro Towhee** *Pipilo socorroensis*

Socorro I. (Revillagigedo Islands off w Mexico)

☐ **Eastern Towhee** *Pipilo erythrophthalmus*

____ *P. e. erythrophthalmus*	E North America (s Canada to Gulf Coast)
____ *P. e. rileyi*	Virginia to Georgia, n Florida and se Alabama
____ *P. e. alleni*	Central peninsular Florida to tip of peninsula
____ *P. e. canaster*	Tennessee to Louisiana, Alabama and nw Florida

☐ **Spotted Towhee** *Pipilo maculatus*

____ *P. m. curtatus*	SE Br. Col. to ne Calif., Nevada and Idaho; > to se Calif.
____ *P. m. oregonus*	Coastal sw Br. Col. to sw Oregon; > to s California
____ *P. m. falcinellus*	Interior sw Oregon to Sierra Nevada and San Joaquin Valley
____ *P. m. arcticus*	Great Plains of North America to sw US; > to ne Mexico
____ *P. m. montanus*	SW US to nw Mexico; > to n Mexico
____ *P. m. falcifer*	Coastal n Calif. (Del Norte to Santa Cruz and San Benito counties)
____ *P. m. megalonyx*	Coastal s Calif. (Monterey) to nw Baja Calif. and Santa Cruz I.
____ *P. m. clementae*	Santa Rosa, Santa Catalina and San Clemente is. (off s California)
____ *P. m. umbraticola*	NW Baja California (latitude 32°N to 30°N)
____ *P. m. consobrinus†*	Formerly Isla Guadalupe (off w Baja California). Extinct
____ *P. m. magnirostris*	Mountains of s Baja California (Sierra de la Laguna)
____ *P. m. gaigei*	Mountains of New Mexico, w Texas and n Mexico (n Coahuila)
____ *P. m. griseipygius*	Mountains of w Mexico (Sierra Madre Occidental)
____ *P. m. orientalis*	Mountains of e Mexico (Sierra Madre Oriental)
____ *P. m. maculatus*	Highlands of e Mexico (Hidalgo to Veracruz and e Puebla)
____ *P. m. sympatricus*	Mountains of e Mexico (Sierra de Tuxtla of Veracruz)
____ *P. m. macronyx*	Mts. of c Mexico (e Michoacán, México, Morelos, Dist. Federal)
____ *P. m. vulcanorum*	Mts. of c Mexico (México, ne Morelos, sw Tlaxcala and w Puebla)
____ *P. m. oaxacae*	Highlands of s Mexico (n and central Oaxaca)
____ *P. m. chiapensis*	Mountains of s Mexico (central Chiapas)
____ *P. m. repetens*	Mountains of s Mexico (se Chiapas) and w Guatemala

☐ **California Towhee** *Pipilo crissalis*

____ *P. c. bullatus*	Semiarid sw Oregon and extreme n-central California
____ *P. c. carolae*	Interior California (Humboldt County to Kern County)
____ *P. c. petulans*	Coastal n California (Humboldt County to Santa Cruz County)
____ *P. c. crissalis*	Coastal cent. Calif. (n Monterey to Kern and Ventura counties)
____ *P. c. eremophilus*	E-central California (Argus Mts. and nw San Bernardino County)
____ *P. c. senicula*	Coastal s California (Los Angeles County) and nw Baja California
____ *P. c. aripolius*	Central Baja California (latitude 29°N to 26°30'N)
____ *P. c. albigula*	Cape district of s Baja California

☐ **Canyon Towhee** *Pipilo fuscus*

____	*P. f. mesoleucus*	Arizona and New Mexico to w Texas, Sonora and nw Chihuahua
____	*P. f. intermedius*	Semiarid nw Mexico (s Sonora to n Sinaloa)
____	*P. f. jamesi*	Isla Tiburón (n Sea of Cortés off Sonora)
____	*P. f. mesatus*	SE Colorado to ne New Mexico and extreme nw Oklahoma
____	*P. f. texanus*	Highlands of w and central Texas to n Mexico (nw Coahuila)
____	*P. f. perpallidus*	Mountains of w Mexico (Sierra Madre Occidental)
____	*P. f. fuscus*	Mts. of w Mexico (Nayarit, Jalisco and Colima to Distrito Federal)
____	*P. f. potosinus*	Highlands of n Mexico (Coahuila to ne Jalisco and sw Tamaulipas)
____	*P. f. campoi*	Highlands of e-cent. Mexico (Hidalgo, adj. Puebla and Veracruz)
____	*P. f. toroi*	S-cent. Mexico (Tlaxcala to Veracruz, Puebla and n Oaxaca)

☐ **Abert's Towhee** *Pipilo aberti*

____	*P. a. aberti*	Deserts of s Utah to sw Nevada and se California
____	*P. a. dumeticolus*	Colorado Desert of ne Baja and nw Mexico (Sonora)
____	*P. a. vorhiesi*	S Arizona (Tucson region) to extreme sw New Mexico

☐ **White-throated Towhee** *Pipilo albicollis*

____	*P. a. marshalli*	Arid oak-pine zone of se Mexico (Puebla)
____	*P. a. albicollis*	Oak-pine zone of s Mexico (s Puebla and e Guerrero to c Oaxaca)

☐ **Bridled Sparrow** *Aimophila mystacalis*

Arid central plateau of Mexico (Puebla to Oaxaca)

☐ **Black-chested Sparrow** *Aimophila humeralis*

Arid highlands of sw Mexico (Jalisco to w Oaxaca)

☐ **Stripe-headed Sparrow** *Aimophila ruficauda*

____	*A. r. acuminata*	Arid nw Mexico (s Durango to se Guerrero and s Puebla)
____	*A. r. lawrencii*	Mexico south of Isthmus of Tehuántepec (Oaxaca and w Chiapas)
____	*A. r. connectens*	E Guatemala (arid valley of Río Motagua)
____	*A. r. ibarrorum*	Guatemala
____	*A. r. ruficauda*	SE Guatemala to nw Costa Rica

☐ **Cinnamon-tailed Sparrow** *Aimophila sumichrasti*

Arid Pacific slope of s Mexico (Oaxaca)

☐ **Stripe-capped Sparrow** *Aimophila strigiceps*

____	*A. s. dabbenei*	NW Argentina (Jujuy, Salta and Tucumán)
____	*A. s. strigiceps*	NE Argentina to sw Paraguay (Presidente Hayes)

☐ **Tumbes Sparrow** *Aimophila stolzmanni*

Arid littoral of sw Ecuador and nw Peru

☐ **Bachman's Sparrow** *Aimophila aestivalis*

____	*A. a. illinoensis*	Indiana and Illinois to Texas and s Louisiana; > to Gulf coast
____	*A. a. bachmani*	Oak-pine woods of Mid-Atlantic states to n Gulf states
____	*A. a. aestivalis*	Oak-pine woods of e South Carolina and e Georgia to s Florida

☐ **Botteri's Sparrow** *Aimophila botterii*

____	*A. b. arizonae*	SE Arizona to n Mexico (s Sonora and n Durango)
____	*A. b. texana*	Extreme s Texas and ne Mexico (e Tamaulipas)
____	*A. b. mexicana*	Central highlands of Mexico
____	*A. b. goldmani*	Coastal w Mexico (Sinaloa to Nayarit)
____	*A. b. botterii*	S highlands of Mexico (s Puebla to Oaxaca and w Chiapas)
____	*A. b. petencia (tabascensis)*	Coastal se Mexico (Veracruz) to Belize, Guatemala and Honduras
____	*A. b. vantynei*	Highlands of central Guatemala
____	*A. b. spadiconigrescens*	Lowland pine savanna of n Honduras and ne Nicaragua
____	*A. b. vulcanica*	Highlands of Nicaragua and n Costa Rica

☐ **Cassin's Sparrow** *Aimophila cassinii*

Arid sw US and adjacent n Mexico; > to central Mexico

☐ **Rufous-crowned Sparrow** *Aimophila ruficeps*

____	*A. r. ruficeps*	Coastal ranges of central California and w slopes of Sierra Nevada

____ *A. r. canescens (lambi)*   SW California and ne Baja (east to base of San Pedro Mártir)
____ *A. r. obscura*   Channel Islands (Santa Cruz, Anacapa and Catalina)
____ *A. r. sanctorum*   Todos Santos Islands (off nw Baja California)
____ *A. r. sororia*   Mountains of s Baja California (Sierra de la Laguna)
____ *A. r. scottii*   N Arizona to New Mexico, ne Sonora and nw Coahuila
____ *A. r. rupicola*   Mountains of sw Arizona
____ *A. r. simulans*   NW Mexico (se Sonora and sw Chihuahua to Nayarit and n Jalisco)
____ *A. r. eremoeca (pallidissima)*   SE Colorado to New Mexico, Texas, n Chihuahua and c Coahuila
____ *A. r. fusca*   W Mexico (s Nayarit to sw Jalisco, n Colima and Michoacán)
____ *A. r. boucardi (extima)*   E Mexico (s Coahuila to San Luis Potosí, n Puebla and s Oaxaca)
____ *A. r. australis*   S Mexico (Guerrero to s Puebla and Oaxaca)

☐ **Rufous-winged Sparrow** *Aimophila carpalis*
____ *A. c. carpalis*   Arid s Arizona to nw Mexico (central Sonora)
____ *A. c. bangsi (distinguenda)*   NW Mexico (se Sonora to n Sinaloa)
____ *A. c. cohaerens*   NW Mexico (central Sinaloa)

☐ **Five-striped Sparrow** *Aimophila quinquestriata*
____ *A. q. septentrionalis*   S Arizona to nw Mexico (Sonora and w Chihuahua to c Sinaloa)
____ *A. q. quinquestriata*   Arid w Mexico (n Jalisco)

☐ **Oaxaca Sparrow** *Aimophila notosticta*
     Arid highlands of s Mexico (central Oaxaca)

☐ **Rusty Sparrow** *Aimophila rufescens*
____ *A. r. antonensis*   Arid nw Mexico (Sierra de San Antonio of n-central Sonora)
____ *A. r. mcleodii*   NW Mexico (e Sonora and w Chihuahua to n Sinaloa, w Durango)
____ *A. r. rufescens (disjuncta, brodkorbi)*   W Mexico (s Sinaloa to Colima, Oaxaca, s Puebla and sw Chiapas)
____ *A. r. pyrgitoides (newmani)*   E Mexico (s Tamaulipas) to Guatemala, Honduras and El Salvador
____ *A. r. pectoralis (gigas)*   S Mexico (Chiapas) to cordillera of w Guatemala and El Salvador
____ *A. r. discolor*   S Belize to ne Guatemala, n Honduras and ne Nicaragua
____ *A. r. hypaethrus*   Pacific slope of nw Costa Rica (Cordillera de Guanacaste)

☐ **Striped Sparrow** *Oriturus superciliosus*
____ *O. s. palliatus*   W Mexico (Sierra Madre Occidental from Sonora to Nayarit)
____ *O. s. superciliosus*   Humid oak-pine forests of southern Central Plateau of Mexico

☐ **Zapata Sparrow** *Torreornis inexpectata*
____ *T. i. inexpectata*   SW Cuba (arid Zapata Peninsula)
____ *T. i. sigmani*   Arid coastal se Cuba
____ *T. i. varonai*   Cayo Coco (off n Cuba)

☐ **American Tree Sparrow** *Spizella arborea*
____ *S. a. arborea*   NE Canada and Labrador; > to e-central US
____ *S. a. ochracae*   N Alaska and n Yukon to n British Columbia; > to sw US

☐ **Chipping Sparrow** *Spizella passerina*
____ *S. p. arizonae (boreophila)*   SE Alaska and w Yukon to n Baja, nw Mexico; > to Oaxaca
____ *S. p. passerina*   SE Canada to s Texas and South Carolina; > to ne Mexico
____ *S. p. atremaeus (comparanda)*   Mexico (Sierra Madre Occidental to pine belt of Nuevo León)
____ *S. p. mexicana (repetens)*   Highlands of w Mexico (Nayarit) to nw Guatemala
____ *S. p. pinetorum*   Pine forests of n Guatemala (Petén) to ne Nicaragua

☐ **Clay-colored Sparrow** *Spizella pallida*
     Breeds w Canada and US; > to Guatemala

☐ **Brewer's Sparrow** *Spizella breweri*
____ *S. b. taverneri*   SW Yukon and nw Br. Col. to se Br. Col. and sw Alberta
____ *S. b. breweri*   Br. Col. and Alberta to sw US; > to Baja and cent. Mexico

☐ **Field Sparrow** *Spizella pusilla*
____ *S. p. pusilla*   SE Canada to se US; > to Gulf Coast and s Florida
____ *S. p. arenacea*   Great Plains of central US; > to Gulf Coast and ne Mexico

☐ **Worthen's Sparrow** *Spizella wortheni*

NE Mexico (Coahuila to Nuevo León, Zacatecas and Veracruz)

☐ **Black-chinned Sparrow** *Spizella atrogularis*
_____ *S. a. evura* — SE Calif. to n Nevada, sw Utah, Arizona, w Texas and n Sonora
_____ *S. a. caurina* — Coastal central California (Contra Costa to San Benito counties)
_____ *S. a. cana* — Interior coastal mountains of California (Monterey) to n Baja
_____ *S. a. atrogularis* — Cent. plateau of Mexico (Durango to w Nuevo Leon and s Oaxaca)

☐ **Vesper Sparrow** *Pooecetes gramineus*
_____ *P. g. gramineus* — SE Canada and e-central US; > to Texas and Gulf Coast
_____ *P. g. confinis* — SW Canada to sw US; > to s Mexico (Chiapas)
_____ *P. g. affinis* — W Washington and w Oregon; > to nw Baja

☐ **Lark Sparrow** *Chondestes grammacus*
_____ *C. g. grammacus* — S-cent. Canada (Ontario) to New York, n Texas and w N Carolina
_____ *C. g. strigatus* — SW Canada and w US to n Mexico; > to s Mexico

☐ **Black-throated Sparrow** *Amphispiza bilineata*
_____ *A. b. bilineata* — N-c Texas to ne Mexico (e Coahuila, Nuevo León and Tamaulipas)
_____ *A. b. opuntia* — SE Colorado to e New Mexico, w Texas and nw Coahuila
_____ *A. b. deserticola* — Arid w-c US to n Baja, islands in Sea of Cortés and nw Chihuahua
_____ *A. b. bangsi* — Cape District of s Baja California and adjacent islands
_____ *A. b. tortugae* — Isla La Tortuga (Gulf of California)
_____ *A. b. carmenae* — Isla Carmen (Gulf of California)
_____ *A. b. belvederei* — Isla Cerralvo (Gulf of California)
_____ *A. b. cana* — Isla San Estéban (Gulf of California)
_____ *A. b. pacifica* — Arid nw Mexico (s Sonora, n Sinaloa and Isla Tiburón)
_____ *A. b. grisea* — W-c Mexico (Chihuahua to s Coahuila, n Jalisco, sw Tamaulipas)

☐ **Sage Sparrow** *Amphispiza belli*
_____ *A. b. nevadensis* — Sagebrush and saltbush of Great Basin and interior California
_____ *A. b. belli* — Coastal sage and chaparral of California, nw Baja, San Clemente I
_____ *A. b. cinerea* — Desert scrub of west-central California

☐ **Lark Bunting** *Calamospiza melanocorys*

Breeds w Canada and w US; > to n Mexico

☐ **Savannah Sparrow** *Passerculus sandwichensis*
_____ *P. s. athinus* — Aleutians and n Alaska to sw Canada; > to Baja, s Mexico
_____ *P. s. sandwichensis* — Amutka I., e Aleutians and w Alaskan Pen.; > to s California
_____ *P. s. crassus* — SE Alaska (Alexander Arch.) and adjacent mainland Alaska
_____ *P. s. brooksi* — Vancouver I. and coastal sw Br. Col. to nw Calif.; > to Baja
_____ *P. s. alaudinus* — Coastal n California (Humboldt to San Luis Obispo County)
_____ *P. s. beldingi* — Coastal s California (Santa Barbara County) to n Baja California
_____ *P. s. anulus* — Bahía Vizcaino area of Baja California
_____ *P. s. sanctorum* — San Benito Islands (off w Baja California)
_____ *P. s. guttatus* — Coastal w Baja California (Laguna San Ignacio region)
_____ *P. s. magdalenae* — Coastal w Baja California (Bahia Magdalena region)
_____ *P. s. rostratus* — NE Baja California (mouth of Colorado River and adjacent Sonora)
_____ *P. s. rufofuscus* — Central Arizona and n New Mexico to n Mexico (Chihuahua)
_____ *P. s. atratus* — Coastal nw Mexico (central Sonora to central Sinaloa)
_____ *P. s. brunnescens* — Mexico (Durango to Jalisco, Puebla, Guerrero and Oaxaca)
_____ *P. s. wetmorei* — Mountains of extreme sw Guatemala
_____ *P. s. nevadensis* — Great Basin and Great Plains of N America; > to s Mexico
_____ *P. s. oblitus* — Central Canada and n-central US; > to ne Mexico
_____ *P. s. mediogriseus* — SE Canada (Ontario to Gaspé Pen.) and ne US; > se US
_____ *P. s. labradorius* — E Quebec, Labrador and Newfoundland; > to se Texas
_____ *P. s. savanna* — Nova Scotia, Prince Edward I. and Magdalen I.; > to Bahamas
_____ *P. s. princeps* — Sable I. (Nova Scotia); > coastal Massachusetts to Georgia

☐ **Seaside Sparrow** *Ammodramus maritimus*
_____ *A. m. maritimus* — Salt marshes from Mass. to n North Carolina; > to ne Florida

____ *A. m. macgillivraii*   Salt marshes from n North Carolina to s Georgia
____ *A. m. pelonotus†*   Salt marshes of n Florida (Georgia border to New Smyrna). Extinct
____ *A. m. mirabilis*   Marshes of sw Florida (Everglades to Cape Sable)
____ *A. m. peninsulae*   Salt marshes of w Florida (Dixie County to Old Tampa Bay)
____ *A. m. junicola*   Gulf Coast of n Florida (Escambie County to Taylor County)
____ *A. m. nigrescens†*   Salt marshes of e coastal and e Florida. Extinct
____ *A. m. fisheri*   Marshes of Gulf Coast (San Antonio Bay to Alabama)
____ *A. m. sennetti*   Marshes of Gulf Coast of s Texas (Aransas County to Nueces Bay)

☐ **Nelson's Sharp-tailed Sparrow** *Ammodramus nelsoni*
____ *A. n. alter*   E Canada (s James Bay and w Quebec); > to n Florida
____ *A. n. subvirgatus*   S Quebec to Nova Scotia and e Maine; > S Car. to n Florida
____ *A. n. nelsoni*   W North America (s Canada to n US; > to nw Mexico)

☐ **Saltmarsh Sharp-tailed Sparrow** *Ammodramus caudacutus*
____ *A. c. caudacutus*   Marshes of s Maine to s New Jersey; > to s Florida
____ *A. c. diversus*   Coastal s New Jersey to N Carolina; > to Florida Gulf Coast

☐ **Le Conte's Sparrow** *Ammodramus leconteii*
    Prairies of central North America; > to Gulf States

☐ **Henslow's Sparrow** *Ammodramus henslowii*
____ *A. h. susurrans*   NY to s N Hamp., e W Virginia and e N Car.; > to c Florida
____ *A. h. henslowii*   N-central US; > to Texas, Louisiana and n Florida

☐ **Baird's Sparrow** *Ammodramus bairdii*
    Great Plains of North America; > sw US and n Mexico

☐ **Grasshopper Sparrow** *Ammodramus savannarum*
____ *A. s. perpallidus*   SE Br. Columbia to w Ontario and sw US; > to El Salvador
____ *A. s. ammolegus*   S Arizona and nw Mexico (n Sonora); > to Guatemala
____ *A. s. pratensis*   SE Canada and e US; > to Guatemala and Cuba
____ *A. s. floridanus*   Central peninsular Florida
____ *A. s. bimaculatus*   S Mexico (Veracruz) to Nicaragua, nw Costa Rica and w Panama
____ *A. s. cracens*   Petén and e Guatemala to Belize, e Honduras and ne Nicaragua
____ *A. s. caucae*   Colombia (upper Cauca Valley) and adjacent n Ecuador
____ *A. s. savannarum*   Jamaica
____ *A. s. intricatus*   Hispaniola
____ *A. s. borinquensis*   Puerto Rico
____ *A. s. caribaeus*   Netherlands Antilles (Bonaire and Curaçao)

☐ **Grassland Sparrow** *Ammodramus humeralis*
____ *A. h. humeralis*   Lowlands of e Colombia to Venezuela, the Guianas and Brazil
____ *A. h. pallidulus*   NE Colombia (Guajira Peninsula)
____ *A. h. xanthornus*   E Bolivia (Beni) to Paraguay, Uruguay, s Brazil and n Argentina
____ *A. h. tarijensis*   E Bolivia (Santa Cruz and Tarija)

☐ **Yellow-browed Sparrow** *Ammodramus aurifrons*
____ *A. a. apurensis*   NE Colombia and w Venezuela
____ *A. a. cherriei*   *Llanos* of e Colombia (Meta)
____ *A. a. tenebrosus*   Tropical e Colombia (Vaupés) and Venezuela (sw Amazonas)
____ *A. a. aurifrons*   Trop. se Colombia to e Ecuador, Peru, Bolivia and w Amaz. Brazil

☐ **Fox Sparrow** *Passerella iliaca*
____ *P. i. chilcatensis*   Alaska
____ *P. i. zaboria*   NW Alaska to sw Canada; > e of Great Plains to c and s US
____ *P. i. altivagans*   Mts. of British Columbia and sw Alberta; > to nw Baja
____ *P. i. unalaschensis (insularis)*   E Aleutian Islands to Alaska Peninsula; > to s California
____ *P. i. ridgwayi*   Alaska (Kodiak Island group); > to s California
____ *P. i. sinuosa*   Kenai Peninsula and Prince William Sound; > to nw Baja
____ *P. i. annectens*   Alaska (Yakutat Bay region); > to s California
____ *P. i. townsendi*   SE Alaska (Glacier Bay to Queen Charlotte Is.); > to c Calif.

____	*P. i. fuliginosa*	Coastal se Alaska to nw Washington; > to s California
____	*P. i. olivacea*	Mts. of sw Br. Col. to central and e Washington; > to n Baja
____	*P. i. schistacea*	SW Br. Col. and sw Alberta to n Nevada; > to w Texas
____	*P. i. swarthi*	Mountains of nw Utah and se Idaho
____	*P. i. fulva*	Oregon e of Cascades to ne California (Modoc); > to n Baja
____	*P. i. megarhyncha*	Mts. of sw Oregon to c Calif. and w-c Nevada; > to nw Baja
____	*P. i. brevicauda*	N and inner coast ranges of California; > to s California
____	*P. i. monoensis*	Mono Lake area of e-c Calif. and adj. Nevada; > to nw Baja
____	*P. i. canescens*	Mts. of Inyo County (Calif.) and adj. Nevada; > to n Baja
____	*P. i. stephensi*	Sierra Nevada and high mts. of s Calif.; > lower elevations
____	*P. i. iliaca*	Labrador and Newfoundland to se Quebec, Ontario; > e US

☐ **Sierra Madre Sparrow** *Xenospiza baileyi*

Locally in montane pine forests of central Mexico

☐ **Song Sparrow** *Melospiza melodia*

____	*M. m. maxima*	Aleutian Islands (Attu to Atka)
____	*M. m. sanaka*	Aleutian Islands (Seguam to Unimak and Sanak to Semidi)
____	*M. m. amaka*	Amak I. (Aleutian Islands)
____	*M. m. insignis*	Kodiak Group (Sitkalidak I. to Barren Is.) and adj. Alaskan Pen.
____	*M. m. kenaiensis*	Coastal s Alaska (Cook Inlet to Copper River)
____	*M. m. caurina*	Coastal se Alaska (Yakutat Bay to Cross Sound); > to n Calif.
____	*M. m. inexspectata*	SE Alaska (Glacier Bay) to interior Br. Col; > to n Oregon
____	*M. m. rufina*	Outer islands of se Alaska to c Br. Col.; > to w Washington
____	*M. m. merrilli*	S Br. Col. to sw Alberta and nw Montana; > to n Mexico
____	*M. m. morphna*	SW British Columbia to sw Oregon; > to n California
____	*M. m. fisherella*	Oregon east of Cascades to e-c Calif., w Nevada and sw Idaho
____	*M. m. cleonensis*	Extreme sw coastal Oregon to n Calif. (w Mendocino County)
____	*M. m. gouldii*	Coastal central California (Mendocino Co. to n San Benito Co.)
____	*M. m. mailliardi*	Central Valley of California (Glenn Co. to Stanislaus Co.)
____	*M. m. samuelis*	Salt marshes of c California (San Pablo and San Francisco bays)
____	*M. m. maxillaris*	Brackish marshes of central California (Suisan Bay)
____	*M. m. pusillula*	Salt marshes of c California (south side of San Francisco Bay)
____	*M. m. heermanni*	S California (Merced Co. to Kern County and Kings Canyon)
____	*M. m. cooperi*	Coastal s Calif. (Santa Cruz) to n Baja, Mojave, Colorado deserts
____	*M. m. micronyx*	San Miguel I. (off coastal s California)
____	*M. m. clementae*	Santa Rosa, Santa Cruz and San Clemente is. (off s California)
____	*M. m. graminea*	Santa Barbara I. (off coastal s California)
____	*M. m. coronatorum*	Los Coronados Islands (off nw Baja California)
____	*M. m. rivularis*	S-central Baja California
____	*M. m. saltonis*	Colorado R. Valley (extreme s Nevada, se California and nw Baja)
____	*M. m. juddi*	NE Br. Columbia to w-central US; > to Texas and se US
____	*M. m. montana*	NE Oregon to w Idaho, e Ariz., and n N. Mex.; > to Sonora
____	*M. m. fallax*	SE Nevada to sw Utah, Arizona and nw Mexico (ne Sonora)
____	*M. m. goldmani*	Sierra Madre Occidental of w Mexico (El Salto area of Durango)
____	*M. m. niceae*	Wetlands of e-central Mexico (Hidalgo)
____	*M. m. mexicana*	Wetlands of s-central Mexico (Tlaxcala and Puebla)
____	*M. m. azteca*	Wetlands of valleys of s-c Mexico (Distrito Federal and México)
____	*M. m. villai*	C Mexico (upper Río Lerma), se Guanajuato and nw Michoacán
____	*M. m. yuriria*	C Mexico (Río Lerma Valley from Lago Yuriria to s Guanajuato)
____	*M. m. adusta*	SW Mexico (Lago Pátzcuaro in Michoacán)
____	*M. m. zacapu*	SW Mexico (Zacapu region of n Michoacán and Laguna Chapala)
____	*M. m. melodia*	SE Canada and ne US; > to e Texas and s Florida
____	*M. m. atlantica*	Coastal New York (Long I.) to N Carolina; > to n Georgia
____	*M. m. euphonia*	N-central US; > to Texas, Alabama, S Carolina and Georgia

☐ **Lincoln's Sparrow** *Melospiza lincolnii*

____	*M. l. lincolnii*	NW Alaska to Canada and n US; > to Baja and Guatemala
____	*M. l. gracilis*	Coastal s Alaska and c Br. Columbia; > to central California
____	*M. l. alticola*	Mts. of Oregon to Ariz. and New Mexico; > to Guatemala

☐ **Swamp Sparrow** *Melospiza georgiana*

____ *M. g. ericrypta*   E and central Canada; > to s US and ne Mexico
____ *M. g. georgiana*   N Dak. to Nova Scotia and ne US; > to s Texas and Florida
____ *M. g. nigrescens*   Atlantic coast from New Jersey to Maryland

☐ **White-crowned Sparrow** *Zonotrichia leucophrys*

____ *Z. l. gambelii*   N Alaska and n Yukon to s-central Canada; > to n Mexico
____ *Z. l. leucophrys*   C and e Canada to Newfoundland; > to se US, Cuba, Jamaica
____ *Z. l. oriantha*   Mts. of sw Canada to sw US; > to s Baja and cent. Mexico
____ *Z. l. pugetensis*   Coastal sw Br. Columbia to nw Calif.; > to sw California
____ *Z. l. nuttalli*   Coastal central California (Mendocino Co. to Santa Barbara Co.)

☐ **White-throated Sparrow** *Zonotrichia albicollis*

Breeds n North America; > to s US and n Mexico

☐ **Golden-crowned Sparrow** *Zonotrichia atricapilla*

Breeds Alaska and w Canada; > to Sonora and Baja Calif.

☐ **Rufous-collared Sparrow** *Zonotrichia capensis*

____ *Z. c. septentrionalis*   Highlands of s Mexico (Chiapas) to Guatemala and Honduras
____ *Z. c. antillarum*   Cordillera Central of Dominican Republic
____ *Z. c. costaricensis*   Mts. of Costa Rica to w Panama; Andes of Colombia, w Venezuela
____ *Z. c. orestera*   Mountains of w Panama (Cerro Campana)
____ *Z. c. insularis*   Netherlands Antilles (Curaçao and Aruba)
____ *Z. c. venezuelae*   Coastal cordillera of n Venezuela
____ *Z. c. roraimae*   S Colombia (Meta) to e Venezuela, w Guyana and adj. n Brazil
____ *Z. c. inaccessibilis*   *Tepuis* of s Venezuela (Cerro de la Neblina)
____ *Z. c. perezchinchillae*   *Tepuis* of s Venezuela (Amazonas)
____ *Z. c. macconelli*   *Tepuis* of s Venezuela (Mt. Roraima)
____ *Z. c. capensis*   French Guiana (lower Oyapock River) and adj. Brazil (Amapá)
____ *Z. c. tocantinsi*   E Brazil (lower Amazonia along Rio Tocantins)
____ *Z. c. novaesi*   E Brazil (Pará)
____ *Z. c. matutina*   NE Brazil (Maranhão to Bahia and Mato Grosso) and adj. e Bolivia
____ *Z. c. huancabambae*   Arid n Peru (Piura, Cajamarca, Amazonas, San Martín and Junín)
____ *Z. c. illescasensis*   N Peru (Cerro Illescas in Piura)
____ *Z. c. peruviensis*   Arid coastal Peru and w slope of Andes (La Libertad to Tacna)
____ *Z. c. carabayae*   E slope of Eastern Andes of Peru (Junín) to w Bolivia
____ *Z. c. pulacayensis*   Andes of Peru (Junín) to w Bolivia and n Argentina
____ *Z. c. subtorquata*   E Brazil (Espírito Santo) to Paraguay, Uruguay and ne Argentina
____ *Z. c. mellea*   Central Paraguay and adjacent n-central Argentina (Formosa)
____ *Z. c. hypoleuca*   E and s Bolivia to ne Argentina
____ *Z. c. antofagastae*   N Chile (Tarapacá and Antofagasta)
____ *Z. c. chilensis*   Chile (Atacama to Islas Guaitecas) and Andes of s Argentina
____ *Z. c. sanborni*   Andes of Chile (Coquimbo, Aconcagua) and Argentina (San Juan)
____ *Z. c. arenalensis*   Andes of n Argentina
____ *Z. c. choraules*   W Argentina (Mendoza, e Neuquén and Río Negro)
____ *Z. c. australis*   S Chile and s Argentina to Cape Horn; > n to Bolivia

☐ **Harris' Sparrow** *Zonotrichia querula*

Breeds n-central Canada; > to s US

☐ **Dark-eyed Junco** *Junco hyemalis*

____ *J. h. hyemalis*   N Alaska and Yukon to n-central US; > to n Mexico
____ *J. h. oreganus*   Coastal se Alaska to cent. Br. Col.; > to cent. California
____ *J. h. cismontanus*   S-c Yukon to w-c Alberta; > to n Baja and central Texas
____ *J. h. montanus*   Interior Br. Col. and sw Alberta to e Oregon, w Montana, c Idaho
____ *J. h. mearnsi*   SE Alberta, sw Saskatchewan to e Idaho, Montana, ne Wyoming
____ *J. h. shufeldti*   W slopes of coastal mts. from sw British Columbia to w Oregon
____ *J. h. thurberi*   S Oregon to mts. of San Diego Co.; > to n Baja, sw N Mex.
____ *J. h. caniceps*   Mts. of s Idaho to Utah and n New Mexico; > to nw Mexico
____ *J. h. dorsalis*   Mts. of New Mexico, n Arizona and extreme w Texas
____ *J. h. aikeni*   SE Montana to w South Dakota, ne Wyoming and nw Nebraska
____ *J. h. pinosus*   Coastal ranges of California (San Francisco to s Monterey Co.)

____	*J. h. pontilus*	Mountains of n Baja California (Sierra Juárez)
____	*J. h. townsendi*	Mountains of n Baja California (San Pedro Mártir)
____	*J. h. carolinensis*	Appalachian Mountains to n Georgia

☐ **Yellow-eyed Junco** *Junco phaeonotus*

____	*J. p. mutabilis*	Mts. of s Nevada and adjacent se California
____	*J. p. palliatus*	Mts. of s Arizona, sw New Mexico and n Mexico
____	*J. p. phaeonotus*	Mts. of c and s Mexico (Jalisco to Hidalgo, Veracruz and Oaxaca)
____	*J. p. bairdi*	Mts. of s Baja California (Sierra de la Laguna)
____	*J. p. fulvescens*	Mts. of s Mexico (interior of Chiapas)
____	*J. p. alticola*	Mts. of s Mexico (se Chiapas) and w Guatemala

☐ **Guadalupe Junco** *Junco insularis*

Oak-pine forests of Guadalupe I. (off w Baja California)

☐ **Volcano Junco** *Junco vulcani*

Volcanic summits of Costa Rica and w Panama (Chiriquí)

☐ **McCown's Longspur** *Calcarius mccownii*

Breeds w Canada to nw US; > to nw Mexico

☐ **Lapland Longspur** *Calcarius lapponicus*

____	*C. l. alascensis*	Aleutian and Pribilof Is., Alaska and w Canada; > to n Texas
____	*C. l. lapponicus*	N Canada across n Siberia to Bering Strait; > n Europe, c US
____	*C. l. coloratus*	E Siberia, Kamchatka Pen. and Komandorskiye Is.; > to Japan

☐ **Smith's Longspur** *Calcarius pictus*

Breeds Alaska and w Canada; > in s-central US

☐ **Chestnut-collared Longspur** *Calcarius ornatus*

Prairies of central North America; > sw US to n Mexico

☐ **Snow Bunting** *Plectrophenax nivalis*

____	*P. n. townsendi*	Komandorskiye, Pribilof and w Aleutian islands
____	*P. n. nivalis*	N North America and n Europe; > to s US and s Europe
____	*P. n. insulae*	Iceland; > to Faeroes, Shetland Islands and n Scotland
____	*P. n. vlasowae*	Tundra of ne Asia; > to c Asia, Manchuria and Japan

☐ **McKay's Bunting** *Plectrophenax hyperboreus*

Islands in Bering Sea; > coastal Alaska and Aleutian Islands

☐ **Large Ground-Finch** *Geospiza magnirostris*

Arid scrub of main Galapagos Islands

☐ **Medium Ground-Finch** *Geospiza fortis*

Arid scrub of main Galapagos Islands

☐ **Small Ground-Finch** *Geospiza fuliginosa*

Arid scrub of main Galapagos Islands

☐ **Sharp-beaked Ground-Finch** *Geospiza difficilis*

____	*G. d. septentrionalis*	Galapagos Islands (Culpepper and Wenman)
____	*G. d. difficilis*	Galapagos Islands (Tower and Abingdon)
____	*G. d. debilirostris*	Galapagos Islands (James, Isabela and Fernandina)

☐ **Common Cactus-Finch** *Geospiza scandens*

____	*G. s. scandens*	Galapagos Islands (James and Jervis)
____	*G. s. intermedia*	Galapagos Is. (Barrington, Floreana, Duncan, Santa Cruz, Isabela)
____	*G. s. abingdoni*	Abingdon I. (Galapagos Islands)
____	*G. s. rothschildi*	Marchena I. (Galapagos Islands)

☐ **Large Cactus-Finch** *Geospiza conirostris*

____	*G. c. darwinii*	Galapagos Islands (Culpepper and Wenman)
____	*G. c. conirostris*	Hood I. (Galapagos Islands)
____	*G. c. propinqua*	Tower I. (Galapagos Islands)

☐ **Vegetarian Finch** *Camarhynchus crassirostris*

Main islands of Galapagos (except for extremely arid islands)

□ **Mangrove Finch** *Camarhynchus heliobates*

Mangrove swamps of Galapagos Islands (Fernandina and Isabela)

□ **Large Tree-Finch** *Camarhynchus psittacula*
_____ *C. p. habeli* — Galapagos Islands (Abingdon and Marchena)
_____ *C. p. affinis* — Galapagos Islands (Isabela and Fernandina)
_____ *C. p. psittacula* — Seymour, Barrington, Santa Cruz, Floreana, Duncan, Jervis, James

□ **Small Tree-Finch** *Camarhynchus parvulus*
_____ *C. p. parvulus* — Humid scrub of main Galapagos Islands (except Chatham I.)
_____ *C. p. salvini* — Chatham I. (Galapagos Islands)

□ **Medium Tree-Finch** *Camarhynchus pauper*

Humid scrub of Floreana I. (Galapagos Islands)

□ **Woodpecker Finch** *Camarhynchus pallidus*
_____ *C. p. pallidus* — Galapagos (James, Jervis, Seymour, Duncan, Santa Cruz, Floreana)
_____ *C. p. productus* — Galapagos Islands (Isabela and Fernandina)
_____ *C. p. striatipectus* — San Cristóbal I. (Galapagos Islands)

□ **Warbler Finch** *Certhidea olivacea*
_____ *C. o. becki* — Galapagos Islands (Culpepper and Wenman)
_____ *C. o. mentalis* — Tower I. (Galapagos Islands)
_____ *C. o. fusca* — Galapagos Islands (Abingdon and Marchena)
_____ *C. o. olivacea* — Galapagos (James, Jervis, Seymour, Duncan, Isabela, Fernandina)
_____ *C. o. bifasciata* — Barrington I. (Galapagos Islands)
_____ *C. o. cinerascens* — Hood I. (Galapagos Islands)
_____ *C. o. ridgwayi* — Floreana I. (Galapagos Islands)

FAMILY: CARDINALIDAE (Saltators, Cardinals and Allies—43)

□ **Red-and-black Grosbeak** *Periporphyrus erythromelas*

S Venezuela to the Guianas and e Amazonian Brazil

□ **Lesser Antillean Saltator** *Saltator albicollis*
_____ *S. a. guadelupensis* — Lesser Antilles (Guadeloupe and Dominica)
_____ *S. a. albicollis* — Lesser Antilles (Martinique and St. Lucia)

□ **Streaked Saltator** *Saltator striatipectus*
_____ *S. s. furax* — Lowlands of sw Costa Rica and w Panama
_____ *S. s. isthmicus* — Panama (except w Chiriquí and Darién)
_____ *S. s. scotinus* — Isla Coiba and Isla Ranchería (off w Panama)
_____ *S. s. melicus* — Isla Taboga (Bay of Panama)
_____ *S. s. speratus* — Pearl Islands (San Miguel, Saboga and Viveros)
_____ *S. s. striatipectus* — E Panama (Darién) and Colombia w of the Andes (south to Cauca)
_____ *S. s. perstriatus* — NE Colombia to mountains of n Venezuela; Trinidad
_____ *S. s. flavidicollis* — SW Colombia (Nariño) to arid w Ecuador and nw Peru (Piura)
_____ *S. s. immaculatus* — Arid coastal Peru (Lambayeque to Ica)
_____ *S. s. peruvianus* — N Peru (upper Marañón Valley in Cajamarca and La Libertad)

□ **Grayish Saltator** *Saltator coerulescens*
_____ *S. c. vigorsii* — NW Mexico (Sinaloa, w Durango, Nayarit and coastal n Jalisco)
_____ *S. c. richardsoni* — W Mexico (Jalisco to Colima, Guerrero and adj. w Oaxaca)
_____ *S. c. grandis* — E Mexico (Tamaulipas) to Guatemala and central Costa Rica
_____ *S. c. yucatanensis* — SE Mexico (Yucatán Pen., adjacent Tabasco and ne Chiapas)
_____ *S. c. hesperis* — Pacific slope of s Mexico (Chiapas) to Nicaragua
_____ *S. c. brevicaudus* — Pacific slope of w Costa Rica (Gulf of Nicoya region)
_____ *S. c. plumbeus* — Caribbean coast of n Colombia (Río Sinú to Magdalena Valley)
_____ *S. c. brewsteri* — Tropical ne Colombia to Venezuela; Trinidad
_____ *S. c. olivascens* — *Tepuis* of s Venezuela to the Guianas and adjacent n Brazil
_____ *S. c. azarae* — Trop. e Colombia to Ecuador, Peru, Bolivia and w Amaz. Brazil
_____ *S. c. mutus* — N Brazil (lower Solimões to Mexiana I., Amapá and n Maranhão)

_____ S. c. *superciliaris*   NE Brazil (s Piauí to n and e Bahia)
_____ S. c. *coerulescens*   E Bolivia to Paraguay, sw Brazil, Uruguay and n Argentina

☐ **Buff-throated Saltator** *Saltator maximus*
_____ S. m. *gigantodes*   Caribbean slope of e Mexico (Veracruz to n Oaxaca and Tabasco)
_____ S. m. *magnoides*   S Mexico (Chiapas and Quintana Roo) to nw Panama
_____ S. m. *intermedius*   SW Costa Rica to w Panama (Canal Zone)
_____ S. m. *iungens*   E Panama and lowlands of nw Colombia
_____ S. m. *maximus*   E Colombia to Venezuela, Guianas, Brazil, e Bolivia and Paraguay

☐ **Black-headed Saltator** *Saltator atriceps*
_____ S. a. *atriceps*   Caribbean slope of Mexico to Guatemala and e Costa Rica
_____ S. a. *suffuscus*   SE Mexico (Sierra de Tuxtla of se Veracruz)
_____ S. a. *flavicrissus*   W Mexico (central Guerrero)
_____ S. a. *peeti*   Pacific slope of s Mexico (Chiapas and adjacent Oaxaca)
_____ S. a. *raptor*   SE Mexico (Yucatán, Quintana Roo and Campeche)
_____ S. a. *lacertosus*   W Costa Rica and Panama (east to Canal Zone)

☐ **Slate-colored Grosbeak** *Saltator grossus*
_____ S. g. *saturatus*   Caribbean slope of Nicaragua to w Ecuador
_____ S. g. *grossus*   E Colombia to Venezuela, the Guianas, Amaz. Brazil and n Bolivia

☐ **Black-throated Grosbeak** *Saltator fuliginosus*
    Humid forests and scrub of e Brazil to Paraguay and ne Argentina

☐ **Black-winged Saltator** *Saltator atripennis*
_____ S. a. *atripennis*   W and Central Andes of Colombia to extreme nw Ecuador
_____ S. a. *caniceps*   W slope of Eastern Andes of Colombia and w Ecuador

☐ **Green-winged Saltator** *Saltator similis*
_____ S. s. *similis*   E Bolivia to se Brazil, Paraguay, Uruguay and ne Argentina
_____ S. s. *ochraceiventris*   SE Brazil (s São Paulo to Paraná and Rio Grande do Sul)

☐ **Orinocan Saltator** *Saltator orenocensis*
_____ S. o. *rufescens*   NE Colombia (Guajira Peninsula) and arid nw Venezuela
_____ S. o. *orenocensis*   *Llanos* of Venezuela (north of the Orinoco River)

☐ **Black-cowled Saltator** *Saltator nigriceps*
    Humid montane forests of s Ecuador and nw Peru

☐ **Golden-billed Saltator** *Saltator aurantiirostris*
_____ S. a. *iteratus*   Andes of n Peru (Cajamarca, Amazonas, La Libertad and Ancash)
_____ S. a. *albociliaris*   Andes of Peru (Ancash and Huánuco) to n Chile (Arica)
_____ S. a. *hellmayri*   Andes of Bolivia (La Paz and Cochabamba to Potosí and n Tarija)
_____ S. a. *aurantiirostris*   S Bolivia to Paraguay, Uruguay, s Brazil and n Argentina
_____ S. a. *parkesi*   S Brazil to Uruguay and ne Argentina
_____ S. a. *nasica*   W-central Argentina (La Rioja, Mendoza and w La Pampa)

☐ **Thick-billed Saltator** *Saltator maxillosus*
    Humid forests of e Paraguay to se Brazil and ne Argentina

☐ **Masked Saltator** *Saltator cinctus*
    Andes of extreme se Colombia to ne Peru

☐ **Black-throated Saltator** *Saltator atricollis*
    Campos of e Bolivia to ne Paraguay and interior ne and c Brazil

☐ **Rufous-bellied Saltator** *Saltator rufiventris*
    Andes of w Bolivia and extreme nw Argentina (Jujuy)

☐ **Black-faced Grosbeak** *Caryothraustes poliogaster*
_____ C. p. *poliogaster*   Lowlands of se Mexico (Veracruz) to Guatemala and n Honduras
_____ C. p. *scapularis*   Caribbean lowlands of Nicaragua to Costa Rica and w Panama

☐ **Yellow-green Grosbeak** *Caryothraustes canadensis*
_____ C. c. *simulans*   Extreme e Panama (Darién)

____	*C. c. canadensis*	SE Colombia (Vaupés) to Venezuela, the Guianas and n Brazil
____	*C. c. frontalis*	NE Brazil (Ceará, Pernambuco and Alagoas)
____	*C. c. brasiliensis*	E-central Brazil (Bahia and e Minas Gerais to Rio de Janeiro)

☐ **Yellow-shouldered Grosbeak** *Parkerthraustes humeralis*

E Colombia to n Bolivia and sw Amazonian Brazil

☐ **Crimson-collared Grosbeak** *Rhodothraupis celaeno*

E Mexico (Nuevo León to n Veracruz and ne Puebla)

☐ **Vermilion Cardinal** *Cardinalis phoeniceus*

Caribbean littoral of ne Colombia and n Venezuela; Isla Margarita

☐ **Northern Cardinal** *Cardinalis cardinalis*

____	*C. c. superbus*	Extremese California to Arizona, sw New Mexico and n Sonora
____	*C. c. seftoni*	Central Baja California (south to latitude 27°N)
____	*C. c. igneus*	S Baja California (north to latitude 27°N)
____	*C. c. clintoni*	Isla Cerralvo (Gulf of California)
____	*C. c. townsendi*	Isla Tiburón (Sea of Cortés) and adjacent coastal Sonora
____	*C. c. affinis*	W Mexico (se Sonora to sw Chihuahua and w Durango)
____	*C. c. sinaloensis*	Coastal w Mexico (Sinaloa and Jalisco)
____	*C. c. mariae*	Tres Marías Islands (off w Mexico)
____	*C. c. carneus*	Coastal w Mexico (Colima to Isthmus of Tehuántepec)
____	*C. c. cardinalis*	E US
____	*C. c. floridanus*	SE Georgia and peninsular Florida
____	*C. c. magnirostris*	SE Texas and s Louisiana
____	*C. c. canicaudus*	W Oklahoma and w Texas to e-central Mexico
____	*C. c. coccineus*	E Mexico (e San Luis Potosí, Veracruz, ne Puebla and n Oaxaca)
____	*C. c. littoralis*	Lowlands of e Mexico (s Veracruz and Tabasco)
____	*C. c. yucatanicus*	SE Mexico (Yucatán Peninsula)
____	*C. c. flammiger*	SE Mexico (s Quintana Roo), Belize and Petén of n Guatemala
____	*C. c. saturatus*	Cozumel I. (off Quintana Roo)

☐ **Pyrrhuloxia** *Cardinalis sinuatus*

____	*C. s. fulvescens*	Arid s Arizona and nw Mexico (Sonora to n Nayarit)
____	*C. s. sinuatus*	Arid s New Mexico to se Texas and ne Mexico
____	*C. s. peninsulae*	Baja California (south of latitude 27°N)

☐ **Yellow Grosbeak** *Pheucticus chrysopeplus*

____	*P. c. dilutus*	Highlands of nw Mexico (s Sonora, sw Chihuahua and n Sinaloa)
____	*P. c. chrysopeplus (rarissimus)*	Highlands of w Mexico (Sinaloa to Guerrero and sw Puebla)
____	*P. c. aurantiacus*	Highlands of s Mexico (Chiapas) to central Guatemala

☐ **Golden-bellied Grosbeak** *Pheucticus chrysogaster*

____	*P. c. laubmanni*	Mountains of n Colombia to coastal cordillera of n Venezuela
____	*P. c. chrysogaster*	Andes of sw Colombia (Nariño) to Ecuador and s Peru

☐ **Black-thighed Grosbeak** *Pheucticus tibialis*

Humid montane forests of Costa Rica and w Panama

☐ **Black-backed Grosbeak** *Pheucticus aureoventris*

____	*P. a. uropygialis*	E and Central Andes of Colombia
____	*P. a. crissalis*	Andes of sw Colombia (Nariño) and Ecuador
____	*P. a. meridensis*	Andes of w Venezuela (Mérida)
____	*P. a. terminalis*	Andes of e Peru (Amazonas and Cuzco)
____	*P. a. aureoventris*	S Peru (Puno) to e Bolivia, n Paraguay, w Brazil and nw Argentina

☐ **Rose-breasted Grosbeak** *Pheucticus ludovicianus*

E Canada and US; > from Mexico to Peru and w Cuba

☐ **Black-headed Grosbeak** *Pheucticus melanocephalus*

____	*P. m. maculatus*	Mountains of sw British Columbia to n Baja California
____	*P. m. melanocephalus*	SE Br. Col. to Rocky Mts., Great Plains and s Mexican Plateau

☐ **Ultramarine Grosbeak** *Cyanocompsa brissonii*

____ *C. b. caucae* — W Colombia (valleys of upper Río Patía, upper Cauca and Dagua)

____ *C. b. minor* — Mountains of n Venezuela (Falcón to Lara, Sucre and Monagas)

____ *C. b. brissonii* — NE Brazil (Piauí and Ceará to Bahia)

____ *C. b. sterea* — E Paraguay to e and s Brazil and ne Argentina

____ *C. b. argentina* — E Bolivia to *chaco* of Paraguay, w Brazil and n Argentina

☐ **Blue Bunting** *Cyanocompsa parellina*

____ *C. p. indigotica* — Pacific slope of Mexico (Sinaloa to Isthmus of Tehuántepec)

____ *C. p. beneplacita (lucida)* — NE Mexico (s Tamaulipas, e San Luis Potosí and s Nuevo León)

____ *C. p. parellina* — E Mexico (Veracruz and e Puebla) to Nicaragua

☐ **Blue-black Grosbeak** *Cyanocompsa cyanoides*

____ *C. c. concreta* — Lowlands of se Mexico (Veracruz) to Guatemala and Honduras

____ *C. c. toddi* — Nicaragua to Costa Rica and w Panama

____ *C. c. cyanoides* — Cent. and e Panama to Colombia, nw Venezuela and w Ecuador

____ *C. c. rothschildii* — E Colombia to Venezuela, the Guianas, Amaz. Brazil and Bolivia

☐ **Glaucous-blue Grosbeak** *Cyanoloxia glaucocaerulea*

E Paraguay to s Brazil, Uruguay and ne Argentina

☐ **Blue Grosbeak** *Passerina caerulea*

____ *P. c. salicaria* — N-cent. Calif. and w Nevada to nw Baja; > to Guerrero

____ *P. c. interfusa* — SW US to ne Baja and nw Mexico; > to Honduras

____ *P. c. deltarhyncha* — Coastal w Mexico (s Sinaloa and Durango to Oaxaca)

____ *P. c. caerulea* — SE US; > to Panama and Cuba

____ *P. c. eurhyncha* — E Mexico (Coahuila to Nuevo León, s Tamaulipas and Oaxaca)

____ *P. c. chiapensis* — S Mexico (Chiapas and adjacent Oaxaca) to Guatemala

____ *P. c. lazula* — Honduras to Nicaragua and Costa Rica

☐ **Lazuli Bunting** *Passerina amoena*

S British Columbia to nw Baja and w Texas; > in Mexico

☐ **Indigo Bunting** *Passerina cyanea*

Canada and US; > to Gr. Antilles, Colombia and Venezuela

☐ **Rose-bellied Bunting** *Passerina rositae*

Pacific slope of sw Mexico (Oaxaca and extreme sw Chiapas)

☐ **Orange-breasted Bunting** *Passerina leclancherii*

____ *P. l. grandior* — SW Mexico (Jalisco to Michoacán, Guerrero, s Puebla and Chiapas)

____ *P. l. leclancherii* — Coastal s Mexico (central Guerrero)

☐ **Varied Bunting** *Passerina versicolor*

____ *P. v. dickeyae* — S Ariz. to w Mexico (Sonora to w Durango, Jalisco and Colima)

____ *P. v. versicolor* — S Texas to s Mexico (Oaxaca and Guerrero)

____ *P. v. pulchra* — S Baja California

____ *P. v. purpurascens* — S Mexico (Chiapas) to central Guatemala

☐ **Painted Bunting** *Passerina ciris*

____ *P. c. pallidior* — SW US and n Mexico; > to w Panama

____ *P. c. ciris* — Coastal se US; > to Bahamas, Cuba, Jamaica and Yucatán

☐ **Yellow-billed Blue Finch** *Porphyrospiza caerulescens*

Cerrado of interior of ne and c Brazil; adjacent Paraguay (?)

☐ **Dickcissel** *Spiza americana*

Breeds e North America; > s Mexico to n South America

FAMILY: ICTERIDAE (Troupials and Allies—101)

☐ **Bobolink** *Dolichonyx oryzivorus*

Grasslands and meadows of N America; > s South America

☐ **Saffron-cowled Blackbird** *Xanthopsar flavus*

Paraguay to se Brazil, Uruguay and ne Argentina

☐ **Tawny-shouldered Blackbird** *Agelaius humeralis*
- ____ *A. h. scopulus* — Cayo Cantiles (off southern Cuba)
- ____ *A. h. humeralis* — Cuba and Haiti

☐ **Yellow-shouldered Blackbird** *Agelaius xanthomus*
- ____ *A. x. monensis* — Mona I. (off w Puerto Rico)
- ____ *A. x. xanthomus* — Lowlands of sw and ne Puerto Rico

☐ **Red-winged Blackbird** *Agelaius phoeniceus*
- ____ *A. p. arctolegus* — SE Alaska and Yukon to n-central US; > to s-central US
- ____ *A. p. fortis* — Montana to se New Mexico (east of Rocky Mts.); > to Texas
- ____ *A. p. nevadensis* — SE Br. Col. to Idaho, se Calif. and s Nevada; > to s Arizona
- ____ *A. p. caurinus* — Coastal sw Br. Col. to nw California; > to central California
- ____ *A. p. mailliardorum* — Coastal central California
- ____ *A. p. californicus* — Central Valley of California
- ____ *A. p. aciculatus* — Mountains of s-central California (e-central Kern County)
- ____ *A. p. neutralis* — Coastal s California (San Luis Obispo County) to nw Baja
- ____ *A. p. sonoriensis* — SE California to ne Baja, s Nevada, cent. Arizona and nw Mexico
- ____ *A. p. nyaritensis* — Coastal plains of sw Mexico (Nayarit)
- ____ *A. p. gubernator* — Mexican Plateau (Durango to Zacatecas, México and Tlaxcala)
- ____ *A. p. pallidulus* — SE Mexico (n Yucatán Peninsula)
- ____ *A. p. nelsoni* — S-c Mexico (Morelos and adj. Guerrero to w Puebla and Chiapas)
- ____ *A. p. arthuralleni* — N Guatemala
- ____ *A. p. grinnelli* — Pacific slope of w Guatemala to nw Costa Rica (Guanacaste)
- ____ *A. p. phoeniceus* — SE Canada to Texas and se US
- ____ *A. p. littoralis* — Gulf Coast of se Texas to nw Florida
- ____ *A. p. mearnsi* — Extreme se Georgia and n Florida
- ____ *A. p. floridanus* — S Florida (Everglades to Key West)
- ____ *A. p. megapotamus* — C Texas and lower Rio Grande Valley to e Mexico (n Veracruz)
- ____ *A. p. richmondi* — Caribbean slope of Mexico (s Veracruz) to Belize and n Guatemala
- ____ *A. p. matudae* — Tropical se Mexico
- ____ *A. p. brevirostris* — Caribbean slope of Honduras and se Nicaragua
- ____ *A. p. bryanti* — NW Bahamas

☐ **Tricolored Blackbird** *Agelaius tricolor*

Marshes and farmlands of sw Oregon to nw Baja California

☐ **Red-shouldered Blackbird** *Agelaius assimilis*
- ____ *A. a. assimilis* — W Cuba
- ____ *A. a. subniger* — Isle of Pines

☐ **Yellow-hooded Blackbird** *Chrysomus icterocephalus*
- ____ *C. i. bogotensis* — E Colombia (Bogotá Plateau)
- ____ *C. i. icterocephalus* — N Colombia to Venezuela, the Guianas, n Brazil and ne Peru

☐ **Chestnut-capped Blackbird** *Chrysomus ruficapillus*
- ____ *C. r. frontalis* — French Guiana and e Brazil
- ____ *C. r. ruficapillus* — SE Bolivia to Paraguay, s Brazil, Uruguay and n Argentina

☐ **Unicolored Blackbird** *Agelasticus cyanopus*
- ____ *A. c. xenicus* — NE Brazil (Amapá to Pará and nw Maranhão)
- ____ *A. c. atroolivaceus* — Coastal se Brazil (Rio de Janeiro)
- ____ *A. c. beniensis* — N Bolivia (Beni)
- ____ *A. c. cyanopus* — E Bolivia (Santa Cruz) to Paraguay and n Argentina

☐ **Pale-eyed Blackbird** *Agelasticus xanthophthalmus*

Locally in marshes of e Ecuador and e Peru

☐ **Yellow-winged Blackbird** *Agelasticus thilius*
- ____ *A. t. alticola* — Andes of se Peru (Cuzco) to nw Bolivia

_____ *A. t. thilius*   Andes of Chile (Atacama to Valdivia) and sw Argentina
_____ *A. t. petersii*   Paraguay to extreme se Brazil, Uruguay and n Argentina

☐ **Jamaican Blackbird** *Nesopsar nigerrimus*

                                         Humid forests of Jamaica

☐ **Red-breasted Blackbird** *Sturnella militaris*

                                         SW Costa Rica to n Bolivia, Guianas and Amaz. Brazil; Trinidad

☐ **White-browed Blackbird** *Sturnella superciliaris*

                                         Extreme se Peru to Paraguay, Uruguay, n Argentina and Brazil

☐ **Peruvian Meadowlark** *Sturnella bellicosa*
_____ *S. b. bellicosa*   Pacific slope of Ecuador to n Peru
_____ *S. b. albipes*   Arid littoral of sw Peru (Ica) to extreme n Chile
_____ *S. b. catamarcanus*   NW Argentina (Jujuy and Catamarca)

☐ **Pampas Meadowlark** *Sturnella defilippii*

                                         Pampas of e Argentina (rarely Uruguay and se Brazil)

☐ **Long-tailed Meadowlark** *Sturnella loyca*
_____ *S. l. loyca*   S Chile and s Argentina to Tierra del Fuego
_____ *S. l. falklandicus*   Falkland Islands

☐ **Eastern Meadowlark** *Sturnella magna*
_____ *S. m. magna*   S Ontario east to Quebec and south to n Texas and ne Georgia
_____ *S. m. argutula*   SE Kansas and Oklahoma to e US (Carolinas to Florida)
_____ *S. m. hoopesi*   S Texas (Eagle Pass) to n Coahuila, Nuevo León and n Tamaulipas
_____ *S. m. lilianae*   N Ariz. to e New Mexico, sw Texas, s Sonora and nw Chihuahua
_____ *S. m. auropectoralis*   Mexico (Durango and Sinaloa to Michoacán, México and n Puebla)
_____ *S. m. saundersi*   S Mexico (Oaxaca)
_____ *S. m. alticola*   Highlands of s Mexico (Guerrero, s Puebla, Veracruz) to Costa Rica
_____ *S. m. mexicana*   Caribbean slope of se Mexico (Veracruz and Tabasco to Chiapas)
_____ *S. m. griscomi*   SE Mexico (arid coastal n Yucatán Peninsula)
_____ *S. m. hippocrepis*   Cuba and Isle of Pines
_____ *S. m. inexpectata*   Pine savanna of Belize, Petén of Guatemala, Honduras, Nicaragua
_____ *S. m. subulata*   Pacific slope of Panama
_____ *S. m. meridionalis*   E Andes of Colombia to Andes of nw Venezuela
_____ *S. m. paralios*   N Colombia and savannas of w Venezuela
_____ *S. m. monticola*   *Tepuis* of s Venezuela (Mt. Roraima)
_____ *S. m. praticola*   *Llanos* of e Colombia to s Venezuela and n Guyana
_____ *S. m. quinta*   Suriname and ne Amazonian Brazil

☐ **Western Meadowlark** *Sturnella neglecta*
_____ *S. n. confluenta*   SW and central British Columbia to w Idaho and s California
_____ *S. n. neglecta*   SE British Columbia to n Baja, Texas and Gulf States

☐ **Yellow-headed Blackbird** *Xanthocephalus xanthocephalus*

                                         S Canada to n Baja and s US; > to central Mexico

☐ **Cuban Blackbird** *Dives atroviolaceus*

                                         Cuba and Isle of Pines

☐ **Melodious Blackbird** *Dives dives*

                                         Gulf slope of e Mexico to Nicaragua and (rarely) nw Costa Rica

☐ **Scrub Blackbird** *Dives warszewiczi*
_____ *D. w. warszewiczi*   Coastal scrub of sw Ecuador and nw Peru (Tumbes and Piura)
_____ *D. w. kalinowskii*   Coastal scrub of w Peru (La Libertad to Ica)

☐ **Rusty Blackbird** *Euphagus carolinus*
_____ *E. c. carolinus*   N Alaska and n Yukon to ne US; > to Gulf States
_____ *E. c. nigrans*   Newfoundland, Magdalen I., and Nova Scotia; > to Georgia

☐ **Brewer's Blackbird** *Euphagus cyanocephalus*

                                         W Canada and w US; > to s Mexico

□ **Boat-tailed Grackle** *Quiscalus major*
_____ *Q. m. torreyi* — E US (Atlantic coast from s New Jersey to extreme ne Florida)
_____ *Q. m. major* — E US (Gulf States from se Texas to Florida Keys)

□ **Common Grackle** *Quiscalus quiscula*
_____ *Q. q. versicolor* — S and se Canada e of Rocky Mts. to c and ne US; > to s US
_____ *Q. q. stonei* — E US (sw Conn. to Alabama and n Georgia); > to Florida
_____ *Q. q. quiscula* — SE US (s Louisiana to e South Carolina and Florida Keys)

□ **Great-tailed Grackle** *Quiscalus mexicanus*
_____ *Q. m. nelsoni* — SE California to s Arizona and w Mexico (ne Baja and s Sonora)
_____ *Q. m. monsoni* — SE Ariz. to w Texas and Mexican Plateau to Jalisco and Guanajuato
_____ *Q. m. prosopidicola* — SE N Mex. to s Texas, Coahuila, San Luis Potosí and s Tamaulipas
_____ *Q. m. graysoni* — Coastal nw Mexico (Sinaloa)
_____ *Q. m. obscurus* — Coastal sw Mexico (Nayarit to Guerrero)
_____ *Q. m. mexicanus* — S Mexico (e Jalisco and San Luis Potosí) to n Nicaragua
_____ *Q. m. loweryi* — Coastal Yucatán Peninsula, Belize and adjacent offshore islands
_____ *Q. m. peruvianus* — Pacific coast of Costa Rica to nw Peru and nw Venezuela

□ **Nicaraguan Grackle** *Quiscalus nicaraguensis*
Lake Nicaragua, Lake Managua and n Costa Rica

□ **Greater Antillean Grackle** *Quiscalus niger*
_____ *Q. n. caribaeus* — W Cuba, Isle of Pines and cays east to Cayos de las Doce Leguas
_____ *Q. n. gundlachii* — Central and e Cuba and inner cays of Jardines de la Reina
_____ *Q. n. caymanensis* — Grand Cayman I.
_____ *Q. n. bangsi* — Little Cayman I. and Cayman Brac
_____ *Q. n. crassirostris* — Jamaica
_____ *Q. n. niger* — Hispaniola, Gonâve, Tortue, Île-a-Vache and Beata islands
_____ *Q. n. brachypterus* — Puerto Rico and Vieques I.

□ **Carib Grackle** *Quiscalus lugubris*
_____ *Q. l. guadeloupensis* — Montserrat, Guadeloupe, Marie Galante, Dominica and Martinique
_____ *Q. l. inflexirostris* — St. Lucia (Lesser Antilles)
_____ *Q. l. contrusus* — St. Vincent (Lesser Antilles)
_____ *Q. l. luminosus* — Lesser Antilles (Grenada, the Grenadines and Islas Los Testigos)
_____ *Q. l. fortirostris* — Barbados; introduced Barbuda and Antigua
_____ *Q. l. orquillensis* — Islas Los Hermanos (off Venezuela)
_____ *Q. l. insularis* — Isla Margarita and Islas Los Frailes (off Venezuela)
_____ *Q. l. lugubris* — NE Colombia to n Venezuela, the Guianas and ne Brazil; Trinidad

□ **Bay-winged Cowbird** *Molothrus badius*
_____ *M. b. fringillarius* — Campos of ne Brazil (Piauí to Pernambuco, Bahia, Minas Gerais)
_____ *M. b. badius* — E Bolivia (Beni and Tarija) to Paraguay, Uruguay and n Argentina
_____ *M. b. bolivianus* — Highlands of central and s Bolivia

□ **Screaming Cowbird** *Molothrus rufoaxillaris*
E Bolivia to Paraguay, s Brazil, Uruguay and n Argentina

□ **Shiny Cowbird** *Molothrus bonariensis*
_____ *M. b. minimus* — L Antilles (n to Martinique), Trinidad, Tobago, Guianas, n Brazil
_____ *M. b. cabanisii* — E Panama (Darién) to Colombia
_____ *M. b. venezuelensis* — Tropical e Colombia and n Venezuela
_____ *M. b. aequatorialis* — Tropical sw Colombia to w Ecuador and Isla Puná
_____ *M. b. occidentalis* — Extreme sw Ecuador (Loja) and w Peru (south to Lima)
_____ *M. b. riparius* — Tropical e Peru (Río Ucayali) to lower Amazon Valley (Pará)
_____ *M. b. bonariensis* — E Bolivia to Paraguay, Brazil, Uruguay and c Argentina

□ **Bronzed Cowbird** *Molothrus aeneus*
_____ *M. a. loyei* — SW US to nw Mexico (Sonora, Chihuahua, Durango and Nayarit)
_____ *M. a. assimilis* — SW Mexico (Jalisco to Colima, Guerrero, Puebla, Oaxaca, Chiapas)
_____ *M. a. aeneus* — S Texas to s Mexico, Yucatán Peninsula and central Panama
_____ *M. a. armenti* — Caribbean coast of n Colombia

☐ **Brown-headed Cowbird** *Molothrus ater*
  ____   *M. a. artemisiae*   S coastal Alaska to sw US; > to Baja and s Mexico
  ____   *M. a. obscurus*   SW US to Guerrero and n Tamaulipas; > to s Baja and Oaxaca
  ____   *M. a. ater*   C and e-central US; > to Gulf Coast, Florida and s Mexico
  ____   *M. a. californicus*   S California to n Baja and Los Coronados Islands (off nw Baja)

☐ **Giant Cowbird** *Molothrus oryzivorus*
  ____   *M. o. impacifa*   Caribbean slope of s Mexico (s Veracruz) to w Panama
  ____   *M. o. oryzivora*   E Panama to Bolivia, Paraguay, se Brazil and ne Argentina

☐ **Moriche Oriole** *Icterus chrysocephalus*
    E Colombia to Venezuela, Guianas, n Brazil and ne Peru; Trinidad

☐ **Epaulet Oriole** *Icterus cayanensis*
  ____   *I. c. cayanensis*   Suriname to French Guiana, Amaz. Brazil, e Peru and e Bolivia
  ____   *I. c. tibialis*   E Brazil (Maranhão to Piauí, Pernambuco, Bahia, Rio de Janeiro)
  ____   *I. c. valenciobuenoi*   SE Brazil (s Goiás to Minas Gerais, São Paulo and se Mato Grosso)
  ____   *I. c. periporphyrus*   NE Bolivia and adjacent w Brazil (w-central Mato Grosso)
  ____   *I. c. pyrrhopterus*   SE Bolivia to Paraguay, se Brazil, Uruguay and n Argentina

☐ **Yellow-backed Oriole** *Icterus chrysater*
  ____   *I. c. chrysater*   S Mexico (s Veracruz) to Belize, Guatemala and n Nicaragua
  ____   *I. c. mayensis*   SE Mexico (n Yucatán Peninsula)
  ____   *I. c. hondae*   Panama (west to Veraguas) to n Colombia
  ____   *I. c. giraudii*   Central Colombia to n Venezuela

☐ **Yellow Oriole** *Icterus nigrogularis*
  ____   *I. n. nigrogularis*   Coastal n Colombia to Venezuela, the Guianas and n Brazil
  ____   *I. n. curasoensis*   Netherlands Antilles (Aruba, Curaçao and Bonaire)
  ____   *I. n. helioeides*   Isla Margarita (off n Venezuela)
  ____   *I. n. trinitatis*   NE Venezuela (e Paria Peninsula); Trinidad and Monos I.

☐ **Jamaican Oriole** *Icterus leucopteryx*
  ____   *I. l. bairdi†*   Grand Cayman I. (extinct ca 1967)
  ____   *I. l. leucopteryx*   Jamaica
  ____   *I. l. lawrencii*   Isla San Andrés (w Caribbean Sea)

☐ **Orange Oriole** *Icterus auratus*
    SE Mexico (arid Yucatán Peninsula)

☐ **Yellow-tailed Oriole** *Icterus mesomelas*
  ____   *I. m. mesomelas*   Trop. se Mexico (Veracruz and Oaxaca) to Belize and Honduras
  ____   *I. m. salvinii*   Caribbean lowlands of Nicaragua to e Panama
  ____   *I. m. carrikeri*   Tropical n and w Colombia to nw Venezuela
  ____   *I. m. taczanowskii*   Tropical w Ecuador and nw Peru
  ____   *I. m. xantholemus*   Ecuador (probable immature *taczanowskii* or unknown hybrid)

☐ **Orange-crowned Oriole** *Icterus auricapillus*
    Lowlands of e Panama to n Colombia and n Venezuela

☐ **White-edged Oriole** *Icterus graceannae*
    Arid scrub of w Ecuador and nw Peru (south to La Libertad)

☐ **Spot-breasted Oriole** *Icterus pectoralis*
  ____   *I. p. pectoralis*   Pacific slope of s Mexico (Colima) to n Nicaragua
  ____   *I. p. espinachi*   Pacific coast of s Nicaragua to nw Costa Rica

☐ **Altamira Oriole** *Icterus gularis*
  ____   *I. g. tamaulipensis*   S Texas (lower Rio Grande Valley) to se Mexico (Campeche)
  ____   *I. g. flavescens*   Coastal sw Mexico (Guerrero)
  ____   *I. g. yucatanensis*   SE Mexico (Yucatán Peninsula), Cozumel I. and extreme n Belize
  ____   *I. g. gularis*   Arid tropical s Mexico (Oaxaca) to Guatemala and El Salvador
  ____   *I. g. troglodytes*   S Mexico (extreme s Chiapas) and Pacific slope of Guatemala
  ____   *I. g. gigas*   Interior s Guatemala to Honduras and w-central Nicaragua

☐ **Streak-backed Oriole** *Icterus pustulatus*

____ *I. p. microstictus*	Tropical w Mexico (n Sonora and Chihuahua)
____ *I. p. yaegeri*	Coastal lowlands of w Mexico (s Sinaloa to s Nayarit)
____ *I. p. graysonii*	Tres Marías Islands (off w Mexico)
____ *I. p. dickermani*	W Mexico (lowlands of sw Jalisco and Colima to s Guerrero)
____ *I. p. pustulatus*	Tropical sw Mexico (Colima to n Oaxaca, Puebla and Veracruz)
____ *I. p. formosus*	Arid tropical s Mexico (Oaxaca and Chiapas) to nw Guatemala
____ *I. p. alticola*	Arid tropical Guatemala and Atlantic slope of Honduras
____ *I. p. sclateri*	Pacific slope of El Salvador to sw Costa Rica (Nicoya Peninsula)

☐ **Hooded Oriole** *Icterus cucullatus*

____ *I. c. nelsoni*	Central California to ne Baja and nw Mexico (s Sonora)
____ *I. c. sennetti*	S Texas (lower Rio Grande Valley) to e Mexico (Tamaulipas)
____ *I. c. cucullatus*	SW Texas (Del Rio) to se Mexico (Veracruz and Oaxaca)
____ *I. c. californicus*	N Baja California
____ *I. c. trochiloides*	S Baja California (latitude 27°N to Cabo San Lucas)
____ *I. c. restrictus*	NW Mexico (s Sonora)
____ *I. c. igneus (cozumelae, duplexus, masoni)*	Yucatán Pen., Cozumel, Contoy, Holbox and Mujeres is. to Belize

☐ **Venezuelan Troupial** *Icterus icterus*

____ *I. i. ridgwayi*	Coastal n Colombia to nw Venezuela; Aruba, Curaçao, Margarita I.
____ *I. i. icterus*	*Llanos* of e Colombia to n Venezuela
____ *I. i. metae*	W Venezuela (extreme sw Apure to Colombian border)

☐ **Campo Troupial** *Icterus jamacaii*

Lowlands of e Brazil

☐ **Orange-backed Troupial** *Icterus croconotus*

____ *I. c. croconotus*	SW Guyana to n Brazil, e Ecuador and e Peru (Madre de Dios)
____ *I. c. strictifrons*	E Bolivia to sw Brazil (Mato Grosso) and *chaco* of Paraguay

☐ **Baltimore Oriole** *Icterus galbula*

E N America (s Canada to s US); > to n South America

☐ **Bullock's Oriole** *Icterus bullockii*

____ *I. b. bullockii*	SW Canada to sw US and n Mexico; > to nw Costa Rica
____ *I. b. parvus*	Extreme sw US to n Baja and nw Sonora; > to Guerrero

☐ **Black-backed Oriole** *Icterus abeillei*

Riparian woodlands of central plateau and eastern Mexico

☐ **Orchard Oriole** *Icterus spurius*

SE Canada to c plain of Mexico; > to Colombia, Venezuela

☐ **Fuertes' Oriole** *Icterus fuertesi*

Caribbean coast of Mexico (s Tamaulipas to s Veracruz)

☐ **Black-cowled Oriole** *Icterus prosthemelas*

____ *I. p. prosthemelas*	Caribbean slope of se Mexico (s Veracruz) to Nicaragua
____ *I. p. praecox*	Caribbean slope of Costa Rica and adjacent w Panama

☐ **Greater Antillean Oriole** *Icterus dominicensis*

____ *I. d. northropi*	Bahamas (Andros, Great Abaco and Little Abaco)
____ *I. d. melanopsis*	Cuba and Isle of Pines
____ *I. d. dominicensis*	Hispaniola, Gonâve I., Tortue I. and Île-a-Vache
____ *I. d. portoricensis*	Puerto Rico

☐ **Black-vented Oriole** *Icterus wagleri*

____ *I. w. castaneopectus*	Arid scrub of nw Mexico (s Sonora to n Sinaloa and Chihuahua)
____ *I. w. wagleri*	Highlands of w Mexico (Sinaloa and Coahuila) to n Nicaragua

☐ **St. Lucia Oriole** *Icterus laudabilis*

St. Lucia (Lesser Antilles)

☐ **Martinique Oriole** *Icterus bonana*

Semiarid hills of Martinique (Lesser Antilles)

☐ **Montserrat Oriole** *Icterus oberi*

Montane forests of Montserrat (Lesser Antilles)

☐ **Audubon's Oriole** *Icterus graduacauda*
- ____ *I. g. audubonii*
- ____ *I. g. nayaritensis*
- ____ *I. g. dickeyae*
- ____ *I. g. graduacauda (richardsoni)*

S Texas (lower Rio Grande Valley) and e Mexico (Tamaulipas)
W-c Mexico (s Nayarit to w Jalisco, n Colima and s Michoacán)
Mountains of sw Mexico (Sierra Madre del Sur of Guerrero)
Southern portion of Mexican Plateau

☐ **Bar-winged Oriole** *Icterus maculialatus*

Oak-pine scrub of s Mexico (se Oaxaca) to El Salvador

☐ **Scott's Oriole** *Icterus parisorum*

Arid sw US to s Mexico

☐ **Yellow-billed Cacique** *Amblycercus holosericeus*
- ____ *A. h. holosericeus*
- ____ *A. h. flavirostris*
- ____ *A. h. australis*

SE Mexico (San Luis Potosí and s Tamaulipas) to n Colombia
W Colombia to w Ecuador and extreme nw Peru (Tumbes)
Colombia to nw Venezuela, e Peru and nw Bolivia

☐ **Yellow-rumped Cacique** *Cacicus cela*
- ____ *C. c. vitellinus*
- ____ *C. c. flavicrissus*
- ____ *C. c. cela*

Tropical e Panama (Canal Zone) to n Colombia
Tropical w Ecuador to extreme nw Peru (Tumbes)
Colombia to Venezuela, the Guianas, Amaz. Brazil and e Bolivia

☐ **Red-rumped Cacique** *Cacicus haemorrhous*
- ____ *C. h. haemorrhous*
- ____ *C. h. affinis*

SE Colombia to e Ecuador, Peru, Bolivia and n Amazonian Brazil
Paraguay to e and s Brazil and ne Argentina

☐ **Scarlet-rumped Cacique** *Cacicus uropygialis*
- ____ *C. u. microrhynchus*
- ____ *C. u. pacificus*
- ____ *C. u. uropygialis*

Extreme ne Honduras to e Panama (except Darién)
E Panama (Darién) to w Colombia and w Ecuador (El Oro)
Andes of Colombia to nw Venezuela, e Ecuador and se Peru

☐ **Selva Cacique** *Cacicus koepckeae*

SE Peru (known from 3 locations in se Peru)

☐ **Golden-winged Cacique** *Cacicus chrysopterus*

E Bolivia to Paraguay, s Brazil, Uruguay and ne Argentina

☐ **Mountain Cacique** *Cacicus chrysonotus*
- ____ *C. c. leucoramphus (peruvianus)*
- ____ *C. c. peruvianus*
- ____ *C. c. chrysonotus*

Andes of Colombia to sw Venezuela (Táchira) and e Ecuador
Andes of n Peru (south to Junín)
Andes of se Peru and nw Bolivia (La Paz and Cochabamba)

☐ **Ecuadorian Cacique** *Cacicus sclateri*

Tropical forests of e Ecuador and extreme n Peru

☐ **Solitary Cacique** *Cacicus solitarius*

Tropical n Venezuela to central Argentina and Amazonian Brazil

☐ **Yellow-winged Cacique** *Cacicus melanicterus*

Pacific lowlands of w Mexico (se Sonora) to se Guatemala

☐ **Russet-backed Oropendola** *Psarocolius angustifrons*
- ____ *P. a. salmoni*
- ____ *P. a. atrocastaneus*
- ____ *P. a. sincipitalis*
- ____ *P. a. neglectus*
- ____ *P. a. oleagineus*
- ____ *P. a. angustifrons*
- ____ *P. a. alfredi*

W and Central Andes of Colombia
Subtropical w Ecuador
W slope of E Andes of Colombia and upper Magdalena Valley
E slope of Eastern Andes of Colombia and nw Venezuela
Coastal cordillera of n Venezuela and interior mountains (Aragua)
Trop. se Colombia to ne Peru and adjacent w Amazonian Brazil
SE Ecuador to e Peru and nw Bolivia (w Santa Cruz)

☐ **Dusky-green Oropendola** *Psarocolius atrovirens*

Andes of se Peru and nw Bolivia

☐ **Chestnut-headed Oropendola** *Psarocolius wagleri*
- ____ *P. w. wagleri*
- ____ *P. w. ridgwayi*

Caribbean slope of se Mexico (s Veracruz) to ne Nicaragua
SE Honduras to Panama, w Colombia and nw Ecuador

697

☐ **Crested Oropendola** *Psarocolius decumanus*
    ____ *P. d. melanterus* — Tropical Panama and n Colombia
    ____ *P. d. insularis* — Trinidad and Tobago
    ____ *P. d. decumanus* — E Colombia to Venezuela, Guianas, n Brazil, Ecuador and n Peru
    ____ *P. d. maculosus* — E Peru to Bolivia, Paraguay and n Argentina

☐ **Casqued Oropendola** *Psarocolius oseryi* — Tropical e Ecuador, e Peru and extreme sw Brazil (Amazonas)

☐ **Green Oropendola** *Psarocolius viridis* — Guianas and s Venezuela to ne Peru and Amazonian Brazil

☐ **Black Oropendola** *Psarocolius guatimozinus* — Humid lowlands of e Panama and n Colombia

☐ **Montezuma Oropendola** *Gymnostinops montezuma* — Gulf-Caribbean lowlands of se Mexico to central Panama

☐ **Amazonian Oropendola** *Gymnostinops bifasciatus*
    ____ *G. b. yuracares* — Trop. se Colombia to s Venezuela, e Bolivia and w Amaz. Brazil
    ____ *G. b. bifasciatus* — N Brazil s of Amazon (Rio Tocantins to Belém and n Mato Grosso)
    ____ *G. b. neivae* — N Brazil south of the Amazon (Rio Tapajós to Rio Xingú)

☐ **Baudo Oropendola** *Gymnostinops cassini* — N Colombia (known from 3 specimens ca 1900 from nw Chocó)

☐ **Band-tailed Oropendola** *Ocyalus latirostris* — Tropical se Colombia to ne Peru and extreme w Brazil

☐ **Oriole Blackbird** *Gymnomystax mexicanus* — E Colombia to Venezuela, the Guianas, Amaz. Brazil and ne Peru

☐ **Yellow-rumped Marshbird** *Pseudoleistes guirahuro* — E Paraguay to s Brazil, Uruguay and n Argentina

☐ **Brown-and-yellow Marshbird** *Pseudoleistes virescens* — Wet pastures of extreme s Brazil, Uruguay and n Argentina

☐ **Scarlet-headed Blackbird** *Amblyramphus holosericeus* — Marshes of n Bolivia, Paraguay, s Brazil and n Argentina

☐ **Red-bellied Grackle** *Hypopyrrhus pyrohypogaster* — Andes of Colombia

☐ **Austral Blackbird** *Curaeus curaeus*
    ____ *C. c. curaeus* — S Argentina and s Chile to Straits of Magellan
    ____ *C. c. recurvirostris* — S Chile (Magellanes)
    ____ *C. c. reynoldsi* — Tierra del Fuego, Navarino and Hoste islands

☐ **Forbes' Blackbird** *Curaeus forbesi* — Known from three localities in extreme e Brazil

☐ **Chopi Blackbird** *Gnorimopsar chopi*
    ____ *G. c. sulcirostris* — E Bolivia to ne Brazil and nw Argentina (n Salta)
    ____ *G. c. chopi* — SE Bolivia to Paraguay, se Brazil, Uruguay and n Argentina

☐ **Bolivian Blackbird** *Agelaioides oreopsar* — Andes of sw Bolivia (Cochabamba, Chuquisaca and Potosí)

☐ **Velvet-fronted Grackle** *Lampropsar tanagrinus*
    ____ *L. t. tanagrinus* — SE Colombia to e Ecuador, n Peru and w Amazonian Brazil
    ____ *L. t. guianensis* — Tropical n Venezuela to nw Guyana and n Brazil (Roraima)
    ____ *L. t. boliviensis* — N Bolivia (along upper Río Beni)
    ____ *L. t. violaceus* — W Brazil (nw Mato Grosso)
    ____ *L. t. macropterus* — W Brazil (upper Rio Juruá)

☐ **Golden-tufted Grackle** *Macroagelaius imthurni* — *Tepuis* of s Venezuela to w Guyana and n Brazil (Roraima)

☐ **Mountain Grackle** *Macroagelaius subalaris* — Locally in Eastern Andes of ne Colombia

698

Extinct Species

The following list of species considered extant since 1600 does not include species that are known only from fossils, paintings or traveler's descriptions, not from "recent" full or partial specimens.

Family	English Name	Scientific Name	Range	Last Seen
Dromaiidae	King Island Emu	*Dromaius ater*	Australia	1850
Podicipedidae	Colombian Grebe	*Podiceps andinus*	Colombia	1977
	Atitlan Grebe	*Podilymbus gigas*	Guatemala	1986
	Alaotra Grebe	*Tachybaptus rufolavatus*	Madagascar	1950
Hydrobatidae	New Zealand Storm-Petrel	*Oceanites maorianus*	New Zealand	1850
	Guadalupe Storm-Petrel	*Oceanodroma macrodactyla*	Mexico	1912
Phalacrocoracidae	Pallas' Cormorant	*Phalacrocorax perspicillatus*	Islands in Bering Sea	1852
Ardeidae	New Zealand Bittern	*Ixobrychus novaezelandiae*	New Zealand	1900
Anatidae	Labrador Duck	*Camptorhynchus labradorius*	NE North America	1975
	Auckland Islands Merganser	*Mergus australis*	Auckland Islands	1902
	Crested Shelduck	*Tadorna cristata*	Siberia and Korea	1916
	Amsterdam Island Wigeon	*Anas marecula*	Amsterdam I.	1780
	Pink-headed Duck	*Rhodonessa caryophyllacea*	India and Myanmar	1935
Falconidae	Guadalupe Caracara	*Caracara lutosa*	Mexico	1900
Phasianidae	New Zealand Quail	*Coturnix novaezelandiae*	New Zealand	1875
	Himalayan Quail	*Ophrysia superciliosa*	West-central Himalayas	1876
Rallidae	Chatham Islands Rail	*Gallirallus modestus*	Chatham Islands	1900
	Wake Island Rail	*Gallirallus wakensis*	Wake Island	1944
	Tahiti Rail	*Gallirallus pacificus*	Tahiti and adj. e Society Is.	1800
	Dieffenbach's Rail	*Gallirallus dieffenbachii*	Chatham Islands	1900
	Sharpe's Rail	*Gallirallus sharpei*	Indonesia	1900
	Red-throated Wood-Rail	*Aramides gutturalis*	Peru	1850
	Laysan Rail	*Porzana palmeri*	Laysan I.	1944
	Bar-winged Rail	*Nesoclopeus poecilopterus*	Fiji	1973
	Hawaiian Rail	*Porzana sandwichensis*	Hawaii	1844
	Kosrae Crake	*Porzana monasa*	Kosrae I. (Micronesia)	1828
	Tristan Moorhen	*Gallinula nesiotis*	Tristan da Cunha	1900
	Samoan Moorhen	*Gallinula pacifica*	Savai'i I. (Western Samoa)	1908
	Lord Howe Swamphen	*Porphyrio albus*	Lord Howe I.	1834

Haematopodidae	Canarian Oystercatcher	*Haematopus meadewaldoi*	Canary Islands	1950
Scolopacidae	White-winged Sandpiper	*Prosobonia leucoptera*	Tahiti and Moorea	1790
Alcidae	Great Auk	*Pinguinus impennis*	Arctic North America	1852
Raphidae	Dodo	*Raphus cucullatus*	Mauritius I.	1662
	Reunion Solitaire	*Raphus solitarius*	Réunion I	1715
	Rodrigues Solitaire	*Pezophaps solitaria*	Rodrigues I.	1770
Columbidae	Mauritius Blue-Pigeon	*Alectroenas nitidissima*	Mauritius I.	1840
	Ryukyu Pigeon	*Columba jouyi*	Okinawa and Daito islands	1936
	Bonin Pigeon	*Columba versicolor*	Bonin Islands	1889
	Passenger Pigeon	*Ectopistes migratorius*	Canada and United States	1914
	Tanna Ground-Dove	*Gallicolumba ferruginea*	Tanna I. (s Vanuatu)	1800
	Choiseul Pigeon	*Microgoura meeki*	Choiseul I. (nw Solomon Is.)	1904
	Red-moustached Fruit-Dove	*Ptilinopus mercierii*	Marquesas Islands	1950
Psittacidae	Cuban Macaw	*Ara cubensis*	Cuba and Isle of Pines	1850
	Hispaniolan Macaw	*Ara tricolor*	Hispaniola	1820
	Carolina Parakeet	*Conuropsis carolinensis*	Eastern United States	1918
	Raiatea Parakeet	*Cyanoramphus ulietanus*	Raiatea I. (e Society Islands)	1773
	Black-fronted Parakeet	*Cyanoramphus zealandicus*	Society Islands	1850
	Mascarene Parrot	*Mascarinus mascarinus*	Réunion I. (w Mascarene Is.)	1834
	Norfolk Island Kaka	*Nestor productus*	Norfolk I. and Phillip I.	1851
	New Caledonian Lorikeet	*Charmosyna diadema*	New Caledonia	1860
	Paradise Parrot	*Psephotus pulcherrimus*	Eastern Australia	1927
	Seychelles Parakeet	*Psittacula wardi*	Seychelles	1900
	Newton's Parakeet	*Psittacula exsul*	Mascarene Islands	1875
	Glaucous Macaw	*Anodorhynchus glaucus*	SE Brazil and adj. Uruguay	1915
Cuculidae	Snail-eating Coua	*Coua delalandei*	Madagascar	1834
Strigidae	Laughing Owl	*Sceloglaux albifacies*	New Zealand	1970
Caprimulgidae	Jamaican Poorwill	*Siphonorhis americana*	Jamaica	1859
Trochilidae	Brace's Emerald	*Chlorostilbon bracei*	Northern Bahamas	1877
	Gould's Emerald	*Chlorostilbon elegans*	Jamaica and Bahamas	1900
	Bogota Sunangel	*Heliangelus zusii*	Colombia	1909
Acanthisittidae	Bush Wren	*Xenicus longipes*	New Zealand	1972
	Stephens Island Wren	*Xenicus lyalli*	Stephens I. (New Zealand)	1894
Corvidae	Banggai Crow	*Corvus unicolor*	Banggai I.	1900
Turdidae	Kittlitz's Thrush	*Zoothera terrestris*	Bonin Islands	1889
	Grand Cayman Thrush	*Turdus ravidus*	Grand Cayman I.	1938
	Amaui	*Myadestes oahensis*	Oahu (Hawaiian Islands)	1825
Sylviidae	Chatham Islands Fernbird	*Megalurus rufescens*	Chatham Islands	1900
Acanthizidae	Lord Howe Gerygone	*Gerygone insularis*	Lord Howe I.	1928
Petroicidae	Piopio	*Turnagra capensis*	New Zealand	1963
Monarchidae	Guam Flycatcher	*Myiagra freycineti*	Guam	1983
	Maupiti Monarch	*Pomarea pomarea*	Society Islands	1850
Timaliidae	Black-browed Babbler	*Malacocincla perspicillata*	Borneo	1850
	Vanderbilt's Babbler	*Malacocincla vanderbilti*	Sumatra	1940
Zosteropidae	Robust White-eye	*Zosterops strenuus*	Lord Howe I.	1928
Meliphagidae	O'ahu Oo	*Moho apicalis*	Hawaiian Islands	1850
	Hawaii Oo	*Moho nobilis*	Hawaiian Islands	1898
	Bishop's Oo	*Moho bishopi*	Hawaiian Islands	1981
	Kaua'i Oo	*Moho braccatus*	Hawaiian Islands	1987
	Kioea	*Chaetoptila angustipluma*	Hawaiian Islands	1900
	Chatham Island Bellbird	*Anthornis melanocephala*	Chatham Islands	1906

Callaeidae	Huia	*Heteralocha acutirostris*	New Zealand	1907
Drepanididae	Greater Akialoa	*Hemignathus ellisianus*	Hawaiian Islands	1969
	Maui Nui Akialoa	*Hemignathus lanaiensis*	Hawaiian Islands	1900
	Lesser Akialoa	*Hemignathus obscurus*	Hawaiian Islands	1940
	Kaua'i Akialoa	*Hemignathus stejnegeri*	Hawaiian Islands	1969
	Greater Amakihi	*Hemignathus sagittirostris*	Hawaiian Islands	1901
	Ula-ai-hawane	*Ciridops anna*	Hawaiian Islands	1937
	Black Mamo	*Drepanis funerea*	Hawaiian Islands	1907
	Hawaii Mamo	*Drepanis pacifica*	Hawaiian Islands	1898
	Lanai Hookbill	*Dysmorodrepanis munroi*	Hawaiian Islands	1920
	Kakawahie	*Paroreomyza flammea*	Hawaiian Islands	1963
	Kona Grosbeak	*Chloridops kona*	Hawaiian Islands	1894
	Lesser Koa-Finch	*Rhodacanthis flaviceps*	Hawaiian Islands	1891
	Greater Koa-Finch	*Rhodacanthis palmeri*	Hawaiian Islands	1896
Sturnidae	Kosrae Starling	*Aplonis corvina*	Kosrae I. (Micronesia)	1880
	Mysterious Starling	*Aplonis mavornata*	Cook Islands	1825
	Norfolk Starling	*Aplonis fusca*	Norfolk Islands	1923
	Reunion Starling	*Fregilupus varius*	Réunion I.	1860
	Rodrigues Starling	*Necropsar rodericanus*	Rodrigues I.	1750
Fringillidae	Bonin Grosbeak	*Chaunoproctus ferreorostris*	Bonin Islands	1900
Icteridae	Slender-billed Grackle	*Quiscalus palustris*	Mexico	1910

Appendix A
Distribution of Bird Species of the World

World regions ranked by the number of bird species present. Adapted with permission from *BirdArea for Windows* © 1996, 2006 Santa Barbara Software Products. Research by Shawneen Finnegan.

An excellent companion to this volume is a pair of Windows software programs that contain the common and scientific names of every order, family, genus, and species; the range of every species; and, optionally, the scientific name of every subspecies. When sightings are recorded, the software automatically updates both annual and life lists for the world and for all faunal zones, ABA reporting areas, nations, major islands or island groups, U.S. states, Canadian provinces, and subdivisions such as counties. Furthermore, it will produce checklists of the birds that can be present in any of these regions except the subdivisions, labeling endemics and species previously sighted in the region, outside it, or both. There is a lexicon file that adds the common and scientific names of each order, family, genus and species, and the scientific name of each subspecies to Windows and Macintosh wordprocessor spelling checkers. More information is available from Santa Barbara Software Products, 1400 Dover Road, Santa Barbara, CA 93103, USA; Phone/fax: 805-963-4886; E-mail: sbsp@aol.com; Web site: http://members.aol.com/sbsp.

Region	Code	Species	Endemics	Region	Code	Species	Endemics
Peru	PE	1793	118	Angola	AO	923	10
Colombia	CO	1757	65	Panama	PA	918	10
Brazil	BR	1719	202	Cameroon	CM	899	8
Ecuador	EC	1534	10	Nigeria	NG	889	4
Bolivia	BO	1347	19	U.S. contiguous 48 states	US	868	11
Venezuela	VE	1346	46	Nepal	NP	868	2
China	CN	1269	50	Costa Rica	CR	852	7
India	IN	1185	36	Ethiopia	ET	840	17
Zaire	ZR	1133	13	South Africa	ZA	825	15
Kenya	KE	1104	6	Vietnam	VN	787	10
Mexico	MX	1075	90	Australia	AU	776	246
Tanzania	TZ	1054	22	Zambia	ZM	766	1
Uganda	UG	1023	1	New Guinea	PG	746	329
Argentina	AR	1005	16	Guyana	GY	743	0
Burma (Myanmar)	BU	1002	4	Ghana	GH	724	0
Sudan	SD	962	2	Ivory Coast	CI	719	0
Thailand	TH	945	2	Russia	RS	712	0

Region	Code	Species	Endemics	Region	Code	Species	Endemics
Guatemala	GT	698	0	Sweden	SE	472	0
Honduras	HN	693	1	Swaziland	SZ	469	0
Bangladesh	BD	688	0	Taiwan (Formosa)	TW	466	15
Mozambique	MZ	685	0	Kampuchea	KH	465	0
Pakistan	PK	682	0	Netherlands	NL	465	0
Suriname	SR	674	0	Egypt	EG	464	0
Malaysia (Peninsular)	MY	672	3	Kazakhstan	KZ	460	0
Somalia	SO	666	10	Hong Kong	HK	459	0
Central African Republic	CF	666	0	Chile	CL	454	8
Nicaragua	NI	659	0	Morocco	MA	454	0
Lao People's Republic	LA	658	0	Oman	OM	449	0
Zimbabwe	ZW	658	0	Trinidad and Tobago	TT	445	1
Paraguay	PY	657	0	Saudi Arabia	SA	445	0
Rwanda	RW	655	0	Turkey	TR	441	0
Gabon	GA	648	0	Burkina-Faso	HV	438	0
Malawi	MW	648	0	Finland	FI	436	0
Canada	CA	645	0	Ireland	IE	436	0
Mali	ML	630	0	Poland	PL	436	0
Namibia	NA	627	1	Denmark	DK	435	0
French Guiana	GF	626	1	France	FR	428	1
Senegal	SN	625	0	Sri Lanka	LK	426	24
Sumatra	TA	614	23	Belgium	BE	426	0
Borneo	IB	608	30	Austria	AT	422	0
Sierra Leone	SL	607	0	United Arab Emirates	AE	412	0
Bhutan	BT	606	0	Ukraine	UR	410	0
Burundi	BI	604	0	Greece	GR	403	0
Philippines	PH	586	192	Gibraltar	GI	402	0
Japan	JP	583	10	Sulawesi (Celebes)	SW	399	107
Liberia	LR	580	1	Lesser Sundas	SS	394	50
United Kingdom	GB	576	1	Portugal	PT	392	0
Togo	TG	572	0	Czech Republic	CZ	389	0
Botswana	BW	570	0	Switzerland	CH	388	0
Congo	CG	569	0	Afghanistan	AF	384	0
Belize	BZ	566	0	Uruguay	UY	383	0
Guinea	GN	552	0	Iraq	IQ	381	0
Gambia	GM	541	0	Hungary	HU	380	0
Mauritania	MR	540	0	Algeria	DZ	378	1
Chad	TD	532	0	Yemen	YE	377	0
Eritrea	ER	532	0	Yugoslavia	YU	375	0
Israel	IL	522	0	Slovenia	SI	374	0
Iran	IR	507	1	Malta	MT	372	0
Java and Bali	ID	504	30	Jordan	JO	371	0
Italy	IT	502	0	Moluccas	CS	370	73
El Salvador	SV	499	0	Mongolia	MN	370	0
Benin	BJ	498	0	Bulgaria	BG	366	0
Spain	ES	494	1	North Korea	KP	365	0
Alaska	ak	493	0	Cuba	CU	361	21
Germany	DE	484	0	Romania	RO	360	0
Niger	NE	479	0	Azerbaijan	ZB	359	0
Norway	NO	473	0	Croatia	RT	359	0

Region	Code	Species	Endemics	Region	Code	Species	Endemics
Bermuda	BM	357	0	Greenland	GL	238	0
Georgia	GG	357	0	U.S. Virgin Islands	VI	224	0
Cyprus	CY	356	0	Leeward Islands	MS	221	2
Tunisia	TN	355	0	Barbados	BB	217	1
Iceland	IS	354	0	Cayman Islands	KY	217	1
South Korea	KR	351	0	Andaman and Nicobar Is.	AI	215	20
Armenia	AM	348	0	Madeira Islands	MD	215	2
Slovakia	SK	348	0	Timor	TP	212	8
Turkmenia	TM	348	0	Seychelles	SC	209	10
Canary Islands	IC	343	6	Western Sahara	EH	195	0
Lebanon	LB	341	0	Guadeloupe	GP	188	0
Estonia	EN	340	0	Socotra	RA	181	6
Uzbekistan	BK	338	0	Martinique	MQ	176	1
Singapore	SG	334	0	British Virgin Islands	VG	176	0
Brunei	BN	332	0	Falkland Islands	FK	171	2
Bahamas	BS	330	1	Dominica	DM	167	2
Guinea-Bassau	GW	329	0	Saint Lucia	LC	165	5
Djibouti	DJ	327	1	Cape Verde	CV	158	3
Syrian Arab Republic	SY	320	0	St. Vincent	VC	157	2
Bahrain	BH	319	0	Andorra	AD	150	0
Latvia	LV	319	0	Maldives	MV	147	0
New Zealand	NZ	317	59	Grenada	GD	144	1
Libyan Arab Jamahiriya	LY	317	0	Galapagos	GS	140	24
Lithuania	LN	316	0	Palau and Caroline Is.	PU	137	12
Puerto Rico	PR	313	13	Guam and Marianas	GU	130	8
St. Pierre and Miquelon	PM	305	0	New Caledonia	NC	125	20
Kirghizia	GZ	304	0	Fiji	FJ	123	26
Macedonia	ME	303	0	Svalbard and Jan Mayen	SJ	121	0
Jamaica	JM	296	29	Comoro Islands	KM	117	18
Hawaii	hi	295	36	Aldabra	DB	116	2
Tasmania	TS	294	12	Monaco	MC	106	0
Albania	AL	294	0	San Marino	SM	105	0
Belarus	BL	294	0	Christmas Island	CX	104	2
Tadzhikistan	ZS	293	0	Micronesia	FM	103	13
Kuwait	KW	287	0	Mauritius	MU	103	7
Moldova	DV	285	0	French Polynesia	PF	94	25
Lesotho	LS	283	0	Admiralty Islands	RL	92	5
Equatorial Guinea	GQ	281	3	Sao Tome and Principe	ST	90	25
Luxembourg	LU	281	0	Reunion	RE	86	5
Dominican Republic	DO	278	1	Marshall Islands	RM	78	0
Madagascar	MG	274	98	Gough/Tristan da Cunha	ZC	76	6
Haiti	HT	270	2	Vanuatu	VU	72	10
Faeroe Islands	FO	263	0	Samoa	WS	64	9
Bosnia and Herzegovina	BA	250	0	Kiribati	KI	64	1
Qatar	QA	248	0	Norfolk Island	NF	60	4
Bismarck Archipelago	MK	247	39	Tonga	TO	60	2
Solomon Islands	SB	244	68	Midway Islands	MI	57	0
Netherlands Antilles	AN	243	0	American Samoa	AS	56	0
Azores	AZ	242	0	Canton and Phoenix Is.	CT	54	0
Liechtenstein	LI	242	0	Cook Islands	CK	53	6

Region	Code	Species	Endemics	Region	Code	Species	Endemics
Cocos (Keeling) Is.	CC	52	0	Niue	NU	28	0
Antarctica	AA	48	1	Tuvalu	TV	27	0
St. Helena/ Ascension Is.	ZH	46	1	Nauru	NR	24	1
Rodriguez	RZ	44	2	Wallis and Futuna Is.	WF	23	0
Macau	MO	40	0	Easter Island	IP	21	0
Pitcairn Island	PN	31	5	Johnston Island	JT	18	0
Wake Island	WK	31	0	Tokelau	TK	15	0

Appendix B
Distribution of Endemic Bird Species of the World

World regions ranked by the number of endemic species present. Adapted with permission from *BirdArea for Windows* © 1996, 2006 Santa Barbara Software Products (http://members.aol.com/sbsp). Research by Shawneen Finnegan.

Region	Code	Species	Endemics	Region	Code	Species	Endemics
New Guinea	PG	746	329	Cuba	CU	361	21
Australia	AU	776	246	Andaman and Nicobar Is.	AI	215	20
Brazil	BR	1719	202	New Caledonia	NC	125	20
Philippines	PH	586	192	Bolivia	BO	1347	19
Peru	PE	1793	118	Comoro Islands	KM	117	18
Sulawesi (Celebes)	SW	399	107	Ethiopia	ET	840	17
Madagascar	MG	274	98	Argentina	AR	1005	16
Mexico	MX	1075	90	South Africa	ZA	825	15
Moluccas	CS	370	73	Taiwan (Formosa)	TW	466	15
Solomon Islands	SB	244	68	Zaire	ZR	1133	13
Colombia	CO	1757	65	Puerto Rico	PR	313	13
New Zealand	NZ	317	59	Micronesia	FM	103	13
China	CN	1269	50	Tasmania	TS	294	12
Lesser Sundas	SS	394	50	Palau and Caroline Is.	PU	137	12
Venezuela	VE	1346	46	U.S. contiguous 48 states	US	868	11
Bismarck Archipelago	MK	247	39	Ecuador	EC	1534	10
India	IN	1185	36	Angola	AO	923	10
Hawaii	hi	295	36	Panama	PA	918	10
Borneo	IB	608	30	Vietnam	VN	787	10
Java and Bali	ID	504	30	Somalia	SO	666	10
Jamaica	JM	296	29	Japan	JP	583	10
Fiji	FJ	123	26	Seychelles	SC	209	10
French Polynesia	PF	94	25	Vanuatu	VU	72	10
Sao Tome and Principe	ST	90	25	Samoa	WS	64	9
Sri Lanka	LK	426	24	Cameroon	CM	899	8
Galapagos	GS	140	24	Chile	CL	454	8
Sumatra	TA	614	23	Timor	TP	212	8
Tanzania	TZ	1054	22	Guam and Marianas	GU	130	8

Region	Code	Species	Endemics	Region	Code	Species	Endemics
Costa Rica	CR	852	7	Antarctica	AA	48	1
Mauritius	MU	103	7	St. Helena/ Ascension Is.	ZH	46	1
Kenya	KE	1104	6	Nauru	NR	24	1
Canary Islands	IC	343	6	Guyana	GY	743	0
Socotra	RA	181	6	Ghana	GH	724	0
Gough/Tristan da Cunha	ZC	76	6	Ivory Coast	CI	719	0
Cook Islands	CK	53	6	Russia	RS	712	0
Saint Lucia	LC	165	5	Guatemala	GT	698	0
Admiralty Islands	RL	92	5	Bangladesh	BD	688	0
Reunion	RE	86	5	Mozambique	MZ	685	0
Pitcairn Island	PN	31	5	Pakistan	PK	682	0
Burma (Myanmar)	BU	1002	4	Suriname	SR	674	0
Nigeria	NG	889	4	Central African Republic	CF	666	0
Norfolk Island	NF	60	4	Nicaragua	NI	659	0
Malaysia (Peninsular)	MY	672	3	Lao People's Republic	LA	658	0
Equatorial Guinea	GQ	281	3	Zimbabwe	ZW	658	0
Cape Verde	CV	158	3	Paraguay	PY	657	0
Sudan	SD	962	2	Rwanda	RW	655	0
Thailand	TH	945	2	Gabon	GA	648	0
Nepal	NP	868	2	Malawi	MW	648	0
Haiti	HT	270	2	Canada	CA	645	0
Leeward Islands	MS	221	2	Mali	ML	630	0
Madeira Islands	MD	215	2	Senegal	SN	625	0
Falkland Islands	FK	171	2	Sierra Leone	SL	607	0
Dominica	DM	167	2	Bhutan	BT	606	0
St. Vincent	VC	157	2	Burundi	BI	604	0
Aldabra	DB	116	2	Togo	TG	572	0
Christmas Island	CX	104	2	Botswana	BW	570	0
Tonga	TO	60	2	Congo	CG	569	0
Rodriguez	RZ	44	2	Belize	BZ	566	0
Uganda	UG	1023	1	Guinea	GN	552	0
Zambia	ZM	766	1	Gambia	GM	541	0
Honduras	HN	693	1	Mauritania	MR	540	0
Namibia	NA	627	1	Chad	TD	532	0
French Guiana	GF	626	1	Eritrea	ER	532	0
Liberia	LR	580	1	Israel	IL	522	0
United Kingdom	GB	576	1	Italy	IT	502	0
Iran	IR	507	1	El Salvador	SV	499	0
Spain	ES	494	1	Benin	BJ	498	0
Trinidad and Tobago	TT	445	1	Alaska	ak	493	0
France	FR	428	1	Germany	DE	484	0
Algeria	DZ	378	1	Niger	NE	479	0
Bahamas	BS	330	1	Norway	NO	473	0
Djibouti	DJ	327	1	Sweden	SE	472	0
Dominican Republic	DO	278	1	Swaziland	SZ	469	0
Barbados	BB	217	1	Kampuchea	KH	465	0
Cayman Islands	KY	217	1	Netherlands	NL	465	0
Martinique	MQ	176	1	Egypt	EG	464	0
Grenada	GD	144	1	Kazakhstan	KZ	460	0
Kiribati	KI	64	1	Hong Kong	HK	459	0

Region	Code	Species	Endemics	Region	Code	Species	Endemics
Morocco	MA	454	0	Brunei	BN	332	0
Oman	OM	449	0	Guinea-Bassau	GW	329	0
Saudi Arabia	SA	445	0	Syrian Arab Republic	SY	320	0
Turkey	TR	441	0	Bahrain	BH	319	0
Burkina-Faso	HV	438	0	Latvia	LV	319	0
Finland	FI	436	0	Libyan Arab Jamahiriya	LY	317	0
Ireland	IE	436	0	Lithuania	LN	316	0
Poland	PL	436	0	St. Pierre and Miquelon	PM	305	0
Denmark	DK	435	0	Kirghizia	GZ	304	0
Belgium	BE	426	0	Macedonia	ME	303	0
Austria	AT	422	0	Albania	AL	294	0
United Arab Emirates	AE	412	0	Belarus	BL	294	0
Ukraine	UR	410	0	Tadzhikistan	ZS	293	0
Greece	GR	403	0	Kuwait	KW	287	0
Gibraltar	GI	402	0	Moldova	DV	285	0
Portugal	PT	392	0	Lesotho	LS	283	0
Czech Republic	CZ	389	0	Luxembourg	LU	281	0
Switzerland	CH	388	0	Faeroe Islands	FO	263	0
Afghanistan	AF	384	0	Bosnia and Herzegovina	BA	250	0
Uruguay	UY	383	0	Qatar	QA	248	0
Iraq	IQ	381	0	Netherlands Antilles	AN	243	0
Hungary	HU	380	0	Azores	AZ	242	0
Yemen	YE	377	0	Liechtenstein	LI	242	0
Yugoslavia	YU	375	0	Greenland	GL	238	0
Slovenia	SI	374	0	U.S. Virgin Islands	VI	224	0
Malta	MT	372	0	Western Sahara	EH	195	0
Jordan	JO	371	0	Guadeloupe	GP	188	0
Mongolia	MN	370	0	British Virgin Islands	VG	176	0
Bulgaria	BG	366	0	Andorra	AD	150	0
North Korea	KP	365	0	Maldives	MV	147	0
Romania	RO	360	0	Svalbard and Jan Mayen	SJ	121	0
Azerbaijan	ZB	359	0	Monaco	MC	106	0
Croatia	RT	359	0	San Marino	SM	105	0
Bermuda	BM	357	0	Marshall Islands	RM	78	0
Georgia	GG	357	0	Midway Islands	MI	57	0
Cyprus	CY	356	0	American Samoa	AS	56	0
Tunisia	TN	355	0	Canton and Phoenix Is.	CT	54	0
Iceland	IS	354	0	Cocos (Keeling) Is.	CC	52	0
South Korea	KR	351	0	Macau	MO	40	0
Armenia	AM	348	0	Wake Island	WK	31	0
Slovakia	SK	348	0	Niue	NU	28	0
Turkmenia	TM	348	0	Tuvalu	TV	27	0
Lebanon	LB	341	0	Wallis and Futuna Is.	WF	23	0
Estonia	EN	340	0	Easter Island	IP	21	0
Uzbekistan	BK	338	0	Johnston Island	JT	18	0
Singapore	SG	334	0	Tokelau	TK	15	0

Major Family References

Complete references to publisher and subtitles are listed in the bibliography

Struthionidae   del Hoyo, J., A. Elliot and J. Sargatal, eds. 1992. *Handbook of Birds of the World*. Vol. 1.

Rheidae   del Hoyo, J., A. Elliot and J. Sargatal, eds. 1992. *Handbook of Birds of the World*. Vol. 1.

Casuariidae   del Hoyo, J., A. Elliot and J. Sargatal, eds. 1992. *Handbook of Birds of the World*. Vol. 1.

Dromaiidae   del Hoyo, J., A. Elliot and J. Sargatal, eds. 1992. *Handbook of Birds of the World*. Vol. 1.

Apterygidae   del Hoyo, J., A. Elliot and J. Sargatal, eds. 1992. *Handbook of Birds of the World*. Vol. 1.

Tinamidae   de Vasconcelos, Marcelo F. 2002. A newly discovered specimen of Kalinowski's Tinamou *Nothoprocta kalinowskii* from the Andean Pacific slope of Peru. *Bull. B. O. C.* 122 (3): 216–218.

    del Hoyo, J., A. Elliot and J. Sargatal, eds. 1992. *Handbook of Birds of the World*. Vol. 1.

Spheniscidae   del Hoyo, J., A. Elliot and J. Sargatal, eds. 1992. *Handbook of Birds of the World*. Vol. 1.

Gaviidae   del Hoyo, J., A. Elliot and J. Sargatal, eds. 1992. *Handbook of Birds of the World*. Vol. 1.

Podicipedidae   del Hoyo, J., A. Elliot and J. Sargatal, eds. 1992. *Handbook of Birds of the World*. Vol. 1.

Diomedeidae   del Hoyo, J., A. Elliot and J. Sargatal, eds. 1992. *Handbook of Birds of the World*. Vol. 1.

Procellariidae   Banks, R. C. *et al.* 2002. Forty-third supplement to the American Ornithologists' Union Check-list of North American Birds. *The Auk* 119 (3): 897–906.

    del Hoyo, J., A. Elliot and J. Sargatal, eds. 1992. *Handbook of Birds of the World*. Vol. 1.

Hydrobatidae   del Hoyo, J., A. Elliot and J. Sargatal, eds. 1992. *Handbook of Birds of the World*. Vol. 1.

Pelecanoididae   del Hoyo, J., A. Elliot and J. Sargatal, eds. 1992. *Handbook of Birds of the World*. Vol. 1.

Phaethontidae   del Hoyo, J., A. Elliot and J. Sargatal, eds. 1992. *Handbook of Birds of the World*. Vol. 1.

Pelecanidae   del Hoyo, J., A. Elliot and J. Sargatal, eds. 1992. *Handbook of Birds of the World*. Vol. 1.

Sulidae   del Hoyo, J., A. Elliot and J. Sargatal, eds. 1992. *Handbook of Birds of the World*. Vol. 1.

Phalacrocoracidae   del Hoyo, J., A. Elliot and J. Sargatal, eds. 1992. *Handbook of Birds of the World*. Vol. 1.

Anhingidae   del Hoyo, J., A. Elliot and J. Sargatal, eds. 1992. *Handbook of Birds of the World*. Vol. 1.

Fregatidae   del Hoyo, J., A. Elliot and J. Sargatal, eds. 1992. *Handbook of Birds of the World*. Vol. 1.

Ardeidae   del Hoyo, J., A. Elliot and J. Sargatal, eds. 1992. *Handbook of Birds of the World*. Vol. 1.

Scopidae   del Hoyo, J., A. Elliot and J. Sargatal, eds. 1992. *Handbook of Birds of the World*. Vol. 1.

Ciconiidae   del Hoyo, J., A. Elliot and J. Sargatal, eds. 1992. *Handbook of Birds of the World*. Vol. 1.

Balaenicipitidae   del Hoyo, J., A. Elliot and J. Sargatal, eds. 1992. *Handbook of Birds of the World*. Vol. 1.

Threskiornithidae   del Hoyo, J., A. Elliot and J. Sargatal, eds. 1992. *Handbook of Birds of the World*. Vol. 1.

Phoenicopteridae   del Hoyo, J., A. Elliot and J. Sargatal, eds. 1992. *Handbook of Birds of the World*. Vol. 1.

    Knox, Alan G., Martin Collinson, Andreas Helbig, David Parkin and George Sangster. 2002. Taxonomic considerations for British Birds. *Ibis* 144 (4): 707–710.

Anhimidae	del Hoyo, J., A. Elliot and J. Sargatal, eds. 1992. *Handbook of Birds of the World*. Vol. 1.
Anatidae	del Hoyo, J., A. Elliot and J. Sargatal, eds. 1992. *Handbook of Birds of the World*. Vol. 1.
	Todd, Frank. 1996. *Natural History of the Waterfowl.*
Cathartidae	del Hoyo, J., A. Elliot and J. Sargatal, eds. 1992. *Handbook of Birds of the World*. Vol. 2.
Pandionidae	del Hoyo, J., A. Elliot and J. Sargatal, eds. 1994. *Handbook of Birds of the World*. Vol. 2.
Accipitridae	del Hoyo, J., A. Elliot and J. Sargatal, eds. 1994. *Handbook of Birds of the World*. Vol. 2.
	Parry, S. J., W. S. Clark and V. Prakash. 2002. On the taxonomic status of the Indian Spotted Eagle *Aquila hastata*. *Ibis* 144 (4): 665–675.
Sagittariidae	del Hoyo, J., A. Elliot and J. Sargatal, eds. 1994. *Handbook of Birds of the World*. Vol. 2.
Falconidae	del Hoyo, J., A. Elliot and J. Sargatal, eds. 1994. *Handbook of Birds of the World*. Vol. 2.
Megapodiidae	Jones, Darryl N., René Dekker and Cees Roselaar. 1995. *The Megapodes.*
	del Hoyo, J., A. Elliot and J. Sargatal, eds. 1994. *Handbook of Birds of the World*. Vol. 2.
Cracidae	del Hoyo, J., A. Elliot and J. Sargatal, eds. 1994. *Handbook of Birds of the World*. Vol. 2.
Meleagrididae	del Hoyo, J., A. Elliot and J. Sargatal, eds. 1994. *Handbook of Birds of the World*. Vol. 2.
	Delacour, Jean and Dean Amadon. 1973. *Curassows and Related Birds.*
Tetraonidae	del Hoyo, J., A. Elliot and J. Sargatal, eds. 1994. *Handbook of Birds of the World*. Vol. 2.
Odontophoridae	del Hoyo, J., A. Elliot and J. Sargatal, eds. 1994. *Handbook of Birds of the World*. Vol. 2.
Phasianidae	del Hoyo, J., A. Elliot and J. Sargatal, eds. 1994. *Handbook of Birds of the World*. Vol. 2.
	Delacour, Jean. 1977. *The Pheasants of the World.*
Numididae	del Hoyo, J., A. Elliot and J. Sargatal, eds. 1994. *Handbook of Birds of the World*. Vol. 2.
Opisthocomidae	del Hoyo, J., A. Elliot and J. Sargatal, eds. 1996. *Handbook of Birds of the World*. Vol. 3.
Mesitornithidae	del Hoyo, J., A. Elliot and J. Sargatal, eds. 1996. *Handbook of Birds of the World*. Vol. 3.
Turnicidae	del Hoyo, J., A. Elliot and J. Sargatal, eds. 1996. *Handbook of Birds of the World*. Vol. 3.
Gruidae	del Hoyo, J., A. Elliot and J. Sargatal, eds. 1996. *Handbook of Birds of the World*. Vol. 3.
Aramidae	del Hoyo, J., A. Elliot and J. Sargatal, eds. 1996. *Handbook of Birds of the World*. Vol. 3.
Psophiidae	del Hoyo, J., A. Elliot and J. Sargatal, eds. 1996. *Handbook of Birds of the World*. Vol. 3.
Rallidae	del Hoyo, J., A. Elliot and J. Sargatal, eds. 1996. *Handbook of Birds of the World*. Vol. 3.
	Taylor, Barry. 1998. *A Guide to the Rails, Crakes, Gallinules and Coots of the World.*
Heliornithidae	del Hoyo, J., A. Elliot and J. Sargatal, eds. 1996. *Handbook of Birds of the World*. Vol. 3.
Rhynochetidae	del Hoyo, J., A. Elliot and J. Sargatal, eds. 1996. *Handbook of Birds of the World*. Vol. 3.
Eurypygidae	del Hoyo, J., A. Elliot and J. Sargatal, eds. 1996. *Handbook of Birds of the World*. Vol. 3.
Cariamidae	del Hoyo, J., A. Elliot and J. Sargatal, eds. 1996. *Handbook of Birds of the World*. Vol. 3.
Otididae	del Hoyo, J., A. Elliot and J. Sargatal, eds. 1996. *Handbook of Birds of the World*. Vol. 3.
	Knox, Alan G., Martin Collinson, Andreas Helbig, David Parkin and George Sangster. 2002. Taxonomic considerations for British Birds. *Ibis* 144 (4): 707–710.
Jacanidae	del Hoyo, J., A. Elliot and J. Sargatal, eds. 1996. *Handbook of Birds of the World*. Vol. 3.
Rostratulidae	del Hoyo, J., A. Elliot and J. Sargatal, eds. 1996. *Handbook of Birds of the World*. Vol. 3.
Dromadidae	del Hoyo, J., A. Elliot and J. Sargatal, eds. 1996. *Handbook of Birds of the World*. Vol. 3.
Haematopodidae	del Hoyo, J., A. Elliot and J. Sargatal, eds. 1996. *Handbook of Birds of the World*. Vol. 3.
Ibidorhynchidae	del Hoyo, J., A. Elliot and J. Sargatal, eds. 1996. *Handbook of Birds of the World*. Vol. 3.
Recurvirostridae	del Hoyo, J., A. Elliot and J. Sargatal, eds. 1996. *Handbook of Birds of the World*. Vol. 3.
Burhinidae	del Hoyo, J., A. Elliot and J. Sargatal, eds. 1996. *Handbook of Birds of the World*. Vol. 3.
Glareolidae	del Hoyo, J., A. Elliot and J. Sargatal, eds. 1996. *Handbook of Birds of the World*. Vol. 3.
Charadriidae	del Hoyo, J., A. Elliot and J. Sargatal, eds. 1996. *Handbook of Birds of the World*. Vol. 3.
Pluvianellidae	del Hoyo, J., A. Elliot and J. Sargatal, eds. 1996. *Handbook of Birds of the World*. Vol. 3.
Scolopacidae	Banks, R. C. *et al.* 2002. Forty-third supplement to the American Ornithologists' Union Check-list of North American Birds. *The Auk* 119 (3): 897–906.
	del Hoyo, J., A. Elliot and J. Sargatal, eds. 1996. *Handbook of Birds of the World*. Vol. 3.
Pedionomidae	del Hoyo, J., A. Elliot and J. Sargatal, eds. 1996. *Handbook of Birds of the World*. Vol. 3.
Thinocoridae	del Hoyo, J., A. Elliot and J. Sargatal, eds. 1996. *Handbook of Birds of the World*. Vol. 3.
Chionididae	del Hoyo, J., A. Elliot and J. Sargatal, eds. 1996. *Handbook of Birds of the World*. Vol. 3.
Stercorariidae	del Hoyo, J., A. Elliot and J. Sargatal, eds. 1996. *Handbook of Birds of the World*. Vol. 3.

Laridae	del Hoyo, J., A. Elliot and J. Sargatal, eds. 1996. *Handbook of Birds of the World*. Vol. 3.
	Yésou, Pierre. 2002. Systematics of *Larus argentatus-cachinnans-fuscus* complex revisited. *Dutch Birding* 24 (5): 271–298.
Sternidae	del Hoyo, J., A. Elliot and J. Sargatal, eds. 1996. *Handbook of Birds of the World*. Vol. 3.
Rynchopidae	del Hoyo, J., A. Elliot and J. Sargatal, eds. 1996. *Handbook of Birds of the World*. Vol. 3.
Alcidae	del Hoyo, J., A. Elliot and J. Sargatal, eds. 1996. *Handbook of Birds of the World*. Vol. 3.
Pteroclidae	del Hoyo, J., A. Elliot and J. Sargatal, eds. 1997. *Handbook of Birds of the World*. Vol. 4.
Columbidae	del Hoyo, J., A. Elliot and J. Sargatal, eds. 1997. *Handbook of Birds of the World*. Vol. 4.
	Garrido, Orlando H., Guy M. Kirwan and David R. Capper. 2002. Species within the Grey-headed Quail-Dove *Geotrygon caniceps* and implications for the conservation of a globally threatened species. *Bird Conservation International* 12 (2): 169–187.
Cacatuidae	del Hoyo, J., A. Elliot and J. Sargatal, eds. 1997. *Handbook of Birds of the World*. Vol. 4.
Psittacidae	del Hoyo, J., A. Elliot and J. Sargatal, eds. 1997. *Handbook of Birds of the World*. Vol. 4.
	Gaban-Lima, Renato, Marcos A. Raposo and Elizabeth Höfling. 2002. Description of a New Species of *Pionopsitta* (Aves: Psittacidae) Endemic to Brazil. *The Auk* 119 (3): 815–819.
Musophagidae	del Hoyo, J., A. Elliot and J. Sargatal, eds. 1997. *Handbook of Birds of the World*. Vol. 4.
Cuculidae	del Hoyo, J., A. Elliot and J. Sargatal, eds. 1997. *Handbook of Birds of the World*. Vol. 4.
Tytonidae	del Hoyo, J., A. Elliot and J. Sargatal, eds. 1999. *Handbook of Birds of the World*. Vol. 5.
	König, Claus, Friedhelm Weick and Jan-Hendrik Becking. 1999. *Owls: A Guide to the Owls of the World*. Sussex: Pica Press.
Strigidae	del Hoyo, J., A. Elliot and J. Sargatal, eds. 1999. *Handbook of Birds of the World*. Vol. 5.
	König, Claus, Friedhelm Weick and Jan-Hendrik Becking. 1999. *Owls: A Guide to the Owls of the World*. Sussex: Pica Press.
	Olsen, Jerry, Michael Wink, Hedi Sauer-Gürth and Susan Trost. 2002. A new *Ninox* owl from Sumba, Indonesia. *Emu* 102: 223–231.
	Warakagoda, Deepal H. and Pamela Rasmussen. 2004. A new species of scops-owl from Sri Lanka. *Bulletin B. O. C.* 124 (2): 85–105.
Steatornithidae	del Hoyo, J., A. Elliot and J. Sargatal, eds. 1999. *Handbook of Birds of the World*. Vol. 5.
Aegothelidae	Cleere, Nigel. 1998. *Nightjars: A Guide to the Nightjars, Nighthawks, and Their Relatives*.
	del Hoyo, J., A. Elliot and J. Sargatal, eds. 1999. *Handbook of Birds of the World*. Vol. 5.
Podargidae	Cleere, Nigel. 1998. *Nightjars: A Guide to the Nightjars, Nighthawks, and Their Relatives*.
	del Hoyo, J., A. Elliot and J. Sargatal, eds. 1999. *Handbook of Birds of the World*. Vol. 5.
Nyctibiidae	Cleere, Nigel. 1998. *Nightjars: A Guide to the Nightjars, Nighthawks, and Their Relatives*.
	del Hoyo, J., A. Elliot and J. Sargatal, eds. 1999. *Handbook of Birds of the World*. Vol. 5.
Caprimulgidae	Cleere, Nigel. 1998. *Nightjars: A Guide to the Nightjars, Nighthawks, and Their Relatives*.
	del Hoyo, J., A. Elliot and J. Sargatal, eds. 1999. *Handbook of Birds of the World*. Vol. 5.
Apodidae	Banks, R. C. *et al.* 2002. Forty-third supplement to the American Ornithologists' Union Check-list of North American Birds. *The Auk* 119 (3): 897–906.
	Chantler, Phil and Gerald Driessens. 1995. *Swifts: A Guide to the Swifts and Treeswifts of the World*.
	del Hoyo, J., A. Elliot and J. Sargatal, eds. 1999. *Handbook of Birds of the World*. Vol. 5.
Hemiprocnidae	del Hoyo, J., A. Elliot and J. Sargatal, eds. 1999. *Handbook of Birds of the World*. Vol. 5.
Trochilidae	Banks, R. C. *et al.* 2002. Forty-third supplement to the American Ornithologists' Union Check-list of North American Birds. *The Auk* 119 (3): 897–906.
	del Hoyo, J., A. Elliot and J. Sargatal, eds. 1999. *Handbook of Birds of the World*. Vol. 5.
Coliidae	Fry, C. Hilary, Stuart Keith and Emil Urban, eds. 1988. *The Birds of Africa*. Volume III.
Trogonidae	Collar, Nigel J. and S. Van Balen. 2002. The Blue-tailed Trogon *Harpactes (Apalharpactes) reinwardtii*: species limits and conservation status. *Forktail* 18: 121–125.
	Peters, James Lee. 1945. *Check-list of Birds of the World*. Volume V.
	Johnsgard, Paul A. 2000. *Trogons and Quetzals of the World*. Smithsonian Institution Press.
Alcedinidae	Fry, C. H., K. Fry and A. Harris. 1992. *Kingfishers, Bee-eaters and Rollers*.
	Peters, James Lee. 1945. *Check-list of Birds of the World*. Volume V.
Todidae	Peters, James Lee. 1945. *Check-list of Birds of the World*. Volume V.

Momotidae	Peters, James Lee. 1945. *Check-list of Birds of the World*. Volume V.
Meropidae	Fry, C. H., K. Fry and A. Harris. 1992. *Kingfishers, Bee-eaters and Rollers*.
	Peters, James Lee. 1945. *Check-list of Birds of the World*. Volume V.
Coraciidae	Fry, C. H., K. Fry and A. Harris. 1992. *Kingfishers, Bee-eaters and Rollers*.
	Peters, James Lee. 1945. *Check-list of Birds of the World*. Volume V.
Brachypteraciidae	Fry, C. H., K. Fry and A. Harris. 1992. *Kingfishers, Bee-eaters and Rollers*.
	Peters, James Lee. 1945. *Check-list of Birds of the World*. Volume V.
Leptosomatidae	Fry, C. H., K. Fry and A. Harris. 1992. *Kingfishers, Bee-eaters and Rollers*.
	Peters, James Lee. 1945. *Check-list of Birds of the World*. Volume V.
Upupidae	Peters, James Lee. 1945. *Check-list of Birds of the World*. Volume V.
Phoeniculidae	Peters, James Lee. 1945. *Check-list of Birds of the World*. Volume V.
Bucerotidae	Kemp, Alan. 1995. *The Hornbills*. New York: Oxford University Press.
Galbulidae	Peters, James Lee. 1948. *Check-list of Birds of the World*. Volume VI.
Bucconidae	Peters, James Lee. 1948. *Check-list of Birds of the World*. Volume VI.
Capitonidae	Peters, James Lee. 1948. *Check-list of Birds of the World*. Volume VI.
Ramphastidae	Peters, James Lee. 1948. *Check-list of Birds of the World*. Volume VI.
	Navarro, Adolfo G., A. Townsend Peterson, Esteban López-Medrano and Hesequio Benítez-Díaz. 2001. Species limits in Mesoamerican *Aulacorhynchus* Toucanets. *Wilson Bulletin* 114 (4): 363–372.
Indicatoridae	Peters, James Lee. 1948. *Check-list of Birds of the World*. Volume VI.
Picidae	Short, Lester. 1982. *Woodpeckers of the World*.
	Winkler, Hans, D. A. Christie and D. Nurney. 1995. *Woodpeckers: An Identification Guide to Woodpeckers of the World*.
Eurylaimidae	Lambert, Frank and Martin Woodcock. 1996. *Pittas, Broadbills and Asities*.
Philepittidae	Lambert, Frank and Martin Woodcock. 1996. *Pittas, Broadbills and Asities*.
Furnariidae	Peters, James Lee. 1957. *Check-list of Birds of the World*. Volume VII.
	Vaurie, Charles. 1980. *Taxonomy and Geographical Distribution of the Furnariidae*.
	Zimmer, Kevin J. 2002. Species limits in Olive-backed Foliage-gleaners (*Automolus*: Furnariidae). *Wilson Bulletin* 114 (1): 20–37.
Dendrocolaptidae	Peters, James Lee. 1957. *Check-list of Birds of the World*. Volume VII.
	Ridgely, Robert and Guy Tudor. 1994. *The Birds of South America*: Vol. II. The suboscine passerines.
	Alexio, Alexandre. 2002. Molecular systematics and the role of the "Várzea"-"Terra-firme" ecotone in the diversification of *Xiphorhynchus* woodcreepers (Aves: Dendrocolaptidae). *The Auk* 119 (3): 621–640.
	da Silva, José Maria C., Fernando C. Novaes and David C. Oren. 2002. Differentiation of *Xiphocolaptes* (Dendrocolaptidae) across the river Xingu, Brazilian Amazonia: recognition of a new phylogenetic species and biogeographic implications. *Bull. B. O. C.* 122 (3): 185–194.
Thamnophilidae	Peters, James Lee. 1957. *Check-list of Birds of the World*. Volume VII.
	Ridgely, Robert and Guy Tudor. 1994. *The Birds of South America*: Vol. II. The suboscine passerines.
Formicariidae	Peters, James Lee. 1957. *Check-list of Birds of the World*. Volume VII.
	Ridgely, Robert and Guy Tudor. 1994. *The Birds of South America*: Vol. II. The suboscine passerines.
Conopophagidae	Peters, James Lee. 1957. *Check-list of Birds of the World*. Volume VII.
	Ridgely, Robert and Guy Tudor. 1994. *The Birds of South America*: Vol. II. The suboscine passerines.
Rhinocryptidae	Peters, James Lee. 1957. *Check-list of Birds of the World*. Volume VII.
	Ridgely, Robert and Guy Tudor. 1994. *The Birds of South America*: Vol. II. The suboscine passerines.
Phytotomidae	Peters, James Lee. 1957. *Check-list of Birds of the World*. Volume VII.
	Ridgely, Robert and Guy Tudor. 1994. *The Birds of South America*: Vol. II. The suboscine passerines.
Cotingidae	del Hoyo, Josep, Andrew Elliot and D. A. Christie, editors. 1992. *Handbook of the Birds of the World*. Volume 9. Cotingas to Pipits and Wagtails. Barcelona: Lynx Edicions.
	Melvin A. Traylor, Jr., ed. 1979. *Check-list of Birds of the World*. Volume VIII. A Continuation of the Work of James L. Peters.
	Ridgely, Robert and Guy Tudor. 1994. *The Birds of South America*: Vol. II. The suboscine passerines.
	Snow, David. 1982. *The Cotingas*.

Pipridae	del Hoyo, Josep, Andrew Elliot and D. A. Christie, editors. 1992. *Handbook of the Birds of the World.* Volume 9. Cotingas to Pipits and Wagtails. Barcelona: Lynx Edicions.
	Melvin A. Traylor, Jr., ed. 1979. *Check-list of Birds of the World.* Volume VIII. A Continuation of the Work of James L. Peters.
	Ridgely, Robert and Guy Tudor. 1994. *The Birds of South America*: Vol. II. The suboscine passerines.
Tyrannidae	del Hoyo, Josep, Andrew Elliot and D. A. Christie, editors. 1992. *Handbook of the Birds of the World.* Volume 9. Cotingas to Pipits and Wagtails. Barcelona: Lynx Edicions.
	Melvin A. Traylor, Jr., ed. 1979. *Check-list of Birds of the World.* Volume VIII. A Continuation of the Work of James L. Peters.
	Ridgely, Robert and Guy Tudor. 1994. *The Birds of South America*: Vol. II. The suboscine passerines.
Oxyruncidae	Melvin A. Traylor, Jr., ed. 1979. *Check-list of Birds of the World.* Volume VIII. A Continuation of the Work of James L. Peters.
Pittidae	Lambert, Frank and Martin Woodcock. 1996. *Pittas, Broadbills and Asities.*
Atrichornithidae	del Hoyo, Josep, Andrew Elliot and D. A. Christie, editors. 1992. *Handbook of the Birds of the World.* Volume 9. Cotingas to Pipits and Wagtails. Barcelona: Lynx Edicions.
	Melvin A. Traylor, Jr., ed. 1979. *Check-list of Birds of the World.* Volume VIII. A Continuation of the Work of James L. Peters.
Menuridae	del Hoyo, Josep, Andrew Elliot and D. A. Christie, editors. 1992. *Handbook of the Birds of the World.* Volume 9. Cotingas to Pipits and Wagtails. Barcelona: Lynx Edicions.
	Melvin A. Traylor, Jr., ed. 1979. *Check-list of Birds of the World.* Volume VIII. A Continuation of the Work of James L. Peters.
Acanthisittidae	del Hoyo, Josep, Andrew Elliot and D. A. Christie, editors. 1992. *Handbook of the Birds of the World.* Volume 9. Cotingas to Pipits and Wagtails. Barcelona: Lynx Edicions.
	Melvin A. Traylor, Jr., ed. 1979. *Check-list of Birds of the World.* Volume VIII. A Continuation of the Work of James L. Peters.
Alaudidae	del Hoyo, Josep, Andrew Elliot and D. A. Christie, editors. 1992. *Handbook of the Birds of the World.* Volume 9. Cotingas to Pipits and Wagtails. Barcelona: Lynx Edicions.
	Mayr, Ernst and James C. Greenway, Jr., eds. 1960. *Check-list of Birds of the World.* Volume IX. A Continuation of the Work of James L. Peters.
Hirundinidae	del Hoyo, Josep, Andrew Elliot and D. A. Christie, editors. 1992. *Handbook of the Birds of the World.* Volume 9. Cotingas to Pipits and Wagtails. Barcelona: Lynx Edicions.
	Turner, Angela and Chris Rose. 1989. *Swallows and Martins*: *An Identification Guide and Handbook.*
Motacillidae	del Hoyo, Josep, Andrew Elliot and D. A. Christie, editors. 1992. *Handbook of the Birds of the World.* Volume 9. Cotingas to Pipits and Wagtails. Barcelona: Lynx Edicions.
	Alström, Per and Krister Mild. 2003. *Pipits and Wagtails.* Princeton University Press.
	Duckworth, J. W., P. Alström, P. Davidson, T. D. Evans, C. M. Poole, T. Setha and R. J. Timmins. 2001. A new species of wagtail from the lower Mekong basin. *Bulletin B. O. C.* 121 (3): 154–182.
	Liversidge, Richard and Gary Voelker. 2002. The Kimberley Pipit: a new African species. *Bulletin B. O. C.* 122 (2): 93–109.
	Mayr, Ernst and James C. Greenway, Jr., eds. 1960. *Check-list of Birds of the World.* Volume IX. A Continuation of the Work of James L. Peters.
Campephagidae	Mayr, Ernst and James C. Greenway, Jr., eds. 1960. *Check-list of Birds of the World.* Volume IX. Continuation of the Work of James L. Peters.
Pycnonotidae	Mayr, Ernst and James C. Greenway, Jr., eds. 1960. *Check-list of Birds of the World.* Volume IX. A Continuation of the Work of James L. Peters.
Regulidae	Mayr, Ernst and G. William Cottrell, eds. 1986. *Check-list of Birds of the World.* Volume XI. A Continuation of the Work of James L. Peters.
Chloropseidae	Mayr, Ernst and James C. Greenway, Jr., eds. 1960. *Check-list of Birds of the World.* Volume IX. A Continuation of the Work of James L. Peters.
Aegithinidae	Mayr, Ernst and James C. Greenway, Jr., eds. 1960. *Check-list of Birds of the World.* Volume IX. A Continuation of the Work of James L. Peters.

Ptilogonatidae	Mayr, Ernst and James C. Greenway, Jr., eds. 1960. *Check-list of Birds of the World.* Volume IX. A Continuation of the Work of James L. Peters.
Bombycillidae	Mayr, Ernst and James C. Greenway, Jr., eds. 1960. *Check-list of Birds of the World.* Volume IX. A Continuation of the Work of James L. Peters.
Hypocoliidae	Mayr, Ernst and James C. Greenway, Jr., eds. 1960. *Check-list of Birds of the World.* Volume IX. A Continuation of the Work of James L. Peters.
Dulidae	Mayr, Ernst and James C. Greenway, Jr., eds. 1960. *Check-list of Birds of the World.* Volume IX. A Continuation of the Work of James L. Peters.
Cinclidae	Mayr, Ernst and James C. Greenway, Jr., eds. 1960. *Check-list of Birds of the World.* Volume IX. A Continuation of the Work of James L. Peters.
Troglodytidae	Mayr, Ernst and James C. Greenway, Jr., eds. 1960. *Check-list of Birds of the World.* Volume IX. A Continuation of the Work of James L. Peters.
Mimidae	Mayr, Ernst and James C. Greenway, Jr., eds. 1960. *Check-list of Birds of the World.* Volume IX. A Continuation of the Work of James L. Peters.
Prunellidae	Mayr, Ernst and Raymond A. Paynter, Jr., eds. 1964. *Check-list of Birds of the World.* Volume X. A Continuation of the Work of James L. Peters.
Turdidae	Clement, Peter. 2000. *Thrushes.* Princeton University Press.
	Mayr, Ernst and Raymond A. Paynter, Jr., eds. 1964. *Check-list of Birds of the World.* Volume X. A Continuation of the Work of James L. Peters.
Cisticolidae	Mayr, Ernst and G. William Cottrell, eds. 1986. *Check-list of Birds of the World.* Volume XI. A Continuation of the Work of James L. Peters.
Sylviidae	Alström, Per and Urban Olsson. 1999. The Golden-spectacled Warbler: a complex of sibling species, including a previously undescribed species. *Ibis* 141 (4): 545–568.
	Baker, Kevin. 1997. *Warblers of Europe, Asia and North Africa.* Princeton University Press.
	Bensch, Staffan and David Pearson. 2002. The Large-billed Reed Warbler *Acrocephalus orinus* revisited. *Ibis* 144 (2) 259–267.
	Mayr, Ernst and G. William Cottrell, eds. 1986. *Check-list of Birds of the World.* Volume XI. A Continuation of the Work of James L. Peters.
	Shirihai, Hadoram, Gabriel Gargallo and Andreas J. Helbig. 2001. *Sylvia Warblers.* Princeton University Press.
Polioptilidae	Mayr, Ernst and Raymond A. Paynter, Jr., eds. 1964. *Check-list of Birds of the World.* Volume X. A Continuation of the Work of James L. Peters.
Muscicapidae	Mayr, Ernst and G. William Cottrell, eds. 1986. *Check-list of Birds of the World.* Volume XI. A Continuation of the Work of James L. Peters.
Platysteiridae	Mayr, Ernst and G. William Cottrell, eds. 1986. *Check-list of Birds of the World.* Volume XI. A Continuation of the Work of James L. Peters.
Rhipiduridae	Mayr, Ernst and G. William Cottrell, eds. 1986. *Check-list of Birds of the World.* Volume XI. A Continuation of the Work of James L. Peters.
Monarchidae	Mayr, Ernst and G. William Cottrell, eds. 1986. *Check-list of Birds of the World.* Volume XI. A Continuation of the Work of James L. Peters.
Petroicidae	Peters, James Lee. 1948. *Check-list of Birds of the World.* Volume VI.
Pachycephalidae	Paynter, Raymond A., Jr., ed., in consultation with Ernst Mayr. 1967. *Check-list of Birds of the World.* Volume XII. A Continuation of the Work of James L. Peters.
Picathartidae	Mayr, Ernst and Raymond A. Paynter, Jr., eds. 1964. *Check-list of Birds of the World.* Volume X. A Continuation of the Work of James L. Peters.
Timaliidae	Mayr, Ernst and Raymond A. Paynter, Jr., eds. 1964. *Check-list of Birds of the World.* Volume X. A Continuation of the Work of James L. Peters.
Pomatostomidae	Mayr, Ernst and Raymond A. Paynter, Jr., eds. 1964. *Check-list of Birds of the World.* Volume X. A Continuation of the Work of James L. Peters.
Paradoxornithidae	Mayr, Ernst and Raymond A. Paynter, Jr., eds. 1964. *Check-list of Birds of the World.* Volume X. A Continuation of the Work of James L. Peters.

Orthonychidae	Mayr, Ernst and Raymond A. Paynter, Jr., eds. 1964. *Check-list of Birds of the World.* Volume X. A Continuation of the Work of James L. Peters.
Eupetidae (Cinclosomatidae)	Mayr, Ernst and Raymond A. Paynter, Jr., eds. 1964. *Check-list of Birds of the World.* Volume X. A Continuation of the Work of James L. Peters.
Aegithalidae	Paynter, Raymond A., Jr., ed., in consultation with Ernst Mayr. 1967. *Check-list of Birds of the World.* Volume XII. A Continuation of the Work of James L. Peters.
Maluridae	Rowley, Ian and Eleanor Russell. 1997. *Fairy-Wrens and Grasswrens.* Schodde, Richard. 1982. *The Fairy Wrens.*
Acanthizidae	Mayr, Ernst and G. William Cottrell, eds. 1986. *Check-list of Birds of the World.* Volume XI. A Continuation of the Work of James L. Peters.
Epthianuridae	Mayr, Ernst and G. William Cottrell, eds. 1986. *Check-list of Birds of the World.* Volume XI. A Continuation of the Work of James L. Peters.
Neosittidae	Paynter, Raymond A., Jr., ed., in consultation with Ernst Mayr. 1967. *Check-list of Birds of the World.* Volume XII. A Continuation of the Work of James L. Peters.
Climacteridae	Paynter, Raymond A., Jr., ed., in consultation with Ernst Mayr. 1967. *Check-list of Birds of the World.* Volume XII. A Continuation of the Work of James L. Peters.
Paridae	Banks, R. C. *et al.* 2002. Forty-third supplement to the American Ornithologists' Union Check-list of North American Birds. *The Auk* 119 (3): 897–906. Harrap, Simon and David Quinn. 1995. *Chickadees, Tits, Nuthatches and Treecreepers.* James, Helen F., Per G. P. Ericson, Beth Slikas, Fu-Min Lei, Frank B. Gill and Storrs L. Olsen. 2003. *Pseudopodoces humilis,* a misclassified terrestrial tit (Paridae) of the Tibetan Plateau: evolutionary consequences of shifting adaptive zones. *Ibis* 145 (2): 185–202.
Sittidae	Harrap, Simon and David Quinn. 1995. *Chickadees, Tits, Nuthatches and Treecreepers.*
Tichidromidae	Harrap, Simon and David Quinn. 1995. *Chickadees, Tits, Nuthatches and Treecreepers.*
Certhiidae	Harrap, Simon and David Quinn. 1995. *Chickadees, Tits, Nuthatches and Treecreepers.* Martens, J., S. Eck and Y. H. Sun. 2002. *Certhia tianquanensis* Li, a treecreeper with relict distribution in Sichuan, China. *Journal Ornithology* 143: 440–456.
Rhabdornithidae	Paynter, Raymond A., Jr., ed., in consultation with Ernst Mayr. 1967. *Check-list of Birds of the World.* Volume XII. A Continuation of the Work of James L. Peters.
Remizidae	Paynter, Raymond A., Jr., ed., in consultation with Ernst Mayr. 1967. *Check-list of Birds of the World.* Volume XII. A Continuation of the Work of James L. Peters.
Nectariniidae	Chekek, R. A. and C. F. Mann. 2001. *Sunbirds: A Guide to the Sunbirds, Flowerpeckers, Spiderhunters and Sugarbirds of the World.* Yale University Press. Fry, C. Hilary and Stuart Keith. *The Birds of Africa,* Vol. VI.
Melanocharitidae	Paynter, Raymond A., Jr., ed., in consultation with Ernst Mayr. 1967. *Check-list of Birds of the World.* Volume XII. A Continuation of the Work of James L. Peters.
Paramythiidae	Paynter, Raymond A., Jr., ed., in consultation with Ernst Mayr. 1967. *Check-list of Birds of the World.* Volume XII. A Continuation of the Work of James L. Peters.
Dicaeidae	Chekek, R. A. and C. F. Mann. 2000. *Sunbirds: A Guide to the Sunbirds, Spiderhunters, Sugarbirds and Flowerpeckers of the World.* Sussex, UK: Pica Press.
Pardalotidae	Paynter, Raymond A., Jr., ed., in consultation with Ernst Mayr. 1967. *Check-list of Birds of the World.* Volume XII. A Continuation of the Work of James L. Peters.
Zosteropidae	Paynter, Raymond A., Jr., ed., in consultation with Ernst Mayr. 1967. *Check-list of Birds of the World.* Volume XII. A Continuation of the Work of James L. Peters.
Promeropidae	Chekek, R. A. and C. F. Mann. 2000. *Sunbirds: A Guide to the Sunbirds, Spiderhunters, Sugarbirds and Flowerpeckers of the World.* Sussex, UK: Pica Press. Fry, C. Hilary and Stuart Keith. *The Birds of Africa,* Vol. VI.
Meliphagidae	Paynter, Raymond A., Jr., ed., in consultation with Ernst Mayr. 1967. *Check-list of Birds of the World.* Volume XII. A Continuation of the Work of James L. Peters.
Oriolidae	Mayr, Ernst and James C. Greenway, Jr., eds. 1962. *Check-list of Birds of the World.* Volume XV. A Continuation of the Work of James L. Peters.

Irenidae	Mayr, Ernst and James C. Greenway, Jr., eds. 1960. *Check-list of Birds of the World*. Volume IX. A Continuation of the Work of James L. Peters.
Laniidae	Lefranc, Norbert. 1997. *Shrikes: A Guide to the Shrikes of the World*.
Malaconotidae	Mayr, Ernst and James C. Greenway, Jr., eds. 1960. *Check-list of Birds of the World*. Volume IX. A Continuation of the Work of James L. Peters.
	Fry, C. Hilary and Stuart Keith. *The Birds of Africa,* Vol. VI.
Prionopidae	Mayr, Ernst and James C. Greenway, Jr., eds. 1960. *Check-list of Birds of the World*. Volume IX. A Continuation of the Work of James L. Peters.
	Fry, C. Hilary and Stuart Keith. *The Birds of Africa,* Vol. VI.
Vangidae	Mayr, Ernst and James C. Greenway, Jr., eds. 1960. *Check-list of Birds of the World*. Volume IX. A Continuation of the Work of James L. Peters.
Dicruridae	Mayr, Ernst and James C. Greenway, Jr., eds. 1962. *Check-list of Birds of the World*. Volume XV. A Continuation of the Work of James L. Peters.
Callaeidae	Mayr, Ernst and James C. Greenway, Jr., eds. 1962. *Check-list of Birds of the World*. Volume XV. A Continuation of the Work of James L. Peters.
Grallinidae	Mayr, Ernst and James C. Greenway, Jr., eds. 1962. *Check-list of Birds of the World*. Volume XV. A Continuation of the Work of James L. Peters.
Corcoracidae	Mayr, Ernst and James C. Greenway, Jr., eds. 1962. *Check-list of Birds of the World*. Volume XV. A Continuation of the Work of James L. Peters.
Artamidae	Mayr, Ernst and James C. Greenway, Jr., eds. 1962. *Check-list of Birds of the World*. Volume XV. A Continuation of the Work of James L. Peters.
Pityriaseidae	Mayr, Ernst and James C. Greenway, Jr., eds. 1960. *Check-list of Birds of the World*. Volume IX. A Continuation of the Work of James L. Peters.
Cracticidae	Mayr, Ernst and James C. Greenway, Jr., eds. 1962. *Check-list of Birds of the World*. Volume XV. A Continuation of the Work of James L. Peters.
Paradisaeidae	Frith, Clifford and Bruce Beehler. 1998. *The Birds of Paradise.*
Ptilonorhynchidae	Mayr, Ernst and James C. Greenway, Jr., eds. 1962. *Check-list of Birds of the World*. Volume XV. A Continuation of the Work of James L. Peters.
Corvidae	Goodwin, Derek. 1976. *The Crows of the World.*
	Knox, Alan G., Martin Collinson, Andreas Helbig, David Parkin and George Sangster. 2002. Taxonomic considerations for British Birds. *Ibis* 144 (4): 707–710.
	Mayr, Ernst and James C. Greenway, Jr., eds. 1962. *Check-list of Birds of the World*. Volume XV. A Continuation of the Work of James L. Peters.
	Saino, Nicola and Simona Villa. 1992. Pair composition and reproductive success across a hybrid zone of Carrion Crows and Hooded Crows. *The Auk* 109 (3): 543–555.
Sturnidae	Feare, Chris and Adrian Craig. 1999. *Starlings and Mynas.*
	Fry, C. Hilary and Stuart Keith. *The Birds of Africa,* Vol. VI.
Passeridae	Mayr, Ernst and James C. Greenway, Jr., eds. 1962. *Check-list of Birds of the World*. Volume XV. A Continuation of the Work of James L. Peters.
Ploceidae	Paynter, Raymond A, Jr., ed., in consultation with Ernst Mayr. 1968. *Check-list of Birds of the World*. Volume XIV. A Continuation of the Work of James L. Peters.
Estrildidae	Goodwin, Derek. 1982. *Estrildid Finches of the World.*
	Restall, Robin. 1997. *Munias and Mannikins.*
	Baptista, Luis F., Robin Lawson, Eleanor Visser and Douglas Bell. 1999. Relationships of some mannikins and waxbills in the estrildidae. *Journal of Ornithology* 140: 179–192.
Viduidae	Paynter, Raymond A, Jr., ed., in consultation with Ernst Mayr. 1968. *Check-list of Birds of the World*. Volume XIV. A Continuation of the Work of James L. Peters.
Vireonidae	Paynter, Raymond A, Jr., ed., in consultation with Ernst Mayr. 1968. *Check-list of Birds of the World*. Volume XIV. A Continuation of the Work of James L. Peters.
Fringillidae	Clement, Paul, Alan Harris and John Davis. 1993. *Finches and Sparrows.*
	Paynter, Raymond A, Jr., ed., in consultation with Ernst Mayr. 1968. *Check-list of Birds of the World*. Volume XIV. A Continuation of the Work of James L. Peters.

Drepanididae   Paynter, Raymond A, Jr., ed., in consultation with Ernst Mayr. 1968. *Check-list of Birds of the World.* Volume XIV. A Continuation of the Work of James L. Peters.

Peucedramidae   Paynter, Raymond A, Jr., ed., in consultation with Ernst Mayr. 1968. *Check-list of Birds of the World.* Volume XIV. A Continuation of the Work of James L. Peters.

Parulidae   Paynter, Raymond A, Jr., ed., in consultation with Ernst Mayr. 1968. *Check-list of Birds of the World.* Volume XIV. A Continuation of the Work of James L. Peters.

Coerebidae   Paynter, Raymond A, Jr., ed., in consultation with Ernst Mayr. 1968. *Check-list of Birds of the World.* Volume XIV. A Continuation of the Work of James L. Peters.

Thraupidae   Isler, Morton L. and Phyllis R. Isler. 1987. *The Tanagers.* Washington: Smithsonian Institution Press.

Paynter, Raymond A, Jr., ed., in consultation with Ernst Mayr. 1970. *Check-list of Birds of the World.* Volume XIII. A Continuation of the Work of James L. Peters.

Emberizidae   Byers, Clive, Jon Curson and Urban Olsson. 1995. *Sparrows and Buntings.*

Paynter, Raymond A, Jr., ed., in consultation with Ernst Mayr. 1970. *Check-list of Birds of the World.* Volume XIII. A Continuation of the Work of James L. Peters.

Cardinalidae   Paynter, Raymond A, Jr., ed., in consultation with Ernst Mayr. 1970. *Check-list of Birds of the World.* Volume XIII. A Continuation of the Work of James L. Peters.

Icteridae   Jaramillo, Alvaro and Peter Burke. 1999. *New World Blackbirds: The Icterids.* Princeton University Press.

Omland, K. E., S. M. Lanyon and S. J. Fritz. 1999. A molecular phylogeny of the New World Orioles (*Icterus*): The importance of dense taxon sampling. *Molecular Phylogenetics and Evolution.* 12: 224–239.

Paynter, Raymond A, Jr., ed., in consultation with Ernst Mayr. 1968. *Check-list of Birds of the World.* Volume XIV. A Continuation of the Work of James L. Peters.

Bibliography

Adams, M. P., J. H. Cooper and N. J. Collar. 2003. Extinct and endangered ('E&E') birds: a proposed list for collection cata-
logues. *Bulletin B. O. C.* 123A: 338–354.

Alexio, Alexandre. 2002. Molecular systematics and the role of the "Várzea"-"Terra-firme" ecotone in the diversification of
Xiphorhynchus woodcreepers (Aves: Dendrocolaptidae). *The Auk* 119 (3): 621–640.

Ali, Salim. 1962. *The Birds of Sikkim.* London: Oxford University Press.

——. 1969. *Birds of Kerala.* London: Oxford University Press.

——. 1972. *Indian Hill Birds.* Bombay Natural History Society.

——. 1977. *Birds of the Eastern Himalayas.* London: Oxford University Press.

Ali, Salim and S. D. Ripley. 1968–1974. *Handbook of the Birds of India and Pakistan* (ten volumes). Bombay: Oxford
University Press.

——. 1983. *A Pictorial Guide to the Birds of the Indian Subcontinent.* Delhi: Oxford University Press.

——. 1983. *Handbook of the Birds of India and Pakistan.* Delhi: Oxford University Press.

Alström, Per and Urban Olsson. 1999. The Golden-spectacled Warbler: a complex of sibling species, including a previously
undescribed species. *Ibis* 141 (4): 545–568.

American Ornithologists Union. 1957. *Check-list of North American Birds.* Fifth Edition.

——. 1997. *Studies in Neotropical Ornithology Honoring Ted Parker.* Ornithological Monographs Number 48.

——. 1983. *Check-list of North American Birds.* Sixth Edition.

——. 1998. *Check-list of North American Birds.* Seventh Edition.

Baker, Kevin. 1997. *Warblers of Europe, Asia and North Africa.* Princeton University Press.

Baker, R. 1951. *Avifauna of Micronesia.* Lawrence: University of Kansas Press.

Banks, Richard C., Carla Cicero, Jon L. Dunn, Andrew W. Kratter, Pamela C. Rasmussen, J. V. Remsen, Jr., James D. Rising
and Douglas F. Stotz. 2002. Forty-third supplement to the American Ornithologists' Union *Check-list of North American
Birds.* *The Auk* 119 (3): 897–906.

——. 2003. Forty-fourth supplement to the American Ornithologists' Union *Check-list of North American Birds.* *The Auk* 120
(3): 923–931.

——. 2004. Forty-fifth supplement to the American Ornithologists' Union *Check-list of North American Birds.* *The Auk* 121
(3): 985–995.

——. 2005. Forty-sixth supplement to the American Ornithologists' Union *Check-list of North American Birds.* *The Auk* 122
(3): 1026–1031.

——. 2006. Forty-seventh supplement to the American Ornithologists' Union *Check-list of North American Birds.* *The Auk* 123
(3): 926–936.

Bannerman, D. A. 1971. *The Birds of West and Equatorial Africa.* Edinburgh: Oliver and Boyd.

Baptista, Luis F., Robin Lawson, Eleanor Visser and Douglas Bell. 1999. Relationships of some mannikins and waxbills in the estrildidae. *Journal of Ornithology* 140: 179–192.

Beehler, Bruce, Thane Pratt and Dale Zimmerman. 1986. *Birds of New Guinea.* Princeton University Press.

Bensch, Staffan and David Pearson. 2002. The Large-billed Reed Warbler *Acrocephalus orinus* revisited. *Ibis* 144 (2) 259–267.

Benson, S. V. 1970. *Birds of Lebanon and the Jordan Area.* Cambridge: International Council for Bird Preservation.

Berger, Andrew. 1981. *Hawaiian Birdlife.* Honolulu: University of Honolulu Press.

Binford, L. C. 1989. *A Distributional Survey of the Birds of the Mexican State of Oaxaca.* Ornithological Monographs. No. 43. Washington: American Ornithologists' Union.

Blake, E. R. 1977. *Manual of Neotropical Birds* (Volume I). Chicago: University of Chicago Press.

Bond, James. 1974. *Birds of the West Indies.* London: Collins.

Borrow, Nik and Ron Demey. 2001. *A Guide to the Birds of Western Africa.* Princeton University Press.

Brewer, David. 2001. *Wrens, Dippers and Thrashers.* New Haven, CT: Yale University Press.

Brown, Leslie, E. K. Urban and K. Newman. 1982. *The Birds of Africa* (Volume 1). New York: Academic Press.

Brudnell-Bruce, P. G. C. 1975. *Birds of the Bahamas.* New York: Taplinger.

Bruner, Phillip. 1972. *Birds of French Polynesia.* Honolulu: Pacific Scientific Information Center.

Burton, John. 1973. *Owls of the World.* New York: Dutton.

Byers, Clive, Jon Curson and Urban Olsson. 1995. *Sparrows and Buntings.* Boston: Houghton Mifflin.

Chantler, Phil and Gerald Driessens. 1995. *Swifts: A Guide to the Swifts and Treeswifts of the World.* South Africa: Russel Friedman Books.

Cheke, Robert A., Clive F. Mann and Richard Allen. 2001. *Sunbirds: A Guide to the Sunbirds, Flowerpeckers, Spiderhunters and Sugarbirds of the World.* New Haven, CT: Yale University Press.

Christidis, Leslie and Walter E. Boles. 1994. *The Taxonomy and Species of Birds of Australia and its Territories.* Royal Australian Ornithologists Union Monograph 2.

Clancy, P. A. 1985. *The Rare Birds of Southern Africa.* Johannesburg: Winchester Press.

Cleere, Nigel. 1998. *Nightjars: A Guide to the Nightjars, Nighthawks, and Their Relatives.* Sussex: Pica Press.

Clement, Paul, Alan Harris and John Davis. 1993. *Finches and Sparrows.* Princeton University Press.

Clement, Peter. 2000. *Thrushes.* Princeton University Press.

Clements, James F. 1991. *Birds of the World: A Checklist.* Fourth Edition. Vista, CA: Ibis Publishing Company.

——. 1998. Supplement to Birds of the World 1992–1998. *Winging It,* Volume 11, No. 8.

Clements, James F. and Noam Shany. 2001. *A Field Guide to the Birds of Peru.* Temecula, CA: Ibis Publishing Company.

Clements, James F. and William Principe, Jr. 1992. *English-Name Index and Supplement to Birds of the World: A Check List.* Vista, CA: Ibis Publishing Company.

Coates, Brian J. and K. David Bishop. 1997. *A Guide to the Birds of Wallacea.* Alderley, Queensland, Australia: Dove Publications.

Collar, Nigel and S. N. Stuart. 1985. *Threatened Birds of Africa and Related Islands.* Cambridge, UK: International Council for Bird Preservation.

Collar, N. J., L. P. Gonzaga, N. Krabbe, A. M. Nieto, L. G. Naranjo, T. A. Parker III and D. C. Wege. 1992. *Threatened Birds of the Americas.* Cambridge: International Council for Bird Preservation.

Collar, Nigel J., Neil Aldrin D. Mallari and Blas R. Tabaranza, Jr. 1999. *Threatened Birds of the Philippines.* Cambridge, UK: BirdLife International.

Collar, Nigel J., A. V. Andreev, S. Chan, M. J. Crosby, S. Subramanya and J. A. Tobias, editors. 2001. *Threatened Birds of Asia:* (two volumes). Cambridge, UK: BirdLife International.

Collar, Nigel J. and S. Van Balen. 2002. The Blue-tailed Trogon *Harpactes (Apalharpactes) reinwardtii*: species limits and conservation status. *Forktail* 18: 121–125.

Coopmans, Paul and Niels Krabbe. 2000. A New Species of Flycatcher (Tyrannidae: *Myiopagis*) from Eastern Ecuador and Eastern Peru. *Wilson Bulletin* 112 (3): 305–312.

Cramp, Stanley, et al. *Birds of Europe, the Middle East and North Africa.* 1977. Volume 1. London: Oxford University Press.

——. 1980. *Birds of Europe, the Middle East and North Africa.* Volume II. London: Oxford University Press.

——. 1983. *Birds of Europe, the Middle East and North Africa.* Volume III. London: Oxford University Press.

——. 1985. *Birds of Europe, the Middle East and North Africa.* Volume IV. London: Oxford University Press.

——. 1988. *Birds of Europe, the Middle East and North Africa.* Volume V. London: Oxford University Press.

——. 1992. *Birds of Europe, the Middle East and North Africa.* Volume VI. London: Oxford University Press.

——. 1993. *Birds of Europe, the Middle East and North Africa.* Volume VII. London: Oxford University Press.

——. 1994. *Birds of Europe, the Middle East and North Africa.* Volume VIII. London: Oxford University Press.

——. 1994. *Birds of Europe, the Middle East and North Africa.* Volume IX. London: Oxford University Press.

Croxall, J. P., P. Evans and R. W. Schreiber. 1982. *Status and Conservation of the World's Seabirds.* Cambridge, UK: International Council for Bird Preservation.

Cuervo, Andrés M., Paul G. W. Salaman, Thomas M. Donegan and José M. Ochoa. 2001. A new species of piha (Cotingidae: *Lipaugus*) from the Cordillera Central of Colombia. *Ibis* 143: 353–368.

da Silva, José Maria C., Fernando C. Novaes and David C. Oren. 2002. Differentiation of *Xiphocolaptes* (Dendrocolaptidae) across the river Xingu, Brazilian Amazonia: recognition of a new phylogenetic species and biogeographic implications. *Bull. B. O. C.* 122 (3): 185–194.

de Vasconcelos, Marcelo F. 2002. A newly discovered specimen of Kalinowski's Tinamou *Nothoprocta kalinowskii* from the Andean Pacific slope of Peru. *Bull. B. O. C.* 122 (3): 216–218.

Dee, T. J. 1986. *The Endemic Birds of Madagascar.* Cambridge, UK: International Council for Bird Preservation.

del Hoyo, Josep, A. Elliot and J. Sargatal, eds. 1992. *Handbook of the Birds of the World.* Volume 1. Ostrich to Ducks. Barcelona: Lynx Edicions.

——. 1994. *Handbook of the Birds of the World.* Volume 2. New World Vultures to Guineafowl. Barcelona: Lynx Edicions.

——. 1996. *Handbook of the Birds of the World.* Volume 3. Hoatzin to Auks. Barcelona: Lynx Edicions.

——. 1997. *Handbook of the Birds of the World.* Volume 4. Sandgrouse to Cuckoos. Barcelona: Lynx Edicions.

——. 1999. *Handbook of the Birds of the World.* Volume 5. Barn Owls to Hummingbirds. Barcelona: Lynx Edicions.

——. 2001. *Handbook of the Birds of the World.* Volume 6. Mousebirds to Hornbills. Barcelona: Lynx Editions.

——. 2002. *Handbook of the Birds of the World.* Volume 7. Jacamars to Woodpeckers. Barcelona: Lynx Editions.

Delacour, Jean and Dean Amadon. 1973. *Curassows and Related Birds.* New York: American Museum of Natural History.

Delacour, Jean. 1977. *The Pheasants of the World.* Surrey, UK: Spur Publications.

Diamond, A. W. 1987. *Studies of Mascarene Island Birds.* Cambridge: Cambridge University Press.

Diamond, Jared M. 1966. *Avifauna of the Eastern Highlands of New Guinea.* Cambridge: Nuttall Club Publication #12.

Dickenson, E. C., Robert S. Kennedy and Kenneth C. Parkes. 1991. *The Birds of the Philippines.* British Ornithologists' Union Checklist No. 12.

Dowsett, R. J. and F. Dowsett-Lemaire. 1993. *A Contribution to the Distribution and Taxonomy of Afrotropical and Malagasy Birds.* Liege, Belgium: Tauraco Press.

Dowsett, R. J. and Alec Forbes-Watson. 1993. *Checklist of Birds of the Afrotropical and Malagasy Regions.* Liege, Belgium: Tauraco Press.

Duckworth, J. W., P. Alström, P. Davidson, T. D. Evans, C. M. Poole, T. Setha and R. J. Timmins. 2001. A new species of wagtail from the lower Mekong basin. *Bulletin B. O. C.* 121 (3): 154–182.

Eames, Jonathan. 2002. Eleven new sub-subspecies of babbler (Passeriformes: Timaliinae) from Kon Tum Province, Vietnam. *Bulletin B. O. C.* 122 (2): 109–141.

Eames, Jonathan, Frank Steinheimer and Ros Bansok. 2002. A collection of birds from the Cardamom Mountains, Cambodia, including a new subspecies of *Arborophila cambodiana. Forktail* 18: 67–86.

Etchécopar, R. and F. Hüe. 1967. *The Birds of North Africa.* Edinburgh: Oliver & Boyd.

——. 1978. *Les Oiseaux de Chine (Non Passerines).* Papeete: Centre National de la Recherche Scientifique.

——. 1983. *Les Oiseaux de Chine (Passerines).* Paris: Centre National de la Recherche Scientifique.

Feare, Chris and Adrian Craig. 1999. *Starlings and Mynas.* Princeton University Press.

Fjeldså, Jon and Niels Krabbe. 1990. *Birds of the High Andes.* Zoological Museum: University of Copenhagen.

Forshaw, Joseph and W. T. Cooper. 1973. *Parrots of the World.* New York: Doubleday.

——. 1983. *Kingfishers and Related Birds* (Volume I). Melbourne: Lansdowne.

——. 1985. *Kingfishers and Related Birds* (Volume II). Melbourne: Lansdowne.

Frith, Clifford and Bruce Beehler. 1998. *The Birds of Paradise.* Oxford: Oxford University Press.

Fry, C. H. 1984. *The Bee-eaters.* Vermillion, SD: Buteo Books.

Fry, C. H., K. Fry and A. Harris. 1992. *Kingfishers, Bee-eaters and Rollers.* London: Christopher Helm.

Fry, C. Hilary and Stuart Keith. 2000. *The Birds of Africa*, Vol. VI. London: Academic Press.

Fuller, E. 1987. *Extinct Birds*. New York: Facts on File.

Gaban-Lima, Renato, Marcos A. Raposo and Elizabeth Höfling. 2002. Description of a new species of *Pionopsitta* (Aves: Psittacidae) endemic to Brazil. *The Auk* 119 (3): 815–819.

Garcia-Moreno, Jaime, Peter Arctander and Jon Fjeldså. 1998. Pre-Pleistocene differentiation among chat-tyrants. *The Condor* 100: 629–640.

Garcia-Moreno, Jaime, J. Ohlson and Jon Fjeldså. 2001. MtDNA Sequences Support Monophyly of *Hemispingus* Tanagers. *Molecular Phylogenetics and Evolution* 21 (3): 424–435.

Garrido, Orlando and Arturo Kirkconnell. 2000. *Field Guide to the Birds of Cuba*. Ithaca, NY: Cornell University Press.

Garrido, Orlando, Guy Kirwan and David Capper. 2002. Species with the Grey-headed Quail-Dove *Geotrygon caniceps* and implications for the conservation of a globally threatened species. *Bird Conservation International* 12 (2): 169–187.

Gill, Frank B. 1995. *Ornithology*. Second edition. New York: W. H. Freeman.

Goodwin, Derek. 1982. *Estrildid Finches of the World*. Ithaca, NY: Cornell University Press.

——. 1976. *The Crows of the World*: Ithaca, NY: Cornell University Press.

——. 1983. *Pigeons and Doves of the World*. Ithaca, NY: Cornell University Press.

Greenway, J. C. 1967. *Extinct and Vanishing Birds of the World*. New York: Dover Publications.

Hadden, D. 1981. *Birds of the North Solomons*. Wau, Papua New Guinea: Wau Ecology Institute.

Hall, B. P. and R. E. Moreau. 1970. *An Atlas of Speciation in African Passerine Birds*. London: Trustees of the British Museum.

Halliday, T. 1978. *Vanishing Birds: Their Natural History and Conservation*. New York: Holt, Rinehart and Winston.

Hancock, James and Hugh Elliot. 1978. *The Herons of the World*. New York: Harper and Row.

Hancock, J., J. Kushlan and M. Kahl. 1992. *Storks, Ibises and Spoonbills of the World*. New York: Academic Press.

Hannecart, F. and Y. Letocart. 1983. *Oiseaux de Nouvelle Caledonia et des Loyautes*. Nouméa: Les Editions Cardinalis.

Harrap, Simon and David Quinn. 1995. *Chickadees, Tits, Nuthatches and Treecreepers*. Princeton University Press.

Harris, M. 1974. *Birds of the Galapagos*. New York: Taplinger.

Harris, Tony. 2000. *Shrikes and Bush-Shrikes*. Princeton University Press.

Harrison, C. J. O. 1978. *Bird Families of the World*. New York: Harry Abrams.

Harrison, Peter. 1983. *Seabirds: An Identification Guide*. Boston: Houghton Mifflin.

Haverschmidt, F. 1971. *The Birds of Surinam*. Edinburgh: Oliver & Boyd.

Hayes, Floyd D. 2001. Geographic variation, hybridization, and the leapfrog pattern of evolution in the Suiriri Flycatcher (*Suiriri suiriri*) complex. *The Auk* 118 (2); 457–471.

Henry, G. M. 1971. *A Guide to the Birds of Ceylon*. London: Oxford University Press.

Hilty, Steven L. and William L. Brown. 1986. *A Guide to the Birds of Colombia*. Princeton University Press.

Howard, Richard and Alick Moore. 1994. *A Complete Checklist of the Birds of the World*. Second Edition. London: Academic Press.

Howell, Steve N. G. and Sophie Webb. 1995. *A Guide to the Birds of Mexico and Northern Central America*. Oxford: Oxford University Press.

Hume, Rob. 1997. *Owls of the World*. London: Parkgate Books.

Inskipp, C. and T. Inskipp. 1985. *A Guide to the Birds of Nepal*. Dover, NH: Tanager Books.

Irwin, M. P. S. 1999. The genus *Nectarinia* and the evolution and diversification of sunbirds: an Afrotropical perspective. *Honeyguide* 45: 45–58.

Isler, Morton L. and Phyllis R. Isler. 1987. *The Tanagers*. Washington: Smithsonian Institution Press.

Isler, Morton L., Phyllis R. Isler, Bret M. Whitney and Barry Walker. 2001. Species limits in antbirds: the *Thamnophilus punctatus* complex continued. *The Condor* 103: 278–286.

Jaramillo, Alvaro and Peter Burke. 1999. *New World Blackbirds: The Icterids*. Princeton University Press.

Johnsgard, Paul. 1981. *The Plovers, Sandpipers and Snipes of the World*. Lincoln: University of Nebraska Press.

——. 1983. *Cranes of the World*. Bloomington: Indiana University Press.

——. 1983. *Grouse of the World*. Lincoln: University of Nebraska Press.

——. 1986. *Pheasants of the World*. Oxford University Press.

——. 1988. *The Quails, Partridges and Francolins of the World*. Oxford University Press.

——. 1991. *Bustards, Hemipodes, and Sandgrouse*. Oxford University Press.

——. 1993. *Cormorants, Darters and Pelicans of the World*. Washington: Smithsonian Institution Press.

——. 2000. *Trogons and Quetzals of the World*. Washington: Smithsonian Institution Press.

Johnson, Ned K. and Robert E. Jones. 2001. A new species of tody-tyrant (Tyrannidae: *Poecilotriccus*) from northern Peru. *The Auk* 118 (2): 334–341.

Jones, Darryl N., René Dekker and Cees Roselaar. 1995. *The Megapodes*. Oxford University Press.

Juniper, Tony and Mike Parr. 1998. *Parrots: A Guide to the Parrots of the World*. New Haven, CT: Yale University Press.

Keith, Stuart, Emil Urban, C. H. Fry. 1992. *The Birds of Africa* (Volume IV). New York: Academic Press.

Kemp, Alan. 1995. *The Hornbills*. New York: Oxford University Press.

King, Ben F. 1997. *Checklist of the Birds of Eurasia*. Vista, CA: Ibis Publishing Company.

——. 2002. Species limits in the Brown Boobook *Ninox scutulata* complex. *Bull. B. O. C.* 124 (4): 250–257.

King, Ben and E. C. Dickenson. 1975. *A Field Guide to the Birds of Southeast Asia*. Boston: Houghton Mifflin.

King, Warren. 1981. *Endangered Birds of the World: The ICBP Red Data Book*. Washington: Smithsonian Institution Press.

Knox, Alan G., Martin Collinson, Andreas Helbig, David Parkin and George Sangster. 2002. Taxonomic considerations for British Birds. *Ibis* 144 (4): 707–710.

König, Claus, Friedhelm Weick and Jan-Hendrik Becking. 1999. *Owls: A Guide to the Owls of the World*. Sussex, UK: Pica Press.

Lambert, Frank and Martin Woodcock. 1996. *Pittas, Broadbills and Asities*. Sussex, UK: Pica Press.

Langrand, Olivier. 1990. *Guide to the Birds of Madagascar*. New Haven, CT: Yale University Press.

Lefranc, Norbert. 1997. *Shrikes: A Guide to the Shrikes of the World*. New Haven, CT: Yale University Press.

Lippens, L. and H. Wille. 1976. *Les Oiseaux de Zaïre*. Brussels: Editions Lannoo Tielt.

Liversidge, Richard and Gary Voelker. 2002. The Kimberley Pipit: a new African species. *Bulletin B. O. C.* 122 (2): 93–109.

Mackworth-Praed, C. W. and C. H. B. Grant. 1957. *Birds of Eastern and North Eastern Africa* (Series 1, Volume 1). London: Longmans.

——. 1960. *Birds of Eastern and North Eastern Africa* (Series 1, Volume 2). London: Longmans.

——. 1962. *Birds of the Southern Third of Africa* (Series 2, Volume 1). London: Longmans.

——. 1963. *Birds of the Southern Third of Africa* (Series 2, Volume 2). London: Longmans.

——. 1970. *Birds of the West Central and Western Africa* (Series 3, Volume 1). London: Longmans.

——. 1973. *Birds of the West Central and Western Africa* (Series 3, Volume 2). London: Longmans.

Madge, Steve and Hilary Burn. 1988. *Waterfowl: An Identification Guide to the Ducks, Geese and Swans of the World*. Boston: Houghton Mifflin.

Martens, J., S. Eck and Y. H. Sun. 2002. *Certhia tianquanensis* Li, a treecreeper with relict distribution in Sichuan, China. *Journal of Ornithology* 143: 440–456.

Mayr, Ernst and Jared Diamond. 2001. *The Birds of Northern Melanesia*. Oxford University Press.

Meyer de Schauensee, Rodolphe. 1982. *A Guide to the Birds of South America*. Pan American Section International Council for Bird Preservation.

——. 1984. *The Birds of China*. Washington: Smithsonian Institution Press.

Monroe, Burt L. and Charles G. Sibley. 1993. *A World Checklist of Birds*. New Haven, CT: Yale University Press.

Navarro, Adolfo G., A. Townsend Peterson, Esteban López-Medrano and Hesequio Benítez-Díaz. 2001. Species limits in Mesoamerican *Aulacorhynchus* Toucanets. *Wilson Bulletin* 114 (4): 363–372.

Olsen, Jerry, Michael Wink, Hedi Sauer-Gürth and Susan Trost. 2002. A new *Ninox* owl from Sumba, Indonesia. *Emu* 102: 223–231.

Omland, K. E., S. M. Lanyon and S. J. Fritz. 1999. A molecular phylogeny of the New World Orioles (*Icterus*): The importance of dense taxon sampling. *Molecular Phylogenetics and Evolution*. 12: 224–239.

Parry, S. J., W. S. Clark and V. Prakash. 2002. On the taxonomic status of the Indian Spotted Eagle *Aquila hastata*. *Ibis* 144 (4): 665–675.

Peters, James L. 1979. *Check-list of Birds of the World*. Volume I, Second Edition. Revision of the Work of James L. Peters. Ernst Mayr and G. William Cottrell, editors. Cambridge, MA: Museum of Comparative Zoology.

——. 1934. *Check-list of Birds of the World*. Volume II. Cambridge, MA: Harvard University Press.

——. 1937. *Check-list of Birds of the World*. Volume III. Cambridge, MA: Harvard University Press.

——. 1940. *Check-list of Birds of the World*. Volume IV. Cambridge, MA: Harvard University Press.

——. 1945. *Check-list of Birds of the World*. Volume V. Cambridge, MA: Harvard University Press.

——. 1948. *Check-list of Birds of the World*. Volume VI. Cambridge, MA: Harvard University Press.

——. 1951. *Check-list of Birds of the World*. Volume VII. Cambridge, MA: Museum of Comparative Zoology.

——. 1979. *Check-list of Birds of the World.* Volume VIII. A Continuation of the Work of James L. Peters. Melvin A. Traylor, Jr., editor. Cambridge, MA: Museum of Comparative Zoology.

——. 1960. *Check-list of Birds of the World.* Volume IX. A Continuation of the Work of James L. Peters. Ernst Mayr and James C. Greenway, Jr., editors. Cambridge, MA: Museum of Comparative Zoology.

——. 1964. *Check-list of Birds of the World.* Volume X. A Continuation of the Work of James L. Peters. Ernst Mayr and Raymond A. Paynter, Jr., editors. Cambridge, MA: Museum of Comparative Zoology.

——. 1986. *Check-list of Birds of the World.* Volume XI. A Continuation of the Work of James L. Peters. Ernst Mayr and G. William Cottrell, editors. Cambridge, MA: Museum of Comparative Zoology.

——. 1967. *Check-list of Birds of the World.* Volume XII. A Continuation of the Work of James L. Peters. Raymond A. Paynter, Jr., editor, in consultation with Ernst Mayr. Cambridge, MA: Museum of Comparative Zoology.

——. 1970. *Check-list of Birds of the World.* Volume XIII. A Continuation of the Work of James L. Peters. Raymond A. Paynter, Jr., editor, in consultation with Ernst Mayr. Cambridge, MA: Museum of Comparative Zoology.

——. 1968. *Check-list of Birds of the World.* Volume XIV. A Continuation of the Work of James L. Peters. Raymond A. Paynter, Jr., editor, in consultation with Ernst Mayr. Cambridge, MA: Museum of Comparative Zoology.

——. 1962. *Check-list of Birds of the World.* Volume XV. A Continuation of the Work of James L. Peters. Ernst Mayr and James C. Greenway, Jr., editors. Cambridge, MA: Museum of Comparative Zoology.

Pratt, H. Douglas, P. Bruner and D. Berrett. 1987. *Birds of Hawaii and the Tropical Pacific.* Princeton University Press.

Price, J. Jordan and Scott M. Lanyon. 2002. A Robust Phylogeny of the Oropendolas: Polyphyly Revealed by Mitochondrial Sequence Data. *The Auk* 119 (2): 335–348.

Puigserver, M., J. D. Rodríguez-Teijero and S. Gallego. 2001. The problem of the subspecies in *Coturnix coturnix* quail. Game and Wildlife Science 18 (3–4): 561–572.

Raffaele, Herbert, James Wiley, Orlando Garrido, Allan Keith and Janis Raffaele. 1998. *A Guide to the Birds of the West Indies.* Princeton University Press.

Rand McNally. 1999. *New Millennium World Atlas Deluxe.* Skokie, Illinois: Rand McNally Company.

Reichel, James D. and Philip O. Glass. 1991. Checklist of the Birds of the Marianas Islands. *Elepaio* 51 (1).

Remsen, J. V., Jr. Editor. 1997. *Studies in Neotropical Ornithology Honoring Ted Parker.* Ornithological Monographs No. 48. Washington, DC: American Ornithologists' Union.

Restall, Robin. 1997. *Munias and Mannikins.* New Haven, CT: Yale University Press.

Ridgely, Robert S. and Paul Greenfield. 2001. *The Birds of Ecuador: Status, Distribution and Taxonomy.* Ithaca, NY: Cornell University Press.

——. 2001. *The Birds of Ecuador: Field Guide.* Ithaca, NY: Cornell University Press.

Ridgely, Robert S. and John Gwynne. 1989. *A Guide to the Birds of Panama with Costa Rica, Nicaragua and Honduras.* Princeton University Press.

Ridgely, Robert S. and Guy Tudor. 1989. *The Birds of South America*: Volume 1. Austin: University of Texas Press.

——. 1994. *The Birds of South America*: Volume 2. Austin: University of Texas Press.

Ripley, S. Dillon. 1977. *Rails of the World.* Boston: David Godine.

Robinson-Dean, J. C., K. R. Willmott, M. J. Catterall, D. J. Kelly, A. Whittington, B. Phalan, N. M. Marples and D. R. S. Boeadi. 2002. A new subspecies of Red-backed Thrush *Zoothera erythronota kabaena* subsp. nov. (Muscicapidae: Turdidae), from Kabaena island, Indonesia. *Forktail* 18: 1–10.

Rowley, Ian and Eleanor Russell. 1997. *Fairy-Wrens and Grasswrens.* Oxford University Press.

Saino, Nicola and Simona Villa. 1992. Pair composition and reproductive success across a hybrid zone of Carrion Crows and Hooded Crows. *The Auk* 109 (3): 543–555.

Salzburger, Walter, Jochen Martens, Alexander Nazarenko, Yua-Hue Sun, Reinhard Dallinger and Christian Sturmbauer. 2002. Phylogeography of the Eurasian Willow Tit (*Parus montanus*) based on DNA sequences of the mitochondrial cytochrome *b* gene. *Molecular Phylogenetics and Evolution* 24 (1) 26–34.

Salzburger, Walter, Jochen Martens and Christian Sturmbauer. 2002. Paraphyly of the Blue Tit (*Parus caeruleus*) suggested from cytochrome *b* sequences. *Molecular Phylogenetics and Evolution* 24 (1) 19–25.

Sætre, Glenn-Peter, Thomas Borge and Truls Moum. 2001. A new bird species? The taxonomic status of 'the Atlas Flycatcher' assessed from DNA sequence analysis. *Ibis* 143: 494–497.

Schodde, Richard. 1982. *The Fairy Wrens.* Melbourne: Lansdowne Editions.

Serle, W., G. J. Morel and W. Hartwig. 1977. *A Field Guide to the Birds of West Africa.* London: Collins.

Shirihai, Hadoram. 1996. *The Birds of Israel.* Academic Press Limited.

Shirihai, Hadoram, Gabriel Gargallo and Andreas J. Helbig. 2001. *Sylvia Warblers*. Princeton University Press.

Short, Lester. 1982. *Woodpeckers of the World*. Wilmington: Delaware Museum of Natural History.

Sibley, Charles G. and J. E. Ahlquist. 1990. *Phylogeny and Classification of Birds*. New Haven, CT: Yale University Press.

Sibley, Charles G. and B. L. Monroe. 1990. *Distribution and Taxonomy of Birds of the World*. New Haven, CT: Yale University Press.

——. 1992. *A Supplement to Distribution and Taxonomy of Birds of the World*. New Haven, CT: Yale University Press.

Sibley, Charles G. and Thayer Birding Software, Ltd. 1994–1996.

Silva, José Maria C., Fernando C. Novaes and David C. Oren. 2002. Differentiation of *Xiphocolaptes* (Dendrocolaptidae) across the river Xingú, Brazilian Amazonia: recognition of a new phylogenetic species and biogeographic implications. *Bull. B. O. C.* 122 (3): 185–194.

Sinclair, Ian and Olivier Langrand. 1998. *Birds of the Indian Ocean Islands*. Cape Town: Struik.

Slud, P. 1964. *The Birds of Costa Rica*. New York: Bulletin of the American Museum of Natural History, Volume 128.

Smythies, B. E. 1960. *The Birds of Borneo*. London: Oliver & Boyd.

——. 1986. *The Birds of Burma*. Pickering, Canada: Silvio Mattacchione & Co.

Snow, David. 1982. *The Cotingas*. Ithaca, NY: Cornell University Press.

Stiles, F. Gary and Alexander F. Skutch. 1989. *A Guide to the Birds of Costa Rica*. Ithaca, NY: Cornell University Press.

Taylor, Barry. 1998. *A Guide to the Rails, Crakes, Gallinules and Coots of the World*. New Haven, CT: Yale University Press.

Tikader, B. K. 1984. *Birds of the Andaman and Nicobar Islands*. Calcutta: Zoological Survey of India.

Todd, Frank. 1996. *Natural History of the Waterfowl*. Vista, CA: Ibis Publishing Company.

Turner, Angela and Chris Rose. 1989. *Swallows and Martins*. Boston: Houghton Mifflin Company.

Urban, E. K., C. Fry and S. Keith. 1986. *The Birds of Africa. Volume II*. London: Academic Press.

Urquhart, Ewan. 2002. *Stonechats: A Guide to the Genus Saxicola*. New Haven, CT: Yale University Press.

Van Marle, J. G. and K. Voous. 1988. *The Birds of Sumatra*. Tring, UK: British Ornithologist's Union.

Vaurie, Charles. 1959. *Birds of the Palearctic Fauna (Passeriformes)*. London: Witherby.

——. 1965. *Birds of the Palearctic Fauna (Non-Passeriformes)*. London: Witherby.

——. 1972. *Tibet and Its Birds*. London: Witherby.

——. 1980. *Taxonomy and Geographical Distribution of the Furnariidae*. New York: American Museum of Natural History, Volume 166: Article 1.

White, C. M. N. 1965. *A Revised Checklist of African Non-Passerine Birds*. Lusaka: Government Printing Office.

White, C. M. N. and M. D. Bruce. 1986. *The Birds of Wallacea*. London: British Ornithologists' Union Check-list No. 7.

Williams, J. G. and N. Arlott. 1980. *Field Guide to the Birds of East Africa*. London: Collins.

Winkler, Hans, D. A. Christie and D. Nurney. 1995. *Woodpeckers: An Identification Guide to Woodpeckers of the World*. Boston: Houghton Mifflin Co.

Yésou, Pierre. 2002. Systematics of *Larus argentatus-cachinnans-fuscus* complex revisited. *Dutch Birding* 24 (5): 271–298.

Zimmer, Kevin J. 2002. Species limits in Olive-backed Foliage-gleaners (*Automolus*: Furnariidae). *Wilson Bulletin* 114 (1): 20–37.

Zimmerman, Dale, Donald Turner and David J. Pearson. 1996. *Birds of Kenya and Northern Tanzania*. Princeton: Princeton University Press.

Index of Scientific Names

aalge, Uria, 107
abbas, Thraupis, 657
abbotti
    Coracina, 370
    Malacocincla, 493
    Sula, 16
abdimii, Ciconia, 23
abeillei
    Abeillia, 196
    Arremon, 679
    Coccothraustes, 637
    Icterus, 696
    Orchesticus, 648
Abeillia, 196
aberdare, Cisticola, 417
aberrans, Cisticola, 415
aberti, Pipilo, 681
abingoni, Campethera, 248
abnormis, Sasia, 246
Abroscopus, 440–441
aburri, Aburria, 56
Aburria, 56
abyssinica
    Cecropis, 361
    Pseudoalcippe, 496
abyssinicus
    Asio, 175
    Bucorvus, 231
    Coracias, 226
    Dendropicos, 249
    Turtur, 115
    Zosterops, 550
acadicus, Aegolius, 173
Acanthagenys, 567
Acanthidops, 673
Acanthisitta, 347
Acanthiza, 519–520
acanthizoides, Cettia, 426
Acanthorhynchus, 566
Acanthornis, 519
accentor, Bradypterus, 428
Accipiter, 40–43
accipitrinus, Deroptyus, 151
Aceros, 230–231
Acridotheres, 598
Acrobatornis, 278
Acrocephalus, 429–432

Acropternis, 309
Acryllium, 74
Actenoides, 222
Actinodura, 506–507
Actitis, 97
Actophilornis, 88
acuminata, Calidris, 99
acunhae, Nesospiza, 668
acuta, Anas, 31
acuticauda
    Apus, 191
    Poephila, 616
    Sterna, 106
acuticaudata, Aratinga, 143
acuticaudus, Lamprotornis, 600
acutipennis
    Chordeiles, 178
    Pseudocolopteryx, 323
acutirostris
    Calandrella, 354
    Formicivora, 297
    Heteralocha, 701
    Scytalopus, 310
adalberti, Aquila, 47
adamsi, Montifringilla, 604
adamsii, Gavia, 7
adansonii, Coturnix, 68
additus, Rhinomyias, 447
adela, Oreotrochilus, 205
adelaidae
    Dendroica, 641
    Platycercus, 136
adelberti, Chalcomitra, 536
adeliae, Pygoscelis, 6
Adelomyia, 203
adolphinae, Myzomela, 558
adorabilis, Lophornis, 196
adscitus, Platycercus, 136
adsimilis, Dicrurus, 578
adspersus, Francolinus, 66
adusta
    Muscicapa, 449
    Roraimia, 278
Aechmophorus, 9
aedon
    Acrocephalus, 431
    Troglodytes, 394

aegithaloides, Leptasthenura, 271
Aegithalos, 514–515
Aegithina, 386–387
Aegolius, 173
Aegotheles, 176–177
Aegypius, 37
aegyptiaca, Alopochen, 28
aegyptius
    Caprimulgus, 181
    Pluvianus, 91
aenea
    Chloroceryle, 223
    Ducula, 127
aeneocauda, Metallura, 208
aeneum, Dicaeum, 548
aeneus
    Dicrurus, 578
    Glaucis, 192
    Molothrus, 694
aenigma, Sapayoa, 265
Aenigmatolimnas, 84
aenobarbus, Pteruthius, 506
Aepypodius, 53
aequatoriale, Apaloderma, 212
aequatorialis
    Androdon, 194
    Momotus, 224
    Sheppardia, 455
    Tachymarptis, 190
aequinoctialis
    Acrocephalus, 431
    Buteogallus, 44
    Geothlypis, 643
    Procellaria, 12
aereus, Ceuthmochares, 155
Aerodramus, 185–187
Aeronautes, 189
aeruginosus, Circus, 39
aestiva, Amazona, 150
aestivalis, Aimophila, 681
aethereus
    Nyctibius, 177
    Phaethon, 15
Aethia, 108
aethiopica, Hirundo, 361

aethiopicus
  Laniarius, 574
  Threskiornis, 24
aethiops
  Myrmecocichla, 465
  Thamnophilus, 292
Aethopyga, 541–543
afer
  Cinnyris, 539
  Euplectes, 610
  Francolinus, 66
  Melaniparus, 527
  Nilaus, 572
  Ptilostomus, 593
  Sphenoeacus, 429
  Turtur, 115
affine, Malacopteron, 495
affinis
  Alophoixus, 383
  Apus, 191
  Arachnothera, 544
  Aythya, 32
  Caprimulgus, 183
  Climacteris, 523
  Cyanocorax, 588
  Emberiza, 666
  Empidonax, 334
  Euphonia, 626
  Garrulax, 492
  Lepidocolaptes, 288
  Megapodius, 54
  Melithreptus, 562
  Mirafra, 348
  Mycerobas, 637
  Ninox, 174
  Penelopides, 230
  Phylloscopus, 437
  Sarothrura, 78
  Scytalopus, 310
  Seicercus, 440
  Turdoides, 504
  Veniliornis, 254
afra
  Eupodotis, 88
  Pytilia, 614
afraoides, Eupodotis, 88
africana
  Mirafra, 349
  Sasia, 246
africanoides, Calendulauda, 350
africanus
  Actophilornis, 88
  Bubo, 167
  Buphagus, 602
  Francolinus, 65
  Gyps, 37
  Phalacrocorax, 18
  Smutsornis, 91
  Spizaetus, 48
Afropavo, 74
agami, Agamia, 21
Agamia, 21
Agapornis, 141
Agelaioides, 698
Agelaius, 692
Agelastes, 74
Agelasticus, 692–693
agile, Dicaeum, 546
agilis
  Amazona, 149
  Anairetes, 322
  Oporornis, 642

Aglaeactis, 204
aglaiae, Pachyramphus, 347
Aglaiocercus, 209
agraphia, Anairetes, 322
agricola, Acrocephalus, 429
Agriornis, 339
aguimp, Motacilla, 367
Agyrtria, 200–201
ahantensis, Francolinus, 66
Ailuroedus, 585–586
Aimophila, 681–682
Aix, 29
ajaja, Platalea, 25
ajax, Cinclosoma, 513
akahige, Erithacus, 456
akool, Amaurornis, 83
Alaemon, 352
alai, Fulica, 86
Alario, 636
alario, Alario, 636
alaschanicus, Phoenicurus,
    460
Alauda, 357
alaudina, Coryphistera, 278
alaudinus, Phrygilus, 668
alaudipes, Alaemon, 352
alba
  Ardea, 19
  Cacatua, 131
  Calidris, 98
  Gygis, 103–104
  Motacilla, 367
  Platalea, 25
  Pterodroma, 10
  Tyto, 161
albatrus, Phoebastria, 9
albellus, Mergellus, 33
albeola, Bucephala, 33
alberti
  Crax, 57
  Menura, 348
  Prionops, 576
  Pteridophora, 584
albertinae, Streptocitta, 598
albertinum, Glaucidium, 172
albertisi
  Aegotheles, 176
  Epimachus, 584
albertisii, Gymnophaps, 129
albescens
  Calendulauda, 351
  Synallaxis, 272
albicapilla
  Cossypha, 458
  Cranioleuca, 275
albicapillus, Spreo, 601
albicauda
  Elminia, 472
  Mirafra, 348
albicaudatus
  Buteo, 46
  Eumyias, 452
albiceps
  Atlapetes, 678
  Cranioleuca, 274
  Elaenia, 320
  Psalidoprocne, 357
  Vanellus, 92
albicilla
  Ficedula, 450
  Haliaeetus, 37
  Mohoua, 483

albicollis
  Corvus, 595
  Ficedula, 450
  Leucochloris, 199
  Leucopternis, 44
  Merops, 226
  Nyctidromus, 179
  Pipilo, 681
  Porzana, 83
  Rhipidura, 468
  Rynchops, 106
  Saltator, 688
  Scelorchilus, 309
  Streptocitta, 597
  Turdus, 412
  Xiphocolaptes, 286
  Zonotrichia, 686
albidinucha, Lorius, 132
albifacies
  Geotrygon, 119
  Myioborus, 644
  Poecilotriccus, 328
  Sceloglaux, 700
albifrons
  Amazona, 149
  Amblyospiza, 611
  Anser, 27
  Conirostrum, 648
  Donacospiza, 668
  Epthianura, 522
  Henicophaps, 115
  Muscisaxicola, 340
  Myioborus, 644
  Myrmecocichla, 465–466
  Phylidonyris, 565
  Pithys, 302
  Platysteira, 466
  Sternula, 104
albigula
  Buteo, 46
  Grallaria, 306
  Myzomela, 557
  Upucerthia, 268
albigularis
  Empidonax, 334
  Hirundo, 360
  Laterallus, 79
  Phyllastrephus, 381
  Rhinomyias, 448
  Sclerurus, 282
  Synallaxis, 272
albilatera, Diglossa, 674
albilinea, Tachycineta, 359
albilineata, Meliphaga, 560
albilora
  Muscisaxicola, 340
  Synallaxis, 273
albiloris
  Oriolus, 568
  Polioptila, 445–446
albinucha
  Actophilornis, 88
  Atlapetes, 676
  Columba, 111
  Ploceus, 608
  Xenopsaris, 346
albipectus
  Illadopsis, 495
  Pyrrhura, 146
albipennis
  Penelope, 55
  Petrophassa, 116

albirostris
　Anthracoceros, 230
　Bubalornis, 604
　Galbula, 232
　Onychognathus, 601
albiscapa, Rhipidura, 470
albispecularis, Heteromyias, 482
albitarsis, Ciccaba, 170
albitorques
　Columba, 110
　Philemon, 563
albiventer
　Fluvicola, 338
　Pnoepyga, 498
　Tachycineta, 359
albiventre, Pellorneum, 494
albiventris
　Dacnis, 662
　Elminia, 472
　Halcyon, 219
　Melaniparus, 526
　Myiagra, 478
　Myiornis, 326
　Pachycephala, 483
albivitta
　Aulacorhynchus, 241
　Procelsterna, 103
alboauricularis, Lichmera, 557
albobrunneus, Campylorhynchus, 388
albocinctus
　Acridotheres, 598
　Turdus, 408
albocoronata, Microchera, 202
albocristata, Sericossypha, 649
albocristatus, Tockus, 229
albofasciata, Chersomanes, 351
albofrenatus, Atlapetes, 677
albofrontata
　Gerygone, 522
　Sturnia, 599
albogriseus, Pachyramphus, 347
albogulare, Malacopteron, 495
albogularis
　Abroscopus, 440–441
　Accipiter, 41
　Anas, 30
　Brachygalba, 232
　Conopophila, 566
　Contopus, 335
　Francolinus, 65
　Garrulax, 489
　Megascops, 166
　Melithreptus, 562
　Phalcoboenus, 49
　Pygarrhichas, 283
　Rhipidura, 468
　Serinus, 635
　Sporophila, 671
　Tyrannus, 342
　Zosterops, 554
albolarvatus, Picoides, 253
albolimbata, Rhipidura, 469
albolimbatus, Megalurus, 442
albolineatus, Lepidocolaptes, 288
alboniger, Spizaetus, 48
albonigra, Oenanthe, 463
albonotata
　Elminia, 472
　Meliphaga, 559
　Poecilodryas, 482
albonotatus
　Buteo, 46

Crocias, 509
　Euplectes, 610
　Todiramphus, 220
alboscapulatus, Malurus, 515
albospecularis, Copsychus, 459
albosquamatus, Picumnus, 245
albostriatus, Chlidonias, 105
alboterminatus, Tockus, 229
albotibialis, Ceratogymna, 231
albovittatus, Conopias, 341
albus
　Chionis, 100
　Corvus, 595
　Eudocimus, 25
　Porphyrio, 699
　Procnias, 314
Alca, 107
Alcedo, 215–217
alchata, Pterocles, 108
alcinus, Macheiramphus, 35
Alcippe, 507–509
alcyon, Ceryle, 223
aldabrana, Nesillas, 428
aldabranus, Dicrurus, 578
Aleadryas, 483
alecto, Myiagra, 478–479
alector, Crax, 57
Alectoris, 64
Alectroenas, 127, 700
Alectrurus, 338
Alectura, 53
Alethe, 414
aleuticus
　Onychoprion, 104
　Ptychoramphus, 108
alexandrae, Polytelis, 139
alexandri
　Apus, 190
　Archilochus, 210
　Psittacula, 140
alexandrinus, Charadrius, 94
alfaroana, Saucerottia, 202
alfredi
　Bradypterus, 427
　Otus, 162
alice, Chlorostilbon, 197
aliciae, Aglaeactis, 204
alienus, Ploceus, 606
alinae
　Cyanomitra, 536
　Eriocnemis, 207
alishanensis, Bradypterus, 428
Alisterus, 139
alius
　Malaconotus, 575
　Otus, 164
alixii, Clytoctantes, 293
Alle, 107
alle, Alle, 107
alleni
　Grallaria, 306
　Porphyrio, 85
Allenia, 399
alligator, Ptilinopus, 123–124
alnorum, Empidonax, 334
alopex
　Calendulauda, 350
　Falco, 51
Alophoixus, 382–383
Alopochelidon, 360
Alopochen, 28
alpestris, Eremophila, 354–355

alphonsianus, Paradoxornis, 512
alpina, Calidris, 99
alpinus
　Anairetes, 322
　Muscisaxicola, 340
altaicus, Tetraogallus, 63
altera, Corapipo, 316
althaea, Sylvia, 443
alticola
　Amazilia, 200
　Apalis, 424
　Aratinga, 143
　Charadrius, 94
　Poospiza, 669
altiloquus, Vireo, 623
altirostre, Chrysomma, 503
altirostris
　Scytalopus, 310
　Turdoides, 503
aluco, Strix, 169
amabilis
　Charmosyna, 133
　Cotinga, 313
　Loriculus, 140
　Malurus, 516
　Pardaliparus, 526
　Polyerata, 201
Amadina, 615
Amalocichla, 479
Amandava, 615
amandava, Amandava, 615
amaurocephala, Nonnula, 234
amaurocephalus
　Hylophilus, 623
　Leptopogon, 325
amaurochalinus, Turdus, 411
Amaurocichla, 436
Amaurolimnas, 82
amauroptera, Pelargopsis, 218
Amaurornis, 83
Amaurospiza, 672
amaurotis
　Anabacerthia, 279
　Ixos, 383–384
Amazilia, 200
amazilia, Amazilia, 200
Amazona, 149–151
amazona, Chloroceryle, 223
Amazonetta, 29
amazonica, Amazona, 150
amazonicus, Thamnophilus, 292
amazonina, Hapalopsittaca, 148
amazonum, Pyrrhura, 145
ambigua
　Carduelis, 631
　Myrmotherula, 294
　Stachyris, 499
ambiguus
　Ara, 142
　Ramphastos, 243
　Thamnophilus, 292
Amblycercus, 697
Amblyornis, 586
Amblyospiza, 611
Amblyramphus, 698
amboimensis, Laniarius, 573
amboinensis
　Alisterus, 139
　Macropygia, 114
ameliae, Macronyx, 367
americana
　Anas, 29

Aythya, 32
Certhia, 532
Chloroceryle, 223
Fulica, 86
Grus, 77
Mycteria, 23
Parula, 639
Recurvirostra, 90
Rhea, 2
Siphonorhis, 700
Spiza, 691
Sporophila, 670
americanus
    Coccyzus, 159
    Ibycter, 49
    Numenius, 97
amethysticollis, Heliangelus,
    206
amethystina
    Calliphlox, 210
    Chalcomitra, 537
amethystinus
    Lampornis, 203
    Phapitreron, 121–122
amherstiae, Chrysolophus, 73
amictus, Nyctyornis, 225
ammodendri, Passer, 602
Ammodramus, 683–684
Ammomanes, 352–353
Ammomanopsis, 351
Ammoperdix, 64
amnicola, Locustella, 429
amoena, Passerina, 691
amoenus, Phylloscopus, 440
Ampeliceps, 598
ampelinus, Hypocolius, 387
Ampelioides, 312
Ampelion, 311
amphichroa, Newtonia, 434
Amphispiza, 683
amurensis, Falco, 51
Amytornis, 516–517
Anabacerthia, 279
Anabathmis, 535–536
anabatina, Dendrocincla, 284
anabatinus, Thamnistes, 293
Anabazenops, 280
anaethetus, Onychoprion,
    104
Anairetes, 322
anais, Mino, 597
analis
    Catamenia, 672
    Coracina, 370
    Formicarius, 303–304
    Iridosornis, 658
analoga, Meliphaga, 560
Anaplectes, 606
Anarhynchus, 95
Anas, 29–31, 699
Anastomus, 23
anchietae
    Anthreptes, 534
    Stactolaema, 236
Ancistrops, 280
andaecola, Upucerthia, 269
andamanensis
    Centropus, 158
    Dicrurus, 579
andamanicus, Caprimulgus, 182
andecola, Haplochelidon, 359
andecolus, Aeronautes, 189

andicola
    Agriornis, 339
    Leptasthenura, 271
andicolus, Grallaria, 306
Andigena, 242
andina
    Gallinago, 96
    Recurvirostra, 90
andinus
    Phoenicopterus, 26
    Podiceps, 699
andrei
    Chaetura, 189
    Taeniotriccus, 328
andrewsi, Fregata, 18
Androdon, 194
andromedae, Zoothera, 403
Andropadus, 378–379
Androphobus, 513
anerythra, Pitta, 267
angelae, Dendroica, 642
angelinae, Otus, 162
angolensis
    Dryoscopus, 573
    Gypohierax, 37
    Hirundo, 360
    Mirafra, 349
    Monticola, 401
    Oryzoborus, 672
    Pitta, 266
    Ploceus, 609
    Uraeginthus, 613
anguitimens, Eurocephalus, 572
angulata, Gallinula, 85
angusticauda, Cisticola, 417
angustifrons, Psarocolius, 697
angustipluma, Chaetoptila, 700
angustirostris
    Lepidocolaptes, 288
    Marmaronetta, 32
    Todus, 223
Anhima, 26
Anhinga, 18
anhinga, Anhinga, 18
ani, Crotophaga, 160
Anisognathus, 657–658
ankoberensis, Carduelis, 633
anna
    Calypte, 210
    Ciridops, 701
annae
    Cettia, 426
    Dicaeum, 545
annamarulae, Melaenornis, 447
anneae, Euphonia, 627
annectans, Dicrurus, 578
annectens, Heterophasia, 509
annumbi, Anumbius, 278
Anodorhynchus, 142, 700
anomala
    Brachycope, 609
    Cossypha, 457
Anomalospiza, 621
anomalus
    Eleothreptus, 184
    Zosterops, 552
anonymus, Cisticola, 414
Anopetia, 192
Anorrhinus, 230
Anous, 103
anoxanthus, Loxipasser, 673
anselli, Centropus, 158

Anser, 27
anser, Anser, 27
Anseranas, 26
ansorgei
    Andropadus, 379
    Nesocharis, 611
    Xenocopsychus, 458
antarctica
    Geositta, 268
    Thalassoica, 10
antarcticus
    Anthus, 365
    Cinclodes, 270
    Lopholaimus, 129
    Pygoscelis, 7
    Rallus, 81
    Stercorarius, 107
Anthobaphes, 536
Anthocephala, 202
Anthochaera, 567
anthoides, Asthenes, 276
anthonyi
    Caprimulgus, 181
    Dicaeum, 546
anthopeplus, Polytelis, 139
anthophilus, Phaethornis, 193
Anthornis, 566–567, 700
Anthoscopus, 533–534
anthracinus, Buteogallus, 44
Anthracoceros, 230
Anthracothorax, 195
Anthreptes, 534–535
Anthropoides, 76
Anthus, 363–366
antigone, Grus, 77
antillarum
    Myiarchus, 344
    Sternula, 104
Antilophia, 316
antipodes, Megadyptes, 7
antiquus, Synthliboramphus,
    108
antisianus, Pharomachrus,
    214
antisiensis, Cranioleuca, 275
antoniae, Carpodectes, 313
Anumbius, 278
Anurolimnas, 79
Anurophasis, 68
Apalis, 422–424
Apaloderma, 212
Apalopteron, 562
Aphanotriccus, 333
Aphelocephala, 522
Aphelocoma, 589–590
Aphrastura, 270
Aphriza, 98
Aphrodroma, 12
apiaster, Merops, 226
apiata, Mirafra, 350
apicalis
    Acanthiza, 520
    Moho, 700
    Myiarchus, 344
apicauda, Treron, 123
apivorus, Pernis, 35
Aplonis, 596–597, 701
apoda, Paradisaea, 585
apolinari, Cistothorus, 395
apperti, Phyllastrephus, 381
approximans, Circus, 39
apricaria, Pluvialis, 93

Aprosmictus, 139
Aptenodytes, 6
Apteryx, 2
Apus, 190–191
apus, Apus, 190
aquatica, Muscicapa, 449
aquaticus, Rallus, 81
Aquila, 47–48
aquila
    Eutoxeres, 192
    Fregata, 18
Ara, 142, 700
arabs, Ardeotis, 87
aracari, Pteroglossus, 242
Arachnothera, 543–544
arada, Cyphorhinus, 397
araea, Falco, 51
Aramides, 82, 699
Aramidopsis, 82
Aramus, 77
ararauna, Ara, 142
Aratinga, 143–144
araucana, Patagioenas, 112
arausiaca, Amazona, 150
arborea
    Dendrocygna, 26
    Lullula, 357
    Spizella, 682
Arborophila, 68–69
arcaei, Bangsia, 657
Arcanator, 493
arcanus, Ptilinopus, 127
archboldi
    Aegotheles, 176
    Eurostopodus, 179
    Newtonia, 435
    Petroica, 480
Archboldia, 586
archeri
    Buteo, 47
    Cossypha, 457
    Heteromirafra, 350
Archilochus, 210
archipelagicus, Indicator, 244
arctica
    Fratercula, 108
    Gavia, 7
arcticus, Picoides, 253
arctitorquis, Pachycephala, 486
arctoa, Leucosticte, 628
arcuata
    Dendrocygna, 26
    Pipreola, 312
Ardea, 19
ardens
    Arborophila, 69
    Euplectes, 610
    Harpactes, 215
    Selasphorus, 211
Ardeola, 20
ardeola, Dromas, 89
Ardeotis, 87
ardesiaca
    Conopophaga, 308
    Egretta, 19
    Fulica, 86
    Sporophila, 670
ardesiacus
    Melaenornis, 447
    Rhopornis, 299
    Thamnomanes, 294
ardosiaceus, Falco, 51

Arenaria, 98
arenarum
    Percnostola, 300
    Sublegatus, 326
arenicola, Toxostoma, 399
arequipae, Asthenes, 277
arfaki
    Oreocharis, 545
    Oreopsittacus, 133–134
arfakiana, Melanocharis, 544
arfakianus
    Aepypodius, 53
    Sericornis, 519
argentata, Alcedo, 216
argentatus, Larus, 101
argentauris
    Leiothrix, 505
    Lichmera, 556
argenteus, Cracticus, 581
argenticeps, Philemon, 563
argentifrons, Scytalopus, 310
argentigula, Cyanolyca, 589
argentina, Columba, 111
argoondah, Perdicula, 68
argus
    Argusianus, 74
    Eurostopodus, 179
Argusianus, 74
argyrofenges, Tangara, 662
argyrotis, Penelope, 55
aricomae, Cinclodes, 269
aridulus, Cisticola, 418
ariel
    Fregata, 18
    Petrochelidon, 362
aristotelis, Phalacrocorax, 17
arizonae, Picoides, 252
armandii, Phylloscopus, 437
armata, Merganetta, 29
armatus, Vanellus, 92
armenicus, Larus, 102
armillaris, Megalaima, 239
armillata
    Cyanolyca, 589
    Fulica, 86
arminjoniana, Pterodroma, 11
arnaudi, Pseudonigrita, 605
arnotti, Myrmecocichla, 466
aroyae, Thamnophilus, 292
arquata
    Cichladusa, 458
    Numenius, 97
    Pitta, 266
arquatrix, Columba, 110
Arremon, 679
Arremonops, 679
arremonops, Oreothraupis, 664
Arses, 477
Artamella, 577
Artamus, 580–581
arthus, Tangara, 659–660
aruensis, Meliphaga, 559
arundinaceus, Acrocephalus, 430
Arundinicola, 338
arvensis, Alauda, 357
ascalaphus, Bubo, 167
Ashbyia, 522
ashi, Mirafra, 349
asiatica
    Megalaima, 239
    Perdicula, 68
    Zenaida, 117

asiaticus
    Caprimulgus, 182
    Charadrius, 95
    Cinnyris, 540
    Ephippiorhynchus, 23
Asio, 175–176
asio, Megascops, 165
Aspatha, 224
assamica, Mirafra, 348
assimilis
    Agelaius, 692
    Chlorostilbon, 197
    Circus, 39
    Dendrocopos, 252
    Haplophaedia, 207
    Myrmotherula, 296
    Puffinus, 13
    Tolmomyias, 330
    Turdus, 412
Asthenes, 276–277
astreans, Chaetocercus, 211
Astrapia, 583
astrild, Estrilda, 612–613
astur, Eutriorchis, 39
atacamensis, Cinclodes, 270
Atalotriccus, 327
Atelornis, 227
ater
    Daptrius, 49
    Dromaius, 699
    Haematopus, 89
    Manucodia, 583
    Melipotes, 565
    Merulaxis, 311
    Molothrus, 695
    Periparus, 525
ateralbus, Centropus, 157
aterrima, Pterodroma, 10
aterrimus
    Knipolegus, 337
    Probosciger, 129
    Rhinopomastus, 228
Athene, 172–173
athertoni, Nyctyornis, 225
atlanticus, Larus, 100
Atlantisia, 82
Atlapetes, 676–678
atra
    Chalcopsitta, 131
    Fulica, 86
    Monasa, 234
    Myiagra, 478
    Porzana, 84
    Pyriglena, 299
    Rhipidura, 470
    Tijuca, 313
atrata
    Carduelis, 632
    Ceratogymna, 231
    Leucosticte, 628
    Pterodroma, 11
atratus
    Coragyps, 34
    Cygnus, 26
    Dendrocopos, 251
    Scytalopus, 309
atricapilla
    Donacobius, 388
    Estrilda, 613
    Heteronetta, 33
    Lonchura, 619
    Megascops, 166

Sylvia, 443
Vireo, 622
Zonotrichia, 686
Zosterops, 551
atricapillus
    Herpsilochmus, 296
    Myiornis, 326
    Philydor, 280
    Poecile, 524
atricaudus, Myiobius, 332–333
atriceps
    Carduelis, 632
    Coracina, 369
    Empidonax, 334
    Hypergerus, 424
    Phalacrocorax, 17
    Phrygilus, 667
    Pycnonotus, 375–376
    Rhopocichla, 502
    Saltator, 689
    Zosterops, 553
Atrichornis, 347–348
atricilla, Larus, 102
atricollis
    Colaptes, 255
    Eremomela, 434
    Ortygospiza, 615
    Saltator, 689
atricristatus, Baeolophus, 529
atrifrons
    Odontophorus, 62
    Zosterops, 552
atrifusca, Aplonis, 597
atrigularis, Napothera, 498
atrimaxillaris, Habia, 654
atrimentalis, Phaethornis, 193
atrinucha, Thamnophilus, 292
atripennis
    Caprimulgus, 182
    Dicrurus, 577
    Phyllanthus, 509
    Saltator, 689
atrirostris, Oryzoborus, 671
atrocaerulea, Hirundo, 361
atrocapillus, Crypturellus, 4
atrocaudata, Terpsiphone, 474
atrochalybeia, Terpsiphone, 473
atrococcineus, Laniarius, 574
atrodorsalis, Puffinus, 13
atroflavus
    Laniarius, 574
    Pogoniulus, 236
atrogularis
    Arborophila, 69
    Aulacorhynchus, 241
    Clytoctantes, 293
    Orthotomus, 433
    Prinia, 419
    Prunella, 400
    Serinus, 634
    Spizella, 683
    Thryothorus, 390
atronitens, Xenopipo, 317
atropileus, Hemispingus, 650
atropurpurea, Xipholena, 313
atrosuperciliaris, Paradoxornis, 512
atrothorax, Myrmeciza, 302
atroviolaceus, Dives, 693
atrovirens
    Lalage, 373
    Psarocolius, 697
Attagis, 100

Atthis, 210
atthis, Alcedo, 215
Atticora, 359
Attila, 345
atyphus, Acrocephalus, 431
aubryana, Gymnomyza, 565
aucklandica
    Anas, 30
    Coenocorypha, 96
audax
    Aphanotriccus, 333
    Aquila, 48
audeberti, Pachycoccyx, 152
audouinii, Larus, 101
Augastes, 209
augur, Buteo, 47
auguralis, Buteo, 47
augusti, Phaethornis, 193
Aulacorhynchus, 241
aura, Cathartes, 34
aurantia
    Sterna, 106
    Tyto, 160
aurantiaca
    Metopothrix, 278
    Pyrrhula, 636
aurantiacus, Manacus, 316
aurantiifrons
    Hylophilus, 624
    Loriculus, 141
    Ptilinopus, 124
aurantiigula, Macronyx, 367
aurantiirostris
    Arremon, 679
    Catharus, 406
    Saltator, 689
aurantiithorax, Sheppardia, 455
aurantiiventris, Trogon, 213
aurantiivertex, Heterocercus, 317
aurantioatrocristatus, Griseotyrannus, 342
aurantiocephala, Pionopsitta, 148
aurantium, Anthreptes, 535
aurantius
    Chaetops, 495
    Lanio, 653
    Ploceus, 607
    Turdus, 413
aurata, Saroglossa, 598
auratus
    Capito, 240
    Colaptes, 255
    Icterus, 695
    Oriolus, 568
aurea
    Aratinga, 144
    Lalage, 372
    Pachycephala, 486
aureliae, Haplophaedia, 207
aureocincta, Bangsia, 657
aureodorsalis, Buthraupis, 657
aureola
    Emberiza, 666
    Pipra, 314
    Rhipidura, 468
aureolimbatum, Dicaeum, 546
aureonucha, Ploceus, 608
aureopectus, Pipreola, 312
aureoventris
    Chlorostilbon, 197
    Pheucticus, 690
aurescens, Heliodoxa, 203

aureus
    Euplectes, 610
    Jacamerops, 232
    Sericulus, 586
auricapillus
    Aratinga, 144
    Icterus, 695
auriceps
    Chlorostilbon, 197
    Cyanoramphus, 135
    Dendrocopos, 251
    Pharomachrus, 214
auricollis, Primolius, 143
auricularis
    Hemispingus, 650
    Heterophasia, 509
    Hylopezus, 304
    Myiornis, 326
    Piculus, 254
    Puffinus, 13
auriculata, Zenaida, 116–117
aurifrons
    Ammodramus, 684
    Chloropsis, 386
    Epthianura, 522
    Melanerpes, 247–248
    Neopelma, 317
    Picumnus, 244
    Psilopsiagon, 146
aurigaster, Pycnonotus, 377
Auriparus, 533
aurita
    Conopophaga, 308
    Zenaida, 117
auritum, Crossoptilon, 72
auritus
    Batrachostomus, 177
    Heliothryx, 209
    Nettapus, 29
    Phalacrocorax, 16
    Podiceps, 8
auriventris, Sicalis, 675
aurocapilla, Seiurus, 642
auropalliata, Amazona, 150
aurorae, Ducula, 128
auroreus, Phoenicurus, 461
aurovirens, Capito, 240
aurulentus, Piculus, 254
austeni
    Anorrhinus, 230
    Garrulax, 491
australasia, Todiramphus, 221
australe, Dicaeum, 546
australis
    Acrocephalus, 430
    Apteryx, 2
    Ardeotis, 87
    Aythya, 32
    Eopsaltria, 481
    Eremopterix, 352
    Gallirallus, 80
    Hyliota, 441
    Lamprotornis, 600
    Leucosticte, 629
    Megalaima, 239
    Mergus, 699
    Oxyura, 33
    Peltohyas, 95
    Petroica, 480
    Phalcoboenus, 49
    Rostratula, 89
    Tchagra, 573

australis (*continued*)
 Treron, 122–123
 Vini, 133
Automolus, 281–282
autumnalis
 Amazona, 150
 Dendrocygna, 26
averano, Procnias, 314
Aviceda, 34–35
Avocettula, 195
avosetta, Recurvirostra, 90
awokera, Picus, 259
axillaris
 Aramides, 82
 Deleornis, 534
 Elanus, 36
 Euplectes, 610
 Herpsilochmus, 296–297
 Monarcha, 475
 Myrmotherula, 295
 Pterodroma, 11
aylmeri, Turdoides, 503
aymara
 Metriopelia, 118
 Psilopsiagon, 146
ayresi, Sarothrura, 78
ayresii
 Aquila, 48
 Cisticola, 418–419
Aythya, 32
azara, Pteroglossus, 242
azarae, Synallaxis, 272
azurea
 Alcedo, 216
 Cochoa, 462
 Coracina, 370
 Hypothymis, 472–473
 Sitta, 531
azureocapilla, Myiagra, 478
azureus, Eurystomus, 227

Babax, 505
baboecala, Bradypterus, 426–427
bacchus, Ardeola, 20
bachmani, Haematopus, 89
bachmanii, Vermivora, 639
badeigularis, Spelaeornis, 499
badia
 Cecropis, 362
 Ducula, 129
 Halcyon, 219
badiceps, Eremomela, 434
badius
 Accipiter, 40
 Caprimulgus, 180
 Molothrus, 694
 Phodilus, 162
 Ploceus, 608
Baeolophus, 528–529
Baeopogon, 380
baeri
 Asthenes, 277
 Aythya, 32
 Leucippus, 200
 Paroaria, 676
 Poospiza, 669
baeticatus, Acrocephalus, 430
baglafecht, Ploceus, 606
bahamensis, Anas, 31
baileyi, Xenospiza, 685
bailleui, Loxioides, 638
bailloni, Baillonius, 242

Baillonius, 242
bairdi
 Oreomystis, 638
 Vireo, 621
bairdii
 Acanthidops, 673
 Ammodramus, 684
 Calidris, 99
 Myiodynastes, 341
 Prinia, 421
 Trogon, 213
bakeri
 Ducula, 128
 Sericulus, 586
 Yuhina, 510
bakkamoena, Otus, 162
balaenarum, Sternula, 104
Balaeniceps, 24
balasiensis, Cypsiurus, 189
Balearica, 76
balearica, Sylvia, 444
balfouri, Chalcomitra, 537
balicassius, Dicrurus, 579
ballarae, Amytornis, 516
balli, Otus, 162
balliviani, Odontophorus, 63
ballmanni, Malimbus, 605
balstoni, Apus, 190
bambla, Microcerculus, 397
Bambusicola, 70
bamendae, Apalis, 423
bangsi, Grallaria, 306
Bangsia, 657
bangwaensis, Bradypterus, 427
banksiana, Neolalage, 475
banksii, Calyptorhynchus, 130
bannermani
 Cyanomitra, 536
 Ploceus, 606
 Tauraco, 151
banyumas, Cyornis, 453
barabensis, Larus, 102
baraui, Pterodroma, 11
barbadensis
 Amazona, 150
 Loxigilla, 673
barbara, Alectoris, 64
barbarus
 Laniarius, 574
 Megascops, 165
barbata
 Carduelis, 632
 Cercotrichas, 458
 Penelope, 55
barbatus
 Amytornis, 517
 Apus, 190
 Criniger, 382
 Dendrortyx, 60
 Gypaetus, 37
 Monarcha, 477
 Myiobius, 332
 Pycnonotus, 376–377
barbirostris, Myiarchus, 343
baritula, Diglossa, 673
barlowi, Calendulauda, 351
barnardi, Barnardius, 136
Barnardius, 136
baroni
 Cranioleuca, 275
 Metallura, 208
barrabandi, Pionopsitta, 148

barratti, Bradypterus, 427
barroti, Heliothryx, 209
bartelsi, Spizaetus, 48
bartletti, Crypturellus, 5
Bartramia, 97
bartschi, Aerodramus, 187
Baryphthengus, 224
basalis, Chrysococcyx, 154
basilanica, Ficedula, 451
Basileuterus, 644–646
basilica, Ducula, 128
basilicus, Basileuterus, 645
Basilornis, 597
bassanus, Morus, 15
Batara, 290
batavicus, Touit, 147
batesi
 Apus, 191
 Caprimulgus, 183
 Cinnyris, 540
 Ploceus, 606
 Terpsiphone, 473
Bathmocercus, 428
Batis, 467–468
Batrachostomus, 177
battyi, Leptotila, 119
baudii, Pitta, 266
baudinii, Calyptorhynchus, 130
baumanni, Phyllastrephus, 381
bayleyi, Dendrocitta, 592
beaudouini, Circaetus, 38
beauharnaesii, Pteroglossus, 242
beccarii
 Cochoa, 462
 Gallicolumba, 121
 Otus, 164
 Sericornis, 518
beckeri, Phylloscartes, 324
bedfordi, Terpsiphone, 473
beecheii, Cyanocorax, 588
beesleyi, Chersomanes, 351
behni, Myrmotherula, 296
beijingnica, Ficedula, 450
belcheri
 Larus, 100
 Pachyptila, 12
beldingi, Geothlypis, 642
belfordi, Melidectes, 564
bella
 Aethopyga, 542
 Goethalsia, 198
 Stagonopleura, 616
belli
 Amphispiza, 683
 Basileuterus, 646
bellicosa, Sturnella, 693
bellicosus, Polemaetus, 48
bellii, Vireo, 622
bellulus, Margarornis, 278
bendirei, Toxostoma, 398
bengalensis
 Bubo, 167
 Centropus, 158
 Graminicola, 442
 Gyps, 37
 Houbaropsis, 88
 Thalasseus, 106
bengalus, Uraeginthus, 613–614
benghalense, Dinopium, 260
benghalensis
 Coracias, 227

Ploceus, 609
Rostratula, 89
benguelensis, Certhilauda, 351
benjamini, Urosticte, 207
bennetti
  Casuarius, 2
  Corvus, 595
bennettii
  Aegotheles, 176
  Campethera, 248
benschi, Monias, 75
bensoni, Pseudocossyphus, 401
bergii, Thalasseus, 106
berigora, Falco, 52
berlepschi
  Aglaiocercus, 209
  Asthenes, 277
  Chaetocercus, 211
  Crypturellus, 3
  Dacnis, 663
  Hylopezus, 305
  Myrmeciza, 301
  Rhegmatorhina, 303
  Thripophaga, 276
Berlepschia, 279
berliozi, Apus, 190
bernardi, Sakesphorus, 290
bernicla, Branta, 27
bernieri
  Anas, 30
  Oriolia, 577
bernsteini
  Centropus, 157
  Thalasseus, 106
bernsteinii
  Megapodius, 54
  Ptilinopus, 124
berthelotii, Anthus, 365
bertrandi, Ploceus, 606
beryllina, Saucerottia, 202
beryllinus, Loriculus, 140
bewickii, Thryomanes, 393
bewsheri, Turdus, 408
biarcuata, Melozone, 680
biarmicus
  Falco, 52
  Panurus, 511
Bias, 466
Biatas, 291
bicalcarata, Galloperdix, 70
bicalcaratum, Polyplectron, 73
bicalcaratus, Francolinus, 66
bichenovii, Taeniopygia, 616
bicinctus
  Charadrius, 94
  Hypnelus, 233
  Pterocles, 109
  Treron, 122
bicknelli, Catharus, 407
bicolor
  Accipiter, 43
  Amaurornis, 83
  Baeolophus, 529
  Conirostrum, 648
  Coracina, 369
  Cyanophaia, 198
  Dendrocygna, 26
  Dicaeum, 546
  Ducula, 129
  Laniarius, 574
  Nigrita, 611
  Ploceus, 608

Rhyacornis, 461
Speculipastor, 601
Spermestes, 617
Spreo, 600
Tachycineta, 358
Tiaris, 672–673
Trichastoma, 493
Turdoides, 504
bicornis, Buceros, 230
biddulphi, Podoces, 592
bidentata, Piranga, 655
bidentatus
  Harpagus, 36
  Lybius, 238
bieti, Garrulax, 490
bifasciatus
  Cinnyris, 539
  Gymnostinops, 698
  Saxicola, 463
bilineata, Amphispiza, 683
bilineatus, Pogoniulus, 236
bilopha, Eremophila, 355
bilophus, Heliactin, 209
bimaculata
  Melanocorypha, 353
  Peneothello, 482
bimaculatus, Pycnonotus, 377
binotata, Apalis, 423
binotatus, Caprimulgus, 181
birostris, Ocyceros, 230
biscutata, Streptoprocne, 185
bishopi
  Catharopeza, 642
  Moho, 565
bistriatus, Burhinus, 90–91
bistrigiceps, Acrocephalus, 429
bitorquata, Streptopelia, 113
bitorquatus
  Pteroglossus, 242
  Rhinoptilus, 91
bivittata
  Buettikoferella, 442
  Petroica, 480
bivittatus, Basileuterus, 644
Biziura, 33
blainvillii, Peltops, 581
blakei, Grallaria, 307
blakistoni, Ketupa, 168
blanchoti, Malaconotus, 575
blanfordi
  Calandrella, 353
  Montifringilla, 604
  Pycnonotus, 378
blasii, Myzomela, 557
Bleda, 381–382
blewitti, Athene, 173
blighi, Myophonus, 402
blissetti, Platysteira, 466
blumenbachii, Crax, 57
blythii
  Onychognathus, 601
  Tragopan, 70
Blythipicus, 260
boanensis, Monarcha, 476
bocagei
  Amaurocichla, 436
  Nectarinia, 538
  Sheppardia, 455
  Telophorus, 575
bodessa, Cisticola, 415
boehmi
  Merops, 226

Muscicapa, 449
Neafrapus, 188
Parisoma, 444–445
Sarothrura, 78
bogotensis, Anthus, 365
boiei, Myzomela, 558
Boissonneaua, 204
boissonneautii, Pseudocolaptes, 279
bojeri, Ploceus, 607
bokermanni, Antilophia, 316
bokharensis, Parus, 527
Bolbopsittacus, 135
Bolborhynchus, 146
boliviana
  Chiroxiphia, 316
  Poospiza, 669
bolivianum, Glaucidium, 171
bolivianus
  Attila, 345
  Scytalopus, 309
  Zimmerius, 324
bollei, Phoeniculus, 228
bollii, Columba, 110
boltoni, Aethopyga, 541
bombus, Chaetocercus, 211
Bombycilla, 387
bonana, Icterus, 696
bonapartei
  Coeligena, 205
  Gymnobucco, 235
  Nothocercus, 3
bonariensis
  Molothrus, 694
  Thraupis, 657
Bonasa, 59–60
bonasia, Bonasa, 59
bonelli, Phylloscopus, 437
bonensis, Hylocitrea, 483
bonthaina, Ficedula, 451
boobook, Ninox, 174
boraquira, Nothura, 5
borbae, Skutchia, 303
borbonica, Phedina, 358
borbonicus
  Hypsipetes, 384
  Zosterops, 550
borealis
  Numenius, 97
  Phylloscopus, 438
  Picoides, 252
borealoides, Phylloscopus, 438
borin, Sylvia, 443
bornea, Eos, 131
borneensis, Myophonus, 402
Bostrychia, 24–25
Botaurus, 22–23
bottae, Oenanthe, 464
botterii, Aimophila, 681
boucardi
  Crypturellus, 4
  Polyerata, 201
bougainvillei
  Actenoides, 222
  Stresemannia, 556
bougainvillii, Phalacrocorax, 17
bougueri, Urochroa, 204
boulboul, Turdus, 408
bourbonnensis, Terpsiphone, 474
bourcieri, Phaethornis, 193
bourcierii, Eubucco, 240
bourkii, Neophema, 136
bouroensis, Oriolus, 567

bouvieri
  Cinnyris, 540
  Scotopelia, 168
bouvreuil, Sporophila, 671
bouvronides, Sporophila, 670
boweri, Colluricincla, 487
boyciana, Ciconia, 23
boyeri, Coracina, 369
braccatus, Moho, 565
bracei, Chlorostilbon, 700
Brachycope, 609
brachydactyla
  Calandrella, 353
  Carpospiza, 604
  Certhia, 532
Brachygalba, 231–232
brachylophus, Lophornis, 196
Brachypteracias, 227
brachypterus
  Buteo, 47
  Cisticola, 417
  Dasyornis, 517
  Tachyeres, 28
Brachypteryx, 413
Brachyramphus, 107–108
brachyrhyncha, Rhipidura, 470
brachyrhynchos, Corvus, 594
brachyrhynchus
  Anser, 27
  Oriolus, 568
brachyura
  Camaroptera, 424
  Chaetura, 189
  Myrmotherula, 294
  Pitta, 267
  Poecilodryas, 481
  Sylvietta, 435
  Synallaxis, 273
brachyurus
  Accipiter, 42
  Anthus, 366
  Buteo, 46
  Celeus, 256
  Graydidascalus, 148
  Idiopsar, 668
  Ramphocinclus, 399
bracteatus
  Dicrurus, 579
  Nyctibius, 178
bradfieldi
  Apus, 190
  Tockus, 229
Bradornis, 446–447
Bradypterus, 426–428
brama, Athene, 173
brandti, Leucosticte, 628
branickii
  Heliodoxa, 204
  Leptosittaca, 144
  Odontorchilus, 389
  Theristicus, 25
bransfieldensis, Phalacrocorax, 17
Branta, 27–28
brasiliana, Cercomacra, 299
brasilianum, Glaucidium, 171
brasilianus, Phalacrocorax, 16
brasiliensis
  Amazona, 150
  Amazonetta, 29
brassi, Philemon, 563
brauni, Laniarius, 573
brazzae, Phedina, 358

brehmeri, Turtur, 115
brehmii
  Monarcha, 476
  Psittacella, 137
brenchleyi, Ducula, 128
bres, Alophoixus, 382
bresilius, Ramphocelus, 656
brevicauda
  Muscigralla, 340
  Paradigalla, 583
brevicaudata
  Napothera, 498
  Nesillas, 428
brevipennis
  Acrocephalus, 431
  Vireo, 621
brevipes
  Accipiter, 40
  Aratinga, 143
  Monticola, 401
  Tringa, 98
brevirostris
  Aerodramus, 186
  Agyrtria, 201
  Aphrodroma, 12
  Brachyramphus, 108
  Certhilauda, 351
  Crypturellus, 5
  Melithreptus, 562
  Pericrocotus, 374
  Rhynchocyclus, 330
  Rissa, 103
  Schoenicola, 442
  Smicrornis, 520
brevis
  Ceratogymna, 231
  Ramphastos, 243
breweri
  Merops, 226
  Spizella, 682
brewsteri, Siphonorhis, 179
bridgesi, Thamnophilus, 291
bridgesii, Drymornis, 285
brigidai, Hylexetastes, 285
brissonii, Cyanocompsa, 691
broadbenti, Dasyornis, 517
brodiei, Glaucidium, 170
brookii, Otus, 162
Brotogeris, 147
browni
  Monarcha, 477
  Reinwardtoena, 115
  Thryorchilus, 396
brucei, Otus, 163
bruijni, Grallina, 580
bruijnii
  Aepypodius, 53
  Epimachus, 584
  Micropsitta, 134
bruniceps, Emberiza, 666
brunnea
  Alcippe, 508
  Luscinia, 456
  Nonnula, 234
brunneatus, Rhinomyias, 447–448
brunneicapillus
  Aplonis, 596
  Campylorhynchus, 389
  Ornithion, 321
brunneicauda
  Alcippe, 508
  Newtonia, 435

brunneiceps
  Hylophilus, 624
  Yuhina, 510
brunneinucha, Buarremon, 678
brunneipectus, Capito, 240
brunneiventris, Diglossa, 674
brunneopectus, Arborophila, 69
brunneopygia, Drymodes, 482
brunnescens
  Cisticola, 418
  Premnoplex, 278
brunneus
  Dromaeocercus, 428
  Melaenornis, 447
  Paradoxornis, 512
  Pycnonotus, 378
  Pyrrholaemus, 519
  Speirops, 549
brunnicephalus, Larus, 102
brunniceps, Myioborus, 644
brunnifrons, Cettia, 426
bryantae, Calliphlox, 210
Buarremon, 678
Bubalornis, 604
Bubo, 166–168
bubo, Bubo, 167
Bubulcus, 20
Bucanetes, 637–638
Buccanodon, 236
buccinator, Cygnus, 26
Bucco, 233
buccoides, Ailuroedus, 585
Bucephala, 33
bucephalus, Lanius, 570
buceroides, Philemon, 563
Buceros, 230
buchanani
  Emberiza, 665
  Prinia, 420
  Serinus, 635
bucinator, Ceratogymna, 231
buckleyi
  Columbina, 117
  Micrastur, 50
Bucorvus, 231
budongoensis, Phylloscopus, 436
budytoides, Stigmatura, 323
buergersi, Erythrotriorchis, 43
Buettikoferella, 442
buettikoferi, Cinnyris, 541
buffoni, Circus, 39
buffonii, Chalybura, 202
Bugeranus, 76
bugunorum, Liocichla, 492
bukidnonensis, Scolopax, 95
bulleri
  Larus, 102
  Puffinus, 12
  Thalassarche, 9
bulliens, Cisticola, 415
bullockii, Icterus, 696
bullockoides, Merops, 225
bulocki, Merops, 225
bulweri, Lophura, 72
Bulweria, 12
bulwerii, Bulweria, 12
Buphagus, 602
burchelli, Pterocles, 109
burhani, Ninox, 174
Burhinus, 90–91
burkii, Seicercus, 440
burmannicus, Acridotheres, 598

burmeisteri
    Chunga, 87
    Microstilbon, 210
    Phyllomyias, 318
burnesii, Prinia, 419
burnieri, Ploceus, 607
burra, Calendulauda, 350
burrovianus, Cathartes, 34
burtoni
    Callacanthis, 637
    Serinus, 635
buruensis
    Ficedula, 451
    Zosterops, 553
buryi, Sylvia, 443
Busarellus, 45
Butastur, 44
Buteo, 45–47
buteo, Buteo, 46
Buteogallus, 44–45
Buthraupis, 657
butleri
    Accipiter, 40
    Strix, 169
Butorides, 20–21

cabanisi
    Emberiza, 666
    Lanius, 571
    Phyllastrephus, 380
    Pseudonigrita, 605
    Synallaxis, 273
    Tangara, 659
caboti, Tragopan, 70
Cacatua, 130–131
cachinnans
    Garrulax, 491
    Herpetotheres, 49–50
    Larus, 102
Cacicus, 697
Cacomantis, 153–154
cactorum
    Aratinga, 144
    Asthenes, 276
    Melanerpes, 247
caerulatus
    Cyornis, 454
    Garrulax, 491
caerulea
    Coua, 157
    Egretta, 20
    Halobaena, 11
    Passerina, 691
    Pitta, 265
    Polioptila, 445
    Urocissa, 591
caeruleiceps, Pyrrhura, 145
caeruleirostris, Loxops, 638
caeruleocephala, Phoenicurus, 460
caeruleogrisea, Coracina, 369
caeruleogularis, Aulacorhynchus, 241
caerulescens
    Chen, 27
    Dendroica, 640
    Dicrurus, 578
    Diglossopis, 674
    Estrilda, 612
    Eupodotis, 87
    Geranospiza, 44
    Melanotis, 399
    Microhierax, 50
    Muscicapa, 449

Porphyrospiza, 691
Ptilorrhoa, 514
Rallus, 81
Sporophila, 670–671
Thamnophilus, 293
Theristicus, 25
caeruleus
    Cyanerpes, 663
    Cyanistes, 528
    Cyanocorax, 588
    Elanus, 36
    Myophonus, 402
caesar, Poospiza, 669
caesia
    Coracina, 370
    Emberiza, 665
caesius, Thamnomanes, 294
cafer
    Promerops, 556
    Pycnonotus, 377
caffer
    Acrocephalus, 431
    Anthus, 366
    Apus, 191
caffra, Cossypha, 457
cahow, Pterodroma, 10
caica, Pionopsitta, 148
cailliautii, Campethera, 248–249
Cairina, 29
cajanea, Aramides, 82
Calamanthus, 519
Calamonastes, 424–425
Calamospiza, 683
calandra
    Emberiza, 667
    Melanocorypha, 353
Calandrella, 353–354
calayanensis, Gallirallus, 80
Calcarius, 687
calcostetha, Leptocoma, 538
caledonica
    Coracina, 368
    Myiagra, 478
    Myzomela, 558
    Pachycephala, 486
caledonicus
    Nycticorax, 21
    Platycercus, 136
calendula, Regulus, 385
Calendulauda, 350–351
Calicalicus, 577
Calidris, 98–99
Caliechthrus, 155
californianus
    Geococcyx, 160
    Gymnogyps, 34
californica
    Aphelocoma, 589
    Callipepla, 61
    Polioptila, 445
californicum, Glaucidium, 170
californicus, Larus, 101
caligata, Hippolais, 432
Callacanthis, 637
Callaeas, 580
calligyna, Heinrichia, 413–414
callinota, Terenura, 298
calliope
    Luscinia, 456
    Stellula, 211
calliparaea, Chlorochrysa, 659
Callipepla, 60–61

Calliphlox, 210
calliptera, Pyrrhura, 146
callizonus, Xenotriccus, 333
Callocephalon, 130
Callonetta, 29
callonotus, Veniliornis, 253
callophrys
    Chlorophonia, 628
    Tangara, 662
callopterus, Piculus, 254
Calochaetes, 655
Calocitta, 587–588
Caloenas, 120
calolaemus, Lampornis, 203
Calonectris, 12
Caloperdix, 70
calophrys, Hemispingus, 650
calopterus
    Aramides, 82
    Mecocerculus, 321
    Poecilotriccus, 329
Calorhamphus, 238
Calothorax, 210
calthropae, Psittacula, 140
calurus, Criniger, 382
calvus
    Geronticus, 24
    Gymnobucco, 235
    Sarcogyps, 38
    Sarcops, 597
    Treron, 123
calyorhynchus, Phaenicophaeus, 156
Calypte, 210
Calyptocichla, 380
Calyptomena, 264
Calyptophilus, 652
Calyptorhynchus, 130
Calyptura, 313
Camarhynchus, 687–688
camaronensis, Zoothera, 403
Camaroptera, 424
cambodiana, Arborophila, 69
camelus, Struthio, 2
cameronensis
    Francolinus, 67
    Vidua, 621
camiguinensis, Loriculus, 140
campanisona
    Chamaeza, 304
    Myrmothera, 305
campbelli, Phalacrocorax, 17
Campephaga, 373–374
Campephilus, 257–258
campestris
    Anthus, 364
    Calamanthus, 519
    Colaptes, 256
    Euneornis, 673
    Uropelia, 118
Campethera, 248–249
Campochaera, 372
Camptorhynchus, 699
Camptostoma, 321
Campylopterus, 194
Campylorhamphus, 289
Campylorhynchus, 388–389
camurus, Tockus, 229
cana, Tadorna, 28
canadensis
    Branta, 27–28
    Caryothraustes, 689–690
    Falcipennis, 57

canadensis (*continued*)
  Grus, 77
  Perisoreus, 587
  Sakesphorus, 290
  Sitta, 530
  Wilsonia, 643
canagica, Chen, 27
canaria, Serinus, 633
canariensis, Phylloscopus, 437
cancellata, Prosobonia, 98
cancrominus, Platyrinchus, 331
candei
  Manacus, 316
  Synallaxis, 274
candicans, Caprimulgus, 180
candida, Agyrtria, 201
candidus, Melanerpes, 246
canente, Hemicircus, 261
canescens, Eremomela, 434
canicapillus
  Bleda, 382
  Dendrocopos, 250
  Nigrita, 611
caniceps
  Geotrygon, 119
  Lonchura, 619
  Myiopagis, 319
  Prionops, 576
  Psittacula, 140
canicollis
  Ortalis, 55
  Serinus, 634
canicularis, Aratinga, 144
canifrons
  Gallicolumba, 121
  Spizixos, 375
canigenis, Atlapetes, 677
canigularis, Chlorospingus, 650
Canirallus, 78
canivetii, Chlorostilbon, 197
cannabina, Carduelis, 633
canningi, Rallina, 79
canorus
  Cuculus, 153
  Garrulax, 491
  Melierax, 39
  Tiaris, 672
cantans
  Cisticola, 414
  Euodice, 618
  Telespiza, 638
cantator
  Hypocnemis, 300
  Phylloscopus, 439
cantillans
  Mirafra, 348
  Sylvia, 444
cantonensis, Pericrocotus, 374
cantoroides, Aplonis, 596
canturians, Cettia, 425
canus
  Agapornis, 141
  Larus, 101
  Picus, 259
  Scytalopus, 310
canutus, Calidris, 98
capellei, Treron, 122
capense
  Daption, 10
  Glaucidium, 172
capensis
  Anas, 30

Asio, 176
Batis, 467
Bubo, 167
Bucco, 233
Burhinus, 90
Corvus, 594
Emberiza, 665
Euplectes, 610
Francolinus, 66
Macronyx, 366
Microparra, 88
Morus, 16
Motacilla, 368
Oena, 115
Pelargopsis, 218
Phalacrocorax, 16
Ploceus, 606
Pycnonotus, 377
Smithornis, 264
Turnagra, 700
Tyto, 161
Zonotrichia, 686
capicola, Streptopelia, 113
capillatus, Phalacrocorax, 17
capistrata
  Heterophasia, 509
  Nesocharis, 612
capistratum, Pellorneum, 494
capistratus
  Muscisaxicola, 340
  Serinus, 634
capitalis
  Aphanotriccus, 333
  Grallaria, 306
  Pezopetes, 676
  Poecilotriccus, 328
  Stachyris, 500
capitata, Paroaria, 676
Capito, 240
capito, Tregellasia, 481
capnodes, Otus, 164
caprata, Saxicola, 463
Caprimulgus, 180–183
caprius, Chrysococcyx, 155
Capsiempis, 323
capueira, Odontophorus, 62
caracae, Scytalopus, 310
Caracara, 49, 699
carajaensis, Xiphocolaptes, 286
carbo
  Cepphus, 107
  Phalacrocorax, 16
  Ramphocelus, 655–656
carbonaria
  Cercomacra, 299
  Diglossa, 674
carbonarius, Phrygilus, 668
Cardellina, 643
Cardinalis, 690
cardinalis
  Cardinalis, 690
  Chalcopsitta, 131
  Myzomela, 558
  Quelea, 609
cardis, Turdus, 408
cardonai, Myioborus, 644
Carduelis, 631–633
carduelis, Carduelis, 633
Cariama, 87
caribaea
  Fulica, 86
  Patagioenas, 112

caribaeus
  Contopus, 335
  Vireo, 621
Caridonax, 221
carinatum, Electron, 224
caripensis, Steatornis, 176
carmioli
  Chlorothraupis, 652
  Vireo, 622
carneipes, Puffinus, 12
carnifex, Phoenicircus, 314
carnipes, Mycerobas, 637
carola, Ducula, 127
carolae, Parotia, 583–584
caroli
  Anthoscopus, 533–534
  Campethera, 249
  Polyonymus, 208
carolina, Porzana, 83
carolinae
  Cettia, 426
  Tanysiptera, 223
carolinensis
  Anas, 30
  Caprimulgus, 180
  Conuropsis, 700
  Dumetella, 397
  Poecile, 524
  Sitta, 530
carolinus
  Euphagus, 693
  Melanerpes, 247
carpalis
  Aimophila, 682
  Bradypterus, 427
carpi, Melaniparus, 526
Carpococcyx, 156
Carpodacus, 629–630
Carpodectes, 313
Carpornis, 312
Carpospiza, 604
carrikeri
  Geotrygon, 119
  Grallaria, 307
carrizalensis, Amaurospiza, 672
carruthersi, Cisticola, 416
carteri, Eremiornis, 442
carunculata
  Acanthagenys, 567
  Bostrychia, 25
  Paradigalla, 583
carunculatus
  Bugeranus, 76
  Foulehaio, 560
  Phalacrocorax, 17
  Phalcoboenus, 49
  Philesturnus, 580
caryocatactes, Nucifraga, 593
caryophyllacea, Rhodonessa, 699
Caryothraustes, 689–690
cashmirensis, Sitta, 530
Casiornis, 343
casiquiare, Crypturellus, 5
caspia, Hydroprogne, 105
caspius, Tetraogallus, 63
cassicus, Cracticus, 581
cassidix, Aceros, 231
cassini
  Gymnostinops, 698
  Leptotila, 119
  Malimbus, 605
  Muscicapa, 449

Neafrapus, 188
Veniliornis, 254
cassinii
 Aimophila, 681
 Carpodacus, 629
 Mitrospingus, 652
 Vireo, 622
castanea
 Anas, 30
 Dendroica, 641
 Hapaloptila, 234
 Myrmeciza, 301
 Philepitta, 265
 Platysteira, 466
 Sitta, 529
 Synallaxis, 273
castaneceps, Alcippe, 507
castaneiceps
 Anurolimnas, 79
 Conopophaga, 308
 Lysurus, 676
 Orthotomus, 433
 Phoeniculus, 228
 Ploceus, 607
castaneicollis, Francolinus, 67
castaneiventris
 Amazilia, 200
 Cacomantis, 154
 Monarcha, 476
 Sporophila, 671
castaneocapilla, Myioborus, 644
castaneocoronata, Tesia, 425
castaneothorax
 Cinclosoma, 513
 Lonchura, 620
castaneoventris
 Delothraupis, 658
 Eulabeornis, 84
 Lampornis, 203
castaneum, Glaucidium, 172
castaneus
 Bradypterus, 428
 Celeus, 256
 Myophonus, 402
 Pachyramphus, 346
 Pithys, 302
 Pteroptochos, 308
castaniceps
 Seicercus, 440
 Yuhina, 510
castanilius, Accipiter, 40
castanonota, Ptilorrhoa, 514
castanonotum, Glaucidium, 172
castanonotus, Hemixos, 384
castanops, Ploceus, 607
castanopterum, Glaucidium, 172
castanopterus, Passer, 602
castanotis
 Pteroglossus, 242
 Taeniopygia, 616
castanotum, Cinclosoma, 513
castanotus
 Colius, 212
 Turnix, 76
castelnau, Picumnus, 246
castelnaudii, Aglaeactis, 204
castro, Oceanodroma, 14
castus, Monarcha, 476
Casuarius, 2
casuarius, Casuarius, 2
Catamblyrhynchus, 663
catamene, Loriculus, 140

Catamenia, 672
Cataponera, 404
Catharopeza, 642
Cathartes, 34
Catharus, 406–407
Catherpes, 390
cathpharius, Dendrocopos, 251
Catreus, 72
caucasicus, Tetraogallus, 63
caudacuta, Culicivora, 323
caudacutus
 Ammodramus, 684
 Hirundapus, 188
 Sclerurus, 282
caudata
 Chiroxiphia, 316
 Drymophila, 298
 Inezia, 326
 Turdoides, 503
caudatus
 Aegithalos, 514
 Bradypterus, 428
 Coracias, 227
 Lamprotornis, 600
 Ptilogonys, 387
 Spelaeornis, 499
 Theristicus, 25
caudifasciatus, Tyrannus, 342
caurensis, Percnostola, 301
caurinus, Corvus, 594
cauta
 Hylacola, 519
 Thalassarche, 9
cayana
 Cotinga, 313
 Dacnis, 662–663
 Piaya, 159
 Tangara, 661
 Tityra, 346
cayanensis
 Icterus, 695
 Leptodon, 35
 Myiozetetes, 340
cayanus
 Cyanocorax, 588
 Vanellus, 93
cayennensis
 Caprimulgus, 181
 Euphonia, 627
 Mesembrinibis, 25
 Panyptila, 189
 Patagioenas, 112
cebuensis
 Copsychus, 460
 Phylloscopus, 439
ceciliae
 Metriopelia, 118
 Phylloscartes, 324
Cecropis, 361–362
cedrorum, Bombycilla, 387
cela, Cacicus, 697
celaeno, Rhodothraupis, 690
celaenops, Turdus, 410
celata, Vermivora, 639
celebense, Trichastoma, 493
celebensis
 Basilornis, 597
 Caprimulgus, 182
 Centropus, 157
 Hirundapus, 188
 Myza, 565

Pernis, 35
Scolopax, 95
celebicum, Dicaeum, 548
Celeus, 256–257
cenchroides, Falco, 51
centralasicus, Caprimulgus, 182
Centrocercus, 60
Centropus, 157–158
Cephalopterus, 314
Cephalopyrus, 534
cephalotes, Myiarchus, 344
Cepphus, 107
ceramensis, Coracina, 371
Ceratogymna, 231
Cercibis, 25
Cercococcyx, 153
Cercomacra, 298–299
Cercomela, 464–465
Cercotrichas, 458–459
Cereopsis, 28
Cerorhinca, 108
Certhia, 531–532
certhia, Dendrocolaptes, 286
Certhiaxis, 275
Certhidea, 688
Certhilauda, 351
certhioides, Ochetorhynchus, 269
certhiola, Locustella, 429
Certhionyx, 559
cerulea
 Dendroica, 642
 Procelsterna, 103
cerverai
 Cyanolimnas, 84
 Ferminia, 393
cervicalis, Pterodroma, 11
cervinicauda, Myiagra, 478
cerviniventris
 Bathmocercus, 428
 Chlamydera, 586
 Phyllastrephus, 380
cervinus, Anthus, 366
Ceryle, 223
cetti, Cettia, 426
Cettia, 425–426
Ceuthmochares, 155
ceylonensis
 Culicicapa, 454
 Zosterops, 551
Ceyx, 217
chabert, Leptopterus, 577
chacoensis
 Anthus, 365
 Nothura, 6
 Strix, 169
chacuru, Nystalus, 233
Chaetocercus, 211
Chaetops, 495
Chaetoptila, 700
Chaetorhynchus, 577
Chaetornis, 442
Chaetura, 188–189
Chaimarrornis, 461
chalcomelas, Cinnyris, 539
Chalcomitra, 536–537
chalconota, Ducula, 128
chalconotus, Phalacrocorax, 17
Chalcoparia, 534
Chalcophaps, 115
Chalcopsitta, 131
chalcoptera, Phaps, 115

chalcopterus
    Pionus, 149
    Rhinoptilus, 91
chalcospilos, Turtur, 115
Chalcostigma, 208–209
chalcothorax, Galbula, 232
chalcurum, Polyplectron, 73
chalcurus
    Lamprotornis, 600
    Ptilinopus, 125
chalybaeus, Lamprotornis, 599
chalybatus, Manucodia, 583
chalybea
    Euphonia, 626
    Platysteira, 466
    Progne, 359
chalybeata, Vidua, 620
chalybeus
    Centropus, 157
    Cinnyris, 538
    Lophornis, 196
Chalybura, 202
Chamaea, 443
Chamaepetes, 56
Chamaeza, 304
chapini
    Apalis, 423
    Kupeornis, 509
chaplini, Lybius, 237
chapmani
    Chaetura, 189
    Pogonotriccus, 325
Charadrius, 93–95
chariessa, Apalis, 423
Charitospiza, 667
charltonii, Arborophila, 69
Charmosyna, 133, 700
charmosyna, Estrilda, 613
Chasiempis, 474
chathamensis, Haematopus, 89
Chauna, 26
Chaunoproctus, 701
chavaria, Chauna, 26
cheela, Spilornis, 38
Chelictinia, 36
chelicuti, Halcyon, 219
Chelidoptera, 235
Chen, 27
cheniana, Mirafra, 349
Chenonetta, 29
Cheramoeca, 358
cherina, Cisticola, 418
cheriway, Caracara, 49
chermesina, Myzomela, 558
cherriei
    Cypseloides, 184
    Myrmotherula, 294
    Synallaxis, 273
    Thripophaga, 275
cherrug, Falco, 53
Chersomanes, 351
Chersophilus, 352
chiapensis, Campylorhynchus, 389
chicquera, Falco, 51
chiguanco, Turdus, 410
chihi, Plegadis, 25
chilensis
    Accipiter, 43
    Phoenicopterus, 26
    Stercorarius, 106
    Tangara, 659
    Vanellus, 93
Chilia, 269

chimachima, Milvago, 49
chimaera, Uratelornis, 227
chimango, Milvago, 49
chimborazo, Oreotrochilus, 204
chinchipensis, Synallaxis, 274
chinensis
    Cissa, 591
    Coturnix, 68
    Garrulax, 490
    Oriolus, 568
    Streptopelia, 113
chiniana, Cisticola, 415
Chionis, 100
chionogaster, Leucippus, 200
chionura, Elvira, 198
chirindensis, Apalis, 424
chiriquensis
    Elaenia, 320
    Geotrygon, 119
chiriri, Brotogeris, 147
Chiroxiphia, 316
chirurgus, Hydrophasianus, 88
Chlamydera, 586
Chlamydochaera, 413
Chlamydotis, 87
Chlidonias, 105
Chloebia, 617
Chloephaga, 28
chloricterus, Orthogonys, 653
Chloridops, 701
chloris
    Acanthisitta, 347
    Carduelis, 631
    Hemimacronyx, 366
    Nicator, 382
    Piprites, 318
    Todiramphus, 220–221
    Zosterops, 552
chlorocephalus, Oriolus, 568
chlorocercus
    Leucippus, 200
    Lorius, 132
Chloroceryle, 223
Chlorocharis, 556
Chlorochrysa, 658–659
Chlorocichla, 380
chlorolepidota, Pipreola, 312
chlorolepidotus, Trichoglossus, 132
chlorolophus, Picus, 258
chloromeros, Pipra, 315
chloronota
    Camaroptera, 424
    Gerygone, 520
chloronothos, Zosterops, 551
chloronotus
    Arremonops, 679
    Criniger, 382
    Phylloscopus, 438
Chloropeta, 432
chloropetoides, Thamnornis, 428
chlorophaeus, Phaenicophaeus, 156
Chlorophanes, 663
Chlorophonia, 627–628
chlorophrys, Basileuterus, 645
Chloropipo, 317
Chloropsis, 386
chloroptera
    Aratinga, 144
    Myzomela, 558
chloropterus
    Alisterus, 139
    Ara, 142
    Lamprotornis, 599

chloropus
    Arborophila, 69
    Gallinula, 85
chloropygius, Cinnyris, 538
chlororhynchos, Thalassarche, 9
chlororhynchus, Centropus, 158
Chlorornis, 649
Chlorospingus, 649–650
Chlorostilbon, 197, 700
Chlorothraupis, 652–653
chlorotica, Euphonia, 626
chlorotis, Anas, 30
chlorurus, Pipilo, 680
chocoensis
    Scytalopus, 310
    Veniliornis, 254
chocolatinus
    Melaenornis, 447
    Spelaeornis, 499
choliba, Megascops, 165
choloensis, Alethe, 414
Chondestes, 683
Chondrohierax, 35
chopi, Gnorimopsar, 698
Chordeiles, 178–179
christinae, Aethopyga, 542
chrysaea, Stachyris, 500
chrysaetos, Aquila, 48
chrysaeus, Tarsiger, 456–457
chrysater, Icterus, 695
chrysauchen, Melanerpes, 247
chrysia, Geotrygon, 120
chrysocephalum, Neopelma, 317
chrysocephalus
    Icterus, 695
    Myiodynastes, 341
    Sericulus, 586
chrysochloros, Piculus, 254
Chrysococcyx, 154–155
Chrysocolaptes, 260
chrysocome, Eudyptes, 7
chrysoconus, Pogoniulus, 236
chrysocrotaphum, Todirostrum,
    329
chrysogaster
    Basileuterus, 644
    Gerygone, 521
    Neophema, 137
    Pheucticus, 690
chrysogenys
    Arachnothera, 543
    Melanerpes, 247
    Oreornis, 562
chrysoides, Colaptes, 255–256
Chrysolampis, 195
chrysolaus, Turdus, 410
chrysolophum, Neopelma, 317
Chrysolophus, 73
chrysolophus, Eudyptes, 7
chrysomela, Monarcha, 477
chrysomelas, Chrysothlypis, 652
Chrysomma, 503
Chrysomus, 692
chrysonotus, Cacicus, 697
chrysoparia, Dendroica, 641
chrysopasta, Euphonia, 627
chrysopeplus, Pheucticus, 690
chrysophrys, Emberiza, 666
chrysopogon, Megalaima, 238
chrysops
    Cyanocorax, 588–589
    Lichenostomus, 560
    Zimmerius, 324

chrysoptera
  Anthochaera, 567
  Brotogeris, 147
  Neositta, 523
  Vermivora, 639
chrysopterus
  Cacicus, 697
  Masius, 316
chrysopterygius, Psephotus, 136
chrysorrheum, Dicaeum, 546
chrysorrhoa, Acanthiza, 520
chrysostoma
  Neophema, 137
  Thalassarche, 9
Chrysothlypis, 652
chrysotis
  Alcippe, 507
  Tangara, 660
chrysura, Hylocharis, 199
Chrysuronia, 199
chthonia, Grallaria, 306
chuana, Certhilauda, 351
chubbi, Cisticola, 415
chukar, Alectoris, 64
chunchotambo, Xiphorhynchus, 287
Chunga, 87
cia, Emberiza, 664
Ciccaba, 169–170
Cichladusa, 458
Cichlherminia, 404
Cichlocolaptes, 281
Cichlopsis, 405
Cicinnurus, 584
Ciconia, 23
ciconia, Ciconia, 23
cinchoneti, Conopias, 341
Cinclidium, 461
Cinclocerthia, 399
Cinclodes, 269–270
Cincloramphus, 442
cinclorhynchus, Monticola, 401
Cinclosoma, 513
Cinclus, 388
cinclus, Cinclus, 388
cincta
  Dichrozona, 296
  Notiomystis, 562
  Poecile, 525
  Poephila, 616
  Riparia, 358
cinctura, Ammomanes, 352
cinctus
  Erythrogonys, 93
  Ptilinopus, 123
  Rhinoptilus, 91
  Rhynchortyx, 63
  Saltator, 689
cineracea
  Ducula, 129
  Emberiza, 665
  Myzomela, 557
cineraceus
  Garrulax, 490
  Sturnus, 599
cinerascens
  Aplonis, 597
  Bubo, 167
  Cercomacra, 299
  Circaetus, 38
  Dendrocitta, 592
  Fraseria, 447
  Melaniparus, 527
  Monarcha, 475

Myiarchus, 344
Nothoprocta, 5
Prinia, 419
Synallaxis, 272
cinerea
  Alcippe, 507
  Apalis, 424
  Ardea, 19
  Batara, 290
  Calandrella, 353
  Coracina, 370
  Creatophora, 599
  Gallicrex, 84
  Gerygone, 520
  Glareola, 92
  Motacilla, 368
  Mycteria, 23
  Piezorhina, 668
  Poospiza, 669
  Porzana, 84
  Procellaria, 12
  Serpophaga, 322
  Struthidea, 580
  Zoothera, 402
cinereicapilla, Zimmerius, 324
cinereicauda, Lampornis, 203
cinereiceps
  Alcippe, 508
  Malacocincla, 493
  Ortalis, 54
  Orthotomus, 433
  Phapitreron, 122
  Phyllastrephus, 381
  Phyllomyias, 319
cinereicollis, Basileuterus, 645
cinereifrons, Garrulax, 488
cinereigulare, Oncostoma, 327
cinereiventris
  Chaetura, 188
  Microbates, 445
cinereocapilla, Prinia, 420
cinereolus, Cisticola, 415
cinereovinacea, Euschistospiza,
    614
cinereum
  Conirostrum, 648
  Malacopteron, 495
  Todirostrum, 329
  Toxostoma, 398
cinereus
  Acridotheres, 598
  Artamus, 581
  Callaeas, 580
  Circaetus, 38
  Circus, 39
  Coccyzus, 159
  Contopus, 335
  Crypturellus, 3
  Muscisaxicola, 339
  Odontorchilus, 389
  Ptilogonys, 387
  Pycnopygius, 563
  Vanellus, 92
  Xenus, 97
  Xolmis, 339
  Zosterops, 555
cinnamomea
  Neopipo, 333
  Sporophila, 671
  Synallaxis, 274
  Terpsiphone, 474
cinnamomeipectus, Hemitriccus,
    328

cinnamomeiventris
  Ochthoeca, 337
  Thamnolaea, 466
cinnamomeum, Cinclosoma, 513
cinnamomeus
  Anthus, 363
  Attila, 345
  Bradypterus, 427
  Certhiaxis, 275
  Cisticola, 418
  Crypturellus, 4
  Hypocryptadius, 556
  Ixobrychus, 22
  Pachyramphus, 346
  Pericrocotus, 374
  Picumnus, 246
  Pyrrhomyias, 333
cinnamominus, Todiramphus, 220
Cinnycerthia, 390
Cinnyricinclus, 600
Cinnyris, 538–541
cioides, Emberiza, 664
Circaetus, 38
circumcincta, Spiziapteryx, 50
Circus, 39
Ciridops, 701
ciris, Passerina, 691
cirlus, Emberiza, 664
cirratus, Picumnus, 245
cirrhata, Fratercula, 108
cirrhatus, Spizaetus, 48
cirrocephalus
  Accipiter, 42
  Larus, 102
cirrochloris, Campylopterus, 194
Cissa, 591
Cissopis, 649
Cisticola, 414–419
Cistothorus, 395–396
citrea, Protonotaria, 642
citreogularis
  Philemon, 563
  Sericornis, 518
citreola, Motacilla, 367
citreolus, Trogon, 212
citrina
  Sicalis, 675
  Wilsonia, 643
  Zoothera, 402–403
citrinella
  Emberiza, 664
  Serinus, 634
  Zosterops, 552
citrinelloides, Serinus, 634
citrinellus, Atlapetes, 678
citrinipectus, Serinus, 635
citriniventris, Attila, 345
Cittura, 218
Cladorhynchus, 90
clamans
  Baeopogon, 380
  Spiloptila, 422
Clamator, 152
clamator, Pseudoscops, 175
clamosus
  Atrichornis, 348
  Cuculus, 153
clanga, Aquila, 47
Clangula, 32
clangula, Bucephala, 33
clappertoni, Francolinus, 66
clara, Motacilla, 368
clarae, Arachnothera, 543

Claravis, 118
clarisse, Heliangelus, 206
clarkii
    Aechmophorus, 9
    Megascops, 165
clarus, Caprimulgus, 183
clathratus, Trogon, 214
cleaveri, Illadopsis, 495
clemenciae, Lampornis, 203
clementsi, Polioptila, 446
Cleptornis, 555
Clibanornis, 278
climacocerca, Hydropsalis, 184
Climacteris, 523
climacurus, Caprimulgus, 183
clotbey, Ramphocoris, 353
clypeata, Anas, 31
Clytoceyx, 218
Clytoctantes, 293
Clytolaema, 203
Clytomyias, 515
Clytorhynchus, 475
Clytospiza, 614
Cnemarchus, 338
Cnemophilus, 582
Cnemoscopus, 650
Cnemotriccus, 333
Cnipodectes, 331
cobanense, Glaucidium, 171
cobbi, Troglodytes, 394
coccinea, Vestiaria, 639
coccineus
    Calochaetes, 655
    Loxops, 638
coccinigastrus, Cinnyris, 540
Coccopygia, 612
Coccothraustes, 637
coccothraustes, Coccothraustes, 637
Coccycua, 159
Coccyzus, 159
cochinchinensis
    Chloropsis, 386
    Hirundapus, 188
Cochlearius, 22
cochlearius, Cochlearius, 22
Cochoa, 462
cockerelli
    Philemon, 563
    Rhipidura, 469
    Trichodere, 557
cocoi, Ardea, 19
codringtoni, Vidua, 621
coelebs, Fringilla, 625–626
coelestis
    Aglaiocercus, 209
    Forpus, 147
    Hypothymis, 473
coelicolor, Grandala, 461
Coeligena, 205–206
coeligena, Coeligena, 205
Coenocorypha, 96
Coereba, 647
coeruleicinctis, Aulacorhynchus, 241
coeruleocapilla, Lepidothrix, 315
coeruleogularis, 199
coerulescens
    Alcedo, 216
    Aphelocoma, 589
    Coracina, 371
    Saltator, 688–689
cognatus, Aulacorhynchus, 241

Colaptes, 255–256
colchicus, Phasianus, 72–73
colensoi, Phalacrocorax, 17
Colibri, 195
Colinus, 61–62
Colius, 211–212
colius, Colius, 212
collari, Otus, 164
collaria, Amazona, 149
collaris
    Accipiter, 42
    Aythya, 32
    Charadrius, 94
    Hedydipna, 535
    Lanius, 571–572
    Microbates, 445
    Mirafra, 350
    Prunella, 400
    Sporophila, 670
    Trogon, 213
colliei, Calocitta, 587
Collocalia, 185
Colluricincla, 486–487
collurio, Lanius, 570
collurioides, Lanius, 570
collybita, Phylloscopus, 437
colma, Formicarius, 303
colombiana, Neocrex, 84
colombianus, Megascops, 165
colombica, Thalurania, 198
Colonia, 340
colonus
    Colonia, 340
    Rhinomyias, 448
coloratus, Myadestes, 405
Colorhamphus, 338
coloria, Erythrura, 617
colubris, Archilochus, 210
Columba, 109–111, 700
columba, Cepphus, 107
columbarius, Falco, 52
columbiana
    Nucifraga, 592
    Sicalis, 675
columbianus
    Cygnus, 27
    Odontophorus, 62
Columbina, 117–118
columboides, Psittacula, 140
comata, Hemiprocne, 191
comatus, Aceros, 230
comechingonus, Cinclodes, 269
comitata, Muscicapa, 449
communis, Sylvia, 443
comorensis, Cinnyris, 541
Compsothraupis, 649
comptus, Trogon, 213
comrii, Manucodia, 583
concinens, Acrocephalus, 430
concinna
    Ducula, 127
    Euphonia, 626
    Glossopsitta, 133
concinnus, Aegithalos, 514–515
concolor
    Amaurolimnas, 82
    Amaurospiza, 672
    Corythaixoides, 152
    Dicaeum, 547
    Falco, 52
    Macrosphenus, 436

Neospiza, 628
Ptyonoprogne, 361
Xenospingus, 668
concreta, Platysteira, 467
concretus
    Actenoides, 222
    Caprimulgus, 183
    Cyornis, 453
    Hemicircus, 261
condamini, Eutoxeres, 192
condita, Tijuca, 313
congensis
    Afropavo, 74
    Cinnyris, 539
congica, Riparia, 358
Conioptilon, 314
conirostris
    Arremonops, 679
    Geospiza, 687
    Indicator, 244
    Spizocorys, 355
Conirostrum, 648
connivens, Ninox, 173–174
Conopias, 341
Conopophaga, 308
Conopophila, 566
Conostoma, 511
Conothraupis, 649
conoveri, Leptotila, 119
consobrinorum, Zosterops, 552
consobrinus, Remiz, 533
conspicillata, Sylvia, 444
conspicillatus
    Basileuterus, 645
    Forpus, 147
    Paradoxornis, 512
    Pelecanus, 15
    Zosterops, 551
constantii, Heliomaster, 209
contaminatus, Heliobletus, 283
Contopus, 334
contra, Gracupica, 599
Conuropsis, 700
conversii, Discosura, 196
cookii
    Cyanoramphus, 135
    Pterodroma, 11
cooperi
    Contopus, 334
    Megascops, 165
cooperii, Accipiter, 43
coprotheres, Gyps, 37
Copsychus, 459–460
coquereli, Coua, 156
coquerellii, Cinnyris, 541
coqui, Francolinus, 65
cora, Thaumastura, 210
Coracias, 226–227
Coracina, 368–372
coracinus, Entomodestes, 406
Coracopsis, 141
Coracornis, 483
Coragyps, 34
coralensis, Ptilinopus, 125
corallirostris, Hypositta, 577
Corapipo, 316
corax, Corvus, 595
coraya, Thryothorus, 391
Corcorax, 580
cordofanica, Mirafra, 348
cordofanicus, Passer, 603

corensis, Patagioenas, 111
Cormobates, 523
corniculata, Fratercula, 108
corniculatus, Philemon, 563
cornix, Corvus, 594–595
cornuta
    Anhima, 26
    Fulica, 86
    Pipra, 315
cornutus
    Batrachostomus, 177
    Eunymphicus, 135
coromanda, Halcyon, 218–219
coromandelianus, Nettapus, 29
coromandelica, Coturnix, 67
coromandelicus, Cursorius, 91
coromandus
    Bubo, 167
    Clamator, 152
coronata
    Dendroica, 640–641
    Hemiprocne, 191
    Lepidothrix, 315
    Paroaria, 676
    Zeledonia, 646
coronatus
    Ampeliceps, 598
    Anthracoceros, 230
    Basileuterus, 645
    Harpyhaliaetus, 45
    Malimbus, 605
    Malurus, 516
    Onychorhynchus, 331
    Phalacrocorax, 18
    Phylloscopus, 438
    Platyrinchus, 331
    Pterocles, 109
    Remiz, 533
    Stephanoaetus, 49
    Tachyphonus, 654
    Vanellus, 92
    Xolmis, 339
corone, Corvus, 594
coronoides, Corvus, 595
coronulatus, Ptilinopus,
    125
correndera, Anthus, 365
corrugatus, Aceros, 230
corruscus, Lamprotornis,
    600
corsicanus, Serinus, 634
coruscans
    Colibri, 195
    Neodrepanis, 265
corvina
    Aplonis, 701
    Corvinella, 572
    Megalaima, 238
    Sporophila, 670
    Terpsiphone, 474
Corvinella, 572
Corvus, 593–595, 700
Corydon, 265
coryi, Schizoeaca, 271
coryphaeus
    Cercotrichas, 459
    Pogoniulus, 236
Coryphaspiza, 667
Coryphistera, 278
Coryphospingus, 667
Corythaeola, 151

corythaix
    Basilornis, 597
    Tauraco, 151
Corythaixoides, 152
Corythopis, 323
Coscoroba, 27
coscoroba, Coscoroba, 27
Cossypha, 457–458
Cossyphicula, 457
costae, Calypte, 210
costaricanum, Glaucidium, 171
costaricensis
    Geotrygon, 119
    Ramphocelus, 656
    Touit, 148
Cotinga, 313
cotinga, Cotinga, 313
cotta, Myiopagis, 319
Coturnicops, 78
Coturnix, 67–68, 699
coturnix, Coturnix, 67
Coua, 156–157, 700
couchii, Tyrannus, 342
couloni, Primolius, 143
courseni, Synallaxis, 272
Cracticus, 581, 582
Cranioleuca, 274–275
crassa
    Aplonis, 596
    Napothera, 498
crassirostris
    Ailuroedus, 585–586
    Arachnothera, 543
    Camarhynchus, 687
    Carduelis, 632
    Corvus, 595
    Cuculus, 152
    Geositta, 268
    Holcia, 555
    Hypsipetes, 384
    Larus, 100
    Lysurus, 676
    Melanocharis, 544
    Oriolus, 568
    Oryzoborus, 671
    Pachyptila, 12
    Reinwardtoena, 115
    Sylvia, 443
    Tyrannus, 342
    Vanellus, 92
    Vireo, 621
crassus, Poicephalus, 142
Crateroscelis, 517
cratitius, Lichenostomus, 561
craveri, Synthliboramphus,
    108
Crax, 57
Creagrus, 103
Creatophora, 599
creatopus, Puffinus, 12
crecca, Anas, 30
Crecopsis, 82
crenatus, Anthus, 365
crepitans, Psophia, 77
Creurgops, 653
Crex, 82
crex, Crex, 82
Crinifer, 152
crinifrons, Aegotheles,
    176
Criniger, 382

criniger
    Setornis, 383
    Tricholestes, 383
crinigera
    Gallicolumba, 120
    Prinia, 419
crinitus, Myiarchus, 344
crispifrons, Napothera, 498
crispus, Pelecanus, 15
crissale, Toxostoma, 399
crissalis
    Pipilo, 680
    Vermivora, 639
cristata
    Alcedo, 216
    Calyptura, 313
    Cariama, 87
    Corythaeola, 151
    Coua, 157
    Cyanocitta, 587
    Elaenia, 320
    Fulica, 85
    Galerida, 355–356
    Goura, 121
    Gubernatrix, 676
    Habia, 654
    Lophostrix, 170
    Lophotibis, 25
    Pseudoseisura, 278
    Rhegmatorhina, 303
    Tadorna, 699
cristatella, Aethia, 108
cristatellus
    Acridotheres, 598
    Cyanocorax, 588
cristatus
    Aegotheles, 177
    Colinus, 61–62
    Furnarius, 270
    Lanius, 570
    Lophophanes, 526
    Orthorhyncus, 196
    Oxyruncus, 311
    Pavo, 74
    Pitohui, 488
    Podiceps, 8
    Psophodes, 513
    Sakesphorus, 290
    Tachyphonus, 653
crocea, Epthianura, 522
croceus, Macronyx, 366
Crocias, 509
croconotus, Icterus, 696
crossleyi
    Atelornis, 227
    Mystacornis, 511
    Zoothera, 403
Crossleyia, 511
Crossoptilon, 72
crossoptilon, Crossoptilon, 72
Crotophaga, 160
crudigularis, Arborophila, 69
cruentata
    Myzomela, 557
    Pyrrhura, 145
cruentatum, Dicaeum, 549
cruentatus, Melanerpes, 247
cruentus
    Ithaginis, 70
    Malaconotus, 575
    Oriolus, 569

cruentus (*continued*)
    Rhodophoneus, 574
    Rhodospingus, 667
crumeniferus, Leptoptilos, 24
cruralis, Cincloramphus, 442
cruziana, Columbina, 118
Crypsirina, 592
crypta, Ficedula, 451
cryptoleuca
    Peneothello, 482
    Progne, 359
cryptoleucus
    Corvus, 594
    Thamnophilus, 291
cryptolophus, Snowornis, 313
Cryptophaps, 129
Cryptospiza, 612
Cryptosylvicola, 435
cryptoxanthus
    Myiophobus, 332
    Poicephalus, 142
Crypturellus, 3–5
cryptus, Cypseloides, 184
cubanensis, Caprimulgus, 180
cubensis
    Ara, 700
    Tyrannus, 342
cubla, Dryoscopus, 572
cuculatus, Orthotomus, 432
cucullata
    Andigena, 242
    Carduelis, 632
    Carpornis, 312
    Cecropis, 361
    Crypsirina, 592
    Cyanolyca, 589
    Grallaricula, 307
    Melanodryas, 481
    Tangara, 661
cucullatus
    Coryphospingus, 667
    Icterus, 696
    Lophodytes, 33
    Ploceus, 607–608
    Raphus, 700
    Spermestes, 617
    Thinornis, 95
cuculoides
    Aviceda, 34
    Glaucidium, 172
Cuculus, 152–153
cujubi, Pipile, 56
Culicicapa, 454
Culicivora, 323
culicivorus, Basileuterus, 645
culik, Selenidera, 241
cumanensis, Pipile, 56
cumingi, Phaenicophaeus, 156
cumingii, Megapodius, 54
cuneata, Geopelia, 116
cunicularia
    Athene, 172–173
    Geositta, 268
cupido, Tympanuchus, 60
cupreicauda, Saucerottia, 202
cupreicaudus, Centropus, 158
cupreiceps, Elvira, 198
cupreocauda, Lamprotornis, 600
cupreoventris, Eriocnemis, 207
cupreus
    Chrysococcyx, 155
    Cinnyris, 540

cupripennis, Aglaeactis, 204
Curaeus, 698
curaeus, Curaeus, 698
curruca, Sylvia, 443
currucoides, Sialia, 405
cursor
    Coua, 157
    Cursorius, 91
Cursorius, 91
curtata, Cranioleuca, 275
curucui, Trogon, 214
curvipennis, Campylopterus, 194
curvirostra
    Loxia, 631
    Treron, 122
curvirostre, Toxostoma, 399
curvirostris
    Andropadus, 379
    Certhilauda, 351
    Limnornis, 270
    Nothoprocta, 5
    Phaenicophaeus, 156
    Vanga, 577
Cutia, 505
cuvieri
    Dryolimnas, 82
    Talegalla, 53
cuvierii
    Falco, 52
    Phaeochroa, 194
cyane, Luscinia, 456
cyanea
    Chlorophonia, 627–628
    Passerina, 691
    Pitta, 266
    Platysteira, 466
cyaneovirens, Erythrura, 617
cyaneoviridis, Tachycineta, 359
Cyanerpes, 663
cyanescens
    Galbula, 232
    Terpsiphone, 474
cyaneus
    Circus, 39
    Cyanerpes, 663
    Diglossopis, 675
    Malurus, 516
cyaniceps, Rhipidura, 468
cyanicollis
    Galbula, 232
    Tangara, 661
Cyanicterus, 657
cyanicterus, Cyanicterus, 657
cyanifrons, Saucerottia, 201
cyanirostris, Knipolegus, 336
Cyanistes, 528
cyaniventer, Tesia, 425
cyaniventris, Pycnonotus, 376
cyanocampter, Cossypha, 457
cyanocephala
    Agyrtria, 201
    Euphonia, 627
    Psittacula, 140
    Starnoenas, 120
    Tangara, 659
    Thraupis, 656–657
cyanocephalus
    Eudynamys, 155
    Euphagus, 693
    Gymnorhinus, 590
    Malurus, 516
    Uraeginthus, 614

Cyanochen, 28
Cyanocitta, 587
Cyanocompsa, 691
Cyanocorax, 588–589
cyanogaster, Coracias, 227
cyanogastra, Irena, 569
cyanogenia, Eos, 131
cyanoides, Cyanocompsa, 691
cyanolaema, Cyanomitra, 536
Cyanolanius, 577
cyanoleuca
    Grallina, 580
    Myiagra, 478
    Notiochelidon, 359
Cyanolimnas, 84
Cyanoliseus, 144
Cyanoloxia, 691
Cyanolyca, 589
cyanomelana, Cyanoptila, 452
cyanomelas
    Cyanocorax, 588
    Rhinopomastus, 229
    Trochocercus, 472
Cyanomitra, 536
cyanopectus
    Alcedo, 216
    Sternoclyta, 204
Cyanophaia, 198
cyanophrys, Eupherusa, 198
Cyanopica, 591
cyanopis, Columbina, 118
cyanopogon
    Chloropsis, 386
    Cyanocorax, 589
Cyanopsitta, 142
cyanoptera
    Anas, 31
    Brotogeris, 147
    Cyanochen, 28
    Tangara, 662
    Thraupis, 656
cyanopterus
    Artamus, 581
    Pterophanes, 206
Cyanoptila, 452
cyanopus, Agelasticus, 692
cyanopygius, Forpus, 146
Cyanoramphus, 135–136, 700
cyanotis
    Cittura, 218
    Entomyzon, 566
    Tangara, 661
cyanouroptera, Minla, 507
cyanoventris
    Halcyon, 219
    Tangara, 659
cyanura, Saucerottia, 202
cyanurus
    Tarsiger, 456
    Psittinus, 137
cyanus
    Cyanistes, 528
    Cyanopica, 591
    Hylocharis, 199
    Peneothello, 482
Cyclarhis, 625
Cyclopsitta, 134–135
cygnoides, Anser, 27
Cygnus, 26–27
cygnus, Cygnus, 26
cylindrica, Ceratogymna, 231
Cymbilaimus, 290

Cymbirhynchus, 264
Cynanthus, 198
Cyornis, 453–454
cyornithopsis, Sheppardia, 455
Cyphorhinus, 397
cypriaca, Oenanthe, 464
Cypseloides, 184
Cypsiurus, 189–190
Cypsnagra, 649
Cyrtonyx, 63

dabbenei, Penelope, 55
Dacelo, 218
dachilleae, Nannopsittaca, 147
Dacnis, 662–663
dacotiae, Saxicola, 462
dactylatra, Sula, 16
Dactylortyx, 63
dahli, Rhipidura, 471
dalhousiae, Psarisomus, 264
damarensis, Phoeniculus, 228
dambo, Cisticola, 418
damii, Xenopirostris, 577
dammermani, Myzomela, 558
Damophila, 199
danae, Tanysiptera, 223
danjoui, Jabouilleia, 497
Daption, 10
Daptrius, 49
darjellensis, Dendrocopos, 251
darnaudii, Trachyphonus, 235
darwinii, Nothura, 6
Dasyornis, 517
dasypus, Delichon, 361
daubentoni, Crax, 57
dauma, Zoothera, 404
daurica, Cecropis, 362
dauurica
    Muscicapa, 448
    Perdix, 67
dauuricus, Corvus, 593
davidi
    Arborophila, 69
    Garrulax, 490
    Niltava, 452
    Poecile, 525
    Strix, 169
davidiana, Montifringilla, 604
davidianus, Paradoxornis, 512
davisoni
    Phylloscopus, 439
    Pseudibis, 24
dayi
    Capito, 240
    Elaenia, 320
dea, Galbula, 232
debilis, Phyllastrephus, 381
decaocto, Streptopelia, 113
decipiens, Streptopelia, 113
deckeni, Tockus, 229
Deconychura, 284
decora
    Paradisaea, 585
    Polyerata, 20
decoratus, Pterocles, 109
decumanus, Psarocolius, 698
decurtatus, Hylophilus, 624
dedemi, Rhipidura, 470
defilippiana, Pterodroma, 11
defilippii, Sturnella, 693
degodiensis, Mirafra, 350
deiroleucus, Falco, 52

delalandei, Coua, 700
delalandi, Corythopis, 323
delatrii, Tachyphonus, 654
delattrei, Lophornis, 196
delawarensis, Larus, 101
delegorguei
    Columba, 110
    Coturnix, 67
Deleornis, 534
delesserti, Garrulax, 490
delicata
    Dendroica, 641
    Gallinago, 96
Delichon, 361
deliciosus, Machaeropterus, 317
Delothraupis, 658
delphinae, Colibri, 195
Deltarhynchus, 345
demersus, Spheniscus, 7
demissa, Cranioleuca, 275
Dendragapus, 57–58
Dendrexetastes, 285
Dendrocincla, 283–284
Dendrocitta, 591, 592
Dendrocolaptes, 286
dendrocolaptoides, Clibanornis, 278
Dendrocopos, 250–252
Dendrocygna, 26
Dendroica, 640–642
Dendronanthus, 367
Dendropicos, 249–250
Dendroplex, 288
Dendrortyx, 60
denhami, Neotis, 87
deningeri, Lichmera, 557
dennistouni, Stachyris, 500
densus, Dicrurus, 579
dentata, Petronia, 604
dentatus, Cnemophilus, 653
denti, Sylvietta, 435
dentirostris, Ailuroedus, 586
derbiana, Psittacula, 140
derbianus
    Aulacorhynchus, 241
    Oreophasis, 56
    Orthotomus, 433
derbyi, Eriocnemis, 206
Deroptyus, 151
deserti
    Ammomanes, 352–353
    Oenanthe, 464
    Sylvia, 443
deserticola, Sylvia, 444
desgodinsi, Heterophasia, 509
desmaresti, Tangara, 659
desmarestii, Psittaculirostris, 135
desmursii, Sylviorthorhynchus,
    271
desolata, Pachyptila, 11
deva, Galerida, 356
devillei
    Drymophila, 298
    Pyrrhura, 145
diabolicus, Eurostopodus, 179
diadema
    Catamblyrhynchus, 663
    Charmosyna, 700
    Ochthoeca, 337
diademata
    Alethe, 414
    Tricholaema, 237
    Yuhina, 510

diadematus
    Euplectes, 610
    Stephanophorus, 658
dialeucos, Odontophorus, 62
diana, Cinclidium, 461
diardi
    Lophura, 72
    Phaenicophaeus, 156
diardii, Harpactes, 215
Dicaeum, 545–549
dichroa
    Aplonis, 596
    Cossypha, 458
dichrocephalus, Ploceus, 608
dichrous
    Lophophanes, 526
    Pitohui, 487
Dichrozona, 296
dickeyi, Cyanocorax, 588
dickinsoni, Falco, 51
dicolorus, Ramphastos, 242
Dicrurus, 577–579
Didunculus, 121
dieffenbachii, Gallirallus, 699
diemenensis, Philemon, 564
difficilis
    Empidonax, 334
    Geospiza, 687
    Phylloscartes, 325
diffusus, Passer, 603
Diglossa, 673–674
Diglossopis, 674–675
dignissima, Grallaria, 306
dignus, Veniliornis, 253
dilectissimus, Touit, 148
diluta
    Rhipidura, 469
    Riparia, 358
dimidiata
    Hirundo, 361
    Pomarea, 474
dimidiatum, Philydor, 280
dimidiatus, Ramphocelus, 655
dimorpha, Uroglaux, 175
dinelliana, Pseudocolopteryx, 323
dinemelli, Dinemellia, 604
Dinemellia, 604
Dinopium, 259–260
diodon, Harpagus, 36
Diomedea, 9
diomedea, Calonectris, 12
diophthalma, Cyclopsitta, 135
diops
    Batis, 467
    Hemitriccus, 327
    Todiramphus, 220
Diopsittaca, 143
diphone, Cettia, 425
discolor
    Certhia, 532
    Dendroica, 641
    Lathamus, 137
    Leptosomus, 228
discors, Anas, 31
Discosura, 196
discurus, Prioniturus, 138
disjuncta, Myrmeciza, 301
dispar, Coracina, 371
disposita, Ficedula, 451
dissimilis
    Psephotus, 136
    Turdus, 408

Diuca, 668
diuca, Diuca, 668
divaricatus, Pericrocotus, 374
Dives, 693
dives
   Dives, 693
   Hylopezus, 305
divisorius, Thamnophilus, 293
Dixiphia, 315
dixoni, Zoothera, 404
dohertyi
   Coracina, 371
   Lophozosterops, 555
   Pitta, 267
   Ptilinopus, 124
   Telophorus, 575
   Zoothera, 402
dohrni, Horizorhinus, 509
dohrnii, Glaucis, 192
dolei, Palmeria, 639
doliatus, Thamnophilus, 291
Dolichonyx, 691
Doliornis, 311
Dolospingus, 672
domesticus, Passer, 602
domicella, Lorius, 132
dominica
   Dendroica, 641
   Nomonyx, 33
   Pluvialis, 93
dominicana, Paroaria, 676
dominicanus
   Larus, 101
   Xolmis, 339
dominicensis
   Carduelis, 632
   Icterus, 696
   Progne, 359
   Spindalis, 656
   Tyrannus, 342
dominicus
   Anthracothorax, 195
   Dulus, 387
   Tachybaptus, 8
Donacobius, 388
Donacospiza, 668
donaldsoni
   Caprimulgus, 182
   Plocepasser, 605
   Serinus, 635
dorae, Dendrocopos, 251
dorbignyanus, Picumnus, 245
dorbignyi, Asthenes, 277
doriae, Megatriorchis, 44
Doricha, 210
dorotheae, Amytornis, 517
dorsale, Ramphomicron, 208
dorsalis
   Anabazenops, 280
   Gerygone, 521
   Lanius, 571
   Mimus, 398
   Phacellodomus, 277
   Phrygilus, 667
   Picoides, 253
dorsimaculatus, Herpsilochmus, 296
dorsomaculatus, Ploceus, 608
dorsostriatus, Serinus, 635
dorsti, Cisticola, 415
Doryfera, 194
dougallii, Sterna, 105
douglasii, Callipepla, 61

dowii, Tangara, 661
Drepanis, 701
Drepanoptila, 127
Drepanorhynchus, 538
Dreptes, 536
Dromaeocercus, 428
Dromaius, 2, 699
Dromas, 89
Dromococcyx, 160
drownei, Rhipidura, 470
dryas
   Catharus, 406
   Rhipidura, 471
Drymocichla, 422
Drymodes, 482
Drymophila, 298
Drymornis, 285
Dryocopus, 257
Dryolimnas, 82
Dryoscopus, 572–573
Dryotriorchis, 38
dubia
   Alcippe, 508
   Cercomela, 465
dubium, Scissirostrum, 598
dubius
   Charadrius, 93
   Leptoptilos, 24
   Lybius, 238
Dubusia, 658
duchaillui, Buccanodon, 236
ducorpsii, Cacatua, 130
Ducula, 127–129
dufresniana, Amazona, 150
dugandi, Herpsilochmus, 296
duidae
   Campylopterus, 194
   Crypturellus, 4
   Diglossa, 674
   Emberizoides, 676
duivenbodei, Chalcopsitta, 131
Dulus, 387
dumasi, Zoothera, 402
Dumetella, 397
dumetaria, Upucerthia, 268
Dumetia, 502
dumetoria, Ficedula, 451
dumetorum, Acrocephalus, 430
dumicola, Polioptila, 446
dumontii, Mino, 597
dunni, Eremalauda, 355
dupetithouarsii, Ptilinopus, 125
duponti, Chersophilus, 352
dupontii, Tilmatura, 210
dussumieri, Cinnyris, 541
duvaucelii
   Harpactes, 215
   Vanellus, 92
duyvenbodei, Aethopyga, 542
dybowskii, Euschistospiza, 614
Dysithamnus, 293–294
Dysmorodrepanis, 701

earlei, Turdoides, 503
eatoni, Anas, 31
eburnea, Pagophila, 103
ecaudatus
   Myiornis, 326
   Terathopius, 38
echo, Psittacula, 139
Eclectus, 139
Ectopistes, 700

edithae, Corvus, 595
edolioides, Melaenornis, 447
eduardi, Tylas, 577
edward, Saucerottia, 202
edwardsi
   Bangsia, 657
   Lophura, 71
edwardsii
   Calonectris, 12
   Carpodacus, 630
   Psittaculirostris, 135
egertoni, Actinodura, 506
egregia
   Chaetura, 188
   Crecopsis, 82
   Pyrrhura, 145
Egretta, 19–20
eichhorni
   Myzomela, 559
   Philemon, 563
eisenmanni
   Pyrrhura, 145
   Thryothorus, 390
eisentrauti, Melignomon, 243
ekmani, Caprimulgus, 180
elachus, Dendropicos, 249
Elaenia, 319–321
Elanoides, 35
Elanus, 36
elaphrus, Aerodramus, 185
elata, Ceratogymna, 231
elatus, Tyrannulus, 319
Electron, 224–225
elegans
   Celeus, 256
   Chlorostilbon, 700
   Emberiza, 666
   Eudromia, 6
   Laniisoma, 311–312
   Leptopoecile, 436
   Malurus, 516
   Melanopareia, 311
   Neophema, 137
   Otus, 164
   Pardaliparus, 525–526
   Phaps, 115–116
   Pitta, 267
   Platycercus, 136
   Progne, 359
   Rallus, 81
   Sarothrura, 78
   Thalasseus, 106
   Trogon, 213
   Xiphorhynchus, 287
elegantissima, Euphonia, 626
eleonorae, Falco, 52
Eleothreptus, 184
elgini, Spilornis, 38
eliciae, Hylocharis, 199
elisabeth, Myadestes, 405
eliza, Doricha, 210
ellioti
   Atthis, 210
   Syrmaticus, 72
   Tanysiptera, 222
elliotii
   Dendropicos, 249
   Garrulax, 492
   Pitta, 266
ellisianus, Hemignathus, 701
Elminia, 472
elphinstonii, Columba, 111

Elseyornis, 95
eludens, Grallaria, 306
Elvira, 198
Emberiza, 664–667
Emberizoides, 675–676
Embernagra, 676
Emblema, 616
emeiensis, Phylloscopus, 439
emiliae, Chlorocharis, 556
emiliana, Macropygia, 114
eminentissima, Foudia, 609
Eminia, 424
eminibey, Passer, 603
Empidonax, 334
Empidonomus, 342
Empidornis, 446
enarratus, Caprimulgus, 183
enca, Corvus, 593
enganensis
    Gracula, 598
    Otus, 164
Enicognathus, 146
enicura, Doricha, 210
Enicurus, 461–462
enigma, Todiramphus, 221
Enodes, 598
Ensifera, 206
ensifera, Ensifera, 206
ensipennis, Campylopterus, 194
Entomodestes, 406
Entomyzon, 566
enucleator, Pinicola, 629
Eolophus, 130
Eophona, 637
Eopsaltria, 481
Eos, 131
eos
    Carpodacus, 629
    Coeligena, 205
epauletta, Pyrrhoplectes, 637
Ephippiorhynchus, 23
epichlorus, Urolais, 422
epilepidota, Napothera, 498
Epimachus, 584
episcopus
    Ciconia, 23
    Thraupis, 656
epomophora, Diomedea, 9
epops, Upupa, 228
Epthianura, 522
epulata, Muscicapa, 449
eques, Myzomela, 557
erckelii, Francolinus, 67
Eremalauda, 355
Eremiornis, 442
eremita
    Geronticus, 24
    Megapodius, 54
    Nesocichla, 404
Eremobius, 269
Eremomela, 434
Eremophila, 354–355
Eremopterix, 352
Ergaticus, 643
Eriocnemis, 206–207
erithaca, Ceyx, 217
Erithacus, 455–456
erithacus, Psittacus, 141
erlangeri, Calandrella, 354
erythaca, Pyrrhula, 636
erythrauchen, Accipiter, 42
erythrinus, Carpodacus, 629

erythrocephala
    Amadina, 615
    Mirafra, 348
    Myzomela, 558
    Pipra, 315
    Piranga, 655
    Pyrrhula, 636
erythrocephalus
    Garrulax, 492
    Harpactes, 215
    Hylocryptus, 282
    Melanerpes, 246
    Trachyphonus, 235
erythrocercum, Philydor, 280
Erythrocercus, 471–472
erythrocercus, Cinnyris, 539
erythrochlamys, Calendulauda, 351
erythrocnemis, Pomatorhinus, 496
erythrogaster
    Laniarius, 574
    Malimbus, 605
    Pitta, 267
erythrogastrus, Phoenicurus, 461
erythrogenys
    Aratinga, 143
    Microhierax, 50
    Pomatorhinus, 496
Erythrogonys, 93
erythroleuca, Grallaria, 306
erythrolophus, Tauraco, 151
erythromelas
    Myzomela, 559
    Periporphyrus, 688
erythronemius, Accipiter, 43
erythronota, Zoothera, 402
erythronotos
    Estrilda, 613
    Formicivora, 297
erythronotus
    Phoenicurus, 460
    Phrygilus, 668
erythrophris, Enodes, 598
erythrophrys, Poospiza, 669
erythrophthalma
    Lophura, 72
    Netta, 32
erythrophthalmus
    Phacellodomus, 277
    Pipilo, 680
erythropleura, Ptiloprora, 564
erythropleurus, Zosterops, 551
erythrops
    Cisticola, 414
    Climacteris, 523
    Cranioleuca, 275
    Myiagra, 477
    Neocrex, 84
    Odontophorus, 62
    Quelea, 609
erythroptera
    Gallicolumba, 120–121
    Mirafra, 348
    Ortalis, 55
    Phlegopsis, 303
    Prinia, 421–422
    Stachyris, 501
erythropterum, Philydor, 280
erythropterus, Aprosmictus, 139
erythropthalmos, Pycnonotus, 378
erythropthalmus, Coccyzus, 159
erythropus
    Accipiter, 42

Anser, 27
Crypturellus, 3
Tringa, 98
erythropygia
    Pinarocorys, 351
    Sturnia, 599
erythropygius
    Cnemarchus, 338
    Pericrocotus, 374
    Picus, 259
    Xiphorhynchus, 288
erythropygus, Morococcyx, 160
erythrorhyncha
    Anas, 31
    Perdicula, 68
    Urocissa, 591
erythrorhynchos
    Dicaeum, 547
    Pelecanus, 15
erythrorhynchus
    Buphagus, 602
    Tockus, 229
erythrostictus, Monarcha, 476
erythrothorax
    Dicaeum, 547
    Stiphrornis, 455
    Synallaxis, 273
erythrotis, Grallaria, 306
Erythrotriorchis, 43
Erythrura, 617
erythrura, Myrmotherula, 295
erythrurus, Terenotriccus, 333
esculenta, Collocalia, 185
estella, Oreotrochilus, 204
estherae, Serinus, 634
Estrilda, 612–613
ethologus, Pericrocotus, 374
Eubucco, 240
euchlorus, Passer, 603
euchrysea, Tachycineta, 358
Eucometis, 653
eucosma, Charitospiza, 667
Eudocimus, 25
Eudromia, 6
Eudynamys, 155
Eudyptes, 7
Eudyptula, 7
Eugenes, 204
eugeniae, Ptilinopus, 126
Eugerygone, 480
Eugralla, 309
Eulabeornis, 84
Eulacestoma, 488
Eulampis, 195
euleri
    Coccyzus, 159
    Lathrotriccus, 333
Eulidia, 211
eulophotes
    Egretta, 20
    Lophotriccus, 327
Eumomota, 225
Eumyias, 452
Euneornis, 673
Eunymphicus, 135
Euodice, 618
euophrys, Thryothorus, 390
euops, Aratinga, 144
eupatria, Psittacula, 139
Eupetes, 514
Euphagus, 693
Eupherusa, 198

Euphonia, 626–627
Euplectes, 610–611
Eupodotis, 87–88
eupogon, Metallura, 208
Euptilotis, 214
eurhythmus, Ixobrychus, 22
eurizonoides, Rallina, 79
Eurocephalus, 572
europaea, Sitta, 529–530
europaeus, Caprimulgus, 181
Eurostopodus, 179
Euryceros, 577
Eurylaimus, 264–265
eurynome, Phaethornis, 193
Eurynorhynchus, 99
euryptera, Opisthoprora, 209
Euryptila, 425
Eurypyga, 86
eurystomina, Pseudochelidon, 357
Eurystomus, 227
euryura, Rhipidura, 468
euryzona, Alcedo, 216
Euscarthmus, 323
Euschistospiza, 614
euteles, Trichoglossus, 132
Euthlypis, 644
eutilotus, Pycnonotus, 377
Eutoxeres, 192
Eutrichomyias, 473
Eutriorchis, 39
evelynae, Calliphlox, 210
everetti
    Aceros, 231
    Dicaeum, 546
    Ixos, 384
    Monarcha, 476
    Tesia, 425
    Turnix, 76
    Yuhina, 510
    Zoothera, 403
    Zosterops, 551–552
eversmanni, Columba,
    110
ewingii, Acanthiza, 520
exarhatus, Penelopides, 230
excellens, Campylopterus, 194
excelsa, Grallaria, 305
excelsior, Cinclodes, 269
excubitor, Lanius, 571
excubitoroides, Lanius, 571
exilis
    Cisticola, 419
    Indicator, 243–244
    Ixobrychus, 22
    Laterallus, 79
    Loriculus, 141
    Picumnus, 245
    Psaltria, 515
eximia
    Aethopyga, 542
    Buthraupis, 657
    Eupherusa, 198
    Megalaima, 239
eximium, Dicaeum, 548
eximius
    Bleda, 381
    Caprimulgus, 182
    Cisticola, 418
    Platycercus, 136
    Pogonotriccus, 325
    Vireolanius, 625
exortis, Heliangelus, 206

explorator
    Monticola, 401
    Zosterops, 554
exquisitus, Coturnicops, 78
exsul
    Myrmeciza, 301
    Psittacula, 700
externa, Pterodroma, 11
exulans, Diomedea, 9
exustus, Pterocles, 108–109
eytoni, Dendrocygna, 26

fabalis, Anser, 27
fagani, Prunella, 400
faiostricta, Megalaima, 238
falcata
    Anas, 29
    Ptilocichla, 498
falcatus, Campylopterus, 194
falcinellus
    Lepidocolaptes, 289
    Limicola, 99
    Plegadis, 25
Falcipennis, 57
falcipennis, Dendragapus, 58
falcirostris
    Sporophila, 669
    Xiphocolaptes, 286
falcklandii, Turdus, 411
Falco, 50–53
falcularius, Campylorhamphus, 289
Falculea, 577
Falcunculus, 483
falkensteini, Chlorocichla, 380
falklandicus, Charadrius, 94
fallax
    Bulweria, 12
    Ceyx, 217
    Elaenia, 320
    Glycichaera, 556
    Leucippus, 200
familiare, Apalopteron, 562
familiaris
    Acrocephalus, 430
    Cercomela, 465
    Certhia, 531–532
    Prinia, 420
famosa, Nectarinia, 538
fanny, Myrtis, 211
fannyi, Thalurania, 198
fanovanae, Newtonia, 435
farinosa, Amazona, 150
farquhari, Todiramphus, 220
fasciata
    Amadina, 615
    Atticora, 359
    Chamaea, 443
    Neothraupis, 648
    Patagioenas, 111–112
    Rallina, 79
fasciatoventris, Thryothorus,
    390
fasciatum, Tigrisoma, 22
fasciatus
    Accipiter, 41
    Anurolimnas, 79
    Aquila, 48
    Campylorhynchus, 389
    Harpactes, 214
    Myiophobus, 332
    Philortyx, 61
    Phyllomyias, 318

Ramsayornis, 566
Tockus, 229
fasciicauda, Pipra, 314–315
fasciinucha, Falco, 53
fasciiventer, Melaniparus, 526
fascinans, Microeca, 479
fasciogularis, Lichenostomus, 561
fasciolata
    Crax, 57
    Locustella, 429
    Mirafra, 350
fasciolatus
    Calamonastes, 425
    Circaetus, 38
fastuosa, Tangara, 659
fastuosus, Epimachus, 584
feadensis, Aplonis, 596
feae
    Pterodroma, 11
    Turdus, 410
featherstoni, Phalacrocorax, 18
fedoa, Limosa, 97
felix, Thryothorus, 391
femoralis
    Falco, 52
    Pholia, 601
    Scytalopus, 309
ferdinandi, Cercomacra, 299
ferina, Aythya, 32
Ferminia, 393
fernandensis, Sephanoides, 206
fernandezianus, Anairetes, 322
fernandinae
    Colaptes, 256
    Teretistris, 643
ferox, Myiarchus, 344
ferreorostris, Chaunoproctus, 701
ferreus, Saxicola, 463
ferrocyanea, Myiagra, 478
ferruginea
    Calidris, 99
    Drymophila, 298
    Gallicolumba, 700
    Hirundinea, 333
    Muscicapa, 449
    Myrmeciza, 301
    Oxyura, 33
    Tadorna, 28
ferrugineifrons, Bolborhynchus,
    146
ferrugineipectus, Grallaricula, 307
ferrugineiventre, Conirostrum, 648
ferrugineus
    Coccyzus, 159
    Enicognathus, 146
    Laniarius, 574
    Pitohui, 487
ferruginosa, Lonchura, 619
ferruginosus, Pomatorhinus, 497
festiva, Amazona, 150
festivus, Chrysocolaptes, 260
Ficedula, 450–452
ficedulinus, Zosterops, 550
figulus, Furnarius, 270
filicauda, Pipra, 315
fimbriata
    Coracina, 372
    Polyerata, 201
fimbriatum, Callocephalon, 130
finlaysoni, Pycnonotus, 378
finschi
    Amazona, 149

Aratinga, 143
Euphonia, 626
Francolinus, 65
Haematopus, 89
finschii
Alophoixus, 382
Ducula, 128
Micropsitta, 134
Neocossyphus, 401
Oenanthe, 464
Psittacula, 140
Zosterops, 555
fischeri
Agapornis, 141
Melaenornis, 447
Phyllastrephus, 380
Ptilinopus, 124
Somateria, 32
Spreo, 601
Tauraco, 151
Vidua, 620
fistulator, Ceratogymna, 231
fjeldsaai, Myrmotherula, 295
flabelliformis, Cacomantis, 154
flagrans, Aethopyga, 542
flammea
Carduelis, 631
Paroreomyza, 701
flammeolus, Otus, 164
flammeus
Asio, 176
Pericrocotus, 374–375
flammiceps, Cephalopyrus, 534
flammigerus, Ramphocelus, 656
flammula, Selasphorus, 211
flammulata, Asthenes, 276
flammulatus
Deltarhynchus, 345
Hemitriccus, 327
Megabyas, 466
Thripadectes, 281
flava
Campephaga, 373
Motacilla, 367
Piranga, 654–655
flavala, Hemixos, 384
flaveola
Capsiempis, 323
Coereba, 647
Sicalis, 675
flaveolus
Alophoixus, 382
Basileuterus, 646
Passer, 602
Platycercus, 136
flavescens
Boissonneaua, 204
Celeus, 256–257
Empidonax, 334
Lichenostomus, 561
Pycnonotus, 378
flavicans
Foudia, 610
Lichmera, 557
Macrosphenus, 436
Myiophobus, 332
Prinia, 421
Prioniturus, 138
flavicapilla, Chloropipo, 317
flaviceps
Atlapetes, 677
Auriparus, 533

Rhodacanthis, 701
flavicollis
Chlorocichla, 380
Hemithraupis, 651–652
Ixobrychus, 22
Lichenostomus, 561
Macronous, 502
Macronyx, 367
Yuhina, 510
flavicrissalis, Eremomela, 434
flavida, Apalis, 423
flavifrons
Amblyornis, 586
Anthoscopus, 533
Megalaima, 238
Melanerpes, 247
Pachycephala, 486
Poicephalus, 142
Vireo, 622
Zosterops, 554–555
flavigaster
Arachnothera, 543
Hyliota, 441
Microeca, 479
Xiphorhynchus, 287
flavigula
Manorina, 566
Piculus, 254
Serinus, 635
flavigularis
Chlorospingus, 650
Chrysococcyx, 155
Platyrinchus, 331
flavinucha
Muscisaxicola, 340
Picus, 258–259
flavipectus, Cyanistes, 528
flavipennis, Chloropsis, 386
flavipes
Hylophilus, 624
Ketupa, 168
Notiochelidon, 359
Platalea, 25
Platycichla, 407
Ploceus, 606
Tringa, 98
flaviprymna, Lonchura, 620
flavirictus, Meliphaga, 560
flavirostra, Amaurornis, 83
flavirostris
Anairetes, 322
Anas, 30
Arremon, 679
Carduelis, 633
Chlorophonia, 627
Grallaricula, 307
Humblotia, 450
Monasa, 234
Paradoxornis, 511
Patagioenas, 112
Phibalura, 312
Porphyrio, 85
Rynchops, 106
Tockus, 229
Urocissa, 591
flaviscapis, Pteruthius, 506
flaviventer
Dacnis, 662
Machaerirhynchus, 479
Porzana, 84
Xanthotis, 561–562

flaviventris
Chlorocichla, 380
Emberiza, 666
Empidonax, 334
Eopsaltria, 481
Motacilla, 368
Phylloscartes, 324
Prinia, 420
Pseudocolopteryx, 323
Serinus, 635
Tolmomyias, 330–331
flavivertex
Heterocercus, 317
Myioborus, 644
Myiopagis, 319
Serinus, 634
flavocinctus, Oriolus, 567
flavogaster, Elaenia, 320
flavogriseum, Pachycare, 483
flavolateralis, Gerygone, 522
flavolivacea, Cettia, 426
flavostriatus, Phyllastrephus, 381
flavotincta, Grallaria, 307
flavovelata, Geothlypis, 642
flavovirens
Chlorospingus, 650
Phylloscartes, 324
flavovirescens, Microeca, 480
flavoviridis
Hartertula, 499
Trichoglossus, 132
Vireo, 623
flavus
Celeus, 257
Hemignathus, 638
Lichenostomus, 561
Xanthopsar, 692
Zosterops, 552
floccosus, Pycnoptilus, 517
florensis, Corvus, 593
floriceps, Anthocephala, 202
florida, Tangara, 659
floris
Spizaetus, 48
Treron, 122
Florisuga, 195
flosculus, Loriculus, 141
fluminea, Porzana, 83
fluminensis, Myrmotherula, 296
fluviatilis
Locustella, 429
Muscisaxicola, 339
Prinia, 421
Fluvicola, 338
fluvicola, Petrochelidon, 362
foersteri
Henicophaps, 115
Melidectes, 564
foetidus, Gymnoderus, 314
fonsecai, Acrobatornis, 278
forbesi
Atlapetes, 678
Charadrius, 94
Curaeus, 698
Cyanoramphus, 135
Leptodon, 35
Lonchura, 619
Rallina, 79
forcipata, Macropsalis, 184
forficatus
Chlorostilbon, 197
Dicrurus, 578

forficatus (*continued*)
   Elanoides, 35
   Tyrannus, 343
Formicarius, 303, 304
Formicivora, 297
formicivora, Myrmecocichla, 465
formicivorus, Melanerpes, 247
formosa
   Anas, 30
   Calocitta, 587–588
   Eudromia, 6
   Pipreola, 312
   Sitta, 531
formosae
   Dendrocitta, 592
   Treron, 123
formosus
   Garrulax, 492
   Oporornis, 642
   Spelaeornis, 499
   Sporaeginthus, 615
fornsi, Teretistris, 643
Forpus, 146–147
forresti, Phylloscopus, 438
forsteni
   Ducula, 127
   Meropogon, 225
   Oriolus, 567
forstenii, Megapodius, 54
forsteri
   Aptenodytes, 6
   Sterna, 106
fortipes, Cettia, 426
fortis
   Coracina, 369
   Geospiza, 687
   Myrmeciza, 302
fossii, Caprimulgus, 183
Foudia, 609–610
Foulehaio, 560
fraenatus, Caprimulgus, 181
francesii, Accipiter, 40
franciae, Agyrtria, 201
francicus, Aerodramus, 185
franciscanus
   Arremon, 679
   Euplectes, 610
   Knipolegus, 337
Francolinus, 65–67
francolinus, Francolinus, 65
franklinii, Megalaima, 239
frantzii
   Catharus, 406
   Elaenia, 320
   Pteroglossus, 242
   Semnornis, 240
fraseri
   Basileuterus, 644
   Deleornis, 534
   Neocossyphus, 401
   Oreomanes, 648
Fraseria, 447
frater
   Monarcha, 476
   Onychognathus, 601
Fratercula, 108
fratrum, Batis, 467
Frederickena, 290
Fregata, 18
Fregetta, 14
Fregilupus, 701
fremantlii, Pseudalaemon, 355

frenata, Geotrygon, 120
frenatus
   Chaetops, 495
   Lichenostomus, 560
freycinet, Megapodius, 54
freycineti, Myiagra, 477
Fringilla, 625–626
fringillaris, Spizocorys, 355
fringillarius, Microhierax, 50
fringillinus, Melaniparus, 526
fringilloides
   Dolospingus, 672
   Spermestes, 618
frontale, Cinclidium, 461
frontalis
   Anarhynchus, 95
   Dendrocitta, 592
   Hemispingus, 651
   Muscisaxicola, 340
   Nonnula, 234
   Ochthoeca, 337
   Orthotomus, 433
   Phoenicurus, 461
   Pipreola, 312
   Pyrrhura, 145
   Sericornis, 518
   Serinus, 634
   Sitta, 531
   Sporophila, 669
   Sporopipes, 605
   Synallaxis, 272
   Veniliornis, 254
frontata, Tricholaema, 237
frontatus, Falcunculus, 483
frugilegus, Corvus, 594
frugivorus, Calyptophilus, 652
fruticeti, Phrygilus, 667
fucata
   Alopochelidon, 360
   Emberiza, 665
fuciphagus, Aerodramus, 187
fucosa, Tangara, 661
fuelleborni
   Alethe, 414
   Laniarius, 574
   Macronyx, 366
fuertesi
   Hapalopsittaca, 148
   Icterus, 696
fugax, Cuculus, 153
fulgens, Eugenes, 204
fulgidus
   Caridonax, 221
   Onychognathus, 601
   Pharomachrus, 214
   Psittrichas, 134
Fulica, 85–86
fulica, Heliornis, 86
fulicarius, Phalaropus, 99
fulicatus, Saxicoloides, 460
fuliginiceps, Leptasthenura, 270
fuliginosa
   Chalcomitra, 536
   Dendrocincla, 283–284
   Geospiza, 687
   Nesofregetta, 14
   Petrochelidon, 362
   Psalidoprocne, 357
   Rhipidura, 470
   Rhyacornis, 461
   Schizoeaca, 271
   Strepera, 582

fuliginosus
   Aegithalos, 515
   Calamanthus, 519
   Calorhamphus, 238
   Dendragapus, 58
   Haematopus, 89
   Larus, 102
   Oreostruthus, 616
   Otus, 163
   Saltator, 689
   Tiaris, 673
fuligiventer, Phylloscopus, 437
fuligula
   Aythya, 32
   Ptyonoprogne, 361
Fulmarus, 10
fulva
   Cinnycerthia, 390
   Petrochelidon, 362–363
   Pluvialis, 93
   Turdoides, 503
fulvescens
   Illadopsis, 495
   Picumnus, 246
   Prunella, 400
   Strix, 169
fulvicapilla, Cisticola, 417
fulvicauda, Basileuterus, 646
fulviceps
   Atlapetes, 678
   Thlypopsis, 651
fulvicollis, Treron, 122
fulvicrissa, Euphonia, 627
fulvifrons
   Empidonax, 334
   Paradoxornis, 512
fulvigula
   Anas, 30
   Timeliopsis, 556
fulvipectus, Rhynchocyclus, 330
fulviventris
   Hylopezus, 305
   Myrmotherula, 295
   Phyllastrephus, 380
   Turdus, 411
fulvogularis, Malacoptila, 233
fulvus
   Gyps, 37
   Lanio, 653
   Mulleripicus, 261
fumicolor, Ochthoeca, 337–338
fumifrons, Poecilotriccus, 329
fumigatus
   Contopus, 335
   Cypseloides, 184
   Melipotes, 565
   Myiotheretes, 338
   Turdus, 412
   Veniliornis, 253
fumosa, Chaetura, 188
funebris
   Laniarius, 574
   Mulleripicus, 261
   Todiramphus, 221
funerea
   Drepanis, 701
   Vidua, 621
funereus
   Aegolius, 173
   Calyptorhynchus, 130
   Melaniparus, 526
   Oryzoborus, 672

furcata
　Oceanodroma, 14
　Tachornis, 189
　Thalurania, 198
furcatus
　Anthus, 365
　Creagrus, 103
　Hemitriccus, 328
furcifer, Heliomaster, 210
Furnarius, 270
fusca
　Allenia, 399
　Aplonis, 701
　Cercomela, 465
　Dendroica, 641
　Gerygone, 522
　Iodopleura, 312
　Malacoptila, 233
　Melanitta, 33
　Phoebetria, 9
　Porzana, 83
fuscans, Lonchura, 618
fuscata
　Padda, 620
　Pseudeos, 131
fuscater
　Catharus, 406
　Turdus, 411
fuscatus, 399
　Cnemotriccus, 333
　Margarops, 399
　Onychoprion, 104
　Phylloscopus, 437
fuscescens
　Catharus, 407
　Dendropicos, 249
　Phalacrocorax, 17
fuscicapilla, Zosterops, 553
fuscicapillus
　Corvus, 594
　Philemon, 563
fuscicauda
　Habia, 654
　Ramphotrigon, 345
　Scytalopus, 310
fusciceps, Thripophaga, 275–276
fuscicollis
　Calidris, 99
　Phalacrocorax, 16
fuscipenne, Philydor, 280
fuscipennis, Dicrurus, 578
fuscirostris, Talegalla, 53
fuscocapillus, Pellorneum, 493
fuscocinereus, Lipaugus, 313
fuscocrissa, Ortygospiza, 616
fusconotus, Nigrita, 611
fuscoolivaceus, Atlapetes, 677
fuscorufa
　Rhipidura, 469
　Synallaxis, 273
fuscorufus, Myiotheretes, 338
fuscus
　Acridotheres, 598
　Anabazenops, 280
　Artamus, 580
　Casiornis, 343
　Cinclodes, 269
　Cinnyris, 540
　Florisuga, 195
　Larus, 101
　Lichenostomus, 561
　Melidectes, 564

Picumnus, 246
Pionus, 149
Pipilo, 681
Scytalopus, 310
Teledromas, 309
Xiphorhynchus, 286
fytchii, Bambusicola, 70

gabar, Micronisus, 40
gabela
　Prionops, 576
　Sheppardia, 455
gabonensis
　Dendropicos, 249
　Ortygospiza, 615
gabonicus, Anthreptes, 535
gaimardi, Phalacrocorax, 17
gaimardii, Myiopagis, 319
galactotes
　Cercotrichas, 459
　Cisticola, 416
galapagoensis
　Buteo, 46
　Zenaida, 117
galatea, Tanysiptera, 222
Galbalcyrhynchus, 231
galbanus, Garrulax, 490
Galbula, 232
galbula
　Galbula, 232
　Icterus, 696
　Ploceus, 607
galeata
　Antilophia, 316
　Ducula, 128
　Myiagra, 478
galeatus
　Basilornis, 597
　Dryocopus, 257
　Lophotriccus, 327
galericulata, Aix, 29
galericulatus, Platylophus, 586–587
Galerida, 355–356
galerita, Cacatua, 130
galeritus, Anorrhinus, 230
galgulus, Loriculus, 140
galinieri, Parophasma, 509
gallardoi, Podiceps, 9
Gallicolumba, 120–121, 700
Gallicrex, 84
gallicus, Circaetus, 38
gallinacea, Irediparra, 88
Gallinago, 96
gallinago, Gallinago, 96
Gallinula, 85, 699
Gallirallus, 80–81, 699
gallopavo, Meleagris, 57
Galloperdix, 70
Gallus, 71
gallus, Gallus, 71
gambagae, Muscicapa, 448
gambeli, Poecile, 524
gambelii, Callipepla, 61
gambensis
　Dryoscopus, 572
　Plectropterus, 28
gambieri, Todiramphus, 221
Gampsonyx, 35–36
Gampsorhynchus, 506
garleppi, Poospiza, 669
garnotii, Pelecanoides, 14
Garrodia, 13

garrula, Ortalis, 54
Garrulax, 488–492
Garrulus, 590–591
garrulus
　Bombycilla, 387
　Coracias, 226
　Lorius, 132
garzetta, Egretta, 20
gaudichaud, Dacelo, 218
Gavia, 7
gavia, Puffinus, 13
gayaquilensis, Campephilus, 258
gayi
　Attagis, 100
　Phrygilus, 667
Gecinulus, 260
geelvinkiana, Micropsitta, 134
geelvinkianum, Dicaeum, 547–548
Gelochelidon, 104–105
genei
　Drymophila, 298
　Larus, 102
genibarbis
　Myadestes, 405
　Thryothorus, 391
gentilis, Accipiter, 43
gentryi, Herpsilochmus, 297
Geobates, 267
Geococcyx, 160
Geocolaptes, 249
geoffroyi
　Augastes, 209
　Geoffroyus, 137–138
　Neomorphus, 160
Geoffroyus, 137–138
Geomalia, 402
Geopelia, 116
Geophaps, 116
Geopsittacus, 137
georgiana
　Eopsaltria, 481
　Melospiza, 686
georgianus, Phalacrocorax, 17
georgica, Anas, 31
georgicus, Pelecanoides, 14
Geositta, 268
Geospiza, 687
Geothlypis, 642–643
Geotrygon, 119–120
Geranoaetus, 45
Geranospiza, 44
germaini, Polyplectron, 73
germani, Aerodramus, 187
Geronticus, 24
Gerygone, 520–522, 700
gibberifrons, Anas, 30
gierowii, Euplectes, 610
gigantea
　Fulica, 86
　Grallaria, 305
　Melampitta, 585
　Pseudibis, 24
giganteus
　Hirundapus, 188
　Macronectes, 10
gigas
　Coua, 156
　Elaenia, 320
　Hydrochous, 185
　Patagona, 206
　Podilymbus, 699
gilberti, Kupeornis, 509

gilletti, Mirafra, 350
gilolensis, Melitograis, 563
gilvicollis, Micrastur, 50
gilviventris, Xenicus, 347
gilvus
  Mimus, 397
  Vireo, 622
gindiana, Eupodotis, 88
gingalensis, Ocyceros, 230
gingica, Arborophila, 68
ginginianus, Acridotheres, 598
githagineus, Bucanetes, 637–638
glabricollis, Cephalopterus, 314
glabrirostris, Melanoptila, 397
glacialis, Fulmarus, 10
glacialoides, Fulmarus, 10
gladiator, Malaconotus, 575
glandarius
  Clamator, 152
  Garrulus, 590
Glareola, 91, 92
glareola, Tringa, 98
glaucescens, Larus, 101
Glaucidium, 170–172
glaucinus, Myophonus, 402
Glaucis, 192
glaucocaerulea, Cyanoloxia, 691
glaucocolpa, Thraupis, 656
glaucogularis, Ara, 142
glaucoides, Larus, 101
glaucopis, Thalurania, 199
glaucopoides, Eriocnemis, 207
glaucops, Tyto, 161
glaucurus, Eurystomus, 227
glaucus
  Anodorhynchus, 700
  Diglossopis, 674
globulosa, Crax, 57
gloriosa, Diglossa, 674
gloriosissima, Diglossa, 674
Glossopsitta, 133
Glycichaera, 556
Glyphorynchus, 285
gnoma, Glaucidium, 171
Gnorimopsar, 698
godeffroyi
  Monarcha, 477
  Todiramphus, 221
godefrida, Claravis, 118
godini, Eriocnemis, 207
godlewskii
  Anthus, 364
  Emberiza, 664
goeldii, Myrmeciza, 301
goeringi
  Brachygalba, 231
  Hemispingus, 650
goertae, Dendropicos, 250
Goethalsia, 198
goffiniana, Cacatua, 130
goiavier, Pycnonotus, 378
goisagi, Gorsachius, 21
golandi, Ploceus, 608
goldiei, Psitteuteles, 132
goldmani, Geotrygon, 119
Goldmania, 198
goliath
  Ardea, 19
  Centropus, 157
  Ducula, 128
gongonensis, Passer, 603

goodenovii, Petroica, 480
goodfellowi
  Lophozosterops, 555
  Regulus, 385
  Rhinomyias, 448
goodsoni, Patagioenas, 112
Gorsachius, 21
goslingi, Apalis, 423
goudoti, Lepidopyga, 199
goudotii, Chamaepetes, 56
goughensis, Rowettia, 668
gouldi, Euphonia, 627
gouldiae
  Aethopyga, 542
  Chloebia, 617
gouldii
  Lophornis, 196
  Selenidera, 241
gounellei, Anopetia, 192
Goura, 121
goyderi, Amytornis, 517
graceannae, Icterus, 695
graciae, Dendroica, 641
gracilipes, Zimmerius, 324
gracilirostris
  Acrocephalus, 431
  Andropadus, 379
  Catharus, 406
  Chloropeta, 432
  Vireo, 623
gracilis
  Anas, 30
  Andropadus, 379
  Heterophasia, 509
  Meliphaga, 560
  Oceanites, 13
  Prinia, 420
Gracula, 598
graculina, Strepera, 582
graculus, Pyrrhocorax, 593
Gracupica, 599
graduacauda, Icterus, 697
graeca, Alectoris, 64
Grafisia, 601
Grallaria, 305–307
grallaria, Fregetta, 14
Grallaricula, 307–308
grallarius, Burhinus, 91
Grallina, 580
gramineus
  Megalurus, 442
  Pooecetes, 683
  Tanygnathus, 138
Graminicola, 442
grammacus, Chondestes, 683
grammiceps
  Seicercus, 440
  Stachyris, 500
grammicus
  Celeus, 256
  Pseudoscops, 175
granadensis
  Hemitriccus, 328
  Myiozetetes, 341
  Picumnus, 246
Granatellus, 647
Granatina, 614
granatina
  Granatina, 614
  Pitta, 266
Grandala, 461
grandidieri, Zoonavena, 187

grandis
  Acridotheres, 598
  Aplonis, 596
  Bradypterus, 427
  Lonchura, 619
  Motacilla, 367
  Niltava, 452
  Nyctibius, 177
  Ploceus, 608
  Rhabdornis, 533
granti
  Ptilopsis, 166
  Sula, 16
grantia, Gecinulus, 260
Grantiella, 566
granulifrons, Ptilinopus, 126
grata, Malia, 488
graueri
  Bradypterus, 427
  Coracina, 370
  Pseudocalyptomena, 265
Graueria, 434
gravis, Puffinus, 12
Graydidascalus, 148
grayi
  Ammomanopsis, 351
  Hylocharis, 199
  Malurus, 515
  Turdus, 412
  Zosterops, 552
grayii, Ardeola, 20
graysoni
  Mimus, 398
  Zenaida, 116
gregalis, Eremomela, 434
gregarius, Vanellus, 93
greyii, Ptilinopus, 125
grillii, Centropus, 158
grimwoodi, Macronyx, 367
grisea
  Formicivora, 297
  Myrmotherula, 296
grisegena, Podiceps, 8
griseicapilla, Odontospiza, 617
griseicapillus, Sittasomus, 284
griseicauda, Treron, 122
griseiceps
  Accipiter, 40
  Basileuterus, 645
  Glaucidium, 171
  Myrmeciza, 301
  Pachycephala, 484
  Phyllomyias, 319
  Piprites, 318
griseicollis, Scytalopus, 310
griseigula, Timeliopsis, 556
griseigularis, Myioparus, 450
griseipectus
  Hemitriccus, 327
  Lathrotriccus, 333
griseisticta, Muscicapa, 448
griseiventris
  Melaniparus, 526
  Taphrolesbia, 209
griseldis, Acrocephalus, 430
griseocapilla, Phyllomyias, 319
griseocephalus, Dendropicos, 250
griseoceps, Microeca, 480
griseocristatus, Lophospingus, 668
griseogularis
  Ammoperdix, 64
  Eopsaltria, 481

Phaethornis, 193–194
griseolus, Phylloscopus, 437
griseomurina, Schizoeaca, 271
griseonota, Pachycephala, 486
griseonucha, Grallaria, 307
griseopyga, Pseudhirundo, 358
griseostriatus, Francolinus, 66
griseotinctus, Zosterops, 553
Griseotyrannus, 342
griseovirescens, Zosterops, 550
griseus
    Campylorhynchus, 389
    Eremopterix, 352
    Limnodromus, 97
    Muscisaxicola, 339
    Nyctibius, 178
    Ocyceros, 229
    Passer, 603
    Puffinus, 12
    Thryothorus, 393
    Vireo, 621
grisola, Pachycephala, 483
grossus, Saltator, 689
grosvenori, Megalurulus, 442
Grus, 76–77
grus, Grus, 77
grylle, Cepphus, 107
gryphus, Vultur, 34
Guadalcanaria, 560
guainumbi, Polytmus, 199
guajana, Pitta, 266
gualaquizae, Phylloscartes, 324
guarauna, Aramus, 77
guarayanus, Thryothorus, 393
Guarouba, 143
guarouba, Guarouba, 143
guatemalae, Megascops, 166
guatemalensis
    Campephilus, 258
    Sclerurus, 282
guatimalensis, Grallaria, 306
guatimozinus, Psarocolius, 698
gubernator, Lanius, 570
Gubernatrix, 676
Gubernetes, 340
guerinii, Oxypogon, 208
guianensis
    Morphnus, 47
    Polioptila, 446
guifsobalito, Lybius, 237
guildingii, Amazona, 151
guilielmi, Paradisaea, 585
guimeti, Klais, 196
guinea, Columba, 110
guineensis, Melaniparus, 526
Guira, 160–651
guira
    Guira, 160
    Hemithraupis, 651
guirahuro, Pseudoleistes, 698
guisei, Ptiloprora, 564
gujanensis
    Cyclarhis, 625
    Odontophorus, 62
    Synallaxis, 273
gularis
    Accipiter, 42
    Aspatha, 224
    Campylorhynchus, 389
    Cuculus, 153
    Egretta, 20
    Eurystomus, 227

Francolinus, 65
Garrulax, 490
Heliodoxa, 204
Hellmayrea, 274
Icterus, 695
Macronous, 502
Melithreptus, 562
Merops, 225
Monticola, 401
Myrmotherula, 295
Nicator, 382
Paradoxornis, 511
Paroaria, 676
Rhinomyias, 448
Serinus, 636
Tephrodornis, 576
Turdoides, 503
Yuhina, 510
gulgula, Alauda, 357
gulielmi, Poicephalus, 142
gulielmitertii, Cyclopsitta, 134–135
gundlachi, Accipiter, 43
gundlachii
    Chordeiles, 179
    Mimus, 397
    Vireo, 621
gunningi, Sheppardia, 455
gurneyi
    Aquila, 47
    Mimizuku, 166
    Pitta, 266
    Promerops, 556
    Zoothera, 403
gustavi, Anthus, 366
guttata
    Chlamydera, 586
    Cichladusa, 458
    Dendrocygna, 26
    Myrmotherula, 294
    Ortalis, 55
    Stagonopleura, 616
    Taeniopygia, 616
    Tangara, 660
    Zoothera, 403
guttaticollis, Paradoxornis, 511
guttatoides, Xiphorhynchus, 287
guttatum, Toxostoma, 398
guttatus
    Catharus, 407
    Hypoedaleus, 290
    Ixonotus, 380
    Odontophorus, 63
    Psilorhamphus, 311
    Tinamus, 3
    Xiphorhynchus, 287
Guttera, 74
guttifer, Tringa, 98
guttulata, Syndactyla, 279
guttuligera, Premnornis, 278
guttulus, Monarcha, 476
gutturalis
    Anthus, 365
    Aramides, 699
    Cinclocerthia, 399
    Corapipo, 316
    Habia, 654
    Irania, 457
    Myrmotherula, 295
    Neocichla, 601
    Oreoica, 483
    Oreoscopus, 517
    Parula, 639

Pseudoseisura, 279
Pterocles, 109
Saxicola, 463
gutturata, Cranioleuca, 275
guy, Phaethornis, 192
Gyalophylax, 274
Gygis, 103–104
Gymnobucco, 235
gymnocephala, Pityriasis, 581
gymnocephalus, Picathartes, 488
Gymnocichla, 300
Gymnocrex, 82–83
Gymnoderus, 314
gymnogenys, Turdoides, 505
Gymnoglaux, 166
Gymnogyps, 34
Gymnomystax, 698
Gymnomyza, 565
Gymnophaps, 129
Gymnopithys, 302–303
gymnops
    Melipotes, 565
    Rhegmatorhina, 303
Gymnorhina, 582
Gymnorhinus, 590
Gymnostinops, 698
Gypaetus, 37
Gypohierax, 37
Gyps, 37
gyrola, Tangara, 660

haastii, Apteryx, 2
habessinicus, Cinnyris, 540
Habia, 654
Habroptila, 84
habroptila, Strigops, 134
haddeni, Cettia, 426
haemacephala, Megalaima, 239
hacmastica, Limosa, 97
haematina, Spermophaga, 613
Haematoderus, 314
haematodus, Trichoglossus, 131
haematogaster
    Campephilus, 258
    Northiella, 136
haematonota, Myrmotherula, 295
haematonotus, Psephotus, 136
Haematopus, 89, 700
haematopus, Himantornis, 78
haematopygus, Aulacorhynchus, 241
Haematortyx, 70
Haematospiza, 638
haematostictum, Dicaeum, 546
haematotis, Pionopsitta, 148
haematuropygia, Cacatua, 130
haemorrhous, Cacicus, 697
haesitatus, Cisticola, 418
hagedash, Bostrychia, 24
hainanus
    Cyornis, 453
    Phylloscopus, 439
Halcyon, 218–219
Haliaeetus, 36–37
haliaetus, Pandion, 34
Haliastur, 36
halli
    Macronectes, 10
    Pomatostomus, 511
Halobaena, 11
hamatus, Rostrhamus, 36
hamertoni, Alaemon, 352
Hamirostra, 35

hamlini, Clytorhynchus, 475
hammondii, Empidonax, 334
Hapalopsittaca, 148
Hapaloptila, 234
Haplochelidon, 359
haplochrous
   Accipiter, 41
   Turdus, 412
haplonota, Grallaria, 306
Haplophaedia, 207
Haplospiza, 673
hardwickii
   Chloropsis, 386
   Gallinago, 96
hardyi, Glaucidium, 171
harmonica, Colluricincla, 487
Harpactes, 214–215
Harpagus, 36
Harpia, 47
Harpyhaliaetus, 45
harpyja, Harpia, 47
Harpyopsis, 47
harrisi, Phalacrocorax, 18
harrisii, Aegolius, 173
harterti
   Batrachostomus, 177
   Ficedula, 451
   Ochetorhynchus, 269
   Phlogophilus, 203
   Schizoeaca, 271
Hartertula, 499
hartlaubi
   Cercotrichas, 458
   Euplectes, 610
   Francolinus, 66
   Otus, 164
   Pseudodacnis, 662
   Tauraco, 151
   Tockus, 229
hartlaubii
   Anabathmis, 536
   Larus, 102
   Lissotis, 88
   Pteronetta, 29
   Turdoides, 504
harwoodi, Francolinus, 66
hasitata, Pterodroma, 10
hastata, Aquila, 47
hatinhensis, Lophura, 71
hattamensis, Pachycephalopsis, 482
hauxwelli
   Myrmotherula, 295
   Turdus, 412
hawaiiensis, Corvus, 594
hebetior, Myiagra, 479
Hedydipna, 535
heermanni, Larus, 100
heilprini, Cyanocorax, 588
heinei
   Tangara, 662
   Zoothera, 404
heinrichi
   Cacomantis, 154
   Cossypha, 458
   Geomalia, 402
Heinrichia, 413–414
heinrothi, Puffinus, 13
Heleia, 555
helenae
   Hypothymis, 472
   Lophornis, 196
   Mellisuga, 210

heliaca, Aquila, 47
Heliactin, 209
Heliangelus, 206, 700
helianthea
   Coeligena, 205
   Culicicapa, 454
helias, Eurypyga, 86
heliobates, Camarhynchus, 688
Heliobletus, 283
heliodor, Chaetocercus, 211
Heliodoxa, 203–204
Heliomaster, 209–210
Heliopais, 86
Heliornis, 86
heliosylus, Zonerodius, 22
Heliothryx, 209
helleri, Schizoeaca, 271
Hellmayrea, 274
hellmayri
   Anthus, 365
   Cranioleuca, 275
   Gyalophylax, 274
   Mecocerculus, 321
Helmitheros, 642
heloisa, Atthis, 210
hemichrysus, Myiodynastes, 341
Hemicircus, 261
Hemignathus, 638, 701
hemilasius, Buteo, 47
hemileucurus
   Campylopterus, 194
   Phlogophilus, 203
hemileucus
   Hypocnemoides, 300
   Lampornis, 203
Hemimacronyx, 366
hemimelaena, Myrmeciza, 301
Hemiphaga, 129
Hemiprocne, 191
Hemipus, 375
Hemispingus, 650–651
Hemitesia, 435
Hemithraupis, 651–652
Hemitriccus, 327–328
hemixantha, Microeca, 479
Hemixos, 384
hemprichii
   Larus, 101
   Tockus, 229
hendersoni, Podoces, 592
henicogrammus, Accipiter, 41
Henicopernis, 35
Henicophaps, 115
Henicorhina, 396
henricae, Cranioleuca, 275
henrici
   Ficedula, 451
   Garrulax, 492
henricii, Megalaima, 239
henslowii, Ammodramus, 684
henstii, Accipiter, 43
herberti
   Phylloscopus, 437
   Stachyris, 501
herbicola, Emberizoides, 675–676
hercules, Alcedo, 215
herero, Namibornis, 459
herioti, Cyornis, 453
herminieri, Melanerpes, 246
herodias, Ardea, 19
Herpetotheres, 49–50
Herpsilochmus, 296–297

herrani, Chalcostigma, 209
Heteralocha, 701
Heterocercus, 317
heteroclitus, Geoffroyus, 138
heterolaemus, Orthotomus, 433
Heteromirafra, 350
Heteromunia, 620
Heteromyias, 482
Heteronetta, 33
Heterophasia, 509–510
heteropogon, Chalcostigma, 209
Heterospingus, 653
heterura, Asthenes, 276
heudei, Paradoxornis, 513
heuglini
   Cossypha, 457
   Larus, 101
   Oenanthe, 464
   Ploceus, 607
heuglinii, Neotis, 87
heyi, Ammoperdix, 64
hiaticula, Charadrius, 93
hildebrandti
   Francolinus, 66
   Lamprotornis, 600
himalayana
   Certhia, 532
   Prunella, 400
   Psittacula, 139
himalayensis
   Dendrocopos, 252
   Gyps, 37
   Sitta, 530
   Tetraogallus, 64
Himantopus, 90
himantopus
   Calidris, 99
   Himantopus, 90
Himantornis, 78
Himatione, 639
hindei, Turdoides, 504
hindwoodi, Lichenostomus, 560
hiogaster, Accipiter, 41
Hippolais, 432
hirsuta, Tricholaema, 236
hirsutus, Glaucis, 192
Hirundapus, 188
hirundinacea
   Cypsnagra, 649
   Euphonia, 626
   Sterna, 105
hirundinaceum, Dicaeum, 548–549
hirundinaceus
   Aerodramus, 186
   Caprimulgus, 180
   Hemipus, 375
Hirundinea, 333
hirundineus, Merops, 225
Hirundo, 360–361
hirundo, Sterna, 105
hispanica, Oenanthe, 464
hispaniolensis
   Contopus, 335
   Passer, 602
   Poospiza, 669
hispidus, Phaethornis, 192
histrio, Eos, 131
histrionica, Phaps, 116
Histrionicus, 32
histrionicus, Histrionicus, 32
Histurgops, 605
hoazin, Opisthocomus, 74

hockingi, Aratinga, 143
hodgei, Dryocopus, 257
hodgsoni
    Anthus, 366
    Batrachostomus, 177
    Muscicapella, 454
    Phoenicurus, 460
    Tickellia, 441
hodgsoniae, Perdix, 67
hodgsonii
    Columba, 111
    Ficedula, 450
    Prinia, 420
Hodgsonius, 461
hoedtii, Gallicolumba, 121
hoematotis, Pyrrhura, 146
hoeschi, Anthus, 363
hoevelli, Cyornis, 453
hoffmanni, Pyrrhura, 146
hoffmannii, Melanerpes, 248
hoffmannsi
    Dendrocolaptes, 286
    Rhegmatorhina, 303
holerythra, Rhytipterna, 343
hollandicus, Nymphicus, 131
holochlora
    Aratinga, 143
    Chloropipo, 317
holochlorus, Erythrocercus, 472
holopolia, Coracina, 372
holosericea, Drepanoptila, 127
holosericeus
    Amblycercus, 697
    Amblyramphus, 698
    Eulampis, 195
holospilus, Spilornis, 38
holostictus, Thripadectes, 281
holsti, Macholophus, 528
hombroni, Actenoides, 222
homeyeri, Pachycephala, 483
homochroa
    Catamenia, 672
    Dendrocincla, 284
    Oceanodroma, 14
homochrous, Pachyramphus, 347
hopkei, Carpodectes, 313
hordeaceus, Euplectes, 610
Horizorhinus, 509
hornbyi, Oceanodroma, 14
hornemanni, Carduelis, 631
horsfieldii
    Myophonus, 402
    Pomatorhinus, 496
hortensis, Sylvia, 443
hortulana, Emberiza, 665
hortulorum, Turdus, 408
horus, Apus, 191
hosii
    Calyptomena, 264
    Oriolus, 569
hoskinsii, Glaucidium, 171
hottentota, Anas, 31
hottentottus
    Dicrurus, 579
    Turnix, 75
Houbaropsis, 88
housei, Amytornis, 517
hova, Mirafra, 348
hoyi, Megascops, 166
huallagae, Aulacorhynchus, 241
huancavelicae, Asthenes, 277

hudsoni
    Asthenes, 276
    Knipolegus, 336
hudsonia, Pica, 592
hudsonica, Poecile, 525
huetii, Touit, 148
huhula, Ciccaba, 170
humbloti
    Ardea, 19
    Cinnyris, 541
Humblotia, 450
humboldti, Spheniscus, 7
humboldtii, Hylocharis, 199
humei
    Phylloscopus, 438
    Sphenocichla, 499
humeralis
    Agelaius, 692
    Aimophila, 681
    Ammodramus, 684
    Cossypha, 457
    Diglossa, 674
    Geopelia, 116
    Parkerthraustes, 690
    Terenura, 298
humiae, Syrmaticus, 72
humicola, Asthenes, 277
humilis
    Asthenes, 276
    Eupodotis, 88
    Ichthyophaga, 37
    Pseudopodoces, 529
    Sericornis, 518
    Yuhina, 510
hunsteini, Lonchura, 619
hunteri
    Chalcomitra, 537
    Cisticola, 415
hutchinsii, Branta, 27
huttoni
    Ptilinopus, 125
    Puffinus, 13
    Vireo, 622
hyacinthinus
    Anodorhynchus, 142
    Cyornis, 453
hybrida
    Chlidonias, 105
    Chloephaga, 28
Hydrobates, 14
hydrocharis, Tanysiptera, 222
Hydrochous, 185
hydrocorax, Buceros, 230
Hydrophasianus, 88
Hydroprogne, 105
Hydropsalis, 184
hyemalis
    Clangula, 32
    Junco, 686–687
Hylacola, 519
Hylexetastes, 285
Hylia, 436
Hyliota, 441
Hylocharis, 199
Hylocichla, 407
Hylocitrea, 483
Hylocryptus, 282
Hyloctistes, 280
Hylomanes, 224
Hylonympha, 204
Hylopezus, 304–305
hylophila, Strix, 169

Hylophilus, 623–624
Hylophylax, 302
Hylorchilus, 390
Hymenolaimus, 29
Hymenops, 336
hyogastrus, Ptilinopus, 126
Hypargos, 614
hyperboreus
    Larus, 101
    Plectrophenax, 687
Hypergerus, 424
hypermelaenus, Poecile, 524
hypermetra, Mirafra, 349
hyperythra
    Arborophila, 69
    Brachypteryx, 413
    Dumetia, 502
    Erythrura, 617
    Ficedula, 450–451
    Myrmeciza, 301
    Pachycephala, 484
    Rhipidura, 469
hyperythrus
    Campylopterus, 194
    Cuculus, 153
    Dendrocopos, 251
    Odontophorus, 62
    Tarsiger, 457
Hypnelus, 233
hypocherina, Vidua, 620
hypochloris, Phyllastrephus, 381
hypochondria, Poospiza, 669
hypochondriaca, Synallaxis, 274
hypochroma, Sporophila, 671
hypochryseus, Vireo, 623
Hypocnemis, 300
Hypocnemoides, 300
Hypocolius, 387
Hypocryptadius, 556
Hypoedaleus, 290
hypoglauca, Andigena, 242
Hypogramma, 535
hypogrammica
    Pytilia, 614
    Stachyris, 500
hypogrammicum, Hedydipna, 535
hypoinochrous, Lorius, 132
hypolais, Zosterops, 551
hypoleuca
    Cissa, 591
    Ficedula, 450
    Grallaria, 307
    Poecilodryas, 481
    Pterodroma, 11
    Serpophaga, 322
    Turdoides, 504
hypoleucos
    Actitis, 97
    Falco, 52
    Pomatorhinus, 496
hypoleucum, Dicaeum, 547
hypoleucus
    Basileuterus, 645
    Capito, 240
    Melanotis, 399
    Sphecotheres, 569
    Synthliboramphus, 108
hypopolius, Melanerpes, 247
hypopyrra, Laniocera, 345
hypopyrrha, Streptopelia, 112–113
Hypopyrrhus, 698
Hypositta, 577

hypospodia, Synallaxis, 272
hypostictus, Leucippus, 200
hyposticutus, Serinus, 634
Hypothymis, 472–473
hypoxantha
    Gerygone, 521
    Hypocnemis, 300
    Neodrepanis, 265
    Pachycephala, 484
    Prinia, 421
    Rhipidura, 468
    Sporophila, 671
hypoxanthus
    Hylophilus, 624
    Ploceus, 609
    Zosterops, 553
Hypsipetes, 384–385
hyrcana, Poecile, 524

iagoensis, Passer, 602
ianthinogaster, Granatina, 614
ibadanensis, Malimbus, 605
ibericus, Phylloscopus, 437
Ibidorhyncha, 90
ibis
    Bubulcus, 20
    Mycteria, 23
Ibycter, 49
ichthyaetus
    Ichthyophaga, 37
    Larus, 102
Ichthyophaga, 37
Icteria, 646
icterina, Hippolais, 432
icterinus, Phyllastrephus, 381
icterioides, Mycerobas, 637
icterocephala, Tangara, 660
icterocephalus, Chrysomus, 692
icterophrys, Satrapa, 338
icteropygialis, Eremomela, 434
icterorhynchus
    Francolinus, 66
    Otus, 162
icterotis
    Ognorhynchus, 143
    Platycercus, 136
Icterus, 695–697
icterus, Icterus, 696
Ictinaetus, 47
Ictinia, 36
idae, Ardeola, 20
idaliae, Phaethornis, 193
Idiopsar, 668
Ifrita, 514
igata, Gerygone, 522
igneus, Pericrocotus, 374
ignicapilla, Regulus, 385
ignicauda, Aethopyga, 543
igniferum, Dicaeum, 548
ignipectum, Dicaeum, 548
ignita, Lophura, 72
igniventris, Anisognathus, 658
ignobilis
    Thripadectes, 281
    Turdus, 411–412
ignota, Myrmotherula, 294
ignotincta, Minla, 507
ignotus, Basileuterus, 646
iheringi
    Myrmotherula, 295
    Neorhopias, 297
ijimae, Phylloscopus, 438

iliaca, Passerella, 684–685
iliacus, Turdus, 410
Ilicura, 316
iliolophus, Toxorhamphus, 545
Illadopsis, 495
imberbe, Camptostoma, 321
imberbis, Anomalospiza, 621
imerinus, Pseudocossyphus, 401
imitans, Euphonia, 627
imitator, Accipiter, 42
immaculata
    Myrmeciza, 302
    Pnoepyga, 499
    Prunella, 400
immaculatus, Enicurus, 461
immer, Gavia, 7
immunda, Rhytipterna, 343
immutabilis, Phoebastria, 9
imparatus, Corvus, 594
impejanus, Lophophorus, 71
impennis, Pinguinus, 700
imperatrix, Heliodoxa, 204
imperialis
    Amazona, 151
    Campephilus, 258
    Gallinago, 96
    Lophura, 71
impetuani, Emberiza, 665
implicata, Pachycephala, 486
importunus, Andropadus, 379
improbus, Zimmerius, 324
imthurni, Macroagelaius, 698
inca
    Coeligena, 205
    Columbina, 118
    Larosterna, 105
Incana, 419
incana
    Drymocichla, 422
    Lichmera, 557
    Tringa, 98
incanus, Incana, 419
Incaspiza, 668–669
incerta
    Amalocichla, 479
    Coracina, 371
    Pterodroma, 10
incertus, Pitohui, 487
incognita, Megalaima, 239
inda, Chloroceryle, 223
indica
    Chalcophaps, 115
    Gracula, 598
    Iole, 383
Indicator, 243–244
indicator
    Baeopogon, 380
    Indicator, 244
indicus
    Anser, 27
    Butastur, 44
    Caprimulgus, 181
    Dendronanthus, 367
    Gyps, 37
    Metopidius, 88
    Pterocles, 109
    Sypheotides, 88
    Tarsiger, 457
    Urocolius, 212
    Vanellus, 93
indigo, Eumyias, 452

indigoticus
    Diglossopis, 674
    Scytalopus, 309
indistincta, Lichmera, 556
indus, Haliastur, 36
inepta, Megacrex, 84
inerme, Ornithion, 321
inexpectata
    Guadalcanaria, 560
    Pterodroma, 10
    Torreornis, 682
inexspectata, Tyto, 161
Inezia, 326
infaustus, Perisoreus, 587
infelix, Monarcha, 477
infuscata
    Muscicapa, 449
    Synallaxis, 272
infuscatus
    Aerodramus, 185
    Automolus, 281
    Bradornis, 446
    Henicopernis, 35
    Phimosus, 25
    Turdus, 411
ingens, Megascops, 165
ingoufi, Tinamotis, 6
innominatus, Picumnus, 244
innotata, Aythya, 32
inopinatum, Polyplectron, 73
inornata
    Acanthiza, 519
    Amblyornis, 586
    Catamenia, 672
    Gerygone, 521
    Inezia, 326
    Lophura, 71
    Pachycephala, 483
    Patagioenas, 112
    Phelpsia, 341
    Pinaroloxias, 673
    Prinia, 421
    Tangara, 659
    Thlypopsis, 651
inornatus
    Baeolophus, 529
    Caprimulgus, 182
    Chlorospingus, 649
    Hemitriccus, 328
    Myiophobus, 332
    Philemon, 563
    Phylloscopus, 438
    Rhabdornis, 533
    Zosterops, 554
inquieta
    Myiagra, 478
    Scotocerca, 419
inquietus, Aerodramus, 187
inquisitor, Tityra, 346
inscriptus, Pteroglossus, 242
insignis
    Aegotheles, 176
    Ardea, 19
    Artamus, 581
    Clytomyias, 515
    Gallirallus, 80
    Panterpe, 197
    Ploceus, 608
    Polihierax, 50
    Prodotiscus, 243
    Rhinomyias, 448

Saxicola, 462
  Thamnophilus, 292
insolitus, Ptilinopus, 126
insularis
  Aphelocoma, 589
  Aplonis, 596
  Arses, 477
  Gerygone, 700
  Junco, 687
  Myophonus, 402
  Otus, 164
  Passer, 602
  Ptilinopus, 125
  Sarothrura, 78
insulata, Sporophila, 671
interjecta, Vidua, 620
intermedia
  Egretta, 20
  Pipreola, 312
  Sporophila, 670
intermedius, Ploceus, 607
internigrans, Perisoreus, 587
interpres
  Arenaria, 98
  Zoothera, 402
involucris, Ixobrychus, 22
Iodopleura, 312–313
iohannis, Hemitriccus, 327
Iole, 383
ios, Ninox, 174
iouschistos, Aegithalos, 515
iozonus, Ptilinopus, 126
iphis, Pomarea, 474
iracunda, Metallura, 208
iraiensis, Scytalopus, 309
Irania, 457
iredalei, Acanthiza, 520
Irediparra, 88
Irena, 569
ireneae, Otus, 162
iriditorques, Columba, 110
Iridophanes, 662
Iridosornis, 658
iris
  Coeligena, 206
  Lamprotornis, 600
  Lepidothrix, 315
  Pitta, 267
  Psitteuteles, 132
irrorata, Phoebastria, 9
irupero, Xolmis, 339
isabella, Stiltia, 91
isabellae
  Cossypha, 457
  Iodopleura, 312
  Oriolus, 568
isabellina
  Amaurornis, 83
  Geositta, 268
  Oenanthe, 464
  Sylvietta, 435
isabellinus, Lanius, 570
isidorei
  Lepidothrix, 315
  Pomatostomus, 511
isidori, Oroaetus, 49
islandica, Bucephala, 33
islerorum, Suiriri, 321
Ispidina, 217
isura, Lophoictinia, 35
Ithaginis, 70
ituriensis, Batis, 467

Ixobrychus, 22, 699
ixoides, Pycnopygius, 562–563
Ixonotus, 380
Ixoreus, 413
Ixos, 383, 384

Jabiru, 24
Jabouilleia, 497
Jacamaralcyon, 232
Jacamerops, 232
Jacana, 88–89
jacana, Jacana, 88–89
jacarina, Volatinia, 669
jacksoni
  Apalis, 422
  Cryptospiza, 612
  Euplectes, 611
  Francolinus, 66
  Ploceus, 608
  Tockus, 229
jacobinus, Clamator, 152
jacquacu, Penelope, 56
jacquinoti
  Ninox, 175
  Pachycephala, 486
jacucaca, Penelope, 56
jacula, Heliodoxa, 204
jacutinga, Pipile, 56
jamacaii, Icterus, 696
jamaica, Euphonia, 626
jamaicensis
  Buteo, 46
  Corvus, 594
  Laterallus, 79
  Leptotila, 119
  Nyctibius, 178
  Oxyura, 33
  Turdus, 412
jambu, Ptilinopus, 124
jamesi
  Phoenicopterus, 26
  Tchagra, 573
jamesoni
  Gallinago, 96
  Parmoptila, 611
  Platysteira, 467
jandaya, Aratinga, 144
jankowskii, Emberiza, 664
janthina, Columba, 111
japonica
  Bombycilla, 387
  Coturnix, 67
  Ninox, 174
japonicus, Zosterops, 551
jardineii, Turdoides, 505
jardini, Boissonneaua, 204
jardinii, Glaucidium, 171
javanense, Dinopium, 259–260
javanica
  Arborophila, 69
  Dendrocygna, 26
  Mirafra, 348
  Rhipidura, 469
javanicus
  Acridotheres, 598
  Charadrius, 94
  Eurylaimus, 264–265
  Leptoptilos, 24
  Lophozosterops, 555
  Phaenicophaeus, 156

javensis
  Batrachostomus, 177
  Coracina, 368
  Dryocopus, 257
  Megalaima, 238
jefferyi
  Chlamydochaera, 413
  Pithecophaga, 47
jelskii
  Iridosornis, 658
  Ochthoeca, 337
  Upucerthia, 268
jerdoni
  Aviceda, 34
  Garrulax, 491
  Saxicola, 463
jobiensis
  Gallicolumba, 120
  Manucodia, 583
  Talegalla, 53
jocosus
  Campylorhynchus, 389
  Pycnonotus, 376
johannae
  Cinnyris, 540
  Doryfera, 194
  Tangara, 659
johannis, Carduelis, 633
johnstoni
  Nectarinia, 538
  Ruwenzorornis, 152
johnstoniae
  Tarsiger, 457
  Trichoglossus, 132
joiceyi, Zoothera, 402
jonquillaceus, Aprosmictus, 139
josefinae, Charmosyna, 133
josephinae, Hemitriccus, 327
jourdanii, Chaetocercus, 211
jouyi, Columba, 700
jubata
  Chenonetta, 29
  Neochen, 28
jubatus, Rhynochetos, 86
Jubula, 170
jucunda, Pipreola, 312
jugger, Falco, 52
jugularis
  Brotogeris, 147
  Cinnyris, 540–541
  Eulampis, 195
  Meiglyptes, 260
  Myzomela, 559
juliae, Arachnothera, 544
julianae, Monarcha, 476
julie, Damophila, 199
julius, Nothocercus, 3
juncidis, Cisticola, 417–418
Junco, 686–687
juninensis, Muscisaxicola, 339
junoniae, Columba, 110
juruanus, Xiphorhynchus, 287
Jynx, 244

kaempferi, Hemitriccus, 328
kaestneri, Grallaria, 306
Kakamega, 496
kalinowskii, Nothoprocta, 5
kandti, Estrilda, 613
kansuensis, Phylloscopus, 438
karamojae, Apalis, 424
kasumba, Harpactes, 214

katangae, Ploceus, 607
katherina, Acanthiza, 520
kauaiensis, Hemignathus, 638
kaupi, Arses, 477
Kaupifalco, 39
kawalli, Amazona, 150
keartlandi, Lichenostomus, 561
keayi, Gallicolumba, 120
keiensis, Micropsitta, 134
kelaarti, Lonchura, 618
kelleyi, Macronous, 502
kempi, Macrosphenus, 436
kennicottii, Megascops, 164–165
Kenopia, 498
kenricki, Poeoptera, 601
keraudrenii, Manucodia, 583
kerearako, Acrocephalus, 431
keri, Sericornis, 518
kerriae, Crypturellus, 4
kessleri, Turdus, 410
Ketupa, 168
ketupu, Ketupa, 168
kieneri, Melozone, 679
kienerii
    Aquila, 48
    Dendroplex, 288
kilimensis, Nectarinia, 538
kinabaluensis, Spilornis, 38
kingi, Aglaiocercus, 209
kioloides, Canirallus, 78
kirhocephalus, Pitohui, 488
kirkii, Veniliornis, 254
kirtlandii, Dendroica, 641
kisserensis, Philemon, 563
kizuki, Dendrocopos, 250
klaas, Chrysococcyx, 155
klagesi, Myrmotherula, 294
Klais, 196
kleinschmidti, Erythrura, 617
klossi, Spilornis, 38
Knipolegus, 336–337
kochi, Pitta, 266
koeniswaldiana, Pulsatrix, 170
koepckeae
    Cacicus, 697
    Megascops, 165
    Phaethornis, 193
koliensis, Serinus, 634
kollari, Synallaxis, 274
komadori, Erithacus, 456
kona, Chloridops, 701
konkakinhensis, Garrulax, 490
kori, Ardeotis, 87
koslowi
    Babax, 505
    Emberiza, 664
    Prunella, 400
kowaldi, Ifrita, 514
krameri, Psittacula, 139
kreffti, Mino, 597
kretschmeri, Macrosphenus, 436
kronei, Phylloscartes, 324
krueperi, Sitta, 530
kubaryi
    Corvus, 594
    Gallicolumba, 120
    Rhipidura, 471
kuehni
    Myzomela, 558
    Zosterops, 553
kuhli, Leucopternis, 44
kuhlii, Vini, 133

kupeensis, Telophorus, 575
Kupeornis, 509

labradorides, Tangara, 661
labradorius, Camptorhynchus, 699
Lacedo, 217
lacernulata, Ducula, 129
lacernulatus, Leucopternis, 44
lacertosa, Woodfordia, 556
lachrymosa, Euthlypis, 644
lachrymosus, Xiphorhynchus, 288
lacrymiger, Lepidocolaptes, 289
lacrymosa, Tricholaema, 237
lacrymosus, Anisognathus, 657–658
lactea
    Glareola, 92
    Polioptila, 446
    Polyerata, 201
lacteus, Bubo, 167
laemosticta, Myrmeciza, 301
laeta
    Cercomacra, 298
    Incaspiza, 669
laetissima, Chlorocichla, 380
laetus, Phylloscopus, 436
lafargei, Myzomela, 559
lafayetii, Gallus, 71
Lafresnaya, 205
lafresnayanus, Gallirallus, 80
lafresnayei, Aegithina, 387
lafresnayi
    Lafresnaya, 205
    Picumnus, 245
lafresnayii, Diglossa, 674
lagdeni, Malaconotus, 575
Lagonosticta, 614–615
Lagopus, 58–59
lagopus
    Buteo, 47
    Lagopus, 58
lagrandieri, Megalaima, 238
lais, Cisticola, 416
Lalage, 372–373
lalandi, Stephanoxis, 196
lamberti, Malurus, 516
lamelligerus, Anastomus, 23
lamellipennis, Xipholena, 313
laminirostris, Andigena, 242
Lampornis, 203
Lamprolaima, 203
Lamprolia, 479
Lampropsar, 698
Lamprospiza, 649
Lamprotornis, 599–600
lanaiensis
    Hemignathus, 701
    Myadestes, 405
lanceolata
    Chiroxiphia, 316
    Locustella, 429
    Micromonacha, 234
    Plectorhyncha, 566
    Rhinocrypta, 309
lanceolatus
    Babax, 505
    Garrulus, 590
    Spizaetus, 48
landanae, Lagonosticta, 615
langbianis, Crocias, 509
langsdorffi, Popelairia, 196
languida, Hippolais, 432
Laniarius, 573–574

laniirostris, Euphonia, 626
Laniisoma, 311–312
Lanio, 653
Laniocera, 345
lanioides
    Lipaugus, 313
    Pachycephala, 486
Lanioturdus, 468
Lanius, 570–572
lansbergei, Pericrocotus, 374
lansbergi, Coccyzus, 159
lanyoni, Pogonotriccus, 325
laperouse, Megapodius, 54
lapponica, Limosa, 97
lapponicus, Calcarius, 687
largipennis, Campylopterus, 194
Larosterna, 105
Larus, 100–103
larvata
    Columba, 110–111
    Coracina, 368
    Lagonosticta, 615
    Tangara, 661
larvaticola, Vidua, 621
larvatus, Oriolus, 569
latebricola, Scytalopus, 310
lateralis
    Cisticola, 414
    Poospiza, 669
    Zosterops, 554
Laterallus, 79–80
lathami
    Alectura, 53
    Calyptorhynchus, 130
    Francolinus, 65
    Melophus, 664
Lathamus, 137
Lathrotriccus, 333
laticincta, Platysteira, 466
latifrons, Microhierax, 50
latimeri, Vireo, 621
latinuchus, Atlapetes, 676–677
latirostris
    Andropadus, 379
    Calyptorhynchus, 130
    Contopus, 335
    Cynanthus, 198
    Ocyalus, 698
    Poecilotriccus, 329
latistriata, Stachyris, 500
latistriatus, Anthus, 363
Latoucheornis, 664
latrans
    Ducula, 128
    Scytalopus, 310–311
laudabilis, Icterus, 696
laurae, Phylloscopus, 436
lauterbachi, Chlamydera, 586
lautus, Aulacorhynchus, 241
lavinia, Tangara, 660
lawesii, Parotia, 584
lawrencei, Carduelis, 632
lawrencii
    Geotrygon, 119
    Gymnoglaux, 166
    Pseudocolaptes, 279
    Turdus, 412
layardi
    Megapodius, 54
    Parisoma, 444
    Ptilinopus, 127
laysanensis, Anas, 30

lazuli, Todiramphus, 220
leachii
    Dacelo, 218
    Mackenziaena, 290
leadbeateri
    Bucorvus, 231
    Cacatua, 130
    Heliodoxa, 203
leari, Anodorhynchus, 142
lebruni, Sicalis, 675
leclancheri, Ptilinopus, 124
leclancherii, Passerina, 691
lecontei
    Ispidina, 217
    Toxostoma, 399
leconteii, Ammodramus, 684
ledanti, Sitta, 530
Legatus, 340
Leiothrix, 505
Leipoa, 53
lembeyei, Polioptila, 445
lemosi, Cypseloides, 184
lempiji, Otus, 163
lemprieri, Cyornis, 454
lendu, Muscicapa, 449
lentiginosus, Botaurus, 22
leontica, Schistolais, 422
lepida
    Eminia, 424
    Pyrrhura, 145
    Rhipidura, 471
Lepidocolaptes, 288–289
Lepidopyga, 199
Lepidothrix, 315
lepidus
    Ceyx, 217
    Cuculus, 153
Leptasthenura, 270–271
Leptocoma, 537–538
Loptodon, 35
leptogrammica, Strix, 168
Leptopoecile, 436
Leptopogon, 325
Leptopterus, 577
Leptoptilos, 24
leptorhynchus, Enicognathus, 146
Leptosittaca, 144
Leptosomus, 228
leptosomus, Brachypteracias, 227
Leptotila, 118–119
lepturus, Phaethon, 15
Lerwa, 63
lerwa, Lerwa, 63
Lesbia, 207
leschenaulti
    Enicurus, 461
    Merops, 226
leschenaultii
    Charadrius, 95
    Phaenicophaeus, 156
lessoni, Mayrornis, 475
Lessonia, 336
lessonii, Pterodroma, 10
letitiae, Popelairia, 196
lettia, Otus, 162–163
lettii, Jubula, 170
leucaspis, Gymnopithys, 302
Leucippus, 200
leucoblepharus, Basileuterus, 646
leucocephala
    Amazona, 149
    Arundinicola, 338

Halcyon, 219
Mycteria, 23
Oxyura, 33
Patagioenas, 111
Turdoides, 504
leucocephalos, Emberiza, 664
leucocephalus
    Aceros, 231
    Chaimarrornis, 461
    Cinclus, 388
    Cladorhynchus, 90
    Colius, 212
    Haliaeetus, 37
    Himantopus, 90
    Hypsipetes, 385
    Lybius, 237
    Tachyeres, 28
Leucochloris, 199
leucogaster
    Agyrtria, 200
    Alcedo, 216
    Centropus, 158
    Cinnyricinclus, 600
    Corythaixoides, 152
    Haliaeetus, 36
    Lepidocolaptes, 288
    Pionites, 148
    Sula, 16
    Thamnophilus, 292
leucogastra
    Dendrocitta, 592
    Galbula, 232
    Lonchura, 618
    Ortalis, 55
    Pachycephala, 486
    Uropsila, 396
leucogastroides, Lonchura, 618
leucogenis, Pyrrhula, 636
leucogenys
    Aegithalos, 514
    Cichlopsis, 405
    Conirostrum, 648
    Pycnonotus, 377
leucogeranus, Grus, 76
leucognaphalus, Corvus, 594
leucogrammica, Ptilocichla, 497
leucogrammicus, Pycnonotus, 375
leucolaema, Zoothera, 402
leucolaemus
    Alario, 636
    Odontophorus, 63
    Piculus, 254
leucolepis, Phyllastrephus, 381
leucolopha, Tigriornis, 22
leucolophus
    Caliechthrus, 155
    Garrulax, 489
    Tauraco, 151
leucomela
    Columba, 111
    Lalage, 372
leucomelaena, Sylvia, 443–444
leucomelanos, Lophura, 71
leucomelas
    Calonectris, 12
    Melaniparus, 526
    Tockus, 229
    Tricholaema, 237
    Turdus, 411
leucometopia, Geotrygon, 119
leucomystax, Pogoniulus, 236

leuconota
    Columba, 110
    Pyriglena, 299
leuconotus
    Gorsachius, 21
    Melaniparus, 526
    Thalassornis, 26
leucopareia, Eremopterix, 352
Leucopeza, 643
leucophaea, Cormobates, 523
leucophaeus
    Aerodramus, 187
    Dicrurus, 578
leucophaius, Legatus, 340
leucophoeus, Speirops, 549
leucophrus, Cichlocolaptes, 281
leucophrys
    Anthus, 364
    Basileuterus, 646
    Brachypteryx, 413
    Callonetta, 29
    Cercotrichas, 458
    Dendrortyx, 60
    Henicorhina, 396
    Mecocerculus, 321
    Myrmoborus, 299
    Ochthoeca, 338
    Rhipidura, 469
    Sylvietta, 435
    Vireo, 623
    Zonotrichia, 686
leucophthalma
    Aratinga, 143
    Myrmotherula, 295
leucophthalmus
    Automolus, 281
    Larus, 101
leucopis, Atlapetes, 677
leucopleura, Thescelocichla, 380
leucopleurus, Oreotrochilus, 205
leucopodus, Haematopus, 89
leucopogon
    Campephilus, 258
    Schistolais, 422
    Thryothorus, 392
leucops
    Platycichla, 407
    Tregellasia, 481
Leucopsar, 598
leucopsis
    Aphelocephala, 522
    Branta, 27
    Sitta, 530
leucoptera
    Fulica, 86
    Henicorhina, 396
    Loxia, 631
    Melanocorypha, 353
    Piranga, 655
    Prosobonia, 700
    Psophia, 77
    Pterodroma, 11
    Pyriglena, 299
    Sporophila, 671
Leucopternis, 44
leucopterus
    Atlapetes, 677
    Chlidonias, 105
    Dendrocopos, 252
    Malurus, 515
    Nyctibius, 178

leucopterus (*continued*)
  Platysmurus, 587
  Serinus, 636
leucopteryx, Icterus, 695
leucopus, Furnarius, 270
leucopyga
  Lalage, 373
  Nyctiprogne, 179
  Oenanthe, 463
leucopygia
  Coracina, 369
  Turdoides, 504
leucopygialis
  Lalage, 372
  Rhaphidura, 188
leucopygius
  Serinus, 634
  Todiramphus, 220
leucopyrrhus, Laterallus, 80
leucorhoa, Oceanodroma, 14
leucorhynchus, Laniarius, 574
leucorodia, Platalea, 25
leucorrhoa, Tachycineta, 359
leucorrhous, Buteo, 45
leucorynchus, Artamus,
    580–581
leucoryphus
  Haliaeetus, 37
  Platyrinchus, 331
Leucosarcia, 116
leucoscepus, Francolinus, 66
leucosoma, Hirundo, 361
leucospila, Rallina, 78
leucospodia, Pseudelaenia, 323
leucostephes, Melidectes, 564
leucosterna, Cheramoeca, 358
leucosticta
  Cercotrichas, 458
  Henicorhina, 396
  Ptilorrhoa, 514
Leucosticte, 628–629
leucostictus
  Bubo, 168
  Dysithamnus, 294
leucostigma
  Percnostola, 300
  Rhagologus, 483
leucothorax
  Lanio, 653
  Rhipidura, 470
leucotis
  Entomodestes, 406
  Eremopterix, 352
  Galbalcyrhynchus, 231
  Hylocharis, 199
  Lichenostomus, 561
  Melozone, 680
  Monarcha, 476
  Phapitreron, 121
  Ptilopsis, 166
  Pycnonotus, 377
  Pyrrhura, 145
  Stachyris, 501
  Stactolaema, 235
  Tauraco, 151
  Thryothorus, 392
  Vireolanius, 625
leucotos, Dendrocopos, 251
leucura
  Lagopus, 58
  Oenanthe, 463
leucurum, Cinclidium, 461

leucurus
  Elanus, 36
  Monarcha, 476
  Saxicola, 462
  Vanellus, 93
leuphotes, Aviceda, 35
levaillantii
  Clamator, 152
  Francolinus, 65
levaillantoides, Francolinus, 66
leverianus, Cissopis, 649
levigaster, Gerygone, 522
levraudi, Laterallus, 79
Lewinia, 81–82
lewinii, Meliphaga, 560
lewis, Melanerpes, 246
leytensis, Micromacronus, 502
lherminieri
  Cichlherminia, 404
  Puffinus, 13
lhuysii, Lophophorus, 71
liberatus, Laniarius, 573
libonyanus, Turdus, 408
Lichenostomus, 560–561
Lichmera, 556–557
lichtensteini, Philydor, 280
lichtensteinii, Pterocles, 109
lictor, Philohydor, 341
lidthi, Garrulus, 591
lignarius, Picoides, 252
lilianae, Agapornis, 141
lilliae, Lepidopyga, 199
limae, Picumnus, 246
limbata, Lichmera, 557
Limicola, 99
limicola, Rallus, 81
Limnodromus, 97
Limnornis, 270
Limnothlypis, 642
Limosa, 97
limosa, Limosa, 97
linaraborae, Aethopyga, 542
linchi, Collocalia, 185
lincolnii, Melospiza, 685
lindsayi, Actenoides, 222
linearis
  Chiroxiphia, 316
  Geotrygon, 119–120
lineata
  Acanthiza, 520
  Conopophaga, 308
  Coracina, 369
  Dacnis, 662
  Megalaima, 238
  Pytilia, 614
lineatum, Tigrisoma, 22
lineatus
  Buteo, 45
  Cymbilaimus, 290
  Dryocopus, 257
  Garrulax, 491
lineifrons, Grallaricula, 308
lineiventris, Anthus, 365
lineola
  Bolborhynchus, 146
  Sporophila, 670
linteatus, Heterocercus, 317
lintoni, Myiophobus, 332
Linurgus, 628
Liocichla, 492–493
Lioptilus, 509
Liosceles, 309

Lipaugus, 313
Lissotis, 88
litae, Piculus, 254
litsipsirupa, Psophocichla, 407
littoralis
  Formicivora, 297
  Ochthornis, 338
livens, Larus, 101
liventer, Butastur, 44
livia, Columba, 109–110
lividus, Agriornis, 339
livingstonei, Erythrocercus, 472
livingstonii, Tauraco, 151
llaneae, Megalurulus, 442
lobata
  Biziura, 33
  Campephaga, 374
lobatus, Phalaropus, 99
Loboparadisea, 582
Lochmias, 283
Locustella, 429
locustella, Paludipasser, 616
Loddigesia, 209
lombokia, Lichmera, 556
lomvia, Uria, 107
Lonchura, 618–620
longicauda
  Bartramia, 97
  Coracina, 370
  Deconychura, 284
  Elminia, 472
  Embernagra, 676
  Henicopernis, 35
  Melanocharis, 544
  Myrmotherula, 294
  Psittacula, 140
longicaudata, Nesillas, 428
longicaudatus
  Anthus, 364
  Mimus, 398
  Spelaeornis, 499
longicaudus
  Discosura, 196
  Stercorarius, 107
longicornis, Otus, 163
longimembris, Tyto, 161
longipennis
  Falco, 52
  Hemiprocne, 191
  Macrodipteryx, 183
  Myrmotherula, 296
longipes
  Myrmeciza, 301
  Xenicus, 347
longirostra
  Arachnothera, 543
  Rukia, 555
longirostre, Toxostoma, 398
longirostris
  Caprimulgus, 180
  Coccyzus, 159
  Dasyornis, 517
  Haematopus, 89
  Heliomaster, 210
  Herpsilochmus, 297
  Nasica, 285
  Phaethornis, 192
  Pterodroma, 11
  Rallus, 81
  Rhizothera, 67
  Thryothorus, 393
  Turdoides, 503

longuemarei, Anthreptes, 535
longuemareus, Phaethornis, 193
lopezi
  Bradypterus, 427
  Poliolais, 434
Lophaetus, 48
Lophodytes, 33
Lophoictinia, 35
Lopholaimus, 129
Lophophanes, 526
Lophophorus, 71
Lophorina, 584
Lophornis, 196
Lophospingus, 668
Lophostrix, 170
lophotes
  Geophaps, 116
  Knipolegus, 337
  Percnostola, 301
  Pseudoseisura, 279
Lophotibis, 25
Lophotriccus, 327
Lophozosterops, 555
Lophura, 71, 72
lorata, Sternula, 104
lorealis, Arses, 477
lorentzi, Pachycephala, 485
lorenzi, Phyllastrephus, 381
loriae, Cnemophilus, 582
loricata
  Compsothraupis, 649
  Grallaricula, 307
  Myrmeciza, 301
loricatus
  Celeus, 256
  Monarcha, 476
Loriculus, 140–141
Lorius, 132
lory, Lorius, 132
lotenius, Cinnyris, 541
louisae, Rhynchostruthus, 628
louisiadensis, Cracticus, 581
lovensis, Ashbyia, 522
loveridgei, Cinnyris, 539
lowei, Sheppardia, 455
loweryi, Xenoglaux, 172
Loxia, 630–631
Loxigilla, 673
Loxioides, 638
Loxipasser, 673
Loxops, 638
loyca, Sturnella, 693
lubomirskii, Pipreola, 312
luchsi, Myiopsitta, 146
luciae
  Polyerata, 201
  Vermivora, 639
luciani, Eriocnemis, 207
lucianii, Pyrrhura, 145
lucida, Hirundo, 360
lucidus
  Chrysococcyx, 154
  Chrysocolaptes, 260
  Cyanerpes, 663
  Hemignathus, 638
lucifer, Calothorax, 210
lucionensis, Tanygnathus, 138
luconensis, Prioniturus, 138
luctuosa
  Ducula, 129
  Sporophila, 670

luctuosus
  Sakesphorus, 290
  Tachyphonus, 654
ludlowi, Alcippe, 508
ludovicae, Doryfera, 194
ludovicensis, Cinnyris, 539
ludoviciana, Piranga, 655
ludovicianus
  Lanius, 570–571
  Pheucticus, 690
  Thryothorus, 392
ludwigii
  Dicrurus, 577
  Neotis, 87
luehderi, Laniarius, 573
lugens
  Haplophaedia, 207
  Oenanthe, 463–464
  Parisoma, 444
  Sarothrura, 78
  Streptopelia, 112
lugubris
  Brachygalba, 232
  Celeus, 256
  Contopus, 335
  Dendropicos, 249
  Garrulax, 489
  Megaceryle, 223
  Melampitta, 585
  Myrmoborus, 299
  Poecile, 523
  Poeoptera, 601
  Quiscalus, 694
  Speirops, 549
  Surniculus, 155
  Vanellus, 92
luizae, Asthenes, 277
Lullula, 357
luluae, Poecilotriccus, 328
lumachella, Augastes, 209
lunatus
  Melithreptus, 562
  Onychoprion, 104
  Serilophus, 264
lunulata
  Anthochaera, 567
  Galloperdix, 70
  Zoothera, 404
lunulatus
  Bolbopsittacus, 135
  Garrulax, 490
  Gymnopithys, 303
Lurocalis, 178
Luscinia, 456
luscinia, Luscinia, 456
luscinioides, Locustella, 429
luscinius, Acrocephalus, 430
lutea
  Leiothrix, 505
  Sicalis, 675
luteicapilla, Euphonia, 626
luteifrons, Nigrita, 611
luteirostris, Zosterops, 554
luteiventris
  Myiodynastes, 341
  Myiozetetes, 341
luteocephala, Sicalis, 675
luteola, Sicalis, 675
luteolus
  Ploceus, 606
  Pycnonotus, 378
luteoschistaceus, Accipiter, 41

luteoventris, Bradypterus, 428
luteovirens, Ptilinopus, 127
luteoviridis
  Basileuterus, 645
  Pselliophorus, 676
lutescens, Anthus, 365
lutetiae, Coeligena, 205
luteus
  Passer, 603
  Zosterops, 553
lutosa, Caracara, 699
luzonica
  Anas, 31
  Gallicolumba, 120
luzoniensis, Copsychus, 460
lyalli, Xenicus, 700
Lybius, 237–238
Lycocorax, 583
Lymnocryptes, 96
lyra, Uropsalis, 183
Lysurus, 676

macao, Ara, 142
maccoa, Oxyura, 33
macconnelli
  Mionectes, 326
  Synallaxis, 273
maccormicki, Stercorarius, 106–107
macdonaldi, Nesomimus, 398
macei
  Coracina, 368
  Dendrocopos, 251
macgillivrayi, Pterodroma, 11
Macgregoria, 583
macgregoriae, Amblyornis, 586
macgregorii, Cnemophilus, 582
macgrigoriae, Niltava, 452
Machaerirhynchus, 479
Machaeropterus, 317
Macheiramphus, 35
Machetornis, 340
machiki, Zoothera, 404
Macholophus, 528
Mackenziaena, 290
mackinlayi, Macropygia, 114
mackinnoni, Lanius, 571
mackloti, Harpactes, 214
macleayanus, Xanthotis, 562
macleayii, Todiramphus, 220
maclovianus, Muscisaxicola, 340
macqueenii, Chlamydotis, 87
Macroagelaius, 698
macrocephala, Petroica, 480
Macrocephalon, 53
macrocerca, Hylonympha, 204
macrocercus, Dicrurus, 578
macrocerus, Eupetes, 514
macrodactyla
  Gallinago, 96
  Napothera, 498
  Oceanodroma, 699
macrodactylus, Bucco, 233
Macrodipteryx, 183
macrolopha, Pucrasia, 71
Macronectes, 10
Macronous, 502
Macronyx, 366–367
macronyx, Remiz, 533
Macropsalis, 184
macroptera, Pterodroma, 10
macropterus, Vanellus, 93
macropus, Scytalopus, 311

Macropygia, 114–115
macrorhynchos
 Corvus, 595
 Cymbirhynchus, 264
 Notharchus, 232–233
macrorhynchus
 Saxicola, 462
 Tauraco, 151
macrorrhina, Melidora, 221–222
Macrosphenus, 436
macrotis, Eurostopodus, 179
macroura
 Dendrortyx, 60
 Euplectes, 610
 Thripophaga, 275
 Vidua, 620
 Zenaida, 116
macrourus
 Campylopterus, 194
 Circus, 39
 Urocolius, 212
 Urotriorchis, 44
macrurus, Caprimulgus, 182
macularius
 Actitis, 97
 Hylopezus, 304
maculata
 Chlamydera, 586
 Cotinga, 313
 Paroreomyza, 638
 Stachyris, 501
 Terenura, 298
maculatum, Todirostrum, 329
maculatus
 Chrysococcyx, 154
 Dendrocopos, 250
 Enicurus, 462
 Indicator, 244
 Myiodynastes, 341–342
 Nystalus, 233
 Pardirallus, 84
 Pipilo, 680
 Prionochilus, 545
maculialatus, Icterus, 697
maculicauda
 Asthenes, 277
 Hypocnemoides, 300
maculicaudus, Caprimulgus,
 181
maculicoronatus, Capito, 240
maculifrons, Veniliornis,
 254
maculipectus
 Phacellodomus, 277
 Rhipidura, 469
 Thryothorus, 391
maculipennis
 Larus, 102
 Phylloscopus, 438
maculirostris
 Muscisaxicola, 339
 Selenidera, 242
 Turdus, 412
maculosa
 Campethera, 248
 Lalage, 373
 Nothura, 6
 Patagioenas, 111
 Prinia, 421
maculosus
 Caprimulgus, 181
 Nyctibius, 178
 Turnix, 75

mada
 Gymnophaps, 129
 Prioniturus, 138
madagascarensis, Margaroperdix, 67
madagascariensis
 Accipiter, 42
 Alectroenas, 127
 Asio, 176
 Aviceda, 34
 Calicalicus, 577
 Caprimulgus, 182
 Foudia, 609
 Hypsipetes, 384
 Ispidina, 217
 Numenius, 97
 Otus, 164
 Oxylabes, 511
 Phyllastrephus, 381
 Rallus, 81
madagascarinus, Cyanolanius, 577
Madanga, 555
madaraspatensis, Motacilla, 367
madaraszi, Psittacella, 137
madeira, Pterodroma, 11
maderaspatanus, Zosterops, 550
maesi, Garrulax, 489–490
magellani, Pelecanoides, 14
magellanica, Carduelis, 632
magellanicus
 Bubo, 167
 Campephilus, 258
 Phalacrocorax, 17
 Scytalopus, 310
 Spheniscus, 7
magentae, Pterodroma, 11
magicus, Otus, 164
magister, Vireo, 623
magna
 Acanthornis, 519
 Alectoris, 64
 Aplonis, 596
 Arachnothera, 544
 Macropygia, 114
 Sitta, 531
 Sturnella, 693
magnificens, Fregata, 18
magnificus
 Cicinnurus, 584
 Gorsachius, 21
 Lophornis, 196
 Ptilinopus, 124
 Ptiloris, 584
magnirostra, Sericornis, 518
magnirostre, Malacopteron,
 494
magnirostris
 Amaurornis, 83
 Burhinus, 91
 Buteo, 45
 Galerida, 356
 Geospiza, 687
 Gerygone, 521
 Myiarchus, 344
 Phylloscopus, 438
magnolia, Dendroica, 640
magnum, Malacopteron, 495
maguari, Ciconia, 23
mahali, Plocepasser, 605
mahrattensis
 Caprimulgus, 182
 Dendrocopos, 251
maillardi, Circus, 39
maja, Lonchura, 619

major
 Brachypteryx, 413
 Bradypterus, 427
 Cettia, 426
 Crotophaga, 160
 Dendrocopos, 251–252
 Diglossa, 674
 Pachyramphus, 347
 Parus, 527
 Podiceps, 8
 Quiscalus, 694
 Schiffornis, 318
 Taraba, 290
 Tinamus, 3
 Xiphocolaptes, 286
makirensis, Phylloscopus, 440
malabarica
 Euodice, 618
 Galerida, 356
 Sturnia, 599
malabaricus
 Copsychus, 459–460
 Vanellus, 92
malacca, Lonchura, 619
malaccensis
 Ixos, 384
 Malacocincla, 493
malacense, Polyplectron, 73
malacensis, Anthreptes, 534–535
malachitacea, Triclaria, 151
malachurus, Stipiturus, 516
Malacocincla, 493, 700
Malaconotus, 575
Malacopteron, 494–495
Malacoptila, 233–234
malacoptilus, Rimator, 497
malacorhynchos, Hymenolaimus, 29
Malacorhynchus, 32
malaitae
 Myzomela, 559
 Rhipidura, 471
malaris, Phaethornis, 192
malayanus, Anthracoceros, 230
malayensis, Ictinaetus, 47
malcolmi, Turdoides, 503
Malcorus, 422
maldivarum, Glareola, 91
maleo, Macrocephalon, 53
malherbi, Cyanoramphus, 136
malherbii, Columba, 110
Malia, 488
malimbica, Halcyon, 219
malimbicus
 Malimbus, 605
 Merops, 226
Malimbus, 605
mallee, Stipiturus, 516
malouinus, Attagis, 100
malura, Drymophila, 298
maluroides, Spartonoica, 271
Malurus, 515–516
mana, Oreomystis, 638
Manacus, 316–317
manacus, Manacus, 316–317
manadensis
 Monarcha, 476
 Otus, 164
 Turacoena, 115
mandellii, Arborophila, 68
Mandingoa, 612
mangle, Aramides, 82
mango, Anthracothorax, 195
manilata, Orthopsittaca, 143

manillae, Penelopides, 230
manillensis, Caprimulgus, 182
manipurensis, Perdicula, 68
manoensis, Cinnyris, 538
Manorina, 566
mantananensis, Otus, 164
mantchuricum, Crossoptilon, 72
mantelli
    Apteryx, 2
    Porphyrio, 85
manu, Cercomacra, 299
Manucodia, 583
manusi, Tyto, 161
manyar, Ploceus, 609
maorianus, Oceanites, 699
maracana, Primolius, 143
marail, Penelope, 55
maranonica
    Melanopareia, 311
    Synallaxis, 273
maranonicus, Turdus, 411
marcapatae, Cranioleuca, 274
marchei
    Anthracoceros, 230
    Cleptornis, 555
    Ptilinopus, 124
marecula, Anas, 699
margaretae, Zoothera, 404
margarethae, Charmosyna, 133
margaritaceiventer, Hemitriccus, 328
margaritae
    Batis, 467
    Conirostrum, 648
margaritatus
    Hypargos, 614
    Megastictus, 293
    Trachyphonus, 235
Margaroperdix, 67
Margarops, 399
Margarornis, 278
margelanica, Sylvia, 443
marginalis, Aenigmatolimnas, 84
marginata
    Upupa, 228
    Zoothera, 404
marginatus
    Charadrius, 94
    Microcerculus, 396
    Pachyramphus, 346
mariae, Nesillas, 428
mariei, Megalurulus, 442
marila, Aythya, 32
marina, Pelagodroma, 13
marinus, Larus, 101
mariquensis
    Bradornis, 447
    Cinnyris, 539
maritima
    Calidris, 99
    Geositta, 268
maritimus, Ammodramus, 683–684
markhami, Oceanodroma, 14
Marmaronetta, 32
marmorata, Napothera, 498
marmoratus, Brachyramphus, 107
marshalli, Megascops, 165
martii, Baryphthengus, 224
martinica
    Chaetura, 188
    Elaenia, 319–320
    Porphyrio, 85
martius, Dryocopus, 257
maryae, Vidua, 621

masafuerae, Aphrastura, 270
Mascarinus, 700
mascarinus, Mascarinus, 700
Masius, 316
massena, Trogon, 214
mastacalis, Myiobius, 332
masteri, Vireo, 623
masukuensis, Andropadus, 379
mathewsi, Cincloramphus, 442
matsudairae, Oceanodroma, 14
matthewsii, Boissonneaua, 204
matthiae, Rhipidura, 471
maugaeus, Chlorostilbon, 197
maugei
    Dicaeum, 548
    Geopelia, 116
mauretanicus, Puffinus, 12
mauri, Calidris, 98
maurus
    Circus, 39
    Saxicola, 462
mavornata, Aplonis, 701
mavors, Heliangelus, 206
maxillosus, Saltator, 689
maxima
    Coracina, 368
    Melanocorypha, 353
    Pitta, 266
maximiliani
    Melanopareia, 311
    Oryzoborus, 672
    Pionus, 149
maximus
    Aerodramus, 187
    Artamus, 580
    Garrulax, 491
    Megaceryle, 223
    Saltator, 689
    Thalasseus, 106
mayeri
    Astrapia, 583
    Nesoenas, 112
maynana, Cotinga, 313
mayottensis
    Otus, 164
    Zosterops, 550
mayri
    Ptiloprora, 564
    Rallina, 79
Mayrornis, 474–475
mccallii, Erythrocercus, 471–472
mcclellandii, Ixos, 384
mccownii, Calcarius, 687
mcgregori, Coracina, 372
mcilhennyi, Conioptilon, 314
mcleannani, Phaenostictus, 303
mcleodii, Nyctiphrynus, 179
meadewaldoi, Haematopus, 700
mearnsi, Aerodramus, 185
Mearnsia, 187
mechowi, Cercococcyx, 153
Mecocerculus, 321–322
media, Gallinago, 96
mediocris, Cinnyris, 539
medius, Dendrocopos, 251
meeki
    Charmosyna, 133
    Corvus, 594
    Microgoura, 700
    Micropsitta, 134
    Ninox, 175
    Zosterops, 553
meekiana, Ptiloprora, 564

meesi, Caprimulgus, 182
Megabyas, 466
megacephalum, Ramphotrigon, 345
Megaceryle, 223
Megacrex, 84
Megadyptes, 7
megaensis, Hirundo, 361
megala, Gallinago, 96
Megalaima, 238–239
megalopterus
    Campylorhynchus, 389
    Phalcoboenus, 49
megalorynchos, Tanygnathus, 138
megalotis, Otus, 163
megalura, Leptotila, 118
Megalurulus, 442
Megalurus, 441–442, 700
megaplaga, Loxia, 631
Megapodius, 54
megapodius, Pteroptochos, 309
megarhyncha
    Colluricincla, 487
    Pitta, 267
    Syma, 222
megarhynchos, Luscinia, 456
megarhynchus
    Dicrurus, 579
    Melilestes, 556
    Ploceus, 609
    Rhamphomantis, 155
Megarynchus, 342
Megascops, 164–166
Megastictus, 293
Megatriorchis, 44
Megaxenops, 283
Megazosterops, 556
meiffrenii, Ortyxelos, 76
Meiglyptes, 260
melacoryphus, Coccyzus, 159
melaena
    Lonchura, 620
    Myrmecocichla, 465
Melaenornis, 447
melambrotus, Cathartes, 34
Melampitta, 585
Melamprosops, 639
melanaria, Cercomacra, 299
melancholicus, Tyrannus, 342
Melanerpes, 246–248
melania, Oceanodroma, 14
melanicterus
    Cacicus, 697
    Pycnonotus, 376
Melaniparus, 526–527
Melanitta, 33
melanocephala
    Alectoris, 64
    Anthornis, 700
    Apalis, 423
    Ardea, 19
    Arenaria, 98
    Carpornis, 312
    Emberiza, 666
    Manorina, 566
    Myzomela, 559
    Sylvia, 444
    Tricholaema, 237
melanocephalus
    Atlapetes, 677
    Larus, 102
    Malurus, 515
    Myioborus, 644
    Pheucticus, 690

melanocephalus (*continued*)
Pionites, 148
Ploceus, 608
Speirops, 549
Threskiornis, 24
Tragopan, 70
Trogon, 212
Vanellus, 92
melanoceps, Myrmeciza, 301
Melanocharis, 544
melanochlamys
Accipiter, 41
Bangsia, 657
Melanochlora, 529
melanochloros, Colaptes, 255
melanochroa, Ducula, 129
melanochrous, Serinus, 636
Melanocorypha, 353
melanocoryphus, Cygnus, 26
melanocorys, Calamospiza, 683
melanocyaneus, Cyanocorax, 588
Melanodera, 668
melanodera, Melanodera, 668
Melanodryas, 481
melanogaster
Anhinga, 18
Conopophaga, 308
Formicivora, 297
Lissotis, 88
Oreotrochilus, 205
Piaya, 159
Ploceus, 606
Ramphocelus, 655
Sporophila, 671
Turnix, 76
melanogenis, Phalacrocorax, 17
melanogenys
Adelomyia, 203
Anisognathus, 657
Basileuterus, 646
melanolaemus, Atlapetes, 677
melanoleuca
Atticora, 359
Corvinella, 572
Heterophasia, 509
Lalage, 372
Lamprospiza, 649
Leucosarcia, 116
Poospiza, 669
Tringa, 98
melanoleucos
Campephilus, 258
Circus, 39
Microhierax, 50
Phalacrocorax, 18
melanoleucus
Accipiter, 43
Geranoaetus, 45
Pycnonotus, 375
Seleucidis, 585
Spizastur, 48
melanolophus
Gorsachius, 21
Periparus, 525
melanonota, Pipraeidea, 658
melanonotus
Odontophorus, 62
Sakesphorus, 290
Touit, 148
Melanopareia, 311
Melanoperdix, 67
melanopezus, Automolus, 281

melanophaius, Laterallus, 79
melanophris, Thalassarche, 9
melanophrys, Manorina, 566
melanopis
Schistochlamys, 648
Theristicus, 25
melanopogon
Acrocephalus, 429
Hypocnemoides, 300
melanops
Centropus, 157
Conopophaga, 308
Elseyornis, 95
Gallinula, 85
Leucopternis, 44
Lichenostomus, 561
Myadestes, 405
Phleocryptes, 270
Phylidonyris, 566
Sporophila, 670
Trichothraupis, 654
Turdoides, 504
melanopsis
Atlapetes, 678
Monarcha, 476
melanoptera
Chloephaga, 28
Coracina, 372
Metriopelia, 118
melanopterus
Acridotheres, 598
Lybius, 237
Vanellus, 92
Melanoptila, 397
melanopygia, Telacanthura, 188
melanorhamphos, Corcorax, 580
melanorhyncha, Pelargopsis, 218
melanorhynchus
Eudynamys, 155
Thripadectes, 281
melanospilus, Ptilinopus, 126
Melanospiza, 673
melanosternon, Hamirostra, 35
melanosticta, Rhegmatorhina, 303
melanota, Pulsatrix, 170
melanothorax
Sakesphorus, 290
Stachyris, 501
Sylvia, 444
Melanotis, 399
melanotis
Ailuroedus, 585
Coccopygia, 612
Coryphaspiza, 667
Hapalopsittaca, 148
Hemispingus, 651
Manorina, 566
Nesomimus, 398
Odontophorus, 62
Oriolus, 567
Pteruthius, 506
melanotos
Calidris, 99
Sarkidiornis, 29
melanoxantha, Phainoptila, 387
melanoxanthum, Dicaeum, 546
melanozanthos, Mycerobas, 637
melanura
Anthornis, 566–567
Cercomela, 465
Chilia, 269
Pachycephala, 485–486

Polioptila, 445
Pyrrhura, 145–146
melanurus
Ceyx, 217
Cisticola, 417
Climacteris, 523
Himantopus, 90
Myophonus, 402
Myrmoborus, 300
Passer, 603
Ramphocaenus, 445
Trogon, 214
melas, Coracina, 371
melaschistos, Coracina, 372
melba
Pytilia, 614
Tachymarptis, 190
meleagrides, Agelastes, 74
Meleagris, 57
meleagris, Numida, 74
Melichneutes, 244
Melidectes, 564–565
Melidora, 221–222
Melierax, 39
Melignomon, 243
Melilestes, 556
melindae, Anthus, 365
Meliphaga, 559–560
meliphilus, Indicator, 243
Melipotes, 565
Melithreptus, 562
Melitograis, 563
melitophrys, Vireolanius, 624–625
melleri, Anas, 30
mellianus, Oriolus, 569
Mellisuga, 210
mellisugus, Chlorostilbon, 197
mellivora, Florisuga, 195
mellori, Corvus, 595
Melocichla, 428–429
meloda, Zenaida, 117
melodia, Melospiza, 685
melodus, Charadrius, 94
Melophus, 664
Melopsittacus, 137
Melopyrrha, 672
meloryphus, Euscarthmus, 323
Melospiza, 685–686
Melozone, 679–680
melpoda, Estrilda, 612
membranaceus, Malacorhynchus, 32
menachensis
Serinus, 636
Turdus, 407
menagei, Gallicolumba, 120
menbeki, Centropus, 157
menckei, Monarcha, 477
mendanae, Acrocephalus, 431
mendeni, Zoothera, 402
mendiculus, Spheniscus, 7
mendozae, Pomarea, 474
menetriesii, Myrmotherula, 296
meninting, Alcedo, 216
mennelli, Serinus, 636
menstruus, Pionus, 149
mentalis
Artamus, 580
Cracticus, 581
Dysithamnus, 293
Melocichla, 428–429
Picus, 259
Pipra, 315

mentawi, Otus, 162
Menura, 348
mercenaria, Amazona, 150
mercierii, Ptilinopus, 700
Merganetta, 29
merganser, Mergus, 33
Mergellus, 33
Mergus, 33, 699
meridae, Cistothorus, 395
meridanus, Scytalopus, 310
meridionalis
    Buteogallus, 45
    Lanius, 57
    Nestor, 134
merlini
    Arborophila, 69
    Coccyzus, 159
Meropogon, 225
Merops, 225–226
merrilli, Ptilinopus, 124
merrotsyi, Amytornis, 517
merula
    Dendrocincla, 284
    Turdus, 408–409
Merulaxis, 311
merulinus
    Cacomantis, 153
    Garrulax, 491
meruloides, Chamaeza, 304
Mesembrinibis, 25
Mesitornis, 75
mesochrysa, Euphonia, 627
mesoleuca
    Conothraupis, 649
    Elaenia, 320
mesomelas, Icterus, 695
metabates, Melierax, 39
Metabolus, 475
metallica
    Aplonis, 596
    Hedydipna, 535
Metallura, 208
metcalfii, Zosterops, 554
metopias, Orthotomus, 432
Metopidius, 88
Metopothrix, 278
Metriopelia, 118
mevesii, Lamprotornis, 600
mexicana
    Sialia, 404–405
    Tangara, 659
mexicanum, Tigrisoma, 22
mexicanus
    Carpodacus, 629
    Catharus, 406
    Catherpes, 390
    Cinclus, 388
    Falco, 53
    Gymnomystax, 698
    Himantopus, 90
    Momotus, 224
    Onychorhynchus, 331
    Quiscalus, 694
    Sclerurus, 282
    Todus, 224
    Trogon, 213
    Xenotriccus, 333
meyeni
    Tachycineta, 359
    Zosterops, 551
meyerdeschauenseei, Tangara, 661
meyeri

Chrysococcyx, 154
Epimachus, 584
Pachycephala, 484
Philemon, 563
Poicephalus, 142
meyerianus, Accipiter, 43
michahellis, Larus, 102
michleri, Pittasoma, 305
micraster, Heliangelus, 206
Micrastur, 50
Micrathene, 172
Microbates, 445
Microcerculus, 396–397
Microchera, 202
Microdynamis, 155
Microeca, 479–480
Microgoura, 700
Microhierax, 50
Microligea, 643
Micromacronus, 502
micromegas, Nesoctites, 246
Micromonacha, 234
Micronisus, 40
Microparra, 88
Micropsitta, 134
microptera
    Mirafra, 348
    Rollandia, 8
micropterus
    Agriornis, 339
    Cuculus, 153
    Scytalopus, 309
Micropygia, 78
Microrhopias, 297
microrhynchum, Ramphomicron, 208
microrhynchus, Bradornis, 447
microsoma, Oceanodroma, 14
Microstilbon, 210
micrura, Myrmia, 211
migrans, Milvus, 36
migratoria, Eophona, 637
migratorius
    Ectopistes, 700
    Turdus, 413
mikado, Syrmaticus, 72
milanjensis, Andropadus, 379
miles, Vanellus, 93
militaris
    Ara, 142
    Haematoderus, 314
    Ilicura, 316
    Sturnella, 693
milleri
    Grallaria, 306
    Polytmus, 199
    Xenops, 283
milleti, Garrulax, 489
milnei, Garrulax, 492
milo, Centropus, 157
Milvago, 49
Milvus, 36
milvus, Milvus, 36
mimikae, Meliphaga, 559
Mimizuku, 166
Mimus, 397–398
mindanensis
    Coracina, 371
    Ptilocichla, 497–498
mindorensis
    Ducula, 127
    Otus, 163
    Penelopides, 230

mineaceus, Picus, 258
miniatus
    Myioborus, 644
    Pericrocotus, 374
minima
    Batis, 467
    Leptocoma, 537
    Mellisuga, 210
minimus
    Catharus, 407
    Centrocercus, 60
    Empidonax, 334
    Hemitriccus, 328
    Lymnocryptes, 96
    Psaltriparus, 515
Minla, 507
minlosi, Xenerpestes, 278
Mino, 597
minor
    Aplonis, 597
    Artamus, 581
    Batis, 467
    Cercotrichas, 459
    Chionis, 100
    Chordeiles, 178
    Coccyzus, 159
    Dendrocopos, 250–251
    Eudyptula, 7
    Fregata, 18
    Furnarius, 270
    Hemitriccus, 327
    Indicator, 244
    Lanius, 571
    Lybius, 237
    Mecocerculus, 322
    Myrmotherula, 295
    Nothura, 6
    Pachyramphus, 347
    Paradisaea, 585
    Phoenicopterus, 26
    Platalea, 25
    Pyrenestes, 613
    Rhinopomastus, 229
    Scolopax, 96
    Zosterops, 553
mintoni, Micrastur, 50
minula, Sylvia, 443
minulla, Batis, 468
minullus
    Accipiter, 42
    Cinnyris, 538
minuta
    Calidris, 99
    Coccycua, 159
    Columbina, 117
    Euphonia, 627
    Sporophila, 671
minutilla, Calidris, 99
minutillus, Chrysococcyx, 154
minutissimum, Glaucidium, 171
minutissimus, Picumnus, 245
minutus
    Anous, 103
    Anthoscopus, 534
    Corvus, 594
    Ixobrychus, 22
    Larus, 103
    Numenius, 97
    Tchagra, 573
    Xenops, 283
    Zosterops, 554
Mionectes, 325–326

mira, Scolopax, 95
mirabilis
  Cyanolyca, 589
  Eriocnemis, 207
  Loddigesia, 209
Mirafra, 348–350
miranda, Neositta, 522
mirandae, Hemitriccus, 328
mirandollei, Micrastur, 50
mirandus, Basilornis, 597
mirificus, Lewinia, 81
mirus, Otus, 163
mississippiensis, Ictinia, 36
mitchellii
  Calliphlox, 210
  Phegornis, 95
mitrata, Aratinga, 143
mitratus, Garrulax, 491
Mitrephanes, 335–336
Mitrospingus, 652
Mitu, 56
mitu, Mitu, 56
mixta, Batis, 467
mixtus, Picoides, 252
mlokosiewiczi, Tetrao, 59
Mniotilta, 642
moabiticus, Passer, 602
mocinno, Pharomachrus, 214
modesta
  Arachnothera, 544
  Asthenes, 276
  Galerida, 356
  Gerygone, 522
  Neochmia, 616
  Pachycephala, 484
  Progne, 359
  Psittacella, 137
  Turacoena, 115
modestus
  Charadrius, 95
  Dicrurus, 578
  Gallirallus, 699
  Larus, 100
  Ramsayornis, 566
  Sublegatus, 326
  Sylviparus, 529
  Thryothorus, 392
  Vireo, 621
  Zosterops, 551
modularis, Prunella, 400
Modulatrix, 493
moesta
  Amaurospiza, 672
  Lalage, 372
  Oenanthe, 464
  Synallaxis, 273
moheliensis, Otus, 164
Moho, 565, 700
Mohoua, 483
molinae, Pyrrhura, 145
molitor, Batis, 467
molleri, Prinia, 421
mollis, Pterodroma, 11
mollissima
  Chamaeza, 304
  Somateria, 32
  Zoothera, 403
Molothrus, 694–695
molucca
  Lonchura, 618
  Threskiornis, 24
moluccana, Amaurornis, 83

moluccensis
  Cacatua, 131
  Dendrocopos, 250
  Falco, 51
  Philemon, 563
  Pitta, 267
mombassica, Campethera, 248
momota, Momotus, 224
momotula, Hylomanes, 224
Momotus, 224
monacha
  Grus, 77
  Oenanthe, 463
  Oriolus, 568
  Pachycephala, 486
  Ptilinopus, 126
Monachella, 479
monachus
  Actenoides, 222
  Aegypius, 37
  Artamus, 580
  Centropus, 158
  Myiopsitta, 146
  Necrosyrtes, 37
Monarcha, 475–477
Monasa, 234
monasa, Porzana, 699
mondetoura, Claravis, 118
monedula, Corvus, 593
moneduloides, Corvus, 593
mongolica, Melanocorypha, 353
mongolicus, Bucanetes, 638
mongolus, Charadrius, 94–95
monguilloti, Carduelis, 631
Monias, 75
monileger
  Ficedula, 451
  Formicarius, 303
  Garrulax, 489
moniliger, Batrachostomus, 177
monocerata, Cerorhinca, 108
monogrammicus, Kaupifalco, 39
monorhis, Oceanodroma, 14
monorthonyx, Anurophasis, 68
montagnii, Penelope, 55
montana
  Brachypteryx, 413
  Buthraupis, 657
  Coracina, 372
  Geotrygon, 120
  Lonchura, 620
  Meliphaga, 559
  Paroreomyza, 638
  Poecile, 524
  Sheppardia, 455
  Xenoligea, 647
montanella, Prunella, 400
montani, Anthracoceros, 230
montanus
  Agriornis, 339
  Andropadus, 378
  Cercococcyx, 153
  Charadrius, 95
  Dicrurus, 579
  Oreoscoptes, 398
  Passer, 603
  Peltops, 581
  Pomatorhinus, 497
  Prioniturus, 138
  Zosterops, 552
monteiri
  Clytospiza, 614

Malaconotus, 575
  Tockus, 229
montezuma, Gymnostinops, 698
montezumae, Cyrtonyx, 63
Monticola, 401
monticola
  Lichmera, 557
  Lonchura, 620
  Megalaima, 239
  Oenanthe, 463
  Troglodytes, 395
  Zoothera, 404
monticolum, Dicaeum, 548
monticolus, Parus, 527
Montifringilla, 604
montifringilla, Fringilla, 626
montis, Seicercus, 440
montium, Paramythia, 545
montivagus, Aeronautes, 189
mooreorum, Glaucidium, 171
moquini, Haematopus, 89
moreaui
  Cinnyris, 539
  Orthotomus, 432
moreirae, Oreophylax, 271
morenoi, Metriopelia, 118
morinellus, Charadrius, 95
morio
  Coracina, 371
  Cyanocorax, 588
  Onychognathus, 601
Morococcyx, 160
morphnoides, Aquila, 48
Morphnus, 47
morphoeus, Monasa, 234
morrisonia, Alcippe, 508
morrisoniana, Actinodura, 507
morrisonianus, Garrulax, 492
mortierii, Gallinula, 85
Morus, 15–16
moschata, Cairina, 29
mosquera, Eriocnemis, 207
mosquitus, Chrysolampis, 195
Motacilla, 367–368
motacilla, Seiurus, 642
motacilloides, Herpsilochmus, 296
motitensis, Passer, 602
motmot, Ortalis, 55
mouki, Gerygone, 521
mouroniensis, Zosterops, 550
moussieri, Phoenicurus, 461
mozambicus, Serinus, 635
muelleri
  Cranioleuca, 275
  Heleia, 555
  Lewinia, 82
  Merops, 225
muelleriana, Monachella, 479
mufumbiri, Laniarius, 574
mugimaki, Ficedula, 450
mullerii, Ducula, 129
Mulleripicus, 261
mulsant, Chaetocercus, 211
multicolor
  Petroica, 480
  Saltatricula, 667
  Telophorus, 575
  Todus, 223
multipunctata, Tyto, 160
multistriata, Charmosyna, 133
multistriatus, Thamnophilus, 291
multostriata, Myrmotherula, 294

munda, Serpophaga, 322
mundus, Monarcha, 476
munroi
    Dysmorodrepanis, 701
    Hemignathus, 638
mupinensis, Turdus, 410
murallae, Sporophila, 670
muraria, Tichodroma, 531
murina
    Acanthiza, 519
    Crateroscelis, 517
    Notiochelidon, 359
    Phaeomyias, 322
murinus
    Agriornis, 339
    Thamnophilus, 292
murphyi
    Progne, 359
    Zosterops, 554
Muscicapa, 448–449
Muscicapella, 454
muscicapinus, Hylophilus, 624
Muscigralla, 340
Muscipipra, 340
Muscisaxicola, 339–340
musculus, Anthoscopus, 533
musica, Euphonia, 627
musicus, Bias, 466
Musophaga, 152
musschenbroekii, Neopsittacus, 134
mustelina, Hylocichla, 407
mustelinus, Certhiaxis, 275
muta, Lagopus, 58–59
mutata, Terpsiphone, 473
muticus, Pavo, 74
muttui, Muscicapa, 449
Myadestes, 405, 700
myadestinus, Myadestes, 405
Mycerobas, 637
Mycteria, 23
mycteria, Jabiru, 24
Myiagra, 477–479
Myiarchus, 343, 344
Myiobius, 332–333
Myioborus, 644
Myiodynastes, 341
Myiopagis, 319
Myioparus, 450
Myiophobus, 332
Myiopsitta, 146
Myiornis, 326
Myiotheretes, 338
Myiotriccus, 326
Myiozetetes, 340–341
Myophonus, 402
myoptilus, Schoutedenapus,
    187
Myornis, 311
myotherinus, Myrmoborus, 299
myristicivora, Ducula, 128
Myrmeciza, 301–302
Myrmecocichla, 465–466
Myrmia, 211
Myrmoborus, 299–300
Myrmochanes, 300
Myrmorchilus, 296
Myrmornis, 302
Myrmothera, 305
Myrmotherula, 294–296
Myrtis, 211
mysolensis, Aplonis, 597
mysorensis, Zosterops, 553

mystacalis
    Aethopyga, 543
    Aimophila, 681
    Cyanocorax, 589
    Diglossa, 674
    Eurostopodus, 179
    Malacoptila, 234
    Thryothorus, 390
mystacea
    Aplonis, 596
    Geotrygon, 120
    Hemiprocne, 191
    Sylvia, 444
mystaceus, Platyrinchus, 331
mystacophanos, Megalaima, 238
Mystacornis, 511
mysticalis, Rhabdornis, 532
Myza, 565
Myzomela, 557–559
Myzornis, 510

nabouroup, Onychognathus,
    601
nacunda, Podager, 179
naevia
    Locustella, 429
    Sclateria, 300
    Tapera, 160
naevioides, Hylophylax, 302
naevius
    Hylophylax, 302
    Ixoreus, 413
    Ramphodon, 191
naevosa, Stictonetta, 28
nagaensis, Sitta, 530
nahani, Francolinus, 66
namaqua, Pterocles, 108
namaquus, Dendropicos, 249
Namibornis, 459
nana
    Acanthiza, 520
    Aratinga, 144
    Cisticola, 417
    Cyanolyca, 589
    Grallaricula, 308
    Lonchura, 618
    Myiagra, 478
    Sylvia, 443
Nandayus, 144
Nannopsittaca, 147
nanum, Glaucidium, 171
nanus
    Accipiter, 42
    Ptilinopus, 126
    Spizaetus, 49
    Taoniscus, 6
    Vireo, 622
napensis
    Megascops, 166
    Stigmatura, 323
napoleonis, Polyplectron, 73
Napothera, 498
narcissina, Ficedula, 450
narcondami, Aceros, 231
narina, Apaloderma, 212
Nasica, 285
nasicus, Corvus, 594
nasutus, Tockus, 229
natalensis
    Caprimulgus, 182
    Chloropeta, 432
    Cisticola, 417

    Cossypha, 458
    Francolinus, 66
natalis
    Ninox, 175
    Zosterops, 552
nationi, Atlapetes, 677
nativitatis, Puffinus, 12
nattereri
    Anthus, 365
    Hylopezus, 305
    Lepidothrix, 315
    Phaethornis, 193
    Selenidera, 241
nattererii, Cotinga, 313
naumanni
    Falco, 50
    Turdus, 410
naungmungensis, Jabouilleia, 497
navai, Hylorchilus, 390
ndussumensis, Criniger, 382
Neafrapus, 188
nebouxii, Sula, 16
nebularia, Tringa, 98
nebulosa, Strix, 169
nebulosus
    Picumnus, 246
    Rhipidura, 470
Necropsar, 701
Necrosyrtes, 37
Nectarinia, 538
nectarinioides, Cinnyris, 539
neergaardi, Cinnyris, 538
neglecta
    Pterodroma, 10
    Sturnella, 693
neglectus
    Anthreptes, 535
    Phalacrocorax, 17
    Phylloscopus, 437
negreti, Henicorhina, 396
nehrkorni
    Dicaeum, 547
    Zosterops, 553
nelicourvi, Ploceus, 609
nelsoni
    Ammodramus, 684
    Geothlypis, 643
    Vireo, 622
nematura, Lochmias, 283
nemoricola
    Gallinago, 96
    Leucosticte, 628
Nemosia, 652
nenday, Nandayus, 144
nengeta, Fluvicola, 338
Neochelidon, 360
Neochen, 28
Neochmia, 616
Neocichla, 601
Neocossyphus, 401
Neocrex, 84
Neoctantes, 293
Neodrepanis, 265
Neolalage, 475
Neolestes, 385
Neomixis, 499
Neomorphus, 160
Neopelma, 317
Neophema, 136–137
Neophron, 37
Neopipo, 333
Neopsittacus, 134

Neorhopias, 297
Neositta, 522–523
Neospiza, 628
Neothraupis, 648
Neotis, 87
neoxena, Hirundo, 360
neoxenus, Euptilotis, 214
Neoxolmis, 338
Nephelornis, 664
nereis
    Garrodia, 13
    Sternula, 104
Nesasio, 176
Nesillas, 428
nesiotis
    Anas, 30
    Gallinula, 85
Nesocharis, 611–612
Nesocichla, 404
Nesoclopeus, 80, 699
Nesoctites, 246
Nesoenas, 113
Nesofregetta, 14
Nesomimus, 398
Nesopsar, 693
Nesospingus, 649
Nesospiza, 668
Nesotriccus, 322
Nestor, 134, 700
Netta, 32
Nettapus, 29
neumanni
    Hemitesia, 435
    Onychognathus, 601
neumayer, Sitta, 530–531
nevermanni, Lonchura, 619
newtoni
    Acrocephalus, 431
    Coracina, 370
    Falco, 51
    Lanius, 572
Newtonia, 434–435
newtoniana, Prionodura, 586
newtonii, Anabathmis, 536
ngoclinhensis, Garrulax, 492
niansae, Apus, 190
nicaraguensis, Quiscalus, 694
Nicator, 382
nicefori, Thryothorus, 392
nicobarica, Caloenas, 120
nicobariensis, Megapodius, 54
nicolli, Ploceus, 608
nidipendulus, Hemitriccus, 328
nieuwenhuisii, Pycnonotus, 377
niger
    Agelastes, 74
    Bubalornis, 604
    Capito, 240
    Certhionyx, 559
    Chlidonias, 105
    Copsychus, 460
    Cypseloides, 184
    Melaniparus, 526
    Melanoperdix, 67
    Neoctantes, 293
    Pachyramphus, 347
    Phalacrocorax, 18
    Phylidonyris, 565
    Quiscalus, 694
    Rynchops, 106
    Threnetes, 192
nigeriae, Vidua, 621

nigerrima, Lonchura, 619
nigerrimus
    Knipolegus, 337
    Nesopsar, 693
    Ploceus, 607
nigra
    Astrapia, 583
    Ciconia, 23
    Coracopsis, 141
    Lalage, 372
    Melanitta, 33
    Melanocharis, 544
    Melopyrrha, 672
    Myrmecocichla, 465
    Penelopina, 56
    Pomarea, 474
nigrescens
    Caprimulgus, 181
    Cercomacra, 298
    Contopus, 335
    Dendroica, 641
    Pitohui, 488
    Turdus, 410
nigricans
    Cercomacra, 299
    Pardirallus, 84
    Petrochelidon, 362
    Pinarocorys, 351
    Pycnonotus, 377
    Sayornis, 336
    Serpophaga, 322
nigricapillus
    Formicarius, 304
    Lioptilus, 509
    Thryothorus, 391
nigricauda, Myrmeciza, 301
nigricephala, Spindalis, 656
nigriceps
    Andropadus, 379
    Apalis, 422
    Ardeotis, 87
    Eremopterix, 352
    Orthotomus, 433
    Polioptila, 445
    Saltator, 689
    Serinus, 634
    Stachyris, 501
    Thamnophilus, 291
    Todirostrum, 330
    Turdus, 411
    Veniliornis, 253
nigricincta, Aphelocephala, 522
nigricollis
    Anthracothorax, 195
    Busarellus, 45
    Gracupica, 599
    Grus, 77
    Phoenicircus, 314
    Ploceus, 606
    Podiceps, 8
    Sporophila, 670
    Stachyris, 501
    Turnix, 76
nigrifrons
    Monasa, 234
    Phylloscartes, 324
    Telophorus, 575
nigrigenis, Agapornis, 141
nigrilore, Dicaeum, 546
nigriloris
    Cisticola, 415
    Estrilda, 613

nigrimenta, Yuhina, 510
nigrimentus, Ploceus, 606
nigripectus, Machaerirhynchus, 479
nigripennis
    Gallinago, 96
    Oriolus, 569
    Pterodroma, 11
nigripes
    Dacnis, 662
    Phoebastria, 9
nigrirostris
    Andigena, 242
    Cyclarhis, 625
    Macropygia, 114
    Patagioenas, 112
nigriscapularis, Caprimulgus, 182
Nigrita, 611
nigrita
    Hirundo, 360
    Myzomela, 557–558
nigriventris, Eupherusa, 198
nigrivestis, Eriocnemis, 206
nigrobrunnea, Tyto, 161
nigrocapillus
    Nothocercus, 3
    Phyllomyias, 319
nigrocapitata, Stachyris, 500
nigrocincta, Tangara, 661
nigrocinereus, Thamnophilus, 291
nigrocinnamomea, Rhipidura, 468
nigrocristatus
    Anairetes, 322
    Basileuterus, 645
nigrocyaneus, Todiramphus, 219
nigrofumosus, Cinclodes, 269
nigrogularis
    Clytorhynchus, 475
    Colinus, 61
    Cracticus, 582
    Icterus, 695
    Phalacrocorax, 17
    Psophodes, 513
    Ramphocelus, 655
nigrolineata, Ciccaba, 170
nigrolutea, Aegithina, 387
nigromaculata, Phlegopsis, 303
nigromitrata, Elminia, 472
nigropectus
    Biatas, 291
    Eulacestoma, 488
nigropunctatus, Picumnus, 245
nigrorufa
    Crateroscelis, 517
    Ficedula, 452
    Hirundo, 361
    Poospiza, 669
    Sporophila, 671
nigrorufus, Centropus, 157
nigrorum
    Stachyris, 500
    Zosterops, 552
nigroventris, Euplectes, 610
nigroviridis, Tangara, 662
Nilaus, 572
nilghiriensis, Anthus, 364
nilotica, Gelochelidon, 104–105
Niltava, 452–453
Ninox, 173–175
nipalense, Delichon, 361
nipalensis
    Aceros, 230
    Actinodura, 506

Aethopyga, 542
Alcippe, 509
Apus, 191
Aquila, 47
Bubo, 167
Carpodacus, 629
Certhia, 532
Cutia, 505
Paradoxornis, 512
Pitta, 265
Pyrrhula, 636
Spizaetus, 48
Turdoides, 503
nippon, Nipponia, 24
Nipponia, 24
nisicolor, Cuculus, 153
nisoria, Sylvia, 443
nisus, Accipiter, 42–43
nitens
    Lamprotornis, 599
    Malimbus, 605
    Phainopepla, 387
    Psalidoprocne, 357
    Trochocercus, 472
nitidissima
    Alectroenas, 700
    Chlorochrysa, 659
nitidula
    Lagonosticta, 614
    Mandingoa, 612
nitidum, Dicaeum, 548
nitidus
    Buteo, 45
    Carpodectes, 313
    Cyanerpes, 663
nivalis
    Montifringilla, 604
    Phalacrocorax, 17
    Plectrophenax, 687
nivea, Pagodroma, 10
niveicapilla, Cossypha, 458
niveigularis, Tyrannus, 342
niveogularis, Aegithalos, 515
niveoguttatus, Hypargos, 614
nivosa, Campethera, 249
njombe, Cisticola, 416
noanamae, Bucco, 233
nobilis
    Chamaeza, 304
    Diopsittaca, 143
    Francolinus, 67
    Gallinago, 96
    Moho, 700
    Oreonympha, 208
    Otidiphaps, 121
noctis, Loxigilla, 673
noctitherus, Caprimulgus, 180
noctivagus, Crypturellus, 4
noctua, Athene, 173
noevius, Coracias, 227
noguchii, Sapheopipo, 260
Nomonyx, 33
Nonnula, 234
nonnula, Estrilda, 613
nordmanni, Glareola, 92
Northiella, 136
notabilis
    Anisognathus, 658
    Lichmera, 557
    Nestor, 134
    Phylidonyris, 566

notata
    Campethera, 248
    Carduelis, 632
    Meliphaga, 560
notatus
    Bleda, 382
    Chlorostilbon, 197
    Cinnyris, 541
    Coturnicops, 78
Notharchus, 232–233
Nothocercus, 3
Nothocrax, 56
Nothoprocta, 5
Nothura, 5–6
Notiochelidon, 359
Notiomystis, 562
notosticta, Aimophila, 682
nouhuysi
    Melidectes, 564
    Sericornis, 518
novacapitalis, Scytalopus, 309
novaeguineae
    Dacelo, 218
    Harpyopsis, 47
    Mearnsia, 187
    Orthonyx, 513
    Toxorhamphus, 544
    Zosterops, 553
novaehollandiae
    Accipiter, 41
    Cereopsis, 28
    Coracina, 368–369
    Dromaius, 2
    Egretta, 20
    Larus, 102
    Menura, 348
    Phylidonyris, 565
    Recurvirostra, 90
    Scythrops, 155
    Tachybaptus, 8
    Tyto, 160
novaeseelandiae
    Anthus, 363
    Aythya, 32
    Falco, 52
    Hemiphaga, 129
    Mohoua, 483
    Ninox, 174
    Prosthemadera, 567
    Thinornis, 95
novaesi, Philydor, 280
novaezelandiae
    Coturnix, 699
    Cyanoramphus, 135
    Himantopus, 90
    Ixobrychus, 699
noveboracensis
    Coturnicops, 78
    Seiurus, 642
nuba, Neotis, 87
nubica, Campethera, 248
nubicoides, Merops, 226
nubicola, Glaucidium, 171
nubicus
    Caprimulgus, 181
    Lanius, 572
    Merops, 226
nuchalis
    Campylorhynchus, 389
    Chlamydera, 586
    Garrulax, 490
    Glareola, 92

Grallaria, 307
Parus, 527
Sphyrapicus, 248
Nucifraga, 592–593
nudiceps, Gymnocichla, 300
nudicollis, Procnias, 314
nudigenis, Turdus, 412
nudigula, Pachycephala, 486
nudipes, Megascops, 166
nuditarsus, Aerodramus, 186
nugator, Myiarchus, 344
Numenius, 97
Numida, 74
nuna, Lesbia, 207
nuttalli, Pica, 592
nuttallii
    Phalaenoptilus, 179
    Picoides, 252
nuttingi
    Myiarchus, 344
    Oryzoborus, 671
nyassae, Anthus, 363
Nyctanassa, 21
nycthemera, Lophura, 71
Nyctibius, 177–178
Nycticorax, 21
nycticorax, Nycticorax, 21
Nyctidromus, 179
Nyctiphrynus, 179–180
Nyctiprogne, 179
Nyctyornis, 225
nympha
    Pitta, 267
    Tanysiptera, 223
Nymphicus, 131
nyroca, Aythya, 32
Nystalus, 233

oahensis, Myadestes, 700
oatesi, Pitta, 265
obbiensis, Spizocorys, 355
oberholseri, Empidonax, 334
oberi
    Icterus, 697
    Myiarchus, 344
oberlaenderi, Zoothera, 403
obscura
    Cyanomitra, 536
    Elaenia, 320
    Luscinia, 456
    Myzomela, 557
    Penelope, 56
    Psalidoprocne, 358
obscurior, Sublegatus, 326
obscurus
    Charadrius, 93
    Dendragapus, 57
    Hemignathus, 701
    Lichenostomus, 560
    Myadestes, 405
    Tetraophasis, 63
    Tiaris, 672
    Turdus, 410
obsoleta
    Cranioleuca, 275
    Rhodospiza, 638
obsoletum, Camptostoma, 321
obsoletus
    Crypturellus, 4
    Dendropicos, 250
    Hemitriccus, 327

obsoletus (*continued*)
  Salpinctes, 390
  Turdus, 412
  Xiphorhynchus, 287
obtusa, Vidua, 620
ocai, Pipilo, 680
occidentalis
  Aechmophorus, 9
  Catharus, 406
  Dendroica, 641
  Dysithamnus, 294
  Empidonax, 334
  Geopsittacus, 137
  Larus, 101
  Leucopternis, 44
  Myadestes, 405
  Onychorhynchus, 331
  Pelecanus, 15
  Psophodes, 513
  Strix, 169
occipitalis
  Chlorophonia, 628
  Dendrocitta, 592
  Lophaetus, 48
  Phylloscopus, 438
  Podiceps, 9
  Ptilinopus, 124
  Trigonoceps, 38
  Yuhina, 510
occulta
  Batis, 468
  Pterodroma, 11
oceanica
  Ducula, 128
  Myiagra, 477
oceanicus, Oceanites, 13
Oceanites, 13, 699
Oceanodroma, 14, 699
ocellata
  Leipoa, 53
  Meleagris, 57
  Rheinardia, 74
  Strix, 168
ocellatum, Toxostoma, 398
ocellatus
  Cyrtonyx, 63
  Garrulax, 490
  Nyctiphrynus, 180
  Podargus, 177
  Turnix, 75
  Xiphorhynchus, 287
Ochetorhynchus, 269
ochotensis, Locustella, 429
ochracea
  Ninox, 174
  Sasia, 246
ochraceiceps
  Hylophilus, 624
  Pomatorhinus, 497
ochraceifrons, Grallaricula, 308
ochraceiventris
  Leptotila, 119
  Myiophobus, 332
ochraceus
  Alophoixus, 382
  Contopus, 335
  Troglodytes, 395
ochrocephala
  Amazona, 150
  Mohoua, 483
ochrogaster, Penelope, 56

ochrolaemus, Automolus, 281
ochroleucus, Hylopezus, 305
ochromalus, Eurylaimus, 265
ochromelas, Melidectes, 564
ochropectus, Francolinus, 67
ochropus, Tringa, 97
ochropyga, Drymophila, 298
ochruros, Phoenicurus, 460
Ochthoeca, 337–338
Ochthornis, 338
ocistus, Aerodramus, 187
ocreata, Fraseria, 447
Ocreatus, 207
octosetaceus, Mergus, 33
ocularis
  Glareola, 92
  Ploceus, 606
  Prunella, 400
oculata, Stagonopleura, 616
oculeus
  Caloperdix, 70
  Canirallus, 78
Oculocincta, 555
Ocyalus, 698
Ocyceros, 229–230
ocypetes, Chaetura, 189
odiosa, Ninox, 175
odomae, Metallura, 208
Odontophorus, 62–63
Odontorchilus, 389
Odontospiza, 617
oedicnemus, Burhinus, 90
oemodium, Conostoma, 511
Oena, 115
Oenanthe, 463–464
oenanthe, Oenanthe, 463
oenanthoides, Ochthoeca, 338
oenas, Columba, 110
oenochlamys, Sitta, 531
oenone, Chrysuronia, 199
oenops, Patagioenas, 112
oglei, Stachyris, 501
Ognorhynchus, 143
okinawae, Gallirallus, 80
olallai, Myiopagis, 319
olax, Treron, 122
oleagineus
  Mionectes, 325–326
  Mitrospingus, 652
  Zosterops, 555
olivacea
  Amaurornis, 83
  Bostrychia, 24
  Carduelis, 632
  Certhidea, 688
  Chlorothraupis, 652
  Cyanomitra, 536
  Gerygone, 521
  Iole, 383
  Pachycephala, 483
  Piranga, 655
  Stactolaema, 236
olivaceiceps, Ploceus, 608
olivaceofuscus, Turdus, 408
olivaceum
  Chalcostigma, 209
  Oncostoma, 327
olivaceus
  Criniger, 382
  Geocolaptes, 249
  Hylophilus, 624

Hypsipetes, 385
Linurgus, 628
Mionectes, 325
Mitrephanes, 336
Phylloscopus, 440
Picumnus, 246
Prionochilus, 545
Psophodes, 513
Rhinomyias, 448
Rhynchocyclus, 330
Telophorus, 575
Tiaris, 672
Turdus, 407–408
Vireo, 623
Zosterops, 550
olivaresi, Chlorostilbon, 197
olivascens
  Cinnycerthia, 390
  Muscicapa, 449
  Sicalis, 675
olivater, Turdus, 411
olivea, Tesia, 425
olivetorum, Hippolais, 432
oliviae, Columba, 110
olivieri, Amaurornis, 83
olivii, Turnix, 76
olivinus, Cercococcyx, 153
olor, Cygnus, 26
olrogi, Cinclodes, 269
omeiensis, Liocichla, 492
omissa, Foudia, 609
omissus, Cyornis, 454
Oncostoma, 327
oneilli, Nephelornis, 664
onocrotalus, Pelecanus, 15
onslowi, Phalacrocorax, 17
Onychognathus, 601
Onychoprion, 104
Onychorhynchus, 331
oorti, Megalaima, 239
opaca
  Aplonis, 597
  Hippolais, 432
Ophrysia, 699
ophthalmica, Cacatua, 131
ophthalmicus
  Chlorospingus, 649
  Pogonotriccus, 325
opistherythra, Rhipidura, 471
Opisthocomus, 74
opisthomelas, Puffinus, 13
Opisthoprora, 209
Opororniris, 642
optatus, Cuculus, 153
oratrix, Amazona, 150
orbignyianus, Thinocorus, 100
orbitalis, Pogonotriccus, 325
orbitatus, Hemitriccus, 327
orbygnesius, Bolborhynchus, 146
orcesi, Pyrrhura, 146
Orchesticus, 648
ordii, Notharchus, 233
oreas
  Lessonia, 336
  Picathartes, 488
orenocensis
  Knipolegus, 337
  Saltator, 689
oreobates, Merops, 225
Oreocharis, 545
Oreoica, 483

Oreomanes, 648
Oreomystis, 638
Oreonympha, 208
Oreophasis, 56
oreophilus, Buteo, 46
Oreopholus, 95
Oreophylax, 271
oreopsar, Agelaioides, 698
Oreopsittacus, 133–134
Oreornis, 562
Oreortyx, 60
Oreoscoptes, 398
Oreoscopus, 517
Oreostruthus, 616
Oreothraupis, 664
Oreotrochilus, 204–205
oreskios, Harpactes, 215
orientalis
    Acrocephalus, 430
    Aerodramus, 186
    Anthreptes, 535
    Arborophila, 69
    Batis, 467
    Eurystomus, 227
    Meliphaga, 559
    Merops, 226
    Phylloscopus, 437
    Pterocles, 109
    Streptopelia, 113
    Vidua, 620
Origma, 517
orinus, Acrocephalus, 430
Oriolia, 577
oriolina, Campephaga, 374
Oriolus, 567–569
oriolus, Oriolus, 568
oritis, Cyanomitra, 536
Oriturus, 682
orix, Euplectes, 610
ornata
    Myrmotherula, 295
    Nothoprocta, 5
    Poospiza, 669
    Thlypopsis, 651
    Thraupis, 656
    Urocissa, 591
ornatus
    Calcarius, 687
    Cephalopterus, 314
    Lamprotornis, 600
    Lichenostomus, 561
    Lophornis, 196
    Merops, 226
    Myioborus, 644
    Myiotriccus, 326
    Ptilinopus, 124
    Spizaetus, 49
    Trichoglossus, 131
Ornithion, 321
Oroaetus, 49
orostruthus, Arcanator, 493
orpheus, Pachycephala, 484
orrhophaeus, Harpactes, 215
orru, Corvus, 595
Ortalis, 54–55
Orthogonys, 653
Orthonyx, 513
orthonyx, Acropternis, 309
Orthopsittaca, 143
Orthorhyncus, 196
Orthotomus, 432–433

ortizi, Incaspiza, 669
ortoni, Penelope, 55
Ortygospiza, 615–616
Ortyxelos, 76
oryzivora, Padda, 620
oryzivorus
    Dolichonyx, 691
    Molothrus, 695
Oryzoborus, 671–672
osburni, Vireo, 622
oscillans, Rhinomyias, 447
oscitans, Anastomus, 23
osculans, Chrysococcyx, 154
osea, Cinnyris, 540
oseryi, Psarocolius, 698
osgoodi, Tinamus, 3
ossifragus, Corvus, 594
ostenta, Coracina, 372
ostralegus, Haematopus, 89
ostrinus, Pyrenestes, 613
Otidiphaps, 121
Otis, 87
ottonis, Asthenes, 276
Otus, 162–164
otus, Asio, 175
oustaleti
    Cinclodes, 269
    Cinnyris, 540
    Phylloscartes, 325
ovampensis, Accipiter, 42
owenii, Apteryx, 2
owstoni, Gallirallus, 80
oxycerca, Cercibis, 25
Oxylabes, 511
Oxypogon, 208
Oxyruncus, 311
Oxyura, 33
oxyurus, Treron, 123

pabsti, Cinclodes, 269
pachecoi, Scytalopus, 309
Pachycare, 483
Pachycephala, 483–486
pachycephaloides, Clytorhynchus, 475
Pachycephalopsis, 482
Pachycoccyx, 152
Pachyphantes, 609
Pachyptila, 11–12
Pachyramphus, 346–347
pachyrhyncha, Rhynchopsitta, 143
pachyrhynchus, Eudyptes, 7
pacifica
    Ardea, 19
    Drepanis, 701
    Ducula, 127
    Gallinula, 699
    Gavia, 7
    Myrmotherula, 294
pacificus
    Apus, 191
    Gallirallus, 699
    Larus, 100
    Puffinus, 12
    Rhynchocyclus, 330
Padda, 620
paena, Cercotrichas, 459
pagodarum, Temenuchus, 599
Pagodroma, 10
Pagophila, 103
palauensis, Megazosterops, 556
palawanense, Malacopteron, 494

palawanensis
    Aerodramus, 186
    Chloropsis, 386
    Ixos, 383
pallasi, Emberiza, 666
pallasii, Cinclus, 388
pallatangae, Elaenia, 321
pallens, Vireo, 621
pallescens, Neopelma, 317
palliata, Falculea, 577
palliatus
    Cinclodes, 270
    Garrulax, 488
    Haematopus, 89
    Thamnophilus, 291
pallida
    Cranioleuca, 275
    Hippolais, 432
    Leptotila, 119
    Lonchura, 619
    Spizella, 682
pallidiceps
    Atlapetes, 678
    Columba, 111
pallidicinctus, Tympanuchus, 60
pallidigaster, Hedydipna, 535
pallidinucha, Atlapetes, 676
pallidipes, Cettia, 425
pallidirostris, Tockus, 229
pallidiventris, Anthus, 364
pallidus
    Alophoixus, 382
    Apus, 190
    Bradornis, 446
    Camarhynchus, 688
    Charadrius, 94
    Contopus, 335
    Cuculus, 153
    Turdus, 410
    Zosterops, 550
pallipes, Cyornis, 453
palliseri, Bradypterus, 428
palmarum
    Charmosyna, 133
    Corvus, 594
    Dendroica, 641
    Glaucidium, 171
    Phaenicophilus, 652
    Thraupis, 657
palmeri
    Myadestes, 405
    Porzana, 699
    Rhodacanthis, 701
    Tangara, 659
Palmeria, 639
palpebralis, Schizoeaca, 271
palpebrata, Phoebetria, 10
palpebrosa, Gerygone, 521
palpebrosus, Zosterops, 551
paludicola
    Acrocephalus, 429
    Estrilda, 612
    Riparia, 358
Paludipasser, 616
palumboides, Columba, 111
palumbus, Columba, 110
palustre, Pellorneum, 493
palustris
    Acrocephalus, 430
    Cistothorus, 395–396
    Megalurus, 442

palustris (*continued*)
  Microligea, 643
  Poecile, 523–524
  Quiscalus, 701
  Sporophila, 671
pamela, Aglaeactis, 204
pammelaena, Myzomela, 559
pammelaina, Melaenornis, 447
panamensis
  Malacoptila, 234
  Myiarchus, 344
  Scytalopus, 310
panayensis
  Aplonis, 596
  Eumyias, 452
panderi, Podoces, 592
Pandion, 34
panini, Penelopides, 230
Panterpe, 197
Panurus, 511
panychlora, Nannopsittaca, 147
Panyptila, 189
papa, Sarcoramphus, 34
papillosa, Pseudibis, 24
papou, Charmosyna, 133
papua, Pygoscelis, 6
papuana
  Erythrura, 617
  Microeca, 480
papuensis
  Aerodramus, 187
  Archboldia, 586
  Chaetorhynchus, 577
  Coracina, 369–370
  Eurostopodus, 179
  Podargus, 177
  Sericornis, 518–519
Parabuteo, 45
Paradigalla, 583
Paradisaea, 585
paradisaea
  Sterna, 105
  Vidua, 620
paradiseus
  Anthropoides, 76
  Dicrurus, 579–580
  Ptiloris, 584
paradisi, Terpsiphone, 474
paradoxa
  Anthochaera, 567
  Eugralla, 309
Paradoxornis, 511–513
paradoxus
  Paradoxornis, 511
  Syrrhaptes, 108
paraensis, Automolus, 281
paraguaiae, Gallinago, 96
Paramythia, 545
parasiticus, Stercorarius, 107
Pardaliparus, 525–526
Pardalotus, 549
pardalotus, Xiphorhynchus, 287
Pardirallus, 84
parellina, Cyanocompsa, 691
parens, Cettia, 426
pareola, Chiroxiphia, 316
pariae, Myioborus, 644
parina, Xenodacnis, 663
Parisoma, 444–445
parisorum, Icterus, 697
parkeri
  Cercomacra, 299

Glaucidium, 171
  Herpsilochmus, 296
  Phylloscartes, 324
  Scytalopus, 310
Parkerthraustes, 690
parkinsoni, Procellaria, 12
Parmoptila, 611
parnaguae, Megaxenops, 283
Paroaria, 676
parodii, Hemispingus, 650
Parophasma, 509
Paroreomyza, 638, 701
Parotia, 583–584
Parula, 639–640
parulus, Anairetes, 322
Parus, 527–528
parva
  Ficedula, 450
  Microdynamis, 155
  Porzana, 83
parvirostris
  Chlorospingus, 650
  Crypturellus, 5
  Elaenia, 320
  Hypsipetes, 384
  Colorhamphus, 338
  Scytalopus, 310
  Tetrao, 59
parvula, Coracina, 370
parvulus
  Anthoscopus, 533
  Camarhynchus, 688
  Caprimulgus, 181
  Nesomimus, 398
parvus
  Conopias, 341
  Cypsiurus, 189–190
  Hemignathus, 638
parzudakii, Tangara, 660
Passer, 602–603
Passerculus, 683–684
Passerella, 684–685
Passerina, 691
passerina
  Columbina, 117
  Mirafra, 349
  Spizella, 682
passerinii, Ramphocelus, 656
passerinum, Glaucidium, 170
passerinus
  Forpus, 146–147
  Veniliornis, 253–254
pastazae, Galbula, 232
pastinator, Cacatua, 130
Pastor, 599
patachonicus, Tachyeres, 28
Patagioenas, 111–112
Patagona, 206
patagonica, Asthenes, 277
patagonicus
  Aptenodytes, 6
  Cinclodes, 269
  Mimus, 398
  Phrygilus, 667
patagonus, Cyanoliseus, 144
pauliani, Otus, 164
paulista, Phylloscartes, 325
pauper, Camarhynchus, 688
Pauxi, 56–57
pauxi, Pauxi, 56
Pavo, 74
pavonina, Balearica, 76

pavoninus
  Dromococcyx, 160
  Lophornis, 196
  Pharomachrus, 214
paykullii, Porzana, 83
pealii, Erythrura, 617
pectardens, Luscinia, 456
pectorale, Dicaeum, 547
pectoralis
  Aphelocephala, 522
  Caprimulgus, 182
  Certhionyx, 559
  Circaetus, 38
  Coracina, 370
  Coturnix, 67
  Cuculus, 153
  Euphonia, 627
  Garrulax, 489
  Herpsilochmus, 297
  Heteromunia, 620
  Hylophilus, 624
  Icterus, 695
  Lewinia, 82
  Luscinia, 456
  Malcorus, 42
  Notharchus, 233
  Pachycephala, 484–485
  Polystictus, 323
  Thlypopsis, 651
pecuarius, Charadrius, 94
Pedionomus, 100
pekinensis, Rhopophilus, 419
pelagica, Chaetura, 188
pelagicus
  Haliaeetus, 37
  Hydrobates, 14
  Phalacrocorax, 17
Pelagodroma, 13
Pelargopsis, 218
Pelecanoides, 14–15
Pelecanus, 15
pelegrinoides, Falco, 53
pelewensis
  Aerodramus, 187
  Ptilinopus, 125
peli
  Gymnobucco, 235
  Scotopelia, 168
pelios, Turdus, 408
pella, Topaza, 195
Pellorneum, 493–494
peltata, Platysteira, 466
Peltohyas, 95
Peltops, 581
pelzelni
  Aplonis, 597
  Elaenia, 320
  Granatellus, 647
  Myrmeciza, 301
  Ploceus, 606
  Pseudotriccus, 323
  Thamnophilus, 292
pelzelnii, Tachybaptus, 8
pembae, Cinnyris, 539
pembaensis
  Otus, 164
  Treron, 123
penduliger, Cephalopterus, 314
pendulinus, Remiz, 533
Penelope, 55–56
penelope, Anas, 29
Penelopides, 230

Penelopina, 56
Peneothello, 482
penicillata, Eucometis, 653
penicillatus
    Lichenostomus, 561
    Phalacrocorax, 17
    Pycnonotus, 378
pennata, Rhea, 2
pennatus, Aquila, 48
pensylvanica, Dendroica, 640
pentlandii
    Nothoprocta, 5
    Tinamotis, 6
peposaca, Netta, 32
peracensis, Alcippe, 509
percivali, Oriolus, 569
percnopterus, Neophron, 37
Percnostola, 300–301
percussus
    Prionochilus, 545
    Xiphidiopicus, 248
perdicaria, Nothoprocta, 5
Perdicula, 68
perdita, Petrochelidon, 362
Perdix, 67
perdix
    Brachyramphus, 107–108
    Perdix, 67
peregrina, Vermivora, 639
peregrinus, Falco, 53
Pericrocotus, 374–375
perijana, Schizoeaca, 271
Periparus, 525
Periporphyrus, 688
Perisoreus, 587
Perissocephalus, 314
perkeo, Batis, 467
perlata
    Pyrrhura, 145
    Rhipidura, 469
perlatum, Glaucidium, 170
perlatus, Ptilinopus, 124
Pernis, 35
pernix, Myiotheretes, 338
peronii
    Charadrius, 94
    Zoothera, 402
perousii, Ptilinopus, 125
perreini, Estrilda, 612
perrotii, Hylexetastes, 285
persa, Tauraco, 151
persicus
    Merops, 226
    Puffinus, 13
personata
    Apalis, 423
    Coracina, 368
    Eophona, 637
    Incaspiza, 668
    Poephila, 616
    Prosopeia, 135
    Rhipidura, 470
    Spizocorys, 355
personatus
    Agapornis, 141
    Artamus, 581
    Atlapetes, 678
    Corythaixoides, 152
    Heliopais, 86
    Pterocles, 109
    Trogon, 213
perspicax, Penelope, 55

perspicillata
    Ducula, 127
    Malacocincla, 700
    Melanitta, 33
    Pulsatrix, 170
perspicillatus
    Garrulax, 489
    Hylopezus, 304–305
    Hymenops, 336
    Phalacrocorax, 699
    Sericornis, 518
perstriata, Ptiloprora, 564
pertinax
    Aratinga, 144
    Contopus, 335
peruana, Cinnycerthia, 390
peruanum, Glaucidium, 171
peruviana
    Conopophaga, 308
    Geositta, 268
    Grallaricula, 308
    Pyrrhura, 145
    Sporophila, 671
    Tangara, 661
    Vini, 133
peruvianus, Rupicola, 314
petechia, Dendroica, 640
petersoni, Megascops, 165
petiti, Campephaga, 373
Petrochelidon, 362–363
Petroica, 480–481
Petronia, 603–604
petronia, Petronia, 604
Petrophassa, 116
petrophila, Neophema, 137
petrosus
    Anthus, 366
    Ptilopachus, 70
Peucedramus, 639
Pezopetes, 676
Pezophaps, 700
Pezoporus, 137
Phacellodomus, 277
Phaenicophaeus, 156
Phaenicophilus, 652
phaenicuroides, Hodgsonius, 461
Phaenostictus, 303
phaeocephalus
    Alophoixus, 383
    Cyphorhinus, 397
    Myiarchus, 344
phaeocercus, Mitrephanes, 335–336
Phaeochroa, 194
phaeochromus, Oriolus, 567
Phaeomyias, 322
phaeonotus, Junco, 687
phaeopus, Numenius, 97
phaeopygia, Pterodroma, 11
phaeosoma, Melamprosops, 639
Phaethon, 15
Phaethornis, 192–194
phaeton, Neochmia, 616
Phaetusa, 104
Phainopepla, 387
phainopeplus, Campylopterus, 194
Phainoptila, 387
phaionota, Pachycephala, 483
Phalacrocorax, 16–18, 699
Phalaenoptilus, 179
Phalaropus, 99
Phalcoboenus, 49
phalerata, Coeligena, 205

Phapitreron, 121–122
Phaps, 115–116
pharetra, Dendroica, 642
Pharomachrus, 214
phasiana, Rhipidura, 470
phasianella, Macropygia, 114
phasianellus
    Dromococcyx, 160
    Tympanuchus, 60
phasianinus, Centropus, 157
Phasianus, 72–73
phayrei, Pitta, 265
Phedina, 358
Phegornis, 95
phelpsi, Cypseloides, 184
Phelpsia, 341
Pheucticus, 690
Phibalura, 312
Phigys, 132
philadelphia
    Larus, 102
    Oporornis, 642
philadelphicus, Vireo, 623
philbyi, Alectoris, 64
Philemon, 563–564
Philentoma, 576
Philepitta, 265
Philesturnus, 580
Philetairus, 605
philippae, Sylvietta, 435
philippensis
    Bubo, 168
    Gallirallus, 80
    Loriculus, 140
    Ninox, 174
    Pelecanus, 15
    Spizaetus, 49
    Sturnia, 599
philippii, Phaethornis, 193
philippinensis, Pachycephala, 484
philippinus
    Ixos, 383
    Merops, 226
    Ploceus, 609
phillipsi
    Oenanthe, 463
    Tangara, 662
Philohydor, 341
Philomachus, 99
philomela, Microcerculus, 396
philomelos, Turdus, 410
Philortyx, 61
Philydor, 280
Phimosus, 25
Phlegopsis, 303
Phleocryptes, 270
Phlogophilus, 203
Phodilus, 162
Phoebastria, 9
phoebe
    Metallura, 208
    Sayornis, 336
Phoebetria, 9–10
phoenicea
    Campephaga, 373
    Liocichla, 492–493
    Petroica, 480
phoeniceus
    Agelaius, 692
    Cardinalis, 690
Phoenicircus, 314
phoenicius, Tachyphonus, 654

phoenicobia, Tachornis, 189
phoenicomitra, Myiophobus, 332
phoenicoptera, Pytilia, 614
Phoenicopterus, 25–26
phoenicopterus, Treron, 122
phoenicotis, Chlorochrysa, 658
Phoeniculus, 228
phoenicura
  Ammomanes, 352
  Rhipidura, 468
Phoenicurus, 460–461
phoenicurus
  Amaurornis, 83
  Attila, 345
  Eremobius, 269
  Phoenicurus, 460
Pholia, 601
Pholidornis, 534
phryganophilus, Schoeniophylax, 271
phrygia, Xanthomyza, 566
Phrygilus, 667–668
Phylidonyris, 565–566
Phyllanthus, 509
Phyllastrephus, 380–381
Phyllolais, 432
Phyllomyias, 318–319
Phylloscartes, 324–325
Phylloscopus, 436–440
Phytotoma, 311
piaggiae, Zoothera, 403
Piaya, 159
Pica, 592
pica
  Fluvicola, 338
  Pica, 592
picaoides, Heterophasia, 510
picata
  Egretta, 19
  Oenanthe, 464
Picathartes, 488
picatus, Hemipus, 375
picazuro, Patagioenas, 111
picina, Mearnsia, 187
pickeringii, Ducula, 128
Picoides, 252–253
picta
  Chloephaga, 28
  Grantiella, 566
  Ispidina, 217
  Psittacella, 137
  Pyrrhura, 145
pictum
  Emblema, 616
  Todirostrum, 329
picturata, Streptopelia, 113
pictus
  Calcarius, 687
  Chrysolophus, 73
  Francolinus, 65
  Myioborus, 644
  Oreortyx, 60
picui, Columbina, 117–118
Piculus, 254–255
Picumnus, 244–246
picumnus
  Climacteris, 523
  Dendrocolaptes, 286
Picus, 258–259
picus, Dendroplex, 288
Piezorhina, 668

pilaris
  Atalotriccus, 327
  Turdus, 410
pileata
  Halcyon, 219
  Leptasthenura, 271
  Nemosia, 652
  Notiochelidon, 359
  Oenanthe, 464
  Penelope, 56
  Pionopsitta, 148
  Piprites, 318
  Timalia, 502–503
pileatus, 677
  Atlapetes, 677
  Chlorospingus, 650
  Coryphospingus, 667
  Dryocopus, 257
  Herpsilochmus, 296
  Lophotriccus, 327
  Monarcha, 476
  Pilherodius, 19
Pilherodius, 19
pinaiae, Lophozosterops, 555
Pinarocorys, 351
Pinaroloxias, 673
Pinarornis, 466
Pinguinus, 700
Pinicola, 629
pinicola, Ridgwayia, 413
pinnatus, Botaurus, 22
pinon, Ducula, 128
pintadeanus, Francolinus, 65
pintoi, Aratinga, 144
pinus
  Carduelis, 631–632
  Dendroica, 641
  Vermivora, 639
Pionites, 148
Pionopsitta, 148
Pionus, 149
pipiens, Cisticola, 416
Pipile, 56
pipile, Pipile, 56
Pipilo, 680–681
pipixcan, Larus, 102
Pipra, 314–315
pipra
  Dixiphia, 315–316
  Iodopleura, 313
Pipraeidea, 658
Pipreola, 312
Piprites, 318
Piranga, 654–655
piscator, Crinifer, 152
pistrinaria, Ducula, 128
pitangua, Megarynchus, 342
Pitangus, 341
Pithecophaga, 47
Pithys, 302
pitiayumi, Parula, 639–640
pitius, Colaptes, 256
Pitohui, 487–488
Pitta, 227–267
Pittasoma, 305
pittoides, Atelornis, 227
pityophila, Dendroica, 641
Pityriasis, 581
piurae
  Hemispingus, 650
  Ochthoeca, 338

placens
  Cormobates, 523
  Poecilodryas, 482
placentis, Charmosyna, 133
placida, Geopelia, 116
placidus, Charadrius, 93
plancus, Caracara, 49
Platalea, 25
platalea, Anas, 31
platenae
  Ficedula, 451
  Gallicolumba, 120
  Prioniturus, 138
plateni
  Aramidopsis, 82
  Prionochilus, 545
  Stachyris, 500
platensis
  Cistothorus, 395
  Embernagra, 676
  Leptasthenura, 270
platura, Hedydipna, 535
platurus, Prioniturus, 138
Platycercus, 136
platycercus, Selasphorus, 211
Platycichla, 407
Platylophus, 586
platypterus, Buteo, 45
platyrhynchos
  Anas, 30
  Platyrinchus, 331
platyrhynchum, Electron, 224–225
Platyrinchus, 331
platyrostris, Dendrocolaptes, 286
Platysmurus, 587
Platysteira, 466–467
platyurus, Schoenicola, 442
plebejus
  Phrygilus, 668
  Turdoides, 505
  Turdus, 411
Plectorhyncha, 566
Plectrophenax, 687
Plectropterus, 28
Plegadis, 25
pleschanka, Oenanthe, 464
pleskei
  Locustella, 429
  Podoces, 592
pleurostictus, Thryothorus, 392
plicatus, Aceros, 231
Plocepasser, 605
Ploceus, 606–609
plumatus, Prionops, 576
plumbea
  Dendroica, 642
  Diglossa, 673
  Euphonia, 626
  Ictinia, 36
  Patagioenas, 112
  Polioptila, 446
  Ptiloprora, 564
  Sporophila, 670
plumbeiceps
  Leptotila, 119
  Phyllomyias, 319
  Poecilotriccus, 329
plumbeiventris, Gymnocrex, 82
plumbeus
  Dysithamnus, 294
  Leucopternis, 44
  Micrastur, 50

Myioparus, 450
Turdus, 410
Vireo, 622
plumifera
Geophaps, 116
Guttera, 74
plumosus
Pinarornis, 466
Pycnonotus, 378
plumulus, Lichenostomus, 561
pluricinctus, Pteroglossus, 242
pluto, Myiagra, 477
Pluvialis, 93
pluvialis, Coccyzus, 159
Pluvianellus, 95
Pluvianus, 91
Pnoepyga, 498–499
Podager, 179
podarginus, Pyrroglaux, 166
Podargus, 177
Podica, 86
Podiceps, 8–9, 699
podiceps, Podilymbus, 8
Podilymbus, 8, 699
podobe, Cercotrichas, 459
Podoces, 592
Poecile, 523–525
poecilinotus, Hylophylax, 302
poecilocercus
Knipolegus, 336
Mecocerculus, 321
poecilochrous, Buteo, 46
Poecilodryas, 481–482
poecilolaemus, Dendropicos, 249
poecilopterus
Geobates, 267
Nesoclopeus, 699
poecilorhyncha, Anas, 30
poecilorhynchus, Garrulax, 491
poecilorrhoa, Cryptophaps, 129
poecilosterna, Calendulauda, 350
poecilotis
Chrysomma, 503
Pogonotriccus, 325
Poecilotriccus, 328–329
poecilurus, Knipolegus, 336
poensis
Batis, 468
Bubo, 167
Laniarius, 574
Neocossyphus, 401
Phyllastrephus, 381
Poeoptera, 601
Poephila, 616
Pogoniulus, 236
Pogonocichla, 454–455
Pogonotriccus, 325
Poicephalus, 141, 142
poiciloptilus, Botaurus, 23
poicilotis, Hylophilus, 623
poioicephala, Alcippe, 508
Polemaetus, 48
Polihierax, 50
poliocephala
Alethe, 414
Chloephaga, 28
Ducula, 127
Geothlypis, 643
Ortalis, 55
Stachyris, 501
poliocephalum, Todirostrum, 329
Poliocephalus, 8

poliocephalus
Accipiter, 42
Caprimulgus, 182
Cuculus, 153
Phaenicophilus, 652
Phyllastrephus, 381
Phylloscopus, 439–440
Poliocephalus, 8
Tolmomyias, 330
Turdus, 409
poliocerca, Eupherusa, 198
poliogaster
Accipiter, 40
Caryothraustes, 689
poliogastrus, Zosterops, 550
poliogenys
Cyornis, 453
Seicercus, 440
Spermophaga, 613
Poliolais, 434
poliolophus
Batrachostomus, 177
Prionops, 576
polionotus, Leucopternis, 44
poliopareia, Estrilda, 612
poliophrys, Alethe, 414
poliopleura, Emberiza, 666
polioptera
Coracina, 372
Cossypha, 457
poliopterus
Melierax, 39
Toxorhamphus, 545
Polioptila, 445–446
poliosoma, Pachycephalopsis, 482
poliothorax, Kakamega, 496
Polioxolmis, 339
polleni, Xenopirostris, 577
pollenii, Columba, 111
pollens, Campephilus, 257
Polyboroides, 39
polychopterus, Pachyramphus, 346
polychroa, Prinia, 419
Polyerata, 201
polyglotta, Hippolais, 432
polyglottos, Mimus, 397
polygrammus, Xanthotis, 562
Polyonymus, 208
polyosoma, Buteo, 46
Polyplectron, 73
Polysticta, 32
Polystictus, 323
Polytelis, 139
Polytmus, 199–200
polytmus, Trochilus, 196
Pomarea, 474, 700
pomarea, Pomarea, 700
pomarina, Aquila, 47
pomarinus, Stercorarius, 107
Pomatorhinus, 496–497
Pomatostomus, 511
pompadora, Treron, 122
pondicerianus
Francolinus, 65
Tephrodornis, 576
Pooecetes, 683
poortmani, Chlorostilbon, 197
Poospiza, 669
Popelairia, 196
popelairii, Popelairia, 196
porphyraceus, Ptilinopus, 125
porphyreolophus, Tauraco, 152

porphyreus, Ptilinopus, 124
Porphyrio, 84–85, 699
porphyrio, Porphyrio, 84–85
porphyrocephala, Glossopsitta, 133
porphyrocephalus, Iridosornis, 658
Porphyrolaema, 313
porphyrolaema
Apalis, 423
Porphyrolaema, 313
Porphyrospiza, 691
portoricensis
Loxigilla, 673
Melanerpes, 246
Spindalis, 656
Porzana, 83–84, 699
porzana, Porzana, 83
praecox, Thamnophilus, 291
prasina
Erythrura, 617
Hylia, 436
prasinus, Aulacorhynchus, 241
pratensis, Anthus, 365
pratincola, Glareola, 91
preciosa, Tangara, 661
Premnoplex, 278
Premnornis, 278
presbytes, Phylloscopus, 439
pretiosa, Claravis, 118
pretrei
Amazona, 149
Phaethornis, 193
preussi
Petrochelidon, 362
Ploceus, 608
prevostii
Anthracothorax, 195
Euryceros, 577
prigoginei
Caprimulgus, 183
Chlorocichla, 380
Cinnyris, 539
Phodilus, 162
primigenia, Aethopyga, 541
Primolius, 143
princei, Zoothera, 403
princeps
Accipiter, 42
Actenoides, 222
Leucopternis, 44
Melidectes, 564
Ploceus, 607
principalis, Campephilus, 258
pringlii, Dryoscopus, 572
Prinia, 419–422
priocephalus, Pycnonotus, 375
Prioniturus, 138
Prionochilus, 545
Prionodura, 586
Prionops, 576
Priotelus, 212
pirit, Batis, 467
pristoptera, Psalidoprocne, 357–358
pritchardii, Megapodius, 54
Probosciger, 129
Procellaria, 12
Procelsterna, 103
Procnias, 314
procurvoides, Campylorhamphus, 289
Prodotiscus, 243
productus, Nestor, 700
Progne, 359
progne, Euplectes, 611

promeropirhynchus, Xiphocolaptes, 285–286
Promerops, 556
propinqua
   Iole, 383
   Synallaxis, 273
proprium, Dicaeum, 546
proregulus, Phylloscopus, 438
Prosobonia, 98, 700
Prosopeia, 135
Prosthemadera, 567
prosthemelas, Icterus, 696
Protonotaria, 642
provocator, Xanthotis, 562
Prunella, 400
prunellei, Coeligena, 205
pryeri, Megalurus, 441
przewalskii
   Grallaria, 307
   Paradoxornis, 512
Psalidoprocne, 357–358
Psaltria, 515
psaltria, Carduelis, 632
Psaltriparus, 515
psammocromius, Euplectes, 611
Psarisomus, 264
Psarocolius, 697–698
Pselliophorus, 676
Psephotus, 136, 700
Pseudalaemon, 355
Pseudelaenia, 323
Pseudeos, 131
Pseudhirundo, 358
Pseudibis, 24
Pseudoalcippe, 496
Pseudobias, 466
Pseudocalyptomena, 265
Pseudochelidon, 357
Pseudochloroptila, 636
Pseudocolaptes, 279
Pseudocolopteryx, 323
Pseudocossyphus, 401
Pseudodacnis, 662
Pseudoleistes, 698
Pseudonestor, 638
Pseudonigrita, 605
Pseudopodoces, 529
Pseudoscops, 175
Pseudoseisura, 278–279
pseudosimilis, Anthus, 364
Pseudotriccus, 323
pseudozosterops, Randia, 434
psilolaemus, Francolinus, 66
Psilopogon, 238
Psilopsiagon, 146
Psilorhamphus, 311
psittacea
   Erythrura, 617
   Psittirostra, 638
Psittacella, 137
psittaceus, Treron, 122
Psittacula, 139–140, 700
psittacula
   Aethia, 108
   Camarhynchus, 688
Psittaculirostris, 135
Psittacus, 141
Psitteuteles, 132
Psittinus, 137
Psittirostra, 638
Psittrichas, 134
Psophia, 77
Psophocichla, 407

Psophodes, 513
psychopompus, Scytalopus, 309
ptaritepui, Crypturellus, 4
pteneres, Tachyeres, 28
Pteridophora, 584
Pterocles, 108–109
Pterodroma, 10–11
Pteroglossus, 242
Pteronetta, 29
Pterophanes, 206
Pteroptochos, 308–309
Pteruthius, 505–506
Ptilinopus, 123–127, 700
Ptilocichla, 497–498
ptilocnemis, Calidris, 99
ptilogenys, Gracula, 598
Ptilogonys, 387
Ptilonorhynchus, 586
Ptilopachus, 70
Ptiloprora, 564
Ptilopsis, 166
ptilorhynchus, Pernis, 35
Ptiloris, 584
Ptilorrhoa, 514
Ptilostomus, 593
ptilosus, Macronous, 502
Ptychoramphus, 108
Ptyonoprogne, 361
Ptyrticus, 496
pubescens, Picoides, 252
pucherani
   Campylorhamphus, 289
   Guttera, 74
   Melanerpes, 247
pucheranii, Neomorphus, 160
Pucrasia, 71
pudibunda, Asthenes, 276
puella
   Hypothymis, 473
   Irena, 569
Puffinus, 12–13
puffinus, Puffinus, 12
pugnax, Philomachus, 99
pulchella
   Charmosyna, 133
   Heterophasia, 510
   Lacedo, 217
   Myzomela, 558
   Neophema, 137
   Ochthoeca, 337
   Phyllolais, 432
pulchellus
   Caprimulgus, 183
   Cinnyris, 539
   Nettapus, 29
   Poecilotriccus, 329
   Ptilinopus, 125–126
   Vireolanius, 625
pulcher
   Calothorax, 210
   Lamprotornis, 600
   Melanerpes, 247
   Myiophobus, 332
   Phylloscopus, 437
pulcherrima
   Aethopyga, 542
   Alectroenas, 127
   Megalaima, 239
pulcherrimus
   Carpodacus, 629
   Iridophanes, 662

   Malurus, 516
   Psephotus, 700
pulchra
   Apalis, 422
   Cyanolyca, 589
   Incaspiza, 668
   Macgregoria, 583
   Pionopsitta, 148
   Pipreola, 312
   Sarothrura, 78
pulchricollis, Columba, 111
pulitzeri, Macrosphenus, 436
pullarius, Agapornis, 141
pullicauda, Neopsittacus, 134
pulpa, Mirafra, 349
Pulsatrix, 170
pulverulenta, Eopsaltria, 481
pulverulentus, Mulleripicus, 261
pumilio, Indicator, 243
pumilo, Cyanolyca, 589
pumilus
   Coccyzus, 159
   Picumnus, 245
puna, Anas, 31
punctata, Tangara, 660
punctatum, Cinclosoma, 513
punctatus
   Falco, 51
   Megalurus, 442
   Pardalotus, 549
   Phalacrocorax, 18
   Thamnophilus, 292
puncticeps, Dysithamnus, 293
punctifrons, Anthoscopus, 533
punctigula, Colaptes, 255
punctulata
   Lonchura, 618
   Ninox, 175
punctulatus, Hylophylax, 302
punctuligera, Campethera, 248
punensis
   Geositta, 268
   Phrygilus, 667
punicea
   Columba, 111
   Xipholena, 313
puniceus
   Carpodacus, 630
   Picus, 258
purnelli, Amytornis, 516
purpurascens
   Penelope, 55
   Phalacrocorax, 17
   Vidua, 621
purpurata, Querula, 314
purpuratus
   Ptilinopus, 125
   Touit, 148
   Trachyphonus, 235
purpurea
   Ardea, 19
   Cochoa, 462
Purpureicephalus, 136
purpureiceps, Lamprotornis, 600
purpureiventris, Nectarinia, 538
purpureus
   Carpodacus, 629
   Lamprotornis, 600
   Phoeniculus, 228
purpuroptera, Lamprotornis, 600
purusianus, Galbalcyrhynchus, 231

pusilla
Acanthiza, 520
Aethia, 108
Alcedo, 216–217
Calidris, 98
Coenocorypha, 96
Emberiza, 666
Eremomela, 434
Glossopsitta, 133
Pnoepyga, 499
Porzana, 83
Sitta, 530
Spizella, 682
Wilsonia, 643
pusillus
Campylorhamphus, 289
Chordeiles, 178
Lophospingus, 668
Loriculus, 141
Merops, 225
Pogoniulus, 236
Serinus, 633
pusio, Micropsitta, 134
pustulatus, Icterus, 696
puveli, Illadopsis, 495
Pycnonotus, 375–378
Pycnoptilus, 517
Pycnopygius, 562–563
pycnopygius, Chaetops, 495
pycrofti, Pterodroma, 11
pygargus, Circus, 39
Pygarrhichas, 283
Pygiptila, 293
pygmaea
Aethia, 108
Sitta, 530
pygmaeum
Dicaeum, 547
Toxorhamphus, 545
pygmaeus
Melanerpes, 247
Phalacrocorax, 18
Picumnus, 245
pygmeus, Eurynorhynchus, 99
Pygoscelis, 6–7
pylzowi, Urocynchramus, 664
pyra, Topaza, 195
Pyrenestes, 613
pyrgita, Petronia, 603
pyrhoptera, Philentoma, 576
Pyriglena, 299
pyrilia, Pionopsitta, 148
Pyrocephalus, 336
pyrocephalus, Machaeropterus, 317
Pyroderus, 314
pyrohypogaster, Hypopyrrhus, 698
pyrolophus, Psilopogon, 238
pyrope, Xolmis, 339
pyrrhocephalus, Phaenicophaeus, 156
Pyrrhocoma, 651
Pyrrhocorax, 593
pyrrhocorax, Pyrrhocorax, 593
pyrrhodes, Philydor, 280
pyrrhogaster, Dendropicos, 249
Pyrrholaemus, 519
pyrrholeuca, Asthenes, 277
Pyrrhomyias, 333
pyrrhonota, Petrochelidon, 362
pyrrhonotus, Passer, 602
pyrrhophia, Cranioleuca, 274
pyrrhophrys, Chlorophonia, 628
Pyrrhoplectes, 637

pyrrhops
Hapalopsittaca, 148
Stachyris, 500
pyrrhoptera
Alcippe, 508
Brotogeris, 147
Illadopsis, 495
pyrrhopterus
Lycocorax, 583
Phylidonyris, 565
pyrrhopygia, Hylacola, 519
pyrrhopygius, Todiramphus, 220
pyrrhothorax, Turnix, 76
pyrrhotis, Blythipicus, 260
pyrrhoura, Myzornis, 510
Pyrrhula, 636–637
pyrrhula, Pyrrhula, 636–637
Pyrrhura, 145–146
pyrrogenys, Pellorneum, 494
Pyrroglaux, 166
pyrropygus, Trichixos, 460
Pytilia, 614
pytyopsittacus, Loxia, 630

quadragintus, Pardalotus, 549
quadribrachys, Alcedo, 216
quadricinctus, Pterocles, 109
quadricolor, Dicaeum, 546
quadrivirgata, Cercotrichas, 458
quartinia, Coccopygia, 612
Quelea, 609
quelea, Quelea, 609
querquedula, Anas, 31
Querula, 314
querula, Zonotrichia, 686
quinquestriata, Aimophila, 682
quinticolor
Capito, 240
Lonchura, 619
quiscalina, Campephaga, 373
Quiscalus, 694, 701
quiscula, Quiscalus, 694
quitensis, Grallaria, 306
quixensis, Microrhopias, 297
quoyi, Cracticus, 582

rabieri, Picus, 259
rabori, Napothera, 498
racheliae, Malimbus, 605
radiata, Ducula, 127
radiatum, Glaucidium, 172
radiatus
Carpococcyx, 156
Erythrotriorchis, 43
Nystalus, 233
Polyboroides, 39
radiolatus, Melanerpes, 247
radiolosus, Neomorphus, 160
radjah, Tadorna, 28
rafflesii
Dinopium, 259
Megalaima, 238
raggiana, Paradisaea, 585
raimondii
Phytotoma, 311
Sicalis, 675
Rallina, 78–79
ralloides
Ardeola, 20
Myadestes, 405
Rallus, 81
rama, Hippolais, 432

ramphastinus, Semnornis, 240
Ramphastos, 242–243
Ramphocaenus, 445
Ramphocelus, 655–656
Ramphocinclus, 399
Ramphocoris, 353
Ramphodon, 191
Ramphomicron, 208
Ramphotrigon, 345
ramsayi, Actinodura, 506
Ramsayornis, 566
randi
Muscicapa, 448
Ninox, 174
Randia, 434
randriansoloi, Cryptosylvicola, 435
ranfurlyi, Phalacrocorax, 17
ranivorus, Circus, 39
rapax, Aquila, 47
Raphus, 700
rara
Bostrychia, 24
Lagonosticta, 615
Phytotoma, 311
raricola, Vidua, 621
rarotongensis, Ptilinopus, 125
raveni, Coracornis, 483
ravidus, Turdus, 700
raytal, Calandrella, 354
razae, Alauda, 357
rectirostris
Anthreptes, 535
Hylocryptus, 282
Limnornis, 270
rectunguis, Centropus, 158
Recurvirostra, 90
recurvirostris
Avocettula, 195
Burhinus, 91
Todiramphus, 220
redivivum, Toxostoma, 399
reevei, Turdus, 411
reevesii, Syrmaticus, 72
regalis
Buteo, 47
Heliangelus, 206
regia
Erythrura, 617
Platalea, 25
Vidua, 620
regina, Ptilinopus, 125
regius
Cicinnurus, 584
Cinnyris, 539
Lamprotornis, 600
reguloides
Acanthiza, 519
Anairetes, 322
Phylloscopus, 438
regulorum, Balearica, 76
Regulus, 385
regulus
Machaeropterus, 317
Prodotiscus, 243
Regulus, 385
rehsei, Acrocephalus, 430
reichardi
Ploceus, 607
Serinus, 636
reichenbachii, Anabathmis, 535
reichenovii, Cryptospiza, 612

reichenowi
    Anthreptes, 534
    Cinnyris, 539
    Drepanorhynchus, 538
    Pitta, 266
    Serinus, 635
    Streptopelia, 113
reinhardti, Iridosornis, 658
reinwardt, Megapodius, 54
reinwardtii
    Harpactes, 214
    Reinwardtoena, 115
    Selenidera, 241
    Turdoides, 504
Reinwardtipicus, 260
Reinwardtoena, 115
reiseri, Phyllomyias, 318
relictus, Larus, 102
religiosa, Gracula, 598
remifer, Dicrurus, 578–579
Remiz, 533
remseni, Doliornis, 311
renauldi, Carpococcyx, 156
rendovae, Zosterops, 554
rennelliana, Rhipidura, 470
rennellianus, Zosterops, 553
repressa, Sterna, 106
resplendens, Vanellus, 93
respublica, Cicinnurus, 584
restrictus, Cisticola, 416
reticulata
    Eos, 131
    Meliphaga, 560
retrocinctum, Dicaeum, 546
retzii, Prionops, 576
revoilii, Merops, 226
rex
    Balaeniceps, 24
    Clytoceyx, 218
reyi, Hemispingus, 651
reynaudii, Coua, 157
Rhabdornis, 532–533
Rhagologus, 483
rhami, Lamprolaima, 203
Rhamphomantis, 155
Rhaphidura, 188
Rhea, 2
Rhegmatorhina, 303
Rheinardia, 74
rhinoceros, Buceros, 230
Rhinocrypta, 309
Rhinomyias, 447–448
Rhinopomastus, 228–229
Rhinoptilus, 91
Rhipidura, 468–471
rhipidurus, Corvus, 595
Rhizothera, 67
Rhodacanthis, 701
Rhodinocichla, 652
rhodocephala, Pyrrhura, 146
rhodochlamys, Carpodacus, 630
rhodocorytha, Amazona, 150
rhodogaster, Accipiter, 42
rhodolaemus, Anthreptes, 535
Rhodonessa, 699
rhodopareia, Lagonosticta, 615
Rhodopechys, 637
rhodopeplus, Carpodacus, 630
Rhodophoneus, 574
Rhodopis, 210
rhodopyga, Estrilda, 612
Rhodospingus, 667

Rhodospiza, 638
Rhodostethia, 103
Rhodothraupis, 690
Rhopocichla, 502
Rhopophilus, 419
Rhopornis, 299
Rhyacornis, 461
Rhynchocyclus, 330
Rhynchopsitta, 143
Rhynchortyx, 63
Rhynchostruthus, 628
rhynchotis, Anas, 31
Rhynchotus, 5
Rhynochetos, 86
Rhytipterna, 343
richardi, Anthus, 363
richardsii
    Monarcha, 476
    Ptilinopus, 125
richardsoni
    Eubucco, 240
    Melanospiza, 673
ricketti, Phylloscopus, 439
ricordii, Chlorostilbon, 197
ridgelyi, Grallaria, 307
ridgwayi
    Aegolius, 173
    Baeolophus, 529
    Buteo, 45
    Caprimulgus, 180
    Cotinga, 313
    Nesotriccus, 322
    Plegadis, 25
    Thalurania, 198
Ridgwayia, 413
ridibundus, Larus, 102
ridleyana, Elaenia, 320
riedelii, Tanysiptera, 222
riefferii
    Chlorornis, 649
    Pipreola, 312
rikeri, Berlepschia, 279
Rimator, 497
rimitarae, Acrocephalus, 431
riocourii, Chelictinia, 36
Riparia, 358
riparia, Riparia, 358
risora, Alectrurus, 338
Rissa, 103
rivoli, Ptilinopus, 126
rivolii, Piculus, 255
rivularis, Basileuterus, 646
rixosa, Machetornis, 340
robbinsi, Scytalopus, 310
roberti
    Conopophaga, 308
    Cossyphicula, 457
robertsi, Prinia, 421
robinsoni, Myophonus, 402
roboratus, Megascops, 165
roborowskii, Carpodacus, 630
robusta
    Arachnothera, 543
    Crateroscelis, 517
    Gracula, 598
robustirostris, Acanthiza, 520
robustus
    Campephilus, 258
    Cisticola, 416–417
    Eudyptes, 7
    Melichneutes, 244
    Poicephalus, 141

rochii, Cuculus, 153
rochussenii, Scolopax, 95
rockefelleri, Cinnyris, 539
rodericanus
    Acrocephalus, 432
    Necropsar, 701
rodinogaster, Petroica, 480
rodochroa, Carpodacus, 630
rodolphei, Stachyris, 499
rodriguezi, Scytalopus, 309
rogersi
    Aerodramus, 186
    Atlantisia, 82
rolland, Rollandia, 8
Rollandia, 8
rolleti, Lybius, 238
Rollulus, 70
rondoniae, Agyrtria, 200
roquettei, Phylloscartes, 324
roraimae
    Automolus, 281
    Herpsilochmus, 296
    Megascops, 166
    Myiophobus, 332
Roraimia, 278
roratus, Eclectus, 139
rosacea, Ducula, 128
rosea
    Petroica, 480
    Rhodinocichla, 652
    Rhodostethia, 103
roseata, Psittacula, 140
roseatus, Anthus, 365
roseicapilla
    Eolophus, 130
    Ptilinopus, 125
roseicollis, Agapornis, 141
roseifrons, Pyrrhura, 145
roseigaster, Priotelus, 212
rosenbergi
    Nyctiphrynus, 180
    Polyerata, 201
rosenbergii
    Gymnocrex, 82
    Myzomela, 559
    Tyto, 161
roseogrisea, Streptopelia, 113
roseogularis, Piranga, 654
roseus
    Carpodacus, 630
    Pastor, 599
    Pericrocotus, 374
    Phoenicopterus, 25
rositae, Passerina, 691
rossae, Musophaga, 152
rossii, Chen, 27
rostrata
    Geothlypis, 643
    Pterodroma, 10
Rostratula, 89
rostratum, Trichastoma, 493
Rostrhamus, 36
rothschildi
    Astrapia, 583
    Bangsia, 657
    Cypseloides, 184
    Leucopsar, 598
    Serinus, 634
rougetii, Rougetius, 82
Rougetius, 82
rouloul, Rollulus, 70
rourei, Nemosia, 652

rovianae, Gallirallus, 80
Rowettia, 668
rowi, Apteryx, 2
rowleyi, Eutrichomyias, 473
rubecula
  Erithacus, 455
  Myiagra, 478
  Nonnula, 234
  Poospiza, 669
  Scelorchilus, 309
rubeculoides
  Cyornis, 453
  Prunella, 400
ruber
  Ergaticus, 643
  Eudocimus, 25
  Laterallus, 79
  Phacellodomus, 277
  Phaethornis, 193
  Phoenicopterus, 25
  Sphyrapicus, 248
rubescens
  Anthus, 366
  Carpodacus, 629
  Chalcomitra, 537
  Gallicolumba, 121
rubetra
  Saxicola, 462
  Xolmis, 339
rubica, Habia, 654
rubicilla, Carpodacus, 630
rubicilloides, Carpodacus, 630
rubicola, Saxicola, 462
rubicunda, Grus, 77
rubida, Prunella, 400
rubidiceps, Chloephaga, 28
rubidiventris, Periparus, 525
rubiensis, Monarcha, 475
rubiginosa, Turdoides, 503
rubiginosus
  Automolus, 282
  Blythipicus, 260
  Margarornis, 278
  Megalurulus, 442
  Piculus, 255
  Ploceus, 608
  Trichoglossus, 132
rubinoides, Heliodoxa, 203
rubinus, Pyrocephalus, 336
rubra
  Crax, 57
  Eugerygone, 480
  Foudia, 609
  Paradisaea, 585
  Piranga, 655
  Rallina, 78
rubratra, Myzomela, 558
rubricapillus
  Megalaima, 239
  Melanerpes, 247
rubricata, Lagonosticta, 615
rubricatus, Pardalotus,
  549
rubricauda
  Clytolaema, 203
  Phaethon, 15
rubriceps
  Anaplectes, 606
  Piranga, 655
rubricera, Ducula, 127
rubricollis
  Campephilus, 258

Drymophila, 298
  Malimbus, 605
rubrifacies, Lybius, 237
rubrifrons
  Cardellina, 643
  Heterospingus, 653
  Parmoptila, 611
rubrigastra, Tachuris, 326
rubrigularis, Charmosyna, 133
rubripes, Anas, 30
rubrirostris
  Arborophila, 69
  Cnemoscopus, 650
rubritorques, Anthreptes, 535
rubritorquis, Aratinga, 143
rubrocanus, Turdus, 410
rubrocapilla, Pipra, 315
rubrocristatus, Ampelion, 311
rubrogenys, Ara, 142
rubronotata, Charmosyna, 133
ruckeri, Threnetes, 192
ruckii, Cyornis, 453
ruddi
  Apalis, 423
  Heteromirafra, 350
rudis, Ceryle, 223
rudolfi, Ninox, 174
rudolphi, Paradisaea, 585
rueppelli
  Eurocephalus, 572
  Sylvia, 444
rueppellii
  Eupodotis, 88
  Gyps, 37
  Poicephalus, 142
rufa
  Alectoris, 64
  Formicivora, 297
  Lessonia, 336
  Malacoptila, 234
  Mirafra, 348–349
  Ninox, 173
  Sarothrura, 78
  Schetba, 577
  Trichocichla, 442
rufalbus, Thryothorus, 392
rufaxilla
  Ampelion, 311
  Leptotila, 118–119
rufescens
  Acrocephalus, 431
  Aimophila, 682
  Atrichornis, 347
  Calandrella, 354
  Egretta, 19
  Illadopsis, 495
  Laniocera, 345
  Megalurus, 700
  Otus, 162
  Pelecanus, 15
  Poecile, 525
  Prinia, 420
  Rhynchotus, 5
  Sericornis, 518
  Sylvietta, 435
  Turdoides, 504
rufibarba, Estrilda, 612
ruficapilla
  Alcippe, 507–508
  Cettia, 426
  Grallaria, 306–307
  Hemithraupis, 651

Nonnula, 234
  Phylloscopus, 436
  Spermophaga, 613
  Sylvietta, 435
  Synallaxis, 272
  Vermivora, 639
ruficapillus
  Baryphthengus, 224
  Charadrius, 94
  Chrysomus, 692
  Enicurus, 461
  Schistochlamys, 648
  Thamnophilus, 291
ruficauda
  Aimophila, 681
  Chamaeza, 304
  Cichladusa, 458
  Cinclocerthia, 399
  Galbula, 232
  Histurgops, 605
  Muscicapa, 449
  Myrmeciza, 301
  Neochmia, 616
  Ortalis, 54
  Ramphotrigon, 345
  Rhinomyias, 448
ruficaudatum, Philydor, 280
ruficaudus, Upucerthia, 269
ruficeps
  Aimophila, 681–682
  Chalcostigma, 208
  Cisticola, 415
  Coua, 157
  Elaenia, 320
  Laniarius, 573
  Luscinia, 456
  Macropygia, 114–115
  Orthotomus, 433
  Paradoxornis, 512
  Pellorneum, 494
  Poecilotriccus, 328
  Pomatostomus, 511
  Pseudotriccus, 323
  Pyrrhocoma, 651
  Stachyris, 500
  Stipiturus, 516
  Thlypopsis, 651
ruficervix, Tangara, 661
ruficollaris, Todiramphus, 221
ruficollis
  Branta, 28
  Calidris, 99
  Caprimulgus, 181
  Chrysococcyx, 154
  Corvus, 595
  Garrulax, 490
  Gerygone, 522
  Hypnelus, 233
  Jynx, 244
  Madanga, 555
  Micrastur, 50
  Montifringilla, 604
  Myiagra, 478
  Oreopholus, 95
  Pomatorhinus, 497
  Sporophila, 671
  Stelgidopteryx, 360
  Syndactyla, 279
  Tachybaptus, 7
  Turdus, 410
ruficrissa, Urosticte, 207
ruficrista, Eupodotis, 88

rufidorsa
  Ceyx, 217
  Rhipidura, 471
rufifrons
  Basileuterus, 645–646
  Formicarius, 304
  Fulica, 86
  Garrulax, 489
  Percnostola, 300
  Phacellodomus, 277
  Rhipidura, 471
  Stachyris, 500
  Urorhipis, 424
rufigaster, Ducula, 128
rufigastra, Cyornis, 454
rufigena, Caprimulgus, 181
rufigenis
  Atlapetes, 678
  Tangara, 661
rufigula
  Dendrexetastes, 285
  Ficedula, 451
  Gallicolumba, 120
  Gymnopithys, 303
  Petrochelidon, 362
  Tangara, 660
rufigularis
  Coccyzus, 159
  Falco, 52
  Hemitriccus, 328
  Ixos, 383
  Sclerurus, 282
rufilatus, Cisticola, 416
rufimarginatus, Herpsilochmus, 297
rufina, Netta, 32
rufinucha
  Aleadryas, 483
  Atlapetes, 677
  Campylorhynchus, 389
rufinus, Buteo, 47
rufipectoralis, Ochthoeca, 337
rufipectus
  Arborophila, 68
  Formicarius, 304
  Leptopogon, 325
  Napothera, 498
  Spilornis, 38
rufipennis
  Butastur, 44
  Cinnyris, 540
  Geositta, 268
  Illadopsis, 495
  Macropygia, 114
  Neomorphus, 160
  Petrophassa, 116
  Polioxolmis, 339
rufipes, Strix, 169
rufipileatus, Automolus, 282
rufitorques
  Accipiter, 41
  Turdus, 413
rufiventer
  Pteruthius, 505
  Tachyphonus, 653
  Terpsiphone, 473
rufiventris
  Accipiter, 43
  Ardeola, 20
  Euphonia, 627
  Lurocalis, 178
  Melaniparus, 526
  Mionectes, 326

  Monticola, 401
  Neoxolmis, 338
  Pachycephala, 486
  Picumnus, 246
  Poicephalus, 142
  Prionops, 576
  Rhipidura, 468–469
  Saltator, 689
  Turdus, 411
rufivertex
  Iridosornis, 658
  Muscisaxicola, 340
rufivirgatus, Arremonops, 679
rufoaxillaris, Molothrus, 694
rufobrunneus
  Serinus, 636
  Thripadectes, 281
rufocarpalis, Calicalicus, 577
rufociliatus, Troglodytes, 394
rufocinctus
  Kupeornis, 509
  Passer, 603
rufocinerea
  Grallaria, 306
  Terpsiphone, 473
rufocinereus, Monticola, 401
rufocinnamomea, Mirafra, 349–350
rufocollaris, Petrochelidon, 363
rufocrissalis, Melidectes, 564
rufofuscus, Buteo, 47
rufogularis
  Acanthagenys, 567
  Alcippe, 508
  Apalis, 423
  Arborophila, 69
  Conopophila, 566
  Garrulax, 490
  Pachycephala, 483
rufolateralis, Smithornis, 264
rufolavatus, Tachybaptus, 699
rufomarginatus, Euscarthmus, 323
rufonuchalis, Periparus, 525
rufopalliatus, Turdus, 412
rufopectus, Poliocephalus, 8
rufopicta, Lagonosticta, 614
rufopictus, Francolinus, 66
rufopileatum, Pittasoma, 305
rufoscapulatus, Plocepasser, 605
rufosuperciliaris, Hemispingus, 650
rufosuperciliata, Syndactyla, 279
rufula, Grallaria, 307
rufulus
  Anthus, 363
  Gampsorhynchus, 506
  Troglodytes, 395
rufum
  Conirostrum, 648
  Philydor, 280
  Toxostoma, 398
rufus
  Attila, 345
  Bathmocercus, 428
  Campylopterus, 194
  Caprimulgus, 180
  Casiornis, 343
  Cisticola, 417
  Climacteris, 523
  Cursorius, 91
  Furnarius, 270
  Neocossyphus, 401
  Pachyramphus, 346
  Selasphorus, 211

  Tachyphonus, 654
  Trogon, 213–214
rugensis, Metabolus, 475
ruki, Rukia, 555
Rukia, 555
rumicivorus, Thinocorus, 100
rupestris
  Chordeiles, 178
  Columba, 110
  Monticola, 401
  Ptyonoprogne, 361
Rupicola, 314
rupicola
  Colaptes, 256
  Pyrrhura, 146
  Rupicola, 314
rupicoloides, Falco, 51
rupurumii, Phaethornis, 193
rushiae, Pholidornis, 534
ruspolii, Tauraco, 151
russatus
  Chlorostilbon, 197
  Poecilotriccus, 328
rustica
  Emberiza, 666
  Haplospiza, 673
  Hirundo, 360
rusticola, Scolopax, 95
rusticolus, Falco, 53
ruticilla, Setophaga, 642
rutila
  Amazilia, 200
  Emberiza, 666
  Phytotoma, 311
  Streptoprocne, 184
rutilans
  Passer, 602
  Synallaxis, 272
  Xenops, 283
rutilus, 391
  Otus, 164
  Thryothorus, 391
ruwenzorii
  Apalis, 422
  Caprimulgus, 182
Ruwenzorornis, 152
ruweti, Ploceus, 607
Rynchops, 106

sabini
  Dryoscopus, 573
  Rhaphidura, 188
  Xema, 103
sabota, Calendulauda, 350
sacerdotum, Monarcha, 476
sacra, Egretta, 20
Sagittarius, 49
sagittatus
  Oriolus, 567
  Otus, 162
  Pyrrholaemus, 519
sagittirostris, Hemignathus, 701
sagrae, Myiarchus, 344
saissetti, Cyanoramphus, 135
sakalava, Ploceus, 608
Sakesphorus, 290
salamonis, Gallicolumba, 121
salangana, Aerodramus, 186–187
salinarum, Xolmis, 339
sallaei, Granatellus, 647

salmoni
  Brachygalba, 231
  Chrysothlypis, 652
Salpinctes, 390
Salpornis, 532
Saltator, 688–689
Saltatricula, 667
salvadorii
  Cryptospiza, 612
  Eremomela, 434
  Onychognathus, 601
  Psittaculirostris, 135
  Zosterops, 551
Salvadorina, 29
salvini
  Caprimulgus, 180
  Gymnopithys, 303
  Mitu, 56
  Pachyptila, 11
  Tumbezia, 337
samarensis
  Eurylaimus, 265
  Orthotomus, 433
  Penelopides, 230
samoensis
  Gymnomyza, 565
  Zosterops, 555
samveasnae, Motacilla, 367
sanblasianus, Cyanocorax, 588
sanchezi, Glaucidium, 171
sanctaecatarinae, Megascops, 166
sanctaecrucis, Gallicolumba, 121
sanctaehelenae, Charadrius, 94
sanctaemariae, Cymbilaimus, 290
sanctaemartae, Scytalopus, 309
sanctihieronymi, Panyptila, 189
sanctithomae
  Brotogeris, 147
  Dendrocolaptes, 286
  Ploceus, 609
  Treron, 123
sanctus, Todiramphus, 221
sandvicensis
  Branta, 28
  Thalasseus, 106
sandwichensis
  Chasiempis, 474
  Passerculus, 683–684
  Porzana, 699
  Pterodroma, 11
sanfordi
  Archboldia, 586
  Cyornis, 453
  Haliaeetus, 37
sanghirensis, Colluricincla, 487
sanguinea
  Cacatua, 130
  Himatione, 639
sanguineus
  Pyrenestes, 613
  Rhodopechys, 637
  Veniliornis, 254
sanguiniceps, Haematortyx, 70
sanguinodorsalis, Lagonosticta, 615
sanguinolenta, Myzomela, 558
sanguinolentum, Dicaeum, 548
sanguinolentus
  Pardirallus, 84
  Ramphocelus, 655
sannio, Garrulax, 491
santaecrucis, Zosterops, 554
santovestris, Aplonis, 596

Sapayoa, 265
Sapheopipo, 260
saphirina, Geotrygon, 119
sapphira, Ficedula, 452
sapphirina, Hylocharis, 199
sapphiropygia, Eriocnemis, 207
Sappho, 208
saracura, Aramides, 82
sarasinorum
  Myza, 565
  Phylloscopus, 439
Sarcogyps, 38
Sarcops, 597
Sarcoramphus, 34
sarda, Sylvia, 444
Sarkidiornis, 29
Saroglossa, 598
Sarothrura, 78
Sasia, 246
sasin, Selasphorus, 211
Satrapa, 338
satrapa, Regulus, 385
saturata
  Aethopyga, 542
  Euphonia, 626
  Percnostola, 300
  Scolopax, 95
saturatus
  Caprimulgus, 180
  Cuculus, 153
  Platyrinchus, 331
saturninus
  Mimus, 398
  Thamnomanes, 294
satyra, Tragopan, 70
Saucerottia, 201–202
saucerrottei, Saucerottia, 201
saularis, Copsychus, 459
saundersi
  Larus, 102
  Sternula, 104
saurophagus, Todiramphus, 221
savana, Tyrannus, 343
savannarum, Ammodramus, 684
savesi, Aegotheles, 176
savilei, Eupodotis, 88
sawtelli, Aerodramus, 187
saxatalis, Aeronautes, 189
saxatilis, Monticola, 401
Saxicola, 462–463
saxicolina, Geositta, 268
Saxicoloides, 460
saya, Sayornis, 336
sayaca, Thraupis, 656
Sayornis, 336
scalaris, Picoides, 252
scandens
  Geospiza, 687
  Phyllastrephus, 380
scandiacus, Bubo, 168
scansor, Sclerurus, 283
scapularis, Alisterus, 139
Sceloglaux, 700
Scelorchilus, 309
schach, Lanius, 570
schalowi, Tauraco, 151
scheepmakeri, Goura, 121
Schetba, 577
Schiffornis, 318
schistacea
  Coracina, 368
  Percnostola, 300

Sporophila, 670
  Zoothera, 402
schistaceigula, Polioptila, 446
schistaceus
  Atlapetes, 677
  Enicurus, 461
  Leucopternis, 44
  Mayrornis, 475
  Thamnophilus, 292
schisticeps
  Abroscopus, 441
  Coracina, 371
  Phoenicurus, 460
  Pomatorhinus, 496
schisticolor, Myrmotherula, 295
schistisagus, Larus, 102
Schistochlamys, 648
schistogynus, Thamnomanes, 294
Schistolais, 422
Schizoeaca, 271
schlegeli
  Arremon, 679
  Eudyptes, 7
  Philepitta, 265
schlegelii
  Cercomela, 464
  Francolinus, 65
  Pachycephala, 486
schleiermacheri, Polyplectron, 73
schneideri, Pitta, 265
schoeniclus, Emberiza, 666–667
Schoenicola, 442
Schoeniophylax, 271
schoenobaenus, Acrocephalus, 429
schomburgkii, Micropygia, 78
Schoutedenapus, 187
schoutedeni, Schoutedenapus, 187
schrankii, Tangara, 659
schreibersii, Heliodoxa, 204
schuettii, Tauraco, 151
schulenbergi, Scytalopus, 310
schulzi
  Cinclus, 388
  Dryocopus, 257
schwarzi, Phylloscopus, 437
scintilla, Selasphorus, 211
scirpaceus, Acrocephalus, 430
Scissirostrum, 598
scita, Stenostira, 450
scitulus, Trochilus, 197
sclateri
  Asthenes, 276
  Cacicus, 697
  Doliornis, 311
  Eudyptes, 7
  Forpus, 147
  Hylophilus, 624
  Lophophorus, 71
  Loriculus, 140
  Melidectes, 565
  Myrmotherula, 294
  Myzomela, 559
  Nonnula, 234
  Phyllomyias, 319
  Picumnus, 245
  Poecile, 524
  Pseudocolopteryx, 323
  Spizocorys, 355
  Thryothorus, 391
Sclateria, 300
sclateriana, Amalocichla, 479
Sclerurus, 282–283

scolopaceus
  Eudynamys, 155
  Limnodromus, 97
  Pogoniulus, 236
Scolopax, 95–96
scopifrons, Prionops, 576
scops, Otus, 163
scopulinus, Larus, 102
Scopus, 23
scoresbii, Larus, 100
scotica, Loxia, 630
Scotocerca, 419
scotocerca, Cercomela, 465
Scotopelia, 168
scotops
  Eremomela, 434
  Serinus, 634
scouleri, Enicurus, 461
scripta, Geophaps, 116
scriptoricauda, Campethera, 248
scriptus, Elanus, 36
scrutator, Thripadectes, 281
scutata, Synallaxis, 273
scutatus
  Augastes, 209
  Malimbus, 605
  Pyroderus, 314
scutulata
  Cairina, 29
  Ninox, 174
Scytalopus, 309–311
Scythrops, 155
sechellarum
  Copsychus, 460
  Foudia, 610
sechellensis, Acrocephalus, 432
seductus, Megascops, 165
seebohmi
  Atlapetes, 677
  Bradypterus, 428
  Cettia, 425
  Dromaeocercus, 428
sefilata, Parotia, 583
segmentata, Uropsalis, 183
segregata, Muscicapa, 448
Seicercus, 440
seimundi
  Anthreptes, 535
  Treron, 123
Seiurus, 642
Selasphorus, 211
seledon, Tangara, 659
Selenidera, 241–142
Seleucidis, 585
sellowi, Herpsilochmus, 296
seloputo, Strix, 168
semibadius, Thryothorus, 391
semibrunneus, Hylophilus, 624
semicincta, Malacoptila, 233
semicinerea, Cranioleuca, 275
semicinereus, Hylophilus, 623–624
semicollaris
  Rostratula, 89
  Streptoprocne, 184
semifasciata, Tityra, 345
semiflava, Geothlypis, 643
semiflavum, Ornithion, 321
semifuscus, Chlorospingus, 650
semilarvata, Eos, 131
semilarvatus, Sittiparus, 528
Semioptera, 584
semipalmata, Anseranas, 26

semipalmatus
  Charadrius, 93
  Limnodromus, 97
  Tringa, 98
semipartitus, Empidornis, 446
semiplumbeus
  Leucopternis, 44
  Rallus, 81
semirubra, Rhipidura, 471
semirufa
  Cecropis, 362
  Cossypha, 457
  Thamnolaea, 466
semirufus
  Atlapetes, 678
  Myiarchus, 343
semitorquata
  Alcedo, 215
  Certhilauda, 351
  Ficedula, 450
  Streptopelia, 113
semitorquatus
  Arremon, 679
  Lurocalis, 178
  Micrastur, 50
  Polihierax, 50
semitorques
  Otus, 163
  Spizixos, 375
Semnornis, 240
semperi
  Leucopeza, 643
  Zosterops, 551
senator, Lanius, 572
senegala, Lagonosticta, 614
senegalensis
  Batis, 467
  Burhinus, 90
  Cecropis, 362
  Centropus, 158
  Chalcomitra, 537
  Dryoscopus, 573
  Ephippiorhynchus, 23
  Eupodotis, 87
  Halcyon, 219
  Otus, 163
  Podica, 86
  Streptopelia, 113–114
  Zosterops, 549–550
senegallus
  Pterocles, 109
  Vanellus, 92
senegaloides, Halcyon, 219
senegalus
  Poicephalus, 142
  Tchagra, 573
senex
  Cypseloides, 184
  Poecilotriccus, 328
senilis
  Myornis, 311
  Pionus, 149
sephaena, Francolinus, 65
sephaniodes, Sephanoides, 206
Sephanoides, 206
sepiaria, Malacocincla, 493
sepium, Orthotomus, 433
septimus, Batrachostomus, 177
serena, Lepidothrix, 315
sericea
  Leptocoma, 537–538
  Loboparadisea, 582

sericeus
  Orthotomus, 433
  Sturnus, 599
sericocaudatus, Caprimulgus, 180
Sericornis, 518–519
Sericossypha, 649
Sericulus, 586
serina, Calyptocichla, 380
Serinus, 633–636
serinus, Serinus, 633
serpentarius, Sagittarius, 49
Serpophaga, 322
serrana
  Formicivora, 297
  Upucerthia, 269
serranus
  Larus, 102
  Turdus, 411
serrator
  Mergus, 33
  Morus, 16
serriana, Coua, 156
serripennis, Stelgidopteryx, 360
serrirostris, Colibri, 195
serva, Cercomacra, 299
setaria, Leptasthenura, 271
sethsmithi, Muscicapa, 449
setifrons, Xenornis, 293
Setophaga, 642
Setornis, 383
severa, Mackenziaena, 290
severus
  Ara, 142
  Falco, 52
sewerzowi, Bonasa, 59
sganzini, Alectroenas, 127
sharpei
  Gallirallus, 699
  Hemimacronyx, 366
  Lalage, 373
  Pseudocossyphus, 401
  Sheppardia, 455
  Smithornis, 264
  Terenura, 298
  Turdoides, 504
sharpii
  Apalis, 423
  Pholia, 601
shelleyi
  Aethopyga, 542
  Bubo, 167
  Cinnyris, 539
  Cryptospiza, 612
  Francolinus, 66
  Lamprotornis, 600
  Nesocharis, 611
  Passer, 603
Sheppardia, 455
shorii, Dinopium, 259
Sialia, 404–405
sialis, Sialia, 404
sibilans, Luscinia, 456
sibilator, Sirystes, 343
sibilatrix
  Anas, 29
  Phacellodomus, 277
  Phylloscopus, 437
  Syrigma, 19
sibirica
  Muscicapa, 448
  Zoothera, 403

sibiricus, Uragus, 638
Sicalis, 675
sicki, Terenura, 298
sidamoensis, Heteromirafra, 350
sieboldii, Treron, 123
siemiradzkii, Carduelis, 632
siemsseni, Latoucheornis, 664
Sigelus, 447
sigillata, Peneothello, 482
signata, Cercotrichas, 458
signatus
    Basileuterus, 645
    Eremopterix, 352
    Knipolegus, 336
siju, Glaucidium, 171
silens, Sigelus, 447
sillemi, Leucosticte, 628
silvestris, Gallinula, 85
silvicola, Otus, 163
similis
    Anthus, 364
    Chloropeta, 432
    Myiozetetes, 340–341
    Saltator, 689
simonsi, Scytalopus, 310
Simoxenops, 280
simplex
    Anthreptes, 534
    Calamonastes, 425
    Chlorocichla, 380
    Columba, 111
    Geoffroyus, 138
    Myrmothera, 305
    Pachycephala, 484
    Passer, 603
    Phaetusa, 104
    Piculus, 254
    Pogoniulus, 236
    Pseudotriccus, 323
    Pycnonotus, 378
    Rhytipterna, 343
    Sporophila, 671
sinaloa, Thryothorus, 392
sinaloae
    Corvus, 594
    Progne, 359
sindianus, Phylloscopus, 437
sinense, Chrysomma, 503
sinensis
    Centropus, 158
    Ixobrychus, 22
    Pycnonotus, 376
    Sturnia, 599
singalensis, Chalcoparia, 534
singularis, Xenerpestes, 278
sinica, Carduelis, 633
sintillata, Chalcopsitta, 131
sinuata, Cercomela, 464
sinuatus, Cardinalis, 690
sipahi, Haematospiza, 638
siparaja, Aethopyga, 543
Siphonorhis, 179, 700
Sipodotus, 515
Siptornis, 278
Siptornopsis, 274
siquijorensis, Ixos, 383
sirintarae, Pseudochelidon, 357
Sirystes, 343
sissonii, Troglodytes, 394
Sitta, 529–531
Sittasomus, 284
sitticolor, Conirostrum, 648

Sittiparus, 528
sittoides, Diglossa, 673–674
sjostedti
    Columba, 110
    Glaucidium, 172
skua, Stercorarius, 107
Skutchia, 303
sladeni, Gymnobucco, 235
sloetii, Campochaera, 372
Smicrornis, 520
smithii
    Anas, 31
    Geophaps, 116
    Hirundo, 360
Smithornis, 264
smithsonianus, Larus, 102
Smutsornis, 91
smyrnensis, Halcyon, 219
snethlage, Pyrrhura, 145
snowi, Myrmotherula, 296
Snowornis, 313
sociabilis, Rostrhamus, 36
socialis
    Pluvianellus, 95
    Prinia, 420
socius, Philetairus, 605
socorroensis, Pipilo, 680
socotrana, Emberiza, 665
socotranus, Rhynchostruthus, 628
sodangorum, Actinodura, 506
soemmerringii, Syrmaticus, 72
sokokensis, Anthus, 366
solala, Caprimulgus, 183
solandri, Pterodroma, 10
solangiae, Sitta, 531
solaris
    Cinnyris, 541
    Pericrocotus, 375
solitaria
    Gallinago, 96
    Origma, 517
    Pezophaps, 700
    Tringa, 98
solitaris, Ficedula, 451
solitarius
    Buteo, 46
    Cacicus, 697
    Cuculus, 153
    Harpyhaliaetus, 45
    Monticola, 401
    Phigys, 132
    Raphus, 700
    Tinamus, 3
    Vireo, 622
soloensis, Accipiter, 40
solomonensis
    Gymnophaps, 129
    Nesasio, 176
    Ptilinopus, 126
solstitialis
    Aratinga, 144
    Troglodytes, 395
somalica
    Calandrella, 354
    Mirafra, 349
    Prinia, 421
somalicus, Lanius, 571
somaliensis, Phoeniculus, 228
Somateria, 32
somptuosus, Anisognathus, 658
songara, Poecile, 524
sonnerati, Chloropsis, 386

sonneratii
    Cacomantis, 153
    Gallus, 71
sophiae, Leptopoecile, 436
sordida
    Cercomela, 465
    Pitta, 266
    Thlypopsis, 651
sordidulus, Contopus, 335
sordidus
    Cynanthus, 198
    Eumyias, 452
    Pionus, 149
sorghophilus, Acrocephalus, 429
soror
    Batis, 467
    Pachycephala, 485
    Pitta, 265
    Seicercus, 440
sororcula, Tyto, 161
soui, Crypturellus, 3
souimanga, Cinnyris, 541
souleyetii, Lepidocolaptes, 289
souliei, Actinodura, 506
soumagnei, Tyto, 161
souzae, Lanius, 570
spadicea, Galloperdix, 70
spadiceus, Attila, 345
spadix, Thryothorus, 390
spaldingii, Orthonyx, 513
sparganura, Sappho, 208
sparsa, Anas, 29
Spartonoica, 271
sparverioides, Cuculus, 152
sparverius, Falco, 51
spatulatus, Coracias, 227
speciosa
    Ardeola, 20
    Geothlypis, 643
    Patagioenas, 111
    Stachyris, 500
speciosum, Conirostrum, 648
speciosus, Odontophorus, 62
spectabilis
    Celeus, 257
    Dryotriorchis, 38
    Elaenia, 320
    Lonchura, 619
    Selenidera, 242
    Somateria, 32
specularioides, Anas, 31
specularis, Anas, 31
speculifera, Diuca, 668
speculiferus, Nesospingus,
    649
speculigera
    Conothraupis, 649
    Ficedula, 450
Speculipastor, 601
Speirops, 549
spekei, Ploceus, 607
spekeoides, Ploceus, 607
Spelaeornis, 499
speluncae, Scytalopus, 309
sperata, Leptocoma, 537
Spermestes, 617–618
Spermophaga, 613
Sphecotheres, 569
Spheniscus, 7
sphenocercus, Lanius, 571
Sphenocichla, 499
Sphenoeacus, 429

sphenurus
Haliastur, 36
Treron, 123
Sphyrapicus, 248
spillmanni, Scytalopus, 310
spilocephalus, Otus, 162
spilodera
Petrochelidon, 362
Rhipidura, 470
Sericornis, 519
spilogaster
Aquila, 48
Picumnus, 245
Veniliornis, 254
spilonotus
Circus, 39
Laterallus, 79
Parus, 528
Salpornis, 532
spiloptera
Porzana, 83
Saroglossa, 598
Zoothera, 403
spilopterus, Centropus, 157
Spiloptila, 422
Spilornis, 38
spilorrhoa, Ducula, 129
Spindalis, 656
spinescens, Carduelis, 632
spinicauda, Aphrastura, 270
spinicaudus, Chaetura, 188
spinicollis, Threskiornis, 24
spinoides, Carduelis, 631
spinoletta, Anthus, 366
spinosa, Jacana, 88
spinosus, Vanellus, 92
spinus, Carduelis, 631
spirurus, Glyphorynchus, 285
spixi, Synallaxis, 272
spixii
Cyanopsitta, 142
Xiphorhynchus, 287
Spiza, 691
spiza, Chlorophanes, 663
Spizaetus, 48–49
Spizastur, 48
Spizella, 682–683
Spiziapteryx, 50
Spizixos, 375
Spizocorys, 355
splendens
Corvus, 593
Malurus, 516
Prosopeia, 135
splendida, Neophema, 137
splendidissima, Astrapia, 583
splendidus
Lamprotornis, 600
Zosterops, 553
spodionota
Myrmotherula, 295
Ochthoeca, 337
spodiops, Hemitriccus, 327
spodioptila, Terenura, 298
spodiopygius, Aerodramus, 186
spodiurus, Pachyramphus, 346
spodocephala, Emberiza, 666
spodocephalus, Dendropicos, 250
sponsa, Aix, 29
Sporaeginthus, 615
Sporophila, 669–671
Sporopipes, 605

spragueii, Anthus, 365
Spreo, 600–601
spurius
Icterus, 696
Purpureicephalus, 136
squalidus, Phaethornis, 193
squamata
Callipepla, 60
Drymophila, 298
Eos, 131
Lichmera, 557
Tachornis, 189
squamatus
Capito, 240
Francolinus, 66
Garrulax, 492
Lepidocolaptes, 289
Mergus, 33
Picus, 259
Pycnonotus, 376
squameiceps, Urosphena, 425
squamiceps
Lophozosterops, 555
Turdoides, 503
squamifrons
Oculocincta, 555
Sporopipes, 605
squamiger
Brachypteracias, 227
Margarornis, 278
Neomorphus, 160
squamigera, Grallaria, 305
squamipila, Ninox, 175
squammata, Columbina, 118
squamosa
Myrmeciza, 301
Patagioenas, 111
squamosus, Heliomaster, 210
squamulata, Turdoides, 504
squamulatus, Picumnus, 245
squatarola, Pluvialis, 93
Stachyris, 499–501
Stactolaema, 235–236
stagnatilis, Tringa, 98
Stagonopleura, 616
stairi, Gallicolumba, 121
stalkeri
Tephrozosterops, 555
Zosterops, 553
stanleyi, Chalcostigma, 209
starki, Spizocorys, 355
Starnoenas, 120
Steatornis, 176
steerii
Centropus, 158
Eurylaimus, 265
Liocichla, 492
Oriolus, 568
Pitta, 266
steinbachi, Asthenes, 277
steindachneri, Picumnus, 245
stejnegeri, Hemignathus, 701
Stelgidopteryx, 360
stellaris
Botaurus, 23
Pygiptila, 293
stellata
Brachypteryx, 413
Gavia, 7
Pogonocichla, 454–455
stellatus
Batrachostomus, 177

Caprimulgus, 183
Margarornis, 278
Odontophorus, 63
stelleri
Cyanocitta, 587
Polysticta, 32
Stellula, 211
Stenostira, 450
stentoreus, Acrocephalus, 430
stenura, Gallinago, 96
stenurus, Chlorostilbon, 197
stephani, Chalcophaps, 115
stephaniae, Astrapia, 583
Stephanoaetus, 49
Stephanophorus, 658
Stephanoxis, 196
stepheni, Vini, 133
Stercorarius, 106–107
Sterna, 105–106
Sternoclyta, 204
Sternula, 104
sterrhopteron, Wetmorethraupis, 657
stewarti, Emberiza, 665
stictigula, Modulatrix, 493
stictocephalus
Herpsilochmus, 297
Pycnopygius, 563
Thamnophilus, 292
stictolaema, Deconychura, 284
stictolophus, Lophornis, 196
Stictonetta, 28
stictopterus
Mecocerculus, 321
Touit, 148
stictothorax
Dysithamnus, 293
Synallaxis, 274
sticturus
Herpsilochmus, 296
Thamnophilus, 292
stierlingi, Dendropicos, 249
Stigmatura, 323
stigmatus, Loriculus, 140
stilesi, Scytalopus, 309
Stiltia, 91
Stiphrornis, 455
Stipiturus, 516
stolidus
Anous, 103
Myiarchus, 344
stolzmanni
Aimophila, 681
Chlorothraupis, 653
Oreotrochilus, 205
Tachycineta, 359
Tyranneutes, 318
Urothraupis, 664
storeri, Cypseloides, 184
stormi, Ciconia, 23
strenua
Aratinga, 143
Ninox, 173
strenuus, Zosterops, 700
Strepera, 582
strepera
Anas, 29
Elaenia, 320
strepitans
Garrulax, 489
Phyllastrephus, 380
Streptocitta, 597–598
Streptopelia, 112–114

streptophorus
  Francolinus, 65
  Lipaugus, 313
Streptoprocne, 184–185
stresemanni
  Hylexetastes, 285
  Merulaxis, 311
  Zaratornis, 311
  Zavattariornis, 592
  Zosterops, 554
Stresemannia, 556
striata
  Aplonis, 596
  Butorides, 20
  Chaetornis, 442
  Coracina, 369
  Dendroica, 642
  Geopelia, 116
  Kenopia, 498
  Leptasthenura, 271
  Lonchura, 618
  Malacoptila, 233
  Muscicapa, 448
  Stachyris, 500
  Sterna, 105
  Turdoides, 504
striaticeps
  Dysithamnus, 293
  Knipolegus, 336
  Macronous, 502
  Phacellodomus, 277
striaticollis
  Alcippe, 507
  Anabacerthia, 279
  Hemitriccus, 327
  Mionectes, 325
  Myiotheretes, 338
  Phacellodomus, 277
  Siptornis, 278
striatigula, Neomixis, 499
striatipectus, Saltator, 688
striativentris, Melanocharis, 544
striatus
  Accipiter, 43
  Amytornis, 517
  Colius, 211–212
  Gallirallus, 80–81
  Garrulax, 489
  Melanerpes, 247
  Pardalotus, 549
  Pycnonotus, 375
  Simoxenops, 280
stricklandi, Picoides, 252
stricklandii, Gallinago, 96
strigiceps, Aimophila, 681
strigilatus
  Ancistrops, 280
  Myrmorchilus, 296
strigirostris, Didunculus, 121
strigoides, Podargus, 177
Strigops, 134
strigula, Minla, 507
strigulosus, Crypturellus, 4
striigularis, Phaethornis, 193
striolata
  Cecropis, 362
  Emberiza, 665
  Leptasthenura, 271
  Stachyris, 501
striolatus
  Machaeropterus, 317

Nystalus, 233
  Serinus, 635
Strix, 168–169
strophianus, Heliangelus, 206
strophiata
  Ficedula, 450
  Prunella, 400
strophium, Odontophorus, 62
struthersii, Ibidorhyncha, 90
Struthidea, 580
Struthio, 2
stuarti, Phaethornis, 193
stuhlmanni
  Cinnyris, 538
  Poeoptera, 601
sturmii, Ixobrychus, 22
Sturnella, 693
Sturnia, 599
sturnina, Sturnia, 599
Sturnus, 599
stygia, Lonchura, 620
stygius, Asio, 175
suahelicus, Passer, 603
suavissima, Lepidothrix, 315
subaffinis, Phylloscopus, 437
subalaris
  Amblyornis, 586
  Macroagelaius, 698
  Snowornis, 313
  Syndactyla, 279
  Turdus, 411
subandina, Pyrrhura, 145
subaureus, Ploceus, 606
subbrunneus, Cnipodectes, 331
subbuteo, Falco, 52
subcaeruleum, Parisoma, 444
subcinnamomea, Euryptila, 425
subcorniculatus, Philemon, 563
subcoronata, Certhilauda, 351
subcristata
  Aviceda, 34–35
  Cranioleuca, 274
  Serpophaga, 322
subflava
  Inezia, 326
  Prinia, 421
subflavus, Sporaeginthus, 615
subfrenatus, Lichenostomus, 560
subgularis, Ptilinopus, 124
subhimachala, Pinicola, 629
subis, Progne, 359
subita, Dendroica, 641
Sublegatus, 326
subminuta, Calidris, 99
subniger, Falco, 52
subochraceus, Phaethornis, 193
subpersonatus, Ploceus, 606
subplacens, Myiopagis, 319
subpudica, Synallaxis, 272
subrubra, Ficedula, 450
subrufa, Turdoides, 504
subruficapilla, Cisticola, 416
subruficollis
  Aceros, 231
  Tryngites, 99
substriata, Prinia, 421
subsulphureus, Pogoniulus, 236
subtilis
  Buteogallus, 44
  Picumnus, 246
subulata, Urosphena, 425

subulatus
  Hyloctistes, 280
  Todus, 223
subunicolor, Garrulax, 491
subvinacea, Patagioenas, 112
subviridis, Phylloscopus, 438
sueurii, Lalage, 372
Suiriri, 321
suiriri, Suiriri, 321
sukatschewi, Garrulax, 490
Sula, 16
sula
  Coracina, 371
  Sula, 16
sulcatus, Aulacorhynchus, 241
sulcirostris
  Crotophaga, 160
  Phalacrocorax, 16
sulfuratus, Ramphastos, 243
sulfureopectus, Telophorus, 575
sulfuriventer, Pachycephala, 484
sulphurata, Emberiza, 666
sulphuratus
  Pitangus, 341
  Serinus, 635
sulphurea
  Cacatua, 130
  Gerygone, 521
  Tyrannopsis, 342
sulphureipygius, Myiobius, 332
sulphureiventer, Neopelma, 317
sulphurescens, Tolmomyias, 330
sulphurifera, Cranioleuca, 274
sultanea, Melanochlora, 529
sumatrana
  Ardea, 19
  Niltava, 453
  Sterna, 105
sumatranus
  Bubo, 167
  Corydon, 265
  Dicrurus, 579
  Phaenicophaeus, 156
  Tanygnathus, 138
sumbaensis, Ninox, 174
sumichrasti
  Aimophila, 681
  Hylorchilus, 390
sundara, Niltava, 452–453
sunensis, Myrmotherula, 295
sunia, Otus, 163
superba
  Lophorina, 584
  Pitta, 266
superbus
  Cinnyris, 540
  Cyornis, 454
  Lamprotornis, 600
  Ptilinopus, 125
superciliaris
  Abroscopus, 441
  Burhinus, 91
  Camaroptera, 424
  Drymodes, 482
  Ficedula, 452
  Hemispingus, 650
  Leptopogon, 325
  Lophozosterops, 555
  Melanerpes, 247
  Ninox, 174
  Ortalis, 55
  Penelope, 55

superciliaris (*continued*)
  Petronia, 603–604
  Phylloscartes, 324
  Polystictus, 323
  Rhipidura, 468
  Scytalopus, 310
  Sternula, 104
  Sturnella, 693
  Tesia, 425
  Thryothorus, 392
  Xiphirhynchus, 497
superciliosa
  Anas, 31
  Eumomota, 225
  Ophrysia, 699
  Parula, 639
  Poecile, 525
  Poecilodryas, 481
  Woodfordia, 556
superciliosus
  Acanthorhynchus, 566
  Accipiter, 42
  Artamus, 581
  Centropus, 158
  Merops, 226
  Oriturus, 682
  Pachyphantes, 609
  Phaenicophaeus, 156
  Phaethornis, 192
  Plocepasser, 605
  Pomatostomus, 511
  Vanellus, 92
superflua, Rhipidura, 470
surdus, Touit, 148
surinamensis, Myrmotherula, 294
surinamus
  Pachyramphus, 347
  Tachyphonus, 653
Surnia, 170
Surniculus, 155
surrucura, Trogon, 214
suscitator, Turnix, 75–76
susurrans, Xiphorhynchus, 287
sutorius, Orthotomus, 433
svecica, Luscinia, 456
swainsoni
  Buteo, 46
  Myiarchus, 343
  Notharchus, 233
swainsonii
  Chlorostilbon, 197
  Francolinus, 66
  Gampsonyx, 35–36
  Limnothlypis, 642
  Passer, 603
  Polytelis, 139
swalesi, Turdus, 413
swierstrai, Francolinus, 67
swindernianus, Agapornis, 141
swinhoii, Lophura, 71
swynnertoni, Swynnertonia, 455
Swynnertonia, 455
sybillae, Lampornis, 203
sylvanus, Anthus, 364
sylvatica
  Prinia, 420
  Zoonavena, 187
sylvaticus
  Bradypterus, 427
  Turnix, 75
sylvestris, Gallirallus, 80
Sylvia, 443–444

sylvia
  Poecilotriccus, 329
  Tanysiptera, 223
Sylvietta, 435
sylviolus, Phylloscartes, 325
Sylviorthorhynchus, 271
Sylviparus, 529
Syma, 222
symonsi, Pseudochloroptila, 636
Synallaxis, 272–274
Syndactyla, 279
syndactylus, Bleda, 381
synoicus, Carpodacus, 630
Synthliboramphus, 108
Sypheotides, 88
syriacus
  Dendrocopos, 252
  Serinus, 633
Syrigma, 19
syrinx, Acrocephalus, 430
Syrmaticus, 72
syrmatophorus, Phaethornis, 193
Syrrhaptes, 108
szalayi, Oriolus, 567
szechenyii, Tetraophasis, 63

tabuensis
  Aplonis, 597
  Porzana, 84
  Prosopeia, 135
tacarcunae, Chlorospingus, 649
tacazze, Nectarinia, 538
tachiro, Accipiter, 40
Tachornis, 189
Tachuris, 326
Tachybaptus, 7–8, 699
Tachycineta, 358–359
Tachyeres, 28
Tachymarptis, 190
Tachyphonus, 653–654
taciturnus, Arremon, 679
tacsanowskius, Bradypterus, 428
taczanowskii
  Cinclodes, 269
  Leptopogon, 325
  Leucippus, 200
  Montifringilla, 604
  Nothoprocta, 5
  Podiceps, 9
  Sicalis, 675
Tadorna, 28, 699
tadorna, Tadorna, 28
tadornoides, Tadorna, 28
taeniata, Dubusia, 658
taeniatus, Peucedramus, 639
taeniopterus, Ploceus, 607
Taeniopygia, 616
Taeniotriccus, 328
tahapisi, Emberiza, 665
tahitica, Hirundo, 360
tahitiensis, Numenius, 97
taitensis, Eudynamys, 155
taiti, Acrocephalus, 431
taivanus, Pycnonotus, 376
takatsukasae, Monarcha, 477
talaseae, Zoothera, 404
talatala, Cinnyris, 540
talaudensis, Gymnocrex, 83
Talegalla, 53
talpacoti, Columbina, 117
tamarugense, Conirostrum, 648
tamatia, Bucco, 233

tanagrinus, Lampropsar, 698
tanganjicae, Zoothera, 403
Tangara, 659–662
tanki, Turnix, 75
tannensis, Ptilinopus, 124
tanneri, Troglodytes, 394
Tanygnathus, 138
Tanysiptera, 222–223
tao, Tinamus, 3
Taoniscus, 6
Tapera, 160
tapera, Progne, 359
Taphrolesbia, 209
Taraba, 290
taranta, Agapornis, 141
tarda, Otis, 87
tarnii, Pteroptochos, 308
Tarsiger, 456–457
tasmanicus, Corvus, 595
tataupa, Crypturellus, 5
tatei
  Aegotheles, 176
  Premnoplex, 278
Tauraco, 151–152
tectes, Saxicola, 462
tectus
  Notharchus, 233
  Vanellus, 92
teerinki, Lonchura, 620
teesa, Butastur, 44
tegimae, Pericrocotus, 374
Telacanthura, 188
telasco, Sporophila, 671
Teledromas, 309
telescophthalmus, Arses, 477
Telespiza, 638
Telophorus, 574–575
Temenuchus, 599
temia, Crypsirina, 592
temminckii
  Aethopyga, 543
  Calidris, 99
  Coracias, 227
  Coracina, 369
  Cursorius, 91
  Dendrocopos, 250
  Eurostopodus, 179
  Orthonyx, 513
  Picumnus, 245
  Tragopan, 70
Temnurus, 592
temnurus
  Priotelus, 212
  Temnurus, 592
temporalis
  Neochmia, 616
  Ploceus, 606
  Pomatostomus, 511
tenebricosa, Tyto, 160
tenebrosa
  Chelidoptera, 235
  Colluricincla, 487
  Gallinula, 85
  Gerygone, 521
  Rhipidura, 470
  Turdoides, 504
tenebrosus, Phyllastrephus, 381
tenella, Neomixis, 499
tenellipes, Phylloscopus, 438
tenellus, Tmetothylacus, 366

tener, Loriculus, 141
teneriffae
    Cyanistes, 528
    Regulus, 385
tenimberensis, Megapodius, 54
tenuepunctatus, Thamnophilus, 291
tenuirostris
    Acanthorhynchus, 566
    Anous, 103
    Cacatua, 130
    Calidris, 98
    Coracina, 370–370
    Geositta, 268
    Gyps, 37
    Inezia, 326
    Macropygia, 114
    Numenius, 97
    Onychognathus, 601
    Oriolus, 568
    Puffinus, 12
    Xenops, 283
    Zosterops, 554
tephrocephalus, Seicercus, 440
tephrocotis, Leucosticte, 628
Tephrodornis, 576
tephrolaemus, Andropadus, 379
tephronota, Sitta, 531
tephronotum, Glaucidium, 171–172
tephronotus
    Lanius, 570
    Turdus, 408
tephropleurus, Zosterops, 554
Tephrozosterops, 555
Terathopius, 38
terborghi, Atlapetes, 677
Terenotriccus, 333
Terenura, 298
Teretistris, 643
Terpsiphone, 473–474
terraereginae, Acrodramus, 186
terrestris
    Phyllastrephus, 380
    Trugon, 121
    Zoothera, 700
terrisi, Rhynchopsitta, 143
Tersina, 663
tertius, Calyptophilus, 652
Tesia, 425
tessmanni, Muscicapa, 449
tethys, Oceanodroma, 14
Tetrao, 59
Tetraogallus, 63–64
Tetraophasis, 63
Tetrax, 88
tetrax, Tetrax, 88
tetrix, Tetrao, 59
textilis, Amytornis, 516
textrix, Cisticola, 418
teydea, Fringilla, 626
teysmanni, Rhipidura, 470
teysmannii, Treron, 122
thagus, Pelecanus, 15
Thalassarche, 9
Thalasseus, 106
thalassina
    Cissa, 591
    Tachycineta, 358
thalassinus
    Colibri, 195
    Eumyias, 452
Thalassoica, 10
Thalassornis, 26

Thalurania, 198–199
Thamnistes, 293
Thamnolaea, 466
Thamnomanes, 294
Thamnophilus, 291–293
Thamnornis, 428
Thaumastura, 210
thayeri, Larus, 101
theklae, Galerida, 356
thenca, Mimus, 398
theomacha, Ninox, 175
theresae, Montifringilla, 604
theresiae
    Metallura, 208
    Polytmus, 200
Theristicus, 25
Thescelocichla, 380
thibetanus, Serinus, 634
thilius, Agelasticus, 692–693
thilohoffmanni, Otus, 162
Thinocorus, 100
Thinornis, 95
Thlypopsis, 651
tholloni, Myrmecocichla, 465
thomensis
    Columba, 110
    Dreptes, 536
    Estrilda, 612
    Zoonavena, 187
thompsoni, Hypsipetes, 385
thoracica
    Apalis, 422
    Ochthoeca, 337
    Poospiza, 669
    Stachyris, 501
thoracicus
    Bambusicola, 70
    Bradypterus, 427
    Charadrius, 94
    Cyphorhinus, 397
    Dactylortyx, 63
    Hylophilus, 623
    Liosceles, 309
    Prionochilus, 545
    Thryothorus, 391
Thraupis, 656–657
Threnetes, 192
threnothorax, Rhipidura, 469
Threskiornis, 24
Thripadectes, 281
Thripophaga, 275–276
thruppi, Melaniparus, 526
Thryomanes, 393
Thryorchilus, 396
Thryothorus, 390–393
thula, Egretta, 20
thura, Carpodacus, 630
thyroideus, Sphyrapicus, 248
tianquanensis, Certhia, 532
Tiaris, 672–673
tibetanus
    Syrrhaptes, 108
    Tetraogallus, 64
tibialis
    Neochelidon, 360
    Pheucticus, 690
    Pselliophorus, 676
tibicen, Gymnorhina, 582
Tichodroma, 531
tickelli
    Anorrhinus, 230
    Pellorneum, 494

Tickellia, 441
tickelliae, Cyornis, 454
tigrina, Dendroica, 640
tigrinus, Lanius, 570
Tigriornis, 22
Tigrisoma, 22
Tijuca, 313
Tilmatura, 210
Timalia, 502–503
Timeliopsis, 556
timorensis, Ficedula, 452
timoriensis, Megalurus,
    441–442
Tinamotis, 6
Tinamus, 3
tinniens, Cisticola, 416
tinnunculus, Falco, 50–51
tiphia, Aegithina, 386
tirica, Brotogeris, 147
tithys, Synallaxis, 273
Tityra, 345–346
Tmetothylacus, 366
tobaci, Saucerottia, 202
Tockus, 229
toco, Ramphastos, 243
tocuyensis, Arremonops,
    679
Todiramphus, 219–221
Todirostrum, 329–330
Todus, 223–224
todus, Todus, 224
togoensis, Vidua, 620
tolmiei, Oporornis, 642
Tolmomyias, 330–331
tombacea, Galbula, 232
tomentosum, Mitu, 56
tonsa, Platysteira, 466
Topaza, 195
torda, Alca, 107
Torgos, 37–38
torotoro, Syma, 222
torquata
    Chauna, 26
    Coeligena, 205
    Grafisia, 601
    Hydropsalis, 184
    Melanopareia, 311
    Myrmornis, 302
    Poospiza, 669
torquatus
    Buarremon, 678
    Celeus, 257
    Ceryle, 223
    Corvus, 595
    Corythopis, 323
    Cracticus, 581
    Gallirallus, 80
    Lanioturdus, 468
    Lybius, 237
    Melidectes, 565
    Myioborus, 644
    Neolestes, 385
    Pedionomus, 100
    Pteroglossus, 242
    Saxicola, 462
    Thamnophilus, 291
    Turdus, 408
torqueola
    Arborophila, 68
    Sporophila, 670
torquilla, Jynx, 244
Torreornis, 682

torridus
 Attila, 345
 Furnarius, 270
torringtoni, Columba, 111
totanus, Tringa, 98
totta, Pseudochloroptila, 636
Touit, 147–148
toulou, Centropus, 158
toussenelii, Accipiter, 40
townsendi
 Dendroica, 641
 Myadestes, 405
toxopei, Charmosyna, 133
Toxorhamphus, 544–545
Toxostoma, 398–399
tracheliotus, Torgos, 37–38
Trachyphonus, 235
tractrac, Cercomela, 465
Tragopan, 70
traillii
 Empidonax, 334
 Oriolus, 569
tranquebarica, Streptopelia, 113
transfasciatus, Crypturellus, 4
traversi, Petroica, 481
traylori, Tolmomyias, 330
Tregellasia, 481
Treron, 122–123
triangularis, Xiphorhynchus, 288
tricarunculatus, Procnias, 314
trichas, Geothlypis, 642
Trichastoma, 493
Trichixos, 460
Trichocichla, 442
Trichodere, 557
Trichoglossus, 131–132
Tricholaema, 236–237
Tricholestes, 383
trichopsis, Megascops, 165
Trichothraupis, 654
trichroa, Erythrura, 617
Triclaria, 151
tricollaris, Charadrius, 94
tricolor
 Agelaius, 692
 Alectrurus, 338
 Ara, 700
 Atlapetes, 677
 Egretta, 20
 Epthianura, 522
 Erythrura, 617
 Ficedula, 452
 Lalage, 372
 Perissocephalus, 314
 Phalaropus, 99
 Ploceus, 608
 Rallina, 79
 Vanellus, 93
tridactyla
 Jacamaralcyon, 232
 Rissa, 103
tridactylus, Picoides, 253
trifasciatus
 Basileuterus, 645
 Carpodacus, 630
 Hemispingus, 650
 Nesomimus, 398
Trigonoceps, 38
trigonostigma, Dicaeum, 546–547
Tringa, 97–98
trinitatis, Euphonia, 626
trinotatus, Accipiter, 41

tristigma, Caprimulgus, 183
tristigmata, Gallicolumba, 120
tristis
 Acridotheres, 598
 Carduelis, 633
 Corvus, 594
 Meiglyptes, 260
 Phaenicophaeus, 156
tristissima, Lonchura, 618
tristrami
 Dicaeum, 548
 Emberiza, 665
 Myzomela, 559
 Oceanodroma, 14
tristramii, Onychognathus, 601
tristriatus
 Basileuterus, 646
 Serinus, 636
triurus, Mimus, 398
trivialis, Anthus, 366
trivirgatus
 Accipiter, 40
 Conopias, 341
 Monarcha, 476
 Phylloscopus, 439
trocaz, Columba, 110
trochileum, Dicaeum, 549
trochilirostris, Campylorhamphus, 289
trochiloides, Phylloscopus, 438
Trochilus, 196–197
trochilus, Phylloscopus, 437
Trochocercus, 472
Troglodytes, 393–395
troglodytes
 Cisticola, 417
 Collocalia, 185
 Estrilda, 612
 Troglodytes, 393–394
troglodytoides, Spelaeornis, 499
Trogon, 212–214
tropica, Fregetta, 14
trudeaui, Sterna, 106
Trugon, 121
Tryngites, 99
tsavoensis, Cinnyris, 539
tschudii, Ampelioides, 312
tschutschensis, Motacilla, 367
tuberculifer, Myiarchus, 343
tuberosum, Mitu, 56
tucanus, Ramphastos, 243
tucinkae, Eubucco, 240
tucumana, Amazona, 149
tucumanum, Glaucidium, 171
tuerosi, Laterallus, 79
tukki, Meiglyptes, 260–261
tullbergi, Campethera, 249
Tumbezia, 337
tumultuosus, Pionus, 149
Turacoena, 115
turatii, Laniarius, 574
turcosa, Cyanolyca, 589
turcosus, Cyornis, 454
turdina
 Chamaeza, 304
 Dendrocincla, 284
 Schiffornis, 318
turdinus
 Campylorhynchus, 389
 Ptyrticus, 496
Turdoides, 503–505
turdoides, Cataponera, 404
Turdus, 407–413, 700

Turnagra, 700
turneri, Eremomela, 434
Turnix, 75–76
Turtur, 115
turtur
 Pachyptila, 12
 Streptopelia, 112
tutus, Todiramphus, 221
Tylas, 577
tympanistria, Turtur, 115
tympanistrigus, Pycnonotus, 375
Tympanuchus, 60
typica
 Coracina, 370
 Nesillas, 428
typicus, Corvus, 593
typus, Polyboroides, 39
Tyranneutes, 318
tyrannina
 Cercomacra, 298
 Dendrocincla, 283
Tyrannopsis, 342
Tyrannulus, 319
tyrannulus, Myiarchus, 344
Tyrannus, 342–343
tyrannus
 Spizaetus, 49
 Tyrannus, 342
tyrianthina, Metallura, 208
tyro, Dacelo, 218
tytleri, Phylloscopus, 438
Tyto, 160–161
tzacatl, Amazilia, 200

ucayalae, Simoxenops, 280
udzungwensis, Xenoperdix, 68
ugiensis, Zosterops, 554
ulietanus, Cyanoramphus, 700
ultima
 Pterodroma, 10
 Telespiza, 638
ultramarina
 Aphelocoma, 589–590
 Vini, 133
ulula, Surnia, 170
umbellus, Bonasa, 59–60
umbra, Otus, 164
umbratilis, Rhinomyias, 448
umbretta, Scopus, 23
umbrina, Colluricincla, 486
umbrinodorsalis, Lagonosticta, 615
umbrovirens, Phylloscopus, 436
unappendiculatus, Casuarius, 2
unchall, Macropygia, 114
uncinatus, Chondrohierax, 35
undata, Sylvia, 444
undatus
 Celeus, 256
 Lybius, 237
underwoodii, Ocreatus, 207
undosus, Calamonastes, 424–425
undulata
 Anas, 30
 Chlamydotis, 87
 Gallinago, 96
undulatus
 Aceros, 231
 Crypturellus, 4
 Melopsittacus, 137
 Phylidonyris, 565
 Zebrilus, 22
unduligera, Frederickena, 290

unicincta, Columba, 110
unicinctus, Parabuteo, 45
unicolor
    Aerodramus, 185
    Aphelocoma, 590
    Apus, 190
    Chamaepetes, 56
    Chloropipo, 317
    Corvus, 700
    Cyanoramphus, 135
    Cyornis, 453
    Haematopus, 89
    Haplospiza, 673
    Lichenostomus, 561
    Mesitornis, 75
    Myadestes, 405
    Myrmotherula, 296
    Paradoxornis, 511
    Phrygilus, 667
    Scytalopus, 310
    Spreo, 601
    Sturnus, 599
    Thamnophilus, 292
    Turdus, 408
unicornis, Pauxi, 57
uniformis
    Chloropipo, 317
    Hylexetastes, 285
unirufa
    Cinnycerthia, 390
    Pseudoseisura, 278
    Synallaxis, 273
unirufus
    Centropus, 157
    Lipaugus, 313
Upucerthia, 268–269
Upupa, 228
Uraeginthus, 613–614
Uragus, 638
uralensis, Strix, 169
Uratelornis, 227
urbicum, Delichon, 361
Uria, 107
urichi, Phyllomyias, 318
urile, Phalacrocorax, 17
urinatrix, Pelecanoides, 14–15
Urochroa, 204
urochrysia, Chalybura, 202
Urocissa, 591
Urocolius, 212
Urocynchramus, 664
urogallus, Tetrao, 59
Uroglaux, 175
Urolais, 422
Uropelia, 118
urophasianus, Centrocercus, 60
Uropsalis, 183
Uropsila, 396
uropygialis
    Acanthiza, 520
    Cacicus, 697
    Carduelis, 632
    Lipaugus, 313
    Melanerpes, 247
    Phyllomyias, 319
    Sicalis, 675
    Zosterops, 552
Urorhipis, 424
Urosphena, 425
urosticta, Myrmotherula, 296
Urosticte, 207
urostictus, Pycnonotus, 377

Urothraupis, 664
Urotriorchis, 44
ursulae, Cinnyris, 540
urubambae, Scytalopus, 310
urubambensis, Asthenes, 276
urubitinga, Buteogallus, 44
urumutum, Nothocrax, 56
usambarae, Hyliota, 441
ussheri
    Muscicapa, 449
    Pitta, 266
    Scotopelia, 168
    Telacanthura, 188
usticollis, Eremomela, 434
ustulatus
    Catharus, 407
    Microcerculus, 397

vaalensis, Anthus, 364
vagabunda, Dendrocitta, 591–592
vagans, Cuculus, 153
vaillantii
    Picus, 259
    Trachyphonus, 235
valentini, Seicercus, 440
validirostris
    Lanius, 570
    Melithreptus, 562
    Upucerthia, 268
validus
    Corvus, 594
    Myiarchus, 344
    Pachyramphus, 347
    Reinwardtipicus, 260
valisineria, Aythya, 32
vana, Lonchura, 619
vanderbilti, Malacocincla, 700
Vanellus, 92–93
vanellus, Vanellus, 92
Vanga, 577
vanikorensis
    Aerodramus, 186
    Myiagra, 478
varia
    Grallaria, 305–306
    Mniotilta, 642
    Strix, 169
    Tangara, 660
variabilis, Emberiza, 666
variegata
    Ninox, 175
    Sula, 16
    Tadorna, 28
variegaticeps
    Alcippe, 507
    Anabacerthia, 279
variegatus
    Certhionyx, 559
    Crypturellus, 4
    Garrulax, 492
    Indicator, 244
    Merops, 225
    Mesitornis, 75
variolosus, Cacomantis, 154
varius
    Cuculus, 152
    Empidonomus, 342
    Fregilupus, 701
    Gallus, 71
    Phalacrocorax, 17
    Psephotus, 136
    Sittiparus, 528

Sphyrapicus, 248
Turnix, 76
varzeae, Picumnus, 245
vasa, Coracopsis, 141
vassali, Garrulax, 490
vassorii, Tangara, 662
vaughani
    Acrocephalus, 431
    Zosterops, 550
vauxi, Chaetura, 188–189
vegae, Larus, 102
velata, Philentoma, 576
velatus
    Enicurus, 461
    Ploceus, 607
    Xolmis, 339
velia, Tangara, 662
vellalavella, Zosterops, 553
velox
    Geococcyx, 160
    Turnix, 76
velutinus, Surniculus, 155
veneratus, Todiramphus, 221
venezuelanus, Pogonotriccus, 325
venezuelensis
    Diglossa, 674
    Myiarchus, 343
Veniliornis, 253–254
ventralis
    Accipiter, 43
    Amazona, 149
    Buteo, 46
    Gallinula, 85
    Phylloscartes, 324
venusta
    Chloropsis, 386
    Dacnis, 662
    Pitta, 266
venustulus, Pardaliparus, 525
venustus
    Cinnyris, 540
    Granatellus, 647
    Platycercus, 136
veraguensis
    Anthracothorax, 195
    Geotrygon, 119
veredus, Charadrius, 95
vermiculatus
    Burhinus, 90
    Megascops, 166
Vermivora, 639
vermivorum, Helmitheros, 642
vernalis, Loriculus, 140
vernans, Treron, 122
veroxii, Cyanomitra, 536
verreauxi
    Coua, 157
    Leptotila, 118
    Paradoxornis, 512
verreauxii, Aquila, 48
verrucosus, Phalacrocorax, 17
versicolor
    Agyrtria, 200
    Amazona, 150
    Anas, 31
    Columba, 700
    Ergaticus, 643
    Eubucco, 240
    Geotrygon, 119
    Lanio, 653
    Lichenostomus, 561
    Mayrornis, 474

versicolor (*continued*)
    Pachyramphus, 346
    Passerina, 691
    Phasianus, 73
    Pitta, 267
    Psitteuteles, 132
    Strepera, 582
versicolurus, Brotogeris, 147
versteri, Melanocharis, 544
verticalis
    Creurgops, 653
    Cyanomitra, 536
    Eremopterix, 352
    Hemispingus, 650
    Monarcha, 477
    Prioniturus, 138
    Tyrannus, 342
vesper, Rhodopis, 210
vespertinus
    Coccothraustes, 637
    Falco, 51
Vestiaria, 639
vestita, Eriocnemis, 206
vetula
    Coccyzus, 159
    Muscipipra, 340
    Ortalis, 54
vexillarius, Macrodipteryx, 183
vicina, Meliphaga, 560
vicinior
    Scytalopus, 309
    Vireo, 622
victor, Ptilinopus, 127
victoria, Goura, 121
victoriae
    Lamprolia, 479
    Lesbia, 207
    Ptiloris, 584
    Sitta, 530
victorini, Bradypterus, 427
Vidua, 620–621
viduata, Dendrocygna, 26
viduus, Monarcha, 477
vieilloti
    Coccyzus, 159
    Lybius, 237
    Sphecotheres, 569
vielliardi, Nyctiprogne, 179
vigil, Buceros, 230
vigorsii
    Aethopyga, 543
    Eupodotis, 87
viguieri, Dacnis, 663
vilasboasi, Lepidothrix, 315
vilcabambae, Schizoeaca, 271
vilissimus, Zimmerius, 324
villarejoi, Zimmerius, 324
villaviscensio, Campylopterus, 194
villosa, Sitta, 530
villosus
    Myiobius, 332
    Picoides, 253
vinacea
    Amazona, 150
    Streptopelia, 113
vinaceigula, Egretta, 19
vinaceus, Carpodacus, 630
vincens, Dicaeum, 546
vincenti, Emberiza, 665
Vini, 133
vinipectus, Alcippe, 507
vintsioides, Alcedo, 216

viola, Heliangelus, 206
violacea
    Anthobaphes, 536
    Euphonia, 626
    Geotrygon, 120
    Hyliota, 441
    Loxigilla, 673
    Musophaga, 152
    Nyctanassa, 21
violaceus
    Centropus, 157
    Cyanocorax, 588
    Ptilonorhynchus, 586
    Trogon, 213
violiceps
    Agyrtria, 201
    Goldmania, 198
violifer, Coeligena, 205
vipio, Grus, 77
virata, Lagonosticta, 615
virens
    Andropadus, 379
    Contopus, 335
    Dendroica, 641
    Hemignathus, 638
    Icteria, 646
    Megalaima, 238
    Sylvietta, 435
virenticeps, Buarremon, 678
Vireo, 621–623
vireo, Nicator, 382
Vireolanius, 624–625
virescens
    Butorides, 21
    Empidonax, 334
    Hypsipetes, 385
    Iole, 383
    Ixos, 384
    Lichenostomus, 561
    Phyllomyias, 318
    Phylloscartes, 324
    Pseudoleistes, 698
    Schiffornis, 318
    Tyranneutes, 318
virgata
    Aphriza, 98
    Asthenes, 276
    Ciccaba, 169–170
    Sterna, 106
virgaticeps, Thripadectes, 281
virgatus
    Accipiter, 42
    Garrulax, 491
    Sericornis, 518
virginiae, Vermivora, 639
virginianus
    Bubo, 166–167
    Colinus, 61
virgo, Anthropoides, 76
viridanum, Todirostrum, 329
viridanus, Picus, 259
viridicata
    Myiopagis, 319
    Pyrrhura, 145
viridicauda, Leucippus, 200
viridicollis, Tangara, 662
viridicyanus, Cyanolyca, 589
viridifacies, Erythrura, 617
viridiflavus, Zimmerius, 324
viridifrons, Agyrtria, 201
viridigaster, Saucerottia, 202
viridigenalis, Amazona, 149

viridigula, Anthracothorax, 195
viridipallens, Lampornis, 203
viridirostris, Phaenicophaeus, 156
viridis
    Androphobus, 513
    Anthracothorax, 195
    Anurolimnas, 79
    Artamella, 577
    Calyptomena, 264
    Carpococcyx, 156
    Centropus, 158
    Cochoa, 462
    Frederickena, 290
    Gecinulus, 260
    Gymnomyza, 565
    Megalaima, 238
    Merops, 226
    Neomixis, 499
    Pachyramphus, 346
    Picus, 259
    Psarocolius, 698
    Psophia, 77
    Pteroglossus, 242
    Ptilinopus, 126
    Sphecotheres, 569
    Telophorus, 575
    Terpsiphone, 473
    Tersina, 663
    Trogon, 213
viridissima, Aegithina, 387
viscivorus, Turdus, 410
vitellina, Dendroica, 641
vitellinus
    Manacus, 316
    Ploceus, 607
    Ramphastos, 243
vitiensis
    Clytorhynchus, 475
    Columba, 111
vitiosus, Lophotriccus, 327
vitriolina, Tangara, 661
vittata
    Amazona, 149
    Graueria, 434
    Melanodryas, 481
    Oxyura, 33
    Pachyptila, 11
    Sterna, 105
vittatum, Apaloderma, 212
vittatus
    Lanius, 570
    Picus, 259
vivida, Niltava, 453
vocifer, Haliaeetus, 37
vociferans
    Lipaugus, 313
    Tyrannus, 342
vociferoides, Haliaeetus, 37
vociferus
    Caprimulgus, 180
    Charadrius, 94
Volatinia, 669
vosseleri, Bubo, 167
vulcani, Junco, 687
vulcania, Cettia, 426
vulcanorum, Aerodramus, 186
vulgaris, Sturnus, 599
vulnerata, Myzomela, 559
vulneratum, Dicaeum, 547
vulpecula, Cranioleuca, 274
vulpina, Cranioleuca, 274
Vultur, 34

vulturina, Pionopsitta, 148
vulturinum, Acryllium, 74

waalia, Treron, 122
waddelli, Babax, 505
wagleri
    Aratinga, 143
    Aulacorhynchus, 241
    Icterus, 696
    Ortalis, 55
    Psarocolius, 697–698
wahlbergi, Aquila, 47
wahnesi, Parotia, 584
waigiuensis, Salvadorina, 29
wakensis, Gallirallus, 699
wakoloensis, Myzomela, 558
waldeni
    Aceros, 231
    Actinodura, 506
waldenii, Dicrurus, 578
wallacei
    Capito, 240
    Megapodius, 54
    Zosterops, 552
wallacii
    Aegotheles, 176
    Habroptila, 84
    Ptilinopus, 124–125
    Semioptera, 584
    Sipodotus, 515
walleri, Onychognathus, 601
wallichi, Catreus, 72
wallicus, Pezoporus, 137
wardi
    Harpactes, 215
    Pseudobias, 466
    Psittacula, 700
wardii, Zoothera, 402
warszewiczi, Dives, 693
watersi, Sarothrura, 78
waterstradti, Prioniturus, 138
watertonii, Thalurania, 198
watkinsi
    Grallaria, 307
    Incaspiza, 669
watsonii, Megascops, 166
webbianus, Paradoxornis, 512
weberi, Lipaugus, 313
websteri, Alcedo, 216
weddellii, Aratinga, 144
wellsi, Leptotila, 119
westermanni, Ficedula, 451
westlandica, Procellaria, 12
wetmorei
    Buthraupis, 657
    Rallus, 81
Wetmorethraupis, 657
weynsi, Ploceus, 608
whartoni, Ducula, 128
whistleri, Seicercus, 440
whiteheadi
    Aerodramus, 186
    Calyptomena, 264
    Harpactes, 215
    Sitta, 530
    Stachyris, 500
    Urocissa, 591
    Urosphena, 425
whitei, Conopophila, 566
whitelyi
    Caprimulgus, 181
    Pipreola, 312

whitemanensis, Melidectes, 564
whitii, Poospiza, 669
whitneyi
    Megalurulus, 442
    Micrathene, 172
    Pomarea, 474
    Synallaxis, 272
whytii
    Serinus, 635
    Stactolaema, 235
    Sylvietta, 435
wilhelminae, Charmosyna, 133
wilkinsi, Nesospiza, 668
willcocksi, Indicator, 243
williami, Metallura, 208
williamsi, Mirafra, 348
williamsoni, Muscicapa, 448
wilsoni
    Coeligena, 205
    Vidua, 621
Wilsonia, 643
wilsonia, Charadrius, 93–94
winchelli, Todiramphus, 219–220
winifredae, Bathmocercus, 428
wolfi, Aramides, 82
wollweberi, Baeolophus, 528
woodfordi
    Corvus, 594
    Nesoclopeus, 80
Woodfordia, 556
woodfordii, Strix, 169
woodhousei, Parmoptila, 611
woodi, Trichastoma, 493
woodwardi
    Amytornis, 517
    Colluricincla, 487
woosnami, Cisticola, 414
worcesteri, Turnix, 76
wortheni, Spizella, 683
wrightii, Empidonax, 334
wumizusume, Synthliboramphus, 108
wyatti, Asthenes, 276
wyvilliana, Anas, 30

xanthocephala, Tangara, 660
Xanthocephalus, 693
xanthocephalus, Xanthocephalus, 693
xanthochlorus, Pteruthius, 506
xanthochroa, Zosterops, 554
xanthocollis, Petronia, 603
xanthogaster, Euphonia, 627
xanthogastra
    Carduelis, 632
    Tangara, 660
xanthogenys
    Pachyramphus, 346
    Parus, 527
xanthogonys, Heliodoxa, 204
xanthogramma, Melanodera, 668
xantholaemus
    Pycnonotus, 378
    Serinus, 635
xantholophus, Dendropicos, 249
xantholora, Amazona, 149
xanthomus, Agelaius, 692
Xanthomyza, 566
xanthonotus
    Indicator, 244
    Oriolus, 567–568
xanthonura, Gallicolumba, 121

xanthophrys
    Crossleyia, 511
    Pseudonestor, 638
xanthophthalmus
    Agelasticus, 692
    Hemispingus, 650
xanthoprymna, Oenanthe, 464
xanthops
    Amazona, 150
    Forpus, 147
    Ploceus, 606
Xanthopsar, 692
xanthopterus
    Dysithamnus, 294
    Ploceus, 607
xanthopterygius, Forpus, 147
xanthopygaeus, Picus, 259
xanthopygius
    Heterospingus, 653
    Prionochilus, 545
    Serinus, 635
xanthopygos, Pycnonotus, 377
xanthorhynchus, Chrysococcyx, 154
xanthornus, Oriolus, 569
xanthorrhous, Pycnonotus, 376
xanthoschistos, Seicercus, 440
Xanthotis, 561–562
xantusii, Hylocharis, 199
xavieri, Phyllastrephus, 381
Xema, 103
Xenerpestes, 278
Xenicus, 347, 700
Xenocopsychus, 458
Xenodacnis, 663
Xenoglaux, 172
Xenoligea, 647
Xenoperdix, 68
Xenopipo, 317
Xenopirostris, 577
xenopirostris, Xenopirostris, 577
Xenops, 283
Xenopsaris, 346
xenopterus, Laterallus, 80
Xenornis, 293
Xenospingus, 668
Xenospiza, 685
xenothorax, Leptasthenura, 271
Xenotriccus, 333
Xenus, 97
Xiphidiopicus, 248
Xiphirhynchus, 497
Xiphocolaptes, 285–286
Xipholena, 313
Xiphorhynchus, 286–288
Xolmis, 339

yanacensis, Leptasthenura, 270
yarrellii
    Carduelis, 632
    Eulidia, 211
yaruqui, Phaethornis, 192
yelkouan, Puffinus, 12
yeltoniensis, Melanocorypha, 353
yemenensis, Carduelis, 633
yersini, Garrulax, 492
yessoensis, Emberiza, 665
yetapa, Gubernetes, 340
yncas, Cyanocorax, 588
ypecaha, Aramides, 82
ypiranganus, Emberizoides, 676
ypsilophora, Coturnix, 67–68

yucatanensis
  Amazilia, 200
  Myiarchus, 343
yucatanicus
  Campylorhynchus, 389
  Cyanocorax, 588
  Nyctiphrynus, 180
Yuhina, 510
yunnanensis
  Phylloscopus, 438
  Sitta, 530

zambesiae, Prodotiscus, 243
zantholeuca, Yuhina, 510
zanthopygia, Ficedula, 450
zappeyi, Paradoxornis, 512
Zaratornis, 311

zarumae, Thamnophilus, 291
Zavattariornis, 592
zealandicus, Cyanoramphus, 700
Zebrilus, 22
zelandica, Aplonis, 596
Zeledonia, 646
zelichi, Sporophila, 671
zena, Spindalis, 656
Zenaida, 116–117
zenkeri, Melignomon, 243
zeylanica, Megalaima, 238
zeylanicus, Pycnonotus, 375
zeylonensis, Ketupa, 168
zeylonica, Leptocoma, 537
zeylonus, Telophorus, 574–575
zimmeri
  Scytalopus, 310

  Synallaxis, 274
Zimmerius, 324
zoeae, Ducula, 129
zonaris, Streptoprocne, 184
zonarius, Barnardius, 136
zonatus, Campylorhynchus, 388
Zonerodius, 22
zoniventris, Falco, 51
Zonotrichia, 686
zonurus, Crinifer, 152
Zoonavena, 187
Zoothera, 402–404, 700
Zosterops, 549–555, 700
zosterops
  Hemitriccus, 327
  Phyllastrephus, 381
zusii, Heliangelus, 700

Index of English Names

Accentor
  Alpine, 400
  Black-throated, 400
  Brown, 400
  Himalayan, 400
  Japanese, 400
  Maroon-backed, 400
  Mongolian, 400
  Radde's, 400
  Robin, 400
  Rufous-breasted, 400
  Siberian, 400
  Yemen, 400
Adjutant
  Greater, 24
  Lesser, 24
Akalat
  Bocage's, 455
  East Coast, 455
  Equatorial, 455
  Gabela, 455
  Iringa, 455
  Lowland, 455
  Rubeho, 455
  Sharpe's, 455
  Usambara, 455
Akekee, 638
Akepa, 638
Akialoa
  Greater, 701
  Kauai, 701
  Lesser, 701
  Maui Nui, 701
Akiapolaau, 638
Akikiki, 638
Akohekohe, 639
Alauahio
  Maui, 638
  Oahu, 638
Albatross
  Black-browed, 9
  Black-footed, 9
  Buller's, 9
  Gray-headed, 9
  Laysan, 9
  Light-mantled, 10

Royal, 9
Short-tailed, 9
Shy, 9
Sooty, 9
Wandering, 9
Waved, 9
Yellow-nosed, 9
Alethe
  Brown-chested, 414
  Cholo, 414
  Fire-crested, 414
  Red-throated, 414
  White-chested, 414
Amakihi
  Greater, 701
  Hawaii, 638
  Kauai, 638
  Oahu, 638
Amaui, 700
Anhinga, 18
Ani
  Greater, 160
  Groove-billed, 160
  Smooth-billed, 160
Anianiau, 638
Ant-Tanager
  Black-cheeked, 654
  Crested, 654
  Red-crowned, 654
  Red-throated, 654
  Sooty, 654
Ant-Thrush
  Red-tailed, 401
  White-tailed, 401
Antbird
  Allpahuayo, 300
  Ash-breasted, 299
  Bananal, 299
  Band-tailed, 300
  Banded, 296
  Bare-crowned, 300
  Bare-eyed, 303
  Bertoni's, 298
  Bicolored, 302
  Black, 299
  Black-and-white, 300

Black-chinned, 300
Black-faced, 299
Black-headed, 300
Black-tailed, 300
Black-throated, 302
Blackish, 298
Caura, 301
Chestnut-backed, 301
Chestnut-crested, 303
Dot-backed, 302
Dull-mantled, 301
Dusky, 298
Dusky-tailed, 298
Esmeraldas, 301
Ferruginous, 298
Ferruginous-backed, 301
Goeldi's, 301
Gray, 299
Gray-bellied, 301
Gray-headed, 301
Hairy-crested, 303
Harlequin, 303
Immaculate, 302
Jet, 299
Long-tailed, 298
Lunulated, 303
Manu, 299
Mato Grosso, 299
Ocellated, 303
Ochre-rumped, 298
Pale-faced, 303
Parker's, 299
Plumbeous, 301
Rio Branco, 299
Rio de Janeiro, 299
Roraiman, 300
Rufous-tailed, 298
Rufous-throated, 303
Scale-backed, 302
Scaled, 298
Scalloped, 301
Silvered, 300
Slate-colored, 300
Slender, 299
Sooty, 302
Southern Chestnut-tailed, 301

Antbird (*continued*)
Spot-backed, 302
Spot-winged, 300
Spotted, 302
Squamate, 301
Striated, 298
Stripe-backed, 296
Stub-tailed, 301
Warbling, 300
White-bellied, 301
White-bibbed, 301
White-breasted, 303
White-browed, 299
White-lined, 301
White-masked, 302
White-plumed, 302
White-shouldered, 301
White-throated, 303
Willis', 298
Wing-banded, 302
Yapacana, 301
Yellow-browed, 300
Zimmer's, 301
Anteater-Chat
Northern, 465
Southern, 465
Antpecker
Jameson's, 611
Red-fronted, 611
Woodhouse's, 611
Antpipit
Ringed, 323
Southern, 323
Antpitta
Amazonian, 305
Bay, 306
Bicolored, 306
Black-crowned, 305
Brown-banded, 306
Chestnut, 307
Chestnut-crowned, 306–307
Chestnut-naped, 307
Crescent-faced, 308
Cundinamarca, 306
Elusive, 306
Fulvous-bellied, 305
Giant, 305
Gray-naped, 307
Great, 305
Hooded, 307
Jocotoco, 307
Masked, 304
Moustached, 306
Ochre-breasted, 307
Ochre-fronted, 308
Ochre-striped, 306
Pale-billed, 307
Peruvian, 308
Plain-backed, 306
Red-and-white, 306
Rufous, 307
Rufous-crowned, 305
Rufous-faced, 306
Rusty-breasted, 307
Rusty-tinged, 307
Santa Marta, 306
Scaled, 306
Scallop-breasted, 307
Slate-crowned, 308
Speckle-breasted, 305
Spotted, 304
Streak-chested, 304–305

Stripe-headed, 306
Tachira, 306
Tawny, 306
Tepui, 305
Thrush-like, 305
Undulated, 305
Variegated, 305–306
Watkins', 307
White-bellied, 307
White-browed, 305
White-lored, 305
White-throated, 306
Yellow-breasted, 307
Antshrike
Acre, 293
Amazonian, 292
Bamboo, 290
Band-tailed, 290
Bar-crested, 291
Barred, 291
Black, 291
Black-backed, 290
Black-crested, 290
Black-hooded, 291
Black-throated, 290
Blackish-gray, 291
Bluish-slate, 294
Castelnau's, 291
Chapman's, 291
Chestnut-backed, 291
Cinereous, 294
Cocha, 291
Collared, 290
Dusky-throated, 294
Fasciated, 290
Giant, 290
Glossy, 290
Great, 290
Large-tailed, 290
Lined, 291
Mouse-colored, 292
Pearly, 293
Plain-winged, 292
Rufous-capped, 291
Rufous-winged, 291
Russet, 293
Saturnine, 294
Silvery-cheeked, 290
Speckled, 293
Spot-backed, 290
Spot-winged, 293
Streak-backed, 292
Tufted, 290
Undulated, 290
Uniform, 292
Upland, 292
Variable, 293
White-bearded, 291
White-shouldered, 292
Antthrush
Barred, 304
Black-faced, 303–304
Black-headed, 304
Brazilian, 304
Mexican, 303
Rufous-breasted, 304
Rufous-capped, 303
Rufous-fronted, 304
Schwartz's, 304
Short-tailed, 304
Striated, 304
Such's, 304

Antvireo
Bicolored, 294
Plain, 293
Plumbeous, 294
Rufous-backed, 294
Spot-breasted, 293
Spot-crowned, 293
Streak-crowned, 293
White-streaked, 294
Antwren
Alagoas, 296
Amazonian, 294
Ancient, 297
Ash-throated, 296
Ash-winged, 298
Ashy, 296
Band-tailed, 296
Banded, 296
Black-bellied, 297
Black-capped, 296
Black-hooded, 297
Brown-backed, 295
Brown-bellied, 295
Caatinga, 296
Checker-throated, 295
Cherrie's, 294
Chestnut-shouldered, 298
Creamy-bellied, 296
Dot-winged, 297
Dugand's, 296
Foothill, 295
Gray, 296
Guianan, 294
Ihering's, 295
Klages', 294
Large-billed, 297
Leaden, 296
Long-winged, 296
Moustached, 294
Narrow-billed, 297
Orange-bellied, 298
Ornate, 295
Pacific, 294
Parana, 297
Pectoral, 297
Pileated, 296
Plain-throated, 295
Plain-winged, 296
Pygmy, 294
Restinga, 297
Rio de Janeiro, 296
Rio Suno, 295
Roraiman, 296
Rufous-bellied, 294
Rufous-rumped, 298
Rufous-tailed, 295
Rufous-winged, 297
Rusty-backed, 297
Salvadori's, 295
Sclater's, 294
Serra, 297
Slaty, 295
Spot-backed, 296
Spot-tailed, 296
Star-throated, 295
Stipple-throated, 295
Streak-capped, 298
Stripe-chested, 294
Todd's, 297
Unicolored, 296
White-eyed, 295
White-flanked, 295

White-fringed, 297
Yellow-breasted, 296–297
Yellow-rumped, 298
Yellow-throated, 294
Apalis
  Bamenda, 423
  Bar-throated, 422
  Black-capped, 422
  Black-collared, 422
  Black-faced, 423
  Black-headed, 423
  Black-throated, 422
  Brown-headed, 424
  Buff-throated, 423
  Chapin's, 423
  Chestnut-throated, 423
  Chirinda, 424
  Gosling's
  Gray, 424
  Karamoja, 424
  Masked, 423
  Rudd's, 423
  Ruwenzori, 422
  Sharpe's, 423
  White-winged, 423
  Yellow-breasted, 423
Apapane, 639
Apostlebird, 580
Araçari
  Black-necked, 242
  Chestnut-eared, 242
  Collared, 242
  Curl-crested, 242
  Fiery-billed, 242
  Green, 242
  Ivory-billed, 242
  Lettered, 242
  Many-banded, 242
  Red-necked, 242
Argus
  Crested, 74
  Great, 74
Asity
  Schlegel's, 265
  Sunbird, 265
  Velvet, 265
  Yellow-bellied, 265
Astrapia
  Arfak, 583
  Huon, 583
  Princess Stephanie's, 583
  Ribbon-tailed, 583
  Splendid, 583
Attila
  Bright-rumped, 345
  Cinnamon, 345
  Citron-bellied, 345
  Dull-capped, 345
  Gray-hooded, 345
  Ochraceous, 345
  Rufous-tailed, 345
Auk, Great, 700
Auklet
  Cassin's, 108
  Crested, 108
  Least, 108
  Parakeet, 108
  Rhinoceros, 108
  Whiskered, 108
Avadavat
  Green, 615
  Red, 615

Avocet
  American, 90
  Andean, 90
  Pied, 90
  Red-necked, 90
Avocetbill, Mountain, 209
Awlbill, Fiery-tailed, 195

Babax
  Chinese, 505
  Giant, 505
  Tibetan, 505
Babbler
  Abbott's, 493
  African Hill, 496
  Arabian, 503
  Arrow-marked, 505
  Ashy-headed, 493
  Bagobo, 493
  Bare-cheeked, 505
  Black-browed, 700
  Black-capped, 494
  Black-chinned, 500
  Black-crowned, 500
  Black-faced, 504
  Black-lored, 504
  Black-throated, 501
  Blackcap, 504
  Brown, 505
  Brown-capped, 493
  Buff-breasted, 494
  Buff-chested, 499
  Capuchin, 509
  Chestnut-capped, 502–503
  Chestnut-crowned, 511
  Chestnut-faced, 500
  Chestnut-rumped, 501
  Chestnut-winged, 501
  Common, 503
  Crescent-chested, 501
  Cretzschmar's, 504
  Crossley's, 511
  Dark-fronted, 502
  Deignan's, 499
  Dusky, 504
  Ferruginous, 493
  Flame-templed, 500
  Golden, 500
  Golden-crowned, 500
  Gray-breasted, 495
  Gray-crowned, 511
  Gray-headed, 501
  Gray-throated, 501
  Hall's, 511
  Hartlaub's, 504
  Horsfield's, 493
  Iraq, 503
  Jerdon's, 503
  Jungle, 504
  Large Gray, 503
  Marsh, 493
  Moustached, 494
  New Guinea, 511
  Orange-billed, 504
  Palawan, 494
  Puff-throated, 494
  Pygmy, 500
  Rufous, 504
  Rufous-capped, 500
  Rufous-crowned, 495
  Rufous-fronted, 500
  Rufous-tailed, 503

Rusty-crowned, 500
Scaly, 504
Scaly-crowned, 495
Short-tailed, 493
Slender-billed, 503
Snowy-throated, 501
Sooty, 501
Sooty-capped, 495
Spiny, 503
Spot-necked, 501
Spot-throated, 494
Striated, 503
Sulawesi, 493
Tawny-bellied, 502
Temminck's, 494
Thrush, 496
Vanderbilt's, 700
White-bibbed, 501
White-breasted, 500
White-browed, 511
White-chested, 493
White-hooded, 506
White-necked, 501
White-rumped, 504
White-throated, 503
Yellow-billed, 504
Yellow-eyed, 503
Balicassiao, 579
Bamboo-Partridge
  Chinese, 70
  Mountain, 70
Bamboo-Tyrant
  Brown-breasted, 327
  Drab-breasted, 327
  Flammulated, 327
Bamboowren, Spotted, 311
Bananaquit, 647
Barbet
  Anchieta's, 236
  Banded, 237
  Bearded, 238
  Black-backed, 237
  Black-banded, 238
  Black-billed, 237
  Black-breasted, 238
  Black-browed, 239
  Black-collared, 237
  Black-girdled, 240
  Black-spotted, 240
  Black-throated, 237
  Blue-eared, 239
  Blue-throated, 239
  Bornean, 239
  Bristle-nosed, 235
  Brown, 238
  Brown-breasted, 237
  Brown-chested, 240
  Brown-headed, 238
  Brown-throated, 238
  Chaplin's, 237
  Coppersmith, 239
  Crested, 235
  Crimson-fronted, 239
  D'Arnaud's, 235
  Double-toothed, 238
  Fire-tufted, 238
  Five-colored, 240
  Flame-fronted, 239
  Gilded, 240
  Gold-whiskered, 238
  Golden-naped, 239
  Golden-throated, 239

Barbet (*continued*)
Gray-throated, 235
Great, 238
Green, 236
Green-eared, 238
Hairy-breasted, 236
Lemon-throated, 240
Lineated, 238
Miombo, 237
Mountain, 239
Moustached, 239
Naked-faced, 235
Orange-fronted, 240
Pied, 237
Prong-billed, 240
Red-and-yellow, 235
Red-crowned, 238
Red-faced, 237
Red-fronted, 237
Red-headed, 240
Red-throated, 238
Red-vented, 238
Scarlet-banded, 240
Scarlet-crowned, 240
Scarlet-hooded, 240
Sladen's, 235
Spot-crowned, 240
Spot-flanked, 237
Toucan, 240
Versicolored, 240
Vieillot's, 237
White-cheeked, 238
White-eared, 235
White-headed, 237
White-mantled, 240
Whyte's, 235
Yellow-billed, 235
Yellow-breasted, 235
Yellow-crowned, 239
Yellow-fronted, 238
Yellow-spotted, 236
Barbtail
Roraiman, 278
Rusty-winged, 278
Spotted, 278
White-throated, 278
Barbthroat
Band-tailed, 192
Pale-tailed, 192
Bare-eye
Black-spotted, 303
Reddish-winged, 303
Barred-Woodcreeper
Amazonian, 286
Northern, 286
Barwing
Black-crowned, 506
Hoary-throated, 506
Rusty-fronted, 506
Spectacled, 506
Streak-throated, 506
Streaked, 506
Taiwan, 507
Bateleur, 38
Batis
Angola, 468
Black-headed, 467
Boulton's, 467
Cape, 467
Chinspot, 467
Fernando Po, 468
Gray-headed, 467

Ituri, 467
Pale, 467
Pririt, 467
Pygmy, 467
Ruwenzori, 467
Senegal, 467
Short-tailed, 467
Verreaux's, 467
West African, 468
Woodward's, 467
Bay-Owl
Congo, 162
Oriental, 162
Baza
Black, 25
Jerdon's, 34
Pacific, 34–35
Bearded-Greenbul
Eastern, 382
Western, 382
Beardless-Tyrannulet
Northern, 321
Southern, 321
Becard
Barred, 346
Black-and-white, 347
Black-capped, 346
Chestnut-crowned, 346
Cinereous, 346
Cinnamon, 346
Crested, 347
Glossy-backed, 347
Gray-collared, 347
Green-backed, 346
Jamaican, 347
One-colored, 347
Pink-throated, 347
Rose-throated, 347
Slaty, 346
White-winged, 346
Yellow-cheeked, 346
Bee-eater
Black, 225
Black-headed, 226
Blue-bearded, 225
Blue-breasted, 225
Blue-cheeked, 226
Blue-headed, 225
Blue-tailed, 226
Blue-throated, 226
Boehm's, 226
Chestnut-headed, 226
Cinnamon-chested, 225
European, 226
Green, 226
Little, 225
Madagascar, 226
Northern Carmine, 226
Purple-bearded, 225
Rainbow, 226
Red-bearded, 225
Red-throated, 225
Rosy, 226
Somali, 226
Southern Carmine, 226
Swallow-tailed, 225
White-fronted, 225
White-throated, 226
Bellbird
Bare-throated, 314
Bearded, 314
Chatham Island, 700

Crested, 483
New Zealand, 566–567
Three-wattled, 314
White, 314
Bentbill
Northern, 327
Southern, 327
Berryeater
Black-headed, 312
Hooded, 312
Berrypecker
Black, 544
Crested, 545
Fan-tailed, 544
Lemon-breasted, 544
Obscure, 544
Spotted, 544
Streaked, 544
Tit, 545
Besra, 42
Bird-of-paradise
Blue, 585
Crested, 582
Emperor, 585
Goldie's, 585
Greater, 585
King, 584
King-of-Saxony, 584
Lesser, 585
Loria's, 582
Macgregor's, 583
Magnificent, 584
Raggiana, 585
Red, 585
Superb, 584
Twelve-wired, 585
Wilson's, 584
Yellow-breasted, 582
Bishop
Black, 610
Black-winged, 610
Fire-fronted, 610
Golden-backed, 610
Orange, 610
Red, 610
Yellow, 610
Yellow-crowned, 610
Zanzibar, 610
Bittern
American, 22
Australasian, 23
Black, 22
Cinnamon, 22
Dwarf, 22
Forest, 22
Great, 23
Least, 22
Little, 22
New Zealand, 699
Pinnated, 22
Schrenck's, 22
Stripe-backed, 22
White-crested, 22
Yellow, 22
Black-Chat
White-fronted,
465–466
White-headed, 466
Black-Cockatoo
Glossy, 130
Red-tailed, 130
Slender-billed, 130

White-tailed, 130
Yellow-tailed, 130
Black-eye, Mountain, 556
Black-Flycatcher
  Northern, 447
  Southern, 447
  Yellow-eyed, 447
Black-Hawk
  Common, 44
  Great, 44
  Mangrove, 44
Black-Tit
  Southern, 526
  White-backed, 526
  White-shouldered, 526
  White-winged, 526
Black-Tyrant
  Amazonian, 336
  Blue-billed, 336
  Caatinga, 337
  Crested, 337
  Hudson's, 336
  Velvety, 337
  White-winged, 337
Blackbird
  Austral, 698
  Bolivian, 698
  Brewer's, 693
  Chestnut-capped, 692
  Chopi, 698
  Cuban, 693
  Eurasian, 408–409
  Forbes', 698
  Gray-winged, 408
  Jamaican, 693
  Melodious, 693
  Oriole, 698
  Pale-eyed, 692
  Red-breasted, 693
  Red-shouldered, 692
  Red-winged, 692
  Rusty, 693
  Saffron-cowled, 692
  Scarlet-headed, 698
  Scrub, 693
  Tawny-shouldered, 692
  Tricolored, 692
  Unicolored, 692
  White-browed, 693
  White-collared, 408
  Yellow-headed, 693
  Yellow-hooded, 692
  Yellow-shouldered, 692
  Yellow-winged, 692–693
Blackcap, 443
  Bush, 509
Blackstart, 465
Bleeding-heart
  Luzon, 120
  Mindanao, 120
  Mindoro, 120
  Negros, 120
  Sulu, 120
Blossomcrown, 202
Blue-Flycatcher
  African, 472
  Bornean, 454
  Hainan, 453
  Hill, 453
  Long-billed, 454
  Malaysian, 454
  Mangrove, 454

Palawan, 454
Pale, 453
Pale-chinned, 453
Pygmy, 454
Rueck's, 453
Sulawesi, 454
Tickell's, 454
Timor, 453
White-bellied, 453
White-tailed, 472
Blue-Pigeon
  Comoro, 127
  Madagascar, 127
  Mauritius, 700
  Seychelles, 127
Bluebill
  Grant's, 613
  Red-headed, 613
  Western, 613
Bluebird
  Eastern, 404
  Mountain, 405
  Western, 404–405
Bluebonnet, 136
Bluetail, Red-flanked, 456
Bluethroat, 456
Boatbill
  Black-breasted, 479
  Yellow-breasted, 479
Bobolink, 691
Bobwhite
  Black-throated, 61
  Crested, 61–62
  Northern, 61
Bokmakierie, 574–575
Boobook
  Chocolate, 174
  Northern, 174
  Southern, 174
  Sumba, 174
Booby
  Abbott's, 16
  Blue-footed, 16
  Brown, 16
  Masked, 16
  Nazca, 16
  Peruvian, 16
  Red-footed, 16
Boubou
  Bulo Burti, 573
  Fuelleborn's, 574
  Gabon, 574
  Mountain Sooty, 574
  Slate-colored, 574
  Sooty, 574
  Southern, 574
  Tropical, 574
  Turati's, 574
  Yellow-breasted, 574
Bowerbird
  Archbold's, 586
  Fawn-breasted, 586
  Fire-maned, 586
  Flame, 586
  Golden, 586
  Golden-fronted, 586
  Great, 586
  Macgregor's, 586
  Regent, 586
  Sanford's, 586
  Satin, 586
  Spotted, 586

Streaked, 586
Vogelkop, 586
Western, 586
Yellow-breasted, 586
Bracken-Warbler, Cinnamon, 427
Brambling, 626
Brant, 27
Brilliant
  Black-throated, 204
  Empress, 204
  Fawn-breasted, 203
  Green-crowned, 204
  Pink-throated, 204
  Rufous-webbed, 204
  Velvet-browed, 204
  Violet-fronted, 203
Bristle-Tyrant
  Antioquia, 325
  Chapman's, 325
  Marble-faced, 325
  Southern, 325
  Spectacled, 325
  Variegated, 325
  Venezuelan, 325
Bristlebill
  Common, 381
  Gray-headed, 382
  Green-tailed, 381
  Lesser, 382
Bristlebird
  Eastern, 517
  Rufous, 517
  Western, 517
Bristlefront
  Slaty, 311
  Stresemann's, 311
Bristlehead, Bornean, 581
Broadbill
  African, 264
  Banded, 264–265
  Black-and-red, 264
  Black-and-yellow, 265
  Dusky, 265
  Grauer's, 265
  Gray-headed, 264
  Green, 264
  Hose's, 264
  Long-tailed, 264
  Rufous-sided, 264
  Silver-breasted, 264
  Visayan, 265
  Wattled, 265
  Whitehead's, 264
Brolga, 77
Bronze-Cuckoo
  Horsfield's, 154
  Little, 154
  Rufous-throated, 154
  Shining, 154
  White-eared, 154
Bronzewing
  Brush, 115–116
  Common, 115
  Flock, 116
  New Britain, 115
  New Guinea, 115
Brownbul
  Northern, 380
  Terrestrial, 380
Brubru, 572

Brush-Finch
  Apurimac, 678
  Bay-crowned, 677
  Black-faced, 677
  Black-spectacled, 678
  Bolivian, 677
  Chestnut-capped, 678
  Cusco, 677
  Dusky-headed, 677
  Fulvous-headed, 678
  Green-striped, 678
  Moustached, 677
  Ochre-breasted, 678
  Pale-headed, 678
  Pale-naped, 676
  Rufous-capped, 677
  Rufous-eared, 678
  Rusty-bellied, 677
  Santa Marta, 677
  Slaty, 677
  Stripe-headed, 678
  Tepui, 678
  Tricolored, 677
  Vilcabamba, 677
  White-headed, 678
  White-naped, 676
  White-rimmed, 677
  White-winged, 677
  Yellow-breasted, 676–677
  Yellow-headed, 677
  Yellow-striped, 678
Brush-turkey
  Australian, 53
  Black-billed, 53
  Brown-collared, 53
  Bruijn's, 53
  Red-billed, 53
  Wattled, 53
Brush-Warbler
  Aldabra, 428
  Anjouan, 428
  Grand Comoro, 428
  Madagascar, 428
  Moheli, 428
  Rodrigues, 432
  Seychelles, 432
Brushrunner, Lark-like, 278
Budgerigar, 137
Buffalo-Weaver
  Red-billed, 604
  White-billed, 604
  White-headed, 604
Bufflehead, 33
Bulbul
  Ashy, 384
  Black, 385
  Black-and-white, 375
  Black-collared, 385
  Black-crested, 376
  Black-fronted, 377
  Black-headed, 375–376
  Blue-wattled, 377
  Brown-breasted, 375–376
  Brown-eared, 383–384
  Buff-vented, 383
  Cape, 377
  Chestnut, 384
  Common, 376–377
  Comoro, 384
  Cream-striped, 375
  Cream-vented, 378
  Finsch's, 382

Flavescent, 378
  Golden, 383
  Gray-bellied, 376
  Gray-cheeked, 382
  Gray-eyed, 383
  Gray-headed, 375
  Hairy-backed, 383
  Hook-billed, 383
  Light-vented, 376
  Madagascar, 384
  Mauritius, 385
  Mountain, 384
  Nicobar, 385
  Ochraceous, 382
  Olive, 383
  Olive-winged, 378
  Orange-spotted, 377
  Philippine, 383
  Puff-backed, 377
  Puff-throated, 382
  Red-eyed, 378
  Red-vented, 377
  Red-whiskered, 376
  Reunion, 384
  Scaly-breasted, 376
  Seychelles, 384
  Sooty-headed, 377
  Spectacled, 378
  Spot-necked, 375
  Straw-headed, 375
  Streak-breasted, 383
  Streak-eared, 378
  Streaked, 384
  Striated, 375
  Stripe-throated, 378
  Styan's, 376
  Sulphur-bellied, 383
  Sunda, 384
  White-browed, 378
  White-cheeked, 377
  White-eared, 377
  White-headed, 385
  White-spectacled, 377
  White-throated, 382
  Yellow-bellied, 383
  Yellow-browed, 383
  Yellow-eared, 378
  Yellow-streaked, 381
  Yellow-throated, 378
  Yellow-vented, 378
  Yellow-wattled, 377
  Yellowish, 384
  Zamboanga, 383
Bullfinch
  Barbados, 673
  Brown, 636
  Cuban, 672
  Eurasian, 636–637
  Gray-headed, 636
  Greater Antillean, 673
  Lesser Antillean, 673
  Orange, 636
  Puerto Rican, 673
  Red-headed, 636
  White-cheeked, 636
Bunting
  Black-faced, 666
  Black-headed, 666
  Blue, 691
  Brown-rumped, 666
  Cabanis', 666
  Cape, 665

Chestnut, 666
  Chestnut-breasted, 665
  Chestnut-eared, 665
  Cinereous, 665
  Cinnamon-breasted, 665
  Cirl, 664
  Corn, 667
  Crested, 664
  Cretzschmar's, 665
  Godlewski's, 664
  Golden-breasted, 666
  Gray, 666
  Gray-hooded, 665
  House, 665
  Indigo, 691
  Lark, 683
  Lark-like, 665
  Lazuli, 691
  Little, 666
  McKay's, 687
  Meadow, 664
  Ochre-rumped, 665
  Orange-breasted, 691
  Ortolan, 665
  Painted, 691
  Pallas', 666
  Pine, 664
  Red-headed, 666
  Reed, 666–667
  Rock, 664
  Rose-bellied, 691
  Rufous-backed, 664
  Rustic, 666
  Slaty, 664
  Snow, 687
  Socotra, 665
  Somali, 666
  Tibetan, 664
  Tristram's, 665
  Varied, 691
  Vincent's, 665
  Yellow, 666
  Yellow-breasted, 666
  Yellow-browed, 666
  Yellow-throated, 666
Bush-Crow, Stresemann's,
    592
Bush-hen
  Isabelline, 83
  Plain, 83
  Rufous-tailed, 83
  Talaud, 83
Bush-Quail
  Jungle, 68
  Manipur, 68
  Painted, 68
  Rock, 68
Bush-Robin
  Collared, 457
  Golden, 456–457
  Rufous-breasted, 457
  White-browed, 457
Bush-Tanager
  Ashy-throated, 650
  Black-backed, 664
  Common, 649
  Dusky, 650
  Gray-hooded, 650
  Pirre, 649
  Short-billed, 650
  Sooty-capped, 650
  Tacarcuna, 649

Yellow-green, 650
Yellow-throated, 650
Bush-Tyrant
Red-rumped, 338
Rufous-bellied, 338
Santa Marta, 338
Smoky, 338
Streak-throated, 338
Bush-Warbler
Aberrant, 426
African, 426–427
Brown, 428
Brownish-flanked, 426
Ceylon, 428
Chestnut-backed, 428
Chestnut-crowned, 426
Chinese, 428
Fiji, 426
Friendly, 428
Gray-sided, 426
Japanese, 425
Long-billed, 427
Long-tailed, 428
Manchurian, 425
Palau, 426
Pale-footed, 425
Philippine, 425
Russet, 428
Spotted, 427
Sunda, 426
Taiwan, 428
Tanimbar, 426
Yellowish-bellied, 426
Bushbird
Black, 293
Buff-banded, 442
Recurve-billed, 293
Rondonia, 293
Bushchat
Buff-streaked, 463
Gray, 463
Jerdon's, 463
Pied, 463
Timor, 463
White-browed, 462
White-throated, 462
Bushlark
Australasian, 348
Bengal, 348
Burmese, 348
Indian, 348
Indochinese, 348
Jerdon's, 348
Singing, 348
Bushshrike
Black-fronted, 575
Braun's, 573
Doherty's, 575
Fiery-breasted, 575
Four-colored, 575
Gabela, 573
Gray-green, 575
Gray-headed, 575
Green-breasted, 575
Lagden's, 575
Luehder's, 573
Many-colored, 575
Monteiro's, 575
Mt. Kupe, 575
Olive, 575
Red-naped, 573
Rosy-patched, 574

Sulphur-breasted, 575
Uluguru, 575
Bushtit, 515
Bustard
Arabian, 87
Australian, 87
Black, 88
Black-bellied, 88
Blue, 87
Buff-crested, 88
Great, 87
Hartlaub's, 88
Heuglin's, 87
Houbara, 87
Indian, 87
Karoo, 87
Kori, 87
Little, 88
Little Brown, 88
Ludwig's, 87
Macqueen's, 87
Nubian, 87
Red-crested, 88
Rueppell's, 88
Savile's, 88
Stanley, 87
White-bellied, 87
White-quilled, 88
Butcherbird
Black, 582
Black-backed, 581
Gray, 581
Hooded, 581
Pied, 582
Silver-backed, 581
Tagula, 581
Buttonquail
Barred, 75–76
Black-breasted, 76
Buff-breasted, 76
Chestnut-backed, 76
Hottentot, 75
Little, 76
Luzon, 76
Madagascar, 76
Painted, 76
Red-backed, 75
Red-chested, 76
Small, 75
Spotted, 75
Sumba, 76
Yellow-legged, 75
Buzzard
Archer's, 47
Augur, 47
Eurasian, 46
Grasshopper, 44
Gray-faced, 44
Jackal, 47
Lizard, 39
Long-legged, 47
Madagascar, 47
Mountain, 46
Red-necked, 47
Rufous-winged, 44
Upland, 47
White-eyed, 44
Buzzard-Eagle, Black-chested, 45

Cacholote
Brown, 279
Caatinga, 278

Gray-crested, 278
White-throated, 279
Cacique
Ecuadorian, 697
Golden-winged, 697
Mountain, 697
Red-rumped, 697
Scarlet-rumped, 697
Selva, 697
Solitary, 697
Yellow-billed, 697
Yellow-rumped, 697
Yellow-winged, 697
Cactus-Finch
Common, 687
Large, 687
Calyptura, Kinglet, 313
Camaroptera
Green-backed, 424
Olive-green, 424
Yellow-browed, 424
Canary
Black-faced, 634
Black-headed, 636
Black-throated, 634
Brimstone, 635
Cape, 634
Damara, 636
Forest, 634
Island, 633
Papyrus, 634
Protea, 636
White-bellied, 635
White-throated, 635
Yellow, 635
Yellow-crowned, 634
Yellow-fronted, 635
Canary-Flycatcher
Citrine, 454
Gray-headed, 454
Canastero
Austral, 276
Berlepsch's, 277
Cactus, 276
Canyon, 276
Cipo, 277
Cordilleran, 276
Creamy-breasted, 277
Dark-winged, 277
Dusky-tailed, 277
Hudson's, 276
Iquico, 276
Junin, 276
Lesser, 277
Line-fronted, 276
Many-striped, 276
Pale-tailed, 277
Patagonian, 277
Puna, 276
Rusty-fronted, 276
Scribble-tailed, 277
Short-billed, 277
Steinbach's, 277
Streak-backed, 276
Streak-throated, 276
Canvasback, 32
Capercaillie
Black-billed, 59
Eurasian, 59
Capuchinbird, 314
Caracara
Black, 49

Caracara (*continued*)
  Carunculated, 49
  Chimango, 49
  Crested, 49
  Guadalupe, 699
  Mountain, 49
  Red-throated, 49
  Southern, 49
  Striated, 49
  White-throated, 49
  Yellow-headed, 49
Cardinal
  Crimson-fronted, 676
  Northern, 690
  Red-capped, 676
  Red-cowled, 676
  Red-crested, 676
  Vermilion, 690
  Yellow, 676
  Yellow-billed, 676
Carib
  Green-throated, 195
  Purple-throated, 195
Casiornis
  Ash-throated, 343
  Rufous, 343
Cassowary
  Dwarf, 2
  Northern, 2
  Southern, 2
Catbird
  Abyssinian, 509
  Black, 397
  Gray, 397
  Green, 585–586
  Spotted, 585
  Tooth-billed, 586
  White-eared, 585
Cave-Chat, Angola, 458
Chachalaca
  Buff-browed, 55
  Chaco, 55
  Chestnut-winged, 54
  Gray-headed, 54
  Little, 55
  Plain, 54
  Rufous-bellied, 55
  Rufous-headed, 55
  Rufous-vented, 54
  Speckled, 55
  West Mexican, 55
  White-bellied, 55
Chaco-Finch, Many-colored, 667
Chaffinch, 625–626
  Blue, 626
Chanting-Goshawk
  Dark, 39
  Eastern, 39
  Pale, 39
Chat
  Boulder, 466
  Brown-tailed, 465
  Crimson, 522
  Desert, 522
  Familiar, 465
  Gray-throated, 647
  Herero, 459
  Indian, 465
  Karoo, 464
  Moorland, 465
  Orange, 522
  Red-breasted, 647

Rose-breasted, 647
Rueppell's, 465
Sicklewing, 464
Sombre, 465
Sooty, 465
Tractrac, 465
White-fronted, 522
Yellow, 522
Yellow-breasted, 646
Chat-Tanager
  Eastern, 652
  Western, 652
Chat-Tyrant
  Brown-backed, 337–338
  Crowned, 337
  D'Orbigny's, 338
  Golden-browed, 337
  Jelski's, 337
  Maroon-chested, 337
  Peruvian, 337
  Piura, 338
  Rufous-breasted, 337
  Slaty-backed, 337
  White-browed, 338
  Yellow-bellied, 337
Chatterer
  Fulvous, 503
  Rufous, 503
  Scaly, 503
Chickadee
  Black-capped, 524
  Boreal, 525
  Carolina, 524
  Chestnut-backed, 525
  Gray-headed, 525
  Mexican, 524
  Mountain, 524
Chiffchaff
  Canary Islands, 437
  Common, 437
  Iberian, 437
  Mountain, 437
Chilia, Crag, 269
Chlorophonia
  Blue-crowned, 628
  Blue-naped, 627–628
  Chestnut-breasted,
    628
  Golden-browed, 628
  Yellow-collared, 627
Chough
  Red-billed, 593
  White-winged, 580
  Yellow-billed, 593
Chowchilla, 513
Chuck-will's-widow, 180
Chukar, 64
Cicadabird, 370–371
Cinclodes
  Bar-winged, 269
  Blackish, 270
  Chilean Seaside, 269
  Comechingones, 269
  Dark-bellied, 269
  Gray-flanked, 269
  Long-tailesd, 269
  Olrog's, 269
  Peruvian Seaside, 269
  Royal, 269
  Stout-billed, 269
  White-bellied, 270
  White-winged, 270

Cisticola
  Aberdare, 417
  Ashy, 415
  Black-lored, 415
  Black-necked, 418
  Boran, 415
  Bubbling, 415
  Carruthers', 416
  Chattering, 414
  Chirping, 416
  Chubb's, 415
  Churring, 416
  Cloud, 418
  Cloud-scraping, 418
  Croaking, 417
  Desert, 418
  Dorst's, 415
  Foxy, 417
  Golden-headed, 419
  Gray, 416
  Hunter's, 415
  Madagascar, 418
  Pale-crowned, 418
  Pectoral-patch, 418
  Piping, 417
  Rattling, 415
  Red-faced, 414
  Red-headed, 416
  Red-pate, 415
  Rock-loving, 415
  Rufous, 417
  Siffling, 417
  Singing, 414
  Slender-tailed, 417
  Socotra, 418
  Stout, 416–417
  Tabora, 417
  Tana River, 416
  Tinkling, 416
  Tiny, 417
  Trilling, 414
  Wailing, 416
  Whistling, 414
  Winding, 416
  Wing-snapping,
    418–419
  Zitting, 417–418
Citril
  African, 634
  Southern, 634
  Western, 634
Cliff-Chat
  Mocking, 466
  White-winged, 466
Cochoa
  Green, 462
  Javan, 462
  Purple, 462
  Sumatran, 462
Cock-of-the-rock
  Andean, 314
  Guianan, 314
Cockatiel, 131
Cockatoo
  Blue-eyed, 131
  Ducorps', 130
  Gang-gang, 130
  Palm, 129
  Philippine, 130
  Pink, 130
  Salmon-crested, 131
  Sulphur-crested, 130

White, 131
Yellow-crested, 130
Coleto, 597
Collared-Dove
  African, 113
  Eurasian, 113
  Island, 113
  Red, 113
  White-winged, 113
Comet
  Bronze-tailed, 208
  Gray-bellied, 209
  Red-tailed, 208
Condor
  Andean, 34
  California, 34
Conebill
  Bicolored, 648
  Blue-backed, 648
  Capped, 648
  Chestnut-vented, 648
  Cinereous, 648
  Giant, 648
  Pearly-breasted, 648
  Rufous-browed, 648
  Tamarugo, 648
  White-browed, 648
  White-eared, 648
Coot
  American, 86
  Caribbean, 86
  Eurasian, 86
  Giant, 86
  Hawaiian, 86
  Horned, 86
  Red-fronted, 86
  Red-gartered, 86
  Red-knobbed, 85
  Slate-colored, 86
  White-winged, 86
Coquette
  Black-crested, 196
  Dot-eared, 196
  Festive, 196
  Frilled, 196
  Peacock, 196
  Racket-tailed, 196
  Rufous-crested, 196
  Short-crested, 196
  Spangled, 196
  Tufted, 196
  White-crested, 196
Cordonbleu
  Blue-breasted, 613
  Blue-capped, 614
  Red-cheeked, 613–614
Corella
  Little, 130
  Long-billed, 130
  Tanimbar, 130
  Western, 130
Cormorant
  Bank, 17
  Black-faced, 17
  Brandt's, 17
  Cape, 16
  Crowned, 18
  Double-crested, 16
  Flightless, 18
  Great, 16
  Guanay, 17
  Indian, 16

Japanese, 17
Little, 18
Little Black, 16
Little Pied, 18
Long-tailed, 18
Neotropic, 16
Pallas', 699
Pelagic, 17
Pied, 17
Pygmy, 18
Red-faced, 17
Red-legged, 17
Socotra, 17
Coronet
  Buff-tailed, 204
  Chestnut-breasted, 204
  Velvet-purple, 204
Cotinga
  Banded, 313
  Bay-vented, 311
  Black-and-gold, 313
  Black-faced, 314
  Black-tipped, 313
  Blue, 313
  Chestnut-bellied, 311
  Chestnut-crested, 311
  Gray-winged, 313
  Lovely, 313
  Plum-throated, 313
  Pompadour, 313
  Purple-breasted, 313
  Purple-throated, 313
  Red-crested, 311
  Shrike-like, 311–312
  Snowy, 313
  Spangled, 313
  Swallow-tailed, 312
  Turquoise, 313
  White-cheeked, 311
  White-tailed, 313
  White-winged, 313
  Yellow-billed, 313
Coua
  Blue, 157
  Coquerel's, 156
  Crested, 157
  Giant, 156
  Red-breasted, 156
  Red-capped, 157
  Red-fronted, 157
  Running, 157
  Snail-eating, 700
  Verreaux's, 157
Coucal
  Andaman, 158
  Bay, 157
  Biak, 157
  Black, 158
  Black-faced, 157
  Black-hooded, 158
  Black-throated, 158
  Blue-headed, 158
  Buff-headed, 157
  Coppery-tailed, 158
  Gabon, 158
  Goliath, 157
  Greater, 158
  Greater Black, 157
  Green-billed, 158
  Kai, 157
  Lesser, 158
  Lesser Black, 157

Madagascar, 158
Pheasant, 157
Philippine, 158
Pied, 157
Rufous, 157
Senegal, 158
Short-toed, 158
Sunda, 157
Violaceous, 157
White-browed, 158
Courser
  Bronze-winged, 91
  Burchell's, 91
  Cream-colored, 91
  Double-banded, 91
  Indian, 91
  Jerdon's, 91
  Temminck's, 91
  Three-banded, 91
Cowbird
  Bay-winged, 694
  Bronzed, 694
  Brown-headed, 695
  Giant, 695
  Screaming, 694
  Shiny, 694
Crab-Hawk, Rufous, 44
Crag-Martin
  Dusky, 361
  Eurasian, 361
Crake
  African, 82
  Andaman, 79
  Ash-throated, 83
  Australian, 83
  Baillon's, 83
  Band-bellied, 83
  Black, 83
  Black-banded, 79
  Black-tailed, 83
  Brown, 83
  Chestnut-headed, 79
  Colombian, 84
  Corn, 82
  Dot-winged, 83
  Gray-breasted, 79
  Henderson Island, 84
  Kosrae, 699
  Little, 83
  Ocellated, 78
  Paint-billed, 84
  Red-and-white, 80
  Red-legged, 79
  Red-necked, 79
  Ruddy, 79
  Ruddy-breasted, 83
  Rufous-faced, 80
  Rufous-sided, 79
  Russet-crowned, 79
  Rusty-flanked, 79
  Slaty-legged, 79
  Spotless, 84
  Spotted, 83
  Striped, 84
  Uniform, 82
  White-browed, 84
  White-throated, 79
  Yellow-breasted, 84
Crane
  Black-necked, 77
  Blue, 76
  Brolga, 77

Crane (*continued*)
Common, 77
Demoiselle, 76
Hooded, 77
Red-crowned, 77
Sandhill, 77
Sarus, 77
Siberian, 76
Wattled, 76
White-naped, 77
Whooping, 77
Creeper
Brown, 532
Hawaii, 638
Spotted, 532
Crescent-chest
Collared, 311
Elegant, 311
Marañón, 311
Olive-crowned, 311
Crested-Flycatcher
African, 472
Blue-headed, 472
Dusky, 472
White-bellied, 472
White-tailed, 472
Crimson-wing
Abyssinian, 612
Dusky, 612
Red-faced, 612
Shelley's, 612
Crocias
Gray-crowned, 509
Spotted, 509
Crombec
Cape, 435
Green, 435
Lemon-bellied, 435
Northern, 435
Red-capped, 435
Red-faced, 435
Short-billed, 435
Somali, 435
White-browed, 435
Crossbill
Hispaniola, 631
Parrot, 630
Red, 631
Scottish, 630
White-winged, 631
Crow
American, 594
Banggai, 700
Bougainville, 594
Brown-headed, 594
Cape, 594
Carrion, 594
Collared, 595
Cuban, 594
Cuban Palm, 594
Fish, 594
Flores, 593
Gray, 594
Guadalcanal, 594
Hawaiian, 594
Hispaniolan Palm, 594
Hooded, 594–595
House, 593
Jamaican, 594
Large-billed, 595
Little, 595
Long-billed, 594

Mariana, 594
New Caledonian, 593
Northwestern, 594
Pied, 595
Piping, 593
Sinaloa, 594
Slender-billed, 593
Somali, 595
Tamaulipas, 594
Torresian, 595
White-necked, 594
Crowned-Crane
Black, 76
Gray, 76
Crowned-Pigeon
Southern, 121
Victoria, 121
Western, 121
Cuckoo
African, 153
African Emerald, 155
Ash-colored, 159
Asian Emerald, 154
Banded Bay, 153
Barred Long-tailed, 153
Bay-breasted, 159
Black, 153
Black-bellied, 159
Black-billed, 159
Black-eared, 154
Brush, 154
Channel-billed, 155
Chestnut-bellied, 159
Chestnut-breasted, 154
Chestnut-winged, 152
Cocos Island, 159
Common, 153
Dark-billed, 159
Dideric, 155
Dusky Long-tailed,
153
Dwarf, 159
Fan-tailed, 154
Gray-capped, 159
Great Spotted, 152
Guira, 160
Himalayan, 153
Indian, 153
Klaas', 155
Lesser, 153
Levaillant's, 152
Little, 159
Long-billed, 155
Madagascar, 153
Mangrove, 159
Moluccan, 154
Olive Long-tailed, 153
Oriental, 153
Pallid, 153
Pavonine, 160
Pearly-breasted, 159
Pheasant, 160
Pied, 152
Plaintive, 153
Red-chested, 153
Squirrel, 159
Striped, 160
Sunda, 153
Thick-billed, 152
Violet, 154
Yellow-billed, 159
Yellow-throated, 155

Cuckoo-Dove
Andaman, 114
Barred, 114
Black-billed, 114
Brown, 114
Crested, 115
Dusky, 114
Great, 115
Little, 114–115
Mackinlay's, 114
Philippine, 114
Pied, 115
Ruddy, 114
Slaty, 115
Slender-billed, 114
White-faced, 115
Cuckoo-Hawk
African, 34
Madagascar, 34
Cuckoo-Roller, 228
Cuckoo-shrike
Ashy, 370
Bar-bellied, 369
Black, 373
Black-bellied, 372
Black-bibbed, 371
Black-faced, 368–369
Black-headed, 372
Black-winged, 372
Blackish, 371
Blue, 370
Boyer's, 369
Buru, 369
Cerulean, 369
Ghana, 374
Golden, 372
Grauer's, 370
Gray, 370
Gray-headed, 371
Ground, 368
Halmahera, 370
Hooded, 370
Indochinese, 372
Javan, 368
Kai, 371
Large, 368
Lesser, 372
Mauritius, 370
McGregor's, 372
Melanesian, 368
Moluccan, 369
New Caledonian, 370
New Guinea, 371
Oriole, 374
Pale-gray, 371
Papuan, 371
Petit's, 373
Pied, 369
Purple-throated, 373
Pygmy, 370
Red-shouldered, 373
Reunion, 370–371
Slaty, 368
Solomon Islands, 372
Stout-billed, 369
Sula, 371
Sulawesi, 371
Sumba, 371
Sunda, 368
Wallacean, 368
White-bellied, 369–370
White-breasted, 370

White-rumped, 369
White-winged, 372
Yellow-eyed, 369
Curassow
Alagoas, 56
Bare-faced, 57
Black, 57
Blue-knobbed, 57
Crestless, 56
Great, 57
Helmeted, 56
Horned, 57
Nocturnal, 56
Razor-billed, 56
Red-billed, 57
Salvin's, 56
Wattled, 57
Yellow-knobbed, 57
Curlew
Bristle-thighed, 97
Eskimo, 97
Eurasian, 97
Far Eastern, 97
Little, 97
Long-billed, 97
Slender-billed, 97
Currawong
Black, 582
Gray, 582
Pied, 582
Cut-throat, 615
Cutia, 505

Dacnis
Black-faced, 662
Black-legged, 662
Blue, 662–663
Scarlet-breasted, 663
Scarlet-thighed, 662
Tit-like, 663
Viridian, 663
White-bellied, 662
Yellow-bellied, 662
Dacnis-Tanager, Turquoise, 662
Dapple-throat, 493
Darter, 18
Dickcissel, 691
Dipper
American, 388
Brown, 388
Rufous-throated, 388
White-capped, 388
White-throated, 388
Diuca-Finch
Common, 668
White-winged, 668
Diucon, Fire-eyed, 339
Diving-Petrel
Common, 14–15
Magellanic, 14
Peruvian, 14
South Georgia, 14
Dodo, 700
Dollarbird, 227
Donacobius, Black-capped, 388
Doradito
Crested, 323
Dinelli's, 323
Subtropical, 323
Warbling, 323

Dotterel
Black-fronted, 94
Eurasian, 94
Inland, 94
Red-breasted, 93
Red-kneed, 93
Rufous-chested, 94
Tawny-throated, 94
Dove
African Mourning, 113
Amethyst, 121–122
Bar-shouldered, 116
Barred, 116
Brown-backed, 119
Caribbean, 119
Cloven-feathered, 127
Dark-eared, 122
Diamond, 116
Eared, 116–117
Emerald, 115
Forest, 111
Galapagos, 117
Golden, 127
Gray-chested, 119
Gray-fronted, 118–119
Gray-headed, 119
Grenada, 119
Inca, 118
Laughing, 113–114
Lemon, 110–111
Mourning, 116
Namaqua, 115
Ochre-bellied, 119
Orange, 127
Pacific, 117
Pallid, 119
Peaceful, 116
Red-eyed, 113
Ring-necked, 113
Scaled, 118
Socorro, 116
Spotted, 113
Stephan's, 115
Stock, 110
Tambourine, 115
Tolima, 119
Velvet, 127
Vinaceous, 113
White-eared, 121
White-faced, 118
White-tipped, 118
White-winged, 117
Zebra, 116
Zenaida, 117
Dovekie, 107
Dowitcher
Asian, 97
Long-billed, 97
Short-billed, 97
Drongo
Aldabra, 578
Andaman, 579
Ashy, 578
Black, 578
Bronzed, 578
Comoro, 578
Crested, 578
Crow-billed, 578
Fork-tailed, 578
Greater Racket-tailed, 579–580
Hair-crested, 579
Lesser Racket-tailed, 578–579

Mayotte, 578
Papuan, 577
Ribbon-tailed, 579
Shining, 577
Spangled, 579
Square-tailed, 577
Sulawesi, 579
Sumatran, 579
Velvet-mantled, 578
Wallacean, 579
White-bellied, 578
Drongo-Cuckoo
Asian, 155
Philippine, 155
Duck
African Black, 29
American Black, 30
Andean, 33
Black-headed, 33
Blue, 29
Blue-billed, 33
Comb, 29
Crested, 31
Falcated, 29
Freckled, 28
Harlequin, 32
Hartlaub's, 29
Hawaiian, 30
Labrador, 699
Lake, 33
Laysan, 30
Long-tailed, 32
Maccoa, 33
Mandarin, 29
Maned, 29
Masked, 33
Meller's, 30
Mottled, 30
Muscovy, 29
Musk, 33
Pacific Black, 31
Philippine, 31
Pink-eared, 32
Pink-headed, 699
Red-billed, 31
Ring-necked, 32
Ruddy, 33
Spectacled, 31
Spot-billed, 30
Torrent, 29
Tufted, 32
White-backed, 26
White-eyed, 32
White-headed, 33
White-winged, 29
Wood, 29
Yellow-billed, 30
Dunlin, 99
Dunnock, 400

Eagle
Bald, 37
Black, 47
Black-and-chestnut, 49
Bonelli's, 48
Booted, 48
Crested, 47
Crowned, 45
Golden, 48
Great Philippine, 47
Greater Spotted, 47
Gurney's, 47

Eagle (*continued*)
Harpy, 47
Imperial, 47
Indian Spotted, 47
Lesser Spotted, 47
Little, 48
Long-crested, 48
Martial, 48
New Guinea, 47
Rufous-bellied, 48
Short-toed, 38
Solitary, 45
Spanish, 47
Steppe, 47
Tawny, 47
Verreaux's, 48
Wahlberg's, 47
Wedge-tailed, 48
White-tailed, 37
Eagle-Owl
Akun, 168
Barred, 167
Cape, 167
Dusky, 167
Eurasian, 167
Fraser's, 167
Grayish, 167
Mindanao, 166
Pharaoh, 167
Philippine, 168
Rock, 167
Shelley's, 167
Spot-bellied, 167
Spotted, 167
Usambara, 167
Verreaux's, 167
Eared-Nightjar, Great, 179
Eared-Pheasant
Blue, 72
Brown, 72
White, 72
Earthcreeper
Band-tailed, 269
Bolivian, 269
Buff-breasted, 268
Chaco, 269
Plain-breasted, 268
Rock, 269
Scale-throated, 268
Straight-billed, 269
Striated, 269
White-throated, 268
Egret
Cattle, 20
Chinese, 20
Great, 19
Intermediate, 20
Little, 20
Reddish, 19
Slaty, 19
Snowy, 20
Eider
Common, 32
King, 32
Spectacled, 32
Steller's, 32
Elaenia
Brownish, 320
Caribbean, 319–320
Foothill, 319
Forest, 319
Gray, 319

Great, 320
Greater Antillean, 320
Greenish, 319
Highland, 320
Jamaican, 319
Large, 320
Lesser, 320
Mottle-backed, 320
Mountain, 320
Noronha, 320
Olivaceous, 320
Pacific, 319
Plain-crested, 320
Rufous-crowned, 320
Sierran, 321
Slaty, 320
Small-billed, 320
White-crested, 320
Yellow-bellied, 320
Yellow-crowned, 319
Elepaio, 474
Emerald
Andean, 201
Blue-tailed, 197
Brace's, 700
Canivet's, 197
Chiribiquete, 197
Coppery, 197
Coppery-headed, 198
Cozumel, 197
Cuban, 197
Garden, 197
Glittering-bellied, 197
Glittering-throated, 201
Golden-crowned, 197
Gould's, 700
Green-tailed, 197
Hispaniolian, 197
Honduran, 201
Narrow-tailed, 197
Plain-bellied, 200
Puerto Rican, 197
Rondonia, 200
Sapphire-spangled, 201
Short-tailed, 197
Versicolored, 200
White-bellied, 201
White-chested, 201
White-tailed, 198
Emu, 2
King Island, 699
Emu-tail
Brown, 428
Gray, 428
Emuwren
Mallee, 516
Rufous-crowned, 516
Southern, 516
Eremomela
Black-necked, 434
Burnt-neck, 434
Green-backed, 434
Greencap, 434
Rufous-crowned, 434
Salvadori's, 434
Senegal, 434
Turner's, 434
Yellow-bellied, 434
Yellow-rumped, 434
Yellow-vented, 434
Euphonia
Antillean, 627

Bronze-green, 627
Chestnut-bellied, 627
Elegant, 626
Finsch's, 626
Fulvous-vented, 627
Golden-rumped, 627
Golden-sided, 627
Green-chinned, 626
Jamaican, 626
Olive-backed, 627
Orange-bellied, 627
Orange-crowned, 626
Plumbeous, 626
Purple-throated, 626
Rufous-bellied, 627
Scrub, 626
Spot-crowned, 627
Tawny-capped, 627
Thick-billed, 626
Trinidad, 626
Velvet-fronted, 626
Violaceous, 626
White-lored, 627
White-vented, 627
Yellow-crowned, 626
Yellow-throated, 626

Fairy
Black-eared, 209
Purple-crowned, 209
Fairy-bluebird
Asian, 569
Philippine, 569
Fairywren
Blue-breasted, 516
Broad-billed, 515
Emperor, 516
Lovely, 516
Orange-crowned, 515
Purple-crowned, 516
Red-backed, 515
Red-winged, 516
Splendid, 516
Superb, 516
Variegated, 516
Wallace's, 515
White-shouldered, 515
White-winged, 515
Falcon
Amur, 51
Aplomado, 52
Barbary, 53
Bat, 52
Black, 52
Brown, 52
Eleonora's, 52
Gray, 52
Laggar, 52
Lanner, 52
Laughing, 49–50
New Zealand, 52
Orange-breasted, 52
Peregrine, 53
Prairie, 53
Pygmy, 50
Red-footed, 51
Red-necked, 51
Saker, 53
Sooty, 52
Taita, 53
White-rumped, 50

Falconet
    Black-thighed, 50
    Collared, 50
    Philippine, 50
    Pied, 50
    Spot-winged, 50
    White-fronted, 50
Fantail
    Arafura, 471
    Bismarck, 471
    Black, 470
    Black-and-cinnamon, 468
    Blue, 468
    Blue-headed, 468
    Brown, 470
    Brown-capped, 469
    Chestnut-bellied, 469
    Cinnamon-backed, 470
    Cinnamon-tailed, 469
    Dimorphic, 470
    Dusky, 470
    Friendly, 469
    Gray, 470
    Kandavu, 470
    Long-tailed, 471
    Malaita, 471
    Mangrove, 470
    Manus, 471
    Matthias, 471
    New Zealand, 470
    Northern, 468–469
    Palau, 471
    Pied, 469
    Pohnpei, 471
    Rennell, 470
    Rufous, 471
    Rufous-backed, 471
    Rufous-tailed, 468
    Rusty-flanked, 470
    Samoan, 470
    Spot-breasted, 468
    Spotted, 469
    Streaked, 470
    Streaky-breasted, 470
    White-bellied, 468
    White-browed, 468
    White-throated, 468
    White-winged, 469
    Yellow-bellied, 468
Fernbird, 442
    Chatham Islands, 700
Fernwren, 517
Field-Tyrant, Short-tailed, 340
Fieldfare, 410
Fieldwren
    Rufous, 519
    Striated, 519
Fig-Parrot
    Double-eyed, 135
    Edwards', 135
    Large, 135
    Orange-breasted, 134–135
    Salvadori's, 135
Figbird
    Australian, 569
    Green, 569
    Wetar, 569
Finch
    Black-crested, 668
    Black-masked, 667
    Black-throated, 616
    Canary-winged, 668

    Cassin's, 629
    Chestnut-eared, 616
    Cinereous, 668
    Citril, 634
    Coal-crested, 667
    Cocos Island, 673
    Corsican, 634
    Crimson, 616
    Crimson-breasted, 667
    Crimson-browed, 629
    Crimson-winged, 637
    Desert, 638
    Double-barred, 616
    Gold-naped, 637
    Gough Island, 668
    Gouldian, 617
    Gray-crested, 668
    House, 629
    Large-footed, 676
    Laysan, 638
    Long-tailed, 616
    Mangrove, 688
    Masked, 616
    Mongolian, 638
    Nightingale, 668
    Nihoa, 638
    Olive, 676
    Oriole, 628
    Peg-billed, 673
    Pileated, 667
    Plum-headed, 616
    Plush-capped, 663
    Purple, 629
    Red-crested, 667
    Red-headed, 615
    Saffron, 675
    Scarlet, 638
    Short-tailed, 668
    Slaty, 673
    Slender-billed, 668
    Sooty-faced, 676
    Spectacled, 637
    St. Lucia Black, 673
    Star, 616
    Sulphur-throated, 675
    Tanager, 664
    Trumpeter, 637
    Uniform, 673
    Vegetarian, 687
    Warbler, 688
    Wilkins', 668
    Woodpecker, 688
    Yellow-billed Blue, 691
    Yellow-bridled, 668
    Yellow-green, 676
    Yellow-thighed, 676
    Zebra, 616
Finchbill
    Collared, 375
    Crested, 375
Finfoot
    African, 86
    Masked, 86
Fire-eye
    Fringe-backed, 299
    White-backed, 299
    White-shouldered, 299
Fireback
    Crested, 72
    Crestless, 72
    Siamese, 72
Firecrest, 385

Firecrown
    Green-backed, 206
    Juan Fernandez, 206
Firefinch
    African, 615
    Bar-breasted, 614
    Black-bellied, 615
    Black-faced, 615
    Brown, 614
    Jameson's, 615
    Mali, 615
    Pale-billed, 615
    Red-billed, 614
    Reichenow's, 615
    Rock, 615
Firetail
    Beautiful, 616
    Diamond, 616
    Mountain, 616
    Painted, 616
    Red-browed, 616
    Red-eared, 616
Firethroat, 456
Firewood-gatherer, 278
Fiscal
    Common, 571–572
    Gray-backed, 571
    Long-tailed, 571
    Newton's, 572
    Somali, 571
    Taita, 571
Fish-Eagle
    African, 37
    Gray-headed, 37
    Lesser, 37
    Madagascar, 37
    Pallas', 37
Fish-Owl
    Blakiston's, 168
    Brown, 168
    Buffy, 168
    Tawny, 168
Fishing-Owl
    Pel's, 168
    Rufous, 168
    Vermiculated, 168
Flameback
    Black-rumped, 260
    Common, 259–260
    Greater, 260
    Himalayan, 259
Flamecrest, 385
Flamingo
    Andean, 26
    Caribbean, 25
    Chilean, 26
    Greater, 25
    Lesser, 26
    Puna, 26
Flatbill
    Dusky-tailed, 345
    Eye-ringed, 330
    Fulvous-breasted, 330
    Large-headed, 345
    Olivaceous, 330
    Pacific, 330
    Rufous-tailed, 345
Flicker
    Andean, 256
    Campo, 256
    Chilean, 256
    Fernandina's, 256

Flicker (*continued*)
  Gilded, 255–256
  Northern, 255
Florican
  Bengal, 88
  Lesser, 88
Flowerpecker
  Ashy, 547
  Bicolored, 546
  Black-fronted, 548
  Black-sided, 548
  Blood-breasted, 548
  Brown-backed, 546
  Cebu, 546
  Crimson-breasted, 545
  Crimson-crowned, 547
  Fire-breasted, 548
  Flame-breasted, 547
  Flame-crowned, 546
  Golden-rumped, 545
  Gray-sided, 548
  Louisiade, 548
  Midget, 548
  Mottled, 548
  Olive-backed, 545
  Olive-capped, 546
  Olive-crowned, 547
  Orange-bellied, 546–547
  Palawan, 545
  Pale-billed, 547
  Plain, 547
  Pygmy, 547
  Red-banded, 548
  Red-capped, 547–548
  Red-chested, 548
  Red-keeled, 546
  Red-striped, 546
  Scarlet-backed, 549
  Scarlet-breasted, 545
  Scarlet-collared, 546
  Scarlet-headed, 549
  Thick-billed, 546
  Whiskered, 546
  White-bellied, 547
  White-throated, 546
  Yellow-bellied, 546
  Yellow-breasted, 545
  Yellow-rumped, 545
  Yellow-sided, 546
  Yellow-vented, 546
Flowerpiercer
  Black, 674
  Black-throated, 674
  Bluish, 674
  Chestnut-bellied, 674
  Cinnamon-bellied, 673
  Deep-blue, 674
  Glossy, 674
  Gray-bellied, 674
  Greater, 674
  Indigo, 674
  Masked, 675
  Merida, 674
  Moustached, 674
  Rusty, 673–674
  Scaled, 674
  Slaty, 673
  Venezuelan, 674
  White-sided, 674
Flufftail
  Buff-spotted, 78
  Chestnut-headed, 78

Madagascar, 78
Red-chested, 78
Slender-billed, 78
Streaky-breasted, 78
Striped, 78
White-spotted, 78
White-winged, 78
Flycatcher
  Acadian, 334
  African Dusky, 449
  African Gray, 447
  Alder, 334
  Amazonian Royal, 331
  Apical, 344
  Ash-throated, 344
  Ashy, 449
  Ashy-breasted, 448
  Asian Brown, 448
  Atlas, 450
  Baird's, 341
  Beijing, 450
  Belted, 333
  Biak, 478
  Black-and-rufous, 452
  Black-banded, 452
  Black-billed, 333
  Black-capped, 334
  Black-tailed, 332–333
  Blue-and-white, 452
  Blue-breasted, 453
  Blue-crested, 478
  Blue-fronted, 453
  Blue-throated, 453
  Boat-billed, 342
  Boehm's, 449
  Bran-colored, 332
  Broad-billed, 478
  Brown-breasted, 449
  Brown-crested, 344
  Brown-streaked, 448
  Brownish, 331
  Buff-breasted, 334
  Canary, 480
  Cassin's, 449
  Chapada, 321
  Chapin's, 449
  Chat, 446
  Chestnut-capped, 471–472
  Cinnamon, 333
  Cinnamon-chested, 451
  Cliff, 333
  Cocos Island, 322
  Collared, 450
  Cordilleran, 334
  Crowned Slaty, 342
  Damar, 451
  Dark-sided, 448
  Dull, 479
  Dull-blue, 452
  Dusky, 334
  Dusky-blue, 449
  Dusky-capped, 343
  Dusky-chested, 341
  Euler's, 333
  European Pied, 450
  Fairy, 450
  Ferruginous, 449
  Fiscal, 447
  Flammulated, 345
  Flavescent, 332
  Fork-tailed, 343
  Furtive, 451

Fuscous, 333
Galapagos, 344
Gambaga, 448
Golden-bellied, 341
Golden-crowned, 341
Grand Comoro, 450
Gray, 334
Gray-breasted, 333
Gray-capped, 341
Gray-crowned, 330
Gray-hooded, 326
Gray-streaked, 448
Great Crested, 344
Grenada, 344
Guam, 477
Hammond's, 334
Handsome, 332
Inca, 325
Indigo, 452
Island, 452
Kashmir, 450
Korean, 450
La Sagra's, 344
Leaden, 478
Least, 334
Lemon-bellied, 479
Lemon-browed, 341
Lesser Antillean, 344
Little Gray, 449
Little Pied, 451
Little Slaty, 451
Livingstone's, 472
Lompobattang, 451
Mariqua, 447
Matinan, 453
McConnell's, 326
Melanesian, 478
Moluccan, 478
Mugimaki, 450
Narcissus, 450
Nilgiri, 452
Nimba, 447
Northern Royal, 331
Nutting's, 344
Oceanic, 477
Ochraceous-breasted, 332
Ochre-bellied, 325–326
Ochre-headed, 478
Olivaceous, 449
Olive-chested, 332
Olive-sided, 334
Olive-striped, 325
Olive-tufted, 336
Orange-banded, 332
Orange-crested, 332
Orange-eyed, 330
Ornate, 326
Pacific Royal, 331
Pacific-slope, 334
Palau, 477
Palawan, 451
Pale, 446
Pale-edged, 344
Panama, 344
Paperbark, 478
Pileated, 333
Pine, 334
Piratic, 340
Pohnpei, 477
Puerto Rican, 344
Red-breasted, 450
Restless, 478

Roraiman, 332
Ruddy-tailed, 333
Rufous, 343
Rufous-breasted, 325
Rufous-browed, 451
Rufous-chested, 451
Rufous-gorgeted, 450
Rufous-tailed, 344
Rufous-throated, 451
Russet-tailed, 451
Rusty-margined, 340
Rusty-tailed, 449
Sad, 343
Samoan, 478
Sapphire, 452
Satin, 478
Scissor-tailed, 343
Semicollared, 450
Sepia-capped, 325
Shining, 478–479
Short-crested, 344
Slaty-backed, 450
Slaty-blue, 452
Slaty-capped, 325
Snowy-browed, 450–451
Social, 340–341
Sooty, 449
Sooty-crowned, 344
Spotted, 448
Steel-blue, 478
Stolid, 344
Streak-necked, 325
Streaked, 341–342
Suiriri, 321
Sulphur-bellied, 341
Sulphur-rumped, 332
Sulphury, 342
Sumba, 451
Sumba Brown, 448
Swainson's, 343
Swamp, 449
Taiga, 450
Tawny-breasted, 332
Tawny-chested, 333
Tessmann's, 449
Three-striped, 341
Torrent, 479
Tufted, 335–336
Ultramarine, 452
Unadorned, 332
Ussher's, 449
Vanikoro, 478
Variegated, 342
Venezuelan, 343
Verditer, 452
Vermilion, 336
Ward's, 466
Whiskered, 332
White-bearded, 341
White-gorgeted, 451
White-ringed, 341
White-tailed, 453
White-throated, 334
Willow, 334
Yellow, 472
Yellow-bellied, 334
Yellow-breasted, 330–331
Yellow-footed, 449
Yellow-legged, 480
Yellow-margined, 330
Yellow-olive, 330
Yellow-rumped, 332

Yellow-throated, 341
Yellowish, 334
Yucatan, 343
Flycatcher-shrike
    Bar-winged, 375
    Black-winged, 375
Flycatcher-Thrush
    Finsch's, 401
    Rufous, 401
Flyrobin
    Golden-bellied, 479
    Olive, 480
Fody
    Forest, 609
    Mauritius, 609
    Red, 609
    Red-headed, 609
    Rodrigues, 610
    Seychelles, 610
Foliage-gleaner
    Alagoas, 280
    Black-capped, 280
    Brown-rumped, 281
    Buff-browed, 279
    Buff-fronted, 280
    Buff-throated, 281
    Chestnut-capped, 282
    Chestnut-crowned, 282
    Chestnut-winged, 280
    Cinnamon-rumped,
        280
    Crested, 280
    Guttulated, 279
    Henna-hooded, 282
    Lineated, 279
    Montane, 279
    Ochre-breasted, 280
    Olive-backed, 281
    Pará, 281
    Ruddy, 282
    Rufous-necked, 279
    Rufous-rumped, 280
    Rufous-tailed, 280
    Russet-mantled, 280
    Scaly-throated, 279
    Slaty-winged, 280
    White-browed, 279
    White-collared, 280
    White-eyed, 281
    White-throated, 281
Forest-Falcon
    Barred, 50
    Buckley's, 50
    Collared, 50
    Cryptic, 50
    Lined, 50
    Plumbeous, 50
    Slaty-backed, 50
Forest-Flycatcher
    African, 447
    White-browed, 447
Forest-Rail
    Chestnut, 78
    White-striped, 78
Forktail
    Black-backed, 461
    Chestnut-naped, 461
    Little, 461
    Slaty-backed, 461
    Spotted, 461
    Sunda, 461
    White-crowned, 461

Francolin
    Ahanta, 66
    Black, 65
    Cameroon, 67
    Cape, 66
    Chestnut-naped, 67
    Chinese, 65
    Clapperton's, 66
    Coqui, 65
    Crested, 65
    Djibouti, 67
    Double-spurred, 66
    Erckel's, 67
    Finsch's, 65
    Forest, 65
    Gray, 65
    Gray-breasted, 66
    Gray-striped, 66
    Gray-winged, 65
    Handsome, 67
    Hartlaub's, 66
    Harwood's, 66
    Heuglin's, 66
    Hildebrandt's, 66
    Jackson's, 66
    Moorland, 66
    Nahan's, 66
    Natal, 66
    Orange River, 66
    Painted, 65
    Red-billed, 66
    Red-necked, 66
    Red-winged, 65
    Ring-necked, 65
    Scaly, 66
    Schlegel's, 65
    Shelley's, 66
    Swainson's, 66
    Swamp, 65
    Swierstra's, 67
    White-throated, 65
    Yellow-necked, 66
Friarbird
    Black-faced, 563
    Brass', 563
    Dusky, 563
    Gray, 563
    Helmeted, 563
    Little, 563
    Meyer's, 563
    New Britain, 563
    New Caledonian, 564
    New Ireland, 563
    Noisy, 563
    Seram, 563
    Silver-crowned, 563
    Timor, 563
    White-naped, 563
    White-streaked, 563
Frigatebird
    Ascension Island, 18
    Christmas Island, 18
    Great, 18
    Lesser, 18
    Magnificent, 18
Frogmouth
    Ceylon, 177
    Dulit, 177
    Gould's, 177
    Hodgson's, 177
    Javan, 177
    Large, 177

Frogmouth (*continued*)
  Marbled, 177
  Papuan, 177
  Philippine, 177
  Short-tailed, 177
  Sunda, 177
  Tawny, 177
Fruit-Dove
  Atoll, 125
  Beautiful, 125–126
  Black-backed, 123
  Black-banded, 123–124
  Black-chinned, 124
  Black-naped, 126
  Blue-capped, 126
  Carunculated, 126
  Claret-breasted, 126
  Cook Islands, 125
  Coroneted, 125
  Cream-breasted, 124
  Crimson-crowned, 125
  Dwarf, 126
  Flame-breasted, 124
  Gray-green, 125
  Gray-headed, 126
  Henderson Island, 125
  Jambu, 124
  Knob-billed, 126
  Makatea, 125
  Many-colored, 125
  Mariana, 125
  Maroon-chinned, 124
  Negros, 127
  Orange-bellied, 126
  Orange-fronted, 124
  Ornate, 124
  Palau, 125
  Pink-headed, 124
  Pink-spotted, 124
  Rapa, 125
  Red-bellied, 125
  Red-eared, 124
  Red-moustached, 700
  Red-naped, 124
  Rose-crowned, 125
  Scarlet-breasted, 124
  Silver-capped, 125
  Superb, 125
  Tanna, 124
  Wallace's, 124–125
  White-breasted, 126
  White-capped, 125
  White-headed, 126
  Wompoo, 124
  Yellow-bibbed, 126
  Yellow-breasted, 124
Fruit-hunter, 413
Fruitcrow
  Bare-necked, 314
  Crimson, 314
  Purple-throated, 314
  Red-ruffed, 314
Fruiteater
  Band-tailed, 312
  Barred, 312
  Black-chested, 312
  Fiery-throated, 312
  Golden-breasted, 312
  Green-and-black, 312
  Handsome, 312
  Masked, 312
  Orange-breasted, 312

Red-banded, 312
  Scaled, 312
  Scarlet-breasted, 312
Fulmar
  Northern, 10
  Southern, 10
Fulvetta
  Brown, 508
  Brown-cheeked, 508
  Chinese, 507
  Dusky, 508
  Gold-fronted, 507
  Golden-breasted, 507
  Gray-cheeked, 508
  Javan, 508
  Ludlow's, 508
  Mountain, 509
  Nepal, 509
  Rufous-throated, 508
  Rufous-winged, 507
  Rusty-capped, 508
  Spectacled, 507–508
  Streak-throated, 508
  White-browed, 507
  Yellow-throated, 507

Gadwall, 29
Galah, 130
Gallinule
  Allen's, 85
  Azure, 85
  Purple, 85
  Spot-flanked, 85
Gallito
  Crested, 309
  Sandy, 309
Gannet
  Australian, 16
  Cape, 16
  Northern, 15
Garganey, 31
Geomalia, 402
Gerygone
  Biak, 521
  Brown, 521
  Brown-breasted, 522
  Chatham Island, 522
  Dusky, 521
  Fairy, 521
  Fan-tailed, 522
  Golden-bellied, 521
  Gray, 522
  Green-backed, 520
  Large-billed, 521
  Lord Howe, 699
  Mangrove, 522
  Mountain, 520
  Norfolk Island, 522
  Plain, 521
  Rufous-sided, 521
  Western, 522
  White-throated, 521
  Yellow-bellied, 521
Glossy-Starling
  Black-bellied, 600
  Bronze-tailed, 599
  Burchell's, 600
  Cape, 599
  Copper-tailed, 600
  Greater Blue-eared, 599
  Lesser Blue-eared, 599
  Long-tailed, 600

Meves', 600
  Principe, 600
  Purple, 600
  Purple-headed, 600
  Rueppell's, 600
  Sharp-tailed, 600
  Splendid, 600
Gnatcatcher
  Black-capped, 445
  Black-tailed, 445
  Blue-gray, 445
  California, 445
  Creamy-bellied, 446
  Cuban, 445
  Guianan, 446
  Iquitos, 446
  Masked, 446
  Slate-throated, 446
  Tropical, 446
  White-lored, 445–446
Gnateater
  Ash-throated, 308
  Black-bellied, 308
  Black-cheeked, 308
  Chestnut-belted, 308
  Chestnut-crowned, 308
  Hooded, 308
  Rufous, 308
  Slaty, 308
Gnatwren
  Collared, 445
  Long-billed, 445
  Tawny-faced, 445
Go-away-bird
  Bare-faced, 152
  Gray, 152
  White-bellied, 152
Godwit
  Bar-tailed, 97
  Black-tailed, 97
  Hudsonian, 97
  Marbled, 97
Goldcrest, 385
Golden-Plover
  American, 93
  Eurasian, 93
  Pacific, 93
Golden-Weaver
  African, 606
  Holub's, 606
  Principe, 607
  Taveta, 607
Goldeneye
  Barrow's, 33
  Common, 33
Goldentail, Blue-throated, 199
Goldenthroat
  Green-tailed, 200
  Tepui, 199
  White-tailed, 199
Goldfinch
  American, 633
  European, 633
  Lawrence's, 632
  Lesser, 632
Gonolek
  Black-headed, 574
  Common, 574
  Crimson-breasted, 574
  Papyrus, 574
Goose
  Andean, 28

Ashy-headed, 28
Bar-headed, 27
Barnacle, 27
Bean, 27
Blue-winged, 28
Cackling, 27
Canada, 27–28
Cape Barren, 28
Egyptian, 28
Emperor, 27
Greater White-fronted,
    27
Greylag, 27
Hawaiian, 28
Kelp, 28
Lesser White-fronted,
    27
Magpie, 26
Orinoco, 28
Pink-footed, 27
Red-breasted, 28
Ross', 27
Ruddy-headed, 28
Snow, 27
Spur-winged, 28
Swan, 27
Upland, 28
Goshawk
    African, 40
    Black, 43
    Black-mantled, 41
    Brown, 41
    Chestnut-shouldered, 43
    Chinese, 40
    Crested, 40
    Doria's, 44
    Fiji, 41
    Frances', 40
    Gabar, 40
    Gray, 41
    Gray-bellied, 40
    Gray-headed, 42
    Henst's, 43
    Meyer's, 43
    Moluccan, 41
    New Britain, 42
    New Caledonia, 41
    Northern, 43
    Pied, 41
    Red, 43
    Red-chested, 40
    Slaty-mantled, 41
    Spot-tailed, 41
    Sulawesi, 40
    Variable, 41
Grackle
    Boat-tailed, 694
    Carib, 694
    Common, 694
    Golden-tufted, 698
    Great-tailed, 694
    Greater Antillean, 694
    Mountain, 698
    Nicaraguan, 694
    Red-bellied, 698
    Slender-billed, 701
    Velvet-fronted, 698
Grandala, 461
Grass-Finch
    Duida, 676
    Lesser, 676
    Wedge-tailed, 675–676

Grass-Owl
    African, 161
    Australasian, 161
Grass-Warbler, Moustached, 428–429
Grassbird
    Bristled, 442
    Broad-tailed, 442
    Cape, 429
    Fan-tailed, 442
    Fly River, 442
    Little, 442
    Marsh, 441
    New Caledonian, 442
    Rufous-rumped, 442
    Striated, 442
    Tawny, 441–442
Grassquit
    Black-faced, 672
    Blue-black, 669
    Cuban, 672
    Dull-colored, 672
    Sooty, 672
    Yellow-faced, 672
    Yellow-shouldered, 672
Grasswren
    Black, 517
    Carpentarian, 517
    Dusky, 516
    Eyrean, 517
    Gray, 517
    Kalkadoon, 516
    Short-tailed, 517
    Striated, 517
    Thick-billed, 516
    White-throated, 517
Graveteiro, Pink-legged, 278
Graytail
    Double-banded, 278
    Equatorial, 278
Grebe
    Alaotra, 699
    Atitlan, 699
    Australasian, 8
    Clark's, 9
    Colombian, 699
    Eared, 8
    Great, 8
    Great Crested, 8
    Hoary-headed, 8
    Hooded, 9
    Horned, 8
    Junin, 9
    Least, 8
    Little, 7–8
    Madagascar, 8
    New Zealand, 8
    Pied-billed, 8
    Red-necked, 8
    Short-winged, 8
    Silvery, 9
    Western, 9
    White-tufted, 8
Green-Pigeon
    African, 123
    Bruce's, 122
    Flores, 122
    Large, 122
    Little, 122
    Madagascar, 122–123
    Pemba, 123
    Pompadour, 122
    São Tomé, 123

Sumba, 122
    Timor, 122
    Whistling, 123
Greenbul
    Ansorge's, 379
    Appert's, 381
    Baumann's, 381
    Cabanis', 380
    Cameroon Mountain,
        378
    Dusky, 381
    Fischer's, 380
    Golden, 380
    Gray, 379
    Gray-crowned, 381
    Gray-headed, 381
    Gray-olive, 380
    Honeyguide, 380
    Icterine, 381
    Joyful, 380
    Liberian, 381
    Little, 379
    Long-billed, 381
    Pale-olive, 380
    Plain, 379
    Prigogine's, 380
    Red-tailed, 382
    Sassi's, 381
    Shelley's, 379
    Simple, 380
    Sjostedt's, 380
    Slender-billed, 379
    Sombre, 379
    Spectacled, 381
    Spotted, 380
    Stripe-cheeked, 379
    Swamp, 380
    Tiny, 381
    White-bearded, 382
    White-throated, 381
    Xavier's, 381
    Yellow-bearded, 382
    Yellow-bellied, 380
    Yellow-necked, 380
    Yellow-throated, 380
    Yellow-whiskered, 379
Greenfinch
    Black-headed, 631
    European, 631
    Oriental, 633
    Vietnamese, 631
    Yellow-breasted, 631
Greenlet
    Ashy-headed, 624
    Brown-headed, 624
    Buff-cheeked, 624
    Dusky-capped, 624
    Golden-fronted, 624
    Gray-chested, 623–624
    Gray-eyed, 623
    Lemon-chested, 623
    Lesser, 624
    Olivaceous, 624
    Rufous-crowned, 623
    Rufous-naped, 624
    Scrub, 624
    Tawny-crowned, 624
    Tepui, 624
Greenshank
    Common, 98
    Nordmann's, 98
Grenadier, Purple, 614

Griffon
  Cape, 37
  Eurasian, 37
  Himalayan, 37
  Rueppell's, 37
Grosbeak
  Black-and-yellow, 637
  Black-backed, 690
  Black-faced, 689
  Black-headed, 690
  Black-thighed, 690
  Black-throated, 689
  Blue, 691
  Blue-black, 691
  Bonin, 701
  Collared, 637
  Crimson-collared, 690
  Evening, 637
  Glaucous-blue, 691
  Golden-bellied, 690
  Golden-winged, 628
  Hooded, 637
  Japanese, 637
  Kona, 701
  Pine, 629
  Red-and-black, 688
  Rose-breasted, 690
  São Tomé, 628
  Slate-colored, 689
  Somali, 628
  Spot-winged, 637
  Ultramarine, 691
  White-winged, 637
  Yellow, 690
  Yellow-billed, 637
  Yellow-green, 689–690
  Yellow-shouldered, 690
Grosbeak-Canary
  Northern, 635
  Southern, 635
Ground-Cuckoo
  Banded, 160
  Bornean, 156
  Coral-billed, 156
  Lesser, 160
  Red-billed, 160
  Rufous-vented, 160
  Rufous-winged, 160
  Scaled, 160
  Sumatran, 156
Ground-Dove
  Bare-eyed, 118
  Bare-faced, 118
  Black-winged, 118
  Blue, 118
  Blue-eyed, 118
  Bronze, 121
  Caroline Islands, 120
  Cinnamon, 120
  Common, 117
  Croaking, 118
  Ecuadorian, 117
  Friendly, 121
  Golden-spotted, 118
  Long-tailed, 118
  Maroon-chested, 118
  Marquesas, 121
  Palau, 121
  Picui, 117–118
  Plain-breasted, 117
  Polynesian, 120–121
  Purple-winged, 118
  Ruddy, 117

Santa Cruz, 121
  Sulawesi, 120
  Tanna, 700
  Thick-billed, 121
  Wetar, 121
  White-bibbed, 120
  White-throated, 121
Ground-Finch
  Large, 687
  Medium, 687
  Sharp-beaked, 687
  Small, 687
Ground-Hornbill
  Abyssinian, 231
  Southern, 231
Ground-Jay
  Iranian, 592
  Mongolian, 592
  Turkestan, 592
  Xinjiang, 592
Ground-Pigeon, Thick-billed, 121
Ground-Robin
  Greater, 479
  Lesser, 479
Ground-Roller
  Long-tailed, 227
  Pitta-like, 227
  Rufous-headed, 227
  Scaly, 227
  Short-legged, 227
Ground-Sparrow
  Prevost's, 680
  Rusty-crowned, 679
  White-eared, 680
Ground-Thrush
  Abyssinian, 403
  Black-eared, 403
  Crossley's, 403
  Gray, 403
  Kivu, 403
  Oberlaender's, 403
  Orange, 403
  Spotted, 403
Ground-Tyrant
  Black-fronted, 340
  Cinereous, 339
  Cinnamon-bellied, 340
  Dark-faced, 340
  Little, 339
  Ochre-naped, 340
  Plain-capped, 340
  Puna, 339
  Rufous-naped, 340
  Spot-billed, 339
  Taczanowski's, 339
  White-browed, 340
  White-fronted, 340
Groundcreeper, Canebrake, 278
Grouse
  Black, 59
  Blue, 57
  Caucasian, 59
  Hazel, 59
  Ruffed, 59–60
  Severtzov's, 59
  Sharp-tailed, 60
  Siberian, 58
  Sooty, 58
  Spruce, 57
Guaiabero, 135
Guan
  Andean, 55
  Band-tailed, 55

Baudo, 55
  Bearded, 55
  Black, 56
  Cauca, 55
  Chestnut-bellied, 56
  Crested, 55
  Dusky-legged, 56
  Highland, 56
  Horned, 56
  Marail, 55
  Red-faced, 55
  Rusty-margined, 55
  Sickle-winged, 56
  Spix's, 56
  Wattled, 56
  White-browed, 56
  White-crested, 56
  White-winged, 55
Guillemot
  Black, 107
  Pigeon, 107
  Spectacled, 107
Guineafowl
  Black, 74
  Crested, 74
  Helmeted, 74
  Plumed, 74
  Vulturine, 74
  White-breasted, 74
Gull
  American Herring, 102
  Andean, 102
  Armenian, 102
  Audouin's, 101
  Belcher's, 100
  Black-billed, 102
  Black-headed, 102
  Black-tailed, 100
  Bonaparte's, 102
  Brown-headed, 102
  Brown-hooded, 102
  California, 101
  Caspian, 102
  Dolphin, 100
  East Siberian, 102
  European Herring, 101
  Franklin's, 102
  Glaucous, 101
  Glaucous-winged, 101
  Gray, 100
  Gray-headed, 102
  Great Black-backed, 101
  Great Black-headed, 102
  Hartlaub's, 102
  Heermann's, 100
  Heuglin's, 101
  Iceland, 101
  Ivory, 103
  Kelp, 101
  Laughing, 102
  Lava, 102
  Lesser Black-backed, 101
  Little, 103
  Mediterranean, 102
  Mew, 101
  Olrog's, 100
  Pacific, 100
  Red-billed, 102
  Relict, 102
  Ring-billed, 101
  Ross', 103
  Sabine's, 103
  Saunders', 102

Silver, 102
Slaty-backed, 102
Slender-billed, 102
Sooty, 101
Steppe, 102
Swallow-tailed, 103
Thayer's, 101
Western, 101
White-eyed, 101
Yellow-footed, 101
Yellow-legged, 102
Gyrfalcon, 53

Hamerkop, 23
Hanging-Parrot
    Blue-crowned, 140
    Camiguin, 140
    Ceylon, 140
    Green-fronted, 141
    Moluccan, 140
    Papuan, 141
    Philippine, 140
    Pygmy, 141
    Sangihe, 140
    Sula, 140
    Sulawesi, 140
    Vernal, 140
    Wallace's, 141
    Yellow-throated, 141
Harrier
    Black, 39
    Cinereous, 39
    Long-winged, 39
    Montagu's, 39
    Northern, 39
    Pallid, 39
    Pied, 39
    Reunion, 39
    Spotted, 39
    Swamp, 39
Harrier-Hawk
    African, 39
    Madagascar, 39
Hawfinch, 637
Hawk
    Barred, 44
    Bat, 35
    Bicolored, 43
    Black-collared, 45
    Black-faced, 44
    Broad-winged, 45
    Chilean, 43
    Cooper's, 43
    Crane, 44
    Ferruginous, 47
    Galapagos, 46
    Gray, 45
    Gray-backed, 44
    Gundlach's, 43
    Harris', 45
    Hawaiian, 46
    Long-tailed, 44
    Mantled, 44
    Plain-breasted, 43
    Plumbeous, 44
    Puna, 46
    Red-backed, 46
    Red-shouldered, 45
    Red-tailed, 46
    Ridgway's, 45
    Roadside, 45
    Rough-legged, 47
    Rufous-tailed, 46

Rufous-thighed, 43
    Savanna, 45
    Semicollared, 43
    Semiplumbeous, 44
    Sharp-shinned, 43
    Short-tailed, 46
    Slate-colored, 44
    Swainson's, 46
    Tiny, 42
    White, 44
    White-browed, 44
    White-necked, 44
    White-rumped, 45
    White-tailed, 46
    White-throated, 46
    Zone-tailed, 46
Hawk-Cuckoo
    Common, 152
    Hodgson's, 153
    Large, 152
    Malaysian, 153
    Moustached, 153
    Northern, 153
    Philippine, 153
    Sulawesi, 152
Hawk-Eagle
    African, 48
    Ayres', 48
    Black, 49
    Black-and-white, 48
    Blyth's, 48
    Cassin's, 48
    Changeable, 48
    Crowned, 49
    Flores, 48
    Javan, 48
    Mountain, 48
    Ornate, 49
    Philippine, 49
    Sulawesi, 48
    Wallace's, 49
Hawk-Owl
    Andaman, 174
    Bismarck, 175
    Brown, 174
    Christmas Island, 175
    Cinnabar, 174
    Jungle, 175
    Little Sumba, 174
    Manus, 175
    Moluccan, 175
    New Britain, 175
    Northern, 170
    Ochre-bellied, 174
    Papuan, 175
    Philippine, 174
    Solomon, 175
    Speckled, 175
    Togian, 174
Heathwren
    Chestnut-rumped, 519
    Shy, 519
Helmetcrest, Bearded,
    208
Helmetshrike
    Angola, 576
    Chestnut-bellied, 576
    Chestnut-fronted, 576
    Gray-crested, 576
    Retz's, 576
    Rufous-bellied, 576
    White, 576
    Yellow-crested, 576

Hemispingus
    Black-capped, 650
    Black-eared, 651
    Black-headed, 650
    Drab, 650
    Gray-capped, 651
    Oleaginous, 651
    Orange-browed, 650
    Parodi's, 650
    Piura, 650
    Rufous-browed, 650
    Slaty-backed, 650
    Superciliaried, 650
    Three-striped, 650
    White-browed, 650
Hermit
    Black-throated, 193
    Broad-tipped, 192
    Bronzy, 192
    Buff-bellied, 193
    Cinnamon-throated, 193
    Dusky-throated, 193
    Eastern Long-tailed, 192
    Gray-chinned, 193
    Great-billed, 192
    Green, 192
    Hook-billed, 192
    Koepcke's, 193
    Little, 193
    Minute, 193
    Needle-billed, 193
    Pale-bellied, 193
    Planalto, 193
    Reddish, 193
    Rufous-breasted, 192
    Saw-billed, 191
    Scale-throated, 193
    Sooty-capped, 193
    Straight-billed, 193
    Streak-throated, 193
    Stripe-throated, 193
    Tawny-bellied, 193
    Western Long-tailed, 192
    White-bearded, 192
    White-browed, 193
    White-whiskered, 192
Heron
    Agami, 21
    Black, 19
    Black-headed, 19
    Boat-billed, 22
    Capped, 19
    Cocoi, 19
    Goliath, 19
    Gray, 19
    Great Blue, 19
    Great-billed, 19
    Green, 21
    Humblot's, 19
    Little Blue, 20
    Pacific, 19
    Pied, 19
    Purple, 19
    Rufous-bellied, 20
    Squacco, 20
    Striated, 20–21
    Tricolored, 20
    Whistling, 19
    White-bellied, 19
    White-faced, 20
    Zigzag, 22
Hillstar
    Andean, 204

Hillstar (*continued*)
Black-breasted, 205
Chimborazo, 204
Green-headed, 205
Wedge-tailed, 205
White-sided, 205
White-tailed, 204
Hoatzin, 74
Hobby
African, 52
Australian, 52
Eurasian, 52
Oriental, 52
Honey-buzzard
Barred, 35
Black, 35
European, 35
Long-tailed, 35
Oriental, 35
Honeycreeper
Golden-collared, 662
Green, 663
Purple, 663
Red-legged, 663
Shining, 663
Short-billed, 663
Honeyeater
Arfak, 565
Banded, 559
Bar-breasted, 566
Barred, 565
Black, 559
Black-backed, 564
Black-chested, 557
Black-chinned, 562
Black-headed, 562
Black-throated, 560
Blue-faced, 566
Bonin, 562
Bougainville, 556
Bridled, 560
Brown, 556
Brown-backed, 566
Brown-headed, 562
Buru, 557
Crescent, 565
Crow, 565
Dark-brown, 557
Dark-eared, 565
Dwarf, 545
Eungella, 560
Forest, 559
Fuscous, 561
Giant, 565
Graceful, 560
Gray, 566
Gray-fronted, 561
Gray-headed, 561
Greater Streaked, 565
Green-backed, 556
Guadalcanal, 560
Indonesian, 557
Kandavu, 562
Leaden, 564
Lewin's, 560
Long-billed, 556
Macleay's, 562
Mangrove, 561
Marbled, 563
Mayr's, 564
Mimic, 560
New Hebrides, 566

New Holland, 565
Obscure, 560
Olive, 556
Olive-streaked, 564
Orange-cheeked, 562
Painted, 566
Pied, 559
Plain, 562–563
Puff-backed, 559
Purple-gaped, 561
Pygmy, 545
Regent, 566
Rufous-backed, 564
Rufous-banded, 566
Rufous-sided, 564
Rufous-throated, 566
Scrub, 559
Seram, 557
Silver-eared, 557
Singing, 561
Smoky, 565
Spangled, 565
Spiny-cheeked, 567
Spotted, 562
Streak-breasted, 560
Streak-headed, 563
Striped, 566
Strong-billed, 562
Sunda, 556
Tagula, 560
Tawny-breasted,
561–562
Tawny-crowned, 566
Varied, 561
Wattled, 560
White-cheeked, 565
White-eared, 561
White-fronted, 565
White-gaped, 561
White-lined, 560
White-naped, 562
White-plumed, 561
White-streaked, 557
White-throated, 562
White-tufted, 557
Yellow, 561
Yellow-eared, 557
Yellow-faced, 560
Yellow-gaped, 560
Yellow-plumed, 561
Yellow-spotted, 560
Yellow-throated, 561
Yellow-tinted, 561
Yellow-tufted, 561
Honeyguide
Cassin's, 243
Dwarf, 243
Greater, 244
Green-backed, 243
Least, 243–244
Lesser, 244
Lyre-tailed, 244
Malaysian, 244
Pallid, 243
Scaly-throated, 244
Spotted, 244
Thick-billed, 244
Wahlberg's, 243
Willcock's, 243
Yellow-footed, 243
Yellow-rumped, 244
Zenker's, 243

Hookbill
Chestnut-winged, 280
Lanai, 701
Hoopoe
Eurasian, 228
Madagascar, 228
Hoopoe-Lark
Greater, 352
Lesser, 352
Hornbill
African Gray, 229
African Pied, 229
Black, 230
Black Dwarf, 229
Black-and-white-casqued,
231
Black-casqued, 231
Blyth's, 231
Bradfield's, 229
Brown, 230
Brown-cheeked, 231
Bushy-crested, 230
Ceylon Gray, 230
Crowned, 229
Eastern Yellow-billed, 229
Great, 230
Helmeted, 230
Hemprich's, 229
Indian Gray, 230
Jackson's, 229
Knobbed, 231
Luzon, 230
Malabar Gray, 229
Mindanao, 230
Mindoro, 230
Monteiro's, 229
Narcondam, 231
Palawan, 230
Pale-billed, 229
Piping, 231
Plain-pouched, 231
Red-billed, 229
Red-billed Dwarf, 229
Rhinoceros, 230
Rufous, 230
Rufous-necked, 230
Rusty-cheeked, 230
Samar, 230
Silvery-cheeked, 231
Southern Yellow-billed,
229
Sulawesi, 230
Sulu, 230
Sumba, 231
Taritic, 230
Trumpeter, 231
Von der Decken's, 229
White-crested, 229
White-crowned, 230
White-thighed, 231
Wreathed, 231
Wrinkled, 230
Writhe-billed, 231
Writhed, 231
Yellow-casqued, 231
Hornero
Bay, 270
Crested, 270
Lesser, 270
Pale-legged, 270
Rufous, 270
Tail-banded, 270

Huet-huet
  Black-throated, 308
  Chestnut-throated, 308
Huia, 701
Hummingbird
  Alfaros, 202
  Allen's, 211
  Amazilia, 200
  Amethyst-throated, 203
  Anna's, 210
  Antillean Crested, 196
  Azure-crowned, 201
  Beautiful, 210
  Bee, 210
  Berylline, 202
  Black-bellied, 198
  Black-chinned, 210
  Blue-capped, 198
  Blue-chested, 201
  Blue-headed, 198
  Blue-tailed, 202
  Blue-throated, 203
  Broad-billed, 198
  Broad-tailed, 211
  Buff-bellied, 200
  Buffy, 200
  Bumblebee, 210
  Calliope, 211
  Charming, 201
  Chestnut-bellied, 200
  Cinnamon, 200
  Copper-rumped, 202
  Copper-tailed, 202
  Costa's, 210
  Dusky, 198
  Emerald-chinned, 196
  Fiery-throated, 197
  Garnet-throated, 203
  Giant, 206
  Glow-throated, 211
  Green-and-white, 200
  Green-bellied, 202
  Green-fronted, 201
  Indigo-capped, 201
  Loja, 200
  Lucifer, 210
  Magnificent, 204
  Mangrove, 201
  Many-spotted, 200
  Oasis, 210
  Olive-spotted, 200
  Purple-chested, 201
  Ruby-throated, 210
  Ruby-topaz, 195
  Rufous, 211
  Rufous-cheeked, 198
  Rufous-tailed, 200
  Sapphire-bellied, 199
  Sapphire-throated, 199
  Scaly-breasted, 194
  Scintillant, 211
  Scissor-tailed, 204
  Shining-green, 199
  Snowy-bellied, 202
  Sombre, 194
  Sparkling-tailed, 210
  Speckled, 203
  Spot-throated, 200
  Steely-vented, 201
  Stripe-tailed, 198
  Swallow-tailed, 194
  Sword-billed, 206

  Tooth-billed, 194
  Tumbes, 200
  Vervain, 210
  Violet-bellied, 199
  Violet-capped, 198
  Violet-chested, 204
  Violet-crowned, 201
  Violet-headed, 196
  Volcano, 211
  Wedge-billed, 209
  White-bellied, 200
  White-eared, 199
  White-tailed, 198
  White-throated, 199
  Wine-throated, 210
  Xantus', 199
Hwamei, 491
Hylia, Green, 436
Hyliota
  Southern, 441
  Usambara, 441
  Violet-backed, 441
  Yellow-bellied, 441
Hypocolius, 387

Ibis
  Andean, 25
  Australian, 24
  Bald, 24
  Bare-faced, 25
  Black-faced, 25
  Black-headed, 24
  Buff-necked, 25
  Crested, 24
  Giant, 24
  Glossy, 25
  Green, 25
  Hadada, 24
  Madagascar, 25
  Olive, 24
  Plumbeous, 25
  Puna, 25
  Red-naped, 24
  Sacred, 24
  Scarlet, 25
  Sharp-tailed, 25
  Spot-breasted, 24
  Straw-necked, 24
  Wattled, 25
  White, 25
  White-faced, 25
  White-shouldered, 24
Ibisbill, 90
Ifrita, Blue-capped, 514
Iiwi, 639
Illadopsis
  Blackcap, 495
  Brown, 495
  Gray-chested, 496
  Mountain, 495
  Pale-breasted, 495
  Puvel's, 495
  Rufous-winged, 495
  Scaly-breasted, 495
Imperial-Pigeon
  Baker's, 128
  Bismarck, 129
  Chestnut-bellied, 128
  Christmas Island, 128
  Cinnamon-bellied, 128
  Collared, 129
  Dark-backed, 129

  Elegant, 127
  Finsch's, 128
  Gray, 128
  Gray-headed, 127
  Green, 127
  Island, 128
  Marquesas, 128
  Micronesian, 128
  Mindoro, 127
  Mountain, 129
  New Caledonian, 128
  Pacific, 127
  Peale's, 128
  Pied, 129
  Pink-bellied, 127
  Pink-headed, 128
  Pinon, 128
  Polynesian, 128
  Purple-tailed, 128
  Red-knobbed, 127
  Rufescent, 128
  Spice, 128
  Spotted, 127
  Timor, 129
  Torresian, 129
  White, 129
  White-bellied, 127
  White-eyed, 127
  Zoe, 129
Inca
  Black, 205
  Bronzy, 205
  Brown, 205
  Collared, 205
  Gould's, 205
Inca-Finch
  Buff-bridled, 669
  Gray-winged, 669
  Great, 668
  Little, 669
  Rufous-backed, 668
Indigobird
  Baka, 621
  Cameroon, 621
  Green, 621
  Jambandu, 621
  Jos Plateau, 621
  Pale-winged, 621
  Purple, 621
  Quailfinch, 621
  Variable, 621
  Village, 620
Iora
  Common, 386
  Great, 387
  Green, 387
  White-tailed, 387

Jabiru, 24
Jacamar
  Blue-cheeked, 232
  Bluish-fronted, 232
  Bronzy, 232
  Brown, 232
  Coppery-chested, 232
  Dusky-backed, 231
  Great, 232
  Green-tailed, 232
  Pale-headed, 231
  Paradise, 232
  Purplish, 232
  Purus, 231

Jacamar (*continued*)
 Rufous-tailed, 232
 Three-toed, 232
 White-chinned, 232
 White-eared, 231
 White-throated, 232
 Yellow-billed, 232
Jacana
 African, 88
 Bronze-winged, 88
 Comb-crested, 88
 Lesser, 88
 Madagascar, 88
 Northern, 88
 Pheasant-tailed, 88
 Wattled, 88–89
Jackdaw
 Daurian, 593
 Eurasian, 593
Jacky-winter, 479
Jacobin
 Black, 195
 White-necked, 195
Jaeger
 Long-tailed, 107
 Parasitic, 107
 Pomarine, 107
Jay
 Azure, 588
 Azure-hooded, 589
 Azure-naped, 588
 Beautiful, 589
 Black-chested, 588
 Black-collared, 589
 Black-headed, 590
 Black-throated, 589
 Blue, 587
 Brown, 588
 Bushy-crested, 588
 Cayenne, 588
 Crested, 586–587
 Curl-crested, 588
 Dwarf, 589
 Eurasian, 590
 Gray, 587
 Green, 588
 Lidth's, 591
 Mexican, 589–590
 Pinyon, 590
 Plush-crested, 588–589
 Purplish, 588
 Purplish-backed, 588
 San Blas, 588
 Siberian, 587
 Sichuan, 587
 Silvery-throated, 589
 Steller's, 587
 Tufted, 588
 Turquoise, 589
 Unicolored, 590
 Violaceous, 588
 White-collared, 589
 White-naped, 589
 White-tailed, 589
 White-throated, 589
 Yucatan, 588
Jery
 Common, 499
 Green, 499
 Stripe-throated, 499
 Wedge-tailed, 499

Jewel-babbler
 Blue, 514
 Chestnut-backed, 514
 Spotted, 514
Jewelfront, Gould's, 203
Junco
 Dark-eyed, 686–687
 Guadalupe, 687
 Volcano, 687
 Yellow-eyed, 687
Jungle-Flycatcher
 Brown-chested, 447–448
 Buru, 447
 Chestnut-tailed, 448
 Eyebrowed, 448
 Flores, 447
 Fulvous-chested, 448
 Gray-chested, 448
 Henna-tailed, 448
 Mindanao, 448
 Negros, 448
 Rusty-flanked, 448
Junglefowl
 Ceylon, 71
 Gray, 71
 Green, 71
 Red, 71

Kagu, 86
Kaka
 New Zealand, 134
 Norfolk Island, 700
Kakapo, 134
Kakawahie, 701
Kamao, 405
Kea, 134
Kestrel
 American, 51
 Australian, 51
 Banded, 51
 Dickinson's, 51
 Eurasian, 50–51
 Fox, 51
 Gray, 51
 Greater, 51
 Lesser, 50
 Madagascar, 51
 Mauritius, 51
 Seychelles, 51
 Spotted, 51
Killdeer, 94
King-Parrot
 Australian, 139
 Moluccan, 139
 Papuan, 139
Kingbird
 Cassin's, 342
 Couch's, 342
 Eastern, 342
 Giant, 342
 Gray, 342
 Loggerhead, 342
 Snowy-throated, 342
 Thick-billed, 342
 Tropical, 342
 Western, 342
 White-throated, 342
Kingfisher
 Amazon, 223
 American Pygmy
 Azure, 216
 Banded, 217

Beach, 221
Belted, 223
Bismarck, 216
Black-backed, 217
Black-billed, 218
Black-capped, 219
Blue-and-white, 220
Blue-banded, 216
Blue-black, 219
Blue-breasted, 219
Blue-capped, 222
Blue-eared, 216
Blyth's, 215
Brown-hooded, 219
Brown-winged, 218
Chattering, 221
Chestnut-bellied, 220
Chocolate-backed, 219
Cinnamon-banded, 221
Collared, 220–221
Common, 215
Crested, 223
Dwarf, 217
Flat-billed, 220
Forest, 220
Giant, 223
Gray-headed, 219
Green, 223
Green-and-rufous, 223
Green-backed, 222
Half-collared, 215
Hook-billed, 221–222
Indigo-banded, 216
Javan, 219
Lazuli, 220
Lilac, 218
Little, 216–217
Malachite, 216
Malagasy, 216
Mangaia, 221
Mangrove, 219
Marquesas, 221
Micronesian, 220
Mountain, 222
Moustached, 222
New Britain, 220
Philippine, 217
Pied, 223
Red-backed, 220
Ringed, 223
Ruddy, 218–219
Rufous-backed, 217
Rufous-collared, 222
Rufous-lored, 219–220
Sacred, 221
Scaly, 222
Shining-blue, 216
Silvery, 216
Small Blue, 216
Sombre, 221
Spotted, 222
Stork-billed, 218
Striped, 219
Sulawesi, 217
Tahiti, 221
Talaud, 221
Tuamotu, 221
Ultramarine, 220
Variable, 217
White-bellied, 216
White-rumped, 221
White-throated, 219

Woodland, 219
Yellow-billed, 222
Kinglet
Canary Islands, 385
Golden-crowned, 385
Ruby-crowned, 385
Kioea, 700
Kiskadee
Great, 341
Lesser, 341
Kite
Australian, 36
Black, 36
Black-breasted, 35
Black-shouldered, 36
Brahminy, 36
Double-toothed, 36
Gray-headed, 35
Hook-billed, 35
Letter-winged, 36
Mississippi, 36
Pearl, 35–36
Plumbeous, 36
Red, 36
Rufous-thighed, 36
Scissor-tailed, 36
Slender-billed, 36
Snail, 36
Square-tailed, 35
Swallow-tailed, 35
Whistling, 36
White-collared, 35
White-tailed, 36
Kittiwake
Black-legged, 103
Red-legged, 103
Kiwi
Great Spotted Kiwi, 2
Little Spotted Kiwi, 2
North Island Brown, 2
Okarito Brown, 2
Southern Brown, 2
Knot
Great, 98
Red, 98
Koa-Finch
Greater, 701
Lesser, 701
Koel
Asian, 155
Australian, 155
Black-billed, 155
Dwarf, 155
Long-tailed, 155
White-crowned, 155
Kokako, 580
Kookaburra
Blue-winged, 218
Laughing, 218
Rufous-bellied, 218
Shovel-billed, 218
Spangled, 218

Lammergeier, 37
Lancebill
Blue-fronted, 194
Green-fronted, 194
Lapwing
Andean, 93
Banded, 93
Black-headed, 92
Black-winged, 92

Brown-chested, 92
Crowned, 92
Gray-headed, 92
Long-toed, 92
Masked, 93
Northern, 92
Pied, 93
Red-wattled, 93
River, 92
Senegal, 92
Sociable, 93
Southern, 93
Spot-breasted, 92
Sunda, 93
Wattled, 92
White-headed, 92
White-tailed, 93
Yellow-wattled, 92
Lark
Algulhas, 351
Angola, 349
Archer's, 350
Ash's, 349
Barlow's, 351
Bar-tailed, 352
Beesley's, 351
Benguela, 351
Bimaculated, 353
Black, 353
Blanford's, 353
Botha's, 355
Calandra, 353
Cape, 351
Cape Clapper, 350
Collared, 350
Crested, 355–356
Degodi, 350
Desert, 352–353
Dune, 351
Dunn's, 355
Dupont's, 352
Dusky, 351
Eastern Clapper, 350
Eastern Long-billed, 351
Erlanger's, 354
Fawn-colored, 350
Ferruginous, 350
Flappet, 349–350
Foxy, 350
Friedmann's, 349
Gillett's, 350
Gray's, 351
Greater Short-toed, 353
Horned, 354–355
Hume's, 354
Karoo, 351
Karoo Long-billed, 351
Kordofan, 348
Large-billed, 356
Latakoo, 349
Lesser Short-toed, 354
Madagascar, 348
Malabar, 356
Masked, 355
Mongolian, 353
Monotonous, 349
Obbia, 355
Pink-billed, 355
Pink-breasted, 350
Red-capped, 353
Red-winged, 349
Rudd's, 350

Rufous-naped, 349
Rufous-rumped, 351
Rufous-tailed, 352
Rusty, 348–349
Sabota, 350
Sand, 354
Sclater's, 355
Short-clawed, 351
Short-tailed, 355
Sidamo, 350
Somali Long-billed, 349
Somali Short-toed, 354
Spike-heeled, 351
Stark's, 355
Sun, 356
Tawny, 356
Temminck's, 355
Thekla, 356
Thick-billed, 353
Tibetan, 353
White-tailed, 348
White-winged, 353
Williams', 348
Wood, 357
Laughingthrush
Ashy-headed, 488
Barred, 490
Biet's, 490
Black, 489
Black-faced, 492
Black-hooded, 489
Black-throated, 490
Blue-winged, 492
Brown-capped, 491
Chestnut-backed, 490
Chestnut-capped, 491
Chestnut-crowned, 492
Chestnut-eared, 490
Collared, 492
Elliot's, 492
Giant, 491
Golden-winged, 492
Gray, 489–490
Gray-breasted, 491
Gray-sided, 491
Greater Necklaced, 489
Lesser Necklaced, 489
Masked, 489
Moustached, 490
Père David's, 490
Prince Henry's, 492
Red-tailed, 492
Red-winged, 492
Rufous-breasted, 491
Rufous-chinned, 490
Rufous-fronted, 489
Rufous-necked, 490
Rufous-vented, 490
Rusty, 491
Scaly, 491
Spot-breasted, 491
Spotted, 490
Streaked, 491
Striated, 489
Striped, 491
Sukatschev's, 490
Sunda, 488
Variegated, 492
White-browed, 491
White-cheeked, 490
White-crested, 489
White-necked, 489

Laughingthrush (*continued*)
  White-throated, 489
  White-whiskered, 492
  Wynaad, 490
  Yellow-throated, 490
Leaf-love, 380
Leaf-Warbler
  Blyth's, 438
  Brooks', 438
  Chinese, 438
  Eastern Crowned, 438
  Emei, 439
  Gansu, 438
  Hainan, 439
  Ijima's, 438
  Island, 439–440
  Kulambangra, 440
  Large-billed, 438
  Pale-legged, 438
  Philippine, 440
  Plain, 437
  Sakhalin, 438
  San Cristobal, 440
  Sichuan, 438
  Sulawesi, 439
  Tickell's, 437
  Timor, 439
  Tytler's, 438
  Western Crowned, 438
  White-tailed, 439
Leafbird
  Blue-masked, 386
  Blue-winged, 386
  Golden-fronted, 386
  Greater Green, 386
  Lesser Green, 386
  Orange-bellied, 386
  Philippine, 386
  Yellow-throated, 386
Leaftosser
  Black-tailed, 282
  Gray-throated, 282
  Rufous-breasted, 283
  Scaly-throated, 282
  Short-billed, 282
  Tawny-throated, 282
Leiothrix, Red-billed, 505
Limpkin, 77
Linnet
  Eurasian, 633
  Warsangli, 633
  Yemen, 633
Liocichla
  Bugun, 492
  Gray-faced, 492
  Red-faced, 492–493
  Steere's, 492
Lizard-Cuckoo
  Great, 159
  Hispaniolan, 159
  Jamaican, 159
  Puerto Rican, 159
Locustfinch, 616
Logrunner
  Northern, 513
  Southern, 513
Longbill
  Bocage's, 436
  Gray, 436
  Kemp's, 436
  Kretschmer's, 436
  Pulitzer's, 436
  Slaty-chinned, 545

  Yellow, 436
  Yellow-bellied, 544
Longclaw
  Abyssinian, 367
  Fuelleborn's, 366
  Grimwood's, 367
  Orange-throated, 366
  Pangani, 367
  Rosy-throated, 367
  Sharpe's, 366
  Yellow-throated, 366
Longspur
  Chestnut-collared, 687
  Lapland, 687
  McCown's, 687
  Smith's, 687
Longtail
  Cricket, 422
  Green, 422
Loon
  Arctic, 7
  Common, 7
  Pacific, 7
  Red-throated, 7
  Yellow-billed, 7
Lorikeet
  Blue, 133
  Blue-crowned, 133
  Blue-fronted, 133
  Duchess, 133
  Fairy, 133
  Goldie's, 132
  Iris, 132
  Josephine's, 133
  Kuhl's, 133
  Little, 133
  Meek's, 133
  Mindanao, 132
  Musk, 133
  New Caledonian, 700
  Olive-headed, 132
  Orange-billed, 134
  Ornate, 131
  Palm, 133
  Papuan, 133
  Plum-faced, 133–134
  Pohnpei, 132
  Purple-crowned, 133
  Pygmy, 133
  Rainbow, 131–132
  Red-chinned, 133
  Red-flanked, 133
  Red-fronted, 133
  Red-throated, 133
  Scaly-breasted, 132
  Stephen's, 133
  Striated, 133
  Ultramarine, 133
  Varied, 132
  Yellow-and-green, 132
  Yellow-billed, 134
Lory
  Black, 131
  Black-capped, 132
  Black-winged, 131
  Blue-eared, 131
  Blue-streaked, 131
  Brown, 131
  Cardinal, 131
  Chattering, 132
  Collared, 132
  Dusky, 131
  Purple-bellied, 132

  Purple-naped, 132
  Red, 131
  Red-and-blue, 131
  Violet-necked, 131
  White-naped, 132
  Yellow-bibbed, 132
  Yellow-streaked, 131
Lovebird
  Black-cheeked, 141
  Black-collared, 141
  Black-winged, 141
  Fischer's, 141
  Gray-headed, 141
  Lilian's, 141
  Red-headed, 141
  Rosy-faced, 141
  Yellow-collared, 141
Lyrebird
  Albert's, 348
  Superb, 348

Macaw
  Blue-and-yellow, 142
  Blue-headed, 143
  Blue-throated, 142
  Blue-winged, 143
  Chestnut-fronted, 142
  Cuban, 700
  Glaucous, 700
  Golden-collared, 143
  Great Green, 142
  Hispaniolan, 700
  Hyacinth, 142
  Lear's, 142
  Military, 142
  Red-and-green, 142
  Red-bellied, 143
  Red-fronted, 142
  Red-shouldered, 143
  Scarlet, 142
  Spix's, 142
Magpie
  Australasian, 582
  Azure-winged, 591
  Black, 587
  Black-billed, 592
  Blue, 591
  Ceylon, 591
  Eurasian, 592
  Formosan, 591
  Gold-billed, 591
  Green, 591
  Short-tailed, 591
  White-winged, 591
  Yellow-billed, 592
  Yellow-breasted, 591
Magpie-Jay
  Black-throated, 587
  White-throated, 587–588
Magpie-lark, 580
Magpie-Robin
  Madagascar, 459
  Oriental, 459
  Seychelles, 460
Maleo, 53
Malia, 488
Malimbe
  Ballman's, 605
  Black-throated, 605
  Crested, 605
  Gray's, 605
  Ibadan, 605
  Rachel's, 605

Red-bellied, 605
Red-crowned, 605
Red-headed, 605
Red-vented, 605
Malkoha
Black-bellied, 156
Blue-faced, 156
Chestnut-bellied, 156
Chestnut-breasted, 156
Green-billed, 156
Raffles', 156
Red-billed, 156
Red-crested, 156
Red-faced, 156
Scale-feathered, 156
Sirkeer, 156
Yellow-billed, 156
Mallard, 30
Malleefowl, 53
Mamo
Black, 701
Hawaii, 701
Manakin
Araripe, 316
Band-tailed, 314
Black, 317
Blue, 316
Blue-backed, 316
Blue-crowned, 315
Blue-rumped, 315
Cerulean-capped, 315
Club-winged, 317
Crimson-hooded, 314
Eastern Striped, 317
Fiery-capped, 317
Flame-crested, 317
Golden-collared, 316
Golden-crowned, 315
Golden-headed, 315
Golden-winged, 316
Green, 317
Helmeted, 316
Jet, 317
Lance-tailed, 316
Long-tailed, 316
Olive, 317
Opal-crowned, 315
Orange-collared, 316
Orange-crested, 317
Pin-tailed, 316
Red-capped, 315
Red-headed, 315
Round-tailed, 315
Scarlet-horned, 315
Snow-capped, 315
Tepui, 315
Western Striped, 317
White-bearded, 316–317
White-collared, 316
White-crowned, 315–316
White-fronted, 315
White-ruffed, 316
White-throated, 316
Wire-tailed, 315
Yellow-crested, 317
Yellow-headed, 317
Yungas, 316
Mango
Antillean, 195
Black-throated, 195
Green, 195
Green-breasted, 195
Green-throated, 195

Jamaican, 195
Veraguan, 195
Mannikin
Black-and-white, 617
Bronze, 617
Magpie, 618
Nutmeg, 618
Manucode
Crinkle-collared, 583
Curl-crested, 583
Glossy-mantled, 583
Jobi, 583
Trumpet, 583
Mao, 565
Marsh-Harrier
African, 39
Eastern, 39
Western, 39
Marsh-Tyrant, White-headed, 338
Marshbird
Brown-and-yellow, 698
Yellow-rumped, 698
Martin
African River, 357
Asian, 361
Banded, 358
Brazza's, 358
Brown-chested, 359
Caribbean, 359
Congo, 358
Cuban, 359
Fairy, 362
Galapagos, 359
Gray-breasted, 359
House, 361
Mascarene, 358
Nepal, 361
Pale Sand, 358
Peruvian, 359
Plain, 358
Purple, 359
Rock, 361
Sinaloa, 359
Southern, 359
Tree, 362
White-eyed River, 357
Masked-Owl
Australian, 160
Lesser, 161
New Britain, 160
Masked-Weaver
Heuglin's, 607
Katanga, 607
Lake Lufira, 607
Lesser, 607
Northern, 607
Southern, 607
Tanganyika, 607
Vitelline, 607
Meadowlark
Eastern, 693
Long-tailed, 693
Pampas, 693
Peruvian, 693
Western, 693
Melampitta
Greater, 585
Lesser, 585
Melidectes
Belford's, 564
Bismarck, 564
Cinnamon-browed, 564
Huon, 564

Long-bearded, 564
Ornate, 565
San Cristobal, 565
Short-bearded, 564
Sooty, 564
Vogelkop, 564
Yellow-browed, 564
Meliphaga
Mountain, 559
Spot-breasted, 559
Merganser
Auckland Islands,
699
Brazilian, 33
Common, 33
Hooded, 33
Red-breasted, 33
Scaly-sided, 33
Merlin, 52
Mesia, Silver-eared, 505
Mesite
Brown, 75
Subdesert, 75
White-breasted, 75
Metaltail
Black, 208
Coppery, 208
Fire-throated, 208
Neblina, 208
Perija, 208
Scaled, 208
Tyrian, 208
Violet-throated, 208
Viridian, 208
Millerbird, 430
Miner
Bell, 566
Black-eared, 566
Campo, 267
Coastal, 268
Common, 268
Creamy-rumped, 268
Dark-winged, 268
Grayish, 268
Noisy, 566
Puna, 268
Rufous-banded, 268
Short-billed, 268
Slender-billed, 268
Thick-billed, 268
Yellow-throated, 566
Minivet
Ashy, 374
Brown-rumped, 374
Fiery, 374
Flores, 374
Gray-chinned, 375
Long-tailed, 374
Rosy, 374
Ryukyu, 374
Scarlet, 374–375
Short-billed, 374
Small, 374
Sunda, 374
White-bellied, 374
Minla
Blue-winged, 507
Chestnut-tailed, 507
Red-tailed, 507
Mistletoebird, 548–549
Mockingbird
Bahama, 397
Blue, 399

Mockingbird (*continued*)
Blue-and-white, 399
Brown-backed, 398
Chalk-browed, 398
Charles, 398
Chilean, 398
Galapagos, 398
Hood, 398
Long-tailed, 398
Northern, 397
Patagonian, 398
San Cristobal, 398
Socorro, 398
Tropical, 397
White-banded, 398
Monal
Chinese, 71
Himalayan, 71
Sclater's, 71
Monarch
Biak, 476
Black, 475
Black-and-white, 477
Black-backed, 476
Black-bibbed, 476
Black-chinned, 476
Black-faced, 476
Black-naped, 472–473
Black-tailed, 477
Black-tipped, 476
Black-winged, 476
Bougainville, 476
Buff-bellied, 475
Celestial, 473
Chestnut-bellied, 476
Fatuhiva, 474
Flores, 476
Frill-necked, 477
Frilled, 477
Golden, 477
Hooded, 476
Iphis, 474
Island, 475
Kulambangra, 477
Loetoe, 476
Manus, 477
Marquesas, 474
Maupiti, 700
Ogea, 474
Pale-blue, 473
Pied, 477
Rarotonga, 474
Rufous, 475
Rufous-collared, 477
Short-crested, 472
Slaty, 475
Spectacled, 476
Spot-winged, 476
Tahiti, 474
Tinian, 477
Truk, 475
Vanikoro, 475
White-breasted, 477
White-capped, 476
White-collared, 477
White-eared, 476
White-naped, 476
White-tailed, 476
White-tipped, 476
Yap, 477
Monjita
Black-and-white, 339

Black-crowned, 339
Gray, 339
Rusty-backed, 339
Salinas, 339
White, 339
White-rumped, 339
Monklet, Lanceolated, 234
Moorchat, Congo, 465
Moorhen
Common, 85
Dusky, 85
Lesser, 85
Samoan, 699
San Cristobal, 85
Tristan, 85
Morepork, 174
Morning-Thrush, Spotted,
458
Morningbird, 487
Motmot
Blue-crowned, 224
Blue-throated, 224
Broad-billed, 224–225
Highland, 224
Keel-billed, 224
Rufous, 224
Rufous-capped, 224
Russet-crowned, 224
Tody, 224
Turquoise-browed, 225
Mountain-Babbler
Chapin's, 509
Red-collared, 509
White-throated, 509
Mountain-Finch
Black-headed, 628
Chestnut-breasted, 669
Cochabamba, 669
Plain, 628
Tawny-headed, 628
Tucuman, 669
Mountain-gem
Gray-tailed, 203
Green-breasted, 203
Green-throated, 203
Purple-throated, 203
White-bellied, 203
White-throated, 203
Mountain-Greenbul
Eastern, 379
Western, 379
Mountain-Pigeon
Long-tailed, 129
Pale, 129
Papuan, 129
Mountain-Tanager
Black-chested, 657
Black-chinned, 658
Blue-winged, 658
Buff-breasted, 658
Chestnut-bellied, 658
Golden-backed, 657
Hooded, 657
Lacrimose, 657–658
Masked, 657
Santa Marta, 657
Scarlet-bellied, 658
Mountain-Toucan
Black-billed, 242
Gray-breasted, 242
Hooded, 242
Plate-billed, 242

Mountaineer, Bearded, 208
Mourner
Cinereous, 345
Grayish, 343
Pale-bellied, 343
Rufous, 343
Speckled, 345
Mouse-Warbler
Bicolored, 517
Mountain, 517
Rusty, 517
Mousebird
Blue-naped, 212
Red-backed, 212
Red-faced, 212
Speckled, 211–212
White-backed, 212
White-headed, 212
Munia
Alpine, 620
Bismarck, 620
Black, 620
Black-breasted, 620
Black-faced, 618
Black-headed, 619
Black-throated, 618
Chestnut, 619
Chestnut-breasted, 620
Dusky, 618
Five-colored, 619
Grand, 619
Gray-banded, 619
Gray-crowned, 619
Gray-headed, 619
Hooded, 619
Javan, 618
Madagascar, 618
Mottled, 619
New Hanover, 619
New Ireland, 619
Pale-headed, 619
Pictorella, 620
Snow Mountain, 620
Streak-headed, 618
White-bellied, 618
White-capped, 619
White-headed, 619
White-rumped, 618
White-throated, 618
Yellow-rumped, 620
Murre
Common, 107
Thick-billed, 107
Murrelet
Ancient, 108
Craveri's, 108
Japanese, 108
Kittlitz's, 108
Long-billed, 107–108
Marbled, 107
Xantus', 108
Myna
Apo, 597
Bali, 598
Bank, 598
Bare-eyed, 598
Ceylon, 598
Collared, 598
Common, 598
Common Hill, 598
Crested, 598
Enggano, 598

Fiery-browed, 598
Finch-billed, 598
Golden, 597
Golden-crested, 598
Helmeted, 597
Javan, 598
Jungle, 598
Long-crested, 597
Long-tailed, 597
Nias, 598
Pale-bellied, 598
Southern Hill, 598
Sulawesi, 597
White-necked, 597
White-vented, 598
Yellow-faced, 597
Myzomela
Ashy, 557
Banda, 558
Black, 557–558
Black-bellied, 559
Black-breasted, 559
Black-headed, 559
Cardinal, 558
Crimson-hooded, 558
Dusky, 557
Ebony, 559
Micronesian, 558
Mountain, 558
New Caledonian, 558
New Ireland, 558
Orange-breasted, 559
Red, 557
Red-bellied, 559
Red-collared, 559
Red-headed, 558
Red-throated, 557
Rotuma, 558
Scarlet, 558
Scarlet-bibbed, 559
Scarlet-naped, 559
Seram, 557
Sooty, 559
Sulawesi, 558
Sumba, 558
Wakolo, 558
White-chinned, 557
Yellow-vented, 559
Myzornis, Fire-tailed, 510

Native-hen
Black-tailed, 85
Tasmanian, 85
Needletail
Brown-backed, 188
Papuan, 187
Philippine, 187
Purple, 188
Silver-backed, 188
Silver-rumped, 188
White-rumped, 187
White-throated, 188
Negrito
Andean, 336
Austral, 336
Negrofinch
Chestnut-breasted, 611
Gray-headed, 611
Pale-fronted, 611
White-breasted, 611

Newtonia
Archbold's, 435
Common, 435
Dark, 434
Red-tailed, 435
Nicator
Eastern, 382
Yellow-spotted, 382
Yellow-throated, 382
Night-Heron
Black-crowned, 21
Japanese, 21
Malayan, 21
Rufous, 21
White-backed, 21
White-eared, 21
Yellow-crowned, 21
Nighthawk
Antillean, 179
Bahia, 179
Band-tailed, 179
Common, 178
Least, 178
Lesser, 178
Nacunda, 179
Rufous-bellied, 178
Sand-colored, 178
Short-tailed, 178
Nightingale
Common, 456
Thrush, 456
Nightingale-Thrush
Black-billed, 406
Black-headed, 406
Orange-billed, 406
Ruddy-capped, 406
Russet, 406
Slaty-backed, 406
Spotted, 406
Nightjar
Abyssinian, 182
Andaman, 182
Archbold's, 179
Band-winged, 180
Bates', 183
Black-shouldered, 182
Blackish, 181
Bonaparte's, 183
Brown, 181
Buff-collared, 180
Cayenne, 181
Collared, 183
Diabolical, 179
Donaldson-Smith's, 182
Dusky, 180
Egyptian, 181
Eurasian, 181
Fiery-necked, 182
Freckled, 183
Golden, 182
Gray, 181
Greater Antillean, 180
Hispaniolan, 180
Indian, 182
Itombwe, 183
Jerdon's, 182
Ladder-tailed, 184
Large-tailed, 182
Little, 181
Long-tailed, 183
Long-trained, 184

Lyre-tailed, 183
Madagascar, 182
Malaysian, 179
Mees's, 182
Montane, 182
Nechisar, 183
Nubian, 181
Papuan, 179
Pennant-winged, 183
Philippine, 182
Plain, 182
Puerto Rican, 180
Pygmy, 180–181
Red-necked, 181
Roraiman, 181
Rufous, 180
Rufous-cheeked, 181
Salvadori's, 183
Savanna, 183
Scissor-tailed, 184
Scrub, 181
Sickle-winged, 184
Silky-tailed, 180
Slender-tailed, 183
Sombre, 181
Spot-tailed, 181
Spotted, 179
Square-tailed, 183
Standard-winged, 183
Star-spotted, 183
Sulawesi, 182
Swallow-tailed, 183
Swamp, 182
Sykes', 182
Tawny-collared, 180
Vaurie's, 182
White-tailed, 181
White-throated, 179
White-winged, 180
Yucatan, 180
Niltava
Fujian, 452
Large, 452
Rufous-bellied, 452–453
Rufous-vented, 453
Small, 452
Vivid, 453
Noddy
Black, 103
Blue, 103
Brown, 103
Gray, 103
Lesser, 103
Nothura
Chaco, 6
Darwin's, 6
Lesser, 6
Spotted, 6
White-bellied, 5
Nukupuu, 638
Nunbird
Black, 234
Black-fronted, 234
White-faced, 234
White-fronted, 234
Yellow-billed, 234
Nunlet
Brown, 234
Chestnut-headed, 234
Fulvous-chinned, 234
Gray-cheeked, 234

Nunlet (*continued*)
 Rufous-capped, 234
 Rusty-breasted, 234
Nutcracker
 Clark's, 592
 Eurasian, 593
Nuthatch
 Algerian, 530
 Beautiful, 531
 Blue, 531
 Brown-headed, 530
 Chestnut-bellied, 529
 Chestnut-vented, 530
 Coral-billed, 577
 Corsican, 530
 Eurasian, 529–530
 Giant, 531
 Kashmir, 530
 Krueper's, 530
 Persian, 531
 Pygmy, 530
 Red-breasted, 530
 Rock, 530–531
 Snowy-browed, 530
 Sulphur-billed, 531
 Velvet-fronted, 531
 White-breasted, 530
 White-browed, 530
 White-cheeked, 530
 White-tailed, 530
 Yellow-billed, 531
 Yunnan, 530

Odeni, 426
Oilbird, 174
Olive-Greenbul
 Cameroon, 381
 Toro, 381
Oliveback
 Fernando Po, 611
 Gray-headed, 612
 White-collared, 611
Olomao, 405
Omao, 405
Oo
 Bishop's, 565
 Hawaii, 700
 Kauai, 565
 Oahu, 700
Openbill
 African, 23
 Asian, 23
Orangequit, 673
Oriole
 African Black-headed, 569
 African Golden, 568
 Altamira, 695
 Audubon's, 697
 Baltimore, 696
 Bar-winged, 697
 Black, 569
 Black-and-crimson, 569
 Black-backed, 696
 Black-cowled, 696
 Black-hooded, 569
 Black-naped, 568
 Black-tailed, 569
 Black-vented, 696
 Black-winged, 569
 Brown, 567
 Bullock's, 696
 Buru, 567

Dark-headed, 568
Dark-throated, 567–568
Epaulet, 695
Eurasian Golden, 568
Fuertes', 696
Greater Antillean, 696
Green, 567
Green-headed, 568
Halmahera, 567
Hooded, 696
Isabela, 568
Jamaican, 695
Maroon, 569
Martinique, 696
Montserrat, 697
Moriche, 695
Olive-backed, 567
Orange, 695
Orange-crowned, 695
Orchard, 696
Philippine, 568
São Tomé 568
Scott's, 697
Seram, 567
Silver, 569
Slender-billed, 568
Spot-breasted, 695
St. Lucia, 696
Streak-backed, 696
Timor, 567
Western Black-headed, 568
White-edged, 695
White-lored, 568
Yellow, 695
Yellow-backed, 695
Yellow-tailed, 695
Oropendola
 Amazonian, 698
 Band-tailed, 698
 Baudo, 698
 Black, 698
 Casqued, 698
 Chestnut-headed, 697
 Crested, 698
 Dusky-green, 697
 Green, 698
 Montezuma, 698
 Russet-backed, 697
Osprey, 34
Ostrich, 2
Ou, 638
Ouzel, Ring, 408
Ovenbird, 642
Owl
 African Long-eared, 175
 Ashy-faced, 161
 Band-bellied, 170
 Bare-legged, 166
 Barking, 173–174
 Barn, 161
 Barred, 169
 Black-and-white, 170
 Black-banded, 170
 Boreal, 173
 Buff-fronted, 173
 Burrowing, 172–173
 Chaco, 169
 Crested, 170
 Elf, 172
 Fearful, 175
 Flammulated, 164
 Fulvous, 169

Great Gray, 169
Great Horned, 166–167
Hume's, 169
Jamaican, 175
Laughing, 700
Little, 173
Madagascar Long-eared, 175
Madagascar Red, 161
Magellanic Horned, 167
Maned, 170
Manus, 161
Marsh, 175
Minahassa, 161
Mottled, 169–170
Northern Hawk, 170
Northern Long-eared, 175
Northern Saw-whet, 173
Northern White-faced, 166
Palau, 166
Père David's, 169
Powerful, 173
Rufous, 173
Rufous-banded, 170
Rufous-legged, 169
Rusty-barred, 169
Short-eared, 175
Snowy, 168
Southern White-faced, 166
Spectacled, 170
Spotted, 169
Striped, 175
Stygian, 175
Sulawesi, 161
Taliabu, 161
Tawny, 169
Tawny-browed, 170
Unspotted Saw-whet, 173
Ural, 169
White-browed, 174
Owlet
 African Barred, 172
 Albertine, 172
 Asian Barred, 172
 Chestnut, 172
 Chestnut-backed, 172
 Collared, 170
 Forest, 173
 Javan, 172
 Jungle, 172
 Long-whiskered, 172
 Pearl-spotted, 170
 Red-chested, 171–172
 Sjostedt's, 172
 Spotted, 173
Owlet-Nightjar
 Archbold's, 176
 Australian, 177
 Barred, 176
 Feline, 176
 Moluccan, 176
 Mountain, 176
 New Caledonian, 176
 Spangled, 176
 Wallace's, 176
Oxpecker
 Red-billed, 602
 Yellow-billed, 602
Oxylabes
 White-throated, 510
 Yellow-browed, 510
Oystercatcher
 African, 89

American, 89
Black, 89
Blackish, 89
Canarian, 700
Chatham, 89
Eurasian, 89
Magellanic, 89
Pied, 89
Sooty, 89
South Island, 89
Variable, 89

Painted-snipe
American, 89
Australian, 89
Greater, 89
Palila, 638
Palm-Swift
African, 189–190
Antillean, 189
Asian, 189
Fork-tailed, 189
Palm-Tanager
Black-crowned, 652
Gray-crowned, 652
Palm-Thrush
Collared, 458
Rufous-tailed, 458
Palmchat, 387
Palmcreeper, Point-tailed, 279
Pampa-Finch, Great, 676
Paradigalla
Long-tailed, 583
Short-tailed, 583
Paradise-crow, 583
Paradise-Flycatcher
African, 473
Asian, 474
Bates', 473
Bedford's, 473
Black-headed, 473
Blue, 474
Cerulean, 473
Japanese, 474
Madagascar, 473
Mascarene, 474
Rufous, 474
Rufous-vented, 473
São Tomé, 473
Seychelles, 474
Paradise-Kingfisher
Biak, 222
Brown-headed, 223
Buff-breasted, 223
Common, 222
Kofiau, 222
Little, 222
Numfor, 223
Red-breasted, 223
Paradise-Whydah
Broad-tailed, 620
Eastern, 620
Long-tailed, 620
Northern, 620
Togo, 620
Parakeet
Alexandrine, 139
Andean, 146
Antipodes, 135
Austral, 146
Azuero, 145

Barred, 146
Black-capped, 146
Black-fronted, 700
Blaze-winged, 145
Blossom-headed, 140
Blue-crowned, 143
Blue-throated, 145
Brown-throated, 144
Caatinga, 144
Canary-winged, 147
Carolina, 700
Chapman's Mitred, 143
Chatham Islands, 135
Cliff, 146
Cobalt-winged, 147
Crimson-bellied, 145
Crimson-fronted, 143
Cuban, 144
Derbyan, 140
Deville's, 145
Dusky-headed, 144
El Oro, 146
Fiery-shouldered, 145
Flame-winged, 146
Golden, 143
Golden-capped, 144
Golden-plumed, 144
Golden-winged, 147
Gray-cheeked, 147
Gray-headed, 140
Gray-hooded, 146
Green, 143
Green-cheeked, 145
Hellmayr's, 145
Hispaniolan, 144
Hocking's, 143
Horned, 135
Jandaya, 144
Layard's, 140
Long-tailed, 140
Madeira, 145
Malabar, 140
Malherbe's, 136
Maroon-bellied, 145
Maroon-tailed, 145–146
Mauritius, 139
Mitred, 143
Monk, 146
Mountain, 146
Nanday, 144
New Caledonian, 135
Newton's, 700
Nicobar, 140
Norfolk Island, 135
Olive-throated, 144
Orange-chinned, 147
Orange-fronted, 144
Pacific, 143
Painted, 145
Peach-fronted, 144
Pearly, 145
Plain, 147
Plum-headed, 140
Raiatea, 700
Red-breasted, 140
Red-crowned, 145
Red-eared, 146
Red-fronted, 135
Red-masked, 143
Red-throated, 143
Rose-headed, 146
Rose-ringed, 139

Rufous-fronted, 146
Santa Marta, 145
Scarlet-fronted, 143
Seychelles, 700
Sinu, 145
Slaty-headed, 139
Slender-billed, 146
Socorro, 143
Sulphur-breasted, 144
Sulphur-winged, 146
Sun, 144
Todd's, 145
Tui, 147
Wavy-breasted, 145
White-eared, 145
White-eyed, 143
White-necked, 146
Yellow-chevroned, 147
Yellow-fronted, 135
Pardalote
Forty-spotted, 549
Red-browed, 549
Spotted, 549
Striated, 549
Pardusco, 664
Parotia
Carola's, 583–584
Lawes', 584
Wahnes', 584
Western, 583
Parrot
Alexandra's, 139
Azure-rumped, 138
Bald, 148
Black, 141
Black-billed, 149
Black-headed, 148
Black-lored, 138
Black-winged, 148
Blue-bellied, 151
Blue-cheeked, 150
Blue-collared, 138
Blue-fronted, 150
Blue-headed, 149
Blue-naped, 138
Blue-rumped, 137
Blue-winged, 137
Bourke's, 136
Bronze-winged, 149
Brown-headed, 142
Brown-hooded, 148
Brown-necked, 141
Burrowing, 144
Caica, 148
Cuban, 149
Dusky, 149
Eclectus, 139
Elegant, 137
Festive, 150
Golden-shouldered, 136
Gray, 141
Great-billed, 138
Ground, 137
Hispaniolan, 149
Hooded, 136
Imperial, 151
Indigo-winged, 148
Kawall's, 150
Lilac-crowned, 149
Maroon-fronted, 143
Mascarene, 700
Mealy, 150

Parrot (*continued*)
Meyer's, 142
Mulga, 136
Niam-Niam, 142
Night, 137
Olive-shouldered, 139
Orange-bellied, 137
Orange-cheeked, 148
Orange-winged, 150
Paradise, 700
Pesquet's, 134
Pileated, 148
Port Lincoln, 136
Puerto Rican, 149
Red-bellied, 142
Red-billed, 149
Red-browed, 150
Red-capped, 136
Red-cheeked, 137–138
Red-crowned, 149
Red-faced, 148
Red-fan, 151
Red-fronted, 142
Red-lored, 150
Red-necked, 150
Red-rumped, 136
Red-spectacled, 149
Red-tailed, 150
Red-winged, 139
Regent, 139
Rock, 137
Rose-faced, 148
Rueppell's, 142
Rusty-faced, 148
Saffron-headed, 148
Scaly-headed, 149
Scaly-naped, 150
Scarlet-chested, 137
Senegal, 142
Short-tailed, 148
Singing, 138
Speckle-faced, 149
St. Lucia, 150
St. Vincent, 151
Superb, 139
Swift, 137
Thick-billed, 143
Tucuman, 149
Turquoise, 137
Vasa, 141
Vinaceous, 150
Vulturine, 148
White-bellied, 148
White-crowned, 149
White-fronted, 149
Yellow-billed, 149
Yellow-crowned, 150
Yellow-eared, 143
Yellow-faced, 150
Yellow-fronted, 142
Yellow-headed, 150
Yellow-lored, 149
Yellow-naped, 150
Yellow-shouldered, 150
Parrotbill
Ashy-throated, 512
Black-breasted, 511
Black-browed, 512
Black-throated, 512
Brown, 511
Brown-winged, 512
Fulvous, 512
Golden, 512

Gray-headed, 511
Gray-hooded, 512
Great, 511
Maui, 638
Reed, 513
Rufous-headed, 512
Rusty-throated, 512
Short-tailed, 512
Spectacled, 512
Spot-breasted, 511
Three-toed, 511
Vinous-throated, 512
Parrotfinch
Blue-faced, 617
Fiji, 617
Green-faced, 617
Papuan, 617
Pin-tailed, 617
Pink-billed, 617
Red-eared, 617
Red-headed, 617
Red-throated, 617
Royal, 617
Tawny-breasted, 617
Tricolored, 617
Parrotlet
Amazonian, 147
Blue-fronted, 148
Blue-winged, 147
Brown-backed, 148
Dusky-billed, 147
Golden-tailed, 148
Green-rumped, 146–147
Lilac-tailed, 147
Mexican, 146
Pacific, 147
Red-fronted, 148
Sapphire-rumped, 148
Scarlet-shouldered, 148
Spectacled, 147
Spot-winged, 148
Tepui, 147
Yellow-faced, 147
Partridge
Arabian, 64
Bar-backed, 69
Barbary, 64
Black, 67
Chestnut-bellied, 69
Chestnut-breasted, 68
Chestnut-headed, 69
Chestnut-necklaced, 69
Crested, 70
Crimson-headed, 70
Daurian, 67
Ferruginous, 70
Gray, 67
Gray-breasted, 69
Hainan, 69
Hill, 68
Long-billed, 67
Madagascar, 67
Orange-necked, 69
Philby's, 64
Przevalski's, 64
Red-billed, 69
Red-breasted, 69
Red-legged, 64
Rock, 64
Rufous-throated, 69
Sand, 64
Scaly-breasted, 69
See-see, 64

Sichuan, 68
Snow, 63
Stone, 70
Szechenyi's, 63
Taiwan, 69
Tibetan, 67
Udzungwa, 68
Verreaux's, 63
Vietnam, 69
White-cheeked, 69
White-necklaced, 68
Parula
Northern, 639
Tropical, 639–640
Pauraque, 179
Peacock, Congo, 74
Peacock-Pheasant
Bornean, 73
Bronze-tailed, 73
Germain's, 73
Gray, 73
Malayan, 73
Mountain, 73
Palawan, 73
Peafowl
Green, 74
Indian, 74
Pelican
American White, 15
Australian, 15
Brown, 15
Dalmatian, 15
Great White, 15
Peruvian, 15
Pink-backed, 15
Spot-billed, 15
Peltops
Lowland, 581
Mountain, 581
Penduline-Tit
African, 533
Black-headed, 533
Chinese, 533
Eurasian, 533
Forest, 533
Mouse-colored, 533
Sennar, 533
Southern, 534
White-crowned, 533
Yellow, 533
Penguin
Adelie, 6
Chinstrap, 7
Emperor, 6
Erect-crested, 7
Fiordland, 7
Galapagos, 7
Gentoo, 6
Humboldt, 7
Jackass, 7
King, 6
Little, 7
Macaroni, 7
Magellanic, 7
Rockhopper, 7
Royal, 7
Snares, 7
Yellow-eyed, 7
Peppershrike
Black-billed, 625
Rufous-browed, 625
Petrel
Antarctic, 10

Antarctic Giant, 10
Atlantic, 10
Barau's, 11
Bermuda, 10
Black-capped, 10
Black-winged, 11
Blue, 11
Bonin, 11
Bulwer's, 12
Cape, 10
Cape Verde, 11
Chatham, 11
Cook's, 11
Defilippe's, 11
Fea's *see* Cape Verde
Fiji, 11
Galapagos, 11
Gould's, 11
Gray, 12
Great-winged, 10
Hall's Giant, 10
Hawaiian, 11
Henderson, 11
Herald, 11
Jouanin's, 12
Juan Fernandez, 11
Kerguelen, 12
Kermadec, 10
Madeira, 11
Magenta, 11
Mascarene, 10
Mottled, 10
Murphy's, 10
Parkinson's, 12
Phoenix, 10
Providence, 10
Pycroft's, 11
Snow, 10
Soft-plumaged, 11
Stejneger's, 11
Tahiti, 10
Vanuatu, 11
Westland, 12
White-chinned, 12
White-headed, 10
White-necked, 11
Zino's *see* Madeira
Petronia
    Bush, 604
    Chestnut-shouldered, 603
    Rock, 604
    Yellow-spotted, 603
    Yellow-throated,
        603–604
Pewee
    Blackish, 335
    Cuban, 335
    Dark, 335
    Greater, 335
    Hispaniolan, 335
    Jamaican, 335
    Lesser Antillean, 335
    Ochraceous, 335
    Smoke-colored, 335
    Tropical, 335
    White-throated, 335
Phainopepla, 387
Phalarope
    Red, 99
    Red-necked, 99
    Wilson's, 99
Pheasant
    Blood, 70

Bulwer's, 72
Cheer, 72
Copper, 72
Edwards', 71
Elliot's, 72
Golden, 73
Green, 73
Hume's, 72
Imperial, 71
Kalij, 71
Koklass, 71
Lady Amherst's, 73
Mikado, 72
Reeves', 72
Ring-necked, 72–73
Salvadori's, 71
Silver, 71–72
Swinhoe's, 71
Vietnamese, 71
Philentoma
    Maroon-breasted,
        576
    Rufous-winged, 576
Phoebe
    Black, 336
    Eastern, 336
    Say's, 336
Piapiac, 593
Piculet
    African, 246
    Antillean, 246
    Bar-breasted, 244
    Black-spotted, 245
    Chestnut, 246
    Ecuadorian, 245
    Fine-barred, 246
    Golden-spangled, 245
    Grayish, 246
    Guianan, 245
    Lafresnaye's, 245
    Mottled, 246
    Ocellated, 245
    Ochraceous, 246
    Ochre-collared, 245
    Olivaceous, 246
    Orinoco, 245
    Plain-breasted, 246
    Rufous, 246
    Rufous-breasted, 246
    Rusty-necked, 246
    Scaled, 245
    Speckle-chested, 245
    Speckled, 244
    Spotted, 245
    Tawny, 246
    Varzea, 245
    White-barred, 245
    White-bellied, 245
    White-browed, 246
    White-wedged, 245
Pied-Babbler
    Hinde's, 504
    Northern, 504
    Southern, 504
Pied-Hornbill
    Malabar, 230
    Oriental, 230
Piedtail
    Ecuadorian, 203
    Peruvian, 203
Pigeon
    Afep, 110
    Band-tailed, 111–112

Bare-eyed, 111
Bolle's, 110
Bonin, 700
Bronze-naped, 110
Cameroon, 110
Chilean, 112
Choiseul, 700
Cinnamon-headed,
    122
Comoro, 111
Crested, 116
Delegorgue's, 110
Dusky, 112
Gray-cheeked, 122
Green-spectacled, 123
Hill, 110
Laurel, 110
Maroon, 110
Metallic, 111
New Zealand, 129
Nicobar, 120
Orange-breasted, 122
Pale-backed, 110
Pale-capped, 111
Pale-vented, 112
Partridge, 116
Passenger, 700
Peruvian, 112
Pheasant, 121
Picazuro, 111
Pink, 112
Pink-necked, 122
Pin-tailed, 123
Plain, 112
Plumbeous, 112
Rameron, 110
Red-billed, 112
Ring-tailed, 112
Rock, 109–110
Ruddy, 112
Ryukyu, 700
São Tomé, 110
Scaled, 111
Scaly-naped, 111
Short-billed, 112
Snow, 110
Somali, 110
Sombre, 129
Speckled, 110
Spinifex, 116
Spot-winged, 111
Squatter, 116
Thick-billed, 122
Tooth-billed, 121
Topknot, 129
Trocaz, 110
Wedge-tailed, 123
White-bellied, 123
White-collared, 110
White-crowned, 111
White-headed, 111
White-naped, 111
Wonga, 116
Yellow-footed, 122
Yellow-legged, 111
Yellow-vented, 123
Piha
    Chestnut-capped, 313
    Cinnamon-vented, 313
    Dusky, 313
    Gray-tailed, 313
    Olivaceous, 313
    Rose-collared, 313

Piha (*continued*)
Rufous, 313
Scimitar-winged, 313
Screaming, 313
Pilotbird, 517
Pintail
Eaton's, 31
Northern, 31
White-cheeked, 31
Yellow-billed, 31
Piopio, 700
Piping-Guan
Black-fronted, 56
Blue-throated, 56
Red-throated, 56
Trinidad, 56
Pipipi, 483
Pipit
African, 363
Alpine, 365
American, 366
Australasian, 363
Berthelot's, 365
Blyth's, 364
Buffy, 364
Bush, 366
Chaco, 365
Correndera, 365
Golden, 366
Hellmayr's, 365
Jackson's, 363
Kimberley, 364
Long-billed, 364
Long-legged, 364
Long-tailed, 364
Malindi, 365
Meadow, 365
Mountain, 363
Nilgiri, 364
Ochre-breasted, 365
Olive-backed, 366
Oriental, 363
Paramo, 365
Pechora, 366
Plain-backed, 364
Red-throated, 366
Richard's, 363
Rock, 366
Rosy, 365
Short-billed, 365
Short-tailed, 366
Sokoke, 366
South Georgia, 365
Sprague's, 365
Striped, 365
Tawny, 364
Tree, 366
Upland, 364
Water, 366
Woodland, 363
Yellow-breasted, 366
Yellow-tufted, 365
Yellowish, 365
Piprites
Black-capped, 318
Gray-headed, 318
Wing-barred, 318
Pitohui
Black, 488
Crested, 488
Hooded, 487
Rusty, 487

Variable, 488
White-bellied, 487
Pitta
African, 266
Azure-breasted, 266
Banded, 266
Bar-bellied, 266
Black-crowned, 266
Black-faced, 267
Black-headed, 266
Blue, 266
Blue-banded, 266
Blue-headed, 266
Blue-naped, 265
Blue-rumped, 265
Blue-winged, 267
Eared, 265
Elegant, 267
Fairy, 267
Garnet, 266
Giant, 265
Green-breasted, 266
Gurney's, 266
Hooded, 266
Indian, 267
Ivory-breasted, 266
Mangrove, 267
Noisy, 267
Rainbow, 267
Red-bellied, 267
Rusty-naped, 265
Schneider's, 265
Sula, 267
Superb, 266
Whiskered, 266
Plains-wanderer, 100
Plantain-eater
Eastern, 152
Western, 152
Plantcutter
Peruvian, 311
Rufous-tailed, 311
White-tipped, 311
Ploughbill, Wattled, 488
Plover
Black-bellied, 93
Blacksmith, 92
Caspian, 94
Chestnut-banded, 94
Collared, 94
Common Ringed, 93
Crab, 89
Double-banded, 94
Egyptian, 91
Forbes', 94
Hooded, 94
Javan, 94
Kittlitz's, 94
Little Ringed, 93
Long-billed, 93
Madagascar, 94
Magellanic, 95
Malaysian, 94
Mountain, 94
Oriental, 94
Piping, 94
Puna, 94
Red-capped, 94
Semipalmated, 93
Shore, 94
Snowy, 94
Spur-winged, 92

St. Helena, 94
Three-banded, 94
Two-banded, 94
White-fronted, 94
Wilson's, 93–94
Plovercrest, 196
Plumeleteer
Bronze-tailed, 202
White-vented, 202
Plushcrown, Orange-fronted, 278
Pochard
Baer's, 32
Common, 32
Ferruginous, 32
Madagascar, 32
Red-crested, 32
Rosy-billed, 32
Southern, 32
Pond-Heron
Chinese, 20
Indian, 20
Javan, 20
Madagascar, 20
Poo-uli, 639
Poorwill
Choco, 180
Common, 179
Eared, 179
Jamaican, 700
Least, 179
Ocellated, 180
Yucatan, 180
Potoo
Andean, 178
Common, 178
Great, 177
Long-tailed, 177
Northern, 178
Rufous, 178
White-winged, 178
Prairie-Chicken
Greater, 60
Lesser, 60
Pratincole
Australian, 91
Black-winged, 92
Collared, 91
Gray, 92
Madagascar, 92
Oriental, 91
Rock, 92
Small, 92
Prickletail, Spectacled, 278
Prinia
Ashy, 420
Banded, 421
Bar-winged, 420
Black-chested, 421
Brown, 419
Drakensberg, 421
Graceful, 420
Gray-breasted, 420
Gray-crowned, 420
Hill, 419
Jungle, 420
Karoo, 421
Namaqua, 421
Pale, 421
Plain, 421
Red-winged, 421–422
River, 421
Roberts', 421

Rufescent, 420
Rufous-fronted, 420
Rufous-vented, 419
São Tomé, 421
Sierra Leone, 422
Striated, 419
Swamp, 419
Tawny-flanked, 421
White-chinned, 422
Yellow-bellied, 420
Prion
Antarctic, 11–12
Broad-billed, 11
Fairy, 12
Fulmar, 12
Salvin's, 11
Slender-billed, 12
Ptarmigan
Rock, 58–59
White-tailed, 58
Willow, 58
Puaiohi, 405
Puffback
Black-backed, 572
Large-billed, 573
Northern, 572
Pink-footed, 573
Pringle's, 572
Red-eye, 573
Puffbird
Barred, 233
Black-breasted, 233
Black-streaked, 233
Brown-banded, 233
Buff-bellied, 233
Chestnut-capped, 233
Collared, 233
Crescent-chested, 233
Moustached, 234
Pied, 233
Rufous-necked, 234
Russet-throated, 233
Semicollared, 233
Sooty-capped, 233
Spot-backed, 233
Spotted, 233
Striolated, 233
Two-banded, 233
White-chested, 233
White-eared, 233
White-necked, 232–233
White-whiskered, 234
Puffin
Atlantic, 108
Horned, 108
Tufted, 108
Puffleg
Black-breasted, 206
Black-thighed, 206
Blue-capped, 207
Buff-thighed, 207
Colorful, 207
Coppery-bellied, 207
Coppery-naped, 207
Emerald-bellied, 207
Glowing, 206
Golden-breasted, 207
Greenish, 207
Hoary, 207
Sapphire-vented, 207
Turquoise-throated,
   207

Purpletuft
Buff-throated, 313
Dusky, 312
White-browed, 312
Pygmy-goose
African, 29
Cotton, 29
Green, 29
Pygmy-Kingfisher
African, 217
Madagascar, 217
Pygmy-Owl
Amazonian, 171
Andean, 171
Austral, 171
Cape, 171
Central American, 171
Cloud-forest, 171
Colima, 171
Costa Rican, 171
Cuban, 171
Eurasian, 170
Ferruginous, 171
Guatemalan, 171
Least, 171
Mountain, 171
Northern, 170
Pernambuco, 171
Peruvian, 171
Subtropical, 171
Tamaulipas, 171
Tucuman, 171
Yungas, 171
Pygmy-Parrot
Buff-faced, 134
Finsch's, 134
Geelvink, 134
Meek's, 134
Red-breasted, 134
Yellow-capped, 134
Pygmy-Tyrant
Black-capped, 326
Bronze-olive, 323
Double-banded, 327
Eared, 326
Hazel-fronted, 323
Helmeted, 327
Long-crested, 327
Pale-eyed, 327
Rufous-headed, 323
Rufous-sided, 323
Scale-crested, 327
Short-tailed, 326
Tawny-crowned, 323
White-bellied, 326
Pyrrhuloxia, 690
Pytilia
Green-winged, 614
Orange-winged, 614
Red-billed, 614
Red-faced, 614
Red-winged, 614

Quail
Banded, 61
Blue, 68
Blue-breasted, 68
Brown, 67–68
California, 61
Common, 67
Elegant, 61
Gambel's, 61

Harlequin, 67
Himalayan, 699
Japanese, 67
Montezuma, 63
Mountain, 60
New Zealand, 699
Ocellated, 63
Rain, 67
Scaled, 60
Singing, 63
Snow Mountain, 68
Stubble, 67
Tawny-faced, 63
Quail-Dove
Blue-headed, 120
Bridled, 120
Buff-fronted, 119
Chiriqui, 119
Crested, 119
Gray-fronted, 119
Key West, 120
Lined, 119–120
Olive-backed, 119
Purplish-backed, 119
Ruddy, 120
Russet-crowned, 119
Sapphire, 119
Tuxtla, 119
Violaceous, 120
White-faced, 119
White-fronted, 119
White-throated, 120
Quail-plover, 76
Quail-thrush
Chestnut, 513
Chestnut-breasted, 513
Cinnamon, 513
Painted, 513
Spotted, 513
Quailfinch
African, 616
Black-faced, 615
Red-billed, 615
Quelea
Cardinal, 609
Red-billed, 609
Red-headed, 609
Quetzal
Crested, 214
Eared, 214
Golden-headed, 214
Pavonine, 214
Resplendent, 214
White-tipped, 214

Racket-tail, Booted, 207
Racquet-tail
Blue-crowned, 138
Blue-headed, 138
Blue-winged, 138
Buru, 138
Golden-mantled, 138
Green, 138
Luzon, 138
Mindanao, 138
Yellowish-breasted, 138
Rail
African, 81
Auckland Islands, 82
Austral, 81
Bar-winged, 699
Bare-eyed, 82

Rail (*continued*)
　Bare-faced, 82
　Barred, 80
　Black, 79
　Blackish, 84
　Bogota, 81
　Buff-banded, 80
　Calayan, 80
　Chatham Islands, 699
　Chestnut, 84
　Clapper, 81
　Dieffenbach's, 699
　Forbes', 79
　Galapagos, 79
　Gray-throated, 78
　Guam, 80
　Hawaiian, 699
　Inaccessible Island, 82
　Invisible, 84
　Junin, 79
　King, 81
　Laysan, 699
　Lewin's, 82
　Lord Howe, 80
　Luzon, 81
　Madagascar, 81
　Mayr's, 79
　New Britain, 80
　New Caledonian, 80
　New Guinea Flightless, 84
　Nkulengu, 78
　Okinawa, 80
　Plain-flanked, 81
　Platen's, 82
　Plumbeous, 84
　Rouget's, 82
　Roviana, 80
　Sakalava, 83
　Sharpe's, 699
　Slaty-breasted, 80–81
　Speckled, 78
　Spotted, 84
　Swinhoe's, 78
　Talaud, 83
　Tahiti, 699
　Virginia, 81
　Wake Island, 699
　Water, 81
　White-throated, 82
　Woodford's, 80
　Yellow, 78
　Zapata, 84
Rail-babbler, Malaysian, 514
Raven
　Australian, 595
　Brown-necked, 595
　Chihuahuan, 594
　Common, 595
　Fan-tailed, 595
　Forest, 595
　Little, 595
　Thick-billed, 595
　White-necked, 595
Rayadito
　Masafuera, 270
　Thorn-tailed, 270
Razorbill, 107
Recurvebill
　Bolivian, 280
　Peruvian, 280
Red-Cotinga
　Black-necked, 314
　Guianan, 314

Redhead, 32
Redpoll
　Common, 631
　Hoary, 631
Redshank
　Common, 98
　Spotted, 98
Redstart
　Ala Shan, 460
　American, 642
　Black, 460
　Blue-capped, 460
　Blue-fronted, 461
　Brown-capped, 644
　Collared, 644
　Common, 460
　Daurian, 461
　Golden-fronted, 644
　Hodgson's, 460
　Luzon, 461
　Moussier's, 461
　Painted, 644
　Plumbeous, 461
　Rufous-backed, 460
　Saffron-breasted, 644
　Slate-throated, 644
　Spectacled, 644
　Tepui, 644
　White-bellied, 461
　White-capped, 461
　White-faced, 644
　White-fronted, 644
　White-throated, 460
　White-winged, 461
　Yellow-crowned, 644
　Yellow-faced, 644
Redthroat, 519
Redwing, 410
Reed-Finch, Long-tailed,
　668
Reed-Warbler
　African, 430
　Australian, 430
　Basra, 430
　Black-browed, 429
　Blyth's, 430
　Caroline, 430
　Clamorous, 430
　Cook Islands, 431
　Eurasian, 430
　Great, 430
　Henderson Island, 431
　Large-billed, 430
　Marquesan, 431
　Nauru, 430
　Nightingale, 430
　Oriental, 430
　Pitcairn, 431
　Rimitara, 431
　Streaked, 429
　Tahiti, 431
　Tuamotu, 431
Reedhaunter
　Curve-billed, 270
　Straight-billed, 270
Reedling, Bearded, 511
Reef-Heron
　Pacific, 20
　Western, 20
Rhabdornis
　Long-billed, 533
　Stripe-sided, 532
　Stripe-breasted, 533

Rhea
　Greater, 2
　Lesser, 2
Riflebird
　Magnificent, 584
　Paradise, 584
　Victoria's, 584
Rifleman, 347
Ringneck, Mallee, 136
Roadrunner
　Greater, 160
　Lesser, 160
Robin
　Alpine, 480
　American, 413
　Black, 411
　Black-chinned, 481
　Black-sided, 481
　Black-throated, 482
　Black-throated Blue, 456
　Blue-fronted, 461
　Blue-gray, 482
　Chatham, 481
　Clay-colored, 412
　Dusky, 481
　European, 455
　Flame, 480
　Forest, 455
　Garnet, 480
　Gray-breasted, 481
　Gray-headed, 482
　Green-backed, 482
　Hooded, 481
　Indian, 460
　Indian Blue, 456
　Japanese, 456
　Mangrove, 481
　Mountain, 411
　New Zealand, 480
　Olive-yellow, 482
　Pale-yellow, 481
　Pink, 480
　Red-capped, 480
　Rose, 480
　Rufous-backed, 412
　Rufous-collared, 413
　Rufous-headed, 456
　Rufous-tailed, 456
　Ryukyu, 456
　Scarlet, 480
　Siberian Blue, 456
　Smoky, 482
　Snow Mountain, 480
　Sooty, 410
　Sunda, 461
　Swynnerton's, 455
　White-breasted, 481
　White-browed, 481
　White-eyed, 482
　White-faced, 481
　White-rumped, 482
　White-starred, 454–455
　White-tailed, 461
　White-throated, 457
　White-winged, 482
　Yellow, 481
　Yellow-bellied, 481
Robin-Chat
　Archer's, 457
　Blue-shouldered, 457
　Cape, 457
　Chorister, 458
　Gray-winged, 457

Mountain, 457
Olive-flanked, 457
Red-capped, 458
Rueppell's, 457
Snowy-crowned, 458
White-bellied, 457
White-browed, 457
White-crowned, 458
White-headed, 458
White-throated, 457
Rock-Pigeon
  Chestnut-quilled, 116
  White-quilled, 116
Rock-Thrush
  Benson's, 401
  Blue, 401
  Blue-capped, 401
  Cape, 401
  Chestnut-bellied, 401
  Forest, 401
  Little, 401
  Littoral, 401
  Miombo, 401
  Rufous-tailed, 401
  Sentinel, 401
  Short-toed, 401
  White-throated, 401
Rockfinch, Pale, 604
Rockfowl
  Gray-necked, 488
  White-necked, 488
Rockjumper
  Damara, 495
  Orange-breasted, 495
  Rufous, 495
Roller
  Abyssinian, 226
  Blue-bellied, 227
  Blue-throated, 227
  Broad-billed, 227
  European, 226
  Indian, 227
  Lilac-breasted, 227
  Purple, 227
  Purple-winged, 227
  Racket-tailed, 227
  Rufous-crowned, 227
Rook, 594
Rosefinch
  Beautiful, 629
  Blanford's, 629
  Common, 629
  Dark-breasted, 629
  Dark-rumped, 630
  Great, 630
  Long-tailed, 638
  Pale, 630
  Pallas', 630
  Pink-browed, 630
  Pink-rumped, 629
  Przewalski's, 664
  Red-fronted, 630
  Red-mantled, 630
  Spot-winged, 630
  Streaked, 630
  Three-banded, 630
  Tibetan, 630
  Vinaceous, 630
  White-browed, 630
Rosella
  Adelaide, 136
  Crimson, 136
  Eastern, 136

Green, 136
Northern, 136
Pale-headed, 136
Western, 136
Yellow, 136
Rosy-Finch
  Asian, 628
  Black, 628
  Brown-capped, 629
  Gray-crowned, 628
Royal Flycatcher
  Amazonian, 331
  Northern, 331
  Pacific, 331
Ruby, Brazilian, 203
Rubythroat
  Siberian, 456
  White-tailed, 456
Ruff and Reeve, 99
Rufous-Warbler
  Black-capped, 428
  Black-faced, 428
Rush-Tyrant, Many-colored,
  326
Rushbird, Wren-like, 270

Sabrewing
  Buff-breasted, 194
  Gray-breasted, 194
  Lazuline, 194
  Long-tailed, 194
  Napo, 194
  Rufous, 194
  Rufous-breasted, 194
  Santa Marta, 194
  Violet, 194
  Wedge-tailed, 194
  White-tailed, 194
Saddleback, 580
Sage-Grouse
  Greater, 60
  Gunnison, 60
Saltator
  Black-cowled, 689
  Black-headed, 689
  Black-throated, 689
  Black-winged, 689
  Buff-throated, 689
  Golden-billed, 689
  Grayish, 688–689
  Green-winged, 689
  Lesser Antillean, 688
  Masked, 689
  Orinocan, 689
  Rufous-bellied, 689
  Streaked, 688
  Thick-billed, 689
Sanderling, 98
Sandgrouse
  Black-bellied, 109
  Black-faced, 109
  Burchell's, 109
  Chestnut-bellied, 108–109
  Crowned, 109
  Double-banded, 109
  Four-banded, 109
  Lichtenstein's, 109
  Madagascar, 109
  Namaqua, 108
  Painted, 109
  Pallas', 108
  Pin-tailed, 108
  Spotted, 109

Tibetan, 108
Yellow-throated, 109
Sandpiper
  Baird's, 99
  Broad-billed, 99
  Buff-breasted, 99
  Common, 97
  Curlew, 99
  Green, 97
  Least, 99
  Marsh, 98
  Pectoral, 99
  Purple, 99
  Rock, 99
  Semipalmated, 98
  Sharp-tailed, 99
  Solitary, 98
  Spoon-billed, 99
  Spotted, 97
  Stilt, 99
  Terek, 97
  Tuamotu, 98
  Upland, 97
  Western, 98
  White-rumped, 99
  White-winged, 700
  Wood, 98
Sandpiper-Plover, Diademed, 95
Sandplover
  Greater, 95
  Lesser, 94–95
Sapayoa, Broad-billed, 265
Sapphire
  Blue-chinned, 197
  Blue-headed, 199
  Gilded, 199
  Golden-tailed, 199
  Humboldt's, 199
  Rufous-throated, 199
  White-chinned, 199
Sapphirewing, Great, 206
Sapsucker
  Red-breasted, 248
  Red-naped, 248
  Williamson's, 248
  Yellow-bellied, 248
Sawwing
  Black, 357–358
  Fanti, 358
  Mountain, 357
  Square-tailed, 357
  White-headed, 357
Scaup
  Greater, 32
  Lesser, 32
  New Zealand, 32
Schiffornis
  Greater, 318
  Greenish, 318
  Thrush-like, 318
Scimitar-Babbler
  Chestnut-backed, 497
  Coral-billed, 497
  Indian, 496
  Large, 496
  Naung Mung, 497
  Red-billed, 497
  Rusty-cheeked, 496
  Short-tailed, 497
  Slender-billed, 497
  Spot-breasted, 496
  Streak-breasted, 497
  White-browed, 496

Scimitar-bill
 Abyssinian, 229
 Black, 228
 Common, 229
Scops-Owl
 African, 163
 Andaman, 162
 Anjouan, 164
 Biak, 164
 Collared, 162–163
 Comoro, 164
 Enggano, 164
 European, 163
 Flores, 162
 Indian, 162
 Japanese, 163
 Javan, 162
 Luzon, 163
 Malagasy, 164
 Mantanani, 164
 Mayotte, 164
 Mentawai, 162
 Mindanao, 163
 Mindoro, 163
 Moheli, 164
 Moluccan, 164
 Mountain, 162
 Nicobar, 164
 Oriental, 163
 Palawan, 163
 Pallid, 163
 Pemba, 164
 Philippine, 163
 Rajah, 162
 Reddish, 162
 Ryukyu, 164
 Sandy, 162
 Sangihe, 164
 São Tomé, 164
 Serendib, 162
 Seychelles, 164
 Simeulue, 164
 Sokoke, 162
 Sulawesi, 164
 Sunda, 163
 Torotoroka, 164
 Wallace's, 163
 White-fronted, 162
Scoter
 Black, 33
 Surf, 33
 White-winged, 33
Screamer
 Horned, 26
 Northern, 26
 Southern, 26
Screech-Owl
 Balsas, 165
 Bare-shanked, 165
 Bearded, 165
 Cinnamon, 165
 Cloud-forest, 165
 Colombian, 165
 Eastern, 165
 Guatemalan, 166
 Hoy's, 166
 Koepcke's, 165
 Long-tufted, 166
 Pacific, 165
 Puerto Rican, 166
 Rio Napo, 166
 Roraima, 166

 Rufescent, 165
 Tawny-bellied, 166
 Tropical, 165
 Variable, 166
 Vermiculated, 166
 Western, 164–165
 West Peruvian, 165
 Whiskered, 165
 White-throated, 166
Scrub-bird
 Noisy, 348
 Rufous, 347
Scrub-Flycatcher
 Amazonian, 326
 Northern, 326
 Southern, 326
Scrub-Jay
 Florida, 589
 Island, 589
 Western, 589
Scrub-Robin
 African, 459
 Bearded, 458
 Black, 459
 Brown, 458
 Brown-backed, 458
 Forest, 458
 Kalahari, 459
 Karoo, 459
 Miombo, 458
 Northern, 482
 Red-backed, 458
 Rufous-tailed, 459
 Southern, 482
Scrub-Warbler
 African, 427
 Bamboo, 427
 Bangwa, 427
 Cameroon, 427
 Grauer's, 427
 Ja River, 427
 Knysna, 427
 Streaked, 419
 Victorin's, 427
 White-winged, 427
Scrubfowl
 Dusky, 54
 Forsten's, 54
 Melanesian, 54
 Micronesian, 54
 Moluccan, 54
 New Guinea, 54
 Niaufoou, 54
 Nicobar, 54
 Orange-footed, 54
 Sula, 54
 Tabon, 54
 Tanimbar, 54
 Vanuatu, 54
Scrubtit, 519
Scrubwren
 Atherton, 518
 Beccari's, 518
 Buff-faced, 518
 Gray-green, 519
 Large, 518
 Large-billed, 518
 Pale-billed, 519
 Papuan, 518–519
 Perplexing, 518
 Tasmanian, 518
 Vogelkop, 518

 White-browed, 518
 Yellow-throated, 518
Scythebill
 Black-billed, 289
 Brown-billed, 289
 Curve-billed, 289
 Greater, 289
 Red-billed, 289
Sea-Eagle
 Solomon, 37
 Steller's, 37
 White-bellied, 36
Secretary-bird, 49
Seed-Finch
 Black-billed, 671
 Chestnut-bellied, 672
 Great-billed, 672
 Large-billed, 671
 Nicaraguan, 671
 Thick-billed, 672
Seedcracker
 Black-bellied, 613
 Crimson, 613
 Lesser, 613
Seedeater
 Band-tailed, 672
 Black-and-tawny, 671
 Black-and-white, 670
 Black-bellied, 671
 Black-eared, 636
 Blackish-blue, 672
 Blue, 672
 Brown-rumped, 636
 Buffy-fronted, 669
 Capped, 671
 Caqueta, 670
 Carrizal, 672
 Chestnut, 671
 Chestnut-bellied, 671
 Chestnut-throated, 671
 Dark-throated, 671
 Double-collared, 670–671
 Drab, 671
 Dubois', 670
 Gray, 670
 Gray-and-chestnut, 671
 Hooded, 670
 Lemon-breasted, 635
 Lesson's, 670
 Lined, 670
 Marsh, 671
 Narosky's, 671
 Paramo, 672
 Parrot-billed, 671
 Plain-colored, 672
 Plumbeous, 670
 Principe, 636
 Reichard's, 636
 Reichenow's, 635
 Ruddy-breasted, 671
 Rusty-collared, 670
 Slate-colored, 670
 Streaky, 635
 Streaky-headed, 636
 Tanzania, 636
 Tawny-bellied, 671
 Temminck's, 669
 Thick-billed, 635
 Tumaco, 671
 Variable, 670
 White-bellied, 671
 White-collared, 670

White-naped, 672
White-rumped, 634
White-throated, 671
Wing-barred, 670
Yellow-bellied, 670
Yellow-browed, 635
Seedsnipe
Gray-breasted, 100
Least, 100
Rufous-bellied, 100
White-bellied, 100
Seriema
Black-legged, 87
Red-legged, 87
Serin
Ankober, 633
European, 633
Fire-fronted, 633
Mountain, 634
Olive-rumped, 634
Salvadori's, 635
Syrian, 633
Tibetan, 634
Yellow-rumped, 635
Yellow-throated, 635
Yemen, 636
Serpent-Eagle
Andaman, 38
Congo, 38
Crested, 38
Madagascar, 39
Mountain, 38
Nicobar, 38
Philippine, 38
Sulawesi, 38
Serra-Finch, Pale-throated, 676
Shag
Antarctic, 17
Auckland Islands, 17
Bounty Islands, 17
Bronze, 17
Campbell Islands, 17
Chatham Islands, 17
Crozet, 17
European, 17
Heard Island, 17
Imperial, 17
Kerguelen, 17
Macquarie, 17
Pitt Island, 18
Rock, 17
Rough-faced, 17
South Georgia, 17
Spotted, 18
Shama
Black, 460
Rufous-tailed, 460
White-browed, 460
White-rumped, 459–460
White-vented, 460
Sharpbill, 311
Sheartail
Mexican, 210
Peruvian, 210
Slender, 210
Shearwater
Audubon's, 13
Balearic, 12
Black-vented, 13
Buller's, 12
Cape Verde, 12
Christmas, 12

Cory's, 12
Flesh-footed, 12
Fluttering, 13
Greater, 12
Heinroth's, 13
Hutton's, 13
Levantine, 12
Little, 13
Manx, 12
Mascarene, 13
Persian, 13
Pink-footed, 12
Short-tailed, 12
Sooty, 12
Streaked, 12
Townsend's, 13
Wedge-tailed, 12
Sheathbill
Black-faced, 100
Snowy, 100
Shelduck
Australian, 28
Common, 28
Crested, 699
Paradise, 28
Radjah, 28
Ruddy, 28
South African, 28
Shikra, 40
Shining-Parrot
Crimson, 135
Masked, 135
Red, 135
Shoebill, 24
Shortwing
Gould's, 413
Great, 413–414
Lesser, 413
Rusty bellied, 413
White-bellied, 413
White-browed, 413
Shoveler
Australian, 31
Cape, 31
Northern, 31
Red, 31
Shrike
Bay-backed, 570
Brown, 570
Bull-headed, 570
Burmese, 570
Chinese Gray, 571
Emin's, 570
Gray-backed, 570
Gray-capped, 570
Lesser Gray, 571
Loggerhead, 570–571
Long-tailed, 570
Mackinnon's, 571
Magpie, 572
Masked, 572
Northern, 571
Red-backed, 570
Rufous-tailed, 570
Southern Gray, 571
Souza's, 570
Tiger, 570
White-crowned, 572
White-rumped, 572
White-tailed, 468
Woodchat, 572
Yellow-billed, 572

Shrike-Babbler
Black-eared, 506
Black-headed, 505
Chestnut-fronted, 506
Green, 506
White-browed, 506
Shrike-flycatcher
African, 466
Black-and-white, 466
Shrike-Tanager
Black-throated, 653
Fulvous, 653
White-throated, 653
White-winged, 653
Shrike-Thrush
Bower's, 487
Gray, 487
Rufous, 487
Sandstone, 487
Sangihe, 487
Sooty, 486
Shrike-tit, Crested, 483
Shrike-Tyrant
Black-billed, 339
Gray-bellied, 339
Great, 339
Lesser, 339
White-tailed, 339
Shrike-Vireo
Chestnut-sided, 624
Green, 625
Slaty-capped, 625
Yellow-browed, 625
Shrikebill
Black-throated, 475
Fiji, 475
Rennell, 475
Southern, 475
Sibia
Beautiful, 510
Black-backed, 509
Black-headed, 509
Gray, 509
Long-tailed, 510
Rufous, 509
Rufous-backed, 509
White-eared, 509
Sicklebill
Black, 584
Black-billed, 584
Brown, 584
Buff-tailed, 192
Pale-billed, 584
White-tipped, 192
Sierra-Finch
Ash-breasted, 668
Band-tailed, 668
Black-hooded, 667
Carbonated, 668
Gray-hooded, 667
Mourning, 667
Patagonian, 667
Peruvian, 667
Plumbeous, 667
Red-backed, 667
White-throated, 668
Silktail, 479
Silky-flycatcher
Black-and-yellow, 387
Gray, 387
Long-tailed, 387
Silver-eye, 554

Silverbill
  African, 618
  Gray-headed, 617
Silverbird, 446
Sirystes, 343
Siskin
  Abyssinian, 634
  Andean, 632
  Antillean, 632
  Black, 632
  Black-capped, 632
  Black-chinned, 632
  Black-headed, 632
  Cape, 636
  Drakensberg, 636
  Eurasian, 631
  Hooded, 632
  Olivaceous, 632
  Pine, 631–632
  Red, 632
  Saffron, 632
  Thick-billed, 632
  Yellow-bellied, 632
  Yellow-faced, 632
  Yellow-rumped, 632
Sittella
  Black, 522
  Varied, 523
Skimmer
  African, 106
  Black, 106
  Indian, 106
Skua
  Brown, 107
  Chilean, 106
  Great, 107
  South Polar, 106–107
Skylark
  Eurasian, 357
  Oriental, 357
  Razo, 357
Slaty-Antshrike
  Bolivian, 292
  Marañó, 292
  Natterer's, 292
  Northern, 292
  Planalto, 292
  Sooretama, 292
  Western, 292
Slaty-Flycatcher
  Abyssinian, 447
  Angola, 447
  White-eyed, 447
Slaty-Thrush
  Andean, 411
  Eastern, 411
Smew, 33
Snake-Eagle
  Banded, 38
  Beaudouin's, 38
  Black-breasted, 38
  Brown, 38
  Fasciated, 38
Snipe
  African, 96
  Andean, 96
  Chatham Islands, 96
  Common, 96
  Fuegian, 96
  Giant, 96
  Great, 96
  Imperial, 96

Jack, 96
Latham's, 96
Madagascar, 96
Noble, 96
Pintail, 96
Puna, 96
Solitary, 96
South American, 96
Subantarctic, 96
Swinhoe's, 96
Wilson's, 96
Wood, 96
Snowcap, 202
Snowcock
  Altai, 63
  Caspian, 63
  Caucasian, 63
  Himalayan, 64
  Tibetan, 64
Snowfinch
  Afghan, 604
  Black-winged, 604
  Blanford's, 604
  Père David's, 604
  Rufous-necked, 604
  White-rumped, 604
  White-winged, 604
Social-Weaver
  Black-capped, 605
  Gray-headed, 605
Softtail
  Orinoco, 275
  Plain, 275–276
  Russet-mantled, 276
  Striated, 275
Solitaire
  Andean, 405
  Black, 406
  Black-faced, 405
  Brown-backed, 405
  Cuban, 405
  Reunion, 700
  Rodrigues, 700
  Rufous-brown, 405
  Rufous-throated, 405
  Slate-colored, 405
  Townsend's, 405
  Varied, 405
  White-eared, 406
Songlark
  Brown, 442
  Rufous, 442
Sooty-Owl
  Greater, 160
  Lesser, 160
Sora, 83
Spadebill
  Cinnamon-crested, 331
  Golden-crowned, 331
  Russet-winged, 331
  Stub-tailed, 331
  White-crested, 331
  White-throated, 331
  Yellow-throated, 331
Sparrow
  American Tree, 682
  Arabian Golden, 603
  Bachman's, 681
  Baird's, 684
  Black-capped, 679
  Black-chested, 681
  Black-chinned, 683

Black-striped, 679
Black-throated, 683
Botteri's, 681
Brewer's, 682
Bridled, 681
Cape, 603
Cape Verde, 602
Cassin's, 681
Chestnut, 603
Chipping, 682
Cinnamon-tailed, 681
Clay-colored, 682
Dead Sea, 602
Desert, 603
Eurasian Tree, 603
Field, 682
Five-striped, 682
Fox, 684–685
Golden-crowned, 686
Golden-winged, 679
Grasshopper, 684
Grassland, 684
Gray-headed, 603
Great Rufous, 602
Green-backed, 679
Half-collared, 679
Harris', 686
Henslow's, 684
House, 602
Java, 620
Kenya Rufous, 603
Kordofan Rufous, 603
Lark, 683
Le Conte's, 684
Lincoln's, 685
Nelson's Sharp-tailed, 684
Oaxaca, 682
Olive, 679
Orange-billed, 679
Parrot-billed, 603
Pectoral, 679
Plain-backed, 602
Rufous-collared, 686
Rufous-crowned, 681–682
Rufous-winged, 682
Russet, 602
Rusty, 682
Saffron-billed, 679
Sage, 683
Saltmarsh Sharp-tailed, 684
San Francisco, 679
Savannah, 683–684
Saxaul, 602
Seaside, 683–684
Shelley's Rufous, 603
Sierra Madre, 685
Sind, 602
Socotra, 602
Somali, 602
Song, 685
Southern Gray-headed, 603
Spanish, 602
Stripe-capped, 681
Stripe-headed, 681
Striped, 682
Sudan Golden, 603
Swahili, 603
Swainson's, 603
Swamp, 686
Timor, 620
Tocuyo, 679
Tumbes, 681

Vesper, 683
White-crowned, 686
White-throated, 686
Worthen's, 683
Yellow-browed, 684
Zapata, 682
Sparrow-Lark
Ashy-crowned, 352
Black-crowned, 352
Black-eared, 352
Chestnut-backed, 352
Chestnut-headed, 352
Fischer's, 352
Gray-backed, 352
Sparrow-Weaver
Chestnut-backed, 605
Chestnut-crowned, 605
Donaldson-Smith's, 605
White-browed, 605
Sparrowhawk
Chestnut-flanked, 40
Collared, 42
Eurasian, 42–43
Imitator, 42
Japanese, 42
Levant, 40
Little, 42
Madagascar, 42
New Britain, 42
Nicobar, 40
Ovampo, 42
Red-thighed, 42
Rufous-chested, 43
Rufous-necked, 42
Small, 42
Vinous-breasted, 42
Spatuletail, Marvelous, 209
Speirops
Black-capped, 549
Cameroon, 549
Fernando Po, 549
Principe, 549
Spiderhunter
Gray-breasted, 544
Little, 543
Long-billed, 543
Naked-faced, 543
Spectacled, 543
Streaked, 544
Streaky-breasted, 544
Thick-billed, 543
Whitehead's, 544
Yellow-eared, 543
Spindalis
Hispaniolan, 656
Jamaican, 656
Puerto Rican, 656
Western, 656
Spinebill
Eastern, 566
Western, 566
Spinetail
Apurimac, 272
Ash-browed, 275
Azara's, 272
Bahia, 272
Baron's, 275
Bat-like, 188
Black, 188
Black-faced, 273
Black-throated, 273
Bolivian, 275

Cabanis', 273
Cassin's, 188
Chestnut-throated, 273
Chicli, 272
Chinchipe, 274
Chotoy, 271
Cinereous-breasted, 272
Creamy-crested, 275
Crested, 274
Dark-breasted, 272
Dusky, 273
Gray-bellied, 272
Gray-headed, 275
Great, 274
Hoary-throated, 274
Light-crowned, 274
Line-cheeked, 275
Malagasy, 187
Marañón, 273
Marcapata, 274
McConnell's, 273
Mottled, 188
Necklaced, 274
Ochre-cheeked, 273
Olive, 275
Pale-breasted, 272
Pallid, 275
Parker's, 274
Pinto's, 272
Plain-crowned, 273
Red-and-white, 275
Red-faced, 275
Red-shouldered, 274
Ruddy, 272
Rufous, 273
Rufous-breasted, 273
Rufous-capped, 272
Russet-bellied, 274
Rusty-backed, 274
Rusty-headed, 273
Sabine's, 188
São Tomé, 187
Scaled, 275
Silvery-throated, 272
Slaty, 273
Sooty-fronted, 272
Speckled, 275
Streak-capped, 275
Stripe-breasted, 274
Stripe-crowned, 274
Sulphur-bearded, 274
Tepui, 275
White-bellied, 273
White-browed, 274
White-lored, 273
White-whiskered, 274
Yellow-chinned, 275
Spinifex-bird, 442
Spoonbill
African, 25
Black-faced, 25
Eurasian, 25
Roseate, 25
Royal, 25
Yellow-billed, 25
Spot-throat, 493
Spurfowl
Ceylon, 70
Painted, 70
Red, 70
Standardwing, Wallace's,
584

Starfrontlet
Blue-throated, 205
Buff-winged, 205
Golden, 205
Golden-bellied, 205
Rainbow, 206
Violet-throated, 205
White-tailed, 205
Starling
Abbott's, 601
African Pied, 600
Ashy, 601
Asian Glossy, 596
Asian Pied, 599
Atoll, 596
Babbling, 601
Black-collared, 599
Black-winged, 598
Brahminy, 599
Bristle-crowned, 601
Brown-winged, 596
Chestnut-bellied, 600
Chestnut-cheeked, 599
Chestnut-tailed, 599
Chestnut-winged, 601
Daurian, 599
Emerald, 600
European, 599
Fischer's, 601
Golden-breasted, 600
Hildebrandt's, 600
Kenrick's, 601
Kosrae, 701
Long-tailed, 596
Madagascar, 598
Magpie, 601
Metallic, 596
Micronesian, 597
Moluccan, 597
Mountain, 596
Mysterious, 701
Norfolk, 701
Narrow-tailed, 601
Neumann's, 601
Pale-winged, 601
Pohnpei, 597
Polynesian, 597
Rarotonga, 597
Red-billed, 599
Red-winged, 601
Rennell, 596
Reunion, 701
Rodrigues, 701
Rosy, 599
Rusty-winged, 596
Samoan, 597
San Cristobal, 596
Sharpe's, 601
Shelley's, 600
Short-tailed, 597
Singing, 596
Slender-billed, 601
Socotra, 601
Somali, 601
Spot-winged, 598
Spotless, 599
Striated, 596
Stuhlmann's, 601
Superb, 600
Tanimbar, 596
Tristram's, 601
Vinous-breasted, 598

Starling (*continued*)
Violet-backed, 600
Waller's, 601
Wattled, 599
White-billed, 601
White-cheeked, 599
White-collared, 601
White-crowned, 601
White-eyed, 596
White-faced, 599
White-headed, 599
White-shouldered, 599
Yellow-eyed, 596
Starthroat
Blue-tufted, 210
Long-billed, 210
Plain-capped, 209
Stripe-breasted, 210
Steamerduck
Falkland, 28
Flightless, 28
Flying, 28
White-headed, 28
Stilt
Banded, 90
Black, 90
Black-necked, 90
Black-winged, 90
Pied, 90
White-backed, 90
Stint
Little, 99
Long-toed, 99
Red-necked, 99
Temminck's, 99
Stitchbird, 562
Stonechat
African, 462
Canary Island, 462
European, 462
Reunion, 462
Siberian, 462
White-tailed, 462
Stork
Abdim's, 23
Black, 23
Black-necked, 23
Maguari, 23
Marabou, 24
Milky, 23
Oriental, 23
Painted, 23
Saddle-billed, 23
Storm's, 23
White, 23
Wood, 23
Woolly-necked, 23
Yellow-billed, 23
Storm-Petrel
Ashy, 14
Band-rumped, 14
Black, 14
Black-bellied, 14
European, 14
Fork-tailed, 14
Gray-backed, 13
Guadalupe, 699
Leach's, 14
Least, 14
Markham's, 14
Matsudaira's, 14
New Zealand, 699

Polynesian, 14
Ringed, 14
Swinhoe's, 14
Tristram's, 14
Wedge-rumped, 14
White-bellied, 14
White-faced, 13
White-vented, 13
Wilson's, 13
Straightbill
Olive, 556
Tawny, 556
Streamcreeper, Sharp-tailed, 283
Streamertail
Black-billed, 197
Red-billed, 196
Striped-Babbler
Luzon, 500
Negros, 500
Palawan, 500
Panay, 500
Striped-Swallow
Greater, 361
Lesser, 361
Stubtail
Asian, 425
Bornean, 425
Timor, 425
Sugarbird
Cape, 556
Gurney's, 556
Sunangel
Amethyst-throated, 206
Bogota, 700
Gorgeted, 206
Little, 206
Longuemare's, 206
Orange-throated, 206
Purple-throated, 206
Royal, 206
Tourmaline, 206
Sunbeam
Black-hooded, 204
Purple-backed, 204
Shining, 204
White-tufted, 204
Sunbird
Amani, 535
Amethyst, 537
Anchieta's, 534
Anjouan, 541
Apricot-breasted, 541
Banded, 535
Bannerman's, 536
Bates', 540
Beautiful, 539
Black, 537–538
Black-bellied, 539
Black-throated, 542
Blue-headed, 536
Blue-throated Brown, 536
Bocage's, 538
Bronze, 538
Buff-throated, 536
Cameroon, 536
Carmelite, 536
Collared, 535
Congo, 539
Copper, 540
Copper-throated, 538
Crimson-backed, 537
Dusky, 540

Eastern Crimson, 543
Eastern Double-collared, 539
Eastern Olive, 536
Elegant, 542
Fire-tailed, 543
Flame-breasted, 541
Flaming, 542
Fork-tailed, 542
Golden-winged, 538
Gould's, 542
Gray-headed, 534
Gray-hooded, 541
Greater Double-collared, 539
Green, 535
Green-headed, 536
Green-tailed, 542
Green-throated, 537
Handsome, 542
Humblot's, 541
Hunter's, 537
Johanna's, 540
Kenya Violet-backed, 535
Lina's, 542
Little Green, 535
Long-billed, 541
Lovely, 542
Loveridge's, 539
Madagascar, 541
Malachite, 538
Mariqua, 539
Mayotte, 541
Metallic-winged, 542
Miombo, 538
Montane Double-collared, 539
Moreau's, 539
Mount Apo, 541
Mouse-brown, 535
Mouse-colored, 536
Neergaard's, 538
Newton's, 536
Nile Valley, 535
Northern Double-collared, 539
Olive-backed, 540–541
Olive-bellied, 538
Orange-breasted, 536
Orange-tufted, 540
Oustalet's, 540
Palestine, 540
Pemba, 539
Plain, 534
Plain-backed, 534
Plain-throated, 534–535
Prigogine's, 539
Principe, 536
Purple, 540
Purple-banded, 539
Purple-breasted, 538
Purple-naped, 535
Purple-rumped, 537
Purple-throated, 537
Pygmy, 535
Red-chested, 539
Red-throated, 535
Red-tufted, 538
Regal, 539
Reichenbach's, 535
Rockefeller's, 539
Ruby-cheeked, 534
Rufous-winged, 540
São Tomé, 536
Scarlet, 543
Scarlet-chested, 537

Scarlet-tufted, 534
Seychelles, 541
Shelley's, 539
Shining, 540
Socotra, 537
Souimanga, 541
Southern Double-collared, 538
Splendid, 540
Stuhlmann's, 538
Superb, 540
Tacazze, 538
Temminck's, 543
Tiny, 538
Tsavo, 539
Uluguru Violet-backed, 535
Ursula's, 540
Variable, 540
Violet-breasted, 539
Violet-tailed, 535
Western Crimson, 543
Western Olive, 536
Western Violet-backed, 535
White-breasted, 540
White-flanked, 542
Sunbittern, 86
Sungem, Horned, 209
Sungrebe, 86
Surfbird, 98
Swallow
Andean, 359
Angola, 360
Bahama, 359
Bank, 358
Barn, 360
Black-and-rufous, 361
Black-capped, 359
Black-collared, 359
Blue, 361
Blue-and-white, 359
Brown-bellied, 359
Cave, 362–363
Chestnut-collared, 363
Chilean, 359
Cliff, 362
Ethiopian, 361
Forest, 362
Golden, 358
Gray-rumped, 358
Mangrove, 359
Mosque, 362
Northern Rough-winged, 360
Pacific, 360
Pale-footed, 359
Pearl-breasted, 361
Pied-winged, 361
Preuss', 362
Red Sea, 362
Red-chested, 360
Red-rumped, 362
Red-throated, 362
Rufous-bellied, 362
Rufous-chested, 362
South African, 362
Southern Rough-winged, 360
Streak-throated, 362
Striated, 362
Tawny-headed, 360
Tree, 358
Tumbes, 359
Violet-green, 358
Welcome, 360
White-backed, 358

White-banded, 359
White-rumped, 359
White-tailed, 361
White-thighed, 360
White-throated, 360
White-throated Blue, 360
White-winged, 359
Wire-tailed, 360
Swallow-Tanager, 663
Swallow-wing, 235
Swamp-Warbler
Cape Verde, 431
Greater, 431
Lesser, 431
Madagascar, 431
Swamphen
Lord Howe, 699
Purple, 84–85
Swan
Black, 26
Black-necked, 26
Coscoroba, 27
Mute, 26
Trumpeter, 26
Tundra, 27
Whooper, 26
Swift
African, 190
Alexander's, 190
Alpine, 190
Andean, 189
Ashy-tailed, 189
Band-rumped, 188
Bates', 191
Biscutate, 185
Black, 184
Bradfield's, 190
Chapman's, 189
Chestnut-collared, 184
Chimney, 188
Common, 190
Costa Rican, 188
Dark-rumped, 191
Forbes-Watson's, 190
Fork-tailed, 191
Gray-rumped, 188
Great Dusky, 184
Great Swallow-tailed, 189
Horus, 191
House, 191
Lesser Antillean, 188
Lesser Swallow-tailed, 189
Little, 191
Madagascar, 190
Mottled, 190
Nyanza, 190
Pale-rumped, 188
Pallid, 190
Plain, 190
Pygmy, 189
Rothschild's, 184
Scarce, 187
Schouteden's, 187
Short-tailed, 189
Sooty, 184
Spot-fronted, 184
Tepui, 184
Tumbes, 189
Vaux's, 188–189
Waterfall, 185
White-chested, 184
White-chinned, 184

White-collared, 184
White-fronted, 184
White-naped, 184
White-rumped, 191
White-throated, 189
White-tipped, 189
Swiftlet
Atiu, 187
Australian, 186
Bare-legged, 186
Black-nest, 187
Caroline Islands, 187
Cave, 185
Edible-nest, 187
German's, 187
Glossy, 185
Himalayan, 186
Indian, 185
Indochinese, 186
Mariana, 187
Marquesan, 187
Mascarene, 185
Mayr's, 186
Moluccan, 185
Mossy-nest, 186–187
Mountain, 186
Palau, 187
Palawan, 186
Papuan, 187
Philippine, 185
Polynesian, 187
Pygmy, 185
Seychelles, 185
Uniform, 186
Volcano, 186
Whitehead's, 186
White-rumped, 186
Sylph
Long-tailed, 209
Venezuelan, 209
Violet-tailed, 209

Tachuri
Bearded, 323
Gray-backed, 323
Tailorbird
African, 432
Ashy, 433
Common, 433
Dark-necked, 433
Gray-backed, 433
Long-billed, 432
Mountain, 432
Olive-backed, 433
Philippine, 433
Rufous-fronted, 433
Rufous-headed, 433
Rufous-tailed, 433
White-browed, 433
White-eared, 433
Yellow-breasted, 433
Takahe, 85
Tanager
Azure-rumped, 659
Azure-shouldered, 656
Bay-headed, 660
Beryl-spangled, 662
Black-and-gold, 657
Black-and-white, 649
Black-and-yellow, 652
Black-backed, 661
Black-capped, 662

Tanager (*continued*)
Black-faced, 648
Black-goggled, 654
Black-headed, 662
Blue-and-black, 662
Blue-and-gold, 657
Blue-and-yellow, 657
Blue-backed, 657
Blue-browed, 661
Blue-capped, 656–657
Blue-gray, 656
Blue-necked, 661
Blue-whiskered, 659
Brassy-breasted, 659
Brazilian, 656
Brown, 648
Brown-flanked, 651
Buff-bellied, 651
Burnished-buff, 661
Cherrie's, 656
Cherry-throated, 652
Chestnut-backed, 661
Chestnut-headed, 651
Cinnamon, 648
Cone-billed, 649
Crimson-backed, 655
Crimson-collared, 655
Diademed, 658
Dotted, 660
Dusky-faced, 652
Emerald, 659
Fawn-breasted, 658
Flame-colored, 655
Flame-crested, 653
Flame-faced, 660
Flame-rumped, 656
Fulvous-crested, 653
Fulvous-headed, 651
Gilt-edged, 659
Glaucous, 656
Glistening-green, 658
Gold-ringed, 657
Golden, 659–660
Golden-chested, 657
Golden-chevroned, 656
Golden-collared, 658
Golden-crowned, 658
Golden-eared, 660
Golden-hooded, 661
Golden-naped, 661
Grass-green, 649
Gray-and-gold, 659
Gray-headed, 653
Green-and-gold, 659
Green-capped, 661
Green-headed, 659
Green-naped, 661
Guira, 651
Hepatic, 654–655
Hooded, 652
Huallaga, 655
Lemon-spectacled, 652
Lesser Antillean, 661
Magpie, 649
Masked, 661
Masked Crimson, 655
Metallic-green, 661
Moss-backed, 657
Multicolored, 659
Ochre-breasted, 653
Olive, 652
Olive-backed, 652

Olive-green, 653
Opal-crowned, 662
Opal-rumped, 662
Orange-eared, 659
Orange-headed, 651
Orange-throated, 657
Palm, 657
Paradise, 659
Passerini's, 656
Plain-colored, 659
Puerto Rican, 649
Purplish-mantled, 658
Red-billed Pied, 649
Red-headed, 655
Red-hooded, 655
Red-necked, 659
Red-shouldered, 654
Rose-throated, 654
Ruby-crowned, 654
Rufous-cheeked, 661
Rufous-chested, 651
Rufous-crested, 653
Rufous-headed, 651
Rufous-throated, 660
Rufous-winged, 660
Rust-and-yellow, 651
Saffron-crowned, 660
Sayaca, 656
Scarlet, 655
Scarlet-and-white, 652
Scarlet-browed, 653
Scarlet-throated, 649
Scrub, 661
Seven-colored, 659
Silver-backed, 662
Silver-beaked, 655–656
Silver-throated, 660
Sira, 662
Slaty, 653
Spangle-cheeked, 661
Speckled, 660
Spotted, 660
Straw-backed, 662
Sulphur-rumped, 653
Summer, 655
Tawny-crested, 654
Turquoise, 659
Vermilion, 655
Western, 655
White-banded, 648
White-capped, 649
White-lined, 654
White-rumped, 649
White-shouldered, 654
White-winged, 655
Yellow-backed, 651–652
Yellow-bellied, 660
Yellow-crested, 653
Yellow-scarfed, 658
Yellow-throated, 658
Yellow-winged, 657
Tapaculo
Ancash, 310
Ash-colored, 311
Bahia, 309
Blackish, 310–311
Bolivian, 309
Brasilia, 309
Brown-rumped, 310
Caracas, 310
Choco, 310
Chucao, 309

Chusquea, 310
Diademed, 310
Dusky, 310
Ecuadorian, 310
Lara, 310
Large-footed, 311
Long-tailed, 309
Magellanic, 310
Matorral, 310
Merida, 310
Mouse-colored, 309
Nariño, 309
Neblina, 310
Ocellated, 309
Ochre-flanked, 309
Pale-throated, 310
Paramo, 310
Planalto, 309
Puna, 310
Rufous-vented, 309
Rusty-belted, 309
Santa Marta, 309
Silvery-fronted, 310
Spillman's, 310
Stiles, 309
Trilling, 310
Tschudi's, 310
Unicolored, 310
Upper Magdalena, 309
Vilcabamba, 310
Wetland, 309
White-breasted, 309
White-browed, 310
White-crowned, 309
White-throated, 309
Zimmer's, 310
Tattler
Gray-tailed, 98
Wandering, 98
Tchagra
Black-crowned, 573
Brown-crowned, 573
Marsh, 573
Southern, 573
Three-streaked, 573
Teal
Andaman, 30
Auckland Islands, 30
Baikal, 30
Bernier's, 30
Blue-winged, 31
Brazilian, 29
Brown, 30
Campbell Islands, 30
Cape, 30
Chestnut, 30
Cinnamon, 31
Eurasian, 30
Gray, 30
Green-winged, 30
Hottentot, 31
Marbled, 32
Puna, 31
Ringed, 29
Salvadori's, 29
Silver, 31
Speckled, 30
Sunda, 30
Tern
Aleutian, 104
Antarctic, 105
Arctic, 105

Black, 105
Black-bellied, 106
Black-fronted, 105
Black-naped, 105
Bridled, 104
Caspian, 105
Chinese Crested, 106
Common, 105
Damara, 104
Elegant, 106
Fairy, 104
Forster's, 106
Gray-backed, 104
Great Crested, 106
Gull-billed, 104–105
Inca, 105
Kerguelen, 106
Large-billed, 104
Least, 104
Lesser Crested, 106
Little, 104
Peruvian, 104
River, 106
Roseate, 105
Royal, 106
Sandwich, 106
Saunders', 104
Snowy-crowned, 106
Sooty, 104
South American, 105
Whiskered, 105
White, 103–104
White-cheeked, 106
White-fronted, 105
White-winged, 105
Yellow-billed, 104
Tesia
　Chestnut-headed, 425
　Gray bellied, 425
　Javan, 425
　Russet-capped, 425
　Slaty-bellied, 425
Thamnornis, 428
Thick-knee
　Beach, 91
　Bush, 91
　Double-striped, 90–91
　Eurasian, 90
　Great, 91
　Peruvian, 91
　Senegal, 90
　Spotted, 90
　Water, 90
Thicket-Fantail
　Black, 469
　Sooty, 469
　White-bellied, 470
Thicketbird
　Bismarck, 442
　Bougainville, 442
　Guadalcanal, 442
　Rusty, 442
Thistletail
　Black-throated, 271
　Eye-ringed, 271
　Itatiaia, 271
　Mouse-colored, 271
　Ochre-browed, 271
　Perija, 271
　Puna, 271
　Vilcabamba, 271
　White-chinned, 271

Thornbill
　Black-backed, 208
　Blue-mantled, 209
　Bronze-tailed, 209
　Brown, 520
　Buff-rumped, 519
　Chestnut-rumped, 520
　Inland, 520
　Mountain, 520
　Olivaceous, 209
　Papuan, 519
　Purple-backed, 208
　Rainbow-bearded, 209
　Rufous-capped, 208
　Slaty-backed, 520
　Slender-billed, 520
　Striated, 520
　Tasmanian, 520
　Western, 519
　Yellow, 520
　Yellow-rumped, 520
Thornbird
　Chestnut-backed, 277
　Common, 277
　Freckle-breasted, 277
　Greater, 277
　Little, 277
　Red-eyed, 277
　Spot-breasted, 277
　Streak-fronted, 277
Thorntail
　Black-bellied, 196
　Coppery, 196
　Green, 196
　Wire-crested, 196
Thrasher
　Bendire's, 398
　Brown, 398
　California, 399
　Cozumel, 398
　Crissal, 399
　Curve-billed, 399
　Gray, 398
　Le Conte's, 399
　Long-billed, 398
　Ocellated, 398
　Pearly-eyed, 399
　Sage, 398
　Scaly-breasted, 399
　Vizcaino, 399
　White-breasted, 399
Thrush
　African, 408
　African Bare-eyed, 408
　Ashy, 402
　Austral, 411
　Aztec, 413
　Bare-eyed, 412
　Bicknell's, 407
　Black-billed, 411–412
　Black-breasted, 408
　Black-hooded, 411
　Brown-headed, 410
　Buru, 402
　Chestnut, 410
　Chestnut-backed, 402
　Chestnut-bellied, 411
　Chestnut-capped, 402
　Chiguanco, 410
　Chinese, 410
　Cocoa, 412
　Comoro, 408

Creamy-bellied, 411
Dark-sided, 404
Dark-throated, 410
Dusky, 410
Ecuadorian, 412
Enggano, 402
Everett's, 403
Eyebrowed, 410
Fawn-breasted, 404
Forest, 404
Glossy-black, 411
Grand Cayman, 700
Gray-backed, 408
Gray-cheeked, 407
Gray-sided, 410
Great, 411
Groundscraper, 407
Hauxwell's, 412
Hermit, 407
Island, 409
Izu, 410
Japanese, 408
Kittlitz's, 700
Kurrichane, 408
La Selle, 413
Lawrence's, 412
Long-billed, 404
Long-tailed, 404
Marañón, 411
Mistle, 410
New Britain, 404
Olivaceous, 408
Olive, 407–408
Olive-tailed, 404
Orange-banded, 402
Orange-headed, 402–403
Pale, 410
Pale-breasted, 411
Pale-eyed, 407
Pale-vented, 412
Pied, 402
Plain-backed, 403
Plumbeous-backed, 411
Red-and-black, 402
Red-legged, 410
Rufous-bellied, 411
Russet-tailed, 404
Rusty-backed, 402
San Cristobal, 404
Scaly, 404
Seram, 402
Siberian, 403
Slaty-backed, 402
Song, 410
Spot-winged, 403
Sulawesi, 404
Sunda, 403
Swainson's, 407
Tickell's, 408
Tristan, 404
Unicolored, 412
Varied, 413
White-backed, 410
White-chinned, 413
White-eyed, 412
White-necked, 412
White-throated, 412
Wood, 407
Yellow-legged, 407
Yemen, 407
Thrush-Babbler, Dohrn's, 509
Thrush-Tanager, Rosy, 652

Tiger-Heron
Bare-throated, 22
Fasciated, 22
Rufescent, 22
Tiger-Parrot
Brehm's, 137
Madarasz's, 137
Modest, 137
Painted, 137
Tinamou
Andean, 5
Barred, 5
Bartlett's, 5
Berlepsch's, 3
Black, 3
Black-capped, 4
Brazilian, 4
Brown, 4
Brushland, 5
Chilean, 5
Choco, 4
Cinereous, 3
Curve-billed, 5
Dwarf, 6
Elegant Crested, 6
Gray, 3
Gray-legged, 4
Great, 3
Highland, 3
Hooded, 3
Kalinowski's, 5
Little, 3–4
Ornate, 5
Pale-browed, 4
Patagonian, 6
Puna, 6
Quebracho Crested, 6
Red-legged, 3
Red-winged, 5
Rusty, 5
Slaty-breasted, 4
Small-billed, 5
Solitary, 3
Taczanowski's, 5
Tataupa, 5
Tawny-breasted, 3
Tepui, 4
Thicket, 4–5
Undulated, 4
Variegated, 4
White-throated, 3
Yellow-legged, 4
Tinkerbird
Green, 236
Moustached, 236
Red-fronted, 236
Red-rumped, 236
Speckled, 236
Western, 236
Yellow-fronted, 236
Yellow-rumped, 236
Yellow-throated, 236
Tit
African Blue, 528
Ashy, 527
Azure, 528
Black-bibbed, 524
Black-breasted, 525
Black-browed, 515
Black-crested, 525
Black-lored, 527
Black-throated, 514–515
Carp's, 526

Caspian, 524
Coal, 525
Crested, 526
Dusky, 526
Elegant, 525–526
Eurasian Blue, 528
Fire-capped, 534
Gray, 527
Gray-crested, 526
Great, 527
Green-backed, 527
Ground, 529
Long-tailed, 514
Marsh, 523–524
Miombo, 526
Palawan, 526
Père David's, 525
Pygmy, 515
Red-throated, 526
Rufous-bellied, 526
Rufous-vented, 525
Somali, 526
Sombre, 523
Songar, 524
Sooty, 515
Stripe-breasted, 526
Sultan, 529
Turkestan, 527
Varied, 528
White-bellied, 526
White-browed, 525
White-cheeked, 514
White-fronted, 528
White-throated, 515
White-winged, 527
Willow, 524
Yellow, 528
Yellow-bellied, 525
Yellow-breasted, 528
Yellow-browed, 529
Yellow-cheeked, 528
Tit-Babbler
Brown, 502
Fluffy-backed, 502
Gray-cheeked, 502
Gray-faced, 502
Miniature, 502
Striped, 502
Tit-Flycatcher
Gray, 450
Gray-throated, 450
Tit-hylia, 534
Tit-Spinetail
Andean, 271
Araucaria, 271
Brown-capped, 270
Plain-mantled,
271
Rusty-crowned, 271
Streaked, 271
Striolated, 271
Tawny, 270
Tufted, 270
White-browed, 271
Tit-Tyrant
Agile, 322
Ash-breasted, 322
Black-crested, 322
Juan Fernandez, 322
Pied-crested, 322
Tufted, 322
Unstreaked, 322
Yellow-billed, 322

Tit-Warbler
Crested, 436
White-browed, 436
Titmouse
Black-crested, 529
Bridled, 528
Juniper, 529
Oak, 529
Tufted, 529
Tityra
Black-crowned, 346
Black-tailed, 346
Masked, 345
Tody
Broad-billed, 223
Cuban, 223
Jamaican, 224
Narrow-billed, 223
Puerto, 224
Tody-Flycatcher
Black-backed, 329
Black-headed, 330
Buff-cheeked, 328
Common, 329
Golden-winged, 329
Ochre-faced, 329
Painted, 329
Ruddy, 328
Rusty-fronted, 329
Short-tailed, 329
Slate-headed, 329
Smoky-fronted, 329
Spotted, 329
Yellow-browed, 329
Yellow-lored, 329
Tody-Tyrant
Black-and-white, 328
Black-throated, 328
Boat-billed, 327
Buff-breasted, 328
Buff-throated, 328
Cinnamon-breasted, 328
Eye-ringed, 327
Fork-tailed, 328
Hangnest, 328
Johannes', 327
Kaempfer's, 328
Lulu's, 328
Pearly-vented, 328
Pelzeln's, 328
Rufous-crowned, 328
Snethlage's, 327
Stripe-necked, 327
White-bellied, 327
White-cheeked, 328
White-eyed, 327
Yungas, 327
Zimmer's, 328
Tomtit, 480
Topaz
Crimson, 195
Fiery, 195
Torrent-lark, 580
Toucan
Black-mandibled, 243
Channel-billed, 243
Choco, 243
Keel-billed, 243
Red-billed, 243
Red-breasted, 242
Toco, 243
Toucanet
Andean, 241

Black-throated, 241
Blue-banded, 241
Blue-throated, 241
Chestnut-tipped, 241
Crimson-rumped, 241
Emerald, 241
Golden-collared, 241
Gould's, 241
Groove-billed, 241
Guianan, 241
Saffron, 242
Santa Marta, 241
Spot-billed, 242
Tawny-tufted, 241
Violet-throated, 241
Wagler's, 241
Yellow-browed, 241
Yellow-eared, 242
Towhee
  Abert's, 681
  California, 680
  Canyon, 681
  Collared, 680
  Eastern, 680
  Green-tailed, 680
  Socorro, 680
  Spotted, 680
  White-throated, 681
Tragopan
  Blyth's, 70
  Cabot's, 70
  Satyr, 70
  Temminck's, 70
  Western, 70
Trainbearer
  Black-tailed, 207
  Green-tailed, 207
Tree-Finch
  Large, 688
  Medium, 688
  Small, 688
Treecreeper
  Bar-tailed, 532
  Black-tailed, 523
  Brown, 523
  Brown-throated, 532
  Eurasian, 531–532
  Papuan, 523
  Red-browed, 523
  Rufous, 523
  Rusty-flanked, 532
  Short-toed, 532
  Sichuan, 532
  White-browed, 523
  White-throated, 523
Treehunter
  Black-billed, 281
  Buff-throated, 281
  Flammulated, 281
  Pale-browed, 281
  Sharp-billed, 283
  Streak-breasted, 281
  Streak-capped, 281
  Striped, 281
  Uniform, 281
Treepie
  Andaman, 592
  Bornean, 592
  Collared, 592
  Gray, 592
  Hooded, 592
  Racket-tailed, 592
  Ratchet-tailed, 592

Rufous, 591–592
Sumatran, 592
White-bellied, 592
Treerunner
  Beautiful, 278
  Fulvous-dotted, 278
  Pearled, 278
  Ruddy, 278
  White-throated, 283
Treeswift
  Crested, 191
  Gray-rumped, 191
  Moustached, 191
  Whiskered, 191
Trembler
  Brown, 399
  Gray, 399
Triller
  Black-and-white, 372
  Black-browed, 373
  Long-tailed, 373
  Pied, 372
  Polynesian, 373
  Rufous-bellied, 372
  Samoan, 373
  Varied, 372
  White-browed, 372
  White-rumped, 372
  White-shouldered, 372
  White-winged, 372
Trogon
  Baird's, 213
  Bar-tailed, 212
  Bare-cheeked, 212
  Black-headed, 212
  Black-tailed, 214
  Black-throated, 213–214
  Blue-crowned, 214
  Cinnamon-rumped, 215
  Citreoline, 212
  Collared, 213
  Cuban, 212
  Diard's, 215
  Elegant, 213
  Hispaniolan, 212
  Javan, 214
  Lattice-tailed, 214
  Malabar, 214
  Masked, 213
  Mountain, 213
  Narina, 212
  Orange-bellied, 213
  Orange-breasted, 215
  Philippine, 215
  Red-headed, 215
  Red-naped, 214
  Scarlet-rumped, 215
  Slaty-tailed, 214
  Sumatran, 214
  Surucua, 214
  Violaceous, 213
  Ward's, 215
  White-eyed, 213
  White-tailed, 213
  Whitehead's, 215
Tropicbird
  Red-billed, 15
  Red-tailed, 15
  White-tailed, 15
Troupial
  Campo, 696
  Orange-backed, 696
  Venezuelan, 696

Trumpeter
  Dark-winged, 77
  Gray-winged, 77
  Pale-winged, 77
Tuftedcheek
  Buffy, 279
  Streaked, 279
Tui, 567
Turaco
  Bannerman's, 151
  Black-billed, 151
  Fischer's, 151
  Great Blue, 151
  Guinea, 151
  Hartlaub's, 151
  Knysna, 151
  Livingstone's, 151
  Prince Ruspoli's, 151
  Purple-crested, 152
  Red-crested, 151
  Ross', 152
  Ruwenzori, 152
  Schalow's, 151
  Violet, 152
  White-cheeked, 151
  White-crested, 151
  Yellow-billed, 151
Turca, Moustached, 309
Turkey
  Ocellated, 57
  Wild, 57
Turnstone
  Black, 98
  Ruddy, 98
Turtle-Dove
  Adamawa, 112
  Dusky, 112
  Eurasian, 112
  Madagascar, 113
  Oriental, 113
Twinspot
  Brown, 614
  Dusky, 614
  Dybowski's, 614
  Green-backed, 612
  Peters', 614
  Pink-throated, 614
Twite, 633
Tyrannulet
  Alagoas, 324
  Amazonian, 326
  Ashy-headed, 319
  Bahia, 324
  Bay-ringed, 325
  Black-capped, 319
  Black-fronted, 324
  Bolivian, 324
  Brown-capped, 321
  Buff-banded, 321
  Cinnamon-faced, 324
  Ecuadorian, 324
  Golden-faced, 324
  Gray-and-white, 323
  Gray-capped, 319
  Greenish, 318
  Minas Gerais, 324
  Mishana, 324
  Mottle-cheeked, 324
  Mouse-colored, 322
  Olive-green, 324
  Oustalet's, 325
  Pale-tipped, 326
  Paltry, 324

Tyrannulet (*continued*)
Peruvian, 324
Plain, 326
Planalto, 318
Plumbeous-crowned, 319
Red-billed, 324
Reiser's, 318
Restinga, 324
River, 322
Rough-legged, 318
Rufous-browed, 324
Rufous-lored, 324
Rufous-winged, 321
São Paulo, 325
Sclater's, 319
Serra do Mar, 325
Slender-billed, 326
Slender-footed, 324
Sooty, 322
Sooty-headed, 319
Sulphur-bellied, 322
Tawny-rumped, 319
Torrent, 322
Urich's, 318
Venezuelan, 324
White-banded, 321
White-bellied, 322
White-crested, 322
White-lored, 321
White-tailed, 321
White-throated, 321
Yellow, 323
Yellow-bellied, 321
Yellow-crowned, 319
Yellow-green, 324
Tyrant
Andean, 336
Black-chested, 328
Cattle, 340
Chocolate-vented, 338
Cinereous, 336
Cinnamon, 333
Cock-tailed, 338
Long-tailed, 340
Patagonian, 338
Riverside, 337
Rufous-tailed, 336–337
Rufous-webbed, 339
Sharp-tailed, 323
Shear-tailed Gray, 340
Spectacled, 336
Strange-tailed, 338
Streamer-tailed, 340
Tumbes, 337
Yellow-browed, 338
Tyrant-Manakin
Dwarf, 318
Pale-bellied, 317
Saffron-crested, 317
Serra, 317
Sulphur-bellied, 317
Tiny, 318
Wied's, 317

Ula-ai-hawane, 701
Umbrellabird
Amazonian, 314
Bare-necked, 314
Long-wattled, 314

Vanga
Bernier's, 577
Blue, 577

Chabert, 577
Helmet, 577
Hook-billed, 577
Lafresnaye's, 577
Pollen's, 577
Red-shouldered, 577
Red-tailed, 577
Rufous, 577
Sickle-billed, 577
Tylas, 577
Van Dam's, 577
White-headed, 577
Veery, 407
Velvetbreast, Mountain,
205
Verdin, 533
Violet-ear
Brown, 195
Green, 195
Sparkling, 195
White-vented, 195
Vireo
Bell's, 622
Black-capped, 622
Black-whiskered, 623
Blue-headed, 622
Blue Mountain, 622
Brown-capped, 623
Cassin's, 622
Choco, 623
Cozumel, 621
Cuban, 621
Dwarf, 622
Flat-billed, 622
Golden, 623
Gray, 622
Hutton's, 622
Jamaican, 621
Mangrove, 621
Noronha, 623
Philadelphia, 623
Plumbeous, 622
Puerto Rican, 621
Red-eyed, 623
Slaty, 621
St. Andrew, 621
Thick-billed, 621
Warbling, 622
White-eyed, 621
Yellow-green, 623
Yellow-throated, 622
Yellow-winged, 622
Yucatan, 623
Visorbearer
Hooded, 209
Hyacinth, 209
Vulture
Black, 34
Cinereous, 37
Egyptian, 37
Greater Yellow-headed, 34
Hooded, 37
Indian, 37
King, 34
Lappet-faced, 37–38
Lesser Yellow-headed, 34
Palm-nut, 37
Red-headed, 38
Slender-billed, 37
Turkey, 34
White-backed, 37
White-headed, 38
White-rumped, 37

Wagtail
African Pied, 367
Cape, 368
Citrine, 367
Eastern Yellow, 367
Forest, 367
Gray, 368
Japanese, 367
Madagascar, 368
Mekong, 367
Mountain, 368
White, 367
White-browed, 367
Yellow, 367
Wagtail-Tyrant
Greater, 323
Lesser, 323
Waldrapp, 24
Wallcreeper, 531
Warbler
Adelaide's, 641
African Desert, 443
African Yellow, 432
Aquatic, 429
Arctic, 438
Arrow-headed, 642
Ashy-throated, 438
Asian Desert, 443
Bachman's, 639
Balearic, 444
Banded, 444–445
Barbuda, 641
Barred, 443
Bay-breasted, 641
Bianchi's, 440
Black-and-white, 642
Black-cheeked, 646
Black-crested, 645
Black-faced, 441
Black-throated Blue, 640
Black-throated Gray, 641
Black-throated Green, 641
Blackburnian, 641
Blackpoll, 642
Blue-winged, 639
Blunt-winged, 430
Booted, 432
Broad-billed, 441
Brown, 444
Buff-barred, 437
Buff-bellied, 432
Buff-rumped, 646
Buff-throated, 437
Canada, 643
Cape May, 640
Cerulean, 642
Cetti's, 426
Chestnut-crowned, 440
Chestnut-sided, 640
Choco, 645
Christmas Island, 431
Citrine, 645
Colima, 639
Connecticut, 642
Crescent-chested, 639
Cryptic, 435
Cyprus, 444
Dartford, 444
Dusky, 437
Eastern Bonelli's, 437
Eastern Olivaceous, 432
Eastern Orphean, 443
Elfin-woods, 642

Eurasian River, 429
Fan-tailed, 644
Flame-throated, 639
Flavescent, 646
Garden, 443
Golden-bellied, 644
Golden-browed, 646
Golden-cheeked, 641
Golden-crowned, 645
Golden-spectacled, 440
Golden-winged, 639
Grace's, 641
Grasshopper, 429
Grauer's, 434
Gray's, 429
Gray-and-gold, 644
Gray-capped, 424
Gray-cheeked, 440
Gray-crowned, 440
Gray-headed, 645
Gray-hooded, 440
Gray-throated, 645
Greenish, 438
Green-tailed, 643
Hermit, 641
Hooded, 643
Hume's, 438
Icterine, 432
Kentucky, 642
Kirtland's, 641
Kopje, 425
Lanceolated, 429
Layard's, 444
Lemon-rumped, 438
Lemon-throated, 439
Long-legged, 442
Lucy's, 639
MacGillivray's, 642
Magnolia, 640
Marmora's, 444
Marsh, 430
Melodious, 432
Ménétries', 444
Middendorff's, 429
Mountain, 439
Mountain Yellow, 432
Mourning, 642
Moustached, 429
Mrs. Moreau's, 428
Nashville, 639
Neotropical River, 646
Neumann's, 435
Olive, 639
Olive-capped, 641
Olive-tree, 432
Orange-crowned, 639
Oriente, 643
Oriole, 424
Paddyfield, 429
Pale-legged, 645
Pale-rumped, 438
Pallas', 429
Palm, 641
Papyrus Yellow, 432
Pine, 641
Pink-headed, 643
Pirre, 646
Plain-tailed, 440
Pleske's, 429
Plumbeous, 642
Prairie, 641
Prothonotary, 642
Radde's, 437

Rand's, 434
Red, 643
Red Sea, 443–444
Red-faced, 643
Red-fronted, 424
Red-winged Gray, 422
Rock, 517
Rueppell's, 444
Rufous-capped, 645–646
Rufous-eared, 422
Rufous-faced, 440–441
Rufous-vented, 444
Russet-crowned, 645
Sakhalin, 429
Santa Marta, 645
Sardinian, 444
Savi's, 429
Sedge, 429
Semper's, 643
Shade, 426
Smoky, 437
Socotra, 419
Speckled, 519
Spectacled, 444
St. Lucia, 641
Subalpine, 444
Sulphur-bellied, 437
Sulphur-breasted, 439
Sunda, 440
Swainson's, 642
Sykes', 432
Tennessee, 639
Thick-billed, 431
Three-banded, 645
Three-striped, 646
Townsend's, 641
Tristram's, 444
Two-banded, 644
Upcher's, 432
Virginia's, 639
Vitelline, 641
Western Bonelli's, 437
Western Olivaceous, 432
Western Orphean, 443
Whistler's, 440
Whistling, 642
White-bellied, 645
White-browed Chinese, 419
White-lored, 645
White-rimmed, 646
White-spectacled, 440
White-striped, 646
White-tailed, 434
White-winged, 647
Willow, 437
Wilson's, 643
Wood, 437
Worm-eating, 642
Yellow, 640
Yellow-bellied, 641
Yellow-breasted, 440
Yellow-browed, 438
Yellow-headed, 643
Yellow-rumped, 640–641
Yellow-streaked, 437
Yellow-throated, 641
Yellow-vented, 439
Yemen, 443
Warbling-Finch
  Bay-chested, 669
  Black-and-chestnut, 669
  Black-and-rufous, 669
  Black-capped, 669

Bolivian, 669
  Cinereous, 669
  Cinnamon, 669
  Collared, 669
  Plain-tailed, 669
  Red-rumped, 669
  Ringed, 669
  Rufous-breasted, 669
  Rufous-sided, 669
  Rusty-browed, 669
Water-Tyrant
  Black-backed, 338
  Drab, 338
  Masked, 338
  Pied, 338
Watercock, 84
Waterhen, White-breasted, 83
Waterthrush
  Louisiana, 642
  Northern, 642
Wattle-eye
  Banded, 466
  Black-necked, 466
  Black-throated, 466
  Brown-throated, 466
  Chestnut, 466
  Jameson's, 467
  Red-cheeked, 466
  White-fronted, 466
  White-spotted, 466
  Yellow-bellied, 467
Wattlebird
  Brush, 567
  Little, 567
  Red, 567
  Yellow, 567
Waxbill
  Anambra, 612
  Arabian, 612
  Black-cheeked, 613
  Black-crowned, 613
  Black-faced, 613
  Black-headed, 613
  Black-rumped, 612
  Black-tailed, 612
  Cinderella, 612
  Common, 612–613
  Crimson-rumped, 612
  Fawn-breasted, 612
  Kandt's, 613
  Lavender, 612
  Orange-cheeked, 612
  Red-rumped, 613
  Swee, 612
  Violet-eared, 614
  Yellow-bellied, 612
  Zebra, 615
Waxwing
  Bohemian, 387
  Cedar, 387
  Japanese, 387
Weaver
  Asian Golden, 609
  Baglafecht, 606
  Bannerman's, 606
  Bar-winged, 609
  Bates', 606
  Baya, 609
  Bengal, 609
  Bertram's, 606
  Black-billed, 606
  Black-chinned, 606
  Black-headed, 608

Weaver (*continued*)
Black-necked, 606
Bob-tailed, 609
Bocage's, 606
Brown-capped, 608
Cape, 606
Chestnut, 608
Cinnamon, 608
Clarke's, 608
Compact, 609
Forest, 608
Fox's, 607
Giant, 608
Golden Palm, 607
Golden-backed, 608
Golden-naped, 608
Grosbeak, 611
Kilombero, 607
Little, 606
Loango, 606
Maxwell's Black, 608
Nelicourvi, 609
Northern Brown-throated, 607
Olive-headed, 608
Orange, 607
Parasitic, 621
Preuss', 608
Red-headed, 606
Rueppell's, 607
Rufous-tailed, 605
Sakalava, 609
Salvadori's, 608
São Tomé, 609
Scaly, 605
Slender-billed, 606
Social, 605
Southern Brown-throated, 607
Speckle-fronted, 605
Spectacled, 606
Speke's, 607
Strange, 606
Streaked, 609
Usambara, 608
Vieillot's, 607
Village, 607–608
Weyns', 608
Yellow, 609
Yellow-capped, 608
Yellow-legged, 606
Yellow-mantled, 608
Wedgebill
Chiming, 513
Chirruping, 513
Weebill, 520
Weka, 80
Wheatear
Black, 463
Black-eared, 464
Capped, 464
Cyprus, 464
Desert, 464
Finsch's, 464
Heuglin's, 464
Hooded, 463
Hume's, 463
Isabelline, 464
Mountain, 463
Mourning, 463–464
Northern, 463
Pied, 464
Red-breasted, 464
Red-rumped, 464

Red-tailed, 464
Somali, 463
Variable, 464
White-tailed, 463
Whimbrel, 97
Whinchat, 462
Whip-poor-will, 180
Whipbird
Eastern, 513
Papuan, 513
Western, 513
Whistler
Bare-throated, 486
Black-headed, 486
Black-tailed, 485–486
Bornean, 484
Brown-backed, 484
Drab, 486
Dwarf, 483
Fawn-breasted, 484
Gilbert's, 483
Golden, 484–485
Golden-backed, 486
Gray, 484
Gray-headed, 484
Green-backed, 483
Hooded, 486
Island, 483
Lorentz's, 485
Mangrove, 483
Maroon-backed, 483
Mottled, 483
New Caledonian, 486
Olive, 483
Olive-flanked, 483
Red-lored, 483
Regent, 486
Rufous, 486
Rufous-naped, 483
Rusty, 484
Samoan, 486
Sclater's, 485
Sulphur-bellied, 484
Tongan, 486
Vogelkop, 484
Wallacean, 486
White-bellied, 486
White-breasted, 486
White-vented, 483
Yellow-bellied, 484
Whistling-Duck
Black-bellied, 26
Fulvous, 26
Lesser, 26
Plumed, 26
Spotted, 26
Wandering, 26
West Indian, 26
White-faced, 26
Whistling-Thrush
Blue, 402
Bornean, 402
Ceylon, 402
Chestnut-winged, 402
Formosan, 402
Javan, 402
Malabar, 402
Malayan, 402
Shiny, 402
White-eye
African Yellow, 549–550
Ambon, 553

Annobon, 550
Ashy-bellied, 552
Australian Yellow, 553
Banded, 553
Bare-eyed, 556
Biak, 553
Black-capped, 551
Black-crowned, 552
Black-fronted, 553
Black-headed, 553
Black-ringed, 552
Bridled, 551
Broad-ringed, 550
Buru, 553
Cape, 550
Capped, 553
Caroline Islands, 551
Ceylon, 551
Chestnut-flanked, 551
Christmas Island, 552
Cinnamon, 556
Comoro, 550
Cream-throated, 553
Dark-crowned, 555
Dusky, 555
Enggano, 551
Everett's, 551–552
Flores, 555
Ganongga, 553
Giant, 556
Golden, 555
Gray, 555
Gray-hooded, 555
Gray-throated, 554
Great Kai, 552
Green-backed, 554
Japanese, 551
Javan, 552
Javan Gray-throated, 555
Kulambangra, 554
Large Lifou, 554
Layard's, 554
Little Kai, 552
Long-billed, 555
Lord Howe, 554
Louisiade, 553
Lowland, 551
Madagascar, 550
Malaita, 554
Mascarene, 550
Mauritius, 551
Mayotte, 550
Mindanao, 555
Mountain, 552
New Guinea, 553
Oriental, 551
Pemba, 550
Plain, 551
Pygmy, 555
Rennell, 553
Reunion, 550
Robust, 700
Rufescent, 555
Rufous-throated, 555
Samoan, 555
Sanford's, 556
Sangihe, 553
Santa Cruz, 554
São Tomé, 550
Seram, 553
Seychelles, 551
Slender-billed, 554

Small Lifou, 554
Solomon Islands, 554
Splendid, 554
Streak-headed, 555
Sulawesi, 552
Timor, 555
Truk, 555
White-breasted, 550
White-browed, 555
White-chested, 554
White-throated, 553
Yap, 555
Yellow-bellied, 552
Yellow-fronted, 554–555
Yellow-spectacled, 552
Yellow-throated, 554
Yellowish, 552
Whiteface
Banded, 522
Chestnut-breasted, 522
Southern, 522
Whitehead, 483
Whitethroat
Greater, 443
Hume's, 443
Lesser, 443
Margelanic, 443
Small, 443
Whitetip
Purple-bibbed, 207
Rufous-vented, 207
Whydah
Pin-tailed, 620
Shaft-tailed, 620
Steel-blue, 620
Straw-tailed, 620
Widowbird
Buff-shouldered, 611
Fan-tailed, 610
Jackson's, 611
Long-tailed, 611
Marsh, 610
Red-collared, 610
White-winged, 610
Yellow-shouldered, 610
Wigeon
American, 29
Amsterdam Island, 699
Chiloe, 29
Eurasian, 29
Willet, 98
Willie-wagtail, 469
Wiretail, Des Murs', 271
Wood-Dove
Black-billed, 115
Blue-headed, 115
Blue-spotted, 115
Emerald-spotted, 115
Wood-Owl
African, 169
Brown, 168
Mottled, 168
Spotted, 168
Wood-Partridge
Bearded, 60
Buffy-crowned, 60
Long-tailed, 60
Wood-Pewee
Eastern, 335
Western, 335
Wood-Pigeon
Andaman, 111

Ashy, 111
Ceylon, 111
Common, 110
Japanese, 111
Nilgiri, 111
Silvery, 111
Speckled, 111
Wood-Quail
Black-breasted, 63
Black-eared, 62
Black-fronted, 62
Chestnut, 62
Dark-backed, 62
Gorgeted, 62
Marbled, 62
Rufous-breasted, 62
Rufous-fronted, 62
Spot-winged, 62
Spotted, 63
Starred, 63
Stripe-faced, 63
Tacarcuna, 62
Venezuelan, 62
Wood-Rail
Brown, 82
Giant, 82
Gray-necked, 82
Little, 82
Madagascar, 78
Red-throated, 699
Red-winged, 82
Rufous-necked, 82
Slaty-breasted, 82
Wood-Warbler
Laura's, 436
Uganda, 436
Yellow-throated, 436
Wood-Wren
Bar-winged, 396
Gray-breasted, 396
Munchique, 396
White-breasted, 396
Woodcock
Amami, 95
American, 96
Bukidnon, 95
Dusky, 95
Eurasian, 95
Moluccan, 95
Sulawesi, 95
Woodcreeper
Bar-bellied, 285
Black-banded, 286
Black-striped, 288
Brigida's, 285
Buff-throated, 287
Carajás, 286
Chestnut-rumped, 287
Cinnamon-throated, 285
Cocoa, 287
Elegant, 287
Great Rufous, 286
Hoffmann's, 286
Ivory-billed, 287
Juruá, 287
Lafresnaye's, 287
Lesser, 286
Lineated, 288
Long-billed, 285
Long-tailed, 284
Montane, 289
Moustached, 286

Narrow-billed, 288
Ocellated, 287
Olivaceous, 284
Olive-backed, 288
Plain-brown, 283–284
Planalto, 286
Red-billed, 285
Ruddy, 284
Scaled, 289
Scalloped, 289
Scimitar-billed, 285
Spix's, 287
Spot-crowned, 288
Spot-throated, 284
Spotted, 288
Straight-billed, 288
Streak-headed, 289
Striped, 287
Strong-billed, 285–286
Tawny-winged, 284
Thrush-like, 284
Tschudi's, 287
Tyrannine, 283
Uniform, 285
Wedge-billed, 285
White-chinned, 284
White-striped, 288
White-throated, 286
Zimmer's, 288
Woodhaunter, Striped, 280
Woodhoopoe
Black-billed, 228
Forest, 228
Green, 228
Violet, 228
White-headed, 228
Woodland-Warbler
Black-capped, 437
Brown, 436
Red-faced, 436
Woodnymph
Fork-tailed, 198
Green-crowned, 198
Long-tailed, 198
Mexican, 198
Violet-capped, 199
Violet-crowned, 198
Woodpecker
Abyssinian, 249
Acorn, 247
American Three-toed, 253
Andaman, 257
Arabian, 251
Arizona, 252
Ashy, 261
Bamboo, 260
Banded, 258
Bar-bellied, 253
Bay, 260
Bearded, 249
Beautiful, 247
Bennett's, 248
Black, 257
Black-and-buff, 260
Black-backed, 253
Black-bodied, 257
Black-cheeked, 247
Black-headed, 259
Black-necked, 255
Blond-crested, 256–257
Blood-colored, 254
Brown-backed, 250

Woodpecker (*continued*)
Brown-capped, 250
Brown-eared, 249
Brown-fronted, 251
Buff-necked, 260–261
Buff-rumped, 260
Buff-spotted, 249
Cardinal, 249
Checker-throated, 259
Checkered, 252
Chestnut, 256
Chestnut-colored, 256
Choco, 254
Cinnamon, 256
Cream-backed, 258
Cream-colored, 257
Crimson-bellied, 258
Crimson-breasted, 251
Crimson-crested, 258
Crimson-mantled, 255
Crimson-winged, 258
Cuban, 248
Darjeeling, 251
Dot-fronted, 254
Downy, 252
Elliot's, 249
Eurasian Three-toed, 253
Fine-spotted, 248
Fire-bellied, 249
Fulvous-breasted, 251
Gabon, 249
Gila, 247
Golden-cheeked, 247
Golden-collared, 254
Golden-crowned, 249
Golden-fronted, 247–248
Golden-green, 254
Golden-naped, 247
Golden-olive, 255
Golden-tailed, 248
Gray, 250
Gray-and-buff, 261
Gray-breasted, 247
Gray-capped, 250
Gray-crowned, 254
Gray-faced, 259
Gray-headed, 250
Great Slaty, 261
Great Spotted, 251–252
Green, 259
Green-backed, 248–249
Green-barred, 255
Ground, 249
Guadeloupe, 246
Guayaquil, 258
Hairy, 253
Heart-spotted, 261
Helmeted, 257
Himalayan, 252
Hispaniolan, 247
Hoffmann's, 248
Imperial, 258
Ivory-billed, 258
Jamaican, 247
Japanese, 259
Knysna, 248
Laced, 259
Ladder-backed, 252
Lesser Spotted, 250–251
Levaillant's, 259
Lewis', 246
Lineated, 257

Lita, 254
Little, 253–254
Little Gray, 249
Little Green, 248
Magellanic, 258
Maroon, 260
Melancholy, 249
Middle Spotted, 251
Mombasa, 248
Nubian, 248
Nuttall's, 252
Okinawa, 260
Olive, 250
Olive-backed, 259
Orange-backed, 260
Pale-billed, 258
Pale-crested, 256
Pale-headed, 260
Philippine, 250
Pileated, 257
Powerful, 257
Puerto Rican, 246
Pygmy, 250
Red-bellied, 247
Red-cockaded, 252
Red-collared, 259
Red-crowned, 247
Red-headed, 246
Red-necked, 258
Red-rumped, 254
Red-stained, 254
Reichenow's, 248
Ringed, 257
Robust, 258
Rufous, 256
Rufous-bellied, 251
Rufous-headed, 257
Rufous-winged, 254
Scaly-bellied, 259
Scaly-breasted, 256
Scarlet-backed, 253
Sind, 252
Smoky-brown, 253
Sooty, 261
Speckle-breasted, 249
Spot-breasted, 255
Stierling's, 249
Streak-breasted, 259
Streak-throated, 259
Strickland's, 252
Stripe-breasted, 251
Stripe-cheeked, 254
Striped, 252
Sulawesi, 250
Syrian, 252
Tullberg's, 249
Waved, 256
West Indian, 247
White, 246
White-backed, 251
White-bellied, 257
White-fronted, 247
White-headed, 253
White-naped, 260
White-spotted, 254
White-throated, 254
White-winged, 252
Yellow-browed, 254
Yellow-crowned, 251
Yellow-eared, 254
Yellow-fronted, 247
Yellow-throated, 254

Yellow-tufted, 247
Yellow-vented, 253
Yucatan, 247
Woodshrike
Common, 576
Large, 576
Woodstar
Amethyst, 210
Bahama, 210
Chilean, 211
Esmeraldas, 211
Gorgeted, 211
Little, 211
Magenta-throated, 210
Purple-collared, 211
Purple-throated, 210
Rufous-shafted, 211
Santa Marta, 211
Short-tailed, 211
Slender-tailed, 210
White-bellied, 211
Woodswallow
Ashy, 580
Bismarck, 581
Black-faced, 581
Dusky, 581
Fiji, 580
Great, 580
Little, 581
Masked, 581
White-backed, 580
White-breasted, 580–581
White-browed, 581
Wren
Apolinar's, 395
Band-backed, 388
Banded, 392
Bay, 391
Bewick's, 393
Bicolored, 389
Black-bellied, 390
Black-throated, 390
Boucard's, 389
Buff-breasted, 392
Bush, 347
Cactus, 389
Canyon, 390
Carolina, 392
Chestnut-breasted, 397
Clarion, 394
Cobb's, 394
Coraya, 391
Fasciated, 389
Fawn-breasted, 393
Flutist, 397
Fulvous, 390
Giant, 389
Gray, 393
Gray-barred, 389
Gray-mantled, 389
Happy, 391
House, 394
Inca, 390
Long-billed, 393
Marsh, 395–396
Mountain, 395
Moustached, 391
Musician, 397
Nava's, 390
Niceforo's, 392
Nightingale, 396
Ochraceous, 395

Paramo, 395
Peruvian, 390
Plain, 392
Plain-tailed, 390
Riverside, 391
Rock, 390
Rufous, 390
Rufous-and-white, 392
Rufous-breasted, 391
Rufous-browed, 394
Rufous-naped, 389
Santa Marta, 395
Scaly-breasted, 396
Sedge, 395
Sharpe's, 390
Sinaloa, 392
Socorro, 394
Song, 397
Sooty-headed, 390
South Island, 347
Speckle-breasted, 391
Spot-breasted, 391
Spotted, 389
Stephens Island, 700
Stripe-backed, 389
Stripe-breasted, 391
Stripe-throated, 392
Sumichrast's, 390
Superciliated, 392
Tepui, 395
Thrush-like, 389
Timberline, 396
Tooth-billed, 389
Whiskered, 390
White-bellied, 396
White-headed, 388
Wing-banded, 397
Winter, 393–394
Yucatan, 389
Zapata, 393
Wren-Babbler
  Bar-winged, 499

Black-throated, 498
Bornean, 497
Eyebrowed, 498
Falcated, 498
Immaculate, 499
Large, 498
Limestone, 498
Long-billed, 497
Long-tailed, 499
Luzon, 498
Marbled, 498
Mishmi, 499
Mountain, 498
Pygmy, 499
Rufous-throated, 499
Rusty-breasted, 498
Scaly-breasted, 498
Spotted, 499
Streaked, 498
Striated, 497–498
Striped, 498
Tawny-breasted, 499
Wedge-billed, 499
Wren-Spinetail, Bay-capped, 271
Wren-Warbler
  Barred, 425
  Gray, 425
  Miombo, 424–425
Wrenthrush, 646
Wrentit, 443
Wrybill, 95
Wryneck
  Eurasian, 244
  Rufous-necked, 244

Xenops
  Great, 283
  Plain, 283
  Rufous-tailed, 283
  Slender-billed, 283
  Streaked, 283
Xenopsaris, White-naped, 346

Yellow-Finch
  Bright-rumped, 675
  Citron-headed, 675
  Grassland, 675
  Greater, 675
  Greenish, 675
  Orange-fronted, 675
  Patagonian, 675
  Puna, 675
  Raimondi's, 675
  Stripe-tailed, 675
Yellowbill, 155
Yellowhammer, 664
Yellowhead, 483
Yellowlegs
  Greater, 98
  Lesser, 98
Yellownape
  Greater, 258–259
  Lesser, 258
Yellowthroat
  Altamira, 642
  Bahama, 643
  Belding's, 642
  Black-polled, 643
  Common, 642
  Gray-crowned, 643
  Hooded, 643
  Masked, 643
  Olive-crowned, 643
Yuhina
  Black-chinned, 510
  Burmese, 510
  Chestnut-crested, 510
  Rufous-vented, 510
  Striated, 510
  Stripe-throated, 510
  Taiwan, 510
  Whiskered, 510
  White-bellied, 510
  White-collared, 510
  White-naped, 510

Distribution of Endemic Bird Species of the World

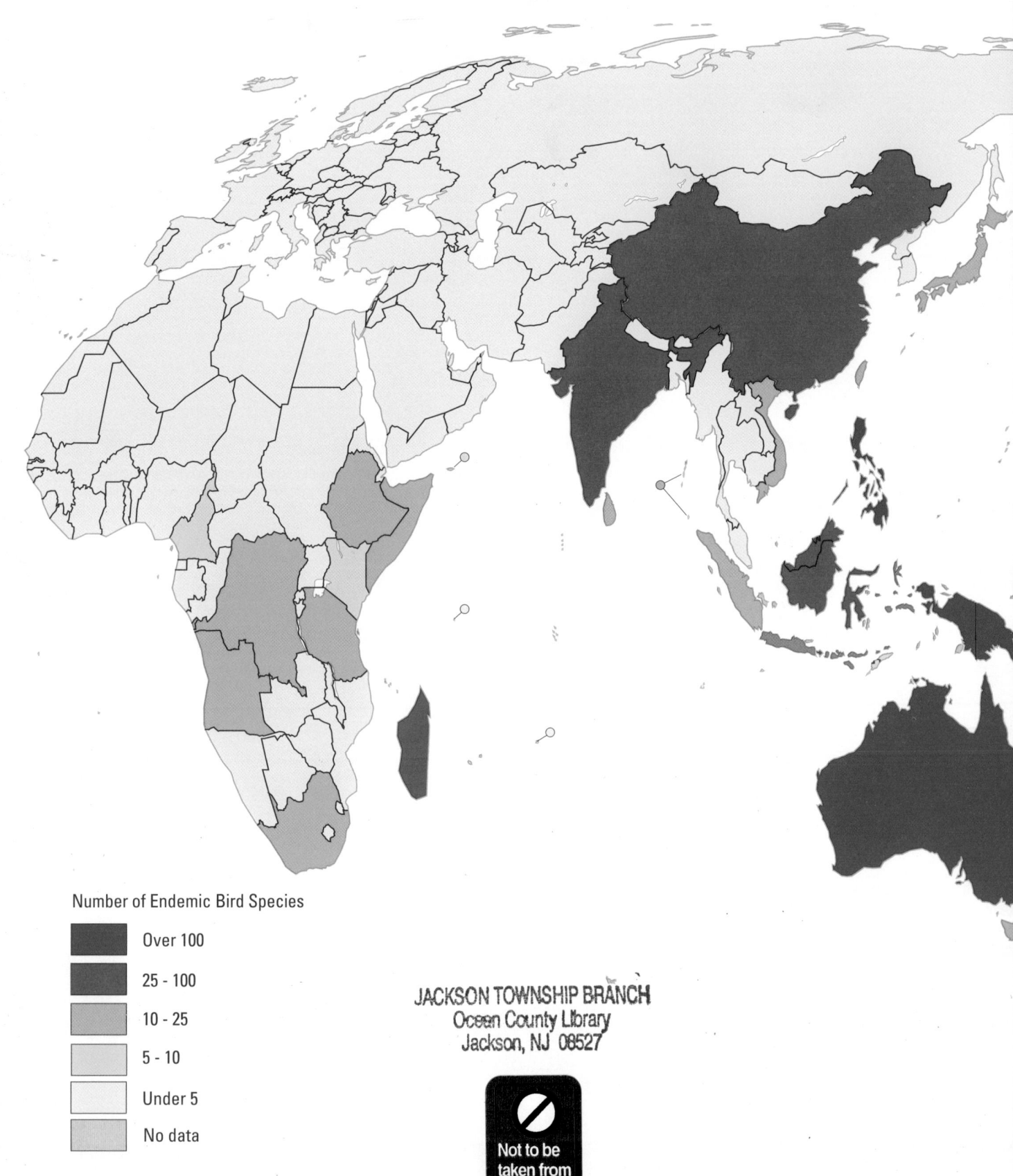

Number of Endemic Bird Species

- Over 100
- 25 - 100
- 10 - 25
- 5 - 10
- Under 5
- No data